The Laboratory Rat

THE LABORATORY RAT

EDITED BY

Mark A. Suckow
University of Notre Dame

Steven H. Weisbroth
Anmed/Biosafe

Craig L. Franklin
University of Missouri-Columbia

ELSEVIER

AMSTERDAM • BOSTON • HEIDELBERG • LONDON • NEW YORK • OXFORD
PARIS • SAN DIEGO • SAN FRANCISCO • SINGAPORE • SYDNEY • TOKYO
Focal Press is an imprint of Elsevier

Elsevier Academic Press
30 Corporate Drive, Suite 400, Burlington, MA 01803, USA
525 B Street, Suite 1900, San Diego, California 92101-4495, USA
84 Theobald's Road, London WC1X 8RR, UK

This book is printed on acid-free paper. ♾

Library of Congress Cataloging-in-Publication Data

The laboratory rat/edited by Mark A. Suckow, Steven H. Weisbroth,
Craig L. Franklin.
  p. cm.
    Includes index.
    ISBN 0-12-074903-3 (alk. paper)
  1. Rats as laboratory animals. I. Suckow, Mark A. II. Weisbroth,
  Steven H. III. Franklin, Craig L.
  SF407.R38L34 2005
  616'.02733–dc22
                                        2005015741

British Library Cataloguing in Publication Data

A catalogue record for this book is available from the British Library

ISBN 13: 978-0-12-074903-3
ISBN 10: 0-12-074903-3

For all information on all Elsevier Academic Press publications
visit our Web site at www.books.elsevier.com

Printed in the United States of America
05 06 07 08 09 10 9 8 7 6 5 4 3 2 1

Contents

List of Contributors

YUKSEL AGCA (7), Comparative Medicine Center, College of Veterinary Medicine, University of Missouri, 1600 East Rollins Road, Columbia, MO 65211

PETER G. ANDERSON (26), Department of Pathology, University of Alabama at Birmingham, 1670 University Blvd., Birmingham, AL 35294-0019

HENRY J. BAKER (1), Scott-Ritchey Research Center, College of Veterinary Medicine, Auburn University, Auburn, AL 36849

DAVID G. BAKER (13), Division of Laboratory Animal Medicine, School of Veterinary Medicine, Louisiana State University, Baton Rouge, LA 70803

DENNIS E. BARNARD (9), NIH/ORS/DIRS/VRP, Building 14G, Room 106A, Bethesda, MD 20892-5590

VALERIE BERGDALL (8), ULAR Department of Veterinary Preventive Medicine, The Ohio State University, 400 W. 12th, Room 115, Columbus, OH 43210

SANFORD P. BISHOP (26), Department of Pathology, University of Alabama at Birmingham, 1670 University Blvd., Birmingham, AL 35294-0019

GARY A. BOORMAN (14), Veterinary Medical Officer, NIEHS NCT TOX/PATH WORKING GROUP P.O. Box 12233, Research Triangle Park, NC 27709

RON BOOT (11), National Institute of Public Health and the Environment, Diagnostic Laboratory for Infectious Diseases and Screening, Section of Laboratory Animal Microbiology, PO Box 1, Bilthoven, 3720 BA, The Netherlands

DENISE I. BOUNOUS (5), Bristol-Myers Squibb, Drug Safety Evaluation and Discovery Toxicology, Province Line Rd. and Route 206, Princeton, NJ 08543-4000

BRUCE D. CAR (5), Bristol-Myers Squibb, Drug Safety Evaluation and Discovery Toxicology, Province Line Rd. and Route 206, Princeton, NJ 08543-4000

PHILIP B. CARTER (22), North Carolina State University, School of Veterinary Medicine, 4700 Hillsborough Street, Raleigh, NC 27606

JOHN K. CRITSER (7), Comparative Medicine Center, College of Veterinary Medicine, University of Missouri, 1600 East Rollins Road, Columbia, MO 65211

DIRCK DILLEHAY (29), Division of Animal Resources, Emory University School of Medicine, Whitehead Biomedical Research Building, 615 Michael Street, Suite GO2, Atlanta, GA 30322

DAVID N. EASTON (17), MPH, CIH, NW Regional Industrial Hygienist Northwest Regional Emergency Preparedness and Response Team, Virginia Department of Health, 671 Berkmar Court, Charlottesville, VA 22901

VICKI M. ENG (5), Princeton University, 1-S-15 Green Hall, Princeton, NJ 08530

CAROL ERB (28), Pfizer, Inc., Global Research and Development, Drug Safety Evaluation, 2800 Plymouth Road, Ann Arbor, MI 48105

NANCY E. EVERDS (5), Stine-Haskell Research Center, DuPont Haskell Laboratory, 1090 Elton Road, PO Box 50, Newark, DE 19714-0050

JEFFREY I. EVERITT (14, 20), GlaxoSmithKline, 5 Moore Drive, Research Triangle Park, NC 27709

ROBERT E. FAITH (10), Medical College of Wisconsin, Animal Resource Center, 8701 Watertown Plank Road, Milwaukee, WI 5322

SANFORD H. FELDMAN (17), Center for Comparative Medicine, University of Virginia, Health Sciences Center, PO 450, Charlottesville, VA 22908

HENRY L. FOSTER (22), Charles River Laboratories, Inc., 251 Ballardville St., Wilmington, MA 01887

CRAIG L. FRANKLIN (16), Department of Veterinary Pathobiology, E111-1 Vet. Med., University of Missouri, 1600 E. Rollins, Columbia, MO 65211

DIANE J. GAERTNER (12), University Laboratory Animal Resources, 423 Guardian Drive Building 30, University of Penssylvania, Philadelphia, PA 19104

BEVERLY J. GNADT (2), Division of Animal Resources, 8611 SUNY, SUNY Stony Brook, Stony Brook, NY 11794-8611

ELIZABETH A. GROSS (20), CIIT Center for Health Research, 6 Davis Drive, P.O. Box 12137, Research Triangle Park, NC 27709

THOMAS E. HAMM (27), 105 Martinique Place, Cary, NC 27511

MARTHA A. HANES (24), Integrated Laboratory Systems, Inc., P.O. Box 13501, Research Triangle Park, NC 27709

FORREST HAUN (8), NeuroDetective Inc., 1757 Wentz Road, Quakertown, PA 18951

HANS J. HEDRICH (3), Department of Laboratory Animal Science and Central Animal Facility, Hannover Medical School, Carl-Neuberg-Strasse 1, Hannover, D-30625, Germany

JACK R. HESSLER (10), 20917 Sunnyacres Dr., Laytonsville, MD 20882-1248

DEBRA L. HICKMAN (4), Portland Veterans Administration Medical Center, 3710 SW US Veterans Hospital Road, Portland, OR 97239

JOHN HOFSTETTER (4), Indiana University School of Medicine, 1120 South Drive, Fesler Hall 302, Indianapolis, IN 46202-5114

MARC HULIN (30), Merck Research Labs - San Diego, MRLSDB1, 3535 General Atomics Ct., San Diego, CA 92121

HOWARD J. JACOB (21), Medical College of Wisconsin, Department of Physiology, 8701 Watertown Plank Road, Milwaukee, WI 53226-0529

ROBERT O. JACOBY (12), Yale University School of Medicine, 375 Congress Avenue, LSOG Rm. 117, Comparative Medicine, New Haven, CT 06510

VERONICA JENNINGS (29), Merck Research Laboratories, Merck & Co., Inc., RY80B-100, PO Box 2000, Rahway, NJ 07065-0900

WILLIAM W. KING (15), Assistant Vice President, Research Services, University of Louisville, School of Medicine, Louisville, KY 40292

ANGELA KING-HERBERT (27), Laboratory of Experimental Pathology, NIEHS/NTP, PO Box 12233, Mail Drop B3-06, Research Triangle Park, NC 27709

JOSEPH J. KNAPKA (9), 19725 Olney Mill Road, Brooville, MD 20833

MICHAEL A. KOCH (18), Covance Laboratories, 3301 Kinsman Boulevard, Madison, WI 53704-2595

DENNIS F. KOHN (11), Institute Of Comparative Medicine, Columbia University, 630 W. 168th St., New York, NY 10032

BRYAN KOLB (8), Department of Psychology and Neuroscience, University of Lethbridge, 4401 University Drive, Lethbridge, Alberta, T1K3M4 Canada

SHERRY M. LEWIS (9), Office of Scientific Coordination National Center for Toxicological Research, Jefferson, AR 72079

J. RUSSELL LINDSEY (1), Department of Comparative Medicine, University of Alabama Medical Center, 402 Volker Hall, 1670 University Blvd., Birmingham, AL 35294-0019

JEFFERY J. LOHMILLER (6), Assistant Director, Genetics and Quality Control, Taconic Farms, 273 Hover Avenue, Germantown, NY 12526

CAROL MORENO-QUINN (21), Medical College of Wisconsin, Human and Molecular Genetics Center, 8701 Watertown Plank Road, Milwaukee, WI 53226

NANCY L. NADON (25), Office of Biological Resources and Resource Development, National Institute on Aging, 7201 Wisconsin Avenue, GW 2C231, Bethesda, MD 20892

GLEN OTTO (16), Department of Comparative Medicine, Stanford University, Quad 7, Building 330, Stanford, CA 94305-5410

DWIGHT R. OWENS (23), Charles River Laboratories, Inc., 251 Ballardville St. WILMINGTON, MA 01887

J. THOMAS PETERSON (26), Cardiovascular Pharmacology, Pfizer Global R & D, 2800 Plymouth Rd., Ann Arbor, MI 48105-2495

ROBERT QUINN (30), Department of Lab Animal Medicine, SUNY Upstate University, 750 E. Adams Street, Syracuse, NY 13210

STEVEN P. RUSSELL (15), Research Resources Center, University of Louisville, Health Sciences Center, Louisville, KY 40292

MARK A. SUCKOW (4), University of Notre Dame, 400 Freiman Life Sciences Center, Notre Dame, IN 46556

SONYA P. SWING (6), Consultant Murine Animal Models, 420 Richmond Dr., SE, Albuquerque, NM 87106

DUANE E. ULLREY (9), 10828 Little Heron Circle, Estero, FL 33928

MARY ANN VASBINDER (27), GlaxoSmithKline, Inc., 5 Moore Drive, P.O. Box 13398, Research Triangle Park, NC 27709-3398

GEORGE A. VOGLER (19), Department of Comparative Medicine, St. Louis University, School of Medicine, 1402 South Grand Blvd., St. Louis, MO 63104-1083

STEVEN H. WEISBROTH (11), AnMed/Biosafe, Inc., 7642 Standish Place, Rockville, MD 20855

IAN Q. WHISHAW (8), Department of Psychology, University of Lethbridge, 4401 University Drive, Lethbridge, Alberta, T1K3M4 Canada

Preface

The Laboratory Rat is one of the texts in the series sponsored by the American College of Laboratory Animal Medicine in furtherance of its educational mission to support the scientific community, in general, and expertise in laboratory animal medicine, in particular. The authoritative series is fully a product of the College by virtue of the editorial leadership for each of the texts and many of the chapters, as is the tradition of assigning all royalty income from the publications to support other continuing education programs of the College.

The Editors of this, the 2nd edition of *The Laboratory Rat* are pleased to present this revision of the original text. More than 26 years have intervened between the two editions and a great deal of new information has accumulated on the biology and experimental utility of this species. As the text goes to press, it is still not certain if the Editors' preference to contain the entire contents to a single volume can be achieved, or if the number of pages dictates two volumes.

There are a number of new chapters, some in subject areas not even imagined in 1979, e.g., transgenics and the rat genome, assisted reproductive technologies and the nude rat. Other new chapters signal subject areas of increasing importance, or represent, in the opinion of the Editors, subjects in the use of laboratory rats that merit compilation and updating of disparate reference material. The chapters on Experimental Modeling and Research Methodology, Metabolic and Traumatic Diseases and Legal and Ethical Perspectives all serve as examples of this concern.

Some of the chapters were meant to fill informational gaps perceived as lacking in the first edition. The chapters on Medical Management and Health Surveillance, Nutrition (with its sections on feeds and feeding), Surgery and Anesthesia, Euthanasia and Necropsy, and Occupational Health represent such subject areas and the Editors believe their inclusion greatly strengthens the utility of the text to readers in the animal care community.

All of the remaining chapters have been extensively revised and updated to include information accumulated since the first edition. For example, in the disease chapters, as the etiologic spectrum of rat pathogens has evolved and been defined with increasing precision over the last twenty-five years, readers will find correspondingly reduced descriptions of older and rarely encountered agents, but heightened attention to newer agents, e.g., the rat parvoviruses and *Helicobacter* — not even known (or delineated) at the time of the first edition.

Finally, the Editors wish to express their thanks to all of the contributors and reviewers for the 2nd edition. While a multiauthored review has the great strengths of capturing the widely separated depth and expertise of eminent individuals, it has the great weakness of the difficulty of getting them all to work within the confines of an arbitrary schedule of deadlines. The patience and support of all involved with this project has earned the gratitude of the Editors. To the extent that the text is found useful as a frequently consulted reference by the intended audience, the Editors will believe their fondest hopes for the book have been realized.

MARK A. SUCKOW
CRAIG L. FRANKLIN
STEVEN H. WEISBROTH

Foreword

Kathryn Bayne, M.S., Ph.D., DV.M., DACLAM, CAAB President,
American College of Laboratory Animal Medicine

It is estimated that between 90–95% of the animals used in biomedical research are mice and rats; thus, the role that they serve in advancing both human and animal health is vital. Their value to the biomedical research enterprise is based in part on the fact that their physiology and genetics closely resemble those of human beings. According to the National Institutes of Health, rats are one of the most important animal models for studying physiological processes and can be used to study virtually every organ system in humans. Indeed, rat models provide more physiological data and some transgenic rats have been found to more closely model certain human diseases than other experimental animals. The rat's short life span makes this animal a valuable contributor to aging studies. Rats are also critical to studies of drug addition, cancer, spinal cord injury, cardiovascular disease, Alzheimer's disease, spontaneous diabetes, and arthritis, to name just a few. The sequencing of the rat genome, estimated to be three billion base pairs in size and comparable in size to that of the human, has escalated the utility of the rat in studying genetically based diseases and using a comparative approach to better understand the mechanisms of the human genome.

One can get a glimpse as to the importance of the rat to biomedical research by reviewing some of the burgeoning resources, specific to the use of rats in research, that are available to investigators. Many of these resources are available online, such as the Rat Resource and Research Center (http://www.nrrrc.missouri.edu), the NIH Autoimmune Rat Model Repository and Development Center (http://www.ors.od.nih.gov/dirs/vrp/ratcenter/index.htm), the Rat Community Forum (http://rgd.mcw.edu/RCF) which includes both discussion and announcements; and the rat genome database (http://www.informatics.jax.org/) sponsored by Jackson Laboratories. This text serves as the definitive print resource for providing an overview of the numerous uses of the rat in research, testing and education while describing methods and considerations to ensure the welfare of the animals. This unique resource is an essential component of the library of veterinarians and investigators engaged in studies of the rat.

The American College of Laboratory Animal Medicine (ACLAM) promotes the humane care and responsible use of laboratory animals through certification of veterinary specialists, professional development, education and research. It is the hope of ACLAM that *The Laboratory Rat* helps fulfill this mission by serving as a resource for those using rats in research and those charged with the care of laboratory rats.

The Editors of the Second Edition of *The Laboratory Rat* wish to acknowledge the Editors and authors of the First Edition for laying a solid foundation for this work.

Editors

Henry J. Baker
J. Russell Lindsey
Steven H. Weisbroth

Authors

Norman H. Altman
Miriam R. Anver
Dennis E.J. Baker
Henry J. Baker
Hervé Bazin
Allan R. Beaudoin
Pravin N. Bhatt
Sanford P. Bishop
W. Sheldon Bivin
N. R. Brewer
G. Bruce Briggs
Joe D. Burek
Philip B. Carter
Gail H. Cassell
Bennett J. Cohen
Donald V. Cramer
M. Pat Crawford
Lyubica Dabich

Jerry K. Davis
Dickson D. Despommier
Michael F. W. Festing
Henry L. Foster
Estelle Hecht Geller
Thomas J. Gill III
Dawn G. Goodman
Carel F. Hollander
Chao-Kuang Hsu
Robert O. Jacoby
Albert M. Jonas
Alan L. Kraus

Sam M. Kruckenberg
Heinz W. Kunz
J. Russell Lindsey
Annie Jo Narkates
Juan M. Navia
Frederick W. Oehme
John C. Peckham
Daniel H. Ringler
Roy Robinson
Adrianne E. Rogers
Steven H. Weisbroth
Christine S. F. Williams

List of Reviewers for Chapters in This Volume

Yuksel Acga — University of Missouri, Columbia, Missouri
Barry Astroff — Eli Lilly and Company, Indianapolis, Indiana
David G. Baker — Louisiana State University, Baton Rouge, Louisiana
Loretta Bober — Schering Plough Corporation, Kenilworth, New Jersey
Paul Boor — University of Texas Medical Branch, Galveston, Texas
Ron Boot — National Institute of Public Health and the Environment, Bilthoven, The Netherlands
Roderick Bronson — Tufts University School of Veterinary Medicine, Grafton, Massachusetts
Fred J. Clubb, Jr. — Texas Heart Institute, Houston, Texas
Melanie Cushion — University of Cincinnati, Cincinnati, Ohio
Peggy J. Danneman — The Jackson Laboratory, Bar Harbor, Maine
Stanley D. Dannemiller — Cleveland Clinic Foundation, Cleveland, Ohio
Jerry K. Davis — Purdue University, West Lafayette, Indiana
Gregory Demas — Indiana University School of Medicine, Indianapolis, Indiana
Bernard J. Doerning — The Procter and Gamble Company, Cincinnati, Ohio
Robert C. Dysko — University of Michigan, Ann Arbor, Michigan
Craig L. Franklin — University of Missouri, Columbia, Missouri
William R. Elkins — National Institute of Allergy and Infectious Diseases, Bethesda, Maryland
Stephen Gammie — University of Wisconsin, Madison, Wisconsin
Thomas J. Gill III — Duxbury, Massachusetts
Robert Gunnels — Pfizer, Inc., Groton, Connecticut
John Hasenau — Baxter Healthcare, Inc., Round Lake, Illinois
C. Terrance Hawk — GlaxoSmithKline, Inc., King of Prussia, Pennsylvania
James D. Henderson, Jr. — Medical College of Wisconsin, Milwaukee, Wisconsin
Gary L. Hofing — Parke-Davis Pharmaceutical Research Division, Ann Arbor, Michigan
Felix Homberger — Weill Medical College of Cornell University and the Memorial Sloan-Kettering Cancer Center, New York, New York
Darrell E. Hoskins — Southern Research Institute, Birmingham, Alabama
Philip Iannaccone — Northwestern University Feinberg School of Medicine, Chicago, Illinois
Heinz Kunz — University of Pittsburgh, Pittsburgh, Pennsylvania
Marie C. LaRegina — Washington University, St. Louis, Missouri
Noel Lehner — Emory University, Atlanta, Georgia
J. Rusell Lindsey — University of Alabama at Birmingham, Birmingham, Alabama
Robert Livingston — University of Missouri, Columbia, Missouri
Brent J. Martin — Medical College of Ohio, Toledo, Ohio
Ronald M. McLaughlin — University of Missouri, Columbia, Missouri
Tore Midtvedt — Karolinska Institute, Stockholm, Sweden
Roger Orcutt — Dunkirk, New York
Scott E. Perkins — Tufts New England Medical Center, Boston, Massachusetts
Morris Pollard — University of Notre Dame, Notre Dame, Indiana
Fred W. Quimby — The Rockefeller University, New York, New York
Robert J. Russell — Harlan, Inc., Indianapolis, Indiana
Marisa Elkins St. Claire — BIOQUAL, Inc., Rockville, Maryland
Valerie A. Schroeder — University of Notre Dame, Notre Dame, Indiana

Patrick Sharp	University of California, Los Angeles, California
Abigail Smith	The Jackson Laboratory, Bar Harbor, Maine
James G. Strake	Southern Illinois University, Carbondale, Illinois
Linda A. Toth	Southern Illinois University School of Medicine, Springfield, Illinois
Philip C. Trexler	Bloomington, Indiana
Huber Warner	National Institute of Aging, Bethesda, Maryland
Joseph Warren	University of Wisconsin, Madison, Wisconsin
Benjamin Weigler	Fred Hutchinson Cancer Research Center, Seattle, Washington
Steven H. Weisbroth	McLean, Virginia
Charles Wiedemeyer	University of Missouri, Columbia, Missouri
Norman Wolf	University of Washington, Seattle, Washington
John D. Young	Cedars-Sinai Medical Center, Los Angeles, California

Chapter 1

Historical Foundations

J. Russell Lindsey and Henry J. Baker

THE LABORATORY RAT, 2ND EDITION

I. ORIGIN OF THE LABORATORY RAT

The purpose of this chapter is to retrace, to around the mid 1970s, one of the most fascinating stories of biomedical history: that of the early events and personalities involved in establishment of *Rattus norvegicus* as a leading laboratory animal. From the outset, three features of the story deserve acknowledgment. First, it will be possible in the brief space of this chapter to touch only the highlights (but these cannot be treated equally because many important details already have been lost). Second, it is predominantly an American story, as most of the strains of rats in use today around the world trace their ancestry to stocks in the United States. Third, the story is literally one of ascendancy from the gutter to a place of nobility, for what creature is more lowly than the rat as a wild pest or more noble than the same species that has contributed so much to the advancement of knowledge as the laboratory rat!

A. Earliest Records

The earliest records of *R. norvegicus* are a bit sketchy, but there appears to be good agreement on the major events concerning the species (Donaldson, 1912a; Donaldson, 1912b; Castle, 1947; Richter, 1954; Robinson, 1965). Its original natural habitat is thought to have been the temperate regions of Asia, specifically the area of the Caspian Sea, Tobolsk and Lake Baykal in Russia, and possibly extending across China and Mongolia (Hedrich, 2000). With the coming of modern civilization, a suitable ecological niche became available to the species as an economic pest, allowing its numbers to increase rapidly and spread over the world in close association with man. It is said to have reached Europe early in the 18th century, England between 1728 and 1730, North America by 1755, and northeastern United States by 1775, arriving on ships with early settlers. That the species did at one time in history spread through Norway is readily accepted, but

the name "Norway rat" or "Norwegian rat" actually has no meaning other than, possibly, to reflect the species name, i.e., *norvegicus* (Hedrich, 2000).

Richter (1954) has summarized the likely sequence of events in the domestication of *R. norvegicus*, as follows:

> It is quite likely that Norway rats come into captivity as albinos. We know that rat-baiting was popular in France and England as early as 1800, and in America soon afterward. This sport flourished for seventy years or more, until it finally was stopped by decree. In this sport 100 to 200 recently trapped wild Norways were placed at one time in a fighting pit. A trained terrier was put into the pit. A keeper measured the time until the last rat was killed. Sportsmen bet on the killing times of their favorite terriers. For this sport many Norway rats had to be trapped and held in pounds in readiness for contests. Records indicate that albinos were moved from such pounds and kept for show purposes and/or breeding. It is thus very likely that these show rats, that probably had been tamed by frequent handling, found their way at one time or another into laboratories.

B. Earliest Experiments

Regardless of the exact details, the Norway rat became the first mammalian species to be domesticated primarily for scientific purposes (Richter, 1959). There is some evidence to suggest that rats probably were used sporadically for nutrition experiments in Europe prior to 1850 (Verzar, 1973). However, the work generally recognized as the first use of the rat for experimental purposes was a study on the effects of adrenalectomy in albino rats by Philipeaux (Philipeaux, 1856) published in France in 1856. Soon afterward in 1863, Savory (Savory, 1863), an English surgeon, published what is thought to be the first attempt to evaluate the nutritional quality of proteins in a mammal. His rats were of mixed coat colors, including black, brown, and white. The first known breeding experiments with rats used both albino and wild animals and were performed in Germany from 1877 to 1885 by Crampe (Crampe, 1877; Crampe, 1883; Crampe, 1884; Crampe, 1885).

The first experiments known to use rats in the United States were neuroanatomical studies performed during the early 1890s by S. Hatai and other faculty members in H. H. Donaldson's Department of Neurology at the University of Chicago (Conklin, 1939). Richter (Richter, 1968) suggested that these albino rats had been brought from the Department of Zoology at the University of Geneva by Adolf Meyer, the Swiss neuropathologist, shortly after he immigrated to the United States and joined Donaldson's faculty around 1890. It is clear that Meyer introduced albino rats into Donaldson's department more than once and was a major stimulus in convincing Donaldson of their value in neurologic investigations (Logan, 1999). On one such occasion Meyer, in 1893, brought albino rats into Donaldson's department for a course in neuroanatomy, after which the surplus rats became a breeding colony, with the progeny being increasingly used by Donaldson and his faculty in neurologic studies (Clause, 1993). In 1894, Colin Stewart at Clark University (in Worcester, Massachusetts, where Donaldson previously had been on the faculty) initiated studies on the effects of alcohol, diet, and barometric phenomena on activity (Stewart, 1898). He began with wild rats, but in 1895 switched to albinos, possibly also provided by Adolf Meyer, who took a job at the mental hospital in Worcester, Massachusetts in 1895 (Warden, 1930). The precise origin of these albino rats remains uncertain, as Donaldson stated explicitly in two separate articles that it was not clear whether the North American albinos were descendants of those stocks that had been used earlier in research laboratories in Europe, or mutants selected from wild rats captured in the United States (Donaldson, 1912a; Donaldson, 1912b). This uncertainty is further supported by a thorough analysis of the scientific publications, correspondences, and memoirs of Donaldson and Meyer (Logan, 1999).

II. THE WISTAR INSTITUTE

A. Background

The Wistar Institute of Philadelphia occupies a special place in the annals of biomedical history. First, it is the oldest independent research institute in the United States. Second, and particularly germane to the subject of this chapter, it provided a major share of the initial foundation on which the rat came to be established as an important laboratory animal.

The Institute was named in honor of a physician, Caspar Wistar (1761–1818). While Professor of Anatomy at the University of Pennsylvania School of Medicine, he authored the first standard text of human anatomy in

America and developed a magnificent collection of anatomic specimens. These specimens were brought together as "The Wistar Museum" in 1808 and given to the University of Pennsylvania after his death. Additional materials subsequently were contributed by his colleagues, particularly Dr. George Horner. Interest in the collection declined over the next 60 years until Dr. James Tyson, Dean of the Medical School, became keenly interested in assuring future preservation of the Museum. He approached General Isaac Wistar (1827–1905), Philadelphia lawyer, great nephew of Caspar Wistar, and former Brigadier General in the Union Army, who initially contributed only $20.00. Later, under the General's direction, however, the Wistar Institute of Anatomy and Biology was incorporated in 1892, and funds were set aside for a "fire-proof museum building" and an endowment. The original building, which still stands today (Fig. 1-1), was built at a cost of $125,000 and formally opened in May 1894. The University of Pennsylvania provided the land (Purcell, 1968–1969; Wistar, 1937).

B. Milton J. Greenman

The first two directors of the Institute served only briefly. A physician, Milton Jay Greenman (1866–1937) (Fig. 1-2) became the third director in 1905, and held that position for 32 years until his death (Annual Reports of Director, 1911–1956; Donaldson, 1937; Wistar, 1937). Although his role was chiefly in administration, Greenman's era was clearly a "golden age" in the life of the Institute. Henry Donaldson (Donaldson, 1927) characterized him as "the Institute's real scientific founder..." and as a "genial, alert man, trained in biology, gifted to an unusual degree with mechanical and inventive abilities, with business capacity and good judgment, based on the imagination needed for an administrator."

Greenman wanted the Institute to be a center of scientific investigation, not merely a museum. Thus, immediately upon being made director, he established a Scientific Advisory Board composed of 10 professors of anatomy and zoology from the nation's leading universities (Fig. 1-3). In April 1905 this Board, at its first meeting, voted unanimously in favor of the Institute being devoted primarily to research, initially focusing its efforts in the areas of "neurology, comparative anatomy and embryology." Further, the Board unanimously recommended one of its own number, Henry Donaldson, to be the Institute's first Scientific Director (Annual Reports of Director, 1911–1956; Wistar, 1937). In 1908, the Institute acquired the rights to five leading biological journals and a printing press that became a well-known publisher, Wistar Press (Purcell, 1968–1969).

Fig. 1-1 Original building of the Wistar Institute of Anatomy and Biology in Philadelphia built at a cost of $125,000 and formally opened in May, 1894. The street (Woodland Avenue) to the right of the building in this photo was subsequently closed and is today a small park. (Courtesy of The Wistar Institute.)

Fig. 1-2 Milton Jay Greenman, Director of The Wistar Institute from 1905 until 1937, and characterized by Henry Donaldson as "the Institute's real scientific founder." (Courtesy of The Wistar Institute.)

C. Henry H. Donaldson

Henry Herbert Donaldson (1857–1938) was one of the truly remarkable men of early American science, undoubtedly one of its greatest heroes (Fig. 1-4). In essence,

Donaldson did for the laboratory rat what Clarence Cook Little (1888–1971) was to do later for the laboratory mouse at the Jackson Laboratory (established in 1929). Donaldson and his team of investigators at Wistar began efforts in 1906 to standardize the albino rat. Initially, the main intent was to produce reliable strains for their studies of growth and development of the nervous system. In reality, the work directly gave the broad foundation for use of the rat in nutrition, biochemistry, endocrinology, genetic, and behavioral research and, indirectly, in many other fields of investigation.

After completion of his B.A. degree at Yale and a year of graduate work in physiology with R. H. Chittenden, Donaldson enrolled in the College of Physicians and Surgeons in New York, but only 1 year of study convinced him that he really wanted a career in basic research rather than clinical medicine. He entered graduate school at the Johns Hopkins University in 1881 and was awarded a Ph.D. in 1885. During the next few years, the work of his dissertation research was continued at different institutions in Europe and the United States, work that culminated in a book, *The Growth of the Brain*, published by Scribners in 1895. From 1889 to 1892 he was Assistant Professor of Neurology at Clark University. Donaldson then became Dean of the Ogden School of Science at the University of Chicago for 6 years and Professor of Neurology there from 1892 until 1906 (during this interval,

Fig. 1-3 General Isaac Wistar and Milton Greenman with the first Scientific Advisory Board of The Wistar Institute during their meeting of April 11 and 12, 1905. On the advice of this group, The Wistar Institute changed its primary mission from that of a museum to scientific research. From left to right: Simon H. Gage, Franklin P. Mall, Isaac J. Wistar, Milton J. Greenman, Edwin G. Conklin, Charles S. Minot, George A. Piersol, Lewellys F. Barker, J. Playfair McMurrich, Henry H. Donaldson (hands on walking cane), Carl Huber, and George S. Huntington (Wistar, 1937). (Courtesy of The Wistar Institute.)

he developed tuberculosis of the hip, which left him permanently crippled) when, at the invitation of Milton Greenman, he became Director of Research at the Wistar Institute and Professor of Neurology at the University of Pennsylvania (Conklin, 1939).

Donaldson brought with him from Chicago one colleague, Dr. Shinkishi Hatai, and four pairs of albino rats. Although Donaldson up to this point in his career had studied primarily the nervous systems of man and frogs, he was convinced (by Adolf Meyer) that the albino rat was the best available animal for laboratory work on problems of growth. Accordingly, he recruited to Wistar a team of scientists who, for their day, launched a colossal multi-disciplinary research program (Conklin, 1939). Donaldson and his group of collaborators carried out the following studies using albino rats [from a summary of Donaldson's life's work (McMurrich and Jackson, 1938)].

1. Determined growth curves for body length and body weight
2. Correlated growth curves of brain and cord with body length and weight
3. Studied various conditions that might affect the nervous system:
   a. Domestication, age, exercise, inbreeding (H. D. King)[1]
   b. Gonadectomy (J. M. Stotsenberg and S. Hatai)

[1]Names of Donaldson's collaborators are given in parenthesis.

   c. Growth of cerebral (N. Sugita) and cerebellar (W. H. F. Addison) cortex
   d. Blood supply to different portions of the brain (E. H. Craigie)
   e. Chemistry of the brain solids (W. and M. L. Koch)
   f. Relation of size of nerve cells to length of their neurites, the spacing of nodes in nerve fibers (S. Hatai)
   g. Regeneration of nerve fibers (M. J. Greenman)
4. Growth curves:
   a. Skull
   b. Skeleton
   c. Viscera (S. Hatai)
   d. Individual organs, the submaxillary glands, the hypophysis (W. H. F. Addison)
   e. Thymus, adrenals (J. C. Donaldson)
   f. Thyroid and effects of thyroidectomy (F. S. Hammett)

Donaldson himself published nearly 100 papers and books, the most famous being *The Rat: Data and Reference Tables for the Albino Rat (Mus norvegicus albinus) and the Norway Rat (Mus norvegicus)* (1st ed., 1915; 2nd ed., 1924). This book (Donaldson, 1915; Donaldson, 1924) was a truly remarkable compendium of data pertaining to the rat; the second edition, published in 1924, contained 469 pages that included 212 tables, 72 charts, and more than 2,050 references! [These references combined with a later list for the period 1924–1929 by Drake and Heron (Drake and Heron, 1930)

Fig. 1-4 Henry Herbert Donaldson, who from 1906 to 1938 led a multidisciplinary research program at the Wistar Institute that laid the initial foundations for use of the rat in research. More than a great scientist, he was a man of exemplary personal qualities, as revealed in the following description by Conklin (Conklin, 1939): "Anyone who had once seen him could never forget his magnificent head, his steady sympathetic eyes, his gentle smile. With these were associated great-hearted kindness, transparent sincerity, genial humor...orderliness, persistence, sincerity. His laboratory and library were always in perfect order, his comings and goings were as timely as the clock, he never seemed hurried and yet he worked *Ohne Hast, Ohne Rast* [no haste, no rest]." (Photo by permission *J. Comp. Neurol.*)

provide a complete bibliography on the rat to 1930.] An appraisal of Donaldson's contribution (McMurrich and Jackson, 1938) reads, "Donaldson's work then, is fundamental: it gives a paradigm of growth with which other investigations may be compared and is an outstanding monument of perseverance, thoroughness, accuracy and cooperation." It remains an all time classic work on the laboratory rat.

Donaldson's choice of the rat as a laboratory animal was no mere accident. The depth of his reasoning can be

seen in the following quotes from his book (Donaldson, 1924):

> In enumerating the qualifications of the rat as a laboratory animal, and in pointing out some of its similarities to man, it is not intended to convey the notion that the rat is a bewitched prince or that man is an overgrown rat, but merely to emphasize the accepted view that the similarities between mammals having the same food habits tend to be close, and that in some instances, at least, by the use of equivalent ages, the results obtained with one form can be very precisely transferred to the other. I selected the albino rat as the animal with which to work. It was found that the nervous system of the rat grows in the same manner as that of man–only some 30 times as fast. Further, the rat of 3 years may be regarded as equivalent in age to a man of 90 years, and this equivalence holds through all portions of the span of life, from birth to maturity. By the use of the equivalent ages observations on the nervous system of the rat can be transferred to man and tested.

It appears that Donaldson's work had its impact particularly on progress in nutrition. Addison (Addison, 1939) gives the following quote from an address presented by H. Gideon Wells [Chairman of the Department of Pathology at the University of Chicago and Director of its affiliated Otha S. A. Sprague Memorial Institute (Chittenden, 1938; Otha, 1911–1941)] in 1938 on the subject "War Time Experience in Nutrition:"

> If it had not been for Henry Donaldson's rats the knowledge of vitamins would not have come in time to be applied when it was most acutely needed. It had taken years for Donaldson to learn the normal curve of growth of white rats and to establish strains of healthy rats that were reliable material for research. These rats and their growth curves were basic factors in nutrition research, and without them this research would have been delayed until nutritionists did what Donaldson had already done for them, and that delay would have been serious for the feeding of starving nations from 1915 to 1919.

D.   Contributions of Donaldson's Colleagues

By the time of Donaldson's death in 1938, his students, assistants, and associates at the Wistar had published more than 360 articles and books, almost exclusively on the rat (Conklin, 1939). Some of them made major contributions to establishment of the rat as a laboratory animal. A few highlights of these contributions follow.

Helen Dean King (1869–1955) (Fig. 1-5) was an extremely productive scientist at the Wistar from 1907 until her retirement in 1949. In 1909 she began inbreeding the albino rats that Donaldson had brought from Chicago. [It is interesting that 1909 was also the year in which Clarence Cook Little began inbreeding the oldest known strain of mouse, the DBA (Heston, 1972)]. This was an enormous undertaking, as some of her many papers amounted to summaries of observations made on 25,000 rats (King, 1918a; King, 1918b; King, 1918c; King, 1919)!

Fig. 1-5 Helen Dean King during the early years of her career as a scientist at The Wistar Institute. In 1909 she initiated the first inbred strain of rats, now known as the PA strain. (Courtesy of The Wistar Institute.)

TABLE 1-1

RAT STRAINS AND STOCKS ORIGINATED AT LEAST IN PART FROM WISTAR GENE POOLS

A. Inbred line began by Helen Dean King in 1909 and now known as the PA strain, one subline: WKA

B. Direct descendants of Wistar commercial stock(s)

AS	K	MNR	SHR	WE/Cpb
B	KYN	MNRA	W	WF
BROFO	LEW	MR	WA	WKY/N
BUF	LOU/C	OKA/Wsl	WAB	WM
GH	LOU/M	R	WAG	WN

C. From crosses between Wistar commercial stock and other rats

BD II	BD VI	BD X	IS	McCollum
BD III	BD VII	BDE	LE	Sprague-Dawley
BD IV	BD VIII	BS	LGE	
BD V	**BD IX**	**GHA**	**MAXX**	

By 1920 these rats were maintained as two separate lines that had reached generation 38 of brother × sister matings, prompting a note by King in the annual report that Wistar was "11 years ahead of all other attempts" to inbreed rats. One of these lines was carried to the 135th generation of inbreeding by King and eventually became known as the King Albino (designated the PA strain). Many other modern strains of rats (Table 1-1) take their origin from this and other stocks at the Wistar (Billingham and Silvers, 1959). For example, the Lewis (LEW) strain was produced from animals selected and inbred at Wistar by Margaret Reed Lewis; in 1956, this strain reached its eighth generation of inbreeding. King also captured some wild Norways in the vicinity of Philadelphia and proceeded to produce an inbred strain of them. By 1934 they had reached generation 35 of inbreeding. This line became the strain known today as the Brown Norway (BN) rat. Several mutations were identified and studied by King (Annual Reports of the Director, 1911–1956; Crampe, 1883; Robinson, 1965).

Another important milestone in the history of the laboratory rat came in 1935 when Eunice Chace Greene (1895–1975) published the first comprehensive treatise on gross anatomy of this species entitled "Anatomy of the Rat" (Greene, 1935). This book (with several reprintings by Hafner Publishing Co.) is unquestionably an all-time classic work on the rat. She also published a chapter in "The Rat in Laboratory Investigation" (Farris and Griffith, 1949; Griffith and Farris, 1942). These two publications comprise her total bibliography. At the time of her death in 1975, she and her husband, Walter F. Greene, formerly Professor of Biology at Springfield College in Massachusetts, lived in Jaffrey Center, New Hampshire (Erikson, 1977; Greene, 1978).

E. The Wistar Rat Colony

It seems impossible to overstate the contribution made by The Wistar Institute to husbandry of the laboratory rat. The Wistar rat colony served as the initial testing ground for developing satisfactory cages and ancillary equipment, diets, breeding practices, and facilities for rats. It also must be emphasized that much of the credit for these innovations and advances goes to none other than the Director of the Institute, Milton Greenman. He gave the colony a great deal of his personal attention and, with the help of an able assistant, Miss F. Louise Duhring, rapidly evolved a highly successful set of husbandry practices. That the two of them took this responsibility seriously is clearly revealed in the first and second editions of their book, *Breeding and Care of the Albino Rat for Research Purposes* (Greenman and Duhring, 1923; Greenman and Duhring, 1931). The preface to the first edition states, "It became more and more evident that clean, healthy, albino rats were essential for accurate research and that their production was a serious, difficult, and worthwhile task."

The first rats brought to the Wistar in 1906 by Donaldson were presumably housed in the museum building (Fig. 1-1). In 1913, however, the colony was moved into a 30 × 90 ft, 3-story building referred as The Annex, which had formerly been a police station (Fig. 1-6). In 1918, there arose a series of circumstances that seriously threatened the colony's productivity and continued existence. Cottonseed meal was introduced into the diet, and many rats died or ceased breeding as result of gossypol toxicity. To meet demands of the Institute's investigators, rats were purchased from outside sources and brought into the facility. Consequently, many further

Fig. 1-6 An early picture of the rat colony at Wistar Institute. The building is thought to be "The Annex," a former police station that housed the Wistar rat colony from 1913 to 1922. (Courtesy of The Wistar Institute.)

Fig. 1-7 The Wistar Colony Building completed in 1922 and referred to by Donaldson in 1927 (Donaldson, 1927) as "a home for the friendless rat." The building is shown here in three perspectives. Figure 1-7 is a view looking across the courtyard from the main Institute building. (Courtesy of The Wistar Institute.)

deaths were incurred because of an epizootic of infectious disease(s). Greenman found himself increasingly involved with the details of operating the colony. As a result of this experience, Miss Duhring was given charge of the colony and Greenman began planning for a new building. Plans were developed over a period of 2 years and, in 1921, a member of the Institute's Board of Managers, Mr. Samuel S. Fels, contributed $30,000 toward its construction. The "Colony Building" (Figs. 1-7, 1-8, and 1-9), with 12 modular rooms measuring 14×22 feet each, for a total of approximately 4635 net feet^2, was built at a cost of $33,458.66! The first rats were moved in on May 5, 1922 and the last entered on August 9, 1922, each rat being treated with tincture of larkspur seed by Miss Duhring to "rid the colony of ectoparasites!" At this time, the colony had a total population of approximately 6,000 rats (Annual Reports of the Director, 1911–1956).

The Wistar Colony Building was remarkable because of its forward-looking design characteristics. It was "constructed of brick, concrete, steel, and glass" because "these materials offer less harbor for dirt and vermin if properly put together." The floor plan (Fig. 1-9) provided for separation of animal facility functions: rooms for housing animals, an office for records, a surgery, a cage-washing room, a diet kitchen, and lavatories for personnel. If one accepts the central hallway as the "clean" corridor and the outside concrete walk as the "dirty" corridor, it qualifies as the first animal facility of the "clean-dirty corridor" concept. It was recognized that "the work of a colony is greatly facilitated if the entire colony is located on one floor level, preferably the ground floor." Also, and possibly the most important principle of all, was the idea

that "no space unfit for human habitation is desirable for an albino rat colony" (Greenman and Duhring, 1923; Greenman and Duhring, 1931).

Husbandry practices for the colony in the early years of the new colony building were documented in detail in the book by Greenman and Duhring (Greenman and Duhring, 1923; Greenman and Duhring, 1931). Most of the cages were constructed of white pine with removable galvanized wire floors (No. 22 wire, 1/8-inch mesh) mounted over steel trays (Figs. 1-10, 1-11, 1-12, and 1-13). (Both editions of Greenman and Duhring's book contained a drawing giving details for construction of these cages

Fig. 1-8 This is a view from the Woodland Avenue side of the Institute's triangular lot looking back toward the main Institute building. (Courtesy of The Wistar Institute.)

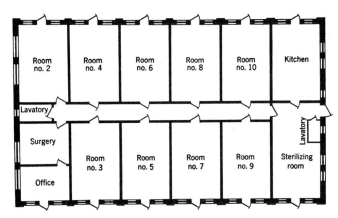

Fig. 1-9 This is the floor plan of the Colony Building. (Courtesy of The Wistar Institute.)

recipes, including "wheat and peas with milk," "barley, salmon, and eggs," "hominy grits, vegetables, and eggs," and many others. Such homemade mixtures probably were continued until 1949 when dry commercial dog and fox diets came into use for the rat colony (Annual Reports of the Director, 1911–1956).

The Annual Reports of the Director (1911–1956) at Wistar give many valuable insights into realities of daily operations of the rat colony. First of all, the preeminence of the colony is revealed in the fact that every report from 1911 to 1956 contains a major section entitled "The Animal Colony." This section is a summary of each year's activities, including rat production, a listing of recipients of animals, progress of the inbreeding program under King, budget information, personnel matters, and comments on health problems encountered in the colony.

Table 1-2 gives the available data on rat production at the Wistar from 1911 through 1928. During this period, albino rats were supplied to many institutions throughout the United States and in many foreign countries. In fact, more animals were supplied to other institutions than

as a large foldout frontispiece.) The floors and pans were washed by hand (Figs. 1-14 and 1-15). The diet, up until about 1920, had been scraps from local restaurants. But, with the diet kitchen in the new colony building, Greenman and Duhring developed a great diversity of

Fig. 1-10 A typical animal room in The Wistar Colony Building. The cages constructed of white pine were suspended from the ceiling on either side. (Courtesy of The Wistar Institute.)

Fig. 1-11   Close-up view of suspended cages (Fig. 1-10) in The Wistar Colony Building. This photo probably was made in the 1940s after pelleted diet came into use.

to investigators in the Wistar. Also, it will be noted that income from rat sales usually covered most of the expenditures for the rat colony except for salaries of personnel. The data after 1928 are less complete.

Parenthetically, it must be emphasized that there has been some confusion about the genetic background of the rats that Wistar sold to users all over the world until 1960 (see below). From its inception, the Wistar apparently maintained a random-bred (heterogeneous) colony,

in addition to those that were maintained by strict brother × sister matings (e.g., the PA, BN, and LEW strains). It was this random-bred stock that served as the commercial colony and presumably gave rise to the LEW strain (Billingham and Silvers, 1959). Whether an albino line(s) other than the one brought from Chicago in 1906 was introduced into the commercial colony is unknown, but this may have happened. It is known that outside breeders were brought in during 1918 to boost production of the breeding colony (Annual Reports of the Director, 1911–1956).

In the early years of the colony, disease problems were common (Annual Reports of the Director, 1911–1956). Some years (excluding 1918 when many died of gossypol toxicity), natural deaths exceeded 10%. Respiratory disease was mentioned frequently and, indeed, the first known report of the disease known today as murine respiratory mycoplasmosis (MRM) was published by the Chicago pathologist, Ludwig Hektoen (Hektoen, 1915–1916), using 328 rats shipped to him from the Wistar in 1915 (Annual Reports of the Director, 1911–1956). In 1931, it was noted that 75% of the rats had middle ear disease and attempts were being made to breed a resistant strain. In 1950 (Farris, 1950) the following was observed:

The incidence of respiratory infections greatly increases when the following conditions exist in an animal colony: (a) the air in the room is stagnant and heavily laden with fumes from the excreta;

Fig. 1-12   "Cage and cage support, one of the older types used in The Wistar Institute Colony" (Greenman and Duhring, 1923). (Courtesy of The Wistar Institute.)

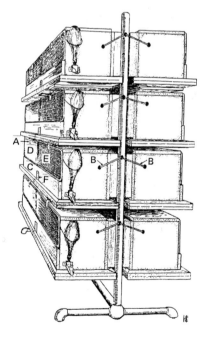

Fig. 1-13   "Cage and cage support, one of the older types used in The Wistar Institute Colony" (Greenman and Duhring, 1923).

Fig. 1-14 The "sterilizing room" in The Wistar Colony Building. The buckets on the floor to the right of the scrub tank had hinged lids and holes in the sides and were used in transporting small groups of rats around the Institute. (Courtesy of The Wistar Institute.)

Fig. 1-15 Hand scrubbing the galvanized wire floors used in cages in The Wistar Colony Building. (Courtesy of The Wistar Institute.)

(b) the debris pans are not changed often enough; (c) the bedding needs changing; (d) the cages need cleaning; (e) the room temperature is unsatisfactory; (f) the atmosphere is dusty.

[These conditions strongly suggest, among other possibilities, what has been demonstrated in recent years,

which is that environmental ammonia exacerbates MRM (Broderson et al., 1976)]. The annual report of 1945 indicates that the disease known as "ringtail" had been eliminated by maintaining the animal rooms at 50% or greater relative humidity (Annual Reports of the Director, 1911–1956).

In 1928, the Wistar Institute was given a 150-acre farm by a relative of General Wistar. The Effingham B. Morris Biological Farm located near Bristol, Pennsylvania, approximately 30 miles from Philadelphia, had facilities for production of rats, amphibians, and opossums (the latter being an interest of Greenman's), and the home of the Director from 1928 until his death. After Greenman's death in 1937, the farm was sold and all rats were returned to the colony building at the Institute (Wistar, 1937).

F. Wistar under Edmund Farris

Edmund J. Farris (1907–1961) became Acting Director in 1937 and was promoted in 1939 to Executive Director, a position he held until 1957. In contrast to the previous sharply focused research program of the Institute, Farris' tenure as Director was marked by diversification, including anatomy, aging, microbiology, cytology, and chemistry. Two editions of the book entitled *The Rat in Laboratory Investigation* were published in 1942 (Griffith and Farris, 1942) and 1949 (Farris and Griffith, 1949). A second text,

TABLE 1-2

RAT PRODUCTION AT THE WISTAR INSTITUTE, 1911–1928[a]

Year	Used in research at Wistar	Supplied to other institutions	Number of institutions supplied	Income from rat sales ($)	Expenditure ($) on rat colony[b]
1911	1,646	0	0	0	645.00
1912	1,737	2,135	—[c]	0	400.00
1913	8,000	3,000	—	0	293.00
1914	12,043	1,197	21	0	919.00
1915	1,091	5,037	21	831.00	1,526.00
1916	1,462	6,025	—	1,218.00	1,021.00
1917	—	~7,000	—	1,348.00	1,213.00
1918	3,313	11,014	—	3,292.00	5,593.00[d]
1919	1,693	5,572	—	5,577.00	5,103.00
1920	1,272	2,517	—	1,594.00	3,815.00
1921	1,450	5,222	59	3,294.00	5,093.00
1922	1,262	2,217	56	2,594.00	2,626.00
1923	1,300	3,093	54	2,819.00	4,809.00
1924	784	1,686	60	2,817.00	3,037.00
1925	522	2,447	55	—	—
1926	727	2,809	~50	4,500.00	—
1927	—	3,443	120	7,231.00	—
1928	787	2,524	121	8,427.00	7,669.00

[a]Compiled from Annual Reports of the Director, The Wistar Institute of Anatomy and Biology, Philadelphia, Pennsylvania.

[b]Amounts shown presumably represented costs of consumable supplies only; salaries (usually slightly more each year than the figures here) and a major cage purchase ($1,596 in 1916) were reported separately.

[c]No data.

[d]Includes cost of mice produced for the U.S. Army to be used "in the diagnosis of pneumonia," no doubt pneumonia secondary to influenza as this was the year of the pandemic.

The Care and Breeding of Laboratory Animals, was published in 1950 (Farris, 1950). During Farris' tenure, the Institute continued to supply rats to other institutions, averaging 4,359 rats to approximately 50 institutions each year. Annual use of rats within the Institute was 6,643, and the average monthly census of rats at the Institute was 2,977 (Annual Reports of the Director, 1911–1956). An effort was made to protect the Institute's interests in commercial production of rats by obtaining, on August 11, 1942 from the U.S. Patent Office, WISTARAT as the registered trademark for its rats (Clause, 1993).

G.   The Modern Era at Wistar

Hilary Koprowski, a virologist, became Director of the Wistar in 1957. Under his leadership, the major thrust of the Institute was redirected toward the viral, degenerative, and neoplastic diseases, fields in which it became recognized as a world leader. Major contributions during this era have included development of improved vaccines against rabies and rubella viruses (Purcell, 1968–1969). The Institute continued to supply the Wistar rats to other laboratories until 1960 when all the breeding stocks and

rights to their perpetuation were sold to a commercial firm (Roosa, 1977).

In 1964, the colony building was renovated and an additional floor placed on top of it (the colony building was stressed for an additional floor when it was built in 1922!). In 1972, the Institute was named one of the first Basic Science Cancer Centers by the National Institutes of Health (Rovera, 1997). Subsequently, in 1975, two additional floors were added to the colony building (using columns of reinforced concrete positioned along the outer walls of the old building) to provide laboratories and animal facilities for a greatly expanded cancer research program. Under the direction of Giovanni Rovera, the Wistar Institute currently has approximately 130 scientists comprising four research programs: molecular genetics, tumor biology, tumor immunology, and structural biology (Rovera, 1997).

III.   BEHAVIORAL RESEARCH

Use of the rat in behavioral research began in the Department of Neurology at Clark University in Worcester, Massachusetts shortly after Donaldson left that department in 1892 for the University of Chicago (Warden, 1930; Conklin, 1939; Munn, 1950).

A. Doctoral Students at Clark University

The first behavioral studies using rats were done by three doctoral students at Clark University: Stewart, Kline, and Small. According to Munn (1950), Colin C. Stewart at Clark University in 1894 began working "on an investigation of the effect of alcohol, diet and barometric changes on animal activity (Stewart, 1898; Logan, 1999). He placed wild gray rats in revolving drums and measured their activity by the number of revolutions of the drums (Fig. 1-16, A). Wild gray rats were difficult to handle and so Stewart changed in 1895 to white rats." The source of the white rats is uncertain, but it seems probable they were given to Stewart by Adolf Meyer who joined the staff of the Worcester State Hospital in 1895 (Miles, 1930). Further contributions from Clark University included Linus W. Kline's work (Kline, 1899) with problem boxes (Fig. 1-16, B) and Willard S. Small's studies (Small, 1900; Small, 1901) with the "Hampton Court Maze" (patterned after the maze at Hampton Court Palace near London, England) (Fig. 1-16, C), in which rats were trained to run through the maze to obtain food (Scott, 1931).

B. Watson and Meyer

After the initial studies using rats at Clark University, the focus of attention in animal behavior quickly shifted to one of Donaldson's doctoral students at the University of Chicago, John Broadus Watson (1878–1958) (Fig. 1-17), who was destined to become one of the great leaders in the field. Munn (Munn, 1950) summarized his contributions made with the rat as follows:

> Although the use of white rats for psychological investigation began at Clark University, research with rats was given its greatest impetus by the Chicago University investigations of Watson and Carr. In 1901 Watson was investigating the relation between myelinization and learning ability in white rats. This work was published in his "Animal Education" in 1903 (Watson, 1903). In 1907 Watson (Watson, 1907) published his very significant work on the role of kinesthetic and organic processes in maze learning by white rats. This work was followed in 1908 by Carr and Watson's study of orientation in the maze (Carr and Watson, 1908). These experiments and others which soon followed from the Chicago Laboratory firmly established the white rat as a subject for psychological investigation.

Watson moved in 1908 to the Johns Hopkins University School of Medicine to become Professor of Experimental and Comparative Psychology, and Director of the new Psychological Laboratory there (Miles, 1930). The next

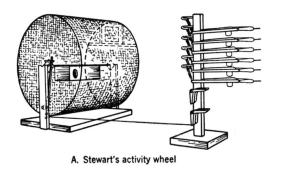 

A. Stewart's activity wheel

B. One of Kline's problem boxes

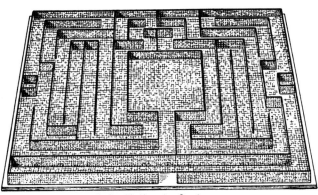

C. Small's Hampton Court maze

Fig. 1-16 "Some historically significant devices used in early research on rat behavior. (**A**) Stewart's activity wheel. The rat lives in the cage at the center of the wheel. Part of the recording mechanism is at the right. (**B**) One of Kline's problem boxes. Escape is effected by burrowing through the sawdust. (**C**) Small's Hampton Court Maze. Food was in center." [From Munn (Munn, 1950), used by permission.]

Fig. 1-17 John Broadus Watson, psychologist of the University of Chicago and The Johns Hopkins University, gave great impetus to early use of the rat in behavioral research. [From Watson (Watson, 1936), used by permission.]

Fig. 1-18 Curt Paul Richter, immediate successor of Watson as Director of the Psychological Laboratory (changed to "Psychobiological Laboratory" in 1919) at The Johns Hopkins University School of Medicine. (Courtesy of the National Library of Medicine, Bethesda, Md; used with permission of Bachrach Studios, Watertown, Mass.)

year, in 1909, Adolf Meyer also moved to the Hopkins as Professor of Psychiatry and Director of the nearby Henry Phipps Psychiatric Clinic. The combined presence of these two great men provided an unusually fertile environment for behavioral research. Watson's work, in particular, was to give further impetus to use of the rat in research. According to Munn (Munn, 1950), the most influential book in bringing about the rapid growth of animal psychology was Watson's *Behavior: An Introduction to Comparative Psychology,* published in 1914. Although Watson believed that animal research would contribute much to the understanding of fundamental processes underlying human behavior, he also believed that animal research had a right to exist for its own sake. He stated, "the range of responses, and the determination of effective stimuli, of habit formation, persistency of habits, interference and reinforcement of habits, must be determined and evaluated in and for themselves, regardless of their generality, or of their bearing upon such laws in other forms, if the phenomena of behavior are ever to be brought within the sphere of scientific control" (Watson, 1914; Watson, 1936).

Watson and Meyer had two graduate students at Hopkins who subsequently proved to be major contributors to the field. Karl Lashley began his work in 1912 and subsequently published on a wide diversity of topics concerning rats, including form and size discrimination and the effects of drugs and experimentally induced brain lesions on learning (Miles, 1930). The second student was Curt Paul Richter (Fig. 1-18), who joined Watson in 1919 shortly after the Psychological Laboratory was transferred

to the Phipps Clinic. Watson resigned from the Hopkins in 1920 as result of a controversy surrounding his divorce. A year later, Curt Richter was appointed director of the laboratory and its name was changed to the "Psychobiological Laboratory" (Miles, 1930).

From 1919 to 1977, Richter conducted a steady stream of research projects on psychobiological phenomena of the rat, including spontaneous activity, biological clocks, physiologic effects of adrenalectomy, self-selection of nutrients, poisoning, stress, and domestication. His laboratory, the direct successor of the laboratory began by Watson, produced approximately 250 research papers and one book of broad fundamental importance to use of the rat in behavioral research (Blass, 1976).

Richter's research, throughout his career, was heavily dependent on a rat colony of 400 to 600 animals he established at Hopkins in 1922 on the third floor of the Phipps Clinic. The foundation stock for this colony was albinos from the Wistar Institute. "A few hooded, tan and black rats from the colony of E. V. McCollum were introduced in 1927." The diet always consisted of graham flour 72.5%, casein 10.0%, butter 5.0%, skim milk powder 10.0%, calcium carbonate 1.5%, and sodium chloride 1.0% (Richter, 1950).

The pioneering work of the group at Clark University, Watson, Richter, and others touched off a major wave of behavioral research using the rat in the 1920s and 1930s. By the 1940s a major share of the basic methodologic approaches used in this field had been developed (Miles, 1930; Warden, 1930; Peterson, 1946; Kreezer, 1949; Richter, 1949; Morse, 1985). With further refinements in methods and widespread use in subsequent years, the laboratory rat has been of inestimable value in behavioral research (Whishaw et al., 1983). Whishaw (Whishaw, 1999) gave a reasoned explanation that compares the behavior of the rat and the mouse (*Mus musculus*): "The brown rat chose complexity as a survival trait whereas the mouse chose simplicity. The rat became social, intelligent, complex, and skilled, all of which are attributes it shares with humans. The answer to the puzzle of how a brain produces mind and action will be found in large part by answering the puzzle of how a brain that weighs just two grams produces the enormous behavioral complexity of the brown rat."

IV. NUTRITION AND BIOCHEMISTRY

The Norway rat probably was first used as an experimental animal in a few nutrition studies conducted in Europe prior to 1850. Its popularity for this purpose continued upward as it was used increasingly in several nutrition laboratories in Europe during the next 50 years (Verzar, 1973).

One of the more notable early studies was an attempt in 1863 to measure the quality of dietary proteins in rats by the English surgeon, Savory (Savory, 1863), who explained, "Rats were chosen as subjects for these experiments because they are omnivorous, and will readily feed on almost any kind of diet. Moreover, from their size, they are very convenient to manage." Many of his rats died on a diet of arrowroot starch, sago, tapioca, and lard or suet. Those on diets of meat with fat and starch survived and experienced some growth. In attempts to perform nitrogen balance studies, the urine volume and nitrogen content were found to be low in animals fed wheat alone but high in those fed meat.

It appears that the majority of early investigators to use rats in nutrition research were interested in dietary proteins and growth (Verzar, 1973). Their choice of the rat as an experimental animal, therefore, takes on special significance as one considers the many possible reasons why rats have excelled in nutrition research over the intervening years. The major reason is not, as commonly thought, that the rat is a good model for man. As pointed out by Hegsted (Hegsted, 1975),

> The very characteristics which have made the rat so attractive for nutritional studies contrast with those found in man. The young rat weighing 50 to 60 gm will grow some 5 to 6 gm/day, a rate

of about 10% of his body weight per day. He continues to grow rapidly until reaching a weight of 200 to 300 gm. This results in a great dilution of the body stores of nutrients and makes it very easy to produce a variety of nutritional deficiencies. In contrast, a child of 4 to 5 years of age also grows about 5 gm/day but weighs 20 kg. The growth rate is negligible compared to body size.

Although probably not fully appreciated by early scientists, these principles were exploited intensively by them.

The advances made through early animal feeding experiments led naturally and increasingly to greater biochemical emphasis. Thus, nutrition and biochemistry are inseparable because of shared information. It may be more revealing, however, to consider that the fields of nutrition and biochemistry and the laboratory rat share much in historical background, including many of the same pioneering personalities.

A. McCollum and the McCollum-Pratt Institute

Elmer Verner McCollum (1879–1967) (Fig. 1-19) has been called "The Abe Lincoln of science" (Snyder and Jones, 1968). As a young man, he was 6 feet tall and weighed only 127 pounds. Furthermore, he was one of the most unusual and distinguished men of early American science–a true pioneer in every sense of the word, including the fact that it was he who pioneered use of the rat in nutrition research.

McCollum was born and reared on a farm in Kansas where he conducted his "first nutrition experiment–collaboratively with Mother." When she discontinued breast feeding him at the age of 7 months, he had developed scurvy and she had treated him successfully by feeding him apples. He received his B.A. degree in 1903 and his M.S. degree in 1904 from the University of Kansas

Fig. 1-19 Elmer Verner McCollum at his desk in Baltimore, 1955. McCollum's pioneering experiments with rats opened a new era of scientific advancement in nutrition and biochemistry. [From McCollum (McCollum, 1964), by permission.]

at Lawrence. He entered graduate school at Yale and obtained a Ph.D. in chemistry under T. B. Johnson in 1906, then remained at Yale an additional year as a postdoctoral fellow in L. B. Mendel's laboratory. For several months in 1906, he worked as an assistant in the laboratory of T. B. Osborne (McCollum, 1964; Rider, 1970).

In 1907, on the advice of Mendel, McCollum accepted a position as instructor in agricultural chemistry at the University of Wisconsin. The salary was $1,200 per year. After a few months' work on a nutrition research project using cows, he became convinced that "the most promising approach to the study of nutritional requirements of animals was through experimenting with small animals fed simplified diets composed of purified nutrients." Despite considerable resistance to this idea from his superiors, McCollum finally obtained their reluctant approval and set about his work by trapping 17 wild rats in the college's horse barn. He next "learned that the rats were too wild, too much alarmed, and too savage to be satisfactory for breeding and experimental work." Undaunted by such problems, he then paid a Chicago pet dealer $6 of his own money for 12 young albino rats. These served as the foundation stock for a colony from which the first experiment was initiated in January 1908. The following year, in 1909, McCollum was joined by a volunteer assistant who took charge of the rat colony, Miss Marguerite Davis, a recent graduate from the University of California at Berkeley. McCollum's idea of experimenting with rats was more than successful; it proved to be a major breakthrough! By the end of his 10-year sojourn in Madison, he and Davis had published numerous papers, including one in 1913 (McCollum and Davis, 1913) announcing the discovery of "fat-soluble A," and another in 1915 (McCollum and Davis, 1915) on the discovery of "water-soluble B." It was probably his early successes, communicated through frequent correspondence with his former mentor, Mendel, that led Osborne and Mendel to establish their rat colony and begin their famous studies in 1909 (McCollum, 1964).

In 1917, McCollum accepted the position of Professor and Chairman of the Department of Chemical Hygiene in the new School of Hygiene and Public Health at The Johns Hopkins University. Fifty female and 10 male rats were forwarded to Baltimore for the establishment of a breeding colony in the new Department. Miss Nina Simmonds was the assistant placed in charge of the colony. Once again, the research program using purified diets was underway, only larger than ever before. During his active career at Hopkins from 1917 to retirement in 1946, approximately 150 papers were published from his laboratory. The topics covered a wide spectrum, including the dietary roles of calcium, phosphorus, potassium, sodium, numerous trace minerals, thiamine, riboflavin, and vitamin E. An illustration of the feeding device he used in his nutrition experiments with rats at Hopkins is shown in Figure 1-20. The most famous work of this period was the discovery of vitamin D as the cause of rickets in 1922 (Day, 1975; McCollum et al., 1922; Proceedings of the Borden Centennial Symposium on Nutrition, 1958). He authored several books, including *A History of Nutrition* (McCollum, 1957) and his autobiography *From Kansas Farmboy to Scientist* (McCollum, 1964). McCollum received many honors during his illustrious career. He served as president of the American Society of Biological Chemists in 1928 and 1929 (Chittenden, 1938).

For several months in 1943, McCollum served as a consultant to the U.S. Lend-Lease Administration in Washington. During some of the committee meetings there he met Mr. John Lee Pratt, a former executive of General Motors and the DuPont Company who had retired to farming at Fredericksburg, Virginia. Their mutual interests in nutrition led to the development of a warm personal friendship. In 1947, Mr. Pratt contributed $500,000 to Hopkins for research on trace elements in nutrition. This led to establishment of the

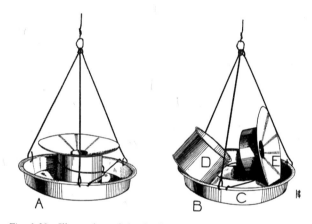

Fig. 1-20 Illustration of the feeding device used by E.V. McCollum, taken from the 1923 edition of Greenman and Duhring's book (Greenman and Duhring, 1923). (**A**) shows the device fully assembled, and (**B**) shows its components. As described by Greenman and Duhring, "It consists of an outer tin pan (**C**)–an ordinary cake pan–8 in. in diameter and $1\frac{1}{4}$ in. deep. In the middle of this pan a plain tin cup (**D**), $3\frac{1}{2}$ in. in diameter and $2\frac{1}{2}$ in deep, is held in place by a wire loop attached at two points to the edge of the pan. This wire loop permits the cup to be removed for cleaning. The cup is covered by a circular disc of tin, 6 in. in diameter, with a central hole $1\frac{1}{8}$ in. in diameter (**E**). This tin disc is slightly convex, dipping toward the central hole, forming a funnel-like top to the cup. A circular flange $1\frac{1}{4}$ in. wide is soldered to the underside of the circular disc. This flange fits accurately into the cup and holds the disc in place. The whole apparatus is suspended by three wires attached to the edge of the pan and brought together in a ring 10 in. above the pan. From this ring a single wire suspends the apparatus from the ceiling of the cage. Food is placed in the cup. The rats may stand on the disc and take food from the cup through the central opening in the disc. Food not eaten tends to fall back into the cup or if scattered over the edge of the disc is caught by the pan below."

McCollum-Pratt Institute on Hopkins' Homewood Campus (McCollum, 1964).

It seems that most assessments of McCollum's life's work emphasize his scientific contributions without recognizing the impact his introduction of the rat had on nutrition research generally. This is a serious mistake, because the new trail he blazed with the rat directly opened a new era of discovery in nutrition and biochemistry, an era of unprecedented advancement spanning the period from 1910 to 1940.

The stock of rats that McCollum began in 1907 at Madison, Wisconsin was maintained until around 1990 in the Department of Biochemistry at The Johns Hopkins University's School of Hygiene and Public Health in Baltimore (Hsueh, 2002; Rider 2002). The initial rats obtained in 1907 were said to be albinos (McCollum, 1964). Other bloodlines were subsequently brought in, including "a dozen young rats from Henry H. Donaldson" (Wistar, 1937) introduced in 1913, and some of Castle's "ruby-eyed, yellow-coated rats," and "pied (black and white) rats" introduced in 1915 (McCollum, 1948; Zucker, 1953). Over the years the colony was random bred and had three coat colors represented: black, white, and albino (Rider, 2002).

B. Osborne and Mendel

Thomas Burr Osborne (1859–1929) (Fig. 1-21) was born in New Haven, Connecticut. Thus, it was only natural that he attended Yale, where he trained in chemistry, receiving

Fig. 1-21 Thomas Burr Osborne, biochemist at the Connecticut Agricultural Experiment Station from 1886 to 1929 and major contributor to understanding the chemistry of vegetable proteins. He and his chief collaborator, Mendel (Fig. 1-22), established the Osborne-Mendel strain of rat. (Courtesy of the National Library of Medicine.)

a B.A. degree in 1881 and a doctorate in 1885. From 1883 to 1886, he was an assistant in analytical chemistry and published several papers on analytic methods for metals and soils. In 1886 he became a member of the scientific staff (and son-in-law of the Director, Dr. S. W. Johnson) of the Connecticut Agricultural Experiment Station, a position he retained until his death. He was a charter member of the American Society of Biological Chemists and served as the fifth president of the Society (Chittenden, 1945). He authored and coauthored over 250 scientific publications during his career. In addition, he was a noted authority on birds and served for many years as director of a New Haven bank (Connecticut Agricultural Experiment Station, 1930; Vickery, 1932).

Osborne's impressive credentials and the innumerable awards he received, however, fail to reveal his real character. The following was written of Osborne by H. B. Vickery and L. B. Mendel, two of his closest associates:

> To those who were privileged to be associated with him in his work he was a rare stimulus, a formidable opponent in argument and an ever genial but just critic. He frequently closed a discussion with the remark that facts were to be found in the laboratory, not in the books. Naturally shy and retiring, the delivery of a public address or of a paper was a severe trial to which he looked forward with trepidation. But among a small group of friends he showed himself as a gifted conversationalist (Connecticut Agricultural Experiment Station, 1930).

In short, he was a brilliant and shy man, a perfectionist in the chemical laboratory where his true genius found its most natural expression.

Osborne's major scientific interest during the early years of his career was the chemistry of vegetable proteins. These studies consumed most of his effort from 1886 to 1909 and culminated in extensive amino acid analyses of many plant proteins, particularly those of major economic importance: gliadin (wheat) and zein (corn). These studies were summarized in his renowned monograph entitled "The Vegetable Proteins" (Osborne, 1909, 1924), first published in 1909, and extensively revised in 1924. This work, which spanned the first half of Osborne's career, set the stage for the remainder of his career and his important collaborative studies with L. B. Mendel.

Lafayette Benedict Mendel (1872–1935) (Fig. 1-22) was a native of Delhi, New York. Like Osborne, he received his undergraduate and graduate training at Yale and spent his entire professional career as a member of the Yale faculty. He received his Ph.D. in Physiological Chemistry at the age of 21 (in 1893) under R. H. Chittenden, one of the most noted early American biochemists. He was first appointed to the Yale faculty in 1892 as Assistant in Physiological Chemistry, and subsequently advanced through the ranks to the position of Professor in 1903. From 1911 to 1935, he held one of the first endowed chairs at Yale, as Sterling Professor of Physiological Chemistry.

Fig. 1-22 Lafayette Benedict Mendel of Yale, a great teacher and early leader in biochemistry and nutrition research. (Courtesy of the National Library of Medicine.)

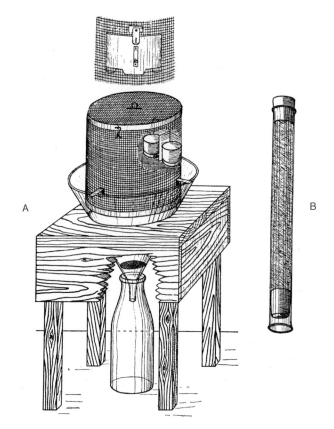

Fig. 1-23 Metabolism cage and food dispenser of Osborne and Mendel, from their publication of 1911, "Feeding Experiments with Isolated Food Substances" (Osborne, 1924). Their description reads as follows. "(A) Sketch of cage used for feeding and collection of urine and feces. Upper figure shows outer view of food-and-water receptacle. (Reduced to one-twelfth natural size). (B) Illustration of tube from which daily ration is discharged during each diet period. (Reduced to one-fourth natural size)."

Mendel was the model university professor. He excelled as an administrator, teacher, and experimentalist and had an enormous capacity for accomplishment. He authored or coauthored a total of 340 publications in his career. Furthermore, he was a most gifted teacher. In the words of one of his students, "He had a wonderful talent for packing thought close, and rendering it portable" (Rose, 1936). During his career, 92 students received their Ph.D. degrees under his mentorship, and 237 graduate students and 96 advanced research fellows received part of their training under his supervision. It is said that Mendel trained more departmental chairmen in biochemistry than any other professor in the United States. Thus, his influence in biochemistry has been profound. He was a charter member of the American Society of Biological Chemists, and at one time or another, held every office of that organization (Chittenden, 1938; Chittenden, 1945; Rose, 1936; Sherman, 1936; Smith, 1956).

In 1909 Osborne ("the shy, retiring servant") and Mendel ("the sociable extrovert scholar") began 20 years of collaborative studies using rats. The work was supported by funds from the Carnegie Institution and yielded an average of eight papers per year from 1911 to 1927 (Smith, 1956). They were convinced that only through the use of purified diets would it be possible to obtain definitive data on the nutritive value of foods. The striking differences in amino acid composition of plant proteins, which had been documented by Osborne, suggested that possible differences might exist in their biological value. The nutritive values of various purified proteins from cereal grains and other plant sources were compared for growth and maintenance in rats. This led to supplementation of "incomplete proteins" with those amino acid(s) limiting each foodstuff's "biologic quality" (e.g., tryptophan and lysine in corn). Casein was found to be a "complete protein," thus paving the way for the use of this protein in modern rat diets. Within a few years, it was possible to list the "essential" and "nonessential" amino acids. These studies necessitated development of a technique (Fig. 1-23, A, B) for feeding individual small animals to permit accurate measurements of food intake (Ferry, 1919; Osborne and Mendel, 1911). They developed one of the first metabolism cages (Fig. 1-23, A) used in animal experiments (Osborne and Mendel, 1911). During the course of these studies, it was noted that coprophagy could be an important variable in use of the rat in nutrition research, and this prompted subsequent investigators in the field to house their rats on wire floors (McCollum, 1957).

Osborne and Mendel's work with rats also provided important contributions in other branches of nutrition. Their independent discovery (Osborne and Mendel,

1913) of vitamin A came only weeks after it was first reported in 1913 by McCollum and Davis (McCollum and Davis, 1913). They were the first to recognize cod liver oil as an important source of vitamin A and to show that it was curative of xerophthalmia in deficient rats and children (Osborne and Mendel, 1914). Along with McCollum's group, they did considerable work with the "water-soluble B" vitamins. In 1918 they showed that phosphorus was an essential nutrient (Osborne and Mendel, 1918). Osborne performed several collaborative studies with H. Gideon Wells of the University of Chicago on the anaphylactogenic properties of plant proteins given to animals (Connecticut Agricultural Experiment Station, 1930).

The rats used by these investigators became known as the Osborne-Mendel strain, although at times they were referred to as the "Yale strain," and were especially noted for their large size (Zucker, 1953). Their origin is uncertain. Presumably, this albino stock was established about 1909 (McCollum, 1964) and maintained as a random-bred colony for all investigators at the Connecticut Agricultural Experiment Station and Mendel's laboratory. It is known to have been maintained as a breeding colony of about 200 animals by Miss Edna L. Ferry (Ferry, 1919) and operated as a closed colony until at least 1935 (Mendel and Hubbell, 1935). Sublines of this stock were established at other institutions around 1915 and used by other famous investigators in the early years of nutrition research, including C. M. McCay and L. A. Maynard at Cornell, and H. C. Sherman, T. F. Zucker, and L. M. Zucker at Columbia University (Heston et al, 1953; Zucker, 1953). A subline established at Vanderbilt University in 1927 by K. E. Mason became known as the "Vanderbilt strain" (Zucker, 1953).

C. Henry C. Sherman

Henry Clapp Sherman (1875–1955) (Fig. 1-24) was born on a farm near Ash Grove, Virginia. He received his B.S. degree from Maryland Agricultural College (now University of Maryland) and remained there two additional years as a graduate student and assistant in chemistry. He then received a fellowship in chemistry from Columbia University, where he was awarded a M.S. degree in 1896 and his Ph.D. in 1897. From his first appointment in 1895 as Fellow in Analytical Chemistry, he rose within the faculty at Columbia to Professor of Organic Analysis in 1907. His title was changed to Professor of Food Chemistry in 1911, and again in 1924 to Mitchell Professor of Chemistry, which remained his title until he retired in 1946. He was executive officer of the Department of Chemistry from 1919 to 1939.

Fig. 1-24 Henry Clapp Sherman of Columbia University, "an analytical chemist, nutritionist, experimental biologist and great humanitarian." He established a subline of the Osborne-Mendel stock of rat, the Sherman strain. (Courtesy of the National Library of Medicine.)

The University awarded him an honorary doctorate in 1929 (Anonymous, 1962; Day, 1957; King, 1975).

According to Day (Day, 1957), Sherman's career progressed through four fairly distinct phases: analytical chemist, nutritionist, experimental biologist, and great humanitarian. The last phase was the natural product of Sherman's real motivations in life: a deep religious faith and a burning desire to achieve long-term improvements in the health of people of all nations. "He was self-effacing because he believed his task transcended self." Thus, he was a quiet, shy person, an introvert whose life was characterized by discipline, courtesy, honesty and kindliness (King, 1975). He served in many capacities dedicated to improving human nutrition, ranging from the local level (e.g., the Association for Improving the Condition of the Poor in New York City) to international (e.g., member of a Red Cross team that evaluated the postrevolution food status of Russia in 1917). He received many honors. He served as president of the American Society of Biological Chemists during 1926 (Chittenden, 1945).

Sherman's career as a scientist was distinguished by its unusual breadth and depth of contribution rather than by an outstanding individual discovery. He and his students often provided the base of knowledge or the analytical techniques that were exploited by many others to advance the field of nutrition. He was by training a chemist, and this background greatly influenced his entire professional career. His early work brought much attention to accurate analysis of foods. During the period from 1895 to 1910, he and his students published about 34 papers on chemical analyses of a wide variety of natural foodstuffs, as well as a few studies dealing with other materials. About 1910, he became interested in digestion and, from 1911 to 1934, published an important series of approximately 50 papers on digestive enzymes, particularly amylases. He and his students clearly established the protein nature of enzymes (Day, 1957; King, 1975).

As the evidence began to accumulate on the importance of dietary deficiencies or imbalances of such things as amino acids, vitamins, iron, and calcium, Sherman began to conduct short-term rat experiments, the first being published in 1919 (Sherman et al., 1919). Within a few years, however, he began emphasizing the two types of lifespan studies that he made most famous: (1) comparisons of experimental diets having definite quantities of ingredients and fed to rats over several generations, and (2) studies of the improvements in longevity from adding known quantities of an ingredient to diets deficient or marginally adequate in that ingredient. Some of these studies using precisely formulated diets continued beyond 40 generations of rats! Sherman also developed many assay methods for food substances, including vitamins A, B_1, B_2, and C. He was a pioneer in the use of statistical methods in the evaluation of biological data. Altogether, Sherman authored or coauthored well over 300 articles. He also published 10 books, many of which went through several editions. One of his collaborators and coauthors was his daughter, Caroline (Mrs. Oscar Lanford, Jr.), who received her Ph.D. in biochemistry at Yale (Anonymous, 1948; Day, 1957; King, 1975).

The first of Sherman's studies to use rats was published in 1919 (Sherman et al., 1919). From that time forward, rats were considered as essential to the accomplishment of his goals as the chemical laboratory itself. When asked or chided about devoting so much effort to animal research, he replied, "These animals are my burettes and balances. They give quantitative answers in chemical terms to many of man's greatest problems!" (King, 1975).

Complete details of the albino rats used by Sherman are not available. However, it appears clear that he obtained his initial breeding stock from Osborne and Mendel shortly before 1919 (Heston et al, 1953; Sherman et al., 1919; 195). Furthermore, the methodology for breeding and maintaining the colony was obtained from the Osborne-Mendel group at Yale and the McCollum group at Hopkins (Sherman et al., 1921). Descendants of this presumably random-bred stock eventually were established in other laboratories.

V. ENDOCRINOLOGY AND REPRODUCTIVE PHYSIOLOGY

The first known experimental use of rats in an endocrinologic study was an assessment of the effects of adrenalectomy in albino rats by the Frenchman, Philipeaux in 1856 (Philipeaux, 1856). This study was prophetic of one of the major uses to be made of rats: experimental ablation of endocrine glands followed by assessment or restoration of altered function(s) as tools in endocrinologic investigation.

Other examples of surgical ablations included studies by J. M. Stotsenburg at the Wistar on the effects of castration (Stotsenberg, 1909) and spaying (Stotsenberg, 1913) on growth of the rat. Frederick S. Hammett (Hammett, 1924) at the Wistar published, in the early 1920s, approximately 20 studies involving thyroidectomies and parathyroidectomies in rats. The most notable achievement in this succession, however, came in 1927 when Phillip E. Smith, an associate of Herbert Evans, succeeded in hypophysectomizing rats by the parapharyngeal approach (Smith, 1927). These techniques, as well as others, served as a pivotal base for the establishment of the rat as the prime animal subject in experimental endocrinology.

A. Long and Evans

Joseph Abraham Long (1879–1953) received his B.S. (1904), M.A. (1905), and Ph.D. (1908) degrees from Harvard. His doctoral dissertation was on maturation of the egg in the mouse (under E. L. Marks, who had previously been advisor on W. E. Castle's doctorate). Immediately upon completion of his doctorate in 1908, he joined the faculty at the University of California in Berkeley as Instructor in Zoology. Through a number of promotions, no doubt largely due to his successful collaborations with Herbert Evans, he became Professor of Embryology in 1939. He became Emeritus Professor of Embryology in 1950. Professor Long was noted as a teacher and as an inventor-master technician in the embryologic research laboratory. For example, he invented the first glass wheel-type sharpener for microtome knives. He did some research in endocrinology, including an evaluation of the effects of hypophysectomy on gestation in the rat (with a student, Richard I. Pencharz). He also contributed to the field of organ culture by inventing several devices to circulate a nutrient medium through excised organs and embryos. In the later years of his career he was an affiliate of the Institute of Experimental Biology of which Herbert Evans was Director (Eakin, 1956; Eakin et al., 1959).

Herbert McLean Evans (1882–1971) was born in Modesto, California, where his father was a well-known physician.

He obtained his B.S. degree at the University of California in 1904 and his M.D. degree at the Johns Hopkins University in 1908. As a medical student, he published seven papers, including one on anatomy of the parathyroid glands with William S. Halsted. He remained at Hopkins as a member of the faculty in the Department of Anatomy from 1908 to 1915, publishing a number of anatomic and embryologic papers, while reaching the rank of Associate Professor. During this interval, as a Research Associate of the Carnegie Institution of Washington, he became interested in vital staining of tissues with benzidine dyes, studies that ultimately led to development of the well-known Evans blue (Bennett, 1975; Evans, 1914). In 1915, Evans became Professor and Chairman of Anatomy at the University of California in Berkeley.

Sometime between 1915 and 1920, these two faculty members at the University of California at Berkeley, Joseph Long of the Department of Zoology and Herbert Evans of the Department of Anatomy (Fig. 1-25), commenced a collaborative research program of major historical and scientific importance, and launched one of the leading strains of rats, Long-Evans (LE). Long had been quietly studying the estrous cycle of the rat for quite a while but had not published any papers. The combined talents of Long and Evans, however, proved catalytic as they authored, during a 3-year period (1920 through 1922), 25 publications dealing with various aspects of reproduction in the rat. These efforts culminated in 1922 with publication of their classic monograph, "The Oestrous Cycle in the Rat and Its Associated Phenomena" (Long and Evans, 1922). Their collaboration was essentially over, but a monumental legacy had been given to endocrinology and reproductive physiology (Bennett, 1975; Friends of H. M. Evans, 1943; Gorski and Whalen, 1963).

Fig. 1-25 Herbert McClean Evans and Joseph Abraham Long, 1945. These two investigators initiated the Long-Evans stock of rats at Berkeley, California, about 1915. (Courtesy of The Bancroft Library, University of California, Berkeley.)

Evans' work with Long on the estrous cycle of the rat yielded more than an understanding of a few basic phenomena of reproduction in the rat. In the words of a later student and colleague of Evans', Leslie Bennett, "It defined clearly for the first time the duration of the estrous cycle of the rat and showed how the changes could be followed clearly and simply by observing the succession of cell types which are thrown off within the vagina. The nature of the vaginal smear was correlated in great detail with histological studies of all parts of the reproductive system." Further, their work "was the launching pad for the chemistry and biology of the anterior lobe hormones" (Bennett, 1975).

Long and Evans (Long and Evans, 1922) gave a brief account of the establishment and maintenance of their rat colony in their monograph, "The Oestrous Cycle in the Rat and Its Associated Phenomena." The rats were "descendants of a cross" made about 1915 "between several white females and a wild gray male caught in Berkeley." The following statement by Leslie Bennett more precisely identifies the origin of the Long-Evans rat. "On many occasions I heard Dr. Evans say that the strain was developed by Dr. Long through crossing the Wistar Institute white rat with the wild Norwegian gray rat which Dr. Long had trapped in the banks of Strawberry Creek as it ran through the Berkeley campus of the University of California" (Bennett, 1978, 1991). The coat colors represented in the colony of Long and Evans (Long and Evans, 1922) were black, gray, and hooded, but there was no "difference to be observed between these different colored rats with respect to the oestrous cycle."

B. Evans and The Institute of Experimental Biology

In 1930, Evans was made Herzstein Professor of Biology and Director of the Institute of Experimental Biology, established on the Berkeley campus for his ever-expanding research group (Fig. 1-26). Evans and his associates (particularly Miriam E. Simpson, Olive Swezy, C. H. Li, and others) went on to characterize the hormones of the anterior pituitary and define the interactions between this gland, the ovary, and the uterus. Many hypophysectomized rats were required in these studies. Initially, they were prepared by Phillip Smith, but he left the department, and this activity was performed by Richard Pencharz, a former student of Long's (Bennett, 1975). By the early 1930s a number of hormones of the anterior pituitary were well known, and a detailed monograph entitled "The Growth and Gonad-Stimulating Hormones of the Anterior Pituitary" was published in 1933. By the end of the 1940s, they had separated the luteotrophic and follicle-stimulating fractions of the anterior pituitary. They next investigated

Fig. 1-26 Professional staff of the Institute of Experimental Biology, posed in 1947 at one of the entrances of the Life Sciences Building (which housed the Institute along with several departments) on the Berkeley campus of the University of California. From left to right: J. A. Long, Hermann Becks, Marjorie Nelson, Alexi A. Knoneff, H.M. Evans, Miriam E. Simpson, Choh H. Li, and Leslie L. Bennett. (Courtesy of The Bancroft Library, University of California, Berkeley.)

adrenocorticotropic and thyrotropic hormones of the anterior pituitary (Friends of H.M. Evans, 1943; Gorski and Whalen, 1963).

During the period from 1922 into the early 1940s, Evans and his colleagues (particularly Katherine Scott Bishop, George Oswald Burr, Gladys A. Emerson, and Oliver H. Emerson) also conducted a major nutrition research program that led to the recognition of vitamin E (Evans and Bishop, 1922), characterization of the muscle and infertility problems in the rat due to vitamin E deficiency, and the chemical isolation of α-tocopherol (Evans, 1962). In 1927, Evans and Burr published their monograph, "The Antisterility Vitamin: Fat Soluble E" (Evans, 1962; Mason, 1977).

Evans also made significant contributions to the husbandry of rats for research purposes (Fig. 1-27). His colony was initially fed table scraps from Berkeley hotels, except for animals on experiment whose diets were supplemented with whole milk and, occasionally, other foods such as raw liver and greens. However, Evans soon identified poor nutrition as a major contributor to biological variability, proceeded to develop a defined diet (adapted from a diet of Osborne and Mendel) for rats in his studies and, as a charter member (in 1928) of the

American Institute of Nutrition, crusaded for the use of standardized rat diets throughout the research community (Raacke, 1983). Other innovations introduced in his rat colony included adequate ventilation, uniform lighting, controlled temperature, easily sanitized cages, and keeping a detailed record on each rat (Raacke, 1983). Crowding of animals in cages was associated with "respiratory inadequacy" (the authors suspect this was murine respiratory mycoplasmosis), necessitating "separation of animals into pairs or at most into fours." Maintenance of room temperature (18° to 20° C) by an electric heater was considered desirable. Experience clearly taught the "inestimable advantages of direct and frequent handling of his stock on the part of the investigator" as opposed to the use of "tongs or other rough devices" for handling rats (Long and Evans, 1922).

Evans retired in 1953, having become one of the most distinguished scientists and scholars of all time. Whereas most individuals are content to make a significant contribution in a single discipline, Evans made highly important contributions in no less than six fields: anatomy, embryology, reproduction, endocrinology, nutrition, and the history of science and medicine (Bennett, 1975; Friends of H.M. Evans, 1943; Gorski and Whalen, 1963)!

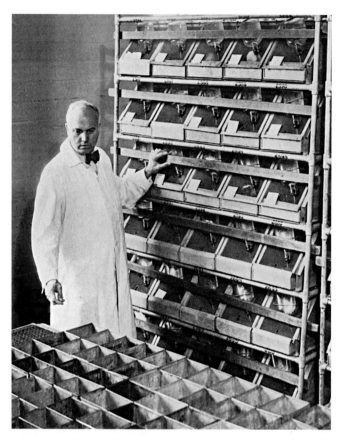

Fig. 1-27 Herbert Evans working in his rat colony, about 1920. (Courtesy of The Bancroft Library, University of California, Berkeley.)

His personal bibliography contains 725 publications, the majority of which concerned studies using rats (Parkes, 1969; Raacke, 1983).

VI. GENETICS

The earliest studies of genetics in the Norway rat were published in Germany by Crampe from 1877 to 1885 (Crampe, 1877, 1883, 1884, 1885). Albino mutants crossed with wild gray (agouti) rats resulted in offspring of three colors (agouti, black, and albino) and two patterns (uniform pigmentation and white spotting). After the rediscovery of Gregor Mendel's laws in 1900, Crampe's data were reexamined by Bateson (Bateson, 1903) and his student Doncaster (Doncaster, 1906) who found the coat color ratios to conform with Mendelian expectations (Castle, 1951; Weisbroth, 1969). Other significant events concerning genetics of the rat came with the establishment of the Wistar Institute and the beginning of an inbred colony by H. D. King in 1909. Over the next few decades, genetics of the rat increasingly gained the attention of one of this nation's first and most distinguished mammalian geneticists, W. E. Castle, who has been called by Morse

the "Father of Mammalian Genetics" (Morse, 1978), and more recently, the "Great-Grandfather of Mammalian Genetics," as he now has at least four generations of intellectual descendants including numerous leaders in the field (Morse, 1981).

A. Castle and The Bussey Institution of Harvard

William Ernest Castle (1867–1962) (Fig. 1-28) was born in 1867 on his father's farm near Alexandria, Ohio. His first college experience was at Denison University in Ohio, a sectarian college emphasizing the classics and ancient languages from which he obtained a B.A. degree. After 3 years of teaching Latin at a college in Kansas, he entered Harvard University and took a second B.A. degree, graduating Phi Beta Kappa in 1893. He then became a laboratory assistant in Zoology at Harvard where he received his M.A. in 1894 and Ph.D. in 1895 (under E. L. Marks). Subsequently, he held brief appointments as Instructor of Zoology at the University of Wisconsin, Knox College, and Harvard. He remained at Harvard, where he was promoted to Assistant Professor in 1903, Professor in 1908, and became Emeritus Professor of Genetics in 1936. Subsequently, he was Research Associate in Mammalian Genetics at the University of California in Berkeley, where

Fig. 1-28 William Ernest Castle of Harvard, first full-time mammalian geneticist in the United States and major contributor to early genetic foundations of the rat. [Courtesy of the National Library of Medicine; see Dunn (Dunn, 1967).]

he continued an active professional career until his death in 1962 (Dunn, 1967).

The rediscovery of Gregor Mendel's principles of inheritance around 1900 gave birth to the new science of genetics. Out of the group of bright young biologists who responded to this new challenge in the United States, there emerged a small nucleus of scientists who were to serve as this nation's first generation of leaders in genetics: T. H. Morgan, C. B. Davenport, W. E. Castle, H. S. Jennings, B. M. Davis, R. A. Emerson, G. H. Schull, and E. M. East. Castle was the first of this illustrious group to devote himself exclusively to genetics, and he had the longest active career–almost 60 years! His first paper on genetics appeared in 1903, and his last (bringing his total bibliography to 246 titles) in 1961 (Dunn, 1967). He published a number of important books, including *Genetics and Eugenics* (Castle, 1924), *The Genetics of Domestic Rabbits* (Castle, 1930), and *Mammalian Genetics* (Castle, 1940).

The early years in Castle's career, 1903 to 1909, were marked by interests in genetics of a wide variety of species, including fruit flies, ants, bees, mice, rats, guinea pigs, rabbits, and many larger domestic animals (Dunn, 1967; Morse, 1985; Snell and Reed, 1993). He was the first to use fruit flies in genetic experiments, and there is good evidence that Castle influenced Thomas Hunt Morgan (Nobel Laureate in 1933) to study fruit flies, and even gave Morgan his collection of inbred flies after he decided to

concentrate his own work on mammalian genetics (Morse, 1985; Snell and Reed, 1993).

After Castle's appointment as Assistant Professor at Harvard in 1903, his career in genetics soared. He was promoted to the rank of Professor in 1908 and elected to membership in the National Academy of Sciences in 1915. One of the key developments in his career came in 1908 with the establishment of the Bussey Institution for Applied Biology. "The Bussey," as it was known, was the graduate school of biology at Harvard and the major base of operations for Castle from 1908 to 1936.

The main building of grey stone (Fig. 1-29) had been built some forty years before in the midst of the fields adjacent to the Arnold Arboretum in that part of Jamaica Plain known as Forest Hills, some ten miles distant from Cambridge. There were in addition greenhouses, barns and outhouses, and a frame dwelling used by the graduate students as a dormitory. Castle moved his rabbits into the basement of the main building, his rats into the large west room on the first floor, and his guinea pigs into an out building known as the pigeon house. Soon a mouse room was established in what had been a greenhouse attached to the front of the main building. Here Castle for the first time had the space and freedom to develop an extensive program in mammalian genetics. He was fortunate in having the help of a foreman who acted also as janitor—Mr. Patch—a former Maine farmer who tilled the fields, grew food for the animals, and invented devices for feeding, watering, and keeping the animals securely in their cages (Fig. 1-30). Here he received as fellow workers the first of the graduate students who took their degrees under his direction (Dunn, 1967).

Fig. 1-29   The Bussey Institution for Applied Biology (The Harvard Graduate School of Biology from 1908 to 1936). It was here that Castle "for the first time had space and freedom to develop an extensive program in mammalian genetics" (Dunn, 1967). (Courtesy of the Harvard Archives.)

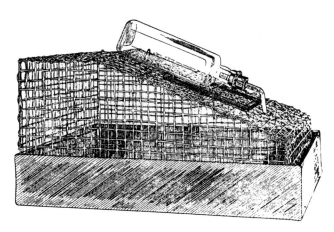

Fig. 1-30 "Rat cage of Professor William E. Castle, of the Bussey Institution, Harvard Institution," from the 1923 edition of the book by Greenman and Duhring (Greenman and Duhring, 1923). The cage almost certainly was designed and built by Mr. Patch, Castle's remarkable janitor, animal technician, and generally versatile assistant. As described by Greenman and Duhring, "The cage consists of a galvanized sheet-metal pan 17 inches long, 15 inches wide and 3 inches deep. In this pan rests the cage proper which is a completely closed box constructed of ½ inch mesh no. 18 galvanized wire cloth. This box is 16 inches long, 14 inches wide, 8½ inches to a height of 4 inches high at the opposite end. For a space of 4 inches wide across the high end, the top is flat, the remaining area of the top slopes from a height of 8½ inches to a height of 4 inches. In the sloping portion of the top is the only entrance to the cage, 5 inches × 5 inches, closed by a wire-cloth door. Upon the door rests the water bottle. The weight of the water bottle is not always sufficient to hold the door closed; an ingenious clip is therefore provided to hold it securely in place. These cages will accommodate five or six adult rats. The cleaning process is reduced to a minimum, as the cage containing the rats may be lifted from the pan which catches the litter falling from the cage."

Genetics as taught at "the Bussey" encompassed both mammalian genetics (under Castle) and plant genetics (under E. M. East). "Both [men] were strong and positive personalities and in temperament and general views they were destined to be often in disagreement." Nevertheless, they spearheaded a cohesive program that by 1937 had prepared 40 doctoral candidates. Of that number, between 1912 and 1937, Castle mentored 21 doctoral students, among them E. C. MacDowell, Clarence Cook Little (Founder of the Jackson Laboratory), Sewall Wright, L. C. Dunn, Clyde E. Keeler, George D. Snell (Nobel Laureate in 1981), Paul B. Sawin, and Sheldon C. Reed (Dunn, 1967; Morse, 1985; Snell and Reed, 1993). The importance of these training activities at "the Bussey" to modern mammalian genetics can be more fully appreciated by studying the Castle "intellectual family tree" described by Morse (Morse, 1978), which documents four generations of his intellectual descendants.

Castle published papers on genetics, particularly inheritance of coat colors, in many animal species. However, the rat proved crucial in answering the question that dominated his early career up to about 1919, that is, "whether Mendelian characters are, as generally assumed, incapable

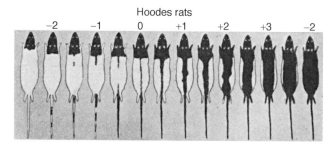

Fig. 1-31 Castle's grading scale for hooded rats, largely worked out by Castle and Phillips in 1914 (Castle, 1951). [From Castle (Castle, 1951).]

of modification by selection." [*Mendelian characters* were referred to by Castle as *unit characters*; they later became known as *genes*.] In order to answer this question, extensive studies were made of coat color inheritance in rats.

Castle, with a succession of collaborators, chiefly J. C. Phillips, proved that a black-and-white spotting pattern, called hooded known to segregate as a "unit-character" in rats, could be modified by selection in both directions toward more black and toward more white and that crossing with wild rats increased the amount of white in the hooded pattern of the spotted animals extracted from the cross. This happened to a conspicuous extent when the line selected toward more white was outcrossed, but there was slight effect of this kind when the dark-selected line was used. This seemed to contradict the dogma of "purity of the gametes," so an extensive study involving some 50,000 rats was carried out between 1907 and 1919. This showed that the character was modifiable, but the crucial test (suggested by Castle's student Sewall Wright) showed that the modifications were due to genes separable in crosses from the hooded gene, which had not itself been changed. The results were shown to be due to the operation of multiple modifying factors (Castle and Phillips, 1914; Dunn, 1967).

Thus, a basic understanding of the hooded pattern in rats had been developed by 1919 (Castle, 1919), laying a firm foundation for further understanding of coat colors in rats (Figs. 1-31, 1-32, 1-33, and 1-34) in subsequent years (Castle, 1951). After about 1920, Castle's main interests shifted to analysis of inheritance of body size in mammals and the construction of chromosome maps for the rabbit, rat, and mouse (Dunn, 1967).

B. Castle's Second Career

The year 1936 was another major turning point in Castle's life. With the completion of the new Biological Laboratories on the main campus, the Bussey closed its doors and Castle went into retirement, both on July 1, 1936. In the words of Dunn (Dunn, 1965), "the rich odors of an old building which housed hundreds or thousands of rats and mice and rabbits and guinea pigs, and the spaciousness (and often the low temperature) of its high-ceilinged rooms failed in the end to compensate for its physical separation from the main center of the University." Despite his disappointment, Castle faced these new

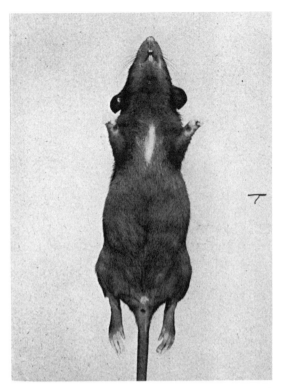

Fig. 1-32 Rats representing three alleles at the *H* locus, from the work of Castle. Fig. 1-32. Irish (hihi) rat of grade +5¾.

Fig. 1-34 Notch (hnhn) rats of grade −3. [From Castle (Castle, 1951), by permission.]

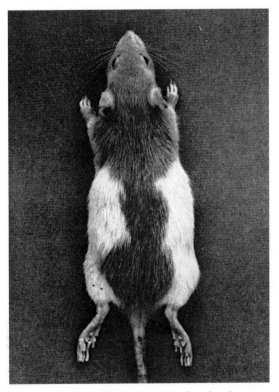

Fig. 1-33 Hooded rat (hh) of grade +2.

realities squarely as expressed in a letter to Dunn in February of 1936. "I am grateful for the long continued opportunities which I have enjoyed for scientific research, which is indeed a privileged status. Now I shall take a vacation and look around for something worth doing while I continue so damned healthy and so am unable to die, as I should." Dunn wrote back, "Why don't you start a second career?" (Dunn, 1936). Castle subsequently became Research Associate in Mammalian Genetics at the University of California in Berkeley and, in essence, launched a second career in genetics that spanned 26 years and produced 55 publications. It was during this interval that he and H. D. King did most of their 10 collaborative linkage studies in the rat (Dunn, 1965). By 1947, he had identified a total of 23 mutations in the rat (Castle, 1947). These and his many other fundamental contributions have served as the foundation for modern genetics of the rat as well as other laboratory animals–indeed for mammalian genetics. Altogether, he published 242 papers, three books and a genetics laboratory manual (Snell and Reed, 1993).

VII.   CANCER RESEARCH INSTITUTES

A.   The Crocker Institute of Cancer Research

A series of key events in the establishment of the rat for research purposes took place in New York City.

In 1912, Mr. George Crocker, a businessman, donated $1,600,000 ("The George Crocker Special Research Fund") to Columbia University for the purpose of studies to find "the cause and cure of cancer" (Eisen, 1954; Wood, 1914–1915). [Elsewhere the figure has been given as $2,500,000 (Dunning, 1951).] Unlike many philanthropists who delight in architectural edifices bearing their own names, Mr. Crocker specifically forbid that any of his contribution be used for a building. For this reason, the cancer program had to be housed in "borrowed space" in the Zoology Department, Schermerhorn Hall, until the University was able to raise the additional support. Eventually, the sum of $40,000 was provided and used to erect, at the corner of Amsterdam Avenue and 116 Street, "the largest possible building, with the sole requirements of the best illumination and the most floor space" for the money. It had a basement and three floors but, because of lack of funds, initially had no partitions. It was opened December 15, 1913 and named "Crocker Research Laboratory" (Fig. 1-35) but, in jest, called "The Workshop of Unrivaled Plainness" (Wood, 1914–1915) or, by Columbia students, "The Canker Fund" (Eisen, 1954). Nevertheless, the name "Institute of Cancer Research" became most widely accepted (Woglom, 1951).

Francis Carter Wood (1869–1951), a pathologist who had authored the text *Clinical Diagnosis* in 1899, was chosen as the first Director of the Crocker Laboratory in 1912.

Wood had major service commitments at surrounding hospitals, particularly St. Luke's located 2 blocks from the Crocker Laboratory. Nevertheless, he and other managers of the Crocker Fund maintained that "In view...of the failure of the medical profession to discover, after two thousand years of observation, the nature and cause of cancer in man, it is considered...inadvisable to expend much energy in investigations along that line" (i.e., clinical studies alone). Thus, Wood was cast throughout most of his 28 years as Director, in the difficult role of conserving the Crocker Fund for "purely experimental purposes" (basic research) against equally strong-willed administrators more interested in clinical medicine. The Fund suffered severe damage in the stock market crash of 1929. Funding for cancer research generally was extremely meager over the next decade, and Wood went into semiretirement in 1936 (his official retirement was announced in 1940), at which time the cancer research program was moved into the College of Physicians and Surgeons, where it became the Department of Cancer Research (Woglom, 1951).

Despite the turbulence of the times, Dr. Wood managed to mount a sizable research program. Early additions to his staff included three other pathologists, Frederick Dabney Bullock (1878–1937) (Fig. 1-36), William H. Woglom (1879–1953), and George Louis Rohdenburg (1883–1967), and a parasitologist, Gary N. Calkins. Bullock and

Fig. 1-35 The Crocker Research Laboratory of Columbia University, built in 1913 at a cost of $40,000. It was here, on the third floor, that inbreeding of six major bloodlines of rats began: August, Copenhagen, Fischer, Marshall, Zimmerman, and Avon. [Courtesy of Ms. Betty Moore, Columbia University; from Wood (Wood, 1914–1915).]

Fig. 1-36 Frederick Dabney Bullock, pathologist and scientist at the Crocker Institute of Cancer Research from 1913 to 1937. (Courtesy of the National Library of Medicine.)

Fig. 1-37 Maynie Rose Curtis, biologist, cancer scientist, and chief developer of the inbred rat colony at the Crocker Institute of Cancer Research. (Courtesy of Dr. W.F. Dunning.)

Rohdenburg held appointments at Lenox Hill Hospital (Eisen, 1954; Moore, 1978).

A major problem confronting cancer research at the time the Crocker Laboratory opened was the need for a reproducible animal model of cancer. It was known that Borrel (Borrel, 1906), a French worker, had reported the production of sarcomas in rats fed tapeworm eggs. Bullock and Rohdenberg attempted to repeat Borrel's work, but failed. At about this time (1916), the group was joined by M. R. Curtis (Dunning, 1977).

Maynie Rose Curtis (1880–1971) (Fig. 1-37) was a native of Mason, Michigan who had obtained undergraduate and advanced degrees (B.S. 1905, M.A. 1908, Ph.D. 1913) in Biology at the University of Michigan, and had subsequently worked for 8 years as a research associate studying poultry reproduction at the Maine Agricultural Experiment Station at Orono. It was her belief that Bullock and Rohdenburg had failed in producing cancers by Borrel's method because their experimentally infected rats simply did not live long enough. Most of their rats had died by 6 months of age. Curtis attributed this to poor housing conditions and diet. The standard practice had been to keep the rats in small boxes and to feed each a piece of dry bread and some vegetable (their sole source of water) every day. Curtis instituted an improved caging system and the practice of dipping the bread in whole milk before it was fed. (These pieces of milk-soaked bread were called *soggies*.) With these improvements, the rats

lived much longer, and Bullock and Curtis reported their first successes in producing sarcomas in 1920 (Bullock and Curtis, 1920). The minimum time for sarcoma induction proved to be 8 months (Dunning, 1977).

Encouraged by their early successes in producing cancers, Bullock and Curtis sought to expand their work. It was soon noted that rats from some vendors developed a higher incidence of sarcomas than those from other sources, when infected with *Taenia* eggs. Curtis then persuaded the group that the best way to find out whether these differences were truly genetic was to develop inbred strains of rats, and further, that these strains should be selected for variations in visible characteristics that might be related (linked?) in some way to the incidence of cancer (Dunning, 1977). In 1919 (Bullock and Curtis, 1930), she purchased a few breeding pairs of rats from each of four local breeders whose names were August, Fischer, Marshall, and Zimmerman. The rats from Marshall were always albinos. Fischer and Zimmerman each had rats that were black nonagouti piebald, but carriers of the albino gene. August had the most varied rats, including some of the pink-eyed dilutes that Thomas Hunt Morgan (Nobel Laureate in Physiology and Medicine, 1931) had given to him (Mr. August) after bringing them back from England some years before (Dunning, 1977). A few rats also were obtained in 1920 from Jacob Rosenstirn (Curtis and Dunning, 1940). This stock he had originally obtained from a breeder in Copenhagen who supplied rats to

TABLE 1-3

INBRED STRAINS OF RATS ORIGINATED AT THE CROCKER RESEARCH INSTITUTE OF COLUMBIA UNIVERSITY

Strain	Date of first mating	Generations of brother × sister matings in 1953[a]
Fischer 230	Sept. 1920	49
Fischer 344	Sept. 1920	51
Zimmerman 61	June 1920	56
Marshall 520	Nov. 1920	51
August 990	Feb. 1921	46
August 7322	Nov. 1925	53
August 28807	Feb. 1936	23
August 35322	Oct. 1942	12
Copenhagen 2331	Aug. 1922	43
A × C 9935	Dec. 1926	38
Avon 34986	Dec. 1941	14

[a]From Heston et al. (1953).

Johannes Fibiger [Nobel Laureate in 1926 for cancer research based on the mistaken claim that gastric cancer in rats was caused by the nematode, *Spiroptera carcinoma*] (Fibiger, 1965; Stolley and Lasky, 1992; Wernstedt, 1965). Many years later, in 1941, an interesting group of rats with "ruby eye" were obtained from a breeder in Avon, Connecticut (Dunning, 1977). These purchases provided the seed stocks for brother × sister matings and the development of several of today's more important inbred strains of rats (Table 1-3). The first litter of pedigreed rats in the Institute was from mating number 344, and this was the beginning of the Fischer 344 (now F344) strain (Dunning, 1977).

Wilhelmina Francis Dunning (1904–1995) (Fig. 1-38) was born in Topsham, Maine. She joined Curtis and Bullock as a part-time student assistant in 1926. She had completed her B.A. degree at the University of Maine and had been admitted as a graduate student with Thomas Hunt Morgan in Zoology at Columbia. She completed the doctoral program at Columbia in Genetics and Cytology in 1932, and elected to join Curtis and Bullock full time. Her major role initially was to analyze the large amount of data that Curtis and Bullock had accumulated on sarcoma induction in the several strains of rats. By 1933, data were available on 3,669 rats in which sarcomas had been induced experimentally. Two major factors emerged as being of paramount importance in sarcoma induction: (1) the relative susceptibility of the strains of rats to the parasite infection, and (2) the longevity of the different strains of rats (Curtis et al., 1933; Dunning and Curtis, 1941). Simultaneously, data were collected on a large number of spontaneous neoplasms from rats of the various lines (Bullock and Curtis, 1930; Curtis et al., 1931), a transplantable lymphosarcoma was described (Curtis and Dunning, 1940), and a number

Fig. 1-38 Wilhelmina Francis Dunning, geneticist, cancer scientist, and chief preserver of the inbred rat strains begun at the Crocker Institute of Cancer Research. (Courtesy of Dr. W.F. Dunning.)

of coat color mutations were documented (Curtis and Dunning, 1937).

By 1940, a great deal of animal research history had been made using the inbred rats at Columbia, which were housed on the third floor of the Crocker Research Laboratory. In a retrospective summary of events for the 20 years from 1920 to 1940 written in 1940 by Dunning (Dunning, 1978), the following recap appeared.

A rat village of some 10,000 individuals was built up, composed of as many divergent strains as was practicable and with the greatest attainable biological uniformity within each line. This colony has been under observation for more than twenty years. The rats were uniformly housed in clean wooden cages on hangers suspended from the ceiling (Fig. 1-39) in a room well suited to such a society. They were uniformly exposed to a constant supply of clean water and an adequate varied diet consisting of bread, cereal, whole milk, fresh vegetables, and raw beef. Pregnant females were removed from breeding cages and reared their young in isolation boxes making it possible to record their birth date and pedigree. Each rat was autopsied and notes were taken on the presence or absence of gross spontaneous or experimentally induced tumors. All gross lesions suggestive of cancer were examined microscopically. To date more than 100,000 rats have been autopsied and there are available a complete pedigree and case history for each of more than 14,000 bearers of experimentally induced and spontaneous tumors.

After the reorganization of the cancer research program at Columbia as the Department of Cancer Research within the Medical School, W. H. Woglom served as Acting

Fig. 1-39 A photograph of the "rat village" at the Crocker Institute of Cancer Research in the late 1930s. Wood cages were kept on wood shelves suspended from the ceiling. (Courtesy of Dr. W.F. Dunning.)

Director from 1940 to 1946 (Eisen, 1954). Subsequently, the program was administered by the Cancer Research Coordinating Committee, with Alfred Gellhorn serving as chairman from 1957 to 1968. Sol Spiegelman became Director in 1968. In 1972, the program was once again reorganized as the Cancer Research Center under Paul A. Marks, Vice President for Health Sciences. It included both the Institute of Cancer Research and the Clinical Cancer Facility in Presbyterian Hospital (Moore, 1978). The Crocker Research Laboratory building (Fig. 1-35) was razed about 1960 to make way for new schools of Law and International Affairs (Moore, 1978).

B. The Detroit Institute for Cancer Research

The reorganization of the Columbia cancer research program, which began in 1936 when the program was made a department in the School of Medicine, continued well into the 1940s. The combined pressures of changing faculty research interests, shortages of laboratory space, the increasing expense of the rat colony, and the general emergency associated with World War II seriously threatened the continued existence of the inbred rat colony. W. C. Rappleye, the Dean of Medicine, made serious efforts to relocate the colony to a farm site, but these attempts proved unsuccessful. Thus, the

environment for continued research at Columbia became unacceptable to Dunning, particularly after Curtis' retirement from that institution in 1941 (Dunning, 1977; Moore, 1978).

Dunning and Curtis soon received an invitation from Rollin Stevens, Radiologist and Chairman of the Cancer Committee of the Wayne County Medical Association, and J. Edgar Norris, Chairman of Pathology and Acting Dean of the College of Medicine at Wayne University (which became Wayne State University in 1956), to move to Detroit and serve as a nucleus of scientists for a new cancer institute. Through the help of Wood, Dunning obtained permission to move the pedigreed rat stocks to Detroit. On Memorial Day of 1941, the entire colony of rats, which had been reduced to approximately 1,000, was transported on the train known as the *Commodore Vanderbilt* from New York City through Canada to Jacksonville, Michigan. From Jacksonville, they were carried by truck approximately 25 miles to the Curtis family farm at Mason, Michigan where Curtis had grown up and which she had now inherited. Work had begun on renovating a swine barn for the rats, but this was incomplete so the rats initially had to be housed in the basement of the farm house. Because of the war, commercial dog food (Carnation's Friskies and Wayne's Lucky brands were used) was not always available, so yellow corn and vegetables grown on the Curtis farm were at times fed to the rats. Initially, the laboratory space given to Dunning at Wayne University was very limited, and it was necessary to maintain the breeding colony at the farm. Dunning spent the weekends at the farm and, as they were needed for experiments, carried a hundred or so rats by car to the laboratory some 84 miles away (Dunning, 1977).

Within a few years, the Southeastern Michigan chapter of the Women's Field Army (now American Cancer Society) had raised approximately $250,000 to be used for a building to house the Detroit Institute for Cancer Research. A 2-story building at 4811 John R (Street) in Detroit, which had been used by the Ford Motor Company's Sales and Service Division, was purchased and renovated. The top floor was used for housing the animal colony and the research laboratories. It was here that Dunning and Curtis (Dunning and Curtis, 1946) demonstrated that the intraperitoneal injection of washed and ground larvae of the tapeworm would produce multiple intraperitoneal sarcomas in rats. Further strain differences in neoplastic diseases of the rats also were elaborated (Dunning and Curtis, 1946b; Dunning et al., 1951).

The Women's Field Army occupied the lower floor of the Institute's building for office space. The frequent (often weekly) visits of up to 200 women in the animal rooms and laboratories proved to be detrimental to the research program, and Dunning once again began searching for a more favorable environment. The Detroit Institute for Cancer Research later became the Michigan Cancer Foundation, which continued a cancer research program at the site (Dunning, 1977; Edward, 1978).

C. The University of Miami and Papanicolaou Cancer Research Institute

For some time the University of Miami had been interested in building a medical school and attracting an affiliated Veterans Administration Hospital to its campus. To be successful, the administration felt that an essential first step was to begin a research program. As part of this effort, the offer of a faculty position was extended to Dunning along with a promise of support for developing laboratory space and renovating building No. 29 (Fig. 1-40) to house the rats on the South Campus (a former naval base for blimps at Perrine, Florida). The sum of $50,000 had been provided for renovations by the Damon Runyon Fund (Dunning, 1977).

After completion of the necessary renovations, in June of 1950, Dunning, accompanied by her close friend and collaborator, Curtis (who sold her farm in Michigan), moved the colony of inbred rats to the University of Miami. The rats were placed in wooden cheese boxes

Fig. 1-40 "Building No. 29," on the South Campus of the University of Miami (Florida). After the inbred rats from Columbia had been moved to Michigan in 1941, they were moved into this building in 1950 and later moved to the Papanicolaou Cancer Research Institute in Miami. (Courtesy of Dr. W.F. Dunning.)

with holes on the sides covered by screen wire, and carried by truck to the Detroit Airport, where they were accepted by a well-known commercial airline and loaded onto an aircraft bound for Miami. Having now moved into the modern era of transportation, about half the rats succumbed on the way. The water bottles arrived separately so that an improvised watering system had to be provided. The Marshall 520 strain was reduced to one pregnant female that later whelped to provide the only breeders for perpetuating that line. The Fischer 230 strain was eventually lost, but this may not have been due to problems in transporting the colony (Dunning, 1977).

At the University of Miami the research program on the larval tapeworm-induced sarcoma in the inbred rats was again very active. Efforts by Dunning and Curtis (Dunning and Curtis, 1953) to isolate the carcinogen of *Taenia sp.* larvae demonstrated that the active agent was associated with the calcareous corpuscles of the parasite. In addition, the inbred rat strains were of crucial importance in studies of transplantable neoplasms (Dunning and Curtis, 1957) and in experimental carcinogenesis (Dunning and Curtis, 1952; Dunning and Curtis, 1952b). Also, many inbred rats (and mice) were produced under contract with the National Cancer Institute. Curtis continued to be active in the research, "working 7 days a week in the laboratory until within 2 days of her death on April 13, 1971 at 91 years of age" (Dunning, 1977).

In 1971, the inbred rat colony was moved to the Papanicolaou Cancer Research Institute. Dunning continued an active research program there until her retirement July 1, 1977. She continued to reside in Miami until her death on January 20, 1995 (Hancock, 1995).

VIII. OTHER STOCKS OF RATS

A. Sprague-Dawley

The precise origin of the so-called Sprague-Dawley rat appears to be lost in uncertainty. The original stock was reportedly established about 1925 by Mr. Robert Worthington Dawley (1897–1949), a physical chemist at the University of Wisconsin (Herrlein et al., 1954). In naming the strain, he simply combined the maiden name (Sprague) of his first wife and his own name to form Sprague-Dawley. He subsequently established in Madison, Wisconsin the commercial firm known as Sprague-Dawley, Inc., dedicated exclusively to the advancement and sale of his rats.

The following excerpt from a letter, dated July 22, 1946, to Mr. Samuel M. Poiley (Poiley, 1953) of the National Institutes of Health from Sprague-Dawley, Inc. summarizes the meager information available on the origin of their rat stock.

> Regarding the origin of the strain as developed by Mr. Dawley, it was started originally with a hybrid hooded male rat of exceptional size and vigor, which genetically was half-white. He was mated to a white female and subsequently to his white female offspring for seven successive generations. The origin of this male is unknown. The original white female was of the Douredoure strain, which probably was from Wistar. After his death, his white offspring were inbred in a number of different lines from which the best ten were combined. Selection was made to retain or acquire characteristics of high lactation, rapid growth, vigor, good temperament, and high resistance to arsenic trioxide.

Mr. Dawley died in 1949. His original company became the ARS/Sprague-Dawley Company, and continues today as Harlan Sprague-Dawley.

Mr. Evan Carl Holtzman, who for many years was an employee of Sprague-Dawley, Inc., left that company in the 1940s and established his own business (also in Madison, Wisconsin) using Sprague-Dawley animals as seed stock. Thus came into being the so-called "Holtzman rat" (Herrlein et al., 1954; Poiley, 1953) used widely in pharmaceutical research.

Many sublines of these stocks exist today and, indeed, Sprague-Dawley–descended animals are among the more popular rats used experimentally. Most of them have been random-bred, perhaps the most distinctive feature of the genealogy of these stocks.

B. Albany Strain

In October 1930, Professor Arthur Knudson of the Department of Biochemistry at the Albany Medical College (Albany, New York) obtained 11 female and five male rats from C. E. Bills of the Mead Johnson Company of Evansville, Indiana. (The origin of Bills' stock is unknown.) Professor Knudson maintained a breeding colony of these animals until 1936, when they were moved to other quarters at the Albany Medical College. Shortly thereafter, they were discovered to have a high incidence of spontaneous mammary tumors and designated the "Albany strain" (Bryan et al., 1938). Consequently, these animals were studied extensively, with comparisons usually being made to the Vanderbilt strain, a subline of Osborne and Mendel's stock that had been established at Vanderbilt by K. E. Mason in 1927 (Wolfe et al., 1938). In 1950, representatives of the Albany stock were transferred from J. M. Wolfe and A. M. Wright at Albany Medical College to the National Institutes of Health, where inbreeding was begun (Festing, 1978).

C. Hunt's Caries-Susceptible and Caries-Resistant Strains

In 1937, at the suggestion of Morris Steggarda of the Carnegie Institution of Washington, H. R. Hunt, C. A. Hoppert, and their associates at Michigan State College (now Michigan State University) initiated a long series of studies of hereditary influences in dental caries of rats. Their strategy and execution of this project are revealed in the following excerpt from a paper by Hunt, Hoppert, and Erwin (Hunt et al., 1944).

> The first objective of the investigation was to determine whether there is an inheritance factor in susceptibility and resistance to dental caries in the albino rat. We therefore undertook to build a caries susceptible line, and a caries resistant line, using phenotypic selection, brother × sister inbreeding, and progeny testing of breeders as genetic techniques to effect this purpose. It was desirable to start with a considerable number of animals from a variety of sources to increase the chance of securing at the outset as large an assortment as possible of genes for susceptibility and resistance. The following rats comprised our first generation: 18 from the colony of the Psychology Department at Michigan State College, 17 from the Nutrition Laboratory of the Home Economics Division, and 84 from the rat laboratory of the Chemistry Department. Selection was fairly stringent at the outset, for only 16 animals (13.4%) chosen as breeders produced young that were themselves mated. These were: susceptibles, 2 males (1 from the Psychology laboratory, 1 from the Chemistry laboratory), 5 females (2 from Psychology, 2 from Chemistry, and 1 from Home Economics); resistants, 3 males (all from the Chemistry colony), and 6 females (1 from Psychology and 5 from Chemistry). The object in selecting and intensively inbreeding by brother × sister matings was not only to build a susceptible and a resistant line, and thereby to demonstrate the inheritance factor, but also to create homozygous susceptible and resistant types which would be crossed, so that genetic segregation could be observed among the grandchildren from such matings, supplying information by which we might determine the number of pairs of genes that are involved.

The studies in Hunt's laboratory were initially based on use of the cariogenic diet of Hoppert, Webber, and Canniff

(Hoppert et al., 1932), which consisted of coarsely ground hulled rice 66%, whole milk powder 30%, alfalfa leaf meal 3%, and NaCl 1%. Although caries-susceptible and caries-resistant strains were developed using this diet (Hunt et al., 1955), the same differences were later demonstrated using a diet containing 57% sucrose (Stewart et al., 1952). Intensive research in this field in recent years by many investigators has led to the belief that the strain differences observed by Hunt and associates are attributable not to Mendelian factors, as originally thought, but to multiple factors, particularly microbial flora of the oral cavity and dietary factors (Larson et al., 1977; Navia, 1977). Nevertheless, these pioneering studies led to the establishment of two inbred strains of rats (Hunt et al., 1955), Hunt's caries-susceptible (CAS) and caries-resistant (CAR).

D. The Oldest Rat Stock and the PAR/Lou Strain

For the sake of completeness, it is necessary to include a discussion of a line of hooded rats that has been bred and maintained since 1856 to feed reptiles at the National Museum of Natural History in Paris, France. In recent years, rats were obtained from that colony and bred at the University of Louvain in Belgium by Herve Bazin to establish an inbred strain, PAR/Lou (Bazin, 1988). Using genetic and biochemical markers, Canzian (Canzian, 1997) showed in 1997 that the PAR/Lou and BN strains are phylogenetically the most divergent of 277 rat strains studied.

IX. GENEALOGY OF MAJOR RAT STOCKS AND STRAINS

Figure 1-41 shows the pedigree relationships of the major established lines of laboratory rats, comprising only about a dozen more or less distinct families, to become the progenitors of hundreds of future strains (inbred, outbred, congenic, consomic, recombinant inbred, and transgenic) (Canzian, 1997; Hedrich, 2000). [The PAR/Lou strain is not included, as it was only very recently discovered and is yet to become established as a laboratory strain (Bazin, 1988). See Heston, et al. (Heston et al., 1953) for additional notes on several of the strains below.]

Donaldson brought an albino stock from the University of Chicago and established it at the Wistar Institute in 1906. These animals gave rise to two lines: the inbred King Albino (now PA) bred by H. D. King and an outbred commercial colony. Other breeding stock from an outside source may have been introduced into the commercial colony in 1918. Rats, referred to by investigators for 50 years as *the Wistar rat,* were disseminated from this

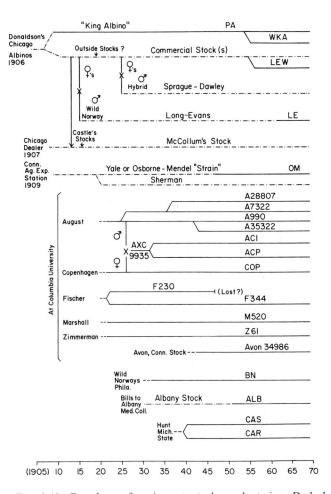

Fig. 1-41 Genealogy of major rat stocks and strains. Dashed lines indicate initiation of b × s matings; solid lines identify inbred strains; lines of dashes and dots designate outbred stocks.

commercial colony all over the world beginning at least as early as 1911 (Table 1-1). Thus, this commercial colony contributed genetically to a major proportion of the established strains of rats. Of the 111 strains listed in the January 1978 issue of *Rat News Letter* (Festing, 1978), at least 45 were known either to have descended directly from this commercial stock or to have otherwise received genetic input from it (Table 1-3). No doubt many others also originated in part or exclusively from commercial Wistars. Regardless, the Wistar gene pool has contributed far more strains of rats than any other single line. In fact, the contribution of the commercial Wistar colony to the total gene pool of all modern strains and stocks of rats may exceed the combined influence of all other gene pools!

McCollum (McCollum, 1964) purchased a stock of albino rats from a Chicago dealer in 1907 and subsequently introduced into his breeding colony Wistars (in 1913) and two strains of animals from Castle (in 1915). It was maintained as an outbred stock (McCollum strain) until around 1990 and may have been discontinued altogether after that time (Hsueh, 2002; Rider, 1970, 2002).

Joseph Long probably began his work on the estrous cycle of the rat with Wistar animals (although the Annual Reports at Wistar do not record a purchase by him) (Purcell, 1978). On the belief that these rats were effete, a number of Wistar females were mated about 1915 to a wild Norway male caught in Berkeley (Bennett, 1975, 1978, 1991). The stock came to be known as "Long-Evans." Muhlbock at the Netherlands Cancer Institute began inbreeding a line of these rats in 1960; representatives were transferred to the National Institutes of Health in 1973 (Festing, 1978).

Robert W. Dawley began the Sprague-Dawley stock about 1925 using females of probable Wistar descent and a hybrid male of unknown origin. There are many Sprague-Dawley–descended stocks in existence today, and practically all are continued as outbred populations (Poiley, 1953).

Possibly as a result of McCollum's early research successes using rats (McCollum, 1964), Osborne and Mendel decided to establish their rat colony at the Connecticut Agricultural Experiment Station about 1909, using stock from an unknown source. Inbreeding of this line was initiated in 1946 by Heston (Festing, 1978). Sherman established his subline of this stock about 1918 (Sherman et al., 1919).

Curtis and Bullock at Columbia University initiated five inbred lines in 1921, using rats from four local vendors (August, Fischer, Marshall, and Zimmerman) and one vendor in Copenhagen, Denmark, from whom rats had been brought to the United States by Rosenstirn (Bullock and Curtis, 1930) (see Section VII.A, above). An accidental mating between an August male with an Irish coat and a COP (originally called Copenhagen 2331) female in 1926 resulted in establishment of the ACI (August × Copenhagen Irish) and ACP (August × Copenhagen Piebald) strains. Inbreeding of a sixth stock, the Avon, began in 1941 with rats obtained from a breeder in Avon, Connecticut. One strain, the F230, was lost in 1945, leaving a total of 11 strains. Wilhelmina F. Dunning joined Curtis and Bullock in 1926 and helped to develop these strains. She continued to maintain most of them at the Papanicolaou Cancer Research Institute in Miami into the 1980s (Dunning, 1977).

The origin of the ALB strain can be traced back only to 1930 when a biochemist, Arthur Knudson, at the Albany Medical College obtained some breeders from C. E. Bills of the Mead Johnson Company of Evansville, Indiana. Inbreeding of this line began in 1950 at the National Institutes of Health (Bryan et al., 1938; Festing, 1978).

The BN strain was developed by Silvers and Billingham from a pen-bred stock maintained by King and Aptekman at Wistar. This stock presumably had been selected out of the wild Norways trapped around Philadelphia about 1930

and inbred for a while by King (Annual Reports of the Director, 1911–1956; Billingham and Silvers, 1959).

Hunt at Michigan State University in 1937 initiated development of a caries-susceptible and a caries-resistant strain of rats, using as foundation stock albino rats from three local departmental (psychology, home economics, and chemistry) colonies. Although these stocks are no longer accepted as being genetically susceptible or resistant (this can now be explained on the basis of oral flora and factors other than genetics), the two strains do represent two branches of a separate and distinct line in the laboratory rat's genealogy (Hunt et al., 1944).

X. GNOTOBIOLOGY: RATS WITHOUT MICROFLORA

In 1885, Pasteur expressed the belief that life without microbes would be impossible, a pronouncement generally credited with stimulating many early scientists to investigate germ-free life (Luckey, 1963; Wostmann, 1996). Approximately 10 years later, around 1895, the first germ-free guinea pig was produced by Nuttal and Thierfelder at the University of Berlin (Wostmann, 1996). Surprisingly, it was another 50 years before germ-free rats were produced in 1946 by the pioneering work of Gustafsson at the University of Lund in Sweden, and Reyniers and colleagues at the Lobund Institute of Notre Dame University in Indiana (Luckey, 1963b). These two scientists established germ-free colonies of rats through hand rearing of cesarean-derived pups and subsequently provided germ-free breeders for foster rearing many of the germ-free rats produced in other laboratories (Newton, 1965).

A. Gustafsson at the University of Lund

Bengt Erik Gustafsson (1916–1986) (Fig. 1-42) was born in Malmo, Sweden. He entered the University of Lund where he completed doctorates in Histology in 1948 and Medicine in 1956 and was a member of the medical faculty from 1948 to 1954. He subsequently served on the Faculty of Dentistry at Malmo, Sweden, then in 1961 became Professor and Chair of the Department of Germfree Research at The Karolinska Institute in Stockholm, where he spent the remainder of his career. Gustafsson's accomplishments in germ-free and dental research earned him a national and international reputation. Among innumerable other prestigious honors and appointments, he served as Secretary General of the Nobel Committee for Physiology and Medicine (1965–1978), and Secretary General of the Swedish Medical Research Council (1963–1977) (Norin, 2002; Pennycuff, 2002).

Fig. 1-42 Bengt Erik Gustafsson of the University of Lund and the Karolinska Institute, pioneer in equipment design and methods for production of germ-free rats. (Courtesy Mr. Tim Pennycuff, Archivist, University of Alabama at Birmingham.)

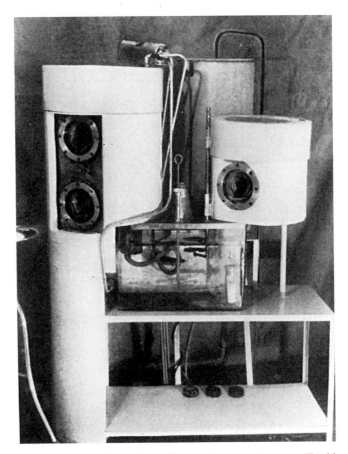

Fig. 1-43 Gustafsson's "germ-free rearing apparatus type 1" with housing capacity of two rats. (From Gustafsson (Gustafsson, 1947); used by permission of S. Karger AG, Basel.)

During his student years, Gustafsson began a career in research as an assistant to Professor Gosta Glimstedt, Head of the Department of Histology at the University of Lund, who in 1932 had derived and maintained guinea pigs germ-free up to 30 days. At Glimstedt's suggestion, Gustafsson began work toward producing germ-free rats in 1939 (Gustafsson, 1948). Inasmuch as there were no published techniques for producing and rearing germ-free mammals born at such an immature stage as rats, the problems that had to be solved were horrendous.

When rearing germ-free experimental animals two main problems must first be solved, one purely technical, *viz.* how the animal shall be isolated in a sterile environment, and the other how, in this environment, the best conditions of life for the animals are to be created. The first problem encompasses a number of minor questions concerning various methods for sterilization of the rearing cage, the air, the food and the vitamins, and how the animal—after caesarean section at full term—shall be brought over into the rearing cage without contamination. The second problem concerns the rearing of the new-born animal without the aid of its mother with everything that implies in the form of food, its composition and administration, cleansing and bedding of the animal, and, finally, the climatological conditions (Gustafsson, 1947).

Accordingly, the first challenge Gustafsson faced was to develop the "apparatus" in which to carry out all of these functions, with the stated aim of "making it simple, easily managed, and cheap." This was accomplished by placing in series a cylindrical chamber for cesarean derivation, a germicidal tank through which pups could be passed (in an enclosed transport container), and a cylindrical chamber (50 cm high $\times$ 27 cm diameter) for rearing two pups continually supplied humidified, heat-sterilized (500° C) air at 37° C (Figs. 1-43 and 1-44). The second challenge was dealt with as follows. Before each experiment, the entire apparatus (with a supply of cow's milk fortified with added minerals and vitamins stored in the rearing cage), was vacuum autoclaved at 120° C. Two female rats would have been mated at the same time 21 days previously. When the first female delivered, the other one was observed for the first sign of labor. When labor began, that female was briefly anesthetized with ether, placed in the surgical chamber, the cervical spinal cord transected and the cesarean performed. Two pups were then placed in a

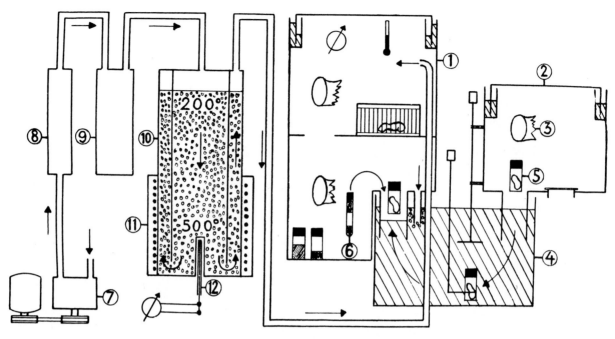

Fig. 1-44 Diagram of germ-free apparatus shown in Fig. 1-43. 1. Rearing cage. 2. Operation cage. 3. Gloves. 4. Sluice tank. 5. Container on its way to rearing cage. 6. Test tube with sterility control. 7. Air compressor. 8. Oil and dust filter. 9. Humidifier. 10. Air sterilization arrangement. 11. Electric element. 12. Automatic Termo-regulator. (From Gustafsson (Gustafsson, 1947); used by permission of S. Karger AG, Basel.)

transfer container and passed through the germicidal trap into the rearing cage. The pups were fed every 4 hours beginning with 0.8 g of milk, increasing to 8.0 g by weaning at 21 days of age. With this apparatus some rats were reared to 28 days of age free of detectable organisms. Beyond that age this initial apparatus (referred to as "type 1") was simply too small to maintain the two pups any longer (Gustafsson, 1947).

Faced with the serious space limitations of the type 1 apparatus, Gustafsson (Gustafsson, 1948) developed a lightweight "type 2 apparatus" capable of housing 10 rats. This was a rectangular tanklike structure measuring 90 cm long × 50 cm high × 45 cm wide and constructed primarily of 2-mm thick sheets stainless steel. He next developed a larger "type 3 apparatus" capable of housing 20 to 30 rats (Figs. 1-45 and 1-46), essentially identical to the type 2 apparatus in design (Gustafsson, 1959). In both types, the top was made of 7-mm thick plate glass fitted to a rim of the steel apparatus with a rubber gasket so as to serve as a lid when being supplied before sterilization prior to use or as a window while housing germ-free rats. Lighting was provided by electric lights mounted above the window. A pair of Neoprene gloves was mounted on one side of the apparatus and a single glove or pair of gloves was mounted on the opposite side for working inside. An air sterilization device and a pass-through germicidal trap were mounted on the ends. The entire apparatus was steam sterilized in a specially built autoclave prior to use. A separate surgical apparatus was used for performing

cesareans, with pups being placed in a transport container and passed through the germicidal trap into the larger apparatus for hand rearing.

Extensive studies were required to understand the composition of rat milk, develop a suitable substitute milk, provide for sterilization of the milk, and work out the details for hand feeding baby pups (Gustafsson, 1948). Having accomplished all these enormous logistical and technical hurdles, Gustafsson successfully weaned germ-free rats in 1946 and achieved normal growth of germ-free rats on an artificial diet by 1947 (Gustafsson, 1947; Luckey, 1963c). The rats were of the Long-Evans strain (Newton, 1965). Some of Gustafsson's original lightweight stainless steel isolators are still in use at the Karolinska Institute (Norin, 2002).

B.   Reyniers and The Lobund Institute

James Arthur Reyniers (1908–1967) (Fig. 1-47) was born in Mishawaka, Indiana on April 16, 1908. After completion of grade school at DePaul Academy of Chicago, he entered Notre Dame University where he earned two degrees in Bacteriology: a B.S. in 1930 and a Masters in 1931. He joined the faculty at Notre Dame as Instructor of Bacteriology in 1931 and subsequently rose through the ranks to become Research Professor of Bacteriology in 1945. He was Director of the Laboratories of Bacteriology from 1931 to 1953. In 1952 he received an honorary

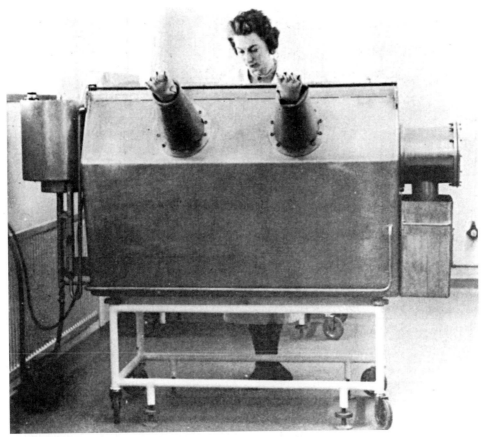

Fig. 1-45 Gustafsson's large germ-free tank or germ-free rearing apparatus type 3, for housing 20 to 30 rats. (From Gustafsson (Gustafsson, 1959), copyright 1959; used by permission New York Academy of Sciences.)

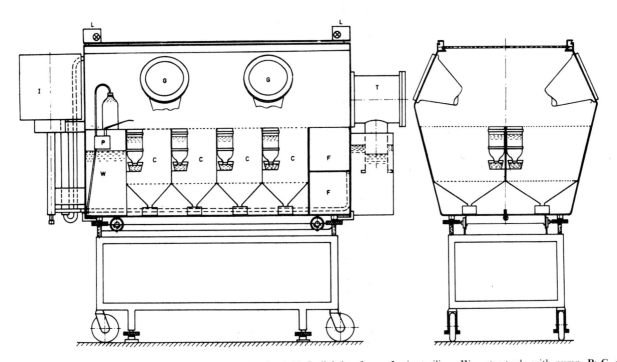

Fig. 1-46 Diagram of large germ-free apparatus shown in Fig. 1-45. **L,** lighting frame; **I,** air sterilizer; **W,** water tank, with pump, **P**; **C,** cages; **G,** glove ports; **F,** food canisters; **T,** food autoclave and transfer unit. (From Gustafsson (Gustafsson, 1959), copyright 1959; used by permission New York Academy of Sciences.)

Fig. 1-47 James Arthur Reyniers, a bacteriologist who was Director of the Laboratories of Bacteriology at the University of Notre Dame (LOBUND) and leader of the germ-free research program at that institution from 1931 to 1953. (Courtesy of University of Notre Dame Archives.)

doctorate from the College of St. Thomas in St. Paul, Minnesota (Lamb, 2002).

As an undergraduate with high ambitions, Reyniers had begun germ-free animal work at Notre Dame in 1928. From a very meager laboratory in Science Hall (built ca. 1890) at the time of his appointment to the faculty, the program expanded so that by 1936 it was moved into 23 labs designed by Reyniers in the new Biology Building. By 1947 it was necessary to add a machine shop, an animal facility mainly for monkeys and rats, and the Reyniers Germfree Life Building (Fig. 1-48). The Reyniers Germfree Life Building was dedicated on June 21, 1950, but the official program read "Program for the Dedication of LOBUND Institute for Research in the Life Sciences." With Reyniers as Director of the Laboratories of Bacteriology at the University of Notre Dame beginning in 1931, the program over the years had become known by the acronym LOBUND. In 1958 Reyniers left LOBUND for other pursuits. Morris Pollard became Director in 1961 (Lamb, 2002; Suckow, 2002).

Reyniers was first and foremost a bacteriologist—actually a bacteriologist with limited scientific training

but unusually gifted as an entrepreneur and fundraiser. According to P. C. Trexler (Fig. 1-49), one of his closest associates beginning in 1931, Reyniers' chief focus was investigation of bacterial variation, and germ-free animals were but one approach to the problem (Trexler, 1985). Trexler characterized the research of Reyniers' career as having three components: "(a) the development of single-cell isolation techniques to make certain that cultures started from a single cell, (b) mechanical methods for increasing the accuracy and reducing the labor required to make viable counts, and (c) germfree animals to make certain that no contamination occurred during the study of bacteria associated with a host" (Trexler, 1985).

Much of the success of LOBUND was the result of Reyniers' very able assistants, Trexler and Ervin (Fig. 1-49), especially the former. Philip Charles Trexler was born in Paradise, California on July 30, 1911. Supported by a scholarship at Notre Dame, he received a B.S. in Biological Sciences (cum laude) in 1934 and an M.S. in Bacteriology (magna cum laude) in 1936. As Reynier's student helper initially and subsequent graduate student, he was involved in the germ-free work at Notre Dame almost from its beginning (Trexler, 2002). He was Associate Director of LOBUND from 1936 to 1961. From Instructor in 1935, he received a number of promotions to become Professor of Bacteriology in 1961. Trexler was clearly the mastermind responsible for both designing and maintaining the germ-free equipment at LOBUND, from the large autoclave-like models to the plastic film types (Luckey, 1963b; Reyniers, 1943; Reyniers, 1959; Trexler, 1959, 1985). After his departure from LOBUND in 1961, he held appointments at many other prestigious institutions, including The Royal College of Veterinary Medicine in London, and pursued a variety of commercial interests. In 1984 he was awarded an honorary Doctor of Science degree by Notre Dame University. His career accomplishments include 76 publications, 18 U.S. patents and nine British patents (Trexler, 2002).

Under the leadership of Reyniers, and with the contributions of the many scientists he attracted to LOBUND over the years (P. C. Trexler, R. F. Ervin, A. W. Phillips, J. F. Reback, M. Wagner, B. A. Teah, J. R. Pleasants, H. A. Gordon, T. D. Luckey, and others), the LOBUND Institute became established as the world's leader in germ-free research. Like Gustafsson, Reyniers' group first reported success in deriving and rearing germ-free rats in 1946 (Reyniers et al., 1946). In the ensuing years, several strains of rats were derived and maintained germ free at LOBUND, including Wistar (now referred to as Lobund Wistar, an inbred Wistar strain that is a model of prostate cancer), Fischer, Holtzman, Long-Evans, Sprague-Dawley, and Buffalo (Pleasants, 1959; Wostmann, 1996a).

Pleasants (Pleasants, 1965) has divided the history of LOBUND from 1928 to 1965 into four periods.

Fig. 1-48 The Reyniers Germfree Life Building dedicated on June 21, 1950 as the LOBUND Institute for Research in the Life Sciences at the University Notre Dame. (Courtesy of University of Notre Dame Archives.)

LOBUND's experiences during the first period, 1928 to 1944, which brought success in producing germ-free guinea pigs, monkeys, and chickens, were strikingly similar to those of Gustafsson (Gustafsson, 1947, 1948).

> During this period, air-tight containers had to be developed which could be sterilized on the inside and then used to maintain and manipulate the germfree animals. Air filtration methods were devised and tested. Food sterilization techniques had to be developed which did not destroy the nutritional value of the diet. Procedures for detecting leaks and contaminations were worked out, as well as the means for preventing them. A practical technique for cesarean delivery of young mammals into the germfree system, and a method of surface sterilization of chicken and turkey eggs were developed.

However, whereas Gustafsson's "apparatus" featured simplicity of design and operation, Reyniers' tended toward heavy, complicated, expensive devices generally resembling cylindrical autoclaves (Fig. 1-50). As described in detail in 1943 (Reyniers, 1943), the complete germ-free system actually consisted of six complicated

devices: rearing cages, storage cages, transfer cage, examination cage, surgical cage, and control panel.

Similarly, during the second period, 1945 to 1954, which brought success in producing germ-free rats and mice, the Reyniers group encountered the same impediments as Gustafsson (Gustafsson, 1947, 1948, 1959):

> In the case of rats and mice, which are very immature at birth, it became necessary for Lobund to pioneer in the development of artificial milk formulas and hand-feeding methods which could be used to rear these germfree mammals from caesarean birth to weaning, since this was the only means by which such species could be obtained germfree. (Pleasants, 1965)

The third period, 1955 to 1958, brought relief from cumbersome, expensive equipment through an innovation that virtually revolutionized the field, as explained by Pleasants (Pleasants, 1965):

> While effective means for transporting germfree animals were being developed, a new type of isolator (Fig. 1-51) was developed by P. C. Trexler, who joined the Lobund staff in 1933 and had done

Fig. 1-49 From left to right: Philip Charles Trexler ("Trex"–Associate Director); James Arthur Reyniers ("Art"–Director); and Robert F. Ervin ("Bob"–Associate Director for Administration), the leadership team of The Lobund Institute during its years of grandeur in gnotobiology. (Photo made in 1949; Courtesy of P.C. Trexler).

Fig. 1-50 Germ-free isolators of the Reyniers type in use at Lobund Institute. (Courtesy of University of Notre Dame Archives.)

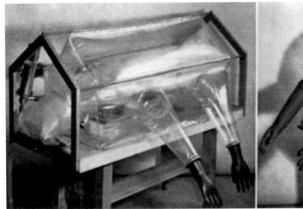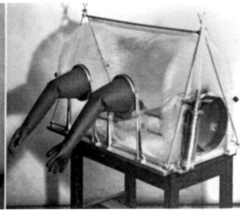

Fig. 1-51 Early examples of flexible film isolators of the type introduced by Philip Trexler. (Suckow, 2002; used by permission of American Soc. Microbiol.).

much of the developmental work on equipment throughout Lobund's history. His innovation was an inexpensive and versatile plastic isolator which could be sterilized by a spray of germicide. This made it possible for many laboratories to venture into germfree research without a large capital investment.

During the fourth period, 1959 to 1965, considerable work was done to develop chemically defined, antigen-free diets for germ-free rodents (Pleasants, 1965). These studies contributed significantly to understanding of the nutrient requirements of rats (Wostmann, 1996b, 1996c).

XI. SECURING THE FOUNDATIONS

The early history of the laboratory rat is replete with accounts of poor reproducibility of research results related to numerous factors, including epizootics of infectious diseases (Greenman and Duhring, 1923, 1931; Hektoen, 1915–1916; Long and Evans, 1922). A prime example is the proceedings of a symposium held at Columbia University on January 31, 1952 (Heston et al., 1953). At that meeting, 32 of the nation's leading rat authorities came together to share their frustrations and experiences with rat quality. Among rat colonies known to them, there were only two "protected colonies" (i.e., partially or completely pathogen free): Nelson's colony at Rockefeller Institute (thought to be free of mycoplasmas) and Reyniers' germ-free rats at LOBUND. One of the symposium participants, Gladys Emerson, observed that, "Apparently, variability in the response of experimental animals cannot be ascribed to a single factor but probably encompasses genetics, diet, general health and environmental conditions" (Heston et al., 1953). The evidence was clear: the prevailing husbandry methods among users and suppliers of rats for research in 1952 were still in a primitive state (Fig. 1-52) (Foster, 1959; Herrlein et al., 1954)!

Significant progress toward resolution of these problems came about through: (1) a major shift in funding of health-related research from private to federal sources coupled with dramatically increased levels of funding; (2) expansion of the number of scientists and scientific organizations; and (3) advances in a wide range of fields, particularly gnotobiology, husbandry, diagnostics, nutrition, and genetics. Thus, a convergence of multiple factors made the period beginning in the 1950s a time of securing the earlier foundations for use of the laboratory rat in research, through achieving a state of definable, quality rat health and an adequate supply of genetically standardized rat models to meet escalating demands.

A. Funding: Role of The National Institutes of Health

Development and use of the rat for research purposes has tracked to a great extent with the public's level of appreciation of research. Harden (Harden, 1986) characterized the public attitude toward research in the United States during the 1920s following World War I (WWI) as an environment of "...confidence in science, scientific medicine, and public health; a positivistic outlook within the scientific community; a rapid rise in college enrollments and in the number of scientists; pressure on the resources of foundations and philanthropists; and increasing emphasis on applied research." In fact, the public's high expectations in response to discoveries early in the 20th century prompted L. B. Mendel of Yale (see Section IV.B, above) to comment in 1927 (Harden, 1986), "Thus it has come about that science has been exposed upon an elevated pedestal to the fullest view of an over-awed public."

The case for federal support of research was clearly emerging but quickly became stymied by the great depression and political wrangling at the national level. Congress finally passed legislation establishing the

Fig. 1-52   Typical mid-20th century commercial rodent breeding facility. Charles River Breeding Laboratories, Wilmington, Massachusetts, 1952. Cooling and ventilation accomplished by open windows and central exhaust fan. Heat provided by fin-type radiator around perimeter walls. (From Foster (Foster, 1980); used by permission American Association of Laboratory Animal Science)

National Institutes of Health (NIH) on April Fool's Day 1930 (Harden, 1986). The first institute, The National Cancer Institute, was authorized in 1937, followed by the establishment of several other institutes and a Research Grants Office at the NIH. However, little happened until after World War II (WWII) when the public's attitudes swung back to those of the post WWI years. The results are reflected in the following representative total budgets for the NIH: 1940: $707,000; 1945: $2.8 million; 1950: $52 million; 1955: $81 million; 1960: $430 million; 1965: $1.1 billion; and 1970: $1.5 billion. Similar dramatic increases in health-related research occurred concurrently in many other federal agencies (NIH Factbook, 1976).

The tremendous upsurge of research using rats as well as other animals in the United States following WWII demanded that measures be taken to improve: (1) laboratory animal health, (2) support and preservation of valuable animal stocks, (3) the nomenclature and genetic standards for various strains and stocks of animals, (4) the dissemination and exchange of information on laboratory animal models, and (5) many other aspects of assuring adequate supplies of high-quality animals for modern research. Numerous organizations–scientific, commercial, and governmental–arose to meet these needs. Historical accounts of some of them were published recently (McPherson and Mattingly, 1999).

B.   Health: Eliminating Pathogens and Nutrients as Variables

The work of Gustafsson and Reyniers in developing the basic methodologies for completely eliminating the microflora and maintaining germ-free breeding colonies of rats were contributions of monumental proportions for many reasons, but largely due to the simple fact that this eliminated pathogens (with possible viral exceptions) and forced meticulous attention to nutritional requirements from birth through senescence (Wostmann, 1996b, 1996c). The importance of their work can be seen in the cascade of events that followed:

1. Germ-free or gnotobiotic rats were available for the first time for stocking breeding colonies for scientists, commercial firms, and governmental agencies, and the cost was reasonable. A germ-free rat purchased from LOBUND in 1951 cost only $20.00 (Suckow, 2002).
2. Philip Trexler's innovation of the inexpensive plastic isolator, generally referred to as *the Trexler isolator*, (Fig. 1-51) made germ-free technology available for widespread use by scientists and suppliers of rats (Trexler, 1959; Trexler and Reynolds, 1957).
3. In June 1960, Trexler held a workshop at LOBUND "to teach lab animal suppliers how to free their stocks of infectious disease agents by utilizing gnotobiotic techniques." It was attended by representatives of ten commercial breeders and one institutional breeder (Trexler and Orcutt, 1999). [More recently, Trexler's methodology was further improved when Alcide™ replaced peracetic acid for sterilizing the interior of plastic isolators (Orcutt et al., 1981).]

4. Trexler isolators fabricated in many sizes and shapes (Fig. 1-53) were used by rat producers for maintaining germ-free rats as well as for exclusion of pathogens (Cumming and Baker, 1968; Hem, 1994; Lattuada et al., 1981), greatly enhancing the availability of pathogen-free rats for research.

5. Programs for pathogen exclusion from rodent facilities (barrier facilities) emerged and became commonplace (Foster, 1959). In retrospect, Henry Foster (Foster, 1980), president of one of the largest commercial firms producing rats, in 1980 summed up No. 1 to No. 5 as follows:

The availability of gnotobiotes coupled with the major break-through and development of the flexible film plastic isolator by P.C. Trexler and the construction of the barrier facility for large-scale rearing of healthy animals, in my opinion, was a quantum jump in the field of laboratory animal production.

6. The innovation by Lisbeth (Liz) M. Kraft (Fig. 1-54) of pathogen exclusion at the cage level, by use of "filter caps" on cages

(Figs. 1-55, 1-56, and 1-57), together with consistently following simple rules to prevent pathogen contamination (Kraft, 1958; Kraft et al., 1964), has been characterized as "a major milestone" and "the mainstay of efforts to maintain rodents free of adventitious pathogens in the research environment" (Hessler, 1999).

Lisbeth M. Kraft (1920–2002) was born in Vienna, Austria. She immigrated with her family to the United States in 1923 and was naturalized as a citizen on March 11, 1929. She attended public schools in Schenectady, New York and subsequently earned degrees at Cornell University: B.S. in Botany in 1942 and D.V.M. in 1945. As a veterinarian, her career was focused on infectious disease research using laboratory animals. During her career, she held appointments at several prestigious institutions (e.g., Harvard, Yale, University of Pennsylvania, and Ames Research Center, National Aeronautics and Space

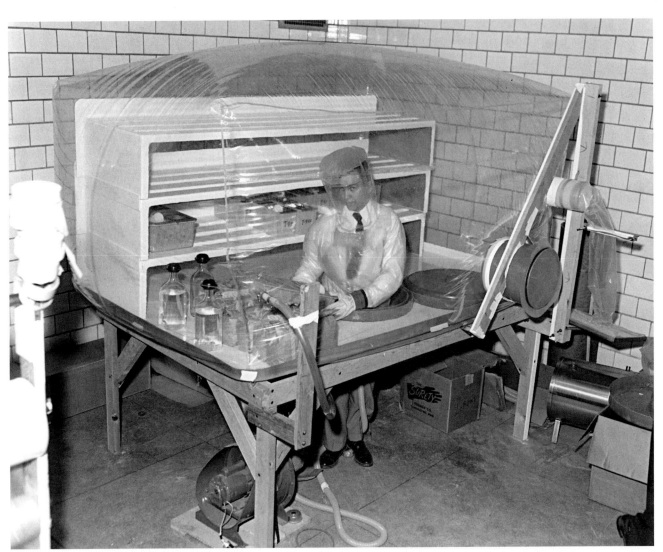

Fig. 1-53 Phil Trexler inside a "half-suit isolator" at Lobund Institute in 1959, demonstrating that very large flexible film isolators could be constructed and used for maintaining germ-free or pathogen-free rodent colonies. The "half-suit" referred to a complete upper body suit of plastic located at the middle of the isolator where the operator worked after entering the suit from beneath. (Courtesy of University of Notre Dame Archives.)

Fig. 1-54 Lisbeth M. Kraft, who introduced and promoted the filter cap cage system for pathogen exclusion. (Courtesy of Lisbeth M. Kraft.)

Agency, Moffett Field, California). She authored and coauthored nearly 60 publications, consistently championing at the national level the cause of improving laboratory animal health. Her painstaking investigations into practical methods for controlling "infant diarrheas" of mice, "lethal intestinal virus of infant mice," and "epizootic diarrhea of infant mice," later identified, respectively, as mouse hepatitis virus and mouse rotavirus (Kraft, 1982), provided

an innovative new methodology with general applicability to preventing the transmission of infectious diseases in laboratory rodents (Hessler, 1999). For these contributions, she received the Griffin Award from The American Association of Laboratory Animal Science in 1972 and the Charles River Prize from the American Veterinary Medical Association in 1981. From 1960 through 1967 she served in a wide variety of leadership roles in the American Association of Laboratory Animal Science. She achieved diplomate status in the American College of Laboratory Animal Medicine in 1961 and served as its President during 1966 and on its Board of Directors during 1965 to 1970 (Anonymous, 2003). At the time of her death on November 29, 2002, she resided in Mountain View, California.

7. The need for rodent pathogen prevention, surveillance, and diagnostic programs in research institutions and breeding facilities began to be recognized, in large measure through the efforts of Wallace P. Rowe (Allen, 1999; Heston et al, 1953; Holdenried, 1966; Kilham, 1966; Rowe et al., 1962).

8. The Animal Resources Branch (ARB) of the Division of Research Facilities and Resources at the NIH was established in 1962 (Allen, 1999; Whitehair, 1999), a time when very little was known about infectious or noninfectious diseases of rats (Heston et al, 1953; King, 1975). The ARB, through its Laboratory Animal Sciences Program, subsequently funded institutional research animal resources that included diagnostic and animal model development laboratories. Activities of these resources, as well as those of other federal agencies, commercial laboratories, and breeders, initiated a major effort leading to the elimination of rodent diseases as research variables (Allen, 1999; Kilham, 1966).

Fig. 1-55   Early design of cage filter caps by Kraft. Views are: **front (A), underside (B),** and **rear (C).** *Arrows* point to support clips. (Kraft et al., 1964); used by permission of American Association of Laboratory Animal Science.)

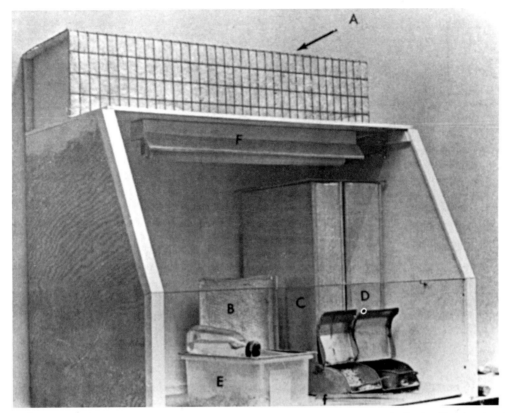

Fig. 1-56 Plywood transfer hood of Kraft for changing filter cap cages, described as follows: "The intake filter housing (**A**) encloses a fan. The sloping portion of the hood front is made of plate glass beneath which the hood is open. **B**, filter cap; **C, D**, feed hoppers; **E**, regular mouse cage; **F**, fluorescent lamp. Forceps (**f**) rest in a protective tube." (Kraft et al., 1964); used by permission of American Association of Laboratory Animal Science.)

Fig. 1-57 Cages demonstrating use of Kraft's "filter caps." (Kraft et al., 1964); used by permission of American Association of Laboratory Animal Science.)

C.   Genetics: Repositories and Standards

Beginning in the 1940s, the Laboratory Aids Branch of the NIH, under the direction of Walter E. Heston, began establishing breeding colonies of selected strains of rodents at the NIH in Bethesda, Maryland (Heston et al., 1953). At the time, George E. Jay, Jr. was geneticist and Samuel E. Poiley was animal husbandman. By 1953, seven inbred strains of rats had been established: A × C 9935, Buffalo, F344, Marshall 520, O'Grady, Osborne-Mendel, and Wistar (Herrlein et al., 1954; Jay, 1953). Random-bred colonies at the NIH at this time included rats of Sprague-Dawley, Holtzman-Rolfsmeyer, and NIH Black stocks (Poiley, 1953). Animal breeding activities at the NIH continued to expand. The name of the unit in charge of the animal colonies was changed to the Veterinary Resources Branch (Division of Research Services). Robert Whitney became chief of this branch. Carl T. Hanson joined the staff of the Veterinary Resources Branch in 1964 and subsequently established within the branch "The NIH Rodent Repository." This repository was "a collection of strains and stocks of rodents and lagomorphs of known genetic characteristics - maintained to provide a defined source of breeders." It included "foundation colonies of inbred and congenic strains and nucleus colonies of outbred and mutant stocks of rodents." The NIH Rodent Repository served as both a national and international resource by virtue of its designation as a World Health Organization (WHO) Collaborating Center for Defined Laboratory Animals. The Repository supplied breeding stocks for scientific purposes to universities, NIH contractors, and commercial breeders (Chase, 1975).

The Laboratory Animals Centre (LAC) in the United Kingdom was founded (as the Laboratory Animals Bureau) in 1947 with the general aim of improving the quality and availability of laboratory animals used in the U.K. Although financed by the Medical Research Council, the Centre was run as a national service to all biomedical research workers in the U.K. and was designated a WHO reference center for the supply of defined laboratory animals. The first director was R. E. Glover (1947–1949), followed by W. Lane-Petter (1949–1965) and John Bleby (from 1965). Breeding colonies were under the control of Michael F.W. Festing, who joined the staff as Geneticist in 1966. The work of the LAC evolved to cover four main areas: information, supply of animals, training, and research (Bleby, 1967; Sparrow, 1976).

These organizations were major participants in the effort that began in the 1950s to establish international standards for the quality, health, and care of laboratory animals. Of crucial importance to their efforts was a "standardized nomenclature" for rat strains and listings of the strains, stocks, and mutants available for research, contributions provided in succession by Billingham and Silvers (Billingham and Silvers, 1959), Jay (Jay, 1963), Robinson (Robinson, 1965), and Festing and Staats (Festing and Staats, 1973), and more recently by others (Gill and Nomura, 1992; Nomura and Potkay, 1991).

D.   Information: Dissemination and Exchange

The LAC in the U.K. also acted as a clearing house for information on all aspects of laboratory animal science. The *Rat News Letter* was established in 1977 after a request from Thomas J. Gill, III of the University of Pittsburgh, who served as an advisor. It was edited initially by M. F. W. Festing with assistance of J. C. Howard (alloantigen news) and Roy Robinson (mutant lists), and was distributed internationally. The *Rat News Letter* subsequently evolved into the *Rat Genome* edited by H. Kunz, T. Natori, and G. Levan (Hedrich, 2000). The LAC also initiated the *International Index of Laboratory Animals* (edited by M. F. W. Festing with the assistance of W. Butler), which by the late 1970s was a world-wide listing of more than 4,000 animal colonies, aimed at assisting people in locating particular stocks of any laboratory species. It subsequently continued to expand (Wolfle, 1999).

Realizing the enormous need for improving the standards and logistics of meeting the national requirements of laboratory animals in the United States, the Division of Biology and Agriculture of the National Research Council—National Academy of Sciences in 1952 established the Institute of Laboratory Animal Resources (ILAR). The first step of that organization toward meeting its charge was the production of the *Handbook of Laboratory Animals* (Herrlein et al., 1954), which summarized the state of the art in terms of genetic standards, nutrition, diseases, and uses of various laboratory animal species. It also gave the first national listing of animal suppliers and users. At that time, there were 25 commercial producers of rats in the U.S. The *Handbook of Laboratory Animals* was the direct forerunner of a plethora of publications and services provided by the ILAR in subsequent years (Wolfle, 1999). For current rat information and links to many relevant Web sites, see the ILAR Home Page (http://www.nas.edu/ilar).

XII.   EPILOGUE

Looking back over the rat's first century as a laboratory animal, first through the careers of those credited in this chapter and second, through our own experiences over the last half century, we (the authors) acknowledge being awed by the contributions of all concerned, including the enormous contributions of the rat to mankind! Inasmuch as the number of scientific publications using rats has exceeded the number using mice for many years (Hedrich, 2000), the rat was arguably the premier laboratory animal of the past century or, at the very least, a close rival to the laboratory mouse. In this context, the statement by Whishaw (Whishaw, 1999) that "The brown rat [*Rattus norvegicus*] is among the great discoveries of twentieth century neuroscience" is no exaggeration! Likewise, "the rat was among the great discoveries of the 20th century" for "advancing the fields of aging, cancer, diabetes, endocrinology, hypertension, immunology, infectious diseases, neurology, nutrition, pharmacology, physiology, reproductive biology, toxicology, transplantation, and countless other fields," (Gill, 1985, 1989) all of which seems to be, at the very best, a major understatement!

ACKNOWLEDGMENTS

Supported in part by research funds of the Veterans Administration, United States Public Health Service Grant RR00463 and NCI Contract N01-CM-6708. The authors gratefully acknowledge the invaluable contributions of all who assisted in preparation and verification of information in both the first and second editions of this chapter. In particular, appreciation is due the following (by institution, historical personality, or rat stock): The Wistar Institute—Dr. Robert Roosa, Mr. William J. Purcell, Mr. Martin Cohn, Mr. Dick Walsh, and Dr. Walter F. Greene; McCollum—Ms. Agatha A. Rider and Dr. Andie M. Hsueh; Sherman—Dr. Paul L. Day and Dr. Charles G. King; Long and Evans—Dr. Leslie L. Bennett, Mr. J. R. K. Kantor, Dr. H. W. Magoun, and Dr. George O. Burr; Castle and The Bussey Institution—Mr. Bill Whalen; Cancer Research Institutes—Dr. Wilhelmina F. Dunning and Ms. Betty R. Moore; Albany Strain—Dr. J. M. Wolfe; Gustafsson—Dr. Elisabeth Norin and Mr. Tim Pennycuff; Reyniers and The Lobund Institute—Dr. Philip C. Trexler, Mr. Charles Lamb, Dr. Mark C. Suckow, and Ms. Sharon Sumpter; and Laboratory Animals Centre—Dr. Michael F. W. Festing. Photographic credits are due Mr. Phil Foster and Mr. David Fisher of The University of Alabama at Birmingham and Ms. Lucinda Keister of The National Library of Medicine. Ms. Hilda Harris of the Lister Hill Library at the University of Alabama at Birmingham provided invaluable assistance in library research.

REFERENCES

Addison, W.H.F. (1939). Henry Herbert Donaldson (1857–1938). *Bios* 10, 4–26.

Allen, A.M. (1999). Evolution of disease monitoring in laboratory rodents. *In* "50 Years of Laboratory Animal Science." (C.W. McPherson and S.F. Mattingly, eds.), pp 136–140. American Association of Laboratory Animal Science, Memphis, Tenn.

Annual Reports of the Director (1911–1956). Wistar Institute of Anatomy and Biology, Philadelphia, Pennsylvania.

Anonymous. (1948). "Selected Works of Henry Clapp Sherman." Macmillan, New York.

Anonymous. (1962). "The National Cyclopaedia of American Biography." Vol. 45, pp 170–171.

Anonymous. (2003). In Memorium–Lisbeth M. Kraft: 1920–2002. *Newsletter* (American College of Laboratory Animal Medicine). **34**, 33–34.

Bateson, W. (1903). The present state of knowledge of colour heredity in mice and rats. *Proc. Zool. Soc. London* **2**, 71–93.

Bazin, H. (1988). A small contribution to the history of the laboratory rat. *Rat News Lett.* **20**, 14.

Bennett, L.L. (1975). Endocrinology and Herbert M. Evans. *In* "Hormonal Proteins and Peptides" (C.H. Li, ed.), Vol. 3, pp 247–272, Academic Press, New York.

Bennett, L.L. (1978). San Francisco, Calif. (personal communication).

Bennett, L.L. (1991). The Long and Evans monograph on the estrous cycle in the rat. *Endocrinol.* **129**, 2812–2814.

Billingham, R.E., and Silvers, W.K. (1959). Inbred animals and tissue transplantation immunity. *Plast. Reconstr. Surg.* **23**, 399–406.

Blass, E.M., ed. (1976). "The Psychobiology of Curt Richter." York Press, Baltimore, Md.

Bleby, J. (1967). The function and work of the United Kingdom Laboratory Animals Centre. *Lab. Anim. Care* **17**, 147–154.

Borrel, A. (1906). Tumeurs cancéreuses et helminthes. *Bull. Acad. Med. Paris* **56**, 141–145.

Broderson, J.R., Lindsey, J.R., and Crawford, J.E. (1976). The role of ammonia in respiratory mycoplasmosis of rats. *Am. J. Pathol.* **85**, 115–130.

Bryan, W.R., Klinck, G.H., Jr., and Wolfe, J.M. (1938). The unusual occurrence of a high incidence of spontaneous mammary tumors in the Albany strain of rats. *Am. J. Cancer* **33**, 370–388.

Bullock, F.D., and Curtis, M.R. (1920). The experimental production of sarcoma of the liver of rats. *Proc. N. Y. Path. Soc.* **20**, 149–175.

Bullock, F.D., and Curtis, M.R. (1930). Spontaneous tumors of the rat. *J. Cancer Res.* **14**, 1–115.

Canzian, F. (1997). Phylogenetics of the laboratory rat *Rattus norvegicus*. *Genome Res.* **7**, 262–267.

Carr, H.A., and Watson, J.B. (1908). Orientation in the white rat. *J. Comp. Neurol. Psychol.* **18**, 27–44.

Castle, W.E. (1919). Piebald rats and selection, a correction. *Am. Nat.* **53**, 370–375.

Castle, W.E. (1924). "Genetics and Eugenics." Harvard Univ. Press, Cambridge, Mass. (1st edition, 1916; revised 1920, 1924, and 1930).

Castle, W.E. (1930). "The Genetics of Domestic Rabbits." Harvard University Press, Cambridge, Mass.

Castle, W.E. (1940). "Mammalian Genetics." Harvard University Press, Cambridge, Mass.

Castle, W.E. (1947). The domestication of the rat. *Proc. Natl. Acad. Sci. U.S.A.* **33**, 109–117.

Castle, W.E. (1951). Variation in the hooded pattern of rats and a new allele of hooded. *Genetics* **36**, 254–266.

Castle, W.E., and Phillips, J.C. (1914). Piebald rats and selection. *Carnegie Inst. Washington Publ.* **195**.

Chase, H.B., Chairman (1975). "NIH Rodent Repository." Report of the Committee on Maintenance of Genetic Stocks. Inst. Lab. Anim. Res., NAS, Washington, D.C.

Chittenden, R.H. (1938). "Biographical Memoir of Lafayette Benedict Mendel, 1872–1935," Biogr. Mem., Vol XVIII, 6th mem., pp 123–155. Natl. Acad. Sci., Washington, D.C.

Chittenden, R.H. (1945). "The First Twenty-five Years of the American Society of Biological Chemists." Waverly Press, Baltimore, Md.

Clause, B.T. (1993). The Wistar rat as a right choice: establishing mammalian standards and the ideal of a standardized mammal. *J. History Biol.* **26**, 329–349.

Conklin, E.G. (1939). "Biographical Memoir of Henry Herbert Donaldson, 1857–1938," Biogr. Mem., Vol. XX, 8th Mem., pp 229–243. Natl. Acad. Sci., Washington, D.C.

Connecticut Agricultural Experiment Station (1930). Thomas B. Osborne, a Memorial. *Conn., Agric. Exp. Stn., New Haven, Bull.* **312**, 281–394.

Crampe, H. (1877). Kreuzungen zwischen wanderratten verschiedener Farbe. *Landwirtsch. Jahrb.* **6**, 385–395.

Crampe, H. (1883). Zuchtversuche mit zahmen Wanderratten. I. Resulte der Zucht in Verwandtschaft. *Landwirtsch. Jahrb.* **12**, 389–449.

Crampe, H. (1884). Zuchtversuche mit zahmen Wanderratten. II. Resultate der Kreuzung der zahmen Ratten mit Wilden. *Landwirtsch. Jahrb.* **13**, 699–754.

Crampe, H. (1885). Die Gesetze der Bererbung der Farbe. *Landwirtsch. Jahrb.* **14**, 539–619.

Cumming, C.N.W., and Baker, D.E.J. (1968). A large flexible film isolator and its use in the production of gnotobiotes. *In* "Advances in Germfree Research and Gnotobiology." (M. Miyakawa and T.D. Luckey, eds.), pp 16–19. CRC Press, Cleveland, Ohio.

Curtis, M.R., Bullock, F.D., and Dunning, W.F. (1931). A statistical study of the occurrence of spontaneous tumors in a large colony of rats. *Am. J. Cancer* **15**, 67–121.

Curtis, M.R., and Dunning, W.F. (1937). Two independent mutations of the hooded or piebald gene of the rat. *J. Hered.* **18**, 382–390.

Curtis, M.R., and Dunning, W.F. (1940). Transplantable lymphosarcomata of the mesenteric lymph nodes of rats. *Am. J. Cancer* **40**, 299–309.

Curtis, M.R., and Dunning, W.F. (1940). An independent recurrence of the blue mutation in the Norway rat. *J. Hered.* **31**, 219–222.

Curtis, M.R., Dunning, W.F., and Bullock, F.D. (1933). Genetic factors in relation to the etiology of malignant tumors. *Am. J. Cancer* **17**, 894–923.

Day, H.G. (1975). Contributions of Elmer Verner McCollum. *In* "Nutrition and Public Health—A Symposium Celebrating the Johns Hopkins University Centennial and Honoring Elmer V. McCollum." (R.M. Herriott, ed.), pp 3–21. The Johns Hopkins School of Hygiene and Public Health, Baltimore, Md.

Day, P.L. (1957). Henry Clapp Sherman. *J. Nutr.* **61**, 3–11.

Donaldson, H.H. (1912). A comparison of the European Norway and albino rats (*Mus norvegicus albinos*) with those of North America in respect to the weight of the central nervous system and to cranial capacity. *J. Comp. Neurol.* **22**, 71–77.

Donaldson, H.H. (1912). The history and zoological position of the albino rat. *J. Acad. Nat. Sci. Philadelphia.* **15**, 365–369.

Donaldson, H.H. (1915). "The Rat. References Tables and Data for the Albino Rat (*Mus norvegicus albinos*) and the Norway Rat (*Mus norvegicus*)." 1st ed., Memoirs No. 6, Wistar Inst. Anat. Biol., Philadelphia, Pa.

Donaldson, H.H. (1924). "The Rat. Reference Tables and Data for the Albino Rat (*Mus norvegicus albinos*) and the Norway Rat (*Mus norvegicus*)." 2nd ed., Memoirs No. 6, Wistar Inst. Anat. Biol., Philadelphia, Pa.

Donaldson, H.H. (1927). The Museum of Caspar Wistar. Address given before Sigma Xi, Philadelphia, Pa., January 19, 1927. *In* "The Donaldson Memoirs." Library of the Wistar Institute, Philadelphia, Pa.

Donaldson, H.H. (1937). Milton Jay Greenman (1866–1937). *Anat Rec.* **68**, 262–265.

Doncaster, L. (1906). On the inheritance of coat color in rats. *Proc. Cambridge Philos. Soc.* **13**, 215–227.

Drake, L.E., and Heron, W.T. (1930). The rat. A bibliography 1924–1929. *Psychol. Bull.* **27**, 141–239.

Dunn, L.C. (1936). Letter From Dunn to William Ernest Castle. History of Medicine Division, National Library of Medicine, Bethesda, Maryland.

Dunn, L.C. (1965). "William Ernest Castle." Biogr. Mem. Natl. Acad. Sci., Washington, D.C., Vol. XXXVIII, pp 33–80.

Dunning, W.F. (1951). In memorium—Francis Carter Wood (1869–1951). *Cancer Res.* **11**, 296.

Dunning, W.F. (1977). The Papanicolaou Cancer Research Institute, Miami, Fla. (personal communication).

Dunning, W.F. (1978). Unpublished manuscript prepared in 1940, made available to J.R. Lindsey in 1978.

Dunning, W.F., and Curtis, M.R. (1941). Longevity and genetic specificity as factors in the occurrence of spontaneous tumors in the hybrids between two inbred lines of rats. *Genetics* **26**, 148–149.

Dunning, W.F., and Curtis, M.R. (1946). Multiple peritoneal sarcoma in rats from intraperitoneal injection of washed, ground *Taenia* larvae. *Cancer Res.* **6**, 668–670.

Dunning, W.F., and Curtis, M.R. (1946). The respective roles of longevity and genetic specificity in the occurrence of spontaneous tumors in the hybrids between two inbred lines of rats. *Cancer Res.* **6**, 61–81.

Dunning, W.F., and Curtis, M.R. (1952). Diethylstilbestrol-induced mammary cancer in reciprocal F1 hybrids between negative and positive inbred lines of rats. *Cancer Res.* **12**, 257–258.

Dunning, W.F., and Curtis, M.R. (1952). The influence of diethylstilbestrol-induced cancer in reciprocal F1 hybrids obtained from crosses between rats of inbred lines that are susceptible and resistant to the induction of mammary cancer by this agent. *Cancer Res.* **12**, 702–706.

Dunning, W.F., and Curtis, M.R. (1953). Attempts to isolate the active agent in *Cysticercus fasciolaris. Cancer Res.* **13**, 838–843.

Dunning, W.F., and Curtis, M.R. (1957). A transplantable acute leukemia in an inbred line of rats. *J. Natl. Cancer Inst.* **19**, 845–852.

Dunning, W.F., Curtis, M.R., and Madsen, M.E. (1951). Diethylstilbestrol-induced mammary gland and bladder cancer in reciprocal F1 hybrids between two inbred lines of rats. *Acta Unio. Int. Cancrum* **7**, 238–244.

Eakin, R.M. (1956). History of zoology at the University of California, Berkeley. *Bios. (Madison, NJ.)* **27**, 67–90.

Eakin, R.M., Evans, H.M., Goldschmidt, R.B., and Lyons, W.R. (1959). Joseph Abraham Long, 1879–1953. "In Memorium," April issue, pp 40–42. University of California.

Edward, A.G. (1978). Department of Comparative Medicine, Wayne State University, Detroit, Michigan (personal communication).

Eisen, M.J. (1954). Obituary—William H. Woglom (1879–1953). *Cancer Res.* **14**, 155–156.

Erikson, G.E. (1977). Eunice Chace Greene, 1895–1975. *Anat. Rec.* **189**, 306–307.

Evans, H.M. (1914). Vital straining of protoplasm. *Science* **39**. 843–844.

Evans, H.M. (1962). The pioneer history of vitamin E. *Vitam. Horm. N.Y.* **20**, 379–387.

Evans, H.M., and Bishop, K.S. (1922). On the existence of a hitherto unrecognized dietary factor essential for reproduction. *Science* **56**, 650–651.

Farris, E.J. (1950). The rat as an experimental animal. *In* "The Care and Breeding of Laboratory Animals." (E. J. Farris, ed.), pp 43–78. Wiley, New York.

Farris, E.J., and Griffith, J.Q., Jr., eds. (1949). "The Rat in Laboratory Investigation." 2nd ed. Lippincott, Philadelphia, Pa.

Ferry, E.L. (1919). Nutrition experiments with rats. A description of methods and technic. *J. Lab. Clin. Med.* **5**, 735–745.

Festing, M.F.W. (1978). "Rat News Letter," No. 3, pp 18–35. Med. Res. Counc. Lab. Anim. Cent., Woodmansterne Rd., Carshalton, Surrey SM5 4EF, England.

Festing, M., and Staats, J. (1973). Standardized nomenclature for inbred strains of rats—fourth listing. *Transplantation.* **16**, 221–245.

Fibiger, J.A.G. (1965). Investigations on *Spiroptera carcinoma* and the experimental induction of cancer: Nobel Lecture, December 12, 1927. *In* "Nobel Lectures Including Presentation Speeches and Laureates' Biographies, Physiology and Medicine, 1922–1941." pp 122–150. Am. Elsevier, New York.

Foster, H.L. (1959). Housing of disease-free vertebrates. *Ann. N.Y. Acad. Sci.* **78**, 80–88.

Foster, H.L. (1980). The history of commercial production of laboratory rodents. *Lab. Anim. Sci.* **30**, 793–798.

Friends of H.M. Evans (1943). Herbert M. Evans—Biographical sketch and Bibliography of Herbert McLean Evans (1804–1902). *In* "Essays in Biology in Honor of Herbert M. Evans." pp ix-x and xii-xxvii. University of California Press, Berkeley.

Gill, T.J., III. (1985). The rat in biomedical research. *Physiologist.* **28**, 9–17.

Gill, T.J., III. (1989). The rat as an experimental animal. *Science.* **245**, 269–276.

Gill, T.J., III, and Nomura, T. (1992). Definition, nomenclature, and conservation of rat strains. *ILAR News.* **34**, S1-S26.

Gorski, R.A., and Whalen, R.E., eds. (1963). "The Brain and Gonadal Function." UCLA Forum for Medical Sciences, No. 3. University of California Press, Berkeley.

Greene, E.C. (1935). "Anatomy of the Rat." Trans. Philos. Soc. Philos. Soc., Philadelphia, Pa.

Greene, W.F. (1978). Jaffrey Center, New Hampshire (personal communication).

Greenman, M.J., and Duhring, F.L. (1923). "Breeding and Care of the Albino Rat for Research Purposes." Wistar Inst. Anat. Biol., Philadelphia, Pa.

Greenman, M.J., and Duhring, F.L. (1931). "Breeding and Care of the Albino Rat for Research Purposes." 2nd edition. Wistar Inst. Anat. Biol, Philadelphia, Pa.

Griffith, J.Q., Jr., and Farris, E.J., eds. (1942). "The Rat in Laboratory Investigation." 1st edition. Lippincott, Philadelphia, Pa.

Gustafsson, B.E. (1947). Germ-free rearing of rats. Preliminary report. *Acta Anat.* **2**, 376–391.

Gustafsson, B.E. (1948). Germ-free rearing of rats. General technique. *Acta Pathol. Microbiol. Scand.*, **73**, 1–130.

Gustafsson, B.E. (1959). Lightweight stainless steel systems for rearing germfree animals. *Ann. N.Y. Acad. Sci.*, **78**, 17–28.

Hammett, F.S. (1924). Studies of the thyroid apparatus. XX. The effect of thyro-parathyroidectomy and parathyroidectomy at 75 days of age on the growth of the brain and spinal cord of male and female albino rats. *J. Comp. Neurol.* **37**, 15–30.

Hancock, D. (1995). Obituary—Wilhelmina Dunning, UM Cancer Researcher. January 24, 1995. Miami Herald, Miami, Fla.

Harden, V.A. (1986). "Inventing the NIH: Federal Biomedical Research Policy, 1887–1927." Johns Hopkins University Press, Baltimore.

Hedrich, H.J. (2000). History, strains and models. *In* "The Laboratory Rat." (George J. Krinke, ed.), pp 3–16. Academic Press, San Diego, Calif.

Hegsted, M. (1975). Relevance of animal studies to human disease. *Cancer Res.* **35**, 3537–3539.

Hektoen, L. (1915–1916). Observations on pulmonary infections in rats. *Trans. Chicago Pathol. Soc.* **10**, 105–109.

Hem, A. (1994). Gnotobiology. *In* "Handbook of Laboratory Animal Science," (P. Svendson and J. Hau, eds.), Vol. I. pp 273–291. CRC Press, Boca Raton, Fla.

Herrlein, H.G., Coursen, G.B., Randall, R., and Slanetz, C.A. (1954). "Handbook of Laboratory Animals." Inst. Anim. Resour., Natl. Acad. Sci.–Natl. Res. Counc., Washington, D.C.

Hessler, J.R. (1999). The history of environmental improvements in laboratory animal science: caging systems, equipment, and facility design. *In* "50 Years of Laboratory Animal Science" (C.W. McPherson and S.F. Mattingly, eds.), pp 92–12. Am. Assoc. Lab. Anim. Sci., Memphis, Tenn.

Heston, W.E. (1972). Obituary—Clarence Cook Little. *Cancer Res.* **32**, 1354–1356.

Heston, W.E., Jay, G.E., Jr., Kaunitz, H., Morris, H.P., Nelson, J.B., Poiley, S.M., Slanetz, C.A., Zucker, L.M., and Zucker, T.F. (1953).

"Rat Quality—A Consideration of Heredity, Diet and Disease." Natl. Vitam. Found., Inc., New York.

Holdenried, R. (ed.). (1966). Symposium on Viruses of Laboratory Rodents. National Cancer Institute Monograph 20, 180 p. U.S. Government Printing Office, Washington, D.C.

Hoppert, C.A., Webber, P.A., and Conniff, T.L. (1932). The production of dental caries in rats fed an adequate diet. *J. Dent. Res.* **12**, 161–173.

Hsueh, A.M. (2002). Texas Womans University, Denton, Texas (personal communication).

Hunt, H.R., Hoppert, C.A., and Erwin, W.G. (1944). Inheritance of susceptibility to caries in albino rats (*Mus norvegicus*). *J. Dent. Res.* **23**, 385–401.

Hunt, H.R., Hoppert, C.A., and Rosen, S. (1955). Genetic factors in experimental rat caries. *In* Sognnaes RF, ed. American Association for the Advancement of Science, Washington, D.C.

Jay, G.E., Jr. (1953). The use of inbred strains in biological research. *In* "Rat Quality—A Consideration of Heredity, Diet and Disease." (W.E. Heston et al., eds.), pp 98–103, Natl. Vitam. Found., New York.

Jay, G.E., Jr. (1963). Genetic strains and stocks. *In* "Methodology in Mammalian Genetics." (Burdette, W.J., ed.), pp 83–126, Holden Day, San Francisco.

Kilham, L. (1966). Viruses of laboratory and wild rats. *In* "Symposium on Viruses of Laboratory Rodents." National Cancer Institute Monograph 20, pp 117–140. U.S. Government Printing Office, Washington, D.C.

King, C.G. (1975). "Biographical Memoir of Henry Clapp Sherman, 1875–1955." Biogr. Mem., Vol. XLVI, pp 397–429. Natl. Acad. Sci., Washington, D.C.

King, H.D. (1918). Studies on inbreeding. I. The effects of inbreeding on the growth and variability in the body weight of the albino rat. *J. Exp. Zool.* **26**, 1–54.

King, H.D. (1918). Studies on inbreeding. II. The effects of inbreeding on the fertility and on the constitutional vigor of the albino rat. *J. Exp. Zool.* **26**, 335–378.

King, H.D. (1918). Studies on inbreeding. III. The effects of inbreeding, with selection, on the sex ratio of the albino rat. *J. Exp. Zool.* **27**, 1–35.

King, H.D. (1919). Studies on inbreeding. IV. A further study of the effects of inbreeding on the growth and variability in the body weight of the albino rat. *J. Exp. Zool.* **29**, 135–175.

Kline, L.W. (1899). Methods in animal psychology. *Am. J. Psychol.* **10**, 256–279.

Kraft, L.M. (1958). Observations on the control and natural history of epidemic diarrhea of infant mice. *Yale J. Biol. Med.* **31**, 121–137.

Kraft, L.M. (1982). Viral diseases of the digestive system. *In* "The Mouse in Biomedical Research." (H.L. Foster, J.D. Small, and J.G. Fox, eds.), Vol. II , pp 159–191. Academic Press, New York.

Kraft, L.M., Pardy, R.F., Pardy, D.A., and Zwickel, H. (1964). Practical control of diarrheal disease in a commercial mouse colony. *Lab. Anim. Care* **14**, 16–19.

Kreezer, G.L. (1949). Technics for the investigation of behavioral phenomena in the rat. *In* "The Rat in Laboratory Investigation." (E.J. Farris and J.Q. Griffith, Jr., eds.), pp 203–277. Lippincott, Philadelphia, Pennsylvania.

Lamb, C. (2002). Archives of the University of Notre Dame, Hesburgh Library, Notre Dame, Indiana (personal communication).

Larson, R.H., Amsbaugh, S.M., Navia, J.M., Rosen, S., Schuster, G.S., and Shaw, J.H. (1977). Collaborative evaluation of a rat caries model in six laboratories. *J. Dent. Res.* **56**, 1007–1012.

Lattuada, C.P., Norberg, R.M., and Harlan, H.P. (1981). Large isolators for rearing rodents. *In* "Recent Advances in Germfree Reseazzrch." (S. Sasaki, ed.), pp 53–58. Tokai University Press, Tokyo, Japan.

Lindsey, J.R., Boorman, G.A., Collins, M.J., Hsu, C.-K., Van Hoosier, G.L., and Wagner, J.E.. (1991). "Infectious Diseases of Mice and Rats." National Academy Press, Washington, D.C.

Logan, C.A. (1999). The altered rationale for the choice of a standard animal in experimental psychology: Henry H. Donaldson, Adolf Meyer, and "the" albino rat. *History Psychol.* **2**, 3–24.

Long, J.A., and Evans, H.M. (1922). "The Oestrous Cycle in the Rat and Its Associated Phenomena." Univ. Calif., No. 6. Univ. of California Press, Berkeley.

Luckey, T.D. (1963). Phylogenetic development of germfree research. *In* "Germfree Life and Gnotobiology." pp 34–98, Academic Press, New York.

Luckey, T.D. (1963). Germfree animal techniques. *In* "Germfree Life and Gnotobiology." pp 99–212, Academic Press, New York.

Luckey, T.D. (1963). Nutrition of germfree animals. *In* "Germfree Life and Gnotobiology." pp 213–282, Academic Press, New York.

Luckey, T.D. (1963). Appendix I—Chronology, Appendix II—Glossary, Appendix III— Diets for germfree animals. *In* "Germfree Life and Gnotobiology." pp 477–497 , Academic Press, New York.

Mason, K.E. (1977). The first two decades of Vitamin E. *Fed. Proc., Fed. Am. Soc. Exp. Biol.* **36**, 1906–1910.

McCollum, E.V. (1948). Letter dated May 4, 1948 from E.V. McCollum to W.D. Salmon of Auburn University, Auburn, Alabama (copy made available by Ms. Agatha Rider, Department of Biochemistry, School of Hygiene and Public Health, The Johns Hopkins University, Baltimore, Md).

McCollum, E.V. (1957). "A History of Nutrition—The Sequence of Ideas in Nutrition Investigation." Houghton, Boston, Mass.

McCollum, E.V. (1964). "From Kansas Farm Boy to Scientist—The Autobiography of Elmer Verner McCollum." University of Kansas Press, Lawrence, Kan.

McCollum, E.V., and Davis, M. (1913). The necessity of certain lipids in the diet during growth. *J. Biol. Chem.* **15**, 167–175.

McCollum, E.V., and Davis, M. (1915). The essential factors in the diet during growth. *J. Biol. Chem.* **23**, 231–246.

McCollum, E.V., Simmonds, N., Becker, J.E., and Shipley, P.G. (1922). Studies on experimental rickets. XXI. An experimental demonstration of the existence of a vitamin which promotes calcium deposition. *J. Biol. Chem.* **53**, 293–312.

McMurrich, J.P., and Jackson, C.M. (1938). Henry Herbert Donaldson (1857–1938). *J. Comp. Neurol.* **69**, 173–179.

McPherson, C.W., and Mattingly, S.F. (eds.). (1999). "Fifty Years of Laboratory Animal Science." Am. Assoc. Lab. Anim. Sci., Memphis, Tenn.

Mendel, L.B., and Hubbell, R.B. (1935). The relation of the rate of growth to diet. *J. Nutr.* **10**, 557–563.

Miles, W.R. (1930). On the history of research with rats and mazes. *J. Gen. Psychol.* **3**, 324–337.

Moore, B.R. (1978). Library Service Coordinator for Cancer Research Center, Columbia University, New York (personal communication).

Morse, H.C., III (1978). Introduction. "Origins of Inbred Mice." (H.C. Morse, III, ed.). Academic Press, New York.

Morse, H.C., III (1981). The laboratory mouse—a historical perspective. *In* "The Mouse in Biomedical Research." (H.L. Foster, J.D. Small, and J.G. Fox, eds.), Vol I, pp 1–16. Academic Press, New York.

Morse, H.C., III (1985). The Bussey Institute and the early days of mammalian genetics. *Immunogenetics* **21**, 109–116.

Munn, N.L. (1950). "Handbook of Psychological Research on the Rat." pp 2–5, Houghton, Boston, Mass.

Navia, J.M. (1977). Experimental dental caries. *In* "Animal Models in Dental Research." pp 257–297, Univ. of Alabama Press, Birmingham.

Newton, W.L. (1965). Methods in germfree animal research. *In* "Methods of Animal Experimentation." Vol. 1, pp 215–271, Academic Press, New York.

NIH Factbook: Guide to National Institutes of Health Programs and Activities. 1st ed. (1976). Marquis Academic Media, Chicago, Ill.

Nomura, T., and Potkay, S. (1991). Establishment and preservation of reference strains of rats for general purpose use. *ILAR News* **33**, 42–44.

Norin, E. (2002). Department of Germfree Research, Karolinska Institute, Stockholm, Sweden (personal communication).

Orcutt, R.P., Otis, A.P., and Alliger, H. (1981). AlcideTM: An alternative sterilant to peracetic acid. *In* "Recent Advances in Germfree Research." (S. Sasaki, ed.), pp 79–81, Tokai University Press, Tokyo, Japan.

Osborne, T.B. (1909). "The Vegetable Proteins." Longmans, Green, New York.

Osborne, T.B. (1924). "The Vegetable Proteins." 2nd ed. Longmans, Green, New York.

Osborne, T.B., and Mendel, L.B. (1911). "Feeding Experiments with Isolated Food-Substances." Parts I and II. Carnegie Institute Washington, Washington, D.C.

Osborne, T.B., and Mendel, L.B. (1913). The influence of butter-fat on growth. *J. Biol. Chem.* **16**, 423–437.

Osborne, T.B., and Mendel, L.B. (1914). The influence of cod liver oil and some other fats on growth. *J. Biol. Chem.* **17**, 401–408.

Osborne, T.B., and Mendel, L.B. (1918). The inorganic elements in nutrition. *J. Biol. Chem.* **34**, 131–139.

Otha, S.A. Sprague Memorial Institute of Chicago (1911–1941). "Collected Papers." Vols. 1–25.

Parkes, A.S. (1969). Herbert McLean Evans. *J. Reprod. Fertil.* **19**, 1–29.

Pennycuff, T. (2002). University Archivist, Lister Hill Library of the Health Sciences, University of Alabama at Birmingham, Birmingham, Ala. (personal communication).

Peterson, G.M. (1946). The rat in animal psychology. *In* "The Encyclopedia of Psychology." (P.L. Harriman, ed.), pp 765–798. Philosophical Library, New York.

Philipeaux, J.M. (1856). Note sur l'extirpation des capsules survenales chez les rats albios (*Mus rattus*). *C. R. Habd. Seances Acad. Sci.* **43**, 904–906.

Pleasants, J.R. (1959). Rearing germfree cesarean-born rats, mice, and rabbits through weaning. *Ann. N.Y. Acad. Sci.* **78**, 116–126.

Pleasants, J.R. (1965). History of germfree animal research at Lobund Laboratory, Biology Department, University of Notre Dame, 1928–1965. *Proc. Indiana Acad. Sci.* **75**, 220–226.

Poiley, S.M. (1953). History and information concerning the rat colonies in the animal section of the National Institutes of Health. *In* "Rat Quality—A Consideration of Heredity, Diet and Disease." (W.E. Heston et al., eds.), pp 86–97. Natl. Vitam. Found., New York.

Proceedings of the Borden Centennial Symposium on Nutrition (1958). "The Nutritional Ages of Man—Nutrition: Past, Present, and Future." pp 120–126. Borden Company Found., Inc., New York.

Purcell, W.L. (1968–1969). "An Outline of the History of the Wistar Institute." Biennial Res. Rep., pp xi-xiii, Wistar Inst. Anat. Biol., Philadelphia, Pa.

Purcell, W.L. (1978). Librarian, Wistar Institute, Philadelphia, Pennsylvania (personal communication).

Raacke, I.D. (1983). Herbert McLean Evans (1882–1971). A biographical sketch. *J. Nutr.* **113**, 929–943.

Reyniers, J.A. (1943). Introduction to the general problem of isolation and elimination of contamination. *In* "Micrurgical and Germ-Free Techniques: Their Application to Experimental Biology and Medicine." (J.A. Reyniers, ed.), pp 95–113, Charles C. Thomas, Springfield, Ill.

Reyniers, J.A. (1959). Design and operation of apparatus for rearing germfree animals. *Ann. N.Y. Acad. Sci.* **78**, 47–79.

Reyniers, J.A., Trexler, P.C., and Ervin, R.F. (1946). Rearing Germ-Free Albino Rats. *Lobund Reports* **1**, 1–84.

Richter, C.P. (1949). The use of the wild Norway rat for psychiatric research. *J. Nerv. Ment. Dis.* **110**, 379–386.

Richter, C.P. (1950). Domestication of the Norway rat and its implications for the problem of stress. *Res. Publ., Assoc. Res. Nerv. Ment. Dis.* **29**, 19–47.

Richter, C.P. (1954). The effects of domestication and selection on the behavior of the Norway rat. *J. Natl. Cancer Inst.* **15**, 727–738.

Richter, C.P. (1959). Rats, man, and the welfare state. *Am. Psychol.* **14**, 18–28.

Richter, C.P. (1968). Experiences of a reluctant rat-catcher. The common Norway rat—Friend or enemy? *Proc. Am. Philos. Soc.* **112**, 403–415.

Rider, A.A. (1970). Elmer Verner McCollum—A biographical sketch. *J. Nutr.* **100**, 1–10.

Rider, A.A. (1978, 2002). Department of Biochemistry, Johns Hopkins University School of Hygiene and Public Health, Baltimore, Md (personal communication).

Robinson, R. (1965). "Genetics of the Norway Rat." Pergamon, Oxford.

Roosa, R. (1977). Wistar Institute, Philadelphia, Pa (personal communication).

Rose, W.C. (1936). Lafayette Benedict Mendel—An appreciation. *J. Nutr.* **11**, 607–613.

Rovera, G. (1997). The Wistar Institute. *Mol. Med.* **3**, 229–230.

Rowe, W.P., Hartley, J.W., and Huebner, R.J. (1962). "The Problems of Laboratory Animal Disease." Academic Press, New York.

Savory, W.S. (1863). Experiments on food; its destination and uses. *Lancet* **2**, 381–383.

Scott, T.R.C. (1931). The Hampton Court maze. *J. Genet. Psychol.* **39**, 287–289.

Sherman, H.C. (1936). Lafayette Benedict Mendel. *Science* **83**, 45–47.

Sherman, H.C., Rouse, M.E., Allen, B., and Woods, E. (1919). Growth and reproduction upon simplified food supply. *Proc. Soc. Exp. Biol. Med.* **17**, 9–10.

Sherman, H. C., Rouse, M. E., Allen, B., and Woods, E. (1921). Growth and reproduction upon simplified food supply. I. *J. Biol. Chem.* **46**, 403–419.

Small, W.S. (1900). An experimental study of the mental processes of the rat. *Am. J. Psychol.* **11**, 133–165.

Small, W.S. (1901). Experimental study of the mental processes of the rat. *Am. J. Psychol.* **12**, 206–239.

Smith, A.H. (1956). Lafayette Benedict Mendel. *J. Nutr.* **60**, 3–12.

Smith, P.E. (1927). The disabilities caused by hypophysectomy and their repair. *J. Am. Med. Assoc.* **88**, 158–161.

Snell, G.D., and Reed, S. (1993). William Ernest Castle, Pioneer Mammalian Geneticist. *Genetics.* **133**, 751–753.

Snyder, E.K., and Jones, E.A. (1968). Elmer Verner McCollum. *J. Am. Diet. Assoc.* **52**, 49.

Sparrow, S. (1976). The microbiological and parasitological status of laboratory animals from accredited breeders in the United Kingdom. *Lab. Anim.* **10**, 365–373.

Stewart, C.C. (1898). Variations in daily activity produced by alcohol and by changes in barometric pressure and diet, with a description of recording methods. *Am. J. Physiol.* **1**, 40–56.

Stewart, W.H., Hoppert, C.A., and Hunt, H.R. (1952). The incidence of dental caries in caries-susceptible and caries-resistant albino rats (*Rattus norvegicus*) when fed diets containing granulated and powdered sucrose. *J. Dent. Res.* **32**, 210–221.

Stolley, P.D., and Lasky, T. (1992). Johannes Fibiger and his Nobel prize for the hypothesis that a worm causes stomach cancer. *Ann. Intern. Med.* **116**, 765–769.

Stotsenberg, J.M. (1909). On the growth of the albino rat (*Mus norvegicus* var. *albus*) after castration. *Anat. Rec.* **3**, 233–244.

Stotsenberg, J.M. (1913). The effect of spaying and semi-spaying young albino rats (*Mus norvegicus albinos*) on the growth in body weight and body length. *Anat. Rec.* **7**, 183–194.

Suckow, M.C. (2002). Notre Dame University, Indiana (personal communication).

Trexler, P.C. (1959). The use of plastics in the design of isolator systems. *Ann. N.Y. Acad. Sci.*, **78**, 29–36.

Trexler, P.C. (1985). The evolution of gnotobiotic technology. *In* "Germfree Research: Microflora Control and Its Application to the Biomedical Sciences." (B.S. Wostmann, ed.), pp 3–7, Alan R. Liss, New York.

Trexler, P.C. (2002). Reyniers, Lobund and Gnotobiotics. *In* "Newsletter of the Association for Gnotobiotics." (R.P. Orcutt, ed.), pp 5–10, Spring 2002.

Trexler, P.C. (2002). Personal communication.

Trexler, P.C., and Orcutt, R.P. (1999). Development of gnotobiotics and contamination control in laboratory animal science. *In* "50 Years of Laboratory Animal Science." (C.W. McPherson and S.F. Mattingly, eds.), pp 121–128, Am. Assoc. Lab. Anim. Sci., Memphis, Tenn.

Trexler, P.C., and Reynolds, L.I. (1957). Flexible film apparatus for the rearing and use of germfree animals. *Appl. Microbiol.* **5**, 406–412.

Verzar, F., ed. (1973). "Clive M. McCay's Notes on the History of Nutrition Research." Huber, Bern.

Vickery, H.B. (1932). "Biographical Memoir of Thomas Burr Osborne, 1859–1929," Biogr. Mem., Vol. XIV, 8th mem., pp 261–304. Natl. Acad. Sci., Washington, D.C.

Warden, C.J. (1930). A note on the early history of experimental methods in comparative psychology. *J. Genet. Psychol.* **38**, 466–471.

Watson, J.B. (1903). "Animal Education. An Experimental Study on the Psychological Development of the White Rat, Correlated with the Growth of its Nervous System." Univ. of Chicago Press, Chicago, Ill.

Watson, J.B. (1907). Kinaesthetic and organic sensations: Their role in the reactions of the white rat. *Psychol. Rev. Monog.* **8**, No. 33, 1–100.

Watson, J.B. (1914). "Behavior: An Introduction to Comparative Psychology." Holt, New York.

Watson, J.B. (1936). John Broadus Watson. *In* "A History of Psychology in Autobiography." (C. Murchison, ed.), Vol. III, pp 271–281, Clark Univ. Press, Worcester, Mass.

Weisbroth, S.H. (1969). The origin of the Long-Evans rat and a review of the inheritance of coat colors in rats (*Rattus norvegicus*). *Lab. Anim. Care* **19**, 733–737.

Weisbroth, S.H. (1999). Evolution of disease patterns in laboratory rodents: the post-indigenous condition. *In* "50 Years of Laboratory Animal Science." (C.W. McPherson, and S.F. Mattingly, eds.), pp 141–146, Am. Assoc. Lab. Anim. Sci., Memphis, Tenn.

Wernstedt, W. (1965). Presentation speech, Nobel prize in physiology or medicine, 1926. *In* "Physiology and Medicine Nobel Lectures Including Presentation Speeches and Laureates Bibliographies, vol. 2, 1922–1941" pp 119–121. Am. Elsevier, New York.

Whishaw, I.Q. (1999). The laboratory rat, the pied piper of twentieth century neuroscience. *Brain Res. Bull.* **50**, 411.

Whishaw, I.Q., Kolb, B., and Sutherland, R.J. (1983). The analysis of behavior in the laboratory rat. *In* "Behavioral Approaches to Brain Research." (T.E. Robinson, ed.), pp 141–211. Oxford University Press, New York.

Whitehair, L.A. (1999). Development and major contributions of the extramural comparative medicine area, NIH (1962–1999). *In* "0 Years of Laboratory Animal Science: 1950–1999." (C.W. McPherson, and S.F. Mattingly, eds.), pp 68–73. Am. Assoc. Lab. Anim. Sci., Memphis, Tenn.

Wistar, I.J. (1937). "Autobiography of Isaac Jones Wistar (1727–1905), Half a Century in War and Peace," with an Appendix entitled "The Wistar Institute of Anatomy and Biology," by Milton J. Greenman (pp 503–516) and an Addendum by Edmond J. Farris (pp 517–518). Wistar Inst. Anat. Biol., Philadelphia, Pa.

Woglom, W.H. (1951). Francis Carter Wood (1869–1951). *Am. J. Roentgenol. Radium Ther.* **65**, 955–959.

Wolfe, J.M., Bryan, W.R., and Wright, A.W. (1938). Histologic observations on the anterior pituitaries of old rats with particular reference to the spontaneous appearance of pituitary adenomata. *Am. J. Cancer* **34**, 352–372.

Wolfle, T.L. (1999). The history of the Institute of Laboratory Animal Resources: 1953–1999. *In* "50 Years of Laboratory Animal Science: 1950–1999." (C.W. McPherson, and S.F. Mattingly, eds.), pp 44–54, Am. Assoc. Lab. Anim. Sci., Memphis, Tenn.

Wood, F.C. (1914–1915). The Crocker Research Laboratory. *Columbia Univ. Quart.,* **17**, 82–86.

Wostmann, B.S. (1996a). "Germfree and Gnotobiotic Animal Models: Background and Applications." CRC Press, Boca Raton, Fla.

Wostmann, B.S. (1996b). Nutrition. *In* "Germfree and Gnotobiotic Animal Models: Background and Applications." pp 71–87, CRC Press, Boca Raton, Fla.

Wostmann, B.S. (1996c). The chemically defined diet. *In* "Germfree and Gnotobiotic Animal Models: Background and Applications." pp 89–99, CRC Press, Boca Raton, Fla.

Zucker, T.F. (1953). Problems in breeding for quality. *In* "Rat Quality—A Consideration of Heredity, Diet and Disease." (W.E. Heston, et al., eds.), pp 48–76, Natl. Vitam. Found., New York.

Chapter 2

Ethical and Legal Perspectives

Beverly J. Gnadt

I. INTRODUCTION

The use of animals in scientific research is a multifaceted and complex issue. The diversity of human views over the millennia as to what constitutes the "appropriate" relationship between humans and nonhuman animals is exemplified by our use of animals as food and clothing sources, beasts of burden, objects of and participants in sport, "pests" to be eradicated, and companions. The use of animals for scientific and medical inquiry for the purpose of advancing basic knowledge and understanding and alleviating disease and suffering of humans and animals is a relatively modern development. When animals are used in research, the humans who use them have a moral obligation to provide humane care and use in the laboratory setting. This obligation encompasses the application of the principles of reduction, refinement, and replacement in experimental design and conduct, not only for ethical and humane reasons, but for scientific integrity as well. To assure animals are used ethically and humanely in quality scientific inquiry, society has deemed it necessary to oversee the conduct of experiments through government agencies and accrediting organizations that provide structure to animal care and use programs, critical oversight, and regulatory enforcement.

II. ETHICAL CONSIDERATIONS

A. The Principles of Reduction, Replacement, and Refinement

In 1954, Dr. William Russell, a zoologist and psychologist, and Mr. Rex Burch, a microbiologist, undertook a systematic study of laboratory techniques from an ethical perspective. The culmination of their study led to the publication of *The Principles of Humane Experimental Technique* in 1959, which outlined three basic ethical tenets: reduction, replacement, and refinement, upon which many of the animal welfare-related laws and regulations have been based over the last 50 years. The stated goals of the "Three Rs" were to:

> ...assist those starting, or about to start work as experimenters on animals, who wish to be as humane as possible to their subject, and seek an orientation not available, to our knowledge, within the covers of any one book. (Russell and Burch, 1959).

The authors sought to provide the scientific community with a systematic classification of techniques to use as a roadmap for the "diminishing or removal of inhumanity." They defined the Three Rs as follows:

> Replacement means the substitution for conscious living higher animals of insentient material. Reduction means reduction in the number of animals used to obtain information of a given amount and precision. Refinement means any decrease in the incident or severity of inhumane procedures applied to those animals which still have to be used. (Russell and Burch, 1959).

Replacement was further defined as being either "relative" or "absolute." In relative replacement, animals were still required to be used but were exposed to no distress during the experimental procedure. Examples of procedures involving relative replacement included non-recovery experiments on living and intact, but completely anesthetized animals, and animals that were humanely euthanized for tissue collection. In absolute replacement, animals were not required in any phase of the experimental procedure; therefore, other nonsentient life forms (metazoan endoparasites, higher plants, microorganisms) and nonliving physical and chemical systems could be used.

The principle of reduction involved three phases. The first phase described two general approach strategies for experimental design: trial and error (random research) versus deductively inspired (guided research). Random research often used tables of random numbers and entailed the large scale use of animals, whereas guided research selected certain experiments to perform and, therefore, used a much smaller number of animals. The second phase in reduction involved the concept of using statistical methods to segregate "controlled" from "uncontrolled" experimental variance, thus being able to determine the minimal number of animals needed for a given experiment. The last phase sought to further reduce experimental variation by controlling the variance between individual animals (i.e., by increasing physiologic uniformity among animals). Russell and Burch concluded that it was possible to vastly decrease the overall number of animals required in most studies if an appropriate experimental design process was selected and variation controlled, and further refined, by statistical methods.

Refinement involved the classification of procedures as either "stressful" or "neutral" investigations. The main objective of stressful investigations was to obtain specific knowledge regarding mechanisms of pain and distress. Neutral investigations, on the other hand, were studies that did not study pain and distress mechanisms directly and, therefore, any degree of stress produced in these studies had the potential to unintentionally affect the study. To refine "neutral" investigations, certain types of procedures could be added into the experimental design of the study, such as antiseptic technique, use of anesthetic agents, use of therapeutic intervention versus surgery, analgesics, euthanasia, improved blood sampling techniques, and selection of needle size for injections. It was also suggested that one should choose specifically one type of procedure over another to reach the given scientific objective. Applying this level of refinement was particularly useful for procedures that were long term or involved large numbers of animals. Scientists, when choosing a specific procedure, were cautioned to avoid elaborate methods, urged to formulate their scientific question carefully, and asked to consider the use of lower vertebrates.

Russell and Burch identified a variety of factors they believed would ultimately influence implementation of their principles of humane techniques. Individual attitudes and cultural mores concerning animals, including personality factors such as "authoritarian" (hostile attitude toward animals) and "revolutionary" (preferential treatment toward certain species) were considered crucial factors. The authors expressed concern that any decrease in experimental efficiency or accuracy, whether real or perceived, would limit acceptance and implementation of these principles by the scientific community. Because of the relative isolation of research laboratories in the 1950s, retrieval and dissemination of information regarding the availability, applicability, and utilization of new advances and techniques, and the lack of educational opportunities, also were considered to be possible deterrents. Russell and Burch viewed specialization within the laboratory environment as an isolating factor, leading to the lack of hybrid vigor, and recommended that specialists be incorporated into generalist scientific teams to provide greater balance in implementing these techniques.

B. Ethical Concepts and Legal Implementation

In retrospect, the universal and all-encompassing nature of Russell and Burch's Three Rs has become one of the essential ethical frameworks upon which U.S. law and subsequent regulations/policies have been formed. The pace of scientific inquiry progressed rapidly in the last half of the 20th century, and associated animal-based legislation was reflective of the evolving cultural and ethical morals. The first animal-related law in the United States, the 28-Hour Law, passed in 1893, focused on the prevention of cruelty to animals. This law ensured that farm animals were provided food, water, and rest at least once every 28 hours during transport. The next major legislation passed for the protection of animals occurred in 1966 in response to public outcry regarding a stolen pet. The Animal Welfare Act was enacted primarily to prohibit the sale of pet animals to research facilities and to require licensing of animal dealers and registration of research facilities. Further amendments to the Animal Welfare Act in 1985 and 1993, and the passing of the Health Research Extension Act (1985) and amendments, have progressively built upon the ethical issues advanced by Russell and Burch by implementing regulations and policies focused on the provision of enriched environments, increased

procedural monitoring, relief of pain and distress, and improved training for research personnel.

Scientists, animal care personnel, regulators, and the public often have conflicting views on what constitutes appropriate care for research animals. At the heart of the issue is the need to protect animals from undue pain and distress while still assuring that scientific advancement is ongoing and research goals are achieved. In designing and carrying out experiments, scientists must be aware of their ethical and regulatory responsibilities toward animals. Likewise, animal care personnel have the responsibility to ensure optimal housing conditions; health monitoring; treatment of experimentally induced and nonexperimentally induced disease; training of animal care and research personnel; and compliance with institutional, state, federal, and accreditation policies and laws. For both scientists and animal care personnel, compliance often can be a daunting task owing to the vast collection of laws, regulations, and recommendations from various oversight agencies that may govern one species, but not another, and that are sometimes contradictory. In addition, there may be disagreement between federal agencies, or between an agency and an institutional animal care and use committee (IACUC), on the most acceptable methods to meet the requirements. Standardization of enforcement between institutions has also been problematic. The scientific community has expressed concern that attempts by agencies to clarify and enhance regulatory compliance or accreditation standards may lead to changes in standards of practice, without the appropriate and required legislative changes or the opportunity to address proposed changes via the federal rule-making process. The public is generally supportive of the use of animals in research but this support is not without qualifications. Animals must be humanely treated, pain relief must be provided during potentially painful procedures, if indicated, and the study must be essential for obtaining new knowledge or medical advances (Fuchs, 2000; Tannenbaum, 1989).

The progressive move toward increased regulation of research animals since the 1980s has met with mixed reaction from the research and animal care community. There is general agreement that protection of animals from undue pain and distress and providing humane care is important, and that basic institutional and regulatory polices and procedures governing their care should be in place. However, investigators and animal care personnel have called for a systematic, scientific, and balanced approach to the creation and subsequent implementation of welfare-related regulations. Scientific data that objectively quantify and correlate the behavioral and physiologic effects of husbandry procedures, environmental enrichment, and/or experimental design changes are contradictory, inconclusive, or lacking (Stark, 2001).

Additional research is urgently needed to fill this void, and scientists must ensure that their data collection is not influenced by anthropomorphism, animal stress, or the introduction of inadvertent variables. Regulations and accrediting requirements based on sound scientific data will help ensure an improved, beneficial environment for laboratory animals, and will provide the proper checks and balances needed for public assurance and support the use of animals for scientific and medical advancement.

III. ENVIRONMENTAL ENRICHMENT FOR LABORATORY RATS

A. Contributions of Russell and Burch

Ethical considerations, regulatory requirements, and scientific findings support the need to investigate what constitutes a beneficial environmental enrichment for laboratory rats, and how that enrichment can be effectively implemented to support and improve animal welfare and research results. In 1959, Russell and Burch addressed the ethical need to consider environmental enrichment of research animals in the laboratory setting, with the goal of increasing refinement in the experimental process. In *The Principles of Humane Experimental Technique*, the authors link the concepts of *neutral* investigations and *inhumanity* with husbandry. They write:

> In neutral studies, the imposition of any degree of distress, however slight, is likely a priori to disturb the efficiency of the investigation . . . Nevertheless, the inadvertent imposition of any degree of distress must always introduce a source of confusion, which will find its ultimate expression in terms of cost, error and wasted effort. In neutral investigations, then, refinement is a major factor for success, and can be simply described as the elimination of contingent inhumanity. It merges into good husbandry, where, as we have seen, the general argument acquires still more cogency. (Russell and Burch, 1959).

In terms of either stressful or neutral investigations, the authors further distinguish procedures as resulting in either direct or contingent inhumanity. *Direct inhumanity* means inflicting unavoidable distress on an animal, even if the procedure is performed without error. *Contingent inhumanity* is the unintentional infliction of distress, either as a direct result of the procedure or as a by-product of the experimental process. Russell and Burch state,

> . . . contingent inhumanity is almost always detrimental to the object of the experiment, since it introduces psychosomatic disturbances likely to confuse almost any biological investigation. The incidence of contingent inhumanity will include the results of every conceivable kind of imperfection in the husbandry of laboratory animals . . . Where chronic experiments over days or

months are concerned, we cannot even in principle separate husbandry from the conduct of the experiment itself. (Russell and Burch, 1959).

In identifying the importance of recognizing and ameliorating the inadvertent effects of husbandry conditions upon the welfare of laboratory animals and experimental results, Russell and Burch laid the foundation for ethical and scientific investigational pursuit of environmental enrichment. The Animal Welfare Act, passed in 1966 and amended in 1970 and 1976, did not specifically address issues pertinent to environmental enrichment, per se, and the 1985 amendment focused on psychological enrichment for nonhuman primates and exercise of dogs. However, 25 years after Russell and Burch's introduction of the concept of refinement of animal procedures in relation to the necessity of good husbandry practice, the Health Research Extension Act of 1985 mandated the Secretary of the Department of Health and Human Services, acting through the Director of the National Institutes of Health (NIH), to establish guidelines for the proper care of animals used in biomedical and behavioral research. In implementing this legislative mandate, the Public Health Services Policy (*PHS Policy or Policy*), which incorporates the *U.S. Government Principles for the Utilization and Care of Vertebrate Animals Used in Testing, Research and Teaching*, refined the definition of proper care by stating, "The living conditions of animals should be appropriate for their species and contribute to their health and comfort." (Public Health Service Policy on Humane Care and Use of Laboratory Animals, 1986; Interagency Research Animal Committee, 1985). The *Guide for the Care and Use of Laboratory Animals*, or "Guide," published by the National Research Council and revised in 1996, further delineates the essential parameters of environmental enrichment for vertebrate animals by addressing both structural and social environments and animal activity. According to the Guide, an animal's structural environment includes any structure or object within their primary enclosure and this environment should:

> . . . include resting boards, shelves or perches, toys, foraging devices, nesting materials, tunnels, swings, or other objects that increase opportunities for the expression of species-typical postures and activities and enhance the animals' well-being; and, Animals should have opportunities to exhibit species-typical activity patterns. and, . . . continuing research into those environments that enhance the well-being of research animals is encouraged. (Committee to Revise the Guide for the Care and Use of Laboratory Animals, 1996).

The Guide clearly states that an animal's social needs should be considered and, when possible and species appropriate, animals should be housed with conspecifics. Singly housed animals must also be afforded the opportunity for positive human interaction and physical enrichment within their standard enclosure.

Since 1959, legislative, regulatory, and policy efforts have slowly progressed toward the recognition of the need to implement appropriate measures to improve animal well-being and to investigate the overall effects of environmental enrichment. Currently, the United States Department of Agriculture (USDA) regulations do not pertain to this issue, because the Animal Welfare Act, as amended in 2002, specifically excludes purpose-bred rats from the definition of "animals" covered by the Act. However, PHS Policy, which regulates PHS-funded activities, and the Association for the Assessment and Accreditation of Laboratory Animal Care, International (AAALAC), for accredited institutions, requires the IACUC to determine environmental enrichment needs for all vertebrate animals (including laboratory rats) housed in their respective institutions and to implement enrichment procedures as part of their animal care and use programs. As most research institutions have PHS-funded activities, AAALAC accreditation, or both, the vast majority of purpose-bred rats are covered either under PHS Policy, or are subject to the guidelines within the Guide. On this legislative and regulatory issue, the animal rights community continues to express its deep-seated concern that legal mandates are necessary to provide appropriate protection for laboratory rats, and that welfare issues, such as minimal cage size and other environmental enrichment issues, should not be left to the discretion of the laboratory animal community (Reinhardt and Reinhardt, 2001).

B.   Enrichment Goals

The primary welfare goal of environmental enrichment is to promote species-typical behavior, minimize or eliminate stress-induced behaviors, and promote physical comfort. In their natural environment, rats display species-typical behaviors, such as burrowing, nesting, and gnawing (Boice, 1997; Brain, 1992). Replication of an environment that meets most, or even some, of these normal behaviors within the nonnatural research setting is challenging and may, in the end, represent an uneven compromise between animal and research needs. In striving to reach these goals, research investigators and animal care personnel have used various methods to enhance environmental conditions within standard caging to enable rats to exert some degree of control over their environment, to allow shelter from light and/or cold, to provide an enclosed area for refuge or nesting activities, and to allow for social interaction.

C.   Testing Methods

Animal welfare-based research is in its infancy. A variety of testing methods have been used to assess whether environmental enrichment, by physical or social means, is truly beneficial. The simplest testing method involves placing inanimate objects in the cage, or increasing the numbers of animals per cage, and anecdotally reporting on observed behavioral effects. Although this method may serve as a starting point in the application of animal welfare principles and may provide subjective information to address in further investigations, it does not supply the objective data required for appropriate enrichment selection.

Preference testing has been used extensively in environmental enrichment studies and measures the frequency of "use" between different experimental options. When applied, the preferred, selected use of one option over another is determined to be the choice that provides the animal with the greatest psychological benefit (Patterson-Kane et al., 2001b). This testing method has been used to investigate whether laboratory rats prefer wire-bottom or solid-bottom cages, small or large cages, cages enriched with various objects, and solitary versus social living conditions. Three basic types of preferential testing include:

1. Demand testing–an animal must "work" in some manner to gain access to a barren cage, an enriched space, or cohabitation with conspecifics;
2. Dwelling time–different configured spaces are joined together, with or without inanimate objects, and the amount of time an animal spends in each space is recorded; and
3. T-maze choice–animals must choose between barren or physically or socially enriched areas before being trapped in one of the areas for a short period. Careful interpretation of preferential testing results is critical and may require additional tests to measure the strength of the preference observed (Duncan, 1992).

Radiotelemetry has recently allowed researchers to continuously record basic physiologic parameters (heart rate, mean arterial pressure, EKG, body temperature, activity) while unrestrained animals undergo behavioral testing. The ability to record a wide variety of parameters, without the confounding artifacts of animal handling and interaction, allows the interpretation and correlation of preferentially determined behavioral choices in relation to physiologic alterations (Rock et al., 1997; Sharp et al., 2003b). The combination of behavioral and physiologic methods allows more accurate documentation of environmental enrichment effects and provides essential baseline data that can be used to assist investigators in initial study design and final interpretation of experimental results.

Accurate assessment of experimental data in enrichment studies is a major concern within the scientific community. Numerous studies, which have analyzed the effects of different cage types, animal group size, human interaction, enrichment devices, and various physiologic parameters on animal well-being, have discovered that the addition of environmental enrichment itself adds new experimental variables and may inadvertently complicate evaluation of results (Augustsson et al., 2002; Brown and Grunberg, 1995). Comparison of results between studies may lead to erroneous conclusions, as scientific papers often do not include housing or enrichment information in their methods sections. Different types of enrichment may also affect the group size needed to achieve statistical significance and, therefore, it may be difficult to estimate the optimal number of animals required for a study, leading to the potential overuse or underuse of animals (Erb, 1990; Mann et al., 1991; Mering et al., 2001). Through careful selection of environmental enrichment testing procedures and accurate interpretation of their results, investigators, animal care personnel, and regulators will be able to make informed decisions regarding enrichment selection. Choices based on anthropomorphic, anecdotal, or preferential judgments alone may, in fact, have no measurable benefit or negatively affect research.

D.   Cage Size, Design and Flooring

The Guide recommends appropriate sizes of rat cages, based on body weight, but also states that other performance measures, such as activity level, behavior, reproductive status, age, social considerations, and health should be used to assess overall space needs. It has been demonstrated that although rats can thrive in relatively small spaces, it is the quality of the living space that has the greatest behavioral and physiologic influence (Patterson-Kane, 2002). Increasing the complexity of the cage design through structural modification has been achieved through the use of platforms, barriers, and walls (Anzaldo et al., 1995; Denny, 1975), and enlarging cage dimensions may lead to group housing. Although potentially beneficial, these environmental changes also can lead to difficulty in identification of individual animals and less opportunity for human interaction (Augustsson et al., 2002).

Radiotelemetry studies have been used recently in the investigation of appropriate cage size for rats, measuring preference for one cage over another, based on activity level. When adult rats were given a choice of larger versus smaller cages, they showed a moderate preference for the larger cage, regardless of whether cagemates were present (Patterson-Kane et al., 2001a; Patterson-Kane, 2002). In a separate study using radiotelemetry, rats housed four per cage were provided 60% (920 cm^2) or 80% (1,250 cm^2) of the recommended floor space stated in the Guide and then subjected to a variety of stressful husbandry or experimental procedures. Physiologic parameters (heart rate, mean arterial blood pressure, and activity) were recorded and compared. Results of this study determined that stress parameters were reduced to a greater degree by

social grouping in a smaller space rather than by providing animals a larger space (Sharp et al., 2003b). Conversely, other studies have documented no significant difference in physiologic parameters between isolated or group-housed rats and cage size (Brown and Grunberg, 1995; Giralt and Armario, 1989).

Laboratory rats are most commonly housed in either solid-bottom plastic cages or hanging cages with wire-bottom floors. Research institutions must weigh regulatory, economic, animal welfare, and experimental factors when selecting one flooring type over another, and experimental results on this issue are inconclusive. The Guide recommends the use of solid-bottom versus wire-bottom flooring for rodent species, and a number of studies have demonstrated that rats appear to avoid cages with wire flooring (Manser et al., 1998; Patterson-Kane, 2003; van de Weerd et al., 1996). Manser et al. (1996) further documented that rats either selected, or worked, to obtain access to solid-bottom floors over wire-bottom floors, irrespective of their previous housing condition. An increased incidence of foot lesions (ulcers and swelling) was reported in rats if they were housed on wire-bottom floors for more than a year (Peace et al., 2001). Conversely, Holson et al. (1991) studied adult rats housed long term in hanging cages and found no hormonal, neurochemical, or anatomical evidence of stress. Manser et al. (1995) also showed no differences in food and water consumption or body weight in rats housed on grid versus solid flooring, and Sherer et al. (2001) reported a preference among Sprague-Dawley rats for wire-bottom cages. A greater activity level was observed in singly housed rats on wire-bottom floors than solid-bottom floors, and multiply housed rats on wire-bottom floors had increased food consumption relative to multiply housed animals on solid-bottom floors (Rock et al., 1997). When the two methods of housing were compared for potential effects upon physical and behavioral functioning, no correlation was observed between the type of flooring and the ease of animal handling (Manser et al., 1995; Rock et al., 1997). Because of conflicting research results involving cage sizes, design, and flooring, further studies are needed to provide additional objective information (Stark, 2001).

E.   Bedding, Nesting Material and Nest Boxes

Bedding, nesting material, and nest boxes placed within standard solid-bottom cages affect air quality, increase the comfort of the cage, and allow rats the opportunity for species-typical behavior, such as digging, chewing, and nest building. Many different factors may contribute to a rat's choice of one bedding material over another, including softness of the material, familiarity with a particular substance, and size of the material. The use of different bedding materials, including hardwood chips, sawdust, shavings, corn cob, and paper strips, have been investigated. Blom et al. (1996) and Ras et al. (2002) found that hardwood chips in solid-bottom cages were preferred over corn cob bedding and proposed this preference was the result of increased comfort or familiarity with the substance. Rats also preferred bedding materials composed of larger pieces, such as shavings or paper strips (Patterson-Kane, 2001b; Ras et al., 2002; van de Weerd et al., 1996).

The use of nest boxes, nesting material, and conduits adds another dimension to cage enrichment and allows animals the opportunity to control their exposure to light and temperature and to escape from dominant or aggressive cagemates. Using preferential testing, rats selected nest boxes, nesting materials, and conduits over empty cages, regardless of whether bedding material was present (Manser, et al., 1998), and nest boxes were preferred over nest material alone (Manser, et al., 1998; Patterson-Kane et al., 2001a; Patterson-Kane, 2003). When offered a choice, rats selected nest boxes that were opaque (vs. transparent or semitransparent), composed of plastic (vs. metal or cardboard), and that had small openings (vs. large openings or open sides) (Manser et al., 1998; Patterson-Kane et al., 2001a; Patterson-Kane, 2003). There is general consensus in the scientific literature that rats prefer some type of nesting material or nest box in their home environment.

F. Inanimate Objects (Toys and Tubes)

Inanimate objects, such as balls, blocks, and tubes, are alternative options for providing enrichment within the cage environment. For ease of use, animal comfort, and health, the selected object(s) should not take up floor space and should be easily sanitized (or disposable), long lasting, nontoxic, and allow species-typical behavior such as gnawing, chewing, and burrowing. When given a choice between 15 different objects, rats showed a strong preference for wood blocks with multiple holes (Chmeil and Noonan, 1996). Wistar rats chewed, and physically moved, nylon balls and blocks throughout their cages with no observed adverse effects on weight, food consumption, or biochemical and hematologic values (Watson, 1993). Adult rats undergoing demand testing, however, preferred the company of cagemates over the opportunity for access to larger cages or toys (Patterson-Kane et al., 2001a). The use of lengths of PVC tubing, placed in the cage to simulate a burrow, has been suggested in the *Guide to the Care and Use of Experimental Animals* (Guide, 1980) and in the Committee to Revise the *Guide for the Care and Use of Laboratory Animals*, 1996). However, when the use of tubing was investigated, both acceptance and use were found to vary widely, depending on the strain of rat, housing conditions (singly vs. multiply), time of day, sex

(some females preferred, some males ignored), and age (increased use by juveniles) (Bradshaw and Poling, 1991; Chmeil and Noonan, 1996; Galef and Sorge, 2000; Patterson-Kane, 2003). In some studies, PVC tubing as an enrichment device was virtually ignored (Chmeil and Noonan, 1996; Galef, 1999). Inanimate object preference in rats has been shown to be extremely selective and, therefore, choosing the object, or objects that will meet the general enrichment needs for all rats housed within an institution will be challenging.

G. Single versus Group Housing

Rats appear to prefer to be housed in groups (Patterson-Kane et al., 2001a; Patterson-Kane et al., 2001b; Sharp et al., 2003b), but there is a divergence of opinion as to whether isolation housing, as opposed to group housing, leads to stress or other physiologic changes. Stress is typically indicated by changes in behavioral (activity, aggression), physiologic (body temperature, heart rate, blood pressure), anatomical (adrenal weight), or biochemical (corticosterone) parameters. In a study of the capacity of male rats to adapt to chronic stress, cage density was not shown to influence any measured physiologic variables, with the exception of elevated corticosterone levels (Giralt and Armario, 1989). Holson et al. (1988) initially reported behavioral and adrenocortical evidence of an "isolation stress syndrome" in adult rats undergoing open field testing, which were reared singly from weaning to adulthood in hanging cages. However, upon re-examination of this study, the authors determined that if the rats housed in hanging cages were handled twice per week, the previously reported isolation syndrome did not occur (Holson et al., 1991). Sharp et al. (2002 and 2003a) used radiotelemetry to evaluate the effects of single versus multiple housing in Sprague-Dawley rats. Male and female rats housed individually or in groups of two to four were subjected to a variety of routine husbandry and experimental procedures, and heart rate, mean arterial blood pressure, and activity were measured. Lower cardiovascular values were recorded for multiply housed rats when stressors were applied than singly housed animals, and the length of the stress response observed was blunted in the group-housed situation. Additionally, pair housing was not shown to decrease physiologic parameters as much as housing rats four per cage. Rock et al. (1997) showed evidence that the diastolic blood pressure of multiply housed adult rats, whether housed on wire-bottom or solid-bottom cages, was significantly reduced. In contrast, Harkin et al. (2002) documented a greater increase in activity, heart rate, and body temperature in pair-housed versus singly housed Sprague-Dawley rats. Male Wistar rats, under crowded conditions, were found to have higher corticosterone levels, whereas female Wistar rats were shown to have elevated corticosterone levels in response to being individually housed (Brown and Grunberg, 1995). Greco et al. (1992) measured changes in the circadian rhythm of norepinephrine, epinephrine, corticosterone, aldosterone, and serotonin in singly housed versus multiply housed adult male rats and determined that isolation only increased circulating norepinephrine, epinephrine, and corticosterone levels. Some studies also have documented that social housing may lead to increased aggressive behavior, predominately in male rats (Galef and Sorge, 2000; Hurst et al., 1996). Work by Hurst et al. (1999) demonstrated that the specific composition of individuals within a group housing situation had the greatest impact on an animal's welfare, rather than the size of the group, *per se*. The authors further suggested that careful observation of certain negative behaviors, such as aggressive grooming and bar-chewing, could be used successfully to group compatible animals.

The scientific literature describes a wide range of potential positive and negative behavioral and physiologic effects and selection preferences associated with modifying housing or social groupings to enrich the environment for laboratory rats. Investigators and animal care personnel have the responsibility to evaluate objectively these data and, if lacking or inconclusive, must support research efforts to provide essential information. With sufficient information, investigators, animal care personnel and regulators can appropriately weigh physiologic and behavioral results, in terms of research and animal welfare goals, and make fully informed decisions regarding enrichment choices that will afford the greatest animal welfare benefit.

IV. LEGAL AND REGULATORY ISSUES

A. U.S. Legislation, Policies, and Guidelines Governing Laboratory Rats

1. The Health Research Extension Act

The Health Research Extension Act of 1985 (Pub.L. 99-158) is the primary piece of U.S. legislation that governs the care and use of animals used in research (Health Research Extension Act of 1985). The law requires the Secretary of the Department of Health and Human Services, via the Director of NIH, to set guidelines for the proper care, treatment, and institutional oversight of live animals used in biomedical and behavioral research. Proper treatment under the law includes the appropriate use of drugs during research procedures (anesthetics, analgesics, tranquilizers, and paralytics), the provision of

perioperative veterinary care, and humane methods of euthanasia. The formation of an IACUC is required at all institutions that receive PHS funding or conduct PHS-sponsored studies, including NIH and the national research institutes. Institutions must file an annual report, including any minority views, with the Director of NIH, which certifies their compliance with the review process and documents any current institutional violation of guidelines or assurances. The institution must assure that all personnel involved with research animals are offered training opportunities in the proper conduct of experimental procedures involving animals, methods to reduce the number of animals used, and ways to minimize animal distress. The law grants the Director of NIH the authority to suspend or revoke funding of PHS projects if the institution has received notification of a violation and has not taken appropriate corrective action within a reasonable time. The Health Research Extension Act specifically excludes any prescription of research methods or the disclosure of trade secrets or privileged or confidential information.

2. Public Health Service Policy

The Public Health Service Policy (*Policy*) was written in 1986 to fulfill the legislative intent of the Health Research Extension Act of 1985. The Policy defines "animal" as "Any live, vertebrate animal used or intended for use in research, research training, experimentation, or biological testing or for related purposes." (Public, 1986). The document specifies the applicability of the Policy, provides basic definitions, and describes implementation by institutions (Animal Welfare Assurances, IACUC composition and function, animal care and use protocol review, PHS application and awards information, recordkeeping, and reporting requirements) and by PHS (OLAW and PHS awarding units responsibilities, special review/site visit conduct, waiver). The Policy is intended to implement and supplement the principles outlined in the *U.S. Government Principles for the Utilization and Care of Vertebrate Animals Used in Teaching, Research and Training*, established by the Interagency Research Animal Committee (IRAC) in 1985. All PHS-supported institutions and entities using animals, including those in the United States, Puerto Rico, or U.S. territory or possession, must comply with the Policy, and institutions in foreign countries involved in PHS-conducted or supported activities must either comply with the Policy or show evidence that equivalent standards are being met. All institutions covered under the PHS Policy are responsible for complying with the Guide, the Animal Welfare Act, and other federal, state, or local laws and standards, if applicable. Federal agencies that require institutional compliance with PHS Policy include: the Agency for Health Care Policy

Research; the Agency for Toxic Substances and Disease Registry; the Centers for Disease Control and Prevention; the Food and Drug Administration; the Health Resources and Services Administration; the Indian Health Services; the National Institutes of Health; and the Substance Abuse and Mental Health Services Administration.

3. NIH Office of Laboratory Animal Welfare

The Office of Laboratory Animal Welfare (OLAW), in the Office of Extramural Research at the NIH, is responsible for administering and implementing the PHS Policy. The Director of NIH has authorized the agency to negotiate and approve Animal Welfare Assurance of Compliance applications (Assurances); evaluate institutional compliance with the requirements of the Policy; and provide information and educational opportunities regarding policy provisions, IACUC procedures and issues, ethical considerations, and procedural techniques.

a) ANIMAL WELFARE ASSURANCE OF COMPLIANCE (ASSURANCE). Institutions that receive PHS support for activities involving the use of live vertebrate animals in research, training, biological testing, or other animal-related activities are required to submit a written Assurance to OLAW for review and approval. The Assurance must describe the process by which animal care and IACUC guideline requirements will be met and provide justification for the experimental use of animals. The institution also must assure that all personnel involved with research animals are offered training opportunities in the proper conduct of experimental procedures using animals, methods to reduce the number of animals used, and ways to minimize animal distress. No animal-related activities, conducted or supported by PHS, may take place at an institution without an approved Assurance on file, and receipt of approval commits the institution to fulfill all requirements as described. After application review, or if an institution is determined to be out of compliance with their approved Assurance, OLAW has the authority to disapprove, restrict, or withdraw the Assurance. The agency maintains lists of domestic and foreign institutions with approved Assurances. These lists are posted on the OLAW Web site (http://grants1.nih.gov/grants/olaw/) and are distributed to the NIH Scientific Review Administrators of initial review and technical evaluation groups.

b) COMPLIANCE WITH PHS POLICY. The IACUC is responsible for achieving and monitoring adherence to the PHS Policy through voluntary compliance with their institution's approved Assurance. OLAW's regulatory role is to provide general oversight and evaluation of Policy compliance by performing routine site visits, conducting site reviews to evaluate alleged noncompliance, and instituting appropriate corrective and preventative

measures for confirmed incidents of noncompliance. Alleged incidents of noncompliance are initially reviewed, addressed, and documented internally by the IACUC, and final outcomes are reported to OLAW. Allegations of noncompliance also may be brought against the institution by anyone, including animal rights groups, private or anonymous individuals, other federal agencies, or the media.

The 1995 Interagency Memorandum of Understanding between PHS, USDA, and the FDA allows these agencies to notify OLAW regarding animal activities discovered during their inspection process that may not be in compliance with PHS Policy (Memorandum of Understanding Among the Animal and Plant Inspection Services, 1995). Once notified by the institution or other entities, OLAW may initially evaluate the incident through an exchange of information with the institution. Evaluations are based on the provisions of the Policy; an institution's approved written Assurance; compliance reports provided by the institution that describe the alleged incident; the internal corrective measures implemented and outcome; and other relevant information.

After comprehensive review, if OLAW determines that the incident has not been fully addressed at the level of the institution, several different actions may be implemented. The institution may be required to provide OLAW with additional information, or OLAW may directly contact other PHS departments, the USDA, or other federal agencies and request information. OLAW may require the institution to institute interim corrective measures or temporarily suspend the Assurance, if the agency deems it necessary for animal welfare concerns.

OLAW also may elect to perform an onsite review at the institution. When the compliance evaluation is complete, OLAW will notify the institution of its findings. Institutions may be found to be in full compliance with PHS Policy and their Assurance. However, OLAW may recommend that certain policies and procedures be added or modified to enhance compliance. OLAW may decide to restrict a portion of the institution's Assurance or suspend specific research projects until corrective measures are implemented. Before restrictions can be lifted, prereview of the amended project, additional training of research personnel, or more rigorous reporting requirements may be required. In serious incidents of noncompliance, evaluation findings may result in the withdrawal of the Assurance, which will restrict the affected research project from receiving PHS support until a new institutional Assurance is filed and approved. If OLAW determines, based on its evaluation, that an official report of findings will be issued to the institution, the report will be forwarded to the institution and to the complainant, who will be allowed to provide written comments to identify errors of fact. Once these documents are received, the final report will be issued to the institution and to the complainant. Final reports, once issued, may be subject to disclosure through the Freedom of Information Act.

c) EDUCATIONAL RESOURCES. OLAW serves as an important resource for individuals and institutions by providing written and Web-based information on PHS Policy interpretation, implementation, and compliance; the structure, function, and maintenance of IACUC's and animal care and use programs; training materials for research and animal care personnel; and ethical issues surrounding the use of laboratory animals. Sample documents, which institutions can use to facilitate Policy implementation and document institutional compliance, can be downloaded from the OLAW Web site, including the Animal Welfare Assurance of Compliance, OLAW Annual Report, and Semi-annual Program and Facility Review Checklists. OLAW also has published, and continues to publish, articles on selected PHS Policy provisions in the *OPRR Reports* (also known as "Dear Colleague" letters), the *NIH Guide for Grants and Contracts*, *Lab Animal*, *Contemporary Topics in Laboratory Animal Science*, and *ILAR News*. These articles are listed under the "Guidance" section of the Web site by general topic ("Published OLAW Guidance" and "Compilation of OLAW Guidance"); by date ("Dear Colleague" letters and NIH Notices); or by journal title ("Published Articles"), which allow easy referencing of materials. The articles, which describe OLAW's past and present interpretation and implementation of Policy provisions, provide a valuable source of information that demonstrates the historical evolution of Policy issues and methods proposed to address implementation and compliance. Additional references, such as the *Institutional Animal Care and Use Committee Guidebook* (2nd ed.), copublished by the Applied Research Ethics National Association (ARENA) and OLAW; "How to Write an Application Involving Research Animals" (National Institute of Allergy and Infectious Disease); the *Institutional Administrator's Manual for Laboratory Animal Care and Use* (NIH publication No. 88-2959); and the "LAMA Disaster Preparedness Resource," cosponsored by the Laboratory Animal Management Association and OLAW, are available on the OLAW Web site and are invaluable resources for institutional officials and support personnel when designing and managing their animal care and use programs. Additional IACUC and animal program-related links and Web resources also are listed as references.

OLAW offers a wide range of educational training opportunities throughout the United States on an annual basis. Training workshops and symposia are offered regionally and are often cosponsored with organizations such as the Scientists Center for Animal Welfare (SCAW), the National Science Foundation (NSF), the Public

Responsibility in Medicine and Research (PRIM&R), ARENA, the Animal Welfare Information Center (AWIC), the Institute for Laboratory Animal Research (ILAR), the American Association for Laboratory Animal Science (AALAS) and the American College of Laboratory Animal Medicine (ACLAM), in conjunction with state biomedical research association organizations and local academic institutions. These programs are open to research investigators, program administrators, IACUC members, veterinarians, and animal care personnel and provide current information in basic and advanced IACUC training, animal welfare and regulatory compliance, development of science-based guidelines in laboratory animal care, information requirements of the Animal Welfare Act, laboratory animal management and technology, the use of nonhuman primates, and applied research ethics. These training opportunities also serve as open forums for discussion of regulatory, compliance, and management issues. Additional training materials accessible on the OLAW Web site include audiovisual materials such as "Working Safely with Nonhuman Primates," "Working with the Laboratory Dog–Training for the Enhancement of Animal Welfare in Research," and "Training in Survival Rodent Surgery." The Web site offers a direct link to the Veterans Administration Office of Research and Development's training program entitled, "ResearchTraining.org" which provides research personnel access to Web-based training modules and associated self-assessment examinations, in topics such as antibody production procedures, use of hazardous materials, preparation of animal care and use protocols for different animal species, responsibilities and procedures for IACUC members and investigators, and postprocedural care of rodents.

4. NIH Revitalization Act

In 1993 the Health Research Extension Act was amended to add the "Plan for Use of Animals in Research" also known as the NIH Revitalization Act (Pub.L. 103-43). This amendment directed NIH to prepare a plan to investigate and validate methods to identify nonanimal alternatives, reduce the number of animals used in research, reduce animal pain and distress, and find alternatives to the use of marine mammals. NIH was given the responsibility to encourage the scientific community to accept viable alternative approaches and provide training necessary to implement new approaches. To meet the legislative objectives, the NIH Director was directed to establish the Interagency Coordinating Committee on the Use of Animals in Research (Committee), whose membership was to include the Directors of each national research institute, the Director of the Center for Research Resources, and representatives from the Environmental Protection Agency (EPA), the Consumer Product Safety

Agency, the National Science Foundation, and the FDA, with at least one representative required to be a laboratory animal medicine veterinarian. The Committee was convened and, in August 2003, released the NIH Plan for the Use of Animals in Research (NIH Plan). The NIH Plan highlighted and described selected ongoing biomedical research initiatives, detailed NIH-supported activities initiated since 1986, and outlined the agency's plan to address the provisions of the NIH Revitalization Act.

Specific research initiatives to be instituted or supported were outlined in the NIH Plan. Previous NIH Program Announcements would be re-evaluated to determine if new areas of research could be supported that would address the legislative objectives. NIH support would be made available for applications that focused on the development of nonmammalian model systems and other methodologies to reduce the number of animals used in biomedical research and testing. Solicitations would be expanded for Small Business Innovation Research (SBIR), Small Business Technology Transfer (STTR) projects, and Academic Research Enhancement Awards to develop technologic approaches and model systems to promote the acquisition of basic biological information to assist in the search for alternatives to animal use. NIH support of ongoing projects that used in vitro model systems for the prescreening of therapeutic compounds and for toxicologic testing of environmental chemicals would continue and new solicitations would be issued. NIH, through the National Center for Research Resources, would issue solicitations to support improvement of research resources through the upgrade of animal facilities and equipment; investigation of methods for disease prevention, detection, and treatment; and cryopreservation of embryos in a broad-based effort to reduce the overall number of animals used. NIH would continue to support resource centers that produce and supply nonmammalian biomaterials, such as cell lines, human tissues, microorganisms, and invertebrates, and support core centers that use nonmammalian aquatic species to study human environmental health issues. Collaboration between computer experts and health researchers would be encouraged to identify and expand the use and availability of nonanimal models.

The NIH plan also outlined specific mechanisms that would be used to disseminate information about the Health Research Extension Act and to encourage acceptance of these new provisions by the scientific community. An Advisory Panel was established to review technology advances and make recommendations to NIH regarding the effectiveness of the new technology to meet the legislative objectives. The National Library of Medicines database, entitled "Alternatives to the Use of Live Vertebrates in Biomedical Research and Testing" was expanded and new resources for the evaluation and comparison of toxicologic test procedures was developed.

Interaction between NIH, other federal agencies, private industry, and national scientific organizations was encouraged to identify alternative-based methodologies for procedures, such as eye irritation testing, dermal irritation and sensitization testing, and to address the policy of reuse of animals in testing procedures. Initial information regarding the NIH Plan, as well as future updates and revisions, would be disseminated by NIH and other scientific organizations. NIH would continue to promote and support national educational opportunities through conferences, workshops, symposia, and publications to encourage acceptance and implementation of new methodologies by the scientific community.

The NIH Plan called for the incorporation of additional training in new technologies, such as mathematics and computer science, for future biomedical and behavioral scientists enrolled in training programs, such as the NIH training grants and predoctoral and postdoctoral fellowships. In addition, training programs and reference materials for research investigators, IACUC members, and animal care personnel in the humane care and use of laboratory animals, alternatives to animal use, and reduction of animal numbers would be provided by the NIH Office for the Protection from Research Risks (now known as the Office for Laboratory Animal Welfare).

5. Good Laboratory Practice for Nonclinical Studies

The Food, Drug, and Cosmetic Act required the Department of Health and Human Services to ensure the proper care and use of laboratory animals used in nonclinical safety testing. In 1979, the FDA published the Good Laboratory Practice (GLP) regulations, which established compliance standards for the acquisition of data (Code of Federal Regulations, 1998). Data collected under GLP standards are provided to the FDA as part of an application to obtain research or marketing permits for drugs, biologics, and medical devices used in humans and animals. Under the Health Research Extension Act of 1985, and the resultant Public Health Services Policy of 1986, the FDA is one of the federal agencies required to comply with the PHS Policy, which includes the requirement to use the *Guide for the Care and Use of Laboratory Animals* as the major standards document for institutional programs using laboratory animals in GLP studies. Specific GLP standards that pertain directly to animal care and use include: review of the animal program to ensure that housing and procedures do not induce stress that could influence research results; inspections to ensure that all husbandry standard operating procedures are being followed and parameters documented; proper maintenance of environmental conditions; isolation and evaluation of new and/or sick animals; and testing of feed and water. The FDA may conduct data audits at institutions to assess compliance with animal-related standard operating procedures, including husbandry, animal identification, food and water consumption, health surveillance reports, and necropsy/pathology findings. If audit reports indicate areas of noncompliance, the FDA will review the findings and, depending on the degree of severity, may issue a warning, reject a study, disqualify an institution as a study site, or revoke a permit.

6. Guide for the Care and Use of Laboratory Animals

The Guide was first published in 1963, as the *Guide for Laboratory Animal Facilities and Care*, and revised six times between 1965 and 1996. The purpose of the Guide, as expressed in the preface of the 1996 edition, is "...to assist institutions in caring for and using animals in ways judged to be scientifically, technically, and humanely appropriate." And, "...recommendations are based on published data, scientific principles, expert opinion, and experience with methods and practices that have proved to be consistent with high-quality humane animal care and use." (Committee to Revise the *Guide for the Care and Use of Laboratory Animals*, 1996). The Guide is widely viewed as the "gold standard" for humane laboratory animal care in the United States, providing recommendations on institutional policies and responsibilities; animal environment, housing, and management; veterinary medical care; and physical plant. Included in the Guide are many of the requirements of the Animal Welfare Act regulations, and the document is used by both the PHS and AAALAC as the primary standards document for animal program evaluation. The Guide encourages performance-based versus engineering-based standards for humane care in that, although goals must be achieved, the methods of reaching the goals are determined by the institution.

B. Reduction of Regulatory Burden in Laboratory Animal Welfare

In 1998 the House Committee on Appropriations instructed the NIH to review current federal regulations governing scientific research in the United States and develop mechanisms for streamlining regulations while maintaining appropriate and mandated levels of protection. One of the five areas of inquiry was animal care and use. In 1999, the "NIH Regulatory Burden: VI. Animal Care and Use–Workgroup Report" (NIH, 1999a) was issued for public comment, followed by the "NIH Regulatory Burden–Three-Month Plan" (NIH, 1999b). The 3-month plan called for OLAW to provide guidance to institutions on synchronizing reporting periods and using AAALAC activities for semiannual program evaluations. Institutions covered under PHS Policy or accredited by

AAALAC must submit annual program reports, but the date of submission for these reports is not specified. The USDA annual report, however, is required by the AWA regulations to be filed on, or before, December 1 for institutions covered under the animal welfare regulations. It was determined that regulatory burden could be reduced if institutions could submit reporting information over the same period for all three entities. In response, both OLAW and AAALAC modified their reporting timeframes to allow institutions to use the USDA reporting period. PHS Policy requires semiannual animal care and use program reviews and facility inspections and allows the institutional IACUC to use *ad hoc* consultants during the evaluation process. Under the regulatory burden reduction plan, both USDA and PHS will allow the use of a recent AAALAC program assessment and report to fulfill the requirement for a semiannual IACUC program evaluation as long as specific criteria are met.

C. U.S. Animal Welfare Act (1966–1990)

In 1966, the Laboratory Animal Welfare Act (Pub.L. 89-544) was enacted in response to public concerns that companion animals were being stolen and sold to research facilities. The law focused primarily on activities of animals in interstate and foreign commerce. Dealers who sold dogs and cats for research purposes were required to be licensed, and the facilities that used them were required to be registered. Mandated standards of care were required for these species; however, this requirement only pertained for the period before and after research use and excluded care to be provided during experimental procedures. The term "animal" was defined as "...any live dog, cat, monkey (nonhuman primate mammal), guinea pig, hamster and rabbit..." Regulatory and enforcement authority was also granted to the USDA pursuant to this law.

In 1970, Congress amended the Laboratory Animal Welfare Act, which was renamed the Animal Welfare Act (Pub.L. 91-579). The definition of "animal" and protection provided based on their intended use was greatly expanded under this amendment. Animals used in research, teaching, exhibition, and the wholesale pet industry were now included and the definition of "animal" included "any live or dead dog, cat, monkey (nonhuman primate mammal), guinea pig, hamster, rabbit, or such other warm-blooded animal, as the Secretary may determine is being used, or is intended for use, for research, testing, experimentation..." This legislation granted discretionary authority to the Secretary of Agriculture to determine which warm-blooded species would be regulated if they were not specifically named. Other legislative changes included the requirement for registration of research facilities, licensure for zoos, redefinition of the term "dealer," expansion of standards of care to include use during the experimental period, and the requirement for submission of detailed annual reports by research facilities.

Between 1970 and 1990, the Animal Welfare Act was amended three more times. The second amendment in 1976 established transportation standards for animals and prohibited transportation of animals for game sporting. In the 1985 amendment, the Food Security Act, legislation specifically focused on issues regarding the experimental use of animals, including the need for further investigation and development of alternatives to the use of live animals, methods to minimize or eliminate duplicative research, and measures to address public concerns regarding the use of animals for research purposes. The third amendment in 1990, the Pet Theft Act, mandated a 5-day holding period for dogs and cats held at municipal pounds, shelters, and research facilities so that owners, or other interested parties, had the opportunity to claim or adopt animals before they were sold or used in research.

D. USDA Regulations Related to Rats, Mice and Birds

1. Regulatory Process

The regulatory process begins at the Congressional level with the passing of legislation. The 1970 Animal Welfare Act amendment gave the Secretary of Agriculture the regulatory authority to implement Congressional intent. The Animal and Plant Health Inspection Service (APHIS), a branch of the USDA, has administrative authority over the subsequent regulatory rule-making process. This process includes publishing proposed regulations in the *Federal Register* with a timeline for public comment, review of all written and oral input, and issuance of a final rule with an effective date for compliance. The Animal Welfare Act amendment in 1970 did not include rats, mice, and birds in the expanded definition of "animal." The Secretary of Agriculture, after closure of the rule-making process, chose to exclude laboratory-bred rats, mice, and birds from the Animal Welfare Act Regulations published in 1971. In subsequent legislative amendments to the Animal Welfare Act (1970–1990), neither the definition of "animal" nor the authority vested in the Secretary of Agriculture to make the final decision regarding covered species was changed.

2. Regulatory Exclusion/Inclusion of Rats, Mice and Birds

a) PETITIONS. In 1989, the Animal Legal Defense Fund and the Humane Society of the United States unsuccessfully petitioned the USDA to remove the exclusion of rats, mice, and birds from the Animal Welfare Act regulations. A subsequent lawsuit was denied on legal technicalities. In 1997 the American Anti-Vivisection Society petitioned the USDA to amend the regulations. The USDA denied the

petition and two lawsuits were filed. The first, the Animal Legal Defense Fund v. Madigan (781 F. Supp. 797 D.D.C. 1992) was vacated *sub nom.* In the second lawsuit, the Animal Legal Defense Fund v. Espy (23 F.3d 496), a decision was vacated because the plaintiffs were found to lack standing to sue.

In April 1998 a petition was filed with the USDA from Alternatives Research and Development Foundation (ARDF), a group affiliated with the American Antivivisection Society, in vitro International, and other private individuals. The petition listed two requests. The first request was to initiate rule-making proceedings to amend the definition of "animal" to eliminate the exclusion of rats, mice, and birds. The proposed regulatory change for the definition of "animal" was, "any live or dead dog, cat, nonhuman primate, guinea pig, hamster, rabbit, or any other warm-blooded animal, which is being used, or is intended for use in research, teaching, testing, experimentation, or exhibition purposes, or as a pet." (Code of Federal Regulations, 1999b). The second request was to "grant such other relief as the Secretary deems just and proper." (Code of Federal Regulations, 1999b). The petition further contended that the decision of the Secretary of Agriculture to exclude rats, mice, and birds was "arbitrary and capricious, an abuse of agency discretion and otherwise not in accordance with law." (Code of Federal Regulations, 1999b). The petition argued that the USDA must extend Animal Welfare Act coverage to these species because the intent of the original language of the law was inclusive of "warm-blooded animals." Their exclusion from the regulations, therefore, resulted in a lack of humane protection as required by law. In response to the petition, the USDA initiated the rule-making process by publishing the Notice of Proposed Rulemaking, as required, in the *Federal Register* in January 1999. In this document, they stated their opposition to extension of coverage for these animals. Comments from the public and the scientific community on this controversial issue were solicited until March 1999.

b) VIEWS ON REGULATORY INCLUSION–USDA AND ARDF. The USDA opposed the proposed regulatory revisions for two major reasons. The agency maintained that Congress had given the Secretary of Agriculture the legal authority to include, or exclude, certain species of animals from regulatory oversight. Appropriate animal care and use protection was already in place for the vast majority of research facilities using rats, mice, and birds either through the PHS Policy or AAALAC. Both entities utilize the Guide as a primary resource for animal program review and evaluation, and Guide standards are generally consistent with, or may exceed, standards in the Animal Welfare Act Regulations. With the large number of AAALAC-accredited facilities, many of whom were also

USDA regulated, an additional layer of regulation would not lead to a greater level of protection. A second justification for exclusion of these species was the obligation of the USDA to manage effectively the limited resources appropriated by Congress for regulatory activities. Annual appropriations from Congress for enforcement remained steady in the 1990s but did not include inflationary adjustments. This shortfall in funds resulted in a progressive decrease in the number of APHIS personnel available for enforcement activities for species currently covered under the Animal Welfare Act. A survey performed by APHIS in 1990 analyzed the potential fiscal effects on regulatory enforcement activity if research rats, mice and birds were added to the regulations. It was determined that the number of regulated facilities would double and 50 additional veterinary staff members would be required to handle the inspection load, at a minimal annual cost of $3.5 million. A similar study, undertaken in 1999 by the Library of Congress, showed that if these species were included, APHIS would be responsible to provide enforcement for an additional 500 million animals. In light of these reports, the USDA expressed deep concern that if additional funds were not appropriated to cover additional species, regulatory enforcement of species already covered would be adversely affected and animals put at risk. New and existing regulated facilities would also incur enormous regulatory-related expenditures due to the expanded documentation requirements, increased personnel needs and potential requirements for new caging and equipment. The USDA, although opposed to inclusion, nevertheless proposed four regulatory options in answer to the petition: (1) regulate all rats, mice, and birds; (2) regulate all rats, mice, and birds at research facilities only; (3) regulate rats and mice only; or (4) maintain the status quo.

The petitioners, Alternative Research and Development Foundation et al., also expressed their viewpoints on the issue of inclusion. They stated that the USDA, in the 1970 Animal Welfare Act Regulations, illegally defined "animal" by excluding rats, mice, and birds and therefore provided a disincentive for the use of alternatives, which impedes the use of nonanimal alternatives. If a species was not designated as an "animal" in the law, then no legal protection was in place to mandate humane care of these animals and the appropriate search to identify possible alternatives to the use of animals and procedures that may cause pain. Legal protection was imperative since between 90% and 95% of research animals were rats and mice and, therefore, the majority of animals used in research were not provided access to care and treatment to minimize pain and distress or the option not be used at all, as alternatives to animal use and procedures were not mandated. The petitioners also asserted that the USDA was required to regulate all warm-blooded animals based on the Supreme Court's judgment in Chevron U.S.A., Inc. v. Natural Resources Defense Council (Chevron, 1984). This judgment set up a two-part process to review an agency's statutes. In summary, if the statute was unambiguous, the agency must comply with the Congressional intent. If the statute was ambiguous, or silent, the Court must determine whether the statute was reasonably interpreted by the agency. The petitioners believed the language in the Animal Welfare Act was unambiguous and, therefore, the intent of Congress was to include all warm-blooded animals. Congress had the opportunity in 1970 to specifically exclude rats, mice, and birds but did not do so. The subsequent exclusion of these species by the Secretary of Agriculture was seen as "arbitrary and capricious," as no reasonable justification was provided for exclusion when the new regulations were enacted in 1971. The petitioners stated the Animal Welfare Act only granted discretion to the Secretary of Agriculture to determine whether certain warm-blooded animals were being used for research, testing, and experimentation, not which species should, or should not, be included.

c) LAWSUITS AND SETTLEMENTS. In March 1999, the USDA rule-making process was in progress and comments from the scientific community, animal welfare organizations, and the public were being received and reviewed. However, less than a month before closure of the comment period, ARDF and other cocomplainants filed a lawsuit in the U.S. District Court in the District of Columbia against Agriculture Secretary Daniel Glickman and USDA/APHIS Administrator W. Ron DeHaven. The lawsuit requested the court to rule on, and compel, the USDA to change the regulatory definition of "animal." In August 2000, the parties engaged in settlement discussions and the District Court agreed to halt all proceedings. In September 2000, the American Association of Medical Colleges sent a letter to the Secretary of Agriculture expressing concern regarding ongoing out-of-court settlement negotiations that might lead to the exclusion of input from the research community and other regulated entities. The organization also pointed out that consultation with other federal agencies and with the Secretary of Health and Human Services on issues concerning the welfare of animals was required by law prior to issuing regulations. During this same period, the National Association for Biomedical Research (NABR) and Johns Hopkins University also petitioned the court and requested intervener status in the lawsuit. In October 2000, a settlement agreement was reached, ARDF and USDA filed a Stipulation of Dismissal with the District Court, and the lawsuit was dismissed.

In the settlement agreement, the USDA agreed to initiate and complete the rule-making process on the regulation of rats, mice, and birds under the Animal Welfare Act in a timely manner. They also agreed to keep the plaintiff's legal counsel informed biannually of the procedural status of the rule-making process and to pay a portion of the plaintiff's legal fees in accordance with the Equal Access to Justice Act. The USDA's position on this issue was that a court settlement would forestall a potential adverse judgment by the U.S. District Court, which might have mandated specific rule-making for the inclusion of rats, mice, and birds as covered species. There was legal concern that such a judgment might preclude the opportunity for input from the research community and immediately lead to the application of engineering-based standards.

d) RESPONSE FROM THE SCIENTIFIC COMMUNITY. Reaction to the ARDF petition and the subsequent 2000 lawsuit settlement was vigorous within the scientific community. National organizations, such as the Federation of American Societies for Experimental Biology (FASEB), the American Physiological Society (APS), and NABR, expressed great concern that additional regulation of these species was redundant and that sufficient standards and guidelines for the provision of quality care were already in place through the PHS Policy, the FDA (GLP), the Guide and peer review evaluation by AAALAC. The recent NIH Animal Care and Use Task Force Report on Reducing Regulatory Burden (NIH, 1999a) had stressed the need for coordination of regulatory efforts between agencies. Adding another layer of regulation on species that are already being monitored countermanded this recommendation. There was also concern that inclusion of these species in the Animal Welfare Act Regulations, without input from scientists and laboratory animal specialists, would result in applying engineering-based instead of performance-based standards. The cost of additional nonproductive, duplicative recordkeeping, increased time in animal use protocol and program review by the IACUC, and tracking of animal numbers would drain resources from crucial ongoing and future research projects. Cost estimates of the paperwork burden alone ranged from $80 to $280 million. This additional regulatory burden would overtax the operational capability of the existing APHIS Animal Welfare Enforcement Program because of lack of funding to provide appropriate oversight for the projected number of covered facilities and not enough trained veterinary medical officers to conduct inspections. As a result, animals currently regulated would be jeopardized. Regulatory inclusion of rats, mice, and birds would place a heavy financial burden on both governmental and research institutions without proof that increased regulation would be beneficial. Regulatory inclusion of mice and rats also would hinder basic science and medical research advances, adversely effect innovative breakthroughs and slow advances in translational research vital for medical advances. Finally, in the 1970 amendment to the Animal Welfare Act, Congress gave the Secretary

of Agriculture discretion to include or exclude regulatory coverage of rats, mice, and birds. In all subsequent Animal Welfare Act amendments, Congress had not objected to their exclusion; therefore, there was no basis for the belief that the intent of Congress was to include these species.

In 1999, responses from two separate National Science Foundation surveys were analyzed and the results combined. The first survey, performed in 1999, surveyed 494 randomly selected IACUC members, and the second survey, completed in 1996, queried 3,811 American Psychology Association members. All members were asked to specify which species of laboratory animals they felt should receive Animal Welfare Act protection. Approximately 74% of the 4,305 respondents, who performed animal research, favored the inclusion of mice and rats into the Animal Welfare Act regulations (Plous and Herzog, 1999).

e) RESPONSE FROM THE LABORATORY ANIMAL COMMUNITY. Reaction within the laboratory animal community was divided. In 2000, AALAS published a statement that expressed the organization's position that all vertebrate animals, including rats, mice, and birds, must be afforded protection under the Animal Welfare Act and that exclusion "of the vast majority of animals used in research is ethically indefensible." (AALAS, 2000). However, while expressing general support for regulatory inclusion, the Association also stated many serious concerns. The position statement stressed that there were insufficient funds for regulatory enforcement of currently covered species, and AALAS was opposed to having any new enforcement costs retrieved through fee-for-service or cost-reimbursement measures. The organization was also opposed to the writing of new standards for rats, mice, and birds, stating that the existing Guide standards provided sufficient oversight and should be directly utilized. The USDA should phase in any new regulatory efforts, and the timeline for implementation should be based on the availability of financial resources and acquisition of additional regulatory personnel. AALAS finally recommended that the entire USDA enforcement process be reviewed and revised before any additional species were regulated.

In September 2000, the American College of Laboratory Animal Medicine (ACLAM) issued a public statement in a letter to the Secretary of Agriculture, stating that a settlement with ARDF and cocomplaintants, without the opportunity for scientific and laboratory animal specialist input, was a mistake. The College further stated that, although they had no philosophical or scientific basis for the regulatory exclusion of rats, mice, and birds, they were opposed to inclusion of these species if it led to deleterious effects upon the species currently covered under the regulations. ACLAM stated their opposition to litigation, called upon the USDA to halt settlement negotiations, and supported the active involvement of the NABR and Johns Hopkins University in the issue. The College also proposed that Congress become involved in the debate to get an interpretation of the intent of the Animal Welfare Act pursuant to inclusion of rats, mice, and birds.

AAALAC issued a statement stating they had no official position on the inclusion issue. The Association, however, strongly recommended that AAALAC-accredited institutions become involved in the national discussion of the issue, determine an official institutional viewpoint, and disseminate that viewpoint as appropriate.

f) CONGRESSIONAL AMENDMENTS AND ACTION. In October 2000, after the USDA settlement with ARDF became public knowledge and serious concerns from the research community were raised, Senator Thad Cochran (R-Miss.) and his colleagues attached an amendment to the 2001 Agriculture, Rural Development, and Related Agencies Appropriations bill. This amendment precluded the USDA from using appropriated funds to initiate the rule-making process on the rats, mice, and birds regulatory issue during the 2001 federal fiscal year. The funding prohibition was for 1 year and rule-making could commence on October 1, 2001. The Agricultural Appropriations Conference Report was passed in the House of Representatives, which allowed the USDA to issue a proposed–but not final–rule on the coverage of rats, mice, and birds during 2002. In February 2002, Senator Jesse Helms (R-N.C.) sponsored two amendments in the Senate. The first amendment, a rider on the Farm Aid Bill (S.1731), sought to change the statutory definition of "animal" to specifically exclude "... birds, rats of the genus *Rattus*, and mice of the genus *Mus*, bred for research." This amendment was the first time the words "birds, rats, and mice" had been included in the definition of "animal." The second amendment called for a study to determine the cost of including these species in the Animal Welfare Act. Both amendments passed in the Senate. The version of the Farm Aid Bill that was presented in the House of Representatives did not include the change in the definition of "animal" and was passed in the House without the change. The bill was therefore moved to a House-Senate Conference committee to work out the differences between the two versions of the bill. The final bill, which included the redefinition of "animal" to exclude birds, and mice and rats bred for research, and a study to examine the cost for potential inclusion, was signed into law by President Bush on May 13, 2002.

In July 2003 two publications were submitted for final approval through the Office of Management and Budget. The first publication was a final rule that would amend the current definition of "animal" so that it was identical to the language passed in the 2002 amended Animal Welfare Act.

The second publication was an Announced Notice of Proposed Rulemaking (ANPR) to define the standards so that these species can be regulated. On June 4, 2004, the USDA published a final ruling amending the Animal Welfare Act regulation's definition of "animal" so that it was consistent with the definition of animal in the AWA. The new definition read as "... birds, rats of the genus *Rattus*, and mice of the genus *Mus*, bred for use in research." The revised definition became effective on June 5, 2004. Purpose-bred rats and mice continue, therefore, to be excluded from coverage by the USDA regulations. An ANPR and Request for Comments was also published by the USDA on June 4, 2004. (Code, 1999c) This notice stated that the USDA was considering whether existing general regulations pertaining to the handling, care, treatment, and transportation of rats and mice covered under the AWA should be changed to establish specific standards and solicited comments from the public. The USDA also requested comments regarding the potential economic impact on institutions if specific standards were established.

REFERENCES

AALAS Position Statement. (2000). Inclusion of Rats, Mice and Birds Under USDA Regulatory Oversight, 2000 AALAS Reference Directory, pp 16–17.

Animal Welfare Act of 1966 (Pub.L. 89-544) and subsequent amendments. (1966). *U.S. Code*, Vol. 7, Secs. 2131–2159 et seq.

Anzaldo, A.J., Harrison, P.C., Riskowski, G.L., et al. (1995). Behavioral evaluation of spatially enhanced caging for laboratory rats at high density. *Contemp. Top. Lab. Anim. Sci.* **34**, 56–50.

Augustsson, H., Lindberg, L., Hoglund, A.U., et al. (2002). Human-animal interactions and animal welfare in conventionally and pen-housed rats. *Lab. Anim.* **36**, 271–281.

Blom, H.J.M., Van Tintelen, G., Van Vorstenbosch C.J.A.H.V., et al. (1996). Preferences of mice and rats for types of bedding material. *Lab. Anim.* **30**, 234–244.

Boice, R. (1997). Burrows of wild and albino rats: effects of domestication, outdoor raising, age, experience, and maternal state. *J. Comp. Physiol. Psychol.* **91**, 99–105.

Bradshaw, A.L., Poling, A. (1991). Choice by rats for enriched versus standard home cages: thermoplastic pipes, wood platforms, wood chips, and paper towels as enrichment items. *J. Exp. Anal. Behav.* **55**, 245–250.

Brain, P.F. (1992). Understanding the behaviors of feral species may facilitate design of optimal living conditions for common laboratory rodents. *Anim. Technol.* **43**, 99–105.

Brown, K.J., Grunberg N.E. (1995). Effects of housing on male and female rats: crowding stresses males but calms females. *Physiol. Behav.* **58**, 1085–1089.

Chevron USA, Inc. v. Natural Resources Council, Inc. 467 U.S. 837 (1984).

Chmeil, D.J. Jr., Noonan, M. (1996). Preference of laboratory rats for potentially enriching stimulus objects. *Lab. Anim.* **30**, 97–101.

Code of Federal Regulations (rev.1998). Title 21: Food and Drugs; Chap. 1: Food and Drug Administration, Department of Health and Human Services; Subchap. A: General; Part 58: Good Laboratory Practice for Nonclinical Laboratory Studies. Office of the Federal Register, Washington, D.C.

Code of Federal Regulations (1999a). Title 9: Animals and Animal Products; Chap.1: Animal and Plant Health Inspection Service, Department of Agriculture; Subchap. A: Animal Welfare; Parts 1, 2, 3, and 4. Office of the Federal Register, Washington, D.C.

Code of Federal Regulations (1999b). Title 9: Animals and Animal Products; Chap.1: Animal and Plant Health Inspection Service, Department of Agriculture; Subchap. A: Animal Welfare; Petition for Rulemaking; Parts 1 and 3. Office of the Federal Register, Washington, D.C.

Code of Federal Regulations (1999c). Title 9: Animals and Animal Products; Chap.1: Animal and Plant Health Inspection Service, Department of Agriculture; Subchap. A: Animal Welfare; Advance Notice of Proposed Rulemaking and Request for Comments; Parts 2 and 3. Office of the Federal Register, Washington, D.C.

Committee to Revise the *Guide for the Care and Use of Laboratory Animals*, (1996). Institute of Laboratory Animal Resources, Commission on Life Sciences, National Research Council. DHHS Publ. (NIH) 96–23. National Academy Press, Washington, D.C.

Denny, M.S. (1975). The rat's long-term preference for complexity in its environment. *Anim. Learning Behav.* 3, 245–249.

Duncan, I.J. (1992). Measuring preference and the strength of preferences. *Poult. Sci.* 71, 658–663.

Erb, H.N. (1990). A statistical approach for calculating the minimum number of animals needed in research. *ILAR News* 32, 11–16.

Fuchs, B.A. (2000). Use of Animals in Biomedical Experimentation. In "Scientific Integrity: An Introductory Text with Cases", (F.L. Macrina, ed.), chap. 6. ASM Press, Washington, D.C.

Gaelf, B.G. Jr., Sorge, R.E. (2000). Use of PVC conduits by rats of various strains and ages housed singly and in pairs. *J. Appl. Anim. Welf. Sci.* 3, 279–292.

Galef, B.G. Jr. (1999). Environmental enrichment for laboratory rodents: Animal Welfare and the methods of science. *J. Appl. Anim. Welf. Sci.* 2, 267–280.

Giralt M., Armario A. (1989). Individual housing does not influence the adaptation of the pituitary-adrenal axis and other physiological variables to chronic stress in adult male rats. *Physiol. Behav.* 45, 477–481.

Greco, A.M., Gambardella P., Sticchi R., et al. (1992). Circadian rhythms of hypothalamic norepinephrine and some circulating substances in individually housed rats. *Physiol. Behav.* 52, 1167–1172.

Guide to the Care and Use of Experimental Animals. (1980). Canadian Council on Animal Care, Ottawa, Ontario, Canada.

Harkin, A.T., Connor T.J., O'Donnell J.M., et al. (2002). Physiological and behavioral responses to stress: what does a rat find stressful? *Lab. Anim.* 31, 42–50.

Health Research Extension Act of 1985 (Pub.L.99-158, Sec. 495) and subsequent amendments.

Holson, R.R., Scallet, A.C., Ali S.F., et al. (1988). Adrenocorticol, B-endorphin and behavioral responses to graded stressors in differentially reared rats. *Physiol. Behav.* 42, 125–130.

Holson, R.R., Scallet, A.C., Ali, S.F., et al. (1991). Isolation stress revisited: isolation rearing effects depend on animal care methods. *Physiol. Behav.* 49, 1107–1118.

Hurst, J.L., Barnard, C.J., Hare, R., et al. (1996). Housing and welfare in laboratory rats: time budgeting and pathophysiology in single-sex groups. *Anim. Behav.* 52, 335–360.

Hurst, J.L., Barnard, C.J., Tolladay, U., et al. (1999). Housing and welfare in laboratory rats: effects of cage stocking density and behavioral predictors of welfare. *Anim. Behav.* 58, 563–586.

Interagency Research Animal Committee. (1985). U.S. government principles for utilization and care of vertebrate animals used in testing, research, and training. *Federal Register* (May 20, 1985).

Mann, M.D., Crouse, D.A., Prentice, E.D. (1991). Appropriate animal numbers in biomedical research in light of animal welfare considerations. *Lab. Anim. Sci.* 41, 6–14.

Manser, C.E., Morris, T.H., Broom, D.M. (1995). An investigation into the effects of solid or grid cage flooring on the welfare of laboratory rats. *Lab. Anim.* 29, 353–363.

Manser, C.E., Morris, T.H., Broom, D.M. (1996). The use of a novel operant test to determine the strength of preference for flooring in laboratory rats. *Lab. Anim.* 30, 1–6.

Manser, C.E., Broom, D.M., Overend, P., et al. (1998). Operant studies to determine the strength of preference in laboratory rats for nest-boxes and nesting materials. *Lab. Anim.* 32, 36–41.

Memorandum of Understanding Among the Animal and Plant Inspection Services. (1995). US Department of Agriculture and the Food and Drug Administration, DHHS and the National Institutes of Health, DHHS Concerning Animal Welfare. December 11, 1995.

Mering, S., Kaliste-Korhonen, E., Nevalainen, T. (2001). Estimates of appropriate number of rats: interaction with housing environment. *Lab. Anim.* 35, 80–90.

NIH Regulatory Burden: VI. Animal Care and Use–Workgroup Report. (1999a). Office of Extramural Research, National Institutes of Health.

NIH Regulatory Burden–Three-month Plan. (1999b) (http://grants.nih.gov/grants/policy/regulatoryburden/regburd3monthplan_09–1999.htm).

Patterson-Kane, E.G., van de Ven, M., Ras, T. (2001a). Enrichment of laboratory rat caging. *Contemp. Top. in Lab. Anim. Sci.* 40, 94.

Patterson-Kane, E.G. (2001b). The cage preferences of laboratory rats. *Lab. Anim.* 35, 74–79.

Patterson-Kane, E.G. (2002). Cage size preference in rats in the laboratory. *J. Appl. Anim. Welf. Sci.* 5, 63–72.

Patterson-Kane, E.G. (2003). Shelter enrichment for rats. *Contemp. Top. Lab. Anim. Sci.* 42, 46–48.

Peace, T.A., Singer, A.W., Niemuth, N.A., et al. (2001). Effects of caging type and animal source on the development of foot lesions in Sprague Dawley rats (rattus norvegicus). *Contemp. Top. Lab. Anim. Sci.* 40, 17–21.

Plous S., Herzog H.A., (1999). Should the AWA cover rats, mice and birds? The results of an IACUC survey. *Lab. Anim.* 28, 38–40.

Public Health Service Policy on Humane Care and Use of Laboratory Animals. (1986). Office for Protection from Research Risks, National Institutes of Health, Public Health Service. (Reprinted by U.S. Department of Health and Human Services, Washington, D.C., 1996).

Ras, T., van de Ven, M., Patterson-Kane, E.G., et al. (2002). Rats' preference for corn versus wood-based bedding and nesting materials. *Lab. Anim.* 36, 420–425.

Reinhardt, V. Reinhardt, A. (2001). Legal space requirement stipulations for animals in the laboratory: are they adequate? *J. Appl. Anim. Welf. Sci.* 4, 143–149.

Rock, F.M., Landi, M.S., Hughes, H.C., et al. (1997). Effects of caging type and group size on selected physiological variables in rats. *Contemp. Top. Lab. Anim. Sci.* 36, 69–72.

Russell, W.M.S., Burch, R.L. (1959). The Principles of Humane Experimental Technique. Methuen, London. (Reprinted 1992, Universities Federation for Animal Welfare, Herts, England).

Sharp, J., Zammit T., Azar T., et al. (2002). Stress-like responses to common procedures in male rats housed alone or with other rats. *Contemp. Top. Lab. Anim. Sci.* 41, 8–14.

Sharp, J., Zammit, T., Azar, T., et al. (2003a). Stress-like responses to common procedures in individually and group-housed female rats. *Contemp. Top. Lab. Anim. Sci.* 42:9–18.

Sharp, J., Azar, T., Lawson D. (2003b). Does cage size affect heart rate and blood pressure of male rats at rest or after procedures that induce stress-like responses? *Contemp. Top. Lab. Anim. Sci.* 42, 8–12.

Sherer, A.D., Rigel, D.F., Iversan, W.O. (2001). Exploratory study to determine the preference of the Sprague-Dawley rat for a solid or wire-cage floor. Abstract. ACLAM forum, Mobile, Ala.

Stark, D.M. (2001). Wire-bottom versus solid-bottom rodent caging issues important to scientists and laboratory animal science specialists. *Contemp. Top. Lab. Anim. Sci.* **40,** 11–14.

Tannenbaum, J. (1989). Animal research. *In,* "Veterinary Ethics–The Veterinarian and Animal Research." (T.S. Satterfield, ed.), Chap. 23. Williams and Wilkins, Baltimore, Md.

van de Weerd, H.A., van den Broek, F.A.R., Baumans, V. (1996). Preference for different types of flooring in two rat strains. *Appl. Anim. Behav. Sci.* **46,** 251–261.

Watson, D.S.B. (1993). Evaluation of inanimate objects on commonly monitored variables in preclinical safety studies for mice and rats. *Lab. Anim. Sci.* **43,** 378–380.

Chapter 3

Taxonomy and Stocks and Strains

Hans J. Hedrich

I. INTRODUCTION

The laboratory or Norway rat (*Rattus norvegicus*) and the laboratory mouse (*Mus musculus laboratorius*) are by far the most commonly used experimental animals in many fields of medical and biological research. The first inbred strains (the PA rat and the DBA mouse) were introduced to the scientific community approximately a century ago. The ease of breeding and short generation times have contributed to the widespread use of these species as experimental mammals. Like their feral ancestors that vied for food, they now compete with respect to their impact on science. Irrespective of this, both species do play a key role in biomedical research, especially as great detail regarding the sequences of both genomes is now

THE LABORATORY RAT, 2ND EDITION

available (Waterston, 2002), permitting comparisons with other mammalian genomes, in particular man, and annotations concerning the functions of the genes studied. Although the laboratory mouse has the great advantage with respect to methods for genetic modification, the rat has advantages over the mouse due to its larger size, which permits greater ease for repeated samplings and surgical modifications. Over time, the number of inbred strains has greatly increased for both species. In addition, innumerable mutations, many of which were shown to be deleterious in their phenotypic effect, have been identified and used in scientific studies to unravel their functional and genetic properties as model systems (see chapter on Spontaneous and Induced Mutants).

The laboratory rat has long been used in experimental physiology and has made significant contributions to several complex areas of mammalian biology. For example, the rat is a highly valuable model organism for the analysis of many complex areas of biomedicine, such as cardiovascular diseases, metabolic disorders (e.g., lipid metabolism, diabetes mellitus), neurologic disorders and behavior (e.g., areas of motor function, hearing, vision, learning, and epilepsy research), autoimmune diseases (e.g., arthritis, experimental autoimmune encephalomyelitis [EAE], etc.), cancer, and renal diseases. The wealth of information and the multiplicity of (inbred, congenic, consomic, and recombinant inbred) strains available with different specific characteristics make the laboratory rat an indispensable tool for biomedical research.

II. TAXONOMY AND GEOGRAPHICAL DISTRIBUTION

Rats are rodents (Table 3-1), and thus members of the largest family of mammals (Fig. 3-1). Approximately 1,325 living species of murid rodents have been described, the members of which are currently placed in 281 genera distributed among 17 subfamilies, and which include most of the familiar rats and mice (Musser and Carlton, 1993). Nevertheless, this family also encompasses an enormously diverse array of other rodents.

The phylogenetic relationships of murids among themselves and with other rodents (Fig. 3-2) have proved to be rather complex, based on comparative morphology and traditional classification, and is somewhat controversial (Musser and Carlton, 1993). This is so because many members of the rodent family are very similar in size and shape. However, understanding of the evolutionary system has advanced due to the molecular analysis of mitochondrial DNA and of long interspersed DNA (LINE) elements (Furano and Usdin, 1995; Verneau et al., 1998), which presumably are homoplasy-free phylogenetic characters (i.e., free from noninherited shared characters). Furano and Usdin (1995) and Verneau et al. (1998) were thus able to determine and date rodent speciation events by using LINE-1 retroposons. Within the very closely related rats, *Rattus sensu stricto*, those authors showed that this lineage arose about 7.5 to 5.5 million years (Myrs) ago

TABLE 3-1

TAXONOMIC CLASSIFICATION OF THE SPECIES *RATTUS*

Kingdom:	*Animalia*	
Phylum:	*Chordata*	
Subphylum:	*Vertebrata*	
Class:	*Mammalia*	
Subclass:	*Theria*	
Infraclass:	*Eutheria*	
Order:	*Rodentia*	
Suborder:	*Myomorpha*	
Family:	*Muridae*	
Superfamily:	*Muroidea*	
Subfamily:	*Murinae*	
Genus:	*Rattus*	
Species:	**Rattus norvegicus (Berkenhout, 1769)**	[Norway or brown rat]
	Rattus exulans (Peale, 1848)	[Polynesian rat]
	Rattus rattus (Linné, 1758)	[house or black rat]
Subspecies:	*R. rattus rattus*	[black rat]
	R. rattus alexandrinus	[Alexandria black rat]
	R. rattus brevicaudatus	[Sawah rat]
	R. rattus diardii	[Malayan black rat]
	R. rattus frugivorous	[fruit rat]

From Grzimek (1968), and Musser and Carlton (1993).

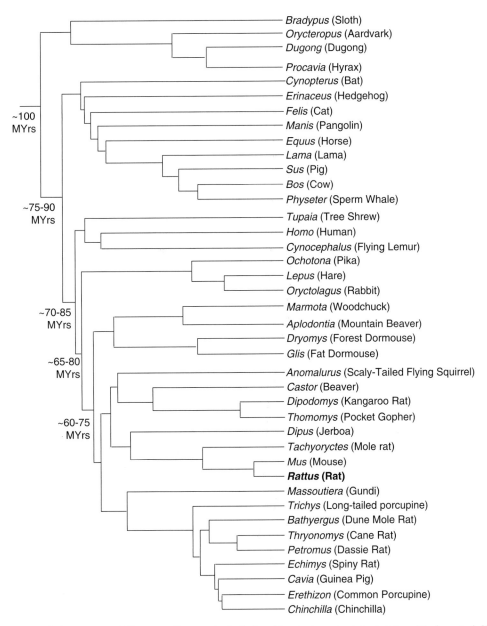

Fig. 3-1 Evolutionary tree of 40 mammalian species including 21 rodent species (adapted from Huchon et al. [2002]).

and apparently underwent two episodes of speciation, an intense one between 2.4 to 2.7 Myrs ago, and a second that began 1.2 Myrs ago, and which may be continuing at present. The divergence between the *Rattus* and *Mus* genera has been estimated as occurring 10 to 12 Myrs ago. Figure 3-2 represents the evolutionary relationship of 26 *Rattus* species, in the broad sense, anchored into the broader phylogeny of the family *Muridae*.

Members of the rodent family can be found on all continents except Antarctica and on many oceanic islands. They occupy ecosystems ranging from dry desert to wet tropical forests. Rats in the broad sense (*Rattus senu lato*) belong to the subfamily *Murinae*. The genus *Rattus* comprises some 50 species (Furano and Usdin, 1995).

Although there are no subspecies to *Rattus norvegicus*, the nearest relative being *Rattus* cf *moluccarius* (Furano and Usdin, 1995), several subspecies have been described for the black rat (*Rattus rattus*) (Table 3-1). These *Rattus rattus* subspecies neither cross-hybridize within the subspecies, nor with the Norway rat (Yoshida, 1980).

According to Meng et al. (1994), "Rodents are first known from many localities of the latest Paleocene to earliest Eocene age in Asia and North America. They are widely considered to have originated in Asia based on the occurrence there of *Eurymylidae*, their perceived nearest relatives." Large parts of the Mediterranean countries, the Middle East, India, China, Japan, the entire Southeast Asia including the Philippines, New Guinea, and Australia

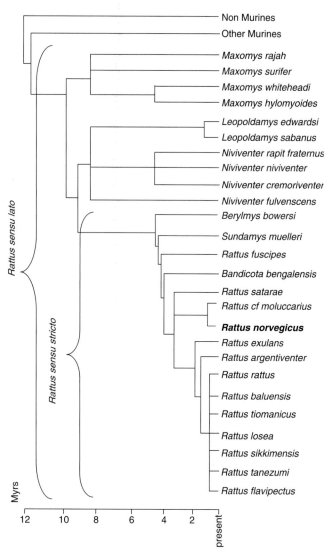

Non Murines
Other Murines
Maxomys rajah
Maxomys surifer
Maxomys whiteheadi
Maxomys hylomyoides
Leopoldamys edwardsi
Leopoldamys sabanus
Niviventer rapit fraternus
Niviventer niviventer
Niviventer cremoriventer
Niviventer fulvenscens
Berylmys bowersi
Sundamys muelleri
Rattus fuscipes
Bandicota bengalensis
Rattus satarae
Rattus cf moluccarius
Rattus norvegicus
Rattus exulans
Rattus argentiventer
Rattus rattus
Rattus baluensis
Rattus tiomanicus
Rattus losea
Rattus sikkimensis
Rattus tanezumi
Rattus flavipectus

Rattus sensu lato

Rattus sensu stricto

Myrs

12   10   8   6   4   2   present

Fig. 3-2 Evolutionary relationship of rats in the broad sense according to Verneau et al. (1998) and following the nomenclature and taxonomy of Musser and Carlton (1993).

can be considered the natural habitat of genus *Rattus*. The original natural habitat of the black rat (*Rattus rattus*) is considered to be the Indo-Malayan region, whereas the Norway rat originates from the vast plains in Asia, probably Northern China and Mongolia. From this region, Norway rats have spread to other parts of the world, somewhere during the Middle Ages according to mammalian taxonomists (Tate, 1936; Silver, 1941; Southern, 1964). They have dispersed from this original natural habitat over the entire world. The Norway rat may have become commensal with man as long as several thousand, or just several hundred years ago; this issue remains still unsettled. Their commensalism with man and their settlement along with human migration is a feature common to both rat species (Yoshida, 1980).

Freye and Thenius (1968) speculate the Norway rat might have been present in Central Europe as early as 1553. They draw this conclusion from an illustration of a presumptive Norway rat depicted in *Historiae animalium* (1551-1558, German edition published in 1669) by the Swiss naturalist Conrad Gesner. This description could fit for the black rat as well. However, Gesner mentions the appearance of a considerable proportion of white rats with red eyes. It is these albino mutations, not uncommon among wild *Rattus norvegicus,* that support the idea of a brown rat being depicted this early. Others suppose that Norway rats crossed the river Volga in South Russia not earlier than 1727 in a huge migratory event and subsequently conquered Western Europe. Little evidence supports the assumption by Castle (1947) that the Norway rat entered Western Europe via the Norwegian peninsula. Historically reliable reports on the presence of Norway rats date back to the 18th century, for example, England 1730, France 1735, Eastern Germany 1750, Spain 1800. It was competing with and displacing the smaller and less aggressive black rat species in most parts of Europe and in the United States. The Norway rat came to North America as early as 1755 via the ships of new settlers (Freye and Thenius, 1968), whereas Lantz (1909) has dated the first arrival on the eastern seaboard of the United States around 1775. However, there are recent reports suggesting that, in Germany, black rats, which have been considered as mostly extinct, are advancing and partly displacing the Norway rat.

III.   ADAPTATION TO THE LABORATORY

As competitors of man, rats were trapped, killed, and sold for food, especially in times of famine in Europe. Moreover, they were used for rat-baiting contests, originally in Europe and later in America, as late as the end of the eighteenth century. In his chapter on rats (*de mure domestico majore*) Gesner (1679) describes that in 1667, tamed rats were on display in Paris dancing on a high wire, or showing other artistry and thus served as a source of income for the poor. A guidebook, *Chingansodategusa*, on breeding of fancy "Daikoku-Nezumi" (Japanese for white or laboratory rats) by Jyo-En-Shi (1787) describes that rats were bred as pets in Japan as early as 1654. Apparently various mutants were maintained, such as rats with a black head, black rats with a crescent shaped white spot, black-eyed white rats as well as dwarf-type rats (Fig. 3-3; published by Chohbei Zeniya in Kyoto, 1787; rediscovered by Prof. T. Serikawa, 2002). In Europe in the eighteenth and nineteenth centuries albino, black, and piebald (hooded) colored Norway rats were captured or selected from offspring of rats entrapped for the baiting

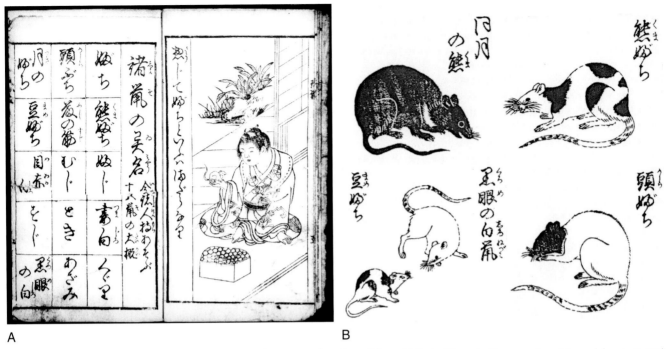

Fig. 3-3 Illustrations from the guidebook Chingansodategusa on breeding of fancy Daikoku-Nezumia (Japanese for white or laboratory rats). Picture kindly provided by Professor T. Serikawa, Kyoto.

contests because of their distinct and more attractive appearance; many of these animals were subsequently tamed. Albino mutants were brought into laboratories early in the nineteenth century, where they then served as subjects for early physiologic studies. These included, for instance fasting studies as early as 1828 (as described by McCay, 1973), attempts to measure the quality of proteins (Savory, 1863), and studies on the importance of the adrenal glands (Philipeaux, 1856). One can therefore state that the Norway rat is the first animal species to be domesticated strictly for scientific purposes (Richter, 1954). The first breeding experiments recorded are those by Crampe (1877; 1883; 1884; 1885) who, shortly after Mendel and long before the rediscovery of Mendel's laws in 1900, reported on the specific patterns of coat color inheritance of the Norway rat. Crampe carefully described the color and ratios of albino, nonagouti black, and hooded offspring obtained from the various breeding experiments.

The Wistar Institute of Philadelphia and its first scientific director, Henry H. Donaldson, paved the ground for standardization of the laboratory rat. The first inbred rat strain ever to be bred is the King Albino strain later designated the PA strain. Helen Dean King, a Ph.D. student of W.E. Castle at the Bussey Institute, Boston (Jamaica Plain), initiated inbreeding of albino rats at the Wistar Institute in 1909 (King, 1918a; 1918b; 1918c; 1919). Interestingly, she started this endeavor at about the same time as Clarence Cook Little, who also was a student of Castle, and who began the selective inbreeding of

dilute (*d*), brown (*b*), nonagouti (*a*) mice, now known as DBA/1 and DBA/2. The Wistar Institute of Anatomy and Biology in Philadelphia served as one of the first laboratory animal breeding and research facilities. Much credit is owed to this institute and its scientists for developing satisfactory cages, ancillary equipment, diets, breeding practices, and facilities for holding and breeding rats. A thorough review by Lindsey (1979) describes the contributions and the impact of the Wistar Institute and other research institutions as well as renowned scientists in the United States toward research in nutrition, biochemistry, physiology, endocrinology, cancer, behavior, and genetics. Most of the rat colonies at these early times were maintained as random-bred stocks. It is the "Wistar rat" and, to a lesser extent, the "Sprague-Dawley rat" that became dominant within animal research laboratories worldwide. Descendants of these two stocks served as founders, or as crossbreeding partners, in the development of most current rat strains. However, a lesser known strain, the PAR rat, has to be considered the eldest rat stock or strain (Bazin, 1988). According to a report in the *Magazin Pitturesque*, a black hooded rat colony existed at the "Jardin des Plantes" as early as 1856. This botanical garden, founded in 1635, was turned into a museum of natural history in 1793 with plants and live animals on exhibit. The rat colony served as a source of food for reptiles and is kept as a closed colony even at present times. Bazin (1988) has established an inbred strain, PAR/Wsl, from this original source.

IV. STOCKS AND STRAINS OF LABORATORY RATS

A. Outbred Stocks

Outbred stocks are closed colonies of genetically more or less heterogeneous animals. New matings are set up either at random or by applying a cyclical mating scheme designed to minimize the increase in the coefficient of inbreeding as much as possible. Random mating means that the animals for breeding are chosen without regard to their relationships and that the stock is closed to introductions from the outside. The purpose of random mating is to avoid the consequences of inbreeding, that is, to maintain the genetic variability and to prevent inbreeding depression. The concept behind the published cyclical mating schemes (Falconer, 1967; Kimura and Crow, 1963; Poiley, 1960; Rapp, 1972) is to preserve a given genetic variability as far as possible with the aim to mimic the genetic make-up of human populations. Nevertheless, animals collected from any outbred stock are genetically undefined, although an enormous body of pathophysiologic data on the various stocks used in toxicologic research has been accumulated. A major advantage is seen in their hybrid vigor and as claimed in various publications by their long lifespans, high disease resistance, early fertility, large and frequent litter sizes, low neonatal mortality, rapid growth, and large size. Factors strongly influencing the heterogeneity of outbred stocks are their effective population size, the sequence of generations, the selection of future breeders, and the breeding system applied (Table 3-2). The stocks derived from Wistar and Sprague-Dawley are among the most popular rats used experimentally, irrespective of the fact that their genetic make-up may be extremely variable owing to founder effects and genetic drift. This must be kept in mind when using rats from such colonies. One should also be aware that animals of a specific outbred stock obtained from a vendor, but maintained at different locations, are likely to differ in their genetic composition, especially in gene frequencies. This holds true despite

attempts of the major vendors to overcome this genetic divergence (e.g. Nomura, 2000; White, 2000). Wistar as well as Sprague-Dawley and others like Osborne-Mendel, Long-Evans, Holtzman, Slonaker, and Albany have served as the source of many of today's inbred strains, which later on showed distinct differences in various parameters despite an identical source. For example, individuals of two inbred strains SPRD/Ztm and SD/Ztm (now extinct), both derived from Han:SPRD, showed different behavioral patterns (Nikkhah et al., 1998).

B. Inbred Strains

1. Standard Inbred Strains

Standard inbred strains are, by definition, developed by at least 20 consecutive generations of brother by sister or parent by offspring matings (Green, 1981a). During inbreeding, the chance of incross increases, whereas that of cross, backcross, and intercross decreases (Green, 1981b). Genetic uniformity (isogenicity) is one of the fundamental properties of an inbred strain. The variety of inbred strains and the possibility for introduction of foreign DNA into the germline make the rat a valuable species. In inbred strains, any phenotypic variability is the result of non-genetic factors, such as environmental and methodologic factors. Therefore, standard inbred, congenic, consomic, or recombinant inbred strains are preferably used. For the analysis of quantitative trait loci (QTL) in multifactorial diseases, such strains play an essential role. A detailed knowledge of the different genetic and phenotypic properties of the various strains would thus be instrumental for biomedical research. Unfortunately there is no database existing that provides an easily retrievable documentation. The genetic variability between strains permits one to characterize a trait in genetic terms. For a variety of pathophysiologic disorders, rat models have been developed (Table 3-3) and permit functional and genetic analyses.

For economic reasons, many holding facilities have abandoned the maintenance of rat strains. However, new repositories have emerged and apparently will continue to expand and maintain valuable strains. Table 3-4 lists most currently known inbred rat strains; Table 3-5 collates standardized strain names with older, nonstandardized strain names; and Table 3-6 refers to internet databases that contain background information on a variety of strains. Inbred strains allow for much better standardization of the test conditions and improve the quality of the results. For this reason, many scientific investigations exclusively use genetically defined inbred strains and their derivatives (congenic, consomic, recombinant inbred strains). Moreover, geneticists can provide not only homogeneous populations, but also, over time, genetically

TABLE 3-2

INBREEDING IN A CLOSED STOCK WITH A RANDOM MATING

No. of breeding pairs	Rate of inbreeding per generation (%)	No. of generations equal to one B × S mating
10	2.5	11
20	1.25	23
40	0.625	46
60	0.42	69
80	0.31	93
100	0.25	116

TABLE 3-3
SELECTED SPONTANEOUS COMPLEX DISEASE MODELS

Disease type	Susceptible or responsive strain	Control or nonresponsive strain
Cardiovascular disorders		
Hypertension	DSS	
	FHH	FHL
	GH	
	LH	LN, LL
	MHS	MNS
	SBH	SBN
	SHR	WKY
	SS	SR
	TGR(mRen2)27	
	WKHT	WKHA
Stroke	SHRSP	
Metabolic disorders		
Type 1 Diabetes Mellitus (T1DM)	BBDP	BBDR
	KDP	KND
	LEW.1AR1-*iddm*	LEW.1AR1
	LETL	LETO
Type 2 Diabetes Mellitus (T2DM)	BBZ	
	CDS	CDR
	CRDH	
	GK	
	LA/N-*Lepr^{fa-cp}*	LA
	OLETF	
	SDT	
	SHHF	
	WBN	
	WDF	
	WOKW	WOKA
	ZDF	
Hypercholesteremia	EXHC	
Hyperlipidemia	FCH	FNL
Behavioral and neurological disorders		
Behavior	IR	INR
	MR	MNR
	RHA	RLA
	TMB	TMD
	TS1	TS3
Alcohol abuse	AA	ANA
	UChB	UChA
	P	NP
Epilepsy	GEPR/3	
	GEPR/9	
	IER	
	NER	
	P77PMC	
	SER	
	WAG/Rij	
	WF/Ztm	
	ZIMY	
ZNS-abnormalities	DMY	
	KZC	
	MD	
	TRM	
	VF	
	ZI	

TABLE 3-3
SELECTED SPONTANEOUS COMPLEX DISEASE MODELS—CONT'D

Disease type	Susceptible or responsive strain	Control or nonresponsive strain
Circling behavior/ syndromal deafness	DOP LEW-*Myo15^{ci2}*	LEW
Spontaneous tumours		
	BUF/Mha	
	BDII	
	Eker rat	
	HMT	
	LEC	
	WN	
Skin and pelage disorders		
	BALD	
	BEG	
	DEBR	
	HWY	
	KHR	

This table summarizes a number of inbred rat strains that are self-functioning as spontaneous complex disease models. For cardiovascular diseases, several strains or sets of strains are available, with SHR and SHRSP being known most widely. Among the group of strains developing metabolic diseases, such as insulin-dependent diabetes mellitus type 1 (T1DM), non-insulin–dependent diabetes mellitus type 2 (T2DM), and metabolic X syndrome, familial dyslipidemic hypertension, disorders of glucose and fatty acid metabolism, the diabetes prone BB (DPBB) rat with T1DM and the GK rat (T2DM) has been intensively studied. Among the strains listed for behavioral research are those specifically selected from a source for a high or low response toward a specific stimulus, and thus must not be considered congenic. The number of strains with seizures is impressive, most being maintained in Japan. With respect to cancer research several strains (see also Table V) have proven valuable in various studies, particularly with respect to precancerous lesions. The Eker rat is a model of dominantly inherited renal cancer, owing to an insertional mutation in the tuberous sclerosis 2 (*Tsc2*) gene. In the homozygous state, the gene acts as an embryonic lethal, whereas in the heterozygous state it leads to renal cell carcinoma, uterine leiomyoma, and hemangioma.

heterogeneous populations with specific characters from isogeneic strains, such as mosaic populations (Cholnoky et al., 1974). Each of the resynthesized populations will be essentially genetically identical to the preceding ones.

2. Coisogenic and Congenic Strains

The use of coisogenic and congenic strains has undergone a recent, rapid increase in biomedical research mainly in conjunction with the identification of genes. As the genomes of inbred strains are highly standardized (F > 40), mutations can be identified and isolated much easier. Mutations either occur spontaneously or may be induced by chemical mutagenesis or through exposure to ionizing radiation. Transgenes can be directly introduced into an

TABLE 3-4

Inbred Rat Strains

Strain	Coat color	Information
AA	albino	Alko Alcohol; derived from outbred Wistar rat with selection for heavy alcohol consumption.
AAW		Atomic Energy Commission, Melbourne.
ABH	brown hooded	Yamada from a cross between BN and outbred Wistar stock, with selection for the non-agouti brown hooded (*abh*) coat colour.
ACH	black hooded	Dunning, 1926, at Columbia University Institute for Cancer Research; high incidence of lymphosarcoma.
ACI	agouti hooded Irish	Derived by Curtis and Dunning, 1926, at Columbia University Institute for Cancer Research, from an August × COP cross; displays high susceptibility to estrogen-induced mammary carcinoma.
ACP	black hooded Irish	Developed by Dunning, to N, to Pit.
AD-1	albino	Albumin deficient rat; developed by Shumiya at Tokyo Metropoliton Institute of Gerontology as inbred strain from Nagase Analbuminemic Rat (NAR) in 1987.
AGA	black	Developed by Nakic, Zagreb.
AGUS	albino	Gustafsson, Karolinska Institute, 1948, from outbred (possibly Sprague-Dawley) stock.
ALB	non-agouti dilute brown	Albany Medical College in an attempt to develop a strain with a high incidence of spontaneous tumours.
ALD		Yoshida and Kuriyama, 1981, from non-inbred Donryu rats with congenital deficiency of lysosomal acid lipase; rats develop genetic lipid storage disease analogous to human Wolman's disease with marked hepatosplenomegaly, lymph node enlargement, and thickened, dilated intestine.
AM	yellow?	Torres, Rio de Janeiro from unknown outbred stock.
AI		Ishibashi at Drug Safety Research Center, Tokyo, from SHRSP rats with hereditary morphological abnormalities of the teeth; mutation leads to amelogenesis imperfecta due to underdevelopment of tooth ameloblasts. Inherited as an autosomal single recessive trait.
AMDIL	dilute yellow?	Torres, Rio de Janeiro from unknown outbred stock.
AN	albino	Originates from outbred Wistar Imamichi strain; carries autosomal recessive mutation, *as*, causing an arrest of spermatogenesis at an early meiotic stage.
ANA	albino	Alko Non-Alcohol; derived from outbred Wistar rat with selection as an alcohol avoiding strain; control strain to AA
AO	albino	From ARC Institute of Animal Diseases, Compton, probably as "WAG" to Gowans, Oxford, 1957. Appears to differ from other WAG sublines in having *A* at the agouti locus.
AR	white spotting	Aganglionosis Rat; developed by Ikadai, 1973, from a cross between an albino female and a wild male. Due to an autosomal recessive mutation in the endothelin-B receptor gene (*Ednrb^{sl}*), homozygous rats show megacolon caused by the absence of myenteric ganglion cells and white coat-color with a small pigmented spot on the head; model of Hirschsprung disease.
APR	albino	Apnea Prone Rat; formerly called MF; developed by Holme and Piechuta, 1979, selective breeding of Sprague-Dawley outbred rats for a prolonged severe dyspnea 14-18 days after challenge with areosolised egg albumin.
AS	albino	University of Otago Medical School, Dunedin, New Zealand, from outbred Wistar rats imported from England in 1930.
AS2	albino	Derived from outbred Wistar colony at the University of Otago Medical School, Dunedin, New Zealand.
AUG	dilute hooded	Derived from one of the US "August" sublines.
AVN	albino	From unknown origin developed at Charles University Prague.
AXC	agouti hooded Irish	Substrain of ACI; aged male rats develop prostate cancer.
A990	agouti hooded	August substrain developed by Curtis and Dunning, 1921, at Columbia University Institute for Cancer Research.
A7322	pink-eyed dilute hooded	August substrain developed by Curtis, 1925, at Columbia University Institute for Cancer Research; spontaneous mammary tumours.
A28807	pink-eyed dilute hooded	August substrain developed by Curtis and Dunning, 1936, at Columbia University Institute for Cancer Research; subline of A7322.
A35322	black hooded	August substrain developed by Curtis and Dunning, 1942, at Columbia University Institute for Cancer Research from a mutation originating in an aunt × nephew cross at F27 of animals of strain A990
B	albino	From Wistar stock known as King B strain.
BALD		Bald rat; from a commercial breeder in Germany to Aventis Japan. Rats lose their hair at about three weeks of age and are devoid of all general body hair except for the vibrissae by five weeks; the bald condition is a simple recessive character (*ba*); aged bald animals show spontaneous tumors of the thymus, pituitary gland, testis, uterus, mammary gland, and skin; incidence of skin tumors is much higher than of other tissue.

TABLE 3-4

INBRED RAT STRAINS—CONT'D

Strain	Coat color	Information
BBDP	albino	Diabetes prone BB rat; mutation causing T1DM in a Wistar rat stock at Bio-Breeding Laboratories, Ontario, Canada; characterized by T lymphopenia (*lyp*). There is a substantial genetic variation among the various substrains.
BBDR	albino	Diabetes-resistant BB rat; developed as non-diabetic control strain to diabetes prone BBDP.
BBZ	albino	Derived at Wor as an obese, noninsulin dependent model of diabetes mellitus (T2DM); morphologically shows oxidative injury and endothelial cell dysfunction in the retina.
BC	agouti hooded	CPB-B renamed BC; susceptible to audiogenic seizures.
BDE	black hooded	Animal model for adenosine-induced bronchoconstriction caused by a direct effect on the airway smooth muscle via activation of NK2 receptors.
BDI	pink-eyed yellow	Druckrey, 1937, from a pair of pink-eyed yellow rats obtained from Kröning, Göttingen; served as progenitor strain for further Berlin-Druckrey strains: BDVI, BDVII, and BDIX; used in experimental oncology.
BDII	albino	Druckrey, 1950, inbred from an outbred colony of Wistar rats obtained from the Unilever Research Institute, Rotterdam; very high incidence (>90%) of endometrial adenocarcinoma in virgin females at about 15-18 months.
BDIII	pink-eyed yellow hooded	Same origin as BDI; BDI and BDIII have been separated after several generations of inbreeding.
BDIV	black hooded	Druckrey from a cross between BDII and BDIII with subsequent brother × sister mating and selection for black-hooded rats.
BDV	pink-eyed yellow, non-agouti hooded	Druckrey from a cross between BDIII and BDIV with subsequent inbreeding and selection for the cream-colored hood.
BDVI	black	Druckrey from a cross between BDI and BDII with subsequent selection of brother-sister pairs for a black selfed coat color.
BDVII	pink-eyed sandy	Druckrey from F2 offspring of a cross between BDVI and BDI with subsequent inbreeding and for a pink-yellow, black-hooded phenotype, resulting in a sandy-colored coat.
BDVIII	agouti hooded	Druckrey from F2 offspring obtained from a cross between BDII and BDIII with subsequent selection of brother-sister pairs for the agouti-hooded phenotype.
BDIX	agouti	Druckrey from a cross between BDI and BDVIII with subsequent selection of brother-sister pairs for agouti coat colour and dark, pigmented eyes. Highly susceptible to transplacental induction of malignant tumors of the central nervous system by ethyl nitrosourea.
BDX	albino	Druckrey from the F2 generation of a cross between BDII and BDV with selection of albino brother-sister pairs carrying the nonagouti, black-hooded, pink-eyed yellow genotype.
BEG	albino	From a cross between SC and TE. Carries rexoid ragged hair gene (*rg*).
BH	black hooded	Developed by Wilson, University of Pennsylvania, from unknown stock; resistant to induction of experimental allergic encephalomyelitis.
BI		Formerly called B3, but now extinct.
BIL-1	dilute cinnamon hooded	University of Pittsburgh from a mutation in a colony of unknown background. BIL/1 carries mutation in the Mhc associated growth and reproduction complex (*grc*).
BIL/2	dilute cinnamon hooded	Same as for BIL/1; control strain to BIL-1.
BIRMA	albino	Mandl, 1952, from albino rats purchased at Birmingham market.
BIRMB	albino	Same as for BIRMA.
BLK	black	Developed at NIH.
BN	brown	Billingham and Silvers, 1958, from a brown mutation maintained by King and Aptekman in a pen-bred colony derived from wild rats captured in the vicinity of Philadelphia; due to its genetic diversity to all other inbred strains widely used in linkage studies.
BP	black hooded Irish	Sekla, Prague; strain carries unusual array of genetic markers selected for resistance to Walker 256 tumor.
BROFO	albino	Developed at TNO (NL) from rats of Wistar origin; selected for large body size.
BS	black	University of Otago Medical School from a cross between wild and outbred Wistar rats, with the F1 hybrids backcrossed to the Wistar stock and selection for a black phenotype.
BUF	albino	Heston from Buffalo stock of Morris; BUF/Mna develop spontaneous thymomas (>98%) >18 months of age.
CAP	albino	Polish Academy of Sciences, Krakow.
CAR	albino	Hunt, 1937; selected for resistance to dental caries; serves as control strain to CAS.
CAS	albino	Developed by Hunt, 1937; selected for a high incidence of dental caries.
CBH	black hooded	Developed at Chester Beatty Institute from rats obtained from Woodruff, Edinburgh.

(Continued)

TABLE 3-4

INBRED RAT STRAINS—CONT'D

Strain	Coat color	Information
CDR	albino	Cohen diabetes resistant rats inbred by Yagil in parallel to CDS as a strain not developing T2DM when fed a high-sucrose low-copper diet. The rate of genetic divergence between CDR and CDS was determined as 43% (polymorphism in 550 microsatellite markers).
CDS	albino	Originally developed by Cohen as a non-inbred colony of rats as a model for T2DM. Animals from the original colony were selectively inbred by Yagil. Cohen diabetes sensitive (CDS) rats that are fed a custom-prepared high-sucrose low-copper diabetogenic diet become overtly diabetic: fasting glucose levels were normal or elevated, and the blood glucose insulin response to glucose loading was markedly abnormal. CDS rats do not become obese or hyperlipidemic, however males have a lower growth rate and more severe glucose intolerance.
CFY		Rat strain with spontaneous mutation causing anophthalmia; identified by Rao in rats of unknown origin.
COP	black hooded	Copenhagen; developed by Curtis, 1921, at Columbia University Institute for Cancer Research; highly resistant to development of estrogen-induced mammary carcinomas.
CPBB	agouti hooded	(BC)
CRDH	albino	Cohen Rosenthal Diabetic Hypertensive rat; derived from a cross between Cohen Diabetic (CDR) and SHR rats with selection for high blood pressure and blood glucose levels on a high sucrose and copper-poor diet.
CWS	albino	From a cross of an outbred Jcl:SD rat with spontaneous cataract and WKAH inbred rats.
DA	agouti	Developed by Odell at Oak Ridge National Laboratory from heterogeneous stocks of unknown origin; named DA because it expresses the $Ag\text{-}D$ $(RT3^a)$ blood group allele of Palm; may be related to COP.
DB	dilute brown	Developed at NIH.
DEBR	black hooded	Dundee experimental bald rat, developed by Oliver, University of Dundee, NZ; possible animal model for human alopecia areata.
DI	black hooded	Schroder from a colony of Long-Evans rats in Brattleboro, Vermont, subsequent substrains developed from these so-called Brattleboro rats. Rats develop hereditary hypothalamic diabetes insipidus due to the absence of circulating vasopressin. Single base deletion in the vasopressin gene (Avp^{di}) causes the diabetes insipidus.
DMY	albino	Autosomal recessive mutation causing demyelination in Sprague Dawley (SD) rats originating from the Universitat Autonoma de Barcelona (Bellaterra Campus).
DON	albino	Formerly DONRYU; Japanese reference strain; developed by Sato, 1950, from Japanese albino rats.
DONRYU	albino	See DON/Kyo.
DOP	dilute	Ohno, 1988, from a breeding colony of Wistar carrying the ataxic mutation dilute-opisthotonus, *dop*.
DRH	albino	Originates from DONRYU rats; carcinogen-resistant strain.
DSS/3N	albino	Developed by Dahl, Brookhaven National Laboratories, from outbred Sprague-Dawley stock with selection for sensitivity to sodium chloride-induced hypertension.
ET	albino	Derived from WKA strain from Taisho Pharmaceutical Co. Ltd.; develops polygenically controlled ectopic scrota in about 70% of males.
EXBH	non-agouti brown red-eyed yellow	Developed at Han as a coat color tester strain (*abCHPr*); may be considered a recombinant inbred strain with E3 and BN as progenitors
EXHC	albino	Exogenously Hypercholesterolemic rat; developed by Imai and Matsumura from SD/Jcl rats with selection for high serum cholesterol.
E3	fawn hooded	Kröning, Göttingen from rats of unknown origin selected for fawn hooded phenotype (*aBhPr*); expresses platelet storage pool defect due to the red-eyed yellow allele *r*.
F344	albino	Curtis and Dunning 1920 at Columbia University Institute for Cancer Research; sublines differ with respect to $Dpp4/Cd26$; Crj and CrlWiga substrains are deficient due to an autosomal semi-dominant mutation which was mapped to RNO3.
FCH		Fice Combined Hyperlipidemic strain; developed from outbred stock by selection for high serum cholesterol.
FH	fawn hooded	Developed by Maier, University of Michigan, Ann Arbor, from a cross between "German brown" and "white Lashley" rats, thereafter inbred.
FHH	fawn hooded	Provoost from fawn hooded rats from Europe; develops focal and segmental glomerular sclerosis, systemic hypertension and proteinuria at a young age.
FHL	fawn hooded	See FHH; however, FHL rats do not develop hypertension or renal damage; used as a control strain to FHH.
FHR	fawn hooded	Substrain to FHH.
FNL		Fice Normolipidemic strain; developed as a control strain for FCH.

TABLE 3-4

INBRED RAT STRAINS—CONT'D

Strain	Coat color	Information
F6R	albino	Mutation in an irradiated F344 strain obtained from the National Institute of Genetics, Misima, Japan. Carries chromosomal translocation (9:14).
G	albino	Developed by Gorter (NL) then transferred to CPB.
GEPR/3	albino	Genetically epilepsy prone; from outbred Sprague-Dawley stock; selected by Jobe, 1971, for moderate susceptibility to audiogenic stimuli-induced seizures.
GEPR/9	albino	Genetically epilepsy prone; see GEPR/3. GEPR-9 is subject to seizures of greater severity; they become sensitive from about 21 days of age post-partum.
GH	albino	Genetically hypertensive; developed by Smirk at University of Otago Medical School, Dunedin, NZ, from rats of Wistar origin imported from England in 1930 with selection for high blood pressure.
GHA	ginger? hooded	Developed at Queen Elizabeth Hospital, Woodville, Australia, from mixed Wistar, LEW and colored stock.
GK	albino	Developed by Goto and Kakizaki from outbred Wistar rats with selection for high glucose levels in and oral glucose tolerance test; model for NIDDM.
HCS		Harvard Caries Susceptible; from a colony of rats at Harvard, then transferred to Liverpool, UK.
HMT	albino	Harwell Mouth Tumor; derived from the Wistar derived outbred Alderley. Park strain 1; inbred since 1964.
HS	black hooded	University of Otago Medical School, probably from the same Wistar × wild rat cross as BS.
HTX	albino	Hydrocephalic rat of unknown origin; hydrocephalus is associated with an abnormality of the cerebral aqueduct and detectable at 1-2 days of age with a maximal survival of 4-5 weeks; 40% penetrance.
HWY	albino	Hairless Wistar Yagi; developed by inbreeding from an undefined Wistar colony; almost complete hairless rat strain; phenotype is caused by an autosomal dominant gene, Hwy, located on RNO15.
IER		Ihara epileptic rat; strain with hereditary neuronal microdysgenesis and aquired lesions in the hippocampal formation leading to abnormal behavior, circling, various degrees of seizure activities, and generalized tonic-clonic convulsions.
IIM	albino	A set of nine inbred strains (IIMa to IIMi) was developed from an outbred colony of albino rats in Buenos Aires, Argentina whose ancestors came from Europe; inbred with selection for large body weight and high fertility. IIMß becomes obese with mild glucose intolerance and glycosurea in older obese rats.
INR	black hooded	Iowa nonreactive; Harrington, 1962, from a stock selected for low open field defecation.
IR	"pale cinnamon beige" hooded	Iowa reactive; Harrington, 1962, from a cross of a "Michigan" and "Berlin" stock; high incidence of congenital abnormalities.
IS	agouti	Ishibashi rat derived from a cross between a wild male and a Wistar female; carries congenital vertebral malformations.
JC	albino	Originates from LEW/Ss transferred to Brisbane, Australia. Because of presumed genetic contamination re-named as JC.
K	albino	Developed by Matthies (Halle-Wittenberg, Germany), 1958, from outbred Wistar stock by selection for resistance to a range of transplantable tumors.
KDP	albino	Diabetes-prone rat developed by Komeda from Long-Evans Tokushima Lean (LETL) rats; frequency of IDDM 70% at 120 d of age, and 82% within 220 d of age; $Cblb$ has been found to be the major susceptibility gene for T1DM in this strain.
KGH	albino	Kunz and Gill, Pittsburgh, from a closed colony of NEDH rats.
KHR	albino	Hairless mutation detected by Kimura in Gunn rats maintained at Kaken Pharmaceuticals, 1987; the khr mutation maps to RNO7.
KIRBY-B	black	Derived from a cross between black hooded rats and outbred CFY rats with selection for resistance to chronic respiratory disease.
KMI	albino	Kyoto Miniature Ishikawa rat; from a breeding colony of Wistar rats at the Ishikawa Animal Laboratory; experimental model for pituitary dwarfism with an autosomal recessive mode of inheritance.
KND	albino	Komeda non-diabetic rats were established simultaneously as controls to KDP from Long-Evans Tokushima Lean (LETL) rats at Tokyo Medical College in 1996.
KX	albino	Formerly designated NEDH; developed from Slonaker colony at University of Chicago. There are substrains that carry infantile ichthyosis (ic) as well as colour genes C and H.
KYN	black hooded notched	Nakata from stock of wild rats from Kyoto crossed with WKAH and subsequent selection for the notched (h^n) pigmentation pattern.
KZC		Komeda Zucker Creeping rat; creeping rats isolated by Komeda as a spontaneous mutation, cre, in a closed colony of the Zucker ($Lepr^{fa}$, fatty) rats; animals show severe ataxia and tremor, and die around 4 weeks of age. The cerebellum of mutants shows severe hypoplasia; abnormal pathological changes are also detected in the cerebrum.
LA	black	From a cross between ALB/N and a hooded stock of unknown origin; serves as control for the congenic strain LA/N-$Lepr^{fa-cp}$.

(Continued)

TABLE 3-4

INBRED RAT STRAINS—CONT'D

Strain	Coat color	Information
LE	black hooded	Mühlbock, Amsterdam, from outbred Orl:LE (Long-Evans) stock.
LEA	agouti	Developed at Hokkaido from outbred Long Evans stock, with selection for agouti coat color (however: Long Evans stock were developed as non-agouti, black hooded rats).
LEC	'cinnamon-like'	Long-Evans cinnamon; originating from a cross between a wild rat and "Castle's black rat" and carrying a mutation that causes fulminant hepatitis and jaundice leading to death within a week after onset (four months of age); survivors develop hepatocellular carcinoma after 1-1.5 years (model for Wilson's disease).
LEJ	black hooded	Long-Evans Japan; derived in Hokkaido, 1956, from an outbred stock of Pacific Farms, USA; susceptible to induction of EAU by interphotoreceptor retinol-binding protein.
LEM	albino	Derived by T.H. Yoshida from LET; carries an inversion of RNO 1.
LEO	black hooded	From National Institute of Genetics Misima, Japan; serves as control strain for LEM and LET; without chromosomal inversions.
LEP	agouti	Long-Evans Praha, developed at Charles University Prague from a cross of outbred animals, including a Long-Evans stock.
LER	albino	Previously designated Le-R or LER and thought to be a mutation within LEW conferring resistance to experimental allergic encephalomyelitis; it was later shown that the phenotype is due to a genetic contamination with BUF genome.
LET	albino	T.H. Yoshida from a cross between LEW and LEJ at National Institute of Genetics, Misima, Japan. Homozygous for a 1:12 chromosomal translocation; see also LEM and LEO.
LETL	black hooded	Develops insulin-dependent T1DM, but no significant T lymphopenia; developed at Tokushima Research Institute by inbreeding of Crl:LE rats that showed polyurea, polyphagia and polydipsia with selection for diabetes mellitus.
LETO	black hooded	Non-diabetic control strain for LETL, developed from the same Crl:LE population.
LEW	albino	Developed by Margaret Lewis from Wistar stock; LEW serves as the genetic background for quite a number of Mhc-haplotypes and other markers/mutations, e.g. Whn^{rnu}, $Myo15^{ci2}$, a model for the Usher type 1 syndrome.
LH	albino	Lyon Hypertensive; from outbred Sprague-Dawley rats selected for high systolic blood pressure.
LL	albino	See LH; LL (Lyon Low blood pressure) was developed as a hypotensive strain.
LN	albino	See LH; LN (Lyon Normotensive) was maintained as a normotensive control.
LOU/C	albino	Bazin and Beckers from rats of presumed Wistar origin; kept at the Université Catholique de Louvain, selected for high incidence of plasmacytomas.
LOU/M	albino	See LOU/C, but selected for low incidence of plasmacytomas.
LUDW	albino	From Wistar stock at Ludwig Institute; susceptible to tumor induction by methyl-nitroso-urea (MNU).
MAXX	black hooded	Developed from a cross of BN and LEW with subsequent inbreeding.
MD	albino	Wistar derived myelin deficient rat carrying X-linked mutation.
MES	albino	Matsumoto Eosinophilia Shinshu; rats with high eosinophilic counts (> 360 cells/μl) were selectively inbred from outbred Slc:SD rats by Matsumoto; locus responsible for onset of eosinophilia was mapped to RNO19.
MF	albino	Developed by Holme and Piechuta, 1979, selective breeding of Sprague-Dawley outbred rats for a prolonged severe dyspnea 14-18 days after challenge with areosolised egg albumin; later named APR (Apnea Prone Rat).
MHS	albino	Milan Hypertensive Strain; derived from outbred Wistar rats with selection for high systolic blood pressure.
MLCS	albino	Derived from a cross between MHS and MNS followed by backcrossing to MNS with selection for low calpastatin activity.
MNR	albino	Maudsley non-reactive rat; Broadhurst, 1954, from a commercial Wistar stock with selection for low defecation response in an open field.
MNRA	albino	Substrain of MNR.
MNS	albino	Milan Normotensive Strain; developed from outbred Wistar rats by selective breeding for low systolic blood pressure; serves as a normotensive control for MHS.
MR	albino	Maudsley reactive; from a commercial Wistar stock with selection for high defecation response in open field.
MSUBL	black	Stroyeva, Institute of Developmental Biology, Moscow from a cross of wild rats × MSU microphthalmic rats; selected for high incidence of microphthalmia.
MW	albino	Munich Wistar stock selected for superficial glomeruli; see also MWF and WMS.
MWF	albino	Frömter, Munich, from outbred Wistar rats selected for large numbers of superficial glomeruli; males develop spontaneous systemic hypertension and proteinuria of glomerular origin, and glomerulosclerosis.
M14	albino	Chapman 1940 from Sprague-Dawley stock selected for low ovarian response to pregnant mare serum gonadotrophin (PMS).

TABLE 3-4

INBRED RAT STRAINS—CONT'D

Strain	Coat color	Information
M17	albino	Chapman 1940 from Sprague-Dawley stock; selected for high ovarian response to PMS.
M520	albino	Curtis, 1920, at Columbia University Institute for Cancer Research.
NAR	albino	Non Albumin Rat, Nagase, 1979, from a random bred Sprague-Dawley colony. The strain carries a recessive mutation of the structural albumin gene causing analbuminemia
NBL	black hooded	Bogden in the mid-1970s from non-inbred Noble (Nb) rats.
NBR	black	NIH black rats; Poiley at NIH from heterogeneous stock.
NEDH	albino	Slonaker (Univ. of Chicago ca. 1928) rats inbred by Warren at the New England Deaconess Hospital; develops pheochromocytomas in about 59% of animals; see also KGH and KX.
NER	albino	Noda from Crj:Wistar rats; animal model of epilepsy; seizures can be induced by pentylenetetrazol, tossing, and transcorneal electroshock, but not by tactile, photic or acoustic stimuli or transauricular electroshock; phenotype controlled by a major autosomal recessive gene for epilepsy.
NIG-III	pink-eyed yellow	From a cross between a wild rat trapped in Misima, Japan, and Castle's black rat.
NP		Alcohol nonpreferring strain; developed by selective breeding for voluntary alcohol consumption; serves as control to P.
NSD	albino	Developed at NIH, 1964, from outbred Sprague-Dawley; develops high systolic blood pressure.
NZR	albino	New Zealand Rat; subline of AS2, develops atriocaval epithelial mesotheliomas of right atrium or inferior vena cava.
ODU	albino	Ito at Osaka Dental University, from outbred Wistar Kyoto stock by selective inbreeding for susceptibility to dental plaque development.
ODUS	albino	Osaka Dental University Rat Susceptible for Plaque Formation. As for ODU heavy plaque formation on the lower incisors; develops both periodontal pocket and gingivitis after being fed with a commercial powdered diet.
OKA	albino	From Okamoto (Kyoto) colony to Bazin, Brussels; hypertensive similar to SHR.
OLETF	black hooded	Otsuka Long-Evans Tokushima fatty rats develop non-insulin-dependent diabetes mellitus (T1DM); late onset of hyperglycemia (at about 18 weeks of age), hyperinsulinemia and mild obesity.
OM	albino	Heston, 1946, from non-inbred Osborne-Mendel colony; becomes obese on a high fat diet and develops moderate hypertension.
OXYR	albino	From Wistar stock at Russian Academy of Sciences, Novosibirsk; in contrast to OXYS resistant to cataractogenic effect of galactose rich diet.
OXYS	albino	See OXYS, but selected for susceptibility to cataractogenic effect of galactose-rich diet.
P		Alcohol preferring strain; developed by selective breeding for voluntary alcohol consumption; NP serves as control.
PA	albino	King, 1909, from Wistar Institute stock; the very first inbred strain of rats; now probably extinct.
PAR	black hooded	Bazin, Brussels, from probably the oldest existing closed colony of rats. According to a report in the *Magazin Pitturesque* (1856) a black hooded colony served as a source of food for reptiles at the Jardin des Plantes, a museum of natural history. This closed colony still serving the same purpose was existent in 1988 when rediscovered by Bazin, and subsequently inbred.
PETH	pink-eyed yellow hooded	Carries gene for retinal dystrophy (Mertk^{rdy}), causing cataracts by about 3 months of age; rapid acetylator due to polymorphism of N-acetyltransferase, in contrast to WKY and NSD.
PKD	albino	Derived from outbred Han:SPRD rats with hereditary dominant polycystic kidney disease (*Pkdr1* or *cy*; RNO5); moderate progression of the disease in heterozygotes; uremia, proteinuria, hyperlipidemia and hypertension in males older than 10 months; clinical signs in males at about 8 months (with death at 14-17 months), and in females clinical signs at 12 months (with death after 17 months of age). Homozygotes show a rapid progression and die by about 3-4 weeks of age.
PVG	black hooded	From a colony at Kings College of Household Science, inbred at Glaxo. PVG serves as genetic background for a number of Mhc-haplotypes and other markers. A substrain PVG/cBkl carries a C6 complement-deficiency, presumably due to a spontaneous mutation, $C6^m$
P77PMC	albino	Beijing Medical College from Wistar colony of unknown descent; rats develop audiogenic seizures (70%) and are susceptible to hyperthermia-induced (45°C) convulsions.
R	albino	Mühlbock, Amsterdam, 1947, from a Wistar stock. A congenic strain, R.Cg-*Ugt1^j*, with UDP-glucuronosyl-transferase deficiency, develops hyperbilirubinemia and acholuric jaundice.
RCS	pink-eyed yellow hooded	Developed before 1965 by Sidman from stock obtained from Sorsby of the Royal College of Surgeons, London; carries allele for retinal dystrophy, Mertk^{rdy}; PETH is presumed to be a subline.
RHA	albino	Roman high avoidance; developed by Bignami with selection for high avoidance conditioning with light as a conditioned stimulus and electric shock as the unconditioned stimulus.
RII/1	albino	Tif from outbred Sprague-Dawley stock from Ivanovas, Germany.
RII/2	albino	Originating from outbred Ico:SD (Sprague-Dawley) stock.
RLA	albino	Roman low avoidance; developed by Bignami with selection for low rates of avoidance conditioning with light as a conditioned stimulus and electric shock as the unconditioned stimulus, see RHA.

(Continued)

TABLE 3-4

INBRED RAT STRAINS—CONT'D

Strain	Coat color	Information
RP	albino	Mühlbock, Amsterdam, 1947, from Wistar stock.
SBH		Sabra Hypertensive; from Sabra outbred rats at Hebrew University by inbreeding and selection for high blood pressure following unilateral nephrectomy and treatment with deoxycorticosterone and sodium chloride.
SBN		Normotensive control strain for SBH selected for low blood pressure as a normotensive control strain for SBH.
SC	albino	Outbred Wistar Imamichi; rats have small eyes and cataract.
SCR		Shumiya cataract rat; the phenotype is due to the interaction of different alleles of two genes *Cats1* (RNO20) and *Cats2* (RNO15); while *Cats1^l* is homozygous lethal.
SD	albino	Sprague-Dawley stock derived (various different strains). "SD" rats have been shown to differ in a number of genetic markers.
SDJ	albino.	Sprague Dawley Japan; at Takeda Chemical Industries from outbred Sprague-Dawley stock.
SDNK	albino	Originates from outbred Sprague-Dawley rats (Japan).
SDT		Spontaneously diabetic Torii; rats develop glucose intolerance, hyperglycemia; T2DM in SDT rats is polygenic in nature.
SER	albino	The spontaneously epileptic rat was developed as a double mutant by Serikawa, Kyoto, by crossing ZI rats (derived from SPRD/Han-*Atrn^{zi}*), carrying an autosomal recessive attractin mutation with TRM rats, carrying a genomic deletion in aspartoacylase (aminoacylase) 2 (*Aspa^{tm}*) that leads to an accumulation of N-acetyl-L-aspartate in the brain.
SHHF	albino	Corpulent gene (*Lepr^{fa-cp}*) partially backcrossed to SHR/N, followed by inconsistent brother × sister mating.
SHR	albino	Okamoto, 1963, from outbred Wistar Kyoto rats, selected for high blood pressure. Several genetically diverse substrains have been developed.
SHRSP	albino	Spontaneously hypertensive stroke-prone rat developed as a substrain of SHR with a tendency to develop cardiovascular lesions and stroke (cerebral hemorrhage or infarction in 82% of males over 100 days of age and 58% of females over 150 days of age), and hypercholesterolemia.
SPRD	albino	From outbred Han:SPRD (Sprague-Dawley) rats carrying a dominant pelage mutation *Cu3* (curly-3); develop mammary adenocarcinoma at almost 100% upon DMBA treatment within 60 days.
SR	albino	Rapp from a Sprague-Dawley outbred colony developed by LK Dahl, Brookhaven National Laboratories, Upton, New York, selected for resistance to salt-induced hypertension.
SS	albino	Rapp from a colony of Sprague-Dawley outbred rats developed by LK Dahl, Brookhaven National Laboratories, Upton, New York, selected for sensitivity to salt-induced hypertension.
S5B	albino	Poiley, 1955, at NIH from a cross of outbred NBR and Sprague-Dawley rats, with five generations of backcrossing of the albino gene.
TA	albino	Derived from outbred Wistar Imamichi. Oligospermia is due to a seminiferous defect caused by an autosomal recessive gene, *ol*.
TE	albino	From outbred Wistar Imamichi rats. Males develop hydrotestes caused by sperm retention cysts in the efferent duct and testicular atrophy.
TF	albino	From outbred Wistar Imamichi rats. Carries an autosomal recessive gene causing pseudohermaphroditism in males due to defect of Leydig cells.
THA	albino	Developed from Jcl:Wistar stock by inbreeding with selection for a high rate of electric shock avoidance by lever pressing.
THE	albino	Tsukuba high emotionality; from Wistar albino rats selected for low ambulation in a bright runway out of a dark starting box (high emotionality); see TLE.
TLE	albino	Tsukuba low emotionality; from Wistar albino rats selected for high ambulation in a bright runway out of a dark starting box (low emotionality); see THE.
TM	red-eyed yellow	Developed as coat color tester strain (Tester Moriyama); inbred at Nagoya University; used as an animal model for platelet storage pool deficiency.
TMB	agouti hooded Irish	Broadhurst from a stock selected by Tryon for good maze learning performance (Tryon maze bright).
TMD	black	Broadhurst from a stock selected by Tryon for poor maze learning performance (Tryon maze dull).
TO	albino	Developed at Hokkaido University; from a commercial breeder in Tokyo, Japan; resistant to the induction of EAU by interphotoreceptor retinol-binding protein.
TOM	albino	Toma Institute, Japan.
TRM	albino	Serikawa from outbred Wistar/Kyo with congenital tremor and curled whiskers and hair; rats develop absence-like seizure (petit mal epilepsy) and spongiform degeneration in the central nervous system; mutant rats (*Aspa^{tm}*) accumulate N-acetyl-L-aspartate in the brain.
TS	albino	WKA derived; develops ectopic scrota in about 70% of males. The defect is controled by multiple genes.

TABLE 3-4

INBRED RAT STRAINS—CONT'D

Strain	Coat color	Information
TS1	agouti hooded Irish	Harrington, 1965, from stock selected by Tryon in 1929 for good maze learning performance.
TS3	black	Harrington, 1965, from stock selected by Tryon for poor maze learning performance.
TT	albino	From outbred Wistar Imamichi strain; carries an autosomal recessive gene causing an arrest of spermatogenesis, *as*.
TU	agouti hooded	From a cross of a wild male and Wistar Imamichi outbred rats; small litter size with malformations of kidneys and vas deferens in about 20% of offspring.
TW	albino	Derived from Wistar Imamichi outbred stock; testicular hypoplasia (unilateral or bilateral) with aplasia of the epididymis and ductus deferens in about 50% of males; female genital organs are normal.
TX	black	From a cross between a wild male and Wistar Imamichi females.
U	hooded	Originates from a colony at the Zootechnical Institute, Utrecht.
UChA	albino	Wistar rats selected for low voluntary 10% ethanol consumption; UChA and UChB do not differ in length of alcohol narcosis induced by 60 mmol/kg ethanol given intraperitoneally.
UChB	albino	Wistar rats selected for high voluntary 10% ethanol consumption.
UPL		The Upjohn Pharmaceuticals Limited rat is a model for cataracts inherited as an autosomal semi-dominant trait.
VF		Derived by a spontaneous mutation, *vf*, that maps on RNO8, and causes abnormal vacuoles in the central nervous system of TRM rats; during subsequent inbreeding $Aspa^{tm}$ was counter selected.
W	albino	Wistar derived; various substrains differing in several genetic markers.
WA	albino	Established at St. Thomas's Hospital from outbred Wistar stock.
WAB	albino	Wistar Albino Boots; established 1926, from same stock as WAG; develops spontaneous thymoma in 23% of individuals over 2 years, with 50% incidence in castrated males and 57% in spayed females.
WAG	albino	Wistar Albino Glaxo; Bacharach, Glaxo Ltd 1924 from Wistar stock; substrains vary in coat color genes hypostatic to Tyr^c and other markers, which could be indicative of unreported outcrosses. WAG/Rij is a model of absence epilepsy.
WBN	albino	Developed in Japan (Shizuoka Laboratory Animal Center, Tokyo) from outbred Wistar rats from Europe; rats ageing >21 weeks develop a diabetic syndrome characterised by impared glucose tolerance and later by glycosurea, hyperlipidaemia and gradual emaciation. The syndrome is characterised as T2DM.
WCF	albino	WKAH/Idr derived with variable incidences for hereditary club-foot.
WDF	albino	Developed by Ikeda, 1981, by backcrossing the fatty gene ($Lepr^{fa}$) to incipient inbred Wistar Kyoto rats. Rats exhibit hyperglycemia, hyperinsulinemia, peripheral and hepatic insulin resistance linked to abnormalities in enzyme levels and activities, diet and sex-dependent diabetes and possible diabetes-linked peripheral neuropathy.
WEC	light beige?	Developed and inbred at TNO from an outcross involving strains B, WAG and others; formerly named WE/Cpb. Rats are hyporesponder to dietary cholesterol.
WEK	brown	Developed at TNO, 1958; formerly known as WEchoc; hyporesponder to dietary cholesterol.
WELS	albino	From outbred Wistar rats in 1976.
WF	albino	Furth, 1945, from a commercial Wistar stock in an attempt to develop a high leukemia rat strain; WF/PmWp carries a pelage mutation (fuzzy, *fz*) that arose spontaneously and has been used in research on dermal toxicity; WF/Ztm are susceptible to polygenically controlled audiogenic seizures (grand mal epilepsy).
WIN	albino	Natori from outbred Wistar rats (Wistar-Imamichi-Natori); carry a unique Mhc class I haplotype, $RT1^s$ ($RT1$-$A^sB^lD^l$).
WIST	albino	From outbred Wistar stock. Several independently derived genetically different strains exist.
WKA	albino	Wistar King Albino; 1909 from Wistar Institute stock to Aptekman; probably genetically identical to PA.
WKAH	albino	WKA substrain maintained at Hokkaido University.
WKAM	albino	Formerly called WKA.
WKHA	albino	From a cross between SHR and WKY with selection for high spontaneous activity and low systolic blood pressure.
WKHT	albino	From a cross between SHR and WKY, with selection for high blood pressure and low spontaneous activity.
WKS	albino	Developed at National Institute of Genetics, Mishima, Japan.
WKY	albino	Hansen at National Institutes of Health, Bethesda, in 1971 from outbred Wistar stock from Kyoto School of Medicine; inbred as a normotensive control strain for SHR. There are considerable genetic differences between WKY and SHR, but also among the different WKY substrains, possibly due to early dichotomy; highly resistant to DMBA-induced mammary-adenocarcinoma. A coisogenic WKY/Ztm-*ter* strain has been reported developing hereditary teratomas of the gonads in both sexes.

(Continued)

TABLE 3-4

INBRED RAT STRAINS—CONT'D

Strain	Coat color	Information
WKYO	albino	Inbred in 1980 from outbred Wistar Kyoto rats. Highly sensitive to the development of experimental glomerulonephritis following injection of nephritogenic antigen from bovine renal basement membrane.
WM	albino	Developed at National Institute of Genetics, Misima in 1951 from outbred Wistar rats.
WMS	albino	From Wistar stock selectively bred for superficial glomeruli and prominent elongated renal papilla at Munich University; see also MW and MWF.
WN	albino	Heston, 1942, from Wistar stock of Nettleship. Female WN rats nearly develop 100% anterior pituitary tumours when >18 months old; in males pheochromocytomas or preneoplastic nodules in adrenal medulla are found at a rate of about 40%.
WNA	albino	Developed at Nagoya University, Institute of Laboratory Animal Research from Wistar rats.
WOKW	albino	W(istar)O(ttawa)K(arlsburg)W(RTl^u), developed by Klöting; rats develop polygenically determined full-blown metabolic syndrome with obesity, hyperinsulinemia, dyslipidemia, and late onset hypertony.
WST	albino	WAG derived.
WTC	albino	Serikawa, 1988. WTC is the coisogenic control strain to TRM ($Aspa^{tm}$), separated at F18.
Y59		Strain developed in Zagreb, Croatia.
YA		Hansen, 1989, at NIH.
YO		Fredrich Cancer Research Facility to Pit; rapid elimination of *Trichinella spiralis* worms.
Z61	albino	Curtis, 1920, at Columbia University Institute for Cancer Research. Susceptible to infection with *Cysticercus*; susceptible to estrogen and 2-acetylaminofluorine-induced tumors.
ZDF		Zucker diabetic fatty developed from rats of undefined outbred background with T2DM by selective inbreeding of diabetic homozygous fatty, $Lepr^{fa}$, males to heterozygous females. Develop quantifiable retinal vascular changes.
ZI	albino	Serikawa from Han SPRD-*zi* rats; the mutation which causes spongiform encephalopathy of the central nervous system with tremors at 15 days of age as well as curly whiskers and hair coat, is due to a 8 bp deletion in intron 12 of the Attractin gene, $Atrn^{zi}$.
ZIMY	albino	Zitter Masao Yamada; developed by Serikawa at Kyoto University by crossing TRM ($Aspa^{tm}$) and ZI ($Atrn^{zi}$) rats.

For further information or references see the strain query form at RGD (http://rgd.mcw.edu/strains/), deposited strains at NBRP (http://www.anim.med.kyoto-u.ac.jp/nbr/cryo.htm), index of major rat strains by M.F.W. Festing at MGD (http://www.informatics.jax.org/external/festing/rat/STRAINS.shtml), or Greenhouse et al. (1990), or retrieve information via an internet search tool.

inbred rat strain by pronuclear microinjection of appropriate DNA constructs, provided it is prolific and allows sufficient numbers of fertilized eggs to be collected; transgenes also can be introduced through infection of preimplantation stages with viral vectors. Congenic strains are typically generated by repeated backcrosses (N) or by a cross-intercross system (M) to the inbred strain of choice, resulting in introgression of a specific character from another strain or stock. A strain developed by this method (Flaherty, 1981) may be regarded as congenic when a minimum number of ten backcross generations or an equivalent thereof to the background strain have been made, counting the first hybrid generation as one (F1 = N1, M1, NE1). Marker-assisted (selection) breeding, also known as "speed congenics" permits the production of congenic strains equivalent to ten backcross generations in as few as five generations (Markel et al., 1997; Wakeland et al., 1997). Provided that the appropriate marker selection has been used, these can be termed *congenic strains*, if the donor strain contribution that is unlinked to the selected locus or chromosomal region is less than 0.1% (i.e., NE10). As congenic strains differ only in a short chromosomal

segment from their background strain, it is possible to investigate the phenotypic effect of this differential locus isolated from distracting effects caused by other loci in the genetic background. On the other hand, it is possible to determine whether background effects modify gene function, and if so, the nature of these effects. These approaches play an important role in experimental immunology and in transplantation biology.

For the rat, a variety of congenic strains had been developed, mostly for alloantigenic markers (Hedrich, 1990). Recent advances in rat genomics have facilitated development of a number of additional sets with the aim to verify the impact of presumed quantitative trait loci on the phenotype of polygenic characters (for examples see Table 3-7).

3. Consomic Strains

Consomic strains are a variation on congenic strains in which a whole chromosome is repeatedly backcrossed onto an inbred strain, resulting in substitution of one recipient chromosome per strain at a time. As with congenic strains,

TABLE 3-5

STANDARDIZED SYNONYMS FOR OLDER OR NONSTANDARDIZED STRAIN NAMES

Nonstandardized strain name	Standardized strain name
ACP 9935 Irish piebald	ACP
Aganglionosis Rat	AR
Albany	ALB
Alko Alcohol	AA
Alko Non-Alcohol	ANA
Apnea Prone Rat	APR
August	AUG
August 990	A990
August 7322	A7322
August 28807	A28807
August 35322	A35322
A×C 9935 Irish	ACI
A×C 9935 Piebald	ACH
Berkeley S1	TMB or TS1
Berkeley S3	TMD or TS3
Birmingham A	BIRMA
Birmingham B	BIRMB
Black hooded	BH
Black Praha	BP
Brattleboro	DI
Brown Norway	BN
Buffalo	BUF
B3	BI
Chester Beatty hooded	CBH
CDF	F344
Cohen diabetes resistant	CDR
Cohen diabetes susceptible	CDS
Cohen Rosenthal diabetic hypertensive	CRDH
CPB-G	G
CPB-WE	WE
CPB-B, CPBB	BC
Copenhagen 2331	COP
Dahl R	SR
Dahl S	SS
DONRYU	DON
Dundee experimental bald	DEBR
Exogenously hypercholesterolemic	EXHC
Experimental 3	E3
Fawn hooded	FH
Fice combined hyperlipidemic	FCH
Fice normolipidemic	FNL
Fischer, Fischer 344	F344
Genetically epilepsy prone	GEPR
Genetic hypertension	GH
Goto-Kakizaki	GK
Hairless Wistar Yagi	HWY
Harvard caries susceptible	HCS
Harwell mouth tumour	HMT
HO	PVG
Hunt's caries resistant	CAR
Hunt's caries susceptible	CAS
Ihara epileptic	IHR
Iowa nonreactive	INR
Iowa reactive	IR
Ishibashi	IS
Kaken hairless	KHR
King albino or P.A.	PA
Komeda diabetes-prone	KDP
Komeda non-diabetic	KND

TABLE 3-5

STANDARDIZED SYNONYMS FOR OLDER OR NONSTANDARDIZED STRAIN NAMES—CONT'D

Nonstandardized strain name	Standardized strain name
Komeda Zucker Creeping	KZC
Kyoto miniature Ishikawa	KMI
Kyoto notched	KYN
Long Evans Tokushima lean	LETL
Long Evans Tokushima Otsuka	LETO
Long-Evans	LE
Long-Evans agouti	LEA
Long-Evans cinnamon	LEC
Long-Evans Japan	LEJ
Long-Evans Praha	LEP
Louvain	LOU
Lewis	LEW
Lyon hypertensive	LH
Lyon low blood pressure	LL
Lyon normotensive	LN
Matsumoto Eosinophilia Shinshu	MES
Maudsley nonreactive	MNR
Maudsley reactive	MR
Marshall 520	M520
MF	APR
Milan hypertensive strain	MHS
Milan normotensive strain	MNS
Munich Wistar Frömter	MWF
NEDH	KX
NIH black	NBR
Noble or Nb	NBL
Noda epilepsy rat	NER
Okamoto (hypertensive)	SHR, OKA/Wsl
Otsuka Long-Evans Tokushima fatty	OLETF
Osborne-Mendel	OM
Roman high avoidance	RHA
Roman low avoidance	RLA
Sabra hypertensive	SBH
Sabra normotensive	SBN
Shumiya cataract rat	SCR
Slonaker	NEDH
Spontaneously diabetic Torii	SDT
Spontaneously epileptic rat	SER
Spontaneously hypertensive rat	SHR
Sprague-Dawley Japan	SDJ
Tester Moriyama	TM
Tokyo	TO
Tryon maze bright	TMB or TS1
Tryon maze dull	TMD or TS3
Tsukuba high emotionality	THE
Tsukuba low emotionality	TLE
WE	WEC
WEchoc	WEK
Wistar albino Boots	WAB
Wistar albino Glaxo	WAG
Wistar Furth	WF
Wistar-Imamichi-Natori	WIN
Wistar King albino	WKA
Wistar Kyoto	WKY
YOS	DONRYU
Zimmerman	Z61
Zitter Masao Yamada	ZIMY

TABLE 3-6

Online Sources Providing Information on Inbred Strains

ILAR's Animal Models & Strains Search Tool (AMSST)
http://dels.nas.edu/ilar/ext_search.asp
Rat Genome Database (RGD)
http://rgd.mcw.edu/strains/
National Bio Resource Project for the Rat in Japan (NBRP)
http://www.anim.med.kyoto-u.ac.jp/nbr/home.htm
Hannover Medical School Rat resource (MHHRR)
http://www.mh-hannover.de/einrichtungen/tierlabor/genearchiv.html
Rat Resource and Research Center
http://www.radil.missouri.edu/rrrc/
Mouse Genome Database at The Jackson Laboratory
http://www.informatics.jax.org/external/festing/rat/STRAINS.shtml
Tokushima strain data
http://www.anex.med.tokushima-u.ac.jp/rat/index-e.html

a minimum of ten backcross generations is required. The Medical College of Wisconsin in Milwaukee has initiated a program to develop novel consomic rat models. These are designed to allow researchers to study the specific function of particular genes that contribute to common polygenetically controlled human diseases, such as hypertension, myocardial infarction, and renal disease. Currently seven strains have been established (http://www.criver.com/products/consomic/index.html) with either FHH or SS as the recipient and BN as the chromosomal donor strain: FHH.1^{BN}/Mcw, SS.2^{BN}/Mcw, SS.6^{BN}/Mcw, SS.8^{BN}/Mcw, SS.13^{BN}/Mcw, SS.18^{BN}/Mcw, SS.Y^{BN}/Mcw. The chromosome substitution is aimed at understanding and treating a particular disease condition from a gene function, or from a pharmacogenomic perspective. In addition, these rat strains allow gene identification much more rapidly.

4. Recombinant Inbred Strains

Recombinant inbred (RI) strains are generated by an initial cross of two inbred strains, followed by F1 intercross and 20 or more generations of strict brother by sister mating (Bailey, 1971; 1981). For an overview on the subject the reader is referred to Silver (1995). Different RI strains

TABLE 3-7

Coisogenic and Congenic Strains on the LEW Background[1]

System	Fucnction/Haplotype	Congenic strain	Donor strain	Recipient strain	Holder
Mhc	$RT1^a$	LEW.1A(AVN)	AVN/Cub	LEW/Cub	Ztm
Mhc	$RT1^{av1}$	LEW.1AV1(DA)	DA/Han	LEW/Han	Ztm
Mhc	$RT1^c$	LEW.1C(WIST)	Han:WIST	LEW/Han	Ztm
Mhc	$RT1^d$	LEW.1D(BDV)	BDV/Max	LEW/Max	Ztm
Mhc	$RT1^f$	LEW.1F(AS2)	AS2/Max	LEW/Max	Ztm
Mhc	$RT1^k$	LEW.1K(SHR)	SHR/Han	LEW/Han	Ztm
Mhc	$RT1^n$	LEW.1N(BN)	LEW/Max	BN/Max	Ztm
Mhc	$RT1^u$	LEW.1W(WP)	WP/Cub	LEW/Cub	Ztm
Mhc	$RT1^{r2}$ ($RT1$-A^aB/D^uC^a)	LEW.1AR1			Ztm
Mhc	$RT1^{r3}$ ($RT1$-A^aB/D^aC^u)	LEW.1AR2			Ztm
Mhc	$RT1^{r4}$ ($RT1$-A^uB/D^uC^a)	LEW.1WR1			Ztm
Mhc	$RT1^{r6}$ ($RT1$-A^uB/D^aC^a)	LEW.1WR2			Ztm
Nkc	NK cell receptor complex	LEW.NKC(BS)	BS/Ztm	LEW/Ztm	Won
Tcra	T cell receptor α chain	LEW.TCR(BH)	BH/Ztm	LEW/Ztm	Won
Tcrb	T cell receptor β chain	LEW.TCRB(BS)	BS/Ztm	LEW/Ztm	Won
RT6	ADP-ribosyltransferase	LEW.6B(BH)	BH/Ztm	LEW/Ztm	Won
Ptprc[2]	Protein-tyrosine-phosphatase	LEW.7B(BH)	BH/Ztm	LEW/Ztm	Won
RT8	blood group locus	LEW.SHR-$RT8^a$	SHR/Han	LEW/Han	Ztm
RT12	blood group locus	LEW.12B(TO)	TO	LEW/Ztm	Won
$Myo15^{ci2}$	Usher type 1	LEW-$Myo15^{ci2}$			Ztm
Iddm	diabetes mellitus type 1	LEW.1AR1-$iddm$			Ztm
Pkdr1[3]	Polycystic kidney disease	LEW-$Pkdr1$	Han:SPRD	LEW/Han	Ztm
Whn[4]	winged helix nude (Rowett)	LEW-Whn^{rnu}	Han:RNU	LEW/Ztm	Ztm
Whn	winged helix nude (NZ)	LEW-Whn^{rnu-nz}	Han:NZNU	LEW/Ztm	Ztm
Whn	winged helix nude (Rowett)	LEW.1AR1-Whn^{rnu}	LEW-Whn^{rnu}	LEW/Ztm	Ztm

[1]These are established strain names (according to ILAR's Definition, Nomenclature and Conservation of Rat Strains), their names have not been adjusted to the joint rules for nomenclature of mouse and rat strains recently set by the International Committee on Standardized Genetic Nomenclature for Mice and the Rat Genome and Nomenclature Committee [http://rgd.mcw.edu/nomen/rules-for-nomen.shtml]; following the latter rules LEW.1A(AVN) would be written as LEW.AVN-$RT1^a$, and LEW.TCRB(BS) as LEW.BS-$Tcrb$

[2]Synonyms: $Cd45$ (leukocyte common antigen), $RT7$

[3]Synonym: cy

[4]Synonyms: rnu (thymus-aplastic nude), $Foxn1^{rnnu}$

TABLE 3-8

LIST OF RI STRAINS AVAILABLE

RI set name	No. of strains maintained per set	Progenitor strain 1	Progenitor strain 2		Developed by
BXH	11	BN-*Lx*/Cub	SHR/OlaIpcv	Cub	Cub, Mpp
DXE	4	DA/Han	E3/Han	Han	Ztm, Rhd
FXLE	15	F344/DuCrj	LE/Stm	Stm	NBRP
HXB	17	SHR/OlaIpcv	BN-Lx/Cub	Ipcv	Cub, Mpp
LEXF	19	LE/Stm	F344/DuCrj	Stm	NBRP[1]
LXB	12	LEW/Han	BN/Han	Han	Ztm
PXO	11	SHR.*Lx*	BXH2	Cub	Cub

[1]NBRP (National Bio Resource Project for the Rat in Japan) collects and distributes strains as well as assembled data, access through http://www.anim.med.kyoto-u.ac.jp/nbr/home.htm.

derived from the same pair of progenitor strains are considered members of a set. For successful usage of such sets, a minimum of 20 strains in a set of RI strains from progenitors differing in as many characters as possible is required (Silver, 1995). RI strains are useful tools primarily to identify the number of genes involved in the expression of a polygenic character and the respective map position (and linkage) of the participating QTLs. Thus, the set of RI strains developed by Pravenec and Kren in Prague (Pravenec et al., 1999; Kren et al., 1996) using the spontaneously hypertensive SHR strain and the congenic BN-*Lx* strain carrying the polydactyly luxate syndrome proved to be extremely useful for identifying genes involved in rat hypertension (Printz et al., 2003). Table 3-8 lists RI strain sets currently available.

V. NOMENCLATURE

A key feature of rat nomenclature is the laboratory registration code, whether it is for the definition of an outbred stock or an inbred strain. This code is usually three to four letters (first letter upper case, followed by all lower case), and identifies a particular producing institute, laboratory, or investigator. Further, these codes are intended to identify outbred stocks and inbred strains, including congenic strains etc. These codes are assigned through the Institute of Laboratory Animal Research, ILAR, at http://dels.nas.edu/ilar_n/ilarhome/labcode.shtml.

A. Nomenclature Rules for Outbred Stocks

Nomenclature rules for outbred stocks have been proposed by Festing et al. (1972) and are based on principles of population genetics. The rules have not been revised

since they were first developed in 1972, although there is considerable uncertainty as to the method for maintaining outbred colonies of the same stock at different locations in a way to keep them genetically stable. To minimize changes due to inbreeding and genetic drift, the population should be maintained in such numbers as to give less than 1% inbreeding per generation. Population size and breeding system (including the method for choosing breeding animals for the next generation and the method of mating the animals that are chosen) are an integral part of the stock characteristics; the stock must be maintained as a closed colony without selection.

Breeding systems that fulfill these requirements have been described elsewhere (Falconer, 1967; Kimura and Crow, 1963; Poiley, 1960; Rapp, 1972); according to Eggenberger (1973) the rotational systems described by Falconer and Rapp provide the lowest increase in inbreeding coefficients. The denomination is based on the same requirements set for inbred strains (see below), a registered laboratory code and a strain name consisting of a unique brief symbol of upper case Roman letters separated by a colon (e.g., Han:WIST, Crl:LE, Hsd:WI, Tac:WKY). This prefixed position of the laboratory (breeder) code, and the separating colon, serve as distinguishing characteristics of outbred stock symbols.

B. Nomenclature Rules for Inbred Strains

For inbred strains of mice and rats, the respective nomenclature committees have agreed to establish a joint set of rules for both species. By this collaboration the former rules for rat strain nomenclature developed by the Committee on Rat Nomenclature (1992) have thus been revised. These rules are accessible through http://rgd.mcw.edu/nomen/nomen.shtml#StrainNomenclature, and the earlier document is also accessible through ILAR at http://dels.nas.edu/ilar_n/ilarjournal/34_4/34_4Definition Backup.shtml or through "Ratmap" at http://rgnc.gen.gu.se/Brief.html.

1. Standard Inbred Strains

In short, inbred strains must be continuously mated brother × sister (B × S; or equivalent; see Green, 1981b). A minimum of 20 B × S generations is required to achieve a level of residual heterozygosity that is on average ≤ 0.01. The denomination of an inbred strain should be by a unique brief symbol made up of upper case Roman letters, or a combination of letters and numerals, beginning with a letter (e.g., IS/Kyo, LEW/Ztm). The level of inbreeding may be added in parenthesis with F followed by the number of generations [e.g., DA/Ztm (F105)]. A genetic divergence of established strains may occur with time either through genetic drift, if substrain dichotomy took place after F20 but before F40, or through mutation, especially if branches have been separated for more than 20 generations from a common ancestor. Genetic alterations owing to an unintentional outcrossing do not form a new substrain; such a contaminated strain should be renamed (e.g., LER) once the segregating markers are fixed again. Substrains are given the root symbol of the original strain, followed by a forward slash and a substrain designation (e.g., SR/JrIpcv; substrain derived at Institute of Physiology, Czech Academy of Sciences [Ipcv] from the colony maintained by John Rapp [Jr]).

2. F1-Hybrids

Rats that are the progeny of two inbred strains, crossed in the same direction, are genetically identical, and can be designated using upper case abbreviations of the two parents (maternal strain listed first), followed by F1. Reciprocal F1 hybrids are genetically not identical, and their designations are, therefore, different. Because there are many strains that are identical in their upper case abbreviations, one should always provide a full designation initially (e.g., [WKY/Ztm × SPRD/Ztm] F1 may then be abbreviated WSF1).

3. Coisogenic Strains

Coisogenic strains are inbred strains that differ at only a single locus through mutation occurring in that strain (e.g., LEW.1AR1-*iddm*).

4. Congenic Strains

Congenic (CR) strains are designated by a symbol consisting of three parts. The full or abbreviated symbol of the recipient strain is separated by a period from an abbreviated symbol of the donor strain. This is the strain in which the allele or mutation originated, and which may or may not be its immediate source in constructing the congenic strain (e.g., DA.KGH-*RT1^g* for which the immediate source has been BN.KGH-*RT1^g*). In cases in which the donor is not inbred or is complex, the symbol Cg should be used to denote congenic. The use of the donor strain symbol or Cg is essential to distinguish congenic from coisogenic strains. A hyphen then separates the strain name from the symbol (in italics) of the differential allele(s) introgressed from the donor strain (e.g., PVG.AO-*RT1^u*; note that a number of established CR strains have maintained their original denomination as proposed by the Committee on Rat Nomenclature in 1992 see [Table 3-7]; according to the newly set rules LEW.1N(BN) should be designated LEW.BN-*RT1^n*; or LEW-*Whn^{rnu}* as LEW.Cg-*Whn^{rnu}*). Moreover, it is advisable to indicate the number of backcross generations as well as the number of brother × sister generations following the intercross (e.g., N10F7), or the equivalent thereof (see Green, 1981b; Silver, 1995; Snell, 1978). Descriptions of speed congenic strains in first publications thereof ideally should include the number and genomic spacing of markers used to define the congenicity of the strain. Because speed congenics depend upon thorough marker analysis and can vary by particular experimental protocols, the inbred status of speed congenics should be regarded with caution.

5. Consomic Strains

Consomic strains (also named chromosome substitution strains, see Nadeau et al., 2000) are produced by repeatedly backcrossing a whole chromosome onto an inbred strain and designated HOST STRAIN-CHROMOSOME$^{DONOR\ STRAIN}$ (e.g., SS.BN2 currently is not denominated in accordance with these rules and should rather be named SS.2^{BN}/Mcw).

6. Recombinant Inbred Strains

Recombinant inbred (RI) strains should be designated by upper case one- or two-letter abbreviations of both parental strain names, with the female strain written first and separated by an upper case letter X with no intervening spaces (e.g., HXB1, HXB2, HXB3; members of the HXB set of RI strains derived from a cross of SHR/OlaIpcv × BN-*Lx*/Cub). Recombinant inbred strains may be intercrossed for mapping complex traits. Such F1 populations are called recombinant inbred intercrosses (RIX) and are symbolized the same as F1 hybrids between standard inbred strains.

VI. SUMMARY

With genomic sequences available for man, mouse, and rat, one can envision that the first few decades of the 21st century are likely to be dominated by assigning function to the complete genomic sequence, particularly with respect to those regions involved in common human diseases. Although the paradigm for ascribing function to the genome is not well defined, it is clear that investigators will use comparative mapping strategies and the available multiple species platforms to accomplish this goal. The laboratory rat, along with the laboratory mouse, will be a lead player in this field.

REFERENCES

Bazin, H. (1988). The PAR Rat. *Rat News Lett.* **20**, 14.

Bailey, D.W. (1971). Recombinant-Inbred Strains. An aid to finding identity, linkage and function of histocompatibility and other genes. *Transplantation* **11**, 325–327.

Bailey, D.W. (1981). Recombinant Inbred Strains and Bilineal Congenic Strains. *In* "The Mouse in Biomedical Research," (H.L. Foster, J.D. Small, and J.G. Fox, eds.), pp 223–239, Academic Press, New York.

Castle, W.E. (1947). The domestication of the rat. *Proc. Natl. Acad. Sci. U S A* **33**, 109–117.

Cholnoky, E., Fischer, J., and Jozsa, S. (1974). Aspects of genetically defined populations in toxicity testing. I. A comparative survey of populations obtained by different breeding systems and "mosaic populations." *Z. Versuchstierkd.* **11**, 298–311.

Committee on Rat Nomenclature. (1992). Definition, nomenclature, and conservation of rat strains. *ILAR News* **34**, S1–S26.

Crampe, H. (1877). Kreuzungen zwischen Wanderratten verschiedener Farbe. *Landwirtsch. Jahrb.* **6**, 385–395.

Crampe, H. (1883). Zuchtversuche mit zahmen Wanderratten. I. Resultate der Zucht in Verwandtschaft. *Landwirtsch. Jahrb.* **12**, 389–449.

Crampe, H. (1884). Zuchtversuche mit zahmen Wanderratten. II. Resultate der Kreuzung der zahmen Ratten mit Wilden. *Landwirtsch. Jahrb.* **13**, 699–754.

Crampe, H. (1885). Die Gesetze der Vererbung der Farbe. *Landwirtsch. Jahrb.* **14**, 539–619.

Eggenberger, E. (1973). Modellpopulationen zur Beurteilung von Rotationssystemen in der Versuchstierzucht. *Z. Versuchstierkd.* **15**, 297–331.

Falconer D.S. (1967). Genetic Aspects of Breeding Methods. *In* "UFAW Handbook on the Care and Management of Laboratory Animals." (W. Lane-Petter, A.N. Worden, A, B.F. Hill, J.S. Paterson, H.G. Vevers, (eds.), pp 72–96, Livingstone, Edinburgh.

Festing, M., Kondo, K., Loosli, R., Poiley, S.M., and Spiegel, A. (1972). International standardized nomenclature for outbred stocks of laboratory animals. *Z. Versuchstierkd.* **14**, 215–224.

Flaherty, L. (1981). Congenic Strains. *In* "The Mouse in Biomedical Research," (H.L. Foster, J.D. Small, and J.G. Fox, eds.), pp 215–222, Academic Press, New York.

Freye, H.A., and Thenius, E. (1968). Die Nagetiere. *In* "Grzimeks Tierleben." (B. Grzimek, ed.), Vol. 11, pp 204–211, Kindler, Zurich.

Furano, A.V., and Usdin, K. (1995). DNA "fossils" and phylogenetic analysis. Using L1 (LINE-1, long interspersed repeated) DNA to determine the evolutionary history of mammals. *J Biol Chem* **270**, 25301–25304.

Gesner, C. (1669). Thier-Buch (partial edition of "Historiae animalium," in German), pp 262–263, reprinted 1980, Schlütersche Verlagsanstalt, Hannover.

Green, E.L. (1981a). Breeding Systems. *In* "The Mouse in Biomedical Research," (H.L. Foster, J.D. Small, and J.G. Fox, eds.), pp 91–104, Academic Press, New York.

Green, E.L. (1981b). "Genetics and Probability." Macmillan Publishers, London.

Greenhouse, D.D., Festing, M.F.W., Hasan, S., and Cohen, A.L. (1990). Catalogue of Inbred Strains of Rats. *In* "Genetic Monitoring of Inbred Strains." (H.J. Hedrich, ed.), pp 1410–480, Gustav Fischer Verlag, Stuttgart.

Hedrich, H.J. (1990). Genetic Monitoring of Inbred Strains. Gustav Fischer Verlag, Stuttgart.

Huchon, D., Madsen, O., Sibbald, M.J., Ament, K., Stanhope, M.J., Catzeflis, F., de Jong, W.W., and Douzery, E.J. (2002). Rodent phylogeny and a timescale for the evolution of Glires: evidence from an extensive taxon sampling using three nuclear genes. *Mol. Biol. Evol.* **19**, 1053–1065.

Jyo-En-Shi. (1787). Chingansodategusa. Chohbei Zeniya, Kyoto.

Kimura, M., and Crow, J.F. (1963). On the maximum avoidance of inbreeding. *Genetical Res. Cambridge* **4**, 399–415.

King, H.D. (1918a). Studies on inbreeding. I. The effects of inbreeding on the growth and variability in the body weight of the albino rat. *J. Exp. Zool.* **26**, 1–54.

King, H.D. (1918b). Studies on inbreeding. II. The effects of inbreeding on the fertility and on the constitutional vigor of the albino rat. *J. Exp. Zool.* **26**, 335–378.

King, H.D. (1918c). Studies on inbreeding. III. The effects of inbreeding, with selection, on the sex ratio of the albino rat. *J. Exp. Zool.* **27**, 1–35.

King, H.D. (1919). Studies on inbreeding. IV. A further study of the effects of inbreeding on the growth and variability in the body weight the of the albino rat. *J. Exp. Zool.* **29**, 135–175.

Kren, V., Krenova, D., Bila, V., Zdobinska, M., Zidek, V., and Pravenec, M. (1996). Recombinant inbred and congenic strains for mapping of genes that are responsible for spontaneous hypertension and other risk factors of cardiovascular disease. *Folia Biol (Praha)* **42**, 155–158.

Lantz, D.E. (1909). *US Dept. of Agriculture, Biol. Survey, Bull.* **33**, 9–54.

Lindsey, J.R. (1979). Historical Foundations. *In* "The Laboratory Rat." (H.J. Baker, J.R. Lindsey, and S.H. Weisbroth, eds.), Vol. 1, pp 1–36. Academic Press, New York.

Markel, P., Shu, P., Ebeling, C., Carlson, G.A., Nagle, D.L., Smutko, J.S., and Moore, K.J. (1997). Theoretical and empirical issues for marker-assisted breeding of congenic mouse strains. *Nat. Genet.* **17**, 280–284.

McCay, C.M. (1973). Notes on the History of Nutrition Research. *In* "Notes on the History of Nutrition Research." (F. Verzár, ed.), pp 109, Hans Huber, Bern.

Meng, J., Wyss, A.R., Dawson, M.R., and Zhai, R. (1994). Primitive fossil rodent from Inner Mongolia and its implications for mammalian phylogeny. *Nature* **370**, 134–136.

Musser, G.G., and Carleton M.D. (1993). Family Muridae. *In* "Mammal Species of the World: A Taxonomic and Geographic Reference." Wilson D.E. and Reeder, D.M. (eds.), 2nd ed., pp 501–756, Smithsonian Institution Press, Washington D.C.

Nadeau, J.H., Singer, J.B., Matin, A., and Lander, E.S. (2000). Analyzing complex genetic traits with chromosome substitution strains. *Nat. Genet.* **24**, 221–225.

Nikkhah, G., Rosenthal, C., Hedrich, H.J., and Samii, M. (1998). Differences in acquisition and full performance in skilled forelimb use as measured by the "staircase test" in five rat strains. *Behav. Brain Res.* **92**, 85–95.

Nomura, T. (2000). Concept for establishment of rat outbred global standard strains. *In* "Microbial Status and Genetic Evaluation of Mice and Rats: Proceedings of the 1999US/Japan Conference." http://books.nap.edu/books/030907195X/html/65.html, pp 65–76, Natl. Acad. Press, Washington, D.C.

Philipeaux, J.M. (1856). Note sur l'extirpation des capsules surénales chez les rats albinos (*Mus rattus*). *C. R. Acad. Sci. (Paris)*. **43**, 904–906.

Poiley, S.M. (1960). A systematic method for breeder rotation for noninbred laboratory animal colonies. *Proc. Anim. Care Panel* **10**, 159–166.

Pravenec, M., Kren, V., Krenova, D., Bila, V., Zidek, V., Simakova, M., Musilova, A., van Lith, H.A., and van Zutphen, L.F. (1999). HXB/Ipcv and BXH/Cub recombinant inbred strains of the rat: strain distribution patterns of 632 alleles. *Folia Biol (Praha)* **45**, 203–215.

Printz, M.P., Jirout, M., Jaworski, R., Alemayehu, A., and Kren, V. (2003). Genetic models in applied physiology. HXB/BXH rat recombinant inbred strain platform: a newly enhanced tool for cardiovascular, behavioral, and developmental genetics and genomics. *Appl. Physiol.* **94**, 2510–2522.

Rapp, K.G. (1972). HAN-rotation, a new system for rigorous outbreeding. *Z. Versuchstierkd.* **14**, 133–142.

Richter, C.P. (1954). The effects of domestication and selection on the behaviour of the Norway rat. *J. Natl. Cancer Inst.* **15**, 727–738.

Savory, W.S. (1863). Experiments on food: its destination and uses. *Lancet* **2**, 381–383.

Silver, J. (1941). The house rat. *Wildlife Circ.* **6**, 1–18.

Silver, L.M. (1995). "Mouse Genetics–Concepts and Applications." Oxford University, Press New York. Available as electronic version at http://www.informatics.jax.org/silver/index.shtml.

Snell, G.D. (1978). Congenic Resistant Strains of Mice. *In* "Origins of Inbred Mice." (H.C. Morse, ed.), pp 1–31, Academic Press, New York.

Southern, H.N. (1964). "The Handbook of the British Mammals." Blackwell Scientific, Oxford.

Tate, G.H.H. (1936). Some muridae of the Indo-Australian region. *Bull. Am. Museum Nat. Hist.* **72**, 501–728.

Verneau, O., Catzeflis, F., and Furano, A.V. (1998). Determining and dating recent rodent speciation events by using L1 (LINE-1) retrotransposons. *Proc Natl Acad Sci U S A* **95**, 11284–11289.

Wakeland, E., Morel, L., Achey, K., Yui, M., and Longmate, J. (1997). Speed congenics: a classic technique in the fast lane (relatively speaking). *Immunol. Today* **18**, 472–477.

Waterston, R.H., Lindblad-Toh, K., Birney, E., et al. (2002). Initial sequencing and comparative analysis of the mouse genome. *Nature* **420**, 520–562.

White, W.J. (2000). Genetic Evaluation of Outbred Rats From the Breeder's Perspective. *In* "Microbial Status and Genetic Evaluation of Mice and Rats: Proceedings of the 1999US/Japan Conference." pp 51–64. Natl. Acad. Press, Washington, D.C.; http://books.nap.edu/books/030907195X/html/51.html.

Yoshida, T.H. (1980). Cytogenetics of the Black Rat. University of Tokyo Press, Tokyo.

Chapter 4

Morphophysiology

John Hofstetter, Mark A. Suckow, and Debra L. Hickman

THE LABORATORY RAT, 2ND EDITION

I. INTRODUCTION

This chapter describes and illustrates the most important anatomical and physiological features of the laboratory rat. It also gives a brief summary of those features that are unique to the rat. An additional goal is to highlight the anatomical and physiologic features that make the rat of particular value as an animal model in research.

II. GENERAL APPEARANCE

A. Head and Body

The body of a normal, healthy Norway rat is long and slender. The tail is hairless and may be as long as 85% of the total body. The tail is proportionately longer in females than in males. Growth rates and maturation times for rats varies by strain (Hughes and Tanner, 1970a; Hughes and Tanner, 1970b). Body weight varies greatly, not only with age but with stock, strain, and source of the rat. Typically adult rats weigh no more than 500 g to 600 g. Interestingly, prolonged exposure to low ambient temperatures causes a reduction in overall growth rate. The tail is most affected, possibly because reduction in its length and surface area may prevent excessive heat loss (Lee et al., 1969).

The integument consists of hair dispersed over the body surface except for the nose, lips, and palmar and plantar surfaces of the feet. Hair growth in the rat is cyclic, with distinct resting and growing phases of approximately 17 days in length each (Fraser, 1931; Butcher, 1934; Greene, 1935). The pelage is composed of two types of hairs: long, stiff guard hairs and shorter, softer, and more numerous underhairs. Females usually have 12 teats, with three pairs in the pectoral and three in the abdominal region.

The ocular globe of the rat normally protrudes from the orbit slightly, the appearance of which is enhanced by a greatly reduced nictitating membrane called the *plica semilunaris*. The eyelids are well developed and have very fine, short eyelashes. Located within the eyelids are the large tarsal or meibomian glands. Hydration of the corneal surface is maintained by secretions from the laterally placed lacrimal and medially placed Harderian glands.

The external nares, shaped like inverted commas, are open on the lateral aspect of the nose. The upper lip is usually cleft in the center by a vertical groove called the philtrum, which ends just below the nares.

Both fore- and hindlimbs have five digits. The first digit ("thumb" or "polex") is greatly reduced on the forelimb and has a flattened nail unlike the more rounded nails of the other digits. Typical pads (tori) are present and likely provide cushion against forces placed on the feet during walking and rest. The forelimb has five apical pads, three interdigital pads, and two basal pads.

B. External Reproductive Genitalia

The vaginal orifice is located approximately 7 mm ventral to the anus, with the clitoris lying in a small prepuce approximately 4 mm cranioventral to the vagina. The urethral orifice is located at the base of the clitoris and is directed ventrocaudally. The sexes can be differentiated based on external genitalia by day 17 of gestation (Inomata et al., 1985).

The scrotum lies caudoventrally and may extend just beyond the abdomen, depending upon the position of the rat. The size of the scrotum increases once the testes descend and may be large enough in the mature male to obscure the anus. Immediately cranial to the scrotum and on the midline is the prepuce containing the penis. The urethral orifice, directed ventrocaudally, opens through the end of the penis. A single cartilaginous or bony process, the os penis, lies on the ventral wall of the penis just proximal to the urethral orifice.

C. Skeletal Structure

The skeleton of the rat is typical for mammals. The vertebral formula is C7 T13 L6 S4 Cy27-30.

Bone maturation progresses more slowly in the rat compared with many other mammals. Ossification is incomplete until after rats are a year old (Strong, 1926). The dorsal segments of the ribs are usually calcified, so the rat lacks true costal cartilages. The clavicle and its associated structures extend as a chain between the sternum and the scapula in the following order: osmosternum, proximal procoracoid piece, clavicle, distal procoracoid piece. Of these, the clavicle and the osmosternum are ossified and the others are cartilaginous (Farris, 1949). The general orientation of the cervical vertebral column when the rat is at rest is vertical and not horizontal as might be suggested by the macroscopic appearance of the neck (Vidal et al., 1986).

The scapula lies somewhat horizontally. The acromion and coracoid processes are large and form a deep socket for the head of the humerus. A large deltoid tuberosity is easily identified on the humerus. Compared with other mammals, the greater, lesser, and third trochanters of the femur are extremely large with a relatively small head and neck. The tibia and fibula are fused in their distal quarter. For more detail concerning the skeleton of the rat, the reader is referred to the works of Greene (1935) and Chiasson (1958).

There is considerable variation in bone structure and bone mineral density among inbred strains of rats (Turner et al., 2001). Interestingly, the variability in skeletal phenotype tends to be both site- and strain-specific. For example, the Copenhagen 2331 has the greatest biomechanical properties in the femoral neck but only modest bone strength at the femoral midshaft compared with other strains. Although the thymus gland is believed to play a role in the regulation of bone metabolism, athymic rats are not significantly different from immunocompetent rats in terms of bone structure, function, or regenerative properties (Kirkeby, 1991).

The development of the skeleton in the rat is in phase with overall body growth, with a peak growth rate at about 7 weeks old (Puche et al., 1988). In the appendicular skeleton, osteogenesis proceeds in a proximo-distal sequence, with pelvic limb bones becoming ossified after those of the pectoral limb (Mohammed and El-Sayad, 1985).

In the rat, androgens help to maintain skeletal integrity. In aged, non-growing male rats, androgen deficiency results in acceleration of bone turnover and bone loss (Vanderschueren, 1996). Although the exact mechanism for bone loss related to androgen deficiency is unclear, it is associated with significant osteopenia (Erben et al., 2000).

Pregnancy and lactation have significant effects on the mechanical properties of bone in the rat. A single pregnancy strengthens the cancellous component of the maternal skeleton, whereas a quick succession of pregnancies weaken it (Shahtaheri et al., 1999). The female rat has excess skeletal mass to accommodate losses associated with the first reproductive cycle; after the first cycle, a new optimal skeletal mass is achieved (Bowman and Miller, 1999; Shahtaheri et al., 1999). In contrast, lactation is associated with a decrease in cortical and cancellous bone strength with a decrease in bone volume, and if reproduction is rapid, the normal process to restore the bone after lactation is impaired (Zeni et al., 1999; Vajda et al., 2001).

The rat has been used extensively as a model for studying the effect of space flight and microgravity on the skeletal system. For example, some of the effects associated with space flight include reduced rate of periosteal bone formation and decreased trabecular bone volume (Wronski et al., 1981); decreased bone formation owing in part to reduced osteoblast function (Carmeliet et al., 2001), which is independent of corticosteroid status (Zerath et al., 2000); and decreased flexural rigidity (Vajda et al., 2001).

D. Musculature

The musculature of the rat is typical of mammals. A comprehensive anatomical description of the muscles of the rat has been published elsewhere (Greene, 1935; Chiasson, 1958), though the terminologies used may be outdated and should be clarified with *nomina anatomica* for more current terms.

A substantial amount of collagen is associated with both cardiac and skeletal muscle in the rat (Borg et al., 1982). The basic organization consists of a perimysium composed of bundles of collagen fibers that connect the epimysium to the endomysium. The endomysium has at least four components: a dense woven network that surrounds the myocytes; myocyte-myocyte collagen struts that connect adjacent myocytes; myocyte-capillary struts that connect myocytes and capillaries; and a complex of single collagen fibers, glycoproteins, and glycosaminoglycans. The amount of collagen in each component varies with the function of the muscle.

Whereas somite-derived skeletal myoblasts are believed to be the prime source of muscle fiber nuclei during pre- and post-natal development, it has been demonstrated in the rat that muscle fiber nuclei may be contributed from other cell types. Primary fibroblasts undergo myogenic conversion when co-cultured with myoblasts, demonstrating the capacity of mesodermal cells to undergo differentiation to myocytes (Salvatori et al., 1995).

As in many species, skeletal muscle function declines as rats age. The isometric twitch duration was prolonged with aging in both fast- and slow-twitch muscles, though the degree of fatigue with contractile activity was not affected by aging (Fitts et al., 1984). In contrast, mitochondrial oxidative capacity decreases with age in rats (Fitts et al., 1984; Fannin et al., 1999; Short and Nair, 2001).

This change appears to be to the result of a decrease in mitochondrial protein, particularly oxidative enzymes (Farrar et al., 1981; Papa, 1996; Rooyackers et al., 1996).

Reactive oxygen intermediates modulate skeletal muscle contraction, and a role for nitric oxide in this process has been demonstrated in rats. In this regard, nitric oxide has been shown to promote relaxation of skeletal muscle through the cGMP pathway and to modulate increases in contractions that are dependent on reactive oxygen intermediates and are thought to occur through reactions with regulatory thiols on the sarcoplasmic reticulum (Kobzik et al., 1994).

E. Superficial Glands of the Head and Neck

The glands in this group lie superficially within the neck and head and can be identified by their location and appearance. They include the orbital glands, lateral nasal glands, salivary and lymphatic glands (lymph nodes) of the neck, and the multilocular adipose tissue or "hibernating gland." Figures 4-1 and 4-2 show the locations and relative sizes of these glands.

The exorbital lacrimal gland lies just ventral and rostral to the ear. Its duct joins with that of the intraorbital lacrimal gland to open onto the conjunctiva in the dorsolateral region of the eye. The intraorbital gland occupies the caudal angle of the orbit and is covered by a connective tissue sheath. The lacrimal duct epithelium is thought to transport sodium, potassium, and chloride ions. The presence of nerve terminals within these ducts suggests that the duct system of the exorbital gland is active in the production of lacrimal fluid.

The Harderian gland is a horseshoe-shaped structure that occupies a large portion of the orbit as it extends medially and caudally to encircle the optic nerve. It is thought that this gland is present in most mammals, except Chiroptera and Simidae. The Harderian gland may have both an apocrine and a holocrine mode of secretion (Muller, 1969).

The glandula nasalis lateralis, or Steno's gland, is the largest of several well-developed rostral nasal glands. This gland is located in the wall of the rostral portion of the maxillary sinus and has an excretory duct that courses to the vestibule along the root of the nasoturbinate. Steno's gland is characterized by cytologic features similar to those described for the major serous salivary glands and is homologous with the salt gland found in marine birds. It produces a watery, nonviscous secretion that is discharged at the nasal airway entrance. Likely, the secretion helps humidify the inspired air and may contribute to the maintenance of the proper viscosity of the mucous blanket covering the ciliated surfaces deeper within the respiratory tract. Because of the large number of autonomic nerves that are found in close contact with the acinar cells, it is believed that Steno's gland is regulated by the nervous system in such a way that rapid adjustment of the secretory activity to changes in the humidity of the inspired air or to airborne irritants is possible (Moe and Bojsen-Moller, 1971).

There are three pairs of salivary glands in the rat: the parotid gland; the submaxillary or submandibular salivary glands; and the sublingual glands. The parotid gland is a diffuse structure extending ventrorostrally beginning behind the ear, coursing along the caudal facial vein, and terminating on the ventrolateral surface of the neck. The caudal border of the gland sometimes extends to the shoulder and the clavicle. The parotid duct is formed by the union of three main branches and crosses the lateral surface of the masseter muscle, in close association with the

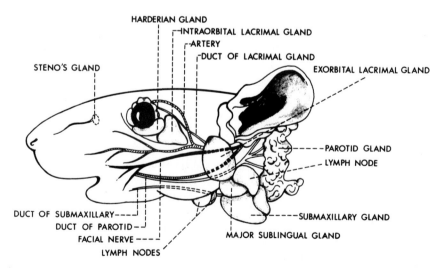

Fig. 4-1 Superficial glands of the head and neck (lateral view). [redrawn from Greene, 1935]

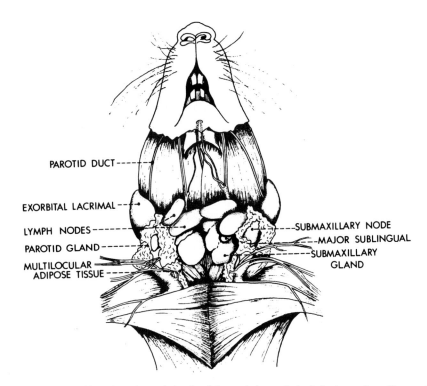

Fig. 4-2 Superficial lymph nodes and glands of the neck (ventral view). [redrawn from Greene, 1935]

buccal and mandibular branches of the facial nerve, to open into the vestibule just opposite the molar teeth. The rat parotid produces saliva with a protein concentration of about 2% (Hall and Schneyer, 1964).

In the rat, the parotid and submaxillary salivary glands are of approximately equal size. The submaxillary or submandibular salivary glands are large, elongated structures that lie just caudal to the angle of the mandible and extend caudally to the cranial aspect of the manubrium. Secretory granules found in the acinar cells of the gland are mucinous, and granules found in the secretory ducts are serous. Sexual dimorphism has been described with respect to the presence of mucous cells in this gland, with such cells occurring in the glands of 100% of males but only 28.5% of sexually mature females (Hatakeyama et al., 1987). With age, there is a relative increase in duct volume and a decrease in the number of acini (Singh et al., 1989). The distal end of the main excretory duct of the submandibular gland has a salivary bladder that is lined by pseudostratified epithelium (Mercurio and Mitchell, 1987). The primary fluid of the gland is modified in the bladder by transepithelial fluid and ion transport.

The sublingual glands are small, rounded structures located at the rostral edge of the submaxillary glands and are sometimes partially embedded in them. The ducts of the sublingual glands run parallel with those of the submaxillary glands to open on the plica sublingualis (Huntington and Schulte H., 1913). With age, the proportion of the gland occupied by ducts increases in

Fisher 344 male rats, and there is an increase in the squamous metaplasia of duct epithelium and periductal lymphocytic foci (Mintz and Mooradian, 1987).

Salivary glands and their ducts secrete digestive enzymes as well as various cell growth factors, including epidermal growth factor, nerve growth factor, transforming growth factor-beta, and basic fibroblast growth factor (Amano and Iseki, 2001). Myoepithelial contractions serve to accelerate the salivary flow and to support the secreting acinar cells and prevent back-flow of fluid from the duct system into the glandular tissue. These contractions occur through a cooperation of sympathetic and parasympathetic innervation (Emmelin, 1981). The moisture content of the diet determines the secreted volume of parotid but not submandibular saliva. Dry diets stimulate a high secreted volume (Ito et al., 2001). Submandibular saliva secretion dominates during grooming.

Additional minor salivary glands are present in most parts of the mouth, and their secretions directly bathe the tissues (Hand et al., 1999). Individual glands are usually in the submucosa or between muscle fibers, and consist of groups of secretory endpieces made up of mucous, serous or seromucous cells. The minor salivary glands contribute approximately 14% of protein and 1% of amylase in whole saliva (Blazsek and Varga, 1999). They also secrete antimicrobial proteins, and the lingual serous (von Ebner's) glands secrete proteins with possible taste perception functions (Gurkan and Bradley, 1988). A similar function has been postulated for Weber's glands, which lie posteriorly in the root of the tongue (Nagato et al., 1997).

Lymph nodes lying superficially in the head and neck are sometimes mistaken for salivary glands. The lymph nodes are irregularly shaped and compact structures that lie in close association with the salivary glands. One lymph node is typically found partially embedded in the parotid salivary gland. Other lymph nodes usually are found rostral to the submaxillary salivary gland.

The thymus gland is located along the ventral aspect of the trachea, dorsal to the sternum at the thoracic inlet. It is composed of two distinct encapsulated masses and extends from the larynx to the heart. Microscopically, the gland contains lobules with cortices of lymphocytes and medullas of epithelial cells (Komarek et al., 2000).

Thymic involution is observed in rats after puberty. This process is believed to be under the control of beta-adrenoceptor receptors, including those specific to glucocorticoids (Pazirandeh et al., 2004; Plecas-Solarovic et al., 2004).

F. Multilocular Adipose Tissue

Multilocular adipose tissue, sometimes referred to as the "hibernating gland," is diffusely distributed throughout the ventral, lateral, and dorsal aspects of the superficial tissues of the neck and between the scapulae. Sometimes referred to as "brown fat" because of the brown color imparted by pigment, the tissue has a glandular appearance microscopically owing to the multilocular appearance imparted by fat cells filled with small lipid droplets. Because the rat does not hibernate, the term "hibernating gland" is a misnomer; however, brown fat plays a critical role in non-shivering thermogenesis during cold exposure, a response likely triggered by the innervation of brown adipose cells by sympathetic nerves (Slavin and Bernick, 1974). Cold acclimation increases the number of brown adipocytes in both young and aged rats, whereas the number of unilocular (white) adipocytes is not affected (Morroni et al., 1995). In contrast to common fat tissue which loses or accumulates neutral fat with changes in the nutritional condition of the animal, multilocular adipose tissue does not.

G. Tissues and Organs Associated with Special Senses

1. Sight and the Eye

In general, the eye of the rat is typical of the mammalian eye, but the rat eye is exophthalmic. The cornea consists of five layers and is approximately 6.80 + 0.2 mm in diameter in the adult male and approximately 6.40 + 0.1 mm in the adult female. The refractive power of the cornea is almost twice as great as that of the lens (Heywood, 1973).

The iris arises from the anterior portion of the ciliary body and is a rostral continuation of the choroids. It is completely devoid of pigment in the albino rat and consists of a loose, highly vascular connective tissue. Few anomalies of the iris have been recorded in the rat (Norrby, 1958).

As in other species, the pupil dilates and contracts to regulate the amount of light entering the eye. Persistence of the pupillary membrane has been seen and is recognized as multiple strands of tissue crossing the pupil.

Bordering the anterior chamber of the eye externally is the cornea and internally is the iris and pupil (Fig. 4-3). The posterior chamber of the eye is bordered rostrally by the iris and caudally by the lens. Both the anterior and posterior chambers are filled with aqueous humor. Normal intraocular pressure in the adult rat has been shown to be 10 to 16.5 mm Hg (Cabrera et al., 1999; Kontiola et al., 2001). Measurements of intraocular pressure using a rebound tonometer have been found to correlate closely with those obtained mannometrically, whereas measurements using an electronic tonometer were found to underestimate intraocular pressure (Goldblum et al., 2002).

The lens is a transparent, biconvex spherical body that occupies approximately two-thirds of the intraocular cavity. It is attached to the ciliary body by suspensory ligaments and changes its shape during the process of visual accommodation. The rate of lens growth in the rat follows an asymptotic curve (Norrby, 1958). With advancing age, the rate of growth of the lens and density of the lens nucleus increases. At birth, the rat lens is relatively flat. However, the lens becomes increasingly spherical in shape after birth. A critical period in lens maturation in the rat occurs about 12 to 16 days after gestation (Carper et al., 1985; Groth-Vasselli and Farnworth, 1986). By day 16 it occupies much of the intraocular space (Sivak and Dovrat, 1984). The refractive components of the rat eye seem similar in appearance and quality to those of a diurnal mammal at birth, but they assume the characteristics associated with nocturnal vision during an early period of postnatal development. The lens of the rat is remarkably

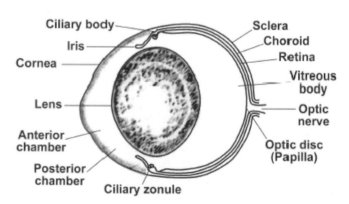

Fig. 4-3 Section of the eye, displaying internal structures.

transparent (Gorgels and van Norren, 1992), though lens opacities and cataracts are occasionally observed. When present, cataracts are usually unilateral, with total cataracts being observed rarely.

The vitreous body fills the vitreal cavity between the lens and the retina. This material consists of a colorless, transparent, amorphous, gelatinous mass. Almost 99% of the vitreous body consists of water, the remainder being hydrophilic polysaccharide, especially hyaluronic acid. Approximately 5% of aged Sprague-Dawley rats have small, globular vitreous opacities that can be observed by routine slit lamp eye examination (Ruttimann and Daicker, 1989).

In the rat, the hyaloid vessels form two distinct entities. There are central hyaloid vessels that consist of three to five arteries running from the optic disc to the caudal pole of the lens. The central vessels can cause posterior capsular damage that is indistinguishable from a cataract when viewed with an ophthalmoscope. Closure of the hyaloid vessels normally occurs between 7 and 22 days old (Cairns, 1959). Remnants of the hyaloid vessels are seen in 60% of rats at 6 weeks old, and in less than 17% of rats 1 year old and older.

The retina is the innermost of the three coats of the eyeball and is the tissue responsible for the reception and transduction of light stimuli and the transmission of these signals in the form of nerve impulses to the brain. The retina has ten distinct layers: pigmented epithelium, photoreceptor cell layer, external limiting membrane, outer nuclear layer, outer plexiform layer, inner nuclear layer, inner plexiform layer, ganglionic cell layer, optic nerve fiber layer, and inner limiting membrane. Because it is derived from the prosencephalon, the retina is usually considered to be part of the brain.

In rats, the peripapillary choroid plays a significant role in both blood supply and venous drainage of the optic nerve head (Sugiyama et al., 1999). Likewise, the central retinal artery also contributes to the optic nerve head circulation. The posterior ciliary artery travels along the inferior side of the optic nerve sheath, directly enters the optic head, and divides into three branches: the central retinal artery, and the medial and lateral long posterior ciliary arteries, which provide several short branches to the choroids.

The rat has an orbital venous plexus retro-orbitally, in contrast to mice and hamsters which have an orbital venous sinus (Timm, 1979). The plexus is formed by the external dorsal ophthalmic vein, external ventral ophthalmic vein, and numerous anastomoses between these veins. Sampling of blood from the rat can be done with relative ease from this site (Sharp and LaRegina, 1998).

Although rats are generally considered to be color-blind, they have two classes of retinal cones: one type contains an ultraviolet-sensitive photopigment and the other contains a pigment maximally sensitive in the middle wavelengths of the visible spectrum (Jacobs et al., 2001). Tests of wavelength discrimination provide evidence that rats may have dichromatic color vision (Jacobs et al., 2001; Koolhass, 2003).

The Harderian gland is located medial to the orbit and, with the lacrimal glands, serves to lubricate the orbit. The ocular surface is covered by a glycocalyx, which allows the eye to resist desiccation. This film contains a glycoprotein that appears on the ocular surface after eyelid opening and is believed to facilitate mucin spread across the ocular surface (Gipson et al., 1992; Watanabe et al., 1993). Rat tear film is made up primarily of mucus, with a lipid layer on the surface, and without a free aqueous layer (Prydal and Campbell, 1992; Chen et al., 1997).

2. Olfaction

Both inhaled and exhaled air is directed laterally by each nare. This feature decreases mixing of odorant intake. The consequence of this is that rats are able to take independent, bilateral samples of the odor environment (Wilson and Sullivan, 1999).

Inhaled air passes over the olfactory epithelium, which lines the ethmoturbinate bones within the nasal cavity. This epithelium secretes a mucous layer that overlies the mucosa, including olfactory receptors, which are neurons interspersed among the epithelial cells. The neurons have short, thick dendrites with expanded ends called *olfactory rods*. From these rods, cilia project to the surface of the mucosa. The axons of the olfactory receptor neurons pierce the cribriform plate of the ethmoid bone and enter the olfactory bulb of the brain. From the olfactory bulb, neurons project to structures of the limbic system, including the amygdala, septal nuclei, pre-pyriform cortex, the entorhinal cortex, hippocampus, and the subiculum.

In addition to the sensory epithelium, there are Bowman's glands in the nasal mucosa; these glands produce a secretion that bathes the surface of the sensory receptors with an aqueous solution containing mucopolysaccharides, immunoglobulins, and enzymes.

As in many mammals, the olfactory epithelium of the rat retains the capacity to recover after damage. Indeed, the posterior zone of epithelium is well preserved in aged rats, suggesting that age-related declines in olfactory capability may be due to the accumulation of inhaled environmental insults (Loo et al., 1996).

The vomeronasal organ is a chemoreceptor organ enclosed in a cartilaginous capsule and separated from the main olfactory epithelium (Keverne, 1999). Vomeronasal sensory neurons resemble the main olfactory epithelium by light microscopy, but they detect specific chemicals, some of which may act as chemical communication signals (pheromones) from other individuals of the same species. When stimulated, sensory neuron receptors activate inositol 1,4,5-triphosphate signaling as opposed to cyclic adenosine monophosphate (cAMP). Stimulation of the vomeronasal organ leads to activation of the hypothalamus by way of the accessory olfactory bulb and amygdala (Keverne, 1999).

3. Taste

Taste buds, the sense organs for taste, are ovoid bodies located on the tongue, the hard and soft palates, and the pharynx in the rat (Miller, Jr. and Spangler, 1982; Iida et al., 1983; Travers and Nicklas, 1990; Harada et al., 2000; El Sharaby et al., 2001). In general, they measure 50 to 70 μm in diameter and 50 to 60 μm in height. Each taste bud is made up of connective tissue and various numbers of hair cells, the gustatory receptors. In addition, basal cells, which are derived from surrounding epithelium, are present to differentiate into new receptor cells. The receptor cells each have a number of hairs projecting into the taste pore, the opening at the epithelial surface of the taste bud.

Primary gustatory fibers synapse in the medulla. Information is subsequently relayed to the somatosensory cortex, hypothalamus, amygdala, and insula. Interestingly, adrenergic transmission and cholecystokinin signaling within the taste bud may play a paracrine role in rat taste physiology (Herness et al., 2002a; Herness et al., 2002b).

It has been postulated that saliva plays a major role in some aspects of taste (Catalanotto and Leffingwell, 1979). Experimental desalivation produced long-term changes in taste preferences in rats. Similarly, the secretions from von Ebner's lingual salivary glands significantly influence taste responses of some taste buds (Gurkan and Bradley, 1988).

4. Hearing and Vestibular Function

Similar to other mammals, the ear of the rat includes an external ear, a middle ear, and an inner ear. The external ear is supported by elastic cartilage and consists of the pinna and the external auditory canal. The pinna serves to funnel sound to the external auditory canal and on to the middle ear. At the entrance to the middle ear is the tympanic membrane, which transmits the sound energy as mechanical energy to three bones: the malleus, incus, and stapes. The stapes transmits the energy to the fluid-filled cochlea of the inner ear and causes waves in the fluid. In the inner ear, the action of the waves on the organ of Corti, which lies within the cochlea, generates action potentials in the nerve fibers via hair cells in the organ of Corti. Nerve impulses are then carried to various regions of the brain, including the inferior colliculi and the auditory cortex. The cortex has been shown to be a necessary component

for normal audition (Moore et al., 2001; Talwar et al., 2001).

The external ear is largely covered by stratified squamous epithelium with interspersed glands. In contrast, the tympanic membrane is covered by simple squamous epithelium. The mucosa of the middle ear in the rat shows striking similarity to that of humans (Albiin et al., 1986); it consists primarily of simple epithelium that transforms into pseudostratified columnar ciliated epithelium near the entrance to the Eustachian tube. A nearly continuous layer of bone-lining cells that separate the general extracellular fluid from bone and its fluid compartment is found in the inner ear (Chole and Tinling, 1994).

The organ of Corti has sensory hair cells that are arranged into a single inner row and three outer rows (Komárek et al., 2000). There is a direct correlation between the hair cell density and the auditory threshold in rats (Burda and Voldrick, 1980). This threshold is diminished by prolonged administration of kanamycin, an aminoglycoside antibiotic (Wu et al., 2001). Loss of hair cells with age has been observed in aging Sprague-Dawley rats (Keithley and Feldman, 1982) and in aging Fischer 344 rats (Nakayama et al., 1994).

The range of hearing in the rat is approximately 250 to 80,000 Hz at 70 dB, although the greatest sensitivity is for sounds between 8,000 and 32,000 Hz (Kelly and Masterton, 1977; Heffner et al., 1994). In contrast, a human being can hear sounds from approximately 16,000 to 20,000 Hz. Although the range of hearing in juvenile and adult rats is the same, rats younger than 5 weeks have a greater susceptibility to noise exposure (Rybalko and Syka, 2001).

The vestibular organ is located in the inner ear in all mammals and functions to maintain equilibrium. Specifically, the vestibule, which contains the utricle and saccule, which sense motion in relation to gravity, and three semicircular canals, which sense rotational motion, are responsible for vestibular function. Briefly, increases or decreases in the velocity of circular movement stimulate a flow of liquid in the semicircular canals. The flow induces lateral movement of the conic cup that covers the sensory area of the vestibular organ. Either bending or tensile forces on the sensory cells elicit nerve impulses.

Differences in the vestibular organs between Wistar rats and Sprague-Dawley rats have been reported (Lindenlaub and Oelschlager, 1999). Specifically, differences in the shape of the lateral semicircular duct and in the cupular mechanical sensitivity were observed.

5. Touch

The main tactile organs in the rat are the vibrissae. The most apparent vibrissae are the mystacial vibrissae, though additional, non-mystacial vibrissae are present in supraorbital, postero-orbital, lateral cervical, median cervical, submental, and carpal forelimb locations. Each vibrissa is sunken in a follicle that is sealed within a blood sinus. When a vibrissa is touched, it bends and pushes the blood against the opposite side of the sinus, triggering a nerve impulse to the barrel cortex in the brain. The barrel cortex occupies a relatively large portion of the somatosensory cortex in rats. Innervation of the mystacial vibrissae is via the trigeminal nerve (Erzurumlu and Killackey, 1983).

Rats use their vibrissae to navigate, orient, and balance. Trimming of vibrissae early in life leads to long-term impairment of sensorimotor function in rats (Carvell and Simons, 1996). The vibrissae of rats have rhythmic motor activity that is used for the tactile localization of objects; this motor activity is under active muscular control (Berg and Kleinfeld, 2003). In rats, the mystacial vibrissae make an average of eight sweeps per second.

The surface of the rat snout pad is ridged in a fashion similar to the dermatoglyphic pattern found in primate digital skin. These epidermal and dermal ridges almost certainly play a role in sensory discrimination. This relationship provides further evidence of the parallel between fingertip function in bipeds and snout tip function in quadrapeds (Macintosh, 1975).

III. DIGESTIVE SYSTEM

We made no attempt to give a highly detailed description of viscera; however, we want to draw attention to areas where the rat is different from the other lab animals. Some important differences include lack of tonsils and gallbladder, an extremely diffuse pancreas, and many visceral accessory glands.

The rat is one of the most popular experimental animals for studying the physiology of digestion. Rats are nocturnal rodents and typically feed semicontinuously when they are active at night. They are coprophagous, that is, some part of the orally ingested food and its metabolites remain in the feces to be reingested (Senoo, 2000). A healthy, 400-g rat eats up to 20 g of dry food per day (about 40 g including liquids) comprising about 50 kcal. A 300-g rat eats up to 4 g of food at each feeding. Water turnover in a 300 g rat is about 24 ml/24 hr (114 ml/ kg/24 hr) under a 25°C ambient temperature, but emotional stress increases drinking. According to Chew (2003), depriving rats of food makes them die sooner than when they are deprived of water.

The following is a description of the digestive system from the oral cavity to the anus and rectum and a discussion of the accessory organs associated with the alimentary tract.

A. Mouth and Buccal Cavity

A cleft upper lip, the philtrum, and an intact lower lip are the rostral boundaries of the mouth. The space between the lips, cheeks, and teeth is the vestibule. Both jaws have a pair of well-developed incisor teeth. The incisors grow continuously and the pulp cavity is open, so it is essential that the teeth are aligned to maintain a sharp cutting edge. Canine teeth are absent, but the open space where they would be is called the diastema. Folds of cheek mucosa fill the diastema and serve to separate the gnawing machinery from the caudal buccal cavity. This dental configuration, characteristic for rodents, is 2(I 1/1, C 0/0, Pm 0/0, M 3/3) = 16.

Along with many mammals, rats have four types of lingual papilla: filiform, fungiform, circumvallate, and foliate, on the dorsal or lateral surface of the tongue (Iwasaki et al., 1997).

Sbarbati and colleagues (1999) suggest that the common description of the relationships between the circumvallate papilla and von Ebner glands must be revised. Formerly textbooks said that the von Ebner glands ancillary to the taste buds were for washing the vallum around the circumvallate papilla or for perireceptorial events. However, recent literature seems to have the structures functioning together. The entity could be called the circumvallata papilla/von Ebner gland complex and seems to be an important enzyme and pheromone production system composed of sensitive (taste buds) and effectory branches linked by feedback control. In this model, the taste buds located in the distal portion of the von Ebner gland's ductal system are analogous to the chemoreceptor cells in other parts of the digestive system, that is, pancreatic and bile ducts. Viewed from this perspective, the complex becomes a rare example of chemoreceptor-secretory organ. (Sbarbati et al., 1999)

Because each fungiform papilla in the rat has a single taste bud, the spatial distribution of fungiform papillae is equivalent to the location of taste buds on the rostral portion of the tongue. A mean total number of 187 fungiform papillae per tongue have been described for an average density of 3.4 papillae/mm. Investigators have long used the rat fungiform papillae as a model system of the mammalian taste receptor (Miller, Jr. and Preslar, 1975). Water taste receptors identified in other animals (dog, cat, pig, etc.) have not been found in the rat, so it is believed rats cannot taste water (Bivin et al., 1979). There is no medial frenum, but two lateral ones extend from near the tip of the band which ties the lower lip to the gum. There are no sublingua; however, a small pair of salivary papillae lie close to the median line behind the incisors.

The hard and soft palates form the roof of the oral cavity. The hard palate has rows of palatine ridges between small, horny, incisive papillae. These ridges are made up of stratified squamous epithelia that are replaced every 6 to 7 days and have, in addition, a most unusual feature. Small outgrowths of hair have been described, sprouting from all sides of the papillae. These hair follicles apparently do not have sebaceous glands. It is assumed that these hairs play some role in the tactile sensitivity of the oral cavity (Jayaraj, 1972).

The soft palate extends from the caudal border of the hard palate to the nasopharyngeal hiatus, where it merges with the ventral wall of the pharynx. Like the hard palate, it is covered by stratified squamous epithelium, which is replaced every 3.25 to 4.5 days (Hamilton and Blackwood, 1974). Laterally the soft palate is continuous with the cheeks and lateral wall of the pharynx. At the caudal extremity, there is no obvious tonsil, no uvula, and no rostral or caudal pillars of the fauces.

An unusual feature is the relationship of the larynx to the soft palate and nasopharyngeal hiatus. The epiglottis lies some 2 to 3 mm rostral to the nasopharyngeal hiatus, whereas the caudal border of the larynx, namely the arytenoid cartilages, actually lies within the nasopharyngeal hiatus rostral to its caudal border. A muscle called the *nasopharyngeal sphincter* surrounds the nasopharyngeal hiatus and is closely related to the cartilage plates on either side of it. There are no levator palati, musculus unulae, or palatoglossus muscles (Cleaton-Jones, 1972).

The pharyngeal region has no other unusual features and is rather typical of most mammalian species. The esophagus begins just dorsal to the glottis and extends caudally to connect the oral cavity with the stomach.

Saliva is water plus amylase, mucin, electrolytes, immunoglobulin, and other trace components. However, secretion from the parotid gland is unique. It has a protein concentration of about 2%. Removing the salivary gland or ligating the salivary duct brings out the many influences that these glands have on numerous endocrine and exocrine functions. For example, parotid hormone stimulates dental fluid transport in teeth and may prevent tooth decay (Senoo, 2000).

B.   Esophagus

The esophagus is accessible caudal to the diaphragm, permitting relatively easy esophageal-intestinal anastomosis and gastrectomy (Bivin et al., 1979). As in all rodents, the epithelium of the esophagus is covered with a layer of keratin. The esophagus enters the stomach at the inner curvature through a fold of the limiting ridge that separates the fore stomach from the glandular stomach. This fold makes it impossible for a rat to vomit.

C.   Stomach

An excellent review of the anatomy and physiology of structures affecting gastrointestinal absorption in rats is in DeSesso and Jacobson (2001).

The stomach lies on the left side of the abdominal cavity in contact with the liver, but attaching it and other organs of the digestive tract to the dorsal body wall is a prominent, highly vascularized mesentery. The sac-like mesentery that attaches to the greater curvature of the stomach and drapes like an apron over the stomach and intestines is called the omentum.

Figure 4-4 shows the recommended name for each part of the stomach. A histologic description of a rat's stomach would not differ significantly from that of most rodents. The nonglandular fore stomach has rumen-like mucosal folds covered with stratified squamous epithelium and serves as a reservoir. A glandular region, the corpus, is characterized by gastric pits with simple columnar epithelium. The gastric glands are composed primarily of parietal cells and chief cells, although argentaffin-like cells have been described on occasion. The pyloric portion of the stomach is again characterized by a mucosa of simple columnar epithelium lining the elongated gastric epithelium, and beneath this lies the pyloric glands (Smith and Calhoun, 1968). Most mammalian gastric mucosae secrete four distinct aspartic proteases: pepsinogen A, pepsinogen C, cathepsin D, and cathepsin E. That of the rat secretes only pepsinogen C and cathepsins D and E (Senoo, 2000).

Gastric mucosal mast cells differ morphologically, histochemically, and pharmacologically from those elsewhere in the body. Histamine is believed to be involved in the secretion of acid by the stomach, and some of this histamine is derived from mast cells. Their true function remains a mystery. However, recent studies indicate that they may be involved in producing local vasodilatation and increased capillary permeability during periods of increased secretory activity (Heap and Kiernan, 1973).

Both the digestive and absorptive functions of the gastrointestinal tract hinge on processes that soften food, propel it along, mix bile with it, and add digestive enzymes secreted by auxiliary glands. Some processes arise from the intrinsic nature of smooth muscle. Others are made of reflexes associated with neurons in the gut or deriving in the central nervous system (Senoo, 2000). Paracrine effects of chemical messengers and gastrointestinal hormones comprise additional processes. Networks of interstitial cells of Cajal embedded in the musculature of the gastrointestinal tract are involved in the generation of electrical pacemaker activity for gastrointestinal motility (Faussone Pellegrini et al., 1977; Senoo, 2000). This pacemaker activity manifests itself as rhythmic slow waves in membrane potential and controls the frequency

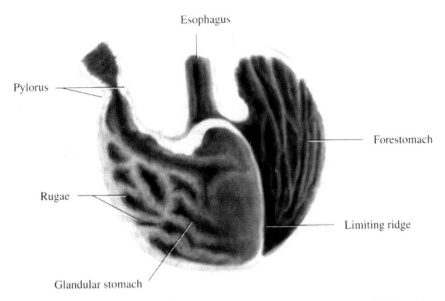

Esophagus

Pylorus

Forestomach

Rugae

Limiting ridge

Glandular stomach

Fig. 4-4 The interior structure of the rat stomach has two distinctive regions separated by a prominent limiting ridge, which precludes vomition. Food entering the stomach goes into the forestomach, a nonsecretory region with a hardy epithelium. The portion of the stomach with an exit to the duodenum is the glandular stomach and is characterized by a delicate secretory epithelium and prominent folds (rugae).

and propagation characteristics of gut contractile activity. The hormones are humoral agents secreted by cells in the mucosa and transported in the circulation to influence the functions of the stomach, the intestine, and the pancreas.

D. Small Intestine

There is a comprehensive review of the ontology of intestinal transport in mammals (Pacha, 2000). A schematic of the rat gastrointestinal tract is in Fig. 4-5. In rats, the small intestinal epithelium differentiates, in a migration-dependent manner, into four lineages. Each produces characteristic gene products and cell structures typical to its function (Senoo, 2000). Enterocytes make brush border hydrolases and Paneth cells lysozyme, criptidins, and defensins. Goblet cells secrete mucins and trefoil peptides, and enterochromaffin cells secrete chromaffins. In the adult, intestinal epithelial crypt cells divide every 10 ± 14 hours. The cell transit time from crypt to villus tip is 48 hours. The Paneth cell is replaced about every 4 weeks but stays at the base of the crypt. This cell produces lysozyme, which may regulate the gut microflora. Normal rat epithelial cells differentiate as they migrate along the villus surface. In the young adult, however, enterocyte cells differentiate slowly by comparison to the aged rat. The mucosal mass is high in the ileum of aged rats. This may be hypertrophy arising from decreasing epithelial efficiency.

Brunner's glands are interesting because only mammals have them. One of their functions is mucus secretion.

The glands secrete an alkaline fluid containing mucin. Mucins protect gastric epithelium and the proximal duodenal mucosa by maintaining a favorable pH gradient and prevent autodigestion. They also secrete bicarbonate ion and epidermal growth factor. Rats have Brunner's glands only in the proximal duodenum (Wolthers et al., 1994).

The rate of cell production in the crypts of Lieberkühn along the length of the small intestine is relatively constant, about 36 cells/crypt. The crypt-to-villi ratio decreases from 27 in the duodenum to 10 in the terminal ileum (Clarke, 1970).

Most nutrients are absorbed through the apices of the epithelial cells in the small intestine (Fig. 4-6). The apical surface faces the lumen. A conspicuous specialization of the apical surface of each epithelial cell is large numbers of microvilli, outward projections from the apical surface that vary in number and density. Microvilli, also known as "brush border," increases the surface area for the important transport and enzymatic activity going on at this surface. Each microvillus is filled with actin filaments that bind at their base to a web of thin filaments called the "terminal web" (http://cellbio.utmb.edu/microanatomy/epithelia/epith_lec.htm, 2002).

The length and surface area of the absorptive gut are well documented. The young rat, most often used in absorption and nutrition experiments, has a mucosal area per unit of gut length ratio of about 2:1 (Boyne et al., 1966; Permezel and Webling, 1971).

Monosaccharides glucose, galactose, and fructose are absorbed by the intestine via specific translocators after the digestion of ingested carbohydrates (Stumpel et al., 2000).

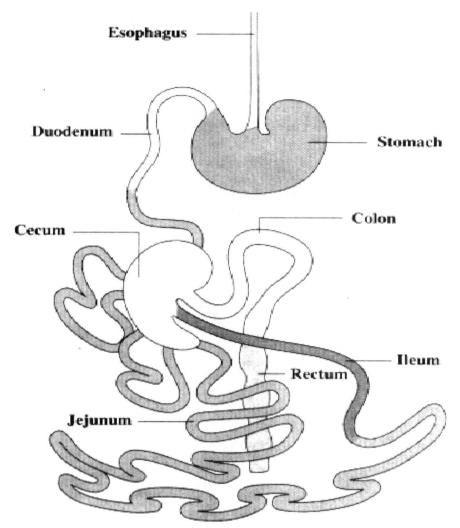

Fig. 4-5 An overview of the rat alimentary canal. The rat gastrointestinal tract shares the same overall organization as that of the human but with a few important differences. The esophagus enters the stomach at a central location and the entry into the duodenum faces cranially. The relative lengths of the small intestine differ in that the rat jejunum makes up nearly the entire small intestine. The cecum of the rat is extremely large and lacks the vermiform appendix. The figure is not drawn to scale.

Glucose and galactose are taken up by sodium-dependent glucose cotransporter-1 and fructose by sodium-independent glucose transporter-5. Both are located in the apical (luminal) plasma membrane of mature enterocytes. On the basolateral membrane of the enterocytes, all three carbohydrates share the same translocator and enter the circulation via the sodium-independent glucose transporter-2.

The absorptive capacity for carbohydrates is not constant. It responds to long-term changes in physiologic conditions. Commonly, changes in intestinal glucose transport come about through modulation of both the sodium-dependent glucose cotransporter-1 and sodium-independent glucose transporter-5 protein concentrations (Stumpel et al., 2000).

Transport of intact peptides and proteins from the intestinal lumen into the blood differs from the regular process of food digestion and absorption. Intestinal absorption of minute amounts of proteins is a routine physiologic process (Ziv and Bendayan, 2000).

The amino acids are absorbed maximally in the jejunal epithelium (Faussone Pellegrini et al., 1977). There are four plasma membrane transport systems for cationic amino acids: one is specific for cationic amino acids and the others also transport zwitterionic amino acids. The remainder of the amino acids is absorbed by the broad-specificity amino acid transporter.

Fat absorption has a maximal absorption rate within the jejunum via the lacteals located within the microvilli of the "brush border" (DeSesso and Jacobson, 2001). Salient features of intestinal absorption of selected vitamins and mineral are reviewed by Thomson et al. (2001).

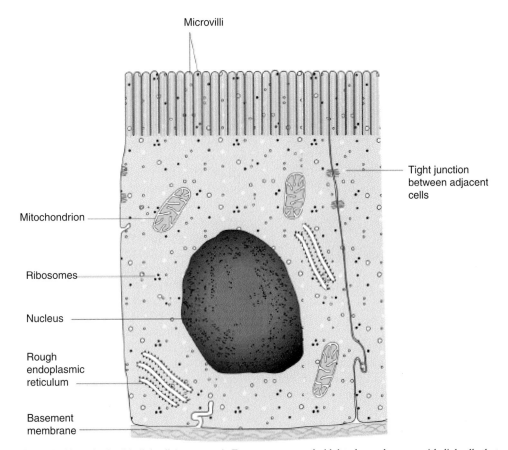

Microvilli

Tight junction between adjacent cells

Mitochondrion

Ribosomes

Nucleus

Rough endoplasmic reticulum

Basement membrane

Fig. 4-6 Diagram of a typical intestinal epithelial cell (enterocyte). Enterocytes are cuboidal to low columnar epithelial cells that rest on a basement membrane. The cells are joined tightly to neighboring cells by tight junctions. The nucleus resides in the basal part of the cell. The apical surface of each cell has microscopic projections (microvilli) that give the appearance of a "brush border" when seen with a light microscope. Microvilli greatly increase the absorptive surface area of the enterocyte.

E. Cecum

The cecum is a large, thin-walled, blind pouch shaped somewhat like a comma. It is lightly constricted about its middle. The rat cecum differs from that of many other rodents; it is devoid of internal septa. Even though the cecum is not divided into septa or cells, it is subdivided into an apical and a basal part. The basal part contains no lymphoid tissue. The apical portion, on the other hand, contains a distinct mass of lymphoid tissue in its lateral wall. This is thought to be analogous with the vermiform appendix of man.

F. Large Intestine

The colon is unremarkable in the rat. From the cecum, the colon runs first cranially to frame the abdominal cavity as the ascending, transverse (across the duodenum), and descending colon until it finally forms the rectum in the pelvic region. The rectum terminates at the anus.

G. Rectum and Anus

Some epithelial cells of the mucous membrane in the rectum of rats have peculiarities. These include glycogen particles forming extended accumulations in young but singly dispersed in adults; a brush border, the microvilli of which are larger and thicker than in the adjoining border cells; and bundles of filaments and rows of vesicles in the apical part of the cytoplasm. These glycogen-containing brush cells are very similar to the sensory cells in the tracheal epithelium; however, no nerve connections have been demonstrated (Luciano et al., 1968).

H. Accessory Organs

The accessory organs are exocrine pancreas, and the liver. The rat pancreas is not a singular mass but is diffuse and extends from the end of the duodenal loop to the left into the gastrosplenic omentum. The number of pancreatic ducts varies from 15 to 40, and all ducts open into the lower common bile duct.

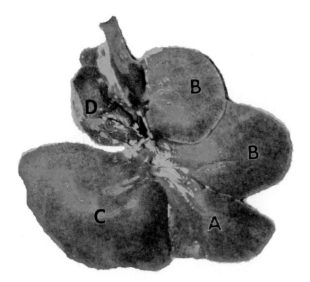

Fig. 4-7 The four major lobes of the liver: **A,** Median or cystic; **B,** Right or lateral; **C,** Left; and **D,** Caudate.

The liver is divided into four major lobes: the median or cystic lobe, which bears a deep fissure; the right lateral lobe, which is partially divided into cranial and caudal lobes; a large left lobe; and a small caudate lobe fitting around the esophagus (Fig. 4-7). The rat has no gallbladder. Bile ducts from each lobe come together to form the common bile duct that is traceable to the descending duodenum. It enters the duodenum about 25 mm from the pyloric sphincter. These ducts are unable to concentrate bile in the way that the gallbladder of other rodents does.

The mesentery in the rat is a double layer of peritoneal membrane extending from the dorsal body wall to the viscera. It is given a specific name according to the particular visceral organ to which it is attached and thus supports: mesogastrium, to the stomach; mesodudenum, to the duodenum; true mesentery or mesentery proper, to all of the small intestine exclusive of the duodenum; mesocolon, to the large intestine; mesorchium covers the testicle of the male; mesovarium, to the ovaries of the female; and broad ligament, to the fallopian tubes and uterus of the female.

IV. RESPIRATORY SYSTEM

A. Upper Respiratory System

The upper respiratory system of the rat does not differ in any great way from that of the typical mammal. The external nares are located rostrolaterally and open caudally through the caudal nares into the nasopharynx. The Eustachian tubes open on the dorsolateral wall of the nasopharynx.

The epiglottis lies in the oropharynx where it covers the opening to the larynx. The beginning of the trachea lies ventral to the esophagus and extends caudally to form the two main branches, the primary bronchi, each of which enters a lung. The diameter of the trachea is approximately 1.6 to 1.7 mm in the adult rat, and the shape is maintained by 18 to 24 rigid C-shaped cartilage structures that form the framework of the trachea. Because of the cartilaginous rings, extension of the head of the rat can result in lengthening the trachea by 50% with no decrease in lumen diameter (Gatto and Houck, 1989).

As in other species, tracheal epithelium is predominantly pseudostratified with cilia, though basal cells and serous cells also are present. Ciliated goblet cells are found in greater numbers in the upper respiratory tract and trachea relative to other areas of the respiratory tract. In the intrapulmonary bronchi, the epithelium becomes shorter and columnar rather than pseudostratified; goblet cells diminish in number. Non-ciliated cells include brush cells, clara cells, and secretory cells (Souma, 1987). Mucosal and submucosal layers are each relatively distinct, and together have a width of approximately 27.5 μm in the adult rat. Seromucous glands are abundant but are limited to the proximal portion of the trachea (Babero et al., 1973).

B. Lower Respiratory System

The lung in the newborn rat is immature. It contains no alveoli or alveolar ducts; instead, gas exchange occurs in smooth-walled channels and saccules, and the prospective alveolar structures. Once the rat reaches 4 days old, a rapid restructuring of lung parenchyma occurs so that by 7 days old, the rat lung is morphologically more mature. Respiratory bronchioles are not present at birth, but by day 10 they are easily identified. After day 13, the primary and secondary septa expand and thin to form true alveolar septa (Burri, 1974). Apoptosis contributes to lung maturation by reducing the number of fibroblasts and type II epithelial cells in the third postnatal week (Schittny et al., 1998; Bruce et al., 1999)

The lung of the average adult rat contains an estimated 975,000,000 cells, of which 74% are in alveolar tissues and 26% in nonalveolar tissues (Stone et al., 1992). The alveoli are lined by large, flat type I pneumocytes and by round type II pneumocytes; both are epithelial cells. Type I pneumocytes constitute part of the extremely thin gaseous diffusion barrier, whereas type II pneumocytes secrete surfactant. With age the ratio of type II cells to type I cells lining the alveolar surfaces decreases (Pinkerton et al., 1982). Epithelial cells lining the alveoli appear to undergo significant deformation with large pulmonary inflations, leading to alveolar basement membrane deformation, which may contribute to lung recoil at high lung pressures (Tschumperlin and Margulies, 1999).

Goblet cells are situated in the epithelium of conducting airways at all levels, but comprise less than 1% of the total cells. In response to inhaled airway insult, a number of products are secreted by goblet cells, including mucins and glycoproteins (Rogers, 1994). Goblet cells serve as the progenitors to ciliated cells, which increase in number toward the periphery. Ciliated cells help to remove mucus and any entrapped material. It is theorized that only the tips of the cilia, which are described as having a claw-like structure, penetrate the periciliary fluid and "claw" the mucus forward. Basal cells decrease and virtually disappear from the small airways and, at all airway levels, the remaining nonciliated cells form a high proportion (about 40%) of the total cells present (Rogers, 1994).

Septal cells in the alveolar interstitium contain contractile filaments containing actin, desmin, and vimentin (Aoki et al., 1995). This feature suggests that septal cells can contract and change the architecture of the air-blood barrier, thereby influencing the regulation of the ventilation/perfusion ratio.

There are five lung lobes in the rat: one left lobe, and four on the right (cranial, middle, accessory, and caudal lobes). The middle lobe lies in contact with the diaphragm and apex of the heart and is notched to accommodate the caudal vena cava. For this reason, the middle lobe is sometimes referred to as the postcaval lobe.

The respiratory bronchioles are relatively short and lead almost immediately into alveolar ducts, each of which subdivides four or five times. The bronchioles in the rat are lined with high cuboidal epithelium on the side adjacent to the accompanying arteriole, whereas the epithelium on the opposite side progressively decreases in height.

The alveolar wall consists of three layers: the epithelial layer lining the alveolar space; a basement membrane; and the endothelial cell lining of the capillary lumen (Kikkawa, 1970; Harrison, 1974). All three layers are relatively thin and continuous, with the thickness of the air-blood barrier similar to that observed in mice and rabbits, approximately 1.5 μm. The alveolar spaces are normally clear of fluid or debris.

Surfactant, a lipid surface tension-lowering agent, coats the alveoli and helps to maintain the patency of conducting airways (Enhorning et al., 1995). Phospholipids, particularly phosphatidylcholine, are essential components of surfactant. Although surfactant is composed of mostly monounsaturated phospholipids in many mammals, rat surfactant has a high content of polyunsaturated phospholipids (Postle et al., 2001).

There are two types of pulmonary arteries in the rat: elastic arteries and muscular arteries (Sasaki et al., 1995). The elastic artery has an abundance of extra-cellular matrix in the media and an oblique arrangement of smooth muscle cells connecting to neighboring elastic laminae. The muscular artery has a paucity of extra-cellular matrix in the media and has a circumferential arrangement of smooth muscle cells enclosing the lumina. In the conscious resting rat, blood flow preferentially distributes to the central, hilar regions of the lung lobes, with less blood flow to the peripheral regions (Kuwahira et al., 1994).

The pulmonary vein of the rat is thicker than in most other species owing to the presence of striated muscle fibers that are contiguous with those of the heart. These muscle fibers are especially apparent in the longer intrapulmonary branches of the vessel. It has been suggested that this relationship facilitates spread of infection from the heart to the lungs (Cotchin, 1967). Precapillary anastomoses between the bronchial and pulmonary arteries have been demonstrated in the rat (Rakshit, 1949), as they have been in man and the guinea pig (Verloop, 1948). These anastomoses are limited to the hilar region in the rat.

Pulmonary lymphatics are critical to clearing lung fluid. A most important site for pulmonary edema formation, the pulmonary capillary is just upstream from small veins which have focal, smooth muscle tufts termed venous sphincters (Schraufnagel and Patel, 1990). Because of their constricting potential, these sphincters may control lung perfusion and cause edema. Increased numbers of lymphatics are adjacent to the venous sphincters, thus helping to moderate edema (Aharinejad et al., 1995).

Innervation of the lung is complex, with wide variation between species. The density of nerves is high in the rat, as it is in the calf, mouse, and guinea pig (Takino and Watanabe, 1937). The rat and the rabbit do not have adrenergic nerve supply to the bronchial muscle (Hebb, 1969). Instead, the bronchoconstriction is controlled by vagal tone. In contrast, the bronchioles are controlled by sympathetic tone in the human, dog, cat, sheep, pig, and the calf.

The lung weight is related to body size, but the surface area of the lung is related to the O_2 consumption. In this regard, the alveolar diameter is related to metabolic rate as measured by the O_2 used per unit of body weight (Tenney, 1963). The mean alveolar diameter for an adult rat is approximately 70 μm, in contrast to an approximate alveolar diameter of 200 to 250 μm for a man. The total surface area of the lung in a 400-g rat is approximately 7.5 m^2, compared with about 75 m^2 in a 70-kg man. This is consistent with the suggestion that smaller rodents have proportionately wider airways than do larger animals (Gomes et al., 2000). Relatively wider airways and a decline in airway resistance with declining body mass results in a relatively high ventilatory dead space that is compensated by a higher breathing frequency compared with larger mammals (Valerius, 1996).

Respiration is regulated, as in other mammals, in response to tissue CO_2 changes in the medullary respiratory center, though the carotid bodies or "glands," as they are sometimes called, also play a role. The carotid bodies are located on each side of the neck, in the bifurcation of the common carotid artery. The carotid bodies are composed of nerve tissue and arise as branches of the glossopharyngeal and vagus trunks and the cranial cervical ganglion, and are accompanied by sympathetic ganglion cells. The carotid bodies function as chemoreceptors and respond when the tissue partial pressure of oxygen decreases below 100 mm Hg. Similar chemoreceptors located in the aorta are called *aortic bodies*; the aortic body afferents travel via the vagi to the brain. In the adult rat, respiration averages about 85 breaths/min, producing an average tidal volume of 1.5 ml (Gay, 1965; Burri, 1974). The minute ventilation is about 100 ml/min.

V. CARDIOVASCULAR SYSTEM

The anatomy of the cardiovascular systems of the human and rat are similar, so we make no attempt in this limited space to describe the system. A schematic of the major blood vessels in the female rat is shown in Fig. 4-8. The figure emphasizes areas frequently targeted in research studies.

Cardiac output, total peripheral resistance, and venous capacity are regulated moment-to-moment to keep blood pressure and volume nearly constant. This regulation is critical to maintaining an ample blood supply to varied and various tissues. To maintain blood pressure and adjust flow efficiently, the central nervous system processes peripheral information on blood pressure, blood gases, pH, volume, and temperature (Dampney, 1994). The information is conveyed by afferent neurons arising from sets of receptors: baroreceptors, chemoreceptors, and cardiopulmonary receptors. This information is integrated by different areas and at different levels of the central nervous system to change the sympathetic and parasympathetic tone to the main effectors of circulatory control, the heart and blood vessels, appropriately. New understanding of control features since the 1990s, such as neurotransmission in the nucleus tractus solitarius, the role of neurons in the most caudal part of the ventrolateral medulla oblongata, and the increasing understanding of the exquisite control of different sympathetic pathways by different neurotransmitter systems can be found in Pilowsky and Goodchild (2002).

There is a sizable genetic component to aerobic exercise capacity, but the genes determining the difference between low and high capacity remain to be identified. Because the rat genome has been fine-mapped, rat models are proving to be a valuable tool for identifying the genetic origin of complex traits, such as aerobic capacity. There are two common types of models. The first is rat lines selectively bred for low and high aerobic capacity. These lines are valuable because alleles underlying the trait variance are associated with the phenotypic extremes. This segregation of alleles increases the power of genetic analysis (Koch et al., 1999). A second type is to use already available inbred strains developed from unselected strains. If the strains differ widely for the trait, comparing the genotype of each strain can lead directly to the underlying quantitative trait genes. The strain of rats with the lowest capacity for endurance running is the Copenhagen (COP). In contrast, the dark agouti (DA) strain of rats has the highest of all the strains tested. Investigators find that in isolated rat heart, working left ventricular function correlates (r = 0.86) with aerobic treadmill running capacity in 11 inbred strains of rats (Barbato et al., 2002). Furthermore, mean α/β-myosin heavy chain isoform ratio for DA rat's hearts is 21-fold higher than buffalo (BUF) hearts, and the DA hearts had a fivefold higher α/β-ratio of steady-state mRNA than the inbred rat strain, BUF, hearts. As a result, the mean rate of myosin heavy chain ATP hydrolysis is 64% more in DA ventricles than in BUF ventricles.

A. Heart

A good description of the physiology of the heart is in d'Uscio et al. (2000). The rat heart is on a midline in the thoracic cavity. Its apex is near the diaphragm and the lungs surround its lateral aspects, but the left lung is small. This means that the heart is for a considerable distance exposed to the thoracic wall. This makes cardiac puncture relatively simple if one inserts the cannula between ribs 3 and 5. The heart is four-chambered: two atria and two ventricles. Aortic and pulmonary valves each have three leaflets, whereas the mitral and tricuspid valves each have two major leaflets and a minor accessory leaflet.

Over the past decade, electrophysiologic studies show transmural heterogeneity, or differences in the action potential waveforms recorded in cells isolated from the epicardial and the endocardial tissues in the left ventricles of mammalian hearts. Recent experiments indicate that the heart morphogenetic information is engraved in the precardiac mesoderm. Nevertheless, specific differentiative signals can be overridden experimentally, which demonstrates cardiac phenotypic instability in the early heart tube stage (Icardo, 1996).

The adult rat is widely used as an experimental model to investigate the electrical heterogeneity in the left ventricle under both normal and pathophysiologic conditions. From patch clamp studies, Pandit et al. (2001) developed

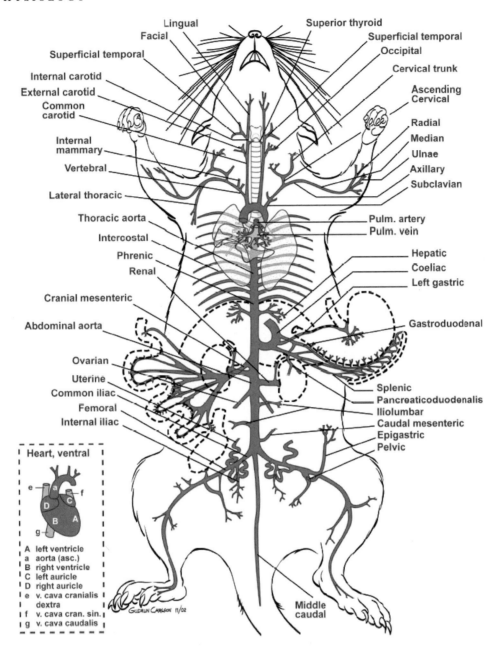

Fig. 4-8 Schematic of the major arteries of the female rat. Redrawn from Bivin et al. 1979, by G. Carlson.

a mathematical model of action potential heterogeneity in adult rat left ventricular myocytes.

The aortic arch bends to the left, as usual in other mammals, and diverges into the brachiocephalic (innominate), the left common carotid, and the left subclavian arteries. The brachiocephalic trunk divides at the sternoclavicular joint into right common carotid and right subclavian arteries. There are good illustrations of aorta and its systemic branches elsewhere (d'Uscio et al., 2000). One notable idiosyncrasy is an extracoronary myocardial blood supply. This makes the blood supply in the rat cardiovascular system more similar to that found in some fish. Each atria gets a large supply of blood from branches of the internal mammary and subclavian arteries. The right cardiac arteries provide a large blood supply to both right and left atria, but the left cardiac arteries supply only a little of what the left atrium needs.

The rat vena cava looks different from that of other mammals as well. There are two precavae. The right precava joins the right atrium directly; however, the left precava extends caudally, joins the azygous vein, unites with the postcava, and finally, joins the right atrium.

B. Peripheral Circulation

The peripheral circulatory system differs little from that of other small mammals. Table 4-1 gives selected hemodynamic parameters for the rat.

Rats are by far the most popular species used in hypertension research (Pinto et al., 1998), and the spontaneously hypertensive rat (SHR) is far and away the most used model. However, the transgenic TGR(mRen2)27 strain is a new rat model that is increasingly used.

Some of the conditions identified that contribute to pathogenesis in the SHR are impaired inhibitory GABA-ergic mechanism, increased activity of both ubiquitous and renal cell-specific isoforms of the Na^+/H^+ exchanger and the Na^+-K^+-2 Cl^--cotransporter (Orlov et al., 1999); increased density of sympathetic innervation; differences in genetic control of neuronal responses to cerebrospinal fluid Na^+ levels; and a genetically defective Cd36 fatty acid transporter. The Cd36 fatty acid transporter affects insulin resistance and disordered fatty acid metabolism; altered fat metabolism promotes disordered carbohydrate metabolism (Pravenec and Kurtz, 2002).

A review of the use of the common animal models for studying drug transport via the intestinal lymphatic system that describes the conscious rat and dog models in detail is in Edwards et al. (2001). In addition, there is a recent review by Dunne et al. (2004) covering the lymphatic system of the head and neck region of the rat. The Harderian gland has the highest density of lymphatics; the major sublingual gland has few lymph vessels (Dunne et al., 2004).

VI. URINARY AND REPRODUCTIVE SYSTEMS

A. Kidneys

The right kidney and the adrenal gland are more cranial than the left. A corresponding difference is present in the branching patterns of the renal vessels, the relation of these vessels to the aorta and caudal vena cava (Fuller and Huelke, 1973). Although it is possible to palpate the kidneys in young animals, it is difficult in mature specimens because the outlines are lost in fat.

The rat kidney enters the ureter directly and is unipapillate with only one calyx, easily accessible for cannulization techniques. The rat kidney also has specialized fornices that are long evaginations of the renal pelvis with epithelium similar to collecting duct epithelium (Schmidt-Nielsen et al., 1980). The fornices are in close association with the thin loops of Henle and likely function to build up the urea concentration in the papilla.

TABLE 4-1

SELECTED HEMODYNAMIC PARAMETERS FOR THE RAT[a]

Blood volume (ml/kg body wt.)	54.3
Arterial blood pressure	
Systolic (mm Hg)	116
Diastolic	90
Heart rate (beats/min)	300
Cardiac output (ml/min)	50

[a]Data are approximations only and vary significantly by strain and on the age, sex, health status, and diet of the rats. Regional blood flow can be measured by using implanted optical fibers and laser-Doppler flowmetry (Tabrizchi and Pugsley, 2000).

The medullary renal pyramid is well developed, including a strong zonation of vascular and tubular elements (Pannabecker et al., 2004). The relative thickness of the kidney regions compared to kidney and overall body weight are illustrated in Table 4-2.

There are both long- and short-looped nephrons in the laboratory rat. The short-looped nephrons arise from the tufts in the middle and outer zone of the cortex (high and middle nephrons). While descending, the thin segment of short loops changes to the broad thick segment, such that the bend of the loop is formed by the latter and the bend occurs in the outmost zone of the medulla or in the medullary ray of the cortex. In some parts of the cortex, usually opposite medullary rays, groups of short nephrons occur which possess no thin segment. The straight descending parts of the proximal tubules join the thick segments directly.

The glomerulus is lobulated, and there are a limited number of connective tissue cells between the vessels near the vascular pole (Fig. 4-9). Within Bowman's capsule, the

TABLE 4-2

RELATION OF BODY WEIGHT, KIDNEY DIMENSIONS, AND THICKNESS OF KIDNEY REGIONS[a]

Body weight (g)	200–300
Weight of left kidney (g)	0.7–2.0
Dimensions (mm)	$15 \times 9 \times 6$
Cortical thickness (mm)	4.0
Outer zone of medulla (mm)	2.2
Inner zone of medulla (mm)	6.0
No. of nephrons counted	846
Cortical (%)	8.2
Short (%)	44.7
Medullary (%)	18.6
Long (%)	25.8

[a]Data are approximations only and vary significantly by strain and on the age, sex, health status, and diet of the rats. For examples, see the manuscripts cited (Abdallah and Tawfik, 1969; Lohr et al., 1991; Christiansen et al., 1997; Obineche et al., 2001; Pannabecker et al., 2004).

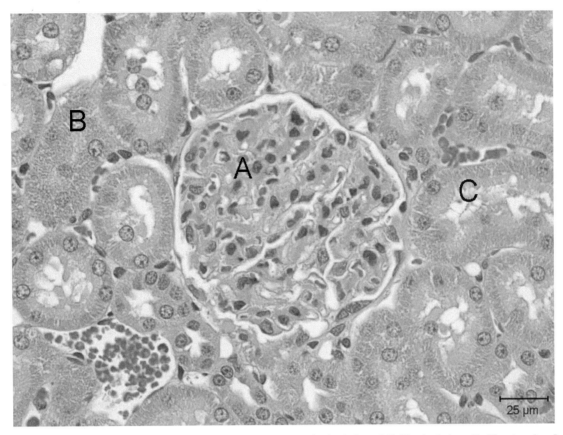

Fig. 4-9 Photomicrograph of rat kidney outer cortex: **A,** Glomerulus; **B,** Proximal tubule; and **C,** Distal tubule. × 40. (Courtesy Ann Lewis, Oregon National Primate Research Center, Beaverton, OR).

wall is thin and lined with squamous epithelium that may become slightly thicker at the urinary pole. The visceral layer covers the glomerulus, whereas the parietal layer covers Bowman's capsule. The visceral cells are characterized by their foot processes, often called podocytes. The glomerular endothelium, basement membrane, and visceral epithelial cell layers compromise the barrier for glomerular ultrafiltration. The capsular free space is usually pronounced.

Normal proteinuria is higher in rats (0.4–1.0 mg/ml urine) (Loeb and Quimby, 1999) than in man (0.03–0.13 mg/ml) (Kasiske and Keane, 2000). In the rat, sodium depletion and local amino acid metabolism control the amount of protein in the urine (Tobian and Nason, 1966; Herrero et al., 1997; Weinstein, 1998). Figure 4-10 shows the pressure gradients responsible for fluid movement from blood to urine as measured by direct puncture (Kallskog et al., 1975). Other renal excretory parameters are presented in Table 4-3.

The proximal tubule is characteristic for most rodents. The cells lining the tubule are broad and of considerable height, each having a rounded central nucleus. The cytoplasm is basophilic and contains granular or irregular mitochondria around the nucleus. Regularly arranged parallel rods occupy the basal part of the cell and are

perpendicular to the basement membrane. The next convoluted part of the proximal tubule, as well as the straight part entering the medullary ray and forming the beginning of Henle's loop, stain less darkly, but still contain numerous parallel rods. The brush border is distinct, and a small intensively stained granule can often be observed at the base of each line in the brush border.

At a somewhat variable position along the loop of Henle, the thick cells of the proximal segment are replaced by a flattened epithelium with a pale staining cytoplasm containing slightly flattened nuclei projecting toward the lumen. There is no brush border on these cells and mitochondria are scarce.

The thick segment of the loop of Henle has about the same diameter as the proximal tubule with no brush border. The thick segment joins the distal convoluted tubule at the macula densa. At this point, the structure of the nephrons changes remarkably and the distal tubule is distinguished by the width of its lumen and light staining of its cells (Pannabecker et al., 2004).

The collecting tubules are lined by cuboidal epithelium with round nuclei and basophilic cytoplasm. The cell borders of the collecting tubules are distinct.

The ureter expands within the renal sinus to create the renal pelvis. The collecting tubules empty into the renal

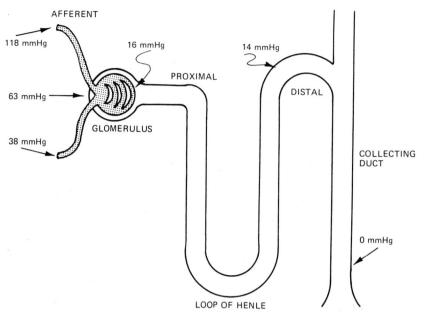

Fig. 4-10 Nephron pressure gradients.

TABLE 4-3

RENAL EXCRETORY PARAMETERS[a]

Blood urea nitrogen (mg%)	21
Urine volume (ml/24 hr/100 g body weight)	5.5–6.2
Na+ excretion (μmol/24 hr/100 g body weight)	191.6
K+ excretion (μmol/24 hr/100 g body weight)	794
Protein (mg/100 ml)	30–100
Urine osmolarity (mOsm/kg of H_2O)	1659
Specific gravity	1.050–1.062
GFR (ml/min/100 g body weight)	1.01–1.236
U/P insulin (mg/ml)	431
Clearance of inulin (μl/min/100 g)	857
Clearance of PAH (ml/min/100 g)	1.341
Filtration fraction	35–45%
Urine flow rate (μl/min/100 g)	4.8-5.2

[a]Data are approximations only and vary significantly by strain and are depend highly on the age, sex, health status, and diet of the rats. For examples, see the manuscripts cited (Flamenbaum et al., 1974; Zabel and Schieber, 1980; Loeb and Quimby, 1999).

pelvis, and the ureter carries the urinary waste products to the bladder for storage prior to excretion via the urethra. In the male, the urethra extends through the penis. In the female, the urethra opens on the perineum separately of the vagina.

In the normal rat, even when allowed an adequate supply of drinking water, the urine osmolarity is in the region of 1,000- to 1,500 mOsm compared with the normal tissue osmolarity of about 300 mOsm. There is a significant osmotic gradient in the medulla, rising from the corticomedullary junction to the tip of the papilla, where it approximates that of the urine (Masilamani et al., 2000).

The ability of the laboratory rat to concentrate its urine is about twice that of man. The osmotic ratio (urine/plasma) for man is 4.2; for the rat it is 8.9 (Christiansen et al., 1997). The specific gravity of urine for humans ranges from 1.003 to 1.030; for the rat it is 1.050 to 1.062 (Kasiske and Keane, 2000). This difference in concentrating ability is to the result of the functional anatomical differences between the kidney anatomy of the two species. The relative medullary thickness of the human kidney is 3.0, with 14% long-looped nephrons, whereas the relative medullary thickness of the rat kidney is 5.8, with approximately 25% long-looped nephrons (Tisher and Madsen, 2000).

The rat kidney contains significant amounts of L-amino acid oxidase, an enzyme that catalyzes the oxidation of 13 amino acids. The enzyme is absent in the kidney of the dog, cat, guinea pig, rabbit, pig, ox, and sheep (Blanchard et al., 1944).

The kidney of the rat (and rabbit, guinea pig, and sheep) also contains glutamine synthetase, an enzyme that converts ammonium glutamate to glutamine (Trevisan et al., 1999). The enzyme is not found in the kidneys of the cat, dog, pig or pigeon, although it is found in the brain of all vertebrates. Glutamine would, therefore, be found in the renal vein of the rat, but not of the dog or cat.

Superficial nephrons in the cortex of the rat kidney provide the model for most micropuncture work examining the individual transport process across the nephrons. This approach offers the most direct method for evaluating tubular function in vivo. A micropipette can be used to collect from single nephrons under conditions of free flow. The collected fluid is analyzed and the tubular collection site is identified by microdissection (Morsing et al., 1988). However, advances in diagnostic imaging techniques, such as diffusion magnetic resonance imaging (MRI), provide alternate methods to evaluate renal physiology (London et al., 1983; Yang et al., 2004).

B.   Male Reproductive System

The rat is one of the most widely used research animals in reproductive physiology. The rat also is useful in assessment of toxicologic insult to the reproductive tract.

The testes of the male lie in two separate thin-walled scrotal sacs located between the anus and the prepuce. They descend between the 30th and 40th day of life and are able to travel between the scrotal sacs and the abdominal cavity throughout life through the open inguinal canal. Prior to testicular descent, measurement of the anogenital distance is necessary for sexing. The anogenital distance is longer in males than in females.

The testicular artery and the pampiniform venus plexus are surrounded by fat as they enter the inguinal canal. Within the scrotum, several epididymal branches are given off from the internal spermatic artery to provide blood supply to the testes and the epididymis.

The epididymis consists of three sections: the enlarged caput epididymis on the proximal end of the testis, nearly completely embedded in fat; the slender corpus epididymis lying along the dorsomedial aspect of the testis; and cauda epididymis on the distal pole. The cauda epididymis doubles upon itself and distally develops into the ductus deferens (Popesko et al., 1990).

The ductus deferens is supplied by the deferential blood vessels and runs proximally through the inguinal canal and crosses the ureter to enter the urethra. The gland of the ductus deferens surrounds the duct close to its opening into the urethra.

The rat has five pairs of accessory sex glands located within the pelvis and surrounding the bladder (Fig. 4-11): the glands of the ductus deferens; two pairs of prostate glands, oriented dorsal and ventral to the ductus deferens; one pair of large scythe-shaped and convoluted vesicular glands; and one pair of coagulating glands in close association to the vesicular glands. The paired bulbourethral glands are in association with the bulboglandular muscle.

The seminal vesicles and coagulating glands are important for fertility in rats. Both organs secrete fluids that are necessary for appropriate formation of a vaginal plug. The role of the vaginal plug is not well understood beyond the observation that pregnancy is rare in the absence of its formation, but it is suspected to act as a reservoir for

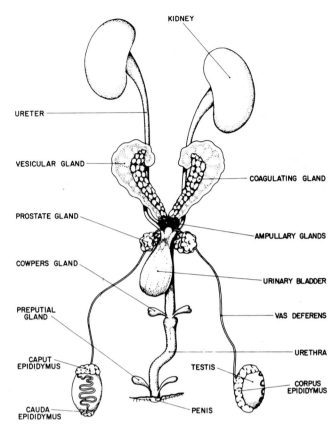

Fig. 4-11   Male urogenital system, ventral view.

the gradual release of sperm or to prevent the outflow of sperm from the vagina (Sofikitis et al., 1992).

The penis lies within a loose prepuce with a single cartilaginous or bony process, the os penis, on the ventral wall. A pair of slender and flattened preputial glands are located beneath the skin of the prepuce (Popesko et al., 1990).

C.   Female Reproductive System

In the female, the ovaries are oriented along the lateral border of psoas muscle and embedded in fat near the kidneys (Fig. 4-12). The mature ovary appears as a mass of follicles. The convoluted oviduct has its open end applied to the ovary and its distal end into the bicornuate uterus. The oviduct is surrounded by the oviductal mesentery (mesosalpinx) forming the ovarian bursa. Although the uterine horns appear to be fused distally, there are two distinct ossa uteri opening into the vagina. Each ossa uteri has an ostium interim and externum and cervical canal (Fig. 4-13).

The only genital structure of the female connected to the urinary system is the clitoris. This organ is located in a prepuce with clitoral glands, analogous to the preputial

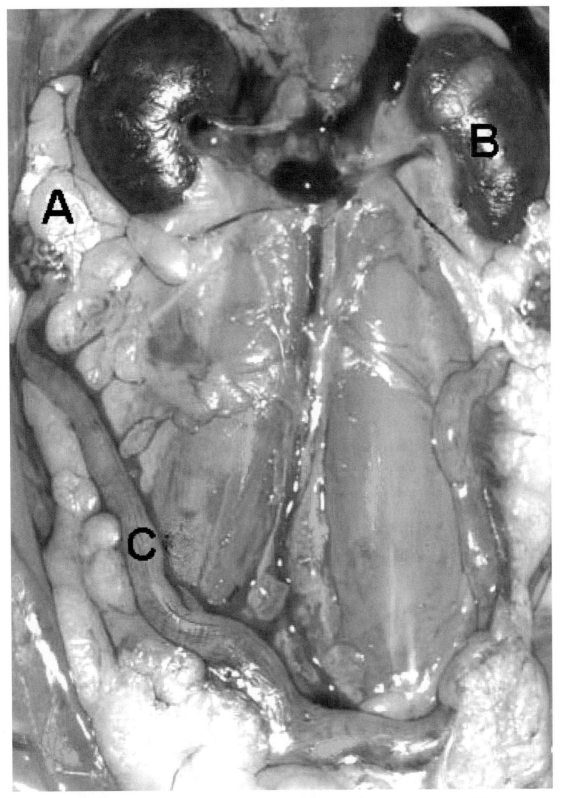

Fig. 4-12 Female reproductive tract: **A,** Right ovary; **B,** Right uterine horn; and **C,** Left kidney.

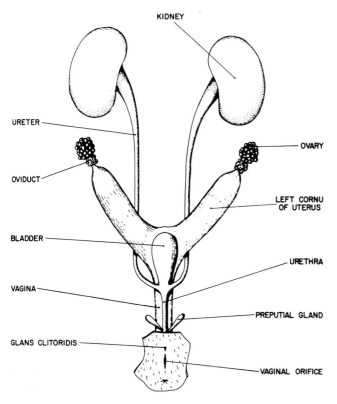

Fig. 4-13 Female urogenital system.

glands of the male. The bladder empties into the urethra that opens to the perineum at the urethral orifice located at the base of the clitoris. The vaginal opening is located dorsal to the urethra. The urethra does not enter the vagina or vestibule, as is seen in other domestic mammals (Popesko et al., 1990).

The estrous cycle of the rat is similar to most other mammals and is approximately 4 to 5 days in duration. During proestrus, serum estrogen and prolactin levels increase until just before ovulation, triggering anatomical changes in the vaginal and uterine linings and behavioral changes consistent with mating behavior acceptance. Estrogen also stimulates the release of follicle stimulating hormone (FSH) to concurrently initiate the maturation of the follicles. The estrogen secretion continues until the luteinizing hormone (LH) surge, is released on estrus, and ovulation occurs within 8 to 14 hours. This LH release is closely tied to the time of day as defined by the light-dark cycle (light on 5:00 AM, off 7:00 PM). The estrus phase of the cycle is characterized by acceptance of the male and superficial squamous cells in the vagina. This LH surge, in addition to causing ovulation, also terminates the ongoing estrogen secretion and starts progesterone secretion by the ovary. Progesterone acts through the nervous system and helps induce mating behavior. As a result of the lowering of estrogen secretion and its negative feedback effect on FSH, the FSH

secretion levels are freed to begin increasing, thus initiating the maturation of the follicles for the next cycle. Hormonal analysis and vaginal cytology are useful when determining the phase of the estrous cycle (Karim et al., 2003).

VII. ENDOCRINE SYSTEM

The endocrine system of the rat has been studied extensively and is a valuable animal model for exploration of multiple endocrine systems. In this chapter, it is not appropriate to describe the many methods that have been used, and the reader is referred to the following papers for examples (Eisenbarth and Jasinski, 2004; Hess and Zimmermann, 2004; Makara et al., 2004; Wong et al., 2004). The endocrine organs include the hypothalamus, pituitary gland, thyroid, parathyroid, adrenal, pancreas, and reproductive organs.

A. Hypothalamus

The hypothalamus is typical for mammals. It surrounds the third ventricle at the base of the diencephalon. The hypothalamus is a senso-neuroendocrine organ that converts neural input into endocrine output by producing hormones to act upon the pituitary gland. The hormones produced by the hypothalamus include gonadotropin-releasing hormone (GnRH), thyrotropin-releasing hormone (TRH), corticotropin-releasing hormone (CRH), growth hormone-releasing hormone (GHRH) somatostatin, and prolactin-releasing and prolactin-inhibiting factors, including dopamine (Ohkura et al., 2000).

B. Pituitary Gland

The pituitary gland, or hypophysis, is situated within a bony cavitation called the *sella turcica* and is surrounded by a meningeal fold. This orientation makes removal of the hypophysis with the brain difficult. At about 40 to 50 days old, the hypophysis of the female becomes heavier than that of the male. This difference tends to increase with age (Stendell, 1913).

The pituitary gland is divided into the anterior lobe, the intermediate lobe, and the posterior lobe. The anterior lobe secretes a number of hormones including LH; FSH; thyroid stimulating hormone (TSH); growth hormone (GH); prolactin; and adrenocorticotropic hormone (ACTH). The intermediate lobe produces melanocyte-stimulating hormone (MSH). The posterior lobe secretes oxytocin and arginine vasopressin (Ohkura et al., 2000).

C. Thyroid and Parathyroid

The two thyroid glands are bilaterally oriented on the trachea, just caudal to the larynx, extending over four to five tracheal rings. The glands are characterized histologically by the presence of colloid in the lumen of the thyroid follicles and by "C" cells that have their origin from the ultimobranchial body (Stendell, 1913).

The thyroid glands secrete hormones important for metabolism, reproduction, and development. The two primary hormones are thyroxine (T4) and triiodothyronine (T3) under stimulation by TSH from the pituitary gland. The C-cells of the thyroid secrete calcitonin, which is important in the regulation of calcium deposition in bone (Yamamoto et al., 1995).

The parathyroid glands are located on the dorsolateral surface of each lobe of the thyroid and secrete parathyroid hormone (PTH), an antagonist of calcitonin (Yamamoto et al., 1995).

D. Adrenals

Adrenal glands in the domesticated laboratory rat are much smaller than those found in wild rats and are located at the craniomedial pole of each kidney. The adrenal cortex of the gland of the female is also consistently larger than that of the male (Dohm et al., 1971). The adrenal gland consists of a cortex and a medulla. The adrenal cortex is further divided into three zones: the zona glomerulosa, zona fasciculata, and zona reticularis (Mulrow, 1986).

Glucocorticoids are secreted by the zona fasciculata and zona reticularis of the adrenal cortex in response to release of ACTH from the anterior pituitary. Elevated levels of ACTH are probably responsible for the negative feedback inhibition of the input stimulus, either in the hypothalamus or anterior pituitary. Figure 4-14 summarizes the feedback control of glucocorticoid secretion. The primary glucocorticoid of the rat is corticosterone (Mizoguchi et al., 2001).

Mineralocorticoids are secreted by the zona glomerulosa of the adrenal cortex. Aldosterone is the most potent of the mineralocorticoids.

The catecholamines, norepinephrine and epinephrine, are produced by the adrenal medulla under direct neurologic and hormonal control (Mulrow, 1986).

E. Pancreas

In addition to its exocrine functions important for digestion, the islets of Langerhans of the pancreas are important in the hormonal control of metabolism. Insulin is secreted by the B cells of the pancreatic islet cells and stimulates the uptake of blood glucose into adipose or

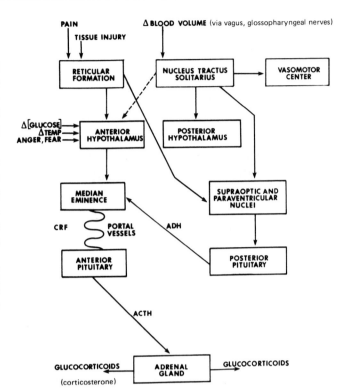

Fig. 4-14 Control scheme for glucocorticoid release by the adrenal gland. *CRF,* Corticotropin releasing factor; *ADH,* Antidiuretic hormone; *ACTH,* Adrenal corticotropin hormone.

muscle cells. Glucagon is secreted by the A cells of the pancreatic islet cells and stimulates glucogenolysis to increase blood glucose concentrations. Somatostatin is secreted by the D cells of the pancreatic islets and is important in metabolic control, but the regulatory mechanisms are not fully identified (Ohkura et al., 2000).

The rat is an important animal model in the study of diabetes, one of the most common human endocrine disorders.

F. Reproductive Organs

The reproductive organs also play a role in the endocrine system. Specifically, the gonads, the uterus, and the placenta secrete hormones important in regulation of physiologic function.

The gonads secrete steroid hormones including the androgens (e.g., testosterone), the estrogens (e.g., estrone) and the progestins (e.g., progesterone). The secretion of these hormones is controlled through feedback by the pituitary gland and aid in the coordination of reproductive events ranging from ovulation to maintenance of pregnancy and formation of sperm. The ovaries also secrete several nonsteroidal hormones: inhibin, activin, and follistatin.

The uterus secretes prostaglandins and oxytocin, which play important roles in luteolysis and parturition. The placenta has been shown to secrete lactogens, which stimulate progesterone and insulin secretions during pregnancy (Ohkura et al., 2000).

G. Miscellaneous

Various other organs, not traditionally classified as endocrine, also are associated with endocrine functions.

Leptin is a hormone secreted by adipose tissue that has been shown to play a role in appetite control and obesity (Pelleymounter et al., 1995).

The kidney secretes two hormones, rennin and angiotensin, which play an important role in the control of blood pressure (Vallotton et al., 1990).

Melatonin, important in regulation of daily and seasonal endocrine activity, is secreted by the pineal gland (Reiter, 1991).

The gastrointestinal tract is associated with four hormones. Gastrin is secreted by the G cells in the antral stomach and stimulates gastric acid secretion. Secretin is secreted by the duodenum and stimulates the release of alkaline juice from the pancreas. Motilin is secreted by the duodenum and stimulates gastrointestinal tract motility. Cholecystokinin (CCK) stimulates pancreatic enzyme and bile release (Ohkura et al., 2000).

VIII. NERVOUS SYSTEM

The rat has a long history of being a useful experimental animal model for studies involving the nervous system. The intent of the discussion presented here is to summarize major anatomical features of the nervous system, not to provide in depth anatomical descriptions. The reader is referred to alternate references (Heidbreder and Groenewegen, 2003; Abrams et al., 2004; Onat and Cavdar, 2004; Schwartzkroin et al., 2004) for additional detailed anatomical descriptions.

A. Peripheral Nervous System

The peripheral nervous system is composed of 34 pairs of spinal nerves: 8 cervical, 13 thoracic, 6 lumbar, 4 sacral, and 3 caudal. The paired spinal nerves arise from the spinal cord through the intervertebral foramina. Each nerve is formed by dorsal and ventral roots. The dorsal root is sensory and carries impulses into the spinal cord. The ventral root carries most fibers away from the spinal cord. The vasculature of the spinal cord includes the median

dorsal spinal artery and vein, two lateral spinal arteries, and ventromedian veins (Scremin, 1995).

The first cervical nerve lacks a ganglion; the ganglion of the second cervical nerve lies dorsal to the bifurcation of the carotid artery. The ganglion of the third cervical nerve and the first few thoracic ganglia contribute to the stellate ganglion, located at the level of the first two thoracic vertebrae. All other spinal ganglia are enclosed within the vertebral canal. From the stellate ganglion to the first sacral nerves, pairs of rami communicantes connect the spinal nerves with the sympathetic ganglia. The rami communicantes are short and cannot be separated easily into white and gray matter (Grant, 1995; Gabella, 1995).

Ventral divisions of the nerves form typical plexuses in the cervical, brachial, and lumbosacral regions. Detailed descriptions of the distribution of these terminal branches are described elsewhere (Gabella, 1995).

The brachial plexus gives rise to five major nerves that supply the forelimb. These include the ulnar nerve that serves the caudal portion of the forearm, the median nerve that serves the medial surface of the forelimb, the axillary nerve that serves the shoulder and lateral surface of the forelimb, the musculocutaneous nerve that serves the shoulder and upper brachium, and the radial nerve that serves the upper caudal muscles of the forelimb and the lower cranial surface of the forelimb.

The lumbar plexus is formed by the thirteenth thoracic and the first four or five lumbar nerves. The major nerves arising from the lumbar plexus include the femoral nerve that serves the medial surface of the thigh and the saphenous nerve that serves the medial surface of the thigh.

The pelvic plexus arises from the last lumbar and first sacral nerves. This plexus provides the neural connections to the genitourinary tract.

The spinal cord terminates in a slender filum terminale. The caudal spinal nerves and the filum terminale are collectively called the *cauda equina* (Popesko et al., 1990).

The cranial nerves also are considered a part of the peripheral nervous system. Their origins and relations to the brain are illustrated in Figs. 4-15 and 4-16. A brief description of each of these nerves is given in Table 4-4.

The autonomic nervous system is composed of two sympathetic trunks bearing 24 ganglia oriented parallel bilaterally on the vertebral column. Each trunk communicates to the ventral root of the spinal nerves. The eighth, ninth, and tenth thoracic and the first lumbar sympathetic ganglia give rise to four branches that unite to form the greater splanchnic nerves on either side. These two nerves pierce the diaphragm to reach the unpaired celiac ganglion supplying branches to the diaphragm, cranial mesenteric artery, and adjacent viscera.

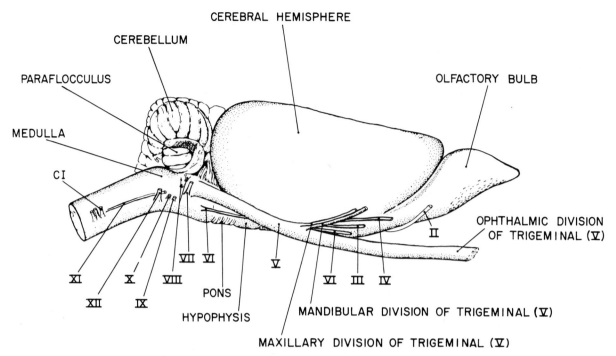

Fig. 4-15   Lateral view of brain.

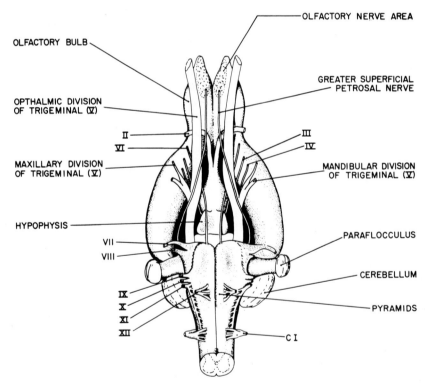

Fig. 4-16   Ventral view of brain.

The lesser splanchnic arises at the level of the third lumbar ganglion and supplies additional branches to the celiac ganglion. The most caudal branches arise from the third and fourth lumbar ganglia and supply the aortic and caudal mesenteric plexuses (Gabella, 1995).

B.   Central Nervous System

The central nervous system is composed of the brain and spinal cord. A three-layered meninges covers both

TABLE 4.4

CHART OF THE CRANIAL NERVES

Name	Superficial origin on brain	Foramen of exit from, or entrance to, cranial cavity	Action	Distribution
I. Olfactory	Olfactory bulb	Cribiform plate of ethmoid	Sensory	Nasal epithelium
II. Optic	Anterior corpora bigemina and thalamus	Optic foramen	Sensory	Retina of eye
III. Oculomotor	Pedunculi cerebri	Anterior lacerated foramen	Motor	Dorsal, ventral and medical recti and ventral oblique muscles
IV. Trochlear	Posterior to posterior corpora bigemina	Anterior lacerated foramen	Motor	Dorsal oblique muscle
V. Trigeminal	Pons	Anterior lacerated foramen (ophthalmic and dorsal maxillary branches) and foramen ovale (ventral maxillary branch)	Motor and sensory	Skin, vibrissae, tongue, teeth, muscles of mastication
VI. Abducens	Anterior medulla oblongata	Anterior lacerated foramen	Motor	Laterall rectus muscle
VII. Facial	Medulla oblogata near V	Facial canal to stylomstoid foramen	Sensory and motor	Facial muscles, tongue, salivary and lacrimal glands
VIII. Vestibulochlear	Medulla oblongata, posterior to VII	Internal acoustic meatus	Sensory	Organs of equilibrium and hearing in inner ear
IX. Glossopharyngeal	Medulla oblongata near X	Posterior lacerated foramen	Sensory and motor	Pharynx, tongue, carotid sinus and parotid gland
X. Vagus	Medulla oblongata posterior to VIII	Posterior lacerated foramen	Sensory and motor	Pharynx, larynx, heart, lungs, diaphragm and stomach
XI. Spinal accessory	Medulla oblongata and anterior end of spinal cord	Posterior lacerated foramen	Motor	Muscles of neck and pharyngeal viscera with vagus
XII. Hypoglossal	Medulla oblongata posterior to X	Hypoglossal canal	Motor	Intrinsic, extrinsic tongue muscles

structures. The outermost layer is the dura mater. On the brain, the dura mater is folded between the cerebral hemispheres as the falx cerebri and between the cerebrum and cerebellum as the tentorium cerebelli. Along the spinal cord, the dura mater is the tough outer fibrous coat that closely invests or surrounds each spinal nerve for a distance as it branches off from the spinal cord. The innermost layer of the meninges is the thin and delicate pia mater that adheres to the brain and the spinal cord. Between the dura and pia is a network of thin fibers called the *arachnoid.*

The brain consists of two large cerebral hemispheres separated by a median fissure. There is no demarcation between the frontal lobes and the larger and more caudal temporal lobes. Two large olfactory bulbs project from the rostral end of the brain and are connected to the hippocampal lobes via the olfactory tracts. The olfactory nerves terminate at the olfactory bulbs after passing through the fenestrated ethmoid plate from the nasal epithelium (Shipley et al., 1995).

The convoluted cerebellum occupies the major portion of the caudal end of the brain. It consists of the vermis and two lateral hemispheres (Voogd, 1995).

Between the cerebral hemispheres and the vermis are four rounded bodies, the corpora quadrigemina. The caudal pair is associated with acute hearing; the rostral pair is associated with vision.

The most caudal portion of the brain is composed of the medulla oblongata as it tapers back into the spinal cord.

On the ventral aspect of the brain are the origins of the cranial nerves (Fig. 4-16). Other structures of importance include the pituitary gland consisting of the stalk, the neurohypophysis and adenohypophysis. On either side of the infundibulum are fiber tracts, the pendunculi cerebri, which connect the cerebrum and the medulla oblongata. The fiber tracts and nuclei of the pons are located caudal to the hypophysis. These tracts serve as relay centers to connect the cerebellar hemispheres and the cerebral cortex. The pyramids extend caudally from the pons and carry information from the cerebral cortex to centers within the spinal cord (Travers, 1995).

A median sagittal plane view of the brain (Fig. 4-17) reveals a prominent corpus callosum or middle commissure. This is a white band of fibers linking the two cerebral hemispheres. A very small white fiber tract lies rostral to

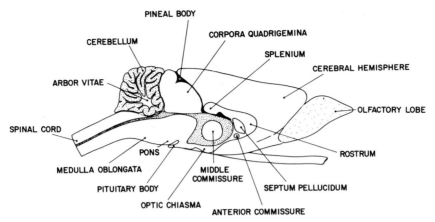

Fig. 4-17 Median sagittal view of brain.

the intermediate mass and is the rostral (anterior) commissure. Also identifiable are the crura cerebri, thickened masses of fibers that lie on the floor of the brain and link the fore- and hindbrains. Other structures of importance include the corpora quadrigemina, the pineal body, the optic chiasm, the medulla oblongata, the arbor vitae of the cerebellum, and the ventricles of the brain.

The ventricular system of the rat consists of one central ventricle (the third) connected with two lateral ones (first and second), and a caudal one (the fourth). Each lateral ventricle communicates with the third ventricle through the interventricular foramen that is on the ventral side of the anterior horn about 1 mm from its caudal end. The cerebral aqueduct is about 3 mm long and connects the third and fourth ventricles. The ventricular system communicates with the cerebral subarachnoid space by two lateral apertures.

The walls of the third ventricle contain the subfornical organ, the vascular organ of the lamina terminalis, the pineal body, the subcommissural organ, and the median eminence (associated with the neurohypophysis). The fourth ventricle contains the area postrema. Within the lumen of the lateral, third, and fourth ventricles is the choroid plexus (Oldfield and McKinley, 1995).

The most studied neurotransmitters of the rat brain are serotonin and tachykinins. Serotonin has been demonstrated to have an effect on multiple functions including blood pressure, sodium and glucose balance, fluid homeostasis, nociception, cognition, and depression and anxiety. Tachykinin peptides have been shown to affect motor, sensory, cardiovascular, respiratory, and gastrointestinal functions. The opioid peptides (dynorphins, enkephalins, and β-endorphins) and cholinergic neurons also provide regulatory effects on brain-controlled functions. For additional information regarding the neurotransmitters, the reader is referred to alternate references (Halliday et al., 1995; Loughlin et al., 1995; Butcher, 1995).

REFERENCES

Abdallah, A., and Tawfik, J. (1969). The anatomy and histology of the kidney of sand rats (*Psammomys obesus*). *Z. Versuchstierk* **11**, 261–275.

Abrams, J.K., Johnson, P.L., Hollis, J.H., and Lowry, C.A. (2004). Anatomic and functional topography of the dorsal raphe nucleus. *Ann. N. Y. Acad. Sci.* **1018**, 46–57.

Aharinejad, S., Bock, P., Firbas, W., and Schraufnagel, D.E. (1995). Pulmonary lymphatics and their spatial relationship to venous sphincters. *Anat. Rec.* **242**, 531–544.

Albiin, N., Hellstrom, S., Stenfors, L.E., and Cerne, A. (1986). Middle ear mucosa in rats and humans. *Ann. Otol. Rhinol. Laryngol. Suppl* **126**, 2–15.

Amano, O., and Iseki, S. (2001). [Expression and localization of cell growth factors in the salivary gland: a review]. *Kaibogaku Zasshi* **76**, 201–212.

Aoki, T., Taira, K., Shibasaki, S., and Fujimoto, T. (1995). The cytological and immunohistochemical study of septal cells in the rat lung. *Acta Histochemica Cytochemica* **28**, 349–355.

Babero, B.B., Yousef, M.K., Wawerna, J.C., and Bradley, W.G. (1973). Comparative histology of the respiratory apparatus of three desert rodents and the albino rat. A view on morphological adaptations. *Comp. Biochem. Physiol. A* **44**, 585–597.

Barbato, J.C., Lee, S.J., Koch, L.G., and Cicila, G.T. (2002). Myocardial function in rat genetic models of low and high aerobic running capacity. *Am. J. Physiol. Regul. Integr. Comp. Physiol.* **282**, R721–R726.

Berg, R.W., and Kleinfeld, D. (2003). Rhythmic whisking by rat: retraction as well as protraction of the vibrissae is under active muscular control. *J. Neurophysiol.* **89**, 104–117.

Bivin, W.S., Crawford, P., and Brewer, N.R. (1979). Morphophysiology. *In* "The Laboratory Rat." pp 73–103, Academic Press Inc., New York.

Blanchard, M., Green, D.E., Nocito, V., and Ratner, S. (1944). L-amino acid oxidase of animal tissue. *J. Biol. Chem.* **155**, 421–440.

Blazsek, J, and Varga, G. (1999). Secretion from minor salivary glands following ablation of the major salivary glands in rats. *Arch. Oral Biol.* **44 Suppl 1**, S45–S48.

Borg, T.K., Sullivan, T., and Ivy, J. (1982). Functional arrangement of connective tissue in striated muscle with emphasis on cardiac muscle. *Scan. Electron. Microsc.*, 1775–1784.

Bowman, B.M., and Miller, S.C. (1999). Skeletal mass, chemistry, and growth during and after multiple reproductive cycles in the rat. *Bone* **25**, 553–559.

Boyne, R., Fell, B.F., and Robb, I. (1966). The surface area of the intestinal mucosa in the lactating rat. *J. Physiol.* **183**, 570–575.

Bruce, M.C., Honaker, C.E., and Cross, R.J. (1999). Lung fibroblasts undergo apoptosis following alveolarization. *Am. J Respir. Cell Mol. Biol.* **20**, 228–236.

Burda, H., and Voldrick, L. (1980). Correlation between the hair cell density and the auditory threshold in the white rat. *Hear. Res.* **3**, 91–93.

Burri, P.H. (1974). The postnatal growth of the rat lung. 3. Morphology. *Anat. Rec.* **180**, 77–98.

Butcher, E.O. (1934). The hair cycles in the albino rat. *Anat. Rec.*, 5–19.

Butcher, L.L. (1995). Cholinergic neurons and networks. *In* "The Rat Nervous System." (G. Paxinos, ed.), Academic Press, San Diego.

Cabrera, C.L., Wagner, L.A., Schork, M.A., Bohr, D.F., and Cohan, B.E. (1999). Intraocular pressure measurement in the conscious rat. *Acta Ophthalmol. Scand.* **77**, 33–36.

Cairns, J.E. (1959). Normal development of the hyaloid and retinal vessels in the rat. *Br J Ophthalmol* **43**, 385–393.

Carmeliet, G., Vico, L., and Bouillon, R. (2001). Space flight: a challenge for normal bone homeostasis. *Crit. Rev. Eukaryot. Gene Expr.* **11**, 131–144.

Carper, D., Russell, P., Shinohara, T., and Kinoshita, J.H. (1985). Differential synthesis of rat lens proteins during development. *Exp. Eye Res.* **40**, 85–94.

Carvell, G.E., and Simons, D.J. (1996). Abnormal tactile experience early in life disrupts active touch. *J Neurosci.* **16**, 2750–2757.

Catalanotto, FA, and Leffingwell, C. (1979). Early effects of desalivation upon fluid consumption and taste acuity in the rat. *Behav. Neural. Biol.* **25**, 190–205.

Chen, H.B., Yamabayashi, S., Ou, B., Tanaka, Y., Ohno, S., and Tsukahara, S. (1997). Structure and composition of rat precorneal tear film. A study by an in vivo cryofixation. *Invest. Ophthalmol. Vis. Sci.* **38**, 381–387.

Chew, R.M. (2003). Water metabolism in mammals. *Physiol. Mammal.* **2**, 143–178.

Chiasson, R.B. (1958). "Laboratory Anatomy of the White Rat." W.C. Brown, Dubuque, Iowa.

Chole, R.A., and Tinling, S.P. (1994). Bone lining cells of the mammalian cochlea. *Hear. Res.* **75**, 233–243.

Christiansen, T., Rasch, R., Stodkilde-Jorgensen, H., and Flyvbjerg, A. (1997). Relationship between MRI and morphometric kidney measurements in diabetic and non-diabetic rats. *Kidney Int.* **51**, 50–56.

Clarke, R.M. (1970). Mucosal architecture and epithelial cell production rate in the small intestine of the albino rat. *J. Anat.* **107**, 519–529.

Cleaton-Jones, P. (1972). Anatomical observations in the soft palate of the albino rat. *Anat. Anz.* **131**, 419–424.

Cotchin, E.R.F.C. (1967). "Pathology of Laboratory Rats and Mice." Blackwell, Oxford.

d'Uscio, L.V., Kilo, J., Lischer, T.F., and Gassmann, M. (2000). Circulation. *In* "The Laboratory Rat." (G.J. Krinke, ed.), pp 345–357. Academic Press, London.

Dampney, R.A. (1994). Functional organization of central pathways regulating the cardiovascular system. *Physiol. Rev.* **74**, 323–364.

DeSesso, J.M., and Jacobson, C.F. (2001). Anatomical and physiological parameters affecting gastrointestinal absorption in humans and rats. *Food Chem. Toxicol.* **39**, 209–228.

Dohm, G., von Seebach, H.V., and Stephan, G. (1971). Der Geschlechtdimorphismus der Nebennierenrinde der Ratte: Lichtomikroskopische und histometrische Untersuchungen. *Z. Zellforsch. Mikrosk. Anat.* **116**, 119–135.

Dunne, A.A., Steinke, L., Teymoortash, A., Kuropkat, C., Folz, B.J., and Werner, J.A. (2004). The lymphatic system of the major head and neck glands in rats. *Otolaryngol. Pol.* **58**, 121–130.

Edwards, G.A., Porter, C.J., Caliph, S.M., Khoo, S.M., and Charman, W.N. (2001). Animal models for the study of intestinal lymphatic drug transport. *Adv. Drug Deliv. Rev.* **50**, 45–60.

Eisenbarth, G.S., and Jasinski, J.M. (2004). Disease prevention with islet autoantigens. *Endocrinol. Metab. Clin. North Am.* **33**, 59–73, viii.

El Sharaby, A., Ueda, K., Kurisu, K., and Wakisaka, S. (2001). Development and maturation of taste buds of the palatal epithelium of the rat: histological and immunohistochemical study. *Anat. Rec.* **263**, 260–268.

Emmelin, N (1981). Nervous control of mammalian salivary glands. *Philos. Trans. R. Soc. Lond B Biol. Sci.* **296**, 27–35.

Enhorning, G., Duffy, L.C., and Welliver, R.C. (1995). Pulmonary surfactant maintains patency of conducting airways in the rat. *Am. J. Respir. Crit. Care Med.* **151**, 554–556.

Erben, R.G., Eberle, J., Stahr, K., and Goldberg, M. (2000). Androgen deficiency induces high turnover osteopenia in aged male rats: a sequential histomorphometric study. *J. Bone Miner. Res.* **15**, 1085–1098.

Erzurumlu, R.S., and Killackey, H.P. (1983). Development of order in the rat trigeminal system. *J. Comp. Neurol.* **213**, 365–380.

Fannin, S.W., Lesnefsky, E.J., Slabe, T.J., Hassan, M.O., and Hoppel, C.L. (1999). Aging selectively decreases oxidative capacity in rat heart interfibrillar mitochondria. *Arch. Biochem Biophys.* **372**, 399–407.

Farrar, R.P., Martin, T.P., and Ardies, C.M. (1981). The interaction of aging and endurance exercise upon the mitochondrial function of skeletal muscle. *J. Gerontol.* **36**, 642–647.

Farris, E.J. and Griffith, J.Q. (1949). "The Rat in Laboratory Investigation." Lippincott, Philadelphia, Pa.

Faussone Pellegrini, M.S., Cortesini, C., and Romagnoli, P. (1977). [Ultrastructure of the tunica muscularis of the cardial portion of the human esophagus and stomach, with special reference to the so-called Cajal's interstitial cells]. *Arch. Ital. Anat. Embriol.* **82**, 157–177.

Fitts, R.H., Troup, J.P., Witzmann, F.A., and Holloszy, J.O. (1984). The effect of ageing and exercise on skeletal muscle function. *Mech. Ageing Dev.* **27**, 161–172.

Flamenbaum, W., Huddleston, M.L., McNeil, J.S., and Hamburger, R.J. (1974). Uranyl nitrate-induced acute renal failure in the rat: micropuncture and renal hemodynamic studies. *Kidney Int.* **6**, 408–418.

Fraser, D.A. (1931). The winter pelage of the adult albino rat. *Am. J. Anat.* **55**–87.

Fuller, P.M., and Huelke, D.F. (1973). Kidney vascular supply in the rat, cat and dog. *Acta Anatomica* **84**, 516–522.

Gabella, G. (1995). Autonomic nervous system. *In* "The Rat Nervous System." (G. Paxinos, ed.), Academic Press, San Diego.

Gatto, L.A., and Houck, B.M. (1989). Mucociliary transport and epithelial morphology with elongation and collapse in rat trachea. *Exp. Lung Res.* **15**, 239–251.

Gay, W.I. (1965). "Methods of Animal Experimentation." Academic Press, New York.

Gipson, I.K., Yankauckas, M., Spurr-Michaud, S.J., Tisdale, A.S., and Rinehart, W. (1992). Characteristics of a glycoprotein in the ocular surface glycocalyx. *Invest. Ophthalmol. Vis. Sci.* **33**, 218–227.

Goldblum, D., Kontiola, A.I., Mittag, T., Chen, B., and Danias, J. (2002). Non-invasive determination of intraocular pressure in the rat eye. Comparison of an electronic tonometer (TonoPen), and a rebound (impact probe) tonometer. *Graefes Arch. Clin. Exp. Ophthalmol.* **240**, 942–946.

Gomes, R.F.M., Shen, X., Ramchandani, R., Tepper, R.S., and Bates, J.H. (2000). Comparative respiratory system mechanics in rodents. *J. Appl. Physiol* **89**, 908–916.

Gorgels, T.G.M.F., and van Norren, D. (1992). Spectral transmittance of the rat lens. *Vision Res.* **32**, 1509–1512.

Grant, G. (1995). Primary afferent projections to the spinal cord. *In* "The Rat Nervous System." (G. Paxinos, ed.), Academic Press, San Diego.

Greene, E.D. (1935). "Anatomy of the Rat." Hafner, New York.

Groth-Vasselli, B., and Farnworth, P.N. (1986). A critical maturation period in neonatal-rat-lens development. *Exp. Eye Res.* **43**, 1057–1066.

Gurkan, S., and Bradley, R.M. (1988). Effects of electrical stimulation of autonomic nervous system on degranulation of von Ebner's gland acini. *Brain Res.* **473**, 127–133.

Hall, H.D., and Schneyer, C.A. (1964). Paper electrophoresis of rat salivary secretions. *Proc. Soc. Exp. Biol. Med.* 1001–1005.

Halliday, G., Harding, A., and Paxinos, G. (1995). Serotonin and tachykinin systems. *In* "The Rat Nervous System." (G. Paxinos, ed.), Academic Press, San Diego.

Hamilton, A.I., and Blackwood, H.J. (1974). Cell renewal of oral mucosal epithelium of the rat. *J. Anat.* **117**, 313–327.

Hand, A.R., Pathmanathan, D., and Field, R.B. (1999). Morphological features of the minor salivary glands. *Arch. Oral Biol.* **44 Suppl 1**, S3–S10.

Harada, S., Yamaguchi, K., Kanemaru, N., and Kasahara, Y. (2000). Maturation of taste buds on the soft palate of the postnatal rat. *Physiol. Behav.* **68**, 333–339.

Harrison, G.A. (1974). Ultrastructural response of rat lung to 90 days' exposure to oxygen at 450 mm Hg. *Aerospace Med.* **45**, 1041–1045.

Hatakeyama, S., Sashima, M., and Suzuki, A. (1987). A sexual dimorphism of mucous cells in the submandibular salivary gland of rat. *Arch. Oral Biol.* **32**, 689–693.

Heap, B.J., and Kiernan, J.A. (1973). Histological, histochemical and pharmacological observations on mast cells in the stomach of the rat. *J. Anat.* **115**, 315–325.

Hebb, C. (1969). Motor innervation of the pulmonary blood vessels of mammals. *In* "The Pulmonary Circulation and Interstitial Space." (A.P. Fishman and H.H. Hecht, eds.), pp 195–222. University of Chicago Press, Chicago.

Heffner, H.E., Heffner, R.S., Contos, C., and Ott, T. (1994). Audiogram of the hooded Norway rat. *Hear. Res.* **73**, 244–247.

Heidbreder, C.A., and Groenewegen, H.J. (2003). The medial prefrontal cortex in the rat: evidence for a dorso-ventral distinction based upon functional and anatomical characteristics. *Neurosci. Biobehav. Rev.* **27**, 555–579.

Herness, S., Zhao, F.L., Kaya, N., Lu, S.G., Shen, T., and Sun, X.D. (2002a). Adrenergic signalling between rat taste receptor cells. *J. Physiol.* **543**, 601–614.

Herness, S, Zhao, F.L., Lu, S.G., Kaya, N., and Shen, T. (2002b). Expression and physiological actions of cholecystokinin in rat taste receptor cells. *J. Neurosci.* **22**, 10018–10029.

Herrero, M.C., Remesar, X., Blade, C., and Arola, L. (1997). Amino acid metabolism in the kidneys of genetic and nutritionally obese rats. *Biochem. Mol. Biol. Int* **42**, 261–269.

Hess, SY, and Zimmermann, M.B. (2004). The effect of micronutrient deficiencies on iodine nutrition and thyroid metabolism. *Int. J. Vitam. Nutr. Res.* **74**, 103–115.

Heywood, R. (1973). Some clinical observations on the eyes of Sprague-Dawley rats. *Lab. Anim.* **7**, 19–27.

Hughes, P.C., and Tanner, J.M. (1970a). A longitudinal study of the growth of the black-hooded rat: methods of measurement and rates of growth for skull, limbs, pelvis, nose-rump and tail lengths. *J. Anat.* **106**, 349–370.

Hughes, P.C., and Tanner, J.M. (1970b). The assessment of skeletal maturity in the growing rat. *J. Anat.* **106**, 371–402.

Huntington, G.S., and Schulte H., v.W. (1913). "Studies in Cancer and Allied Subjects." pp 315–324, Columbia University Press, New York.

Icardo, J.M. (1996). Developmental biology of the vertebrate heart. *J. Exp. Zool.* **275**, 144–161.

Iida, M.I.Y., Yoshioka, I., and Muto, H. (1983). Taste bud papillae on the retromolar mucosa of the rat, mouse and golden hamster. *Acta Anat. (Basel)* **117**, 374–381.

Inomata, T., Eguchi, Y., and Nakamura, T. (1985). Development of the external genitalia in rat fetuses. *Jikken Dobutsu* **34**, 439–444.

Ito, K., Morikawa, M., and Inenaga, K. (2001). The effect of food consistency and dehydration on reflex parotid and submandibular salivary secretion in conscious rats. *Arch. Oral Biol.* **46**, 353–363.

Iwasaki, S., Yoshizawa, H., and Kawahara, I. (1997). Study by scanning electron microscopy of the morphogenesis of three types of lingual papilla in the rat. *Anat. Rec.* **247**, 528–541.

Jacobs, G.H., Fenwick, J.A., and Williams, G.A. (2001). Cone-based vision of rats for ultraviolet and visible lights. *J. Exp. Biol.* **204**, 2439–2446.

Jayaraj, A.P. (1972). Hair growth in oral cavity. *Acta Anatomica* **83**, 367–371.

Kallskog, O., Lindbom, L.O., Ulfendahl, H.R., and Wolgast, M. (1975). Kinetics of the glomerular ultrafiltration in the rat kidney. An experimental study. *Acta Physiol. Scand.* **95**, 293–300.

Karim, B.O., Landolfi, J.A., Christian, A., Ricart-Arbona, R., Qiu, W., McAlonis, M., Eyabi, P.O., Khan, K.A., Dicello, J.F., Mann, J.F., and Huso, D.L. (2003). Estrous cycle and ovarian changes in a rat mammary carcinogenesis model after irradiation, tamoxifen chemoprevention, and aging. *Comp. Med.* **53**, 532–538.

Kasiske, B.L., and Keane, W.F. (2000). Laboratory assessment of renal disease: clearance, urinalysis, and renal biopsy. *In* "Brenner & Rector's The Kidney." (B.M. Brenner, ed.), W.B. Saunders, Philadelphia, Pa.

Keithley, E.M., and Feldman, M.L. (1982). Hair cell counts in an age-graded series of rat cochleas. *Hear. Res.* **8**, 249–262.

Kelly, J.B., and Masterton, B. (1977). Auditory sensitivity of the albino rat. *J. Comp. Physiol. Psychol.* **91**, 930–936.

Keverne, E.B. (1999). The vomeronasal organ. *Science* **286**, 716–720.

Kikkawa, Y. (1970). Morphology of alveolar lining layer. *Anat. Rec.* **167**, 389–400.

Kirkeby, O.J. (1991). Bone metabolism and repair are normal in athymic rats. *Acta Orthop. Scand.* **62**, 253–256.

Kobzik, L., Reid, M.B., Bredt, D.S., and Stamler, J.S. (1994). Nitric oxide in skeletal muscle. *Nature* **372**, 546–548.

Koch, L.G., Britton, S.L., Barbato, J.C., Rodenbaugh, D.W., and DiCarlo, S.E. (1999). Phenotypic differences in cardiovascular regulation in inbred rat models of aerobic capacity. *Physiol Genomics* **1**, 63–69.

Komárek, V., Gembart, C., Krinke, A., Mahrous, T.A., and Schaetti, P. (2000). Synopsis of the organ anatomy. *In* "The Laboratory Rat." (G.J. Krinke, ed.), pp 283–319, Academic Press, London.

Kontiola, A.I., Goldblum, D., Mittag, T., and Danias, J. (2001). The induction/impact tonometer: a new instrument to measure intraocular pressure in the rat. *Exp. Eye Res.* **73**, 781–785.

Koolhass, J.M. (2003). The laboratory rat. *In* "The UFAW Handbook on the Care and Management of Laboratory Animals," pp 313–330. Blackwell Science Ltd., Malden, Mass.

Kuwahira, I., Moue, Y., Ohta, Y., Mori, H., and Gonzalez, N.C. (1994). Distribution of pulmonary blood flow in conscious resting rats. *Respir. Physiol.* **97**, 309–321.

Lee, M.M., Chu, P.C., and Chan, H.C. (1969). Effects of cold on the skeletal growth of albino rats. *Am. J. Anat.* **124**, 239–249.

Lindenlaub, T., and Oelschläger, H.A. (1999). Morphological, morphometric, and functional differences in the vestibular organ of different breeds of the rat (Rattus norvegicus). *Anat. Rec.* **255**, 15–19.

Loeb, W.F., and Quimby, F. (1999). "The Clinical Chemistry of Laboratory Animals." Taylor & Francis, London.

Lohr, J., Mazurchuk, R.J., Acara, M.A., Nickerson, P.A., and Fiel, R.J. (1991). Magnetic resonance imaging (MRI) and pathophysiology of the rat kidney in streptozotocin-induced diabetes. *Magn. Res. Imaging* **9**, 93–100.

London, D.A., Davis, P.L., Williams, R.D., Crooks, L.E., Sheldon, P.E., and Gooding, C.A. (1983). Nuclear magnetic resonance imaging of induced renal lesions. *Radiology* **148**, 167–172.

Loo, A.T., Youngentob, S.L., Kent, P.F., and Schwob, J.E. (1996). The aging olfactory epithelium: neurogenesis, response to damage, and odorant-induced activity. *Int. J. Dev. Neurosci.* **14**, 881–900.

Loughlin, S.E., Leslie, F.M., and Fallon, J.H. (1995). Endogenous opioid systems. *In* "The Rat Nervous System." (G. Paxinos, ed.), Academic Press, San Diego.

Luciano, L., Reale, E., and Ruska, H. (1968). [On a glycogen containing brush cell in the rectum of the rat]. [German]. *Zeitschrift fur Zellforschung und Mikroskopische Anatomie* 91, 153–158.

Macintosh, S.R. (1975). Observations on the structural and innervation of the rat snout. *J. Anat.* 119, 537–546.

Makara, G.B., Mergl, Z., and Zelena, D. (2004). The role of vasopressin in hypothalamo-pituitary-adrenal axis activation during stress: an assessment of the evidence. *Ann. N. Y. Acad. Sci.* 1018, 151–161.

Masilamani, S., Knepper, M.A., and Burg, M.B. (2000). Urine concentration and dilution. *In* "Brenner & Rector's The Kidney." (B.M. Brenner, ed.), W. B. Saunders, Philadelphia, Pa.

Mercurio, A.R., and Mitchell, O.G. (1987). Submandibular duct salivary bladder of the rat. *J. Morphol.* 192, 247–256.

Miller, I.J., Jr., and Preslar, A.J. (1975). Spatial distribution of rat fungiform papillae. *Anat. Rec.* 181, 679–684.

Miller, I.J., Jr., and Spangler, K.M. (1982). Taste bud distribution and innervation on the palate of the rat. *Chem. Senses* 7, 99–108.

Mintz, G.A., and Mooradian, A.D. (1987). Age-related changes in rat sublingual salivary gland morphology. *Gerodontology* 6, 137–144.

Mizoguchi, K., Yuzurihara, M., Ishige, A., Sasaki, H., Chui, D.H., and Tabira, T. (2001). Chronic stress differentially regulates glucocorticoid negative feedback response in rats. *Psychoneuroendocrinology* 26, 443–459.

Moe, H., and Bojsen-Moller, F. (1971). The fine structure of the lateral nasal gland (Steno's gland) of the rat. *J. Ultrastructure Res.* 36, 127–148.

Mohammed, M.B.H., and El-Sayad, F. I. (1985). Time and order of appearance of ossification centres of the rat skeleton. *Qatar Univ. Sci. Bull.* 5, 255-266.

Moore, D.R., Rothholtz, V., and King, A.J. (2001). Hearing: cortical activation does matter. *Curr. Biol.* 11, R782—R784.

Morroni, M., Barbatelli, G., Zingaretti, M.C., and Cinti, S. (1995). Immunohistochemical, ultrastructural and morphometric evidence for brown adipose-tissue recruitment due to cold-acclimation in old rats. *Int. J. Obes. Relat. Metab. Disord.* 19, 126–131.

Morsing, P., Persson, A.E., and Boberg, U. (1988). A micropuncture assessment of the effects of contrast media of different osmolality. *Invest. Radiol.* 23, 767–771.

Muller, H.B. (1969). [The postnatal development of the rat harderian gland. I. Light microscopic observations]. [German]. *Zeitschrift fur Zellforschung und Mikroskopische Anatomie* 100, 421–438.

Mulrow, P.J. (1986). "The Adrenal Gland." Elsevier, New York.

Nagato, T., Ren, X.Z., Toh, H., and Tandler, B. (1997). Ultrastructure of Weber's salivary glands of the root of the tongue in the rat. *Anat. Rec.* 249, 435–440.

Nakayama, M., Helfert, R.H., Konrad, H.R., and Caspary, D.M. (1994). Scanning electron microscopic evaluation of age-related changes in the rat vestibular epithelium. *Otolaryngol. Head Neck Surg.* 111, 799–806.

Norrby, A. (1958). On the growth of the crystalline lens, the eyeball and cornea in the rat. *Acta Ophthalmol (Copenh)* 49, 5–5.

Obineche, E.N., Mensah-Brown, E., Chandranath, S.I., Ahmed, I., Naseer, O., and Adem, A. (2001). Morphological changes in the rat kidney following long-term diabetes. *Arch. Physiol. Biochem.* 109, 245.

Ohkura, S., Tsukamura, H., and Maeda, K. (2000). Endocrinology. *In* "The Laboratory Rat." (G.J. Krinke, ed.), pp 283–319, Academic Press, London.

Oldfield, B.J., and McKinley, M.J. (1995). Circumventricular organs. *In* "The Rat Nervous System." (G. Paxinos, ed.), Academic Press, San Diego.

Onat, F., and Cavdar, S. (2004). Cerebellar connections: hypothalamus. *Cerebellum* 2, 263–269.

Orlov, S.N., Adragna, N.C., Adarichev, V.A., and Hamet, P. (1999). Genetic and biochemical determinants of abnormal monovalent ion transport in primary hypertension. *Am. J. Physiol.* 276, C511–C536.

Pacha, J. (2000). Development of intestinal transport function in mammals. *Physiol. Rev.* 80, 1633–1667.

Pandit, S.V., Clark, R.B., Giles, W.R., and Demir, S.S. (2001). A mathematical model of action potential heterogeneity in adult rat left ventricular myocytes. *Biophysical J.* 81, 3029–3051.

Pannabecker, T.L., Abbott, D.E., and Dantzler, W.H. (2004). Three-dimensional functional reconstruction of inner medullary thin limbs of Henle's loop. *Am. J. Physiol. Renal Physiol.* 286, F38–F45.

Papa, S. (1996). Mitochondrial oxidative phosphorylation changes in the life span. Molecular aspects and physiopathological implications. *Biochim. Biophys. Acta* 1276, 87–105.

Pazirandeh, A., Jondal, M., and Okret, S. (2004). Glucocorticoids delay age-associated thymic involution through directly affecting the thymocytes. *Endocrinology* 145, 2392–2401.

Pelleymounter, M.A., Cullen, M.J., Baker, M.B., Hecht, R., Winters, D., Boone, T., and Collins, F. (1995). Effects of the obese gene product on body weight regulation in ob/ob mice. *Science* 269, 540–543.

Permezel, N.C., and Webling, D.D. (1971). The length and mucosal surface area of the small and large gut in young rats. *J. Anat* 108, 295–296.

Pilowsky, P.M., and Goodchild, A.K. (2002). Baroreceptor reflex pathways and neurotransmitters: 10 years on. *J. Hypertens* 20, 1675–1688.

Pinkerton, K.E., Barry, B.E., O'Neil, J.J., Raub, J.A., Pratt, P.C., and Crapo, J.D. (1982). Morphologic changes in the lung during the lifespan of Fischer 344 rats. *Am. J. Anat.* 164, 155–174.

Pinto, Y.M., Paul, M., and Ganten, D. (1998). Lessons from rat models of hypertension: from Goldblatt to genetic engineering. *Cardiovasc. Res.* 39, 77–88.

Plecas-Solarovic, B., Lalic, L., and Leposavic, G. (2004). Age-dependent morphometrical changes in the thymus of male propranolol-treated rats. *Ann. Anat.* 186, 141–147.

Popesko, P., Rajitova, V., and Horak, J. (1990). "A Colour Atlas of Anatomy of Small Laboratory Animals, Volume Two: Rat, Mouse, Hamster." W.B. Saunders, Philadelphia, Pa.

Postle, A.D., Heeley, E.L., and Wilton, D.C. (2001). A comparison of the molecular species compositions of mammalian lung surfactant phospholipids. *Comp. Biochem. Physiol. A. Mol. Integr. Physiol.* 129, 65–73.

Pravenec, M, and Kurtz, T.W. (2002). Genetics of Cd36 and the hypertension metabolic syndrome. *Semin. Nephrol.* 22, 148–153.

Prydal, J.I., and Campbell, F.W. (1992). Study of precorneal tear film thickness and structure by interferometry and confocal microscopy. *Invest. Ophthalmol. Vis. Sci.* 33, 1996–2005.

Puche, R.C., Alloatti, R., and Ledesma, S. (1988). Growth and development of the bone mass of two strains of inbred rats. *Bone Miner.* 4, 341–353.

Rakshit, P. (1949). Communicating blood vessels between bronchial and pulmonary circulations in the guinea pig and rat. *Q. J. Exp. Physiol. Cogn. Med. Sci.* 47, 53–.

Reiter, R.J. (1991). Neuroendocrine effects of light. *Int. J. Biometeorol.* 35, 169–175.

Rogers, D.F. (1994). Airway goblet cells: responsive and adaptable front-line defenders. *Eur. Respir. J.* 7, 1690–1706.

Rooyackers, O.E., Adey, D.B., Ades, P.A., and Nair, K.S. (1996). Effect of age on in vivo rates of mitochondrial protein synthesis in human skeletal muscle. *Proc. Natl. Acad. Sci. U. S. A.* 93, 15364–15369.

Ruttimann, G., and Daicker, B. (1989). [Retrolental cloudiness of the vitreous humor in the rat eye]. *Zentralbl. Veterinarmed. A.* 36, 340–347.

Rybalko, N., and Syka, J. (2001). Susceptibility to noise exposure during postnatal development in rats. *Hear. Res.* 155, 32–40.

Salvatori, G., Lattanzi, L., Coletta, M., et al. (1995). Myogenic conversion of mammalian fibroblasts induced by differentiating muscle cells. *J. Cell Sci.* **108 (Pt 8)**, 2733–2739.

Sasaki, S., Kobayashi, N., Dambara, T., Kira, S., and Sakai, T. (1995). Structural organization of pulmonary arteries in the rat lung. *Anat. Embryol. (Berl)* **191**, 477–489.

Sbarbati, A., Crescimanno, C., and Osculati, F. (1999). The anatomy and functional role of the circumvallate papilla/von Ebner gland complex. *Med. Hypotheses* **53**, 40–44.

Schittny, J.C., Djonov, V., Fine, A., and Burri, P.H. (1998). Programmed cell death contributes to postnatal lung development. *Am. J. Respir. Cell Mol. Biol.* **18**, 786–793.

Schmidt-Nielsen, B., Churchill, M., and Reinking, L.N. (1980). Occurrence of renal pelvic refluxes during rising urine flow rate in rats and hamsters. *Kidney Int.* **18**, 419–431.

Schraufnagel, D.E., and Patel, K.R. (1990). Sphincters in pulmonary veins. An anatomic study in rats. *Am. Rev. Respir. Dis.* **141**, 721–726.

Schwartzkroin, P.A., Roper, S.N., and Wenzel, H.J. (2004). Cortical dysplasia and epilepsy: animal models. *Adv. Exp. Med. Biol.* **548**, 145–174.

Scremin, O.U. (1995). Cerebral vascular system. *In* "The Rat Nervous System." (G. Paxinos, ed.), Academic Press, San Diego.

Senoo, H. (2000). Digestion, metabolism. *In* "The Laboratory Rat." (G.J. Krinke, ed.), pp 359–383, Academic Press, London.

Shahtaheri, S.M., Aaron, J.E., Johnson, D.R., and Paxton, S.K. (1999). The impact of mammalian reproduction on cancellous bone architecture. *J. Anat.* **194 (Pt 3)**, 407–421.

Sharp, P.E., and LaRegina, M.C. (1998). "The Laboratory Rat." CRC Press, Boca Raton, Fl.

Shipley, M.T., McLean, J.H., and Ennis, M. (1995). Olfactory system. *In* "The Rat Nervous System." (G. Paxinos, ed.), Academic Press, San Diego.

Short, K.R., and Nair, K.S. (2001). Does aging adversely affect muscle mitochondrial function? *Exerc. Sport Sci. Rev.* **29**, 118–123.

Singh, S.K., Dhall, U., and Dhall, J.C. (1989). Age-related changes in rat submandibular salivary gland: a stereological study. *J. Anat. Soc. (India)* **38**, 22–28.

Sivak, J.G., and Dovrat, A. (1984). Early postnatal development of the rat lens. *Exp. Biol.* **43**, 57–65.

Slavin, B.G., and Bernick, S. (1974). Morphological studies on denervated brown adipose tissue. *Anat. Rec.* **179**, 497–506.

Smith E.M., and Calhoun, M.L. (1968). "The Microscopic Anatomy of the White Rat." Iowa State University Press, Ames, Iowa.

Sofikitis, N, Takahashi, C., Kadowaki, H., Okazaki, T., Shimamoto, T., Nakamura, I., and Miyagawa, I. (1992). The role of the seminal vesicles and coagulating glands in fertilization in the rat. *Int J Androl* **15**, 54–61.

Souma, T. (1987). The distribution and surface ultrastructure of airway epithelial cells in the rat lung: a scanning electron microscopic study. *Arch. Histol. Jpn.* **50**, 419–436.

Stendell, W. (1913). Zur vergleichenden Anatomie und Histologie der Hypophysis cerebri. *Arch. Mikrosk. Anat. Entwicklungsmech.* **82**, 289–332.

Stone, K.C., Mercer, R.R., Freeman, B.A., Chang, L.Y., and Crapo, J.D. (1992). Distribution of lung cell numbers and volumes between alveolar and nonalveolar tissue. *Am. Rev. Respir. Dis.* **146**, 454–456.

Strong, R.M. (1926). The order, time and rate of ossification of the albino rat skeleton. *Am. J. Anat.* 313–355.

Stumpel, F., Scholtka, B., and Jungermann, K. (2000). Stimulation by portal insulin of intestinal glucose absorption via hepatoenteral nerves and prostaglandin E2 in the isolated, jointly perfused small intestine and liver of the rat. *Ann. N. Y. Acad. Sci.* **915**, 111–116.

Sugiyama, K., Gu, Z.B., Kawase, C., Yamamoto, T., and Kitazawa, Y. (1999). Optic nerve and peripapillary choroidal microvasculature of the rat eye. *Invest. Ophthalmol. Vis. Sci.* **40**, 3084–3090.

Tabrizchi, R., and Pugsley, M.K. (2000). Methods of blood flow measurement in the arterial circulatory system. *J. Pharmacol. Toxicol. Methods* **44**, 375–384.

Takino, M., and Watanabe, W. (1937). Uber das Vovkommen der Ganglienzellen von unipolaren Typus in der Lunge des Mehschen und Schweiner. *Acta Sch. Med. Univ. Imp. Kioto* 317–320.

Talwar, S.K., Musial, P.G., and Gerstein, G.L. (2001). Role of mammalian auditory cortex in the perception of elementary sound properties. *J. Neurophysiol.* **85**, 2350–2358.

Tenney, S.M., Remmers, J.E. (1963). Comparative quantitative morphology of mammalian lungs. *Nature* 54–56.

Thomson, A.B.R., Keelan, M., Thiesen, A., Clandinin, M.T., Ropeleski, M., and Wild, G.E. (2001). Small bowel review—Normal physiology part 1. *Dig. Dis. Sci.* **46**, 2567–2587.

Timm, K.I. (1979). Orbital venous anatomy of the rat. *Lab. Anim. Sci.* **29**, 636–638.

Tisher, C.C., and Madsen, K.M. (2000). Anatomy of the kidney. *In* "Brenner & Rector's The Kidney." (B.M. Brenner, ed.), W. B. Saunders, Philadelphia.

Tobian, L., and Nason, P. (1966). The augmentation of proteinuria after acute sodium depletion in the rat. *J. Lab. Clin. Med.* **67**, 224–228.

Travers, J.B. (1995). Oromotor nuclei. *In* "The Rat Nervous System." (G. Paxinos, ed.), Academic Press, San Diego.

Travers, S.P., and Nicklas, K. (1990). Taste bud distribution in the rat pharynx and larynx. *Anat. Rec.* **227**, 373–379.

Trevisan, A., Cristofori, P., and Fanelli, G. (1999). Glutamine synthetase activity in rat urine as sensitive marker to detect S3 segment-specific injury of proximal tubule induced by xenobiotics. *Arch. Toxicol.* **73**, 255–262.

Tschumperlin, D.J., and Margulies, S.S. (1999). Alveolar epithelial surface area-volume relationship in isolated rat lungs. *J. Appl. Physiol.* **86**, 2026–2033.

Turner, C.H., Roeder, R.K., Wieczorek, A., Foroud, T., Liu, G., and Peacock, M. (2001). Variability in skeletal mass, structure, and biomechanical properties among inbred strains of rats. *J. Bone Miner. Res.* **16**, 1532–1539.

Vajda, E.G., Wronski, T.J., Halloran, B.P., Bachus, K.N., and Miller, S.C. (2001). Spaceflight alters bone mechanics and modeling drifts in growing rats. *Aviat. Space Environ. Med.* **72**, 720–726.

Valerius, K.P. (1996). Size-dependent morphology of the conductive bronchial tree in four species of myomorph rodents. *J. Morphol.* **230**, 291–297.

Vallotton, M.B., Capponi, A.M., Johnson, E.I., and Lang, U. (1990). Mode of action of angiotensin II and vasopressin on their target cells. *Horm. Res.* **34**, 105–110.

Vanderschueren, D. (1996). Androgens and their role in skeletal homeostasis. *Horm. Res.* **46**, 95–98.

Verloop, M. (1948). The arterial bronchioles and their anastomoses with the arteria pulmonalis in the human lung. *Acta Anatomica* 171–205.

Vidal, P.P., Graf, W., and Berthoz, A. (1986). The orientation of the cervical vertebral column in unrestrained awake animals. I. Resting position. *Exp. Brain Res.* **61**, 549–559.

Voogd, J. (1995). Cerebellum. *In* "The Rat Nervous System." (G. Paxinos, ed.), Academic Press, San Diego.

Watanabe, H., Tisdale, A.S., and Gipson, I.K. (1993). Eyelid opening induces expression of a glycocalyx glycoprotein of rat ocular surface epithelium. *Invest. Ophthalmol. Vis. Sci.* **34**, 327–338.

Weinstein, A.M. (1998). A mathematical model of the inner medullary collecting duct of the rat: pathways for Na and K transport. *Am. J. Physiol.* **274**, F841—F855.

Wilson, D.A., and Sullivan, R.M. (1999). Respiratory airflow pattern at the rat's snout and an hypothesis regarding its role in olfaction. *Physiol. Behav.* **66**, 41–44.

Wolthers, T., Grofte, T., Flyvbjerg, A., Frystyk, J., Vilstrup, H., Orskov, H., and Foegh, M. (1994). Dose-dependent stimulation of insulin-like growth factor-binding protein-1 by lanreotide, a somatostatin analog. *J. Clin. Endocrinol. Metab.* **78,** 141–144.

Wong, D.L., Tai, T.C., Wong-Faull, D.C., Claycomb, R., and Kvetnansky, R. (2004). Genetic mechanisms for adrenergic control during stress. *Ann. N. Y. Acad. Sci.* **1018,** 387–397.

Wronski, T.J., Morey-Holton, E., and Jee, W.S. (1981). Skeletal alterations in rats during space flight. *Adv. Space Res.* **1,** 135–140.

Wu, W.J., Sha, S.H., McLaren, J.D., Kawamoto, K., Raphael, Y., and Schacht, J. (2001). Aminoglycoside ototoxicity in adult CBA, C57BL and BALB mice and the Sprague-Dawley rat. *Hear. Res.* **158,** 165–178.

Yamamoto, M., Seedor, J.G., Rodan, G.A., and Balena, R. (1995). Endogenous calcitonin attenuates parathyroid hormone-induced cancellous bone loss in the rat. *Endocrinology* **136,** 788–795.

Yang, D., Ye, Q., Williams, D.S., Hitchens, T.K., and Ho, C. (2004). Normal and transplanted rat kidneys: diffusion MR imaging at 7 T. *Radiology* **231,** 702–709.

Zabel, M., and Schieber T.H. (1980). Histochemical, autoradiographic and electron microscopic investigations of the renal proximal tubule of male and female rats after castration. *Histochemistry* **69,** 276.

Zeni, S.N., Di Gregorio, S., and Mautalen, C. (1999). Bone mass changes during pregnancy and lactation in the rat. *Bone* **25,** 681–685.

Zerath, E., Holy, X., Roberts, S.G., Andre, C., Renault, S., Hott, M., and Marie, P.J. (2000). Spaceflight inhibits bone formation independent of corticosteroid status in growing rats. *J. Bone Miner. Res.* **15,** 1310–1320.

Ziv, E., and Bendayan, M. (2000). Intestinal absorption of peptides through the enterocytes. *Microsc. Res. Tech.* **49,** 346–352.

Chapter 5

Clinical Pathology of the Rat

Bruce D. Car, Vicki M. Eng, Nancy E. Everds, and Denise I. Bounous

I. INTRODUCTION

An extensive literature and experience base surrounds the clinical pathology of the laboratory rat, developed largely through the use of the rat in pharmaceutical and chemical industries and university research. The rat is the most widely utilized species for determining the potential toxicity of xenobiotics. A wide variety of readily measurable analytes combined with an in-depth understanding of the pathophysiology of the rat permits insightful interpretation of alterations in hematology, clinical chemistry, special chemistries, endocrinology, urinalysis, and blood coagulation. Although limited in comparison to the mouse, the investigation of transgenic and spontaneous genetic mutations of the rat also have contributed to the discipline of comparative pathophysiology. The relationship of alterations in clinical pathology parameters in the rat to human pathophysiology has been extensively explored; the rat manifests alterations both relevant to human pathophysiology and unique to itself.

General considerations relating to effects of environment, conditions of collection, storage, and quality control will be included in each section, as appropriate. Approaches to the interpretation of clinical pathology datasets, including use of individual animal controls, reference intervals, concurrent control groups, and pattern recognition will be discussed. The combined evaluation of alterations in clinical biochemistry, special chemistries, hematology, blood coagulation, and urinalysis of the rat are central to the understanding of the pharmacologic and toxicologic effects of xenobiotics. Considerable experience in the pharmaceutical and chemical industries has led to a significant enhancement in the ability of scientists to interpret the clinical pathology data of rats and use these interpretations in the risk assessment of novel chemicals.

The application of these skills to evaluation of rodent pathophysiology is the phenotypes of rat transgenics are key to providing greater insights in these parallel disciplines. Toxicologic and pharmacologic evaluations in rats are increasingly accompanied by systems biology datasets that include the products of metabonomic, proteomic, and transcriptomic evaluations. A thorough understanding of the interpretation of traditional clinical pathology parameters is required to support these more exploratory analyses.

Given the limited scope of this chapter, an extensive literature review on the background of specific clinical pathology analytes will not be provided. These may be obtained in more specialized publications. In general, responses of the rat are similar to those of other species. Changes unique to the rat and its responses to toxic xenobiotics will be highlighted. Readers requiring further specific information concerning rodent clinical chemistry and the interpretation of clinical pathology data are directed to Clinical Chemistry of Domestic Animals (Kaneko et al., 1997), Schalm's Hematology of Domestic Animals (Jain, 1986; Moore, 2000) and Clinical Pathology of Laboratory Animals (Loeb and Quimby, 1999). The mean historic values for a variety of commonly used rat strains are given in Chapter 5 of the previous edition of this text (Ringler and Dabich, 1979). Values presented in this chapter will be those obtained from the outbred Sprague-Dawley (SD) rat, as it is the most widely used research strain, followed by Wistar and inbred F344 rats. Tables in this chapter represent historic values obtained from International Gold Standard, diet-restricted (approximately 25% less than *ad libitum* feeding), SD rats with data generated by the following instruments: Roche Hitachi 917 Chemistry Analyzer, Bayer Advia hematology analyzer, Bayer Atlas urinalysis analyzer and BCG Coagulation analyzer. In-house, instrument-specific and strain-specific analyte ranges always should be used in preference to historic information. Comparisons to other strains, transgenic rats, humans, and mouse will be drawn as appropriate.

This chapter provides a basis for how clinical pathology parameters are used to correctly diagnose toxicity and naturally occurring disease in the rat.

II. CONSIDERATIONS OF BLOOD, URINE COLLECTION AND SAMPLE HANDLING

A. Collection of Blood

To ensure the correct values are obtained for clinical pathology analytes, appropriate quality control of sample collection handling is required. Overnight fasting, consistent sample collection times, and attention to the quality of procedure for blood collection are all critical to the integrity of clinical pathology datasets. This is particularly important for small animals such as the rat, for which venipuncture of smaller vessels is technically more demanding, and the collection of urine in metabolism cages is fraught with potential artifact. For consistency, samples should be collected from the same anatomical site, using the same restraint or anesthetic, and following the same procedure for processing. Values from abnormal animals should be compared to adequately sized control groups. If at all possible, single animals from different groups should be selected on a rotating basis for sample collection. A progressively increasing serum Cl^- concentration is noted when large numbers of samples are processed for clinical chemistry, owing to a slight increase in resistance in the Cl^- ion-specific electrode. If samples are run from control through high dose in order, an apparent but false drug-related increase in Cl^- concentration may be observed. Periodic insertion of quality control sera should be included through long sample runs, in addition to at the beginning and ends of sample runs.

The preferred sites for blood collection in rats are venipuncture (jugular vein, tail vein, saphenous vein, abdominal aorta, or vena cava at necropsy), cardiac puncture, or paraorbital sinus puncture under appropriate anesthesia. Subtle differences in analyte values [e.g., white (WBC) and red blood cell (RBC) counts, phosphate, bicarbonate, glucose, CK, AST, K^+] may occur following different routes of collection (Mahl et al., 2000; Katein et al., 2002). Generally, the small differences that occur from differing routes of collection do not interfere with the determination of significant alterations in clinical pathology analytes. Puncture by needle or collection device must be clean to avoid vascular damage or introduction of tissue into collected blood, which may activate coagulation, and

to prevent hemolysis of erythrocytes. For hematology, blood is collected using K_2EDTA as the anticoagulant. For the preparation of plasma, blood is collected into appropriately anticoagulated tubes (including but not limited to sodium citrate, K_2EDTA, heparin, hirudin or other anticoagulants for other specialized analyses), and plasma is separated after centrifugation. Blood for serum is collected into serum tubes (without anticoagulant), allowed to clot, and serum is separated after centrifugation. Both serum and plasma are suitable substrates for clinical chemistry, with serum more widely used in North America and plasma in Japan and Europe. If EDTA is used to obtain plasma for serum chemistry, chelation of K+ and Ca++ will result in low concentrations of these analytes. When serum sodium levels well exceed those normal for a rat, sodium salts of citrate or heparin may have been inadvertently used to collect plasma for serum analysis.

B. Effects of Freezing or Hemolysis

Methods used for collection and processing of blood may result in hemolysis (rupture of red blood cells). Hemolysis is recognized by a reddish tint of plasma or serum, or increased mean cell hemoglobin concentration on hematology results. If hemolysis is substantial, the samples are probably not appropriate for analysis.

Freezing of serum or plasma may affect the determination of certain analytes. In addition, it is important to adequately mix previously frozen samples. Inadequately mixed samples frequently have much lower than expected levels of albumin and other analytes. Samples for hematology must not be frozen.

Table 5-1 lists analytes most seriously affected by significant hemolysis of blood or freezing of samples, using common methods of analysis. Analytes absent from this table are generally not seriously affected by freezing or hemolysis. However, it is important to realize that the effect of hemolysis or freezing may be method dependent. The application sheets provided for tests with the clinical pathology instruments in a given laboratory should be carefully reviewed. Instrument application sheets with indications regarding freezing and the stability of samples are applicable to human samples and are not necessarily correct for the rat. Hence, it is important for each laboratory to determine analyte stability under its own conditions of handling and projected storage.

C. Collection of Urine

The collection of rat urine is generally performed in metabolism cages. Metabolism cages should be scrupulously cleaned. In addition, rat urine allowed to stand for prolonged periods at room temperature may become more

TABLE 5-1

EFFECTS OF FREEZING AND HEMOLYSIS ON CLINICAL CHEMISTRY, HEMATOLOGY AND BLOOD COAGULATION

Clinical chemistry tests	Effect of hemolysis	Effect of freezing
AST	↑	none
ALT	↑	none
SDH	↑	↓
LDH	↑	↓
ALP	↓	none
CK	↑	↓
K$^+$	↑	none
Bilirubin	↑	none
Albumin	↑	none
Glucose	↑	none
Hematology tests		
MCH	↑	–
MCHC	↑	–
Platelets	↓	–
RBC	↓	–
PCV	↓	
CHCM*	none	–
Hemoglobin	none	–
Blood coagulation tests		
APTT	↑	↑ (slight over time)
PT	↑	↑ (slight over time)
Fibrinogen	↓	none
Clotting time	↑	↓ (slight over time)
Coagulation factors	↓	↓ (slight over time)

*Cellular hemoglobin concentration mean – A direct rather than calculated determination, the CHCM is numerically identical to MCHC in the absence of hemolysis. (Guder and Ross, 1986)

alkaline owing to bacterial metabolism. Collection by cystocentesis under anesthesia or at necropsy is also possible, producing the highest quality sample, but in limited volume. Urine can also be collected "free-catch" from rats. To do this, rats are picked up and held over a weigh boat to collect urine. To collect urine for metabonomic analysis, any bacterial contamination voids the validity of the sample. To this end, refrigerated collection of urine into containers spiked with sodium azide (final concentration of approximately 1% v/v) is recommended.

D. Estrous Cycle

When collection of analytes from female rats, such as sex steroid hormones, gonadotrophins, and some other endogenous metabolites, is to be undertaken, one must be cognizant of the phase within the 5-day estrous cycle during which samples are collected. This is generally achieved by vaginal cytologic evaluation for estrus, proestrus, and

diestrus, but may also be determined from endocrine data and vaginal histology, if available.

III. USE OF REFERENCE INTERVALS IN DATA INTERPRETATION

To interpret alterations in clinical pathology of the rat, reference must be made to either concurrent control groups or reference intervals. It is ideal to have access to both sets of information. Control groups must be of sufficient sample size and run concurrently with samples from diseased or treated rats. Control data are particularly useful for interpretation of data from rat strains for which limited reference intervals are available. When there are few or no control rats, or when only affected animals are studied, reference intervals are invaluable to data interpretation. Reference intervals should be annotated with age range, sex, strain of rat and should include analyses from 50 or more rats per analyte. Often, ranges are established by using either 95 percentile units or means ± 2 standard deviations for parameters with normal distribution.

Reference ranges should be used as a tool, but generally should not be used as the sole guide to determine if values are normal or abnormal, or if there are treatment-related effects. Robust changes in clinical pathology analytes occurring *within* the reference interval may still be consistent with significant treatment-related effects. Pairwise statistical tests also can be used in conjunction with reference ranges, to put analyte changes into perspective. However, a result should not necessarily be interpreted to mean there is a significant treatment-related effect just because it is statistically different from control. Likewise, the lack of statistical significance does not imply that there are no treatment-related effects, particularly when important effects are observed in individual rats. Determination of whether changes are related to treatment requires thorough evaluation of individual data points, in the context of other changes observed in the same animal and its group.

Although it is ideal to have control data from individual rats prior to testing, the small size of young rats (generally 5–8 weeks of age at study start) precludes collection of pretest blood in most cases. A new branch of clinical chemistry, *metabonomics,* has reached the point of maturity whereby urine may be collected from pretest rats and analyzed for hundreds to thousands of small molecular-weight endogenous constituents by nuclear magnetic resonance (NMR) or mass spectroscopy (MS). Models of normalcy using thousands of metabolites readily identify outlier rats prior to initiation of study without a requirement for invasive blood collection procedures.

IV. HEMATOLOGY, CLINICAL CHEMISTRY, URINALYSIS AND BLOOD COAGULATION

A. Hematology Tests

Hematology data are generally collected by highly automated instrumentation that differentiates and quantifies the formed elements in blood based on their size, internal structures, staining, and/or biochemical characteristics. Such analyses then use species-specific algorithms to classify cells according to type. The interested reader is directed to a recent review of normal rat hematology (Moore, 2000) in the 5th edition of Schalm's Veterinary Hematology. Confirmation of these measurements can be made by routine approaches, including manual counting of diluted specimens and enumeration of differentiated cells on Romanowsky-stained blood smears. The evaluation of morphologic alterations of leukocytes, such as toxic change of neutrophils, platelet aggregation, the morphology of erythrocyte, and bone marrow cytology must be performed by experienced individuals. Large unstained cells (LUCs) are recorded by instruments capable of performing automated differential counts, when leukocytes are observed but not classified by instrument software. In humans, these cells are occasionally shown to be leukemic blasts in circulation, although in blood collected from rats this is rarely the case. It is recommended that instrument-generated white cell counts be replaced with manual differential counts when LUCs exceed 4% of the automated count. When platelet aggregation is detected microscopically during evaluation of the blood smear, an instrument-generated platelet count using the same blood is likely to underestimate the actual count.

An evaluation of bone marrow cells is generally not necessary for the interpretation of peripheral blood changes. .However, it is important to prepare bone marrow smears at the time of sacrifice for potential future examination. Brushed bone marrow smears are generally made from the femoral marrow. An examination of bone marrow alterations may provide additional information to the interpretation made from peripheral blood analyses, particularly in cases of unexplained cytopenias or malignancies. When bone marrow evaluation is undertaken, the proportions of different hematopoietic precursors are quantified by cytologic evaluation of smears stained with a Romanowsky-type stain. Complementing this morphologic evaluation of bone marrow are histology and flow cytometric (FACS) analyses. Histology of hematopoietic tissues, including bone marrow, and extramedullary hematopoiesis in spleen and liver provides the best indication of cellularity (quantity of productive hematopoietic elements). In addition, bone marrow fibrosis (myelofibrosis) and necrosis of hematopoietic tissues is most easily demonstrated in formalin-fixed sections of bone marrow. FACS of rat bone marrow has been developed to differentiate hematopoietic lineages (Criswell et al., 1998; Saad et al., 2001), thereby rapidly providing quantitative information that would otherwise require many hours of microscopic cytologic evaluation.

B. Coagulation Tests

Routine coagulation tests including the prothrombin time (PT) and activated partial thromboplastin time (APTT) are determined on coagulation analyzers using citrated plasma. Ca^{++} is an essential cofactor for certain clotting enzymes in the coagulation cascade and is sequestered by the citrate in the anticoagulant. To initiate the clotting assays, Ca^{++} is spiked into plasma. The PT is an evaluation of extrinsic coagulation initiated by the interaction of tissue factor, which is present outside of the blood circulation, with Factor VII. When these components bind, Factor VII becomes activated (VIIa), activating factors X and IX, and ultimately prothrombin to thrombin, which cleaves fibrinogen to allow fibrin clot formation. The source of tissue factor most frequently used in the PT assay is from rabbit brain. The endpoint of routine coagulation assays is formation of a clot, which is measured either mechanically or optically. The PT most closely recapitulates coagulation as it occurs *in vivo*.

The APTT assay is initiated by activation of Factor XII or the contact system of plasma by activating agents, including kaolin and ellagic acid. The order of activation of clotting factors in the APTT is XII, XI, IX, X, then II. The APTT provides a convenient assessment for the activity of IX, VIII, and the common pathway factors (X and II). Factors XII and XI are largely redundant in coagulation; their deficiencies generally do not manifest in increased bleeding tendency.

C. Clinical Chemistry Tests

Routine clinical chemistry tests are run on automated clinical chemistry analyzers. These instruments use a variety of reagents to determine concentrations or activities of analytes in serum or plasma.

D. Urinalysis Tests

There are several different steps involved in urinalysis. Urine appearance and volume is recorded. Urine constituents are semiquantitatively measured using dipstick technology. Urine is centrifuged and examined microscopically for cells and crystals. Frozen urine may be stable

TABLE 5-2

LIFE-SPAN OF CELLS AND COMPONENTS IN PERIPHERAL BLOOD

Cells in peripheral blood	Life-span or half-life
Blood cells	
Mature erythrocyte	Life-span 42–65 days
Reticulocyte	Life-span 1–1.5 days
Neutrophil	Life-span 6 hours
Monocyte	Life-span ~18 hours*
Lymphocytes	Life-span 60–100 days
Eosinophil	Life-span 18–23 hours
Platelet	Life-span 3.7–8 days
Serum chemistry components	
Albumin	Half-life 2.5 days
Immunoglobulin G (IgG)	Half-life 5 days
ALT	Half-life 4.5–8 hours
AST	Half-life 1.6 hours
Liver ALP	Half-life 28 hours
Intestinal ALP	Half-life 3 minutes
CK	Half-life 34 minutes

(Allison, 1960; Boyd, 1983; Cohen, 1957; Foot, 1961; Friedel et al., 1976; Jain, 1986; Little et al., 1961; Loeffler et al., 1989; Hoffman et al., 1999; Kekki and Eisalo, 1964; Moore, 2000; Osim et al., 1999; Shi et al., 2001; van Furth, 1989)

*approximation only. Value is from the mouse.

for some parameters but is not acceptable for most microscopic analyses.

The lifespan of hematologic elements of blood and half-lives of commonly analyzed serum biochemistry analytes is provided in Table 5-2. The numbers collectively provide information that may be useful in understanding alterations in peripheral blood parameters.

Reference interval data for diet-restricted, SD rats are provided under hematology (Table 5-3), blood coagulation (Table 5-4), clinical chemistry (Table 5-5), and urinalysis (Table 5-6). Separate ranges should be established for older rats, different strains, sexes, and different instrumentation and for each unique laboratory environment.

V. INTERPRETATION OF DISEASE STATES IN THE RAT

A. Hematopoietic Neoplasia

The interested reader is directed to several detailed reviews of hematopoietic neoplasia, including lymphomas and leukemias in the rat (Sanderson and Philips, 1981;

TABLE 5-3

95% REFERENCE INTERVALS FOR HEMATOLOGIC PARAMETERS IN DIET-RESTRICTED 7–11 WEEK OLD SPRAGUE-DAWLEY RATS COLLECTED UNDER ISOFLURANE ANESTHESIA

Parameter	Unit	Male range (2.5–97.5%)	Mean (males)	Mean (females)
Red Blood Cells (RBC)	$\times 10^6$/uL	7.34–8.85	8.14	8.19
Hemoglobin (HGB)	g/dL	14.7–17.3	15.9	15.9
Hematocrit (HCT)	%	44.9–51.7	48.5	46.5
Mean Cell Volume (MCV)	fL	55.1–64.2	59.7	56.9
Mean Cell Hemoglobin (MCH)	pg	18.6–20.7	19.6	19.5
Mean Cell Hemoglobin Concentration (MCHC)	g/dL	31.3–34.4	32.8	34.3
Red Cell Distribution Width (RDW)	%	11.3–14.2	12.4	11.5
Absolute reticulocytes	$\times 10^6$/uL	0.114–0.399	0.236	0.195
Reticulocytes	%	1.3–4.94	2.81	2.28
Platelets	$\times 10^3$/uL	903–1594	1159	1146
White Blood Cells (WBC)	$\times 10^3$/uL	6.63–20.35	12.43	12.02
Neutrophils	$\times 10^3$/uL	0.37–2.63	0.95	0.72
Lymphocytes	$\times 10^3$/uL	6.10–18.45	10.85	10.79
Monocytes	$\times 10^3$/uL	0.04–0.50	0.20	0.16
Eosinophils	$\times 10^3$/uL	0.02–0.27	0.11	0.15
Basophils	$\times 10^3$/uL	0.01–0.12	0.05	0.05
Large unstained cells	$\times 10^3$/uL	0.04–0.35	0.14	0.12
Neutrophils	%	3.5–18.7	7.9	6.0
Lymphocytes	%	75.8–92.9	88.0	89.9
Monocytes	%	0.5–3.4	1.5	1.3
Eosinophils	%	0.0–1.9	0.9	1.4
Basophils	%	0.2–0.6	0.4	0.4
Large unstained cells	%	0.5–2.4	1.1	1.0

Data generated with a Bayer Advia hematology analyzer.

TABLE 5-4

95% Reference Intervals For Coagulation Parameters in 8–12 Week Old Sprague Dawley Rats Collected Under CO_2 Narcosis

Parameter	Unit	Male range (2.5–97.5%)	Mean (males)	Mean (females)
Prothrombin time (PT)	sec	13.6–16.6	15.5	15.5
Activated Partial Thromboplastin Time (APTT)	sec	10.4–16.3	13.5	13.0
Fibrinogen	mg/dL	210–267	237	195
Bleeding time	sec		120	

Data generated on a BCG Coagulation analyzer.

TABLE 5-5

95% Reference Intervals for Clinical Chemistry Parameters in 7–10 Week Old Sprague Dawley Rats Collected Under Isoflurane Anesthesia

Parameter	Unit	Male range (2.5–97.5%)	Mean (males)	Mean (females)
Aspartate aminotransferase (AST)	IU/L	77–157	111	115
Alanine aminotransferase (ALT)	IU/L	24–53	35	31
Alkaline phosphatase (ALP)	IU/L	132–312	210	155
γ-glutamyl transpeptidase (γGT)	IU/L	0–3	0	0
Bilirubin	mg/dL	0.10–0.21	0.14	0.17
Amylase	IU/L	766–1850	1252	771
Total bile acids	μmol/L	10–68	30	20
Non-esterified fatty acids*	meq/L	0.4–1.0	0.64	0.68
Free glycerol*	mg/dL	1.08–2.65	1.84	1.53
Haptoglobin	mg/dL	10–35	22.3	7.2
Gastrin	pg/mL	12.5–40.1	23.3	17.7
Urea Nitrogen	mg/dL	9–17	13	15
Corticosterone**	ng/dL	33–252	139	334
Creatinine	mg/dL	0.15–0.35	0.26	0.29
Cholesterol	mg/dL	47–88	65	83
Triglycerides	mg/dL	25–145	64	49
Glucose	mg/dL	82–187	140	126
Total Protein	g/dL	5.5–6.6	6.0	6.4
Albumin	g/dL	4.0–4.8	4.4	4.7
Globulin	g/dL	1.2–2.0	1.6	1.7
A:G ratio		2.1–3.8	2.8	2.9
Calcium	mg/dL	9.1–10.9	10.0	10.1
Na^+	mmol/L	138.3–146.9	142.2	142.3
K^+	mmol/L	4.31–5.53	4.84	4.44
Cl^-	mmol/L	100.4–107.5	103.6	105.4
HCO_3^-	mmol/L	17.7–25.7	21.0	20.8

*Rats 33 weeks of age; **2 PM sampling.
Data generated on a Roche Hitachi Hitachi 917 Chemistry Analyzer.

Frith, 1988; Ward et al., 1990; Stromberg, 1992). These lymphomas and leukemias have similar counterparts in humans (Ward et al., 1990), although currently little is known of the molecular basis of hematopoietic neoplasia in rats.

F344/N rats have a high incidence of large granular cell leukemia (LGL) in 24-month carcinogenicity studies. These lymphocytes are identified by the surface antigen OX-8 and contain prominent azurophilic granules. In nonlymphomatous rats, they may comprise up to 10% of circulating lymphocytes (Ward, et al., 1990). Affected rats frequently exhibit a Coombs'-positive, immune-mediated hemolytic anemia, with spherocytosis, reticulocytosis, anisocytosis, and polychromasia and show evidence of monocytic erythrophagocytosis in peripheral blood (Stromberg et al., 1983a; Stromberg et al., 1983b; Stromberg, 1985; MacKenzie and Alison, 1990; Stromberg, 1992). In addition, there is a significant neutrophilia with left shift, mild lymphopenia, and moderate to severe thrombocytopenia WBC counts ranging from $5.0–370 \times 10^3$ cells/μl.

TABLE 5-6

95% Reference Intervals For Urinalysis Parameters in 9–11 Week Old Sprague Dawley Rats from 16 hour Urine Collection in Metabolism Cages

Parameter	Unit	95% range (males)	Mean (males)	Mean (females)
Specific Gravity (SG)		1.016–1.066	1.036	1.034
pH		6.5–8.5	7.7	7.4
Volume (16 hours)	mL	3.0–18.0	9.3	8.5
Volume (24 hours)	mL	9–37	20	18
Urea Nitrogen	mg/24h	600–1920	1245	1156
Creatinine	mg/24h	32–96	61	44
Na^+	mmol/24h	28–190	96	77
K^+	mmol/24h	76–421	246	184
Cl^-	mmol/24h	43–286	148	118

Data generated on a Atlas urinalysis and Roche Hitachi Hitachi 917 Chemistry Analyzer (urine electrolytes).

Altered hemostasis is reflected by significantly prolonged prothrombin time, hypofibrinogenemia, and slightly increased to normal partial thromboplastin time (Stromberg et al., 1983a), likely a combined effect of hepatic dysfunction and consumptive coagulopathy. Characteristic changes in clinical chemistry include marked increases in conjugated (late) and unconjugated (early) serum bilirubin, alanine aminotransferase, aspartate aminotransferase, lactate dehydrogenase, and alkaline phosphatase. These changes are likely all secondary to hepatic infiltration in advanced leukemic disease. Spontaneous myelofibrosis and osteosclerosis are reported to occur in approximately one quarter of rats with LGL leukemia (Stromberg et al., 1983b).

SD rats less than 5 months of age infrequently develop widespread lymphoma. Lymphoblastic lymphoma is the most common hematopoietic neoplasm of SD rats (0.65% incidence), followed closely by an LGL leukemia (0.60%) similar to that of the F344 rat (Frith, 1988). Myelogenous leukemia is rare in rats (0.1–0.3% of SD rats; Frith, 1988), and is stained histochemically positive for chloracetate esterase and alkaline phosphatase, unlike in humans (Ward et al., 1990).

Usually diagnosed by histologic evaluation of lymphopoietic tissues, hematopoietic neoplasia also is readily identified by evaluation of Romanowsky-stained blood smears or bone marrow cytology. For the pharmaceutical industry, hematologic information generated by hematology analyzers is inconsistently obtained in carcinogenicity studies, although peripheral blood smears are frequently obtained at the end of study. Both sets of information are consistently obtained in chemical industry studies. For these chronic rat bioassays, naturally occurring lymphomas or leukemias should be differentiated from induced tumors (Ward et al., 1990). Antigenic markers, including the cluster differentiation antigens used to qualify hematopoietic neoplasia in the rat, are detailed in Ward et al. (1990).

B. Bone Marrow Toxicity

Following the administration of a cytotoxic agent, rapidly divided cells are targeted for destruction by apoptosis as they pass through their cell-cycle checkpoints. Examples of such agents include nucleotide analogues (e.g., azidothymidine), DNA alkalating agents (e.g., cyclophosphamide), cyclin-dependent kinase inhibitors, and microtubule disrupting agents (e.g., taxol), or other agents, which target bone marrow such as chloramphenicol and oxazolidonone antibiotics. These latter agents inhibit mitochondrial protein synthesis at high tissue concentrations. Rapidly dividing cells for which the earliest cytotoxic agent-induced toxicity is noted in rats include cells of bone marrow (hematopoietic progenitors and precursors) and intestine (epithelial crypt cells). The earliest change observed in the peripheral blood of the rat is absolute reticulocytopenia. At 24 hours post-injection, reticulocytes may be substantially increased, likely from release from bone marrow stores; however, as early as 48 hours after injection of a cytotoxic agent, are markedly reduced, consistent with their lifespan (1–1.5 days) in the rat (Table 5-1). Unless red blood cells are also destroyed (hemolysis) by the compound, the attrition of red cells is gradual with continued administration of cytotoxic agent. With a lifespan of 42 to 65 days (Table 5-1), after 7 days, a decrease in the red count of up to 11% to 17% is calculated and observed in the face of complete marrow ablation. Given their short lifespan, neutrophils disappear rapidly from the peripheral blood, followed by platelets. Eosinophils also disappear rapidly, but because eosinophils also decrease with stress, their blood counts should not be used in the evaluation of hematopoiesis. Regeneration of bone marrow and peripheral blood, even in the face of mild bone marrow injury with myelofibrosis, is generally rapid and complete within 2 weeks after withdrawal of the cytotoxic agent. Although overshoot thrombocytosis to levels that are two to three times greater than controls commonly occurs in mice during recovery from administered cytotoxics, it is less characteristic of the response in the rat and generally less than 1.5 times control values. A variety of techniques available to assess the contribution of altered bone marrow function to altered hematologic indices are detailed in Car and Blue (2000).

C. Decreased Red Cell Mass

Most conditions resulting in decreased red cell mass are similar in the rat and other species. The generation of detailed sets of information from hematology analyzers provides the tools with which red cell changes may be characterized and understood. When required, standard hematology tests may be complemented by analysis of investigative hematologic parameters, bone marrow cytology, or bone marrow histology to determine the mechanisms of reduced mass.

1. "Anemia" of Chronic Disease

This condition is typically normochromic and normocytic, with slight reduction from expected hemoglobin and red cell concentrations. In many cases, rats are not truly anemic but have mild decreases in red cell mass well within reference intervals for RBCs and hemoglobin (HGB) (Table 5-3). Mean corpuscular hemoglobin (MCH) is generally normal, mean corpuscular volume (MCV) is normal to slightly reduced, and red blood cell distribution width (RDW) is unchanged. There may be histologic evidence of chronic inflammation. Minor statistically significant effects consistent with the "anemia of chronic disease" are frequently observed at high doses in toxicology studies with rats and are generally considered to lack toxicologic significance.

2. Immune-Mediated Hemolytic Anemia

Immune-mediated hemolytic anemia (IMHA) is extremely rare as a primary disease in rats. However, immune-mediated disease may be observed secondary to compound administration or associated with large granular lymphocyte leukemia of F344 rats or other neoplasias (Stromberg et al., 1983a). The morphology of IMHA is typical of that for other species with spherocytes, other poikilocytes, and rouleaux or agglutination on smears of peripheral blood. The presence of spherocytes, although highly suggestive, is not diagnostic as in dogs or humans; other red cell shape changes may also occur. Because immune-mediated destruction of red cells is highly regenerative, the RDW is generally increased. The large lysis-resistant reticulocytes sometimes present in IMHA may result in falsely high white blood cell counts, although IMHA may be associated with real, absolute leukocytosis. Erythrophagocytosis may rarely be observed in blood smears, although it is more readily identified in cytologic preparations of spleen or bone marrow.

3. Oxidant Injury

Many xenobiotic agents (such as aryl amines or hydrazines) either cause oxidant injury directly or are metabolized in the liver by p450 mixed function oxidases to more reactive intermediates that cause oxidant injury to hemoglobin. Oxidant injury is manifested by either Heinz body or methemoglobin formation, or both. Heinz bodies result from oxidized hemoglobin. The denatured hemoglobin molecules aggregate subjacent to the plasma membrane of the erythrocyte. Heinz bodies are best visualized in methylene blue-stained preparations but can also be observed in Wright's stained smears. The presence of Heinz bodies results in a shortened lifespan of the red cells through enhanced susceptibility to hemolysis; thus, Heinz body hemolytic processes are regenerative (increased reticulocytes) with an increased RDW. Occasionally, compounds that are oxidants cause increases in blood methemoglobin, either alone or in addition to Heinz body formation. Methemoglobin is reversibly oxidized hemoglobin (Fe is in the ferric [3+] state). Percent methemoglobin can be measured in the laboratory. It is important to note that the absence of methemoglobinemia in the face of Heinz body hemolysis does not confound the interpretation that administration of the test substance has caused oxidation of hemoglobin (Eyer, 1983). Normal blood concentrations of methemoglobin in the rat (0.1–0.4 g/dl vs. 16 g/L for oxyhemoglobin), as in the mouse (Hejtmancik et al., 2002), are much lower than in humans. This may result from the higher erythrocyte levels of methemoglobin reductase in rats and mice than humans (Stolk and Smith, 1966). Because methemoglobin reductase activity is higher in rodents, care must be taken to analyze methemoglobin immediately after sample collection. Generation of sulfhemoglobin is another possible sequela to administration of xenobiotics, such as sodium thiosulfate or N-(2-mercaptopropionyl)-glycine (Nomura, 1980). Unlike methemoglobin, sulfhemoglobin formation is irreversible in red cells and, therefore, persists for the lifespan of the cell (42 to 65 days in the rat).

4. *Plasmodium berghei*-Induced Anemia

Young Wistar rats are highly susceptible to the malarial parasite, *Plasmodium berghei* (Holloway et al., 1995). Unlike its human counterparts with trophism for mature erythrocytes, *Plasmodium berghei* preferentially parasitizes reticulocytes. In addition to exhibiting a marked regenerative anemia with clearly demonstrable parasitemia, infected rats have marked lactic acidosis and hypoglycemia that contribute to the morbidity of the disease.

5. Hereditary Anemias

Unlike the mouse for which a large number of naturally occurring and experimentally generated genetic mutants manifest a diverse array of hereditary anemias (Car and Eng, 2001), relatively few such models have been identified in the rat. Most extensively studied is the Belgrade rat, which develops a hypochromic, microcytic anemia with the phenotypic abnormality in iron metabolism associated with a mutation in the Nramp2 gene (Wood and Han, 1998; Zunic et al., 1993).

6. Howell-Jolly Bodies

A low proportion of the reticulocytes (0–0.5%) of normal rats and mice contain usually single basophilic-stained rounded DNA fragments left over from erythropoiesis. It was recognized that an increase in the proportion of these fragments (>0.5%) is observed with clastogenic (genotoxic) xenobiotics. Known as "micronuclei" in the field of genetic toxicology, the assessment in rats and mice of the effects of pharmaceutical and chemical agents on the frequency of reticulocyte Howell-Jolly bodies is a key part of the genotoxicity assessment required by regulatory authorities. The mouse and rat *in vivo* micronucleus test, in which doses of xenobiotics up to 2 g/kg are administered once and bone marrow smears prepared generally 24 or 48 or hours postdosing, is an *in vivo* test to evaluate genetic toxicity. Micronuclei are either microscopically enumerated in reticulocytes (1000–5000) in bone marrow smears or are counted in peripheral blood by flow cytometry (Torous et al., 2003). Fewer Howell-Jolly bodies are present in higher species that have more efficient splenic removal of red cell inclusions than rodents (Blue and Weiss, 1981).

7. Poikilocytosis

Poikilocytosis refers to the abnormality in erythrocyte morphology documented in Romanowsky-stained blood smears or as observed by scanning electron microscopy. Most poikilocytes observed in higher species are also noted in the rat. Specific poikilocytes are associated with diseases, but may also be observed during the evaluation of rat blood in toxicity studies. Many compounds are designed to be cationic, hydrophobic, and lipophilic to enhance their ability to penetrate plasma membranes (Isomaa et al., 1987; Mark et al., 2001). These qualities also allow the intercalation of xenobiotics into the erythrocyte lipid bilayer; this may be responsible for the observation of echinocytosis and acanthocytosis in some in rat toxicity

studies. Acanthocytes are erythrocytes with short membrane protrusions (spikes) occurring with irregular shape and incidence around the perimeter of the cell (Biemer, 1980). They should not be confused with echinocytes, which have more regularly crenated plasma membranes and frequently occur as artifacts of drying blood smears, but may also be secondary to renal toxicity (Gojer, 1992). Acanthocytes also are commonly observed in the blood of rats with severe liver disease as they are in other species.

Given the frequency with which poikilocytosis is documented in rat toxicity studies, relatively few publications document the specific pathophysiology associated with these changes in rats. Red blood cell fragments (helmet cells, schistocytes) are observed in hemolytic processes such as those associated with Heinz body formation, osmotic shock, and disseminated intravascular coagulation. Rarely, red cell shapes such as stomatocytes, target cells, ghost cells, spherocytes, and others are observed on study (Vittori et al., 1999; Hamada et al., 1998; Mark et al., 2001). Reference should be made to the interpretation of these forms in other species.

D. Hepatobiliary Disease

Hepatocellular and/or biliary toxicity is a common finding after the administration of xenobiotics to rats. In this regard, tests of the hepatobiliary system have been studied extensively, and a large body of literature documents biomarkers of hepatotoxicity and hepatic disease in rats.

1. Hepatocellular Disease

In addition to the acanthocytosis mentioned above, hepatocellular necrosis in rats is associated with increased serum activities for aspartate aminotransferase (AST), alanine aminotransferase (ALT), and sorbitol dehydrogenase (SDH). AST is sensitive but lacks specificity, whereas both ALT and SDH provide specificity. SDH is frequently the most sensitive and specific indicator of hepatocellular toxicity in rats (Travlos et al., 1996). The determination of SDH activity requires rapid processing of samples from blood collection to chemistry analyzer, unlike the more stable analytes, AST and ALT. Handling of SDH should be determined by stability testing at the local laboratory. In addition, SDH interpretation is not confounded by compound interference owing to modification of pyridoxal phosphate cofactor interaction, as occurs with transaminases. Recently, glutamate dehydrogenase (GLDH) has been studied as a potentially superior biomarker of hepatocellular toxicity in rats than ALT, AST, or SDH (O'Brien et al., 2002). The combination of AST, ALT, and SDH is recommended for the evaluation of hepatocellular toxicity in the rat. Frequently, increases up to twofold the upper limit of normal for these analytes are not associated with histologically detectable injury to hepatocytes.

Hepatocellular hypertrophy is a frequent finding in rats and is generally associated with liver weights that are 25% to 30% greater than normal, although increases as great as 100% may be observed. Though not experimentally established, compound-induced hepatocellular hypertrophy has been suggested as a cause for increases in activities of hepatocellular enzymes above the normal levels. Theoretically, if one assumes a steady-state turnover of hepatocytes, this hepatic hypertrophy would lead to similar increases in the activity of enzymes used to identify hepatocellular injury.

2. Cholestatic Disease

Increased serum activities of gamma-glutamyl transpeptidase (γGT), alkaline phosphatase, total bile acids, 5′-nucleotidase (5′NT) and bilirubin have been associated with cholestatic disease in rats. ALP, γGT, and bilirubin are recommended for routine use. ALP is highly sensitive, although isoforms exist in several tissues, so this analyte may lack specificity. Although the highest concentrations of γGT are present in the renal proximal tubular brush borders, increased serum γGT activity is very specific for cholestatic disease, yet is very insensitive in the rat. Xenobiotics may induce γGT and/or ALP through noncholestatic mechanisms; this induction can complicate interpretation of these parameters. Although touted for its utility in the evaluation of cholestatic disease in rodents (Carakostas et al., 1990), 5′NT levels are highly variable over time, and unlike other biomarkers of hepatobiliary injury, exhibit marked sexual dimorphism in serum activity. The highest specific activity of 5′NT in the liver is in the Kupffer cells (Haugen et al., 1986; De Valck et al., 1988). Toxicities targeting Kupffer cells, such as systemic phospholipidosis, are not uncommon in toxicology studies and further contribute to the high variability in this parameter. Total serum bile acids also are highly variable in rats. Compounds affecting excretion or conjugation of bile acids affect the utility of this measurement as a surrogate for hepatobiliary function. Bilirubin, although not very sensitive or specific, is a readily available test and is also used for determination of cholestatic processes. Bilirubin may also be increased under conditions of hemolysis; this decreases the specificity of this parameter.

3. Inherited Models of Cholestasis in Rats

Several hereditary models of hyperbilirubinemia (>0.2 mg/dl of serum bilirubin) have been identified and studied extensively in the rat (Muraca et al., 1987). These include the Gunn rat, which develops severe unconjugated hyperbilirubinemia due to defective bilirubin conjugation. Gunn rats are deficient in bilirubin UDP-glucuronosyltransferase (UGT1A1) (Chowdhury et al., 1993; Seppen et al., 1997; Kren et al., 1999) and serve as a rodent model for Crigler-Najjar syndrome type I in humans. Less severe phenotypes associated with mutations of this gene are referred to as Gilbert's syndrome. Hyperbilirubinemic SD and Wistar strains have been described, which lack the MRP-2 transport protein (TR-/- or Eisai rats). This protein transports organic anions from the hepatocyte into the biliary canaliculus (Kitamura et al., 1990; Tanaka et al., 2003). The hyperbilirubinemia of these rats is predominantly of conjugated bilirubin and serves as a model of Dubin-Johnson syndrome in humans.

4. Miscellaneous Hepatic Disease

Xenobiotics have been reported to target Kupffer cells, leading to severe multifocal granulomatous liver disease with only minor or no increases in routine hepatocellular biomarkers. The search for more sensitive markers of liver injury is, therefore, still important.

5. Metabonomics and Biomarkers

Metabonomic evaluation of urine from rats dosed with hepatotoxicants has shown changes in the urinary analytes, such as taurine, to be highly sensitive for hepatic injury (Holmes et al., 2000; Robertson et al., 2000; Nicholson et al., 2002). With increasing sophistication of this technology and the broadening of its technologic base to include mass spectroscopic (MS) approaches, this branch of chemistry is expected to deliver topographically and mechanistically specific biomarkers of liver disease with equal or greater sensitivity and localization to traditional endpoints. Biomarkers identified by this technology yield diagnostic information, in much the same manner that analysis of ALT, AST, and SDH are much superior in combination to the diagnosis of hepatocellular injury than as individual analytes.

E. Renal Disease

Following the liver, the kidney represents the target organ most likely damaged by xenobiotic agents. Toxicants specific to renal papilla, medulla, cortex, and glomeruli have been described and used in the validation of sensitive methodologies for detecting alterations in renal function. In rats, the kidney is the organ that declines in function the earliest with age (Keenan et al., 1997), especially when rats are fed *ad libitum*. The urine chemistry of the normal SD rat was reviewed by Shevock et al. (1993).

The male rat is uniquely susceptible to a nephropathy that derives from hepatic production of α2-urinary globulin, which is readily filtered by the renal glomerulus. This protein, which contributes to the peculiar odor of rat and mouse urine, is readsorbed by proximal tubular epithelial cells and degraded. Alpha-2 urinary globulin of male rats is unique in its ability to bind xenobiotic agents, such as D-limonene, which in turn slows the degradation of the globulin. Persistence and excessive build-up of α2-urinary globulin in proximal tubular epithelial cells ultimately results in cytotoxicity to tubular epithelium, which in its most severe state causes significant renal pathology and can predispose rats to renal carcinoma (Lehman-McKeeman et al., 1998; Swenberg and Lehman-McKeeman, 1999).

Please refer to "Considerations of Blood and Urine Collection and Sample Handling" for appropriate urine collection in rats. The renal pathophysiology of the rat closely mimics that of larger animal species. The interpretation of analytes that measure renal function may be interpolated to a large extent from those of other species. Evaluation of renal function is described in detail by Finco (1997), Loeb (1998), Ragan and Weller (1999), and Gregory (2003).

Traditionally, as with other species, blood urea nitrogen (BUN) and creatinine have been used to assess renal function in rats. Elevated serum creatinine concentration is more specific and more consistently elevated in glomerular disease. Increased serum BUN concentration (azotemia) generally results from compromised tubular function, although it is also increased when glomerular filtration is impaired. Prerenal azotemia, in which BUN and/or creatinine are increased, occurs secondary to fluid loss or decreased fluid intake. Prerenal and renal azotemia sometimes coexist, confounding interpretation. Azotemia may also occur with increased protein catabolism or with gastrointestinal bleeding. BUN and creatinine are relatively insensitive tests when conducted outside of a research laboratory. Because of the consistent environment provided by a typical toxicology facility, BUN and creatinine are more sensitive for biomedical researchers than under clinical conditions. Nevertheless, renal injury must be fairly extensive (less than one third of nephron function maintained) to result in detectable increases of UN and creatinine. Therefore, compared with biomarkers of liver injury, detection of renal injury by traditional analytes is insensitive. Uric acid is not useful in assessing the renal function of rats.

A variety of specific assays and experimental approaches are used to enhance the sensitivity of traditional assays. These approaches and their advantages are listed in Table 5-7.

TABLE 5-7

ASSAYS AND EXPERIMENTAL APPROACHES TO EVALUATE RENAL DISEASE IN THE RAT

Assay	Renal function evaluated	Comments
Urea nitrogen (serum)	Tubular disease and severe glomerular disease	Neither specific, nor sensitive. Moderately increased in dehydration.
Creatinine (serum)	Tubular and glomerular disease	Neither specific, nor sensitive. Mildly increased in dehydration.
Urine cytology	Tubular integrity	Presence of casts, renal epithelial cells, crystals highly sensitive to cortical injury
Urinalysis – routine	Integrated renal function	
–Urinary protein		Sensitive to glomerular and tubular injury
–Volume and specific gravity		Renal and extra renal causes.
		Tubular readsorption of H_2O, extrarenal causes of polyuria or polydipsia
–pH		Renal and extrarenal regulation. Highly alkaline = bacterial contamination
–Glucose		Proximal tubular/extra renal
–Ketones		Monitors extrarenal function
–Bilirubin, urobilinogen		Monitors extrarenal function
Urinalysis – special		
–fractional excretion of Na^+, K^+, Cl^-	Proximal tubular and renal papillary function	Overnight collection or forced diuresis (4 hour urine collection after oral gavage with 10 mL/kg of water). Highly sensitive to functional derangements of cortex. Requires serum and urinary creatinine concentrations.
–β2-microglobulin, other proteins	Glomerular injury	Requires immunoassay (1)
–ALP, γGT, N-acetyl-β-D-glucosaminidase	Cortical tubular injury	Requires enzyme assay, highly sensitive for detecting brush border injury (2)
Para-amino hippurate uptake	Renal blood flow	Measured by colorimetry.
Inulin clearance	Glomerular filtration	Measured by colorimetry, fluorimetry (FITC-Inulin) (3) or Technetium-99m DTPA dimethyl ester (4)
ADH challenge	Urine concentrating ability	

(1. Ivandic et al., 2000; 2. Guder and Ross, 1984; 3. Qi et al., 2003; 4. Bhowal et al., 2003.)

Metabonomic evaluation of urine is particularly sensitive to nephrotoxicity and provides noninvasively obtained information permitting the specific topography of renal injury to be interpolated (Holmes et al., 2000; Robertson et al., 2000; Nicholson et al., 2002). This methodology will likely become more routine in the evaluation of renal disease of rats in the future.

F. Cardiac and Skeletal Muscle Disease

The condition termed *cardiomyopathy* in male SD rats (focal inflammation with focal fiber degeneration) is a relatively common naturally occurring disease, although restricted feeding reduces its incidence in aging animals (Van Vleet and Ferrans, 1986; Keenan et al., 1997). Xenobiotic injury to the heart, however, is fairly uncommon. Anthracycline (e.g., adriamycin)-induced myocardial necrosis and peroxisome proliferators-activated receptor-gamma (PPARγ) agonist-induced cardiomyopathy epitomize compound-related effects. Although the adriamycin-induced pathology is predictive of human cardiac pathology, the cardiomyopathy of PPARγ agonists appears to be rat specific. Rats also develop severe inflammatory cardiomyopathy after the administration of triiodothyronine; however, this syndrome is unlike the hypertrophic cardiomyopathy observed in hyperthyroid cats. Rats have proven a valuable species for the development of heart-specific biomarkers.

Skeletal muscle lesions occur in rats given high doses of HMG CoA reductase inhibitors (e.g., cerivastatin), but this is a relatively unusual finding in toxicity studies or naturally occurring diseases of the rat.

Following skeletal or cardiac muscle injury, serum AST and creatinine kinase (CK) activities increase. If ALT and SDH—two enzymes more specific for liver disease—are within their reference intervals, one must suspect skeletal or cardiac muscle injury. Unlike the situation in humans and higher domestic species, CK and lactate dehydrogenase (LDH) isozyme analysis is generally not useful in the specific diagnosis or assessment of cardiac pathology in the rat. However, troponin-T concentrations as measured by enzyme-linked immunosorbent assay (ELISA) in plasma are reported to be relatively specific for cardiac injury in rats (O'Brien et al., 1997; Bertsch et al., 1999). Troponin-T is accepted by regulatory authorities as a sensitive surrogate for cardiac muscle injury. Although multiple troponin assay platforms are currently available, none has been optimally qualified for the rat and recognized by regulatory agencies. Therefore, determination of this analyte is currently best utilized as a second-tier test rather than a cardiotoxicity screening assay for rats, unless careful validation of serum concentrations and cardiac histopathology for a compound series has been undertaken.

G. Pancreatic Disease

1. Endocrine Disease

As the rat is the principal research animal used to study diabetes mellitus, considerable experience surrounds the assessment of its pancreatic endocrine function. Commonly used in pancreatic endocrine research are Wistar BB rats (Nahkooda et al., 1977), and the fa/fa or "Zucker" rat (Zucker and Zucker, 1961). The Wistar BB rat develops type I diabetes mellitus following severe islet cell destruction and demonstrates hyperglycemia, glycosuria, insulinopenia, and ketonemia. Type I diabetes also can be induced experimentally in rats by administration of streptozotocin. The fa/fa rat is a model of type II diabetes mellitus. In addition to the characteristics described for type I diabetes, this strain demonstrates marked insulin resistance. For a detailed overview of carbohydrate metabolism in laboratory animals, the interested reader is referred to Kaneko (1999).

The principal biomarkers for pancreatic endocrine function in the rat are serum and urinary glucose, insulin, for which there are two isoforms in the rat, IGF-1, and glycated hemoglobin and albumin as biomarkers of chronic hyperglycemia (Neuman et al., 1994; Wan Nazaimoon and Khalid, 2002).

2. Exocrine Disease

The exocrine pancreas is targeted by a limited range of xenobiotics; however, it is particularly susceptible to surface-active agents administered by intraperitoneal injection. Long-term administration of inhibitors of trypsin to rats results in an increased incidence of pancreatic adenocarcinoma. Exocrine pancreatic inflammation, necrosis, and insufficiency have been extensively modeled in the rat. Infusion of taurocholate (Marks et al., 1984) or oleic acid (Henry and Steinberg, 1993) into the pancreatic duct reliably produces severe pancreatic exocrine inflammation and necrosis followed by insufficiency. Serum lipase and amylase activities are both used as sensitive biomarkers of exocrine pancreatic disease in rats. Because amylase isoforms exist in the salivary glands of rats, which are occasionally targeted by xenobiotics excreted through the saliva, amylase lacks specificity for pancreatic injury in rats. Trypsin and chymotrypsin inhibition, and pancreatic insufficiency are readily detected by the bentiromide test in rats. Bentiromide is cleaved by chymotrypsin to para-amino benzoic acid (PABA), which is absorbed by the small intestine. Blood collection should be at 30, 60, and 90 minutes after oral gavage dosing with bentiromide, as PABA is more rapidly cleared from the rat circulation than higher species, such as the dog. Analysis of fecal chymotrypsin content is among the most sensitive indicators of exocrine pancreatic insufficiency in the rat (Henry and Steinberg, 1993). Experimentally, exocrine pancreatic function may be assessed in the rat after administration of cholecystokinin (CCK-8; 26 pmol/kg/h), followed by evaluation of pancreatic juice volume, amylase, and trypsin output (Shimuzu et al., 2000). Immunoreactive trypsin-associated proteins in serum are sensitive to pancreatic injury in the rat, as in other species (Marks et al., 1984).

H. Gastrointestinal Disease

A variety of naturally occurring and xenobiotic-induced pathologies of the rat alimentary tract have been documented.

1. Stomach

In rats, agents that ablate gastric parietal cells, including infection with *Helicobacter pylori* (Neu et al., 2002) and certain xenobiotics (Goldenring et al., 2000), or that inhibit gastric acid secretion such as H^+K^+ ATPase inhibitors and histamine-1 receptor antagonists, produce severe hypergastrinemia, which is readily detected in rat serum and plasma by immunoassay. Occasionally, xenobiotics cause a delay in gastric emptying and subsequent distention of the stomach with gastric contents. As with gastric stasis in other species, this results in stimulation of acid secretion, metabolic alkalosis, and hypochloremia in the rat.

2. Small and Large Intestine

Crypt cells of the small intestine provide stem cells for renewal of the intestinal epithelium, which turns over each 3 to 4 days. Xenobiotics that target rapidly dividing cells result in epithelial villus atrophy. Specific biomarkers for small intestinal mucosal injury are limited. The use of 3-methylglucose, D-xylose, folate, and cobalamin absorption tests are detailed in Hornbuckle and Tennant (1997), Bounous (2003), and Martin et al. (2003). Experimental resection of the small intestine, "short bowel syndrome," is used to model human malabsorption syndromes in rats (Martin et al., 2003). Accompanying small and large intestinal mucosal injury are alterations in serum electrolytes, with hyponatremia most commonly observed. Inflammatory diseases of the small intestine are best evaluated histologically and by culture of intestinal contents. Specific biomarkers of inflammatory small intestinal disease are limited.

I. Skeletal Disease

More routinely evaluated by imaging technologies, skeletal turnover in the rat may also be evaluated with biomarkers used experimentally, and in the evaluation of osteoporosis and its progression in humans. In the rat, sensitive serum biomarkers to osteoblast function are the specific bone isoform of ALP and osteocalcin. Indicators of osteoclast function include serum calcium, phosphate, tartrate-resistant acid phosphatase, and urinary and serum collagen type II fragments including pyridinium cross-links, and N-telopeptides (Allen, 2003; Seed, 2003). In addition, monitoring of plasma concentrations of calcitonin and parathyroid hormone are performed in the rat. Given the availability of mouse knockout models of osteoporosis, osteopetrosis, and cartilage metabolism, the mouse is more widely used in the modeling of skeletal disease than the rat (Ammann et al., 2000).

J. Acid-Base Abnormalities

Acid-base abnormalities are infrequent changes in toxicity studies. In general, they occur secondary to high dose toxicities. Typically, assessments from which alterations in acid-base metabolism may be interpolated include changes in potassium (increased with metabolic acidosis), HCO_3^- or total carbon dioxide (TCO$_2$) (decreased with metabolic acidosis, increased with alkalosis) and urine pH (reduced in metabolic acidosis, reduced in metabolic alkalosis with paradoxic aciduria, present in renal tubular acidosis leading to aciduria). Most frequent are acid-base abnormalities relating to severe systemic disease or toxicity, altered or sequestered gastric acid or alterations in renal function, followed by lactic acidosis, which occurs in severe diabetes mellitus, mitochondrial and other metabolic intoxications. Upon observation of an abnormality consistent with a systemic or renal affect on acid-base metabolism, further tests may be conducted to dissect the mechanism and/or severity of the abnormality. Such tests that may be used to determine mechanism include blood pH, blood lactate, lactate, β-hydroxybutyrate and acetoacetate, PO$_2$, calculation of anion gap, administration of buffers of gastric pH (HCO_3^-) or coadministration of urinary acidifiers (NH$_4$Cl). Acid-base abnormalities are considered together with water balance and electrolytes. For more detailed information, readers are referred to Carlson (1997), Riley and Cornelius (1999), DeBartola (2000), Rose (2002), and George (2003).

K. Responses of the Rat to Stress

The term "stress" applies to two different physiologic phenomena. Stress related to epinephrine release has a relatively short duration, whereas stress related to corticosteroids (mostly corticosterone in the rat) is a longer lasting process. These two processes may occur simultaneously but have differing effects on peripheral blood constituents. In addition, stress-related changes may be difficult to separate from primary disease processes. In rats, the major alterations observed as a result of epinephrine are increased glucose and increased numbers of lymphocytes, whereas those observed as a result of corticosteroids are decreased numbers of lymphocytes and eosinophils.

To document endocrine alterations in chronically stressed rats, serum adrenocorticotrophic hormone (ACTH) and corticosterone may be measured (Bamberg et al., 2001). These parameters are so sensitive to perceived stress and rapidly regulated that true baselines are difficult to obtain by routes of blood sampling other than decapitation. The evaluation of the 24-hour urinary excretion of corticosterone serves as an alternate robust surrogate for these serum assays.

Alterations documented as secondary to stress are listed in Table 5-8. Many of these are not consistently altered in the experience of the authors; therefore, parameters most consistently affected are indicated with an asterisk. Alterations of serum cholesterol and triglycerides, as per the experience of the authors, are unlikely in rats.

The neutrophilia associated with increased plasma glucocorticoids results from displacement of the marginating pool into the central compartment of the blood vessels and release of neutrophils from bone marrow stores. With fewer neutrophils extravasated over time, the average lifespan (6 hours) of the neutrophil increases and is reflected in an increased proportion of neutrophils with multisegmented (>3) nuclear lobes (Fehr and Grossman, 1979; Jilma et al., 1998).

L. Effects of Fasting

The effects of fasting on rats have been documented in several studies. Overnight fasting of rats is undertaken to ensure uniformity of samples for clinical chemistry and hematologic analyses. Because rats are nocturnal feeders, overnight deprivation of food has significant impact on daily caloric intake; therefore, reference intervals should specify whether rats are fasted and, if so, for what period. Due to the significant hyperlipemia that occurs in 30% of overnight fasted rats (Waner and Nyksa, 1994), fasting is essential to ensure accurate determination of analytes restricted to either fat or water-soluble fractions of blood (e.g., electrolytes). The effects of fasting of rats on various serum analytes are detailed below in Table 5-9.

Despite alterations in other serum lipids, cholesterol is not changed by overnight fasting in rats. Fasting increases unconjugated serum bilirubin in both humans and rats (Kotal et al., 1996). This finding is not uniform across laboratory and domestic animal species. Following the negative catabolic effects of 2 weeks of feed restriction in the rat, serum glucose, bilirubin, urea, sodium, and chloride may increase, whereas total protein, globulin, potassium, and calcium decrease (Levin et al., 1993).

M. Compound-Induced Hypothyroidism

Many exogenously administered agents alter thyroid hormone metabolism in rodents, leading to marked changes in clinical pathology profiles. It is thought that the relative deficiency in thyroid-binding globulin in rats results in a much shorter plasma half-life of T_4 than in higher species, which dictates the rat's greater susceptibility to flux in thyroid hormone status. Rats depend 'to a greater extent' on transthyretin for T_4 transport (Vranckx et al., 1994). The most common xenobiotic-induced effect is that typified by phenobarbital administration to rats, which is thought to increase the clearance of T_4 by

TABLE 5-8

REPORTED RESPONSES OF THE RAT TO STRESS

Increased parameters	Decreased parameters
Neutrophil count	Eosinophil count*
Monocyte count	Basophil count
Lymphocytes count (immediate response)*	Lymphocyte count (all types except NK cells – chronic response)
Cholesterol	Testosterone
Potassium	Prolactin
Glucose*	Triglycerides
Adrenalin *	
Corticosterone *	
Adrenocorticotrophic hormone*	
Arginine vasopression	

(Balasubramanian et al., 1982; Bean-Knudsen and Wagner, 1987; Flaherty et al., 1993; Meno-Tetang et al., 1999; Moore, 2000; Perez et al., 1997).

*Indicates those parameters that are most consistently affected.

TABLE 5-9

RESPONSES OF THE RAT TO FASTING

Increased parameters	Decreased parameters
Unconjugated (direct) bilirubin	Free fatty acids
Erythrocytes	Triglycerides
Hemoglobin	T_4, T_3, TSH
PCV	Transthyretin
	Glucose
	Alkaline phosphatase

(Kast and Nishikawa, 1981; Kotal et al., 1996; Spear et al., 1994; Waner and Nyksa, 1994).

enhancing UDP-glucuronysyl transferase activity in hepatocytes (De Sandro et al., 1991). Increased glucuronidated T_4 is then excreted into the bile. Serum T_4 concentrations rapidly decrease, sometimes to undetectable levels. T_3 levels are reduced, but to a lesser extent, and TSH is markedly increased. Other agents such as propylthiouracil, which inhibits thyroid peroxidase, also decreases T_3, and T_4, and increases TSH. Reductions in T_4 are generally less significant than those observed due to increased clearance of T_4. Rarely, compounds may interfere with T_4 deiodination to T_3, in which case serum T_3 concentrations are relatively more severely affected than T_4, which may be normal or increased based on TSH stimulation of the thyroid. Marked increase in liver mass, a common effect of many xenobiotics that induce the cytochromes P450 enzymes, may also directly increase T_4 clearance (De Sandro et al., 1991), and decrease serum T_4 concentrations.

In the rat, increased T_4 clearance may be associated with hypotriglyceridemia and hypercholesterolemia. In most other species, decreased T_4 results in hypertriglyceridemia. Urine volume also may increase due to impaired renal concentration in hypothyroid rats (Cadnapaphornchai et al., 2003) and lower serum sodium concentrations. Although nonregenerative anemia may be observed with hypothyroidism in other species, no clear association has been demonstrated in the rat.

N. Acute-Phase Responses in the Rat

The observation of an acute phase response in the rat is relatively common and may be heralded by a mild nonregenerative decrease in red cell mass. These changes often overlay primary disease or direct compound-related alterations and need to be assessed independently of them. Thrombocytosis results from the direct stimulatory effects of interleukin (IL)-6 on megakaryocytopoiesis; however, it is more commonly observed with exogenously administered IL-6 than in endogenous acute-phase reactions. The nonregenerative decreased red cell mass is thought to result from decreased availability of Fe^{++} for erythropoiesis, decreased survival time of red blood cells, and decreased bone marrow production. The cytokines, IL-6, IL-1β, involved in the induction of acute-phase response in the rat derive largely from macrophages involved in inflammatory reactions (van Gool et al., 1984; Gabay and Kushner, 1999; Jinbo et al., 2002), and from corticosterone from the adrenal cortex. Alpha-2 macroglobulin is considered the most consistently elevated acute phase reactant of the rat (Jinbo et al., 2002), although the diagnosis of an acute phase reaction should derive from an observation of a constellation of effects (French, 1989; Schreiber et al., 1989), including many of those listed in Table 5-10.

O. Abnormalities of Blood Coagulation in the Rat

The rat has been used extensively in the study of blood coagulation and platelet-mediated thrombosis. No models of coagulation factor-deficient rats were identified on Medline search (2004). Genetically modified mice and inbred dog breeds are used more extensively as models of coagulation factor deficiencies than the rat (Car and Eng, 2001).

1. Acquired Hemorrhagic Diathesis

Several models of disseminated intravascular coagulation and the Shwartzmann reaction have been developed and are widely used in the rat (Carthew, 1991; Hara et al., 1997; Asakura et al., 2002). In each of these models, thrombocytopenia, hypofibrinogenemia, increased fibrin D-dimer fragment, and consumption of coagulation factors with prolongation of activated partial thromboplastin time (APTT) and prothrombin time (PT) are observed. Although D-dimer tests function well with rat plasma, standard tests for fibrin degradation products do not cross-react sufficiently well to be of diagnostic utility in the rat (Nieuwenhuizen et al., 1982). The use of human factor-deficient plasmas in evaluation of coagulation factor activities in the rat is well established.

2. Coagulation Factors and Vitamin K

Vitamin K is required for the activity of the epoxide reductase enzyme complex in hepatocytes. This enzyme complex is involved in an essential posttranslational event leading to functional coagulation factors. Deficiency in vitamin K and administration of agents inhibitory to this enzyme complex, such as warfarin, lead to defective hepatic

---

TABLE 5-10

REPORTED ACUTE PHASE RESPONSE IN THE RAT

Increased parameters	Decreased parameters
Platelets	Erthrocytes* and reticulocytes
Fibrinogen*	Albumin*
α-2 macroglobulin*	Serum Fe+++*
C-reactive protein	
α-1 acidic glycoprotein	
α-1 proteinase inhibitor	
Ceruloplasmin	
Haptoglobin	
Transferrin	
Total alpha and beta globulins*	

(Marinkovic et al., 1989; Myrset et al., 1993; Ruot et al., 2000; Schreiber et al., 1989).

*Indicates those parameters that are most consistently affected.

production of coagulation factors II, VII, IX, and X and a bleeding diathesis that is typically more severe in male than female rats (Hara et al., 1994). A mutation in the vitamin K 2,3-epoxide reductase enzyme complex renders rats relatively resistant to the effect of vitamin K-antagonizing rodenticides (Cain et al., 1998). Xenobiotic-induced prolongation of PT and APTT may indicate compound or metabolite interference with bile acid secretion leading secondarily to impaired bile-acid facilitated intestinal uptake of vitamin K (Kambayashi et al., 1985). This toxicity is correctable by parenteral administration of vitamin K. Tabular comparisons of rat coagulation factor activities and hemostatic parameters to those of other species appear in Dodds (1997).

3. Inherited Platelet Dysfunction

Several models of inherited platelet dysfunction have been identified and studied in rats (Jackson, 1989). These include the hypertensive Fawn Hooded rat, which has macroplatelets with abnormal dense granules (Datta et al., 2003), TM rats with platelet storage pool deficiency (Hamada et al., 1997), and rat models of Chediak-Higashi syndrome, which have abnormal lysosomes in multiple cell types including platelets (Nishikawa and Nishimura, 2000).

VI. SUMMARY

The clinical pathology of the rat is a discipline with a current level of sophistication that is largely derived from its extensive use in toxicology in the chemical and pharmaceutical industry. As specific clinical pathology information for this species grows, and with highly promising areas of technology such as proteomics and metabonomics, large biological datasets from rats will become available for analysis and greatly enhance our understanding of the pathophysiology of this species.

REFERENCES

Allen, M.J. (2003). Biochemical markers of bone metabolism in animals: uses and limitations. *Vet. Clin. Pathol.* **32**, 101–113.

Allison, A.C. (1960). Turnovers of erythrocytes and plasma proteins in mammals. *Nature* **188**, 37–40.

Ammann, P., Rizzoli, R., and Bonjour, J.P. (2000). Modèles animaux et développement pré-clinique des médicaments contre l'ostéoporose. *Ann. Med. Interne.* (Paris) **151**, 380–384.

Asakura, H., Okudaira, M., Yoshida, T., Ontachi, Y., Yamazaki, M., Morishita, E., Miyamoto, K., and Nakao, S. (2002). Induction of vasoactive substances differs in LPS-induced and TF-induced DIC models in rats. *Thromb. Haemost.* **88**, 663–667.

Asakura, H., Suga, Y., Aoshima, K., Ontachi, Y., Mizutani, T., Kato, M., Saito, M., Morishita, E., Yamazaki, M., Takami, A., Miyamoto, K., and Nakao, S. (2002). Marked difference in pathophysiology between tissue factor and lipopolysaccharide-induced disseminated intravascular coagulation models in rats. *Crit. Care Med.* **30**, 161–164.

Balasubramanian, K., Pereira, B.M., and Govindarajulu, P. (1982). Epididymal carbohydrate metabolism in experimental hypercorticosteronism: studies on mature male rats. *Int. J. Androl.* **5**, 534–544.

Bamberg, E., Palme, R., and Meingassner, J.G. (2001). Excretion of corticosteroid metabolites in urine and faeces of rats. *Lab. Anim.* **35**, 307–314.

Bean-Knudsen, D.E., and Wagner, J.E. (1987). Effect of shipping stress on clinicopathologic indicators in F344/N rats. *Am. J. Vet. Res.* **48**, 306–308.

Bertsch, T., Bleuel, H., Deschl, U., and Rebel, W. (1999). A new sensitive cardiac Troponin T rapid test (TROPT) for the detection of experimental acute myocardial damage in rats. *Exp. Toxicol. Pathol.* **51**, 565–569.

Bhowal, K., Bhattacharyya, S., Majumdar, A., Giri, C., Vanaja, R., Ramamoorthy, N., Ganguly, S., Sarkar, B.R., Banerjee, S., and Chatterjee Debnath, M. (2003). Technetium-99m DTPA dimethyl ester: a renal function imaging agent. Comparative studies in animals with technetium-99m mercaptoacetyl triglycine and ^{131}I-ortho-iodohippurate. *Nucl. Med. Commun.* **24**, 583–595.

Biemer, J.J. (1980). Acanthocytosis–biochemical and physiological considerations *Ann. Clin. Lab.* **10**, 238–249.

Blue, J., and Weiss, L. (1981). Electron microscopy of the red pulp of the dog spleen including vascular arrangements, periarterial macrophage sheaths (ellipsoids), and the contractile, innervated reticular meshwork. *Am. J. Anat.* **161**, 189–218.

Bounous, D.I. (2003). Digestive system. *In* "Duncan & Prasse's Veterinary Laboratory Medicine, 4th edition." (K.S. Latimer, E.A. Mahaffey, and K.W. Prasse, eds.), pp 215–230, Iowa State Press, Ames, Ia.

Boyd, J.W. (1983). The mechanisms relating to increases in plasma enzymes and isoenzymes in disease of animals. *Vet. Clin. Pathol.* **12**, 9–24.

Cadnapaphornchai, M.A., Kim, Y.W., Gurevich, A.K., Summer, S.N., Falk, S., Thurman, J.M., and Schrier, R.W. (2003). Urinary concentrating defect in hypothyroid rats: role of sodium, potassium, 2–chloride co-transporter, and aquaporins. *J. Am. Soc. Nephrol.* **14**, 566–574.

Cain, D., Hutson, S.M., and Wallin, R. (1998). Warfarin resistance is associated with a protein component of the vitamin K 2,3–epoxide reductase enzyme complex in rat liver. *Thromb. Haemost.* **80**, 128–133.

Car, B.D., and Blue, J.T. (2000). Approaches to evaluation of bone marrow function. *In* "Schalm's Veterinary Hematology." (B.F. Feldman, J.G. Zinkl, and N.C. Jain, eds.), pp 33–37, Williams & Wilkins, Philadelphia, Pa.

Car, B.D., and Eng, V.M. (2001). Special considerations in the evaluation of the hematology and hemostasis of mutant mice. *Vet. Pathol.* **38**, 20–30.

Carakostas, M.C., Power, R.J., and Banerjee, A.K. (1990). Serum 5'nucleotidase activity in rats: a method for automated analysis and criteria for interpretation. *Vet. Clin. Pathol.* **19**, 109–113.

Carlson, G.P. (1997). Fluid, electrolyte, and acid-base balance. *In* "Clinical Biochemistry of Domestic Animals." (J.J. Kaneko, J.W. Harvey, and M.L. Bruss, eds.), pp 485–516, Academic Press, San Diego, Calif.

Carthew, P., Dorman, B.M., and Edwards, R.E. (1991). Increased susceptibility of aged rats to haemorrhage and intravascular hypercoagulation following endotoxin administered in a generalized Shwartzman regime. *J. Comp. Pathol.* **105**, 323–330.

Chowdhury, J.R., Kondapalli, R., and Chowdhury, N.R. (1993). Gunn rat: a model for inherited deficiency of bilirubin glucuronidation. *Adv. Vet. Sci. Comp. Med.* **37**, 149–173.

Cohen, S. (1957). Turnover of some chromatographically separated serum protein fractions in the rat. *South Af. J. Med. Sci.* **23**, 245–256.

Criswell, K.A., Bleavins, M.R., Zielinski, D., Zandee, J.C., and Walsh, K.M. (1998). Flow cytometric evaluation of bone marrow differentials in rats with pharmacologically induced hematologic abnormalities. *Cytometry* **32**, 18–27.

Datta, Y.H., Wu, F.C., Dumas, P.C., Rangel-Filho, A., Datta, M.W., Ning, G., Cooley, B.C., Majewski, R.R., Provoost, A.P., and Jacob, H.J. (2003). Genetic mapping and characterization of the bleeding disorder in the fawn-hooded hypertensive rat. *Thromb. Haemost.* **89**, 1031–1042.

DeBartola, S.P. (2000) "Fluid Therapy in Small Animal Practice, 2nd edition," W B Saunders, Philadelphia, Pa.

Delaunay, J. (2002). Molecular basis of red cell membrane disorders. *Acta Haematol.* **108**, 210–218.

De Sandro, V., Chevrier, M., Boddaert, A., Melcion, C., Cordier, A., and Richert, L. (1991). Comparison of the effects of propylthiouracil, amiodarone, diphenylhydantoin, phenobarbital, and 3–methylcholan-threne on hepatic and renal T_4 metabolism and thyroid gland function in rats. *Toxicol. Appl. Pharmacol.* **111**, 263–278.

De Valck, V., Geerts, A., Schellinck, P., and Wisse, E. (1988). Localization of four phosphatases in rat liver sinusoidal cells. An enzyme cytochemical study. *Histochemistry* **89**, 357–363.

Dodds, W.J. (1997). Hemostasis. *In* "Clinical Biochemistry of Domestic Animals, 5th edition." (J J. Kaneko, JW. Harvey, and M.L. Bruss, eds.), pp 246–247, Academic Press, San Diego, Calif.

Eyer, P. (1983). The red cell as a sensitive target for activated toxic arylamines. *Arch. Toxicol. Suppl.* **6**, 3–12.

Fehr, J., and Grossmann, H.C. (1979). Disparity between circulating and marginated neutrophils: evidence from studies on the granulocyte alkaline phosphatase, a marker of cell maturity. *Am. J. Hematol.* **7**, 369–379.

Finco, D.R. (1997). Urinalysis. *In* "Clinical Biochemistry of Domestic Animals, 5th edition." (J J. Kaneko, J.W. Harvey, and M.L. Bruss, eds.), pp. 460–467. Academic Press, San Diego, Calif.

Flaherty, D.K., McGarity, K.L., Winzenburger, P., and Panyik, M. (1993). The effect of continuous corticosterone administration on lymphocyte subpopulations in the peripheral blood of the Fischer 344 rat as determined by two color flow cytometric analyses. *Immunopharmacol. Immunotoxicol.* **15**, 583–604.

Foot, E.C. (1961). Eosinophil turnover in the normal rat. *Brit. J. Haemat.* **11**, 439–445.

French., T. (1989). Specific proteins D. Acute phase proteins. *In* "The Clinical Chemistry of Laboratory Animals." (W.F. Loeb, and F.W. Quimby, eds.), pp 201–235. Pergamon Press, New York.

Friedel, R., Bode, R., Trautshold, I., and Mattenheimer, H. (1976). Verteilung heterologer, homologer, und autologer Enzyme nach intravenoser Injektions-Verteilung und Transport von Zellenzymen im extrazellularen Raum. III. Mitteilung. *J. Clin. Chem. Clin. Biochem.* **14**, 129–136.

Frith, C.H. (1988). Morphologic classification and incidence of hemato-poietic neoplasia in the Sprague-Dawley rat. *Toxicol. Pathol.* **16**, 451–457.

Gabay, C., and Kushner, I. (1999). Acute-phase proteins and other systemic responses to inflammation. *N. Engl. J. Med.* **340**, 448–454.

George, J.W. (2003). Water, electrolytes, and acid base. *In* "Duncan & Prasse's Veterinary Laboratory Medicine, 4th edition." (K.S. Latimer, E.A. Mahaffey, and K.W. Prasse, eds.), pp 136–161. Iowa State Press, Ames, Ia.

Gojer, M., and Sawant, V. (1992) . Uranyl nitrate induced corpuscular derangement: an early indication of induced acute renal failure. *Indian J. Exp. Biol.* **30**, 119–121.

Goldenring, J.R, Ray, G.S., Coffey, R.J., Meunier, P.C., Haley, P.J., Barnes, T.B., and Car, B.D. (2000). Reversible drug-induced oxyntic atrophy in rats. *Gastroenterology* **118**, 1080–1093.

Gregory, C.R., (2003). Urinary system, *In* "Duncan & Prasse's Veterinary Laboratory Medicine, 4th edition." (K.S. Latimer, E.A. Mahaffey, and K.W. Prasse, eds.), pp 231–259. Iowa State Press, Ames, Ia.

Guder, W.G. (1986). Haemolysis as an influence and interference factor in clinical chemistry. *J. Clin. Chem. Clin. Biochem.* **24**, 125–126.

Guder, W.G., and Ross, B.D. (1984). Enzyme distribution along the nephron. *Kid. Int.* **26**, 101–111.

Gurer, H., Ozgunes, H., Neal, R., Spitz, D.R., and Ercal, N. (1998). Antioxidant effects of N-acetylcysteine and succimer in red blood cells from lead-exposed rats. *Toxicology* **128**, 181–189.

Hamada, S., Nishikawa, T., Yokoi, N., and Serikawa T. (1997). TM rats: a model for platelet storage pool deficiency. *Exp. Anim.* **46**, 235–239.

Hamada, T, Tanimoto, A., Arima, N., Ide, Y., Sasaguri, T., Shimajiri, S., and Sasaguri, Y. (1998). Altered membrane skeleton of red blood cells participates in cadmium-induced anemia. *Biochem. Mol. Biol. Int.* **45**, 841–847.

Hara, K., Akiyama, Y., and Tajima, T. (1994). Sex differences in the anticoagulant effects of warfarin. *Jpn. J. Pharmacol.* **66**, 387–392.

Hara, S., Asada, Y., Hatakeyama, K., Marutsuka, K., Sato, Y., Kisanuki, A., and Sumiyoshi, A. (1997). Expression of tissue factor and tissue factor pathway inhibitor in rats lungs with lipopolysaccharide-induced disseminated intravascular coagulation. *Lab. Invest.* **77**, 581–589.

Haugen, T.B., and Fritzson, P. (1986). Measurement and activity of cytosolic deoxyribonucleoside-activated nucleotidase in various cell types from rat liver and spleen. *Int. J. Biochem.* **18**, 167–170.

Hejtmancik, M.R., Trela, B.A., Kurtz, P.J., Persing, R.L., Ryan, M.J., Yarrington, J.T., and Chabra, R.S. (2002). Comparative gavage subchronic toxicity studies of o-chloroaniline and m-chloroaniline in F344 rats and B6C3F1 mice. *Toxicol. Sci.* **69**, 234–243.

Henry, J.P., and Steinberg, W.M. (1993). Pancreatic function tests in the rat model of chronic pancreatic insufficiency. *Pancreas* **8**, 622–626.

Hoffman, W.E., Solter, P.F., and Wilson, W.W. (1999). Clinical enzymology. *In* "The Clinical Chemistry of Laboratory Animals, 2nd edition." (W.F. Loeb, and F.W. Quimby, eds.), pp 399–454. Taylor & Francis, Ann Arbor, Mich.

Holloway, P.A., Knox, K., Bajaj, N., Chapman, D., White, N.J., O'Brien, R., Stacpoole, P.W., and Krishna, S. (1995). Plasmodium berghei infection: dichloroacetate improves survival in rats with lactic acidosis. *Exp. Parasitol.* **80**, 624–632.

Holmes, E., Nicholls, A.W., Lindon, J.C., Connor, S.C., Connelly, J.C., Haselden, J.N., Damment, S.J., Spraul, M., Neidig, P., and Nicholson, J.K. (2000). Chemometric models for toxicity classification based on NMR spectra of biofluids. *Chem. Res. Toxicol.* **13**, 471–478.

Hornbuckle, W.E., and Tennant, B.C. (1997). Gastrointestinal function. *In* "Clinical Biochemistry of Domestic Animals." (J.J. Kaneko, J.W. Harvey, M.L. Bruss, eds), pp 367–406, Academic Press, San Diego, Calif.

Isomaa, B., Hagerstrand, H., and Paatero, G. (1987). Shape transforma-tions induced by amphiphiles in erythrocytes. *Biochim. Biophys. Acta* **899**, 93–103.

Ivandic, M., Ogurol, Y., Hofmann, W., and Guder, W.G. (2000). From a urinalysis strategy to an evaluated urine protein expert system. *Methods Inf. Med.* **39**, 93–98.

Jackson, C.W. (1989). Animal models with inherited hematopoietic abnormalities as tools to study thrombopoiesis. *Blood Cells* **15**, 237–253.

Jain N. C. (1986). "Schalm's Veterinary Hematology, 4th edition." pp 288–298, Lea and Febiger, Philadelphia, Pa.

Jilma, B., Stohlawetz, P., Pernerstorfer, T., Eichler, H.G., Mullner, C., and Kapiotis, S. (1998). Glucocorticoids dose-dependently increase plasma levels of granulocyte colony stimulating factor in man. *J. Clin. Endocrinol. Metab.* **83**, 1037–1040.

Jinbo, T., Sakamoto, T., and Yamamoto, S. (2002). Serum alpha2–macroglobulin and cytokine measurements in an acute inflammation model in rats. *Lab. Anim.* **36**, 153–157.

Kambayashi, J., Ohshiro, T., Mori, T., and Kosaki, G. (1985). Hemostatic defects in experimental obstructive jaundice. *Jpn. J. Surg.* **15**, 75–80.

Kaneko, J.J. (1999). Carbohydrate metabolism. *In* "The Clinical Chemistry of Laboratory Animals, 2nd edition." (W.F. Loeb and F.W. Quimby, eds.), pp 165–179. Taylor and Francis, Philadelphia, Pa.

Kaneko, J.J., Harvey, J.W., and Bruss, M.L. (1997). "Clinical Biochemistry of Domestic Animals, 5th edition". Academic Press, San Diego, Calif.

Kast, A., and Nishikawa, J. (1981). The effect of fasting on oral acute toxicity of drugs on rats and mice. *Lab. Anim.* **15**, 359–364.

Katein, A.M., O'Bryan, S.M., and Bounous, D.I. (2002). Specimen collection comparison for clinical pathology analysis. Abstract, American College of Veterinary Pathology Meeting, New Orleans, La.

Keenan, K.P., Ballam, G.C., Dixit, R., Soper, K.A., Laroque, P., Mattson, B.A., Adams, S.P., and Coleman, J.B. (1997). The effects of diet, overfeeding and moderate dietary restriction on Sprague-Dawley rat survival, disease and toxicology. *J. Nutr.* **127**, 851S–856S.

Kekki, M., and Eisalo, A. (1964). Turnover of ³⁵S-labeled serum albumin and gamma globulin in the rat: Comparison of the resolution of plasma radioactivity curve by graphic means (manually) and by computer. *Ann. Med. Exp. Finn.* **42**, 196–208.

Kitamura, T., Jansen, P., Hardenbrook, C., Kamimoto, Y., Gatmaitan, Z., and Arias, I.M. (1990). Defective ATP-dependent bile canalicular transport of organic anions in mutant (TR-) rats with conjugated hyperbilirubinemia. *Proc. Natl. Acad. Sci. U S A* **87**, 3557–3561.

Kotal, P., Vitek, L., and Fevery, J. (1996). Fasting-related hyperbilirubinemia in rats: the effect of decreased intestinal motility. *Gastroenterology* **111**, 217–223.

Kren, B.T., Parashar, B., Bandyopadhyay, P., Chowdhury, N.R., Chowdhury, J.R., and Steer, C.J. (1999). Correction of the UDP-glucuronosyltransferase gene defect in the gunn rat model of Crigler-Najjar syndrome type I with a chimeric oligonucleotide *Proc. Natl. Acad. Sci. U S A* **96**, 10349–10354.

Lehman-McKeeman, L.D., Caudill, D., Rodriguez, P.A., and Eddy, C. (1998). 2–sec-butyl-4,5–dihydrothiazole is a ligand for mouse urinary protein and rat alpha 2u-globulin: physiological and toxicological relevance. *Toxicol. Appl. Pharmacol.* **149**, 32–40.

Levin, S., Semler, D., and Ruben, Z. (1993). Effects of two weeks of feed restriction on some common toxicologic parameters in Sprague-Dawley rats. *Toxicol. Pathol.* **21**, 1–14.

Little, J.R., Brecher, G., Bradly, T.R., and Rose, S. (1961). Determination of lymphocyte turnover by continuous infusion of H³ thymidine. *Blood* **19**, 236–242.

Loeb, W.F. (1998). The measurement of renal injury. *Toxicol. Pathol.* **26**, 26–28.

Loeb, W.F., and Quimby, F.W. (1999). "The Clinical Chemistry of Laboratory Animals." Taylor and Francis, Ann Arbor, Mich.

Loeffler, M., Pantel, K., Wulff, H., and Wichmann, H.E. (1989). A mathematical model of erythropoiesis in mice and rats. Part 1: Structure of the model. *Cell Tissue Kinet.* **22**, 13–30.

MacKenzie, W.F., and Alison, R.H. (1990). Heart. *In* "Pathology of the Fischer Rat." (G.A. Boorman, S.L. Eustis, M.R. Elwell, C.A. Montgomery, and W.F. MacKenzie, eds.), pp 461–471, Academic Press, Inc., San Diego, Calif.

Mahl, A., Heining, P., Ulrich, P., Jakubowski, J., Bobadilla, M., Zeller, W., Bergmann, R., Singer, T., and Meister, L. (2000). Comparison of clinical pathology parameters with two different blood sampling techniques in rats: retrobulbar plexus versus sublingual vein. *Lab. Anim.* **34**, 351–361.

Marinkovic, S., Jahreis, G.P., Wong, G.G., and Baumann, H. (1989). IL-6 modulates the synthesis of a specific set of acute phase plasma proteins in vivo. *J. Immunol.* **142**, 808–812.

Mark, M., Walter, R., Meredith, D.O., and Reinhart, W.H. (2001). Commercial taxane formulations induce stomatocytosis and increase blood viscosity. *Br. J. Pharmacol.* **134**, 1207–1214.

Marks, W.H., Genell, S., Hjelmqvist, B., and Ohlsson, K. (1984). Pancreatic secretory proteins in sera from rats before and after induction of experimental pancreatitis. *Scand. J. Gastroenterol.* **19**, 552–560.

Martin, G.R., Meddings, J.B., and Sigalet, D.L. (2003). 3–0 methylglucose absorption in vivo correlates with nutrient absorption and intestinal surface area in experimental short bowel syndrome. *J. Parenteral Nutr.* **27**, 65–70.

Meno-Tetang, G.M., Hon, Y.Y., Van Wart, S., and Jusko, W.J. (1999). Pharmacokinetic and pharmacodynamic interactions between dehydroepiandrosterone and prednisolone in the rat. *Drug Metabol. Drug Interact.* **15**, 51–70.

Moore, D.M. (2000). Hematology of the rat (Rattus novegicus). *In* "Schalm's Veterinary Hematology." (B.F. Feldman, J.G. Zinkl, and N.C. Jain, eds.), pp 1210–1218, Williams & Wilkins, Philadelphia, Pa.

Muraca, M., Fevery, J., and Blanckaert, N. (1987). Relationships between serum bilirubins and production and conjugation of bilirubin. Studies in Gilbert's syndrome, Crigler-Najjar disease, hemolytic disorders, and rat models. *Gastroenterology* **92**, 309–317.

Myrset, A.H., Halvorsen, B., Ording, E., and Helgeland, L. (1993). The time courses of intracellular transport of some secretory proteins of rat liver are not affected by an induced acute phase response. *Eur. J. Cell Biol.* **60**, 108–114.

Nakhooda, A.F., Like, A.A., Chappel, C.I., Murray, F.T., and Marliss, E.B. (1977). The spontaneously diabetic Wistar rat. Metabolic and morphologic studies. *Diabetes* **26**, 100–112.

Neu, B., Randlkofer, P., Neuhofer, M., Voland, P., Mayerhofer, A., Gerhard, M., Schepp, W., and Prinz, C. (2002). Helicobacter pylori induces apoptosis of rat gastric parietal cells. *Am. J. Physiol. Gastrointest. Liver Physiol.* **283**, G309–318.

Neuman, R.G., Hud, E., and Cohen, M.P. (1994). Glycated albumin: a marker of glycaemic status in rats with experimental diabetes. *Lab. Anim.* **28**, 63–99.

Nicholson, J.K., Connelly, J., Lindon, J.C., and Holmes, E. (2002). Metabonomics: a platform for studying drug toxicity and gene function. *Nat. Rev. Drug Discov.* **1**, 153–161.

Nieuwenhuizen, W., Emeis, J.J., and Vermond, A. (1982). Catabolism of purified rat fibrin(ogen) plasmin degradation products in rats. *Thromb. Haemost.* **48**, 59–61.

Nishikawa, T., and Nishimura, M. (2000). Mapping of the beige (bg) gene on rat chromosome 17. *Exp. Anim.* **49**, 43–45.

Nomura, A. (1980). Studies of sulfhemoglobin formation by various drugs (4). Influences of various antidotes on chemically induced methemoglobinemia and sulfhemoglobinemia. *Nippon Yakurigaku Zasshi* **76**, 435–446.

O'Brien, P.J., Dameron, G.W., Beck, M.L., Kang, Y.J., Erickson, B.K., Di Battista, T.H., Miller, K.E., Jackson, K.N., and Mittelstadt, S. (1997). Cardiac troponin T is a sensitive, specific biomarker of cardiac injury in laboratory animals. *Lab. Anim. Sci.* **47**, 486–495.

O'Brien, P.J., Slaughter, M.R., Polley, S.R., and Kramer, K. (2002) Advantages of glutamate dehydrogenase as a blood biomarker of acute hepatic injury in rats. *Lab. Anim.* **36**, 313–321.

Osim, E., Mudzingwa, S., Musabayane, C., Mbajiorgu, F., and Munjeri, O. (1999). The effect of chloroquine on circulating platelet survival in the rat. *J. Cardiovasc. Pharmacol. Ther.* **4**, 97–102.

Pérez, C., Canal, J.R., Dominuez, E., Campillo, J.E., Guillén, M., and Torres M.D. (1997). Individual housing influences certain biochemical parameters in the rat. *Lab. Anim.* **31**, 357–361.

Qi, Z., Whitt, I., Mehta, A., Jianping, J., Zhao, M., Harris, R.C., Fogo, A.B., and Breyer, M.D. (2004). Serial determination of glomerular filtration rate in conscious mice using FITC-inulin clearance. *Am. J. Physiol. Renal Physiol.* **286**, F590–F596.

Ragan, H.A., and Weller, R.E. (1999). Markers of renal function and injury. *In* "The Clinical Chemistry of Laboratory Animals, 2nd edition." (W.F. Loeb, and F.W. Quimby, eds.), pp 549–642, Taylor & Francis, Philadelphia, Pa.

Riley, J.H., and Cornelius, L.M. (1999). Electrolytes, blood gases, and acid-base balance. *In* "The Clinical Chemistry of Laboratory Animals, 2nd edition." (W.F. Loeb, and F.W. Quimby, eds.), pp 549–642, Taylor & Francis, Philadelphia, Pa.

Ringler, D.H., and Dabich, L. (1979). Hematology and clinical biochemistry. *In* "The Laboratory Rat, Volume I: Biology and Diseases." (H.J. Baker, J.R. Lindsey, and S.H. Weisbroth, eds.), pp 105–121. Academic Press, San Diego, Calif.

Robertson, D.G., Reily, M.D., Sigler, R.E., Wells, D.F., Paterson, D.A., and Braden, T.K. (2000). Metabonomics: evaluation of nuclear magnetic resonance (NMR) and pattern recognition technology for rapid in vivo screening of liver and kidney toxicants. *Toxicol. Sci.* **57**, 326–337.

Rose, B.D. (2000). "Clinical Physiology of Acid-Base and Electrolyte Disorders, 5th edition." McGraw-Hill Professional, New York.

Ruot, B., Breuille, D., Rambourdin, F., Bayle, G., Capitan, P., and Obled, C. (2000). Synthesis rate of plasma albumin is a good indicator of liver albumin synthesis in sepsis *Am. J. Physiol. Endocrinol. Metab.* **279**, E244–E251.

Saad, A., Palm, M., Widell, S., and Reiland, S. (2001). Differential analysis of rat bone marrow by flow cytometry. *Comp. Haematol. Int.* **10**, 97–101.

Sanderson, J.H., and Philips, C.E. (1981). Rats. *In* "An Atlas of Laboratory Animal Haematology." pp 38–87, Clarendon Press, Oxford.

Schreiber, G., Tsykin, A., Aldred, A.R., Thomas, T., Fung, W.P., Dickson, P.W., Cole, T., Birch, H., De Jong, F.A., Milland, J. (1989). The acute phase response in the rodent. *Ann. NY Acad. Sci.* **557**, 61–85.

Seed, M.P. (2003). The assessment of inflammation, cartilage matrix, and bone loss in experimental monoarticular arthritis of the rat. *Methods Mol. Biol.* **225**, 161–174.

Seppen, J., Tada, K., Ottenhoff, R., Sengupta, K., Chowdhury, N.R., Chowdhury, J.R., Bosma, P.J., Oude Elferink, R.P. (1997). Transplantation of Gunn rats with autologous fibroblasts expressing bilirubin UDP-glucuronosyltransferase: correction of genetic deficiency and tumor formation *Hum. Gene Ther.* **8**, 27–36.

Shevock, P.N., Khan, S.R., and Hackett, R.L. (1993). Urinary chemistry of the normal Sprague-Dawley rat. *Urol. Res.* **21**, 309–312.

Shi, J., Gilbert, G.E., Kokubo, Y., and Ohashi, T. (2001). Role of the liver in regulating numbers of circulating neutrophils. *Blood* **98**, 1226–1230.

Shimizu, K., Shiratori, K., Hayashi, N., Fujiwara, T., Horikoshi, H. (2000). Effect of troglitazone on exocrine pancreas in rats with streptozotocin-induced diabetes mellitus. *Pancreas* **21**, 421–426.

Smith, P.F., Grossman, S.J., Gerson, R.J., Gordon, L.R., Deluca, J.G., Majka, J.A., Wang, R.W., Germershausen, J.I., and MacDonald, J.S. (1991). Studies on the mechanism of simvastatin-induced thyroid hypertrophy and follicular cell adenoma in the rat. *Toxicol. Pathol.* **19**, 197–205.

Spear, P.A., Higueret, P., and Garcin, H.J. (1994). Effects of fasting and 3,3′,4,4′,5,5′-hexabromobiphenyl on plasma transport of thyroxine and retinol: fasting reverses elevation of retinol. *Toxicol. Environ. Health* **42**, 173–183.

Stolk, J.M., and Smith, R.P. (1966). Species differences in methemoglobin reductase activity *Biochem. Pharmacol.* **15**, 343–351.

Stromberg, P.C. (1992). Changes in the hematologic system. *In* "Pathobiology of the Aging Rat." (U. Mohr, D.L. Dungworth, and C.C. Capen, eds.), pp 15–24, ILSI Press, Washington, DC.

Stromberg, P.C. (1985). Large granular lymphocyte leukemia in F344 rats. Model for human T gamma lymphoma, malignant histiocytosis, and T-cell chronic lymphocytic leukemia. *Am. J. Pathol.* **119**, 517–519.

Stromberg, P.C., Vogtsberger, L.M., and Marsh, L.R. (1983a). Pathology of the mononuclear cell leukemia of Fischer rats. III. Clinical chemistry. *Vet. Pathol.* **20**, 718–726.

Stromberg, P.C., Vogtsberger, L.M., Marsh, L.R., and Wilson, F.D. (1983b). Pathology of the mononuclear cell leukemia of Fischer rats. II. Hematology. *Vet. Pathol.* **20**, 709–717.

Swenberg, J.A., and Lehman-McKeeman, L.D. (1999) Alpha 2–urinary globulin-associated nephropathy as a mechanism of renal tubule cell carcinogenesis in male rats. *IARC Sci. Publ.* **147**, 95–118.

Tanaka, H., Sano, N., and Takikawa, H. (2003). Biliary excretion of phenolphthalein sulfate in rats. *Pharmacology* **68**, 177–182.

Torous, D.K., Hall, N.E., Murante, F.G., Gleason, S.E., Tometsko, C.R., and Dertinger, S.D. (2003). Comparative scoring of micronucleated reticulocytes in rat peripheral blood by flow cytometry and microscopy. *Toxicol. Sci.* **74**, 309–314.

Travlos, G.S., Morris, R.W., Elwell, M.R., Duke, A., Rosenblum, S., and Thompson, M.B. (1996). Frequency and relationships of clinical chemistry and liver and kidney histopathology findings in 13–week toxicity studies in rats. *Toxicology* **107**, 17–29.

van Furth, R. (1989). Origin and turnover of monocytes and macrophages. *Curr. Top. Pathol.* **79**, 125–150.

van Gool, J., Boers, W., Sala, M., and Ladiges, W.C. (1984). Glucocorticoids and catecholamines as mediators of acute-phase proteins, especially rat alpha-macrofoetoprotein. *Biochem. J.* **220**, 125–132.

Van Vleet, J.F., and Ferrans, V.J. (1986). Myocardial diseases of animals *Am. J. Pathol.* **124**, 98–178.

Vittori, D., Nesse, A., Perez, G., and Garbossa, G. (1999). Morphologic and functional alterations of erythroid cells induced by long-term ingestion of aluminium. *J. Inorg. Biochem.* **76**, 113–120.

Vranckx, R., Rouaze-Romet, M., Savu, L., Mechighel, P., Maya, M., and Nunez, E.A. (1994). Regulation of rat thyroxine-binding globulin and transthyretin: studies in thyroidectomized and hypophysectomized rats given tri-iodothyronine or/and growth hormone. *J. Endocrinol.* **142**, 77–84.

Wan Nazaimoon, W.M., and Khalid, B.A. (2002). Tocotrienols-rich diet decreases advanced glycosylation end-products in non-diabetic rats and improves glycemic control in streptozotocin-induced diabetic rats. *Malays. J. Pathol.* **24**, 77–82.

Waner, T., and Nyska, A. (1994). The influence of fasting on blood glucose, triglycerides, cholesterol, and alkaline phosphatase in rats. *Vet. Clin. Pathol.* **23**, 78–80.

Ward, K.M., Rehm, S., and Reynolds, C.W. (1990). Tumours of the haematopoietic system. *In* "Pathology of Tumours in Laboratory Animals, Vol. 1–Tumors of the Rat, 2nd edition." (V. Turusov, and U. Mohr, eds.), pp 625–645. Oxford University Press, Oxford.

Wood, R.J., and Han, O. (1998). Recently identified molecular aspects of intestinal iron absorption. *J. Nutr.* **128**, 1841–1844.

Zucker, L.M., and Zucker, T.F. (1961). Fatty, a new mutation in the rat. *J. Hered.* **52**, 275–278.

Zunic, G., Rolovic, Z., Basara, N., Simovic, M., and Vasiljevski, M. (1993). Decreased plasma proteins, increased total plasma-free amino acids, and disturbed amino acid metabolism in the hereditary severe anemia of the Belgrade laboratory (b/b) rat. *Proc. Soc. Exp. Biol. Med.* **203**, 366–371.

Chapter 6

Reproduction and Breeding

Jeffrey J. Lohmiller and Sonya P. Swing

I.   INTRODUCTION

Rats have been used extensively in various areas of reproductive research, including fecundity and fertility studies, behavioral aspects of reproduction, and screening of compounds for teratogenic effects. Familiarity with and understanding of normal reproductive traits and behavior of rats are imperative for these areas of scientific investigation. The increasing production of unique genetically modified rat strains and stocks necessitates a greater understanding of reproduction by those maintaining these valuable lines.

II.   NORMATIVE BIOLOGY

A.   Determination of Sex

Determination of sex is most easily performed in adult rats by evaluating the anatomical structures in the perineal region. Males normally have a pronounced scrotum located between the anus and urethral opening where none is present in the female. However, it should be noted that the male retains the ability to retract the testes into the abdomen. Alternatively, comparison of the anogenital distance, the distance between the anus and the urethral opening, can be used to differentiate the sexes. The anogenital distance is larger in the male compared to the female (Figure 6-1). In rat pups less than two weeks of age, sex determination may be more challenging due to the relatively small size of the pups. However, sex can be accurately determined based on the greater anogenital distance in male compared to female pups. The presence of testes in pre-weanling pups cannot reliably be used to aid in sex determination of rat pups because the testes normally descend into the scrotum at approximately 15 days of age (Russell, 1992). Observation of mammae can be used to positively identify females, as males do not possess nipples (Cowie, 1984).

B.   Puberty and Gonadotropins

Puberty is defined as sexual maturity with the ability to produce viable young (Fox and Laird, 1970). The age of pubertal onset varies greatly between strains or stocks and is influenced by a number of other factors. For example, rats on a high nutritional plane and rats from smaller litters have earlier onset of puberty (Bennett and Vickery, 1970). Conversely, female rats born from an intrauterine

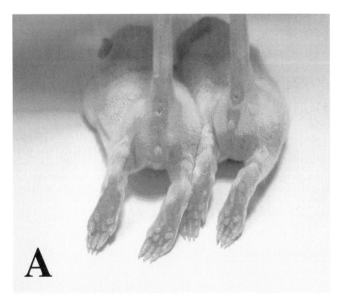 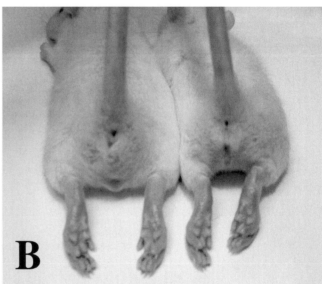

Fig. 6-1   Sex differentiation of SD rats. The photograph in panel A depicts the relatively longer anogenital distance in the male (left) versus the female (right) in two-week-old rats. Panel B depicts the anogenital distance in six-week-old male (left) and female (right) rats.

position between two males are slightly delayed in reaching puberty (Vandebergh, 2003). A number of measures have been proposed to predict the onset of puberty. Bennett and Vickery (1970) observed that puberty typically occurs at approximately 50% of the mature body weight, but that body length of 148 to 150 mm is a more reliable indicator of puberty than age or weight. The exact stimulus for pubertal onset has not been fully identified. Leptin has been proposed as a trigger for puberty, though the influence may be due to leptin's ability to influence the availability of metabolic fuels (Cunningham et al., 1999). Nazian and Cameron (1999) suggested a correlation between increased concentrations of leptin and testosterone at puberty. Thus, leptin may be involved not only with pubertal onset but with secondary sex organ development.

In the male rat, puberty is accompanied by descent of the testes into the scrotum and the onset of spermatogenesis. Sperm are first produced in the testes about day 45 of age, but optimal production does not occur until about 75 days (Russell, 1992). Gonadotropin-releasing hormone (GnRH) secretion results in increased secretion of both luteinizing hormone (LH) and follicle-stimulating hormone (FSH) and corresponds to increasing testosterone levels during puberty. LH stimulates Leydig cells to increase testosterone production. The exact role of FSH remains unclear in the male (Setchell, 1982). Hypophysectomy prior to puberty prevents pubertal changes. Administration of LH and FSH or testosterone and FSH after hypophysectomy will restore spermatogenesis (Setchell, 1982).

Puberty in the female rat corresponds with vaginal opening and the first proestrus. The opening of the vaginal orifice occurs from 33 to 42 days after birth, at a body weight of approximately 100 grams (Maeda et al., 2000). Regular estrous cycles begin about one week after the vaginal opening. Ojeda et al. (1986) proposed four phases of prepubertal ovarian development based on morphology and gonadotropin levels: a neonatal period from birth to 1 week, an infantile period from 1 to 3 weeks, a juvenile period from 3 to 4.5 weeks, and a final peripubertal period lasting approximately three days. During the neonatal period, the ovary begins to convert testosterone to estadiol-17β in response to FSH.

The infantile period is one of follicular development in response to both FSH and LH. During the juvenile stage, the ovary goes through a transition, with follicles capable of estrogen secretion following appropriate hormonal stimulation. During the three-day peripubertal phase, the uterus fills with fluid, follicles secrete large amounts of estrogen, and gonadotropin spikes lead to the first ovulations. Thus, ovarian development is dependent on the gonadotropins, FSH, LH, GnRH, prolactin, and growth hormone (GH). FSH levels increase from birth, peak by day 12, and then gradually decrease. LH concentrations elevate moderately and rise dramatically during hormone release surges. GnRH in the hypothalamus increases from birth until puberty. Prolactin release occurs approximately every three hours during the juvenile stage, resulting in estradiol and progesterone secretion from the ovary (Advis and Ojeda, 1979). GH pulses follow a pattern similar to that of prolactin and are associated with ovarian steroidogenesis.

C. Estrous Cycle

1. Cycle Description

The rat estrous cycle averages four to five days in length, occurs throughout the year without seasonal influence in laboratory colonies, and occurs from pubertal onset until senescence, including during the postpartum period (Bennett and Vickery, 1970; Kohn and Clifford, 2002). The cycle consists of four stages: proestrus, estrus, metestrus, and diestrus (see Table 6-1). Hormonal fluctuations regulated by gonadotropins secreted by the anterior pituitary result in ovarian and follicular changes as well as detectable changes in vaginal cytology. LH profiles show a pulsatile fluctuation, with highest frequencies during proestrus and lowest during estrus. Each LH pulse corresponds to a GnRH pulse (Gallo, 1981a, 1981b; Dluzen and Raimirea, 1986; Levine and Duffy, 1988; Levine and Powell, 1989; Levine et al., 1991). LH surges during proestrus stimulate the preovulatory follicles to ovulate and form corpora lutea.

Two periovulatory FSH surges stimulate growth of small follicles (Peluso, 1992). The first surge correlates to the LH surge, whereas the second pulse is associated with

TABLE 6-1

BEHAVIOR AND VAGINAL CYTOLOGY WITH ESTROUS CYCLE PHASES

Cycle phase	Duration (hrs.)	Behavior	Vaginal smear morphology
Proestrus	12	Male acceptance at end of phase	Nucleated epithelial cells
Estrus	12	Lordosis; male acceptance	75% nucleated cells; 25% cornified cells
Metestrus	21	No male acceptance	Many leukocytes with nucleated and cornified cells
Diestrus	57	No male acceptance	Leukocytes

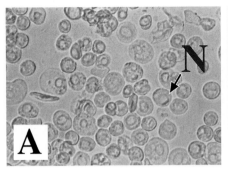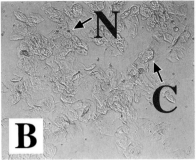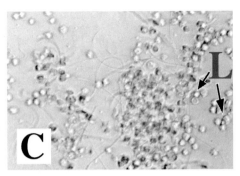

Fig. 6-2 Estrus detection in vaginal smears. Panel A shows an unstained vaginal swab during proestrus, panel B shows an unstained vaginal swab during estrus, and panel C shows an unstained vaginal swabs during diestrus. Magnification of all images is 20×. N = nucleated cell; C = cornified cell; L = leukocyte. (Photomicrographs courtesy of Dr. Yuksel Agca.)

a decrease in inhibin after ovulation (Ackland et al., 1990; Haisenleder et al., 1990; Watanabe et al., 1990). During metestrus, the corpora lutea secrete progesterone. The progesterone levels decline during diestrus, and follicular development is associated with an increase in estradiol-17β. The cycle is completed as estrogen peaks during proestrus stimulate gonadotropin release that triggers ovulation (Freeman, 1988).

Pseudopregnancy lasting 12 to 14 days occurs when the female receives cervical stimulation by a vasectomized male or mechanical stimulation such as with a glass rod (Maeda et al., 2000). The cervical stimulation results in both diurnal and nocturnal serum prolactin surges that continue throughout the pseudopregnancy (Smith et al., 1975). The prolactin surges increase the number of luteal LH receptors and thus stimulate the corpora lutea of the estrous cycle to produce progesterone, which maintains the pseudo-pregnant state (Niswender and Nett, 1988). The leutolytic mechanism that terminates pseudopregnancy is still unknown. However, splenic macrophages have been proposed in regulation (Takahashi et al., 1989; Matsuyama et al., 1990, 1992; Matsuyama and Takahashi, 1995).

2. Estrous Cycle Stage Identification

A number of methods have been used to detect the phase of the estrous cycle to maximize breeding success in the rat. Techniques range from observation of behavioral changes to examination of vaginal cytology to measurement of electrical impedence in the vagina.

The duration of the estrus phase is 9 to 15 hours and is defined as the period when the female is sexually receptive and will allow copulation. Behavioral changes that indicate acceptance of the male include increased running activity, ear quivering resulting from stroking of the head or back, and pronounced lordosis (dorsiflexion of the vertebral column) upon pelvic stimulation (Fox and Laird, 1970; Pfaff, 1980). The vaginal wall appears dry during estrus and the vulva is swollen (Baker, 1979).

Examination of the cellular morphology of vaginal smears (Figure 6-2) is a method widely used to detect the phase of the estrous cycle. Accurate phase identification is dependent on smears taken at a fixed time during the day, as cell populations vary throughout a 24-hour period. Nucleated epithelial cells are characteristic of proestrus, which lasts approximately 12 hours (Sharp and LaRegina, 1998). It should also be noted that during proestrus, particularly in the prepubertal rat, the uterus may appear fluid-filled. This should not be confused with hydrometra (Kohn and Clifford, 2002). During the estrous phase, cytology consists initially of approximately 75% nucleated cells and 25% cornified cells, which then progresses to a majority of swollen cornified cells without nuclei. In the later hours of estrus, the cornified cells appear degenerate and may appear to form an epithelial pavement. Metestrus follows shortly after ovulation and lasts approximately 21 hours. The vaginal cytology during metestrus consists of many leukocytes along with nucleated cells and corni-fied cells. Diestrus is the longest phase, lasting roughly 57 hours, and the vaginal smear consists primarily of leukocytes (Fox and Laird, 1970; Maeda et al., 2000).

Bartos first described (1977) the use of an impedence meter to detect electrical resistance of the vaginal mucous membrane by insertion of an electrical probe into the vagina. A peak in impedence occurs during proestrus, with the lowest resistance during estrus. Many investi-gations have been conducted to refine this method of estrus detection (Koto et al., 1987a, 1987b; Ramos et al., 2001).

3. Estrous Synchronization

A study to improve the accuracy of timed mating compared the use of dry and pre-moistened swabs for vaginal cytology. During this study, the estrous cycle was synchronized in 50 to 75% of female SD rats by means of vaginal swabbing over two to three consecutive days and by acclimating female rats to the light cycle and to males within the room (Harris and Kesel, 1990). A high percentage of female rats will synchronize their estrous cycle three days after being withdrawn from daily injections of progesterone (Shirley, 1978). Stimuli associated with introduction of rats to a new environment caused the rats to begin new estrous cycles. Rats with four- and five-day cycles responded differently, but within both groups partial synchronization of estrous cycles occurred (Shirley, 1978).

D. Mating and Reproductive Behavior

Three classical reproductive phenomena of mice are thought to either not exist or to be much weaker in rats. Rats do not show the Bruce effect, in which implantation of embryos is delayed with the presence of pheromones from newly introduced males. Likewise, the Whitten effect, in which female mice undergo estrus synchronization when a male is introduced, is lacking in rats. The Lee-Boot effect, in which groups of females housed together become anestrous, is thought to exist in rats, although not as markedly as in mice (Sharp and LaRegina, 1998).

As stated previously, lordosis is a characteristic mating behavior of the female rat. Although both estrogen and progesterone are involved in the reflex, estrogen alone is sufficient to induce the behavior (Maeda et al., 2000). Testosterone is essential to the exhibition of mating behavior in male rats. Castrated males show no sexual behavior (Meisel and Sachs, 1994). Olfactory cues from pheromones are also critical to male sexual behavior (Nelson, 1995). Auditory stimuli play important roles in the reproductive behavior of both sexes. A 50-kHz vocalization is produced by both males and females during copulation. This call is produced by males in response to females in estrus, and by the female to solicit male attention (Maeda et al., 2000). The male rat also produces a 22-kHz vocalization in the post-ejaculatory refractory period (Barfield and Thomas, 1986).

Copulation in rats occurs most often during the latter portion of the dark cycle (Mercier et al., 1987). The male initiates mating behavior with genital sniffing of a female in estrus. The receptive female displays hopping and ear-quivering, resulting in male mounting, which in turns solicits lordosis from the female. Mounting behavior consists of combinations of intromission and ejaculation, including mounting without intromission; mounting with intromission but without ejaculation, frequently with a backward lunge; and mounting with intromission and ejaculation. An intromission typically consists of two to nine pelvic thrusts and lasts 0.3 to 0.6 seconds (Bennett and Vickery, 1970). Ejaculation is typically preceded by 3 to 44 intromissions, and a non-responsive refractory period occurs after ejaculation. Mating continues intermittently for up to three hours, until 3 to 10 ejaculations have been achieved.

Mating can be confirmed by the presence of sperm in a vaginal smear, by observation of a vaginal plug, or by direct observation of the mating behavior. The vaginal plug does not persist as long in the rat as it does in the mouse, and thus the lack of a plug is not a reliable indicator that mating did not occur. Vaginal smears that detect sperm-positive animals are commonly used to confirm mating, and a 90 to 94% correlation between sperm-positive vaginal smears and pregnancy has been reported (Baker, 1979).

E. Pregnancy and Pregnancy Detection

Conception rates of greater than 85% are reported in outbred stocks of rats, with slightly lower rates for inbred strains (Kohn and Clifford, 2002). Thus, detection of sperm in the vaginal smear following coitus is an excellent predictor of pregnancy. Fetuses may be palpated as early as 10 days of gestation, but palpation is more accurate after 12 days. Abdominal enlargement is usually visible by day 13 of gestation (Baker, 1979). By day 14, mammary development and nipple enlargement can be observed (Bennett and Vickery, 1970).

F. Gestation

Gestation averages 21 to 23 days from copulation to parturition, but may be longer if the dam is suckling a litter (Bennett and Vickery, 1970). Implantation of the hatched blastocyst occurs on day 5 of gestation (Maeda et al., 2000). Unlike other animal species, the rat embryo does not produce estrogen and maternal estrogen regulates implantation (Dey and Johnson, 1986). During the first half of pregnancy, progesterone is produced by the ovary, stimulated by prolactin surges induced by coitus (Smith et al., 1975). During the second half of gestation, progesterone is also produced by the placenta (Gibori et al., 1988). The ovary continues to produce estrogen throughout gestation to activate and sustain the corpus luteum of pregnancy. Estrogen is not produced by the rat placenta, and progesterone is not produced in amounts sufficient to maintain pregnancy (Gibori et al., 1988).

Placentation in the rat is discoidal and hemochorial (Enders 1965; Kaufmann and Burton, 1994). Discoid placentation indicates a circular area of attachment between the fetal and maternal tissues. Hemochorial placentation indicates that the fetal trophoblasts invade the maternal vessels and are in direct contact with the maternal blood. The placenta secretes polypeptides known as placental lactogens, which serve to stimulate mammary development and stimulate the corpora lutea to produce progesterone (Soares et al., 1991; Sores et al., 1998).

G. Parturition

Nest-building behavior in the female rat begins about five days prior to parturition and is maintained throughout lactation (Bennett and Vickery, 1970). Relaxation of the pubic symphysis begins by day 17 of gestation and is complete prior to parturition. The hormone relaxin, produced by the corpus luteum during the second half of pregnancy, is responsible for cervical extensibility as well as relaxation of the pubic symphysis (Samuel et al., 1986).

The exact physiological stimulus for parturition is not fully understood. It has been proposed that estrogen could play a role as ovarian estradiol secretion increases during the final 24-hour period of gestation (Thorburn and Challis, 1979). Higuchi et al. (1986) have demonstrated an elevation in oxytocin secretion by the pituitary that correlates with uterine contraction. The Ferguson reflex is a neuroendocrine reflex in which the fetal distension of the cervix stimulates a series of neuroendocrine responses, leading to oxytocin production.

Vaginal discharge is evident 1.5 to 4 hours prior to birth of the first pup (Bennett and Vickery, 1970). During labor, the female extends her body while moving about the cage at intervals decreasing from 2 minutes to 15 seconds. She eventually rests her abdomen on the cage floor with her hind limbs extending off the floor. Licking of the vulva precedes delivery. The female assumes a semi-crouched position during delivery.

The entire delivery process varies with litter size and ranges from 55 minutes to nearly 4 hours, with 1.5 hours as an average (Baker, 1979). Fetal evacuation begins at the cervical end of the uterine horn and proceeds until the horns are entirely empty. Presentation tends to be alternatively breech and head-first. Dystocia is very uncommon in the rat. As the pups are delivered, the dam pulls the placenta from the birth canal and eats it (Kohn and Clifford, 2002). After placental consumption, the dam licks the young and removes the amniotic covering. Pups are typically not nursed until the entire litter is born.

Many variables affect litter size, including stock, strain, and maternal age. The second litter is typically the largest, and reproductive function declines after nine months (Bennett and Vickery, 1970; Niggeschulze and Kast, 1994). Rosen et al. (1987) examined the effect of the age at first mating on litter size and weaning weight of Wistar rats. Maternal ages of 70 or 105 days resulted in larger litters and higher weaned weights, but also increased pup mortality, compared to dams bred at 35 days of age. Sprague-Dawley rats have been shown to produce litters reduced by three or more pups when bred at first estrus in comparison to breeding of more mature females (Evans, 1986). The sex of littermates also influences the subsequent litter size of female rats. A study by Sharpe et al. (1973) showed that litter size was larger in female rats raised with both male and female littermates when compared to all female littermates. Cannibalism is rare in the rat and indicates maternal stress. Gonzalez and Deis (1990) showed that maternal behavior is enhanced in older rats of 19 to 20 months when compared to 3 to 4-month-old dams.

H. Lactation and Pup Development

Six pairs of mammary glands and nipples are present in the female rat: three pair in the thoracic region and three pair in the abdominal-inguinal region. The mammary glands and nipples enlarge during lactation, extending from the parotid glands to the anal region.

Lactation is induced without a suckling stimulus, but requires suckling for maintenance. The typical lactation period is three weeks, but lactation can be maintained for up to 70 days if suckling pups are continually provided (Bennett and Vickery, 1970). Initiation of lactation is dependent on prolactin and glucocorticoids, with plasma corticosterone increasing during the last few days of gestation (Yokahama and Ota, 1978). The estrous cycle and ovulation are delayed during lactation and resume following cessation of nursing.

Gestational development of the mammary gland and milk production are influenced by placental hormones and lactogens (Tucker, 1994). The role of placental lactogen has been demonstrated through maintenance of milk production even with removal of the anterior pituitary during pregnancy. Rat milk consists (on a percentage weight/volume basis) of 10.3% fat, 6.4% casein, 2.0% whey protein, 2.6% lactose, and 1.3% minerals (Nagasawa et al., 1983). Milk yield peaks about day 17 of lactation and then gradually declines.

A neuroendocrine loop (known as the "milk ejection reflex") is triggered by the suckling stimulus of the pups and results in oxytocin release from the posterior pituitary, which causes contraction of the myoepithelial cells and ejection of milk (Lincoln, 1984). As the milk is ejected, which may be up to 10 minutes following attachment of pups to the nipple, the suckling pups stretch all their legs in a typical nursing behavior. Pups suckle 50 to 80 times a day, with the pups remaining attached to the nipple for up to a total of 18 hours a day. Antibodies contained in the maternal milk can be transferred across the intestinal mucosa in the pup until approximately 21 days of age (Martin et al., 1997).

Ovulation ceases in the dam during vigorous lactation due to high prolactin levels. However, an LH surge and postpartum ovulation occur within 24 hours of parturition (Maeda et al., 2000). If successful fertilization occurs, implantation may be blocked and the blastocysts remain in the uterus until the end of lactation. Implantation is delayed due to the suppression of progesterone from the corpora lutea as well as ovarian estrogen. Food restriction in lactating dams during the first two weeks of lactation will prolong lactational diestrus (Woodside and Jans, 1995).

Rat pups are born hairless, with closed eyes and ear canals, and with poorly developed limbs and short tails (Baker, 1979). An inverse relationship exists between litter size and birth weight (Romero et al., 1992), an important factor for reproductive toxicologists to consider in evaluating compounds. The ear canal opens between 2.5 and 3.5 days of age, with mature morphology of the cochlea and organ of Corti by weaning. Although vocalization begins at birth, the ability to hear develops around nine days of age (Feldman, 1992). The eyelids open within 14 to 17 days, but the retina is not fully developed until 30 to 40 days of age. Pups are fully haired within 7 to 10 days of age. Incisors begin to erupt at six to eight days of age, but molars erupt from 16 to 34 days (Weisse, 1992).

I. Weaning

Weaning typically occurs at about 21 days of age, when pups are able to eat and drink. Weaning earlier than 17 days may cause urinary disease due to lack of maternal stimulation to induce urination (Kohn and Clifford, 2002). The suckling stimulus is required for continued lactation. Removal of rat pups from the dam results in a drop in plasma prolactin levels within hours (Maeda et al., 2000).

J. Male Reproductive Development

Large numbers of primary germ cells are present at birth, with approximately 75% degenerating during the first week of life. Each testicle has approximately 20 to 30 seminiferous tubules tightly coiled within, with no divisions by fibrous tissue (Russell, 1992). Spermatogenesis occurs in the seminiferous tubule and is characterized by a cycle of recognizable cell types and cell associations, termed the spermatogenic cycle. The spermatogenic cycle, based on cellular changes over time, is separated into distinct phases defined by cellular morphology and associations and by relative position of the developing spermatocyte within the tubule. All cells within a confined region of the seminiferous tubule are synchronized in the same phase of the spermatogenic cycle, with adjacent regions in either previous or subsequent phases.

The spatial developmental change in spermatozoa seen over the seminiferous tubule is called the spermatic wave. The spermatogenic cycle lasts approximately 12 days, and must be completed four times from the first division of a spermatogonia to produce spermatozoa in the tubule lumen (Setchell, 1982). The time from spermatogonia to spermatozoa is relatively consistent and requires 48 to 52 days in the rat (de Kretser, 1982). Spermatogenic cycles begin in very young rats, but the cycles are incomplete and irregular until puberty, when spermatogenesis becomes more coordinated and regular (Russell, 1992). From the lumen of the seminiferous tubules, spermatozoa collect in the tail of the epididymis and take approximately eight days to traverse the head of the epididymis (Russell, 1992).

K. Reproductive Senescence

Fertility in female rats begins to drop by seven to nine months of age. The drop in fertility is not related to decreased ovulation or fertilization rate but to increased losses at pre- and post-implantation due to reduction of oocyte viability with maternal age (Mattheij and Swarts, 1991). Proestral LH surges are lower in middle-age rats, but the use of hormone supplements to increase the LH surge does not increase the fecundity of aging female rats. Rats with regular four- or five-day estrous cycles have increased viability of ovulated ova compared to those with prolonged cycles (Mattheij and Swarts, 1991).

In male rats, there appears to be prolonged reproductive capacity compared to females. Although two-year-old male rats are able to sire offspring, their fertility is suboptimal. Seminiferous tubular atrophy was reported in 20% of Sprague-Dawley rats by 24 months of age, and in 100% of F344 by 18 months of age (Takahashi, 1992). Longitudinal reproductive studies carried out in male Long-Evans rats comparing plasma hormone levels and mating behavior showed that LH and testosterone declined steadily with age. LH was significantly lower by 13 months of age, and testosterone dropped by 15 months of age compared to seven-month-old males (Smith et al., 1992). Sexual dysfunction was noted by 11 months of age, with significant increases in mount latency, intromission latency, ejaculatory latency, and post-ejaculatory interval. Reproductive behavioral decline was not directly correlated with decline in testosterone and LH.

To maximize the production of offspring, Gridgeman and Taylor (1972) mathematically determined that SD breeder cages should be maximally replaced at intervals of 27 to 28 weeks. Most commercial breeders routinely retire breeder rats by about 10 months of age to maintain optimum reproductive performance.

III. FACTORS AFFECTING BREEDING PERFORMANCE

A. Light

Rats are continual breeders in the laboratory setting, without any indication of seasonality. Constant light exposure has been demonstrated to result in vaginal opening six days earlier than females held in a 12-hour-light/12-hour-dark cycle (Fiske, 1941). Constant light has also been demonstrated to induce cystic ovaries and persistent estrus (Shirley 1978; Maeda et al., 2000). Continual exposure of SD rats to low levels of indirect light (30 lux), as would occur by a light left on in an anteroom during the dark cycle, leads to decreased numbers and size of corpora lutea and increased numbers of tertiary follicles and ovarian cysts, indicative of ovarian atrophy (Beys et al., 1995). However, exposure of rats to light levels of 0.2 lux during the dark period were insufficient to produce ovarian changes (Beys et al., 1995).

Changes in photoperiod can alter reproductive characteristics in rats. Extending the light cycle from 12 to 16 hours a day changed the length of the estrous cycle in a majority of SD rats from four days to five days. However, no effect was noted on the estrous cycle of Blue Spruce rats (Clough, 1982). Timing of ovulation in rats occurs approximately two hours after the induction of the dark cycle. With complete reversal of the light cycle, synchronization of the ovulation cycle required about two weeks in rats (Clough, 1982).

B. Noise

Equipment emitting 40 Hz, a frequency within the range to which rats are most sensitive, has been noted to disrupt maternal behavior (Clough, 1982). Fire alarms in the most sensitive hearing range of rats (i.e., those in the ultrasound range near 40 kHz) will alter the estrous cycle. Alarms (92 dB for 1 minute, three times a day for four consecutive days) within the audible range of rats, but not in the most sensitive range, did not alter the estrous cycle based on vaginal cytology results (Gamble, 1976).

C. Vibration

Female SD rats naturally exposed to a 7.1-magnitude earthquake on days 7, 8, or 9 of gestation demonstrated no teratogenic effects on their pups, but there was an increase in fetal resorptions when compared to rats not exposed to an earthquake (Fujinaga et al., 1992). Mechanically induced whole-body vibration of 9 to 11 day gestation female Wistar rats resulted in decreased uterine blood flow, increases in plasma corticosterone, and decreases in plasma progesterone and prostaglandin E2–but no changes in plasma estradiol and prostaglandin F2–compared to control rats (Nakamura et al., 1996).

Wistar rats exposed to a cage environment with a vibrating floor at varying ages were preference tested to determine whether they would choose a stable or vibrating environment. Rats exposed to the vibrating environment all chose the stable environment over the vibrating environment. However, rats exposed at 1 to 21 days of age avoid the vibrating environment significantly less than rats exposed when older than 21 days of age (Soskin, 1963).

D. Pheromones and Odors

The potential exists for pheromones to interfere with estrous cycles. Sharp et al. (2002) evaluated the effects of presentation of dried blood, feces, and urine from stressed rats to other rats. They reported induction of transient stress as measured by blood pressure and heart rate, but the stress was not sufficient to alter the estrous cycle, regardless of the phase of the cycle the females were in at the time of exposure.

E. Litter Size and Characteristics

Male rat pups from large litters (14 pups/dam) had lower body weights at weaning and had delayed hypothalamic and testicular maturation compared to male pups from small litters (6/dam). However, pups from large litters that were allowed to increase food intake prior to weaning did not exhibit the same hypothalamic and testicular changes (Bourguignon et al., 1992). This information suggests that fostering some pups from very large litters may positively impact the later fertility of males in the litter.

F. Diet and Food Intake, Caloric Restriction, and Phytoestrogens

Modifications in nutritional status of rats during pre- and postnatal development, as well as during their reproductive life, may modify future reproductive success. Engelbregt et al. (2001) studied pre- and postnatal nutritional restrictions using uterine artery ligation to mimic retarded intrauterine growth and enlargement of litter size to 20 pups two days postnatal to mimic food restriction. They reported that in both simulations of undernutrition by pubertal onset both males and females had similar body mass indices, body composition, and leptin levels compared to controls. Bourguignon et al. (1992) similarly restricted feed by increasing litter size. They noted that feed-restricted male pups gained more weight post weaning and had accelerated hypothalamic and testicular development compared to control male pups. By puberty, the feed-restricted male rats were similar to control male rats, suggesting that a period of early food restriction could be potentially reversed.

Despite the aforementioned studies demonstrating negligible effects of undernutrition pre- and postnatally, effects on rat pups from undernutrition during pregnancy and lactation altered future reproductive behavior in Wistar versus Lister outbred rats (Tonkiss et al., 1984). The behavioral changes in male Lister rats included reduced numbers of non-penetrative mounts and a shortened latency to ejaculation compared to age-matched pups from *ad libitum* fed dams. Mount latency was reduced in Wistar rats but not in Lister rats.

Simulation of undernutrition via feed restriction of breeding female rats to 50% of *ad libitum* resulted in delays in conception and decreased number of pups and milk production (McGuire et al., 1995). Decreased breeding performance of chronic feed-restricted female rats (60 to 75% of *ad libitum*) was less pronounced in second-litter pups compared to controls presumably due to weight gains in the dams between litters (Fischbeck and Rasmussen, 1987). Young and Rasmussen (1985) reported that the impact to pups born to chronically feed-restricted dams was proportional to the degree of feed restriction. Feed restriction of lactating dams for either the first two weeks or the last two weeks of lactation resulted in undernourishment of the pups as well as prolongation of lactational diestrus (Woodside and Jans, 1995; Woodside et al., 1998).

Restriction of feed access opportunities can have consequences on future female reproductive performance. Parshad (1990) limited feed access of 21-day-old weanling rats for nine weeks. Groups of female rats restricted to feeding for 2, 4, and 8 hours had delayed onset of puberty, smaller ovaries, lower body weights, and greater ovarian follicular atresia compared to rats limited to 12 hours or continual feed access. The lowered reproductive status was believed to be due to nutritional deficiency (Parshad, 1990). Other studies of feed restriction have noted that providing an equivalent of 50% of *ad libitum* calories to female rats resulted in cessation of the estrous cycle due to disruption of gonadotropin secretion (Spranger and Piacsek, 1997). A fast of 48 hours duration at the start of estrus decreases LH secretions and prevents ovulation in rats (Meada et al., 2000). Thus, the extent of resulting reproductive changes varies with modification of nutritional plane and depends on the age at onset, chronicity, and severity of undernutrition.

The effects of phytoestrogens on reproduction in laboratory rats are currently being debated. Yang and Bittner (2002) reviewed the varied results from several studies and concluded inconsistencies to be in part due to the complex interactions between, and sources of, compounds with estrogen-like activities. Santti et al. (1998) investigated the effects of phytoestrogens on male rats. No effects were noted at doses similar to those in soybean feed. However, artificially high doses induced DES-like lesions (Santti et al., 1998). Exposure of male rat pups to phytoestrogens during lactation decreased testicular but not serum testosterone near weaning and caused altered sexual behavior in adulthood (Santti et al., 1998). Further studies on the effects of phytoestrogens will be necessary to more clearly elucidate the exact effects on reproduction and breeding in rats.

G. Health

Chemically induced colitis resulted in delayed onset of puberty in males, smaller accessory sex glands, and lower plasma testosterone levels compared to controls. Colitis also delayed the onset of puberty in females and disrupted the estrous cycle (Azooz et al., 2001). Colitis potentiated undernutrition in rats and exacerbated the reproductive abnormalities described previously.

H. Caging (Micro and Macro Environments)

In vivo exposure of rat testes to 43° C for 30 minutes results in a 50% weight loss in the testes and disrupts spermatogenesis (Setchell, 1982). High temperature also increases embryonic deaths, reduces litter size and growth rate, and negatively impacts lactating females (Clough, 1982). Exposure of male SD rats age 4 to 66 days to elevated environmental temperatures (74 to 100° F) due to HVAC failure resulted in testicular changes. Approximately three weeks after the temperature insult, 25% of the surviving males were identified with irreversible bilateral testicular atrophy characterized by degeneration of seminiferous tubules and spermatocyte maturation failure. Males in utero during this episode did not have reproductive deficiencies (Pucak et al., 1977).

I. Handling and Experimental Stress

Minimally invasive research procedures (restraint and SC and IV tail injections) induced stress responses based on mean arterial pressure and heart rate but did not alter the estrous cycle, regardless of the phase of the cycle the female was in. Similarly, exposure of female rats to cage changing, entry into rooms, and decapitation of conspecifics resulted in stress based on elevation in heart rate and mean arterial pressure, but did not result in alterations of estrous cycle (Sharp et al., 2002). Immobilization or induction of acute stress in female rats resulted in suppression of LH pulses independent of estrogen (Maeda et al., 2000).

IV. BREEDING

A. Breeding Schemes

1. Breeding

a. BREEDING AGE.   Consideration must be given to the age of onset of puberty for determination of optimal mating age. Male rat fertility begins to increase at puberty and is generally maximal at approximately 75 days of age. Commercial breeders strive for reproductive efficiency. Therefore, many begin to mate rats at approximately 10 weeks of age. Rats younger than 10 weeks of age may not obtain maximal reproductive performance until later, but are fertile and capable of successful breeding. Certain rat strains or stocks, such as genetically modified rats with abbreviated reproductive lives, may require more intensive mating at early ages to maintain the line. In these instances, greater numbers of rats may need to be mated to overcome reproductive inefficiencies seen in young rats, especially for young male rats.

b. MATING FORMAT.   For rat breeding colonies, the mating format used is an important consideration for determining size of the colony, production of progeny, and the required levels of caretaker involvement.

i. Monogamous Stable-pair Mating.   Stable-pair mating involves placement of one male and one female into a breeding cage where the pair remains together for their reproductive life. Continual presence of the male allows mating to occur during post-partum estrus, thereby maximizing the breeding output of the female with an interval between litters of three to four weeks. Stable pair mating does require more males per female in the breeding colony than other mating formats. From the labor standpoint, this is the least labor intensive method. The primary activity for stable-pair mating cages is to record new litters and to wean rat pups at the appropriate age. This mating system is the easiest for record keeping purposes because all progeny produced by this cage will have the same parents.

ii. Polygamous Stable Trio Mating.   Trio mating is accomplished by placing one male and two females into a single breeding cage. The male will breed the females as they come into estrus. The trio can remain as a stable group if the breeder cage size is appropriate to accommodate the adults and subsequent litters up to weaning. Stable trio

mating has the advantage of requiring half as many males as females to produce the same number of pups compared to stable-pair mating. Like stable-pair mating, this system allows for mating at postpartum estrus, maximizing breeding efficiency. This breeding system also affords the pups of lactating females the advantage of multiple lactating dams from which to nurse, sometimes referred to as the "aunting phenomenon." However, the presence of more than one female rat with a litter may cause aggressive behavior in the females of some stocks or strains and decrease the reproductive performance of the cage. In such cases, one of the females must be removed prior to parturition to prevent aggression.

Cage size is most often insufficient to accommodate both gestating and lactating dams and their pups, and pregnant females must be removed prior to parturition and allowed to deliver in a separate cage. The female is not returned to the male's cage until her litter has been weaned. When pregnant females are removed, it does not allow the male to mate with the female at post-partum estrus, thus increasing the interval between litters to seven to eight weeks. This mating format requires one male for every two females in the breeding colony, and may increase the overall number of cages in the breeding colony to accommodate gestating and lactating females. Compared to the stable-pair format, trio mating is more labor intensive because pregnant females must be removed prior to parturition and returned to the male after the pups are weaned. Record keeping may be more complicated, especially if the female is not returned to the same male for subsequent breeding.

iii. Polygamous Harem Mating.   Harem mating involves placement of three or more females with a male in a breeding cage. The male will breed the females as they come into estrus. Pregnant females are removed prior to parturition and additional non-pregnant females may be placed in the male's cage. This system is labor intensive because it requires that close attention be paid to the cages to remove pregnant females and to return the females into mating after litters are weaned. This system also requires greater attention to record keeping due to the frequent animal movements, especially if the females are not mated to the same male.

Fewer males are required for mating with a greater number of females compared to the other mating systems—an advantage from the colony manager's standpoint. However, in small colonies each male's genome may be overrepresented, thus potentially decreasing the genetic diversity in outbred or randomly bred colonies. Utilization of this mating system in small colonies of genetically modified rats may also pose a risk because a genetic mutation may be fixed rapidly within the colony due to the limited number of breeder males.

c. CROSS FOSTERING OF PUPS. Cross fostering is the act of removing pups from one dam and transferring them to another lactating dam with pups of the same approximate age. Female rats are generally excellent mothers and accept pups readily. This practice is used for many purposes, including by colony managers to maximize production within a breeding colony. Pups from small litters (six or fewer) can be cross fostered onto another lactating dam, thereby allowing her to reenter the breeding pool and not postponing her next mating until after her litter has been weaned. Similarly, some pups from litters of more than 12 pups can be removed and fostered onto another dam, limiting the number of pups per dam to 12 to 14. This practice will also minimize the variability of pup weaning weight due to the effects of litter size. The practice of cross fostering is most successful when performed within the first several days of life.

To perform cross fostering, the pups to be fostered are removed from their dam. Great care must be taken to hasten the process to maintain normal body temperature in the pups, in that hypothermic pups are less likely to be accepted by the new dam. The dam that will receive the pups is temporarily removed from her cage and the new pups are added to her existing litter. To increase the scent on the pups to be fostered, a technician may gently roll the new pups through soiled bedding in the new cage before placing the pups with the female's existing litter. The female is then returned to her litter and observed.

Cross fostering may also be used try to save valuable litters in instances of acute death of a female. When the pups to be cross fostered must be readily identified, it is vital to be able to visibly distinguish the pups. This is most commonly performed by cross fostering pups onto a dam and litter of a different coat color due to the challenges in permanently identifying very young pups.

d. EVALUATION OF REPRODUCTIVE PERFORMANCE. To effectively monitor a breeding colony, the performance (or productivity) of the "breeding unit" must be evaluated. The breeding unit evaluated can be defined either as production by breeder female or as production by cage (in breeding formats that may have more that one female per cage). Although most colony managers focus on breeding efficiency of females, it must be understood that reproductive failures may be the result of the female, male, or both.

The primary reason for evaluating reproductive performance is to ensure maximum breeding efficiency for the colony. However, continual or periodic evaluation of reproductive parameters will allow colony managers to monitor strains and/or stocks for expected reproductive characteristics. Examples of reproductive traits are provided in Table 6-2. Sudden or dramatic changes in reproductive characteristics or gradual changes over time may indicate significant genetic changes such as genetic contamination, genetic mutation, and/or genetic drift within a line. Reproductive monitoring also allows evaluation of breeders to select replacement breeder stock from parents that meet the expected reproductive characteristics for that rat stock or strain. These reproductive evaluations can only be meaningful if the expected reproductive characteristics are known.

Using various measures in evaluating the reproductive efficiency of a breeding colony allows colony managers to readily identify and remove animals or cages with poor or aberrant reproductive characteristics. For large breeding colonies, this results in replacement of the abnormal animals or breeder cages with new, younger breeder stock. Consistent evaluation of even the smallest breeding colonies will identify reproductive problems before the colony is in reproductive crisis. A situation characterized

TABLE 6-2

REPRODUCTIVE CHARACTERISTICS OF SELECT RAT STOCKS AND STRAIN

Strain/Stock	Mating format	Age at setup (wks.)	Days to first litter[1]	# Newborn (all litters)[1]	% Weaned (all litters)[2]	% Female pups weaned[3]	Inter-litter interval (days)[1]	PEI
NTac:SD	Monogamous pair	10	26.8 ± 2.3	11.2 ± 3.3	98.8	52.0	34.6 ± 6.8	2.00
Hsd:Sprague Dawley SD	Harem	9–10	25.2 ± 3.4	10.9 ± 2.3	96	50.0	25.3 ± 3.5	1.46
Hsd:WI	Monogamous pair	9–10	29.6 ± 6.8	10.9 ± 3.9	99	52.1	30.5 ± 8.4	1.52
HsdBlu:LE	Monogamous pair	9–10	27.6 ± 6.7	10.7 ± 2.1	98	50.0	26.5 ± 3.7	2.21
SimTac:LE	Monogamous pair	10	28.2 ± 4.9	12.3 ± 2.6	99.5	50.8	27.0 ± 4.4	2.00
Tac:SHR	Monogamous pair	10	34.4 ± 10.6	8.9 ± 3.9	74.2	50.4	27.8 ± 11.9	1.50
F344/NTac	Monogamous pair	10	29.2 ± 9.2	8.8 ± 3.7	90.4	51.7	45.3 ± 16.2	1.10
F344/NHsd	Monogamous pair and harem	9–10	29.1 ± 5.5	8.7 ± 2.9	94	50.0	36.6 ± 11.3	.78–1.16
LEW/SsNHsd	Monogamous pair and harem	9–10	32.3 ± 9.2	9.2 ± 2.5	91	50.3	43.3 ± 7.5	.58–.77

1 = Data presented as Mean ± Standard Deviation; 2 = (Total pups weaned/total pups born) × 100; 3 = (Total number of females weaned/Total number of pups weaned) × 100; PEI = Production Efficiency Index = number of pups weaned per female per week. The data provided by Harlan on the F344 and Lewis rats is combined information from the foundation, pedigreed expansion, and production colonies.

by a few reproductively active or many geriatric animals indicates the risk of loss of the colony.

Colony managers should set an upper limit on the time any breeder cage will remain active in the colony regardless of productivity. In the authors' experience the most common cause of reproductive failure or breeding inefficiencies in colonies is failure to rotate breeder stock and replace the oldest breeder cages on a regular basis. This forced timely rotation will continue to introduce young breeder stock into the colony. Rotating breeder stock on a periodic basis (weekly, biweekly, or monthly) will prevent production peaks and valleys that may occur with replacement of large numbers of breeder cages at one time. This breeder rotation also minimizes the production effects on the colony of smaller litter size commonly associated with a female's first litter.

2. Inbred Strains

Inbred rat strains are produced by breeding related individuals, primarily brother by sister, for at least 20 consecutive generations. This breeding process results in allelic segregation such that after 20 generations rats are homozygous at 98.7% of all alleles and after 40 generations are homozygous at 99.98% alleles (Silvers, 1995). Residual heterozygosity refers to those alleles that have not yet been fixed in the genome. To minimize genetic differences among inbred rats, colony managers must continue to utilize inbreeding. As the strain is continually inbred, the amount of residual heterozygosity decreases. Inbreeding may also fix in the genome genetic changes that result from genetic drift.

Commercial breeders that produce large numbers of inbred rats for sale generally utilize a production system that begins with a foundation colony from which all subsequent animals of the strain are derived (Figure 6-3). The foundation colony is a closed colony (no new animals are introduced) maintained by strict inbreeding. It serves as the genetic repository for all future rats of that strain. To prevent line divergence, great attention paid to breeding performance in the selection of breeders and replacement breeders in the foundation colony. The foundation colony is the only colony in the production scheme that is self-replacing (i.e., saves replacement breeders for itself).

Progeny from the foundation colony are used as breeder stock for the pedigreed expansion colony. The pedigreed expansion colony is maintained by brother/sister mating, and serves to expand breeder stock and supply breeder stock to the expansion colony. The expansion colony is maintained by strict inbreeding and serves to further expand the number of available breeders. Progeny from the expansion colony supply breeders to the production colony, which then produces large numbers of animals for use. Based on the numbers of animals required in the production colony, there may be several consecutive expansion colonies to ensure adequate numbers of breeders to meet the required production colony needs. In this production scheme, a small foundation colony supplies a larger expansion colony, which in turn provides even greater numbers of breeders for the production colony (Figure 6-3).

Research colonies are often small. They may need one colony only for maintenance, and may not require the multi-generational approach to meet the numbers necessary for experimental use. However, strict genetic control

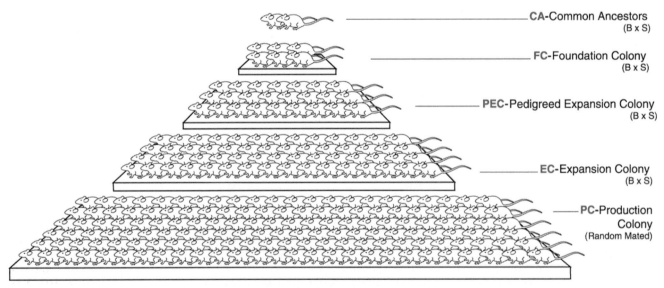

Fig. 6-3   Inbred colony scheme. Depiction of colony animal flow in an inbred or congenic rat colony to produce large numbers of progeny. Strict adherence to inbreeding and expansion of available breeding animals increases with each colony type. B × S = brother/sister mating.

must be maintained to prevent genetic drift and unwanted research variability.

3. Outbred Stocks

Outbred rat stocks are genetically diverse within the population, and there is likely genetic variability between any two animals within the population. Outbred stocks are primarily used in studies designed to evaluate or model the human population. Accordingly, mating systems aimed at maintaining outbred stocks must ensure that the genetic diversity in the population is maintained (i.e., must prevent or minimize the breeding of related individuals). Such populations must be started with as many individuals as possible; minimally 25 breeder pairs. Several systems have been described using random mating tables or circular or rotational mating systems (Poiley, 1960; Kimura and Crow, 1963; Falconer, 1967; Rapp, 1972). These rotational mating schemes are based on a predetermined number of breeding units, wherein each breeder unit has roughly equal numbers of breeding cages in the colony.

In setting up new breeder cages, males are selected from a specific breeder unit, females are selected from a different breeder unit, and the breeder cage created is assigned a different breeder unit number. These systems eliminate the risk of mating closely related individuals, in that male and female progeny from any given breeder unit are set up in separate breeder units. The probability of mating related individuals is inversely related to the number of breeder cages within the breeder unit and to the total number of breeder units. These mating systems are easily employed and can be instituted fairly easily in research colonies. Further, these mating systems require the use of stable trios or pairs. Table 6-3 outlines several examples of rotational mating schemes as described by Poiley (1960).

In large outbred colonies, there may be a need to maintain the rats in two or more locations. In this case, there must be a plan in place to provide consistency in genetic diversity among the separately maintained outbred colonies. This can be accomplished by periodic exchange of breeder stock among the various locations. Alternatively, population genetic analysis will reveal whether the frequency distribution of screened alleles or markers has changed significantly to warrant transfer of breeder stock among locations.

Random mating is frequently chosen by holders of small colonies due to ease of colony maintenance. This does, however, increase the likelihood that genetic diversity will decrease with each generation due to the limited number of breeders (as described previously).

4. Consideration for Genetically Manipulated Lines

The breeding system chosen in maintaining genetically modified rats is most commonly driven by the background

TABLE 6-3

ROTATIONAL MATING SCHEME FOR MAINTENANCE OF OUTBRED COLONIES USING DIFFERENT NUMBERS OF BREEDER UNITS

Males from breeder unit	×	Females from breeder unit	=	Create cage in breeder unit
Rotational Mating Scheme Based on Four Breeder Units				
1		2		3
2		3		4
3		4		1
4		1		2
Rotational Mating Scheme Based on Eight Breeder Units				
1		8		2
2		3		5
3		4		6
4		5		7
5		6		8
6		7		1
7		1		3
8		2		4
Rotational Mating Scheme Based on Twelve Breeder Units				
1		3		4
2		4		5
3		5		6
4		6		7
5		7		8
6		8		9
7		9		10
8		10		11
9		11		12
10		12		1
11		1		2
12		2		3

Example of a rotational mating scheme for outbred rats to maintain maximal genetic diversity. Equal numbers of breeder rat cages are assigned into groups. To set up new breeder cages, the males are selected from a specific breeder unit listed above, the females from a different breeder unit listed above, and the breeder cage created is assigned the breeder unit number defined above. Examples using different numbers of groups is provided and taken from Poiley, 1960.

upon which the genetic mutation(s) are founded. When the genetic modification is created on an inbred background, the inbreeding system described previously is frequently followed. However, in instances in which certain zygosities for the gene(s) of interest must be selected and the colonies are fairly small, colony managers may elect to mate rats that are not related to maximize the number of available animals as breeders. Long-term mating of nonrelated individuals based on a presumed stable inbred background increases the likelihood of genetic differences and maximizes the effects of subtle genetic changes due to genetic drift or persistent residual heterozygosity.

Maintaining genetically modified rats based on an outbred or mixed background requires mating of animals to retain genetic diversity. Genetic diversity is often difficult to maintain because colonies commonly start out with a small number of genetically modified animals and thus with a limited gene pool.

B. Record Keeping

Maintenance of breeding records is a critical component of colony management that allows for the evaluation of breeding colony performance and that provides essential documentation of inbred and congenic lines. The definition of specific record keeping requirements is diverse and is often tailored to specific research needs.

1. Animal Identification

Regardless of the record keeping tools employed, identification of animals is essential, primarily in inbred or congenic colonies. Individual identification may not be required in outbred colonies, unless there are genetic modifications in the line requiring that each rat be individually identified. Identification can occur at either the level of individual animals or by breeding cage. In the case of individual animal identification, all animals must be provided a unique number by means of tattoos, ear tags, subcutaneous chips/transponders, ear punches, or other approved methods. Ensuring accurate and easily readable numbering is essential to the accuracy of breeding records and to the maintenance of genetic integrity in the breeding colony. The requisite longevity of the identification method should be a strong consideration when determining the method.

Individually identifying animals is beneficial when animals may be mated to one or more rats and positive identification of all rats is imperative, such as may occur when back-crossing a line or intercrossing two different lines. However, individual identification requires more labor and record keeping and may not be as easily justified in colonies using a stable-pair mating format.

An alternative means of identification is to assign each breeder cage a unique number and track all animals by cage number. This identification method is only feasible when stable-pair mating is used.

2. Record Keeping Tools

The level of detail and items of information required for any breeding colony records will vary depending on the intended animal use and/or investigator needs. Differences in the level of detail of breeding information will depend on which methods of record keeping are employed, including cage cards, logbooks, and custom and/or commercially available programs or systems.

a. CAGE CARDS. Use of cage cards as a major component of breeding colony record keeping is one of the most basic methods, and has the advantage of being readily viewable at the cage level. Individuals maintaining the colonies readily utilize cage cards as a primary means of record keeping. An example of a breeding colony cage card is shown in Figure 6-4. Most cage cards in use are custom made to fit the needs of the colony and/or investigative group. The cage card shown in Figure 6-4 is used to record at the cage level all breeding information for inbred and congenic lines using individual animal identification. Different cage cards may be required for weaned offspring, depending on colony size and animal use. This particularly applies to offspring used as experimental animals under conditions in which specific facility cage identification requirements such as investigator name, date of birth, etc., may be necessary for cage card identification.

Benefits of cage cards include accessibility, visibility, and an increase in the amount of information provided directly at the animal's cage. Cage cards permit a quick assessment of breeder performance and allow for identification of both nonproductive cages and those that should be retired due to advanced reproductive age. Cage cards are also easily customized to include the information required for use in the institution or investigator's laboratory. Disadvantages of cage cards are that they are more difficult to maintain/archive long term, are easy to misplace/lose, and (like any hard-copy record) do not have a convenient means of backup for informational purposes. Cage cards are generally inadequate as a sole means of documentation of breeding records for an inbred or congenic line.

b. LOGBOOKS. Logbooks or other handwritten tracking systems are frequently used to document breeding records. This record keeping tool allows for continual tracking of breeder performance and provides key pieces of information for the line. Specific pieces of information are discussed in material following. Advantages of logbooks are that they are technologically easy to maintain and relatively quick to implement. They are readily transportable and may be maintained in the animal room, will track breeding records longitudinally, and may not require retention of cage cards if all information is duplicated in the logbook. Like cage cards, they are readily adapted to any required format to capture key pieces of information for the specific institution or laboratory's needs. Disadvantages to logbooks are that the information contained within are unlikely to exist in another form, and thus a backup for irreplaceable data is lacking. Further, because they are readily transportable they may be lost or misplaced.

c. COMPUTERIZED SYSTEMS. Use of computerized systems to document breeding information continues to

ID:						Strain:				
Generation#	N#		F#			DOB	Parent Reg # or Group #	Breeder #	Genotype	
Cage # :		Set Up Date						F#		
Genotyping Sample(s)			Ear	Blood	Tail			F#		
Gene(s) of Interest								M#		
						Backcross?	No	Yes	To Line #	
DOB	# NB	F wnd.	M wnd.	Total wnd.	Wean Date	Missing/ Dead	Comments			
Colony Type		Technician:					Tear Down Date:			

BRED 28

Fig. 6-4 Example of a rat breeding cage card. Key: Generation field (N# = number of generations of back-crossing, F# = number of generations of inbreeding, and DOB = date of birth), and breeder # field (F# = female identification number, M# = male identification number, #NB = total number of newborn pups, F wnd = number of female pups weaned, and M wnd = number of male pups weaned).

increase the use of computerized record keeping. Several commercially available computer systems exist that have as a component of the program or as its major focus the maintenance of breeding records. Evaluation and comparison of such systems must be made based on individual needs.

Creation of customized computer systems using existing/available programs continues to be very common in smaller institutions. This approach allows customization of the system for recording data but is limited to the abilities of the developer for the reporting and tracking capabilities.

Advantages to computerized systems are the ease of data back-up as well as the ability to transition data from or even eliminate paper records. The ability to customize and/or design a program is directly tied to cost and skilled technical support. Disadvantages of computerized records are cost, potential limitation for customization of existing computer programs, and the resources required to create a computerized breeding program de novo.

3. Colony Record Keeping Requirements

Regardless of the record keeping system or tools being employed, there is essential information that must be maintained on each animal or cage of animals. This information includes:

- Identification number of individual animal or cage unit
- Date of birth
- Identification number of dam and sire
- Generation number
- Genotype (if line has a genetic manipulation)
- Date set up as a breeder and with whom
- For each litter born
  - Date of birth
  - Number pups born
  - Date weaned
  - Number weaned
  - Sex of each pup
  - Genotype of pups (if line has a genetic manipulation)
  - Comment field for abnormal or notable findings or events
  - Final disposition

With the aforementioned pieces of information, whether stored on paper or in a computerized format, evaluation of breeding performance and maintenance of genetic integrity can be successfully achieved in rat breeding colonies. Additional information—such as phenotypic data, experimental results, and so on—provides a more comprehensive record system, if required.

REFERENCES

Ackland, J. F., D'Agostino, J., Ringstrom, S. J., Hostetler, J. P., Mann, B. G., and Scwartz, N. B. (1990). Circulating radioimmunoassayable inhibin during periods of transient follicle-stimulating hormone rise: secondary surge and unilateral ovariectomy. *Biol. Reprod.* **43**, 347–352.

Advis, J. P., and Ojeda, S. R. (1979). Acute and delayed effects of anterior pituitary transplants in inducing precocious puberty in female rat. *Biol. Reprod.* **20**, 879–887.

Azooz, O. G., Farthing, M. J., Savage, M. O., and Ballinger, A. B. (2001). Delayed puberty and response to testosterone in a rat model of colitis. *Am. J. Physiol. Regul. Integr. Comp. Physiol.* **281**, R1483–R1491.

Baker, D. E. J. (1979). Reproduction and breeding. *In* H. J. Baker, J. R. Lindsey, and S. H. Weisbroth (eds.), The Laboratory Rat, Volume 1 , pp. 153–168, New York: Academic Press.

Barfield, R. J., and Thomas, D. A. (1986). The role of ultrasonic vocalizations in the regulation of reproduction in rats. *Annals N.Y. Acad. Sci.* **474**, 33–43.

Bartos, L. (1977). Vaginal impedance measurement used for mating in the rat. *Lab. Anim.* **11**, 53–55.

Bennett, J. P., and Vickery, B. H. (1970). Rats and mice. *In* E. S. E. Hafez (ed.), Reproduction and Breeding Techniques for Laboratory Animals, pp. 299–315, Philadelphia: Lea and Febiger.

Beys, E., Hodge, T., and Nohynek, G. J. (1995). Ovarian changes in Sprague-Dawley rats produced by nocturnal exposure to low intensity light. *Lab. Anim.* **29**, 335–338.

Bourguignon, J. P., Gérard, A., Alvarez Gonzalez, M. L., Fawe, L., and Franchimont, P. (1992). Effects of changes in nutritional conditions on timing of puberty: Clinical evidence from adopted children and experimental studies in the male rat. *Horm. Res.* **38 Suppl. 1**, Los Angeles, 97–105.

Clough, G. (1982). Environmental effects on animals used in biomedical research. *Ind. Rev.* **57**, 487–523.

Cowie, A. T. (1984). Lactation. *In* C. R. Austin and R. V. Short (eds.), Reproduction in Animals, Volume 3, pp. 195–231, New York: Cambridge University Press.

Cunningham, M. J., Clifton, D. K., and Steiner, R. A. (1999). Leptin's actions on the reproductive axis: Perspectives and mechanisms. *Biol. Reprod.* **60**, 216–222.

de Kretser, D. M. (1982). The testis. *In* C. R. Austin and R. V. Short (eds.), Reproduction in Animals, Volume 3, pp. 76–90, New York: Cambridge University Press.

Dey, S. K., and Johnson, D. C. (1986). Embryonic signals in pregnancy. *Annals N.Y. Acad. Sci.* **476**, 49–62.

Dluzen, D. E., and Ramirez, V. D. (1986). Transient changes in the in vitro activity of the luteinizing hormone-releasing hormone pulse generator after ovariectomy in rats. *Endocrinology* **118**, 1110–1113.

Enders, A. (1965). A comparative study of the fine structure of the trophoblast in several hemochorial placentas. *Am. J. Anat.* **116**, 29–68.

Engelbregt, M. J., van Weissenbruch, M. M., Popp-Snijders, C., Lips, P., Delmarre-van de Waal, H. A. (1992). Body mass index, body composition, and leptin at onset in male and female rats after intrauterine growth retardation and after postnatal food restriction. *Pediatr. Res.* **50**, 474–478.

Evans, A. M. (1986). Age at puberty and first litter size in early and late paired rats. *Biol. Reprod.* **34**, 322–326.

Falconer, D. S. (1989). Genetic aspects of breeding methods. *In* W. Lane-Petter (ed.), UFAW Handbook on the Care of and Maintenance of Laboratory Animals, pp. 72–96. New York: Churchill Livingstone.

Feldman, M. L. (1992). Changes in the ear. *In* U. Mohr, D. L. Dungworth, and C. C. Capen (eds.), Pathobiology in the Again Rat, Volume 2, pp. 121–147, Washington, D.C.: ILSI Press.

Fischbeck, K. L., and Rasmussen, K. M. (1987). Effect of repeated reproductive cycles on maternal nutritional status, lactational performance, and litter growth in *ad libitum*—fed and chronically food restricted rats. *J. Nutr.* **117**, 1967–1975.

Fiske, V. M. (1941). Effect of light on sexual maturation, estrous cycles, and anterior pituitary of the rat. *Endocrin.* **29**, 189–196.

Freeman, M. E. (1988). The ovarian ocycle in the rat. *In* E. Knobil and J. D. Neill (eds.), The Physiology of Reproduction, pp. 1983, New York: Raven Press.

Fox, R. R., and Laird, C. W. (1970). Sexual cycles. *In* E.S.E. Hafez (ed.), Reproduction and Breeding Techniques for Laboratory Animals, Los Angeles, pp. 107–122, Philadelphia: Lea and Febiger.

Fujinaga, M., Baden, J. M., and Mazze, R. I. (1992). Reproduction and teratogenic effects of a major earthquake in rats. *Lab. Anim. Sci.* **42**, 209–210.

Gallo, R. V. (1981a). Pulsatile LH release during the ovulatory LH surge on proestrus in the rat. *Biol. Reprod.* **24**, 100–104.

Gallo, R. V. (1981b). Pulsatile LH release during periods of low level LH secretion in the rat estrous cycle. *Biol. Reprod.* **24**, 771–777.

Gamble, M. R. (1976). Fire alarms and oestrus in rats. *Lab. Anim.* **10**, 161–163.

Gibori, G., Khan, I., Warshaw, M. I., McLean, M. P., Puryear, T. K., Nelson, S., Durkee, T. J., Azhar, S., Steinschneider, A., and Rao, M. C. (1988). Placental-derived regulators and the complex control of luteal cell function. *Recent Prog. Horm. Res.* **44**, 377–429.

Gonzalez, D. E., and Deis, R. P. (1990). The capacity to develop maternal behavior is enhanced during aging in rats. *Neurobiol. Aging* **11**, 237–241.

Gridgeman, N. T., and Taylor, J. M. (1972). Maximization of long-term productivity in a rat colony. *Lab. Anim.* **6**, 203–206.

Haisenleder, D. J., Ortolano, G. A., Jolly, D., Dalkin, A. C., Landefeld, T. D., Vale, W. W., and Marshall, J. C. (1990). Inhibin secretion during the rat estrous cycle: Relationships to FSH secretion and FSH beta subunit mRNA concentrations. *Life Sci.* **47**, 1769–1773.

Harris, M. A., and Kesel, M. L. (1990). An improved method for accurately timed mating in rats. *Lab. Anim. Sci.* **40**, 424–425.

Higuchi, T., Honda, K., and Negoro, H. (1986). Detailed analysis of blood oxytocin levels during suckling and parturition in the rat. *J. Endocrinol.* **110**, 251–256.

Kaufmann, P., and Burton, G. (1994). *In* E. Knobil and J. D. Neill (eds.), The Physiology of Reproduction (2d ed.), pp. 441–484, New York: Raven Press.

Kimura, M., and Crow, J. F. (1963). On the maximum avoidance of inbreeding. *Genet. Res.* **4**, 399–415.

Kohn, D. F., and Clifford, C. B. (2002). Biology and diseases of rats. *In* Fox, L. C Anderson, F. M. Leow, and F. W. Quimby (eds.), Laboratory Animal Medicine, pp. 121–165, San Diego: Academic Press.

Koto, M., Miwa, M., Togashi, M., Tsuji, K., Okamoto, M., and Adachi, J. (1987a). A method for detecting the optimum for mating during the 4-day estrous cycle in the rat: Measuring the value of electrical impedance of the vagina. *Jikken Dobotsu* **36**, 195–198.

Koto, M., Miwa, M., Tsuji, K., Okamoto, M., and Adachi, J. (1987b). Change in the electrical impedance caused by cornification of the epithelial cell layer of the vaginal mucosa in the rat. *Jikken Dobotsu* **36**, 151–156.

Levine, J. E., and Duffy, M. T. (1988). Simultaneous measurement of luteinizing hormone (LH)-releasing hormone, LH, and follicle-stimulating hormone release in intact and short-term castrate rats. *Endocrin.* **122**, 2211–2221.

Levine, J. E., and Powell, K. D. (1989). Microdialysis for measurement of neuroendocrine peptides. *Methods Enzymol.* **168**, 166–181.

Levine, J. E., Bauer-Dantoin, A. C., Besecke, L. M., Congahan, L. A., Legan, S. J., Meredith, J. M., Strobl, F. J., Urban, J. H., Vogelsong, K. M., and Wolfe, A. M. (1991). Neuroendocrine regulation of the luteinizing hormone-releasing hormone pulse generator in the rat. *Recent Prog. Horm. Res.* **47**, 97–153.

Lincoln, D. W. (1984). The posterior pituitary. *In* C. R. Austin and R. V. Short (eds.), Hormonal Control of Reproduction: Reproduction in Mammals, Volume 3, pp. 21–51. Cambridge, UK: Cambridge University Press.

McGuire, M. K., Littleton, A. W., Schulze, K. J., and Rasmussen, K. M. (1995). Pre- and postweaning food restrictions interact to determine reproductive success and milk volume in rats. J. Nutr. **125**, 2400–2406.

Maeda, K-I., Ohkura, S., and Tsukamura, H. (2000). Physiology of reproduction. *In* G. J. Krinke (ed.), The Laboratory Rat, pp. 145–176, New York: Academic Press.

Martin, M. G., Wu, S. V., and Walsh, J. H. (1997). Ontogenetic development and distribution of antibody transport and Fc receptor mRNA expression in the rat intestine. *Digest. Dis. Sci.* **42**, 1062–1069.

Matsuyama, S., and Takahashi, M. (1995). Immunoreactive (ir)-transforming growth factor (TGF)-beta in rat corpus luteum: ir-TGF beta is expressed by luteal macrophages. *Endocrin. J.* **42**, 203–217.

Matsuyama S., Shiota, K., and Takahashi, M. (1990). Possible role of transforming growth factor-beta as a mediator of luteotropic action of prolactin in rat luteal cell cultures. *Endocrin.* **127**, 1561–1567.

Matsuyama, S., Shiota, K., Tachi, C., Nishihara, M., and Takahashi, M. (1992). Splenic macrophages enhance prolactin and luteinizing hormone action in rat luteal cell cultures. *Endocrinol. Jpn.* **39**, 51–57.

Mattheij, J. A. M., and Swarts, J. J. M. (1991). Quantification and classification of pregnancy wastage in 5-day cyclic young and middle-aged rats. *Lab. Anim.* **25**, 30–34.

Meisel, R. L., and Sachs, B. D. (1994). *In* E. Knobil and J. D. Neill (eds.), The Physiology of Reproduction, pp. 3–105, New York: Raven Press.

Mercier, O., Perraud, J., and Stadler, J. (1987). A method for routine observation of sexual behaviour in rats. *Lab. Anim.* **21**, 125–130.

Nagasawa, H., Fujiwara, K., Maejima, K., Matsushita, H., Yamada, J., and Yokoyama, A. (1983). *Jikken Dobotsu Handbook of Experimental Animals*. Tokyo: Yokendo.

Nakamura, H., Ohsu, Y., Nagase, H., Okazawa, T., Yoshida, M., and Okada, A. (1996). Uterine dysfunction induced by whole-body vibration and its endocrine pathogenesis. *Eur. J. Applied Physiol.* **72**, 292–296.

Nazian, S. J., and Camerson, D. F. (1999). Temporal relation between leptin and various indices of sexual maturation in the male rat. *J. Androl.* **20**, 487–491.

Nazian, S. J., and Mahesh, V. B. (1980). Hypothalamic, pituitary, testicular, and secondary organ functions and interactions during the sexual maturation of the male rat. *Arch. Androl.* **4**, 283–303.

Nelson, R. J. (1995). An Introduction to Behavioral Endocrinology. Sunderland, MA: Sinauer Associates.

Niggeschulze, A., and Kast, A. (1994). Maternal age, reproduction, and chromosomal aberrations in Wistar derived rats. *Lab. Anim.* **28**, 55–62.

Nisender, G. D., and Nett, T. M. (1988). The corporus luteum and its control. *In* E. Knobil and J. D. Neill (eds.), The Physiology of Reproduction, pp. 489, New York: Raven Press.

Ojeda, S. R., Urbanski, H. F., and Ahmed, C. E. (1986). The onset of female puberty: studies in the rat. *Recent Prog. Horm. Res.* **42**, 385–441.

Parshad, R. K. (1990). Effect of restriction in daily feeding periods on reproduction in female rats. *Acta Physiol. Hung.* **76**, 205–209.

Peluso, J. J. (1992). Morphologic and physiologic features of the ovary. *In* U. Mohr, D. L. Dungworth, and C. C. Capen (eds.), Pathobiology in the Aging Rat, Volume 1, pp. 337–349, Washington, D.C.: ILSI Press.

Pfaff, D. W. (1980). Estrogens and Brain Function. New York: Springer.

Poiley, S. (1960). A systematic method of breeder rotation for non-inbred laboratory animal colonies. *Proc. of the Anim. Care Panel* **10**, 159–166.

Pucak, G. J., Lee, C. S., and Zaino, A. S. (1977). Effects of high temperature on testicular development and fertility in the male rat. *Lab. Anim. Sci.* **27**, 76–77.

Ramos, S. D., Lee, J. M., and Peuler, J. D. (2001). An inexpensive meter to measure differences in electrical resistance in the rat vagina during the ovarian cycle. *J. Appl. Physiol.* **91**, 667–670.

Rapp, K. G. (1972). HAN-rotation, a new system for rigorous outbreeding. *Z. Versuchstierkd.* **14**, 133–142.

Romero, A., Villamayor, F., Grau, M. T., Sacristan, A., and Ortiz, J. A. (1992). Relationship between fetal weight and litter size in rats: Application to reproductive toxicology studies. *Reprod. Toxicol.* **6**, 453–456.

Rosen, M., Kahan, E., and Derazne, E. (1987). The influence of the first-mating age of rats on the number of pups born, their weights, and their mortality. *Lab. Anim.* **21**, 348–352.

Russell, L. D. (1992). Normal development of the testis. *In* U. Mohr, D. L. Dungworth, and C. C. Capen (eds.), Pathobiology in the Aging Rat, Volume 1, pp. 395–405, Washington, D.C.: ILSI Press.

Samuel, C. S., Butkus, A., Coghlan, J. P., and Bateman, J. F. (1996). The effect of relaxin on collagen metabolism in the nonpregnant rat pubic symphysis: The influence of estrogen and progesterone in regulating relaxin activity. *Endocrin.* **137**, 3884–3890.

Santti, R., Mäkelä, S., Strauss, L., Korkman, J., and Kostian, M. L. (1998). Phytoestrogens: Potential endocrine disruptors in males. *Toxicol. Ind. Health* **14**, 223–237.

Setchell, B. P. (1982). Spermatogenesis and spermatozoa. *In* C. R. Austin and R. V. Short (eds.), Reproduction in Mammals, Volume 1, pp. 63–101. Cambridge University Press, New York.

Sharp, P. E., and LaRegina, M. C. (1998). The Laboratory Rat, pp. 15–19, New York: CRC Press.

Sharp, J. L., Zammit, T. G., and Lawson, D. M. (2002). Stress-like responses to common procedures in rats: Effect of the estrous cycle. *Cont. Topics* **41**, 15–22.

Sharpe, R. M., Morris, A., and Wyatt, A. C. (1973). The effect of the sex of litter-mates on the subsequent behavior and breeding performance of cross-fostered rats. *Lab. Anim.* **7**, 51–59.

Shirley, B. J. (1978). Partial synchrony of the oestrous cycles of rats introduced to a new environment. *Endocrin.* **77**, 195–202.

Silvers, L. M. (1995). Laboratory mice. *In* Silvers, L.M., Mouse Genetics, pp. 32–61, New York: Oxford University Press.

Smith, E. R., Stefanick, M. L., Clark, J. T., and Davidson, J. M. (1992) Hormones and sexual behavior in relationship to aging in male rats. *Horm. Behav.* **26**, 110–135.

Smith, M. S., Freeman, M. E., and Neili, J. D. (1975). The control of progesterone secretion during the estrous cycle and early pseudopregnancy in the rat: Prolactin, gonadotropin, and steroid levels associated with rescue of the corpus luteum of pseudopregnancy. *Endocrin.* **96**, 219–226.

Soares, M. J., Faria, T. N., Roby, K. F., and Deb, S. (1991). Pregnancy and the prolactin family of hormones: Coordination of anterior pituitary, uterine, and placental expression. *Endocr. Rev.* **12**, 402–423.

Sores, M. J., Muller, H., Orwing, K. E., Peters, T. J., and Dai, G. (1998). The uteroplacental prolactin family and pregnancy. *Biol. Reprod.* **58**, 273–284.

Soskin R. A. (1963). The effect of early experience upon the formation of environmental preferences in rat. *J. Comp. Physiol. Psych.* **56**, 303–306.

Sprangers, S. A., and Paicsek, B. E. (1997). Chronic underfeeding increases the positive feedback efficacy of estrogen on gonadotropin secretion. *Proc. Soc. Exp. Biol. Med.* **216**, 398–403.

Takahashi, M., Kasuga, F., Saito, S., Matsuyama, S., Yamanouchi, K., Murata, T., and Shiota, K. (1989). Role of germ cells and splenocytes in the steroidogenesis of ovarian endocrine cells. *Prog. Clin. Biol. Res.* **294**, 101–115.

Takahashi, M., Shinoda, K., and Hayashi, Y. (1992). Nonneoplastic changes in the testis. *In* U. Mohr, D. L. Dungworth, and C. C. Capen (eds.), Pathobiology in the Aging Rat, Volume 1, pp. 407–411, Washington, D.C.: ILSI Press.

Tang, C. Z., and Bittner, G. D. (2002). Effects of some dietary phytoestrogens in animal studies: Review of a confusing landscape. *Lab. Anim.* **31,** 43–48.

Thorburn, G. D., and Challis, R. G. (1979). Endocrine control of parturition. *Physiol. Rev.* **59,** 863–918.

Tonkiss, J., Smart, J. L., and Griffiths, E. C. (1984). Mating behavior of male rats following pre- and postnatal under nutrition: A comparison of two outbred stocks. *Physiol. Behav.* **32,** 397–401.

Tucker, H. A. (1994). Lactation and its hormonal control. *In* E. Knobil and J. D. Neill (eds.), The Physiology of Reproduction (2d ed.), pp. 1065–1098, New York: Raven Press.

Vandenbergh, J. G. (2003). Prenatal hormone exposure and sexual variation. *Amer. Sci.* **91,** 218–225.

Watanabe, G., Taya, K., and Sasamoto, S. (1990). Dynamics of ovarian inhibin secretion during the oestrous cycle of the rat. *Endocrin.* **126,** 151–157.

Weisse, I. (1992). Aging and ocular changes. *In* U. Mohr, D. L. Dungworth, and C. C. Capen (eds.), Pathobiology in the Aging Rat, Volume 2, pp. 65–119, Washington, D.C.: ILSI Press.

Woodside, B., Abizaid, A., and Caporale, M. (1998). The role of specific macronutrient availability in the effect of food restriction on length of lactational diestrus in rats. *Physiol. Behav.* **64,** 409–414.

Woodside, C., and Jans, J. E. (1995). Role of the nutritional status of the litter and length and frequency of mother litter contact bouts in prolonging lactational diestrus in rats. *Horm. Behav.* **29,** 154–176.

Yokoyama, A., and Ota, K. (1978). *In* A. Yokoyama, H. Mizuno, and H. Nagasawa (eds.), Physiology of Mammary Glands, pp. 266–284, Tokyo/Baltimore: Japan Scientific Societies Press/University Park Press.

Young, C. M., and Rasmussen, K. M. (1985). Effects of varying degrees of chronic dietary restriction in rat dams on reproductive and lactational performance and body composition in dams and their pups. *Am. J. Clin. Nutr.* **41,** 979–987.

Chapter 7

Assisted Reproductive Technologies and Genetic Modifications in Rats

Yuksel Agca and John K. Critser

I. INTRODUCTION

To date, rats as models of human health and disease have made valuable contributions in almost every field of human medicine (Jacob, 1999; Bohlender et al., 2000; Jacob and Kwitek, 2002). Rats are important models in a wide range of disciplines, including physiology, behavior, biochemistry, neurobiology, endocrinology, toxicology, drug metabolism, cancer, pharmacology, and genetic etiology of complex diseases. During the last two decades there have been significant advancements in the field of assisted reproductive technology (ART), such as in vitro oocyte maturation, fertilization, embryo culture, embryo cryopreservation, embryo transfer, microinsemination, in vitro spermatogonia culture, and transplantation. Furthermore, in parallel with improvements made in ART advances in molecular and cellular biology techniques have allowed isolation, identification, and mutation of genes in the study of the molecular basis of particular diseases or disorders.

The synergism among these fields is crucial to the development of reagents, genomic information, and critical germ line modification techniques, including transgenesis, embryonic stem cell methodology, and nuclear transfer. With these ART and molecular biology techniques in place, genetically modified rats (GMRs) can make unprecedented contributions toward unraveling the molecular basis of human disease and the development of new therapeutic strategies. However, the efficient management of such rat lines requires extensive expertise in several disciplines, including reconstitution, breeding, genome banking, genotyping, phenotyping, infectious disease screening, and the distribution of these animal models.

The rapid increase in the production of GMRs created a need for a rat genome resource that meets the needs of the biomedical research community. The Rat Research and Resource Center (www.nrrrc.missouri.edu) has developed the infrastructure for distributing high-quality well-characterized inbred, hybrid, and GMRs to investigators, either as live rats or as cryopreserved embryos or gametes. In this chapter we describe methodologies for rapidly advancing ART in the rat as well as its potential applications in the genetic modification in this species.

II. ASSISTED REPRODUCTIVE TECHNOLOGY

A. Gamete, Embryo, and Gonad Recovery

1. Recovery from Female Rats

a. SUPEROVULATION. A large number of oocytes can be collected from immature superovulated rats when abundant numbers of oocytes and/or embryos are needed for experimental purposes, genome banking, or genetic

modification studies (i.e., transgenics and cloning). For example, one of the first steps in creating a transgenic rat is the ability to obtain a large number of zygotes (700 to 800) so that the gene of interest can be inserted in a relatively short period of time. Obtainment of such high numbers of zygotes from naturally cycling mature female rats is very inefficient because such rats yield 7 to 15 oocytes per reproductive cycle depending on genetic background. Larger numbers can be achieved via superovulation through administration of exogenous gonadotropins. Compared to mice, the response of female rats to exogenous gonadotropins is poorly understood (Hamilton and Armstrong, 1991). Therefore, superovulation protocols that have been successful for the mouse do not work well when applied to the rat. The light cycle of the rat and hormone administration need to be considered to obtain an optimal oocyte yield. The authors use a light cycle of 14 hours of light and 10 hours of darkness. Lights are on from 05:00 to 19:00 and off between 19:00 and 05:00. The superovulation method described in the following and all other animal procedures described in this chapter are based on this light cycle.

Outbred (i.e., Sprague-Dawley and Wistar) or inbred (Fischer 344, Lewis, PVG, and Wistar-Furth) rats can be used to obtain oocytes. There are two commonly used methods of superovulating immature (28 to 35-day-old) rats. The first method involves an intraperitoneal (IP) injection of 15 to 30 IU of pregnant mare serum gonadotropin (PMSG) at 10:00 and a subsequent injection (54 hours later) of 15 to 25 IU of human chorionic gonadotropin (hCG) at 16:00. The second method utilizes follicle-stimulating hormone (FSH) delivered by osmotic pump followed by a single IP injection of luteinizing hormone (LH), or hCG (Armstrong and Opavsky, 1988). A lyophilized powder form of FSH is diluted in 0.9% saline and loaded into an ALZET® osmotic pump and equilibrated in a 37°C water bath overnight to initiate flow before implantation (eight units per pump) (Figure 7-1).

Pumps are then inserted subcutaneously (8:00 to 9:00) through an incision in the skin over the back of the neck under light isoflurane anesthesia and the wound closed. Approximately 50 to 52 hours post FSH implantation,

Fig. 7-1 An ALZET® osmotic pump.

(13:00 to 14:00), female rats are injected with 15 units LH IP and ovulation takes place about 01:00. The method of intraperitoneal injection of PMSG + hCG is cheaper and easier to perform, whereas the FSH + LH method is more expensive and labor intensive due to osmotic pump preparation and subcutaneous implantation into the donor females.

The number of oocytes/zygotes collected per donor female will largely depend on the genetic background and method of superovulation (Corbin and McCabe, 2002). Outbred stocks generally exhibit a consistent superovulatory response with the use of FSH + LH or PMSG + hCG, yielding 30 to 50 oocyte per immature female (Walton and Armstrong, 1983; Jiang et al., 1999). Corbin and McCabe (2002) suggested an optimal PMSG (30 IU) and HCG (25 IU) amount for immature SD rats and found that an increase in PMSG causes poor-quality oocytes. With inbred strains, the PMSG + hCG protocol generally yields 30 to 40 oocytes per donor, depending on the genetic background (Mukumoto et al., 1995; Iannaccone et al., 2001). Dorsch and Hedrich (1998) reported a retrospective rat superovulation study in which a total of 68 sexually immature (28 to 30 days) and mature (> older than 60 days) inbred donor females were used for superovulation using the method described by Armstrong and Opavsky (1988). In their report, synchronized mature females yielded 17 to 20 embryos per donor, whereas immature females yielded about seven embryos per donor.

b. Oocyte Isolation (Metaphase II). The recovery of rat oocytes is a vital step for optimal in vitro fertilization, somatic cell nuclear transfer, and oocyte cryopreservation procedures. Superovulated or naturally cycling female rats are euthanized by CO_2 inhalation. Both oviducts and a small portion of the uterus are carefully removed and placed in a petri dish containing pre-warmed (37° C) Hepes-buffered Tyrode's lactate (TL-Hepes; Bavister et al., 1983) or M2 media containing 3 mg/ml bovine serum albumin (BSA). The oviducts are washed twice in this solution, each oviduct is stabilized by fine tweezers, and the transparent bulge containing cumulus oocyte complexes (COCs) is torn using a 27-g needle to release the COCs under a stereomicroscope (Figure 7-2). In some cases (e.g., microinsemination) it may be desirable to remove cumulus cells and/or the zona pellucida. To remove cumulus cells, COCs are exposed to 1 mg/ml bovine testicular hyaluronidase in TL-Hepes at 37°C for 5 to10 minutes, and as soon as the cumulus cells disperse the naked oocytes are transferred to fresh media using a 120 to 150 μm inner-diameter pulled borosilicate glass pipette (Figure 7-2). To remove the zona pellucida, the oocytes or embryos are further treated with 0.04% pronase prepared in TL-Hepes until the zona pellucida is digested.

c. Embryo Isolation. Pre-implantation rat embryos, at a desired stage, can be used in basic research, embryo culture studies, transgenesis, genome banking, and conceivably establishment of embryonic stem cells (Jiang et al., 1999; Rall et al., 2000; Iannaccone and Galat, 2002). To collect embryos, superovulated or naturally cycling outbred or inbred female rats are mated with intact males and checked for the presence of a vaginal plug and/or sperm in a vaginal smear. Those that have mated are euthanized by CO_2 inhalation and their oviducts and uterus are removed and placed in TL-Hepes media containing 3 mg/ml BSA. The oviducts and uterus are placed on a sterile filter paper to remove blood and fat, and washed twice in TL-Hepes to remove contaminants. The time of embryo collection depends on the embryo stage needed (Figure 7-3). For example, if one wishes to collect pronuclear-stage embryos for pronuclear DNA injection to create transgenic rats embryos should be collected 24 hours post LH/hCG injection (12:00 to 2:00). In this case, embryos need to be released under a stereomicroscope from the oviduct into the collection media described previously. Most of the embryos will be partially free of cumulus cells, but some may require hyaluronidase treatment to completely remove the cumulus cells in order to perform optimal microinjection.

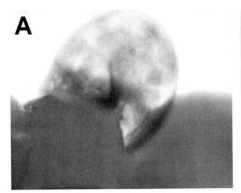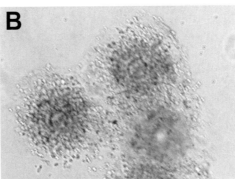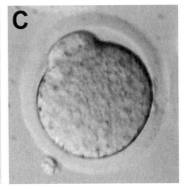

Fig. 7-2 Oviduct of superovulated rat containing cumulus oocyte complexes (*A*), cumulus oocyte complexes (*B*), and cumulus cell free metaphase II oocyte after hyaluronidase treatment (*C*).

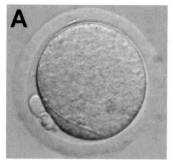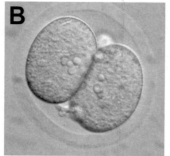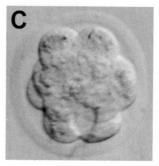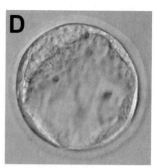

Fig. 7-3 Rat 1-cell (*A*), 2-cell (*B*), morulae (*C*), and blastocyst (*D*) stage embryos.

It should be noted that rat embryos demonstrate highly irregular morphology and thus it is often difficult to identify their developmental stage. However, using the superovulation protocol and the light cycle described previously, fertilization (day = 0) takes place in the early morning (04:00 to 05:00). One can then obtain two-cell to four-cell embryos on the first day, eight-cell to 16-cell on the second and third days, and morula- and blastocyst-stage embryos on days 4 and 5 (respectively) of fertilization. To collect two-cell to morula-stage embryos, the infundibulum is gently held using fine tweezers and each horn is flushed with 100 μl solution to expel embryos. Blastocysts can be collected by puncturing the uterus close to the utero-tubal junction and flushing the uterine horn. Collected embryos are then transferred into a clean petri dish containing TL-Hepes and washed three times in TL-Hepes prior to use.

d. OVARY AND FOLLICLE ISOLATION.   The ability to develop methodologies for the removal of ovarian tissue and the optimal isolation of various stages of follicles is a valuable research tool studying studying the molecular and cellular nature of folliculogenesis in vitro and following transplantation in vivo. The knowledge gained from these types of studies will ultimately allow for the development of protocols for in vitro growth of oocytes, which can in turn be fertilized to obtain developmentally competent embryos (Cain et al., 1995). Ovaries and surrounding tissues are removed from the desired age of anesthetized or euthanized rats. Ovaries are then released from their surrounding ovarian bursa and fat pad under a steriomicroscope and washed twice in TL-Hepes solution under sterile conditions at 20°C. Fully grown (70 to 80 μm diameter) germinal vesicle stage oocytes arrested at Prophase I of meiosis are isolated by puncturing the antral follicles of the whole ovary using fine forceps and a 27-g needle in appropriate tissue culture medium between 40 and 44 hours post FSH implantation.

If one desires to collect preantral follicles (i.e., primordial follicles), ovarian remnants from the previous procedure are incubated with collagenase (Type 1, 200 IU) and

DNase I (100 IU) at 37°C for 20 minutes with gentle shaking in a water bath. The cell suspension is then gently agitated using a Pasteur pipette, resuspended with TCM 199 and incubated overnight at 4°C in KRB. The digest is then centrifuged at 54 × g for 3 minutes at 4°C. The supernatant is discarded and 5 ml of Ca^{++} and Mg^{++} free KRB containing 0.5% BSA is added. The suspension is then filtered through Teflon or nylon mesh to separate the isolated follicles according to their size (Kishi and Greenwald, 1999).

2. Recovery from Male Rats

a. SPERM ISOLATION.   Careful sperm isolation is crucial because rat sperm are known to have extreme sensitivity to suboptimal conditions such as centrifugation, pipetting, and chilling (Nakatsukasa et al., 2001). Spermatozoa can be obtained from sexually mature male rats that have been euthanized by CO_2 inhalation. The scrotal region is sprayed with 70% ethanol and then dissected to expose the cauda epididymidis. The cauda epididymis is removed using fine scissors, taking extreme care to remove excess fat and blood vessels (Figure 7-4).

The tissue is then blotted on sterile filter paper to remove blood and epididymides are aseptically incised. Dense masses of spermatozoa are squeezed out and scooped up using a fine glass rod (Toyoda and Chang, 1974), submerged into an appropriate pre-equilibrated modified-Krebs-Ringers bicarbonate (mKRB) medium at 37°C, and covered with mineral oil in order to allow the sperm cells to diffuse. Alternatively, the cauda epididymis is punctured in several places and held in mKRB medium to allow the sperm to diffuse. This will generally yield a total of $20–25 × 10^6$ spermatozoa/ml. If the male is irreplaceable, the recovery of ejaculated sperm from a female's genital tract just after coitus is also possible. This procedure generally yields about 50 to 60 million sperm per male rat. If the purpose of sperm collection is to obtain a sperm-rich sample that is free of uterine cells, a simple sperm swim-up procedure can be performed in a 1.5-ml Eppendorf tube by gently diluting the semen in 1 ml mKRB medium,

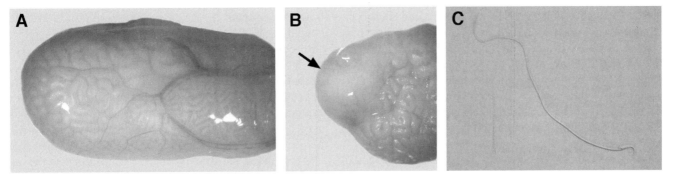

Fig. 7-4 Cauda epididymis (*A*), arrow indicates dense masses of spermatozoa (*B*), and a single rat spermatozoa (*C*).

followed by removal of the upper third portion of the media containing the highly motile sperm fraction.

b. SPERM EVALUATION. Measures of sperm production, motility characteristics, and morphology are important predictors for the assessment of male fertility. Although some subjective semen examination techniques are used to give information about the fertility potential of a semen sample, there are several more objective means of semen evaluation currently available.

i. Computer-assisted Sperm Analysis. Computer-assisted sperm analysis (CASA) is a technique that has been successfully used to determine sperm motion characteristics (e.g., percentage of progressively motile spermatozoa) in men (Shibahara et al., 2004). This device captures about 30 to 60 images of sperm per second and re-constructs those images to determine sperm path and velocity. Recently, with some modifications CASA has been successfully adapted to analyze rat sperm motility in toxicology studies to test the effects of certain chemical compounds on sperm motility characteristics and fertilizing ability (Working and Hurtt, 1987; Perreault, 1998; Higuchi et al., 2001).

ii. Flow Cytometric Sperm Analysis. Although sperm motility and morphology are the most commonly used parameters for the assessment of sperm viability, neither is an accurate indicator of function. Spermatozoa gross morphology is insufficient to detect damage to the plasma membrane, mitochondria, and acrosome. Flow cytometric analysis using specific fluorescent probes is a valuable method of determining these structural changes that may hamper normal fertilization and subsequent embryo development. This method provides a rapid and accurate means of determining the functional status of large numbers of spermatozoa.

Gravance et al. (2001) used flow cytometric analysis of JC-1 staining patterns of rat spermatozoa to detect chemical-induced alterations of sperm mitochondrial membrane potential. They found that the mid-piece (mitochondrial location) of live, highly motile spermatozoa stained

bright orange, whereas the mid-piece of live, non-motile spermatozoa stained green. The mid-piece of slightly or non-progressively motile spermatozoa stained a faint orange-green. Gravance et al. (2003) were able to assess frozen-thawed rat sperm viability using SYBR-14 and propidium iodide. Sperm mitochondria were differentially labeled with JC-1. Motile sperm stained with JC-1 appeared orange in the mid-piece, indicating a high mitochondrial membrane potential, whereas immotile sperm with a low membrane potential stained green.

c. TESTICULAR TISSUE, SPERMATID, AND SPERMATOGONIA ISOLATION. Isolation of testicular tissue containing fully grown (i.e., spermatozoa) and developing (i.e., spermatogonia, round and elongated spermatids) male germ cells offers great potential for the reconstitution of valuable rat strains and the study of mechanisms underlying spermatogonia differentiation and proliferation and in vitro spermatogenesis (Ogura et al., 1996; Brinster and Nagano, 1998). Although complete spermatogenesis under entirely in vitro conditions is the ultimate goal of the reproductive biologist, this field of study is still in its infancy. Currently, isolation of developing male germ cells with high purity is possible and has proven crucial to genome banking and the aforementioned areas of study.

For complete removal of testicles, the rat is euthanized, an incision is made in the scrotum and testis, and the cauda and caput epididymides are removed. If the purpose is to collect seminiferous tubules, an incision is made in the tunica albuginea and the seminiferous tubules are gently removed and kept in Dulbecco's Modified Eagle medium containing 50% rat serum supplemented with 2 mM L-glutamine, 100 milligram/milliliter streptomycin, and 100 U/ml penicillin (Ogura et al., 1996). To isolate round spermatids, seminiferous tubules are placed in erythrocyte-lysing buffer. They are then transferred into the GL-PBS (Dulbecco's phosphate buffered saline supplemented with glucose, sodium lactate, and polyvinyl pyrrolidone [PVP]) at 4° C, cut into small pieces using a pair of fine scissors, and shaken gently to release spermatogenic cells into the medium (Ogura et al., 1996). The cell suspension is then filtered through a 40 μm nylon mesh filter and then centrifuged at 200 g for 5 minutes at 4° C. The cells are then re-suspended in GL-PBS containing 0.2 mg/ml pronase E. After centrifugation at 400 g for 5 minutes at 4° C, elongated spermatids and testicular spermatozoa that have agglutinated into sticky masses are discarded from the suspension. The remaining suspension, which is rich in round spermatids, is washed twice with GL-PBS by centrifugation at 200 g for 5 minutes each at 4° C.

Optimal methods of isolating type A spermatogonia from prepubertal and adult rat testis have been previously described (Morena et al., 1996; van Pelt et al., 1996). Briefly, decapsulated testis are placed in modified Eagle's medium (MEM) containing 5 milligram/ml DNase, 1 mg/ml hyaluronidase, 1 mg/ml trypsin, and 1 mg/ml collagenase for 15 minutes at 32° C and repeatedly pipetted. The tissue is further incubated in the same media in the absence of trypsin for another 30 minutes at 32° C. After repeated pipetting, tubular fragments are centrifuged at 30 g for 2 minutes. The supernatant is then collected and filtered through a 70- and then 55-μm mesh nylon filter to eliminate cell aggregates. Following enzymatic digestion, the cell suspension is plated on lectin-coated dishes and incubated for 1 hour at 5% CO_2 in air at 32° C. Cells not adhering to the lectin are then subjected to further purification (75 to 85%) using a discontinuous density gradient.

B.   Generation of Rat Embryos

1.   In Vitro Fertilization

In vitro fertilization (IVF) is one of the most widely used ART to study sperm/egg interaction and basic molecular and cellular mechanisms of mammalian fertilization (Figure 7-5). IVF and subsequent transfer of resulting embryos to foster mothers are also powerful techniques to reconstitute scientifically valuable rat strains. This technique requires a relatively low number (0.5 to 1×10^6 motile sperm/ml) of sperm to achieve optimal fertilization as compared to natural mating (50 to 60×10^6 per ejaculate) and artificial insemination (100 to 200×10^6 per ml). In vitro fertilization is performed as previously described (Niwa and Chang, 1974; Toyoda and Chang, 1974; Miyoshi et al., 1997).

Epididymal spermatozoa are obtained from proven males as described previously. Dense creamy sperm are introduced into 400 μl mKRB media (pre-equilibrated in 5% CO_2 in air at 37° C under mineral oil) and held for 10 minutes so that the sperm swim out and evenly disperse into the media. The sperm concentration is then determined using a hemocytometer so that sufficient numbers (1.0×10^6 motile spermatozoa/ml) of spermatozoa are aged

to achieve optimal fertilization. The spermatozoa are then incubated in IVF drops of modified rat 1-cell embryo culture medium (mR1ECM) for 5 hours under 5% CO_2 in air at 37° C for the sperm to capacitate before co-incubation with oocytes. It should be noted that capacitation is essential to obtain optimal fertilization. Oviducts containing cumulus oocyte complexes are submerged in mineral oil and pulled into the IVF drops with the capacitated live sperm and co-incubated for 5 to 7 hours under paraffin oil at 37° C in 5% CO_2 in air. Fertilization can be confirmed by observation of two pronuclei and the presence of the sperm tail (Toyoda and Chang, 1974).

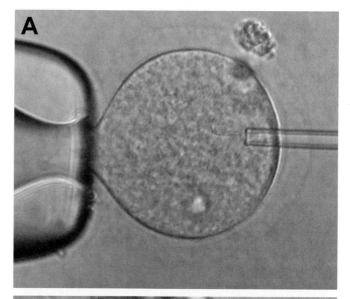

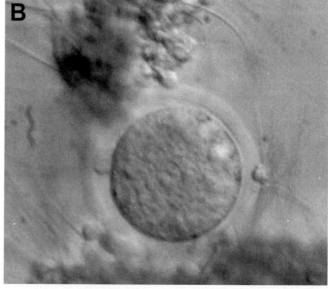

Fig. 7-5 Micro insemination by intracytoplasmic sperm injection (*A*), and in vitro fertilization (*B*).

2. Microinsemination

Microinsemination is a procedure by which a single relatively advanced (spermatocyte, round spermatid, spermatozoa) male germ cell is directly injected into an ooplasm via a fine pulled (5 to 6 μm ID) glass injection pipette using micromanipulation systems. This technique has been widely used to circumvent male infertility due to poor sperm quantity and/or quality. However, it must be noted that there are several technical issues that make micro-manipulation of rodent gametes technically more difficult than in other mammalian species (e.g., hook-shaped sperm head, unusually long tail, and elastic oolemma). Successful micro-insemination requires a short procedure time with minimal damage to the oolemma to avoid cytoplasmic leakage, and minimal injection of the holding media PVP into the ooplasm to avoid the toxic effect of PVP.

With the aid of a Piezzo-driven micro-manipulator and skillful manipulation, it is possible to separate the sperm tail from the head, drill the zona pellucida, penetrate the oolemma, and deposit the gamete with minimal damage (Figure 7-5) (Wakayama et al., 1998). A standard micro-manipulation setup is shown in Figure 7-6.

To date, there have been two microinsemination techniques introduced in ART. Whereas the intracytoplasmic sperm injection (ICSI) method uses mature sperm head, the round spermatid injection (ROSI) method uses immature male germ cells for injection. Although ICSI attempts in rats were made as early as 1979 (Thadani, 1979; Dozortsev et al., 1998), Hirabayashi et al. (2002) were the first to obtain live offspring using this technique. In this report, rat sperm were first sonicated for 20 seconds to separate the sperm heads and then stored at −20° C. The sperm heads were then aspirated into a 2 to 4 μm inner diameter injection pipette in media containing 12% PVP and injected into MII rat oocytes with the aid of a Piezzo-driven micromanipulator. Using the ROSI technique and chemical activation by strontium chloride, Hirabayashi et al. (2002) were then able to produce live pups. However, the overall efficiency (obtaining live births) using either of the microinsemination techniques was very low (2 to 10%). Recently, ICSI has been used very effectively in mice (Szczygiel et al., 2002). Further improvement of ICSI in the rat will have a significant impact on genome banking due to the simplicity and low cost of sperm collection procedures compared to embryos.

3. Artificial Insemination

Artificial insemination (AI) is a procedure by which one manually deposits a sperm suspension, fresh or frozen-thawed, into the female reproductive tract to overcome logistical problems associated with natural mating.

There are several situations in which AI can be valuable in restoring fertility in rats, such as when the number of sperm or the proportion of motile sperm in the semen is not sufficient to achieve fertility through natural mating. Scientifically important (i.e., muscular disorders, arthritis) but reproductively impaired (unable to mate) rat strains can also be restored through AI (Zhou et al., 1998). Furthermore, the use of rat models in combination with AI is a powerful tool to test the effects of known environmental toxins (i.e., ethane dimethanesulfonate) on male fertility (Klinefelter et al., 1994; Woods and Garside, 1996; Hsu et al., 1999; Perreault and Cancel, 2001; Klinefelter et al., 2002). In the case of sperm freezing, the number of motile sperm significantly decreases post-thaw in rats and may also require AI.

Artificial insemination can be performed non-surgically via the cervix under general anesthesia with the use of a speculum. It is estimated that rat ejaculate contains 50 to 60×10^6 spermatozoa based on the spermatozoa recovered from the female tract following natural mating (Blandau and Odor, 1949). On the other hand, sperm isolated from dense masses of spermatozoa, obtained from the cauda epididymides contains 20 to 25×10^6 motile sperm/ml and will yield greater than an 80% fertilization rate. Prior to AI, the recipient female needs to be mated with a proven vasectomized male between 16:00 and 23:00 hours to induce pseudopregnancy using the light cycle specified previously. Females with a copulatory plug can be inseminated after ovulation (01:00 to 02:00).

Artificial insemination can also be performed using a surgical technique. Recipient rats are anesthetized, the dorsal and lateral abdomen is shaved, and the surgery site is prepared with three alternating scrubs of betadine and alcohol. The skin of the dorsum is incised (10 to 15 mm in length) on the midline at the level of the paralumbar fossa and then undermined by blunt dissection. The skin incision is then rolled laterally over the paralumbar fossa. The muscle layers and the peritoneal wall are opened via an incision and spread to create a window about 8 to 12 mm in length, which permits the retraction of the ovarian fat pad. The ovary is gently grasped with tissue forceps and retracted until the ovary, ovarian bursa, and oviduct are exposed. These are viewed using a stereomicroscope under an illuminator at 10 to 15× magnification. For AI, sperm is first suspended in mKRB and loaded into a fine glass pipette with an inner diameter of approximately 200 μm. A 50 to 100 μl sperm suspension is then gently deposited into the oviductal end of each uterine horn that has been punctured with a 27-g needle. The hole can be cauterized to prevent any back flow of the sperm suspension. The uteri are then gently returned to the abdominal cavity and the musculature and skin closed. After repeating the procedure on the contralateral side, the skin is closed using zwound clips.

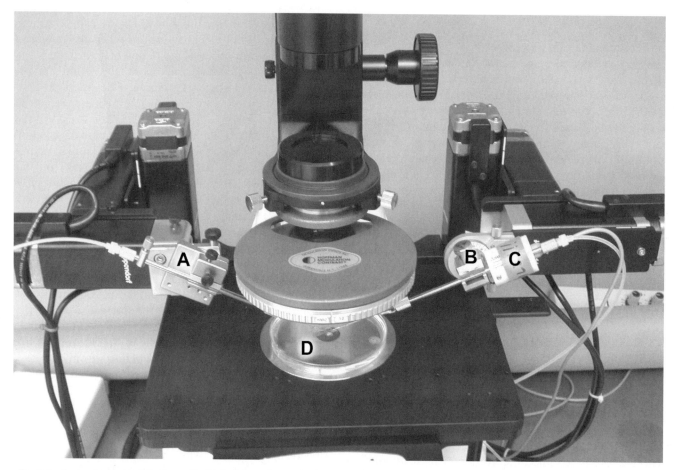

Fig. 7-6 A standard microinjection unit attached to an inverted microscope for gamete and embryo manipulation: left arm to stabilize oocytes or embryos via holding pipette (*A*), right arm to control micro-microinjection pipette (*B*), piezzo injection unit (*C*), and petri dish for micro-injection (*D*).

4. In Vitro Embryo Culture

The ability to grow pre-implantation mammalian embryos under entirely in vitro conditions is crucial for biomedical research as well as for rapid advancement of embryo biotechnology such as transgenesis, cloning, and gene knockout technologies (Y. Zhou et al., 2003). Although in vitro development of pre-implantation embryos from one-cell to blastocyst has been very successful for many outbred stocks and some inbred mouse strains, to date limited success has been achieved with the in vitro development of rat embryos. Rat embryo culture systems such as R1ECM are somewhat efficient in outbred Sprague-Dawley and Wistar rats but are highly variable. One of the main reasons for this inefficiency is the failure of cultured rat embryos to progress from two to four cells into later embryonic stages (Toyoda and Chang, 1974; Whittingham, 1975). Although the exact nature of this developmental block is not fully understood, it has been attributed to detrimental effects of glucose, inorganic phosphate, and proteins in the culture media (Bavister, 1995; Matsumoto and Sugawara, 1998).

Chemically defined hamster embryo culture medium (HECM-1), which does not contain glucose and inorganic phosphate (Schini and Bavister, 1988), was adapted to culture rat embryos in vitro (Kishi et al., 1991; Miyoshi et al., 1994). It has been reported that adjustment of inorganic phosphate concentration in the culture media based on the embryonic stage significantly increased developmental potential (80 to 90%), particularly for the embryos collected from outbred stocks (Miyoshi and Niwa, 1997; Nishikimi et al., 2000). However, the success rate varies from lab to lab and therefore there is an urgent need to develop more reliable and strain-specific rat embryo culture conditions.

Most recently, a two-phase chemically defined culture system for pre-implantation rat embryos was suggested (Y. Zhou et al., 2003). In this report, successful in vitro culture of young one-cell rat embryos was achieved in a commonly used chemically-defined culture media such as potassium simplex optimized medium (KSOM), Brinster's medium for ovum culture (BMOC), or human tubal fluid (HTF) for 18 to 22 hours followed by in vitro culture in mR1ECM medium. This approach allowed successful development of blastocyst-stage embryos from both outbred (SD) and inbred strains of rat (Wistar-Furth, LEW, F344, and PVG).

C.   Rat Embryo Transfer

Embryo transfer is an indispensable technique that can be useful in several areas, such as (1) Reconstitution of frozen embryos to live colonies, (2) Rederivation of rats contaminated with specific pathogens, (3) Development of various ART such as gamete cryopreservation, in vitro fertilization, and micro-insemination methods, and (4) Transferring transgenic embryos to recipient females. The following sections explore components of the rat embryo transfer procedure.

1.   Vasectomy

Vasectomized male rats are required to induce pseudopregnancy in female recipients of transferred embryos. Seven to eight-week-old SD males are generally used because of their high plug frequency and abundant availability. The vasectomy procedure is performed under general anesthesia. The lower abdominal region is shaved and wiped with 70% ethanol to clean and remove loose hairs. A 1.5-cm-long transverse incision is made in the skin and muscle. The testicle is gently exposed by pulling on the attached fat pad and the vas deferens located. Vasculature accompanying the vas deferens is separated and two silk ligatures about 1 cm apart are placed around the vas deferens. A 5 to 6-mm section of the vas deferens is removed and the cut ends are cauterized. The testicle is returned to the abdominal cavity and the procedure is repeated on the opposite side. The abdominal wall is closed using absorbable suture, followed by closure of the skin with an auto-clip to prevent post-operative chewing/self-mutilation. Animals are allowed to recover for two weeks before they are used again, and should be retired at eight to nine months of age.

2.   Establishment of Pseudopregnant Females

To perform embryo transfer (ET) or AI in female rats it is essential to achieve a state of pseudopregnancy. Pseudopregnancy is the lengthening of diestrus to approximately 13 days between consecutive estrous cycles, which allows for the formation of functional progesterone-producing corpora lutea that facilitate embryo implantation and subsequent gestation. To meet this requirement, one needs to first determine the stage of estrous cycle. This can be done by monitoring vaginal cytology of the recipients (see the reproduction chapter in this book). Females in the proestrus stage can be induced into

pseudopregnancy by mating to a vasectomized male. The following morning, at the beginning of the light cycle, females are inspected for a copulatory plug (waxy substance) by spreading the vaginal opening with blunt forceps under a strong light. Otoscopic inspection can also be used. Alternatively, one can mechanically stimulate the cervix and/or vagina with a glass rod to achieve pseudopregnancy. For pseudopregnant females, the authors generally use 8 to 10-week-old Sprague-Dawley female rats because of their docile behavior and good maternal behavior.

3.   Recipient Synchronization

When a large number of recipient females are needed, sexually mature females can be synchronized with an IP injection of a 40-milligram analog of gonadotropin-releasing hormone, LH-RH (des-Gly^{10} [D-Ala^6] ethylamide), between 08:00 and 09:00 a.m. Synchronized females are expected to be in proestrus in the afternoon (01:00 to 03:00 p.m.) four days following LH-RH injection. LH-RH synchronization can also be used to produce time-mated rats, for example when large numbers of fetuses with exact gestational length are needed. The authors have achieved a 70 to 80% success rate with this LH-RH synchronization.

4.   Surgical Embryo Transfer

Embryo transfer to surrogate mothers is an extremely useful technique extensively used for the re-derivation of cryopreserved embryos, elimination of infectious diseases, and production of genetically modified rats (Rouleau et al., 1993; Rall et al., 2000). To achieve optimal pregnancy, and to ensure good motherhood following birth, it is important to use females of an appropriate strain and age. To obtain pseudopregnant recipient rats, 8 to 10-week-old SD female rats are synchronized via an injection of LHRH and mated 1:1 with mature vasectomized males. For verification of mating and hence pseudopregnancy, the female rats are removed the following day between 8:00 and 9:30 a.m. and checked for the presence of a copulatory plug.

For ET, the surgical and anesthetic procedures are the same as those described for AI. The rat ovarian bursa is highly vascular and thus bleeding is virtually unavoidable even with minor trauma. In addition, the rat infundibulum is obscured between the ovary and oviduct, which provides very little space to work. Therefore, to perform optimal ET a few drops (approximately 25 μl) of epinephrine are sprinkled over the bursa to reduce bleeding and a 30 to 40-degree angled transfer pipette is used to gain access to the infundibulum.

Embryos are loaded into a small column consisting of a few microliters of PBS within a 150 μm inner diameter transfer pipette. The ovarian bursa is gently dissected using watchmaker forceps and the transfer pipette is inserted into the infundibulum. Embryos are then discharged into the oviduct, and as soon as air bubbles are observed in the infundibulum the transfer pipette is carefully pulled out and the infundibulum crushed (for 10 to 15 seconds) with fine tweezers. This should prevent back-flow of the fluid into the surgical field or abdominal cavity. PBS is used to moisten the tissue before gently placing it back into the abdominal cavity. The abdominal wall is closed with absorbable suture (optional). After repeating the procedure on the contralateral side, the skin is closed using wound clips. Typically, one-cell to morulae-stage embryos (2-cell, 4-cell, 8-cell, 16-cell) can be transferred to the oviducts of day 0.5 to day 1 pseudopregnant recipients, whereas blastocyst-stage embryos can be transferred to the uterine horn of 3.5 day pseudopregnant recipients. If one wishes to transfer blastocyst-stage embryos, a glass pipette containing the embryos is used to expel them into the uterine horn close to the utero-tubal junction.

III. GENOME BANKING

There are several major impediments to the realization of the full potential of rats as models of human health and disease. Many inbred spontaneous mutant and genetically modified rat strains are not readily available to researchers. Although individual investigators develop and characterize many mutant rat models, they lack the physical and financial resources to maintain all populations of potentially important strains. To date, many mutant rat strains have been lost because of inadequate resources to maintain breeding populations. Loss of important models wastes both scientific and financial resources used to initially create or identify them.

Application of methods to cryopreserve embryos, reproductive cells (spermatozoa, oocytes), and tissues (ovarian tissue, testicular tissue) can be used to prevent the loss of mutant strains while safeguarding unique genotypes from several potential problems, including (a) genetic changes (genetic drift, instability, and contamination), (b) pathogen contamination, and (c) loss due to disease or catastrophic disasters to housing facilities. In addition, preservation of germ cells and embryos enhances management efficiencies by (a) saving animal room space by allowing the discontinuation of breeding colonies of strains not currently in use, (b) reducing workload for staff, (c) facilitating shipment of genotypes by allowing distribution of frozen germ cells rather than live animals, and (d) supporting ownership and patent claims.

The ability to cryopreserve each of these sources of germ cells permits greater overall flexibility in the management of this process (Glenister and Thornton, 2000). Cryopreservation of these cells or tissue types provides greater assurance that reconstitution of a given strain will be possible. For example, if females of a particular strain have a poor ovulatory response or have other reproductive limitations, sperm and/or ovarian tissue provide alternative "routes" to reconstitution. Cryopreservation of embryos and germ cells has been widely used to store genetically and scientifically important livestock and lab animals over the years and has allowed free distribution worldwide, stress-free transportation, and avoidance of disease transmission among animal colonies. With the increasing numbers of GMR, rat embryos, and germ cells, cryopreservation becomes even more crucial. Adventitious infections of rat models can compromise or invalidate research studies, resulting in devastating scientific and economic losses (Rouleau et al., 1993). The use of cryopreserved embryos and germ cells from rats free of pathogenic microorganisms facilitates easy distribution and ensures that distributed rats are of the highest quality.

In general, embryo, cell, and tissue cryopreservation procedures involve (1) initial exposure of the sample to permeating (glycerol, dimethylsulfoxide [DMSO], propylene glycol [PG]) and non-permeating (egg yolk, raffinose, sucrose) cryoprotective agents (CPA), (2) cooling to subzero temperatures and storage, and (3) thawing, dilution, and removal of the CPA, with return to physiological conditions to allow further development. Depending on the cell type, structural and functional integrity of the cells must be either completely or partially (at least nuclear integrity) maintained throughout the cryopreservation procedure. The major factors that determine the success of cell and tissue cryopreservation are (1) cell type, tissue type, or embryo developmental stages, (2) species of origin, and (3) surface-to-volume ratio. To date, two general methods have been introduced to cryopreserve embryos, reproductive cells, and tissues: equilibrium freezing and non-equilibrium freezing.

Equilibrium freezing involves (1) placement of the material to be cryopreserved into (approximately 1 to 2 M) CPA, followed by loading into 0.25-ml plastic straws/cryovials, (2) cooling (1- to $2°$ C per minute) to $-7°$ C and seeding (induction of ice crystallization) by touching precooled forceps to induce ice formation, (3) dehydrating by further cooling (0.3 to $0.5°$ C per minute) to -40 to $-80°$ C in the presence of extracellular ice, and (4) plunging into LN_2 at $-196°$ C for long-term storage. This method has been successfully used to cryopreserve embryos from many mammalian species, including the rat and mouse (Whittingham et al., 1972). During the slow freezing process, however, damage may occur to the cells due to formation of intracellular ice because of trapped intracellular water, solute damage due to extreme high solute concentration, so-called solution effects, and chilling injury (Mazur et al., 1972; Mazur et al., 1992).

The second method is the non-equilibrium freezing (or so-called ultra-rapid) cooling (2,500 to 5,000° C per minute). Vitrification is the transition of aqueous solutions from the liquid state to the glass state (solid), bypassing the crystalline solid state. To achieve optimal vitrification, the solution in which the cells or tissue are suspended must have a relatively high CPA concentration (6 to 7 M) (Fahy et al., 1984). In practice, for the vitrification of biological samples there is a practical limit to achievable cooling rates and there is a biological limit to the concentration of CPA the cells will tolerate. Therefore, a balance must be sought to maximize the cooling rate and minimize the CPA concentration to avoid damage caused by chemical toxicity and osmotic stress.

A. Embryo Cryopreservation

Embryo cryopreservation techniques have been successfully used to reconstitute mutant, transgenic, knock-out, and congenic mouse lines for decades (Whittingham et al., 1972). Similarly, pre-implantation rat embryos at various developmental stages have also been successfully cryopreserved using both equilibrium (Whittingham, 1975; Hirabayashi et al., 1997; Pfaff et al., 2000; Rall et al., 2000) and ultra-rapid (Isachenko et al., 1997; Takahashi et al., 1999) cooling methods, with post-thaw survival rates ranging between 40 and 90%. Furthermore, the ability to cryopreserve pronuclear-stage rat embryos has overcome logistical problems associated with transgenic rat production (Takahashi et al., 1999). The most reliable method of evaluating the developmental potential of thawed embryos remains the demonstration of live young following ET. However, one report (in which about 700 rat embryos from 33 genotypes were used) showed that post-thaw developmental potential of these embryos in vivo (ratio of number of pups born to number of embryos

thawed) varies significantly (10 to 58%) among strains. Therefore, the number of embryos to be banked should be adjusted based on genetic background.

For genome banking purposes, the authors cryopreserve rat embryos at the morulae stage using the equilibrium freezing procedure. Embryos are exposed to 1.5 M DMSO for 10 minutes. After equilibration, embryos (20 per straw) are loaded into the CPA-containing column within 0.25-ml plastic straws, and the straws are then sealed (Figure 7-7). The proximal 60% of each straw contains a column of fluid consisting of 0.5 M sucrose in TL-Hepes, and the distal 40% of the straw contains three small columns consisting of a permeable CPA in TL-Hepes solution. The embryos are placed into the middle column of the three permeable CPA-containing columns using a disposable pulled-and-polished micro-pipette. The straws are then placed into a programmable freezer, cooled to the seeding temperature ($-7°$ C), and held at $-7°$ C for 10 minutes. Seeding is induced by pinching the straw containing the sucrose/TL-Hepes column with LN_2-cooled forceps. Ten minutes after seeding, cooling is continued at a rate of $0.5°$ C per minute until reaching $-35°$ C, at which time the straws are quickly removed from the freezing apparatus and immediately submerged in LN_2.

For ET, straws containing the embryos are held in air for 10 seconds and then plunged into a 20° C water bath for another 10 seconds. The content of the straws is emptied into a petri dish. Embryos are maintained for approximately five minutes in the thawed CPA-sucrose-TL-Hepes solution, and then transferred into a fresh TL-Hepes solution. The morphology of the embryos is evaluated under a stereo-dissecting microscope (50×) to assess post-thaw survival rates, and only those embryos that have an intact zona pellucida and an intact plasma membrane and cytoplasm are counted as viable and suitable for ET.

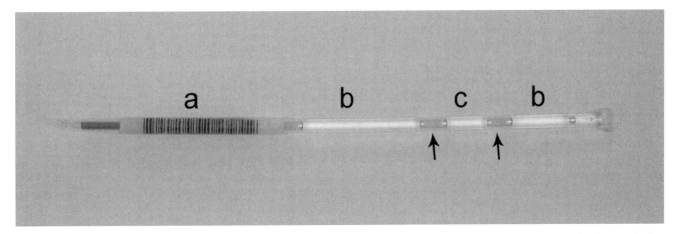

Fig. 7-7 A security plastic straw used for cryopreserving rat embryos (Cryo Bio System, France). Both ends are pressure sealed and arrows indicate air space. The bar code information (*A*), 0.5 molar sucrose solution (*B*), and 1.5 M DMSO solution containing embryos (*C*).

B.   Sperm, Spermatid, and Spermatogonia Cryopreservation

Cyopreservation of male germ cells (i.e., spermatozoa, spermatids) is an alternative to embryo banking and an integral part of the establishment of a rat genome resource bank (Thornton et al., 1999). Embryo production and isolation is a laborious and significantly more expensive procedure than sperm isolation from several males, due to the cost associated with hormone stimulation and additional animal housing. Although effective sperm cryopreservation methods using a raffinose and skim milk mixture have been reported for a number of mouse strains (Critser and Mobraaten, 2000), to date there has been only one successful report of rat sperm cryopreservation (Nakatsukasa et al., 2001). This is in part due to the extreme sensitivity of epididymal rat sperm to subphysiologic conditions. In this report, epididymal rat sperm was cryopreserved in 23% egg yolk, 8% lactose monohydrate, and 10% Tris (hydroxymethyl) aminomethane. The sperm sample was first held at 15° C for 30 minutes and then at 5° C for another 30 minutes. The sperm suspension is then loaded into 0.25-ml straws and exposed to LN_2 vapor at approximately −170° C for 10 minutes, followed by plunging into LN_2. Following intrauterine insemination using a 100 μl frozen-thawed sperm suspension containing 1 to 3×10^6 sperm with 9.6% motility, 5 out of 12 (41.7%) recipients became pregnant and yielded 13 pups. Although these results are encouraging, more investigation is needed to improve pup yield.

Recently, significant progress has been made in the long-term storage of mouse sperm in the freeze-dried state. Live young have been obtained by injecting freeze-dried sperm heads into mature oocytes (Wakayama and Yanagimachi, 1998). Although the long-term effects of this procedure on mouse genotype and phenotype are unknown, this technique is very promising and presents a cost-effective alternative for the genome banking of rat strains. Round rat spermatids have been successfully cryopreserved (Hirabayashi et al., 2002) using GL-PBS supplemented with 7.5% glycerol and 7.5% FBS as originally described by Ogura et al. (1996), who cryopreserved mouse round spermatids at 1° C per minute to −80° C followed by plunging into LN_2 24 hours later. Frozen thawed rat round spermatids gave rise to live offspring after being injected into MII rat oocytes (Hirabayashi et al., 2002).

One of the most promising recent studies involves long-term culture of spermatogonia and their transplantation. This approach has been successful in terms of producing mature sperm after transplanting cryopreserved mouse or fresh rat spermatogonia into seminiferous tubules of busulfan-treated recipient mice (Clouthier et al., 1996; Brinster and Nagano, 1998). In the Clouthier report, mouse spermatogonia were initially exposed to a freezing solution consisting of FBS, Dulbecco's modified Eagles medium, and DMSO in a ratio of 1:3:1 and then placed into an insulated container at −70° C for 12 hours followed by plunging into LN_2 for long-term storage. The procedures used to cryopreserve both spermatogonia and round spermatids in those studies are quite simple to perform and are essentially the same as those used for cryopreserving somatic cells. These reports are overall very encouraging for the utilization of cryopreserved rat male germ cells virtually at any stage for genome banking or as readily available sources of germ cells for gene modification studies.

C.   Oocyte Cryopreservation

Although the availability of working cryopreservation protocols for rat oocytes would be an important addition to overall ARTs in the rat, the low cryosurvival rate of oocytes makes this a less attractive means of genome banking than embryo and sperm cryopreservation. The labor and cost involved in the superovulation procedure is another consideration. Equilibrium freezing protocols that were used to cryopreserve metaphase II (MII) stage rat oocytes (Kasai et al., 1979) resulted in poor post-thaw survival and fertilization rates. A vitrification solution consisting of 2 M acetamine + 4 M PG + 2 M DMSO was used to cryopreserve MII rat oocytes. Although the post-thaw survival rate (65%) was relatively high, fetal development into live pups was low (3%) (Nakagata, 1992). These limited numbers of investigations are not sufficient enough to draw conclusions on their cryopreservation efficiency. Intrinsic factors such as high spontaneous activation and fragmentation of rat oocytes are major limitations for the development of better cryopreservation protocols for rat oocytes (Zernicka-Goetz, 1991; Ben-Yosef et al., 1995).

D.   Ovarian and Testicular Tissue Cryopreservation

Cryopreservation of ovarian tissue (OT) in the rat offers an easy way of banking female germ cells because they contain thousands of primordial follicles, which are exceptionally resistant to cryopreservation. In practice, frozen-thawed OT from mice has been effectively used in the reconstitution of specific mouse strains following orthotopic autologous transplantation (Krohn, 1958). Promising results have also been reported for in vitro maturation and fertilization of primordial follicles isolated from frozen-thawed mouse OT. In that optimized follicle isolation (Morena et al., 1996; van Pelt et al., 1996) and OT cryopreservation protocols (Aubard et al., 1998) are available for the rat, these technologies may soon be part of ARTs in rat strains. Before cryopreservation, the removed ovary is cut into small pieces the size of a post-pubertal mouse ovary (1 mm × 1 to 1.5 mm) using a scalpel and stereo-microscopy. Ovarian tissue pieces are then randomly distributed to the fresh or frozen-thawed transplantation groups and either transplanted immediately to recipients fresh or cryopreserved for further experimentation.

The authors use a slow cooling procedure to cryopreserve rat ovarian tissue similar to the procedure used in mouse ovarian tissue cryopreservation (Harp et al., 1994; Gunasena et al., 1997). Ovarian tissue pieces are placed into 1-ml cryo-tubes containing 1.5 M DMSO prepared in TL-Hepes media and held for 10 minutes before they are then placed into a programmable freezer at 20° C. The samples are initially cooled to 10° C using a cooling rate of 1° C per minute, and then to −55° C using a cooling rate of 0.5° C per minute before plunging into LN_2. Ice nucleation is induced manually by using pre-cooled forceps at −7° C. For thawing, cryo-tubes are placed at 20° C, and tissue is expelled into a fresh TL-Hepes solution and washed to remove DMSO.

The extraction of sperm, spermatids, and spermatogonia from cryopreserved testicular tissue has great potential when successfully combined with microinsemination into mature oocytes in ICSI and ROSI methods for embryo production. Jezek et al. (2001) cryopreserved rat testicular tissue in 10% DMSO in Dulbecco's modified Eagle medium using a cooling rate of 2° C per minute followed by rapid thawing by plunging into a 37° C water bath. The effects of the cryopreservation procedure on the morphology of the cellular structures in the testicular biopsies were analyzed post-thaw. This study demonstrated that although spermatogonia and spermatocytes were detrimentally affected during the procedure the vast majority of round and elongated spermatids and spermatozoa maintained their normal structure. These results suggest the potential usage of simple testicular tissue cryopreservation when it is not possible to perform optimal germ cell extraction procedures such as in the case of an unexpected death of a valuable male rat.

IV. GENETIC MODIFICATIONS

Over the last two decades, methods developed to mature, grow, and micromanipulate mammalian gametes and embryos have opened many avenues to the investigation of genes and their functional products during fetal development and adulthood. For example, creation of transgenic animals by random genome insertion, the modification of existing genes by targeted mutagenesis, and the inactivation of endogenous gene(s) continue to play major roles in advancing our understanding of the relationship between phenotype and genotype. Having both molecular biology techniques (i.e., recombinant DNA technology, polymerase chain reaction, fluorescence in situ hybridization, southern blot) and ARTs described in this chapter, research opportunities are presently limited only by one's imagination. The creation of germ-line modified lab animals has significance in the understanding of gene regulation in terms of loss- (i.e., gene knockout) or gain-of-function (i.e., transgenic) studies.

Recent advances in genome characterization have rapidly increased the ability to create and utilize new animal models for biomedical research. Scientists now have the ability to create GMR using techniques such as pronuclear microinjection or viral transfection (Mullins and Mullins 1996; Brenin et al., 1997; Lois et al., 2002). Using these techniques, a rapidly expanding number of GMR have been developed for use in studies ranging from basic gene function analysis to toxicological testing of therapeutic agents to comparative studies of numerous human disorders. These GMR have become invaluable in the study of human diseases such as hypertension, cardiovascular diseases (Bader et al., 2000), endocrinopathies (Rocco et al., 1994), autoimmunity (Taurog et al., 1994), and carcinogenesis (Hully et al., 1994), as well as in research in the field of pharmacology (Hochi et al., 1992; Hirabayashi et al., 1996). This is in part due to the fact that rat models provide specific advantages over the mouse for some areas of investigation, for reasons including their larger size. This is important in experiments for which greater blood volumes need to be collected (for multiple analyses), when repeated samples are needed, or in cases requiring long-term monitoring involving surgical implantation (e.g., blood pressure measurement via telemetry) (Mullins et al., 2002).

In some instances, the production of transgenic rats has provided data that are new and relevant compared to data obtained from mice bearing the same transgene (Charreau et al., 1996). Transgenic rat technology also opens up important perspectives for transplantation research, in which microsurgery is an essential procedure (Charreau et al., 1996; Mullins et al., 2002). The development of GMR is still in its infancy. Only a small percentage of rat genes have well-characterized phenotypic profiles. Studies using systematic inactivation of each rat gene or random mutagenesis to identify new genes are ongoing, and the technologies for assessing sequence variation on a genome-wide scale will prompt comprehensive studies of comparative genomic diversity. These studies will vastly improve our understanding of the genetic basis for complex diseases and traits.

A. Transgenic Technology

The essential steps involved in the creation of transgenic rats as they relate to reproductive biology are reviewed here. Transgenesis is a technique by which foreign cloned DNA is introduced into a living organism using molecular tools and micro-tools to achieve genomic alterations in the living organism. Transgenic animals remain crucial to our understanding of the action of a single gene in the context of the whole organism. Over the past two decades, many techniques have been introduced for the purpose of delivering foreign DNA into germ cells (spermatogonia, spermatozoa, oocytes), embryos, and somatic or embryonic stem cells (nuclear cloning) to create transgenic animals (Chan et al., 1999).

1. Pronuclear Injection (PI)

To date, direct microinjection of a recombinant DNA solution into the pronucleus of a fertilized oocyte has been the most widely used method for creating thousands of transgenic mouse lines (Gordon et al., 1980). Current technology used to create transgenic rats has been primarily adapted from the mouse and requires a substantial amount of time and technical effort (Mullins et al., 2002). The transgenic concept was first demonstrated by injection of the metallothionein-growth hormone fusion gene to alter the phenotype of the mice by dramatic overgrowth (Palmiter et al., 1982). This paved the way for the exponential development of gene transfer technology using PI. However, the implementation of this technology in the rat species took another decade. Mullins et al. (1990) were able to create transgenic rats carrying the mouse Ren-2 gene to study the genetic basis of hypertension.

Although efficiency is significantly lower than mice, PI has been successfully used to create hundreds of transgenic rat lines (Mullins et al., 1990). With this procedure, insertion of exogenous DNA occurs randomly in the host chromosome. It generally results in a single integration site, but multiple copies of the transgene are usually present at this single locus. With PI, the site of integration and number of copies cannot be controlled. Although linear DNA fragments integrate more effectively than supercoiled DNA, the size and length of the DNA fragments does not affect integration frequency. In addition, time of DNA integration is a crucial issue in terms of ubiquitous expression of the transgene. If transgene integration occurs prior to DNA replication, all developing cells (including germ cells) contain the transgene. However, if integration occurs after DNA replication some cells or tissues contain the transgene, causing mosaic integration of the transgene (that may prevent germ-line transmission of the transgene).

Another critical issue associated with creating transgenic rats via PI is the possibility of unintended insertional mutagenesis to a functionally important locus of the host genome. The loss of an appropriate function at the locus of insertion may confound interpretation of the significance of transgene expression. For these reasons, multiple founder animals are usually produced to help elucidate the insertional effects and copy number effects on the phenotype of the transgenic rat.

DNA construct design and preparation are complex procedures described in detail elsewhere (Polites and Pinkert, 2002). Generally, a typical transgenic construct used for PI contains three distinct elements (Figure 7-8): (1) a gene's introns, (2) promoter, including proximal regulatory element, and (3) a polyadenylation sequence to enhance the stability of transgene messenger RNA. Promoters usually contain the proximal region, which is the major unit providing cell-specific expression. Enhancers located upstream increase the frequency level of transcriptional activity. With increasing understanding of promoter structure and function, more promoter regions containing regulatory elements are available for optimal construct design. The structure of the transgene totally depends on the purpose of the experiment. Depending on the question of interest, ubiquitous, inducible, developmental stage-specific, cell-specific, or tissue-specific promoters may be used.

There are many essential steps that need to be followed to produce transgenic rats before actual injection of transgene construct (Mullins et al., 2002). These are (1) isolation of DNA and transgene construct design, (2) purification of DNA, and (3) quantification of DNA. After these steps are carefully performed, the gene construct is injected into a pronucleus of fertilized oocytes with the aid of a microinjection unit. The injected embryos are then transferred to surrogate mothers. Typically, the determination of the presence of transgene and evaluation of transgene expression using appropriate molecular biology tools (e.g., polymerase chain reaction, southern blot) is performed as early as two weeks postpartum. DNA for microinjection can be prepared from plasmid or cosmid clones by any of the standard protocols (Polites and Pinkert, 2002). It is imperative that DNA fragments for microinjection are free of strands or contaminants that may be cytotoxic to the embryo or obstruct the injection needle during PI.

Delivery of a transgene into an embryo requires a micromanipulation apparatus, as previously described for the ICSI procedure. Briefly, a rat zygote having two visible pronuclei is secured by a holding pipette (10 to 15 micrometer ID) using negative pressure and an injection needle (approximately 1 µm ID) containing a transgene construct is inserted into the larger PN after puncturing the zona pellucida, plasma membrane, and nuclear membrane. A few

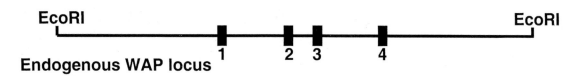

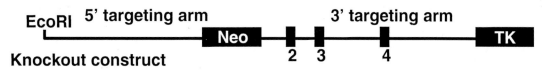

Fig. 7-8 Schematic of transgenic and knockout vectors for the whey acidic protein (WAP) locus. **Top**: Endogenous WAP locus consisting of four exons. **Middle**: Representative design for a transgenic vector for overexpression of a foreign gene via pronuclear injection. The important components of this vector are (1) a promoter for tissue, cell, inducible, or developmental-specific expression of the foreign introduced gene, (2) the gene of interest to be expressed (green fluorescent protein or GFP), (3) intronic sequence for increased mRNA expression and stability (can be provided from endogenous gene sequences or synthetic introns), (4) a polyadenylation sequence for mRNA stability, and (5) flanking restriction enzyme sites (EcoR1) used to liberate the transgene from the parental vector. **Bottom**: Representative design for a replacement-type knockout construct for deletion of an endogenous gene in embryonic stem (ES) cells. The construct consists of (1) a 5′ targeting arm having homology to the endogenous gene, (2) a neomycin (Neo) cassette for positive selection, (3) a 3′ targeting arm having homology to the endogenous gene, (4) a thymidine kinase (TK) cassette for negative selection, and (5) a restriction enzyme site (EcoR1) for linearizing the targeting vector prior to introduction into ES cells. (Courtesy of Dr. Ed Rucker, University of Missouri.)

picoliters of solution containing (1 microgram/milliliter) 100 to 200 copies of the transgene construct are then injected into PN by positive pressure generated by a transjector and confirmed by nuclear membrane swelling (Figure 7-9). The major impediment in creating a transgenic rat by PI is the difficulty in penetrating the pronuclear membrane due to its extreme elasticity as well as obscurity. This results in a lower (40 to 60% depending on the strain) post-injection survival rate when compared to the mouse and ultimately lowers the overall efficiency of transgenic rat production to 0.1 to 1% (Dycaico et al., 1994; Charreau et al., 1996; Hirabayashi et al., 2001).

2. Viral Vector Mediated Gene Transfer

Gene delivery into gametes, pre-implantation embryos and embryonic carcinoma cells can also be achieved using retroviral vectors such as moloney murine leukemia (MLV), and more recently using lentiviral vectors (Jaenisch and Mintz, 1974; Teich et al., 1977; Jahner et al., 1982; Chan et al., 1998; Lois et al., 2002; Orwig et al., 2002). Retroviruses are a class of enveloped viruses containing a

double-stranded RNA molecule as the genome. Following infection, the viral genome is reverse transcribed into two identical copies of DNA and subsequently integrates into the host cell genome. For most retroviruses, DNA integration only occurs in dividing cells. However, lentiviruses (a subclass of retroviruses) are able to integrate into both proliferating and nonproliferating cells.

Wider application of viral-mediated gene delivery has become more attractive following the development of replication-defective retroviral vectors due to reduced biosafety risk associated with vectors incapable of producing viral particles (Shimotohno and Temin, 1981; Miyoshi et al., 1998). A similar strategy has been achieved for lentiviruses, with some modifications to prevent viral particle replication in the host genome. The lentiviral vector was first used to create transgenic rats by infecting pronuclear-stage rat embryos (Lois et al., 2002). The basic design of the vector consists of a cassette in which a promoter drives the expression of the gene of interest. This promoter and gene are then inserted into a lentiviral vector flanked by 5-foot and 3-foot-long terminal repeat LTR.

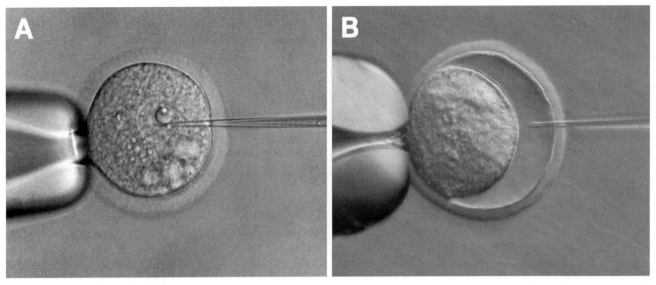

Fig. 7-9 Injection of plasmid DNA into a pronucleus of rat zygote (*A*), and injection of lentiviral gene construct into perivitelline space of rat zygote (*B*).

a. Construction and Production of Lentiviral Vectors. The lentiviral vectors contain self-inactivating long terminal repeats, but the viral genes have been deleted and a multicloning site has been added. The transgene of interest and promoter to control the transgene are inserted between the long terminal repeats. Figure 7-10 illustrates the lentiviral vector used to create a transgenic rat model with ubiquitous expression of the green fluorescence protein gene. Once the transgene has been inserted into the cloning vector, viral stocks are made by transfecting cell lines with the cloning vector. The cell lines have previously been stably transfected with plasmids that allow packaging and envelope production of the lentiviral vector (Kafir et al., 1999). The viral stock is then collected from culture supernatant by ultracentrifugation. The titer of the concentrated viral stock is determined by performing immune staining of cell cultures infected with 10-fold serial dilutions of the viral stock. The titered viral vector is then used for zygote injections.

b. Perivitelline Space Injection. Perivitelline space injection is a simple procedure compared to pronuclear injection and yields a higher post-injection survival rate (>90%). A zygote is first restrained by a holding pipette, and an injection needle containing viral vector punctures the zona pellucida to allow deposition of the viral vector into the perivitelline space (Figure 7-9). Injected zygotes are transferred to surrogate mothers as described previously.

B. Gene Knockout Technology

One of the most powerful methods of studying gene function is to look at the phenotype in the absence of a specific gene product. In the context of studying the genetic origin of human diseases, gene knockout technology in animals offers great insight into the function of conserved genes. Disruption of the normal coding sequence of a specific gene to interfere with expression provides a powerful tool for understanding its functional importance (Doetschman, 2002; Rucker et al., 2002). Recently there has been significant interest in implementing gene knockout techniques in rats. The following sections review existing and developing methodologies for creating knockout rats.

1. Embryonic Stem Cell Technology

The preimplantation blastocyst-stage embryo has two distinct domains, the inner cell mass (ICM) and the trophectoderm. Whereas the ICM gives rise to the fetus itself and extraembryonic tissues (i.e., allantois, amnion), trophectoderm contributes only to trophoblast layers of the placenta. Establishment of immortal embryonic stem (ES) cells that have the ability to transform into many tissues has been interest to scientists for decades. One of the most significant events in mammalian developmental biology was the isolation and long-term culture of cells from the inner cell mass of preimplantation blastocysts.

It was observed early in the 1960s that a 129 mouse strain (LT129) had a high incidence of developing ovarian teratocarcinomas due to spontaneous activation (parthenogenetic) of unovulated oocytes. This finding led to an experiment to determine if embryonic cells derived from a 129-strain tumor had the ability to incorporate into an embryo derived from a genetically different mouse strain (Papaioannou et al., 1978). These research efforts resulted in the production of chimeric offspring with the

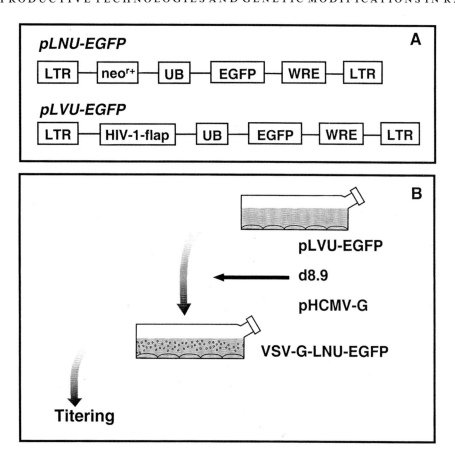

Fig. 7-10 Schematic of lentiviral vector (LV) for the production of a transgenic rat harboring enhanced green fluorescent protein (EGFP) gene. The basic design of the vector consists of a cassette in which a promoter (UB, ubiquitin) drives the expression of a gene. The promoter and gene are inserted into a lentiviral vector flanked by 5' and 3'-long terminal repeats (LTRs). WRE: woodchuck hepatitis virus post-transcriptional regulatory element (A). Generation of VSV-G pseudotyped lentiviral vector: Three DNA plasmids (including pLVU-EGFP, d8.9, and pHCMV-G) were co-transfected in 293FT packaging cells. Supernatant is collected at 48 hours post-transfection followed by titering and ultracentrifugation (B). (Courtesy of Dr. Anthony Chan, Emory University, Atlanta, Ga.)

genetic makeup of teratocarcinoma cells in addition to host genome. Although these mice did not transmit the genetic material of the teratocarcinoma to their offspring (Papaioannou et al., 1978; Fujii and Martin, 1980), these studies ultimately led to the development of 129 ES cells.

ES cells are pluripotent and can be maintained by culturing the ICM on feeder layers in Dulbecco's modified Eagle's medium supplemented with heat-inactivated fetal calf serum. ES cells can be propagated indefinitely in an undifferentiated state by repeated passages in the presence of leukemia-inhibiting factor to prevent spontaneous differentiation (Evans and Kaufman, 1981). It was then quite logical to assume that because ICM cells are derived from the blastocyst they are more likely to incorporate into the host genome when reintroduced to the blastocyst cavity using the appropriate tool. Bradley et al. (1984) elegantly demonstrated that ES cells injected into a blastocyst from a genetically distinguishable mouse strain are capable of contributing to multiple tissue types (pluripotency), including germ cells.

This achievement was followed by a second logical step, which is the ability to deliver a gene through ES cell manipulation (Robertson et al., 1986). Mammalian genetics was revolutionized by the finding that homologous recombination of DNA occurs in ES cells, thus allowing for the development of gene targeting (Doetschman et al., 1987). Gene targeting is a powerful technology that can be employed to investigate the developmental or functional significance of a gene by deleting it. The principle behind ES cell-based gene targeting is that modified cloned DNA sequences are delivered to ES cells and homologous recombination occurs, resulting in ES cells carrying the targeted genetic modification of interest. In the course of homologous recombination, the inserted DNA replaces the locus of endogenous DNA that has the complimentary nucleotide homology (Figure 7-8). These ES cell lines are then injected into a developing blastocyst and the resulting mice that develop are chimeric for the genetic modification and the wild type gene.

Gametes of the chimera containing the targeted mutation are passed on to the offspring, a process referred to as germ line transmission. The position of the homologous regions in vector and chromosomal DNA will ultimately determine if the engineered gene replaces the endogenous one or if the entire vector is inserted into the genome. Based on the mouse, there are several considerations for the preparation of a gene knockout construct for ES cells: (1) the incorporation of positive selectable marker(s) (i.e., neomycin), (2) the inclusion of thymidine kinase negative selectable markers for enrichment, (3) the length of homology, (4) use of isogenic DNA, and (5) the type of targeting vector. Figure 7-8 illustrates a replacement-type vector (Doetschman, 2002).

There is no doubt that hundreds of knockout mouse models have greatly improved our understanding of gene function. However, in some cases the rat may be a more appropriate animal model for studying certain human diseases. Therefore, the establishment of rat ES cell lines and their subsequent usage to produce knockout rats is essential. Unfortunately, research efforts with rat ES cells have not yet yielded true pluripotent cell lines and germ-line chimeras following injection of these lines into the rat blastocyst (Iannaccone et al., 1994; Brenin et al., 1997).

2. *N*-ethyl-*N*-nitrosourea (ENU) Induced Mutagenesis

Although ENU-induced mutagenesis was discovered more than 20 years ago, only recently have genome-wide mutagenesis protocols using ENU been successfully demonstrated in rats. ENU-induced mutagenesis is an alternative strategy for the creation of rat models that contain disrupted genes. ENU is a highly mutagenic and toxic agent widely used for gene disruption studies to obtain heritable phenotype in the mouse and to study loss-of-function phenotypes (Russell et al., 1979). Treating male mice with ENU is an efficient way of mutagenizing spermatogonial stem cells, and this treatment has proven effective in a variety of screens for both dominant and recessive mutations. ENU is an alkylating agent that mainly causes nucleic acid base substitutions and point mutations (Favor et al., 1988). Because the mutagenic action of ENU occurs via alkylation of nucleic acids, metabolic activation is not necessary for mutagenesis. However, because ENU is a potent carcinogen—which causes various types of tumor development as well as altered phenotypes—a balance needs to be sought between mutation frequency and toxicity. Thus, initial careful titration of the ENU doses per strain studied is recommended before production of large-scale mutants (Justice et al., 2000).

A shortcoming of this method is that the efficiency of mutagenesis is assessed most often by the empirical determination of a per-locus mutation frequency by using the specific locus test, which is expensive, time-consuming,

and logistically difficult (Favor et al., 1991; Davis et al., 1999). It is also important to note that for any individual ENU-treated animal many different mutations are simultaneously induced. The major concern with this high incidence of multiple mutations in a single animal is that it may lead to the demonstration of phenotypes that are the results of a multigenic trait, confounding the identification of the responsible gene(s).

It has been shown in mouse studies that whereas F1 hybrid lines are quite resistant many inbred lines cannot tolerate high doses of ENU without significant risk of death and/or permanent sterility. Generally, a dose of 100 to 200 mg/kg body weight is administered IP either as a single injection or as multiple injections of the same dose. The dose can then be optimized to yield approximate mutation rates of 1 in 1,000 gametes (Favor et al., 1997). It has also been shown that ENU is an efficient mutagen of spermatogonial cells. The rate of mutation significantly drops in advanced male germ cells (i.e., spermatids) and female germ cells. ENU is a very ineffective mutagen of mature spermatozoa. Administration of ENU to males causes a temporary period (14 to 15 weeks) of sterility due to a depletion of spermatogonial stem cells. Therefore, the spermatocytes need to be repopulated from the remaining stem cells so that treated males can be used for the production of F1 progeny.

Recently, an elegant study combined ENU mutagenesis and yeast-based screening assay to create knockout rats for the breast cancer suppressor genes *Brca1* and *Brca2* (Zan et al., 2003) (Figure 7-11). In this study, sexually mature rats of one outbred stock (SD) and two inbred strains (F344 and Wistar-Furth) were injected with ENU to induce mutations. Whereas F344 and Wistar-Furth received single doses of 100-mg/kg and 25-mg/kg ENU, respectively, SD rats received a split dose of 120 mg/kg at 9 and 10 weeks of age. Following mating, preweaning F1 rat pups from mutagenized males were screened for functional mutagenesis for *Brca1* and *Brca2* genes using the yeast truncation assay. One of the exons for *Brca1* (from genomic DNA) and cDNA for *Brca2* (from mRNA) were PCR-amplified from F1 pups.

To determine if the mutation disrupted gene function, a yeast base assay was developed. Gap-repair vectors containing approximately 100-bp fragments from both 5′ and 3′ ends of the clone of interest were constructed. The vector contained ADH1 upstream of the 5′ fragment and ADE2 downstream of the 3′ fragment. The vector was linearized, mixed with the PCR reaction, and transformated into yeast. Thus, the gene-specific fragment was cloned in vivo into the gap-repair vector. If the gene of interest (i.e., *Brca1* or *Brca2*) had a functional mutation, translation of ADE2 was prevented and the yeast produced red colonies. In the absence of a functional mutation, yeast produced white colonies.

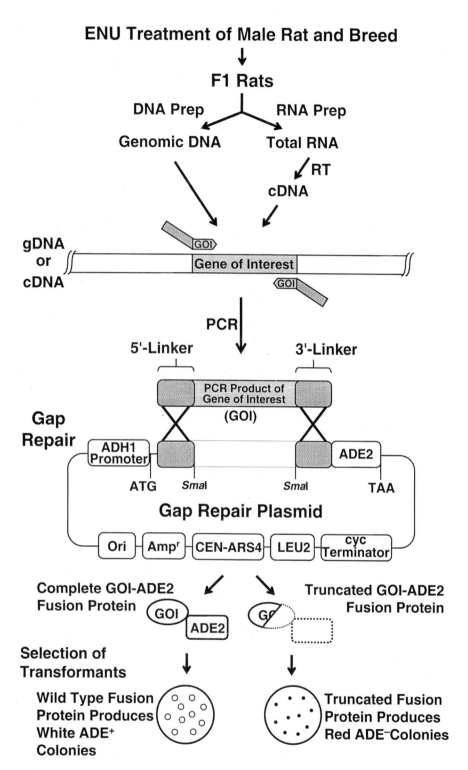

Fig. 7-11 Schematic diagram of ENU-mediated mutagenesis and gDNA/cDNA truncation assays. Male rats are treated with ENU and bred to produce F1 pups. DNA and RNA are isolated from tail clips of one-week-old F1 rats. Total RNA is reverse-transcribed to cDNA. Genomic DNA or cDNA fragments of interest are PCR amplified with chimeric oligonucleotide primers. The 3′-termini of chimeric primers are sequences derived from a gene of interest (GOI) that will specify the PCR product. The 5′-termini of the chimeric primers are homologous to the linkers in the gap repair vector. Following co-transformation into the yeast cells, the GOI fragment is cloned into the gap repair vector by homologous recombination. A wild-type gene fragment will produce a functional fusion protein with the ADE2 gene of the vector and form large white yeast colonies when plated on a selective medium. A gene fragment with truncation mutation will not form a functional fusion protein, and approximately 50% of the yeast colonies will be small and red. (Courtesy of Dr. Michael Gould, University of Wisconsin, Madison, WI.)

3. Somatic Cell Nuclear Transfer

Nuclear transfer is powerful technique for studying genomic imprinting, nuclear-cytoplasmic interaction, totipotency, and the contribution of paternal and maternal genomes to developing embryos. Somatic cell nuclear transfer (SCNT) is a procedure by which a nucleus from a fully differentiated cell (e.g., fibroblast) undergoes complete genetic reprogramming when it is introduced into an enucleated oocyte. The prime advantage of SCNT is the possibility of creating multiple genetically identical animals. By minimizing genetic variation, more conclusive results may be obtained in a shorter time period using fewer experimental animals. Cloning also contributes to basic research efforts in understanding the mechanisms underlying the molecular/cellular basis of donor nucleus reprogramming by the recipient oocyte and studies involving gene function and differentiation (Latham, 2004).

One of the earliest studies simply demonstrated totipotency of the blastomere of two-cell-stage rat embryos by obtaining rat pups following destruction of one blastomere (Nicholas and Hall, 1942). Kono et al. (1988) produced live births by transferring the donor zygote nuclei into a recipient enucleated zygote. Later cloning studies in mammals utilized totipotent blastomere nuclei from pre-implantation-stage embryos as nuclear donors. The limited availability of donor cells is one of the shortcomings of the production of large numbers of cloned animals (Willadsen, 1986; Tsunoda et al., 1987; Tsunoda and Kato, 1997). In this regard, the successful production of live offspring with somatic cells, derived from mammary gland of adult sheep, was a major milestone in developmental biology (Wilmut et al., 1997). Despite the low efficiency in production of live births (2 to 5% of manipulated oocytes) and various serious health problems, many mammalian species (including sheep, cattle, mouse, pig, goat, rabbit, cat, and rat) have been successfully cloned using the nucleus of somatic cells (i.e., fibroblast, cumulus, tail tip, sertoli cells, myoblast) (Rideout et al., 2000). These studies overall suggested the importance of the state and nature of recipient oocytes, cell cycle status and developmental potential of donor nucleus, method of activation, and genetic background on the SCNT outcome.

ZIt has recently been demonstrated that it is possible to manipulate genes in the donor nucleus for cloning to produce both transgenic and knockout pigs (Lai et al., 2002). Although cloned mice were produced from gene-targeted embryonic stem cells, this approach is not yet possible in the rat due to lack of embryonic stem cell technology (Ono et al., 2001). However, these advances collectively demonstrate the future potential of nuclear transfer in combination with gene-targeting technology in the biomedical field. Surprisingly, despite its common usage in many fields of medical research the rat only recently joined the long list of cloned mammalian species. This opened a new era for the possibility for gene knockout technology via SCNT in the rat (Fitchev et al., 1999; Hayes et al., 2001; Q. Zhou et al., 2003). The procedure for somatic cell nuclear transfer (SCNT) in the rat includes (1) preparation of recipient oocytes and donor cells, (2) introduction of donor nuclei by intracytoplasmic injection and subsequent removal of MII chromosomes from the recipient oocytes (enucleation), (3) chemical activation of reconstructed oocytes, and (4) development to term in a foster mother. The following sections provide a detailed description of each of these procedures.

a. PREPARATION OF RECIPIENT OOCYTES AND DONOR CELLS. Compared to oocytes from other species, rat oocytes present a unique challenge due to their tendency to activate spontaneously. This phenomenon was one of the most significant factors preventing successful SCNT in the rat. Zhou Q overcame this problem and produced the first cloned rat by isolating the rat cumulus oocyte complexes in M2 medium containing proteasomal inhibitor MG132, which reversibly blocks the first meiotic metaphase-anaphase transition in the rat (Zhou Q et al., 2003). The impact of the origin (embryonic, fetal, and adult) and type of donor cells on the efficiency of mammalian cloning is a current topic of discussion due to subtle variations in their development potential. Currently, fetal somatic cells such as fibroblasts are the most popular choice due to their abundance, genetic stability during long-term culture, and high efficiency with genetic manipulations. Primary rat embryonic fibroblasts can be prepared from 12.5-d pregnant Sprague-Dawley rats. Donor fibroblasts, used for NT, are synchronized in metaphase by treatment with 0.05 µg/ml Demecolcin for 2 hours.

b. NUCLEAR TRANSFER AND ENUCLEATION PROCEDURE. The nuclear transfer procedure requires complete removal of MII chromosomes from the recipient oocyte so that the donor nucleus can take over control of embryonic development and ensure appropriate reprogramming. The genetic material remaining in the oocyte is mostly maternal mitochondrial DNA scattered through the ooplasm. The SCNT is performed at room temperature in MG132-free M2 medium. A single fibroblast cell is aspirated into a 10-µm inner diameter glass injection pipette used to break the plasma membrane. The donor nucleus is then injected into cytoplasm after penetrating the zona pellucida and oolemma with the aid of piezzo actuation. The injection pipette is then gently withdrawn until it is close to the MII chromosomes, which are then sucked into the pipette allowing closure of the oolemma (Figure 7-12).

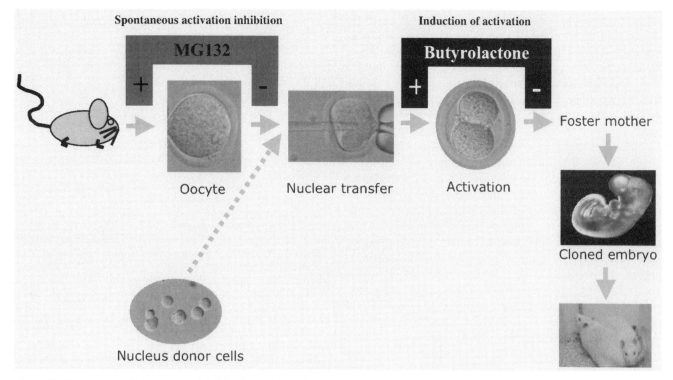

Fig. 7-12   Rat somatic cell nuclear transfer (SCNT) procedure. Oocytes are maintained at metaphase II stage until SCNT by incubation in MG132 supplemented medium to suppress spontaneous activation. The SCNT is performed in MG132-free medium. Activation is initiated by transfer of reconstructed rat oocytes into medium containing butyrolactone. (Courtesy of Dr. Jean Cozzi, Lyon, France.)

c. ACTIVATION PROCEDURE. Exit from MII arrest is naturally initiated by sperm incorporation into the oocyte following fertilization, generally referred to as oocyte activation. This event consists of a series of cellular changes that includes transient intracellular Ca^{++} oscillation, which then triggers sets of other events of activation such as cortical granule exocytosis to prevent polyspermy, recruitment of maternal mRNAs, resumption of meiosis with extrusion of a second polar body (PB2), formation of male and female pronuclei, rapid DNA synthesis, and ultimately the first mitotic cleavage (Wassarman, 1999). It is also well documented that the MII arrest of the mammalian oocyte is mediated by a cytostatic factor (CSF), which contains several molecules, including maturation-promoting factor (MPF)—a complex formed by cdk-1 and cyclin B.

Inactivation of MPF, mediated by degradation of cyclin B1 by the proteasomes, makes possible the exit from meiosis. Because cloning procedures do not involve sperm, reconstructed embryos require artificial activation to initiate the aforementioned cellular events. In addition to fertilization, oocyte activation can be induced by serial electrical pulses (1.3 kV/cm for 20 μsec); by chemicals such as calcium ionophore, ethanol, and strontium chloride ($SrCl_2$); or by protein synthesis inhibitors such as dimethylaminopurine (DMAP). The methods and time of activation of the reconstructed oocytes significantly affect the success of NT efficiency. For example, it was suggested that the donor nucleus needs to be introduced into recipient mouse oocytes before egg activation while the oocyte still posseses a high level of MPF activity (Rideout et al., 2000). This is thought to be one of the crucial factors for proper donor nucleus reprogramming. It is well known that rat oocytes undergo rapid spontaneous activation after ovulation if they are not fertilized (Zernicka-Goetz, 1991). This phenomenon presents a significant drawback because the NT procedure takes a few hours to complete. Activation is induced by incubating the reconstructed embryos in R1ECM medium containing 150 μM cdc2-specific kinase inhibitor butyrolactone I for two hours. Activated oocytes are then transferred to surrogate mothers to establish pregnancy.

4. Alternative Gene Knockout Technologies

Although gene disruption via gene-targeted ES cells and ENU mutagenesis are both powerful methods, they are hampered by the laborious construct design, positive and/or negative selection steps, and the cost associated with breeding and screening large numbers of animals. Therefore, alternative gene-targeting technologies are urgently needed. In this regard, RNA interference (RNAi) is a promising new technology to silence genes. The RNAi technique is a simple technique to reduce or abolish translation of a particular protein. The RNAi is based on introduction of small interfering RNA (siRNA) into cells to silence RNA expression. The vectors used for RNAi contain a promoter sequence upstream on the short hairpin RNA and downstream thymidine residues to terminate transcription. The siRNA contains approximately 19 to 23 bp of sense and antisense RNA separated by a 7-bp spacer.

A promoter sequence enables transcription of the short hairpin RNA. The proposed mechanism of action of siRNA is that the double-stranded RNA is degraded to 19 bp siRNA by an enzyme called dicer. The siRNAs are unwound by RNA-induced silencing complex, and antisense RNA binds to complementary sequence of the mRNA of interest. The newly formed double-stranded RNA (dsRNA) is degraded and protein translation is blocked. To date, it has been demonstrated that RNAi is a powerful way of interfering with gene expression in a range of organisms, including *Caenorhabditis elegans* and *Drosophila melanogaster* (Kennerdell and Carthew 2000; Timmons et al., 2003). Recently, RNAi interference of gene expression has been successfully used to inhibit c-*mos* and E-cadherin gene expression following microinjection of appropriate RNAi into mouse oocytes and zygotes, respectively (Wianny and Zernicka-Goetz, 2000). This report was followed by successful silencing of ubiquitously expressed EGFP using siRNA in the rat (Hasuwa et al., 2002).

The advantage of RNAi is that it offers an alternative to study loss-of-function phenotypes in specific cells and at specific stages of development of preimplantation embryos. Moreover, this technique will allow for study of loss-of-function of not only a single gene but multiple genes. These reports collectively suggest that RNAi could function as an alternative method of gene silencing in rats.

V. CONCLUSIONS

In conclusion, rats have been used in scientific studies for almost two centuries to understand mammalian anatomy and physiology, as well as various human disorders. Implementation of ART and genomic modification in the rat have less than three decades of history, and their development lags behind that of the mouse and livestock species. Several areas, including ART and genetic modification, require urgent attention in the rat so that the full potential of this species in animal model development can be realized. It is also important to note that the rat genome has recently been sequenced and extraordinary genomic and proteomic tools are becoming available. Therefore, the biomedical research community will greatly benefit from the synergism between reproductive and genomic advancements in the near future.

ACKNOWLEDGEMENTS

We thank Dr. Beth Bauer for her thoughtful comments and critical reading, and Howard Wilson for his technical assistance with the images.

REFERENCES

Armstrong, D.T., and Opavsky, M.A. (1988). Superovulation of immature rats by continuous infusion of follicle-stimulating hormone. *Biol. Reprod.* **39,** 511–518.

Aubard, Y., Newton, H., Scheffer, G., and Gosden, R. (1998). Conservation of the follicular population in irradiated rats by the cryopreservation and orthotopic autografting of ovarian tissue. *Eur. J. Obstet. Gynecol. Reprod. Biol.* **79,** 83–87.

Bader, M., Bohnemeier, H., Zollmann, F.S., Lockley-Jones, O.E., and Ganten, D. (2000). Transgenic animals in cardiovascular disease research. *Exp. Physiol.* **85,** 713–731.

Bavister, B.D., Leibfred, M.L., and Lieberman, G. (1983). Development of preimplantation embryos of the golden hamster in a defined medium. *Biol. Reprod.* **28,** 235–247.

Bavister, B.D. (1995). Culture of pre implantation embryos. Facts and artifacts *Hum. Reprod.* Update **1,** 91–148.

Ben-Yosef, D., Oron, Y., and Shalgi R. (1995). Low temperature and fertilization-induced Ca^{+2} changes in rat eggs. *Mol. Reprod. Dev.* **42,** 122–129.

Blandau, R.J., and Odor, D.L. (1949). The total number of spermatozoa reaching various segments of the reproductive tract in the female albino rat at intervals after insemination *Anat. Rec.* **103,** 93–94.

Bohlender, J., Ganten, D., and Luft, F.C. (2000). Rats transgenic for human renin and human angiotensinogen as a model for gestational hypertension. *J. Am. Soc. Nephrol.* **11,** 2056–2061.

Bradley, A., Evans, M., Kaufman, M.H., and Robertson, E. (1984). Formation of germ-line chimaeras from embryo-derived teratocarcinoma cell lines. *Nature* **309,** 255–256.

Brenin, D., Look, J., Bader, M., Hubner, N., Levan, G., and Iannaccone, P. (1997). Rat embryonic stem cells: a progress report. *Transplant. Proc.* **29,** 1761–1765.

Brinster, R.L., and Nagano, M. (1998). Spermatogonial stem cell transplantation, cryopreservation and culture. *Semin. Cell. Dev. Biol.* **9,** 401–409.

Cain, L., Chatterjee, S., and Collins, T.J. (1995). In vitro folliculogenesis of rat preantral follicles. *Endocrinology* **136,** 3369–3377.

Charreau, B., Tesson, L., Soulillou, J.P., Pourcel, C., and Anegon, I. (1996). Transgenesis in rats: technical aspects and models. *Transgenic Res.* **5,** 223–234.

Chan, A.W., Homan, E.J., Ballou, L.U., Burns, J.C., and Bremel, R.D. (1998). Transgenic cattle produced by reverse-transcribed gene transfer in oocytes. *Proc. Natl. Acad. Sci. USA* **95,** 14028–14033.

Chan, A.W.S. (1999). Transgenic animals: Current and alternative strategies. *Cloning* **1,** 25–46.

Clouthier, D.E., Avarbock, M.R., Maika SD, Hammer R.E., and Brinster, R.L. (1996). Rat spermatogenesis in mouse testis. *Nature* **381,** 418–219.

Corbin, T.J., and McCabe, J.G. (2002). Strain variation of immature female rats in response to various superovulatory hormone preparations and routes of administration. Contemp Top *Lab. Anim. Sci.* **41,** 18–23.

Critser, J.K., and Mobraaten, L.E. (2000). Cryopreservation of murine spermatozoa. *ILAR J.* **41,** 197–206.

Davis, A.P., Woychik, R.P., and Justice, M.J. (1999). Effective chemical mutagenesis in FVB/N mice requires low doses of ethylnitrosourea. *Mamm. Genome.* **10,** 308–310.

Doetschman, T., Gregg, R.G., Maeda, N., Hooper, M.L., Melton, D.W., Thompson, S., and Smithies, O. (1987). Targetted correction of a mutant HPRT gene in mouse embryonic stem cells. *Nature.* **330**, 576–578.

Doetschman, T. (2002). Gene targeting in embryonic stem cells: history and methodology. *In* "Transgenic Animal Technology" (C. Pinkert, Ed) pp. 113–141. Academic Press, San Diego.

Dorsch, M.M., and Hedrich, H.J. (1998). Effective superovulation for rat inbred strains and in vitro culture of preimplantation embryos: A retrospective study. *J. Exp. Anim Sci.* **39**, 99–108.

Dozortsev, D., Wakaiama, T., Ermilov, A., and Yanagimachi, R. (1998). Intracytoplasmic sperm injection in the rat. *Zygote* **6**, 143–147.

Dycaico, M.J, Provost G.S., Kretz P.L., Ransom S.L., Moores J.C., and Short J.M. (1994). The use of shuttle vectors for mutation analysis in transgenic mice and rats. *Mutat. Res.* **307**, 461–478.

Evans, M.J., and Kaufman, M.H. (1981). Establishment in culture of pluripotential cells from mouse embryos. *Nature.* **292**, 154–156.

Fahy, G.M., MacFarlane, D.R., Angell, C.A., and Meryman, H.T. (1984). Vitrification as an approach to cryopreservation. *Cryobiology* **21**, 407–426.

Favor, J., Neuhauser-Klaus, A., and Ehling, U.H. (1988). The effect of dose fractionation on the frequency of ethylnitrosourea-induced dominant cataract and recessive specific locus mutations in germ cells of the mouse. *Mutat. Res.* **198**, 269–275.

Favor, J., Neuhauser-Klaus A., and Ehling, U.H. (1991). The induction of forward and reverse specific-locus mutations and dominant cataract mutations in spermatogonia of treated strain DBA/2 mice by ethylnitrosourea. *Mutat. Res.* **249**, 293–300.

Favor, J., Neuhauser-Klaus, A., Ehling, U.H., Wulff, A., and van Zeeland, A.A. (1997). The effect of the interval between dose applications on the observed specific-locus mutation rate in the mouse following fractionated treatments of spermatogonia with ethylnitrosourea. *Mutat. Res.* **374**, 193–199.

Fitchev, P., Taborn, G., Garton, R., and Iannaccone, P. (1999). Nuclear transfer in the rat: potential access to the germline. *Transplant. Proc.* **31**, 1525–1530.

Fujii, J.T., and Martin, G.R. (1980). Incorporation of teratocarcinoma stem cells into blastocysts by aggregation with cleavage-stage embryos. *Dev. Biol.* **74**, 239–244.

Glenister, P.H., and Thornton, C.E. (2000) Cryoconservation-archiving for the future. *Mamm. Genome* **11**, 565–571.

Gordon, J.W., Scangos, G.A., Plotkin, D.J., Barbosa, J.A., and Ruddle, F.H. (1980). Genetic transformation of mouse embryos by microinjection of purified DNA. *Proc. Natl. Acad. Sci. USA* **77**, 7380–7384.

Gravance, C.G., Garner, D.L., Miller, M.G., and Berger, T. (2001). Fluorescent probes and flow cytometry to assess rat sperm integrity and mitochondrial function. *Reprod. Toxicol.* **15**, 5–10.

Gravance, C.G., Garner, D.L., Miller, M.G., and Berger, T. (2003). Flow cytometric assessment of changes in rat sperm mitochondrial function after treatment with pentachlorophenol. *Toxicol. In Vitro* **17**, 253–257.

Gunasena, K.T., Villines, P.M., Critser, E.S., and Critser, J.K. (1997). Live births after autologous transplant of cryopreserved mouse ovaries. *Hum. Reprod.* **12**, 101–106.

Hamilton, G.S. and Armstrong, D.T. (1991). The superovulation of synchronous adult rats using follicle-stimulating hormone delivered by continuous infusion. *Biol. Reprod.* **44**, 851–856.

Harp, R., Leibach, J., Black, J., Keldahl, C., and Karow, A. (1994). Cryopreservation of murine ovarian tissue. *Cryobiology.* **31**, 336–343.

Hasuwa, H., Kaseda, K., Einarsdottir, T., and Okabe, M. (2002). Small interfering RNA and gene silencing in transgenic mice and rats. *FEBS Lett.* **532**, 227–230.

Hayes, E., Galea, S., Verkuylen, A., Pera, M., Morrison, J., Lacham-Kaplan, O., and Trounson, A. (2001). Nuclear transfer of adult and genetically modified fetal cells of the rat. *Physiol. Genomics* **5**, 193–204.

Higuchi, H., Nakaoka, M., Kawamura, S., Kamita, Y., Kohda, A., and Seki, T. (2001). Application of computer-assisted sperm analysis system to elucidate lack of effects of cyclophosphamide on rat epididymal sperm motion. *J Toxicol Sci.* **26**, 75–83.

Hirabayashi, M., Kodaira, K., Takahashi, R., Sagara, J., Suzuki, T., and Ueda, M. (1996). Transgene expression in mammary glands of newborn rats. *Mol. Reprod. Dev.* **43**, 145–149.

Hirabayashi, M., Takahashi, R., Sekiguchi, J., and Ueda M. (1997). Viability of transgenic rat embryos after freezing and thawing. *Exp. Anim.* **46**, 111–115.

Hirabayashi, M., Ito, K., Sekimoto, A., Hochi, S., and Ueda, M. (2001). Production of transgenic rats using young Sprague-Dawley females treated with PMSG and hCG. *Exp. Anim.* **50**, 365–369.

Hirabayashi, M., Kato, M., Aoto, T., Ueda, M., and Hochi, S. (2002). Rescue of infertile transgenic rat lines by intracytoplasmic injection of cryopreserved round spermatids. *Mol. Reprod. Dev.* **62**, 295–299.

Hochi, S., Ninomiya, T., Waga-Homma, M., Sagara, J., and Yuki, A. (1992). Secretion of bovine alpha-lactalbumin into the milk of transgenic rats. *Mol. Reprod. Dev.* **33**, 160–164.

Hsu, P.C., Hsu, C.C., and Guo, Y.L. (1999). Hydrogen peroxide induces premature acrosome reaction in rat sperm and reduces their penetration of the zona pellucida. *Toxicology* **139**, 93–101.

Hully, J.R., Su, Y., Lohse, J.K., Griep, A.E., Sattler, C.A., Haas, M.J., Dragan, Y., Peterson, J., Neveu, M., and Pitot, H.C. (1994). Transgenic hepatocarcinogenesis in the rat. *Am. J. Pathol.* **145**, 386–397.

Iannaccone, P.M., Taborn, G., and Garton, R. (2001). Preimplantation and postimplantation development of rat embryos cloned with cumulus cells and fibroblasts. *Zygote* **9**, 135–143.

Iannaccone, P.M., Taborn, G.U., Garton, R.L., Caplice, M.D., Brenin, D.R., and Isachenko, V.V. (1994). Pluripotent embryonic stem cells from the rat are capable of producing chimeras. *Dev. Biol.* **163**, 288–292.

Iannaccone, P.M., and Galat, V. (2002). Production of Transgenic Rats. *In* "Transgenic Animal Technology" (C. Pinkert, Ed) pp 235–250. Academic Press, San Diego.

Isachenko, E.F., Ostashko, F.I., and Grishchenko, V.I. (1997). Ultrarapid freezing of rat embryos with rapid dilution of permeable cryoprotectants. *Cryobiology* **34**, 157–164.

Jacob, H.J. (1999). Functional genomics and rat models. *Genome Res.* **9**, 1013–1016.

Jacob, H.J., and Kwitek, A.E. (2002). Rat genetics: attaching physiology and pharmacology to the genome. *Nat. Rev. Genet.* **3**, 33–42.

Jaenisch, R., and Mintz, B. (1974). Simian virus 40 DNA sequences in DNA of healthy adult mice derived from preimplantation blastocysts injected with viral DNA. *Proc. Natl. Acad. Sci. USA* **71**, 1250–1254.

Jahner, D., Stuhlmann, H., Stewart, C.L., Harbers, K., Lohler, J., Simon, I., and Jaenisch, R. (1982). De novo methylation and expression of retroviral genomes during mouse embryogenesis. *Nature* **12**, 623–628.

Jezek, D., Schulze, W., Kalanj-Bognar, S., Vukelic, Z., Milavec-Puretic, V., Krhen, I. (2001). Effects of various cryopreservation media and freezing-thawing on the morphology of rat testicular biopsies. *Andrologia* **33**, 368–378.

Jiang, J.Y., Miyoshi, K., Umezu, M., and Sato, E. (1999). Superovulation of immature hypothyroid rdw rats by thyroxine therapy and the development of eggs after in vitro fertilization. *J. Reprod. Fertil.* **116**, 19–24.

Justice, M.J., Carpenter, D.A., Favor, J., Neuhauser-Klaus, A., Hrabe de Angelis, M., Soewarto, D., Moser, A., Cordes, S., Miller, D., Chapman, V., Weber, J.S., Rinchik, E.M., Hunsicker, P.R., Russell, W.L., and Bode, V.C. (2000). Effects of ENU dosage on mouse strains. *Mamm. Genome.* **11**, 484–488.

Kafri, T., van Praag, H., Ouyang, L., Gage, F.H., and Verma, I.M. (1999). A packaging cell line for lentivirus vectors. *J. Virol.* **73**, 576–584.

Kasai, M., Iritani, A., and Chang, M.C. (1979). Fertilization in vitro of rat ovarian oocytes after freezing and thawing. *Biol. Reprod.* **21**, 839–844.

Kennerdell, J.R., and Carthew, R.W. (2000). Heritable gene silencing in Drosophila using double-stranded RNA. *Nat. Biotechnol.* **18**, 896–898.

Kishi, H., and Greenwald, G.S. (1999). In vitro steroidogenesis by dissociated rat follicles, primary to antral, before and after injection of equine chorionic gonadotropin. *Biol. Reprod.* **61**, 1177–1183.

Kishi, J., Noda, Y., Narimoto, K., Umaoka, Y., and Mori, T. (1991). Block to development in cultured rat 1-cell embryos is overcome using medium HECM-1. *Hum. Reprod.* **6**, 1445–1448.

Klinefelter, G.R., Laskey, J.W., Perreault, S.D., Ferrell, J., Jeffay, S., Suarez, J., and Roberts, N. (1994). The ethane dimethanesulfonate-induced decrease in the fertilizing ability of cauda epididymal sperm is independent of the testis. *J. Androl.* **15**, 318–327.

Klinefelter, G.R., Strader, L.F., Suarez, J.D., and Roberts, N.L. (2002). Bromochloroacetic acid exerts qualitative effects on rat sperm: implications for a novel biomarker. *Toxicol. Sci.* **68**, 164–173.

Krohn, P.L. (1958). Litters from CSH and CBA ovaries orthotopically transplanted into tolerant A strain mice. *Nature* **181**, 1671–1672.

Kono, T., Shioda, Y., and Tsunoda, Y. (1988). Nuclear transplantation of ovulated rat oocytes during in vitro culture *J. Exp. Zool.* **224**, 371–377.

Lai, L., Kolber-Simonds, D., Park, K.W., Cheong, H.T., Greenstein, J.L., Im, G.S., Samuel, M., Bonk, A., Rieke, A., Day, B.N., Murphy, C.N., Carter, D.B., Hawley, R.J., and Prather, R.S. (2002). Production of alpha-1,3—galactosyltransferase knockout pigs by nuclear transfer cloning. *Science* **295**, 1089–1092.

Latham, K.E. (2004). Cloning: questions answered and unsolved. *Differentiation* **72**, 11–22.

Lois, C., Hong, E.J., Pease, S., Brown, E.J., and Baltimore, D. (2002). Germline transmission and tissue-specific expression of transgenes delivered by lentiviral vectors. *Science* **295**, 868–872.

Matsumoto, H., and Sugawara, S. (1998). Effect of phosphate on the second cleavage division of the rat embryo. *Hum. Reprod.* **13**, 398–402.

Mazur, P., Leib, S.P., and Chu, E.H.Y. (1972). Two factor hypothesis of freezing injury, evidence from Chinese tissue-culture cells. *Exp. Cell. Res.* **71**, 345–355.

Mazur, P., Schneider U., and Mahowald A.P. (1992). Characteristics and kinetics of subzero chilling injury in drosophila embryos. *Cryobiology* **29**, 39–68.

Miyoshi, K., Funahashi, H., Okuda, K., and Niwa, K. (1994). Development of rat one-cell embryos in a chemically defined medium: effects of glucose, phosphate and osmolarity. *J. Reprod. Fertil.* **100**, 21–26.

Miyoshi, K., Kono, T., and Niwa, K. (1997). Stage-dependent development of rat 1-cell embryos in a chemically defined medium after fertilization in vivo and in vitro. *Biol. Reprod.* **56**, 180–185.

Miyoshi, K., and Niwa, K. (1997). Stage-specific requirement of phosphate for development of rat 1-cell embryos in a chemically defined medium. *Zygote* **5**, 67–73.

Miyoshi, H., Blomer, U., Takahashi, M., Gage, F.H., and Verma, I.M. (1998). Development of a self-inactivating lentivirus vector. *J. Virol.* **72**, 8150–8157.

Morena, A.R., Boitani, C., Pesce, M., De Felici, M., and Stefanini, M. (1996). Isolation of highly purified type A spermatogonia from prepubertal rat testis. *J. Androl.* **17**, 708–717.

Mukumoto, S., Mori, K., and Ishikawa, H. (1995). Efficient induction of superovulation in adult rats by PMSG and hCG. *Exp. Anim.* **44**, 111–118.

Mullins, J.J., Peters, J., and Ganten, D. (1990). Fulminant hypertension in transgenic rats harbouring the mouse Ren-2 gene. *Nature* **344**, 541–544.

Mullins, L.J., and Mullins, J.J. (1996). Transgenesis in the rat and larger mammals. *J. Clin. Invest.* **97**, 1557–1560.

Mullins, L.J., Brooker, G., and Mullins J.J. (2002). Transgenesis in the rat. *Methods Mol. Biol.* **180**, 255–270.

Nakagata, N. (1992). Cryopreservation of unfertilized rat oocytes by ultrarapid freezing. *Exp. Anim.* **41**, 443–447.

Nakatsukasa, E., Inomata, T., Ikeda, T., Shino, M., and Kashiwazaki, N. (2001). Generation of live rat offspring by intrauterine insemination with epididymal spermatozoa cryopreserved at -196 degrees C. *Reproduction* **122**, 463–467.

Nicholas, J.S., and Hall, B.V. (1942). Experiments on developing rats II. The development of isolated blastomeres and fused eggs. *J. Exp. Zool.* **90**, 441–459.

Nishikimi, A., Uekawa, N., and Yamada, M. (2000). Involvement of glycolytic metabolism in developmental inhibition of rat two-cell embryos by phosphate. *J. Exp. Zool.* **287**, 503–509.

Niwa, K., and Chang, M.C. (1974). Optimal sperm concentration and minimal number of spermatozoa for fertilization in vitro of rat eggs. *J. Reprod. Fertil.* **40**, 471–474.

Ogura, A., Matsuda, J., Asano, T., Suzuki, O., and Yanagimachi, R. (1996). Mouse oocytes injected with cryopreserved round spermatids can develop into normal offspring. *J. Assist. Reprod. Genet.* **13**, 431–434.

Ono, Y., Shimozawa, N., Muguruma, K., Kimoto, S., Hioki, K., Tachibana, M., Shinkai, Y., Ito, M., and Kono, T. (2001). Production of cloned mice from embryonic stem cells arrested at metaphase. *Reproduction* **122**, 731–736.

Orwig, K.E., Avarbock, M.R., and Brinster, R.L. (2002). Retrovirus-mediated modification of male germline stem cells in rats. *Biol. Reprod.* **67**, 874–879.

Palmiter, R.D., Brinster, R.L, Hammer, R.E, Trumbauer, M.E., Rosenfeld, M.G., Birnberg, N.C., and Evans, R.M. (1982). Dramatic growth of mice that develop from eggs microinjected with metallothionein-growth hormone fusion genes. *Nature* **300**, 611–615.

Papaioannou, V.E., Gardner, R.L., McBurney, M.W., Babinet, C., and Evans, M.J. (1978). Participation of cultured teratocarcinoma cells in mouse embryogenesis. *J. Embryol. Exp. Morphol.* **44**, 93–104.

Perreault, S.D. (1998). Gamete Toxicology, The impact of new technologies in Reproductive Development and Toxicology. (K.S. Korach, Ed). pp. 635–654. Marcel Dekker, New York.

Perreault, S.D., and Cancel, A.M. (2001). Significance of incorporating measures of sperm production and function into rat toxicology studies. *Reproduction* **121**, 207–216.

Pfaff, R.T., Agca, Y., Liu, J., Woods, E.J., Peter, A.T., and Critser, J.K. (2000). Cryobiology of rat embryos I: determination of zygote membrane permeability coefficients for water and cryoprotectants, their activation energies, and the development of improved cryopreservation methods. *Biol. Reprod.* **63**, 1294–1302.

Polites, H.G., and Pinkert, C.A. (2002). DNA microinjection and transgenic animal production. *In* "Transgenic Animal Technology" (C. Pinkert, Ed) pp.15–65. Academic Press, San Diego.

Rall, W.F., Schmidt, P.M., Lin, X., Brown, S.S., Ward, A.C., and Hansen, C.T. (2000). Factors affecting the efficiency of embryo cryopreservation and rederivation of rat and mouse models. *ILAR J.* **41**, 221–227.

Rideout, W.M., Wakayama, T., Wutz, A., Eggan, K., Jackson-Grusby, L., Dausman J., Yanagimachi R., and Jaenisch, R. (2000). Generation of mice from wild-type and targeted ES cells by nuclear cloning. *Nat. Genet.* **24**, 109–110.

Rocco, S., Rebuffat, P., Cimolato, M., Opocher, G., Peters, J., Mazzocchi, G., Ganten, D., Mantero, F., and Nussdorfer, G.G. (1994). Zona glomerulosa of the adrenal gland in a transgenic strain of rat: a morphologic and functional study. *Cell Tissue Res.* **278**, 21–28.

Rouleau, A.M., Kovacs, P.R., Kunz, H.W., and Armstrong, D.T. (1993). Decontamination of rat embryos and transfer to specific pathogen-free recipients for the production of a breeding colony. *Lab. Anim. Sci.* **43**, 611–615.

Robertson, E., Bradley, A., Kuehn, M., and Evans, M. (1986). Germ-line transmission of genes introduced into cultured pluripotential cells by retroviral vector. *Nature* **323**, 445–448.

Rucker, E.B., Thomson, J.G., and Piedrahita, J.A. (2002). Gene targeting in embryonic stem cells: conditional technologies. *In* "Transgenic Animal Technology" (C. Pinkert, Ed) pp. 143–171. Academic Press, San Diego.

Russell, W.L., Kelly, E.M., Hunsicker, P.R., Bangham, J.W., Maddux, S.C., and Phipps, E.L. (1979). Specific-locus test shows ethylnitrosourea to be the most potent mutagen in the mouse. *Proc. Natl. Acad. Sci. U S A.* **76**, 5818–5819.

Schini, S.A., and Bavister, B.D. (1988). Two-cell block to development of cultured hamster embryos is caused by phosphate and glucose. *Biol. Reprod.* **39**, 1183–1192.

Shibahara, H., Obara, H., Ayustawati, Hirano, Y., Suzuki. T., Ohno, A., Takamizawa, S., and Suzuki, M. (2004). Prediction of pregnancy by intrauterine insemination using CASA estimates and strict criteria in patients with male factor infertility. *Int. J. Androl.* **27**, 63–68.

Shimotohno., K., and Temin, H.M. (1981). Formation of infectious progeny virus after insertion of herpes simplex thymidine kinase gene into DNA of an avian retrovirus. *Cell* **26**, 67–77.

Szczygiel, M.A., Kusakabe, H., Yanagimachi, R., and Whittingham, D.G. (2002). Intracytoplasmic sperm injection is more efficient than in vitro fertilization for generating mouse embryos from cryopreserved spermatozoa. *Biol. Reprod.* **67**, 1278–1284.

Takahashi, R., Hirabayashi, M., and Ueda, M. (1999). Production of transgenic rats using cryopreserved pronuclear-stage zygotes. *Transgenic Res.* **8**, 397–400.

Teich, N.M., Weiss, R.A., Martin, G.R., and Lowy, D.R. (1977). Virus infection of murine teratocarcinoma stem cell lines. *Cell* **12**, 973–982.

Thadani, V.M. (1979). Injection of sperm heads into immature rat oocytes. *J. Exp. Zool.* **210**, 161–168.

Thornton, C.E., Brown, S.D., and Glenister, P.H. (1999). Large numbers of mice established by in vitro fertilization with cryopreserved spermatozoa: implications and applications for genetic resource banks, mutagenesis screens, and mouse backcrosses. *Mamm. Genom.* **10**, 987–992.

Timmons, L., Tabara, H., Mello, C.C., and Fire, A.Z. (2003). Inducible systemic RNA silencing in Caenorhabditis elegans. *Mol. Biol. Cell.* **14**, 2972–2983.

Taurog, J.D., Richardson, J.A., Croft, J.T., Simmons, W.A., Zhou, M., Fernandez-Sueiro, J.L., Balish, E., and Hammer, R.E. (1994). The germfree state prevents development of gut and joint inflammatory disease in HLA-B27 transgenic rats. *J. Exp. Med.* **180**, 2359–2364.

Toyoda, Y., and Chang, M.C. (1974). Fertilization of rat eggs in vitro by epididymal spermatozoa and the development of eggs following transfer. *J. Reprod. Fertil.* **36**, 9–22.

Tsunoda, Y., Yasui, T., Shioda, Y., Nakamura, K., Uchida, T., and Sugie, T. (1987). Full-term development of mouse blastomere nuclei transplanted into enucleated two-cell embryos. *J. Exp. Zool.* **242**, 147–151.

Tsunoda, Y., and Kato, Y. (1997). Full-term development after transfer of nuclei from 4-cell and compacted morula stage embryos to enucleated oocytes in the mouse. *J. Exp. Zool.* **278**, 250–254.

van Pelt, A.M., Morena, A.R., van Dissel-Emiliani, F.M., Boitani, C., Gaemers, I.C., de Rooij, D.G., and Stefanini, M. (1996). Isolation of the synchronized A spermatogonia from adult vitamin A-deficient rat testes. *Biol. Reprod.* **55**, 439–444.

Wakayama, T., Whittingham, D.G., and Yanagimachi, R. (1998). Production of normal offspring from mouse oocytes injected with spermatozoa cryopreserved with or without cryoprotection. *J. Reprod. Fertil.* **112**, 11–17.

Wakayama, T., and Yanagimachi, R. (1998). Development of normal mice from oocytes injected with freeze-dried spermatozoa. *Nat, Biotechnol.* **16**, 639–641.

Walton, E.A., and Armstrong, D.T. (1983). Oocyte normality after superovulation in immature rats. *J. Reprod. Fertil.* **67**, 309–314.

Wassarman, P.M. (1999). Fertilization in animals. *Dev. Genet.* **25**, 83–86.

Whittingham, D.G., Leibo, S.P., and Mazur, P. (1972). Survival of mouse embryos frozen to −196 and −169°C. *Science* **178**, 411–414.

Whittingham, D.G. (1975). Survival of rat embryos after freezing and thawing *J. Reprod. Fert.* **43**, 575–578.

Wianny, F., and Zernicka-Goetz, M. (2000). Specific interference with gene function by double-stranded RNA in early mouse development. *Nat. Cell. Biol.* **2**, 70–75.

Willadsen, S.M. (1986). Nuclear transplantation in sheep embryos. *Nature* **320**, 63–65.

Wilmut, I., Schnieke, A.E., McWhir, J., Kind, A.J., and Campbell, K.H. (1997). Viable offspring derived from fetal and adult mammalian cells. *Nature* **385**, 810–813.

Woods, J., and Garside, D.A. (1996). An in vivo and in vitro investigation into the effects of alpha-chlorohydrin on sperm motility and correlation with fertility in the Han Wistar rat. *Reprod. Toxicol.* **10**, 199–207.

Working, P.K., and Hurtt, M.E. (1987). Computerized videomicrographic analysis of rat sperm motility. *J. Androl.* **8**, 330–337.

Zernicka-Goetz, M. (1991). Spontaneous and induced activation of rat oocytes. *Mol. Reprod. Dev.* **28**, 169–176.

Zan, Y., Haag, J.D., Chen, K.S., Shepel, L.A., Wigington, D., Wang, Y.R., Hu, R., Lopez-Guajardo, C.C., Brose, H.L., Porter, K.I., Leonard, R.A., Hitt, A.A., Schommer, S.L., Elegbede, A.F., and Gould, M.N. (2003). Production of knockout rats using ENU mutagenesis and a yeast-based screening assay. *Nat. Biotechnol.* **21**, 645–651.

Zhou, M., Sayad, A., Simmons, W.A., Jones, R.C., Maika, S.D., Satumtira, N., Dorris, M.L., Gaskell, S.J., Bordoli, R.S., Sartor, R.B., Slaughter, C.A., Richardson, J.A., Hammer, R.E., and Taurog, J.D. (1998). The specificity of peptides bound to human histocompatibility leukocyte antigen (HLA)-B27 influences the prevalence of arthritis in HLA-B27 transgenic rats. *J. Exp. Med.* **188**, 877–886.

Zhou, Q., Renard, J.P., Le Friec, G., Brochard, V., Beaujean, N., Cherifi, Y., Fraichard, A., and Cozzi, J. (2003). Generation of fertile cloned rats by regulating oocyte activation. *Science* **302**, 1179.

Zhou, Y., Galat, V., Garton, R., Taborn, G., Niwa, K., and Iannaccone, P. (2003). Two-phase chemically defined culture system for preimplantation rat embryos. *Genesis* **36**, 129–133.

Chapter 8

Analysis of Behavior in Laboratory Rats[1]

Ian Q. Whishaw, Valerie Bergdall, and Bryan Kolb

[1]A version of this chapter previously appeared in: Whishaw, I.Q. et al. (1999). Analysis of behavior in Laboratory Rodents. In Windhorst, U., Johansson, H. (eds), Modern techniques in neuroscience research, pp. 1244–1275, Heidelberg: Springer.

I. INTRODUCTION

To see the world in a grain of sand
And a heaven in a wildflower
Hold infinity in the palm of your hand
And eternity in an hour
–William Blake

The nervous system is intimately involved in the production of behavior, and so behavioral analysis is the ultimate assay of neural function. This chapter provides an overview of the behavior of rodents, and references for details of behavioral testing (Whishaw and Kolb, 2005). Most of the behavioral methodology comes from research on rats, but the ethograms of rodents are similar enough to allow for generalization of the methods, if not many aspects of behavior, to other species. The testing method can be conceived of as having a number of stages, sequentially involving the description of: (1) general appearance, (2) sensorimotor behavior, (3) immobility and its reflexes, (4) locomotion, (5) skilled movement, (6) species-specific behaviors, and (7) learning. For convenience, tables summarizing each class of behavior are found in the respective sections that follow.

The thoughts expressed in the opening line of William Blake's poem provide advice for behavioral neuroscientists, in both the surface and deep meaning of the words. The surface meaning is that the observation of details can provide insights into the larger structure of behavior. The behaviorist in the neuroscience laboratory who is attempting to diagnose the effects of a drug, a neurotoxin, or a genetic manipulation can heed the advice that it is often subtle cues that provide the insights into the effects of the treatment (Hutt and Hutt, 1970). The deeper meaning is quite simply that one should believe what one sees and

not be biased by theory to the extent that observed behavior is ignored, even when the particular behaviors seem at odds with theory.

To make this point in another way, the beginning student and even the seasoned worker may have been taught that the proper way to do science is to state a theory consisting of a number of postulates, logically deduce predictions about behavioral outcomes from the theory, and then compare the predictions with the obtained results of carefully controlled experiments, which leads to a revision or a confirmation of the theory. This way of doing science is potent, but unfortunately this has not been an especially productive way of conducting behavioral neuroscience. This is because our current understanding of how the brain produces behavior is not sufficiently advanced to permit the generation of non-trivial and readily testable theories. Put another way, there is no one-to-one congruence between behavioral effects and brain function (Vanderwolf and Cain, 1994).

Consider the following example. Suppose that Professor Alpha believes he has discovered a gene for learning. He predicts that if the gene is knocked out in an experimental animal the animal, although otherwise normal, will no longer be able to learn. He develops a knockout mouse that does not have the gene and then examines the learning ability of the mouse in an apparatus widely used for testing learning. Sure enough, the mouse is unable to solve the task and Professor Alpha publishes to much acclaim. Some time later, Professor Alpha's knockout mouse is examined in another laboratory, where it is discovered that it has a defect in its retina rendering it functionally blind. Of course, the reader may argue that Professor Alpha is unlikely to be so naive, but in actuality errors of this sort are common (see Huerta et al., 1996, for an example of avoiding such an error). Even when the more obvious sensorimotor functions that could affect learning are examined, there are potentially dozens of other subtle problems that might keep the animal from learning.

A different way of proceeding in behavioral neuroscience is to use an empirical and inductive approach (Whishaw et al., 1983). Empirical means that an animal's behavior is carefully assessed, without regard to theories, in order to describe its condition. Inductive means that from the description generalizations and conclusions are drawn about the effects of the treatment. Inductive science has been criticized, because, it has been argued, there is no way to tell which conclusions are correct and which are incorrect. We argue, however, that for behavioral research in general and behavioral neuroscience in particular conclusions made through induction can be subject to rigorous evaluation using the theoretical method.

The inductive technique is widely used as a first analytical step in clinical medicine (Denny-Brown et al., 1982) and in neuropsychology (Kolb and Whishaw, 1996). For example, when a patient goes to see a physician with a specific complaint, the careful physician will administer a physical examination in which sensory processes, motor status, circulatory function, and so on are examined. Only after such an examination does the physician venture a conclusion about the cause of the patient's symptoms. In neuropsychology, a wide-ranging battery of behavioral and cognitive tests is given to a patient and then the outcome of the tests is compared to the results obtained from patients who have known brain damage. Similar clinical tests have been developed for rodents (Whishaw et al., 1983).

Had Professor Alpha administered a physical and neuropsychology examination to his knockout mouse, he may have noticed that the mouse was blind and therefore tested it in conditions in which vision would not be essential for performance. The testing protocol given here is designed to provide an "ethogram" that becomes the foundation for subsequent detailed testing. For studies of genetically-manipulated rodents in particular, this comprehensive behavioral evaluation is intended to cast as wide a net as possible to capture multiple brain functions that may have been altered by even a single gene manipulation.

II. GENERAL METHODOLOGICAL APPROACHES

The three main ways of evaluating behavior are:

- End-point measures
- Kinematics
- Movement description

End-point measures are measures of the consequences of actions; for example, a bar was pressed, an arm of the maze was entered, or a photobeam was broken (Ossenkopp et al., 1996). Kinematics provides Cartesian representations of action, including measures of distance, velocity, and trajectories (Fentress and Bolivar, 1996; Fish, 1996; Whishaw and Miklyaeva, 1996). Movements can be described using formal languages that have been adapted to the study of behavior, such as Eshkol-Wachman movement notation (Eshkol and Wachmann, 1958). This system has been used for describing behaviors as different as social behavior (Golani, 1976), solitary play (Pellis, 1983), skilled forelimb use in reaching (Whishaw and Pellis, 1990), walking (Ganor and Golani, 1980), and recovery from brain injury (Golani et al., 1979; Whishaw et al., 1993). For a comprehensive description of behavior, all three methods are recommended (see the example following). Endpoint measures provide an excellent way of quantifying behavior, but animals are extremely versatile and can display compensatory behavior after almost any treatment.

There are many ways that they can press a bar, enter an alley, or intersect a photobeam. Kinematics provides excellent quantification of movement, but unless every body segment is described ambiguity can exist about which body part produced a movement. Movement notation provides an excellent description of behavior, but quantification is difficult.

A. Video Recording

As a prelude to behavioral analysis, it is recommended that behavior be video-recorded. Regardless of the type of experiment, equipment for video recording is relatively inexpensive, as off-the-shelf camcorders and VCRs are suitable for most behavioral studies (Whishaw and Miklyaeva, 1996). Equipment required is a hand-held video recorder that has an adjustable shutter speed. Most of the filming of human movements uses a shutter speed of about 1/100 of a second, but in order to capture blur-free pictures of the rapid movements made by rodents (who have a respiratory cycle, lick rate, and whisker brushing rate of about seven times per second), shutter speeds of 1/1,000 and higher are required. Fast shutter speeds require fairly bright lights, but with habituation rodents do not appear distressed by lights.

B. Video Analysis

To analyze the film, a videocassette recorder that has a frame-by-frame video advance option is necessary. A VHS model cassette recorder can be used, and the record acquired on the recorder cassette can be copied over to the VHS tape. Optional, but not essential, equipment includes a computer with a frame grabber that allows individual

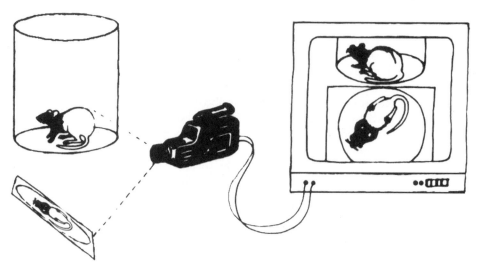

Fig. 8-1 Video-recording method. The video camcorder is placed so that it simultaneously records the rat from a lateral view and, through an inclined mirror, from a ventral view. Thus, the rat can be seen from two perspectives on the monitor (After Pinel et al., 1992).

frames of behavior to be captured for computer manipulation. Some camcorders direct digital filming onto disc, but the resolution may be less than that obtained with digital tape. Once a videotape of behavior is made, the behavior can undergo frame-by-frame analysis. Each video frame provides a 1/30 second snapshot of behavior. If rat licking is being studied, a single lick cycle would be represented on three successive video frames, a resolution that is adequate for most purposes.

One important feature of video frames, however, is that they are actually made up of two fields that are superimposed. Because computer-based frame grabbers can capture each field, a computer image of the video record increases resolution to 1/60 second. The analysis of some behaviors may require still greater resolution and higher-speed cameras are available, but they are presently very expensive. For most studies, it is helpful to use a mirror to film the animal from below (Pinel et al., 1992) or so that the surface on the animal's body that faces away from the camera can also be seen (Golani et al., 1979). A typical recording set-up is illustrated in Figure 8-1.

1. Example of Video-based Behavioral Analysis

The following example illustrates the complementary role of the different types of video-based behavioral analysis. Reaching for food by rats with unilateral motor cortex injury was studied using an end-point video recording and movement notation analysis (Whishaw et al., 1991). The study began with an end-point measure. The animal was allowed to obtain a piece of food on a tray by reaching through a slot in its cage. To force the rat to use its non-preferred limb, a light bracelet was placed on the normal limb, thus preventing it from going between the bars. The end-point measure of behavior was

the success in reaching for food with the limb contralateral to the lesion.

The end-point measure revealed that motor cortex lesions impaired the grasping of food. Normal control animals have a success rate of about 70%, and rats with motor cortex lesions had success rates that varied from near 20% to about 50%, the extent of impairment varying directly with the lesion size. Thus, this type of analysis shows that the forelimb representation of the motor cortex of the rat plays a role in skilled reaching movements. Note, however, that although this analysis identified that the motor cortex lesion interfered with reaching for food it did not identify the reason for the poor reaching. This question was addressed by the movement notation and kinematic analyses (Figure 8-2).

Analysis with the Eshkol-Wachman movement notation system, which is designed to express relations and changes of relations between parts of the body, revealed that the motor impairments were attributable to an inability to pronate the paw over the food in order to grasp it, as well as an inability to supinate the paw at the wrist to bring the food to the mouth. Finally, once a movement description was established with movement notation a number of other aspects of the movement were measured and documented using a Cartesian coordinate system, with initial and terminating components of the movements serving as reference points. For this analysis, points on the body were digitized and the trajectory of the movements of the limb reconstructed. This analysis revealed that in addition to the inability to supinate and pronate properly the animals with motor cortex injury also had an abnormal movement trajectory of the limb so that the aiming of the limb to the food was impaired.

Taken together, these three analyses provide a description of the various motor components that are affected

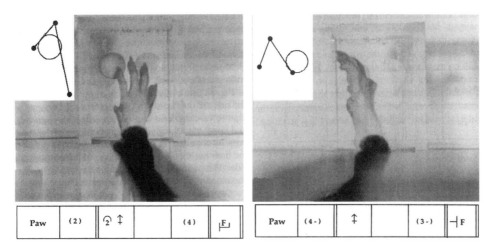

Fig. 8-2 Three methods of describing behavior: (1) Endpoint measure. The photographs are of a rat reaching through a slot in order to obtain a food pellet located on the shelf. In the left figure a control rat is about to grasp the food while in the right photo a rat with a motor cortex lesion has knocked the food pellet off the shelf and lost it. (2) Movement notation. On the bottom of each figure a movement notation score describes the movement. The first box indicates that it is the paw that is being described, the second box indicates the starting position of the paw and the last box indicates the end position of the paw. The notation in the three middle boxes indicates movement in three video frames (1/30 sec). Right: The paw advances and pronates and grasps the food pellet (F). Right: The paw advances and turns sideways without pronation to swipe the food away. (3) Cartesian reconstruction. The insert shows the trajectory of the tip of the third digit of the paw relative to the food pellet on the same three video frames. The three forms of analysis are necessary for a complete description of the movement and its results (After Whishaw et al., 1991).

by the motor cortex injury, as well as the subsequent effect the impairments have on the behavior. Such an analysis is necessary not only to understand the role of a neural structure in the control of movement but to investigate the effects of different treatments on the observed impairments. Thus, it might be shown that a particular drug influenced recovery of function, which could be documented by an improvement in an end-point measure. The subsequent movement notation and kinematic analyses would be required, however, to determine which aspects of the movement had been affected. For example, it has been shown that rats with motor cortex lesions show some recovery of skilled limb reaching, but this is largely due to changes in a variety of whole-body movements that compensate for the deficits in aiming (Whishaw et al., 1991). The impairments in pronation and supination remain.

III. THE NEUROBEHAVIORAL EXAMINATION

Many tests can be administered quite quickly while an animal is in its home cage, whereas others are given when the animal is removed from its cage. Most tests require no special equipment and are intended to be simple, rapid, and inexpensive ways of evaluating an animal's condition. In administering the clinical examination, the standard is that a healthy laboratory animal be clean, lively, and inquisitive, but not aggressive. The tests we describe may require innovative and liberal use of paraphernalia found around

the laboratory. For all of the tests described, it is assumed that control animals are also examined to provide the standard against which an experimental group is compared. In studies of genetically altered animals, often several classes of control animals are needed to be able to attribute behaviors to particular genetic alterations (Crawley et al., 1997; Upchurch and Wehner, 1989).

A. Appearance

The main features of the physical examination are outlined in Table 8-1. Animals should be examined in the home cage and removed for individual inspection. The following should be made points of inspection.

TABLE 8-1

EXAMINATION OF APPEARANCE AND RESPONSIVENESS

Appearance	Inspect body shape, eyes, vibrissae, limbs, fur and tail, and coloring.
Cage Examination	Examine the animal's cage, including bedding material, nest, food storage, and droppings.
Handling	Remove the animal from its cage and evaluate its response to handling, including movements and body tone, and vocalization. Lift the lips to examine teeth, especially the incisors, and inspect digits and toenails. Inspect the genitals and rectum.
Body Measurements	Weigh the animal and measure its body proportions, e.g., head, trunk, limbs, and tail. Measure core temperature with a rectal or aural thermometer.

- The appearance of the fur and its color should be noted.
- The proportions of body parts should be examined, including the length of the snout, head and body, limb, and tail.
- An examination of the eyes should include using a small flashlight to test pupillary response. The pupils should constrict to light and dilate when the light is removed, which indicates intact mid-brain function. Rodents secrete a reddish fluid, called Hardarian fluid, from the lacrimal glands. This fluid is collected on their paws as the paws are rubbed across the eyes during grooming. It is then mixed with saliva when the animal licks its paws, and the mixture is spread onto the fur during subsequent grooming. Chromodacryorrhea (reddening of the fur around the eyes) due to the accumulation of Hardarian secretions indicates that the animal is not grooming. Rodents are fastidious groomers, and a rich rough appearance to the fur also indicates normal grooming.
- The teeth should be inspected by gently retracting the lips. The incisor teeth of rodents grow continuously and are kept at an appropriate length by chewing. Excessively long or crooked teeth indicate an absence of chewing or a jaw malformation. Overgrown teeth should be appropriately trimmed. If the teeth are inadequate for chewing, an animal can be fed a liquid diet.
- During grooming, rats trim their toenails, especially the toenails of the hind feet, and thus these should be short with rough tips, indicative of daily care. Long, unkempt toenails may indicate poor grooming habits, or problems with teeth and mouth, which are used for making the fine nail cutting movements. (We must note, however, that many inbred strains of animals display less than propitious toenail care.)

B. Example of Analysis of Appearance

The power of a simple analysis of appearance is illustrated in Figure 8-3. In the course of studying the embryological development of neurons in the rat cortex, we noticed a dramatic effect of our treatments on coat coloration pattern. Pregnant rats were injected with a standard dose of bromodeoxyuridine (BrdU, 60 mg/kg) on different embryonic days, ranging from E11 to E21. In the rat, neocortical neurons are generated during this period and we were interested in the effects of different postnatal treatments on the numbers of neurons generated at different ages. The BrdU labels cells that are mitotic during the two or so hours following the injection. Labeled cells can later be identified using immunohistological techniques. Interestingly, Long-Evans hooded rats with BrdU injections on days E11 to about E15 had a dramatically altered pattern of black-and-white coloration in their fur

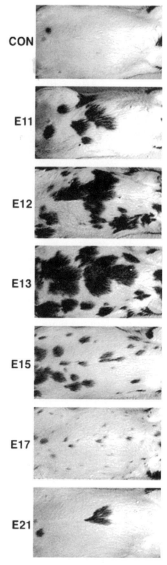

Fig. 8-3 Coat color on the ventrum of Long-Evans rats whose mother had received bromodeoxyuridine at different times of gestation. The changes in coat color are paralleled by changes in brain and behavior (After Kolb et al., 1997).

coats, and this pattern varied with the precise age at which rats were treated with BrdU.

The injections produced spots, much like those on a Dalmatian dog, the size of the spots varying with injection age (see Figure 8-3). Because we knew that the skin cells (specifically the melanocytes in this case) and the brain cells were derived from the same embryonic source, we immediately became suspicious that the BrdU was altering not just skin cells but the pattern of migration of neurons (Kolb et al., 1997). This led us to analyze the behavior of the animals in more detail and it turned out that the BrdU treatment was producing marked changes in many behaviors. This discovery was only possible because we had observed the general appearance of the animals.

C. Body Weight

Animal suppliers provide weight curves for the strains of animals they sell, and thus it is a simple matter to weigh an animal and compare its weight to a standard weight curve. Growth in rodents is extremely sensitive to nutrition, being accelerated or retarded by changes in nutritional status at any age. It is also influenced by the behavior of conspecifics, as subordinate animals are typically smaller than dominant conspecifics. Rats, especially males, continue to increase in size and weight throughout life, but size and rate of growth vary appreciably across strains. Differences in actual weight from expected weight may signal malnutrition, overfeeding, developmental disorders, or any of a variety of peripheral and central problems.

D. Body Temperature

At the time of weighing, the animal's temperature can be recorded with a rectal or aural thermometer. Rodent core temperature is quite variable and can fall to about 35° C when an animal is resting in its home cage and can increase to 41° C when the animal is aroused upon removal from its home cage. Temperatures lower or higher than this range indicate hypothermia or fever. Animals display a variety of postures, reflexes, and complex behaviors in order to maintain temperature, and these are described in detail by Satinoff (1983).

E. Response to Handling

During handling, an animal may typically make soft vocalizations. Excessive squeaking may indicate distress or sickness. Rodents maintained in group housing are usually unaggressive when handled by an experimenter. Animals raised in isolation (e.g., housed individually) may be very sensitive to handling and squeak and struggle or even display rage responses. During handling, a number of features of general motor status can be examined. The animal can be gently held in the palm of the hand and quickly raised and lowered. Limb muscles should tense and relax as the rat adjusts itself to the movements of the hand. Absence of muscle tone or excessive rigidity are both indicative of problems with motor status (e.g., drugs that stimulate dopamine function produce flaccid muscle tone, whereas drugs that block dopamine function produce rigidity).

IV. SENSORIMOTOR BEHAVIOR

The objective of sensorimotor tests is to evaluate the sensory and motor abilities of animals (see Table 8-2).

TABLE 8-2

SENSORY AND SENSORIMOTOR BEHAVIOR

Home Cage	Response to auditory, olfactory, somatosensory, taste, vestibular, and visual stimuli. The home cage should provide easy viewing of an animal. Holes in the sides and bottom of the cage provide entries for probes to touch the animal or to present objects to the animal or to present food items. Animals are extremely responsive to inserted objects and treat capturing the objects as a "game". Slightly opening an animal's cage can attenuate its responses to introduced stimuli, showing that it notices the change.
Open Field	Response to auditory, olfactory, somatosensory, taste, vestibular, and visual stimuli. The same tests are administered. Generally animals taken out of their home cage are more interested in exploring and so ignore objects that they responded to when in the home cage.

The tests evaluate the ability of animals to orient to objects in the environment in each sensory modality. The term *sensorimotor* derives from the recognition that it is ordinarily very difficult to determine whether the absence of a response is related to an inability to detect a stimulus or to an inability to respond to a detected stimulus. For the purposes of the present overview, such distinctions are not necessary, but it is worthwhile pointing out that some theoretical positions suggest that such distinctions are not possible (Teitelbaum et al., 1983).

A. Home Cage

Tests of sensory and motor behavior are best administered to an animal in its home cage, preferably a hanging wire mesh cage because the holes in the mesh allow easy access to the animal. Sensorimotor behavior of animals is radically different if they are assessed in an open area where even neurologically intact animals will act as though they are neglecting sensory information (see material following). For the cage examination, food pellets are placed into the cage and paper towels placed beneath the home cage to catch residue. If the tray is slid out from beneath the animal's cage a day to two later, residue can be examined. Rodents are fastidious in their eating and toilet habits, and thus feces and urine will be found in one location on the tray and food droppings will be found in another location, an indication that the animal has compartmentalized its home spatially. Residue from food should be quite fine, indicating that the animal is chewing its food. Many rodents are central place foragers; they carry food to their home territory and store it for subsequent use. An examination of the inside of the cage should indicate that the food is piled in one corner of the cage.

B. Orienting

While the rat is in its home cage, its sensory responsiveness can be examined. In an analysis of animals recovering from lateral hypothalamic damage, Marshall et al. (1971) observed a rostrocaudal recovery of sensory responsiveness. Generally, normal animals are much more responsive to rostrally than to caudally applied stimulation. Schallert and Whishaw (1978) reported that a syndrome of hyperresponsiveness and hyporesponsiveness can occur after hypothalamic damage. Procedures for evaluation of sensory responsiveness include the following.

- Placement of a cotton-tipped applicator, of the type used in surgery, into the animal's cage, gently touching different parts of the animal's body, including its vibrissae, body, paws, and tail. The animal should perceive this as a "game" and vigorously pursue and bite the applicator, thus allowing assessment of the sensitivity of different body parts. Rubbing the applicator gently against the cage can be used to test the rat's auditory acuity, as it will orient to the sound. Placing objects on the rat provides an additional assessment of its sensory responsiveness, as a rat will quickly remove the object (Figure 8-4). Pieces of sticky tape of various sizes placed on the ulnar surface of the forearm or bracelets tied with single or double knots provide tests of detection, obscuration, or neglect (Schallert and Whishaw, 1984).
- An animal's responsiveness to odors can be tested by placing a small drop of an odorous substance on the tip of the applicator. Animals will investigate food smells, recoil from noxious odors such as ammonia, and recoil from the odors of predators such as stoats and foxes (Heale et al., 1996).
- Simple ingestive responses can be investigated by placing a drop of food substances on the blade of a spatula. In their home cage, rodents, unless water deprived, are usually indifferent to water, but they enthusiastically ingest sweet foods such as sugar water, milk, or a mash of sweet chocolate-flavored cookie. Lip licking indicates that the animal's sensitivity to sweet food is normal. If the spatula is held adjacent to the cage, the animal will stick its tongue out to lick up the food, providing an indication of the motor status of the tongue. Bitter-tasting food, such as quinine, elicits a series of rejection responses, including wiping of the snout with the paws, wiping of the chin on the floor of the cage, and tongue protrusion to remove the food. Grill and Norgren (1978) have described taste-responsive reflexes in rats that have subsequently been widely used to assess gustatory responses.
- An animal's ability to eat and chew may be further examined by giving the animal a food pellet, a piece of cheese, or some other food substance of a standard size. Rodents sniff the food, grasp it in their incisors, sit back on their haunches while transferring the food to the paws, and eat the food from the paws while in a sitting position. Observing the processes of food identification, food handling, and eating speed all provide insights into the function of the front end of the animal's body. More detailed analysis of rat eating speed shows that the animals eat more quickly when exposed than when in a secure environment, and that they eat more quickly at normal meal times than at other times (Whishaw et al., 1992a).

Fig. 8-4 Left: A sticky dot placed on the ulnar pad of the forearm provides a simple test for orienting because the rat will quickly remove the tape with its mouth. Sensitivity can be tested by reducing the size of the dot. Right: Competition between stimuli can be tested with bracelets tied with one (right wrist) or two (left wrist) knots. If a stimulus on the good side obscures the stimulus on the bad side, a rat will persist in attempting to untie the difficult right knot while ignoring the easy left knot (After Schallert and Whishaw, 1984).

C. Open Field Behavior

Sensory tests may also be given to an animal that is removed from its home cage, but here the meaning of the responses changes. Normally, an animal in a novel environment ignores food in favor of making exploratory movements. Ingestion of food outside the home means it has habituated to that location or is insensitive to novelty. Generally, it may take a number of days or weeks to habituate an animal to an environment that provides no hiding place, as is the case with most mazes. Animals also display a number of defensive responses when they find food, and will turn or dodge away from other animals with the food, or run or "hoard" the food to a secure location for eating. These behaviors can be used as "natural" tests of orienting and defensive behavior (Whishaw, 1988).

V. IMMOBILITY AND ASSOCIATED REFLEXES

Posture and locomotion are supported by independent neural subsystems (see Table 8-3). A condition of immobility in which posture is supported against gravity is the objective of a large number of local and whole-body reflexes. Thus, immobility should be viewed as a behavior with complex allied reflexes. Even animals that are catatonic and appear completely unresponsive may move quickly to regain postural support if they are placed in a condition of unstable equilibrium (Figure 8-5). Postural and righting reflexes are mediated by the visual system, the vestibular system, surface body senses, and proprioceptive senses. Although responses mediated by each system are allied they are frequently independent (Pellis, 1996).

A. Postural Reflexes

If the animal is placed on a flat surface and gently lifted by the tail, it should display a number of postural reflexes. The head should be raised and the forelimbs and hindlimbs extended outward, while the forequarters are twisted from side to side. When mice are raised and then quickly lowered, their digits should extend, a response not seen in rats. The posture and movements are typical of an animal searching for a surface on which to obtain support.

Asymmetries in movement are used as tests of brain asymmetries that might be produced by unilateral injury (Kolb and Whishaw, 1985). For example, when suspended by the tail adult animals with unilateral cortical lesions usually turn contralateral to the lesion, whereas animals with unilateral dopamine depletions turn ipsilateral to the lesion. The posture of the limbs provides a sensitive measure of central motor status. Flexion of the forelimbs toward the body, including grasping of the fur of the ventrum and flexure of the hindlimbs toward each other (including grasping of each other), can indicate abnormalities in descending pyramidal or extrapyramidal systems (Whishaw et al., 1981b). On a flat surface, animals typically rotate toward the side of injury, and when placed on irregular surfaces they may favor a limb contralateral to an injury.

B. Postural Support

Placing an animal on a flat surface allows inspection of its postural support. An animal may maintain a condition of immobility in which it has posture or adopt a position of immobility in which it lacks posture. The two types of

TABLE 8-3

POSTURE AND IMMOBILITY

Immobility and Movement with Posture	Animals usually have postural support when they move about and they maintain posture when they stand still and remain still while rearing. Posture and movement can be dissociated: in states of catalepsy postural support is retained while movement is lost.
Immobility and Movement without Posture	An animal has posture only with limb movement. When a limb is still, the animal collapses, unable to maintain posture when still. When still, the animal remains alert but has no posture, a condition termed cataplexy.
Movement and Immobility of Body Parts	Mobility and immobility of body parts can be examined by placing a limb in an awkward posture or placing it on an object such as a bottle stopper and timing how long it takes an animal to move it.
Restraint-induced Immobility	Restraint-induced immobility, also called tonic immobility or hypnosis, is induced by placing an animal in an awkward position, e.g., on its back. The time it remains in such a position is typically measured. Animals will maintain awkward positions while maintaining body tone or when body tone is absent. During tonic immobility animals are usually awake.
Righting Responses	Supporting, righting, placing, hopping reactions are used to maintain a quadrupetal posture. When placed on side or back or dropped in a supine or prone position, adjustments are made to regain a quadrupetal position. Righting responses are mediated by tactile, proprioceptive, vestibular, and visual reflexes.
Environmental Influences on Immobility	Feeding fatigue potentiates immobility. Warming induces heat loss postures, e.g., sprawling, and thus potentiates immobility without tone. Cooling induces heat gain posture with shivering and thus potentiates immobility with muscle tone.

Male   Female

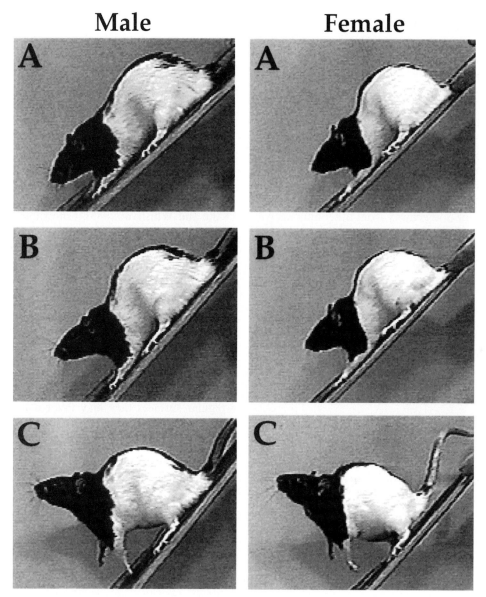

Fig. 8-5 Animals defend immobility when placed in a condition of instability. Rats treated with haloperidol (5 mg/kg) display immobility with postural support. When made unstable by tilting the substrate, they first brace (A) but eventually jump to regain a new supporting position when postural instability becomes too great (B,C). The tactics for maintaining postural stability prior to jumping are sexually dimorphic (After Field et al., 2000).

immobility are independent. An animal without posture while immobile may achieve posture as a part of locomotion. DeRyck et al. (1980) have demonstrated that morphine (an opioid agonist) produces a condition of immobility without postural support, whereas haloperidol (a dopamine antagonist) produces immobility with postural support. The two types of immobility are part of normal behavior, as an animal that is cold and shivering has postural support while otherwise remaining immobile, whereas an animal that is hot will sprawl without postural support as part of a heat loss strategy. Immobility with postural support is characteristic of an animal that pauses during a bout of exploratory behavior or an animal that rears and stands against a wall. Immobility without postural support is typical of an animal that is resting or sleeping.

Although difficult to achieve in normal rodents, if an animal is gently restrained in almost any position it may remain in that position when released. This form of immobility, sometimes called animal hypnosis or tonic immobility, is usually accompanied by a good deal of body tone, as the latter term implies (Gallup and Maser, 1977). If an animal is frightened, it may "freeze" in place in a condition of immobility with postural support. When hiding or attempting to escape, it frequently crouches into immobility without postural support. (The suggestion that animal hypnosis can be used as a condition of anesthesia is not supported by research, which indicates on the contrary that hypnosis is but one of many forms of adaptive immobility.)

C. Placing Responses

Placing responses are movements of the head, body, or limbs directed toward regaining a quadrupedal posture. If an animal is lifted by the tail, as the animal is lowered toward a surface contact of its long whiskers with the surface will trigger a placing reaction of extending the forelimbs to contact the surface. Placing reactions of each of the limbs can be tested by holding the rat in both hands while touching the dorsal surface of each paw against the edge of a table. Upon contact, the paw should be lifted and placed on the substrate (Wolgin and Bonner, 1985). Placing responses are sensitive to damage to corticospinal systems.

D. Bracing Responses

If an immobile animal is gently pushed, it will often push back against displacement to maintain static equilibrium, a behavior termed a bracing response. If the push begins to make the animal unstable, it will step or turn away to relieve pressure. Animals that have been rendered cataleptic by a treatment (i.e., immobile with postural support) may be unable to step and thus are reduced to bracing to maintain stability (Schallert et al., 1979). Bracing can be examined in a single limb. Animals that are made hemi-Parkinsonian with a unilateral injection of 6-hydroxydopamine, a dopamine-depleting neurotoxin, can be held so that they are standing on a single forelimb. When gently pushed forward, they will step to gain postural support with their good forelimb while being

reduced to bracing with their bad forelimb (Schallert et al., 1992; Olsson et al., 1995).

E. Righting Response

When placed in a condition of unstable equilibrium, animals will attempt to regain an upright posture in relation to gravity, a behavior termed righting. Tests of righting examine visual, vestibular, tactile, and proprioceptive reflexes. If an animal is dropped from the height of less than a meter, onto a cushion, it will adjust its posture so that it lands "on its feet." If released with feet facing down, it arches its back, and extends its limbs to parachute to the surface. If it is released in a prone position, it will right itself by first turning the forequarters of the body and then the hindquarters, a response mediated by vestibular receptors. Video recordings of righting responses show that righting is also visually modulated. Sighted animals right proximal to the landing surface, whereas in the absence of visual cues animals initiate righting immediately upon release. If a rat is placed on its side or back, on the surface of a table, it will right itself so that it returns to its feet. The righting responses of parts of the body can be tested by holding the head, forequarters, or hindquarters. Details of righting reflexes and their sensory control are provided by Pellis (1996).

VI. LOCOMOTION

Locomotor behavior includes all of the acts in which an animal moves from one place to another (see Table 8-4). It includes the acts of initiating movement (often referred to as warm-up), turning behavior, exploratory behavior,

TABLE 8-4

Locomotion

General Activity	Video-recording, movement sensors, activity wheels, open field tests.
Movement Initiation	The warm-up effects: Movements are initiated in a rostral-caudal sequence, small movements precede large movements, and lateral movements precede forward movements, which precede vertical movements.
Turning and Climbing	Components of movements can be captured by placing animals in cages, alleys, tunnels, etc.
Walking and Swimming	Rodents have distinctive walking and swimming patterns. Rats and mice walk by moving limbs in diagonal couplets with a forelimb leading a contralateral hindlimb. They swim using the hindlimbs with the forelimbs held beneath the chin to assist in steering.
Exploratory Activity	Rodents select a home base as their center of exploration, where they turn and groom, and make excursions of increasing distance from the home base. Outward trips are slow and involve numerous pauses and rears while return trips are more rapid.
Circadian Activity	Most rodents are nocturnal and are more active in the night portion of their cycle. Peak activity typically occurs at the beginning and end of the night portion of the cycle. Embedded within the circadian cycle are more rapid cycles of eating and drinking, especially during the night portion of the circadian cycle.

and a variety of movement patterns on dry land, water, or vertical substrates.

A. Warm-up

Movement is conceived of as being organized along three dimensions as illustrated by warm-up. The initiation of locomotion, or warm-up, may be observed in an animal gently placed in the center of an open field (Golani et al., 1979). There are four principles of warm-up.

- Lateral, forward, and vertical movements are independent dimensions of movement.
- During warm-up, an animal can be observed to alternate between lateral, forward, and vertical movements.
- Small movements recruit larger movements. For example, a small lateral head turn will be shortly followed by a larger head turn and so forth until the animal turns in a complete circle.
- Rostral movements precede caudal movements. That is, a head turn will precede movements of the front limbs, which will precede movements of the hind limbs.
- The relationship between the movements is such that lateral movements precede forward movements, which precede vertical movements.

In a novel environment, warm-up may be lengthy, whereas in a familiar environment it may precede quickly (e.g., an animal may simply turn and walk away). Almost any nervous system treatment or pharmacological treatment may affect warm-up. Conceptually, warm-up is thought to be a reflection of evolutionary and ontogenetic processes, and brain structural organization (Golani, 1992). Functionally, warm-up allows an animal to systematically examine an environment into which it is moving.

B. Turning

Rodents have a variety of strategies for turning. Turning may be incorporated into patterns of forward locomotion, in which case an animal turns its head and then "follows" using a normal walking pattern. Incorporated into this pattern, or used independently, it may make most of the turn with the hindquarters (thus pivoting in part or in whole around its hindquarters) or most of the turn with its forequarters, thus pivoting around its forequarters. These two patterns of turning are incorporated into a variety of other behaviors, including locomotion, aggression, play, and sexual behavior. In rats, there is sexual dimorphism in the extent to which the patterns are used (Field et al., 1997). Females make greater use of forequarter turning, whereas males make greater use of hindquarter turning. Dimorphism, in turn, may be related

to the way the animals turn during sex and aggression, respectively. Animals may also turn by first rearing and then using the potential energy of the rear to pivot and fall in one direction or the other.

The incidence of turning as well as the form of turning is widely used as an index of asymmetrical brain function (Miklyaeva et al., 1995). For example, animals with unilateral dopamine depletions turn ipsilateral to their lesion when given amphetamine and contralateral to their lesion when given apomorphine. Some papers have suggested that direction of rotation can be used as an index of recovery after therapeutic treatments (Freed, 1983). Miklyaeva and colleagues' analysis shows, however, that the depleted rats tend not to exert force with the limbs contralateral to their lesion and the impairment persists irrespective of turning direction or drug treatment (Figure 8-6). Thus, it is more appropriate to use analyses of limb use rather than turning direction in assessing functional recovery (Schallert et al., 1992; Olsson et al., 1995). The causes of turning direction induced by drugs remain enigmatic.

C. Walking and Running

Although this may be a little difficult for the novice to observe, a rodent's major source of propulsion comes from its hindlimbs. During slow locomotion, the forelimbs are used for contacting and exploring the substrate and walls (Clarke, 1992). Limb contact with irregular surfaces or the wall of a cage can be used as a test of normal forelimb function. At its simplest, the number of times a limb contacts a wall when an animal rears can be a sensitive measure of forelimb function (Kozlowski et al., 1996). When a rodent walks, it moves with diagonal limb couplets. One forelimb and the contralateral hindlimb move together followed by the other forelimb and hindlimb. Rodents also have species typical movement patterns. For example, rats seldom walk. They move either hesitantly with turns and pauses or they trot. Patterns of locomotion are difficult to analyze by eye unless they are grossly abnormal, but a number of simple video-recording techniques have been used for detailed locomotor analysis (Clarke, 1992; Miklyaeva et al., 1995). The structure of walking movements has been described by Ganor and Golani (1980).

D. Exploration

The movements of an animal can be observed by removing it from its cage and placing it in a small open environment or field. Golani and co-workers have described some of the geometric aspects of rodent exploratory behavior (Eliam and Golani, 1989; Golani et al., 1993;

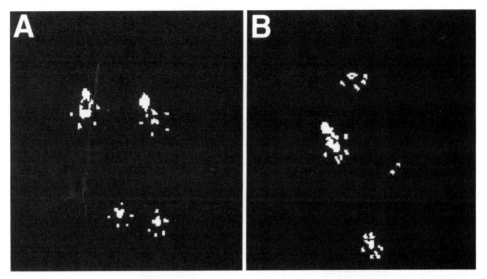

Fig. 8-6 Limb use can be measured by using the reflectance technique, in which a light shone through the edge of a glass table top reflects from the paw pads of an animal standing on the table surface. The technique illustrates that a control animal distributes its weight evenly when standing while a rat with unilateral dopamine depletion rests its weight on its good (ipsilateral to lesion) limbs (After Miklyaeva et al., 1959).

Tchernichovski and Golani, 1995). An animal will usually treat the place at which it is first placed as a "home base." It will pause, rear, turn, and groom at this location, often before exploring the rest of the open field. When it does rear, it will support itself by touching the wall of the field with its forepaws. As it begins to explore, it extends its forequarters and head and examines the area surrounding its home base. It will eventually begin to make trips away from its home base, usually along the edge of the walls of the enclosure. Exploration will proceed with brief and slow outward excursions followed by more rapid returns to the home base.

The outward excursions become gradually longer until the surrounding area is explored. During the course of its exploration, the animal may choose another location as its home base, and this can be identified because it circles and grooms at this location and uses the location as the home base for its outbound trips. Typically, return trips are more rapid than outbound trips. A 10-minute exploratory test can provide a wealth of behavior to analyze, including number of home bases, number of trips, kinematics of excursions and returns, number of stops, number of rears, incidence of grooming, duration of trips, and so on (Golani et al., 1993; Whishaw et al., 1994).

Another feature of open field behavior is habituation. Over time, animals will normally show a reduction in open field activity. Furthermore, they will show a shift in behavior and may spend more time grooming or sitting immobile once they are familiar with the environment. Animals with various forebrain injuries, such as frontal cortex or hippocampal lesions, may display slower habituation, even with extended exposure to the open field (Kolb, 1974).

E. Swimming

If a swimming pool is available, movements of swimming can be observed. Rats are semiaquatic, as their natural environments are usually along the margins of streams and rivers. Rats are proficient swimmers and propulsion comes entirely from the hindlimbs. Other rodents may be less proficient in water than rats, and some may use quite idiosyncratic swimming strategies. Hamsters, for example, inflate their cheek pouches and use them as "water-wings." In typical rodent swimming, the forelimbs are tucked up under the chin and the open palms of the paws are used for steering (Salis, 1972; Fish, 1996). Changes in the way animals swim may occur with development and aging, under the effects of drugs and brain damage, but swimming itself is quite resistant to central nervous system damage (Whishaw et al., 1981b).

F. Circadian Activity

Tests of circadian activity involve recording the general activity of animals across a day/night cycle. Usually the activity cycle is entrained by having lights come on at 0800 hrs and go off at 2000 hrs. Rodents are typically active during the dark portion of the cycle. The test requires a dedicated room in which lighting can be regulated with a timer. A test apparatus consists of a cage that has a photocell at each end. The photocells are connected to a microcomputer, which records instances of beam breaks. The computer is programmed so that it records beam breaks at each photocell, as well as instances in which beam breaks occur at successive photocells.

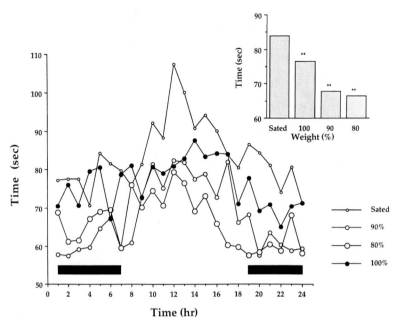

Fig. 8-7 Time taken to eat a one-gram food pellet as a function of time of day and food deprivation. Sated rats were never food-deprived, 100% represents rats previously food-deprived to 90 and 80% and returned to free feeding. Note that rats eat more quickly after having been subject to food deprivation, when hungry, and at usual feeding times (after Whishaw et al., 1992).

Beam breaks at a single photocell provide a measure of stereotyped movements, such as head bobbing, grooming, and circling. Successive beam breaks provide a measure of locomotion (i.e., walking from one end of the cage to the other). If an animal is placed in the activity cages at 1200, it is initially very active in exploring the apparatus but habituation occurs across the first hour or two. At 2000 hrs, when the light turns off, there is a burst of activity followed by bouts of increased activity across the dark cycle.

Just prior to lights-on at 0800 hrs, animals show another burst of activity. During subsequent lights-on periods, animals are typically inactive. These features of circadian activity can be recorded in a single 24-hour recording period, and one recording session can frequently reveal distinctive differences between control and experimental groups. More detailed analysis of circadian activity can be examined in which the effects of light, sound, feeding, and so forth are assessed (Mistleburger and Mumby, 1992). Figure 8-7 illustrates circadian activity in eating speed in rats on different levels of food deprivation. Rats eat more quickly at their usual feeding times at the onset and offset of the lights-off period.

VII. SKILLED MOVEMENT

The term skilled movement is somewhat arbitrary in that it refers to movement in which the mouth or paws

TABLE 8-5

SKILLED MOVEMENTS

Limb Movements	Bar-pressing, reaching and retrieving food through a slot, spontaneous food handling of objects or nesting material and limb movements used in fur grooming and social behavior. Rodents use limb movements that are order-typical and species-typical.
Climbing and Jumping	Movements of climbing up a screen, rope, ladder, etc. and jumping from one base of support to another.
Oral Movements	Mouth and tongue movements in acceptance or rejection of food such as spitting food out or grasping and ingesting food. Movements used in grooming, cleaning pups, nest building, teeth cutting.

are used to manipulate objects. The term may also be used to include movements used to traverse difficult terrain, such as walking on a narrow beam or climbing a rope, and swimming (see Table 8-5). The cohesive feature of the movements is that they seem much more disrupted by cortical lesions than are species-typical movements or movements of locomotion on a flat surface. The distinguishing feature of the movements is that they require rotatory movements, irregular patterns of movement, selective movements of a limb, and movements that break up the patterns of normal antigravity support (Figure 8-8). Skilled movements in rodents and primates are quite comparable, which makes rodent models quite

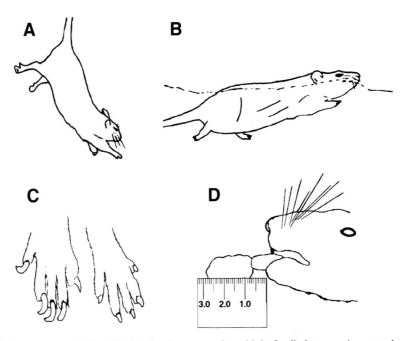

Fig. 8-8 Examples of skilled movements. A: When lifted in the air, a rat reaches with its forelimbs to regain postural support. B: When swimming, a rat tucks the forelimbs under the chin and tilts the paws to steer. C: Toenails are regularly clipped (right) and if skilled movements of chewing are impaired, they grow long (left). D: Skilled movements of the tongue are used to reach for food as illustrated by a rat licking through the bars of a cage to obtain food from a ruler. Maximum tongue extension is about 11 mm.

generalizable to humans (Whishaw et al., 1992b). Two commonly used tests of skilled movement are beam walking and skilled reaching.

A. Beam Walking

When a normal rat walks a narrow beam, it has the surprising ability to move along rapidly with its feet placed on the dorsal surface of the beam. A sign of motor incoordination is that it grasps the edge of the beam with its digits as it walks. Following unilateral motor system damage, only the contralateral paws are likely to be used for grasping. Grasping can be measured by video-recording the placement of the paws on the beam or by painting the feet of the animals so that paw placement can be visualized (Becker et al., 1987). The overall body posture of the animal also may be abnormal on a narrow beam. This can be quantified by measuring the angle of the back and head relative to the beam (Gentile et al., 1978). Another formal test related to beam walking is balancing on a rotating rod, the "rotorod" test. The test consists of a rotating rod upon which the animal balances. As the animals learn to balance, the rod is turned increasingly faster. The measure of motor skill is the time the animal spends on the rotorod as a function of speed with which the rod rotates (Le Marec et al., 1997).

B. Skilled Forelimb Movement

Rodents use their paws and digits to reach for, hold, and manipulate food. Tests of skilled forelimb use have an animal reach through a slot to obtain a food pellet. One form of the test has animals reach into a tray to retrieve food (Whishaw and Miklyaeva, 1996). Limb preference and success (number of pellets per reach) are used as performance measures. Limb use by an animal can also be controlled by restricting the use of one limb, thus forcing the animal to use the other limb. If the animal is reaching through a slot, a bracelet placed on a limb will prevent the animal from inserting its limb through the slot while otherwise leaving use of the limb unimpaired. A second form of the test has an animal reach for a single food pellet while its movements are filmed (Whishaw and Pellis, 1990). Video analysis of skilled reaching shows that rodents use a variety of whole-body preparatory movements and the reach itself consists of a number of subcomponents such as limb lifting, aiming, pronating, grasping, and supination upon withdrawal.

Animals with central nervous system damage may learn to compensate for their impairments and regain presurgical performance as judged by success measures, but movement analysis will indicate that the movements used in reaching are permanently changed (see Figure 8-2). Use of the forelimbs can also be evaluated by watching

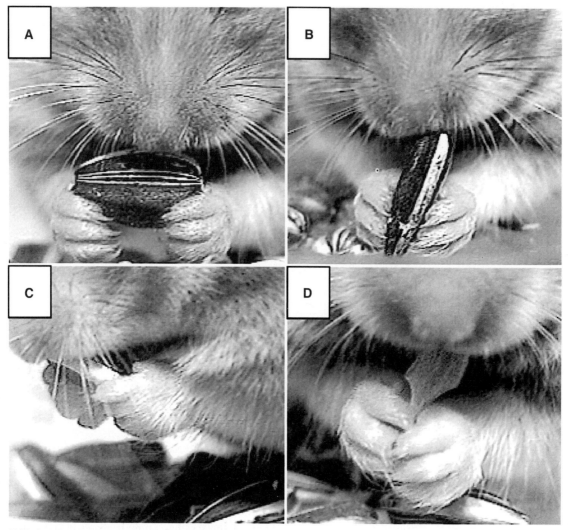

Fig. 8-9 Skilled paw and digit movements in a hamster. **A,C:** Food is held between digit 1 (thumb) and digit 2. **B:** Food is held with all digits. **D:** Food is held bilaterally with digit 1.

spontaneous food retrieval and manipulation. With the exception of guinea pigs, most rodent species have five "order-typical" movements in spontaneous eating (Whishaw et al., 1998). They (1) identify food by sniffing, (2) grasp food in the incisors, (3) sit back on their haunches to eat, (4) take the food from the mouth with an inward movement of the forelimbs, and (5) grasp and handle the food with the digits. Each of these movements has its characteristic features that can be subject to further analysis, and there are species-typical features of the movements in different rodents (Figure 8-9).

VIII. SPECIES-TYPICAL BEHAVIORS

Most movements performed by rodents are sufficiently stereotyped that they are recognizable from occurrence to occurrence and from animal to animal within a species. A variety of complex actions, however, are referred to as species-typical movements (see Table 8-6). Examples of species-typical behaviors include grooming, nest building, play, sexual behavior, social behavior, and care of young. With the resolution provided by a video record, it is possible to document successive behavioral acts in species-typical behavior in order to produce a description of their order or syntax.

A. Grooming

Berridge (1990) provides a comparative description of the grooming behavior of a number of species of rodents, including most rodents commonly used in laboratories. His method involved filming the animal through a mirror placed beneath the animal's holding cage. Grooming was

TABLE 8-6

SPECIES-SPECIFIC BEHAVIORS

Grooming	Grooming movements are species-distinctive and are used for cleaning and temperature regulation. Begin with movements of paw cleaning and proceed through face washing, body cleaning, and limb and tail cleaning.
Food Foraging/ Hoarding	Food carrying movements are species-distinctive and used for transporting food to shelter for eating, scattering food throughout a territory, or storing food in depots. Size of food, time required to eat, difficulty of terrain, and presence of predators influence carrying behavior. Both mouth-carrying or cheek-pounching are used by different species. Rodents also engage in food wrenching, in which food is stolen from a conspecific, and dodging, in which the victim protects food by evading the robber.
Eating	Incisors are used for grasping and biting, rear teeth are used for chewing, tongue is used for food manipulation and drinking.
Exploration/ Neophobia	Species vary in responses to novel territories and objects. Objects are explored visually or with olfaction, avoided, or buried. Spaces are explored by slow excursions into space and quick returns to a starting point. Spaces are subdivided into home bases, familiar territories, and boundaries.
Foraging and Diet Selection	Food preferences are based on size and eating time of food, nutritive value, taste, and familiarity. For colony species the colony is an information source with acceptable foods identified by smelling and licking snout of conspecifics.
Sleep	Rodents display all typical aspects of sleep including resting, napping, quiet sleep and rapid eye movement sleep. Most rodents are nocturnal, thus sleeping during the day with major activity periods occurring at sun up and sun down. Cycles in natural habitats vary widely with seasons.
Nest-building	Different species are nest builders, tunnel builders, and build nests for small family groups or large colonies. All kinds of objects are carried, manipulated, and shredded for nesting material.
Maternal Behavior	Laboratory rodents typically have large litters that are immature when born. Pups are fed for the first two to three weeks of life and thereafter become independent.
Social Behavior	Colony or family rodents have rich social relations including territorial defense, social hierarchies, family groupings, and greeting behaviors. Solitary rodents may have simplified social patterns. Defensive and attack behavior in males and females is distinctive.
Sexual Behavior	Characteristic sexual behavior displayed by males and females. Males display territorial control or territory invasion, and engage in courtship and often group sexual behavior. Sexual behavior is often long-lasting with many bouts of chasing, mounting and intromission, and incidents of ejaculation. Mounting is followed by genital grooming and intromission is followed by immobility and high frequency vocalizations. Females engage in soliciting including approaches and darting, pauses and ear wiggling, and dodging and lordosis to facilitate male mounting.
Play Behavior	Many rodents have rich play behavior with the highest incidence in the juvenile period. Play typically consists of attack in which snout-to-neck contacts is the objective and defense in which the neck is protected.

elicited by spraying a little water onto an animal's fur. A typical grooming sequence consists of an animal walking forward and making a few body shakes to remove the water from its fur. It then sits back onto its haunches, in which posture it performs a number of grooming acts in a relatively fixed sequence. The animal first licks its paws and then wipes its nose with rotatory movements of the paws. This is followed by face washing, which consists of wiping the paws down across the face, with the successive wiping movements becoming larger until the paws reach behind the ears and then wash downward across the face. Once an animal has finished a sequence of head grooming, it turns to one side, grasps its fur with a paw and then proceeds to groom its body (Figure 8-10).

A single grooming bout thus begins at the snout and moves caudally down the body and may consist of more than one hundred individual grooming acts. Berridge has examined the internal consistency of grooming (i.e., the extent to which one grooming act predicts another) to derive a grooming syntax. The syntax, in turn, provides the baseline against which central nervous system manipulations are contrasted.

The grooming syntax of rats and mice, which are each other's closest relatives, is slightly different in that mice make fewer asymmetrical limb movements when face washing. The grooming of other less closely related rodents

is different still. For instance, an animal may use only a single limb to wipe the face. Grooming syntax in turn becomes a powerful tool for the analysis of neural control of action patterns. For example, to answer the question of whether grooming is produced by a grooming center or has its control distributed across a number of neural systems Berridge and co-workers have sectioned the brain at various rostrocaudal levels to find that different features of grooming control are represented at many different nervous system levels (Berridge, 1989).

B. Food Hoarding

Rodents have a number of food-handling patterns depending on the time required to eat food (Figure 8-11). Rodents are usually cautious in leaving the home base and cautious in approaching a food source. Small pieces of food that take little time to eat are consumed in situ. Items that take a little longer to eat are consumed from a sitting posture, whereas items that take a long time to eat are carried to a secure location or to the home base. Rodents that cheek-pouch carry items of all sizes, whereas animals that do not cheek-pouch eat smaller items at the location at which they are found and carry larger objects. The decision

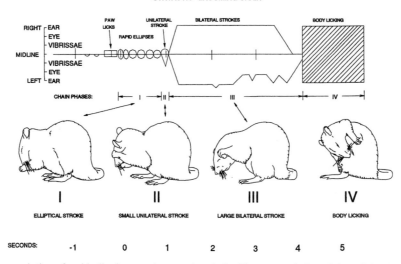

Fig. 8-10 Choreographic transcription of an idealized syntactic grooming chain. Time proceeds from left to right. The horizontal axis represents the center of the nose. The line above the horizontal axis denotes movement of the left forepaw. Small rectangles denote paw licks. Large rectangle denotes body licks. Chain phases are: (I) 5 to 8 rapid elliptical strokes around the nose (6.5 Hz); (II) unilateral strokes of small amplitude; (III) symmetrically bilateral strokes of large amplitude; (IV) licking the ventrolateral torso (After Berridge and Whishaw, 1992).

on whether to carry an object is based on an estimate of eating time.

Food-carrying animals may store food in a single central location, such as a nesting area, hide it at a number of locations, or hide individual items throughout a territory. Colony-living rodents may carry food to a home area, but here it is likely to be stolen and eaten quickly by other colony members (Whishaw and Whishaw, 1996). Tests of food carrying are used to evaluate exploration, spatial abilities, time estimates, and social competition for food (Whishaw et al., 1990).

C. Foraging and Diet Selection

A rodent colony is frequently an information center, especially regarding novel foods (Galef, 1993). Rodents, especially those that have been subjected to attempts at eradication, are extremely wary of novel objects and foods, a trait called neophobia. Thus, rodents in a colony may share information about types of foods, food safety, and food locations. An animal will sniff and lick the snout of a conspecific and so gain information about food types and sources. Sulfur dioxide on the breath of the conspecific indicates to the inquirer that the food is palpable. Tests of neophobia and information sharing can be used to study both learning and social behavior in animals (Galef et al., 1997). Pairing food items with a sickness-inducing agent is commonly used to study food aversion responses (Perks and Clifton, 1997). Conditioned aversion is considered a special form of learning because the ingestion of food and subsequent illness may be

separated by hours or days but will still be strongly acquired.

D. Nest Building

If rodents are provided with appropriate material, they will build nests (Kinder, 1927). The nests contribute to thermoregulation and provide a place to gather young. The quality of a nest built by female rats will also wax and wane over the four-day estrous cycle. If an animal is given strips of paper about 2 cm wide and 20 cm long, each behavior involved in nest building can be recorded as it occurs (e.g., pick up material, carry, push, chew, and so on). The quality of the nest constructed can be rated on a four-point scale over three to four days, the time required to fashion an optimal nest. Limbic system and medial frontal cortex lesions have been reported to disrupt nest building (Shipley and Kolb, 1977).

E. Social Behavior

Social behavior can be defined as all behavior that influences, or is influenced by, other members of the same species. The term thus covers all sexual and reproductive activities and all behavior that tends to bring individuals together as well as all forms of aggressive behavior (Grant, 1963). It is traditional to describe sexual behavior separately (see material following), and in recent years aggressive behavior has also come to be seen as a separate form of social behavior as well (see material following).

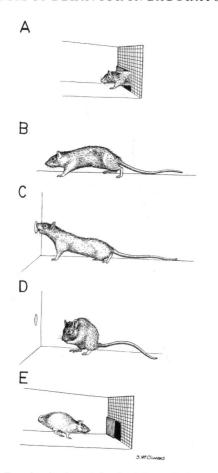

A

B

C

D

E

Fig. 8-11 Foraging in the rat for food items that are passed to it through a hole in the test apparatus. **A:** The animal "stops, looks and listens" before leaving its home base. **B:** The food location is approached cautiously. **C:** Food items that can be eaten quickly are swallowed. **D:** Larger items are eaten from a sitting position. **E:** Very large items are carried "home", usually at a gallop. (After Whishaw et al., 1990).

It is generally recognized that social behavior is not a unitary behavior with a unitary neurological basis. Rather, different aspects of social behavior have different neural and endocrine bases (Moyer, 1968). It is therefore necessary to examine social behavior in a number of different situations before concluding that a particular treatment has produced a general change in social behavior. Behavioral changes may be situation-specific.

Most studies of social behavior in rodents usually utilize some type of test of free interaction in which a group of subjects is housed in a large group cage, often with several interconnected chambers (Lubar et al., 1973). A less natural version has animals paired in a novel situation, often repeatedly over days (Latane, 1970). In both situations the behavior is videotaped and can either be analyzed by calculating specific behaviors, such as time in contact, or by writing a detailed description of the behaviors such as described by Grant and Mackintosh (1963). Other behaviors such as vocalizations

(Francis, 1977) or urine marking (Brown, 1975) can also be recorded.

F. Aggression

Aggressive behavior is used to establish social hierarchies and to defend territories. As rodents vary widely in their tendency to live in colonies or in relative isolation, their tendency to engage in aggressive behavior also varies. Patterns of aggressive behavior are also usually distinctive in male and female animals. The targets of aggressive contact (e.g., the location on the body on which an animal attempts to inflicts bites) are usually quite distinctive from the targets of play contact (Pellis, 1997). In rats, bites are typically directed to the back and rear. Fellow residents and strangers are usually identified by their odor. Aggressive behavior is widely used as an animal model of human aggression (Blanchard et al., 1989).

G. Sexual Behavior

Sexual behavior requires the integrity of hormonal systems and neural systems, developmental experiences, learning, context, and an appropriate partner. Sexual behavior consists of at least two phases, courtship and consummation, and both are extremely complex requiring complex independent and interdependent actions by both the male and female. Dewsbury (1973) has described the social behaviors associated with sexual activity, including exploration, sniffing, grooming, soliciting, ear wiggling, hopping and darting in females, genital and nongenital grooming, as well as mounting, pelvic thrusting, ejaculation, lordosis, immobility, ultrasonic songs, and other behaviors. The patterning of movements of sexual behavior has been described by Sachs and Barfield (1976) for male rats and Carter et al. (1982) for female rats. Paradigms in which the female is given the opportunity to pace sexual activity are described by Mermelstein and Becker (1995). Paradigms in which interest in access to sexual partners and interest in sexual intercourse are described are presented by Everitt (1990). Michal (1973) provided a detailed discussion of the behavioral patterns of rats with limbic lesions.

H. Care of Young

Rodents are immature when born and are hairless with immature sensory and motor systems and receive extensive parental care (Grota and Ader, 1969). Pups are cared for mainly by the mother with elaborate patterns of pup cleaning and feeding under hormonal and thermoregulatory control (Leon et al., 1978). Thermoregulation influences much of the pups' social behavior as they have elaborate huddling strategies (Alberts, 1978).

I. Play

Rodents may display play behavior at any stage of development, but play is particularly prominent during the juvenile stage of life. Play is highly ritualized but incorporates many of the movements used by animals in other aspects of life, including sex, aggression, and skilled manipulation. In rodent play, the rich array of movements used by the participants appears to be orchestrated around the attempt of an initiator to thrust the tip of its snout into the neck of the recipient while the recipient attempts to avoid the contact (Figure 8-12). The patterns of thrust and parry are distinctive in different rodent species (Pellis et al., 1996).

IX. LEARNING

Research on the neural basis of learning suggests that there are a number of at least partially independent learning and memory systems. These include short-term memory (thought to be a frontal lobe function), object memory (thought to be a function of rhinal cortex), emotional memory (thought to be functions of the amygdala and related circuits), and implicit and explicit spatial memory (thought to be a function of the neocortex and hippocampal formation, respectively). A quick overview of an animal's learning ability therefore requires a number of tests, in order to tap into all of these systems (see Table 8-7). Widely used tests for rodents include the following.

- Passive avoidance
- Defensive burying

Fig. 8-12 A sequence of play fighting between two 30-day-old male rats. Note the repeated attack and defense of the neck (After Pellis et al., 1996).

TABLE 8.7

LEARNING

Classical Conditioning or Conditional Learning	Unconditioned stimuli are paired with conditioned stimuli and the strength of an unconditioned response to the conditioned stimuli is measured. Almost any arrangement of stimuli, environments, treatments, or behavior can be used.
Instrumental Conditioning	Animals are reinforced for performing motor acts such as running, jumping, sitting still, lever pressing, or opening puzzle latches.
Avoidance Learning	Passive responses including avoiding preferred places or objects which have been associated with noxious stimuli such as electric shock. Active responses including moving away from noxious items or burying noxious items.
Object Recognition	Including simple recognition of one or more objects, matching to sample, and nonmatching to sample in any sensory modality. Tasks are formal in which an animal makes an instrumental response or inferential in which recognition is inferred from exploratory behavior.
Spatial Learning	Dry land- and water-based tasks are used. Spatial tasks can be solved using *allothetic cues*, which are external and relatively independent to movements, or *idiothetic cues,* which include cues from vestibular or proprioceptive systems, reafference from movement commands, or sensory flow produced by movements themselves. Animals are required to move to/away from locations. *Cue tasks* require responding to a detectable cue. *Place tasks* require moving using the relationships between a number of cues, no one of which is essential. *Matching tasks* require learning a response based on a single information trial.
Memory	Memory includes *procedural memory* in which response and cues remain constant from trial to trial and *working memory* in which response or cues change from trial to trial. Tasks are constructed to measure one or both types of learning. Memory is typically divided into object, emotional, and spatial and each category can be further subdivided into sensory and motor memory.

- Conditioned place preference
- Conditioned emotional response
- Object recognition task
- Swimming pool place and matching-to-place tasks

Although these tests provide a rapid way of screening for learning and memory deficits, all of the tests are sensitive to quite a number of different functions and brain regions. At this point, we emphasize that the enormous number of tests in all their variations and the diversity of opinion concerning what they actually measure requires consultation with more comprehensive sources than provided here (see Olton et al., 1985; Vanderwolf and Cain, 1994).

A. Memory

Memory is also often described as being either procedural or working memory. *Procedural memory* is memory for the rules of task solution. For example, the rule may be that food is found at the end of an alley or that an escape platform is located somewhere in the swimming pool. *Working memory* is trial unique memory; that is, "on the last trial I found food here." It is thought that each sensory or motor system may be involved in system-specific procedural or working memory. For example, the visual pathway underlying the perception of objects is likely to be involved in the storage of object information. Memory is also described as being either *short-term*, to be used only for the moment, or *long-term*, to be used for long durations. That is, for each of the types of memory described previously there are procedural and working memory, short-term, and long-term memory. Detailed discussion concerning terminology, tests, and their significance can be found in comprehensive sources (Dudai, 1989; Martinez and Kesner, 1991).

B. Passive Avoidance

The passive avoidance apparatus consists of a box with two compartments and a connecting door. One of the boxes is white and the other is black, and the floor of both boxes is constructed of grids that can pass a small electric shock. An animal is placed in the white compartment each day for three days and is removed after it has crossed over into the dark compartment. As most rodents display a strong preference for the dark, by the third day the animal's passage between the boxes is quite quick. Now, however, the animal is given a brief electric shock when it enters the dark compartment. After one hour to 24 hours, the animal is once again placed in the white compartment of the box and the measure of learning is the time taken to again enter the now noxious dark compartment. Usually, a five-minute cutoff for entry is used. Passive avoidance has

been found to be a very sensitive measure of the effects of drugs that affect memory, such as the muscarinic blocker atropine (or scopolamine). Certain types of brain damage, including damage to the limbic system and globus pallidus and their transmitters, are similarly sensitive to the passive avoidance task (Bammer, 1982; Slagen et al., 1990).

C. Defensive Burying

The strength of the defensive burying test is that it provides a natural test of an animal's response to a threatening or noxious object. A large number of modifications of the test and their uses have been described by Pinel and Treit (1983). Traditionally, animals have been thought to have two primary defensive strategies, flight or fight. The defensive burying paradigm reveals that the responses of rats and mice (but not gerbils and hamsters) to threat are much more complex than has been thought and include investigating the object, removing it, or burying it so that it is no longer threatening. At its simplest, an animal is placed in and briefly habituated to a box that contains sawdust on the floor. After habituation, a probe that can deliver a shock is inserted into the box through a hole in the wall. When the animal investigates the object, it receives a brief electric shock from the probe. The response of the animal is to first withdraw from the object, then investigate it cautiously, and finally to use the forelimbs to cover the object with sawdust.

Measures of the strength of learning about the probe include the number of times the animal investigates the object, the length of time that it spends burying the object, and the depth of the sawdust that eventually covers the object. A variety of variations of the task have been used, including burying objects that deliver noxious sounds or odors. Animals will also bury other objects that are proximal to the offending object, indicating that the burying response can be secondarily conditioned to other objects. Defensive burying has been used to examine the effects of aging on behavior and to examine the effects of potential anti-anxiety agents (Pinel and Treit, 1983).

D. Object Recognition

Object recognition can be tested in a three-compartment box, also called the Mumby box (Mumby et al., 1989; Mumby and Pinel, 1994), or other similar test situations (Ennaceur and Aggleton, 1997). The central compartment of the box is a waiting area and is connected to two side boxes, choice box 1 and choice box 2, by sliding doors. A door is opened and the animal is allowed to enter choice box 1, where it finds two food wells, one of which is covered by a "sample" object. When it displaces the sample, it receives a food reward. The rat then shuttles into the waiting area until it is allowed access to choice box 2. Here it again finds two food wells, one covered by an object identical to the previously encountered sample in choice box 1, and a novel object. To obtain food reward, it must displace the novel object. New sample and novel trials are given with new sample and novel objects. The measure of success is the animal's memory that sample objects will not provide reinforcement on two successive trials.

The rodent object recognition test is similar to non-matching-to-sample tests previously developed for primates. Both short-term and long-term memory for objects can be measured by introducing a delay of variable duration after the sample trial. Object recognition in more natural environments uses very similar methodology to that described previously, but the time spent sniffing and examining objects or animals placed in the animal's home environment is used as the measure of recognition.

animal finds pleasant or noxious become conditioned to the location in which the object, event, or substance is encountered (Schechter and Calcagnetti, 1993; Cabib et al., 1996). Typically, a two-compartment box is used and the measure of behavior is the time spent in the compartments. For example, if the experimenter wishes to determine whether a drug treatment is perceived as being pleasant the animal is exposed to one of the compartments of the box while under the effects of the drug. At some later date, while undrugged, the animal is given access to the original box or to a different box. If the animal spends more time in the original "conditioned" box, that can be taken as evidence that the animal perceived the treatment as positively rewarding. If it shows a preference for the other box, it perceived the treatment as negatively rewarding. Conditioned place preferences can be modified to measure the strength of memories and their duration by varying intervals between sample and test trials.

E.   Conditioned Place Preferences

The conditioned place preference task takes advantage of the fact that objects, events, or substances that an

F.   Spatial Navigation

A large number of maze tests have been used to measure spatial navigation (Figure 8-13). The central idea for

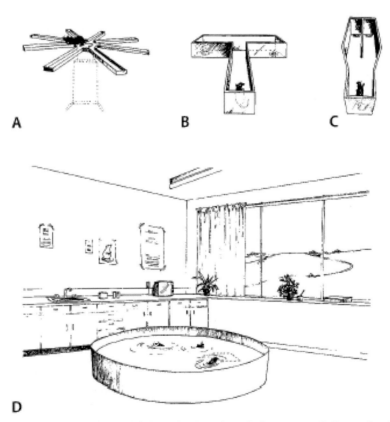

Fig. 8-13   Spatial tasks: **A:** radial arm maze in which food is located only at the end of some arms, **B:** T-maze in which food is located in one arm, **C:** Grice box in which food is located on one side, **D:** swimming pool task, in which an animal searches for a platform hidden just beneath the surface of the water. The tests are usually conducted in a open room that provides many spatial cues.

all of the tests consists of having an animal (1) learn to find food at one or more locations or (2) escape to a refuge from different locations. Most of the tests are administered on dry land, but because of the excellent swimming ability of the rat tests have been developed in a swimming pool. The two most widely used tests are the radial arm maze and the swimming pool place task (also commonly referred to as the Morris water maze task). The radial arm maze consists of a central box or platform from which protrude a number of arms (Olton et al., 1979; Jarrard, 1983). The location of the arms is either fixed or marked by a cue on the arm (e.g., roughness of the surface of the arm, color of the arm, a light at its end, and so on). Food is located at the end of one or more arms. The task of the animal is to learn the location of the food over a number of test days. This evaluates its ability to form a procedural memory for the task. The animal's performance can be interrupted to see if it can "pick up where it left off" in order to evaluate its working memory.

The swimming pool task has become extremely popular, mainly because the animals do not have to be food- or water-deprived to motivate them to perform (Morris, 1984; Sutherland and Dyck, 1984; Whishaw, 1985; McNamara and Skelton, 1993). Although the task is excellent for the rat, which is semiaquatic, it may be less useful for other species (Whishaw, 1995). The apparatus consists of a circular round pool about 1.5 m in diameter, filled with tepid water made opaque with powdered milk, paint, sawdust, or floating beads. A platform about 10 cm sq is placed in the pool with its surface either visible or hidden about 1 cm beneath the surface of the water. The animal is placed into the water facing the wall of the pool, and in order to escape from the water must reach the platform. On successive trials, the animal is placed into the water at new locations, and its response time decreases until it escapes by swimming directly to the hidden platform. Its ability to learn to escape to a platform hidden at a fixed location is thought to be a measure of spatial procedural memory. If the platform is moved repeatedly to new locations, the task becomes mainly a test of spatial working memory (Whishaw, 1985). That is, the animal has to match on its second trial the location at which the platform was found on the first trial. At their simplest, spatial tasks attempt to measure three aspects of spatial behavior.

- Place tasks measure whether an animal can find a food item or an escape platform using the relational properties of ambient cues, usually visual cues.
- Cue tasks measure whether an animal can find a food item or an escape platform using a visible cue marking the location of the target.
- Response tasks measure whether an animal can use body cues (e.g., turn left or right) to locate an object.

It is widely assumed that these different types of spatial learning are mediated by different neural systems. Thus, administration of more than one task can be used to dissociate spatial functions.

X. INDUCED-BEHAVIOR ANALYSIS

In addition to analysis of naturally-occurring behaviors, various interventions have been utilized to induce behavior. For example, stereotaxic surgery can be utilized to "lesion" or chemically destroy specific brain areas. A neurotoxic agent such as 6-hydroxydopamine administered during stereotaxic surgery directly into the basal ganglia will cause dopaminergic deficits resembling those seen in patients with Parkinson's disease. Stereotaxic surgical intervention requires the use of aseptic technique and appropriate anesthesia selection. For lengthy surgeries, the use of an inhalant anesthetic delivered via stereotaxic nose cone is optimal at providing a steady plane of anesthesia.

Surgical preparation of the rat for this type of surgery involves placement of the rat in a stereotaxic frame device to stabilize the head. The frame has microcalibrated arms to precisely measure coordinates in three dimensions. Brain atlases such as that by Paxinos and Watson (1998) are consulted to determine exact coordinates for the area of interest. Correct placement of the rat in this device is critical, which can be difficult for the beginner. Ear bars are placed into the anesthetized rat's external ear canal, and the bars are then affixed to the frame. Correct placement of the ear bars can be confirmed by free up/down rotation of the head and lack of side to side movement. In addition, the head should appear to be symmetrical.

Using aseptic technique, a midline incision is made over the calvarium to expose both bregma and lambda landmarks. Using the stereotaxic arm, the head is adjusted in a vertical plane using the nose bar so that the two landmarks are at the same vertical coordinate. Once this is confirmed, the surgeon can proceed to access the brain area of interest via a craniotomy overlying the site coordinates. Postoperative care of the rat should include supportive care in response to the created deficits, as well as general postoperative care such as that described by Waynforth (1992).

Pharmacologic-induced behaviors are also utilized to investigate brain function. For example, amphetamine given parenterally (intraperitoneally or subcutaneously) induces rotational behavior in an animal with a unilateral basal ganglia lesion. The rotation frequency is recorded over a defined period after the amphetamine injection by placing the rat in a round chamber with an electronic swivel to record the number and direction of rotations over time. A reduction in rotation frequency subsequently serves as one measure of treatment effectiveness in this model.

Various "incidental" behaviors may also be induced with a pharmacologic intervention even though they may not be of primary interest. In the case of amphetamine, rats are sensitive to overheating, hyper-excitable, and with chronic use will have weight loss. Analysis of induced behavior is a powerful tool. However, it is very important to observe the animal's entire behavioral repertoire as we described in this chapter in order to accurately assess behavior.

XI.   COMMENTS ON GENERALIZING FROM BEHAVIORAL ANALYSIS

Because the ultimate goal of studies on rodents is to understand the brain-behavior relationships in humans, it is reasonable to ask to what extent the behavior of rodents is useful in understanding human brain-behavior relationships. Indeed, one difficulty with choosing any mammal or mammalian order to use as a model of brain function is that each species has a unique behavioral repertoire that permits the animal to survive in its particular environmental niche. There is, therefore, the danger that neural organization is uniquely patterned in different species in a way that reflects the unique behavioral adaptations of the different species. Stated differently, it is possible that the brain-behavior relationships of rodents are not representative of other mammalian species, especially primates.

We have emphasized elsewhere that although the details of behavior may differ somewhat mammals share many similar behavioral traits and capacities (Kolb and Whishaw, 1983). For example, all mammals must detect and interpret sensory stimuli, relate this information to past experience, and act appropriately. Similarly, all mammals appear to be capable of learning complex tasks under various conditions of reinforcement (Warren, 1977), and all mammals are mobile and have developed mechanisms for navigating in space. The details and complexity of these behaviors clearly vary, but the general capacities are common to all mammals. Warren and Kolb (1978) proposed that behaviors and behavioral capacities demonstrable in all mammals could be designated as *class-common* behaviors. In contrast, behaviors that are unique to a species and that have been selected to promote survival in a particular niche are designated as *species typical* behaviors.

The distinction between these two types of behavior can be illustrated by the manner in which different mammals use their forelimbs to manipulate food objects. Monkeys will grasp objects with a single forepaw and often sit upright, holding the food item to consume the food. Rats too will grasp objects with one forepaw and then typically transfer the food to their mouth, assume a sitting posture, transfer the food back to both forepaws, and then eat.

Thus, many mammals use the forepaws to manipulate food (or other items) that it can be considered a class-common behavior. Nonetheless, the details vary from species to species. Some species-typical differences are large indeed, such as the use of the forelimbs in bats or carnivores versus rodents or primates.

It would seem foolish indeed to use dogs as a model for studying the details of the neural control of object manipulation by humans, as their limb use is relatively rudimentary. But what about rodents? There are clearly species-typical differences among rodent species and between rodents and primates. The question is whether these species-typical differences necessarily reflect fundamental differences in the neural control of skilled forelimb use. One way to address this question is to examine the order-specific characteristics of forelimb use. That is, one can study the similarity in forelimb use in different species within an order. This type of analysis would allow us to determine the commonalities in behavior across species of an order, which would give us a better basis for comparing across orders (Whishaw et al., 1992b).

XII.   L'ENVOIE

The following example from our laboratory illustrates how behavioral analysis can provide insights into behavior. Aged Fischer 344 rats are widely used to examine the effects of aging on memory (Lindner, 1997). We were interested in using these animals for studies that examine the ameliorative effects of exogenously supplied compounds thought to have neurotrophic properties. Since previous studies had used swimming pool spatial tasks, we first sought to confirm that 24-month-old Fischer rats were impaired relative to six-month-old rats. We found that when the rats were given eight trials a day on a task that required them to find a hidden platform at a fixed location in a swimming pool the 24-month-old rats were indeed severely impaired relative to the six-month-old rats. When given just one trial a day, however, they learned the task much more quickly, reaching the hidden platform as rapidly as the young rats after only 14 to 16 trials, indicating that learning per se was not impaired.

Finally, when given a matching-to-place task in which they were given two trials a day with the platform in a new location each day, the aged rats were severely impaired and showed no improvement from the first (test) trial to the second (matching) trial, whereas the young rats showed a marked reduction in latency. Results from these three tests seemed to support the idea that the animals had a selective spatial deficit, in that a very similar pattern of results is obtained from rats that have selective hippocampal lesions. This conclusion was severely compromised by the results of further tests. In open field tests, the old rats were less active than the young rats as they walked, and they reared less. When given locomotion tests, they were slower swimmers and walked more slowly for food in a straight alley. When required to climb out of a 9-inch-deep cage to obtain food, they were extremely impaired.

Further tests showed that the motor impairments of old rats were quite selective. The aged animals could protrude their tongue normally, eat one gram food pellets as quickly as the young rats, and they reached for food in a skilled reaching task as well as did the young rats. When given tests of righting, they were impaired relative to the young rats but when their forequarters and hindquarters were tested separately, forequarter righting was unimpaired whereas hindquarter righting was impaired. The results of these neurologic tests suggested that the old rats were selectively impaired in using their hindlimbs, and this result was confirmed by kinematic analysis of hindlimb movements used in swimming and walking. Thus, it is unclear whether their "spatial" deficit was due to a learning impairment or related to impaired use of the hindlimbs.

These results are relevant to the discussion of methodology given in the opening of this chapter. There is a general expectation that aged rats will be impaired in spatial memory tasks. Our tests confirmed this expectation. The comprehensive follow-up analysis showed, however, that the animals had a selective motor impairment in use of the hindlimbs to move themselves. The selectivity of the deficit was completely unexpected. Because commonly used spatial tasks require the animals to use their hindlimbs to move themselves, the results of the spatial tests are confounded by the rats' motor impairment. The particular result from this study suggests two new hypotheses. Aged Fischer 344 rats may have only a motor deficit that impairs their swimming performance, or the animals may have both a learning deficit and a motor deficit. Subsequent testing (e.g., spatial tests that do not require movement) could be used to assess these hypotheses.

Fortuitously, however, the finding of a selective motor deficit provides a very good model for studying motor impairments associated with aging. The lesson from this example is, therefore, that a careful examination of behavior can provide insights into the specific impairments of an animal, can provide new models for behavioral analysis, and can assist in evaluating the specificity of animal models of functional disorders.

REFERENCES

Alberts, J. R. (1978). Huddling by rat pups: Multisensory control of contact behavior. *J. C. Physiol. Psychol.* **92,** 220–230.

Bammer, G. (1982). Pharmacological investigations of neurotransmitter involvement in passive avoidance responding: a review and some new results. *Neurosci. Biobehav. Rev.* **6,** 247–296.

Becker, J. B., Snyder, P. J., Miller, M. M., Westgate, S. A., Jenuwine, M. J. (1987). The influence of esterous cycle and intrastriatal estradiol on sensorimotor performance in the female rat. *Pharm. Biochem. Behav.* **27,** 53–59.

Berridge, K. C. (1989). Progressive degradation of serial grooming chains by descending decerebration. *Behav. Brain Res.* **33,** 241–253.

Berridge, K. C. (1990). Comparative fine structure of action: rules of form and sequence in the grooming patterns of six rodent species. *Behavior* **113,** 21–56.

Berridge, K. C., and Whishaw, I. Q. (1992). Cortex, striatum and cerebellum: Control of a syntactic grooming sequence. *Exp. Brain Res.* **90,** 275–290.

Blake, W. (1905). Auguries of innocence [microform] / by WilliamOpie collection of children's literature, E.V. Lucas.

Blanchard, B. J., Blanchard, D. C., and Hori, K. (1989). An ethoexperimental approach to the study of defense. In R. J. Blanchard, P. F. Brain, D. C. Blanchard, and S. Parmigiani (eds.), Ethoexperimental Approaches to the Study of Behavior, pp. 114–137, London: Kluwer Academic Publishers.

Brown, R. E. (1975). Object-directed urine marking by male rats (Rattus norvegicus). *Behavioral Biology* **15,** 251–254.

Cabib, S., Puglisi-Allegra, S., Genua, C., Simon, H., Le Moal, M., and Pizza, P.V. (1996). Dose-dependent aversive and rewarding effects of amphetamine as revealed by a new place conditioning apparatus. *Psychopharmacology* **125,** 92–96.

Carter, C. S., Witt, D. M., Kolb, B., and Whishaw, I.Q. (1982). Neonatal decortication and adult female sexual behavior. *Physiol. Behav.* **29,** 763–766.

Clarke, K. A. (1992). A technique for the study of spatiotemporal aspects of paw contact patterns applied to rats treated with a TRH analogue. *Behav. Res. Meth. Instrum. Comput.* **24,** 407–411.

Crawley, J. N., Belknap, J. K., Collins, A., Crabbe, J. C., Frankel, W., Henderson, N., Hitzemann, R. J., Maxson, S. C., Miner, L. L., Silva, A. J., Wehner, J. M., Wynshaw-Boris, A., and Paylor, R. (1997). Behavioral phenotypes of inbred mouse strains: Implications and recommendations for molecular studies. *Psychopharmacology* **132,** 107–124.

Denny-Brown, D., Dawson, D. M., and Tyler, H. R. (1982). Handbook of Neurological Examination and Case Recording. Cambridge, MA: Harvard University Press.

DeRyck, M., Schallert, T., and Teitelbaum, P. (1980). Morphine versus Haloperidol catalepsy in the rat: A behavioral analysis of postural support mechanisms. *Brain Res.* **201,** 143–172.

Dewsbury, D. A. (1973). A quantitative description of the behavior of rats during copulation. *Behaviour* **29,** 154–178.

Dudai, Y. (1989). The Neurobiology of Memory. Oxford, UK: Oxford University Press.

Eliam, D., and Golani, I. (1989). Home base behavior of rats (Rattus norvegicus) exploring a novel environment. *Behav. Brain Res.* **34,** 199–211.

Ennaceur, A., and Aggleton, J. P. (1997). The effects of neurotoxic lesions of the perirhinal cortex compared to fornix transection on object recognition memory in the rat. *Behav. Brain Res.* **88,** 181–193.

Eshkol, N., and Wachmann, A. (1958). Movement Notation. London: Weidenfeld and Nicholson.

Everitt, B. J. (1990). Sexual motivation: A neural and behavioral analysis of the mechanisms underlying appetitive and copulatory responses of male rats. *Neuroscience and Biobehavioral Reviews* **14,** 217–232.

Fentress, J. C., and Bolivar, V. J. (1996). Developmental Aspects of Movement Sequences in Mammals. In K-P. Ossenkopp, M. Kavaliers, and P. R. Sanberg (eds.), Measuring Movement and Locomotion: From Invertebrates to Humans, pp. 95–114, Austin, TX: R. G. Landes.

Field, E. F., Whishaw, I. Q., and Pellis, S. M. (1997). A kinematic analysis of sex-typical movement patterns used during evasive dodging to protect a food item: The role of testicular hormones. *Behav. Neurosci.* **111**, 808–815.

Field, E. F., Whishaw, I. Q., Pellis, S. M. (2000). Sex differences in catalepsy: evidence for hormone-dependent postural mechanisms in haloperidol-treated rats. *Behav Brain Res*, **109**, 207–212.

Fish, F. F. (1996). Measurement of swimming kinematics in small terrestrial mammals. In K-P. Ossenkopp, M. Kavaliers, P. R. Sanberg (eds.), Measuring Movement and Locomotion: From Invertebrates to Humans, pp. 171–188, Austin, TX: R. G. Landes.

Francis, R. L. (1977). 22-kHz calls by isolated rats. *Nature* **265**, 236–238.

Freed, W. (1983). Functional brain tissue transplantation: Reversal of lesion-induced rotation by intraventricular substantia nigra and adrenal medulla grafts, with a note on intracranial retinal grafts. *Biol. Psychiat.* **18**, 1205–1267.

Galef, B. G. (1993). Functions of social learning about food: A causal analysis of effects of diet novelty on preference transmission. *Anim. Behav.* **46**, 257–265.

Galef, B. G., Whiskin, E. E., and Bielavska, E. (1997). Interaction with demonstrator rats changes observer rats' affective responses to flavors. *J. Comp. Psychol.* **111**, 393–398.

Gallup, G. G., and Maser, J. D. (1977). Tonic immobility: Evolutionary underpinnings of human catalepsy and catatonia. In J. D. Maser and M. E. P. Seligman (eds.), Psychopathology: Experimental Models, pp. 334–462, San Francisco: W. H. Freeman.

Ganor, I., and Golani, I. (1980). Coordination and integration in the hindleg step cycle of the rat: Kinematic synergies. *Brain Res.* **195**, 57–67.

Gentile, A. M., Green, S., Nieburgs, A., Schmelzer, W., and Stein, D. G. (1978). Disruption and recovery of locomotor and manipulatory behavior following cortical lesions in rats. *Behav. Biol.* **22**, 417–455.

Golani, I. (1976). Homeostatic motor processes in mammalian interactions: A choreography of display. In P. G. Bateson and P. H. Klopfer (eds.), Prospectives in Ethology, Volume 2, pp. 237–334, New York: Plenum Press.

Golani, I. (1992). A mobility gradient in the organization of vertebrate movement: The perception of movement through symbolic language. *Behavioral and Brain Sciences* **15**, 249–266.

Golani, I., Benjamini, Y., and Eilam, D. (1993). Stopping behavior: Constraints on exploration in rats (Rattus norvegicus). *Behav. Brain Res.* **26**, 21–33.

Golani, I., Wolgin, D. L., and Teitelbaum, P. (1979). A proposed natural geometry of recovery from akinesia in the lateral hypothalmic rat. *Brain Res.* **164**, 237–267.

Grant, E. C. (1963). An analysis of the social behaviour of the male laboratory rat. *Behaviour* **21**, 260–281.

Grant, E. C., and Mackintosh, J. H. (1963). A comparison of the social postures of some common laboratory rodents. *Behaviour* **21**, 246–259.

Grill, H. J., and Norgren, R. (1978). The taste reactivity test: I. Mimetic responses to gustatory stimuli in neurologically normal rats. *Brain Res.* **13**, 63–279.

Grota, L. J., and Ader, R. (1969). Continuous recording of maternal behavior in Rattus norvegicus. *Anim. Behav.* **21**, 78–82.

Heale, V. R., Petersen, K., and Vanderwolf, C. H. (1996). Effect of colchicine-induced cell loss in the dentate gyrus and Ammon's horn on the olfactory control of feeding in rats. *Brain Res.* **712**, 213–220.

Huerta, P. T., Scearce, K. A., Farris, S. M., Empson, R. M., and Prusky, G. T. (1996). Preservation of spatial learning in fyn tyrosine kinase knockout mice. *Neuroreport* **10**, 1685–1689.

Hutt, S. J., and Hutt, C. (1970). Direct Observation and Measurement of Behavior. Springfield, IL: Charles C. Thomas.

Jarrard, L. E. (1983). Selective hippocampal lesions and behavior: Effects of kainic acid lesions on performance of place and cue tasks. *Behav. Neurosci.* **97**, 873–889.

Kinder, E. F. (1927). A study of the nest-building activity of the albino rat. *J. Comp. Physiol. Psychol.* **47**, 117–161.

Kolb, B. (1974). Some tests of response habituation in rats with prefrontal lesions. *Can. J. Psychol.* **28**, 260–267.

Kolb, B., and Whishaw, I. Q. (1983). Problems and principles underlying interspecies comparisons. In T. E. Robinson (ed.), Behavioral Approaches to Brain Research, pp. 237–265, Oxford, UK: Oxford University Press.

Kolb, B., and Whishaw, I. Q. (1985). An observer's view of locomotor asymmetry in the rat. *Neurobehav. Toxicol. Teratolog.* **7**, 71–78.

Kolb, B., and Whishaw, I. Q. (1996). Fundamentals of Human Neuropsychology. New York: W. H. Freeman.

Kolb, B., Gibb, R., Pedersen, B., and Whishaw, I. Q. (1997). Embryonic injection of BrdU blocks later cerebral plasticity. *Society for Neuroscience Abstracts* **23**, 677.16.

Kozlowski, D. A., James, D. C., and Schallert, T. J. (1996). Use-dependent exaggeration of neuronal injury after unilateral sensorimotor cortex lesions. *Neurosci.* **16**, 4776–4786.

Lassek, A. M. (1954). The Pyramidal Tract. Springfield, IL: Charles C. Thomas.

Latane, B. (1970). Gregariousness and fear in laboratory rats. *J. Exp. Soc. Psychol.* **5**, 61–69.

Le Marec, N., Stelz, T., Delhaye-Bouchaud, N., Mariani, J., and Caston, J. (1997). Effect of cerebellar granule cell depletion on learning of the equilibrium behaviour: Study in postnatally X-irradiated rats. *Eur. J. Neurosci.* **9**, 2472–2478.

Leon, M., Croskerry, P. G., and Smith, G. K. (1978). Thermal control of mother-young contact in rats. *Physiol. Behav.* **21**, 793–811.

Lindner, M. D. (1997). Reliability, distribution, and validity of age-related cognitive deficits in the Morris water maze. *Neurobiol. Learn. Mem.* **68**, 203–220.

Lubar, J. F., Herrmann, T. F., Moore, D. R., and Shouse, M. N. (1973). Effect of septal and frontal ablations on species-typical behavior in the rat. *J. Comp. Physiol. Psychol.* **83**, 260–270.

McNamara, R. K., and Skelton, R. W. (1993). The neuropharmacological and neurochemical basis of place learning in the Morris water maze. *Brain Res. Rev.* **18**, 33–49.

Marshall, J., Turner, B. H., and Teitelbaum, P. (1971). Further analysis of sensory inattention following lateral hypothalamic damage in rats. *J. Comp. Physiol. Psychol.* **86**, 808–830.

Martinez, J. L., and Kesner, R. P. (1991). Learning and Memory: A Biological View (2d ed.), New York: Academic Press.

Mermelstein, P. G., and Becker, J. B. (1995). Increased extracellular dopamine in the nucleus accumbens and striatum of the female rat during paced copulatory behavior. *Behav. Neurosci.* **109**, 354–365.

Michal, E. K. (1973). Effects of limbic lesions on behavior sequences and courtship behavior of male rats (Rattus norvegicus). *Anim. Behav.* **244**, 264–285.

Miklyaeva, E. I., Martens, D. J., and Whishaw, I. Q. (1995). Impairments and compensatory adjustments in spontaneous movement after unilateral dopamine depletion in rats. *Brain Res.* **681**, 23–40.

Mistleburger, R. E., and Mumby, D. G. (1992). The limbic system and food-anticipatory circadian rhythms in the rat: ablation and dopamine blocking studies. *Behav. Brain Res.* **47**, 159–168.

Morris, R. (1984). Developments of a water-maze procedure for studying spatial learning in the rat. *Neurosci. Meth.* **11**, 47–60.

Moyer, K. E. (1968). Kinds of aggression and their physiological basis. *Com. Behav. Biol.* **2**, 65–87.

Mumby, D. G., and Pinel, J. P. J. (1994). Rhinal cortex lesions impair object recognition in rats. *Behav. Neurosci.* **108**, 11–18.

Mumby, D. G., Pinel, J. P. J., and Wood, E. R. (1989). Nonrecurring items delayed nonmatching-to-sample in rats: A new paradigm for testing nonspatial working memory. *Psychobiology* **18**, 321–326.

Olsson, M., Nikkhah, G., Bentlage, C., and Bjorklund, A. (1995). Forelimb akinesia in the rat Parkinson model: Differential effects of dopamine. *Neurosci.* **15**, 3863–3875.

Olton, D. S., Becker, J. T., and Handlemann, G. E. (1979). Hippocampus, space and memory. *Behav. Brain Sci.* **2**, 313–365.

Olton, D. S., Gamzu, E., and Corkin, S. (1985). Memory dysfunctions: An integration of animal and human research from preclinical and clinical perspectives. *Ann. NY Acad. Sci.* 444.

Ossenkopp, K-P., Kavaliers, M., and Sanberg, P. R. (1996). Measuring Movement and Locomotion: From Invertebrates to Humans. Austin, TX: R. G. Landes.

Paxinos, G., and Watson, C. (1998). The Rat Brain in Stereotaxic Coordinates. San Diego: Academic Press.

Pellis, S. M. (1983). Development of head and food coordination in the Australian magpie Gymnorhina tibicen, and the function of play. *Bird Behav.* **4**, 57–62.

Pellis, S. M. (1996). Righting and the modular organization of motor programs. In K-P. Ossenkopp, M. Kavaliers, and P. R. Sanberg (eds.), Measuring Movement and Locomotion: From Invertebrates to Humans, pp. 115–133, Austin, TX: R. G. Landes.

Pellis, S. M. (1997). Targets and tactics: The analysis of moment-to-moment decision making in animal combat. *Agg. Behav.* **23**, 107–129.

Pellis, S. M., Field, E. F., Smith, L. K., and Pellis, V. (1996). Multiple differences in the play fighting of male and female rats: Implications for the causes and functions of play. *Neurosci. Biobehav. Rev.* **21**, 105–120.

Perks, S. M., and Clifton, P. G. (1997). Reinforcer revaluation and conditioned place reference. *Physiol. Behav.* **61**, 1–5.

Pinel, J. P. J., and Treit, D. (1983). The conditioned defensive burying paradigm and behavioral neuroscience. In T. E. Robinson (ed.), Behavioral Contributions to Brain Research, pp. 212–234, Oxford, UK: Oxford University Press.

Pinel, J. P. J., Hones, C. H., and Whishaw, I. Q. (1992). Behavior from the ground up: rat behavior from the ventral perspective. *Psychobiology* **20**, 185–188.

Sachs, B. D., and Barfield, R. J. (1976). Functional analysis of masculine copulatory behavior in the rat. In J. S. Rosenblatt, R. A. Hinde, E. Shaw, and C. Beer (eds.), Advances in the Study of Behavior, Volume 7, New York: Academic Press.

Salis, M. (1972). Effects of early malnutrition on the development of swimming ability in the rat. *Physiol. Behav.* **8**, 119–122.

Satinoff, E. (1983). A reevaluation of the concept of the homeostatic organization of temperature regulation. In E. Satinoff and P. Teitelbaum (eds.), Handbook of Behavioral Neurobiology, Volume 6, pp. 443–467. New York: Plenum Press.

Schallert, T., and Whishaw, I. Q. (1978). Two types of aphagia and two types of sensorimotor impairment after lateral hypothalamic lesions: Observations in normal weight, dieted, and fattened rats. *J. Comp. Physiol. Psychol.* **92**, 720–741.

Schallert, T., and Whishaw, I. Q. (1984). Bilateral cutaneous stimulation of the somatosensory system in hemidecorticate rats. *Behav. Neurosci.* **98**, 375–393.

Schallert, T., DeRyck, M., Whishaw, I. Q., Ramirez, V. D., and Teitelbaum, P. (1979). Excessive bracing reactions and their control by atropine and l-dopa in an animal analog of Parkinsonism. *Exp. Neurol.* **64**, 33–43.

Schallert, T., Norton, D., and Jones, T. A. (1992). A clinically relevant unilateral rat model of Parkinsonian akinesia. *J. Neural. Transpl. Plast.* **3**, 332–333.

Schechter, M. D., and Calcagnetti, D. J. (1993). Trends in place preference conditioning with a cross-indexed bibliography: 1957–1991. *Neurosci. Biobehav. Rev.* **17**, 21–41.

Shipley, J., and Kolb, B. (1977). Neural correlates of species-typical behavior in the Syrian golden hamster. *J. Comp. Physiol. Psychol.* **91**, 1056–1073.

Slagen, J. L., Earley, B., Jaffard, R., Richelle, M., and Olton, D. S. (1990). Behavioral models of memory and amnesia. *Pharmacopsychiatry* Suppl. **2**, 81–83.

Sutherland, R. J., and Dyck, R. H. (1984). Place navigation by rats in a swimming pool. *Can. J. Psychol.* **38**, 322–347.

Tchernichovski, O., and Golani, I. (1995). A phase plane representation of rat exploratory behavior. *J. Neurosci. Method.* **62**, 21–27.

Teitelbaum, P., Schallert, T., and Whishaw, I. Q. (1983). Sources of spontaneity in motivated behavior. In E. Satinoff and P. Teitelbaum (eds.), Handbook of Behavioral Neurobiology, Volume 6, pp. 23–61, New York: Plenum Press.

Vanderwolf, C. H., and Cain, D. P. (1994). The behavioral neurobiology of learning and memory: a conceptual reorientation. *Brain Res.* **19**, 264–297.

Upchurch, M., and Wehner, J. M. (1989). Inheritance of spatial learning ability in inbred mice: A classical genetic analysis. *Behav. Neurosci.* **103**, 1251–1258.

Warren, J. M. (1977). A phylogenetic approach to learning and intelligence. In A. Oliverio (ed.), Genetics, Environment, and Intelligence, pp. 37–56, Amsterdam: Elsevier.

Warren, J. M., and Kolb, B. (1978). Generalizations in neuropsychology. In S. Finger (ed.), Recovery from Brain Damage, pp. 36–49. New York: Plenum Press.

Waynforth, H. B., and Flecknell, P. A. (1992). Experimental and Surgical Technique in the Rat (2d ed.), pp. 100–152, London: Academic Press.

Whishaw, I. Q. (1985). Formation of a place learning-set in the rat: A new procedure for neuro *behavioural* studies. *Physiol. Behav.* **35**, 139–143.

Whishaw, I. Q. (1988). Food wrenching and dodging: use of action patterns for the analysis of sensorimotor and social behavior in the rat. *J. Neurosci. Method.* **24**, 169–178.

Whishaw, I. Q. (1995). A comparison of rats and mice in a swimming pool place task and matching to place task: Some surprising differences. *Physiol. Behav.* **58**, 687–693.

Whishaw, I. Q., Kolb, B. (2005). The behavior of the laboratory rat: A handbook with tests. Oxford UK: Oxford University Press.

Whishaw, I. Q., and Miklyaeva, E. (1996). A rat's reach should exceed its grasp: Analysis of independent limb and digit use in the laboratory rat. In K-P. Ossenkopp, M. Kavaliers, and P. R. Sanberg (eds.), Measuring Movement and Locomotion: From Invertebrates to Humans, pp. 135–164, Austin, TX: R. G. Landes.

Whishaw, I. Q., and Pellis, S. M. (1990). The structure of skilled forelimb reaching in the rat: a proximally driven movement with a single distal rotatory component. *Behav. Brain Res.* **41**, 49–59.

Whishaw, I. Q., and Whishaw, G. E. (1996). Conspecific aggression influences food carrying: Studies on a wild population of Rattus norvegicus. *Agg. Behav.* **22**, 47–66.

Whishaw, I. Q., Cassel, J-C., Majchrazak, M., Cassel, S., and Will, B. (1994). Short-stops in rats with fimbriafornix lesions: Evidence for change in the mobility gradient. *Hippocampus* **5**, 577–582.

Whishaw, I. Q., Dringenberg, H. C., and Comery, T. A. (1992a). Rats (Rattus norvegicus) modulate eating speed and vigilance to optimize food consumption: Effects of cover, circadian rhythm, food deprivation, and individual differences. *J. Comp. Psychol.* **106**, 411–419.

Whishaw, I. Q., Kolb, B., Sutherland, R. J. (1983). The analysis of behavior in the laboratory rat. In T. E. Robinson (ed.), Behavioral Approaches to Brain Research, pp. 237–264, Oxford, UK: Oxford University Press.

Whishaw, I. Q., Nonneman, A. J., and Kolb, B. (1981a). Environmental constraints on motor abilities used in grooming, swimming, and eating by decorticate rats. *J. Comp. Physiol. Psychol.* **95**, 792–804.

Whishaw, I. Q., Oddie, S. D., McNamara, R. K., Harris, T. W., and Perry, B. (1990). Psychophysical methods for study of sensory-motor behavior using a food-carrying (hoarding) task in rodents. *J. Neurosci. Method.* **32**, 123–133.

Whishaw, I. Q., Pellis, S. M., and Gorny, B. P. (1992b). Skilled reaching in rats and humans: Parallel development of homology. *Behav. Brain Res.* **47**, 59–70.

Whishaw, I. Q., Pellis, S. M., Gorny, B. P., Kolb, B., and Tetzlaff, W. (1993). Proximal and distal impairments in rat forelimb use in reaching follow unilateral pyramidal tract lesions. *Behav. Brain Res.* **56,** 59–76.

Whishaw, I. Q., Pellis, S. M., Gorny, B. P., and Pellis, V. C. (1991). The impairments in reaching and the movements of compensation in rats with motor cortex lesions: An endpoint, videorecording, and movement notation analysis. *Behav. Brain Res.* **42,** 77–91.

Whishaw, I. Q., Sarna, J. R., and Pellis, S. M. (1998). Evidence for rodent-common and species-typical limb and digit use in eating, derived from a comparative analysis of ten rodent species. *Behav. Brain Res.* **96,** 79–91.

Whishaw, I. Q., Schallert, T., and Kolb, B. (1981b). An analysis of feeding and sensorimotor abilities of rats after decortication. *J. Comp. Physiol. Psychol.* **95,** 85–103.

Wolgin, D. L., and Bonner, R. (1985). A simple, computer-assisted system for measuring latencies of contact placing. *Physiol. Behav.* **34,** 315–317.

Chapter 9

Nutrition

Sherry M. Lewis, Duane E. Ullrey, Dennis E. Barnard, and Joseph J. Knapka

THE LABORATORY RAT, 2ND EDITION

I. INTRODUCTION

"Nutrition involves various chemical and physiological activities which transform food elements into body elements." This simple definition (Maynard et al., 1979) describes the science of nutrition, a chemistry-based discipline interacting to varying degrees with many of the other physical and biological sciences. This definition also implicates nutrition as one of the environmental factors that influences the ability of animals to attain their genetic potential for growth, reproduction, longevity, or response to stimuli. Therefore, the nutritional status of animals involved in biomedical research has a profound effect on the quality of experimental results. The process of supplying adequate nutrition for laboratory animals involves establishing requirements for approximately 50 essential nutrients, formulating and manufacturing diets with the required nutrient concentrations, and managing numerous factors related to diet quality. Factors potentially affecting diet quality include the bioavailability of nutrients, palatability or acceptance by animals, procedures involved in preparation or storage, and the concentration of chemical contaminants.

II. NUTRIENT REQUIREMENTS

An estimate of the nutrient requirements of rats must be obtained in order to provide an adequate diet. The report "Nutrient Requirements of Laboratory Animals" published by the National Research Council (NRC) contains a rat chapter and is the most reliable source for the estimated nutrient requirements for this species. Revised editions of this report are published when there is sufficient new information in the literature to justify revisions in the estimated nutrient requirements. Therefore, readers are referred to the latest report (NRC, 1995) in the series for the most current estimates of the nutrient requirements for rats. Some are provided in the minerals section. Multiple reports of estimated nutrients for rats, which may contain different values for some nutrients, would be confusing to individuals attempting to formulate diets with adequate nutrient concentrations. The nutrient requirements of the laboratory rat are dynamic in that they are influenced by genetic and environmental factors. Therefore, the objective of the NRC report is to provide guidelines to adequate nutrition and not to describe the requirements of a single rat or rat colony. Factors that influence nutrient requirements must be identified and considered in selecting the nutrient composition of diets for specific rat colonies. Genetic factors to consider include body size or growth potential, reproductive performance, and metabolic changes caused by genetic manipulation in a specific stock or strain of rat. Environmental factors that influence the required nutrient concentrations in rat diets include the stage of the life cycle of the animals of interest, the microbiological status of their environment, experimentally induced stress, dietary nutrient interactions, and the addition of test substances to the diet (Knapka, 1983).

III. REQUIRED NUTRIENTS

The nutrients required for laboratory rats do not differ from those required by other mammalian species. The quantitative requirement for each nutrient may be different from other species, however, because of the relatively small body size and greater metabolic rate of rats. Inconsistencies in reports on nutrient requirements of rats may arise because the research was conducted under different environmental conditions or with different rat stocks or strains.

A. Protein and Amino Acids

Protein and amino acid requirements of healthy rats are influenced by their physiologic status (e.g., age, growth rate, pregnancy, or lactation), energy concentration of the diet, amino acid composition of the protein, and bioavailability of the amino acids. Protein requirements decline with age after weaning, and requirements of male and nonbreeding female rats older than 3 months of age are considerably lower than is the amount required during the female rat's most active growth period. The requirements discussed in this chapter, unless stated otherwise, are based on studies in which the fat content of the diet was approximately 5% by weight, and metabolizable energy (ME) concentrations were approximately 4 kcal/g. For diets that contain a greater amount of fat, increasing protein to maintain a constant amino acid-to-calorie ratio should be considered. Mixtures of proteins can be used to supply the necessary essential amino acids; specific supplementation of deficient proteins can be made by using L-amino acids. Animals depend on an exogenous source for approximately half the 20 amino acids commonly found in dietary proteins.

Protein deficiency results in decreased food intake and weight gain, hypoproteinemia, anemia, edema, depletion of body protein, muscle wasting, rough haircoat, irregularity or cessation of estrus, and poor reproduction with fetal resorption or delivery of weak or dead neonates. Protein deficiency in mothers during lactation results in poor growth of offspring. Diets supplying 13% protein from wheat gluten fed to six dams from conception through 15 days of lactation resulted in 48 pups on day 1 of lactation, but only two pups survived to lactation day 15. Post-weaning survival was significantly better when wheat gluten was supplemented with lysine and threonine or when 13% protein from casein plus methionine was fed. However, research suggested that impaired mammary development owing to amino acid deficiencies during pregnancy was not completely overcome by a nutritionally adequate diet during lactation (Jansen et al., 1987). Deficiencies of single amino acids may, in some cases, cause specific abnormalities in addition to those cited. Deletion of a single dietary essential amino acid commonly results in an immediate reduction of food intake, with a return to normal within a day or so of the amino acid replacement. A deficiency of tryptophan may result in cataracts, corneal vascularization, and alopecia (Cannon, 1948; Mesiter, 1957). A deficiency of lysine may result in a hunched stance, ataxia, dental caries, blackened teeth, and impaired bone calcification (Harris et al., 1943; Cannon, 1948; Kligler and Krehl, 1952; Bavetta and McClure, 1957; Likins et al., 1957; Meister, 1957). Fatty liver has been seen in a deficiency of methionine (Follis, 1958). A deficiency of arginine has resulted in increased plasma and liver concentrations of glutamate and glutamine (Gross et al., 1991) and increased urinary excretion of urea, citrate, and orotate (Milner et al., 1974).

Protein deficiency may have significant effects on experimental results in studies of chemical toxicity and carcinogenesis, immunologic processes, infection, and many other areas of interest. The activity of hepatic and probably other tissue microsomal oxidases (enzymes responsible for metabolism of steroid hormones and many exogenous chemicals) decrease markedly and rapidly in protein deficiency (Campbell and Hayes, 1975). Therefore, the response of rats to toxins and carcinogens may be altered. Rats that are protein deficient may respond abnormally to infectious agents because both cellular and humoral immunologic responses are depressed (Newberne, 1974).

As rats grow, the metabolically active visceral organs form an increasingly smaller fraction of total body weight (BW); however, the skeletal muscle fraction remains constant except in obese animals. The most rapid increase in total body protein occurs during the suckling period when both body mass and percentage of protein increase.

Protein synthesis during that time is not uniform but follows a cyclical pattern that varies from organ to organ. Several techniques are available for the study of nutritional requirements in newborn rats; alteration of litter weight is used extensively. Individual variations in intake may be large, and animals that die during the study must be replaced because litter size must be maintained throughout the experiment. A second approach is to restrict energy intake of the dam, which reduces the variation in intake among the young to some extent. The primary deficiency induced by these techniques is likely of protein rather than of energy. Better control of intake can be achieved by hand-feeding neonates (Miller, 1969).

Several methods are used to evaluate nutritional value of proteins. The protein efficiency ratio (PER) is the grams of weight gain per gram of protein consumed; it may be different at different dietary concentrations of test protein. For example, for casein the highest PER value is found at 7% protein in the diet; for plant proteins, the highest PER is at about 15%. Ten percent protein is used in most studies. Strain, age, sex, and duration of the feeding study influence PER. The final evaluation of proteins requires an accounting of both quality and quantity, as well as concentration, of other dietary components. Protein can be spared by increasing dietary concentrations of fat or carbohydrate in cases in which some protein is being used for energy.

Other methods of protein evaluation include chemical, microbiological, and whole-animal bioassays. Net protein utilization (NPU) is a measure of that proportion of food nitrogen retained by test animals and is defined as the body nitrogen of the test group minus body nitrogen of a nonprotein group divided by nitrogen consumed by the test group. NPU decreases with an increasing protein-to-energy ratio because of utilization of protein for energy. The relationship between NPU and the percentage of protein in the diet depends also on whether protein is being used for growth or maintenance and on the specific protein examined, because the proportions of amino acids required for maintenance appear to be different from the proportions required for growth (McLaughlan and Campbell, 1969; Said and Hegsted, 1970).

Weanling rats fed diets containing 3.6% to 25% protein (lactalbumin) for 3 weeks exhibited in a steep rise in weight gain as dietary protein increased from 3.6% to 12%; 12% protein and above gave equivalent weight gain. Feed efficiency (grams weight gained per gram of food consumed) was maximum at 12% protein. PER was maximal at 8% protein and decreased as protein content of the diet increased above that level. Fecal and urinary nitrogen increased with increasing dietary protein; urinary nitrogen rose markedly at levels greater than 12%, that is, after maximal growth rate was reached. Retention efficiency for most essential amino acids reached a maximum at 6% dietary protein (at which point BW gain was approximately one-half maximum), plateaued with diets that contained 6% to 10% protein, and then declined thereafter (Bunce and King, 1969a).

Carcass retention efficiency is the ratio of the increase in content of an amino acid in the carcass to the amount of that amino acid consumed. Dietary amino acids can be (1) incorporated into protein; (2) metabolized to carbon dioxide (CO_2), water, and urea; or (3) synthesized into nonessential amino acids or other compounds such as creatine. If the major factor operating in utilization of dietary amino acids is requirement for protein synthesis, retention efficiency for each amino acid is inversely related to its dietary supply as that supply approaches the minimal requirement. Maximum retention efficiency is used to determine dietary requirement. In rats fed 6% to 10% protein, 17% of total ingested amino acids were catabolized; in rats fed diets that contained lower concentrations of protein, which were therefore deficient in some amino acids, about 50% of amino acids ingested were catabolized because they could not be used for protein synthesis. At low dietary protein intake, there was a change in carcass amino acid composition, apparently representing alterations in the proportions of body proteins (Bunce and King, 1969a). If rats were fed casein rather than lactalbumin, maximum weight gain and feed efficiency occurred at 18% to 20% dietary protein; the highest PER, at 10% to 12% protein. Fecal nitrogen excretion was the same as in rats fed lactalbumin, and urinary nitrogen rose before maximum weight gain was achieved, indicating that amino acids from casein were not used as efficiently as were amino acids from lactalbumin (Bunce and King, 1969b).

It has been assumed that the nutritive value of protein is determined by the most limiting essential amino acid, and that nutritional quality of proteins decreases linearly as the limiting essential amino acid(s) decrease below the quantity found in an "ideal" protein. Protein utilization should decrease regardless of the essential amino acid that is deficient and stop when an essential amino acid is absent from the protein being tested. However, in rats fed diets lacking a single essential amino acid, protein utilization dropped to 20% to 40% of the value for rats fed an adequate diet, but protein utilization did not decrease to zero except in rats fed diets that lacked threonine, isoleucine, or the sulfur amino acids. Rats may be able to adapt to diets deficient in some essential amino acids by increased coprophagy, by alteration of catabolism, or possibly by limiting protein synthesis (Said and Hegsted, 1970).

In a review of data on protein requirements of rats, the NRC (1978) concluded that growing, pregnant, or lactating rats required 12% ideal protein in the diet or 14% casein supplemented with 0.2% DL-methionine. For maintenance of adult rats, the corresponding values were 4.2% and 4.8%, respectively. Subsequent examination of these issues (NRC, 1995) found that 17% and 23% dietary crude

protein from unsupplemented casein were required to support 95% and 100%, respectively, of the maximum growth response. Based on a need of 1.11 times more casein than lactalbumin for support of maximum nitrogen gain, requirements from lactalbumin for 95% and 100% of the maximum growth response were estimated to be about 15% and 21%, respectively, of dietary protein. The NRC (1995) observed that, in practice, diets made up of natural ingredients with 18% to 25% crude protein support high rates of postweaning growth. Protein requirements for gestation and lactation do not appear to differ significantly from these (NRC, 1995). Adult maintenance requirements, when a high-quality protein was used, were estimated to be about 5% compared with about 7% in diets made up of natural ingredients (NRC, 1995).

If rats are fed diets that contain amino acids to replace whole proteins, mixtures of essential and nonessential amino acids give a greater growth rate than do mixtures of only essential amino acids. The range of acceptable ratios of nonessential to essential amino acids is wide; satisfactory results have been obtained by using ratios of 0.5 to 4.0 in diets that contained 10% to 15% total amino acids. In addition to amino acids known to be essential from earlier studies, an apparent need for other animo acids has been shown by growing rats for arginine, asparagine, glutamic acid, and probably proline, presumably because they cannot be synthesized at a rate needed for rapid growth (Brookes et al., 1972; Milner et al., 1974; NRC, 1978). These amino acids may not be required for maintenance of adult rats. Estimated protein and amino acid requirements for maintenance, growth, and reproduction of rats have been published by the NRC (1995).

In early studies in which dietary protein was replaced by amino acids in amounts equal to or greater than the known requirements of rats, growth rate was not equal to that of rats fed casein. Addition of casein up to 12% in diets that already contained 22% amino acids was required to achieve maximal growth. It became necessary to define concentrations of essential amino acids and to establish relative or absolute concentrations of nonessential or partially dispensable amino acids such as arginine, proline, and glutamic acid. A low intake of one increased the quantity of the others needed for normal growth. Also, rats fed amino acid diets had decreased food intake, but if diets were mixed 1-to-1 with 3% agar, intake and weight gain were normal. Possibly the water in agar gel diets mediated the osmotic effects of amino acids in the stomach and upper small intestine. Rats fed amino acid diets have been carried through four generations with good, but not maximal, weight gain and with normal reproduction (Rogers and Harper, 1965; Schultz, 1956, 1957).

The maintenance requirement of young female adult rats was met by lactalbumin at 3.2% of the diet. The rats were then fed amino acid diets deficient in a single amino acid or diets that contained no protein or amino acids. Growth was depressed in all rats fed deficient diets, but only diets that lacked threonine, isoleucine or methionine, and cystine depressed growth as markedly as did protein-free diets. Discrepancies were explained in part by the observation that the cornstarch fed contained leucine, methionine, and tyrosine in amounts to provide 15% to 21% of the maintenance requirement, and other essential amino acids were present in amounts that supplied 2% to 10% of maintenance needs (Said and Hegsted, 1970).

Deficiencies of single amino acids have been studied in rats fed *ad libitum* (AL) or force-fed. Young rats force-fed amino acid diets deficient in a single essential amino acid were studied at intervals of 3 to 7 days. They had poor haircoats with areas of alopecia and pigmentation from porphyrins around the mouth and nose; they were weak and lethargic and lost weight. Rats fed diets deficient in threonine or histidine became hyper-reactive to external stimuli. Threonine, histidine, or methionine deficiency resulted in livers that were enlarged and yellow and that had periportal hepatocytes distended with fat. The pancreas was edematous in threonine- or histidine-deficient rats, and the acinar cells had decreased zymogen granules. In methionine-deficient animals the pancreas was normal (Sidransky and Farber, 1958a,b; Lyman and Wilcox, 1963; Sidransky and Verney, 1969; Harper et al., 1970). If rats were force-fed diets deficient in valine, tryptophan, leucine, isoleucine, or lysine, there was significant reduction in activities of pancreatic enzymes (Lyman and Wilcox, 1963).

Measurement of plasma-free amino acids can be used to study amino acid requirements. If rats, fed a diet that contains all but one of the essential amino acids in adequate quantities, are supplemented with increasing amounts of the deficient amino acid, the blood content of the supplemented amino acid shows an inflection point and then rises sharply when the requirement is reached. This method was used to determine the tryptophan requirement of growing and adult male rats. In young rats the requirement for maximal growth was 0.14%; older rats (300 g) required 0.05% to 0.09%. On the basis of plasma amino acid content, the requirement was 0.11% for young rats and 0.04% for older rats (Young and Munro, 1973).

Another method for determination of amino acid requirement is measurement of metabolism or oxidation of specific amino acids. The latter method assumes, as does the carcass retention method, that as dietary supply approaches and exceeds the requirement, amino acids will be used increasingly for lipogenesis, gluconeogenesis, excretion, and oxidation. For example, rats were fed a sesame seed protein diet supplemented with lysine. Average daily gain and food efficiency ratio indicated a lysine requirement of approximately 0.7%; serum lysine analysis indicated a requirement of 0.8%. Rats were then injected with radioactive lysine, and their expired CO_2 was collected

and measured. At low lysine intakes, little was oxidized, but oxidation rose as dietary lysine increased beyond 0.6% (Brookes et al., 1972).

Interactions between amino acids or between amino acids and other nutrients influence requirements. One-third to one-half the requirement for phenylalanine can be met by tyrosine. The tryptophan requirement (0.15%) assumes an adequate dietary content of niacin. Of the total requirement for methionine and cystine (0.6%), up to one-half can be supplied by L-cystine. Both total sulfur amino acids and the relative proportions supplied by methionine and cystine affect food intake, growth, and body composition of rats. Weight gain and food efficiency ratios were equivalent in rats fed 0.4% to 0.7% methionine in diets that contained 0% to 0.6% cystine, respectively (Salmon and Newberne, 1962). Rats fed 0.2% to 0.8% methionine and no cystine had increased weight gain up to 0.6%, with no further improvement at 0.8% methionine. Addition of cystine improved food intake and weight gain at all levels, but the effect was not marked at 0.6% or 0.8% methionine (Shannon et al., 1972).

Methionine supplies methyl groups for synthesis of choline, creatine, carnitine, nucleic acid, and histones. It is metabolized more extensively than are other essential amino acids; almost twice as much methionine carbon is recovered as CO_2 as is carbon from other essential amino acids. Turnover of the methyl carbon is high before methionine is incorporated into protein. Cystine supplementation affects several enzymes of methionine metabolism in rats fed a diet low in methionine (Aguilar et al., 1974; Benevenga, 1974).

Arginine, in addition to being a constituent of protein, is required for transport, storage, and excretion of nitrogen. It is required for growth of young rats but may not be required for maintenance of adult rats. Weanling rats, fed amino acid diets complete except for arginine, had depressed food intake and growth, although a small positive nitrogen balance was maintained. Urinary excretion of amino acids and urea was increased, whereas excretion of ammonia and creatinine was decreased. Older male rats (150 to 175 g) fed the arginine-deficient diet had slightly decreased food intake and no significant alteration of weight gain but had increased excretion of urea, orotic acid, and citric acid. Activities of hepatic urea cycle enzymes were increased in arginine deficiency, presumably because of increased catabolism of other amino acids. Increased urinary orotic acid may be derived from shunting of ammonia into carbamyl phosphate and pyrimidine synthesis (Milner et al., 1974).

Specific effects of tryptophan deficiency on eye development have been reported, but may have resulted from combined deficiencies. Female rats fed diets deficient in several amino acids produced young with a 30% incidence of cataracts. Supplementation with either L-tryptophan or

DL-α-tocopherol prevented the abnormality. In subsequent studies, rats were fed an amino acid diet deficient only in tryptophan. Dams had decreased weight gain and feed efficiency, and offspring had decreased weight gain but did not have cataracts unless the diet was also deficient in α-tocopherol. Unilateral or bilateral lenticular opacities occurred in 33% of offspring of mothers deficient in both nutrients and in 6% of offspring of mothers deficient only in α-tocopherol, but opacities did not occur in offspring of mothers deficient only in tryptophan (Bunce and Hess, 1976).

Amino acids may have specific functions other than those related directly to protein synthesis. For example, intake and plasma content of tryptophan are closely related to brain content of serotonin and therefore to neurotransmission in the brain. This relationship may be responsible for integration of information about metabolic state and food intake. Administration or endogenous secretion of insulin lowers blood concentrations of glucose and most amino acids, but plasma tryptophan is increased, as are brain tryptophan and serotonin concentrations (Fernstrom, 1976). Elevated brain tryptophan and serotonin depend not only on plasma tryptophan but also on the relative content of other neutral amino acids that share the brain transport system with tryptophan: tyrosine, phenylalanine, leucine, isoleucine, and valine. Histidinemia produced by intraperitoneal injections of histidine into rats inhibited the transport of tryptophan and the production of serotonin in the brain, resulting in mental retardation (Aono, 1985). With protein intake there is no increase in brain tryptophan or serotonin because most proteins contain the neutral amino acids in addition to tryptophan, and all compete for entry into the brain (Fernstrom, 1976). Administration of tryptophan-enriched diets to female rats throughout pregnancy and lactation retarded serotonergic innervation in the cortex and brain stem of their offspring (Huether et al., 1992).

Amino acids fed at concentrations greater than required may be toxic (Harper et al., 1970). Methionine, which is required at slightly less than 1% of the diet, can be toxic at concentrations as low as 2%; toxicity is manifested by growth depression and tissue damage. Adaptation occurs with prolonged intake; expiration of labeled CO_2 from labeled methionine is decreased in rats fed 3% methionine for several days compared with rats fed the diet for the first time. Supplementation of the high-methionine diet with glycine or serine is protective; the amino acids may increase metabolism of the methyl carbon. Incorporation of the methyl carbon into phospholipid choline is most important at low methionine concentration, but at higher concentrations conversion to *S*-methyl-L-cysteine may occur. *S*-Methyl-L-cysteine is toxic and results in growth depression and anemia, signs associated with methionine toxicity (Benevenga, 1974). Some adaptation to methionine excess appears to involve increases in synthesis of cystathionine, increased flow through the betaine reaction, and markedly increased metabolism of homocysteine (Finkelstein and Martin, 1986). Long-term consumption (6 to 9 months) of excess methionine (1.6% of the diet) slowed growth rate of male rats and increased hepatic concentrations of iron, ferritin, and thiobarbituric-reactive substances (Mori and Hirayama, 2000).

B. Energy

Food energy is expressed in calories or joules. A calorie is the amount of energy required at one atmosphere of pressure to raise the temperature of 1 g of water from 14.5°C to 15.5°C (NRC, 1981). The joule (J), the preferred unit of the Le Système International d'Unites (SI; International System of Units), is often used alternatively. One calorie is equal to 4.184 J.

When a foodstuff is completely oxidized in a bomb calorimeter to CO_2 and water, the energy released is known as gross energy (GE). The GE of a foodstuff can be expressed in kilocalories per gram. Average GE concentrations in carbohydrates, proteins, and fats have been estimated to be 4.1, 5.6, and 9.4 kcal/g, respectively (Mayes, 1996a).

However, not all of the GE in food is available to the animal owing to losses in digestion and metabolism. The GE of a foodstuff minus the GE contained in the feces equals the apparent digestible energy (DE). Apparent DE is a function of diet composition, the amount of diet consumed, and the degree to which the diet is digested in the gastrointestinal tract. High-fiber dietary components (essentially plant cellulose) are better utilized by animals with significant gastrointestinal microbial fermentation capacity. As rodents depend on endogenous gastrointestinal digestive enzymes for most digestive processes, diets high in fiber are not well utilized by rats and fiber is frequently used to dilute the energy density of rodent diets.

Apparent ME of a foodstuff is equal to GE, minus fecal GE, minus urine GE, and minus GE of combustible gases (as a consequence of digestive processes). In most cases, gaseous GE loss is largely in the form of methane from microbial fermentation in the foregut or hindgut. Gaseous energy loss is not significant for rats and is often disregarded (Lloyd et al., 1978).

The system most widely applied to estimate the apparent ME of foodstuffs has been calculated by use of physiological fuel values. This system makes use of Atwater constants to apply kilocalorie values to the energy providing components in diets, carbohydrates, protein, and fat (Merrill and Watt, 1955). Under this system, the physiological fuel values, 4 kcal/g for carbohydrates and protein and 9 kcal/g for fat, give reasonable approximations of apparent ME in the composition of rodent diets. The fiber components, cellulose and hemicellulose, are generally considered to provide no available energy for rodents.

The rat requirements for dietary energy to support the most basic life functions (i.e., vital cell activity, respiration, cardiovascular distribution of blood) can be expressed in terms of metabolic body size, which is commonly referred to as basal metabolic rate (BMR).

Basal requirements may be considered equivalent to the maintenance energy requirement when animals are in metabolic equilibrium. Ideally, BMR defines basal energy requirement when the animal is in a post-absorptive state and housed in a thermoneutral environment (Curtis, 1983).

Kleiber (1975) established the concept of metabolic body size as a power function of BW (BW^n) and determined that BMR of fasting adult animals, varying in BW from mice (0.021 kg) to cattle (600 kg), could be expressed as: BMR in kcal/day = $70 \times BW_{kg}^{0.75}$. When species and sex classifications of animals are disregarded, then Kleiber's $BW_{kg}^{0.75}$ is useful although no term describes all physiological states (Thonney et al., 1976). The $BW_{kg}^{0.75}$ value, however, is suitable for use with male and female rats. An accurate prediction of the maintenance energy requirement of the rat, however, requires consideration of sex, age, reproductive status, the amount of dietary intake, health status, and body composition. The data suggest that maintenance energy requirements of the rat will be met in most cases by dietary intake of 112 kcal $ME/BW_{kg}^{0.75}$/day (470 kJ/$BW_{kg}^{0.75}$/day) (NRC, 1995).

In rats, approximately 60% to 75% of the ME supplied by the diet is used to meet maintenance (BMR) requirements (Lloyd et al., 1978; Curtis, 1983). About 5% to 10% is used to support events associated with diet digestion (Forsum et al., 1981; Mayes, 1996). Caged animals generally would require additions of only 13% to 35% to the maintenance requirement for activity (Lloyd et al., 1978; Scott, 1986).

Because dietary energy requirements may differ among various rat strains, daily requirements for growth are difficult to estimate accurately. Body composition during growth and weight gain significantly influences required dietary energy inputs. The mass-specific BMRs of rapidly growing animals are higher than are those of adults, and as a consequence, the energy to support rapid growth and development may reach three to four times that of the adult (Clarke et al., 1977; Scott, 1986). During the 4-week growth period after weaning at 21 days of age, the average daily energy requirement is at least 227 kcal $ME/BW_{kg}^{0.75}$/day (950 kJ/$BW_{kg}^{0.75}$/day) (NRC, 1995).

The gestational energy requirement may be 10% to 30% greater than that of mature but non-reproductive female rats (Rogers, 1979; NRC, 1995). Approximately one-third of the 100 to 201 kcal (420 to 840 kJ) stored during gestation is deposited in fetal tissues. The daily ME requirement of female rats is about 143 kcal $ME/BW_{kg}^{0.75}$/day

(600 kJ/$BW_{kg}^{0.75}$/day) in early gestation and may increase to 265 kcal $ME/BW_{kg}^{0.75}$/day (1,110 kJ/$BW_{kg}^{0.75}$/day) during late gestation (NRC, 1995).

Despite the increased demand for energy during late gestation, rats are generally in negative energy balance during peak lactation and maternal adipose stores are mobilized to meet the energy requirement of lactation. Lactating rats may have an energy requirement two to four times that of nonlactating females (Rogers, 1979). Although lactation demands will vary owing to litter size, during peak lactation the dam's daily ME requirement will be at least 311 kcal $ME/BW_{kg}^{0.75}$/day (1300 kJ/$BW_{kg}^{0.75}$/day) (NRC, 1995).

The energy for maintenance and growth can be met by diets with a wide range of energy densities, and rats can adjust their intake to meet their energy requirements when they are fed AL. However, rapidly growing weanlings require an energy density of at least 3 kcal ME/g (Rogers, 1979). Generally, purified or chemically defined diets that contain 10% fat contain 4 to 4.5 kcals GE/gram. Of this total, 90% to 95% is DE; ME varies from 90% to 95% of DE (NRC, 1995). In purified diets to which cellulose is added or in natural ingredient (chow) diets, the digestibility may be somewhat lower, between 75% and 80% (Roe et al., 1995; Duffy et al., 2001).

The primary consequence of energy excess is obesity, which is associated with decreased life span, increased incidence and severity of degenerative diseases, earlier onset and incidence of neoplasia (McCay, 1935; Yu et al., 1985; Weindruch and Walford, 1988; Keenan et al., 1994; Masoro, 1996), and increased variability in statistical interpretation of research results (Seng et al., 1998). The strain of rat can influence maintenance, growth and aging rate, feed utilization efficiency, metabolic rate, and tendency for obesity. The daily ME intakes of several rat strains fed AL or diet restricted and compared with the predicted maintenance energy requirement for adult animals, 112 kcal $ME/W_{kg}^{0.75}$/d (NRC, 1995), are presented in Table 9-1. Data from various sources were adapted for this presentation by (1) adjusting purified diets to 95% digestibility based on ingredient composition (Watt and Merrill, 1963); (2) adjusting natural ingredient diets to 80% digestibility (Duffy et al., 2001) except for the diet of Roe et al. (1995), which was 75% digestible; and/or (3) calculating dietary ME by physiological fuel values, also based on ingredient composition. Generally, average daily ME intake, either AL or diet restricted, was lower for all strains of male and female rats fed purified diets compared with natural ingredient diets. Daily ME intake for female rats fed AL was slightly greater than was the predicted maintenance energy requirement; ME intake of female rats fed 40% diet restricted was slightly lower than was the predicted requirement (NRC, 1995). Conversely, daily ME intake of male rats fed AL or diet restricted was at or slightly below the predicted requirement; these data are in

TABLE 9-1

Daily Metabolizable Energy Intake (kcal $ME/BW_{kg}^{0.75}$) of Rats Fed *Ad Libitum* (AL) or Diet Restricted (DR) Compared to the Estimated Adult Maintenance Energy Requirement of 112 kcal $ME/BW_{kg}^{0.75}$/day (NRC, 1995)

Intake level	Rat strain	Sex	Diet type[a]	Energy intake (kcal $ME/BW_{kg}^{0.75}$/day)[2] Age, 6 mo	Age, 24 mo	Citation
AL	Fischer 344	M	NI	119	106	Duffy et al., 2001
	Fischer 344	M	NI	142	115	Turturro et al., 1999
	Sprague-Dawley	M	NI	125	111	Duffy et al., 2001, 2003
	Brown Norway	M	NI	120	103	Turturro et al., 1999
	BN × F344[b]	M	NI	124	101	Turturro et al., 1999
	Wistar	M	NI	145	128[c]	Roe et al., 1995
	Sprague-Dawley	M	SP	110	83	Lewis et al., 2003
	Fischer 344	M	SP	106	102	Yu et al., 1985
	Fischer 344	M	SP	124	99	Turturro et al., 1999
	Fischer 344	F	NI	154	123	Turturro et al., 1999
	Brown Norway	F	NI	144	133	Turturro et al., 1999
	BN × F344	F	NI	137	122	Turturro et al., 1999
	Wistar	F	NI	157	150[c]	Roe et al., 1995
	Fischer 344	F	SP	136	107	Turturro et al., 1999
DR[d]						
10%	Fischer 344	M	NI	121	108	Duffy et al., 2001
10%	Sprague-Dawley	M	NI	121	109	Duffy et al., 2001, 2002
25%	Fischer 344	M	NI	107	101	Duffy et al., 2001
25%	Sprague-Dawley	M	NI	111	103	Duffy et al., 2001
31%	Sprague-Dawley	M	SP	94	82	Duffy et al., 2002
40%	Fischer 344	M	NI	96	87	Turturro et al., 1999
40%	Fischer 344	M	NI	96	98	Duffy et al., 2001
40%	Fischer 344	M	NI	102	100	Turturro et al., 1999
40%	Sprague-Dawley	M	NI	99	95	Duffy et al., 2001
40%	Brown Norway	M	NI	94	93	Turturro et al., 1999
40%	BN × F344	M	NI	98	99	Turturro et al., 1999
40%	Fischer 344	M	SP	91	92	Yu et al., 1985
40%	Fischer 344	F	NI	101	98	Turturro et al., 1999
40%	Fischer 344	F	NI	102	106	Turturro et al., 1999
40%	Brown Norway	F	NI	111	108	Turturro et al., 1999
40%	BN × F344	F	NI	99	102	Turturro et al., 1999

[a]NI = naturalingredient diet; SP = semipurified diet.
[b]Brown Norway × Fischer 344 hybrid.
[c]Data for 15 months of age.
[d]Percentage diet restriction compared to *ad libitum* controls.

agreement with those of Keenan et al. (1997) who found similar patterns of intake for Sprague-Dawley rats fed AL or a restricted diet.

C. Fat

Carbohydrates and lipids serve as the primary energy sources in the diet in order to spare protein. Dietary fats are a significant source of calories in the diet and supply twice the energy provided by carbohydrates and protein; that is, the comparative physiological fuel values are 4 kcal/g for carbohydrates and protein, and 9 kcal/g for fat. Diets for rats generally contain between 5% and 15% fat by weight (Rogers, 1979). A level of 5% to 6% dietary lipid is recommended for both males and females during rapid growth and for adult females during reproduction

and lactation (NRC, 1995). This amount is also satisfactory for absorption of carotene and vitamin A. Many lipids provide sufficient essential fatty acids (EFAs) when included in the diet at the 5% concentration.

Dietary lipids provide EFAs that function to (1) promote growth; (2) prevent or alleviate skin abnormalities; (3) maintain a normal ratio of phospholipids to triglycerides in tissues; (4) promote prostaglandin formation; (5) maintain normal ratios of polyunsaturated fatty acids (PUFAs), which are required for synthesis of tissue lipids and cellular membranes; and (6) provide for the normal absorption and utilization of the fat-soluble vitamins. Signs of EFA deficiency in rats are reduced growth rate with a growth plateau 12 to 18 weeks after weaning onto a deficient diet, scaling and reduced thickness of the skin, roughening and thinning of hair, necrosis of the tail, fatty liver, and renal damage (Rogers, 1979; NRC, 1995). Failure of reproductive function occurs in both sexes, although males have a greater EFA requirement than do females. The classical signs of EFA deficiency appear to be ameliorated by n-6 PUFA (NRC, 1995).

Monounsaturated fatty acids contain only one double bond. Of these, oleic acid is the most common in foods. Olive oil, canola oil, peanut oil, peanuts, pecans, almonds, and avocados have concentrated amounts of monounsaturated fatty acids.

The PUFA contain two or more double bonds. There are two main families of PUFA, n-3 (linolenic acid, 18:3 n-3) and n-6 (linoleic acid, 18:2 n-6), which have different biochemical roles. In fatty acid nomenclature, using the example 18:2 n-6, the number before the colon indicates the number of carbon atoms in the chain (18), the number following the colon indicates the number of double bonds (2), and the number following the "n" indicates the position of the first double bond from the methyl end (between the sixth and seventh carbon atoms) (Mayes, 1996b).

Of the three PUFA that have been considered essential (linoleic, n-6; linolenic, n-3; arachidonic, n-6), linoleic is the most readily available in foods and can be converted by rats to arachidonic acid, which is the major EFA in membranes. The rat requires fatty acids from the n-6 family as a component of membranes, for optimal membrane-bound enzyme function, and for prostaglandin formation (NRC, 1995). The presence of n-3 fatty acids in specific tissues (retina, cerebral cortex, testis, and sperm) and the tendency for those tissues to sequester these fatty acids have led to the speculation that n-3 fatty acids are required for some functions and may be required in the diet of rats (NRC, 1995). Based on tissue saturation of 20:4 n-6 and 22:6 n-3, it was determined that 12 g of linoleic acid and 2 g of α-linolenic acid per kilogram of diet were the minimal requirements for the rat (Bourre et al., 1989, 1990).

Oils derived from a variety of seeds (corn, cottonseed, soybeans, and peanuts) contain 50% or more linoleic acid. Corn oil can contain as much as 63% linoleate; hydrogenated vegetable fat, approximately 30%. Lard, butter, and beef tallow contain between 2% and 10% linoleic acid (Rogers, 1979). Soybean and cottonseed meals and starches from several sources may contain significant levels of EFAs. The recommended amounts of n-3, n-6, and their appropriate EFA ratio can be met by the addition of soybean oil in the diet (NRC, 1995). Soybean oil contains about 14% saturated fatty acids, 23% monounsaturated fatty acids, 51% linoleic acid, and 7% linolenic acid.

D. Carbohydrates

Carbohydrates are important dietary sources of energy, and glucose or glucose precursors (e.g., other sugars, glycerol, glucogenic amino acids) in small amounts may be required for optimal energy metabolism. Growth of young male rats was not supported by diets containing 90% of ME from fatty acids and 10% from protein, but growth was allowed by substitution of soybean oil for fatty acids or by addition of glycerol equivalent to that in the soybean oil. However, growth rates were still not equal to those achieved on a 78% starch diet (Konijn et al., 1970; Carmel et al., 1975). When rats were fed carbohydrate-free diets with 80% of ME from fatty acids and 20% from protein, weight gain occurred; however, weight gain was greater when the diet was supplemented with glucose or neutral fat (Goldberg, 1971; Akrabawi and Salji, 1973). Low-protein (10% of ME), carbohydrate-free diets resulted in rats that were hypoglycemic with abnormal glucose tolerance curves (Konijn et al., 1970; Carmel et al., 1975). When fed higher-protein (18% of ME), carbohydrate-free diets, blood glucose concentrations were normal, but glucose tolerance curves were still somewhat abnormal (Goldberg, 1971). Replacement of fatty acids with neutral fats in carbohydrate-free diets containing 20% of ME from protein resulted in increased weight gain in meal-fed (once per day) but not in AL-fed rats (Akrabawi and Salji, 1973). When dietary fat-to-carbohydrate ratios (ME basis) ranged from 0.2 to 1.4, heat increment was constant at 47.5% of ME, suggesting that carbohydrate and fat may be used with equal efficiency under these circumstances (Hartsook et al., 1973).

The rat can use a number of dietary carbohydrates for growth, including glucose, fructose, sucrose, maltose, dextrins, and starch (NRC, 1995). However, the metabolism of fructose (preformed or from sucrose) is mediated by fructokinase and aldolase B, thus bypassing phosphofructokinase control of glycolysis and increasing the glycolytic flux. Use of fructose or sucrose as the principal dietary carbohydrate increases liver weight, concentrations of liver lipid and glycogen, and activities of the liver lipogenic enzymes glucose-6-phosphate dehydrogenase, malic enzyme, adenosine triphosphate (ATP) citrate lyase, and fatty acid synthetase (Worcester et al., 1979; Narayan and McMullen, 1980; Michaelis et al., 1981; Cha and Randall, 1982; Herzberg and Rogerson, 1988a,b). Increased hepatic synthesis (Herzberg and Rogerson, 1988b) and/or decreased peripheral clearance (Hirano et al., 1988) of triglyceride may account for the hyperlipidemia associated with fructose feeding. Nephrocalcinosis and increases in kidney weight have been seen when rats were fed diets containing 55% sucrose (Kang et al., 1979) or 63% fructose (Koh et al., 1989). Starch was more easily metabolized than was sucrose when rats were fed low-protein (12.5% casein) (Khan and Munira, 1978) or protein-free diets (Yokogoshi et al., 1980). High-sucrose diets also may exacerbate EFA deficiencies (Trugnan et al., 1985).

Rats fed lactose or galactose have exhibited poor performance and cataracts (Day and Pigman, 1957). When fed either α- or β-lactose, weanling rats exhibited diarrhea (Baker et al., 1967). Rats fed 15% or more xylose exhibited diarrhea and lens opacities (Booth et al., 1953). Sorbose in the diet may supply significant energy, perhaps from hindgut fermentation, but tends to decrease feed intake and growth rate (Furuse et al., 1989). Rats fed a carbohydrate-free diet had improved growth when up to 8% mannose was added, suggesting that at least low concentrations can be metabolized (Keymer et al., 1983). A bond isomer of sucrose, leucrose (D-glycosyl-α[1–5]D-fructopyranose), appears to be metabolized as well as is sucrose (Ziesenitz et al., 1989). Sorbitol can be metabolized by rat liver (Ertel et al., 1983). Lactitol and xylotol, when added to the diet at 16% of dry matter, decreased feed intake and growth rate, although some adaptation appeared to occur within 2 weeks (Grenby and Phillips, 1989).

The qualitative requirement for carbohydrate for successful reproduction has been studied by using a carbohydrate-free, low-protein diet (4.25 kcal ME/g; 12% of ME from protein). Although 78% of embryos were normal after 6 days, only 25% and 0.6% were normal after 8 and 10 days, respectively. All embryos had been absorbed by 12 days (Taylor et al., 1983). A carbohydrate-free diet (4.11 kcal ME/g) containing 10% of ME from protein did not maintain pregnancy to term unless supplemented with 4% glucose (or glycerol equivalent). Six percent to 8% glucose was required to support normal maternal weight gain and normal fetal weight. Twelve percent glucose was required to produce fetal liver glycogen concentrations half as large as those in controls fed a 62% carbohydrate diet (Koski et al., 1986a). When pregnant rats were fed a low-protein (10% of ME) diet, containing 6% or less glucose from gestation day 9 through lactation day 7, no pups survived. When fed 8% or 12% glucose, pup survival at lactation day 7 was 6% and 30%, respectively. Pups from control dams fed 62% glucose had a 93% survival rate (Koski and Hill, 1986a). When a high-carbohydrate diet was fed from the final two gestation days through lactation to dams previously fed a low-protein (10% of ME), 4% glucose diet, pup survival was much improved (Koski and Hill, 1990b). Lower concentrations of lipid and carbohydrate were found in the milk produced by lactating rats fed 6% glucose diets, and there was an association with retarded postnatal growth of the pups (Koski et al., 1990a,b). Four percent fructose diets may support pregnancy, but fetal liver glycogen concentrations during lactation were not supported as well by 4% fructose, or equivalent glycerol, as by 4% glucose (Fergusson and Koski, 1990). Pregnant rats fed sucrose-based diets (61.5%) were more likely to exhibit EFA deficiencies than were those fed glucose-based diets (Cardot et al., 1987).

E. Fiber

Fiber has not been shown to be a dietary essential for the rat as are the amino acids, fatty acids, minerals, vitamins, and water discussed in this chapter. However, depending on the properties of the fiber source (i.e., solubility, fermentability, viscosity), digesta transit time, fecal bulk, and gastrointestinal health may be influenced. Analytical procedures for fiber are still under development despite a long history (DeVries et al., 1999). This continuing search for more meaningful analyses is driven by the variability and complexity of fiber and recognition that certain compounds have unique physiological significance. Reviews of methodology and their limitations and advantages have been published by Englyst and Cummings (1990), Spiller (1992), and Van Soest (1994).

Crude fiber is the insoluble organic residue remaining after sequential treatment of samples with acid and alkali to mimic digestion in the stomach and intestine. Thus, crude fiber was intended to represent that fibrous fraction of the plant cell that was indigestible. Unfortunately, the procedure results in significant solubilization of hemicelluloses and lignin, thus seriously underestimating the fiber content (Englyst and Cummings, 1990; Spiller, 1992; Van Soest, 1994). As a consequence, variable proportions of these substances appear in the nonstructural carbohydrate fraction (or nitrogen-free extract) by difference. Hemicelluloses, although carbohydrates, cannot be digested by endogenous enzymes and yield energy to the host only after gastrointestinal fermentation. Lignin is a noncarbohydrate phenolic polymer that cannot be digested by endogenous mammalian enzymes or fermented by gastrointestinal microbes. Thus, its placement in nitrogen-free extract is a serious error.

DeVries et al. (1999) suggested that Hipsley in 1953 may have been the first to use the term "dietary fiber" for the indigestible constituents that make up the plant cell wall. These indigestible constituents were known to include cellulose, hemicelluloses, and lignin, and the term dietary fiber was intended to distinguish more clearly between these indigestible components and components measured as crude fiber. The definition of dietary fiber was subsequently broadened to include "remnants of edible plant cells, polysaccharides, lignin, and associated substances resistant to (hydrolysis) digestion by the alimentary enzymes of humans." Included in dietary fiber were cellulose, hemicelluloses, lignin, gums, modified celluloses, mucilages, oligosaccharides, and pectins and associated minor substances, such as waxes, cutin, and suberin. After successful collaborative studies, AOAC Method 985.29 (1995a) and AACC Method 32-05 (1995a) were officially declared defining procedures for quantifying dietary fiber. Further modifications to separate total, soluble, and insoluble fiber were adopted as AOAC Method 991.43 (1995b) and AACC Method 32-07 (1995b). A reference standard with analytical values for these fractions is now available (Caldwell and Nelson, 1999).

Despite this progress, analytical problems in defining dietary carbohydrates and fiber persist (Delcour and Erlingen, 1996). Starch that is resistant to hydrolysis by digestive enzymes has physiological effects in rats that make it comparable to dietary fiber. The formation, structure, and properties of enzyme-resistant starch have been reviewed (Erlingen and Delcour, 1995), and its physiological properties have been described (Annison and Topping, 1994). Type I resistant starch is physically trapped within the food matrix. For example, starch granules within cell contents may be physically separated from amylolytic digestive enzymes by an unbroken cell wall. Enzymatic digestion will proceed if the cell wall is ruptured by chewing or by food processing such as grinding. Type II resistant starch is native granular starch that is resistant to enzymatic digestion because of its compactness and partial crystalline structure. This resistance can be overcome by gelatinization (heating in the presence of water to disrupt hydrogen bonding and destroy crystallinity). Type III resistant starch is formed during retrogradation (recrystallization), primarily of amylose, although retrogradation of amylopectin also may be involved.

The implications of the above for "accuracy" of the current AOAC/AACC methods depend on the intent to include or not include resistant starch in the dietary fiber residue. Type I resistant starch generally *would not* be included in the dietary fiber residue because of destruction of type I enzyme resistance during grinding of the sample in preparation for the analysis. Type II resistant starch *would not* appear in the dietary fiber residue because the temperature (100°C) to which it is exposed during the analysis results in gelatinization, and it would be hydrolyzed by the added heat-stable α-amylase. Type III resistant starch, consisting of retrograded amylopectin, generally *would not* be included in the dietary fiber residue because heating to 100°C would destroy most or all of its enzyme resistance. Retrograded amylose *would* be included in the dietary fiber residue because its enzyme resistance would not be destroyed until it reached a temperature of about 150°C, which is above the temperature used in the analysis.

It is apparent that progress is being made in defining the physiologically functional components of dietary fiber in human foods, but few total dietary fiber determinations have been made on the foods consumed by other animals. Except for the crude fiber values required by regulatory agencies on commercial feed labels, most measurements of fiber in these foods have been expressed as neutral-detergent fiber, acid-detergent fiber, and acid-detergent lignin, commonly using the procedures described by Van Soest et al. (1991) with the modifications described by Robertson and Horvath (1992). Although this detergent system of analysis does not quantify soluble fibers, quantification of insoluble fibers is comparable to that of the total dietary fiber system just described (Lee et al., 1992; Popovich et al., 1997).

Feeding rats fiber increases fecal bulk and decreases gastrointestinal transit time, and decreases in transit time are more pronounced with insoluble fibers (Fleming and Lee, 1983). Increases in the weight of the cecum and colon are observed when fiber is included in rat diets, with inclusion of cellulose (insoluble fiber) in the diet leading to greater enlargement of the colon. Glucomannan (soluble fiber) led to greater enlargement of the cecum (Konishi et al., 1984).

Increases in weight of the cecal wall occur in rats fed lactulose, a disaccharide fermented in the cecum, suggesting that microbial fermentation plays a role in stimulating this hypertrophy (Remesy and Demigne, 1989). The viscosity of fiber sources also may influence cecal hypertrophy (Ikegami et al., 1990). Fiber as an energy source depends on fermentation in the hindgut. Microbial fermentation results in the production of volatile fatty acids, predominantly acetate, propionate, and butyrate, which are absorbed and can be used as energy sources by the rat. The DE values of cellulose, a relatively unfermentable fiber, and guar, a highly fermentable fiber, were 0 and 2.4 kcal/g, respectively, for the rat. Consumption of guar-containing diets, however, increased heat production by rats such that, despite additional energy supply from guar, there was no additional gain of body energy (i.e., NE = 0) (Davies et al., 1991). It is unknown if this thermogenic effect of guar applies to other fermentable fibers.

Additions of insoluble, undegradable sources of fiber such as cellulose, oat hulls, wheat bran, and corn bran to rat diets at concentrations up to 20% do not affect growth, even though these nonfermentable fiber sources dilute the nutrient density of the diet (Schneeman and Gallaher, 1980; Fleming and Lee, 1983; Lopez-Guisa et al., 1988; Nishina et al., 1991). At high concentrations, viscous polysaccharides such as pectin, guar, and carboxymethyl-cellulose may decrease weight gain, perhaps because of decreased feed intake, especially during initial adaptation (Davies et al., 1991). The effects of pectin in particular are difficult to assess because its properties can vary greatly among sources because of molecular weight and degree of esterification. The more viscous pectins (high-molecular weight and degree of esterification) tend to cause greater decreases in feed intake than do less viscous pectins (Atallah and Melnik, 1982). Delorme and Gordon (1983) observed a 30% decrease in growth of rats when 4.8% pectin was added to diets and a 50% mortality when 28.6% pectin was added. Fleming and Lee (1983) observed a

35% decrease in weight gain when 10% pectin was added to the diet, but Nishina et al. (1991), Thomsen et al. (1983), and Track et al. (1982) found no differences in growth when 5% to 8% pectin was added to purified fiber-free diets. Guar added to diets at 5% of dry matter had no effect on BW (Ikegami et al., 1990), but 8% guar depressed gain (Cannon et al., 1980; Track et al., 1982).

Nitrogen metabolism can be altered by dietary additions of fermentable fiber sources. Fecal nitrogen excretion increases and urinary nitrogen excretion decreases as a result of microbial fermentation in the hindgut. Remesy and Demigne (1989) demonstrated that absorption of ammonia from the hindgut increased when fermentable fiber sources (pectin and guar) were added to the diet, but transfer of urea to the gut was stimulated to a greater extent such that net fecal excretion of nitrogen was increased. The addition of fermentable fiber sources to diets deficient in arginine may improve growth by decreasing the need for arginine for hepatic urea synthesis (Ulman and Fisher, 1983).

Many fiber sources have been used in rat diets, including soybean fiber (Levrat et al., 1991), carrageenan, xanthan, alginates (Ikegami et al., 1990), and gum arabic (Tulung et al., 1987). The effects of these fibers can generally be predicted based on their physical properties and fermentabilities. Some carbohydrates that cannot be properly called fiber also elicit some responses similar to those observed for true fibers. Lactulose (disaccharide), raffinose (trisaccharide), and fructooligosaccharides are not absorbed in the small intestine but are rapidly fermented in the hindgut (Fleming and Lee, 1983; Remesy and Demigne, 1989; Tokunaga et al., 1989). Some starches, particularly raw potato starch, escape small intestinal digestion, are fermented in the cecum, and exert effects similar to true fibers (Calvert et al., 1989).

F. Minerals

The total mineral content of a diet is expressed as "ash" and consists of the residue remaining after a diet sample has been subjected to complete oxidation (AOAC, 1995a,b) The concentrations of individual minerals are not identified in the ash fraction of a diet, but ash content is an indicator of diet quality. A good quality natural ingredient rat diet will contain from 7% to 8.5% ash, whereas a purified diet will contain 3.5% to 4.5%. Estimated mineral requirements established in the NRC reports (NRC, 1995) are for maximum rat growth or reproduction. However, the mineral concentrations in most natural ingredient rat diets are substantially higher than are the published estimated requirements to compensate for the low bioavailability of the chemical forms of minerals in feed ingredients, or to respond to phytates (Maynard et al., 1979) associated with plant products.

Minerals are found in every cell, tissue, and organ. Their functions include serving as obligatory cofactors in metalloenzyme activity, maintaining pH, providing a medium essential for normal cellular activity, conducting nerves, determining the osmotic properties of body fluids, contracting muscles, producing energy, and bestowing hardness to bones and teeth.

1. Macrominerals

The minerals occurring in living tissues in substantial concentrations are commonly referred to as the "macro minerals" to distinguish them from minerals appearing in smaller concentrations and designated as "trace minerals." Historically this distinction was used because of difficulties in the accurate analysis of many of the trace minerals (Underwood and Mertz, 1987). This distinction is still of some practical value because the trace minerals are generally added to diets in premixes and the macro minerals are added as primary ingredients.

CALCIUM.   Calcium is the most abundant divalent cation in the animal body averaging about 1.5% of total BW. More than 99% of the calcium is found in the skeleton and teeth. The remaining calcium is distributed in both the extra- and intracellular-fluids and serves as an intracellular/intercellular messenger or regulator. Calcium mobilization and deposition are influenced by age, diet, hormonal status, and physiological state (Arnaud and Sanchez, 1990).

Calcium absorption occurs in the small intestine and involves two principle transport processes. The calcitriol-dependent calcium transport system requires energy and involves calbindin, a calcium binding protein, which is regulated by calcitriol $(1,25[OH]_2D_3)$. This system is stimulated by ingestion of low-calcium diets and during times of increased calcium requirement such as growth, pregnancy, and lactation. Renal calbindin concentrations were highest in rats fed a diet containing 0.1% calcium and lowest in rats fed a 2.5% calcium diet (Bogden et al., 1992). The second mechanism is nonsaturable, passive, and paracellular. This process is activated when the calcium intake is increased (Groff and Gropper, 2000a). Findings show that in Sprague-Dawley rats intestinal transit time, calcium solubility, and mucosal permeability to calcium determine the rate of passive (paracellular) calcium absorption (Duflos et al., 1995). Calcium absorption is influenced by protein (Orwall et al., 1992), vitamin D (Brown et al., 2002), dietary fiber (Watkins et al., 1992), phytate (Gueguen et al., 2000; Kamao et al., 2000), oxalate (Peterson et al., 1992; Morozumi and Ogawa, 2000), competing divalent cations such as magnesium and zinc (Chonan et al., 1997), unabsorbed fatty acids (Brink et al.,

1995), and type of calcium salt (Pansu et al., 1993; Chonan et al., 1997; Tsugawa et al., 1999). In rats, insoluble calcium carbonate is minimally absorbed in the small intestine, but the large intestine compensates for this insufficient absorption (Shiga et al., 1998). Prebiotics such as inulin, oligofructose, glucooligosaccharide, and galactooligosaccharide have been shown to stimulate calcium absorption in the rat (Scholz-Ahrens et al., 2001).

Three hormones are involved in calcium homeostasis in the blood (extracellular fluid): parathyroid hormone (PTH), calcitriol (1,25[OH]$_2$D$_3$) (Brown et al., 1995), and calcitonin (Tordoff et al., 1998). PTH acts to increase extracellular fluid calcium concentrations through interactions with the kidney and bone. Calcitriol accelerates absorption of calcium from the gastrointestinal tract. Calcitonin counteracts PTH, lowering serum calcium by inhibiting osteoclast activity and preventing mobilization of calcium from bone (Weaver and Heaney, 1999). Rats fed a low-calcium diet (0.03%) had raised circulating concentrations of 1,25(OH)$_2$D and PTH and lowered 25(OH)D$_3$ and Ca^{2+}, but a high-calcium diet (5.46%) raised calcitonin, Ca^{2+}, 25(OH)D$_3$, and 1,25(OH)$_2$D (Persson et al., 1993). Feeding a diet deficient in calcium and vitamin D to weanling rats did not affect calcitonin mRNA concentration, but did result in significant increases in PTH mRNA concentration (Naveh-Many et al., 1992).

Bone mineralization is one of the many functions of calcium. Bone formation continues throughout life; it continues to lose and gain mineral matter via remodeling through the action of osteoblasts, which synthesize the collagen matrix, and osteoclasts, which are stimulated by PTH to reabsorb calcium when needed (Calvo, 1993). Geng and Wright (2001) showed there is an increased sensitivity to dietary calcium deficiency in female rats, involving a significant loss in axial bone mass compared with male rats fed the same diet. Rats fed diets containing 0.2% calcium had 20% lower mineralized bone area and 20% larger medullary cavity area than did rats fed diets containing 0.4% or 0.8% calcium (Kunkel et al., 1990). Low calcium intake (0.25%) by rats through adolescence had a nonreversible, detrimental effect on peak bone mass, whereas higher intakes (0.5% to 1.0%) promoted greater bone mass, providing potential protection from age-related bone loss (Peterson et al., 1995). Increased calcium intake (20 g/kg diet) had no effect on bone mineral composition or bone resorption in growing female Wistar rats, but reduced intake resulted in bone resorption (Creedon and Cashman, 2001). Pregnancy accelerates intestinal calcium absorption and calcium accumulation in Sprague-Dawley rats, but bone density of the lumbar spine decreased during pregnancy (Omi and Ezawa, 2001). Young rats fed 0.1 g/kg dietary calcium for 8 days exhibited slow growth, anorexia, increased BMR, reduced activity and sensitivity, male infertility, poor lactation, osteoporosis, internal hemorrhage, and rear leg paralysis (Boelter and Greenberg, 1941; Boelter and Greenberg, 1943). Estimated calcium requirements for growth and maintenance of nonlactating rats are 5.0 and 3.0 g/kg, respectively (NRC, 1995). It is recommended that during lactation dietary calcium should be increased by 25%.

Skeletal changes were examined in female rats fed diets containing 0.02%, 0.5%, and 1.0% calcium during gestation and lactation. Dams fed the 0.02% calcium diet showed decreased bone area in the vertebrae, increased osteoid surface, and increased osteoblasts. Offspring from these dams incurred spontaneous fractures and increased mortality. Dams maintained on 0.5% calcium showed bone mineral depletion during lactation which was not resolved until 28 days after weaning. Control dams (1% dietary calcium) showed no significant bone mineral depletion or postweaning repletion. Offspring from dams fed the 0.5% calcium diet had a bone mineral content deficit despite being fed the 1.0% calcium diet after weaning (Gruber and Stover, 1994).

The calcium not associated with bone is essential for blood clotting, nerve conduction, muscle contraction, enzyme regulation, and membrane permeability (Groff and Gropper, 2000a). Calcium is an integral part of cell signaling. Ionized calcium is the most common signal transduction element in cells because of its ability to reversibly bind to proteins. Movement of calcium across cell membranes regulates the actions of hormones and neurotransmitters to affect intracellular processes. Increased free Ca^{2+} concentrations in the cell affect cell functions directly or through calcium-binding proteins (Weaver and Heaney, 1999). Increased free Ca^{2+} can activate neutrophils and platelet phospholipase A$_2$. Phospholipase A$_2$ removes fatty acids such as arachidonic acid from phospholipids. Arachidonic acid can be metabolized to form thromboxanes, prostaglandins, or leukotrienes (Berdanier, 1998). Alanko et al. (2003) fed Wistar rats a high-calcium (3%) diet, which decreased prostacyclin production by 30% and increased thromboxane A$_2$ production two-fold.

Findings show that the activities of delta-5, delta-6, and delta-9 desaturases were significantly decreased in rats fed a calcium-deficient diet (Marra and de Alaniz, 2000). Examination of the phospholipase A$_2$ activity in calcium-deficient rats showed the decreased unsaturated fatty acid synthesis was due to alterations in the acylation/deacylation cycles via inhibition of phospholipase A$_2$ (Marra et al., 2002).

The flux of calcium across membranes is facilitated by the calcium binding protein, calmodulin. Calmodulin mediates many of the effects of calcium such as activation of phosphodiesterase, a component of the cyclic adenosine monophosphate (cAMP) second messenger system, and stimulation of renal Ca^{2+} and Mg^{2+} adenosine triphosphatases (ATPases). Calcium is required to activate protein kinase C in another cellular second messenger system called the phosphatidylinositol system. Protein kinase C catalyzes the phosphorylation of proteins involved in glucose transport, gastric acid release, hormone release, and energy-driven processes (Rasmussen, 1986a,b). The association between calcium, cAMP, and phosphatidylinositol signaling systems explains the actions of many hormones and cell regulators such as angiotensin, cholecystokinin, acetylcholine, insulin-like growth factors (IGFs), insulin, and glucagon. Calcium flux between the cytoplasm and mitochondria regulates mitochondrial activity (Berdanier, 1998).

Troponin C is a calcium binding protein found in skeletal muscle. Skeletal muscle, stimulated by the neurotransmitter acetylcholine, triggers increased calcium concentration which binds to troponin C and results in muscle contraction (Rasmussen, 1986a,b).

Calcium interacts with many nutrients both at the absorptive surface of the intestinal cell and within the body. Urinary calcium excretion is decreased with phosphorus, potassium (Grases et al., 2004), magnesium, and boron, but urinary calcium excretion increased with sodium (Faqi et al., 2001), protein (Amanzadeh et al., 2003), and boron plus magnesium (Nielsen and Shuler, 1992). Dietary calcium supplementation has been reported to decrease heme and non-heme iron absorption (Cook et al., 1991). In magnesium-deficient rats, calcium deficiency provided a significant protection against the pro-inflammatory effect of magnesium deficiency (Bussiere et al., 2002). Iron deficiency in weanling Long-Evans rats resulted in decreased total tibia and femur widths, decreased cortical widths, reduced cortical bone area, and decreased bone density. These effects were exacerbated in rats fed a diet deficient in iron and low in calcium (Medeiros et al., 2002). Calcium deficiency during gestation and lactation in Wistar rats produced an increase in zinc utilization that was reflected in the increase of maternal tissue zinc levels and in femur zinc concentration (Weisstaub et al., 2003). In ovariectomized rats, the combination of both isoflavones and calcium supplementation is more protective against the loss of femur and vertebra bone mass density than are isoflavones or high-calcium diet alone (Breitman et al., 2003).

Calcium affects the absorption of fatty acids and thus influences serum lipid concentrations and the fatty acid composition of bile (Groff and Gropper, 2000a,b). Increased calcium in diets fed to rats significantly decreased the solubility of bile acid in the colon and feces, as well as fatty acids in the ileum, colon, and feces (Govers and Van der Meet, 1993). Data show rats fed a high-fat diet containing 1% calcium had total bile acid concentrations similar to that of rats fed low-fat diets (Lupton et al., 1994). Male Wistar rats fed a diet containing 2.5% calcium had increased fecal excretion of dietary lipids in association with decreased weight gain and 29% less carcass fat

compared with controls (Papakonstantinou et al., 2002). Serum lipids were measured in rats fed diets containing 1% cholesterol and calcium concentrations ranging from 0.2% to 2.1%. The findings show there was a dose-dependent decrease in serum total cholesterol, low-density lipoprotein (LDL)-cholesterol, and trigylceride concentrations, as well as increased high-density lipoprotein (HDL)-cholesterol and HDL/LDL cholesterol ratio (Vaskonen et al., 2002). Examination of the effects of feeding an "atherogenic" diet on calcium homeostasis showed male Sprague-Dawley rats developed dyslipidemia, hyperoxaluria, hypercalciuria, dysproteinuria, loss of bone calcium, and nephrocalcinosis (Schmiedl et al., 2000).

Dietary calcium can alter blood pressure in rat models of hypertension. High-calcium diets lower blood pressure, and low-calcium diets elevate blood pressure (Hatton et al., 1993a; Schleiffer and Gairard, 1995). The mechanisms responsible for the effect of calcium on hypertension have not been determined, but a few potential mechanisms have been identified. Spontaneously hypertensive rats (SHR) fed a high-calcium diet (2%) had decreased blood pressure and decreased pressor responses to exogenous norepinephrine compared with that of rats fed the low-calcium (0.1%) diet. The difference in blood pressure was eliminated by α_1–adrenergic receptor blockers, indicating the diet-induced effects were related to α-$_1$-adrenergic activity (Hatton et al., 1993, 1995). Abnormal intestinal regulation of calbindin-D9K by calcitriol and doudenal calmodulin by dietary calcium was observed in SHRs but not normotensive Wistar-Kyoto rats (Roullet et al., 1991). A high-calcium (3%) diet fed to NaCl-hypertensive rats normalized blood pressure and endothelium-dependent and endothelium-independent vasorelaxation, but did not cause hypercalcemia (Kahonen et al., 2003). Dietary calcium (2%) reduced blood pressure, PTH, and platelet cytosolic calcium responses in SHRs (Rao et al., 1994).

Calcium supplementation decreased chenodeoxycholic acid concentration in bile and the lithocholate-to-deoxycholate ratio in feces. This may lower the risk of colon cancer (Lui et al., 2001; Lupton et al., 1996). A low dietary calcium supplement (3.2 g/L of water) fed to Sprague-Dawley rats reduced the number of cancer tumors induced by dimethylhydrazine, increased the number of tumor-free rats, and changed tumor location toward the distal colon (Vinas-Salas et al., 1998).

It has been shown that dietary calcium phosphate can enhance host resistance to intestinal pathogenic bacteria. Fecal excretion of *Salmonella* and the translocation of the pathogen across the intestine to the systemic circulation were decreased in infected rats fed a diet containing a high calcium concentration (Bovee-Oudenhoven et al., 1997a,b). Supplemental calcium phosphate precipitated fatty acids and bile acids in the intestinal lumen and reduced the cytotoxicity of ileal bile acids, providing a trophic effect for endogenous lactobacilli and increasing the rat's resistance to *Salmonella* (Bovee-Oudenhoven et al., 1999).

Nephrocalcinosis in rats, particularly females, is a long-standing health problem that has a dietary etiology. The Sprague-Dawley and Wistar strains appear to be more susceptible than are other strains (NRC, 1995). Rats fed purified diets are more likely to develop nephrocalcinosis than are rats fed natural ingredient diets (Ritskes-Hoitinga et al., 1991). Research shows that a dietary calcium-to-phosphorus molar ratio below 1.3 is associated with nephrocalcinosis (Hoek et al., 1988; Ritskes-Hoitinga et al., 1991; Reeves et al., 1993). There was an increase in the incidence and severity of nephrocalcinosis in weanling female rats fed modified AIN-93 diets containing calcium and phosphorus at the same ratio (1.3) and at increasing concentrations. However, rats fed the AIN-76A diet with a calcium-to-phosphorus ratio of 0.78 had more severe nephrocalcinosis with increasing calcium and phosphorus concentrations (Cockell et al., 2002). Peterson et al. (1996) showed that dietary-induced nephrocalcinosis in female rats is irreversible and is induced predominantly before the completion of adolescence. Adding excess calcium and phosphorus to the diet decreased magnesium absorption and increased the deposition of calcium in the kidney. Dietary calcium gluconate decreased the accumulation of calcium in the kidney and increased the serum magnesium concentration, resulting in less severe nephrocalcinosis compared with that of rats fed a calcium carbonate diet (Chonan et al., 1996). Examination of the short-term effects of feeding a diet with a low calcium-to-phosphorus molar ratio to female rats showed, that after 2 weeks, there was a significant increase in the severity and incidence of nephrocalcinosis that could not be reversed by switching to the control diet (Cockell and Belonje, 2004).

PHOSPHORUS. Approximately 85% of the body's phosphorus is contained in bone, 1% is found in the blood and body fluids, and the remaining 14% is associated with soft tissue. Phosphorus and calcium metabolism are very closely related because they are both involved in bone mineralization (Shapiro and Heaney, 2003). Zeni et al. (2003) showed that regardless of dietary calcium content, the maternal skeleton is affected very little by pregnancy but is significantly affected by lactation. The influence of the response depends on both calcium and the calcium-to-phosphorus molar ratio. There is an optimal calcium-to-phosphorus molar ratio (1.3) for healthy bone formation and the prevention of nephrocalcinosis in the rat (Reeves et al., 1993). Rats fed high-phosphorus diets using polyphosphate salts as the phosphorus source had more severe nephrocalcinosis than did the rats fed the monophosphate salts. These findings show that the phosphorus source as well as concentration affects the development of nephrocalcinosis (Matsuzaki et al., 1999).

Phosphorus is absorbed in its inorganic form. Organically bound phosphorus is hydrolyzed in the small intestine by phospholipase C or alkaline phosphatase before it is absorbed. Phosphorus absorption occurs predominantly in the duodenum and jejunum. Vitamin D (calcitriol) stimulates phosphorus absorption, particularly if the intake is low (Groff and Gropper, 2000a,b). Phytic acid can interfere with phosphorus absorption. Other minerals that can impair absorption include magnesium, aluminum, and calcium.

Phosphorus is an essential component in every cell of the body. Phosphorus is the anion in hydroxyapatite used in bone mineralization, and is a critical element of DNA and RNA, phospholipids, phosphoproteins, adenine nucleotides, guanine nucleotides, and the second messenger systems (Berdanier, 1998). Dietary phosphorus depletion in rats caused increased nitrogen retention, increased urea concentrations in plasma, kidney, and liver; depletion also caused reduced weight gain, decreased food conversion ratio, decreased renal glutamate dehydrogenase activity, and decreased fructose diphosphatase activity, indicating impaired nitrogen and carbohydrate metabolism (Huber and Breves, 1999). Within cells, phosphate functions in acid-base balance as the primary intracellular buffer. The dietary phosphorus requirement for rats is 3.0 g/kg for growth and maintenance, but the requirement may be increased during lactation (NRC, 1995).

MAGNESIUM. Magnesium is the most abundant intracellular divalent cation found in living organisms. Approximately 60% of magnesium in the body is found in bone. Bone magnesium is associated with either phosphorus and calcium as part of the crystal lattice (about 70%) or it is found on the surface (about 30%). Magnesium that is not part of the bone is found in extracellular fluids and in soft tissue, primarily muscle (Shils, 1999).

In the rat, magnesium is absorbed throughout the small and large intestine (Hardwick et al., 1991; Kayne and Lee, 1993). Absorption of magnesium can be affected by dietary phytate, fiber, lactose, and high quantities of unabsorbed fatty acids, as found with steatorrhea (NRC, 1989; Rude, 1993; Rimbach and Pallauf, 1999; Groff and Gropper, 2000a,c). Coudray et al. (2002) showed that magnesium absorption and fecal endogenous excretion were proportional to dietary intake. These findings indicate magnesium absorption is a process of passive diffusion.

Magnesium acts as an allosteric activator or structural cofactor of enzyme activity for more than 300 fundamental metabolic reactions (Shils, 1999). Some of the essential roles for magnesium include oxidative decarboxylation in the Krebs cycle, cofactor for hexokinase and phosphofructokinase in glycolysis, nucleic acid synthesis, protein synthesis, cardiac and smooth muscle contraction, β-oxidation in lipid metabolism, vascular reactivity and coagulation, amino acid activation, nucleic acid synthesis, DNA and RNA transcription, and synthesis of glutathione (GSH) (Groff and Gropper, 2000a). Magnesium is involved in the production of cAMP from adenylyl cyclase, and is involved in the mediation of many hormones, including PTH and the hydroxylation of vitamin D in the liver (Wester, 1987).

In magnesium-deficient rats, enterocytes from the upper jejunum showed increased calcium, iron, zinc, copper, and manganese concentrations (Planells et al., 2000). Magnesium is required for the synthesis of PTH, which maintains calcium homeostasis, and both high calcium and magnesium concentrations will inhibit PTH secretion. Magnesium and calcium use overlapping transport systems in the kidney, which allows competition for reabsorption sites. These minerals work as antagonists in blood coagulation, with magnesium as inhibitor and calcium as promotor. The ratio of magnesium-to-calcium affects muscle contraction (Iseri and French, 1984). Calcium deficiency in magnesium-deficient rats induced hypocalcemia and protected against the pro-inflammatory response to magnesium deficiency (Bussiere et al., 2002). Magnesium inhibits phosphorus absorption (Fine et al., 1991) and influences the balance between extracellular and intracellular potassium (Wester, 1987). Magnesium deficiency results in decreased manganese concentration in plasma, lung, liver, spleen, kidney, testis, and bone (Kimura et al., 1996). Magnesium deficiency decreases selenium absorption, altering selenium tissue concentrations and GSH peroxidase (GSH-Px) activity (Jimenez et al., 1997; Zhu et al., 1993). The estimated requirement for rat growth and reproduction is 0.5 and 0.6 g/kg dietary magnesium, respectively (NRC, 1995).

In the rat, acute magnesium deficiency signs are different from other species. The signs include hyperemia of ears and feet owing to histamine release from basophils, hyperirritability associated with typically fatal tonic-clonic convulsions, high serum calcium concentration with low inorganic phosphate, and decreased PTH in response to hypercalcemia (Alcock and Shils, 1974; Shils, 1999). Magnesium deficiency in young rats results in hypomagnesemia, slow growth rate, alopecia, and skin lesions. Signs of chronic deficiency in the rats are edema, hypertrophic gums, leukocytosis, and splenomegaly (Alcock et al., 1973). Crystalluria in the kidney and degenerative changes in the kidney, muscle, heart, and aorta are found in magnesium-deficient rats (Heggtveit, 1969; Whang et al., 1969). Hypomagnesemic, hypercalcemic rats were resistant to the calcemic effect of vitamin D and its metabolites calcidiol and calcitriol (Carpenter et al., 1987). Magnesium deficiency in rats impaired pyridoxine status by inhibiting alkaline phosphatase, which is required for the uptake of pyridoxal phosphate (PLP) by tissues (Planells et al., 1997).

Impaired bone growth, decreased bone formation, increased bone resorption, osteoporosis, and increased skeletal fragility have been reported in magnesium-deficient rats (Carpenter et al., 1992; Kenney et al., 1994; Rude et al., 1999). Bone loss as a result of magnesium deficiency in rats is associated with increased release of substance P and tumor necrosis factor-α (Rude et al., 2004). Chronic excessive magnesium supplementation is harmful, but suboptimal levels, provided to produce a moderate magnesium deficiency, were reported to benefit some parameters of bone health in rats (Riond et al., 2000).

Magnesium deficiency in rats results in increased plasma/serum triglyceride and phospholipid concentrations, altered cholesterol levels, changes in fatty acid concentrations, and modifications in lipoprotein levels (Cunnane et al., 1985; Geuex et al., 1991). Hyperlipidemia in magnesium-deficient rats was associated with increased oxidation of LDLs and very low density lipoproteins (VLDLs). Magnesium-deficient rats also had increased lipid oxidation in the muscle, liver, and heart (Rayssiguier et al., 1993). Dietary magnesium deficiency in rats caused decreases in myocardial phospholipid and carbohydrate concentrations (Altura et al., 1996). Magnesium deficiency resulted in significant decreases in superoxide dismutase (SOD) and catalase in rat heart (Kumar and Shivakumar, 1997). Recent findings suggest that the oxidative stress and hypertriglyceridemia reported in magnesium-deficient rats are caused by the inflammatory response that occurs during magnesium deficiency (Rayssiguier et al., 2001).

Sucrose feeding in magnesium-deficient rats was associated with higher concentrations of plasma triglycerides and higher susceptibility to lipid peroxidation of heart and liver tissue compared with levels for magnesium-deficient rats fed a starch diet (Busserolles et al., 2003). Hans et al. (2003) showed that alloxanic diabetes is associated with decreased magnesium status and increased oxidative stress, and that magnesium supplementation can partially restore the antioxidant parameters and decrease the oxidative stress in experimental diabetic rats.

Plasma and tissue prostanoid concentrations were significantly higher in magnesium-deficient rats compared with controls. Geuex et al. (1991) suggested these findings indicate that magnesium depletion inhibited adenylyl cyclase activity, thus lowering cAMP levels and allowing increased cyclooxygenase activity and stimulation of prostanoid synthesis.

Research shows that magnesium deficiency in rats increases blood pressure (Laurant et al., 2000; Touyz et al., 2002; Schooley and Franz, 2002), and magnesium supplementation prevents the development and severity of hypertension (Touyz and Milne, 1999; Kh et al., 2000). Magnesium depletion in stroke-prone spontaneously hypertensive rats (SPSHR) was associated with increased vascular superoxide anion, and phosphorylation of mitogen-activated protein kinases was increased significantly (Touyz et al., 2002). Research findings show magnesium supplementation of deoxycorticosterone-acetate salt to hypertensive rats prevents hypertension by inhibiting tissue endothelin-1 activity and/or production (Berthon et al., 2003).

POTASSIUM. Potassium is the principle intracellular cation. Approximately 98% of potassium resides in the cell. However, potassium concentration in the extracellular fluid is a critical determinant of neuromuscular excitability. Acute changes in extracellular potassium concentration alter vascular smooth muscle potential and tension development as a result of altered vascular smooth muscle sodium pump activity. In the rat, it was shown that chronically decreased extracellular potassium caused chronically decreased sodium-pump activity (Songu-Mize et al., 1987); however, an adaptive change occurs in rats fed high potassium, such that sodium-pump activity remains normal despite elevated extracellular potassium.

Because potassium is an intracellular ion, the number of cells in the body can be determined by using an infusion of the heavy isotope, potassium-^{40}K. The Na^+ K^+-ATPase activity in the inner medullary collecting tube was shown to be modulated by potassium intake (Helou et al., 1994). Almost all potassium is excreted in the urine, with very little in the feces. However, diarrhea can result in significant loss of potassium, affecting the animal's health. Rats fed a diet supplemented with KCl absorbed potassium more efficiently than did those fed diets containing potassium salts of HCO_3 and HSO_4 (Kaup et al., 1991).

Potassium is involved with the contractility of smooth, skeletal, and cardiac muscle, as well as the excitability of nerve tissue. Potassium is also important in maintaining electrolyte and pH balance (Girard et al., 1985). The estimated potassium requirement is 3.6 g/kg diet (NRC, 1995).

Dietary potassium decreases calcium excretion. NaCl can be replaced with KCl because it is not as calciuric as sodium and reduces the excretory rate of calcium. Hyperkalemia is a toxemic condition that can result in cardiac arrhythmias and possibly cardiac arrest. It is almost impossible to produce dietary hyperkalemia in an animal with normal circulation and renal function. Hypokalemia is associated with muscular weakness, nervous irritability, and mental disorientation (Groff and Gropper, 2000a).

Potassium depletion in fasted rats resulted in metabolic alkalosis, reduced plasma insulin, and increased creatine phosphokinase activity, indicating impaired carbohydrate metabolism (Schaefer et al., 1985). Chronic hypokalemia in young growing rats resulted in growth retardation, increased renin-angiotensin system activity, high plasma renin activity, recruitment of renin-producing cells along the afferent arterioles, and downregulation of angiotensin II receptors in renal glomeruli (Ray et al., 2001).

Ong and Sabatini (1999) examined the effects of dietary-induced hypokalemia in young (4 months) and aged nonobese Fisher 344 × Brown Norway F(1) rats (30 months). In the senescent rats, but not among the young rats, hypokalemia caused hyperbicarbonatemia, hyperglycemia, and azotemia. Hypokalemia significantly increased renal brush border and basolateral membrane protein concentration in the young but not the aged rats. In both age groups, hypokalemia resulted in a significant reduction in plasma aldosterone and an increase in sodium concentration. Hypokalemia significantly increased the K^+-ATPase activity in both the cortical basolateral membrane vesicles and in the microdissected proximal convoluted tubule in both age groups.

Feeding SHRs a diet containing 42 g/kg dietary potassium attenuated the elevation of blood pressure induced by ingestion of 8% salt solution (Sato et al., 1991). In Wistar-Kyoto rats, the potassium supplemented diet did not affect blood pressure. A combination of dietary potassium and magnesium supplementation fed to SHRs ingesting a high-sodium diet had beneficial effects against cyclosporine-A induced hypertension and nephrotoxicity (Pere et al., 2000). Nephrocalcinosis was more severe in rats fed a high-phosphorus diet when potassium tripolyphosphate rather than potassium dihydrogenphosphate was the source of the dietary phosphorus (Matsuzaki et al., 2001).

Pamnani et al. (2003) showed that potassium and magnesium have additive effects in preventing the blood pressure increase in reduced renal mass-salt hypertensive rats. Dietary potassium supplementation reduces the incidence of stroke in Dahl rats independently of blood pressure, which may be associated with its enhancing effect on endothelium-dependent relaxations (Raij et al., 1988). Potassium supplementation in Dahl salt-sensitive (DS) rats resulted in increased endothelial nitric oxide production, which appears to be responsible for the improvement in endothelial-dependent relaxations associated with attenuation of hypertension (Zhou et al., 2000). DS and Dahl salt-resistant (DR) rats were fed a high-salt, low-potassium diet which resulted in significantly higher serum levels of $1,25(OH)_2D_3$ and lower serum levels of $25(OH)D_3$ in the DS rats compared with the DR rats. These results indicate a genetic difference in vitamin D metabolism between DR and DS rats, which may directly contribute to hypertension (Wu et al., 2000).

SHRs are resistant to NaCl-induced increase of blood pressure when fed a diet containing 2.1% potassium, but they were susceptible to hypertension when the diet contained 0.5% potassium (Ganguli and Tobian, 1991). DiBona and Jones (1992) demonstrated that dietary KCl supplementation in borderline hypertensive rats does not protect against hypertension and exaggerated responses to acute environmental stress as seen in SHRs and Dahl rats. Stroke-prone SHRs fed a diet supplemented with potassium (2.1%) and high in NaCl (5%) had reduced oxidative stress on the endothelium independently of blood pressure changes (Ishimitsu et al., 1996).

SODIUM.   Sodium is the major extracellular electrolyte, constituting 93% of the cations in the body. Approximately 70% is found in extracellular fluid, nerve, and muscle tissue. The remaining sodium, located on the surface of bone crystals, can be released into the bloodstream if hyponatremia develops (Berdanier, 1998).

Almost 95% of ingested sodium is absorbed. There are three pathways for absorption of sodium across the intestinal mucosa: (1) the Na^+/glucose cotransport systems, (2) the electroneutral Na^+ and Cl^- cotransport system, and (3) the electrogenic sodium absorption mechanism (Groff and Gropper, 2000a). Both hormones and physical/chemical factors regulate sodium levels in the blood; these systems are also involved in the regulation of water balance, pH, and osmotic pressure (Oh and Uribarri, 1999). Hormones involved in sodium metabolism include vasopressin, natriuretic hormone, renin, angiotensin II, and aldosterone. Most of these hormones are released in response to sodium concentration or to osmoreceptors located in the anterolateral hypothalamus (Schmid et al., 1997; Grove and Deschepper, 1999; Hettinger et al., 2002; Ingert et al., 2002). The sodium ion is the most potent of the solutes activating the osmoreceptors, which in turn signal the release of hormones regulating osmolality (Brody, 1999).

Rats fed a high (4%) NaCl diet had decreased plasma renin activity and renal renin mRNA (Holmer et al., 1993). Hodge et al. (2002) reported that hypertension in SHRs is associated with salt sensitivity, which is related to a loss of the normal regulatory effect of dietary sodium on angiotensin II and angiotensinogen synthesis. In Sprague-Dawley rats, cyclooxygenase I derived prostanoids were shown to play a role in the regulation of the renin system by salt intake (Hocherl et al., 2002).

Early research showed that sodium deficiency (20 to 70 mg/kg diet) in rats caused reduced appetite, decreased body fat and protein, growth retardation, corneal lesions, soft bones, male infertility, delayed sexual maturity in females, and death (Kahlenberg et al., 1937; Orent-Keiles et al., 1937). Salt hunger is a behavior of rats suffering from sodium deficiency, and perinatal sodium deprivation may result in increased salt intake over a lifetime (Leshem, 1999).

Consumption of excess sodium as NaCl is associated with elevated blood pressure in various rat strains, including salt-sensitive Sprague-Dawley rats, DS rats, DR rats (treated with deoxycorticosterone acetate), obese

Zucker rats, and SHRs (Kaup et al., 1991a,b; Tobin, 1991; Reddy and Kotchen, 1992b). In the DS rat, a high-sodium without chloride diet and a chloride without sodium diet do not increase arterial pressure as significantly as does a high-NaCl diet (Reddy and Kotchen, 1992a). The estimated sodium requirement for rats is 0.5 g/kg diet (NRC, 1995).

The Dahl/Rapp rat is a model for salt-sensitive hypertension. It has been shown that the L-arginine/nitric oxide (NO) pathway is integrally involved in the production of hypertension in response to high dietary NaCl intake (Sanders, 1996). Data from Sprague-Dawley rats fed a high-salt diet indicate it impairs endothelium-dependent relaxation via reduced NO levels and increased superoxide production (Zhu et al., 2003). Chronic consumption of a high-fat, refined-carbohydrate diet by female Fischer rats resulted in increased blood pressure and salt sensitivity in association with reactive oxygen species-mediated NO inactivation and depressed renal-neuronal NO synthase protein expression (Roberts et al., 2003). Nishikawa et al. (2003) showed high vitamin C intake by SHR resulted in antihypertensive effects. Ettarh et al. (2002) demonstrated that the antihypertensive effect of vitamin C is associated with a reduction in vascular sensitivity to noradrenaline and enhancement of endothelium-dependent relaxation owing to increased NO bioavailability.

A comparison of the salt-sensitivity of male and female F2 progeny obtained from crosses between Wistar-Kyoto/Izumo rats and stroke-prone SHRs after salt loading showed the resulting blood pressure responses were different. This indicates that a possible hormonal difference between sexes may influence salt sensitivity in stroke-prone SHR rats (Nara et al., 1994). Sympathetic nervous system activity and decreased vascular reactivity may contribute to elevated arterial pressure in type-2 diabetic, obese Zucker rats, but the sympathetic nervous system does not appear to contribute to the dietary salt-sensitive hypertension in this model (Carlson et al., 2000).

High-salt diets contribute to the pathogenesis of hypertension; data further suggest an association with insulin resistance. DS rats fed a high-salt diet had increased insulin resistance and hypertension; supplementation with potassium (8%) ameliorated the changes in insulin sensitivity and decreased blood pressure (Ogihara et al., 2002). Dietary potassium depletion in young, rapidly growing Sprague-Dawley rats induced salt sensitivity and resulted in increased renin-angiotensin system activity, increased blood pressure, tubulointerstitial injury, and kidney fibrosis (Ray et al., 2001). Salt-loading in Wistar fat rats elevated blood pressure and increased insulin resistance in association with increased Na^+-H^+ exchanger activity (Hayashida et al., 2001). Prada et al. (2000) reported that a high-salt diet did not affect insulin sensitivity.

Sodium also functions in nerve conduction, active transport both by enterocytes and other cell types, and is involved in the formation of the bone mineral apatite. Sodium involvement in nerve conduction, active transport, and water balance is related to its function in the Na^+K^+-ATPase enzyme. This enzyme is called the Na^+K^+ pump because it pumps sodium out and, as potassium returns to the cell, there is a concurrent hydrolysis of ATP (Berdanier, 1998). DS rats fed a high-salt diet (8% NaCl) had decreased Na^+K^+-ATPase activity in the hypothalamus compared with that of the DR rats (Abdelrahman et al., 1995).

CHLORIDE. Chloride along with sodium and potassium are responsible for osmotic pressure and acid-base balance. Chloride is the most abundant anion in the extracellular fluid. As an electronegative element, Cl^- is an oxidizing agent. In addition to its passive role in electrolyte balance, chloride is required for the production of gastric hydrochloric acid secreted from the parietal cells of the gastric mucosa in the stomach (Groff and Gropper, 2000a). This mucosa also releases pepsinogen, which is activated by HCl and is the intrinsic factor needed for vitamin B_{12} absorption and mucus production. Mucus protects the organ from being digested by the HCl and proteases. HCl acts as a bacteriocide preventing bacterial overgrowth of the gastrointestinal tract (Berdanier, 1998). It also functions as the exchange anion in the red blood cell for HCO_3^-, known as the chloride shift. This process allows the transfer of CO_2 derived from the tissues back to the lungs (Groff and Gropper, 2000).

The estimated dietary chloride requirement for rats is 0.5 g/kg diet (NRC, 1995). Chloride deficiency in rats develops slowly owing to their ability to conserve the electrolyte by significantly reducing urinary excretion during depletion. Signs of chloride deficiency are poor growth, decreased feed efficiency, decreased blood chloride, reduced chloride excretion, and increased blood CO_2 (NRC, 1995). Rats fed diets containing high chloride concentrations grew normally and had normal muscle and kidney chloride concentrations (Whitescarver et al., 1986; Kaup et al., 1991a,b). However, Sprague-Dawley rats fed 15.6 g/kg dietary chloride had elevated blood pressure and enlarged kidneys (Kaup et al., 1991a,b). Kotchen et al. (1983) fed salt-sensitive Dahl rats 4.86 g/kg dietary chloride, resulting in increased blood pressure.

2. Microminerals

ZINC. Zinc is the most abundant intracellular trace element and is involved in diverse catalytic, structural, and regulatory functions (King and Keen, 1999). In biologic systems, zinc is in a divalent state and is not involved in redox activity. Zinc absorption is poor, and it easily complexes to amino acids, peptides, proteins, and nucleotides. It also binds to ligands that contain sulfur, nitrogen, or oxygen. Small-molecular-weight ligands such as amino acids improve absorption, whereas large-molecular-weight compounds such as phytic acid reduce absorption (Zhou et al., 1992; Rimbach et al., 1995). Rats fed a phytate-free soybean protein-based diet resulted in improved zinc, calcium, magnesium, iron, and phosphorus absorption compared with levels for rats fed soybean protein-isolate and casein diets (Kamao et al., 2000). The estimated zinc requirement is 12 mg/kg of diet for growth and 25 mg/kg of diet for reproduction (NRC, 1995). Higher levels may be needed when diets contain high phytate ingredients such as soybean meal.

Zinc serves as an essential cofactor for more than 300 enzymes. Zinc can have a structural, catalytic, or regulatory role in its involvement with metalloenzymes. However, it appears that zinc deficiency does not impair the activity of these enzymes (Berdanier, 1998) because they are intracellular and retain zinc. Increased absorption efficiency also protects zinc-dependent enzymes during zinc deficiency.

Zinc serves as a required structural component of DNA-binding proteins (transcription factors). These proteins with zinc attached are called zinc fingers because of their configuration. DNA protein containing zinc fingers can bind retinoic acid, thyroxine (T_4), vitamin D, estrogen, androgens, IGF-I, and growth hormone. With zinc attached to the DNA protein, the protein-hormone complex binds to DNA to affect gene expression (Groff and Gropper, 2000a). Zinc-regulated genes have been identified in small intestine, thymus, and monocytes. Most of the genes regulated by zinc are involved with signal transduction, responses to stress and oxidation, and growth and energy (Cousins et al., 2003).

Zinc has a critical role in the structure and function of biomembranes (Avery and Bettger, 1992; Kraus et al., 1997). Zinc deficiency results in increased oxidative damage, structural stresses, and alterations in specific receptor sites and transport systems. The effect of zinc on cell membranes may be through (1) direct effects on membrane protein conformation and/or protein-to-protein interactions; (2) effects on plasma membrane enzymes such as alkaline phosphatase, carbonic anhydrase, and SOD (Bettger and O'Dell, 1993); (3) stabilizing and maintaining phospholipid and thiol (SH) groups in a reduced state; and (4) protection from peroxidative damage (Sato and Bremmer, 1993). Zinc is instrumental in the association between skeletal and cytoskeletal membrane proteins. Zinc binds to tubulin, a protein component of microtubules that acts as a framework for structural support of the cell. The rate of tubulin polymerization is decreased in brain extracts of zinc-deficient rats (Oteiza et al., 1990). Kraus et al. (1997) showed that dietary supplementation with vitamin C, vitamin E, or β-carotene decreased the osmotic fragility and oxidative damage of erythrocytes in zinc-deficient rats. Vitamin E levels in the liver, testis, brain, heart, and kidney were lowered in rats consuming diets low in zinc (Noh and Koo, 2001).

In the rat, zinc deficiency significantly reduces plasma zinc concentration within 24 hours; anorexia is demonstrated within 3 days (Hambidge et al., 1986). Zinc-deficient rats develop fissures at the corners of the mouth, anorexia, an unthrifty haircoat, scaly feet, and a kangaroo-like posture (Brody, 1999). Prolonged deficiency can cause growth retardation; decreased efficiency of food utilization; abnormalities in platelet aggregation and hemostasis (Emery et al., 1990; O'Dell, 2000); alopecia; membrane lipid peroxidation (Hammermueller et al., 1987; Yousef et al., 2002); hyperirritability; and impairment of lipid (Reaves et al., 1999), carbohydrate (Tobia et al., 1998), and protein metabolism (Groff and Gropper, 2000a). Zinc deficiency impairs the immune system, resulting in a small thymus and decreased T- and B-lymphocyte production (Keen and Gershwin, 1990).

Many similarities exist between EFA and zinc deficiencies, and zinc deficiency intensifies the effects of EFA deficiency in rats (Bettger et al., 1979). Dietary zinc deficiency alters the fatty acid composition of phospholipids of the liver and red blood cells (Cunnane, 1988; Kudo et al., 1990; Eder and Kirchgessner 1994a, 1994b). Zinc deficiency resulted in higher concentrations of linoleic acid and lower concentrations of arachidonic acid in tissue phopholipids (Cunnane, 1988). Increased omega-3 long-chain PUFAs were found in liver phopholipids of zinc-deficient young rats (Eder and Kirchgessner, 1994b). Kudo et al., (1990), using a fat-free diet, showed zinc deficiency resulted in decreased liver delta-9 desaturase activity. Eder and Kirchgessner (1995) showed that zinc-deficient rats fed a diet containing coconut oil as the fat source developed fatty liver characterized by elevated levels of triglycerides with saturated and monounsaturated fatty acids and increased lipogenic enzyme activity. Zinc-deficient rats fed a diet with linseed oil as the fat source did not have the elevated lipogenic enzyme activity, increased triglyceride levels, or fatty liver.

Zinc deficiency during gestation is highly teratogenic, resulting in defects of the central nervous system, soft tissue, skeletal system, lung, heart, and pancreas. The teratogenicity of zinc deficiency is related to abnormal nucleic acid and protein synthesis, alterations in differential rates of cellular growth required for normal morphogenesis, impaired tubulin polymerization, chromosomal defects, abnormal apoptosis, and increased lipid peroxidation of cell membranes (Keen and Hurley, 1989). Reduced testicular size, atrophied seminiferous epithelium resulting in impaired spermatogenesis, and reduced testosterone levels have been observed in zinc-deficient males. McClain

et al. (1984) concluded that testicular dysfunction relating to zinc deficiency is the result of impaired Leydig cell function and a secondary effect of the pituitary-gonadal axis. Testicular androgen content was significantly decreased in zinc-deficient males, indicating that hypogonadism is related directly to impaired testicular steroidogenesis or indirectly to Leydig cell failure (Hamdi et al., 1997). Zinc-deficient dams are characterized by difficult deliveries, including a delay in parturition and excessive bleeding (Keen and Hurley, 1989). The delay in delivery in zinc-deficient rats is possibly related to significantly low ovarian 20-α-hydroxysteroid dehydrogenase, an enzyme that catalyzes the catabolism of progesterone, which is inhibitory to parturition (Bunce, 1989).

Offspring of zinc-deficient dams have anorexia, low growth rate, increased incidence of neonatal death, and behavioral abnormalities (Apgar, 1985; Bunce, 1989; Keen and Hurley, 1989). It has been shown that zinc-deficient rats had significant decreases in expression of the IGF-I and growth hormone receptor genes. This indicates that growth retardation due to zinc deficiency is associated with defects in the growth hormone receptor signaling pathway (McNall et al., 1995; Ninh et al., 1995).

An investigation into the effect of zinc deficiency in Sprague-Dawley rats on skeletal metabolism revealed significantly decreased values in ponderal growth rate, femur weight and length, circulating IGF-I, the thickness of the overall growth plate, and hypertrophic cartilage (Rossi et al., 2001). The investigators concluded the effects of zinc deficiency on bone growth are the result of reduced activity of the growth plate owing to impairment of the IGF-I system. In the presence of different intakes of copper, the effects of low dietary zinc were more significant on the trabecular bones of the spine than on long bones (Roughead and Lukaski, 2003).

Reduced food intake is one of the first signs of zinc deficiency in rats (Chesters and Quarterman, 1970; Chesters and Will, 1973). Early research indicated zinc-deprived rats appeared to develop an aversion to protein (Chester, 1975; Reeves and O'Dell, 1981). However, Rains and Shay (1995) showed that zinc-deprived rats selected more protein and fat and less carbohydrate than did the controls. Zinc-deficient rats were reported to prefer a high-carbohydrate diet during the beginning of zinc deficiency, but 25% switched to a higher-fat diet toward the end of a zinc-deficient dietary regimen (Kennedy et al., 1998). Recent findings by Reeves (2003) show a reduced food intake but not a change in feeding pattern among zinc-deficient rats within 3 to 4 days after initiation of a zinc-deficient diet regimen. Rats reduced their selection of protein from 12% to 8% of the diet and increased carbohydrate intake from 69% to 73%. At dietary protein concentrations of 20% and 25%, zinc-deficient rats developed marked signs of zinc deficiency and had reduced feed intake, growth, serum alkaline phosphatase activity, and zinc concentrations in serum and femur compared with levels for controls (Roth, 2003). These findings may be owing to disturbed protein synthesis as demonstrated by the increased activities of alanine amino-transferase, glutamate dehydrogenase, and carbamoylphosphate synthetase in the liver.

The possibility that zinc deficiency-induced anorexia in the rat is related to impaired leptin metabolism has been investigated. Lower plasma leptin levels were reported in zinc-deficient rats (Mangian et al., 1998; Gaetke et al., 2002). Lee et al. (2003) showed that serum leptin and inguinal adipocyte leptin mRNA levels increased with zinc depletion as expected with decreased appetite. However, there was a decrease in abdominal adipocyte leptin mRNA level, which is inconsistent with the increases in serum leptin and inguinal adipocyte leptin mRNA levels.

Donaldson (1973) identified zinc-containing neurons and nerve terminals in the hippocampus, cortex, and cerebellum of the rat brain. Zinc was able to modulate the activity of glutamate and γ-aminobutyric acid (GABA) receptors in the brain, indicating an important role in neurotransmission (Vallee and Falchuk, 1994; Slomianka, 1992; Xie and Smart, 1991). Feeding rats zinc-deficient diets at a critical age resulted in impaired spatial memory in adult rats and induced behavioral changes, including impaired short-term memory processing (Halas et al., 1983; Keller et al., 2001; Chu et al., 2003).

Zinc interacts with vitamin A, copper, iron, calcium, folate, cadmium, and lead (Groff and Gropper, 2000a,c). Increased hepatic methionine synthase and decreased plasma folate concentrations have been reported in zinc-deficient rats (Hong et al., 2000; Tamura et al., 1987). Zinc deficiency in pregnant rats decreases folate bioavailability, and folate supplementation does not prevent fetal growth retardation (Favier et al., 1993). Zinc deficiency results in decreased serum copper and iron concentrations (Hendy et al., 2001). Copper and manganese concentrations in the liver, kidney, and femur were significantly higher in zinc-deficient rats compared with controls. Iron concentration in zinc-deficient rats was increased in liver, kidney, and muscle (Sakai et al., 2004).

Chemically similar metal ions such as iron and zinc have been shown to have biological interactions. Rats fed an iron-deficient diet had decreased zinc absorption and increased levels of zinc in the plasma, liver, spleen, kidney, and femur (Kaganda et al., 2003). Interactions between zinc and iron affect ceruloplasmin activity, red blood cells, mean corpuscular volume, and monocyte count in rats (Uthus and Zaslavsky, 2001). Rats fed diets containing 10.44, 388, and 827 mg/kg iron demonstrated that low dietary iron increased zinc absorption and zinc concentration in the brain and liver (Dursen and Aydogan, 1995). Conversely, zinc concentrations decreased in the brain,

liver, ileum, and duodenum of rats fed diets containing the higher iron concentrations. These findings indicate that dietary iron can influence zinc metabolism at the intestinal and cellular transport levels. Bougle et al. (1999) also showed that zinc absorption was affected by dietary iron levels; there was a positive correlation between growth and liver zinc content, but a negative correlation with liver iron content.

Tanaka et al. (1995) reported rats fed a zinc-deficient and cadmium-supplemented (100 ppm) diet had diminished bone growth, cortical thinning of the femur, and enhanced renal toxicity. Naturally occurring zinc in sunflower seeds minimized cadmium absorption in rats (Reeves and Chaney, 2001).

Zinc, through its involvement in protein synthesis and cellular enzyme functions, participates in the absorption, mobilization, and transport of vitamin A (Christian and West, 1998). Vitamin A supplementation increased metallothionein concentration, whereas vitamin A deficiency in rats decreased intestinal zinc absorption and altered tissue mineral concentrations (Rahman et al., 1995, 1999; Sundaresan et al., 1996; Christian and West, 1998). Kelleher and Lonnerdal (2002) documented an interaction between marginal zinc and vitamin A intake during lactation on zinc transporter mRNA expression in the rat mammary gland.

Emerick and Kayongo-Male (1990) reported an apparent antagonism between zinc and silica in rats. Zinc deficiency contributed to silica urolith formation among rats fed a diet containing 2700 mg/kg silicon (Stewart et al., 1993).

COPPER. Copper, zinc, and iron help regulate the expression of the genes for the metallothioneins, the metal-binding proteins. These genes have metal response elements specific for each of these minerals. The gene affecting copper response encodes metallothionein, which binds copper and other heavy metals such as zinc and cadmium. Copper influences gene expression by binding to specific transcription factors called binding proteins. Data indicate that transcription of the copper-responsive metallothionein is a function of both copper and zinc because this metal-binding protein can only be synthesized in the presence of both metals (Berdanier, 1998). Ten other genes have been shown to have copper-responsive elements required for expression. Many of these have considerable homology with ferritin mRNA, fetuin mRNA, mitochondrial 12S and 16S rRNA, and mitochondrial tRNA for phenylalanine, valine, and leucine. This suggests that copper plays a role in mitochondrial gene expression, which relates to the reported decrease in oxidative phosphorylation among copper-deficient rats (Berdanier, 1998).

Copper is essential as an enzyme cofactor and allosteric component of enzymes (Prohaska, 1988). The activities of these enzymes are normally depressed during copper deprivation (Prohaska and Baily, 1995). The decreased oxidative phosphorylation in copper-deficient rats is related to the requirement for copper by cytochrome c oxidase, which functions in the terminal oxidative step in mitochondrial electron transport (Davies and Lawrence, 1986).

Ceruloplasmin, also called ferroxidase I, contains six copper atoms. Ceruloplasmin, a multifaceted oxidative enzyme and antioxidant, transports copper in the blood and is found on cell surface receptors on plasma membranes of cells. As ferroxidase I it is responsible for the oxidation of ferrous (Fe^{2+}) iron and manganese (Mn^{2+}). It is involved with the transfer of iron from storage sites to sites of hemoglobin synthesis (Turnlund, 1999). The sex of the rat and level of dietary iron affect the hematologic response to copper deficiency (Cohen et al., 1985; Johnson and Kramer, 1987).

SOD is found in the cytosol of cells and depends on copper and zinc (Prohaska, 1991). The function of SOD is to scavenge superoxide radicals and protect cell membranes from oxidative damage. Increased peroxidation of cell membranes is found with copper deficiency (Sukalski et al., 1997; Groff and Gropper, 2000a).

Copper-dependent amine oxidases are found in the blood and body tissues, where they inactivate and catalyze the oxidation of physiologically active amines such as histidine, tyramine, and polyamines. The activity of amine oxidases is increased when connective tissue activation and deposition occurs during normal growth and during liver fibrosis, congestive heart failure, and hyperthyroidism (Turnlund, 1999). Diamine oxidase inactivates histamine and polyamines involved in cell proliferation. Lysyl oxidase is found in connective tissue cells and is essential in the cross-linking between collagen and elastin (Farquharson et al., 1989). Lysyl oxidase functions in the formation of connective tissue, including bone, blood vessels, vasculature, skin, lungs, and teeth. Decreased lysyl oxidase activity owing to copper deficiency results in vascular disease, spontaneous rupture of major blood vessels, defective bone matrices, and osteoporosis (O'Dell, 1990). In copper-deficient weanling rats, decreased lysyl oxidase activity and elevated soluble and total cardiac collagen concentrations affect the cardiac system integrity and bone formation (Werman et al., 1995).

Tyrosinase, which is involved in the synthesis of melanin, depends on copper. Deficiency of tryrosinase in skin results in achromotrichia of hair (Turnlund, 1999). Copper has other physiologic functions. Some are not well understood; these include angiogenesis, immunity, nerve myelination, thermal regulation, cholesterol metabolism, and glucose metabolism. Copper is required for formation and maintenance of myelin which is composed primarily of phospholipids. Phospholipid synthesis depends on the copper-dependent enzyme, cytochrome c oxidase, which explains poor myelination and necrosis of nerve tissue with copper deficiency (O'Dell and Prohaska, 1983). In copper-deficient rats, decreased norepinephrine and striatal dopamine levels have been observed (Prohaska, 1987). The depressed striatal dopamine is not reversed by copper repletion. The low striatal dopamine, along with neurological signs owing to copper deficiency, do not develop in all copper-deficient rats, and it has been suggested that a genetic component is involved (Miller and O'Dell, 1987).

Copper is involved in many roles in the central nervous system ranging from antioxidant activity to oxidative phosphorylation, neurotransmission, and neuropeptide maturation (Prohaska, 1990). Early research by Carlton and Kelly (1969) showed copper deficiency in rats caused gross neural focal lesions in the occipital and parietal parts of the cerebral cortex and abnormalities in the corpus striatum. Prohaska and Baily (1995) studied perinatal copper deficiency in Sprague-Dawley rat offspring to investigate regional changes in brain cuproenzymes. The copper-deficient rats had decreased activities of the cuproenzymes peptidylglycine α-amidating monoxygenase, cytochrome c oxygenase, and copper- and zinc-SOD in all six brain regions studied. After 4 months of copper repletion, the rats continued to have low copper in all brain regions and low cytochrome c oxidase activity. Diminished auditory startle response was observed after 4 months of copper repletion in rats made copper deficient during the perinatal period (Prohaska and Hoffman, 1996). These data indicate long-term neurochemical and behavioral abnormalities persist after perinatal copper defieincy. To study the mechanisms for the observed neuropathy resulting from copper deficiency, perinatal copper deficiency was induced in Holtzman rats. Analysis of the brains showed that cytochrome c and mitochondrial mass were significantly increased (Gybina and Prohaska, 2004).

Copper interacts with many organic and inorganic components. Dietary constituents that facilitate copper absorption include amino acids such as histidine, methionine, and cysteine (Du et al., 1996; Aoyama, 2001). Organic acids such as citric, gluconic, lactic, acetic, and malic acids act as binding ligands to improve solubilization and absorption of copper (Groff and Gropper, 2000a). Inhibitors of copper absorption include zinc (Abdel-Mageed and Oehme, 1991; Barone et al., 1998), iron (Yu et al., 1995; Reeves et al., 2004), molybdenum (Groff and Gropper, 2000a), and ascorbic acid (Johnson and Murphy, 1988; Van den Berg et al., 1994).

Anemia, neutropenia, and osteoporosis are universal signs of copper deficiency. The reduced survival time of erythrocytes in copper-deficient rats was shown to be related to changes in membrane fluidity and increased susceptibility to peroxidation (Rock et al., 1995). Reeves and DeMars (2004) reported that iron deficiency anemia in copper-deficient rats is caused in part by reductions in iron absorption and retention. Other manifestations of copper deficiency are reduced growth rate (Allen, 1994); alterations in platelet function (Johnson and Dufault, 1989; Lominadze et al., 1996); skeletal abnormalities, fractures, and spinal deformities (Strause et al., 1986; Smith et al., 2002); alterations in thromboxane and prostaglandin synthesis (Allen et al., 1991; Saari, 1992); ataxia (Prohaska, 1987); depigmentation and impaired keratinization of hair; and reproductive failure, including low fertility, fetal death, and resorption (Davis and Mertz 1987). Other effects include cardiovascular disorders such as myocardial degeneration, cardiac hypertrophy and failure, rupture of blood vessels, and electrocardiographic changes (Medeiros et al., 1992); impaired immune function (Koller et al., 1987; Kramer et al., 1988; Babu and Failla, 1990a,b); changes in lipid and cholesterol metabolism (Lei, 1991; al-Othman et al., 1994; Fields et al., 1999); increased lipid peroxidation (Rayssiguier et al., 1993; Sukalski et al., 1997); and impaired pancreatic function (Dubick et al., 1989; Fields and Lewis, 1997). Food restriction in copper-deficient rats abated many of the signs of deficiency including cardiac hypertrophy, red blood cell defects, reduction in SOD activity, and decreasing mortality associated with copper deficiency (Saari et al., 1993).

Hypercholesterolemia and hypertriglyceridemia can be induced in copper-deficient rats with increased iron concentrations (Klevay, 1973; Davis and Mertz, 1986; Lei, 1991; al-Othman et al., 1994; Fields and Lewis, 1977, 1999; Bureau et al., 1998). In copper-deficient rats, both the HDL cholesterol and HDL apolipoprotein E were significantly increased compared with controls (Lei, 1983). Hassel et al. (1988) showed a greater total specific binding of ^{125}iodine apolipoprotein E-rich HDL to hepatic membranes from copper-deficient rats. Tang et al. (2000) showed copper deficiency in rats resulted in an increase in the nuclear transcription factor, sterol regulatory element binding protein-1 (SREBP-1), a strong enhancer of fatty acid synthase promoter activity. The investigators suggest that copper deficiency stimulates hepatic lipogenic gene expression by increasing the hepatic translocation of mature SREBP-1. Copper deficiency in rats resulted in reduced activity of hepatic methionine synthase which caused increases in plasma homocysteine concentration, a risk factor for cardiovascular disease (Tamura et al., 1999).

Cardiac myopathy is reported in copper-deficient weanling rats (NRC, 1995). Ventricular aneurysms and decreased norepinephrine concentrations are also common in copper-deficient rats (Prohaska and Heller, 1982). Electrocardiography of copper-deficient rats, both normal and SHHS/Mcc-cp, a hypertensive cardiomyopathic strain, showed abnormalities in the ST segment, QRS height, and occurrence of bundle-branch block (Viestenz and Klevay, 1982; Medeiros et al., 1991; Jalili et al., 1996).

Abnormalities in cardiac ultrastructure were reported in rats fed diets marginally low in copper, despite minimal changes in conventional biochemical indicators of copper status (Wildman et al., 1995). It has been demonstrated that cardiac adaptations in copper-deficient rats are influenced by type and amount of dietary fat and cholesterol (Jenkins and Medeiros, 1993; Jalli et al., 1996).

Copper metabolism is adversely affected by simple sugars (Reiser et al., 1983; Fields et al., 1984). Copper-deficient rats fed a diet with sucrose as its carbohydrate source had significantly lower hematocrit, lower hemoglobin, and depressed copper absorption compared with levels for rats fed a diet containing starch as the carbohydrate source (Johnson and Gratzek, 1986). Xu et al. (2001) reported dietary galactose and fructose exacerbate effects of chronic marginal copper intake by rats, including hypertrophy of liver, heart, and kidney, hyperlipidemia, and increased mortality. Copper deficiency reduced carbohydrate and increased utilization of fat as a substrate for energy, and reduced fat mass in rats (Hoogeveen et al., 1994).

Copper deficiency impairs both humoral and cell mediated immunity in rats (Failla et al., 1988; Prohaska and Failla, 1993; Lominadze et al., 2004). Phenotypic profiles of mononuclear cells from rat spleen and blood are altered by dietary copper deficiency (Bala et al., 1991). Thymus weights are decreased, antibody titers after immunization with sheep erythrocytes are decreased, natural killer (NK) cell cytotoxicity is suppressed, and helper T cells are decreased (Failla et al., 1988). Bala and Failla (1993) showed dietary copper deficiency reversibly suppresses the maturation and function of splenic T helper cells. Chronic intake of diet marginally low in copper suppressed the *in vitro* activities of T lymphocytes and neutrophils without affecting the tissue copper levels or the activities of cuproenzymes in serum and most tissues (Hopkins and Failla, 1995).

In humans, Wilson's disease is the result of a defect in copper metabolism. The Long-Evans Cinnamon (LEC) rats have excessive accumulation of copper in the liver due to a gene mutation homologous to the human Wilson's disease gene, ATP7B, which encodes a copper-transporting P-type ATPase. Du et al. (2004) showed that PUFAs suppress the development of acute hepatitis and prolong survival in female LEC rats.

The estimated copper requirements for rat growth and reproduction are 5 and 8 mg/kg of diet, respectively (NRC, 1995). Weanling male rats fed diets with copper concentrations less than 3 to 4 mg/kg had depressed functional measures of copper status, including platelet cytochrome c oxidase activity, serum ceruloplasmin activity, plasma copper, heart, and liver copper, and copper and zinc-superoxide mutase activity (Johnson et al., 1993; Klevay and Saari, 1993). Feeding diets containing 8 mg/kg copper during pregnancy and lactation improved serum copper, serum ceruloplasmin activity, and liver copper concentrations in dams and their offspring (Spoerl and Kirchgessner, 1975a,b).

Rats are more tolerant to excessive levels of dietary copper than are other species. Aburto et al. (2001) reported hepatic necrosis, portal inflammation, hyaline remnants, and reduced growth in Fischer 344 rats fed 1250 mg/kg of diet. Young Fischer 344 rats were observed to be more susceptible to copper-induced liver injury than were adults (Fuentealba et al., 2000).

IRON. Iron is a pivotal element in the metabolism of all living organisms. The essentiality of iron is due to its relationship with the prosthetic group of hemoglobin, heme, which is the active site of electron transport in cytochromes and cytochrome oxidase. Heme is also involved in the transport of oxygen to tissues and within muscle cells. Another essential function of iron is as the iron-sulfur active center for enzymes. Aconitase, an important enzyme in the tricarboxylic acid cycle has an iron-sulfur active site and links cellular iron content with energy production via oxidative phosphorylation both in carbohydrate and lipid metabolism (Chen et al., 1998; Stangl and Kirchgessner, 1998; Fairbanks, 1999). Iron deficiency in rats results in a significant reduction in the activities of iron-sulfur enzymes succinate-ubiquinone oxidoreductase and nicotinamide adenine dinucleotide (NADH)-ubiquinone oxidoreductase in skeletal muscle mitochondria (Ackrell et al., 1984). In iron-deficient rats, hepatocyte total iron regulatory protein (IRP) activity was increased and mitochondrial aconitase was decreased without altering the tricarboxylic acid cycle capacity (Ross and Eisenstein, 2002).

Monooxygenases and dioxygenases function to insert oxygen into substrates. These enzymes are also involved in amino acid metabolism and synthesis of carnitine (Bartholmey and Sherman, 1986), procollagen, nitric oxide, and vitamin A (During et al., 1999). Catalase contains four heme groups and helps prevent cellular peroxidation (Miret et al., 2003). Myeloperoxidase, another heme containing enzyme, is involved in the production of hypochlorite, a strong cytotoxic oxidant important in the destruction of foreign substances such as bacteria. Iron deficiency can impair myeloperoxidase activity and increase susceptibility or severity of infection (Mackler et al., 1984; Sunder-Plassmann et al., 1999). Ribonucleotide reductase, an enzyme that does not depend on heme-iron, is involved in DNA synthesis and, thus, cell replication. Thyroperoxidase, a heme-iron-dependent enzyme, is necessary for the synthesis of thyroid hormones T_3 and T_4.

The ferric (Fe^{3+}) and the ferrous (Fe^{2+}) forms are the only oxidative states of iron stable in the aqueous environment of the animal body and in food. Absorption of iron is more efficient in the ferrous state because ferric iron is less soluble at the alkaline pH of intestinal fluid. Normal iron balance is maintained to a large extent by absorption, and is primarily dependent upon whether the source is heme or non-heme iron (Groff and Gropper, 2000a). Iron is absorbed by proteins in the mucosal epithelium of the luminar surface of the duodenum. Nramp2 is one protein responsible for absorption of iron by the intestinal epithelial cells, and it is the intracellular transport protein in the erythroblast. Lack of this protein in Belgrade rat erythroblasts causes hereditary microcytic anemia (Fairbanks, 1999). Gomez-Ayala et al. (1997) reported that iron-deficient anemic rats have impaired heme iron absorption via duodenal active transport but that non-heme iron absorption was not affected.

Chelators or ligands can bind with non-heme iron to either inhibit or enhance its absorption. Examples of non-heme iron absorption enhancers include ascorbic acid, citric acid, tartaric acid, fructose, sorbitol, and amino acids. Perez-Lamas et al. (2001) showed that the apparent digestibility coefficients for amino acids were greater in rats fed a diet containing ferrous sulfate compared with ferrous lactate. Iron retention was greater in rats fed a diet containing casein (animal protein) as compared with soybean protein. Dietary ascorbic acid raised iron absorption in iron-deficient anemic rats by enhancing mucosal iron uptake independent of iron solubility in digesta (Wienk et al., 1997). Inhibitors of iron absorption are phytates, polyphenols, oxalic acid, EDTA, calcium, manganese (Rodriguez-Matas et al., 1998), calcium phosphate salts, and zinc (Groff and Gropper, 2000a). In iron-deficient rats, bone morphology, strength, and density are compromised and calcium restriction exacerbates the condition (Medeiros et al., 2002). Rats fed a diet containing phytate-free soybean protein had significantly higher mineral absorption and retention ratios, including iron, than those in rats fed soy protein isolate and casein (Kamao et al., 2000). Iron status of the animal also affects iron absorption.

There is an association between vitamin A and iron. In rats with a marginal vitamin A deficiency, increased hepatic iron accumulation along with decreased plasma iron, blood hemoglobin, and hematocrit have been observed (Houwelingen et al., 1993). Lead inhibits aminolevulinic acid dehydratase which is required for heme synthesis, and lead also inhibits ferrochetalase, the enzyme that incorporates iron into heme. Iron deficiency will decrease selenium concentrations and GSH-Px synthesis and activity.

Iron depends on the copper-containing enzyme ceruloplasmin for its mobilization from ferritin stores. Yu et al. (1993) showed that increased dietary intake of iron depressed copper absorption and biliary excretion, producing a decrease in plasma and organ copper concentrations.

These findings show that increased iron intake interferes with mobilization of copper stores. During pregnancy, iron deficiency resulted in increased maternal but decreased fetal hepatic copper concentration. This demonstrates that iron deficiency during pregnancy has a differential effect on copper metabolism in the mother and fetus (Gambling et al., 2004). Crowe and Morgan (1996) reported that iron- and copper-loading in developing rats caused increased non-heme iron concentration in the brain and liver compared with that in rats loaded with iron alone. An iron-deficient diet enhanced the onset of hepatitis C owing to increased hepatic copper deposition in the LEC rat (Sugawara and Sugawara, 1999).

In rats, iron deficiency during pregnancy results in growth retardation, decreased serum triglyceride, cardiac hypertrophy, and high blood pressure in the adult offspring (Lewis et al., 2001, 2002). Lisle et al. (2003) showed that maternal iron restriction during pregnancy caused a decrease in nephron number in the adult rat offspring. Iron deficiency in male weanling Sprague-Dawley rats causes cardiac eccentric hypertrophy (Medeiros and Beard, 1998). In rats, postweaning iron deficiency produces a decrease in the total amount of iron in the brain that is reversible with iron repletion (Chen et al., 1995a; Erickson et al., 1997; Pinero et al., 2000). Decreased concentration of iron in the brain has been linked to altered dopaminergic functioning (Nelson et al., 1997). Examination of dopamine function during iron deficiency showed decreased D_1 and D_2 receptor density and increased extracellular dopamine concentrations (Beard et al., 1994; Chen et al., 1995b; Nelson et al., 1997; Erickson et al., 2001). Iron-deficient and iron-supplemented rat pups showed decreased activity and stereotypic behavior. Iron repletion of the deficient pups did not reverse these functional variables (Pinero et al., 2001).

Iron deficiency in male Wistar rats resulted in the following mineral alterations: calcium concentrations in blood and liver increased; calcium concentration in the lung decreased; magnesium concentration in blood increased; copper concentrations in blood, liver, spleen, and tibia increased; zinc concentration in blood decreased; and manganese concentrations in brain, heart, kidney, testis, femoral muscle, and tibia increased (Yokoi et al., 1991). Iron deficiency in rats resulted in a decrease in zinc absorption but increased concentration of zinc in plasma, liver, spleen, kidney, and femur (Kaganda et al., 2003).

Rao and Jagadeesan (1995) showed that Wistar and Fischer strains of rats appear to be more susceptible and developed clinical and biochemical signs of iron deficiency earlier than did the Sprague-Dawley strain. A more recent study reports that the Fischer 344 strain was less sensitive to an iron-deficient diet than were the Sprague-Dawley and Wistar strains (Kasaoka et al., 1999).

Stangl and Kirchgessner (1998) reported moderate iron deficiency in rats caused decreased hepatic cholesterol concentration, decreased serum lipoproteins, and depressed serum phospholipid levels. The findings also showed significant differences in phosphotidylcholine and phosphatidyl-ethanolamine fatty acid compositions, indicating impaired desaturation by delta-9 and delta-6 desaturases of saturated and EFAs.

Iron supplementation can induce oxidative stress and inflammation in rats predisposed to colitis and in normal rats (Knutson et al., 2000; Carrier et al., 2002; Uritski et al., 2004). Knutson et al. (2000) showed that iron deficiency, as well as iron supplementation results in lipid peroxidation. Supplementation of rat pups during early infancy resulted in increased small intestine and liver iron concentrations, and the pups were unable to down-regulate intestinal iron transporters, divalent metal transporter 1, and ferroportin 1 (Leong et al., 2003).

The liver is a primary site for deposition in iron overload, a condition that leads to hepatic cirrhosis and/or fibrosis (Britton et al., 1994; Olynyk and Clarke, 2001). Hepatic iron accumulation has been shown to cause increased lipid peroxidation, believed to be the initial step by which excess iron may cause cellular injury (Dabbagh et al., 1994; Khan et al., 2002; Brown et al., 2003). Dietary iron overload depleted hepatic antioxidants, decreased carbon tetrachloride-induced necrosis and cell proliferation, enhanced apoptosis, but did not facilitate fibrogenesis (Wang et al., 1999). Stal et al. (1999) reported that dietary iron overload increased the number of preneoplastic foci but did not enhance their progression to hepatocellular carcinomas. Dietary iron overload-induced lipid peroxidation was apparently correlated with significant perturbations in plasma lipid transport and with hepatic sterol metabolism (Brunet et al., 1999; Whittaker and Chanderbhan, 2001).

SELENIUM. Schwarz and Foltz (1957) showed that traces of selenium prevented liver necrosis in vitamin E-deficient rats. Selenium deficiency, without vitamin E deficiency, has been seen only in laboratory animals under experimental conditions. Throughout the world, selenium concentration of soils varies more than any other essential trace element and affects the selenium concentration in food plants. Most selenium in biologic systems occurs in amino acids as components of proteins. Selenium is present naturally in food almost entirely as selenomethionine, selenocystine, selenocysteine, and selenium-methyl selenomethionine.

The organic and inorganic forms of selenium are efficiently absorbed primarily in the duodenum. Selenomethionine is absorbed better than are the inorganic forms of selenium (Vendeland, 1992). Smith and Picciano (1987) showed feeding a diet containing 500 μg/kg selenium as sodium selenite resulted in a maximum GSH-Px activity in dams and pups. However, when the source of selenium was selenomethionine, only 250 μg/kg diet was required for maximum activity. Selenium absorption is enhanced by vitamins C, A, and E, and reduced by GSH in the intestinal lumen. Phytates and heavy metals, such as mercury, inhibit selenium absorption (Groff and Gropper, 2000a).

The source of selenium in the new AIN-93 diet is selenate not selenite, which was used in the AIN-76 diet (Reeves, 1993). Because of concerns that this change might increase research variability, a study, planned to compare dietary lipid oxidation and hepatic peroxidation in rats, found no apparent differences in oxidation of dietary components between the selenate and selenite diets. Livers isolated from the rats fed both diets showed no differences in thiobutaric acid-reactive substances or lipid hydroperoxides (Moak and Christensen, 2001).

Selenium is transported from the intestine via blood by transport proteins to the liver and other tissues. Selenoprotein P and extracellular GSH-Px have been identified in the plasma. Selenoprotein-P, a selenocystine-containing plasma protein, has been isolated in the rat and appears to function as a selenium transport and storage protein (Kato et al., 1992). Selenium is incorporated into many different proteins that provide transport and storage roles. In rats, selenium apparently controls the synthesis of these proteins which are important in the transfer of selenium among tissues (Evenson and Sunde, 1988).

Eleven selenoproteins have been characterized in animals. Enzymatic functions have been established for most, but the biochemical functions of some remain unknown. Selenium is a cofactor for GSH-Px. There are four selenium-dependent GSH peroxidase enzymes found in different tissues. Selenium deficiency results in less mRNA as less liver GSH-Px is produced and enzyme activity is decreased (Burk and Hill, 1993). GSH-Px catalyzes the reduction of organic peroxides derived from unsaturated fatty acids and nucleic acids as well as hydrogen peroxide. Turan et al. (2001) showed that feeding rats diets deficient in both selenium and vitamin E, or diets containing excess selenium, resulted in significantly lower liver and brain GSH-Px and GSH reductase actitivities when compared to rats fed a control diet. GSH depletion in selenium-deficient rats results in lipid peroxidation, centrilobular hepatic necrosis, and renal tubular necrosis (Burk et al., 1995a). Hyperlipidemic rats fed diets supplemented with vitamin E and selenium did not have the glomerular and tubulointerstitial damage observed in controls. The synergistic antioxidant activities of vitamin E and selenium prevented renal damage owing to oxidized lipids (Gonca et al., 2000).

There is an interdependent relationship among selenium, vitamin E, iron, zinc, and copper in their role as antioxidants preventing free radical-induced cell damage (Groff and Gropper, 2000a). GSH-Pxs also have regulatory functions because they affect the concentrations of oxidant molecules, which have functions in metabolism and signaling pathways (Burk and Levander, 1999). Selenium can protect against the toxicities of other metals such as silver, mercury, and cadmium (Levander and Cheng, 1980).

Selenium is involved in iodine metabolism because iodothyronine deiodinases are selenoproteins. These enzymes regulate the concentration of the hormone triiodothyronine (T_3) by catalyzing the deiodination of T_4, T_3, and reverse T_3 (Burk and Levander, 1999).

Selenoprotein-P is an extracellular glycoprotein located in the plasma and associated with endothelial cells. Selenoprotein-P function has not been completely characterized, but it has been associated with oxidant defense properties and the transport of selenium (Burk et al., 2003). Burk et al. (1995b) showed selenium-deficient rats susceptible to diquat-induced lipid peroxidation and liver necrosis were protected by selenium repletion which correlated with selenoprotein-P plasma concentration. Atkinson et al. (2001) reported that liver lesions caused by selenium deficiency are a sign of injury to endothelial cells in the centrilobular region. Protection against this injury by selenium correlates to the presence of the extracellular selenoprotein, selenoprotein-P, which associates with endothelial cells.

Christensen et al. (1995) investigated the tissue-specific effects of selenium intake on selenoprotein gene expression and enzyme activity. In the liver, selenium intake did not affect transcription of genes for cellular glutathionine peroxidase, type-1 iodothyronine 5'-deiodinase, and selenoprotein-P. In the liver and kidney, selenium deficiency significantly reduced the activities of both GSH-Px mRNA and iodothyronine 5'-deiodinase mRNA, with a greater effect on GSH-Px. Selenoprotein-P mRNA was reduced significantly more in the kidney than the liver. These results suggest that translation and protein turnover may determine the level of enzyme activity attained in response to dietary selenium intake.

Selenium interacts with heavy metals and other nutrients. Lead will lower tissue selenium concentration, and iron deficiency decreases the synthesis of hepatic GSH-Px and decreases liver selenium concentrations (Moriarty et al., 1993). It has been shown that selenium deficiency in rats resulted in up-regulation of transferrin mRNA, transferrin receptor, and IRP-1 genes involved in iron metabolism (Christensen et al., 2000). Copper deficiency decreases the activity of GSH-Px and 5'-deiodinase (Olin et al., 1994). Copper has been shown to provide protection from selenium toxicity in the Fischer 344 rat (Tatum, 2000). Yu and Beynan (2001) showed that the amount of selenium in the diet determines whether or not increased dietary copper concentration affects selenium metabolism.

Rats fed a selenium-deficient and vitamin E-adequate diet for two generations had alopecia, growth retardation, and reproductive failure. Selenium deficiency causes significant changes in many biochemical systems. GSH S-transferase activities in rat lung, liver, and kidney increase in selenium deficiency. In the selenium-deficient rat, increased hepatic synthesis of GSH led to increased plasma GSH (Hill et al., 1987). Olsen et al. (2004) reported that the caudal epididymal spermatozoa of selenium-deficient rats exhibited flagellar defects. The data suggest loss of male fertility in selenium deficiency is owing to sequential development of sperm defects expressed during both spermatogenesis and maturation in the epididymis. Selenium deficiency decreased homocysteine transferase activity, reduced homocysteine concentration in the heart and kidney, and increased the ratio of plasma-free reduced homocysteine to free oxidized homocysteine (Uthus et al., 2002). Selenium deficiency in rats can result in calcium ionophore A23187-stimulated lymphocytes producing less prostaglandins and lower phopholipase-D activation by 12-O-tetradecanoylphorbol-13-acetate. From these findings the investigators concluded that dietary selenium plays an important role in the regulation of arachidonic acid metabolism that affects phopholipase-D activation (Cao et al., 2002). The activity and expression of GSH-Px, SOD activity, and total antioxidant capacity were significantly decreased in selenium-deficient rats (Wu et al., 2003).

The nutritional requirement for selenium has been assessed using various criteria. GSH-Px activities in tissues other than red blood cells were maximized at 200 µg/kg dietary selenium when the selenium source was sodium selenite (Whangler and Butler, 1988). Pence (1991) reported that the GSH-Px activities in liver and colon of rats fed a diet containing 120 µg/kg dietary selenium were only 59% of the activities in those fed 520 µg/kg dietary selenium. Yang et al. (1989), using sodium selenate as the selenium source, reported GSH-Px activities in plasma and liver were maximized at 500 µg/kg dietary selenium. The liver GSH-Px activity of rats fed 1000 µg/kg dietary selenium for 25 weeks was 20% higher than in those fed 100 µg/kg dietary selenium (L'Abbe et al., 1991). Eckhert et al. (1993) reported that a diet containing 200 µg/kg dietary selenium protected retinal microvasculature from sucrose-induced damage compared to rats fed a diet containing 100 µg/kg dietary selenium. Incorporation of selenium-75 into selenoproteins and GSH-Px mRNA concentrations were used to determine the amount of dietary selenium required to maximize GSH-Px in growing rats. The results showed that 100 µg/kg dietary selenium was required for all three assessments (Evenson et al., 1992; Sunde et al., 1992). Plasma selenoprotein-P concentrations reached a plateau between 100 and 500 µg/kg dietary selenium (Yang et al., 1987). Rats were fed diets containing

between 10 and 500 µg/kg dietary selenium for 20 weeks, and the liver and thyroidal 5'-deiodinase activities were significantly decreased only in those fed the 10 µg/kg dietary selenium (Vadhanavikit and Ganther, 1993).

Based on the previous data, the NRC (1995) estimated minimal selenium requirement for growth and maintenance at 150 µg/kg dietary selenium. However, data collected from other investigators indicate that the minimal selenium requirement in the form of selenite for pregnant and lactating rats may be at least 400 µg/kg dietary selenium (Smith and Picciano, 1986, 1987). If the dietary source of selenium is selenate or selenomethionine, the requirement may be less (Whanger and Butler, 1988; Lane et al., 1991; Vendeland et al., 1992).

CHROMIUM. Chromium is a ubiquitous metal found in several oxidation states, with the trivalent form being the most important for animals. Brewer's yeast, a common ingredient in laboratory animal diets, contains a significant amount of glucose tolerance factor, a biologically active organically-complexed form of chromium. Glucose tolerance factor, which has not been purified, is thought to contain chromium attached to nicotinic acid and the amino acids glycine, glutamic acid, and cysteine (Stoecker, 1999).

The primary function of chromium, as part of glucose tolerance factor, is to potentiate insulin action that affects glucose uptake, intracellular carbohydrate, and lipid metabolism. In the rat it has been shown that chromium is involved in pancreatic insulin secretion (Striffler et al., 1993). There is evidence that chromium may improve glucose intolerance and affect lipoprotein lipase activity (Thomas and Gropper, 1996; Anderson, 1997). Research findings indicate that another role for chromium is in the maintenance of the structural integrity of nuclear strands and in the regulation of gene expression (Stoecker, 1999).

Amino acids such as methionine and histidine acting as ligands to chromium improve its solubility, which enhances its absorption (Stoecker, 1999). Rats dosed with [51]CrCl$_3$ and supplemented with ascorbic acid had increased ^{51}Cr concentration in the urine without reducing tissue levels, thus indicating ascorbate enhanced chromium absorption (Seaborn and Stoecker, 1990). Absorption of ^{51}Cr was increased in zinc-deficient rats, but zinc repletion decreased chromium absorption, indicating competition for absorption (Offenbacher et al., 1997).

Early research showing an enhancing effect on insulin and glucose metabolism in the rat provided evidence for the essentiality of chromium (Schwartz and Mertz, 1959; Schroeder et al., 1963; Schroeder, 1966; Roginski and Mertz, 1969). Chromium deficiency in rats results in retarded growth, insulin hyper-responsiveness to glucose, increased insulin resistance, decreased cAMP-dependent phosphodiesterase activity, decreased glycogen reserves, increased incidence of aortic lesions, and disturbances in amino acid utilization for protein synthesis (Mertz, 1969; Striffler et al., 1995, 1999). Striffler et al. (1999) showed chromium deficiency resulted in elevated insulin secretory response to glucose. Insulin resistance induced by feeding a high-fat, low-chromium diet to Wistar rats was improved by chromium supplementation (Striffler et al., 1998). However, findings showing vanadium has similar effects on insulin as chromium question its specificity (Fagin et al., 1987; Pederson et al., 1989). Other studies do not confirm the positive effects of chromium on glucose tolerance or glucose utilization in rats (Woolliscroft and Barbosa, 1977; Flatt et al., 1989; Holdsworth and Neville, 1990).

Spicer et al. (1998) examined the effect of chromium depletion and streptozotocin-induced diabetes on maternal and fetal insulin, glucose, IGF binding protein, IGF-I and IGF-II concentrations, pregnancy outcome, and fetal and placental protein and hydroxyproline content. Female rats were fed a low-chromium diet (70 µg/kg diet) starting at 21 days of age and were then fed a diet containing 40 µg/kg chromium starting at day 1 of pregnancy. Chromium depletion increased urinary hydoxyproline excretion, percentage of protein per fetus, and fetal IGF-I and IGF-II concentrations. Chromium depletion had no effect on maternal hormones, IGF binding protein, glucose, or placental and fetal hydroxyproline concentrations.

Roginski and Mertz (1969) showed rats fed a low-chromium diet (100 µg/kg diet) had a decrease in glycogen formation in liver and heart after an injection of insulin compared with control rats. Campbell et al. (1989) studied the effects of low dietary chromium and exercise on liver and muscle glycogen, glycogen synthase, and phosphorylase. The results of the study showed that after 18 weeks, liver glycogen phosphorylase activity for rats fed the chromium supplemented diet was higher than was the activity for the rats fed the non-supplemented diet. Dietary chromium increased total protein concentration in the liver but decreased it in the gastrocnemius muscle. There was also a chromium/exercise interaction on glycogen synthase activities in the liver and gastrocnemius muscle.

Kim et al. (2002) examined the effects of chromium on insulin sensitivity and glucose intolerance in insulin-resistant rats induced by dexamethasone treatment. Their findings show that chromium supplementation in dexamethasone-treated rats can decrease serum trigylcerides, increase insulin sensitivity, and reverse a catabolic state.

Evaluations of the toxicity of chromium chloride and chromium tripicolinate were studied at 100 mg/kg diet (Anderson et al., 1997). The results showed there were no differences in blood glucose, cholesterol, triglycerides, blood urea nitrogen, and lactic acid dehydrogenase total protein, as well as no histological differences in liver and kidney between controls and chromium supplemented rats.

MANGANESE. Manganese is biologically active in two oxidative states Mn^{2+} and Mn^{3+}. Mn^{2+} is the form in solution, metal-enzyme complexes, and in metalloenzymes. Mn^{3+} is the form found in the enzyme manganese SOD, the oxidative state that binds to transferrin, and the form that may interact with Fe^{3+}. The chemistries of Mn^{2+} and Mg^{2+} are similar; thus, most enzymatic reactions activated by Mn^{2+} are nonspecific because they can be activated by Mg^{2+} (Nielsen, 1999). Rats fed a diet low in calcium and magnesium had increased manganese concentration in central nervous system tissues and visceral organs (Yasui et al., 1995). Among rats fed a magnesium-deficient diet for 2 weeks, manganese concentrations in plasma, brain, spinal cord, lung, spleen, kidney, and bone were depleted (Kimura et al., 1996). Wistar rats fed a magnesium-deficient diet had significantly increased manganese absorption with increased blood, skeletal muscle, and kidney manganese concentrations (Sanchez-Morito et al., 1999).

Absorption efficiency apparently declines with increased dietary manganese intake, and increases with decreased manganese status; however, the mechanism of manganese absorption remains unknown. Findings indicate absorption could be through an active process involving a high-affinity, low-capacity, active transport system or by a nonsaturable, simple diffusion process (Wiegand et al., 1986; Davis et al., 1992). Research findings show that fiber, oxalic acid, calcium, and phosphorus can precipitate manganese in the rat gastrointestinal tract and prevent absorption (Garcia-Aranda et al., 1983). Iron competes with manganese for common binding sites, thus influencing manganese absorption (Johnson and Korynta, 1992). Davis et al. (1992) reported that iron depressed manganese absorption by inhibiting manganese uptake into mucosal cells. It was concluded that control of gut absorption is the primary means to maintain manganese homeostatsis. Rodriguez-Mataz et al. (1998) showed manganese absorption was increased during iron deficiency, but did not reflect manganese concentrations in the organ tissues. Manganese availability is also influenced by protein concentration and source (Johnson and Korynta, 1990; Takeda et al., 1996).

On absorption, the Mn^{3+} is bound to transferrin and taken up by extrahepatic tissue (Chua and Morgan, 1996). Manganese and iron interact during transfer from the plasma to the brain, liver, and kidneys in a synergistic nature. Manganese is found primarily in mitochrondria-rich organs such as liver, kidney, and pancreas. A high concentration of manganese is found in bone as part of the apatite moiety (Nielsen, 1999).

Manganese functions as an enzyme activator and as a constituent of metalloenzymes. Enzymes activated by manganese include oxidoreductases, lyases, ligases, hydrolases, kinases, and transferases (Groff and

Gropper, 2000a). Most enzymes activated by manganese can be activated by magnesium. Exceptions are the glyocsyltransferases which are involved in syntheses of mucopolysaccharides, important components of connective tissue. A manganese-deficient diet fed to rats resulted in increased plasma ammonia concentration and a decreased plasma urea concentration that was associated with decreased arginase activity (Brock et al., 1994). Phosphoenolpyruvate carboxykinase is important in gluconeogenesis, and its activity is decreased with manganese deficiency (Groff and Gropper, 2000).

The manganese-dependent metalloenzyme, SOD, prevents lipid peroxidation by superoxide radicals as do the copper- and zinc-SOD. However, manganese-SOD is located in the mitochondria, and copper- and zinc-SOD are found in the cytoplasm. Manganese-deficient rats have been shown to have depressed manganese-SOD activity in the heart and liver (Zidenberg-Cherr et al., 1983; Davis et al., 1990, 1992; Malecki et al., 1994). Dietary manganese protects against *in vivo* lipid peroxidation of heart mitochondrial membranes from Sprague-Dawley rats (Malecki and Greger, 1996). The heart mitochondrial manganese-SOD activities for rats fed diets containing high concentrations of γ-linolenic acid were significantly increased compared with the manganese-SOD activities for rats fed a diet high in linoleic acid (Phylactos et al., 1994). Finley and Davis (2001) observed decreased manganese absorption and retention in rats fed a diet high in saturated fat as compared with control rats. Increasing dietary lipid and iron content decreases manganese-SOD activity in colonic mucosa (Kuratko, 1997). Aberrant crypt foci, preneoplastic lesions of colon cancer, were increased, and heart SOD activity was decreased in rats fed a low-manganese diet (Davis and Feng, 1999). The investigators suggested that dietary alterations affecting SOD activity may affect cancer susceptibility.

Hurley and Keen (1987) reported litters from dams fed a diet containing 1 mg/kg manganese were characterized by ataxia, skeletal defects, and a high incidence of early postnatal death. Litters from dams fed 3 mg/kg dietary manganese were normal except that certain rat strains still had a high incidence of ataxia owing to inner ear defects (Baly et al., 1986; Hurley and Keen, 1987). Rats fed diets containing less than 1 mg/kg dietary manganese had reduced food consumption, decreased growth, bone abnormalities, and early mortality. Manganese deficiency also results in decreased pancreatic insulin synthesis (Baly et al., 1985). Manganese-deficient weanling male rats have been reported to have elevated pancreatic amylase activity which was not reversed by manganese repletion (Brannon et al., 1987; Werner et al., 1987). The pancreatic lipase was significantly increased in weanling rats fed a manganese-deficient and high-fat diet (Werner et al., 1987). Clegg et al. (1998) reported manganese-deficient rats had reduced BW compared with controls, but daily diet intake was not decreased. The manganese-deficient rats also had lower insulin and circulating concentrations of IGF-I and an elevated growth hormone status. The investigators suggested that these alterations in IGF metabolism are responsible for the growth and bone abnormalities observed in manganese-deficient rats. Eder et al. (1996) showed feeding offspring of manganese-depleted dams a diet containing 0.1 mg/kg dietary manganese affects growth and thyroid hormone metabolism, but a dietary concentration of 0.5 mg/kg dietary Mn is sufficient for growth and normal thyroid hormone metabolism.

Decreases in HDL protein, cholesterol, and apolipoprotein E owing to manganese deficiency were more pronounced in Sprague-Dawley rats than Wistar rats (Kawano et al., 1987). Davis et al. (1990) reported lowered plasma cholesterol, HDL cholesterol, HDL protein, and HDL apolipoprotein E in manganese-deficient Sprague-Dawley rats, confirming Kawano's findings. Klimis-Tavantzis et al. (1983) reported manganese deficiency in Wistar rats caused decreased hepatic fatty acid synthesis but had no effect on cholesterol and lipid metabolism in the hypercholesterolemic rat. Examination of the ultrastructural architecture of the optic nerves from manganese-deficient rats showed decreased diameters and lamellae of myelinated axons and abnormal mitochondria in the axons (Gong and Amemiya, 1999). Femur calcium concentration was decreased in manganese-deficient rats (Strause et al., 1986). Manganese supplementation was an effective inhibitor of loss of bone mass after ovariectomy in rats (Rico et al., 2000).

There are conflicting data regarding the rat dietary manganese requirement. Early research findings indicated the optimal manganese intake for growth was between 2 and 50 mg/kg diet (Holtkamp and Hill, 1950; Anderson and Parker, 1955). More recent data indicate that 5 mg/kg of diet is adequate for normal development and growth (Baly et al., 1986; Hurley and Keen, 1987). Because there are reports indicating different strains of rats respond differently to dietary manganese intake (Hurley and Bell, 1974; Kawano et al., 1987) the NRC (1995) estimated the manganese requirement at 10 mg/kg diet. The lack of data supporting the NRC (1978) manganese dietary requirement of 50 mg/kg diet and the potential negative effects of excess manganese on iron metabolism (Davis et al., 1990) resulted in the reduced manganese requirement.

An inverse relationship between manganese supplementation level and striatal dopamine concentration has been reported and indicates that intake of high dietary manganese during infancy can be neurotoxic to rat pups, resulting in developmental deficits (Tran et al., 2002).

IODINE. Iodine, a nonmetal, functions in its ionic form iodide, I^-. Iodine is involved in the synthesis of, and is an essential component of, thyroid hormones, T_4 and T_3. The primary functions of the thyroid hormones are growth and development (Hetzel and Clugston, 1999).

Iodide is absorbed rapidly throughout the gastrointestinal tract. After absorption, iodide is transported via the blood to tissues. The thyroid gland aggressively traps iodide by using a sodium-dependent, active transport mechanism called the iodine pump which is regulated by thyroid-stimulating hormone released from the pituitary to regulate thyroid secretion. The nucleus is the site from which thyroid hormone regulates metabolic events (Clugston and Hetzel, 1994).

Early research indicates that the iodine requirement in the rat is between 100 and 200 µg/kg diet (Levine et al., 1933; Remington and Remington, 1938; Halverston et al., 1945; Parker et al., 1951). The NRC (1995) estimated iodine requirement for growth and reproduction in the rat is 150 µg/kg diet. Signs of iodine deficiency in the rat are goiter (Taylor and Poulson, 1956), impaired reproduction (Feldman, 1960), decreased serum T_4 (Abrams and Larsen, 1973), increased serum thyrotropin (Pazos-Moura et al., 1991), and increased Type I iodothyronine 5'-deiodinase activity (Arthur et al., 1991). Chronic dietary iodine deficiency in rats results in thyroid follicular adenomas and follicular carcinomas (Ohshima and Ward, 1986; Ward and Ohshima, 1986). Krupp and Lee (1988) reported there is a redistribution of organelles in thyroid cells owing to iodine deficiency and iodine supplementation, which may be related to alterations in intracellular iodine metabolism. Iodine deficiency induced thyroid autoimmune reactivity in Wistar rats (Mooij et al., 1993).

Both iodine and selenium are required for optimal thyroid hormone metabolism. Type I iodothyronine 5'-deiodinase, which is involved in the synthesis of T_3, is a selenoenzyme. Beckett et al. (1993) reported rats deficient in selenium and iodine had significantly lower thyroid T_4, T_3, total iodine, and hepatic and plasma T_4, but plasma thyroid-stimulating hormone and thyroid weight were significantly higher than in rats deficient in iodine alone. In second-generation iodine-deficient rats, the thyroid gland mRNA levels for iodothyronine deiodinase, cytosolic GSH-Px, and phospholipid hydroperoxide GSH-Px increased (Mitchell et al., 1996). Contempre et al. (1993) demonstrated that rats deficient in both iodine and selenium had more thyroid tissue damage than did rats deficient in iodine alone. Thyroidal selenium-dependent GSH-Px was decreased in Sprague-Dawley rats fed a diet containing high iodine and low selenium concentrations (Hotz et al., 1997). The findings suggest the high iodine intake when selenium is deficient may permit thyroid tissue damage owing to low thyroidal GSH-Px activity during thyroidal stimulation.

Ammerman et al. (1964) showed rats fed 500 to 2000 mg/kg dietary iodine during pregnancy had increased neonatal mortality, and rats fed 500 mg/kg dietary iodine had decreased milk production. The fertility of male rats fed 2500 mg/kg dietary iodine was not affected. Pregnant rats fed 500 to 1000 mg/kg dietary iodine had increased number of stillbirths and decreased survival rates of liveborn pups. Most pup deaths were owing to iodine-induced anemia and/or agalactia (Stowe et al., 1980). Feeding excess iodine to the Buffalo strain of rat, which is genetically susceptible to autoimmune thyroiditis, resulted in enhanced thyroglubulin antibody production and an increase in the severity of the thyroiditis (Cohen and Weetman, 1988). Fischer et al. (1989) showed that feeding BB/Wistar rats diets containing 2 to 3 mg/kg dietary iodine have increased thyroglobulin antibodies and lymphocytic thyroiditis. Excessive dietary iodine intake results in swollen and disrupted mitochondria and extreme dilation of rough endoplasmic reticulum as well as accelerates development of lymphatic thyroiditis in the BB/Wistar but not the Wistar rat (Li and Boyages, 1994).

MOLYBDENUM. Molybdenum is a transition metal that is usually found in the body in either the Mo^{4+} or Mo^{6+} valence state bound to sulfur or oxygen. As a transition element that easily changes oxidation state, it functions as an electron transfer agent in oxidation-reduction reactions. Molybdenum is found in tissues as molybdate, molybdopterin, or bound to enzymes. The liver, kidney, and bone contain the largest amounts of molybdenum (Groff and Gropper, 2000a).

Molybdenum, as molybdopterin, is a cofactor for three metalloenzymes that catalyze oxidation reduction reactions (Nielsen, 1999). Sulfite oxidase catalyzes the terminal step in the metabolism of sulfur-containing amino acids such as methionine and cysteine. Aldehyde oxidase functions in the liver as a true oxidase, oxidizing and detoxifying pyrimidines, purines, and pteridines. Xanthine dehyrogenase and oxidase enzymes hydoxylate purines, pteridines, and pyrimidines. Purine catabolism results in the production of hypoxanthine which is oxidized by xanthine dehydrogenase, producing uric acid (Rajagopalan, 1988). Rats fed diets with very low molybdenum concentrations have decreased activities for these enzymes but do not exhibit signs of deficiency. Feeding the molybdenum antagonist, tungsten, to rats will almost eliminate the molybdenum-dependent enzymes activities, but there is no effect on growth (NRC, 1995).

A study examining the effect of low dietary molybdenum on intestinal and hepatic xanthine oxidase/dehydrogenase indicated 20 µg/kg dietary molybdenum maintained normal growth and reproduction, but xanthine oxidase activity was low (Higgins et al., 1956). Results from titration experiments showed intestinal xanthine oxidase/dehydrogenase activity was maximized at 100 µg/kg dietary molybdenum.

A more recent experiment studied the criteria for the assessment of nutritional status of molybdenum and to determine the requirement for female rats fed the AIN-76A diet (Wang et al., 1992). Molybdenum concentration in the brain and liver increased linearly in rats fed diets containing 50, 100, and 200 µg/kg dietary molybdenum. There was no further increase in brain and liver molybdenum concentration when the rats were fed diets containing higher levels of molybdenum. Maximal activities for hepatic xanthine dehydrogenase/oxidase, hepatic sulfite oxidase, and small-intestinal mucosa xanthine dehydrogenase/oxidase were attained at 50, 50, and 100 µg/kg dietary molybdenum, respectively. The investigators estimated the molybdenum requirement for rats fed the AIN-76 A diet was 200 µg/kg dietary molybdenum. The estimated molybdenum requirement is 150 µg/kg dietary molybdenum (NRC, 1995).

Parry et al. (1993) reported that excessive dietary molybdenum intake (6 mg/kg dietary molybdenum) in rats caused a significant reduction in longitudinal bone growth associated with decreased glucose-6-phosphate dehydrogenase activity and cell proliferation.

FLUORINE. Fluorine exists as the fluoride ion or as hydrofluoric acid in body fluids. Almost 99% of total body fluoride is found in mineralized tissues as fluoride apatite. A major function of fluoride is its ability to promote mineral precipitation from meta-stable solutions of calcium and phosphate, resulting in the formation of apatite (Groff and Gropper, 2000a). This does not meet the definition of an essential function. Rao (1984) reported that aluminum, calcium, magnesium, and chloride reduce fluoride uptake, but phosphate and sulfate increased fluoride uptake. Rats fed a chloride-deficient diet had significant fluoride retention and skeletal uptake (Cerklewski et al., 1986). Metabolic studies showed low dietary magnesium enhanced fluoride absorption and high dietary magnesium reduced fluoride absorption (Cerklewski, 1987).

Neither dietary iron nor zinc affected skeletal uptake of fluoride in the rat (Cerklewski, 1985). High dietary protein can enhance fluoride absorption but reduce fluoride retention in rat femurs (Boyd and Cerklewski, 1987). Consumption of excessive fluoride facilitates calcium oxalate crystalluria, promoting bladder stones in rats (Anasuya, 1982). Fluoride supplementation reduces caries development in rats exposed to cariogenic challenges (Tabchoury et al., 1998). Significant reductions in plasma cholesterol, plasma and VLDL-esterified cholesterol, and HDL cholesterol were observed in RICO rats fed a diet supplemented with fluoride and magnesium (Luoma et al., 1998). Fluoride deficiency in laboratory animals results in infertility, anemia, and slow growth (Groff and Gropper, 2000a).

ARSENIC. In food, arsenic is found in both organic and inorganic forms. In biologic material it exists in both trivalent and pentavalent ionic forms (Vahter, 1983). The most biochemically important organoarsenicals are methylated which are the least toxic and most readily absorbed. The form of the organic arsenic determines its absorption rate. In rats it has been shown that 70% to 80% of arsenocholine was recovered in the urine (Yamauchi et al., 1986).

The metabolic function of arsenic is not positively defined. Recent research suggests that arsenic is involved in methionine metabolism to taurine and arginine (Nielsen, 1993). Taurine production from methionine is decreased in arsenic-deficient rats (Uthus and Nielsen, 1993). Growth deficits and decreased cystathionase and ornithine decarboxylase activities were observed in arsenic-deficient rats fed guanidoacetate, an arginine metabolite, compared with arsenic-supplemented rats (Uthus, 1992). Phosphatidylcholine (PC) synthesis was altered in arsenic-deprived rats indicating arsenic may play a role in phospholipid synthesis (Cornatzer et al., 1983). Arsenic appears to play a role in gene expression at the transcription level by affecting methylation of core histones (Desrosiers and Tanguay, 1986). The most consistent signs of arsenic deprivation in rats are depressed growth, impaired fertility, and increased prenatal mortality (Uthus, 1994). Factors that interact with arsenic and affect the animal's response to arsenic deprivation include zinc, arginine, choline, methionine, taurine, and guanidoacetic acid.

Uthus and Poellot (1991) showed that rats fed a diet deficient in arsenic, pyridoxine, and methionine had impaired growth and abnormal plasma amino acids. These findings indicated an interaction between arsenic and pyridoxine that could be affected by methionine status. Effects of arsenic deprivation can be influenced by nutritional stressors that affect sulfur amino acid or labile-methyl-group metabolism (Nielsen, 1999).

BORON. Boron in food, sodium borate, and boric acid are rapidly absorbed and excreted mostly in the urine. Boron complexes with sugars, adenosine-5-phosphate, pyridoxine, riboflavin, dehydroascorbic acid, and pyridine nucleotides (Nielsen, 1999). Boron is distributed throughout the body tissues, but it is most concentrated in bone, fingernails, and teeth (Ward, 1993). Although a biochemical function for boron has not been clearly defined, it is involved with the composition, structure, and strength of bones. Rats fed diets containing 200 to 9000 ppm boric acid had reduced serum phosphorus and magnesium; serum chloride was increased and bone strength was increased (Chapin et al., 1997). The mechanism is unknown, but it appears that boron affects the metabolism of magnesium, calcium, phosphorus, and vitamin D (Nielsen and Shuler, 1992; Hunt, 1994). Dietary boron and estrogen treatment in ovariectomized rats increased calcium, phosphorus, and magnesium absorption as well as improved trabecular bone quality (Sheng et al., 2001a,b). Animal responses to boron-depleted diets are most significant when dietary calcium, vitamin D, or magnesium are also low (Nielsen et al., 1988; Nielsen, 1993; Hunt, 1994). It is hypothesized that boron is involved in cell membrane function or stability such that it influences the response to hormone action, transmembrane signaling, or transmembrane movement of regulatory cations or anions (Nielsen, 1991). This theory is supported by data showing that boron influences the transport of extracellular calcium and the release of intracellular calcium in rat platelets activated by thrombin (Nielsen, 1994).

NICKEL.   A specific function of nickel in animal nutrition has not been clearly defined. Nickel may function as a cofactor or structural component in metalloenzymes. In several enzyme systems, nickel can be substituted for other minerals such as magnesium (Fishelson et al., 1982). There appears to be a synergistic relation between iron and nickel; signs of nickel deprivation were more severe when dietary iron was low, and signs of iron deficiency were more severe when dietary nickel was deficient (Nielsen et al., 1980). It appears that nickel is absorbed by the absorptive mechanism for iron in the intestinal epithelium (Tallkvist and Tjalve, 1997). Vitamin B_{12} and folic acid affect signs of nickel deprivation in rats (Nielsen et al., 1989). The nickel and folic acid interaction affects the vitamin B_{12}- and folate-dependent pathway of methionine synthesis (Nielsen et al., 1993; Uthus and Poellot, 1996, 1997). Nickel deficiency in rats resulted in reduced sperm count and reduced sperm motility, decreased the weight of the seminal vesicles and prostate glands, and decreased testicular nucleic acids (Das and Dasgupta, 2000; Yokoi et al., 2003).

SILICON.   The chemistry of silicon is similar to carbon. Silicon is involved in the growth and development of bone, connective tissue, and cartilage. Examination of the biochemical changes in bone during silicon deficiency indicates silicon influences bone formation by affecting the process of cartilage calcification (Carlisle, 1981, 1988; Nielsen, 1999). Most of the signs of silicon deficiency in rats indicate abnormal metabolism of connective tissue and bone (Nielsen, 1999). Rats fed a diet high in aluminum and low in calcium and silicon accumulated significant amounts of aluminum in the brain. However, silicon supplements prevented accumulation of aluminum in the brain (Carlisle and Curran, 1987). Thyroidectomized rats fed a diet containing high-aluminum and low-silicon concentrations had low brain zinc concentrations. Silicon supplementation prevented the depression in brain zinc levels (Carlisle et al., 1991).

G. Vitamins

VITAMIN A.   The period of vitamin discovery started with thiamin in 1912 and ended with vitamin B_{12} in 1948. As other nutrients, vitamins are required for growth, maintenance, and reproduction. The vitamin classification defines a group of organic compounds required in minute amounts that are not catabolized to satisfy the energy requirement and are not used for structural purposes. Vitamin functions include regulation of metabolism by gene expression and cofactors for enzymes, facilitating the conversion of fat and carbohydrate to energy and helping in the formation of bones and tissues. Vitamins are classified according to their solubility in fat or water.

1. Fat Soluble Vitamins

Vitamin A is a fat-soluble vitamin in the form of retinyl esters when the dietary source is of animal origin. Natural vitamin A is usually found as retinyl ester, a molecule of retinol esterified with a fatty acid. Vitamin A is required by rats for vision, differentiation of epithelial cells, and reproductive functions. All three functions can be supported by dietary retinyl esters, retinol and retinal, which are interconvertible. Retinal can be oxidized to form retinoic acid, but the reaction is irreversible. Dietary retinoic acid supports growth and epithelial differentiation but not vision and reproduction (Ross, 1999).

Plants do not contain vitamin A; however, some plants have high concentrations of provitamin A-active compounds, such as beta-carotene, which is the most active of the carotenoids. Beta-carotene is cleaved and converted to retinol in the small intestine. The biological activity of vitamin A and carotenoids is not equivalent on a per-weight basis and is affected by many factors, including the presence of fat in the diet, diet preparation, and the binding of carotenoids to other components of the diet. Rats do not readily absorb carotenoids and are known as a "white fat animal" (Brody, 1999).

The primary source of vitamin A in rat diets is in the form of a retinyl ester in the vitamin premix. Vitamin A is prone to oxidation when exposed to oxygen, light, and heat in a humid environment. Therefore, retinyl acetate, proprionate, or palmitate in the presence of antioxidants are stabilized in the form of a gelatin-carbohydrate matrix and used in laboratory rodent diets. The beadlets can retain 90% of their activity for 6 months under good storage conditions (Olson, 1984). Based on kinetic studies (Green et al., 1987; Green and Green, 1991) and a lactation study (Gerber and Erdman, 1982) the NRC (1995) recommended vitamin A requirement is 2300 IU/kg diet (2.4 µmol/kg). Increased dietary concentrations are recommended owing to the instability of vitamin A and to the evidence that animals exposed to stress respond better at higher dietary concentrations (Gerber and Erdman, 1982; Demetriou et al., 1984). The typical vitamin A concentration in laboratory rodent diets is over 20,000 IU/kg diet of diet, much higher than the NRC requirement.

For a healthy animal, the overall absorption of dietary vitamin A is between 80% and 90%. Retinol is esterified with long-chain fatty acids in the intestinal mucosa and the retinyl esters are released into the lymphatic system as chylomicrons where they deliver retinyl esters, some unesterified retinol, and carotenoids to extrahepatic tissues and the liver (Blomhoff et al., 1992). In healthy animals, 90% of the vitamin A is stored in the liver from where it is mobilized to other tissues as retinol linked to retinol binding protein (RBP). Factors such as nutritional status, intestinal mucosal integrity, and dietary protein and fat concentrations influence bioavailability and digestion of vitamin A (Olson, 2001). Vitamin A metabolism is affected by protein status as transport and utilization depends upon vitamin A binding proteins (Olson, 1991). Rosales et al. (1999) showed that iron deficiency causes a reduction of plasma retinol and an accumulation of hepatic retinyl esters that are refractory to vitamin A intake. In a search for an appropriate rat strain to mimic vitamin A metabolism in human diabetics, Tuitoek et al. (1994) determined that vitamin A metabolism may be strain dependent.

Vitamin A is essential for vision, cellular differentiation, growth, reproduction, bone development, immune function, development of the cardiovascular and central nervous systems, and the integrity of epithelial tissues (Wolbach and Howe, 1925) Tissues that are normally composed of columnar or cuboidal mucus secreting cells, were squamous, dry, and keratinized in vitamin A-deficient rats. Tissues particularly sensitive to vitamin A deficiency were trachea, skin, salivary gland, cornea, and testes. Retinoid-induced cell differentiation can be accompanied by an inhibition of cell proliferation (Pfhal, 1993; Wang et al., 1997), or initiation of apoptosis (Lotan, 1995). Induction of differentiation, inhibition of proliferation, and induction of apoptosis have been shown to be related to the actions of retinoids as anti-cancer agents and in normal embryonic development (Olson, 2001). Evidence shows nuclear retinoic acid receptors appear in different cells during different times of development, suggesting that retinoic acid acts as a morphogen in embryonic development (Wolf, 1991; Hoffman et al., 1994; Smith et al., 1998). Offspring of vitamin A-deficient rats show many abnormalities, including craniofacial defects, microphthalmia, umbilical hernia, edema, and spongy tissue structures of the liver, heart, and thymus (Morriss-Kay and Sokolava, 1996).

UDP-glucuronosyltransferases play a major role in detoxification and elimination of endogenous substrates such as steroids, bile acids, bilirubin, and exogenous compounds, including food additives, therapeutic drugs, and environmental pollutants. Haberkorn et al. (2002) has shown in rats that the expression of family 1-UDP-glucuronosyltransferase mRNAs are co-regulated by both vitamin A and thyroid hormones.

Most animals, including rats, are born with very low liver vitamin A content because of limited placental transfer of fat-soluble vitamins (Ross, 1999). Therefore, the status of neonates depends on the vitamin A concentration of the dam's milk and the postweaning diet. Vitamin A deficiency in rats is characterized by anorexia (NRC, 1978), retarded growth (Rogers et al., 1971), epithelial metaplasia and keratinization (Underwood, 1984), xerophthalmia (Wald, 1968), bone defects (Underwood, 1984), increased cerebral spinal fluid pressure (Corey and Hayes, 1972), reproductive failure (Wilson et al., 1953), and compromised immune system (Ross, 1994). The required retinol intake for repletion of vitamin A-deficient rats varies with criteria being examined such as epithelial keratinization, hepatic storage, or retinol kinetics. The retinol intake for repletion of vitamin A-deficient rats based on growth rate, positive vitamin A balance, and hepatic storage criteria are 14, 100, and 200 nmol/kg BW/day, respectively (NRC, 1995).

Hypervitaminosis A, the result of an acute overdose or chronic exposure to excessive levels, is indicated by the presence of esterified retinol with normal holo-RBP concentration in fasting plasma (Smith and Goodman, 1976). Vitamin A toxicity varies according to its chemical form, retinoic acid being the most toxic followed by retinol, and retinyl esters (NRC, 1995). Rats fed 25,000 to 75,000 IU/day vitamin A for 16 days exhibited weight and hair loss, and loss of balance. Osteoporosis was observed at necropsy. Renal and cardiac calcification were present in rats fed 75,000 IU/day vitamin A. Rats fed 180 μmol retinol/kg BW/day showed signs of acute toxicity (Leelaprute et al., 1973). Vitamin A toxicity was observed in rats fed 47 μmol retinoic acid/kg BW/day (Kurtz et al., 1984). Carotenoids are not considered toxic up to levels of 1800 μmol/kg BW/day (Heywood et al., 1985). Research shows that vitamin E can significantly reduce the toxicity of vitamin A (Jenkins and Mitchell, 1975).

Vitamin A interacts with iron (Van Houwelingen et al., 1992; Sijtsma et al., 1993;), zinc, copper, and vitamins E and K (Smith et al., 1973; Olson, 1991). Vitamin E provides protection against oxidation during the conversion of β-carotene to retinal (Erdman et al., 1988). Excess vitamin A, however, may interfere with vitamin K absorption (Groff and Gropper, 2000c).

It has been shown that long-term dietary restriction in rats results in increased intestinal retinol absorption and hepatic retinoid concentrations (Ferland et al., 1992; Chevalier et al., 1996). A follow-up study investigating retinoid and RBP metabolism, as influenced by dietary restriction, showed low plasma retinol-RBP levels, which were associated with decreased hepatic RBP, but unchanged RBP mRNA levels. Dietary restriction did not reduce the retinoid content in extrahepatic tissues, indicating alternative pathways of retinol delivery other than that involving RBP (Chevalier et al., 1996).

Elements of the immune system function are influenced by vitamin A. In vitamin A deficiency, both cell-mediated and antibody-mediated immunity responses are depressed. Research shows vitamin A appears to be required for T-lymphocyte function and for antibody response to viral, parasitic, and bacterial infections (Carman et al., 1992, Cantora et al., 1994). Vitamin A deficiency can also result in impaired phagocytosis and reduced natural-killer (NK) cell activity in young rats (Naus et al., 1985; Ross and Hammerling, 1994). The reduced activity is attributed to a decrease in both the number of NK cells as well as lytic efficiency (Zhao et al., 1994). Dawson et al. (1999) studied the effects of a wide range of vitamin A intakes from marginal (0.35 mg RE/kg diet) to supplemented (50 mg retinal equivalents/kg diet) on NK cells and found that vitamin A status affected the number of NK cells. Chronic marginal vitamin A deficiency reduces circulating NK cells, and vitamin A supplementation increased the number of NK cells. It was also shown that rats fed the marginal and supplemented vitamin A diets for a lifetime may have impaired maintenance of T cell-dependent and/or NK T cell-dependent immune responses, leading to increased risk of infectious and neoplastic disease in older animals (Dawson and Ross, 1999). Aging results in decreased immunocompetence, particularly in the differentiation and function of T cells which increases the susceptiblitiy to autoimmunity, infectious diseases, and cancer (Miller, 1995; Pawelec et al., 1998).

Neutrophils serve as the first line of defense against infections. The formation of neutrophils from myeloblasts depends on retinoic acid bound to the nuclear retinoic acid receptor (Gratas et al., 1993). Vitamin A deficiency in the rat disrupts normal neutrophil development and can result in decreased chemotaxis, adhesion, and phagocytosis (Twining et al., 1997). These defects may lead to decreased clearance of bacteria from the bloodstream of vitamin A-deficient rats reported by Ongsakul et al. (1985). However, vitamin A-deficient rats have normal numbers of neutrophils in the blood and inflamed tissues. This is due to the sequestration of retinol in bone marrow of vitamin A-deficient rats (Twining et al., 1996).

VITAMIN D.   Vitamin D is a fat-soluble vitamin which has two provitamin forms, ergosterol found in plants and 7-dehydrocholesterol found in animals. Exposure of the provitamins to sunlight results in the conversion to ergocalciferol (vitamin D_2) and cholecalciferol (vitamin D_3) in plants and animals, respectively. Cholecalciferol can be synthesized in the skin of rats when it is exposed to ultraviolet light in the range of 280 to 320 nm (NRC, 1995). In rats, vitamin D_2 is more potent than is vitamin D_3 (Brody, 1999). Because the body is able to produce cholecalciferol, vitamin D does not agree with the classic definition of a vitamin and should be referred to as a prohormone. Vitamins D_2 and D_3 are not biologically active and require successive hydroxylations in the liver and kidney to form 1,25-dihydroxy-ergocalciferol (ercalcitriol) and 1,25-dihydroxy-cholecalciferol (calcitriol), respectively, the biologically active forms of vitamin D (Reichel et al., 1989). Calcitriol functions similar to a steroid hormone because it is synthesized in one organ and acts on other target organs (Norman et al., 1994).

In the rat, vitamin D is absorbed via the lymphatic system and its associated chylomicrons (Schachter et al., 1964; Rosenstreich et al., 1971), approximately 50% of dietary vitamin D is absorbed (Norman and DeLuca, 1963; Schachter et al., 1964); however, exposure to sunlight provides the major source of vitamin D via production in the skin from 7-dehydrocholesterol. Vitamin D is transferred from the chylomicrons to vitamin D-binding protein (Mallon et al., 1980) which is involved in the cellular internalization of vitamin D sterols (Bouillon et al., 1981). The concentration of vitamin D-binding protein in plasma affects the proportion of bound to free vitamin D levels, which influences the biological activity of the hormone (van Baelen et al., 1988; Cooke and Haddad, 1989).

Vitamin D is transported by vitamin D-binding protein to the liver where it is hydroxylated, forming the major circulating form of vitamin D, 25-hydroxyvitamin D (25-OH-D). It has been shown that blood has the highest concentration of vitamin D compared with other tissues. Results from a rat study show that no tissue can store vitamin D against a concentration gradient (Rosenstreich et al., 1971).

The major function of vitamin D, with help from calcitonin and PTH, is to act on bone, kidney, parathyroid gland, and intestine to maintain intracellular and extracellular calcium and phosphate concentrations within a physiologically acceptable range (Lee et al.,1980; Haussler, 1986). Maintenance of plasma calcium concentration is important for normal functioning of the nervous system, muscle contraction, blood clotting, membrane structure, growth of bones, and preservation of bone mass. Phosphorus is an important element of DNA, RNA, membrane lipids, and ATP.

Vitamin D is required for a healthy skeleton. However, studies using vitamin D-deficient rats have shown that the rats had the ability to mineralize their bones in a manner similar to rats fed a vitamin D control diet

(Underwood and DeLucca, 1984; Dutta-Roy et al., 1993a,b). This indicates that vitamin D is not absolutely required for bone ossification, but it is responsible for maintaining extracellular calcium and phosphorus levels in a saturated state that results in bone mineralization (Nalecz et al., 1992).

Recent studies indicate that the hormonally active form of vitamin D, 1-α, 25-OH$_2$-D$_3$, has receptors and activities in tissues not related to calcium homeostasis such as brain, pancreas, immune cells, pituitary, and muscle. Findings show vitamin D is also involved in muscle function, immune and stress response, insulin and prolactin secretion, and cellular differentiation of skin and blood cells. Experiments show that with or without normal serum calcium concentrations, vitamin D increases insulin release from isolated perfused rat pancreas (Kadowski and Norman, 1984; Cade and Norman, 1986, 1987).

Vitamin D requirements are difficult to determine because cholecalciferol is produced in the skin on exposure to sunlight. The daily requirement for vitamin D in animals can also be affected by dietary calcium/phosphorus ratio, physiological stage of development, sex, amount of fur or hair, color, and strain. The estimated NRC (1995) requirement for the rat is 1000 IU vitamin D/kg of diet.

Vitamin D deficiency results in impaired intestinal absorption and renal reabsorption of calcium and phosphate. During vitamin D deficiency, serum calcium and phosphate levels decrease and serum alkaline phosphatase activity increases (Jones, 1971). Without sufficient serum calcium and phosphorus, bone mineralization cannot occur. Deficiency signs in animals include a decline in plasma concentrations of calcium and phosphorus, irritablity, low growth rate, tetany, and demineralization of the bones (Steenbock and Herting, 1955; Mathews et al., 1986; Uhland et al.,1992). Vitamin D deficiency in rats results in poor reproductive performance (Halloran and DeLucca, 1980; Kwiecinski et al., 1989); however, calcium and phosphorus supplementation has been reported to improve the reproductive performance in vitamin-deficient rats (Halloran and DeLucca, 1980; Mathews et al., 1986; Uhland et al., 1992).

There are several disease states that can affect vitamin D status (Collins and Norman, 2001). Fat malabsorption and gastric surgical procedures can impair vitamin D absorption. Liver disorders such as obstructive jaundice and primary biliary cirrhosis can result in malabsorption of calcium and bone disease. Renal failure can cause skeletal abnormalities, including growth retardation, osteitis fibrosa, osteomalacia, and osteosclerosis. Hyperparathyroidism results in a bone disease resembling osteomalacia, and hypoparathyroidism causes hypocalcemia.

Vitamin D toxicity can result in irreversible calcification of the heart, lungs, kidneys, and other soft tissues; early signs of hypercalcemia intoxication are followed by kidney calcification. Other signs of toxicity include hypercalciuria, anorexia, bone demineralization, uremia, kidney failure, and calcification of the arteries, liver, and heart (Potvliege, 1962; Bajwa et al., 1971). When pregnant rats were treated with 2500 µmol of ergocalciferol/day, pups had impaired ossification of long bones and reduced growth rate. These results indicate vitamin D toxicity can have a teratogenic effect (Ornoy et al., 1968).

VITAMIN E. The fat-soluble vitamin E was discovered during a series of studies on the influence of nutrition on rat reproduction (Evans and Bishop, 1922). Vitamin E deficiency results in dead fetuses, spontaneous abortions, and fetal resorption. Based on this deficiency effect, the rat fertility test was developed to measure the potency of various forms of vitamin E (Brody, 1999).

Vitamin E refers to two classes of compounds, the tocopherols and tocotrienols. Vitamin E activity varies among these compounds, with α-tocopherol being the greatest as determined by the rat fertility test. Vegetable oils are the best sources of vitamin E.

As an antioxidant, the principal function of vitamin E is for the maintenance of biological membrane integrity (Chow, 1991; Groff and Gropper, 2000c) and for development and maintenance of the nerves and skeletal muscle (Sokol, 1988; MacEvilly et al., 1996). Vitamin E has also been shown to regulate immune response or cell-mediated immunity by modulating the generation of prostaglandins and the metabolism of arachidonic acid (Goetzel, 1981; Douglas et al., 1986).

There are a number of species-dependent and tissue-specific signs of vitamin E deficiency exhibited by rats which include reproductive failure (Ames, 1974), kyphoscoliosis (humped back), erythrocyte destruction (Jager, 1972), depigmentation, abnormal behavior (Sarter and Van der Linde, 1987), kidney degeneration, (Gabriel et al., 1980), eosinophilia, skin ulcers, neurological abnormalities, and liver necrosis (Evans and Emerson, 1943; Maclin et al., 1977; Chow, 2001). The development and severity of vitamin E deficiency is related to the status of other nutrients, including vitamin A, β-carotene (Blakely et al., 1990), PUFAs (Jager and Hjoutsmuller, 1970; Buckingham, 1985), selenium (Hong et al., 1988), and sulfur amino acids (Hakkarauinen et al., 1986).

Vitamin E dietary requirement is difficult to determine because it is influenced by other nutrients such as PUFAs, ascorbic acid, selenium, vitamin A, and sulfur amino acids (Leedle and Aust, 1990; Chow, 1991). There are strain differences within a species. The SHR is more sensitive to a deficiency than are the Wistar-Kyoto and Sprague-Dawley rats (Bendich et al., 1983a,b; Bendich et al., 1986). By genetic selection, inbred strains of rats have been developed that are susceptible or resistant to vitamin E deficiency (Gabriel and Maclin, 1982; Bendich et al., 1983). Evans and Emerson (1943) showed that age is another factor influencing the rat vitamin E requirement. The daily vitamin E requirement to maintain fertility in male mice greater than 9 months of age increased from 0.18 to 0.57 mg/day. The vitamin E requirement to maintain 50% fetal viability increased sevenfold from the first pregnancy to the fourth (Ames, 1974).

The NRC (1995) recommends the vitamin E requirement for the most frequently used strains of rats to be 18 mg RRR-α-tocopherol/kg diet or 27 mg all-rac-α-tocopheryl acetate/kg diet when dietary fat concentration is not more than 10%.

VITAMIN K. Vitamin K is a cofactor for an enzyme necessary for the posttranslational carboxylation of specific glutamic acid residues to form γ-carboxyglutamate on prothrombin and factors VII, IX, and X of 13 factors required for normal coagulation of blood (Suttie, 2001). Four other vitamin K-dependent plasma proteins that inhibit coagulation have been found (Brinkley and Suttie, 1995). In addition, vitamin K-dependent proteins are found in calcified tissue, including bone, dentine, atherosclerotic tissue, and renal stones (Olson, 1984).

There are several compounds that have vitamin K activity; the naturally occurring forms of vitamin K are phylloquinone (K_1), isolated from plants, and menaquinones (K_2) synthesized by bacteria. The gut flora produce menaquinones which cannot be absorbed in the large intestine, making them available in the feces to the coprophagous rats (Mameesh and Johnson, 1960; Wostmann et al., 1963; Giovanetti, 1982; Mathers et al., 1990). Menadione (K_3) is a synthetic form of vitamin K that must be alkylated by tissue enzymes for activity. Menadione and the menadione sodium bisulfite complex are the sources of vitamin K activity in commercial laboratory rodent diets. Menadione possesses high biological activity, but its absorption depends on the amount of fat in the diet. Presence of bile salts and pancreatic juice enhance its absorption (Suttie, 2001). Impaired fat absorption may reduce vitamin K absorption to 20% or 30% of the ingested vitamin (Suttie, 1985). The menadione sodium bisulfite complex is more readily absorbed and is almost as active on a molar basis as phylloquinone.

The dietary vitamin K requirement in animals is difficult to determine owing to intestinal flora synthesis and the degree of coprophagy practiced (Mameesh and Johnson, 1960; Mathers et al., 1990; Ichihashi et al., 1992). Gustafsson et al. (1962) reported that the conventional daily requirement is 10 μg/kg BW vitamin K compared with 25 μg/kg BW for germ-free rats. Other factors influencing the requirement are age, sex, strain, and lipid absorption (Greaves and Ayres, 1973; Matschiner and Bell, 1973, 1974; Will et al., 1992). Kindberg and Suttie (1989) measured plasma prothrombin concentration and protein carboxylase activities to determine the vitamin K status of male Holtzman and Sprague-Dawley rats. They showed that the optimal concentration was 2.22 μmol (1 mg) phylloquinone/kg diet; further, the rats depend on a continuous dietary supply of vitamin K to maintain optimal levels. The NRC (1995) estimated vitamin K requirement is 1 mg/kg diet. The American Institute of Nutrition recommends that the vitamin K requirement for purified rodent diets, such as the AIN-93 rodent diet, should be 750 μg phylloquinone/kg diet (Reeves et al., 1993).

It has been shown that excess intake of vitamins A and E antagonize vitamin K. Vitamin A appears to interfere with the absorption of vitamin K (Elliott et al., 1940; Matschiner and Doisy, 1962). Vitamin E in excess has been shown to affect coagulation mechanisms and to increase vitamin K requirements in some species (Rao and Mason, 1975; Uotila, 1988).

VITAMIN B_6. Vitamin B_6 (pyridoxine) is a water-soluble vitamin that exists as six vitamers with similar activities and are interchangeable (Leklem, 2001). The active coenzyme forms of the vitamin are pyridoxal phosphate (PLP) and pyridoxamine phosphate. PLP serves as a coenzyme for more than 100 enzymes primarily involved in amino acid metabolism (Ink and Henderson, 1984). The involvement of PLP with many enzymes explains its effect on growth, cognitive development, immunity, and steroid hormone activity.

The three major vitamers of vitamin B_6, pyridoxine, pyridoxal, and pyridoxamine are absorbed by a nonsaturable passive process (Serebro et al., 1966; Hamm et al., 1979; Mehanso et al., 1979; Middleton, 1982, 1985). During digestion, amino acids and oligopeptides have been shown to inhibit hydrolysis of pyridoxal-5′-phosphate, the first step in intestinal absorption (Middleton, 1990). Dietary fiber did not influence the *in vitro* jejunal absorption rates of pyridoxine, pyridoxal, or pyridoxamine (Nguyen et al., 1983). Coburn et al. (1989) concluded that similar urinary pyridoxic acid concentrations between germ-free and conventional rats fed nutritionally complete diets, indicated that vitamin B_6 synthesized in the intestinal tract was not readily absorbed and metabolized. Therefore, coprophagy did not make a detectable contribution to vitamin B_6 requirements. Research in rats has shown that the liver is the primary organ responsible for metabolism of vitamin B_6 and supplies the active form, PLP, to the circulation and other tissues (Lument and Li, 1980).

The NRC (1995) estimated vitamin B_6 requirement is 6 mg/kg diet. Male weanling rats fed diets containing between 1 and 8 mg vitamin B_6/kg diet maintained liver, serum, and red blood cell aminotransferase activity only at concentrations above 4 mg/kg (Chen and Marlatt, 1975). The red blood cell alanine and aspartate aminotransferase activities for vitamin B_6-depleted female Long-Evans rats were fully restored when fed diets containing 7 mg vitamin B_6/kg diet (Skala et al., 1989). Pregnant rats fed diets containing 8 or 40 mg/kg diet had similar hepatic aspartate aminotransferase, erythrocyte alanine aminotransferase, and muscle glycogen phosphorylase activities (Shibuya et al., 1990). Vitamin B_6-deficient pregnant rats produced vitamin B_6-deficient progeny with abnormal cerebral lipid composition, increased tissue and urinary concentrations of cystathionine, and retarded renal differentiation (Kurtz et al., 1972; DiPaolo et al., 1974; Pang and Kirksey, 1974).

Vitamin B_6-deficient rats develop symmetrical scaling dermatitis (acrodynia) on the tail, paws, nose, chin, ears, and upper thorax; hyperirritability; microcytic anemia; convulsions; and muscular weakness (NRC, 1995). The involvement of vitamin B_6 with so many enzymes results in many different deficiency signs. Neurological signs include depressed amplitude of response to acoustic and tactile stimuli, as well as differences in angle and width of the rear leg gait (Guilarte and Wagner, 1987; Schaeffer, 1987; Schaeffer and Kretsch, 1987; Schaeffer et al., 1990). Deficient rats had deficits in active and passive avoidance learning (Stewart et al., 1975). Vitamin B_6 deficiency in rats can result in decreased cerebroside and ganglioside content, as well as decreased fatty acids (Thomas and Kirksey, 1976).

Pharmacological doses of vitamin B_6 have been used to treat a variety of disease states, including autism, gestational diabetes, carpal tunnel syndrome, depression, atherosclerotic heart disease, muscular fatigue, and diabetic neuropathy (Cohen et al., 1989; Leklem, 2001). In rats there is the risk of toxicity from megadoses of vitamin B_6 (Schaeffer et al., 1990; NRC, 1995). Rats injected intraperitoneally with 200 mg/kg BW developed an unsteady gait and peripheral neuropathy (Windebank et al., 1985). The size of testis epididymis and prostate gland were decreased and spermatid counts decreased in rats given oral doses of 500 and 1000 mg pyridoxine hydrochloride for 6 weeks (Mori et al., 1992).

Research has shown that vitamin B_6 deficiency affects both humoral and cell-mediated immune responses (Rall and Meydani, 1993). Vitamin B_6 can affect lymphocyte production and antibody response to antigens (Axelrod and Trakatelles, 1964; Chandra and Puri, 1985). Repletion of the vitamin will restore lymphocyte differentiation and maturation, increase delayed-type hypersensitivity responses, and improve antibody production.

By acting as a coenzyme for transaminase and glycogen phosphorylase, PLP is involved in gluconeogenesis.

Rat studies have shown that a vitamin B_6 deficiency results in decreased activities of both liver and muscle glycogen phosphorylase (Angel and Mellor, 1974; Black et al., 1978). Studies indicate that a vitamin B_6 deficiency does not lead to mobilization of stored vitamin B_6 from the muscle (Black et al., 1977). However, caloric restriction in rats does result in a decrease in muscle phosphorylase concentration (Black et al., 1978). These findings show that the vitamin B_6 stored in muscle is primarily used for gluconeogenesis. Injections of different vitamin B_6 vitamers resulted in increased liver glucogenolysis which was mediated by adrenal catecholamines (Lau-Cam et al., 1991).

PLP is a cofactor for γ-aminolevulinate synthase, which catalyzes the first and rate-limiting step in heme synthesis (Kikuchi et al., 1958). Therefore, vitamin B_6 performs a fundamental role in erythropoiesis. A deficiency in vitamin B_6 can result in hypochromic microcytic anemia (Bottomley, 1983). Another function vitamin B_6 has in erythrocyte metabolism is both PLP and PL bind to hemoglobin, affecting oxygen binding affinity, and may be important in sickle cell anemia (Maeda et al., 1976; Reynolds and Natta, 1985).

PLP-dependent enzymes are involved in the synthesis of neurotransmitters serotonin, epinephrine, norepinephrine, dopamine, histamine, and GABA (Dakshinamurti, 1982). Neurological abnormalities in rats fed vitamin B_6-deficient diets show the vitamin plays an important role in nervous system function (Stephens et al., 1971; Alton-Mackey and Walker, 1973; Chang et al., 1981; Wasynczuk et al., 1983a,b; Groziak et al., 1984). Research has shown that vitamin B_6 restriction in rat dams was associated with decreased alanine aminotransferase and glutamic acid decarboxylase activity, and low brain weights in the progeny (Aycock and Kirksey, 1976) and reduced myelination (Morre et al., 1978). Alterations in myelin fatty acid concentrations in the cerebellum and cerebrum were reported in progeny from dams fed a diet providing 1.2 mg/kg vitamin B_6 daily (Thomas and Kirksey, 1976). Cerebral sphingolipids were decreased up to 50% in progeny of dams fed vitamin B_6-deficient diets (Kurtz et al., 1972).

Studies conducted on rats more than 60 years ago showed that vitamin B_6 deficiency resulted in decreased body fats (McHenry and Gauvin, 1938). More recent findings showed that liver lipid concentrations were significantly lower in vitamin B_6-deficient versus pair-fed rats (Audet and Lupien, 1974). Rats fed a vitamin B_6-deficient but high-protein (70%) diet developed fatty livers as a result of impaired lysosomal degradation of lipid (Abe and Kishino, 1982). Conflicting findings have shown the synthesis of fat in vitamin B_6-deficient rats to be depressed (Angel and Song, 1973) normal (Desikachar and McHenry, 1954; Angel, 1975), or increased (Sabo et al., 1971).

These differences may be related to the feeding regimen (Witten and Holman, 1952).

Vitamin B_6 deficiency affects fatty acid metabolism. Pyridoxine deficiency in rats impairs the conversion of linoleic acid to arachidonic acid by the inhibition of linoleic acid desaturation and γ-linoleic acid elongation (Audet and Lupien, 1974; Cunnane et al., 1984; She et al., 1994). Delorme and Lupien (1976) reported decreased arachidonic acid in vitamin B_6-deficient rat liver phospholipids and increased linoleic acid. Loo and Smith (1986) showed this change was the result in decreased phospholipid methylation in the liver.

Pyridoxine can act as a modulator of steroid hormone receptors (Cidlowski and Thanassi, 1981; Bender et al., 1989; Tully et al., 1994). Research using rats, has shown that when PLP is at physiological concentrations, reversible reactions occur with receptors for estrogen (Muldoon and Cidlowski, 1980) and androgen (Hiipakka and Liao, 1980). Injection of vitamin B_6-deficient female rats with [^{3}H]-estradiol resulted in more of the isotope accumulating in the uterine tissues of the deficient rats than in the tissues of the control rats (Holley et al., 1983). Bunce and Vessal (1987) showed that there is an increased uptake of estrogen in both vitamin B_6- and zinc-deficient rats, but the number of receptors was not increased. This suggests that there is increased sensitivity of the uterus to steroids when vitamin B_6 is deficient.

Studies (Oka et al., 1995a,b, 1997) have shown that albumin and cystosolic aminotransferase mRNA are significantly increased in vitamin B_6-deficient rats compared with controls. These findings indicate that PLP may be a modulator of gene expression in animals.

Vitamin B_6 is involved with four enzyme reactions in the conversion of tryptophan to niacin. A vitamin B_6 deficiency has a negative effect on niacin formation from tryptophan. Vitamin B_6-dependent enzyme reactions are also involved in the synthesis of histamine, taurine, and dopamine (Leklem, 2001).

NIACIN. Nicotinamide adenine dinucleotide (NAD$^+$) and nicotinamide adenine dinucleotide phosphate (NADP$^+$) are formed in the body from the vitamin niacin. NAD and NADP are required for oxidation reduction reactions by approximately 200 enzymes, mostly dehydrogenases (Kirkland and Rawlings, 2001). These coenzymes are required for the catabolism of glucose, fatty acids, ketone bodies, and amino acids. The major role of NADH is to transfer its electrons during ADP-ribosylation to produce ATP. NADP is essential for the biosynthetic reactions involved in energy storage. NADPH acts as a reducing agent in biosynthesis of fatty acids, cholesterol, steroid

hormones, and deoxyribonucleotides (Groff and Gropper, 2000b).

NADH has a non-redox function; it acts as a donor of adenosine phosphate ribose during protein synthesis (Miro et al., 1989; Cervantes-Laurean et al., 1999). NAD is a substrate for poly (ADP-ribose) polymerase, which is involved in DNA transcription, replication, and repair. Research on the relationship between poly (ADP-ribose) polymerase activity and carcinogenesis has been conducted (Yamagami et al., 1985; Rawling et al., 1993). Rats treated with diethylnitrosamine and 3-aminobenzamide, a competitive inhibitor of poly (ADP-ribose) polymerase, resulted in a significant increase in the formation of precancerous lesions in the liver (Takahashi et al., 1984).

Microorganisms and plants can synthesize the pyridine ring of NAD *de novo*, but animals cannot; therefore, niacin is a vitamin (Hankes, 1984). However, rats practicing coprophagy can take advantage of colonic synthesis of niacin by microflora. NAD can be synthesized in the liver from tryptophan with the help of vitamin B_6 and the riboflavin coenzyme derivative flavin adenine dinucleotide (FAD). Species vary in their ability to convert tryptophan to niacin; and rats are efficient in the synthesis of niacin from tryptophan (Hankes et al., 1948). For optimal growth, rats fed a diet containing 20% casein do not require niacin in the diet (Hundley, 1947). The source of dietary carbohydrate can affect the niacin requirement. Hundley (1949) showed in rats, high sucrose and fructose diets increase the niacin requirement compared with diets containing starch and glucose. Research has shown that niacin deficiency can be induced in rats fed niacin-free diets containing low tryptophan and fructose, and 15 mg niacin/kg diet is required to restore normal growth (Krehl et al., 1946; Henderson et al., 1947).

In rats, niacin deficiency results in signs such as retarded growth, diarrhea, behavioral abnormalities, convulsions, rough haircoat, and alopecia (Brookes et al., 1972; Rawling et al., 1994; NRC, 1995). A histopathological study of the peripheral nervous tissue of niacin-deficient rats showed degenerative changes in motor, sensory, and automotive neurons (Hankes, 1984). Rats have been used to determine if niacin deficiency plays a causal role in the process of carcinogenesis. Tumor incidence, size, and progression were reduced in rats with esophageal cancer when a diet containing 20 mg/kg nicotinic acid was fed compared with rats fed a niacin-deficient diet (Van Rensberg et al., 1986). It has been shown that lymphocytes from niacin-deficient rats are more susceptible to oxygen radical-induced DNA damage (Zhang et al., 1993). Research on DNA damage to the bone marrow during chemotherapy, using niacin-deficient rats, showed niacin deficiency resulted in increased severity of acute anemia and leukopenia as well as increased rate of development of nitrosourea-induced leukemias (Boyonoski et al., 1999, 2000).

BIOTIN. The toxic properties of feeding raw egg white to animals were first observed in 1916. When rats were fed egg-white protein containing avidin, a glycoprotein that binds to biotin, the biotin was biologically unavailable. This resulted in a syndrome of dermatitis, hair loss, and neuromuscular dysfunction known as "egg-white injury" (Mock, 2001). Biotin is an essential cofactor for four carboxylases involved in critical steps in gluconeogenesis, fatty acid synthesis, and amino acid metabolism (Bai et al., 1989; Shriver et al., 1993; Mock, 1999).

There are conflicting data regarding the intestinal transport of biotin. Early studies indicated that intestinal and renal transport were by simple diffusion. Research by Leon-Del-Rio et al. (1990) supported simple diffusion transport in rats. However, a biotin transporter in the brush border has been described (Bowman and Rosenberg, 1987; Said et al., 1989, 1993). Intestinal absorption of biotin is upregulated as rats age, and the site of maximal transport by the biotin transporter shifts from the ileum to the jejunum (Said et al., 1993). Bowman and Rosenberg (1987) concluded that absorption of biotin from the proximal colon suggests nutritional significance of biotin synthesized by enteric flora.

Rats do not require dietary biotin, because it is provided by intestinal microorganisms through coprophagy. There are four ways to produce biotin deficiency in rats fed a biotin-deficient diet (1) use germ-free animals; (2) prevent coprophagy; (3) feed sulfa drugs; and (4) feed raw egg whites (NRC, 1995). Deficiency signs include loss of appetite, decreased growth, seborrheic dermatitis, exfoliative dermatitis and hyperkeratosis in advanced cases, achromatrichia, alopecia, "spectacle eye," "kangaroo gait," resorption of fetuses, stillbirths, and depressed immunity (Rabin, 1983; Bonjour, 1984).

Research has shown biotin deficiency will result in abnormal fatty acid metabolism, which may be responsible for the pathogenesis of dermatitis and alopecia (Kramer et al., 1984). Rat studies have shown that odd-chain fatty acid accumulation in hepatic, cardiac, and serum phospholipids, as well as abnormal PUFA metabolism, are a result of biotin deficiency (Suchy et al., 1986; Mock et al., 1988a). Mock et al. (1988a) suggested that the abnormal PUFA composition affects prostaglandin metabolism. Supplementation of biotin-deficient rats with n-6 PUFA prevented the development of dermatitis, which indicated that an abnormality in n-6 PUFA metabolism is involved in biotin deficiency-related dermatitis (Mock, 1988b).

The assessment of biotin requirement is difficult owing to biotin's enteric synthesis by intestinal microflora. Dietary composition and the requirement for biotin by some gut microbes may affect overall biotin biosynthesis. The NRC (1995) recommends 0.2 mg biotin/kg diet as the requirement for rats, but it is dependent on the dietary protein. When a diet using 20% raw egg whites as the source of protein was fed, 2 ppm biotin was required to obtain optimal growth (Klevay, 1976). Purified diets AIN-76, AIN-93M, and AIN-93G contain 0.2 mg d-biotin/kg diet when the protein source is casein (AIN, 1977; Reeves et al., 1993).

FOLATE. Folate is composed of three parts, pterin, para-aminobenzoic acid, and glutamate, all of which must be present for vitamin activity. The metabolically active form is tetrahydrofolic acid, a constituent of a coenzyme involved in the transfer of one-carbon units in reactions involved in amino acid and nucleotide metabolism. Folate is involved in the metabolism of serine, methionine, glycine, and histidine (Brody and Shane, 2001). The synthesis of pyrimidine and purine nucleotides requires folate, making it essential for cell division (Wagner, 1995).

Coprophagy and the ability of the rat to absorb folate produced by gut flora make it difficult to achieve a deficiency without supplementing the folate-deficient diet with antibacterial drugs (Clifford et al., 1989; Ward and Nixon, 1990; Rong et al., 1991). Early research estimated the folate requirement to be between 5 and 20 µg/day (Asenjo, 1948). The NRC (1995) based its estimated folate requirement on more recent research that shows 2 mg/kg diet is no more effective than is 1 mg/kg diet for the rat (Clifford et al., 1989). Deficiency may be owing to malabsorption syndromes; drug treatment; increased requirement, such as pregnancy; increased excretion due to liver disease; and increased oxidative destruction (Herbert, 1999). Signs of folate deficiency in the rat are decreased growth rate, leukopenia, anemia, and excretion of formiminoglutamate (Clifford et al., 1989; Ward and Nixon, 1990; Varela-Moreiras and Selhub, 1992). Two reports on the folate distribution among tissues in folate-deficient rats show that, in liver, kidneys, and spleen, folate concentration decreased by 60% of the control animals, and there was an elongation of the glutamate chains (Ward and Nixon, 1990; Varela-Moreiras and Selhub, 1992). The investigators also reported there was no depletion of brain folate concentrations in either long-term or short-term folate deficiency. Feeding young and old rats either a folate-deficient or folate-repleted diet showed that a deficient diet resulted in significant decreases in serum and hepatic folate. Aging was reported to result in a 50% reduction in serum folate but no change in liver folate concentration (Varela-Moreiras et al., 1994).

Hyperhomocysteinemia is a risk factor for cerebrovascular and occlusive vascular disease (Zhang et al., 2004). Fasting homocysteine levels are inversely correlated with plasma folate concentrations and folate intake

(Varela-Moreiras et al., 1994; Hankey and Eikelboom, 1999). In rats, folate deficiency induced hyperhomocystein-emia, resulting in prothrombotic effects on platelets and macrophages that were related to increased plasma lipid peroxidation (Durand et al., 1996). It appears oxidative stress induced by folate depletion and elevated homo-cysteine is involved in the pathogenesis of cardiovascular disease (Durand et al., 1996; Zhang et al., 2004).

In rats, the liver contains the most folate and is susceptible to folate depletion (Clifford et al., 1990; Varela-Moreiras and Selhub, 1992). Miller et al. (1994) demonstrated folate deficiency in rats disrupts hepatic one-carbon metabolism, increasing homocysteine levels by impairing enyzmes methylenetetrahydrofolate reductase and cystathionine synthase. Huang et al. (2001) fed rats folate-deficient diets to determine if folate depletion would result in oxidative stress in the liver as it is reported to do in the cardiovascular system. In folate-depleted rats, elevated plasma homocysteine and decreased plasma and liver folate concentrations were significantly correlated with increased hepatic lipid peroxidation.

The effects of dietary folate supplementation are being studied because of the role of folates in the prevention of neural tube defects. Improved liver morphology and increased hepatocyte division were reported when 18-month-old male Wistar rats were fed a diet containing 40 mg/kg folic acid for a month (Roncales et al., 2004). High dietary supplementation (40 mg/kg dietary folate) had a negative effect on dietary protein utilization in pregnant and weanling Wistar rats (Achon et al., 1999). It was also shown there were significant reductions in BW and vertex-coccyx length in fetuses from dams fed a folate-supplemented diet (40 mg/kg).

Folic acid interacts with vitamin B_{12} and ascorbic acid. As an antioxidant vitamin C can protect folate from oxidative destruction. The megaloblastic anemia produced by folic acid deficiency is identical to that observed in vitamin B_{12} deficiency. The interrelationship between these two vitamins is explained by the methyl trap hypothesis (Shane and Stokstad, 1985; Stokstad et al., 1988). The two vitamins are cofactors for the methionine synthase reaction. If vitamin B_{12} was deficient, this results in trapping folate in a nonfunctional form with an associated reduction in the level of other folate coenzymes required for one-carbon metabolism. In the rat, this results in gross reduction in tissue folate owing to an inability to retain folate, increased plasma folate concentrations, and an increase in the proportion of hepatic folate in the 5-methyl-H_4PteGlu non functional short chain poly-glutamate form (Brody and Stokstad, 1991). There are conflicting results from rat studies about whether high concentrations of dietary folate (350 μg/d) can result in zinc/folate complexes inhibiting zinc absorption (Grishan et al., 1986; Keating et al., 1987).

VITAMIN B_{12}. There are four cobalamins (vitamin B_{12}) that play a significant role in animal cell metabolism (Beck, 2001). Cyanocobalamin and its analog hydroxycobalamin are equivalents of the natural vitamin. Adenosylcobalamin and methylcobalamin are alkyl derivatives synthesized from the vitamin serving as coenzymes. Adenosylcobala-min and methylcobalamin function in only two enzyme systems in animal cells, adenosylcobalamin-dependent methylmalonyl coenzyme A (CoA) mutase (Lardy, 1956; Gurnani, 1959) and methylcobalamin-dependent methionine synthase (Beck, 1990). Methylmalonyl CoA mutase is involved in fatty acid synthesis, and methionine synthase controls nucleic acid synthesis and methylation reactions.

Gastric juice contains several cobalamin binding pro-teins, but only intrinsic factor facilitates absorption of cobalamins in the ileum. Cobalamin is transferred to portal blood. The vitamin is transported in the blood by two proteins, haptocorrin and transcobalamin (Begley, 1983).

Increased metabolic requirements such as growth and pregnancy increase cobalamin requirements (Beck, 1983). Cobalamin deficiency can result from (1) a deficient diet; (2) hypermetabolic states; (3) drug interactions; or (4) defective vitamin metabolism, including malabsorption and utilization in the tissues (Ellenbogen, 1984). Cobala-min deficiency is not induced easily in rats and does not produce the typical megaloblastic anemia and neuropathy reported in humans. The neuropathy related to cobalamin deficiency may be determined by abnormalities in the ratio of S-adenosylmethionine to S-adenosylhomocysteine, the methylation ratio (McKeever et al., 1995a). The methylation ratio controls the activity of tissue methyl-transferases. The inactivation of cobalamin-dependent methionine synthase reduces the methylation ratio in rats and pigs. Abnormal methylation ratios found in nitrous oxide-induced, cobalamin-inactivated methionine synthase from pigs and humans significantly inhibit their methyl-transferases. However, McKeever et al. (1995b) determined the altered methylation ratio in deficient rats only minimally affected their methyltransferases. This may explain the myelopathy produced by impairment of methionine synthase in pigs and humans but not rats (Beck, 2001).

Both cobalamin-dependent enzymes, methylmalonyl CoA mutase and methionine synthase, are involved in amino acid metabolism. Ebara (2001) studied alterations in plasma concentrations of amino acids in cobalamin-deficient Wistar rats. Dietary cobalamin deficiency resulted in significant increases in plasma serine, threonine, glycine, alanine, tyrosine, lysine, and histidine.

Woodward and Newberne (1966) reported that cobala-min deficiency can be induced in rats fed vegetable rather than animal protein. Their study showed a 10% incidence in hydrocephalopathy, decreased birth weights, and decreased growth in offspring born to rats fed a diet containing soy protein but deficient in vitamin B_{12}. Liver cobalamin concentrations were significantly decreased in both the dams and pups. Data from Doi et al. (1989) showed that 0.5% DL-methionine in a cobalamin-deficient diet fed to rats born from cobalamin-deficient dams would prevent growth retardation. Abortions, cannibalization, and short-lived pups were observed when germ-free female rats were fed a soy protein diet deficient in cobalamin (Valencia and Sacquet, 1968).

Cullen and Oace (1989a) showed pectin added to a fiber-free diet significantly increases urinary methylmalonic acid in cobalamin-deficient rats, accelerating cobalamin loss. Administering neomycin, a cobalamin sparing antibiotic, did not prevent the loss of cobalmin associated with dietary pectin (Cullen and Oace, 1989b). It was concluded that pectin directly interferes with cobalamin absorption or may stimulate cobalamin uptake by neomycin-resistant gut flora.

Cobalamin deficiency impairs methylmalonyl CoA mutase enzyme activity involved in fatty acid synthesis, resulting in abnormal odd-chain fatty acids. Rats fed cobalamin-deficient diets accumulated abnormal levels of odd-chain fatty acids in the cerebrum and liver, as well as other signs of abnormal fatty acid metabolism (Peifer and Lewis, 1979). Increased urinary methylmalonate from rats fed a diet containing 10 µg B_{12}/kg of diet indicated the dietary vitamin B_{12} level was inadequate (Thenen, 1989). The current vitamin B_{12} requirement is 50 µ/kg diet (NRC, 1995). The requirement is based on the report that 50 µg B_{12}/kg diet supports normal growth and reproduction (Woodward and Newberne, 1969).

Megaloblastic anemia is the result of a defect in DNA synthesis without impaired RNA synthesis. Norohna and Siverman (1962) developed the "methylfolate trap" hypothesis to explain the role of cobalamin in DNA synthesis and account for abnormalities of folate metabolism in cobalamin deficient rats. Clinical signs of megaloblastic anemia in humans and most animals include macrocytic anemia, low serum cobalamin concentration, neuropathy, weakness, dyspnea, weight loss, alopecia, and pancytopenia (Beck, 2001). Treatment with folic acid will reverse the anemia but not the neurological abnormalities.

THIAMINE. Thiamin plays a critical role regulating carbohydrate metabolism. The three most important thiamin esters are thiamin monophosphate, thiamin pyrophosphate (TPP), and thiamin triphosphate. Thiamin monophosphate is hydrolyzed to form free thiamin that is phosphorylated to produce TPP, the active coenzyme form of thiamin. TPP functions as the Mg^{2+} coordinated enzyme for the active aldehyde transfers in oxidative decarboxylation of α-keto acids and the transketolase reaction. The α-keto acid enzyme complexes play key roles in energy generating pathways. The metabolic significance of the transketolase reaction is its role in producing NADPH for fatty acid synthesis and ribonucleotides (Tanphaichitr, 2001). Blair et al. (1999), showed that thiamin deficiency and excess result in significant changes in rat mitochondrial TPP levels that have major but opposite effects on α-ketoglutarate dehydrogenase and branched-chain α-ketoglutarate dehydrogenase activities.

Plasma thiamin concentration is regulated by both the intestine and kidney. Thiamin plasma concentration in rats does not exceed 240 nmol/L. In the rat, thiamin is not bound to a protein in plasma, and typically, it is reabsorbed rather than secreted by the kidney (Rindi, 1968; Verri, 2002). The small intestine absorbs thiamin by passive diffusion when the concentration exceeds 1.25 µmol/L but via an active process when concentrations are below 1.25 µmol/L (Rindi, 1984).

Jejunoileal bypass surgery on rats resulted in significant decrease in the hepatic concentrations of thiamin, folate, riboflavin, vitamin B_{12}, β-carotene, and vitamin E (Baker, 1992).

Thiamin transport across the blood-brain barrier involves saturable and nonsaturable transport mechanisms (Greenwood, 1985). Thiamin binding protein is involved in the transfer of thiamin across membranes. The significance of the transfer of thiamin across the placenta was demonstrated by Adiga (1978) and Subramanian (1996) when they immunized pregnant rats with chicken thiamin binding protein antibodies, resulting in fetal resorption. Kirchgessner (1996) examined the effect of dietary thiamin supply during gestation on body thiamin status of lactating rats and their pups, as well as thiamin in milk. The dams and their offspring were thiamin-deficient based on reduced thiamin concentration of the liver and red blood cell transketolase activity.

Bioavailability of thiamin depends on processing (e.g., autoclaving and irradiation), antithiamin factors, and folate and protein status (Tanphaichitr, 2001). In folate-deficient rats, thiamin absorption was significantly decreased compared with pair-fed controls (Howard, 1977). However, a more recent study showed there were no differences in thiamin absorption or excretion as a result of folate deficiency (Walzem, 1988). Rats fed a diet containing bracken fern, which contains the antithiamin factor thiaminase I, became thiamin deficient (Evans, 1975).

Thiamin is a cofactor for key enzymes in the tricarboxylic acid cycle and pentose pathway; both are involved in carbohydrate metabolism. Therefore, the requirement for thiamin can be influenced by the level of carbohydrate in the diet relative to other energy sources. A rat model of glucose-induced Wernicke encephalopathy has been developed in which glucose loading (10 g/kg BW, intraperitoneal) significantly increased the rate of progression of neurological dysfunction in thiamin-deficient rats (Zimitat, 2001).

SHR/NCrj rats have a defective fatty acid translocase, CD36, which results in an impaired uptake of myocardial long-chain fatty acids but is offset by increased glucose utilization. Tanaka (2003) used this rat model to determine if the heart becomes vulnerable to thiamin deficiency when the substrate shifts from long-chain fatty acids to glucose. Two weeks of feeding a thiamin-deficient diet resulted in (1) increased weight of body, liver, and lungs; (2) lactic acidemia; (3) peripheral vasal dilation with high cardiac output and hyperdynamic state of the heart; and (4) fluid retention of the heart and lungs. The investigators concluded these are characteristics of cardiovascular beriberi and are evidence of a gene-environment interaction involved in its etiology.

Thiamin functions as a cofactor for pyruvate dehydrogenase and α-ketoglutarate dehydrogenase critical enzymes in the regulation of carbohydrate metabolism. Perturbations in carbohydrate metabolism such as found in diabetes can affect thiamin status. Sprague-Dawley rats fed a thiamin-deficient diet (0.35 mg/kg) for 6 weeks had significantly reduced plasma insulin and increased glucagon and corticosterone concentrations compared with those of controls (Molina, 1994). Reddi (1993) examined the changes in tissue distribution and concentrations of several water-soluble vitamins in short-term diabetic Wistar rats. The results showed that thiamin is significantly decreased in heart and liver but not affected in the brain of short-term diabetic rats. Thiamin status of the offspring of diabetic rats was studied by Berant (1988). The offspring of diabetic dams had increased weight, significantly lower glucose levels, significantly higher insulin concentrations, increased transketolase activity, and higher TPP. The investigators concluded a fetal thiamin deficiency evolved owing to enhanced fetal glucose turnover.

To determine the thiamin requirement for rats, feed efficiency (Mackerer et al., 1973), growth rates (Itokawa and Fujiwara, 1973; Mercer et. al., 1986), plasma and erythrocyte thiamin concentrations (Chen et al., 1984), urinary thiamin excretion (Leclerc, 1991), and organ thiamin retention (Roth-Maier, 1990) have been studied. From the results of these studies the NRC (1995) estimated the requirement for thiamin to be 4 mg/kg diet for maintenance, growth, pregnancy, and lactation. Rains et al. (1997) fed a chemically defined diet containing graded levels of thiamin to weanling male Sprague-Dawley rats. Based on diet consumption, weight gain, and hepatic transketolase activity, it was concluded the minimum thiamin requirement was 0.55 mg/kg diet. Roth-Maier

(1990) examined retention and utilization of thiamin by pregnant and non-pregnant rats by varying dietary thiamin. Based on the best utilization of thiamin, thiamin retention to thiamin intake, it was suggested that thiamin requirement for adult rats is 7 to 8 mg/kg diet.

Signs of thiamin deficiency in the rat include anorexia, weight loss, roughed appearance, hypertrophy of the heart, ataxia, opisthotonus, and neuropathy. Bruce (2003) showed rats fed a marginally thiamin-deficient diet (1.6 mg/kg) had decreased plasma thiamin, red blood cell transketolase, and an increased number of aberrant crypt foci, a biomarker for colon cancer. Molina (1994) reported multiple changes in whole-body carbohydrate metabolism and glucoregulatory hormone concentrations in thiamin-deficient Sprague-Dawley rats.

Thiamin deficiency affects central nervous system neurotransmitter systems. Studies using thiamin-deficient rats have shown reductions in acetylcholine turnover and utilization in the cortex, midbrain, diencephalon, and brain stem, indicating depressed central cholinergic mechanisms (Gibson, 1975; Butterworth, 1982; Kulkarni, 1983). Iwata (1970) demonstrated decreased synthesis of catecholamines in the thiamin-deficient rat brain. Behavioral deficits in rats after reversal of thiamin deficiency have been shown to be linked to significant reductions in norepinephrine content of cortex, hippocampus, and olfactory bulbs (Mair, 1985). Glutamate, aspartate, GABA, and glutamine, amino acids with neurotransmitter functions, are decreased in thiamin-deficient rat cerebellum and other areas of the brain (Butterworth, 1982; Gaitonde, 1982).

Matsuda (1989) reported thiamin metabolism in rat brain changes during postnatal development in a different way from that in liver, and the development of thiamin metabolism differs among brain regions. Fournier (1990) studied the effects of thiamin deficiency on thiamin-dependent enyzmes in regions of the brain of pregnant rats and their offspring. The findings demonstrated no effect on the enzymes in the brain among dams. However, the offspring had significantly reduced activity in all three enzymes from the cerebral cortex.

Pyrithiamine-induced thiamin deficiency in rats is used to model the etiology, diencephalic neuropathology, and memory deficits of Korsakoff amnesia. Mumby (1995) demonstrated pyrithiamine-induced thiamin deficiency impairs object recognition in rats. An investigation of the relationship between thiamin status and learning a task in rats was investigated by Terasawa (1999). The results showed that rats fed a thiamin-deficient diet had a slower response time to an electrical impulse than did controls. Thiamin is important in nerve conduction, and thiamin triphosphate appears to be involved in nerve membrane function (Haas, 1988). Peripheral neuropathy with axonopathy is caused by thiamin deficiency (Takahashi, 1981).

McLane (1987) showed reduced conduction and increased axonal protein transport in aural nerves from thiamin-deficient rats.

RIBOFLAVIN. Riboflavin serves as the precursor of the flavin coenzymes, flavin mononucleotide and flavin adenine dinucleotide (FAD). FAD and flavin mononucleotide are widely distributed in intermediary metabolism and function as cofactors for many different oxidative enzyme systems. The ability of FAD to accept a pair of hydrogen atoms in the electron transport chain makes riboflavin a principal factor in energy production. Riboflavin is also involved in drug and steroid metabolism, in conjunction with cytochrome P450 enzymes, and lipid metabolism (Rivlin and Pinto, 2001).

The flavoproteins are involved in fatty acid synthesis as well as their catabolism. A riboflavin-deficient (1 mg/kg diet), high-fat (40% of calories) diet fed to rats resulted in significant increases in liver total lipids, triglycerides, cholesterol, and lipid peroxides compared with levels in pair-fed controls (Liao and Huang, 1987). The fatty acid composition of liver phospholipids from riboflavin-deficient rats is altered compared with controls (Taniguchi and Nakamura, 1976; Taniguchi et al., 1978; Olpin and Bates, 1982a). The most significant change is the increase in linoleic acid (18:2) and the decrease in arachidonic acid (20:4), a precursor of the prostaglandins. Olpin and Bates (1982b) demonstrated that riboflavin deficiency resulted in decreased acyl coenzyme A dehydrogenase activity, indicating a depression in mitochondrial β-oxidation may be responsible for the fatty acid changes.

Pelliccione et al. (1985) investigated the effects of riboflavin deficiency on prostaglandin synthesis in the rat kidney. The findings showed riboflavin deficiency increased the rates of synthesis of both prostaglandin E_2 and prostaglandin F_2, indicating a role for riboflavin in the regulation of renal prostaglandin synthesis.

Riboflavin coenzymes are required to convert vitamin B_6 and folate to their active forms, pyridoxal phosphate and N-5-methyl tetrahydrofolate, respectively. Bates and Fuller (1985) showed riboflavin deficiency in Norwegian hooded rats resulted in a significant decreases in methylenetetrafolate reductase activity but not dihydrofolate reductase activity. Because flavoproteins are involved in the activation and transformations of pyridoxine, folic acid, niacin, and vitamin K, riboflavin deficiency can be observed in conjunction with deficiencies of these vitamins (Rivlin, 1984).

Riboflavin deficiency in rats has been reported to induce eye signs varying in severity from an inflamed condition of the cornea to its complete opacity. The data are conflicting regarding the association between riboflavin deficiency and cataract formation (Bhat, 1982, 1987; Yagi et al., 1989; Dutta et al., 1990). Takami et al. (2004) examined the conjunctiva and cornea of riboflavin-deficient rats. Riboflavin deficiency resulted in a decrease of microvilli and microplicae in the cornea and conjunctiva epithelium. From these findings the investigators concluded riboflavin is essential in the development, maintenance, and function of the ocular surface.

As precursor to FAD, riboflavin has antioxidant activity (Miyazawa, 1983). GSH-Px destroys reactive lipid peroxides via the GSH redox cycle. To function, this enzyme requires the FAD-containing GSH reductase to reduce oxidized GSH to GSH. This demonstrates that riboflavin nutrition is critical for regulating the inactivation of lipid peroxides. Feeding riboflavin-deficient diets to rats is associated with increased lipid peroxidation, and supplementation limits the reaction (Taniguchi and Hara, 1983; Rivlin and Dutta, 1995).

A study of the relationship between riboflavin and protein utilization in Sprague-Dawley rats showed the effect of protein on riboflavin requirement is related to the rate of growth and not to protein intake (Turkki, 1982). Results from investigations on the effects of energy restriction on tissue riboflavin depletion suggest energy restriction impairs flavoprotein synthesis in muscle but not the liver. The investigators concluded not all tissues are equally efficient in utilizing dietary riboflavin to correct deficiencies during energy restriction (Turrki and Degruccio, 1983; Turkki, 1989).

Riboflavin via the oxidation-reduction potential of the flavoproteins is involved in iron metabolism. Powers (1986) showed iron mobilization in epithilial mucosa from the upper part of the small intestine of riboflavin-deficient weanling and adult rats was significantly decreased compared with that of controls. Hepatic ferritin-iron concentration in young riboflavin-deficient rats was 36% of that in livers from control rats. Low hepatic iron is a characteristic of riboflavin deficiency (Powers, 1985; Adelekan and Thurnham, 1986). These results indicate flavin mononucleotide-oxidoreductase activity may be involved in iron absorption and that riboflavin deficiency, therefore, can affect ferritin-iron mobilization.

Rat studies have shown that riboflavin deficiency decreases absorption of iron and increases gastrointestinal loss (Powers et al., 1991). Butler and Topham (1993) demonstrated a reduced uptake of iron by brush-border-membrane vesicles from riboflavin-deficient rats, confirming reduced iron absorption. To determine if enhanced rate of endogenous iron loss due to riboflavin deficiency was caused by an increased rate of turnover in the small intestine, epithelial crypt cell proliferation and crypt cell production rate were measured. Riboflavin deficiency was associated with hyperproliferation of crypt cells from the upper small intestine (Powers et al., 1993).

Villi morphometry and the kinetics of cell movement on the villi from riboflavin-deficient female Wistar rats were studied. Feeding a riboflavin-deficient diet to weanling rats resulted in a significantly lower number of villi, a significant increase in villus length, and an increased rate of transit of enterocytes along the villi compared with that of controls (Williams et al., 1995, 1996a). The morphological and cytokinetic changes in duodenums from weanling rats fed a riboflavin-deficient diet for 5 weeks could not be reversed by a 21-day riboflavin-repletion period (Williams et al., 1996b). The earliest point at which riboflavin deficiency affects post-weaning bowel development in rats has been identified to be 96 hours after initiating the deficient diet. The changes affected duodenal crypt cell proliferation and bifurcation with no reduction in villus number (Yates et al., 2001).

Riboflavin deficiency in the rat during pregnancy causes congenital abnormalities (Sheperd et al., 1968) and fetal resorption (Duerden and Bates, 1985). The possibility that these findings are due to the existence of a flavin-dependent step in placental iron transfer has been investigated. Powers and Bates (1984) studied the effect of riboflavin deficiency on hepatic iron stores in pregnant Norwegian hooded rats. The results demonstrated that pregnancy and rapid growth increased the demand for iron turnover, depleted ferritin stores, and riboflavin deficiency impaired iron mobilization for these purposes. Rats were fed a riboflavin-deficient diet (0.25 mg/kg diet) from 10 weeks of age through gestation. Riboflavin deficiency was associated with a reduction in fetal mass, which limited maternal iron depletion and maternal-fetal iron transfer (Powers, 1987).

Clinical signs of riboflavin deficiency in rats include unthrifty appearance with areas of alopecia on the skin, seborrheic inflammation, cheilosis, angular stomatitis, glossitis, anemia, hyperkeratosis of the epidermis, neuropathy, blepharitis, conjunctivitis, corneal opacity and vascularization, anestrus, and birth defects (Cooperman and Lopez, 1984; NRC, 1995). The NRC estimated riboflavin requirements for growth and reproduction are 3 and 4 mg/kg diet, respectively (NRC, 1995).

PANTOTHENIC ACID. One of the functions of pantothenic acid is its role as a component of coenzyme A (CoA) (Branca et al., 1984; Plesofsky, 2001). In the form of CoA, pantothenic acid is fundamental to the energy-yielding oxidation of glycolytic products and other metabolites through the Krebs cycle. The synthesis of fatty acids, phospholipids, and sphingolipids requires CoA. The synthesis of amino acids, methionine, leucine, and arginine requires a pantothenate-dependent step. The β-oxidation of fatty acids and the oxidative catabolism of amino acids also depend on CoA.

Pantothenic acid, as a component of CoA, is involved in the synthesis of 3-hydroxy-3-methylglutaryl-CoA (HMG-CoA) required for the production of isoprenoid-derived compounds, such as cholesterol, steroid hormones, vitamin A, vitamin D, and heme A (Wells and Hogan, 1968).

Riebel et al. (1982) reported the pantothenic acid content of heart, kidney, gastrocnemius muscle, and testes of rats fed a pantothenic acid-deficient diet was reduced by more than 90% compared with controls. However, the CoA levels did not have a corresponding decrease. Weanling rats fed a pantothenic acid-free diet for 11 days did not show signs of deficiency, but concentrations of hepatic total and free CoA, dephospho-CoA, and 4'-phosphopantethiene were decreased. The concentrations of long-chain acyl-CoA, the ratios of free CoA/total CoA, and long chain-acyl CoA/total CoA did not change (Moiseenok et al., 1986). The investigators from these studies suggested the stability of the tissue CoA pools is due to pantothenate-protein complexes that serve as a reserve to meet requirements of CoA biosynthesis during pantothenic acid deficiency.

Youssef et al. (1997) studied the effects of CoA depletion, in pantothenic acid-deficient rats, on the mitochondrial and peroxisomal pathways of fatty acid oxidation. The findings show that peroxisomal β-oxidation is inhibited in livers from CoA-deficient rats; hepatic mitochondrial β-oxidation is not affected. The investigators suggest that because the role of hepatic mitochondrial β-oxidation is energy production, whereas peroxisomal β-oxidation is responsible for detoxification, the mitochondrial pathway of β-oxidation is spared at the expense of the peroxisomal pathway when hepatic CoA concentrations are low.

Pantothenic acid also functions as the prosthetic group for acyl carrier protein, an important component of the fatty acid synthase complex that is involved in the synthesis of fatty acids (Groff and Gropper, 2000b). Wittwer et al. (1990) demonstrated that mild pantothenate deficiency in rats caused increased serum and free fatty acid levels.

Pantothenic acid is involved in the protein acetylation process, which in turn affects protein functions. Acetylation of proteins may protect them from catabolism and may determine activity, location, and function in the cell (Plesofsky-Vig et al., 1988).

Early research showed 80 μg/day D-pantothenate was required for optimal growth in the rat (Unna 1940; Barboriak et al., 1957). Nelson and Evans (1961) reported normal growth in suckling pups from dams being fed a diet containing 10 mg/kg calcium D-pantothenate. The AIN-76 (American Institute of Nutrition, 1977), AIN-93G, and AIN-93M (Reeves et al., 1993) purified rodent diets provide 15 mg/kg calcium D-pantothenate. The NRC

(1995) estimated pantothenic acid requirement for growth and reproduction is 10 mg/kg diet.

Deficiency signs of pantothenic acid vary among the different animal species. Pantothenic acid deficiency in rats results in exfoliative dermatitis, achromotrichia, oral hyperkeratosis, necrosis of the adrenals, and porphyrin-caked whiskers (Fox, 1984; NRC, 1995). Pantothenic acid deficiency in pregnant rats resulted in congenital defects, growth retardation, and adrenal necrosis of the offspring (Lederer et al., 1975). By studying the distribution of ^{14}C-pantothenate is rat tissues, Pietrzik and Hornig (1980) demonstrated that adrenal function is a sensitive indicator of the nutritional status of pantothenic acid.

CHOLINE. Dietary choline is essential for normal growth and functioning of all mammalian cells (Garner et al., 1995; Zeisel, 1999). Choline is a precursor to the neurotransmitter acetylcholine; a precursor to cell membrane constituents phosphatidylcholine and sphingo-myelin, and; a precursor to signaling lipids sphingosylphosphocholine and platelet-activating factor (Zeisel and Holmes-McNary, 2001).

Choline and its metabolites are involved in cholinergic neurotransmission, phospholipid synthesis, transmembrane signaling, methyl metabolism, and lipid-cholesterol transport and metabolism (Blusztajn, 1998). Research shows that choline plays an important role in development. Alteration of choline concentrations in the diet fed to pregnant rats caused changes in the embryonic brain chemistry (Holler et al., 1996; Cermack et al., 1998), which resulted in permanent changes in learning and memory in the offspring (Meck et al., 1989; Meck and Williams, 1997a,b).

Recent research using rats has shown that prenatal choline supplementation produces permanent enhancements of central nervous system function. Holler et al. (1996) showed prenatal choline supplementation increased phopholipase-D activity in the hippocampus of offspring at maturity. Cermak et al. (1999) reported a decreased acetylcholine esterase activity in the hippocampus of juvenile offspring from dams supplemented with choline. These findings are consistent with the improvements in spatial (Meck et al., 1989; Williams et al., 1998) and temporal memory (Meck and Williams, 1997a) observed in adults after prenatal choline supplementation. Enhancement of N-methyl-D-aspartate receptor-mediated neuro-transmission (Montoya and Swartzwelder, 2000) and long-term-potentiation (Payapali et al., 1998) in the hippocampus from rats supplemented with choline prenatally also supports the conclusion that choline supplementation during embryonic development facilitates cognitive function and visuospatial memory in adult rats.

To understand the relationship among maternal choline intake, brain cytoarchitecture, and behavior, the effect of choline availability on cell proliferation, apoptosis and differentiation in the fetal rat brain septum was investigated (Albright et al., 1999a,b). The findings demonstrated that choline availability during pregnancy alters the timing of mitosis, apoptosis, and early commitment to neuronal differentiation by progenitor cells in regions of the fetal brain septum and hippocampus, two brain regions known to be associated with learning and memory.

Li et al. (2003) investigated the possibility that the improvements in spatial and temporal memory in the offspring of rats supplemented with choline are due to morphological or neurophysiological alterations in hippocampal pyramidal cells. Pregnant Sprague-Dawley rats were fed a choline-supplemented diet (7.9 mmol/kg) during a 6-day period of gestation (days 12 through 17). The findings show that dietary supplementation with choline during a critical period of prenatal development alters the structure and function of hippocampal pyramidal cells.

The NRC (1995) estimated requirement for choline is 750 mg/kg diet. Choline status can be influenced by dietary conditions (carbohydrate overloading) that increase hepatic triglyceride synthesis because choline is needed for export of triglyceride from the liver (Carroll and Williams, 1982). The choline requirement also depends on the methionine and fat content of the diet (NRC, 1995). Choline, methionine, and methyl folate are closely interrelated in methyl-donor metabolism (Zeisel et al., 1989; Selhub et al., 1991). Disturbing the metabolism of one of the methyl donors results in compensatory changes in the others. The use of choline as a methyl donor is a major factor determining how quickly a choline-deficient diet will induce pathology (Newberne and Rogers, 1986).

The primary lesion of choline deficiency is fatty liver. Choline deficiency results in impaired triglyceride transport from the liver (Yao and Vance, 1988). Prolonged deficiency can result in cirrhosis (Zaki et al., 1963; Rogers and MacDonald, 1965). Hemorrhagic kidney degeneration, myocardial necrosis, and atheromatous changes in arteries have also been reported in choline-deficient rats (Chan, 1984; NRC, 1995).

Choline is the only nutrient for which dietary deficiency causes the development of hepatocarcinomas in the absence of a known carcinogen (Newberne and Rogers, 1986). Not only do choline-deficient rats have a high incidence of spontaneous heptatocarcinoma, but they are significantly more sensitized to the effects of carcinogens. This indicates that choline deficiency has both cancer-initiating and cancer-promoting activities (Zeisel and Holmes-McNary, 2001). Investigations into the mechanisms for the cancer-promoting effect in rats fed a choline-deficient diet show (1) there is a progressive increase in cell proliferation, related to regeneration after parenchymal cell death (Chandar and Lombardi, 1988); (2) cell proliferation with increased rate of DNA synthesis may increase sensitivity to carcinogens (Ghoshal et al., 1983); (3) undermethylation of DNA may perturb regulation of genetic information (Locker et al., 1986; Dizik et al., 1991); (4) nuclear lipid peroxidation modifies DNA (Rushmore et al., 1984); and (5) accumulation of diacylglycerol produces abnormal protein kinase C-mediated signal transduction triggering carcinogenesis (da Costa et al., 1993).

H. Water

Water is an essential nutrient for the health and well being of all animals, including rats. It is the medium within which the chemical reactions of the body take place (Harris and Van Horn, 1992). Water is involved in hydrolytic processes; transport of hormones, nutrients, and metabolites; lubrication of joints; transmission of light in the eyes and sound in the ears; and excretion of waste (Robinson, 1957). It also gives form to the body and provides protective cushioning for the nervous system (Askew, 1996).

The role of water in thermoregulation is particularly vital. It absorbs heat where generated, with little temperature rise, and dissipates it throughout the fluids in the body. Thus, enzyme and structural protein damage is minimized, and heat-bearing blood plasma is routed to the skin, where heat is transferred to the environment through conduction, radiation, convection, and evaporation. Body heat also is transferred to the environment via moisture carried out by each exhalation. For every liter of water vaporized at 20°C, 585 kcal of heat are lost (Kleiber, 1975; Askew, 1996).

Water comprises about 99% of all the molecules in the animal body (McFarlane and Howard, 1972) as a consequence of its high percentage of body mass and the small size of the water molecule, compared with molecules of carbohydrate, protein, and fat. On a fat-free body mass basis, the adult animal is said to be a relatively constant 68% to 72% water (NRC, 1974). Water balance in 60-day-old Wistar, Zucker obese, and Zucker lean rats has been studied by measuring liquid water intake; water in food consumed; water lost in urine, feces, and water vapor; metabolic water production; and net water accretion with increases in BW. Daily water accretion was 1.2% to 1.4% of total body water mass. The contribution of metabolic water to the daily water budget was 23.6% for obese Zucker rats, 22.5% for lean Zucker rats, and 15.9% for Wistar rats (Rafecas et al., 1993). Studies of water and electrolyte metabolism in rats revealed that the high water concentrations found in various organs at 2 weeks of age declined remarkably in the liver, spleen, and testes with aging (Hiraide, 1981).

Laboratory rats usually get their drinking water from municipal water systems. Although the composition of water varies among municipal systems, all are required, as a minimum, to meet national primary drinking water standards (National Primary Drinking Water Regulations) established by the U.S. Environmental Protection Agency (EPA). These primary standards set limits on levels of specific contaminants that may adversely affect public health and are known to occur or may be anticipated to occur in public water systems. These contaminants are categorized as inorganic chemicals, organic chemicals, radionuclides, and microorganisms. The EPA has included two levels in the National Primary Drinking Water Regulations for each contaminant. The first is known as the Maximum Contaminant Level Goal (MCLG), defined as the maximum level in drinking water at which no known or anticipated adverse effect on human health would occur. MCLG goals are non-enforceable public health goals. The second is the Maximum Contaminant Level (MCL), defined as the maximum permissible level of a contaminant in water delivered to any user of a public water system. MCLs are enforceable standards. MCLG goal levels are equal to or lower than MCLs, with margins of safety that ensure that exceeding the MCL slightly does not pose significant risk to public health.

The EPA also has established secondary drinking water standards (National Secondary Drinking Water Regulations) that are non-enforceable guidelines regulating contaminants that may cause cosmetic effects (such as skin or tooth discoloration) or aesthetic effects (such as taste, odor, or color) in drinking water. Although the EPA does not require compliance with these secondary regulations, some states do. Detailed information on the EPA drinking water standards for U.S. public water systems may be obtained at the EPA Web site, http://www.epa.gov/safewater/mcl.html. Composition of specific municipal water supplies can usually be obtained from public works departments or state departments of health.

Requirements for liquid water intake are dictated by the need to balance water loss when metabolic water and water from food are inadequate for that purpose. Thus, liquid water requirements will vary with food composition, intake, and metabolism, as well as with activity levels and the need to dissipate body heat, although adult rats typically consume 10 to 12 mL of water per 100 g BW each day. The efficiency of this latter process varies with environmental circumstances, particularly ambient temperature and relative humidity, which in turn affect food intake.

Thirst is the clue that water balance needs attention, encouraging the thirsty subject to seek and consume water. Lick sensors have been used as tools in studying drinking behavior in rats (Weijnen, 1989), and it is apparent that, unless research protocols dictate otherwise, rats receiving pelleted or extruded diets should have AL access to water.

IV. NATURAL-OCCURRING CONTAMINANTS

Plants are the major source of naturally occurring chemicals; and, their composition is complex and highly variable. The great majority of these constitutive chemicals found in food plants are present in the diet at levels that pose no toxicological consequence to the consumer. Most naturally occurring toxicants, except phytic acid, are essentially absent from the major cereal grains used to formulate rodent diets. However, there are some natural, environmentally-acquired chemicals that are known potent rodent carcinogens such as various mycotoxins resulting from feed contamination (NRC, 1996a). Plants have also evolved bioactive compounds that serve as defensive agents against predators. Of 50 such natural compounds evaluated in animal cancer tests, about half have demonstrated carcinogenic properties (Ames et al., 1990).

When plant biocides are considered as natural pesticides, the amount of such naturally occurring compounds exceeds that of synthetic pesticide residues used in agricultural production. There are approximately 70 constitutive naturally occurring chemicals from dietary plants that are reported to possess both mutagenic and antimutagenic and, in some cases, antioxidant properties. Most of these fall under the following classes: flavonoids, phenolic acids, phenylpropanoids, coumarins, depsides, cyclitols, isothiocyanates, catechins, simple phenols, monoterpenes, sesquiterpenes, amino acids, and anthraquinones (NRC, 1996a). Although many of these compounds are not considered to be nutrients, several are metabolized with significant biological effects.

The concentration and toxicity of naturally occurring chemicals may be exacerbated by harvest, processing, and storage conditions (Rao and Knapka, 1987). Natural ingredient diets often contain endogenous metal contaminants derived from growth conditions–that is, arsenic, cadmium, lead, mercury, and selenium–when it is above the requirement concentration (Rogers, 1979). Nitrates, which can be converted to nitrite and form carcinogenic nitrosamines, may be present in variable amounts from fertilizers, and pesticide and herbicide residues may be present as well. Manufacturers take care to ensure that natural toxicants remain below acceptable levels in dietary products. And, although natural ingredient dietary components vary widely in nutrient and non-nutrient content, efforts by manufacturers to maintain a homogeneous product by component selection practices ameliorate extensive variability. Purified diets, the components of which are refined products, do not contain toxic endogenous dietary or microbial contaminants.

Synthetic dietary compounds such as butylated hydroxyanisole, added as an antioxidant preservative, have been found to both inhibit carcinogenesis (Ito et al., 1989) and induce oncogene expression in Fischer 344 male rats (Ito et al., 1993). Forestomach cancer was induced by butylated hydroxyanisole and high doses of several naturally occurring plant antioxidants. However, these same antioxidants given at low doses, with more potent carcinogens, were effective in inhibiting cancer at a number of sites. When the antioxidants α-tocopherol, t-butylhydroquinone, propyl gallate, and butylated hydroxytoluene were examined by using a multi-organ carcinogenesis model, none of the agents studied was unequivocal in exerting either positive or negative influence (Hirose et al., 1993).

A. Non-Nutrient Constituents of Diets that have Biological Consequences

There are several non-nutrient endogenous constituents of plants that are of particular biological interest. The non-nutrient plant antioxidants such as the flavone derivatives, isoflavones, catechins, coumarins, phenylpropanoids, polyfunctional organic acids, phosphatides, as well as the nutrient antioxidants the tocopherols, ascorbic acid, and carotenes have clear roles in dietary plants. They act as reducing compounds, as free radical chain interrupters, as quenchers or inhibitors of the formation of singlet oxygen, and as inactivators of pro-oxidant metals (NRC, 1996a). The biologically active phytoestrogens, plant compounds structurally and/or functionally similar to ovarian and placental estrogens and their active metabolites, include members of the flavonoid family: isoflavonoids, lignans and also some of the flavones, flavanones, chalcones, coumestans, and stilbenes (Wiseman, 2000; Whitten and Patisaul, 2001). Phytoestrogens bind to estrogen receptors (ERs), an action complicated by the recent discovery that ERs are divided into two distinct subtypes: the originally sequenced ER, ER-α, and new variant, ER-β (Whitten and Patisaul, 2001). Phytoestrogens may be important modulators in rat uterotropic response (Boettger-Tong et al., 1998), induction of uterine c-*fos* and estrogenic effects in mammary and pituitary glands (Odum et al. 2001), hormone-mediated tumor development (Wiseman, 2000), and pup growth parameters (Odum et al., 2001). However, inconsistent reports of the effects of phytoestrogens on female reproductive organs, induction of cancers, and blood cholesterol levels have also been reported (Yang and Bittner, 2002).

Soybean meal, alfalfa meal, and other legumes contribute phytoestrogens to rodent diets. Soybean meal is the primary contributor of the glycosides, genistin and daidzin, which are hydrolyzed by large intestinal bacteria to release the isoflavone phytoestrogens genistein (4′,5,7-trihydroxyisoflavone) and daidzein (4′,7-dihydroxyisoflavone), respectively (Wiseman, 2000; Yang and Bittner, 2002). Rat studies using [14C] genistein (4 mg/kg) revealed the highest concentration of radioactivity was found in the gut; significant levels of radioactivity were also found in a variety of other organs, particularly the liver and reproductive organs but also the brain, heart, lungs, and kidneys (Coldham and Sauer, 2000). Retention in the liver was sexually dimorphic, with females showing nearly 2.5 times the radioactivity as did males 2 and 7 hours after the initial dose was given.

The most significant isoflavan is equol, a metabolite of daidzein. Coumestrol is the best-known coumestan and the isoflavonoid with the highest estrogenic potency. Alfalfa is one of the richest sources of coumestans (Boettger-Tong et al., 1998).

The plant-derived phytoestrogens are structurally and/or functionally similar to estrogens (mimics) that exhibit estrogenic or antiestrogenic effects (Whitten and Patisaul, 2001); doses generally reported to be active in rodents range between 10 and 100 mg/kg BW/day. Total and/or individual concentrations of daidzein and genistein in select natural ingredient commercial rodent diets can range between 6 to 491 µg/g diet (Boettger-Tong et al., 1998; Thigpen et al., 1999; Odum et al., 2001); purified diets are generally phytoestrogen-free. Isoflavonoids found in legumes, especially in soybeans and soybean-based products, can contain as much as 0.2 to 1.6 mg of isoflavones per gram dry weight (Whitten and Patisaul, 2001). Total isoflavonoid concentrations of soy protein isolate can range between 0.62 and 0.99 mg/g (Anderson and Wolf, 1995).

Lignans, minor components of cell walls and cereal seed fibers, are found in a number of whole grains. Secoisolariciresinol and matairesinol are lignan glycosides that give rise to the mammalian phytoestrogens enterodiol and enterolactone in the colon by bacterial action (Wiseman, 2000). Lignan concentration of cereals range between 100 and 700 mg/100 g; soy products can contain 900 mg/100 g (Whitten and Patisaul, 2001).

B. Mycotoxins, Aflatoxins, and Fumonisins

Mycotoxins are acquired, naturally occurring substances that result from fungal growth on foodstuffs either in the field or during harvest and storage. Dietary contamination by one or more mycotoxin is common; the presence of one generally implies co-contamination by others as a single fungus can generate several mycotoxins or several mycotoxin-producing fungi may infect the same plant (NRC, 1996a). Contamination by two toxigenic species of *Aspergillus, A. flavus,* and *A. parasiticus,* both known to produce hepatocarcinogenic aflatoxins, appears ubiquitous. *A. flavus* produces aflatoxins B_1 and B_2, whereas *A. parasiticus* produces aflatoxins B_1, B_2, G_1, and G_2 (Pitt et al., 1993). Although all four aflatoxins are toxic and believed to be

carcinogenic in animals, B_1 is the most prevalent and the most potent, causing hepatocellular adenomas and carcinomas and colon tumors in rats (NRC, 1996). Corn and other grains, peanuts, and cottonseed meal are common to aflatoxin-producing fungi. Concentrations of the toxin depend not only on infection but also on pre- and post-harvest conditions and can vary geographically with the southeastern United States affected most frequently. Environmental stressors such as drought or insect attack cause corn crops to become particularly susceptible to *A. flavus* growth. The median levels of aflatoxins in corn range from less than 0.1 to 80 ng/kg (NRC, 1996a). Currently, aflatoxin B_1 is the only mycotoxin for which the Food and Drug Administration has set action levels, the level of contamination at which the foodstuff is regarded adulterated, in corn (Riley et al., 1993). Ochratoxin A, produced predominantly by *A. ochraceus* and *Penicillium verrucosum,* occurs worldwide in many grains (barley and wheat) and is implicated in urinary tumorigenesis in humans and rodents (NRC, 1996a).

Aflatoxins are currently the mycotoxins of greatest concern in the United States; however, the fumonisins have become the fastest growing area of mycotoxin research.

Cereal grains can be infected by the fungal genus *Fusarium*, which produces estrogenic compounds (mycoestrogens), highly stable compounds that can be ingested, inhaled, or absorbed through the skin (Whitten and Patisaul, 2001; Yang and Bittner, 2002). Zearalenone, a mycotoxin with estrogenic properties, is produced by the molds *F. graminearum* and *F. culmorum,* which are commonly found in plants, soil, and stored grains such as barley, corn, rice, oats, and rye (Park et al., 1996; Boettger-Tong et al., 1998). Zearalenone has been associated with mammary tumorigenesis (Shier et al., 2001) and may be uterotropic (Sheehan et al., 1984). When 500 cereal samples from 19 countries were analyzed for zearalenone, more than 40% had concentrations as high as 0.045 µg/g cereal (Tanaka et al., 1988), whereas concentrations of 21 μg/g have been reported in moldy corn in the United States (Park et al., 1996).

Another widely distributed toxigenic fungus, *F. moniliforme* is a ubiquitous, seed-borne (Riley et al., 1993) contaminant in corn, which produces the water-soluble fumonisins B_1 (FB$_1$) and B_2 (FB$_2$) and fusarin C (NRC, 1996a). The T_2 toxin, also produced by *Fusarium* and other species, has been reported to be carcinogenic. Although fumonisins are poorly absorbed and rapidly eliminated, small amounts do accumulate in the liver and kidney (Voss et al., 2001). Among Sprague-Dawley rats dosed orally for three consecutive days with [^{14}C] FB$_1$ (0.045 µCi), 80% of the label was excreted in feces within 48 h and less than 3% from urine within 96 h of the final dose (Norred et al., 1993). Liver- and kidney-specific activities peaked at 24 h (after final dose) and persisted for 48 h. Similar to FB$_1$,

FB$_2$ is also rapidly cleared from plasma and excreted (82% within 72 h, primarily during the first 24 h) (Voss et al., 2001). FB$_1$ does not cross the placenta and is not teratogenic *in vivo* in rats, mice, or rabbits but is embryotoxic at maternally toxic doses.

A number of studies have demonstrated the carcinogenicity of *F. moniliforme* to rats (Marasas et al., 1984). Rats developed liver tumors when fed 50 ppm purified FB$_1$ (Gelderblom et al., 1991). Tumorigenic effects were not found at diet concentrations of less than 15 ppm FB$_1$ (Howard et al., 2001a); however, FB$_1$ caused kidney adenomas and carcinomas in male rats. Fumonisin intake levels of between 0.08 and 0.16 mg FB$_1$/100 g BW/day for 2-years produced liver cancer in male BD IX rats (diet concentration was not given in this article, but over a narrow range of BW and intakes, this diet could have contained about 16 to 25 ppm FB$_1$), thus suggesting that rat strain differences do occur to FB$_1$ exposure (Gelderblom et al., 2001). Exposure levels less than 0.08 ppm mg FB$_1$/100 g BW/day failed to produce cancer but did induce mild toxicity and preneoplastic lesions. The diets used in the long-term experiments were marginally deficient in lipotropes and vitamins and could have played an important modulating role in fumonisin-induced hepatocarcinogenesis.

FB$_1$ is a rodent carcinogen that induces renal tubule tumors in male Fischer 344 (Howard et al., 2001b). The FB$_1$ concentration of the control NIH-31 rodent diet was less than 0.06 ppm. Male and female Fischer 344 rats were fed varying concentrations of a purified extract of FB$_1$. Female Fischer 344 rats lost BW when fed 100 ppm FB$_1$ compared with control rats. There were no FB$_1$-dependent BW changes in male Fischer 344 rats fed up to 150 ppm. There were no dose-related differences in female or male Fischer 344 survival at 104 weeks. Renal tubule adenomas and carcinomas became evident among male rats fed 50 ppm FB$_1$ (but not 0, 5, or 15 ppm) but were more pronounced in rats fed 150 ppm. There were no apparent FB$_1$-dependent changes in tumor incidence among female rats, even when fed 100 ppm FB$_1$.

In rats, FB$_1$ induces apoptosis of hepatocytes and of proximal tubule epithelial cells (Tolleson et al., 1996). More advanced lesions in both organs are characterized by simultaneous cell loss (apoptosis and necrosis) and proliferation (mitosis); when there is an imbalance between cell loss and replacement, the potential for carcinogenesis increases (Voss et al., 2001).

Fumonisins are specific inhibitors of the enzyme ceramide synthase (sphinganine and sphingosine N-acyltransferase), which is a key enzyme in the pathway leading to formation of sphingomyelin and complex sphingolipids (Wang et al., 1991). Sphingolipids, formed *de novo* from sphinganine, rich in brain tissue, perform a wide variety of biological functions: participation in cellular communication, modulation of behavior of cellular proteins and receptors, and signal transduction both as extracellular agonists and intracellular mediators. Fumonisins alter sphingolipid biosynthesis, induce hepatotoxicity, and elevate serum cholesterol concentration in all species studied (Haschek et al., 2001). Greatest sphingolipid alterations occur for sphingosine and sphinganine concentrations in kidney, liver, lung, and heart.

FB$_1$-disrupted lipid biosynthsis was reported for Fischer 344 rats. Major changes were seen in phosphatidylcholine, phosphatidylethanolamine, and cholesterol in serum and liver (Gelderblom et al., 2001). In short-term studies, phosphatidylethanolamine increased, but sphingomyelin decreased when rats were fed 250 ppm in the diet. However in long-term studies with male BD IX rats, dietary levels of 1, 10, and 25 ppm FB$_1$ increased phosphatidylethanolamine in liver tissues. Changes in saturation of biological fatty acids may determine the responsiveness of cells to transformation, or the expression of certain cell types associated with neoplastic development. When PE is markedly increased in the membrane fractions of hepatocytes, there is an increase in the absolute concentration of C20:4ω6 (arachidonate metabolized to prostaglandin E$_2$) within the cell. This latter event, together with an increase in C18:1ω9, implies a lower oxidative status and likely favors cell proliferation, especially in the initiated hepatocyte cell population. The disruption of the phospholipid and n-6 fatty acid metabolic pathway, producing changes in the level of C20:4ω6, appears to be critical with respect to cancer promotion, particularly with low dietary FB$_1$ concentration, in which cancer promotion is effected in the absence of apoptosis and disruption of the sphingolipid metabolic pathway.

Nutrient-compound metabolic interaction, drug toxicity, genetic response, survivability, and various pathology endpoints are all influenced by diet. Selection of an appropriate dietary formulation and understanding of the performance characteristics of the rat when fed a particular diet are of primary importance during protocol development.

V. DIET RESTRICTION

The concept of restricted-feeding, without malnutrition, to control BW, decrease premature death, slow the incidence and severity of degenerative diseases, and delay the onset and incidence of neoplasia while increasing life expectancy and life span has been firmly established (McCay et al., 1935, Tannenbaum 1945, Silberberg and Silberberg, 1955; Ross, 1972; Weindruch and Walford, 1988; Hart et al., 1995; Roe et al., 1995). Dietary restriction (also food or calorie restriction, diet optimization) inhibits age-related hyperparathyroidism and senile bone loss

(Kalu et al., 1988), supports reproductive capacity in older rodents (Merry and Holehan, 1979), and protects rats from the toxic actions of drugs (Duffy et al., 1995). Other reported benefits of dietary restriction include the modulation of endogenous processes such as apoptosis, gene expression, cell turnover, and free radical-induced DNA damage (Grasl-Kraupp et al., 1994; Fernandes et al., 1995; Fischer and Lutz, 1998). Recently, dietary restriction has proven to be a powerful control for statistical variation among rodents assigned to gerontology and toxicology research (Seng et al., 1998; Keenan et al., 1999). Although increased BW is influenced by breeding selection and consequent genetic drift, the practice of AL feeding, after weaning, is the most important factor associated with obesity and early mortality among research animals. In research, use of rodents of specific genetic makeup is desired because of uniformity and dependability. However, one important disadvantage inherent in the use of inbred animals is that negative genetic characteristics may accumulate, resulting in life span changes. Barring outbreak of serious life-threatening disease, research rats during the 1920s and 1930s often reached 36 or 40 months of age, but by the mid-1950s, the oldest ages reached by individuals of several inbred strains had decreased by one-fourth or one-third compared with the ages reached by rats of mixed stocks (Silberburg and Silberburg, 1955). Consequently, conclusions about the actual life spans of currently used stocks maintained to 24 months of age must be made with caution.

The level of dietary restriction used to achieve beneficial health effects has varied widely, and dietary restriction levels between 10% and 40% of AL intake of natural ingredient diets have avoided the effects of malnutrition (Weindruch and Walford, 1988; Hart et al., 1995; Roe et al., 1995; Christian et al., 1998; Keenan et al., 1999; Duffy, 2001) (Table 9-1). Extension of life span is not apparently linked to early growth mechanisms (Bellamy, 1990; Masoro, 1996). Some report that metabolic rate, usually expressed as oxygen consumption per unit of lean body mass, is decreased by long-term dietary restriction in rats (Dulloo and Girardier, 1993; Roe et al., 1995), although others found no apparent alteration of metabolic rate (McCarter et al., 1985; Duffy et al., 1989). Because rates of oxygen consumption differ among organs, metabolic rates between AL and dietary-restricted rats based on total lean body mass may obscure organ-specific variations (Roe et al., 1995, Weindruch and Sohal, 1997). Dietary restriction may be governed by energy intake per rat, rather than per unit metabolic mass, whereas protein and mineral restrictions are, comparatively, of less importance (Masoro and Yu, 1989; Bellamy, 1990). Although dietary restriction at 40%, the most commonly reported limit, has increased the average life span and the maximal life span (the mean survival of the longest-lived decile), and retarded the age-associated progression of disease among male Fischer 344 rats (Yu et al., 1985, Weindruch and Walford, 1988; Masoro et al., 1989b; Turturro et al., 1999) and Sprague-Dawley rats (Duffy, 2001), this degree of dietary restriction produces rats of very different size and body composition compared with the bodies of their AL control-fed counterparts (Duffy et al., 1989; Weindruch and Sohal, 1997).

The longer an animal is on a regimen of dietary restriction, the greater the apparent benefit. The age of dietary restriction initiation was not as important a determinant as the duration of dietary restriction in regard to incidence and age of onset of leukemia in male Fischer 344 rats (Higami et al., 1994). Dietary restriction is effective and protective when initiated in both young and adult animals (Roe et al., 1995; Masoro, 1996; Weindruch and Sohal, 1997), and its mode of action is apparently linked to a reduction of energy intake, which alters the characteristics of fuel use via alteration in nervous and/or diet-related endocrine functions (Yu et al., 1985; Masoro, 1996). Dietary restriction protects against long-term damage of such fuel use from glycation or oxidative damage. Although antioxidant defense systems deteriorate with age, even moderate dietary restriction maintains the function of antioxidant defense systems and proteolytic and lipolytic enzymes involved with removal of reactive-oxygen damaged molecules (Hart et al., 1995; Masoro, 1996; Weindruch and Sohol, 1997).

No single diet can be optimal for all physiological life stages, and between the extremes of undernutrition and dietary enrichments are those regimens that are optimal for longevity (Silberberg and Silberberg, 1955). Optimal diets vary with age and sex, and diet optimization is altered by reproductive and lactational stressors. In addition, dietary restriction may act on longevity by different mechanisms during the developmental phase of life compared with adult life (Yu et al., 1985). The complexity increases when one considers the nutrient requirements of different strains within one and the same species, or the multitude of transgenic and immunodeficient rodent models that have become increasingly available.

Long-term dietary restriction programs usually involve meal eating rather than the nibbling pattern of food intake typical of AL-fed rats. When male Fischer 344 rats were fed their 60% of AL aliquot at either one or two meal periods, median life span did not change between the groups; however, the diurnal pattern of plasma corticosterone and glucose concentrations differed between the groups (Masoro et al., 1995). Similarly, night-time normal feeding behavior was found to better synchronize physiological performance between AL and dietary-restricted rats than did day-time feeding, thus allowing more precise evaluation of qualitative changes in metabolism and energy expenditure as they relate to feeding behaviors (Duffy et al., 1989).

Dietary restriction can be accomplished by reducing the quality of the diet (e.g., decreased caloric density by modifying energy components) or by reducing the amount of diet consumed (restriction of all components). The present view is that the beneficial effects of dietary restriction can be attributed solely to caloric intake irrespective of diet composition. Excess energy, more than any specific nutrient, is the most important dietary factor contributing to obesity, early senescence, premature appearance of age-related disease and spontaneous tumors, and endocrine disruption (Tannenbaum, 1945; Roe et al., 1995; Keenan et al., 1999). Restriction of energy intake delayed death owing to neoplasms and chronic nephropathy; restriction of dietary fat or protein without energy restriction did not provide the same response (Masoro et al., 1989a,b; Masoro, 1992). Protein restriction in the absence of food (caloric) restriction did slow the progression of chronic nephropathy and cardiomyopathy, although less effectively than did calorie restriction (Maeda et al., 1985, Yu et al., 1985). However, when Fischer 344 male rats were AL-fed 21% protein as soy in a diet isocaloric to a 21%-casein diet, severe renal lesions were reduced by 60% at time of death, thus demonstrating that the nature of dietary protein does influence the age-associated progression of nephropathy by mechanisms that are not secondary to dietary restriction intake (Masoro and Yu, 1989).

Data from several studies indicate that controlling energy, rather than protein, intake results in increased rodent survival and prevention of renal disease during long-term studies with Fischer 344, Sprague-Dawley, and Wistar rats (Maeda et al., 1985, Roe et al., 1995; Keenan et al., 1999). Reduction of specific dietary components (energy, fat, protein, and mineral) or different sources of dietary protein (casein, soy, and lactalbumin) without energy restriction affected mortality (Shimokawa et al., 1996). Neither restricted intake of dietary fat nor minerals influenced the median or maximum life span of male Fischer 344 rats (Iwasaki et al., 1988). Restricted fat intake did retard the development of chronic nephropathy, but it also increased the prevalence of lymphoma and leukemia. In the case of leukemia/lymphoma, dietary restriction delayed the age of occurrence (Masoro, 1992) but did not retard its progression, that is, the time between occurrence and death (Shimokawa et al., 1993). The incidence and age of leukemia onset appear to relate to the total cumulative energy intake of the rat so that duration of dietary restriction is a significant factor (Higami et al., 1994).

Dietary restriction supports greater spontaneous loco-motive activity into old age (Yu et al., 1985; Duffy et al., 1989). Several studies reviewed by Goodrick (1980), have shown that exercise can increase longevity of rodents, but the effect is small compared with that of dietary restriction. However, dietary restriction does not apparently reduce caloric expenditure per gram lean body mass over a significant portion of the life span (McCarter et al., 1985), nor does dietary restriction appear to reduce arterial blood pressure in male Fischer 344 rats (Yu et al., 1985). Dietary restriction also substantially reduced the incidences of endocrine-mediated tumors in Sprague-Dawley and Fischer 344 rats (Christian et al., 1998).

The beneficial effects of dietary restriction on the control of plasma glucose and insulin efficiency may result from lower sustained concentrations of glucose, conditions that are less damaging and endocrine disruptive throughout the life span (Hart et al., 1995; Masoro et al., 1995).

Dietary restriction reduced muscle fiber loss and decreased the accumulation of mtDNA deletions in aged rat muscle (Aspnes et al., 1997).

Cost-associated increases with dietary restriction are exceeded by advantages of increased survival, reduced disease and tumor incidence, increased ease of histopathological processing and evaluation (Christian et al., 1998; Lewis et al., 1999), and increased bioassay sensitivity by reducing variability (Seng et al., 1998).

Questions remain about the mechanisms of dietary or caloric restriction, which influence aging and disease processes in rodents. In the search for single nutritional factors other than total dietary intake control that may play a role in determining the life span, the factors that promote or inhibit disease, that accelerate or retard the rate of aging, and that accelerate or retard the rate of growth and development would need to be investigated (Silberberg and Silberberg, 1955).

VI. CLASSIFICATION AND SELECTION OF DIETS

The type and composition of diets used in production and experimental rat colonies are major considerations for maintaining animals in good health or obtaining consistent experimental results. The best diet for a particular rat colony depends on production or experimental objectives. A detailed evaluation not only of nutrient requirements but also of the dietary requirements of specific rat colonies is essential in order to obtain optimal results. Diets for laboratory rats are classified according to the degree of ingredient refinement (NRC, 1995).

A. Natural Ingredient Diets

Diets formulated with appropriately processed whole grains, such as wheat, corn, or oats, and commodities that have been subjected to limited refinement, such as fishmeal, soybean meal, or wheat bran, are referred to as natural ingredient diets. These types of diets also have been referred to in the literature as "cereal based," "chow," "unrefined," "non-purified," or "stock" diets. Knapka et al. (1974) published the formulation of a natural ingredient diet that has proven to be satisfactory for rat growth and reproduction in conventional environments. Autoclavable diet formulations for rats maintained in germfree or specific pathogen-free environments have been published by Kellogg and Wostmann (1969), Knapka et al. (1983), and NRC (1995).

Natural ingredient diets are the most readily available of the three types of diets because they are economical to manufacture and are generally palatable and well accepted by rats. Therefore, the natural ingredient diets are the most widely used diets in rat colonies associated with biomedical research. Several factors associated with natural ingredient diets limit their use in rat colonies maintained for biomedical research.

Because of variations that occur in natural ingredients, it is not possible to completely control the nutrient concentrations among production batches of a product. The amount of variation in nutrient concentrations occurring in shipments of a particular diet will depend on factors such as the number, kind, and quality of ingredients; ingredient manufacturing procedures; and environmental control during diet warehousing and transporting. Knapka (1983) reported the variation in nutrient concentrations among production batches of two rodent diets during a 3- to 4-year period. Representative samples of diets manufactured for the National Institutes of Health were sent to an independent laboratory for analyses of the proximate nutrients, calcium, and phosphorus. The results indicated that nutrient concentrations in a large percentage of the samples were slightly in excess of the planned levels; however, occasional batches of diet contained either excess or deficient concentrations of nutrients. The observed variations would have little effect on rats under most practical conditions but may justify concern if rats are subjected to stress or in experiments in which the dietary concentration of a specific nutrient is critical to the quality of experimental results. It should be recognized, however, that the variation observed in analytical results from a series of feed samples includes inherent errors associated with product sampling, subsampling, and analytical procedures.

A factor also restricting the use of natural ingredient diets for research relates to the difficulty of making changes in the concentration of single nutrients. Each natural ingredient may contain some percentage of all the approximately 50 required nutrients. Therefore, it is not possible to change the concentration of a single nutrient by altering the amount of any one ingredient in a formulation without changing the concentration of practically all other nutrients. This is an undesirable characteristic,

particularly for experimental designs requiring diets with graded concentrations of a specific nutrient.

The potential for residual concentrations of pesticides, heavy metals, phytoestrogens, and other agents that might alter responses to experimental treatments may limit the use of natural ingredient diets in toxicology-related studies. These diets may become contaminated with man-made or naturally occurring compounds (Newberne, 1975; Fox et al., 1976; Greenman et al., 1980; Rao and Knapka, 1987). Although dietary chemical concentrations are generally below the levels that produce clinical signs of toxicity, they may influence biochemical or physiological processes in test animals and alter experimental results. Procedures for decontaminating diets are difficult or nonexistent, but the concentrations of chemical contaminants can be controlled to a degree by manufacturing diets from ingredients that have low potentials for contamination.

B. Purified Diets

Diets formulated with only refined ingredients are referred to as purified diets (NRC, 1995). In these diets, casein or isolated soy protein are examples of protein sources; sugar or starch is a source of carbohydrates; and vegetable oil or animal fat is added as a source of energy and essential fatty acids. Cellulose is used for crude fiber, and inorganic salts and pure vitamins are added for the minerals and vitamins, respectively. These diets also have been described in the literature as "semipurified," "synthetic," or "semi-synthetic." Formulations of purified diets that have resulted in acceptable growth and reproduction in experimental rat colonies have been published (Hurley and Bell, 1974; AIN, 1977; Reeves et al., 1993; Lewis et al., 2003).

Planned nutrient concentrations in purified diets can be readily obtained with minimal variation among production batches of diet provided that ingredient quality is maintained. Nutrient concentrations can be readily reproduced or altered for inducing nutritional deficiencies or excesses, and there is a low potential for even residual concentrations of chemical contaminants in diets manufactured with purified ingredients.

Purified diets are more expensive than are natural ingredient diets because of higher ingredient costs, and their acceptance by some strains of rats might be marginal. Purified diets have been used in experiments involving rats for many years, but almost all such studies have been of relatively short duration. The performance of rats fed purified diets for short and long terms may differ because purified ingredients may not contain adequate amounts of nutrients that are required in trace concentrations. In natural ingredient formulations, this is of little concern because relatively unrefined feed ingredients usually contain ample amounts. Errors of omission in the formulation of purified diets are critical because each ingredient may be the only source of an essential nutrient.

C. Chemically Defined Diets

Diets formulated entirely with chemically pure compounds are designated as chemically defined diets (NRC, 1995). Amino acids, sugars, triglycerides, EFAs, inorganic salts, and vitamins are used to provide the required nutrients. Pleasants et al. (1970) published the formulation for a chemically defined diet for rats. These diets are useful in studies in which strict control of the concentration of specific nutrients is essential. However, their use even in experimental rat colonies has been very limited because of the high ingredient costs, their instability, and the experience required to formulate and prepare these diets.

D. Closed Formula Diets

Diets manufactured and marketed by commercial institutions that consider the quantitative ingredient composition of the diet as privileged information are referred to as closed formula diets (Knapka et al. 1974). The list of ingredients used to formulate the diet and a "guaranteed" analysis are readily available, as well as the mean nutrient analysis. However, the amount of ingredients used in the manufacture of the diet may be changed without the users' knowledge.

E. Open Formula Diets

Diets manufactured in accordance with a readily available quantitative ingredient formulation are referred to as open formula diets (Knapka et al. 1974). A change in the amount of ingredients used in an open formula diet may be appropriate when there is a change in the nutrient composition of ingredients; however, these changes are made with the knowledge and consent of the customers. Open formula diets have been recommended for experimental laboratory animal colonies in publications originating from the National Institutes of Health (Knapka et al.,1974), the American Institute of Nutrition (1977), and the NRC (1978). Diets manufactured in accordance with readily available quantitative ingredient composition may be essential for evaluating and interpreting experimental results.

VII. DIET STERILIZATION

As many research and commercial breeding facilities house animals under specific pathogen-free barrier or isolater-maintained conditions, diet sterilization is an integral part of rodent maintenance protocol. Diet sterilization also is required by many facilities that conventionally house rodents. The presence of microbial agents in diets presents a risk for potential specific pathogen-free-barrier contamination, introduction of potentially infectious agents or heavy microbial burdens to immune-compromised animal resources, and economic loss. Diets can be sterilized or decontaminated to remove pathogenic bacteria, molds, viruses, and insect pests (NRC, 1996b). Autoclave processing is the most commonly used method to achieve diet sterilization. Autoclavable diets are vitamin fortified to offset vitamin loss owing to heat effects and are recommended if diets are to be autoclaved. After steam autoclave sterilization or pasteurization treatment of the diets, post-autoclave quality should be monitored for adequate vitamin concentrations. Diets should be stored in temperature-controlled environments before and after autoclave processing.

Although steam autoclaving at 121°C for 15 to 20 minutes is frequently recommended for diet sterilization, some diet formulations (i.e., purified, high casein or sugar components) are less successfully autoclaved at these temperatures due to heat effects (e.g., Maillard reaction formation and pellet hardening). Pasteurization at 100°C for 5 minutes is successful in significantly reducing microbial contamination but does not equal high temperature autoclaving in elimination of bacterial bioload. Pasteurization at 80°C for 5 to 10 minutes will remove vegetative forms of mold but not spores (Clarke et al, 1977); therefore, pasteurized diets are not sterile but microbial contaminants are reduced. Pelleted and meal forms of semipurified and vitamin-fortified, autoclavable natural ingredient diets were either pasteurized at 105°C for 6 minutes, or steam autoclaved at 135°C for 5 minutes to control microbial contaminants (Table 9-2) (Lewis et al., 1999). In all cases, ground meal diet was higher in numbers of microbial contaminants. Steam autoclaving effectively removed all microbial contaminants from both the pelleted and meal forms of the diet. Pasteurization eliminated 92% of the microbial contaminants from the pelleted diet, but eliminated only 14% of the total bacteria from the meal form of the diet. Gram-negative bacteria, mold, and coliforms were effectively removed by pasteurization.

Laboratory animal diets may also be sterilized by ethylene oxide fumigation (NRC, 1996b) and ionizing radiation. There were no differences in intake, nitrogen retention, or dry matter, protein, and fat digestibilities among male germ-free Wistar rats fed autoclaved or irradiated diets (Yamanaka et al., 1981).

Treatment with high-energy (ionizing) radiation is a cold, nonchemical process for preserving food that has been studied extensively for more than 40 years. Irradiation is

TABLE 9-2

EFFECT OF AUTOCLAVING OR PASTEURIZATION ON MICROBIAL CONTAMINATION OF DIETS

Diet type	Diet form	Treatment[a]	CFU/g[b]		
			Total bacteria	Gram negative bacteria	Mold
Natural-ingredient	Pellet	None	30,000	0	0
		Autoclaved	0	0	0
Natural-ingredient	Meal	None	52,867	2	0
		Autoclaved	0	0	0
Semi-purified	Pellet	None	6,600	60	0
		Pasteurized	540	0	0
Semi-purified	Meal	None	56,000	2,040	7,400
		Pasteurized	48,000	0	0

[a]Autoclaved at 135°C for 5 min; pasteurized at 105°C for 6 min.
[b]Mean colony forming units per gram, 3 diet lots per sample.
Adapted from Lewis et al., 1999.

usually accomplished with gamma radiation from a radio-isotope source; most commercial facilities use cobalt-60. The technology is well developed, and irradiated composite research diets are rapidly gaining acceptance (Thayer, 1990; Steele and Engel, 1992; Woods, 1994). Yet, questions remain about diet irradiation and breakdown products with potential toxicities, that is, free radicals, thiobarbituric acid reactants, antivitamin effects, benzo(α)pyrene quinones (Wills, 1980; CAST, 1986, Gower and Wills, 1986, Tritsch, 1993), the potential for bacterial or viral mutations (WHO, 1977; CAST, 1986; Hoekstra, 1993), and evidence of peripheral lymphocytic polyploidy among research animals and malnourished children fed irradiated wheat (Bhaskram and Sadasivan, 1975). Although many studies, some multi-generation protocols, have accumulated adequate knowledge to ensure that product safety is well managed within the food irradiation industry, the debate continues.

Most of the radiolytic products identified in irradiated foods can also be found in nonirradiated foods, and many are generated in foods by other processing procedures, including cooking (WHO, 1977). The greatest concentrations of identified radiolytic products, produced with radiation doses up to 60 kilogray (kGy), are only in the milligram per kilogram range. The low concentrations of these products suggest that health hazards are negligible. Further, methods of testing the functional properties of packaging materials and detecting migrating compounds are well established and are applied to nonirradiated as well as irradiated packaging materials (WHO, 1977).

In radiation processing the absorbed dose, generally measured in units of kGy where one gray (symbol Gy) is an energy absorption of 1 J per kilogram (Woods, 1994), determines the degree of chemical and physical changes produced within a feedstuff. To effectively sterilize

laboratory animal feeds, an absorbed dose of 30 to 50 kGy is required. The FDA (Title 21, CFR579) recommends irradiation of animal diets not exceed KGy.

Specialized terms used to describe irradiation applications in processing of foods are, briefly: (1) *radicidation*, doses of 0.1 to 8 kGy, eliminates pathogenic organisms and microorganisms other than viruses and decreases the numbers of viable non-spore-forming pathogenic microorganisms (requiring 2 to 8 kGy); (2) *radurization*, doses of 0.4 to 10 kGy, improves product shelf-life by substantial reduction of microorganisms that cause spoilage; and (3) *radappertization*, doses of 10 to 50 kGy, brings about complete sterilization (Woods, 1994).

In 1980, the Joint Food and Agriculture Organization/International Atomic Energy Agency/World Health Organization Expert Committee on the Wholesomeness of Irradiated Food concluded that irradiation of any food commodity with an absorbed dose of up to 10 kGy causes no toxicological hazard and induces no special nutritional or microbiological problems (Woods, 1994). The 10-kGy level is not an upper safe limit but rather a level at which, or below which, safety has been proven.

Although ionizing energy can modify the physical and chemical properties of individual dietary components (CAST, 1986), little change in nutritive value of animal feeds was noted when diets were irradiated with doses of 5 and 10 kGy (Kennedy, 1965). The biological value of proteins and the ME value of rodent diets were unaltered by sterilization with 56 kGy of ionizing energy (CAST, 1986). Germ-free rats were maintained for 5 years on diets that had been sterilized with ionization energy (Ley, 1975). When a diet composed of 35% whole-milk powder and 65% standard, natural ingredient rat diet was irradiated and fed to five subsequent generations of rats, there were no observable changes in fertility, duration of pregnancy, litter size, weaning index, sex ratio, or number of fetal malformations (Renner et al., 1973). In addition, the high content of free radicals produced no observed harmful effects on the characteristics investigated.

Diets containing high concentrations of refined sugars do not respond as favorably as do natural ingredient diets to high doses of ionizing energy in terms of palatability and nutritional quality owing to the caramelizing effect seen with both heat and ionizing sterilization methods (CAST, 1986). Vitamins may be disproportionately affected by irradiation as with heat sterilization. Supplementation with additional vitamins, particularly those most heat labile, may be required by this sterilization method, as with autoclaving. Irradiation does not prevent subsequent recontamination, and appropriate post-irradiation measures should be taken: proper packaging, temperature limits, moisture controls, and inventory turnover (Woods, 1994). The health risks associated with consumption of irradiated diets are still unconfirmed but may be inconsequential compared with the implications of microbiologically borne diseases.

Although irradiation can cause bacteria to mutate, heat can also generate mutations. Commercial and laboratory experimentation has not proven such mutations confer any properties to the bacteria that would be detrimental to research animals, humans, or the environment (WHO, 1977; Maxcy, 1983; Hoekstra, 1993). Ionizing energy in sufficient doses detoxifies aflatoxin (Temcharoen and Thilly, 1982). Others report that the spores of *Clostridium botulinum* are resistant to the permitted doses of radiation and thus would not be eliminated (Tritsch, 1993). With the dose required to eliminate the important bacterial pathogens, that is, *Salmonella* and *Campylobacter* sp., surviving organisms are weakened and present no apparent special public health risk (Maxcy, 1983).

VIII. DIET FORMULATION

Diet formulation is a process to determine the amount of various feed ingredients required in a feed formulation that will produce a diet with the planned concentrations of nutrients. The process of formulating natural ingredient diets is complex in that each ingredient may contribute a percentage of the approximately 50 required nutrients to the total dietary concentration. Therefore, it is necessary to account for the nutrients in each ingredient to determine the total dietary nutrient concentrations.

The initial step in the process of diet formulation is to establish the planned dietary nutrient concentrations. This can be based on published guidelines for the dietary requirements of rats (NRC, 1995) or on experimental objectives. Dietary nutrient concentrations must be adjusted for estimated losses occurring during the manufacturing process, feed storage, feed sterilization, and

factors that may affect feed consumption. The amount of heat-labile vitamins lost during the manufacturing and sterilization process will mostly depend on the amount of heat applied in the process. The amount of nutrients lost during storage depends on the time between manufacture and use, and loss is increased with increased humidity and temperature. Factors that alter feed consumption and may require compensatory adjustments include dietary energy levels or the addition of foul-tasting test articles.

The second step in diet formulation is to select the major ingredients to be used. These are selected primarily on the basis of nutrient composition, but factors such as availability in the marketplace, palatability, and the potential for biological or chemical contaminants are also important considerations. In general, diet quality increases as the number of ingredients used in the formulation increases. Large numbers of ingredients used in a formulation tend to minimize the effect of variation in nutrient concentrations that any one ingredient has on the total nutrient concentration.

A typical natural ingredient diet will include one or two ingredients as the primary source of each nutrient class and supplemental vitamins and minerals. Crude protein is supplied by a combination of ingredients of animal and plant origin, such as fish meal, dairy products, soybean meal, or corn gluten meal. The primary source of carbohydrates is a combination of whole feed grains—such as wheat, barley, corn, or oats—or by-products of these ingredients such as wheat middlings, wheat bran, or oat groats. Alfalfa meal, dried beet pulp, or oat hulls are used as a source of fiber. Dietary energy concentrations are increased and EFAs are provided by adding soybean or corn oil. An ingredient such as brewer's dried yeast may be used as a source of water-soluble vitamins. Limestone, dicalcium phosphate, calcium carbonate, and NaCl are used as sources of the major minerals. The trace minerals and vitamins are added in premixes that are formulated to provide the differences in concentrations between the total amounts supplied by the ingredients and the planned dietary concentrations. Separate vitamin and mineral premixes are formulated, prepared, and stored separately to minimize the oxidation of vitamins through mineral-catalyzed reactions.

The final step in the process is to determine the amount and ratios of each of the selected ingredients that will be required to produce the diet. This involves the use of nutrient data from ingredient tables (NRC, 1971) to calculate the concentrations of each nutrient in a specific combination of ingredients. The amount of each ingredient is expressed as a percentage by weight, and the diet formula must total 100%. A systematic and detailed procedure for formulating natural ingredient diets was previously published by Knapka and Morin (1979).

Computer programs for calculating diet formulations are commercially available.

The process of formulating purified diets is similar to that for natural ingredient diets in that planned dietary nutrient concentrations must be established and an ingredient selected to supply each nutrient. Each purified ingredient essentially contains a single nutrient or nutrient class. Therefore, the process of calculating the amount of each ingredient to be used is less complex than for natural ingredient diets.

IX.   DIET MANUFACTURE

The manufacture of diets involves a process in which ingredients are ground into fine particles, blended in the amounts specified in the formulation, mixed, made into an acceptable physical form, and packaged for protection until use. The efficient manufacture of natural ingredient diets requires a large capital investment for facilities, milling apparatus, and inventories of ingredients that are least costly when purchased in large bulk quantities. Natural ingredient diets should be purchased only from manufacturers with facilities that produce only laboratory animal diets and do not use feed additives such as rodenticides, insecticides, hormones, antibiotics, or fumigants. Areas in which ingredients and diets are stored and processed should be kept clean and enclosed to prevent entry of domestic or wild animals, bird, or insects.

Purified diets are routinely manufactured by commercial manufacturers according to formulations provided by clients or from various catalogs. The nutrient composition of "catalog diets" should be carefully checked because the original formulations may have been designed to meet the requirements of specific research programs, and their use for other projects may not be valid. Purified diets can be prepared in laboratories or diet kitchens with minimal amounts of special equipment. Essential apparatus includes a grinder, an analytical balance or scale, a mixer, and perhaps a pellet mill. Ingredients for purified diets are readily available from various biochemical suppliers. Navia (1977) published guidelines for purified diet preparation.

Frequently, the final step of diet manufacture is completed in the laboratory by the incorporation of various test compounds, which involves combining a relatively small amount of test compound with a large amount of an otherwise complete diet in a mechanical mixer. The length of time required to mix dietary ingredients and obtain maximum distribution of all the constituents depends on factors such as particle size and density, as well as mixer size and speed. Overmixing results in particle separation, depending on factors such as particle density and physical form and the susceptibility of particles to static electrical charges that can develop in mixers. The text on feed

manufacturing written by Pfast (1976) is an excellent source of information for individuals involved in diet preparation.

X.   PHYSICAL FORMS OF RAT DIETS

Diets for laboratory rats can be provided in various physical forms. The criteria for selecting particular forms are usually related to specific program or experimental requirements. Pelleted diets are the most widely used. Feed pellets are formed by adding heat and moisture to the diet in meal form and then forcing the meal through a die. Hot air is used to dry the pellets, resulting in a relatively dense product that is the most efficient form of feed for laboratory rats. Pelleted diets are easy to store, handle, and feed. However, test articles or feed additives cannot be added after the pelleting process is complete.

Diets in meal form are frequently the most inefficient to use because large amounts of feed are generally wasted unless specially designed feeders are used. Diets in meal form will cake in less than ideal storage conditions. Meal may be required, however, if test articles are to be added to otherwise complete diets. Dust from meal diets may be hazardous if toxic compounds are involved. This problem may be overcome by adding water, agar, gelatin, or other gelling agents to the meal. A mixture of equal parts of a 3% agar solution and the complete meal diet will solidify into a gel mass that can be easily cut into blocks for weighing and feeding. Unconsumed agar diet can be readily collected and weighed to determine food consumption accurately. Agar contains minerals and gelatin contains amino acids that should be accounted for when dietary mineral or amino acid concentrations are critical to experiments. Gel diets are also susceptible to microbial growth, and they must be stored under refrigeration before use.

Liquid diets have been developed for laboratory rodents (Pleasants et al., 1970) to accommodate specific program requirements such as filter sterilization or studies requiring the administration of large amounts of alcohol. Diets manufactured by baking or extrusion have also been fed to rats, but because of the low density of these products, animals tend to waste large amounts of diets in these physical forms.

XI.   DIET STORAGE

Nutrient stability in complete diets generally increases as environmental temperature and humidity decrease. The shelf life of any particular lot of feed depends on the environmental conditions during diet manufacture and storage. The quality of feeds stored at high temperatures and humidity may deteriorate within weeks, whereas the same diet stored in a freezer may maintain its original quality for a year or more. In general, it is not economical to store large amounts of natural ingredient or purified diets under refrigeration. As a rule of thumb, natural ingredient diets stored in conventional areas should be used within 180 days of manufacture; purified diets, within 40 days. Diets formulated without antioxidants or with large quantities of perishable ingredients such as fat may require special storage procedures. For instance, purified diets that do not contain antioxidants should be stored under refrigeration. Because the most heat labile nutrients are thiamin and vitamin A, diets stored for long periods or under unusual environmental conditions should be assayed for at least these nutrients before use.

REFERENCES

Abdel-Mageed, A.B., and Oehme, F.W. (1991). The effect of various dietary zinc concentrations on the biological interactions of zinc, copper, and iron in rats. *Biol. Trace Elem. Res.* **29**, 239–256.

Abdelrahman, A.M., Harmsen, E., and Leeman, F.H. (1995). Dietary sodium and Na, K-ATPase activity in Dahl salt-sensitive versus salt-resistant rats. *J. Hypertens.* **13**, 517–522.

Abe, M., and Kishino, Y. (1982). Pathogenesis of fatty liver in rats fed a high protein diet without pyridoxine. *J. Nutr.* **112**, 205–210.

Abrams, G.M., and Larsen, P.R. (1973). Triiodothyronine and thyroxine in the serum and thyroid glands of iodine-deficient rats. *J. Clin. Invest.* **52**, 2522–2531.

Aburto, E.M., Cribb, A.E., Fuentealba, I.C., Ikede, B.O., Kibenge, F.S., and Markham, F. (2001). Morphological and biochemical assessment of the liver response to excess dietary copper in Fischer 344 rats. *Can. J. Vet. Res.* **65**, 97–103.

Achon, M., Reyes, L., Alonso-Aperte, E., Ubeda, N., and Varela-Moreiras, G. (1999). High dietary folate supplementation affects gestational development and dietary protein utilization in rats. *J. Nutr.* **129**, 1204–1208.

Ackrell, B.A., Maguire, J.J., Dallman, P.R., and Kearney, E.B. (1984). Effect of iron deficiency on succinate- and NADH-ubiquinone oxidoreductases in skeletal muscle mitochondria. *J. Biol. Chem.* **259**, 10053–10059.

Adelekan, D.A., and Thurnham, D.I. (1986). Effects of combined riboflavin and iron deficiency on the hematological status and tissue iron concentrations of the rat. *J. Nutr.* **116**, 1257–1265.

Adiga, P.R., and Muniyappa, K. (1978). Estrogen induction and functional importance of carrier proteins for riboflavin and thiamin in the rad during gestation. *J. Steriod Biochem.* **9**, 829.

Aguilar, T.S., Benevenga, N.J., and Harper, A.E. (1974). Effect of dietary methionine level on its metabolism in rats. *J. Nutr.* **104**, 761–771.

Akrabawi, S., and Salji, J.P. (1973). Influence of meal-feeding on some of the effects of dietary carbohydrate deficiency in rats. *Br. J. Nutr.* **30**, 37–43.

Alanko, J., Jolma, P., Koobi, P., Riutta, A., Kalliovalkama, J., Tolvanen, J.P., and Porsti, I. (2003). Prostacyclin and thromboxane A_2 production in nitric oxide-deficient hypertension *in vivo*. Effects of high calcium diet and angiotensin receptor blockade. *Prostaglandins Leukot. Essent. Fatty Acids* **69**, 345–350.

Albright, C.D., Friedrich, C.B., Brown, E.C., Mar, M.H., and Zeisel, S.H. (1999a). Maternal dietary choline availability alters mitosis, apoptosis and the localization of TOAD-64 protein in the developing fetal rat septum. *Brain Res. Dev. Brain Res.* **115**, 123–129.

Albright, C.D., Tsai, A.Y., Friedrich, C.B., Mar, M.H., and Zeisel, S.H. (1999b). Choline availability alters embryonic development of the hippocampus and septum in the rat. *Brain Res. Dev. Brain Res.* **113**, 13–20.

Alcock, N.W., and Shils, M.E. (1974). Serum immunoglobulin G in the magnesium-depleted rat. *Proc. Soc. Exp. Biol. Med.* **145**, 855–858.

Alcock, N.W., Shils, M.E., Lieberman, P.H., and Erlandson, R.A. (1973). Thymic changes in the magnesium-depleted rat. *Cancer Res.* **33**, 2196–2204.

Allen, C.B. (1994). Effects of dietary copper deficiency on relative food intake and growth efficiency in rats. *Physiol. Behav.* **59**, 247.

Allen, K.G.D., Lampi, K.J., et al. (1991). Increased thromboxane production in recalcified challenged whole blood from copper-deficient rats. *Nutr. Res.* **11**, 61.

al-Othman, A.A., Rosenstein, F., and Lei, K.Y. (1994). Pool size and concentration of plasma cholesterol are increased and tissue copper levels are reduced during early stages of copper deficiency in rats. *J. Nutr.* **124**, 628–635.

Alton-Mackey, M.G., and Walker, B.L. (1973). Graded levels of pyridoxine in the rat diet during gestation and the physical and neuromotor development of offspring. *Am. J. Clin. Nutr.* **26**, 420–428.

Altura, B.M., Gebrewold, A., Altura, B.T., and Brautbar, N. (1996). Magnesium depletion impairs myocardial carbohydrate and lipid metabolism and cardiac bioenergetics and raises myocardial calcium content in-vivo: relationship to etiology of cardiac diseases. *Biochem. Mol. Biol. Int.* **40**, 1183–1190.

Amanzadeh, J., Gitomer, W.L., Zerwekh, J.E., Preisig, P.A., Moe, O.W., Pak, C.Y., and Levi, M. (2003). Effect of high protein diet on stone-forming propensity and bone loss in rats. *Kidney Int.* **64**, 2142–2149.

American Association of Cereal Chemists [AACC]. (1995a). Total dietary fiber, Method 32-05. Approved Methods of the AACC., 9th Ed. St. Paul, MN.

American Association of Cereal Chemists [AACC]. (1995b). Determination of soluble, insoluble and total dietary fiber in foods and food products, Method 32-07. *In* "Approved Methods of the AACC," 9th ed. AACC, St. Paul, MN.

American Institute of Nutrition [AIN]. (1977). Ad Hoc Committee on standards for Nutritional Studies. Report of the Committee. *J. Nutr.* **107**, 1340–1348.

Ames, B.N., Profet, M., and Gold, L.S. (1990). Dietary pesticides (99.99% all natural). *Proc. Natl. Acad. Sci. USA* **87**, 7777–7781.

Ames, S.R. (1974). Age, parity, and vitamin A supplementation and the vitamin E requirement of female rats. *Am. J. Clin. Nutr.* **27**, 1017–1025.

Ammerman, C.B., Arrington, L.R., Warnick, A.C., Edwards, J.L., Shirley, R.L. and Davis, G.K. (1964). Reproduction and lactation in rats fed excessive iodine. *J. Nutr.* **84**, 107–112.

Anasuya, A. (1982). Role of fluoride in formation of urinary calculi: studies in rats. *J. Nutr.* **112**, 1787–1795.

Anderson, B.M., and Parker, H.E. (1955). The effects of dietary manganese and thiamine levels on growth rate and manganese concentration in tissues of rats. *J. Nutr.* **57**, 55–59.

Anderson, R.A. (1997). Nutritional factors influencing the glucose/insulin system: chromium. *J. Am. Coll. Nutr.* **16**, 404–410.

Anderson, R.A., Bryden, N.A., and Polansky, M.M. (1997). Lack of toxicity of chromium chloride and chromium picolinate in rats. *J. Am. Coll. Nutr.* **16**, 273–279.

Anderson, R.L., and Wolf, W.J (1995). Compositional changes in trypsin inhibitors, phytic acid, saponins and isoflavones related to soybean processing. *J. Nutr.* **125**, 581S—588S.

Angel, J.F. (1975). Lipogenesis by hepatic and adipose tissues from meal-fed pyridoxine deprived rats. *Nutr. Rep. Int.* **11**, 369.

Angel, J.F., and Mellor, R.M (1974). Glycogenesis and glucogenesis in meal-fed pyridoxine-deprived rats. *Nutr. Rep. Int.* **9**, 97.

Angel, J.F., and Song, G.W. (1973). Lipogenesis in pyroxidine deficient nibbling and meal-fed rats. *Nutr. Rep. Int.* **8**, 393.

Annison, G., and Topping, D.L. (1994). Nutritional role of resistant starch: chemical structure vs physiological function. *Annu. Rev. Nutr.* **14**, 297–320.

Aono, S. (1985). Studies on tryptophan metabolism of histidinemia. *J. Osaka City Med. Center* **34**, 155–168.

Aoyama, Y., Kato, C., and Sakakibara, S. (2001). Expression of metallothionein in the liver and kidney of rats is influenced by excess dietary histidine. *Comp. Biochem. Physiol. C Toxicol. Pharmacol.* **128**, 339–347.

Apgar, J. (1985). Zinc and reproduction. *Annu. Rev. Nutr.* **5**, 43–68.

Arnaud, C.D., and Sanchez, S.D. (1990). Calcium and phosphorus *In* "Present Knowledge in Nutrition." (M.L. Brown, ed), p. 212. International Life Sciences Institute Nutrition Foundation, Washington, DC.

Arthur, J.R., Nicol, F., Grant, E., and Beckett, G.J. (1991). The effects of selenium deficiency on hepatic type-I iodothyronine deiodinase and protein disulphide-isomerase assessed by activity measurements and affinity labelling. *Biochem. J.* **274**, 297–300.

Asenjo, C.F. (1948). Pteroylglutamic acid requirement of the rat and a characteristic lesion observed in the spleen of the deficient animal. *J. Nutr.* **36**, 601.

Askew, E.W. (1996). Water. *In* "Present Knowledge in Nutrition." (E.E. Ziegler and L.J. Filer, eds), pp. 98–108. ILSI Press, Washington, DC.

Aspnes, L.E., Lee, C.M., Weindruch, R., Chung, S.S., Roecker, E.B., and Aiken, J.M. (1997). Caloric restriction reduces fiber loss and mitochondrial abnormalities in aged rat muscle. *FASEB J.* **11**, 573–581.

Association of Official Analytical Chemists [AOAC]. (1995a). Total dietary fiber in foods - enzymatic-gravimetric method, Method 985.29. *In* "Official Methods of Analysis," 16th ed. Association of Official Analytical Chemists, Gaithersburg, MD.

Association of Official Analytical Chemists [AOAC]. (1995b). Total, insoluble and soluble dietary fiber in food: enzymatic-gravimetric method, MES-TRIS buffer, Method 991.43. *In* "Official Methods of Analysis," 16th ed. Association of Official Analytical Chemists, Gaithersburg, MD.

Atallah, M.T., and Melnik, T.A. (1982). Effect of pectin structure on protein utilization by growing rats. *J. Nutr.* **112**, 2027–2032.

Atkinson, J.B., Hill, K.E., and Burk, R.F. (2001). Centrilobular endothelial cell injury by diquat in the selenium-deficient rat liver. *Lab. Invest.* **81**, 193–200.

Audet, A., and Lupien, P.J. (1974). Triglyceride metabolism in pyridoxine-deficient rats. *J. Nutr.* **104**, 91–100.

Avery, R.A., and Bettger, W.J. (1992). Zinc deficiency alters the protein composition of the membrane skeleton but not the extractability or oligomeric form of spectrin in rat erythrocyte membranes. *J. Nutr.* **122**, 428–434.

Axelrod, A.E., and Trakatelles, A.C. (1964). Relationship of pyridoxine to immunological phenomena. *Vitamin Horm.* **22**, 591–607.

Aycock, J.E., and Kirksey, A. (1976). Influence of different levels of dietary pyridoxine on certain parameters of developing and mature brains in rats. *J. Nutr.* **106**, 680–688.

Babu, U., and Failla, M.L. (1990a). Respiratory burst and candidacidal activity of peritoneal macrophages are impaired in copper-deficient rats. *J. Nutr.* **120**, 1692–1699.

Babu, U., and Failla, M.L. (1990b). Copper status and function of neutrophils are reversibly depressed in marginally and severely copper-deficient rats. *J. Nutr.* **120**, 1700–1709.

Bai, D.H., Moon, T.W., Lopez-Casillas, F., Andrews, P.C., and Kim, K.H. (1989). Analysis of the biotin-binding site on acetyl-CoA carboxylase from rat. *Eur. J. Biochem.* **182**, 239–245.

Bajwa, G.S., Morrison, L.M., and Ershoff, B.H. (1971). Induction of aortic and coronary athero-arteriosclerosis in rats fed a hypervitaminosis D, cholesterol-containing diet. *Proc. Soc. Exp. Biol. Med.* **138**, 975–982.

Berdon, W.E., Baker, D.H., Becker, J., De Sanctis, P. (1967). Response of the weanling rat to α- or β-lactose with or without an excess of dietary phosphorus. *J. Dairy Sci.* **50**, 1314–1318.

Baker, H., Vanderhoof, J.A., Tuma, D.J., Frank, O., Baker, E.R., and Sorrell, M.F. (1992). A jejunoileal bypass rat model for rapid study of the effects of vitamin malabsorption. *Int. J. Vitam. Nutr. Res.* **62**, 43–46.

Bala, S., and Failla, M.L. (1993). Copper repletion restores the number and function of CD4 cells in copper-deficient rats. *J. Nutr.* **123**, 991–996.

Bala, S., Failla, M.L., and Lunney, J.K. (1991). Alterations in splenic lymphoid cell subsets and activation antigens in copper-deficient rats. *J. Nutr.* **121**, 745–753.

Baly, D.L., Curry, D.L., Keen, C.L., and Hurley, L.S. (1985). Dynamics of insulin and glucagon release in rats: influence of dietary manganese. *Endocrinology* **116**, 1734–1740.

Barboriak, J.J., Krehl, W.A., and Cowgill, G.R. (1957). Pantothenic acid requirement of the growing and adult rat. *J. Nutr.* **61**, 13–21.

Barone, A., Ebesh, O., Harper, R.G., and Wapnir, R.A. (1998). Placental copper transport in rats: effects of elevated dietary zinc on fetal copper, iron and metallothionein. *J. Nutr.* **128**, 1037–1041.

Bartholmey, S.J., and Sherman, A.R. (1986). Impaired ketogenesis in iron-deficient rat pups. *J. Nutr.* **116**, 2180–2189.

Bates, C.J., and Fuller, N.J. (1986). The effect of riboflavin deficiency on methylenetetrahydrofolate reductase (NADPH) (EC 1.5.1.20) and folate metabolism in the rat. *Br. J. Nutr.* **55**, 455–464.

Bavetta, L.A., and McClure, E.J. (1957). Protein factors and experimental rat caries. *J. Nutr.* **63**, 107–117.

Beard, J.L., Chen, Q., Connor, J., and Jones, B.C. (1994). Altered monamine metabolism in caudate-putamen of iron-deficient rats. *Pharmacol. Biochem. Behav.* **48**, 621–624.

Beck, W.S. (1983). The megaloblastic anemias. *In* "Hematology." (W.J. Williams, E. Beutler, A.J. Erslev, and M.A. Lichtman, ed), p. 434. McGraw Hill, New York.

Beck, W.S. (1990). Cobalamin as coenzyme: a twisting trail of research. *Am. J. Hematol.* **34**, 83–89.

Beck, W.S. (2001). Cobalamin (vitamin B12). *In* "Handbook of Vitamins." (R.B. Rucker, J.W. Suttie, D.B. McCormick, and L.J. Machlin, eds), p. 427. Marcel Dekker, New York.

Beckett, G.J., Nicol, F., Rae, P.W., Beech, S., Guo, Y., and Arthur, J.R. (1993). Effects of combined iodine and selenium deficiency on thyroid hormone metabolism in rats. *Am. J. Clin. Nutr.* **57**, 240S–243S.

Begley, J.A. (1983). The materials and process of plasma transport. *In* "The Cobalamins." (C.A. Hall, ed), p. 109. Churchill Livingstone, Edinburgh.

Bellamy, D. (1990). Dietary restriction–a procedure with an undefined goal: a critical review of two recent books. *Gerontology* **36**, 99–103.

Bender, D.A., Ghartey-Sam, K., and Singh, A. (1989). Effects of vitamin B6 deficiency and repletion on the uptake of steroid hormones into uterus slices and isolated liver cells of rats. *Br. J. Nutr.* **61**, 619–628.

Bendich, A., Gabriel, E, and Machlin, L.J. (1983a). Differences in vitamin E levels in tissues of the spontaneously hypertensive and Wistar-Kyoto rats. *Proc. Soc. Exp. Biol. Med.* **172**, 297–300.

Bendich, A., Gabriel, E, and Machlin, L.J. (1983b). Effect of dietary level of vitamin E on the immune system of the spontaneously hypertensive (SHR) and normotensive Wistar Kyoto (WKY) rat. *J. Nutr.* **113**, 1920–1926.

Bendich, A., Gabriel, E, and Machlin, L.J. (1986). Dietary vitamin E requirement for optimum immune responses in the rat. *J. Nutr.* **116**, 675–681.

Benevenga, N.J. (1974). Toxicities of methionine and other amino acids. *J. Agric. Food Chem.* **22**, 2–9.

Berant, M., Berkovitz, D., Mandel, H., Zinder, O., and Mordohovich, D. (1988). Thiamin status of the offspring of diabetic rats. *Pediatr. Res.* **23**, 574–575.

Berdanier, C.D. (1998). Trace minerals. *In* "Advanced Nutrition of Micronutrients." (I. Wolinsky and J.F. Hickson, eds), p. 194. CRC Press, Boca Raton.

Bert, B.N., and Simms, H.S. (1961). Nutrition and longevity in the rat, III: food restriction beyond 800 days. *J. Nutr.* **74**, 23–32.

Berthon, N., Laurant, P., Fellmann, D., and Berthelot, A. (2003). Effect of magnesium on mRNA expression and production of endothelin-1 in DOCA-salt hypertensive rats. *J. Cardiovasc. Pharmacol.* **42**, 24–31.

Bettger, W., and O'Dell, B. (1993). Physiological roles of zinc in the plasma membrane of mammalian cells. *J. Nutr. Biochem.* **4**, 194.

Bettger, W.J., Reeves, P.G., Moscatelli, E.A., Reynolds, G., and O'Dell, B.L. (1979). Interaction of zinc and essential fatty acids in the rat. *J. Nutr.* **109**, 480–488.

Bhaskram, C., and Sadasivan, G. (1975). Effects of feeding irradiated wheat to malnourished children. *Am. J. Clin. Nutr.* **28**, 130–135.

Bhat, K.S. (1982). Alterations in the lenticular proteins of rats on riboflavin deficient diet. *Curr. Eye Res.* **2**, 829–834.

Bhat, K.S. (1987). Changes in lens and erythrocyte glutathione reductase in response to endogenous flavin adenine dinucleotide and liver riboflavin content of rat on riboflavin deficient diet. *Nutr. Res.* **7**, 1203.

Binkley, N.C., and Suttie, J.W. (1995). Vitamin K nutrition and osteoporosis. *J. Nutr.* **125**, 1812–1821.

Black, A.L., Guirard, B.M., and Snell, E.E. (1977). Increased muscle phosphorylase in rats fed high levels of vitamin B6. *J. Nutr.* **107**, 1962–1968.

Black, A.L., Guirard, B.M., and Snell, E.E. (1978). The behavior of muscle phosphorylase as a reservoir for vitamin B6 in the rat. *J. Nutr.* **108**, 670–677.

Blair, P.V., Kobayashi, R., Edwards, H.M. 3rd, Shay, N.F., Baker, D.H., and Harris, R.A. (1999). Dietary thiamin level influences levels of its diphosphate form and thiamin-dependent enzymic activities of rat liver. *J. Nutr.* **129**, 641–648.

Blakely, S.R., Grundel, E., Jenkins, M.Y., and Mitchel, G.V. (1990). Alterations in beta-carotene and vitamin E stress in rats fed beta-carotene and excess vitamin A. *Nutr. Res.* **10**, 1035.

Blomhoff, R., Green, M.H., and Norum, K.R. (1992). Vitamin A: physiological and biochemical processing. *Annu. Rev. Nutr.* **12**, 37–57.

Blusztajn, J.K. (1998). Choline, a vital amine. *Science* **281**, 794–795.

Boelter, M.D.D., and Greenberg, D.M. (1941). Severe calcium deficiency in growing rats, I: symptoms and pathology. *J. Nutr.* **21**, 61.

Boelter, M.D.D., and Greenberg, D.M. (1943). Effect of severe calcium deficiency on pregnancy and lactation in the rat. *J. Nutr.* **26**, 105.

Boettger-Tong, H., Murthy, L., Chiappetta, C., Kirkland, J.L., Goodwin, B., Adlercreutz, H., Stancel, G.M., and Makela, S. (1998). A case of a laboratory animal feed with high estrogenic activity and its impact on *in vivo* responses to exogenously administered estrogens. *Environ. Health Perspect.* **106**, 369–373.

Bogden, J.D., Gertner, S.B., Christakos, S., Kemp, F.W., Yang, Z., Katz, S.R., and Chu, C. (1992). Dietary calcium modifies concentrations of lead and other metals and renal calbindin in rats. *J. Nutr.* **122**, 1351–1360.

Bonjour, J.P. (1984). Biotin. *In* "Handbook of Vitamins." (L.J. Machlin, ed), p. 403. Marcel Dekker, New York.

Booth, A., Wilson, R., and Deeds, F. (1953). Effects of prolonged ingestion of xylose on rats. *J. Nutr.* **49**, 347–355.

Bottomley, S.S. (1983). Iron and vitamin B6 metabolism in the sideroblastic anemias. *In* "Nutrition in Hematology." (J.L. Lindenbaum, ed), p. 203. Churchill Livingstone, New York.

Bougle, D., Isfaoun, A., Bureau, F., Neuville, D., Jauzac, P., and Arhan, P. (1999). Long-term effects of iron:zinc interactions on growth in rats. *Biol. Trace Elem. Res.* **67**, 37–48.

Bouillon, R., Van Assche, F.A., Van Baelen, H., Heyns, W., and De Moor, P. (1981). Influence of the vitamin D-binding protein on the serum concentration of 1,25-dihydroxyvitamin D3: significance of the free 1,25-dihydroxyvitamin D3 concentration. *J. Clin. Invest.* **67**, 589–596.

Bourre, J.M., Francois, M., Youyou, A., Dumont, O., Piciotti, M., Pascal, G., and Durand, G. (1989). The effects of dietary α-linolenic acid on the composition of nerve membranes, enzymatic activity, amplitude of electrophysiological parameters, resistance to poisons and performance of learning tasks in rats. *J. Nutr.* **119**, 1880–1892.

Bourre, J.M., Piciotti, M., Dumont, O., Pascal, G., Durand, G. (1990). Dietary linoleic acid and polyunsaturated fatty acids in rat brain and other organs: minimal requirements of linoleic acid. *Lipids* **25**, 465–472.

Bovee-Oudenhoven, I.M., Termont, D.S., Heidt, P.J., and Van der Meer, R. (1997a). Increasing the intestinal resistance of rats to the invasive pathogen *Salmonella enteritidis*: additive effects of dietary lactulose and calcium. *Gut* **40**, 497–504.

Bovee-Oudenhoven, I.M., Termont, D.S., Weerkamp, A.H., Faassen-Peters, M.A., and Van der Meer, R. (1997b). Dietary calcium inhibits the intestinal colonization and translocation of Salmonella in rats. *Gastroenterology* **113**, 550–557.

Bovee-Oudenhoven, I.M., Wissink, M.L., Wouters, J.T., and Van der Meer, R. (1999). Dietary calcium phosphate stimulates intestinal lactobacilli and decreases the severity of a salmonella infection in rats. *J. Nutr.* **129**, 607–612.

Bowman, B.B., and Rosenberg, I.H. (1987). Biotin absorption by distal rat intestine. *J. Nutr.* **117**, 2121–2126.

Boyde, C.D., and Cerklewski, F.L. (1987). Influence of type and level of dietary protein on fluoride bioavailability in the rat. *J. Nutr.* **117**, 2086–2090.

Boyonoski, A.C., Gallacher, L.M., ApSimon, M.M., Jacobs, R.M., Shah, G.M., Poirier, G.G., and Kirkland, J.B. (1999). Niacin deficiency increases the sensitivity of rats to the short and long term effects of ethylnitrosourea treatment. *Mol Cell Biochem* **193**, 83–87.

Boyonoski, A.C., Gallacher, L.M., ApSimon, M.M., Jacobs, R.M., Shah, G.M., Poirier, G.G., and Kirkland, J.B. (2000). Niacin deficiency in rats increases the severity of ethylnitrosourea-induced anemia and leukopenia. *J. Nutr.* **130**, 1102–1107.

Branca, D., Scutari, G., and Siliprandi, N. (1984). Pantethine and pantothenate effect on the CoA content of rat liver. *Int. J. Vitamin Nutr. Res.* **54**, 211–216.

Brannon, P.M., Collins, V.P., and Korc, M. (1987). Alterations of pancreatic digestive enzyme content in the manganese-deficient rat. *J. Nutr.* **117**, 305–311.

Breitman, P.L., Fonseca, D., Cheung, A.M., and Ward, W.E. (2003). Isoflavones with supplemental calcium provide greater protection against the loss of bone mass and strength after ovariectomy compared to isoflavones alone. *Bone* **33**, 597–605.

Brink, E.J., Haddeman, E., de Fouw, N.J., and Weststrate, J.A. (1995). Positional distribution of stearic acid and oleic acid in a triacylglycerol and dietary calcium concentration determines the apparent absorption of these fatty acids in rats. *J. Nutr.* **125**, 2379–2387.

Brinkley, N.C, and Suttie, J.W. (1995). Vitamin K nutrition and osteoporosis. J. Nutr. **125**, 1812-1821.

Britton, R.S., Ramm, G.A., Olynyk, J., Singh, R., O'Neill, R., and Bacon, B.R. (1994). Pathophysiology of iron toxicity. *Adv. Exp. Med. Biol.* **356**, 239–253.

Brock, A.A., Chapman, S.A., Ulman, E.A., and Wu, G. (1994). Dietary manganese deficiency decreases rat hepatic arginase activity. *J. Nutr.* **124**, 340–344.

Brody, T. (1999). Vitamins. *In* "Nutritional Biochemistry." (T. Brody, ed), p. 491, 638. Academic Press, San Diego.

Brody, T. (1999). Inorganic nutrients. *In* "Nutritional Biochemistry." (T. Brody, ed), p. 693. Academic Press, San Diego.

Brody, T., and Shane, B. (2001). Folic acid. *In* "Handbook of Vitamins." (R.B. Rucker, J.W. Suttie, D.B. McCormick, and L.J. Machlin, eds), p. 427. Marcel Dekker, New York.

Brody, T. and Stokstad E.L.R. (1991). Incorporation of the 2-ring carbon of histidine into folylpolyglutamate coenzymes. *J. Nutr. Biochem.* **2**, 492.

Brookes, I.M., Owens, F.N., and Garrigus, U.S. (1972). Influence of amino acid level in the diet upon amino acid oxidation by the rat. *J. Nutr.* **102**, 27–35.

Brown, A.J., Finch, J., and Slatopolsky, E. (2002). Differential effects of 19-nor-1,25-dihydroxyvitamin D_2 and 1,25-dihydroxyvitamin D_3 on intestinal calcium and phosphate transport. *J. Lab. Clin. Med.* **139**, 279–284.

Brown, A.J., Zhong, M., Finch, J., Ritter, C., and Slatopolsky, E. (1995). The roles of calcium and 1,25-dihydroxyvitamin D_3 in the regulation of vitamin D receptor expression by rat parathyroid glands. *Endocrinology* **136**, 1419–1425.

Brown, K.E., Dennery, P.A., Ridnour, L.A., Fimmel, C.J., Kladney, R.D., Brunt, E.M., and Spitz, D.R. (2003). Effect of iron overload and dietary fat on indices of oxidative stress and hepatic fibrogenesis in rats. *Liver Int.* **23**, 232–242.

Bruce, W.R., Furrer, R., Shangari, N., O'Brien, P.J., Medline, A., and Wang, Y. (2003). Marginal dietary thiamin deficiency induces the formation of colonic aberrant crypt foci (ACF) in rats. *Cancer Lett.* **202**, 125–129.

Brunet, S., Thibault, L., Delvin, E., Yotov, W., Bendayan, M., and Levy, E. (1999). Dietary iron overload and induced lipid peroxidation are associated with impaired plasma lipid transport and hepatic sterol metabolism in rats. *Hepatology* **29**, 1809–1817.

Buckingham, K.W. (1985). Effect of dietary polyunsaturated/saturated fatty acid ratio and dietary vitamin E on lipid peroxidation in the rat. *J. Nutr.* **115**, 1425–1435.

Bunce, G.E. (1989). Zinc in endocrine function. *In* "Human Biology." (C.F. Mills, ed), p. 249. Springer-Verlag, New York.

Bunce, G.E., and Hess, J.L. (1976). Lenticular opacities in young rats as a consequence of maternal diets low in tryptopham and/or vitamin E. *J. Nutr.* **106**, 222–229.

Bunce, G.E., and King, K.W. (1969a). Amino acid retention and balance in the young rat fed varying levels of lactalbumin. *J. Nutr.* **98**, 159–167.

Bunce, G.E., and King, K.W. (1969b). Amino acid retention and balance in the young rat fed varying levels of casein. *J. Nutr.* **98**, 168–176.

Bunce, G.E., and Vessal, M. (1987). Effect of zinc and/or pyridoxine deficiency upon oestrogen retention and oestrogen receptor distribution in the rat uterus. *J. Steriod Biochem.* **26**, 303–308.

Bureau, I., Lewis, C.G., and Fields, M. (1998). Effect of hepatic iron on hypercholesterolemia and hypertriacylglycerolemia in copper-deficient fructose-fed rats. *Nutrition* **14**, 366–371.

Burk, R., and Levander, O. (1999). Selenium. *In* "Modern Nutrition in Health and Disease." (M.E. Shils, J.A. Olson, M. Shike, and A.C. Ross, eds), p. 265. Williams and Wilkins, Baltimore.

Burk, R.F., and Hill, K.E. (1993). Regulation of selenoproteins. *Annu. Rev. Nutr.* **13**, 65–81.

Burk, R.F., Hill, K.E., Awad, J.A., Morrow, J.D., Kato, T., Cockell, K.A., and Lyons, P.R. (1995a). Pathogenesis of diquat-induced liver necrosis in selenium-deficient rats: assessment of the roles of lipid peroxidation and selenoprotein P. *Hepatology* **21**, 561–569.

Burk, R.F., Hill, K.E., Awad, J.A., Morrow, J.D., and Lyons, P.R. (1995b). Liver and kidney necrosis in selenium-deficient rats depleted of glutathione. *Lab. Invest.* **72**, 723–730.

Burk, R.F., Hill, K.E., and Motley, A.K. (2003). Selenoprotein metabolism and function: evidence for more than one function for selenoprotein P. *J. Nutr.* **133**, 1517S—1520S.

Busserolles, J., Gueux, E., Rock, E., Mazur, A., and Rayssiguier, Y. (2003). High fructose feeding of magnesium deficient rats is associated with increased plasma triglyceride concentration and increased oxidative stress. *Magnes. Res.* **16**, 7–12.

Bussiere, F.I., Gueux, E., Rock, E., Mazur, A., and Rayssiguier, Y. (2002). Protective effect of calcium deficiency on the inflammatory response in magnesium-deficient rats. *Eur. J. Nutr.* **41**, 197–202.

Butler, B.F., and Topham, R.W. (1993). Comparison of changes in the uptake and mucosal processing of iron in riboflavin-deficient rats. *Biochem. Mol. Biol. Int.* **30**, 53–61.

Butterworth, R.F. (1982). Neurotransmitter function in thiamin deficiency encephalopathy. *Neurochem. Int.* **4**, 449.

Butterworth, R.F., Merkel, A.D., and Landreville, F. (1982). Regional amino acid distribution in relation to function in insulin hypoglycaemia. *J. Neurochem.* **38**, 1483–1489.

Cade, C., and Norman, A.W. (1986). Vitamin D3 improves impaired glucose tolerance and insulin secretion in the vitamin D-deficient rat *in vivo*. *Endocrinology* **119**, 84–90.

Cade, C., and Norman, A.W. (1987). Rapid normalization/stimulation by 1,25-dihydroxyvitamin D3 of insulin secretion and glucose tolerance in the vitamin D-deficient rat. *Endocrinology* **120**, 1490–1497.

Caldwell, E.F., and Nelson, T.C. (1999). Development of an analytical reference standard for total, insoluble, and soluble dietary fiber. *Cereal Foods World* **44**, 360–382.

Calvert, R.J., Otsuka, M., and Satchithanandam, S. (1989). Consumption of raw potato starch alters intestinal function and colonic cell proliferation in the rat. *J. Nutr.* **119**, 1610–1616.

Calvo, M.S. (1993). Dietary phosphorus, calcium metabolism and bone. *J. Nutr.* **123**, 1627–1633.

Campbell, T.C., and Hayes, J.R. (1975). Role of nutrition in the drug-metabolizing enzyme system. *Pharmacol. Rev.* **26**, 171–197.

Campbell, W.W., Polansky, M.M., Bryden, N.A., Soares, J.H. Jr, and Anderson, R.A. (1989). Exercise training and dietary chromium effects on glycogen, glycogen synthase, phosphorylase and total protein in rats. *J. Nutr.* **119**, 653–660.

Cannon, M., Flenniken, A., and Track, N.S. (1980). Demonstration of acute and chronic effects of dietary fibre upon carbohydrate metabolism. *Life Sci.* **27**, 1397–1401.

Cannon, P. R. (1948). "Some Pathologic Consequences of Protein and Amino Acid Deficiencies." Charles C. Thomas, Springfield, IL.

Cantorna, M.T., Nashold, F.E., and Hayes, C.E. (1994). In vitamin A deficiency multiple mechanisms establish a regulatory T helper cell imbalance with excess Th1 and insufficient Th2 function. *J. Immunol.* **152**, 1515–1522.

Cao, Y.Z., Weaver, J.A., Reddy, C.C., and Sordillo, L.M. (2002). Selenium deficiency alters the formation of eicosanoids and signal transduction in rat lymphocytes. *Prostaglandins Other Lipid Mediat.* **70**, 131–143.

Cardot, P., Chambaz, J., Thomas, G., Rayssiguier, Y., and Bereziat, G. (1987). Essential fatty acid deficiency during pregnancy in the rat: influence of dietary carbohydrates. *J. Nutr.* **117**, 1504–1513.

Carlisle, E.M. (1984). Silicon. *In* "Biochemistry of the Essential Ultratrace Elements." (E. Frieden, ed), p. 257. Plenum, New York.

Carlisle, E.M. (1988). Silicon as a trace nutrient. *Sci. Total Environ.* **73**, 95–106.

Carlisle, E.M., and Curran, M.J. (1987). Effect of dietary silicon and aluminum on silicon and aluminum levels in rat brain. *Alzheimer Dis. Assoc. Disord.* **1**, 83–89.

Carlson, S.H., Shelton, J., White, C.R., and Wyss, J.M. (2000). Elevated sympathetic activity contributes to hypertension and salt sensitivity in diabetic obese Zucker rats. *Hypertension* **35**, 403–408.

Carlton, W.W., and Kelly, W.A. (1969). Neural lesions in the offspring of female rats fed a copper-deficient diet. *J. Nutr.* **97**, 42–52.

Carman, J.A., Pond, L., Nashold, F., Wassom, D.L., and Hayes, C.E. (1992). Immunity to *Trichinella spiralis* infection in vitamin A-deficient mice. *J. Exp. Med.* **175**, 111–120.

Carmel, N., Konijn, A.M., Kaufman, N.A., and Guggenheim, K. (1975). Effects of carbohydrate-free diets on the insulin-carbohydrate relationships in rats. *J. Nutr.* **105**, 1141–1149.

Carpenter, T.O., Carnes, D.L. Jr, and Anast, C.S. (1987). Effect of magnesium depletion on metabolism of 25-hydroxyvitamin D in rats. *Am. J. Physiol.* **253**, E106–E113.

Carpenter, T.O., Mackowiak, S.J., Troiano, N., and Gundberg, C.M. (1992). Osteocalcin and its message: relationship to bone histology in magnesium-deprived rats. *Am. J. Physiol.* **263**, E107—E114.

Carrier, J., Aghdassi, E., Cullen, J., and Allard, J.P. (2002). Iron supplementation increases disease activity and vitamin E ameliorates the effect in rats with dextran sulfate sodium-induced colitis. *J. Nutr.* **132**, 3146–3150.

Carroll, C., and Williams, L. (1982). Choline deficiency in rats as influenced by dietary energy. *Nutr. Rep. Int.* **25**, 773.

Cerklewski, F. L. (1987). Influence of dietary magnesium on fluoride bioavailability in the rat. *J. Nutr.* **117**, 496–500.

Cerklewski, F.L., and Ridlington, J.W. (1985). Influence of zinc and iron on dietary fluoride utilization in the rat. *J. Nutr.* **115**, 1162–1167.

Cerklewski, F.L., Ridlington, J.W., and Bills, N.D. (1986). Influence of dietary chloride on fluoride bioavailability in the rat. *J. Nutr.* **116**, 618–624.

Cermak, J.M., Blusztajn, J.K., Meck, W.H., Williams, C.L., Fitzgerald, C.M., Rosene, D.L., and Loy, R. (1999). Prenatal availability of choline alters the development of acetylcholinesterase in the rat hippocampus. *Dev. Neurosci.* **21**, 94–104.

Cermak, J.M., Holler, T., Jackson, D.A., and Blusztajn, J.K. (1998). Prenatal availability of choline modifies development of the hippocampal cholinergic system. *FASEB J.* **12**, 349–357.

Cervantes-Laurean, D., McElvaney, G., et al. (1999). Niacin. *In* "Modern Nutrition in Health and Disease." (M.E. Shils, J.A. Olson, M. Shike, and A.C. Ross, eds), p. 401. Williams and Wilkins, Baltimore.

Cha, C.J., and Randall, H.T. (1982). Effects os substitution of glucose-oligosaccharides by sucrose in a defined formula diet on intestinal disaccharidases, hepatic lipogenic enzymes and carbohydrate metabolism in young rats. *Metabolism* **31**, 57–66.

Chan, M.M. (1984). Choline and carnitine. *In* "Handbook of Vitamins." (R.B. Rucker, J.W. Suttie, D.B. McCormick, and L.J. Machlin, eds), p. 549. Marcel Dekker, New York.

Chandar, N., and Lombardi, B. (1988). Liver cell proliferation and incidence of hepatocellular carcinomas in rats fed consecutively a choline-devoid and a choline-supplemented diet. *Carcinogenesis* **9**, 259–263.

Chandra, R.K., and Puri, S. (1985). Vitamin B6 modulation of immune responses and infection. *In* "Vitamin B6: Its Role in Health and Disease." (R.D. Reynolds and J.E. Leklem, eds), p. 163. Alan R. Liss, New York.

Chang, S.J., Kirksey, A., and Morre, D.M. (1981). Effects of vitamin B-6 deficiency on morphological changes in dendritic trees of Purkinje cells in developing cerebellum of rats. *J. Nutr.* **111**, 848–857.

Chapin, R.E., Ku, W.W., Kenney, M.A., McCoy, H., Gladen, B., Wine, R.N., Wilson, R., and Elwell, M.R. (1997). The effects of dietary boron on bone strength in rats. *Fundam. Appl. Toxicol.* **35**, 205–215.

Chen, L.H., and Marlatt, A.L. (1975). Effects of dietary vitamin B-6 levels and exercise on glutamic-pyruvic transaminase activity in rat tissues. *J. Nutr.* **105**, 401–407.

Chen, L.T., Bowen, C.H., et al. (1984). Vitamin B1 and B12 blood levels in rats during pregnancy. *Nutr. Rep. Int.* **30**, 433.

Chen, O.S., Blemings, K.P., Schalinske, K.L., and Eisenstein, R.S. (1998). Dietary iron intake rapidly influences iron regulatory proteins, ferritin subunits and mitochondrial aconitase in rat liver. *J. Nutr.* **128**, 525–535.

Chen, Q., Beard, J.L., et al. (1995a). Abnormal brain monoamine metabolism in iron deficiency anemia. *J. Nutr. Biochem.* **6**, 486.

Chen, Q., Connor, J.R., and Beard, J.L. (1995b). Brain iron, transferrin and ferritin concentrations are altered in developing iron-deficient rats. *J. Nutr.* **125**, 1529–1535.

Chesters, J.K. (1975). Food intake control in zinc-deficient rats of the Zucker-Zucker strain. *Proc. Nutr. Soc.* **34**, 103A–104A.

Chesters, J.K., and Quarterman, J. (1970). Effects of zinc deficiency on food intake and feeding patterns of rats. *Br. J. Nutr.* **24**, 1061–1069.

Chesters, J.K., and Will, M. (1973). Some factors controlling food intake by zinc-deficient rats. *Br. J. Nutr.* **30**, 555–566.

Chevalier, S., Blaner, W.S., Azais-Braesco, V., and Tuchweber, B. (1999). Dietary restriction alters retinol and retinol-binding protein metabolism in aging rats. *J. Gerontol. A Biol. Sci. Med. Sci.* **54**, B384–B392.

Chevalier, S., Ferland, G., and Tuchweber, B. (1996). Lymphatic absorption of retinol in young, mature, and old rats: influence of dietary restriction. *FASEB J.* **10**, 1085–1090.

Chonan, O., Takahashi, R., Kado, S., Nagata, Y., Kimura, H., and Uchida, K, Watanuki, M. (1996). Effects of calcium gluconate on the utilization of magnesium and the nephrocalcinosis in rats fed excess dietary phosphorus and calcium. *J. Nutr. Sci. Vitaminol. (Tokyo)* **42**, 313–323.

Chonan, O., Takahashi, R., Yasui, H., and Watanuki, M. (1997). The effect of calcium gluconate and other calcium supplements as a dietary calcium source on magnesium absorption in rats. *Int. J. Vitam. Nutr. Res.* **67**, 201–206.

Chow, C. (2001). Vitamin E. *In* "Handbook of Vitamins." (R.B. Rucker, J.W. Suttie, D.B. McCormick, and L.J. Machlin, eds), p. 165. Marcel Dekker, New York.

Chow, C.K. (1991). Vitamin E and oxidative stress. *Free Radic. Biol. Med.* **11**, 215–232.

Christensen, M.J., Cammack, P.M., and Wray, C.D. (1995). Tissue specificity of selenoprotein gene expression in rats. *J. Nutr. Biochem.* **6**, 367–372.

Christensen, M.J., Olsen, C.A., Hansen, D.V., and Ballif, B.C. (2000). Selenium regulates expression in rat liver of genes for proteins involved in iron metabolism. *Biol. Trace Elem. Res.* **74**, 55–70.

Christian, M.S., Hoberman, A.M., Johnson, M.D., Brown, W.R., and Bucci, T.J. (1998). Effect of dietary optimization on growth, survival, tumor incidences and clinical pathology parameters in CD Sprague-Dawley and Fischer-344 rats: a 104-week study. *Drug Chem. Toxicol.* **21**, 97–117.

Christian, P., and West, K.P., Jr. (1998). Interactions between zinc and vitamin A: an update. *Am. J. Clin. Nutr.* **68**, 435S–441S.

Chu, Y., Mouat, M.F., Harris, R.B., Coffield, J.A., and Grider, A. (2003). Water maze performance and changes in serum corticosterone levels in zinc-deprived and pair-fed rats. *Physiol. Behav.* **78**, 569.

Chua, A.C., and Morgan, E.H. (1996). Effects of iron deficiency and iron overload on manganese uptake and deposition in the brain and other organs of the rat. *Biol. Trace Elem. Res.* **55**, 39–54.

Cidlowski, J.A., and Thanassi, J.W. (1981). Pyridoxal phosphate: a possible cofactor in steroid hormone action. *J. Steriod Biochem.* **15**, 11–16.

Clark, S.A., Boass, A., and Toverud, S.U. (1987). Effects of high dietary contents of calcium and phosphorus on mineral metabolism and growth of vitamin D—deficient suckling and weaned rats. *Bone Miner.* **2**, 257–270.

Clarke, H.E., Coates, M.E., Eva, J.K., Ford, D.J., Milner, C.K., O'Donoghue, P.N., Scott, P.P., and Ward, R.J. (1977). Dietary standards for laboratory animals: report of the Laboratory Animals Center Diets Advisory Committee. *Lab. Anim.* **11**, 1–38.

Clegg, M.S., Donovan, S.M., Monaco, M.H., Baly, D.L., Ensunsa, J.L., and Keen, C.L. (1998). The influence of manganese deficiency on serum IGF-I and IGF binding proteins in the male rat. *Proc. Soc. Exp. Biol. Med.* **219**, 41–47.

Clifford, A.J., Heid, M.K., Muller, H.G., and Bills, N.D. (1990). Tissue distribution and prediction of total body folate of rats. *J. Nutr.* **120**, 1633–1639.

Clifford, A.J., Wilson, D.S., and Bills, N.D. (1989). Repletion of folate-depleted rats with an amino acid-based diet supplemented with folic acid. *J. Nutr.* **119**, 1956–1961.

Clugston, G., and Hetzel, B. (1994). Iodine. *In* "Modern Nutrition in Health and Disease." (M.E. Shils, J.A. Olson, and M. Shike, eds), p. 252. Lea and Febiger, Baltimore.

Coburn, S.P., Mahuren, J.D., Wostmann, B.S., Snyder, D.L., and Townsend, D.W. (1989). Role of intestinal microflora in the metabolism of vitamin B-6 and 4'-deoxypyridoxine examined using germfree guinea pigs and rats. *J. Nutr.* **119**, 181–188.

Cockell, K.A., and Belonje, B. (2004). Nephrocalcinosis caused by dietary calcium:phosphorus imbalance in female rats develops rapidly and is irreversible. *J. Nutr.* **134**, 637–640.

Cockell, K.A., L'Abbe, M.R., and Belonje, B. (2002). The concentrations and ratio of dietary calcium and phosphorus influence development of nephrocalcinosis in female rats. *J. Nutr.* **132**, 252–256.

Cohen, N.L., Keen, C.L., Hurley, L.S., and Lonnerdal, B. (1985). Determinants of copper-deficiency anemia in rats. *J. Nutr.* **115**, 710–725.

Cohen, P.A., Schneidman, K., Ginsberg-Fellner, F., Sturman, J.A., Knittle, J., and Gaull, G.E. (1973). High pyridoxine diet in the rat: possible implications for megavitamin therapy. *J. Nutr.* **103**, 143–151.

Cohen, S.B., and Weetman, A.P. (1988). The effect of iodide depletion and supplementation in the Buffalo strain rat. *J. Endocrinol. Invest.* **11**, 625–627.

Coldham, N.G., and Sauer, M.J. (2000). Pharmacokinetics of [(^{14})C]Genistein in the rat: gender-related differences, potential mechanisms of biological action, and implications for human health. *Toxicol. Appl. Pharmacol.* **164**, 206–215.

Collins, E.D., and Norman, A.W. (2001). Vitamin D. *In* "Handbook of Vitamins." (R.B. Rucker, J.W. Suttie, D.B. McCormick, and L.J. Machlin, eds), p. 51. Marcel Dekker, New York.

Contempre, B., Denef, J.F., Dumont, J.E., and Many, M.C. (1993). Selenium deficiency aggravates the necrotizing effects of a high iodide dose in iodine deficient rats. *Endocrinology* **132**, 1866–1868.

Cook, J.D., Dassenko, S.A., and Whittaker, P. (1991). Calcium supplementation: effect on iron absorption. *Am. J. Clin. Nutr.* **53**, 106–111.

Cooke, N.E., and Haddad, J.G. (1989). Vitamin D binding protein (Gc-globulin). *Endocr. Rev.* **10**, 294–307.

Cooperman, J.M., and Lopez, R. (1984). Riboflavin. *In* "Handbook of Vitamins." (R.B. Rucker, J.W. Suttie, D.B. McCormick, and L.J. Machlin, eds), p. 299. Marcel Dekker, New York.

Corey, J.E., and Hayes, K.C. (1972). Cerebrospinal fluid pressure, growth, and hematology in relation to retinal status of the rat in acute vitamin A deficiency. *J. Nutr.* **102**, 1585–1593.

Cornatzer, W., Uthus, E., et al. (1983). Effect of arsenic deprivation on phosphatidylcholine biosynthesis on liver microsomes in the rat. *Nutr. Rep. Int.* **27**, 821.

Coudray, C., Feillet-Coudray, C., Grizard, D., Tressol, J.C., Gueux, E., and Rayssiguier, Y. (2002). Fractional intestinal absorption of magnesium is directly proportional to dietary magnesium intake in rats. *J. Nutr.* **132**, 2043–2047.

Council for Agricultural Science and Technology [CAST]. (1986). Ionizing Energy in Food Processing and Pest Control, I: Wholesomeness of Food Treated with Ionizing Energy, Report No. 109. Council for Agricultural Science and Technology, Washington, DC.

Cousins, R.J., Blanchard, R.K., Moore, J.B., Cui, L., Green, C.L., Liuzzi, J.P., Cao, J., and Bobo, J.A. (2003). Regulation of zinc metabolism and genomic outcomes. *J. Nutr.* **133**, 1521S–1526S.

Creedon, A., and Cashman, K.D. (2001). The effect of calcium intake on bone composition and bone resorption in the young growing rat. *Br. J. Nutr.* **86**, 453–459.

Crowe, A., and Morgan, E.H. (1996). Iron and copper interact during their uptake and deposition in the brain and other organs of developing rats exposed to dietary excess of the two metals. *J. Nutr.* **126**, 183–194.

Cullen, R.W., and Oace, S.M. (1989a). Neomycin has no persistent sparing effect on vitamin B-12 status in pectin-fed rats. *J. Nutr.* **119**, 1399–1403.

Cullen, R.W., and Oace, S.M. (1989b). Dietary pectin shortens the biologic half-life of vitamin B-12 in rats by increasing fecal and urinary losses. *J. Nutr.* **119**, 1121–1127.

Cunnane, S.C. (1988). Evidence that adverse effects of zinc deficiency on essential fatty acid composition in rats are independent of food intake. *Br. J. Nutr.* **59**, 273–278.

Cunnane, S.C., Manku, M.S., and Horrobin, D.F. (1984). Accumulation of linoleic and gamma-linolenic acids in tissue lipids of pyridoxine-deficient rats. *J. Nutr.* **114**, 1754–1761.

Cunnane, S.C., Soma, M., McAdoo, K.R., and Horrobin, D.F. (1985). Magnesium deficiency in the rat increases tissue levels of docosahexaenoic acid. *J. Nutr.* **115**, 1498–1503.

Curtis, S.E. (1983). Animal energetics and thermal environment. *In* "Environmental Management in Animal Agriculture." p. 79–96. Iowa State University Press, Ames.

da Costa, K.A., Cochary, E.F., Blusztajn, J.K., Garner, S.C., and Zeisel, S.H. (1993). Accumulation of 1,2-sn-diradylglycerol with increased membrane-associated protein kinase C may be the mechanism for spontaneous hepatocarcinogenesis in choline-deficient rats. *J. Biol. Chem.* **268**, 2100–2105.

Dabbagh, A.J., Mannion, T., Lynch, S.M., and Frei, B. (1994). The effect of iron overload on rat plasma and liver oxidant status *in vivo*. *Biochem. J.* **300**, 799–803.

Dakshinamurti, K. (1982). Neurobiology of pyridoxine. *In* "Advances in Nutritional Research." (H.H. Draper, ed), p. 143. Plenum Press, New York.

Das, K.K., and Dasgupta, S. (2000). Effect of nickel on testicular nucleic acid concentrations of rats on protein restriction. *Biol. Trace Elem. Res.* **73**, 175–180.

Davies, I.R., Brown, J.C., and Livesey, G. (1991). Energy values and energy balance in rats fed on supplements of guar gum or cellulose. *Br. J. Nutr.* **65**, 415–433.

Davies, N.T., and Lawrence, C.B. (1986). Studies on the effect of copper deficiency on rat liver mitochondria, III: effects on adenine nucleotide translocase. *Biochim. Biophys. Acta* **848**, 294–304.

Davis, C.D., and Feng, Y. (1999). Dietary copper, manganese and iron affect the formation of aberrant crypts in colon of rats administered 3,2'-dimethyl-4-aminobiphenyl. *J. Nutr.* **129**, 1060–1067.

Davis, C.D., Ney, D.M., and Greger, J.L. (1990). Manganese, iron and lipid interactions in rats. *J. Nutr.* **120**, 507–513.

Davis, C.D., Wolf, T.L., and Greger, J.L. (1992). Varying levels of manganese and iron affect absorption and gut endogenous losses of manganese by rats. *J. Nutr.* **122**, 1300–1308.

Davis, G.K., and Mertz, W. (1987). Copper. *In* "Trace Elements in Human and Animal Nutrition." (W. Mertz, ed), p. 301. Academic Press, Orlando, FL.

Dawson, H.D., Li, N.Q., DeCicco, K.L., Nibert, J.A., and Ross, A.C. (1999). Chronic marginal vitamin A status reduces natural killer cell number and function in aging Lewis rats. *J. Nutr.* **129**, 1510–1517.

Dawson, H.D., and Ross, A.C. (1999). Chronic marginal vitamin A status affects the distribution and function of T cells and natural T cells in aging Lewis rats. *J. Nutr.* **129**, 1782–1790.

Day, H.G., and Pigman, W. (1957). Carbohydrates in nutrition. *In* "The Carbohydrates." (W. Pigman, ed), pp. 779–806. Academic Press, New York.

Delcour, J.A., and Erlingen, R.C. (1996). Analytical implications of the classification of resistant starch as dietary fiber. *Cereal Foods World* **41**, 85–86.

Delorme, C.B., and Gordon, C.I. (1983). The effect of pectin on the utilization of marginal levels of dietary protein by weanling rats. *J. Nutr.* **113**, 2432–2441.

Delorme, C.B., and Lupien, P.J. (1976). The effect of a long-term excess of pyridoxine on the fatty acid composition of the major phospholipids in the rat. *J. Nutr.* **106**, 976–984.

Demetriou, A.A., Franco, I., Bark, S., Rettura, G., Seifter, E., and Levenson, S.M. (1984). Effects of vitamin A and beta carotene on intra-abdominal sepsis. *Arch. Surg.* **119**, 161–165.

Desikachar, H.S., and McHenry, E.W. (1954). Some effects of vitamin B6 deficiency on fat metabolism in rats. *Biochem. J.* **56**, 544–547.

Desrosiers, R., and Tanguay, R.M. (1986). Further characterization of the posttranslational modifications of core histones in response to heat and arsenite stress in *Drosophila. Biochem. Cell Biol.* **64**, 750–757.

DeVries, J.W., Prosky, L., Li, B., and Cho, S. (1999). A historical perspective on defining dietary fiber. *Cereal Foods World* **44**, 367–369.

DiBona, G.F., and Jones, S.Y. (1992). Effect of dietary potassium chloride in borderline hypertensive rats. *J. Am. Soc. Nephrol.* **3**, 188–195.

DiPaolo, R.V., Caviness, V.S. Jr, and Kanfer, J.N. (1974). Delayed maturation of the renal cortex in the vitamin B_6-deficient newborn rat. *Pediatr. Res.* **8**, 546–552.

Dizik, M., Christman, J.K., and Wainfan, E. (1991). Alterations in expression and methylation of specific genes in livers of rats fed a cancer promoting methyl-deficient diet. *Carcinogenesis* **12**, 1307–1312.

Doi, T., Kawata, T., Tadano, N., Iijima, T., and Maekawa, A. (1989). Effect of vitamin B_{12}-deficiency on the activity of hepatic cystathionine beta-synthase in rats. *J. Nutr. Sci. Vitaminol. (Tokyo)* **35**, 101–110.

Donaldson, J. (1973). Determination of Na^+, K^+, Mg^{2+}, Cu^{2+}, Zn^{2+} and Mn^{2+} in rat brain regions. *Can. J. Biochem.* **51**, 85.

Douglas, C.E., Chan, A.C., and Choy, P.C. (1986). Vitamin E inhibits platelet phospholipase A2. *Biochim. Biophys. Acta* **876**, 639–645.

Du, C., Fujii, Y., Ito, M., Harada, M., Moriyama, E., Shimada, R., Ikemoto, A., and Okuyama, H. (2004). Dietary polyunsaturated fatty acids suppress acute hepatitis, alter gene expression and prolong survival of female Long-Evans Cinnamon rats, a model of Wilson disease. *J. Nutr. Biochem.* **15**, 273–280.

Du, Z., Hemken, R.W., Jackson, J.A., and Trammell, D.S. (1996). Utilization of copper in copper proteinate, copper lysine, and cupric sulfate using the rat as an experimental model. *J. Anim. Sci.* **74**, 1657–1663.

Dubick, M.A., Yu, G.S., and Majumdar, A.P. (1989). Morphological and biochemical changes in the pancreas of the copper-deficient female rat. *J. Nutr.* **119**, 1165–1172.

Duerden, J.M., and Bates, C.J. (1985). Effect of riboflavin deficiency on reproductive performance and on biochemical indices of riboflavin status in the rat. *Br. J. Nutr.* **53**, 97–105.

Duffy, P.H., Feuers, R.J., Leakey, J.A., Nakamura, K., Turturro, A., Hart, R.W. (1989). Effect of chronic caloric restriction on physiological variables related to energy metabolism in the male Fischer 344 rat. *Mech. Ageing Dev.* **48**, 117–133.

Duffy, P.H., Feuers, R., Nakamura, K.D., Leakey, J., and Hart, R.W. (1990). Effect of chronic caloric restriction on the synchronization of various physiological measures in old female Fischer 344 rats. *Chronobiol. Int.* **7**, 113–124.

Duffy, P.H., Feuers, R.J., et al. (1995). The effect of dietary restriction and aging on the physiological response of rodents to drugs. *In* "Dietary Restriction: Implications for the Design and Interpretation of Toxicity and Carcinogenicity Studies." (R.W. Hart, D.A. Neuman, and R.T. Robertson, eds), pp. 125–140. ILSI Press, Washington, DC.

Duffy, P.H., Seng, J.E., Lewis, S.M., Mayhugh, M.A., Aidoo, A., Hattan, D.G., Casciano, D.A., and Feuers, R.J. (2001). The effects of

different levels of dietary restriction on aging and survival in the Sprague-Dawley rat: implications for chronic studies. *Aging (Milano)* **13**, 263–272.

Duffy, P.H., Lewis, S.M., Mayhugh, M.A., McCracken, A., Thorn, B.T., Reeves, P.G., Blakely, S.A., Casciano, D.A., and Feuers, R.J. (2002). Effect of the AIN-93M purified diet and dietary restriction on survival in Sprague-Dawley rats: implications for chronic studies. *J. Nutr.* **132**, 101–107.

Duflos, C., Bellaton, C., Pansu, D., and Bronner, F. (1995). Calcium solubility, intestinal sojourn time and paracellular permeability codetermine passive calcium absorption in rats. *J. Nutr.* **125**, 2348–2355.

Dulloo, A.G., and Girardier, L. (1993). Twenty-four hour energy expenditure several months after weight loss in the underfed rat: evidence for a chronic increase in whole-body metabolic efficiency. *Int. J. Obes. Relat. Metab. Disord.* **17**, 115–123.

Durand, P., Prost, M., and Blache, D. (1996). Pro-thrombotic effects of a folic acid deficient diet in rat platelets and macrophages related to elevated homocysteine and decreased n-3 polyunsaturated fatty acids. *Atherosclerosis* **121**, 231–243.

During, A., Fields, M., Lewis, C.G., and Smith, J.C. (1999). Beta-carotene 15,15′-dioxygenase activity is responsive to copper and iron concentrations in rat small intestine. *J. Am. Coll. Nutr.* **18**, 309–315.

Dursen, N., and Aydogan, S. (1995). The influence of dietary iron on zinc in rat. *Biol. Trace Element Res.* **48**, 161.

Dutta, P., Rivlin, R.S., and Pinto, J. (1990). Enhanced depletion of lens reduced glutathione adriamycin in riboflavin-deficient rats. *Biochem. Pharmacol.* **40**, 1111–1115.

Dutta-Roy, A.K., Gordon, M.J., Leishman, D.J., Paterson, B.J., Duthie, G.G., and James, W.P. (1993a). Purification and partial characterisation of an α-tocopherol-binding protein from rabbit heart cytosol. *Mol. Cell Biochem.* **123**, 139–144.

Dutta-Roy, A.K., Leishman, D.J., Gordon, M.J., Campbell, F.M., and Duthie, G.G. (1993b). Identification of a low molecular mass (14.2 kDa) α-tocopherol-binding protein in the cytosol of rat liver and heart. *Biochem. Biophys. Res. Commun.* **196**, 1108–1112.

Ebara, S., Toyoshima, S., Matsumura, T., Adachi, S., Takenaka, S., Yamaji, R., Watanabe, F., Miyatake, K., Inui, H., and Nakano, Y. (2001). Cobalamin deficiency results in severe metabolic disorder of serine and threonine in rats. *Biochim. Biophys. Acta* **1568**, 111–117.

Eckhert, C.D., Lockwood, M.K., and Shen, B. (1993). Influence of selenium on the microvasculature of the retina. *Microvasc. Res.* **45**, 74–82.

Eder, K., and Kirchgessner, M. (1994a). Levels of polyunsaturated fatty acids in tissues from zinc-deficient rats fed a linseed oil diet. *Lipids* **29**, 839–844.

Eder, K., and Kirchgessner, M. (1994b). Dietary fat influences the effect of zinc deficiency on liver lipids and fatty acids in rats force-fed equal quantities of diet. *J. Nutr.* **124**, 1917–1926.

Eder, K., Kralik, A., and Kirchgessner, M. (1996). The effect of manganese supply on thyroid hormone metabolism in the offspring of manganese-depleted dams. *Biol. Trace Elem. Res.* **55**, 137–145.

Eiam-Ong, S., and Sabatini, S. (1999). Age-related changes in renal function, membrane protein metabolism, and Na, K-ATPase activity and abundance in hypokalemic F344 x BNF(1) rats. *Gerontology* **45**, 254–264.

El Hendy, H.A., Yousef, M.I., and Abo El-Naga, N.I. (2001). Effect of dietary zinc deficiency on hematological and biochemical parameters and concentrations of zinc, copper, and iron in growing rats. *Toxicology* **167**, 163–170.

Ellenbogen, L. (1984). Vitamin B_{12}. *In* "Handbook of Vitamins." (R.B. Rucker, J.W. Suttie, D.B. McCormick, and L.J. Machlin, eds), pp. 497–546. Marcel Dekker, New York.

Emerick, R.J., and Kayongo-Male, H. (1990). Interactive effects of dietary silicon, copper, and zinc in the rat. *J. Nutr. Biochem.* **1**, 35.

Emery, M.P., Browning, J.D., and O'Dell, B.L. (1990). Impaired hemostasis and platelet function in rats fed low zinc diets based on egg white protein. *J. Nutr.* **120**, 1062–1067.

Englyst, H., and Cummings, J. (1990). Dietary fibre and starch: definition, classification and measurement. *In* "Dietary Fibre Perspectives: Reviews and Bibliography." (A.R. Leeds, ed), pp. 3–26. John Libby & Co., London.

Erdman, J., Poor, C., and Dietz, J.M. (1988). Factors affecting the bioavailability of vitamin A, carotenoids, and vitamin E. *Food Tech.* **42**, 214.

Erikson, K.M., Jones, B.C., Hess, E.J., Zhang, Q., and Beard, J.L. (2001). Iron deficiency decreases dopamine D1 and D2 receptors in rat brain. *Pharmacol. Biochem. Behav.* **69**, 409–418.

Erikson, K.M., Pinero, D.J., Connor, J.R., and Beard, J.L. (1997). Regional brain iron, ferritin and transferrin concentrations during iron deficiency and iron repletion in developing rats. *J. Nutr.* **127**, 2030–2038.

Erlingen, R.C., and Delcour, J.A. (1995). Formation, analysis, structure and properties of type III enzyme resistant starch. *J. Cereal Sci.* **22**, 129–138.

Ertel, N.H., Akgun, S., Kemp, F.W., and Mittler, J.C. (1983). The metabolic fate of exogenous sorbitol in the rat. *J. Nutr.* **113**, 566–573.

Ettarh, R.R., Odigie, I.P., and Adigun, S.A. (2002). Vitamin C lowers blood pressure and alters vascular responsiveness in salt-induced hypertension. *Can. J. Physiol. Pharmacol.* **80**, 1199–1202.

Evans, H.M., and Bishop, K.S. (1922). On the existence of a hitherto unrecognized dietary factor essential for reproduction. *Science* **56**, 650.

Evans, H.M., and Emerson, F.A. (1943). The prophylactic requirement of the rat for α-tocopherol. *J. Nutr.* **26**, 555.

Evans, W.C. (1975). Thiaminases and their effects on animals. *Vitam. Horm.* **33**, 467–504.

Evenson, J.K., and Sunde, R.A. (1988). Selenium incorporation into selenoproteins in the Se-adequate and Se-deficient rat. *Proc. Soc. Exp. Biol. Med.* **187**, 169–180.

Fagin, J.A., Ikejiri, K., and Levin, S.R. (1987). Insulinotropic effects of vanadate. *Diabetes* **36**, 1448–1452.

Fairbanks, V.F. (1999). Iron in medicine and nutrition. *In* "Modern Nutrition in Health and Disease." (M.E. Shils, J.A. Olson, M. Shike, and A.C. Ross, eds), p. 193. Williams and Wilkins, Baltimore.

Faqi, A.S., Sherman, D.D., Wang, M., Pasquali, M., Bayorh, M.A., Thierry-Palmer, M. (2001). The calciuric response to dietary salt of Dahl salt-sensitive and salt-resistant female rats. *Am. J. Med. Sci.* **322**, 333–338.

Farquharson, C., Duncan, A., and Robins, S.P. (1989). The effects of copper deficiency on the pyridinium crosslinks of mature collagen in the rat skeleton and cardiovascular system. *Proc. Soc. Exp. Biol. Med.* **192**, 166–171.

Favier, M., Faure, P., Roussel, A.M., Coudray, C., Blache, D., and Favier, A. (1993). Zinc deficiency and dietary folate metabolism in pregnant rats. *J. Trace Elem. Electrolytes Health. Dis.* **7**, 19–24.

Feldman, D. (1960). Iodine deficiency in newborn rats. *Am. J. Physiol.* **199**, 1081.

Fergusson, M.A., and Koski, K.G. (1990). Comparison of effects of dietary glucose versus fructose during pregnancy on fetal growth and development in rats. *J. Nutr.* **120**, 1312–1319.

Ferland, G., Tuchweber, B., Bhat, P.V., and Lacroix, A. (1992). Effect of dietary restriction on hepatic vitamin A content in aging rats. *J. Gerontol.* **47**, B3–B8.

Fernandes, G., Chandrasekar, B., Troyer, D.A., Venkatraman, J.T., and Good, R.A. (1995). Dietary lipids and calorie restriction affect mammary tumor incidence and gene expression in mouse mammary tumor virus/v-Ha-ras transgenic mice. *Proc. Natl. Acad. Sci. USA* **92**, 6494–6498.

Fernstrom, J.D. (1976). The effect of nutritional factors on brain amino acid levels and monoamine synthesis. *Fed. Proc.* **35**, 1151–1156.

Fields, M., Ferretti, R.J., Smith, J.C. Jr, and Reiser, S. (1984). The interaction of type of dietary carbohydrates with copper deficiency. *Am. J. Clin. Nutr.* **39**, 289–295.

Fields, M., and Lewis, C.G. (1997). Impaired endocrine and exocrine pancreatic functions in copper-deficient rats: the effect of gender. *J. Am. Coll. Nutr.* **16**, 346–351.

Fields, M., and Lewis, C.G. (1999). Level of dietary iron, not type of dietary fat, is hyperlipidemic in copper-deficient rats. *J. Am. Coll. Nutr.* **18**, 353–357.

Fine, K.D., Santa Ana, C.A., Porter, J.L., and Fordtran, J.S. (1991). Intestinal absorption of magnesium from food and supplements. *J. Clin. Invest.* **88**, 396–402.

Finkelstein, J.D., and Martin, J.J. (1986). Methionine metabolism in mammals: adaptation to methionine excess. *J. Biol. Chem.* **261**, 1582–1587.

Finley, J.W. and Davis, C.D. (2001). Manganese absorption and retention in rats is affected by the type of dietary fat. *Biol. Trace Elem. Res.* **82**, 143–158.

Fischer, P.W., Campbell, J.S., and Giroux, A. (1989). Effect of dietary iodine on autoimmune thyroiditis in the BB Wistar rats. *J. Nutr.* **119**, 502–507.

Fischer, W.H., and Lutz, W.K. (1998). Influence of diet restriction and tumor promoter dose on cell proliferation, oxidative DNA damage and rate of papilloma appearance in the mouse skin after initiation with DMBA and promotion with TPA. *Toxicol. Lett.* **98**, 59–69.

Fishelson, Z., and Muller-Eberhard, H.J. (1982). C3 convertase of human complement: enhanced formation and stability of the enzyme generated with nickel instead of magnesium. *J. Immunol.* **129**, 2603–2607.

Flatt, P.R., Juntti-Berggren, L., Berggren, P.O., Gould, B.J., and Swanston-Flatt, S.K. (1989). Effects of dietary inorganic trivalent chromium (Cr3+) on the development of glucose homeostasis in rats. *Diabetes Metab.* **15**, 93–97.

Fleming, S.E,. and Lee, B. (1983). Growth performance and intestinal transit time of rats fed purified and natural dietary fibers. *J. Nutr.* **113**, 592–601.

Follis, R.H. (1958). "Deficiency Disease." Charles C. Thomas, Springfield, IL.

Forsum, E., Hillman, P.E., and Nesheim, M.C. (1981). Effect of energy restriction on total heat production, basal metabolic rate, and specific dynamic action of food in rats. *J. Nutr.* **111**, 1691–1697.

Fox, H.M. (1984). Pantothenic acid. *In* "Handbook of Vitamins." (R.B. Rucker, J.W. Suttie, D.B. McCormick, and L.J. Machlin, eds), p. 437. Marcel Dekker, New York.

Fox, J.G., Aldrich, F.D., and Boylen, G.W. Jr. (1976). Lead in animal foods. *J. Toxicol. Environ. Health* **1**, 461–467.

Fuentealba, I.C., Mullins, J.E., Aburto, E.M., Lau, J.C., and Cherian, G.M. (2000). Effect of age and sex on liver damage due to excess dietary copper in Fischer 344 rats. *J. Toxicol. Clin. Toxicol.* **38**, 709–717.

Furuse, M., Tamura, Y., Matsuda, S., Shimizu, T., and Okumura, J. (1989). Lower fat deposition and energy utilization of growing rats fed diets containing sorbose. *Comp. Biochem. Physiol. A* **94**, 813–817.

Gabriel, E., Machlin, L.J., Filipski, R., and Nelson, J. (1980). Influence of age on the vitamin E requirement for resolution of necrotizing myopathy. *J. Nutr.* **110**, 1372–1379.

Gaetke, L.M., Frederich, R.C., Oz, H.S., and McClain, C.J. (2002). Decreased food intake rather than zinc deficiency is associated with changes in plasma leptin, metabolic rate, and activity levels in zinc deficient rats (small star, filled). *J. Nutr. Biochem* **13**, 237–244.

Gaitonde, M.K. (1982). Neurotransmitter function in thiamine-deficiency encephalopathy. *Neurochem. Int.* **4**, 465.

Gambling, L., Dunford, S., and McArdle, H.J. (2004). Iron deficiency in the pregnant rat has differential effects on maternal and fetal copper levels. *J. Nutr. Biochem.* **15**, 366–372.

Ganguli, M., and Tobian, L. (1991). In SHR rats, dietary potassium determines NaCl sensitivity in NaCl-induced rises of blood pressure. *Clin. Exp. Hypertens. A* **13**, 677–685.

Garcia-Aranda, J.A., Wapnir, R.A., and Lifshitz, F. (1983). In vivo intestinal absorption of manganese in the rat. *J. Nutr.* **113**, 2601–2607.

Garner, S.C., Mar, M.H., and Zeisel, S.H. (1995). Choline distribution and metabolism in pregnant rats and fetuses are influenced by the choline content of the maternal diet. *J. Nutr.* **125**, 2851–2858.

Gelderblom, W.C., Abel, S., Smuts, C.M., Marnewick, J., Marasas, W.F., Lemmer, E.R., and Ramljak, D. (2001). Fumonisin-induced hepato-carcinogenesis: mechanisms related to cancer initiation and promotion. *Environ. Health Perspect.* **109**, 291–300.

Gelderblom, W.C., Kriek, N.P., Marasas, W.F., and Thiel, P.G. (1991). Toxicity and carcinogenicity of the *Fusarium moniliforme* metabolite, fumonisin B1, in rats. *Carcinogenesis* **12**, 1247–1251.

Gerber, L.E., and Erdman, J.W. Jr. (1982). Effect of dietary retinyl acetate, beta-carotene and retinoic acid on wound healing in rats. *J. Nutr.* **112**, 1555–1564.

Ghishan, F.K., Said, H.M., Wilson, P.C., Murrell, J.E., and Greene, H.L. (1986). Intestinal transport of zinc and folic acid: a mutual inhibitory effect. *Am. J. Clin. Nutr.* **43**, 258–262.

Ghoshal, A.K., Ahluwalia, M., and Farber, E. (1983). The rapid induction of liver cell death in rats fed a chorine-deficient methionine-low diet. *Am. J. Pathol.* **113**, 309–314.

Gibson, G.E., Jope, R., and Blass, J.P. (1975). Decreased synthesis of acetylcholine accompanying impaired oxidation of pyruvic acid in rat-brain minces. *Biochem. J.* **148**, 17–23.

Giovanetti, P.M. (1982). Effect of coprophagy on nutrition. *Nutr. Res.* **2**, 335.

Girard, P., Brun-Pascaud, M., and Paillard, M. (1985). Selective dietary potassium depletion and acid-base equilibrium in the rat. *Clin. Sci. (Lond)* **68**, 301–309.

Goetzel, E.J. (1981). Vitamin E modulates the lipoxygenation of arachidonic acid in leukocytes. *Nature* **288**, 193.

Goldberg, A. (1971). Carbohydrate metabolism in rats fed carbohydrate-free diets. *J. Nutr.* **101**, 693–697.

Gomez-Ayala, A.E., Campos, M.S., Lopez-Aliaga, I., Pallares, I., Hartiti, S., Barrionuevo, M., Alferez, M.J., Rodriguez-Matas, M.C., and Lisbona, F. (1997). Effect of source of iron on duodenal absorption of iron, calcium, phosphorus, magnesium, copper and zinc in rats with ferropoenic anaemia. *Int. J. Vitamin Nutr. Res.* **67**, 106–114.

Gonca, S., Ceylan, S., Yardimoglu, M., Dalcik, H., Yumbul, Z., Kokturk, S., and Filiz, S. (2000). Protective effects of vitamin E and selenium on the renal morphology in rats fed high-cholesterol diets. *Pathobiology* **68**, 258–263.

Gong, H., and Amemiya, T. (1999). Optic nerve changes in manganese-deficient rats. *Exp. Eye Res.* **68**, 313–320.

Goodrick, C.L. (1980). Effects of long-term voluntary wheel exercise on male and female Wistar rats, I: longevity, body weight, and metabolic rate. *Gerontology* **26**, 22–33.

Govers, M.J., and Van der Meet, R. (1993). Effects of dietary calcium and phosphate on the intestinal interactions between calcium, phosphate, fatty acids, and bile acids. *Gut* **34**, 365–370.

Gower, J.D., and Wills, E.D. (1986). The oxidation of benzo[a]pyrene mediated by lipid peroxidation in irradiated synthetic diets. *Int. J. Radiat. Biol. Relat. Stud. Phys. Chem. Med.* **49**, 471–484.

Grases, F., Perello, J., Simonet, B.M., Prieto, R.M., and Garcia-Raja, A. (2004). Study of potassium phytate effects on decreasing urinary calcium in rats. *Urol. Int.* **72**, 237–243.

Grasl-Kraupp, B., Bursch, W., Ruttkay-Nedecky, B., Wagner, A., Lauer, B., and Schulte-Hermann, R. (1994). Food restriction eliminates preneoplastic cells through apoptosis and antagonizes carcinogenesis in rat liver. *Proc. Natl. Acad. Sci. USA* **91**, 9995–9999.

Gratas, C., Menot, M.L., Dresch, C., and Chomienne, C. (1993). Retinoid acid supports granulocytic but not erythroid differentiation of myeloid progenitors in normal bone marrow cells. *Leukemia* **7**, 1156–1162.

Greaves, J.H., and Ayres, P. (1973). Warfarin resistance and vitamin K requirement in the rat. *Lab. Anim.* **7**, 141–148.

Green, M.H., and Green, J.B. (1991). Influence of vitamin A intake on retinol (ROH) balance, utilization, and dynamics. *FASEB J.* **5**, A718.

Green, M.H., Green, J.B., and Lewis, K.C. (1987). Variation in retinol utilization rate with vitamin A status in the rat. *J. Nutr.* **117**, 694–703.

Greenman, D.L., Oller, W.L., Littlefield, N.A., and Nelson, C.J. (1980). Commercial laboratory animal diets: toxicant and nutrient variability. *J. Toxicol. Environ. Health* **6**, 235–246.

Greenwood, J., and Pratt, O.E. (1985). Comparison of the effects of some thiamine analogues upon thiamine transport across the blood-brain barrier of the rat. *J. Physiol.* **369**, 79–91.

Grenby, T.H., and Phillips, A. (1989). Dental and metabolic effects of lactitol in the diet of laboratory rats. *Br. J. Nutr.* **61**, 17–24.

Groff, J.L., and Gropper, S.S. (2000a). Microminerals. *In* "Advanced Nutrition and Human Metabolism." (L. Graham, ed), p. 401. Wadsworth/Thomson, Belmont, CA.

Groff, J.L., and Gropper, S.S. (2000b). Water soluble vitamins. *In* "Advanced Nutrition and Human Metabolism." (L. Graham, ed), p. 279. Wadsworth/Thomson, Belmont, CA.

Groff, J.L., and Gropper, S.S. (2000c). Fat soluble vitamins. *In* "Advanced Nutrition and Human Metabolism." (L. Graham, ed), p. 343. Wadsworth/Thomson, Belmont, CA.

Gross, K.L., Hartman, W.J., Ronnenberg, A., and Prior, R.L. (1991). Arginine-deficient diets alter plasma and tissue amino acids in young and aged rats. *J. Nutr.* **121**, 1591–1599.

Grove, K.L., and Deschepper, C.F. (1999). High salt intake differentially regulates kidney angiotensin IV AT4 receptors in Wistar-Kyoto and spontaneously hypertensive rats. *Life Sci.* **64**, 1811–1818.

Groziak, S., Kirksey, A., and Hamaker, B. (1984). Effect of maternal vitamin B-6 restriction on pyridoxal phosphate concentrations in developing regions of the central nervous system in rats. *J. Nutr.* **114**, 727–732.

Gruber, H.E., and Stover, S.J. (1994). Maternal and weanling bone: the influence of lowered calcium intake and maternal dietary history. *Bone* **15**, 167–176.

Gueguen, L., and Pointillart, A. (2000). The bioavailability of dietary calcium. *J. Am. Coll. Nutr.* **19**, 119S–136S.

Gueux, E., Mazur, A., Cardot, P., and Rayssiguier, Y. (1991). Magnesium deficiency affects plasma lipoprotein composition in rats. *J. Nutr.* **121**, 1222–1227.

Guilarte, T.R., and Wagner, H.N. Jr. (1987). Increased concentrations of 3-hydroxykynurenine in vitamin B$_6$ deficient neonatal rat brain. *J. Neurochem.* **49**, 1918–1926.

Gurnani, S., Mistry, S.P., and Johnson, B.C. (1959). Function of vitamin B12 in methylmalonate metabolism, I: effect of a cofactor form of B12 on the activity of methylmalony CoA isomerase. *Biochim. Biophys. Acta* **38**, 187–188.

Gustafsson, B.E., Daft, F.S., McDaniel, E.G., Smith, J.C., and Fitzgerald, R.J. (1962). Effects of vitamin K-active compounds and intestinal microorganisms in vitamin K-deficient germfree rats. *J. Nutr.* **78**, 461–468.

Haas, R.H. (1988). Thiamin and the brain. *Annu. Rev. Nutr.* **8**, 483–515.

Haberkorn, V., Heydel, J.M., Mounie, J., Artur, Y., and Goudonnet, H. (2002). Vitamin A modulates the effects of thyroid hormone on UDP-glucuronosyltransferase expression and activity in rat liver. *Mol. Cell Endocrinol.* **190**, 167–175.

Hakkarainen, J., Tyopponen, J., and Jonsson, L. (1986). Vitamin E requirement of the growing rat during selenium deficiency with special reference to selenium dependent–and selenium independent glutathione peroxidase. *Zentralbl. Veterinarmed. A* **33**, 247–258.

Halas, E.S., Eberhardt, M.J., Diers, M.A., and Sandstead, H.H. (1983). Learning and memory impairment in adult rats due to severe zinc deficiency during lactation. *Physiol. Behav.* **30**, 371–381.

Halloran, B.P., and DeLuca, H.F. (1980). Effect of vitamin D deficiency on fertility and reproductive capacity in the female rat. *J. Nutr.* **110**, 1573–1580.

Halverston, A.W., Shaw, J.H., et al. (1945). Goiter studies in the rat. *J. Nutr.* **30**, 59.

Hambidge, K.M., Casey, C.E., et al. (1986). Zinc. *In* "Trace Elements in Human and Animal Nutrition." (W. Mertz, ed), p. 1. Academic Press, Orlando.

Hamdi, S.A., Nassif, O.I., and Ardawi, M.S. (1997). Effect of marginal or severe dietary zinc deficiency on testicular development and functions of the rat. *Arch. Androl.* **38**, 243–253.

Hamm, M.W., Mehansho, H., and Henderson, L.M. (1979). Transport and metabolism of pyridoxamine and pyridoxamine phosphate in the small intestine of the rat. *J. Nutr.* **109**, 1552–1559.

Hammermueller, J.D., Bray, T.M., and Bettger, W.J. (1987). Effect of zinc and copper deficiency on microsomal NADPH-dependent active oxygen generation in rat lung and liver. *J. Nutr.* **117**, 894–901.

Hankes, L.V. (1984). Nicotinic acid and nicotinamide. *In* "Handbook of Vitamins." (L.J. Machlin, ed), p. 329. Marcel Dekker, New York.

Hankes, L.V., Henderson, L.M., et al. (1948). Effect of amino acids on the growth rate of rats on niacin-tryptophan deficient rations. *J. Biol. Chem.* **174**, 873.

Hankey, G.J., and Eikelboom, J.W. (1999). Homocysteine and vascular disease. *Lancet* **354**, 407–413.

Hans, C.P., Chaudhary, D.P., and Bansal, D.D. (2003). Effect of magnesium supplementation on oxidative stress in alloxanic diabetic rats. *Magnes. Res.* **16**, 13–19.

Hardwick, L.L., Jones, M.R., Brautbar, N., and Lee, D.B. (1991). Magnesium absorption: mechanisms and the influence of vitamin D, calcium and phosphate. *J. Nutr.* **121**, 13–23.

Harman, D. (1971). Free radical theory of aging: Effect of the amount and degree of unsaturation of dietary fat on mortality rate. *J. Gerontol.* **26**, 451–457.

Harper, A.E., Benevenga, N.J., and Wohlhueter, R.M. (1970). Effects of ingestion of disproportionate amounts of amino acids. *Physiol. Rev.* **50**, 428–558.

Harris, H.A., Neuberger, A., and Sanger, F. (1943). Lysine deficiency in young rats. *Biochem. J.* **37**, 508–513.

Harris, J.B., and Van Horn, H.H. (1992). Water and its importance to animals, Circular 1017, Dairy Production Guide. Florida Cooperative Extension Service.

Hartsook, E.W., Hershberger, T.V., and Nee, J.C. (1973). Effects of dietary protein content and ratio of fat to carbohydrate calories on energy metabolism and body composition of growing rats. *J. Nutr.* **103**, 167–178.

Haschek, W.M., Gumprecht, L.A., Smith, G., Tumbleson, M.E., and Constable, P.D. (2001). Fumonisin toxicosis in swine: an overview of porcine pulmonary edema and current perspectives. *Environ. Health Perspect.* **109**, 251–257.

Hassel, C.A., Carr, T.P., Marchello, J.A., and Lei, K.Y. (1988). Apolipoprotein E-rich HDL binding to liver plasma membranes in copper-deficient rats. *Proc. Soc. Exp. Biol. Med.* **187**, 296–308.

Hatton, D.C., Scrogin, K.E., Levine, D., Feller, D., and McCarron, D.A. (1993). Dietary calcium modulates blood pressure through α_1-adrenergic receptors. *Am. J. Physiol.* **264**, F234–F238.

Hatton, D.C., Yue, Q., and McCarron, D.A. (1995). Mechanisms of calcium's effects on blood pressure. *Semin. Nephrol.* **15**, 593–602.

Haussler, M.R. (1986). Vitamin D receptors: nature and function. *Annu. Rev. Nutr.* **6**, 527–562.

Hayashida, T., Ohno, Y., Otsuka, K., Suzawa, T., Shibagaki, K., Suzuki, H., Ikeda, H., and Saruta, T. (2001). Salt-loading elevates blood pressure and aggravates insulin resistance in Wistar fatty rats: a possible role for enhanced Na^+-H^+ exchanger activity. *J. Hypertens.* **19**, 1643–1650.

Heggtveit, H.A. (1969). Myopathy in experimental magnesium deficiency. *Ann. NY Acad. Sci.* **162**, 758–765.

Helou, C.M., de Araujo, M., and Seguro, A.C. (1994). Effect of low and high potassium diets on H(+)-K(+)-ATPase and Na(+)-K(+)-ATPase activities in the rat inner medullary collecting duct cells. *Ren. Physiol. Biochem.* **17**, 21–26.

Henderson, L.M., Deodhar, T., Krehl, W.A., and Elvehjem, C.A. (1947). Factors affecting the growth of rats receiving niacin-tryptophanpdeficient diets. *J. Biol. Chem.* **170**, 261.

Herbert, V. (1999). Folic Acid. *In* "Modern Nutrition in Health and Disease." (M.E. Shils, J.A. Olson, M. Shike, and C.A. Ross, eds), p. 433. Williams and Wilkins, Baltimore.

Herzberg, G.R., and Rogerson, M. (1988a). Hepatic fatty acid synthesis and triglyceride secretion in rats fed fructose- or glucose-based diets containing corn oil, tallow or marine oil. *J. Nutr.* **118**, 1061–1067.

Herzberg, G.R., and Rogerson, M. (1988b). Interaction of dietary carbohydrate and fat in the regulation of hepatic and extrahepatic lipogenesis in the rat. *Br. J. Nutr.* **59**, 233–241.

Hettinger, U., Lukasova, M., Lewicka, S., and Hilgenfeldt, U. (2002). Regulatory effects of salt diet on renal renin-angiotensin-aldosterone, and kallikrein-kinin systems. *Int. Immunopharmacol.* **2**, 1975–1980.

Hetzel, B.S., and Clugston, G.A. (1999). Iodine. *In* "Modern Nutrition in Health and Disease." (M.E. Shils, J.A. Olson, M. Shike, and C.A. Ross, eds), p. 241. Williams and Wilkins, Baltimore.

Heywood, R., Palmer, A.K., Gregson, R.L., and Hummler, H. (1985). The toxicity of beta-carotene. *Toxicology* **36**, 91–100.

Higami, Y., Yu, B.P., Shimokawa, I., Masoro, E.J., and Ikeda, T. (1994). Duration of dietary restriction: an important determinant for the incidence and age of onset of leukemia in male F344 rats. *J. Gerontol.* **49**, B239–B244.

Higgins, E.S., Richert, D.A., and Westerfeld, W.W. (1956). Molybdenum deficiency and tungstate inhibition studies. *J. Nutr.* **59**, 539–559.

Hiipakka, R.A., and Liao, S. (1980). Effect of pyridoxal phosphate on the androgen receptor from rat prostate: inhibition of receptor aggregation and receptor binding to nuclei and to DNA-cellulose. *J. Steriod Biochem.* **13**, 841–846.

Hill, K.E., Burk, R.F., and Lane, J.M. (1987). Effect of selenium depletion and repletion on plasma glutathione and glutathione-dependent enzymes in the rat. *J. Nutr.* **117**, 99–104.

Hipsley, E.H. (1953). Dietary fibre and pregnancy toxaemia. *Br. Med. J.* **2**, 420–422.

Hiraide, K. (1981). Alterations in electrolytes with aging. *Nihon Univ. J. Med.* **23**, 21–32.

Hirano, T., Mamo, J., Poapst, M., and Steiner, G. (1988). Very-low-density lipoprotein triglyceride kinetics in acute and chronic carbohydrate-fed rats. *Am. J. Physiol.* **255**, E236–E240.

Hirose, M., Yada, H., Hakoi, K., Takahashi, S., and Ito, N. (1993). Modification of carcinogenesis by α-tocopherol, t-butylhydroquinone, propyl gallate and butylated hydroxytoluene in a rat multi-organ carcinogenesis model. *Carcinogenesis* **14**, 2359–2364.

Hocherl, K., Kammerl, M.C., Schumacher, K., Endemann, D., Grobecker, H.F., and Kurtz, A. (2002). Role of prostanoids in regulation of the renin-angiotensin-aldosterone system by salt intake. *Am. J. Physiol. Renal Physiol* **283**, F294–F301.

Hodge, G., Ye, V.Z., and Duggan, K.A. (2002). Dysregulation of angiotensin II synthesis is associated with salt sensitivity in the spontaneous hypertensive rat. *Acta Physiol. Scand.* **174**, 209–215.

Hoek, A.C., Lemmens, A.G., Mullink, J.W., and Beynen, A.C. (1988). Influence of dietary calcium:phosphorus ratio on mineral excretion and nephrocalcinosis in female rats. *J. Nutr.* **118**, 1210–1216.

Hoekstra, B. (1993). Questioning the safety of food irradiation. *Public Health Rep.* **108**, 402.

Hoffman, C., and Eichele, G. (1994). Retinoids in development. *In* "The Retinoids: Biology, Chemistry, and Medicine." (M.B. Sporn, A.B. Roberts, and D.S. Goodman, eds), p. 387. Raven Press, New York.

Holdsworth, E.S., and Neville, E. (1990). Effects of extracts of high- and low-chromium brewer's yeast on metabolism of glucose by hepatocytes from rats fed on high- or low-Cr diets. *Br. J. Nutr.* **63**, 623–630.

Holler, T., Cermak, J.M., and Blusztajn, J.K. (1996). Dietary choline supplementation in pregnant rats increases hippocampal phospholipase D activity of the offspring. *FASEB J.* **10**, 1653–1659.

Holley, J., Bender, D.A., Coulson, W.F., and Symes, E.K. (1983). Effects of vitamin B6 nutritional status on the uptake of [3H]-oestradiol into the uterus, liver and hypothalamus of the rat. *J. Steriod Biochem.* **18**, 161–165.

Holmer, S., Eckardt, K.U., LeHir, M., Schricker, K., Riegger, G., and Kurtz, A. (1993). Influence of dietary NaCl intake on renin gene expression in the kidneys and adrenal glands of rats. *Pflugers Arch.* **425**, 62–67.

Holtkamp, D.E., and Hill, R.M. (1950). The effect on growth of the level of manganese in the diet of rats, with some observations on the manganese-thiamine relationships. *J. Nutr.* **41**, 307–316.

Hong, C.B., and Chow, C.K. (1988). Induction of eosinophilic enteritis and eosinophilia in rats by vitamin E and selenium deficiency. *Exp. Mol. Pathol.* **48**, 182–192.

Hong, K.H., Keen, C.L., Mizuno, Y., Johnston, K.E., and Tamura, T. (2000). Effects of dietary zinc deficiency on homocysteine and folate metabolism in rats(1). *J. Nutr. Biochem.* **11**, 165–169.

Hoogeveen, R.C., Reaves, S.K., Reid, P.M., Reid, B.L., and Lei, K.Y. (1994). Copper deficiency shifts energy substrate utilization from carbohydrate to fat and reduces fat mass in rats. *J. Nutr.* **124**, 1660–1666.

Hopkins, R.G., and Failla, M.L. (1995). Chronic intake of a marginally low copper diet impairs *in vitro* activities of lymphocytes and neutrophils from male rats despite minimal impact on conventional indicators of copper status. *J. Nutr.* **125**, 2658–2668.

Hotz, C.S., Fitzpatrick, D.W., Trick, K.D., and L'Abbe, MR. (1997). Dietary Iodine and selenium interact to affect thyroid hormone metabolism of rats. *J. Nutr.* **127**, 1214–1218.

Howard, L., Wagner, C., and Schenker, C. (1977). Malabsorption of thiamine in folate deficient rats. *J. Nutr.* **107**, 775.

Howard, P.C., Eppley, R.M., Stack, M.E., Warbritton, A., Voss, K.A., Lorentzen, R.J., Kovach, R.M., and Bucci, T.J. (2001a). Fumonisin B1 carcinogenicity in a two-year feeding study using F344 rats and B6C3F1 mice. *Environ. Health Perspect.* **109**, 277–282.

Howard, P.C., Warbritton, A., Voss, K.A., Lorentzen, R.J., Thurman, J.D., Kovach, R.M., and Bucci, T.J. (2001b). Compensatory regeneration as a mechanism for renal tubule carcinogenesis of fumonisin B1 in the F344/N/Nctr BR rat. *Environ. Health Perspect.* **109 Suppl 2**, 309–314.

Huang, R.F., Hsu, Y.C., Lin, H.L., and Yang, F.L. (2001). Folate depletion and elevated plasma homocysteine promote oxidative stress in rat livers. *J. Nutr.* **131**, 33–38.

Huber, K., and Breves, G. (1999). Influence of dietary phosphorus depletion on central pathways of intermediary metabolism in rats. *Arch. Tierernahr.* **52**, 299–309.

Huether, G., Thomke, F., and Adler, L. (1992). Administration of tryptophan-enriched diets to pregnant rats retards the development of the serotonergic system in their offspring. *Brain Res. Dev. Brain Res.* **68**, 175–181.

Hundley, J.M. (1947). Production of niacin deficiency in rats. *J. Nutr.* **34**, 253.

Hundley, J.M. (1949). Influence of fructose and other carbohydrates on the niacin requirement of the rat. *J. Biol. Chem.* **181**, 1.

Hunt, C.D. (1994). The biochemical effects of physiologic amounts of dietary boron in animal nutrition models. *Environ. Health Perspect.* **102**, 35–43.

Hurley, L.S., and Bell, L.T. (1974). Genetic influence on response to dietary manganese deficiency in mice. *J. Nutr.* **104**, 133–137.

Hurley, L.S., and Keen, C.L. (1987). Manganese. *In* "Trace Elements in Human and Animal Nutrition." (W. Mertz, ed), p. 185. Academic Press, Orlando.

Ichihashi, T., Takagishi, Y., Uchida, K., and Yamada, H. (1992). Colonic absorption of menaquinone-4 and menaquinone-9 in rats. *J. Nutr.* **122**, 506–512.

Ikegami, S., Tsuchihashi, F., Harada, H., Tsuchihashi, N., Nishide, E., and Innami, S. (1990). Effect of viscous indigestible polysaccharides on pancreatic-biliary secretion and digestive organs in rats. *J. Nutr.* **120**, 353–360.

Ingert, C., Grima, M., Coquard, C., Barthelmebs, M., and Imbs, J.L. (2002). Effects of dietary salt changes on renal renin-angiotensin system in rats. *Am. J. Physiol. Renal Physiol.* **283**, F995–F1002.

Ink, S.L., and Henderson, L.M (1984). Vitamin B6 metabolism. *Annu. Rev. Nutr.* **4**, 455–470.

Iseri, L.T., and French, J.H. (1984). Magnesium: nature's physiologic calcium blocker. *Am. Heart J.* **108**, 188–193.

Ishimitsu, T., Tobian, L., Sugimoto, K., and Everson, T. (1996). High potassium diets reduce vascular and plasma lipid peroxides in stroke-prone spontaneously hypertensive rats. *Clin. Exp. Hypertens.* **18**, 659–673.

Ito, N., and Hirose, M. (1989). Antioxidants–carcinogenic and chemopreventive properties. *Adv. Cancer Res.* **53**, 247–302.

Ito, N., Hirose, M., and Takahashi, S. (1993). Cell proliferation and forestomach carcinogenesis. *Environ. Health Perspect.* **101**, 107–10.

Itokawa, Y., and Fujiwara, M. (1973). Lead and vitamin effect on heme synthesis. *Arch. Environ. Health* **27**, 31.

Iwasaki, K., Gleiser, C.A., Mascro, E.J., McMahan, C.A., Seo, E.J., and Yu, B.P. (1988). Influence of the restriction of individual dietary components on longevity and age-related disease of Fischer rats: the fat component and the mineral component. *J. Gerontol.* **43**, B13–B21.

Iwata, H., Nishikawa, T., and Baba, A. (1970). Catecholamine accumulation in tissues of thiamine-deficient rats after inhibition of monoamine oxidase. *Eur. J. Pharmacol.* **12**, 253–256.

Jager, F.C. (1972). Long-term dose-response effects of vitamin E in rats: significance of the *in vitro* haemolysis test. *Nutr. Metab.* **14**, 1–7.

Jager, F.C., and Houtsmuller, U.M. (1970). Effect of dietary linoleic acid on vitamin E requirement and fatty acid composition of erythrocyte lipids in rats. *Nutr. Metab.* **12**, 3–12.

Jalili, T., Medeiros, D.M., and Wildman, R.E. (1996). Aspects of cardiomyopathy are exacerbated by elevated dietary fat in copper-restricted rats. *J. Nutr.* **126**, 807–816.

Jansen, G.R., Grayson, C., and Hunsaker, H. (1987). Wheat gluten during pregnancy and lactation: effects on mammary gland development and pup viability. *Am. J. Clin. Nutr.* **46**, 250–257.

Jenkins, J.E., and Medeiros, D.M. (1993). Diets containing corn oil, coconut oil and cholesterol alter ventricular hypertrophy, dilatation and function in hearts of rats fed copper-deficient diets. *J. Nutr.* **123**, 1150–1160.

Jimenez, A., Planells, E., Aranda, P., Sanchez-Vinas, M., and Llopis, J. (1997). Changes in bioavailability and tissue distribution of selenium caused by magnesium deficiency in rats. *J. Am. Coll. Nutr.* **16**, 175–180.

Johnson, M.A., and Gratzek, J.M. (1986). Influence of sucrose and starch on the development of anemia in copper- and iron-deficient rats. *J. Nutr.* **116**, 2443–2452.

Johnson, M.A., and Murphy, C.L. (1988). Adverse effects of high dietary iron and ascorbic acid on copper status in copper-deficient and copper-adequate rats. *Am. J. Clin. Nutr.* **47**, 96–101.

Johnson, P.E., and Korynta, E.D. (1990). The effect of dietary protein source on manganese bioavailability to the rat. *Proc. Soc. Exp. Biol. Med.* **195**, 230–236.

Johnson, P.E., and Korynta, E.D. (1992). Effects of copper, iron, and ascorbic acid on manganese availability to rats. *Proc. Soc. Exp. Biol. Med.* **199**, 470–480.

Johnson, W.T., and Dufault, S.N. (1989). Altered cytoskeletal organization and secretory response of thrombin-activated platelets from copper-deficient rats. *J. Nutr.* **119**, 1404–1410.

Johnson, W.T., Dufault, S.N., and Thomas, A.C. (1993). Platelet cytochrome c oxidase activity is an indicator of copper status in rats. *Nutr. Res.* **13**, 1153.

Johnson, W.T., and Kramer, T.R. (1987). Effect of copper deficiency on erythrocyte membrane proteins of rats. *J. Nutr.* **117**, 1085–1090.

Jones, J.H. (1971). Vitamin D requirement for animals. *In* "The Vitamins, "Vol III. (W.H. Sebrell and R.S. Harris, eds) p. 285. Academic Press, New York.

Kadowaki, S., and Norman, A.W. (1984). Dietary vitamin D is essential for normal insulin secretion from the perfused rat pancreas. *J. Clin. Invest.* **73**, 759–766.

Kaganda, J., Matsuo, T., and Suzuki, H. (2003). Development of iron deficiency decreases zinc requirement of rats. *J. Nutr. Sci. Vitaminol. (Tokyo)* **49**, 234–240.

Kahlenberg, O.J., Black, A., Batzler, J.W., and Forbes, E.B. (1937). The utilization of energy producing nutriment and protein as affected by sodium deficiency. *J. Nutr.* **13**, 97.

Kahonen, M., Nappi, S., Jolma, P., Hutri-Kahonen, N., Tolvanen, J.P., Saha, H., Koivisto, P., Krogerus, L., Kalliovalkama, J., and Porsti, I. (2003). Vascular influences of calcium supplementation and vitamin D-induced hypercalcemia in NaCl-hypertensive rats. *J. Cardiovasc. Pharmacol.* **42**, 319–328.

Kalu, D.N., Masoro, E.J., Yu, B.P., Hardin, R.R., and Hollis, B.W. (1988). Modulation of age-related hyperparathyroidism and senile bone loss in Fischer rats by soy protein and food restriction. *Endocrinology* **122**, 1847–1854.

Kamao, M., Tsugawa, N., Nakagawa, K., Kawamoto, Y., Fukui, K., Takamatsu, K., Kuwata, G., Imai, M., and Okano, T. (2000). Absorption of calcium, magnesium, phosphorus, iron and zinc in growing male rats fed diets containing either phytate-free soybean protein or soybean protein isolate or casein. *J. Nutr. Sci. Vitaminol. (Tokyo)* **46**, 34–41.

Kang, S.S., Price, R.G., Yudkin, J., Worcester, N.A., and Bruckdorfer, K.R. (1979). The influence of dietary carbohydrate and fat on kidney calcification and the urinary excretion of N-acetyl-β-glucosaminidase (EC 3.2.1.30). *Br. J. Nutr.* **41**, 65–71.

Kasaoka, S., Yamagishi, H., and Kitano, T. (1999). Differences in the effect of iron-deficient diet on tissue weight, hemoglobin concentration and serum triglycerides in Fischer-344, Sprague-Dawley and Wistar rats. *J. Nutr. Sci. Vitaminol. (Tokyo)* **45**, 359–366.

Kato, T., Read, R., Rozga, J., and Burk, R.F. (1992). Evidence for intestinal release of absorbed selenium in a form with high hepatic extraction. *Am. J. Physiol.* **262**, G854–G858.

Kaup, S.M., Behling, A.R., and Greger, J.L. (1991a). Sodium, potassium and chloride utilization by rats given various inorganic anions. *Br. J. Nutr.* **66**, 523–532.

Kaup, S.M., Greger, J.L., Marcus, M.S., and Lewis, N.M. (1991b). Blood pressure, fluid compartments and utilization of chloride in rats fed various chloride diets. *J. Nutr.* **121**, 330–337.

Kawano, J., Ney, D.M., Keen, C.L., and Schneeman, B.O. (1987). Altered high density lipoprotein composition in manganese-deficient Sprague-Dawley and Wistar rats. *J. Nutr.* **117**, 902–906.

Kayne, L.H., and Lee, D.B. (1993). Intestinal magnesium absorption. *Miner. Electrolyte Metab.* **19**, 210–217.

Keating, J.N., Wada, L., Stokstad, E.L., and King, J.C. (1987). Folic acid: effect on zinc absorption in humans and in the rat. *Am. J. Clin. Nutr.* **46**, 835–839.

Keen, C.L., and Gershwin, M.E. (1990). Zinc deficiency and immune function. *Annu. Rev. Nutr.* **10**, 415–431.

Keen, C.L., and Hurley, L.S. (1989). Zinc and reproduction: effects of deficiency on foetal and postnatal development. *In* "Zinc in Human Biology. " (C.F. Mills, ed), p. 183. Springer-Verlag, New York.

Keenan, K.P., Ballam, G.C., Dixit, R., Soper, K.A., Laroque, P., Mattson, B.A., Adams, S.P., and Coleman, J.B. (1997). The effects of diet, overfeeding and moderate dietary restriction on Sprague-Dawley rat survival, disease and toxicology. *J. Nutr.* **127**, 851S–856S.

Keenan, K.P., Ballam, G.C., Soper, K.A., Laroque, P., Coleman, J.B., and Dixit, R. (1999). Diet, caloric restriction, and the rodent bioassay. *Toxicol. Sci.* **52**, 24–34.

Keenan, K.P., Smith, P.F., Hertzog, P., Soper, K., Ballam, G.C., and Clark, R.L. (1994). The effects of overfeeding and dietary restriction on Sprague-Dawley rat survival and early pathology biomarkers of aging. *Toxicol. Pathol.* **22**, 300–315.

Kelleher, S.L., and Lonnerdal, B. (2002). Zinc transporters in the rat mammary gland respond to marginal zinc and vitamin A intakes during lactation. *J. Nutr.* **132**, 3280–3285.

Keller, K.A., Grider, A., and Coffield, J.A. (2001). Age-dependent influence of dietary zinc restriction on short-term memory in male rats. *Physiol. Behav.* **72**, 339–348.

Kellogg, T.F., and Wostmann, B.S. (1969). Stock diet for colony production of germfree rats and mice. *Lab. Anim. Care* **19**, 812–814.

Kennedy, K.J., Rains, T.M., and Shay, N.F. (1998). Zinc deficiency changes preferred macronutrient intake in subpopulations of Sprague-Dawley outbred rats and reduces hepatic pyruvate kinase gene expression. *J. Nutr.* **128**, 43–49.

Kennedy, T.S. (1965). Studies on the nutritional value of foods treated with beta-radiation II: Effects on the protein in some animal feeds, egg and wheat. *J. Sci. Food Agric.* **16**, 433–437.

Kenney, M.A., McCoy, H., and Williams, L. (1994). Effects of magnesium deficiency on strength, mass, and composition of rat femur. *Calcif. Tissue Int.* **54**, 44–49.

Keymer, A., Crompton, D.W., and Singhvi, A. (1983). Mannose and the 'crowding effect' of Hymenolepis in rats. *Int. J. Parasitol.* **13**, 561–570.

Kh, R., Khullar, M., Kashyap, M., Pandhi, P., and Uppal, R. (2000). Effect of oral magnesium supplementation on blood pressure, platelet aggregation and calcium handling in deoxycorticosterone acetate induced hypertension in rats. *J. Hypertens.* **18**, 919–926.

Khan, M.A., and Munira, B. (1978). Biological utilization of protein as influenced by dietary carbohydrates. *Acta Agr. Scand.* **28**, 282–284.

Khan, M.F., Wu, X., Tipnis, U.R., Ansari, G.A., and Boor, P.J. (2002). Protein adducts of malondialdehyde and 4-hydroxynonenal in livers of iron loaded rats: quantitation and localization. *Toxicology* **173**, 193–201.

Kikuchi, G., Kumar, A., Talmage, P., and Shemin, D. (1958). The enzymatic synthesis of g-aminolevulinic acid. *J. Biol. Chem.* **233**, 1214–1219.

Kim, D.S., Kim, T.W., Park, I.K., Kang, J.S., and Om, A.S. (2002). Effects of chromium picolinate supplementation on insulin sensitivity, serum lipids, and body weight in dexamethasone-treated rats. *Metabolism* **51**, 589–594.

Kimura, M., Ujihara, M., and Yokoi, K. (1996). Tissue manganese levels and liver pyruvate carboxylase activity in magnesium-deficient rats. *Biol. Trace Elem. Res.* **52**, 171–179.

Kindberg, C.G., and Suttie, J.W. (1989). Effect of various intakes of phylloquinone on signs of vitamin K deficiency and serum and liver phylloquinone concentrations in the rat. *J. Nutr.* **119**, 175–180.

King, J.C., and Keen, C.L. (1999). Zinc. *In* "Modern Nutrition in Health and Disease." (M.E. Shils, J.A. Olson, M. Shike, and C.A. Ross, eds), p. 223. Williams and Wilkins, Baltimore.

Kirchgessner, M., Trubswetter, N., Stangl, G.I., and Roth-Maier, D.A. (1997). Dietary thiamin supply during gestation effects thiamin status of lactating rats and their suckling offspring. *Int. J. Vitam. Nutr. Res.* **67**, 248–254.

Kirkland, J.B., and Rawlings, J.M. (2001). Vitamin B6. *In* "Handbook of Vitamins." (R.B. Rucker, J.W. Suttie, D.B. McCormick, and L.J. Machlin, eds), p. 339. Marcel Dekker, New York.

Kleiber, M. (1975). Life as a combustion process. *In* "The Fire of Life: An Introduction to Animal Energetics," 2nd rev. ed. p. 3-8. R.E. Krieger Publishing Co., New York.

Klevay, L.M. (1976). The biotin requirement of rats fed 20% egg white. *J. Nutr.* **106**, 1643.

Klevay, L.M., and Saari, J.T. (1993). Comparative responses of rats to different copper intakes and modes of supplementation. *Proc. Soc. Exp. Biol. Med.* **203**, 214–220.

Kligler, D., and Krehl, W.A. (1952). Lysine deficiency in rats II: Studies with amino acid diets. *J. Nutr.* **46**, 61–74.

Klimis-Tavantzis, D.J., Leach, R.M. Jr, and Kris-Etherton, P.M. (1983). The effect of dietary manganese deficiency on cholesterol and lipid metabolism in the Wistar rat and in the genetically hypercholesterolemic RICO rat. *J. Nutr.* **113**, 328–336.

Knapka, J.J. (1983). Nutrition. *In* "The Mouse in Biomedical Research." (J.D.S.H.L. Foster and J.G. Fox, eds), pp. 52–67. Academic Press, New York.

Knapka, J.J., and Morin, L.M. (1979). Open formula natural ingredient diets for nonhuman primates. *In* "Primates in Nutritional Research." (K.C. Hayes, ed), p. 121. Academic Press, New York

Knapka, J.J., Smith, K.P., and Judge, F.J. (1974). Effect of open and closed formula rations on the performance of three strains of laboratory mice. *Lab. Anim. Sci.* **24**, 480–487.

Knutson, M.D., Walter, P.B., Ames, B.N., and Viteri, F.E. (2000). Both iron deficiency and daily iron supplements increase lipid peroxidation in rats. *J. Nutr.* **130**, 621–628.

Koh, E.T., Reiser, S., and Fields, M. (1989). Dietary fructose as compared to glucose and starch increases the calcium content of kidney of magnesium-deficient rats. *J. Nutr.* **119**, 1173–1178.

Koller, L.D., Mulhern, S.A., Frankel, N.C., Steven, M.G., and Williams, J.R. (1987). Immune dysfunction in rats fed a diet deficient in copper. *Am. J. Clin. Nutr.* **45**, 997–1006.

Konijn, A.M., Muogbo, D.N., and Guggenheim, K. (1970). Metabolic effects of carbohydrate-free diets. *Isr. J. Med. Sci.* **6**, 498–505.

Konishi, F., Oku, T., and Hosoya, N. (1984). Hypertrophic effect of unavailable carbohydrate on cecum and colon in rats. *J. Nutr. Sci. Vitaminol. (Tokyo)* **30**, 373–379.

Koski, K.G., and Hill, F.W. (1986a). Effect of low carbohydrate diets during pregnancy on parturition and postnatal survival of the newborn rat pup. *J. Nutr.* **116**, 1938–1948.

Koski, K.G., Hill, F.W., and Hurley, L.S. (1986b). Effect of low carbohydrate diets during pregnancy on embryogenesis and fetal growth and development in rats. *J. Nutr.* **116**, 1922–1937.

Koski, K.G., and Hill, F.W. (1990a). Evidence for a critical period during late gestation when maternal dietary carbohydrate is essential for survival of newborn rats. *J. Nutr.* **120**, 1016–1027.

Koski, K.G., Hill, F.W., and Lonnerdal, B. (1990b). Altered lactational performance in rats fed low carbohydrate diets and its effect on growth of neonatal rat pups. *J. Nutr.* **120**, 1028–1036.

Kotchen, T.A., Luke, R.G., Ott, C.E., Galla, J.H., and Whitescarver, S. (1983). Effect of chloride on renin and blood pressure responses to sodium chloride. *Ann. Intern. Med.* **98**, 817–822.

Kramer, T.R., Briske-Anderson, M., Johnson, S.B., and Holman, R.T. (1984). Effects of biotin deficiency on polyunsaturated fatty acid metabolism in rats. *J. Nutr.* **114**, 2047–2052.

Kramer, T.R., Johnson, W.T., and Briske-Anderson, M. (1988). Influence of iron and the sex of rats on hematological, biochemical and immunological changes during copper deficiency. *J. Nutr.* **118**, 214–221.

Kraus, A., Roth, H.P., and Kirchgessner, M. (1997). Supplementation with vitamin C, vitamin E or beta-carotene influences osmotic fragility and oxidative damage of erythrocytes of zinc-deficient rats. *J. Nutr.* **127**, 1290–1296.

Krehl, W.A., Sarma, P.S., and Teply, L.I. (1946). Factors affecting the dietary niacin and tryptophane requirement of the growing rat. *J. Nutr.* **31**, 85.

Krupp, P.P., and Lee, K.P. (1988). The effects of dietary iodine on thyroid ultrastructure. *Tissue Cell* **20**, 79–88.

Kudo, N., Nakagawa, Y., and Waku, K. (1990). Effects of zinc deficiency on the fatty acid composition and metabolism in rats fed a fat-free diet. *Biol. Trace Elem. Res.* **24**, 49–60.

Kulkarni, A.B., and Gaitonde, B.B. (1983). Effects of early thiamin deficiency and subsequent rehabilitation on the cholinergic system in developing rat brain. *J. Nutr. Sci. Vitaminol. (Tokyo)* **29**, 217–225.

Kumar, B.P., and Shivakumar, K. (1997). Depressed antioxidant defense in rat heart in experimental magnesium deficiency. Implications for the pathogenesis of myocardial lesions. *Biol. Trace Elem. Res.* **60**, 139–144.

Kunkel, M.E., Powers, D.L., and Hord, N.G. (1990). Comparison of chemical, histomorphometric, and absorptiometric analyses of bones of growing rats subjected to dietary calcium stress. *J. Am. Coll. Nutr.* **9**, 633–640.

Kuratko, C.N. (1997). Increasing dietary lipid and iron content decreases manganese superoxide dismutase activity in colonic mucosa. *Nutr. Cancer* **28**, 36–40.

Kurtz, D.J., Levy, H., and Kanfer, J.N. (1972). Cerebral lipids and amino acids in the vitamin B6-deficient suckling rat. *J. Nutr.* **102**, 291–298.

Kurtz, P.J., Emmerling, D.C., and Donofrio, D.J. (1984). Subchronic toxicity of all-trans-retinoic acid and retinylidene dimedone in Sprague-Dawley rats. *Toxicology* **.30**, 115–124.

Kwiecinski, G.G., Petrie, G.I., and DeLuca, H.F. (1989). Vitamin D is necessary for reproductive functions of the male rat. *J. Nutr.* **119**, 741–744.

L'Abbe, M.R., Fischer, P.W.F., et al. (1991). Dietary Se and tumor glutathione peroxidase and superoxide dismutase activity. *J. Nutr. Biochem.* **2**, 430.

Lardy, H.A., and Adler, J. (1956). Synthesis of succinate from propionate and bicarbonate by soluble enzymes from liver mitochondria. *J. Biol. Chem.* **219**, 933–942.

Lau-Cam, C.A., Thadikonda, K.P., and Kendall, B.F. (1991). Stimulation of rat liver glycogenolysis by vitamin B6: a role for adrenal catecholamines. *Res Commun. Chem. Pathol. Pharmacol.* **73**, 197–207.

Laurant, P., Dalle, M., Berthelot, A., and Rayssiguier, Y. (1999). Time-course of the change in blood pressure level in magnesium-deficient Wistar rats. *Br. J. Nutr.* **82**, 243–251.

Leclerc, J. (1991). Study of vitamin B1 nutrition in pregnant rat and its litter as a function of dietary thiamin supply. *Int. J. Vitamin Nutr. Res.* **57**, 45.

Lederer, H., Kumar, M., and Axelrod, A.E. (1975). Effects of pantothenic acid deficiency on cellular antibody synthesis in rats. *J. Nutr.* **105**, 17.

Lee D.B, Walling M.W., Gafter U., Silis V., and Coburn J.W. (1980). Calcium and inorganic phosphate transport in rat colon: dissociated response to 1,25-dihydroxyvitamin D3. *J. Clin. Invest.* **65**, 1326–1331.

Lee, S., Prosky, L., and DeVries, J. (1992). Determination of total, soluble, and insoluble dietary fiber in foods. Enzymatic-gravimetric method, MEW-TRIS buffer: collaborative study. *J. Assoc. Off. Anal. Chem.* **75**, 395–416.

Leedle, R.A., and Aust, S.D. (1990). The effect of glutathione on the vitamin E requirement for inhibition of liver microsomal lipid peroxidation. *Lipids* **25**, 241–245.

Leelaprute, V., Boonpucknavig, V., Bhamarapravati, N., and Weerapradist, W. (1973). Hypervitaminosis A in rats: varying responses due to different forms, doses, and routes of administration. *Arch. Pathol.* **96**, 5–9.

Lei, K.Y. (1991). Dietary copper: cholesterol and lipoprotein metabolism. *Annu. Rev. Nutr.* **11**, 265–283.

Leon-Del-Rio, A., Velazquez, A., Vizcaino, G., Robles-Diaz, G., and Gonzalez-Noriega, A. (1990). Association of pancreatic biotinidase activity and intestinal uptake of biotin and biocytin in hamster and rat. *Ann. Nutr. Metab.* **34**, 266–272.

Leong, W.I., Bowlus, C.L., Tallkvist, J., and Lonnerdal, B. (2003). Iron supplementation during infancy: effects on expression of iron transporters, iron absorption, and iron utilization in rat pups. *Am. J. Clin. Nutr.* **78**, 1203–1211.

Leshem, M. (1999). The ontogeny of salt hunger in the rat. *Neurosci. Biobehav. Rev.* **23**, 649–659.

Levander, O.A., Welsh, S.O., and Morris, V.C. (1980). Erythrocyte deformability as affected by vitamin E deficiency and lead toxicity. *Ann. NY Acad. Sci.* **355**, 227–239.

Levine, H., Remington, R.E., and von Klinitz, H. (1933). Studies on the relation to goiter, II.: the iodine requirement of the rat. *J. Nutr.* **6**, 347.

Levrat, M.A., Behr, S.R., Remesy, C., and Demigne, C. (1991). Effects of soybean fiber on cecal digestion in rats previously adapted to a fiber-free diet. *J. Nutr.* **121**, 672–678.

Lewis, R.M., Forhead, A.J., Petry, C.J., Ozanne, S.E., and Hales, C.N. (2002). Long-term programming of blood pressure by maternal dietary iron restriction in the rat. *Br. J. Nutr.* **88**, 283–290.

Lewis, R.M., Petry, C.J., Ozanne, S.E., and Hales, C.N. (2001). Effects of maternal iron restriction in the rat on blood pressure, glucose tolerance, and serum lipids in the 3-month-old offspring. *Metabolism* **50**, 562–567.

Lewis, S.M., Johnson, Z.J., Mayhugh, M.A., and Duffy, P.H. (2003). Nutrient intake and growth characteristics of male Sprague-Dawley rats fed AIN-93M purified diet or NIH-31 natural ingredient diet in a chronic two-year study. *Aging Clin. Exp. Res.* **15**, 460–468.

Lewis, S.M., Leard, B.L., Turtorro, A., and Hart, R.W. (1999). Long-term housing of rodents under specific pathogen-free barrier conditions. *In* "Methods in Aging Research." (B.P. Yu, ed) pp. 217–235. CRC Press, Boca Raton, FL.

Ley, F.J. (1975). Radiation sterilization: an industrial process. *In* "Radiation Research: Biomedical, Chemical and Physical Perspectives, Proceedings of the 5th International Congress of Radiation Research." Academic Press, New York.

Li, M., and Boyages, S.C. (1994). Iodide induced lymphocytic thyroiditis in the BB/W rat: evidence of direct toxic effects of iodide on thyroid subcellular structure. *Autoimmunity* **18**, 31–40.

Li, Q., Guo-Ross, S., Lewis, D.V., Turner, D., White, A.M., Wilson, W.A., and Swartzwelder, H.S. (2004). Dietary prenatal choline supplementation alters postnatal hippocampal structure and function. *J. Neurophysiol.* **91**, 1545–1555.

Liao, F., and Huang, P.C. (1987). Effects of moderate riboflavin deficiency on lipid metabolism in rats. *Proc. Natl. Sci. Counc. Repub. China B* **11**, 128–132.

Likins, R.C., L. A. Bavetta, L.A., and Posner, A.S. (1957). Calcification in lysine deficiency. *Arch. Biochem. Biophys.* **70**, 401–412.

Lisle, S.J., Lewis, R.M., Petry, C.J., Ozanne, S.E., Hales, C.N., and Forhead, A.J. (2003). Effect of maternal iron restriction during pregnancy on renal morphology in the adult rat offspring. *Br. J. Nutr.* **90**, 33–39.

Lloyd, L.E., McDonald, B.E., and Crampton, E.W. (1978). Energy requirements of the body. *In* "Fundamentals of Nutrition." pp. 396–438. W.H. Freeman and Co., San Francisco.

Locker, J., Reddy, T.V., and Lombardi, B. (1986). DNA methylation and hepatocarcinogenesis in rats fed a choline-devoid diet. *Carcinogenesis* **7**, 1309–1312.

Lominadze, D., Saari, J.T., Percival, S.S., and Schuschke, D.A. (2004). Proinflammatory effects of copper deficiency on neutrophils and lung endothelial cells. *Immunol. Cell Biol.* **82**, 231–238.

Lominadze, D.G., Saari, J.T., Miller, F.N., Catalfamo, J.L., Justus, D.E., and Schuschke, D.A. (1996). Platelet aggregation and adhesion during dietary copper deficiency in rats. *Thromb. Haemost.* **75**, 630–634.

Loo, G., and Smith, J.T. (1986). Effect of pyridoxine deficiency on phospholipid methylation in rat liver microsomes. *Lipids* **21**, 409–412.

Lopez-Guisa, J.M., Harned, M.C., Dubielzig, R., Rao, S.C., and Marlett, J.A. (1988). Processed oat hulls as potential dietary fiber sources in rats. *J. Nutr.* **118**, 953–962.

Lotan, R. (1995). Retinoids and apoptosis: implications for cancer chemoprevention and therapy. *J. Natl. Cancer Inst.* **87**, 1655–1657.

Liu, Z., Tomotake, H., Wan, G., Watanabe, H., and Kato, N. (2001). Combined effect of dietary calcium and iron on colonic aberrant crypt cell foci, cell proliferation and apoptosis, and fecal bile acids in 1,2-dimethylhydrazine-treated rats. *Oncol. Rep.* **8**, 893–897.

Lumment, L., and Li, T.K. (1980). Mammalian vitamin B6 metabolism: regulatory role of protein binding and the hydrolysis of pyridoxal 5′-phosphate in storage and transport. *In* "Vitamin B_6 Metabolism and Role in Growth. " (G.P. Tryflates, ed), p. 27. Food and Nutrition Press, Westport, CT.

Lupton, J.R., Chen, X.Q., Frolich, W., Schoeffler, G.L., and Peterson, M.L. (1994). Rats fed high fat diets with increased calcium levels have fecal bile acid concentrations similar to those of rats fed a low fat diet. *J. Nutr.* **124**, 188–195.

Lupton, J.R., Steinbach, G., Chang, W.C., O'Brien, B.C., Wiese, S., Stoltzfus, C.L., Glober, G.A., Wargovich, M.J., McPherson, R.S., and Winn. R.J. (1996). Calcium supplementation modifies the relative amounts of bile acids in bile and affects key aspects of human colon physiology. *J. Nutr.* **126**, 1421–1428.

Lyman, R.L., and Wilcox, S.S. (1963). Effect of acute amino acid deficiencies on carcass composition and pancreatic function in the force-fed rat, II: deficiencies of valine, lysine, tryptophan, leucine and isoleucine. *J. Nutr.* **79**, 37–44.

MacEvilly, C.J., and Muller, D.P. (1996). Lipid peroxidation in neural tissues and fractions from vitamin E-deficient rats. *Free Radic. Biol. Med.* **20**, 639–648.

MacFarlane, W.V., and Howard, B. (1972). Comparative water and energy economy of wild and domestic mammals. *Symp. Zool. Soc. Lond.* **31**, 261–296.

Machlin, L.J., Filipski, R., Nelson, J., Horn, L.R., and Brin, M. (1977). Effects of a prolonged vitamin E deficiency in the rat. *J. Nutr.* **107**, 1200–1208.

Machlin, L.J., and Gabriel, E. (1982). Kinetics of tissue α-tocopherol uptake and depletion following administration of high levels of vitamin E. *Ann. NY Acad. Sci.* **393**, 48–60.

Mackerer, C.R., Mehlman, M.A., and Tobin, R.B. (1973). Effects of chronic acetylsalicylate administration on several nutritional and biochemical parameters in rats fed diets of varied thiamin content. *Biochem. Med.* **8**, 51–60.

Mackler, B., Person, R., Ochs, H., and Finch, C.A. (1984). Iron deficiency in the rat: effects on neutrophil activation and metabolism. *Pediatr. Res.* **18**, 549–551.

Maeda, H., Gleiser, C.A., Masoro, E.J., Murata, I., McMahan, C.A., and Yu, B.P. (1985). Nutritional influences on aging of Fischer 344 rats, II: pathology. *J. Gerontol.* **40**, 671–688.

Maeda, N., Takahashi, K., Aono, K., and Shiga, T. (1976). Effect of pyridoxal 5'-phosphate on the oxygen affinity of human erythrocytes. *Br. J. Haematol.* **34**, 501–509.

Mair, R.G., Anderson, C.D., Langlais, P.J., and McEntee, W.J. (1985). Thiamine deficiency depletes cortical norepinephrine and impairs learning processes in the rat. *Brain Res.* **360**, .273–284.

Malecki, E.A., and Greger, J.L. (1996). Manganese protects against heart mitochondrial lipid peroxidation in rats fed high levels of polyunsaturated fatty acids. *J. Nutr.* **126**, 27–33.

Malecki, E.A., Huttner, D.L., and Greger, J.L. (1994). Manganese status, gut endogenous losses of manganese, and antioxidant enzyme activity in rats fed varying levels of manganese and fat. *Biol. Trace Elem. Res.* **42**, 17–29.

Mallon, J.P., Matuszewski, D., and Sheppard, H. (1980). Binding specificity of the rat serum vitamin D transport protein. *J. Steriod Biochem.* **13**, 409–413.

Mameesh, M.S., and Johnson, B.C. (1960). Dietary vitamin K requirement of the rat. *Proc. Soc. Exp. Biol. Med.* **103**, 378–380.

Mangian, H.F., Lee, R.G., Paul, G.L., Emmert, J.L., and Shay, N.F. (1998). Zinc deficiency supresses plasma leptin concentration in rats. *J. Nutr. Biochem.* **9**, 47.

Marasas, W.F., Kriek, N.P., Fincham, J.E., and van Rensburg, S.J. (1984). Primary liver cancer and oesophageal basal cell hyperplasia in rats caused by *Fusarium moniliforme. Int. J. Cancer* **34**, 383–387.

Marra, C.A., and de Alaniz, M.J. (2000). Calcium deficiency modifies polyunsaturated fatty acid metabolism in growing rats. *Lipids* **35**, 983–990.

Marra, C.A., Rimoldi, O., and de Alaniz, M.J. (2002). Correlation between fatty acyl composition in neutral and polar lipids and enzyme activities from various tissues of calcium-deficient rats. *Lipids* **37**, 701–714.

Masoro, E.J. (1992). Aging and proliferative homeostasis: modulation by food restriction in rodents. *Lab. Anim. Sci.* **42**, 132–137.

Masoro, E.J. (1996). Possible mechanisms underlying the antiaging actions of caloric restriction. *Toxicol. Pathol.* **24**, 738–741.

Masoro, E.J. (2000). Caloric restriction and aging: an update. *Exp. Gerontol.* **35**, 299–305.

Masoro, E.J., Iwasaki, K., Gleiser, C.A., McMahan, C.A., Seo, E.J., and Yu, B.P. (1989). Dietary modulation of the progression of nephropathy in aging rats: an evaluation of the importance of protein. *Am. J. Clin. Nutr.* **49**, 1217–1227.

Masoro, E.J., and Yu, B.P. (1989). Diet and nephropathy. *Lab. Invest.* **60**, 165–167.

Masoro, E.J., Katz, M.S., and McMahan, C.A. (1989). Evidence for the glycation hypothesis of aging from the food-restricted rodent model. *J. Gerontol.* **44**, B20–B22.

Masoro, E.J., Shimokawa, I., Higami, Y., McMahan, C.A., and Yu, B.P. (1995). Temporal pattern of food intake not a factor in the retardation of aging processes by dietary restriction. *J. Gerontol. A Biol. Sci. Med. Sci.* **50A**, B48–B53.

Mathers, J.C., Fernandez, F., Hill, M.J., McCarthy, P.T., Shearer, M.J., and Oxley, A. (1990). Dietary modification of potential vitamin K supply from enteric bacterial menaquinones in rats. *Br. J. Nutr.* **63**, 639–652.

Mathews, C.H., Brommage, R., and DeLuca, H.F. (1986). Role of vitamin D in neonatal skeletal development in rats. *Am. J. Physiol.* **250**, E725–E230.

Matschiner, J.T., and Bell, R.G. (1973). Effect of sex and sex hormones on plasma prothrombin and vitamin K deficiency. *Proc. Soc. Exp. Biol. Med.* **144**, 316–320.

Matschiner, J.T., and Doisy, E.A. Jr. (1962). Role of vitamin A in induction of vitamin K deficiency in the rat. *Proc. Soc. Exp. Biol. Med.* **109**, 139–42.

Matschiner, J.T., and Willingham, A.K. (1974). Influence of sex hormones on vitamin K deficiency and epoxidation of vitamin K in the rat. *J. Nutr.* **104**, 660–665.

Matsuda, T., Doi, T., Tonomura, H., Baba, A., and Iwata, H. (1989). Postnatal development of thiamine metabolism in rat brain. *J. Neurochem.* **52**, 842–846.

Matsuzaki, H., Kikuchi, T., Kajita, Y., Masuyama, R., Uehara, M., Goto, S., and Suzuki, K. (1999). Comparison of various phosphate salts as the dietary phosphorus source on nephrocalcinosis and kidney function in rats. *J. Nutr. Sci. Vitaminol. (Tokyo)* **45**, 595–608.

Matsuzaki, H., Masuyama, R., Uehara, M., Nakamura, K., and Suzuki, K. (2001). Greater effect of dietary potassium tripolyphosphate than of potassium dihydrogenphosphate on the nephrocalcinosis and proximal tubular function in female rats from the intake of a high-phosphorus diet. *Biosci. Biotechnol. Biochem.* **65**, 928–934.

Maxcy, R.B. (1983). Significance of residual organisms in foods after substerilizing doses of gamma radiation: a review. *J. Food Safety* **5**, 203–211.

Mayes, P.A. (1996a). Nutrition. *In* "Harper's Biochemistry." (R.K. Murray, D.K. Granner, P.A. Mayes, and V.W. Rodwell, eds). pp. 625–634. Appleton & Lange, Stamford, CT.

Mayes, P.A. (1996b). Lipids of physiologic significance. *In* "Harper's Biochemistry." (R.K. Murray, D.K. Granner, P.A. Mayes, and V.W. Rodwell, eds). pp. 146–157. Appleton & Lange, Stamford, CT.

Maynard, L.A., Loosli, J.K., Fintz, H.F., and Warner, R.S. (1979). "Animal Nutrition," 7th. ed. McGraw-Hill, New York.

McCarter, R., Masoro, E.J., and Yu, B.P. (1985). Does food restriction retard aging by reducing the metabolic rate? *Am. J. Physiol.* **248**, E488–E490.

McCay, C., Crowell, M., and Maynard, L.A. (1935). The effect of retarded growth upon the length of the life span and upon the ultimate size. *J. Nutr.* **10**, 63–79.

McClain, C.J., Gavaler, J.S., and Van Thiel, D.H. (1984). Hypogonadism in the zinc-deficient rat: localization of the functional abnormalities. *J. Lab. Clin. Med.* **104**, 1007–1015.

McHenry, E.W., and Gauvin, G. (1938). The B vitamins and fat metabolism, 1: effects of thiamine, riboflavin and rice polish concentrate upon body fat. *J. Biol. Chem.* **125**, 653.

McKeever, M., Molloy, A., Young, P., Kennedy, S., Kennedy, D.G., Scott, J.M., and Weir, D.G. (1995a). An abnormal methylation ratio induces hypomethylation *in vitro* in the brain of pig and man, but not in rat. *Clin. Sci. (Lond)* **88**, 73–79.

McKeever, M., Molloy, A., Weir, D.G., Young, P.B., Kennedy, D.G., Kennedy, S., and Scott, J.M. (1995b). Demonstration of hypomethylation of proteins in the brain of pigs (but not in rats) associated with chronic vitamin B12 inactivation. *Clin. Sci. (Lond)* **88**, 471–477.

McLane, J.A., Khan, T., and Held, I.R. (1987). Increased axonal transport in peripheral nerves of thiamine-deficient rats. *Exp. Neurol.* **95**, 482–491.

McLaughlan, J.M., and Campbell, J.A. (1969). Methodology of protein evaluation. *In* "Mammalian Protein Metabolism." (H.N. Munro and J.B. Allison, eds), pp. 391–418. Academic Press, New York..

McNall, A.D., Etherton, T.D., and Fosmire, G.J. (1995). The impaired growth induced by zinc deficiency in rats is associated with decreased expression of the hepatic insulin-like growth factor I and growth hormone receptor genes. *J. Nutr.* **125**, 874–879.

Meck, W.H., Smith, R.A., and Williams, C.L. (1989). Organizational changes in cholinergic activity and enhanced visuospatial memory as a function of choline administered prenatally or postnatally or both. *Behav. Neurosci.* **103**, 1234–1241.

Meck, W.H., and Williams, C.L. (1997a). Characterization of the facilitative effects of perinatal choline supplementation on timing and temporal memory. *Neuroreport* **8**, 2831–2835.

Meck, W.H., and Williams, C.L. (1997b). Perinatal choline supplementation increases the threshold for chunking in spatial memory. *Neuroreport* **8**, 3053–3059.

Medeiros, D.M., and Beard, J.L. (1998). Dietary iron deficiency results in cardiac eccentric hypertrophy in rats. *Proc. Soc. Exp. Biol. Med.* **218**, 370–375.

Medeiros, D.M., Liao, Z., and Hamlin, R.L. (1991). Copper deficiency in a genetically hypertensive cardiomyopathic rat: electrocardiogram, functional and ultrastructural aspects. *J. Nutr.* **121**, 1026–1034.

Medeiros, D.M., Liao, Z., and Hamlin, R.L. (1992). Electrocardiographic activity and cardiac function in copper-restricted rats. *Proc. Soc. Exp. Biol. Med.* **200**, 78–84.

Medeiros, D.M., Plattner, A., Jennings, D., and Stoecker, B. (2002). Bone morphology, strength and density are compromised in iron-deficient rats and exacerbated by calcium restriction. *J. Nutr.* **132**, 3135–3141.

Mehansho, H., Hamm, M.W., and Henderson, L.M. (1979). Transport and metabolism of pyridoxal and pyridoxal phosphate in the small intestine of the rat. *J. Nutr.* **109**, 1542–1551.

Meister, A. (1957). "Biochemistry of the Amino Acids." Academic Press, New York.

Mercer, L.P., Dodds, S.J., et al. (1986). The determination of nutritional requirements: a modeling approach. *Nutr. Rep. Int.* **34**, 337.

Merrill, A.L., and Watt, B.K. (1955). Derivation of current calorie factors. Energy Value of Foods, Basis and Derivation, Handbook. No. 74, Agriculture Research Service, Washington, DC.

Merry, B.J., and Holehan, A.M. (1979). Onset of puberty and duration of fertility in rats fed a restricted diet. *J. Reprod. Fert.* **57**, 253–259.

Mertz, W. (1969). Chromium occurrence and function in biological systems. *Physiol. Rev.* **49**, 163–239.

Mertz, W., and Roginski, E.E. (1969). Effects of chromium 3E supplementation on growth and survival under stress in rats fed low protein diets. *J. Nutr.* **97**, 531–536.

Michaelis, O.E., Martin, R.E., Gardener, L.B., and Ellwood, K.C. (1981). Effect of simple and complex carbohydrate on lipogenic parameters of spontaneously hypertensive rats. *Nutr. Rep. Int.* **24**, 313–321.

Middleton, H.M. III. (1977). Uptake of pyridoxine hydrochloride by the rat jejunal mucosa *in vitro*. *J. Nutr.* **107**, 126–131.

Middleton, H.M. III. (1982). Characterization of pyridoxal 5′-phosphate disappearance from *in vivo* perfused segments of rat jejunum. *J. Nutr.* **112**, 269–275.

Middleton, H.M. III. (1985). Uptake of pyridoxine by *in vivo* perfused segments of rat small intestine: a possible role for intracellular vitamin metabolism. *J. Nutr.* **115**, 1079–1088.

Middleton, H.M. III. (1990). Intestinal hydrolysis of pyridoxal 5′-phosphate *in vitro* and *in vivo* in the rat. Effect of amino acids and oligopeptides. *Dig. Dis. Sci.* **35**, 113–120.

Miller, D.S., and O'Dell, B.L. (1987). Milk and casein-based diets for the study of brain catecholamines in copper-deficient rats. *J. Nutr.* **117**, 1890–1897.

Miller, J.W., Nadeau, M.R., Smith, J., Smith, D., and Selhub, J. (1994). Folate-deficiency-induced homocysteinaemia in rats: disruption of *S*-adenosylmethionine's co-ordinate regulation of homocysteine metabolism. *Biochem. J.* **298**, 415–419.

Miller, R.A. (1995). Immune system. *In* "Handbook of Physiology." (E.J. Masoro, ed), p. 555. Oxford University Press, New York.

Miller, S.A. (1969). Protein metabolism during growth and development. *In* "Mammalian Protein Metabolism." (H.N.M.a.J.B. Allison, ed), pp. 183–227. Academic Press, New York.

Milner, J.A., Wakeling, A.E., and Visek, W.J. (1974). Effect of arginine deficiency on growth and intermediary metabolism in rats. *J. Nutr.* **104**, 1681–1689.

Miret, S., McKie, A.T., Saiz, M.P., Bomford, A., and Mitjavila, M.T. (2003). IRP1 activity and expression are increased in the liver and the spleen of rats fed fish oil-rich diets and are related to oxidative stress. *J. Nutr.* **133**, 999–1003.

Miro, A., Costas, M.J., Garcia-Diaz, M., Hernandez, M.T., and Cameselle, J.C. (1989). A specific, low Km ADP-ribose pyrophosphatase from rat liver. *FEBS Lett.* **244**, 123–126.

Mitchell, J.H., Nicol, F., Beckett, G.J., and Arthur, J.R. (1996). Selenoenzyme expression in thyroid and liver of second generation selenium- and iodine-deficient rats. *J. Mol. Endocrinol.* **16**, 259–267.

Miyazawa, T., Sato, C., and Kaneda, T. (1983). Antioxidative effects of α-tocopherol and riboflavin-butyrate in rats dosed with methyl linoleate hydroperoxide. *Agric. Biol. Chem.* **47**, 1577.

Moak, M.A., and Christensen, M.J. (2001). Promotion of lipid oxidation by selenate and selenite and indicators of lipid peroxidation in the rat. *Biol. Trace Elem. Res.* **79**, 257–269.

Mock, D.M., Johnson, S.B., and Holman, R.T. (1988a). Effects of biotin deficiency on serum fatty acid composition: evidence for abnormalities in humans. *J. Nutr.* **118**, 342–348.

Mock, D.M. (1988b). Evidence for a pathogenetic role of fatty acid (FA) abnormalities in the cutaneous manisfestations of biotin deficiency. *FASEB J.* **2**, A1204.

Mock, D.M. (1999). Biotin. *In* "Modern Nutrition in Health and Disease." (M.E. Shils, J.A. Olson, M. Shike, and C.A. Ross, eds), p. 459. Williams and Wilkins, Baltimore.

Mock, D.M. (2001). Biotin. *In* "Handbook of Vitamins." (R.B. Rucker, J.W. Suttie, D.B. McCormick, and L.J. Machlin, eds), p. 397. Marcel Dekker, New York.

Moiseenok, A.G., Sheibak, V.M., and Gurinovich, V.A. (1987). Hepatic CoA, S-acyl-CoA, biosynthetic precursors of the coenzyme and pantothenate-protein complexes in dietary pantothenic acid deficiency. *Int. J. Vitam. Nutr. Res.* **57**, 71–77.

Molina, A., Oka, T., Munoz, S.M., Chikamori-Aoyama, M., Kuwahata, M., and Natori, Y. (1997). Modulation of albumin gene expression by amino acid supply in rat liver is mediated through intracellular concentration of pyridoxal 5′-phosphate. *J. Nutr. Biochem.* **8**, 211.

Molina, P.E., Myers, N., Smith, R.M., Lang, C.H., Yousef, K.A., Tepper, P.G., and Abumrad, N.N. (1994). Nutritional and metabolic characterization of a thiamine-deficient rat model. *JPEN J. Parenter. Enteral. Nutr.* **18**, 104–111.

Montoya, D., and Swartzwelder, H.S. (2000). Prenatal choline supplementation alters hippocampal *N*-methyl-D-aspartate receptor-mediated neurotransmission in adult rats. *Neurosci. Lett.* **296**, 85–88.

Mooij, P., de Wit, H.J., Bloot, A.M., Wilders-Truschnig, M.M., and Drexhage, H.A. (1993). Iodine deficiency induces thyroid autoimmune reactivity in Wistar rats. *Endocrinology* **133**, 1197–1204.

Mori, K., Kaido, M., Fujishiro, K., Inoue, N., and Koide, O. (1992). Effects of megadoses of pyridoxine on spermatogenesis and male reproductive organs in rats. *Arch. Toxicol.* **66**, 198–203.

Mori, N., and Hirayama, K. (2000). Long-term consumption of a methionine-supplemented diet increases iron and lipid peroxide levels in rat liver. *J. Nutr.* **130**, 2349–2355.

Moriarty, P., Picciano, M., Beard, J., et al. (1993). Iron deficiency decreases Se-GPX mRNA level in the liver and impairs selenium utilization in other tissues. *FASEB J.* **7**, A277.

Morozumi, M., and Ogawa, Y. (2000). Impact of dietary calcium and oxalate ratio on urinary stone formation in rats. *Mol. Urol.* **4**, 313–320.

Morre, D.M., Kirksey, A., and Das, G.D. (1978). Effects of vitamin B-6 deficiency on the developing central nervous system of the rat. Myelination. *J. Nutr.* **108**, 1260–1265.

Morriss-Kay, G.M., and Sokolova, N. (1996). Embryonic development and pattern formation. *FASEB J.* **10**, 961–968.

Muldoon, T.G., and Cidlowski, J.A. (1980). Specific modification of rat uterine estrogen receptor by pyridoxal 5′-phosphate. *J. Biol. Chem.* **255**, 3100–3107.

Mumby, D.G., Mana, M.J., Pinel, J.P., David, E., and Banks, K. (1995). Pyrithiamine-induced thiamine deficiency impairs object recognition in rats. *Behav. Neurosci.* **109**, 1209–1214.

Nalecz, K.A., Nalecz, M.J., and Azzi, A. (1992). Isolation of tocopherol-binding proteins from the cytosol of smooth muscle A7r5 cells. *Eur. J. Biochem.* **209**, 37–42.

Nara, Y., Ikeda, K., Nabika, T., Sawamura, M., Mano, M., Endo, J., and Yamori, Y. (1994). Comparison of salt sensitivity of male and female F2 progeny from crosses between WKY and SHRSP rats. *Clin. Exp. Pharmacol. Physiol.* **21**, 899–902.

Narayan, K.A., and McMullen, J.J. (1980). Accelerated induction of fatty livers in rats fed fat-free diets containing sucrose or glycerol. *Nutr. Rep. Int.* **21**, 689–697.

National Research Council [NRC]. (1971). "Atlas of Nutritional Data on United States and Canadian Feed." National Academy Press, Washington, DC.

National Research Council [NRC]. (1974). "Nutrients and Toxic Substances in Water for Livestock and Poultry." National Academy Press, Washington, DC.

National Research Council [NRC]. (1978). "Nutrient Requirements of Laboratory Animals," 3rd ed. National Academy Press, Washington, DC.

National Research Council [NRC]. (1978). "Control of Diets in Laboratory Animal Experiments." National Academy Press, Washington, DC.

National Research Council [NRC]. (1981). "Nutritional Energetics of Domestic Animals and Glossary of Energy Terms," 2nd ed. National Academy Press, Washington, DC.

National Research Council [NRC]. (1987). "Vitamin Tolerance of Animals. "National Academy Press, Washington, DC.

National Research Council [NRC]. (1989). "Recommended Dietary Allowances," 10th ed. National Academy Press, Washington, DC.

National Research Council [NRC]. (1995). "Nutrient Requirements of the Laboratory Rat: Nutrient Requirements of Laboratory Animals," 4th ed. National Academy Press, Washington, DC.

National Research Council [NRC]. (1996a). "Carcinogens and Anticarcinogens in the Human Diet: A Comparison of Naturally Occurring and Synthetic Substances." National Academy Press, Washington, DC.

National Research Council [NRC]. (1996b). "Institute of Laboratory Animal Resources, Committee on Rodents, Laboratory Animal Management. National Academy Press, Washington, DC.

Nauss, K.M., and Newberne, P.M. (1985). Local and regional immune function of vitamin A-deficient rats with ocular herpes simplex virus (HSV) infections. *J. Nutr.* **115**, 1316–1324.

Naveh-Many, T., Raue, F., Grauer, A., and Silver, J. (1992). Regulation of calcitonin gene expression by hypocalcemia, hypercalcemia, and vitamin D in the rat. *J. Bone Miner. Res.* **7**, 1233–1237.

Navia, J.M. (1977). "Animal Models in Dental Research." University of Alabama Press, Birminham.

Nelson, C., Erikson, K., Pinero, D.J., and Beard, J.L. (1997). In vivo dopamine metabolism is altered in iron-deficient anemic rats. *J. Nutr.* **127**, 2282–2288.

Nelson, M.M., and Evans, H.M. (1961). Dietary requirements for lactation in rats and other laboratory animals. *In* "Milk: The Mammary Gland and Its Secretion," Vol 2. (S.K. Kon and A.T. Cowie, eds). pp. 137–191. Academic Press, New York.

Newberne, P.M. (1974). The influence of nutrition on response to infectious disease. *Adv. Vet. Sci. Comp. Med.* **17**, 265–289.

Newberne, P.M. (1975). Influence of pharmacological experiments of chemicals and other factors in diets of laboratory animals. *Fed. Proc.* **34**, 209–218.

Newberne, P.M., and Rogers, A.E. (1986). Labile methyl groups and the promotion of cancer. *Annu. Rev. Nutr.* **6**, 407–432.

Nguyen, L.B., Gregory, J.F. 3rd, and Cerda, J.J. (1983). Effect of dietary fiber on absorption of B-6 vitamers in a rat jejunal perfusion study. *Proc. Soc. Exp. Biol. Med.* **173**, 568–573.

Nielsen, F.H. (1989). New essential trace elements for the life sciences. *Biol. Trace Elem. Res.* **26–27**, 599.

Nielsen, F.H. (1991). Nutritional requirements for boron, silicon, vanadium, nickel, and arsenic: current knowledge and speculation. *FASEB J.* **5**, 2661–2667.

Nielsen, F.H. (1993). Ultratrace elements of possible importance in human health: an update. *In* "Essential and Toxic Trace Elements in Human Health." (A.S. Prasad, ed), p. 355. Wiley-Liss, New York.

Nielsen, F.H. (1994). Biochemical and physiologic consequences of boron deprivation in humans. *Environ. Health Perspect.* **102**, 59–63.

Nielsen, F.H. (1999). Ultratrace minerals: manganese. *In* "Modern Nutrition in Health and Disease." (M.E. Shils, J.A. Olson, M. Shike, and C.A. Ross, eds), p. 289. Williams and Wilkins, Baltimore.

Nielsen, F.H., Hunt, C.D., and Uthus, E.O. (1980). Interactions between essential trace and ultratrace elements. *Ann. NY Acad. Sci.* **355**, 152–164.

Nielsen, F.H., and Shuler, T.R. (1992). Studies of the interaction between boron and calcium, and its modification by magnesium and potassium, in rats: effects on growth, blood variables, and bone mineral composition. *Biol. Trace Elem. Res.* **35**, 225–237.

Nielsen, F.H., Shuler, T.R., Zimmerman, T.J., and Uthus, E.O. (1988). Magnesium and methionine deprivation affect the response of rats to boron deprivation. *Biol. Trace Elem. Res.* **17**, 91–107.

Nielsen, F.H., Uthus, E.O., Poellot, R.A., and Shuler, T.R. (1993). Dietary vitamin B_{12}, sulfur amino acids, and odd-chain fatty acids affect the responses of rats to nickel deprivation. *Biol. Trace Elem. Res.* **37**, 1–15.

Nielsen, F.H., Zimmerman, T.J., Shuler, T.R., Brossart, B., and Uthus, E.O. (1989). Evidence for a cooperative metabolic relationship between nickel and Vitamin B_{12} in rats. *J. Trace Elem. Res. Exp. Med.* **2**, 21.

Ninh, N.X., Thissen, J.P., Maiter, D., Adam, E., Mulumba, N., and Ketelslegers, J.M. (1995). Reduced liver insulin-like growth factor-I gene expression in young zinc-deprived rats is associated with a decrease in liver growth hormone (GH) receptors and serum GH-binding protein. *J. Endocrinol.* **144**, 449–456.

Nishikawa, Y., Tatsumi, K., Matsuura, T., Yamamoto, A., Nadamoto, T., and Urabe, K. (2003). Effects of vitamin C on high blood pressure induced by salt in spontaneously hypertensive rats. *J. Nutr. Sci. Vitaminol. (Tokyo)* **49**, 301–309.

Nishina, P.M., Schneeman, B.O., and Freedland, R.A. (1991). Effects of dietary fibers on nonfasting plasma lipoprotein and apolipoprotein levels in rats. *J. Nutr.* **121**, 431–437.

Noh, S.K., and Koo, S.I. (2001). Feeding of a low-zinc diet lowers the tissue concentrations of α-tocopherol in adult rats. *Biol. Trace Elem. Res.* **81**, 153–168.

Norman, A.W., Bouillon, R., et al. (1994). Vitamin D, a pluripotent steroid hormone: structural studies. *In* "Molecular Endocrinology and Clinical Applications," 9th ed. Walter deGruyter, Berlin.

Norman, A.W., and Deluca, H.F. (1963). The preparation of H3—vitamins D2 and D3—their localization in the rat. *Biochemistry* **13**, 1160–1168.

Norohna, J.M., and Sibverman, M. (1962). On folic acid, vitamin B12, methionine and formiminoglutamic acid metabolism. *In* "Vitamin B12 und Intrinsic Faktor." (H.C.F. Heinrich, ed), p. 728. Stuttgart, Enka, Stuttgart.

Norred, W.P., Plattner, R.D., and Chamberlain, W.J. (1993). Distribution and excretion of [^{14}C]fumonisin B1 in male Sprague-Dawley rats. *Nat. Toxins* **1**, 341–346.

O'Dell, B.L. (1990). Copper. *In* "Present Knowledge of Nutrition." (M.L. Brown, ed), p. 261. ILSI Press, Washington, DC.

O'Dell, B.L. (2000). Role of zinc in plasma membrane function. *J. Nutr.* **130**, 1432S–1436S.

O'Dell, B.L., and Prohaska, J.R. (1983). Biochemical aspects of copper deficiency in the nervous system. *In* "Neurobiology of the Trace Elements." (I.E. Dreosti and R.M. Smith, eds), p. 41. Humana Press, Clifton, NJ.

Odum, J., Tinwell, H., Jones, K., Van Miller, J.P., Joiner, R.L., Tobin, G., Kawasaki, H., Deghenghi, R., and Ashby, J. (2001). Effect of rodent diets on the sexual development of the rat. *Toxicol. Sci.* **61**, 115–127.

Offenbacher, E.G., Pi-Sunyer, F.X., et al. (1997). Chromium. *In* "Handbook of Nutritionally Essential Mineral Elements." (B.L. O'Dell and R.A. Sunde, eds), p. 389. Marcel Dekker, New York.

Ogihara, T., Asano, T., Ando, K., Sakoda, H., Anai, M., Shojima, N., Ono, H., Onishi, Y., Fujishiro, M., Abe, M., Fukushima, Y., Kikuchi, M., and Fujita, T. (2002). High-salt diet enhances insulin signaling and induces insulin resistance in Dahl salt-sensitive rats. *Hypertension* **40**, 83–89.

Oh, M.S., and Uribarri, J. (1999). Electrolytes, water, and acid-base balance. *In* "Modern Nutrition in Health and Disease." (M.E. Shils, J.A. Olson, M. Shike, and C.A. Ross, eds), p. 241. Williams and Wilkins, Baltimore, MD.

Ohshima, M., and Ward, J.M (1986). Dietary iodine deficiency as a tumor promoter and carcinogen in male F344/NCr rats. *Cancer Res.* **46**, 877–883.

Oka, T., Komori, N., Kuwahata, M., Hiroi, Y., Shimoda, T., Okada, M., and Natori, Y. (1995a). Pyridoxal 5′-phosphate modulates expression of cytosolic aspartate aminotransferase gene by inactivation of glucocorticoid receptor. *J. Nutr. Sci. Vitaminol. (Tokyo)* **41**, 363–375.

Oka, T., Komori, N., Kuwahata, M., Okada, M., and Natori, Y. (1995b). Vitamin B6 modulates expression of albumin gene by inactivating tissue-specific DNA-binding protein in rat liver. *Biochem. J.* **309 (Pt 1)**, 243–248.

Olin, K.L., Walter, R.M., and Keen, C.L. (1994). Copper deficiency affects selenoglutathione peroxidase and selenodeiodinase activities and antioxidant defense in weanling rats. *Am. J. Clin. Nutr.* **59**, 654–658.

Olpin, S.E., and Bates, C.J. (1982a). Lipid metabolism in riboflavin-deficient rats, 1: effect of dietary lipids on riboflavin status and fatty acid profiles. *Br. J. Nutr.* **47**, 577–588.

Olpin, S.E., and Bates, C.J. (1982b). Lipid metabolism in riboflavin-deficient rats, 2: mitochondrial fatty acid oxidation and the microsomal desaturation pathway. *Br. J. Nutr.* **47**, 589–596.

Olson, G.E., Winfrey, V.P., Hill, K.E., and Burk, R.F. (2004). Sequential development of flagellar defects in spermatids and epididymal spermatozoa of selenium-deficient rats. *Reproduction* **127**, 335–342.

Olson, J.A. (1984). Vitamin A. *In* "Handbook of Vitamins." (L.J. Machlin, ed), p. 1. Marcel Dekker, New York.

Olson, R. (1984). The function and metabolism of vitamin K. *Ann. Rev. Nutr.* **4**, 281.

Olynyk, J.K., and Clarke, S.L. (2001). Iron overload impairs pro-inflammatory cytokine responses by Kupffer cells. *J. Gastroenterol. Hepatol.* **16**, 438–444.

Omi, N., and Ezawa, I. (2001). Change in calcium balance and bone mineral density during pregnancy in female rats. *J. Nutr. Sci. Vitaminol. (Tokyo)* **47**, 195–200.

Ongsakul, M., Sirisinha, S., and Lamb, A.J. (1985). Impaired blood clearance of bacteria and phagocytic activity in vitamin A-deficient rats. *Proc. Soc. Exp. Biol. Med.* **178**, 204–208.

Orent-Keiles, E., Robinson, A., and McCollum E.V. (1937). The effects of sodium deprivation on the animal organism. *Am. J. Physiol.* **119**, 651.

Ornoy, A., Menczel, J., and Nebel, L. (1968). Alterations in the mineral composition and metabolism of rat fetuses and their fetuses and their placentas induced by maternal hypervitaminosis D2. *Isr. J. Med. Sci.* **4**, 827.

Orwoll, E., Ware, M., Stribrska, L., Bikle, D., Sanchez, T., Andon, M., and Li, H. (1992). Effects of dietary protein deficiency on mineral metabolism and bone mineral density. *Am. J. Clin. Nutr.* **56**, 314–319.

Oteiza, P.I., Cuellar, S., Lonnerdal, B., Hurley, L.S., and Keen, C.L. (1990). Influence of maternal dietary zinc intake on *in vitro* tubulin polymerization in fetal rat brain. *Teratology* **41**, 97–104.

Pamnani, M.B., Bryant, H.J., Clough, D.L., and Schooley, J.F. (2003). Increased dietary potassium and magnesium attenuate experimental volume dependent hypertension possibly through endogenous sodium-potassium pump inhibitor. *Clin. Exp. Hypertens.* **25**, 103–115.

Pang, R.L., and Kirksey, A. (1974). Early postnatal changes in brain composition in progeny of rats fed different levels of dietary pyridoxine. *J. Nutr.* **104**, 111–117.

Pansu, D., Duflos, C., Bellaton, C., and Bronner, F. (1993). Solubility and intestinal transit time limit calcium absorption in rats. *J. Nutr.* **123**, 1396–1404.

Papakonstantinou, E., Flatt, W.P., Huth, P.J., and Harris, R.B. (2002). High dietary calcium reduces body fat content, digestibility of fat, and serum vitamin D in rats. *Obesity Res.* **11**, 387–394.

Park, J.J., Smalley, E.B., and Chu, F.S. (1996). Natural occurrence of Fusarium mycotoxins in field samples from the 1992 Wisconsin corn crop. *Appl. Environ. Microbiol.* **62**, 1642–1648.

Parker, H.E., Andrews, F.N., Hauge, S.M., and Quackenbush, F.W. (1951). Studies on the iodine requirements of white rats during growth, pregnancy and lactation. *J. Nutr.* **44**, 501–511.

Parry, N.M., Phillippo, M., Reid, M.D., McGaw, B.A., Flint, D.J., and Loveridge, N. (1993). Molybdenum-induced changes in the epiphyseal growth plate. *Calcif. Tissue Int.* **53**, 180–186.

Pawelec, G., Remarque, E., Barnett, Y., and Solana, R. (1998). T cells and aging. *Front. Biosci.* **3**, 59–99.

Pazos-Moura, C.C., Moura, E.G., Dorris, M.L., Rehnmark, S., Melendez, L., Silva, J.E., and Taurog, A. (1991). Effect of iodine deficiency and cold exposure on thyroxine 5′-deiodinase activity in various rat tissues. *Am. J. Physiol.* **260**, E175–E182.

Pederson, R.A., Ramanadham, S., Buchan, A.M., and McNeill, J.H. (1989). Long-term effects of vanadyl treatment on streptozocin-induced diabetes in rats. *Diabetes* **38**, 1390–1395.

Peifer, J.J., and Lewis, R.D. (1979). Effects of vitamin B-12 deprivation on phospholipid fatty acid patterns in liver and brain of rats fed high and low levels of linoleate in low methionine diets. *J. Nutr.* **109**, 2160–2172.

Pelliccione, N.J., Karmali, R., Rivlin, R.S., and Pinto, J. (1985). Effects of riboflavin deficiency upon prostaglandin biosynthesis in rat kidney. *Prostaglandins Leukot. Med.* **17**, 349–358.

Pence, B.C. (1991). Dietary selenium and antioxidant status: toxic effects of 1,2-dimethylhydrazine in rats. *J. Nutr.* **121**, 138–144.

Pere, A.K., Lindgren, L., Tuomainen, P., Krogerus, L., Rauhala, P., Laakso, J., Karppanen, H., Vapaatalo, H., Ahonen, J., and Mervaala, E.M. (2000). Dietary potassium and magnesium supplementation in cyclosporine-induced hypertension and nephrotoxicity. *Kidney Int.* **58**, 2462–2472.

Perez-Llamas, F., Garaulet, M., Martinez, J.A., Marin, J.F., Larque, E., and Zamora, S. (2001). Influence of dietary protein type and iron source on the absorption of amino acids and minerals. *J. Physiol. Biochem.* **57**, 321–328.

Persson. P., Gagnemo-Persson, R., and Hakanson, R. (1993). The effect of high or low dietary calcium on bone and calcium homeostasis in young male rats. *Calcif. Tissue Int.* **52**, 460–464.

Peterson, C.A., Baker, D.H., and Erdman, J.W., Jr. (1996). Diet-induced nephrocalcinosis in female rats is irreversible and is induced primarily before the completion of adolescence. *J. Nutr.* **126**, 259–265.

Peterson, C.A., Eurell, J.A., and Erdman, J.W., Jr. (1992). Bone composition and histology of young growing rats fed diets of varied calcium bioavailability: spinach, nonfat dry milk, or calcium carbonate added to casein. *J. Nutr.* **122**, 137–144.

Peterson, C.A., Eurell, J.A., and Erdman, J.W. Jr. (1995). Alterations in calcium intake on peak bone mass in the female rat. *J. Bone Miner. Res.* **10**, 81–95.

Pfahl, M. (1993). Nuclear receptor/AP-1 interaction. *Endocr. Rev.* **14**, 651–8.

Pfast, H.B. (1976). Feed Manufacturing Technology. Feed Production Council, American Feed Manufacturing Association, Inc.

Pietrzik, K., and Hornig, D. (1980). Studies on the distribution of (1–14C) pantothenic acid in rats. *Int. J. Vitamin Nutr. Res.* **50**, 283–293.

Pinero, D., Jones, B., and Beard, J. (2001). Variations in dietary iron alter behavior in developing rats. *J. Nutr.* **131**, 311–318.

Pinero, D.J., Li, N.Q., Connor, J.R., and Beard, J.L. (2000). Variations in dietary iron alter brain iron metabolism in developing rats. *J. Nutr.* **130**, 254–263.

Pitt, J.I., Hocking, A.D., Bhudhasamai, K., Miscamble, B.F., Wheeler, K.A., and Tanboon-Ek, P. (1993). The normal mycoflora of commodities from Thailand, 1: nuts and oilseeds. *Int. J. Food Microbiol.* **20**, 211–226.

Planells, E., Lerma, A., Sanchez-Morito, N., Aranda, P., and Llopis, J. (1997). Effect of magnesium deficiency on vitamin B2 and B6 status in the rat. *J. Am. Coll. Nutr.* **16**, 352–356.

Planells, E., Sanchez-Morito, N., Montellano, M.A., Aranda, P., and Llopis, J. (2000). Effect of magnesium deficiency on enterocyte Ca, Fe, Cu, Zn, Mn and Se content. *J. Physiol. Biochem.* **56**, 217–222.

Pleasants, J.R., Reddy, B.S., and Wostmann, B.S. (1970). Qualitative adequacy of a chemically defined diet for reproducing germfree mice. *J. Nutr.* **100**, 498.

Pleasants, J.R., Wostmann, B.S., and Reddy, B.S. (1973). Improved lactation in germfree mice following changes in the amino acid and fat components of chemically defined diets. Germfree Research. (I.B. Heneghan ed.). pp. 245. Academic Press, New York.

Plesofsky-Vig, N. (1999). Pantothenic acid. *In* In Modern Nutrition in Health and Disease." (M.E. Shils, J.A. Olson, M. Shike, and C.A. Ross, eds), p. 423. Williams and Wilkins, Baltimore.

Plesofsky-Vig, N., and Brambl, R. (1988). Pantothenic acid and coenzyme A in cellular modification of proteins. *Annu. Rev. Nutr.* **8**, 461–482.

Popovich, D.G., Jenkins, D.J., Kendall, C.W., Dierenfeld, E.S., Carroll, R,W., Tariq, N., and Vidgen, E. (1997). The western lowland gorilla diet has implications for the health of humans and other hominoids. *J. Nutr.* **127**, 2000–2005.

Potvliege, P.R. (1962). Hypervitaminosis D2 in gravid rats. *Arch. Pathol.* **73**, 371.

Powers, H.J. (1986). Investigation into the relative effects of riboflavin deprivation on iron economy in the weanling rat and the adult. *Ann. Nutr. Metab.* **30**, 308–315.

Powers, H.J. (1987). A study of maternofetal iron transfer in the riboflavin-deficient rat. *J. Nutr.* **117**, 852–856.

Powers, H.J., and Bates, C.J. (1984). Effects of pregnancy and riboflavin deficiency on some aspects of iron metabolism in rats. *Int. J. Vitamin Nutr. Res.* **54**, .179–183.

Powers, H.J., Wright, A.J., and Fairweather-Tait, S.J. (1988). The effect of riboflavin deficiency in rats on the absorption and distribution of iron. *Br. J. Nutr.* **59**, 381–387.

Prada, P., Okamoto, M.M., Furukawa, L.N., Machado, U.F., Heimann, J.C., and Dolnikoff, M.S. (2000). High- or low-salt diet from weaning to adulthood: effect on insulin sensitivity in Wistar rats. *Hypertension* **35**, 424–429.

Prohaska, J.R. (1987). Functions of trace elements in brain metabolism. *Physiol. Rev.* **67**, 858–901.

Prohaska, J.R. (1988). Biochemical functions of copper in animals. *In* "Essential and Toxic Trace Elements in Human and Disease." (A.S. Prasad, ed), p. 105. Alan R. Liss, New York.

Prohaska, J.R. (1990). Biochemical changes in copper deficiency. *J. Nutr. Biochem.* **1**, 452.

Prohaska, J.R. (1991). Changes in Cu, Zn-superoxide dismutase, cytochrome c oxidase, glutathione peroxidase and glutathione transferase activities in copper-deficient mice and rats. *J. Nutr.* **121**, 355–363.

Prohaska, J.R., and Bailey, W.R. (1995). Alterations of rat brain peptidylglycine α-amidating monooxygenase and other cuproenzyme activities following perinatal copper deficiency. *Proc. Soc. Exp. Biol. Med.* **210**, 107–116.

Prohaska, J.R., and Failla, M.L. (1993). Copper and Immunity. *In* "Nutrition and Immunology. Human Nutrition: A Comprehensive Treatise," Vol. 8. (D.E.M. Klurfeld, ed), p. 309. Plenum Press, New York.

Prohaska, J.R., and Heller, L.J. (1982). Mechanical properties of the copper-deficient rat heart. *J. Nutr.* **112**, 2142–2150.

Pyapali, G.K., Turner, D.A., Williams, C.L., Meck, W.H., and Swartzwelder, H.S. (1998). Prenatal dietary choline supplementation decreases the threshold for induction of long-term potentiation in young adult rats. *J. Neurophysiol.* **79**, 1790–1796.

Rabin, B.S. (1983). Inhibition of experimentally induced autoimmunity in rats by biotin deficiency. *J. Nutr.* **113**, 2316–2322.

Rafecas, I., Esteve, M., Fernandez-Lopez, J.A., Remesar, X., and Alemany, M. (1993). Water balance in Zucker obese rats. *Comp. Biochem. Physiol. Comp. Physiol.* **104**, 813–818.

Rahman, A.S., Kimura, M., and Itokawa, Y. (1999). Testicular atrophy, zinc concentration, and angiotensin-converting enzyme activity in the testes of vitamin A-deficient rats. *Biol. Trace Elem. Res.* **67**, 29–36.

Rahman, A.S., Kimura, M., Yokoi, K., Tanvir, E.N., and Itokawa, Y. (1995). Iron, zinc, and copper levels in different tissues of clinically vitamin A-deficient rats. *Biol. Trace Elem. Res.* **49**, 75–84.

Rains, T.M., Emmert, J.L., Baker, D.H., and Shay, N.F. (1997). Minimum thiamin requirement of weanling Sprague-Dawley outbred rats. *J. Nutr.* **127**, 167–70.

Rains, T.M., and Shay, N.F. (1995). Zinc status specifically changes preferences for carbohydrate and protein in rats selecting from separate carbohydrate-, protein-, and fat-containing diets. *J. Nutr.* **125**, 2874–2879.

Rajagopalan, K.V. (1988). Molybdenum: an essential trace element in human nutrition. *Annu. Rev. Nutr.* **8**, 401–427.

Rall, L.C., and Meydani, S.N. (1993). Vitamin B6 and immune competence. *Nutr. Rev.* **51**, 217–225.

Rao, G.H., and Mason, K.E. (1975). Antisterility and antivitamin K activity of d-α-tocopheryl hydroquinone in the vitamin E-deficient female rat. *J. Nutr.* **105**, 495–498.

Rao, G.N., and Knapka, J.J. (1987). Contaminant and nutrient concentrations of natural ingredient rat and mouse diet used in chemical toxicology studies. *Fundam Appl Toxicol.* **9**, 329–338.

Rao, G.S. (1984). Dietary intake and bioavailability of fluoride. *Annu. Rev. Nutr.* **4**, 115–136.

Rao, J., and Jagadeesan, V. (1995). Development of a rat model for iron deficiency and toxicological studies: comparison among Fischer 344, Wistar, and Sprague Dawley strains. *Lab. Anim. Sci.* **45**, 393–397.

Rao, R.M., Yan, Y., and Wu, Y. (1994). Dietary calcium reduces blood pressure, parathyroid hormone, and platelet cytosolic calcium responses in spontaneously hypertensive rats. *Am. J. Hypertens.* **7**, 1052–1057.

Rasmussen, H. (1986a). The calcium messenger system (2). *N. Engl. J. Med.* **314**, 1164–1170.

Rasmussen, H. (1986b). The calcium messenger system (1). *N. Engl. J. Med.* **314**, 1094–1101.

Rawling, J.M., Driscoll, E.R., Poirier, G.G., and Kirkland, J.B. (1993). Diethylnitrosamine administration *in vivo* increases hepatic poly(ADP-ribose) levels in rats: results of a modified technique for poly(ADP-ribose) measurement. *Carcinogenesis* **14**, 2513–2516.

Rawling, J.M., Jackson, T.M., Driscoll, E.R., Kirkland, J.B. (1994). Dietary niacin deficiency lowers tissue poly(ADP-ribose) and NAD+ concentrations in Fischer-344 rats. *J. Nutr.* **124**, 1597–1603.

Ray, P.E., Suga, S., Liu, X.H., Huang, X., Johnson, R.J. (2001). Chronic potassium depletion induces renal injury, salt sensitivity, and hypertension in young rats. *Kidney Int.* **59**, 1850–1858.

Rayssiguier, Y., Bussiere, F., et al. (2001). Acute phase response to magnesium deficiency: possible relevance to atherosclerosis. *In* "Advances in Magnesium Research: Nutrition and Health." (Y. Rayssiguier, A. Mazur, and J. Durlach, eds), p. 277. John Libbey and Co., London.

Rayssiguier, Y., Gueux, E., Bussiere, L., and Mazur, A. (1993). Copper deficiency increases the susceptibility of lipoproteins and tissues to peroxidation in rats. *J. Nutr.* **123**, 1343–1348.

Reaves, S.K., Fanzo, J.C., Wu, J.Y., Wang, Y.R., Wu, Y.W., Zhu, L., and Lei, K.Y. (1999). Plasma apolipoprotein B-48, hepatic apolipoprotein B mRNA editing and apolipoprotein B mRNA editing catalytic subunit-1 mRNA levels are altered in zinc-deficient rats. *J. Nutr.* **129**, 1855–1861.

Reddi, A.S., Jyothirmayi, G.N., DeAngelis, B., Frank, O., and Baker, H. (1993). Tissue concentrations of water-soluble vitamins in normal and diabetic rats. *Int. J. Vitamin Nutr. Res.* **63**, 140.

Reddy, S.R., and Kotchen, T.A. (1992a). Hemodynamic effects of high dietary intakes of sodium or chloride in the Dahl salt-sensitive rat. *J. Lab. Clin. Med.* **120**, 476–482.

Reddy, S.R., and Kotchen, T.A. (1992b). Dietary sodium chloride increases blood pressure in obese Zucker rats. *Hypertension* **20**, 389–393.

Reeves, P.G. (2003). Patterns of food intake and self-selection of macronutrients in rats during short-term deprivation of dietary zinc. *J. Nutr. Biochem.* **14**, 232–243.

Reeves, P.G., and Chaney, R.L. (2001). Mineral status of female rats affects the absorption and organ distribution of dietary cadmium derived from edible sunflower kernels (*Helianthus annuus* L.). *Environ. Res.* **85**, 215–225.

Reeves, P.G., and DeMars, L.C. (2004). Copper deficiency reduces iron absorption and biological half-life in male rats. *J. Nutr.* **134**, 1953–1957.

Reeves, P.G., Nielsen, F.H., and Fahey, G.C. Jr. (1993). AIN-93 purified diets for laboratory rodents: final report of the American Institute of Nutrition ad hoc writing committee on the reformulation of the AIN-76A rodent diet. *J. Nutr.* **123**, 1939–1951.

Reeves, P.G., and O'Dell, B.L. (1981). Short-term zinc deficiency in the rat and self-section of dietary protein level. *J. Nutr.* **111**, 375–383.

Reeves, P.G., Ralston, N.V., Idso, J.P., and Lukaski, H.C. (2004). Contrasting and cooperative effects of copper and iron deficiencies in male rats fed different concentrations of manganese and different sources of sulfur amino acids in an AIN-93G–based diet. *J. Nutr.* **134**, 416–425.

Reichel, H., Koeffler, H.P., and Norman, A.W. (1989). The role of the vitamin D endocrine system in health and disease. *N. Engl. J. Med.* **320**, 980–991.

Reiser, S., Ferretti, R.J., Fields, M., and Smith, J.C. Jr. (1983). Role of dietary fructose in the enhancement of mortality and biochemical changes associated with copper deficiency in rats. *Am. J. Clin. Nutr.* **38**, 214–222.

Remesy, C., and Demigne, C. (1989). Specific effects of fermentable carbohydrates on blood urea flux and ammonia absorption in the rat cecum. *J. Nutr.* **119**, 560–565.

Remington, R.E., and Remington, J.W. (1938). The effect of enhanced iodine intake on growth and on the thyroid glands of normal goitrous rats. *J. Nutr.* **15**, 539.

Renner, H.W., Grunewald, T., and Ehrenberg-Kieckebusch, W. (1973). Mutagenicity testing of irradiated food using the dominant lethal test. *Humangenetik* **18**, 155–164.

Reynolds, R.D., and Natta, C.L. (1985). Vitamin B6 and sickle cell anemia. *In* "Vitamin B6: Its Role in Health and Disease." (R.D. Reynolds and J.E. Leklem, eds), p. 301. Alan R. Liss, New York.

Rico, H., Gomez-Raso, N., Revilla, M., Hernandez, E.R., Seco, C., Paez, E., and Crespo, E. (2000). Effects on bone loss of manganese alone or with copper supplement in ovariectomized rats. a morphometric and densitomeric study. *Eur. J. Obstet. Gynecol. Reprod. Biol.* **90**, 97–101.

Riley, R.T., Norred, W.P., and Bacon, C.W. (1993). Fungal toxins in foods: recent concerns. *Annu. Rev. Nutr.* **13**, 167–189.

Rimbach, G., and Pallauf, J. (1999). Effect of dietary phytate on magnesium bioavailability and liver oxidant status in growing rats. *Food Chem. Toxicol.* **37**, 37–45.

Rimbach, G., Pallauf, J., Brandt, K., and Most, E. (1995). Effect of phytic acid and microbial phytase on Cd accumulation, Zn status, and apparent absorption of Ca, P, Mg, Fe, Zn, Cu, and Mn in growing rats. *Ann. Nutr. Metab.* **39**, 361–370.

Rindi, G. (1984). Thiamin absorption by small intestine. *Acta Vitaminol. Enzymol.* **6**, 47–55.

Rindi, G., De Giuseppe, L., and Sciorelli, G. (1968). Thiamine monophosphate, a normal constituent of rat plasma. *J. Nutr.* **94**, 447–454.

Riond, J.L., Hartmann, P., Steiner, P., Ursprung, R., Wanner, M., Forrer, R., Spichiger, U.E., Thomsen, J.S., and Mosekilde, L. (2000). Long-term excessive magnesium supplementation is deleterious whereas suboptimal supply is beneficial for bones in rats. *Magnes. Res.* **13**, 249–64.

Ritskes-Hoitinga, J., Mathot, J.N., Danse, L.H., and Beynen, A.C. (1991). Commercial rodent diets and nephrocalcinosis in weanling female rats. *Lab. Anim.* **25**, 126–132.

Rivlin, R.S. (1984). Riboflavin. *In* "Present Knowledge of Nutrition." (R.E. Olsen, H.P. Broquist, C.O. Chichester, et al., eds), p. 285. Nutrition Foundation, Washington, DC.

Rivlin, R.S., and Dutta, P. (1995). Vitamin B$_2$ (riboflavin): relevance to malaria and antioxidant activity. *Nutr. Today* **30**, 62.

Rivlin, R.S., and Pinto, J.T. (2001). Riboflavin (vitamin B$_2$). *In* "Handbook of Vitamins." (R.B. Rucker, J.W. Suttie, D.B. McCormick, and L.J. Machlin, eds), p. 255. Marcel Dekker, New York.

Roberts, C.K., Vaziri, N.D., Sindhu, R.K., and Barnard, R.J. (2003). A high-fat, refined-carbohydrate diet affects renal NO synthase protein expression and salt sensitivity. *J. Appl. Physiol.* **94**, 941–946.

Robertson, J.B., and Horvath, P.J. (1992). Detergent analysis of foods. *In* "CRC Handbook of Dietary Fiber in Human Nutrition," 2nd ed. (G.A. Spiller, ed), pp. 49–52. CRC Press., Boca Raton.

Robinson, J.R. (1957). Functions of water in the body. *Proc. Nutr. Soc.* **16**, 108–112.

Rock, E., Gueux, E., Mazur, A., Motta, C., and Rayssiguier, Y. (1995). Anemia in copper-deficient rats: role of alterations in erythrocyte membrane fluidity and oxidative damage. *Am. J. Physiol.* **269**, C1245–C1249.

Rodriguez-Matas, M.C., Campos, M.S., Lopez-Aliaga, I., Gomez-Ayala, A.E., and Lisbona, F. (1998). Iron-manganese interactions in the evolution of iron deficiency. *Ann. Nutr. Metab.* **42**, 96–109.

Roe, F.J., Lee, P.N., Conybeare, G., Kelly, D., Matter, B., Prentice, D., and Tobin, G. (1995). The Biosure Study: influence of composition of diet and food consumption on longevity, degenerative diseases and neoplasia in Wistar rats studied for up to 30 months post weaning. *Food Chem. Toxicol.* **33 Suppl 1**, 1S–100S.

Rogers, A. (1979). Nutrition. *In* "The Laboratory Rat," Vol I. Academic Press, San Diego.

Rogers, A.E., and MacDonald, R.A. (1965). Hepatic vasculature and cell proliferation in experimental cirrhosis. *Lab. Invest.* **14**, 1710–1726.

Rogers, Q.R., and Harper, A.E. (1965). Amino acid diets and maximal growth in the rat. *J. Nutr.* **87**, 267–273.

Rogers, W.E. Jr., Bieri, J.G., and McDaniel, E.G. (1971). Vitamin A deficiency in the germfree state. *Fed. Proc.* **30**, 1773–1778.

Roginski, E.E., and Mertz, W. (1969). Effects of chromium 3+ supplementation on glucose and amino acid metabolism in rats fed a low protein diet. *J. Nutr.* **97**, 525–530.

Roncales, M., Achon, M., Manzarbeitia, F., Maestro de las Casas, C., Ramirez, C., Varela-Moreiras, G., and Perez-Miguelsanz, J. (2004). Folic acid supplementation for 4 weeks affects liver morphology in aged rats. *J. Nutr.* **134**, 1130–1133.

Rong, N., Selhub, J., Goldin, B.R., and Rosenberg, I.H. (1991). Bacterially synthesized folate in rat large intestine is incorporated into host tissue folyl polyglutamates. *J. Nutr.* **121**, 1955–1959.

Rosales, F.J., Jang, J.T., Pinero, D.J., Erikson, K.M., Beard, J.L., and Ross, A.C. (1999). Iron deficiency in young rats alters the distribution of vitamin A between plasma and liver and between hepatic retinol and retinyl esters. *J. Nutr.* **129**, 1223–1228.

Rosenstreich, S.J., Rich, C., and Volwiler, W. (1971). Deposition in and release of vitamin D3 from body fat: evidence for a storage site in the rat. *J. Clin. Invest.* **50**, 679–687.

Ross, A.C., and Hammerling, U.G. (1994). Retinoids and the immune system. *In* "The Retinoids: Biology, Chemistry, and Medicine." (M.B. Sporn, A.B. Roberts, and D.S. Goodman, eds), p. 521. Raven Press, New York.

Ross, M.H. (1972). Length of life and caloric intake. *Am. J. Clin. Nutr.* **25**, 834–838.

Rossi, L., Migliaccio, S., Corsi, A., Marzia, M., Bianco, P., Teti, A., Gambelli, L., Cianfarani, S., Paoletti, F., and Branca, F. (2001). Reduced growth and skeletal changes in zinc-deficient growing rats are due to impaired growth plate activity and inanition. *J. Nutr.* **131**, 1142–1146.

Roth, H.P. (2003). Development of alimentary zinc deficiency in growing rats is retarded at low dietary protein levels. *J. Nutr.* **133**, 2294–2301.

Roth-Maier, D.A., Kirchgessner, M., and Rajtek, S. (1990). Retention and utilization of thiamin by gravid and non gravid rats with varying dietary thiamin supply. *Int. J. Vitamin Nutr. Res.* **60**, 343–350.

Roughead, Z.K., and Lukaski, H.C. (2003). Inadequate copper intake reduces serum insulin-like growth factor-I and bone strength in growing rats fed graded amounts of copper and zinc. *J. Nutr.* **133**, 442–448.

Roullet, C.M., Roullet, J.B., Duchambon, P., Thomasset, M., Lacour, B., McCarron, D.A., and Drueke, T. (1991). Abnormal intestinal regulation of calbindin-D9K and calmodulin by dietary calcium in genetic hypertension. *Am. J. Physiol.* **261**, F474–F480.

Rude, R.K. (1993). Magnesium metabolism and deficiency. *Endocrinol. Metab. Clin. North Am.* **22**, 377–395.

Rude, R.K., Gruber, H.E., Norton, H.J., Wei, L.Y., Frausto, A., and Mills, B.G. (2004). Bone loss induced by dietary magnesium reduction to 10% of the nutrient requirement in rats is associated with increased release of substance P and tumor necrosis factor-α. *J. Nutr.* **134**, 79–85.

Rude, R.K., Kirchen, M.E., Gruber, H.E., Meyer, M.H., Luck, J.S., and Crawford, D.L. (1999). Magnesium deficiency-induced osteoporosis in the rat: uncoupling of bone formation and bone resorption. *Magnes. Res.* **12**, 257–267.

Rushmore, T.H., Lim, Y.P., Farber, E., and Ghoshal, A.K. (1984). Rapid lipid peroxidation in the nuclear fraction of rat liver induced by a diet deficient in choline and methionine. *Cancer Lett.* **24**, 251–5.

Saari, J.T. (1992). Dietary copper deficiency and endothelium-dependent relaxation of rat aorta. *Proc. Soc. Exp. Biol. Med.* **200**, 19–24.

Saari, J.T., Johnson, W.T., Reeves, P.G., and Johnson, L.K. (1993). Amelioration of effects of severe dietary copper deficiency by food restriction in rats. *Am. J. Clin. Nutr.* **58**, 891–896.

Sabo, D.J., Francesconi, R.P., and Gershoff, S.N. (1971). Effect of vitamin B6 deficiency on tissue dehydrogenases and fat synthesis in rats. *J. Nutr.* **101**, 29–34.

Said, A.K., and Hegsted, D.M. (1970). Response of adult rats to low dietary levels of essential amino acids. *J. Nutr.* **100**, 1363–1375.

Said, H.M., Horne, D.W., and Mock, D.M. (1990). Effect of aging on intestinal biotin transport in the rat. *Exp. Gerontol.* **25**, 67–73.

Said, H.M., Mock, D.M., and Collins, J.C. (1989). Regulation of intestinal biotin transport in the rat: effect of biotin deficiency and supplementation. *Am. J. Physiol.* **256**, G306–G311.

Said, H.M., Thuy, L.P., Sweetman, L., and Schatzman, B. (1993). Transport of the biotin dietary derivative biocytin (N-biotinyl-L-lysine) in rat small intestine. *Gastroenterology* **104**, 75–80.

Sakai, T., Miki, F., Wariishi, M., and Yamamoto, S. (2004). Comparative study of zinc, copper, manganese, and iron concentrations in organs of zinc-deficient rats and rats treated neonatally with l-monosodium glutamate. *Biol. Trace Elem. Res.* **97**, 163–182.

Salmon, W.D., and Newberne, P.M. (1962). Cardiovascular disease in choline-deficient rats. *Arch. Pathol.* **73**, 190–209.

Sanchez-Morito, N., Planells, E., Aranda, P., and Llopis, J. (1999). Magnesium-manganese interactions caused by magnesium deficiency in rats. *J. Am. Coll. Nutr.* **18**, 475–480.

Sanders, P.W. (1996). Salt-sensitive hypertension: lessons from animal models. *Am. J. Kidney. Dis.* **28**, 775–782.

Sarter, M., and van der Linde, A. (1987). Vitamin E deprivation in rats: some behavioral and histochemical observations. *Neurobiol. Aging* **8**, 297–307.

Sato, M., and Bremner, I. (1993). Oxygen free radicals and metallothionein. *Medicine* **14**, 325.

Schachter, D., Finkelstein, J.D., and Kowarski, S. (1964). Metabolism of vitamin D, I: preparation of radioactive vitamin D and its intestinal absorption in the rat. *J. Clin. Invest.* **43**, 787–796.

Schaefer, R.M., Heidland, A., and Horl, W.H. (1985). Carbohydrate metabolism in potassium-depleted rats. *Nephron* **41**, 100–109.

Schaeffer, M.C. (1987). Attenuation of acoustic and tactile startle responses of vitamin B6 deficient rats. *Physiol. Behav.* **40**, 473.

Schaeffer, M.C., Cochary, E.F., and Sadowski, J.A. (1990). Subtle abnormalities of gait detected early in vitamin B6 deficiency in aged and weanling rats with hind leg gait analysis. *J. Am. Coll. Nutr.* **9**, 120–127.

Schaeffer, M.C., and Kretsch, M.J. (1987). Quantitative assessment of motor and sensory function in vitamin B6 deficient rats. *Nutr. Res.* **7**, 851.

Schaeffer, M.C., Sampson, D.A., Skala, J.H., Gietzen, D.W., and Grier, R.E. (1989). Evaluation of vitamin B-6 status and function of rats fed excess pyridoxine. *J. Nutr.* **119**, 1392–1398.

Schleiffer, R., and Gairard, A. (1995). Blood pressure effects of calcium intake in experimental models of hypertension. *Semin. Nephrol.* **15**, 526–535.

Schmid, C., Castrop, H., Reitbauer, J., Della Bruna, R., and Kurtz, A. (1997). Dietary salt intake modulates angiotensin II type 1 receptor gene expression. *Hypertension* **29**, 923–929.

Schmiedl, A., Schwille, P.O., Bonucci, E., Erben, R.G., Grayczyk, A., and Sharma, V. (2000). Nephrocalcinosis and hyperlipidemia in rats fed a cholesterol- and fat-rich diet: association with hyperoxaluria, altered kidney and bone minerals, and renal tissue phospholipid-calcium interaction. *Urol. Res.* **28**, 404–415.

Schneeman, B.O., and Gallaher, D. (1980). Changes in small intestinal digestive enzyme activity and bile acids with dietary cellulose in rats. *J. Nutr.* **110**, 584–590.

Scholz-Ahrens, K.E., Schaafsma, G., van den Heuvel, E.G., and Schrezenmeir, J.- (2001). Effects of prebiotics on mineral metabolism. *Am. J. Clin. Nutr.* **73**, 459S–464S.

Schooley, C., and Franz, K.B. (2002). Magnesium during pregnancy in rats increases systolic blood pressure and plasma nitrite. *Am. J. Hypertens.* **15**, 1081.

Schroeder, H.A. (1966). Chromium deficiency in rats: a syndrome simulating diabetes mellitus with retarded growth. *J. Nutr.* **88**, 439–445.

Schroeder, H.A., Vinton, W.H. Jr., and Balassa, J.J. (1963). Effects of chromium, cadmium and lead on the growth and survival of rats. *J. Nutr.* **80**, 48–54.

Schultz, M.O. (1956). Reproduction of rats fed protein-free amino acid rations. *J. Nutr.* **60**, 35–45.

Schultz, M.O. (1957). Nutrition of rats with compounds of known chemical structure. *J. Nutr.* **61**, 585–596.

Schwartz, K., and Mertz, W. (1959). Chromium (III) and the glucose tolerance factor. *Arch. Biochim. Biophys.* **85**, 292.

Schwarz, K., and Foltz, C.M. (1957). Selenium as an integral part of factor 3 against dietary necrotic liver degeneration. *J. Am. Chem. Soc.* **79**, 3292.

Scott, M.L. (1986). Energy: requirements, sources, and metabolism. *In* "Nutrition in Humans and Selected Animal Species." pp. 12–78. John Wiley & Sons, New York.

Seaborn, C.D., and Stoecker, B.J. (1990). Effects of antacid and ascorbic acid on tissue accumulation and urinary excretion of 51 chromium. *Nutr. Res.* **10**, 1401.

Selhub, J., Seyoum, E., Pomfret, E.A., and Zeisel, S.H. (1991). Effects of choline deficiency and methotrexate treatment upon liver folate content and distribution. *Cancer Res.* **51**, 16–21.

Seng, J.E., Allaben, W.T., Nichols, M.I., et al. (1998). Putting dietary control to the test: increasing biosassay sensitivity by reducing variability. *Lab. Anim.* **27**, 35–38.

Serebro, H.A., Solomon, H.M., Johnson, J.H., Hendrix, T.R. (1966). The intestinal absorption of vitamin B6 compounds by the rat and hamster. *Bull. Johns Hopkins Hosp.* **119**, 166.

Shane, B., and Stokstad, E.L. (1985). Vitamin B_{12}-folate interrelationships. *Annu. Rev. Nutr.* **5**, 115–141.

Shannon, B.M., Howe, J.M., Clark, H.E. (1972). Interrelationships between dietary methionine and cystine as reflected by growth, certain hepatic enzymes and liver composition of weanling rats. *J. Nutr.* **102**, 557–562.

Shapiro, R., and Heaney, R.P. (2003). Co-dependence of calcium and phosphorus for growth and bone development under conditions of varying deficiency. *Bone* **32**, 532–540.

She, Q.B., Hayakawa, T., and Tsuge, H. (1994). Effect of vitamin B6 deficiency on linoleic acid desaturation in arachidonic acid biosynthesis of rat liver microsomes. *Biosci. Biochem.* **58**, 459.

Sheehan, D.M., Branham, W.S., Medlock, K.L., and Shanmugasundaram, E.R. (1984). Estrogenic activity of zearalenone and zearalanol in the neonatal rat uterus. *Teratology* **29**, 383–392.

Sheng, M.H., Taper, L.J., Veit, H., Qian, H., Ritchey, S.J., and Lau, K.H. (2001a). Dietary boron supplementation enhanced the action of estrogen, but not that of parathyroid hormone, to improve trabecular bone quality in ovariectomized rats. *Biol. Trace Elem. Res.* **82**, 109–123.

Sheng, M.H., Taper, L.J., Veit, H., Thomas, E.A., Ritchey, S.J., and Lau, K.H. (2001b). Dietary boron supplementation enhances the effects of estrogen on bone mineral balance in ovariectomized rats. *Biol. Trace Elem. Res.* **81**, 29–45.

Shepard, T.H., Lemire, R.J., Aksu, O., and Mackler, B. (1968). Studies of the development of congenital anomalies in embryos of riboflavin-deficient, galactoflavin fed rats. I. Growth and embryologic pathology. *Teratology* **1**, 74–92.

Shibuya, M., Hisaoka, F., et al. (1990). Effects of pregnancy on vitamin B6-dependent enzymes and B6 content in tissues of rats fed diets containing two levels of pyridoxine-hydrochloride. *Nippon Eiyo Shokuryo Gakkaishi* **43**, 189.

Shier, W.T., Shier, A.C., Xie, W., and Mirocha, C.J. (2001). Structure-activity relationships for human estrogenic activity in zearalenone mycotoxins. *Toxicon* **39**, 1435–1438.

Shiga, K., Hara, H., and Kasai, T. (1998). The large intestine compensates for insufficient calcium absorption in the small intestine in rats. *J. Nutr. Sci. Vitaminol. (Tokyo)* **44**, 737–744.

Shils, M.E. (1999). Magnesium. *In* "Modern Nutrition in Health and Disease." (M.E. Shils, J.A. Olson, M. Shike, and C.A. Ross, eds), p. 160. Williams and Wilkins, Baltimore.

Shimokawa, I., Higami, Y., Yu, B.P., Masoro, E.J., and Ikeda, T. (1996). Influence of dietary components on occurrence of and mortality due to neoplasms in male F344 rats. *Aging (Milano)* **8**, 254–262.

Shimokawa, I., Yu, B.P., Higami, Y., Ikeda, T., and Masoro, E.J. (1993). Dietary restriction retards onset but not progression of leukemia in male F344 rats. *J. Gerontol.* **48**, B68–B73.

Shriver, B.J., Roman-Shriver, C., and Allred, J.B. (1993). Depletion and repletion of biotinyl enzymes in liver of biotin-deficient rats: evidence of a biotin storage system. *J. Nutr.* **123**, 1140–1149.

Sidransky, H., and Farber, E. (1958a). Chemical pathology of acute amino acid deficiencies. *Arch. Pathol.* **66**, 119–134.

Sidransky, H., and Farber, E. (1958b). Chemical pathology of acute amino acid deficiencies, II: biochemical changes in rats fed threonine- or methionine-devoid diets. *Arch. Pathol.* **66**, 135–149.

Sidransky, H., and Verney, E. (1969). Chemical pathology of acute amino acid deficiencies: morphologic and biochemical changes in young rats force-fed a threonine-deficient diet. *J. Nutr.* **96**, 349–358.

Sijtsma, K.W., Van Den Berg, G.J., Lemmens, A.G., West, C.E., and Beynen, A.C. (1993). Iron status in rats fed on diets containing marginal amounts of vitamin A. *Br. J. Nutr.* **70**, 777–785.

Silberberg, M., and Silberberg, R. (1955). Diet and life span. *Physiol. Rev.* **35**, 347–362.

Skala, J.H., Schaeffer, M.C., et al. (1989). Effects of various levels of pyridoxine on erythrocyte aminotransferase activities in the rat. *Nutr. Res.* **9**, 195.

Slomianka, L. (1992). Neurons of origin of zinc-containing pathways and the distribution of zinc-containing boutons in the hippocampal region of the rat. *Neuroscience* **48**, 325–352.

Smith, A.M., and Picciano, M.F. (1986). Evidence for increased selenium requirement for the rat during pregnancy and lactation. *J. Nutr.* **116**, 1068–1079.

Smith, A.M., and Picciano, M.F. (1987). Relative bioavailability of seleno-compounds in the lactating rat. *J. Nutr.* **117**, 725–731.

Smith, B.J., King, J.B., Lucas, E.A., Akhter, M.P., Arjmandi, B.H., and Stoecker, B.J. (2002). Skeletal unloading and dietary copper depletion are detrimental to bone quality of mature rats. *J. Nutr.* **132**, 190–196.

Smith, F.R., and Goodman, D.S. (1976). Vitamin A transport in human vitamin A toxicity. *N. Engl. J. Med.* **294**, 805–808.

Smith, J.C. Jr., McDaniel, E.G., Fan, F.F., and Halsted, J.A. (1973). Zinc: a trace element essential in vitamin A metabolism. *Science* **181**, 954–955.

Smith, S.M., Dickman, E.D., Power, S.C., and Lancman, J. (1998). Retinoids and their receptors in vertebrate embryogenesis. *J. Nutr.* **128**, 467S–470S.

Sokol, R.J. (1988). Vitamin E deficiency and neurologic disease. *Annu. Rev. Nutr.* **8**, 351–373.

Songu-Mize, E., Caldwell, R.W., and Baer, P.G. (1987). High and low dietary potassium effects on rat vascular sodium pump activity. *Proc. Soc. Exp. Biol. Med.* **186**, 280–287.

Spicer, M.T., Stoecker, B.J., Chen, T., and Spicer, L.J. (1998). Maternal and fetal insulin-like growth factor system and embryonic survival during pregnancy in rats: interaction between dietary chromium and diabetes. *J. Nutr.* **128**, 2341–2347.

Spiller, G.A. (1992). Definition of dietary fiber. *In* "CRC Handbook of Dietary Fiber in Human Nutrition," 2nd ed. (G.A. Spiller, ed), pp. 15–18. CRC Press, Boca Raton.

Spoerl, R., and Kirchgessner, M. (1975a). Effect of Cu-deficiency in rats during rearing and pregnancy on reproduction. *Z. Tierphysiol. Tierernahr. Futtermittelkd.* **35**, 321–328.

Spoerl, R., and Kirchgessner, M. (1975b). Changes of Cu status and ceruloplasmin activity in mother and suckling rats during a gradual increase of copper supply. *Z. Tierphysiol. Tierernahr. Futtermittelkd.* **35,** 113–127.

Stal, P., Wang, G.S., Olsson, J.M., and Eriksson, L.C. (1999). Effects of dietary iron overload on progression in chemical hepatocarcinogenesis. *Liver* **19,** 326–334.

Stangl, G.I., and Kirchgessner, M. (1998). Effect of different degrees of moderate iron deficiency on the activities of tricarboxylic acid cycle enzymes, and the cytochrome oxidase, and the iron, copper, and zinc concentrations in rat tissues. *Z. Ernahrungswiss.* **37,** 260–268.

Steele, J.H., and Engel, R.E. (1992). Radiation processing of food. *J. Am. Vet. Med. Assoc.* **201,** 1522–1529.

Steenbock, H., and Herting, D.C. (1955). Vitamin D and growth. *J. Nutr.* **57,** 449–468.

Stephens, M.C., Havlicek, V., and Dakshinamurti, K. (1971). Pyridoxine deficiency and development of the central nervous system in the rat. *J. Neurochem.* **18,** 2407–2416.

Stewart, C.N., Coursin, D.B., and Bhagavan, H.N. (1975). Avoidance behavior in vitamin B-6 deficient rats. *J. Nutr.* **105,** 1363–1370.

Stewart, S.R., Emerick, R.J., and Kayongo-Male, H. (1993). Silicon-zinc interactions and potential roles for dietary zinc and copper in minimizing silica urolithiasis in rats. *J. Anim. Sci.* **71,** 946–954.

Stoecker, B.J. (1999). Chromium. *In* "Modern Nutrition in Health and Disease." (M.E. Shils, J.A. Olson, M. Shike, and C.A. Ross, eds), p. 277. Williams and Wilkins, Baltimore.

Stokstad, E.L., Reisenauer, A., Kusano, G., and Keating, J.N. (1988). Effect of high levels of dietary folic acid on folate metabolism in vitamin B12 deficiency. *Arch. Biochem. Biophys.* **265,** 407–414.

Stowe, H.D., Rangel, F., Anstead, C., and Goelling, B. (1980). Influence of supplemental dietary vitamin A on the reproductive performance of iodine-toxic rats. *J. Nutr.* **110,** 1947–1957.

Strause, L.G., Hegenauer, J., Saltman, P., Cone, R., and Resnick, D. (1986). Effects of long-term dietary manganese and copper deficiency on rat skeleton. *J. Nutr.* **116,** 135–141.

Striffler, J., Polansky, M., and Anderson, R. (1993). Dietary chromium enhances insulin secretion in perfused rat pancreas. *J. Trace Elem. Exp. Med.* **6,** 75.

Striffler, J.S., Law, J.S., Polansky, M.M., Bhathena, S.J., and Anderson, R.A. (1995). Chromium improves insulin response to glucose in rats. *Metabolism* **44,** 1314–1320.

Striffler, J.S., Polansky, M.M., and Anderson, R.A. (1998). Dietary chromium decreases insulin resistance in rats fed a high-fat, mineral-imbalanced diet. *Metabolism* **47,** 396–400.

Striffler, J.S., Polansky, M.M., and Anderson, R.A. (1999). Overproduction of insulin in the chromium-deficient rat. *Metabolism* **48,** 1063–1068.

Subramanian, S., Karande, A.A., and Adiga, P.R. (1996). Establishment of the functional importance of thiamin carrier protein in pregnant rats by using monoclonal antibodies. *Indian J. Biochem. Biophys.* **33,** 111–115.

Suchy, S.F., Rizzo, W.B., and Wolf, B. (1986). Effect of biotin deficiency and supplementation on lipid metabolism in rats: saturated fatty acids. *Am. J. Clin. Nutr.* **44,** 475–480.

Sugawara, N., and Sugawara, C. (1999). An iron-deficient diet stimulates the onset of the hepatitis due to hepatic copper deposition in the Long-Evans Cinnamon (LEC) rat. *Arch. Toxicol.* **73,** 353–358.

Sukalski, K.A., LaBerge, T.P., and Johnson, W.T. (1997). In vivo oxidative modification of erythrocyte membrane proteins in copper deficiency. *Free. Radic. Biol. Med.* **22,** 835–842.

Sundaresan, P.R., Kaup, S.M., Wiesenfeld, P.W., Chirtel, S.J., Hight, S.C., and Rader, J.I. (1996). Interactions in indices of vitamin A, zinc and copper status when these nutrients are fed to rats at adequate and increased levels. *Br. J. Nutr.* **75,** 915–928.

Sunde, R.A., Weiss, S.L., et al. (1992). Dietary selenium regulation of glutathione peroxidase mRNA-: implications for the selenium requirement. *FASEB J.* **6**(part1), A1365.

Sunder-Plassmann, G., Patruta, S.I., and Horl, W.H.. (1999). Pathobiology of the role of iron in infection. *Am. J. Kidney. Dis.* **34,** S25–S29.

Suttie, J. (1985). Vitamin K. *In* "Fat Soluble Vitamins." (A.T. Diplock, ed), p. 225. Technomic, Lancaster, PA.

Suttie, J. (2001). Vitamin K. *In* "Handbook of Vitamins." (R.B. Rucker, J.W. Suttie, D.B. McCormick, and L.J. Machlin, eds), p. 115. Marcel Dekker, New York.

Tabchoury, C.M., Holt, T., Pearson, S.K., and Bowen, W.H. (1998). The effects of fluoride concentration and the level of cariogenic challenge on caries development in desalivated rats. *Arch. Oral. Biol.* **43,** 917–924.

Takahashi, K. (1981). Thiamine deficiency neuropathy, a reappraisal. *Int. J. Neurol.* **15,** 245.

Takahashi, S., Nakae, D., Yokose, Y., Emi, Y., Denda, A., Mikami, S., Ohnishi, T., and Konishi, Y. (1984). Enhancement of DEN initiation of liver carcinogenesis by inhibitors of NAD+ ADP ribosyl transferase in rats. *Carcinogenesis* **5,** 901–906.

Takami, Y., Gong, H., and Amemiya, T. (2004). Riboflavin deficiency induces ocular surface damage. *Opthalmic Res.* **36,** 156.

Takeda, T., Kimura, M., Yokoi, K., and Itokawa, Y. (1996). Effect of age and dietary protein level on tissue mineral levels in female rats. *Biol. Trace Elem. Res.* **54,** 55–74.

Tallkvist, J., and Tjalve, H. (1997). Effect of dietary iron-deficiency on the disposition of nickel in rats. *Toxicol. Lett.* **92,** 131–138.

Tamura, T., Hong, K.H., Mizuno, Y., Johnston, K.E., and Keen, C.L. (1999). Folate and homocysteine metabolism in copper-deficient rats. *Biochim. Biophys. Acta* **1427,** 351–356.

Tamura, T., Kaiser, L.L., Watson, J.E., Halsted, C.H., Hurley, L.S., and Stokstad, E.L. (1987). Increased methionine synthetase activity in zinc-deficient rat liver. *Arch. Biochem. Biophys.* **256,** 311–316.

Tanaka, M., Yanagi, M., Shirota, K., Une, Y., Nomura, Y., Masaoka, T., and Akahori, F. (1995). Effect of cadmium in the zinc deficient rat. *Vet. Hum. Toxicol.* **37,** 203–208.

Tanaka, T., Kono, T., Terasaki, F., Kintaka, T., Sohmiya, K., Mishima, T., and Kitaura, Y. (2003). Gene-environment interactions in wet beriberi: effects of thiamine depletion in CD36-defect rats. *Am. J. Physiol. Heart Circ. Physiol.* **285,** H1546–H1553.

Tanaka, T.A., Hasegawa, A., Yamamoto, S., et al. (1988). Worldwide contamination of cereals by the Fusarium mycotoxins nivalenol, deoxynivalenol, and zearalenone, 1: survey of 19 countries. *J. Agric. Food Chem.* **36,** 979–983.

Tang, Z., Gasperkova, D., Xu, J., Baillie, R., Lee, J.H., and Clarke, S.D. (2000). Copper deficiency induces hepatic fatty acid synthase gene transcription in rats by increasing the nuclear content of mature sterol regulatory element binding protein 1. *J. Nutr.* **130,** 2915–2921.

Taniguchi, M., and Hara, T. (1983). Effects of riboflavin and selenium deficiencies on glutathione and its relating enzyme activities with respect to lipid peroxide content of rat livers. *J. Nutr. Sci. Vitaminol. (Tokyo)* **29,** 283–292.

Taniguchi, M., and Nakamura, M. (1976). Effects of riboflavin deficiency on the lipids of rat liver. *J. Nutr. Sci. Vitaminol. (Tokyo)* **22,** 135–146.

Taniguchi, M., Yamamoto, T., and Nakamura, M. (1978). Effects of riboflavin deficiency on the lipids of rat liver mitochondria and microsomes. *J. Nutr. Sci. Vitaminol. (Tokyo)* **24,** 363–381.

Tannenbaum, A. (1945). The dependence of tumour formation on the composition or the calorie-restricted diet as well as the degree of restriction. *Cancer Res.* **5,** 616–625.

Tanphaichitr, V. (2001). Thiamine. *In* "Handbook of Vitamins." (R.B. Rucker, J.W. Suttie, D.B. McCormick, and L.J. Machlin, eds), p. 275. Marcel Dekker, New York.

Tatum, L., Shankar, P., Boylan, L.M., and Spallholz, J.E. (2000). Effect of dietary copper on selenium toxicity in Fischer 344 rats. *Biol. Trace Elem. Res.* **77,** 241–249.

Taylor, S., and Poulson, E. (1956). Long-term iodine deficiency in the rat. *J. Endocrinol.* **13**, 439–444.

Taylor, S.A., Shrader, R.E., Koski, K.G., and Zeman, F.J. (1983). Maternal and embryonic response to a carbohydrate-free diet fed to rats. *J. Nutr.* **113**, 253–267.

Temcharoen, P., and Thilly, W.G. (1982). Removal of aflatoxin B1 toxicity but not mutagenicity by 1 megarad gamma radiation of peanut meal. *J. Food Safety* **4**, 199–205.

Terasawa, M., Nakahara, T., Tsukada, N., Sugawara, A., and Itokawa, Y. (1999). The relationship between thiamine deficiency and performance of a learning task in rats. *Metab. Brain Dis.* **14**, 137–148.

Thayer, D.W. (1990). Food irradiation: benefits and concerns. *J. Food Quality* **13**, 147–169.

Thenen, S.W. (1989). Megadose effects of vitamin C on vitamin B-12 status in the rat. *J. Nutr.* **119**, 1107–1114.

Thigpen, J.E., Setchell, K.D., Ahlmark, K.B., Locklear, J., Spahr, T., Caviness, G.F., Goelz, M.F., Haseman, J.K., Newbold, R.R., and Forsythe, D.B. (1999). Phytoestrogen content of purified, open- and closed-formula laboratory animal diets. *Lab. Anim. Sci.* **49**, 530–536.

Thomas, M.R., and Kirksey, A. (1976). Postnatal patterns of brain lipids in progeny of vitamin B-6 deficient rats before and after pyridoxine supplementation. *J. Nutr.* **106**, 1404–1414.

Thomas, V.L., and Gropper, S.S. (1996). Effect of chromium nicotinic acid supplementation on selected cardiovascular disease risk factors. *Biol. Trace Elem. Res.* **55**, 297–305.

Thomsen, L.L., Tasman-Jones, C., and Maher, C. (1983). Effects of dietary fat and gel-forming substances on rat jejunal disaccharidase levels. *Digestion* **26**, 124–130.

Thonney, M.L., Touchberry, R.W., Goodrich, R.D., and Meiske, J.C. (1976). Intraspecies relationship between fasting heat production and body weight: a reevaluation of W. *J. Anim. Sci.* **43**, 692–704.

Tobia, M.H., Zdanowicz, M.M., Wingertzahn, M.A., McHeffey-Atkinson, B., Slonim, A.E., and Wapnir, R.A. (1998). The role of dietary zinc in modifying the onset and severity of spontaneous diabetes in the BB Wistar rat. *Mol. Genet. Metab.* **63**, 205–213.

Tobian, L. (1991). Salt and hypertension: lessons from animal models that relate to human hypertension. *Hypertension* **17**, I52–I58.

Tokunaga, T., Oku, T., and Hosoya, N. (1989). Utilization and excretion of a new sweetener, fructooligosaccharide (Neosugar), in rats. *J. Nutr.* **119**, 553–559.

Tolleson, W.H., Dooley, K.L., Sheldon, W.G., Thurman, J.D., Bucci, T.J., and Howard, P.C. (1996). The mycotoxin fumonisin induces apoptosis in cultured human cells and in livers and kidneys of rats. *Adv. Exp. Med. Biol.* **392**, 237–250.

Tordoff, M.G., Hughes, R.L., and Pilchak, D.M. (1998). Calcium intake by rats: influence of parathyroid hormone, calcitonin, and 1,25-dihydroxyvitamin D. *Am. J. Physiol.* **274**, R214–R231.

Touyz, R.M., and Milne, F.J. (1999). Magnesium supplementation attenuates, but does not prevent, development of hypertension in spontaneously hypertensive rats. *Am. J. Hypertens.* **12**, 757–765.

Touyz, R.M., Pu, Q., He, G., Chen, X., Yao, G., Neves, M.F., and Viel, E. (2002). Effects of low dietary magnesium intake on development of hypertension in stroke-prone spontaneously hypertensive rats: role of reactive oxygen species. *J. Hypertens.* **20**, 2221–2232.

Track, N.S., Cannon, M.M., Flenniken, A., Katamay, S., and Woods, E.F. (1982). Improved carbohydrate tolerance in fibre-fed rats: studies of the chronic effect. *Can. J. Physiol. Pharmacol.* **60**, 769–776.

Tran, T.T., Chowanadisai, W., Crinella, F.M., Chicz-DeMet, A., and Lonnerdal, B. (2002). Effect of high dietary manganese intake of neonatal rats on tissue mineral accumulation, striatal dopamine levels, and neurodevelopmental status. *Neurotoxicology* **23**, 635–643.

Tritsch, G.L. (1993). The safety of irradiated foods. *JAMA* **270**, 575–576.

Trugnan, G., Thomas-Benhamou, G., Cardot, P., Rayssiguier, Y., and Bereziat, G. (1985). Short term essential fatty acid deficiency in rats. Influence of dietary carbohydrates. *Lipids* **20**, 862–868.

Tsugawa, N., Yamabe, T., Takeuchi, A., Kamao, M., Nakagawa, K., Nishijima, K., and Okano. T. (1999). Intestinal absorption of calcium from calcium ascorbate in rats. *J. Bone Miner. Metab.* **17**, 30–36.

Tuitoek, P.J., Thomson, A.B., Rajotte, R.V., and Basu, T.K. (1994). Intestinal absorption of vitamin A in streptozotocin-induced diabetic rats. *Diabetes Res.* **25**, 151–158.

Tully, D.B., Allgood, V.E., and Cidlowski, J.A. (1994). Modulation of steroid receptor-mediated gene expression by vitamin B6. *FASEB J.* **8**, 343–349.

Turan, B., Acan, N.L., Ulusu, N.N., and Tezcan, E.F. (2001). A comparative study on effect of dietary selenium and vitamin E on some antioxidant enzyme activities of liver and brain tissues. *Biol. Trace Elem. Res.* **81**, 141–152.

Turkki, P.R., and Degruccio, G.D. (1983). Riboflavin status of rats fed two levels of protein during energy deprivation and subsequent repletion. *J. Nutr.* **113**, 282–292.

Turkki, P.R., and Holtzapple, P.G. (1982). Growth and riboflavin status of rats fed different levels of protein and riboflavin. *J. Nutr.* **112**, 1940–1952.

Turkki, P.R., Ingerman, L., Kurlandsky, S.B., Yang, C., and Chung, R.S. (1989). Effect of energy restriction on riboflavin retention in normal and deficient tissues of the rat. *Nutrition* **5**, 331–337.

Turnlund, J.R. (1999). Copper. *In* "Modern Nutrition in Health and Disease." (M.E. Shils, J.A. Olson, M. Shike, and C.A. Ross, eds), p. 241. Williams and Wilkins, Baltimore.

Turturro, A., Witt, W.W., Lewis, S., Hass, B.S., Lipman, R.D., and Hart, R.W. (1999). Growth curves and survival characteristics of the animals used in the Biomarkers of Aging Program. *J. Gerontol. A Biol. Sci. Med. Sci.* **54**, B492–B501.

Twining, S.S., Schulte, D.P., Wilson, P.M., Fish, B.L., and Moulder, J.E. (1996). Retinol is sequestered in the bone marrow of vitamin A-deficient rats. *J. Nutr.* **126**, 1618–1626.

Twining, S.S., Schulte, D.P., Wilson, P.M., Fish, B.L., and Moulder, J.E. (1997). Vitamin A deficiency alters rat neutrophil function. *J. Nutr.* **127**, 558–565.

Uhland, A.M., Kwiecinski, G.G., and DeLuca, H.F. (1992). Normalization of serum calcium restores fertility in vitamin D-deficient male rats. *J. Nutr.* **122**, 1338–1344.

Ulman, E.A., and Fisher, H. (1983). Arginine utilization of young rats fed diets with simple versus complex carbohydrates. *J. Nutr.* **113**, 131–137.

Underwood, B.A. (1984). Vitamin A in animal and human nutrition. *In* "The Retinoids," Vol I. (M.B. Spron, A.B. Roberts, and D.S. Goodman, eds), p. 281. Academic Press, New York.

Underwood, J.L., and DeLuca, H.F. (1984). Vitamin D is not directly necessary for bone growth and mineralization. *Am. J. Physiol.* **246**, E493–E498.

Unna, K. (1940). Pantothenic acid requirement of the rat. *J. Nutr.* **20**, 565.

Uotila, L. (1988). Inhibition of vitamin K group, XI: pharmacology and toxicology. *In* "Current Advances in Vitamin K Research" (J.W. Suttie, ed), p. 59. Elsevier, New York.

Uritski, R., Barshack, I., Bilkis, I., Ghebremeskel, K., and Reifen, R. (2004). Dietary iron affects inflammatory status in a rat model of colitis. *J. Nutr.* **134**, 2251–2255.

Uthus, E., and Poellot, R. (1991). Effect of dietary pyridoxine on arsenic deprivation in rats. *Magnes. Trace Elem.* **10**, .339–347.

Uthus, E.O. (1994). Arsenic essentiality and factors affecting its importance. *In* "Arsenic: Exposure and Health." (W.R. Chappell, C.O. Abernathy, and C.R. Cothern, eds), p. 1999. Science and Technology Letters, Northwood, UK.

Uthus, E.O., and Nielsen, F.H. (1993). Determination of the possible requirement and reference dose levels for arsenic in humans. *Scand. J. Work Environ. Health* **19**, 137–138.

Uthus, E.O., and Poellot, R.A. (1996). Dietary folate affects the response of rats to nickel deprivation. *Biol. Trace Elem. Res.* **52**, 23–35.

Uthus, E.O., and Poellot, R.A. (1997). Dietary nickel and folic acid interact to affect folate and methionine metabolism in the rat. *Biol. Trace Elem. Res.* **58**, 25–33.

Uthus, E.O., Yokoi, K., and Davis, C.D. (2002). Selenium deficiency in Fisher-344 rats decreases plasma and tissue homocysteine concentrations and alters plasma homocysteine and cysteine redox status. *J. Nutr.* **132**, 1122–1128.

Uthus, E.O., and Zaslavsky, B. (2001). Interaction between zinc and iron in rats: experimental results and mathematical analysis of blood parameters. *Biol. Trace Elem. Res.* **82**, 167–183.

Vadhanavikit, S., and Ganther, H.E. (1993). Selenium requirements of rats for normal hepatic and thyroidal 5′-deiodinase (type I) activities. *J. Nutr.* **123**, 1124–1128.

Vahter, M. (1983). Metabolism of arsenic. *In* "Biological and Environmental Effects of Arsenic." (B.A. Fowler, ed), p. 171. Elsevier, Amsterdam.

Valencia, R., and Raibaud, P. (1968). Influence of culture conditions on the production of cobalamins and on their fermentation by the bacteria of the intestinal flora of the rat. *Ann. Nutr. Aliment.* **22**, 77–81.

Vallee, B.L., and Falchuk, K.H. (1994). The biochemical basis of zinc physiology. *Physiol. Rev.* **73**, 79.

Van Baelen, H., Allewaert, K., and Bouillon, R. (1988). New aspects of the plasma carrier protein for 25-hydroxycholecalciferol in vertebrates. *Ann. NY Acad. Sci.* **538**, 60–68.

Van den Berg, G.J., Yu, S., Lemmens, A.G., and Beynen, A.C. (1994). Dietary ascorbic acid lowers the concentration of soluble copper in the small intestinal lumen of rats. *Br. J. Nutr.* **71**, 701–707.

Van Houwelingen, F., Van den Berg, G.J., Lemmens, A.G., Sijtsma, K.W., and Beynen, A.C. (1993). Iron and zinc status in rats with diet-induced marginal deficiency of vitamin A and/or copper. *Biol. Trace Elem. Res.* **38**, 83–95.

van Rensburg, S.J., Hall, J.M., and Gathercole, P.S. (1986). Inhibition of esophageal carcinogenesis in corn-fed rats by riboflavin, nicotinic acid, selenium, molybdenum, zinc, and magnesium. *Nutr. Cancer* **8**, 163–170.

VanSoest, P.J. (1994). "Nutritional Ecology of the Ruminant," 2nd ed. Cornell University Press, Ithaca, NY.

Varela-Moreiras, G., Perez-Olleros, L., Garcia-Cuevas, M., and Ruiz-Roso, B. (1994). Effects of ageing on folate metabolism in rats fed a long-term folate deficient diet. *Int. J. Vitamin Nutr. Res.* **64**, 294–299.

Varela-Moreiras, G., and Selhub, J. (1992). Long-term folate deficiency alters folate content and distribution differentially in rat tissues. *J. Nutr.* **122**, 986–991.

Vaskonen, T., Mervaala, E., Sumuvuori, V., Seppanen-Laakso, T., and Karppanen, H. (2002). Effects of calcium and plant sterols on serum lipids in obese Zucker rats on a low-fat diet. *Br. J. Nutr.* **87**, 239–245.

Vendeland, S.C., Butler, J.A., and Whanger, P.D. (1992). Intestinal absorption of selenite, selenate, and selenomethionine in the rat. *J. Nutr. Biochem.* **3**, 359.

Verri, A., Laforenza, U., Gastaldi, G., Tosco, M., and Rindi, G. (2002). Molecular characteristics of small intestinal and renal brush border thiamin transporters in rats. *Biochim. Biophys. Acta* **1558**, 187–197.

Viestenz, K.E., and Klevay, L.M. (1982). A randomized trial of copper therapy in rats with electrocardiographic abnormalities due to copper deficiency. *Am. J. Clin. Nutr.* **35**, 258–266.

Vinas-Salas, J., Biendicho-Palau, P., Pinol-Felis, C., Miguelsanz-Garcia, S., and Perez-Holanda, S. (1998). Calcium inhibits colon carcinogenesis in an experimental model in the rat. *Eur. J. Cancer* **34**, 1941–1945.

Voss, K.A., Riley, R.T., Norred, W., et al. (2002). An overview of rodent toxicities: liver and kidney effects of fumonisins and Fusarium moniliforme. *Environ. Health Perspect.* **109**, 259–266.

Wagner, C. (1995). Biochemical role of folate in cellular metabolism. *In* "Folate in Health and Disease." (L.B. Baily, ed), p. 23. Marcel Dekker, New York.

Wald, G. (1968). The molecular basis of visual excitation. *Nature* **219**, 800–807.

Walzem, R.L., and Clifford, A.J. (1988). Thiamin absorption is not compromised in folate-deficient rats. *J. Nutr.* **118**, 1343–1348.

Wang, E., Norred, W.P., Bacon, C.W., Riley, R.T., and Merrill, A.H. Jr. (1991). Inhibition of sphingolipid biosynthesis by fumonisins. Implications for diseases associated with Fusarium moniliforme. *J. Biol. Chem.* **266**, 14486–14490.

Wang, G.S., Eriksson, L.C., Xia, L., Olsson, J., and Stal, P. (1999). Dietary iron overload inhibits carbon tetrachloride-induced promotion in chemical hepatocarcinogenesis: effects on cell proliferation, apoptosis, and antioxidation. *J. Hepatol.* **30**, 689–698.

Wang, J.L., Swartz-Basile, D.A., Rubin, D.C., and Levin, M.S. (1997). Retinoic acid stimulates early cellular proliferation in the adapting remnant rat small intestine after partial resection. *J. Nutr.* **127**, 1297–1303.

Ward, G.J., and Nixon, P.F. (1990). Modulation of pteroylpolyglutamate concentration and length in response to altered folate nutrition in a comprehensive range of rat tissues. *J. Nutr.* **120**, 476–484.

Ward, J.M., and Ohshima, M. (1986). The role of iodine in carcinogenesis. *Adv. Exp. Med. Biol.* **206**, 529–542.

Ward, N.I. (1993). Boron levels in human tissues and fluids. *In* "Trace Elements in Man and Animals." (M. Anke, D. Meissner, and C.F. Mills, eds), p. 724. Verlag Media Touristik, Gersdorf.

Wasynczuk, A., Kirksey, A., and Morre, D.M. (1983a). Effect of maternal vitamin B-6 deficiency on specific regions of developing rat brain: amino acid metabolism. *J. Nutr.* **113**, 735–745.

Wasynczuk, A., Kirksey, A., and Morre, D.M. (1983b). Effects of maternal vitamin B-6 deficiency on specific regions of developing rat brain: the extrapyramidal motor system. *J. Nutr.* **113**, 746–754.

Watkins, D.W., Jahangeer, S., Floor, M.K., and Alabaster, O. (1992). Magnesium and calcium absorption in Fischer-344 rats influenced by changes in dietary fibre (wheat bran), fat and calcium. *Magnes. Res.* **5**, 15–21.

Watt, B.K., and Merrill, A.L. (1963). Composition of Foods, Agriculture Handbook No. 8. U.S. Government. Printing Office, Washington, DC.

Weaver, C.M., and Heaney, R.P. (1999). Calcium. *In* "Modern Nutrition in Health and Disease." (M.E. Shils, J.A. Olson, M. Shike, and C.A. Ross, eds), p. 141. Williams and Wilkins, Baltimore.

Weijnen, J.A. (1989). Lick sensors as tools in behavioral and neuroscience research. *Physiol. Behav.* **46**, 923–928.

Weindruch, R., and Sohal, R.S. (1997). Caloric intake and aging: seminars in medicine of the Beth Israel Deaconess Medical Center. *N. Engl. J. Med.* **337**, 986–994.

Weindruch, R., and Walford, R.L. (1988). "The Retardation of Aging and Disease by Dietary Restriction." Charles C. Thomas, Springfield, IL.

Weisstaub, A., de Ferrer, P.R., Zeni, S., and de Portela, M.L. (2003). Influence of low dietary calcium during pregnancy and lactation on zinc levels in maternal blood and bone in rats. *J. Trace Elem. Med. Biol.* **17**, 27–32.

Wells, I.C., and Hogan, J.M. (1968). Effects of dietary deficiencies of lipotropic factors on plasma cholesterol esterification and tissue cholesterol in rats. *J. Nutr.* **95**, 55–62.

Werman, M.J., Barat, E., and Bhathena, S.J. (1995). Gender, dietary copper and carbohydrate source influence cardiac collagen and lysyl oxidase in weanling rats. *J. Nutr.* **125**, 857–863.

Werner, L., Korc, M., and Brannon, P.M. (1987). Effects of manganese deficiency and dietary composition on rat pancreatic enzyme content. *J. Nutr.* **117**, 2079–2085.

Wester, P.O. (1987). Magnesium. *Am. J. Clin. Nutr.* **45**, 1305–1312.

Whang, R., Oliver, J., Welt, L.G., and MacDowell, M. (1969). Renal lesions and disturbance of renal function in rats with magnesium deficiency. *Ann. NY Acad. Sci.* **162**, 766–774.

Whanger, P.D., and Butler, J.A. (1988). Effects of various dietary levels of selenium as selenite or selenomethionine on tissue selenium levels and glutathione peroxidase activity in rats. *J. Nutr.* **118**, 846–852.

Whitescarver, S.A., Holtzclaw, B.J., Downs, J.H., Ott, C.E., Sowers, J.R., and Kotchen, T.A. (1986). Effect of dietary chloride on salt-sensitive and renin-dependent hypertension. *Hypertension* **8**, 56–61.

Whittaker, P., and Chanderbhan, R.F. (2001). Effect of increasing iron supplementation on blood lipids in rats. *Br. J. Nutr.* **86**, 587–592.

Whitten, P.L., and Patisaul, H.B. (2001). Cross-species and interassay comparisons of phytoestrogen action. *Environ. Health Perspect.* **109**, 5–20.

Wiegand, E., Kirchgessner, M., et al. (1986). Manganese. *Biol. Trace Elem. Res.* **10**, 265.

Wienk, K.J., Marx, J.J., Santos, M., Lemmens, A.G., Brink, E.J., Van der Meer, R., and Beynen, A.C. (1997). Dietary ascorbic acid raises iron absorption in anaemic rats through enhancing mucosal iron uptake independent of iron solubility in the digesta. *Br. J. Nutr.* **77**, 123–131.

Wildman, R.E., Hopkins, R., Failla, M.L., and Medeiros, D.M. (1995). Marginal copper-restricted diets produce altered cardiac ultrastructure in the rat. *Proc. Soc. Exp. Biol. Med.* **210**, 43–49.

Will, B.H., Usui, Y., and Suttie, J.W. (1992). Comparative metabolism and requirement of vitamin K in chicks and rats. *J. Nutr.* **122**, 2354–2360.

Williams, C.L., Meck, W.H., Heyer, D.D., and Loy, R. (1998). Hypertrophy of basal forebrain neurons and enhanced visuospatial memory in perinatally choline-supplemented rats. *Brain Res.* **794**, 225–238.

Williams, E.A., Powers, H.J., and Rumsey, R.D. (1995). Morphological changes in the rat small intestine in response to riboflavin depletion. *Br. J. Nutr.* **73**, 141–146.

Williams, E.A., Rumsey, R.D., and Powers, H.J. (1996a). An investigation into the reversibility of the morphological and cytokinetic changes seen in the small intestine of riboflavin deficient rats. *Gut* **39**, 220–225.

Williams, E.A., Rumsey, R.D., and Powers, H.J. (1996b). Cytokinetic and structural responses of the rat small intestine to riboflavin depletion. *Br. J. Nutr.* **75**, 315–324.

Wills, E.D. (1980). Studies of lipid peroxide formation in irradiated synthetic diets and the effects of storage after irradiation. *Int. J. Radiat. Biol. Relat. Stud. Phys. Chem. Med.* **37**, 383–401.

Wilson, J.G., Roth, C.B., and Warkany, J. (1953). An analysis of the syndrome of malformations induced by maternal vitamin A deficiency: effects of restoration of vitamin A at various times during gestation. *Am. J. Anat.* **92**, 189.

Windebank, A.J., Low, P.A., Blexrud, M.D., Schmelzer, J.D., and Schaumburg, H.H. (1985). Pyridoxine neuropathy in rats: specific degeneration of sensory axons. *Neurology* **35**, 1617–1622.

Wiseman, H. (2000). Dietary phytoestrogens, oestrogens and tamoxifen: mechanisms of action in modulation of breast cancer risk and in heart disease prevention. *In* "Biomolecular Free Radical Toxicity: Causes and Prevention." (H. Wiseman, P. Goldfarb, T. Ridgway, and A. Wiseman, eds). John Wiley & Sons, Ltd, West Sussex, England.

Witten, P.W., and Holman, R.T. (1952). Polyethenoid fatty acid metabolism, VI: effect of pyridoxine on essential fatty acid conversions. *Arch. Biochem. Biophys.* **41**, 266–273.

Wolbach, S.B., and Howe, P.R. (1925). Tissue changes following deprivation of fat-soluble A vitamin. *J. Exp. Med.* **42**, 753–777.

Wolf, G. (1991). The intracellular vitamin A-binding proteins: an overview of their functions. *Nutr. Rev.* **49**, 1–12.

Woodard, J.C., and Newberne, P.M. (1966). Relation of vitamin B12 and one-carbon metabolism to hydrocephalus in the rat. *J. Nutr.* **88**, 375–381.

Woods, R.J. (1994). Food irradiation. *Endeavour* **18**, 104–108.

Woolliscroft, J., and Barbosa, J. (1977). Analysis of chromium induced carbohydrate intolerance in the rat. *J. Nutr.* **107**, 1702–1706.

Worcester, N.A., Bruckdorfer, K.R., Hallinan, T., Wilkins, A.J., Mann, J.A., and Yudkins, J. (1979). The influence of diet and diabetes on stearoyl conenzyme A desaturase (EC 1.14.99.5) activity and fatty acid composition in rat tissues. *Br. J. Nutr.* **41**, 239–252.

World Health Organization [WHO]. (1977). "Wholesomeness of irradiated food." WHO, Geneva.

Wostmann, B.S., Knight, P.L., Keeley, L.L., and Kan, D.F. (1963). Metabolism and function of thiamine and naphthoquinones in germfree and conventional rats. *Fed. Proc.* **22**, 120–124.

Wu, X., Vieth, R., Milojevic, S., Sonnenberg, H., and Melo, L.G. (2000). Regulation of sodium, calcium and vitamin D metabolism in Dahl rats on a high-salt/low-potassium diet: genetic and neural influences. *Clin. Exp. Pharmacol. Physiol.* **27**, 378–383.

Xie, X.M., and Smart, T.G. (1991). A physiological role for endogenous zinc in rat hippocampal synaptic neurotransmission. *Nature* **349**, 521–524.

Xue, Q., Aliabadi, H., and Hallfrisch, J. (2001). Effects of dietary galactose and fructose on rats fed diets marginal or adequate in copper for 9- to 21 months. *Nutr. Res.* **21**, 1078–1087.

Yagi, K., Komuro, S., et al. (1989). Serum lipid peroxides and cataractogenesis in riboflavin deficiency. *J. Clin. Biochem. Nutr.* **6**, 39.

Yamagami, T., Miwa, A., Takasawa, S., Yamamoto, H., and Okamoto, H. (1985). Induction of rat pancreatic B-cell tumors by the combined administration of streptozotocin or alloxan and poly(adenosine diphosphate ribose) synthetase inhibitors. *Cancer Res.* **45**, 1845–1849.

Yamanaka, M., Saito, M., and Nomura, T. (1981). A comparison of the nutritional evaluation of irradiated and autoclaved diets in germfree rats. *Jikken Dobutsu* **30**, 299–302.

Yamauchi, H., Kaise, T., and Yamamura, Y. (1986). Metabolism and excretion of orally administered arsenobetaine in the hamster. *Bull. Environ. Contam. Toxicol.* **36**, 350–355.

Yang, C.Z., and Bittner, G.D. (2002). Effects of some dietary phytoestrogens in animal studies: review of a confusing landscape. *Lab. Anim.* **31**, 43–48.

Yang, J.G., Hill, K.E., and Burk, R.F. (1989). Dietary selenium intake controls rat plasma selenoprotein P concentration. *J. Nutr.* **119**, 1010–1012.

Yang, J.G., Morrison-Plummer, J., and Burk, R.F. (1987). Purification and quantitation of a rat plasma selenoprotein distinct from glutathione peroxidase using monoclonal antibodies. *J. Biol. Chem.* **262**, 13372–13375.

Yao, Z.M., and Vance, D.E. (1988). The active synthesis of phosphatidylcholine is required for very low density lipoprotein secretion from rat hepatocytes. *J. Biol. Chem.* **263**, 2998–3004.

Yasui, M., Ota, K., and Garruto, R.M. (1995). Effects of calcium-deficient diets on manganese deposition in the central nervous system and bones of rats. *Neurotoxicology* **16**, 511–517.

Yates, C.A., Evans, G.S, and Powers, H.J. (2001). Riboflavin deficiency: early effects on post-weaning development of the duodenum in rats. *Br. J. Nutr.* **86**, 593–599.

Yokogoshi, H., Hayase, K., and Yoshida, A. (1980). Effect of carbohydrates and starvation on nitrogen sparing action of methionine and threonine in rats. *Agr. Biol. Chem.* **44**, 2503–2506.

Yokoi, K., Kimura, M., and Itokawa, Y. (1991). Effect of dietary iron deficiency on mineral levels in tissues of rats. *Biol. Trace Elem. Res.* **29**, 257–265.

Yokoi, K., Uthus, E.O., and Nielsen, F.H. (2003). Nickel deficiency diminishes sperm quantity and movement in rats. *Biol. Trace Elem. Res.* **93**, 141–154.

Young, V.R., and Munro, H.N. (1973). Plasma and tissue tryptophan levels in relation to tryptophan requirements of weanling and adult rats. *J. Nutr.* **103**, 1756–1763.

Yousef, M.I., El-Hendy, H.A., El-Demerdash, F.M., and Elagamy, E.I. (2002). Dietary zinc deficiency induced-changes in the activity of enzymes and the levels of free radicals, lipids and protein electrophoretic behavior in growing rats. *Toxicology* **175**, 223–234.

Youssef, J.A., Song, W.O., and Badr, M.Z. (1997). Mitochondrial, but not peroxisomal, beta-oxidation of fatty acids is conserved in coenzyme A-deficient rat liver. *Mol. Cell Biochem.* **175**, 37–42.

Yu, B.P., Masoro, E.J., and McMahan, C.A. (1985). Nutritional influences on aging of Fischer 344 rats, I: physical, metabolic, and longevity characteristics. *J. Gerontol.* **40**, 657–670.

Yu, S., and Beynen, A.C. (2001). The lowering effect of high copper intake on selenium retention in weanling rats depends on the selenium concentration of the diet. *J. Anim. Physiol. Anim. Nutr. (Berl)* **85**, .29–37.

Yu S, West CE, Beynen AC. (1994). Increasing intakes of iron reduce status, absorption and biliary excretion of copper in rats. *Br. J. Nutr.* **71**, 887–895.

Zaki, F.G., C. Bandt, C., and Hoffbauer, F.W. (1963). Fatty cirrhosis in the rat, III: liver lipid and collagen content in various stages. *Arch. Pathol.* **75**, 648–653.

Zeisel, S.H. (1999). Choline and phosphatidylcholine. *In* "Modern Nutrition in Health and Disease." (M.E. Shils, J.A. Olson, M. Shike, and C.A. Ross, eds), p. 513. Williams and Wilkins, Baltimore.

Zeisel, S.H., and Holmes-McNary, M. (2001). Choline. *In* "Handbook of Vitamins." (R.B. Rucker, J.W. Suttie, D.B. McCormick, and L.J. Machlin, eds), p. 513. Marcel Dekker, New York.

Zeisel, S.H., Zola, T., daCosta, K.A., and Pomfret, E.A. (1989). Effect of choline deficiency on S-adenosylmethionine and methionine concentrations in rat liver. *Biochem. J.* **259**, 725–729.

Zeni, S., Weisstaub, A., Di Gregorio, S., Ronanre De Ferrer, P., and Portela, M.L. (2003). Bone mass changes *in vivo* during the entire reproductive cycle in rats feeding different dietary calcium and calcium/phosphorus ratio content. *Calcif. Tissue Int.* **73**, 594–600.

Zhang, J.Z., Henning, S.M., and Swendseid, M.E. (1993). Poly(ADP-ribose) polymerase activity and DNA strand breaks are affected in tissues of niacin-deficient rats. *J. Nutr.* **123**, 1349–1355.

Zhang, R., Ma, J., Xia, M., Zhu, H., and Ling, W. (2004). Mild hyperhomocysteinemia induced by feeding rats diets rich in methionine or deficient in folate promotes early atherosclerotic inflammatory processes. *J. Nutr.* **134**, 825–830.

Zhao, Z., Murasko, D.M., and Ross, A.C. (1994). The role of vitamin A in natural killer cell cytotoxicity, number and activation in the rat. *Nat. Immun.* **13**, 29–41.

Zhou, J.R., Fordyce, E.J., Raboy, V., Dickinson, D.B., Wong, M.S., Burns, R.A., and Erdman, J.W. Jr. (1992). Reduction of phytic acid in soybean products improves zinc bioavailability in rats. *J. Nutr.* **122**, 2466–2473.

Zhou, M.S., Kosaka, H., and Yoneyama, H. (2000). Potassium augments vascular relaxation mediated by nitric oxide in the carotid arteries of hypertensive Dahl rats. *Am. J. Hypertens.* **13**, 666–672.

Zhu, J., Mori, T., Huang, T., and Lombard, J.H. (2004). Effect of high-salt diet on NO release and superoxide production in rat aorta. *Am. J. Physiol. Heart Circ. Physiol.* **286**, H575–H583.

Zhu, Z., Kimura, M., and Itokawa, Y. (1993). Selenium concentration and glutathione peroxidase activity in selenium and magnesium deficient rats. *Biol. Trace Elem. Res.* **37**, 209–217.

Zidenberg-Cherr, S., Keen, C.L., Lonnerdal, B., and Hurley, L.S. (1983). Superoxide dismutase activity and lipid peroxidation in the rat: developmental correlations affected by manganese deficiency. *J. Nutr.* **113**, 2498–2504.

Ziesenitz, S.C., Siebert, G., Schwengers, D., and Lemmes, R. (1989). Nutritional assessment in humans and rats of leucrose [D-glucopyranosyl-α(1–5)-D-fructopyranose] as a sugar substitute. *J. Nutr.* **119**, 971–978.

Zimitat, C., and Nixon, P.F. (2001). Glucose induced IEG expression in the thiamin-deficient rat brain. *Brain Res.* **892**, 218–227.

Chapter 10

Housing and Environment

Robert E. Faith and Jack R. Hessler

THE LABORATORY RAT, 2ND EDITION

I. INTRODUCTION

The methods and techniques for animal care and research have reached a high level of sophistication. Since the end of World War II, there have been two major expansions in the use of rodents in biomedical research. Immediately after World War II through the 1960s, there was rapid expansion of animal research with rapidly increasing sophistication (Baker et al., 1979). There was a movement to standardize research animals. The scientific community exerted pressure to reduce infectious diseases (Rowe et al., 1962; Holdenreid, R., 1966). Application of gnotobiotic principles to large-scale rodent production led to major advances in quality (Foster, 1958; Foster et al., 1963; Foster and Pfan, 1963; Simons and Brick, 1970). Husbandry practices and facilities capable of maintaining rats free of common infectious diseases were developed during this time (Foster, 1962; Jonas, 1965; Christie et al., 1968; Brick et al., 1969; Beall et al., 1971; Lang and Harrell, 1971; Serrano, 1971b).

In the early 1980s, techniques were developed that allowed for the consistent genetic manipulation of mammals, and these have fueled a second expansion (Gordon et al., 1985; Ledermann, 1985; Connelly et al., 1989; Evans, 1989; Robertson, 1991; Sigmund, 1993; Wight and Wagner, 1994). These techniques have led to the development of large numbers of transgenic, gene knockout, and targeted mutagenic rodents, especially mice (Wilder and Rizzino, 1993; Moreadith and Radford, 1997; Erickson, 1999; Jacob, 1999; Babinet, 2000; Justice, 2000; Richa, 2001; Jacob and Kwitek, 2002). These genetically manipulated rodents have been shown to be useful for a wide variety of scientific investigations, including *in vivo* gene function and regulation, mammalian development and biology, the development of models of human disease, and models for the study of gene therapy (Cline, 1986; Connelly et al., 1989; Wilder and Rizzino, 1993; Babinet, 2000; Ledermann, 2000; Hamilton and Frankel, 2001; Hemminki, 2002; Izaraeli and Rechavi, 2002; Van Damme et al., 2002); this has led to an explosion of rodent populations in many research institutions. Although much of the rodent population growth has been caused by increases in numbers of mice, the rat lends itself well to the study of human medical problems and is the major focus at some institutions (Kwitek-Black and Jacob, 2001). The rat genome project and the rat genome database may add impetus to the growth in laboratory rat populations, especially as the rat genome database includes Virtual Comparative Mapping (VCMap), an essential tool for comparative genomics, which provides a dynamic sequence-based homology tool that allows researchers of rat, mouse, and human to view mapped genes and sequences and their locations in the other two organisms (Jacob and Kwitek, 2002; Twigger et al., 2002).

Over the years, compelling evidence has accumulated showing that numerous environmental variables can have profound effects on biologic responses of research animals (Jonas, 1976; Vesell et al., 1976; Lindsey et al., 1978). The biologic response of the laboratory rat is the result of multiple genetic and environmental effects experienced by the animal during the continuum from zygote to death (Lindsey et al., 1978) (Fig. 10-1). The environmental factors that affect animal health and well being include cage design and construction materials, bedding material, food and water, available living area, air exchange and air quality (relative humidity [RH], ammonia [NH_3], carbon dioxide [CO_2], dust), temperature, light (intensity, photoperiod, and wavelength), vibration, noise level (including ultrasound), electrical and magnetic fields of force, pheromones, microorganisms, parasites, and pollutants (Clough, 1982, 1984). The effects of environmental and husbandry factors on animal physiological and behavioral function may be subtle, to the extent that effects are not observed but may still cause perturbations in research results. This makes it extremely important that all reasonable steps be taken to control research variables, and emphasizes the need to define laboratory rats in terms of both genetics and environment (physical, chemical, and microbial factors) and to report these crucial data in scientific publications.

This chapter will review environmental factors, including primary and secondary housing enclosures, and their potential effects on animal research.

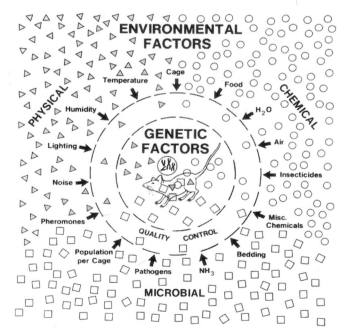

Fig. 10-1 The biological response of the laboratory rat to experimental interventions may be profoundly influenced by multiple genetic and environmental variables that are illustrated in this figure from the first volume of *The Laboratory Rat*. (From Baker et al., 1979.)

II. HOUSING

A. Facility Design

The design, construction, and maintenance of the animal research facility play a major role in the type and quality of the husbandry of laboratory animals. Rat housing space should be located in structurally sound, vermin-proof buildings (especially wild rodent–proof), with the animal facility physically separated from offices, laboratories, and other areas used primarily for human occupation. In addition to animal housing, there are a number of other functions that must be accommodated in the animal research facility. Typically, these include all, or some mix of, the following: receiving; quarantine; cage wash and sanitation; diet storage and preparation; supply and equipment storage; waste disposal; personnel lockers, showers, sinks, toilets, and break rooms; specialized laboratories (such as surgery, necropsy, diagnostic procedures, imaging, and infectious disease and biohazard containment) (Hessler, 1991a, 1995; Hessler et al., 1999); and administrative activities and training activities. Increasingly, rodent facilities are operated as clean facilities where animals are not allowed to leave the facility and return. In these cases the facility should contain a sufficient number of animal procedure rooms to allow for all of the research manipulations that must be carried out in the animal facility. In effect, animal rooms and animal procedure rooms are extensions of the research laboratory. The space assigned to the various functions should be arranged in a way to optimize traffic flow and minimize cross-contamination between clean and dirty functions. A high degree of security against unnecessary traffic must be accomplished.

Room dimensions should accommodate standard racks and caging units, or specialty caging used by the institution. Animal rooms should also be of sufficient size to accommodate specialty equipment such as laminar flow work stations or hoods, and laboratory carts brought into the room by investigative and animal care staff. Larger rodent rooms are more efficient in terms of the number of cages that can be placed per square foot of facility. Large rooms that house multiple projects, be they from different laboratories or the same laboratory, may lead to unacceptable competition for space and equipment in the room from a number of individuals wanting to work with their animals at the same time. This also may result in competition between husbandry and investigative staff needing to work in the room at the same time. A mix of large and small rooms is probably best in an academic setting. Animal housing areas should be constructed with monolithic wall, floor, and ceiling finishes that minimize accumulation of debris, facilitate microbial decontamination, and discourage harborage for vermin. Animal rooms should be windowless with light provided by fluorescent fixtures controlled by automatic timers at the room or preferably by a central computerized system. Electrical timers tend to provide more reliable service than do mechanical timers. The facility should provide good control of environmental parameters and have redundancy of essential mechanical systems such as air handlers, steam, chilled and hot water supply, water and vacuum pumps, and electrical supply. Corridor systems, single or dual, can have a major impact on the functionality of an animal facility. Each system has significant advantages and disadvantages (Hessler, 1991b); however, the current trend in animal facility layout tends more toward a single corridor system. Facilities incorporating multiple suites of animal holding rooms and procedure rooms provide a good degree of flexibility in terms of disease control and quality assurance, and separation. Additional principles of facilities design can be found in the *Guide for Care and Use of Laboratory Animals* (1996) and in several other publications (Jonas, 1965; McSheehy, 1976; Institute of Laboratory Animal Resources, 1978; Ruys, 1991; Sorensen, 1994; Hessler and Leary, 2002; Hessler and Hoglund, 2002).

B. Primary Enclosures

Rodents have traditionally been housed in either solid-bottom or wire-bottom cages (Fig. 10-2). Solid-bottom cages are usually referred to as shoebox cages. They are rectangular boxes with solid sides and bottoms, usually constructed of plastic or metal, although metal cages have mostly fallen out of use. Plastic cages are generally preferred over metal cages because of the enhanced visibility of animals in the cage, thermal insulation, economy, seamless construction, durability, and supposed chemical inertness. Traditionally these cages have been constructed of polycarbonate or polypropylene. Today many cages are being molded of more heat and chemical-resistant plastics such as polysulfone to better withstand the rigors of repeated autoclaving. Cages molded of less durable, inexpensive plastics such as polystyrene may be used in instances in which contamination makes disposable cages desirable. Shoebox cages are topped by a cover that may be constructed of perforated sheet metal or wire bars. In some systems shoebox cages are suspended from a rack so that a shelf forms the top of the cage. Cage lids should be locked in place to prevent displacement and escape by larger rats.

The plastics, polycarbonate and polysulfone, used to manufacture most rodent cages may be of concern. This is because bisphenol A (BPA), a chemical that has been shown to be an endocrine disrupter with estrogenic activity, is one of the materials used to make these plastics.

 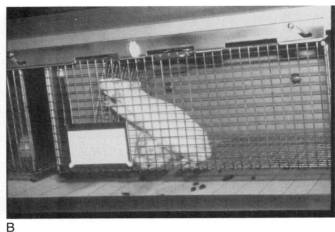

A   B

Fig. 10-2   (**A**) A rat in a solid bottom cage. (**B**) A rat in a wire-bottom cage.

There is considerable interest and debate regarding environmental estrogens, naturally occurring or man-made chemicals with estrogen-like activity found in the environment of man and animals. These chemicals have biologic activity, and there is concern that they may affect developmental biology and play a role in cancer causation (Turner and Sharpe, 1997; Roy et al., 1998; Degen and Bolt, 2000; Fenner-Crisp, 2000; Markey et al., 2002, 2003; Melnick et al., 2002). The polycarbonate used for animal cages is a polymer of BPA molecules linked together through carbonate groups, and polysulfone is a polymer produced from the polycondensation reaction between BPA and 4,4'-dichlorodiphenylsulfone (Solvay Advanced Polymers, 2003). Products made from both plastics have small amounts of monomeric BPA on their surfaces when new, and this BPA can leach into aqueous solutions. A significant difference between the two plastics is the susceptibility of the carbonate linkage in polycarbonate to hydrolysis; the bonds of polysulfone are not subject to hydrolysis (Solvay Advanced Polymers, 2003). Polycarbonate has been shown to release biologically active amounts of BPA into aqueous solution under normal use conditions (Krishnan et al., 1993; Feldman and Krishnan, 1995). Used polycarbonate animal cages release more BPA into aqueous solution than do new cages (Howdeshell et al., 2003). The hydrolysis of polycarbonate is accelerated in the presence of alkaline solutions. When immersed in a 10% aqueous solution of sodium hydroxide at 96°C for 480 hours, polycarbonate loses more than 50% of its mass, but there is no loss of polysulfone (Solvay Advanced Polymers, 2003). Thus, as discussed in the later section on "Cleaners, Deodorizers, and Chemical Sterilants," washing polycarbonate cages or water bottles in the presence of alkaline detergents can potentially have negative impacts on research.

Studies in both rats and mice have shown exposure to BPA to result in a variety of biologic effects. In adult mice, BPA exposure has been shown to cause meiotic aneuploidy (Hunt et al., 2003), and *in utero* exposure results in advanced postnatal development (Takai et al., 2001), a significantly lowered age at which vaginal opening occurs and a lowered age at which first vaginal estrus occurs (Honma et al., 2002), altered maternal behavior (Palanza et al., 2002), and alteration in the development and tissue organization of the mammary gland in ways that are consistent with changes associated with carcinogenesis in both rodents and humans (Markey et al., 2001). Pre- and/or perinatal exposure of rats to low levels of BPA results in morphological changes in the vagina of postpubertal offspring of exposed dams and lack of expression of full-length estrogen receptor alpha in the vaginas of these females during estrus (Schonfelder et al., 2002); increases in body weights in offspring from treated dams that is apparent soon after birth and continues into adulthood, and altered patterns of estrous cyclicity and decreased levels of leuteinizing hormone in adulthood (Rubin et al., 2001); changes in nonsocial behavior of adults as exemplified by a reduction in the motivation to explore and anxiety in males, and depression of motor activity and motivation to explore in females (Farabollini et al., 1999); a potentiation of female behavior in females and a depotentiation of male behavior in males (Farabollini et al., 2002); a masculinization of female behavior in two behavioral categories (play with females and social-sexual exploration) (Dessi-Fulgheri et al., 2002); modified activity of neuronal pathways and/or centers involved in nociception and pain in a sex-related manner (Alosi et al., 2002); increased fibroblastic/smooth muscle cell ratio, decreased the number of androgen receptor positive cells of the periductal stroma of the ventral prostate, and diminished expression of prostatic acid phosphatase in prostate ductal secretory cells (Ramos et al., 2001); and increased plasma and pituitary prolactin levels, increased plasma concentrations of luteinizing hormone, decreased plasma concentrations of testosterone, and decreased testicular levels of inhibin in male rats (Tohei et al., 2001). Although the studies cited above clearly demonstrate that BPA exposure may result in a variety of biologic changes, multigenerational studies have shown BPA exposure has no effect on estrous cyclicity, copulation index, fertility index, number of implantations, gestation length, litter size, pup/sex ratio, pup viability, or other functional reproductive measures in rats (Ema et al., 2001; Tyl et al., 2002).

Generally, shoebox cages are placed on mobile racks approximately 5 feet long by 5 to 6 feet high with multiple shelves of sheet metal or wire-bar construction. Cages can be arranged in one or two rows, depending on the depth of the shelves. The position of the cage on the rack and the location of the rack in the room have significant influence on intracage ventilation, temperature, and light intensity (Weihe et al., 1969; Woods et al., 1974; Woods, 1978; Clough, 1976). Assignment of cage positions by standard methods of randomization and/or rotation of cages in a systematic manner can be used to control this variable.

Wire-bottom cages are usually constructed entirely of metal but may have plastic sides with wire floors. These cages are suspended from racks to permit excrement to fall through the wire floor and collect on absorbent material below. Absorbent cellulose products of paper, corncob, or wood are generally used for this purpose. Wire-bottom cages allow for frequent removal of excreta without having to manipulate the animals. The cages exchange air readily with the room, which helps keep intracage NH_3 levels low. Wire-bottom cages are not suitable for whelping and raising young. Rodents are frequently housed in wire-bottom cages in pharmaceutical and contract toxicology laboratories owing to economic and experimental considerations (Stark, 2001). There is a concern about the use of wire-bottom cages for housing rodents (*Guide for the Care and Use of Laboratory Animals*, 1996). Pathological lesions associated with wire-bottom housing of rodents include pressure neuropathies of the hind foot, peripheral nerve abnormalities, urologic syndrome, nodular swellings, and ulcerations (Fullerton and Gilliatt, 1967, Grover-Johnson and Spencer, 1981, Ortman et al., 1983; Everitt et al., 1988; Peace et al., 2001). It must be noted that some of these lesions also occur in rats housed in solid-bottom cages (Grover-Johnson and Spencer, 1981; Peace et al., 2001). In a retrospective study, foot lesions were shown to be more common in rats housed in wire cages than in rats housed in solid-bottom cages, but despite differences in animal weight or cage type, lesions were not found until the rats had been housed for more than 1 year (Peace et al., 2001). However, preference testing has clearly shown that rats prefer solid-bottom cages with bedding, especially when they are resting

(Manser et al., 1995; Manser et al., 1996; Stauffacher, 1966; Rock et al., 1997). In addition, no differences in body weight gain, water consumption, physiologic data, or ease of handling have been reported associated with either wire- or solid-bottom cages (Stauffacher, 1966; Manser et al., 1995; Rock et al., 1997). Based on reported preference studies, rats should be housed in solid-bottom cages with bedding whenever possible, especially for long-term studies.

1. Isolation Caging

As efforts to improve the quality of rodents used in research have progressed, methods to control disease and disease transmission in rodent colonies have been sought. Caging systems have evolved as part of this effort. In efforts to reduce cross-contamination between cages and between the cage and the environment, various types of filter tops were devised for shoebox cages. The most popular of the early renditions of filtered cage lids was the molded polyester filter lid. In the early 1980s, Sedlacek and coworkers (Sedlacek et al., 1981) designed a rigid filter lid that has become widely adopted in rodent caging systems. This has evolved into what we now know as the microisolation caging system, a formed plastic lid with a filter medium insert that fits over the top of the shoebox cage. An important component of the isolation "system" is a high efficiency particulate air (HEPA) filtered laminar air flow work station in which microisolation cages are opened one at a time for performing routine husbandry or animal procedures (Fig. 10-3). The microisolation cage functions much similar to a Petri dish, provides a barrier at the cage level, and is extremely cost-effective when compared with more traditional methods of containment such as flexible film isolators and laminar flow equipment. The environment (microenvironment) in the cage is often quite different from that of the room (macroenvironment). The caging type plays a major role in this difference. Open cages, especially those with wire bottoms, exchange air with the room much more than do shoebox cages with filter or microisolation tops. Depending on cage type, there can be marked differences between the macroenvironment and the microenvironment, including temperature, humidity, and the levels of waste gases such as CO_2 and NH_3 (Serrano, 1971a; Woods, 1980; Hirsjarvi and Valiaho, 1987; Corning and Lipman, 1991; Lipman, 1992). Although temperature inside the covered cage will vary from room temperature, the variance is generally not significant. Heat exchange occurs primarily through the cage surface area (Besch, 1975). The majority of air exchange in microisolation cage systems takes place not through the filter on top of the lid but at the interface of the cage lid and the cage base (Keller et al., 1989). In studies with mice, RH in microisolation cages tends

Fig. 10-3 This figure illustrates the microisolation cage system consisting of microisolation cages (in this case, static mouse cages) and a high efficiency particulate air (HEPA)-filtered laminar flow air work station (in this case, a type II-A biosafety cabinet) in which microisolation cages are opened one at a time for performing routine husbandry or animal procedures.

to be at least 20% greater than the RH in the room (Corning and Lipman, 1991). To maintain microenvironmental RH within the recommended ranges, the macroenvironmental humidity must be kept below 40% RH, which would be very difficult and expensive for most facilities located within the temperate climate regions of the United States (Corning and Lipman, 1991). High intracage humidity prevents desiccation of urine and feces and provides a good environment for bacterial growth and subsequent NH_3 production. At 50% macroenvironmental RH, NH_3 levels measured at the level of the wire-bar cover may be in excess of 450 ppm by the seventh day in mouse cages (Corning and Lipman, 1991). In these cage systems, waste gases can reach and exceed levels that are documented to impact physiological parameters of tracheal epithelium, inhibit phagosome–lysosome fusion in macrophages, enhance pathogen growth in the respiratory tract, and impair immune function (Broderson et al., 1976; Gamble and Clough, 1976; Gordon et al., 1980; Schoeb et al., 1982; Lindsey et al., 1985; Targowski et al., 1984). The population density and the type of bedding material used have significant effects on the rate of NH_3 production in the cage (Serano, 1971; Perkins and Lipman, 1995; Lipman, 1999), and bedding can be selected to reduce NH_3 production. As discussed below, ground corncob tends to control NH_3 levels better than do most other bedding materials.

Individually ventilated caging systems (VCSs) are a significant improvement over the static microisolation cage (Fig. 10-4). The improvement in microenvironmental air quality obtained by the use of VCS may be the principle reason for their growth in popularity. There were several early attempts to ventilate rodent cages. The earliest successful use appears to have been at the Jackson Laboratory, where individually ventilated cages have been in use since the late 1960s (Les, 1983). The first VCS used at Jackson Laboratory supplied and exhausted each cage individually and could be operated in a neutral, positive, or negative mode at desired pressures. The system reduced cross-contamination between cages and provided a high air exchange rate, which reduced NH_3 build-up in the cage. Other early attempts at ventilating rodent cages include the suspension of specialized plastic cages in mass air flow racks (Nielsen and Bailey, 1979), the delivery of filtered air into the top chamber of microisolation cages through a polyvinylchloride piping system (Keller et al., 1983), and a forced air ventilation system in which air was delivered to cages through a distribution tube running through a microisolation type top just above the wire-bar lid (Wu et al., 1985). Each of these systems functioned to control NH_3 and humidity levels in the cage. Jackson Laboratory entered into collaborations with Thoren Caging Systems in the late 1970s, and the original system developed for Jackson Laboratory was modified

Fig. 10-4 Four ventilated rodent cage racks showing different ventilation patterns. (**A**) Above each cage rack on wall-mounted shelves are two fan/filter units. One supplies high efficiency particulate air (HEPA)-filtered room air to each cage on the rack, and the other HEPA filters the air exhausted from each cage before returning it to the room. (**B**) A fan/filter unit on top of the rack supplies HEPA-filtered room air to all the cages on the rack, and a second fan/filter unit directs air from the cages into the building exhaust. (**C**) A fan/filter unit on top of the rack supplies HEPA-filtered room air to all the cages on the rack, and the rack is directly exhausted to building exhaust. (**D**) A rodent cage rack with individually ventilated microisolation cages. A single fan/filter unit located in an interstitial space above the ceiling supplies HEPA-filtered room air to all the cages on all the racks in the room. The air from the cages on each rack is exhausted directly into the facility exhaust duct.

in the early 1980s into the first commercially available VCS (Lipman, 1999). The VCSs have gained widespread popularity, and there are a number of commercial systems available from which to choose (Lipman, 1999).

There are a number of advantages of VCS over static systems. Marked improvement of microenvironmental air quality has been clearly documented in these systems (Keller et al., 1983; Iwarsson and Noren, 1992; Lipman et al., 1992; Huerkamp, and Lehner, 1994; Yoshida et al., 1995; Perkins and Lipman, 1996; Hasegawa et al., 1997; Rivard et al., 2000). When compared with static cages housed under the same macroenvironmental conditions with the same strain and biomass of mice, the intracage concentrations of NH_3 and CO_2 are considerably lower in VCSs. The day on which NH_3 is first detected is delayed

in VCSs compared with static cages. Reduced NH_3 production improves macroenvironmental air quality for personnel working in animal rooms and cage wash.

The improvement in the microenvironmental air quality allows cage changing frequency to be reduced to once a week or once every 2 weeks or, in some cases, longer periods (Reeb et al., 1998; Reeb-Whitaker et al., 2001). Increasing the period between cage changes results in a reduction of labor cost in the animal room and cage wash, a reduction in the amount of bedding used, and a reduction in the use of chemicals associated with husbandry procedures and cage sanitation. The savings in labor and supplies can result in considerable cost savings to the institution. Provided there is sufficient cooling capacity to dissipate heat loads, VCS may be used in areas or facilities with inadequate ventilation instead of up-grading the existing HVAC (heating, ventilating, air conditioning) system to accommodate the use of other types of caging. VCSs may be integrated into facilities by a variety of methods. Integration approaches include supply and exhaust air provided by the building central system, single supply fans for all racks in a room with exhaust through the building system, or individual rack supply fans with exhaust through the building system. These have been reviewed in detail elsewhere (Lipman, 1993; Hessler and Leary, 2002). VCSs that pressurize the cage with HEPA-filtered air can provide an additional level of barrier to protect animals from infection (Cunliffe-Beamer and Les, 1983; Lipman et al., 1993; Clough et al., 1995). VCSs have been shown not to impair the breeding performance of rats (Chaguri et al., 2001). In fact, certain air-speed levels led to a greater number of, and more uniform, litters, with decreased mortality rates.

Clearly, there are advantages to housing animals in VCSs. When selecting or using this system, the user must clearly understand the operating principals of the system they intend to use. The method of introduction, and the quantity and quality of the air supplied to each cage differs between various VCSs. The ideal intracage ventilation rate has not been determined, but it probably depends on several factors, including species, strain or stock housed, population density, and bedding material used. An ideal ventilation rate in one situation may be excessive or insufficient in another. It has been recommended that ventilation rates be established in these caging systems so that, before changing cages, microenvironmental NH_3 and CO_2 concentrations are less than 25 and 5000 ppm, respectively, and temperature and RH fall within prescribed limits (Lipman, 1999). The 25-ppm level for NH_3 is based on a regulatory standard for humans, which stipulates an 8-hour weighted average of 25 ppm as the maximum exposure for humans. The equivalent 24-hour weighted average would be 8.33 ppm, which might reasonably be rounded to 10 ppm. This might not be a scientifically documented standard, but it has a rational basis and is not difficult to maintain with a VCS. In some systems the cages are not tightly sealed and simply become static isolators if there is a loss of power or other failure of the ventilation system. Although microenvironmental conditions will deteriorate in these systems, the animals are not in immediate danger. However, in those systems in which the cages are tightly sealed, if there is a loss of power or other ventilation failure, the microenvironmental conditions rapidly deteriorate and animals may die in periods as short as 17 minutes after ventilation failure (Kanzaki et al., 2001). Deaths in rats housed in a VCSs occurring within 60 minutes of an inadvertent power failure have recently been reported (Huerkamp et al., 2003). Loss of power to a VCS creates a potentially life-threatening condition to the animals housed in them. Emergency back-up power or ventilation is required to prevent loss of animals owing to a power failure.

One factor that has driven the development of new caging systems is the control of aeroallergens produced by rodents. These are powerful allergens and can cause significant allergic disease in animal care and research staff. The Association for the Assessment and Accreditation of Laboratory Animal Care, International (AAALAC), expects institutions to include risk assessment of allergen exposure as a component of any animal care and use program. A comprehensive review of the issues involved in personnel exposure to rodent allergens has recently been published (Feldman, 2003). Several approaches have been used to reduce the exposure of workers to these aeroallergens. These include increase in room ventilation rates, the use of filter tops on static cages, the use of negatively pressurized ventilated cages, and the use of ventilated cage-changing stations. Increase of room ventilation from 6 to 10, 15, and 20 air changes per hour (ac/hour) had little effect on aeroallergen levels and no impact on airborne particulate matter (Reeb-Whitaker et al., 1999). Covering open static cages with simple filter sheet tops reduced aeroallergens from 5.1 to 1.3 ng/m^3, and a fitted filter bonnet reduced the level to 0.8 ng/m^3. When ventilated cages were used, ambient aeroallergen levels were 1.1 ng/m^3 when the cages were positive and only 0.3 ng/m^3 when the cages were negative in relative pressure. The use of ventilated cage changing stations in combination with the use of negative cage pressure reduced breathing zone allergen from 28 ng/m^3 with neither control strategy in place to 9 ng/m^3 (Reeb-Whitaker et al., 1999). Many of the VCSs either exhaust directly into building exhaust or filter the exhaust air before releasing it into the room. This reduces the concentration of allergenic particulates in the macroenvironment, providing personnel protection from allergen exposure (Clough et al., 1995; Gordon et al., 1997; Renstrom et al., 2001). Leakage of cage effluent into the macroenvironment has been detected in those VCSs that operate by pressurizing the cage and attempting to capture cage effluent after it escapes from the cage (Tu et al., 1997; Renstrom et al., 2001). To minimize personnel allergen exposure, these cages must be changed or worked with in a laminar flow work station or hood.

C. Population Density

Changing the cage type that animals are housed in may significantly affect research results if the animals are not allowed to acclimate to the new cages (Damon et al., 1982; Damon et al., 1986). It was reported that a minimum of 3 to 5 days were required for food and water consumption to return to normal after placing rats in a different cage type. Cage size (living area) can have influence in several areas of health and well-being. For example, living area influences the development of foot lesions in rats (Shaw and Gallerger, 1984). Cage size has been shown to influence sexual behavior of male rats (Saito et al., 1996). Sexually inexperienced male rats were placed in circular or rectangular cages of increasing floor area shortly before introduction of a receptive female. Mounting frequency decreased markedly as the floor area increased, whereas intromission frequency tended to increase slightly and ejaculatory frequency increased as floor area increased.

Population density refers to the number of animals per area of floor space in the primary enclosure, or the number of animals per cage in a standard size cage. Current standards provide recommendations on the amount of floor space individual animals require. Population density can have marked effects on reproduction (Christian and LeMunyan, 1958) and behavior (Davis, 1978) and may have important metabolic effects (Armario et al., 1984a,c; Klir et al., 1984). Crowded rats show changes in emotional reactivity and are more responsive to acute noise stress (Armario et al., 1984b). Immune responsiveness is influenced by population density (Johnson et al., 1963; Vessey, 1964; Fritz et al., 1968; Plant et al., 1969; Rasmussen, 1969; Solomon, 1969; Joasoo and McKenzie, 1976; Grewal et al., 1997). A number of studies have been performed in both mice and rats (the preponderance in mice) that have shown population density can influence multiple components of the host defense mechanism. In studies comparing mice housed five per cage with mice housed individually, it was shown that mononuclear cells from individually housed mice had a greater capacity to phagocytize dead cells of *Candida albicans*; had spleens that produced more macrophage colony stimulating factor (M-CSF); were more responsive to M-CSF; had peritoneal macrophages that released greater quantities of interleukin-1 *in vitro* into the surrounding medium and that had a greater capacity to migrate toward a chemotactic stimulus; had higher titers of IgM hemagglutination antibody to sheep erythrocytes; and had greater lymphocyte reactivity to the mitogen concanavalin A (Rabin and Salvin, 1987; Salvin et al., 1990). At least some of these effects appear to be transient (Rabin and Salvin, 1987). Grewal et al. (1997a) found that male mice housed 6 per cage had markedly lower responsiveness of lymph node T cells to antigen stimulation than did those of male mice housed individually for 4 to 15 weeks. This was shown to result from a reduction in the efficiency of the splenic antigen presenting cells in the group-housed animals (Grewal et al., 1997a). When mice were analyzed for their ability to synthesize autoantibodies against autologous bromelain-treated erythrocytes it was found that males housed six per cage had a

significantly lower number of autoimmune plaque forming cells (APFCs) in their spleens than did age-matched females housed under similar conditions. When the male mice were housed individually for 4 to 44 weeks, a marked increase in numbers of APFCs was found in their spleens, approaching those of female control mice (Grewal et al., 1997b). Karp et al., (1993) demonstrated that the secondary antibody response is higher in mice housed individually than it is in group-housed animals. Individually housed mice have been shown to have higher natural killer cell activity than did group-housed mice (Hoffman-Goetz et al., 1992). The production of a number of cytokines by lymphocytes has been shown to be higher in cells from individually housed mice than in cells from group-housed mice (Karp et al., 1994, 1997). Rabin et al. (1987) have shown that genetic factors related to the major histocompatibility complex apparently do not influence alteration of the immune response, which occurs with differential housing conditions in mice.

Similar results have been obtained in studies with rats. Individual housing of rats for 5 weeks resulted in significantly enhanced splenic lymphocyte proliferative responses to phytohemagglutinin compared with those of group-housed animals (Jessop et al., 1987). In following the time course of the observed changes, it was found that there was no difference, or suppression of B- and T-lymphocyte responses, in cells from individually housed animals compared with group-housed animals during the first week of individual housing; however, after 2 weeks of individual housing, splenic and blood T lymphocytes from individually housed animals demonstrated increased proliferative responses to suboptimum and maximum concentrations of phytohemagglutinin (Jessop et al., 1988; Jessop and Bayer, 1989). In addition, splenic B lymphocyte responses to lipopolysaccharide from individually housed animals were also increased by two to threefold compared with that of group-housed animals (Jessop et al., 1988). The increased lymphocyte responsiveness was still present after 35 days of individual housing (Jessop and Bayer, 1989). The increased responsiveness could not be explained by changes in total white blood cell numbers, the proportion of splenic T or T-helper lymphocytes, or a significant change in the sensitivity of lymphocytes from individually housed animals to phytohemagglutinin (Jessop et al., 1987, 1988). In contrast, Joasoo and McKenzie (1976) reported that the *in vitro* response of sensitized splenic lymphocytes to antigen was increased by crowding and decreased by isolation in female rats. Both isolated and crowded male rats responded by a decrease in the *in vitro* reactivity of lymphocytes to antigen (Joasoo and McKenzie, 1976). Stefanski (2001) provides data indicating that the early social environment of rats affects numbers and proportions of many blood immune cell subsets in later life.

D. Individual Versus Group Housing

Rodents may be individually housed because of experimental considerations; however, studies have shown that individual housing results in physiological, neurochemical, and behavioral changes of animals (Brain and Benton, 1979; 1983). The magnitude of the effects of individual housing is influenced by the strain, sex, and age of the animals when they are individually housed. In addition, cage type and husbandry procedures may vary between studies, and this may affect the resulting response of the animal. It appears that the most critical factor in social deprivation is the age or developmental stage during the period of deprivation (Hall, 1998). The effects may be separated into three major developmental stages, with each stage related to deprivation of specific types of social interaction: preweaning/neonatal, postweaning/adolescent, and adult. Social deprivation during each of these stages appears to have effects that are neurochemically and behaviorally specific. When animals are individually housed, they are confronted with an environment devoid of the most basic aspects of social interaction. For example, changes in environment may result in temporary changes in eating behavior. In this regard, studies of weight gain in rodents housed individually housed or in groups have shown mixed results; some studies have demonstrated reduction of weight gain in individually housed animals (Wiberg et al., 1966); some studies have demonstrated no difference in weight gains between individually housed or group-housed rodents (Sokel et al., 1979; Morinan and Leonard, 1980; Niesink and van Ree, 1982); and some studies have shown increased weight gain in individually housed animals (Les, 1968; Hughes and Syme, 1972; Chvedoff et al., 1980; File, 1982).

Individual housing of rats has been reported to have both no effect on basal corticosterone levels (Gentsch et al., 1981), and to result in lower corticosterone levels compared with that of group-housed rats (Holson et al., 1988).

Cage type may play a role in the effects of individual housing. For example, an increase in mean systolic blood pressure was found when rats were housed individually in glass metabolism cages (Gardiner and Bennett, 1977), and the blood pressure remained elevated during the 3 weeks of the study. Heart rate concomitantly increased from 342 to 386 beats/min but returned to normal by the 13th day of individual housing. When housed individually in standard cages, hypertension developed in some but not all rats. Hypertension associated with individual housing has also been reported by other investigators (Mills and Ward, 1984; Naranjo and Fuentes, 1985): the hypertension developed within 24 hours of individual housing, although blood pressure reverts to normal shortly after the animals are returned to group housing (Mills and Ward, 1984).

Hypertension has been reported to be increased (Connolly et al., 1983) or attenuated (Hallback, 1975) in spontaneously hypertensive rats when they are socially isolated. Renal function may also be affected by social isolation. One study showed that some rats have significantly decreased urinary flow and decreased creatinine clearance during four consecutive days of individual housing in metabolism cages (Vadiei et al., 1969). Housing rats individually may result in a decrease in the LD_{50} for a given drug, and even control animals may develop pathophysiological changes indicative of Selye's stress syndrome (Hatch et al., 1965).

There have been numerous reports of individually housed and group-housed animals having differences in responses to pharmacological challenges. Often, conflicting results have been obtained, most likely owing to differences in animal characteristics and housing procedures. Individual housing has been shown to reduce the potency of central nervous system depressant drugs (Baumel et al., 1969; Dairman and Balazs, 1970; Einon et al., 1976). Individual housing markedly influenced the aggression inducing effect of Δ^9-tetrahydrocannabinol in rats (Fujiwara and Ueki, 1978, 1979). A single dose of tetrahydrocannabinol induced muricidal (mouse killing) and rod-attack behavior in rats when they were individually housed before or immediately after they were administered the drug. The aggressive behavior persisted as long as the animals were individually housed. Group-housed animals did not display muricidal behavior when dosed with tetrahydrocannabinol. In rats, the pain threshold for noxious stimuli was not affected by individual housing and was independent of aggressiveness (Ader et al., 1975), was not affected in nonkillers but increased in killing rats (Kostowski et al., 1977), and decreased after 1 day of individual housing (Panksepp, 1980). Morphine-induced analgesia was increased by individual housing in the studies of Katz and Steinberg (1972) and was not affected in muricidal and nonmuricidal rats according to Adler et al. (1975); and morphine analgesic efficiency was lowered in the test procedures of Kostowski et al. (1977). The sensitivity for morphine in young rats was reported to be decreased after a short period of individual housing (Panksepp, 1980).

In general, group-housed rodents develop a hierarchical structure. The social structure highly depends on the number of animals in the group. Rank position can be scored on the basis of the observation of agonistic behavioral categories in the home cage (Militzer and Reinhard, 1982; Lundberg, 1986; Schuhr, 1987). In rats, the social structure is often linear, with as many ranks as there are animals in the group. Each animal is dominant to all other animals with a lower status. The responsiveness to environmental stimuli may vary with the relative rank position of the various group members in group-housed animals.

In general, dominant animals and individually housed animals share several physiological and behavioral characteristics. In male albino rats, it has been found that dominance was correlated with high relative weights of thymus and testes and with low relative weights of adrenal glands (Militzer and Reinhard, 1982). Rearing rats in isolation modified their behavioral response (Karim and Arslan, 2000). Furthermore, rats reared in isolation were more exploratory than were socially reared rats in open field testing. Compared with socially reared rats, isolation reared rats are nervous, aggressive, and hyperactive. Isolation rearing in early life may modify a variety of behaviors in adults. Biologic responsiveness is impacted by many variables that may act in concert. However, there is variance in results between some studies that show a factor may or may not have an effect, and many studies may have been inadequately controlled or unrecognized variables may have influenced the results. Nonetheless, it is clear that the housing and environmental conditions of research animals may significantly influence study outcomes.

III. ENVIRONMENTAL FACTORS

Animals used for biomedical research should be kept under conditions that permit as standardized a response to experimental parameters as possible. Much evidence has accumulated over the years that indicate that many environmental conditions can influence research results to a greater extent than is recognized by many investigators (Clough, 1982). Changes in the animal's external environment are perceived by exteroreceptors and relayed to the brain. If an environmental condition is sufficient to unbalance homeostasis, the neuroendocrine system is stimulated to restore homeostasis (Clough, 1982). Thus, control of environmental factors can lead to a significant reduction in the variability of experimental results often seen between laboratories, or even within the same laboratory (Roe, 1965; Magee, 1970; Golberg, 1974; van der Touw et al., 1978; Chvedoff et al., 1980). Although the majority of investigators clearly recognize the need to control variables in their work, such as with biochemical assays, many do not appreciate the potential influence of environmental variables and, hence, fail to give adequate environmental descriptions in published works (Vesell and Lang, 1976). Innumerable factors can influence the response of animals to experimental procedures, including ambient temperature, humidity, air movement, light, sound, vibration, air pressure, gravity, electrical and magnetic fields, ionizing radiation, odors, microorganisms, parasites, dust, chemicals, and other pollutants (Clough, 1982). Of these, temperature, humidity, and ventilation are especially important environmental variables. Together, they determine relative heat loss or retention and ultimately contribute heavily to metabolic rate.

A. Temperature, Humidity, and Ventilation

Current environmental standards for rodent housing are based on room conditions, rather than conditions in the cage (primary enclosure). The conditions in the primary enclosure (microenvironment) may differ significantly from the room (macroenvironment) conditions in factors such as dry-bulb temperature, RH, gaseous content, and particulate concentrations (Woods, 1978; Baker et al., 1979; Clough, 1982, 1984). The differences between microenvironment and macroenvironment will vary markedly between facilities. Complex interactions exist between production of heat and water inside the cage and their dissipation into the macroenvironment. Many factors may influence these interactions, including the population density (Yamauchi et al., 1965), cage design (Serrano, 1971a), presence or absence of filter tops (Simmons et al., 1968; Besch, 1975), and the amount and velocity of air flowing over the cage (Woods, 1975). The extent to which differences in microenvironment affect the responsiveness of animals to research stimuli depends on the extent of the variations in microenvironmental factors, and on the ability of the animals to adapt to these environmental changes. At present, there is very little information available on the variation in experimental results caused by differences in microenvironmental conditions.

The relationship between the microenvironment and the macroenvironment is not a simple one. A number of factors affect exchange between these environments. These include thermic isolation and air exchange, which are in turn determined by room ventilation intensity and pattern, and cage type (plastic shoebox or wire grid), and cage dimensions and by whether the cage is covered with a filter bonnet or not. When considering non-VCSs, cage ventilation rates are generally less than are room ventilation rates. Cage type greatly influences this difference; cage ventilation rates have been shown to be 82% to 92% of room ventilation rates for cages with wire-grid floors and 20% to 58% of room ventilation rates for shoebox type cages (Murakani, 1971; Clough, 1984). Static microisolation cages have been shown to have ventilation rates as low as 0.68 ac/hour (Keller et al., 1989). The microenvironment is not solely determined by the room ventilation rate but is also a function of the cage type and size, the animals housed in the cage and their activity level, population density, bedding material, and cage changing frequency. One study investigated the differences between room temperature and the temperature in cages as a function of the cage location on the rack.

In-cage temperatures varied with the height of the cage in the room, with cages on the top shelf being 4.9°C warmer than those on the bottom shelf, and in-cage temperatures being up to 5.7°C warmer than the "recommended environmental temperature" for rats (Clough, 1984). Other studies reported that the in-cage temperature varied not only with the population density but also with the construction material of the cage (Yamauchi et al., 1965; Murakami, 1971; Hirsjarvi and Valiaho, 1987).

Heat is one of the most important environmental factors affecting living organisms and its effects have been extensively studied (Rose, 1967). Rodents are homeothermic mammals and must therefore maintain a fine degree of control over both heat production and heat loss in order to achieve homeothermy. Rats can adapt to a wide range of environmental temperatures (10°C to 30°C) (Weihe, 1965; 1971). Metabolic adaptation can commence within minutes (Clough, 1982) but continues for varying periods depending on the animal's previous experience and the extent of the temperature change. Complete adaptation commonly takes 2 to 3 weeks in rats (Gelineo, 1934), but it can take 5 to 6 weeks (Sellers, 1957) or even 7 to 12 weeks (Depocas, 1960); therefore, sudden fluctuations in temperature should be avoided. Temperatures of rodent housing rooms are typically maintained below the thermoneutral zone of rats (Romanovsky et al., 2002). Consequently, changes in the ambient temperature cause changes in the metabolic rate of the animals (Svendsen, 1994) and can affect enzyme activity (Shysh and Noujaim, 1972), as well as the toxicity of or response to drugs and chemicals (Balazs et al., 1962; Weihe, 1973; Sanvordecker and Lambert, 1974; Clough, 1982).

In the controlled environment of the research animal facility, the core body temperature of the rat will be higher than that of the air around the animal. Thus, there is a temperature gradient from the central parts of the body, through the superficial layers, to the surrounding air. Homeotherms are generally considered to consist of a core and a shell (Verbiest, 1956). Obviously, ambient temperature significantly affects this gradient. Changes in the temperature gradient and thickness of the shell are more significant in hairless animals, and are particularly noticeable in the limbs and tail. There is an increase or decrease in the rate of many tissue and cellular functions, up to 20% to 30%, for each 1°C change in ambient temperature (Irving, 1964). Thus, the absorption and/or activity of any substance given epicutaneously, subcutaneously, or intramuscularly is likely to vary with ambient temperature.

Changes in ambient temperature are compensated for by one of several methods, including increases or decreases in metabolic rate, activity (e.g., shivering and nonshivering thermogenesis), peripheral circulation, insulation (e.g., fat or fur), and evaporative loss (e.g., sweating and changes in respiratory rate such as by panting). In addition, animals may use so-called behavioral thermoregulatory activities, such as changes in the ratio of surface area to mass (e.g., huddling or extension of limbs including the tail), voluntary variation in the extent of external insulation (e.g., nest building), and selection or creation of a less thermally stressful habitat (e.g., shelter and shade-seeking, nest building). In nonsweating species such as rodents, increases in ambient temperature result in increases in respiratory rate (Clough, 1982). The main response of rodents to changes in ambient temperature is alteration of their metabolic rate (Weihe, 1971).

The temperature range of 20°C to 26°C has been shown to be the optimal temperature range for the macroenvironment of rat rooms (Yamauchi et al., 1981). Exposure of male rats to high ambient temperatures (31.6°C to 32.5°C), such as may happen with air conditioning failure in hot weather, may result in irreversible testicular atrophy or even death (Pucak et al., 1977). Lactation is impaired in rats exposed to high temperatures (Benson and Morris, 1971; Yagil et al., 1976). Changes in ambient temperature can modulate immune function. Cold stress has been shown to decrease antibody production (Sabiston et al., 1978), and heat stress resulted in elevated corticosteroid levels, changes in lymphocyte migratory patterns, decrease in thymus and spleen weights, increase in phagocytic index, and decreased antibody response (Krynicki and Olszewski, 1989; Joseph et al., 1991; Yamamoto et al., 1999). Lowered temperature (20°C versus 25°C) inhibited hepatic microsomal enzymes (HME) and prolonged hexobarbital sleep times in mice (Vesell, 1968). It appears that the temperature to which rats are acclimated affects the thermoneutral zone of the animal and alters the set-point temperature around which thermal responses are regulated (Gwosdow and Besch, 1985).

RH inside the cage tends to be higher that in the room owing to water output by the animals in the cage. RH affects the thermoregulatory capacity of animals and the control and management of airborne diseases (Clough, 1982). High RH in the cage encourages the production of NH_3 by urease-positive bacteria. It has been shown that room air exchange rates of 12 to 16 ac/hour are required to keep in-cage (in open cages) RH from rising above 70% when the RH of the supply air is 45% and when the calculation was based on respiratory water only. When the calculations are based on total water turnover, the room ventilation rates need to be between 44 and 87 ac/hour to keep in-cage RH from rising above 70%. If the RH of the supply air is 60% the room ventilation rates may need to be as high as 200 ac/h to control in-cage RH (Clough, 1984). Current standards for room ventilation rates appear not to be excessive in light of these calculations, and in fact may be low. These data also indicate that in high humidity climates, dehumidifiers may be required in conjunction with the HVAC system to control humidity.

Room ventilation functions to supply adequate oxygen; remove heat loads caused by animal respiration, lights, and equipment; dilute gaseous and particulate contaminants; adjust room RH; and create static-pressure differentials between adjoining spaces where appropriate. Before the 1996 revision, the *Guide for the Care and Use of Laboratory Animals (Guide)* recommended room ventilation rates of 10 to 15 fresh ac/hour. In the 1996 revision, the recommendation became a performance standard recommending that the ventilation rate be the minimal required to control the heat load expected to be generated by the largest number of animals to be housed in the room in question plus any heat expected to be generated by nonanimal sources and heat transfer through room surfaces (Committee to Revise the *Guide for the Care and Use of Laboratory Animals*, 1996). This allows for flexibility in the use of existing space and in the design of new space. This type of approach can be used to determine the maximum number of animals that can be housed in an existing space with a fixed rate of ventilation. In addition to controlling heat loads, ventilation rates must be sufficient to control odors, allergens, airborne particulates, and metabolically generated gases (*Guide*, 1996). This may require ventilation rates beyond the calculated minimum based on heat load. The ventilation rates for various areas within the facility should be adjusted so that clean spaces are positive in air pressure relative to potentially contaminated areas.

B. Light

Light is an important part of an animal's environment, and intensity, quality, and photoperiod are variables that can influence biological response. One of the best known responses to light is the retinal degeneration that occurs in albino rodents, especially rats, as a consequence of exposure to light (O'Steen and Anderson, 1972; Bellhorn, 1980; Semple-Rowland and Dawson, 1987a,b). Sprague-Dawley rats raised for 15 weeks in a 12-hour light/12-hour dark cycle at a light intensity of 6 lux and then exposed to a light intensity of 270 lux in a 12-hour light/12-hour dark cycle developed severely damaged retinas within 3 to 7 days. It has been shown that Lewis and Buffalo rats are more susceptible to retinal photic injury than are Wistar and Fischer rats (Borges et al., 1990). Pigmented strains appear to be less susceptible (Reiter, 1973) but may experience photic injury (Williams et al., 1985). Chronic stress increases susceptibility to retinal damage (O'Steen and Brodish, 1985).

The material of which the cage is constructed plays a major role in light exposure, with clear, translucent cages allowing the most light into the cage. The position of the cage on the rack is important because light intensity decreases with the square of the distance from the light source, and upper cages and shelves block light from lower cages. Light at the top shelf of a rack may be 80 times more intense than that at the bottom shelf (Weihe, 1976).

Light is a powerful stimulant and synchronizer of rhythms related to reproductive biology. In rats, the estrous cycle lasted 4 days with a 12-hour light/12-hour dark cycle, and 5 days with a 16-hour light/8-hour dark cycle; with a 22-hour light/2-hour dark cycle, the estrous cycle became irregular and the animals did not reproduce (Svendsen, 1994). Female rats housed in continuous light displayed persistent vaginal estrus and perturbations of 17β-estradiol and estrone levels (Takeo, 1984). Very short periods of light occurring during the dark phase can have significant effects on physiology (Clough, 1982). Therefore, lights should not be turned on during the dark phase of the light cycle unless it is absolutely necessary. Room light timers should be checked on a regular basis to ensure that lights are cycling correctly. It has been said that rodents do not see red light, but continuous red light induced persistent estrus in female rats (Lambert, 1975).

Photoperiod has a profound regulatory effect on circadian rhythms (Hastings and Menaker, 1976). Photoperiod influences redox enzyme activity in the central nervous system, drug metabolism, and drug toxicity (Radzialowski and Bousquet, 1968; Nair and Casper, 1969; Jori et al., 1971; LeBouton and Handler, 1971; Chedid and Nair, 1972; Baran et al., 2000; Pozdeyev and Lavrikova, 2000). The light/dark cycle also influences immune function, exemplified by the fact that humoral and cellular immune responses to thymus-dependent antigens were shown to have a circadian rhythm (Hayashi and Kikuchi, 1982). Phase shift in the light/dark cycle resulted in suppression of the immune response to thymus-dependent antigens (Hayashi and Kikuchi, 1985).

There have not been many studies on the effects of various colors of light on biological processes, but the limited data in the literature indicate that varying light spectra may have significant effects. The voluntary wheel-running activity of mice was strongly influenced by differently colored lights (Spalding et al., 1969a,b). Wavelength of light may affect body weight and organ weights in mice (Saltarelli and Coppola, 1979). Blue or white light has been shown to provide protection against bilirubin toxicity and to stunt the growth of infant rats (Heller et al., 1969; Ballowitz, 1971). In addition, different colored lights influenced human biological rhythms (Morita and Tokura, 1998).

C. Noise

There are numerous sources of noise in research animal facilities. They originate from ventilation systems, personnel and equipment movement, vocalization and activity of animals, husbandry and cleaning procedures, and the operation of equipment. The noise levels in animal facilities may vary between 30 to 102 dB throughout the day (Pfaff, 1974; Peterson, 1980). Milligan et al. (1993) monitored sound in animal facilities in both low and high frequency ranges and found that during the work day, sound levels commonly reached values of 80 to 95 dB in the low frequency range (0.01 to 12.5 kHz) and 50 to 75 dB in the high frequency range (12.5 to 70 kHz). Man has the lowest upper cut-off hearing frequency of all species so far examined, and therefore, sounds that are inaudible to a human might be stressful to rodents (Clough, 1982). A concern has been raised that ambient ultrasound may be common in animal facilities; its effect on laboratory animals should be investigated and guidelines on acceptable levels be formulated (Sales et al., 1988). Hearing in the rat ranges from 500 Hz to 60 to 80 kHz (Sharp and LaRegina, 1998), so ultrasound in the animal facility may be of concern. Decibel levels that exceed the usual background noise in animal facilities caused various degrees of destruction of sensory hairs and supporting cells of several animal species (Fletcher, 1976). Similar to man, rats experienced mechanical damage at 160 db, pain at about 140 db, and signs of inner ear damage after prolonged exposures to about 100 db (Anthony, 1962).

Noise has been shown to have effects on a variety of experimental parameters. Noise stress reduced fertility of rodents (Zakem and Alliston, 1974; Fletcher, 1976; Gamble, 1976); affected components of blood, including plasma lipids, corticosterone, total cholesterol, serum glutamic oxaloacetic transaminase, serum glutamic pyruvic transaminase, and triglyceride levels (Geber et al., 1966; Friedman et al., 1967; Prabhakaran et al., 1988); affected components of the immune system as evidenced by a significant increase in thymus weight and cell count, a significant decrease in antibody titer and spleen weight and cell count, a reduction in the migration of prethymic stem cells to the thymus, and time-dependent suppression and enhancement of splenic lymphocyte proliferation in response to mitogen stimulation (Bomberger and Haar, 1992; Van Raaij et al., 1996; Archana and Namasivayam, 1999, 2000); impaired wound healing (McCarthy et al., 1992; Wysocki, 1996); affected the function of the adrenal glands resulting in increases in dopamine, nora-drenaline, adrenaline, and their metabolites (Gesi et al., 2001, 2002); increased durations of exploring, grooming, and resting behaviors (Krebs et al., 1996); reduced food intake (Nayfield and Besch, 1981); and affected cardiovascular function, with resultant increases in systolic blood pressure and pulse pressure (Gao and Zhang, 1992; Baudrie et al., 2001).

There are sex, strain, and age differences in the response to noise stress (Glowa and Hansen, 1994; Blaszezyk and

Tajchert, 1996; Pryce et al., 2001; Faraday, 2002; Maslova et al., 2002). The banging of cages in an animal room can cause a 100% to 200% increase in plasma corticosterone in rats, which persists for 2 to 4 hours (Barrett and Stockham, 1963). It has been shown that the noise of oxygen rushing into a hyperbaric chamber contributed to the incidence of convulsions of rats being treated with hyperbaric oxygen (Boyle and Villanueva, 1976). Exposure of pregnant rats to an 85 to 90 dB fire alarm bell results in alteration of immune function in the offspring (Sobrian et al., 1997). Interestingly, rats appear to adapt to chronic noise stress (Armario et al., 1984b, 1985). The idea of playing background music as white noise to mask sudden noises in a facility is one that has been around for quite some time. Music has been shown to reduce the effects of noise stress (Nunez et al., 2002), and therefore, a source of white noise in an animal facility may be beneficial.

When considering noise and its sources in animal facilities, one must consider noise generated by ventilation systems such as those used with VCSs. The supply and/or exhaust blowers both produce macroenvironmental and microenvironmental noise. The volume and frequency of the noise generated depend on the system type and the number of units per holding room. In an investigation of three commercially available VCSs, all three were shown to produce macro- and microenvironmental noise significantly greater than room background noise. Macroenvironmental noise ranged between 74 and 80 dB, whereas microenvironmental noise ranged between 79 and 89 dB (Perkins and Lipman, 1996). Another group reported no detection of ultrasonic frequencies produced by a VCS (Clough et al., 1995). To fully understand the potential effect of noise generated by VCSs on rodents, these units need to be evaluated for sounds over the complete hearing range of rodents. Sound levels for multiple VCSs in a room are determined by a logarithmic equation. In a room containing four units generating 80 dB each, the room noise level would be 86 dB, which is above the 8-hour exposure level established by the American Conference of Governmental Industrial Hygienists (Lipman, 1999).

IV. FEED, WATER, AND BEDDING

A. Feed

One of the most important environmental factors influencing the well being of rodent colonies is access to a diet that provides adequate nutrition (Knapka, 1999). Diets are discussed in detail elsewhere in this volume and therefore will not be emphasized here.

Variations in the quantity and quality of nutrients, as well as the presence of extraneous materials, in animal diets are not infrequent. Because these variations may interact with experimental variables in biomedical research to produce unwanted results, diets of the best possible quality should be used.

B. Water

The water provided to research animals is subject to considerable variation depending on the geographic location of the facility and area geology; the use of surface versus well water; proximity to industrial, agricultural, or major urban centers; and the type of water treatment used. Many of the agents found in water occur naturally and enter water after exposure to rock, soil, and air. Chemical contaminants found in water are generally classified as suspended solids and as organic and inorganic solutes (Shapiro, 1980). The suspended solids are generally harmless, but they may act as carriers of biological agents. More than 700 organic chemicals have been identified in drinking water in the United States (Shapiro, 1980). Of these, over 90% are natural decomposition products of animal and plant origin. Drinking water also often contains low levels of many synthetic organic compounds, such as pesticides and cyclic aromatic and halogenated hydrocarbons. Trihalomethanes make up the largest portion of identifiable synthetic organic contaminants. These compounds are derived from the interaction of a halogen, usually chlorine or bromine, with methane groups from natural organic materials, or they may enter the water supply as contaminants of the chlorine used for water treatment. Trihalomethanes are found in virtually all drinking water that has been disinfected with chlorine. One of these compounds, chloroform, has biological impact (Vessel et al., 1976) and may be found in high relative concentrations in chlorine-treated water. Water is often contaminated with nitrates, which are of particular concern because of their potential as procarcinogens.

The drinking water provided to research animals is often treated at the facility to inhibit microbial growth in the water. This treatment often takes the form of the addition of chlorine in the form of sodium hypochlorite or acidification with hydrogen chloride. The treatment of drinking water by chlorination may result in the formation of mutagenic compounds, and chlorinated water may cause induction of HMEs (Douglas et al., 1986; Liimatainen et al., 1988). Hyperchlorination of water has been shown to result in a reduction in recoverable numbers of macrophages and in a reduction in macrophage-mediated cytotoxicity in mice (Fidler, 1977). Because research rodents are fed diets that provide 100% of their mineral needs, they can be given drinking water that is pure. Given the highly variable content of community water supplies and the potential for water content to alter an animal's biology

and the data derived from the animal, providing purified drinking water for research animals is strongly recommended. The water purification process most often used in contemporary animal facilities is reverse osmosis. Often, the reverse osmosis water is acidified to a pH of 2.5 to 3.5 to control microbial contamination.

C. Bedding

Various absorbent materials termed "bedding" are normally placed in shoebox cages to absorb urine and other moisture and to provide nesting material. To perform well, bedding materials should have a high capacity for moisture absorption (without desiccation or other injury of neonatal animals); have a high capacity for binding NH_3; be free from contamination with excreta of wild rodents, pathogenic agents, and harmful residues such as preservatives and pesticides; be as dust free as possible; and be nonabrasive, economical, and conveniently disposed (Hastings, 1970; Burkhart and Robinson, 1978; Kraft, 1980). Rats appear to prefer bedding materials consisting of large fibrous particles (Blom et al., 1996). In preference tests, they generally avoid relatively small particles (1.2×1.6 mm^2).

Bedding materials can have effects on research animals as a function of the material itself or through trace contaminants in the bedding material. Most bedding materials in use currently are natural products that are byproducts of other industries, and therefore, the potential for contamination with a variety of chemicals certainly exists. A variety of wood and numerous other products are widely used as bedding materials. Most of these materials are natural products, and there may be large variability of the quality between batches. Because of the compounds that may be present in bedding materials and their potential affect on the health and responsiveness of research animals, bedding material must be considered as a potential variable between experiments (Potgieter and Wilke, 1992). Laboratory rodents are known to consume significant quantities of bedding (Weisbroth, 1979); therefore, it has been recommended that contact beddings be considered equivalent to animal diets in terms of likely chemical and microbial contaminants, and the same standards for permissible levels of various contaminants should be applied to bedding materials as are applied to feeds (Weisbroth, 1979).

Bedding materials may contain endogenous constituents that directly affect the responsiveness of animals to pharmacological challenges (Pick and Little, 1965). Untreated softwood bedding materials have been show to have constituent volatile hydrocarbons that have ability to induce hepatic enzymes (Weisbroth, 1979). These induced enzymes may alter biologic responses to pharmacological compounds. Use of white pine or red cedar chips as bedding reduced hexobarbitone-induced sleeping time in mice, and these animals showed increases in aniline hydroxylase and ethylmorphine N-demethylase in liver microsomes compared with levels of control mice housed on hardwood bedding (Vesell, 1967, 1968). In one study, cedar chip bedding was shown to cause an 18% increase in liver cytochrome P-450, a 46% increase in liver ethylmorphine N-demethylase, and a 49% increase in liver benzo(a) pyrene activities. In contrast, animals housed on heat-treated pine shavings showed a 21% decrease in liver ethylmorphine N-demethylase activity with no significant change in liver cytochrome P-450 or benzo(a) pyrene hydroxylase activity compared with levels of control animals housed in wire-bottom cages with no bedding (Weichbrod et al., 1988). *In vitro* studies using a mouse hepatoma cell line (Hepa-1) have corroborated and expanded these findings. By use of cytotoxicity and induction of cytochrome P450IA1 (aryl hydrocarbon hydroxylase) and aldehyde dehydrogenase as measures, corncob was shown to be practically nontoxic, whereas the softwoods and alder were more cytotoxic than wasaspen, and both softwood and hardwood extracts were shown to contain enzyme inducers, with pine extracts being the most potent enzyme inducers and hardwood extracts being much less active (Torronen et al., 1989; Pelkonen and Hanninen, 1997). Probable enzyme-inducing substances in softwood beddings are cedrene and a-pinene (Wade et al., 1968; Nielsen et al., 1986). Autoclaving these bedding materials appeared not to inactivate the enzyme inducing properties (Cunliffe-Beamer et al., 1981).

As mentioned earlier, bedding materials may contain trace contaminants. Wood shavings may come from trees that were clear cut and may therefore be contaminated with defoliants, whereas corncob may come from fields that have been sprayed with insecticides. In addition, bedding materials may be contaminated with chemicals during the manufacturing process. Acute lethal effects in Long-Evans rats caused by incidental contamination of bedding material with an organophosphate pesticide (terbufos) were reported by Gibson et al. (1987). Contamination of bedding materials with dichlorodiphenyltrichloroethane and parathion has also been reported (Foley, 1979). In 1985, Faith et al., reported a reduction in breeding efficiency of Sprague-Dawley rats housed on pine shaving bedding. Pups from affected litters were runted and died at about 10 days of age, but the dams remained healthy. No infectious agents could be identified as the cause of the problem. Although chemical analysis was not performed on the bedding, the investigators speculated that the bedding contained chemical contaminants that caused the runting and death of the pups.

In addition to chemical contaminants, animal beddings may have high levels of microbial contaminants. Fungal contamination of an animal facility from spores in bedding has recently been reported. In the course of studies involving the colonization of mice with *C. albicans*, it was discovered that the animals were being exposed to large numbers of *Aspergillus fumigatus* spores, which interfered with the *C. albicans* colonization (Mayeux et al., 1995). In the course of investigation, it became clear that the source of the contamination was the untreated hardwood chip bedding that was being used in the mouse colony. Cultures of untreated bedding revealed high levels of contamination with *A. fumigatus*, as well as contamination with *Mucor*, *Rhizopus*, and *Penicillium* spp. When the bedding was obtained from the manufacturer as a heat-treated product, it was uncontaminated with fungal spores. The investigators also cultured corncob and aspen beddings and found untreated cob to contain *Fusarium*, *Cladosporium*, and *Penicillium* spp. and untreated aspen to have various zygomycetes. Heat treatment of these materials eliminated the contamination. In another report, bedding materials were analyzed for the presence of fungal spores after two rats were diagnosed with fungal rhinitis. Corncob bedding from two manufacturers was found to be contaminated with six species of fungus (*Cladosporidium*, *Acremonium*, *Penicillium*, *Aspergillus*, *Fusarium*, and *Scolobasidium*) (Royals et al., 1999). These investigators found aspen chip bedding to be free of significant fungal contamination.

Ground corncob is widely used as a rodent bedding material. Perhaps the primary reason for its widespread use is its ability to diminish intracage NH_3 build-up (Ras et al., 2002), which allows for longer periods between cage changes and therefore a reduction in labor cost. Although corncob controls NH_3 levels better than do most other bedding materials, it too is not biologically inert. Depression in reproductive efficiency and protein anabolism has been reported for mice housed on corncob bedding compared with levels for animals on pine bedding (Port and Kaltenbach, 1969). The cause of the observed depression was not investigated, but it was theorized that it could be owing to either a reduction in nutrient intake caused by ingestion of the corncob, or owing to mycotoxins that might have been present in the corncob. Recently, it has been shown that there are factors in corncob bedding that impede the mating behavior of male and female rats and cause estrous acyclicity in females (Markaverich et al., 2002a,b). This activity can be extracted from the corncob bedding and has been found in fresh corn (both kernels and cob) and corn products such as corn tortillas, leading to the belief that it is a natural product in the corn and not a contaminant. Corncob extracts are mitogenic for estrogen receptor–positive (MCF-7) and estrogen receptor–negative (MDA-MD-231) breast cancer cells, and PC-3 human prostatic cancer cells *in vitro*. In addition, the growth rate of PC-3 cell xenografts is accelerated in nude male mice housed on ground corncob as opposed to pure cellulose bedding. The active component is not a phytoestrogen, bioflavonoid, mycotoxin, or other known endocrine-disrupting agent that modifies cell growth via estrogen receptor or type II [^{3}H] estradiol binding sites. This endocrine disrupting agent in ground corncob bedding may influence behavioral and physiologic reproductive response profiles and malignant cell proliferation in experimental animals. Corncob bedding should not be used for rats involved in reproductive studies or in studies in which reproductive hormones play a role in the experimental outcome. The mitogenic activity has been shown to reside in an isomeric mixture of linoleic acid derivatives with a tetrahydrofuran ring and two hydroxyl groups (THF-diols) (Markaverich et al., 2002b). Exposure to THF-diols may disrupt endocrine function in experimental animals at doses about 200 times lower than that of classical phytoestrogens.

V. CHEMICAL FACTORS IN THE ENVIRONMENT

Research animals may be intentionally exposed to chemicals through the use of pharmacologic agents or test compounds, or unintentionally exposed through the use of chemicals in cleaning and sanitation of the facility or by chemicals found in air, water, food, and bedding and on caging or other equipment. Many elements of the research animal's environment may be inadvertently contaminated by a great variety of man-made and/or naturally occurring chemicals that may be deleterious to the animal's health or may subtly affect experimental results (Cass, 1970; Newberne, 1975; Fouts, 1976; Lindsey et al., 1978; Baker et al., 1979; Newberne and McConnell, 1980).

Environmental chemicals may affect a wide variety of biological responses. A chemical contaminant may exert its effect at various sites in the body. It may have a topical effect at the portal of entry (gastrointestinal tract, skin, or respiratory system) or be absorbed across one or more body surfaces into the general circulation. Once absorbed, it may be translocated to storage sites, biotransformation sites, or sites of excretion. It may be inherently toxic, or metabolism by the host may form a toxic product. Subtle effects of chemical exposure may have special significance in chronic studies such as assessment of toxicity, cancer research, or studies on aging. The ultimate response to exogenous chemicals is determined by interplay of host and environmental factors. Age, state of health, nutritional status, sex, genetic constitution, and immune function are important host factors. Environmental variables include the concentration of the agent, its physicochemical properties, potential interactions with other agents in the exposure environment, and the duration, frequency, and route of exposure. There are a number of examples of chemicals affecting the HMEs and the immune system (Lindsey et al., 1978). HMEs play a principal role in metabolism and elimination of exogenous substances and are of particular interest in pharmacological and biochemical studies that use rats. A great many environmental chemicals and drugs have been shown to enhance or impair HME activity. The effects of chemicals that modulate HME activity may be subtle. For example, unknown exposure of rats in a carcinogen study to low levels of a chemical affecting HME activity might render the rats far more or less susceptible to the induction of cancer by the agent being studied.

A. Gaseous Pollutants

In the past, a variety of volatile organic solvents, such as acetone, ether, and chloroform, were used commonly in animal facilities. Fortunately, the use of such chemicals in animal housing areas has decreased greatly, and people have become aware of the importance of working with volatile compounds in fume hoods. The use of volatile chemicals, such as insecticides and odor counteractants, in routine animal husbandry is poor practice. Whenever possible, nonvolatile materials should be used (Newberne and Fox, 1978).

The main gaseous pollutants occurring from animal wastes in the cage are CO_2 and NH_3 (Briel et al., 1972). NH_3 is generated by urease positive bacteria typically found in feces, breaking down each urea molecule from urine into two NH_4 molecules. The generation of gaseous pollutants from animal wastes depends on the cage type, population density, strain of animal, bedding material, frequency of cage change, and temperature and humidity in the cage. NH_3 and CO_2 levels may become significantly elevated in shoebox cages with wire-bar lids and in suspended cages with wire-grid floors. In shoebox cages with wire-bar lids housing mice, CO_2 may approach levels three times greater than that of room air; in suspended cages with wire-grid floors, four times that of room air (Serrano, 1971a). NH_3 in the cage becomes detectable several days after cage change and may reach levels in excess of 11 ppm by the fourth day in shoebox cages housing mice, or 45 ppm by the seventh day in suspended cages housing mice (Murakami, 1971; Serrano, 1971a). When the cage is covered with any type of top that restricts air exchange between the cage and the room, NH_3 levels in the cage become detectable in 1 to 2 days and may reach levels in excess of 700 ppm by the seventh day (Flynn, 1968; Gamble and Clough, 1976; Hasenau et al., 1993). HME activity has been shown to be impaired in such an environment (Vesell et al., 1973, 1976). NH_3 levels as low as 25 ppm have been shown to play a contributory role in the pathogenesis of respiratory mycoplasmosis in rats (Broderson et al., 1976), and levels as low as 2 ppm have been shown to enhance the growth of *Mycoplasma pulmonis* in the respiratory tracts of rats (Schoeb et al., 1982). There is a direct correlation between the concentration of NH_3 in the cage and the development of lung lesions in rats infected with *M. pulmonis* (Lindsey et al., 1978). Very low levels of NH_3 (about 3 ppm) cause the rapid cessation of ciliary activity in the tracheal epithelium of rats (Dalhamn, 1956). Clearly, it is desirable to control NH_3 levels in the cage. This can be done by frequent cage changes, selection of a bedding material that controls NH_3 generation (Perkins and Lipman, 1995), placement of cages in mass air displacement units (Corning and Lipman, 1992), the use of VCSs as discussed above, or a combination of these.

B. Pesticides

Exposure to pesticides may result in induction or inhibition of enzymes and increased or decreased toxicity of other compounds. The function of endocrine organs and the metabolism of their products may be altered by pesticides. These factors, as well as the various types of pesticides and their impact on the environment, have been reviewed elsewhere (Hayes, 1975; Matsumura, 1975; Quraishi, 1977; Wagner, 1983). Chlorinated hydrocarbons have essentially been banned from general use because of their profound effects on biological systems and their environmental persistence. Organophosphates and carbamates are now the most widely used insecticides. HMEs are stimulated by halogenated hydrocarbon insecticides and inhibited by organophosphate insecticides or pesticide synergists of the methylene-dioxyphenyl type (Hart and Fouts, 1965; Kolmodin et al., 1969; Fouts, 1970; Campbell et al., 1983; Haake et al., 1987). Carbamate fungicides inhibit HME activity (Periquet and Derache, 1981; Borin et al., 1985), whereas substituted urea herbicides induce HME activity (Schoket and Vincze, 1985). Fungicides have also been shown to affect sexual differentiation in rats (Gray et al., 1994; Gray and Kelce, 1996; Gray and Ostby, 1998; Ostby et al., 1999). Pesticides have been shown to alter immune responses (Casale et al., 1984; Bernier et al., 1987; Banerjee et al., 1996, 1998; Koner et al., 1998).

It is generally recommended that nonchemical means of vermin control be used to their maximum potential in animal facilities owing to the high risk of unwanted complications associated with pesticide use. Pesticides should be used only when absolutely necessary and then used only sparingly and with the utmost discretion and control. It is preferable to use nonvolatile forms of pesticides and limit their application to areas outside of animal housing rooms. If insecticides must be used, pyrethroids may be the agents of choice as they are considered to be relatively nontoxic and only slightly induce HME activity (Krechniak and Wrzesniowska, 1991). However, at least one pyrethroid insecticide has been shown to be immunomodulatory (Punareewattana et al., 2001; Prater et al., 2002). Investigators should be informed about the intended use of pesticides, and only specifically approved compounds should be used.

C. Drugs

Pharmaceuticals may be used in the research setting for treatment or control of disease, anesthesia, or euthanasia and for the altering physiological status for specific research objectives, such as the induction of diabetes with alloxan. A wide variety of drugs and anesthetics affect HMEs and liver function and thus may affect research results (Dale et al., 1983; Dale and Nilsen, 1984; Daniel et al., 1984; Govindwar et al., 1984; Belanger and Atitse-Gbeassor, 1985; Knights et al., 1987; Rikans and Moore, 1988; Akita et al., 1989; Dalvi and Terse, 1990; Ishikawa et al., 1991; Ozaki et al., 1993; Ohishi et al., 1994; Kodam et al., 1996; Kodam and Govindwar, 1997; Thompson et al., 2002). The effect that drugs may have on research results depends on the characteristics of the active ingredient, the characteristics of the vehicles or diluents, the route of administration, frequency of treatment, and dosage. Animals should normally not be treated with drugs without the knowledge and consent of the principal investigator.

The addition of drugs or chemicals to food or water is a useful method of administration of therapeutic or experimental materials. However, many factors influence the suitability of this approach for specific agents. Consideration must be given to the extent to which the material is absorbed via the gastrointestinal tract, the stability of the material in the presence of nutrients or other chemicals (e.g., chlorine in water), and degradation by light or bacteria. It may be necessary to use distilled water, brown glass receptacles, and/or frequent replacement of treated drinking water or food. Dosage of a drug or chemical administered in the food or water varies with the amount of food or water consumed, and usually cannot be controlled precisely because of wastage and/or competition among occupants of a cage. If the material imparts an unpleasant flavor, the consumption of food or water may be reduced below expected levels. Direct effects of the drug or chemical on the gastrointestinal tract (including gut flora) may affect the health of the animal as well as alter the rate of absorption. Unabsorbed portions or metabolites of the material may be present in excreta. Coprophagy may result in repeated exposure to the material or its metabolites in an unexpected way. The use of cages with wire mesh floors will not eliminate this complication.

D. Cleaners, Deodorizers, and Chemical Sterilants

A wide variety of compounds, including soaps, detergents, wetting agents, disinfectants, and solvents, are used for cleaning and sanitizing interior surfaces and equipment in animal facilities. Many of the cleaning compounds available contain volatile substances that often have pleasant odors and the ability to modify HME activity (Conney and Burns, 1972; Lang and Vesell, 1976). Deodorants or odor counteractants should not be used in the animal facility because their use may mask poor sanitation, and they can induce HME activity (Jori et al., 1969; Cinti et al., 1976; Fouts, 1976). It is not possible to select a single agent that meets all cleaning and sanitation needs. However, it is prudent to minimize as much as possible the number of different agents being used in the facility. The halogens, especially the chlorine-based compounds, are effective against the majority of viruses but are rapidly inactivated by the presence of organic matter. There are four main classes of detergents: anionic, cationic, non-ionic, and amphoteric.

The bulk of detergents in use today are linear alkyl sulfonates. These compounds and the detergents containing them have a wide margin of safety and do not appear to be a threat to human or animal safety. They have generally tested negative in mutagenicity tests and reproductive tests, and teratogenesis studies have yielded positive effects only at levels great enough to produce maternal toxicity (Burek and Schwetz, 1980; Robinson and Nair, 1992; Robinson and Schroeder, 1992). Two-year feeding and chronic skin painting studies have not yielded any significant effects, and the linear alkyl sulfonates appear to be rapidly cleared from the body after oral exposure (Burek and Schwetz, 1980). These compounds are relatively safe for use in the research animal environment.

There are many disinfectants that may be used in the animal facility, including derivatives of phenols, cresols, resorcinols, alcohols, aldehydes, acids, halogens, oxidizing agents, or heavy metals. The toxicity of these compounds has been reviewed elsewhere (Hardman et al., 2001). When used properly, there is relatively little chance of animal exposure or toxicity. However, they all have a potential for animal toxicity and altering research data if present in sufficiently large amounts. Paraformaldehyde, peracetic acid, and ethylene oxide have all been used as chemical sterilants in animal research facilities. Paraformaldehyde has been used for the disinfection or sterilization of animal rooms, especially in barrier facilities. Because of its high toxicity and explosive nature at higher concentrations,

there has been a trend toward using much safer fogged chlorine dioxide (ClO_2) for this purpose in recent years. Peracetic acid is an effective bactericide, fungicide, and virucide, which is used in the sterilization of gnotobiotic equipment (Kline and Hull, 1960; Baldry, 1983). Personnel working with peracetic acid must wear protective clothing, gas masks, and gloves (Pleasants, 1974). Animals housed under gnotobiotic conditions are almost invariably exposed to peracetic acid when materials are passed in and out of the isolator in which they are housed. This may be a serious problem to research using these animals, as peracetic acid has been shown to be a carcinogen and a cancer promoter (Bock et al., 1975).

ClO_2 has become widely used in rodent colonies as a disinfectant for surfaces, equipment, and hands, especially where isolator caging systems are in use. There has been interest in the use of ClO_2 as a disinfectant for treatment of water and waste water for a number of years (Aston and Synan, 1948). ClO_2 is a stronger disinfectant than is chlorine and chloramine. Ozone has greater antimicrobial effects, but limited residual disinfection capability (Gomella and Musquere, 1980; Hoff and Geldreich, 1981). Recent research demonstrates that ClO_2 is an effective agent against a number of microorganisms—capable of destroying enteroviruses, *E. coli*, amoebae, and a variety of fungal agents—and is effective against cryptosporidium cysts (Terleckyj and Axler, 1987; Peeters et al., 1989; Tanner, 1989; Junli et al., 1997a,b). Although ClO_2 is an effective disinfectant/sterilant, it appears to be a relatively safe chemical to use in the animal environment. Studies with ClO_2 have shown that it does not affect litter size, pup viability, or pup weight in rats (Carlton et al., 1991), nor does it affect fetal development or cause teratology in rats and mice (Gerges et al., 1985; Skowronski et al., 1985); cause chromosomal damage or sperm-head abnormalities in mice (Meier et al., 1997); or affect distribution of dietary iodide in rats (Harrington et al., 1985). Even though it appears to be relatively safe, ClO_2 has been shown to have some negative effects. There are indications that ClO_2 affects thyroid function (Orme et al., 1985; Harrington et al., 1986) and may have central neurotoxic potential (Toth et al., 1990). When given to mice in the drinking water, chlorite (a product formed from ClO_2 disinfection) produced increases in red blood cell mean corpuscular volume, osmotic fragility, glucose-6-phosphate dehydrogenase activity, and the number of acanthocytes (all changes consistent with red cell damage) (Moore and Calabrese, 1980).

Indiscriminate or inappropriate use of chemicals may lead to erroneous and nonreproducible research results. All chemicals must be used properly in accordance with the label directions in order to protect personnel and animals from either exposure or overexposure. All warning and caution notices should be heeded. Whenever possible, the use of chemicals should be minimized. Because any chemical may cause experimental changes, investigators should be made aware of the chemicals used in the facility and their potential impact on research. The chemicals used and their schedule of use should be documented.

A recently reported incident exemplifies how inappropriate use of chemicals can profoundly affect research (Hunt et al., 2003; Koehler et al., 2003). Investigators in two laboratories studying meiotic chromosome behavior experienced a sudden, spontaneous increase in meiotic disturbances, including aneuploidy, in studies of oocytes from control female mice. The investigators noticed that the polycarbonate cages that the mice were housed in appeared to be melting and subsequently discovered that the cages had mistakenly been washed with a highly alkaline detergent that caused breakdown of the plastic. The process was purposely repeated. Again the cages were damaged, and female mice housed in them exhibited meiotic disturbances. Further investigation led to the determination that BPA released from the damaged plastic was responsible for the meiotic disturbances (Hunt et al., 2003). BPA is a weakly estrogenic compound used in the manufacture of plastics and resins. Recently, it has been shown that BPA is released in low levels from new cages or water bottles made of polycarbonate and polysulfone at room temperature, and that the level of leaching is markedly increased under conditions of normal wear (Howdeshell et al., 2003). Autoclaving polycarbonate bottles has been shown to result in the release of BPA (Krishman et al., 1993; Feldman and Krishman, 1995).

VI. MISCELLANEOUS ENVIRONMENTAL FACTORS

A. Stress and Stressors

Many factors in the environment of laboratory animals may be stressful to the animal. This includes things as routine and simple as transportation, either between facilities or within the facility. The simple act of transferring the animals from a laboratory cage to a shipping box results in weight loss, even under the best conditions (Wallace, 1976). Mice taken from a breeding unit with the cage wrapped in black plastic and carried to an experimental room (about a 12-minute trip, a 10-minute walk, and 2 minutes in elevators) had significantly elevated corticosterone levels when tested immediately after being moved (Tuli et al., 1994). Corticosterone levels returned to baseline within 24 hours. In a similar study, mice were transported within a facility in a pattern designed to simulate that commonly used by investigators before experimental manipulation (Drozdowicz et al., 1990). A significant increase was seen in

plasma corticosterone levels immediately after transport. The normal circadian rhythm of plasma corticosterone levels was altered for the subsequent 24 hours, and there was a corresponding significant decrease in total white blood cell numbers, lymphocyte count, and thymus weight. Some behavioral parameters do not return to baseline within 4 days after transportation of animals within a facility (Tuli et al., 1994).

The question of how long animals should be allowed to equilibrate after transit before being placed on study is a difficult one. Data in the literature indicate that acclimation after shipping may require periods as short as 1 to 2 days or as long as several weeks. A number of studies have shown that animals lose weight during shipment (Stanetz et al., 1956; Dymsza et al., 1963; Foster et al., 1967; Manning and Banks, 1968; Flynn et al., 1971; Grant et al., 1971; Weisbroth et al., 1977). Weight loss appears to be more severe in warm weather than in cold (Slanetz et al., 1956), and air shipment seems to be more stressful than is surface transport (Slanetz et al., 1956; Dymsza et al., 1963). In a controlled study of wild and laboratory mice subjected to a 28-hour journey, the following results were found. Transportation by a combination of van and rail alone caused weight loss that was exacerbated by food or food and water deprivation during transit. Water deprivation caused retardation in the recovery of the weight loss. Wild mice were more severely affected than were laboratory mice (Wallace, 1976). Hooded Lister rats were shipped by rail for a journey of 5 hours. On arrival at the facility, they were housed in soundproof rooms at a constant temperature of 21°C with a 12-hour light/12-hour dark cycle. The animals were cared for by the same attendant at the same time each day and given food and water *ad libitum*, and water consumption was measured daily. Based on water consumption, acclimation took 7 to 25 days for individual rats, with half of the animals requiring more than 14 days to acclimate (Grant et al., 1971). The investigators suggest that 4 weeks in a constant environment might be an appropriate adjustment period for rats after transit. In mice shipped by truck or plane, plasma corticosterone levels were markedly elevated on arrival and remained at high values for a 48-hour period (Landi et al., 1982). These animals also exhibited significantly depressed immune function on arrival. Immune function returned to baseline within 48 hours. Similarly natural killer cell activity was shown to be depressed in mice after transport by air and truck (Aguila et al., 1988), but the activity returned to normal by 24 hours after arrival at the facility. Aged rats and mice are more susceptible to shipping stress than are young animals (Manning and Banks, 1968; Flynn et al., 1971). Lack of water intake during shipment resulting in dehydration has been identified as one of the important components in shipping stress (Weisbroth et al., 1977).

If the equilibration period is defined as that amount of time required at the destination for a young, growing animal to regain lost weight to within one standard deviation of the unshipped growth rate curve, then animals shipped without water or with water substitutes, such as vegetables or gelled diets, require 2 to 4 days to equilibrate for short duration trips (up to 24 hours) and 4 to 7 days for long duration trips (Dymsza et al., 1963; Foster et al., 1967; Weisbroth et al., 1977). The provision of a source of drinking water to animals during transit greatly reduces the equilibration period on arrival (Salvia et al., 1979). The severity of the stress response appears to depend on a number of factors, including the mode and duration of transport, the age of the animal, the ambient temperature, whether or not drinking water is provided during transit, and other factors such as the level of noise and vibration experienced by the animal during transit. Therefore, the equilibration period after arrival at the facility should be determined from the transportation parameters and the type of research for which the animals will be used.

B. Personnel Factors

An unavoidable part of every research animal's life is the contact with and handling by the animal husbandry technician during husbandry procedures. A study was performed to determine if blood pressure and heart rate of adult male Sprague-Dawley rats are affected by the routine animal husbandry procedure of moving animals to clean cages (Duke et al., 2001). The animals were implanted with telemetry devices that were used to obtain the cardiovascular responses. Animals were moved to clean polycarbonate shoebox cages with fresh wood chip bedding. After being placed in the clean cages, the animals exhibited prompt and significant increases in systolic, diastolic, and mean arterial blood pressures; heart rate; and cage behavior (movement, rearing, and grooming). The cardiovascular and behavioral perturbations lasted for 45 to 60 minutes. Animals that observed the cage change but were not moved themselves did not show significant increases in blood pressure, heart rate, or activity. Responses of animals at the fourth weekly change were no different than were responses on the first weekly change. The changes observed were very similar to those observed after exposure of rats to novel environments that are quite different from a normal animal cage. Thus, the routine act of cage changing caused a transient stress response in rats.

In many studies, animals must be handled to have procedures preformed on them. The simple act of catching and removing an animal from its cage has significant effects on plasma corticosterone levels. It has been found that reliable baseline measurements of plasma corticosterone can only be obtained within 3 to 5 minutes after the initial disturbance of rodents by handling (Riley, 1981). In addition, corticosterone levels in animals remaining in the cage during the sequential capture of their cage mates are affected by what is termed "contagious anxiety." The seemingly innocuous act of moving a cage containing rats from the rack to a table or to the floor had significant impact on a number of blood characteristics (Gartner et al., 1980). Adult male Han:Sprague rats weighing 230 to 300 g were housed in groups of 2 or 4 in shoebox cages with sterile wooden bedding. The animal room temperature was maintained at 21°C and 55% RH with a 12-hour light/12-hour dark cycle. Food and water were provided *ad libitum*. On the day of the experiment, the animal room was opened and four cages were removed. The rats in these cages were decapitated immediately with a guillotine, or blood was collected by cardiac puncture without anesthesia, simultaneously by eight trained persons. All samples were obtained 50 to 100 seconds after the animal room door was opened. These samples provided baseline data. Animals in other cages were stressed by moving the cages from the racks to the floor or to a table. Plasma concentrations of 25 substances linked with stress and shock reactions were measured. Five minutes after the stressor, serum prolactin, corticosterone, thyroid-stimulating hormone, luteinizing hormone, triiodothyronine, and thyroxin levels were elevated 150% to 500% compared with baseline. Heart rate (telemetrically recorded), packed cell volume, hemoglobin, and plasma protein content were 10% to 20% elevated 2 to 10 minutes after cage movement, indicating circulatory and microcirculatory perturbations. Serum glucose, pyruvate, and lactate concentrations rose by 20% to 100% 1 to 5 minutes after cage movement. Phosphate, calcium, urea, aspartate and alanine transferases, alkaline phosphatase, and leucine arylamidase were not altered significantly by the stressor. The presence of a familiar animal attendant working in the room without touching the cages did not markedly affect the blood characteristics studied. Increased corticoid concentrations cause injury to elements of the immunological apparatus, which may leave the subject vulnerable to the action of latent oncogenic viruses, newly transformed cancer cells, or other incipient pathological processes that are normally held in check by an intact immune system (Fauci, 1985; Elenkov and Chrousos, 1999; Elenkov et al., 1999).

It is well known that animals habituated to a handler, or that are gentled early in life, show less handling stress in later life and react only to the particular experimental stimuli used in the study, whereas nonhandled animals are much more likely to react to a new handler and to the test situation (Williams and Wells, 1970; Hirsjarvi and Junnila, 1986). If an experimental protocol calls for the frequent handling of animals during the course of a study, the animals should be trained by regular handling to accustom them to being caught and restrained before the investigation begins. This preconditioning will reduce experimental variability and habituation effects during the course of the investigation. Similarly, it is considered ill-advised to begin a study on recently acquired animals until they have adjusted to their cages and new surroundings.

The way in which animals are handled influences the effect of the handling. Hyperbaric oxygen exposure has long been associated with convulsive seizures in rats. In one laboratory studying hyperbaric oxygen exposure in rats, seizures in the exposed rats were common. Midway through the experiments, the animal handlers were changed. The original handlers had lifted the rats by their tails when transferring from chambers and cages. The new handlers were gentler in their treatment of the rats, and especially avoided dangling the animals by their tails. After the change in animal handlers, the frequency of convulsions in the rats decreased until they ceased altogether (Boyle and Villanueva, 1976). This is in agreement with findings that rough handling, specifically lifting by their tails, induces convulsions in mice treated with certain central nervous system drugs (Goldstein, 1973).

C. Environmental Enrichment

Rodents used in biomedical research are raised and maintained in environments that are markedly different from feral rodent natural habitats. They are typically reared and maintained in small cages that lack key features of natural environments. Research animal housing necessitates, in many cases, that we restrict the animal's ability for social interaction and for problem-solving activities such as searching for food, avoiding predators, and finding environments with favorable climatic conditions. These barren housing conditions impose constraints on behavior and brain development that result in altered brain function (Wurbel, 2001). The barren environment of the animal facility interferes with brain development in three ways: early environmental deprivation, thwarting of behavioral response rules, and disruption of habitat-dependent adaptation processes have all been shown to result in aberrant or maladaptive brain functions (Wurbel, 2001). Housing conditions consistent with current standards could therefore compromise the utility of rodents for research, especially in behavioral neuroscience. A better understanding of the environmental factors involved in the control of behavior, and of animal needs, could offer a biological basis for refinement of housing methodologies. Not all investigators agree that barren housing conditions are a detriment (Benefiel and Greenough, 1998). An animal's brain structure and behavior are molded by

experience. Experience that plays a critical role in early organization of the brain may be encoded via a process of overproduction of synaptic connections followed by the loss of those that are underutilized during a critical period. As the individual animal is exposed to new environmental stimuli, the novel information may be encoded throughout life by the formation of new synapses. Many laboratory species will exhibit an increase in the number of synapses per neuron as well as other anatomical differences when reared in complex environments or when trained to perform complex tasks; these changes often occur independent of the age at which the animal is exposed to the complexity in environment or task. Young rats raised in an environment enriched with toys have larger forebrains than do rats raised in isolation (Bennett et al., 1969). Rats raised in isolation and later maze-trained develop larger brains than do rats raised and maintained in isolation (Cummins et al., 1973). There are changes in brain RNA/DNA ratios in rats raised in enriched versus impoverished environments (Vivitskaya et al., 1982). Learning in mice occurs more quickly in enriched groups than in impoverished groups (Bouchon and Will, 1982), and enriched groups of rodents have more well-developed brain structure than do impoverished groups (Wahlstein, 1982). Benefiel and Greenough (1998) state, "even though increased environmental stimulation may result in more 'normal' anatomical and physiological development for that species, there is no conclusive evidence that enriched caging is essential or even that it increases well-being in laboratory rodents."

The typical rodent cage is a rather sterile environment, providing few environmental stimuli. The small size of a cage relative to an animal's natural range restricts the animals' ability for social interaction and for problem-solving activities such as foraging, avoiding predators, and finding environments with favorable climatic conditions. The goal of environmental enrichment is to alter the living environment of the animal in order to provide opportunities to express more of its natural behavioral repertoire. This is likely to be beneficial for the animal (Van de Weerd et al., 1994). The feelings on environmental enrichment for laboratory animals runs the entire spectrum, from those who feel there is no need for enrichment to those who are certain that enrichment for all animals is necessary. There is currently a movement to enrich the environment of a number of laboratory species, including laboratory rodents. Rodent caging systems are typically designed and constructed more with the needs of the facility and caregiver in mind, than with concern for emulating the animals' natural environment. These factors are generally related to equipment costs, space in the animal room, workload of animal caretakers and technicians, hygienic aspects for man and animal, and control/observation of the animal (Baumans and van de Weerd,

1996). The requirements of the animal are related to the possibility of expressing species-specific behavior such as resting, exploring, grooming, foraging, eating, drinking, nest-building, digging, urination/defecation, social behavior, and reaction to environmental stimuli.

Environmental enrichment may focus on several aspects of the environment, including the social environment, the nutritional environment, the sensory environment, the psychological environment, and/or the physical environment (Baumans and van de Weerd, 1996). The social environment can be enriched by group housing. Most laboratory rodents are social animals and benefit from group housing. The benefit of group housing depends on stable social relationships and environmental conditions, such as the opportunity to hide (Baumans and van de Weerd, 1996). Group housing in the laboratory results in groups of age and sex composition that will always be different from groups in nature. This may cause problems, such as aggression in group-housed males of some strains. Husbandry and research may play a role in social enrichment of rodents.

Nutritional enrichment can be provided by giving the animals an opportunity to forage for food, scattering food in the bedding rather than feeding *ad libitum*. Rats have been shown to prefer earned food, even when free food was available, if the work demand is not too high (Baumans and van de Weerd, 1996). In nature, rodents can control their environment by moving from one set of environmental conditions to another. The inability to control their environment in this manner in the laboratory setting may be stressful. Providing shelter or refuge in the cage gives the animal the opportunity to withdraw from frightening stimuli within or outside the cage and also provides a place to hide from too much light. Physical enrichment can be accomplished by enlarging the cage or functionally enlarging the cage space by providing structure, such as climbing accessories, shelters, or exercise devices in the cage.

If environmental enrichment is used, it is important that it be evaluated to ensure that it is beneficial. There are several methods that can be used to evaluate the enrichment program. These include things such as preference tests, home-cage behavior, behavioral tests, and physiological parameters (Baumans and van de Weerd, 1996). The animals' reaction to enrichment should be monitored and compared with baseline behavior that was assessed before the introduction of enrichment. An increase in species-typical behavior or a decrease in abnormal behavior may be seen as an indicator that the animals are responding positively to the enrichment. Assessment of the effects of enrichment should include determination of whether the changes are long term or short term. In addition, the animals may use the enrichment in a different way after some period of time. Cage enrichment generally benefits

the animal (more stimuli, increased alertness, less boredom, and fewer sterotypies), but the consequences for the facility may be higher costs and greater workloads, sanitation issues, and reduced control of the animals. The costs and the practical use of enrichment items are important factors. Enrichment items introduced into the cage should be stimulating for the animals, but they should also be easy to remove, easily sanitizable, and easy to replace (Baumans and van de Weerd, 1996). There has been published a very nice description of the development of a rodent enrichment program (Smith and Hargaden, 2001). A wide variety of items were used as enrichment devices, ranging from the cardboard cores and toilet paper rolls and used plastic soda bottles to items specifically manufactured as enrichment devices. Ideally, enrichment devices should be inexpensive, easy to place in and retrieve out of the cage, easy to clean and sanitize or be disposable, and free of any contaminants that might negatively impact research parameters.

VII. DISASTER PLANNING

It is unfortunate that personnel often learn the value of emergency preparedness through hard experience, but an emergency does not have to become a full-fledged disaster. Hazards can often be mitigated or avoided altogether by a comprehensive, systematic emergency-preparedness program. Such programs provide a means for recognizing and preventing risks and for responding effectively to emergencies. Small-scale emergencies can be contained if staff members are prepared to react quickly. Damage can be limited even in the face of a large-scale disaster. As an example, cultural institutions in Charleston, South Carolina, formed a consortium that focused on disaster preparedness several years before they were hit by hurricane Hugo in 1989. Many of those institutions sustained only minor damage because they were able to put their early warning procedures into operation (Lindblom and Motylewski, 1993).

Disaster planning is complex. The planning process should culminate in a written plan that is the result of a wide range of preliminary activities. The disaster plan should guide individuals during emergencies, inform individuals of potential emergency situations before they occur, and help to anticipate and avoid dangerous situations. The disaster plan for the animal facility should be a part of the institution's overall disaster plan. Disaster planning must be supported at the highest level of the organization if the resulting plan is to be effective. Early in the planning process, a timetable for the project should be established and the scope and goals of the plan should be defined. The scope and goals will depend largely on the potential risks faced by the institution. In identifying risks, a logical first step is to list geographic and climatic hazards and other risks that could jeopardize the facility or its programs. These might include the institution's susceptibility to hurricanes, tornadoes, flash flooding, earthquakes, or forest fires and even the possibility of unusual hazards such as volcanic eruptions. One should consider man-made disasters such as power outages, civil disturbances, break-ins, sprinkler discharges, fuel or water supply failures, chemical spills, arson, bomb threats, or other problems such as terrorist activity. Further, it is important to take note of the environmental risks that surround the institution. Chemical industries, shipping routes for hazardous materials, and adjacent construction projects all may expose the institution to damage. Although all institutions are not vulnerable to all disasters, any event that is a real possibility should be covered under the emergency plan.

Once the institution's hazards are specified, the disaster planner should devise a program with concrete goals, identifiable resources, and a schedule of activities for eliminating as many risks as possible. Disaster planning should not take place in a vacuum; to be effective, the plan must be integrated into the routine operating procedures of the institution. Three important characteristics of an effective disaster plan are comprehensiveness, simplicity, and flexibility. The plan should address all types of emergencies and disasters that the institution is likely to face. It should include schemes for immediate response and long-term salvage and recovery efforts. In the case of animal facilities, the plan should include schemes for moving animals when necessary, with prioritization as to which animals are moved first. The plan should take into account the fact that normal services may be disrupted and contingencies made for proceeding without electrical power, water, or other services for some period of time.

The disaster plan must be easy to follow. In disaster situations, people often have trouble thinking clearly, so concise instructions and training are critical to the success of the plan. It is impossible to anticipate every detail, so one should be sure that the plan provides basic instructions but allows for some on-the-spot creativity. Further, it is important to decide who will be responsible for various activities when responding to an emergency. Who will be the senior decisionmaker? Who will interact with fire, police, or civil defense authorities? Who will talk to the press? Who will serve as back-up if any of the team members are unable to get to the site? The plan should identify a location for a central command post. It is also critical to set up a system for relaying information to members of the recovery team. Written information is less susceptible to misunderstanding and should be used whenever possible. Good communication is essential to avoid confusion and duplication of effort in an emergency.

Before writing the plan, one should identify potential sources of assistance in a disaster. Supplies that will be needed for disaster response and recovery efforts, (flash lights, rubber gloves, etc) should be purchased and kept on hand. These supplies should be kept in a clearly marked location, inventoried periodically, and replaced as needed. If the storage location is locked, it is important to make sure the keys will be available in an emergency.

The first priority in any disaster is human safety.

The following is an outline that can be used for writing a disaster response plan.

1. Introduction stating the lines of authority and the possible events covered by the plan.
2. Actions to be taken if advance warning is available.
3. First response procedures, including who should be contacted first in each type of emergency, what immediate steps should be taken, and how staff or teams will be notified.
4. Emergency procedures with sections devoted to each emergency event covered by the plan. This will include what is to be done during the event and the appropriate recovery procedures to be followed once the initial excitement is over. Floor plans should be included.
5. Recovery plans for returning the institution back to normal operations.
6. Appendices, which may include evacuation/floor plans, listing of emergency services, listing of emergency response team members and responsibilities, telephone tree, location of keys, fire/intrusion alarm procedures, listing of animal priorities, arrangements for relocation of animals if necessary, listing of in-house supplies, listing of outside suppliers and services, listing of volunteers, and detailed recovery procedures.

No matter how much effort has gone into creating the perfect disaster plan, it will be largely ineffective if the staff is not aware of it, if it is outdated, or if it cannot be found during a disaster. Effort must be expended to educate and train staff in emergency procedures. Each staff member should be made aware of his/her responsibilities, and regular drills should be conducted if possible. Copies of the plan should be distributed in various locations, including off-site (ideally in waterproof containers). Each copy of the plan should indicate where other copies may be found. The plan must be updated periodically, as names, addresses, phone numbers, and personnel frequently change.

The animal facility disaster plan should be part of a comprehensive institutional disaster plan. The institution's plan must be in concert with the community's disaster plan. Many disasters will require coordination of the institutional response with city, county, or state emergency response efforts. The institutional plan should be developed with knowledge of the local governmental plans. The Incident Command System (ICS) is recognized as an effective system for managing emergencies. Several states have adopted ICS as their standard for emergency management, and others are considering adopting ICS. A number of institutions are adopting ICS as the basis for developing their disaster plans. As ICS gains wider use, there is a need to provide training for those who are not first responders (i.e., law enforcement, fire, or emergency medical services personnel) who may be called upon to function in an ICS environment. The Federal Emergency Management Agency through its Emergency Management Institute provides independent training programs in ICS and a number of other disaster preparedness areas. These may be accessed on its Web site (*http://training.fema.gov/EMIWeb/IS/*).

VIII. CONCLUSIONS

Research animals respond to many factors, or changes in their environment. These responses may affect experimental results. The greater the understanding of the effects that environmental elements can have on biological processes, the better the variables can be controlled, or at least the better experimental results can be discussed in light of them. A constant, reproducible environment (to within prescribed limits) is desirable to minimize the physiological variations associated with environmental changes; however, the effective control of all environmental variables at all times is a difficult goal. However, all reasonable attempts should be made to control those environmental factors most likely to interfere with the work in progress. Records should be kept of environmental variables relevant to the research programs in the facility. Planned significant environmental changes should not be made without prior consultation with the investigators to ensure that minimal impact occurs to research programs. A cardinal rule of laboratory animal science and medicine is to always keep in mind that procedures carried out in health and husbandry programs may have significant effects on research results.

REFERENCES

Adler, M.W., Bendotti, C., Ghezzi, D., Samanin, R., and Valzelli, L. (1975). Morphine analgesia in grouped and isolated rats. *Psychopharmacol.* **41**, 15–18.

Aguila, H.N., Pakes, S.P., Lai, W.C., and Lu, Y.S. (1988). The effect of transportation stress on splenic natural killer cell activity in C57BL/6J mice. *Lab. Anim. Sci.* **38**, 148–151.

Aloisi, A.M., Seta, D.D., Rendo, C., Ceccarelli, I., Scaramuzzino, A., and Farabollini, F. (2002). Exposure to the estrogen pollutant bisphenol A affects pain behavior induced by subcutaneous formalin injection in male and female rats. *Brain Res.* **937**, 1–7.

Akita, S., Kawahara, M., Takeshita, T., Morio, T., and Fujii, K. (1989). Halothane-induced hepatic microsomal lipid peroxidation in guinea pigs and rats. *J. Appl. Toxicol.* **9**, 9–14.

Anthony, A. (1962). Criteria for acoustics in animal housing. *Lab. Anim. Care* **13**, 340–347.

Archana, R. and Namasivayam, A. (1999). The effect of acute noise stress on neutrophil functions. *Indian J. Physiol. Pharmacol.* **43**, 491–495.

Archana, R. and Namasivayam, A. (2000). Acute noise-induced alterations in the immune status of albino rats. *Indian J. Physiol. Pharmacol.* **44**, 105–108.

Armario, A., Castellanos, J.M., and Balasch, J. (1984a). Effect of crowding on emotional reactivity in male rats. *Nuroendocrin* **39**, 330–333.

Armario, A., Castellanos, J.M., and Balasch, J. (1984b). Adaptation of anterior pituitary hormones to chronic noise stress in male rats. *Behav. Neural. Biol.* **41**, 71–76.

Armario, A., Ortiz, R., and Balasch, J. (1984c). Effect of crowding on some physiological and behavioral variables in adult male rats. *Physiol. Behav.* **32**, 35–37.

Armario, A., Castellanos, J.M., and Balasch, J. (1985). Chronic noise stress and insulin secretion in male rats. *Physiol. Behav.* **34**, 359–361.

Aston, R. and Synan, J. (1948). Chlorine dioxide as a bactericide in waterworks operation. *J. N. Eng. Water Works Assoc.* **62**, 80.

Babinet, C. (2000). Transgenic mice: an irreplaceable tool for the study of mammalian development and biology. *J. Am. Soc. Nephrol.* **11(Suppl 16)**, 88–94.

Baker, H.J., Lindsey, J.R., and Weisbroth, S.H. (1979). Housing to control research variables. *In* "The Laboratory Rat," Vol. 1. pp. 169–192. Academic Press, New York.

Balazs, T., Murphy, J.B., and Grice, H.C. (1962). The influence of environmental changes on the cardiotoxicity of isoprenaline in rats. *J. Pharm. Pharmacol.* **14**, 750–755.

Baldry, M.G. (1983). The bactericidal, fungicidal and sporicidal properties of hydrogen peroxide and peracetic acid. *J. Appl. Bacteriol.* **54**, 417–423.

Ballowitz, L. (1971). Effects of blue and white light on infant (Gunn) rats and on lactating mother rats. *Biol. Neonate* **19**, 409–425.

Banerjee, B.D., Koner, B.C., Ray, A., and Pasha, S.T. (1996). Influence of subchronic exposure to lindane on humoral immunity in mice. *Indian J. Exp. Biol.* **34**, 1109–1113.

Banerjee, B.D., Pasha, S.T., Hussain, Q.Z., Koner, B.C., and Ray, A. (1998). A comparative evaluation of immunotoxicity of malathion after subchronic exposure in experimental animals. *Indian J. Exp. Biol.* **36**, 273–282.

Baran, D., Paduraru, I., Saramet, A., Petrescu, E., and Haulica, I. (2000). Influence of light-dark cycle alteration on free radical level in rat cns. *Rom. J. Physiol.* **37**, 23–38.

Barrett, A.M. and Stockham, M.A. (1963). The effect of housing conditions and simple experimental procedures upon the corticosterone level in the plasma of rats. *J. Endocrinol.* **26**, 97–105.

Baudrie, V., Laude, D., Chaouloff, F., and Elghozi, J.L. (2001). Genetic influences on cardiovascular responses to an acoustic startle stimulus in rats. *Clin. Exp. Pharmacol. Physiol.* **28**, 1096–1099.

Baumans, V. and van de Weerd, H.A. (1996). Enrichment of laboratory animal housing: basic need or luxury? *Scand. Lab. Anim. Sci.* **23(Suppl 1)**, 93–95.

Baumel, I., de Fao, J.J., and Lal, H. (1969). Decreased potency of CNS depressants after prolonged isolation in mice. *Psychopharmacol.* **15**, 153–159.

Beall, J.R., Torning, F.E., and Runkle, R.S. (1971). A laminar flow system for animal maintenance. *Lab. Anim. Sci.* **21**, 206–212.

Belanger, P.M. and Atitse-Gbeassor, A. (1985). Effect of nonsteroidal anti-inflammatory drugs on the microsomal monooxygenase system of rat liver. *Can. J. Physiol. Pharmacol.* **63**, 798–803.

Bellhorn, R.W. (1980). Lighting in the animal environment. *Lab. Anim. Sci.* **30**, 440–450.

Benefiel, A.C. and Greenough, W.T. (1998). Effects of experience and environment on the developing and mature brain: implications for laboratory animal housing. *ILAR J.* **39**, 5–11.

Bennett, E.L., Rosenzweig, M.R., and Diamond, M.C. (1969). Rat brain: effects of environmental enrichment on wet and dry weights. *Science* **163**, 825–826.

Benson, G.K. and Morris, L.R. (1971). Foetal growth and lactation in rats exposed to high temperatures during pregnancy. *J. Reprod. Fertil.* **27**, 369–384.

Bernier, J., Hugo, P., Krzystyniak, K., and Fournier, M. (1987). Suppression of humoral immunity in inbred mice by dieldrin. *Toxicol. Lett.* **35**, 231–240.

Besch, E.L. (1975). Animal cage-room dry-bulb and dew-point temperature differentials. *ASHRAE* **81**, 549–557.

Blaszezyk, J. and Tajchert, K. (1996). Sex and strain differences of acoustic startle reaction development in adolescent albino Wistar and hooded rats. *Acta Neurobiol. Exp.* **56**, 919–925.

Blom, H.J., Van Tintelen, G., Van Vorstenbosch, C.J., Baumans, V., and Beynen, A.C. (1996). Preferences of mice and rats for types of bedding material. *Lab. Anim.* **30**, 234–244.

Bock, F.G., Myers, H.K., and Fox, H.W. (1975). Cocarcinogenic activity of peroxy compounds. *J. Natl. Cancer Inst.* **55**, 1359–1361.

Bomberger, C.E. and Haar, J.L. (1992). Restraint and sound stress reduce the in vitro migration of prethymic stem cells to thymus supernatant. *Thymus* **19**, 111–115.

Borges, J.M., Edward, D.P., and Tso, M.O. (1990). A comparative study of photic injury in four inbred strains of albino rats. *Curr. Eye Res.* **9**, 799–803.

Borin, C., Periquet, A., and Mitjavila, S. (1985). Studies on the mechanism of nabam- and zineb-induced inhibition of the hepatic microsomal monooxygenases of the male rat. *Toxicol. Appl. Pharmacol.* **81**, 460–468.

Bouchon, R. and Will, B. (1982). Effects of post-weaning rearing conditions on learning performance in "dwarf mice." *Physiol. Behav.* **28**, 971–978.

Boyle, E. and Villanueva, P.A. (1976). Hyperbaric oxygen seizures in rats; effects of handling and chamber noise. *Lab. Anim. Sci.* **26**, 100–101.

Brain, P. and Benton, D. (1979). The interpretation of physiological correlates of differential housing in laboratory rats. *Life Sci.* **24**, 99–116.

Brain, P. and Benton, D. (1983). Conditions of housing, hormones, and aggressive behavior. *In* "Hormones and Aggressive Behavior." (B.B. Svare, ed.), pp. 351–372, Plenum Press, New York.

Brick, J.O., Newell, R.R., and Doherty, D.G. (1969). A barrier system for a breeding and experimental rodent colony: Description and operation. *Lab. Anim. Care* **19**, 93–97.

Briel, J.E., Kruckenberg, S.M., and Besch, E.L. (1972). "Observations of Ammonia Generation in Laboratory Animal Quarters." Publication no. 72-03, Institute of Environmental Research, Kansas State University, Manhattan.

Broderson, J.R., Lindsey, J.R., and Crawford, J.E. (1976). The role of environmental ammonia in respiratory mycoplasmosis of rats. *Am. J. Pathol.* **85**, 115–127.

Burek, J.D. and Schwetz, B.A. (1980). Considerations in the selection and use of chemicals within the animal facility. *Lab. Anim. Sci.* **30**, 414–421.

Burkhart, C.A. and Robinson, J.L. (1978). High rat pup mortality attributed to the use of cedar-wood shavings as bedding. *Lab. Anim.* **12**, 221–222.

Campbell, M.A., Gyorkos, J., Leece, B., Homonko, K., and Safe, S. (1983). The effects of twenty-two organochlorine pesticides as inducers of the hepatic drug-metabolizing enzymes. *Gen. Pharmacol.* **14**, 445–454.

Carlton, B.D., Basaran, A.H., Mezza, L.E., George, E.L., and Smith, M.K. (1991). Reproductive effects in Long-Evans rats exposed to chlorine dioxide. *Environ. Res.* **56**, 170–177.

Casale, G.P., Cohen, S.D., and DiCapua, R.A. (1984). Parathion-induced suppression of humoral immunity in inbred mice. *Toxicol. Lett.* **23**, 239–247.

Cass, J.S. (1970). Chemical factors in laboratory animal surroundings. *Bio-Sci* **20**, 658–662.

Chaguri, L.C.A.G., Souza, N.L., Teixeira, M.A., Mori, C.M.C., Carissimi, A.S., and Merusse, J.L.B. (2001). Evaluation of reproductive indices in rats (*Rattus norvegicus*) housed under an intracage ventilation system. *Contemp. Top. Lab. Anim. Sci.* **40**, 25–30.

Chedid, A. and Nair, V. (1972). Diural rhythm in endoplasmic reticulum of rat liver: electron microscopic study. *Science* **175**, 176–179.

Christian, J.J. and LeMunyan, C.D. (1958). Adverse effects of crowding on lactation and reproduction of mice and two generations of their progeny. *Endocrinology* **63**, 517–529.

Christie, R.J., Williams, F.P., Whitney Jr., R.A., and Johnson, D.J. (1968). Techniques used in the establishment and maintenance of a barrier mouse breeding colony. *Lab. Anim. Care* **18**, 544–549.

Chvedoff, M., Clarke, M.R., Irisarri, E., Faccini, J.M., and Monro, A.M. (1980). Effects of housing conditions on food intake, body weight and spontaneous lesions in mice: a review of the literature and results of an 18-month study. *Food Cosmest. Toxicol.* **18**, 517–522.

Cinti, H.E., Lemelin, M.A., and Christian, J., (1976). Induction of liver microsomal mixed-function oxidases by volatile hydrocarbons. *Biochem. Pharmacol.* **25**, 100–103.

Cline, J. (1986). Gene therapy: current status and future directions. *Schweiz Med. Wochenschr.* **116**, 1459–1464.

Clough, G. (1976). The immediate environment of the laboratory animal. *In* "Control of the Animal House Environment." (T. McSheehy, ed.), pp. 77–94, Lab. Anim. Ltd., London.

Clough, G. (1982). Environmental effects on animals used in biomedical research. *Biol. Rev.* **57**, 487–523.

Clough, G. (1984). Environmental factors in relation to the comfort and well-being of laboratory rats and mice. *In* "Standards In Laboratory Animal Management: Part 1." pp. 7–24. Universities Federation for Animal Welfare, Herfordshire, United Kingdom.

Clough, G., Wallace, J., Gamble, M.R., Merryweather, E.R., and Bailey, E. (1995). A positive, individually ventilated caging system: a local barrier system to protect both animals and personnel. *Lab. Anim.* **29**, 139–151.

Committee to Revise the Guide for the Care and Use of Laboratory Animals (1996). *Guide for the Care and Use of Laboratory Animals.* National Academy Press, Washington, DC.

Connelly, C.S., Fahl, W.E., and Iannaccone, P.M. (1989). The role of transgenic animals in the analysis of various biological aspects of normal and pathologic states. *Exp. Cell Res.* **183**, 257–276.

Conney, A.H. and Burns, J.J. (1972). Metabolic interactions among environmental chemicals and drugs. *Science* **178**, 576–586.

Connolly, M.S., Suman, M.A., Halberg, E., and Halberg, F. (1983). Lighting cycle and social isolation affect development of elevated blood pressure in spontaneously hypertensive rats. *Med. Biol.* **61**, 113–119.

Corning B.F. and Lipman, N.S. (1991). A comparison of rodent caging systems based on microenvironmental parameters. *Lab. Anim. Sci.* **41**, 498–503.

Corning, B.F. and Lipman, N.S. (1992). The effects of a mass air displacement unit on the microenvironmental parameters within isolator cages. *Lab. Anim. Sci.* **42**, 91–93.

Cummins, R.A., Walsh, R.N., Budtz-Olsen, O.E., Kontantinos, T., and Horsfall, C.R. (1973). Environmentally-enduced changes in the brains of elderly rats. *Nature* **243**, 516–518.

Cunliffe-Beamer, T.L., Freeman, L.C., and Myers, D.D. (1981). Barbiturate sleeptime in mice exposed to autoclaved or unautoclaved wood beddings. *Lab. Anim. Sci.* **31**, 672–675.

Cunliffe-Beamer, T.L. and Les, E.P. (1983). Effectiveness of pressurized individually ventilated (PIV) cages in reducing transmission of pneumonia virus of mice (PVM). *Lab. Anim. Sci.* **33**, 495.

Dairman, W. and Balazs, T. (1970). Comparison of liver microsome enzyme systems and barbiturate sleep times in rats caged individually or communally. *Biochem. Pharmacol.* **19**, 951–955.

Dale, O. and Nilsen, O.G. (1984). Glutathione and glutathione *S*-transferases in rat liver after inhalation of halothane and enflurane. *Toxicol. Lett.* **23**, 61–66.

Dale, O., Nielsen, K., Westgaard, G., and Nilsen, O.G. (1983). Drug metabolizing enzymes in the rat after inhalation of halothane and enflurane: different pattern of response in liver, kidney and lung and possible implications for toxicity. *Br. J. Anaesth.* **55**, 1217–1224.

Dalhamn, T. (1956). Mucous flow and ciliary activity in the trachea of healthy rats and rats exposed to respiratory irritant gases. *Acta Physiol. Sci.* **36(Suppl 123)**, 1–158.

Dalvi, R.R. and Terse, P.S. (1990). Induction of hepatic microsomal drug metabolizing enzyme system by levamisole in male rats. *J. Pharm. Pharmacol.* **42**, 58–59.

Damon, E.G., Eidson, A.F., and Hahn, F.F. (1982). Acute uranium toxicity resulting from subcutaneous implantation of soluble yellow-cake powder in Fischer-344 rats. *In* "Biological Characterization of Radiation Exposure and Dose Estimates for Inhaled Uranium Milling Effluents," Annual Progress Report, April 1980–March 1981, pp. 32–37. NUREG/CR-2539, LMF-94, Springfield, VA.

Damon, E.G., Eidson, A.F., Hobbs, C.H., and Hahn, F.F. (1986). Effect of acclimation to caging on nephrotoxic response of rats to uranium. *Lab. Anim. Sci.* **36**, 24–27.

Daniel, W., Friebertshauser, J., and Steffen, C. (1984). The effect of imipramine and desipramine on mixed function oxidase in rats. *Naunyn Schmiedebergs Arch. Pharmacol.* **328**, 83–86.

Davis, D.E. (1978). Social behavior in a laboratory environment. *In* "Laboratory Animal Housing." pp. 44–63. Institute of Laboratory Animal Resoures, National Academy of Science, Washington, DC.

Degen, G.H. and Bolt, H.M. (2000). Endocrine disruptors: update on xenoestrogens. *Int. Arch Occup. Environ. Health* **73**, 433–441.

Depocas, F. (1960). The calorigenic response of cold-acclimated white rats to infused nor-adrenalin. *Canadian J. Biochem. Physiol.* **38**, 107–114.

Dessi-Fulgheri, F., Porrini, S., and Farabollini, F. (2002). Effects of perinatal exposure to bisphenol A on play behavior of female and male juvenile rats. *Environ. Health Persp.* **110(Suppl 3)**, 403–407.

Douglas, G.R., Nestmann, E.R., and Lebel, G. (1986). Contribution of chlorination to the mutagenic activity of drinking water extracts in *Salmonella* and Chinese hamster ovary cells. *Environ. Health Perspect.* **69**, 81–87.

Drozdowicz, C.K., Bowman, T.A., Webb, M.L., and Lang, C.M. (1990). Effect of in-house transport on murine plasma corticosterone concentration and blood lymphocyte populations. *Am. J. Vet. Res.* **51**, 1841–1846.

Duke, J.L., Zammit, T.G., and Lawson, D.M. (2001). The effects of routine cage-changing on cardiovascular and behavioral parameters in male Sprague-Dawley rats. *Contemp. Top. Lab. Anim. Sci.* **40**, 17–20.

Dymsza, H.A., Miller, S.A., Maloney, J.F., and Foster, H.L. (1963). Equilibration of the laboratory rat following exposure to shipping stresses. *Lab. Anim. Care* **13**, 60–65.

Einon, D., Stewart, J., Atkinson, S., and Morgan, M. (1976). Effect of isolation on barbiturate anesthesia in the rat. *Psychopharmacol.* **50**, 85–88.

Elenkov, I.J. and Chrousos, G.P. (1999). Stress, cytokine patterns and susceptibility to disease. *Baillieres Best Pract. Res. Clin. Endocrinol. Metab.* **13**, 583–595.

Elenkov, I.J., Webster, E.L., Torpy, D.J., and Chrousos, G.P. (1999). Stress, corticotrophin-releasing hormone, glucocorticoids, and the immune/inflammatory response: acute and chronic effects. *Ann. NY Acad. Sci.* **876**, 1–11.

Ema, M., Fujii, S., Furukawa, M., Kiguchi, M., Ikka, T., and Harazono, A. (2001). Rat two-generation reproductive toxicity study of bisphenol A. *Reprod. Toxicol.* **15**, 505–523.

Erickson, R.P. (1999). Antisense transgenics in animals. *Methods* **18**, 304–310.

Evans, M.J. (1989). Potential for generic manipulation of mammals. *Mol. Biol. Med.* **6**, 557–565.

Everitt, J.I., Paul, W.R., and Davis, T.W. (1988). Urologic syndrome associated with wire caging in AKR mice. *Lab. Anim. Sci.* **38**, 609–611.

Faith, R.E., Henning, S.J., McCarty, D.R., and McKenzie, W.F. (1985). Reduction of reproductive efficiency in Sprague-Dawley rats by soft wood bedding. *Lab. Anim. Sci.* **35**, 555.

Farabollini, F., Porrini, S., and Dessi-Fulgheri, F. (1999). Perinatal exposure to the estrogenic pollutant bisphenol A affects behavior in male and female rats. *Pharmacol. Biochem. Behav.* **64**, 687–694.

Farabollini, F., Porrini, S., Seta, D.D., Bianchi, F., and Dessi-Fulgheri, F. (2002). Effects of perinatal exposure to bisphenol A on sociosexual behavior of female and male rats. *Environ. Health. Persp.* **110(Suppl 3)**, 409–414.

Faraday, M.M. (2002). Rat sex and strain differences in responses to stress. *Physiol. Behav.* **75**, 507–522.

Fauci, A.S. (1985). Clinical aspects of immunosuppression: use of cytotoxic agents and corticosteroids. *In* "Immunology III." (J.A. Bellanti, ed.), pp. 546–557, W.B. Saunders, Philadelphia.

Feldman, D. and Krishnan, A. (1995). Estrogens in unexpected places: possible implications for researchers and consumers. *Environ. Health Perspect.* **103(Suppl 7)**, 129–133.

Feldman, S.H. (2003). Rodent allergens and occupational health programs in research facilities: the complexity of political and immune responses. *Contemp. Topics Lab. Anim. Sci.* **42**, 144–147.

Fenner-Crisp, P.A. (2000). Endocrine modulators: risk characterization and assessment. *Toxicol. Pathol.* **28**, 438–440.

Ferstrom, J.D. and Wurtman, R.J. (1971). Brain serotonin content: physiological dependence on plasma tryptophan levels. *Science* **173**, 149.

Fidler, I.J. (1977). Depression of macrophages in mice drinking hyperchlorinated water. *Nature* **270**, 735–736.

File, S.E. (1978). Exploration, distraction and habituation in rats, reared in isolation. *Dev. Psychobiol.* **11**, 73–81.

Fletcher, J.I. (1976). Influence of noise on animals. *In* "Control of the Animal House Environment." (T. McSheehy, ed.), pp. 51–62, Laboratory Animal Limited, London.

Flynn, R.J. (1968). A new cage cover as an aid to laboratory rodent disease control. *Proc. Soc. Exp. Biol. Med.* **129**, 714–717.

Flynn, R.J., Poole, C.M., and Tyler, S.A. (1971). Long distance air transport of aged laboratory mice. *J. Gerontol.* **26**, 201–203.

Foley, K.M. (1979). A comparison of the pesticide residues in corn cob and wood beddings. *Abstr.* **22**, 29th Annual Session, American Association for Laboratory Animal Sciences, New York.

Foster, C.H.L., Trexler, P.C., and Rumsey, G. (1967). A canned sterile shipping diet for small laboratory rodents. *Lab. Anim. Care* **17**, 400–405.

Foster, H.L. (1958). Large scale production of rats free of commonly occurring pathogens and parasites. *Proc. Anim. Care Panel* **8**, 92–100.

Foster, H.L. (1962). Establishment and operation of SPF colonies. *In* "Problems of Laboratory Animal Disease." (R.J.C. Harris, ed.), pp. 249–259, Academic Press, New York.

Foster, H.L. and Pfan, E.S. (1963). Gnotobiotic animal production at the Charles River Breeding Laboratories, Inc. *Lab. Anim. Care* **13**, 609–632.

Foster, H.L., Foster, S.J., and Pfan, E.S. (1963). The large scale production of caesarian originated barrier sustained mice. *Lab. Anim. Care* **13**, 711–718.

Fouts, J.R. (1970). Some effects of insecticides on hepatic microsomal enzymes in various animal species. *Rev. Can. Biol.* **29**, 377–389.

Fouts, J.R. (1976). Overview of the field: environmental factors affecting chemical or drug effects in animals. *Fed. Proc. Fed. Am. Soc. Exp. Biol.* **35**, 1162–1165.

Friedman, M., Byers, S.O., and Brown, A.E. (1967). Plasma lipid responses of rats and rabbits to an auditory stimulus. *Am. J. Physiol.* **212,** 1174–1178.

Fritz, T.E., Tolle, D.V., and Flynn, R.J. (1968). Hemorrhagic diathesis in laboratory rodents. *Proc. Soc. Exp. Biol. Med.* **128,** 228–234.

Fujiwara, M. and Ueki, S. (1978). Muricide induced by single injections of Δ^9-tetrahydrocannibol. *Physiol. Behav.* **21,** 581–585.

Fujiwara, M. and Ueki, S. (1979). The course of aggressive behavior induced by a single injection of Δ^9-tetrahydrocannabinol and its characteristics. *Physiol. Behav.* **22,** 535–540.

Fullerton, P.M. and Gilliatt, R.W. (1967). Pressure neuropathy in the hind foot of the guinea pig. *J. Neurol. Neurosurg. Psychiatry* **30,** 18–25.

Gamble, M.R. (1976). Fire alarms and oestrus in rats. *Lab. Anim.* **10,** 161–163.

Gamble, M.R. and Clough, G. (1976). Ammonia build-up in animal boxes and its effect on rat tracheal epithelium. *Lab. Anim.* **10,** 93–104.

Gao, H. and Zhang, S.Z. (1992). Effect of noise on blood pressure of various types of rats. *Zhonghua Yu Fang Yi Xue Za Zhi* **26,** 275–277.

Gardiner, S.M. and Bennet, T. (1977). The effects of short-term isolation on systolic blood pressure and heart rate in rats. *Med. Biol.* **55,** 325–329.

Gartner, K., Buttner, D., Dohler, B.K., Friedel, R., Lindena, J., and Trautschold, I. (1980). Stress response of rats to handling and experimental procedures. *Lab. Anim.* **14,** 267–274.

Geber, W.F., Anderson, T.A., and Van Dyne, V. (1966). Physiologic responses of the albino rat to chronic noise stress. *Arch. Environ. Health* **12,** 751–754.

Gelineo, S. (1934). Influence du milieu thermique d'adaptation sur la thermogenese des homeothermes. *Annales de Physiologie et de Physiocochimie Biologique* **10,** 1083–1115.

Gentsch, C., Lichtsteiner, M., and Feer, H. (1981). Locomotor activity, defecation score and corticosterone levels during an open field exposure: a comparison among individually and group housed rats, and genetically selected rat lines. *Physiol. Behav.* **27,** 183–186.

Gerges, S.E., Abdel-Rahman, M.S., Skowronski, G.A., and Von Hagen, S. (1985). Effects of Alcide gel on fetal development in rats and mice, II. *J. Appl. Toxicol.* **5,** 104–109.

Gesi, M., Fornai, F., Lenzi, P., Natale, G., Soldani, P., and Paparelli, A. (2001). Time-dependent changes in adrenal cortex ultrastructure and corticosterone levels after noise exposure in male rats. *Eur. J. Morphol.* **39,** 129–135.

Gesi, M., Lenzi, P., Alessandri, M.G., Ferrucci, M., Fornai, F., and Paparelli, A. (2002). Brief and repeated noise exposure produces different morphological and biochemical effects in noradrenaline and adrenaline cells of adrenal medulla. *J. Anat.* **200,** 159–168.

Gibson, S.V., Besch-Williford, C., Raisbeck, M.F., Wagner, J.E., and McLaughlin, R.M. (1987). Organophosphate toxicity in rats associated with contaminated bedding. *Lab. Anim. Sci.* **37,** 59–62.

Glowa, J.R. and Hansen, C.T. (1994). Differences in response to an acoustic startle stimulus among 46 rat strains. *Behav. Genet.* **24,** 79–84.

Golberg, L. (1974). "Carcinogenesis Testing of Chemicals." CRC Press, Cleveland, OH.

Goldstein, D. (1973). Convulsions elicited by handling: a sensitive method of measuring CNS excitation in mice treated with reserpine or convulsant drugs. *Psychopharm.* **32,** 27–32.

Gomella, C. and Musquere, P. (1980). Disinfection of water by chlorine, ozone, and chlorine dioxide. *Int. Water Supply Assoc., Congr. [Proc.]* **13,** k1–k12.

Gordon, A.H., D'Arcy Hart, P., and Young, M.R. (1980). Ammonia inhibits phagosome-lysosome fusion in macrophages. *Nature* **286,** 79–80.

Gordon, J.W. (1983). Studies of foreign genes transmitted through the germ lines of transgenic mice. *J. Exp. Zool.* **228,** 313–324.

Gordon, J.W. and Ruddle, F.H. (1985). DNA-mediated genetic transformation of mouse embryos and bone marrow: a review. *Gene* **33,** 121–136.

Gordon, S., Wallace, J., Cook, A., Tee, R.D., and Newman-Taylor, A.J. (1997). Reduction of exposure to laboratory animal allergens in the workplace. *Clin. Exp. Allergy* **27,** 744–751.

Govindwar, S.P., Siddiqui, A.M., Hashmi, R.S., Kachole, M.S., and Pawar, S.S. (1984). Effect of ampicillin on hepatic microsomal mixed-function oxidase system in male rats. *Toxicol. Lett* **23,** 201–204.

Grant, L., Hopkinson, P., Jennings, G., and Jenner, F.A. (1971). Period of adjustment of rats used for experimental studies. *Nature* **232,** 135.

Gray Jr., L.E. and Kelce, W.R. (1996). Latent effects of pesticides and toxic substances on sexual differentiation of rodents. *Toxicol. Ind. Health* **12,** 515–531.

Gray Jr., L.E. and Ostby, J. (1998). Effects of pesticides and toxic substances on behavioral and morphological reproductive development: endocrine versus nonendocrine mechanisms. *Toxicol. Ind. Health* **14,** 159–184.

Gray Jr., L.E., Ostby, J.S., and Kelce, W.R. (1994). Developmental effects of an environmental antiandrogen: the fungicide vinclozolin alters sex differentiation of the male rat. *Toxicol. Appl. Pharmacol.* **129,** 46–52.

Grewal, I.S., Heilig, M., Miller, A., and Sercarz, E.E. (1997). Environmental regulation of T-cell function in mice: group housing of males affects accessory cell function. *Immunol.* **90,** 165–168.

Grewal, I.S., Miller, A., and Sercarz, E.E. (1997). The influence of mouse housing density on autoimmune reactivity. *Autoimmunity* **26,** 209–214.

Grover-Johnson, N. and Spencer, P.S. (1981). Peripheral nerve abnormalities in aging rats, *J. Neuropath. Exp. Neurol.* **40,** 155–165.

[Guide] 1996. "Guide for the Care and Use of Laboratory Animals." National Academy Press, Washington, DC.

Gwosdow, A.R. and Besch, E.L. (1985). Effect of thermal history on the rat's response to varying environmental temperature. *J. Appl. Physiol.* **59,** 413–419.

Haake, J., Kelley, M., Keys, B., and Safe, S. (1987). The effects of organochlorine pesticides as inducers of testosterone and benzo [a] pyrene hydroxylases. *Gen. Pharmacol.* **18,** 165–169.

Hall, F.S. (1998). Social deprivation of neonatal, adolescent, and adult rats has distinct neurochemical and behavioral consequences. *Crit. Rev. Neurobiol.* **12,** 129–162.

Hallback, M. (1975). Consequence of social isolation on blood pressure, cardiovascular reactivity and design in spontaneously hypertensive rats. *Acta Physiol. Scand.* **93,** 455–465.

Hamilton, B.A. and Frankel, W.N. (2001). Of mice and genome sequence. *Cell* **107,** 13–16.

Hardman, J.G., Limbird, L.E., and Gilman, A.G. (eds.) (2001). "Goodman and Gilman's: the Pharmacological Basis of Therapeutics." McGraw-Hill, New York.

Harrington, R.M., Shertzer, H.G., and Bercz, J.P. (1985). Effects of ClO_2 on the absorption and distribution of dietary iodide in the rat. *Fundam. Appl. Toxicol.* **5,** 672–678.

Harrington, R.M., Shertzer, H.G., and Bercz, J.P. (1986). Effects of chlorine dioxide on thyroid function in the African green monkey and the rat. *J. Toxicol. Environ. Health* **19,** 235–242.

Hart, L. and Fouts, J. (1965). Further studies on the stimulation of hepatic microsomal drug metabolizing enzymes by DDT and its analogs. *Arch. Exp. Pathol. Pharmacol.* **249,** 486–500.

Hasegawa, M., Kurabayashi, Y., Ishii, T., Yoshida, K., Uebayashi, N., Sato, N., and Kurosawa, T. (1997). Intra-cage air change rate on forced-air-ventilated microisolation environment within cages: carbon dioxide and oxygen concentration. *Exp. Anim.* **46,** 251–257.

Hasenau, J.J., Baggs, R.B., and Kraus, A.L. (1993). Microenvironments in microisolation cages using BALB/c and CD-1 mice. *Contemp. Top. Lab. Anim. Sci.* **32,** 11–16.

Hastings, J.S. (1967). Long term use of vermiculite. *J. Inst. Anim. Tech.* **18**, 184–190.

Hastings, J.W. and Menaker, M. (1976). Physiological and biochemical aspects of circadian rhythms. *Fed. Proc., Fed. Am. Soc. Exp. Biol.* **35**, 2325–2357.

Hatch, A.M., Wibert, G.S., and Zawidzka, Z. (1965). Isolation syndrome in the rat. *Toxicol. Appl. Pharmacol.* **7**, 737–745.

Hayashi, O. and Kikuchi, M. (1982). The effects of the light-dark cycle on humoral and cell-mediated immune response of mice. *Chronobiologia* **9**, 291–300.

Hayashi, O. and Kikuchi, M. (1985). The influence of phase shift in the light-dark cycle on humoral immune responses of mice to sheep red blood cells and polyvinylpyrrolidone. *J. Immunol.* **134**, 1455–1461.

Hayes Jr., W.J. (1975). "Toxicology of Pesticides." Williams and Wilkins, Baltimore.

Heller, R., Ballowitz, L., and Natszchka, J. (1969). Wirkungen von blaulicht auf junge Gunn-ratten; bietrag zur frage der phototherapie bei hyperbilirubinamie. *Monatsshrift fur Kinderheilkunde* **117**, 437–440.

Hemminki, A. (2002). From molecular changes to customized therapy. *Eur. J. Cancer* **38**, 333–338.

Hessler, J.R. (1991a). Facilities to support research. *In* "Handbook of Facilities Planning: Vol. 2, Laboratory Animal Facilities." (T. Ruys, ed.), pp. 35–54, Van Nostrand Reinhold, New York.

Hessler, J.R. (1991b). Single versus dual-corridor systems: advantages, disadvantages, limitations and alternatives for effective contamination control. *In* "Handbook of Facilities Planning: Vol. 2, Laboratory Animal Facilities." (T. Ruys, ed.), pp. 59–67, Van Nostrand Reinhold, New York.

Hessler, J.R. (1995). Methods of biocontainment. *In* "Current Issues and New Frontiers in Animal Research." (K.A. Bayne, L.M. Greene, and E.D. Prentice, eds), pp. 61–68, Scientist Center for Animal Welfare, Greenbelt, MD.

Hessler, J.R., Broderson, R., and King, C. (1999). Animal research facilities and equipment. *In* "Anthology of Biosafety 1. Perspectives on Laboratory Design." (J.Y. Richmond, ed.), pp. 191–217, American Biological Safety Association, Mundelein, IL.

Hessler, J.R. and Leary, S.L. (2002). Design and management of animal facilities. *In* "Laboratory Animal Medicine," 2nd Ed. (J. Fox, L.C. Anderson, F. Lowe, and F.W. Quimby eds) pp. 909-933. Academic Press, New York, NY.

Hisjarvi, P. and Junnila, M. (1986). Happy rats: reliable results. *Acta Physiol. Sc*and. **128(Suppl 554)**, 32.

Hirsjarvi, P.A. and Valiaho, T.V. (1987). Microclimate in two types of rat cages. *Lab. Anim.* **21**, 95–98.

Hoff, J.C. and Geldreich, E.E. (1981). Comparison of the biocidal efficiency of alternative disinfectants. *J. Amer. Water Works Assoc.* 73, 40–49.

Holdenreid, R., ed. (1966). "Viruses of Laboratory Rodents," Monograph no. 20. pp. XIII–XIV. National Cancer Institute, Bethesda, MD.

Hoffman-Goetz, L., MacNeil, B., and Arumugam, Y. (1992). Effect of differential housing in mice on natural killer cell activity, tumor growth, and plasma corticosterone. *Proc. Soc. Exp. Biol. Med.* **199**, 337–344.

Honma, S., Suzuki, A., Buchannan, D.L., Katsu, Y., Watanabe, H., and Iguchi, T. (2002). Low dose effect of in utero exposure to bisphenol A and diethylstilbestrol on female mouse reproduction. *Reprod. Toxicol.* **16**, 117–122.

Howdeshell, K.L., Peterman, P.H., Judy, B.M., Taylor, J.A., Orazio, C.E., Ruhlen, R.L., vom Saal, F.S., and Welshons, W.V. (2003). Bisphenol A is released from used polycarbonate animal cages into water at room temperature. *Environ. Health Perspect.* **111**, 1180–1187.

Huerkamp, M.J. and Lehner, N.D.M. (1994). Comparative effects of forced-air, individual cage ventilation or an absorbent bedding additive on mouse isolator cage microenvironment. *Contemp. Top. Lab. Anim. Sci.* **33**, 58–61.

Huerkamp, M.J., Thompson, W.D., and Lehner, N.D.M. (2003). Failed air supply to individually ventilated caging system causes acute hypoxia and mortality of rats. *Contemp. Top. Lab. Anim. Sci.* **42**, 454–445.

Hughes, R.N. and Syme, L.A. (1972). The role of social isolation and sex in determining effects of chlordiazepoxide and methylphenidate on exploratory activity. *Psychopharmacology* **27**, 359–366.

Hunt, P.A., Koehler, K.E., Susiarjo, M., Hodges, C.A., Ilagan, A., Voigt, R.C., Thomas, S., Thomas, B.F., and Hassold, T.J. (2003). Bisphenol a exposure causes meiotic aneuploidy in the female mouse. *Curr. Biol.* **13**, 546–553.

Institute of Laboratory Animal Resources (1978). "Laboratory Animal Housing." Institue of Laboratory Animal Resources, National Academy of Science, Washington, DC.

Irving, L. (1964). Terrestrial animals in cold: birds and mammals. *In* "Adaptation to the Environment," Handbook of Physiology, Section 4. pp. 361–377, American Physiological Society, Washington, DC.

Ishikawa, M., Ozaki, M., Takayanagi, Y., and Sasaki, K. (1991). Induction of hepatic cytochrome P-450 and drug metabolism by doxapram in the mouse. *Res. Commun. Chem. Pathol. Pharmacol.* **72**, 109–112.

Iwarsson, K. and Noren, L. (1992). Comparison of microenvironmental conditions in standard versus forced-air ventilated rodent filter-top cages. *Scand. J. Lab. Anim. Sci.* **19**, 167–173.

Izaraeli, S. and Rechavi, G. (2002). Molecular medicine: an overview. *Isr. Med. Assoc. J.* **4**, 638–640

Jacob, H.J. (1999). Functional genomics and rat models. *Genome Res.* **9**, 1013–1016.

Jacob, H.J. and Kwitek, A.E. (2002). Rat genetics: attaching physiology and pharmacology to the genome. *Nat. Rev. Genet.* **3**, 33–42.

Jessop, J.J. and Bayer, B.M. (1989). Time-dependent effects of isolation on lymphocyte and adrenal activity. *J. Neuroimmunol.* **23**, 143–147.

Jessop, J.J., Gale, K., and Bayer, B.M. (1987). Enhancement of rat lymphocyte proliferation after prolonged exposure to stress. *J. Neuroimmunol.* **16**, 261–271.

Jessop, J.J., Gale, K., and Bayer, B.M. (1988). Time-dependent enhancement of lymphocyte activation by mitogens after exposure to isolation or water scheduling. *Life Sci.* **43**, 1133–1140.

Joasoo, A. and McKenzie, J.M. (1976). Stress and immune response in rats. *Int. Arch. Allergy Appl. Immunol.* **50**, 659–663.

Johnson, T., Lavender, J.F., Hultin, E., and Rasmussen Jr., A.F. (1963). The influence of avoidance-learning stress on resistance to cocksackie B virus in mice. *J. Immunol.* **91**, 569–575.

Jonas, A.M. (1965). Laboratory animal facilities. *J. Am. Vet. Med. Assoc.* **146**, 600–606.

Jonas, A.M. (1976). Long-term holding of laboratory rodents. *ILAR News* **19**, 1–25.

Jori, A., Bianchetti, A., and Prestini, P.E. (1969). Effect of essential oils on drug metabolism. *Biochem. Pharmacol.* **18**, 2081–2085.

Jori, A., DiSalle, E., and Santini, V. (1971). Daily rhythmic variation and liver drug metabolism in rats. *Biochem. Pharmacol.* **29**, 2965–2969.

Joseph, I.M., Suthanthirarajan, N., and Namasivayam, A. (1991). Effect of heat stress on certain immunological parameters in albino rats. *Indian J. Physiol. Pharmacol.* **35**, 269–271.

Junli, H., Li, W., Nenqi, R., Fang, M., and Li, J. (1997a). Disinfection effect of chlorine dioxide on bacteria in water. *Water Res.* **31**, 607–613.

Junli, H., Li, W., Nenqi, R., Li, L.X., Fun, S.R., and Guanle, Y. (1997b). Disinfection effect of chlorine dioxide on viruses, algae, and animal planktons in water. *Water Res.* **31**, 455–460.

Justice, M.J. (2000). Capitalizing on large-scale mouse mutagenesis screens. *Nat. Rev. Genet.* **1**, 109–115.

Kanzaki, M., Fujieda, M., and Furukawa, T. (2001). Effects of suspension of air-conditioning on airtight-type racks. *Exp. Anim.* **50**, 379–385.

Karim, A. and Arslan, M.I. (2000). Isolation modifies the behavioral response in rats. *Bangladesh Med. Res. Counc. Bull.* **26**, 27–32.

Karp, J.D., Cohen, N., and Moynihan, J.A. (1994). Quantitative differences in interleukin-2 and interleukin-4 production by antigen-stimulated splenocytes from individually- and group-housed mice. *Life Sci.* **55,** 789–795.

Karp, J.D., Moynihan, J.A., and Ader, R. (1993). Effects of differential housing on the primary and secondary antibody responses of male C57BL/6 and BALB/c mice. *Brain Behav. Immun.* **7,** 326–333.

Karp, J.D., Moynihan, J.A., and Ader, R. (1997). Psychosocial influences on immune responses to HSV-1 infection in BALB/c mice. *Brain Behav. Immun.* **11,** 47–62.

Katz, D.M. and Steinberg, H. (1972). Factors which might modify morphine dependence in rats. *In* "Biochemical and Pharmacological Aspects of Dependence and Reports on Marihuana Research." (H.M. Prang, ed.). pp. 57–66. De Erven Bohn, Haarlem.

Keller, G.L., Mattingly, S.F., and Knapke Jr., F.B. (1983). A forced-air individually ventilated caging system for rodents. *Lab. Anim. Sci.* **33,** 580–582.

Keller, L.S.F., White, W.J., Snider, M.T., and Lang, C.M. (1989). An evaluation of intra-cage ventilation in three animal caging systems. *Lab. Anim. Sci.* **39,** 237–242.

Kline, L.B. and Hull, R.N. (1960). The virucidal properties of peracetic acid. *Am. J. Clin. Pathol.* **33,** 30–33.

Klir, P., Bondy, R., Lachout, J., and Hanis, T. (1984). Physiological changes in laboratory rats caused by different housing. *Physiol. Bohemoslov.* **33,** 111–121.

Knapka, J.J. (1999). Nutrition of rodents. *Vet. Clin. N. Am. Exot. Anim. Pract.* **2,** 153–167.

Knights, K.M., Gourlay, G.K., and Cousins, M.J. (1987). Changes in rat hepatic microsomal mixed function oxidase activity following exposure to halothane under various oxygen concentrations. *Biochem. Pharmacol.* **36,** 897–906.

Kodam, K.M., Adav, S.S., and Govindwar, S.P. (1996). Effect of sulfamethazine on phenobarbitol and benzo[a]pyrene induced hepatic microsomal mixed function oxidase system in rats. *Toxicol. Lett.* **87,** 25–30.

Kodam, K.M. and Govindwar, S.P. (1997). *In vivo* and *in vitro* effect of sulfamethazine on hepatic mixed function oxidases in rats. *Vet. Hum. Toxicol.* **39,** 141–146.

Koehler, K.E., Voigt, R.C., Thomas, S., Lamb, B., Urban, C., Hassold, T., and Hunt, P.A. (2003). When disaster strikes: rethinking caging materials. *Lab. Anim.* **32,** 24–27.

Kolmodin, B., Azarnoff, D.L., and Sjoqvist, F. (1969). Effect of environmental factors in drug metabolism: decreased plasma half-life of antipyrine in workers exposed to chlorinated hydrocarbon insecticides. *Clin. Pharmacol. Ther.* **10,** 638–642.

Koner, B.C., Banerjee, B.D., and Ray, A. (1998). Organochlorine pesticide-induced oxidative stress and immune suppression in rats. *Indian J. Exp. Biol.* **36,** 395–398.

Kostowski, W., Czlonkowski, A., Rewerski, W., and Piechocki, T. (1977). Morphine action in grouped and isolated rats and mice. *Psychopharmacology* **53,** 191–193.

Kraft, L.M. (1980). The manufacturing, shipping and receiving and quality control of rodent bedding materials. *Lab. Anim. Sci.* **30,** 366–374.

Krebs, H., Macht, M., Weyers, P., Weijers, H.G., and Janke, W. (1996). Effects of stressful noise on eating and non-eating behavior in rats. *Appetite* **26,** 193–202.

Krechniak, J. and Wrzesniowska, K. (1991). Effects of pyrethroid insecticides on hepatic microsomal enzymes in rats. *Environ. Res.* **55,** 129–134.

Krishnan, A.V., Stathis, P., Permuth, S.F., Tokes, L., and Feldman, D. (1993). Bisphenol-A: an estrogenic substance is released from polycarbonate flasks during autoclaving. *Endocrinology* **132,** 2279–2286.

Krynicki, M. and Olszewski, W.L. (1989). Influence of thermal stress on lymphocyte migration pattern in rats. *Arch. Immunol. Ther. Exp. (Warsz)* **37,** 601–607.

Kwitek-Black, A.E. and Jacob, H.J. (2001). The use of designer rats in the genetic dissection of hypertension. *Curr. Hypertens. Rep.* **3,** 12–18.

Lambert. H.H. (1975). Continuous red light induces persistent estrus without retinal degeneration in the albino rat. *Endocrinology* **97,** 208–210.

Landi, M.S., Kreider, J.W., Lang, C.M., and Bullock, L.P. (1982). Effects of shipping on the immune function in mice. *Am. J. Vet. Res.* **43,** 1654–1657.

Lang, C. and Harrell Jr., G.T. (1971) "A Comprehensive Animal Facility for a College of Medicine." USHEW, PHS, NIH, Bethesda, MD.

Lang, C. and Vesell, E.S. (1976). Environmental and genetic factors affecting laboratory animals: impact on biomedical research. *Fed. Proc., Fed. Am. Soc. Exp. Biol.* **35,** 1123–1124.

LeBouton, A.V. and Handler, S.D. (1971). Persistent circadian rhythmicity of protein synthesis in liver of starved rats. *Experientia* **27,** 1031–1032.

Ledermann, B. (2000). Embryonic stem cells and gene targeting. *Exp. Physiol.* **85,** 603–613.

Les, E.P. (1968). Cage population density and efficiency of feed utilization in inbred mice. *Lab. Anim. Care* **18,** 305–313.

Les, E.P. (1983). Pressurized, individually ventilated (PIV) and individually exhausted caging for laboratory mice. *Lab. Anim. Sci.* **33,** 495.

Liimatainen, A., Muller, D., Vartiainen, T., Jahn, F., Kleeberg, U., Klinger, W., and Hanninen, O. (1988). Chlorinated drinking water is mutagenic and causes 3-methylcholanthrene type induction of hepatic monooxygenase. *Toxicology* **51,** 281–289.

Lindblom, B. and Motylewski, K. (1993). "Disaster Planning for Cultural Institutions." technical leaflet 183, American Association for State and Local History, Nashville, TN.

Lindsey, J.R., Conner, M.W., and Baker, H.J. (1978). Physical, chemical and microbial factors affecting biologic responses. *In* "Laboratory Animal Housing." pp. 37–43. Institute for Laboratory Animal Resources, National Academy of Science, Washington, DC.

Lindsey, J.R., Davidson, M.K., Schoeb, T.R., and Cassell, G.H. (1985). *Mycoplasma pulmonis*–host relationships in a breeding colony of Sprague-Dawley rats with enzootic murine respiratory mycoplasmosis. *Lab. Anim. Sci.* **35,** 597–608.

Lipman, N.S. (1992). Microenvironmental conditions in isolator cages: an important research variable. *Lab. Anim.* **21,** 23–26.

Lipman, N.S. (1993). Strategies for architectural integration of ventilated caging systems. *Contemp. Top. Lab. Anim. Sci.* **32,** 7–12.

Lipman, N.S. (1999). Isolator rodent caging systems (state of the art): a critical view. *Contemp. Top. Lab. Anim. Sci.* **38,** 9–17.

Lipman, N.S., Corning, B.F., and Coiro, M.A. (1992). The effects of intracage ventilation on microenvironmental conditions in filter-top cages. *Lab. Anim.* **26,** 206–210.

Lipman, N.S., Corning, B.F., and Saifuddin, M. (1993). Evaluation of isolator caging systems for protection of mice against challenge with mouse hepatitis virus. *Lab. Anim.* **27,** 134–140.

Lundberg, V. (1986). Dominanzstrukturen mannlicher laborratten. *Zeitschr. Versuchstierk.* **28,** 257–268.

Magee, P.N. (1970). Tests for carcinogenic potential. *In* "Methods in Toxicology." (G.E. Paget ed.), pp. 158–196. Blackwells, Oxford.

Manning, P.J. and Banks, K.L. (1968). The effects of transportation on the aged rat. *In* "Laboratory Animals in Gerontological Research." pp. 98–103. Publication 1591, National Academy of Sciences, Washington, DC.

Manser, C.E., Morris, T.H., and Broom, D.M. (1995). An investigation into the effects of solid or grid cage flooring on the welfare of laboratory rats. *Lab. Anim.* **2(9),** 353–363.

Manser, C.E., Morris, T.H., and Broom, D.M. (1996). The use of a novel operant test to determine the strength of preference for flooring in laboratory rats. *Lab. Anim.* **30**, 1–6.

Markaverich, B., Mani, S., Alejandro, M.A., Mitchell, A., Markaverich, D., Brown, T., Velez-Trippe, C., Murchison, C., O'Malley, B., and Faith, R. (2002). A novel endocrine-disrupting agent in corn with mitogenic activity in human breast and prostatic cancer cells. *Environ. Health Persp.* **110**, 169–177.

Markaverich, B.M., Alejandro, M.A., Markaverich, D., Zitzow, L., Casajuna, N., Camarao, N., Hill, J., Bhirdo, K., Faith, R., Turk, J., and Crowley, J.R. (2002). Identification of an endocrine disrupting agent from corn with mitogenic activity. *Biochem. Biophys. Res. Comm.* **291**, 692–700.

Markey, C.M., Coombs, M.A., Sonnenschein, C., and Soto, A.M. (2003). Mammalian development in a changing environment: exposure to endocrine disruptors reveals the developmental plasticity of steroid-hormone target organs. *Evol. Dev.* **5**, 67–75.

Markey, C.M., Luque, E.H., de Toro, M.M., Sonnenschein, C., and Soto, A.M. (2001). In utero exposure to bisphenol A alters the development and tissue organization of the mouse mammary gland. *Biol. Reprod.* **65**, 1215–1223.

Markey, C.M., Rubin, B.S., Soto, A.M., and Sonnenschein, C. (2002). Endocrine disruptors: from Wingspread to environmental developmental biology. *J. Steriod Biochem. Mol. Biol.* **83**, 235–244.

Maslova, L.N., Bulygina, V.V., and Markel, A.L. (2002). Chronic stress during prepubertal development: immediate and long lasting effects on arterial blood pressure and anxiety-related behavior. *Psychoneuroendocrinology* **27**, 549–561.

Matsumura, F. (1975). "Toxicology of Insecticides." Plenum Press, New York.

Mayeux, P., Dupepe, L., Dunn, K., Balsmo, J., and Domer, J. (1995). Massive fungal contamination in animal care facilities traced to bedding supply. *Appl. Environ. Microbiol.* **61**, 2297–2301.

McCarthy, D.O., Ouimet, M.E., and Daun, J.M. (1992). The effects of noise stress on leukocyte function in rats. *Res. Nurs. Health* **15**, 131–137.

McSheehy, T., ed. (1976). "Control of the Animal House Environment." Laboratory Animals Ltd., London.

Meier, J.R., Bull, R.J., Stober, J.A., and Cimino, M.C. (1985). Evaluation of chemicals used for drinking water disinfection for production of chromosomal damage and sperm-head abnormalities in mice. *Environ. Mutagen.* **7**, 201–211.

Melnick, R., Lucier, G., Wolfe, M., Hall, R., Stancel, G., Prins, G., Gallo, M., Reuhl, K., Ho, S.M., Brown, T., Moore, J., Leakey, J., Haseman, J., and Kohn, M. (2002). Summary of the National Toxicology Program's report of the endocrine disruptors low-dose peer review. *Environ. Health Perspect.* **110**, 427–431.

Militzer, K. and Reinhard, H.J. (1982). Rank position in rats and their relations to tissue parameters. *Physiol. Psychol.* **10**, 251–260.

Milligan, S.R., Sales, G.D., and Khirnykh, K. (1993). Sound levels in rooms housing laboratory animals: an uncontrolled daily variable. *Physiol. Behav.* **53**, 1067–1076.

Mills, D.E. and Ward, R. (1984). Attenuation of psychosocial stress-induced hypertension by gamma-linolenic acid (GLA) administration in rats. *Proc. Soc. Exp. Biol.* **176**, 32–37.

Moore, G.S. and Calabrese, E.J. (1980). The effects of chlorine dioxide and sodium chlorite on erythrocytes of A/J and C57L/J mice. *J. Environ. Pathol. Toxicol.* **4**, 513–524.

Moreadith, R.W. and Radford, N.B. (1997). Gene targeting in embryonic stem cells: the new physiology and metabolism. *J. Mol. Med.* **75**, 208–216.

Morinan, S. and Leonard, B.E. (1980). Some anatomical and physiological correlates of social isolation in the young rat. *Physiol. Behav.* **24**, 637–640.

Morita, T. and Tokura, H. (1998). The influence of different wavelengths of light on human biological rhythms. *Appl. Human Sci.* **17**, 91–96.

Murakami, H. (1971). Difference between internal and external environment of the mouse cage. *Lab. Anim. Sci.* **21**, 680–684.

Nair, V. and Casper, R. (1969). The influence of light on daily rhythm in hepatic drug metabolizing enzymes in rat. *Life Sci.* **21**, 680–684.

Naranjo, J.R. and Fuentes, J.A. (1985). Association between hypoalgesia and hypertension in rats after short-term isolation. *Neuropharmacology* **24**, 167–171.

Nayfield, K.C. and Besch, E.L. (1981). Comparative responses of rabbits and rats to elevated noise. *Lab. Anim. Sci.* **31**, 386–390.

Newberne, P.M. (1975). Influence on pharmacological experiments of chemicals and other factors in diets of laboratory animals. *Fed. Proc., Fed. Am. Soc.Exp. Biol.* **34**, 209–218.

Newberne, P.M. and Fox, J.G. (1978). Chemicals and toxins in the animal facility. *In* "Laboratory Animal Housing." pp. 118–138, Institute of Laboratory Animal Resources, National Academy of Sciences, Washington, DC.

Newberne, P.M. and McConnell, R.G. (1980). Dietary nutrients and contaminants in laboratory animal experimentation. *J. Environ. Pathol. Toxicol.* **4**, 105–122.

Nielsen, F.H. and Bailey, B. (1979). The fabrication of plastic cages for suspension in mass air flow racks. *Lab. Anim. Sci.* **29**, 502–506.

Nielsen, J.B., Andersen, O., and Svendsen, P. (1986). Hepatic O-demethylase activity in mice on different types of bedding. *Zeitschr. Versuchstierk.* **28**, 69–75.

Niesink, R.J.M. and van Ree, J.M. (1982). Short-term isolation increases social interactions of male rats: A parametric analysis. *Physiol. Behav.* **29**, 819–825.

Nunez, M.J., Mana, P., Linares, D., Riveiro, M.P., Balboa, J., Suarez-Quintanilla, J., Maracchi, M., Mendez, M.R., Lopez, J.M., and Freire-Garabal, M. (2002). Music, immunity and cancer. *Life Sci.* **71**, 1047–1057.

Ohishi, N., Imaoka, S., and Funae, Y. (1994). Changes in content of P-450 isozymes in hepatic and renal microsomes of the male rat treated with cis-diamminedichloroplatinum. *Xenobiotica* **24**, 873–880.

Orme, J., Taylor, D.H., Laurie, R.D., and Bull, R.J. (1985). Effects of chlorine dioxide on thyroid function in neonatal rats. *J. Toxicol. Envoiron. Health* **15**, 315–322.

Ortman, J.A., Sahenk, Z., and Mendell, J.R. (1983). The experimental production of Renaut bodies. *J. Neurol. Sci.* **62**, 233–241.

Ostby, J., Kelce, W.R., Lambright, C., Wolf, C.J., Mann, P., and Gray Jr., L.E. (1999). The fungicide procymidone alters sexual differentiation in the male rat by acting as an androgen-receptor antagonist *in vivo* and *in vitro*. *Toxicol. Ind. Health* **15**, 80–93.

O'Steen, W.K. and Anderson, K.V. (1972). Photoreceptor degeneration after exposure of rats to incandescent illumination. *Z. Zellforsch. Mikrosk. Anat.* **127**, 306–313.

O'Steen W.K. and Brodish, A. (1985). Neuronal damage in the rat retina after chronic stress. *Brain. Res.* **344**, 231–239.

Ozaki, M., Ishikawa, M., Takayanagi, Y., and Sasaki, K. (1993). Sex-related differences in rat liver microsomal enzymes and their induction by doxapram. *J. Pharm. Pharmacol.* **45**, 975–978.

Palanza, P., Howdeshell, K.L., Parmigiani, S., and vom Saal, F.S. (2002). Exposure to a low dose of bisphenol A during fetal life or in adulthood alters maternal behavior in mice. *Environ. Health Perspect.* **110(Suppl 3)**, 415–422.

Panksepp, J. (1980). Brief social isolation, pain responsivity, and morphine analgesia in young rats. *Psychopharmacol.* **72**, 111–112.

Peace, T. A., Singer, A.W., Niemuth, N.A., and Shaw, M.E. (2001). Effects of caging type and animal source on the development of foot lesions in Sprague Dawley rats (*Rattus norvegicus*). *Contemp. Top. Lab. Anim. Sci.* **40**, 17–21.

Peeters, J.E., Mazas, E.A., Masschelein, W.J., Martinez de Maturana, I.V., and Debacker, E. (1989). Effect of disinfection of drinking water with ozone or chlorine dioxide on survival of cryptosporidium parvum oocysts. *Appl. Environ. Microbiol.* **55**, 1519–1522.

Pelkonen, K.H. and Hanninen, O.O. (1997). Cytotoxicity and biotransformation inducing activity of rodent beddings: a global survey using the Hepa-1 assay. *Toxicol.* **122**, 73–80.

Periquet, A. and Derache, R. (1981). Hepatic microsomal monoxygenase inhibition by nabam in the rat. *Toxicol. Eur. Res.* **3**, 285–291.

Perkins, S.E. and Lipman, N.S. (1995). Characterization and quantification of microenvironmental contaminants in isolator cages with a variety of contact beddings. *Contemp. Top. Lab. Anim. Sci.* **34**, 93–97.

Perkins, S.E. and Lipman, N.S. (1996). Evaluation of microenvironmental conditions and noise generation in three individually ventilated rodent caging systems and static isolator cages. *Contemp. Top. Lab. Anim. Sci.* **35**, 61–65.

Peterson, E.A. (1980). Noise and laboratory animals. *Lab. Anim. Sci.* **30**, 422–439.

Pfaff, J. (1974). Noise as an environmental problem in the animal house. *Lab Anim.* **8**, 347–354.

Pick, J.R. and Little, J.M. (1965). Effect of type of bedding material on thresholds of penthylenetetrazole convulsions in mice. *Lab. Anim. Care* **15**, 28–33.

Plant, S.M., Alder, R., Friedman, S.B., and Ritterson, A.L. (1969). Social factors and resistance to malaria in the mouse: effects of group vs. individual housing on resistance to *Plasmodium berghei* infection. *Psychosom. Med.* **31**, 536–540.

Pleasants, J.R. (1974). Gnotobiotics. *In* "Handbook of Laboratory Animal Science," Vol. 1. (E.C. Melby Jr. and N.H. Altman, eds.), pp. 119–174, CRC Press, Cleveland, OH.

Port, C.D. and Kaltenbach, J.P. (1969). The effect of corncob bedding on reproductivity and leucine incorporation in mice. *Lab. Anim. Care* **10**, 46–49.

Potgieter, F.J. and Wilke, P.J. (1992). Laboratory animal bedding: a review of wood and wood constituents as a possible source of external variables that could influence experimental results. *Anim. Technol.* **43**, 65–88.

Pozdeyev, N.V. and Lavrikova, E.V. (2000). Diurnal changes of tyrosine, dopamine, and dopamine metabolites content in the retina of rats maintained at different lighting conditions. *J. Mol. Neurosci.* **15**, 1–9.

Prabhakaran, K., Suthanthirarajan, N., and Namasivayam, A. (1988). Biochemical changes in acute noise stress in rats. *Indian J. Physiol. Pharmacol.* **32**, 100–104.

Prater, M.R., Gogal Jr., R.M., Blaylock, B.L., Longstreth, J., and Holladay, S.D. (2002). Single-dose topical exposure to the pyrethroid insecticide, permethrin in C57BL/6N mice: effects on thymus and spleen. *Food Chem. Toxicol.* **40**, 1863–1873.

Pryce, C.R., Bettschen, D., Bahr, N.I., and Feldon, J. (2001). Comparison of the effects of infant handling, isolation, and nonhandling on acoustic startle, prepulse inhibition, locomotion, and HPA activity in the adult rat. *Behav. Neurosci.* **115**, 71–83.

Pucak, G.J., Lee, C.S., and Zaino, A.S. (1977). Effects of prolonged high temperature on testicular development and fertility in the male rat. *Lab. Anim. Sci.* **27**, 76–77.

Punareewattana, K., Smith, B.J., Blaylock, B.L., Longstreth, J., Snodgrass, H.L., Gogal Jr., R.M., Prater, R.M., and Holladay, S.D. (2001). Topical permethrin exposure inhibits antibody production and macrophage function in C57BL/6N mice. *Food Chem. Toxicol.* **39**, 133–139.

Quraishi, M.S. (1977). "Biochemical Insect Control: Its Impact on Economy, Environment and Natural Selection." John Wiley and Sons, New York.

Rabin, B.S., Lyte, M., and Hamill, E. (1987). The influence of mouse strain and housing on the immune response. *J. Neuroimmunol.* **17**, 11–16.

Rabin, B.S. and Salvin, S.B. (1987). Effect of differential housing and time on immune reactivity to sheep erythrocytes and Candida. *Brain Behav. Immun.* **1**, 267–275.

Radzialowski, F.M. and Bousquet, W.F. (1968). Daily rhythmic variation in hepatic drug metabolism in the rat and mouse. *J. Pharmacol. Exp. Ther.* **163**, 229–238.

Ramos, J.G., Varayoud, J., Sonnenschein, C., Soto, A.M., de Toro, M.M., and Luque, E. H. (2001). Prenatal exposure to low doses of bisphenol A alters the periductal stroma and glandular cell function in the rat ventral prostate. *Biol. Reprod.* **65**, 1271–1277.

Ras, T., Van De Ven, M., Patterson-Kane, E.G., and Nelson, K. (2002). Rats' preferences for corn versus wood-based bedding and nesting materials. *Lab. Anim.* **36**, 420–425.

Rasmussen Jr., A.F. (1969). Emotions and immunity. *Ann. NY Acad. Sci.* **164**, 458–461.

Reeb, C.K., Jones, R.B., Bearg, D.W., Cih, P.E., Bedigian, H., Myers, D.D., and Paigen, B. (1998). Microenvironment in ventilated animal cages with differing ventilation rates, mice populations, and frequency of bedding changes. *Contemp. Top. Lab. Anim. Sci.* **37**, 43–49.

Reeb-Whitaker, C.K., Harrison, D.J., Jopnes, R.B., Kacergia, J.B., Myers, D.D., and Paigen, B. (1999). Control strategies for aeroallergens in an animal facility. *J. Allergy Clin. Immunol.* **103**, 139–146.

Reeb-Whitaker, C.K., Paigen, B., Beamer, W.G., Bronson, R.T., Churchill, G.A., Schweitzer, I.B., and Myers, D.D. (2001). The impact of reduced frequency of cage changes on the health of mice housed in ventilated cages. *Lab. Anim.* **35**, 58–73.

Reiter, R.J. (1973). Comparative effects of continual lighting and pinealectomy on the eyes, the Harderian glands and reproduction in pigmented and albino rats. *Comp. Biochem. Physiol.* **44**, 503–509.

Renstrom, A., Bjoring, G., and Hoglund, A.U. (2001). Evaluation of individually ventilated cage systems for laboratory rodents: occupational health aspects. *Lab. Anim.* **35**, 42–50.

Richa, J. (2001). Production of transgenic mice. *Mol. Biotechnol.* **17**, 261–268.

Rikans, L.E. and Moore, D.R. (1988). Acetaminophen hepatotoxicity in aging rats. *Drug Chem. Toxicol.* **11**, 237–247.

Riley, V. (1981). Psychoneuroendocrine influence on immunocompetence and neoplasia. *Science* **212**, 1100–1109.

Rivard, G.F., Neff, D.E., Cullen, J.F., and Welch, S.W. (2000). A novel vented microisolation container for caging animals: microenvironmental comfort in a closed-system filter cage. *Contemp. Top. Lab. Anim. Sci.* **39**, 22–27.

Robertson, E.J. (1991). Using stem cells to introduce mutations into the mouse germ line. *Biol. Reprod.* **44**, 238–245.

Robinson, E.C. and Nair, R.S. (1992). The genotoxic potential of linear alkylbenzene mixtures in a short-term test battery. *Fundam. Appl. Toxicol.* **18**, 540–548.

Robinson, E.C. and Schroeder, R.E. (1992). Reproductive and developmental toxicity studies of a linear alkylbenzene mixture in rats. *Fundam. Appl. Toxicol.* **18**, 549–556.

Rock, F.M., Landi, M.S., Hughes, H.C., and Gagnon, R.C. (1997). Effects of caging type and group size on selected physiologic variables in rats. *Contemp. Top. Lab. Anim. Sci.* **36**, 69–72.

Roe, F.J.C. (1965). Spontaneous tumors in rats and mice. *Food Cosmet. Toxicol.* **3**, 707–720.

Romanovsky, A.A., Ivanov, A.I., and Shimansky, Y.P. (2002). Selected contribution: ambient temperature for experiments in rats: a new method for determining the zone of thermal neutrality. *J. Appl. Physiol.* **92**, 2667–2679.

Rose, A.H. (ed.) (1967). "Thermobiology." Academic Press, New York.

Rowe, W.P., Hartley, J.W., and Huebner, R.J. (1962). Polyoma and other indigenous mouse viruses. *In* "The Problems of Laboratory Animal Diseases." (R.J.C. Harris, ed.), p. 131, Academic Press, New York.

Roy, D., Colerangle, J.B., and Singh, K.P. (1998). Is exposure to environmental or industrial endocrine disrupting estrogen-like chemicals able to cause genomic instability? *Front. Biosci.* **3**, 913–921.

Royals, M., Getzy, D.M., and Vandewoude, S. (1999). High fungal spore load in corncob bedding associated with fungal-induced rhinitis in two rats. *Contemp. Top. Lab. Anim. Sci.* **38**, 64–66.

Rubin, B.S., Murray, M.K., Damassa, D.A., King, J.C., and Soto, A.M. (2001). Perinatal exposure to low doses of bisphenol A affects body weight, patterns of estrous cyclicity, and plasma LH levels. *Environ. Health Perspect.* **109**, 675–680.

Ruys, T., ed. (1991). "Handbook of Facilities Planning, Volume 2: Laboratory Animal Facilities," Van Nostrand Reinhold, New York.

Sabiston, B.H., Rose, J.E., and Cinader, B. (1978). Temperature stress and immunity in mice: effects of environmental temperature on the antibody response to human immunoglobulin of mice, differing in age and strain. *J. Immunogenet.* **5**, 197–212.

Saito, T.R., Motomura, N., Taniguchi, K., Hokao, R., Arkin, A., Takahashi, K.W., and Sato, N.L. (1996). Effect of cage size on sexual behavior pattern in male rats. *Contemp. Top. Lab. Anim. Sci.* **35**, 80–82.

Sales, G.D., Wilson, K.J., Spencer, K.E., and Milligan, S.R. (1988). Environmental ultrasound in laboratories and animal houses: a possible cause for concern in the welfare and use of laboratory animals. *Lab. Anim.* **22**, 369–375.

Saltarelli, C.G. and Coppola, C.P. (1979). Influence of visible light on organ weights of mice. *Lab. Anim. Sci.* **29**, 319–322.

Salvia, M., Weisbroth, S.H., and Paganelli, R.G. (1979). Lab animals show less "shipment stress" given drinking water. *Lab. Anim.* **8**, 38–42.

Salvin, S.B., Rabin, B.S., and Neta, R. (1990). Evaluation of immunologic assays to determine the effects of differential housing on immune reactivity. *Brain Behav. Immun.* **4**, 180–188.

Sanvordecker, D.R. and Lambert, H.J. (1974). Environmental modification of mammalian drug metabolism and biological response. *Drug Metab. Rev.* **3**, 201–229.

Schoeb, T.R., Davidson, M.K., and Lindsey, J.R. (1982). Intracage ammonia promotes growth of *Mycoplasma pulmonis* in the respiratory tract of rats. *Infect. Immun.* **38**, 212–217.

Schoket, B. and Vincze, I. (1985). Induction of rat hepatic drug metabolizing enzymes by substituted urea herbicides. *Acta Pharmacol. Toxicol.* **56**, 283–288.

Schonfelder, G., Flick, B., Mayr, E., Talsness, C., Paul, M., and Chahoud, I. (2002). In utero exposure to low doses of bisphenol A lead to long-term deleterious effects in the vagina. *Neoplasia* **4**, 98–102.

Schuhr, B. (1987). Social structure and plasma corticosterone level in female albino mice. *Physiol. Behav.* **40**, 689–693.

Sedlacek, R.S., Orcutt, R.P., Suit, H.D., and Rose, E.F. (1981). A flexible barrier at cage level for existing colonies: Production and maintenance of a limited stable anaerobic flora in a closed inbred mouse colony. *In* "Recent Advances in Germfree Research: Proceedings of the VIIth International Symposium on Gnotobiology." (S. Sasaki, A. Ozawa, and K. Hashimoto eds.), pp. 65–69, Tokai University Press, Tokyo.

Sellers, E.A. (1957). Adaptive and related phenomena in rats exposed to cold: a review. *Revue Canadienne de Biologie* **16**, 175–188.

Semple-Rowland, S.L. and Dawson, W.W. (1987a). Cyclic light intensity threshold for retinal damage in albino rats raised under 6 lx. *Exp. Eye Res.* **44**, 643–661.

Semple-Rowland, S.L. and Dawson, W.W. (1987b). Retinal cyclic light damage threshold for albino rats. *Lab. Anim. Sci.* **37**, 289–298.

Serrano, L.J. (1971a). Carbon dioxide and ammonia in mouse cages: effect of cage covers, population, and activity. *Lab. Anim. Sci.* **21**, 75–85.

Serrano, L.J. (1971b). Defined mice in a radiobiological experiment. *In* "Defining the Laboratory Animal." pp. 13–41, Institues of Laboratory Animal Resources, National Academy of Science, Washington, DC.

Sharp, P.E. and LaRegina, M.C. (1998). "The Laboratory Rat." p. 30, CRC Press, Boca Raton.

Shaw, D.C. and Gallaggher, R.H. (1984). Group or singly housed rats? *In* "Standards in Laboratory Animal Management, Part 1" pp. 65–70. Universities Federation for Animal Welfare, Herferdshire, United Kingdom.

Shapiro, R. (1980). Chemical contamination of drinking water: what it is and where it comes from. *Lab Anim.* **9**, 45–51.

Shysh, A. and Noujaim, A.A. (1972). Alterations in hepatic microsomal drug metabolizing systems in cold stressed mice. *Can. J. Pharm. Sci.* **7**, 23

Sigmund, C.D. (1993). Major approaches for generating and analyzing transgenic mice. An overview. *Hypertension* **22**, 599–607.

Simmons, M.L. and Brick, J.O. (1970). "The Laboratory Mouse Selection and Management," pp. 153–156. Prentice-Hall, Englewood Cliffs, NJ.

Simmons, M.L., Robie, D.M., Jones, J.B., and Serrano, L.J. (1968). Effect of a filter cover on temperature and humidity in a mouse cage. *Lab. Anim.* **2**, 113–120.

Skowronski, G.A., Abdel-Rahman, M.S., Gerges, S.E., and Klein, K.M. (1985). Teratologic evaluation of Alcide liquid in rats and mice, I. *J.Appl. Toxicol.* **5**, 97–103.

Slanetz, C.A., Fratta, I., Crouse, C.W., and Jones, S.C. (1956). Stress and transportation of animals. *Proc. Anim. Care Panel* **7**, 278–289.

Smith, M.M. and Hargaden, M. (2001). Developing a rodent enrichment program. *Lab Anim.* **30**, 36–41.

Solomon, G.F. (1969). Stress and antibody response in rats. *Int. Arch. Allergy Appl. Immunol.* **35**, 97–104.

Sobrian, S.K., Vaughn, V.T., Ashe, W.K., Markovic, B., Djuric, V., and Jankovic, B.D. (1997). Gestational exposure to loud noise alters the developmental and postnatal responsiveness of humoral and cellular components of the immune system in offspring. *Environ. Res.* **73**, 227–241.

Sokel, E.H., Zuppinger, K.A., and Joss, E.E. (1979). Effects of isolation, handling and restraint on growth of rats. *Growth* **43**, 1–6.

Solvay Advanced Polymers (2003). Questions and answers about bisphenol A in lab animal caging materials. pp. 1–9. Information Sheet Provided by Solvay Advanced Polymers to Interested Individuals.

Sorensen, D.R. (1994). Laboratory animal facilities. *In* "Handbook of Laboratory Animal Science, Vol. I. Selection and Handling of Animals in Biomedical Research." (P. Svendssen and J. Hau, eds.), pp. 79–87, CRC Press, Boca Raton, FL.

Spalding, J.F., Archuleta, R.F., and Holland, L.M. (1969a). Influence of visible color spectrum on activity in mice. *Lab. Anim. Care* **19**, 50–54.

Spalding, J.F., Holland, L.M., and Tietjen, G.L. (1969b). Influence of the visible color spectrum on activity in mice, II: influence of sex, color and age on activity. *Lab. Anim. Care* **19**, 209–213.

Stauffacher, M. (1966). Comparative studies on housing conditions. *In* "Harmonization of Laboratory Animal Husbandry." (P.N. O'Donoghue, ed.), pp. 5–9, Royal Society of Medicine Press, London.

Stefanski, V. (2001). Social rearing conditions before weaning influence numbers and proportions of blood immune cells in laboratory rats. *Dev. Psychobiol.* **39**, 46–52.

Svendsen, P. (1994). Environmental impact on animal experiments. *In* "Handbook of Laboratory Animal Science," Vol. 1. pp. 191–202, CRC Press, Boca Raton, FL.

Takai, Y., Tsutsumi, O., Ikezuki, Y., Kamei, Y., Osuga, Y., Yano, T., and Taketan, Y., (2001). Preimplantation exposure to bisphenol A advances postnatal development. *Reprod. Toxicol.* **15**, 71–74.

Takeo, Y. (1984). Influence of continuous illumination on estrus cycle of rats:time course of changes in levels of gonadotropins and ovarian steroids until occurrence of persistent estrus. *Neuroendocrinology* **39**, 97–104.

Tanner, R.S. (1989). Comparative testing and evaluation of hard-surface biocides. *J. Ind. Microbiol.* **4**, 145.

Targowski, S.P., Klucinski, W., Babiker, S., and Nonnecke, B.J. (1984). Effect of ammonia on *in vivo* and *in vitro* immune responses. *Infect Immun.* **43**, 289–293.

Terleckyj, B. and Axler, D.A. (1987). Quantitative neutralization assay of fungicidal activity of disinfectants. *Antimicrob. Agents Chemother.* **31**, 794–798.

Thompson, J.S., Brown, S.A., Khurdayan, V., Zeynalzadedan, A., Sullivan, P.G., and Scheff, S.W. (2002). Early effects of tribromo-ethanol, ketamine/xylazine, pentobarbital, and isoflurane anesthesia on hepatic and lymphoid tissue in ICR mice. *Comp. Med.* **52**, 63–67.

Tohei, A., Suda, S., Taya, K., Hashimoto, T., and Kogo, H. (2001). Bisphenol A inhibits testicular functions and increases luteinizing hormone secretion in adult male rats. *Exp. Biol. Med.* **226**, 216–221.

Torronen, R., Pelkonen, K., and Karenlampi, S. (1989). Enzyme-inducing and cytotoxic effects of wood-based materials used as bedding for laboratory animals: comparison by a cell culture study. *Life Sci.* **45**, 559–565.

Toth, G.P., Long, R.E., Mills, T.S., and Smith, M.K. (1990). Effects of chlorine dioxide on the developing rat brain. *J. Toxicol. Environ. Health* **31**, 29–44.

Tu, H., Diberadinis, J., and Lipman, N.S. (1997). Determination of air distribution, exchange, velocity, and leakage in three individually ventilated rodent caging systems. *Contemp. Top. Lab. Anim. Sci.* **36**, 69–73.

Tuli, J.S., Smith, J.A., and Morton, D.B. (1994). Stress measurements in mice after transportation. *Lab. Anim.* **29**, 132–138.

Turner, K. and Sharpe, R.M. (1997). Environmental oestrogens–present understanding. *Rev. Reprod.* **2**, 69–73.

Twigger, S., Lu, J., Shimoyama, M., Chen, D., Pasko, D., Long, H., Ginster, J., Chen, C.F., Nigam, R., Kwitek, A., Eppig, J., Maltais, L., Maglott, D., Schuler, G., Jacob, H., and Tonellato, P.J. (2002). Rat genome database (RGD): mapping disease onto the genome. *Nucleic Acids Res.* **30**, 125–128.

Tyl, R.W., Myers, C.B., Marr, M.C., Thomas, B.F., Keimowitz, A.R., Brine, D.R., Veselica, M.M., Fail, P.A., Chang, T.Y., Seely, J.C., Joiner, R.L., Butala, J.H., Dimond, S.S., Cagen, S.Z., Shiotsuka, R.N., Stropp, G.D., and Waechter, J.M. (2002). Three-generation reproductive toxicity study of dietary bisphenol A in CD Sprague-Dawley rats. *Toxicol. Sci.* **68**, 121–146.

Vadiei, K., Berend, K.L., and Luke, D.R. (1969). Isolation-induced renal functional changes in rats from four breeders. *Lab. Anim. Sci.* **40**, 56–59.

Van Damme, A., Vanden Driessche, T., Collen, D., and Chush, M.K. (2002). Bone marrow stromal cells as targets for gene therapy. *Curr. Gene Ther.* **2**, 195–209.

Van der Touw, J., Thrower, S.J., and Olley, J. (1978). Non-specific neural stimuli and metabolic rhythms in rats. *Physiol. Bohemos.* **27**, 501–504.

Van de Weerd, H.A., Baumans, V., Koolhaas, J.M., and Van Zutphen, L.F.M. (1994). Strain specific behavioural response to environmental enrichment in the mouse. *J. Exp. Anim. Sci.* **36**, 117–127.

Van Raaij, M.T., Oortgiesen, M., Timmerman, H.H., Dobbe, C.J., and Van Loveren, H. (1996). Time-dependent differential changes of immune function in rats exposed to chronic intermittent noise. *Physiol. Behav.* **60**, 1527–1533.

Verbiest, H. (1956). Temperature and heat regulation. *Folia Psychiatrica, Neurologica et Neurochirurgica Neerl* **59**, 363–407.

Vesell, E.S. (1967). Induction of drug-metabolizing enzymes in liver microsomes of mice and rats by soft wood bedding. *Science* **157**, 1057–1058.

Vesell, E.S. (1968). Genetic and environmental factors affecting hexobarbital metabolism in mice. *Ann. N. Y. Acad. Sci.* **151**, 900–912.

Vessell, E.S. and Lang, C.M. (1976). Environmental and genetic factors affecting laboratory animals: impact on biomedical research. *Fed. Proc.* **35**, 1123–1165.

Vessel, E.S., Lang, C.M., White, W.J., Passananti, G.T., Hill, R.N., Clemens, T.L., Liu, D.K., and Johnson, W.D. (1976). Environmental and genetic factors affecting the response of laboratory animals to drugs. *Fed. Proc., Fed. Am. Soc. Exp. Biol.* **35**, 1125–1132.

Vesell, E.S., Lang, C.M., White, W.J., Passananti, G.T., and Tripp, S.L. (1973). Hepatic drug metabolism in rats: impairment in a dirty environment. *Science* **179**, 896–897.

Vessey, S.H. (1964). Effects of grouping on levels of circulating antibodies in mice. *Proc. Soc. Exp. Biol. Med.* **115**, 225–255.

Vivitskaya, L.V., Bikbulatova, I.S., and Vivitsky, V. (1982). Changes in the content of nucleic acids and proteins in different brain parts of rats raised in enriched and impoverished media. *Zhurnal Vysshei Deyatel Nosti* **32**, 455–462.

Wade, A.E., Holl, J.E., Hilliard, C.C., Molton, E., and Greene, F.E. (1968). Alterations of drug metabolism in rats and mice by environment of cedar wood. *Pharmacol.* **1**, 317–328.

Wagner, S.L. (1983). "Clinical Toxicology of Agricultural Chemicals." Noyes Data Corporation, Park Ridge.

Wahlstein, D. (1982). Deficiency of corpus callosum varies with strain and supplier of the mice. *Brain Res.* **239**, 329–347.

Wallace, M.E. (1976). Effects of stress due to deprivation and transport in different genotypes of house mouse. *Lab. Anim.* **10**, 335–347.

Weichbrod, R.H., Cisar, C.F., Miller, J.G., Simmonds, R.C., Alvares, A.P., and Ueng, T.H. (1988). Effects of cage beddings on microsomal oxidative enzymes in rat liver. *Lab. Anim. Sci.* **38**, 296–298.

Weihe, W.H. (1965). Temperature and humidity climatographs for rats and mice. *Lab. Anim. Care* **15**, 18–28.

Weihe, W.H. (1971). The significance of the physical environment for the health and state of adaptation of laboratory animals. *In* "Defining the Laboratory Animal." IVth Symposium of the International Committee on Laboratory Animals, National Academy of Sciences, Washington, DC.

Weihe, W.H. (1973). The effect of temperature on the action of drugs. *Annu, Rev. Pharmacol.* **13**, 409–425.

Weihe, W.H. (1976). Influence of light on animals. *In* "Control of the Animal House Environment." (T. McSheehy, ed.), pp. 63–76, Laboratory Animal, Ltd., London.

Weihe, W.H., Schidlow, J., and Strittmatter, J. (1969). The effect of light intensity on the breeding and development of rats and golden hamsters. *Int. J. Biometeorol.* **13**, 69–79.

Weisbroth, S.H. (1979). Chemical contamination of lab animal beddings: problems and recommendations. *Lab. Anim.* **8**, 24–34.

Weisbroth, S.H., Paganelli, R.G., and Salvia, M. (1977). Evaluation of a disposable water system during shipment of laboratory rats and mice. *Lab. Anim. Sci.* **27**, 186–194.

Wiberg, G.S., Airth, J.M., and Grive, H.C. (1966). Methodology in long-term toxicity tests: A comparison of individual versus community housing. *Fd. Cosmet. Toxicol.* **4**, 47–55.

Wight, D.C. and Wagner, T.E. (1994). Transgenic mice: a decade of progress in technology and research. *Mutat. Res.* **307**, 429–440.

Wilder, P.J. and Rizzino, A. (1993). Mouse genetics in the 21st century: using gene targeting to create a cornucopia of mouse mutants possessing precise genetic modifications. *Cytotechnology* **11**, 79–99.

Williams, D.I. and Wells, P.A. (1970). Differences in home-cage emergence in the rat in relation to infantile handling. *Psychonom. Sci.* **18**, 168–169.

Williams, R.A., Howard, A.G., and Williams, T.P. (1985). Retinal damage in pigmented and albino rats exposed to low levels of cyclic light following a single mydriatic treatment. *Curr. Eye Res.* **4**, 97–102.

Woods, J.E. (1978). Interactions between primary (cage) and secondary (room) enclosures. *In* "Laboratory Animal Housing." pp. 65–83, National Academy of Science, Washington D.C.

Woods, J.E. (1980). The animal enclosure: a microenvironment. *Lab. Anim. Sci.* **30**, 407–413.

Woods, J.E., Besch, E.L., and Nevins, R.G. (1974). Heat and moisture transfer in filter-top rodent cages. *AALAS Publ.* **74-3**, Abstr. 77.

Wu, D., Joiner, G.N., and McFarland, A.R. (1985). A forced-air ventilation system for rodent cages. *Lab. Anim. Sci.* **35**, 499–504.

Wurbel, H. (2001). Ideal homes? Housing effects on rodent brain and behavior. *Trends Neurosci.* **24**, 207–211.

Wysocki, A.B. (1996). The effect of intermittent noise exposure on wound healing. *Adv. Wound Care* **9**, 35–39.

Yagil, R., Etzion, Z., and Berlyne, G.M. (1976). Changes in rat milk quantity and quality due to variations in litter size and high ambient temperature. *Lab. Anim. Sci.* **26**, 33–37.

Yamamoto, S., Ando, M., and Suzuki, E. (1999). High-temperature effects on antibody response to viral antigen in mice. *Exp. Anim.* **48**, 9–14.

Yamauchi, C., Fujita, S., Obara, T., and Ueda, T. (1981). Effects of room temperature on reproduction, body and organ weights, food and water intake, and hematology in rats. *Lab. Anim. Sci.* **31**, 251–258.

Yamauchi, C., Takahashi, H., and Ando, A. (1965). Effect of environmental temperature on physiological events in mice, 1: relationship between environmental temperature and number of caged mice. *Jap. J. Vet. Sci.* **27**, 471–478.

Yoshida, K., Okamoto, M., Tajima, M., and Kurosawa, T. (1995). Invention of a forced-air-ventilated micro-isolation cage and rack system: environment within cages: temperature and ammonia concentration. *Exp. Anim.* **43**, 703–710.

Zakem, H.B. and Alliston, C.W. (1974). The effects of noise level and elevated ambient temperatures upon selected reproductive traits in female Swiss-Webster mice. *Lab. Anim. Sci.* **24**, 469–475.

Chapter 11

Bacterial, Mycoplasmal and Mycotic Infections

Steven H. Weisbroth, Dennis F. Kohn and Ron Boot

I. INTRODUCTION

This chapter will review the pertinent microbiological, pathological, diagnostic, and medical features of naturally occurring bacterial, mycoplasmal, and mycotic infections of the laboratory rat. Viewed from the perspective of time, it is important to recognize that an evolutionary process has just as surely been at work on the microbial flora, as on the host itself, since domestication of the rat more than a century ago (Robinson, 1965). What the veterinary history of laboratory rodents seems to demonstrate is that the spectrum of pathogenic agents the rats may carry is not a list frozen for all time but, rather, resembles a moving boundary in which old pathogens are eradicated, creating invasive opportunities for new pathogens and thus periodic reconstitution of the list (Weisbroth, 1999a). The emphasis of the chapter will, of course, be on the pathogenic agents of greatest current significance.

As the ecological niche of the rat has been sequentially modified by refinements in both breeder production and laboratory environments, the pathogenic microbial species accompanying these changes have shifted to forms largely unrecognized as rat pathogens at the beginning of this process. As a general statement, it may be said that the primary bacterial pathogens posing major epizootic hazards were among the first to be eliminated from laboratory environments by these ecologic changes. Examples include *Salmonella* sp., *Leptospira* sp. and *Streptobacillus moniliformis,* which figure prominently as agents encountered in the decades just before and after the turn of the 20th century. These organisms are still common in wild rats, and the concept that laboratory rats differed little in a microbial sense from their wild counterparts in those earlier times is a valid one. Under the generally improved circumstances of laboratory environments that obtained between the turn of the 20th century and the mid-1950s, the main accomplishments of disease control programs were to reduce the quantitative incidence of primary pathogens but not to effect a qualitative change in the distribution of bacterial pathogens carried by rat hosts.

As this pattern applies to the laboratory rat, by the early 1950s bacterial diseases, as a practical matter, had been reduced to a small number of extremely important diseases of high morbidity when they occurred. The bacterial components of this spectrum were now seen to include *Streptococcus pneumoniae, Mycoplasma pulmonis, Pasteurella pneumotropica,* and *Corynebacterium kutscheri,* entities of minor or unrecognized significance at the turn of the century. Moreover, although exposure alone is generally sufficient to produce infection with clinical signs and lesions in the case of such frank primary pathogens as *Salmonella* sp. and *S. pneumoniae,* in modern times the increasing significance of inapparent infection with bacterial forms of lower pathogenicity has been recognized. By inapparent infection is meant the carriage of infectious organisms by the host, whose presence is unrecognized or masked until a suitable stress sufficiently impairs the defensive competence of the host in a direction favoring multiplication and invasion of the microbe accompanied by clinical signs and lesions. Such microbial forms as *C. kutscheri, P. pneumotropica, Pneumocystis carinii, Clostridium piliforme,* and *Pseudomonas aeruginosa* characteristically occur in the rat as inapparent infections. Exogenous stressors that transiently impair immunocompetence—such as relocation and shipment, other intercurrent infectious diseases, vitamin deficiency, X irradiation, administration of adrenal corticosteroids, immunosuppressive drugs, thymectomy, and others—are documented examples of the types of incitant stimuli that often are known to precede or trigger clinical signs and/or microscopic lesions associated with particular bacterial and mycotic diseases.

To further confound the issue of inapparence, it needs to be recognized that intrinsic factors of rat host immunocompetency likewise regulate the pathogenicity of microbial infection. These latter factors prevent universally applicable guidelines to usefully categorize microbes as being "primary pathogens," "opportunists," or "commensal" flora because it is recognized that the issue is as largely determined by heritable constitutional factors of relative resistance or susceptibility of the host to given microbial forms, as much as it is to microbial properties of pathogenicity (Weisbroth, 1999a). Since the early 1980s, a substantial base of literature has accumulated to document episodes of disease in immunodeficient rodents with infectious agents that are usually clinically silent in their immunocompetent counterparts.

II. BACTERIAL INFECTIONS

A. Streptobacillosis (*Streptobacillus moniliformis*)

Considered only within the context of its status as a pathogen of laboratory rats, *S. moniliformis* occupies a somewhat ambiguous role. It has been circumstantially associated with various aspects of respiratory disease in laboratory rats since early times; however, two other features of the organism have received intense scrutiny, tending to obscure or deemphasize the limited potential for pathogenicity of the organism in its natural reservoir host. These other features include the etiological significance of *S. moniliformis* in rat bite and Haverhill fevers of man, and the cultural instability of the organism leading to L-phase variants that for a certain period of time were confused with *Mycoplasma* of rat origin.

The binomial name *S. moniliformis*, first suggested by Levaditi et al. (1925) will be used throughout this text in accordance with the ninth edition of *Bergey's Manual* (Holt et al., 1994), although based on relationship to the actinobacilli, Wilson and Miles (Collier et al., 1998) have used the name *Actinobacillus moniliformis*, which is here considered to be a synonym (for a review, see Wullenweber, 1995). The above citations (and Nelson, 1930a) describe the evolution of synonymy with this organism and the following list of names may be found there and in the older literature: *Streptothrix muris ratti, Nocardia muris, Actinomyces muris ratti, Haverhilia multiformis, Actinomyces muris, Asterococcus muris, Proactinomyces muris, Haverhilia moniliformis, Actinobacillus muris,* and *Clostridium actinoides* var. *muris*.

Although a streptothrix-like organism had been recognized in the blood of some human patients with recurrent fever after a rat bite, the organism was not isolated in pure culture and characterized until 1914 (Schottmuller, 1914). Blake (1916) described the first etiologically confirmed case of human streptobacillosis in America. The association of the organism with human disease was clear from the beginning, but many years were to elapse before an understanding of the relationship was established between *S. moniliformis* and its natural reservoir host, the rat. The earliest associations of *S. moniliformis* and its rat host were in the context of respiratory disease, particularly pneumonia (Tunnicliff, 1916a,b; Jones, 1922; Turner, 1929), and, to this day, in cases of otitis media (Nelson, 1930a,b; Koopman et al., 1991; Wullenwebber et al., 1992). It was this respiratory context that was contributory to the confusion (discussed later in this chapter) between *S. moniliformis* L-phase variants and *M. pulmonis*. Although Tunicliff's original isolates were from pneumonic tissues and, in addition, she was to claim that intraperitoneal injection of *S. moniliformis* caused pneumonias (Tunnicliff, 1916a,b), critical examination of her work casts doubt on whether the organism caused pneumonia or merely was reisolated from (preexisting) pneumonic tissues. Moreover, the weight of experimental and circumstantial evidence since has, to the contrary, established the inability of this organism to cause respiratory disease in rats (Nelson, 1930; Strangeways, 1933; Boot et al., 1993a, 1996).

It was not until 1933 that Strangeways (1933) reported the asymptomatic occurrence of *S. moniliformis* in both wild and laboratory rats as a commensal of the nasopharynx. Streptobacillosis had inadvertently been caused in mice receiving *Trypanosoma equiperdum*–infected blood from rat donors. Investigation revealed that the manner of collecting blood from the rat donors, that is, by "a blow on the head of sufficient violence to cause blood to flow freely

from the head and nose" was instrumental in directing her attention to the nasopharynx because heart blood collected from the rats did not (except on two occasions) transfer *S. moniliformis* to the mice. Subsequent investigations established the nasopharynx (and saliva) as the usual location of this organism in the rat and provided the ecological link to explain the injection of *S. moniliformis* into humans after a rat bite, at the time a puzzle of some significance. Swallowing of the organism by rat carriers results in contamination of the feces with *S. moniliformis* and forms the basis for understanding how Haverhill fever (in man) can result from milk or water contaminated with rat feces (McEvoy et al., 1987).

At the present time, *S. moniliformis* is believed to be a commensal with low potential for pathogenicity for the rat. It has not been etiologically associated with naturally occurring disease processes in this species except as a secondary invader. Experimental infections in rats with certain isolates ("strain C") given intravenously have been used as a model of the infective arthritides (Lerner and Silverstein, 1957; Lerner and Sokoloff, 1959), but these bone and joint lesions have not been seen in the rat under natural circumstances. Although *S. moniliformis* occurred commonly in laboratory rats during the first half of the 20th century, variably as high as 50% (Nelson and Gowen, 1930; Strangeways, 1933; Bhandari, 1976), the incidence of *S. moniliformis* in the laboratory has markedly decreased in modern times under the influence of higher animal care standards and the general use of laboratory rats produced by gnotobiotic derivation. It continues, however, to be occasionally isolated from both wild (Boot et al., 1993a) and conventional laboratory rats (Matheson et al., 1955; Koopman et al., 1991) and, rarely, from hysterectomy-derived barrier-maintained laboratory mice (Wullenwebber et al., 1990a, Glastonbury et al., 1996). Rat-bite fever remains an occasional occupational hazard for laboratory personnel and the general public who have contact with pet or wild rats (Rake, 1936; Dawson and Hobby, 1939; Larson, 1941; Thjotta and Jonsen, 1947; Borgen and Gaustad, 1948; Hamburger and Knowles, 1953; Smith and Sampson, 1960; Holden and MacKay, 1964; Roughgarden, 1965; McGill et al., 1966; Gledhill, 1967; Gilbert et al., 1971; Anonymous, 1975; Biberstein, 1975; Anderson et al., 1983; McEvoy et al., 1987; Rumley et al., 1987; Cunningham et al., 1998; Hagelskjaer et al., 1998; Downing et al., 2001).

Laboratory rats may act as zoonotic reservoirs for *S. moniliformis* infections in laboratory mice (Freundt, 1956, 1959; Glastonbury et al., 1996) (and in nature, as reservoirs for wild mouse infections) (Taylor et al., 1994). Unlike rat infections, mouse infections with *S. moniliformis* ordinarily are not lesionless and asymptomatic. More recently, heritable variability in expression of disease has been observed with different inbred mouse strains (Wullenwebber et al., 1990). The mouse is not thought to be a carrier of commensal streptobacilli. Infections in mice may be acute with generalized septicemia and lymphadenitis (erythema multiforme) or may be more chronic and characterized by osteomyelitis and polyarthritis, especially of the lower hind extremity and tail vertebrae. (Mackie et al., 1933; Van Rooyen, 1936; Freundt, 1956a,b, 1959; Kaspareit-Rittinghausen et al., 1990; Savage et al., 1991). Natural infection of pregnant mice may also result in abortions (Mackie et al., 1933; Sawicki et al., 1962). The mode of transmission from rat to mouse has been investigated but remains uncertain (Levaditi et al., 1932; Van Rooyen, 1936). Infections do not persist in mouse colonies rigidly separated from rats (Gledhill, 1967). In this connection, mouse infections appear to mimic human infections in terms of the ecology of *S. moniliformis*. In the mouse, as in man, septicemia and polyarthritis may follow rat bites (Levaditi et al., 1932) or may be unrelated to this mode of transmission (Mackie et al., 1933; Van Rooyen, 1936). Similarly, Haverhill fever in man (erythema arthriticum epidemicum) occurred as epidemics in Haverhill, Massachusetts, in 1926 and in Chester, Pennsylvania, in 1925, in which the source of infection was presumptively oral and eventually traced to infected milk to which rats had access (Parker and Hudson, 1926; Place et al., 1926; Place and Sutton, 1936; Editorial, 1939). Even in more recent times, human *S. moniliformis* infections may be unrelated to rat bites (Hazard and Goodkind, 1932; Sprecher and Copeland, 1947; Osimani et al., 1972; Lambe et al., 1973; McEvoy et al., 1987; Rumley et al., 1987), although direct inoculation by rat bite appears to be the most common mode of transmission to man. There appear to be very few human cases on record of streptobacillary rat-bite fever after a mouse bite (Arkless, 1970; Gilbert et al., 1971).

The morphology of the organism is variable and pleomorphic, depending on environmental factors and age of the culture (Nelson, 1931; Heilman, 1941). Smears from blood or other clinical materials most commonly show small (less than 1 μm wide and 1 to 5 μm long) Gram-negative rods and filaments (Francis, 1932). Blood, serum, or ascetic fluid is required for growth; however, growth on Loeffler's serum agar slants is poor. Growth does not occur in nutrient or infusion broths or agars unless they are supplemented with one of these body fluids. After subculture on 20% horse serum infusion agar, slender interwoven masses of filaments 0.4 to 0.6 μm develop. Occasional filaments show spherical, oval, fusiform, or club-shaped swellings that may occur terminally, subterminally, or irregularly (hence the name "moniliformis") by 6 to 12 hours after subculture (Collier et al., 1998). After 12 to 18 hours, the filaments fragment into chains of bacillary, coccoid, or fusiform bodies. Some 24 to 30 hours after subculture on agar, tiny coccoid

or ring forms are seen as well, accompanied by masses of bubbles and irregularly shaped cholesterol-containing droplets. Gram negativity is variable, and staining is more intense in the monilia-like swellings than in the filaments. In 20% horse serum infusion broth cultures, the morphology is similar, but aggregate filamentous masses are characteristically seen, giving the appearance of a coarsely granular sediment. The supernatant is typically clear, without turbidity, and contains small flocculent masses of organisms, giving the tube a snowflake or cotton ball appearance when the tube is agitated. Although described as a facultative anaerobe, ordinary aerobic incubator conditions suffice for isolation and propagation. Growth may be enhanced in 8% to 10% carbon dioxide (CO_2) environments, which are recommended (Smith and Sampson, 1960; Hansen, 2000).

Colonies on serum or ascetic fluid infusion agar are 1 to 3 mm in diameter after 48 to 72 hours of growth, circular, and low convex; may vary from colorless to grayish white; be translucent or opaque; and have a smooth, glistening appearance with butyrous consistency (Heilman, 1951). Best results have been obtained by using trypticase soy agar with 20% horse serum (Smith and Sampson, 1960). There is little or no increase in size of colonies with prolonged incubation. On 5% sheep blood agar, colonies are barely visible at 24 hours, are 1 to 3 mm by 48 to 72 hours, and are smooth, gray, translucent, and essentially nonhemolytic. Gram-stained preparations of the growth show the characteristic morphology of the bacterium. Cystine trypticase broth, 10% horse serum, and 1% of various carbohydrates may be used for study of fermentative reactions (Smith and Sampson, 1960; Hansen, 2000). Other cultural and biochemical characteristics are summarized in Table 11-1. The organism is viable in serum broth and agar cultures for 2 to 4 days at 37°C, 7 to 10 days at 3°C to 4°C, 14 to 15 days at 3° to 4°C in sealed cultures, and for several years in infected tissues, body fluids, or cultures maintained at from −25°C to −70°C (Holt et al., 1994). The organism may be difficult to culturally retrieve from resistant and lightly infected carriers.

It is believed that all isolates of *S. moniliformis* may undergo reversible conversion to transitional-phase variants. The term L1 was first used by Klieneberger (1935, 1936, 1942) to designate the occurrence in *S. moniliformis* cultures of small, filterable (Berkfeld V) variants morphologically resembling the bovine pleuropneumonia organism. Indeed, a sensation was caused when Klieneberger affirmed that the L1 variant of *S. moniliformis* belonged to the pleuropneumonia-like organisms (PPLOs). In this designation the "L" (for Lister Institute) was later applied to any bacterial variant reproducing in the form of very small cells lacking rigid cell walls. Individual isolates of *S. moniliformis* vary in degree of stability, and the

TABLE 11-1

CHARACTERISTIC BIOCHEMICAL REACTIONS AND OTHER PROPERTIES OF *STREPTOBACILLUS MONILIFORMIS*

Characteristic	Reaction or property
Morphology	Gram-negative, pleomorphic, filaments or fragmented filaments (bacillary) with moniliform swellings
Carbon dioxide requirement	Enhances growth but not required
Catalase	(−)
Oxidase	(−)
Growth on nutrient agar or broth	(−), requires serum or blood
Indole	(−)
MR/VP	(−)/(−)
Glucose fermentation	(+), acid
Growth in 10% serum nutrient broth	(+), whitish flocculent aggregates, clear supernate
Growth on blood agar	1- to 2-mm colonies in 48 to 72 hours, smooth, gray, nonhemolytic
Gelatin liquefaction	(−)
Nitrate reduction	(−)
Hydrogen sulfide production	(−)
Arginine hydrolysis	(+)

MR/VP, Methyl Red/Voges Proskauer.
(Data from Buchanan and Gibbons, 1974; Cowan, 1974; Wilson and Miles, 1975.)

L phase is typically unstable. L-phase variants differ from the bacillus in cellular and colonial morphology and in susceptibility to antibiotics, for example, penicillin, that act primarily on the cell wall. Occasionally, stable L variants may be achieved by manipulation, which do not undergo reversion to the parental type. One such, LI Rat 30, was stabilized and has never since reversed (Klieneberger, 1938). The individual cells of LI Rat 30 vary from 0.3 to 3 μm, with a mean of 1.0 ± 0.5 μm. On serum or ascetic fluid infusion agar, colonies of the LI variant average about 300 to 500 μm in diameter (compared with 1000 to 3000 μm in the parental form), will show oil-like droplets, and the central portion will penetrate the agar to a depth of 30 to 50 μm. Bacterial L variants may not be distinguished from true *Mycoplasma* colonies on the basis of size, "fried egg" appearance, or deep brown color in the central area (Freundt, 1956; McGee and Wittler, 1969). However, the usual reversion of L variants to parental forms on supportive media without inhibitors will ordinarily serve to identify bacteria (and is a useful proof to sustain diagnostic isolation of *Mycoplasma*). The L-phase variant with an incomplete cell wall is resistant to penicillin concentrations as much as 10,000 times higher than is the normal bacillus with cell wall (Razin and Boschwitz, 1968; McGee and Wittler, 1969), and is ordinarily only encountered in the laboratory after subculture, although it has been isolated directly from pneumonic rats (Klieneberger, 1938) and from the blood of penicillin-treated rat-bite fever patients (Dolman et al., 1951). When inoculated into mice, the L-phase variant is nonpathogenic; however, it frequently reverts *in vivo* to the bacillary form with full recovery of the pathogenic properties of the original *S. moniliformis* isolate (Freundt, 1956).

Great confusion was inadvertently introduced into the emerging understanding of the etiology of respiratory disease of laboratory rats when Klieneberger and Steabben (1937, 1940) isolated from pneumonic rats hitherto undescribed organisms that they termed L3 and believed to be L-phase variants of *S. moniliformis*. Also in 1937, Nelson (1937) described "coccobacilliform bodies" isolated from mice with respiratory disease; several years later, from rats (Nelson, 1940). Accumulated findings in these and other laboratories led to the conclusion that the L3 and coccobacilliform bodies were identical, and they have since been classified within the Mycoplasmatales as *M. pulmonis* (Freundt, 1957). It is now generally understood that L1 and L3 are not associated, that the L1 variant (and other L variants) of *S. moniliformis* are unrelated to the mycoplasmas (PPLOs), and that *S. moniliformis* and/or its L variants play no necessary role in murine respiratory mycoplasmosis. More recently, partial immunologic identity of the antigens of *Acholeplasma laidlawii* with those of *S. moniliformis* has been demonstrated (Boot et al., 1993).

Opsonins, agglutinins, and complement fixing antibodies have been demonstrated in the postinfective sera of both experimentally and naturally infected rats (Tunnicliff, 1916b; Heilman 1941, 1956), naturally and experimentally infected mice (Nelson, 1933; Van Rooyen, 1936; Merrikin and Terry, 1972), and human *S. moniliforms* infections (Brown and Nunemaker, 1942; Hamburger and Knowles, 1953; Robinson, 1963; Gilbert et al., 1971; Lambe et al., 1973). Use of Tween 80 in media (1.5%) used to propagate *S. moniliformis* for agglutination antigen appears to prevent the tendency of the organism to clump (Lambe et al., 1973). Agglutinins may not appear in promptly treated human cases (Holden and MacKay, 1964). Both direct (Lambe et al., 1973) and indirect (Holmgren and Tunevall, 1970; Lambe et al., 1973) immunofluorescence have been used to confirm *S. moniliformis* isolates and antibodies from rat-bite fever patients. The L-phase variant shares at least one common antigen with the bacillus, but it lacks at least one antigen present in the bacterial form (Klieneberger, 1942). An enzyme-linked immunosorbent assay (ELISA) procedure has been developed and recommended for use as a screening test in rats (Boot et al., 1993), especially with lightly infected carriers where cultural retrieval is difficult and likely to result in a false negative (Koopman et al., 1991; Boot et al., 1996).

The antigenic relatedness of *S. moniliformis* isolates from various species has been investigated to some extent. Isolates of *S. moniliformis* from mouse, rat, and human infections were cross-reactive by ELISA (Boot et al., 1993a). Varying degrees of electrophoretic protein patterns were shown among human, rodent, and avian isolates (Costas and Owen, 1987). Such isolates may have varying biological characteristics. Guinea pig isolates of *S. moniliformis,* for example, are strict anaerobes, are biochemically inert, and, when injected intraperitoneally into mice, are pyogenic with abscess formation rather than septicemic and polyarthritic as is usually the case with rat, mouse, and human isolates (Smith, 1941; Aldred and Young, 1974). For these reasons, some doubt exists as to whether guinea pig isolates are truly *S. moniliformis* or, rather, taxonomically removed. *Bergey's Manual* classifies such organisms as *Sphaerophorus caviae* (Kirchner et al., 1992; Holt et al., 1994).

Diagnosis is established by isolation and cultural characterization of *S. moniliformis*. Important confirmatory information may be obtained by polymerase chain reaction (PCR) with samples derived from multiple sites, for example, pharynx, trachea, and lymph nodes (Boot et al., 2002). Older diagnostic strategies also used intraperitoneal and foot pad inoculation of test mice with causation of septicemia and embolically distributed lesions, including polyarthritis from which *S. moniliformis* could be reisolated. Sufficient evidence is not available as to whether or not antibodies invariably occur in asymptomatic carrier rats so as to make serological screening useful in monitoring laboratory rat colonies, but conventional rats were almost invariably found to be ELISA positive (Boot et al., 1993a).

With regard to treatment, little is known about the efficacy of antibiotic treatments in abolishing latent infections of the nasopharynx in rats, nor has this strategy been of interest. Treatment of colonies with important zoonotic infection should be discouraged. Considerable information is available, of course, with regard to antibiotic treatment of systemic *S. moniliformis*

infections in man (Roughgarden, 1965; Lambe et al., 1973; Wullenweber et al., 1995; Cunningham et al., 1998; Downing et al., 2001). However, information about systemic treatment of embolic foci (in man) is of little use in predicting treatment methods for the rat, in which the ecology of the organism is quite different. There is no evidence that the organism may hematogenously traverse the rat uterus or reside as a commensal in rat reproductive tract. Infections have not been reported to occur in rats derived by gnotobiotic means and shielded from proximity to conventional rats.

B. Spirochetosis (*Spirillum minus*)

The status of *Spirillum minus* as a bacterial pathogen of rats, as with *S. moniliformis,* is somewhat uncertain. Similar to *S. moniliformis,* the pathogenicity of *S. minus* for its natural hosts has been overshadowed by the interest attending this organism as a cause of rat-bite fever in man. Although illness after rat bites had been known since ancient times in India, where it is believed to have originated (Rowe, 1918), and had been known for centuries in Japan as sodoku (*so* meaning rat; *doku* meaning poisoning), it was not until 1902 that the disease was described by Japanese workers in European journals (Miyake, 1902) and given the name "Rattenbisskrankheit," or rat-bite fever. The article by Miyake (1902) led to the recognition of rat-bite fever as a clinical entity and to the realization that a number of previous accounts of "poisoning" after rat bites had been made from America and the European continent (for early reviews, see McDermott, 1928; Robertson, 1924, 1930). Before Miyake's report, it had been recognized that wild *Ratti norvegicus (Mus decumanus)* could carry spirillar organisms in the blood (Carter, 1888); slightly later (1906), mice (Breinl and Kinghorn, 1906; Wenyon, 1906). The association was left, however, until 1915, when Futaki and his coworkers discovered this cause of rat-bite fever in man (Futaki et al., 1916). Readers will recall that only 1 year before, *S. moniliformis* also had been associated with rat-bite fever in man (Schottmuller, 1914), and a controversy was initiated that continued for almost 3 decades as to the "true" cause of rat-bite fever. As so often happens in science, the passage of time and the careful documentation of clinical cases and laboratory data were to reveal that both of these organisms would be recognized as causing similar, but nonetheless different, syndromes of "fever" after rat bites (Brown and Nunemaker, 1942). The evidence was summarized at about the same time by Allbritten et al. (1940) and Farrell et al. (1939), who carefully distinguished three diseases: sodoku and streptobacillosis after rat bite, and Haverhill fever after (presumed) ingestion of

S. moniliformis (for reviews, see Adams et al., 1972; Watkins, 1946). Sodoku and streptobacillosis were differentiated on the basis of (1) longer incubation period (14 days); (2) the formation of an indurated "chancre" with regional lymphadenitis related to the bite site; (3) more regularly periodic recurrent fever; (4) maculopapular, erythematous rash spreading from the initial lesion; (5) the absence or rare occurrence of petechiae; and (6) polyarthritis in the case of sodoku (*S. minus* infection). Streptobacillosis, on the other hand, was usually shorter in onset (within 10 days) and septicemic, with morbilliform or petechial hemorrhages, endocarditis, irregularly recurrent fever, and polyarthritis. All known cases of rat-bite fever contracted from laboratory rats and mice have been of the streptobacillary type (Gledhill, 1967). Human rat-bite fever of the *S. minus* type, in India and Brazil, has a seasonal epidemiology (Chopra et al., 1939; Cole et al., 1969; Bhatt et al., 1992; Hinrichsen et al., 1992).

The organism has been identified in a number of other hosts, including dog, cat, weasel, ferret, although it is believed that carnivore infections represent zoonoses contracted from rodent carriers in the course of predation (Hata, 1912). At present only one species, *S. minus*, is taxonomically recognized (Holt et al., 1994). The older literature, however, contains references to *Spirochaeta morsus muris*, *Spirillum minor*, *Spirochaeta laverani*, *Spironema minor*, *Leptospira morsus minor*, *Spirochaeta muris*, and *Spirochaeta petit*, which are all regarded as synonyms. The frequent early references to this organism as a spirochete were based not only on morphologic criteria: indeed, it is a matter of some historic interest that Ehrlich used the mouse–*S. minus* system in experiments leading to development of salvarsan for chemotherapy of human syphilis (McDermott, 1928), that early cases of *S. minus* rat-bite fever in man were successfully treated with salvarsan (Hata, 1912; Surveyor, 1913; Dalal, 1914; Spaar, 1923), and that the organism has been advocated as a model system for the study of syphilitic infections (Stuhmer, 1929).

S. minus is classified by *Bergey's Manual* within the Spirillaceae, and it appears to be the only species in the genus that is both parasitic and, at present, incultivable on artificial media (Holt et al., 1994). As identified in fixed and stained blood smears from infected humans or laboratory rodents, the cells are Gram negative, appear short and thick but are approximately 0.5 μm in width and 1.7 to 5 μm in length, and have two to six spirals (for illustrations, see Bayne-Jones, 1931). The organism has tufts of flagellae (MacNeal, 1907; Adachi, 1921) at each end and is actively motile.

The little that is known about the pathogenicity of *S. minus* in its natural hosts (rats and mice) is related to the diagnostic inoculation of laboratory rats and mice with blood from rat-bite fever patients (or strains of

S. minus initially derived from such patients). Although some surveys of both wild and laboratory rats and mice have indicated the prevalence of the organism in the decades surrounding World War II (de Araujo, 1931; Francis, 1936; Knowles et al., 1936; Brown and Nunemaker, 1942; Jellison et al., 1949; Humphreys et al., 1950; Dolman et al., 1951), these surveys have not reported the occurrence of lesions referable to *S. minus*. At the same time, the presence of unrecognized intercurrent disease was so prevalent at the time studies were made in experimental laboratory hosts, the interpretation of clinical signs and lesions said to be caused by *S. minus* must be viewed with circumspection. In modern times, it still occurs as an infrequent cause of rat-bite fever.

The experimental disease has been studied by the intraperitoneal injection of blood into laboratory rats, mice, guinea pigs, rabbits, cats, ferrets, and rhesus monkeys. These studies achieve a fair consensus on placing the incubation period in experimental hosts at about 3 to 5 days or longer (Wenyon, 1906; Dalal, 1914; Leadingham, 1928; McDermott, 1928; Stuhmer, 1929; Leadingham, 1938). Indeed, it has been pointed out that the observation of *S. minus* in the blood of experimental hosts before 3 days postinoculation should be regarded as intercurrent (preexisting) infection of the host with *S. minus* (Robertoson, 1930; Francis, 1932, 1936; Knowles et al., 1936; DasGupta, 1938). The height of infection in terms of demonstrable organisms by dark-field examination of thick films occurs 2 to 3 weeks after inoculation of experimental rats, mice, and guinea pigs. Experimental animals may remain infective 6 to 12 months after inoculation (Leadingham, 1938). Few, if any, signs other than an initial bacteremia have been reported in mouse hosts (MacNeal, 1907; McDermott, 1928). In the rat and guinea pig, however, subcutaneous *S. minus* infection may typically also cause indurated or ulcered chancroid lesions at the inoculation site, with regional lymphadenopathy and fever reminiscent of human sodoku-type rat-bite fever (McDermott, 1928). Spirochetes frequently are demonstrable in the serous discharge from primary lesions (McDermott, 1928).

In considering the essential features of the experimental disease in rats, McDermott (1928) described the following successive phases: incubation, development of primary inflammatory lesion, and lymphadenopathy with organisms initially limited to these tissues; a secondary septicemic stage with organisms demonstrable in the blood; a latent stage in which the blood was free of (demonstrable) spirochetes and there were no obvious lesions; and, finally after many months, a tertiary stage with gummatoid lesions (abscesses and granulomas) of the lungs, spleen lymph nodes, and liver. Gummas have been described as long-term lesions by others as well (Stuhmer, 1929; Leadingham, 1938). The gummatoid lesions particularly are suspect of being otherwise induced and merit reexamination. It is worth restating, however, that none of these lesions has been reported as occurring in naturally infected rat hosts.

Diagnosis of *S. minus* infection may be made by demonstration of the actively motile organisms in wet mounts made of primary lesion exudates or peripheral blood by dark-field or phase-contrast microscopy. Blood films stained with Wright's, Giemsa, or silver impregnation stains also should be examined. Because it is recognized that septicemic phases may be brief, failure to demonstrate the organism in peripheral blood should always be supported by culture (to isolate *S. moniliformis,* if present) and animal inoculation. It is recommended that four mice and two guinea pigs be used to screen each suspect blood sample and that preinoculation blood samples be examined from the test animals to preclude use of intercurrently infected mice or rats (Brown and Nunemaker, 1942; Rogosa, 1974). Suspect blood should be inoculated subcutaneously, intraperitoneally, and intradermally (Wilson and Miles, 1975). Blood and peritoneal fluid from the test animals should be examined microscopically at weekly intervals for 4 weeks. Peritoneal fluid is usually richer in organisms than is blood (Jellison et al., 1949). An equal number of animals should be identically inoculated and examined by using an aliquot of suspect blood heat-treated to 52°C for 1 hour (Rogosa, 1974). The heart and tongue of test animals at the conclusion of the test period (4 weeks) should be used to make Giemsa-stained impression smears. Organisms often are present in fair numbers in tongue and cardiac tissue when scarce (as in latent phases) in blood or peritoneal fluid (Jellison et al., 1949; Rogosa, 1974; Wilson and Miles, 1975).

Live organisms in fresh preparation have been immobilized by immune sera (McDermott, 1928), but serologic tests have not been advocated for diagnosis because of the technical problems encountered in making antigenic suspensions. Approximately one-half of human cases are positive by *Treponema pallidum* serodiagnostic methods, and the importance of taking a serum sample early in the course for this purpose has been stressed (Woolley, 1936). Rabbits infected with *S. minus* blood suspensions have been reported to develop positive Weil-Felix reactions with *Proteus* OXK strains (Savoor and Lewthwaite, 1941).

The mechanism of zoonotic transfer of *S. minus* from infected rats via bite lesions is unknown. The organism has not been demonstrated in saliva, although this question has been investigated (McDermott, 1928; Leadingham, 1938). It has been observed in urine of human rat-bite fever cases (Leadingham, 1938), as well as isolated (by inoculation) from the urine of naturally and experimentally infected rats (Humphreys et al., 1950). The current hypothesis suggests that bleeding oral abrasions at the time of the bite may result in bite inoculation. There have been only two human cases of record after a mouse bite despite the documented widespread distribution of *S. minus* in both wild and laboratory mouse populations (Jellison et al., 1949).

The organism has not been reported in laboratory animals in modern times, and it is unknown, therefore, whether it can be vertically transmitted to gnotobiotically derived animals. That it may not be is suggested by McDermott's experiment in which *S. minus* could not be demonstrated in the progeny of infected female mice (McDermott, 1928). Chemotherapy of naturally acquired or experimental *S. minus* infection has not been reported for the rat in recent times, although human cases have been reported as successfully treated with penicillin or streptomycin (Jellison et al., 1949; Roughgarden, 1985). A battery of antibiotics has been assessed for efficacy in prevention of transmission of *S. minus* from infected donor mice to treated mouse recipients (Tani and Takano, 1958).

C. *Streptococcus pneumoniae* Infection

In the Introduction to this chapter, we explained that *S. pneumoniae* was unrecognized as a pathogen of laboratory rats during the first half of the 20th century. The organism appears not to have been encountered even in microbial surveys of respiratory flora from laboratory rats during this period (Nelson, 1930a,b; Block and Baldock, 1937), unlike similar surveys undertaken later (Matheson et al., 1955; Weisbroth and Freimer, 1969). As will be described in more detail later, in recent times *S. pneumoniae* has been observed in the context of acute respiratory episodes in rats most commonly in, but not limited to, adolescent and young adult age groups. The evidence does not suggest that these infections occurred and were unrecognized earlier but rather that fundamental changes in the ecology of the host occurred later, rendering it more susceptible to this organism.

The first report of respiratory epizootics in laboratory rats owing to *S. pneumoniae* was published in 1950 (Mirick et al., 1950). However, as late as 1963 it was written that acute respiratory episodes in rats of any age were rare (Tuffery and Innes, 1963); as late as 1969, that bacterial pneumonias, including those caused by *S. pneumoniae,* were unlikely to be encountered in properly managed facilities (Brennan et al., 1969b). It was regarded as a widespread and major primary bacterial pathogen of laboratory rats 30 to 40 years ago (Weisbroth and Freimer, 1969), but the incidence of infection of laboratory rats by *S. pneumoniae* has been much reduced in more recent years, principally as a result of common availability of barrier-reared disease-free rat stocks. The pathogenicity of certain serotypes of the organism is sufficient to serve as a primary cause of disease in the absence of physiologic stress or other infectious incitants, although it has been recognized as frequently associated with *Mycoplasma* infections of rats. Certain serotypes appear to cause essentially asymptomatic infections (Fallon et al., 1988). In conformity with the ninth edition of *Bergey's Manual* (Holt et al., 1994), the name *S. pneumoniae* has been adopted for this organism, and the former names *Diplococcus* and *Pneumococcus* are regarded as synonyms.

That the organism is widespread among commercial rat breeding stocks was documented in 1969 (Weisbroth and Freimer, 1969). In the study by Weisbroth and Freimer, 19 of 22 commercial sources were found to be offering for sale adolescent rat carriers of *S. pneumoniae* with varying degrees of clinical signs. The majority of animals from which pneumococci were isolated were asymptomatic carriers of the organism in the nasoturbinate mucosa. It was concluded that the organism is generally carried in the upper respiratory tract, particularly the nasoturbinates and middle ear in the absence of clinical signs. Of 20 rats without clinical signs, no pneumococci were isolated from pulmonary tissues, whereas *S. pneumoniae* was isolated from 17 in nasoturbinate washings (Weisbroth and Freimer, 1969).

The infection is not ordinarily inapparent, however, and clinical signs of respiratory disease with varying degrees of mortality are more typical (Kelemen and Sargent, 1946; Innes et al., 1956a; Ford, 1965; Baer, 1967; Baer and Preiser, 1969; Mitruka, 1971; Tucek, 1971; Adams et al., 1972). Serosanguinous to mucopurulent nasal exudates are frequently the first signs of clinical illness and ordinarily precede pulmonary involvement. Rhinitis, sinusitis, conjunctivitis, and otitis media are common gross lesions of upper respiratory infection. Outward bulging of the tympanic membrane may be noted as a sign of pus under pressure in the middle ear. Histologically, mucopurulent exudates are seen to overlay the respiratory mucosa of the turbinates, sinuses, Eustachian tube, nasolacrimal duct, and tympanic cavity. Acute inflammatory cells, primarily neutrophils, and more chronic inflammatory cells, including plasma cells and lymphocytes, infiltrate the mucosa and underlying submucosa of these tissues. Abundant numbers of organisms may be observed in the exudates and superficial levels of the mucosa by means of tissue Gram stains or smears (Fig. 11-1). Concomitant clinical signs include postural changes (e.g., hunching), abdominal breathing as pneumonia supervenes, dyspnea, conjunctival exudation, anorexia with loss of weight, depression, and gurgling or snuffling respiratory sounds (rales). The onset of clinical signs and lesions is more often acute or subchronic than chronic and affects rats of all ages but particularly younger age groups.

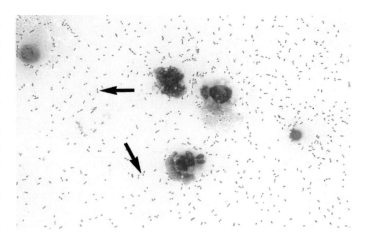

Fig. 11-1 *Streptococcus pneumoniae*: impression smear with Gram-positive diplococci occurring in pairs (arrows). Gram, 1000×. (Courtesy of Dr. Shari R. Hamilton.)

The ordinary route of progression of upper respiratory infections is by descent to pulmonary tissues. Frequently, the right intermediate lobe of the lung is the first to be affected. Initially, fibrinous bronchopneumonias in certain lobes develop rapidly into classical confluent fibrinous lobar pneumonia (Fig. 11-2). Frothy, serosanguinous fluid exudes from the cut trachea. Microscopically, the mucosa of the trachea, bronchi, and bronchioles is necrotic and eroded, with fluid and purulent exudates, often with blood, in the lumina. The alveolar capillaries are congested, and alveoli themselves are filled with blood, proteinaceous fluids, neutrophils, or variable combinations of these. Tissue Gram stains reveal abundant Gram-positive cocci in affected tissues, in microcolonies, as freely distributed single cells, and intracellularly within phagocytic cells, including neutrophils. Death may supervene at this point depending on the degree of pulmonary involvement. Frequently, however, the organism escapes to the thoracic cavity, and lesions of fibrinous pleuritis,

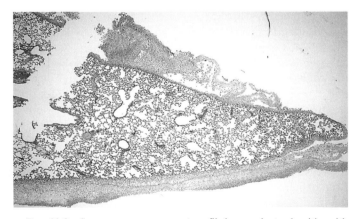

Fig. 11-2 *Streptococcus pneumoniae*: fibrinopurulent pleuritis with necrotizing pneumonia. Hematoxylin and eosin, 25×. (Courtesy of Dr. Shari R. Hamilton.)

pleural effusion, and fibrinous pericarditis are common. Organisms may ascend from the thoracic organs to become septicemic, and septicemia is a frequent terminating event. Less commonly, the organism is embolically distributed to any of a variety of organs and locations, and such complications as purulent arthritis; focal necrosis and/or infarction of the liver, spleen and kidneys; and fibrinopurulent peritonitis, orchitis, and meningitis have been observed (Baer and Preiser, 1969; Weisbroth and Freimer, 1969; Mitruka, 1971). It is known that the serum biochemistry of naturally infected rats is disturbed with increases in several serum enzymes (including glutamic-pyruvic transaminase, glutamic-oxalecetic transaminase, and lactate dehydrogenase) and α- and β-globulins (Mitruka, 1971).

Diagnosis is established by retrieval of *S. pneumoniae* from affected tissues and is supported by characteristic gross and microscopic lesions. Rats may be screened for *S. pneumoniae* by the culture of nasopharyngeal swabs (Mirick et al., 1950; Accreditation, 1971) or saline washings of the tympanic cavity and nasoturbinates (Weisbroth and Freimer, 1969), as these upper respiratory locations in carrier rats are more likely to yield the organism than are pulmonary tissues. Both quantitative nasopharyngeal cultures and quantitative PCR (compared with nasopharyngeal swabs) have been established as yielding somewhat higher numbers of positive rat carriers, but neither of the former quantitative methods was compared with nasopharyngeal wash cultures (Kontiokari et al., 2000). During diagnostic necropsy, the nares should be prepared for examination by snipping off the tip of the nose with sterile scissors; the middle ear, by reflecting the skin and ear from the external auditory meatus. Sterile saline may be introduced and withdrawn from either location by means of sterile plugged Pasteur pipettes or automatic pipette plastic tips. Both techniques follow the method of Nelson (1963). The pipette should puncture the tympanic membrane to enter the bulla. Primary clinical samples (swabs, washings, exudates, heart blood, etc.) are plated directly onto 5% blood agar. Growth is facilitated by 10% CO_2, and some isolates require microaerophilic environments (Wilson and Miles, 1975). Colonies of capsulated pneumococci on blood agar are raised, circular, and 1 to 2 mm in diameter with steeply shelving sides and an entire edge. Early in incubation, the colonies are dome-shaped and glistening (especially type 3), but by 24 to 48 hours, the center collapses because of autolysis, giving a typical concave umbilication to the top of the colony. *S. pneumoniae* colonies are surrounded by a small (alpha) zone of hemolysis with greenish (viridans) discoloration in the media. The organism is grouped with the α-hemolytic streptococci.

The morphology of *S. pneumoniae* is consistent both from clinical materials—for example, smears of exudates or tissues—and from subcultivation on artificial media. The individual cells are ovoid or lanceolate and typically occur as pairs, with occasional short chains, surrounded by polysaccharide capsules. The capsule may be demonstrated by Gin's method (Baer and Preiser, 1969). Gram-staining reactions may be inconsistent from clinical materials; however, the majority of smears and isolates are distinctly Gram positive with clear nonstaining capsular material surrounding the cells and are of value in presumptive identification.

Suspicious colonies from primary isolates should be individually picked and established in isolation by subculture on blood agar. Identity of the isolate should be confirmed by characteristic reactions in differential media, sensitivity to optochin and bile solubility, and serological typing (Nakagawa and Weyant, 1994). Twenty-four–hour glucose-enriched (Todd-Hewitt) broth subcultures may be used for the serological typing. Characteristic biochemical reactions are summarized in Table 11-2, although for convenience, the API 20 Strep kit and similar identification systems can be used for identification of primary isolates (Hansen, 2000). Optochin (hydrocuprein hydrochloride) sensitivity is established by a zone of

TABLE 11-2

CHARACTERISTIC BIOCHEMICAL REACTIONS AND OTHER PROPERTIES OF *STREPTOCOCCUS PNEUMONIAE*

Characteristic	Reaction or property
Growth on blood agar	α–Hemolysis with viridans discoloration, colony surface frequently concave after 24 to 48 hours
Morphology	Gram-positive diplococci
Carbohydrate acid from:	
Glucose	(+)
Glycerol	(−)
Lactose	(+)
Maltose	(+)
Mannitol	(−)
Raffinose	(+)
Salicin	(−)
Sorbitol	(−)
Sucrose	(+)
Trehalose	(+)
Voges-Proskauer (VP)	(−)
Litmus milk	Acid
Gelatin liquefaction	(−)
Aesculin hydrolysis	(−)
Arginine hydrolysis	(−)
Hippurate hydrolysis	(−)
Bile solubility	(+)
Optochin sensitivity	(+)
Growth in 6% sodium chloride	(−)

(Data from Buchanan and Gibbons, 1974; Cowan, 1974; Wilson and Miles, 1975.)

inhibited growth surrounding optochin (5 µg)-impregnated paper disks placed on the surface of evenly streaked blood agar plates (Bowers and Jeffries, 1955). Sensitive isolates are inhibited for a zone of about 5 mm from the edge of the disk. Occasional viridans streptococci are inhibited by optochin; however, pneumococci are rarely resistant (Accreditation, 1971; Nakagawa and Weyant, 1994). Bile solubility may be ascertained by resuspension of the organisms from 5 ml of a centrifuged overnight Todd-Hewitt broth culture in 0.5 ml of the supernate. To each of two small tubes (3 × 0.5 inches) is added 2 ml phosphate buffer (pH 7.8) and one drop of concentrated organism resuspension. Two drops of 10% sodium deoxycholate are added to one of the tubes, and both are mixed and incubated for 1 hour. Isolations that are bile soluble will clear (i.e., autolyze) during this period, and it is generally understood that bile solubility simply acts to accelerate the natural autolytic process that causes central collapse of colonies on blood agar (Wilson and Miles, 1975). Bile solubility of S. pneumoniae isolates is more variable than is optochin sensitivity (Accreditation, 1971; Wilson and Miles, 1975).

Biochemically confirmed isolates of S. pneumoniae may be typed by the use of type-specific antisera in the Neufeld-Quellung reaction (Austrian, 1974; Austrian, 1976), although serotyping is not required for the diagnosis of S. pneumoniae infection. Quellung reactions are based on swelling of the capsules surrounding individual organisms in the presence of specific antisera and may be performed by the microscopic examination of cover-slipped slides with a mixture of one loopful each of an overnight Todd-Hewitt broth culture, specific antiserum, and methylene blue (Weisbroth and Freimer, 1969). Typing of the isolate may help determine the source of the infection. Weisbroth and Freimer (1969) found that rat isolates from various colonies were usually monotypic and that the same type appeared to prevail in distinct geographical patterns. Most commonly, types 2, 3, and 19 (Mirick et al., 1950; Innes et al., 1956b; Baer and Preiser, 1969; Weisbroth and Freimer, 1969; Mitruka, 1971; Adams etal., 1972)—and less commonly types 8 and 16 (Ford, 1965; Baer, 1967)—have been clinically encountered in rats on the American continent. The pneumococcal type may also bear on pathogenicity for the rat. Type 35, for example, has been documented as being carried by rats (and even mice) as an inapparent agent in the absence of disease (Fallon et al., 1988). Virulence of S. pneumoniae is known to be highly correlated with capsular material (Nakagawa and Weyant, 1994).

Although it is recognized that rats respond with antibodies to natural infections with S. pneumoniae, this property has not received a great deal of attention as a diagnostic method, perhaps because of the ease with which the organism may be culturally retrieved. Several serologic tests have been reported (Lund et al., 1978,

Schiffman et al., 1980; Jalonen et al., 1989), but none have been recommended for use in health surveillance of laboratory rats. It has been established that certain rat stocks are more resistant to certain pneumococcal types and that young animals are more susceptible than are mature rats (Ross, 1934). Natural resistance occurs to experimental infection with pneumococcal types commonly found in rats (e.g., types 2 and 3) (Ross, 1931, 1934). As with serologic tests, although both direct and indirect immunofluorescence have been used to locate S. pneumoniae in infected tissues (Coons et al., 1942; Kaplan et al., 1950), and pneumococcal antigens have been detected by counter electrophoresis of clinical specimens (Spencer and Savage, 1976), these methods have not received attention as routine diagnostic methods in rats.

It is recognized that laboratory mice are very sensitive to the intraperitoneal injection of clinical samples or broth cultures containing S. pneumoniae (Rake, 1936; Morch, 1947; Cowan, 1974; Lennette et al., 1974), with death often occurring in 24 to 72 hours with dilutions containing as few as five microorganisms (Ford, 1965). Although animal inoculation has been advocated as a biological sieve for human diagnostic specimens, especially where speed may be of the essence (Rathbun and Govani, 1967), it has not generally been advocated as a diagnostic method in the rat (Accreditation, 1971; Nakagawa and Weyant, 1994).

Antibiograms of isolates of S. pneumoniae from the rat have generally established the susceptibility of the organism to antibiotics effective against streptococci (Osimani et al., 1972). Antibiotic sensitivity should be established before treatment, because penicillin-resistant isolates have been retrieved from naturally infected rats (Mirick et al., 1950). Most isolates are sensitive to penicillin, however, and benzathine-based penicillin has been used for effected treatments in the course of epizootics (Baer, 1967; Weisbroth and Freimer, 1969; Adams, 1972). Rats also have been effectively treated by intraperitoneal injection of type-specific (rabbit) antiserum (Mirick et al., 1950). It has been generally agreed that antibiotics are useful in the amelioration or prevention of systemic infection (septicemia, pneumonia); however, antibiotics are thought to be ineffective in eradication of organism from infected rat colonies, with rebound after withdrawal of treatment. It was concluded (Weisbroth and Freimer, 1969) that the bacteria in and on the mucosa of the upper respiratory tract are shielded from antibiotic access. The organism may be excluded by establishment of barrier colonies of gnotobiotically derived rats. No pneumococci were isolated from commercial sources of rats thus derived in the course of a microbiologic survey (Weisbroth and Freimer, 1969).

Rats have been used as the host in model systems employing S. pneumoniae in experimental infections.

Such studies have established a number of factors bearing on susceptibility, including splenectomy (Biggar et al., 1972; Boart et al., 1972; Leung et al., 1972), age (Ross, 1931, 1934), genetic stock (Ross, 1931), iron deficiency (Shu-heh et al., 1976), and pulmonary edema (Johanson et al., 1974). Other reports have explored the effect of *S. pneumoniae* infections on such host systems as nitrogen metabolism and protein synthesis (Powanda et al., 1972), serum proteins and enzymes (Mitruka, 1971), blood pH and electrolytes (Elwell et al., 1975), hepatic enzymes (DeRubertis and Woeber, 1972; Canonico et al., 1975), thyroid physiology (Shambaugh and Beisel, 1966), and the effect of antihyperlipidemic agents (e.g., clofibrate) (Powanda and Canonico, 1976).

D. *Enterococcus* sp. Infection

Although traditionally regarded as a streptococcal genus, the fecal streptococci are now classified as a separate genus, *Enterococcus*. The enterococci are catalase-negative and microaerophilic, form pinpoint colonies, usually exhibit α-hemoloysis, and are grouped in chains. Isolates are mainly serologically typed as Lancefield group D and fall into two groups, those able and those unable to grow on tellurite-containing media. Because the most common species in rodents, *E. faecium,* will not grow on tellurite-containing media and the latter should not be used for isolation of *Enterococcus,* primary isolation may be easily accomplished by incubation of fecal samples (or swabs) directly onto eosin methylene blue agar. Suspicious colonies can be picked for isolation and propagation into pure culture in typticase soy broth (Hoover et al., 1985). The enterococci may be speciated by commercial kits (e.g., API Strep or similar).

Enterococci are ubiquitous in nature and commonly carried as bacterial commensals in the human gastrointestinal tract. The most common species in humans, *E. faecium* and *E. faecalis,* are likewise most commonly carried by laboratory rats (and the origins of contamination by *Enterococcus* in rederived rats, as with *S. pneumoniae,* are thereby suggested). Most clinical episodes have been reported as occurring in essentially conventional animal care environments. In rats also, enterococci are ordinarily carried in the gastrointestinal tract as nonpathogenic commensals and may even be regarded as having a probiotic function in that their abundance competes with other forms to act as an inhibitor on ascendancy of bacterial pathogens (Hansen, 2000). In dogs, *E. faecium* may favor adhesion and colonization of *C. jejuni* in the intestine (Rinkinon et al., 2003).

In a number of instances, enterococci have been diagnosed as causal in an enteropathic syndrome of diarrhea in neonatal rats. The indicted species include *E. durans* (formerly *S. faecium durans*-2) (Hoover et al., 1985) and *E. hirae* (Etheridge et al., 1988), as well as enterococci and *Enterococcus*-like agents not further identified (Etheridge and Vonderfecht, 1992; Gades et al., 1999). Clinical episodes have tended to occur in the 6- to 12-day age group. The highest morbidity and mortality have occurred in outbreaks associated with *E. durans,* whereas milder clinical effects have been reported with *E. hirae* and other enterococcal species.

Clinical signs typically have included abdominal distension and rough, yellowish matted haircoats, with perineal scalding and perhaps ulceration of the skin of the hindfeet (Gades et al., 1999). The cages of affected litters have wet bedding and frequently are devoid of formed feces. Dams with affected litters may also have soft feces (Hoover et al., 1985), and in the case reported by Gades et al. (1999), the dam was clinically affected as well. The affected pups appear undersized and dehydrated, and those surviving the clinical phase have delayed growth and pelage development (Hoover et al., 1985). At necropsy, affected pups consistently have stomachs filled with clotted milk. Typically, the small and large intestines are distended with gas and yellowish fluid, and the other viscera are usually normal in appearance. Microscopic changes are quite minimal. Large numbers of cocci can be seen as aggregates on the surface of villi in the small intestine, but usually without a significant inflammatory component.

All investigators have emphasized that isolation of an *Enterococcus* species alone is not diagnostic for neonatal enteropathy. This is so because so many clinically normal rats (perhaps all, once removed from gnotobiotic isolators) carry commensal enterococci, and because biochemically identical isolates may differ markedly in pathogenic potential (Etheridge and Vonderfecht, 1992; Gades et al., 1999). Although virulence factors such as adhesins and cytolysins have received increasing attention, little is known about how such factors are triggered to occur in pathogenic strains, and none have been characterized sufficiently to act as diagnostic criteria (Forbes et al., 1998). It is known that once pathogenic strains are isolated in pure culture, Koch's postulates can be fulfilled by oral or intragastric inoculation of neonatal test rats (Hoover et al., 1985, Etheridge et al., 1988; Etheridge and Vonderfecht, 1992). The possible contribution to this syndrome by rat rotavirus seems not to have been considered in most reports.

E. Pseudomoniasis (*Pseudomonas aeruginosa*)

The association of *P. aeruginosa* with disease (pseudomoniasis) of laboratory rats and mice has only rarely been encountered as a naturally occurring phenomenon in immunologically competent animals. It may be fairly stated that *Pseudomonas* is a frequently occurring bacterial contaminant with only weak pathogenic potential for unstressed immunocompetent laboratory rodents. It assumes the status of a significant and serious pathogen in the course of experimental treatments that impair normal host defensive responses. The occurrence of pseudomoniasis as a consequence of such treatments is intensified by the ubiquitous and almost universal distribution of the organism in conventional stocks of rodents, other laboratory species, and man, as well as the ability of the organism to thrive in a variety of inanimate materials outside of mammalian systems that include untreated drinking water, dirty bedding, feces, water bottles, and disinfectant and detergent solutions. In the discussion that follows, information derived from experience with mouse infections has been borrowed, where appropriate for application to the rat, because of the similarity of epizootiology of the infection in these two species and comparatively greater availability of information reported with the mouse.

The notoriety of pseudomoniasis in research animals has been primarily established in connection with "early mortality" (3 to 8 days postirradiation) consequent to lethal whole-body X irradiation in the 750 to 1000 rad (Gy) range. Experience has indicated that lower dose ranges of X irradiation court infection as well. After irradiation, rats and mice infected with *Pseudomonsas* die earlier and in larger numbers than do uninfected animals (Gundel et al., 1932; Miller et al., 1950, 1951; Gonsherry et al., 1953; Hammond, 1954, 1963; Hammond et al., 1954a,b, 1955; Wensinck et al., 1957; Ainsworth and Forbes, 1961; Flynn 1963b,c; Woodward, 1963; Hightower, 1966; Taffs, 1974). Similarly, supervening pseudomoniasis has complicated burn research in laboratory animals (Ross, 1931; Millican, and Rust, 1960; Millican 1963; McRipley and Garrison, 1964; Millican et al., 1966) and has, indeed, been used as a model for burn wound contamination in man (Miyake, 1902; Flynn, 1960; Flynn et al., 1961; Millican, 1963; Teplitz et al., 1964a,b; Lemperle, 1967; Stone et al., 1967; Grogan, 1969; Fox et al., 1970; Nance et al., 1970; Neely et al., 1994; Stevens et al., 1994). *Pseudomonas* infections in laboratory rodents have complicated or been provoked by other experimental stressors, for example, cortisone injections (Millican et al., 1957), infections with other microorganisms (Flynn et al., 1961; Tsuchimori et al., 1994), antilymphocyte serum (Grogan, 1969), tumors (Flynn, 1960), cold stress (Halkett et al., 1968), neonatal thymectomy (Taffs, 1974), and injection abscesses or surgical procedures (Flynn, 1963b; Wyand and Jonas, 1967; Taffs, 1974; Yamaguchi and Yamada, 1991; Bradfield et al., 1992; Heggers et al., 1992; Dixon and Misfeldt, 1994; Wang et al., 1994). *P. aeruginosa*

infection has been used as a model system for the study of cystic fibrosis (Mahenthiralingam et al., 1994; Johansen et al., 1996), modulation of resistance by diabetes (Kitahara et al., 1995), mucosal colonization (Pier et al., 1992), and oral vaccines (Cripps et al., 1995). With the exception of a single report of otitis media in a closed mouse colony from which *P. aeruginosa* was incriminated (Ediger et al., 1971; Olson and Ediger, 1972), no convincing evidence exits to suggest that clinically overt disease syndromes may be induced by *Pseudomonas* in unstressed immunocompetent laboratory rats and mice, although clinical pseudomoniasis should be recognized as an infectious hazard for immunolgically impaired mouse and rat mutants (Johansen et al., 1993; Dietrich et al., 1996).

The ecology of *P. aeruginosa* within animal care facilities is complex and does not depend on animal infection. The organism grows well at ordinary room temperatures and may be introduced by a variety of nonsterile supply fomites, including food, bedding, inadequately sanitized cages, food hoppers, water bottles, bottle stoppers, and sipper tubes (Wensinck et al., 1957; Beck, 1963; Van Der Waay et al., 1963; Hoag et al., 1965). Considerable evidence has accumulated that the organism may be introduced by human contacts (animal care technicians, investigators) even within barrier facilities, primarily by contact with ungloved hands (Wensinck et al., 1957; Van Der Waay et al., 1963). Human carriers appear to be relatively frequent (10 to 15% of fecal samples and 90% of sewage cultures) (Hunter and Ensign, 1947; Ringen and Drake, 1952; Lowbury and Fox, 1954; Wilson et al., 1961). The organism has been cultivated from the floor, water faucets, water pipe lines, tap water, cages, water bottles, and sipper tubes within animal rooms (Wensinck et al., 1957; Beck, 1963; Van Der Waay, 1963; Woodward, 1963; Hoag et al., 1965; Trentin et al., 1966). *P. aeruginosa* thrives in quarternary ammonium disinfectant solutions (Pyrah et al., 1955, Plotkin and Austrian, 1958; Lowbury, 1959) and has been cultured from a number of assumedly sterile parenteral materials (for review, see Flynn, 1963c).

It is generally believed that the other floral components of the oropharynx and gastrointestinal tract of untreated conventional animals are effective in regulating the population dynamics of *Pseudomonas in vivo*. Repeated animal colony survey data indicate that the ordinary incidence of *Pseudomonas* carriers is no higher than 5% to 20% (Beck, 1963; Hoag et al., 1965; Hightower et al., 1966), in comparison to the near 100% incidence in lethally irradiated rats and mice from the same colonies. Ironically, reduction of these other microbial forms by antibiotic treatments or by establishment of the gnotobiotic state, in effect, removes their inhibitory effect and facilitates colonization by *Pseudomonas* to a much higher degree when the animals are exposed to it (Flynn, 1963b;

Van Der Waay et al., 1963; Hoag et al., 1965). *Pseudomonas* appears to localize in the oropharynx, upper respiratory, and rectocolonic tissues in colonized normal animals. It is thought that isolations of *P. aeruginosa* from deeper reaches of the gastrointestinal tract, that is, duodenum or ileum, represent organisms being transported through the tract, rather than localization or colonization of these sites (Hightower et al., 1966).

The organism ascends from the normal areas of localization *in vivo* to become septicemic after treatments that deprive the host of competence to inhibit invasion. It has been shown that bradykinin and proteases generated in infective foci facilitate septicemic ascension of the organism (Sakata et al., 1996). Septicemia and hematogenous distribution to such organs as the spleen, lungs, and liver generally describe the pathogenesis of pseudomoniasis, and there is some evidence that even peripheral localizations, for example, abscesses of middle and inner ears, pyelonephritis, and skin burn wound sepsis may be hematogenously initiated (Teplitz et al., 1964a,b). The gross pathology, therefore, is generally that of a septicemia, that is, congestion of splanchnic and thoracic viscera with occasional abscesses of organs with capillary nets. The usual circumstance, however, is that of fulminating infection with onset of death before pyogenic lesion development.

Experimentally, a number of factors have been shown to bear on or to modulate the host's reaction to *P. aeruginosa* (for review, see Baker, 1998). Heritable differences in susceptibility have been shown with inbred mice, for example, mice of the BALB/c strain are resistant to pulmonary infection but those of the DBA/2 strain are susceptible (Morrisette et al., 1995, 1996). Both humoral and cellular immunity arise and act to modulate the course of experimental infection (Dunkley et al., 1994; Pier et al., 1995; Stevenson et al., 1995) and are enhanced by treatments with vitamin B_2 and interleukin (IL) (Araki et al., 1995; Vogels et al., 1995). Cellular components of host resistance include T-helper type 1 cells (Fruh et al., 1995), macrophages and neutrophils (Qin et al., 1996), and tumor necrosis factor (TNF) (Gosselin et al., 1995; Morrisette, 1996).

Virulence factors associated with *P. aeruginosa* have been investigated, and it is recognized that *Pseudomonas* strains vary in virulence (Furuya et al., 1993). Such factors as the bacterial flagellum (Mahenthiralingam and Speert, 1995), pyoverdin (Meyer et al., 1996) and pyocyanin production (Shellito et al., 1992), elastase (Tamura et al., 1992), and exotoxins (Hirikata et al., 1995; Miyazaki et al., 1995; Gupta et al., 1996; O'Callaghan, 1996; Pittet et al., 1996; Tang et al., 1996) have been shown to be important elements that determine virulence and have been characterized in the course of model systems of surgical wound infection, respiratory infection, or surgical sepsis.

Diagnosis is established by the characteristic case history that usually involves immunosuppressive pretreatment or treatments that induce a shocklike state, and by isolation of *P. aeruginosa*. It may be isolated from heart blood or spleen of septicemic cadavers. The organism is aerobic and grows readily on blood agar, usually producing pigments that darken the agar initially and also a powerful hemolysin that may clear the entire plate. The colonial morphology is irregularly round, 2 to 3 mm in diameter, usually with a smooth matt surface, butyrous consistency, and floccular internal structure. Colonial variants that include mucoid, rough umbonate, rugose, or smaller coliform-like forms are not uncommon. The organism is a typical Gram-negative rod, 1.5 to 3 μm in length and 0.5 μm wide. Suspect colonies should be isolated in pure culture and confirmed by a number of tests that characterize the pigment-producing and biochemical properties of *P. aeruginosa* (Table 11-3). Although taxonomy of *P. aeruginosa* has not been in recent flux, it may be found

TABLE 11-3

CHARACTERISTIC BIOCHEMICAL REACTIONS AND OTHER PROPERTIES OF
PSEUDOMONAS AERUGINOSA

Characteristic	Reaction or property
Morphology	Gram-negative rods
Motility	Motile
Pigments	Pyocyanin (blue-green), fluorescein (yellow)
Growth at 42°C	(+)
Growth at 5°C	(−)
Oxidase/catalase	(+/+)
Growth on blood agar	2- to 3-mm colonies, early discoloration, later hemolysis
Nitrate reduction	(+)
Growth on brilliant green	(+), reddish colonies, alkaline (red) agar
Growth on glycerol-peptone	(+), 2- to 3-mm colonies, blue-green pigment in medium
Gelatin liquefaction	(+)
Urease	(+)
Arginine hydrolase	(+)
Lysine decarboxylase	(−)
Carbohydrates, acid from:	
Glucose	(+)
Lactose	(−)
Maltose	(−)
Mannitol	(+)
Salicin	(−)
Sucrose	(−)
Xylose	(+)
Dulcitol	(−)
Inulin	(−)
Raffinose	(−)

(Data from Buchanan and Gibbons, 1974; Cowan, 1974; Lennette, 1974; Wilson and Miles, 1975.)

in the older literature under the names *P. pyocyanea, Bacterium aeruginosum, B. pyocyaneum, B. aerugineum, Clostridium aeruginosis,* and *B. pyocyaneus.*

More typically, however, the organism is suspected in dying animals, or monitoring programs are used to screen laboratory animals or component equipment. A number of specialized isolation or detection systems for presumptive cultural identification of *P. aeruginosa* oriented to its biochemical properties have been developed for such purposes. Such studies have emphasized the desirability of enrichment broths for primary isolation followed by subculture to selective or differential semisolid media. Direct inoculation of agar plates for primary isolation from test materials (e.g., water, swabs, feces, or tissue samples)—even when using such recommended media as brilliant green agar, *Pseudomonas* agar (King's medium B-Flo agar), glycerol-peptone agar (King's medium A), or Wensinck's glycerol agar—have been shown to yield fewer isolations than when the same test materials are selectively enriched in broths before subculture on agar (Hoag et al., 1965). A number of broths for primary culture have been explored, including Wensinck's glycerol enrichment broth (Wensinck et al., 1957), tetrathionate broth, and Koser's citrate medium, which may be enriched by addition (10% by volume) of brain-heart infusion broth (Beck, 1963; Woodward, 1963). There is some evidence favoring Koser's citrate medium as the broth of choice for this purpose (Hoag et al., 1965). King's medium B slants may be used to demonstrate Wood's light fluorescence of fluorescein pigments diffusing from surface growth of several *Pseudomonas* species, including *P. aeruginosa.* There is no convincing evidence to consider *Pseudomonas* species (other than *P. aeruginosa*) as pathogens of laboratory rodents, even though they are commonly isolated in the course of health surveillance testing.

Presumptive determination of the presence of *P. aeruginosa* may be made by the observation of water-soluble greenish-blue pigments (pyocyanin) in the medium, which may be extracted in chloroform. The latter method, without subculture, has been used for routine screening of water and swabs of water bottles and bottle parts (Flynn and Greco, 1962; Beck, 1963; McPherson, 1963). Other test materials, including oropharyngeal swabs, rectal swabs, feces, and tissue samples, should be subcultured from 24-hour broth primary cultures onto agar plates (e.g. brilliant green or glycerol-peptone agars). On brilliant green, *P. aeruginosa* colonies are reddish and a pronounced red color (non-lactose-fermenting, basic pH) develops in the agar. On glycerol-peptone agar, the greenish-blue pigment may be seen diffusing through the medium. Colonies should be confirmed as oxidase positive by the development of a purple color within 15 seconds when inoculating loop samples are applied to filter paper disks soaked in Kovac's reagent (Hightower et al., 1966).

Final confirmation of individual isolates should be made by establishment of a biochemical profile consistent with the summary in Table 11-4. Several systems have been established for the serological (O and H antigens) and pyocin typing of *P. aeruginosa* (for reviews, see Cowan, 1974; Lenette et al., 1974; Wilson and Miles, 1975; Urano, 1994); however, little correlation between type and pathogenicity or between type and origin of animal isolate has emerged from these studies (Matsumoto et al., 1968; Maejima et al., 1973). These systems and bacteriophage typing are being replaced by pulsed field gel electrophoresis typing (Speert, 2002).

Investigations of the method of choice for screening rodent populations have generally established a higher rate of success from culture of oropharyngeal swabs than fecal samples, and the former is recommended for monitoring programs designed to detect shedder or carrier animals (Trentin et al., 1966). Similarly, screening of water bottle contents after 2 to 4 days and of water bottle parts have generally yielded higher percentages or indications of carrier animals than do fecal cultures (Beck, 1963; Flynn, 1963a,c; Woodward, 1963; Hoag et al., 1965), although it has been shown that water bottles may frequently fail to yield *Pseudomonas* from cages with mice having positive oropharyngeal cultures, particularly when the bottles are well sanitized (Trentin et al., 1966). Serological tests have been developed to detect antibodies to *P. aeruginosa* (Johansen et al., 1993), but their utility for diagnostic surveillance programs has not been explored.

A variety of approaches has been explored to cope with the problems of pseudomoniasis in immunosuppressed or traumatized animals. Although there are some antibiotics (e.g., gentamacin) (Summerlin and Artz, 1966; McDougall et al., 1967) and bacterocidins (e.g., pyocin) (Merrikin and Terry, 1972) that have been used to extend the life of X-irradiated animals (Wolf et al., 1965), the approach of mass treatments using antibiotics for control of *Pseudomonas* within rodent colonies has been generally concluded to be ineffective (Hammond, 1963; Hoag et al., 1965; Lusis and Soltys, 1971). Similarly, the use of bacterins for active immunization and passively administered anti-*P. aeruginosa* hyperimmune sera have been explored and found to have limited value for control of pseudomoniasis in immunosuppressed animals (McDougall et al., 1967; Lusis and Soltys, 1971).

The treatment of drinking water with sodium hypochlorite to provide 10 to 12 ppm chlorine, adjusting water to pH 2.5 to 2.8 by the addition of concentrated hydrochloric acid, or, preferably, both has been found to suppress *Pseudomonas* sufficiently to prevent its effects in irradiated or burned animals (Beck, 1963; Woodward, 1963; Hoag et al., 1965). Potable tap water chlorinated to 2 ppm at the source has been found to drop to less

TABLE 11-4

CHARACTERISTIC BIOCHEMICAL REACTIONS AND OTHER DIFFERENTIAL PROPERTIES OF RODENT ENTEROBACTERIACEA

Characteristic	EC	ET	YE	CF	S	EH	P	H	SM	EA	KA
Catalase	+	+	+	+	+	+	+	+	+	+	+
Oxidase	−	−	−	−	−	−	−	−	−	−	−
Motility	+	+	+	+	+	+	+	+	+	+	−
Growth in potassium cyanide	−	−	−	+	−	−	+	+	+	+	+
Simmons citrate	−	−	−	+	+	+	+	+	+	+	+
MR	+	+	+	+	+	−	+	−	+	−	−
VP	−	−	−	−	−	+	−	+	+	+	+
Indole	+	+	−	−	−	−	+	−	−	−	−
Gelatin liquefaction	−	−	−	−	−	+	+	−	+	+	−
Urease	−	−	+	−	−	−	+	−	−	−	+
Phenylalanine	−	−	−	−	−	+	+	−	−	−	−
Hydrogen sulfide on triple sugar iron	−	+	−	+	+	−	+	−	−	−	−
Lysine decarbosylase	+	+	−	−	+	−	−	+	+	+	−
Ornithine decarboxylase	+	+	+	+	+	−	−	+	+	+	−
Carbohydrates, acid from:											
Adonitol	−	−	−	−	−	−	−	−	v	+	+
Arabinose	+	−	+	+	+	+	−	+	−	v	v
Dulcitol	v	−	−	v	+	−	−	−	−	+	+
Glycerol				v	v	+	+	+	+	+	+
Inositol	−			−	v	v	−	−	v	+	+
Lactose	+	−	−	v	−	v	−	−	−	+	+
Maltose	+	+	+	+	+	+	+	+	+	+	+
Mannitol	+	−	+	+	+	+	−	+	+	+	+
Raffinose					+			−	−	+	+
Rhamnose	v	−	−	+	+	+	−	+	−	+	+
Salicin	v	−	−	v	−	v	v	−	+	+	=
Sorbitol	v	−	+	+				v	v	+	+
Sucrose	v	−	+	v	−	+	+	v	+	+	+
Trehalose	+	−	+	+	+	+	+	+	+	+	+
Xylose	v	−	+	+	+	+	v	v	v	+	+

EC indicates *Escherichia coli*; ET, *Edwardsiella tarda*; YE, *Yersinia enterocolitica*; CF, *Citrobacter freundii*; S, Salmonella, Arzona; EH, *Erwinia herbicola*; P, Proteus; H, *Hafnia alvei*; SM, *Serratia marcesens*; EA, *Enterobacter aerogenes*; KA, *Klebsiella aerogenes*; MR, methyl red; VP, voges proskauer; v, variable reaction depending on isolate. (Data from Cowan, 1974; Lennette et al., 1974.)

than 1 ppm at the tap and to have little inhibitory effect on *Pseudomonas* (Trentin et al., 1966; Les, 1973). Similarly, chlorine at the 10-ppm level initially is effective in sanitizing water bottles to eliminate *Pseudomonas* but decreases to 6 ppm by 24 hours, 2.5 ppm by 48 hours, and 0.25 ppm by 72 hours and resembles tap water (McPherson, 1963). Such bottles, with sodium hypochlorite alone, may become recontaminated by 48 to 72 hours if the mice are oropharyngeal carriers of *Pseudomonas*.

It was found earlier that *Pseudomonas* is quite sensitive to low pH (2.5 to 2.8) and that chlorinated-acidified drinking water was effective in eliminating the *Pseudomonas* carrier state (and substantially suppressing coliform and *Proteus* populations) in laboratory mice (Schaedler and Dubos, 1962). The combination of acidified and chlorinated water is recommended for mass treatment as a standard method for control of pseudomoniasis in immunocompromised animals. Palability studies have indicated little, if any, preference for untreated tap water

compared with chlorine-acid-treated drinking water (McPherson, 1963; Thunert and Heine, 1975). However, at least one study has indicated retarded growth on the basis of reduced intake in some mouse strains (Les, 1968a). Several studies through reproductive lifetime cycles have failed to demonstrate deleterious effects of such treatments (Mullink and Rumke, 1971); indeed, most parameters of reproductive performance have increased (Schaedler and Dubos, 1962; McPherson, 1963; Les, 1968b). The low pH has no demonstrable effect on rat tooth enamel (Tolo and Erichsen, 1969). It is known that a small proportion of rats and mice on treated water may continue to shed *Pseudomonas* in feces (Hoag et al., 1965; Trentin et al., 1966). If it is desirable to do so, such animals may be detected by fecal or oropharyngeal screening and removed from the colony. The decision to recommend initiation of acidification-chlorination treatment of drinking water in rat colonies resolves itself to an assessment of the risk of pseudomoniasis in the course of experimentation and should be determined by the nature of the research program. No specific disease ameliorative or preventive effects with other defined rat pathogens have been shown for chlorinated-acidified drinking water.

F. *Helicobacter* Infection

The index cases of *Helicobacter* infections in mice were reported as incidental findings of hepatitis in 1992 (Fox et al., 1994; Ward et al., 1994a). It was determined that the lesions were caused by a newly recognized helical bacterium, later designated *Helicobacter hepaticus,* isolated from affected livers. The agent could be passed to satisfy Koch's postulates as the incitant of the hepatitis and, as was learned later, of the hepatomas induced as late-stage sequelae in certain mouse strains (Ward et al., 1994a). These cases opened a new chapter in diseases of laboratory animals and served as the introduction to an intensive diagnostic and research effort to characterize the nature and biology of this new genus of gastrointestinal pathogens (for reviews, see Fox and Lee, 1997; Weisbroth, 1999).

Based only on morphologic criteria, *Helicobacter*-like spiral bacteria had been noted earlier in rat cecum (Davis et al., 1972; Phillips and Lee, 1983; also see references in Solnick and Schaner, 2001). *Helicobacter trogontum* isolated from the rat (Mendes et al., 1996) has not been reported as a natural infection of other rodents nor has it been associated with either overt or clinically inapparent disease in the rat. Experimentally, the agent can be used to infect mice (Mendes et al., 1998, 1999). The rat has been reported as being not susceptible to infection by *H. hepaticus* (Ward et al., 1994b), although several other *Helicobacter* species usually associated with mice, for example, *H. muridarum* and *H. bilis* have been reported from rats (Phillips and Lee, 1983; Lee et al., 1992). In fact, the only report of clinically apparent disease in rats owing to *Helicobacter* occurred as typhylocolitis in athymic rats infected with *H. bilis* (Haines et al., 1998).

The original diagnosis of murine helicobacteriosis was based on histopathologic demonstration of silver-staining spiral bacteria in clinically incidental cases of hepatitis by use of Warthin Starry or Gomori's methenamine silver (GMS) stains (Ward et al., 1994a). Histopathology remains an important tool for evaluation of lesioned tissues, but it has not been adopted as a useful screening method because it is not sensitive: even in susceptible mouse strains, liver lesion development is variable and may not exceed 10% of infected individuals (Ward et al., 1994a). Resistant strains do not develop lesions to focus the microscopic search for organisms; in addition, the method does not allow for discrimination between *Helicobacter* species and cannot discriminate between spiroform commensals and *Helicobacter* forms. On balance, histopathology is useful in describing morphologic changes attributable to *Helicobacter* and for confirming clinical diagnosis in lesioned animals, but not as a screening test in clinically normative rodents.

Microbiologic isolation of murine *Helicobacter* species involves use of Brucella agar plates with trimethoprin, vancomycin, and polymyxin (TVP) to inhibit contaminating organisms from clinical materials (tissue or fecal suspensions) and a specialized microaerophilic incubation environment (nitrogen, hydrogen, CO_2 in a ratio of 90:5:5) (Shames et al., 1995). It was determined that unwanted overgrowing microbial contaminants can be further reduced by passing the clinical suspension through a 0.65-μm filter, which retains many of the contaminants but allows passage of *H. hepaticus* as well as the larger helicobacters, for example, *H. bilis, H. rappini,* and *H. trogontum* (which are retained by a 0.45-μm filter) (Fox and Lee, 1997). Isolated *Helicobacter* cultures are identified by their spreading growth films with spiroform Gram-negative bacteria that are catalase, oxidase, and urease positive. Some species (*H. rodentium*) are urease negative, but urease production has not proven as useful a feature for the diagnosis of murine helicobacteriosis as it has for human *H. pylori* infections. Isolation in pure culture is required to enable speciation of isolates on the basis of biochemical criteria (see tables in Mendes et al., 1996; Fox and Lee, 1997). When strains are isolated that do not match existing profiles, a new species may have been encountered (Dewhirst et al., 2000b). In summary, culture has been widely used as a screening method, but important limitations include the obscuring presence of contaminating microbial overgrowth and low *Helicobacter* populations, both of which bear on the reliability of negative results (Weisbroth, 1999b).

Most of what is known about the immunological response to *Helicobacter* infection is based on investigative experience with mice, which is used here as a model for immune response in the rat. The immune response to *Helicobacter* is prompt and appears proportional to the intensity of infection or degree of tissue invasiveness but may not confer protection (Fox and Lee, 1997). Unlesioned and naturally resistant mouse strains may be colonized by *H. hepaticus* without development of a detectable antibody response (Ward et al., 1996). Both whole-cell sonicates (Fox and Lee, 1997) and outer cell membrane proteins (Livingston et al., 1997; Wharry et al., 1998) have been used to prepare ELISA antigens. In parallel with lesion development and serum alanine transaminase levels, antibodies have first been detected in naturally infected animals (of susceptible strains) from enzootically infected colonies at about 6 months of age. The intensity of the optical density of the ELISA titers increases as the animals age, reaching maximal levels at about 12 to 18 months of age (Fox and Lee, 1997). The ELISA test using *H. hepaticus* antigens appeared quite specific, recognizing antibodies to *H. hepaticus* but not to *H. bilis* or *H. muridarum* (Livingston et al., 1997).

Serology has received some attention as a screening method (Whary et al., 2000a) but, similar to cultural isolation, has important limitations that inhibit its endorsement as a generally useful diagnostic tool. Because antibody levels to *Helicobacter* are generally proportional to the intensity of microbial challenge, positive serology is probably reliable as an indicator of exposure to antigens used in the test. The problem is that negative serology is not a reliable result for any of a number of reasons, including (1) innate host resistance that prevents lesion development or significant antigenic challenge; (2) initially low populations of *Helicobacter* in the gastrointestinal reservoir; (3) serum samples that may be drawn early in the course before development of detectable antibody levels; and, perhaps most important, (4) the posibility of false negative owing to infection by *Helicobacter* species that stimulate antibodies not recognizing antigens used in the test (i.e., issues of specificity). One practical problem is that validated antigens are not generally available for sustenance of testing programs.

The PCR has been developed (Battles et al., 1995) as the most useful single screening test for detection of murine helicobacters from clinical materials, including tissue and fecal specimens (Shames et al., 1995; Beckwith et al., 1997; Riley et al., 1997; Hodzic et al., 2001) and retrospective material, including wet, fixed tissues, paraffin embedded tissues, and even stained histologic specimens. The method is based on the detection of unique and subsequently amplified 16S rRNA gene sequences of the agent extracted from the sample material. The test is rapid (1 day) and, by use of restriction endonucleases, can be used to speciate *Helicobacter* DNA detected in generic tests (Riley et al., 1996). The procedure is specific to the one or more *Helicobacter* sequences specified by the primers used for amplification and is not complicated by the presence of contaminating microorganisms in the sample material. One important property of the PCR test is extreme sensitivity for detection of low numbers of *Helicobacter* in the sample specimens. In direct comparisons, PCR tests detected as positive 30% to 50% more specimens found negative from animals also tested by culture and/or histopathology or electron microscopy (Shames et al., 1998). Furthermore, the method permits sample pooling up to dilutions of 1:5 to 1:10 of positive material without significant loss of sensitivity, thus economizing on the cost of testing (Whary et al., 2000a,b). The combination of PCR and serology for

screening programs appears to offer superior detection rates (Whary et al., 2000a).

Because of concern about the potential of *Helicobacter* infections to spread within both breeding and user facilities and to adversely impact research programs, treatment modalities have been explored as a means of eradicating *Helicobacter* from asymptomatic infected carriers and of reducing the clinical manifestation in susceptible mouse strains. No direct experience with the effect of treatment has been reported on rat infections. It has been shown that 2-week treatments of amoxicillin alone, dosed via the drinking water, were effective in eliminating (or preventing) *H. hepaticus* infection in weanlings but not in older mice with established enteric colonization (Russell et al., 1995). Triple treatment with combinations of amoxicillin, metronidazole, and bismuth (known to eradicate *H. pylori* in humans and *H. mustelae* in ferrets) given by gavage 3 times daily for 2 weeks appeared effective in eradicating *H. hepaticus* infections in 6- to 8-month-old susceptible inbred mice (Foltz et al., 1995), but the labor-intensive nature of the dosing regimen limits usefulness of this individual treatment. More promising was the *ad libitum* oral treatment with amoxicillin triple treatment formulated into flavored dietary wafers use to eradicate *H. hepaticus* in 6- to 10-month-old naturally infected susceptible inbred mice (Foltz et al., 1996). As with the gavage method, the dietary wafer modality appears most useful to clear small, defined groups of the infection. Both modalities share the general problem with antibiotic treatments to suppress rather than to fully eradicate, thus setting the stage for eventual rebound of residual *Helicobacter* populations after withdrawal of treatment. An additional area of concern is the uncertain susceptibility of other *Helicobacter* species to the treatments discussed above for *H. hepaticus*. It has been reported (Shomer et al., 1998) that although diarrheas owing to dual infection by *H. bilis* and *H. rodentium* (in *scid* mice) could be clinically ameliorated, the *Helicobacter* infections were not eradicated after 2-week treatments with dietary wafers.

Not a great deal is known about other means to eradicate *Helicobacter* infections. Successful eradication of *Helicobacter* infections should be oriented to principles familiar with eradication of other highly infectious murine pathogens, such as depopulation (or rederivation) of affected units, sanitization of empty rooms and repopulation with same strain stocks rederived by caesarean section or embryo transfer, or use of stocks free of *Helicobacter* brought in from the outside. Stringent maintenance regimens for admission of personnel and reprocessed equipment and supplies should be put in place to prevent reinfection of the clean stock.

A number of qualifiers bear on the pathogenic significance of rodent *Helicobacter* infections. It would seem warranted to regard the rodent helicobacters as essentially opportunistic enteric bacteria with only low-grade potential to cause disease. That said, reports are steadily accumulating that detail enterohepatic disease syndromes in variably immunocompetent but susceptible inbred mice, as well as in immunodeficient mouse and rat strains. Complicating the issue is the proliferating diversity of microbial species and strains within the genus *Helicobacter*. In general, the *Helicobacter* species are fairly host specific, but not entirely. For example, both *H. bilis* and *H. muridarum* of the mouse have been isolated from rats (Fox and Lee, 1997), and *H. bilis* has been reported as a cause of enteric disease in athymic rats (Haines et al., 1998). Most laboratories working with these agents find occasional isolates with profiles that do not neatly match one of the characterized species, thus not all of the variants have been carefully speciated as yet (see Dewhirst, 2000b). Conversely, the entire taxon of *Flexispira (Helicobacter) rappini* appears to be undergoing dissolution by reassignment of its isolates to other, established *Helicobacter* species (Dewhirst et al., 2000a).

The *Helicobacter* species also vary in pathogenicity, even in susceptible hosts. Thus, *H. bilis* and *H. hepaticus* have been documented as pathogens on a number of occasions in a wide number mouse strains, as well as *H. bilis* as a pathogen of immunodeficient rats, and should therefore be regarded as frank pathogens of susceptible hosts. On at least two occasions (Phillips and Lee, 1983; Lee et al., 1992), *H. muridarum* (a species noted above as isolated from rats) has been microscopically associated with mild gastritis in naturally infected mice. There is no strong, convincing evidence, at present, to indict *H. trogontum* as a pathogen of laboratory rats under circumstances of natural infection. Nonetheless, the (tenuous) designation of "nonpathogen" should be cautiously applied. Given the propensity of this genus to cause enterohepatic disease in such a widely diverse range of mammalian hosts, it may well only be that the right mix of agent/host/milieu has not been encountered to eventuate and be recognized as a cause of clinical disease. A good example of this potential was reported in which *scid* mice developed hypertrophic colitis with diarrhea when dually infected with *H. bilis* and *H. rodentium*. When infected with *H. rodentium* alone (a species previously not known to be pathogenic), typhylocolitis was caused experimentally in same strain host mice (Shomer et al., 1998).

G. *Clostridium piliforme* Infection (Tyzzer's Disease)

In the interval between the first and second editions of this text, greater advances have been made in understanding the pathobiology *C. piliforme,* and in the technology of its diagnosis, than in any other infection in the rat. For example, development of generally available antigen systems to support diagnostic serology programs have enabled the distinction to be made between (asymptomatic) infection with the causative agent, *C. piliforme,* and the clinical and pathologic consequences that it may cause (Tyzzer's disease). Tyzzer's disease of the rat occurs as an acute (or latent) generally fatal epizootic with cardinal clinicopathologic features of megaloileitis, focal necrosis of the liver, myocarditis, and diarrhea. It has been observed preeminently in the laboratory mouse (Gard, 1944; Rights et al., 1947; Porter, 1952; Saunders, 1958; Carda et al., 1959; Tuffery, 1966; Friedmann et al., 1969; Kaneko, 1969; Peace and Soave, 1969; Francis, 1970; Meshorer, 1970), but naturally occurring disease has been described in the laboratory rat (Fujiwara et al., 1963a; Takagaki et al., 1968; Stedham and Bucci, 1969; Jonas et al., 1970; Stedham and Bucci, 1970; Yamada et al., 1970), gerbil (Carter et al., 1969; White and Waldron, 1969; Veazey et al., 1992), hamster (Fujiwara et al., 1968; Takagaki et al., 1974; White and Waldron, 1969), guinea pig (Boot and Walvoort, 1994), rabbit (Allen et al., 1965; Cutlip et al., 1971; Ganaway et al., 1971b; Van Kruiningen and Blodgett, 1971) and larger animals such as primates (Niven, 1968), cats (Kovatch and Zebaith, 1973; Ikegami et al., 1999a), red pandas (Langan et al., 1990), calves (Ikegami et al., 1999b), horses (Hall and Van Kruiningen, 1974; Thomson et al., 1977; Hook et al., 1995; Fosgate et al., 2002), and immuncompromised humans (Smith et al., 1996). The condition is named after Ernest Tyzzer of Harvard, who initially reported the disease as a fatal epizootic of Japanese waltzing mice, *Mus bactrianus* (Tyzzer, 1917; Kerrin, 1928). Tyzzer's original report of the disease, with descriptions of its histopathology, still stands as the fundamental study. The disease has emerged, with worldwide distribution, as one of the major infectious diseases of the rat.

Ganaway et al. (1971a) pointed out that only 2 articles had been reported on the disease between 1917 and 1933 and 6 more articles between 1933 and 1959. Of 35 articles between 1959 and 1971, 23 had appeared since 1965. In that same context, the disease was recognized only in *Mus musculus, Mus bactrianus,* and their crosses before 1965, whereas naturally occurring Tyzzer's disease has since been recognized in a wide range of mammalian species. It is uncertain if the disease is emerging or, rather, being subjected to more accurate surveillance. Certainly, the advent of generally available serologic reagents has demonstrated the common occurrence of serologic positivity in rodent health surveillance programs, especially the seeming occurrence of *C. piliforme* infection in the absence of Tyzzer's disease (Hansen et al., 1990; Motzel et al., 1991; Hansen and Skovgaard-Hansen, 1995). The degree to which this disease has interfered with the interpretation of experimental results is difficult to assess for several reasons; however, it is thought to be substantial. On the one hand, it is likely that the disease frequently has gone unrecognized by investigators, and on the other, it is a general fact that ruined experiments are seldom published (for documented experimental complications, see references in Lindsey, 1991). In England, Tyzzer's disease is claimed to have ruined more carcinogenesis experiments than has any other single disease (Wilson and Miles, 1975).

The taxonomic status of *C. piliforme,* the Tyzzer's disease agent, has been greatly clarified since the first edition of this text. Wilson and Miles (1975) proposed the name *Actinobacillus piliformis* based on the morphologic analogy of monilial thickenings at irregular locations in the vegetative rods seen in *Streptobacillus (Actinobacillus) moniliformis (muris)*. The analogy appears to end there, however, and the name originally proposed by Tyzzer *(Bacillus piliformis)* has been retained until recently. On the basis of 16S rRNA sequence analysis, the agent has been assigned to the clostridia and given the name *C. piliforme* (Duncan et al., 1993). The organism stains poorly with hematoxylin, although it is generally recognizable, with some experience, in hematoxylin and eosin–stained tissue sections. Methylene blue or Giemsa stains are satisfactory for demonstration of the organism in smears and tissue sections. The silver impregnation stains (e.g., Warthin-Starry or GMS) similarly are satisfactory and may be the stains of choice, for demonstration of the organism in tissue sections. The organism is periodic acid–Schiff (PAS) positive and Gram (Brown and Brenn) negative. Excellent color illustrations may be found in the publications by Jonas et al. (1970) and Fujiwara et al. (1963a).

The morphology of the organism is pleomorphic, but general agreement has been reached that the vegetative phase of long slender rods, approximately 0.5×8 to 10 μm (or more) are the obligately intracellular forms seen in large numbers (sheaves or bundles) within infected cells. Sporelike monilial thickenings are seen less frequently in tissue sections from liver and intestine, but more frequently in yolk sac preparations from infected embryonating chicken eggs (Craigie, 1966a). Although Tyzzer believed the organism to be nonmotile, observation of motility by phase-contrast microscopy (Allen et al., 1965; Craigie, 1966a), supported by demonstration of peritrichous flagellae using electron microscopy (Fujiwara et al., 1968; Jonas et al., 1970; Kurashina and Fujiwara, 1972), has led to its being accepted as motile.

It is generally concluded that *C. piliforme* may not be successfully cultivated in cell-free media (Thunert, 1984). The two reports to the contrary (Kanazawa and Imai, 1959; Simon, 1977) have not been supported by other investigators. The organism may be propagated in embryonating eggs by yolk sac inoculations of 10- to 12-day chick embryos (Allen et al., 1965; Craigie, 1966a; Fries, 1977b) without loss of pathogenicity. Preferentially, the organism may be cultivated in a number of primary and continuous cell lines, for example, mouse fibroblast 3T3, mouse L-929, mouse hepatocyte NCTC 1469, and human carcinoma Caco-2 (Franklin et al., 1993; Kawamura, 1983a,b; Riley et al., 1990; Spencer et al., 1990).

The viability of the vegetative phase, that is, the bundles of long (8 to 40 μm) slender rods seen in infected cells, is extremely unstable (Fujiwara, 1967). Infective preparations made from heavily infected livers, even when injected into cortisone-prepared substrate hosts, are likely to fail if made from a donor animal that has been dead more than 2 hours. Preparations made from infected yolk sacs lose substantial activity after 15 to 20 minutes at room temperature and even more at 37°C. The vegetative phase is killed by heating to 45°C for 15 minutes. Infectivity is completely lost after 24 hours at 4°C. Preparations of the vegetative form may be stabilized by freezing at −70°C.

The spore form is considerably more stable. Spores are about 0.5 wide $\times$ 3 μm long and occur as terminal endospores of vegetative cells or may be free (Riley et al., 1994). They will survive 56°C temperatures for 1 hour and repeated cycles of freeze and thaw, but they will not survive 80°C or greater for 30 minutes or more (Ganaway, 1980). Spores are resistant to ethanol, phenolic germicidal detergents, aldehydes, and quarternary ammonium disinfectants, but spores are inactivated by iodophores, formalin, peracetic acid, and sodium hypochlorite solutions (Ganaway, 1980; Itoh et al., 1987; Hansen et al., 1990, 1992b; Riley et al., 1994). Infectivity of spore forms may be retained in dead yolk sac-inoculated chicken eggs for 1 year at room temperature (Craigie, 1966a) and for similar periods in infected mouse and rabbit bedding at room temperature (Tyzzer, 1917; Allen et al., 1965; Christensen, 1968).

The consensus is that natural infections are initiated by oral ingestion of spore forms from the bedding contaminated by infected animals shedding spores in the feces. Both Tyzzer (1917) and Gard (1944) used contaminated cages for transmission studies. Transmission studies directly exploring the oral route have produced varying results. Oral infections in the young rat (Jonas et al., 1970; Kurashina and Fujiwara, 1972), rabbit (Allen et al., 1965), and gerbil (Carter et al., 1969) have generally been successful, with characteristic production of both intestinal and liver lesions. In the mouse, Gorer (1947) used dried liver blocks as a means of stabilizing the sporulated form, which were ground up and placed in cooked cereal for oral transmission studies. Interestingly, Fujiwara et al. (1973a) reported higher rates of success with raw liver blocks than with ground-up liver suspensions and proposed that cannibalism might serve for transmission and maintenance of the cycle in breeder colonies. Tyzzer's disease was induced via gastric intubation of untreated suckling mice up to 6 to 7 days old, with enhanced transmission in cortisone-treated weanlings and adults starved 24 to 48 hours before infection (Fujiwara et al., 1973). It has been suggested that variability in transmission studies may relate to fragility of the vegetative form *in vitro* and/or failure to administer sufficient spores for oral transmission (Ganaway et al., 1971b). Support for this view was shown in the higher infectivity rate obtained in nonfasted adult mice inoculated with 10^7 organisms directly into the duodenum or cecum (38% and 50%, respectively), than intragastrically (13%), and also by the demonstration that the infectivity rate was dose related (Fujiwara et al., 1973; Takagaki and Fujiwara, 1974). Experimentally, it has been demonstrated that spores are shed for 1 to 2 weeks after oral infection and that rodents exposed to contaminated feces can contract Tyzzer's disease (Waggie et al., 1984; Itoh et al., 1988; Franklin et al., 1994).

Little is known about antecedent factors predisposing the pathogenesis and history of natural infections. Clinically apparent Tyzzer's disease may appear in epizootic proportions with seemingly little in the way of predisposing stress, but more commonly the disease appears to be precipitated into overt, but sporadic, clinically apparent disease by a number of factors that impair immunocompetence of carrier hosts. The serologic evidence suggests that the dominant mode of infection may be a clinically inapparent, immunizing exposure with only brief duration of infectivity, as may be suggested by the frequent failure to induce clinical Tyzzer's disease by immunosuppressive stress tests on serologically positive rats. Precedent experimental treatments that include tumor transplantation (Tyzzer, 1917; Craigie, 1966b; Takenaka and Fujiwara, 1975), administration of steroid or other immunosuppressive drugs (Takagaki et al., 1963; Fujiwara et al., 1964b; Niven, 1968; Yamada et al., 1970; Taffs, 1974; Fries, 1979a), leukocyte injections (Taffs, 1974), and X irradiation (Takagaki et al., 1963; Takagaki et al., 1966; Taffs, 1974) have been either observed in the course of natural outbreaks or used to facilitate transmission studies with the Tyzzer's agent. Other factors bearing on susceptibility are known (or believed) to include poor sanitation and overcrowding (Gard, 1944; Rights et al., 1947; Peace and Soave, 1969), transportation stress (Peace and Soave, 1969), nutritional status (Maejima et al., 1965), dietary form (Gard, 1944), and carbon tetrachloride liver injury (Takenaka and Fujiwara, 1975).

In assessing the issue of genetic susceptibility and resistance, Gowen and Schott (Gowen and Schott, 1933; Takagaki et al., 1966) were unable to relate resistance in *M. musculus, M. bactrianus,* and their crosses to the waltzing factor, dominant white, or sex; they concluded that resistance was best interpreted as depending on a major dominant factor. More recently, it has been reported that mouse strains with impaired immunocompetence have heightened susceptibility and more severe lesions than do their normative counterparts (Waggie et al., 1981; Livingston et al., 1996). Resistant mouse strains (C57Bl) depleted of natural killer cells, neutrophils, and macrophages have increased susceptibility that is comparable to susceptible strains (DBA/2) (Van Andel et al., 1997). Mice experimentally infected with either toxigenic or nontoxigenic *C. piliforme* isolates respond within a day with elevation of hepatic TNA (TNF-α), interferon (IFN)-γ, and serum proteins (Van Andel et al., 2000a) and with elevation of IL-6 (Van Andel et al., 2000b) and IL-12 (Van Andel et al., 1998). Immunologic neutralization of reactive IL-12, *in vivo,* was associated with increased severity of experimental infections. The latter result was interpreted as indicating a protective role for IL-12 as a mediator of infection. In the rat, it has been shown epidemiologically that certain haplotypes were associated with resistance to natural infection, whereas others were more susceptible (Hansen et al., 1990, 1992a)

The balance of evidence now suggests that the pathogenesis of natural infections is in accord with the following sequence: oral ingestion of infectious spores from contaminated litter with establishment of primary infection in tissues of the jejunum, ileum, and cecum. The initial stage is followed by ascension of organisms via the portal vein to the liver and by bacteremic embolization to other tissues, for example, the liver and myocardium. Natural and experimental oral infections in the rabbit, rat, gerbil, and hamster characteristically include intestinal localization of the organism with lesion production in the intestine, as well as liver and (more variably) myocardial lesions (White and Waldron, 1969; Jonas et al., 1970; Yamada et al., 1970; Ganaway et al., 1971b; Van Kruiningen and Blodgett, 1971; Carter et al., 1975; Nakayama et al., 1975; Lee et al., 1976; Waggie et al., 1984; Yokomori et al., 1989; Veazey et al., 1992). In natural and experimental oral infections in the mouse, intestinal lesions are more variable and liver lesions do not appear to depend on prior development of intestinal lesions (Fujiwara et al., 1973). The naturally heightened susceptibility of gerbils may account for more extensive lesions in that species, including encephalitis (Veazey et al., 1992), although factors (e.g., variable susceptibility of gerbil stocks and/or pathogenicity of *C. piliforme* strains) have been suggested as explanations for why gerbils do not always contract the infection (Motzel and Riley, 1992).

Virulence determinants have been studied as a function of diversity between *C. piliforme* isolates. Experimentally, it has been shown (in Caco-2 cultures) that isolates vary in their ability to exit from intracellular entry phagosomes to replicate in the cytoplasm (Franklin et al., 1993).

Similarly, isolates vary in their capacity to produce cytotoxins, and toxigenic isolates have been shown to exert increased pathogenicity in infected hosts (Riley et al., 1992). Thus, the severity of outbreaks may depend on virulence of the isolate, as well as on innate resistance of the host. Bacteremia occurs routinely subsequent to induction of liver lesions (Takagaki and Fujiwara, 1968) and elevation of serum transaminases (ALAT, ASAT) indicative of liver destruction (Hoag et al., 1965; Naiki et al., 1965). Experimental infections of any host species given by intravenous or intraperitoneal routes typically induce liver and myocardial lesions but not intestinal lesions. It was originally shown by Rights et al. (1947) that the organism could be propagated in brain tissue inoculated intracerebrally, and several studies have extended this observation (Fujiwara et al., 1964; Onodera and Fujiwara, 1970, 1972). Similarly, induction via the intracerebral route does not induce intestinal lesions.

Few clinical signs specific to naturally occurring Tyzzer's disease in the mouse and rat have been reported. Even in the presence of enteritic lesions, signs of diarrhea have been variable and not often observed in the rat, although diarrhea is more frequent in other species. Affected rats are depressed, with ruffled haircoat and short period of illness. Infected adolescents may have a potbellied appearance. The few outbreaks that have been studied (before development of high-quality serologic reagents) have indicated low morbidity but high mortality in rat groups. It has been mentioned previously that the serum transaminases are elevated in sick animals. In the mouse, both morbidity and mortality may be high; Tyzzer lost his entire colony of 79 animals (Tyzzer, 1917).

Gross lesions in the rat are suggestive and include flaccid segmental dilatation of affected portions of the intestines (up to three to four times normal diameter) with an edematous, atonic ileum (Jonas et al., 1970). This gross lesion (megaloileitis), with its *in vivo* potbellied appearance, may be seen as one of the earliest signs of an outbreak in newly contaminated rat breeding colonies (Hansen et al., 1992a, 1994; Hansen, 1996). Ileal lesions may extend to adjacent cecum and jejunum. On the basis of these lesions, Jonas et al. (1970) have suggested that several earlier reports of megaloileitis or segmental ileitis in the rat (Geil et al., 1961; Hottendorf et al., 1969) may have actually have represented unrecognized instances of Tyzzer's disease (although it has been pointed out et al., that the lesion is not pathognomonic and may be induced by chloral hydrate alone) (Fleischman et al., 1977). The liver (Fig. 11-3) may have too many disseminated pale foci, up to several millimeters in diameter, scattered throughout the parenchyma (Jonas et al., 1970; Yamada et al., 1970), but frequently it is not lesioned. Circumscribed, grayish foci may be seen in the myocardium in

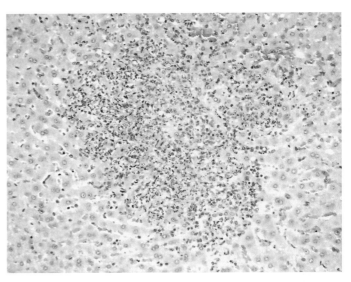

Fig. 11-3 Tyzzer's disease: rabbit liver with early stage, focal necrotizing hepatitis. Hematoxylin and eosin, 200×. (Courtesy of Dr. Shari R. Hamilton.)

some cases. The mesenteric lymph nodes are usually swollen. No other lesions are seen at necropsy. Excellent color illustrations of gross lesions may be seen in the report by Jonas et al. (1970).

Definitive diagnosis of Tyzzer's disease in the rat continues to rest on histopathologic demonstration of the organism in infected tissues; although it has been pointed out that *C. piliforme* may be isolated from infected tissues, propagated via yolk sac inoculations in chick embryo or permissive cell lines, and these isolates may be used to satisfy Koch's postulates (Allen et al., 1965; Craigie, 1966a; Franklin et al., 1994). Intestinal lesions are characterized microscopically by transmural involvement of the intestinal wall, particularly the ileum, in a process varying from segmental sharply demarcated necrosis to acute and chronic inflammatory infiltration of ball levels that include the mucosa, submucosa, and muscular and serosal layers. Edema and hemorrhage are common in the subserosal and muscular layers. *C. piliforme* may be found, often profusely, within epithelial cells of the villi and their crypts. The organism has not been seen deep to the mucosa (Jonas et al., 1970). Although Jonas et al. (1970) described villi as intact but blunted, Japanese investigators have (without detailed microscopic descriptions) described the intestinal process in the rat as an ulcerative enterocolitis (Yamada et al., 1970).

Myocardial lesions, more common in the rat than hepatic, vary in size from only several myofibers to complete transmural involvement. Inflammatory infiltrates are variable, and the myocardial process has been described as degenerative rather than necrotic. *C. piliforme* may be demonstrated by special stains within affected myocardial muscle cells.

Microscopic lesions in the liver are initiated by foci of coagulative necrosis varying in size from several micrometers several millimeters, with a tendency to centrolobular orientation in association with central veins and distinct demarcation between necrotic and adjacent normal hepatic parenchyma. *C. piliforme* typically is found in viable hepatocytes on the periphery of necrotic foci (Fig. 11-4). Necrotic foci thought to be early developing are relatively acellular and barely eosinophilic, with hepatic cell ghosts and karyolitic debris (Jonas et al., 1970). Lesions interpreted to be later in development demonstrate moderate to marked infiltration with neutrophils, Kupfer cells, macrophages, and fibroblasts, which may almost obliterate previously acellular foci of necrosis (Ganaway et al., 1971a). Mineralized material similar to that seen in rabbit lesions has been seen in necrotic foci of rat livers (Jonas et al., 1970). Giant cells and histiocytes may aggregate peripheral to infiltrated necrotic foci (Jonas et al., 1970). In any given liver, relatively acellular "early" foci may be seen, along with infiltrated "later" foci; others, with fibroplasias and giant cells. These observations are suggestive of successive showers of organisms from the intestines. The ultrastructure of hepatic lesions has been described (Fujiwara et al., 1963a, 1968; Jonas et al., 1970; Kurashina and Fujiwara, 1972).

The advent of a more powerful armamentarium of diagnostic methodology has enabled a fuller understanding of the biology of natural infections. Principally, the wide use of serologic screening techniques in health surveillance programs (Fries, 1978, 1979b; Motzel and Riley, 1991; Motzel et al., 1991; Hansen et al., 1992b, 1994;

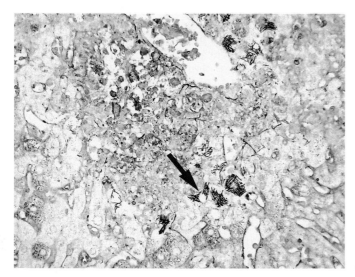

Fig. 11-4 Tyzzer's disease: focal lesion in mouse liver with argyrophilic (silver-staining) vegetative *Clostridium piliforme* rods within hepatocytes at lesion periphery (arrow). Steiner, 400×. (Courtesy of Dr. Shari R. Hamilton.)

Boivin et al., 1994; Hansen, 1996) and the introduction of molecular methods (PCR) for genomic detection of the organism (Duncan et al., 1993; Goto and Itoh, 1996; Furakawa et al., 2002) have greatly enlarged the dimensions of infection and enabled the distinction to be made between asymptomatic *C. piliforme* infection and overt Tyzzer's disease with clinical signs and lesions. Certainly, experimental infections of immunocompetent rat and mouse hosts support the concept of an acute, transient, immunogenic, but self-limiting infection that may last no more than 1 to 2 weeks (Waggie et al., 1984; Itoh et al., 1988; Motzel and Riley, 1992; Franklin et al., 1994).

At the same time, inapparent infection is commonly observed and may be inferred from several lines of evidence; chief among them the frequent reports of intercurrent epizootics subsequent to stressful or immunosuppressive treatments and the documented results of immunosuppressive diagnostic screening techniques (Maejima et al., 1965; Takagaki et al., 1967; Karasek, 1970; Yamada et al., 1970; Fries, 1979a; Thunert, 1980; Nakayama et al., 1984; Gibson et al., 1987). By use of antigens derived from infected liver suspensions, it has been established by fluorescent antibody techniques, complement fixation, and agar gel immunodiffusion that the organism is immunogenic and that the antibodies thus induced are protective and usually present in higher titer in younger age groups than in retired breeders (Fujiwara et al., 1965, 1969; Fujiwara, 1967; Onodera and Fujiwara, 1970). Fluorescent antibody techniques, both direct and indirect, have been used to locate *C. piliforme* antigen in tissues (Fujiwara et al., 1968, 1969; Savage and Lewis, 1972; Fries, 1977). One technique used to detect inapparence in the earlier era of insensitive serology tests (complement fixation) was the anamnestic challenge assay. This assay was used to detect latent infections (and previous exposure to the agent) in adult rodents that carried antibody levels below the detectable level. The basis of the test is the anamnestic antibody response to a defined challenge with killed bacterial or viral antigens (Fujiwara, 1971). The principle has been shown to be equally useful for serologic monitoring of rodent colonies for other pathogens, especially mouse hepatitis virus but also Sendai virus, *C. kutscheri,* and *Salmonella enteritidis*. As a rule, antibody levels are determined on sera from the sixth day after antigen injection. The working assumption has been made that primary antibody responses are detectable only subsequent to the ninth day, whereas those that occur before day 9 are truly anamnestic and reflect sensitization from previous experience with the antigen (Fujiwara, 1971). The assay has fallen into disuse because of general availability of adequately sensitive immunofluorescence assay and ELISA serologic reagents.

Nonetheless, results with immunosuppressive stress tests for confirmation of infection in serologically positive rats have had mixed results and frequently fail to provoke a histopathologically definitive diagnosis. Explanations for failed provocation tests include the following: the detected antibodies were indicators of previous (and transient) infection; the organism was no longer carried by the animals in the stress test; and the particular *C. piliforme* isolate, although immunogenic, was insufficiently toxigenic to produce lesions. Finally, it needs to be recognized that one of the currently most difficult and confounding diagnostic issues with Tyzzer's disease of the rat is the not infrequent occurrence of positivity in colonies monitored by serologic health surveillance programs in which it has not been possible to confirm serodiagnosis by a combination of immunosuppressive provocation, traditional histopathology, and molecular techniques. This circumstance suggests the additional possibility of immunologic cross-reaction with antibodies stimulated by other commensal flora carried by the rat. Although the foregoing explanations for the inability to confirm *C. piliforme* infection in serologically positive rats are only conjectural, the phenomenon has been widely recognized (Hansen, 1996; Riley et al., 1994; Weisbroth, 1999b). Even the Federation of European Laboratory Animal Science Associations (FELASA) has suggested caution by stating that positive serologic reactions occur frequently in the absence of clinical signs (Nicklas et al., 2002).

Despite morphologic similarity, isolates from different host species may have substantial biochemical and biologic diversity (Fujiwara et al., 1973a; Fries, 1980; Manning, 1993; Riley et al., 1994). The degree of host specificity of *C. piliforme* isolates and the hazards of interspecific transmission are of great interest and have been studied in some detail. These studies have shown that heterologous transmission (i.e., rat isolate to mouse and vice versa) is less virulent than is homologous transmission, particularly when inoculated by the oral and intravenous routes; that immunity to infective challenge was higher in homologous systems; and that although rat and mouse isolates had certain antigens in common, they also had other unshared antigens (Fujiwara et al., 1971, 1973, 1974; Waggie et al., 1987; Motzel and Riley, 1991; Franklin et al., 1994). Based on the greater stability of vegetative suspensions from rat liver, Fujiwara et al. (1971) suggested that *B. piliforme* may originally have been a pathogen of the rat, being forced to mutate to a form with surface changes in heterologous hosts. The general view at present is that regardless of host origin, all isolates must be considered as potentially causative agents of *C. piliforme* infection and Tyzzer's disease in heterologous hosts.

The sensitivity of *B. piliforme* to various antibiotics has been investigated primarily from the standpoint of treatment and has shed light on the taxonomic position of the organism by development of antibiograms. These studies have established the effectiveness of cephaloridine, tetracycline, and chlorampenicol for protection against experimental infection (Craigie, 1966a,b; Takagaki et al., 1966; Yokoiyama and Fujiwara, 1971).

The emphasis has been on prevention of infection, however, rather than treatment. The importance of good sanitary standards was emphasized previously in the discussion on transmission and factors known to predispose infection. The effectiveness of cage lid filters in preventing infection in enzootic environments has been demonstrated (Boot et al., 1996). The disease may be eradicated and excluded by gnotobiotic principles and maintenance within barriers, although the potential for reinfection of gnotobiotically rederived rat colonies is considerable (Hunter, 1971; Hansen and Mollegaard-Hansen, 1990; Hansen et al., 1992; Hansen, 1996). Although there is no evidence that *C. piliforme* may traverse the rat placental barrier in natural infections (to infect gnotobiotic neonates during rederivation), the possibility exists that it may because it has been demonstrated experimentally in immunosuppressed rats (Fries, 1978, 1979c) and as a natural occurrence in the guinea pig (Boot and Walvoort, 1984). An additional possibility for reinfection of rederived rodent stocks may follow reuse of inadequately sanitized contaminated barriers owing to the extreme viability of spores (Hansen and Mollegaard-Hansen, 1990).

H. *Corynebacterium kutscheri* Infection (Pseudotuberculosis)

Pseudotuberculosis in laboratory rats and mice is caused by *C. kutscheri,* first isolated and described from the mouse in 1894 (Kutscher, 1894). Synonymy with this organism is confusing, both because of the descriptive name for gross lesion morphology (pseudotuberculosis) and because of other similarly named but distinct bacterial entities, for example, *Corynebacterium pseudotuberculosis* (the Preisz-Nocard bacterium) and *Yersinia pseudotuberculosis.* The synonymy with *Y. pseudotuberculosis* is additionally confusing, because it includes *Bacterium pseudotuberculosis rodentium* (Schutze, 1928, 1932; Hass, 1938). Both of these latter organisms are culturally and biochemically distinct from *C. kutscheri,* and there is no convincing evidence that either may cause natural or intercurrent disease syndromes in laboratory rats and mice. The older literature contains references to *Clostridium pseudotuberculosis murium* (Reed, 1902; Hojo, 1938; Ford and Joiner, 1968), *Corynethrix pseudotuberculosis murium* (Schechmeister and Adler, 1953), and *Corynebacterium pseudotuberculosis murium* (Schechmeister and Adler, 1953), which are regarded as synonyms of *C. kutscheri*. In addition, as pointed out by Giddens et al. (1968), older references to *Clostridium muris* (Klein, 1903; Mitchell, 1912) have been cited in reviews on murine pneumonia that describe isolation of Gram-positive diphtheroids from rat pulmonary abscesses consistent with those associated with *C. kutscheri*. Wilson and Miles (1975) have continued *C. murium* as the current name for this organism; however, *Bergey's Manual* retains the name *C. kutscheri*.

Naturally occurring syndromes of clinically apparent disease occur in laboratory rats and mice, and the latent carrier state has been recognized in both of these species. The organism has been isolated from wild rats (Boot et al., 1995a) and from humans (Fitter et al., 1979; Messina et al., 1989; Sixl et al., 1989); thus, both of these carrier hosts represent potential reservoirs. Similarly, the agent has been isolated from hamsters and guinea pigs, which can act as aymptomatic carriers (Vallee et al., 1969; Nelson, 1973; Amao et al., 1991).

C. kutscheri has been encountered as a primary pathogen in unprovoked infections of the rat (Vallee and Levaditi, 1957; Pestana de Castro et al., 1964; Ford and Joiner, 1968; Giddens et al., 1968; Nelson, 1973; Bhandari, 1976; McEwen and Percy, 1985) as well as the mouse (Kutscher, 1894; Bonger, 1901; Hojo, 1939; Polak, 1944; Wolff, 1950; Bicks, 1957; Weisbroth and Scher, 1968a,b; Savage, 1972). More commonly, however, in both species, inapparent or latent infections have been unmasked in the course of research protocols by experimental treatments lowering host resistance or impairing immunocompetence. Such treatments have included cortisone (Antopol et al., 1950, 1951, 1953; Berlin et al., 1952; LeMaistre and Thompsett, 1952; Speirs, 1956; Pierce-Chase et al., 1964; Caren and Rosenberg, 1966; Fauve and Pierce-Chase, 1967; Robinson et al., 1968; Yamada et al., 1970), X irradiation (Schechmeister and Adler, 1953; Schechmeister, 1956) and vitamin deficiency (particularly biotin [Gundel et al., 1932; Giddens et al., 1968] and pantothenic acid [Seronde, 1954; Zucker and Zucker, 1954; Seronde et al., 1955, 1956; Zucker et al., 1956; Zucker, 1957]). Some infections with other organisms— for example, ectromelia virus (Lawrence, 1957) and *Salmonella* (Topley and Wilson, 1922–1923)—have been thought to provoke overt expression of disease, whereas others (e.g., Sendai, rat virus, and sialodacryoadentitis virus) have not had that effect (Barthold and Brownstein, 1988).

Indeed, the rat was long considered relatively insusceptible. Rats were frequently found refractory to unconditioned transmission attempts (Kutscher, 1894; Bonger, 1901; Seronde, 1954; Ford and Joiner, 1968), and clinical expression was thought to require conditioning by immunosuppressive or nutritionally deficient treatments (Kutscher, 1894; Bonger, 1901; Seronde, 1954).

Other modulators of infection are known to include anti-inflammatory agents such as L18-MDP, which can restore rat resistance ordinarily ablated by cortisone treatment (Ishihara, 1984). A tumorlytic factor, distinct from TNF-α or -β, has been isolated from the serum of mice injected with T-cell mitogens extracted from *C. kutscheri* (Kita et al., 1995). The stimulatory factor also induced cytokine production, for example, interleukins (IL-1 and IL-2) and TNF-α, which contributes to nonspecific resistance of rodent hosts (Kita et al., 1992). It is known that heritable resistance and susceptibility occurs among various mouse strains (Pierce-Chase et al., 1964; Weisbroth and Scher, 1968a; Amao, 1993) and rat (Seronde et al., 1955; Seronde, 1956), and that rat stocks and age groups vary in their capacity for immunologic response to infection (Suzuki et al., 1986, 1988).

The concept of latency in the mouse has been investigated in some detail by Pierce-Chase and Fauve (Fauve et al., 1964, 1966; Pierce-Chase et al., 1964; Fauve and Pierce-Chase, 1967). Their initial observation was that randomly chosen mice from their colony could be placed into one of two categories based on the mice's susceptibility or resistance to an intravenous challenge with *C. kutscheri*. They concluded that the resistant mice were actually resistant to superinfection because a state of premunition (i.e., infection immunity) existed in these stocks. The resistant strains, they thought, were immunologically sensitized (and protected) by virtue of preexisting latent infections. They believed the latent organism was an avirulent variant of *C. kutscheri* carried in the same organs as the virulent form and that shortly after a suitable stress, for example, cortisone administration, these avirulent "A" bacterial populations shifted to the virulent "K" form. These results were not supported by subsequent investigation (Bruce et al., 1969; Hirst and Olds, 1978), and in the light of experience since, "resistance" and "susceptibility" are thought to relate to more to host factors of relative immunocompetence rather than to variants of the agent. It is assumed that the occasional isolation of *C. kutscheri* from the oral cavity, middle ear, lungs, and (abscessed) preputial glands in the course of routine diagnostic monitoring of essentially normal conventional rats (Matheson et al., 1955; Amao et al., 1986, 1988, 2002) actually represents cultural confirmation of latent infection in resistant hosts.

Presumptive diagnosis may be made at necropsy by smears of affected tissues (or later, in sectioned material), in which typical aggregates of *C. kutscheri* are demonstrated by Giemsa or Gram stains. Although grayish-blue amorphous bacterial colonies may be seen in tissue sections, typical organisms are not readily discernible in material stained with hematoxylin and eosin. The organisms may easily be demonstrated by Giemsa or tissue Gram (Brown and Brenn or

Gram-Weigert) stains as irregular palisades or "Chinese letter" arrangements of Gram-positive diphtheritic rods with metachromatic granules. Gram reactions of *C. kutscheri* may be variable in clinical material. Definitive diagnosis requires cultural isolation and biochemical characterization of the organism.

Triturates of aseptically collected suspect tissues, swabs, and aspirated washings of the nasal turbinates or middle ears should be cultured directly on 5% blood agar plates and incubated aerobically at 37°C. A selective FNC (furazolidone–nalidixic acid–colimycin) medium may be used to facilitate recovery of low numbers of organisms from nonlesioned or contaminated clinical material (Amao et al., 1990, 1995b; Brownstein et al., 1985). Colonies of *C. kutscheri* on blood agar are 1 to 2 mm in size after 24 hours, circular, entire, dome-shaped, grayish-white in color, smooth, and usually nonhemolytic (whereas *C. pseudotuberculosis* is hemolytic on blood agar) (Wilson and Miles, 1975). Growth and biochemical characteristics are summarized in Table 11-5. Isolates of *C. kutscheri* are facultatively aerobic and nonmotile. Cultures may be

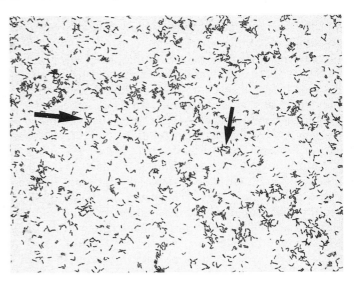

Fig. 11-5 Corynebacterium kutscheri: broth culture. Note irregular palisades with "Chinese letter" formations (arrows). Gram, 1000×. (Courtesy of Dr. Shari R. Hamilton.)

propagated on trypticase soy broth or agar (Fig. 11-5). Although *C. kutscheri* does not form spores, it is quite stable at room temperature in exudates, urine, and feces. It will survive for at least 8 days at 4 and −20°C in phosphate-buffered saline, which was recommended as a transport medium (Shimoda et al., 1991). As a Gram-positive rod, *C. kutscheri* should be regarded as requiring washing temperatures of at least 180°F to inactivate.

The means of spread of *C. kutscheri* within a colony is not known with certainty, but given that the organism is quite consistently isolated from submaxillary lymph nodes of experimentally infected rats (Brownstein et al., 1985) and from cervical lymph node and oral structures of subclinically infected conventional rats and mice (Amao et al., 1990, 1995a,b, 2002; Fox et al., 1987), it would seem reasonable to conclude that oral transmission by direct contact is probably the dominant mode. A higher rate of organism recovery has been reported from oral sites than from nasopharyngeal washings (Amao et al., 1995b; Fox et al., 1987), even though rat lungs are the primary site of overt infection. Likewise, given the predilection for kidney abscesses in both host species and high recovery rates from large bowel contents (Amao et al., 1995a), both urine and feces from subclinical carriers should be regarded as infective. Although vertical transmission has been demonstrated experimentally in the mouse (Juhr and Horn, 1975), the marked decline in prevalence of this condition in recent years is testimony to the success of its eradication in rederived rodents. Because of the chronicity of inapparent infections, the origin of epizootics, when detected, is often obscure.

TABLE 11-5

CHARACTERISTIC BIOCHEMICAL REACTIONS AND OTHER PROPERTIES OF *CORYNEBACTERIUM KUTSCHERI*

Characteristic	Reaction or property
Growth on blood agar	Grayish- or yellowish-white, 1- to 3-mm colonies; no hemolysis
Morphology	Gram-positive rods with metachromasia in Chinese letter arrangement
Carbohydrate, acid from:	
Glucose	(+)
Xylose	(±)
Lactose	(−)
Sucrose	(+)
Maltose	(+)
Salicin	(+)
Mannitol	(−)
Dulcitol	(−)
Fructose	(+)
Mannose	(+)
Aesculin hydrolysis	(+)
Growth on MacConkey	(−)
Hydrogen sulfide production	(±)
Indol	(−)
Urease	(+)
Nitrate reduction	(+)
Gelatin liquefaction	(−)
Litmus milk	No change
Catalase	(+)
Oxidase	(−)
MR/VP	(−)/(−)

MR/VP, Methyl Red/Voges Proskauer.
(Data from Buchanan and Gibbons, 1974; Weisbroth and Scher, 1968a,b; Cowan, 1974; Wilson and Miles, 1975.)

Both morbidity and mortality are extremely variable and depend for expression on the interaction of husbandry, heritable factors of resistance or susceptibility, and modulators (e.g., virus infections that lower resistance). Clinical signs in unprovoked rats usually are those associated with subchronic respiratory disease. Variable degrees of upper respiratory signs may be seen, including reddish to mucopurulent discharges from the nares and medial canthus. Affected rats are depressed and anorexic, with humped posture and ruffled haircoat. Respiratory rales are common. Untreated animals typically die 1 to 7 days after the onset of signs. Any age group may be affected, and a pattern of ongoing but sporadic, low-intensity mortality is typical of enzootically infected colonies.

Gross lesions in the rat may be confined to pulmonary tissues and consist of numerous pale foci 1 to 5 mm in size scattered throughout the parenchyma of the lung (Fig. 11-6). Typically, the lesions are liquefied centrally but may be caseous. Occasional areas of coalescence

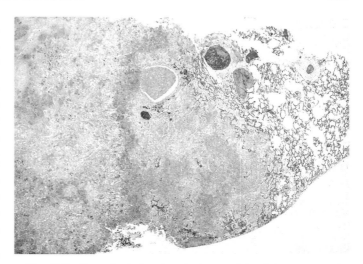

Fig. 11-7 *Corynebacterium kutscheri*: necrotizing lobar pneumonia. Hematoxylin and eosin, 25×. (Courtesy of Dr. Shari R. Hamilton.)

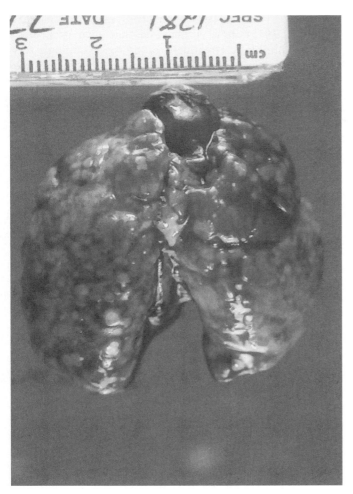

Fig. 11-6 *Corynebacterium kutscheri*: rat lung with multifocal coalescing abscesses. (Courtesy of Dr. Shari R. Hamilton.)

of foci may be quite large, up to 1.5 cm, and involve substantial portions of a given lobe. The tissue surrounding foci is congested, and entire lobes may be consolidated (Fig. 11-7). Fibrinous adhesions to the thoracic wall are not uncommon and occasionally make retraction of the lung difficult. Such adhesions become fibrous in more chronic cases. Similar foci may less frequently be seen in the liver and kidney of infected rats than in the mouse, in which hematogenous septic embolization and abscessation of organs with capillary nets (e.g., the liver and kidney) are the rule (Weisbroth and Scher, 1968a). Infrequently, arthritic lesions of the pedal extremities and subcuticular abscesses may be seen (Fischl et al., 1931; Nelson, 1973). It was pointed out earlier that abscesses of the preputial glands and middle ears in the rat frequently yield *C. kutscheri* in the absence of pulmonary lesions. Under conditions of cortisone preconditioning (or provocation), the appearance of hematogenous distribution may be more apparent. In considering the issue of how organisms get to the lungs, Giddens et al. (1969), reasoned that inasmuch as the pulmonary lesions of rats were interstitial rather than bronchial, hematogenous embolization to the lung was more consistent with the observations than was inhalation. They also pointed out that pulmonary infections in their own experiments followed oral or intraperitoneal inocula, as did those provoked by the intracardiac injection of *C. kutscheri* by other investigators (LeMaistre and Thompsett, 1952). Fauve et al. recovered inapparent *C. kutscheri* from the lungs of asymptomatic lesionless mice (1964).

Microscopic lesions in any of the affected tissues consist of abscesses that vary in size from several micrometers up to the size seen grossly at necropsy. The centers of such lesions are necrotized and, depending on the stage, may be coagulative and relatively acellular early, liquifac-tive (later) as acute inflammatory cells infiltrate, and mature as inspissated or caseous abscesses. The inflamma-tory infiltrate surrounding and invading necrotized areas consists of neutrophils early in the course and includes macrophages, lymphocytes, and plasma cells as the lesions become subacute. Fibroplasia is not pro-nounced in pulmonary lesions; epitheloid cells, calcifi-cation, or giant cells are not seen. Thus, although the lesions are spoken of in the older literature as being "granulomatous," Giddens et al. (1968) pointed out that there is little histopathologic support for this description, and the term "pseudotuberculosis" would not really seem to apply. In the lung, suppurative exudates are seen in the brochioles and bronchi. Interstitial tissues peripheral to pulmonary abscesses are congested, alveoli are filled with proteinaceous fluids, and leukocytes may be seen as perivascular cuffs around blood vessels. Foci are believed to enlarge and coalesce by radial expansion (Weisbroth and Scher, 1968).

Both subclinically and overtly infected animals develop antibodies that may be detected by tube or microplate agglutination and indirect immunofluorescence (Brownstein et al., 1985; Suzuki et al., 1986, 1988) or, more effectively, by the ELISA method (Ackerman et al., 1984; Fox et al., 1987; Boot et al., 1995a). Although Weisbroth and Scher (1968b) were unable to detect antibodies in lesionless mice from an enzootically infected colony by tube agglutination, the more sensitive ELISA technology has established the use of this technique in health surveil-lance programs to serologically detect asymptomatic, inapparently infected rats and mice (Ackerman et al., 1984; Boot et al., 1995a). In enzootically infected breeding colonies, colostral antibodies to C. kutscheri may be detected for the first 3 to 4 weeks of life, declining to undetectable by the eighth week, rising slowly thereafter to become detectable by the fifth to sixth month and having the highest levels after the eighth month of life (Suzuki et al., 1986, 1988). Although minor differences may exist in homology between isolates from various host species or geographic origin, the great majority of C. kutscheri isolates form a culturally, morphologically, ecologically, and serologically homogeneous group (Yokoiyama et al., 1977; Boot et al., 1995a).

Several other methods to detect latent infection have been explored. Cortisone provocation has been exten-sively used as a means of expressing latent infections and has been advocated for diagnostic surveillance of rat and mouse colonies (Utsumi et al., 1969; Karasek, 1970; Yamada et al., 1970). Similarly, Fujiwara (1971) and his colleagues have explored the use of the anamnestic response test (ART) with this organism and greatly in-creased the rate of serologic detection of rodents with, at the time, otherwise indetectable asymptomatic infections.

Killed suspensions of organisms are injected and the test animals bled on the sixth day. Antibody responses, if any, by the sixth day are held to be anamnestic and reflective of previous experience (sensitization) with C. kutscheri. The practical need for both of these tests, that is, immunosup-pressive provocation and the ART, has declined in recent years, supplanted by the more practical and cost-effective ELISA serology. The use of molecular methods for in situ detection of the agent in latent infections has been explored (Saltzgraber-Muller and Stone, 1986) but not widely reported.

Visibly ill animals can be treated by medication of the water or individual injections of suitable antibiotics after antibiotic sensitivity in vitro has been determined. The organism has been shown to be sensitive in vitro to a variety of narrow- and broad-spectrum antibiotics (Weisbroth and Scher, 1968a), and antibiograms have been developed for C. kutscheri rat isolates (Owens et al., 1975). If the morbidity is low, a number of unprovoked epizootics have been brought under control by the expedient of simply removing visibly ill animals from the colony (Georg, 1960; Weisbroth and Scher, 1968b; Bhandari, 1976). It is emphasized that such procedures, that is, antibiotic treatment and rigid culling, serve to remove the visibly ill tip of the iceberg but leave behind the great bulk of inapparently infected reservoir hosts. The disease may be prevented or excluded from rat colonies by maintenance of the animals inside micro-biologic barrier systems.

I. Cilia Associated Respiratory Bacillus Infection

Cilia-associated respiratory (CAR) bacillus infection was originally described as a chronic respiratory disease of aging rats, with lesions resembling those of murine mycoplasmosis (van Zwieten et al., 1980a,b) (Fig. 11-8), although intimations of a similar disease had been seen earlier (see NRC, 1991). The condition was accom-panied by clinical signs of respiratory distress and was characterized microscopically by bronchitis, bronchiecta-sis, mucopurulent exudation, lymphoid follicular hyper-plasia (Fig. 11-9), and basophilic striation of airway epithelial cilia. Silver impregnation stains revealed the striations to consist of densely packed bacteria among the cilia bordering the airways (Fig. 11-10). The argyro-philic bacteria were seen from the upper respiratory tract, including the nasoturbinates and middle ears, to the larynx, trachea, and various levels in the lung. The etiological significance of the argyrophilic bacteria was uncertain, because the rats involved were concurrently infected with other respiratory pathogens, including Sendai virus, pneumonia virus of mice, and M. pulmonis. Within several years, similar conditions, with morphologically

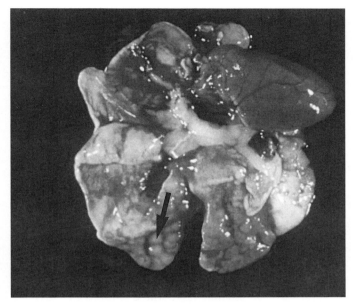

Fig. 11-8 Cilia-associated respiratory (CAR) bacillus: multiple foci of bronchiectasis (arrow) and mucopurulent bronchopneumonia. (Courtesy of Dr. Shari R. Hamilton.)

The condition was seen initially in the Netherlands, but within a short period of time reports came from the United States, Italy, Sweden, Spain, Japan, and Australia (Ganaway et al., 1985; Matsushita, 1986; Matsushita et al., 1987; France, 1994; Oros et al., 1997; Caniatti et al., 1998). In many of these earlier reports, other respiratory pathogens (or serologic evidence of infection) were simultaneously present, thus CAR bacillus, at the time, was relegated to the status of a coinfecting opportunist rather than that of a primary pathogen. This interpretation was supported by PCR analysis of a number of isolates, including that of Ganaway's (1985), indicating contamination by *Mycoplasma* species, including *M. pulmonis* (Schoeb et al., 1997a). Nonetheless, outbreaks were reported in which careful diagnostic searches could not support concurrent infection by copathogens and in which lesions of chronic respiratory disease could only be attributed to CAR bacillus (Medina et al., 1994). There appears little doubt that CAR bacillus should be regarded as a primary respiratory pathogen for laboratory rats, mice, and rabbits and as an opportunistic copathogen that may complicate any of a number of rat respiratory diseases, including mycoplasmosis.

The name cilia-associated respiratory bacillus and its acronym CAR bacillus were proposed by Ganaway et al. (1985) for use until such time as the organism could be taxonomically classified. The CAR bacillus is difficult to cultivate, thus impeding biochemical characterization and taxonomic assignment. Isolation was initially successful in embryonating hens eggs (Ganaway et al., 1985), then in 3T3 tissue cultures (Cundiff et al., 1994a,b), and finally on supplemented (Shoji et al., 1992) and unsupplemented artificial media (Schoeb et al., 1993a). The CAR bacillus is a Gram-negative, argyrophilic, and

consistent bacterial involvement were reported in a wider range of host species, including *Mystromys* (MacKenzie et al., 1981), laboratory mice (Griffith et al., 1988) and rabbits (Kurisu et al., 1987; Waggie et al., 1990), pigs (Nietfeld et al., 1999), and cattle (Hastie et al., 1993; Nietfeld et al., 1999), and additional episodes of infection were seen in both wild (Brogden et al., 1993) and laboratory (Matsushita, 1986; Itoh et al., 1987) rats. Experimental infections (with rat isolates) demonstrated that CAR bacillus could asymptomatically colonize guinea pigs and hamsters (Shoji-Darkye et al., 1991).

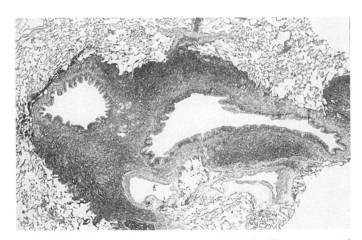

Fig. 11-9 Cilia-associated respiratory (CAR) bacillus: exaggerated peribronchial lymphoid aggregates. Hematoxylin and eosin, 40×. (Courtesy of Dr. Shari R. Hamilton.)

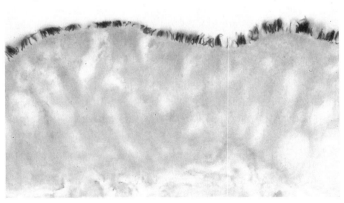

Fig. 11-10 Cilia-associated respiratory (CAR) bacillus: bronchiolar epithelium demonstrating argyrophilic (silver staining) palisades of the adherent bacillis. Steiner, 1000×. (Courtesy of Dr. Shari R. Hamilton.)

filamentous rod measuring approximately 0.2 μm × 6.0 μm with a triple layer cell wall and bulbous ends. It stains poorly with aniline dyes (and may be missed in tissues stained with hematoxylin and eosin), is not acid fast, is PAS negative and does not form spores. The CAR bacillus is motile but lacks apparent flagellae, pili, or axial filaments and has been classified with the "gliding bacteria" (Ganaway et al., 1985). It is heat labile (56°C for 30 minutes) but survives successive freezing and thawing and remains viable at room temperature for at least 1 week in allantoic fluid preparations (Waggie et al., 1994).

Despite morphologic similarity leading to their common (and temporary) nomenclature as CAR bacillus, rat and rabbit isolates are different and somewhat species specific. The cells of rat isolates in culture are more motile, are larger and thicker than rabbit isolates, and form multicell aggregates (Schoeb et al., 1993; Cundiff et al., 1994a). Rat isolates have been shown to vary in pathogenicity (Schoeb et al., 1997b). Mice used for infectivity studies with rat isolates demonstrated colonization of respiratory epithelium, seroconverted, and developed pulmonary lesions, whereas mice inoculated with rabbit isolates may seroconvert but were refractory to infection (Shoji-Darkye et al., 1991; Cundiff et al., 1994a). Rabbits inoculated with rabbit origin CAR bacillus developed nasal discharge and microscopic lesions of the respiratory tract and seroconverted, but they were not colonized and did not seroconvert when inoculated with rat origin CAR bacillus (Matsushita et al., 1989; Cundiff et al., 1995a). Sequence analysis of the 16S rRNA gene from rabbit and rat isolates indicated these isolates were quite different (with only 48.8% homology), suggesting they probably should be classified in different genera. Rabbit isolates identified most closely with *Helicobacter* (about 90% homology), and it was concluded the isolates were host specific and belong to separate genera (Cundiff et al., 1995a). The genetic and antigenic relationships remain quite complicated, at present, as all isolates share certain epitopes (Hook et al., 1998). Analysis of rat isolate 16S rRNA sequences indicated that rat origin CAR bacillus most is closely related to *Flavobacterium* and *Flexibacter* (Schoeb et al., 1993; Kawano, 2000). Because it has been shown experimentally that transmission can occur between rodent species, it must be assumed that rodent carriers of CAR bacillus may be infectious for other rodent species (Shoji et al., 1988b; Matsushita et al., 1989).

Infection appears to be most effectively transmitted by direct contact (Matsushita et al., 1989) and, in enzootically infected breeding colonies, may be transmitted by dams to neonates as early as 1 to 2 weeks after birth (Ganaway et al., 1985). Dirty bedding from infected animals fails to transmit the infection to exposed sentinels (Cundiff et al., 1995b). The clinical signs and corresponding pathology of CAR bacillus infection are quite variable, and expression depends on the contending interplay of modulators such as isolate qualities, host age, degree of heritable resistance, and the effects of resistance-lowering copathogens. Infected rat carriers may be asymptomatic (Schoeb et al., 1987; Shoji et al., 1988a; Cundiff and Besch-Williford, 1992). Nonspecific signs of respiratory disease—including labored breathing, persistent rales, nasal discharge, periocular porphyria, head tilt, hunched posture, roughened haircoat, and weight loss—have been observed (Matsushita and Joshima, 1989; NRC, 1991; Cundiff and Besch-Williford, 1992). Similarly, the gross lesions of CAR bacillus infection seen at necropsy are not distinctive. They can include the presence of exudates in the nasoturbinates and middle ears (tympanic bulla) and serous to mucopurulent exudates in the trachea. The lungs may fail to deflate owing to mucopurulent plugging, and there may be areas of consolidation or scattered foci on the pleural surface (Fig. 11-8). These lesions, of course, are quite similar to those of murine respiratory mycoplasmosis, in the past a frequent copathogen, and the question is open as to how much of the pathology ascribed in the past to mycoplasmosis (alone) may really have been owing to complicated infections with both agents.

Experiments monitored to exclude contribution by copathogens have clarified and confirmed the ability of monospecific CAR bacillus infections to cause the lesions of chronic respiratory disease (Ganaway et al., 1985; Ganaway, 1986; Matsushita and Joshima, 1989). These studies have shown that CAR bacillus infections initiate as an upper respiratory condition. The earliest lesions (rhinitis, tympanitis, and tracheitis) of experimental infection in rats occur in the ciliated epithelium of the nasoturbinates, Eustachian tube, middle ears, and trachea. Microscopically, chronic inflammatory infiltrates are seen in the lamina propria underlying the ciliated epithelium, and with silver stains, argyrophilic bacteria can be observed on the surface of the ciliated border. Later, the infection descends to the pulmonary tree and microscopic lesions of mucopurulent bronchitis and bronchiolitis develop, along with prominent peribronchial cuffs of chronic inflammatory cells (Fig. 11-9). Multifocal necrosis of bronchiolar and bronchial epithelium may occur, and acute inflammation with abscess formation may be seen in these foci (NRC, 1991). The ciliated epithelium lining the airways is markedly infiltrated with chronic inflammatory cells, predominantly lymphocytes and plasma cells. Silver impregnation stains, for example, Warthin Starry or GMS, of pulmonary sections demonstrate palisading masses of argyrophilic, rod-shaped bacteria aligned to the apices of cells of the ciliated respiratory epithelium (Fig. 11-10). The apical insertions of CAR bacilli have been demonstrated by electron microscopy (see van Zwieten et al., 1980a,b). Excellent light and electron microscopy photomicrographs of CAR bacillus may also be seen work by Ganaway et al. (1985).

Health surveillance screening tests to detect CAR bacillus infections in rats are principally oriented to serologic tests for antibodies indicative of exposure and PCR tests to detect genomic material resident in rat respiratory system. Both immunofluorescence assay (Matsushita et al., 1987) and ELISA (Ganaway et al., 1985; Lukas et al., 1987; Shoji et al., 1988a; Thuis and Boot, 1999) are commonly used as serologic screening tests. Although the potential exists for cross-reactive false positives owing to shared antigenic epitopes with other bacterial forms (Hook et al., 1998), this has not proven to be a practical limitation to the use of serodiagnostics. The optimum age for serologic detection of rats from infected breeding colonies is 10 to 14 weeks of age, based on the sequence that rat pups born of antibody-positive dams have colostral titers for the first 3 to 4 weeks of life, which will wane to become low or negative by 5 to 7 weeks. These adolescents will then contract direct infection within the enzootic environment and seroconvert to positive by 8 to 10 weeks after birth (Ganaway et al., 1985; Matsushita et al., 1987; Shoji et al., 1988a). Experimental infectivity studies have shown that rats will seroconvert by about 2 weeks after inoculation and that symptomatic animals should be serologically positive (Ganaway et al., 1985; Matsushita et al., 1987, 1989). Screening tests for detection of CAR bacillus infection also include PCR and reverse transcription–PCR, which offer means for overcoming some of the disadvantages of serology, for example, the potential for false positives and the problem that rats sampled for testing, although exposed and colonized, may not yet have developed maximal titers (Cundiff et al., 1994b; Goto et al., 1995). Antemortem (or necropsy) samples for PCR screening tests may be taken by swab with a higher rate of recovery from the nasoturbinates (Franklin et al., 1999) than that of samples from the oral cavity (Goto et al., 1995).

Diagnostic confirmation of screening test positives from either asymptomatic or clinically affected rats requires demonstration of the organism in infected tissues. Confirmation may be obtained by light microscopy with special stains, by immunofluorescent or immunoperoxidase methods to demonstrate the CAR bacillus in tissues (Oros et al., 1996), by isolation of the organism in culture, or by some combination of these approaches. One caveat here is that resistant rodent strains may only have brief periods of a few weeks when the organism is in residence (Shoji-Darkye et al., 1991), thus older rats may be serologically positive but no longer infected. Because it has been shown that upper respiratory involvement precedes pulmonary involvement (Itoh et al., 1987), necropsy tissue selection for diagnosis of CAR bacillus infection should include tracheal sections or smears. An improved method for demonstrating the organism in tracheal scrapings has been reported (Medina et al., 1996).

Spread of the organism within contaminated facilities proceeds principally by direct contact or by contact with contaminated fomites, for example, gloved hands or unsanitzed cages and ancillary cage parts, although dirty bedding sentinels fail to contract the infection (Cundiff et al., 1995b). Other sentinel studies have indicated probable aerosol transmission within rooms of infected rats (Medina et al., 1994). Ordinary quarantine procedures should be regarded as sufficient to contain the infection to contaminated rooms in a facility. Treatment studies of experimentally infected rats (Ganaway et al., 1985) and mice (Matsushita and Suzuki, 1995) have shown the effectiveness of sulfonamides, especially sulfamerazine (compared with chlortetracycline and ampicillin), in preventing colonization by the organism or lesion development, but the role of treatment modalities to achieve eradication of natural infections has not been reported.

J. *Salmonella* sp. Infection (Salmonellosis)

Salmonella infections were among the first to be recognized as naturally occurring enzootic or epizootic diseases of laboratory rats and mice. Indeed, the first isolation of *Salmonella* typhimurium was made by Loeffler in the course of an epizootic of laboratory mice in 1892 (Pappenheimer and von Wedel, 1914). The scientific literature of the next 3 decades (1895–1925) was particularly rich on the subject of rat salmonellosis and dealt with perhaps three major themes: incidence and significance of salmonellosis in wild rats as vectors of human food contamination, the use of "rat viruses" (*Salmonella* cultures) for biological control of wild rat populations, and the documentation and description of intercurrent salmonellosis in laboratory rat colonies.

The genus *Salmonella* contains two serotypes, *S. enterica* and *S. bongori*. Based on DNA–DNA hybridization, *S. enterica* is divided into seven subgroups that are all designated as subspecies. Of these, *S. enterica* spp *enterica* is the most important, as it contains all serotypes relevant to warm-blooded animals. *Salmonella enterica* spp *enterica* has numerous serotypes, including typhimurium and enteritidis (which are not italicized as serotype designations have no standing in nomenclature), which are frequently abbreviated for convenience as *S.* typhimurium and *S.* enteritidis, respectively. Both have been recovered with sufficient frequency as to be regarded as typical for murine salmonellosis. The older literature contains a confusing synonomic array for both serotypes, as would be expected for a large bacterial group undergoing taxonomic study for almost 100 years. The synonymy for *S. typhimurium* includes *Clostridium typhimurium, Bacterium aertrycke, S. pestis caviae, S. aertrycke,* and *S. psittacosis.* In like manner, the following are regarded as synonymous with *S. enteritidis: Bacterium enteritidis, Clostridium enteritidis, Danysz Clostridium, S. enteritidis* var. *danysz,* and Gaertner's (or Gärtner's) *Clostridium.* Both *Salmonella* species have essentially identical epizootiology in rats, and the discussion that follows applies to both, except where noted. Salmonellas isolated from both wild and laboratory rats before the 1950s most commonly included *S. typhimurium* and *S. enteritidis.* Although many rodent infections with these agents are subclinical, they are quite pathogenic for rats (and mice) and the pathologic descriptions given below are based on infections caused by these two serotypes.

As mentioned previously, surveys of wild rat populations have established that these animals were, and remain, enzootic vectors of human *Salmonella* food poisoning, particularly of dairy, bakery, and meat products (Macalister and Brooks, 1914; Savage and Reid, 1914; Savage and White, 1923; Salthe and Krumweide, 1924; Jordan, 1925; Meyer and Matsumura, 1927; Verder, 1927; Kerrin, 1928; Khalil, 1934; Nafiz, 1935; Staff and Grover, 1936; Bartram et al., 1940; Welch et al., 1941). Although the incidence of survey reports has declined in recent decades, such studies indicate the continuing association of wild rats with *Salmonella* (Varela et al., 1949; Lee, 1955; Lee and Mackeras, 1955; Brown and Parker, 1957; Bruce et al., 1969; McKeil et al., 1970). For this reason, wild rats must still be regarded as significant hazards for the introduction of salmonellas to laboratory rat colonies. A wild rat source was shown to be the cause of a waterborne contamination that survived (faulty) chlorination in one of the few reported episodes of salmonellosis in laboratory rats since 1939 (Steffens and Wagner, 1983).

S. typhimurium and *S. enteritidis* are still occasionally encountered in laboratory rodents (usually with a wild rodent source in the history), but in recent years, most salmonellas isolated from laboratory animals are species more suggestive of originating from some other source. These sources include dietary components, drinking water, bedding materials, and avian or human contacts. The *Salmonella* serotypes encountered include the following: Agona, Amsterdam, Anatum, Binza, Blockley Bredeney, California Cerro, Infantis, Kentucky, Livingstone, Montevideo Oranienburg, Poona, Senftenberg, Tennessee, Mbamdaka, and others (Weisbroth, 1979; Simmons and Simpson, 1980, 1981; Kirchner et al., 1982;

Lentsch et al., 1983; Seps et al., 1987). These ordinarily nonrodent salmonellas tend to be less pathogenic for rats and mice and typically occur as essentially subclinical infections. Sporadic but quite current isolation of these nonrodent serotypes underscores the continuing importance of maintaining the high sanitary standards associated with rodent barrier facilities, including treatment of diets, beddings, and drinking water and personnel hygiene practices that include use of protective outerwear.

Considerable early interest attended the practical use of *Salmonella* cultures as rodenticides. They were prepared for this purpose on a large scale by commercial and public health organizations. The idea was first suggested in 1893 by the success observed by Loeffler with the use of *S. typhimurium* in combating a plague of field mice in Thessaly (Pappenheimer and von Wedel, 1914). Another of the early isolates was the *Clostridium* (*S. enteritidis*) isolated by Danysz (1914) from an epizootic of field mice in Charny-en-Seine, France. The isolate was observed to have considerable pathogenicity for rats. These agents were known as rat viruses and used as popular rodenticides for about 15 years (1895–1910). Such cultures as the Liverpool virus, the Azoa virus of Wherry, the Danysz and Issatschenko viruses, and Ratin I and II (all isolates of *S. enteritidis*) were widely used on the European continent, England, and in the United States as rat "poisons" (Bahr, 1905; Mallory and Ordway, 1909; Bahr et al., 1910; Pappenheimer and von Wedel, 1914; Jordan, 1925). They were still used until recently for this purpose in eastern Europe (Trakhanov and Kadirov, 1972). The predictable curtailment of this usage came about when it was recognized that the salmonellas could not be safely contained in the field, and they were, in fact, indicted in numerous human epidemics (Savage and White, 1923; Jordan, 1925; Bartram et al., 1940).

Salmonella infections have traditionally been internationally regarded as among the most important of laboratory animal diseases. The renown was not misplaced since it was a serious and common disease of laboratory rats for about 4 decades (1890–1930), as the frequent reports during that period indicate (Bainbridge and Boycott, 1909; Boycott, 1911; Pappenheimer and von Wedel, 1914; Cannon, 1920; Ball and Price-Jones, 1926; Price-Jones, 1926–1927; Friesleben, 1927; Bloomfield and Lew, 1943a,b,c). Much of the earlier literature, before 1914, has been summarized by Pappenheimer and von Wedel (1914). The naturally occurring disease in rats has been reported on several occasions subsequently (Buchbinder et al., 1935; Jones and Stewart, 1939; Ratcliffe, 1949; Taylor and Atkinson, 1956; Saquet, 1958; Kunze and Marwitz, 1967; Potel, 1967; Ray and Mallick, 1970; Lentsch and Wagner, 1983). The decreasing prevalence of this condition is worth emphasizing as it has only been reported in laboratory rats in the

United States once (Lentsch and Wagner, 1983) since 1939. The epizootiology of this disease has undergone a radical transition in American rodent stocks, attributable perhaps in equal measure to the concepts of gnotobiotic technology with the principle of disease exclusion, to the elevation of standards of animal care generally, and as a consequence of U.S. federal laws (viz., 91:579, the Animal Welfare Act and amendments) and the accreditation scheme of the American Association for Accreditation of Laboratory Animal Care. In most countries with similar standards, the disease is comparatively rare.

The clinical form of the disease in rats may vary from that of an acute epizootic with substantial mortality to that of a clinically latent disease that may be silent until the animals are placed under investigatory stress. Many factors are known to modulate the clinical expression of *Salmonella* infections in laboratory rats and mice. They include the level of acquired immunity, genetic resistance, virulence and dosage of the organism, age of the host, nutritional balance, and any of a variety of stressors that may act to lessen the immunologic competence or general vitality of the host (Topley and Wilson, 1922–1923; Webster, 1924; Guggenheim and Buechler, 1947; Bohme et al., 1954; Haberman and Williams, 1958; Blanden et al., 1966; Mackaness et al., 1966; McGuire et al., 1968; Newberne et al., 1968; Kampelmacher et al., 1969; Groschl, 1970; Robson and Vas, 1972; Tannock and Smith, 1972; Marecki et al., 1975). It is also pointed out that the pathologic consequences of *Salmonella* infection as a monocontaminant of the germfree rats and mice (Margard and Peters, 1964; Wostman, 1970; Ruitenberg et al., 1971) are considerably attenuated in comparison to microbially associated animals, thus the floral constitution of the host (and the issue of colonization resistance) comes into play as well. After the acute phase, natural infections in laboratory rats from enzootically infected colonies tend to moderate thereafter with little subsequent clinical signs or mortality.

Clinical signs of salmonellosis include reduced activity, thirst, ruffled fur, hunched posture, anorexia, weight loss, and softer lighter stools (Pappenheimer and von Wedel, 1914; Buchbinder et al., 1935; Ratcliffe, 1949; Maenza et al., 1970). Diarrhea has been an inconstant sign under conditions of natural infection, seen more commonly during acute episodes, and less frequently in enzootically infected colonies (Buchbinder et al., 1935). Experimental infections of rats with *S.* typhimurium 0 by Maenza et al. (1970) have established the brief incubation period of 2 to 3 days before development of clinical signs, including diarrhea. Diarrhea incidence averaged 15% to 20% of infected rats but was increased in rats that fasted before infection and by overcrowding. Maenza et al.

confirmed the results of others that the diarrheal sign was dose dependent (Bloomfield and Lew, 1942; Maenza et al., 1970). The incidence of diarrhea was maximal 4 days after infection and ranged in persistence from 1 to 21 days. The character of the stools in rats with diarrhea ranged from formless but normally colored, to mucoid and occasionally bloody bowel movements. The latter have been seen by others as well (Pappenheimer and von Wedel, 1914). Experimental *Salmonella* diarrhea has been extensively investigated as a diarrhea model in rats (Wostman, 1970; Powell et al., 1971a,b).

The development of methemoglobinemia during septicemic phases was first observed by Boycott (1911). This transitory sign causes browning of the normally pink eyes and an anemic, bluish discoloration of the tail, feet, and ears in albino rats. The observation of methemoglobinemia is not specific, disappears shortly after death, and is of little diagnostic value (Boycott, 1911; Pappenheimer and von Wedel, 1914; Ball and Price-Jones, 1926; Price-Jones, 1926–1927). Research results have documented nonspecific changes in serum enzyme concentrations (Woodward et al., 1969). Several reports have described sanguinous discharges accumulating about the conjunctiva and nares as terminal events (Pappenheimer and von Wedel, 1914). The clinical signs of gastrointestinal disturbance and methemoglobinemia may be suggestive, but neither is sufficiently unique or constant to be of diagnostic value.

The gross findings in fulminating and acute cases are limited to those consistent with septicemia. Most abdominal organs, particularly the liver and spleen, are congested, and lesions may be thus limited. Subchronic and chronic cases, with longer duration before death, generally have specific lesions. The disease is preeminently a classic enteric infection of the ileum and cecum; however, intestinal lesions may extend on either side of this location. Bacteremic or septicemic ascension to the liver and spleen occurs commonly. It is doubtful if the cardiac and pulmonary lesions described in the older literature (Pappenheimer and von Wedel, 1914) are *Salmonella* related, because they have not been described by more recent investigators (Buchbinder et al., 1935; Naughton et al., 1996).

Gross lesions of the intestines include thickening of the ileum and cecum with intraluminal accumulations of semiliquid feces, gas, and flecks of blood. The mesenteric and serosal vessels are congested, and the mesenteries edematous (Buchbinder et al., 1935). In animals with diarrhea, the intestinal lesions typically extend into the jejunum and include as key features large and small ulcers of the cecal epithelium and, less commonly, the terminal ileum. The ulcers are usually covered with fibrinous exudates. Microscopically, the ileal and cecal lesions are seen to consist of diffuse enteritis with infiltration of the lamina propria with macrophages, neutrophils, and, later, lymphocytes. The crypt epithelium is typically hyperplastic; however, the villous epithelium shows degenerative changes. Ulcers of the ileal mucosa, infrequently present, tend to be microscopic in size, whereas those of the cecum are larger (up to 20 mm with irregular outline) and grossly apparent. Cecal ulcers were not always accepted as occurring as direct consequences of salmonellosis in rats, and the relationship had been seriously questioned (Jones and Stewart, 1939, Stewart and Jones, 1941; Bloomfield and Lew, 1941, 1942, 1943a,b,c, 1948; Bloomfield et al., 1949). They are now recognized, however, as lesions attributable to *Salmonella* infection (Buchbinder et al., 1935; Maenza et al., 1970). It must be emphasized that subclinically infected animals without diarrhea have similar but less extensive lesions on both gross and microscopic levels of observation. A positive correlation has been shown to exist between the signs of diarrhea and the severity of ileal lesions (Maenza et al., 1970). Similarly, the persistence of diarrhea is related to the severity of ileal lesions, with sign subsiding as the tissues heal. The intestines of surviving animals were histologically healed by 4 to 5 weeks after infection (Maenza et al., 1970).

Grossly apparent indications of reticuloendothelial involvement are initiated early in the course and include enlargement of the Peyer's patches of the ileum, the mesenteric lymph nodes, and spleen. Microscopically, the Peyer's patches and lymph nodes are edematous, have inflammatory infiltrates, and may show focal necrosis or granuloma formation. Splenomegaly is a common feature (Thygessen et al., 2000). Such spleens are dark red or have a grayish capsule and may approximate 4 cm in length by 1 cm width (Pappenheimer and von Wedel, 1914; Buchbinder et al., 1935). Microscopically, the splenomegaly is seen to be owing to intense congestion of splenic sinuses, with spilling over of erythrocytes into the white pulp. Extensive inflammatory infiltrates, particularly of phagocytic cells, are seen in splenic tissues. Focal lesions varying from necrosis to granulomata are commonly seen.

The liver is variably congested and dark red or pale and friable. Focal lesions (1 to 2 mm) may be seen grossly under the capsule and on cut surface but more commonly are not. On microscopic examination, however, small areas of coagulative, acellular necrosis are seen, as has been described in mice (Tuffery and Innes, 1963). Granuloma formation in the rat liver has been described (Macalister and Brooks, 1914). In the original histopathologic description provided by Pappenheimer and von Wedel (1914), small microthrombi (cell fragment thrombi)—composed of mononuclear phagocytes containing cell fragments and bacteria, degenerating phagocytes, cellular

debris, and fibrin—were observed in organs with focal necrosis, for example, spleen, liver, and lymph nodes. The investigators concluded that the focal necroses were actually sequella of embolically occluded (thrombosed) microvasculature. One is tempted to recall, in support of this concept, the pulmonary thrombi in hamster salmonellosis described as pathognomonic by Innes et al. (1956). Certain lesions seen experimentally by using the rat *Salmonella* model—for example, cholangitis (Arora and Sangal, 1973), placentitis (Hall, 1974), and polyarthritis (Volkman and Collins, 1973)—have not been observed in natural infections.

Diagnosis of salmonellosis in the rat is based on a consistent case history, isolation of the organism in pure culture, and cultural and immunologic confirmation of the identified isolate as one of the *Salmonella* serotypes. It has been pointed out, in the mouse, that difficulty should not be experienced in the retrieval of *Salmonella* species encountered during acute epizootics (Morello et al., 1964). Large numbers of animals die of the infection, and the many ill animals shed high ratios of *Salmonella* organisms into the feces. More problems have been encountered in the diagnosis of salmonellosis in asymptomatic carrier animals from enzootically infected colonies. The asymptomatic carrier animal (mouse) has been investigated in some detail (Jenkin et al., 1964). It is understood from such studies that the relationship with the host is one of intracellular multiplication by the organism within a mileu of humoral antibody. Exacerbation of the disease in recovered carriers is prevented by specific antibody, but complete elimination is inhibited by smoldering replication of the bacteria within macrophages. Such animals may shed *Salmonella* intermittently in the feces over long periods of time and represent infective reservoirs of the disease. Difficulty of detection owing to episodic shedding is compounded by the problem that the prevalence of asymptomatic carriers may range from a high 50% of the animals to as low as 5%, thus introducing a substantial risk of fecal sampling error (NRC, 1991). That essentially the same factors are involved in the rat is generally accepted, and the asymptomatic carrier rat has been amply documented (Buchbinder et al., 1935; Naughton, 1996).

Although clearly of great diagnostic importance, the focus of infection in recovered carriers is not well understood and probably is not the same in each carrier. Several studies have attempted to localize the most likely organs or other locations in carrier animals for more efficient diagnostic retrieval (Buchbinder et al., 1935; Slanetz, 1948; Hoag and Rogers, 1961; Margard et al., 1963; Hoag et al., 1964; Morello et al., 1964, 1965). The consensus is that although a greater number of isolates may be retrieved from intestines—compared with liver, spleen, heart, or other organs—none of these tissue locations results in higher percentage of isolations than from feces alone. For this reason, diagnostic monitoring programs for murine salmonellosis have been oriented to fecal culture.

Salmonella is quite stable in dried feces and on environmental surfaces. Strains inoculated in media can survive at room temperature (23°C) for more than 1 year (Kagiyama and Wagner, 1994). As a Gram-negative rod, it is quite sensitive to heat and is killed by exposure to temperatures of 55°C for 1 hour or of 63°C (145°F) for 20 minutes. Note that the pasteurizing temperatures at which rodent diets are pelleted are sufficiently border-line in terms of time and temperature as to variably inactivate *Salmonella* (NRC, 1991). Cage washing machine temperatures of 82°C (180°F) are sufficient to inactivate *Salmonella*. Ordinary disinfectants, including sodium hypochlorite, iodophore detergents, ethanol, phenol, and formaldehyde, are effective (Kagiyama and Wagner, 1994).

The factors relating to efficacy of retrieval of *Salmonella* from rat and mouse feces under varying conditions have been studied. It is known that recovery from individual animal samples is higher than that from pooled samples (Slanetz, 1948; Margard et al., 1963), that the highest number of isolates will be obtained from the first six fecal pellets in any given animal (Margard et al., 1963), and that storage up to 8 days in fluid media does not result in greater degree of isolations than dry feces stored for 8 days (Margard et al., 1963). Similarly, storage temperatures (whether frozen, at room temperature [20°C] or at 40°C) did not affect the percentage of isolations (Margard et al., 1963). It is recommended that feces shipped for diagnostic culture be kept cool and dry in order to minimize growth of species other than *Salmonella* that may interfere with isolation.

In like manner, primary culture media used for isolation of the organism before biochemical characterization have been explored in great detail. The consensus is that the highest numbers of isolations follow the sequence of overnight enrichment in a broth medium to permit preferential multiplication of *Salmonella* (compared with coliforms) followed by plating on differential solid agar media. The media have been thoroughly reviewed by Margard et al. (1963), but they and others (Haberman and Williams, 1958; McPherson, 1963; Morello et al., 1964; Black, 1975) recommend the sequence of incubating fecal pellets or tissue macerates in selenite-F plus cystine broth overnight followed by streaking onto brilliant green agar as the method of choice. Suspicious colonies are picked from the brilliant green plates and inoculated on triple sugar iron. Isolates consistent with *Salmonella* sp. are then confirmed by agglutination in polyvalent antisera and characterized biochemically according to the profile in Table 11-4 (see also identification

tables in Hansen, 2000). The serotyping of isolates is beyond the competence of most laboratories, and characterized isolates should be forwarded to a recognized diagnostic center for complete identification.

Both rats and mice respond immunologically to infective challenge by *Salmonella* (Boycott, 1919; Topley and Ayrton, 1924; Ball and Price-Jones, 1926; Buchbinder et al., 1935; Slanetz, 1948; Hobson, 1957; Kent et al., 1967; Collins, 1968, 1970). Traditionally, it has been thought that postrecovery antibody levels, especially in immune asymptomatic carriers, are too low and inconsistent to be of great value in diagnosis. Numerous studies have documented the two- to threefold higher number of successful isolations of *Salmonella* from feces compared with detection of antibodies in the same animals (Slanetz, 1948; Hobson, 1957; Sacquet, 1958, 1959). For this reason, immunodiagnosis has not been generally recommended, other than as an adjunct to cultural methods for diagnosis. Of the methods explored, complement fixation was found to be more sensitive than was tube agglutination (Morello et al., 1964). More recently, application of ELISA methodology appears to offer promise as a means of detecting low postrecovery antibody levels in carriers found negative by less sensitive serologic tests (Lentsch et al., 1981). *Salmonella* can be detected by PCR and may detect as few as 10 to 50 bacterial cells in the sample (Soumet et al., 1995; Stone et al., 1994).

Vaccination has not been endorsed as a method of prevention, although the concept has been seriously explored. Vaccination with bacterins or bacteriophage has not been completely effective in either preventing infection in vaccinated individuals or in preventing spread when vaccinated and nonvaccinated mice were placed in close proximity (Topley and Wilson, 1922–1923, 1925; Topley et al., 1925; Ibraham and Schutze, 1928; Walker et al., 1964; Badakhsh and Herzxberg, 1969; Rauss and Kalovich, 1971; Cameron and Fuls, 1974).

The concept of treatment of infected individuals has been discouraged because of the incomplete effectiveness of various treatment regimens (Haberman and Williams, 1958). The problem has been that although the majority of animals may be apparently freed of *Salmonella,* a residue of infection always remains to act as a reservoir for rebound of infection in susceptible animals (Bruner and Moran, 1949). Similarly, although fecal testing programs oriented to detection and culling of infected animals may seem to be statistically effective, it is almost impossible, with certainty, to eradicate the infection from a given colony by these means (Steffen and Wagner, 1983). Current recommendations favor destruction of colonies in which the disease has been detected, followed by replacement with gnotobiotically derived stock known to be free of *Salmonella*. There is no evidence that the organism may traverse the placental barrier.

K. Pasteurellaceae Infection

Based on accumulating information, the 2nd edition of this text recognizes that any discussion of *P. pneumotropica* needs to take place within its context as a member of the family Pasteurellaceae. As a group, the Pasteurellaceae are Gram-negative, fermentative, nonmotile coccobacilli that are oxidase and catalase positive and that variably fail to grow on MacConkey agar. As closer scrutiny and more precise methods of taxonomic analysis have been applied to the Pasteurellaceae, to the three classic genera—viz. *Pasteurella* (1897), *Actinobacillus* (1910), and *Haemophilus* (1917)—four new genera have recently been added, namely, *Lonepinella* (1995), *Mannheimia* (1999), *Phocoenobacter* (2001), and *Gallibacterium* (2003). The Pasteurellaceae may prove to be as taxonomically complex as the Enterobacteriaceae and might contain as many as 20 genera (Mutters et al., 1989; de Ley et al., 1990; Bisgaard, 1995). There are many poorly characterized Pasteurellaceae and several members whose taxonomic position is unclear. For health monitoring programs, FELASA has taken the position of recommending surveillance for all members of the Pasteurellaceae as a health monitoring strategy, rather than selecting the genera and species associated with disease, a policy that recognizes the current difficulties with taxonomy and speciation, the impossibility of separating pathogenically relevant species from irrelevant commensals, and the differences in testing outcomes in different laboratories (Nicklas et al., 2002).

1. *Pasteurella pneumotropica*

The most frequently reported rodent pathogen of the Pasteurellaceae is *P. pneumotropica*. However, in many of these reports, when examined critically, it needs to be recognized that the bacteriologic characteristics presented really do not allow definitive identification to the species level (see Mutters et al., 1989 for cultural criteria). Based on biochemical differences (see Table 11-6), *P. pneumotropica* is divided into two biotypes, viz. *P. pneumotropica* type Heyl and *P. pneumotropica* type Jawetz. The Heyl and Jawetz biotypes are genetically distinct (less than 30% related by DNA–DNA hybridization), and both biotypes appear related more to the the genus *Actinobacillus* than to *Pasteurella* (Ryll et al., 1991; Bisgaard, 1995). On the basis of cultural criteria, DNA–DNA hybridization and randomly amplified polymorphic DNA analysis by PCR, biotype Heyl isolates are genetically homogeneous, but biotype Jawetz isolates form a genus-like cluster that even contains some V factor (NAD)-dependent strains (Ryll et al., 1991; Weigler et al.,

TABLE 11-6

CHARACTERISTIC BIOCHEMICAL REACTIONS AND OTHER PROPERTIES OF *PASTEURELLA PNEUMOTROPICA*

First stage Pasteurellaceae		Second stage *Pasteurella pneumotropica*		Third stage biotypes:	Heyl	Jawetz
carb F/O/−	F	adonitol	−	arabinose	+	−
Catalase	+	aesculine	−	LDC	v	−
coccus/rod	rod	cellobiose	−	melibiose	+	−
gr. in oxygen	+	dulcitol	−	raffinose	+	−
gr. Anaerobic	+	glucose	+			
Gram stain	−	glucose gas	v			
Motility	−	indole	+			
Nitrate	+	gr. MacConkey	v			
Oxidase	+	mannitol	−			
		galactose	+			
		mannose	+			
		urease	+			
		ODC	+			
		ONPG	+			
		phosphatase	+			
		fructose	+			
		salicine	−			
		sorbitol	−			
		sorbose	−			
		sucrose	+			
		trehalose	(+)			

F/O/− indicates carbohydrates fermented/oxidized/unutilized, gr., growth; v, variable reaction depending on isolate; (+), delayed reaction; ODC, ornithine decarboxylase; ONPG, 2-nitrophenyl 6-D-galactopyranoside.

1996; Schultz, 1997; Kodjo et al., 1999). Mouse, but not rat, isolates of *P. pneumotropica* have been shown to possess a hemagglutinating property for human type O Rh+ cells (Boot et al., 1993b). A previously included third biotype, *P. pneumotropica* type Henriksen, that fails to ferment xylose has been reclassified as *P. dagmatis* (Mutters et al., 1985). Although occasionally isolated from rats (Boot and Bisgaard, 1995), *P. dagmatis* is encountered primarily in carnivore bite wounds of humans and does not appear to play a role in rodent clinical infections.

Since its original description (Jawetz, 1948), *P. pneumotropica* has been associated mainly with the respiratory tract of laboratory rodents, a predilection emphasized in the species name (Jawetz, 1950; Jawetz and Baker, 1950). It is uncertain if the organism has emerged since the late 1940s or, more likely, has only been recognized as a distinct entity since that time. Sporadic reports of pasteurellosis in rats occur before that time, notably those of Schipper (1947), and Meyer and Batchelder (1926). Similarly, the older literature contains references to respiratory infection in rats with organisms that on close examination resemble what are now recognized as Pasteurellaceae (Hoskins and Stout, 1920; Matheson et al., 1955; Harr et al., 1969).

P. pneumotropica has emerged since the 1950s as an important and globally distributed infectious disease of laboratory rats. It has been described as a clinical entity in laboratory mice on a number of occasions (Jawetz, 1950; Gray and Campbell, 1953; Heyl, 1963; Brennan et al., 1965; Wheater, 1967; Weisbroth et al., 1969; Black, 1975; Needham and Cooper, 1976; Wilson, 1976; McGinn et al., 1992; Grossman et al., 1993, Dickie et al., 1996; Marcotte et al., 1996; Artwohl et al., 2000; Macy et al., 2000). *P. pneumotropica* may be carried asymptomatically as an inapparent organism in respiratory and other tissues of a number of host species other than laboratory rats and mice, including hamsters (Brennan et al., 1965; McKenna et al., 1970; Sparrow, 1976; Lesher et al., 1985), guinea pigs (Sparrow, 1976; Boot et al., 1995b), and rabbits (Sparrow, 1976). The organism has been reported from carnivores (Olson and Meadows, 1969) and man (Henriksen and Jyssum, 1960, 1961; Henriksen, 1962; Jones, 1962), although most, if not all, carnivore and human isolates would now be reclassified as *P. dagmatis*. *P. pneumotropica* is commonly isolated from wild rats and mice (Van der Schaaf et al., 1970; Boot et al., 1986). These other host species should be regarded as reservoirs of potential *P. pneumotropica* infectivity for laboratory rats, and this potential strengthens the accepted principle of separation of species having different pathogen profiles in animal care facilities.

Although encountered less commonly in recent years, a number of surveys have established the widespread distribution of *P. pneumotropica* in breeding stocks of rats and mice (Tregier and Homburger, 1961; Hoag et al, 1962; Flynn et al., 1965; Wheater, 1967; Casillo and Blackmore, 1972; Hill, 1974; Hooper and Sebesteny, 1974; Rehbinder and Tschappat, 1974; Young and Hill, 1974; Sparrow, 1976), frequently in the absence of clinically overt disease. Indeed, inapparent infection has remained one of characteristics of the organism. Even under conditions of experimental infection, the organism may be clinically silent (Van der Schaff et al., 1970; Jakab and Dick, 1973; Hill, 1974; Burek et al., 1976). Alternatively, *P. pneumotropica* may occupy the role of secondary opportunist or copathogen, synergistically enhancing the pathogenicity of a number of respiratory agents, including Sendai virus (Jakab and Dick, 1973; Jakab, 1974), *M. pulmonis* (Lutsky and Organick, 1966; Brennan et al., 1969a; Eamens, 1984), and *P. carinii* (Macy et al., 2000). Finally, it may appear to act as a primary pathogen associated with production of disease and otherwise meeting Koch's postulates for cause and effect (Brennan et al., 1969a,b; Van der Schaaf et al., 1970), although the resistance-lowering potential of viral coinfection in these reports is not known. Little is known about the triggering mechanisms promoting progression from inapparence to pathogenicity with this organism, although such factors as impairment of pulmonary phagocytic defense mechanisms are known to favor multiplication of *P. pneumotropica* (Goldstein and Green, 1967; Jakab and Dick, 1973; Jakab, 1974). This factor, at least, may be used to interpret the synergism of infections coupled with other pulmonary pathogens. An increasing body of clinical experience indicates that immunologic impairment of the rodent host predisposes to clinically overt expression of infection (Kent et al., 1976; Moore and Aldred, 1978; Hajjar, 1991; Grossman et al., 1993; Hansen, 1995; Dickie et al., 1996; Marcotte et al., 1996; Artwohl et al., 2000; Macy et al., 2000). Thus, immunocompetence of the host would seem to be as important as is pathogenicity of the organism in determining outcome of infection.

Under conditions of natural infection, *P. pneumotropica* has been isolated most frequently among the microflora of the nasoturbinates, pharynx, conjunctiva, trachea, lungs, and uterus. Less commonly, infection of deeper organs of laboratory rats has been observed, for example, liver, spleen, and kidney (Wheater, 1967; Needham and Cooper, 1976), and *P. pneumotropica* has been encountered as a contaminant of Morse Antibody Production (MAP) test tissues (Nakai et al., 2000). The earliest primary site of colonization with *P. pneumotropica,* is the laryngopharynx (Saito et al., 1981; Mikazuki et al., 1994). Colonization of the laryngopharynx may approach an incidence of 100% in rat colonies (Saito et al., 1981), and this nidus underlies common isolation of the organism from the nasoturbinates, lower intestine, and feces. This primary localization also accords well with the analysis of common clinical presentations outlined by Besch-Williford and Wagner (1982): (1) conjunctivitis, blepharitis, and panophthalmitis (Gray and Campbell, 1953; Brennan et al., 1969; Wagner et al., 1969; Weisbroth et al., 1969; Rehbinder and Tschappat, 1974; Young and Hill, 1974; Kent et al., 1976; Moore et al., 1978; Roberts and Gregory, 1980; Artwohl, 2000); (2) skin furunculosis, superficial lymph node abscesses, and mastitis (Brennan et al., 1965; Wheater, 1967; Weisbroth et al., 1969; Van der Schaaf et al., 1970; Sebesteny, 1973; Wilson, 1976; Hong and Ediger, 1978a; Moore and Aldred, 1978); (3) otitis media (Wheater, 1967; Eamens et al., 1984; Harkness and Wagner, 1975; McGinn et al., 1992); and (4) uterovaginal infections (see citations below). Besch-Williford and Wagner (1982) suggested that *P. pneumotropica* could ascend via the Eustachian tube and nasolacrimal duct to extend infection to the middle ears and orbital structures, via oronasal contact with the genitalia to infect the preputial glands, vagina, and uterus, or by bites and suckling to infect the skin or mammaries, respectively.

Reports involving inapparent infection in axenic rats indicate that the organism readily colonizes the intestines, where it may be carried asymptomatically for long periods of time (Moore et al., 1973), although enteric colonization of rats may not have epizootiologic significance beyond that of a peculiarity of gnotobiosis. In neonatal hamsters, *P. pneumotropica* was associated with enteritis (Lesher et al., 1985). Because of the more common association of *P. pneumotropica* with the respiratory system and of the difficulties of isolating *P. pneumotropica* from the profusion of vigorous enteric forms encountered in rat feces, it is possible that the epizootiological role of intestinal colonization has not been adequately investigated. In defense of a more common enteric occurrence, it has been pointed out that intranasal inoculation of fecal extracts (from infected carriers) results in respiratory infection by *P. pneumotropica* (Wheater, 1967).

The vagina and uterus have been frequently detected as infected with *P. pneumotropica* (Blackmore and Casillo, 1972; Burek et al., 1972; Ackerman and Fox, 1981). Reproductive tract involvement may be either asymptomatic or clinically apparent. Uterine infections appear to ascend from the vagina (Blackmore and Casillo, 1972), where *P. pneumotropica* may be the dominant microbial form (Larsen et al., 1976; Yamada et al., 1983). Studies have established that *P. pneumotropica* colonization of rat uterus may approach an incidence of 60% to 70% under conditions of natural infection (Graham, 1963; Casillo and Blackmore, 1972; Larsen et al., 1976). Vaginal populations of the organism have been shown to wax

and wane according to stages in the estrus cycle, rising to highest levels at stages with greatest numbers of cornified, nonnucleated cells, and cellular debris (Yamada et al., 1986).

Transmission, as pointed out above, may take place via uterovaginal contamination (venereally), but transmission more commonly occurs by direct oronasal contact. Same cage sentinels were shown to be culture positive after 2 weeks and to have seroconverted by 3 weeks after contact exposure (Scharmann and Heller, 2000). Spread via contaminated vectors has only limited potential to spread infection. The isolate they studied (NCTC 8254) was quite fragile on inanimate surfaces, surviving no more than 30 minutes on laboratory coat fabric, about 1.5 hours on plastic surfaces, and up to 2 hours on mouse hair (Scharmann and Heller, 2000). Although freshly passed feces may act as a contaminated vector substance (Hong and Ediger, 1978a), the practice of exposing sentinels to dirty bedding from infected carrier rodents over a period of 12 weeks did not result in infection (Scharmann and Heller, 2000).

The pathology of *P. pneumotropica* infection is not distinctive and resembles that caused by any of a variety of pyogenic bacteria in similar sites. Subcutaneous abscesses of the skin, adnexal organs, or orbital structures are generally encapsulated and filled with liquid exudates. Microscopically, the lesions are suppurative with liquifactive necrosis centrally surrounded by a zone of granulomatous inflammation. Clinically silent infections in the lungs, upper respiratory tract, uterus, and intestines frequently occur without histopathologic indication of epithelial inflammation. In the lungs of mice, areas of consolidation with perivascular and peribronchial infiltration by acute and chronic inflammatory cells may be produced (Brennan et al., 1969a), although this lesion is not distinctive. It has been pointed out that the pathologic architecture of experimental pulmonary disease caused by *P. pneumotropica* and *M. pulmonis* in combination more resembles naturally occurring chronic respiratory conditions in rodents than either organism used singly (Brennan et al., 1969a). In this connection, it is worth emphasizing that disease conditions with *P. pneumotropica* in immunocompetent hosts have been more commonly encountered in conventional rats and mice than they have been in gnotobiotes or rodents free of common pathogens. Isolator or barrier breaks involving gnotobiotes have more frequently tended to be clinically silent, perhaps reinforcing the status of the organism as an organism of ordinarily low pathogenicity requiring the resistance-lowering presence of either a primary copathogen or immunologic impairment for infection to result in disease.

Diagnosis of *P. pneumotropica* infection is established (or confirmed) by cultural retrieval of the organism from infected tissues and isolation in pure culture.

Primary isolation of the organism is accomplished readily on blood agar by using swabs from epithelial surfaces or lesion contents or aspirates from the nasoturbinates, larynx, or trachea. Isolation from feces or intestinal contents is facilitated by use of a selective medium (Garlinghouse et al., 1981; Mikazuki et al., 1987). Blood agar cultures should be incubated in a moist, micro-aerophilic environment. After 24 to 48 hours, colonies are 1 to 2 mm in diameter, nonhemolytic (although greening of the agar deep to the colonies may occur), entire, smooth, convex, and grayish. Colonies with organisms that are Gram-negative, fermentative coccobacilli, nonmotile and catalase, oxidase and nitrate positive should be regarded as presumptive for Pasteurellaceae and further identified as provisional *P. pneumotropica* isolates according to the profile in Table 11-6. Many isolates, especially *P. pneumotropica* biotype Jawetz, fail to grow on McConkey agar. Biotype Heyl isolates are positive for fermentation of melibiose and raffinose, whereas biotype Jawetz isolates fail to use those sugars (Nicklas et al., 1995). Mutters et al. (1995) used the reactions to lysine decarboxylase, melibiose, and arabinose to separate Heyl and Jawets biotypes. Investigators of rodent Pasteurellaceae stress that definitive speciation of provisional *P. pneumotropica* isolates requires analysis of a panel of about 33 to 40 biochemical traits (Kunstyr and Hartmann, 1983; Boot and Bisgaard, 1995), and biochemical reactions may be delayed and weak (Table 11-6). Commercially available nonfermenter identification kits; for example, API20 NE and Remel RapID NF Plus, can be used to establish isolates as belonging to the Pasteurellaceae, but not for speciation. Biochemical variants are common and suggestive of other, closely related Pasteurellaceae (see ahead under "Actinobacillus").

PCR methodology has been applied both as a screening tool for detection of *P. pneumotropica* infection (Wang et al., 1996; Bootz et al., 1998; Kodjo et al., 1999) and as confirmation of cultural isolates (Weigler et al., 1996, Nozu et al., 1999). In testing samples from a known positive colony, PCR was more sensitive than was culture and detected about 30% more samples to be PCR positive (Bootz et al., 1998). The problem has been to balance the goals of the screening program against the range of Pasteurellaceae detected by the PCR. The primer sequences developed by Wang et al. (1996) were based on the 16S rDNA sequence of the *P. pneumotropica* Jawetz biotype, but were not validated against other known *P. pneumotropica* biotypes. The PIN-1 and PIN-2 primers used by Wang et al. were evaluated by Weigler et al. (1999) and found to detect all biotypes of *P. pneumotropica* and the closely related *A. muris* but not other Pasteurellaceae. PCR tests reported by Kodjo et al. (1999), were similarly structured to narrowly limit detection to *P. pneumotropica* biotypes. A different strategy was used by Bootz et al.,

(1997, 1998), who developed primer sets meant to detect all Pasteurellaceae known (at that time) to infect rats and mice. With the use of these primers (and an adequately representative sampling), a negative result would mean the colony the sample was drawn from could be declared free of all Pasteurellaceae. A positive result was more complicated and, if speciation was required, would mean extensive follow-up cultural and molecular testing to either identify the isolate or to rule out specific taxons.

The advent of more sensitive serologic methods, for example, ELISA, has done much to dispel the limitation cited in earlier reports (based on complement fixation and agglutination tests) that although sera from lesioned animals were usually positive, even with autologous antigens, these test methods would not consistently detect the lower antibody levels deriving from inapparent infection (Hoag et al., 1962; Weisbroth et al., 1969). The use of ELISA methodology has been demonstrated to be useful (and more sensitive than culture) in detection of clinically inapparent carriers from selected, known infected colonies of rats and mice (Wullenwebber-Schmidt et al., 1988; Manning et al., 1991; Boot et al., 1995a). In this connection, both the sensitivity and specificity of detection have been improved through the use of cleaner cell wall antigens (lipooligosaccharides) than whole-cell preparations (Manning et al., 1989). The problem in recommending general use of ELISA in testing serologic unknowns derives from the issue of antigenic variability among the Pasteurellaceae generally and *P. pneumotropica* biotypes more specifically. There is some evidence that antigenic divergence (and some degree of host specificity) occurs among rat biotypes of *P. pneumotropica*, compared with those occurring in mice (Nakagawa and Saito, 1984; Boot et al., 1994a,b). Similarly, antibodies to guinea pig Pasteurellaceae, including guinea pig *P. pneumotropica* biotypes, do not cross-react in ELISA tests using mouse origin *P. pneumotropica* antigens (Boot et al., 1995b,d). The problem that results for the diagnostic laboratory is that there would appear to be no single (monovalent) serotype with antigens that would detect antibodies to all rodent biotypes, thus the potential for false negative is substantial. At the same time, the issue of common antigens among the Pasteurellaceae is widespread and underlies the potential for false positives if the objective is to serologically detect only *P. pneumotropica* (Wullenwebber-Schmidt et al., 1988). The latter potential is sufficiently prevalent that it has been recommended that rodents older than 12 weeks not be serologically tested (Bootz et al., 1998).

Treatment modalities, especially for amelioration of clinical expressions of disease have greatly improved in recent years. Earlier experience with treatment, using an earlier armamentarium of antibiotics, commonly resulted in clinical amelioration during treatment phases, but incomplete elimination of the organism and its rebound after cessation of treatment (Gray and Campbell, 1953; Hoag et al., 1962; Wheater, 1967; Spencer and Savage, 1976; Hansen, 1995). Oral dosing with enrofloxacin, a fluoroquinolone antibiotic, has been shown effective in apparently eliminating *P. pneumotropica* without recurrence after withdrawal of the drug (Goelz et al., 1996; Ueno et al., 2002).

The organism may be excluded from rat stocks by gnotobiotic rederivation, with the proviso that only culturally negative uteri be used, or by embryo transfer (Reetz et al., 1988; Suzuki et al., 1996). Rodents with infected uteri may conceive and produce full-term fetuses as effectively as with uteri devoid of *P. pneumotropica* (Flynn et al., 1968). Uterovaginal infection has been observed to result in contamination of neonates and fetuses derived by hysterectomy (Casillo and Blackmore, 1972; Hong and Ediger, 1978; Ward et al., 1978), and this potential needs to be carefully considered by those conducting rederivation programs (Reetz et al., 1988). Lack of appreciation of the clinically inapparent colonization of rat (and mouse) uterus by *P. pneumotropica* more than any other single factor has been responsible for the reintroduction of this organism to isolators or barriers of gnotobiotic rodents, and has been a frequent cause of barrier "breaks" (Blackmore, 1972; Casillo and Blackmore, 1972). Enrofloxacin has been used preparatory to rederivation to clear or reduce uterine populations of *P. pneumotropica* to enhance likelihood of a successful outcome (Macy et al., 2000).

2. *Actinobacillus*

Increasingly, the rodent actinobacilli are being diagnostically encountered, usually in asymptomatic carriers, but occasionally, as *P. pneumotropica,* in association with clinical conditions of the respiratory and reproductive systems. The actinobacilli are included with the Pasteurellaceae in the profile for required health surveillance recommended by FELASA (Nicklas et al., 2002). Several *Actinobacillus* species have been reported from laboratory rodents, often in coinfection with *P. pneumotropica*. In a survey of feral rodents, many were found to be reservoir hosts for *Actinobacillus* species, *P. pneumotropica,* or both (Boot et al., 1986).

The earliest reference to rodent actinobacilli is that of Simpson and Simmons (1980), who reported isolation of two species, *A. equuli* and one not further identified, from nasopharyngeal washes and ceca of laboratory rats and hamsters undergoing health surveillance screening tests. One of the hamster isolates originated from a cheek pouch abscess, but the rats and other hamsters were asymptomatic. In the same year, Lentsch and Wagner (1980) reported isolation of *A. equuli* and *A. lignierseii* from

oropharynx, conjunctiva, and middle ears of laboratory rats, mice, and guinea pigs. Shortly after, *P. ureae* was reported as a copathogen with *P. pneumotropica* in an epizootic of mouse reproductive tract disease (Ackerman and Fox, 1981). The Ackerman and Fox isolate was later reclassified by Bisgaard (1986) as *A. muris,* in the original description of this new species. The species name *A. ureae* is retained to identify a different organism whose natural distribution appears limited to humans (Mutters et al., 1984). Also in 1981, an organism later reclassified as *A. muris* was isolated from instances of mastitis in laboratory mice (Gialamas, 1981).

The organism previously named *Haemophilus influenzaemurium* (or *H. influenzae murium*) has similarly undergone revision. The taxonomic position of *H. influenzaemurium* is unsettled, but it does not depend on V or X for growth and belongs with the *P. pneumotropica* Jawetz biotype group. The name *H. influenzaemurium* is no longer listed as a valid name (Euzby, 2003). Based on DNA homology, Nicklas et al., (1997) have classified this organism with *A. muris,* and for that reason, it is included in this section. *H. influenzaemurium* was first isolated from mice with respiratory disease (Kairies and Schwartzer, 1936; Ivanovics and Ivanovics, 1937) and was not reported again for 40 years until encountered during health surveillance tests on healthy laboratory mice in Hungary, reisolated, and characterized (Czukas, 1976).

The rodent actinobacilli are closely related phylogenetically and often confused with *P. pneumotropica* because of the similarity of locations in the host, DNA homology and, when isolated, biochemical profiles. The actinobacilli differ from *P. pneumotropica* by lacking ornithine decarboxylase and urease and fail to use indole (see Table 11-6 and also biochemical profiles tabulated in Bisgaard 1986; Boot and Bisgaard, 1995; Bootz et al., 1998). Isolates of *A. muris* may also be distinguished from *P. pneumotropica* (and other actinobacilli) by genetic fingerprinting via randomly amplified polymorphic DNA analysis (Weigler et al., 1986). The rodent actinobacilli may be detected by diagnostic PCR of swab samples. The primers used by Bootz et al. (1997, 1998) to detect all members of the Pasteurellaceae were demonstrated to detect all reference strains of rodent actinobacilli. The primers used by Wang et al. (1996) and designed to detect *P. pneumotropica* would not detect *A. muris* if a more stringent annealing temperature (57°C) were used (Weigler et al., 1998).

3. *Haemophilus*

Both the incidence and importance of rat *Haemophilus* species as pathogens of rats are difficult to assess. On the one hand, the X factor– and V factor–dependent Pasteurellaceae (including *Haemophilus*) are in the profile of organisms recommended by FELASA as important for routine health surveillance of laboratory rats and mice (Nicklas, 2002). They appear to be only weakly pathogenic and are usually clinically inapparent. The *Haemophilus* species are fastidious in their growth requirements, and microbiologic screening for them requires an additional sequence of media. Their isolation and identification pose a significant challenge for clinical microbiologists.

On the other hand, the incidence of *Haemophilus* in laboratory rats may be substantially underestimated because the usual diagnostic regimen does not use the media required for their isolation. One recent survey (Nicklas et al., 1994) indicated an incidence of 21% in rat diagnostic specimens and 2.4% in those from mice. Barrier-reared guinea pigs seem to rarely be free of infection, as indicated by the frequency of serologic positivity (Boot et al., 1999). The V factor–dependent Pasteurellaceae have been encountered in episodes of respiratory disease in rats often under circumstances where a diagnostic search for X factor– or V factor– dependent pathogens was prompted by failure to isolate an etiologic agent on blood agar (Nicklas et al., 1989) or by failure to isolate *Pasteurella* in rat sentinels that had seroconverted to positive against *P. pneumotropica* antigens (Boot et al., 1995c). Moreover, intercurrent infection by *Haemophilus* can modulate certain research protocols. It has been demonstrated that infection has a depressive effect on development of adjuvant arthritis in Lewis (LEW) rats (Boot et al., 1995a).

Clinical specimens for isolation are essentially the same as those recommended for *P. pneumotropica,* that is, nasopharyngeal washes or swabs of the oropharynx or vagina. The rodent *Haemophilus* species are V factor– dependent for their growth and will not be isolated from clinical inocula cultured on (unsupplemented) blood agar. The required V (NAD) and X (hemin) factors are both provided by the gentle heating of blood cells in the preparation of chocolate agar, the medium recommended for clinical isolation. Supplementation of the medium with clindamycin can be used to inhibit overgrowing microflora (Garlinghouse et al., 1981; Boot et al., 1995c). Inoculated chocolate agar plates should be incubated in a moist, microaerophilic (5% to 10% CO_2) environment for 24 to 48 hours. Dependence on V or X factors, or both, for growth can be determined by the growth patterns of lawns of the isolate streaked onto Mueller-Hinton agar plates, onto which discs with V or X factors or both (V-X) are placed. Radial growth, for example, surrounding the V and X-V discs are indicative of V factor dependence. Caution should be exercised in assigning V factor–dependent Pasteurellaceae to the genus *Haemophilus* however, because there are X factor– and V factor–dependent strains of both *Pasteurella* and *Actinobacillus* (Nicklas et al., 1993; Nicklas and Benner, 1994; Boot et al., 1995a). Although one outbreak of respiratory disease in rats involving X-factor dependency has been reported (Harr et al., 1969), the X-factor dependency of the isolate was not definitively established. The transmissibility of V factor–dependent Pasteurellaceae between laboratory rodents, including rats, mice, and guinea pigs has been demonstrated (Boot et al., 2000); thus, the separation of host species having disparate pathogen profiles is recommended.

Presumptive *Haemophilus* isolates may be further characterized by use of commercial kits, for example, API NH (bioMerieux France), with the proviso that it be recognized that profiles or software supplied by the kit are oriented to identification of human isolates and are generally inadequate for rodent isolates (Brands and Mannheim, 1996). Rat *Haemophilus* isolates will usually be classified by the API NH system as *H. parainfluenzae* (Boot et al., 1995a, 1997). Biochemical profiles for rodent *Haemophilus* species are tabulated in the following references: Brands and Mannheim (1996). Boot et al. (1997, 1999), and Hansen (2000).

Rodent *Haemophilus* isolates species may be detected in swab samples from infected rats by PCR; however, the primers reported by Bootz et al. (1997, 1998) detect all members of the Pasteurellaceae and thus cultural criteria would be needed to confirm PCR positives. PCR negatives would be acceptable to demonstrate that the samples were free of Pasteurellaceae, including *Haemophilus* sp.

Because of a one-way cross-reaction, antibodies to many *Haemophilus* strains sufficiently react with *P. pneumotropica* ELISA antigen as to enable detection of *Haemophilus* infection, although the converse is not necessarily true (Boot et al., 1995c). As mentioned earlier, this predilection also limits specificity of serologic screening tests to detect *P. pneumotropica.* In their study of *Haemophilus* strains, Boot et al. (1997) demonstrated that a panel of at least two antigenic types (*P. pneumotropica* and the H21 strain of *Haemophilus*) would be needed to reliably detect all strains of *Haemophilus.* Thus, PCR may be a simpler and more cost-effective way of screening rat populations for presumptive *Haemophilus* status.

L. Potential Pathogens

There are several organisms isolated occasionally in the course of health surveillance with rats that are difficult to categorize as to pathogenicity and that do not recur often enough either as commensals or in periodic association with disease states as to be regarded as natural pathogens of the rat. Clinical conditions caused by such organisms as *Bordetella bronchiseptica, Klebsiella pneumoniae, K. oxytoca,* and *Staphylococcus aureus* fall within this group and might, more properly, be regarded as idiosyncratic infections. Although infrequently associated with disease states in the rat, the latter bacteria are more commonly associated with disease in other species. When isolated, the question becomes one of significance as to whether one has isolated a "pathogen" or a "commensal," and whether one is dealing with an "infection" or a "floral resident." For health surveillance purposes, all of this group should be considered reportable. The presence of these agents in laboratory rats reinforces the principle of separation of species having disparate microbial profiles within animal care facilities.

1. *Klebsiella pneumoniae*

K. pneumoniae, the Friedlander bacillus, has long been associated both as a common commensal and as an infrequent opportunist of laboratory rats and mice. *Klebsiella* species are ubiquitous in nature and commonly carried as bacterial commensals in the human and rodent intestinal tract. The epizootiology of the species of interest, *K. pneumoniae* and *K. oxytoca*, are quite similar and in the discussion that follows, they are considered together. The early European literature on *Klebsiella* is reviewed by Cohrs et al. (1958). Surveys taken in the 1930s have reported low incidence recovery of *Klebsiella* among the bacteria isolated from rat middle ear (Needham and Cooper, 1976), and a similar low incidence was reported among the isolates from rats undergoing health surveillance tests in conjunction with the (then-operational) British animal accreditation program (Sparrow, 1976). *Klebsiella* is presently not in the profile of organisms recommended by FELASA as important for routine health surveillance of laboratory rats (Nicklas et al., 2002).

Klebsiella species are Gram-negative, nonmotile, capsulated, facultatively anaerobic members of the Enterobacteriaceae. They are readily isolated from enteric sequence cultures made from feces or oropharyngeal swabs and less frequently from nasopharyngeal washes. Primary isolation media include Gram-negative broth, MacConkey agar, or blood agar. A selective medium has been developed to facilitate clinical isolation of this organism (Kregten et al., 1984). The identifying features of *Klebsiella* isolates are outlined in standard works on clinical microbiology (Wilson and Miles, 1975; Forbes, 1998; Hansen, 2000). The IMViC reactions may be used to distinguish *K. pneumoniae* ($- - + +$) from *K. oxytoca* ($+ - + +$), although, in practice, clinical microbiologists simply assign indole positive rodent *Klebsiella* isolates to *K. oxytoca*. Commercial identification kits, for example, API 20E, are widely used to identify presumptive *Klebsiella* isolates to species.

Klebsiella has been somewhat more frequently encountered clinically in mice than rats (Webster, 1930; Ehrenworth and Baer, 1956; Flamm, 1957; Hartwick and Shouman, 1965; Schneemilch, 1976; Davis et al., 1987; Rao et al., 1987). Typical clinical presentations in mice include abscess formations in lymph nodes or internal organs (e.g., ovary) and as opportunists isolated in mixed flora from upper respiratory or pulmonary conditions initiated by primary pathogens. In the single reported clinical incidents in laboratory rats, the condition presented as abscessed superficial lymph nodes, occasionally with fistulous tracts to the skin (Jackson et al., 1980; Arseculeratne et al., 1981). Despite the ubiquitous distribution and reputation of *Klebsiella* as an opportunist, it seems not to have arisen as a pathogenic hazard of immunologically impaired rats and mice.

The polysaccharide capsular material of *Klebsiella* can be typed into one of six types and is known to be an important modulator of virulence (Held et al., 1992), but it appears to be of limited value as an epizootiologic tool (Davis et al., 1987). Both rats and mice are used extensively as research models for *Klebsiella* infection (for summaries, see NRC, 1991; Baker, 1998).

2. *Bordetella bronchiseptica*

B. bronchiseptica is the name retained for this organism in the ninth edition of *Bergey's manual* (Holt et al., 1994). The organism is readily isolated from nasopharyngeal washes or respiratory tissues onto blood agar and may be identified according to criteria in the standard works cited above (for *Klebsiella*). It has been reported as an isolate of low incidence from rats submitted for microbiologic examination in the English accreditation program (Sparrow, 1976). *B. bronchiseptica* is occasionally isolated or detected serologically (Fujiwara et al., 1979; Boot et al., 1996; Bemis et al., 2003) in clinically normal rats during health surveillance tests. Although there is some evidence of antigenic divergence in *B. bronchiseptica* isolates from different host species (Wullenweber and Boot, 1993), it would seem prudent to regard carriers of one rodent species to be potential infectious reservoirs for others. *B. bronchiseptica* may be detected by PCR. Despite lack of evidence for pathogenicity, *Bordetella* is in the profile of organisms recommended by FELASA as important for routine health surveillance of laboratory rats (Nicklas, 2002).

Although the organism is recognized as a frank pathogen of guinea pigs (for review, see Wagner and Manning, 1976), there is no critical evidence (and no original references) indicting *B. bronchiseptica* in naturally occurring disease processes in the laboratory rat. In his review of this organism, Winsser (1960) described lesions in rats from which *B. bronchiseptica* was recovered; however, the rats in those studies were conventional, and the role assumed for *B. bronchiseptica* cannot with confidence be attributed solely to that organism. In recognition of the ambiguity of the Winsser study, Burek et al. (1972) explored the experimental pathogenicity of *B. bronchiseptica* for laboratory rats. The organism induced acute to subchronic bronchopneumonias in both germfree and conventional rats. The pathology of experimental infections is described (Burek et al., 1972; Percy and Barthold, 2001).

3. *Staphylococcus aureus*

S. aureus has for many years been associated with a multifactorial ulcerative dermatitis condition of laboratory rats (Ash, 1971; Fox et al., 1977; Wagner et al., 1977; Galler et al., 1979; Kunstyr et al., 1995). Illustrations of the gross and microscopic appearance of such lesions may be found in works by Fox et al. (1977), Percy and Barthold (2001), and Wagner et al. (1977). The condition occurs episodically, does not depend on age or strain and usually does not have an incidence of more than 1% to 2% of the animals at risk, but the incidence may be as high as 20%. Irregularly segmental and bilaterally symmetrical reddish, weeping, or crusty ulcerating skin lesions, perhaps 1 × 3 cm in size, develop dorsally over the shoulders, dorsolaterally on the neck, and ventrally under the chin (Fig. 11-11 and 11-12). In all cases, the lesions appear pruritic and are self-inflicted by staphylococcal contamination of skin abraded by scratching or biting. *S. aureus* is consistently isolated from such lesions, and characteristic microcolonies of staphylococci are observed microscopically in the superficial debris (Fig. 11-13).

Staphylococcus may be isolated from nasopharyngeal washes, the oropharynx, and skin of asymptomatic carrier rats and is frequently encountered during routine health surveillance tests. It is readily recovered from ulcerative skin lesions on blood agar. Isolates of *S. aureus* are Gram-positive cocci and occur as grapelike clusters, are coagulase positive, produce a yellow pigmented colonies, and are hemolytic on blood agar. Isolated staphylococci may be identified to species by traditional biochemical criteria using the standard references given above for *Klebsiella,* or by using identification kits, for example, API STAPH (bioMerieux France). Staphylococci are quite resistant to environmental temperature extremes and to drying, and are easily vectored by contaminated clothing, ungloved hands, and respiratory droplets (Blackmore and Francis, 1970).

Because nonlesioned laboratory rats so commonly seem to carry *S. aureus* as a commensal of the skin and mucous membranes, the question arises as to whether the staphylococci isolated in association with lesions

Fig. 11-11 Staphylococcus aureus: rat with contaminated skin ulcer overlying lateral right thorax. (Courtesy of Dr. Shari R. Hamilton.)

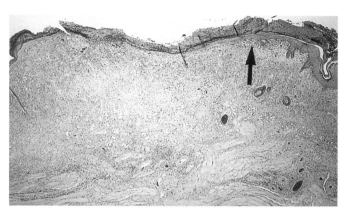

Fig. 11-12 Staphylococcus aureus: skin with fibrinopurulent debris overlying ulcered area. Note margin of ulcer (arrow) with epithelial hyperplasia to the right and necrotic slough of epithelium to the left. Hematoxylin and eosin, 100×. (Courtesy of Dr. Shari R. Hamilton.)

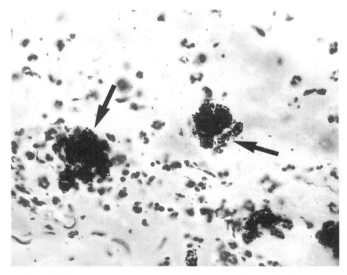

Fig. 11-13 Staphylococcus aureus: tissue section of ulcered area showing staphylococcal microcolonies (arrows). Gram, 1000×. (Courtesy of Dr. Shari R. Hamilton.)

are simply opportunistic saprophytes of devitalized tissue or rather incitants of the condition. With both rats and mice, several factors have been shown to bear on the origin of staphylococcal contamination of barriered rodent colonies and the potential pathologic effects of such contamination. The case for understanding that contamination of barriered rodent colonies by their human animal care and research staff contacts has been persuasively summarized by Besch-Williford and Wagner (1982). *S. aureus* is not regarded as a primary pathogen of laboratory rodents, nor is it on the list of agents recommended by FELASA for detection and reporting in conjunction with rodent health surveillance programs (Nicklas et al., 2002). This indeterminate status as an opportunistic pathogen contributes to a climate of indifference about human carriers and faulty managerial practices that may permit introduction and persistence of staphylococcal contamination. That this occurs has been demonstrated on a number of occasions (Blackmore and Francis, 1970; Ash, 1971; Shults et al., 1973; Lenz et al., 1978; Wullenweber et al., 1990).

At least in mice, a heritable component to ulcerative dermatitis has been shown in that Swiss Webster mice and certain inbred strains, those with normative immunocompetence (e.g., fC57Bl/6 and VM/Dk mice) as well as those with impaired immunocompetence (e.g., BSVS, C3H, and A strains), seem to have a predilection for clinical *S. aureus*–associated lesion development (Festing and Blackmore, 1971; Shults et al., 1973; Clarke et al., 1978; Hong and Ediger, 1978b; Taylor and Neal, 1980; Wardrip et al., 1994; Slattum et al., 1998). Certain transgenic (Shapiro et al., 1997) and athymic nude mice have been shown to have special susceptibility to staphylococcal syndromes, and this potential is an important and frequently limiting aspect of how they are cared for at user institutions (Cooper, 1971; McBride et al., 1981; Bradfield et al., 1993).

Ulcerative dermatitis may also develop secondarily to certain factors affecting integrity of the epidermis, for example, arthropod parasitism (Weisbroth et al., 1976). Even in C57Bl/6 strain mice, ulcerative dermatitis may occur secondarily to immune complex vasculitis (Andrews et al., 1994) or to cage mate hair chewing or barbering (Thornburg et al., 1973). A nutritional component has been shown to arise as a complication of certain experimental and ostensibly adequate diets in both rats (Galler et al., 1979) and mice (Nutini and Berberich, 1965; Nutini et al., 1968), but the nutritional basis for lesion development remains obscure. It was shown that the incidence of ulcerative dermatitis in C57Bl6N mice had a lower incidence and delayed onset in mice fed the NIH-31 diet compared with those on another ostensibly adequate commercial diet (Witt, 1989).

III. MYCOTIC INFECTIONS

A. Dermatomycosis (*Trichophyton mentagrophytes*)

Dermatomycosis (dermatophytosis, ringworm, or favus) has long been associated with wild and laboratory rodents. Although the literature, particularly older reports, is rife with synonymy, it is now generally concluded that the etiology of the majority of naturally occurring rodent ringworm is caused by *Trichophyton mentagrophytes*. This species, however, is one of the most polymorphic of the dermatophytes, and failure to recognize its range of forms has led to confusion in taxonomy. Two principal forms of the organism are recognized: (1) a zoophilic variant with granular colonial surface and red pigmentation named *T. mentagrophytes* var. *T. mentagrophytes* and (2) an anthropophilic form with white, fluffy colonial surface and no pigmentation designated *T. mentagrophytes* var. *inter-digitalis* (Ajello, 1974). Molecular approaches have been used to classify *T. mentagrophytes* substrains on the basis of DNA sequences (Makimura et al., 1998; Kim, 2001). Many reports, particularly in the mouse, refer to *T. quinckeanum* as a separate and distinct species, but most modern mycologists consider *T. quinckeanum* to be synonymous with *T. mentagrophytes* var. *mentagrophytes* (Ajello et al., 1968). Both variants may infect laboratory rodents. Much less frequently, several other dermatophytic species have been encountered in both wild and laboratory rodents (Feuerman et al., 1975; Kunstyr, 1980; Papini et al., 1997; Connole et al., 2001).

Dermatomycosis is more common as a disease of laboratory mice (Parish and Craddock, 1931; Booth, 1952; Brown and Parker, 1957; Menges et al., 1957; Dolan et al., 1958; Mackenzie, 1961; Cetin et al., 1965; Davies and Shewell, 1965; Reith, 1968) and guinea pigs (Menges and Georg, 1956; Kaffka and Reith, 1960; Mohapatra et al., 1964; Otcenasek et al., 1974; Owens and Wagner, 1975; Pombier and Kim, 1975; Kunstyr et al., 1980) than of rats (Dolan et al., 1958; Dolan and Fendrick, 1959; Georg, 1960; Povar, 1965; Mizoguchi et al., 1986), as the volume of literature indicates. Periodic survey work indicates that *T. mentagrophytes* is not uncommon in wild rats (Smith et al., 1957; Georg, 1960; Thierman and Jeffries, 1980) and mice (Brown and Suter, 1969; Chmel et al., 1975), although, as will be pointed out below, the asymptomatic carrier state may be more common than is realized. The disease in rats may assume an epizootic form, with many of the animals showing lesions, or may be insidious and characterized by lesionless carriers. In both presentation modes, substantial hazard to human contacts is present; indeed, human infection by persons handling the animals is frequently the first indication that the infection is in the colony. Most human infections occur on the exposed, relatively hairless parts of the body; especially the hands and arm.

As mentioned above, the infection assumes variable form in rats and is thought to be influenced by a number of factors that include those directly bearing on susceptibility or resistance, for example, age, genetic constitution, immunologic competence, and phase of hair growth cycle, as well as other less understood factors. Cortisone injections to test this hypothesis experimentally failed to influence degree of infection compared with that in untreated guinea pigs (Fisher and Sher, 1972). The few reported epizootics with lesioned rats all occurred in nonutilized animals before experimentation (Dolan et al., 1958; Povar, 1965; Mizoguchi et al., 1986). Lesions, when present, may occur in the skin of any area but are most common on the neck, back, and base of the tail. Lesions are not as are classically described, that is, uniformly discoid with alopecia and raised margins, but rather, may have a scurfy or erythematous papular-pustular appearance with irregular, patchy hair loss. Lesions on the tail (typically seen in the mouse) were not seen in rats by Povar (1965) in the outbreak he described.

Diagnosis of dermatophytosis is established by demonstration of fungal elements in skin scrapings and isolation of the causative organism by culture. Histopathology of affected skin is supportive in the attribution of lesion development to isolated dermatophytes if invasion of epidermal structures can be demonstrated. Histologic sections stained with Gridley fungus stain reveal fungal elements in the superficial epithelium and invasion of hair follicles. Secondary invasion of fungal lesions by bacteria with suppurative inflammation is commonly observed and is the cause of kerion-like lesions in both animals and man. The differential diagnosis for dermatophytosis should also consider other causes of similarly appearing skin lesions, including staphylococcal ulcerative dermatitis, fighting bite trauma, hair chewing or barbering by cage mates, and ectoparasitic hypersensitivity (Kunstyr, 1980).

Skin scrapings should be carefully taken from the lesion periphery, mounted in 10% potassium hydroxide under a vaseline-ringed cover slip, and examined under the microscope immediately and again in 30 minutes. When present, septate mycelia are observed in squamous cells. Small spore (2 to 3 µm) ectothrix invasion of hairs, especially near the base, are seen in *T. mentagrophytes* infections. Scrapings, similarly collected, should be inoculated onto the surface of a suitable agar medium and cultivated aerobically at room temperature for at least 10 days before discarding as negative. Suitable media include DTM (dermatophyte test medium with color indicators) (Carroll, 1974) or Sabouraud's medium with

cycloheximide and chloramphenical to inhibit nondermatophytic contaminants (Rosenthal and Furnari, 1957; Rebell and Taplin, 1970). Typical microscopic feature of *T. mentagrophytes* (macroconidia, spiral coils) should be demonstrated (Rebell and Taplin, 1970).

MacKenzie's hair brush technique is used to evaluate the incidence of lesionless, asymptomatic carrier rats (Mackenzie, 1963; Rosenthal and Wapnick, 1963; Papaini et al., 1997). This technique may be used to screen sample groups of rats to establish their status as asymptomatic carriers of dermatophytes, although in the absence of suspicious lesions, scheduled health surveillance testing for dermatophytes is not recommended by FELASA (Nicklas et al., 2002). There is some evidence to indicate that this state may occur frequently in the rat (Dolan et al., 1958; Dolan and Fendrick, 1959; Gugnani et al., 1971; Balsari et al., 1981; Papini et al., 1997) as is more commonly recognized in the guinea pig, mouse, and cat (Fuentes and Aboulafia, 1955; Fuentes et al., 1956; Menges et al., 1957; Dolan et al., 1958; Rosenthal and Wapnick, 1963; Gip and Martin, 1964; Feuerman et al., 1975). In this technique, the animal to be screened is held over an opened Petri dish of suitable agar medium and the hair brushed with a sterile surgical scrub brush so that hairs, flakes, and desquamated cellular debris fall directly down on the medium surface. The plate is incubated as described above.

Eradication and control in research colonies usually entails destruction of the affected groups and sterilization or disinfection of equipment and environmental surfaces. A modified rederivation approach to eradication in a breeding colony was described by Mizoguchi et al. (1986). This program entailed removal of all rats from the colony, disinfection of the premises with formalin and sodium propionate, and subsequent restocking of the colony with weanlings from nonlesioned dams dipped in sodium propionate before reintroduction. Although the clinical efficacy of feeding griseofulvin to treat dermatomycosis has had mixed results in other laboratory species (Cetin et al., 1965; Pombier and Kim, 1975), its effectiveness for this purpose has not been evaluated in rats. The organism is not known to traverse the placenta and has not been recovered from barrier-reared laboratory rats.

B. Deep Mycoses

The deep or systemic mycoses (fungal infections other than dermatomycosis) are extremely rare in rats; indeed, two comprehensive reviews—one covering the period up to 1954 (Schwarz, 1954) and another to 1967 (Smith and Austwick, 1967)—failed to document a single episode in unimpaired laboratory rats. Wild *R. norvegicus,* and other rat species have been encountered as both asymptomatic and lesioned carriers of several fungal species and are regarded as playing some role as natural reservoir hosts for certain organisms. The reviews cited above may be consulted for information related to wild rats.

Aspergillosis has been reported in naturally occurring outbreaks of pulmonary (Singh and Chawla, 1974; Gupta, 1978; Hubbs et al., 1991) and nasal (Nyska and Kuttin, 1988; Rehm et al., 1988) infections of laboratory rats. In their review of systemic mycoses as experimental diseases in rat hosts, Smith and Austwick (1967) concluded that rats were quite refractory to most fungal infections; however, a number of reports have documented aspergillosis in cortisone-treated laboratory rats and mice as unanticipated outbreaks (Sagi and Lapis, 1956; Sidransky and Friedman, 1959; Sethi et al., 1964; Sanhu et al., 1970). Experimentally, lesions of aspergillosis were more severe in rats treated with cortisone than those immunosuppressed with immuran (Tuner et al., 1976). That immunity is important in resistance is suggested both by the lethal aspergillosis induced experimentally in mice with antilymphocyte serum (Swenberg et al., 1969) and by the protection induced by immunization (Corbel and Eades, 1977). In these reports, as in the natural outbreaks cited above, pulmonary aspergillosis presented as granulomatous pneumonia with miliary distribution in that organ primarily, although hematogenous distribution to other organs with capillary nets, for example, liver and kidney, was observed to a lesser degree. In the two reports of *Aspergillus* rhinitis, the lesions involved granulomatous inflammation of the maxilloturbinates.

Diagnosis of aspergillosis is established by cultural retrieval of the organism and supportive histopathologic lesions. Sabouraud's agar with bacterial inhibitors (chloramphenicol, penicillin, and streptomycin) is satisfactory for this purpose, but care must be used to avoid modern media designed for isolation of dermatophytes, because the cycloheximide (Actidione) used as inhibitor will prevent growth by fungal "contaminants" such as *Aspergillus*. Isolated *Aspergillus* cultures should be speciated according to the criteria outlined by Austwick (1974), which emphasizes subculture on Czapek-Dox medium for optimal growth of differential features useful in speciation.

Granuloma formation is typical of pulmonary and upper respiratory aspergillosis in all species, including the rat. The lesions are seen microscopically to consist of macrophage and epitheloid cells centrally, as well as peripheral aggregation with both acute and chronic inflammatory cells, eosinophils, and giant cells of the Langhans type (Singh and Chawla, 1974). Congestion, microscopic hemorrhage, and septal thickening occurs in surrounding parenchyma. *Aspergillus* is seen quite distinctly in material stained with hematoxylin and eosin, but Gridley's fungus stain or Gomori-Grocott should

be used for careful study. The *Aspergillus* species may be differentiated in tissue from those causing phycomycosis by the more uniform and smaller hyphae (2.5 μm in diameter), regular branching at 45° angles, and prominent septae. In contrast, phycomycotic fungi in tissue have thicker hyphae (10 μm in diameter) and irregular branches and are nonseptate.

Phycomycosis (or mucormycosis) is the general term used to describe tissue infection by nonseptate fungi of the genera *Mucor, Absidia,* and *Rhizopus.* The infection is quite rare in all species and has most oftenly been reported in association with primary conditions lowering resistance or otherwise predisposing to phycomycosis, for example, diabetis mellitus, cortisone treatment, and immunosuppressive therapy. Mycotic encephalitis may be induced experimentally in mice by cortisone treatment (Swenberg et al., 1969a,b). Phycomycotic encephalitis has been reported in juvenile rats as a naturally occurring disease before experimentation (Rapp and McGrath, 1975; Moody et al., 1986). The lesions presented as purulent, necrotizing foci with acute inflammatory responses. Phycomycotic nonseptate hyphae up to 10 to 15 μm diameter were demonstrated with various stains. The investigators emphasized that the immaturity of these rats (2 to 4 weeks) was predisposing, because neither their dams nor older rats in the same room developed signs or lesions.

C. Pneumocystosis (*Pneumocystis carinii*)

Pneumocystis infection was originally described as a pneumonia of laboratory guinea pigs by Chagas (1909), who mistook the cysts he saw as forms of *Trypanosoma cruzi.* Again, in 1911, he saw the cyst forms (which he reported as trypanosomes) in lung tissue from a human case of Chagas disease (Chagas, 1911). Shortly afterwards, similar pneumonias in rats were described as caused by a new organism named *P. carinii* by Delanoe and Delanoe (1912). The organism was reported several years later in English laboratory mice (Porter, 1915). There are, however, few references to rodent pneumocystosis in the intervening years before the era of radiation and anti-inflammatory (cortisone) research in the 1950s and 1960s (Frenkel et al., 1966). At that time, scientists began to encounter clinically overt *Pneumocystis* infections, principally in rats, as complications of research protocols now recognized as inducing a state of immunosuppression. Interestingly, most references to unanticipated, "spontaneous" *C. piliforme* and *C. kutscheri* infections of rats date to the same era, for the same reasons.

Animals, especially rats and mice and, to a lesser degree rabbits and ferrets, have been an indispensable element of *Pneumocystis* research dating from Frenkel's 1966 paper (Dei-Cas et al., 1998). The histologic features of pulmonary *Pneumocystis* pneumonia (PCP) in animals (see description and illustrations in Percy and Barthold, 2001) are nearly identical to those of the disease in humans. Animal models have provided the basis for most of our current concepts of the epizootiology, taxonomy, diagnosis, genetics, and therapy of pneumocystosis (Hughes, 1989; Walzer, 1991). In the past, rats and mice from conventional commercial breeding colonies of immunocompetent stocks have acted as convenient, but erratic and unreliable sources of carrier *Pneumocystis* infections. The term "conventional" describes a rodent population in which one or more murine viruses cycle as enzootic infections and may also be infected with a range of helminth, protozoan, arthropod, and bacterial pathogens. The pathogen load is emblematic of unshielded husbandry environments and the correspondingly high likelihood that such populations also latently (actually, inapparently) carry *Pneumocystis*. Investigators needing *Pneumocystis*-infected rodents for research programs favored procurement from such colonies in the past because it was assumed they were likely to be "latent" carriers (Russell and McGinley, 1991) although it has since been shown that *Pneumocystis* infection dynamics are quite independent of murine virus status (Pesanti and Shanley, 1984; Cushion and Linke, 1991). After arrival at the user institution, these animals would be deliberately subjected to an immunosuppressive protocol that would culminate, after an inductive phase of about 4 to 8 weeks, with development of severe lesions of PCP with abundant *Pneumocystis*. Faced with impending disappearance of conventional *Pneumocystis* carrier sources as commercial rodent breeders mounted intensive rederivation programs to eradicate murine viruses (Bartlett et al., 1987), investigators have turned to the more reliable method of inducing PCP in pathogen-free rodents. In this procedure, animals are first rendered susceptible to overt infection by immunosuppressive treatments and then infected by intranasal or intratracheal installation of *Pneumocystis* inocula (Bartlett et al., 1987, 1988, 1990, 1991; Shellito et al., 1990; Boylan and Current, 1992) or by short-term housing with infected seed rats or mice (McFadden et al., 1991). Additional models of pneumocystosis include the athymic (nude) rat, athymic (nude), and SCID mice, which offer physiologic analogs comparable to the HIV-infected human with PCP. Neonatal rabbits commonly develop transient lesions of pneumocystosis, without immunosuppressive treatment, but these infections resolve naturally as immunocompetence develops in the weeks after birth (Soulez et al., 1989).

Over the years, until the 1980s, informed opinion had wavered on the kingdom in which to place *Pneumocystis*, but historically, the organism was widely thought to be a protozoan. Uncertainty still surrounds details of the life cycle of *Pneumocystis*. The issue is complicated, and even the terms used to describe morphologic life forms of *Pneumocystis* derive from protozoology. The mature form is termed a cyst. Within the cyst, visible more or less depending on the stain and stage of maturation are eight intracystic bodies or "sporozoites." When excystation takes place, the sporozoites enter a trophic stage. The vegetative, trophic forms in turn become precysts, crescent-shaped cysts, and finally cysts (Kim et al., 1972; Yoneda et al., 1982; Cushion et al., 1988). Cyst production involves a sexually reproductive phase (Cushion et al., 1997; Cushion, 2003). An ste3-like pheromone receptor has been identified in the organism (Smulian et al., 2001). Although the morphologic stage variants described above have been recognized microscopically, the lack of a continuous *in vitro* culture system for propagation of *Pneumocystis* has hampered an understanding of the progression of the life cycle and definition of the infective form. *Pneumocystis* may be carried for several passages in tissue culture, with an increase of up to 10- to 20-fold in numbers, but a continuous *in vitro* culture method has so far proven elusive (Bartlett et al., 1979; Cushion and Walzer, 1984a,b; Cushion et al., 1985; Cushion and Ebbets, 1990; Aliouat et al., 1995; Beck et al., 1996; Atzori et al., 1998). At least one phase, the most studied, takes place in the mammalian lung, although tracing methods using PCR commonly detect *Pneumocystis* DNA in extrapulmonic locations, for example, bone marrow, liver, blood stream (antigenemia), and spleen (Reddy and Zammit, 1991; Schluger et al., 1991; Chary-Reddy and Graves, 1996; Rabdonirina et al., 1997). Lesions in spontaneous rodent PCP are usually confined to the lung, but there are reports of SCID mice developing extrapulmonic foci in the heart and spleen (Reddy and Zammit, 1991).

The cyst walls stain best with GMS, PAS, or toluidine blue O stains (Thompson et al., 1982; Sundberg et al., 1989). Giemsa stain does not stain the cyst wall. The best single stain for diagnostic visualization of *Pneumocystis* forms in concentrated smears of lung homogenates or impression smears of cut lung surface appears to be the Diff-Quik and similar stains because all stages are stained. Although several lines of evidence, for example, histologic tinctorial qualities (argyrophilia), were consistent with relatedness to fungi (Walkeden, 1990), antifungal drugs were shown to be ineffective against PCP, although *Pneumocystis* has since been shown, *in vitro*, susceptible to some antifungals (Kaneshiro et al., 2000). Conversely, drugs with confirmed activity against *Pneumocystis*, for example, trimethoprim-sulfamethoxazole, fansidar, and pentamidine, have an antiprotozoan spectrum. Molecular and biochemical analyses have since established that *Pneumocystis* is a genus of unusual eukaryotic single-celled and genetically complex fungi, most likely an ascomycete but lacking in ergosterol (Edman et al., 1988; Cushion, 1993; Stringer et al., 1992, 1993, 2002).

Genetic studies with *Pneumocystis* isolates derived from various mammalian hosts have shown that these organisms are quite different. The accumulating results of DNA studies directed to taxonomic analysis have established *P. jiroveci* and *P. carinii* as separate species. Based on genetic divergence studies (Stringer et al., 1993) and infective isolation for its human host, the human strains of this organism were designated *P. jiroveci.* (Frenkel 1999), as earlier suggested by Frenkel (1976). Although DNA sequence polymorphism has been shown with human *Pneumocystis* isolates (Lee et al., 1993), suggesting that numerous strains of *P. jiroveci* exist, this diversity has been thought insufficient to designate either special forms of *P. jiroveci* or additional human species (Wakefield, 1998; Stringer, 2002). *Pneumocystis jiroveci* has not been found in the lungs of any other mammal, including non-human primates (Wakefield et al., 1990; Stringer, 2002), nor may it establish productive infection even in SCID mice (Durand-Joly et al., 2002). Accordingly, it has been concluded that humans do not carry rodent *Pneumocystis* and, conversely, that humans do not contract pneumocystosis from animal contacts.

Experience from several lines of inquiry supports the concept that *Pneumocystis* strains have diverged in genetic heterogeneity, progressing to speciation according to homology with a given mammalian host species (Shah et al., 1996; Stringer, 2001). Infectivity studies have demonstrated that heterologous hosts, even if immunodeficient, are refractory to replication of *Pneumocystis* derived from other host species (Gigliotti et al., 1993; Aliouat et al., 1994). Similarly, the ultrastructural morphology and isoenzyme diversity among *Pneumocystis* isolates from different host species is host specific and supports evidence of genetic diversity demonstrated by molecular analysis (Mazars et al., 1997; Nielsen et al., 1998). For this reason, under natural circumstances, rats should not be regarded as vectors of *Pneumocystis* infectious for other rodent species, and conversely, other rodent species, including mice, are not carriers of *Pneumocystis* infectious for rats.

Isolates of *Pneumocystis* from rats are taxonomically quite complex. A trinomial system has been adopted (Wakefield, 1998; Cushion, 1998) in which the various strains of *Pneumocystis* have been designated as special forms (formae speciales, or f. sp.). At least five species of rat *Pneumocystis* have been delineated by DNA analysis, but only two of them occur in laboratory rats, that is, *P. carinii* and *P. wakefieldiae (formerly, P. carinii f. sp. ratti)* (Cushion, 2003). The binomial *P. carinii* is reserved for the more common *Pneumocystis* species occurring in rats and the former designation *P. carinii*

f. sp. carinii is now shortened) (Stringer et al., 2001). Three additional species have been detected in wild rats, viz., *P. carinii* f. sp. ratti-secundi, *P. carinii* f. sp. ratti-tertii, and *P. carinii* f. sp. ratti-quarti (Cushion et al., 1993; Palmer et al., 2000). These forms may be discriminated by PCR (Palmer et al., 1999, 2000; Schaffazin et al., 1999; Nahimana et al., 2001) or by immunoblotting (Vasquez et al., 1996), and investigators should do so because the two species vary in responses to drug therapy trials, the life cycles differ, and the use of mixed isolates from lung preparations (not uncommon) could lead to a confusing mixture of gene sequences in gene libraries. At present, *P. carinii* f. sp. carinii is believed to be the most prevalent strain in commercial rat producer stocks (Icenhour et al., 2001), although coinfection by more than one special form in the same rat is recognized as not uncommon (Cushion et al., 1993; Nahimana et al., 2001). The name newly given to the mouse strain is *P. murina.*

The common denominator of immunodeficiency that underlies PCP has been linked with the common occurrence of intercurrent or "spontaneous" PCP in immunodeficient (athymic and *scid*) mutant and transgenic rat and mouse stocks, with induced rodent models for human PCP, and of course, with human PCP, itself a consequence of impaired immunocompetence resulting from HIV infection. Progressive wasting (emaciation, or cachexia), cough, dyspnea, and cyanosis are clinical signs of PCP in rats. Clinically overt signs of pneumocystosis occur in rats and mice under the following circumstances:

1. Immunologically competent rodents that have been rendered immunodeficient as a consequence of experimental treatments in which in which PCP develops as an unplanned complication of a research protocol, as cited above during earlier eras of radiation and anti-inflammatory research. Such animals may have been infected at the source, before arrival at the user institution or first become exposed to same host species *Pneumocystis* shedders at the institution and infected after arrival.

2. Unplanned outbreaks in immunologically deficient mutants lacking cell-mediated immune capability and T cell–assisted antibody responses. Such mutants include the athymic (nude) mouse and rat, the beige triple-deficient mouse, the *scid* mouse, and a range of similarly affected heritably immunodeficient mutant and transgenic rodent models (Veda et al., 1977; VanHooft et al., 1986; Weir et al., 1986; Walzer et al., 1989; Sundberg et al., 1989; Gordon et al., 1992; Deerberg et al., 1993; Furuta et al., 1993; Percy and Barta, 1993; Pohlmeyer and Deerberg, 1993). These animals need only adequate exposure to *Pneumocystis* to initiate development of PCP (Soulez et al., 1991). On this basis, athymic mice have been used in mouse colonies as sentinels for detection of *Pneumocystis* (Serikawa et al., 1991). The clinical course of PCP in immunodeficient animals may be fulminant but is more likely to be chronic. The microscopic qualities of the lesions that develop in immunodeficient mutants are the same as in immunocompetent rodents that develop PCP as a result of immunosuppressive treatments. There is no support in the literature for the view that immunodeficient mutants may carry *Pneumocystis* in a latent form, and further, it is established that immunosuppressive treatments of athymic mutants do not

increase the incidence or severity of PCP during experimental induction (Veda et al., 1977; Walzer and Powell, 1982; Soulez et al., 1991; McFadden et al., 1991).

3. Immunologically competent carrier rats and mice rendered immunodeficient by chemical immunosuppressants, for example, corticosteroids or antimetabolites such as cyclophosphamide, as a planned process to develop PCP for some further experimental purpose; to propagate *Pneumocystis* for inoculation or antigen material; or to detect *Pneumocystis* carriers through the outcome of a diagnostic stress test.

Clinically overt pneumocystosis (PCP) should be viewed in all species as an unusual outcome requiring special circumstances of infection (immunodeficiency) as outlined above. More commonly, pneumocystosis occurs as a clinically inapparent infection of immunocompetent rodent stocks and strains that, because of its symptomalogic silence and difficulty of diagnostic detection (until recently), seems to have provided little incentive to commercial and institutional rodent breeders for eradication. Indeed, as recently as 2001, most rats from good commercial producers in the United States were detected as contaminated with DNA of *P. carinii* (Icenhour et al., 2001).

It is gradually becoming clear that the natural mode for transmission and maintenance of rodent *Pneumocystis* infection is by exposure via aerosol of susceptible immunocompetent hosts (Walzer et al., 1977; Hughes, 1982; Hughes et al., 1983). Even germfree rats were shown to be insusceptible to oral dosing via the diet or by ingestion of infected lung (Hughes, 1982). In enzootically infected breeding colonies, neonates are born to variably immune dams, exposed within hours of birth (Icenhour et al., 2002) but, it may assumed, are probably protected from productive infection for the first few weeks of life by colostral antibody. Thereafter, they become susceptible and cage-to-cage transmission occurs via aerosol from infected shedder animals. In enzootically infected, closed, barriered colonies, weanling (3- to 4-week-old [100 to 125 g]) rats can be envisioned as living in a cloud of *Pneumocystis* spores, with spore contamination (at least as denoted by PCR detection of *Pneumocystis* DNA) on all environmental surfaces (Icenhour et al., 2002), thus with ample opportunity for exposure. The prime infective (shedder) phase would thus appear to be the 6- to 12-week-old rat (200 to 250 g), which corresponds to the age at which most rats are procured and shipped to user institutions. Thereafter, the immunologic response to these light (and clinically inapparent) infections supervenes, and *Pneumocystis* populations decline until eliminated by the host (An et al., 2003). Thus, the infection in the colony would seem to be propagated by successive cohorts of the productively infected, spore-shedding 4- to 12-week age group, infectious only to the next cohort of 3- to 4-week-old rats that are in the process of losing their colostral immunity. It would appear that older rats may be immune to reinfection and younger rats protected by colostral antibody.

As more has been learned about the pathogenesis of pneumocystosis in immunocompetent rat and mouse (and human) hosts, doubt has been cast as to whether the term "latent" ever actually applies to *Pneumocystis* infections. Even after PCP induced by immunsuppression, most rat hosts allowed to immunologically recover eliminate *Pneumocystis* from the lungs (Vargas et al., 1995), as do infected SCID mice immunologically reconstituted with normative spleen cells (Chen et al., 1993). The basis for inadvertent introduction of asymptomatic and lightly infected shedder rodents into user institutions has been outlined above as reflecting the enzootic infective cycle in breeding colonies in which it can be envisioned that essentially all rodents in the 4- to 6-week to 12-week adolescent age group become infected for a brief period of time, after which they become immune noncarriers (An et al., 2003). During this period (the shedder phase), they are infective hazards for other *Pneumocystis*-free rodents of the same species and have been a frequent cause of clinically overt infective "breaks" in immunodeficient stocks at user institutions (Dumoulin et al., 2000). This same infectivity model is believed to hold with humans as well, who become almost universally exposed and immune at a very young age (Stringer et al., 2002). In the same way, it is now understood that human PCP derives not from activation of endogenous "latent" infections but more probably from chance or nosocomial aerosol exposure to spores during a receptive phase of immunodeficiency (Chen et al., 1993; Vargas et al., 1995). Although the infective form of *Pneumocystis* in aerosols has not been defined, there is ample evidence of *Pneumocystis* DNA detected by PCR of spore trap air samples (Wakefield, 1994, 1996; Olsson et al., 1995). Immunosuppression, as by HIV infection or preparation for organ transplantation in humans or by immunosuppressive treatment of immunocompetent rodents, can effectively override naturally acquired or experimental immunity and render even immune hosts susceptible to PCP (Shellito et al., 1990; Harmsen et al., 1995). Antibodies, whether derived actively after infection or passively by injection of immune sera, adequately protect against experimental homologous *Pneumocystis* infection (Gigliotti and Harmsen, 1997; Bartlett et al., 1998).

Definitive diagnosis of PCP in rats continues to be based on morphologic criteria of consistent gross and microscopic lesions of lobar pneumonia, with microscopic correlates of paravascular cuffing and alveolar distension with foamy, pinkish material in hematoxylin and eosin–stained material (Fig. 11-14). With use of special stains (e.g., GMS), multiple discoid blackish staining *Pneumocystis* cysts may be demonstrated in the alveoli (Fig. 11-15). Smears of lung tissue prepared at necropsy may be stained

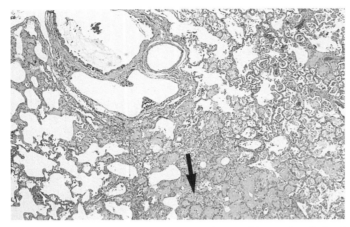

Fig. 11-14 Pneumocystis carinii: rat lung with patchy foci of histioalveolitis. Note foamy alveolar contents (arrow). Hematoxylin and eosin, 40×. (Courtesy of Dr. Shari R. Hamilton.)

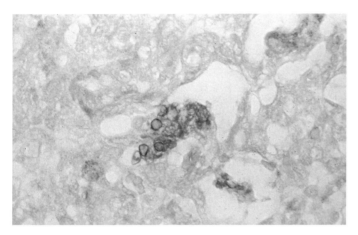

Fig. 11-15 Pneumocystis carinii: lung section demonstrating silver staining (black) cyst forms in alveolar foam. Gomori's methenamine silver, 1000×. (Courtesy of Dr. Shari R. Hamilton.)

with toluidine blue O or GMS to detect cyst forms of *Pneumocystis* in postmortem material. Morphologic diagnosis of PCP can be confirmed by PCR tests (Kitada et al., 1991; Schluger et al., 1992; O'Leary et al., 1995), and for this purpose, individual or pooled lung tissue, bronchioalveolar lavage, or oral swab samples may be used.

In considering the use of serology as a diagnostic screening tool, the many practical hurdles and potential sources of false-negative results impose important limitations to adoption of this method. It has been shown that the host species origin of *Pneumocystis* antigen determines the serologic specificity of anti-*Pneumocystis* antibodies (Walzer and Rutledge, 1980; Kovacs et al., 1989; Bauer et al., 1991, 1993). Although many investigators use serology to detect prior exposures to *Pneumocystis* (Peglow et al., 1990), rats are quite variable in the age

when detectable levels of antibody appear (Walzer et al., 1987). Antibody levels in sera after homologous experimental infections, that is, rat origin *Pneumocystis* used to infect rats, have a much higher titer against homologous *Pneumocystis* antigen than do the same sera tested against *Pneumocystis* antigens derived from other host species. Scientists recognized as early as 1966 (Frenkel) that *Pneumocystis* antigens derived from either rat or human would fix complement only with serum from homologous hosts, although when tested by immunoblotting, sufficient cross-reactivity of human sera against rat trophozoite antigens has been shown as to be diagnostically useful (Chatterton et al., 1999). A further potential cause of false negative could accrue to the point that, at least in rats, there are multiple special forms of *P. carinii* with an unestablished degree of cross-reactivity. These problems combine to make unlikely the development of a universal antigen that would detect antibodies to *Pneumocystis* in all laboratory species, or even in just rats and mice. Rather, it would seem that homologous antigen/antibody systems would need to be used. Because of the practical issues related to the present inability for large-scale propagation of *Pneumocystis*, rat origin antigens are not commercially available and serodiagnosis is not commonly used for diagnostic screening.

Diagnosis of light and clinically inapparent rat infections, as is the case in immunocompetent rat stocks, in the past depended on use of a diagnostic stress test (for a review of procedural parameters used in conducting rodent stress tests, see Weisbroth, 1995). The traditional stress test required histologic demonstration of *Pneumocystis* in stained lung sections after use of chemical immunosuppressants and a low-protein diet during a 2- to 4-week induction period (Frenkel et al., 1966; Walzer et al., 1979, 1980; Milder et al., 1980). The test was also used, when negative, to provide evidence that the institutional or commercial source of the animals in the test were free of contamination by *Pneumocystis*. Stress tests are lengthy, cumbersome, and expensive and have been shown to be no more effective in detection of inapparent infection than the are same rats tested by PCR at the start of the procedure before induction (Weisbroth et al., 1999b). For this reason, the stress test as a screening method for diagnosis of *Pneumocystis* infection has largely been discarded in favor of PCR. As mentioned above, for this purpose individual or pooled lung tissue, bronchioalveolar lavage, or oral swab samples can be used for screening purposes (Feldman et al., 1996; Icenhour et al., 2001). Because of the sensitivity of PCR, immunocompetent (but *Pneumocystis* free) rats may be used as sentinels to monitor *Pneumocystis* status of given rat environments (Icenhour et al., 2001).

There are very few options for control and eradication of *Pneumocystis* from rat populations. Owing almost entirely to the role of rats as prime models for exploration of treatment modalities for human PCP, there is an extensive body of experience in treatment of clinically apparent PCP in rats. The generality is that although a fair number of treatments—including trimethoprim-sulfamethoxazole, dapsone, or pentamidine, as well as an impressive array of newer therapeutics (beyond the scope of this text for review)—have been shown effective in amelioration of clinical signs (and lung counts), none have been shown effective in eradication of *Pneumocystis* (Hughes, 1979, 1988). For that reason, treatment options would appear limited to those few instances in which the life of a small population of immunodeficient animals needed to be prolonged for some purpose.

Gnotobiotic rederivation into isolators followed by introduction to a rodent barrier stringently managed for exclusion of rodent pathogens is recommended as effective for eradication and restart of rat populations free of *Pneumocystis*. There is no convincing evidence that *P. carinii* can traverse the rodent placenta to effect vertical transmission (Icenhour et al., 2000), rather, there is a good deal of experience demonstrating efficacy of gnotobiotic rederivation (Wagner, 1985; Ito et al., 1991).

IV. MYCOPLASMAL AND RICKETTSIAL INFECTIONS

A. Introduction

The genus *Mycoplasma* is one of eight genera within the Class *Mollicutes*. *Mycoplasma* are bacteria that lack a cell wall and, with a 0.2-μm diameter, they represent the smallest free-living procaryotes known. Their genome may be as small as 600 kb, one-tenth that of *Escherichia coli*. *Mollicutes* are often models for studying the minimal metabolism necessary to sustain independent life. They lack cytochromes or the tricarboxylic acid cycle except for malate dehydrogenase activity and are unable to synthesize pyrimidines or purines *de novo* (Pollack et al., 1997). The first *Mycoplasma* species identified was the agent of contagious bovine pleuropneumonia, which was cultivated on artificial media by Nocard and Roux in 1898. As other isolates with similar characteristics were isolated from a wide range of animal species, they were called PPLOs (Klieneberger, 1938; Klieneberger and Steabben, 1940). This term was commonly used in the literature until the mid 1960s.

Mycoplasma are known to be significant pathogens in most species of animals as well as in humans. Generally, they have tropisms for the respiratory tract, urogenital tract, or articular cartilage/synovial tissue in the animal or human host. Their pathogenicity is typically associated with a chronic disease course rather than a fulminant course. In laboratory rats, two *Mycoplasma* species, *M. pulmonis* and *M. arthritidis,* have been shown to be pathogenic; however, *M. pulmonis,* is the only significant pathogen of the two.

B. Respiratory Mycoplasmosis (*M. pulmonis*)

Chronic respiratory disease (murine respiratory mycoplasmosis) was arguably one of the most significant infectious diseases in laboratory rats until the 1970s, at which time commercial sources of rats began offering stock free of *M. pulmonis*. Today, *M. pulmonis* infections are essentially limited to colonies that are produced by noncommercial institutions, rats that become infected through contact with such breeding colonies, or rats that have been inoculated with contaminated biological materials. Personal communication with directors of several large diagnostic laboratories indicated that *M. pulmonis*–positive ELISA tests occur at a rate of about 3 per 1000 rats and that these positive tests are from conventional colonies. Until 1969, chronic respiratory disease was thought to be associated with two agents; a virus that produced bronchiectasis and *M. pulmonis* that induced an upper respiratory tract catarrh infection (Nelson, 1951). Kohn and Kirk (1969) first demonstrated that *M. pulmonis* was independently responsible for both the upper and lower respiratory tract disease syndrome. In the next several years, other reports confirmed this finding and significantly expanded information on the pathogenesis of this *Mycoplasma* (Kohn, 1971a,b; Lindsey et al., 1971; Jersey et al., 1973; Cassell et al., 1973).

Clinical signs in rats infected with *M. pulmonis* are quite variable; in many cases, no signs are present even though significant pulmonary lesions may exist. The prevalence and severity of signs typically increases with the age of the rat and the presence of experimental or environmental stresses placed upon the animal. Clinical signs include sniffling, rales, roughened haircoat, dyspnea, weight loss, and red staining of the eyes peri-orbitally and at the nares owing to increased porphyrin secretion of the Harderian glands. Not uncommon is torticollis associated with extension of the infection from the nasal mucosa through the Eustachian tube to the middle and inner ear.

Mucopurulent exudation occurs in the nasal cavity and trachea; however, it may not be readily evident. The pulmonary lesions vary from small surface areas that are dark red or gray to entire lobes that are affected. The apical lobe and cranial portion of the left lobe tend to be more often affected than are the other lobes, however, distribution varies among animals. The typical end-stage lung lesion is a cobble stone–like surface, reflecting

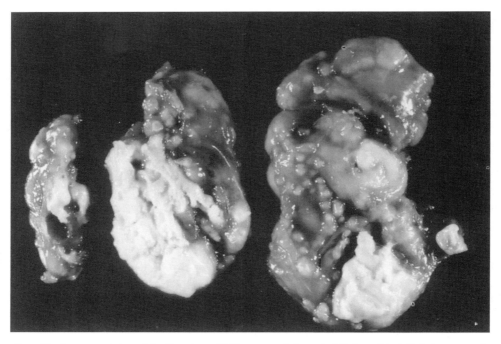

Fig. 11-16 Lung sectioned to show caseated exudates from bronchial lumens and characteristic bronchiectatic lesions on surface of lung infected with *Mycoplasma pulmonis*. (Courtesy of Dr. Trenton R. Schoeb.)

marked bronchiectasis. The size of these abscessed bronchi vary from several mm to coalesced abscesses up to 1 cm in diameter (Fig. 11-16). The lung surface surrounding these multiple yellowish lesions is dark because of consolidation and atelectasis. Cross-sections of affected lungs reveal markedly dilated bronchi impacted with caseated exudates and consolidation of varying amounts of parenchyma.

The inflammatory response to *M. pulmonis* infection is variable but can be intense, particularly after an extended period of time. The response at all sites in the respiratory tract and middle ear is characterized by a submucosal lymphocytic infiltrate and primarily a neutrophilic leukocyte exudate at the epithelial cell surface. Within the lung, peribronchial lymphoid hyperplasia is the initial dominant lesion, and as it increases, leukocyte infiltration of the bronchial and bronchiolar lumens leads to bronchiectasis and bronchiolectasis. Mucosal epithelial cells often become cuboidal or squamoid in bronchiectatic airways, and hyperplasia of the epithelial mucosa may occur.

Another more recently described outcome of *M. pulmonis* infection is angiogenesis and blood vessel remodeling associated with microvascular leakiness owing, in part, to formation of endothelial gaps. This is accompanied by an influx of inflammatory cells, epithelial thickening, mucous gland hypertrophy, fibrosis, and an abnormal sensitivity to substance P (Kwan et al., 2001).

M. pulmonis is transmitted vertically and horizontally within rat colonies. It is readily transmitted from infected to noninfected rats by direct and indirect contact.

However, the rapidity of transmission may vary within colonies owing to differences in room ventilation rates, husbandry practices, and animal density. The importance of fomites in transmission is not well defined. The organism does not survive long periods outside the host; however, personnel and contaminated equipment may be of importance in some instances. Within breeding colonies, dams may infect their litters postpartum or *in utero* because the female genital tract is frequently a site of infection. Most significant in the epizootiology of *M. pulmonis* infection is that infected rats typically remain as a life-long source for transmission to naive rats of any age.

The most important factors that modulate the pathogenesis *M. pulmonis* are (1) those effecting cytadherence, (2) genotype of the rat, (3) coinfection with a virus pathogen, and (4) environmental influences. A summary of studies that have dealt with elucidation of these modulating factors is given below.

Mycoplasma that are pathogens in animals and humans typically colonize the epithelial cell surfaces of the respiratory and urogenital tracts. Unlike most other bacteria, *Mycoplasma* lack fimbriae that are involved in adhesion to the host cell membrane. Cytadherence of *Mycoplasma* is associated with adhesins located within the cell membrane that attach to specific receptors on the host cell membrane (Razin and Jacobs, 1992). *M. pulmonis* has a tropism for the ciliated epithelial cells of the respiratory tract and the genital tract in rats (Fig. 11-17). Under experimental conditions, it also cytadsorbs to the ciliated

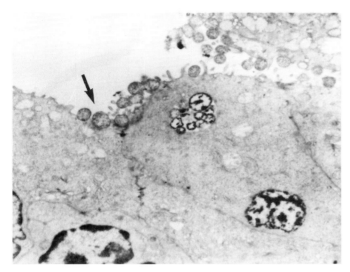

Fig. 11-17 Electron micrograph showing cytadherence of *Mycoplasma pulmonis* to bronchial epithelial cells. Note the single organism being partially surrounded by two microvilli, and the intracytoplasmic vacuole containing numerous degenerating organisms. This latter location is rarely observed.

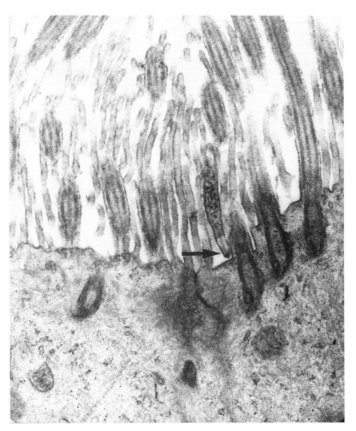

Fig. 11-18 Electron micrograph of a ciliated bronchial epithelial cell from a hamster experimentally infected with *Mycoplasma pneumoniae*. Electron dense attachment organelle of the organism indicated by arrow.

epithelial surface of ependymal cells (Kohn and Chinookoswong, 1981). Absence of a cell wall imparts a unique cell membrane to cell membrane relationship, allowing for transfer of needed host-metabolites to the microorganism and exposure of the host cell to *Mycoplasma* generated cytopathological metabolites such as peroxide and superoxide radicals (Razin and Jacobs, 1992). Some *Mycoplasma,* most notably the human pathogen *M. pneumoniae* have an attachment organelle that aligns the organism with the host cell receptor site (Fig. 11-18) (Collier and Clyde, 1971).

Adherence of *M. pulmonis* to the host cell membrane is key to the pathogenesis of respiratory disease, and the degree to which the organism successfully adheres to the nasal mucosa is likely of major significance to the outcome of infection in the lower respiratory tract (Schoeb et al., 1993). Using tracheal organ culture, Schoeb et al. (1993) studied the modulating effects that sialodacryo-adenitis virus (SDAV) and Sendai virus infections, age of the host, vitamin A deficiency, and exposure to ammonia may have on adherence owing to alteration of the respiratory mucosal surface. In their *in vitro* model, the investigators found that prior exposure of rats to SDAV and to vitamin A deficiency substantially increased the number of organisms adhering to the tracheal explants derived from the animals.

It is known that mouse genotype influences respiratory disease susceptibility to *M. pulmonis* (Lai et al., 1993; Cartner et al., 1998). In rats, marked susceptibility differences exist between LEW and Fischer 344 (F344) strains. After experimental infection with *M. pulmonis,* LEW rats mount a significantly increased inflammatory response to

the organism in comparison to F344 rats (Davis et al., 1982a; Davis and Cassell, 1982). Lung lesions in LEW rats developed earlier after infection, becoming more severe, and were characterized by larger areas of bronchus-associated lymphoid tissue hyperplasia. Proliferation of all classes of lymphoid cells, T and B lymphocytes, and plasma cells continued for the observation period of 4 months. In contrast, lymphoid proliferation in F344 rats reached a plateau at 28 days, with lesions tending to resolve by 4 months. The investigators suggested that these differences in severity and progression are related to differences in the degree of nonspecific lymphocyte activation and/or an imbalance in regulation of lymphocyte proliferation in Lewis rats.

In both Lewis and F344 rats, the upper respiratory nodes were determined to be the initial site of antibody production after experimental infection (Simecka et al., 1991), and a major site of antibody forming cells through a 28-day study. By 14 days after infection, antibody forming cells were found in other tissues, suggesting that the upper respiratory tract may be a significant source of primed or activated B cells that migrate to the infected tissues. In the same study, there was a major difference in the number of *M. pulmonis*–specific B cells in the draining lymph nodes of the lungs, with LEW rats having a larger proportion. There was also a greater antibody response in the lungs and upper respiratory nodes of LEW rats. The investigators noted that this may indicate that antibody production in the upper and lower respiratory nodes contributes significantly to the differences in response to infection between the two strains. Serum antibody levels alone do not resolve lesions and are not responsible for differences in lesion severity between the two rat strains (Simecka et al., 1989).

The pathogenicity of *M. pulmonis* is modulated by coinfection with viruses. Schoeb et al. (1985) found that gnotobiotic F344/N rats experimentally infected with Sendai virus and *M. pulmonis* had more severe lesions and a greater proliferation of *M. pulmonis* than did rats infected with only *M. pulmonis*. Tracheal mucosa in the coinfected rats was nearly destroyed by severe lymphocyte and plasmacyte accumulations; whereas, mono-infections induced mild inflammatory changes. At 3 to 4 weeks after infection, lungs from coinfected rats had suppurative pneumonia, atelectasis, and, in some cases, bronchiectasis, whereas lungs from mono-infected rats had essentially normal lungs. Similarly, Schoeb and Lindsey (1987) observed that F344/N rats coinfected with SDAV developed more severe lesions throughout the respiratory tract than animals mono-infected. Although the mechanism for this enhanced pathogenicity is not known, the investigators suggested that coinfection with a virus such as Sendai virus or SDAV, could induce immunosuppression, inhibition of macrophage function, or increased cytadhesion of *M. pulmonis*. However, it has been shown that coinfection with either Sendai virus or SDAV does not alter intrapulmonary killing or physical clearance of *M. pulmonis* in either LEW or F344 rats (Nichols et al., 1992).

Broderson et al. (1976) first reported the direct association of increased intracage ammonia levels on severity of rhinitis, otitis media, tracheitis, and pneumonia in rats infected with *M. pulmonis*. Pinson et al. (1986) examined the nasal mucosa of rats infected with *M. pulmonis* exposed to an ammonia level of 100 ppm. Within 9 days of ammonia exposure, more severe degenerative changes occurred in the mucosa infected with *Mycoplasma* than in the mucosa of noninfected rats. Exposure to ammonia in the absence of *Mycoplasma* infection induced only mild inflammatory changes, whereas widely distributed hyperplastic, metaplastic, and inflammatory changes were induced in *Mycoplasma*-infected rats. In another report (Schoeb et al., 1982), F344 rats were exposed to 100 ppm of ammonia and inoculated intranasally with *M. pulmonis*. This resulted in an increase in *M. pulmonis* titers and an exacerbation of lung lesions. In addition, it was determined that ammonia is almost entirely absorbed in the nasal

passages, not in the trachea or lungs. This finding indicated that the increased numbers of *Mycoplasma* in the trachea and lung are probably a secondary effect, rather than a direct effect, of ammonia levels at either site.

Today in the United States, *M. pulmonis* infections in laboratory rats are primarily limited to a small proportion of conventional institutional colonies. This was not the case before 1980, when a survey showed that the organism was present in 11 of 13 conventional rat facilities and in 6 of 9 barrier-maintained facilities (Cassell et al., 1981a). ELISA testing is the most frequently used method to determine if the organism is present within a rat colony, and it is part of the pathogen repertoire of commercial diagnostic laboratories that provide serological testing services to institutions. The ELISA tests reflect immunoglobulin G detection. A presumptive diagnosis is difficult to make based only on characteristic gross lung lesions associated with *M. pulmonis* because other pathogens such as the CAR bacillus and *Corynebacterium kutscheri* induce similar lesions. Culturing the organism from sites such as the nasopharynx, middle ear, trachea, lung, and oviduct is a reliable means to define the *M. pulmonis* status of an animal or a colony (Cassell et al., 1983). However, the nutrient requirements for *Mycoplasma* are more complex than for most other bacteria because *Mycoplasma* require a source of sterol, usually supplied by addition of either sterile horse or swine serum to the broth and agar media. Accordingly, ELISA testing has replaced culturing as the principal means to diagnose infections because it is less time consuming and costly and is comparable in reliability (Mia et al., 1981; Cassell et al., 1986, Lindsey et al., 1991).

PCR is a sensitive means to diagnose *M. pulmonis* infections that may be below the threshold for detection by culture and ELISA testing, and detect *M. pulmonis* in either nonfixed or paraffin-embedded tissues (van Kuppeveld et al., 1993; Brunnert et al., 1994; Sanchez et al., 1994; Schoeb et al., 1997). An excellent review of laboratory methods to diagnose mycoplasmosis in laboratory rodents is presented by Davidson et al. (1994).

Although *M. pulmonis* infrequently occurs in laboratory rat colonies today, diligent and frequent testing for this pathogen must be part of any screening program because the organism continues to exist in a small subset of conventional colonies of rats and mice and in wild or feral rodents. Once *M. pulmonis* is introduced into a colony, there is no effective means to eliminate it other than through rederivation by cesarean section or embryo transfer, and removal of all animals shown to be infected. If cesarean section is the method chosen, the uterus and oviducts of the donor dam must be tested for *M. pulmonis* by culturing and/or PCR and the rederived litter maintained in quarantine until determined to be *M. pulmonis*-free by repeated ELISA and PCR testing. Antibiotic therapy is not effective in eliminating the organism from infected animals. Immunization protocols have been investigated in which either live or formalin-inactivated *M. pulmonis* vaccines were given intravenously or intranasally. However, none of the protocols resulted in protective outcomes that would warrant their use (Cassell et al., 1978).

C. Genital Mycoplasmosis *(Mycoplasma pulmonis)*

Although *M. pulmonis* is most often associated with respiratory infections, the organism has a strong tropism for the epithelia of the female genital tract. In female rats that have naturally occurring respiratory mycoplasmosis, up to 40% will also have *M. pulmonis*–infected genital tracts. Of these, it is estimated that about 30% will have grossly evident oophoritis and salpingitis (Cassell et al., 1979). Infection of the female genital tract, in most instances, is owing to colonization of the vagina with subsequent ascending spread to the uterus and oviducts. Hematogenous spread to the female genital tract may also occur in naturally occurring infections because intravenous inoculation of *M. pulmonis* results in a very high rate of oviductal and uterine infection. Infection of the genital tract in male rats has been reported. *M. pulmonis* has been isolated from the testes, epididymides, and vas deferens of rats that were bred to genital tract–infected females. Infection of the male genital tract may be a factor associated with infertility (Steiner and Brown, 1993). It has also been shown *in vitro* that the organism adheres to spermatozoa (Cassell et al., 1981b).

Clinical signs associated with genital mycoplasmosis are limited to infertility and to fetal and neonatal deaths. The most frequently occurring lesions of the female genital tract are purulent exudates within enlarged oviducts and distended ovarian bursas owing to exudation. Uteri may not show any gross abnormalities; however, in some cases pyometra occurs (Graham, 1963; Cassell et al., 1979).

The inflammatory response in the oviduct is characterized by a polymorphonuclear cell exudate within the lumen, hyperplasia, and metaplasia of the ciliated columnar epithelium and a submucosal lymphoid infiltrate. This response quite closely mirrors that seen with *M. pulmonis* infected trachea and bronchi. Within the uterus, a polymorphonuclear cell inflammatory response occurs with varying intensity. Similar to infection in the respiratory tract, the organism adheres to the cell surface without migration into the submucosa. In the gravid uterus, there are foci of polymorphonuclear cells in the decidual and basal layers of the placenta (Steiner and Brown, 1993). *M. pulmonis* induces a placentitis similar to that described in deciduitis of humans and represents a model system for investigations into the pathogenicity of uterine infection (Peltier et al., 2003).

A number of studies have been done to determine factors involved in outcomes associated with genital tract infections. The pathogenesis associated with fetal death has been studied by experimentally inoculating rats intravaginally with *M. pulmonis* before breeding. Subsequent to inoculation, the organism infects the uterus, invading the placenta and establishing an infection in the amnion cavity at day-14 gestation. At gestation day 18, it can be isolated from the oropharynx and lungs of fetuses (Steiner et al., 1993). Under experimental conditions, the time of initial infection is a major factor in the pregnancy outcome and spread from the genital tract to the respiratory tract (Brown and Steiner, 1996). These investigators infected female rats at three time points (10 days before breeding, and day 11 and day 14 of gestation) and found that the most severe loss of pups occurred in dams that had been infected before breeding. In this group, one-third of the fetuses at day-14 gestation, and one-half of the fetuses at day-18 gestation were infected; whereas, in the two groups of dams inoculated at either 11 or 14 days of gestation, *M. pulmonis* was not isolated from the placentas, amniotic fluid or fetuses. Approximately, one-third of the female rats inoculated 10 days before breeding were found to be infertile.

Similarly to respiratory mycoplasmosis, rat stocks/strains differ in their susceptibility to genital tract infection and pregnancy outcome. A study comparing infection outcome of Sprague-Dawley, Wistar, F344, and Lewis rats found that Sprague-Dawley rats had a greater prevalence of infertility, decreased litter size, decreased pup birth weight, increased resorptions and stillbirths, and the highest rate of pulmonary infection in neonatal pups. Lewis dams, at 1-day postpartum, had the lowest genital tract infection rate, and their litters lacked evidence of vertical transmission of *M. pulmonis*. In this study, intravaginal inoculation of the organism resulted in self-limiting genital tract infections in the Lewis and F344 strains (Reyes et al., 2000).

The hormonal influence of testosterone treatment on increasing the prevalence and severity of *M. pulmonis*–induced genital mycoplasmosis has been reported (Leader et al., 1970).

Caesarean section is commonly used to rederive a rat line to render it free of *M. pulmonis*. If the genital tract or the fetuses of the rat undergoing caesarean section were infected with the organism, it is obvious that the success of the process to rederive a litter free of *Mycoplasma* would very likely be compromised. Graham (1963) was perhaps the first to caution that vertical transfer of *M. pulmonis* via infected uteri of hysterectomized rats could lead to unsuccessful rederivation of litters. Vertical transmission of the organism has been reported in germfree rats that were contaminated by hysterectomized dams that were infected with *M. pulmonis* (Ganaway et al., 1973).

Accordingly, it is essential to use ELISA and PCR testing to ensure that rederived litters are, in fact, free of *Mycoplasma*. Rederivation by early embryo transfer is an alternative to caesarean section. Trypsin treatment of embryos has been successfully used to help ensure that *Mycoplasma* and viruses are not contaminants of transferred embryos (Rouleau et al., 1993).

D. Articular Mycoplasmosis
(*Mycoplasma arthritidis*)

Klieneberger (1938) isolated a PPLO that she termed L4 from the middle ear and oropharynx of laboratory rats. The L4 organism was subsequently reclassified as *Mycoplasma arthritidis*. Although rarely a cause of naturally occurring arthritis in laboratory rats, the organism has been reported in earlier surveys to infect conventionally maintained colonies and has been reported in barrier-maintained rats (Cox et al., 1988). Today, all commercial sources of barrier-maintained rats in the United States are free of *M. arthritidis* and based on personal communication with directors of several large diagnostic laboratories, sera samples from conventionally maintained rats are rarely *M. arthritidis*–antibody positive. The organism has been isolated, typically without pathological implications, in various nonarticular sites such as the submaxillary gland (Klieneberger, 1938); the oropharynx; middle ear; lung, usually with coinfection by *M. pulmonis* or other bacteria (Cole et al., 1967; Stewart and Buck, 1975); the large intestine; lower female genital tract; upper male genital tract; and the Harderian gland (Cox et al., 1988). Experimental inoculation in athymic nude rats and euthymic (rnu/+) littermates resulted in more severe arthritis and progressive disease leading to deaths in the athymic (rnu/rnu) rats (Binder et al., 1993). A number of *Mycoplasma* induce arthritis in mammalian and avian species, and there has been much interest in *Mycoplasma* as models of infectious and rheumatoid arthritis in humans. The disease course after intravenous inoculation of *M. arthritidis* into mice and rats differs significantly in these two species. In the rat, the arthritis is acute and self-limiting with resolution in less than 2 months; whereas in the mouse, the arthritis subsides after 70 to 150 days but then may undergo a pronounced resurgence (Cole and Ward, 1979).

Naturally occurring *M. arthritidis*–induced arthritis has been rarely reported in rats. Nearly all the information associated with *M. arthritidis*–induced arthritis reflects experimental infections; however, the clinical signs seen in both naturally occurring and induced arthritis are similar. Typically, multiple joints are swollen, inflamed, and tender, with rats tending to avoid weight-bearing on the affected limbs. One of the first reports (Preston,

1942) described arthritis in a female rat that was in continuous oestrus induced by testosterone treatment, along with another rat that had not been treated with testosterone. Other rats in the colony had intra-abdominal and para-ovarian abscesses from which the L4 organism (*M. arthritidis*) was isolated. In another report (Ito, 1957), the L4 organism was associated with arthritis in more than 100 rats in which the joints of the forelimbs were involved in nearly all cases. Isolation of the PPLO strain from the oropharynx in all arthritic rats led the investigator to the hypothesis that the preponderance of forelimb involvement was a function of L4 transfer from the oral cavity to abrasions associated with licking and biting. However, attempts by Ito to reproduce this scenario experimentally were unsuccessful.

The articular and para-articular tissues are swollen, and the synovial cavity contains a thick, purulent exudate. After intravenous inoculation of *M. arthritidis*, polyarthritis becomes evident in 2 to 9 days, with the arthritis resolving in about 2 months. *Mycoplasma* can be easily recovered during the peak stage of the disease but not after 6 to 8 weeks. Microscopically, there is an intense polymorphonuclear infiltration of the synovial space and membrane and of the surrounding articular tissue (Fig. 11-19). This is followed by periostial osteoneogenesis and destruction of articular cartilage (Cole and Ward, 1979). Hindlimb paralysis may follow perineural inflammation and edema that compresses nerve roots within the spinal foramina (Cassell et al., 1979).

The rat has been widely used to study the immunologically based pathogenesis of articular disease induced by intravenous inoculation of *M. arthritidis* (Cole and Ward, 1979). Although the mechanisms contributing to the pathogenesis of *M. arthritidis* are not well defined, of considerable interest is the *M. arthritidis* mitogen (MAM) superantigen, which is a potent T-cell mitogen and inducer of IFN-γ for murine lymphocytes. Superantigens, as a class, can induce autoimmunity and act as potent immuno-modulatory molecules. Although the MAM superantigen probably has a role in the arthritogenicity of *M. arthritidis*, all strains of the organism release MAM during lytic cell senescence (Cole and Atkin, 1991; Anh-Hue et al., 2002).

A lysogenic bacteriophage, MAV1, is considered to be a virulence factor associated with arthritis in rats and mice because MAV1 lysogens are more arthritogenic than are nonlysogens. One of the MAV genes transcribed is thought to code for a membrane surface-exposed lipoprotein. This lipoprotein, MAA1, may mediate adherence of *M. arthritidis* to joint tissues and thus enhance its pathogenicity (Washburn et al., 2000; Anh-Hue et al., 2002). An excellent reiview of the pathogenic mechanisms involved in *M. arthritidis*–induced murine arthritis is presented by Washburn and Cole (2002).

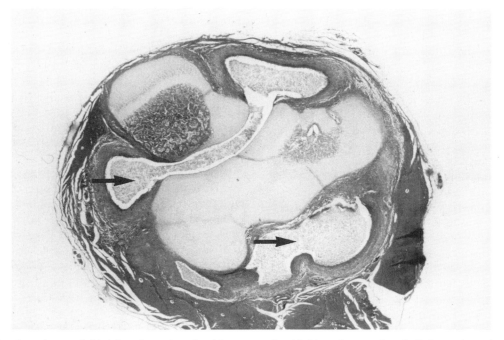

Fig. 11-19 Section through an arthritic joint of a rat inoculated intravenously with *Mycoplasma arthritidis*. Polymorphonuclear cell exudate within the synovial spaces indicated by arrows.

ELISA testing is the most frequent means to survey for *M. arthritidis*; however, because of one-way cross-reacting antigens with *M. pulmonis*, immunofluorescence assay or a Western blot must be done to confirm if a positive ELISA serum sample reflects *M. arthritidis* infection.

E. Research Complications Owing To *M. pulmonis* and *M. arthritidis*

Before the elimination of *M. pulmonis* from commercial sources of laboratory rats, this organism was responsible for catastrophic effects on many long-term toxicologic, gerontologic, and cancer studies. Today, *M. pulmonis* remains in some conventional rat and mouse colonies; these sources and contamination of *M. pulmonis*–free colonies by feral or wild rodents still pose risks to long-term studies if appropriate measures are not taken to exclude the organism from barrier facilities. *M. pulmonis* infection is incompatible with essentially any investigation in which the respiratory or female genital tract is involved. A number of studies have shown its effect on the immune system. These include suppression of humoral antibody response to sheep erythrocytes, modulation of collagen-induced arthritis, and induction of nonspecific mitogen effects on lymphocytes (Cassell et al., 1986; Lindsey et al., 1991). The effects of *M. arthritidis* on research are negligible compared with those of *M. pulmonis*. Naturally occurring latent infections have

been shown to interfere with experimentally induced models of arthritis, and experimental infections have been shown to alter immunologic responsiveness (Cassell et al., 1986).

F. Mycoplasma of Negligible Pathogenicity in the Rat

1. *Mycoplasma neurolyticum*

Mycoplasma neurolyticum has been associated with conjunctivitis in mice. In the rat, there are no reports of naturally occurring disease induced by the organism, and there is essentially no evidence that it infects rats even as a commensal (Cassell et al., 1979). Most notably, the organism induces "rolling disease" in mice after intravenous, intraperitoneal, or intracerebral inoculation of the organism or cellfree culture filtrates (Findlay et al., 1938; Thomas and Bitensky, 1966). Rats, after intravenous inoculation of *M. neurolyticum*, develop neurological signs in less than an hour. Initially, animals become weak, ataxic, and prostrate; this is followed by clonic convulsive seizures and usually death in 3 to 4 hours. In a few rats, there is a prolonged survival of several days. The pathogenicity of *M. neurolyticum* is owing to an exotoxin (Tully, 1969; Kaklamanis and Thomas, 1970), and its action induces severe distension of astrocytes and a generalized edema of the brain, causing compression of neurons. Eight hours after inoculation, there are focal areas of spongiform degeneration, most prominently in the deep layers of the frontoparietal cortex and underlying white matter and in the molecular layer of the cerebrum (Thomas and Bitensky, 1966; Thomas et al., 1966).

2. *Mycoplasma volis*

This mycoplasma species was isolated from the respiratory tract of clinically normal prairie voles *(Microtus ochrogaster)*. To test the possible pathogenicity of this newly defined species in rats, 10 Sprague-Dawley rats were inoculated intranasally with *M. volis* (Evans-Davis et al., 1998). No clinical signs of infection were induced in any of the rats during the 6-week observation period, although all animals seroconverted to *M. volis*. The only histopathological changes were mild to severe hyperplasia of the bronchial-associated lymphoid tissue. *M. volis* was isolated from the nasopharynx of 9 rats and the lungs of 10 rats. The degree of peribronchial lymphoid hyperplasia was less at 6 weeks compared with that observed at 4 weeks after inoculaton. This would tend to suggest that the rat is able to clear *M. volis* without long-term sequelae.

G. Rickettsial Infection (*Hemobartonella muris*)

Hemobartonella muris is an extracellular parasite of erythrocytes in *Rattus norvegicus*. This Gram-negative rickettsia is pleomorphic varying from a 350-nm coccus to a 700-nm rod. It is an obligate parasite that is not cultivable outside the host (Lindsey et al., 1991). The blood-sucking rat louse, *Polyplax spinulosa,* is the vector necessary for transmission of *H. muris* between rats. Today, because of the elimination of *P. spinulosa* from all reasonably well managed colonies, this rickettsial infection would rarely be present in laboratory rats.

If the reticuloendothelial system is intact, natural infections of *H. muris* are invariably latent. However, if the host has been splenectomized, there is a rapid increase in the number of organisms, leading to hemolytic episodes within 3 to 10 days, dyspnea, hemoglobinuria, and often death.

Splenomegaly is essentially the only change seen upon gross examination of rats subsequent to natural infection with the agent. Red blood cell smears show *H. muris* as coccoid and rod bodies arranged singly, and in chains or clusters on erythrocytes in splenectomized rats, but parasitemia would be infrequently detected in nonsplenectomized rats (Cassell et al., 1979).

As noted, *P. spinulosa*–transmitted *H. muris* infections have been essentially eliminated in laboratory rat colonies. Exceptions to this would be colonies contaminated by wild or feral rats infected with both H. muris and the vector, as well as institutional colonies in which husbandry

and veterinary practices do not ensure the absence of *P. spinulosa*.

Transmission has been reported from contaminated blood, tumors, and other blood-laden biologic material (Cassell et al., 1979; Lindsey et al., 1991). Today, receipt of these biological materials from unreliable sources poses the greatest risk of infection to a naïve group of recipient rats. Although transplacental transmission has been suggested, there is negligible evidence to support vertical transfer of the agent (Owen, 1982).

Splenectomy, as a means to activate a latent infection, is used to detect the organism's presence within a colony. Older rats develop more severe parasitemias than do young rats after splenectomy and, accordingly, should be used as subjects for activation. Positive identification of *H. muris* in blood smears is achieved by use of Giemsa or Romanowsky stains; however, care must be taken to not confuse normal basophilic stippling of erythrocytes with the organism (Cassell et al., 1979).

The most likely research complications associated with *H. muris* would be through accidental transmission by contaminated blood, plasma, tumors, or other biological material. It is extremely important that investigators and veterinarians are aware that not all sources of biologic materials are derived from pathogen-free rats and mice, and that these materials could be contaminated with viral, bacterial, *Mycoplasma*, or ricketsial agents. *H. muris* infection has been shown to affect transplantable tumor kinetics, reticuloendothelial function, erythrocyte half-life, IFN production, and host response to experimental infections with viral and protozoal agents. (Sacks and Egdahl, 1960; Stansly, 1965; Cassell et al., 1979; Lindsey et al., 1991).

REFERENCES

Accreditation Microbiological Advisory Committee [Accreditation] (1971). "Microbiological Examination of Laboratory Animals for Purposes of Accreditation." Laboratory Aminal Center, Medical Research Council, Carshalton, United Kingdom.

Ackerman, J.I. and Fox, J. (1981). Isolation of *Pasteurella ureae* from reproductive tracts of congenic mice. *J. Clin. Microbiol.* **13,** 1049–1053.

Ackerman, J.I., Fox, J.G., and Murphy, J.C. (1984). An enzyme linked immunosorbent assay for detection of antibodies to *Corynebacterium kutscheri* in experimentally infected rats. *Lab. Anim. Sci.* **38,** 38–43.

Adachi, K. (1921). Flagellum of the microorganism of rat-bite fever. *J. Exp. Med.* **33,** 647–652.

Adams, L.E., Yamauchi, Y., Carleton, J., Townsend, L., and Kirn, O.J. (1972). An epizootic of respiratory tract disease in Sprague-Dawley rats. *J. Am. Vet. Med. Assoc.* **161,** 656–660.

Ainsworth, E.J. and Forbes, P.D. (1961). The effect of *Pseudomonas* pyrogen on survival of irradiated mice. *Radiat. Res.* **14,** 767–774.

Ajello, L. (1974). Natural history of the dermatophytes and related fungi. *Mycopathol. Mycol. Appl.* **53,** 93–110.

Ajello, L., Bostick, L., and Cheng, S.L. (1968). The relationship of *Trichophyton quinckeanum* to *Trichophyton mentagrophytes*. *Mycologia* **60**, 1185–1189.

Aldred, P., Hill, A.C., and Young, C. (1974). The isolation of *Streptobacillus moniliformis* from cervical abscesses of guinea-pigs. *Lab. Anim.* **8**, 275–277.

Aliouat, E.M., Dei-Cas, E., Billaut, P., DujarClin, L., and Camus, D. (1995). *Pneumocystis carinii* organisms from in vitro culture are highly infectious to the nude rat. *Parasitol. Res.* **81**, 82–85.

Aliouat, E.M., Mazars, E., Dei-Cas, E., Delcourt, P., Billaut, P., and Camus, D. (1994). Pneumocystis cross infection experiments using SCID mice and nude rats as recipient host showed strong host species specificity. *J. Eukaryotic Microbiol.* **41**, 715S

Allbritten, F.F., Sheely, R.F., and Jeffers, W.A. (1940). *Haverhilia multiformis* septicemia. *JAMA* **114**, 2360–2363.

Allen, A.M., Ganaway, J.R., Moore, T.D., and Kinard, R.F. (1965). Tyzzer's disease syndrome in laboratory rabbits. *Am. J. Pathol.* **4**, 859–882.

Amao, H., Akimoto, T., Komukai, Y., Sawada, T., Saito, M., and Takahashi, K.W. (2002). Detection of *Corynebacterium kutscheri* from the oral cavity of rats. *Exp. Anim.* **51**, 99–102.

Amano, H., Akimoto, T., and Takahashi, K.W. (1991). Isolation of *Corynebacterium kutscheri* from aged Syrian hamsters *(Mesocricetus auratus)*. *Lab. Anim. Sci.* **41**, 265–268.

Amao, H., Komukai, Y., Akimoto, T., Sugiyama, M., Takahashi, K.W., Sawada, T., and Saito, M. (1995a). Natural and subclinical *Corynebacterium kutscheri* infection in rats. *Lab. Anim. Sci.* **45**, 11–14.

Amao, H., Komukai, Y., Sugiyama, M., Saito, T.R., Takahashi, K.W., and Saito, M. (1993). Differences in susceptibility of mice among various strains to oral infection with *Corynebacterium kutscheri*. *Exp. Anim.* **42**, 539–545.

Amao, H., Komukai, Y., Sugiyama, M., Takghashi, K.W., Sawada, T., and Saito, M. (1995b). Natural habitats of *Corynebacterium kutscheri* in subclinically infected ICGN and DBA/2 strains of mice. *Lab. Anim. Sci.* **45**, 6–10.

Amao, H., Saito, M., Takahashi, K.W., and Nakagawa, M. (1990). Selective medium for *Corynebacterium kutscheri* and localization of the organism in mice and rats. *Exp. Anim.* **39**, 519–529.

An, C.L., Cigliotti, F., and Harmsen, A.G. (2003). Exposure of immunocompetent adult mice to *Pneumocystis carinii* f. sp. muris by cohousing: growth of *P. carinii f. sp. muris* and host immune response. *Infect. Immun.* **71**, 2065–2070.

Andrews, A.G., Dysko, R.C., Spillman, S.C., Kunkel, R.G., Brammer, D.W., and Johnson, K.J. (1994). Immune complex vasculitis with secondary ulcerative dermatitis in aged C57Bl/6NNia mice. *Vet. Pathol.* **31**, 293–300.

Anh-Hue, T., Lindsey, J.R., Schoeb, T.R., Elgavish, A., Huilan, Y., and Dybvig, K. (2002). Role of bacteriophage MAV1 as a mycoplasmal virulence factor for the development of arthritis in mice and rats. *J. Infect. Dis.* **186**, 432–435.

Anonymous (1975). Rat-bite fever a lab job risk. *Hosp. Pract.* **10**, 53.

Antopol, W. (1950). Anatomic changes in mice treated with excessive doses of cortisone. *Proc. Soc. Exp. Biol. Med.* **73**, 262–265.

Antopol, W., Glaubach, S., and Quittner, H. (1951). Experimental observations with massive doses of cortisone. *Rheumatism* **7**, 187–197.

Antopol, W., Quittner, H., and Saphra, I. (1953). "Spontaneous" infections after the administration of cortisone and ACTH. *Am. J. Pathol.* **29**, 599–600.

Araki, S., Suzuki, M., Fujimoto, M., and Kimura, M. (1995). Enhancement of resistance to bacterial infection in mice by vitamin B_2. *J. Vet. Med. Sci.* **57**, 599–602.

Arkless, H.A. (1970). Rat-bite fever at Albert Einstein Medical Center. *Pa. Med.* **73**, 49.

Arora, H.L. and Sangal, B.C. (1973). *Salmonella typhi* cholangitis in rats: an experimental study. *Indian J. Pathol. Bacteriol.* **16**, 28–34.

Arseculeratne, S.N., Panabokke, R.G., Navaratnam, C., and Weliange, L.V. (1981). An epizootic of *Klebsiella aerogenes* infection in laboratory rats. *Lab. Anim.* **15**, 333–337.

Artwohl, J.E., Flynn, J.C., Bunte, R.M., Angen, O., and Herold, K.C. (2000). Outbreak of *Pasteurella pneumotropica* in a closed colony of stock-Cd28(tmlMak) mice. *Contemp. Top. Lab. Anim. Sci.* **39**, 39–41.

Ash, G.W. (1971). An epidemic of chronic skin ulceration in rats. *Lab. Anim.* **5**, 115–122.

Atzori, C., Aliouat, E.M., Bartlett, M.S., DujarClin, L., Cargnel, A., and Dei-Cas, E. (1998). Current in vitro culture systems for *Pneumocystis*. *FEMS Immunol. Med. Microbiol.* **22**, 169–172.

Austrian, R. (1974). *Streptococcus pneumonias (Pneumococcus)*. *In* "Manual of Clinical Microbiology," 2nd Ed. (E.H. Lennette, E.H. Spaulding, and J.P. Truant, eds.), pp 109–115. American Society of Microbiology, Washington, DC.

Austrian, R. (1976). The Quellung reaction: a neglected microbiologic technique. *Mt. Sinai J. Med.* **43**, 699–709.

Austwick, P.K.C. (1974). Medically important *Aspergillus* species. *In* "Manual of Medical Microbiology," 2nd Ed. (E.H. Lennette, E.H. Spaulding, and J.P. Truant, eds.), pp 550–556. American Society of Microbiology, Washington, DC.

Badakhsh, F.F. and Herzberg, M. (1969). Deoxycholate-treated, non-toxic, whole-cell vaccine protective against experimental salmonellosis of mice. *J. Bacteriol.* **100**, 738–827.

Baer, H. (1967). *Diplococcus pneumoniae* type 16 in laboratory rats. *Can. J. Comp. Med. Vet. Sci.* **31**, 216–218.

Baer, H. and Preiser, A. (1969). Type 3 *Diplococcus pneumonia* in laboratory rats. *Can. J. Comp. Med.* **33**, 113–117.

Bahr, L. (1905). Uber die zur Vertilgung von Ratten und Mausen benutzten Bakterien. *Zentralbl. Bakteriol. Parasitenkd. Infektionskr. Hyg. Abt. I Orig.* **39**, 263–274.

Bahr, L., Raebiger, M., and Grosso, G. (1910). Ratin I und II, sowie uber die Stellung der Ratinbazillus zur Gartnergruppe. *Zentralbl. Bakteriol. Parasitenkd. Infektionskr. Hyg. Abt. 1 Orig.* **54**, 231–234.

Bainbridge, F.A. and Boycott, A.E. (1909). The occurrence of spontaneous rat epidemics due to Gaertner's bacillus. *J. Pathol. Bacteriol.* **13**, 342.

Baker, D.G. (1998). Natural pathogens of laboratory rats, mice, and rabbits and their effects on research. *Clin. Microbiol. Rev.* **11**, 231–266.

Baker, H.J., Cassell, G.H., and Lindsey, J.R. (1971). Research complications due to *Hemobartonelia* and *Eperythrozoon* infections in experimental animals. *Am. J. Pathol.* **64**, 625–656.

Ball, N.D. and Price-Jones, C. (1926). A laboratory epidemic in rats due to Gaertner's bacillus. *J. Pathol. Bacteriol.* **29**, 27–30.

Balows, A. (ed) (1998). "Topley and Wilson's Microbiology and Microbial Infections, Vol. 2: Systematic bacteriology," 9th Ed. Chapter 52, Hodder Arnold, London.

Balsari, Bianchi, C., Cocilova, A., Dragoni, I., Poli, G., and Ponti, W. (1981). Dermatophytes in clinically healthy laboratory animals. *Lab. Anim.* **15**, 75–77.

Barthold, S.W. and Brownstein, D.G. (1988). The effect of selected viruses on *Corynebacterium kutscheri* in rats. *Lab. Anim. Sci.* **38**, 580–583.

Bartlett, M.S., Angus, W.C., Shaw, M.M., Durant, P.J., Lee, C., Pascle, J.M., and Smith, J.W. (1998). Antibody to *Pneumocystis carinii* protects rats and mice from developing pneumonia. *Clin. Diag. Lab. Immunol.* **5**, 74–77.

Bartlett, M.S., Durkin, M.M., and Jay, M.A. (1987). Sources of rats free of latent *Pneumocystis carinii*. *J. Clin. Microbiol.* **25**, 1794–1795.

Bartlett, M.S., Fishman, J.A., Durkin, M.M., Queener, S.F., and Smith, J.W. (1990). *Pneumocystis carinii*: Improved models to study efficacy of drugs for treatment or prophylaxis of *Pneumocystis* pneumonia in the rat (*Rattus* sp.). *Exp. Parasitol.* **70**, 100–106.

Bartlett, M.S., Fishman, J.A., Queener, S.F., Durkin, M.M., Jay, M.A., and Smith, J.W. (1988). New rat model of *Pneumocystis carinii* infection. *J. Clin. Microbiol.* **26**, 1100–1102.

Bartlett, M.S., Queenre, S.F., Durkin, M.M., Shaw, M.M., and Smith, J.W. (1991). Inoculated mouse model of *Pneumocystis carinii* pneumonia. *J. Protozool.* **38**, 130S–131S.

Bartlett, M.S., Verbanac, P.A., and Smith, J.W. (1979). Cultivation of *Pneumocystis carinii* with WI-38 cells. *J. Clin. Microbiol.* **10**, 796–799.

Bartram, J.T., Welch, H., and Ostrolenk, M. (1940). Incidence of members of the *Salmonella* group in rats. *J. Infect. Dis.* **67**, 222–226.

Battles, J.K., Williamson, J.C., Pike, K.M., Gorelick, P.L., Ward, J.M., and Gonda, M.A. (1995). Diagnostic assay for *Helicobacter hepaticus* based on nucleotide sequence of its 16S rRNA gene. *J. Clin. Microbiol.* **33**, 1344–1347.

Bauer, N.L., Paulsiud, J.R., Bartlett, M.S., Smith, J.W., and Wilde III, C.E. (1991). Immunologic comparisons of *Pneumocystis carinii* strains obtained from rats, ferrets and mice using convalescent sera from the same sources. *J. Protozool.* **38**, 1665S–1685S.

Bauer, N.L., Paulsrod, J.R., Bartlett, M.S., Smith, J.W., and Wilde III, C.E. (1993). *Pneumocystis carinii* organisms obtained from rats, ferrets, and mice are antigenically different. *Infect. Immun.* **61**, 1315–1319.

Bayne-Jones, S. (1931). Rat-bite fever in the United States. *Int. Clin.* **3**, 235–253.

Beck, J.M., Newbury, R.L., and Palmer, B.E. (1996). *Pneumocystis carinii* pneumonia in SCID mice induced by viable organisms produced in-vitro. *Infect. Immun.* **64**, 4643–4647.

Beck, R.W. (1963). A study in the control of *Pseudomonas aeruginosa* in a mouse breeding colony by the use of chlorine in the drinking water. *Lab. Anim. Care* **13**, 41–46.

Beckwith, C.A., Franklin, C.L, Hook Jr., R.R., Besch-Williford, C.L., and Riley, L.K. (1997). Fecal PCR assay for diagnosis of Helicobacter infection in laboratory rodents. *J. Clin. Microbiol.* **35**, 1620–1623.

Beeson, P.B. (1943). Problem of etiology of rat-bite fever: report of two cases due to *Spirillum minus*. *JAMA* **123**, 332–334.

Bemis, D.A., Shek, W.R., and Clifford, C.B. (2003). *Bordetella bronchiseptica* infection of rats and mice. *Comp. Med.* **53**, 11–20.

Berlin, B.S., Johnson, G., Hawke, W.D., and Lawrence, A.G. (1952). The occurrence of bacteremia and death in cortisone treated mice. *J. Lab. Clin. Med.* **40**, 82–89.

Besch-Williford, C.B. and Wagner, J.E. (1982). Bacterial and mycotic diseases of the integumentary system. *In* "The Mouse in Biomedical Research, Vol. 2: Disease." (H.L. Foster, J.D. Small, and J.G. Fox, eds.), pp 55–72. Academic Press, New York.

Besch-Williford, C.B. and Wagner, J.E. (1992). Bacterial and mycotic disease of the integumentary system. *In* "The Mouse in Biomedical Research, Vol. 2: Disease." (H.L. Foster, J.D. Small, and J.G. Fox, eds.), pp 58–60. Academic Press, New York.

Bhandari, J.C. (1976). A spontaneous outbreak of pseudotuberculosis in rats. *27th Annu. Sess., Am. Assoc. Lab. Anim. Sci.* Abstract No. 73.

Biberstein, E.L. (1975). Rat bite fever. *In* "Diseases Transmitted from Animals to Man," 6th Ed. (W.T. Hubbert, W.F. McCulloch, and P.R. Schnurrenberger, eds.), pp 186–188. Thomas, Springfield, IL.

Bicks, V.A. (1957). Infection of laboratory mice with *Corynebacterium murium*. *Aust. J. Sci.* **20**, 20–23.

Biggar, W.D., Bogart, D., Holmes, B., and Good, R.A. (1972). Impaired phagocytosis of *Pneumococcus* type 3 in splenectomized rats. *Proc. Soc. Exp. Biol. Med.* **139**, 903–908.

Binder, A., Hedrich, H.J., Wonigeit, K., and Kirchhoff, H. (1993). The *Mycoplasma arthritidis* infection in congenitally athymic nude rats. *J. Exp. Anim. Sci.* **35**, 177–185.

Bisgaard, M. (1986). *Actinobacillus muris* sp. nov. isolated from mice. *Acta Path. Microbiol. Immunol. Scand. B* **94**, 1–8.

Bisgaard, M. (1995). Taxonomy of the family Pasteurellaceae Pohl 1991. *In* "Haemophilus, Actinobacillus and Pasteurella." (W. Donachie, F.A. Lainson, and J.C. Hodgson, eds.), pp 1–7, Plenum Press, New York.

Black, D.J. (1975). *Pasteurella pneumotropica* isolated from a Harderian gland abscess of a mouse. *Lab. Anim. Dig.* **9**, 74–77.

Blackmore, D.K. (1972). Accidental contamination of a specified-pathogen-free unit. *Lab. Anim.* **6**, 257–271.

Blackmore, D.K. and Francis, R.A. (1970). The apparent transmission of staphylococci of human origin to laboratory animals. *J. Comp. Pathol.* **80**, 645–651.

Blackmore, D.L. and Casillo, S. (1972). Experimental investigation of uterine infections of mice due to *Pasteurella pneumotropica*. *J. Comp. Pathol.* **82**, 471–475.

Blake, F.G. (1916). The etiology of rat-bite fever. *J. Exp. Med.* **23**, 39–60.

Blanden, R.V., Mackaness, G.B., and Collins, F.M. (1966). Mechanisms of acquired resistance in mouse typhoid. *J. Exp. Med.* **124**, 585–599.

Block Jr., O. and Baldock, H. (1937). A case of rat-bite fever with demonstration of *Spirillum minus*. *J. Pediatr.* **10**, 358–360.

Bloomfield, A.L. and Lew, W. (1941). Protective effect of sulfaguanadine against ulcerative cecitis in rats. *Proc. Soc. Exp. Biol. Med.* **48**, 363–368.

Bloomfield, A.L. and Lew, W. (1942). Significance of Salmonella in ulcerative cecitis of rats. *Proc. Soc. Exp. Biol. Med.* **51**, 179–182.

Bloomfield, A.L. and Lew, W. (1943a). Increased resistance to ulcerative cecitis of rats on a diet deficient in the vitamin B complex. *J. Nutr.* **25**, 427–431.

Bloomfield, A.L. and Lew, W. (1943b). Prevention of infectious ulcerative cecitis in the young of rats by chemotherapy of the mother. *Am. J. Med. Sci.* **205**, 383–388.

Bloomfield, A.L. and Lew, W. (1943c). Prevention by succinylsulfathiazole of ulcerative cecitis in rats. *Proc. Soc. Exp. Biol. Med.* **51**, 28–29.

Bloomfield, A. L., and Lew, W. (1948). Cure of ulcerative cecitis of rats by streptomycin. *Proc. Soc. Exp. Biol. Med.* **69**, 11–14.

Bloomfield, A.L., Rantz, L.A., Lew, W., and Zuckerman, A. (1949). Relation of a specific strain of Salmonella to ulcerative cecitis of rats. *Proc. Soc. Exp. Biol. Med.* **71**, 457–461.

Boart, D., Biggar, W.D., and Good, R.A. (1972). Impaired intravascular clearance of *Pneumococcus* type-3 following splenectomy. *Res. J. Reticuloendothel. Soc.* **11**, 77–87.

Bohme, D.H., Schneider, H.A., and Lee, J.M. (1954). Some pathophysiological parameters of natural resistance to infection in murine salmonellosis. *J. Exp. Med.* **110**, 9–26.

Boivin, G.P., Hook Jr., R.R., and Riley, L.K. (1993). Antigenic diversity in flagellar epitopes among *Bacillus piliformis* isolates. *J. Med. Microbiol.* **38**, 177–182.

Bonger, J. (1901). *Corynethrix pseudotuberculosis murium*, ein neuer pathogener Bacillus fur Mause. *Z. Hyg. Infektionskr.* **37**, 449–475.

Boot, R., Bakker, R.W.G., Thuis, H., Veenema, J.L., and DeHoog, H. (1993a). An enzyme-linked immunosorbent assay (ELISA) for monitoring rodent colonies for *Streptobacillus moniliformis*. *Lab. Anim.* **27**, 350–357.

Boot, R. and Bisgaard, M. (1995). Reclassification of 30 Pasteurellaceae strains isolated from rodents. *Lab. Anim.* **29**, 314–319.

Boot, R., Lammers, R.M., and Busschbach, A.E. (1986). Isolation of members of the Hemophilus-Pasteurella-Actinobacillus group from feral rodents. *Lab. Anim.* **20**, 36–40.

Boot, R., Oosterhuis, A., and Thuis, H.C.W. (2002). PCR for the detection of *Streptobacillus moniliformis*. *Lab Anim.* **36**, 200–208.

Boot, R., Thuis, M., Bakker, R., and Veenema, J.L. (1995a). Serological studies of *Corynebacterium kutscheri* and coryneform bacteria using an enzyme-linked immunosorbent assay (ELISA). *Lab. Anim.* **29**, 294–299.

Boot, R., Thuis, H., Bakker, R.H., and Veenema, J.L. (1995b). An enzyme-linked immunosorbent assay (ELISA) for monitoring antibodies to SP group Pasteurellaceae in guinea pigs. *Lab. Anim.* **29**, 59–65.

Boot, R., Thuis, H., and Koedam, M.A. (1995c). Infection by V factor dependent Pasteurellaceae (*Haemophilus*) in rats. *J. Exp. Anim. Sci.* **37**, 7–14.

Boot, R., Thuis, H., and Teppema, J.S. (1993b). Hemagglutination by Pasteurellaceae isolated from rodents. *Zentrabl. Bakteriol.* **279**, 259–273.

Boot, R., Thuis, H.C., and Van de Berg, L. (1999). Most European SPF "Pasteurella free" guinea pig colonies are *Haemophilus* spp positive. *Scand. J. Lab. Anim. Sci.* **26**, 148–152.

Boot, R., Thuis, H.C., and Veenema, J.L. (1994a). Comparison of mouse and rat Pasteurellaceae by micro-agglutination assay. *J. Exp. Anim. Sci.* **37**, 158–165.

Boot, R., Thuis, H.C., and Veenema, J.L. (1997). Serological relationship of some V-factor dependent Pasteurellaceae (*Haemophilus*) from rats. *J. Exp. Anim. Sci.* **38**, 147–152.

Boot, R., Thuis, H.C., and Veenema, J.L. (1998). Serological relationships of some V-factor dependent Pasteurellaceae (*Haemophilus* sp.) from guinea pigs and rabbits. *Lab. Anim.* **33**, 91–94.

Boot, R., Thuis, H.C., and Veenema, J.L. (2000). Transmission of rat and guinea pig *Haemophilus spp* to mice and rats. *Lab. Anim.* **34**, 409–412.

Boot, R., Thuis, C.W., Veenema, J.L., and Bakker, R.G.H. (1995d). An enzyme-linked immunosorbent assay (ELISA) for monitoring rodent colonies for *Pasteurella pneumotropica* antibodies. *Lab. Anim.* **29**, 307–313.

Boot, R., Thuis, H.C., Veenema, J.L., Bakker, R.H., and Walvoort, H.C. (1994b). Colonization and antibody response in mice and rats experimentally infected with Pasteurellaceae from different rodent species. *Lab. Anim.* **28**, 130–137.

Boot, R., van Herck, H., and van der Logt, J. (1996). Mutual viral and bacterial infections after housing rats of various breeders within an experimental unit. *Lab. Anim.* **30**, 42–45.

Boot, R. and Walvoort, H.C. (1994). Vertical transmission of *Bacillus piliformis* infection (Tyzzer's disease) in a guinea pig: case report. *Lab Anim.* **18**, 195–199.

Booth, B.H. (1952). Mouse ringworm. *Arch. Dermatol. Syphilol.* **66**, 65–69.

Bootz, F., Kirschnek, S., and Nicklas, W. (1997). Detection of Pasteurellaceae in rodents by polymerase chain reaction. *In* "Proceedings of the sixth FELASA Symposium Harmonization of Laboratory Animal Husbandry." (P.N. O'Donoghue, ed), pp 76–78, Royal Society of Medicine Press, London.

Bootz, F., Kirschnek, S., Nicklas, W., Wyss, S.K., and Homberger, F.R. (1998). Detection of Pasteurellaceae in rodents by polymerase chain reaction analysis. *Lab. Anim. Sci.* **48**, 542–546.

Borgen, L.O. and Gaustad, V. (1948). Infection with *Actinomyces muris ratti (Streptobacillus moniliformis)* after bite of a laboratory rat. *Acta Med. Scand.* **130**, 189–198.

Bowers, E.F. and Jeffries, L.R. (1955). Optochin in the identification of *Str. pneumoniae*. *J. Clin. Pathol.* **8**, 58–60.

Boycott, A.E. (1911). Infective methaemoglobinaemia in rats caused by Gaertner's bacillus. *J. Hyg.* **11**, 443–472.

Boycott, A.E. (1919). Note on the production of agglutinins by mice. *J. Pathol. Bacteriol.* **23**, 126.

Boylan, D.J. and Current, W.L. (1992). Improved rat model of *Pneumocystis carinii* pneumonia: induced laboratory infections in *Pneumocystis*-free animals. *Infect. Immun.* **60**, 1589–1597.

Bradfield, J.F., Schaetman, T.R., McLaughlin, R.M., and Steffen, E.K. (1992). Behavioral and physiologic effects of inapparent wound infection in rats. *Lab. Anim. Sci.* **42**, 572–578.

Bradfield, J.F., Wagner, J.E., Boivin, G.P., Steffer, E.K., and Russel, R.J. (1993). Epizootic of fatal dermatitis in athymic nude mice due to *Staphylococcus xylosus*. *Lab. Anim. Sci.* **432**, 111–113.

Brands, U. and Mannheim, W. (1996). Use of two commercial rapid test kits for the identification of *Haemophilus* and related organisms: reactions of authentic strains. *Med. Micrbiol. Lett.* **5**, 133–144.

Breini, A. and Kinghom, A. (1906). A preliminary note on a new *Spirochaeta* found in a mouse. *Lancet* **2**, 651–652.

Brennan, P.C., Fritz, T.E., and Flynn, R.J. (1965). *Pasteurella pneumotropica*: Cultural and biochemical characteristics and its association with disease in laboratory animals. *Lab. Anim. Care* **15**, 307–312.

Brennan, P.C., Fritz, T.E., and Flynn, R.J. (1969a). Murine pneumonia: a review of the etiologic agents. *Lab. Anim. Care* **19**, 360–371.

Brennan, P.C., Fritz, T.E., and Flynn, R.J. (1969b). The role of *Pasteurella pneumotropica* and *Mycoplasma pulmonis* in murine pneumonia. *J. Bacteriol.* **97**, 337–349.

Broderson, J.R., Lindsey, J.R., and Crawford, J.E. (1976). The role of environmental ammonia in respiratory mycoplasmosis of rats. *Am. J. Pathol.* **85**, 115–130.

Brogden, K.A., Cutlip, R.C., and Lehmkuhl, H.D. (1993). Cilia-associated respiratory bacillus in wild rats in central Iowa. *J. Wildl. Dis.* **29**, 123–126.

Brown, C.M. and Parker, M.T. (1957). *Salmonella* infections in rodents in Manchester with special reference to *Salmonella enteritidis* var *danysz*. *Lancet* **2**, 1277–1279.

Brown, G.W. and Suter, I.I. (1969). Human infections with mouse favus in a rural area of South Australia. *Med. J. Aust.* **2**, 541–542.

Brown, M.B. and Steiner, D.A. (1996). Experimental genital mycoplasmosis: time of infection influences pregnancy outcome. *Infect. Immun.* **64**, 2315–2321.

Brown, T.M. and Nunemaker, J.C. (1942). Rat-bite fever: a review of the American cases with reevaluation of etiology: report of cases. *Bull. Johns Hopkins Hosp.* **70**, 201–328.

Brownstein, D.G., Barthold, S.W., Adams, R.L., Termillgon, G.A., and Aftosmis, J.G. (1985). Experimental *Corynebacterium kutscheri* infection in rats: bacteriology and serology. *Lab. Anim. Sci.* **35**, 135–138.

Bruce, D.L., Bismanis, J.E., and Vickerstaff, J.M. (1969). Comparative examinations of virulent *Corynebacterium kutscheri* and its presumed a virulent variant. *Can. J. Microbiol.* **15**, 817–818.

Bruner, D.W. and Moran, A. (1949). *Salmonella* infections of domestic animals. *Cornell Vet.* **39**, 3–63.

Brunnert, S.R., Dai, Y., and Kohn, D.F. (1994). Comparison of polymerase chain reaction and immunohistochemistry for the detection of *Mycoplasma pulmonis* in paraffin-embedded tissue. *Lab. Anim. Sci.* **44**, 257–260.

Buchanan, R.E. and Gibbons, N.E., eds. (1974). "Bergey's Manual of Determinative Bacteriology", 8th ed. Williams and Wilkins, Baltimore, Maryland.

Buchbinder, L., Hall, L., Wilens, S.L., and Slanetz, C.A. (1935). Observations on enzootic paratyphoid infection in a rat colony. *Am. J. Hyg.* **22**, 199–213.

Burek, J.D., Jersey, G.C., Whitehair, C.K., and Carter, G.R. (1972). The pathology and pathogenesis of *Bordetella bronchiseptica* and *Pasteurella pneumotropica* in conventional and germfree rats. *Lab. Anim. Sci.* **22**, 844–849.

Cahill, J.F., Cole, B.C., Wiley, B.B., and Ward, J.R. (1971). Role of biologic mimicry in the pathogenesis of rat arthritis induced by *Mycoplasma arthritidis*. *Infect. Immun.* **3**, 24–35.

Cameron, C.M. and Fuls, W.J.P. (1974). A comparative study on the immunogenicity of live and inactivated *Salmonella typhimurium* vaccines in mice. *Onderstepoort J. Vet. Res.* **41**, 81–92.

Caniatti, M., Crippa, L., Giusti, M., Mattiello, S., Grilli, G., Orsenigo, R., and Scanziani, E. (1998). Cilia-associated respiratory (CAR) bacillus infection in conventionally reared rabbits. *Zentrabl. Veterinarmed.* **45**, 363–371.

Cannon, T.R. (1920). Bacillus enteritidis infection in laboratory rats. *J. Infect. Dis.* **26**, 402–404.

Canonico, P.G., White, J.D., and Powanda, M.C. (1975). Peroxisome depletion in rat liver during pneumococcal sepsis. *Lab. Invest.* **33**, 147–150.

Carda, P., Barros, C., and Salamanka, M.E. (1959). Epidemic of *Bacillus piliformis* infection in a colony of white mice. *Ann. Inst. Invest. Vet. (Madrid)* **9**, 153–160.

Caren, L.D. and Rosenberg, L.T. (1966). The role of complement in resistance to endogenous and exogenous infection with a common mouse pathogen *Corynebacterium kutscheri*. *J. Exp. Med.* **124**, 689–699.

Carroll, H.W. (1974). Evaluation of dermatophyte-test medium for diagnosis of dermatophytosis. *J. Am. Vet. Med. Assoc.* **165**, 192–195.

Carter, G.R., Whitenack, D.L., and Julius, L.A. (1969). Natural Tyzzer's disease in Mongolian gerbils (*Meriones unguiculatus*). *Lab. Anim. Care* **19**, 648–651.

Carter, H.V. (1888). Note on the occurrence of a minute blood-spirillum in an Indian rat. *Sci. Mem. Med. Officers Army India* **3**, 45–48.

Carthew, P. and Gannon, J. (1981). Secondary infection of rat lungs with *Pasteurella pneumotropica* after Kilham rat virus infection. *Lab. Anim.* **15**, 219–221.

Cartner, S.C., Lindsey, J.R., Gibbs-Erwin, J., Cassell, G.H., and Lindsey, J.R. (1996). Roles of innate and adaptive immunity in respiratory mycoplasmosis. *Infect. Immun.* **66**, 3485–3491.

Casebolt, D.B. and Schoeb, T.R. (1988). An outbreak in mice of salmonellosis caused by *Salmonella enteritidis* serotype enteritidis. *Lab. Anim. Sci.* **38**, 190–192.

Casillo, S. and Blackmore, D.K. (1972). Uterine infection caused by bacteria and mycoplasma in mice and rats. *J. Comp. Pathol.* **82**, 477–482.

Cassell, G.H., Davidson, M.K., Davis, J.K., and Lindsey, J R. (1983). Recovery and identification of murine mycoplasmas. *In* "Methods of Mycoplasmology: Diagnostic Mycoplasmology", Vol. 2. (J.G. Tully and S. Razin, eds.), pp 129–142, Academic Press, New York.

Cassell, G.H. and Davis, J.K. (1978). Active immunization of rats against *Mycoplasma pulmonis* respiratory disease. *Infect. Immun.* **21**, 69–75.

Cassell, G.H., Davis, J.K., Simecka, J.W., Lindsey, J.R., Cox, N.R., Ross, S., and Fallon, M. (1986). Mycoplasmal infections: disease pathogenesis, implications for biomedical research, and control. *In* "Viral and Mycoplasmal Infections of Laboratory Rodents: Effects on Biomedical Research." (P.N. Bhatt, R.O. Jacoby, H.C. Morse III, and A.E. New, eds.), pp 87–130, Academic Press, Orlando, FL.

Cassell, G.H. and Hill, A. (1979). Murine and other small-animal mycoplasmas. *In* "The Mycoplasmas II: Human and Animal Mycoplasmas." (J. Tully and R. Whitcomb, ed.), pp 235–273, Academic Press, New York.

Cassell, G.H., Lindsey, J.R., Baker, H.J., and Davis, J.K. (1979). Mycoplasmal and Rickettsial Diseases. *In:* "The Laboratory Rat, Vol. I: Biology and Diseases." (Baker H.J., Lindsey J.R., and Weisbroth S.H., eds.), pp 243–269, Academic Press, New York.

Cassell, G.H., Lindsey, J.R., Davis, J.K., Davidson, M.K., Brown, M.B., and Mayo, J.G. (1981a). Detection of natural *Mycoplasma pulmonis* infection in rats and mice by an enzyme linked immunosorbent assay (ELISA). *Lab. Anim. Sci.* **31**, 676–682.

Cassell, G.H., Lindsey, J.R., Overcash, R.G., and Baker, H.J. (1973). Murine *Mycoplasma* respiratory disease. *Ann. N.Y. Acad. Sci.* **225**, 395–412.

Cassell, G.H., Wilborn, W.H., Silvers, S.H., and Minion, F.C. (1981b). Adherance and colonization of *Mycoplasma pulmonis* to genital epithelium and spermatozoa in rats. *Israel J. Med. Sci.* **17**, 593–598.

Cetin, E.T., Tashinoglu, M., and Volkan, S. (1965). Epizootic of *Trichophyton* mentagrophytes in white mice. *Pathol. Microbiol.* **28**, 839–846.

Chagas, C. (1909). Nova tripanosomiaze humana. *Mem. Istit. Oswaldo Cruz.* **1**, 159–218.

Chagas, C. (1911). Nova entidade morbider do homen. *Mem. Inst. Oswaldo Cruz.* **3**, 219–275.

Chary-Reddy, S. and Graves, D.C. (1996). Identification of extrapulmonary *Pneumocystis carinii* in immunocompromised rats by PCR. *J. Clin. Microbiol.* **34**, 1660–1665.

Chatterton, J.M., Joss, A.W., Pennington, T.H., and Ho-Yen, D.O. (1999). Usefulness of rat-derived antigen in the serodiagnosis of *Pneumocystis carinii* infection. *J. Med. Microbiol.* **48**, 681–687.

Chen, W., Gigliotti, F., and Harmsen, A.G. (1993). Latency is not an inevitable outcome of infection with *Pneumocystis carinii*. *Infect. Immun.* **61**, 5406–5409.

Chmel, L., Buchwald, J., and Valentova, M. (1975). Spread of *Trichophyton mentagrophytes* var. gran. infection to man. *Int. J. Dermatol.* **14**, 269–272.

Chopra, R.N., Basu, B.C., and Sen, J. (1939). Rat-bite fever in Calcutta. *Indian Med. Gat.* **74**, 449–451.

Christensen, L.R. (1968). Commentary following presentation by J.S. Niven, *Z. Versuchstierkd.* **10**, 174.

Clarke, M.C., Taylor, R.J., Hall, G.A., and Jones, P.W. (1978). The occurrence of facial and mandibular abscesses associated with *Staphylococcus aureus*. *Lab. Anim.* **12**, 121–123.

Cohrs, P., Jaffe, R., and Meessen, H. (1958). "Pathologie der Laboratoriumstiere." Springer-Verlag, New York.

Cole, B.C. and Atkin, C.L. (1991). The *Mycoplasma arthritidis* T-cell mitogen, MAM: a model superantigen. *Immunol. Today* **12**, 271–274.

Cole, B.C., Miller, M.L., and Ward, J.R. (1967). A comparative study on the virulence of *Mycoplasma hominis,* type II strains in rats. *Proc. Soc. Exp. Biol. Med.* **124**, 103–107.

Cole, B.C. and Ward, J.R. (1979). Mycoplasmas as arthritogenic agents. *In* "The Mycoplasmas II: Human and Animal Mycoplasmas." (J.G. Tully and R.F. Whitcomb, eds.), pp 367–398, Academic Press, Orlando, FL.

Cole, J.S., Stole, R.W., and Bulger, R.J. (1969). Rat-bite fever: report of three cases. *Ann. Intern. Med.* **71**, 979–981.

Collier, A.M. and Clyde, W.A. (1971). Relationships between *Mycoplasma pneumoniae* and human respiratory epithelium. *Infect. Immun.* **3**, 694–701.

Collier, L., Mahy, B.W., Hausler, W.J., and Sussman, M., eds. (1998). "Topley and Wilson's Microbiology and Microbial Infections, Vol. 3: Bacterial Infections," 9th Ed., Chapter 43, Hodder Arnold, London.

Collins, F.M. (1968). Recall of immunity in mice vaccinated with *Salmonella enteritidis* or *Salmonella typhimurium*. *J. Bacteriol.* **95**, 2014–2021.

Collins, F.M. (1970). Immunity to enteric infection in mice. *Infect. Immun.* **1**, 243–250.

Connole, M.D., Yamaguchi, W., Elad, D., Hasegawa, A., Segal, E., and Torres-Rodriguez, J.M. (2000). Natural pathogens of laboratory animals and their effects on research. *Med. Mycol.* **38(Suppl 1)**, 59–65.

Coons, A.H., Creech, H.S., Jones, R.N., and Berliner, E. (1942). The demonstration of pneumococcal antigens in tissues by the use of fluorescent antibody. *J. Immunol.* **45**, 159–170.

Cooper, J.E. (1977). Furunculosis in the mouse. *Vet. Rec.* **101**, 433.

Corbel, M.J. and Eades, S.M. (1977). Examination of the effect of age and acquired immunity on the susceptibility of mice to infection with *Aspergillus fumigatus*. *Mycologia* **60**, 79–85.

Costas, M. and Owen, J. (1987). Numerical analysis of electrophoretic patterns of *Streptobacillus moniliformis* strains from human, murine and asvian infections. *Med. Microbiol.* **25**, 303–311.

Cowan, S.T. (1974). "Cowan and Steel's Manual for the Identification of Medical Bacteria." Cambridge University Press, London.

Cox, N.R., Davidson, M.K., Davis, J.K., Lindsey, J.R., and Cassell, G.H. (1988). Natural mycoplasmal infections in isolator-maintained LEW/Tru rats. *Lab. Anim. Sci.* **38**, 381–388.

Craigie, J. (1966a). *Bacillus piliformis* (Tyzzer) and Tyzzer's disease of the laboratory mouse, I: propagation of the organism in embryonated eggs. *Proc. R. Soc. Lond. B* **165**, 35–61.

Craigie, J. (1966b). *Bacillus piliformis* (Tyzzer) and Tyzzer's disease of the laboratory mouse, II: mouse pathogenicity of *B. piliformis* grown in embryonated eggs. *Proc. R. Soc. Lond. B* **165**, 61–78.

Cripps, A.W., Dunkley, M.L, Clancy, R.L, Wallace, F., Buret, A., and Taylor, D.C. (1995). An animal model to study the mechanisms of immunity induced in the respiratory tract by intestinal immunization. *Adv. Exp. Med. Biol.* **371P,** 749–753.

Csukas, Z. (1976). Reisolation and characterization of *Haemophilus influenzaemurium. Acta Microbiol. Acad. Sci. Hung.* **23,** 89–96.

Cundiff, D.D. and Besch-Williford, C.L. (1992). Respiratory disease in a colony of rats. *Lab. Anim.* **21,** 16–18.

Cundiff, D.D., Besch-Williford, D.L., Hook, R.R., Franklin, C.L., and Riley, L.K. (1994a). Characterization of cilia-associated respiratory bacillus isolates from rats and rabbits. *Lab. Anim. Sci.* **44,** 305–312.

Cundiff, D.D., Besch-Williford, C.L., Hook, R.R., Franklin, C.L., and Riley, L.K. (1994b). Detection of cilia-associated respiratory bacillus by PCR. *J. Clin. Microbiol.* **32,** 1930–1934.

Cundiff, D.D., Besch-Williford, C.L., Hook, R.R., Franklin, C.L., and Riley, L.K. (1995a). Characterization of cilia-associated respiratory bacillus in rabbits and analysis of the 16S rRNA gene sequence. *Lab. Anim. Sci.* **45,** 22–26.

Cundiff, D.D., Riley, L.K., Franklin, C.L., Hook, R.R., and Besch-Williford, C.L. (1995b). Failure of a soiled bedding sentinel system to detect cilia-associated respiratory bacillus infection in rats. *Lab. Anim. Sci.* **45,** 219–221.

Cunningham, B.B., Paller, A., and Katz, B.Z. (1998). Rat bite fever in a pet lover. *J. Am. Acad. Dermatol.* **38,** 330–332.

Cushion, M.T. (1993). Genetic stability and diversity of *Pneumocystis carinii* infecting rat colonies. *Infect. Immun.* **61,** 4801–4813.

Cushion, M.T. (1998). Taxonomy, genetic organization, and life cycle of *Pneumocystis carinii. Semin. Respir. Infect.* **13,** 304–312.

Cushion, M.T. (2003). Pneumocystosis. *In* "Manual of Clinical Microbiology." (P.R. Murray, E.J. Baron, J.H. Jorgensen, M.A. Pfaller, and R.H. Yolken, eds.), pp 1712–1725, ASM Press, Washington, DC.

Cushion, M.T. and Ebbets, D. (1990). Growth and metabolism of *Pneumocystis carinii* in axenic culture. *J. Clin. Microbiol.* **28,** 1385–1394.

Cushion, M.T., Keely, S., and Stringer, J.R. (2004). Molecular and phenotypic description of *Pneumocystis wakefieldiae* sp. nov., a new species in rats. *Mycologia* (in press).

Cushion, M.T. and Linke, M.J. (1991). Factors influencing *Pneumocystis* infection in the immunocompromised rat. *J. Protozool.* **38,** 133S–135S.

Cushion, M.T., Ruffolo, J.J., Linke, M.J., and Walzer, P.D. (1985). *Pneumocystis carinii:* growth variables and estimates in the A549 and WI-38VA13 human cell lines. *Exp. Parasitol.* **60,** 43–54.

Cushion, M.T., Ruffolo, J.J., and Walzer, P. (1988). Analysis of the developmental stages of *Pneumocystis carinii* in-vivo. *Lab. Invest.* **58,** 324–331.

Cushion, M.T. and Walzer, P.D. (1984a). Cultivation of *Pneumocystis carinii* in lung-derived cell lines. *J. Infect. Dis.* **149,** 644.

Cushion, M.T. and Walzer, P.D. (1984b). Growth and serial passage of *Pneumocystis carinii* in the A549 cell line. *Infect. Immun.* **44,** 245–251.

Cushion, M.T, Walzer, P.D., Smulian, A.G., Linke, M.J., Ruffolo, J.J., Kaneshiro, E.S. (1997). Terminology for the life cycle of *Pneumocystis carinii. Pneumocystis carinii. Infect. Immun.* **65,** 4365.

Cushion, M.T., Zhang, J., Kaselis, M., Gluntoli, D., Stringer, S.I., and Stringer, J.R. (1993). Evidence for two genetic variants of *Pneumocystis carinii* coinfecting laboratory rats. *J. Clin. Microbiol.* **31,** 1217–1223.

Cutlip, R.C., Amtower, W.C., Beall, C.W., and Matthews, P.J. (1971). An epizootic of Tyzzer's disease in rabbits. *Lab. Anim. Sci.* **21,** 356–361.

Dalal, A.K. (1914). Case of rat-bite fever, treated with intravenous injection of neo-salvarsan. *Practitioner* **92,** 449.

Danysz, J. (1914). Un microbe pathogene pour les rats (*Mus decumanus* et *Mus ratus*) et son application à la destruction de ces animaux. *Ann Inst. Pasteur, Paris* **28,** 193–201.

Das Gupta, B.M. (1938). Spontaneous infection of guinea pigs with a spirillum, presumably s*pirillum minus.* Carter, 1887. *Indian Med. Gaz.* **73,** 140–141.

Davidson, M.K., Davis, J.K., Gambill, G.P., Cassell, G.H., and Lindsey, J.R. (1994). Mycoplasmas of laboratory rodents. *In* "Mycoplasmosis in Animals:Laboratory Diagnosis." (H.W. Whitford, R.F. Rosenbusch, and L.H. Lauerman, eds.), pp 97–133, Iowa State University Press, Ames, IO.

Davies, R.R. and Shewell, J. (1965). Ringworm carriage and its control in mice. *J. Hyg.* **63,** 507–515.

Davis, C.P., Mulcahy, D., Takeguchi, A., and Savage, D.C. (1972). Location and description of spiral-shaped microorganisms in the normal rat cecum. *Infect. Immun.* **6,** 184–192

Davis, J.K. and Cassell, G.H. (1982). Murine respiratory mycoplasmosis in LEW and F344 rats strain differences in lesion severity. *Vet. Pathol.* **19,** 280–293.

Davis, J.K., Gaertner, D.J., Cox, N.R., Lindsey, J.R., Cassell, G.H., Davidson, M.K., Kervin, C.C., and Rao, G.N. (1987). The role of *Klebsiella oxytoca* in utero-ovarian infection in B6C3F1 mice. *Lab. Anim. Sci.* **37,** 159–166.

Davis, J.K., Parker, R.F., White, H., Dziedzic, D., Taylor, G., Davidson, M.K., Cox, N.R., and Cassell, G.H. (1982). Murine respiratory mycoplasmosis in F344 and LEW rats: evolution of lesions and lung lymphoid cell populations. *Infect. Immun.* **36,** 720–729.

Dawson, M.H. and Hobby, G.L. (1939). Rat-bite fever. *Trans. Assoc. Am. Physicians* **54,** 329–332.

de Araujo, E. (1931). Diagnostico experimenta de sodoco. *Bras.-Med.* **45,** 924–928.

Deerberg, F., Pohlmeyer, G., Wullenweber, M., and Hedrich, H.J. (1993). History and pathology of an enzootic *Pneumocystis carinii* pneumonia in athymic Han:RNU and Han:NZNU rats. *J. Exp. Anim. Sci.* **36,** 1–11.

Dei-Cas E., Brun-Pascaud, M., Bile-Hansen, V., Allaert, A., and Aliouat, E.M. (1998). Animal models of pneumocystosis. *FEMS Immunol. Med. Microbiol.* **22,** 163–168.

Delanoe, P. and Delanoe, Mme. (1912). Sur les rapports des kystes des carinii du poumon des rats avec le *Trypanosoma lewisi. C.R. Acad-Sci* (Paris). **155,** 658–660.

De Ley, J., Mannheim, W., Mutters, R., Piechula, K., Tytgat, R., Segers, P., Bisgaard, M., Fredrickson, W., Hinz, K.-H., and Vanhouke, M. (1990). Inter- and intrafamilial similarities of rRNA cistrons of the Pasteurellaceae as determined by comparison of 16S rRNA sequences. *Int. J. Syst. Bacteriol.* **174,** 2002–2013.

De Rubertis, F.R. and Woeber, K.A. (1972). The effect of acute infection with *Diplococcus pneumoniae* on hepatic mitochondrial α-glycerophosphate dehydrogenase activity. *Endocrinology* **90,** 1384–1387.

Dewhirst, F.E., Fox, J.G., Mendes, E.N., Paster, B.J., Gates, C.E., Kirkbride, C.A., and Eaton, A.K. (2000a). "*Flexispira rappini*" strains represent at least 10 *Helicobacter* taxa. *Int. J. Syst. Microbiol.* **50,** 1781–1787.

Dewhirst, F.E., Fox, J.G., and On, S.L. (2000b). Recommended minimal standards for describing new species of the genus *Helicobacter. J. Syst. Evol. Microbiol.* **50,** 2231–2237.

Dewhirst, F.E., Paster, B.J., Olsen, I., and Fraser, G.J. (1992). Phylogeny of 54 representative strains of species in the family Pasteurellaceae as determined by comparison of 16S rRNA sequences. *J. Bacteriol.* **174,** 2002–2013.

Dickie, P., Mounts, P., Purcell, D., Miller, G., Fredrickson, T., Chang, L.J., and Martin, M.A. (1996). Myopathy and spontaneous *Pasteurella pneumotropica*-induced abscess formation in an HIV-1 ransgenic mouse model. *J. Acquired. Immune Defic. Syndromes Human Retrovirol.* **13,** 101–116.

Dietrich, H.M., Khaschabi, D., and Albini, B. (1996). Isolation of *Enterococcus durans* and *Pseudomonas aeruginosa* in a SCID mouse colony. *Lab. Anim.* **30,** 102–107.

Dixon, D.M. and Misfeldt, M.L. (1994). Proliferation of immature T cells within the splenocytes of athymic mice by *Pseudomonas* exotoxin A. *Cell. Immunol.* **158,** 71–82.

Dolan, M.M. and Fendrick, A.J. (1959). Incidence of *Trichophyton mentagrophytes* in laboratory rats. *Proc. Anim. Care Panel* **9**, 161–164.

Dolan, M.M., Kligman, A.M., Kobylinski, P.G., and Motsavage, M.A. (1958). Ringworm epizootics in laboratory mice and rats: experimental and accidental transmission of infection. *J. Invest. Dermatol.* **30**, 23–35.

Dolman, C.E., Kerr, D.E., Chang, H., and Shearer, A.R. (1951). Two cases of rat-bite fever due to *Streptobacillus moniliformis*. *Can. J. Public Health* **42**, 228–241.

Downing, N.D., Dewnany, G.D., and Radford, P.J. (2001). A rare and serious consequence of a rat bite. *Ann. R. Coll. Surg. Engl.* **83**, 279–280.

Dumoulin, A., Mazara, E., Seguy, N., Gargallo-Viola, D., Vargas, S., Cailliez, J.C., Aliouat, E.M., Wakefield, A.E., and Dei-Cas, E. (2000). Transmission of *Pneumocystis carinii* disease from immunocompetent contacts of infected hosts to susceptible hosts. *Eur. J. Clin. Microbiol. Infect. Dis.* **19**, 671–678.

Duncan, A.J., Carman, R.J., Olsen, G.J., and Wilson, K.H. (1993). Assignment of the agent of Tyzzer's disease to *Clostridium piliforme* comb. Nov. on the basis of 16S rRNA sequence analysis. *Int. J. Syst. Bacteriol.* **43**, 314–318.

Dunkley, M.L., Clancy, R.L., and Cripps, A.W. (1994). A role for CD4+ T cells from orally immunized rats in enhanced clearance of *Pseudomonas aeruginosa* from the lung. *Immunology.* **83**, 362–369.

Durand-Joly, I., el Aliouat, M., Recourt, D., Guyot, K., Francois, N., Wauguier, M., Camus, D., and Dei-Cas, E. (2002). *Pneumocystis carinii f. sp. hominis* is not infectious for SCID mice. *J. Clin. Microbiol.* **40**, 1862–1865.

Eamens, G.J. (1984). Bacterial and mycoplasmal flora of the middle ear of laboratory rats with otitis media. *Lab. Anim. Sci.* **34**, 480–483.

Ediger, R.D., Rabstein, M.M., and Olson, L.D. (1971). Circling in mice caused by *Pseudomonas aeruginosa*. *Lab. Anim. Sci.* **21**, 845–848.

Editorial (1939). Haverhill fever (erythema arthriticum epidemicum). *JAMA* **113**, 941.

Edman, J.D., Kovacs, J.A., Masur, H., Santi, D.V., Elwood, H.J., and Sogin, M.L. (1988). Ribosomal RNA sequence shows *Pneumocystis carinii* to be a member of the fungi. *Nature.* **334**, 519–522.

Ehrenworth, L. and Baer, H. (1956). The pathogenicity of *Klebsiella pneumoniae* for mice: the relationship to the quantity and rate of production of type-specific capsular polysaccharide. *J. Bacteriol.* **2**, 713–717.

Eisenberg Jr., G.H.G. and Cavanaugh, D.C. (1974). Pasteurella. *In* "Manual of Clinical Microbiology," 2nd Ed. (E.H. Lennette, E.H. Spaulding, and J.P. Truant, eds.), pp 246–249, American Society of Microbiology, Washington, DC.

Elwell, M.R., Sammons, M.L., Liu, C.T., and Beisel, W.R. (1975). Changes in blood pH in rats after infection with *Streptococcus pneumoniae*. *Infect. Immun.* **11**, 724–726.

Etheridge, M.E. and Vonderfecht, S.L. (1992). Diarrhea caused by a slow-growing *Enterococcus*-like agent in neonatal rats. *Lab. Anim. Sci.* **42**, 548–550.

Etheridge, M.E., Yolken, R.H., and Vonderfecht, S.L. (1988). *Enterococcus hirae* implicated as a cause of diarrhea in suckling rats. *J. Clin. Microbiol.* **26**, 1741–1744.

Euzby, P.J. (2003). List of Bacterial Names with Standing in Nomenclature. Societe de Bacteriologie Systematique et Veterinaire. www//bacterio.cict.fr/

Evans-Davis, K.D., Dillehay, D.L., Wargo, D.N., Webb, S.K., Talkington, D.F., Thacker, W.L., Small, L.S., and Brown, M.B. (1998). Pathogenicity of *Mycoplasma volis* in mice and rats. *Lab. Anim. Sci.* **48**, 38–44.

Fallon, M.T., Reinhard, B.M., Gray, M., Davis, T.W., and Lindsey, J.R. (1988). Inapparent *Streptococcus pneumoniae* Type 35 infections in commercial rats and mice. *Lab. Anim. Sci.* **38**, 129–132.

Farrell, E., Lord, G.H., and Vogel, J. (1939). Haverhill fever: report of a case with review of the literature. *Arch. Intern. Med.* **64**, 1.

Fauve, R.M., Bouanchaud, D., and Delaunay, A. (1966). Resistance cellulaire a l'infection bacterienne. *Ann. Inst. Pasteur, Paris* **110**, 106–117.

Fauve, R.M. and Pierce-Chase, C.H. (1967). Comparative effects of corticosteroids on host resistance to infection in relation to chemical structure. *J. Exp. Med.* **125**, 807–821.

Fauve, R.M., Pierce-Chase, C.H., and Dubos, R. (1964). Corynebacterial pseudotuberculosis in mice, II: activation of natural and experimental latent infections. *J. Exp. Med.* **120**, 283–304.

Feldman, S.H., Weisbroth, S.P., and Weisbroth, S.H. (1996). Detection of *Pneumocystis carinii* in rats by polymerase chain reaction: comparasion of lung tissue and bronchioalveolar lavage specimens. *Lab. Anim. Sci.* **46**, 628–634.

Festing, M.R.W. and Blackmore, D.K. (1971). Lifespan of specified pathogen free (MRC category 4) mice and rats. *Lab. Anim. Sci.* **5**, 19–192.

Feuerman, E., Alteras, I., Honig, E., and Lehrer, N. (1975). Saprophytic occurrence of *Trichophyton mentagrophytes* and *Microsporum gypseum* in the coats of healthy laboratory animals: preliminary report. *Mycopathologia* **55**, 13–16.

Findlay, G.M., Klieneberger, E., MacCallum, F.O., and MacKenzie, R.D. (1938). Rolling disease: new syndrome in mice associated with a pleuropneumonia-like organism. *Lancet* **ii**, 1511–1513.

Fischi, V., Koech, M., and Kussat, E. (1931). Infektarthritis bei Meriden. *Z. Hyg. Infektionskr.* **112**, 421–425.

Fisher, M. and Sher, A.M. (1972). Virulence of *Trichophyton mentagrophytes* infecting steroid-treated guinea pigs. *Mycopathol. Mycol. Appl.* **47**, 121–127.

Fitter, W.F., de Sa, D.J., and Richardson, H. (1979). Chorioamnionitis and funsitis due to *Corynebacterium kutscheri*. *Exp. Anim.* **42**, 539–545.

Flamm, H. (1957). *Klebsiella*-Enzootic in einer Mausezucht. *Schweiz Z. Pathol. Vet.* **7**, 492–497.

Fleischman, R.W., McCracken, D., and Forces, W. (1977). Adynamic ileus in the rat induced by chloral hydrate. *Lab. Anim. Sci.* **27**, 238–243.

Flynn, R.J. (1960). "Progress Report: Diseases of Laboratory Animals. Pseudomonas Infection of Mice," Semi-Annual Report, ANL-6264, pp 155–157. Biology Medicine Research Division, Argonne National Laboratory, Lemont, IL.

Flynn, R.J. (1963a). The diagnosis of *Pseudomonas aeruginosa* infection in mice. *Lab. Anim. Care* **13**, 126–130.

Flynn, R.J. (1963b). Introduction: *Pseudomonas aeruginosa* infection and its effects on biological and medical research. *Lab. Anim. Care* **13**, 1–6.

Flynn, R.J. (1963c). *Pseudomonas aeruginosa* infection and radiobiological research at the Argonne National Laboratory: effects, diagnosis, epizootiology, and control. *Lab. Anim. Care* **13**, 25–35.

Flynn, R.J., Ainsworth, E.J., and Greco, I. (1961). Progress Report: Diseases and Care of Laboratory Animals. I. Effects of *Pseudomonas* Infection of Mice, Summary Report, ANL-6368, pp 35–37. Biology Medicine Research Division, Argonne National Laboratory, Lemont, IL.

Flynn, R.J., Brennan, P.C., and Fritz, T.E. (1965). Pathogen status of commercially produced laboratory mice. *Lab. Anim. Care* **15**, 440–448.

Flynn, R.J. and Greco, I. (1962). "Progress Report: Disease and Care of Laboratory Animals. Further Studies on the Diagnosis of *Pseudomonas aeruginosa* Infection of Mice," Semi-annual Report. ANL 6535. Biology Medicine Research Division, Argonne National Laboratory, Lemont, IL.

Flynn, R.J., Simkins, R.C., Brennan, P.C., and Fritz, T.E. (1968). Uterine infection in mice. *Z. Versuchstierkd.* **10**, 131–136.

Foltz, C.J., Fox, J.G., Yan, L., and Shames, B. (1995). Evaluation of antibiotic therapies for eradication of *Helicobacter hepaticus*. *Antimicrobial Agents Chemother.* **39**, 1292–1294.

Foltz, C.J., Fox, J.G., Yan, L., and Shames, B. (1996). Evaluation of various antimicrobial formulations for eradication of *Helicobacter hepaticus*. *Lab. Anim. Sci.* **46**, 193–197.

Forbes, B.A., Sahm, D.F., and Weissfeld, A.S. (1998). Streptococcus, Enterococcus and similar organisms. *In* "Bailey and Scott's Diagnostic Microbiology," 106th ed. Mosby, St. Louis, pp. 619–635.

Ford, T.M. (1965). An outbreak of pneumonia in laboratory rats associated with *Diplococcus pneumoniae* type 8. *Lab. Anim. Care* **15**, 448–451.

Ford, T.M. and Joiner, G.N. (1968). Pneumonia in a rat associated with *Corynebacterium pseudotuberculosis*, a case report and literature survey. *Lab. Anim. Care* **18**, 220–223.

Fosgate, G.T., Hird, D.W., Read, D.H., and Walker, R.L. (2002). Risk factors for *Clostridium piliforme* infection in foals. *J. Am. Vet. Med. Assoc.* **220**, 785–790.

Fox Jr., C.L., Sampath, A.C., and Stanford, J.W. (1970). Virulence of *Pseudomonas* infection in burned rats and mice: comparative efficacy of silver sulfadiazine and mafenide. *Arch. Surg.* **101**, 508–512.

Fox, J.G., Dewhirst, F.E., Tully, J.G., Paster, B.J., Yan, L., Taylor, N.S., Collins, M.J., Gorelick, P.L., and Ward, J.M. (1994). *Helicobacter hepaticus*, sp. nov.: a microaerophilic bacterium isolated from the livers and intestinal mucosal scrapings from mice. *J. Clin. Microbiol.* **32**, 1238–1245.

Fox, J.G. and Lee, A. (1997). The role of *Helicobacter species* in newly recognized gastrointestinal tract disease of animals. *Lab. Anim. Sci.* **47**, 222–255.

Fox, J.G., Niemi, S.M., Ackerman, J., and Murphy, J.C. (1987). Comparison of methods to diagnose an epizootic of *Corynebacterium kutsheri* pneumonia in rats. *Lab. Anim. Sci.* **37**, 72–75.

Fox, J.G., Niemi, S.M., Murphy, J.C., and Quimby, F.W. (1977). Ulcerative dermatitis in the rat. *Lab. Anim. Sci.* **27**, 671–678.

France, M.P. (1994). Cilia-associated respiratory bacillus infection in laboratory rats with chronic respiratory disease. *Australian Vet. J.* **71**, 350–351.

Francis, E. (1932). Rat-bite fever and relapsing fever in the United States. *Trans. Assoc. Am. Physicians* **47**, 143–151.

Francis, E. (1936). Rat-bite fever spirochetes in naturally infected white mice. *Mus musculus. Public Health Rep.* **51**, 976–977.

Francis, R.A. (1970). Tyzzer's disease in laboratory animals. *J. Inst. Anim. Tech.* **21**, 167–171.

Franklin, C.L., Kinden, D.A., Stogsdill, P.L., and Riley, L.K. (1993). In-vitro model of adhesion and invasion by *Bacillus piliformis*. *Infect. Immun.* **61**, 876–883.

Franklin, C.L., Motzel, S.L., Besch-Williford, C.L., Hook Jr., R.R., and Riley, L.K. (1994). Tyzzer's infection: host specificity of *Clostridium piliforme* isolates. *Lab. Anim. Sci.* **44**, 568–572.

Franklin, C.L., Pletz, J.D., Riley, L.K., Livingston, B.A., Hook, R.R., and Besch-Wiliford, C.L. (1999). Detection of cilia-associated respiratory (CAR) bacillus in nasa-swab specimens from infected rats by use of polymerase chain reaction. *Lab. Anim. Sci.* **49**, 114–117.

Frenkel, J.K. (1976). *Pneumocystis jiroveci* n. sp. from man: morphology, physiology and immunology in relation to pathology. *NCI Monogr.* **43**, 13–27.

Frenkel, J.K., Good, J.T., and Schultz, J.A. (1966). Latent pneumocystis infection of rats, relapse, and chemotherapy. *Lab. Invest.* **15**, 1553–1557.

Freundt, D.E. (1957). Mycoplasmatales. *In* "Bergey's Manual of Determinative Bacteriology," 7th Ed. (R.S. Breed, E.G.D. Murray, and N.R. Smith, eds.), pp 914–926, Williams & Wilkins, Baltimore, MD.

Freundt, E.A. (1959). Arthritis caused by *Streptobacillus moniliformis* and pleuropneumonia-like organisms in small rodents. *Lab. Invest.* **8**, 1358–1366.

Friedmann, J.C., Guillon, J.C., Vassor, M.J., and Mahouy, G. (1969). La maladie de Tyzzer de la souris. *Exp. Anim.* **2**, 195–213.

Fries, A.S. (1977a). Studies on Tyzzer's disease: application of immunofluorescence for detection of *Bacillus piliformis* and for demonstration and determination of antibodies to it in sera from mice and rabbits. *Lab. Anim.* **11**, 69–73.

Fries, A.S. (1977b). Studies on Tyzzer's disease: isolations and propagation of *Bacillus piliformis*. *Lab. Anim.* **11**, 75–78.

Fries, A.S. (1978). Demonstration of antibodies to *Bacillus piliformis* in SPF colonies and experimental transplacental infection by *Bacillus piliformis* in mice. *Lab. Anim.* **12**, 23–26.

Fries, A.S. (1979a). Studies on Tyzzer's disease: acquired immunity against infection and activation of infection by immunosuppressive treatment. *Lab. Anim.* **13**, 143–147.

Fries, A.S. (1979b). Studies on Tyzzer's disease: a long term study of the humoral antibody response in mice, rats and rabbits. *Lab. Anim.* **13**, 37–41.

Fries, A.S. (1979c). Studies on Tyzzer's disease: transplacental transmission of *Bacillus piliformis* in rats. *Lab. Anim.* **13**, 43–46.

Fries, A.S. (1980). Studies on Tyzzer's disease: comparison between *Bacillus piliformis* strains from rat, mouse and rabbit. *Lab. Anim.* **14**, 61–63.

Friesleben, M. (1927). Das Vorkommen von Bazillen der Paratyphys-B-Gruppe in gesunden Schlachttieren sowie in Ratten und Mausen. *Dtsch. Med. Wochenschr.* **53**, 1589–1591.

Fruh, R., Blum, B., Mossman, H., Domdey, H., and von Specht, B.U. (1995). TH1 cells trigger tumor necrosis factor α–mediated hypersensitivity to *Pseudomonas aeruginosa* after adoptive transfer into SCID mice. *Infect. Immun.* **63**, 1107–1112.

Fuentes, C.A. and Aboulafia, R. (1955). *Trichophyton mentagrophytes* from apparently healthy guinea pigs. *AMA Arch. Dermatol.* **71**, 478–480.

Fuentes, C.A., Bosch, Z.E., and Boudet, C.C. (1956). Occurrence of *Trichophyton mentagrophytes* and *Microsporum gypseum* on hairs of healthy cats. *J. Invest. Dermatol.* **23**, 311–313.

Fujiwara, K. (1967). Complement fixation reaction and agar-gel double diffusion test in Tyzzer's disease of mice. *Jpn. J. Microbiol.* **11**, 103–117.

Fujiwara, K. (1971). Problems in checking in apparent infections in laboratory mouse colonies: an attempt at serological checking by anamnestic response. *In* "Defining the Laboratory Animal," pp 77–92. International Committee on Laboratory Animals, National Academy of Science, Washington, DC.

Fujiwara, K., Fukuda, S., Takagaki, Y., and Tajima, Y. (1963a). Tyzzer's disease in mice. Electron microscopy of the liver lesions. *Jpn. J. Exp. Med.* **33**, 203–212.

Fujiwara, K., Hirano, N., Takenaka, S., and Sato, K. (1973a). Peroral infection in Tyzzer's disease of mice. *Jpn. J. Exp. Med.* **43**, 33–42.

Fujiwara, K., Kurashina, H., Maejima, K., Tajima, Y., Takagaki, Y., and Naiki, M. (1965). Actively induced immune resistance to experimental Tyzzer's disease of mice. *Jpn. J. Exp. Med.* **35**, 259–275.

Fujiwara, K., Kurashina, H., Magaribuchi, T., Takenaka, S., and Yokoiyama, S. (1973b). Further observation on the difference between Tyzzer's organisms from mice and those from rats. *Jpn. J. Exp. Med.* **43**, 307–315.

Fujiwara, K., Kurashina, H., Matsunuma, N., and Takahashi, R. (1968). Demonstration of peritrichous flagella of Tyzzer's disease organism. *Jpn. J. Microbiol.* **12**, 361–363.

Fujiwara, K., Maejima, K., Takagaki, Y., Maiki, M., Tajima, Y., and Takahashi, R. (1964a). Multiplication des organismes de Tyzzer dans les tissus cerebraux de la souris experimentalement infectee. *C. R. S. Soc. Biol.* **158**, 407–413.

Fujiwara, K., Nakayama, M., Nakayama, H., Toriumi, W., Oguihara, S., and Thunert, A. (1985). Antigenic relatedness of *Bacillus piliformis* from Tyzzer's disease occurring in Japan and other regions. *Jpn. J. Vet. Sci.* **47**, 9–16.

Fujiwara, K., Takagaki, Y., Maejima, K., Kato, K., Naiki, M., and Tajima, Y. (1963b). Tyzzer's disease in mice: pathologic studies on experimentally infected animals. *Jpn. J. Exp. Med.* **33**, 183–202.

Fujiwara, K., Takagaki, Y., Naiki, M., Maejima, K., and Tajima, Y. (1964b). Tyzzer's disease in mice: effects of cortico steroids on the formation of liver lesions and the level of blood transaminases in experimentally infected animals. *Jpn. J. Exp. Med.* **34**, 59–75.

Fujiwara, K., Takahashi, R., Kurashina, H., and Matsunuma, N. (1969). Protective serum antibodies in Tyzzer's disease of mice. *Jpn. J. Exp. Med.* **39**, 491–504.

Fujiwara, K., Takasaki, Y., Kubokawa, K., Takenaka, S., Kubo, M., and Sato, K. (1974). Pathogenic and antigenic properties of the Tyzzer's organisms from feline and hamster cases. *Jpn. J. Exp. Med.* **44**, 365–372.

Fujiwara, K., Tanishima, Y., and Tanaka, M. (1979). Seromonitoring of laboratory mouse and rat colonies for common murine pathogens. *Jikken Dobutsu.* **28**, 297–306.

Fujiwara, K., Yamada, H., Ogawa, H., and Oshima, Y. (1971). Comparative studies on the Tyzzer's organisms from rats and mice. *Jpn. J. Exp. Med.* **41**, 125–134.

Furukawa, T., Furumoto, K., Fujieda, M., and Okada, E. (2002). Detection by PCR of the Tyzzer's disease organism (*Clostridium piliforme*) in feces. *Exp. Anim.* **51**, 513–516.

Furuta, T., Fugita, M., Machii, R., Kobayashi, K., Kojima, S., and Veda, K. (1993). Fatal spontaneous pneumocystosis in nude rats. *Lab. Anim. Sci.* **43**, 551–556.

Furuya, N., Hirakata, Y., Tomono, K., Matsumoto, T., Tateda, K., Kaku, M., and Yamaguchi, K. (1993). Mortality rates amongst mice with endogenous septicaemia caused by *Pseudomonas aeruginosa* isolates from various clinical sources. *J. Med. Microbiol.* **39**, 141–146.

Futaki, K., Takaki, S., Taniguchi, T., and Osumi, S. (1916). The cause of rat-bite fever. *J. Exp. Med.* **23**, 249–250.

Gades, N.M., Mandrell, T.D., and Rogers, W.P. (1999). Diarrhea in neonatal rats. *Contemp. Topics Lab. Anim. Sci.* **38**, 44–46.

Galler, J.R., Fox, J.G., Murphy, J.C., and Melanson, D.E. (1979). Ulcerative dermatitis in rats with over fifteen generation of protein malnutrition. *Brit. J. Nutr.* **41**, 611–618.

Ganaway, J.F. (1980). Effect of heat and selected chemical disinfectants upon infectivity of spores of *Bacillus piliformis* (Tyzzer's disease). *Lab. Anim. Sci.* **30**, 192–196.

Ganaway, J.R. (1986). Isolation of a newly recognized respiratory pathogen of laboratory rats: the CAR bacillus. *In* "Viral and Mycoplasma Infections of Laboratory Rodents, Effects on Biomedical Research." (P. Bhatt, R.O. Jacoby, W.C. Morse, and A.E. New, eds), pp 131–136, Academic Press, Orlando, FL.

Ganaway, J.R., Allen, A.M., and Moore, T.D. (1971a). Tyzzer's disease. *Am. J. Pathol.* **64**, 717–732.

Ganaway, J.R., Allen, A.M., and Moore, T.D. (1971b). Tyzzer's disease of rabbits: Isolation and propagation of *Bacillus piliformis* (Tyzzer) in embryonated eggs. *Infect. Immun.* **3**, 429–437.

Ganaway, J.R., Allen, A.M., Moore, T., and Bohner, H.J. (1973). Natural infection of germfree rats with *Mycoplasma pulmonis. J. Infect. Dis.* **127**, 529–537.

Ganaway, J.R., Spencer, T.R., Moore, T.D., and Allen, A.M. (1985). Isolation, propagation and characterization of a newly recognized pathogen, cilia-associated respiratory bacillus of rats, an etiological agent of chronic respiratory disease. *Infect. Immun.* **47**, 472–479.

Gard, S. (1944). *Bacillus piliformis* infection in mice, and its prevention. *Acta Pathol. Microbiol. Scand., Suppl.* **54**, 123–134.

Garlinghouse, L.E., DiGiacomo, R.F., Van Hoosier, G.L., and Condon, J. (1981). Selective media for *Pasteurella multocida* and *Bordetella bronchiseptica. Lab. Anim. Sci.* **31**, 39–42.

Garrity, G.M., ed. (2001). "Bergey's Manual of Systematic Bacteriology," 2nd Ed. Springer-Verlag, New York.

Geil, R.G., Davis, D.L., and Thomson, S.W. (1961). Spontaneous ileitis in rats: a report of 64 cases. *Am. J. Vet. Res.* **22**, 932–936.

Georg, L.K. (1960). "Animal Ringworm in Public Health," Public Health Service Publication no. 727, pp 9–17. U.S. Government Printing Office, Washington, DC.

Gialamas, J. (1981). Mastitis in einer SPF-Mausezucht. *Z. Versuchstierkd.* **23**, 274–277.

Gibson, S.V., Waggie, K.S., Wagner, J.E., and Ganaway, J.R. (1987). Diagnosis of subclinical *Bacillus piliformis* infection in a barrier-maintained mouse production colony. *Lab. Anim. Sci.* **37**, 786–788.

Giddens, W.E., Keahey, K.K., Carter, G.R., and Whitehair, C.K. (1968). Pneumonia in rats due to infection with *Corynebacterium kutscheri. Pathol. Vet.* **5**, 227–237.

Gigliotti, F. and Harmsen, A.G. (1997). *Pneumocystis carinii* host origin defines the antibody specificity and protective response induced by immunization. *J. Infect. Dis.* **176**, 1322–1326.

Gigliotti, F., Harmsen, A.G., Haidaris, C.G., and Haidaris, P.J. (1993). *Pneumocystis carinii* is not universally transmissible between mammalian species. *Infect. Immun.* **61**, 2886–2890.

Gilardi, G.L. (1976). *Pseudomonas* species in clinical microbiology. *Mt. Sinai J. Med. N.Y.* **43**, 710–726.

Gilbert, G.L., Cassidy, J.F., and Bennett, N.M. (1971). Rat-bite fever. *Med. J. Aust.* **2**, 1131–1134.

Gip, L. and Martin, B. (1964). Occurrence of *Trichophyton mentagrophytes asteroid* on hairs of guinea pigs without ringworm lesions. *Acta Derm.-Venereol.* **44**, 208–210.

Glastonbury, J.R., Morton, J.G., and Matthews, L.M. (1996). *Streptobacillus moniliformis* infection in Swiss white mice. *J. Vet. Diag. Invest.* **8**, 202–209.

Gledhill, A.W. (1967). Rat-bite fever in laboratory personnel. *Lab. Anim.* **1**, 73–76.

Goelz, M.F., Thigpen, J.E., Mahler, J., Rogers, W.P., Locklear, J., Weigler, B.J., and Forsythe, D.B. (1996). Efficacy of various therapeutic regimens in eliminating *Pasteurella pneumotropica* from the mouse. *Lab. Anim. Sci.* **46**, 280–285.

Goldstein, E. and Green, G.M. (1967). Alteration of the pathogenicity of *Pasteurella pneumotropica* for the murine lung caused by changes in pulmonary antibacterial activity. *J. Bacteriol.* **93**, 1651–1656.

Gonsherry, L., Marston, R.Q., and Smith, W.W. (1953). Naturally occurring infections in untreated and streptomycin-treated X-irradiated mice. *Am. J. Physiol.* **172**, 359–364.

Gordon, B.E., Durfee, W.J., Feldman, S.H., and Richardson, J.A. (1992). Diagnostic exercise: Pneumonia in a congenic immunodeficient mouse. *Lab. Anim. Sci.* **42**, 76–77.

Gorer, P.A. (1947). Some observations on the diseases of mice. *In* "The Care and Management of Laboratory Animals." (A.N. Worden, ed.), Williams & Wilkins, Baltimore, Maryland.

Gosselin, D., DeSanctis, J., Boule, M., Skamené, E., Matouk, C., and Radzioch, D. (1995). Role of tumor necrosis factor α in innate resistance to mouse pulmonary infection with *Pseudomonas aeruginosa. Infect. Immun.* **63**, 3272–3278.

Goto, K. and Itoh, T. (1996). Detection of *Clostridium piliforme* by enzymatic assay of amplified cDNA segment in microtitration plates. *Lab. Anim. Sci.* **46**, 493–496.

Goto, L., Nozu, R., Takakura, A., Matsushita, S., and Itoh, T. (1995). Detection of cilia-associated respiratory bacillus in experimentally and naturally infected mice and rats by the polymerase chain reaction. *Exp. Anim.* **44**, 333–336.

Gowen, J.W., and Schott, R.G. (1933). Genetic predisposition to *Bacillus piliformis* among mice. *J. Hyg.* **33**, 370–378.

Graham, W.R. (1963). Recovery of a pleuropneumonia-like organism (PPLO) from the genitalia of the female albino rat. *Lab. Anim. Care.* **13**, 719–724.

Gray, D.F. and Campbell, A.L. (1953). The use of chloramphenicol and foster mothers in the control of natural pasteurellosis in experimental mice. *Aust. J. Exp. Biol.* **31**, 161–166.

Griffith, J.W., White, W.J., Danneman, P.J., and Lang, C.M. (1988). Cilia-associated respiratory bacillus infection of obese mice. *Vet. Pathol.* **25**, 72–76.

Grogan, J.B. (1969). Effect of antilymphocyte serum on mortality of *Pseudomonas aeruginosa*-infected rats. *Arch. Surg.* **99**, 382–384.

Groschel, D. (1970). Genetic resistance of inbred mice to *Salmonella*. *Zentralbl. Bakteriol. Parasitenkd. Infektionskr. Hyg. Abt. 1 Orig.* **15**, 441–444.

Grossman, A., Maggio-Price, L., Shiota, F.M., and Liggitt, D.V. (1993). Pathologic features associated with decreased longevity of mutant sppha/sppha mice with chronic hemolytic anemia: similarities to sequelae of sickle cell anemia in humans. *Lab. Anim. Sci.* **43**, 217–221.

Guggenheim, K. and Buechler, E. (1947). Nutritional deficiency and resistance to infection: the effect of caloric and protein deficiency on the susceptibility of rats and mice to infection with *Salmonella typhimurium. J. Hyg.* **45**, 103–109.

Gugnani, H.C., Randhawa, H.S., and Shrivastar, J.B. (1971). Isolation of dermatophytes and other keratinophilic fungi from apparently healthy skin coats of domestic animals. *J. Med. Res.* **59**, 1699–1702.

Gundel, J., Gyorgy, P., and Pager, W. (1932). Experimentelle Beobachtungen zu der Frage der Resistenzverminderung und Infektion *Z. Hyg. Infektionskrankh.* **113**, 629–644.

Gupta, B.N. (1978). Pulmonary aspergilloma in a rat. *J. Am. Vet. Med. Assoc.* **173**, 1196–1197.

Gupta, S.K., Masinick, S.A., Hobden, J.A., Berk, R.S., and Hazlett, L.D. (1996). Bacterial proteases and adherence of *Pseudomonas aeruginosa* to mouse cornea. *Exp. Eye Res.* **62**, 641–650.

Habermann, R.T. and Williams, F.P. (1958). Salmonellosis in laboratory animals. *J. Natl. Cancer Inst.* **20**, 933–947.

Hagelskjaer, L., Sorensen, I., and Randers, E. (1998). *Streptobacillus moniliformis* infection: two cases and a literature review. *Scand. J. Infect. Dis.* **30**, 309–311.

Haines, D.C., Goerlick, P.L., Battles, J.K., Pike, K.M., Anderson, R.J., Fox, J.G., Taylor, N.S., Shen, Z., Dewhirst, F.E., Anver, M.R., and Ward, G.M. (1998). Natural and experimental inflammatory large bowel disease in immunodeficient rats infected with *Helicobacter bilis. Vet. Pathol.* **35**, 202–208.

Hajjar, A.M., DiGiacomo, R.F., Carpenter, J.K., Bingel, S.A., and Moazed, T.C. (1991). Chronic sialodacryoadenitis virus (SDAV) infection in athymic rats. *Lab. Anim. Sci.* **41**, 22–25.

Halkett, J.A.E., Davis, A.J., and Natsios, G.A. (1968). The effect of cold stress and *Pseudomonas aeruginosa* gavage on the survival of three-week-old swiss mice. *Lab. Anim. Care* **18**, 94–96.

Hall, G.A. (1974). An investigation into the mechanism of placental damage in rats inoculated with *Salmonella dublin. Am. J. Pathol.* **77**, 299–312.

Hall, W.C. and Van Kruiningen, H.T. (1974). Tyzzer's disease in a horse. *J. Am. Vet. Med. Assoc.* **164**, 1187–1189.

Hamburger, M. and Knowles Jr., H.C. (1953). *Streptobacillus moniliformis* infection complicated by acute bacterial endocarditis. *Arch. Intern. Med.* **92**, 216–220.

Hammond, C.W. (1954). The treatment of post-irradiation infection. *Radial. Res.* **1**, 448–458.

Hammond, C.W., Colling, M., Cooper, D.B., and Miller, C.P. (1954a). Studies on susceptibility to infection following ionizing radiation, II: its estimation by oral inoculation at different times post-irradiation. *J. Exp. Med.* **99**, 411–418.

Hammond, C.W., Rumi, D., Cooper, C.B., and Miller, C.P. (1955). Studies on susceptibility to infection following ionizing radiation, III: susceptibility of the intestinal tract to oral inoculation with Pseudomonas aeroginosa. *J. Exp. Med.* **102**, 403–411.

Hammond, C.W., Tompkins, M., and Miller, C.P. (1954b). Studies on susceptibility to infection following ionizing radiation, I: the time of onset and duration of the endogenous bacteremias in mice. *J. Exp. Med.* **99**, 405–410.

Hammond, W. (1963). *Pseudomonas aeruginosa* infection and its effects on radiobiological research. *Lab. Anim. Care* **13**, 6–11.

Hansen, A.K. (1995). Antibiotic treatment of nude rats and its impact on the aerobic bacterial flora. *Lab. Anim.* **29**, 37–44.

Hansen, A.K. (1996). Improvement of health monitoring and the microbiological quality of laboratory rats. *Scand. J. Lab. Anim. Sci.* **23(Suppl 2)**, 1–70.

Hansen, A.K. (2000). "Handbook of Laboratory Animal Bacteriology." CRC Press, Boca Raton, FL.

Hansen, A.K., Andersen, H.V., and Svendsen, O. (1994). Studies on the diagnosis of Tyzzer's disease in laboratory rat colonies with antibodies against *Bacillus piliformis (Clostridium piliforme). Lab. Anim.* Sci. **44**, 424–429.

Hansen, A.K., Dagnaes-Hansen, F., Mollegaard-Hansen, K.E. (1992a). Correlation between megaloileitis and antibodies to *Bacillus piliformis* in laboratory rat colonies. *Lab. Anim. Sci.* **42**, 449–453.

Hansen, A.K. and Skovgaard-Jensen, H.J. (1995). Experiences from sentinel health monitoring in units containing rats and mice in experiments. *Scand. J. Lab. Anim. Sci.* **22**, 1–9.

Hansen, A.K., Skovgaard-Jensen, H.J., Thomsen, P., Svendsen, O., Dagnaes-Hansen, F., and Mollegaard-Hansen, K.E. (1992b). Rederivation of rat colonies seropositive to *Bacillus piliformis* and the subsequent screening for antibodies. *Lab. Anim. Sci.* **42**, 444–448.

Hansen, A.K., Svendsen, O., and Mollegaard-Hansen, K.E. (1990). Epidemiological studies of *Bacillus piliformis* infection and Tyzzer's disease in laboratory rats. *Z. Versuchskd.* **33**, 163–169.

Hansen, A.K. and Velschow, S. (2000). Antibiotic resistance in bacterial isolates from laboratory animal colonies naïve to antibiotic treatment. *Lab. Anim.* **34**, 413–422.

Hansen, K.E. and Mollegaard-Hansen, K.E. (1990). An improved technique for the decontamination of barrier units contaminated with *Bacillus piliformis* strains of rat origin. *Scand. J. Lab. Anim. Sci.* **17**, 66–71.

Harkness, J.E. and Wagner, J.E. (1975). Self-mutilation in mice associated with otitis media. *Lab. Anim. Sci.* **25**, 315–318.

Harmsen, A.G., Chen, W., and Gigliotti, F. (1995). Active immunity to *Pneumocystis carinii* reinfection in T-cell–depleted mice. *Infect. Immun.* **63**, 2591–2395.

Harr, J.R., Tinsley, I.J., and Weswig, P.H. (1969). *Haemophilus* isolated from a rat respiratory epizootic. *J. Am. Vet. Med. Assoc.* **155**, 1126–1130.

Hartwich, J. and Shouman, M.T. (1965). Untersuchungen uber gehauft auftende *Klebsiella*-Infectionen bei Versuchsratten. *Z. Versuchstierkd.* **6**, 114–146.

Hass, V.A. (1938). A study of pseudotuberculosis rodentium recovered from a rat. *U.S. Public Health Rep.* **53**, 1033–1038.

Hastie, A.T., Evans, L.P., and Allen, A.M. (1993). Two types of bacteria adherent to bovine respiratory tract ciliated epithelium. *Vet. Pathol.* **30**, 12–19.

Hata, S. (1912). Salvarsantherapie der Rattenbisskrankheit in Japan. *Munch. Med. Wochenschr.* **59**, 854.

Hazard, J.B. and Goodkind, R. (1932). Haverhill fever (erythema arthriticum epidemicum) a case report and bacteriologic study. *JAMA* **99**, 534–538.

Heggers, J.P., Haydon, S., Ko, F., Hayward, P.G., Carp, S., and Robson, M.C. (1992). *Pseudomonas aeruginosa* Exotoxin A: its role in retardation of wound healing: the 1992 Lindberg Award. *J. Burn Care Rehabil.* **13**, 512–518.

Heilman, F.R. (1941). A study of *Asterococcus muris (Streptobacillus moniliformis)*, I: morphologic aspects and nomenclature. *J. Infect. Dis.* **69**, 32–44.

Heilman, F.R. (1951). A study of *Asterococcus muris (Streptobacillus moniliformis)*, II: cultivation and biochemical activities. *J. Infect. Dis.* **69**, 45–51.

Held, T.K., Trautman, M., Mielke, M.E., Neudeck, H., Cryz, S.J., and Cross, A.S. (1992). Monoclonal antibody against *Klebsiella* capsular polysaccharide reduces severity and hematogenic spread of experimental *Klebsiella pneumoniae* infection. *Infect. Immun.* **60**, 1771–1778.

Henriksen, S.D. (1962). Some *Pasteurella* strains from the human respiratory tract: a correction and supplement. *Acta Pathol. Microbiol. Scand.* **55**, 355–356.

Henriksen, S.D. and Jyssum, K. (1960). A new variety of *Pasteurella hemolytica* from the human respiratory tract. *Acta Pathol. Microbiol. Scand.* **50**, 443.

Henriksen, S.D. and Jyssum, K. (1961). A study of some *Pasteurella* strains from the human respiratory tract. *Acta Pathol. Microbiol. Scand.* **51**, 354–368.

Heyl, J.G. (1963). A study of *Pasteurella* strains from animal sources. *Antonie van Leeuwenhoek* **29**, 79–83.

Hightower, D., Uhrig, H.T., and Davis, J.I. (1966). *Pseudomonas aeruginosa* infection in rats used in radiobiology research. *Lab. Anim. Care* **16**, 85–93.

Hill, A. (1974). Experimental and natural infection of the conjunctiva of rats. *Lab. Anim.* **8**, 305–310.

Hirakata, Y., Furuya, N., Tateda, K., Matsumoto, T., and Yamaguchi, K. (1995). The influence of exo-enzyme S and proteases on endogenous *Pseudomonas aeruginosa* bacteraemia in mice. *J. Med. Microbiol.* **43**, 258–261.

Hirst, R.G. and Olds, R.J. (1978). *Corynebacterium kutscheri* and its alleged avirulent variant in mice. *J. Hyg.* **80**, 349–356.

Ho, M., Tsugane, T., Kobayashi, K., Kuramochi, T., Hioki, K., Furuta, T., and Nomura, T. (1991). Study on placental transmission of *Pneumocystis carinii* in mice using immunodeficient SCID mice as a new animal model. *J. Protozool.* **38**, 218S–219S.

Hoag, W.G. and Rogers, J. (1961). Techniques for the isolation of *Salmonella typhimurium* from laboratory mice. *J. Bacteriol.* **82**, 153–154.

Hoag, W.G., Strout, J., and Meier, H. (1964). Isolation of *Salmonella* spp from laboratory mice and from diet supplements. *J. Bacteriol.* **88**, 534–536.

Hoag, W.G., Strout, J., and Meier, H. (1965). Epidemiological aspects of the control of *Pseudomonas* infection in mouse colonies. *Lab. Anim. Care* **15**, 217–225.

Hoag, W.G., Wetmore, P.W., Rogers, J., and Meier, H. (1962). A study of latent *Pasteurella* infection in a mouse colony. *J. Infect. Dis.* **11**, 135–140.

Hobson, D. (1957). Chronic bacterial carriage in survivors of mouse typhoid. *J. Pathol. Bacteriol.* **73**, 399–410.

Hodzic, E., McKisic, M., Feng, S., and Barthold, S.W. (2001). Evaluation of diagnostic methods for *Helicobacter bilis* infection in laboratory mice. *Comp. Med.* **51**, 406–412.

Hojo, E. (1939). On the *Bacillus pseudo-tuberculosis murium* prevalent in mouse. *Jpn. J. Exp. Med.* **10**, 113.

Holden, F.A. and MacKay, J.C. (1964). Rat-bite fever: an occupational hazard. *Can. Med. Assoc. J.* **91**, 78.

Holmgren, E.B. and Tunevall, G. (1970). Rat-bite fever. *Scand. J. Infect. Dis.* **2**, 71–74.

Holt, J.G., Krieg, N.R., Sneath, P.H.A., Staley, J.T., and Williams, S.T. (1994). "Bergey's Manual of Determinative Bacteriology," 9th Ed. Williams & Wilkins, Baltimore, MD.

Hong, C.C. and Ediger, R.D. (1978a). Chronic necrotizing mastitis in rats caused by *Pasteurella pneumotropica*. *Lab. Anim. Sci.* **28**, 317–320.

Hong, C.C. and Ediger, R.D. (1978b). Self-mutilation of the penis in C57Bl/N mice. *Lab. Anim.* **12**, 55–57.

Hook, R.R., Franklin, C.L., Riley, L.K., Livingston, B.A., and Besch-Williford, C.L. (1998). Antigenic analyses of cilia-associated respiratory (CAR) bacillus isolates by use of monoclonal antibodies. *Lab. Anim. Sci.* **48**, 234–239.

Hook, R.R., Riley, L.K., Franklin, C.L., and Besch-Williford, C.L. (1995). Seroanalysis of Tyzzer's disease in horses: implications that multiple strains can infect Equidae. *Equine Vet. J.* **27**, 8–12.

Hooper, A. and Sebesteny, A. (1974). Variation in *Pasteurella pneumotropica*. *J. Med. Microbiol.* **7**, 137–140.

Hoover, D., Bendele, S.A., Wightman, S.R., Thompson, C.Z., and Hoyt, J.A. (1985). Streptococcal enteropathy in infant rats. *Lab. Anim. Sci.* **35**, 635–641.

Hoskins, H.P. and Stout, A.L. (1920). *Bacillus bronchisepticus* as the cause of an infectious respiratory disease of the white rat. *J. Lab. Clin. Med.* **5**, 307–310.

Hottendorf, G.H., Hirth, R.S., and Peer, R.L. (1969). Megaloileitis in rats. *J. Am. Vet. Med. Assoc.* **155**, 1131–1135.

Hozbor, D., Fougue, F., and Guiso, N. (1999). Detection of *Bordetella bronchiseptica* by the polymerase chain reaction. *Res. Microbiol.* **150**, 333–341.

Hubbs, A.F., Hahn, F.F., and Lundgren, D.L. (1991). Invasive tracheobronchial aspergillosis in an F344/N rat. *Lab. Anim. Sci.* **41**, 521–524.

Hughes, W.T. (1979). Limited effect of trimethoprim-sulfamethoxazile prophylaxis on *Pneumocystis carinii*. *Antimicrob. Agents Chemother.* **16**, 333–335.

Hughes, W.T. (1982). Natural mode of acquisition for denovo infection with *Pneumocystis carinii*. *J. Infect. Dis.* **145**, 842–848.

Hughes, W.T. (1988). Comparison of dosages, intervals and drugs in the prevention of *Pneumocystis carinii* pneumonia. *Antimicrob. Agents Chemother.* **32**, 623–625.

Hughes, W.T. (1989). Animal models for *Pneumocystis carinii* pneumonia. *J. Protozool.* **36**, 41–45.

Hughes, W.T., Bartley, D.L., and Smith, B.M. (1983). A natural source of infection due to *Pneumocystis carinii*. *J. Infect. Dis.* **147**, 595.

Humphreys, F.A., Campbell, A.G., Driver, M.W., and Hatton, G.N. (1950). Rat-bite fever. *Can. J. Public Health* **41**, 66–71.

Hunter, B. (1971). Eradication of Tyzzer's disease in a colony of barrier-maintained mice. *Lab. Anim.* **5**, 271–276.

Hunter, C.A. and Ensign, P.R. (1947). An epidemic of diarrhea in a newborn nursery caused by *Pseudomonas aeruginosa*. *Am. J. Public Health* **37**, 1166–1169.

Ibraham, H.M. and Schutze, H. (1928). A comparison of the prophylactic value of the H, O and R antigens of *Salmonella aertrycke*, together with some observations on the toxicity of its smooth and rough variants. *Br. J. Exp. Pathol.* **9**, 353–360.

Icenhour, C.R., Rebholz, S.L., Collins, M.S., and Cushion, M.T. (2001). Widespread occurrence of *Pneumocystis carinii* in commercial rat colonies detected using targeted PCR and oral swabs. *J. Clin. Microbiol.* **39**, 3437–3441.

Icenhour, C.R., Rebholz, S.L., Collins, M.S., and Cushion, M.T. (2002). Early acquisition of *Pneumocystis carinii* in neonatal rats as evidenced by PCR and oral swabs. *Eukaryotic Cell.* **1**, 414–419.

Ikegami, T., Shirota, K., Goto, K., Takakura, A., Itoh, T., Kawamura, S., Une, Y., Nomura, Y., and Fujiwara, K. (1999a). Enterocolitis associated with dual infection by *Clostridium piliforme* and feline panleukopenia virus in three kittens. *Vet. Pathol.* **36**, 613–615.

Ikegami, T., Shirota, K., Une, Y., Nomura, Y., Wada, Y., Goto, K., Takakura, A., Itoh, T., and Fujiwara, K. (1999b). Naturally occurring Tyzzer's disease in a calf. *Vet. Pathol.* **36**, 253–255.

NRC [National Research Council] (1991). "Infectious Diseases of Mice and Rats," pp 48–30, 54–58, 132–139, 211–214, Institute of Laboratory Animal Resources, National Academy Press, Washington, DC.

Innes, J.R.M., McAdams, A.J., and Yevich, P. (1956a). Pulmonary disease in rats: a survey with comments on "chronic murine pneumonia." *Am. J. Pathol.* **32**, 141–159.

Innes, J.R.M., Wilson, C., and Ross, M.A. (1956b). Epizootic *Salmonella enteritidis* infection causing septic pulmonary phlebothrombosis in hamsters. *J. Infect. Dis.* **98**, 133–141.

Ishihara, C., Yamamoto, K., Hamada, N., and Azuma, T. (1984). Effect of steroyl-N-acetylmuramyl-*h*-atanyl-D-isoglutamine on host resistance to *Corynebacterium kutscheri* infection in cortisone-treated mice. *Vaccine* **2**, 261–264.

Ito, M., Tsugane, T., Kobayashi, K., Kuramochi, T., Hioki, K., Furuta, T., and Nomura, T. (1991). Study on placental transmission of *Pneumocystis carinii* in mice using immunodeficient SCID mice as a new animal model. *J. Protozool.* **38**, 218S–219S.

Ito, S. (1957). A disease of rats caused by a pleuropneumonia-like organism (PPLO). *Jpn. J. Exp. Med.* **27**, 243–248.

Itoh, T., Eburkuro, M., and Ragiyama, N. (1987). Inactivation of *Bacillus piliformis* spores by heat and certain chemical disinfectants. *Exp. Anim.* **36**, 239–244.

Itoh, T., Kagiyama, N., and Fujiwara, K. (1988). Production of Tyzzer's disease in rats by ingestion of bacterial spores. *Jpn. J. Exp. Med.* **59**, 9–15.

Itoh, T., Kohyama, K., Takakura, A., Takenouchi, T., and Kagiyama, N. (1987). Naturally occurring CAR bacillus infection in a laboratory rat colony and epizootiological observations. *Exp. Anim.* **36**, 387–393.

Ivanovics, G. and Ivanovics, C. (1937). Beitrage zur Kenntnis des Bacterium Influenzae murium (Kairies und Schwartzer). *Zentrabl Bakteriol. Parasitenk Infektionskr. Abt. I Orig.* **139**, 184–188.

Jackson, N.N., Wall, H.G., Miller, C.A., and Rogul, M. (1980). Naturally acquired infections of *Klebsiella pneumoniae* in Wistar rats. *Lab. Anim.* **14**, 357–361.

Jakab, G.J. (1974). Effect of sequential inoculation of Sendai virus and *Pasteurella pneumotropica* in mice. *J. Am. Vet. Med. Assoc.* **164**, 723–728.

Jakob, G.J. and Dick, E.C. (1973). Synergistic effect in viral-bacterial infection: combined infection of the murine respiratory tract with Sendai virus and *Pasteurella pneumotropica*. *Infect. Immun.* **8**, 762–768.

Jalonen, E., Paton, J.C., Koskela, M., Kertula, Y., and Leionen, M. (1989). Measurement of antibody responses to pneumolysin: a promising method for the presumptive aetiological diagnosis of pneumococcal pneumonia. *J. Infect.* **19**, 127–134.

Jawetz, E. (1948). A latent pneumotropic *Pasteurella* of laboratory animals. *Proc. Soc. Exp. Biol. Med.* **68**, 46–48.

Jawetz, E. (1950). A pneumotropic *Pasteurella* of laboratory animals, 1: bacteriological and serological characteristics of the organism. *J. Infect. Dis.* **86**, 172–183.

Jawetz, E. and Baker, W.H. (1950). A pneumotropic *Pasteurella* of laboratory animals, II: pathological and immunological studies with the organism. *J. Infect. Dis.* **86**, 184–196.

Jellison, W.L., Eneboe, P.L., Parker, R.R., and Hughes, L.E. (1949). Rat-bite fever in Montana. *Public Health Rep.* **64**, 1661–1665.

Jenkin, C.R., Rowley, D., and Auzins, I. (1964). The basis for immunity in mouse typhoid. I. The carrier state. *Aust. J. Exp. Biol. Meet. Sci.* **42**, 215–228.

Jersey, G., Whitehair, C.K., and Carter, G.R. (1973). *Mycoplasma pulmonis* as the primary cause of chronic respiratory disease in rats. *J. Am. Vet. Med. Assoc.* **163**, 599–604.

Johansen, H.K., Espersen, F., Pedersen, S.S., Hougen, H.P., Rygaard, J., and Hoiby, N. (1993). Chronic *Pseudomonas aeruginosa* lung infection in normal and athymic rats. *APMIS* **1**, 207–225.

Johansen, H.K., Hougon, H.P., Rygaard, J., and Hoiby, N. (1996). Interferon-γ (IFN-γ) treatment decreases the inflammatory response in chronic *Pseudomonas aeruginosa* pneumonia in rats. *Clin. Exp. Immunol.* **103**, 212–218.

Johanson Jr., W.G., Jay, S.J., and Pierce, A.K. (1974). Bacterial growth *in vivo*: an important determinant of the pulmonary clearance of *Diplococcus pneumoniae* in rats. *J. Clin. Invest.* **53**, 1320–1325.

Jonas, A.M., Percey, D., and Craft, J. (1970). Tyzzer's disease in the rat. *Arch. Pathol.* **90**, 561–566.

Jones, B.F. and Stewart, H.L. (1939). Chronic ulcerative cecitis in the rat. *Public Health Rep.* **54**, 172–175.

Jones, D.M. (1962). A *Pasteurella*-like organism from the human respiratory tract. *J. Pathol. Bacteriol.* **83**, 143–151.

Jones, F.S. (1922). An organism resembling *Bacillus actinoides* isolated from pneumonic lungs of white rats. *J. Exp. Med.* **35**, 361–366.

Jordan, E.O. (1925). The differentiation of the paratyphoid enteritidis group, IX: strains from various mammalian hosts. *J. Infect. Dis.* **36**, 309–329.

Juhr, N.C. and Horn, J. (1975). Modellinfektion mit Corynebacterium kutscheri bei der Maus. *Z. Versuchstierkd.* **17**, 129–141.

Kaffka, A. and Reith, H. (1960). *Trichophyton mentagrophytes*: Varianten bei Laboratoriumstieren. *Zentralbl. Bakteriol. Parasitenkd. Infektionskr. Hyg. Abt. 1 Orig.* **177**, 96–106.

Kagiyama, N. and Wagner, J.E. (1994). *Salmonella* sp. *In* "Manual of Microbiologic Monitoring of Laboratory Animals," 2nd Ed. (K. Waggie, N. Kagiyama, A.M. Allen and T. Nomura, eds.), pp 155–157. National Institutes of Health [NIH] Publication no. 94–2498, PHS, NIH, Bethesda, MD.

Kairies, A. and Schwartzer, K. (1956). Studien zu einer bakteriellen Influenza der Mause und Beschreibung eines "Bacterium influenzae murium." *Zentrabl. Bakteriol. Parasitenk. Infektionsk. Abt. I Orig.* **137**, 357–359.

Kaklamanis, E. and Thomas, L. (1970). *In* "Microbial Toxins," Vol. 3. (T.C. Montie, S. Kadis, and S. Ajl, eds.), pp 493–505, Academic Press, New York.

Kampelmacher, E.H., Guinee, P.A.M., and van Noorle Jansen, L.M. (1969). Artificial *Salmonella* infections in rats. *Zentralbl. Veterinaermed. Reihe B* **16**, 173–182.

Kanazawa, K. and Imai, A. (1959). Pure culture of the pathogenic agent of Tyzzer's disease of mice. *Nature (London)* **184**, 1810.

Kaneko, J., Fujita, H., Matsuyama, S., Kojima, H., Asakura, H., Nakamura, Y., and Kodama, T. (1969). An outbreak of Tyzzer's disease among the colonies of mice. *Bull. Exp. Anim.* **9**, 148–156.

Kaneshiro, E.S., Collins, M.S., and Cushion, M.T. (2000). Inhibitors of sterolbiosynthesis and amphotericin B reduce the viability of *Pneumocystis carinii* f. sp. carinii. *Antimicrob. Agents Chemother.* **44**, 1630–1638.

Kaplan, M.H., Coons, A.H., and Deanne, H.W. (1950). Localization of antigen in tissue cells, III: cellular distribution of penumococcal polysaccharides types II and III in the mouse. *J. Exp. Med.* **91**, 15–30.

Karasek, E. (1970). Der Kortisontest zum Nachweis latenter infektionen bei Versuchsmausen. *Z. Versuchstierkd.* **12**, 155–161.

Kaspareit-Rittinghausen, J., Wullenweber, M., Deerberg, F., and Farouq, M. (1990). Pathological changes in *Streptobacillus moniliformis* infection of C57Bl/6J mice. *J. Berl. Munch. Tierartzl. Wochenschr.* **103**, 84–87.

Kawamura, S., Taguchi, F., and Fujiwara, K. (1983a). Plaque formation by Tyzzer's organism in primary monolayer cultures of mouse hepatocytes. *Microbiol. Immunol.* **27**, 415–424.

Kawamura, S., Taguchi, F., Ishida, T., Nakayama, M., Fujiwara, K. (1983b). Growth of Tyzzer's organism in primary monolayer cultures of adult mouse hepatocytes. *J. Gen. Microbiol.* **129**, 277–283.

Kawano, A., Nenoi, M., Matsushita, S., Matsumoto, T., and Mita, K. (2000). Sequence of 16S rRNA gene of rat-origin cilia associated respiratory (CAR) bacillus SMR strain. *J. Vet. Med. Sci.* **62**, 797–800.

Keely, S.P., Fischer, J.M., Cushion, M.T., Stringer, J.R. (2004). Phylogenetic identification of Pneumocystis murina sp. nov., a new species in laboratory mice. Microbiology **150**, 1153–1165.

Kelemen, G. and Sargent, F. (1946). Nonexperimental pathologic nasal findings in laboratory rats. *Arch. Otolaryngol.* **44**, 24–42.

Kent, R.L., Lutzner, M.A., and Hansen, C.T. (1976). The masked rat: an x-ray–induced mutant with chronic blepharitis, alopecia and pasteurellosis. *J. Hered.* **67**, 3–5.

Kerrin, J.C. (1928). *Bacillus enteriditis* infection in wild rats. *J. Pathol. Bacteriol.* **31**, 588–589.

Khalil, A.M. (1934). The incidence of organisms of *Salmonella* group in wild rats and mice in Liverpool. *J. Hyg.* **38**, 75–78.

Kim, H.K., Hughes, W.T., and Feldman, S. (1972). Studies of morphology and immunofluorescence of *Pneumocystis carinii*. *Proc. Soc. Exp. Biol. Med.* **141**, 304–309.

Kim, J.A., Takahashi, Y., Tanaka, R., Fukushima, K., Nishimura, K., and Miyaji, M. (2001). Identification and subtyping of *Trichophyton mentagrophytes* by random amplified polymorphic DNA. *Mycoses* **44**, 157–165.

Kirchner, B.K., Dixon, L.W., Lentsch, R.H., and Wagner, J.E. (1982). Recovery and pathogenicity of several *Salmonella* species isolated from mice. *Lab. Anim. Sci.* **32**, 506–507.

Kirchner, B.K., Lake, S.G., and Wightman, S.R. (1992). Isolation of *Streptobacillus moniliformis* from a guinea pig with granulomatous pneumonia. *Lab. Anim. Sci.* **42**, 519–521.

Kirschnek, S., Ryll, M., Busse, J., and Nicklas, W. (1987). Identification and characterization of *Haemophilus influenzae murium* isolated from laboratory mice. *In* "Proceeding of the sixth FELASA Symposium Harmonization of Laboratory Animal Husbandry." (P.N. O'Donoghue, ed.), pp 66–68, Royal Society of Medicine Press, London.

Kita, E., Kamikaidu, N., Oku, D., Nakano, A., Katsui, N., and Kashiba, S. (1992). Nonspecific stimulation of host defense by *Coynebacterium kutscheri,* III: enhanced cytokine induction by the active moiety of *C. kutscheri. Nat. Immun.* **11**, 46–55.

Kita, E., Matsui, N., Sawaki, M., Mikasa, K., and Katsui, N. (1995). Murine tumorlytic factor, immunologically distinct from tumor necrosis factor-α and -β, induced in the serum of mice treated with a T-cell mitogen of *Corynebacterium kutscheri. Immunol. Lett.* **46**, 101–106.

Kitada, K., Oka, S., Kimura, S., Shimada, K., Serikawa, T., Yamada, J., Tsunoo, H., Egawa, K., and Nakamura, Y. (1991). Detection of *Pneumocystis carinii* sequences by polymerase chain reaction: animal models and clinical application to noninvasive specimen. *J. Clin. Microbiol.* **29**, 1985–1990.

Kitahara, Y., Ishibashi, T., Harada, Y., Takamoto, M., and Tanaka, K. (1981). Reduced resistance to *Pseudomonas* septicemia in diabetic mice. *Clin. Exp. Immunol.* **43**, 590–598.

Klein, E. (1903). Discussion of the Pathologic Society of London. *Lancet* **1**, 238–239.

Klieneberger, E. (1935). The natural occurrence of pleuropneumonia-like organisms in apparent symbiosis with *Streptobacillus moniliformis* and other bacteria. *J. Pathol. Bacteriol.* **40**, 93–105.

Klieneberger, E. (1936). Further studies on *Streptobacillus moniliformis* and its symbiont. *J. Pathol. Bacteriol.* **42**, 587–598.

Klieneberger, E. (1938). Pleuropneumonia-like organisms of diverse provenance: some results of an enquiry into methods of differentiation. *J. Hyg.* **38**, 458–476.

Klieneberger, E. (1942). Some new observations bearing on the nature of the pleuropneumonia-like organism known as LI associated with *Streptobacillus moniliformis. J. Hyg.* **42**, 485–497.

Klieneberger, E. and Steabben, D.B. (1937). On a pleuropneumonia-like organism in lung lesions or rats with notes on the clinical and pathological features of the underlying condition. *J. Hyg.* **37**, 143–152.

Klieneberger, E. and Steabben, D.B. (1940). On the association of the pleuropneumonia-like organism L3 with bronchiectatic lesions in rats. *J. Hyg.* **40**, 223–227.

Knowles, R., Das Gupta, B.M., and Sen, S. (1936). Natural *Spirillum minus* infection in white mice. *Indian Med. Gaz.* **71**, 210–212.

Kodjo, A., Villard, L., Veillet, P., Escande, P., Borges, E., Maurin, F., Bonnod, J., and Richard, Y. (1999). Identification by 16s rDNA fragment amplification and determination of genetic diversity by random amplified polymorphic DNA analysis of *Pasteurella pneumotropica* isolated from laboratory rodents. *Lab. Anim. Sci.* **49**, 49–53.

Kohn, D.F. (1971a). Bronchiectasis in rats infected with *Mycoplasma pulmonis*: an electron microscopy study. *Lab. Anim. Sci.* **21**, 856–861.

Kohn, D.F. (1971b). Sequential pathogenicity of *Mycoplasma pulmonis* in laboratory rats. *Lab. Anim. Sci.* **21**, 849–855.

Kohn, D.F. and Chinookoswong, N. (1981). Cytadsorption of *Mycoplasma pulmonis* to the rat ependyma. *Infect. Immun.* **34**, 292–295.

Kohn, D.F. and Kirk, B.E. (1969). Pathogenicity of *Mycoplasma pulmonis* in laboratory rats. *Lab. Anim. Care* **19**, 321–330.

Kontiokari, T., Renko, M., Kaijalainen, T., Kuisma, L., and Leinonen, M. (2000). Comparison of nasal swab culture, quantitative culture of nasal mucosal tissue and PCR in detecting *Streptococcus pneumoniae* carriage in rats. *APMIS* **108**, 734–738.

Koopman, J.P. and Janssen, F.G.J. (1975). The occurrence of salmonellas in laboratory animals and a comparison of three enrichment methods used in their isolation. *Z. Versuchstierkd.* **17**, 155–158.

Koopman, J.P., Van-den-Brink, M.E., Vennix, P.P., Kypers, W., Bost, R., and Bakker, R.H. (1991). Isolation of *Streptobacillus moniliformis* from the middle ear of rats. *Lab. Anim.* **25**, 35–39.

Kovacs, J.A., Halpern, J.L., Lundgren, B., Swan, J.C., Parrillo, J.E., and Masur, H. (1989). Monoclonal antibodies to *Pneumocystis carinii*: identification of specific antigens and characterization of antigenic differences between rat and human isolates. *J. Infect. Dis.* **159**, 60–70.

Kovatch, R.M., and Zebaith, G. (1973). Naturally occurring Tyzzer's disease in a cat. *J. Am. Vet. Med. Assoc.* **162**, 136–138.

Kregten, E.V., Westerdaal, N.A.C., and Willers, J.M.N. (1984). New simple medium for selective recovery of *Klebsiella pneumoniae* and *K. oxytoca* from human feces. *J. Clin. Microbiol.* **20**, 936–941.

Kunstyr, I. (1980). Laboratory animals as a possible source of dermatophytic infections in humans. *Zentrabl. Bakteriol Orig. A* **245(Suppl 8)**, 361–367.

Kunstyr, I., Ernst, H., and Lenz, W. (1995). Granulomatous dermatitis and mastitis in two SPF rats associated with a slowly growing *Staphylococcus aureus*: a case report. *Lab. Anim.* **29**, 177–179.

Kunstyr, I., Hackbarth, H., Naumann, S., Rode, B., and Muller, H. (1980). Zur *Pasteurella-pneumotropica* infection in einer Barrier-ratten-Zuchteinheit. *Z. Versuchstierkd.* **22**, 303–308.

Kunstyr, I. and Hartman, D. (1983). *Pasteurella pneumotropica* and the prevalence of the AHP (*Actinobacillus, Haemophilus, Pasteurella*) group in laboratory animals. *Lab. Anim.* **17**, 156–160.

Kunstyr, I., Heimann, W., Matthiesen, T., and Militzer, K. (1980). Dermatomykosen bei Versuchstieren unter der besonderen berucksichtigung von differential-diagnose, prophylaxe und therapie. *Berl. Munch. Tierartzl. Wschr.* **93**, 347–350.

Kunze, B. and Marwitz, T. (1967). Tilgung einer *Salmonella enteritidis*–Infektionen in einem Rattenzuchtbestand. *Z. Versuchstierkd.* **9**, 257–259.

Kurashina, H. and Fujiwara, K. (1972). Fine structure of the mouse liver infected with the Tyzzer's organism. *Jpn. J. Exp. Med.* **42**, 139–154.

Kurisu, K., Kyo, S., Shiomoto, Y., and Matsushita, S. (1990). Cilia-associated respiratory bacillus infection in rabbits. *Lab. Anim. Sci.* **40**, 413–415.

Kutscher, D. (1894). Ein Beitrag zur Kentniss det bacillaren Pseudotuberculose der Nagethiere. *Z. Hyg. Infektionskr.* **18**, 327–342.

Kwan, M.L., Gomez, A.D., Baluk, P., Hashizume, H., and McDonald, D.M. (2001). Airway vasculature after mycoplasma infection: chronic leakiness and selective hypersensitivity to substance P. *Am. J. Physiol Lung Cell Mol. Physiol.* **280**, L286–L297.

Kwiatkowsa-Patzer, B., Patzer, J.A., and Heller, L.J. (1993). *Pseudomonas aeruginosa* exotoxin A enhances automaticity and potentiates hypoxic depression of isolated rat hearts. *Proc. Soc. Exp. Biol. Med.* **202**, 377–383.

Lai, W.C., Linton, G., Bennett, M., and Pakes, S.P. (1993). Genetic control of resistance to *Mycoplasma pulmonis* infection in mice. *Infect. Immun.* **61**, 4615–4621.

Lambe Jr., D.W., McPhedran, A.M., Mertz, J.A., and Stewart, S. (1973). *Streptobacillus moniliformis* isolated from a case of Haverhill fever: biochemical characterization and inhibitory effect of sodium polyanethol sulfonate. *Am. J. dm. Pathol.* **60,** 854–860.

Langan, J., Bemis, D., Harbo, S., Pollock, C., and Schumacher, J. (2000). Tyzzer's disease in a red panda (*Ailurus fulgens fulgens*). *J. Zoo Wildl. Med.* **31,** 558–562.

Larsen, B., Markovetz, A.J., and Galask, R.P. (1976). The bacterial flora of the female rat genital tract. *Proc. Soc. Exp. Biol. Med.* **151,** 571–574.

Larson, C.L. (1941). Rat-bite fever in Washington D.C. due to *Spirillum minus* and *Streptobacillus moniliformis*. *Public Health Rep.* **56,** 1961–1969.

Lawrence, J.J. (1957). Infection of laboratory mice with *Corynebacterium murium*. *Aust. J. Sci.* **20,** 147.

Leader, R.W., Leader, I., and Witschi, E. (1970). Genital mycoplasmosis in rats treated with testosterone propionate to produce constant estrus. *J. Am. Vet. Med. Assoc.* **157,** 1923–25.

Leadingham, R.S. (1928). Rat-bite fever. *J. Med. Assoc. Ga.* **17,** 16–19.

Leadingham, R.S. (1938). Rat-bite fever (sodoku): report of five cases. *Am. J. Clin. Pathol.* **8,** 333–344.

Lee, A., Phillips, M.H., O'Rourke, B.J., Paster, B.J., Dewhirst, F.E., Fraser, G.J., Fox, J.G., Sly, L.I., Romaniuk, L.I., Trust, T.J., and Kouprach, S. (1992). *Helicobacter muridarum*, sp. nov.: a microaerophilie helical bacterium with a novel ultrastructure isolated from the intestinal mucosa of rodents. *Int. J. Syst. Bacteriol.* **42,** 27–36.

Lee, C.H., Lu, J.J., Bartlett, M.S., Durkin, M.M., Lin, T.H., Wang, J., Jiang, B., and Smith, J.W. (1993). Nucleotide sequence variation in *Pneumocystis carinii* strains that infect humans. *J. Clin. Microbiol.* **31,** 754–757.

Lee, P.E. (1955). *Salmonella* infections of urban rats in Brisbane, Queensland. *Aust. J. Exp. Biol. Med. Sci.* **33,** 113–115.

Lee, P.E. and Mackerras, I.M. (1955). *Salmonella* infections of Australian native animals. *Aust. J. Exp. Biol. Med. Sci.* **33,** 117–125.

Lee, Y.S., Hirose, H., Ogisho, Y., Goto, N., Takahashi, R., and Fujiwara, K. (1976). Myocardiopathy in rabbits experimentally infected with the Tyzzer's organism. *Jpn. J. Exp. Med.* **46,** 371–382.

LeMaistre, C. and Thompsett, R. (1952). The emergence of pseudotuberculosis in rats given cortisone. *J. Exp. Med.* **95,** 393–407.

Lemer, E.M. and Silverstein, E. (1957). Experimental infection of rats with *Streptobacillus moniliformis*. *Science* **126,** 208–209.

Lemer, E.M. and Sokoloff, L. (1959). The pathogenesis of bone and joint infection produced in rats by *Streptobacillus moniliformis*. *Arch. Pathol.* **67,** 364–372.

Lemperle, G. (1967). Stimulation of the reticuloendothelial system in burned rats infected with *Pseudomonas aeruginosa*. *J. Infect. Dis.* **117,** 7–14.

Lennette, E.H., Spaulding, E.H., and Truant, J.P., eds. (1974). "Manual of Clinical Microbiology," 2nd Ed. American Society of Microbiology, Washington, DC.

Lentsch, R.H., Batema, R.P., and Wagner, J.E. (1981). Detection of *Salmonella* infections by polyvalent enzyme-linked immunosorbent assay. *J. Clin. Microbiol.* **14,** 281–287.

Lentsch, R.H., Kirchner, B.K., Dixon, L.W., and Wagner, J.E. (1983). A report of an outbreak of *Salmonella oranienburg* in a hybrid mouse colony. *Vet. Microbiol.* **8,** 105–109.

Lentsch, R.H. and Wagner, J. (1980). Isolation of *Actinobacillus ligniersii* and *Actinobacillus equuli* from laboratory rodents. *J. Clin. Microbiol.* **12,** 351–354.

Lenz, W., Thunert, A., and Brandis, H. (1978). Untersuchungen zur Epidemiologie von Staphylokokken-Infektionen in SPF-Versuchstierbestanden. *Zentrabl. Bakteriol. Hyg. I Abt. Orig. A.* **240,** 447–465.

Lerner, E.M. and Silverstein, E. (1957). Experimental infection of rats with *Streptobacillus moniliformis*. *Science* **126,** 208–209.

Lerner, E.M. and Sokoloff, L. (1959). The pathogenesis of bone and joint infection produced in rats by *Streptobacillus moniliformis*. *Arch. Pathol.* **67,** 364–372.

Les, E.P. (1968a). Effect of acidified-chlorinated water on reproduction in C3H/HeJ and C57BL/6J mice. *Lab. Anim. Care* **18,** 210–213.

Les, E.P. (1968b). Environmental factors influencing body weight of C57BL/6J and DBA/2J mice. *Lab. Anim. Care* **18,** 623–625.

Les, E.P. (1973). "Acidified-chlorinated Drinking Water for Mice," Jax Notes, no. 415 (August). Jackson Laboratory, Bar Harbor, ME.

Lesher, R.J., Jeszenka, E.V., and Swan, E. (1985). Enteritis caused by *Pasteurella pneumotropica* infection in hamsters. *J. Clin. Microbiol.* **22,** 48.

Leung, L.S.E., Szal, G.J., and Drachman, R.H. (1972). Increased susceptibility of splenectomized rats to infection with *Diplococcus pneumoniae*. *J. Infect. Dis.* **126,** 507–513.

Levaditi, C., Nicolau, S., and Poincloux, P. (1925). Sur le role étiologique de *Streptobacillus moniliformis* (nov. spec.) dans l'érythème polymorphe aigu septicemique. *C. R. Hebd. Seances Acad Sci.* **180,** 1188–1190.

Levaditi, C., Selbie, R.F., and Schoen, R. (1932). Le rheumatisme infectieux spontane de la souris provogue par le *Streptobacillus moniliformis*. *Ann. Inst. Pasteur, Paris* **48,** 308–343.

Lindsey, J.R., Baker, H.J., Overcash, R.G., Cassell, G.H., and Hunt, C.E. (1971). Murine chronic respiratory disease: significance as a research complication and experimental production with *Mycoplasma pulmonis*. *Am. J. Pathol.* **64,** 675–716.

Livingston, R.S., Franklin, C.L., Besch-Williford, C.L., Hook Jr., R.R., and Riley, L.K. (1996). A novel presentation of *Clostridium piliforme* infection (Tyzzer's disease) in nude mice. *Lab. Anim. Sci.* **46,** 21–25.

Livingston, R.S., Riley, L.K., Besch-Williford, C.L., and Besch-Williford, C. (1998). Transmission of *Helicobacter hepaticus* infection to sentinel mice by contaminated bedding. *Lab. Anim. Sci.* **48,** 291–293.

Livingston, R.S., Riley, L.K., Steffen, E., Besch-Williford, C.L., Hook Jr., R.P., and Franklin, C.L. (1997). Serodiagnosis of *Helicobacter hepaticus* infection in mice by an enzyme-linked immunosorbent assay. *J. Clin. Microbiol.* **35,** 1236–1238.

Lowbury, E.J.L. (1951). Contamination of cetrimide and other fluids with *Pseudomonas pyocyanea*. *Br. J. Ind. Med.* **8,** 22–25.

Lowbury, E.J.L. and Fox, J. (1954). The epidemiology of infection with *Pseudomonas pyocyanea* in a burns unit. *J. Hyg.* **52,** 403–416.

Lund, E. and Henrichsen, J. (1978). Laboratory diagnosis, serology and epidemiology of *Streptococcus pneumoniae In* "Methods in Microbiology," Vol. 12. (T. Bergan and N.J. Norris, eds.), pp 193–241, Academic Press, New York.

Lusis, P.I. and Soltys, M.A. (1971). Immunization of mice and chinchillas against *Pseudomonas aeruginosa*. *Can. J. Comp. Med.* **35,** 60–66.

Lutsky, I. and Organick, A.B. (1966). Pneumonia due to Mycoplasma in gnotobiotic mice, I: pathogenicity of *Mycoplasma pneumoniae*, *Mycoplasma salivarium*, and *Mycoplasma pulmonis* for the lungs of conventional and gnotobiotic mice. *J. Bacteriol.* **92,** 1154–1163.

MacAlister, G.H., and Brooks, R.S.J. (1914). Report upon the postmortem examination of rats at Ipswich. *J. Hyg.* **14,** 316–330.

Mackaness, G.B., Blanden, R.V., and Collins, F.M. (1966). Host-parasite relations in mouse typhoid. *J. Exp. Med.* **124,** 573.

Mackenzie, D.W.R. (1961). *Trichophyton mentagrophytes* in mice; infections of humans and incidence amongst laboratory animals. *Sabouraudia* **1,** 178–182.

Mackenzie, D.W.R. (1963). "Hairbrush diagnosis" in the detection and eradication of non-fluorescent scalp ringworm. *Br. Med. J.* **2,** 363–365.

MacKenzie, W.F., Magill, L.S., and Hulse, M. (1981). A filamentous bacterium associated with respiratory disease in wild rats. *Vet. Pathol.* **18,** 836–837.

Mackie, T.J., VanRooyen, C.E., and Gilroy, E. (1933). An epizootic disease occurring in a breeding stock of mice: bacteriological and experimental observations. *Br. J. Exp. Pathol.* **14,** 132–136.

MacNeal, W.J. (1907). A spirochaete found in the blood of a wild rat. *Proc. Soc. Exp. Biol. Med.* **4**, 125.

Macy, J.D., Weir, E.C., Compton, S.R., Shlomchik, M.J., and Brownstein, D.G. (2000). Dual infection with *Pneumocystis carinii* and *Pasteurella pneumotropica* in B cell–deficient mice: diagnosis and therapy. *J. Comp. Med.* **50**, 49–55.

Maejima, K., Fujiwara, K., Takagaki, Y., Naiki, M., Kucashin, H., and Tajima, Y. (1965). Dietetic effects on experimental Tyzzer's disease of mice. *Jpn. J. Exp. Med.* **35**, 1–10.

Maejima, K., Urano, T., Itoh, K., Fujiwara, K., Homma, J.Y., Suzuki, K., and Yanabe, M. (1973). Serological typing *Pseudomonas aeruginosa* from mouse breeding colonies. *Jpn. J. Exp. Med.* **43**, 179–184.

Maenza, R.M., Powell, D.W., Plotkin, G.R., Focmal, S.B., Jervis, H.R., and Sprinz, H. (1970). Experimental diarrhea: *Salmonella* enterocolitis in the rat. *J. Infect. Dis.* **121**, 475–485.

Mahenthiralingam, E., Campbell, M.E., and Speert, D.P. (1994). Nonmotility and phagocytic resistance of *Pseudomonas aeruginosa* isolates from chronically colonized patients with cystic fibrosis. *Infect. Immun.* **62**, 596–605.

Mahenthiralingam, E. and Speert, D.P. (1995). Nonopsonic phagocytosis of *Pseudomonas aeruginosa* by macrophages and polymorphonuclear leukocytes requires the presence of the bacterial flagellum. *Infect. Immun.* **63**, 4519–4523.

Mahler, M., Bedigian, H.G., Burgett, B.L., Bates, R.J., Hogan, M.E., and Sundberg, J.P. (1998). Comparison of four diagnostic methods for detection of *Helicobacter* species in laboratory mice. *Lab. Anim. Sci.* **48**, 85–91.

Makamura, K., Mochizuki, T., Hasegawa, A., Uchida, K., Saito, H., and Yamaguchi, H. (1998). Phylogenetic classification of *Trichophyton mentagrophytes* complex strains based on DNA sequences of nuclear ribosomal internal transcribed spacer 1 regions. *Clin. Microbiol.* **36**, 2629–2633.

Makazuki, K., Takahashi, K., Nanba, K., and Hayashi, Y. (1987). (Selective media for *Pasteurella pneumotropica*). *Jikken Dobutsu.* **36**, 229–237.

Mallory, F.B. and Ordway, T. (1909). Lesions produced in the rat by a typhoid-like organism (Danysz virus). *JAMA* **52**, 1455.

Manning, P.J. (1993). Two rat isolates of *B. piliformis* are different based on protein banding patterns in acrylamide gels and immunoblots. *Lab. Anim. Sci.* **43**, 208–209.

Manning, P.J., DeLong, D., Gunther, R., and Swanson, D. (1991). An enzyme-linked immunosorbent assay for detection of chronic subclinical *Pasteurella pneumotropica* infection in mice. *Lab. Anim. Sci.* **41**, 162–165.

Manning, P.J., Gaibor, J., DeLong, D., and Gunther, R. (1989). Enzyme-linked immunosorbent assay and immunoblot analysis of the immunoglobulin G response to whole cell and lipooligosaccharide antigens of *Pasteurella pneumotropica* in laboratory mice with latent pasteurellosis. *J. Clin. Microbiol.* **27**, 2190–2194.

Marcotte, H., Levesque, D., Delanay, K., Bourgeaulat, A., de la Durantaye, R., Brochu, S., and Lavoie, M.C. (1996). *Pneumocystis carinii* infection in transgenic B cell–deficient mice. *J. Infect. Dis.* **173**, 1034–1037.

Marecki, N.M., Hsu, H.S., and Mayo, D.R. (1975). Cellular and humoral aspects of host resistance in murine salmonellosis. *Br. J. Exp. Pathol.* **56**, 231–243.

Margard, W.L. and Peters, A.C. (1964). A study of gnotobiotic mice monocontaminated with *Salmonella typhimurium*. *Lab. Anim. Care* **14**, 200–207.

Margard, W.L., Peters, A.C., Dorko, N., Litchfield, J.H., Davidson, R.S., and Rheins, M.S. (1963). Salmonellosis in mice: diagnostic procedures. *Lab. Anim. Care* **13**, 144–165.

Matheson, B.H., Grice, H.C., and Connell, M.R.E. (1955). Studies of middle ear disease in rats, I: age of infection and infecting organisms. *Can. J. Comp. Med. Vet. Sci.* **19**, 91–97.

Matsumoto, H., Tazaki, T., and Kato, T. (1968). Serological and pyocine types of *Pseudomonas aeruginosa* from various sources. *Jpn. J. Microbiol.* **12**, 111–119.

Matsushita, S. (1986). Spontaneous respiratory disease associated with cilia-associated respiratory (CAR) bacillus in a rat. *Jpn. J. Vet. Sci.* **48**, 437–440.

Matsushita, S. and Joshima, H. (1989). Pathology of rats intranasally inoculated with the cilia-associated respiratory bacillus. *Lab. Anim.* **23**, 89–95.

Matsushita, S., Joshima, H., Matumoto, T., and Fukutsu, K. (1989). Transmission experiments of cilia-associated respiratory bacillus in mice, rabbits and guinea pigs. *Lab. Anim.* **23**, 96–102.

Matsushita, S., Kashima, M., and Joshima, N. (1987). Serodiagnosis of cilia-associated respiratory bacillus infection by the indirect immunofluorescence assay technique. *Lab. Anim.* **21**, 356–359.

Matsushita, S. and Suzuki, E. (1995). Prevention and treatment of cilia-associated respiratory bacillus in mice by use of antibiotics. *Lab. Anim. Sci.* **45**, 503–507.

Maura, S.B., Mendes, E.N., Queiroz, D.M., Camargos, E.S., Evangelina, M., Fonseca, F., Rocha, G.A., and Nicoli, J.R. (1998). Ultrastructure of *Helicobacter trogontum* in culture and in the gastrointestinal tract of gnotobiotic mice. *J. Med. Microbiol.* **47**, 513–520.

Mazars, E., Guyot, K., Durand, I., Dei-Cas, E., Boucher, S., Abderrazak, S.B., Banuls, A.L., Tibayrene, M., and Camus, D. (1997). Isoenzyme diversity in *Pneumocystis carinii* from rats, mice and rabbits. *J. Infect. Dis.* **175**, 655–660.

McBride, D.F., Stark, D.M., and Walberg, J.A. (1981). An outbreak of staphylococcal furunculosis in nude mice. *Lab. Anim. Sci.* **31**, 270–272.

McDermott, E.N. (1928). Rat-bite fever: study of the experimental disease, with a critical review of the literature. *Q. J. Med.* **21**, 433–458.

McDougall, P.T., Wolf, N.S., Steinbach, W.A., and Trentin, J.J. (1967). Control of *Pseudomonas aeruginosa* in an experimental mouse colony. *Lab. Anim. Care* **17**, 204–215.

McEvoy, M.B., Noah, N.D., and Pilsworth, R. (1987). Outbreak of fever caused by *Streptobacillus moniliformis*. *Lancet.* **2**, 1361–1363.

McEwen, S.A. and Percy, D.H. (1985). Diagnostic exercise: pneumonia in a rat. *Lab. Anim. Sci.* **35**, 485–487.

McFadden, D.C., Powles, M.A., Pittarelli, L.A., and Schmatz, D.M. (1991). Establishment of *Pneumocystis carinii* in various mouse strains using natural transmission to effect infection. *J. Protozool.* **38**, 126S–127S.

McGee, A.A. and Wittler, R.G. (1969). The role of L phase and other wall-defective microbial variants in disease. *In* "The Mycoplasmatales and the L-phase of Bacteria." (L. Hayflick, ed.), pp 697–720, Appleton, New York.

McGill, R.C., Martin, A.M., and Edmunds, P.N. (1966). Rat-bite fever due to *Streptobacillus moniliformis*. *Br. Med. J.* **1**, 1213–1214.

McGinn, M.D., Bean-Knudsen, D., and Ermel, R.W. (1992). Incidence of otitis media in CBA/J and CBA/CaJ mice. *Hear. Res.* **59**, 1–6.

McGuire, E.A., Young, V.R., Newbeme, P.M., and Payne, B.J. (1968). Effects of *Salmonella typhimurium* infection in rats fed varying protein intakes. *Arch. Pathol.* **86**, 60–68.

McKeil, J.A., Rappay, D.E., Cousineau, J.G., Hall, R.R., and McKenna, H.E. (1970). Domestic rats as carriers of leptospires and salmonellae in eastern Canada. *Can. J. Public Health* **61**, 336–340.

McKenna, J.M., South, F.E., and Mussachia, X.J. (1970). *Pasteurella* infection in irradiated hamsters. *Lab. Anim. Care* **20**, 443–446.

McPherson, C.W. (1963). Reduction of *Pseudomonas aeruginosa* and coliform bacteria in mouse drinking water following treatment with hydrochloric acid or chlorine. *Lab. Anim. Care* **13**, 737–745.

McRipley, R.J. and Garrison, D.W. (1964). Increased susceptibility of burned rats to *Pseudomonas aeruginosa*. *Proc. Soc. Exp. Biol. Med.* **115**, 336–338.

Medina, L.V., Chladny, J., Fortman, J.D., Artwohl, J.E., Bunte, R.M., and Bennett, B.T. (1996). Rapid way to identify the cilia-associated respiratory bacillus: tracheal mucosal scraping with a modified microwave Steiner silver impregnation. *Lab. Anim. Sci.* **46**, 113–115.

Medina, L.V., Fortman, J.D., Bunte, R.M., and Bennett, B.T. (1994). Respiratory disease in a rat colony: identification of CAR bacillus without other respiratory pathogens by standard diagnostic screening methods. *Lab. Anim. Sci.* **44**, 521–525.

Mendes, E.N., Quieroz, D.M.M., Dewhirst, F.E., Paster, B.J., Moura, S.B., and Fox, J.G. (1996). *Helicobacter trogontum*, sp. nov., isolated from the rat intestine. *Int. J. Syst. Bacteriol.* **46**, 916–921.

Menges, R.W. and Georg, L.K. (1956). An epizootic of ringworm among guinea pigs caused by *Trichophyton mentagrophytes*. *J. Am. Vet. Med. Assoc.* **128**, 395–398.

Menges, R.W., Georg, L.K., and Habermann, R.T. (1957). Therapeutic studies on ringworm-infected guinea pigs. *J. Invest. Dermatol.* **28**, 233–237.

Merrikin, D.J. and Terry, C.S. (1972). Use of pyocin 78-C2 in the treatment of *Pseudomonas aeruginosa* infection in mice. *Appl. Microbiol.* **23**, 164–169.

Meshorer, A. (1974). Tyzzer's disease in laboratory mice. *Refuah Vet.* **31**, 41–42.

Messina, O.D., Maldonado-Cocco, J.A., Pescio, A., Farinati, A., and Garcia-Morteo, O. (1989). *Corynebacterium kutscheri* septic arthritis. *Arthritis Rheum.* **32**, 1053.

Meyer, J.M., Neely, A., Stintzi, A., Georges, C., and Holder, I.A. (1996). PyoverClin is essential for virulence of *Pseudonomas aeruginosa*. *Infect. Immun.* **64**, 518–523.

Meyer, K.F. and Batchelder, A.P. (1926). A disease in wild rats caused by *Pasteurella muricida* n. sp. *J. Infect. Dis.* **39**, 386–412.

Meyer, K.F. and Matsumura, K. (1927). The incidence of carriers of *B. aertrycke* (*B. pestiscaviae*) and *B. enteritidis* in wild rats of San Francisco. *J. Infect. Dis.* **41**, 395–404.

Mia, A.S., Kravcak, D.M., and Cassell, G.H. (1981). Detection of *Mycoplasma pulmonis* antibody in rats and mice by a rapid micro enzyme linked immunosorbent assay. *Lab. Anim. Sci.* **31**, 356–359.

Mikazuki, K., Hirasawa, T., Chiba, H., Takahashi, K., Sakai, Y., Ohhara, S., and Nenui, H. (1994). Colonization pattern of *Pasteurella pneumotropica* in mice with latent pasteurellosis. *Exp. Anim.* **43**, 375–379.

Milder, J.E., Walzer, P.D., Coonrod, J.D., and Rutledge, M.E. (1980). Comparison of histological and immunological techniques for detection of *Pneumocystis carinii* in rat bronchial lavage fluid. *J. Clin. Microbiol.* **11**, 409–417.

Miller, C.P., Hammond, C.W., and Tompkins, M.A. (1950). The incidence of bacteremia in mice subjected to total body X-irradiation. *Science* **11**, 540–541.

Miller, C.P., Hammond, C.W., and Tompkins, M.A. (1951). The role of infection in radiation injury. *J. Lab. Clin. Med.* **38**, 331–343.

Millican, R.C. (1963). *Pseudomonas aeruginosa* infection and its effects in non-radiation stress. *Lab. Anim. Care* **13**, 11–19.

Millican, R.C., Evans, G., and Markley, K. (1966). Susceptibility of burned mice to *Pseudomonas aeruginosa* and protection by vaccination. *Ann. Surg.* **163**, 603–610.

Millican, R.C. and Rust, J.D. (1960). Efficacy of rabbit *Pseudomonas* antiserum in experimental *Pseudomonas aeruginosa* infection. *J. Infect. Dis.* **107**, 389–396.

Millican, R.C., Rust, J.D., Verder E., and Rosenthal, S.M. (1957). Experimental chemotherapy of *Pseudomonas* infections, I: production of fatal infections in cortisone-infected mice. *Antibiot. Annu.* 486–493.

Minion, F.C., Brown, M.B., and Cassell, G.H. (1984). Identification of cross-reactive antigens between *M. pulmonis* and *M. arthritidis* screening with ELISA. *Infect. Immun.* **43**, 115–121.

Mirick, G.S., Richter, C.P., Schaub, I.G., Franklin, R., MacCleary, R., Schipper, G., and Spitznager, J. (1950). An epizootic due to *Pneumococcus* type 11 in laboratory rats. *Am. J. Hyg.* **52**, 48–53.

Mitchell, D.W.H. (1912). Bacillus muris as the etiologic agent of pneumonitis in white rats and its pathogenicity for laboratory animals. *J. Infect. Dis.* **10**, 17–23.

Mitruka, B.M. (1971). Biochemical aspects of *Diplococcus pneumoniae* infections in laboratory rats. *Yale J. Biol. Med.* **44**, 253–264.

Miyake, H. (1902). Ueber die Rattenbissenkrankheit. *Mitt. Gremgeb. Meet. Chir.* **5**, 231–262.

Miyazaki, S., Matsumoto, T., Tateda, K., Ohno, A., and Yamaguchi, K. (1995). Role of exotoxin A in inducing severe *Pseudomonas aeruginosa* infections in mice. *J. Med. Microbiol.* **43**, 169–175.

Mizoguchi, J., Hokao, R., Sano, J., Kagiyama, N., and Imamichi, T. (1986). An outbreak of *Trichophyton mentagrophytes* infection in a rat breeding stock and its successful control. *Exp. Anim.* **35**, 125–130.

Mohapatra, L.N., Gugnani, H.C., and Shivrajan, K. (1964). Natural infection in laboratory animals due to *Trichophyton mentagrophytes* in India. *Mycopathol. Mycol. Appl.* **24**, 275–280.

Moody, K.D., Griffity, J.W., and Lang, C.M. (1986). Fungal meningoencephalitis in a laboratory rat. *J. Am. Vet. Med. Assoc.* **189**, 1152–1153.

Moore, G.J. (1979). Conjunctivitis in the nude rat (nu/nu). *Lab. Anim.* **18**, 35.

Moore, G.J. and Aldred, P. (1978). Treatment of *Pasteurella pneumotropica* abscesses in nude mice (nu/nu). *Lab. Anim.* **22**, 227–228.

Moore, T.D., Alien, A.M., and Ganaway, J.R. (1973). Latent *Pasteurella pneumotropica* infection of the intestine of gnotobiotic and barrier-held rats. *Lab. Anim. Sci.* **5**, 657–661.

Morch, E. (1947). Mechanism of *Pneumococcus* infection in mice. *Acta Pathol. Microbiol. Scand.* **24**, 169–180.

Morello, J.A., Digenio, T.A., and Baker, E.E. (1964). Evaluation of serological and cultural methods for the diagnosis of chronic salmonellosis in mice. *J. Bacteriol.* **88**, 1277–1282.

Morello, J.A., Digenio, T.A., and Baker, E.E. (1965). Significance of salmonellae isolated from apparently healthy mice. *J. Bacteriol.* **89**, 1460–1464.

Morrissette, C., Francoeur, C., Darmond, Zweig, C., and Gerrais, F. (1996). Lung phagocyte bactericidal function in strains of mice resistant and susceptible to *Pseudomonas aeruginosa*. *Infect. Immun.* **64**, 4984–4992.

Morrissette, C., Skamene, E., and Gervais, F. (1995). Endobronchial inflammation following *Pseudomonas aeruginosa* infection on resistant and susceptible strains of mice. *Infect. Immun.* **63**, 1718–1724.

Motzel, S.L., Meyer, J.K., and Riley, L.K. (1991). Detection of serum antibodies to *Bacillus piliformis* in mice and rats using an enzyme-linked immunosorbent assay. *Lab. Anim. Sci.* **41**, 26–30.

Motzel, S.L. and Riley, L.K. (1991). *Bacillus piliformis* flagellar antigens for serodiagnosis of Tyzzer's disease. *J. Clin. Microbiol.* **29**, 2566–2570.

Motzel, S.H. and Riley, L.K. (1992). Subclinical infection and transmission of Tyzzer's disease in rats. *Lab. Anim. Sci.* **42**, 439–443.

Moura, S.B., Mendes, E.N., Quieroz, D.M., Nicoli, J.R., Cabral, M.M., Magalhaes, P.P., Rocha, B.A., and Vieria, E.C. (1999). Microbiological and histological study of the gastrointestinal tract of germfree mice infected with *Helicobacter trogontium*. *Res. Microbiol.* **150**, 205–212.

Mullink, J.W.M.A. and Rumke, C.L. (1971). Reaction on hexobarbital and pathological control of mice given acidified drinking water. *Z. Versuchstierkd.* **13**, 196–200.

Mutters, R., Frederiksen, W., and Mannheim, W. (1984). Lack of evidence for the occurrence of Pasteurella ureae in rodents. *Vet. Microbiol.* **9**, 83–93.

Mutters, R., Ihm, P., Polh, S., Fredriksen, W., and Mannheim, W. (1985). Reclassification of the genus *Pasteurella* Trevisan 1887 on the basis of deoxyribonucleic acid homology, with proposals for the new species *Pasteurella dagmatis, Pasteurella canis, Pasteurella stomatis, Pasteurella anatis* and *Pasteurella langaa. Int. J. Sys. Bacteriol.* **35,** 309–322.

Mutters, R., Mannheim, W., and Bisgaard, M. (1989). Taxonomy of the group. *In* "*Pasteurella* and Pasteurellosis." (C. Adlam and J.M. Rutter, eds.), pp 3–34, Academic Press, New York.

Nafiz, M. (1935). Untersuchungen an wilden Ratten in Mïnchen auf das Vorkommen von Typhus und Paratyphuskeimen. *Arch. Hyg.* **113,** 245–246.

Nahimana, A., Cushion, M.T., Blanc, D.S., and Hauser, P.M. (2001). Rapid PCR-single-strand conformation polymorphism method to differentiate and estimate relative abundance of *Pneumocystis carinii* special forms infecting rats. *J. Clin. Microbiol.* **39,** 4563–4565.

Naiki, M., Takagaki, Y., and Fujiwara, K. (1965). Note on the change of transaminases in the liver and the significance of the transaminase ratio in experimental Tyzzer's disease of mice. *Jpn. J. Exp. Med.* **35,** 305–309.

Nakagawa, M. and Saito, M. (1984). Antigenic characterization of *Pasteurella pneumotropica* isolated from mice and rats. *Exp. Anim.* **33,** 187–192.

Nakagawa, M. and Weyant, R.S. (1994). *Streptococcus pneumoniae. In* "Manual of Microbiologic Monitoring of Laboratory Animals," 2nd Ed. (K. Waggie, N. Kagiyama, A.M. Allen, and T. Nomura, eds.). National Institutes of Health [NIH], Bethesda, MD, NIH Publication no. 94-2498.

Nakai, N., Kawaguchi, C., Nawa, K., Kobayashi, S., Katasuta, Y., and Watanabe, M. (2000). Detection and elimination of contaminating microorganisms in transplantable tumors and cell lines. *Exp. Anim.* **49,** 309–313.

Nakayama, H., Oguihara, S., Osaki, K., Toriumi, W., and Fujiwara, K. (1984). Effect of cyclophosphamide on Tyzzer's disease of mice. *Jpn. J. Vet. Sci.* **47,** 81–88.

Nakayama, M., Saegusa, J., Itoh, K., Kiuch, Y., Tamura, T., Veda, K., and Fujiwara, K. (1975). Transmissible enterocolitis in hamsters caused by Tyzzer's organisms. *Jpn. J. Exp. Med.* **45,** 33–41.

Nance, F.C., Lewis Jr., V.L., and Hines, J.L. (1970). *Pseudomonas* sepsis in gnotobiotic rats. *Surg. Forum* **21,** 234–235.

Naughton, P.J., Gran, R., Spencer, R.J., Bardocz, S., and Pusztai, A. (1996). A rat model of infection by *Salmonella typhimurium* or *Salmonella enteritidis. J. Appl. Bacteriol.* **81,** 651–656.

Needham, J.R. and Cooper, J.E. (1976). An eye infection in laboratory mice associated with *Pasteurella pneumotropica. Lab. Anim.* **9,** 197–200.

Neely, A.N., Miller, R.G., and Holder, I.A. (1994). Proteolytic activity and fatal Gram negative sepsis in burned mice: effect of exogenous proteinase inhibition. *Infect. Immun.* **62,** 2158–2164.

Nelson, J.B. (1930a). The bacteria of the infected middle ear in adult and young albino rats. *J. Infect. Dis.* **46,** 64–75.

Nelson, J.B. (1930b). The reaction of the albino rat to the intraaural administration of certain bacteria associated with middle ear disease. *J. Exp. Med.* **52,** 873–883.

Nelson, J.B. (1931). The biological characters of *B. actinoides* variety muris. *J. Bacteriol.* **21,** 183–195.

Nelson, J.B. (1933). The reaction of antisera for *B. actinoides. J. Bacteriol.* **26,** 321–327.

Nelson, J.B. (1937). Infectious catarrh of mice. *J. Exp. Med.* **65,** 833–860.

Nelson, J.B. (1940). Infectious catarrh of the albino rat, I: experimental transmission in relation to the role of Actinobacillus muris. *J. Exp. Med.* **72,** 645–654.

Nelson, J.B. (1951). Development of a rat colony free from respiratory infections. *J. Exp. Med.* **94,** 377–384.

Nelson, J.B. (1963). Chronic respiratory disease in mice and rats. *Lab. Anim. Care* **13,** 137–143.

Nelson, J.B. (1973). Response of mice to *Corynebacterium kutscheri* on footpad injection. *Lab. Anim. Sci.* **23,** 370–372.

Nelson, J.B. and Gowen, J.W. (1930). The incidence of middle ear infection and pneumonia in albino rats at different ages. *J. Infect. Dis.* **46,** 53–63.

Newbeme, P.M., Hunt, C.E., and Young, V.R. (1968). The role of diet and the reticuloendothelial system in the response of rats to *Salmonella typhimurium* infection. *Br. J. Exp. Pathol.* **49,** 448–457.

Nichols, P.W., Schoeb, T.R., Davis, J.K., Davidson, M.K., and Lindsey, J.R. (1992). Pulmonary clearance of *Mycoplasma pulmonis* in rats with respiratory viral infections or of susceptible genotype. *Lab. Anim. Sci.* **42:** 454–457.

Nicklas, W. (1989). *Haemophilus* infection in a colony of laboratory rats. *J. Clin. Microbiol.* **27,** 1636–1639.

Nicklas, W., Baneux, P., Boot, R., Decelle, T., Deeny, A.A., Fumanelli, M., and Illgen-Wilcke, B. (2002). Recommendations for the health monitoring of rodent and rabbit colonies in breeding and experimental units. *Lab. Anim.* **36,** 20–42.

Nicklas, W. and Benner, A. (1994). Prevalence of V factor-dependent Pasteurellaceae in laboratory rodents and their phenotypic classication. *In* "Proceedings of the Fifth FELASA Symposium: Welfare and Science." (I. Bunyan, ed.), pp 375–377, Royal Society of Medicine, London.

Nicklas, W., Benner, A., and Mauter, P. (1995). Computer assisted classification of *Pasteurella pneumotropica* biotypes on the basis of biochemical criteria. *Contemp. Topics Lab. Anim. Sci.* **34,** 68.

Nicklas, W., Staut, M., and Benner, A. (1993). Prevalence and biochemical properties of V factor-dependent Pasteurellaccae from rodents. *Zentrabl. Bakteriol.* **279,** 114–124.

Nietfeld, J.C., Fickbohm, B.L., Rogers, D.G., Franklin, C.L., and Riley, L.K. (1999). Isolation of cilia-associated respiratory (CAR) bacillus from pigs and calves and experimental infection of gnotobiotic pigs and rodents. *J. Vet. Diag. Invest.* **11,** 252–258.

Nielsen, M.H., Settnes, O.P., Aliouat, E.M., Cailliez, J.C., and Dei-Cas, E. (1998). Different ultrastructural morphology of *Pneumocystis carinii* derived from mice, rats and rabbits. *APMIS* **106,** 771–779.

Niven, J.S.F. (1968). Tyzzer's disease in laboratory animals. *Z. Versuchstierkd.* **10,** 168–174.

Nozu, R., Goto, K., Ohashi, H., Takakua, A., and Itch, T. (1999). Evaluation of PCR as a means of identification of *Pasteurella pneumotropica. Exp. Anim.* **48,** 51–54.

Nutini, L.G. and Berberich, J. (1965). Effect of diet and strain difference on virulence of *Staphylococcus aureus* for mice. *Appl. Microbiol.* **13,** 614–617.

Nutini, L.G., Mukkada, A.T., and Cook, E.S. (1968). Susceptibility of Swiss albino mice to *Staphylococcus aureus*: diet and sex factors. *Appl. Microbiol.* **16,** 815–816.

Nyska, A. and Kuttin, E.S. (1988). Upper respiratory tract infection in mouse and rats. *Mycopathologia* **101,** 95–98.

O'Callaghan, R.J., Engel, L.L., Hobden, A., Callegan, M.C., Green, L.C., and Hill, J.M. (1996). *Pseudomonas* keratitis: the role of an uncharacterized exoprotein, protease IV, in corneal virulence. *Investig. Ophthalmol. Visual Sci.* **37,** 534–543.

O'Leary, T.L., Tsai, M.M., Wright, C.F., and Cushion, M.T. (1995). Use of semiquantitative PCR to assess onset and treatment of *Pneumocystis carinii* infection in rat model. *J. Clin. Microbiol.* **33,** 718–724.

Olson, L.D. and Ediger, R.D. (1972). Histopathologic study of the heads of circling mice infected with *Pseudomonas aeruginosa. Lab. Anim. Sci.* **22,** 522–527.

Olson, J.R. and Meadows, T.R. (1969). *Pasteurella pneumotropica* infection resulting from a cat bite. *Am. J. Clin. Pathol.* **51,** 709–710.

Olsson, M., Sukura, A., Lindberg, L.A., and Lindner, E. (1996). Detection of *Pneumocystis carinii* DNA by filtration of air. *Scand. J. Infect. Dis.* **28,** 279–282.

Onodera, T. and Fujiwara, K. (1970). Experimental encephalopathy in Tyzzer's disease of mice. *Jpn. J. Exp. Med.* **40**, 295–323.

Onodera, T. and Fujiwara, K. (1972). Neuropathology of intraspinal infection in mice with the Tyzzer's organism. *Jpn. J. Exp. Med.* **42**, 263–282.

Oros, J., Matsushita, S., Rodriguez, J.L., Rodriguez, F., and Fernandez, A. (1996). Demonstration of rat CAR bacillus using a labeled strepaviClin biotin (LSAB) method. *J. Vet. Med. Sci.* **58**, 1219–1221.

Oros, J., Poveda, J.B., Rodriguez, J.L., Franklin, C.L., and Fernandez, A. (1997). Natural cilia-associated respiratory bacillus infection in rabbits used for elaboration of hyperimmune serum against *Mycoplasma* sp. *Zentrabl. Veterinarmed.* **44**, 313–317.

Osimani, J.R., Nairac, R., Hormaeche, C., Fonseca, D., de Hormaeche, R.D., and Portillo, M. (1972). Fiebre estreptobacNRC por mordedura de rata. *Rev. Latinoam. Microbiol.* **14**, 197–201.

Otcenasek, M., Stros, K., Krivanek, K., and Komarek, J. (1974). Epizoocie dermatofytozy ve velkochovu morcat. *Vet. Med. (Prague)* **19**, 277–281.

Owen, D.G. (1982). *Haemobartonella muris:* case report and investigation of transplacental transmission. *Lab. Anim.* **16**, 17–19.

Owens, D.R. and Wagner, J.E. (1975). *Trichophyton mentagrophytes* infection in guinea pigs. *Lab. Anim. Dig.* **10**, 14–18.

Owens, D.R., Wagner, J.E., and Addison, J.B. (1975). Antibiograms of pathogenic bacteria isolated from laboratory animals. *J. Am. Vet. Med. Assoc.* **167**, 605–609.

Palmer, R.J., Cushion, M.T., and Wakefield, A.E. (1999). Discrimination of rat-derived *Pneumocystis carinii f. sp. carinii* and *Pneumocystis carinii f. sp. ratti* using the polymerase chain reaction. *J. Mol. Cell Probes* **13**, 147–155.

Palmer, R.J., Settnes, O.P., Lodal, J., and Wakefield, A.E. (2000). Population structure of rat-derived *Pneumocystis carinii* in Danish wild rats. *Appl. Environ. Microbiol.* **66**, 4954–4961.

Papini, R., Gazzano, A., and Mancianti, F. (1997). Survey of dermatophytes isolated from the coats of laboratory animals in Italy. *Lab. Anim. Sci.* **47**, 75–77.

Pappenheimer, A.M. and von Wedel, H. (1914). Observations on a spontaneous typhoid-like epidemic of white rats. *J. Infect. Dis.* **14**, 180–215.

Parish, H.J. and Craddock, J. (1931). Ringworm epizootic in mice. *Br. J. Exp. Pathol.* **12**, 209–213.

Parker Jr., F. and Hudson, N.P. (1926). The etiology of Haverhill fever (erythema arthriticum epidemicum). *Am. J. Pathol.* **2**, 357–379.

Parmely, M., Gale, A., Clabaugh, M., Horvat, R., and Zhou, W.W. (1990). Proteolytic inactivation of cytokines by *Pseudomonas aeruginosa. Infect. Immun.* **58**, 3009–3014.

Peace, T. and Soave, O.A. (1969). Tyzzer's disease in a group of newly purchased mice. *Lab. Anim. Dig.* **5**, 8–10.

Peglow, S.L., Smulian, A.G., Linke, M.J., Pogue, C.L.N., Crisler, S.J., Phair, J., Gold, J W., Armstrong, D., and Walzer, P.D. (1990). Serologic responses to *Pneumocystis carinii* antigens in health and disease. *J. Infect. Dis.* **161**, 296.

Peltier, M.R., Richey, L.J., and Brown, M.B. (2003). Placental lesions caused by experimental infection of Sprague-Dawley rats with *Mycoplasma pulmonis. Am. J. Reprod. Immun.* **50**, 254–262.

Percy, D.H. and Barta, J.R. (1993). Spontaneous and experimental infections in SCID and SCID/Beige mice. *Lab. Anim. Sci.* **43**, 127–132.

Pesanti, E.L. and Shanley, J.D. (1984). Murine cytomegalovirus and *Pneumocystis carinii. J. Infect. Dis.* **149**, 643.

Pestana de Castro, A.F., Giorgi, W., and Ribeiro, W.B. (1964). Estudo de umna amostra de *Corynebacterium kutscheri* isolada de ratos e camundongos. *Arq. Inst. Biol. (Sao Paulo)* **31**, 91–99.

Phillips, M.W. and Lee, A. (1983). Isolation and characterization of a spiral bacterium from the crypts of rodent gastrointestinal tracts. *Appl. Environ. Microbiol.* **45**, 675–683.

Pier, G.B., Meluleni, G., and Goldberg, J.B. (1995). Clearance of *Pseudomonas aeruginosa* from the murine gastrointestinal tract is effectively mediated by O-antigen–specific circulating antibodies. *Infect. Immun.* **63**, 2818–2825.

Pier, G.B., Meluleni, G., and Neuger, E. (1992). A murine model of chronic mucosal colonization by *Pseudomonas aeruginosa. Infect. Immun.* **60**, 4668–4672.

Pierce-Chase, C.H., Fauve, M., and Dubos, R. (1964). Corynebacterial pseudotuberculosis in mice, I: comparative susceptibility of mouse strains to experimental infection with *Corynebacterium kutscheri. J. Exp. Med.* **120**, 267–281.

Pinson, D.M., Schoeb, T.R., Lindsey, J.R., and Davis, J.K. (1986). Evaluation by scoring and computerized morphometry of lesions of early *Mycoplasma pulmonis* infection and ammonia exposure in F344/N rats. *Vet. Pathol.* **23**, 550–555.

Pittet, J.F., Hashimoto, S., Pian, M., McElroy, M.C., Nitenberg, G., and Wiener-Kronish, J.P. (1996). Exotoxin A stimulates fluid reabsorption from distal airspaces of lung in anesthetized rats. *Am. J. Physiol.* **270**, L232–L241.

Place, E.H. and Sutton, L.E. (1934). Erythema arthriticum epidemicum. *Arch. Intern. Med.* **54**, 659–684.

Place, E.H., Sutton, L.E., and Willner, O. (1926). Erythema arthriticum: preliminary report. *Boston Med. Surg. J.* **194**, 285–287.

Plotkin, S.A. and Austrian, R. (1958). Bacteremia caused by *Pseudomonas* sp. following the use of materials stored in solutions of a cationic surface-active agent. *Am. J. Med. Sci.* **235**, 621–627.

Pohlmeyer, G. and Deerberg, F. (1993). Nude rats as a model of natural *Pneumocystis carinii* pneumonia: a sequential morphologic study of long lesions. *J. Comp. Pathol.* **109**, 217–230.

Polak, M.F. (1944). Epidemic survenue parmi des souris blanches à la suite d'une infection par le *Corynebacterium pseudotuberculosis murium. Antonie van Leeuwenhoek* **10**, 23–27.

Pollack, J.D., Williams, M.V., and McElhaney, R.N. (1997). The comparative metabolism of the Mollicutes (*Mycoplasma*): the utility for taxonomic classification and the relationship of putative gene annotation and phylogeny to enzymatic function in the smallest free-living cells. *Critical Rev. Microbiol.* **23**, 269–354.

Pombier, E.C. and Kirn, J.C.S. (1975). An epizootic outbreak of ringworm in a guinea pig colony caused by *Trichophyton mentagrophytes. Lab. Anim.* **9**, 215–221.

Port, C.D., Richter, W.R., and Moise, S.M. (1970). Tyzzer's disease in the gerbil (*Meriones unguiculatus*). *Lab. Anim. Care* **20**, 109–111.

Porter, A. (1915–16). The occurrence of *Pneumocystis carinii* in mice in England. *Parasitol.* **8**, 255–259.

Porter, G. (1952). An experience of Tyzzer's disease in mice. *J. Inst. Anim. Tech.* **2**, 17–18.

Potel, J. (1967). (Latent *Salmonella* infection in laboratory animals.) *Zentralbl. Bakteriol., Parasitenkd., Infektionskr. Hyg., Abt. 1: Orig.* **203**, 292–295.

Povar, M.L. (1965). Ringworm (*Trichophyton mentagrophytes*): infection in a colony of albino Norway rats. *Lab. Anim. Care* **15**, 264–265.

Powanda, M.C. and Canonico, P.G. (1976). Protective effect of clofibrate against *S. pneumoniae* infection in rats. *Proc. Soc. Exp. Biol. Med.* **152**, 437–430.

Powanda, M.C., Wannemacher Jr., R.W., and Cockerell, G.L. (1972). Nitrogen metabolism and protein synthesis during pneumococcal sepsis in rats. *Infect. Immun.* **6**, 266–271.

Powell, D.W., Plotkin, G.R., Maenza, R.M., Solberg, L.I., Catlin, D.H., and Formal, S.B. (1971a). Experimental diarrheam 1: intestinal water and electrolyte transport in rat *Salmonella* enterocolitis. *Gastroenterology* **60**, 1053–1064.

Powell, D.W., Plotkin, G.R., Solberg, L.I., Catlin, D.H., Maenza, R.M., and Formal, S.B. (1971b). Experimental diarrhea, 2: glucose-stimulated sodium and water transport in rat *Salmonella* enterocolitis. *Gastroenterology* **60**, 1065–1075.

Powell, D.W., Solberg, L.I., Plotkin, G.R., Catlin, D.H., Maenza, R.M., and Formal, S.B. (1971c). Experimental diarrhea, 3: bicarbonate transport in rat *Salmonella* enterocolitis. *Gastroenterology* **60**, 1076–1086.

Preston, W.S. (1942). Arthritis in rats caused by pleuropneumonia-like microorganisms and the relationship of similar organisms to human rheumatism. *J. Infect. Dis.* **70**, 180–184.

Price-Jones, C. (1926–1927). Infection of rats by Gartner's bacillus. *J. Pathol. Bacteriol.* **30**, 45–54.

Pyrah, L.N., Goldie, W., Parsons, F.M., and Raper, F.P. (1955). Control of *Pseudomonas pyocanea* infection in a urological ward. *Lancet* **2**, 314–317.

Qin, L., Quinlan, W.M., Doyle, N.A., Graham, L., Sligh, J.E., Takei, F., Beaudet, A.L., and Doerschuk, C.M. (1995). The roles of CD11/CD18 and ICAM-1 in acute *Pseudomonas aeruginosa*-induced pneumonia in mice. *J. Immunol.* **157**, 5016–5021.

Rabodonirina, M., Wilmotte, R., Dannaoui, E., Persat, F., Bayle, G., and Mojon, M. (1997). Detection of *Pneumocystis carinii* DNA by PCR amplification in various rat organs in experimental pneumocystosis. *J. Med. Microbiol.* **46**, 665–668.

Rake, G. (1936). Pathology of *Pneumococcus* infection in mice following intranasal installation. *J. Exp. Med.* **63**, 17–31.

Rao, G.N., Hickman, R.L., Seilkop, S.K., and Boorman, G.A. (1981). Utero-ovarian infection in aged B63CF1 mice. *Lab. Anim. Sci.* **37**, 153–158.

Rapp, J.P. and McGrath, J.T. (1975). Mycotic encephalitis in weanling rats. *Lab. Anim. Sci.* **25**, 477–480.

Ratcliffe, H.L. (1949). Spontaneous diseases of laboratory rats. *In* "The Rat in Laboratory Investigation," 2nd Ed. (E.J. Fan-is and J.Q. Griffith, eds.), pp 515–530, Lippincott, Philadelphia, PA.

Rathbun, H.K. and Govani, I. (1967). Mouse inoculation as a means of identifying pneumococci in the sputum. *Johns Hopkins Med. J.* **120**, 46–48.

Rauss, K. and Kalovich, B. (1971). *Salmonella* immunity in mice. *Zentralbl. Bakteriol., Parasitenkd., Infektionskr. Hyg., Abt. 1: Orig.* **216**, 32–53.

Ray, J.P. and Mallick, B.B. (1970). Public Health significance of *Salmonella* infections in laboratory animals. *Indian Vet. J.* **47**, 1033–1037.

Razin, S. and Boschwitz, C. (1968). The membrane of the *Streptobacillus moniliformis* L phase. *J. Gen. Microbiol.* **54**, 21–32.

Razin, S. and Jacobs, E. (1992). Review article: mycoplasma adhesion. *J. Gen. Microbiol.* **138**, 407–422.

Rebell, G. and Taplin, D. (1970). "Dermatophytes: Their Recognition and Identification," 2nd. rev. pp 10–19, University of Miami Press, Coral Gables, FL.

Reddy, L.F. and Zammit, C. (1991). Proliferation patterns of latent *Pneumocystis carinii* in rat organs during progressive stages of immunosuppression. *J. Protozool.* **38**, 455–475.

Reed, D.M. (1902). The *Bacillus pseudotuberculosis murium* its streptothrix forms and pathogenic action. *Johns Hopkins Hosp. Rep.* **9**, 525–541.

Reetz, I.C., Wullenweber-Schidt, M., Kraft, V., and Hedrich, H.J. (1988). Rederivation of inbred strains of mice by means of embryotransfer. *Lab. Anim. Sci.* **38**, 696–701.

Rehbinder, C. and Tschappat, V. (1974). *Pasteurella pneumotropica,* isoliert von der Konjunktivalschleimhaut gesunder Laboratoriumsmause. *Z. Versuchstierkd.* **16**, 359–365.

Rehm, S., Waalkes, M.P., and Ward, J.M. (1988). *Aspergillus* rhinitis in Wistar (CRL:(WI) BR) rats. *Lab. Anim. Sci.* **38**, 162–166.

Reith, von H. (1968). Spontane und experimentelle Pilzerkrankungen bei Mausen. *Z. Versuchstierkd.* **10**, 75–81.

Reyes, L., Steiner, D.A., Hutchison, J., Crenshaw, B., and Brown, M. B. (2000). *Mycoplasma pulmonis* genital disease: effect of rat strain on pregnancy outcome. *Comp. Med.* **50**, 622–627.

Rights, F.L., Jackson, E.B., and Smadel, F.E. (1947). Observations on Tyzzer's disease in mice. *Am. J. Pathol.* **23**, 627–635.

Riley, L.K., Besch-Williford, C., and Waggie, K.S. (1990). Protein and antigenic heterogenecity among isolates of *Bacillus piliformis*. *Infect. Immun.* **58**, 1010–1016.

Riley, L.K., Caffrey, C.J., Musille, V.S., and Meyer, J.K. (1992). Cytotoxicity of *Bacillus piliformis*. *J. Med. Microbiol.* **37**, 77–80.

Riley, L.K., Franklin, C.L., Hook Jr., R.R., and Besch-Wiliford, C. (1994). "Tyzzer's disease: an update of current information." Charles River Laboratories, Wilmington, MA.

Riley, L.K., Franklin, C.H., Hook Jr., R.R., and Besch-Williford, C. (1996). Identification of murine *Helicobacters* by PCR and restriction enzyme analyses. *J. Clin. Microbiol.* **34**, 942–946.

Ringen, L.M. and Drake, C.H. (1952). A study of the incidence of *Pseudomonas aeruginosa* from various natural sources. *J. Bacteriol.* **64**, 841–845.

Rinkinon, M., Jalava, K., Westermarck, E., Salminen, S., and Ouwehand, A.C. (2003). Interaction between probiotic lactic acid bateria and canmine pathogens: a risk factor for intestinal *Enterococcus faecium* colonization? *Vet. Microbiol.* **92**, 111–119.

Roberts, S.A. and Gregory, B.J. (1980). Facultative *Pasteurella pneumotropica* ophthalmitis in hooded Lister rats. *Lab. Anim.* **14**, 323–324.

Robertson, A. (1924). Causal organism of rat-bite fever in man. *Ann. Trop. Med. Parasitol.* **18**, 157–175.

Robertson, A. (1930). *Spirillum minus* Carter 1887, the etiological agent of rat-bite fever: A review. *Ann. Trop. Med. Parasitol.* **24**, 367–410.

Robinson, H.J., Phares, H.F., and Graessle, O.E. (1968). Effects of indomethacin on acute, subacute, and latent infections in mice and rats. *J. Bacteriol.* **96**, 6–13.

Robinson, L.B. (1963). *Streptobacillus moniliformis* infections. *In* "Diagnostic Procedures and Reagents," 4th Ed. (A.H. Harris and M.B. Coleman, eds.), pp 642–651, American Public Health Association, New York.

Robinson, R. (1965). "Genetics of the Norway Rat," pp 5–6. Pergamon, Oxford.

Robson, H.G. and Vas, S.I. (1972). Resistance of inbred mice to *Salmonella typhimurium*. *J. Infect. Dis.* **126**, 378–386.

Rogosa, M. (1974). *Streptobacillus moniliformis* and *Spirillum minus*. *In* "Manual of Clinical Microbiology," 2nd Ed. (E.H. Lennette, E.H. Spaulding, and J.P. Truant, eds.), pp 326–332, American Society of Microbiology, Washington, DC.

Rosenthal, S.A. and Fumari, D. (1957). The use of a cyclohexamide-chloramphenicol medium in routine culture for fungi. *J. Invest. Dermatol.* **28**, 367–371.

Rosenthal, S.A. and Wapnick, H. (1963). The value of MacKenzie's hair brush technic in the isolation of *Trichophyton mentagrophytes* from clinically normal guinea pigs. *J. Invest. Dermatol.* **41**, 5–6.

Rosenthal, S.M., Millican, R.C., and Rust, J. (1957). A factor in human γ globulin preparations active against *Pseudomonas aeruginosa* infections. *Proc. Soc. Exp. Biol. Med.* **94**, 214–217.

Ross, V. (1931). Oral immunization against *Pneumococcus* types II and III and the normal variation in resistance to these types among rats. *J. Exp. Med.* **54**, 875–898.

Ross, V. (1934). Protective antibodies following oral administration of *Pneumococcus* types 2 and 3 to rats, with some data for types 4, 5 and 6. *J. Immunol.* **27**, 273–306.

Roughgarden, J.W. (1965). Antimicrobial therapy of rat-bite fever; a review. *Arch. Intern. Med.* **116**, 39–54.

Rouleau, A.M.J., Kovacs, P.R., Kunz, H.W., and Armstrong, D.T. (1993). Decontamination of rat embryos and transfer to specific pathogen-free recipients for the production of a breeding colony. *Lab. Anim. Sci.* **43**, 611–615.

Row, R. (1918). Cutaneous spirochetosis produced by rat-bite in India. *Bull. Soc. Pathol. Exot.* **11**, 88–195.

Rozenqurt, N. and Sanchez, S. (1993). *Aspergillus niger* isolated from an outbreak of rhinitis in rats. *Vet. Rec.* **132**, 656–657.

Ruitenberg, E.J., Guinee, P.A.M., Kruyt, B.C., and Berkvens, J.M. (1971). *Salmonella* pathogenesis in germ-free mice. *Br. J. Exp. Pathol.* **52**, 192–197.

Rumlely, R.L., Patrone, N.A., and White, L. (1987). Rat-bite fever as a cause of septic arthritis: a diagnostic dilemma. *J. Ann. Rheum. Dis.* **46**, 793–795.

Russell, R.J., Haines, D.C., Anver, M.R., Battles, J.K., Gorelick, P.L., Blumenauer, L.L., Gonda, M.A., and Ward, G.M. (1995). Use of antibiotics to prevent hepatitis and typhlitis in male SCID mice spontaneously infected with *Helicobacter hepaticus*. *Lab. Anim. Sci.* **45**, 373–379.

Russell, R.J. and McGinley, J.R. (1991). *Pneumocystis carinii*-animal production perspective. *J. Protozool.* **38**, 128S-129S.

Ryll, M., Mutters, R., and Mannheim, W. (1991). Untersuchungen zur genetischen klassifikation der *Pasteurella pneumotropica* lemplexes. *Berl. Munch. Tierarztl. Wschr.* **104**, 243–245.

Sacks, J.H. and Egdahl, R.H. (1960). Protective effects of immunity and immue serum on the development of hemolytic anemia and cancer in rats. *Surg. Forum* **10**, 22–25.

Sacquet, E. (1958). La salmonellose du rat de laboratoire: detection et traitement des porteurs sains. *Rev. Fr. Etud. Clin. Biol.* **3**, 1075–1078.

Sacquet, E. (1959). Valeur de la sero-agglutination dans le diagnostic des salmonellosis du rat et de la souris. *Rev. Fr. Etud. Clin. Biol.* **4**, 930–932.

Sagi, T. and Lapis, L. (1956). Unter dem Enfluss der Cortison behandlung entstehende Lungenaspergillose bei Ratten. *Acta Microbiol. Acad. Sci. Hung.* **3**, 337–340.

Saito, M., Kohjima, K., Sano, J., Nakayama, K., and Nakagawa, M. (1981). Carrier state of *Pasteurella pneumotropica* in mice and rats. *Jikken Dobutsu.* **30**, 313–316.

Sakata, Y., Akaike, T., Suga, M., Ijiri, S., Ando, M., and Maeda, H. (1996). Bradykinin triggered by Pseudomonas proteases facilitates invasion of the systemic circulation by *Pseudomonas aeruginosa*. *Microbiol. Immunol.* **40**, 415–423.

Salthe, O. and Krumweide, C. (1924). Studies on the paratyphoid-enteriditis group, VIII: an epidemic of food infection due to a paratyphoid bacillus of rodent origin. *Am. J. Hyg.* **4**, 23–32.

Saltzgaber-Muller, J. and Stone, B. A. (1986). Detection of *Corynebacterium kutscheri* in animal tissues by DNA–DNA hybridization. *J. Clin. Microbiol.* **24**, 759–763.

Sanchez, S., Tyler, K., Rozengurt, N., and Lida, J. (1994). Comparison of a PCR-based diagnostic assay for *Mycoplasma pulmonis* with traditional detection techniques. *Lab. Anim.* **28**, 249–256.

Sandhu, K.K., Sandhu, R.S., Damodaran, V.N., and Randhawa, H.S. (1970). Effect of cortisone on brochopulmonary aspergillosis in mice exposed to spores of various *Aspergillus* species. *Sabouraudia* **8**, 32–38.

Saunders, L.Z. (1958). Tyzzer's disease. *J. Natl. Cancer Inst.* **20**, 893–897.

Savage, N.L. (1972). Host-parasite relationships in experimental *Streptobacillus moniliformis* arthritis in mice. *Infect. Immun.* **5**, 183–190.

Savage, N.L., Joiner, G.N., and Florey, D.W. (1984). Clinical, microbiological and histological manifestations of *Streptobacillus moniliformis*–induced arthritis in mice. *Infect. Immun.* **34**, 605–609.

Savage, N.L. and Lewis, D.H. (1972). Application of immunofluorescence to detection of Tyzzer's disease agent (*Bacillus piliformis*) in experimentally infected mice. *Am. J. Vet. Res.* **33**, 1007–1012.

Savage, W.D. and Read, W.J. (1914). Gaertner group bacilli in rats and mice. *J. Hyg.* **13**, 343–352.

Savage, W.G. and White, P.B. (1923). Rats and *Salmonella* group bacilli. *J. Hyg.* **21**, 258–261.

Savoor, S.R. and Lewthwaite, R. (1941). The Weil-Felix reaction in experimental rat-bite fever. *Br. J. Exp. Pathol.* **22**, 274–292.

Sawicki, L., Bruce, H.M., and Andrewes, C.H. (1962). *Streptobacillus moniliformis* infection as a probable cause of arrested pregnancy and abortion in laboratory mice. *Br. J. Exp. Pathol.* **43**, 194–197.

Schaedler, R.W. and Dubos, R.J. (1962). The fecal flora of various strains of mice. Its bearing on their susceptibility to endotoxin. *J. Exp. Med.* **115**, 1149–1160.

Schaffzin, J.K., Garbe, R.R., and Stringer, J.R. (1999). Major surface glycoprotein genes from *Pneumocystis carinii f. sp. ratti. J. Fungal Genet. Biol.* **28**, 214–226.

Scharman, W. and Heller, A. (2001). Survival and transmissibility of *Pasteurella pneumotropica. Lab. Anim.* **35**, 163–166.

Schechmeister, J.L. (1956). Pseudotuberculosis in experimental animals. *Science* **123**, 463–464.

Schechmeister, J.L. and Adler, F.L. (1953). Activation of pseudo-tuberculosis in mice exposed to sublethal total body radiation. *J. Infect. Dis.* **92**, 228–239.

Schiffman, G., Douglas, R.M., Bonner, M.J., Roberts, M., and Austrian, R. (1980). A radioimmunoassay for immunologic phenomena in pneumococcal disease and for the antibody response to pneumococcal vaccines, I: method for the radioimmunassay of anticapsular antibodies and comparison with other techniques. *J. Immunol. Meth.* **33**, 133–144.

Schipper, G.J. (1947). Unusual pathogenicity of *Pasteurella multocida* isolated from the throats of common wild rats. *Bull. Johns Hopkins Hosp.* **81**, 333–356.

Schluger, N., Godwin, T., Sepkowitz, K., Armstrong, D., Bernard, E., Rifkin, M., Cerami, A., and Bucala, R. (1992). Application of DNA amplification to pneumocystosis: presence of serum *Pneumocystis carinii* DNA during human and experimentally induced *Pneumocystis carinii* pneumonia. *J. Exp. Med.* **176**, 1327–1333.

Schluger, N., Sepkonitz, K., Armstrong, D., Bernard, E., Rifkin, M., Cerami, A., and Bucala, R. (1991). Detection of *Pneumocystis carinii* in serum of AIDS patients with *Pneumocystis pneumonia* by the polymerase chain reaction. *J. Protozool.* **38**, 123S-125S.

Schneemilch, H.D. (1976). A naturally acquired infection of laboratory mice with *Klebsiella* capsule type 6. *Lab. Anim.* **10**, 305–310.

Schoeb, T.R., Davidson, M.K., and Davis, J.K. (1997a). Pathogenicity of cilia-associated respiratory (CAR) bacillus isolates for F344, LEW and SD rats. *Vet. Pathol.* **34**, 263–270.

Schoeb, T.R., Davidson, M.K., and Lindsey, J.R. (1982). Intracage ammonia promotes growth of *Mycoplasma pulmonis* in the respiratory tract of rats. *Infect. Immun.* **38**, 212–217.

Schoeb, T.R., Dybvig, K., Davidson, M.K., and Davis, J.K. (1993a). Cultivation of cilia-associated respiratory bacillus in artificial medium and determination of the 16S rRNA gene sequence. *J. Clin. Microbiol.* **31**, 2751–2757.

Schoeb, T.R., Dybvig, K., Keisling, K.F., Davidson, M.K., and Davis, J.K. (1997b). Detection of *Mycoplasma pulmonis* in cilia-associated respiratory bacillus isolates and in respiratory tracts of rats by nested PCR. *J. Clin. Microbiol.* **35**, 1667–1670.

Schoeb, T.R., Juliana, M.M., Nichols, P.W., Davis, J.K., and Lindsey, J.R. (1993b). Effects of viral and mycoplasmal infections, ammonia exposure, vitamin A deficiency, host age, and organism strain on adherence of *Mycoplasma pulmonis* in cultured rat tracheas. *Lab. Anim. Sci.* **43**, 417–424.

Schoeb, T.R., Kervin, K.C., and Lindsey, J.R. (1985). Exacerbation of murine respiratory mycoplasmosis in gnotobiotic F344/N rats by Sendai virus infection. *Vet. Pathol.* **22**, 272–282.

Schoeb, T.R. and Lindsey, J.R. (1987). Exacerbation of murine respiratory mycoplasmosis by sialodacryoadenitis virus infection in gnotobiotic F344 rats. *Vet. Pathol.* **24**, 392–399.

Schottmuller, H. (1914). Zur Atiologie und Klinik der Bisskrankheit Rattin, Katzen-, Eichhornchen-Bisskrankheit). *Dermatol. Wochenschr.* **58(Suppl),** 77–103.

Schultz, S., Pohl, S., and Mannheim, W. (1977). Mischinfektion von Albinoratten durch *Pasteurella pneumotropica* und eine neue pneumotropie *Pasteurella* species. *Zentrabl. Veterinarmed.* B. **24,** 476–485.

Schutze, H. (1928). Bacterium pseudotuberculosis rodentium. *Arch. Hyg.* **100,** 181.

Schutze, H. (1932). Studies in *B. pestis* antigens, II: the antigenic relationship of *B. pestis* and *B. pseudotuberculosis rodentium.* *Br. J. Exp. Pathol.* **13,** 289–292.

Schwarz, J. (1954). The deep mycoses in laboratory animals. *Proc. Anim. Care Panel* **5,** 37–70.

Sebesteny, A. (1973). Abscesses of the bulbourethral glands of mice due to *Pasteurella pneumotropica. Lab. Anim.* **7,** 315–317.

Seps, S.L., Cera, L.M., Terese Jr., S.C., and Vosioky, J. (1987). Investigations of the pathogenicity of *Salmonella enteritidis* Amsterdam following a naturally occurring infection in rats. *Lab. Anim. Sci.* **37,** 326–330.

Serikawa, T., Kitada, K., Muraguchi, T., and Yamada, J. (1991). A survey of *Pneumocystis carinii* infection in research mouse colonies in Japan. *Lab. Anim. Sci.* **41,** 411–414.

Seronde, J. (1954). Resistance of rats to inoculation with *Corynebacterium* pathogenic in pantothenate deficiency. *Proc. Soc. Exp. Biol. Med.* **85,** 521–524.

Seronde, J., Zucker, L.M., and Zucker, T.F. (1955). The influence of duration of pantothenate deprivation upon natural resistance of rats to a *Corynebacterium. J. Infect. Dis.* **97,** 35–38.

Seronde, J., Zucker, T.F., and Zucker, L.M. (1956). Thiamine, pyridozine and pantothenic acid in the natural resistance of the rat to a *Corynebacterium* infection. *J. Nutr.* **59,** 287–298.

Sethi, K.K., Salfelder, K., and Schwarz, J. (1964). Pulmonary fungal flora in experimental pulmocystosis of cortisone treated rats. *Mycopathol. Mycol. Appl.* **24,** 121–129.

Shah, J.S., Pieciak, W., Liu, J., Buharin, A., and Lane, D.J. (1996). Diversity of host species and strains of *Pneumocystis carinii* is based on rRNA sequences. *Clin. Diagn. Lab. Immunol.* **3,** 119–127.

Shambaugh III, G.E. and Beisel, W.R. (1966). Alterations in thyroid physiology during pneumococcal septicemia in the rat. *Endocrinology* **79,** 511–523.

Shames, B., Fox, J.G., Dewhirst, F., Yan, L., Shen, Z., and Taylor, N.S. (1995). Identification of widespread *Helicobacter hepaticus* infection in feces in commercial mouse colonies by culture and PCR assay. *J. Clin. Microbiol.* **33,** 2968–2972.

Shapiro, R.L., Duguette, J.G., Nunes, I., Rosees, D.F., Harris, M.N., Wilson, E.L., and Rifkin, D.B. (1997). Urokinase-type plasminogen activator deficient mice are predisposed to staphylococcal botryomycosis, pleuritis and effacement of lymphoid follicles. *Amer. J. Pathol.* **150,** 359–369.

Shellito, J., Nelson, S., and Sorensen, R.V. (1992). Effect of pyocyanine, a pigment of *Pseudomonas aeruginosa* on production of reactive nitrogen intermediates by murine alveolar macrophages. *Infect. Immun.* **60,** 3913–3915.

Shellito, J., Suzara, V.V., Blumenfeld, W., Beck, J.M., Steger, H.J., and Ermak, T.H. (1990). A new model of *Pneumocystis carinii* infection in mice selectively depleted of helper T lymphocytes. *J. Clin. Invest.* **85,** 1686–1693.

Shimoda, K., Maejima, K., Kuhara, T., and Nakagawa, M. (1991). Stability of pathogenic bacteria from laboratory animals in various transport media. *Lab. Anim.* **25,** 228–231.

Shoji, Y., Itoh, T., and Kagiyama, N. (1988a). Enzyme-linked immunosorbent assay for detection of serum antibody to CAR bacillus. *Exp. Anim.* **37,** 67–72.

Shoji, Y., Itoh, T., and Kagiyama, N. (1988b). Pathogenicities of two CAR bacillus strains in mice. *Exp. Anim.* **37,** 447–453.

Shoji, Y., Itoh, T., and Kagiyama, N. (1992). Propagation of CAR bacillus in artificial media. *Exp. Anim.* **41,** 231–234.

Shoji-Darkye, Y., Itoh, T., and Kagiyama, N. (1991). Pathogenesis of CAR bacillus in rabbits, guinea pigs, Syrian hamsters and mice. *Lab. Anim. Sci.* **41,** 567–571.

Shomer, N.H., Dangler, C.A., Marini, C.P., and Fox, J.G. (1998). *Helicobacter bilis/Helicobacter rodentium* co-infection associated with diarrhea in a colony of SCID mice. *Lab. Anim. Sci.* **48,** 455–459.

Shu-heh, W., Chu, Welch, K.J., Murray, E.S., and Hagsted, D.M. (1976). Effect of iron deficiency on the susceptibility to *Streptococcus pneumoniae* infection in the rat. *Nutr. Rep. Int.* **14,** 605–609.

Shults, F.S., Estes, P.C., Franklin, J.A., and Richter, C.S. (1973). Staphylococcal botryomycosis in a specified-pathogen-free mouse colony. *Lab. Anim. Sci.* **23,** 36–42.

Sidransky, H. and Friedman, L. (1959). The effect of cortisone and antibiotic agents on experimental pulmonary asperigillosis. *Am. J. Pathol.* **35,** 169–184.

Simecka, J.W., Davis, J.K., and Cassell, G.H. (1989). Serum antibody does not account for differences in the severity of chronic respiratory disease caused by *Mycoplasma pulmonis* in LEW and F344 rats. *Infect. Immun.* **57,** 3570–3575.

Simecka, J.W., Patel, P., Davis, J.K., Ross, S.E., Otwelll, P., and Cassell, G.H. (1991). Specific and nonspecific antibody responses in different segments of the respiratory tract on rats infected with *M. pulmonis. Infect. Immun.* **59,** 3715–3721.

Simmons, D.J.C. and Simpson, W. (1977). The biochemical and cultural characteristics of *Pasteurella pneumotropica. Med. Lab. Sci.* **34,** 145–148.

Simmons, D.J.C. and Simpson, W. (1980). *Salmonella monterideo* salmonellosis in laboratory mice: successful treatment of the disease by oral oxytetracycline. *Lab. Anim.* **14,** 217–219.

Simon, P.C. (1977). Isolation of *Bacillus piliformis* from rabbits. *Can. Vet. J.* **18,** 46–48.

Simpson, W. and Simmons, D. J. C. (1980). Two *Actinobacillus* isolated from laboratory rodents. *Lab. Anim.* **14,** 15–16.

Simpson, W. and Simmons, D.J.C. (1981). *Salmonella livingstone* salmonellosis in laboratory mice: successful containment and treatment of the disease. *Lab. Anim.* **15,** 261.262.

Singh, B. and Chawla, R.S. (1974). A note on an outbreak of pulmonary aspergillosis in albino rat colony. *Indian J. Anim. Sci.* **44,** 804–807.

Sixl, W., Sixl-Voigt, B., Stogerer, M., Kock, M., Marth, E., Schuhmann, G., and Pichler-Semmelrock, F. (1989). Different species of corynebacteria in human investigations in Cairo. *Georg. Med. Suppl.* **5,** 117–134.

Slanetz, C.A. (1948). The control of *Salmonella* infections in colonies of mice. *J. Bacteriol.* **56,** 771–775.

Slattum, M.M., Stein, S., Singleton, W.L., and Decelle, T. (1998). Progressive necrosing dermatitis of the pinna in outbread mice: an institutional survey. *Lab. Anim. Sci.* **48,** 95–98.

Smith, C.D. and Sampson, C.C. (1960). Studies of *Streptobacillus moniliformis* from case of human rat-bite fever. *Am. J. Med. Technol.* **26,** 47–50.

Smith, J.M.B., and Austwick, P.K.C. (1967). Fungal diseases of rats and mice. *In* "Pathology of Laboratory Rats and Mice." (E. Cotchin and F.J.C. Roe, eds.), pp 681–719, Davis, Philadelphia, PA.

Smith, K.J., Skelton, H.G., Hilyard, E.J., Hadfield, T., Moeller, R.S., Tuur, S., Decker, C., Wagner, K.F., and Angritt, P. (1996). *Bacillus piliformis* infection (Tyzzer's disease) in a patient infected with HIV-1: confirmation with 16S ribosomal RNA sequence analysis. *J. Am. Acad. Dermatol.* **34,** 343–348.

Smith, W. (1941). Cervical abscesses of guinea pigs. *J. Pathol. Bacteriol.* **53,** 29–37.

Smith, W.W., Menges, R.W., and Georg, L.I. (1957). Ecology of ringworm fungi on commensal rats from rural premises in southwestern Georgia. *Am. J. Trop. Med. Hyg.* **6**, 81–85.

Smulian, A.G., Sesterhenn, T., Tanaka, R., and Cushion, M.T. (2001). The ste3 pheromone receptor gene of *Pneumocystis carinii* is surrounded by a cluster of signal transduction genes. *Genetics* **157**, 991–1002.

Solnick, J.V. and Schauer, D.B. (2001). Emergence of diverse *Helicobacter* species in the pathogenesis of gastric and enterohepatic diseases. *Clin. Microbiol. Rev.* **134**, 59–97.

Soulez, B., Dri-Cas, E., Charet, P., Mougeot, G., Caillaux, M., and Camus, D. (1989). The young rabbit: a nonimmunosuppressed model for *Pneumocystis carinii* pneumonia. *J. Infect. Dis.* **160**, 355–356.

Soulez, B., Pailuaull, F., Cesbron, J. Y., Dei-Cas, W., Capron, A., and Carous, D. (1991). Introduction of *Pneumocystis carinii* in a colony of SCID mice. *J. Protozool.* **38**, 455–475.

Soumet, C., Ermel, G., Boutin, P., Boscher, E., and Colin, P. (1994). Evaluation of different DNA extraction procedures for the detection of *Salmonella* from chicken products by polymerase chain reaction. *Lett. Appl. Microbiol.* **19**, 294–298.

Spaar, R.C. (1923). Two cases of rat-bite fever: rapid cure by the intravenous injection of neo-salvarsan. *J. Trop. Med. Hyg.* **26**, 239.

Sparrow, S. (1976). The microbiological and parasitological status of laboratory animals from accredited breeders in the United Kingdom. *Lab. Anim.* **10**, 365–373.

Speert, D.P. (2002). Molecular epidemiology of *Pseudomonas aeruginosa*. *Front. Biosci.* **7**, e354–e361.

Speirs, R.S. (1956). Effect of oxytetracycline upon cortisone induced pseudotuberculosis in mice. *Antibiot. Chemother. (Washington, DC)* **6**, 395–399.

Spencer, R.C. and Savage, M.A. (1976). Use of counter and rocket immunoelectrophoresis in acute respiratory infections due to *Streptococcus pneumoniae*. *J. Clin. Pathol.* **29**, 187–190.

Spencer, T.H., Ganaway, J.R., and Waggie, K.S. (1990). Cultivation of *Bacillus piliformis* (Tyzzer) in mouse fibroblasts (373 cells). *Vet. Microbiol.* **22**, 291–297.

Sprecher, M.W. and Copeland, J.R. (1947). Haverhill fever due to *Streptobacillus moniliformis*. *JAMA* **134**, 1014–1016.

Staff, E.J. and Grover, M.L. (1936). An outbreak of *Salmonella* food infection caused by filled bakery products. *Food Res.* **1**, 465–479.

Stansly, P.G. (1965). Nononcogenic infectious agents associated with experimental tumors. *Prog. Exp. Tumor Res.* **7**, 224–258.

Stedham, M.A. and Bucci, T.J. (1969). Tyzzer's disease in the rat. *Lab. Invest.* **20**, 604.

Stedham, M.A. and Bucci, T.J. (1970). Spontaneous Tyzzer's disease in a rat. *Lab. Anim. Care* **20**, 743–746.

Steffen, E.K. and Wagner, J.E. (1983). *Salmonella enteritidis* serotype Amsterdam in a commercial rat colony. *Lab. Anim. Sci.* **33**, 454–456.

Steiner, D.A. and Brown, M.B. (1993). Impact of experimental genital mycoplasmosis on pregnancy outcome in Sprague-Dawley rats. *Infect. Immun.* **61**, 633–639.

Steiner, D.A., Uhl, E.W., and Brown, M.B. (1993). In utero transmission of *Mycoplasma pulmonis* in experimentally infected Sprague-Dawley rats. *Infect. Immun.* **61**, 2985–2990.

Stevens, E.J., Ryan, C.M., Friedberg, J.S., Barnhill, R.L., Yarmush, M.L., and Tompkins R.G. (1994). A quantitative model of invasive *Pseudomonas* infection in burn injury. *J. Burn Care Rehabil.* **15**, 232–235.

Stevenson, M.M., Kondratieva, T.K, Apt, A.S., Tam, M.F., and Skamene, E. (1995). In vitro and in vivo T cell responses in mice during bronch-pulmonary infection with mucoid *Pseudomonas aeruginosa*. *Clin. Exp. Immunol.* **99**, 98–105.

Stewart, D.D. and Buck, G.E. 1975. The occurrence of *Mycoplasma arthritidis* in the throat and middle ear of rats with chronic respiratory disease. *Lab. Anim. Sci.* **25**, 769–773.

Stewart, H.L. and Jones, B.F. (1941). Pathologic anatomy of chronic ulcerative cecitis: A spontaneous disease of the rat. *Arch. Pathol.* **31**, 37–54.

Stone, G.G., Oberst, R.D., Hays, M.P., McVey, S., and Chengappa, M.M. (1995). Combined PCR-oligonucleotide ligation assay for rapid detection of *Salmonella* serovars. *J. Clin. Microbiol.* **33**, 2888–2893.

Stone, H.H., Given, K.S., and Martin Jr., J.D. (1967). Delayed rejection of skin homografts in *Pseudomonas* sepis. *Surg. Gynecol. Obstet.* **124**, 1067–1070.

Strangeways, W.L. (1933). Rats as carriers of *Streptobacillus*. *J. Pathol. Bacteriol.* **37**, 45–51.

Stringer, J.R. (1993). The identity of *Pneumocystis carinii*: not a single protozoan, but a diverse group of exotic fungi. *Infect. Agents Dis.* **2**, 109–117.

Stringer, J.R. (2002). *Pneumocystis*. *Int. J. Med. Microbiol.* **292**, 391–404.

Stringer, J.R., Beard, C. B., Miller, R.F., and Wakefield, A.L. (2002). A new name (*Pneumocystis jiroveci*) for *Pneumocystis* from humans. *Emerg. Infect. Dis.* **8**, 891–896.

Stringer, J.R., Cushion, M.T., and Wakefield, A.E. (2001). New nomenclature for the genus *Pneumocystis*. *J. Eukaryot. Microbiol.* **(Suppl)**, 184S–189S.

Stringer, J.R., Erdman, J.C., Cushion, M.T., Richards, F.F., and Watanabe, J. (1992). The fungal nature of *Pneumocystis*. *J. Med. Vet. Mycol. Suppl.* **1**, 271–278.

Stringer, J.R., Stringer, S.L., Zhang, J., Baughman, R., Smulian, A.G., and Cushion, M.T. (1993). Molecular genetic distinction of *Pneumocystis carinii* from rats and humans. *J. Eukaryot. Microbiol.* **40**, 733–741.

Stuhmer, A. (1929). Die Rattenbiszerkrankung (sodoku) als Modellinfektion fur Syphilisstudien. *Arch. Dermatol. Syph.* **158**, 98–110.

Summerlin, W.T. and Artz, C.P. (1966). Gentamicin sulfate therapy of experimentally induced *Pseudomonas* septicemia. *J. Trauma* **6**, 233–238.

Sundberg, J.P., Burnstien, T., Schultz, L.D., and Bedigian, H. (1989). Identification of *Pneumocystis carinii* in immunodeficient mice. *Lab. Anim. Sci.* **39**, 213–218.

Surveyor, N.F. (1913). A case of rat-bite fever treated with neosalvarsan. *Lancet* **2**, 1764.

Suzuki, E., Mochida, K., and Nakagawa, M. (1988). Naturally occurring subclinical *Corynebacterium kutscheri* infection in laboratory rats: strain and age-related antibody response. *Lab. Anim. Sci.* **38**, 42–45.

Suzuki, E., Mochida, K., Takayama, S., Saitoh, M., Orikasa, M., and Nakagawa, M. (1986). Serological survey of *Corynebacterium kutscheri* infection in mice and rats. *Exp. Anim.* **35**, 45–49.

Suzuki, H., Yorozu, K., Watanabe, T., Nakura, M., and Adachi, J. (1996). Rederivation of mice by means of in vitro fertilization and embryo transfer. *Exp. Anim.* **45**, 33–38.

Swenberg, J.A., Koestner, A., and Tewari, R.P. (1969a). The pathogenesis of experimental mycotic encephalitis. *Lab. Invest.* **21**, 365–373.

Swenberg, J.A., Koestner, A., and Tewari, R.P. (1969b). Experimental mycotic encephalitis. *Acta Neuropathol.* **13**, 75–90.

Taffs, L.F. (1974). Some diseases in normal and immunosuppressed experimental animals. *Lab. Anim.* **8**, 149–154.

Takagaki, Y. and Fujiwara, K. (1968). Bacteremia in experimental Tyzzer's disease of mice. *Jpn. J. Microbiol.* **12**, 129–143.

Takagaki, Y., Ito, M., Naiki, W., Fujiwara, K., Okugi, M., Maejima, K., and Tajima, Y. (1966). Experimental Tyzzer's disease in different species of laboratory animals. *Jpn. Exp. Med.* **36**, 519–534.

Takagaki, Y., Naiki, M., Fujiwara, K., and Tajima, Y. (1963) Maladie de Tyzzer experimentale de la souris traitee avec la cortisone. *C. R. Seances Soc. Biol. Ses Fil.* **157**, 438–441.

Takagaki, Y., Naiki, M., Ito, M., Noguchi, and Fujiwara, K. (1967). Checking of infections due to *Corynebacterium* and Tyzzer's organism among mouse breeding colonies by cortisone injection. *Bull. Exp. Anim.* **16**, 12–19.

Takagaki, Y., Ogisho, Y., Sato, K., and Fujiwara, K. (1974). Tyzzer's disease in hamsters. *Jpn. J. Exp. Med.* **44**, 267–270.

Takagaki, Y., Tsuji, K., and Fujiwara, K. (1968). Tyzzer's disease-like liver lesions observed in young rats. *Exp. Anim.* **17**, 67–69.

Takenaka, J. and Fujiwara, K. (1975). Effect of carbon tetra-chloride on experimental Tyzzer's disease of mice. *Jpn. J. Exp. Med.* **45**, 393–402.

Tamura, Y., Suzuki, S., and Saweda, T. (1992). Role of elastase as a virulence factor in experimental *Pseudomonas aeruginosa* infection in mice. *Microb. Pathog.* **12**, 237–244.

Tani, T. and Takano, S. (1958). Prevention of *Borrelia duttoni, Trypanosoma gambiense, Spirillum minus* and *Treponema pallidum* infections conveyable through transmission. *Jpn. J. Med. Sci. Biol.* **11**, 407–413.

Tank, H.B., DiMango, E., Bryan, R., Gambello, M., Iglewski, B.H., Goldberg, J.B., and Prince, A. Contribution of specific *Pseudomonas aeruginosa* virulence factors to pathogenesis of pneumonia in a neonatal mouse model of infection. *Infect. Immun.* **64**, 37–43.

Tannock, G.W. and Smith, J.M.B. (1972). The effect of food and water deprivation (stress) on *Salmonella*-carrier mice. *J. Med. Microbiol.* **5**, 283–289.

Taylor, D.M. and Neal, D.L. (1980). An infected eczematous condition in mice: Methods of treatment. *Lab. Anim.* **14**, 325–328.

Taylor, J. and Atkinson, J.D. (1956). *Salmonella* in laboratory animals. *Lab. Anim. Bur. Collect. Pap. (Carshalton)* **4**, 57–66.

Taylor, J.D., Stephens, C.P., Duncan, R.G., and Singleton, G.R. (1994). Polyarthritis in wild mice (*Mus musculus*) caused by *Streptobacillus moniliformis*. *Aust. Vet. J.* **71**, 143–145.

Teplitz, C., Davis, D., Mason Jr., A.D., and Moncrief, J.A. (1964a). *Pseudomonas* burn wound sepsis, I: pathogenesis of experimental *Pseudomonas* burn wound sepsis. *J. Surg. Res.* **4**, 200–216.

Teplitz, C., Davis, D., Walker, H.L., Raulston, G.L., Mason Jr., A.D., and Moncrief, J.A. (1964b). *Pseudomonas* burn wound sepsis, II: hematogenous infection at the junction of the burn wound and the unburned hypodennis. *J. Surg. Res.* **5**, 217–222.

Thierman, A.B. and Jeffries, C.D. (1980). Opportunistic fungi in Detroit's rats and opossum. *Mycopathologia.* **71**, 39–43.

Thjotta, T. and Jonsen, J. (1947). *Streptothrix (Actinomyces) muris rani (Streptobacillus moniliformis)* isolated from a human infection and studied as to its relation to Emmy Klieneberger's LI. *Acta Pathol. Microbiol. Scand.* **24**, 336–351.

Thomas, L., Aleu, F., Bitensky, M.W., Davidson, M., and Gesner, B. (1966). Studies of PPLO infection, II: the neurotoxin of *Mycoplasma neurolyticum*. *J. Exp. Med.* **124**, 1067–1082.

Thomas, L. and Bitensky, M.W. (1966). Studies of PPLO infection, IV: the neurotoxicity of intact mycoplasmas, and their production of toxin in vivo and in vitro. *J. Exp. Med.* **124**, 1089–1098.

Thompson Jr., R.E., Smith, T.F., and Wilson, W.R. (1982). Comparison of two methods used to prepare smears of mouse lung tissue for detection of *Pneumocystis carinii*. *J. Clin. Microbiol.* **16**, 303–306.

Thomson, G.W., Wilson, R.W., Hall, E.A., and Physick-Sheard, P. (1977). Tyzzer's disease in the foal: Case reports and review. *Can. Vet. J.* **18**, 41–43.

Thornburg, L.P., Stowe, H.D., and Pick, J.R. (1973). The pathogenesis of alopecia due to hair chowing in mice. *Lab. Anim. Sci.* **23**, 843–850.

Thuis, H.C.W. and Boot, R. (1999). Monitoring of rat colonies for antibodies to CAR bacillus. *Scand. J. Lab. Anim. Sci.* **25**, 198–201.

Thunert, A. (1980). Investigations into an agent causing Tyzzer's disease in mice (*Bacillus piliformis*). *Z. Versuchstierkd.* **22**, 323–333.

Thunert, A. (1984). Is it possible to cultivate the agent of Tyzzer's disease (*Bacillus piliformis*) in cell free media? *Z. Versuchstierkd.* **26**, 145–150.

Thunert, A. and Heine, W. (1975). Zur Trinkwasserversorgung von SPF-Tieranlagen. III. Erhitzung und Ansauerung von Trinkwasser. *Z. Versuchstierkd.* **17**, 50–52.

Thygessen, P., Martinsen, C. Hongen, H.P., Hattori, R., Stenvang, J.P., and Rygaard, J. (2000). Histologic, cytologic and bacteriologic examinations of experimentally induced *Salmonella typhinurium* infection in Lewis rats. *Comp. Med.* **50**, 124–132.

Tolo, K.J. and Erichsen, S. (1969). Acidified drinking water and dental enamel in rats. *Z. Versuchstierkd.* **11**, 229–233.

Topley, W.W.C. and Ayrton, J. (1924). Biologic characteristics of *B. enteritidis (aertrycke)*. *J. Hyg.* **23**, 198–222.

Topley, W.W.C. and Wilson, G.S. (1922–1923). The spread of bacterial infection: the problem of herd immunity. *J. Hyg.* **21**, 243–249.

Topley, W.W.C. and Wilson, J. (1925). Further observations on the role of the Twort-d'Herelle phenomenon in the epidemic spread of mouse typhoid. *J. Hyg.* **24**, 295–300.

Topley, W.W.C., Wilson, J., and Lewis, E.R. (1925). Immunization and selection as factors in herd-resistance. *J. Hyg.* **23**, 421–436.

Trakhanov, D.F. and Kadirov, A.F. (1972). Izuchenie vospriim-chivosti porosyat k bakteriyam Isachenko. *Probl. Vet. Sanit.* **42**, 248–253.

Tregier, A. and Homburger, F. (1961). Bacterial flora of the mouse uterus. *Proc. Soc. Exp. Biol. Med.* **108**, 152–154.

Trentin, J.J., Van Hoosier Jr., G.L., Shields, J., Stepens, K., and Stenback, W.A. (1966). Establishment of a caesarean-derived, gnoto-biote foster nursed inbred mouse colony with observations on the control of *Pseudomonas*. *Lab. Anim. Care* **16**, 109–118.

Tsuchimori, N., Hayashi, R., Shino, A., Yamazaki, T., and Okonogi, K. (1994). *Enterococcus faecalis* aggravates pyelonephritis caused by *Pseudomonas aeruginosa* in experimental ascending mixed urinary tract infection in mice. *Infect. Immun.* **62**, 4534–4541.

Tucek, P.C. (1971). Diplococcal pneumonia in the laboratory rat. *Lab. Anim. Dig.* **7**, 32–35.

Tuffery, A.A. (1956). The laboratory mouse in Great Britain. IV. Intercurrent infection (Tyzzer's disease). *Vet. Rec.* **68**, 511–515.

Tuffery, A.A. and Innes, J.R.M. (1963). Diseases of laboratory mice and rats. *In* "Animals for Research: Principles of Breeding and Management" (W. Lane-Petter, ed.), pp 47–107, Academic Press, New York.

Tully, J.G. (1969). *In* "The Mycoplasmatales and the L-Phase of Bacteria." (L. Hayflick, ed.), pp 581–583, North-Holland Publishers, Amsterdam.

Tunnicliff, R. (1916a). *Streptothrix* in bronchopneumonia of rats similar to that in rat-bite fever. A preliminary report. *JAMA* **66**, 1606.

Tunnicliff, R. (1916b). *Streptothrix* in bronchopneumonia of rats similar to that in rat-bite fever. *J. Infect. Dis.* **19**, 767–771.

Turner, K.J., Hackshaw, R., Papadimitriou, J., and Perrott, J. (1976). The pathogenesis of experimental pulmonary aspergillosis in normal and cortisone-treated rats. *J. Pathol.* **118**, 65–73.

Turner, R.G. (1929). Bacteria isolated from infections of the nasal cavities and middle ear of rats deprived of vitamin A. *J. Infect. Dis.* **45**, 208–213.

Tyzzer, E.E. (1917). A fatal disease of the Japanese waltzing mouse caused by a sport-bearing bacillus (*Bacillus piliformis* N. sp.). *J. Med. Res.* **37**, 307–338.

Urano, T. (1994). *Pseudomonas aeruginosa. In* "Manual of Microbiologic Monitoring of Laboratory Animals, 2nd ed. (K. Waggie, N. Kagiyama, A.M. Allen, and T. Nomura, eds.), National Institutes of Health, Bethesda, MD. NIH Publ. No. 94–2498.

Utsumi, K., Matsui, Y., Ishikawa, T., Fukagawa, S., Tatsumi, H., Fujimoto, K., and Fujiwara, K. (1969). Checking of corynebacterial infection in rats by cortisone treatment. *Bull. Exp. Anim.* **18**, 59–67.

Vallee, A., Guillon, J.C., and Cayeux, R. (1969). Isolement d'une souche de *Corynebacterium kutscheri* chez un cobaye. *Bull. Acad. Vet. Fr.* **42**, 797–800.

Vallee, A. and Levaditi, J.C. (1957). Abcès miliaires des reins observes chex le rat blanc et provoque par un Corynebacterium aerobié voisin du type *C. kutscheri*. *Ann. Inst. Pasteur, Paris* **93**, 468–474.

Van Andel, R.A., Franklin, C.L., Besch-Williford, C.L., Hook, R.R., and Riley, L.K. (2000a). Prolonged perturbations of tumour necrosis factor-α and interferon-γ in mice inoculated with *Clostridium piliforme*. *J. Med. Microbiol.* **49**, 557–563.

Van Andel, R.A., Franklin, C.L., Besch-Williford, C.L, Hook, R.R., and Riley, L.K. (2000b). Role of interleukin-6 in determining the course of murine Tyzzer's disease. *J. Med. Microbiol.* **49**, 171–176.

Van Andel, R.A., Hook Jr., R.R., Franklin, C.L, Beach-Williford, C.L., and Riley L.K. (1998). Interleukin-12 has a role in mediating resistance of murine strains to Tyzzer's disease. *Infect. Immun.* **66**, 4942–4946.

Van Andel, R.A., Hook Jr., R.R., Franklin, C.L, Besch-Williford, C.L., van Rooijen, N., and Riley, L.K. (1997). Effects of neutrophil, natural killer cell, and macrophage depletion on murine *Clostridium piliforme* infection. *Infect. Immun.* **65**, 2725–2731.

Van der Schaaf, A., Mullink, J.W.M.A., Nikkels, R.J., and Goudswaard, J. (1970). *Pasteurella pneumotropica* as a causal microorganism of multiple subcutaneous abscesses in a colony of Wistar rats. *Z. Versuchstierkd.* **12**, 356–362.

Van Der Waay, D., Zimmerman, W.M.T., and Van Bekkum, D.W. (1963). An outbreak of *Pseudomonas aeruginosa* infection in a colony previously free of this infection. *Lab. Anim. Care* **13**, 46–53.

Van Hooft, J.I.M., Van Zwieten, M.J., and Solleveld, H.A. (1986). Spontaneous pneumocystosis in athymic nude rat. *Lab. Anim. Sci.* **36**, 588.

Vargas, S.L., Hughess, W.T., Wakefield, A.E., and Oz, H.S. (1995). Limited persistence in and subsequent elimation of *Pneumocystis carinii* from the lungs after P. carinii pneumonia. *J. Infect. Dis.* **172**, 506–510.

Van Kruiningen, H.J. and Blodgett, S.B. (1971). Tyzzer's disease in a Connecticut rabbitry. *J. Am. Vet. Med. Assoc.* **158**, 1205–1212.

Van Kuppeveld, F.J., Melchers, W.J., Willemse, H.F., Kissing, J., Galama, J.M., and van der Logt, J.T. (1993). Detection of *Mycoplasma pulmonis* in experimentally infected laboratory rats by 16S rRNA amplification. *J. Clin. Microbiol.* **31**, 524–527.

Van Rooyen, C.R. (1936). The biology pathogenesis and classification of *Streptobacillus moniliformis*. *J. Pathol. Bacteriol.* **43**, 455–472.

Van Zwieten, M.J., Solleveld, H.A., Lindsey, J.R., deGroot, F.G., Zurcher, C., and Hollander, C.F. (1980b). Respiratory disease in rats associated with a long rod-shaped bacterium: preliminary results of a morphologic study. *In* "Animal Quality and Models in Research." (A. Spiegel, S. Erichsen, and H.A. Solleveld, eds.), pp 243–247, Gustav Fischer Verlag, New York.

Van Zwieten, M.J., Sulleveld, H.A., Lindsey, J.R., de Groot, F.G., Zurcher, C., and Hollander, C.F. (1980a). Respiratory disease in rats associated with a filamentous bacterium; a preliminary report. *Lab. Anim. Sci.* **30**, 215–221.

Varela, G., Olarte, J., and Mata, F. (1948). Salmonellas en las ratas de la Ciudad de Mexico estudio de 1927 *Ratas norvegicus*. *Rev. Inst. Salubr. Enferm. Trap., Mexico City* **9**, 239–243.

Vasquez, J., Smulian, A.J., Linke, J., and Cushion, M.T. (1996). Antigenic differences associated with genetically distinct *Pneumocystis carinii* from rats. *Infect. Immun.* **64**, 290–297.

Veazey II, R.S., Paulsen, D.B., and Schaeffer, D.O. (1992). Encephalitis in gerbils due to naturally occurring infection with *Bacillus piliformis* (Tyzzer's disease). *Lab. Anim. Sci.* **42**, 516–518.

Veda, K., Goto, Y., Yamazaki, S., and Fugiwara, K. (1977). Chronic fatal pneumocystosis in nude mice. *Jpn. J. Exp. Med.* **47**, 475–482.

Veno, Y., Shimizu, R., Nozu, R., Takahashi, S., Yamamoto, M., Sugiyama, F., Takakura, A., Itah, T., and Yagami, K. (2002). Elimination of *Pasteurella pneumotropica* from a contaminated mouse colony by oral administration of enrofloxacin. *Exp. Anim.* **51**, 401–405.

Verder, E. (1927). The wild rat as a carrier of organisms of the paratyphoid-enteriditis group. *Am. J. Public Health* **17**, 1007.

Volgels, M.T., Eling, W.M., Otten, A., and van der Meer, J.W. (1995). Interleukin-1 (IL-1)-induced resistance to bacterial infection: role of the type I IL-1 receptor. *Antimicrob. Agents Chemother.* **39**, 1744–1747.

Volkman, A. and Collins, F.M. (1973). Polyarthritis associated with *Salmonella*. *Infect. Immun.* **8**, 814–827.

Waggie, K.S., Ganaway, J.R., Wagner, J.E., and Spencer, T.H. (1984). Experimentally induced Tyzzer's disease in Mongolian gerbils (*Meriones ungniculatus*). *Lab. Anim. Sci.* **34**, 53–57.

Waggie, K.S., Hansen, C.T., Ganaway, J.R., and Spencer, T.S. (1981). A study of mouse strain susceptibility to *Bacillus piliformis* (Tyzzer's disease): the association of B-cell function and resistance. *Lab. Anim. Sci.* **31**, 139–142.

Waggie, K., Kagiyama, N., and Itoh, T. "Cilia Associated Respiratory (CAR) Bacillus (1994). *In* "Manual of Microbiologic Monitoring of Laboratory Animals," 2nd Ed. (K. Waggie, N. Kagiyama, A.M. Allen and T. Nomura, eds). National Institutes of Health, Bethesda, MD. NIH Publ. No. 94–2498.

Wagner, J.E. and Manning, P.J., eds. (1976). "The Biology of the Guinea Pig." Adacemic Press, New York.

Wagner, J.E., Owens, D.S., LaRegina, M.C., and Vogler, G.A. (1977). Self trauma and *Staphylococcus aureus* in ulcerative dermatitis of rats. *J. Am. Vet. Med. Assoc.* **171**, 839–841.

Wagner, M. (1985). Absence of *Pneumocystis carinii* in Lobund germfree and conventional rat colonies. *In* "Germfree Research: Microflora Control and its Application to the Biomedical Sciences." (B.S. Wostmann, ed.), pp 51–54, Alan R. Liss, New York.

Wakefield, A.E. (1994). Detection of DNA sequences identical to *Pneumocystis carinii* in samples of ambient air. *J. Eukaryotic Microbiol.* **41**, 116S.

Wakefield, A.E. (1996). DNA sequences identical to *Pneumocystis carinii f. sp. carinii* and *Pneumocystis carinii f. sp. hominis* in samples of air spora. *J. Clin. Microbiol.* **34**, 1754–1759.

Wakefield, A.E. (1998). Genetic heterogeneity in *Pneumocystis carinii*: an introduction. *FEMS. Immunol. Med. Microbiol.* **22**, 5–13.

Wakefield, A.E., Banerji, S., Pixley, F.J., and Hopkin, J.M. (1990). Molecular probes for the detection of *Pneumocystis carinii*. *Trans. R. Soc. Trop. Med. Hyg.* **84(Suppl 1)** 17–18.

Walkeden, P.J. (1990). A rapid method for fungi and *Pneumocystis carinii*. *Histo-Logic.* **20**, 188–192.

Walker, H.L., Mason Jr., A.D., and Raulston, G.L. (1964). Surface infection with *Pseudomonas aeruginosa*. *Ann. Surg.* **160**, 297–305.

Walzer, P.D. (1991). Overview of animal models of *Pneumocystis carinii* pneumonia. *J. Protozool.* **38**, 1225–1235.

Walzer, P.D., Kim, C.K., Linke, M.J., Pogue, C.L., Huerkamp, M.J., Chrisp, C.E., Lerro, A.V., Wixson, S.K., Hall, E., Schultz, L.D. (1989). Outbreaks of *Pneumocystis carinii* pneumonia in colonies of immunodeficient mice. *Infect. Immun.* **57**, 62–70.

Walzer, P.D. and Powell, R.D. (1982). Experimental *Pneumocystis carinii* infection in nude and steroid-treated normal mice. *In* "Proceedings Third International Workshop on Nude Mice." pp 123–132. Gustav Fisher, New York.

Walzer, P.D., Powell, R.D., and Yoneda, K. (1979). Experimental *Pneumocystis carinii* pneumonia in different strains of cortisonized mice. *Infect. Immun.* **24**, 939–947.

Walzer, P.D., Powell, R., Yoneda, K., Rutledge, M.E., and Milder, J.E. (1980). Growth characteristics and pathogenesis of experimental *Pneumocystis carinii* pneumonia. *Infect. Immun.* **27**, 928–937.

Walzer, P.D. and Rutledge, M.E. (1980). Comparison of rat, mouse, and human *Pneumocystis carinii* by immunofluorescence. *J. Infect. Dis.* **142**, 449.

Walzer, P.D., Schnelle, V., Armstrong, D., and Rosen, P.P. (1977). Nude mouse: a new model for *Pneumocystis carinii* infection. *Science* **197**, 177–179.

Walzer, P.D., Stanforth, M.J., Linke, M.J., and Cushion, M.T. (1987). *Pneumocystis carinii*: immunoblotting and immunofluorescent analyses of serum antibodies during experimental rat infections and recovery. *Exp. Parasitol.* **63**, 319–328.

Wang, R.F., Campbell, W., Cao, W.W., Summage, C., Steele, R.S., and Cerniglia, C.E. (1996). Detection of *Pasteurella pneumotropica* in laboratory mice and rats by polymerase chain reaction. *Lab. Anim. Sci.* **46**, 81–85.

Wang, S.D., Huang, J., Lin, Y.S., and Lei, W.Y. (1994). Sepsis-induced apoptosis of the thymocytes in mice. *J. Immunol.* **152**, 5014–5021.

Ward, G.E., Moffatt, R., and Olfert, E. (1978). Abortion in mice associated with *Pasteurella pneumotropica*. *J. Clin. Microbiol.* **8**, 177–180.

Ward, J.M., Fox, J.G., Anver, M.R., Haines, D.C., George, C.V., Collins Jr., M.J., Gorelick, P.L., Nagashima, K., Ganda, M.A., Gilden, R.V., Tully, J.G., Russell, R.J., Beneveniste, R.E., and Paster, B.J. (1994). Chronic active hepatitis and associated liver tumors in mice caused by a persistent bacterial infection with a novel *Helicobacter species*. *J. Natl. Cancer Inst.* **86**, 1222–1227.

Ward, J.M., Anver, M.R., Haines, D.C., and Beneveniste, R.E. (1994a). Chronic active hepatitis in mice caused by *Helicobacter hepaticus*. *Am. J. Pathol.* **145**, 959–968.

Ward, J.M., Anver, M.R., Haines, D.C., Melhorn, J.M., Gorelick, P., Yan, L., and Fox, J.G. (1996). Inflammatory large bowel disease in immundeficient mice naturally infected with *Helicobacter hepaticus* *Lab. Anim. Sci.* **46**, 15–20.

Ward, J.R., and Jones, R.S. (1962). The pathogenesis of mycoplasmal arthritis in rats. *Arthritis Rheum.* **5**, 163–175.

Wardrip, C.L., Artwohl, J.E., Bunte, R.M., and Bennett, T.B. (1994). Diagnostic exercise: Head and neck swelling in A/JCr mice. *Lab. Anim. Sci.* **44**, 280–282.

Washburn, L.R. and Cole, B.C. (2002). *Mycoplasma arthritidis* pathogenicity: membranes, MAM, and MAV1. *In* "Molecular Biology and Pathogenicity of Mycoplasmas." (S. Razin and R. Herrmann, eds.), chapter 21, Kluwer Academic, New York.

Washburn, L.R., Miller, E.J., and Weaver, K.E. (2000). Molecular characterization of *Mycoplasma arthritidis* membrane lipoprotein MAA1. *Infect. Immun.* **68**, 437–442.

Watkins, C.G. (1946). Rat-bite fever. *J. Pediatr.* **28**, 429–448.

Webster, L.T. (1924). Microbic virulence and host susceptibility in paratyphoid-enteriditis infection of white mice. *J. Exp. Med.* **39**, 129–135.

Webster, L.T. (1930). The role of microbic virulence, dosage, and host resistance in determining the spread of bacterial infections among mice, II: B. Friedlaenderi-like infection. *J. Exp. Med.* **52**, 909–929.

Weigler, B.J., Thigpen, J.E., Goelz, M.F., Babineau, C.A., and Forsythe, D.B. (1996). Randomly amplified polymorphic DNA polymerase chain reaction assay for molecular epidemiologic investigation of *Pasteurella pneumotropica* in laboratory rodent colonies. *Lab. Anim. Sci.* **46**, 386–392.

Weigler, B.J., Witron, L.A., Hancock, S.I., Thigpen, J.E., Goelz, M.F., and Forsythe, D.B. (1998). Further evaluation of a diagnostic polymerase chain reaction assay for *Pasteurella pneumotropica*. *Lab. Anim. Sci.* **48**, 193–196.

Weisbroth, S.H. (1995). *Pneumocystis carinii*: Review of diagnostic issues in laboratory rodents. *Lab. Animal.* **24**, 36–39.

Weisbroth, S.H. (1999a). Development of rodent pathogen profiles and adequacy of detection technology. *In* "Microbial and Phenotypic Definition of Rats and Mice: Proceedings 1998 US/Japan Conference."

pp 141–146, National Research Committee, National Academy of Science Press, Washington, DC.

Weisbroth, S.H. (1999b). The rodent helicobacters: Present status of diagnostic detection and guidelines for institutional procurement standards. *Lab. Animal.* **28**, 41–45.

Weisbroth, S.H. and Freimer, E.H. (1969). Laboratory rats from commercial breeders as carriers of pathogenic pneumococci. *Lab. Anim. Care* **19**, 473–478.

Weisbroth, S.H., Geistfeld, G., Weisbroth, S.P., Williams, B., Feldman, S.H., Linke, M.J., Orr, S., and Cushion, M.T. (1999). Latent *Pneumocystis carinii* infection in commercial rat colonies: comparison of inductive immunosupressants plus histopathology, PCR and serology as detection methods. *J. Clin. Microbiol.* **37**, 1441–1446.

Weisbroth, S.H., Peters, R., Riley, L.K., and Shek, W. (1999). Microbial assessment of laboratory rats and mice. *NRC J.* **39**, 272–290.

Weisbroth, S.H. and Scher, S. (1968a). *Corynebacterium kutscheri* infection in the mouse, I: report of an outbreak, bacteriology, and pathology of spontaneous infections. *Lab. Anim. Care* **18**, 451–458.

Weisbroth, S.H. and Scher, S. (1968b). *Corynebacterium kutscheri* infection in the mouse, II: diagnostic serology. *Lab. Anim. Care* **18**, 459–468.

Weisbroth, S.H., Scher, S., and Boman, I. (1969). *Pasteurella pneumotropica* abscess syndrome in a mouse colony. *J. Am. Vet. Med. Assoc.* **155**, 1206–1210.

Welch, H., Ostrolenk, M.. and Bartram, M.T. (1941). Role of rats in the spread of food poisoning bacteria of the *Salmonella* group. *Am. J. Public Health* **31**, 332–340.

Well, E.C., Brownstein, D.G., and Barthold, S.W. (1986). Spontaneous wasting disease in nude mice associated with *Pneumocystis carinii* infection. *Lab. Anim. Sci.* **36**, 140–144.

Wensinck, F.D., Van Bekkum, D.W., and Renaud, H. (1957). The prevention of *Pseudomonas aeruginosa* infections in irradiated mice and rats. *Radiol. Res.* **7**, 491–499.

Wenyon, C.M. (1906). Spirochetosis of mice. N. sp. in the blood. *J. Hyg.* **6**, 580–585.

Whary, M.T., Cline, J.H., King, A.E., Corcoran, C.A., Xu, S., and Fox, J.G. (2000b). Containment of *Helicobacter hepaticus* by husbandry practices. *Comp. Med.* **50**, 78–81.

Whary, M.T., Cline, J.H., King, A.E., Hewes, K.M., Chojnacky, D., Salvarrey, A., and Fox, J.G. (2000a). Monitoring sentinel mice for *Helicobacter hepaticus, H. rodentium* and *H. bilis* infection by use of polymerase chain reaction analysis and serological testing. *Comp. Med.* **50**, 436–443.

Whary, M.T., Morgan, T.J., Dangler, C.A., Gaudes, K.J., Taylor, N.S., and Fox, J.G. (1998). Chronic active hepatitis induced by *Helicobacter hepaticus* in the A/JCr mouse is associated with Th1 cell-mediated immune response. *Infect. Immun.* **66**, 1342–1348.

Wheater, D.F.W. (1967). The bacterial flora of an S.P.F. colony of mice, rats, and guinea pigs. *In* "Husbandry of Laboratory Animals." (M.L. Conalty, ed.), pp 343–360, Academic Press, New York.

White, D.J. and Waldron, M.M. (1969). Naturally-occurring Tyzzer's disease in the gerbil. *Vet. Rec.* **85**, 111–114.

Wilson, G.S. and Miles, A.A., eds. (1975). "Topley and Wilson's Principles of Bacteriology, Virology, and Immunity," 6th ed., Vols. I and II. Williams & Wilkins, Baltimore, MD.

Wilson, M.G., Nelson, R.C., Phillips, L.H., and Boak, R.A. (1961). New source of *Pseudomonas aeruginosa* in a nursery. *JAMA* **175**, 1146–1148.

Wilson, P. (1976). *Pasteurella pneumotropica* as the causal organism of abscesses in the masseter muscle of mice. *Lab. Anim.* **10**, 171–172.

Winsser, J. (1960). A study of *Bordetella bronchiseptica*. *Proc. Anim. Care Panel* **10**, 87–101.

Witt, W.M. (1989). An idiopathic dermatitis in C57Bl/6N mice effectively modulated by dietary restriction. *Lab. Anim. Sci.* **39,** 470.

Witt, W.M. (1989). An idiopathic dermatitis in C57Bl/6N mice effectively modulated by dietary restriction. *Lab. Anim. Sci.* **39,** 470.

Wolf, N., Stenback, W., Taylor, P., Graber, C., and Trentin, J. (1965). Antibiotic control of post-irradiation deaths in mice due to *Pseudomonas aeruginosa. Transplantation* **3,** 585–589.

Wolff, H.L. (1950). On some spontaneous infections observed in mice. I. *C. kutscheri* and *C. pseudotuberculosis. Antonie van Leeuwenhoek* **16,** 105–110.

Woodward, J.M. (1963). *Pseudomonas aeruginosa* infection and its control in the radiobiological research program at Oak Ridge National Laboratory. *Lab. Anim. Care* **13,** 20–25.

Woodward, J.M., Camblin, M.L., and Jobe, M.H. (1969). Influence of bacterial infection on serum enzymes of white rats. *Appl. Microbiol.* **17,** 145–149.

Woolley Jr., P.V. (1936). Rat-bite fever. Report of a case with serologic observations. *J. Pediatr.* **8,** 693–696.

Wostman, B.S. (1970). Antimicrobial defense mechanisms in the *Salmonella typhimurium* associated ex-germfree rat. *Proc. Soc. Exp. Biol. Med.* **134,** 294–299.

Wullenweber, M. (1995). *Streptobacillus moniliformis*: a zoonotic pathogen. Taxonomic considerations, host species, diagnosis, therapy, geographical distribution. *Lab. Anim.* **29,** 1–15.

Wullenweber, M., Jonas, C., and Kunstyr, I. (1992) *Streptobacillus monliformis* isolated from otitis media of conventionally kept laboratory rats. *J. Exp. Anim. Sci.* **35,** 49–57.

Wullenweber, M., Kaspareit-Rittinghausen, J., and Farouq, M. (1990a). *Streptobacillus moniliformis* epizootic in barrier-maintained C57BL/6J mice and susceptibility to infection of different strains of mice. *Lab. Anim. Sci.* **40,** 608–612.

Wullenweber, M., Lenz, W., and Werhan, K. (1990b). *Staphylococcus aureus* phage types in barrier-maintained colonies of SPF mice and rats. *Z. Versuchstierkd.* **33,** 57–61.

Wullenweber-Schmidt, M., Meyer, B., Kraft, V., and Kaspareit, J. (1988). An enzyme-linked immunosorbent assay (ELISA) for the detection of antibodies to *Pasteurella pneumotropica* in murine colonies. *Lab. Anim. Sci.* **38,** 37–41.

Wyand, D.S. and Jonas, A.M. (1967). *Pseudomonas aeruginosa* infection in rats following implantation of an indwelling jugular catheter. *Lab. Anim. Care* **17,** 261–267.

Yamada, A., Osada, Y., Takayama, S., Akimoto, T., Ogawa, H., Oshima, Y., and Fujiwara, K. (1970). Tyzzer's disease syndrome in laboratory rats treated with adrenocorticotropic hormone. *Jpn. J. Exp. Med.* **39,** 505–518.

Yamada, S., Baba, E., and Arakawa, A. (1983). Proliferation of *Pasteurella pneumotropica* at oestrus in the vagina of rats. *Lab. Anim.* **17,** 261–266.

Yamada, S., Mizoguchi, J., and Ohtaki, T. (1986). Effect of oestrogen on *Pasteurella pneumotropica* in rat vagina. *Lab. Anim.* **20,** 185–188.

Yamaguchi, T. and Yamada, H. (1991). Role of mechanical injury on airway surface in the pathogenesis of *Pseudomonas aeruginosa. Am. Rev. Respir. Dis.* **144,** 1147–1152.

Yokoiyama, S. and Fujiwara, K. (1971). Effects of antibiotics on Tyzzer's disease. *Jpn. J. Exp. Med.* **41,** 49–58.

Yokoiyama, S., Mizuno, K., and Fujiwara, K. (1977). Antigenic heterogeneity of *Corynebacterium kutscheri* from mice and rats. *Exp. Anim.* **26,** 263–266.

Yokomori, K., Okada, N., Murai, Y., Goto, N., and Fujiwara, K. (1989). Enterohepatitis in Mongolian gerbils (*Meriones unguiculatus*) inoculated perorally with Tyzzer's organism (*Bacillus piliformis*). *Lab. Anim. Sci.* **39,** 16–20.

Yoneda, K., Walzer, P.D., Richey, C.S., and Birk, M.G. (1982). *Pneumocystis carinii*: freeze fracture study of stages of the organism. *Exp. Parasitol.* **53,** 68–76.

Young, C. and Hill, A. (1974). Conjunctivitis in a colony of rats. *Lab. Anim.* **8,** 301–304.

Zucker, T.F. (1957). Pantothenate deficiency in rats. *Proc. Anim. Care Panel* **7,** 193–202.

Zucker, T.F. and Zucker, L.M. (1954). Pantothenic acid deficiency and loss of natural resistance to a bacterial infection in the rat. *Proc. Soc. Exp. Biol. Med.* **85,** 517–521.

Zucker, T.F., Zucker, L.M., and Seronde, J. (1956). Antibody formation and natural resistance in nutritional deficiencies. *J. Nutr.* **59,** 299–308.

Chapter 12

Viral Disease

Robert O. Jacoby and Diane J. Gaertner

I. INTRODUCTION

The previous edition of this chapter began with an observation that is still largely true: "The number of viruses naturally infectious for rats is small (Table 12-1), and most cause inapparent infections which usually are detected by serological monitoring." This statement was not meant either then or now to imply that viral infections of rats are trivial. Their prevalence, and their capacity or potential for disruption of research, belies their modest number. A national survey (Jacoby and Lindsey, 1998) found that two virus families in particular—parvoviruses and coronaviruses—remain widely distributed in the United States: with a recent prevalence of more than 30% among rat colonies at 72 major biomedical research centers. Viruses of laboratory rats also are highly infectious and, at least for parvoviruses, can persist in animals and in the environment, thereby extending their impact on research.

TABLE 12-1

VIRUSES OF LABORATORY RATS

	Virus	Signs	Duration	Transmission	Lesions
DNA	Adenovirus	None	Presumed acute	Feces	Intranuclear inclusions in enterocytes
	Papovavirus	*Euthymic:* none *Athymic:* wasting	Unknown	Urine, saliva	*Euthymic:* Intranuclear inclusions in lung *Athymic:* Intranuclear inclusions in salivary gland, respiratory tract, kidney
	Poxvirus	None through dermal pox and deaths	Unknown	Skin contact, respiratory aerosol	*Severe cases:* Pox lesions in skin; necrotizing, rhinotracheitis and pleuropneumonia
	Rat cytomegalovirus	None	Persistent	Saliva	Intranuclear inclusions and cytomegaly in salivary glands; variable sialoadenitis
	Rat parvovirus	None	Persistent	Presumed urine, possibly feces	None
	Rat virus/H-1 virus	Usually none *Dams:* rarely abortion, fetal resorption *Infants:* rarely icterus, diarrhea, ataxia *All ages:* rarely sudden death	Acute after exposure as adults; Persistent after pre- or peri-natal exposure, and in athymic rats	Presumed urine, possibly feces, intrauterine	Necrosis/hemorrhage in liver, CNS, lymphoid tissue; intranuclear inclusions, fetal deaths
RNA	Coronavirus	Photophobia, lacrimation, sneezing, cervical swelling	Acute in immunocompetent rats Persistent in athymic rats	Respiratory aerosol, saliva, potentially tears; potentially urine in athymic rats	Necrotizing, rhinitis, sialodacryoadenitis, keratoconjunctivitis; interstitial pneumonia
	Hantavirus	None	Persistent	Bite wounds, skin, urine	None, mild multisystemic inflammation, insulitis
	Rat respiratory virus (presumed hanta-like virus)	None	Unknown	Presumed respiratory aerosol	Interstitial pneumonia
	Picornavirus	None	Unknown	Presumed feces	None
	Pneumonia virus of mice	None	Acute	Respiratory aerosol	Rhinitis, pulmonary perivasculitis, interstitial pneumonia
	Rotavirus	*Adults:* none *Infants:* diarrhea	Acute	Feces	Fluid and gas in bowels, enterocytic syncytia and necrosis, occasional intracytoplasmic inclusions
	Sendai virus	None, respiratory distress	Acute	Respiratory aerosol	Rhinitis, bronchiolitis, pneumonia

	Virus	Tissues to save for virus isolation	Tissues to save for pathology	Elimination	Interference with research
DNA	Adenovirus	Small intestine (based on pathology)	Small intestine	Not determined	None documented
	Papovavirus	Salivary glands, lung, kidney (based on pathology)	Lung, salivary gland, kidney	Not determined	None documented
	Poxvirus	Skin, lung	Skin, lung	Euthanasia, chlorine dioxide disinfection	Clinical disease, potential zoonotic hazard
	Rat cytomegalovirus	Salivary glands, lacrimal glands	Salivary glands, lacrimal glands	Cesarean rederivation, embryo transfer	None documented, but potentially disruptive to studies of salivary or lacrimal glands
	Rat parvovirus	Mesenteric lymph nodes	Small intestine, mesenteric lymph nodes—both for ISH[a]	Quarantine, cesarean rederivation, embryo transfer, chlorine dioxide disinfection	Potential to disrupt biological responses that depend on cell proliferation *in vivo* and *in vitro*
	Rat virus/H-1 virus	Liver, brain, spleen, lymph nodes, mesenteric vessels	Liver, brain, spleen, lymph nodes, mesenteric vessels—and for ISH or IHC[b]	As for rat parvovirus	As for rat parvovirus; poor breeding performance, sudden death
RNA	Coronavirus	Submandibular salivary glands, Harderian glands, nasal washes, lung	Salivary glands, lacrimal glands, nasal turbinates, lung, eye	Quarantine, cessation of breeding, chlorine dioxide disinfection	Disrupted respiratory studies, anesthetic risks, disrupted ophthalmology studies, poor reproductive performance, reduced food intake, retarded growth
	Hantavirus	Lung, vessels, kidney	Lung, vessels, kidney	Euthanasia, chlorine dioxide disinfection	Zoonotic hazard
	Rat respiratory virus (presumed hantalike virus)	Not determined	Lung	Not determined	None documented
	Picornavirus	Small intestine	Small intestine	As for rat coronavirus	None documented
	Pneumonia virus of mice	Lung	Lung	As for rat coronavirus	None documented
	Rotavirus	Small intestine	Small intestine	As for rat coronavirus	Clinical disease, retarded growth
	Sendai virus	Lung	Lung	As for rat coronavirus	Clinical disease, poor reproductive performance, disrupted respiratory studies, disrupted immune responses

Notes: Signs, duration, and lesions listed are typical for immunocompetent rats, except as noted. Immunodeficient rats can be expected to have infection with increased severity and duration.
Acute infection: Termination of infection correlates with onset of immunity, which usually occurs within 10–14 day after exposure.
Persistent infection: Infection persists for varying periods after onset of immunity.
Transmission: Transmission by direct contact should be assumed in addition to routes listed.
Major lesions: Represents acute infection, especially prior to the onset of immunity.
a = *in situ* hybridization.
b = immunohistochemistry.

Laboratory rats also are susceptible to less prevalent viruses. These include Sendai virus, pneumonia virus of mice, rat cytomegalovirus, rat rotavirus, and hantaviruses. Finally, poxviruses, papovaviruses, and picornaviruses can induce natural infections in rats, but are rarely found. In this edition, we have reduced or omitted discussion of agents that were reported transiently in older literature or were misconstrued as viral because of incomplete diagnostic assessment. These include rat salivary gland virus, MHG virus, Novy virus, and virus-like pneumotropic agents such as "gray lung virus" and wild rat pneumonia agent.

The major sections of this chapter are divided between DNA viruses (Section II) and RNA viruses (Section III). However, please note that Section IV provides a general discussion of detection, diagnosis, risk assessment, and decision-making regarding viral infections.

II. INFECTION CAUSED BY DNA VIRUSES

A. Parvoviruses

The Parvoviridae are small (18–30 nm), nonenveloped, single-stranded, negative-sense DNA viruses with a genome of approximately 5 kb. Productive replication requires cellular factors that are expressed only during cell differentiation and division (Tattersall and Cotmore, 1986; Cotmore and Tattersall, 1987). They account for the predilection of parvoviruses for mitotically active cells and the pathogenicity of several serotypes. Infection begins with binding of virions to cell receptors, internalization by endocytosis, and transport to the nucleus where they replicate and from which they are released during apoptic or lytic cell death. Recent evidence from *in vivo* studies also suggests that productive replication may not lead irrevocably to cell lysis (Jacoby, et al., 2000) a possibility that requires confirmation. In this context, nonproductive infection (abortive, cryptic, or restrictive) has been demonstrated *in vitro* for the murine parvovirus, minute virus of mice (MVM) but not for the parvoviruses of rats (Tattersall and Cotmore, 1986; Cotmore and Tattersall, 1987; Jacoby and Ball-Goodrich, 1995; Jacoby et al., 1996).

Current knowledge, based partially on extrapolation from studies of parvoviruses of mice, indicate that parvoviruses of rats replicate autonomously; that is, they do not require a helper virus. They encode two nonstructural regulatory proteins, NS1 and NS2, which are conserved and account for immunologic cross-reactivity among serotypes (Cross and Parker, 1972). They also encode two capsid proteins, VP1 and VP2, which are serotype specific. VP2 is the major capsid protein, although its sequence is contained within the VP1 coding region. A third capsid protein, VP3, results from the proteolytic cleavage of VP2 but is present in only minute quantities in mature virions.

Three established serotypes of parvoviruses infect laboratory rats. The prototype agent is rat virus (RV), which was isolated from a transplantable neoplasm of rats by Kilham and Olivier (Kilham and Olivier,1959). Therefore, it also has been called Kilham rat virus (KRV). An antigenically distinct virus (H-1 virus) was isolated shortly afterward by Toolan from a human tumor cell line that had been passaged through rats (Toolan et al., 1960). A third serotype, rat parvovirus (RPV), was isolated recently from naturally infected rats (Ball-Goodrich, et al., 1998). A similar agent has been identified by Japanese workers who initially referred to it as "rat orphan parvovirus" (Ueno et al., 1996). Most recently, three strains of a putative fourth serotype have been identified by Wan and co-workers (Wan et al, 2002), which they have named, collectively, minute virus of rats. The genomic and amino acid sequences for these isolates are closely related to KRV and H-1 virus but are significantly different from RPV. Because parvovirus infections remain prevalent among laboratory rats and have the potential to distort rat-based research results, efforts to detect, eliminate, and exclude them should receive justifiably high priority.

Many early descriptions of parvoviruses of rats were based on experimentally induced infections (Kilham, 1960; Kilham, 1961; Kilham and Margolis, 1964; Kilham and Margolis, 1966; Kilham and Margolis, 1969), often induced by parenteral inoculation. This reflected interest in using these agents for developing rodent models of virus-induced birth defects. Contemporary studies have emphasized inoculation by natural (oronasal) routes at various ages in both immunocompetent and immunodeficient rats to resemble more closely exposure during natural outbreaks (Jacoby et al., 1987, 1988, 1991; Gaertner et al., 1989, 1993, 1995, 1996).

Agents. The three documented serogroups of parvoviruses infective for rats (RV, H-1 virus, and RPV) each contain multiple isolates. Because they are small, nonenveloped viruses, they are highly resistant to environmental inactivation (Yang et. al., 1995), a characteristic that complicates control and elimination. Early work with RV and H-1 virus showed that they retain infectivity after exposure to lipid solvents and are stable over a pH range of 1 to 12 (Siegl, 1976). They also are highly heat resistant. For example, RV can remain infectious after exposure to 80°C for up to 2 hours and for up to 60 days at 40°C. Rodent parvoviruses are resistant to ultrasonication, treatment with RNAse, DNAse, papain, and trypsin—properties which, however, also facilitate preparation of purified virus from infected cell cultures. RV and H-1 agglutinate guinea pig erythrocytes and aggregate, to variable degrees, erythrocytes of other species (Tattersall and Cotmore, 1986). This property was useful for serotyping viral isolates by hemagglutination inhibition (HAI), a method that has been superseded by more modern serodiagnostic tests, discussed below. There is no evidence that strains within a given serogroup are antigenically distinguishable, even though they may differ in nucleic acid sequence, replication kinetics, and pathogenicity. For example, RV-UMass and RV-Y belong to the same serogroup, but RV-UMass appears to replicate faster *in vitro* and to be more pathogenic *in vivo* (Ball-Goodrich et al., 2001).

RV and H-1 virus grow well in primary rat embryo cells and C6 rat glial cells. RV also replicates in continuous cell lines such as 324K (human embryonic kidney) and BHK21 (hamster kidney). H-1 virus also can be propagated in rat nephroma cells (Tattersall and Cotmore, 1986; Paturzo et al., 1987) whereas RPV replicates well in 324K cells (Ball-Goodrich, et al., 1998). Productively infected cells usually develop cytopathic effects (CPE), beginning with intranuclear inclusions and proceeding rapidly to cytolysis. The kinetics and severity of CPE depend on virus strain and dose and the level of confluence of the cell substrate. High concentrations of input virus can induce intranuclear antigen indicative of replication within 12 hours and CPE within several days, whereas low concentrations may not induce measurable effects for a week or more. Cultures that contain actively dividing cells are more susceptible to viral infection and replication than confluent cultures.

Clinical signs. Natural parvoviral infections in rats are usually asymptomatic and first detected by seroconversion or suspected from distorted research results. The prevalence of clinical signs depends on virus strain and dose, host age, and route of exposure. Host genotype is not known to affect susceptibility to infection or disease. In the context of these variables, RV and H-1 serotypes are demonstrably pathogenic after experimental inoculation, whereas RPV is not. Clinical signs of RV or H-1 virus infection are most readily elicited in pregnant or infant rats. Resistance to clinical illness appears to develop by 1 week postpartum (Gaertner et al., 1996). The onset of resistance has been attributed to a decline in mitotic activity among target tissues and postnatal development of immune competence.

Prenatal infection with pathogenic serotypes can cause fetal deaths resulting in partial or complete loss of litters in dams that appear otherwise clinically normal (Kilham and Ferm, 1964; Kilham and Margolis, 1966; Kilham and Margolis, 1969; Jacoby et al., 1988; Gaertner et al., 1996). It is unclear whether intrauterine mortality reduces a dam's subsequent reproductive performance. Infection in late pregnancy or within several days after parturition can cause severe or fatal disease in infants, particularly

due to hemorrhage and necrosis in the liver and central nervous system (Kilham and Margolis, 1966; Cole et al., 1970; Margolis and Kilham, 1970). Signs in affected infants can include ataxia, icterus, and diarrhea, or sudden death (Figs. 12-1 and 12-2). Rats that survive acute disease may have locomotion deficits or may die later from chronic active hepatitis and progressive hepatic fibrosis. Although clinical signs are rare in rats exposed to virus beyond 1 week after birth, fatal hemorrhagic lesions have been reported in weanlings and immunosuppressed adults (El Dadah et al., 1967; Coleman et al., 1983). They also have been elicited in infected rats with induced increases in hepatic mitotic activity (Ruffolo et al., 1966).

Epidemiology. The relative prevalence of the three known serotypes has been complicated by the use of generic antigens for serologic testing. This raises the possibility that RPV has contributed to seroconversions attributed historically to RV and/or H-1 virus. The recent availability of serotype-specific antigens (see below) should make differentiation of parvoviral infections more precise.

Fig. 12-1 Acute rat virus infection. Ataxia (splayed feet) from RV-induced cerebellar hypoplasia.

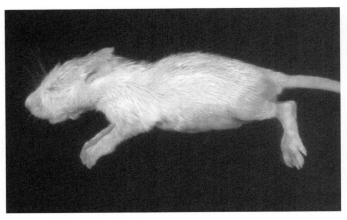

Fig. 12-2   Acute rat virus infection. Diarrhea in a rat pup with severe hepatic necrosis.

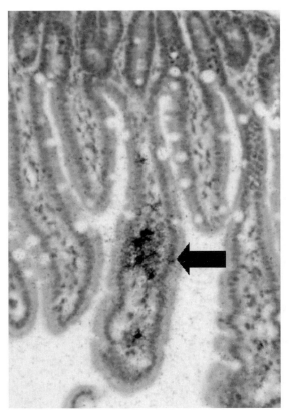

Fig. 12-4   Acute rat parvovirus infection. Viral DNA in renal tubular epithelium detected by in situ hybridization.

Rattus norvegicus is the only known natural host for RV, H-1 virus, and RPV. Based largely on studies with RV, infections should be considered highly contagious, rat-to-rat transmission occurring primarily by contact with infectious animals per se or particulate contaminants, such as animal bedding (Jacoby et al., 1988; Yang et al., 1995). Both RV and RPV (and presumably H-1 virus) infect renal tubular epithelium (Figs. 12-3 and 12-4) which facilitates excretion in urine. There is some evidence that RV can infect intestinal mucosa and lead to fecal excretion (Lipton et al., 1973), but intestine does not appear to be a prominent target tissue. Intestinal infection is prominent with RPV, although affecting predominantly the lamina propria (Fig. 12-5). Therefore, its contribution to fecal excretion of virus is unresolved. Natural postnatal exposure to parvoviruses of rats appears due to inhalation and/or ingestion of virus or virus-contaminated fomites. Prenatal transmission is secondary to maternal infection. However, it appears to require exposure of dams to large doses of virus and/or virulent strains (of RV or H-1 virus) (Kilham and Margolis, 1966; Jacoby et al., 1988;

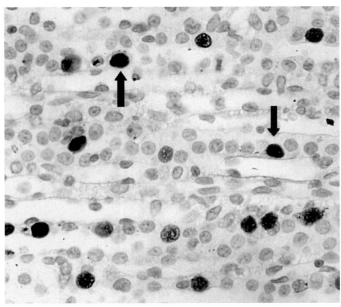

Fig. 12-3   Acute rat virus infection. Intranuclear rat virus antigen (arrows) in renal tubular epithelial cells detected by immunoperoxidase staining. (Modified from Jacoby et al., 1987 with permission from Archives of Virology.)

Fig. 12-5   Acute rat parvovirus infection. Viral DNA in an intestinal villus detected by *in situ* hybridization (arrow).

Gaertner et al., 1996; Kajiwara et al., 1996). This requirement helps to explain why prenatal infection is uncommon during natural outbreaks in breeding colonies.

The risk of transmission is prolonged by persistent infection in rats (Jacoby et al., 1991) and protracted environmental contamination. These properties have been clearly established for RV but should be assumed for all parvoviruses of rats. Susceptibility to persistent RV infection depends heavily on age and immunologic status. Infection in adult immunocompetent rats is usually eliminated within 4 weeks, whereas pre- or perinatal infection can lead to persistent infection lasting at least 6 months, despite the onset of antiviral humoral immunity (Gaertner et al., 1989, 1991, 1995; Jacoby et al., 1991). Persistently infected rats also can excrete virus for up to 3 months, and episodic excretion over longer periods cannot be ruled out (Jacoby et al., 1988). Passive immunization with RV immune serum protects rats from acute and persistent infection (Gaertner et al., 1991). Similarly, maternally acquired immunity appears to protect rats born to persistently infected dams (Jacoby et al., 1988). However, passive immunity is most effective if it is established prior to infection. Antibody administered even 1 day after exposure to virus provides only partial protection against infection. Thus, the potential for persistent infection develops soon after virus reaches its cellular targets. Immunodeficient (athymic) rats, unsurprisingly, are susceptible to persistent RV infection at any age (Gaertner et al., 1989). Administration of RV-immune serum to infected athymic adults

suppresses but does not eliminate virus, which can re-emerge after immunity decays (Gaertner et al., 1991). The environmental stability of RV has been demonstrated by Yang and co-workers, who found that virus remained infectious on plastic surfaces for at least 5 weeks (Yang et al., 1995).

Pathology and pathogenesis. Pioneering studies by Kilham, Margolis, Toolan, Cole, Nathanson, and others indicated that the predilection of RV and H-1 virus for mitotically active cells was responsible for pathogenic infections in fetal and infant rats, especially in the liver, central nervous system, and lymphopoietic tissues (Toolan et al., 1960; Kilham and Ferm, 1964; Kilham and Margolis, 1966; Margolis et al., 1968; Cole et al., 1970; Novotny and Hetrick, 1970). Virus infection in these tissues has been documented more recently by immunohistochemistry and *in situ* hybridization (Jacoby et al., 1987; Gaertner et al., 1993), indicating that oronasal exposure leads to viremic dissemination. Lesions caused by pathogenic strains are often widespread but most readily detected in liver, which is both susceptible to infection and mitotically active during infancy. Viral replication causes intranuclear inclusions (Fig. 12-6), followed rapidly by degenerative changes, including ballooning degeneration, intensified cytoplasmic eosinophilia, dissociation of histoarchitecture, nuclear pyknosis and karyorrhexis, and cell lysis. Associated lesions may include focal hemorrhage, bile stasis, and formation of blood-filled spaces resembling peliosis hepatis. These changes are reflected by gross lesions that

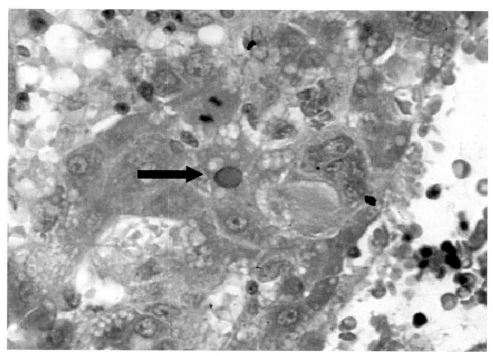

Fig. 12-6 Acute rat virus infection. Hepatocellular necrosis, hemorrhage, and an intranuclear hepatocytic inclusion (arrow).

may include icterus and yellow-red mottling of the liver accompanied by distorted lobular contours caused by necrosis. Attempts at repair result in increased mitotic activity, which may include the formation of polyploid giant cells. Persistent infection can incite chronic active hepatitis. Portal triads in affected livers may sustain chronic inflammation with infiltration by mononuclear cells, fibrosis, nodular hyperplasia, and biliary hyperplasia (Fig. 12-7). Corresponding gross lesions can include fibrosis and nodular distortion of one or more lobes.

Necrosis and hemorrhage can occur virtually anywhere in the central nervous system but often affect the cerebellum. Segmental or pancerebellar destruction of the external germinal layer (Fig. 12-8) leads to granuloprival cerebellar hypoplasia (Fig. 12-9). Hemorrhage may be significant in the cerebellum or other sites and may result in infarction and malacia. Lymphoreticular lesions also are typified by necrosis, which can affect thymus, lymph nodes, and spleen. Intrauterine uterine infection often results in necrosis of fetuses, placentas, and degrees of fetal resorption. Corresponding gross lesions may vary from a reduction to a total loss of viable fetuses and include segmental reddish-black discoloration of the uterus and accumulation of necrotic debris at placentation sites.

The prevalence of hemorrhagic lesions illustrates the importance of vasculotropism in pathogenic parvovirus infections. Hemorrhage during acute infection appears to result from viral-induced endothelial injury (Cole et al., 1970). Inclusion bodies, viral DNA, and viral antigens have been demonstrated in endothelium (Fig. 12-10) (Margolis and Kilham, 1970; Jacoby et al., 1987; Gaertner et al., 1993; Jacoby et al., 2000). Infection may cause cells to swell or lyse, thereby disrupting vascular integrity. Baringer and Nathanson (1972) found that platelet-fibrin aggregates attached preferentially to RV-infected cells and suggested that endothelial infection activated the clotting mechanism, resulting in infarction and hemorrhage. Additionally, RV infection of megakaryocytes with resulting thrombocytopenia can occur during acute infection (Margolis and Kilham, 1972). Endothelial infection may also exacerbate viremia and facilitate fetal infection through involvement of placental and fetal vasculature.

RV infects vascular and intestinal smooth muscle cells (SMC), which appear to be major sites of persistent infection (Fig. 12-11) (Jacoby et al., 2000). Persistent vascular SMC infection may be accompanied by perivascular mononuclear cell infiltrates, which most often are found in kidney and liver. Viral DNA can be demonstrated in adjacent vessels by *in situ* hybridization (Fig. 12-12) (Jacoby et al., 2000). The mechanisms of viral persistence are unknown, but recent studies suggest that SMC sequester non-replicating RV. RV replication also may continue among mitotically quiescent SMC (an observation that requires confirmation)

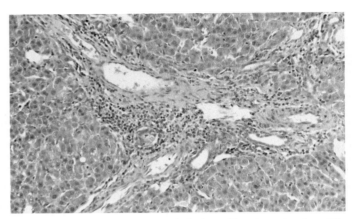

Fig. 12-7 Persistent rat virus infection. Chronic active hepatitis, fibrosis and biliary hyperplasia.

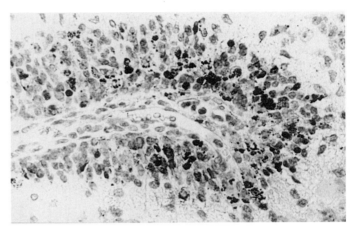

Fig. 12-8 Acute rat virus infection. Necrosis of the external germinal layer of the cerebellum. (Modified from Jacoby et al., 1987 with permission from Archives of Virology.)

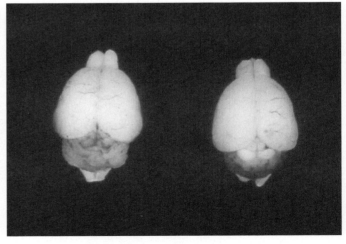

Fig. 12-9 Acute rat virus infection. Hypoplastic cerebellum **(right)** compared with normal cerebellum **(left)**. Note hemorrhage in affected cerebellum.

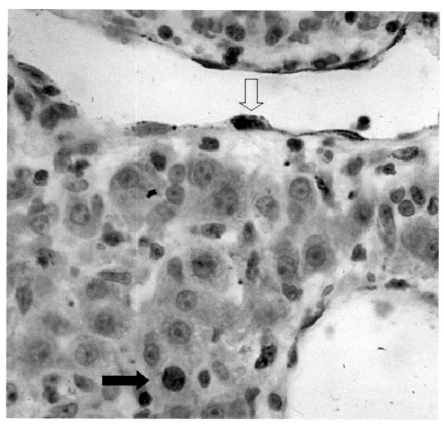

Fig. 12-10   Acute rat virus infection. Rat virus antigen in swollen hepatic endothelial cell (open arrow) and hepatocyte (solid arrow) detected by immunoperoxidase staining. (Modified from Jacoby et al., 1987 with permission from Archives of Virology.)

(Jacoby et al., 2000). Persistent infection of SMC does not cause prominent necrosis, but may provoke focal myolysis. Focal SMC infection also could exacerbate mitotic activity in attempts at repair and may provide additional targets for viral infection.

The role of host immunity in parvoviral infection is only partially understood. Although pre-existing humoral immunity can prevent infection, anti-viral antibody reduces but does not eliminate infection (Robey et al., 1968; Gaertner et al., 1991, 1995). The increased susceptibility of athymic rats confirms an important role for T lymphocytes in eliminating infection (Gaertner et al., 1989, 1991), whereas mononuclear cell infiltrates that develop during convalescent or persistent infection suggest at least an ancillary role for cell-mediated immunity.

In contrast to RV and H-1 virus, RPV is non-pathogenic even in infant rats. However, it does express similar tissue tropisms. In particular, it infects vascular endothelium and renal tubular epithelium (Fig. 12-4) as well as regional lymph nodes (Fig. 12-13) (Ball-Goodrich et al., 1998). However, unlike RV and H-1 virus, it also infects the lamina propria of intestinal mucosa (Fig. 12-5), a tropism comparable to that of MVM and mouse parvovirus (MPV). Therefore, mechanisms of productive RPV

replication and excretion may differ from those of the cytolytic parvoviruses. Further, there is no evidence that RPV is transmitted *in utero*. The duration of RPV infection and factors that may influence duration have not been studied adequately. Virus has been detected after seroconversion in both infant and adult rats for as long as 8 weeks (Ueno et al., 1997; Ball-Goodrich et al., 1998; Ueno et al., 1998), a finding consistent with persistent infection.

Diagnosis. The diagnosis of parvovirus infection depends on the complementary use of several methods. Clinical signs can be dramatic, but occur rarely. Standard pathologic examination can be useful, especially to detect inclusions and/or lesions compatible with pathogenic infection. However, diagnostic sensitivity and specificity is best achieved with tissue sections by immunostaining or *in situ* hybridization. Fixation of tissues in freshly prepared paraformaldehyde-lysine-periodate for not more than 16 hours preserves cytoarchitecture and minimizes denaturation of parvoviral antigens and DNA (Jacoby et al., 1987; Gaertner et al., 1993). Therefore, this method should be used prior to paraffin embedding to permit maximum flexibility in primary and follow-up examination of tissues or cells if parvovirus infection is suspected.

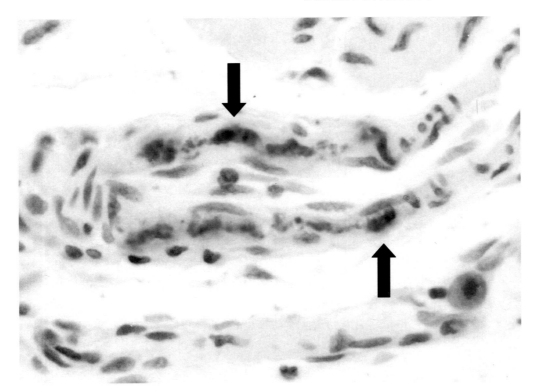

Fig. 12-11 Persistent rat virus infection. Viral DNA detected by *in situ* hybridization in nuclei of arteriolar smooth muscle cells (arrows). (Modified from Jacoby et al., 2000 with permission of Journal of Virology.)

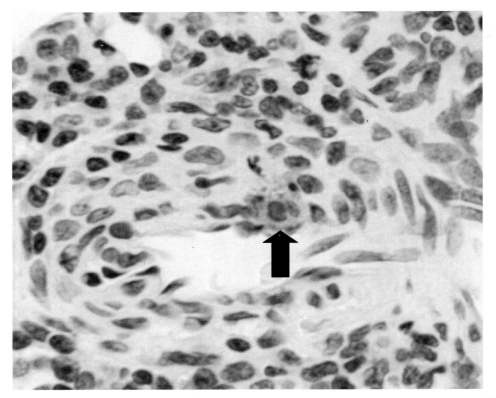

Fig. 12-12 Persistent rat virus infection. DNA detected by *in situ* hybridization in the nucleus of a smooth muscle cell of a renal arteriole (arrow). Note adjacent accumulations of mononuclear cells. (Modified from Jacoby et al., 2000 with permission of Journal of Virology.)

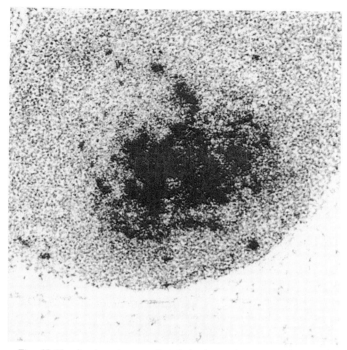

Fig. 12-13 Acute rat parvovirus infection. Viral DNA detected by *in situ* hybridization in the germinal center of a mesenteric lymph node.

Because infections are frequently asymptomatic, serology is essential for primary detection. Enzyme-linked immunosorbent assays (ELISAs) or immunofluorescence assays (IFAs) using virions or infected cells as generic antigens can be used to detect all known serotypes. Alternatively an ELISA using rNS-1 from MPV can detect infection caused by parvoviruses of rats because NS-1 is highly conserved among rodent parvoviruses (Riley et al., 1996). Follow-up testing can employ HAI to directly distinguish RV and H-1 virus infection. Additionally, sera positive by ELISA and/or IFA and negative by HAI would suggest RPV infection. However, the sensitivity of the generic assays is higher than that of HAI, so testing of sera with low concentrations of antibody could produce a false-negative HAI result. Neutralization serology also can be used to distinguish among serotypes, as all replicate in at least one established cell line. This option is comparatively expensive, time consuming, and technically more demanding, making it best suited to use in reference laboratories. The most promising emerging strategy for specific, sensitive cost-effective serologic detection employs recombinant, serotype-specific (capsid) ELISA antigens. They have been produced for RV and RPV as recombinant VP-2 (Ball-Goodrich et al., 2002). Addition of a corresponding H-1 antigen should be straightforward.

Although serologic testing indicates historical or contemporary exposure to virus, immunostaining and molecular diagnostics may be required to obtain evidence for active infection. Antisera against rat parvoviruses are available and can be applied to snap-frozen or aldehyde-fixed tissues. Serotype-specific antibodies are not yet generally available, and immunostaining results from experimental studies indicate that antibodies to NS-1 proteins are more sensitive than those against VP2 proteins. Thus, immunostaining alone does not currently provide the means to confirm serotype specificity. *In situ* hybridization can employ strand-specific probes to detect not only virus genome, but also replicative forms and mRNA indicative of active infection.

The rat antibody production (RAP) test, adapted from a method developed for detection of mouse viruses, can detect viral antigens in clinical specimens. Test tissues are homogenized and inoculated into non-immune pathogen-free rats to elicit antibodies against suspected viral agents. Testing requires that live rats be held in quarantine for several weeks to allow antibodies to develop, making it relatively laborious and expensive. Although, in practical terms, it also requires that infectious virus be present in the test inoculum, it does not inherently rule out seroconversions elicited by viral antigen that may be present in non-infectious samples.

Four polymerase chain reaction (PCR) assays are currently available to detect parvoviruses of rats. They are very sensitive and improved primers make them increasingly specific. One assay amplifies a region of the NS-1 gene, thereby detecting all serotypes, but not distinguishing among them. The other three amplify VP region sequences for RV, H-1 virus, or RPV, making them serotype-specific (Besselsen et al. 1995a,b; Yagami et al., 1995). They can be used to test tissues, excreta, or cells (e.g., tumor lines) for viral DNA.

Virus isolation can be accomplished by inoculation of the cell lines listed above or by explant culture, which is an amplification technique especially conducive to detecting small quantities of infectious virus, such as those occurring during persistent infection (Paturzo et al., 1987). Briefly, it entails culture of multiple small fragments of suspect tissue for about 3 weeks. Cultures are harvested and aliquots are inoculated into indicator cells, which are examined for viral antigen and/or CPE.

Differential diagnosis. Pathogenic parvoviral infections should be differentiated from chemical toxicity, neoplasia, trauma, genetically determined developmental abnormalities, or rotavirus infection. Reproductive failures or reduced litter size also can occur during Sendai virus or coronavirus infection or environmental insults.

Control and prevention. Control and prevention must take into account the ability of parvoviruses to persist in rats and in the environment, and their capacity for both pre- and postnatal transmission. Initial assessment of an outbreak should begin with quarantine of seropositive rooms, extensive serologic testing in each room, in adjacent or regional facilities, and sites to which potentially affected rats or rat tissues may have been transported, including research laboratories. Cell lines, transplantable tumors, and other biological products, which may be either a source of or exposed to infection, should be tested rapidly by molecular diagnostics, or the RAP method and contaminated items should be discarded. Strategies to eliminate infection depend on multiple factors, including the value and immunologic status of infected or exposed animals and the ability to replace them from specific pathogen-free sources. Selective culling can be attempted if animals are housed in isolation or individually ventilated cages, but one must consider risks posed by persistent infection and environmental contamination. Attempts to minimize further transmission should include use of barrier caging and change stations, limiting personnel traffic, barrier garb, and strict adherence to room order entry and sanitation during cage changing, including heat sterilization of soiled equipment and supplies. Directional air flow should be tested and affected rooms placed under negative pressure.

Rederivation can be accomplished by embryo transfer or cesarean section under stringently aseptic conditions. However, offspring should be tested for prenatal infection by contact transmission (which may include testing of foster dams) or pathologic examination of strategically chosen animals (e.g., at least one per litter). It is important to remember that the prenatal transfer of maternal antibody may render derived progeny at least transiently seropositive, yet protected. Therefore, they should be tested to ensure that antibody titers decay completely by 3 months after birth. A reversal of decreasing titer would suggest active infection.

Because pre-existing humoral immunity protects rats from infection, it can be exploited to help rederive breeding colonies. For example, progeny weaned from seropositive dams can be segregated from virus-free rats until they lose maternally derived immunity. However their dams should remain segregated if they will be used for further matings or be discarded as suspects for persistent infection. The use of contact sentinels to confirm the absence of infection in rederived rats is an optional step. This immunologically based strategy may take 3- to 4 months to complete, but research "down time" can be minimized by re-initiation of breeding among rederived rats while the decay of maternal immunity proceeds. Incremental serologic surveillance of facilities housing or previously housing infected rats is wise until confidence in elimination of infection is secure. Refer to Sections III A. and IV for additional information about the prevention and control of viral agents in rats.

Previously infected rooms should be evacuated and thoroughly disinfected, including physical removal of any debris or fomites from floors, walls, and ceilings. Detergent washes should be followed by disinfection with an oxidizing agent, such as chlorine dioxide, and a drying period of 2- to 3 days (Saknimit et al., 1988). Additional checks on the thoroughness of decontamination may include placement of sentinel rats and/or swabbing of surfaces for PCR analysis.

Interference with research. Although infection with pathogenic strains can cause illness and death, the more likely effects of parvovirus infection will be disruption of biological responses that rely on cell proliferation. RV is lymphocytotropic and can induce functional distortions in immune function. These include: interference with T cell responses to transplantable neoplasms (Campbell et al., 1977), provocation of autoimmune diabetes in diabetes-resistant rat strains by disruption of Th1-like T lymphocyte responses (Guberski et al., 1991; Brown et al., 1993; Ellerman et al., 1996), and diminished proliferative and cytolytic responses to alloantigens in mixed lymphocyte cultures (McKisic et al., 1995). Further, parvoviruses of rats are oncotropic, raising the possibility for disruption of tumor kinetics (Bergs, 1969). Finally, contamination of cell cultures or other biological products must be considered (Nicklas et al., 1993). It can lead to cytopathic effects or inadvertent transmission to rats if such products are used *in vivo*.

B. Rat Cytomegalovirus

Rat cytomegalovirus (RCMV) is the only known herpesvirus of rats. It has physical, chemical, and biological characteristics typical of the Herpesviridae, including the capacity for persistent infection of salivary glands (Priscott and Tyrell, 1982; Bruggeman et al., 1983; Bruggeman et al., 1985). Virus can be propagated in primary rat embryo fibroblasts, rat kidney cells, and hamster kidney cells. Infection of immunocompetent rats is asymptomatic. Experimental infection of rats immunosuppressed by whole body irradiation induced systemic disease with high mortality, but no studies of infection in naturally immunodeficient rats have been reported. RCMV appears to be infectious only for rats. There is no published evidence for natural cross-infection of rats with cytomegaloviruses of other species or for infection of other species with RCMV. Although infection has been detected in wild rats, the prevalence of RCMV in contemporary rat colonies is believed to be very low. However, epidemiologic data are inadequate to confirm this perception because serologic surveillance is performed sporadically. Nevertheless, prevention of infection in vivaria should focus on exclusion of wild rats, especially from support areas such as those used for supply and equipment storage.

Although acute infection may be generalized, persistent infection occurs in the salivary glands and may affect the lacrimal glands (Lyon et al., 1959). Virus can be excreted

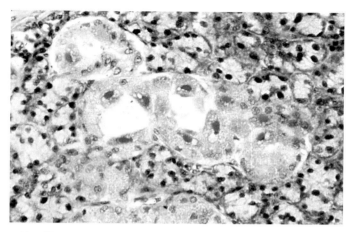

Fig. 12-14 Rat cytomegalovirus infection. Intranuclear inclusions in acinar epithelial cells of a submandibular salivary gland accompanied by mild interstitial inflammation.

in saliva for at least several months. Typical lesions include cytomegaly with formation of intranuclear inclusions, primarily in ductal epithelium (Fig. 12-14), accompanied by mild, non-suppurative interstitial inflammation. Infection can be detected from typical histopathologic changes, which are most apparent during the first few weeks of infection. An ELISA is available for serologic testing (Bruggeman et al., 1983). Virus can be isolated by inoculation of or cultivation with rat embryo fibroblasts (Rabson et al., 1969).

Effects on research. Reported effects on research include altered macrophage function, transient suppression of humoral immunity, transfer of infection during organ transplant studies, and exacerbation of collagen-induced arthritis (Bruggeman et al., 1985). Latent infection can be activated by immunosuppression. Conversely, RCMV replication can be augmented by allogeneic immune responses (Tamura et al., 1999). No significant effects on research have been reported from natural infections.

C. Poxviruses

An agent called Turkemia rodent poxvirus, distinct from ectromelia virus, has been reported in rats from Eastern Europe and the former Soviet Union (Iftimovici et al., 1976; Marrenikova and Shelukhina, 1976; Krikun, 1977; Marenikova et al., 1978; Kraft et al., 1982). Although infection can be asymptomatic, if clinical signs occur they resemble mousepox including the development of dermal lesions, tail amputation, and high mortality. Rats also develop lesions of the upper and lower respiratory tract. The latter can include interstitial pneumonia, pulmonary edema and hemorrhage, and pleural effusions.

D. Adenovirus

There is serologic and histologic evidence that rats are susceptible to infection with one or more adenoviruses antigenically related to mouse adenovirus-2 (Ward and Young, 1976). However, the causative agent(s) has not been isolated. Infection, to the best of available knowledge, is asymptomatic. Histologic changes are minimal and resemble those found in adenoviral infections of mice, with the sporadic development of intranuclear inclusions in enterocytes of the small intestine.

E. Papovavirus

A polyoma virus antigenically distinct from polyoma virus of mice has been detected in rats. It was found initially in athymic rats that developed a wasting disease accompanied, in a minority of animals, by pneumonia and sialoadenitis (Ward et al., 1984). Intranuclear inclusions consistent with papovavirus infection occurred in parotid salivary glands, and viral antigen was detected in salivary glands, respiratory airway mucosa, and kidney. Immunocompetent rats did not develop disease. More recent studies in athymic rats have detected intranuclear inclusions in alveolar lining cells and associated with interstitial pneumonia (Percy and Barthold, 2001).

III. INFECTION AND DISEASE CAUSED BY RNA VIRUSES

A. Coronaviruses

Agents. The Coronaviridae are enveloped, pleomorphic single-stranded RNA viruses with club-like projections (peplomers) radially arranged on their capsids. They replicate in the cytoplasm and are released by budding through the endoplasmic reticulum (Compton et al., 1993). They are widespread among mammals and birds, including laboratory rats and mice, but individual viruses are largely species-specific. Classification of the coronaviruses of rats has been somewhat confused by the context in which the two prototype strains were initially identified. In 1964, Hartley and co-workers detected antibodies in rat sera to a common coronavirus of mice, mouse hepatitis virus (MHV), which suggested that rats were susceptible to an antigenically related agent (Hartley et al., 1964). Parker and associates subsequently isolated a coronavirus from lungs of clinically normal rats which became widely known as *Parker's rat coronavirus* (Parker et al., 1970). Within several years, a second coronavirus was isolated by Bhatt and co-workers from rats with sialodacryoadenitis and was named *sialodacryoadenitis virus* (Bhatt et al., 1972). Subsequent studies showed that RCV and SDAV were similar antigenically and physicochemically, and that both induced sialodacryoadenitis and inflammation of the respiratory tract. Therefore, RCV and SDAV should be perceived as different strains of a coronavirus indigenous to rats and called, most inclusively, RCV (Percy and Williams, 1990).

Replication of single-stranded RNA viruses is more error-prone than that of DNA viruses and leads to emergence of divergent strains during natural infections. Thus, additional RCV isolates have been reported since the initial isolations by Parker and Bhatt. These include Japanese isolates—CARS (Kojima et al., 1980; Maru and Sato, 1982)—and U.S. isolates RCV-BCMM, RCV-W (Compton et al., 1999a), and RCV-NJ (Compton et al., 1999b).

RCV has a non-segmented, positive-sense RNA genome (Compton et al., 1993). It is sensitive to lipid solvents and relatively stable at acid pH. It can be stored at $-60°$ C for at least 7 years, but loses infectivity rapidly if stored at $-20°$ C (Bhatt et al., 1972; Jacoby et al., 1979). It replicates *in vitro* in primary rat kidney (PRK) cells in which most virus strains form multinucleate syncytia typical of coronaviral CPE (Bhatt et al., 1972). Virus can be detected antigenically in infected cells within 12 hours and CPE can develop within 24 hours. Established cell lines such as BHK-21, VERO, Hep-2, and NCTC 1469 resist RCV infection, but mouse L2 fibroblasts and sublines, as well as rat LBC cells and intestinal RCV-9 cells, also can be infected (Hirano et al., 1986; Percy et al., 1989; Percy et al., 1990b; Gaertner et al., 1996; Ohsawa et al., 1996). L2 cell sublines and LBC cells also have been used for plaque assays and plaque purification (Gaertner et al., 1993c).

RCV antigens appear to be similar among all isolates, although some also have a hemagglutinating surface protein (Gaertner et al., 1996a; Gagneten et al., 1996). Because RCV is antigenically related to MHV (Barker et al., 1994), MHV antigens are routinely used in serologic testing to detect RCV infection (Peters and Collins, 1981; Smith and Winograd, 1986; Percy et al., 1991).

Clinical signs. Clinical signs may vary with RCV strain, but their overall expression is characteristic of coronavirus infection. Mild infection can be clinically silent and revealed first by seroconversion of sentinel or index rats, but clinical signs are common in many outbreaks. They can occur in non-immune rats of any age, usually last about a week, and occur in various combinations and with varying severity (Jacoby et al., 1975; Bhatt and Jacoby, 1985). Early signs, especially in sucklings, often include squinting, photophobia, and lacrimation (Fig. 12-15). Older rats frequently develop audible sneezing and palpable enlargement of cervical salivary glands. Closer examination may also reveal serous nasal discharge and pawing at the

Fig. 12-15 Rat coronavirus infection. Photophobia and lacrimation due to keratoconjunctivitis and secondary to RCV infection of the lacrimal glands.

external nares resulting in wet forefeet. Red-tinged discharges containing porphyrin (chromodacryorrhea) may stain periorbital and perinasal skin (Fig. 12-16). Severe ocular inflammation is expressed initially as keratoconjunctivitis, which may resolve quickly or become chronic. Chronic ocular inflammation can lead to corneal opacities and ulcers, pannus, hypopyon, hyphema, and megaloglobus (Lai et al., 1976) (Figs. 12.17 and 12.18). Ancillary effects attributable to the discomfort of infection include transient anorexia and weight loss (Sato et al., 2001) and disruption of estrus (Utsumi et al., 1978, 1980). Additionally, birth rates and fertility in breeding

Fig. 12-16 Rat coronavirus infection. Periocular porphyrin staining (chromodacryorrhea).

Fig. 12-17 Rat coronavirus infection. Opaque cornea indicating keratitis.

Fig. 12-18 Rat coronavirus infection. Postinfectious hyphema and megaloglobus.

colonies have been observed to markedly decline during acute epizootics. Although morbidity may be high, RCV infection rarely causes mortality. Most rats recover, with the most frequent long term sequelae associated with ocular disease.

Epidemiology. RCV spreads rapidly among rats housed in "open" cages and among rooms in conventional facilities. Transmission appears to be primarily by direct contact with infected rats or by aerosol. However, limited tests have shown that RCV can remain infectious for at least 2 days when dried on plastic surfaces (Gaertner et al., 1993a). This suggests that fomite transmission also

may occur. Experimental studies have demonstrated the ease of horizontal transmission of RCV (LaRegina et. al., 1992), but there is no evidence for intrauterine transmission. Morbidity frequently reaches 100% among conventionally housed rats.

Colony infections often occur in either of two patterns: explosive epizootics among non-immune animals or endemic infection in rooms where young rats are protected transiently by maternal antibody. Non-immune, immunocompetent rats excrete virus in oronasal and lacrimal discharges for about 1 week (Bhatt and Jacoby, 1985). The onset of antiviral immunity terminates infection and is most easily detected by seroconversion. Immunity offers protection against recurrence of clinical disease, but protection is relatively short–lived (<6 months) and does not preclude transient re-infection, virus shedding, and a resultant increase in antibody titer (Percy et al., 1990a; Weir et al., 1990a; Percy et al., 1991; Percy and Scott, 1991). Re-infection of seropositive animals implies exposure to antigenically variant strains. This risk is inherent to coronavirus infections because of spontaneous mutations leading to the emergence of variant strains, especially during endemic infection. Infection of immunodeficient (e.g., athymic) rats is persistent (Weir et al., 1990b; Hajjar et al., 1991). Such rats can sustain chronic necrosis and inflammation in the salivary and lacrimal glands and the respiratory tract, as described below. Additionally, they can excrete virus for prolonged periods because their capacity for immune elimination of infection is impaired. Apart from oronasal and lacrimal excretion, immunodeficient rats can develop urinary tract infection, which makes transmission by contaminated urine likely. In contrast to the prolonged course of RCV infection in athymic rats, Hanna and co-workers (Hanna et al., 1984) found that immunosuppression with prednisolone had a negligible effect on the course of infection, and treatment with cyclophosphamide resulted in only a modest prolongation.

Although there is no firm evidence for genetic resistance and susceptibility to RCV infection (Percy et al., 1984), recent studies suggested that Lewis rats developed more severe clinical signs than other rat strains tested (Liang et al., 1995). There also is evidence that RCV strains may vary in virulence, but corresponding antigenic differences detected by plaque neutralization are not useful to predict the severity of clinical disease (Compton et al., 1999b).

Rats appear to be the sole hosts for naturally occurring infection, although there is experimental evidence that mice develop transient infection and interstitial pneumonia after inoculation with virus or contact exposure to infected rats (Bhatt et al., 1977; Barthold et al., 1990; La Regina et al., 1992). Mice experimentally inoculated with large doses of RCV can transmit infection to mice in close contact; however, RCV infection does not appear to spread mouse-to-mouse under natural conditions.

A national survey found that RCV infection was prevalent among institutional rat colonies in the United States (Jacoby and Lindsey, 1998) and similar results were reported from France (Zenner and Regnault, 2000). Commercial breeders have made significant strides in eliminating infection. Nevertheless, a recent break in preventive procedures at a commercial vendor resulted in widespread contamination of client colonies. This episode underscores the highly contagious character of RCV infection and the need for effective vigilance.

Pathology and pathogenesis. The lesions of RCV reflect acute inflammation of the salivary and lacrimal glands, lymphoid tissue, respiratory tract, and eye (Jacoby et al., 1975; Bhatt and Jacoby, 1977). They occur in various combinations and with varying severity. The submandibular and parotid salivary glands and adjacent cervical connective tissue are often edematous and the glands proper may be pale yellow to white, with red spots caused by vascular congestion (Fig. 12-19). The cervical lymph nodes also may be enlarged, edematous and contain multiple red foci. The exorbital and infraorbital lacrimal glands also may be pale and enlarged, whereas the Harderian lacrimal gland located on the dorso-lateral and posterior aspects of the eye globe may be flecked with yellow-gray foci. Reddish-brown mottling of the Harderian gland, however, indicative of porphyrin pigment, is a normal finding. The thymus may be small (as a result of stress atrophy) and the lungs may contain small grey foci. The external nares and eyes may reveal serous exudate and staining of adjacent skin by porphyrin pigment. Acute ocular inflammation includes corneal opacities and swollen conjunctiva, whereas advanced ocular disease may, as noted above, include pannus, hypopyon, hyphema, and megaloglobus (Lai et al., 1976).

Histologic lesions of RCV infection develop initially in the nasopharynx. They include necrosis of respiratory epithelium (Fig. 12-20) and submucosal glands with inflammatory edema of the lamina propria. Similar changes also affect the vomeronasal organ. The nasal meatus often contains neutrophils, cell debris, and mucus (Jacoby et al., 1975; Bhatt and Jacoby, 1977; Bihun and Percy, 1995). Mild, nonsuppurative tracheitis with focal necrosis of respiratory epithelium also may occur and hyperplastic peribronchial lymphoid nodules may develop. Pulmonary lesions are less frequent and are characterized by focal, interstitial pneumonia presenting as infiltration of alveolar septae by mononuclear cells and neutrophils (Fig. 12-21). More severe lesions may be accompanied by alveolar exudates containing exfoliated pneumocytes, macrophages, and edema fluid. Although such lesions are typically asymptomatic, Parker and colleagues demonstrated that RCV can induce lethal interstitial pneumonia in neonatal rats (Parker et al., 1970).

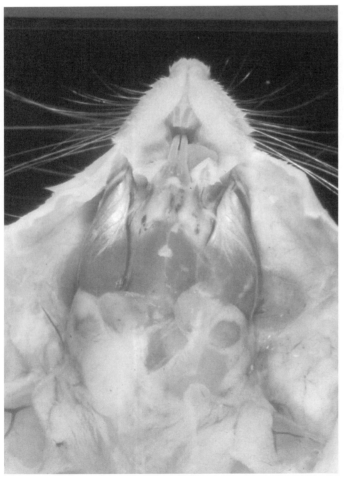

Fig. 12-19   Rat coronavirus infection. Pale, enlarged salivary glands and interstitial inflammatory edema.

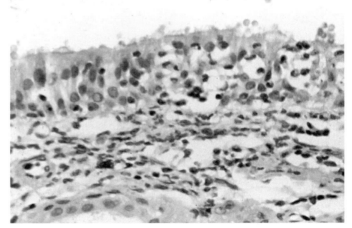

Fig. 12-20   Rat coronavirus infection. Necrosis of respiratory epithelium and inflammation of nasal mucosa.

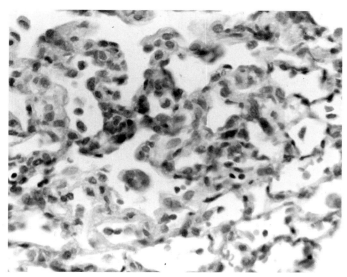

Fig. 12-21   Rat coronavirus infection. Interstitial pneumonia.

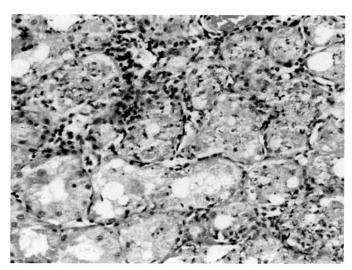

Fig. 12-23   Rat coronavirus infection. Necrosis in a Harderian gland.

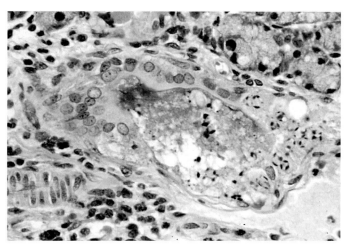

Fig. 12-22   Rat coronavirus infection. Necrosis of a salivary duct during early stages of infection. (Modified from Jacoby et al., 1975 with permission from Veterinary Pathology.)

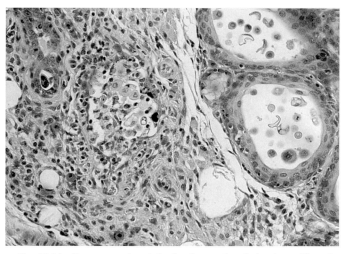

Fig. 12-24   Rat coronavirus infection in an athymic (rnu) rat. Chronic, active sialoadenitis with fibrosis in a submandibular salivary gland. (From Weir et al., 1990b, with permission from the American Association for Laboratory Animal Science.)

Salivary gland lesions occur in serous or mixed parenchyma and are, therefore, most easily visualized in the submandibular and parotid salivary glands, although smaller serous glands lining the oral cavity also are susceptible. Mucous salivary glands are not affected. Lesions begin as necrosis of ductular epithelium (Fig. 12-22), which progresses rapidly to diffuse acinar necrosis. Moderate to severe interstitial inflammatory edema also develops within glandular parenchyma and in periglandular connective tissues. The combined effects of necrosis and inflammation often lead to regional or panacinar effacement of affected glands. Lacrimal gland lesions develop in a similar pattern (Fig. 12-23).

Histologic lesions in immunodeficient (athymic) rats are characterized by chronic active inflammation in the respiratory tract, salivary glands (Fig. 12-24), and lacrimal glands. They represent chronic tissue damage caused by persistent infection concomitant with ineffective attempts by the host to eliminate virus and effect tissue repair. As noted above, viral antigen has been detected in the urinary tract of athymic rats, indicating that defective immunity leads to altered tissue tropism and the likelihood of prolonged urinary excretion (Fig. 12-25) (Weir et al., 1990b).

Immunohistochemistry has shown that the foregoing lesions, irrespective of immune status, result from initial

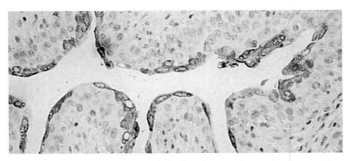

Fig. 12-25 Rat coronavirus infection in an athymic (rnu) rat. Viral antigen in epithelium of urinary bladder mucosa detected by immunoperoxidase staining.

infection of the respiratory tract, which extends to the salivary and lacrimal glands. It is not clear how this transition occurs. No viremic phase has been detected during experimental infections, but detection of viral antigen in salivary excretory ducts suggests that retrograde infection from the pharynx may occur. Retrograde infection of nasolacrimal ducts also could account for lacrimal gland infection.

Cervical lymph nodes often sustain focal necrosis and inflammatory edema succeeded by hyperplasia as immunity develops. Mild thymic necrosis with widening of interlobular septae is viewed as a non-specific response to the stress of infection.

Tissue repair in immunocompetent rats commences 5 to 7 days after infection begins. Reconstitution of respiratory mucosa and alveolar parenchyma is rapid and uneventful, although transient squamous metaplasia may occur in respiratory epithelium. Healing in salivary and lacrimal glands is characterized by prominent squamous

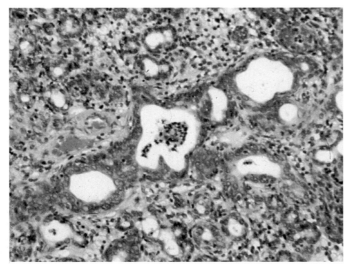

Fig. 12-26 Rat coronavirus infection. Squamous metaplasia in a submandibular salivary gland.

metaplasia of ductular epithelium, including the tubuloalveolar epithelium of the Harderian gland, and proliferation of hyperchromatic regenerating acinar cells (Fig. 12.26). Restoration of cytoarchitecture is substantially complete in 4 to 6 weeks, facilitated by survival of basement membranes, which provide an effective framework for parenchymal regeneration. Residual lesions include focal lymphocytic infiltrates and mild fibrosis.

Keratoconjunctivitis does not result from direct RCV infection of the eye. It is attributed to impeded tear production by compromised lacrimal glands resulting in keratitis sicca. Changes are characterized by focal or diffuse interstitial keratitis with superficial corneal ulceration and associated conjunctivitis. These lesions may resolve within 4 to 6 weeks without further complication or progress to permanent scarring or more severe sequelae in a small number of rats. Severe outcomes include transmural corneal ulceration, hypopyon, hyphema, synechia, and glaucoma with lenticular and retinal degeneration. Keratitis also may facilitate secondary bacterial infection and increase the severity of ocular lesions.

Diagnosis. Clinical signs, serology, lesions, immunohistochemistry, and virus isolation can all aid the diagnosis of RCV infection. However, the severity and prevalence of clinical signs can vary enough to caution against using them as the sole basis for diagnosis. Gross and microscopic lesions, particularly necrosis and inflammation of the salivary and lacrimal glands, are virtually pathognomonic for RCV infection, as they are not duplicated in other known diseases of rats. The course of infection and tissue repair is sufficiently prolonged (4–6 weeks) to facilitate detection of acute or reparative stages, provided the number of rats sampled is adequate. Because acute lesions develop within 5 days of exposure to virus and are often reflected in clinical signs, it is best to select animals for pathologic study that have active clinical signs or have been in recent contact with clinically affected animals. Immunohistochemistry to detect RCV antigens in aldehyde-fixed, paraffin-embedded tissues is useful to visualize virus in tissues collected during acute infection, as virus is eliminated rapidly with the onset of host immunity.

Serology is used routinely to detect or confirm infection. However, it generally takes 7 to 10 days after initial exposure to virus before an individual rat becomes seropositive. The preferred serologic tests are an immunofluorescent assay (IFA) using virus-infected cells or an ELISA (Peters and Collins, 1981; Smith and Winograd, 1986; Machii et al., 1988; Percy et al., 1991), both of which use MHV as antigen to exploit its cross-reactivity with RCV. Virus isolation is rarely used or needed but can be accomplished by inoculation of test specimens, such as clarified homogenates of cervical salivary gland into PRK cells or L2 or LBC cell lines (Bhatt et al., 1972;

Hirano et al., 1986; Hirano, 1990; Gaertner et al., 1996a; Hirano et al., 1995). Molecular probes for RCV have been developed and can be used to detect infection *in situ* or in tissue extracts. Reverse-transcriptase PCR techniques also have been used to identify cages housing infected rats rapidly–and prior to seroconversion (Compton et al., 1999b).

Differential diagnosis. Salivary gland lesions are virtually pathognomonic for RCV infection. Rat cytomegalovirus (II.B.) can cause mild, asymptomatic lesions of the salivary glands, but corresponding lesions are characterized by enlargement of ductal epithelial cells that may contain herpetic, intranuclear inclusions. Porphyrin-tinged naso-ocular discharges can occur when rats are subject to other stressors, or during other infectious diseases of the respiratory tract such as mycoplasmosis. Irritants such as ammonia can cause photophobia and lacrimation. Respiratory tract lesions due to RCV should be differentiated from those caused by Sendai virusa recently recognized and putative hantavirus-like respiratory virus of rats (Simmons and Riley, 2002), pneumonia virus of mice, or pathogenic bacteria such as *Mycoplasma pulmonis*. One cannot exclude, in this regard, the potential for RCV to facilitate or exacerbate secondary bacterial infection (Michaels and Myerowitz, 1983). Bacterial conjunctivitis can occur in rats, but RCV must be ruled out as an underlying cause.

Control and prevention. Effective use of barrier equipment and procedures is essential to control and prevent RCV infection, because it spreads rapidly and can remain endemic in any colony where non-immune rats are regularly added. However, rapid spread and limited duration also can be exploited toward control; that is, it is reasonable to expect that, in a given colony room, all rats should be infected and develop immunity within 3 to 5 weeks. This latter interval may be longer among rats housed in static isolation caging or in individually ventilated caging. Thus, strict quarantine of infected rooms is essential for at least 5 weeks, and preferably 6 to 8 weeks. Additionally, quarantined rooms should be kept under negative air pressure with respect to shared corridors and should be serviced last. Personnel entering barrier or quarantined animal rooms should wear protective garb and should not enter rooms housing uninfected rats without thorough personal sanitation steps, as fomite transmission of RCV can occur (La Regina et al., 1992). Cages should be serviced in HEPA-filtered stations and soiled equipment should be sent for sanitation by a predetermined and secure route, to prevent cross-infection. Finally, serologic testing should be intensified in adjacent colonies to determine if quarantine has been effective in containing the spread of infection. The quarantine period can be reduced by purposefully exposing all rats to infection, which will accelerate the development of immunity, provided that it will not interfere with research objectives. Production should cease in breeding colonies for about 6 weeks. After this interval, seropositive rats can be moved to a different room to resume breeding, because they should no longer be contagious (Brammer et al., 1993). Here, as well, purposeful exposure of breeding stock will accelerate colony-wide immunity, which also will lead to protection of normally susceptible litters by maternally acquired immunity. The foregoing strategies assume that attempts will be made to salvage infected animals. Under certain conditions, such as those applicable during pharmacologic testing, even transient infection and RCV seropositivity may not be tolerable and may require euthanasia of entire colonies.

Because RCV is relatively labile, routine disinfection will inactivate virus. Sanitation of caging and equipment should include thorough washing at 83°C, but need not require autoclaving. Similarly, room surfaces can be sanitized with standard disinfectants, such as chlorine dioxide, followed by enforced vacancy for 2 to 3 days (Saknimit et al., 1988).

Prevention of infection depends on effective surveillance as outlined in Chapter 16. Because infected rats are thought to be the major source of contamination, procurement from vendors with sound barrier housing and serologic monitoring programs is essential. This includes demonstrated attention to separation of functions, such as surgical manipulations, within vendor facilities. An effective quarantine program for biological materials for use in rats and for rats arriving from non-commercial vendors is also essential (Nicklas et al., 1993; Shek and Gaertner, 2002). Irrespective of source, it is prudent to decontaminate the external surfaces of shipping containers before unpacking animals.

Interference with research. RCV infection may hamper studies involving the respiratory system, salivary glands, lacrimal glands, the immune system, and the eye. This includes the distortion of physiologic responses or the interpretation of histologic changes. Furthermore, acutely infected rats may be at increased risk for death during inhalation anesthesia, as inflammatory exudate in the airways and enlarged salivary glands may impede normal respiration. Eye research can be compromised by the ocular manifestations of infection, a problem especially relevant to long-term toxicologic studies. Published reports include other diverse effects upon research, including altered feeding behavior (Sato et al., 2001), inhibition of phagocytosis and interleukin-1 production by pulmonary macrophages (Boschert et al., 1988), enhancement of nasal colonization with *Haemophilus influenzae* type b (Michaels and Myerowitz, 1983), depletion of salivary gland epidermal growth factor (Percy et al., 1988), and graft-versus-host disease in rats with transplanted bone marrow (Rossie et al., 1988). Finally, it is worth remembering that pneumotropic viruses can facilitate co-infection by opportunistic bacteria, as illustrated in the following description of Sendai virus infection. Thus, bacterial pneumonias of rats should be assessed for underlying coronaviral involvement, especially during active RCV infection.

B. Paramyxoviruses (Sendai Virus, Pneumonia Virus of Mice)

1. Sendai Virus

Sendai virus infection, with resultant necrosis and inflammation in the respiratory tract, was a major impediment to rodent-based research for many years. Its prevalence has abated owing to improved housing, husbandry, and serologic surveillance. Mice were the primary targets of Sendai virus infection. There are, however, several reports of natural outbreaks in rats. Because Sendai virus also is infectious for hamsters and guinea pigs, infection would place multiple species at risk.

Agent. Sendai virus is an enveloped, pleomorphic, spherical, or filamentous virus of the family Paramyxoviridae. It has a single-stranded RNA genome that encodes six proteins including a surface HN glycoprotein with hemagglutinin and neuraminidase moieties that are responsible for adsorption to host cells, and a surface F glycoprotein that facilitates cell fusion important to virus entry, expressed cytologically as syncytia formation (Brownstein, 1986). Sendai virus is also known as parainfluenza virus 1 and can be serologically differentiated from strains 2, 3 and 4. It grows well in embryonated eggs and in multiple established cell lines, including BHK-21 cells. Other biological and biochemical characteristics of Sendai virus have been described elsewhere (Brownstein, 1986).

Clinical signs. Infection in rats, in contrast to mice, is primarily asymptomatic. However, clinical signs have been observed during natural (Makino et al., 1972) and experimental (Castleman, 1983; Giddens et al., 1987) infection. These include dyspnea, anorexia, and ruffled pelage. Indirect clinical effects have been attributed to the general stress of infection and include fetal resorption, reduced litter size, and retarded growth (Coid and Wardman, 1971).

Epidemiology. Rats of all ages are susceptible to infection, but it is acute and self-limiting in immunocompetent animals. Elimination of virus coincides with the onset of antiviral immunity at about 1 week after initial exposure. Because infection is highly contagious and transmitted by aerosol, rapid spread is common. This effect has been documented by Makino and co-workers (et al., 1972) who observed infection in a colony of 500 breeding rats with subsequent outbreaks re-occurring at 8- to 10-month

intervals as newly susceptible rats emerged from the breeding program. Signs of respiratory disease and retarded growth among suckling rats occurred during each episode. Endemic infection is punctuated by fresh clinical outbreaks, typical for acute viral infections in which recovery leads to transient maternally derived immunity followed by renewed susceptibility. Antibody responses to virus remain high for at least several months after acute infection, but decline to low or undetectable levels within 9 months. In these contexts, the epidemiologic pattern and risks from Sendai virus infection in rats resemble those described for RCV infection. However, there is no specific evidence regarding long-term risks for re-infection with Sendai virus. Infection has not been described in immunodeficient rats, but one should assume, as for immunodeficient mice, that it would persist in individual animals and that the risks for transmission would be high. There is no evidence for prenatal infection.

Pathology and pathogenesis. Patterns of infection and lesions are similar in infant, weanling, and young adult rats, but viral clearance is slower in infants (Castleman, 1983; Castleman et al., 1987; Giddens et al., 1987). Thus, virus can be detected beyond the first week of infection, whereas older rats generally clear virus by 1 week. This difference is reflected in a slower onset of humoral anti-Sendai virus immunity in infant rats. Infection causes rhinitis with subsequent development of bronchiolitis and pneumonia. Grossly, affected portions of lung are dark red and consolidated. Otitis media may develop in some rats. Upper respiratory lesions begin within 4 days of exposure to virus, but viral antigen can be detected in respiratory epithelium within 1- to 2 days. Early lesions are characterized by epithelial necrosis with mucosal infiltration by neutrophils and lymphoid cells, all of which increase in severity during the ensuing week. Bronchiolitis and pneumonia may begin as early as 2 days after exposure to virus and feature necrosis and inflammation (Fig. 12-27). Bronchiolar epithelium undergoes necrosis and erosion, which may be accompanied by epithelial hyperplasia. Viral replication occurs in type I and type II pneumocytes and in macrophages, resulting in necrosis and interstitial infiltration by neutrophils, macrophages, and lymphocytes. Pneumonic lesions are most severe at 5 to 8 days, depending on the age of affected rats, with lesions peaking earlier in weanling or young adult rats than in infant rats. Resolution of pneumonia in immunocompetent rats begins within 2 weeks after exposure and is largely complete within 3 weeks. Residual lesions include perivascular and peribronchial accumulations of mononuclear cells, which may last up to several weeks. Some rats may develop interstitial fibrosis.

Diagnosis. Because clinical signs occur sporadically, they are not a reliable indicator of infection. However, Sendai virus infection should be considered if unexpected

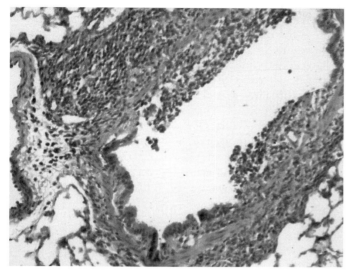

Fig. 12-27 Sendai virus infection. Bronchopneumonia with necrosis of bronchial epithelium.

morbidity or deaths occur during procedures that impact the respiratory tract, or if unexplained distortions occur in immune responses. Serologic testing, accompanied or followed by pathologic assessment, is the primary means for detecting infection. Sensitive IFA and ELISA tests are available for detection of antiviral antibodies (Smith, 1983; Wan et al., 1995). Histopathologic examination of the upper and lower respiratory tracts should be performed in clinically affected animals, on those that have recently seroconverted, or even in asymptomatic rats that have been exposed to affected animals. Immunohistochemistry to detect viral antigen can be performed on aldehyde-fixed, paraffin-embedded tissues harvested during acute infection. Respiratory epithelial cells and alveolar pneumocytes should be examined for intracytoplasmic viral antigen. Virus can be isolated by inoculation of nasobronchial washes or clarified lung homogenates into embryonated eggs, or preferably, BHK-21 cells. Cultures should be checked for viral antigen by 1 week after inoculation, whether or not CPE has developed.

Differential diagnosis. Sendai virus infections must be differentiated from those caused by RCV, PVM, and rat respiratory virus (RRV) (putative hanta-like virus). RCV pneumonia is primarily interstitial and infection also causes sialodacryoadenitis. RRV also appears to target the alveolar parenchyma, so rhinotracheitis and bronchitis are not characteristic features. PVM causes pulmonary vasculitis and interstitial pneumonia. Sendai virus pneumonia can be confused with early stages of murine respiratory mycoplasmosis, which it also can exacerbate. However, the latter often progress to chronic respiratory disease with bronchiectasis and bronchiolectasis. Other bacterial pneumonias also can be considered in the

differential diagnosis and are described in the chapter on bacterial disease by Weisbroth, Kohn, and Boot. Because Sendai virus infection has been associated with fetal resorption and retarded growth, it must be differentiated from parvovirus infection.

Control and prevention. Strategies for control and prevention are equivalent to those described for RCV infection. Because mice, rats, hamsters, and guinea pigs are susceptible to Sendai virus, precautions should be taken to prevent cross-infection. The essential components of containment are quarantine of infected rooms, strict sanitary measures, and tightly controlled room servicing procedures and schedules. Because Sendai virus is highly labile in the environment, clean up and disinfection should follow recommendations described for RCV. Transport of rats to laboratories should be discouraged. As with RCV, the primary source of infection is likely to be infected animals, so strict control of animal entry, including vendor surveillance and quarantine and testing of animals from non-commercial sources is essential.

Interference with research. Because Sendai virus infects the respiratory tract, experimental procedures ranging from anesthesia to inhalation toxicology can be affected. This includes the distortion of physiologic responses or the interpretation of histologic changes. There also is evidence that systemic immune responses can be transiently impaired (Garlinghouse and van Hoosier, 1978). Finally, Sendai virus should be considered a potential copathogen or aggravating factor in other respiratory infections of rats, as demonstrated by its aforementioned relationship with murine mycoplasmosis (Schoeb et al., 1985).

2. Pneumonia Virus of Mice

Agent. Pneumonia virus of mice (PVM) is a member of the genus *Pneumovirus* in the family Paramyxoviridae. It is most widely known as a cause of rhinotracheitis and interstitial pneumonia in mice, but also is naturally infectious for rats.

Clinical signs. Infection in rats is asymptomatic and thus detected primarily by serology, for which sensitive ELISA and IFA tests are available (Descoteaux et al., 1981; London, et al., 1983; Smith, 1983). PVM infection is mildly pathogenic, so diagnosis can be supplemented by pathologic assessment during acute infection.

Epidemiology. Rats, mice, hamsters, gerbils, and possibly guinea pigs and rabbits are susceptible to infection. A recent survey of major biomedical research centers reported that approximately 20% of rat colonies had some evidence of PVM infection (Jacoby and Lindsey, 1998).

Pathology. The following picture has emerged from anecdotal observations and a brief report of experimental infection (Votsberger et al., 1982; Brownstein, 1985). Although rhinotracheitis may occur, the most prominent lesions are perivascular mononuclear cell infiltrates, non-suppurative interstitial pneumonia and hyperplasia of bronchial-associated lymphoid tissue, all of which may last up to several weeks. The vascular orientation is a point of differentiation from pneumonia caused by Sendai virus or rat coronavirus.

Control and prevention. Prevention, control, and elimination follow principles described above for rat coronaviruses.

Interference with research. The risks to research pertain to PVM's potential capacity as a co-pathogen and to transmission to other susceptible species, notably mice. It is worth noting, in these contexts, that PVM can exacerbate pneumocystosis in immunocompromised (SCID) mice (Bray et al., 1993). It is not known if immunodeficient rats are susceptible to this effect.

C.   Hantaviruses and Rat Respiratory (Hanta-like) Virus

Agents. Hantaviruses are tri-segmented, negative-sense, single-stranded RNA viruses of the genus *Hantavirus* in the family Bunyaviridae (Nichol, 2001). There are more than 20 known strains, each of which is closely associated with a rodent or insectivorous reservoir host (Schmaljohn and Hjelle, 1997; Simmons and Riley, 2002). Several hantavirus strains, including Seoul virus, are indigenous to *Rattus norvegicus*. Black Creek Canal virus, a strain that is highly virulent for humans, has been isolated from cotton rats (*Sigmodon hispidus*) and implies that extreme caution should be employed if wild-caught cotton rats are used for research. In fact virtually all hantaviruses are known or have the potential to be zoonotic for man with varying pathogenicity ranging from asymptomatic to lethal infection. A newly recognized agent, named rat respiratory virus (RRV), has been detected in laboratory rats with interstitial pneumonia (Simmons and Riley, 2002). Serologic tests suggest that it is related to the hantaviruses.

Clinical signs. Rodent hantavirus infections are typically asymptomatic but persistent, a condition attributed to co-evolution of hantaviruses with their respective hosts.

Epidemiology. Rat-to-rat transmission occurs through direct contact and bite wounds, although indirect contact by fomites contaminated with excreta or secreta is possible. Recent *in situ* hybridization studies have detected persistent infection in kidney, exocrine pancreas, salivary gland, and skin—sites conducive to secretion/excretion of infectious virus (Compton et al., 2003). Further, one must consider the possibility for transmission by contaminated transplantable rat tumors. Hantavirus infection should be considered persistent in individual rats

despite antiviral immunity (Meyer and Schmaljohn, 2000). The mechanism(s) of persistence is not well-understood, but prolongs risks for transmission. The distribution of rat-associated infection is virtually worldwide and has included zoonotic infections among laboratory animal workers (Desmyter et al., 1983; Lloyd et al., 1984; Lloyd and Jones, 1986; LeDuc, 1987; Tsai, 1987). As noted above, human infection has a variety of clinical outcomes ranging from asymptomatic infection to lethal hemorrhagic fevers (Hart and Bennett, 1999). The prevalence of virulent and/or zoonotic hantaviruses in laboratory rats is extremely low. Early assessments of RRV infection indicate a prevalence of approximately 8% among laboratory rats and more than 20% among laboratory mice (Simmons and Riley, 2002).

Pathology. Most studies of hantavirus pathology in rats have utilized experimental infection with Asian isolates such as Seoul virus (Lee et al., 1986; Compton et al., 2003). Acute hantavirus infection has a widespread tissue distribution and a predilection for endothelial cells, as well as glandular and renal epithelium. Infection does not cause significant tissue damage, but can provoke mononuclear cell infiltrations in affected tissues, including lung, salivary gland, pancreas, and kidney, especially during persistent infection. Pancreatic lesions may include insulitis with at least transient elevations in blood glucose (Compton et al., 2003). As noted above, RRV infection appears to cause interstitial pneumonia.

Diagnosis. Diagnosis relies primarily on detection of specific anti-hantavirus antibodies by IFA, with an ELISA serving primarily as a back-up test (Lee et al., 1999). Immunostaining of tissue sections or molecular techniques (*in situ* hybridization or RT-PCR) also can be used; however, primer sets are not readily available for all strains. Further, RT-PCR is complicated by heterogeneity inherent to hantavirus genomes. Hantaviruses can be propagated in cell culture, with Vero-6 cells being the preferred substrate, but replicate slowly and with low yield.

Attempts at virus isolation should be reserved for only the most experienced and well-equipped laboratories, because of technical and safety concerns.

Differential diagnosis. Interstitial pneumonia, which may be caused be RRV, must be differentiated from other causes of similar lesions including rat coronavirus, Sendai virus, and PVM.

Control and prevention. Serologic monitoring is the key to prevent entry and spread of hantavirus infection. Infected or exposed rats should be placed under strict quarantine, and attempts to speciate the causative agent by a qualified laboratory should be undertaken as a means to assess zoonotic risk. Thus, the Asian hantaviruses present a clear zoonotic hazard, whereas there is currently no evidence for zoonotic transmission of RRV. Confirmation of a zoonotic strain should provoke depopulation of affected and exposed animals and thorough decontamination of facilities, equipment, and supplies. Personnel working in affected or potentially affected areas should don full protective gear, including ventilated hoods. If the potential for infection has been identified, the safest action is to notify and engage institutional safety officers. It also is prudent to assess rat-derived biological materials obtained from countries where hantavirus infection is endemic.

Effects on research. Zoonotic threats from virulent strains are the major impediment to research. As noted above, infection of rats with Seoul virus can cause insulitis and hyperglycemia. There is no evidence for or against the possibility that interstitial pneumonia associated with RRV alters pulmonary function or susceptibility to opportunistic respiratory infections.

D. Rotavirus and Reovirus

A rotavirus has been incriminated as the cause of infectious diarrhea of infant rats (IDIR), a non-lethal condition that resembles rotavirus-caused epidemic diarrhea of infant mice (EDIM) (Vonderfecht et al., 1984; Vonderfecht, 1986). The causative agent is an antigenically distinct group B rotavirus that may be of human origin. Infected suckling rats develop clinical signs within a day or two after initial exposure. These include diarrhea for up to 1 week accompanied by cracking and bleeding of the perineal region, dry, flaky skin, and transient growth retardation. Although rats of all ages are susceptible to infection, resistance to clinical disease develops at about 2 weeks of age. Gross lesions are limited to the gastrointestinal tract. The stomach usually contains milk. The ileum and colon may be distended with yellow-green fluid and gas. Histologic lesions are most prominent in the ileum and are characterized by enterocytic syncytia, which may contain intracytoplasmic inclusions, enterocytic necrosis, and attenuation of villi. Viral antigen and rotaviral particles may be found in affected enterocytes. Diagnosis rests on characteristic clinical and pathologic changes accompanied by detection of rotaviral antigens and/or particles. Serologic assays have not been developed. Differential diagnoses should include clinically apparent infection by pathogenic enterotropic bacteria. The significance of IDIR pertains primarily to transient growth retardation and the possibility that it originated as a human infection.

Margolis and Kilham have induced prenatal reovirus 3 infection in fetal rats (Margolis and Kilham, 1973) and there is anecdotal evidence that rats can develop antibodies to reoviruses, but there are no documented reports of natural reovirus infection. Because reovirus antigens are fairly widespread in nature, it is decidedly unclear whether such seroconversions are due to active infection or exposure to inert reovirus antigens from sources such as plant matter.

E. Picornaviruses

Two types of picornavirus infection have been linked to rats. Theiler's mouse encephalomyelitis virus (TMEV) is a member of the *cardiovirus* genus of the family Picornaviridae, the natural host for which is the mouse (Jacoby et al., 2002). Anecdotal observations indicate that rats can develop antibodies to TMEV, but no associated disease or lesions have been found. In historical perspective, a TMEV-like agent called MHG virus was isolated from rats and infected intestine, lung, and brain after experimental inoculation. A second type of cardiovirus genetically distinct from TMEV has been isolated from rats and designated rat cardiovirus (Ohsawa et al., 2003). It does not appear to be pathogenic after experimental inoculation.

IV. COMMENTS ON DETECTION, DIAGNOSIS, RISK-ASSESSMENT AND DECISION-MAKING

Contemporary biomedical research should use rats free of adventitious viruses with the potential to cause illness or distort research results. Validation is based primarily on the absence of antibodies to common viruses. Broad agreement among experts in the field regarding antigen panels, sampling protocols, and serologic methods has not yet materialized. Therefore, definitions of health status vary and must be established for each set of housing and husbandry conditions. This step is essential to protect the integrity of individual colonies, to reduce risks for spread of infection during transport or exchange of rats among laboratories, and to document the virologic status of rats in research grant applications and in scientific reports.

A. Detection and Diagnosis

Monitoring for viral infections should cover at least three venues: animals in established breeding and experimental colonies, animals held in entry quarantine, and animal tissues and products destined for *in vivo* use. For established colonies, monitoring should be tailored to local conditions. Effective monitoring should encompass sampling on a pre-arranged schedule, which can be intensified if evidence or suspicion of viral infection emerges. Although sampling should utilize index animals, such as retired breeders, whenever possible, the most widely accepted approach to assess contemporary conditions in a colony is serologic monitoring of strategically placed and exposed sentinels. Sentinels must be immunocompetent, virus-antibody free, and protected from exposure to infectious agents during shipment. The number of sentinels used should reflect assessment of risks to a colony, as illustrated in Chapter 16. Cages of sentinels should be placed, by rack and room, to facilitate controlled exposure to potential sources of infection. One popular configuration for rats housed in conventional wire-top caging is to place sentinels on the lowest shelf of a rack to maximize natural exposure to bedding expelled from overlying cages. Additionally, samples of soiled bedding collected during routine cage changing (e.g., weekly) can be placed in sentinel cages to increase the sensitivity of detection. This strategy also helps to compensate for the use of isolation and individually ventilated caging, which limits airborne transmission of viral agents and by doing so, also may inhibit early detection of infection. However, limited durability of excreted virus in soiled bedding and the potential that small amounts of virus could be diluted to non-immunogenic levels in pooled bedding samples also must be considered. Targeted selection of soiled bedding samples also can be helpful to trace infection to a sector of a suspect rack or shelf. For this procedure to work satisfactorily, cage positions should not be changed during the testing interval.

Sampling intervals for sentinel rats generally range from 30 to 90 days, but, minimally, must be long enough to complete desired exposure and allow opportunity for seroconversion, which typically occurs by 2 weeks after an immunogenic exposure. Sentinel animals can be sampled repeatedly if they remain seronegative, but should be replaced if they develop antibodies. As a general rule, sentinels, regardless of serologic status, should be replaced every 6 to 12 months to permit incremental assessment, such as histopathology.

Molecular techniques can be used selectively to confirm serologic results. For example, PCR testing of excreta, soiled bedding, equipment, or surfaces can be used in lieu of virus isolation to help determine if infection initially detected by serology is currently active. Additionally, there is preliminary evidence from studies with mouse coronavirus that PCR sampling of exhaust air manifolds can provide a rapid, "realtime" overview of infection among individually ventilated cage racks (SR Compton, personal communication); however, it remains to be seen whether this technology is applicable to viruses of rats.

Sampling of quarantine arrivals, which is likely to increase with the wider use and exchange of genetically altered rats, should be based on the same sampling principles outlined above. Depending on source, age, gender, animal number and behavioral considerations, sentinel exposures can use soiled bedding or direct placement with index animals. Because cohorts of imported rats are generally small, sentinel exposure for 30 days should be adequate to elicit potential seroconversion. Sampling of animal products destined for *in vivo* use, such as transplantable tumors or blood products, is essential for a sound monitoring program. It also is important, in this regard, to counsel investigators about the hazards of introducing untested biological products into rats. Aliquots can be tested by inoculation of susceptible cell lines or by molecular methods. The aforementioned RAP test, although apparently sensitive, tends to be costly, time consuming, and likely to be replaced by molecular methods.

Because viral infections of rats can spread insidiously, early detection and epidemiologic "staging" should employ a detection matrix that includes clinical observation, appropriate sampling, and sensitive and specific diagnostic testing. The value of serology to detect infection and determine prevalence and incidence is obvious. As noted in previous sections of this chapter, sensitive and specific tests are available for the most common viruses of laboratory rats. We re-emphasize, however, that the number of test animals should be large enough and the duration of exposure long enough to maximize the likelihood of detecting infection in a room or colony. Further, the clinician/pathologist should have high confidence that the testing laboratory has the testing conditions and expertise to produce incontrovertible results.

Clinical signs are helpful if they are comparatively specific, such as the cervical and ocular manifestations of coronavirus infection. However, non-specific signs, such as inappetence, altered reproduction, problematic anesthesia, distorted immunity, also may signal underlying viral infection.

Exposed rats, and especially those with clinical morbidity, should be subjected to pathologic examination, beginning with a thorough necropsy and including proper preservation of tissues for histopathology, collection of serum for antibody testing, and selection of tissues for potential molecular and/or virologic analysis. Tissue selection and preservation should be guided by a differential diagnosis emerging from clinical and gross pathology examinations complemented by a sound working knowledge of viral infections of rats. Such knowledge should account for the possibility that signs and gross lesions will be absent. Table 12-1 summarizes current recommendations for tissue collection, preservation, and examination. As noted earlier, aldehyde fixation followed by embedding in paraffin wax is generally suitable for routine histopathology. However, we recommend fixation for approximately 16 hours in freshly prepared paraformaldehyde-lysine-periodate (McClean and Nakane, 1974) if immunohistochemistry and/or *in situ* hybridization are contemplated, because it provides improved stabilization of viral antigens and nucleic acids. We also recommend freezing selected tissues at −60° C. These samples can be used for virus isolation, molecular testing, or antibody production testing, as initial results warrant. Careful attention to minimize contamination during collection of such samples is critical, especially if the option for PCR testing is to be preserved. It also is prudent to complement examinations for potential viral infection with other microbiological testing to assess the possibility for dual infections or emergence of opportunistic bacterial infections secondary to viral infection.

B. Risk-Assessment and Decision-Making

If evidence of viral infection is obtained, the decision to take corrective action should be based on several key questions: Is the infection a threat to animal or human health? Is the infection a threat to research at the affected or collaborating sites? What are the options for control and elimination? Are the options epidemiologically effective, minimally disruptive to research, and fiscally sound? What are the options to prevent reinfection? These questions are best addressed through a thorough understanding of the biology and pathobiology of the offending agent and the functional and structural characteristics of the affected vivarium. Characteristics such as mode of spread, environmental stability, duration, and virulence in immunocompetent compared with immunodeficient hosts and known effects on research can lead to well-established actions for containment and elimination that have been illustrated for specific agents in this chapter. Further, working knowledge of viral infections can facilitate extrapolation of what is known to what may be conjectural due to lack of specific data. For example, the predilection of parvoviruses for mitotically active cells should be a generic factor in considering the potential impact of parvovirus infection on research. However, decision-making is likely to become more problematic as the use of genetically altered rats increases. Rats with either targeted or serendipitous alterations in resistance to infection may have correspondingly altered responses to viral infections. This potential is illustrated by cryptic immunodeficiencies in laboratory mice that can occur as unintended sequelae of genetic manipulation. Therefore, it will be important to consider phenotype as integral to risk assessment, and to broaden clinical, epidemiologic, and pathologic conception of the "textbook" spectra of infection.

REFERENCES

Ball-Goodrich, L.J., Johnson, E.A., and Jacoby, R.O. (2001). Divergent replication kinetics in two phenotypically different parvoviruses of rats. *J. Gen. Virol.* **82**, 537–546.

Ball-Goodrich, L.J., Leland, S.E., Johnson, E.A., Paturzo, F.X., and Jacoby, R.O. (1998). Rat parvovirus type 1: the prototype for a new rodent parvovirus serogroup. *J. Virol.* **72**, 3289–3299.

Ball-Goodrich, L.J., Paturzo, F.X., Johnson, E.A., Steger, K., and Jacoby, R.O. (2002). Immune responses to the major capsid proteins during parvovirus infection of rats. *J. Virol.* **76**, 10044–10049.

Baringer, J.R., and Nathanson, N. (1972). Parvovirus hemorrhagic encephalopathy of rats: electron microscopic observations of the vascular lesions. *Lab. Invest.* **27**, 514–522.

Barker, M.G., Percy, D.H., Hovland, D.J., and MacInnes, J.I. (1994). Preliminary characterization of the structural proteins of the coronaviruses, sialodacryoadenitis virus and Parker's rat coronavirus. *Can. J. Vet. Res.* **58**, 99–103.

Barthold, S.W., deSouza, M.S., and Smith, A.L. (1990). Susceptibility of laboratory mice to intranasal and contact infection with coronaviruses of other species. *Lab. Anim. Sci.* **40**, 481–485.

Bergs, V.V. (1969). Rat virus-mediated suppression of leukemia induction by Moloney virus in rats. *Cancer Res.* **29**, 1669–1672.

Besselsen, D.G., Besch-Williford, C.L., Pintel, D.J., Franklin, C.L., Hook, R.R., Jr., and Riley, L.K. (1995a). Detection of newly recognized rodent parvoviruses by PCR. *J. Clin. Microbiol.* **33**, 2859–2863.

Besselsen, D.G., Besch-Williford, C.L., Pintel, D.J., Franklin, C.L., Hook, R.R., Jr., and Riley, L.K. (1995b). Detection of H-1 parvovirus and Kilham rat virus by PCR. *J. Clin. Microbiol.* **33**, 1699–1703.

Bhatt, P.N., and Jacoby, R.O. (1985). Epizootiological observations of natural and experimental infection with sialodacryoadenitis virus in rats. *Lab. Anim. Sci.* **35**, 129–134.

Bhatt, P.N., and Jacoby, R.O. (1977). Experimental infection of adult axenic rats with Parker's rat coronavirus. *Arch. Virol.* **54**, 345–352.

Bhatt, P.N., Jacoby, R.O., and Jonas, A.M. (1977). Respiratory infection in mice with sialodacryoadenitis virus, a coronavirus of rats. *Infect. Immun.* **18**, 823–827.

Bhatt, P.N., Percy, D.H., and Jonas, A.M. (1972). Characterization of the virus of sialodacryoadenitis of rats: a member of the coronavirus group. *J. Infect. Dis.* **126**, 123–130.

Bihun, C.G., and Percy, D.H. (1995). Morphologic changes in the nasal cavity associated with sialodacryoadenitis virus infection in the Wistar rat. *Vet. Pathol.* **32**, 1–10.

Boschert, K.R., Schoeb, T.R., Chandler, D.B., and Dillehay, D.L. (1988). Inhibition of phagocytosis and interleukin-1 production in pulmonary macrophages from rats with sialodacryoadenitis virus infection. *J. Leukoc. Biol.* **44**, 87–92.

Brammer, D.W., Dysko, R.C., Spilman, S.C., and Oskar, P.A. (1993). Elimination of sialodacryoadenitis virus from a rat production colony by using seropositive breeding animals. *Lab. Anim. Sci.* **43**, 633–634.

Bray, M.V., Barthold, S.W., Sidman, C.L., Roths, J., and Smith, A.L. (1993). Exacerbation of *Pneumocystis carinii* pneumonia in immunodeficient (SCID) mice by concurrent infection with a pneumovirus. *Infect. Immun.* **61**, 158–1588.

Brown, D.W., Welsh, R.M., and Like, A.A. (1993). Infection of peripancreatic lymph nodes but not islets precedes Kilham rat virus-induced diabetes in BB/Wor rats. *J. Virol.* **67**, 5873–5878.

Brownstein, D.G. (1985). Pneumonia virus of mice infection, lung, mouse and rat. *In* "Monographs on Pathology of Laboratory Animals: Respiratory System." (T.C. Jones, U. Mohr, and R.D. Hunt, eds.), pp 206–210, Springer-Verlag, New York.

Brownstein, D.G. (1986). Sendai virus. *In* "Viral and Mycoplasmal Infections of Rodents: Effects on Biomedical Research." (P.N. Bhatt, R.O. Jacoby, H.C. Morse, and A.E. New, eds.), pp.37–61. Academic Press, Orlando, Fl.

Bruggeman, C.A., Debie, W.M., Grauls, G., Majoor, G., and van Boven, C.P. (1983). Infection of laboratory rats with a new cytomegalo-like virus. *Arch. Virol.* **76**, 189–199.

Bruggeman, C.A., Meijer, H., Bosman, F., van Boven, C.P.A. (1985). Biology of rat cytomegalovirus infection. *Intervirol.* **24**, 1–9.

Campbell, D.A., Staal, S.P., Manders, E.K., Bonnard, G.D., Oldham, R.K., Salzman, R.K., and Herberman, R.B. (1977). Inhibition of *in vitro* lymphoproliferative responses by *in vivo* passaged rat 13762 mammary adenocarcinoma cells. II. Evidence that Kilham rat virus is responsible for the inhibitory effect. *Cell. Immunol.* **33**, 378–391.

Castleman, W.L. (1983). Respiratory tract lesions in weanling outbred rats infected with Sendai virus. *Am. J. Vet. Res.* **44**, 1024–1031.

Castleman, W.L., Brudnage-Anguish, L.J., Kreitzer, L., and Neuenschwander, S.B. (1987). Pathogenesis of bronchiolitis and pneumonia induced in neonatal and weanling rats by parainfluenza (Sendai) virus. *Am. J. Pathol.* **129**, 277–296.

Cole, G.A., Nathanson, N., and Rivet, H. (1970). Viral hemorrhagic encephalopathy of rats II. Pathogenesis of central nervous system lesions. *Am. J. Epidemiol.* **91**, 339–350.

Coid, R., and Wardman, G. (1971). The effect of parainfluenza type 1 (Sendai) virus infection on early pregnancy in the rat. *J. Reprod. Fertil.* **29**, 39–43.

Coleman, G.L., Jacoby, R.O., Bhatt, P.N., and Jonas, A.M. (1983). Naturally occurring lethal parvovirus infection of juvenile and young adult rats. *Vet. Pathol.* **20**, 49–56.

Compton, S.R., Barthold, S.W., and Smith, A.L. (1993). The cellular and molecular pathogenesis of coronaviruses. *Lab. Anim. Sci.* **43**, 15–28.

Compton, S.R., Jacoby, R.O., Paturzo, F.X., and Smith, A.L. (2003). Persistent Seoul virus infection in Lewis rats. Submitted for publication.

Compton, S.R., Smith, A.L., and Gaertner, D.J. (1999a). Comparison of the pathogenicity in rats of rat coronaviruses of different neutralization groups. *Lab. Anim. Sci.* **49**, 514–518.

Compton, S.R., Vivas-Gonzales, B.E., and Macy, J.D. (1999b). Reverse transcriptase chain reaction-based diagnosis and molecular characterization of a new rat coronavirus strain. *Lab. Anim. Sci.* **49**, 506–513.

Cotmore, S.F., and Tattersall, P. (1987). The autonomously replicating parvoviruses of vertebrates. *Adv. Virus Res.* **33**, 91–174.

Cross, S.S., and Parker, J.C. (1972). Some antigenic relationships of the murine parvoviruses: minute virus of mice, rat virus, H-1 virus. *Proc. Soc. Exp. Biol. Med.* **139**, 105–108.

Descoteaux, J.P., and Payment, P. (1981). Comparison of hemagglutination inhibition, serum neutralization and ELISA PVM tests for the detection of antibodies to pneumonia virus of mice in rat sera. *J. Virol. Meth.* **3**, 83–87.

Desmyter, J., LeDuc, J.W., Johnson, K.M., Brasseur, F., Deckers, C., and van Ypersele de Strihou. (1983). Laboratory rat associated outbreak of hemorrhagic fever with renal syndrome due to Hantaan-like virus in Belgium. *Lancet* **2**, 1445–1448.

El Dadah, A.H., Nathanson, N., Smith, K.O., Squire, R.A., Santos, G.W., and Melby, E.C. (1967). Viral hemorrhagic encephalopathy in rats. *Science* **156**, 392–394.

Ellerman, K.E., Richards, C.A., Guberski, D.L., Shek, W.R., and Like, A.A. (1996). Kilham rat virus triggers, T-cell-dependent autoimmune diabetes in multiple strains of rat. *Diabetes* **45**, 557–562.

Gaertner, D.J., Compton, S.R., and Winograd, D.F. (1993a). Environmental stability of rat coronaviruses (RCVs). *Lab. Anim. Sci.* **43**, 403–404.

Gaertner, D.J., Compton, S.R., Winograd, D.F., and Smith, A.L. (1996a). Growth characteristics and protein profiles of prototype and wild-type rat coronavirus isolates grown in a cloned subline of mouse fibroblasts (L2p.176 cells). *Virus Res.* **41**, 55–68.

Gaertner, D.J., Jacoby, R.O., Johnson, E.A., Paturzo, F.X., and Smith, A.L. (1995). Persistent rat virus infection in juvenile athymic rats and its modulation by antiserum. *Lab. Anim. Sci.* **45**, 249–253.

Gaertner, D.J., Jacoby, R.O., Johnson, E.A., Paturzo, F.X., Smith, A.L., and Brandsma, J.L. (1993b). Characterization of acute rat parvovirus infection by *in situ* hybridization. *Virus Res.* **28**, 1–18.

Gaertner, D.J., Jacoby, R.O., Paturzo, F.X., Johnson, E.A., Brandsma, J.L., and Smith, A.L. (1991). Modulation of lethal and persistent rat parvovirus infection by antibody. *Arch. Virol.* **118**, 1–9.

Gaertner, D.J., Jacoby, R.O., Smith, A.L., Ardito, R.B., and Paturzo, F.X. (1989). Persistence of rat virus in athymic rats. *Arch. Virol.* **105**, 259–268.

Gaertner, D.G., Smith A.L., and R.O. Jacoby. (1996b). Efficient induction of persistent and prenatal infection with a parvovirus of rats. *Virus Res.* **44**, 67–78.

Gaertner, D.J., Winograd, D.F., Compton, S.R., Patruzo, F.X., and Smith, A.L. (1993c). Development and optimization of plaque assays for rat coronaviruses. *J. Virol. Meth.* **43**, 53–64.

Gagneten, S., Scanga, C.A., Dveksler, G.S., Beauchemin, N., Percy, D.H., and Holmes, K.V. (1996). Attachment glycoproteins and receptor specificity of rat coronaviruses. *Lab. Anim. Sci.* **46**, 159–166.

Garlinghouse, L.E., and van Hooosier, G.L., Jr. (1978). Studies on adjuvant-induced arthritis, tumor transplantability, and serologic response to bovine serum albumin in Sendai virus-infected rats. *Am. J. Vet. Res.* **39**, 297–300.

Giddens, W.E., van Hoosier, G.L., and Garlinghouse, L.E. (1987). Experimental Sendai virus infection in laboratory rats. II. Pathology and histochemistry. *Lab. Anim. Sci.* **37**, 442–448.

Guberski, D.L., Thomas, V.A., Shek, W.R., Like, A.A., Handler, E.S., Rossini, A.A., Wallace, J.E., and Welsh, R.M. (1991). Induction of type I diabetes by Kilham rat virus in diabetes-resistant BB/Wor rats. *Science* **254**, 1010–1013.

Hajjar, A.M., DiGiacomo, R.F., Carpenter, J.K., Bingel, S.A., and Moazed, T.C. (1991). Chronic sialodacryoadenitis virus (SDAV) infection in athymic rats. *Lab. Anim. Sci.* **41**, 22–25.

Hanna, P.E., Percy, D.H., Paturzo, F.X., and Bhatt, P.N. (1984). Sialodacryoadenitis in the rat: effects of immunosuppression on the course of the disease. *Am. J. Vet. Res.* **45**, 2077–2083.

Hart, C.A., and Bennett, M. (1999). Hantavirus infections: epidemiology and pathogenesis. *Microbe Infect.* **1**, 1229–1237.

Hartley, J.W., Rowe, W.P., Bloom, J.J., and Turner, H.C. (1964). Antibodies to mouse hepatitis virus in human sera. *Proc. Soc. Exp. Biol. Med.* **115**, 414–418.

Hirano, N., Suzuki, Y., Ono, K., Murakami, T. and Fujiwara, K. (1986). Growth of sialodacryoadenitis virus in LBC cell culture. *Nippon Juigaku Zasshi* **48**, 193–195.

Hirano, N. (1990). Plaque assay and propagation in rat cell line LBC cells of rat coronaviruses and 5 strains of sialodacryoadenitis virus. *Zentralbl. Veterinarmed.* **37**, 91–96.

Hirano, N., Ono, K., Nomura, R., and Tawara, T. (1995). Isolation and characterization of sialodacryoadenitis virus (coronavirus) from rats by established cell line LBC. *Zentralbl. Veterinarmed.* **42**, 147–154.

Iftimovici, R., Iacobescu, V. Mutui, A., and Puca, D. (1976). Enzootic with ectromelia symptomatology in Sprague-Dawley rats. *Rev. Roum. Med. Virol.* **27**, 65–66.

Jacoby, R.O., and Ball-Goodrich, L.B. (1995). Parvovirus infections of mice and rats. *Sem. Virol.* **6**, 329–33.

Jacoby, R.O., Ball-Goodrich, L.J., Besselsen, D.G., McKisic, M.D., Riley, L.K., and Smith, A.L. (1996). Rodent parvovirus infections. *Lab. Anim. Sci.* **46**, 370–380.

Jacoby, R.O., Bhatt, P.N., Gaertner, D.J., Smith, A.L., and Johnson, E.A. (1987). The pathogenesis of rat virus infection in infant and juvenile rats after oronasal inoculation. *Arch. Virol.* **95**, 251–270.

Jacoby, R.O., Bhatt, P.N., and Jonas, A.M. (1975). Pathogenesis of sialodacryoadenitis in gnotobiotic rats. *Vet. Pathol.* **12**, 196–209.

Jacoby, R.O., Bhatt, P.N., and Jonas, A.M. (1979). Viral diseases. *In* "The Laboratory Rat." First edition. (H. Baker, J.R. Lindsey, and S.H. Weisbroth, eds.), Volume 1, pp. 271–306, Academic Press, San Diego.

Jacoby, R.O., Fox, J.G., and Davisson, M. (2002). Biology and diseases of mice. *In* "Laboratory Animal Medicine." Second edition. (J.G. Fox, L.C. Anderson, F.M. Loew, and F.W Quimby, eds.), pp. 35–120, Academic Press, San Diego.

Jacoby, R.O., Gaertner, D.J., Bhatt, P.N., Paturzo, F.X., and Smith. A.L. (1988). Transmission of experimentally-induced rat virus infection. *Lab. Anim. Sci.* **38**, 11–14.

Jacoby, R.O., Johnson, E.A., Paturzo, F.X., and Ball-Goodrich, L.J. (2000). Persistent rat virus infection in smooth muscle of euthymic and athymic rats. *J. Virol.* **74**, 11841–11848.

Jacoby, R.O., Johnson, E.A., Paturzo, F.X., Gaertner, D.J., Brandsma, J.L., and Smith, A.L. (1991). Persistent rat parvovirus infection in individually housed rats. *Arch. Virol.* **117**, 193–205.

Jacoby, R.O., and Lindsey, J.R. (1998). Risks of infection among laboratory rats and mice at major biomedical research institutions. *ILAR J.* **39**, 266–271.

Kajiwara, N., Yutaka U., Takahashi, A., Sugiyama, F., and Yagami, K. (1996). Vertical transmission to embryo and fetus in maternal infection with rat virus (RV). *Exp. Anim.* **45**, 239–244.

Kilham, L., and Olivier, L. (1959). A latent virus of rats isolated in tissue culture. *Virology* **7**, 428–437.

Kilham, L. (1960). Mongolism associated with rat virus (RV) infection in hamsters. *Virology* **13**, 141–143.

Kilham, L. (1961). Rat virus (RV) infections in hamsters. *Proc. Soc. Exp. Biol. Med.* **106**, 825–829.

Kilham, L., and Ferm, V.H. (1964). Rat virus (RV) infections of pregnant, fetal and newborn rats. *Proc. Soc. Exp. Biol. Med.* **117**, 874–879.

Kilham, L., and Margolis, G. (1964). Cerebellar ataxia in hamsters inoculated with rat virus. *Science* **143**, 1047–1048.

Kilham, L., and Margolis, G. (1966). Spontaneous hepatitis and cerebellar hypoplasia in suckling rats due to congenital infection with rat virus. *Am. J. Pathol.* **49**, 457–475.

Kilham, L., and Margolis, G. (1969). Transplacental infection of rats and hamsters induced by oral and parenteral inoculations of H-1 and rat viruses (RV). *Teratology* **2**, 111–23.

Kojima, A., Fujimani, F., Doi, K., Yasoshima, A., and Okaniwa, A. (1980). Isolation and properties of sialodacryoadenitis virus of rats. *Jikken Dobutsu* **29**, 409–418.

Kraft, L.M., D'Amelio, E.D., and D'Amelio, F.E. (1982). Morphological evidence for natural poxvirus infection in rats. *Lab. Anim. Sci.* **32**, 648–654.

Krikun, V.A. (1977). Pox in rats: isolation and identification of poxvirus. *Vopr. Virusol.* **22**, 371–373.

Lai, Y.L., Jacoby, R.O., Bhatt, P.N., and Jonas, A.M. (1976). Keratoconjunctivitis associated with sialodacryoadenitis in rats. *Invest. Ophthalmol.* **15**, 538–541.

La Regina, M., Woods, L., Klender, P., Gaertner, D.J., and Paturzo, F.X. (1992). Transmission of sialodacryoadenitis virus (SDAV) from infected rats to rats and mice through handling, close contact, and soiled bedding. *Lab. Anim. Sci.* **42**, 344–346.

LeDuc, J.W. (1987). Epidemiology of Hantaan and related viruses. *Lab. Anim. Sci.* **37**, 413–418.

Lee, H.W., Callisher, C.H., and Schmaljohn, C. (eds.). (1999). Manual of hemorrhagic fever with renal syndrome and hantavirus pulmonary syndrome. WHO Collaborating Center for Virus Reference and Research (*Hantaviruses*) Asian Institute for Life Sciences, Seoul, Korea.

Lee, H.W., French, G.R., Lee, P.W., Baek, L.J., Tsuchiya, K., and Foulke, R.S. (1981). Observations on natural and laboratory infections with the etiologic agent of Korean hemorrhagic fever. *Am. J. Trop. Med. Hyg.* **30**, 477–482.

Lee, P.W., Yanagihara, R., Gibbs, Jr., C.J., and Gadjusek, D.C. (1986). Pathogenesis of experimental Hantaan virus infection in laboratory rats. *Arch. Virol.* **88**, 57–66.

Liang, S.C., Schoeb, T.R., Davis, J.K., Simecka, J.W., Cassell, G.H., and Lindsey, J.R. (1995). Comparison severity of respiratory lesions of sialodacryoadenitis virus and Sendai virus infections in LEW and F344 rats. *Vet. Pathol.* **32**, 661–667.

Lipton, H.G., Nathanson, N., and Hodous, J. (1973). Enteric transmission of parvoviruses: pathogenesis of rat virus infection in adult rats. *Am. J. Epidemiol.* **6**, 443–446.

Lloyd, G., Bowen, E.T., Jones, N., and Pendry, A. (1984). HFRS outbreak associated with laboratory rats in UK. *Lancet* **1**, 1175–1176.

Lloyd, G., and Jones, N. (1986). Infection of laboratory workers with hantavirus acquired from immunocytomas propagated in laboratory rats. *J. Infect.* **12**, 117–125.

London, B.A., Kern, J., and Gehle, W.D. (1983). Detection of murine virus antibody by ELISA. *Lab. Anim.* **12**, 40–47.

Lyon, H.W., Christian, J.J., and Mitler, C.W. (1959). Cytomegalic inclusion disease of lacrimal glands in male laboratory rats. *Proc. Soc. Exp. Biol. Med.* **101**, 164–166.

Machii, K., Iwai, H., Otsuka, Y., Ueda, K., and Hirano, N. (1988). Reactivities of 4 murine coronavirus antigens with immunized or naturally infected rat sera by enzyme linked immunosorbent assay. *Jikken Dobutsu* **37**, 251–255.

Makino, S., Seko, S., Nakao, H., and Midazuki, K. (1972). An epizootic of Sendai virus infection in a rat colony. *Exp. Anim.* **22**, 275–280.

Margolis, G., and Kilham, L. (1970). Parvovirus infections, vascular endothelium and hemorrhagic encephalopathy. *Lab. Invest.* **22**, 478–488.

Margolis, G., and Kilham, L. (1973). Pathogenesis of intrauterine infection in rats due to reovirus type 3. II. Pathologic and fluorescent antibody studies. *Lab. Invest.* **28**, 605–613.

Margolis, G., and Kilham, L. (1972). Rat virus infection of megakaryocytes: a factor in hemorrhagic encephalopathy. *Exp. Molec. Pathol.* **16**, 326–340.

Margolis, G., Kilham, L., and Ruffulo, P.R. (1968). Rat virus disease, an experimental model of neonatal hepatitis. *Exp. Molec. Pathol.* **8**, 1–20.

Marrenikova, S.S., and Shelukhina, E.M. (1976). White rats as a source of pox infection in carnivora of the family Felidae. *Acta Virol.* **2**, 422.

Marrenikova, S.S., Shalukhina, E.M., and Fimina, V.A. (1978). Pox infection in white rats. *Lab. Anim.* **12**, 33–36.

Maru, M., and Sato, K. (1982). Characterization of a coronavirus isolated from rats with sialodacryoadenitis. *Arch. Virol.* **73**, 33–43.

McClean, I.W., and Nakane, P.K. (1974). Periodate-lysine-paraformaldehyde fixative: a new fixative for immunoelectron microscopy. *J. Histochem. Cytochem.* **22**, 1077–1083.

McKisic, M.D., Paturzo, F.X., Gaertner, D.J., Jacoby, R.O., and Smith, A.L. (1995). A nonlethal rat parvovirus infection suppresses rat T lymphocyte effector functions. *J. Immunol.* **155**, 3979–3986.

Meyer, B.J., and Schmaljohn, C.S. (2000). Persistent hantavirus infections: characteristics and mechanisms. *Trends Microbiol.* **8**, 61–67.

Michaels, R.H., and Myerowitz, R.L. (1983). Viral enhancement of nasal colonization with *Haemophilus influenzae* type b in the infant rat. *J. Pediatr. Res.* **17**, 472–473.

Nichol S.N. (2001). Bunyaviruses. *In* "Field's Virology," (D. Knipe, and P. Howley, eds.), pp.1603–1633, Lippincott Williams & Wilkins, Philadelphia, Pa.

Nicklas, W., Kraft, V., and Meyer, B. (1993). Contamination of transplantable tumors, cell lines, and monoclonal antibodies with rodent viruses. *Lab. Anim. Sci.* **43**, 296–300.

Novotny, J.F., and Hetrick, F.M. (1970). Pathogenesis and transmission of Kilham rat virus infection in rats. *Infect. Immun.* **2**, 298–303.

Ohsawa, K., Watanabe, Y., Takakura, A., Itoh, T., and Sato, H. (1996). Replication of rat coronaviruses in intestinal cell line, RCN-9, derived from F344 rats. *J. Exp. Anim.* **45**, 389–393.

Ohsawa, K., Watanabe, Y., Miyata, H., and Sato, H. (2003). Genetic analysis of a Theiler-like virus isolated from rats. *Comp. Med.* **53,** 191–196.

Parker, J.C., Cross, S.S., and Rowe, W.P. (1970). Rat coronavirus (RCV): a prevalent, naturally occurring pneumotropic virus of rats. *Arch. Gesamte Virusforsch.* **31,** 293–302.

Paturzo, F.X., Jacoby, R.O., Bhatt, P.N., Smith, A.L., Gaertner, D.G., and Ardito, R.B. (1987). Persistence of rat virus in seropositive rats as detected by explant culture. *Arch. Virol.* **95,** 137–142.

Percy, D.H., and Barthold, S.W. (2001). "Pathology of Laboratory Rodents and Rabbits." Second edition, p. 109, Iowa State University Press, Ames, Iowa.

Percy, D.H., Bond, S., and MacInnes, J. (1989). Replication of sialodacryoadenitis virus in mouse L-2 cells. *Arch. Virol.* **104,** 323–333.

Percy, D.H., Bond, S.J., Paturzo, F.X., and Bhatt, P.N. (1990a). Duration of protection from reinfection following exposure to sialodacryoadenitis virus in Wistar rats. *Lab. Anim. Sci.* **40,** 144–149.

Percy, D.H., Hanna, P.E., Paturzo, F.X., and Bhatt, P.N. (1984). Comparison of strain susceptibility to experimental sialodacryo-adenitis in rats. *Lab. Anim. Sci.* **34,** 255–260.

Percy, D.H., Hayes, M.A., Kocal, T.E., and Wojcinski, Z.W. (1988). Depletion of salivary gland epidermal growth factor by sialodacryo-adenitis virus infection in the Wistar rat. *Vet. Pathol.* **25,** 183–192.

Percy, D.H., and Scott, R.A. (1991). Coronavirus infection in the laboratory rat: immunization trials using attenuated virus replicated in L-2 cells. *Can. J. Vet. Res.* **55,** 60–66.

Percy, D.H., and Williams, K.L. (1990). Experimental Parker's rat coronavirus infection in Wistar rats. *Lab. Anim. Sci.* **40,** 603–607.

Percy, D.H., Williams, K.L., Bond, S.J., and MacInnes, J.L. (1990b). Characteristics of Parker's rat coronavirus (PRC) replicated in L-2 cells. *Arch. Virol.* **112,** 195–202.

Percy, D.H., Williams, K.L., and Paturzo, F.X. (1991). A comparison of the sensitivity and specificity of sialodacryoadenitis virus, Parker's rat coronavirus, and mouse hepatitis virus-infected cells as a source of antigen for the detection of antibody to rat coronaviruses. *Arch. Virol.* **119,** 175–180.

Peters, R.L., and Collins, M..J. (1981). Use of mouse hepatitis virus antigen in an enzyme-linked immunosorbent assay for rat corona-viruses. *Lab. Anim. Sci.* **31,** 472–475.

Priscott, P.K., and Tyrell, D.A.J. (1982). The isolation and partial characterization of a cytomegalovirus from the brown rat. *Rattus norvegicus. Arch. Virol.* **73,** 145–150.

Rabson, A.S., Edgcomb, J.H., Legallais, F.Y., and Tyrell, S.A. (1969). Isolation and growth of rat cytomegalovirus in vitro. *Proc. Soc. Exp. Biol. Med.* **131,** 923–927.

Riley, L.K., Knowles, R., Purdy, G., Salome, N., Pintel, D., Hook, R.R., Jr., Franklin, C.L., and Besch-Williford, C.L. (1996). Expression of recombinant parvovirus NS1 protein by a baculovirus and appli-cation to serologic testing of rodents. *J. Clin. Microbiol.* **34,** 440–444.

Robey, R.E., Woodman, D.R., and Hetrick, F.M. (1968). Studies on the natural infection of rats with the Kilham rat virus. *Am. J. Epidemiol.* **88,**139_143.

Rossie, K.M., Sheridan, J.F., Barthold, S.W., and Tutschka, P.J. (1988). Graft-versus-host disease and sialodacryoadenitis viral infection in bone marrow transplanted rats. *Transplantation* **45,** 1012–1016.

Ruffolo, P.R., Margolis, G., and Kilham, L. (1966). The induction of hepatitis by partial hepatectomy in resistant adult rats injected with H-1 virus. *Am. J. Pathol.* **49,** 795–824.

Saknimit, M., Inatsuki, I., Sugiyama, Y., and Yagami, K. (1988). Virucidal efficacy of physico-chemical treatments against coronaviruses and parvoviruses of laboratory animals. *Jikken Dobutsu* **37,** 341–345.

Sato, T., Mequid, M.M., Quinn, R.H., Zhang, L., and Chen, C. (2001). Feeding behavior during sialodacryoadenitis viral infection in rats. *Physiol. Behav.* **72,** 721–726.

Schmaljohn C., and Hjelle, B. (1997). Hantaviruses: a global disease problem. *Emerg. Infect. Dis.* **3,** 95–104.

Schoeb, T.R., Kervin, K.C., and Lindsey, J.R. (1985). Exacerbation of murine respiratory mycoplasmosis in gnotobiotic F344/N rats by Sendai virus infection. *Vet. Pathol.* **22,** 272–282.

Shek, W.R., and Gaertner, D.J. (2002). Microbiological quality control for laboratory rodents and lagomorphs. *In* "Laboratory Animal Medicine." Second Edition. (J.G. Fox, L.C. Andersen, F.W. Loew, and F.W. Quimby, eds.). pp. 365–393, Academic Press, San Diego.

Siegl, G. (1976). The parvoviruses. *In* "Virology Monographs." (S. Gard and G. Hallauer, eds.). Vol. 15, Springer-Verlag, Berlin.

Simmons, J.H., and Riley, L.K. (2002). Hantaviruses: an overview. *Comp. Med.* **52,** 97–110.

Singh, S.B., and Lang, C.M. (1984). Enzyme-linked immunosor-bent assay in the diagnosis of Kilham rat virus infection in rats. *Lab. Anim. Sci.* **18,** 364–370.

Smith, A.L. (1983). An immunofluorescence test for detection of serum antibody to rodent coronaviruses. *Lab. Anim. Sci.* **33,** 157–160.

Smith, A.L., and Winograd, D.F. (1986). Two enzyme immunoassays for the detection of antibody to rodent coronaviruses. *Virol. Methods* **14,** 335–343.

Tamura, K., Ohsawa, K., Koji, T., Watanabe, Y., Katamine, S., Sato, H. and Ayabe, H. Allogeneic immune responses augment rat cytomegalo-virus replication in rats. *Transplant. Proc.* **31,** 1376–1377.

Tattersall, P., and Cotmore, S.F. (1986). The rodent parvoviruses. *In* "Viral and Mycoplasmal Infections of Rodents: Effects on Biomedical Research," (P.N. Bhatt, R.O. Jacoby, H.C. Morse, and A.E. New, eds.), pp. 305–348, Academic Press, Orlando, FL.

Toolan, H.W., Dalldorf, G., Barclay, M., Chandra, S., and Moore, A.E. (1960). An unidentified filterable agent isolated from transplanted human tumors. *Proc. Nat. Acad. Sci. U.S.A.* **46,** 1256–1259.

Ueno, Y., Iwama, M., Ohshima, T., Sugiyama, Y., Takakura, A., Itoh, T., and Sugiyama, K. (1998). Prevalence of "orphan" parvovirus infections in rats and mice. *Exp. Anim.* **47,** 207–210.

Ueno, Y., Sugiyama, F., Sugiyama, Y., Ohsawa, K., Sato, H., and Yagami, K. (1997). Epidemiological characterization of newly recog-nized rat parvovirus, "rat orphan parvovirus." *J. Vet. Med. Sci.* **59,** 265–269.

Ueno, Y., Sugiyama, F., and Yagami, K. (1996). Detection and *in vivo* transmission of rat orphan parvovirus (ROPV). *Lab. Anim.* **30,** 114–119.

Utsumi, K., Ishikawa, T., Maeda, T., Shimizu, S., Tatsumi, H., and Fujiwars, K. (1980). Infectious sialodacryoadenitis and rat breeding. *Lab. Anim.* **14,** 303–307.

Utsumi, K., Maeda, T., Tatsumi, H. and Fujiwara, K. (1978). Some clinical and epizootiological observations of infectious sialoade-nitis in rats. *Jikken Dobutsu* **27,** 283–287.

Vonderfecht, S.L. (1986). Infectious diarrhea of infant rats (IDIR) induced by an antigenically distinct rotavirus. *In* "Viral and Myco-plasmal Infections of Rodents: Effects on Biomedical Research." (P.N. Bhatt, R.O. Jacoby, H.C. Morse, and A.E. New, eds.), pp. 245–252. Academic Press, Orlando, Fl.

Vonderfecht, S.L., Huber, A.C., Eiden, J., Mader, L.C., and Yolken, R.H. (1984). Infectious diarrhea of infant rats produced by a rotavirus-like agent. *J. Virol.* **52,** 94–98.

Votsberger, L.M., Stromberg, P.C., and Rice, J.M. (1982). Histological and serological response of B6C3F1 mice and F344 rats to experimental pneumonia virus of mice infection. *Lab. Anim. Sci.* **32,** 419.

Wan, C.H., Riley, M.I., Hook, R.R. Jr., Franklin, C.L., Besch-Williford, C.L., and Riley, L.K. (1995). Expression of Sendai virus nucleo-capsid protein in a baculovirus expression system and applica-tion to diagnostic assays for Sendai virus. *J. Clin. Microbiol.* **33,** 2007–2011.

Wan, C.H., Soderlund-Venermo, M., Pintel, D.J., and Riley, L.K. (2002). Molecular characterization of three newly recognized rat parvoviruses. *J. Gen. Virol.* **83**, 2075–2083.

Ward, J.M., Lock, A., Collins, M.J., Gonda, M.A., and Reynolds, C.W. (1984). Papovaviral sialoadenitis in athymic nude rats. *Lab. Anim.* **18**, 84–89.

Ward, J.M., and Young, D.M. (1976). Latent adenoviral infection of rats: intranuclear inclusions induced by treatment with a cancer chemotherapeutic agent. *J. Am. Vet. Med. Assoc.* **169**, 952–953.

Weir, E.C., Jacoby, R.O., Paturzo, F.X., and Johnson, E.A. (1990a). Infection of SDAV-immune rats with SDAV and rat coronavirus. *Lab. Anim. Sci.* **40**, 363–366.

Weir, E.C., Jacoby, R.O., Paturzo, F.X., Johnson, E.A., and Ardito, R.B. (1990b). Persistence of sialodacryoadenitis virus in athymic rats. *Lab. Anim. Sci.* **40**, 138–143.

Yagami, K., Goto, Y., Ishida, J., Ueno, Y., Kajiwara, N., and Sugiyama, F. (1995). Polymerase chain reaction for detection of rodent parvoviral contamination in cell lines and transplantable tumors. *Lab. Anim. Sci.* **45**, 326–328.

Yang, F-C., Paturzo, F.X., and Jacoby, R.O. (1995). Environmental stability and transmission of rat virus. *Lab. Anim. Sci.* **45**, 140–144.

Zenner, L., and Regnault, J.P. (2000). Ten-year long monitoring of laboratory mouse and rat colonies in French facilities: a retrospective study. *Lab. Anim.* **34**, 76–83.

Chapter 13

Parasitic Diseases

David G. Baker

I. INTRODUCTION

Laboratory rats may serve as hosts for a wide variety of parasites. Although most of these infections are clinically silent, some may result in severe disease. At the least, infections may affect host physiology, and therefore serve as unwanted variables in research. Consequently, only rats free of parasitic infections are entirely suitable as research subjects. Fortunately, advances in animal husbandry have resulted in a dramatic decline in parasitic infections in laboratory rats. This is particularly true for parasites requiring multiple hosts for completion of the lifecycle. In fact, parasites other than pinworms are uncommonly found associated with rats in the modern vivarium. However, wild rats may at times gain access to and contaminate animal facilities, as well as feed and bedding storage areas. Also, animal care personnel may privately and unknowingly purchase pet rats infected with parasites. Thus, the laboratory animal professional should be aware of those parasites capable of infecting the laboratory rat. This chapter reviews those parasites that may be found infecting laboratory rats. The information has been arranged primarily according to organ system(s) involved, and secondarily by parasite phylogeny. Each parasite is presented with a brief morphologic description, lifecycle, pathobiology, diagnosis, treatment, prevention, and control.

Treatment of parasite infections deserves a special word of caution. It is well known that many drugs used to treat parasitic infections or infestations have themselves the potential to alter host physiology. In so doing, these drugs may compromise the usefulness of research animals, including rats. Fortunately, the decline in prevalence of parasitic infections in laboratory rats has been accompanied by a decline in the use of parasiticides.

However, the discovery of parasitic infection in laboratory rats will require action. At times, parasites may be removed chemically without affecting the usefulness of rats as research subjects. At other times, the needs of the research study will preclude the use of parasiticides, and necessitate non-chemical methods of parasite elimination, such as cesarean rederivation. To determine whether parasiticides are safe to use in specific situations, the laboratory animal professional should be aware of the physiologic effects of commonly used anti-parasitic compounds. Relatively few such agents are used to treat rats. Among those in use, thiabendazole has nonimmunosuppressive anti-inflammatory properties in rodents (Van Arman et al., 1975). Levamisole has been used as a broad-spectrum anthelminthic in both human and veterinary medicine. It has been shown to have numerous *in vivo* and/or *in vitro* effects on rats or rat cells. These effects include superoxide scavenging (Schinetti et al., 1984), antagonism of gastric ulcers caused by necrotizing and anti-inflammatory agents (Evangelista et al., 1984), enhancement of gastrointestinal absorption of some compounds (Utsumi et al., 1987), inhibition of platelet aggregation (Pinto et al., 1990), upregulation of pyruvate dehydrogenase and glycogen synthase activity in rat adipose tissue (Thomaskutty et al., 1993; Basi et al., 1994), inhibition of skeletal tissue mineralization (Klein et al., 1993), inhibition of fibroblast collagen synthesis (de Waard et al., 1998), decreased post-surgical translocation of bacteria (Cetinkaya et al., 2002), immune stimulation (Trabert et al., 1976), and others. Fenbendazole causes far fewer effects. In rats, these appear to be limited to subtle changes in behavioral performance in young rats from dams fed fenbendazole during pregnancy (Barron et al., 2000), although other minor effects have been demonstrated in other species. Albendazole, another benzimidazole with anthelminthic properties, is embryotoxic in rats, induces drug-metabolizing enzymes in the liver (Souhaili-el Amri et al., 1988), and delays rat brain microtubule assembly (Solana et al., 1998). Lastly, ivermectin, a macrocyclic lactone disaccharide and member of the avermectin anthelminthic family, is not without its effects on rat physiology. This is important considering the common use of ivermectin in the treatment of parasitic infections in rats and other host species. In rats, ivermectin and doramectin, another avermectin, are known to possess anxiolytic effects (de Souza Spinosa et al., 2000, 2002). Still other members of the avermectin family may be selectively cytostatic against tumor cells (Driniaev et al., 2001). Lastly, the avermectins may cause developmental neurotoxicity in rats (Poul, 1988; Wise et al., 1997). Each of these reports highlight the potential for commonly used anthelminthics to alter host physiology.

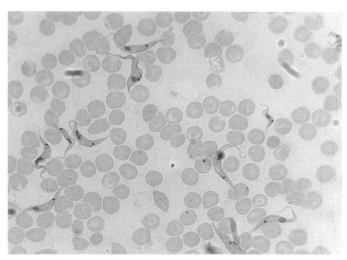

Fig. 13-1 Trypanosoma lewisi. Several trypomastigotes in a blood smear. Romanowski stain. ×1000.

II. PARASITES OF PARENTERAL SYSTEMS

A. Flagellates

Both *Trypanosoma lewisi* and *T. cruzi* are hemoflagellates commonly found in wild rats. Natural infection of the trypanosomes in animal hosts requires the presence of suitable arthropod vectors. Commercially purchased rats should arrive free of trypanosome infection, and the lack of suitable arthropod hosts in the well-managed animal facility should assure that rats remain trypanosome-free. However, experimental infection of rats with *T. lewisi* has been extensively used as an animal model for studying human and ruminant trypanosomiasis.

1. *T. lewisi*

Description and lifecycle. In host blood the trypanosome has a long, slender body tapering to a point at the posterior end, typically measuring from 25 to 36 μm in total length (Fig. 13-1). However, a longer form has been identified in *Rattus norvegicus* in India. This form measures 35 to 39 μm in total length (Saxena and Miyata, 1993). A simple flagellum arising from the kinetoplast runs along the body and is free at the anterior end.

The rat fleas *Nosopsyllus fusciatus*, and possibly *Xenopsylla cheopis*, serve as natural arthropod hosts. *T. lewisi* multiplies rapidly in the alimentary tract of the flea. Rats become infected by ingestion of fleas or their feces. The rat louse *Polyplax spinulosa* may also serve as a mechanical vector (Khachoian and Arakelian, 1978).

Host. *T. lewisi* is host specific, occurring almost exclusively in the genus *Rattus*. There is one report of natural infection with *T. lewisi* in *Gerbilus pyramidous* (Sakla and Monib, 1984). *T. lewisi* has been experimentally transmitted to laboratory mice, Mongolian gerbils, and guinea pigs, but these infections are short-lived (Smith, 1972). Human infections with *T. lewisi*-like trypanosomes have been reported twice (Johnson, 1933; Shrivastava and Shrivastava, 1974).

Pathobiology. *T. lewisi* infection is generally asymptomatic, occurring as a short-term infection in the blood. Clinical disease may develop in experimentally infected rats following manipulations such as irradiation, splenectomy, immune suppression, and adrenalectomy. In addition, *T. lewisi* may cause a wide range of potentially unrecognized physiologic changes in the rat host. These include immunoregulation (Ndarathi, 1992), lipopolysaccharide hyperreactivity (Ferrante et al., 1984), autoimmune hemolytic anemia with splenomegaly and glomerulonephritis (Thoongsuwan and Cox, 1978), increased susceptibility to other pathogens such as *Toxoplasma gondii* (Catarinella Arrea et al., 1998) and *Salmonella typhimurium* (Nielsen et al., 1978), decreased total iron-binding capacity (Lee et al., 1977), and altered liver enzyme activities (Lee and Boone, 1983).

Clinical symptoms. Even in heavily infected rats, symptoms usually are not observed, in spite of the physiologic changes noted above. However, rats infected early in gestation abort (Shaw and Quadagno, 1975). Interestingly, rats infected with *T. lewisi* gain up to 31% more weight than uninfected controls, depending upon the age of rat examined. The rate of weight gain is affected to a greater extent in juvenile versus nearly mature rats (Lincicome et al., 1963). Those authors suggest several possible explanations for their observation, but the underlying mechanism remains unknown.

Diagnosis. Diagnosis of infection is established by demonstrating trypanosomes in blood smears (Fig. 13-1).

Inoculation of suspected blood into uninfected young rats is a useful confirmatory step.

Treatment. Rats have been cleared of *T. lewisi* infection by daily treatment with 54 mg/kg rifampicin for 30 days (el-Ridi et al., 1985).

Prevention and control. Infection with *T. lewisi* is unlikely to occur in a modern animal facility. Infection is prevented by excluding fleas, lice, and wild rodents.

B. Sporozoa

The class Sporozoa belongs to the phylum Apicomplexa. Parasites in this class reproduce both sexually and asexually. Oocysts contain infective sporozoites which arise from sporogony. Locomotion is primarily by body flexion and gliding. In addition to the species covered here, *Isospora felis* and *I. rivolta*, both parasites of cats, have been shown experimentally to infect rats (Frenkel and Dubey, 1972). This is biologically important because *Isospora* spp. were long regarded as single-host parasites. However, natural infection of rats with these parasites is likely so rare that further discussion of them in this text is unwarranted.

1. *Hepatozoon muris*

Description and lifecycle. *Hepatozoon muris*, formerly known as *H. perniciosum*, is an apicomplexan parasite in the family Haemogregarinidae. Miller first described this parasite from laboratory rats in Washington, D.C. (Miller, 1908). Schizogony takes place in the parenchymal cells of the liver. The schizonts measure 30 to 35 by 25 to 28 μm, and give rise to 20 to 30 merozoites each. Gametogony occurs in lymphocytes, monocytes or, occasionally, in granulocytes. Further development requires the spiny rat mite, *Laelaps echidninus*, which ingests the parasite while taking a blood meal. Fertilization and sporogony then take place in the gut and hemocoele of the mite. Rats become infected by ingesting the mite. Sporozoites released in the intestine penetrate the gut wall and pass via the blood to the liver.

Host. The parasite is presumed common in wild Norway and black rats throughout the world. Historically, wild rats in the Washington, D.C. area have been reported to be naturally infected (Price and Chitwood, 1931). The current distribution in wild rats is unknown. *H. muris* occurs only in rats of the genus *Rattus*.

Pathobiology. *H. muris* may be pathogenic, although most infections are mild and asymptomatic. After natural infection, parasite burdens increase gradually over the course of a few months. The infection induces profound leukocytosis and monocytosis and may also cause splenomegaly, hepatic degeneration, and anemia.

Grossly, affected rats are pale, the liver and spleen are enlarged, and the lungs may have minute surface hemorrhages. Other organs may show degenerative changes (Miller, 1908).

Clinical symptoms. Mild parasite burdens are asymptomatic. In contrast, heavily infected rats develop anorexia, lethargy, emaciation, terminal diarrhea, and death. A mortality of 50% has been reported in rats experimentally infected by the parasite. Growing rats are more susceptible to overwhelming infection than are mature rats (Miller, 1908).

Diagnosis. Gametocytes can be demonstrated within leukocytes on stained blood smears, whereas merozoites can be observed in liver sections.

Treatment. No effective treatment is known. Macintire and coworkers report on the treatment of dogs infected with *H. americanum*. Treatment with anticoccidial agents ameliorated clinical disease but did not eliminate tissue stages (Macintire et al., 2001). Likewise, Krampitz and Haberkorn (1988) administered toltrazuril subcutaneously or in the drinking water or feed of bank voles (*Clethrionomys glareolus*) infected with *H. erhardovae*. Although clinical recovery occurred following nearly all routes and dosages, infection was not reproducibly eliminated with any of the regimens. This approach has not been tested in rats infected with *H. muris*. However, it is unlikely that antiprotozoal treatment would be preferred to rederivation and elimination of infected laboratory rats.

Prevention and control. Infection with *H. muris* is unlikely to occur in a modern animal facility. Infection is prevented by excluding mites and wild rodents.

2. *Toxoplasma gondii*

Description and lifecycle. *Toxoplasma gondii* is a common coccidian parasite. Although infections are generally asymptomatic in mature animals, fatal infections occur, depending on parasite strain and host factors, including species, age, and immune status. The lifecycle of *T. gondii* has been thoroughly described. Domestic cats and other felids are the only definitive hosts of *T. gondii*. Cats become infected through ingestion of infected intermediate hosts or by ingestion of sporulated oocysts excreted from another cat. Following infection, asexual, and later, sexual development occurs in the feline intestinal tract. Ultimately, unsporulated oocysts are released into the environment. Oocysts can survive for long periods under moderate conditions of temperature and humidity. In this way, oocysts may come to contaminate feed or bedding intended for use in the laboratory environment. Rats become infected following ingestion of feed or bedding contaminated with oocysts. In rats, asexual multiplication results in the formation of bradyzoite

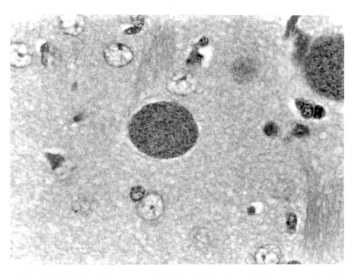

Fig. 13-2 *Toxoplasma gondii.* A well-developed tissue cyst in the brain of a rat 29 days after infection. Hematoxylin and eosin. ×650. (Courtesy of Dr. J.P. Dubey.)

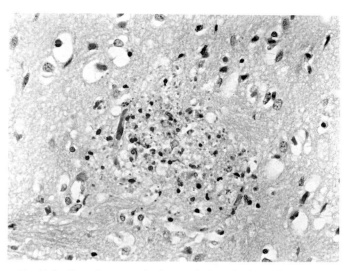

Fig. 13-3 *Toxoplasma gondii.* A necrotic focus in the cerebrum of a rat that died 7 days after being fed oocysts. Hematoxylin and eosin. ×750. (Courtesy of Dr. J.P. Dubey.)

tissue cysts in multiple tissues (Fig. 13-2), there awaiting ingestion by a cat. Once established in a rat colony, offspring may also become infected through transplacental transmission (Dubey and Frenkel, 1998).

Host. The definitive host of *T. gondii* is the cat or other felids. Intermediate animal hosts include virtually all terrestrial mammalian and avian species living in proximity to cats. The zoonotic nature of *T. gondii*, and the occasionally serious consequences of human infection, make this parasite of special concern for animal care personnel. Human fetuses and immune-compromised persons are at increased risk of potentially life-threatening illness with *T. gondii*. Globally, rats of the genus *Rattus* have frequently been found naturally infected with *T. gondii* (Dubey and Frenkel, 1998).

Pathobiology. *T. gondii* is an intracellular parasite affecting all organs, with particular affinity for the central nervous system and tissues of the mononuclear phagocytic system. In rats ingesting sporulated oocysts, sporozoites penetrate and multiply within the lamina propria of the small intestine, causing enteritis. There-after, parasites disseminate and form bradyzoite cysts in brain, lung, heart, muscle, liver, spleen, kidneys, uterus, intestine, lymph nodes, and other organs (Dubey, 1997). Cysts are most numerous in the brain, and are often up to 50 μm in diameter (Fig. 13-2) (Dubey, 1996). Tissue cysts are formed inside the cell as the host develops both cellular and humoral immunity. Tissue cysts may persist for many months, years, or for the life of the host. Ordinarily, they do not provoke inflammatory host responses. However, ruptured cysts elicit a lymphocytic inflammatory response (Fig. 13-3).

Clinical symptoms. Rats are among the most resistant hosts of *T. gondii* infection and rarely show clinical signs. However, clinical outcome of infection is highly dependent on parasite strain, stage inoculated, and route of inoculation. Infection of *T. gondii* in newborn or young rats usually results in acute clinical manifestations with fatal pneumonia (Lewis and Markell, 1958). Older rats experimentally inoculated with few parasites become chronically infected but rarely show clinical signs of infection. However, rats administered large numbers of sporulated oocysts may experience fatal infection (Dubey, 1996). In acute cases, death may occur owing to enteritis prior to dissemination to extraintestinal organs. Rats surviving longer may die of encephalitis or compromise of other vital organs. In addition, *T. gondii* alters cognitive function in the rat so that the rat's innate aversion to cats is diminished, thereby facilitating predation and completion of the lifecycle (Berdoy et al., 2000).

Diagnosis. Spontaneous toxoplasmosis in laboratory rats is rare in modern animal facilities. Diagnosis of *T. gondii* infection is based on the demonstration of parasites in tissue section or in stained cell preparations; serologic tests (a commercial modified agglutination test kit is available); or by bioassay of suspect material in mice. The latter is considered the only reliable method of determining the persistence of chronic *T. gondii* infection in rats (Dubey and Frenkel, 1998).

Treatment. No antiprotozoal treatments are known to completely and reliably eradicate *T. gondii* infection in rats. Infected rats should be culled, as they are likely unfit for most research applications. Valuable stocks or strains may be rederived free of *T. gondii* by embryo transfer.

Prevention and control. Natural infection of laboratory rats with *T. gondii* is unlikely to occur in the modern animal facility. Clearly, cats should not be housed near laboratory rodents. Occasionally though, rodent feed or bedding may become contaminated with cat feces containing *T. gondii* oocysts. This most commonly occurs when these products are warehoused, though contamination of materials may also occur during production. It should also be noted that cockroaches may serve as mechanical vectors of *T. gondii*, through ingestion of oocyst-containing cat feces. Rodents become infected when they consume contaminated cockroaches (Smith and Frenkel, 1978). Lastly, caretakers may transfer oocysts from their own personal pet cats, or from laboratory cats located elsewhere in the facility, to laboratory rodents. To prevent these occurrences, strict attention must be paid to storage of feed and bedding materials, and caretakers and other personnel working in the facility must practice excellent personal hygiene.

3. *Hammondia hammondi*

Description and lifecycle. *Hammondia hammondi* is a coccidian parasite very similar to *T. gondii* (Frenkel and Dubey, 1975). The lifecycles of the two parasites are also similar, although *H. hammondi* has an obligatory two-host lifecycle, whereas *T. gondii* may be passed from rat to rat, or from cat to cat. This means that only sporulated oocysts are infective to rats, and only bradyzoite cysts are infective to cats (Frenkel and Dubey, 2000).

Host. The definitive host of *H. hammondi* is the cat or other felids. Intermediate hosts include mice (*Mus musculus*), rats (*Rattus norvegicus*), deer mice (*Peromyscus* sp.), multimammate rats (*Mastomys natalensis*), black and red tamarins (*Saguinus nigricollis*), rabbits, hamsters, goats, dogs, and pigs. *H. hammondi* is not infective to humans. The prevalence of *H. hammondi* is unknown. Occasionally, cats initially thought to be infected with *T. gondii* have later been determined to have actually been infected with *H. hammondi* (Wallace, 1975). In at least one large study involving cats euthanized at an animal control facility, only 2 of 1000 cats were found actively shedding *H. hammondi* oocysts (Christie et al., 1977). The prevalence of *H. hammondi* in wild rats is unknown, but reports of natural infection are rare (Mason, 1978).

Pathobiology. *H. hammondi* is apparently nonpathogenic in rats. Elongated intracellular bradyzoite cysts form primarily in skeletal and cardiac muscle, yet are incidental.

Clinical symptoms. *H. hammondi* is not known to cause clinical disease in rats. However, most studies of intermediate hosts have focused on mice. Additional studies are needed to verify that *H. hammondi* is truly nonpathogenic in rats.

Diagnosis. Spontaneous hammondiosis has not been reported in laboratory rats. Diagnosis of *H. hammondi* infection is similar to that of *T. gondii*. Rats infected with *H. hammondi* may give cross-reacting positive results to *T. gondii* serology, depending on serologic test used. Therefore, serology alone is inadequate for distinguishing *H. hammondi* infection from *T. gondii* infection (Dubey and Frenkel, 1998). The two infections can be distinguished only via bioassay in mice. *H. hammondi* bradyzoites and tachyzoites are not infective to mice. Only sporulated oocysts are infective. In contrast, both *T. gondii* bradyzoites and tachyzoites are readily transferred between mice (Frenkel and Dubey, 1975). Also, *H. hammondi* tissue cysts are more frequent in striated muscle than in the brain of mice used for bioassay, whereas the opposite is true for *T. gondii*. This feature alone however, is inadequate for differentiating the two infections (Dubey and Frenkel, 1998).

Treatment. No antiprotozoal treatments are known to eradicate *H. hammondi* infection in rats. Affected rat colonies should be culled.

Prevention and control. Natural infection of laboratory rats with *H. hammondi* is unlikely to occur in the modern animal facility. Preventive measures are similar to those for *T. gondii*. That is, cats should not be housed near laboratory rodents, and rodent feed and bedding supplies should be protected from accidental contamination with cat feces.

4. *Sarcocystis* spp.

Description and lifecycle. *Sarcocystis* spp. are apicomplexan protozoan parasites in the family Sarcocystidae. Several *Sarcocystis* spp. have been reported from wild rats. These include *S. singaporensis*, *S. villivillosi*, *S. zamani*, *S. cymruensis*, *S. sulawesiensis*, *S. dirumpens*, and *S. murinotechis* (Ashford, 1978; Beaver and Maleckar, 1981; Hafner and Frank, 1986; Jakel et al., 1997). The elongated cysts (or Miescher's tubes) of *Sarcocystis* spp. occur in cardiac and skeletal muscle tissues throughout the body, or in vascular endothelium, depending on parasite species. Cysts are bounded by a parasitophorous vacuole membrane. Zoites liberated from cysts are crescent shaped and differ in length and width by species. Many years ago, *Sarcocystis* spp. infections were common in laboratory rodents (Twort and Twort, 1932). However, that is no longer the case.

Typical of the family, the lifecycles of *Sarcocystis* spp. involve predator-prey relationships. The definitive hosts of *S. singaporensis*, *S. villivillosi*, and *S. zamani* are snakes. The definitive host of *S. cymruensis* is the cat, whereas the definitive hosts of *S. sulawesiensis*, *S. dirumpens*, and *S. murinotechis* are not known. Predators become infected while consuming rodents bearing tissue cysts.

Rats become infected by ingestion of sporulated oocysts or sporocysts shed in predator feces.

Hosts. In the wild, the *Sarcocystis* spp. described above may utilize a range of intermediate hosts, including *Rattus* sp., *Mastomys natalensis*, *Mus musculus*, *Meriones unguiculatus*, *Phodopus sungorus*, and *Bandicota indica* (Hafner and Frank, 1986; Jakel et al., 1997). In contrast, laboratory rats are rarely infected. The prevalence of natural infection in wild snakes is high, whereas cats are uncommonly infected.

Pathobiology. Of the species found in rats, *S. singaporensis* is the most pathogenic and has been proposed as a means of wild rat population control because rat populations collapse rapidly following baited introduction of the parasite. Although many rats develop fatal protozoal pneumonia, surviving rats mount a rapid and specific immune response (Jakel et al., 1996, 1999, 2001).

Clinical symptoms. Infections with most *Sarcocystis* spp. are asymptomatic. *S. singaporensis* causes terminal anorexia, labored breathing, and death (Jakel et al., 1996). Although it is possible that infection may result in myositis, muscle pain, and decreased motor activity, this has yet to be demonstrated.

Diagnosis. Diagnosis of sarcocystosis is based on the demonstration of the characteristic tissue cysts. The cyst wall may be smooth or bear projections. Cyst wall morphology is one of the primary features used to differentiate parasite species. Serologic tests have also been developed, but these are not likely to be practical.

Treatment. There is no information available on treatment of sarcocystosis in rats. In mice, *S. muris* infection can be eliminated by combined treatment with sulfaquinoxaline plus pyrimethamine (Rommel et al., 1981). It is not known whether this combination will also clear rats of *Sarcocystis* spp.

Prevention and control. Given the heteroxenous lifecycle of *Sarcocystis* spp., infection of laboratory rats is extremely unlikely. Clearly, snakes and cats should not be housed near laboratory rodents, and rodent feed and bedding stocks should be protected from contamination with snake and cat feces. Lastly, caretakers should practice adequate hygiene to prevent spread of contamination from house cats and pet snakes.

5. *Frenkelia* spp.

Description and lifecycle. *Frenkelia* spp. are apicomplexan parasites in the family Sarcocystidae and are genetically and biologically similar to *Sarcocystis* spp. (Mugridge et al., 1999). Bradyzoite cysts of *Frenkelia spp.* were first reported in the brain of laboratory-reared Fischer 344 rats (Hayden et al., 1976). The cysts are thin-walled and multilobulated, with many compartments outlined by fine interlacing septae. Each compartment contains numerous crescent-shaped, periodic acid-Shiff (PAS)-positive bradyzoites.

The lifecycles of *Frenkelia* spp. are typical of the family and involve predator-prey relationships. The definitive hosts of *Frenkelia* spp. are raptorial birds in the families Falconiformes and Strigiformes. These include hawks, eagles, vultures, and owls. Intermediate hosts, including rats, become infected by ingesting sporocysts or oocysts shed in the feces of definitive hosts. Only asexual reproduction occurs in the rat.

Hosts. *Frenkelia* spp. occur in many species of rodents, including rats, voles (*Microtus agrestis*), chinchillas (*Chinchilla laniger*), meadow mice (*Microtus modestus*), muskrats (*Ondatra zibethica*), lemmings (*Lemmus lemmus*), red-backed mice (*Clethrionomys glareolus*), and others. Relatively little is known of the host distribution and specificity of *Frenkelia* in raptorial birds. Whereas infections in raptorial birds are routinely diagnosed (Baker et al., 1996), infection in a laboratory rat has been reported only once (Hayden et al., 1976).

Pathobiology. After infection, merozoite formation (merogony) occurs in the rodent liver. Merogony may result in hepatic necrosis, and perivascular cellular infiltration in several organs. Tissue cysts containing bradyzoites are found in the cerebrum, cerebellum, brainstem, and cervical spinal cord. Usually no local reaction is noted (Hayden et al., 1976). However, cyst growth may occasionally cause pressure necrosis, followed by inflammatory changes characterized by granulomatous encephalitis with giant cell formation; perivascular and meningeal infiltrates; and gliosis. Usually however, no host local reaction is noted (Hayden et al., 1976).

Clinical symptoms. Infections with *Frenkelia* spp. are asymptomatic.

Diagnosis. In wild rats, infection with *Frenkelia* spp. is usually an incidental finding. Although many bradyzoite cysts are macroscopic, diagnosis is based on the histologic demonstration of cysts in the brain.

Treatment. Nothing is known about treatment of laboratory rodents for *Frenkelia* spp. infection. Infected rodents should be culled and the source of infection determined.

Prevention and control. Because *Frenkelia* spp. have an obligate two-host lifecycle, infection of laboratory rats is extremely unlikely. Raptorial birds should not be housed near laboratory rodents, and rodent feed and bedding stocks should be protected from contamination with raptor feces, including that introduced by filth-bearing insect transport hosts.

6. *Encephalitozoon cuniculi*

Description and lifecycle. *Encephalitozoon* (formerly *Nosema*) *cuniculi* is a microsporidian parasite capable of infecting a range of mammalian species, including rats (Muller-Doblies et al., 2002). The microsporidia are obligate intracellular parasitic fungi that lack mitochondria and reproduce asexually by binary fission. Sporoblasts within the host divide to produce spores, which are oval, about 1.5 by 2.5 μm in size, and secrete a thick, resistant spore wall. The spore contains a nucleus near one end, and a polar filament forming five to six coils at the other end. Spores usually form clusters in the brain, kidney, liver, macrophages, and less commonly, peritoneal exudates, heart muscle, pancreas, spleen, and other organs. *E. cuniculi* is disseminated via the bloodstream, either free or within macrophages.

The lifecycle is direct. The primary mode of transmission is through the ingestion of spores excreted in urine. Intrauterine transmission has been suggested in gnotobiotic mice, rabbits, and dogs, but has not been demonstrated in rats.

Host. *E. cuniculi* infections have been reported in several species of insects and mammals. Mammalian hosts include rats, mice, muskrat, hamsters, guinea pigs, multimammate rats, rabbits, cottontails, dogs, monkeys, and humans. At least three antigenically distinct strains of *E. cuniculi* have been recognized (Didier et al., 1995). The extent to which these strains are cross-infective and/or represent distinct species is unknown.

Pathobiology. In rats, infection with *E. cuniculi* has been primarily associated with lesions in the brain, kidneys, and to a lesser extent, liver. In one study, investigators found granulomatous encephalitis throughout the brain in 21% of 365 adult laboratory rats (Attwood and Sutton, 1965). The granulomas ("glial nodules" of tumor cells of a Yoshida transplantable ascites sarcoma (Petri, 1966). The infected tumor cells became less pathogenic than usual and did not give rise to solid tumors after subcutaneous inoculation of rats.

Clinical symptoms. Rats infected with *E. cuniculi* are asymptomatic. The most serious concern with *E. cuniculi* infection involves the potential compromise of research studies as a result of induced histologic and physiologic changes.

Diagnosis. Exudates from suspected animals should be air-dried, fixed with methanol, stained with Giemsa stain, and examined for spores. Normally, only a small portion of peritoneal mononuclear cells are infected, although the infection rate can be increased by immunosuppressive treatments. In tissue sections, one can demonstrate spores or granulomatous inflammation using an appropriate special stain. *E. cuniculi* stains positively with Giemsa, Goodpasture's carbol fuchsin, iron hematoxylin, and Gram's stain, whereas *T. gondii* does not. Conversely, *T. gondii* stains well with hematoxylin and eosin, whereas *E. cuniculi* stains poorly. Serologic assays also have been developed for the detection of *E. cuniculi*

infection. These are useful for colony surveillance, particularly in rabbit colonies. Lastly, newer methods will likely assume a greater role in diagnostic efforts. These include polymerase chain reaction assays (Didier et al., 1995; Kock et al., 1997) and single-strand conformation polymorphism analysis (Fedorko et al., 2001).

Treatment. No effective treatment is known. Some have reported the successful use of fenbendazole for eliminating *E. cuniculi* in rabbits (Suter et al., 2001). Similar studies have not been performed with infected rats. Infected rat colonies should be culled.

Prevention and control. Because of improvements in husbandry standards, infection of laboratory rat colonies is much less common than in the past. In contrast, some rabbit colonies continue to experience high prevalence rates. Laboratory rabbits should not be housed near rat colonies. Likewise, animal care personnel working with potentially infected rabbits should not also care for laboratory rats, or should care for rats first. *E. cuniculi* may cause severe disease in immunocompromised persons (Mohindra et al., 2002). Therefore, such persons should exercise caution when working with potentially infected laboratory animals, or preferably, should avoid them altogether.

C. Nematodes

1. *Trichinella spiralis*

Description and lifecycle. *Trichinella spiralis* is a member of the order Enoplida, and is therefore phylogenetically related to *Trichuris* spp., *Calodium* spp., and *Trichosomoides crassicauda*, the latter being another historically important parasite of rats.

Following ingestion of encysted infective larvae in undercooked muscle meat, worms rapidly develop to the adult stage in the intestine within 4 days. After copulation, the adult male worm dies and the adult female worms penetrate into the intestinal mucosa, where they produce eggs that hatch within the uterus of the worm. Adult female worms live for up to about 6 weeks and then die. However, the lifespan of adult but not larval worms is influenced by rat strain. Rats of the BUF and YO strains rapidly expel adult worms. In contrast, rats of the BI and WKA strains expel worms more slowly (Bell, 1992). Larvae released within the mucosa are taken up by the lymphatic system and disseminate throughout the body. Those reaching voluntary muscles such as the tongue, diaphragm, larynx, eye, and the masticatory and intercostal muscles enter striated muscle fibers and become surrounded by a capsule formed by the muscle fiber (Fig. 13-4). Larvae can remain infectious for several months in this "nurse cell" while awaiting consumption by another

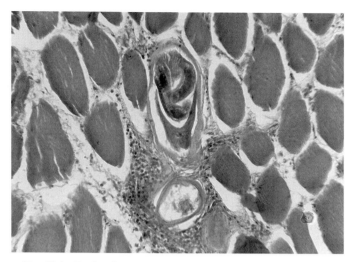

Fig. 13-4 *Trichinella spiralis*. Worm larva in the diaphragm of a rat. Granulomatous and eosinophilic inflammation is centered primarily around a degenerating larva. Hematoxylin and eosin. ×250.

suitable host. Transplacental transmission also has been demonstrated (Cosoroaba and Orjanu, 1998).

Hosts. *T. spiralis* infection may occur in several species of carnivorous and omnivorous mammals, including rats. Other species more commonly infected include wild pigs, bears, walruses, cats, dogs, and humans. Infection of wild rats may be high in areas where natural sylvatic cycles are established (Oivanen et al., 2000). In contrast, natural infection of laboratory rats is rare.

Pathobiology. Infected rats may develop enteritis due to the presence of large numbers of adult worms. Intestinal inflammation has been associated with functional motor changes and morphologic alterations (Tanovic et al., 2002). Heavy infection of the muscles of the diaphragm may result in respiratory compromise and death. Larval dissemination to the kidney, liver, pancreas, and other organs results in additional inflammatory responses in those sites. Lastly, heavy infection typically results in a marked eosinophilia and other cellular and humoral indicators of an immediate hypersensitivity reaction (Stewart et al., 1999).

Clinical symptoms. Infection with *T. spiralis* is usually asymptomatic. When adult stages are present in the intestine, rats may develop diarrhea and weakness. During larval migration and invasion, rats may experience muscular pain, weakness, anorexia, and respiratory distress.

Diagnosis. Definitive diagnosis is made by finding the larvae by microscopic examination of diaphragm or other skeletal muscle. Squash preparations of unfixed muscle are satisfactory for this purpose.

Treatment. Mebendazole eliminates >95% of larval stages of *T. spiralis*. Efficacy against adult worms in the intestine is less. Other benzimidazole anthelmintics are less effective (McCracken, 1978). Regardless, naturally infected rat colonies should be culled.

Prevention and control. While the laboratory rat has been extensively used as a model of *T. spiralis* infection, natural infections are rare. Laboratory rodent feeds should be protected from contamination with body parts of potentially infected hosts, including wild rodents.

2. *Calodium hepaticum*

Description and lifecycle. *Calodium hepaticum*, formerly known as *Capillaria hepatica*, is a member of the order Enoplida, and is therefore phylogenetically related to *Trichinella* spiralis, *Trichuris* sp., and *Trichosomoides crassicauda*. Adult worms are slender. The eggs are brownish, barrel-shaped, and possess a thick double wall, of which the outer one is distinctly pitted. At each end of the egg there is a plug (operculum) that does not bulge beyond the outline of the outer wall.

Calodium hepaticum has a direct lifecycle but may involve a transport host. The natural host acquires the infection by ingesting infective embryonated eggs in the environment. These hatch in the cecum, and the larvae penetrate the intestinal mucosa and enter the portal vein. Larvae reach the liver, where they mature within 3 weeks. The adult worms live for short periods but deposit large numbers of unembryonated eggs in the liver (Figs. 13-5 and 13-6). These eggs do not develop but may remain viable in the liver for many months. When the infected mammal is eaten by a carnivorous animal, eggs are then discharged and passed out in the feces. Alternatively, cannibalism and carcass decomposition may release eggs into the environment (Farhang-Azad, 1977).

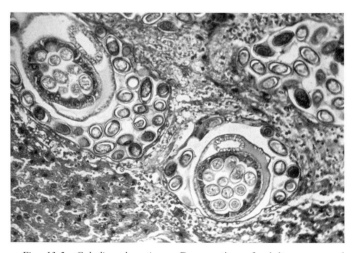

Fig. 13-5 *Calodium hepaticum*. Cross-section of adult worms and eggs, surrounded by inflammatory cells in the liver of an infected rat. Hematoxylin and eosin. ×100.

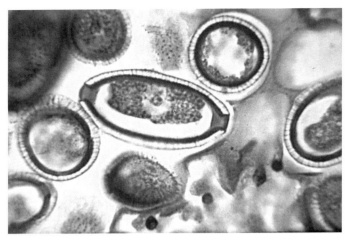

Fig. 13-6 Calodium hepaticum. Higher power magnification of eggs in the liver of an infected rat. Hematoxylin and eosin. ×400.

Eggs embryonate in the environment, and become infective in 2 to 6 weeks under suitable conditions.

Hosts. C. hepaticum is commonly found in the liver of wild rats and, less frequently, in mice, squirrels, muskrats, hares, beavers, dogs, pigs, nonhuman primates, and rarely, in man. The infection rate is often very high in wild rats. Ceruti and coworkers (2001) reported a prevalence of 36% of 47 wild rats, Conlogue and coworkers (1979) found 82% of 86 wild rats infected, and lastly, Luttermoser (1936) found 86% of 2,500 wild rats infected.

Pathobiology. Although the parasite has low pathogenicity in rats, infection can result in hepatomegaly and liver damage. Accumulations of eggs in the liver are usually visible as irregular yellow-gray spots or streaks on the surface. Histologically, the liver architecture is distorted by granulomatous foci. Adult parasites and/or eggs may be observed in the central portion of the lesions, which consist of an amorphous center with parasites surrounded by eosinophils, plasma cells, macrophages, epithelioid cells, and multinucleated giant cells. In more advanced stages, there are extensive areas of septal fibrosis, containing large numbers of eggs (Santos et al., 2001). In fact, the C. hepaticum-rat system has been developed as a valuable model for testing antifibrotic drugs (de Souza et al., 2000).

Clinical symptoms. Infection with C. hepaticum is generally asymptomatic. It can be anticipated that heavily infected rats may experience anorexia, malaise, and other signs compatible with hepatic dysfunction.

Diagnosis. Final diagnosis is based on the demonstration of parasites or eggs in the liver. Parasite eggs are only found in the rat feces after cannibalism of other infected rats. However, this is uncommon.

Treatment. Both mebendazole and ivermectin have been shown to eliminate early larval stages of C. hepaticum and to reduce fecundity of adult worms (El Gebaly et al., 1996). However, unless infected colonies are highly valuable, they should be culled.

Prevention and control. C. hepaticum infection should not occur in well-managed laboratory rat colonies. Rodent feed and bedding stocks should be protected from contamination with feces from other animals, particularly that of feral dogs.

3. *Trichosomoides crassicauda*

Description and lifecycle. Trichosomoides crassicauda is a member of the order Enoplida and is therefore phylogenetically related to *Trichuris* sp., *Trichinella spiralis*, and *Calodium hepaticum*. The small male worms, 1.3 to 3.5 mm long, live permanently in the uterus or vagina of the adult female worm. The latter is 9 to 10 mm long. The eggs are oval, brown, and thick-shelled with an operculum at each end, measuring 55 to 70 by 30 to 35 μm.

The lifecycle of *T. crassicauda* is direct. Infection is by ingestion of embryonated eggs passed in the urine. Infection most often occurs shortly after birth, from adult females to offspring (Weisbroth and Scher, 1971). Ingested eggs hatch in the stomach. Newly hatched larvae penetrate the stomach wall to migrate via the bloodstream to the lungs, kidney, and ureter, by which they arrive in the urinary bladder, where they mature into adult worms. The prepatent period is 50 to 60 days.

Host. T. crassicauda occurs in the urinary bladder, ureter, and renal pelvis of wild and laboratory rats. Historically, this parasite was frequently found in conventional laboratory rat colonies (Bone and Harr, 1967; Chapman, 1964; Bell, 1968; Weisbroth and Scher, 1971). More recently however, natural infection of laboratory rats is rare.

Pathobiology. The anterior end of the adult female embeds within tunnels burrowed into the transitional epithelium of the urinary bladder. The presence of the worms causes the epithelium to become hyperplastic but not inflamed (Antonakopoulos et al., 1991). Occasionally, worms also can be found in the lower part of the medullary collecting tubules or in the urinary pelvic cavity, with associated papillitis and pyelonephritis (Zubaidy et al., 1981). In the bladder, dead parasites may act as a nidus for calculus formation (Smith, 1946). Infection also has been associated with bladder tumors (Chapman, 1964). In addition, migrating larvae in the lung may produce multifocal granulomas (Zubaidy et al., 1981) and eosinophilia (Ahlquist et al., 1962).

Clinical symptoms. Infections with *T. crassicauda* are asymptomatic.

Diagnosis. Diagnosis depends on the demonstration of characteristic eggs in the urine, or female worms in the bladder.

Treatment. *T. crassicauda* infection can be eliminated by treatment of infected rats with 3 mg/kg ivermectin orally (Summa et al., 1992). Treatment of female rats prior to entry into a breeding program will prevent transmission of infection to offspring (Summa et al., 1992).

Prevention and control. Natural infection of laboratory rats with *T. crassicauda* is extremely uncommon. Purchase of parasite-free rats from reputable sources and exclusion of feral rats from the animal facility should prevent entry of this parasite.

4. *Gongylonema neoplasticum*

Description and lifecycle. *Gongylonema neoplasticum* is a spirurid nematode parasite in the family Thelaziidae. Male worms in this family have pre- and post-anal papillae and unequal spicules. *G. neoplasticum* is characterized by having the cephalic and esophageal regions ornamented with numerous cuticular plaques irregularly arranged in longitudinal rows on the dorsal and ventral parts of the body.

G. neoplasticum occurs in the tongue and in the squamous portion of the rat stomach. The lifecycle involves intermediate hosts, such as cockroaches, mealworms, and fleas. In these hosts, eggs hatch and the larvae develop and encyst in muscle. Infection occurs by ingestion of the infected intermediate host.

Hosts. *G. neoplasticum* utilizes rats and mice as definitive hosts. *G. neoplasticum* is uncommon even in wild rodents (Sahin, 1979; Krishnasamy et al., 1980) and is rarely found in laboratory rats. This is primarily due to exclusion of suitable intermediate hosts.

Pathobiology. The nematode parasitizes the wall of the stomach and can cause gastric ulcers. The parasitic lesions resemble a carcinoma-like proliferation of the gastric mucosa. Although *G. neoplasticum* was initially thought to contribute to the induction of stomach tumors, later studies failed to reproduce neoplastic lesions by the parasites. It was found that the earlier reported neoplastic lesions were due to the original rats having unwittingly been fed a vitamin A-deficient diet (Hitchcock and Bell, 1952).

Clinical symptoms. No clinical symptoms have been reported in rats infected with *G. neoplasticum.*

Diagnosis. Infection with *G. neoplasticum* can be diagnosed by finding characteristic eggs in fecal flotation examination or by finding the worms in the stomach at necropsy.

Treatment. Typical of this order of helminth parasites, there is no satisfactory anthelmintic treatment for *G. neoplasticum* infection. Infected rats should be culled.

Prevention and control. Natural infection of laboratory rats with *G. neoplasticum* is extremely uncommon. Rats should be purchased from reputable sources.

Laboratory rodent colonies should be protected from exposure to feral rodents and potentially infected arthropod intermediate hosts such as cockroaches.

D. Cestodes

1. *Taenia taeniaeformis (Cysticercus fasciolaris)*

Description and lifecycle. *Taenia taeniaeformis* is a cyclophyllidean tapeworm. Formerly, the larval, or metacestode form was named *Cysticercus fasciolaris.* Because the cysticercus transforms into a strobilocercus during larval development in the intermediate host, this parasite also has been referred to as *Strobilocercus fasciolaris.* The strobilocercus can reach a considerable size, up to 6 to 12 cm, and is infective for the definitive host roughly 60 days after infection. Larval forms of *T. taeniaeformis* produce an eosinophil chemotactic factor (Potter and Leid, 1986). The infection of rats occurs by ingestion of the embryonated eggs, which are shed in the feces of the definitive host, a cat. Oncospheres emerge in the small intestine and make their way to the liver where they grow rapidly and develop into the cysticercus stage by day 30 (Fig. 13-7). By day 42, the scolex evaginates and becomes connected to the bladder by a segmented strobila. In this way it resembles a small tapeworm (Soulsby, 1982). Transmission to the definitive host is by ingestion of infected liver. After ingestion, the strobila and bladder of the larva are digested away and the scolex, which is all that remains, attaches itself to the wall of the upper small intestine, and strobilation proceeds. The prepatent period in the cat is 36 to 42 days, and the adult worms can survive for up to 3 years.

Hosts. *T. taeniaeformis* occurs in the liver of wild rats, mice, and other rodents. Infection of laboratory

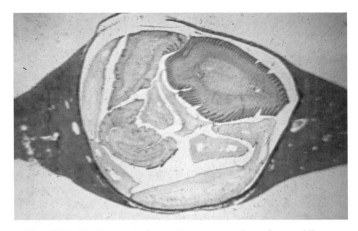

Fig. 13-7 *Taenia taeniaeformis.* Low-power view of a strobilocercus in the liver of a rat.

rats is rare. Laboratory rat strains have been shown to differ in susceptibility to infection with *T. taeniaeformis* (Williams et al., 1981). In addition, strains of *T. taeniaeformis* differ in infectivity (Brandt and Sewell, 1981).

Pathobiology. Infection of rats with *T. taeniaeformis* can cause both lesion development and physiologic alterations. Infection is recognized by the development of the white to clear cysticercus and/or strobilocercus in the liver. Microscopically, the parasite induces proliferation of connective tissue, which forms a thin circular fibrous capsule. A chronic inflammatory reaction to the parasite can result in formation of multinodular fibrosarcomas (Hanes and Stribling, 1995). Other pathologic findings include hypertrophic gastritis and mucous cell proliferation (Cook and Williams, 1981; Konno et al., 2000), thymic atrophy, and splenic changes (Cook et al., 1981). Roughly 2 months after heavy experimental infection, circulating levels of specific hepatic enzymes, as well as blood urea nitrogen, are elevated, whereas serum glucose and total protein levels are lower (Konno et al., 2000). There is a reduced response to histamine, leading to suppression of gastric acid secretion and hypergastrinemia (Oku et al., 2002). Production of interleukin-2 by splenic lymphocytes is reduced, as are proliferative responses to mitogens (Burger et al., 1986).

Clinical symptoms. Infection of rats with *T. taeniaeformis* reduces both male and female fertility. The decrease in male reproductive parameters appears to be due to reduced testosterone production (Lin et al., 1990). Additional clinical signs accompanying infection are limited to abdominal enlargement associated with tumor formation.

Diagnosis. Definitive diagnosis of infection in rats is based on demonstrating the metacestode stages in the liver.

Treatment. Elimination of infection has been achieved with the use of substituted benzimidazole anthelmintics (Gupta et al., 1992). However, infected rats are unsuitable for most research uses and should be culled.

Prevention and control. Effective prevention and control include proper management of the animal facility to prevent contamination of rodent feed and bedding stocks with feces from feral cats.

III. PARASITES OF THE ALIMENTARY SYSTEM

A. Flagellates

1. *Giardia muris*

Description and lifecycle. Giardia muris is a flagellated protozoan in the phylum Sarcomastigophora. *G. muris*

occurs in two lifecycle stages. The trophozoites are 7 to 13 by 5 to 10 μm, pear-shaped, and bilaterally symmetrical. Distinguishing features include an attachment disk, two anterior nuclei, four pair of flagellae, and a "claw hammer"-shaped median body. The median body appears to function as a repository of cytoskeletal elements. Cysts are ellipsoidal (15 by 7 μm), and may have up to two median bodies and four nuclei.

The lifecycle of *G. muris* is direct. Trophozoites live in the anterior small intestine, where they colonize the mucosal surface of the proximal small intestine, adhering to columnar cells near the bases of intestinal villi. Upon entering the stream of ingesta, trophozoites transform into cysts, which are environmentally resistant. Infection occurs after ingestion of infective cysts and excystment in the small intestine. Shortly after excystment, trophozoites divide longitudinally and colonize the small intestine.

Hosts. G. muris infects rats, mice, hamsters, and many other rodents. However, Kunstyr and coworkers (1992) failed to infect rats with *G. muris* recovered from mice and hamsters. It is uncertain whether this was due to differences at the host strain or species level. Currently, the prevalence of infection in laboratory rats is unknown. There is no evidence that *G. muris* is zoonotic.

Pathobiology. Remarkably little is known of the pathologic changes that occur in rats infected with *G. muris*. In mice, pathologic changes include villous blunting; increased numbers of intraepithelial lymphocytes, goblet cells, and mast cells; and alterations in intestinal disaccharidase content (Scott et al., 2000).

Clinical symptoms. Infection with *G. muris* is typically asymptomatic. Heavy infections may potentially cause some degree of inappetence and lethargy, but these have not been adequately documented.

Diagnosis. Trophozoites can be recognized easily in a wet preparation of intestinal contents by their characteristic rolling and tumbling movements. Alternatively, a small amount of fresh feces can be mixed with saline and examined after addition of a small amount of iodine, which stains internal structures. Lastly, routine fecal flotation procedures reveal cysts, though these may be distorted by the flotation medium.

Treatment. Treatment is usually not attempted due to the generally nonpathogenic nature of the infection. For studies where *Giardia*-free rats are needed, eradication can be accomplished by rederivation and barrier maintenance. Metronidazole may be useful for controlling clinical infections but will not eradicate the infection (Cruz et al., 1997).

Prevention and control. Control depends on proper sanitization. Cysts can be killed by 2.5% phenol or Lysol, temperatures at or above 50°C (Červa, 1955), and by ultraviolet light (Hayes et al., 2003).

2. Spironucleus (Hexamita) muris

Description and lifecycle. *Spironucleus muris,* formerly *Hexamita muris,* is a flagellated protozoan in the Phylum Sarcomastigophora. Trophozoites are 7 to 9 by 2 to 3 μm, pear-shaped, bilaterally symmetrical, and contain two longitudinal axostyles, and four pairs of flagellae arising from a pair of blepharoplasts situated between the paired nuclei (Fig. 13-8). There is disagreement concerning the existence of a true cyst stage (Kunstyr, 1977). Cyst-like structures are observed, and are roughly 4 by 3 μm, oval, and contain up to four nuclei (Baker et al., 1998).

The lifecycle of *S. muris* is direct and is similar to *Giardia muris.* Trophozoites live in the distal small intestine and cecum, where they colonize the mucosal surface, especially the crypts of Lieberkuhn. Infectious stages are released with the fecal stream.

Hosts. Susceptible hosts include rats, mice, hamsters, and possibly other rodents. Little information is available on cross-infection of *S. muris* isolates between rodent species. Currently, the prevalence of infection in laboratory rats is unknown.

Pathobiology. Little is known of the pathobiology of *S. muris* infection in rats. In mice, pathologic changes may include intestinal mucosal epithelial desquamation, subepithelial edema, crypt hyperplasia, variable villous atrophy, and enteritis (Schagemann et al., 1990; Whitehouse et al., 1993). Mouse strains differ in severity of infection based on major histocompatibility complex (Baker et al., 1998). It is unknown whether similar strain differences occur in rats.

Clinical symptoms. Many infections with *S. muris* are asymptomatic. Clinical signs are more likely to be observed in stressed or very young animals, versus immunologically normal adults. When present, clinical signs may include rough hair coat, depression, weight loss, listlessness, distended abdomen, diarrhea, and occasionally, death.

Diagnosis. *S. muris* trophozoites may be demonstrated in a fresh wet mount of small intestinal contents. Alternatively, feces or intestinal contents can be preserved in sodium acetate-acetic acid-formalin, followed by trichrome staining for trophozoites or cyst-like structures (Baker et al., 1998).

Treatment. To date, antiprotozoal agents have been ineffective in eliminating the organism (Meshorer, 1969; Oxberry et al., 1994).

Prevention and control. Prevention of disease is based on obtaining *S. muris*-free rats, strict adherence to good hygiene, and stress reduction. Because infection is easily overlooked, preshipment health records may not accurately reflect the infection status of commercially available rodents (Baker et al., 1998).

3. Hexamastix muris

Hexamastix muris occurs as a nonpathogenic protozoan of the cecum in the rat, hamster, and other rodents. The prevalence rate of this infection in laboratory rats is unknown. *H. muris* has a piriform body (9 by 7 μm) with an anterior nucleus and cytostome, a pelta, a conspicuous axostyle, a prominent parabasal body, five anterior flagella, and a trailing flagellum (Levine, 1985). Cysts are not formed.

4. Chilomastix bettencourti

Chilomastix bettencourti occurs in the cecum of the rat, mouse, hamster, and wild rodents. Current prevalence rates in laboratory rats are unknown. The trophozoites of this nonpathogenic flagellate are piriform, with an

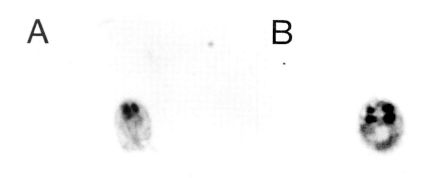

A   B

Fig. 13-8 Spironucleus muris. Both cysts have undergone internal division. **(A)** Two nuclei are in focus and a third is in partial focus in one cyst. **(B)** All four nuclei are seen in the other cyst. **Bar = 5μ.** Trichrome stain. ×3400.

anterior nucleus, a large cytostomal groove, three anterior flagella, a short fourth flagellum that undulates within the cytostomal cleft, and a cytoplasmic fibril along the anterior end and sides of the cytostomal groove (Levine, 1985). The cysts are usually lemon shaped and contain a nucleus and all organelles of the trophozoites.

5. *Tritrichomonas muris*

Tritrichomonas muris is nonpathogenic and occurs in the cecum, colon, and small intestine of the rat, mouse, hamster, and wild rodents (Levine, 1985). It is pear shaped and measures 16 to 26 by 10 to 14 μm. It has an anterior vesicular nucleus. Anterior to the nucleus is the blepharoplast from which arise three anterior flagella and a posterior flagellum that is attached to the body by means of an undulating membrane that continues posteriorly as a free flagellum. It has a large oval-shaped nucleus, slitlike cytostome, and a sausage-shaped parabasal body in the anterior region. Reproduction is by binary fission. The organism swims with a characteristic wobbly movement. Although some report the existence of a true cyst stage (Chinchilla et al., 1987), more recent studies indicate that the undulating membrane and flagella become internalized without forming a cyst wall. This stage is termed a "pseudocyst" (Lipman et al., 1999). Transmission is by ingestion of pseudocysts in feces.

6. *Tritrichomonas minuta*

This nonpathogenic trichomonad occurs in the cecum and colon of the rat, mouse, hamster, and other rodents. It measures 4 to 9 by 2 to 5 μm. Little is known of the biology of *T. minuta*. It is assumed to be similar to *T. muris* (Levine, 1985).

7. *Tetratrichomonas microti*

This nonpathogenic flagellate is common in the cecum of the rat, mouse, hamster, vole, and other wild rodents in North America. It is 4 to 9 μm long. The organism has four anterior flagellae and a posterior free-trailing flagellum. It has been experimentally transmitted from the mouse to several other species of mammals, including the guinea pig, ground squirrel (*Citellus citellus*), dog, and cat (Levine, 1985).

8. *Pentatrichomonas hominis*

Pentatrichomonas hominis occurs in the cecum and colon of the rat, mouse, hamster, dog, cat, guinea pig, ground squirrel, several species of nonhuman primates, and man (Levine, 1985; Romatowski, 2000). This generally nonpathogenic organism can be cross-transmitted among these species (Fukushima et al., 1990). Although disease has not been reported in laboratory rats, *P. hominis* has been associated with diarrhea in kittens (Romatowski, 2000). *P. hominis* is piriform, measuring 8 to 20 by 3 to 14 μm, and ordinarily has five anterior flagellae plus one posterior flagellum.

B. Sporozoa

1. *Eimeria nieschulzi*

Description and lifecycle. *Eimeria nieschulzi* is a true coccidian parasite in the phylum Apicomplexa and family Eimeriidae. The biology of *E. nieschulzi* has been well-characterized as a model for coccidiosis. In addition, *E. nieschulzi* affects host physiology and in so doing, interferes with concurrent infection by other parasites (Upton et al., 1987). Oocysts of *E. nieschulzi* are ellipsoidal to ovoid, taper at both ends, smooth-walled, and measure 16 to 26 by 13 to 21 μm (Soulsby, 1982).

After ingestion of sporulated oocysts, four generations of schizogony, followed by gametogony, occur in the small intestine. The prepatent period is 7 days (Duszynski, 1972). Oocysts are shed in the feces for only 4 to 7 days. Shed oocysts sporulate within 3 days. Sporulated oocysts contain four sporocysts, each with two sporozoites (Soulsby, 1982; Smith et al., 1995).

Host. *Eimeria nieschulzi* is common in the small intestine of wild rats but uncommon in the laboratory rat. In experimental studies, sex of rat did not influence the severity of infection (Liburd, 1973). *E. nieschulzi* does not develop beyond the early schizont stage in mice (Marquardt, 1966).

Pathobiology. *E. nieschulzi* is considered mildly pathogenic in small numbers. However, after primary infection with large numbers of sporulated oocysts, pathologic changes may be severe, and include crypt hyperplasia, villous atrophy, increased villous to crypt ratio, cellular infiltration, edema, hyperemic to hemorrhagic bowel, and increased mass of the small intestine (Becker, 1932; Smith et al., 1995; Duszynski, 1972; Duszynski et al., 1978). Intestinal disaccharidase and peroxidase levels decline during the first week post–inoculation in a dose-dependent manner before returning to normal by day 16 (Duszynski et al., 1978). In addition to inducing pathologic lesions, *E. nieschulzi* infection reduces digestibility of dry matter, organic matter, and nitrogen (Frandsen, 1983). Also, *E. nieschulzi* may alter the host response to concurrent or subsequent infection with nematode parasites. For example, *E. nieschulzi* exacerbates anorexia, prevents eosinophil release from bone marrow, and lengthens patency in *Nippostrongylus brasiliensis*-infected rats (Frandsen, 1983; Upton et al., 1987), and reduces worm burden in *Trichinella spiralis*-infected rats (Stewart et al., 1980). Typical of coccidial infections, the intensity of immunity following infection is dose-dependent (Liburd, 1973).

Clinical symptoms. Clinical signs are not observed after infection with up to 5,000 oocysts. In contrast, infection with 30,000 or more oocysts causes loss of appetite, weight loss or reduced weight gain, and transient diarrhea or reduced fecal output (Becker, 1934; Liburd, 1973; Duszynski, 1972). Infection is self-limiting and induces strong humoral immunity (Liburd, 1973; Smith et al., 1995).

Diagnosis. Diagnosis is based on the histologic demonstration of the organism in the intestine or the identification of oocysts in the feces.

Treatment. Although commonly used anticoccidial drugs will likely ameliorate clinical signs, laboratory rats infected with *E. nieschulzi* are unsuitable as research subjects. Infected rats should be eliminated or caesarean rederived.

Prevention and control. Because coccidiosis is self-limiting, control primarily depends on good sanitization and purchase of parasite-free rats.

2. *Eimeria* spp.

Wild and laboratory rats may become infected with several *Eimeria* spp., in addition to *E. nieschulzi*. These include *E. miyairii, E. separata, E. nochti, E. hasei, E. ratti, E. carinii,* and the closely related, *Isospora ratti* (Soulsby, 1982). These are generally nonpathogenic, and should rarely be found in laboratory rats. Lifecycles and other biological features are essentially similar to *E. nieschulzi*.

C. Amebae

1. *Entamoeba muris*

Entamoeba muris is common in wild rats, mice, hamsters, and many wild rodents (Levine, 1985; Abd el-Wahed et al., 1999). Prevalence rates in laboratory rodent colonies are unknown, but have historically been high (Mudrow-Reichenow, 1956; Higgins-Opitz, 1990). *E. muris* trophozoites are 8 to 30 μm long. Oocysts are 9 to 20 μm and contain up to eight nuclei. A food vacuole often is present in the cyst. The cyst produces eight amebae after ingestion by a new host. Trophozoites multiply by binary fission. Normally these amebae live in the lumen of the cecum and less commonly, the upper colon (Lin, 1971), where they feed on food particles and bacteria. *E. muris* is nonpathogenic but is undesirable in research animals.

D. Ciliates

1. *Balantidium coli*

Description and lifecycle. *Balantidium coli* is a large ciliate in the phylum Ciliophora. Trophozoites are ovoid, 30 to 150 by 150 by 25 μm, and with a subterminal tubular mouth. They are covered with cilia and with these they move rapidly. Cysts are spherical to ovoid, 40 to 50 μm in diameter, with many starch-containing food vacuoles. Multiplication is by transverse binary fission and conjugation. Transmission is by ingestion of cysts.

Hosts. *B. coli* commonly occurs in the cecum and colon of pigs, nonhuman primates, and humans. Less commonly, *B. coli* may be found in the rat, hamster, guinea pig, and dog (Hankinson et al., 1982; Nakauchi, 1999).

Pathobiology. *B. coli* is usually nonpathogenic but may sometimes secondarily invade lesions initiated by other pathogens.

Clinical symptoms. *B. coli* has not been associated with clinical disease in rats.

Diagnosis. Balantidosis can be recognized easily by microscopic examination of intestinal contents or by histologic examination of intestine.

Treatment. Metronidazole and tetracyclines have been used successfully to eliminate *B. coli* infection in humans (Garcia-Laverde and de Bonilla, 1975). Treatment of laboratory rats is generally not attempted. Infected rats should be rederived or culled.

Prevention and control. Commercial vendors of laboratory rats routinely test for infection with *B. coli*. Therefore, commercially purchased rats should not be infected with *B. coli*. Purchase of *B. coli*-free rats, along with strict adherence to standard operating procedures for sanitation, should prevent infection with this organism.

E. Nematodes

1. *Syphacia muris*

Description and lifecycle. *Syphacia muris* is the common pinworm of rats. *S. muris* closely resembles the common mouse pinworm, *S. obvelata*, with which it is superficially confused (Fig. 13-9) (Hussey, 1957). The two can be differentiated by the size of the egg, positions of excretory pore and vulva or mamelons, and length of the tail (Hussey, 1957; Tafts, 1976). The adult worm is cylindrical, with a rounded anterior region and a tapered posterior that ends in a sharply pointed tail. The mouth is surrounded by three distinct lips. The male is about 1.2 to 1.3 mm long and 100 μm wide. The anterior mamelon is near the middle of the body. The tail of the male is bent ventrally. The male worm has a single, long,

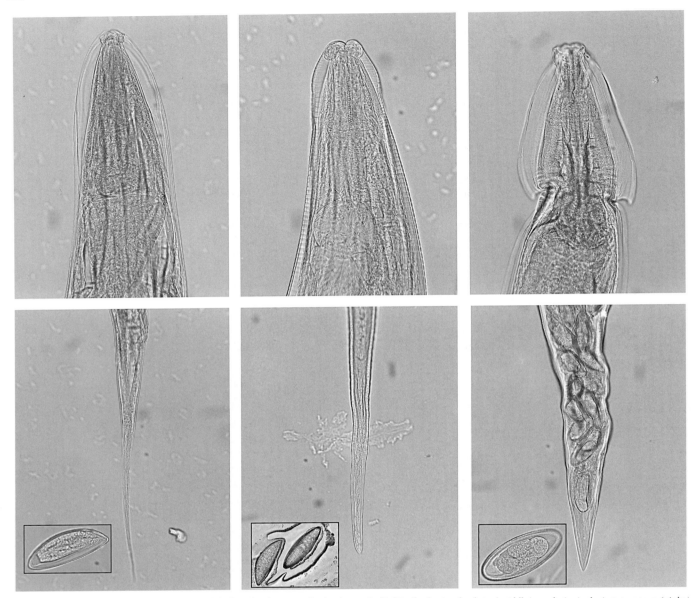

Fig. 13-9 Heads, tails, and eggs of the adult female pinworm *Syphacia muris (left)*, *Syphacia obvelata (middle)*, and *Aspiculuris tetraptera (right)*. Heads and tails photographed at ×250. Eggs photographed at ×400.

prominent spicule and a gubernaculum. The female is 2.8 to 4.0 mm long with its vulva in the anterior quarter of the body. The egg is vermiform, slightly flattened on one side and measures 72 to 82 by 25 to 36 µm.

The lifecycle of *S. muris* is direct (Stahl, 1963). Adult worms inhabit the cecum and colon. Eggs are deposited by the female on the perianal area of the host or in the colon. The eggs embryonate within several hours, thereby becoming infective. Infection of the rat occurs by ingestion of embryonated eggs on the perianal area, or in the cage environment. Retroinfection via the anus, by newly hatched larvae, also occurs, and may account for the extremely high parasite burdens

observed with this infection. The prepatent period is 7 to 8 days (Lewis and D'Silva, 1986).

Hosts. *S. muris* infects rats, and occasionally, laboratory mice, gerbils, and the Mongolian hamster (Hussey, 1957; Ross et al., 1980).

Pathobiology. Like other pinworm infections of rodents, *S. muris* is generally considered mildly pathogenic. Lesions are not observed in infected rats. However, infection with *S. muris* is known to alter host physiology and thereby confound research. For example, rats infected with *S. muris* have impaired intestinal electrolyte transport (Lïbcke et al., 1992) and decreased weight gain (Wagner, 1988).

Clinical symptoms. *S. muris* is not known to induce clinical disease other than reduced weight gain (Wagner, 1988).

Diagnosis. Infection with *S. muris* is readily diagnosed by finding eggs on clear tape that has been applied to the perianal region. Tape tests should be performed in the afternoon, as perianal egg counts are higher during this time than in the morning, due to the periodicity of egg production (Tafts, 1976). Additionally, adult worms can be observed in the large intestine. This is readily accomplished by removing and opening the cecum, and placing it in a petri dish containing a small amount of warm water or saline. Within minutes the worms will migrate away from the fecal mass and into the liquid, where they may be observed with a dissecting scope or hand–held magnifying glass.

Treatment. Relatively few anthelmintics are consistently effective at eliminating all lifecycle stages of pinworms from rats and other rodents. Although several treatment strategies have been reported (Pritchett and Johnston, 2002), the most efficacious and cost effective means is through provision of feed containing fenbendazole at a rate of 150 mg/kg of feed, given for 2 to several weeks of treatment (Pritchett and Johnston, 2002). Fenbendazole not only has larvicidal and adulticidal efficacy, but is also ovicidal (Kirsch, 1978; Lacey et al., 1987). Although most advocate simultaneous disinfection of all surfaces within the animal room, others have found feeding of medicated feed adequate to eliminate the infection (Huerkamp et al., 2000).

Prevention and control. It is desirable to prevent entry of pinworms into the animal facility. This is accomplished by only purchasing parasite-free animals. Alternatively, suspect or known positive rats can be treated with fenbendazole-medicated feed while in quarantine and released to the animal facility when determined to be cleared of infection. Control depends on colony surveillance and strict attention to cleaning protocols that remove and disinfect potentially contaminated fomites.

2. *Syphacia obvelata*

Laboratory rats are susceptible to infection with the common mouse pinworm, *Syphacia obvelata* (Hussey, 1957; Kellogg and Wagner, 1982). The two parasites can be distinguished morphologically (Fig. 13-9) (Hussey, 1957; Tafts, 1976). Adult worms are found in the cecum and colon. Male pinworms are 1.1 to 1.5 mm long and 120 to 140 μm wide, with a long tail bearing a distinct spicule and gubernaculum. The mouth is surrounded by three simple lips. The anterior end has small cervical alae. The posterior end is bent ventrally. The females are 3.4 to 5.8 mm long and 0.24 to 0.40 mm wide.

The eggs are thin-shelled, crescent shaped, flattened on one side, and contain an undifferentiated embryo *in utero*. Eggs measure 111 to 153 by 33 to 55 μm.

The lifecycle of *S. obvelata* is direct and similar to that of *S. muris*. Eggs are deposited on the perianal region, where they embryonate within several hours. Infection is either by ingestion of infective eggs or by retroinfection. The prepatent period is 11 to 15 days (Tafts, 1976).

S. obvelata is known to reduce the incidence of experimentally induced adjuvant arthritis in the rat (Pearson and Taylor, 1975). Other aspects of the host-parasite relationship, diagnosis, treatment, prevention, and control, are as described for *S. muris*.

3. *Aspiculuris tetraptera*

Like *S. obvelata*, *Aspiculuris tetraptera* more commonly infects mice. Occasionally, laboratory rats may also become infected. *A. tetraptera* is morphologically similar to, but readily distinguished from *S. obvelata* (Fig. 13-9) (Tafts, 1976). The males are 2 to 4 mm long and 120 to 190 μm wide with a short conical tail measuring 117 to 169 μm long. Both spicule and gubernaculum are absent. The females are 3 to 4 mm long and 215 to 275 μm wide with a conical tail of 445 to 605 μm long. The eggs are symmetrically ellipsoidal and measure 70 to 98 by 29 to 50 μm.

The lifecycle of *A. tetraptera* is direct but distinct from that of *Syphacia* spp. in that unembryonated eggs are passed entirely in the feces and are not found in the perianal region of the host. *A. tetraptera* eggs develop and become infective in 5 to 8 days at 27°C (Chan, 1955). The eggs are resistant to desiccation and many disinfectants but are sensitive to high temperatures. Infection is by ingestion of infective eggs. Larvae hatch and develop in the posterior colon and then migrate anteriorly and develop to maturity in the proximal colon. They remain in the lumen of the intestine and do not invade the mucosa. The prepatent period is 23 days.

A. tetraptera is generally regarded as harmless in rats. Diagnosis is by finding the characteristic eggs in fecal floatation preparations or adult worms in the colon. *A. tetraptera* may be eliminated by feeding fenbendazole-medicated feed, as described for the *Syphacia* spp. (Boivin et al., 1996). Likewise, prevention and control are as described for *Syphacia* spp.

4. *Heterakis spumosa*

Heterakis spumosa is a member of the superfamily Subuluroidea and family Heterakidae (Soulsby, 1982).

Members of this superfamily resemble members of the superfamily Ascaridoidea. The male worms are 3.5 to 8 mm long with a distinct spicule. The males have a preanal sucker with a chitinous rim. The females are 6.8 to 8 mm long. The eggs have a thick-walled, mammillated shell and measure 55 to 60 by 40 to 55 μm.

The lifecycle is direct. Unembryonated eggs are passed in the feces and become infectious in 14 days under optimal conditions. Infection is by ingestion of infective eggs that hatch in the stomach. The larvae then migrate to the cecum and colon where they develop and mature. Eggs appear in the feces 26 to 47 days after infection (Smith, 1953). Modern methods of rodent colony management have made infection with *H. spumosa* extremely rare.

H. spumosa is considered nonpathogenic in rats and mice, causing neither clinical disease nor lesion development. Diagnosis of infection is based on demonstrating eggs in the feces or adult worms in the large intestine. *H. spumosa* has been eliminated from mice by combination therapy with febantal and pyrantel, but not with either drug alone (Mehlhorn and Harder, 1997). It is likely that similar results would follow treatment of infected rats. Prevention and control are by purchase of parasite-free rats, and through strict adherence to good sanitation and facility management practices.

F. Cestodes

1. *Rodentolepis nana*

Description and lifecycle. *Rodentolepis nana* (dwarf tapeworm), formerly referred to as *Hymenolepis nana* or *Vampirolepis nana*, is a member of the class Eucestoda and family Hymenolepididae (Soulsby, 1982). Tapeworms of this genus are usually narrow and threadlike. Adult *R. nana* vary greatly in size, but usually range from 25 to 40 mm in length by 1 mm wide. The scolex bears four unarmed suckers and a retractable muscular rostellum encircled at the anterior end by a single ring of 20 to 27 small hooklets (Fig. 13-10). The strobila consists of a chain of segments. Gravid segments contain 100 to 200 eggs that are liberated by disintegration of the terminal segments and are passed out in the feces. The egg is oval, colorless, and measures 44 to 62 by 30 to 55 μm. The embryo is spherical and thin walled with a knob at each pole, from which six fine filaments emerge. The oncosphere (hexacanth embryo) possesses three pairs of small hooks (Soulsby, 1982).

The lifecycle of *R. nana* may be direct or indirect (without or with an intermediate host) as well as by autoinfection. In the direct lifecycle, embryonated eggs are ingested by the definitive host and hatch in the

small intestine, where the oncosphere emerges and penetrates into villi to develop into a cysticercoid larva in 4 to 5 days. The larva then re-enters the lumen of the gut, and the scolex evaginates and attaches to the mucosa. Mature worms then develop in 10 to 11 days. Therefore the direct lifecycle requires 14 to 16 days. The adults live for only a few weeks in the host. The direct cycle induces a certain degree of immunity, which can prevent autoinfection. In the indirect lifecycle, proper intermediate hosts such as grain beetles (*Tenebrio molitor* and *T. obscurus*) and fleas (*Pulex irritans*, *Ctenocephalides canis*, and *Xenopsylla cheopis*) are required. The embryonated eggs are ingested by the arthropod intermediate hosts. In the gut, the egg hatches, and the oncosphere penetrates the hemocoele and develops into a cysticercoid larva. This process is completed at metamorphosis. The definitive host becomes infected by ingesting an infected arthropod. The indirect route of transmission confers little host immunity and permits autoinfection to occur. During autoinfection, the eggs hatch within the small intestine of the same host in which they are produced. The embryos develop into mature worms without leaving the original host (Heyneman, 1961).

Hosts. *R. nana* occurs in the small intestine of mice, rats, and hamsters (Katiyar et al., 1983). Ito and Kamiyama (1984) were unable to produce adult tapeworms when they infected rats with mouse-derived cysticercoids and suggested that the rat is an unnatural host for this parasite. Also, although many consider *R. nana* of rodents to be zoonotic, others have reported inability to infect rats with human isolates of *R. nana*, suggesting that rodent and human isolates of *R. nana* may differ between and among host species (Macnish et al., 2002). In earlier periods, infections

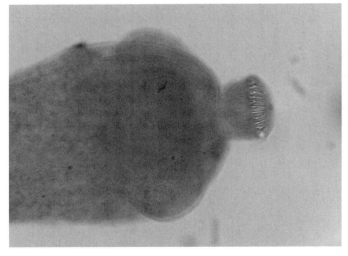

Fig. 13-10 Rodentolepis nana. Scolex-bearing rostellum armed with hooklets. ×250.

were common in laboratory animal colonies. More recently however, improvements in husbandry practices have rendered the infection extremely uncommon.

Pathobiology. R. nana is pathogenic only when large numbers of worms are present. Infection results in catarrhal enteritis. Chronically, abscesses and focal granulomatous lymphadenitis of mesenteric lymph nodes have been reported (Simmons et al., 1967). Serum albumin levels decline for the first 20 days post-infection in rats experimentally infected with R. nana. This is accompanied by increased levels of λ-globulins, indicative of a developing immune response. Thereafter, albumin levels return toward normal, coincident with worm expulsion (Katiyar et al., 1973). Lastly, increased amounts of histamine or a histamine-like compound are found in the intestine of R. nana-infected rats (Katiyar and Sen, 1970).

Clinical symptoms. Heavy infections of young rodents with R. nana may result in retarded growth and weight loss (Soulsby, 1982).

Diagnosis. Diagnosis is based on the demonstration of the characteristic eggs in the feces or the adults in the intestine at necropsy.

Treatment. Infection may be eliminated with a single oral dose of praziquantel (12.5 mg/kg) (Gupta et al., 1980). Other compounds are likewise effective. These include niclosamide (Hughes et al., 1973), nitroscanate (Gonenc and Sarimehmetoglu, 2001), and others (Soulsby, 1982). In general however, infected colonies should be culled.

Prevention and control. An effective prevention and control program should include purchase of parasite-free animals, vermin control, and high standards of animal husbandry and facility management.

2. *Hymenolepis diminuta*

Description and lifecycle. Hymenolepis diminuta is another member of the tapeworm family Hymenolepididae (Soulsby, 1982). The parasite is commonly known as the rat tapeworm. H. diminuta may be distinguished from R. nana by its greater size, being 20 to 60 mm in length and 3 to 4 mm in width; and its having a small pear-shaped scolex bearing four deep suckers with an unarmed rostellum (Fig. 13-11). The proglottids are similar to those of R. nana, being broader than they are long and containing three oval testes. The gravid segments break away from the strobila and are passed in the feces. The eggs are spherical, measure 60 to 88 μm by 52 to 81 μm and contain a hexacanth embryo that also possesses three pairs of small hooks. Unlike R. nana, the embryo does not have polar filaments. The eggs are more resistant to the environment than those of R. nana and can survive in the feces for up to 6 months.

The lifecycle of H. diminuta is always indirect and requires arthropods such as mealworm beetle (*Tenebrio molitor*), confused flour beetle (*Tribolium confusum*), fleas (*Nosopsyllus fasciatus*), or moths as an intermediate host. The eggs are ingested by intermediate hosts and hatch, and the oncosphere migrates to the hemocoele, where it develops into a cysticercoid. Infection of the definitive host is by ingestion of the infected arthropod. The larva is liberated from the cysticercoid, and the scolex evaginates and attaches itself to the intestinal mucosa where they develop into adult worms in 19 to 21 days. Unlike R. nana, the cysticercoid of H. diminuta cannot develop within the definitive host.

Hosts. H. diminuta occurs in the upper small intestine of rats, mice, hamsters, and primates, including humans. Rat strains differ in susceptibility to infection. In one study, inbred rats of the TM and DA strains developed 60% and 30% fewer total adult worms, respectively, versus F344/N, JAR-2, LOU/M, and outbred Wistar rats (Ishih et al., 1992).

Pathobiology. Light infections with H. diminuta are nonpathogenic (Insler and Roberts, 1976). Heavy infections are rare because primary infection limits the size of the enteric worm population, and results in strong but short-lived immune resistance to reinfection (Andreassen and Hopkins, 1980). Heavy infection may cause acute catarrhal enteritis or chronic enterocolitis with lymphoid hyperplasia. However, even mild infections may alter host physiology, including increased intestinal permeability (Zimmerman et al., 2001).

Clinical symptoms. Infection with H. diminuta is often asymptomatic. Because the lifecycle includes an obligatory intermediate host, and worm populations

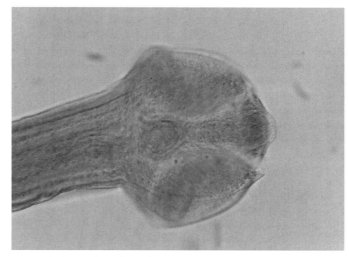

Fig. 13-11 Hymenolepis diminuta. Scolex with unarmed rostellum. ×250.

are density-dependent, heavy worm burdens are rare. It is possible that young rats might experience poor growth and development.

Diagnosis. Diagnosis is based on finding eggs or proglottids in the feces or the adult worms in the intestine.

Treatment. Treatments recommended for *R. nana* will likely also be effective against *H. diminuta*. However, infected colonies should be culled.

Prevention and control. An effective prevention and control program should include purchase of parasite-free animals, insect control, and high standards of animal husbandry and facility management.

IV. PARASITES OF THE INTEGUMENT

A. Lice

1. *Polyplax spinulosa*

Description and lifecycle. *Polyplax spinulosa* (the spined rat louse) is a member of the order Phthiraptera and suborder Anoplura (sucking lice). *P. spinulosa* is the common louse of both laboratory and wild rats. It occurs in the fur of the midbody, shoulders, and neck. It is slender, yellow-brown, and 0.6- to 1.5 mm long with a brown tinge. The head is rounded with two five-segmented antennae. Eyes are absent. The body is thickset. The ventral thoracic plate is pentagonal. The abdomen has about seven lateral plates on each side and seven to 13 dorsal plates. The third segment of the male antenna is provided with a pointed apophysis. The eggs are elongated and are laid and fastened to the hair near the skin. They have a conelike operculum with a row of pores along the cone.

Eggs hatch in 5 to 6 days. Nymphs emerge from the operculum and undergo three molts to become adults. Nymphs resemble adults but are paler in color and lack reproductive organs. The entire lifecycle takes place on the body fur and is completed in 26 days. Adults survive 28 to 35 days. Transmission between animals is by direct contact.

Hosts. *P. spinulosa* infests laboratory and wild rats and guinea pigs. Although infestation rates were once high, improved husbandry standards have greatly reduced the incidence of pediculosis. As a group, lice tend to be more host-specific than other ectoparasites and so are less likely to be found on other hosts or in the environment.

Pathobiology. *P. spinulosa* is a blood-sucking louse. Heavy infestation may result in dermatitis and anemia. In addition, *P. spinulosa* may serve as a vector of *Mycoplasma haemomuris* (formerly *Haemobartonella muris*), *Rickettsia typhi*, *Trypanosoma lewisi*, *Borrellia duttoni*, and *Brucella brucei*.

Clinical symptoms. Rats differ in the degree and duration of pediculosis, depending on rat strain, age, and sex (Volf, 1991). Infested animals usually show unthrifty appearance, constant scratching and irritation, restlessness, and debilitation. Chronic infestation usually induces partial immunity and diminished louse burdens (Volf, 1991).

Diagnosis. Diagnosis depends on finding and identifying adult lice, nymphs, or eggs on the fur.

Treatment. Several insecticides have been used to treat pediculosis. Formulations have included dusts, sprays, and dips. Insecticides should be applied both to the animals and to their bedding. Virtually all insecticides formulated for treatment of pediculosis in veterinary species should be effective. In addition to these, ivermectin administered by subcutaneous injection at a dosage of 600 μg/kg, is also effective (Uhlir and Volf, 1992).

Prevention and control. Pediculosis is rarer than it once was. Entry into the colony can be prevented by purchase of parasite-free rats, proper quarantine procedures, and vermin exclusion. If infestation occurs, insecticide treatment and sanitation are effective control measures.

B. Mites

1. *Radfordia ensifera*

Description and lifecycle. *Radfordia ensifera*, the fur mite of rats, is closely related to *Myobia musculi* and *Radfordia affinis* of mice. Morphologically, it can be readily distinguished by the claws on the tarsi of the second pair of legs of *R. ensifera*. They are paired and equal, whereas those of *R. affinis* are paired and subequal; *M. musculi* has only a single empodial claw on these segments. In addition, the humoral setae are broader than those found on mouse fur mites (Tenorio and Goff, 1980). The lifecycle is poorly known, but is probably similar to that of *Myobia musculi* of the mice.

Host. *R. ensifera* occurs on wild and laboratory rats. Males and adult rats are more commonly and heavily infested than their female or juvenile counterparts, respectively (Soliman et al., 2001). Improved husbandry and disease control measures have resulted in greatly reduced prevalence of infestation.

Pathobiology. Lesions are secondary to the host response to infestation. Infested rats scratch aggressively, and therefore may excoriate the skin. This facilitates development of dermatitis, hair loss, and secondary bacterial infections.

Clinical symptoms. Infestations with *R. ensifera* are sometimes asymptomatic. When signs occur, they include pruritus, patchy alopecia, and dermal ulcers (Kondo et al., 1998).

Diagnosis. Infestation with *R. ensifera* may be diagnosed by examining skin scrapings, carcass washings (Kondo et al., 1998), the pelt, or tape applied to the dorsum and then transferred to a microscope slide (West et al., 1992). The latter appears to be the most reliable.

Treatment. Infestations may be eradicated by three biweekly applications of 1% ivermectin to the dorsum of all colony rats (Kondo et al., 1998). Success also has been reported using topically applied insecticides such as malathion, chlorpyrifos, and others (Kondo et al., 1998).

Prevention and control. Infestation of laboratory rats with *R. ensifera* is infrequent. In contrast, feral rats are commonly infested (Soliman et al., 2001). Prevention of infestation is through purchase of parasite-free rats, and through exclusion of feral rats. Control is through vigilance, husbandry practices, and prompt treatment or elimination of affected rats.

2. *Notoedres muris*

Description and lifecycle. *Notoedres muris* is a burrowing mite resembling *Sarcoptes scabiei*. Adult females measure 300×200 μm. The dorsal position of the anal opening, somewhat smaller body size, and absence of heavy dorsal spines, cones, and triangular scales distinguish *N. muris* from *S. scabiei* (Soulsby, 1982). The males and immature forms resemble those of *S. scabiei*. Both sexes bear suckers on unjointed pedicles on the first two pairs of legs. The third and fourth pairs of legs are short and do not protrude beyond the edge of the body. The male also bears suckers on the fourth pair of legs.

The female mites burrow into the deeper layers of the epidermis, laying eggs at regular intervals in the stratum corneum. The larvae hatch in 4 to 5 days and either remain in the maternal burrow or emerge to excavate a pocket on the surface in which to molt. Two nymphal stages follow, each of which may form its own molting pocket. The second-stage nymphs molt and develop into adults 16 to 18 days after oviposition. The males emerge from their pockets to search for the females; copulation takes place in the burrow. The whole lifecycle from egg to adult takes 19 to 21 days (Soulsby, 1982). Transmission is by direct contact.

Host. *N. muris* infests rats, hamsters, and some other rodent host species (Flynn, 1960; Klompen and Nachman, 1990; Beco et al., 2001). Infestation may be common in wild rats.

Pathobiology. *N. muris* causes a scabies-like mange in the rat. The lesions are characterized by red papules, reddened and thickened skin, thick crusts, and erythematous vesicles on the hairless or sparsely haired regions of the body, such as ear pinnae, nose, face, head, tail,

and sometimes external genitalia and legs. In severe cases, the outer ear may be completely eroded. The mites are usually restricted to the stratum corneum but occasionally they penetrate this layer, giving rise to a more severe skin reaction and serum exudation. Epidermal cells proliferate and gradually cornify (Watson, 1962). The inflammation is localized and characterized by an influx of neutrophils and lymphocytes.

Clinical symptoms. Although some infestations are asymptomatic, clinical signs include intense pruritus, patchy alopecia, and yellow crusts of dried serum at the lesion site. Secondary bacterial infections may follow self-excoriation.

Diagnosis. Diagnosis is by finding and identifying the mites in deep skin scrapings of lesions.

Treatment. Historically, effective treatments included topical application of 0.1 to 0.25% lindane solution once or twice, or dipping in a 2% aqueous suspension of a wettable powder containing 15% butylphenoxyisopropyl chlorethyl sulfite (Aramite-15W). Recently, single subcutaneous doses of doramectin cleared cats of infestation with *N. cati* (Delucchi and Castro, 2000), and topically applied selamectin eliminated canine infestations with *Sarcoptes scabiei*, a closely related burrowing mite (Shanks et al., 2000). Ivermectin (400 μg/kg subcutaneously once per week for 8 weeks) and moxidectin (400 μg/kg orally once or twice per week for 8 weeks) have been used to treat *N. muris* infestation in hamsters (Beco et al., 2001). At the end of that study, skin scrapings were negative in only 60% to 70% of the animals treated with either compound. Where *Notoedres muris* infestations are severe, crusts and papules should be softened and removed before treatment.

Prevention and control. The prevention and control of *N. muris* infestation includes vermin exclusion, purchase of parasite-free rats, quarantine and examination of all newly acquired rats, and strict adherence to excellent husbandry standards.

3. *Laelaps (Echinolaelaps) echidninus*

Description and lifecycle. *Laelaps echidninus* is the "spiny rat mite." *L. echidninus* is also referred to as *Echinolaelaps echidninus*. However, *Echinolaelaps* is considered a subgenus of *Laelaps* (Strandtmann and Mitchell, 1963) and by convention, organisms are not typically referred to by subgenus. These mites are non-burrowing, long-legged, and globular in shape. The females are about 1.1 mm long with sclerotized, reddish brown shields. The males average 0.88 mm in length (Strandtmann and Mitchell, 1963).

Owen (1956) studied the lifecycle of *L. echidninus*. Female mites are viviparous and produce hexapod larvae that do not feed, but molt into protonymphs in

10 to 13 hours. The protonymphs feed and molt in 3 to 11 days to deutonymphs, which also feed and molt in 3 to 9 days. The females feed shortly after molting and produce larvae within 5 to 17 days, sometimes by parthenogenesis. The whole lifecycle requires at least 16 days. Females can live 2 to 3 months if they feed. The mites live in the bedding and come out at night to feed on abraded skin and body fluids.

Host. L. echidninus is common on wild rats, and under modern husbandry conditions, only rarely found on laboratory rats and mice.

Pathobiology. Although *L. echidninus* is a blood-sucking arthropod, it is generally considered nonpathogenic. *L. echidninus* serves as the natural vector of the hemogregarine protozoon *Hepatozoon muris*.

Clinical symptoms. Infestation with *L. echidninus* has not been associated with clinical signs.

Diagnosis. Diagnosis is by finding and identifying the mites on the rat or in the bedding or cage crevices.

Treatment. There are no guidelines for treatment of *L. echidninus*. However, treatments described for other mite infestations would likely be successful. Because *L. echidninus* spends much of its life off the host, control programs should address that portion of the mite population found in the environment.

Prevention and control. Prevention and control involve exclusion of wild rodents, and high standards in hygienic practices.

REFERENCES

Abd el-Wahed, M.M., Salem, G.H., and el-Assaly, T.M. (1999). The role of wild rats as a reservoir of some internal parasites in Qalyobia governorate. *J. Egypt. Soc. Parasitol.* **29**, 495–503.

Ahlquist, J., Rytömaa, T., and Borgmästar, H. (1962). Blood eosinophilia caused by a common parasite in laboratory rats. *Acta Haematol.* **28**, 306–312.

Andreassen, J., and Hopkins, C.A. (1980). Immunologically mediated rejection of *Hymenolepis diminuta* by its normal host, the rat. *J. Parasitol.* **66**, 898–903.

Antonakopoulos, G.N., Turton, J., Whitfield, P., and Newman, J. (1991). Host-parasite interface of the urinary bladder-inhabiting nematode *Trichosomoides crassicauda*: changes induced in the urothelium of infected rats. *Int. J. Parasitol.* **21**, 187–193.

Ashford, R.W. (1978). *Sarcocystis cymruensis* n. sp., a parasite of rats *Rattus norvegicus* and cats *Felis catus. Ann. Trop. Med. Parasitol.* **72**, 37–43.

Attwood, H.D., and Sutton, R.D. (1965). *Encephalitozoon* granuloma in rats. *J. Pathol. Bacteriol.* **89**, 735–738.

Baker, D.G., Malineni, S., and Taylor, H.W. (1998). Experimental infection of inbred mouse strains with *Spironucleus muris. Vet. Parasitol.* **77**, 305–310.

Baker, D.G., Morishita, T.Y., Bartlett, J.L., and Brooks, D.L. (1996). Coprologic survey of internal parasites of Northern California raptors. *J. Zoo Wildl. Med.* **27**, 358–363.

Barron, S., Baseheart, B.J., Segar, T.M., Deveraux, T., and Willford, J.A. (2000). The behavioral teratogenic potential of fenbendazole: a medication for pinworm infestation. *Neurotoxicol. Teratol.* **22**, 871–877.

Basi, N.S., George, M., and Pointer, R.H. (1994). Regulation of glycogen synthase activity in isolated rat adipocytes by levamisole. *Life Sci.* **54**, 1027–1034.

Beaver, P.C., and Maleckar, J.R. (1981). *Sarcocystis singaporensis* Zaman and Colley, (1975) 1976, *Sarcocystis villivilliso* sp. n., and *Sarcocystis zamani* sp. n.: development, morphology, and persistence in the laboratory rat, *Rattus norvegicus. J. Parasitol.* **67**, 241–256.

Becker, E.R. (1934). "Coccidia and Coccidiosis of Domestic, Game, and Laboratory Animals and of Man." Collegiate Press, Ames, p. 147.

Becker, E.R., Hall, P.R., and Hager, A. (1932). Quantitative, biometric, and host-parasite studies on *Eimeria miyairii* and *Eimeria separata* in rats. *Iowa St. Coll. J. Sci.* **6**, 299–316.

Beco, L., Petite, A., and Olivry, T. (2001). Comparison of subcutaneous ivermectin and oral moxidectin for the treatment of notoedric acariasis in hamsters. *Vet. Record* **149**, 324–327.

Bell, E.P. (1968). Disease in a caesarian-derived albino rat colony in a conventional colony. *Lab. Anim.* **2**, 1–17.

Bell, R.G. (1992). Variation in responsiveness to *Trichinella spiralis* infection in inbred rat strains. *Parasitology* **105**, 125–130.

Berdoy, M., Webster, J.P., and Macdonald, D.W. (2000). Fatal attraction in rats infected with *Toxoplasma gondii. Proc. R. Soc. Lond. B Biol. Sci.* **267**, 1591–1594.

Boivin, G.P., Ormsby, I., and Hall, J.E. (1996). Eradication of *Aspiculuris tetraptera*, using fenbendazole-medicated food. *Contemp. Top. Lab. Anim. Sci.* **35**, 69–70.

Bone, J.F., and Harr, J.R. (1967). *Trichosomoides crassicauda* infection in laboratory rats. *Lab. Anim. Care* **17**, 321–326.

Brandt, J.R., and Sewell, M.M. (1981). Varying infectivity of *Taenia taeniaeformis* for rats and mice. *Vet. Res. Commun.* **5**, 187–191.

Burger, C.J., Rikihisa, Y., and Lin, Y.C. (1986). *Taenia taeniaeformis*: inhibition of mitogen induced proliferation and interleukin-2 production in rat splenocytes by larval in vitro product. *Exp. Parasitol.* **62**, 216–222.

Catarinella Arrea, G., Chinchilla Carmona, M., Guerrero Bermudez, O.M., and Abrahams, E. (1998). Effect of *Trypanosoma lewisi* (Kinetoplastida: Trypanosomatidae) on the infection of white rats with *Toxoplasma gondii* (Eucoccidia: Sarcocystidae) oocysts. *Rev. Biol. Trop.* **46**, 1121–1123.

Ceruti, R., Sonzogni, O., Origgi, F., Vezzoli, F., Cammarata, S., Giusti, A.M, and Scanziani, E. (2001). *Capillaria hepatica* infection in wild brown rats (*Rattus norvegicus*) from the urban area of Milan, Italy. *J. Vet. Med. B Infect. Dis. Vet. Public Health* **48**, 235–240.

Červa, L. (1955). Resistance cyst *Lamblia intestinalis* vuci zevnim faktorum. *Cest. Parazitol.* **2**, 17–21.

Cetinkaya, Z., Ulger, H., Akkus, M.A., Dogru, O., Cifter, C., Doymaz, M.Z., and Ozercan, I.H. (2002). Influence of some substances on bacterial translocation in the rat. *World J. Surg.* **26**, 9–12.

Chan, K.-F. (1955). The distribution of larval stages of *Aspiculuris tetraptera* in the intestine of mice. *J. Parasitol.* **41**, 529–532.

Chapman, W.H. (1964). The incidence of a nematode, *Trichosomoides crassicauda*, in the bladder of laboratory rats. Treatment with nitrofurantoin and preliminary report of their influence on urinary calculi and experimental bladder tumors. *Invest. Urol.* **2**, 52–57.

Chinchilla, M., Portilla, E., Guerrero, O.M., and Marin, R. (1987). The presence of cysts in *Tritrichomonas muris. Rev. Biol. Trop.* **35**, 21–24.

Christie, E., Dubey, J.P., and Pappas, P.W. (1977). Prevalence of *Hammondia hammondi* in the feces of cats in Ohio. *J. Parasitol.* **63**, 929–931.

Conlogue, G., Foreyt, W., Adess, M., and Levine, H. (1979). *Capillaria hepatica* (Bancroft) in select rat populations of Hartford, Connecticut, with possible public health implications. *J. Parasitol.* **65**, 105–108.

Cook, R.W., Trapp, A.L., and Williams, J.F. (1981). Pathology of *Taenia taeniaeformis* in the rat: hepatic, lymph node and thymic changes. *J. Comp. Pathol.* **91**, 219–226.

Cook, R.W., and Williams, J.F. (1981). Pathology of *Taenia taeniaeformis* infection in the rat: gastrointestinal changes. *J. Comp. Pathol.* **91**, 205–217.

Cosoroaba, I., and Orjanu, N. (1998). Congenital trichinellosis in the rat. *Vet. Parasitol.* **77**, 147–151.

Cruz, C.C., Ferrari, L., and Sogayar, R. (1997). A therapeutic trial in *Giardia muris* infection in the mouse with metronidazole, tinidazole, secnidazole and furazolidone. *Rev. Soc. Bras. Med. Trop.* **30**, 223–228.

de Souza, M.M., Silva, L.M., Barbosa, A.A. Jr., de Oliveira, I.R., Parana, R., and Andrade, Z.A. (2000). Hepatic capillariasis in rats: a new model for testing antifibrotic drugs. *Braz. J. Med. Biol. Res.* **33**, 1329–1334.

de Souza Spinosa, H., Gerenutti, M., and Bernardi, M.M. (2000). Anxiolytic and anticonvulsant properties of doramectin in rats: behavioral and neurochemistric evaluations. *Comp. Biochem. Physiol. C Toxicol. Pharmacol.* **127**, 359–366.

de Souza Spinosa, H., Stilck, S.R.A.N., and Bernardi, M.M. (2002). Possible anxiolytic effects of ivermectin in rats. *Vet. Res. Commun.* **26**, 309–321.

de Waard, J.W., de Man, B.M., Wobbes, T., van der Linden, C.J., and Hendriks, T. (1998). Inhibition of fibroblast collagen synthesis and proliferation by levamisole and 5–fluorouracil. *Eur. J. Cancer* **34**, 162–167.

Delucchi, L., and Castro, E. (2000). Use of doramectin for treatment of notoedric mange in five cats. *J. Am. Vet. Med. Assoc.* **216**, 215–216.

Didier, E.S., Vossbrinck, C.R., Baker, M.D., Rogers, L.B., Bertucci, D.C., and Shadduck, J.A. (1995). Identification and characterization of three *Encephalitozoon cuniculi* strains. *Parasitology* **111**, 411–421.

Driniaev, V.A., Mosin, V.A., Krugliak, E.B., Sterlina, T.S., Viktorov, A.V., Tsyganova, V.G., Korystova, A.F., Grichenko, A.S., Zenchenko, K.I., and Kokoz, I.M. (2001). Selective cytostatic and cytotoxic effects of avermectins. *Antibiot. Khimioter.* **46**, 13–16.

Dubey, J.P. (1996). Pathogenicity and infectivity of *Toxoplasma gondii* oocysts for rats. *J. Parasitol.* **82**, 951–956.

Dubey, J.P. (1997). Distribution of tissue cysts in organs of rats fed *Toxoplasma gondii* oocysts. *J. Parasitol.* **83**, 755–757.

Dubey, J.P., and Frenkel, J.K. (1998). Toxoplasmosis of rats: a review, with considerations of their value as an animal model and their possible role in epidemiology. *Vet. Parasitol.* **77**, 1–32.

Duszynski, D.W. (1972). Host and parasite interactions during single and concurrent infections with *Eimeria nieschulzi* and *E. separata* in the rat. *J. Protozool.* **19**, 82–88.

Duszynski, D.W., Roy, S.A., and Castro, G.A. (1978). Intestinal disaccharidase and peroxidase deficiencies during *Eimeria nieschulzi* infections in rats. *J. Protozool.* **25**, 226–231.

El Gebaly, M.W., Nassery, S.F., El Azzouni, M.Z., Hammouda, N.A., and Allam, S.R. (1996). Effect of mebendazole and ivermectin in experimental hepatic capillariasis: parasitological, scanning electron microscopy and immunological studies. *J. Egypt. Soc. Parasitol.* **26**, 261–272.

el-Ridi, A.M.S., Hamdy, E.I., Nasr, N.T., Gerges, Z.I., and Sobhy, M.M.K. (1985). Effect of some antibiotics and chemotherapeutics on *Trypanosoma lewisi* infection in albino rats. *J. Egypt. Soc. Parasitol.* **15**, 119–123.

Evangelista, S., Maggi, C.A., and Meli, A. (1984). The effect of levamisole on gastric ulcers induced in the rat by anti-inflammatory and necrotizing agents. *J. Pharm. Pharmacol.* **36**, 270–272.

Farhang-Azad, A. (1977). Ecology of *Capillaria hepatica* (Bancroft 1893). II. Egg-releasing mechanisms and transmission. *J. Parasitol.* **63**, 701–706.

Fedorko, D.P., Nelson, N.A., Didier, E.S., Bertucci, D., Delgado, R.M., and Hruszkewycz, A.M. (2001). Speciation of human microsporidia by polymerase chain reaction single-strand conformation polymorphism. *Am. J. Trop. Med. Hyg.* **65**, 397–401.

Ferrante, A., Carter, R.F., Ferluga, J., and Allison, A.C. (1984). Lipopolysaccharide hyperreactivity of animals infected with *Trypanosoma lewisi* or *Trypanosoma musculi*. *Infect. Immun.* **46**, 501–506.

Flynn, R.J. (1960). *Notoedres muris* infestation of rats. *Proc. Anim. Care Panel* **10**, 69–70.

Frandsen, J.C. (1983). Effects of low-level infections by coccidia and roundworms on the nutritional status of rats fed an adequate diet. *J. Anim. Sci.* **57**, 1487–1497.

Frenkel, J.K., and Dubey, J.P. (1972). Rodents as vectors for feline coccidia, *Isospora felis* and *Isospora rivolta*. *J. Infect. Dis.* **125**, 69–72.

Frenkel, J.K., and Dubey, J.P. (1975). *Hammondia hammondi* gen. nov., sp. nov., from domestic cats, a new coccidian related to *Toxoplasma* and *Sarcocystis*. *Z. Parasitenkd.* **46**, 3–12.

Frenkel, J.K., and Dubey, J.P. (2000). The taxonomic importance of obligate heteroxeny: distinction of *Hammondia hammondi* from *Toxoplasma gondii*–another opinion. *Parasitol. Res.* **86**, 783–786.

Fukushima, T., Mochizuki, K., Yamazaki, H., Watanabe, Y., Yamada, S., Aoyama, T., Sakurai, Y., Mori, H., and Nakazawa, M. (1990). *Pentatrichomonas hominis* from beagle dogs–detection method, characteristics and route of infection. *Jikken Dobutsu* **39**, 187–192.

Garcia-Laverde, A., and de Bonilla, L. (1975). Clinical trials with metronidazole in human balantidiasis. *Am. J. Trop. Med. Hyg.* **24**, 781–783.

Gonenc, B., and Sarimehmetoglu, H.O. (2001). Continuous feed medication with nitroscanate for the removal of *Hymenolepis nana* in naturally infected mice and rats. *Dtsch. Tierarztl. Wochenschr.* **108**, 434–436.

Gupta, S., Jain, M.K., Katiyar, J.C., and Maitra, S.C. (1992). Substituted methyl benzimidazole carbamate: efficacy against experimental cysticercosis. *Ann. Trop. Med. Parasitol.* **86**, 51–57.

Gupta, S., Katiyar, J.C., Sen, A.B., Dubey, S.K., Singh, H., Sharma, S., and Iyer, R.N. (1980). Anticestode activity of 3,5 dibromo-2′-chlorosalicylanilide-4′-isothiocyanate- a preliminary report. *J. Helminthol.* **54**, 271–273.

Hafner, U., and Frank, W. (1986). Morphological studies on the muscle cysts of *Sarcocystis dirumpens* (Hoare 1933) in several host species revealing endopolygeny in metrocytes. *Z. Parasitenkd.* **72**, 453–461.

Hanes, M.A., and Stribling, L.J. (1995). Fibrosarcomas in two rats arising from hepatic cysts of *Cysticercus fasciolaris*. *Vet. Pathol.* **32**, 441–444.

Hankinson, G.J., Murphy, J.C., and Fox, J.G. (1982). Diagnostic exercise. *Eimeria caviae* infection with concurrent *Balantidium coli* infection. *Lab. Anim. Sci.* **32**, 35–36.

Hayden, D.W., King, N.W., and Murthy, A.S.K. (1976). Spontaneous Frenkelia infection in a laboratory-reared rat. *Vet. Pathol.* **13**, 337–342.

Hayes, S.L., Rice, E.W., Ware, M.W., and Schaefer, F.W. III. (2003). Low pressure ultraviolet studies for inactivation of *Giardia muris* cysts. *J. Appl. Microbiol.* **94**, 54–59.

Heyneman, D. (1961). Studies on helminth immunity. III. Experimental verification of autoinfection from cysticercoids of *Hymenolepis nana* in the white mouse. *J. Infect. Dis.* **109**, 10–18.

Higgins-Opitz, S.B., Dettman, C.D., Dingle, C.E., Anderson, C.B., and Becker, P.J. (1990). Intestinal parasites of conventionally maintained BALB/c mice and *Mastomys coucha* and the effects of a concomitant schistosome infection. *Lab. Anim.* **24**, 246–252.

Hitchcock, C.R., and Bell, E.T. (1952). Studies on the nematode parasite, *Gongylonema neoplasticum*, and avitaminosis A in the forestomach of rats: Comparison with Fibiger's results. *J. Natl. Cancer Inst.* **12**, 1345–1387.

Huerkamp, M.J., Benjamin, K.A., Zitzow, L.A., Pullium, J.K., Lloyd, J.A., Thompson, W.D., Webb, S.K., and Lehner, N.D.M. (2000). Fenbendazole treatment without environmental decontamination eradicates *Syphacia muris* from all rats in a large, complex research institution. *Contemp. Top. Lab. Anim. Sci.* **39**, 9–12.

Hughes, H.C. Jr., Barthel, C.H., Lang, C.M. (1973). Niclosamide as a treatment for *Hymenolepis nana* and *Hymenolepis diminuta* in rats. *Lab. Anim. Sci.* **23**, 72–73.

Hussey, K.L. (1957). *Syphacia muris* vs. *S. obvelata* in laboratory rats and mice. *J. Parasitol.* **43**, 555–559.

Insler, G.D., and Roberts, L.S. (1976). *Hymenolepis diminuta*: lack of pathogenicity in the healthy rat host. *Exp. Parasitol.* **39**, 351–357.

Ishih, A., Nishimura, M., and Sano, M. (1992). Differential establishment and survival of *Hymenolepis diminuta* in syngeneic and outbred rat strains. *J. Helminthol.* **66**, 132–136.

Ito, A., and Kamiyama, T. (1984). *Hymenolepis nana*: worm recovery from congenitally athymic nude and phenotypically normal rats and mice. *Exp. Parasitol.* **58**, 132–137.

Jakel, T., Burgstaller, H., and Frank, W. (1996). *Sarcocystis singaporensis*: studies on host specificity, pathogenicity, and potential use as a biocontrol agent of wild rats. *J. Parasitol.* **82**, 280–287.

Jakel, T., Khoprasert, Y., Endepols, S., Archer-Baumann, C., Suasaard, K., Promkerd, P., Kliemt, D., Boonsong, P., and Hongnark, S. (1999). Biological control of rodents using *Sarcocystis singaporensis*. *Int. J. Parasitol.* **29**, 1321–1330.

Jakel, T., Khoprasert, Y., Sorger, I., Kliemt, D., Seehabutr, V., Suasaard, K., and Hongnark, S. (1997). Sarcosporidiasis in rodents from Thailand. *J. Wildl. Dis.* **33**, 860–867.

Jakel, T., Scharpfenecker, M., Jitrawang, P., Ruckle, J., Kliemt, D., Mackenstedt, U., Hongnark, S., and Khoprasert, Y. (2001). Reduction of transmission stages concomitant with increased host immune responses to hypervirulent *Sarcocystis singaporensis*, and natural selection for intermediate virulence. *Int. J. Parasitol.* **31**, 1639–1647.

Johnson, P.D. (1933). A case of infection by *Trypanosoma lewisi* in a child. *Trans. Roy. Soc. Trop. Med. Hyg.* **26**, 467–468.

Katiyar, J.C., Gupta, S., and Sen, A.B. (1983). Susceptibility, chemotherapeutic reaction and immunological response of rat and mouse to *Hymenolepis nana*—a comparative study. *Indian J. Exp. Biol.* **21**, 371–374.

Katiyar, J.C., and Sen, A.B. (1970). Occurrence of histamine in the intestine of rats harboring cysticercoids of *Hymenolepis nana*. *Indian J. Exp. Biol.* **8**, 191–193.

Katiyar, J.C., Tangri, A.N., Ghatak, S., and Sen, A.B. (1973). Serum protein pattern of rats during infection with *Hymenolepis nana*. *Indian J. Exp. Biol.* **11**, 188–190.

Kellogg, H.S., and Wagner, J.E. (1982). Experimental transmission of *Syphacia obvelata* among mice, rats, hamsters, and gerbils. *Lab. Anim. Sci.* **32**, 500–501.

Khachoian, V.I., and Arakelian, L.A. (1978). Case of the transmission of the rat trypanosome by lice. *Parazitologiia* **12**, 451–453.

Kirsch, R. (1978). *In vitro* and *in vivo* studies on the ovicidal activity of fenbendazole. *Res. Vet. Sci.* **25**, 263–265.

Klein, B.Y., Gal, I., and Segal, D. (1993). Studies of the levamisole inhibitory effect on rat stromal-cell commitment to mineralization. *J. Cell. Biochem.* **53**, 114–121.

Klompen, J.S., and Nachman, M.W. (1980). Occurrence and treatment of the mange mite *Notoedres muris* in marsh rats from South America. *J. Wildl. Dis.* **26**, 135–136.

Kock, N.P., Petersen, H., Fenner, T., Sobottka, I., Schmetz, C., Deplazes, P., Pieniazek, N.J., Albrecht, H., and Schottelius, J. (1997). Species-specific identification of microsporidia in stool and intestinal biopsy specimens by the polymerase chain reaction. *Eur. J. Clin. Microbiol. Infect. Dis.* **16**, 369–376.

Kondo, S., Taylor, A., and Chun, S. (1998). Elimination of an infestation of rat fur mites (*Radfordia ensifera*) from a colony of Long Evans rats, using the micro-dot technique for topical administration of 1% Ivermectin. *Contemp. Top. Lab. Anim. Sci.* **37**, 58–61.

Konno, K., Abella, J.A., Oku, Y., Nonaka, N., and Kamiya, M. (2000). Histopathology and physiopathology of gastric mucous hyperplasia in rats heavily infected with *Taenia taeniaeformis*. *J. Vet. Med. Sci.* **61**, 317–324.

Krampitz, H.E., and Haberkorn, A. (1988). Experimental treatment of *Hepatozoon* infections with the anticoccidial agent toltrazuril. *Zentralbl. Veterinarmed. B* **35**, 131–137.

Krishnasamy, M., Singh, K.I., Ambu, S., and Ramachandran, P. (1980). Seasonal prevalence of the helminth fauna of the wood rat *Rattus tiomanicus* (Miller) in West Malaysia. *Folia Parasitol.* **27**, 231–235.

Kunstyr, I. (1977). Infectious form of *Spironucleus* (*Hexamita*) *muris*: banded cysts. *Lab. Anim.* **11**, 185–188.

Kunstyr, I., Schoeneberg, U., and Friedhoff, K.T. (1992). Host specificity of *Giardia muris* isolates from mouse and golden hamster. *Parasitol. Res.* **78**, 621–622.

Lacey, E., Brady, R.L., Prichard, R.K., and Watson, T.R. (1987). Comparison of inhibition of polymerisation of mammalian tubulin and helminth ovicidal activity by benzimidazole carbamates. *Vet. Parasitol.* **23**, 105–119.

Lee, C.M., and Boone, L.Y. (1983). Rat liver glucose-6–phosphate dehydrogenase isozymes: influence of infection with *Trypanosoma*. *Comp. Biochem. Physiol. B* **75**, 505–508.

Lee, C.M., George, Y.G., and Aboko-Cole, G. (1977). Iron metabolism in *Trypanosoma lewisi* infection: serum iron and serum iron-binding capacity. *Z. Parasitenkd.* **53**, 1–6.

Levine, N.D. (1985). "Veterinary Protozoology." Iowa State University Press, Ames.

Lewis, J.W., and D'Silva, J. (1986). The life-cycle of *Syphacia muris* Yamaguti (Nematoda: Oxyuroidea) in the laboratory rat. *J. Helminthol.* **60**, 39–46.

Lewis, W.P., and Markell, E.K. (1958). Acquisition of immunity to toxoplasmosis by the newborn rat. *Exp. Parasitol.* **7**, 463–467.

Liburd, E.M. (1973). *Eimeria nieschulzi* infections in inbred and outbred rats: infective dose, route of infection, and host resistance. *Can. J. Zool.* **51**, 273–279.

Lin, T.M. (1971). Colonization and encystation of *Entamoeba muris* in the rat and the mouse. *J. Parasitol.* **57**, 375–382.

Lin, Y.C., Rikihisa, Y., Kono, H., and Gu, Y. (1990). Effects of larval tapeworm (*Taenia taeniaeformis*) infection on reproductive functions in male and female host rats. *Exp. Parasitol.* **70**, 344–352.

Lincicome, D.R., Rossan, R.N., and Jones, W.C. (1963). Growth of rats infected with *Trypanosoma lewisi*. *Exp. Parasitol.* **14**, 54–65.

Lipman, N.S., Lampen, N., and Nguyen, H.T. (1999). Identification of pseudocysts of *Tritrichomonas muris* in Armenian hamsters and their transmission to mice. *Lab. Anim. Sci.* **49**, 313–315.

Lübcke, R., Hutcheson, F.A.R., and Barbezat, G.O. (1992). Impaired intestinal electrolyte transport in rats infested with the common parasite *Syphacia muris*. *Dig. Dis. Sci.* **37**, 60–64.

Luttermoser, G.W. (1936). A helminthological survey of Baltimore house rats (*Rattus norvegicus*). *Am. J. Hyg.* **24**, 350.

Macintire, D.K., Vincent-Johnson, N.A., Kane, C.W., Lindsay, D.S., Blagburn, B.L., and Dillon, A.R. (2001). Treatment of dogs with *Hepatozoon americanum*: 53 cases. *J. Am. Vet. Med. Assoc.* **218**, 77–82.

Macnish, M.G., Morgan, U.M., Behnke, J.M., and Thompson, R.C. (2002). Failure to infect laboratory rodent hosts with human isolates of *Rodentolepis* (= *Hymenolepis*) *nana*. *J. Helminthol.* **76**, 37–43.

Majeed, S.K., and Zubaidy, A.J. (1982). Histopathological lesions associated with *Encephalitozoon cuniculi* (nosematosis) infection in a colony of Wistar rats. *Lab. Anim.* **16**, 244–247.

Marquardt, W.C. (1966). Attempted transmission of the rat coccidium *Eimeria nieschulzi* to mice. *J. Parasitol.* **52**, 691–694.

Mason, R.W. (1978). The detection of *Hammondia hammondi* in Australia and the identification of a free-living intermediate host. *Z. Parasitenkd.* **57**, 101–106.

McCracken, R.O. (1978). Efficacy of mebendazole and albendazole against *Trichinella spiralis* in mice. *J. Parasitol.* **64**, 214–219.

Mehlhorn, H., and Harder A. (1997). Effects of the synergistic action of febantel and pyrantel on the nematode *Heterakis spumosa*: a light and transmission electron microscopy study. *Parasitol. Res.* **83**, 419–434.

Meshorer, A. (1969). Hexamitiasis in laboratory mice. *Lab. Anim. Care* **19**, 33–37.

Miller, W.W. (1908). *Hepatozoon perniciosum* (n.g., n.s.); a hemogregarine pathogenic for white rats; with a description of the sexual cycle in the intermediate host, a mite (*Laelaps echidninus*). *Bull. U.S. Hyg. Lab.* **46**, 7–51.

Mohindra, A.R., Lee, M.W., Visvesvara, G., Moura, H., Parasuraman, R., Leitch, G.J., Xiao, L., Yee, J., and del Busto, R. (2002). Disseminated microsporidiosis in a renal transplant recipient. *Transpl. Infect. Dis.* **4**, 102–107.

Mudrow-Reichenow, L., (1956). Spontanes vorkommen von amöben und ciliaten bei laboratoriumstieren. *Z. Tropenmed. Parasitol.* **7**, 198–211.

Mugridge, N.B., Morrision, D.A., Johnson, A.M., Luton, K., Dubey, J.P., Votypka, J., and Tenter, A.M. (1999). Phylogenetic relationships of the genus *Frenkelia*: a review of its history and new knowledge gained from comparison of large subunit ribosomal ribonucleic acid gene sequences. *Int. J. Parasitol.* **29**, 957–972.

Muller-Doblies, U.U., Herzog, K., Tanner, I., Mathis, A., and Deplazes, P. (2002). First isolation and characterization of *Encephalitozoon cuniculi* from a free-ranging rat (*Rattus norvegicus*). *Vet. Parasitol.* **107**, 279–285.

Nakauchi, K. (1999). The prevalence of *Balantidium coli* infection in fifty-six mammalian species. *J. Vet. Med. Sci.* **61**, 63–65.

Ndarathi, C.M. (1992). Cellular responses to culture-derived soluble exoantigens of *Trypanosoma lewisi*. *Parasitol. Res.* **78**, 324–328.

Nielsen, K., Sheppard, J., Holmes, W., and Tizard, I. (1978). Increased susceptibility of *Trypanosoma lewisi* infected, or decomplemented rats to *Salmonella typhimurium*. *Experimentia* **34**, 118–119.

Oivanen, L., Mikkonen, T., and Sukura, A. (2000). An outbreak of trichinellosis in farmed wild boar in Finland. *APMIS* **108**, 814–818.

Oku, Y., Yamanouchi, T., Matsuda, K., Abella, J.A., Ooi, H.K., Ohtsubo, R., Goto, Y., and Kamiya, M. (2002). Retarded gastric acid secretion in rats infected with larval *Taenia taeniaeformis*. *Parasitol. Res.* **88**, 872–873.

Owen, B.L. (1956). Life history of the spiny rat mite under artificial conditions. *J. Econ. Entomol.* **49**, 702–703.

Oxberry, M.E., Thompson, R.C., and Reynoldson, J.A. (1994). Evaluation of the effects of albendazole and metronidazole on the ultrastructure of *Giardia duodenalis*, *Trichomonas vaginalis* and *Spironucleus muris* using transmission electron microscopy. *Int. J. Parasitol.* **24**, 695–703.

Pearson, D.J., and Taylor, G. (1975). The influence of the nematode *Syphacia obvelata* on adjuvant arthritis in the rat. *Immunology* **29**, 391–396.

Petri, M. (1966). The occurrence of *Nosema cuniculi* (*Encephalitozoon cuniculi*) in the cells of transplantable, malignant ascites tumours and its effect upon tumour and host. *Acta Pathol. Microbiol. Scand.* **66**, 13–30.

Pinto, A., Sorrentino, R., and Sorrentino, L. (1990). Levamisole inhibits in vivo rat platelet aggregation by a release of prostacycline-like factor. *Gen. Pharmacol.* **21**, 255–259.

Potter, K., and Leid, R.W. (1986). A review of eosinophil chemotaxis and function in *Taenia taeniaeformis* infections in the laboratory rat. *Vet. Parasitol.* **20**, 103–116.

Poul, J.M. (1988). Effects of perinatal ivermectin exposure on behavioral development of rats. *Neurotoxicol. Teratol.* **10**, 267–272.

Price, E.W., and Chitwood, B.G. (1931). Incidence of internal parasites in wild rats in Washington, D.C. *J. Parasitol.* **18**, 55.

Pritchett, K., and Johnston, N.A. (2002). A review of treatments for the eradication of pinworm infections from laboratory rodent colonies. *Contemp. Topics* **41**, 36–46.

Romatowski, J. (2000). *Pentatrichomonas hominis* infection in four kittens. *J. Am. Vet. Med. Assoc.* **216**, 1270–1272.

Rommel, M., Schwerdtfeger, A., and Blewaska, S. (1981). The *Sarcocystis muris*-infection as a model for research on the chemotherapy of acute sarcocystosis of domestic animals. *Zentralbl. Bakteriol. Microbiol. Hyg. A.* **250**, 268–276.

Ross, C.R., Wagner, J.E., Wightman, S.R., and Dill, S.E. (1980). Experimental transmission of *Syphacia muris* among rats, mice, hamsters and gerbils. *Lab. Anim. Sci.* **30**, 35–37.

Sahin, I. (1979). Parasitosis and zoonosis in mice and rats caught in and around Beytepe Village. *Mikrobiyol. Bul.* **13**, 283–290.

Sakla, A.A., and Monib, M.E.M. (1984). Redescription of the life cycle of *Trypanosoma* (*Herpetosoma*) *lewisi* from upper Egyptian rats. *J. Egypt. Soc. Parasit.* **14**, 367–376.

Santos, A.B., Tolentino, M. Jr., and Andrade, Z.A. (2001). Pathogenesis of hepatic septal fibrosis associated with *Capillaria hepatica* infection of rats. *Rev. Soc. Bras. Med. Trop.* **34**, 503–506.

Saxena, V.K., and Miyata, A. (1993). An unusual morphological type of *Trypanosoma* (*Herpetosoma*) *lewisi* (Kent, 1880) detected in the blood of *Rattus norvegicus* in India. *J. Commun. Dis.* **25**, 15–17.

Schagemann, G., Bohnet, W., Kunstyr, I., and Friedhoff, T. (1990). Host specificity of cloned *Spironucleus muris* in laboratory rodents. *Lab. Animals* **24**, 234–239.

Schinetti, M.L., Mazzini, A., Greco, R., and Bertelli, A. (1984). Inhibiting effect of levamisole on superoxide production from rat mast cells. *Pharmacol. Res. Commun.* **16**, 101–107.

Scott, K.G., Logan, M.R., Klammer, G.M., Teoh, D.A., and Buret, A.G. (2000). Jejunal brush border microvillous alterations in *Giardia muris*-infected mice: role of T lymphocytes and interleukin-6. *Infect. Immun.* **68**, 3412–3418.

Shanks, D.J., McTier, T.L., Behan, S., Pengo, G., Genchi, C., Bowman, D.D., Holbert, M.S., Smith, D.G., Jernigan, A.D., and Rowan, T.G. (2000). The efficacy of selamectin in the treatment of naturally acquired infestations of *Sarcoptes scabiei* on dogs. *Vet. Parasitol.* **91**, 269–281.

Shaw, G.L., and Quadagno, D. (1975). *Trypanosoma lewisi* and *T. cruzi*: effect of infection on gestation in the rat. *Exp. Parasitol.* **37**, 211–217.

Shrivastava, K.K., and Shrivastava, G.P. (1974). Two cases of *Trypanosoma* (*Herpetosoma*) species infection of man in India. *Trans. Roy. Soc. Trop. Med. Hyg.* **68**, 143–144.

Simmons, M.L., Richter, C.B., Franklin, J.A., and Tennant, R.W. (1967). Prevention of infectious diseases in experimental mice. *Proc. Soc. Exp. Biol. Med.* **126**, 830–837.

Smith, D.D., and Frenkel, J.K. (1978). Cockroaches as vectors of *Sarcocystis muris* and of other coccidia in the laboratory. *J. Parasitol.* **64**, 315–319.

Smith, N.C., Ovington, K.S., Deplazes, P., and Eckert, J. (1995). Cytokine and immunoglobulin subclass responses of rats to infection with *Eimeria nieschulzi*. *Parasitology* **111**, 51–57.

Smith, O. (1972). Survival and growth pattern of *Trypanosoma lewisi* in two heterologous hosts: albino mouse and guinea-pig. *Acta Zool. Pathol. Antverp.* **55**, 3–18.

Smith, P.E. (1953). Life history and host parasite relations of *Heterakis spumosa*, a nematode parasite in the colon or the rat. *Am. J. Hyg.* **57**, 194–221.

Smith, V.S. (1946). Are vesical calculi associated with *Trichosomoides crassicauda*, the common bladder nematode of rats. *J. Parasitol.* **32**, 142–149.

Solana, H.D., Teruel, M.T., Najle, R., Lanusse, C.E., and Rodriguez, J.A. (1998). The anthelmintic albendazole affects in vivo the dynamics and the detyrosination-tyrosination cycle of rat brain microtubules. *Acta. Physiol. Pharmacol. Ther. Latinoam.* **48**, 199–205.

Soliman, S., Marzouk, A.S., Main, A.J., and Montasser, A.A. (2001). Effects of sex, size, and age of commensal rat hosts on the infestation parameters of their ectoparasites in a rural area of Egypt. *J. Parasitol.* **87**, 1308–1316.

Souhaili-el Amri, H., Fargetton, X., Benoit, E., Totis, M., and Batt, A.M. (1988). Inducing effect of albendazole on rat liver drug-metabolizing enzymes and metabolite pharmacokinetics. *Toxicol. Appl. Pharmacol.* **92**, 141–149.

Soulsby, E.J.L. (1982). "Helminths, Arthropods and Protozoa of Domesticated Animals." Seventh edition, Lea & Febiger, Philadelphia, Pa.

Stahl, W.B. (1963). Studies on the life cycle of *Syphacia muris*, the rat pinworm. *Keio J. Med.* **12**, 55–60.

Stewart, G.L., Na, H., Smart, L., and Seelig, L.L. Jr. (1999). The temporal relationship among anti-parasite immune elements expressed during the early phase of infection of the rat with *Trichinella spiralis*. *Parasitol. Res.* **85**, 672–677.

Stewart, G.L., Reddington, J.J., and Hamilton, A.M. (1980). *Eimeria nieschulzi* and *Trichinella spiralis*: analysis of concurrent infection in the rat. *Exp. Parasitol.* **50**, 115–122.

Strandtmann, R.W., and Mitchell, C.J. (1963). The laelaptine mites of the *Echinolaelaps* complex from the Southwest Pacific area (Acarina: Mesostigmata). *Pac. Insects* **5**, 541–576.

Summa, M.E., Ebisui, L., Osaka, J.T., and de Tolosa, E.M. (1992). Efficacy of oral ivermectin against *Trichosomoides crassicauda* in naturally infected laboratory rats. *Lab. Anim. Sci.* **42**, 620–622.

Suter, C., Muller-Doblies, U.U., Hatt, J.M., and Deplazes, P. (2001). Prevention and treatment of *Encephalitozoon cuniculi* infection in rabbits with fenbendazole. *Vet. Rec.* **148**, 478–480.

Tafts, L.F. (1976). Pinworm infections in laboratory rodents: a review. *Lab. Anim.* **10**, 1–13.

Tanovic, A., Jimenez, M., and Fernandez, E. (2002). Changes in the inhibitory responses to electrical field stimulation of intestinal smooth muscle from *Trichinella spiralis* infected rats. *Life Sci.* **71**, 3121–3136.

Tenorio, J.M., and Goff, M.L. (1980). "Ectoparasites of Hawaiian rodents." Allen Press, Lawrence, Kansas.

Thomaskutty, K.G., Basi, N.S., McKenzie, M.L., and Pointer, R.H. (1993). Regulation of pyruvate dehydrogenase activity in rat fat pads and isolated hepatocytes by levamisole. *Pharmacol. Res.* **27**, 263–271.

Thoongsuwan, S., and Cox, H.W. (1978). Anemia, splenomegaly, and glomerulonephritis associated with autoantibody in *Trypanosoma lewisi* infections. *J. Parasitol.* **64**, 669–673.

Trabert, U., Rosenthal, M., and Muller, W. (1976). The effect of levamisole on adjuvant arthritis in the rat. *J. Rheumatol.* **3**, 165–174.

Twort, J.M., and Twort, C.C. (1932). Disease in relation to carcinogenic agents among 60,000 experimental mice. *J. Pathol. Bacteriol.* **35**, 219–242.

Uhlir, J., and Volf, P. (1992). Ivermectin: its effect on the immune system of rabbits and rats infested with ectoparasites. *Vet. Immunol. Immunopathol.* **34**, 325–336.

Upton, S.J., Mayberry, L.F., Bristol, J.R., Favela, S.H., and Sambrano, G.R. (1987). Suppression of peripheral eosinophilia by the coccidium *Eimeria nieschulzi* (Apicomplexa: Eimeriidae) in experimentally infected rats. *J. Parasitol.* **73**, 300–308.

Utsumi, E., Yamamoto, A., Kawaratani, T., Sakane, T., Hashida, M., and Sezaki, H. (1987). Enhanced gastrointestinal absorption of drugs in rats pretreated with the synthetic immunomodulator, levamisole. *J. Pharm. Pharmacol.* **39**, 307–309.

Van Arman, G.G., and Campbell, W. (1975). Anti-inflammatory activity of thiabendazole and its relation to parasitic disease. *Tex. Rep. Biol. Med.* **33**, 303–311.

Volf, P. (1991). *Polyplax spinulosa* infestation and antibody response in various strains of laboratory rats. *Folia Parasitol.* **38**, 355–362.

Wagner, M. (1988). The effect of infection with the pinworm (*Syphacia muris*) on rat growth. *Lab. Anim. Sci.* **38**, 476–478.

Wallace, G.D. (1975). Observations on a feline coccidium with some characteristics of *Toxoplasma* and *Sarcocystis*. *Z. Parasitenkd.* **46**, 167–178.

Watson, D.P. (1962). On the immature and adult stages of *Notoedres alepis* (Railliet and Lucet, 1893) and its effect on the skin of the rat. *Acarologia* **4**, 64–77.

Weisbroth, S.H., and Scher, S. (1971). *Trichosomoides crassicauda* infections of a commercial rat breeding colony. I. Observations on the life cycle and propagation. *Lab. Anim. Sci.* **21**, 54–61.

West, W.L., Schofield, J.C., and Bennett, B.T. (1992). Efficacy of the "micro-dot" technique for administering topical 1% Ivermectin for the control of pinworms and fur mites in mice. *Contemp. Top. Lab. Anim. Sci.* **31**, 7–10.

Whitehouse, A., France, M.P., Pope, S.E., Lloyd, J.E., and Ratcliffe, R.C. (1993). *Spironucleus muris* in laboratory mice. *Aus. Vet. J.* **70**, 193.

Williams, J.F., Shearer, A.M., and Ravitch, M.M. (1981). Differences in susceptibility of rat strains to experimental infection with *Taenia taeniaeformis*. *J. Parasitol.* **67**, 540–547.

Wise, L.D., Allen, R.L., Hoe, C.M., Verbeke, D.R., and Gerson, R.J. (1997). Developmental neurotoxicity evaluation of the avermectin pesticide, emamectin benzoate, in Sprague-Dawley rats. *Neurotoxicol. Teratol.* **19**, 315–326.

Zimmerman, N.P., Bass, P., and Oaks, J.A. (2001). Modulation of caudal intestinal permeability in the rat during infection by the tapeworm *Hymenolepis diminuta*. *J. Parasitol.* **87**, 1260–1263.

Zubaidy, A.J., and Majeed, S.K. (1981). Pathology of the nematode *Trichosomoides crassicauda* in the urinary bladder of laboratory rats. *Lab. Anim.* **15**, 381–384.

Chapter 14

Neoplastic Disease

Gary A. Boorman and Jeffrey I. Everitt

THE LABORATORY RAT, 2ND EDITION

I. INTRODUCTION

A. General Considerations

Cancer is a genetic disease believed to result from the progressive accumulation of mutations, causing a loss of normal mechanisms of cellular growth control. It is a disease of disordered proliferation of somatic cells, and as such often manifests clinically as a mass. Current thinking is that the pathways to neoplasia involve mutation of caretaker or gatekeeper tumor-suppressor genes. Gatekeepers are genes that directly regulate the growth of tumors by inhibiting growth or promoting cell death. Caretakers are responsible for maintaining overall genomic integrity (Vogelstein and Kinsler, 1993). The neoplastic transformation process is a multistep event involving alterations in two major classes of genes: the cellular oncogenes and the tumor-suppressor genes. The multistep nature of the cancer process can be considered as three phases, including initiation, promotion, and tumor progression. Rats have been useful animal models in which to study all phases of the carcinogenic process.

Our understanding of cancer in rats comes from two general sources. First is the use of the rat as a bioassay model to test chemicals for potential carcinogenicity. Thousands of studies have been conducted by pharmaceutical firms, chemical firms, and government agencies.

The largest government program is the National Toxicology Program (NTP), which has conducted more than 500 two-year studies over the past 25 years (Bucher et al., 1996; Bucher, 2002). The bioassay has evolved over the years to a fairly standard protocol involving rats and mice 6 to 8 weeks of age being dosed for 2 years, at which time a complete necropsy is performed with the collection and examination of approximately 40 tissues. There are generally 50 animals per sex and per species, occasionally with interim sacrifices of smaller groups at 6 or 12 months (Goodman et al., 1985; Hamm, 1985; Prejean, 1985). Increased rates of cancer, especially when unexpected or different results occur between different laboratories, prompted further mechanistic studies, as well as effects to improve tumor classification (Maronpot et al., 1986; Maronpot, 1990; Haseman and Hailey, 1997; Cohen, 2002). The NTP also instituted a procedure whereby all tumor diagnoses by the study pathologist are reviewed by a second pathologist, with discrepancies resolved by a pathology working group (Boorman and Eustis, 1986; Maronpot, 1990; Goodman and Sauer, 1992; Boorman et al., 1997). This procedure has improved accuracy, consistency, and standardized terminology between and within studies (Ward et al., 1995). The other major source of cancer information in the rat is in its use as a model in oncology. This aspect is discussed in more detail later in this chapter in the following section.

It should be recognized that neoplastic disease in the rat, as in most species, increases dramatically with increasing age. Many genetic and environmental factors influence the development of neoplasia, and attaining an understanding of these factors and their control is critical for scientists who use the rat in the laboratory setting. This chapter will discuss spontaneous and induced neoplastic lesions in the rat, but the separation of these entities becomes increasingly difficult as we gain greater understanding of the genetic and epigenetic effects and the role of hormones, diet, and environment on cancer development. The increasing use of genetically modified rats in carcinogenesis studies will also blur the lines that were previously used to differentiate between spontaneous and induced neoplastic tumor models.

B. Diagnosis of Neoplasia

Cancer is a proliferative disease and manifests most commonly as an abnormal cellular response or as a mass. For most rat neoplasms there is relatively little information concerning the genetic and phenotypic characteristics that correlate with biological behavior. This often makes the distinction between the preneoplastic state and early neoplasia difficult and sometimes arbitrary, and further creates uncertainty in the separation of benign and malignant tumors. The pathologist must characterize and differentiate proliferative states that appear as a biological continuum of lesions ranging from the preneoplastic to overt malignancy. The classification of lesions is often based upon subjective morphologic criteria. Fortunately, the widespread use of the rat in chronic toxicity/oncogenicity bioassays has resulted in extensive publications describing the morphology of rat proliferative lesions. In rat cancer studies, it is important to understand the histogenic relationship of various tumors so that they can be properly grouped together for the determination of tumor incidence. Excessive diagnostic "splitting" and categorizing by tumor type can artificially create or obscure potential treatment-related increases in cancer incidence in rodent bioassays. Published guidelines are available to aid in this challenging effort (McConnell et al., 1986). The reader is directed to the Web sites of the National Toxicology Program (NTP) (http://ntp-server. niehs.nih.gov) and the Society of Toxicologic Pathology (www.toxpath.org) for linkages to information on rodent tumor guidelines, incidences, and other pertinent information concerning this topic.

C. Dietary Factors

The incidence of spontaneous and induced cancer in the rat is heavily influenced by both genetic and environmental factors. Diet is one of the more important environmental factors that contribute to variability in cancer incidence. The composition of the diet is known to significantly influence cancer development (Rogers et al., 1993). Dietary factors such as vitamin or fat content can directly influence tumor development in certain target organs. One example is the association of fat intake and acinar cell neoplasia of the pancreas in the rat (Eustis and Boorman, 1985). Chronic feeding of diets deficient in choline and methionine are examples of models to induce hepatic neoplasia in rats (Nakae, 1999). In addition to a primary modulation of tumor incidence rates, dietary influences can induce disease and physiologic states that have secondary effects on cancer development in the rat. One example is the high dietary protein-enhancing chronic progressive nephropathy in the rat that in turn modulates the development of renal cell neoplasia (Seely et al., 2002).

Caloric intake also affects tumor incidence in rats (Keenan et al., 1995). Rats fed *ad libitum* have lower survival and a higher incidence of pancreatic, mammary, and pituitary tumors than rats subjected to moderate dietary restriction of identical diets. In addition to increasing tumor incidence, high caloric intake decreases latency and enhances rate of tumor growth (Keenan et al., 1995). Tumors with endocrine influence, such as the pituitary and mammary glands, have been most severely altered by *ad libitum* feeding.

D. Hormonal Factors

Hormonal factors are known to be a major environmental influence on the incidence of neoplasia in the rat. One example is the significant negative correlation between testicular interstitial cell tumors and pituitary tumors in control male Fischer 344 rats (Nyska et al., 2002). It appears that single versus group housing alters hormonal physiology and cellular kinetic changes in the pituitary and testes, which influence the incidence of tumors in these organs. Thus the incidence of neoplasia in group-housed retired breeder animals can differ significantly from control animals that are singly housed in long-term toxicity/oncogenicity bioassays. Parity also is associated with a dramatic reduction in rat uterine leiomyoma incidence (Walker et al., 2001).

E. Microbial Effects

There is general agreement that rats should be maintained free of known murine pathogens during cancer studies but the influence of microbial factors on rat carcinogenic responses is not as well described as that for mice. In mice there are well-known associations of natural pathogens with changes in incidence of sponta-neous neoplasia rates (Baker, 1998). In one report, groups of female rats with Sendai virus infection had a greater survival and a higher prevalence of lung tumors than groups without Sendai infection. However, none of the tumor prevalence and survival differences was statistically significant when interlaboratory variability and time-related effects were accounted for (Rao et al., 1989). It should be noted, though, that in experimental rat carcinogenesis studies with nitrosamines, respiratory infec-tion with *M. pulmonis* has been associated with enhance-ment of the pulmonary neoplastic response (Schreiber et al., 1972). Murine pathogens can modulate tumor incidence rates via a number of important processes, including cellular kinetic and metabolism effects (Everitt and Richter, 1990). Questions of synergism between chemical and infectious agents have arisen in rat oncogenicity studies in which infections were present. These include the finding of nasal proliferative lesions in a rat study infected with *M. pulmonis* (Lynch et al., 1984) and the association of salivary gland tumors in a study infected with sialodacryoadenitis virus (SDAV) (Burek et al., 1984).

F. Miscellaneous Environmental Factors That Influence Tumor Rates

Numerous aspects of the rodent environment can impact tumor development and incidence, including hous-ing type, temperature, bedding, altitude, light levels, and caging practices (Fox, 1983; Everitt, 1984). Housing and other environmental effects on tumor incidence are well described in rat studies and must be controlled for via cage rotation schemes and proper randomization of animals and cage allocation during the course of oncogenicity bioassays (Steplewski et al., 1987; Young, 1989; Herzberg and Lagakos, 1991).

II. THE RAT AS AN EXPERIMENTAL MODEL IN ONCOLOGY

Although cancer is generally considered to be a disease of aged animals, it is occasionally noted in young animals as a hereditary condition. One example is here-ditary renal carcinoma in the rat, first described by the pathologist Reidar Eker and thus termed the *Eker rat* (Eker and Mossige, 1961). This condition is an example of a Mendelian dominantly inherited predisposition to the development of a specific cancer. In Eker rats, 100% of heterozygote animals develop bilateral, multiple renal cell tumors at a young age. Hereditary or familial forms

of neoplasia often develop in litter mates, may manifest at a young age, or may be found as multiple tumors or present bilaterally. Because familial forms of cancer exist, proper randomization and assignment to experi-mental study groups is critical when conducting a rodent oncogenicity bioassay, especially when outbred rats are used. One procedure is to obtain more animals than necessary, weigh all the animals after acclimation and eliminate the lightest and heaviest animals, and then randomize to allow equal body weights in each of the dose and control groups (Hamm, 1985).

Hereditary cancers have been important in the under-standing of neoplasia and have led to the cloning of the first tumor suppressor genes (Knudson, 1971). The genetic changes responsible for familial cancers are often the same as those responsible for the more common spontaneous sporadic forms of neoplasia. Heredi-tary renal carcinoma in the Eker rat is caused by a germ line retrotransposon insertion in the rat homologue of the human tuberous sclerosis (*TSC2*) gene (Kobayashi et al., 1995). This leads to a "natural knockout" of the tumor suppressor function of the tuberous sclerosis 2 gene and the development of a cancer syndrome leading to hereditary renal cell tumors as well as leiomyomas and hemangiomas (Everitt et al., 1992). Eker rats have proven to be useful models for studying the molecular under-pinnings of both renal and smooth muscle tumorigenesis. Studies in these animals have led to the finding that the Tsc2 tumor suppressor function is important in rodent and potentially in human renal cell tumors.

Spontaneous mutations in cancer susceptibility genes and genetic predispositions to chemically-induced neo-plasia have provided important tools for the study of cancer in rats; however, it will soon be possible to mani-pulate the rat genome and create oncogenesis models in a manner very analogous to that used for genetically-engineered mice. The National Institutes of Health (NIH) has established a National Rat Genetic Resource Center (*http://www.nih.gov/science/models/rat/*) and launched the Rat Genome Program in 1995. Rat models provide important strengths for the study of human health and disease, as the vast amount of physiologic, biochemical, cellular, pharmacologic, and toxicologic data provide a superb platform on which to build the genetic and genomic tools to elucidate the connections between genes and biology. It is anticipated that there will be active research interactions to develop mutagenesis programs and the phenotyping of rat models in parallel with the more mature and established NIH mouse programs.

Spontaneous genetic and chemically induced rat cancer models are used to study major cancers, and the use of rats has contributed much useful information. A discussion of individual models and cancers is beyond the scope of this chapter, as the literature is voluminous. A recent search of PubMed publications revealed over 5,000 review articles of cancer studies in rats. Despite the many well-studied models available, the advent of technology to allow the manipulation of the rat genome and the creation of genetically-modified rats is revolutionizing how this species contributes to our knowledge of the carcinogenic process.

III. INTEGUMENTARY SYSTEM

A. Skin

Proliferative lesions of the skin are usually visible clinically or at necropsy. The most common neoplasms that appear in the skin have specific anatomic locations, such as the mammary gland, preputial gland, clitoral gland, or Zymbal glands but are not actually skin tumors (Elwell et al., 1990). The diagnosis of the specific skin tumor type and separation from non-neoplastic lesions, such as squamous cell hyperplasia or epidermal cyst, requires microscopic examination. One of the more common skin tumors (occurring at 2% to 3% incidence, Tables 14-1, 14-2, and 14-3) is squamous cell papilloma often appearing as a papillary structure in the facial area (Fig. 14-1). The papilloma is comprised of a fibro-vascular core covered by usually well-differentiated epithelium (Fig. 14-2). Its malignant counterpart, the

TABLE 14-1

FISCHER 344/N RATS

Common Tumors of the F344/N Rats			
Site	Tumor	Males (%)	Females (%)
Testes	Adenoma	89.1	—
Hematopoietic	Leukemia (MCL)	50.5	28.1
Pituitary, pars distalis	Adenoma	30.0	53.1
Mammary gland	Fibroadenoma	4.3	41.2
Adrenal medulla	Pheochromocytoma	31.9	5.1
Uterus	Stromal polyp	—	14.2
Thyroid	C-cell adenoma	13.0	11.7
Preputial/clitoral gland	Adenoma	7.4	9.2
Subcutis	Fibroma	5.1	1.6
Pancreatic islet	Adenoma	4.0	1.5
Skin	Keratoacanthoma	4.0	0.4
Preputial/clitoral gland	Carcinoma	3.3	3.0
Lung	Adenoma	2.6	1.6
Mammary gland	Carcinoma	0.1	3.1
Liver	Cholangioma	2.8	0.7
Liver	Adenoma	2.3	0.6
Skin	Papilloma	2.1	0.8

Table based on 1,338 male and 1,336 female F344/N rats from 2-year feeding studies (Haseman et al., 1998). Includes all tumors that occur at greater than 2% incidence in either sex.

TABLE 14-2

SPRAGUE-DAWLEY RATS

Common Tumors of Sprague-Dawley Rats			
Site	Tumor	Males (%)	Females (%)
Mammary Gland	Fibroadenoma	0.0	19.0
Mammary gland	Adenocarcinoma	0.0	8.8
Mammary gland	Adenoma	0.0	3.5
Adrenal medulla	Pheochromocytoma	4.0	0.0
Skin	Keratoacanthoma	4.0	0.0
Thyroid	C-cell adenoma	3.6	2.7
Pancreatic islet	Adenoma	3.7	1.0
Uterus	Stromal polyp	—	2.6
Skin	Fibroma	2.0	1.3
Skin	Papilloma	2.0	0.0
Liver	Adenoma	2.0	0.0

Table based on 1,340 male and 1,329 female Sprague-Dawley rats used as controls in 17 carcinogenicity studies (Chandra et al., 1992). Includes all tumors that occur at greater than 2% incidence in either sex.

squamous cell carcinoma, is uncommon as a spontaneous tumor but can be induced in dermal carcinogenicity studies. Cellular atypia and local invasion are usually the hallmark of malignancy for the squamous cell carcinoma. Tumors that arise from the basal layer of the skin

TABLE 14-3

WISTAR RATS

Common Tumors of Wistar Rats			
Site	Tumor	Males (%)	Females (%)
Pituitary, pars distalis	Adenoma	31.6	50.3
Mammary gland	Fibroadenoma	2.6	36.1
Uterus	Stromal polyp	—	15.5
Exocrine pancreas	Adenoma	13.3	0.7
Skin	Keratoacanthoma	11.2	0.7
Testes	Adenoma	10.5	—
Adrenal medulla	Pheochromocytoma	9.7	1.3
Thyroid	C-cell adenoma	5.8	8.4
Thymus	Thymoma	3.0	7.5
Mammary gland	Carcinoma	0.0	6.7
Pancreatic islet	Adenoma	5.4	1.7
Subcutis	Fibroma	4.5	3.0
Adrenal cortex	Adenoma	3.4	4.1
Thyroid	Follicular cell adenoma	3.9	2.8
Exocrine pancreas	Carcinoma	3.4	0.0
All sites	Hemangioma	3.4	0.7
Subcutis	Fibrosarcoma	3.2	0.0
Skin	Papilloma	3.2	0.4
Liver	Adenoma	2.8	2.2
Brain	Granular cell tumor	2.4	1.5
Hematopoietic	Lymphoma	1.3	2.4

Table based on 465 male and 465 female Wistar rats that served as controls for carcinogenicity studies (Poteracki and Walsh, 1998). Includes all tumors that occur at greater than 2% incidence in either sex.

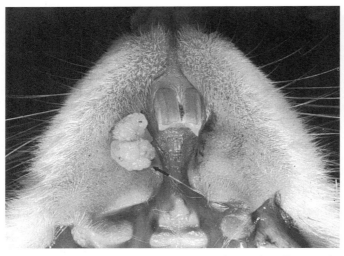

Fig. 14-1 Typical macroscopic appearance of an oral papilloma at the mucocutaneous junction.

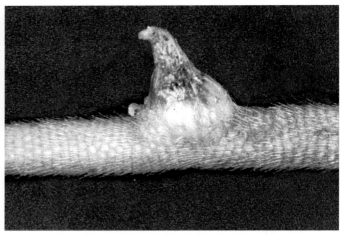

Fig. 14-3 Raised proliferative keratin "horn" noted with a trichoepithelioma on the skin of an F344 rat.

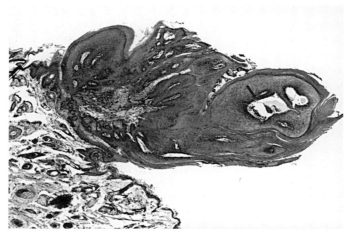

Fig. 14-2 Photomicrograph of papilloma depicting overlying thickened squamous epithelium forming deep papillary projections into a proliferative fibrovascular core *(arrow)*.

are known as basal cell adenomas and basal cell carcinomas. These tumors are very uncommon as spontaneous lesions and are recognized microscopically by their tendency to grow in ribbons, cords, or to form nests of darkly basophilic cells with prominent nuclei and scant cytoplasm (Elwell et al., 1990).

Several types of tumors arise in the adnexa. The most common is the keratoacanthoma; a crateriform neoplasm that often forms a keratin horn, giving the tumor a characteristic appearance (Fig. 14-3). The tumor is thought to arise from the hair follicle. One report suggests a high incidence of keratoacanthoma in a small group of Long-Evans rats (Esfandiari et al., 2002), but a larger study found one keratoacanthoma in 490 control males and none in 490 control females from carcinogenicity studies

(Sommer, 1997). Keratoacanthoma occurred at about a 3% incidence in aged male Sprague-Dawley rats but less than 1% in the females (Zwicker et al., 1992). The trichoepithelioma is an uncommon tumor that arises from the hair follicle with sufficient differentiation that hair follicle formation is seen within the tumor (Elwell et al., 1990). Sebaceous gland adenomas and carcinomas arise from the sebaceous glands surrounding the hair follicle. Both benign and malignant tumors contain characteristic sebaceous cells with foamy cytoplasm and a peripheral layer of basal cells. The carcinomas are more poorly differentiated and usually show invasion into the adjacent structures.

B. Subcutis

Fibromas are the most common tumor of the subcutis, occurring at about a 5% incidence in male Fischer 344 (Table 14-1) and Wistar rats (Table 14-3) and at a lower incidence in other strains and in the female Fischer 344 and Wistar rats (Maekawa et al., 1986; Chandra et al., 1992; Haseman et al., 1998; Poteracki and Walsh, 1998). Fibromas appear at necropsy as firm white subcutaneous masses. Grossly they appear similar to fibroadenomas of the mammary gland. Histologically, fibromas are comprised only of spindle cells in a whorled pattern (Fig. 14-4). The cells are well differentiated and mitotic figures are rare. The malignant counterpart, fibrosarcoma, is quite uncommon, shows more atypia, and because it is invasive, tends to be attached to surrounding tissue. Special stains or ultrastructural examination can demonstrate, in some cases, that an occasional spindle cell tumor is of Schwann or nerve cell origin. Lipomas are

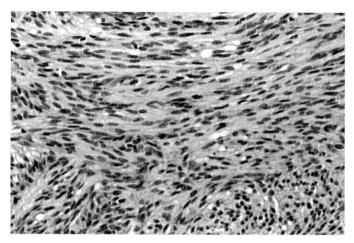

Fig. 14-4 High-magnification photomicrograph showing the typical appearance of a fibroma with long fusiform neoplastic mesenchymal cells arranged in interlacing fascicles.

subcutaneous tumors that appear grossly to be comprised of fatty tissue and, histologically, they resemble mature adipose cells. In Fischer 344 rats, neural crest tumors have been described that appear most commonly on the pinna (Fig. 14-5). The tumor is observed as a thickening or nodule on the pinna of the ear. The tumor is not common, but one or two may occasionally be observed in a control group of 50 (Elwell et al., 1990). The tumors stain strongly positive for S-100, indicating either Schwann cell or melanocytic origin (Elwell et al., 1990). The tumor cells on ultrastructural examination contain premelanosomes, suggesting to some that these tumors on the pinnae of the ear are amelanotic melanomas (Yoshitomi et al., 1995; Nakashima et al., 1996) rather than neural crest tumors.

C. Specialized Sebaceous Glands

Specialized sebaceous glands are associated with the ear, prepuce, clitoris, and anus in the rat. Zymbal's gland is an auditory sebaceous gland ventral to the orifice of the external ear. Although spontaneous tumors are rare, this neoplasm gained notoriety when several important environmental chemicals were shown to cause Zymbal gland tumors (Eustis et al., 1995; Dunnick et al., 1997; Quast, 2002). The incidence of this tumor is probably underestimated because the gland is not examined unless a mass is seen grossly (Fig. 14-6) (Copeland-Haines and Eustis, 1990). Further, a Zymbal gland tumor could be mistaken as a squamous cell carcinoma unless sufficient sections are taken to show the sebaceous gland origin (Fig. 14-7). Zymbal gland tumors usually appear as firm subcutaneous masses ventral to the ear. Although chemicals that induce Zymbal gland tumors also often induce hyperplasia of the gland, most tumors are diagnosed as malignant, (i.e., Zymbal gland carcinoma) when they are found. The tumors are characterized by abundant keratin production and papillary projections of epithelial cells comprised of a mixture of squamous cells and sebaceous cells. Intense scirrhous response and inflammation are often associated with Zymbal gland carcinomas. The amount of squamous and sebaceous components varies between tumors, but there seems to be little justification for subclassifying Zymbal gland tumors based on predominant cell type (Copeland-Haines and Eustis, 1990).

Spontaneous preputial and clitoral gland tumors are not uncommon, occurring at about a 3% incidence in male and female Fischer 344 rats, respectively (Haseman et al., 1998). The tumors appear as firm subcutaneous pale pink to orange masses lateral to the penis or ventral to the

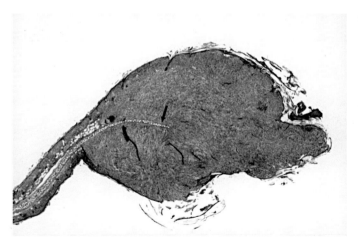

Fig. 14-5 Low-magnification photomicrograph of an amelanotic melanoma on the pinna of the ear of an F344 rat. Remnants of aural cartilage form a central core (arrows) through the mass.

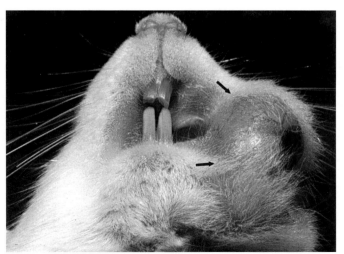

Fig. 14-6 An ulcerated firm mass on the side of the face of an F344 rat that is typical of the appearance of a Zymbal's gland tumor.

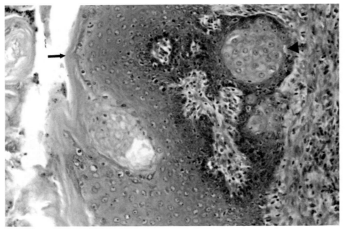

Fig. 14-7 Photomicrograph of Zymbal's gland tumor showing the typical mixed squamous *(arrow)* and sebaceous differentiation. The arrowhead depicts a nest of plump neoplastic sebaceous epithelial cells within an area of squamous epithelial proliferation.

vaginal orifice. Occasionally these tumors become ulcerated, resulting in removal of the animal from the study. The tumors have varied histologic patterns ranging from papillary to cystic to solid growths with varying components of sebaceous and squamous cells. Sub-classification based on cell type or pattern does not appear warranted (Copeland-Haines and Eustis, 1990). The tumors appear to represent a continuum and classification into benign and malignant categories is based on cellular atypia, size, and invasion. These tumors rarely metastasize (Reznik and Ward, 1981). As with the Zymbal glands some chemicals will induce tumors of the preputial and clitoral glands. These specialized sebaceous glands do not always respond uniformly to chemical exposure. In some studies, only the preputial or clitoral glands have increased cancer rates (Fujii et al., 1991; Irwin et al., 1995), whereas with other chemicals only Zymbal glands may be involved (Dunnick et al., 1997; Quast, 2002), and with other chemicals both preputial/clitoral and Zymbal glands cancer rates are increased (Huff et al., 1989; Eustis et al., 1995).

IV.   MUSCULOSKELETAL SYSTEM

A.   Skeletal Muscle

Primary tumors of the skeletal muscle in rats are rare. Rhabdomyomas, the benign tumors, have not been reported in the literature to our knowledge. The malignant counterpart, rhabdomyosarcomas occurs at less than 1% incidence in the Fischer 344/N rats (McDonald and Hamilton, 1990). Rhabdomyosarcomas can be induced by local injection of some carcinogenic metals, such as nickel sulfide (Shibata et al., 1989). The hallmark for

diagnosis of skeletal muscle tumors is the presence of cross-striations in the cancer cells. Special stains such as PTAH can help demonstrate the cross-striations and ultrastructure can provide confirmation. The diagnosis is not always easy, as highly pleomorphic cancer cells may have little resemblance to muscle cells. When cancers invade skeletal muscle, one must be careful not to misinterpret preexisting entrapped muscle fibers as neoplastic cells. The presence of striated cells at multiple sites or in metastases can be helpful. Occasionally rhabdomyosarcomas occur in unusual sites, such as the nasal cavity of rats exposed to 2, 6-xylidine (National, 1986).

B.   Bones and Joints

Primary bone tumors are uncommon, occurring at less than 1% in Fischer 344/N rats (Leininger and Riley, 1990). Osteosarcomas have been reported to occur with the injection of iron oxide in athymic rats (Axmann et al., 1998). In the rat, the tumors appear to be more common in males than females and occur more commonly in the flat bones of the head and vertebrae than in the long bones. Unless the skeleton is subject to special scrutiny, these tumors can be easily overlooked. In a sodium fluoride study in which there was concern about the potential for induction of skeletal tumors, the rats were subject to whole body x-ray at the time of necropsy to be sure that no bone tumors were missed (National, 1990). A high incidence of osteosarcoma was found in both male and female Fischer 344 rats after daily subcutaneous injections of recombinant human parathyroid hormone PTH (1-34) (Vahle et al., 2002).

The benign bone tumor, osteoma, may appear at necropsy as a hard dense pale mass attached to the surface of a bone (Leininger and Riley, 1990). Histologically osteomas are comprised of osteoid with few cells. The malignant counterpart, osteosarcomas, are more common. The appearance at necropsy varies from firm, dense, pale masses to more hemorrhagic tumors. Histologically, osteosarcomas may have a diverse appearance, ranging from those comprised mainly of osteoid with few cells, to highly cellular tumors comprised mainly of osteoblast cells. Some osteosarcomas may show a fibroblastic appearance, whereas telangiectatic osteosarcomas are filled with dilated blood vessels (Leininger and Riley, 1990). As one might predict, tumors of fibroblast origin (fibrosarcoma) invading bone and inciting osteoblast reaction are difficult to distinguish from the fibroblastic variant of osteosarcoma. It is helpful to find bone formation in tumor areas away from the bone, such as occurs with pulmonary metastases. Occasionally large tumors of various types may contain areas of necrosis and/or mineralization. Although the tumors may appear gritty when cut, suggesting the presence of bone, secondary mineralization of tumors rarely provides diagnostic problems. Rats have been used to study the development of sarcomas after radiation, usually of a hind limb. Osteosarcomas, malignant fibrous histiocytomas, and fibrosarcomas are often seen after irradiation (Tinkey et al., 1998).

Tumors of cartilage may arise in bone or at articular surfaces and are even less common than bone tumors in the rat. Benign tumors are known as chondromas, whereas malignant tumors of cartilage are chondrosarcomas. When both cartilage and bone formation are present in the same tumor, the convention is to classify the tumors as compound osteosarcomas.

V. RESPIRATORY SYSTEM

A. Nasal Cavity

Primary tumors of the nasal cavity in rats are rare. Tumors induced by chemicals are also uncommon. Only 12 chemicals of nearly 500 NTP studies were associated with tumors of the nasal cavity (Haseman and Hailey, 1997). This is, however, an important target organ of toxicity in inhalation and other types of oncogenicity bioassays due to the fact that rats are obligate nasal breathers and this organ has extremely high metabolic activity. It is worth noting that only five of the 12 nasal carcinogens in NTP studies were administered by inhalation (Haseman and Hailey, 1997). Squamous cell carcinomas have been induced in the rat respiratory epithelium of the nasal cavity (Swenberg et al., 1980), and tumors of neuroepithelial origin have been reported in the olfactory region (Reznik-Schuller, 1983). Nasal squamous cell carcinomas have been reported to have a relationship to the malocclusion syndrome in aging rats (Feron and Woutersen, 1989). Perhaps the most unexpected nasal tumors were the rhabdomyosarcomas of the nasal cavity found with 2, 6-xylidine exposure (Haseman and Hailey, 1997). A high incidence of inflammation, hyperplasia, and squamous metaplasia often was found in association with the nasal neoplasms in NTP studies (Haseman and Hailey, 1997). However, a high incidence of these nonneoplastic lesions was often found in inhalation studies showing no evidence of nasal carcinogenicity (Haseman and Hailey, 1997). This suggests that chronic toxicity does not necessarily lead to carcinogenicity in the nasal cavity.

B. Lung

Primary lung tumors are not a common spontaneous lesion in any common strain or stock of rat (Goodman et al., 1980; Kroes et al., 1981; Sher et al., 1982;

Haseman et al., 1998). When they occur, these tumors are usually of bronchiolar/alveolar origin (alveolar type II cell and/or Clara cell), although squamous cell carcinomas also have been reported (Mohr et al., 1990). Most pulmonary neoplasms in the rat arise in the distal lung and not within large airways, as in man and certain other species. The incidence of primary lung tumors in rats varies from 0 to 2% in most stocks and strains in chronic studies. In Fischer 344 rats that have been commonly used by the NTP in oncogenicity bioassays there is a slightly higher incidence in males than in females and virtually all spontaneous tumors are of the bronchiolar/alveolar type. They usually appear as solitary, firm, pale white nodules that project above the pleural surface. In advanced cases, such as carcinomas, an entire lung lobe may be replaced by tumor tissue. Most spontaneous pulmonary neoplasms in rats are incidental findings and are not necessarily related to the cause of death. They do not generally metastasize widely and little is known of their biological behavior (Reznik-Schuller and Reznik, 1982; Boorman, 1985).

Similar to other neoplasms of the rat, primary pulmonary carcinogenesis is a biological continuum of lesions without clear demarcation of stages. Pulmonary lesions range from the pre-neoplastic hyperplasia to the benign bronchiolar/alveolar adenoma (Fig. 14-8) and then the more advanced malignant bronchiolar/alveolar adenocarcinoma. The lung also is a common site of metastasis of epithelial neoplasms. Thus the bronchiolar/alveolar carcinoma must be differentiated also from metastatic carcinomas arising from the mammary glands, reproductive or gastrointestinal tract, or hepatic bile ducts. Metastatic carcinomas are often multiple, tend to occur in lymphatics or blood vessels, and generally have discrete margins. Systemic mesenchymal neoplasms, such as

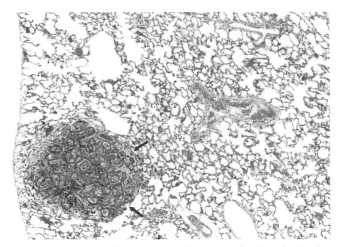

Fig. 14-8 Low-magnification photomicrograph of a bronchoalveolar adenoma *(arrows)* showing the circumscribed proliferation of neoplastic alveolar epithelial cells causing local compression of the pulmonary parenchyma.

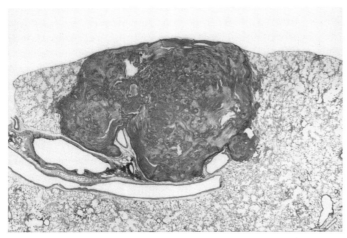

Fig. 14-9 Low-magnification photomicrograph of proliferative keratinizing cyst in the lung of a rat. This is a unique benign squamous epithelial lesion that has been induced in inhalation studies.

histiocytic sarcomas, lymphoid tumors, and hematopoietic malignancies also commonly invade the pulmonary parenchyma and cause macroscopic and microscopic lesions in the rat lung. These lesions are often more diffuse than epithelial tumors and can resemble inflammatory and other conditions. In most stocks and strains of rats they are more common than primary lung tumors.

Rats are frequently used in chronic inhalation bioassays to assess oncogenicity because they have a low incidence of spontaneous lung cancer, but respond with lung tumors when exposed to a variety of inhaled carcinogens (Maronpot, 1990). Squamous metaplasia is common in inhalation studies, as are keratinizing lesions ranging from simple cysts to larger cystic keratinizing lesions (Boorman et al., 1996). An unusual feature of pulmonary oncogenesis in the rat is the propensity of this species to develop primary pulmonary neoplasms when exposed to low toxicity insoluble particulates at extremely high concentrations that overwhelm the ability of alveolar macrophage-mediated lung clearance (so-called "pulmonary overload" tumorigenesis). These rats also often develop a benign keratinizing cyst (Figs. 14-9 and 14-10) filled with keratin that may be mistaken for neoplasia (Carlton, 1994; Boorman et al., 1996). Thus in judging the significance of cancer in the rat lung following chemical exposure, one must keep in mind not only the chemical, but if a particulate, whether the dose exceeded the pulmonary clearance.

VI. CARDIOVASCULAR SYSTEM

A. Heart

Proliferative lesions of the heart are uncommon but several very characteristic tumors are found in the heart.

The most common tumor for many strains of rats begins as endocardial hyperplasia that then appears to progress to endocardial schwannoma (MacKenzie and Alison, 1990). The diagnosis of the tumor and separation from the non-neoplastic endocardial hyperplasia is difficult, as these lesions appear to represent a spectrum of the same disease process (MacKenzie and Alison, 1990). The lesion appears to begin in the left ventricle as a proliferation of cells between the endocardium and the underlying myocardium and, when only a few cell layers thick, appears to be a hyperplastic lesion (Boorman et al., 1973). Endocardial hyperplasia occurs at less than 0.5% in Fischer 344, Wistar, or Sprague-Dawley rats (Novilla et al., 1991) and endocardial schwannomas occur at less than 0.1% incidence (Alison et al., 1987). As the tumors become larger, they can fill the left ventricle and extend along the major vessels that are attached to the heart. Some tumors will show palisading of nuclei that is characteristic for schwannomas and some tumors stain weakly for S-100 (MacKenzie and Alison, 1990). Although these tumors are uncommon, they can be induced by intravenous injections of nitrosamines (Berman et al., 1980). These induced tumors show ultrastructural features characteristic of Schwann cells (Berman et al., 1980).

Intramural schwannomas have morphologic features similar to endocardial schwannomas, except that they occur within the myocardium (MacKenzie and Alison, 1990). They usually appear as a small focal mass of spindle cells within the left ventricle or interventricular septum. Larger lesions may show palisading of nuclei. Their relationship to the endocardial lesions is not known (MacKenzie and Alison, 1990).

An unusual but very characteristic tumor is an atriocaval mesothelioma that occurs in the right atrium of the rat.

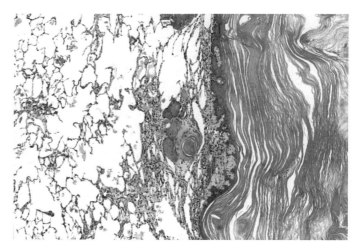

Fig. 14-10 High-magnification photomicrograph of proliferative keratinizing cyst in a rat, showing the proliferation of acellular squamous debris forming a central core with a thin benign squamous epithelial lining.

This tumor is common in the NZR/Gd inbred albino rat (Goodall et al., 1975), but very uncommon in other strains (MacKenzie and Alison, 1990). The tumor is typically a mass involving the wall of the right atrium. Histologically the tumor is comprised of glandular structures that contain erythrocytes, eosinophilic material, cell debris, hematoidin crystals, and macrophages. Larger neoplasms may show necrosis, invasion, and, rarely, pulmonary metastases (MacKenzie and Alison, 1990). Ultrastructurally these tumors have features of mesotheliomas and thus, *atriocaval mesothelioma* is the most appropriate term for these rare and interesting tumors (Chandra et al., 1993; Peano et al., 1998).

B. Vessels

Hemangiomas are benign, well-demarcated vascular tumors that do not exhibit atypia or local invasion (Mitsumori, 1990). The tumors consist of well-formed vessels lined by a single layer of endothelial cells. Vessels packed with erythrocytes often are the most noticeable feature of this neoplasm. Vascular tumors must be distinguished from angiectasis and other non-neoplastic proliferations of dilated vessels. Hemangiomas may occur in any organ but in the rat are often found in the spleen, kidney, subcutaneous tissue, and liver (Zwicker et al., 1995). A relatively high incidence of hemangiomas has been reported in the mesenteric lymph nodes of Wistar rats (Reindel et al., 1992). One has to be cautious diagnosing hemangiomas in mesenteric lymph nodes, because hemorrhage, hemosiderin deposition, and angiectasis are common features in the mesenteric lymph nodes of aging rats and can be confused with benign tumors.

Hemangiosarcomas are the malignant counterpart to hemangiomas and generally are found in the same organs. Hemangiosarcomas are locally invasive, the cells are much more prominent, and usually show cellular atypia. Some hemangiosarcomas are highly cellular, and areas of vessel formation can be found only upon careful examination. The occurrence of hepatic hemangiosarcomas in rats after vinyl chloride exposure (Maltoni and Cotti, 1988) is of interest, because vinyl chloride causes similar tumors with occupational exposure (Lee and Harry, 1974). The bioassay results of vinyl chloride in rats were available (Maltoni and Lefemine, 1975) as the understanding of the potential occupational risk was unfolding and played an important role in the risk assessment process for this important human chemical exposure.

Hemangiopericytomas are rare tumors of pericytes that have a characteristic whorled pattern of spindle cells around capillaries. These lesions can be difficult to distinguish from fibromas and fibrosarcomas that also may arise from the vessel wall. The differential diagnoses of these various soft tissue neoplasms provides a diagnostic challenge that may still be open to question even after special stains and ultrastructural examination. The diagnostic difficulties are due in part to the pluripotent nature of mesenchymal cells and the lack of distinguishing features of spindle cells.

VII. DIGESTIVE SYSTEM

A. Oral Cavity

Primary tumors of the oral cavity in rats are uncommon. Squamous cell papillomas of the oral cavity are most likely to occur on the dorsum of the tongue as a papillary projection (Fig. 14-11). Occasionally they occur on the hard or soft palate. Papillomas have an appearance similar to papillomas at other sites, with a squamous epithelium covering a fibrovascular core. The malignant counterpart, squamous cell carcinomas, have been described as spontaneous lesions but are very rare (Brown and Hardisty, 1990). Both squamous cell papillomas and squamous cell carcinomas may be induced by exposure to some chemicals. Although the oral exposure to multisite carcinogens may be expected to result in oral neoplasia (Dunnick et al., 1997), inhalation exposure also may cause oral cancer because of carcinogen deposition on the fur and grooming behavior (Melnick et al., 1999).

Painting of 4-nitroquinoline 1-oxide (4NQO) on the tongue or oral palate is used as a model for the induction of oral cancer with squamous cell carcinomas developing within 6 months (Nauta et al., 1996). This model has been

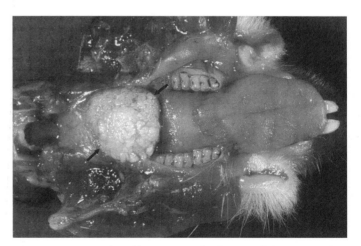

Fig. 14-11 Irregular white mass on the dorsum of the tongue of an F344 rat that is typical for a papilloma on the tongue. Microscopic examination is necessary to determine whether this lesion is a squamous cell papilloma or a squamous cell carcinoma. This lesion may be missed if the skull is not properly disarticulated to allow complete examination of the tongue.

used to test the modifying effects of various compounds (Tanaka et al., 2002; Yanaida et al., 2002).

Malocclusions and fractures may give rise to dysplastic lesions that need to be distinguished from odontomas of the teeth. Generally the growth pattern is most helpful with the tumor giving rise to tooth structures with variable amounts of dentin, enamel, and cementum (Brown and Hardisty, 1990). Although several morphologic types have been seen and subclassifications could be made on predominant cell type and matrix (Ernst et al., 1998; Jang et al., 2002), generally these tumors are simply referred to as odontomas (Brown and Hardisty, 1990).

B. Salivary Glands

Primary salivary gland tumors are rare in Fischer 344 rats (Neuenschwander and Elwell, 1990; Hosokawa et al., 2000) but may occur following chemical exposure (Takegawa et al., 2000). Adenomas of ductular or acinar origin may form expansive structures within the gland. Tubular adenomas consist of well-formed tubules, whereas acinar adenomas consist of well-formed acini resembling the normal gland. The growth pattern and compression of the adjacent tissue is helpful in separating adenoma from focal hyperplasia. Adenocarcinomas are less well differentiated, show local invasion, and, in some cases, incite a significant desmoplastic response. Squamous cell carcinomas of the salivary glands are rare in Fischer 344 rats but have been induced in other strains (Neuenschwander and Elwell, 1990; Sumitomo et al., 1996; Takegawa et al., 1998).

Malignant schwannomas, fibrosarcomas, or undifferentiated sarcomas may occur adjacent to or within the salivary gland (Neuenschwander and Elwell, 1990). These tumors infiltrate between the salivary ducts and acini that may undergo atrophic changes. Unless one is aware of the propensity of these mesenchymal tumors to occur within salivary glands, they may be mistaken for anaplastic carcinomas in rats (Neuenschwander and Elwell, 1990).

C. Esophagus

Primary tumors of the esophagus are rare in rats, but an occasional squamous cell papilloma has been reported (Neuenschwander and Elwell, 1990).

D. Forestomach

The stomach in the rat is divided into a forestomach lined by squamous epithelium and a glandular stomach separated by a limiting ridge (Fig. 14-12). Hyperplasia along the limiting ridge is common, especially in gavage studies (Neuenschwander and Elwell, 1990), and one must

Fig. 14-12 Squamous epithelial hyperplasia on the margo plicata of the rat stomach, the normal ridge that separates the squamous and epithelial portions of the gastric mucosa. This is a common site for proliferative lesions in the stomach of the rat and should be carefully examined in all rat oncology studies.

be cautious in diagnosing papillomas along the limiting ridge. Squamous cell papillomas are not uncommon in the forestomach, especially with gavage studies. Grossly the lesions appear as papillary structures or a cauliflower-shaped mass on the mucosal surface. These lesions are essentially diagnosed at necropsy and a careful examination results in more tumors being found. Squamous cell carcinomas are very rare as a spontaneous tumor but are not infrequently found after administration of carcinogens in the diet. Chemicals that cause irritation also appear to have a propensity to cause squamous cell carcinomas (Ghanayem et al., 1991). It is not uncommon for the male rats to have a higher incidence of forestomach tumors than the females (Tamano et al., 1992).

E. Glandular Stomach

Naturally occurring tumors of the glandular stomach are rare in the rat (Neuenschwander and Elwell, 1990). Adenomas may appear grossly as plaques or small polypoid nodules. The tumors are well differentiated, with cells forming glands with little or no atypia or invasion. Squamous cell carcinomas may show extensive invasion with epithelial cells growing through the wall of the stomach. Rat stocks and strains vary in their sensitivity to carcinogenicity of the glandular stomach with Lewis, Wistar, and Sprague-Dawley rats being similar and WKY rats less sensitive (Tanaka et al., 1995).

Spontaneous gastric carcinoid is extremely rare in the rat (Neuenschwander and Elwell, 1990). However, gastric carcinoids are important in humans (Gilligan et al., 1995) and have been induced in rats by chemicals that stimulate gastrin secretion (Hakanson and Sundler, 1990; Havu et al., 1990). The typical pattern is the appearance of nests of endocrine cells in the gastric mucosa. By immuno-histochemistry, these neuroendocrine tumors stain for neuron-specific enolase. The Sevier-Munger stain is specific for the enterochromaffin-like cells that form the carcinoids (Neuenschwander and Elwell, 1990).

The significance of carcinoids in rats gained considerable attention with the demonstration that exposure to several herbicides resulted in increased rates of these tumors (Hard et al., 1995; Heydens et al., 1999). Exposure to methyleu-genol, which is widely found in foods and beverages, also is associated with increased rates of gastric carcinoids in rats (Johnson et al., 2000). In many cases the carcinoids may be secondary to increased gastric pH, increased serum gastrin, and gastric-induced tropic effects on the enterochromaffin-like cells (Heydens et al., 1999; Abdo et al., 2001). Naturally occurring gastric carcinoids develop commonly with age in cotton rats (Waldum et al., 1999).

A variety of other tumors, such as leiomyomas, fibromas, hemangiomas, and their malignant counterparts, may also be found in the stomach. Lymphomas can involve the wall of the stomach and provide a diagnostic challenge. The morphology of the neoplastic cells and the tendency of lymphocytes to occur as individual cells as they spread into the adipose tissue are helpful clues as to the nature of the neoplasm.

F. Small Intestine

Primary tumors of the small intestine are rare in the rat. None were seen in nearly 4,000 control female Fischer 344 rats (Haseman et al., 1990) and only 10 in more than 3,700 control male Fischer 344 rats from NTP carcino-genicity studies. It is interesting that seven of the 10 small intestine tumors found in the NTP historical controls were malignant, suggesting that perhaps they were detected because of their size. Small polyploid lesions in the small intestine are difficult to detect unless the intestine is opened and examined carefully. Azoxymethane and 1,2-dimethylhydrazine are two of the more widely used chemicals in the induction of small intestinal neoplasia in the rat (Thorling et al., 1994; Hirose et al., 2002).

G. Colon and Cecum

There is considerable interest in animal models of colon cancer because of the frequency of this cancer in humans. Azoxymethane and its parent compound, dimethylhydra-zine, have been used to induce tumors in the large intestine in Fischer 344 rats (Elwell and McConnell, 1990). In the NTP studies only a few chemicals have produced tumors of the large intestine, mainly colon, and tumor rates are

generally higher in male rats than female rats. Spontaneous tumors of the colon are very uncommon in the Fischer 344 rat in 2-year studies, found in approximately one out of every 1,000 male rats and even less common in female rats (Haseman et al., 1998).

Almost all gastrointestinal tumors are recognized at necropsy and, thus, opening the bowel and careful examination of the luminal surface is required. The tumors appear as polyps that protrude into the lumen (Fig. 14-13). Invasion into the wall of the intestine is useful in distinguishing benign tumors from carcinomas. Metastases to distant tissues are very uncommon.

H. Liver

Unlike the situation in mice, most stocks and strains of rats do not have a high spontaneous incidence of primary liver cancer. Nevertheless, the hepatic parenchyma is one of the most frequently identified target organs in toxicity studies, and liver tumors are readily induced in the rat by a number of chemical carcinogens and other experimental regimens, such as dietary manipulation. The most common liver tumors in the rat are hepatocellular adenomas and hepatocellular carcinomas. They are more frequent in males and are found at an incidence of app-roximately 4% to 6% in Fischer 344 rats at 2 years of age (Haseman et al., 1998). Hepatocellular adenomas are benign nodular masses that are noted at necropsy to be lighter or darker than surrounding liver parenchyma, depending on the degree of vascular congestion or the amount of lipid and glycogen within neoplastic cells.

The distinction between focal hyperplasia, foci of cellular alteration, and hepatocellular adenoma can pose diagnostic difficulties. Hepatodiaphragmatic nodule, a rounded mass that protrudes from the diaphragmatic surface of

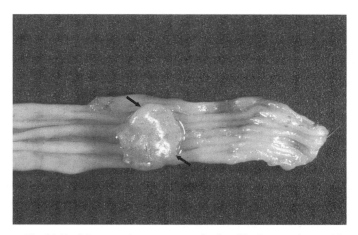

Fig. 14-13 Macroscopic appearance of polypoid adenoma *(arrows)* in the colon of a rat. These types of lesions are easily missed if the entire gastrointestinal tract is not opened and examined.

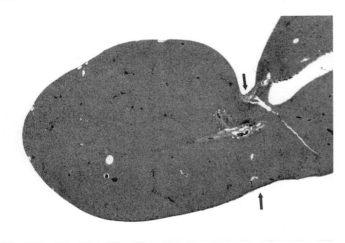

Fig. 14-14 Low-magnification photomicrograph of a rat liver that has herniated through the diaphragm forming a hepatodiaphragmatic nodule (*arrows* depict where the liver was entrapped). This relatively common lesion is often misdiagnosed as a hepatocellular adenoma.

the median lobe of the liver, must not be misdiagnosed as a hepatocellular adenoma (Fig. 14-14). This is a congenital lesion that may occur at any age in the rat (Eustis et al., 1990). The incidence may vary from 1% to 11% in the F344 rat but is rare in other strains. There is no demarcation between the nodule and the remainder of the liver parenchyma. In contrast, hepatocellular adenomas consist of cells resembling relatively normal hepatocytes, but there is loss of the normal lobular architecture and compression of surrounding parenchyma (Fig. 14-15).

Focal areas of hepatocytes within the liver of aging rats or in livers of rats exposed to carcinogens will show different tinctorial characteristics. These lesions have been classified as foci of alteration. The five types of alteration

include basophilic, eosinophilic, clear cell, vacuolated, and mixed cell-type foci (Eustis et al., 1990). Although these foci are excellent markers for early neoplastic change following carcinogen exposure (Bannasch et al., 2003), they can also be found in nearly 100% of F344/N rats by 2 years of age (Maronpot et al., 1989; Eustis et al., 1990). In rodent bioassays, basophilic foci are the most common, with the hepatocytes within the foci appearing generally smaller with more basophilic cytoplasm. Spontaneous eosinophilic foci are less common, with the hepatocytes within the foci generally appearing larger with a homogeneous eosinophilic cytoplasm (Eustis et al., 1990). Clear cell foci are characterized by enlarged hepatocytes with a clear perinuclear zone of accumulated glycogen. Vacuolated foci contain hepatocytes with lipid vacuoles. Mixtures of the above types may be classified as mixed cell foci (Eustis et al., 1990). In some cases, carcinogen exposure will give rise to foci that appear distinct from the naturally occurring foci (Bannasch et al., 1989; Maronpot et al., 1989).

There has been much controversy concerning the classification and nomenclature of rat hepatic lesions, and the accepted diagnostic criteria have changed over the years (Squire and Levitt, 1975; ILAR, 1980; Maronpot et al., 1986). Histologically, hepatocellular carcinomas have a more abnormal growth pattern than do adenomas. Carcinomas are characterized by cytologic atypia and marked disruption of parenchymal architecture. Hepatocellular carcinomas can have trabecular, glandular, solid, or mixed patterns and frequently metastasize to the lung.

Tumors of biliary origin also arise within the rat liver but are rare in control rats. Tumors of biliary origin have been diagnosed as cholangiomas and cholangiocarcinomas (Fig. 14-16). Other rare neoplastic lesions that have been

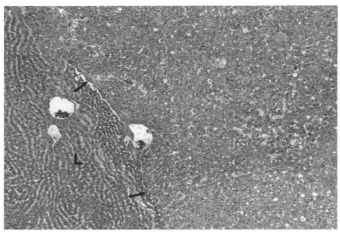

Fig. 14-15 Photomicrograph of the junction *(arrows)* between normal liver (L) and area of hepatocellular adenoma characterized by disorganization and loss of normal trabecular architecture.

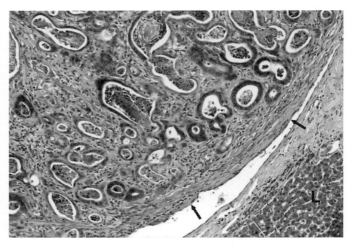

Fig. 14-16 Low-magnification photomicrograph of cholangiocarcinoma *(arrows)* in a rat liver showing neoplastic epithelial bile ductules separated by mesenchymal proliferation. Remnants of normal hepatic parenchyma (L) can be seen.

reported to arise within the liver of rats include vascular neoplasms and histiocytic tumors. Interestingly for laboratory animal scientists, rats have been reported to develop hepatic fibrosarcomas after infection with *Cysticercus fasciolaris*—the larval stage of the cat tapeworm (*Taenia taeniformis*)—but this finding is of historical note and should rarely occur in the modern laboratory. It should be recognized that the liver is one of the most frequent sites of metastatic tumors. The early stages of mononuclear cell leukemia are often first recognized in the liver sections.

I. Pancreas

Spontaneous neoplasms of the exocrine pancreas have been reported to occur infrequently in most stocks and strains of rats, although the incidence has been shown to depend on the sampling scheme used and the number of sections examined microscopically (Boorman and Eustis, 1984). In general, rats have pancreatic tumor incidences of less than 1% at 2 years of age. More recently, there have been reports of relatively high (up to 40% in aged Wistar rats) incidence (Dominick et al., 1990). Dietary factors, as well as pathology issues, such as diagnostic criteria and sectioning methods, can dramatically affect the incidence of acinar cell tumors. Acinar cell hyperplasia and neoplasia have been shown to be related to dietary fat levels and are more frequently noted in control rats receiving corn oil vehicles in chronic oncogenicity bioassays (Eustis and Boorman, 1985). Experimental data indicate that high levels of dietary fat (primarily unsaturated fat) enhance chemically induced carcinogenesis of the pancreas. Caloric restriction or feeding a diet containing retinoids (vitamin A analogues) inhibits the formation of pancreatic cancer in the rat (Longnecker, 1990).

There is a clear morphologic continuum from focal acinar cell hyperplasia to adenoma and to carcinoma. These proliferative lesions range in size from a millimeter to large multilobulated masses 1 cm or more in diameter. The criteria for differentiating hyperplastic lesions from overt neoplasia is difficult, as the biological potential of these pancreatic lesions has not been fully characterized. There is a higher incidence and wider distribution of exocrine pancreatic neoplasms in male rats as compared with females. This is consistent with androgenic modulation of exocrine pancreatic growth and development.

Rats have provided an important animal model for studies of human pancreatic tumorigenesis, because a number of chemical treatments can reproducibly induce these tumors. Azaserine treatment of Wistar/Lewis rats comprises a well-studied model of chemically-induced pancreatic cancer (Longnecker, 1981). Rats generally develop acinar cell neoplasms that are rare in humans, whereas humans generally develop ductal tumors.

VIII. ENDOCRINE SYSTEM

Spontaneous tumors of the endocrine glands are common in most rat stains and quite variable between strains and sex. For example, Fischer 344 (Table 14-1) and Wistar rats (Table 14-3) generally have a very high incidence of pituitary adenomas compared with Sprague-Dawley rats (Table 14-3). Also male Fischer 344 rats have a greater than 30% incidence of pheochromocytomas, whereas female Fischer 344 rats have an approximately 5% incidence (Table 14-1). Because endocrine tumors are small and easily missed, the incidence is directly related to the histologic techniques used. Multiple sections of endocrine tissues result in a greater incidence. Studies that require both sections of adrenal medulla to be examined for each rat will have a higher incidence of adrenal medullary tumors than from studies in which the protocols require only adrenal (cortex or medulla) to be present on the slide.

Endocrine glands respond to hormonal stimuli and constant prolonged stimulation will result in hyperplasia and eventually, in most cases, neoplasia in target tissue. Hyperplasia, when diffuse, is easily separated from benign tumors, but focal hyperplasia can provide a diagnostic challenge for the pathologist. Lesions that even appear to be tumors can regress when the hormonal stimulus is removed. This has lead to some unusual terminology, such as benign metastasizing goiter, in the older human literature. The separation of hyperplasia from neoplasia and the complex interplay of hormonal stimulation on cell growth are beyond the scope of this chapter. However, one must be cautious when dealing with endocrine tumors. The diagnoses of endocrine tumors can be more accurate if the hormonal status of the rat, the potential effect of the chemical on the endocrine system, and the chemical effect on the target organs are understood.

A. Pituitary

Primary tumors of the pituitary gland in rats are extremely common in the Fischer 344 and Wistar rats but uncommon in Sprague-Dawley rats. The incidence increases with age and is a significant factor in limiting the lifespan of Fischer 344 and Wistar rats. Pituitary tumors can become very large, causing compression of the brain (Figs. 14-17 and 14-18). The tumor in many rats secretes prolactin. Lactation in an aging rat is often a sign of a functional pituitary tumor. Growth hormone, adrenocorticotropic hormone (ACTH), thyroid-stimulating

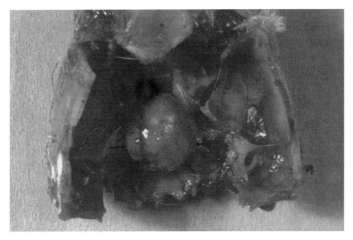

Fig. 14-17 Large pituitary nodule *(arrows)* projecting from the base of the skull. This is the typical appearance of a pituitary adenoma.

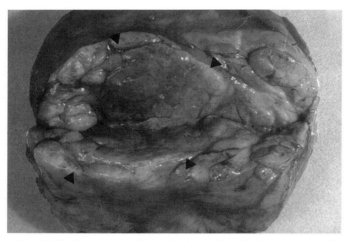

Fig. 14-18 Deep indentation in the base of the brain from a rat with a large pituitary adenoma *(arrowheads)*. This lesion is the cause of the frequent neurologic manifestations associated with this tumor.

hormone (TSH), and follicle-stimulating hormone (FSH) also have been found in rat pituitary adenomas (Friedrich et al., 2000). Prolactin-secreting pituitary adenomas can be induced in rats by chronic diethylstilbestrol administration (Piroli et al., 2000). This estrogen effect on pituitary tumors can be abolished by dietary restriction in the Fischer 344 rat but not the ACI rat, suggesting that genetic background is also important (Spady et al., 1999).

Many of the pituitary adenomas are diagnosed as chromophobe adenoma based on hematoxylin and eosin (H&E) stained sections. Because the diagnosis of "chromophobe adenoma" provides no information as to the endocrine status of the tumor and H&E sections do not allow for differentiation of hormonal status, more recent terminology is to simply diagnose as adenoma or carcinoma (MacKenzie and Boorman, 1990). Although adenomas are

common, carcinomas are rare, perhaps because compression of the brain caused death before tumor progression can occur. Generally clear evidence of invasion or distant metastases are required for the diagnosis of carcinoma (MacKenzie and Boorman, 1990).

Some rare pituitary tumors include adenomas of the pars intermedia and craniopharyngioma of the pars nervosa. Craniopharyngioma arises from remnants of Rathke's pouch and contain glandular and tubular structures (MacKenzie and Boorman, 1990). Primary astrocytic tumors of the pars nervosa also have been described in the aging rat (Satoh et al., 2000).

B. Thyroid

The thyroid consists of two very different endocrine cell populations, follicular cells and C-cells (Hardisty and Boorman, 1990). Follicular cells secrete thyroid hormones responsible for metabolism, differentiation, and growth. C-cells or calcitonin-producing cells help regulate serum calcium levels. Naturally occurring C-cell tumors are commonly found in aging rats but are rarely associated with chemical exposure. Follicular cell tumor response following chemical exposure is very common, especially for chemicals that alter the hormonal status of the thyroid.

Numerous anti-thyroid compounds inhibit thyroid hormones resulting in increasing TSH secretion by the pituitary, leading to follicular cell hyperplasia, adenomas, and, eventually, carcinomas (Hardisty and Boorman, 1990). Because reversing TSH secretion can halt the progression of the lesions and even cause reversal of lesions that morphologically appear to be adenomas, the diagnosis of thyroid follicular cell tumors can be controversial. The significance of follicular cell tumors in rat studies for risk assessment for humans also has been subject to considerable debate (Hill et al., 1998).

Follicular cell adenomas may be cystic or solid, have a papillary or glandular pattern, and cause compression of the adjacent thyroid (Fig. 14-19). The cells resemble thyroid follicular cells and often colloid is found within the tumor. Follicular cell carcinomas resemble adenoma but tend to have a more solid growth and more cellular atypia, but invasion of the capsule and distant metastases are hallmarks of malignancy. Different growth patterns within the neoplasm are more often a feature of the carcinomas.

C-cell adenoma and C-cell hyperplasia are common age-related lesions in many strains of rats. The lesions appear as focal collections of pale staining cells with abundant cytoplasm, poorly defined cell boundaries, and central nucleus with finely stippled chromatin (Fig. 14-20). The C-cells appear beneath the follicular cells and the basement membrane lining the follicle. There is a continuum

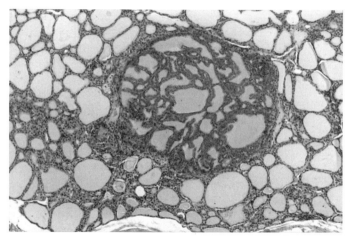

Fig. 14-19 Proliferative nodule of thyroid follicular cells representing a small adenoma within background of normal thyroid follicles.

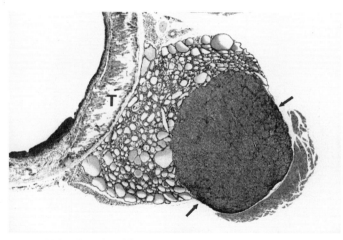

Fig. 14-21 Parathyroid mass *(arrows)* projecting from the thyroid gland ("T" labels tracheal cartilage). A diagnosis of parathyroid adenoma was made based on the fact that the lesion was unilateral and showed compression of normal thyroid architecture.

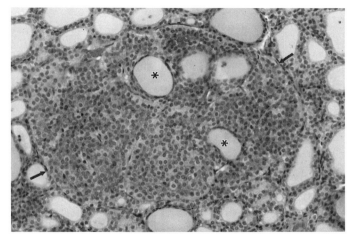

Fig. 14-20 C-cell adenoma *(arrows)* in the thyroid of a rat. Remnants of thyroid follicles are entrapped (*) within the proliferative C-cells.

between focal hyperplasia and larger lesions that can be considered tumors. Thus, the convention has been to use size as a criterion. Generally, focal lesions larger than five follicles in diameter are considered C-cell adenomas (Hardisty and Boorman, 1990). As the lesions become larger and invade the capsule or blood vessels, the C-cell lesions are diagnosed as carcinomas. There is almost no cellular atypia, even in C-cell carcinomas with distant metastases.

C. Parathyroid

Parathyroid hyperplasia is common with severe chronic renal disease in which both parathyroid glands may be enlarged and diffusely hyperplastic. Parathyroid adenomas are very uncommon (Fig. 14-21) and reserved for unilateral lesions (Seely and Hildebrandt, 1990). When the parathyroid is found within the thyroid, one must be careful to distinguish a focal lesion from a C-cell tumor, as C cells and parathyroid cells have morphologic similarities.

D. Adrenal

The adrenal gland is comprised of cells of very different origin. The adrenal cortex arises from the urogenital ridge. Adrenal cortical cells under the influence of ACTH secrete glucocorticoids that have broad effects essential for life, and mineralocorticoids that regulate electrolytes and water balance. The adrenal medulla is of neural crest origin, acting as a neuroendrocrine organ releasing epinephrine and norepinephrine in response to stress. The adrenal is one of the small organs for which historical tumor rates can vary widely, depending on histologic technique. The cortical adenoma usually is not apparent grossly. Similarly, the adrenal medullary pheochromocytoma usually is microscopic and unilateral. Whether the study protocol requires one adrenal medulla or both to be examined histologically will affect tumor rates.

1. Adrenal Cortex

Adrenal cortical tumors (Fig. 14-22) appear to be more common in Wistar (Table 14-3) and Donryu rats (Table 14-4) than other common strains (Tables 1 and 2). Adrenal cortical tumors often are not increased in response to chemical exposure (Rosol et al., 2001) and have

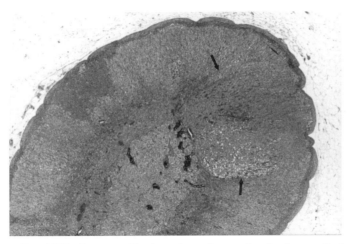

Fig. 14-22 Low-magnification photomicrograph of adrenocortical adenoma *(arrows)* in an F344 rat. This lesion was not noted macroscopically, thus demonstrating the need for careful dissection and preparative technique.

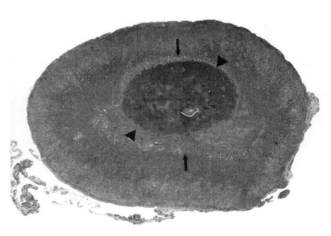

Fig. 14-23 Low-magnification photomicrograph of proliferation of atypical medullary cells representing a pheochromocytoma *(arrowheads)*. *Arrows* depict the normal corticomedullary junction.

engendered less research than adrenal medullary tumors in rats. Adenomas appear to arise near the capsule in the zona fasciculata or reticularis (Hamlin and Banas, 1990). The normal arrangement of cords in the cortex is disrupted and there is compression of the adjacent parenchyma. Dilated vessels may make the lesion more prominent. There usually

is little cellular atypia. Carcinomas usually show invasion of the capsule or vessels and in some cases replace the whole adrenal. Because adrenal cells even in hyperplasia may show considerable variability, atypia is not a good diagnostic criterion (Hamlin and Banas, 1990).

TABLE 14-4

DONRYU RATS

Common Tumors of the Donryu Rats			
Site	Tumor	Males (%)	Females (%)
Pituitary, pars distalis	Adenoma	35.8	33.3
Mammary gland	Fibroadenoma	1.0	35.4
Uterus	Adenocarcinoma	—	35.4
Adrenal medulla	Pheochromocytoma	16.8	3.1
Pancreatic islet	Adenoma	11.6	1.0
Adrenal cortex	Adenoma	6.3	7.3
Thyroid	C-cell adenoma	2.1	7.3
Hematopoietic	Histiocytic sarcoma	5.3	4.2
Hematopoietic	Leukemia/lymphoma	4.2	4.2
Mammary gland	Adenoma	0.0	5.2
Liver	Neoplastic cell tumor	4.2	2.1
Testes	Adenoma	4.2	—
Brain	Granular cell tumor	3.2	2.1
Pituitary, pars distalis	Carcinoma	2.1	1.0
Urinary bladder	Papillomatosis	2.1	0.0
Ovary	Granulosa cell tumor	—	2.1
Uterus	Sarcoma	—	2.1
Uterus	Stromal polyp	—	2.1
Parathyroid	Adenoma	2.1	0.0

Table based on 95 male and 96 female Donryu rats observed up to 120 weeks of age (Maekawa et al., 1986). Includes all tumors that occur at greater than 2% incidence in either sex.

2. Adrenal Medulla

Adrenal medullary tumors or pheochromocytomas (Fig. 14-23) occur commonly with aging in males of most rat strains (Tables 14-1 through 14-4). Pheochromocytomas are often increased in response to chemical exposure and have been extensively studied in the rat (Rosol et al., 2001). Excess growth hormone, stimulation of cholinergic nerves and diet-induced hypercalcemia may cause proliferation of adrenal medullary cells (Rosol et al., 2001). Pheochromocytomas arise as nests of deeply basophilic cells in the medulla (Hamlin and Banas, 1990). There are variable growth patterns but nests of small cells with small hyperchromatic nuclei are common. The lesion may fill the medulla and compression is a variable feature. Malignant pheochromocytomas may show necrosis, hemorrhage and replace the whole adrenal gland. Malignant tumors may show considerable cellular atypia but invasion and distant metastases are hallmarks of malignancy. Some pheochromocytomas contain a mixture of ganglion cells, neuroblasts as well as the pheochromocytes. These are usually diagnosed as complex pheochromocytoma (Hamlin and Banas, 1990). Medullary tumors consisting entirely of ganglion and Schwann cells may be diagnosed as ganglioneuromas (Hamlin and Banas, 1990).

E. Pancreatic Islets

Tumors of the pancreatic islets occur at a less than 5% incidence in Fischer 344 and Sprague-Dawley rats. Pancreatic islet cell tumors can be induced by repeated exposure to 4-hydroxyaminoquinoline 1-oxide (Imazawa et al., 2001) or by X-irradiation of the gastric region in rats (Kido et al., 2000). Islet cell lesions are usually not visible at necropsy and the incidence of islet cell tumors is dependent on the amount of pancreatic tissue examined histologically. Islet cell adenomas are discrete, well-circumscribed, solitary nodules that may compress the adjacent pancreatic tissue (Riley and Boorman, 1990). Pancreatic islet cell adenomas may incorporate acinar cells near the margin of the lesion, and this should not be interpreted as invasion or a sign of malignancy. The cells are usually uniform and grow in rows or cords. Most islet cell adenomas are comprised of beta cells and stain positive for insulin (Riley and Boorman, 1990). Islet cell carcinomas show more variability in growth pattern and may incite a scirrhous response. Metastases are uncommon and when found are usually in the lung. In the Fischer 344 rat, a mixed acinar-islet cell adenoma has been described (Riley and Boorman, 1990). These lesions consist of large nodules comprised of both acinar and islet cells throughout the lesion (Fig. 14-24).

IX. HEMATOPOIETIC AND LYMPHATIC SYSTEM

Lymphomas and leukemias are the most common tumors arising in the lymph nodes, thymus, spleen, and

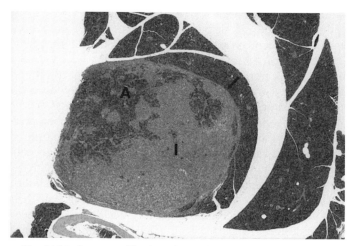

Fig. 14-24 Low-magnification photomicrograph of mixed pancreatic tumor *(arrow)* in an F344 rat. The mass is comprised of an admixture of islet (I) and acinar (A) cells with compression of surrounding normal pancreatic parenchyma.

bone marrow. Lymphomas are generally solid tumors of the lymph nodes and thymus and are very rare in the rat. This is in contrast to many mouse strains in which lymphomas arising in either the thymus or lymph nodes are very common. The late stage of lymphoma when neoplastic cells appear in the peripheral blood is known also as the leukemic phase, but these tumors are still classified as lymphomas. In contrast, in leukemias, neoplastic cells appear very early in the peripheral blood when often it is difficult to recognize the primary neoplastic process in the spleen or bone marrow where these tumors arise. In humans, leukemias often arise in bone marrow, but in rats most leukemias appear to arise in the spleen.

The occurrence of hematopoietic neoplasia varies significantly by rat strain. It is extremely common in Fischer 344 (Moloney et al., 1970; Moloney et al., 1971), Wistar (Bulay et al., 1979) and Lew/Han (Baum et al., 1995) rats and relatively uncommon in other stocks and strains. Most leukemias in rats are either of myeloid origin or of large granular lymphocyte origin. The latter disease is by far the most common hematopoietic neoplasm of Fischer 344 rats. The classification of myeloid and lymphoid neoplasia is complex and difficult. The characterization of these tumors in the rat has not been as well developed using immunophenotyping and molecular markers as it has in humans and mice. These neoplasms are generally late-onset diseases that are most common in aged rats with relatively few cases noted in animals younger than 20 months of age.

A. Leukemia/Lymphoma

By far the most common leukemia of rats is the large granular lymphocyte (LGL) leukemia. LGL lymphoma is very common in some strains of rats, such as the Fischer 344 (Moloney et al., 1970; Moloney et al., 1971), Wistar/Furth (Moloney et al., 1969), and Wistar (Bulay et al., 1979), and has been reported in much lower incidence in aged Sprague-Dawley rats (Abbott et al., 1983; Frith et al., 1993). It also has been termed mononuclear cell leukemia (MCL) in these animals due to the early occurrence of atypical cells in the circulation. The disease has been best characterized and studied in the Fischer 344 rat, in which its incidence has been more than 50% in aged animals. The average incidence in untreated control animals within oncogenicity bioassays has been approximately 34% in males and 20% in females (Stromberg and Vogtberger, 1983; Ward and Reynolds, 1983; Losco and Ward, 1984; Frith, 1988).

In the MCL of Fischer 344 rats, immunophenotyping and histochemistry of the cell of origin has shown the cells to resemble large granular lymphocytes containing esterase, acid phosphatase, and beta-glucuronidase.

The MCL cells also have Fc receptors, are capable of phagocytosis, and exhibit NK activity. Thy 1.1 and OX-8 antigens characteristic of suppressor/cytotoxic T lymphocytes are present in MCL cells (Ward and Reynolds, 1983). Clinically rats become depressed, pale, and icteric with palpably enlarged spleens. Detection of atypical mononuclear cells in the peripheral blood is an early manifestation of the disease (Stromberg et al., 1983). Macroscopic lesions include splenomegaly, enlarged mesenteric lymph nodes, and hepatomegaly with infiltrates (Fig. 14-25). The disease appears to originate in the spleen. Splenectomy significantly reduces the incidence of leukemia in rats (Moloney and King, 1973), as do chemicals causing splenic toxicity or atrophy (Elwell et al., 1996). Histologic evidence of neoplastic infiltrates is common in liver (nearly all), lungs (66%), bone marrow (43%), mesenteric lymph nodes (43%), adrenal glands (23%), mandibular lymph node (20%), and kidneys (19%) (Stephanski et al., 1990).

A variety of other lymphomas can arise in rats, albeit at much lower incidence than MCL (LGL) leukemia. Less than 1% of aged Sprague-Dawley rats developed various non-LGL lymphomas, including lymphoblastic, immunoblastic, or follicular center cell subtypes (Frith, 1988). Immunoblastic lymphomas have been reported to arise from the specialized lymphoid tissue known as Peyer's patches at the ileocecal junction in LOU rats (Rehm et al., 1990).

B. Non-lymphoid Neoplasia

Occasional cases of myeloid leukemias (granulocytic and erythroid) and mast cell tumors have been reported in rats either as rare spontaneous lesions or induced by chemicals and viruses (Frith et al., 1993). A tumor termed histiocytic sarcoma is believed to arise from cells of the

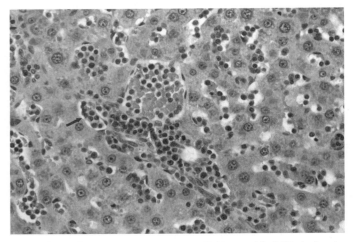

Fig. 14-25 Section of liver from an F344 rat with LGL leukemia. Hepatic sinusoids are filled with circulating neoplastic mononuclear cells (arrow).

mononuclear-phagocyte system and occurs in low incidence of some commonly utilized stocks and strains such as the Sprague-Dawley (Squire et al., 1981; Frith et al., 1993). These tumors have been alternatively termed malignant fibrous histiocytomas and histiocytic lymphomas. The morphology in this tumor can vary from animal to animal, resembling either sarcomatous neoplasia or having granulomatous features with multinucleate giant cells. The liver is the most commonly affected organ, although lung, lymph nodes, and spleen also can be involved, as can mediastinum and retroperitoneal sites.

X. URINARY SYSTEM

A. Kidney

Kidney tumors in the rat represent a number of distinct entities that arise from either renal tubular epithelium (adenomas and carcinomas), transitional epithelium of the renal pelvis (papillomas, transitional cell or metaplastic squamous cell carcinomas), renal connective tissue (renal mesenchymal tumors and lipomatous tumors), or embryonic primordia (nephroblastomas) (Hard, 1990). The incidence of spontaneous kidney tumors is very low compared with other tumor sites, with observed incidences of each type of the above tumors in most stocks and strains not exceeding 1% in either males or females (Tsuda and Krieg, 1992). In certain instances, hereditary rat kidney tumors have been noted to arise spontaneously in high incidence with genetic predisposition (Eker et al., 1981; Everitt et al., 1992; Thurman et al., 1995). Hereditary cancers in rats, as in other species, usually share genotypic and phenotypic similarity with their spontaneous sporadic counterparts. Kidney tumors of all types can be induced in the rat through various modes of induction, including chemical carcinogens, irradiation, and oncogenic viruses (Hard, 1998).

Renal cell tumors of tubular origin are primarily found within the cortex and appear macroscopically as well-demarcated grayish-white or grayish yellow masses. These neoplasms often contain cystic or hemorrhagic components (Bannasch et al., 1997). Microscopically these tumors can be classified as acidophilic, basophilic, clear cell, chromophobic, oncocytic, and mixed types on the basis of their cytologic characteristics and as solid, tubular, cystic, papillary, and mixed types depending on their histologic patterns (Bannasch et al., 1997). Small preneoplastic lesions (hyperplasias), adenomas, and malignant adenocarcinomas of tubular epithelial cell origin represent a biological spectrum of the same disease and are difficult to differentiate and stage into separate entities (Williams et al., 2001). These tumors are more common in the male rat and appear to be influenced to some degree by the occurrence of chronic progressive nephropathy (Seely et al., 2002). However, there appears to be a threshold, and minimal nephropathy does not appear to contribute to the development of renal tumors (Williams et al., 2001).

Tumors of transitional cell urothelium in the rat kidney consist of transitional cell papillomas and carcinomas, and by metaplasia of the urothelium into squamous cells, squamous cell papillomas, and carcinomas (Hard, 1990). Often, especially with carcinomas, these different tumor phenotypes are admixed. These tumors can occur anywhere in the area of the urothelium and can be single or multiple, papillary or nodular. Most transitional cell tumors in the rat undergo exophytic growth within the renal pelvis and, in cases of invasive growth, into the medullary region. Tumors of transitional cell origin often obstruct or interfere with urine drainage.

Renal mesenchymal tumors arise from pluripotent connective tissue elements in the interstitial tissue of rat kidney and can present with a complex histologic spectrum of lesions. These tumors tend to be neoplasms of the young rat without a gender preference and have been confused with nephroblastomas, although the histogenesis of these two lesions is very different (Hard, 1997). Renal mesenchymal tumors grow rapidly, but rarely metastasize. A specific type of mesenchymal tumor also arising in the interstitium of rats has been termed the lipomatous tumor (Gordon, 1986). These neoplasms are usually found in aged animals and have been noted in a variety of stocks and strains. Although considered rare, they are regarded among the most common spontaneous renal tumors in certain strains, such as the Osborne-Mendel (Goodman et al., 1980). These neoplasms can range in size from 1 mm to several centimeters in diameter, often being found at the corticomedullary region arising in the interstitium of the outer medulla.

Nephroblastomas (also known as embryonal nephromas, or Wilm's tumor) are generally but not always observed in young rats (Fig. 14-26), and are believed to arise from remnants of embryonal metanephric blastema (Hard, 1990; Cardesa and Ribalta, 1997). They are unilateral focal lesions that seem to arise within the renal cortex and grow as expansile multinodular masses. Microscopically, they usually present with a triphasic pattern, consisting of blastemal, epithelial, and stromal elements. The histologic appearance can resemble other tumor types, such as renal mesenchymal tumors, from which they need to be differentiated.

B. Ureter

Development of spontaneous tumors of the ureter in most commonly used stocks and strains has not

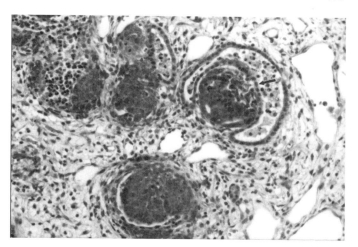

Fig. 14-26 Photomicrograph of section of embryonal nephroma demonstrating primordial glomerular structures *(arrow)* as part of tumor.

been reported. An exception is the Brown-Norway rat in which an unusually high frequency of naturally occurring ureter neoplasms of epithelial origin (22% of females and 6% of males) was found (Boorman and Hollander, 1974). In these rats, spontaneous neoplasms arise in both the ureter and urinary bladder in high incidence.

C. Urinary Bladder

Although the rat serves as an important animal model of experimental and chemically-induced urinary bladder carcinogenesis, spontaneous neoplasms of the urinary bladder rarely occur in most stocks and strains (Kunze, 1992). Exceptions are in DA/Han and Brown-Norway rats in which an exceptionally high incidence (14%–54% DA and 3%–35% BN) of naturally occurring epithelial bladder tumors has been reported (Boorman and Hollander, 1974; Deerberg et al., 1985). As in man, spontaneous bladder tumors occur more frequently in male than in female rats and occur at a younger age. Most spontaneous epithelial bladder tumors show an exophytic pattern of growth with a polypoid intraluminal appearance. Microscopically, most urinary bladder neoplasms manifest as either transitional cell papillomas or carcinomas that can develop varying degrees of squamous differentiation.

In addition to the genetic predisposition noted in certain rat strains, a variety of environmental factors has been noted to affect the development of urinary bladder carcinogenesis. Urolithiasis has been reported to modify spontaneous and experimental bladder carcinogenesis in the rat (Cohen, 1998). The presence of the parasite *Trichosomoides crassicauda* also may affect bladder cancer rates (Chapman, 1964). A variety of nutritional factors has been reported to effect incidence and latency of bladder tumors in rats. These include both caloric intake and protein content of diet (Ross and Bras, 1973).

XI. FEMALE REPRODUCTIVE SYSTEM

A. Ovary

Ovarian neoplasms are uncommon spontaneous and induced lesions in most stocks and strains of rats. They occur with an overall incidence of 0.5% to −3%, and most are found as incidental lesions at necropsy of old rats (Gregson et al., 1984; Maekawa et al., 1987). The exceptions include a single report in Osborne-Mendel rats in which approximately 33% of rats over 18 months of age developed granulosa cell tumors (Snell, 1961) and a report of approximately 30% incidence of ovarian carcinomas (not further described) in female Lewis rats (Feldman and Woda, 1980). In rats the basis for classification of ovarian tumors is morphologic, as there is little information on the biological behavior or endocrine effects of these neoplasms. The current recommended grouping of ovarian tumors is to aggregate the various histologic subtypes into the categories of epithelial, sex cord/stromal, and germ cell (McConnell et al., 1986; Alison and Morgan, 1987).

Ovarian epithelial tumors, including adenoma/adenocarcinoma and mesotheliomas, are derived from the surface epithelium and are described by morphologic subtypes based on histology, such as tubular, papillary, and cyst papillary. These tumors are either bilateral or unilateral and are often found only on microscopic examination of the ovary. They comprise less than 10% of ovarian tumors in Fischer 344 and Sprague-Dawley rats (Alison and Morgan, 1987; Lewis, 1987). Sex cord/stromal tumors comprise the majority of ovarian tumors in Fischer 344, Sprague-Dawley, and Wistar rats. Granulosa cell and granulosa-derived tumors, including those of theca and luteal origin, make up the bulk of this group in Fischer 344 rats (Alison and Morgan, 1987), whereas sertoliform tubular adenomas comprise the majority of sex cord/stromal tumors in Sprague-Dawley rats (Lewis, 1987). These tumors are usually unilateral, although they can be bilateral. Occasionally sex cord/stromal tumors are associated with hormonal activity as evidenced by smooth muscle hyperplasia of the genital tract or mammary gland hyperplasia (Maekawa and Hayashi, 1987). Germ cell tumors in rats are exceedingly rare, although yolk sac carcinoma has been described (Sobis, 1987).

B. Uterus

Endometrial stromal polyps (Fig. 14-27) are the most commonly diagnosed neoplasm in the uterus of the rat of most stocks and strains (Goodman and Hildebrandt, 1987) and have been found in up to 37% of control Fischer 344 rats (Leininger and Jokinen, 1990). This benign proliferation of stromal cells extends into the uterine lumen on either a broad-based or narrow stalk (Fig. 14-28). The increased density of the stromal cells and capillary growth characterizes polyps. Another relatively common uterine neoplasm in most stocks and strains is the endometrial adenoma and carcinoma. In some strains, such as the LEW/Han rat, the endometrial carcinoma is among the most common neoplasms of the female with an incidence of 45% in aged animals on lifetime studies

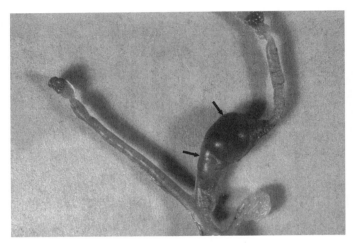

Fig. 14-27 Segmental dilatation *(arrows)* of uterine horn depicting the typical macroscopic appearance of a uterine polyp. This is one of the most common uterine proliferative lesions in the rat.

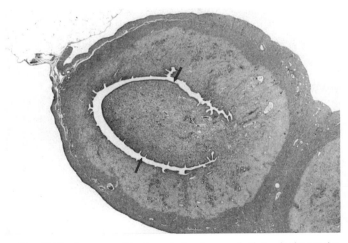

Fig. 14-28 Low-magnification photomicrograph through the uterine horn of an F344 rat depicting a uterine poly *(arrows)* consisting of an epithelial-lined stromal proliferation projecting into the uterine horn.

(Baum et al., 1995). The adenoma is usually solitary, causing local compression without invasion into the myometrium, whereas endometrial carcinomas are characterized by cellular pleomorphism and atypia or evidence of invasion. Metastatic spread, when it occurs, is often to the lung. Occasionally squamous cell lesions occur in the rat uterus. These can arise spontaneously or be induced by long-term exposure to chemicals and pharmaceutical agents. They need to be differentiated from squamous cell carcinoma arising in the cervix or vagina and invading the uterus. Rarely embryonal carcinomas, choriocarcinomas, and teratomas arise within the uterus of rats (Sobis, 1987).

The incidence of uterine proliferative lesions increases with age in most stocks and strains and is highly dependent on the influence of hormonal and breeding status for both mesenchymal and epithelial tumors (Tang and Tang, 1981). In the Han:Wistar rat, a high incidence in endometrial carcinoma (40%) has been reported in virgin females after 24 months, whereas only 5% was reported in multiparous animals (Deerberg et al., 1981; Rehm et al., 1984). Endometrial proliferative lesions in the Fischer 344 rats may be directly related to an anovulatory, noncyclic, reproductive tract (Leininger and Jokinen, 1990). Smooth muscle tumors of the uterus are rare, with the exception of those that occur in a hereditary cancer syndrome that occurs in Eker rats that bear a mutation in the tuberous sclerosis-2 allele (Everitt et al., 1995). These animals exhibit both a typical smooth muscle cytologic appearance and an epithelioid variant of smooth muscle tumor similar to that noted in cervical tumors of BN/Bi rats (Burek et al., 1976).

Deciduomas are uncommonly seen in young adult female rats, often in short-term studies (Leininger and Jokinen, 1990). The deciduoma is not a tumor but rather a decidual response that mimics an implantation site. The deciduoma usually occurs in pseudopregnant rats but occasionally may be found in an aging rat as a nodular mass (Leininger and Jokinen, 1990). The characteristic epithelial-like decidual cells with large eosinophilic cytoplasmic granules provide ample clues as to the nature of the lesion.

C. Vagina and Cervix

Neoplasms arising in the vagina and cervix are rare with the exception of an unusual variant of smooth muscle tumor that has been reported in the cervix of BN/Bi rats (Burek et al., 1976). Stromal tumors, sarcoma, and polyp occur occasionally in the cervix. Spontaneous squamous cell papillomas and carcinomas occasionally arise from the surface epithelium of the vagina or cervix and histologically resemble those of other sites.

XII. MAMMARY GLAND

Primary tumors of the mammary gland are some of the most common tumors in rats occurring in up to a third of the females by 2 years of age. For Fischer 344 and Wistar rats, the vast majority of tumors are fibroadenomas, but in Sprague-Dawley rats adenocarcinomas may also approach 10% incidence (Table 14-2). The Fischer 344 rat also is less susceptible to DMBA-induced breast cancer than is the Sprague-Dawley rat (Boorman et al., 1990). Mammary tumors are rare before 1 year of age and are generally found after 18 months of age. Exposure to estrogen (Russo and Russo, 1998) and prolonged exposure to prolactin increase mammary gland tumors in rats, whereas parity significantly decreased the incidence of mammary tumors. Increased mammary gland tumors are found in rats with pituitary tumors, but a causal effect is difficult to determine, as they tend to occur together. Estrogens induce both pituitary and mammary tumors in the rat (Leung et al., 2002). Also both mammary gland and pituitary tumor incidence correlates with body weight (Haseman et al., 1997).

The mammary gland is one of the more common sites at which carcinogenic effects (see NTP Web site at http://ntp-server.niehs.nih.gov) are observed following chemical exposure (Boorman et al., 1990). The rat is also frequently used as a model to study mammary gland cancer. Generally a single dose of a potent carcinogen is given at about 50 days of age and factors modifying cancer development are evaluated (Rowlands et al., 2001). Another model to study development of breast cancer is sex hormone-induced breast cancer in the female Noble rat (Leung et al., 2003).

Rats have six pairs of mammary glands extending anteriorly to the salivary glands and posteriorly to the perianal region. Thus, unless great care is taken at necropsy, small tumors can be missed. In studies in which the mammary gland is the tissue of interest, it is common to remove the entire ventral anterior skin with accompanying mammary glands and examine for the presence of masses using a light box. Using this procedure. both the number and location—for example, L1, R1, etc. of the tumor can be recorded.

Fibroadenomas are the most common benign neoplasms of the mammary gland. They can range in size from several millimeters up to several centimeters (Fig. 14-29). Very large tumors may become ulcerated and necessitate the removal of the animal from the study. Fibroadenomas are firm, glistening masses on cut surface. Histologically the tumors are comprised of a mixture of glandular tissue, connective tissue, and ducts (Fig. 14-30). The proportions of glandular to connective tissue vary widely between and within fibroadenomas. Some of the fibromas diagnosed in the region of the mammary gland may represent fibroadenomas with very little glandular tissue. It does not

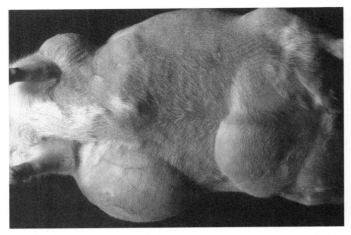

Fig. 14-29 Typical appearance of multiple mammary masses in the rat. Microscopically, these can be benign fibroadenomas or malignant adenocarcinomas.

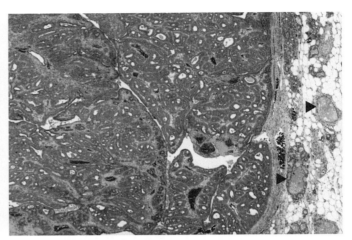

Fig. 14-31 Low-magnification photomicrograph of rat mammary adenocarcinoma depicting solid compact arrangement of neoplastic mammary glandular epithelial cells. Remnants of unaffected glandular epithelium reside within the normal subcutis *(arrow heads)*.

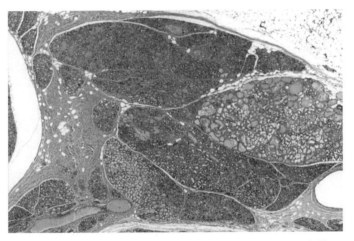

Fig. 14-30 Low-magnification photomicrograph of a typical fibro-adenoma of the rat. These tumors commonly show a mixed pattern with differentiation along both epithelial and mesenchymal lines.

appear useful to subclassify tumors based on the amount of fibrous versus glandular tissue but in some literature, the term *fibroadenoma* is used for tumors with more connective tissue and *adenofibroma* for those with a preponderance of glandular tissue.

Adenomas are masses consisting entirely of glandular tissue. Many of the adenomas are small, suggesting that further growth may result in the development of connective tissue and the diagnosis of fibroadenomas (Boorman et al., 1990). The administration of potent mammary gland carcinogens has induced small focal lesions that have been diagnosed as adenomas but that may represent incipient carcinomas.

Naturally occurring adenocarcinomas (Fig. 14-31) of the mammary gland are most common in the Sprague-Dawley

rat (Table 14-2) and uncommon in other rat stains (Tables 1, 3, 4). Approximately half the chemicals that induce mammary gland tumors cause adenocarcinomas at this site (Boorman et al., 1990). Adenocarcinomas exhibit a broad range of histologic patterns, ranging from solid sheets of pleomorphic cells to more glandular patterns of growth. Occasionally an adenocarcinoma will arise in an existing fibroadenoma, but this is uncommon. Areas of necrosis and hemorrhage are suggestive of malignancy. Naturally occurring adenocarcinomas and those induced in most carcinogenicity studies generally do not metastasize. However in breast cancer models, such as induction by DMBA, the tumor cells are highly pleomorphic and pulmonary metastases are common. Some adenocarcinomas contain areas of squamous differentiation but in our experience, this does not warrant a separate classification.

The relevance of mammary gland tumors in the rat for indicating a potential human hazard of a chemical exposure will continue to be debated when only an increased incidence of fibroadenomas is found because the human counterpart of fibroadenoma, although common in the rat, is rare in humans. However, several chemicals have been shown to promote the development of mammary gland adenocarcinomas. Thus, the rat has become a useful model to study factors that modify the development of mammary gland cancer.

XIII. MALE REPRODUCTIVE SYSTEM

A. Testes

Primary tumors of the testes in rats vary widely between strains. The most common tumor is interstitial cell tumor and the tumor incidence approaches 90% in the Fischer 344 rat (Table 14-1), 10% in the Wistar rat (Table 14-3), and is uncommon in the Sprague-Dawley rat (Table 14-2). Diet restriction will reduce the incidence of interstitial tumors in the Fischer 344 rat (Thurman et al., 1995).

The interstitial, or Leydig cell tumor, begins as a proliferation of interstitial cells between the seminiferous tubules (Boorman et al., 1990). Interstitial cell hyperplasia is uncommon in the Fischer 344 rat before 9 months of age but by 15 months of age, one third of the rats have interstitial cell hyperplasia and 80% have interstitial cell adenoma (Boorman et al., 1990). The tumors appear as yellow or white masses within the testes and are often bilateral and multiple (Fig. 14-32). Histologically the lesions appear to represent a continuum from small interstitial collections of cells to masses that completely efface the testes. Therefore a collection of interstitial cells less than the diameter of one seminiferous tubule is considered hyperplasia, and those larger are considered to be interstitial cell tumors (Boorman et al., 1990). Microscopically, interstitial cells may have abundant eosinophilic cytoplasm, pale vacuolated cytoplasm, or appear as small basophilic cells. Often all three microscopic patterns are found within the same tumor. These tumors are almost always benign.

Because of the high incidence of interstitial cell tumors in Fischer 344 rats, it is nearly impossible to demonstrate an increased incidence following chemical exposure. This is in contrast to the Sprague-Dawley rat, in which numerous pharmaceutical preparations have produced interstitial cell proliferations, including tumors, in 2-year studies (Prentice and Meikle, 1995). The interstitial cells are regulated by leuteinizing hormone (LH) secreted by the pituitary. Many of the compounds causing tumors interfere with a

Fig. 14-33 Low-magnification photomicrograph depicting the typical papillar appearance of testicular mesothelioma in the rat.

feed-back loop, resulting in increased levels of LH (Prentice and Meikle, 1995). Cadmium causes increased incidence of interstitial cell tumors in Wistar rats but not Fischer 344 rats (Waalkes et al., 1991).

Mesotheliomas, most commonly arising from the mesothelial lining of the tunica vaginalis testis of the testes (Fig. 14-33) and epididymis, are common tumors involving the testes of rats (Boorman et al., 1990). This is a tumor that is easily missed unless careful gross and histologic examination is performed. Other tumors of the testis, such as seminoma (Kerlin et al., 1998) and Sertoli cell tumor, have been described but are exceedingly rare in the rat (Boorman et al., 1990). Acrylamide exposure for 2 years causes mesotheliomas in the Fischer 344 rat (Friedman et al., 1995; Damjanov and Friedman, 1998).

B. Epididymis

Primary tumors unique to the epididymis have not been described in the rat (Boorman et al., 1990). Other tumors, such as hemangiomas and hemangiosarcomas, may occur but are similar to vascular tumors at other sites. Mesotheliomas of the testis also may involve the epididymis.

C. Accessory Sex Glands

With the exception of the prostate and preputial glands, neoplastic lesions of the accessory sex glands are very uncommon in the rat (Boorman et al., 1990). This section will only address lesions of the prostate and preputial gland.

The common occurrence of prostatic cancer in humans has prompted interest in these tumors in rats. The rat has

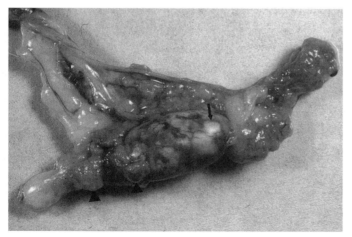

Fig. 14-32 Macroscopic appearance of atrophic testis of F344 rat showing pale irregular testicular masses typical of interstitial cell tumors *(arrow)* and irregular thickening of tunica vaginalis testis from mesothelioma *(arrowheads)*.

three lobes of the prostate of different embryologic origin and not completely analogous to the prostate of man (Suwa et al., 2001). The human prostate appears homologous to the dorsal area of the dorsolateral lobe in the rat (Boorman et al., 1990). In many rat carcinogenicity studies, the ventral lobe is examined because it is more prominent. The rat ventral lobe appears to be homologous to the paraurethral ducts of women (Boorman et al., 1990). In the Fischer 344 rat, focal hyperplasia and adenoma are common lesions but malignant lesions are rarely found (Reznik et al., 1981). Adenomas are glandular tumors that cause compression of the adjacent architecture. The tumor cells grow in complex papillary or cribriform patterns (Boorman et al., 1990). The tumor cells often stain more intensely than the surrounding gland. In contrast to the ventral prostate, the only tumors found in the dorsolateral lobe are carcinomas (Mitsumori and Elwell, 1988; Boorman et al., 1990). The tumor cells tend to be pleomorphic and invasion is a usual feature (Boorman et al., 1990).

Tumors of the preputial gland occur at 2% to 5% incidence in the male Fischer 344 rat (Table 14-1). Preputial gland tumors appear as firm subcutaneous masses in the perineum lateral to the penis (Copeland-Haines and Eustis, 1990). The masses often appear white or pale yellow. Larger lesions may be ulcerated. Proliferative lesions of the preputial gland form a morphologic continuum of hyperplasia, adenoma, and carcinoma (Copeland-Haines and Eustis, 1990). Preputial gland tumors may have cystic, papillary, or solid patterns (Reznik and Ward, 1981) of growth but sub-classification is usually not done because mixed patterns occur in most lesions (Copeland-Haines and Eustis, 1990). Adenomas are circumscribed masses that tend to have expansile growth. Carcinomas show more cellular atypia and tend to be invasive. Metastases are uncommon (Copeland-Haines and Eustis, 1990), but these tumors grow on transplantation (Maronpot et al., 1988). Chemicals that induce Zymbal gland tumors also frequently cause preputial gland tumors (Ward, 1975; Copeland-Haines and Eustis, 1990).

XIV. NERVOUS SYSTEM

Primary tumors of the central or peripheral nervous system are uncommon in the rat (Sills et al., 1999). Routine histologic sections of the brain in most carcinogenicity studies include three levels of brain to allow visualization of cerebral cortex, cerebellum, medulla, thalamus, corpus callosum, and the ventricles (Solleveld and Boorman, 1990). Although three levels of examination are sufficient for routine studies, when brain tumors are expected to occur, a more through examination of the brain may be indicated. The adult rat is also quite resistant to the

chemical induction of central nervous system tumors, with only 10 of 500 NTP carcinogenicity studies showing a possible neoplastic effect in the brain (Sills et al., 1999). Most brain cancer models require intrauterine exposure at a sensitive age for the pups (Vaquero et al., 1994; Mandeville et al., 2000). Nitrosoureas are common inducing agents for brain tumors in rats (Tanigawa et al., 1995; Kish et al., 2001).

A. Brain

Primary tumors of the brain in rats are uncommon, but the most frequently reported tumor is glioma (Gopinath, 1986). Gliomas usually are not visible at necropsy, and even larger tumors appear to blend into the adjacent nervous tissue. Histologically, gliomas may present diagnostic problems. They can vary from a focal increase in glial cells, which may be termed gliosis, to an obvious malignancy (Fig. 14-34). The glial tumors can be sub-classified into astrocytomas or oligodendrogliomas, each with specific cellular morphology (Solleveld and Boorman, 1990). The oligodendroglioma cells can be distinctive, with clear cytoplasm and dark central nuclei. Vessels also may be prominent in oligodendrogliomas (Solleveld and Boorman, 1990). Mixed glial cell variants occur and some glial tumors are very undifferentiated (Solleveld and Boorman, 1990). Medulloblastomas are neoplasms of primitive neural cells (Krinke et al., 2000). The tumors are very uncommon, but when found, usually occur in the cerebellum. The presence of tubular or rosette-like structures is a feature of this uncommon neoplasm (Solleveld and Boorman, 1990). Another poorly defined proliferation of the central nervous system of the rat is termed malignant reticulosis (Solleveld and Boorman, 1990). The cells may resemble lymphocytes, histiocytes, or microglia. Reactive astrocytes are present

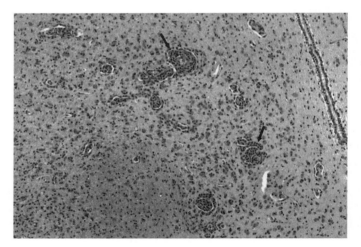

Fig. 14-34 Photomicrograph of a highly cellular glioma in a rat brain showing whorls of neoplastic cells around blood vessels (arrows).

throughout the lesion. Whether this lesion should be considered separately from glial neoplasms is still subject to debate (Solleveld and Boorman, 1990).

B. Meninges

Meningeal tumors are the second most frequent tumors involving the central nervous system. Meningeal tumors often are detected at necropsy as pale plaque-like lesions on the brain surface or meninges (Solleveld and Boorman, 1990). Meningeal tumors may be classified as fibroblastic, meningothelial, and granular cell. As might be expected from the names, fibroblastic meningiomas consist of spindle-shaped cells; the meningeal type consists of lobules of epithelial cells and granular meningioma consists of cells with granular cytoplasm (Solleveld and Boorman, 1990). The first two types have a malignant counterpart, and meningeal sarcomas often invade the brain. Granular cell tumors rarely if ever appear malignant. Granular cell tumors may be of meningeal origin, perhaps from a meningeal arachnoid cell (Yoshida et al., 1997). The tumor cells have a granular cytoplasm (Fig. 14-35) that stains intensely with a periodic acid-Schiff (PAS) stain.

C. Spinal Cord and Peripheral Nerves

Primary tumors of the spinal cord are extremely rare and have not been identified in the NTP historical control database of more than 500 carcinogenicity studies conducted in F344/N rats (Solleveld and Boorman, 1990). They may be more common in Sprague-Dawley rats, with three cases reported in approximately 1,400 controls (Zwicker et al., 1992). Tumors of the spinal cord include

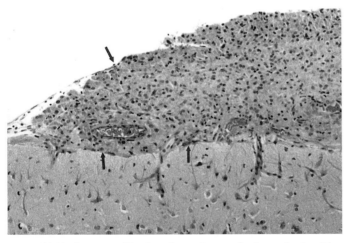

Fig. 14-35 Low-magnification photomicrograph of a rat brain with a granular cell tumor manifested by marked thickening and infiltration of the meninges by neoplastic granular cells *(arrows)*.

both gliomas and schwannomas (Kaspareit-Rittinghausen and Deerberg, 1989; Zwicker et al., 1992). The low incidence of spinal cord tumors may be due to the spinal cord's inaccessibility and because it is rarely removed and examined in its entirety.

Tumors of the peripheral nerves are also uncommon. Benign and malignant schwannomas occur at less than 1% incidence in the Fischer 344 rat. The most common site for schwannomas in the Fischer 344 rat is by the salivary gland, most likely arising from the trigeminal nerve (Solleveld and Boorman, 1990). Schwannomas may occur in the subcutaneous tissue and on the endocardium of the heart (see Section VI.A in this chapter). Many tumors exhibit a spindle pattern that makes them difficult to distinguish from other spindle cell tumors but occasionally a distinct pattern of palisading neoplastic cells is seen (Solleveld and Boorman, 1990). Tumors may arise in the paraganglia that are distributed throughout the body, mainly along major vessels. Paragangliomas are rare in most strains of rats except for a sub-strain of Wistar rats (Wag/Rij), in which the incidence exceeds 10% in females (van Zwieten et al., 1979). Most gangliomas in the rat occur in the aortic body or the abdominal cavity. The tumors are characterized by ganglion and neuron-like cells embedded in an eosinophilic matrix (Pace et al., 2002). The paragangliomas may be misdiagnosed as ectopic adrenal glands or metastatic pheochromocytomas (Solleveld and Boorman, 1990). The location of the tumors near the aorta or large arteries is helpful, coupled with a lack of tumors in the adrenal. Tumors of the spinal cord and peripheral nerve are rare in the rat and may present a diagnostic challenge.

REFERENCES

Abbott, D.P., Prentice, D.E., and Cherry, C.P. (1983). Mononuclear cell leukemia in aged Sprague-Dawley rats. *Vet. Pathol.* **20**, 434–439.

Abdo, K.M., Cunningham, M.L., Snell, M.L., Herbert, R.A., Travlos, G.S., Eldridge, S.R., and Bucher, J.R. (2001). 14-week toxicity and cell proliferation of methyleugenol administered by gavage to F344 rats and B6C3F1 mice. *Food Chem. Toxicol.* **39**, 303–316.

Alison, R., and Morgan, K.T. (1987). Ovarian neoplasms in F344 rats and B6C3F1 mice. *Environ. Health Perspect.* **73**, 91–106.

Alison, R.H., Elwell, M.R., Jokinen, M.P., Dittrich, K.L., and Boorman, G.A. (1987). Morphology and classification of 96 primary cardiac neoplasms in Fischer 344 rats. *Vet. Pathol.* **24**, 488–496.

Axmann, C., Bohndorf, K., Gellissen, J., Prescher, A., and Lodemann, K.P. (1998). Mechanisms of accumulation of small particles of iron oxide in experimentally induced osteosarcomas of rats: a correlation of magnetic resonance imaging and histology. Preliminary results. *Invest. Radiol.* **33**, 236–245.

Baker, D.G. (1998). Natural pathogens of laboratory mice, rats, and rabbits and their effects on research. *Clin. Microb. Rev.* **11**, 231–266.

Bannasch, P., Enzmann, H., Klimek, F., Weber, E., and Zerban, H. (1989). Significance of sequential cellular changes inside and outside foci of altered hepatocytes during hepatocarcinogenesis. *Toxicol. Pathol.* **17**, 617–628.

Bannasch, P., Haertel, T., and Su, Q. (2003). Significance of hepatic preneoplasia in risk identification and early detection of neoplasia. *Toxicol. Pathol.* **31**, 134–139.

Bannasch, P., Zerban, H., and Ahn, Y.S. (1997). Renal cell adenoma and carcinoma, rat. *In* "Urinary System." (T.C. Jones, G.C. Hard and U. Mohr, eds.), pp. 79–118, Springer-Verlag, Heidelberg.

Baum, A., Pohlmeyer, G., Rapp, K.G., and Deerberg, F. (1995). Lewis rats of the inbred strain LEW/Han: life expectancy, spectrum and incidence of spontaneous neoplasms. *Exp. Toxicol. Pathol.* **47**, 11–18.

Berman, J.J., Rice, J.M., and Reddick, R. (1980). Endocardial schwannomas in rats. Their characterization by light and electron microscopy. *Arch. Pathol. Lab. Med.* **104**, 187–191.

Boorman, G.A. (1985). Bronchiolar/alveolar hyperplasia, lung, rat. *In* "Monographs on Pathology of Laboratory Animals: Respiratory System." (T.C. Jones, U. Mohr, and R.D. Hunt, eds.), pp. 177–179, Springer-Verlag, Heidelberg.

Boorman, G.A., Botts, S., Bunton, T.E., Fournie, J.W., Harshbarger, J.C., Hawkins, W.E., Hinton, D.E., Jokinen, M.P., Okihiro, M.S., and Wolfe, M.J. (1997). Diagnostic criteria for degenerative, inflammatory, proliferative nonneoplastic and neoplastic liver lesions in medaka (Oryzias latipes): consensus of a National Toxicology Program pathology working group. *Toxicol. Pathol.* **25**, 202–210.

Boorman, G.A., Brockmann, M., Carlton, W.W., Davis, J.M., Dungworth, D.L., Hahn, F.F., Mohr, U., Reichhelm, H.B., Turusov, V.S., and Wagner, B.M. (1996). Classification of cystic keratinizing squamous cell lesions of the rat lung: report of a workshop. *Toxicol. Pathol.* **24**, 564–572.

Boorman, G.A., Chapin, R.E., and Mitsumori, K. (1990). Testis and epididymis. *In* "Pathology of the Fischer Rat." (G.A. Boorman, S.L. Eustis, M.R. Elwell, C.A.J. Montgomery and W.F. MacKenzie, eds.), pp. 405–418, Academic Press, London.

Boorman, G.A., Elwell, M.R., and Mitsumori, K. (1990). Male accessory sex glands, penis and scrotum. *In* "Pathology of the Fischer Rat." (G.A. Boorman, S.L. Eustis, M.R. Elwell, C.A.J. Montgomery, and W.F. MacKenzie, eds.), 1, 419–428. Academic Press, London.

Boorman, G.A., and Eustis, S.L. (1984). Proliferative lesions of the exocrine pancreas in male F344/N rats. *Environ. Health Perspect.* **56**, 213–217.

Boorman, G.A., and Eustis S.L. (1986). The pathology working group as a means for assuring pathology quality in toxicological studies. *In* "Managing Conduct and Data Quality of Toxicological Studies." (B.K. Hoover, J.K. Baldwin, A.F. Uelner, C.E. Whitmire, C.L. Davies, and D.W. Bristol, eds.), pp. 271–275, Princeton Scientific Publishing Co., Inc., Princeton, NJ.

Boorman, G.A., and Hollander, C.F. (1974). High incidence of spontaneous urinary bladder and ureter tumors in the Brown Norway rat. *J. Natl. Cancer Inst.* **52**, 1005–1008.

Boorman, G.A., Wilson, J.T.H., van Zwieten, M.J., and Eustis, S.L. (1990). Mammary gland. *In* "Pathology of the Fischer Rat." (G.A. Boorman, S.L. Eustis, M.R. Elwell, C.A.J. Montgomery, and W.F. MacKenzie, eds.), pp. 295–337, Academic Press, London.

Boorman, G.A., Zurcher, C., Hollander, C.F., and Feron, V.J. (1973). Naturally occurring endocardial disease in the rat. *Arch. Pathol.* **96**, 39–45.

Brown, H.R., and Hardisty, J.F. (1990). Oral cavity, esophagus, and stomach. *In* "Pathology of the Fischer Rat." (G.A. Boorman, S.L. Eustis, M.R. Elwell, C.A.J. Montgomery, and W.F. MacKenzie, eds.), pp. 9–30, Academic Press, London.

Bucher, J.R. (2002). The National Toxicology Program rodent bioassay: design, interpretations, and scientific contributions. *Ann. N. Y. Acad. Sci.* **982**, 198–207.

Bucher, J.R., Portier, C.J., Goodman, J.I., Faustman, E.M., and Lucier, G.W. (1996). Workshop overview. National Toxicology Program Studies: principles of dose selection and applications to mechanistic based risk assessment. *Fundam. Appl. Toxicol.* **31**, 1–8.

Bulay, O., Mirvash, S.S., Garcia, H., Pelfrene, A.F., Gold, B., and Eagen, M. (1979). Carcinogenicity test of six nitrosamides and a nitrosocyanamide administered orally to rats. *J. Natl. Cancer Inst.* **62**, 1523–1528.

Burek, J.D., Nitschke, K.D., Bell, T.J., Wackerle, D.L., Childs, R.C., Beyer, J.E., Dittenber, D.A., Rampy, L.W., and McKenna, M.J. (1984). Methylene chloride: a two year inhalation toxicity and oncogenicity study in rats and hamsters. *Fundam. Appl. Toxicol.* **4**, 30–47.

Burek, J.D., Zurcher, C., and Hollander, C.F. (1976). High incidence of spontaneous cervical and vaginal tumors in an inbred strain of Brown Norway rats (BN/Bi). *J. Natl. Cancer Inst.* **57**, 549–554.

Cardesa, A., and Ribalta, T. (1997). Nephroblastoma, kidney, rat. *In* "Urinary System." (T.C. Jones, G.C. Hard, and U. Mohr, eds.), pp. 129–138, Springer-Verlag, Heidelberg.

Carlton, W.W. (1994). "Proliferative keratinizing cyst," a lesion in the lungs of rats following chronic exposure to para-aramid fibrils. *Fundam. Appl. Toxicol.* **23**, 304–307.

Chandra, M., Davis, H., and Carlton, W.W. (1993). Naturally occurring atriocaval mesotheliomas in rats. *J. Comp. Pathol.* **109**, 433–437.

Chandra, M., Riley, M.G.I., and Johnson, D.E. (1992). Spontaneous neoplasms in aged Sprague-Dawley rats. *Arch. Toxicol.* **66**, 496–502.

Chapman, W.H. (1964). The incidence of a nematode, Trichosomoides crassicauda, in the bladder of laboratory rats. Treatment with nitro-furantoin and preliminary report of the influence on urinary calculi and experimental bladder tumors. *Invest. Urol.* **2**, 52–57.

Cohen, S.M. (1998). Urinary bladder carcinogenesis. *Toxicol. Pathol.* **26**, 121–127.

Cohen, S.M. (2002). Comparative pathology of proliferative lesions of the urinary bladder. *Toxicol. Pathol.* **30**, 663–671.

Copeland-Haines, D., and Eustis, S.L. (1990). Specialized sebaceous glands. *In* "Pathology of the Fischer Rat." (G.A. Boorman, S.L. Eustis, M.R. Elwell, C.A.J Montgomery, and W.F. MacKenzie, eds.), pp. 279–293, Academic Press, London.

Damjanov, I., and Friedman, M.A. (1998). Mesotheliomas of tunica vaginalis testis of Fischer 344 (F344) rats treated with acrylamide: a light and electron microscopy study. *In Vivo* **12**, 495–502.

Deerberg, F., Rehm, S., and Jostmeyer, H.H. (1985). Spontaneous urinary bladder tumors in DA/Han rats: a feasible model of human bladder cancer. *J. Natl. Cancer Inst.* **75**, 1113–1121.

Deerberg, F., Rehm, S., and Pittermann, W. (1981). Uncommon frequency of adenocarcinomas of the uterus in virgin Han:Wistar rats. *Vet. Pathol.* **18**, 707–713.

Dominick, M.A., Bobrowski, W.F., and Metz, A.L. (1990). Proliferative exocrine pancreatic lesions in aged Wistar rats. *Toxicol. Pathol.* **18**, 423–426.

Dunnick, J.K., Heath, J.E., Farnell, D.R., Prejean, J.D., Haseman, J.K., and Elwell, M.R. (1997). Carcinogenic activity of the flame retardant, 2, 2–bis(bromomethyl)-1, 3–propanediol in rodents and comparison with the carcinogenicity of other NTP brominated chemicals. *Toxicol. Pathol.* **25**, 541–548.

Eker, R., and Mossige, J. (1961). A dominant gene for renal adenomas in the rat. *Nature* **189**, 858–859.

Eker, R., Mossige, J., Johannessen, J.V., and Aars, H. (1981). Hereditary renal adenomas and adenocarcinomas in rats. *Diagn. Histopathol.* **4**, 99–110.

Elwell, M.R., Dunnick, J.K., Hailey, J.R., and Haseman, J. (1996). Chemicals associated with decreases in the incidence of mononuclear cell leukemia in the Fischer rat. *Toxicol. Path.* **24**, 238–245.

Elwell, M.R., and McConnell, E.E. (1990). Small and large intestine. *In* "Pathology of the Fischer Rat." (G.A. Boorman, S.L. Eustis, M.R. Elwell, C.A.J. Montgomery, and W.F. MacKenzie, eds.), pp. 43–61, Academic Press, London.

Elwell, M.R., Stedham, M.A., and Kovatch, R.M. (1990). Skin and subcutis. *In* "Pathology of the Fischer Rat." (A. Boorman, S.L. Eustis, M.R. Elwell, C.A.J. Montgomery, and W.F. MacKenzie, eds.), pp. 261–277, Academic Press, London.

Ernst, H., Scampini, G., Durchfeld-Meyer, B., Brander-Weber, P., and Rittinghausen, S. (1998). Odontogenic fibroma in Sprague-Dawley rats: a report of 2 cases. *Exp. Toxicol. Pathol.* **50**, 384–388.

Esfandiari, A., Loya, T., and Lee, J.L. (2002). Skin tumors in aging Long-Evans rats. *J. Natl. Med. Assoc.* **94**, 506–510.

Eustis, S.L., and Boorman, G.A. (1985). Proliferative lesions of the exocrine pancreas. Relationship to corn oil gavage in the NTP. *J. Natl. Cancer Inst.* **75**, 1067–1073.

Eustis, S.L., Boorman, G.A., Harada, T., and Popp, J.A. (1990). Liver. *In* "Pathology of the Fischer Rat." (A. Boorman, S.L. Eustis, M.R. Elwell, C.A.J. Montgomery, and W.F. MacKenzie, eds.), pp. 71–94, Academic Press, London.

Eustis, S.L., Haseman, J., MacKenzie, W.F., and Abdo, K.M. (1995). Toxicity and carcinogenicity of 2,3–dibromo-1–propanol in F344/N rats and B6C3F1 mice. *Fundam. Appl. Toxicol.* **26**, 41–50.

Everitt, J.I., Goldsworthy, T.L., Wolf, D.C., and Walker, C.L. (1992). Hereditary renal cell carcinoma in the Eker rat: a rodent familiar cancer syndrome. *J. Urol.* **148**, 1932–1936.

Everitt, J.I., and Richter, C.B. (1990). Infectious diseases of the upper respiratory tract: implications for toxicology studies. *Environ. Health Perspect.* **85**, 239–247.

Everitt, J.I., Wolf, D.C., Howe, S.R., Goldsworthy, T.L., and Walker, C. (1995). Rodent model of reproductive tract leiomyomata: clinical and pathologic features. *Am. J. Pathol.* **146**, 1568–1579.

Everitt, R. (1984). Factors affecting spontaneous tumor incidence rates in mice: a literature review. *CRC Crit. Rev. Toxicol.* **13**, 235–251.

Feldman, J.D., and Woda, B.A. (1980). Pathology and tumor incidence in aged Lewis and BN rats. *Clin. Immunol. Immunopathol.* **15**, 331–334.

Feron, V.J., and Woutersen, H. (1989). Role of tissue damage in nasal carcinogenesis. *In* "Nasal Carcinogenesis in Rodents: Relevance to Human Health Risk." (V.J. Feron, and M.C. Bosland, eds.), pp. 76–84, Pudoc, Wageningen.

Fox, J.G. (1983). Intercurrent disease and environmental variables in rodent toxicology studies. *Prog. Exp. Tumor Res.* **26**, 208–240.

Friedman, M.A., Dulak, L.H., and Stedham, M.A. (1995). A lifetime oncogenicity study in rats with acrylamide. *Fundam. Appl. Toxicol.* **27**, 95–105.

Friedrich, R.E., Saeger, W., Laas, R., and Friedrich, S.B. (2000). Hormone production in pituitary adenomas following external irradiation: an experimental study in rats. *Anticancer Res.* **20**, 5165–5170.

Frith, C.H. (1988). Morphologic classification and incidence of hematopoietic neoplasms in the Sprague-Dawley rats. *Vet. Pathol.* **16**, 451–457.

Frith, C.H., Ward, J.W., and Chandra, M. (1993). The morphology, immunohistochemistry, and incidence of hematopoietic neoplasms in mice and rats. *Toxicol. Path.* **21**, 206–218.

Fujii, T., Mikuriya, H., and Sasaki, M. (1991). Chronic toxicity and carcinogenicity study of thiabendazole in rats. *Food Chem. Toxicol.* **29**, 771–775.

Ghanayem, B.I., Matthews, H.B., and Maronpot, R.R. (1991). Sustainability of forestomach hyperplasia in rats treated with ethyl acrylate for 13 weeks and regression after cessation of dosing. *Toxicol. Pathol.* **19**, 273–279.

Gilligan, C.J., Lawton, G.P., Tang, L.H., West, A.B., and Modlin, I.M. (1995). Gastric carcinoid tumors: the biology and therapy of an enigmatic and controversial lesion. *Am. J. Gastroenterol.* **90**, 338–352.

Goodall, C.M., Christie, G.S., and Hurley, J.V. (1975). Primary epithelial tumour in the right atrium of the heart and inferior vena cava in NZR/Gd inbred rats; pathology of 18 cases. *J. Pathol.* **116**, 239–251.

Goodman, D.G., Boorman, G.A., and Strandberg, J.D. (1985). Selection and use of the B6C3F1 mouse and F344 rat in long-term bioassays for carcinogenicity. *In* "Handbook of Carcinogen Testing." (H.A. Milman, and E.K. Weisburger, eds.), pp. 282–325, Noyes publications, Park Ridge, NJ.

Goodman, D.G., and Hildebrandt, P.K. (1987). Stromal polyp, endometrium, rat. *In* "Genital System." (T.C. Jones, U. Mohr, and R.D. Hunt, eds.), pp. 146–148, Springer Verlag, Heidelberg, Germany.

Goodman, D.G., and Sauer, R.M. (1992). Hepatotoxicity and carcinogenicity in female Sprague-Dawley rats treated with 2,3,7,8-tetrachlorodibenzo-p-dioxin (TCDD): a pathology working group reevaluation. *Regul. Toxicol. Pharmacol.* **15**, 245–252.

Goodman, D.G., Ward, J.W., Squire, R.A., Paxton, M.B., Reichardt, W.D., Chu, K.C., and Linhart, M.S. (1980). Neoplastic and nonneoplastic lesions in aging Osborne-Mendel rats. *Toxicol. Appl. Pharmacol.* **55**, 433–447.

Gopinath, C. (1986). Spontaneous brain tumours in Sprague-Dawley rats. *Food Chem. Toxicol.* **24**, 113–120.

Gordon, L.R. (1986). Spontaneous lipomatous tumors in the kidney of the Crl:CD(SD)BR rat. *Toxicol. Path.* **14**, 175–182.

Gregson, R.L., Lewis, D.J., and Abbott, D.P. (1984). Spontaneous ovarian neoplasms of the laboratory rat. *Vet. Pathol.* **21**, 292–299.

Hakanson, R., and Sundler, F. (1990). Proposed mechanism of induction of gastric carcinoids: the gastrin hypothesis. *Eur. J. Clin. Invest.* **20**, S65–S71.

Hamlin, M. H. I., and Banas, D.A. (1990). Adrenal Gland. *In* "Pathology of the Fischer Rat." (G.A. Boorman, S.L. Eustis, M.R. Elwell, C.A.J. Montgomery, and W.F. MacKenzie, eds.), pp. 501–518, Academic Press, London.

Hamm, T.E. (1985). Design of a long-term animal bioassay for carcinogenicity. *In* "Handbook of Carcinogen Testing." (H.A Milman., and E.K. Weisburger, eds.), pp. 252–267, Noyes Publications, Park Ridge, NJ.

Hard, G.C. (1990). Tumours of the kidney, renal pelvis, and ureter. *In* "Pathology of Laboratory Animals. I. Tumours of the Rat." (V.S. Turusov, and U. Mohr, eds.), pp. 301–344, IARC Scientific Pubs, Lyon.

Hard, G.C. (1997). Mesenchymal tumor, kidney, rat. *In* "Urinary System." (T.C. Jones, G.C. Hard, and U. Mohr, eds.), pp. 118–129, Springer-Verlag, Heidelberg, Germany.

Hard, G.C. (1998). Mechanisms of chemically-induced renal carcinogenesis in the laboratory rodent. *Toxicol. Path.* **26**, 104–112.

Hard, G.C., Iatropoulos, M.J., Thake, D.C., Wheeler, D., Tatematsu, M., Hagiwara, A., Williams, G.M., and Wilson, A.G. (1995). Identity and pathogenesis of stomach tumors in Sprague-Dawley rats associated with the dietary administration of butachlor. *Exp. Toxicol. Pathol.* **47**, 95–105.

Hardisty, J.F., and Boorman, G.A. (1990). Thyroid gland. *In* "Pathology of the Fischer Rat." (G.A. Boorman, S.L. Eustis, M.R. Elwell, C.A.J. Montgomery, and W.F. MacKenzie, eds.), pp. 519–536, Academic Press, London.

Haseman, J., and Hailey, J.R. (1997). An update of the National Toxicology Program database on nasal carcinogens. *Mutat. Res.* **380**, 3–11.

Haseman, J.K., Arnold, J.E., and Eustis, S.L. (1990). Tumor incidence in Fischer 344 rats: NTP historical data. *In* "Pathology of the Fischer Rat." (G.A. Boorman, S.L. Eustis, M.R. Elwell, C.A.J. Montgomery, and W.F. MacKenzie, eds.), pp. 555–564, Academic Press, London.

Haseman, J.K., Hailey, J.R., and Morris, R.W. (1998). Spontaneous neoplasm incidences in Fischer 344 rats and B6C3F1 mice in two-year carcinogenicity studies: a National Toxicology Program update. *Toxicol. Pathol.* **26**, 428–441.

Haseman, J.K., Young, E., Eustis, S.L., and Hailey, J.R. (1997). Body weight-tumor correlations in long-term rodent carcinogenicity studies. *Toxicol. Pathol.* **25**, 256–263.

Havu, N., Mattsson, H., Ekman, L., and Carlsson, E. (1990). Enterochromaffin-like cell carcinoids in the rat gastric mucosa following long-term administration of ranitidine. *Digestion* **45**, 189–195.

Herzberg, A.M., and Lagakos, S.W. (1991). Cage allocation designs for rodent cacinogenicity experiments. *Environ. Health Perspect.* **96**, 199–202.

Heydens, W.F., Wilson, A.G., Kier, L.D., Lau, H., Thake, D.C., and Martens, M.A. (1999). An evaluation of the carcinogenic potential of the herbicide alachlor to man. *Hum. Exp. Toxicol.* **18**, 363–391.

Hill, R.N., Chrisp, T.M., Hurley, P.M., Rosenthal, S.L., and Singh, D.V. (1998). Risk assessment of thyroid follicular cell tumors. *Environ. Health Perspect.* **106**, 447–457.

Hirose, M., Yamaguchi, T., Mizoguchi, Y., Akagi, K., Futakuchi, M., and Shirai, T. (2002). Lack of inhibitory effect of green tea catechins in 1, 2–dimethyl hydrazine induced intestinal carcinogenesis model: comparison of the different formulations, administration routes and doses. *Cancer Lett.* **188**, 163–170.

Hosokawa, S., Imai, T., Hayakawa, K., Fukuta, T., and Sagami, F. (2000). Parotid gland papillary cystadenocarcinoma in a Fischer 344 rat. *Contemp. Top. Lab. Anim. Sci.* **39**, 31–33.

Huff, J., Haseman, J., DeMarini, D.M., Eustis, S.L., Maronpot, R.R., Peters, A.C., Persing, R.L., Chrisp, C.E., and Jacobs, A.C. (1989). Multiple-site carcinogenicity of benzene in Fischer 344 rats and B6C3F1 mice. *Environ. Health Perspect.* **82**, 125–163.

ILAR (1980). Institute of Laboratory Animal Resources. National Academy of Sciences. Histological typing of liver tumors of the rat. *J. Natl. Cancer Inst.* **64**, 179–206.

Imazawa, T., Nishikawa, A., Shibutani, M., Ogasawara, H., Furukawa, F., Ikeda, T., Suda, K., and Hirose, M. (2001). Induction of pancreatic islet cell tumors in rats by repeated intravenous administration of 4–hydroxyaminoquinoline 1–oxide. *Toxicol. Path.* **29**, 320–327.

Irwin, R.D., Haseman, J., and Eustis, S.L. (1995). 1,2,3–trichloropropane: a multisite carcinogen in rats and mice. *Fundam. Appl. Toxicol.* **25**, 241–252.

Jang, D.D., Kim, C.K., Ahn, B., Kang, J.S., Nam, K.T., Kim, D.J., Han, D.U., Jung, K., Chung, H.K., Ha, S.K., Choi, C., Cho, W.S., Kim, J., and Chae, C. (2002). Spontaneous complex odontoma in a Sprague-Dawley rat. *J. Vet. Med. Sci.* **64**, 289–291.

Johnson, J.D., Ryan, M.J., Tuft, J.D., Graves, S.W., Hejmancik, M.R., Cunningham, M.L., Herbert, R., and Abdo, K.M. (2000). Two-year toxicity and carcinogenicity study of methyleugenol in F344/N rats and B6C3F1 mice. *J. Agric. Food Chem.* **48**, 3620–3632.

Kaspareit-Rittinghausen, J., and Deerberg, F. (1989). Spontaneous tumours of the spinal cord in laboratory rats. *J. Comp. Pathol.* **100**, 209–215.

Keenan, K.P., Soper, K.A., Hertzog, P.R., Gumprecht, L.A., Smith, P.F., Mattson, B.A., Ballam, G.C., and Clark, R.L. (1995). Diet, overfeeding, and moderate dietary restriction in control Sprague-Dawley rats: II. Effects on age-related proliferative and degenerative lesions. *Toxicol. Pathol.* **23**, 287–302.

Keenan, K.P., Soper, K.A., Smith, P.F., Ballam, G.C., and Clark, R.L. (1995). Diet, overfeeding, and moderate dietary restriction in control Sprague-Dawley rats: I. Effects on spontaneous neoplasms. *Toxicol. Pathol.* **23**, 269–286.

Kerlin, R.L., Roesler, A.R., Jakowski, A.B., Boucher, G.G., Krull, D.L., and Appel, W.H. (1998). A poorly differentiated germ cell tumor (seminoma) in a Long Evans rat. *Toxicol. Path.* **26**, 691–694.

Kido, S., Haruma, K., Kitadai, Y., Yoshihara, M., Sumii, K., Kajiyama, G., and Watanabe, H. (2000). Enhanced tumorigenicity of insulinoma by X-irradiation of the gastric regions in Sprague-Dawley male rats. *J. Gastroenterol. Hepatol.* **15**, 766–770.

Kish, P.E., Blaivas, M., Strawderman, M., Muraszko, K.M., Ross, D.A., Ross, B.D., and McMahon, G. (2001). Magnetic resonance imaging of ethyl-nitrosourea-induced rat gliomas: a model for experimental therapeutics of low-grade gliomas. *J. Neurooncol.* **53**, 243–257.

Knudson, A.G. (1971). Mutation and cancer: statistical study of retinoblastoma. *Proc. Natl. Acad. Sci.* **68**, 820–823.

Kobayashi, T., Hirayama, Y., Kobayashi, E., Kubo, Y., and Hino, O. (1995). A germline insertion in the tuberous sclerosis (*Tsc*2) gene gives rise to the Eker rat model of dominantly inherited cancer. *Nature Genetics* **9**, 70–74.

Krinke, G.J., Kaufmann, W., Mahrous, A.T., and Schaetti, P. (2000). Morphological characterization of spontaneous nervous system tumors in mice and rats. *Toxicol. Path.* **28**, 178–192.

Kroes, R., Garbis-Berkvens, J.M., deVries, T., and van Nesselrooy, J.H.J. (1981). Histopathological profile of a Wistar rat stock including a survey of the literature. *J. Gerontol.* **36**, 259–279.

Kunze, E. (1992). Nonneoplastic and neoplastic lesions of the urinary bladder, ureter and renal pelvis. *In* "Pathobiology of the Aging Rat." (U. Mohr, D.L. Dungworth, and C.C. Capen, eds.), pp. 259–285, ILSI Press, Washington, DC.

Lee, F.I., and Harry, D.S. (1974). Angiosarcoma of the liver in a vinyl-chloride worker. *Lancet* **1**, 1316–1318.

Leininger, J.R., and Jokinen, M.P. (1990). Oviduct, uterus and vagina. *In* "Pathology of the Fischer Rat." (G.A. Boorman, S.L. Eustis, M.R. Elwell, C.A.J. Montgomery, and W.F. MacKenzie, eds.), pp. 443–459, Academic Press, London.

Leininger, J.R., and Riley, M.G.I. (1990). Bones, joints, and synovia. *In* "Pathology of the Fischer Rat." (G.A. Boorman, S.L. Eustis, M.R. Elwell, C.A.J. Montgomery, and W.F. MacKenzie, eds.), pp. 209–226, Academic Press, London.

Leung, G., Tsao, S.W., and Wong, Y.C. (2002). The effect of flutamide and tamoxifen on sex hormone-induced mammary carcinogenesis and pituitary adenoma. *Breast Cancer Res. Treat.* **72**, 153–162.

Leung, G., Tsao, S.W., and Wong, Y.C. (2003). Sex hormone-induced mammary carcinogenesis in female Noble rats: detection of differentially expressed genes. *Breast Cancer Res. Treat.* **77**, 49–63.

Lewis, D.J. (1987). Ovarian neoplasia in the Sprague-Dawley rat. *Environ. Health Perspect.* **73**, 77–90.

Longnecker, D.S. (1981). Animal model of human disease: carcinoma of the pancreas in azaserine-treated rats. *Am. J. Pathol.* **105**, 94–96.

Longnecker, D.S. (1990). Experimental pancreatic cancer. Role of species, sex and diet. *Bull. Cancer* **77**, 27–37.

Losco, P.E., and Ward, J.W. (1984). The early stage of large granular lymphocyte leukemia in the F344 rat. *Vet. Pathol.* **21**, 286–291.

Lynch, D.W., Lewis, T.R., Moorman, W.J., Burg, J.R., Groth, D.H., Khan, A., Ackerman, L.J., and Cockrell, B.Y. (1984). Carcinogenic and toxicologic effects of inhaled ethylene oxide and propylene oxide in F344 rats. *Toxicol. Appl. Pharmacol.* **76**, 69–84.

MacKenzie, W.F., and Alison, R. (1990). Heart. *In* "Pathology of the Fischer Rat." (G.A. Boorman, S.L. Eustis, M.R. Elwell, C.A.J. Montgomery, and W.F. MacKenzie, eds.), pp. 461–472, Academic Press, London.

MacKenzie, W.F., and Boorman, G.A. (1990). Pituitary Gland. *In* "Pathology of the Fischer Rat." (G.A. Boorman, S.L. Eustis, M.R. Elwell, C.A.J. Montgomery, and W.F. MacKenzie, eds.), pp. 485–500, Academic Press, London.

Maekawa, A., and Hayashi, Y. (1987). Granulosa/theca cell tumor, ovary, rat. *In* "Genital System." (T.C. Jones, U. Mohr, and R.D. Hunt, eds.), pp. 15–22, Springer Verlag, Heidelberg, Germany.

Maekawa, A., Onodera, H., Tanigawa, H., Furuta, K., Matsuoka, C., Kanno, J., Ogiu, T., and Hayashi, Y. (1986). Spontaneous neoplastic and non-neoplastic lesions in aging Donryu rats. *Jpn. J. Cancer Res.* **77**, 882–890.

Maekawa, A., Onodera, H., Tanigawa, H., Furuta, K., Matsuoka, C., Kanno, J., Ogiu, T., and Hayashi, Y. (1987). Experimental induction of ovarian Sertoli cell tumors in rats by N-nitrosoureas. *Environ. Health Perspect.* **73**, 115–123.

Maltoni, C., and Cotti, G. (1988). Carcinogenicity of vinyl chloride in Sprague-Dawley rats after prenatal and postnatal exposure. *Ann. N. Y. Acad. Sci.* **534**, 145–159.

Maltoni, C., and Lefemine, G. (1975). Carcinogenicity bioassays of vinyl chloride: current results. *Ann. N. Y. Acad. Sci.* **246**, 195–218.

Mandeville, R., Franco, E., Sidrac-Ghali, S., Paris-Nadon, L., Rocheleau, N., Mercier, M., Desy, M., Devaux, C., and Gaboury, L. (2000). Evaluation of the potential promoting effect of 60 Hz magnetic fields on N-ethyl-N-nitrosourea induced neurogenic tumors in female F344 rats. *Bioelectromagnetics* **21**, 84–93.

Maronpot, R.R. (1990). Pathology working group review of selected upper respiratory tract lesions in rats and mice. *Environ. Health Perspect.* **85**, 331–352.

Maronpot, R.R., Harada, T., Murthy, A.S., and Boorman, G.A. (1989). Documenting foci of hepatocellular alteration in two-year carcinogenicity studies: current practices of the National Toxicology Program. *Toxicol. Pathol.* **17**, 675–683.

Maronpot, R.R., Montgomery, C.A.J., and McConnell, E.E. (1986). National Toxicology Program nomenclature for hepatoproliferative lesions of rats. *Toxicol. Pathol.* **14**, 263–273.

Maronpot, R.R., Ulland, B., and Mennear, J. (1988). Transplantation characteristics, morphologic features, and interpretation of preputial gland neoplasia in the Fischer 344 rat. *Environ. Health Perspect.* **77**.

McConnell, E.E., Solleveld, H.A., Swenberg, J.A., and Boorman, G.A. (1986). Guidelines for combining neoplasms for evaluation of rodent carcinogenicity studies. *J. Natl. Cancer Inst.* **76**, 283–289.

McDonald, M.M., and Hamilton, B.F. (1990). Skeletal muscle. *In* "Pathology of the Fischer Rat." (G.A. Boorman, S.L. Eustis, M.R. Elwell, C.A.J. Montgomery, and W.F. MacKenzie, eds.), pp. 193–207, Academic Press, London.

Melnick, R.L., Sills, R.C., Portier, C.J., Roycroft, J.H., Chou, B.J., Grumbein, S.L., and Miller, R.A. (1999). Multiple organ carcinogenicity of inhaled chloroprene (2–chloro-1,3–butadiene) in F344/N rats and B6C3F1 mice and comparison of dose-response with 1,3–butadiene in mice. *Carcinogenesis* **20**, 867–878.

Mitsumori, K. (1990). Blood and lymphatic vessels. *In* "Pathology of the Fischer Rat." (G.A. Boorman, S.L. Eustis, M.R. Elwell, C.A.J. Montgomery, and W.F. MacKenzie, eds.), pp. 473–484, Academic Press, London.

Mitsumori, K., and Elwell, M.R. (1988). Proliferative lesions in the male reproductive system of F344 rats and B6C3F1 mice: incidence and classification. *Environ. Health Perspect.* **77**, 11–21.

Mohr, U., Rittinghausen, S., Takenaka, S., Ernst, H., Dungworth, D.L., and Pylev, L.N. (1990). Tumors of the lower respiratory tract and pleura. *In* "Pathology of Tumours in Laboratory Animals Volume 1: Tumours of the Rat." (V. S. Turusov, and U. Mohr, eds.), pp. 275–299, IARC Scientific Pubs, Lyon, France.

Moloney, W.C., Batata, M., and King, V. (1971). Leukemogenesis in the rat: further observations. *J. Natl. Cancer Inst.* **46**, 1139–1143.

Moloney, W.C., Boscheti, A.E., and King, V. (1969). Observations on leukemia in Wistar-Furth rats. *Cancer Res.* **29**, 938–946.

Moloney, W.C., Boscheti, A.E., and King, V. (1970). Spontaneous leukemia in Fischer rats. *Cancer Res.* **30**, 41–43.

Moloney, W.C., and King, V.P. (1973). Reduction of leukemia incidence following splenectomy in the rat. *Cancer Res.* **33**, 573–574.

Nakae, D. (1999). Endogenous liver carcinogenesis in the rat. *Pathol. Int.* **49**, 1028–1042.

Nakashima, N., Takahashi, K., Harada, T., and Maita, K. (1996). An epithelioid cell type of amelanotic melanoma of the pinna in a Fischer-344 rat: a case report. *Toxicol. Path.* **24**, 258–261.

National, T.P. (1986). Toxicology and carcinogenesis studies of 2,6–xylidine (2, 6–dimethylaniline) (CAS NO. 87–62–7) in Charles River CD rats (feed studies). NTP Technical Report No. 278. Research Triangle Park, NC 27709, U.S. Department of Health and Human Services, Public Health Service, National Institutes of Health.

National, T.P. (1990). Toxicology and carcinogenesis studies of sodium fluoride (CAS NO. 7681–49–4) in F344/N rats and B6C3F1 mice (drinking water studies). NTP Technical Report No. 393. Research Triangle Park, NC 27709, U.S. Department of Health and Human Services, Public Health Service, National Institutes of Health.

Nauta, J.M., Roodenburg, J.L., Nikkels, P.G., Witjes, M.J., and Vermey, A. (1996). Epithelial dysplasia and squamous cell carcinoma of the Wistar rat palatal mucosa: 4NQO model. *Head Neck* **18**, 441–449.

Neuenschwander, S.B., and Elwell, M.R. (1990). Salivary glands. *In* "Pathology of the Fischer Rat." (G.A. Boorman, S.L. Eustis, M.R. Elwell, C.A.J. Montgomery, and W.F. MacKenzie, eds.), pp. 31–42, Academic Press, London.

Novilla, M.N., Sandusky, G.E., Hoover, D.M., Ray, S.E., and Wightman, K.A. (1991). A retrospective survey of endocardial proliferative lesions in rats. *Vet. Pathol.* **28**, 156–165.

Nyska, A., Hester, S.D., Cooper, R.L., Goldman, J.M., Stoker, T.E., House, D., and Wolf, D.C. (2002). Single or group housing altered hormonal physiology and cell kinetics. *J. Toxicol. Sci.* **27**, 449–457.

Pace, V., Perentes, E., and Germann, P.G. (2002). Pheochromocytomas and ganglioneuromas in the aging rats: morphological and immuno-histochemical characterization. *Toxicol. Path.* **30**, 492–500.

Peano, S., Conz, A., Carbonatto, M., Goldstein, J., and Nyska, A. (1998). Atriocaval mesothelioma in a male Sprague-Dawley rat. *Toxicol. Path.* **26**, 695–698.

Piroli, G.G., Torres, A., Pietranera, L., Grillo, C.A., Ferrini, M.G., Lux-Lantos, V., Aoki, A., and De Nicola, A.F. (2000). Sexual dimorphism in diethylstilbestrol-induced prolactin pituitary tumors in F344 rats. *Neuroendocrinology* **72**, 80–90.

Poteracki, J., and Walsh, K.M. (1998). Spontaneous neoplasms in control Wistar rats: a comparison of reviews. *Toxicol. Sci.* **45**, 1–8.

Prejean, J.D. (1985). Conduct of long-term animal bioassays. *In* "Handbook of Carcinogen Testing." (H.A. Milman, and E.K. Weisburger, eds.), pp. 268–281, Noyes Publications, Park Ridge, NJ.

Prentice, D.E., and Meikle, A.W. (1995). A review of drug-induced Leydig cell hyperplasia and neoplasia in the rat and some comparisons with man. *Hum. Exp. Toxicol.* **14**, 562–572.

Quast, J.F. (2002). Two-year toxicity and oncogenicity study with acrylonitrile incorporated in the drinking water of rats. *Toxicol. Lett.* **132**, 153–196.

Rao, G.N., Haseman, J., and Edmonson, J. (1989). Influence of viral infections on body weight, and tumor prevalence in Fischer F344NCr rats on two year studies. *Lab. Anim. Sci.* **39**, 389–393.

Rehm, S., Deerberg, F., and Rapp, K.G. (1984). A comparison of life-span and spontaneous tumor incidence of male and female Han:Wistar virgin and retired breeder rats. *Lab. Anim. Sci.* **34**, 458–464.

Rehm, S., Eberly, K., and Pollard, M. (1990). Immunoblastic lymphoma, ileocecal lymph nodes, LOU/C rat. *In* "Hematopoietic System." (T.C. Jones, J.W. Ward, U. Mohr, and R.D Hunt, eds.), pp. 201–204, Springer-Verlag, Berlin.

Reindel, J.F., Dominick, M.A., and Gaugh, A.W. (1992). Mesenteric lymph node hemangiomas of Wistar rats. *Toxicol. Path.* **20**, 268–273.

Reznik, G., Hamlin, M.H.I., Ward, J.W., and Stinson, S.F. (1981). Prostatic hyperplasia and neoplasia in aging F344 rats. *Prostate* **2**, 261–268.

Reznik, G., and Ward, J.W. (1981). Morphology of hyperplastic and neoplastic lesions in the clitoral and preputial gland of the F344 rat. *Vet. Pathol.* **18**, 228–238.

Reznik-Schuller, H.M. (1983). Pathogenesis of tumors induced with N-nitrosomethylpiperzine in the olfactory region of the rat nasal cavity. *J. Natl. Cancer Inst.* **71**, 165–172.

Reznik-Schuller, H.M., and Reznik, G. (1982). Morphology of spontaneous and induced tumors in the bronchiolo-alveolar region of F344 rats. *Anticancer Res.* **2**, 53–58.

Riley, M.G.I., and Boorman, G.A. (1990). Endocrine Pancreas. *In* "Pathology of the Fischer Rat." (G.A. Boorman, S.L. Eustis, M.R. Elwell, C.A.J. Montgomery, and W.F. MacKenzie, eds.), pp. 545–553, Academic Press, London.

Rogers, A.E., Zeisel, S.H., and Groopman, J. (1993). Diet and carcinogenesis. *Carcinogenesis* **14**, 2205–2217.

Rosol, T.J., Yarrington, J.T., Lantendresse, J., and Capen, C.C. (2001). Adrenal gland: structure, function and mechanism of toxicity. *Toxicol. Path.* **29**, 41–48.

Ross, M.H., and Bras, G. (1973). Influence of protein under- and overnutrition on spontaneous tumor prevalence in the rat. *J. Nutr.* **103**, 944–963.

Rowlands, J.C., He, L., Hakkak, R., Ronis, M.J., and Badger, T.M. (2001). Soy and whey proteins downregulate DMBA-induced liver and mammary gland CYP1 expression in female rats. *J. Nutr.* **131**, 3281–3287.

Russo, I.H., and Russo, J. (1998). Role of hormones in mammary cancer initiation and progression. *J. Mammary Gland Biol. Neoplasia* **3**, 49–61.

Satoh, H., Iwata, H., Furuhama, K., and Enomoto, M. (2000). Pituicytoma: primary astrocytic tumor of the pars nervosa in aging Fischer 344 rats. *Toxicol. Path.* **28**, 836–838.

Schreiber, H., Nettesheim, P., Lijinsky, W., Richter, C.B., and Walburg, H.E.J. (1972). Induction of lung cancer in germfree, specific-pathogen-free, and infected rats by N-nitrosoheptamethylenei- mine: enhancement by respiratory infection. *J. Natl. Cancer Inst.* **49**, 1107–1114.

Seely, J.C., Haseman, J., Nyska, A., Wolf, D.C., Everitt, J.I., and Hailey, J.R. (2002). The effect of chronic progressive nephropathy on the incidence of renal tubule neoplasms in the F344 rat. *Toxicol. Path.* **30**, 681–686.

Seely, J.C., and Hildebrandt, P.K. (1990). Parathyroid Gland. *In* "Pathology of the Fischer Rat." (G.A. Boorman, S.L. Eustis, M.R. Elwell, C.A.J. Montgomery, and W.F. MacKenzie, eds.), pp. 37–543, Academic Press, London.

Sher, S.P., Jensen, R.D., and Bokelman, D.L. (1982). Spontaneous tumors in control F344 and Charles-River-CD rats and Charles River CD-1 and B6C3F1 mice. *Toxicol. Lett.* **11**, 103–110.

Shibata, M., Izumi, K., Sano, N., Akagi, A., and Otsuka, H. (1989). Induction of soft tissue tumours in F344 rats by subcutaneous, intramuscular, intra-articular, and retroperitoneal injection of nickel sulphide (Ni3S2). *J. Pathol.* **157**, 263–274.

Sills, R.C., Hailey, J.R., Neal, J., Boorman, G.A., Haseman, J., and Melnick, R.L. (1999). Examination of low-incidence brain tumor responses in F344 rats following chemical exposures in National Toxicology Program carcinogenicity studies. *Toxicol. Path.* **27**, 589–599.

Snell, K. (1961). Spontaneous lesions of the rat. *In* "The Pathology of Laboratory Animals." (W.E. Ribelin, and J.T. McCoy, eds.), pp. 241–300, CC Thomas, Springfield, Illinois.

Sobis, H. (1987). Choriocarcinoma, uterus, rat. *In* "Genital System." (T.C. Jones, U. Mohr, and R.D. Hunt, eds.), pp. 138–140, Springer Verlag, Heidelberg, Germany.

Sobis, H. (1987). Embryonal carcinoma, uterus, rat. *In* "Genital System." (T.C. Jones, U. Mohr, and R.D. Hunt, eds.), pp. 134–137, Springer Verlag, Heidelberg, Germany.

Sobis, H. (1987). Teratoma, uterus, rat. *In* "Genital System." (T.C. Jones, U. Mohr, and R.D. Hunt, eds.), pp. 120–126, Springer Verlag, Heidelberg, Germany.

Sobis, H. (1987). Yolk sac carcinoma, uterus, rat. *In* "Genital System." (T.C. Jones, U. Mohr, and R.D. Hunt, eds.), pp. 127–133, Springer Verlag, Heidelberg, Germany.

Solleveld, H.A., and Boorman, G.A. (1990). Brain. *In* "Pathology of the Fischer Rat." (G.A. Boorman, S.L. Eustis, M.R. Elwell, C.A.J. Montgomery, and W.F. MacKenzie, eds.), pp. 155–177, Academic Press, London.

Sommer, M.M. (1997). Spontaneous skin neoplasms in Long-Evans rats. *Toxicol. Path.* **25**, 506–510.

Spady, T.J., Pennington, K.L., McComb, R.D., Birt, D.F., and Shull, J.D. (1999). Estrogen-induced pituitary tumor development in the ACI rat not inhibited by dietary energy restriction. *Mol. Carcino.* **26**, 239–253.

Squire, R.A., Brinkhous, K.M., Peiper, S.C., Firminger, H.I., Mann, R.B., and Strandberg, J.D. (1981). Histiocytic sarcoma with a granuloma- like component occurring in a large colony of Sprague-Dawley rats. *Am. J. Pathol.* **105**, 21–30.

Squire, R.A., and Levitt, M.H. (1975). Report of a workshop on classification of specific hepatocellular lesions in rats. *Cancer Res.* **35**, 3214–3223.

Stephanski, S.A., Elwell, M.R., and Stromberg, P.C. (1990). Spleen, lymph node and thymus. *In* "Pathology of the Fischer Rat." (G.A. Boorman, S.L. Eustis, M.R. Elwell, C.A.J. Montgomery, and W.F. MacKenzie, eds.), pp. 369–393, Academic Press, London.

Steplewski, Z., Golman, P.R., and Vogel, W.H. (1987). Effect of housing stress on the formation and development of tumors in rats. *Cancer Lett.* **34**, 257–261.

Stromberg, P.C., and Vogtberger, L.M. (1983). Pathology of the mononuclear cell leukemia of Fischer rats. I. Morphologic studies. *Vet. Pathol.* **20**, 698–708.

Stromberg, P.C., Vogtberger, L.M., Marsh, L.R., and Wilson, F.D. (1983). Pathology of the mononuclear cell leukemia of Fischer rats. II. Hematology. *Vet. Pathol.* **20**, 709–717.

Sumitomo, S., Hashimura, K., and Mori, M. (1996). Growth pattern of experimental squamous cell carcinoma in rat submandibular glands– an immunohistochemical evaluation. *Eur. J. Cancer B Oral Oncol.* **32B**, 97–105.

Suwa, T., Nyska, A., Peckham, J.C., Hailey, J.R., Mahler, J.F., Haseman, J., and Maronpot, R.R. (2001). A retrospective analysis of background lesions and tissue accountability for male accessory sex organs in Fischer-344 rats. *Toxicol. Path.* **29**, 467–478.

Swenberg, J.A., Kerns, W.D., Mitchell, R.I., Gralla, E.J., and Pavkow, K.L. (1980). Induction of squamous cell carcinomas of the rat nasal cavity by inhalation exposure to formaldehyde vapor. *Cancer Res.* **40**, 3398–3902.

Takegawa, K., Mitsumori, K., Onodera, H., Shimo, T., Kitaura, K., Yasuhara, K., Hirose, M., and Takahashi, M. (2000). Studies on the carcinogenicity of potassium iodide in F344 rats. *Food Chem. Toxicol.* **38**, 773–781.

Takegawa, K., Mitsumori, K., Onodera, H., Yasuhara, K., Kitaura, K., Shimo, T., and Takahashi, K. (1998). Induction of squamous cell carcinomas in the salivary glands of rats by potassium iodide. *Jpn. J. Cancer Res.* **89**, 105–109.

Tamano, S., Hirose, M., Tanaka, H., Asakawa, E., Ogawa, K., and Ito, N. (1992). Forestomach neoplasm induction in F344?DuCrj rats and B6C3F1 mice exposed to sesamol. *Jpn. J. Cancer Res.* **83**, 1279–1285.

Tanaka, H., Hirose, M., Hagiwara, A., Imaida, K., Shirai, T., and Ito, N. (1995). Rat strain differences in catechol carcinogenicity to the stomach. *Food Chem. Toxicol.* **33**, 93–98.

Tanaka, T., Kohno, H., Sakata, K., Yamada, Y., Hirose, Y., Sugie, S., and Mori, H. (2002). Modifying effects of dietary capsaicin and rotenone on 4–nitroquinoline 1–oxide-induced rat tongue carcino- genesis. *Carcinogenesis* **23**, 1361–1367.

Tang, F.Y., and Tang, L.K. (1981). Association of endometrial tumors with reproductive tract abnormalities in the aged rat. *Gynecol. Oncol.* **12**, 51–63.

Tanigawa, H., Onodera, H., and Maekawa, A. (1995). Effects of barbital on neuro-oncogenesis in a transplacental carcinogenicity model using F344 rats. *J. Toxicol. Sci.* **20**, 55–65.

Thorling, E.B., Jacobsen, N.O., and Overvad, K. (1994). The effect of treadmill exercise on azoxymethane-induced intestinal neoplasia in the male Fischer rat on two different high fat diets. *Nutr. Cancer* **22**, 31–41.

Thurman, J.D., Hailey, J.R., Turturro, A., and Gaylor, D.W. (1995). Spontaneous renal tubular carcinoma in Fischer-344 rat littermates. *Vet. Pathol.* **32**, 419–422.

Thurman, J.D., Moeller, R.B.J., and Turturro, A. (1995). Proliferative lesions of the testis in ad libitum-fed and food-restricted Fischer-344 and FBNF1 rats. *Lab. Anim. Sci.* **45**, 635–640.

Tinkey, P.T., Lembo, T.M., Evans, G.R., Cundiff, J.H., Gray, K.N., and Price, R.E. (1998). Postirradiation sarcomas in Sprague-Dawley rats. *Rad. Research* **149**, 401–404.

Tsuda, H., and Krieg, K. (1992). Neoplastic lesions in the kidney. *In* "Pathobiology of the Aging Rat." (U. Mohr, D.L. Dungworth, and C.C. Capen, eds.), pp. 227–241, ILSI Press, Washington, DC.

Vahle, J.L., Sato, M., Long, G.G., Young, J.K., Francis, P.C., Engelhardt, J.A., Westmore, M.S., Linda Y., and Nold, J.B. (2002). Skeletal changes in rats given daily subcutaneous injections of recombinant human parathyroid hormone (1–34) for 2 years and relevance to human safety. *Toxicol. Path.* **30**, 312–321.

van Zwieten, M.J., Burek, J.D., Zurcher, C., and Hollander, C.F. (1979). Aortic body tumors and hyperplasia in the rat. *J. Pathol.* **128**, 99–112.

Vaquero, J., Oya, S., Coca, S., and Zurita, M. (1994). Experimental induction of primitive neuro-ectodermal tumours in rats: a re-appraisement of the ENU-model of neurocarcinogenesis. *Acta Neurochir. (Wein.)* **131**, 294–301.

Vogelstein, B., and Kinsler, K.W. (1993). The multistep nature of cancer. *Trends Genet.* **9**, 138–141.

Waalkes, M.P., Rehm, S., Sass, B., Konishi, N., and Ward, J.W. (1991). Chronic carcinogenic and toxic effects of a single subcutaneous dose of cadmium in the male Fischer rat. *Environ. Res.* **55**, 40–50.

Waldum, H.L., Rorvik, H., Falkmer, S., and Kawase, S. (1999). Neuroendocrine (ECL cell) differentiation of spontaneous gastric carcinomas of cotton rats (Sigmodon hispidus). *Lab. Anim. Sci.* **49**, 241–247.

Walker, C.L., Cesen-Cummings, K., Houle, C., Baird, D., Barrett, J.C., and Davis, B. (2001). Protective effect of pregnancy for development of uterine leiomyoma. *Carcinogenesis* **22**, 2049–2052.

Ward, J.W. (1975). Dose response to a single injection of azoxymethane in rats: induction of tumors in the gastrointestinal tract, auditory sebaceous glands, kidney, liver and preputial gland. *Vet. Pathol.* **12**, 165–177.

Ward, J.W., Hardisty, J.F., Hailey, J.R. and Street, C.S. (1995). Peer review in toxicologic pathology. *Toxicol. Pathol.* **23**, 226–234.

Ward, J.W., and Reynolds, C.W. (1983). Large granular lymphocyte leukemia. A heterogenous lymphocytic leukemia in F344 rats. *Am. J. Pathol.* **111**, 1–10.

Williams, K.D., Dunnick, J.K., Horton, J., Greenwell, A., Eldridge, S.R., Elwell, M.R., and Sills, R.C. (2001). p-Nitrobenzoic acid alpha 2u nephropathy in 13–week studies is not associated with renal carcinogenesis in 2–year feed studies. *Toxicol. Pathol.* **29**, 507–513.

Yanaida, Y., Kohno, H., Yoshida, K., Hirose, Y., Yamada, Y., Mori, H., and Tanaka, T. (2002). Dietary silymarin suppresses 4–nitroquinoline 1–oxide-induced tongue carcinogenesis in male F344 rats. *Carcinogenesis* **23**, 787–794.

Yoshida, T., Mitsumori, K., Harada, T., and Maita, K. (1997). Morphological and ultrastructural study of the histogenesis of meningeal granular cell tumors in rats. *Toxicol. Path.* **25**, 211–216.

Yoshitomi, K., Elwell, M.R., and Boorman, G.A. (1995). Pathology and incidence of amelanotic melanomas of the skin in F-344/N rats. *Toxicol. Path.* **23**, 16–25.

Young, S.S. (1989). What is the proper experimental unit for long-term rodent studies? An examination of the NTP benzyl acetate study. *Toxicology* **54**, 233–239.

Zwicker, G.M., Eyster, R.C., Sells, D.M., and Gass J.H. (1992). Spontaneous brain and spinal cord/nerve neoplasms in aged Sprague-Dawley rats. *Toxicol. Path.* **20**, 576–584.

Zwicker, G.M., Eyster, R.C., Sells, D.M., and Gass, J.H. (1992). Spontaneous skin neoplasms in aged Sprague-Dawley rats. *Toxicol. Path.* **20**, 327–340.

Zwicker, G.M., Eyster, R.C., Sells, D.M. and Gass, J.H. (1995). Spontaneous vascular neoplasms in aged Sprague-Dawley rats. *Toxicol. Path.* **23**, 518–526.

Chapter 15

Metabolic, Traumatic, and Miscellaneous Diseases

William W. King and Steven P. Russell

THE LABORATORY RAT, 2ND EDITION

I. INTRODUCTION

Techniques developed from germfree and specific-pathogen-free colony production and management have reduced the potential for inadvertent introduction of infectious agents to contemporary rodent colonies. A great deal has been learned regarding the nutritional requirements for optimal rodent growth and reproduction as well. Although much remains to be discovered in these disciplines, one of the most significant challenges remains further determining noninfectious diseases of laboratory rats. To further remove these sources of pain, distress, and model system variability, recognition of congenital, degenerative, traumatic, and other spontaneous diseases is paramount.

Because of the "miscellaneous" designation of its title, this chapter is charged with discussing any noninduced disease entity not associated with infectious agents or neoplasia. A work of this scope could diverge in many directions, including a general description of all pathological findings described in laboratory rats. Given inherent limitations, this work focuses primarily on those conditions that either result in recognized clinical signs or are likely to be seen in the postmortem examination of rats. The term "recognized," however, is an admitted shortcoming—an attempt at separating "incidental" from "clinically relevant" conditions may simply represent an inability to appreciate subtle clinical signs in laboratory rodents. Nonetheless, for a more encyclopedic description of lesions that have been discovered in rats and their significance, the reader is referred to one of the many texts devoted to the pathology of this species.

In companion animal practice, rats suffering from nonneoplastic, noninfectious diseases are likely to present to the clinician with ulcerative dermatitis, "ringtail," malocclusion, or renal failure (Donnelly, 1997). In the laboratory, the incidence of noninfectious, nonneoplastic causes of spontaneous death has been estimated at 29%

in males and 12% of females (Ettlin et al., 1994), mostly associated with age-related diseases. In specific-pathogen-free rats, the most significant sources of morbidity are neoplasia, chronic progressive nephropathy (Coleman et al., 1977), myocardial degeneration (Maeda et al., 1985; Keenan et al., 1995b), and polyarteritis nodosa (Berg and Simms, 1960; Ettlin et al., 1994). Frequently cited lesions include radiculoneuropathy, skeletal muscle degeneration, reproductive senescence, and bile duct hyperplasia (Coleman et al., 1977; Anver et al., 1982). Interestingly, many of these diseases may be coincidental, that is, appearing the same time, e.g. hypertension, chronic progressive nephropathy, polyarteritis nodosa, and myocardial degeneration (Wilens and Sproul, 1938b; Rapp, 1973; Wexler et al., 1981; Bishop, 1989; Weber et al., 1990; Weber et al., 1993; Saito and Kawamura, 1999; Percy and Barthold, 2001). As additional information elucidates the aging process and associated cellular degeneration, these and other relationships should be clarified.

A concerted effort to summarize documented incidence rates of various disorders has been made in this chapter. A vital component of appreciating a condition's significance involves understanding its possible genetic predisposition. Highlighting stock and strain differences may also yield important clues to pathogenesis and prevention. Interpreting incidence rates has risks, however, depending on the intent of the referenced report and attention to what might be considered "incidental" by the author—for example, a summary of all lesions regardless of severity versus those thought to be causing morbidity/mortality. In an attempt to control some of this variability, the incidence rates in this chapter's tables cite, when available, only those animal groups with disease of moderate or greater severity and therefore more likely to be of clinical significance. Other causes of variability include genetic drift, the use of different diets, the criteria and statistical methods used by the pathologist to identify lesions (Roe, 1994), environmental conditions, husbandry practices, specific health status, and even individual animal differences. Differences accordant with gender, which may be illustrated when strain, diet, health status, and age are controlled, may be attributable to the hormonal milieu, but might also be indirectly related to differences in body weight and/or conformation (Berg and Simms, 1960). These incidence rates should therefore be interpreted cautiously.

Additionally, the incidence tables reflect only rats fed *ad libitum*. A note should be made, however, regarding the impact of feeding practices on degenerative diseases. The beneficial effects of dietary restriction have been touted for over 65 years (Cornwell et al., 1991). Commercial rodent diet formulations optimize early development, growth, and the increased metabolic demands of reproduction and lactation (Goldstein et al., 1988). Therefore, older and more sedentary rats may suffer the consequences of overfeeding afforded by the common practice of *ad libitum* feeding (Keenan et al., 1995a). Even moderate dietary restriction in rats has a significantly positive effect on life span, primarily through the diet's impact on spontaneous diseases (Masoro, 1991; Masoro et al., 1991; Roe et al., 1991; Keenan et al., 1994; Keenan et al., 1995b). The contribution of protein and calories to overfeeding continues to be debated, and may be organ or disease specific. While reducing dietary protein reduces the incidence and severity of degenerative lesions, it is less effective than calorie restriction (Maeda et al., 1985; Masoro et al., 1989; Keenan et al., 1995a). Calorie restriction may circumvent the course of aging (Stern et al., 2001) by optimizing energy utilization through cellular mechanisms such as glucocorticoid or glucose metabolism, or free oxygen radical production (Masoro et al., 1991). Unlike neoplasia, where it appears to reduce only the age of onset (Keenan et al., 1995b), dietary restriction reduces both the age of onset and progression of degenerative diseases such as chronic progressive nephropathy, cardiomyopathy, and reproductive deterioration (Keenan et al., 1994; Keenan et al., 1995a). Automated methods of dispensing limited feed amounts to rats are being explored (Petruska et al., 2001).

II. CARDIOVASCULAR SYSTEM

A. Congenital Heart Defects

Descriptions of congenital heart defects in rats are uncommon. A survey of Sprague-Dawley (Hsd:Sprague-Dawley SD) fetuses revealed cardiac defects in approximately 2.3% of the fetal hearts, including dextrocardia, atrial and ventricular septal defects, an endocardial cushion defect, and an aortic valve defect (Johnson et al., 1993). The Olson-Goss subline of the Long-Evans rat has been reported to have a 30% incidence of congenital membranous interventricular septal defects, compared with 4% in the California subline (Fox, 1969; Jimenez-Marin, 1971).

B. Myocardial Degeneration and Fibrosis

Some of the earliest literature describing lesions in older rats involves the cardiovascular system. In a pair of surveys of spontaneous cardiovascular diseases, Wilens and Sproul (1938a, 1938b) discussed intracardiac thrombosis, chronic auriculitis, endocarditis, coronary artery sclerosis, cardiac hypertrophy, pericarditis, and calcification of various arteries. Noting the relative paucity of literature on cardiovascular lesions in laboratory rats, these authors also recognized that the significance of myocardial degeneration and fibrosis was far greater than other spontaneous diseases of this system. Others have noted this disorder ranks third behind neoplasia and chronic renal disease for causes of rat mortality (Keenan et al., 1995b).

The onset of myocardial degeneration varies with age, gender, and strain. Lesions are commonly found in laboratory rats beginning at 12 months, with the incidence and severity increasing with age (Anver and Cohen, 1979; Lewis, 1992). Myocardial degeneration occurs more commonly and at an earlier age in males (Keenan et al., 1995b; Percy and Barthold, 2001), which also develop more severe, multifocal lesions, contrasting to the focal lesions seen in females (Dixon et al., 1995). Germ-free rats may be resistant to myocardial degeneration (Pollard and Kajima, 1970). The incidence has been estimated at 80.7% in Fischer-344 males (Maeda et al., 1985), and 73.2% to 85.9% in Sprague-Dawley stocks (Cohen et al., 1978; Anver et al., 1982). Table 15-1 summarizes various reported age, gender, and stock/strain incidences of myocardial degeneration and fibrosis.

In most cases, lesions are discovered during routine or scheduled necropsy with no appreciable clinical signs (Anver et al., 1982; Keenan et al., 1995b). Although slight variations in electrocardiogram findings have been noted (Berg, 1955a), microscopic evidence of congestive heart failure is usually absent (Cornwell et al., 1991). Differences in most physiological indices of cardiac function in Wistar rats between 6 and 12 months of age are insignificant (Lee et al., 1972); however, the rate of myocardial degeneration is typically low in this strain. A correlation between the onset and progression of myocardial fibrosis in Fischer-344 rats between 20 and 29 months of age and increased end-diastolic pressure suggests that subtle cardiac functional changes may be present although clinically inapparent (Anversa and Capasso, 1991).

Lesions may be difficult to discern grossly, although foci of pale to grayish tissue may be present in severe cases

TABLE 15-1

Stock/Strain, Age, and Gender Distribution of Laboratory Rats with Myocardial Degeneration and Fibrosis

| | Incidence (%)* of Myocardial Degeneration and Fibrosis | | | | | | |
| | 0–52 Weeks of Age | | 52–104 Weeks of Age | | >104 Weeks of Age | | |
Stock/Strain	Male	Female	Male	Female	Male	Female	Reference
Fischer-344	79.1	22.8	—	—	—	—	Dixon et al., 1995
	0.0–46.1	—	35.0–75.0	—	76.6–86.7	—	Coleman et al., 1977
	—	—	—	—	33.1	17.6	Goodman et al., 1979
	10.9	0.0	19.0–34.7	5.2–12.5	31.1	15.0	Biology Databook Editorial Board, 1985
	18.0–33.3	—	51.7–74.6	—	75.0–100.0	—	Maeda et al., 1985
	32.1 (between 0–9 mo.) 67.9 (by 6–9 mo)	11.1 (between 0–9 mo.) 22.2 (by 6–9 mo)	—	—	—	—	Hall et al., 1992
Osborne-Mendel	—	—	—	—	40.2	18.9	Goodman et al., 1980
	—	—	65.6	—	—	55.3	Wilens and Sproul, 1938a
	0.0	0.0	9.4–19.2	0.0–4.5	16.9	7.7	Biology Databook Editorial Board, 1985
Sprague-Dawley	2.4	0.0	2.1–3.8	0.0–1.9	21.8	8.3	Biology Databook Editorial Board, 1985
Lobund-Wistar	0.4	—	1.2	—	2.2–3.4	—	Cornwell et al., 1991

*Age distributions were not consistent in all references cited. Some references may have stated average age rather than age ranges; such values were approximated in the divisions employed this table. It should also be noted that some references may have used smaller age ranges, or even more divisions among age ranges; these were included with similar age groups for ease of comparison.

(Anver and Cohen, 1979; Percy and Barthold, 2001). Foci are often distributed randomly (Dixon et al., 1995), but with a predilection for the left ventricular papillary muscles, their sites of attachment, and the interventricular septum (Anver and Cohen, 1979; Percy and Barthold, 2001). Histological evidence follows a progression of myocardial atrophy, necrosis, and focal interstitial myocarditis composed primarily of mononuclear cells, followed by interstitial fibrosis and accumulation of fibrous connective tissue (Coleman et al., 1977; Anver et al., 1982; Dixon et al., 1995; Keenan et al., 1995b; Fig. 15-1). As severity increases, lesions coalesce to form large areas of fibrous tissue with mineralization (Masoro et al., 1989).

Numerous predisposing factors have been proposed in the pathogenesis of myocardial degeneration and fibrosis in rats. The lesions have been suggested to be primarily of inflammatory origin as a sequela of chronic myocarditis or myocardial ischemia (Fairweather, 1967; Coleman et al., 1977). A strong correlation exists between the onset and severity of cardiomyopathy and chronic progressive nephropathy, suggesting a developmental relationship between these two significant age-related diseases (Maeda et al., 1985). Dietary restriction reduces both the incidence and severity of cardiomyopathy in several rat stocks and strains (Maeda et al., 1985; Masoro et al., 1989; Keenan et al., 1994), with differences evident by 30 weeks of age (Cornwell et al., 1991). This may be related to calorie

restriction's beneficial effects on chronic progressive nephropathy development, although some diets affecting chronic progressive nephropathy do not prevent myocardial fibrosis (Iwasaki et al., 1988). Certain forms of renovascular hypertension lead to cardiac myocyte necrosis and fibrosis, although the mechanisms are not clear (Weber et al., 1990). The relationships between hypertension, chronic progressive nephropathy, and myocardial fibrosis warrant further investigation (Weber et al., 1993).

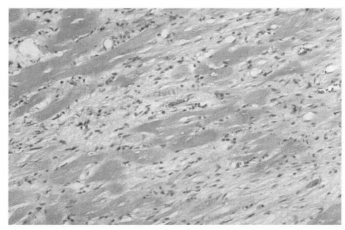

Fig. 15-1 Myocardial degeneration and fibrosis, photomicrograph (100×). (*Image courtesy of Dr. S.W. Barthold*)

C. Intracardiac Thrombi

Intracardiac thrombi in one or more heart chambers are not uncommon. The incidence of this age-associated condition varies between stocks and strains, ranging from 0.5% in Sprague-Dawley, 2.1% in Fischer-344 (Biology Databook Editorial Board, 1985), and from 0% to 6.4% in Osborne-Mendel rats (Wilens and Sproul, 1938a; Biology Databook Editorial Board, 1985). Although thrombi may be encountered in any chamber, the left atrium is most frequently affected (Lewis, 1992). Grossly, thrombi are characterized as firm, laminated, gray-red masses that progress to well-organized accumulations of fibrous connective tissue seen in histological sections (Ayers and Jones, 1978). Older thrombi may involve the endo- and myocardium (Anver and Cohen, 1979). Large thrombi may induce pulmonary congestion and ventricular hypertrophy (Lewis, 1992).

D. Valvular Endocardiosis

Heart valve thickening with myxomatous connective tissue in aging rats has been referred to as valvular endocardiosis, chronic valvular fibrosis, chronic valvular disease, valvular myxoma, myxomatous degeneration, mucinous degeneration, and valvular endocarditis (Ayers and Jones, 1978). This lesion affects up to 10% of rats in some Sprague-Dawley colonies, usually involving the atrioventricular valves (Anver et al., 1982). Although there are no known clinical effects, these masses microscopically resemble those associated with systolic murmurs and congestive heart failure in dogs (Ayers and Jones, 1978; Anver and Cohen, 1979).

E. Endocardial Proliferative Lesions

Endocardial and subendothelial proliferative lesions have been noted in many stocks and strains of rats. Microscopically, they are composed of endocardial accumulations of fibroblast-like cells, with occasional lymphocytic infiltration, expanding into grossly evident tumors penetrating the myocardium (Boorman et al., 1973; Zaidi et al., 1982). The nature of these lesions is controversial (Boorman et al., 1973; Frith et al., 1977), with some suggesting these masses are intermediate forms of developing Schwann cell malignancies (Alison et al., 1987), and others making distinctions between neoplasia and fibroproliferative masses based on the discrete morphological features (Novilla et al., 1991).

The stock-/strain-related differences in the occurrence of these lesions range from 1% in the Wistar-derived CIVO rat, 4% in the Wistar-derived WAG/Rij rat, and 7% in the Brown Norway (BN/Rij) rat (Boorman et al., 1973). An overall incidence of 0.2% has been described in Fischer-344, Wistar, Sprague-Dawley, and Long Evans rats (Novilla et al., 1991); one report documented a 1.25% incidence in a Sprague-Dawley colony (Zaidi et al., 1982). The condition, also described as (sub)endocardial fibrosis, fibroelastosis, endocardiosis, or endocardial fibromatous proliferation, is without gender predilection, but is generally seen in animals older than 71 weeks of age (Boorman et al., 1973; Novilla et al., 1991), although it has also been reported in younger rats (Frith et al., 1977).

F. Polyarteritis Nodosa

Although described in dogs, cats, and various other mammals, polyarteritis nodosa (PAN) has been principally characterized in laboratory rats (Bishop, 1989). The first survey of inflammatory vascular wall lesions in rats, provided by Wilens and Sproul (1938b), described a 9.7% incidence of periarteritis in an Osborn-Mendel colony. PAN was attributed to the cause of death (or euthanasia in extremis) in 19.2% and 7.5% male and female Sprague-Dawley-derived OFA Sandoz rats, respectively (Ettlin et al., 1994).

PAN is considered a degenerative disease of aging rats (Skold, 1961). Early reports discussed preponderance in females (Wilens and Sproul, 1938b; Cutts, 1966), although more recent references describe a predisposition in males (Yang, 1965; Bishop, 1989; Percy and Barthold, 2001). Incidence rates in various arteries differ greatly (Table 15-2). PAN occurs with increased frequency in various hooded strains such as August (Opie et al., 1970) and H/C (Cutts, 1966). Essentially all male spontaneously hypertensive rats (SHR) develop PAN by 15 months of age (Suzuki et al., 1979; Bishop, 1989), although there are differences in the histological appearance in this strain (Wexler et al., 1981).

Stock and strain impacts the site affected, with medium-sized muscular arteries such as the mesenteric, splenic, testicular, and pancreatic arteries being commonly affected (Skold, 1961; Anver et al., 1982; Bishop, 1989; Saito and Kawamura, 1999). Other organs with PAN-associated lesions include the kidney, tongue, urinary bladder, brain, salivary glands, liver, thymus, seminal vesicles, lymph nodes, cecum, skeletal muscle, adrenal glands, stomach, intestine, heart, ovary, and uterus (Yang, 1965; Anver and Cohen, 1979; Anver et al., 1982). As in humans, lung tissue remains uninvolved (Bishop, 1989). Grossly, affected arteries display a firm, segmental thickening, often described as nodular and tortuous, with thrombi and aneurysmal dilations that occasionally rupture (Opie et al., 1970; Kohn and Barthold, 1984; Bishop, 1989) (Fig. 15-2). Early changes include fibrinoid necrosis of the intima

TABLE 15-2

STOCK/STRAIN AND AGE DISTRIBUTION OF LABORATORY RATS WITH LESIONS ASSOCIATED WITH POLYARTERITIS NODOSA

| Stock/Strain | Incidence (%)* of Polyarteritis Nodosa | | | | Reference |
	0–52 Weeks of Age	52–110 Weeks of Age	110–128 Weeks of Age	>128 Weeks of Age	
F-344	0.0	0.0	0.2	0.0	Biology Databook Editorial Board, 1985
	—	1.6 p 2.4 s		—	Coleman, et al., 1977
Osborne-Mendel	0.0	1.0	0.0	0.0	Biology Databook Editorial Board, 1985
	0.0	3.0	13.1	15.7	Wilens and Sproul, 1938b
August	0.0	36.4	75.0	85.7	Opie et al., 1970
Sprague-Dawley[1]	0.0	0.0	25.9 t 15.4 s 17.6 m 9.4 p	48.8 t 28.6 s 22.2 m 16.3 p	Anver et al., 1982
Sprague-Dawley[2]	0.0	9.1 t 9.1 s 18.2 m 18.2 p	35.3 t 15.2 s 14.7 m 2.9 p	—	Anver et al., 1982
Sprague-Dawley[3]	—	9.2 p 13.1 s		—	Cohen et al., 1978
Sprague-Dawley[4]	0.6	3.1	1.8	6.5	Biology Databook Editorial Board, 1985
	—	14.6	—	—	Yang, 1965
	—	—	60.0	—	Berg, 1967
Long-Evans	0.0	0.0	0.0	1.6	Opie et al., 1970
Stroke-Prone SHR	57.1 t 28.6 m	100.0 t 60.0 m	—	—	Saito and Kawamura, 1999
Stroke-Resistant SHR	—	42.9 t 28.6 m	—	—	Saito and Kawamura, 1999
Wistar	0.0	7.1	0.0	4.3	Opie et al., 1970
Wistar-Kyoto	0.0	0.0	—	—	Saito and Kawamura, 1999

t = testicular artery; s = splenic artery; SHR = spontaneously hypertensive rat; m = mesenteric artery; p = pancreatic artery.

[1]Crl:COBS[R]CD[R] (SD).

[2]Hap:(SD).

[3]Crl:CD(SD)BR.

[4]Designation not specified.

*Age distributions were not consistent in all references cited. Some references may have stated average age rather than age ranges; such values were approximated in the divisions employed this table. Some references may have used smaller, more division, or slightly different age ranges or divisions; these were included with similar age groups for ease of comparison.

and media leading to inflammatory cellular infiltration and disruption of the internal elastic laminae (Bishop, 1989; Percy and Barthold, 2001). Chronic lesions are characterized by intimal proliferation and a marked fibrotic reaction within the adventitia and media; both acute and chronic changes may be appreciated during microscopic evaluation (Skold, 1961; Anver and Cohen, 1979). Subsequent features include stenosis (with or without recanalization), thromboses, and aneurysm formation (Anver and Cohen, 1979; Bishop, 1989; Percy and Barthold, 2001). Aneurysm formation occurring subsequent to elastic laminar disruption without further evidence of PAN has also been reported in the anterior cerebral artery of a 35-week-old Sprague-Dawley rat (Kim and Cervós-Navarro, 1991).

The differences in onset and severity related to age, strain, and gender indicate multifactorial influences impacting the development of PAN. An immune-mediated pathogenesis has been proffered (Yang, 1965). In humans, PAN lesions respond to immunosuppressive therapy (Bishop, 1989), and may be seen concurrently with other collagen-associated vascular diseases such as rheumatoid arthritis, polymyositis, myocarditis, and dermatitis (Yamazaki et al., 1997). PAN-like lesions may be experimentally induced in rats by administering chemical agents such as 4'-fluoro-10-methyl-1,2-benzanthracene (Hartmann et al., 1959) and dopaminergic agonists (Kerns et al., 1989).

A relationship between hypertension and PAN has been previously noted. PAN lesions in SHR affect testicular

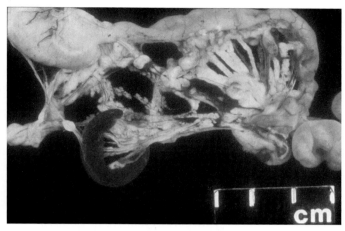

Fig. 15-2 Polyarteritis nodosa in mesenteric vasculature. Note tortuous and nodular thickenings of the vessel walls. (*Image courtesy of Dr. S.W. Barthold*)

arterioles at an earlier age and become more severe in mesenteric arteries, especially in stroke-prone (as compared to stroke-resistant) strains, but are absent in the normotensive Wistar-Kyoto rat (Saito and Kawamura, 1999). Various surgical and mechanical renal insults producing renal secondary hypertension can induce PAN (Yang, 1965; Cutts, 1966).

Several other factors influence the progression of PAN, including nonspecific stress and reproduction history (Bishop, 1989). A positive correlation between the incidence and number of pregnancies has been demonstrated in a colony of Sprague-Dawley rats (Wexler, 1970), although supplemental estrogen may decrease the age of onset in female hooded (H/C strain) rats (Cutts, 1966). Lesion development adjacent to a pancreatic endocrine tumor in a male Holtzmann rat led to the speculation for a role of pancreatic hormones in the pathogenesis of PAN (Baczako and Dolderer, 1997). Like myocardial and skeletal muscle degeneration, germfree Wistar rats are reportedly resistant to PAN development (Pollard and Kajima, 1970).

G. Arteriosclerosis/Atherosclerosis

Reviews of age-associated changes in the arteries of rats have included arteriosclerosis and atherosclerosis of the aorta, carotid, coronary, and cerebral arteries; aortic and carotid lesions resembling Mönkeberg's medial sclerosis; and other changes seen in the arterial wall (Anver and Cohen, 1979; Lewis, 1992). Rats are generally regarded as resistant to atherosclerosis except following relatively invasive experimental treatments and in those strains specifically bred for studying related cardiovascular diseases, such as the SHR, the obese SHR, the La/N-cp rat, the stroke-prone rat, the arteriolipidosis-prone rat (ALR),

the normotensive atherogenic rat (NAR), and the myocardial infarction rat (MR) (Ritskes-Hoitinga and Beynen, 1988). Nonetheless, spontaneous lesions have been described in common stocks and strains. In Fischer-344, arteriosclerosis was reported in 0.06% of males over 104 weeks of age (Biology Databook Editorial Board, 1985). Gender-related differences have been shown in Osborne-Mendel rats, affecting 10.9% of males and 2.7% of females between 109 and 126 weeks of age (Goodman et al., 1980). Others estimate the incidence in this strain to be approximately 3.3% in animals over 52 weeks of age (Biology Databook Editorial Board, 1985). A difference in the incidence of arteriosclerosis between retired breeders of one line of Sprague-Dawley–derived stocks (Crl:COBS[R]CD[R](SD)) and virgins of another (Hap:(SD)) may be related to either subtle genetic differences or variations induced by reproductive status (Anver et al., 1982). Similar to PAN, repeated breeding appears to increase the likelihood of arteriosclerosis (Wexler, 1970). Medial calcification tends to occur primarily in the pulmonary artery (Coleman et al., 1977; Cohen et al., 1978; Goodman et al., 1979).

H. Hypertension

Hypertension is usually associated with strains bred specifically to develop this condition, such as the SHR, and substrains that are susceptible or resistant to the effects of salt such as the Dahl and Sabra rat (Yagil and Yagil, 1998). Nonetheless, spontaneous hypertension can develop in aged rats of various traditional stocks and strains (Anver and Cohen, 1979). A 46% incidence was reported in male Sprague-Dawley rats over 120 weeks of age, which, as with other spontaneous cardiovascular diseases, are more prone to hypertension than females (Berg and Harmison, 1955).

Of particular interest is the relationship between hypertension and other age-related, degenerative diseases, including chronic progressive nephropathy (Rapp, 1973; Wexler et al., 1981; Magro and Rudofsky, 1982; Rudofsky and Magro, 1982; Alden and Frith, 1991), myocardial degeneration (Weber et al., 1990; Weber et al., 1993), and polyarteritis nodosa (Rapp, 1973; Wexler et al., 1981; Bishop, 1989; Saito and Kawamura, 1999). Decreased nociperception in hypertensive rats, like humans, may have clinical relevance to their care and use (Ghione, 1996).

III. HEMOLYMPHATIC SYSTEM

A. Hematopoiesis and Coagulation

Extramedullary hematopoiesis (EMH), which is characterized by foci of erythroid precursors, megakaryocytes, and hemosiderosis, or hemosiderin-laden macrophages (Eustis et al., 1990) typically follows hemorrhage or phlebotomy (Scipioni et al., 1997). There are conflicting opinions as to the significance of spontaneous EMH and hemosiderosis in rats. Reports range from over 50% in some Sprague-Dawley-derived colonies less than 1 year of age (Anver et al., 1982) to 100% in 19- to 21-week-old Fischer-344 rats (Dixon et al., 1995), suggesting an incidental occurrence. Others have stated that spontaneous EMH is rare without an underlying pathological condition (Eustis, et al., 1990; Percy and Barthold, 2001). When present, EMH is found primarily in the spleen, although other sites such as the perirenal adipose tissue and liver can also be involved (Eustis et al., 1990). The incidence of EMH appears to be greater in breeding females (Dixon et al., 1995; Percy and Barthold, 2001).

Spontaneous hemorrhage may occur secondary to hypovitaminosis K in germfree or certain specific pathogen-free rats. Such colonies are predisposed due to the absence of vitamin-synthesizing gut microflora and the destruction of dietary vitamin K in autoclaved feeds (Kohn and Barthold, 1984). Certain anesthetic agents administered to rats, notably urethane and ketamine-xylazine combinations, can alter coagulation assays (Stringer and Seligmann, 1996).

Apostolou and colleagues (1976) noted that overnight fasting, even with free access to water, resulted in hemoconcentration, restricting the authors' ability to obtain useful clinical samples. The overnight fast resulted in a significant body weight loss in both 37- and 71-day-old rats, and markedly greater hematocrit and erythrocyte counts (as well as a decreased leukocyte concentration, serum glucose, blood urea nitrogen (BUN), alanine aminotransferase (ALT), and alkaline phosphatase (ALP) in the older rats, increased creatinine in older females, and increased SGOT in older male rats) due to reduced water consumption.

B. Thymic Involution

Immune system senescence, especially in the thymus, has been the focus of many investigations. This process is characterized morphologically by an age-associated proliferation of thymic epithelial components, including epithelial cords, tubules, and cysts (Cherry et al., 1967; Meihuizen and Burek, 1978). Thymic involution appears to advance more rapidly in males than females (Kuper et al., 1986). Although some cellular responses from aging thymus tissue are generally uncompromised in aging animals (Cherry et al., 1967), thymic involution has been associated with reduced T-cell–dependent function and immune competence (Kuper et al., 1986; Mazzeo, 1994; Shinkai et al., 1997; Aspinall and Andrew, 2000).

IV. RESPIRATORY SYSTEM

A. Inhalation Pneumonia

Aspiration pneumonia is an occasional finding in older rats (Maeda et al., 1985). Inhaled foreign bodies, including dusty bedding and food, result in an inflammatory response (Percy and Barthold, 2001); microscopic evaluation may reveal traces of plant material and mineral deposits (Kohn and Barthold, 1984; Fig. 15-3). Although such findings are not uncommon, most are focal and rarely associated with significant disease (Hollander, 1976). Residual endotoxin and other microbial cell wall elements in bedding, even if autoclaved, may induce airway responses in rats, including moderate, multifocal, perivascular, and interstitial inflammatory infiltration (Ewaldsson et al., 2002).

The morphology of the gastro-esophageal junction essentially eliminates the potential for regurgitation in rats (Bilhun, 1997), therefore pneumonia following inhalation of vomitus (e.g., during anesthesia) is improbable. However, esophageal disorders may cause secondary inhalation pneumonia (see Megaesophagus and Gastrointestinal Impaction section of this chapter).

Fischer-344 rats appear to be particularly susceptible to developing age-related tracheal cartilage degeneration and seromucinous adenitis. The use of a rigid, metal gavage cannula and the irritant properties of the gavaged material may also play a role (Germann and Ockert, 1994), although spontaneous lesions occur in untreated rats and may be seen as early as 6 weeks of age (Germann et al., 1995). These inflammatory lesions are thought to lead to impaired salivation; food and bedding may become lodged in the oropharyngeal cavity, resulting in asphyxia (Germann and Ockert, 1994). Microscopically, lesions

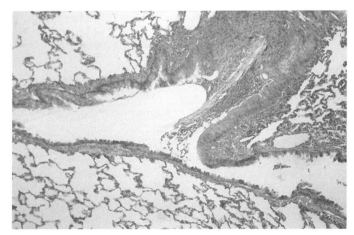

Fig. 15-3 Inhalation pneumonia, photomicrograph (100×). Note inflammatory response around plant material. (*Image courtesy of Dr. S.W. Barthold*)

begin as diffuse lymphohistiocytic inflammation of the tracheal seromucinous glands and nodular granuloma formation. This can become extensive at the epiglottis-arytenoid junction and lead to osseous metaplasia of the tracheal chondrocytes (Germann et al., 1995). Gavage-associated complications such as aspiration and airway injury are more frequent when gavage volumes exceed 10 ml/kg body weight (Brown et al., 2000).

B. Agent-Induced Pulmonary Edema

Pulmonary edema has been noted to occur more frequently in rats dying spontaneously versus those undergoing euthanasia (Maeda et al., 1985). Euthanasia technique, however, has also been shown to have an impact on pulmonary lesions. Danneman and colleagues (1997) described mild perivascular edema and intra-alveolar hemorrhage following euthanasia with pentobarbital, and mild to marked perivascular edema, perivascular hemorrhage, and intra-alveolar hemorrhage following euthanasia with concentrations of carbon dioxide ranging from 60% to 100%. Feldman and Gupta (1976) noted that most common methods of euthanasia, including inhalant anesthetic and sodium pentobarbital overdose, may all result in some level of pulmonary vascular change, ranging from mild capillary congestion to significant pulmonary edema. Carbon dioxide asphyxiation and physical methods such as decapitation and cervical dislocation may result in frank blood within alveolar spaces (Feldman and Gupta, 1976).

In addition to its clinically useful effects of sedation, muscle relaxation, and analgesia, and its negative impact on glucose metabolism (Koppel et al., 1982), the α2-adrenergic agonist xylazine induces pulmonary edema in rats. When administered intramuscularly at doses of greater than 20 mg/kg, xylazine results in pleural effusion and alveolar edema (Amouzadeh et al., 1991). The optimal "edemagenic" dose is approximately 42 mg/kg, which has been used to study the progression of increased pulmonary vascular permeability (Amouzadeh et al., 1993).

C. Effects of Environmental Conditions

The impact of environmental contaminants present in the housing environment has been the subject of many investigations. For example, softwood organic compounds are known to influence hepatic enzyme function (Vesell, 1967). The sanitation level of primary enclosures has also been cited as a major contributor to hepatic microsome dysfunction (Vesell et al., 1973). Although such changes might result from metabolic toxins other than ammonia (Schaerdel et al., 1983), many have discussed the potential

for ammonia-induced toxicity and its effects on pulmonary function.

The concentration of ammonia, which is a byproduct of bacterial urease breakdown of urine-soiled bedding material, can range from 0 to over 200 ppm in rodent and rabbit primary enclosures, with approximately 25 ppm occurring with regular frequency (Gamble and Clough, 1976). Using filtered micro-isolator cage covers facilitates ammonia accumulation if soiled bedding is not removed within 5 or 6 days (Serrano, 1971). Controversy remains, however, on the minimal concentration of ammonia that is deleterious to laboratory rats and its mechanism of toxicity.

Some studies have demonstrated ammonia-induced lesions in nasal passages even at low levels (Broderson et al., 1976). Exposure of rats to ammonia approximately 200 ppm resulted in histological changes in the tracheal epithelium in 4 days and epithelial hyperplasia with loss of cilia within 8 days (Gamble and Clough, 1976). Exposure to ammonia 500 ppm for 3 weeks induced evidence of nasal and upper respiratory epithelial inflammation; these changes were absent after 8 weeks, suggesting a compensatory mechanism (Richard et al., 1978). However, Schaerdel and colleagues (1983) found that following a peak in blood ammonia concentration at 8 hours, exposure to concentrations over 700 ppm for 7 days failed to incite changes in blood pH, pCO_2 and minimal lesions within the lung or trachea. These findings indicated the presence of compensatory metabolic mechanisms (Manninen et al., 1988) and suggested that ammonia concentrations in contemporary rat facilities may have little direct toxicity (Schaerdel et al., 1983).

What is clear is the potentiation of environmental ammonia on murine respiratory mycoplasmosis (MRM). Ammonia levels as low as 25 ppm increase the severity of MRM-induced lesions (Broderson et al., 1976). Exposure to 50 to 100 ppm augments the growth of *Mycoplasma pulmonis* in the respiratory tract of rats and mice (Saito, et al., 1982). The enhancement of MRM in the lower respiratory tract is likely a secondary consequence, as most gaseous ammonia is absorbed in the nasal passages (Schoeb, et al., 1982).

D. Alveolar Histiocytosis

An age-related change of questionable clinical significance is the subpleural accumulation of pulmonary foam cells (PFC), also know as alveolar histiocytes, foamy macrophages, and xanthoma cells, which may be appreciated grossly as pale stippling (Beaver et al., 1963; Anver and Cohen, 1979). These lipid-laden phagocytes, which were first thought to result from cirrhogenic, and protein- or pantothenic acid-deficient diets (Beaver et al., 1963), are considered spontaneous, incidental findings in laboratory rats (Yang et al., 1966). The incidence has been estimated to be as high as 55.3% and 52.8% of male and female 19- to 21-week-old Fischer-344 rats (Dixon et al., 1995).

The origin of PFC remains an enigma. Studies of rats with dietary- and carbon tetrachloride-induced hyper β-lipoproteinemia suggest PFCs are lipid-laden monocytes that migrate into pulmonary alveoli (Shibuya et al., 1991; Shibuya et al., 1995; Shibuya et al., 1997). Increased concentration of PFCs in conjunction with eosinophilic infiltration has been associated with biotin-deficient diets (Tanaka et al., 1995). Their numbers are greater in lobes with other lesions such as pneumonia and neoplasia, suggesting a relationship with liberation of adjacent cellular lipids (Shibuya et al., 1986). Their distribution has led to speculation that their accumulation occurs from the lymphatic system following chronic hypoxic conditions (Wolman et al., 1993). PFCs amass in response to cryptococcal and *Pneumocystis* sp. infection (Goldman et al., 1994; el-Nassery et al., 1994). Significant PFC aggregation in suspected pathogen-free rats may then represent a subclinical response to an undetected infectious agent.

V. GASTROINTESTINAL SYSTEM

A. Malocclusion and Periodontitis

A characteristic of the family Muridae is hypsodontic (continually-erupting) incisors to compensate for coarse dietary materials; overgrowth follows misalignment (Emily, 1991). Fracture of one incisor is often associated with overgrowth of the opposing one (Sharp and LaRegina, 1998; Percy and Barthold, 2001), but any genotypic, dietary, infectious, or traumatic condition interrupting the normal apposition and thereby the natural wearing of the occlusal surfaces of the upper and lower incisors can lead to overgrowth (Harkness and Wagner, 1995; Bilhun, 1997).

Malocclusion may lead to a mechanical obstruction of prehension and dysphagia, ptyalism, anorexia, weight loss, and, if untreated, death (Anver and Cohen, 1979; Harkness and Wagner, 1995). Incisor overgrowth may continue until the soft tissues of the mandible or maxilla are penetrated, leading to inflammatory lesions and secondary bacterial infections, abscess formation, and cellulitis (Donnelly, 1997; Percy and Barthold, 2001). Excess salivation associated with overgrown incisors may cause moist dermatitis (Harkness and Wagner, 1995). Chronic inflammation induced by overgrown incisors was implicated in the development of a rhabdomyosarcoma (Brockus et al., 1999).

Overgrown incisors are treated by physically reducing the length of the tooth or teeth involved. Clippers or rongeurs may be used to quickly treat conscious rats, but must be used with caution as the teeth tend to split longitudinally, producing jagged edges and predisposing to apical abscess formation (Emily, 1991). This can be prevented by using a dental drill, bur, small hacksaw blade, or embryotomy wire under sedation or general anesthesia to bring the teeth into occlusion (Emily, 1991; Donnelly, 1997). A speculum, otoscope, or small tongue depressor may assist in visualizing the site (Harkness and Wagner, 1995). Rats with severe malocclusion will likely require re-trimming every 4 to 6 weeks (Emily, 1991). Incisors without apposition may be removed, although the extraction procedure may prove challenging due to the length of the tooth roots (Donnelly, 1997).

Personnel working with rats should be trained to recognize the 1:3 ratio of upper to lower incisor crown length to avoid an inappropriate diagnosis of malocclusion (Bilhun, 1997). Because spontaneous malocclusion without a history of trauma may have a genetic component, affected animals should not be used for breeding (Harkness and Wagner, 1995; Sharp and LaRegina, 1998). It should be noted that yellowing on the rostral aspect of rodent incisors follows iron deposition and is a normal physiological phenomenon (Donnelly, 1997), not to be confused with discoloration associated with tooth death (Harkness and Wagner, 1995).

Periodontitis is not uncommon in older rats. Fibers and chaff from oat and barley have been associated with increased inflammatory oral disease severity, especially at the level of the first molar (Robinson et al., 1991). Madsen (1989) reported that such chaff predispose rats to foreign-body reactions that can lead to tooth loss, oropharyngeal/oronasal fistulation, and subsequent gastrointestinal air distension, as well as to epithelial hyperplasia, metaplasia, and squamous-cell carcinoma.

B. Megaesophagus and Gastrointestinal Impaction

Sporadic cases of idiopathic megaesophagus have been reported in several stocks and strains (Harkness and Ferguson, 1979). This neuromuscular disorder is characterized by flaccid esophageal enlargement, frequently presenting with feed or bedding accumulation and leading to anorexia or choke (Harkness and Wagner, 1995). Some reports suggest a heritable etiology (Ruben et al., 1983; Baiocco et al., 1993).

Fatal esophageal impaction has been implicated in several strains, including Wistar (Hla:(WI)BR; Harkness and Ferguson, 1979), BDE/HAN (Deerberg and Pittermann, 1972), Fischer-344 (COBS CD F/CrlBR; Will et al., 1979), Long-Evans (Baiocco et al., 1993), and thought to be the cause of 21% mortality in an Srl:BHE rat colony (Ruben et al., 1983). The condition is associated with various feed formulations, including powdered (Harkness and Ferguson, 1979) and high-bulk diets (Will et al., 1979). Will and colleagues (1979) described a female predilection for acquiring esophageal impaction on standard laboratory diet in Fischer-344 rats. In one case report, terminal pneumonia reportedly followed pharyngeal paresis/paralysis-induced dysphagia, a sequela to a large pituitary tumor (Dixon and Jure, 1988).

Clinical signs depend on the chronicity of the lesion. In cases with esophageal impaction, rats may display ptyalism, anorexia, dehydration, perioral dermatitis, dyspnea, weakness, emaciation, asphyxiation, or sudden death (Ruben et al., 1983; Will et al., 1979; Harkness and Wagner, 1995). Contrast esophagram reveals thoracic esophageal enlargement (Baiocco et al., 1993). At necropsy, the esophagus may be greatly distended and impacted with material; evidence of bronchopneumonia is frequently present due to aspiration of ingesta (Harkness and Ferguson, 1979). The esophageal musculature is generally thin and atrophied, with inflammation, necrosis, and myenteric ganglion cell loss (Ruben et al., 1983; Baiocco et al., 1993).

Younger rats are prone to intestinal impaction following ingestion of softwood bedding (Kohn and Barthold, 1984). In one report, death of 42.9% of suckling rats was attributed to fine particles in sawdust bedding causing obstruction at the ileocecal junction (Smith et al., 1968).

A dose-related gastric distension secondary to ingested bedding following oral or systemic administration of the analgesic buprenorphine has been described (Clark et al., 1995; Clark et al., 1997; Bender, 1998). The incidence of such pica has been estimated at 2% to 3%; the effects may commence within 30 minutes of administration and may be lethal (Jacobson, 2000). This behavior in rats may be a response to gastric irritation and nausea, similar to emesis in other species (Mitchell et al., 1976; Mitchell et al., 1977; Takeda et al., 1993). Use of the lowest effective dose and limiting access to readily ingestible bedding following administration of this opioid has been suggested (Flecknell et al., 1999; Jacobson, 2000; Jablonski et al., 2001; Roughan and Flecknell, 2002).

C. Trichobezoars

Trichobezoars, or hair balls, are generally less common in laboratory rats than in other species such as rabbits. Feeding a casein-based semi-purified diet to a Wistar-Kyoto (WKY/NHsd) rat colony resulted in a gastric trichobezoar prevalence of 100% (Krugner-Higby et al., 1996). These rats presented with nonspecific signs of illness, anorexia, and evidence of abdominal pain resolving after conversion to a standard laboratory feed. Gastric trichobezoar development in this colony may have had a genetic component, as Sprague-Dawley rats on the same feed failed to develop clinically significant hair balls. However, others have also reported an increased incidence of trichobezoars in Sprague-Dawley rats fed semi-synthetic diets compared to commercial feeds, perhaps related to ulcerative gastritis development (Anastasia et al., 1990). Calorie restriction increases the incidence of intestinal obstruction by trichobezoars, possibly due to increased grooming (Masoro et al., 1989), reaching an incidence of 8.4% in a Fischer-344 rat colony (Maeda et al., 1985).

D. Gastrointestinal Ulcerative Diseases

Punctate erosions and ulcerations are occasionally seen in the glandular stomach. More often, the nonglandular stomach is involved, and may present with larger, occasionally penetrating ulcerations. Gastric ulcer formation appears to be an age-related phenomenon (Anver and Cohen, 1979), with the incidence varying greatly with strain. Rates for gastric erosions and/or ulcerations have been documented to range between 0.6% (Goodman et al., 1979) and 1.7% (Biology Databook Editorial Board, 1985) for Fischer-344 rats, 4.1% (Goodman et al., 1980) and 4.9% (Biology Databook Editorial Board, 1985) for Osborne-Mendel rats, and 1.9% in Sprague-Dawley rats (Biology Databook Editorial Board, 1985). Such lesions have also been reported in WAG/Rij, BN/Bi/Rij, and their F1 hybrids (Burek, 1978).

Nonspecific stressors, such as those mimicked by immobilization, may result in gastric ulceration (Mikhail and Holland, 1966). Wright and colleagues (1981) reported a significant increase in multifocal gastric ulcers in diabetic BB/Wistar (Bb1:(WI)) rats—32.1% versus 9.7% in nondiabetic litter mates and none in Wistar controls. The incidence was greater in rats dying spontaneously, suggesting that uncontrolled diabetes led to the formation of stress ulcers.

Gastric ulceration is associated with both incidence and severity of chronic progressive nephropathy (Iwasaki et al., 1988). In one survey, 79% of the cases with gastric ulceration also had severe nephropathy; rats with less severe nephropathy had fewer gastric ulcers, suggesting a relationship with nephropathy-induced uremia (Maeda et al., 1985). Anastasia and colleagues (1990) described a positive correlation between gastric ulceration and feeding semi-synthetic diets, noting rats on such diets consumed more calories than those fed standard laboratory chow. Reduced gastric inflammation and ulceration was seen in dietary-restricted rats (Masoro et al., 1989; Shimokawa et al., 1993), which was likely due to reduced nephropathy associated with calorie restriction.

The rat is uniquely susceptible among rodents to gastrointestinal ulceration following nonsteroidal anti-inflammatory drug (NSAID) administration (Wilhelmi, 1974), leading to the species' use in determining the "ulcerogenic dose-50 (UD50)" of anti-inflammatory compounds (Liles and Flecknell, 1992). For example, a single oral dose of indomethacin results in jejunal epithelial injury within 3 to 6 hours (Nygård et al., 1994). Inadvertent overdose of rats with flunixin meglamine resulted in fatal perforating ulcers in the jejunum, ileum, and cecum within 4 days after administration (King and Miller, 1997). Rats administered nonphysiological or irritating solutions may also develop erosive or ulcerative conditions of the gastrointestinal tract. These inflammatory conditions can lead to scarring and strictures followed by obstructive lesions and death (Brors, 1987). In addition to its predilection for inducing peritonitis and adynamic ileus (see Injection-Induced Ileus and Peritonitis section in this chapter), the intraperitoneal or subcutaneous administration of chloral hydrate can also cause gastric ulceration (Ogino et al., 1990).

Multifocal ulcerative typhlocolitis has been reported in Chester Beatty Hooded Rowett nude rats (CBH-*rnu/rnu*) presenting with intermittent, mild diarrhea and subclinical hematochezia (Thomas and Pass, 1997). Although, the familial history suggested a heritable predisposition, an unidentified infectious etiology could not be excluded because the disease surfaced only after removal of the colony from microbiological isolator housing.

E. Intestinal Dilatation

The ceca in germfree rats can reach up to 5 times the normal size. This is due to the increased luminal osmotic pressure, disruption in the mucosal solute-water resorption mechanisms due to reduced ion availability and vasoactive compounds, and subsequent smooth muscle atony (Kohn and Barthold, 1984). Increased fluid and semi-solid ingesta buildup within the lumen leads to cecal wall thinning (Pollard and Kajima, 1970), predisposing older animals to volvulus and torsion (Kohn and Barthold, 1984). A 630° torsion at the ileo-cecal-colic junction was reported in two male, nonaxenic, Sprague-Dawley (Caw:CFE(SD)SPF) rats (Pollock and Hagan, 1972). Lipman and colleagues (1998) described a spontaneous mutation resulting in dilatation of the cecum and proximal colon (Familial Megacecum and Colon, FMC) heritable as an autosomal recessive disorder.

F. Injection-Induced Ileus and Peritonitis

Administrating various agents to rats via the intraperitoneal route is common, but not without complications (Lee et al., 1994). Intraperitoneal injection of several anesthetic agents can result in life-threatening sequelae. Chloral hydrate concentrations between 125 and 275 mg/ml can result in adynamic ileus, peritonitis, and gastric ulceration (Silverman and Muir, 1993). Clinical signs, including lethargy, anorexia, rough hair coat, constipation, or death, can occur up to 5 weeks after administration, and mortality may exceed 50% of the treated animals (Fleischman et al., 1977). Marked abdominal distension must be differentiated from Tyzzer's disease (Kohn and Barthold, 1984; Davis et al., 1985). Gross findings include segmental atony and distention of the small intestine and cecum, increased intestinal fragility, moderate ascites, and mild to moderate omental, mesenteric, and intestinal adhesions (Fleischman et al., 1977; Davis et al., 1985; Fig. 15-4). Although use of lower concentrations can reduce ileus and inflammatory response to chloral hydrate (Vachon et al., 2000), even 40 mg/ml may produce mild to moderate serositis (Spikes et al., 1996).

Intraperitoneal (IP) tribromoethanol administration, even at recommended levels of 300 mg/kg, can yield intestinal distension, hepatomegaly, peritonitis, and hepatic capsular fibrosis up to 8 days following injection (Reid et al., 1999). Fibrous adhesions and ileus from tribromoethanol can result in volvulus (Spikes et al., 1996). Although to a lesser extent, the IP pentobarbital injection can also result in splenomegaly and moderate peritonitis (Feldman and Gupta, 1976; Spikes et al., 1996).

G. Hepatic and Exocrine Pancreatic Anomalies

Hepatodiaphragmatic nodules (HDN) are congenital anomalies resulting from a small mass of liver protruding through, but not perforating, the thin central diaphragmatic

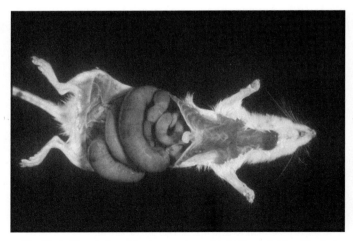

Fig. 15-4 Ileus induced by intraperitoneal injection of chloral hydrate. (*Image courtesy of Dr. S.W. Barthold*)

tendon into the thoracic cavity (Eustis et al., 1990). Displacement usually involves the median or left lateral hepatic lobes (Hall et al., 1992). Although HDN is uncommon in most rats, it may occur in 1% to 11% of Fischer-344 rat colonies (Eustis et al., 1990).

Several degenerative hepatic lesions have been described in laboratory rats. Few, if any, are likely of clinical significance, but may be important in the recognition of experimentally induced changes. These lesions include telangiectasia, periportal hepatocellular vacuolation, hepatocellular necrosis, polyploidy, intranuclear cytoplasmic invagination, intracytoplasmic inclusions, focal sinusoidal dilation, spongiosis, peliosis, areas of hepatocellular alteration, extramedullary hematopoiesis, lipidosis, hepatic cysts, and focal granulomas (Eustis et al., 1990; Keenan et al., 1995a; Percy and Barthold, 2001). Calorie restriction decreases the incidence and severity of many degenerative hepatic lesions (Masoro et al., 1989; Keenan et al., 1994).

Bile duct hyperplasia and hepatic cholangiofibrosis are common findings in aging rats (Squire and Levitt, 1975; Eustis et al., 1990). Incidence rates vary with age, strain or stock, gender, and source, and have been reported in 24.5% to 98.3% and 12.5% of male and female Fischer-344 (Coleman et al., 1977; Goodman et al., 1979), 13.4% and 13.1% of male and female Osborne-Mendel (Goodman et al., 1980), and 51.9% of male Sprague-Dawley (Crl:CD(SD)BR) rats (Cohen et al., 1978). Cholangiofibrosis is uncommon in WAG/Rij and BN/Bi/Rij strains (Burek, 1978). Hyperplastic ductules may become dilated and associated with marked fibrosis, often resulting in the appearance of cirrhosis (Percy and Barthold, 2001).

Several degenerative changes are also seen in the pancreas. One curious finding is the presence of pancreatic hepatocytes. These cells, which are microscopically and ultrastructurally identical to hepatocytes, occur spontaneously in rats (Reddy et al., 1984). Increased pancreatic hepatocyte production can be induced by chemical compounds such as 2,6-dichloro-p-phenylenediamine (McDonald and Boorman, 1989), ciprofibrate (Reddy et al., 1984; Rao et al., 1986), and by dietary copper depletion (Rao et al., 1986).

Exocrine pancreatic acinar atrophy and fibrosis is also a common finding in aging rats. As with most degenerative rat lesions, the incidence varies with age, gender, and strain or stock. Rates in Osborne-Mendel rats 109 to 126 weeks of age have been reported as 2.1% and 0.2% in males and females, respectively (Goodman et al., 1980). In Fischer-344 rats, the incidence has ranged from 4.9% to 6.9% in male and 3.8% to 5.0% in females of various ages (Dixon et al., 1995; Goodman et al., 1979; Coleman et al., 1977). Up to 40% of 12- to 39-month-old male Sprague-Dawley (Crl:CD(SD)BR) rats may be affected (Cohen et al., 1978).

VI. URINARY SYSTEM

A. Chronic Progressive Nephropathy

The development of end-stage renal disease is the most common age-related, degenerative condition of laboratory rats and is considered the primary etiology of non-neoplastic mortality (Kohn and Barthold, 1984; Ettlin et al., 1994; Donnelly, 1997; Percy and Barthold, 2001). It has been noted that other pathological findings are of minor significance by comparison (Owen and Heywood, 1986). Its prevalence in rat colonies has garnished numerous descriptions and terms for this condition, including old rat nephropathy, protein overload nephropathy, chronic renal disease, chronic nephritis, dietary nephritis, glomerulosclerosis, glomerulonephritis, progressive glomerulonephrosis, glomerular hyalinosis, progressive renal disease, chronic progressive glomerulonephropathy, spontaneous nephrosis, and chronic progressive nephrosis, with some of the terms alluding to the various lesions that develop and others to the condition's progression (Coleman et al., 1977; Barthold, 1979; Solleveld and Boorman, 1986; Rao et al., 1993; Percy and Barthold, 2001). The term *chronic progressive nephropathy* (CPN) has received widespread use, and will be used in this description.

Although lesions can be evident as early as 3 months (Harkness and Wagner, 1995), they do not usually become severe until over 52 weeks of age (Percy and Barthold, 2001). CPN is more frequently associated with males than females (Kohn and Barthold, 1984; Owen and Heywood, 1986; Ettlin et al., 1994; Percy and Barthold, 2001). In one study, CPN-associated mortality occurred from 82 to 188 weeks of age in males, and 85 to 124 weeks in females (Ettlin et al., 1994). Although the progression of lesions are similar, the reported age of onset, overall incidence rates, and severity varies significantly with stock/strain (Gray et al., 1982b; Goldstein et al., 1988; Donnelly, 1997; Table 15-3). Albino stocks and strains may be more inclined to develop CPN (Kohn and Barthold, 1984); axenic rats are relatively resistant (Kohn and Barthold, 1984; Harkness and Wagner, 1995).

Although numerous pathogenic factors are implicated, essentially any nephrotoxic insult eventually results in CPN. Subtle nephrotoxins can be identified by their ability to augment the onset or severity of the condition (Alden and Frith, 1991). Prolactin administration increases CPN incidence and severity; prolactin secretion by or in association with adrenal or pituitary tumors has been suggested as the rationale for increased CPN seen in these cases (Richardson and Luginbïhl, 1976). Facets of the humoral (Couser and Stilmant, 1975; Elema and Arends,

TABLE 15-3

STOCK/STRAIN, GENDER, AND AGE DISTRIBUTION OF LABORATORY RATS WITH MODERATE OR SEVERE CHRONIC PROGRESSIVE NEPHROPATHY

| | Incidence (%)* of Chronic Progressive Nephropathy | | | | | | | | | | | |
| | 0–26 Weeks of Age | | 26–51 Weeks of Age | | 52–77 Weeks of Age | | 78–104 Weeks of Age | | 104–129 Weeks of Age | | >130 Weeks of Age | | |
Stock/ Strain	Male	Female	Male	Female	Male	Female	Male	Female	Male	Female	Male	Female	Reference
Fischer-344	90.6	19.1	—	—	—	—	—	—	—	—	—	—	Dixon et al., 1995
	0.0		7.7		25.0		82.5		97.9		100.0		Coleman et al., 1977
	—	—	—	—	—	—	—	—	66.5	38.9	—	—	Goodman et al., 1979
	—	—	38.7	10.8	—	—	—	—	—	—	—	—	Hall et al., 1992
	0.0	—	0.0	—	0.0	—	88.6	—	100.0	—	100.0	—	Maeda et al., 1985[1]
	0.0	—	20.0	—	80.0	—	100.0	—	100.0	—	—	—	Maeda et al., 1985[2]
	—	—	—	—	—	—	—	—	99.0	86.0	—	—	Solleveld and Boorman, 1986
Osborne- Mendel	—	—	—	—	—	—	—	—	100.0	87.0	—	—	Solleveld and Boorman, 1986
	—	—	—	—	—	—	—	—	75.7	36.0	—	—	Goodman et al., 1980
ACI	—	—	—	—	—	—	—	—	99.0	99.0	—	—	Solleveld and Boorman, 1986
A28807	—	—	—	—	—	—	—	—	93.0	92.0	—	—	Solleveld and Boorman, 1986
M520	—	—	—	—	—	—	—	—	96.0	100.0	—	—	Solleveld and Boorman, 1986
Sprague- Dawley	7.0	3.0	33.0	11.0	60.0	18.5	75.0	38.0	100.0	44.0	—	—	Owen and Heywood, 1986
	—	—	—	—	—	—	—	—	91.4	55.7	—	—	Keenan et al., 1995b

ACI = August Copenhagen Irish; CPN = chronic progressive nephropathy.

[1]CPN listed as "cause of death or moribund state."

[2]Animals submitted for scheduled necropsy.

*Age distributions were not consistent in all references cited. Some references may have stated average age rather than age ranges; such values were approximated in the divisions employed this table. It should also be noted that some references may have used smaller age ranges, or even more divisions among age ranges; these were included with similar age groups for ease of comparison.

1975; Bolton et al., 1976) and cellular (Bolton et al., 1976) immunological system have also been considered.

Dietary factors significantly impact CPN progression. While the effects of some components such as mineral, lipid, and carbohydrate content appear to conflict or play a minor role (Klahr et al., 1983; Iwasaki et al., 1988; Tapp et al., 1989; Shackelford and Jokinen, 1991), others such as protein composition and total caloric content are more significant (Bras, 1969; Alden and Frith, 1991). The relative importance of these two constituents has been debated for over 3 decades.

Diets relatively high in protein contribute to the severity of CPN (Lalich and Allen, 1971; Rao et al., 1993; Saxton and Kimball, 1941; Blatherwick and Medlar, 1937). Rats consuming a diet of 35% protein suffer more marked nephron insult than those fed 20% protein; further protein reduction to 6% essentially prevented CPN (Bertani et al., 1989). Protein content reduction from 23% to 15% significantly reduces the severity of lesions, with or without calorie restriction (Rao et al., 1993). Such findings suggest limiting total dietary protein content in rats to 4% to 7% (Donnelly, 1997). Reducing dietary protein content in aging Fischer-344 rats by time-restricted feeding also decreases CPN development over *ad libitum* feeding (Maeda et al., 1985). The protein source is also important, as diets using casein or lactalbumin as the sole protein source are more nephropathogenic than others, such as soy (Saxton and Kimball, 1941; Klahr et al., 1983; Iwasaki et al., 1988; Donnelly, 1997; Shimokawa et al., 1993).

Similarly, calorie restriction greatly reduces the incidence and severity of CPN (Saxton and Kimball, 1941; Bras and Ross, 1964; Shimokawa et al., 1993; Keenan et al., 2000). Several investigations indicate that calorie restriction has more impact than protein restriction alone (Masoro and Yu, 1989; Masoro et al., 1989; Gumprecht et al., 1993). Regardless of protein intake, feed restriction delays the onset and reduces the severity and development rate of CPN in aging rats and in experimental models in which renal compromise is induced by renal ablation (Tapp et al., 1989; Maeda et al., 1985). Masoro and colleagues (1989) demonstrated that CPN is less severe in diet-restricted rats even though total protein intake was shown to be 70% greater than groups fed *ad libitum*. Keenan et al. (1995a) found that protein-restricted diets fed *ad libitum* has no effect on the incidence of CPN, and nephropathy decreases only when food intake was restricted as well. Dietary restriction initiated at weaning or in young adult rats must continue throughout adulthood to effectively reduce the incidence of CPN; when ceased at 6 months of age, dietary restriction has no effect (Maeda et al., 1985).

A current hypothesis on CPN pathogenesis identifies glomerular hypertrophy as the initial pathological consequence of *ad libitum* feeding (Gumprecht et al., 1993; Keenan et al., 2000). Caloric overload produces hemodynamic and/or glomerular hydrostatic pressure alterations, leading to mesangial and glomerular epithelial trauma and proliferation (Fries et al., 1989; Keenan et al., 1995a). This is followed by expansion of the mesangial matrix, basement membrane thickening, and further endothelial damage, leading to glomerulosclerosis, nephron loss, and progressive protein hyperfiltration and tubular damage – the hallmarks of CPN (Brenner, 1985; Fogo and Ichikawa, 1989; Gumprecht et al., 1993; Keenan et al., 1995a).

Clinical signs are usually inapparent until renal decompensation and uremia cause weight loss and lethargy (Percy and Barthold, 2001). Polydipsia and palpably enlarged kidneys have been used as indicators of CPN (Spangler and Ingram, 1996). Serum urea nitrogen and creatinine values are usually unchanged until disease nears end stage (Everitt, 1958; Coleman et al., 1977; Maeda et al., 1985; Gumprecht et al., 1993). Proteinuria has been considered the classical indicator of CPN (Peter et al., 1986; Harkness and Wagner, 1995; Palm, 1998). This finding, which has lead to the moniker "protein leakage disease" (Gray, 1986), likely results from an altered composition of the glomerular basement membrane (Abrass, 2000). Perry (1965) reported the change in protein excretion from 25 to 100 mg/dL at 8 weeks of age to 1000- to 3000 mg/dL at 12 to 24 months in Wistar rats. In Sprague-Dawley rats, proteinuria levels can rise from 10 mg/day in early stages to 137 mg/day in males and 76 mg/day in females at 18 months of age (Short and Goldstein, 1992), and may exceed a loss of 280 mg/day (Neuhaus and Flory, 1978). Urine protein concentrations measured in potential sires at 4 months of age can be predictive of CPN and CPN-related lifespan of offspring; this index may be used to select breeding stock for producing progeny with less severe disease (Gray et al., 1982a). In late stages, the urinary protein profile transforms from the normal α-globulinuria towards a semblance of serum (Perry, 1965; Kohn and Barthold, 1984). Sequential examination of urine specific gravity reveals decreased concentrating ability (Owen and Heywood, 1986); urine concentration must be considered if protein levels are measured to account for diuresis (Barthold, 1979). When disease approaches end stage, rats may develop signs consistent with nephrotic syndrome: hypoproteinemia, azotemia, and hypercholesterolemia (Barthold, 1979; Percy and Barthold, 2001). Compensatory mechanisms persist until very late stage disease with continued nephron loss, eventually resulting in rapid decompensation and death (Harkness and Wagner, 1995).

Postmortem examination typically reveals enlarged kidneys with an irregular, often pitted surface (Short and Goldstein, 1992; Fig. 15-5). Coloration may vary from pale to yellow with mottling and varying amounts of brownish pigmentation (Owen and Heywood, 1986; Harkness and Wagner, 1995). Grossly visible streaks or striations may be appreciated on cut surface (Percy and Barthold, 2001). Small cystic structures may be present in the renal cortex (Anver and Cohen, 1979).

The histological appearance of CPN depends on the disease's chronicity, with severity judged by the proportion of nephrons affected (Gray et al., 1982b; Fig. 15-6). Fig. 15-7 provides three scales used to grade lesions according to severity. Mild CPN is characterized by one

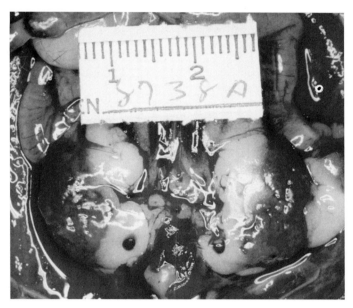

Fig. 15-5 Chronic progressive nephropathy. Note pale, mottled, and dimpled surface of renal capsule. (*Image courtesy of Dr. J.R. Lindsey*)

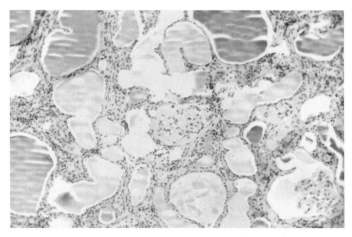

Fig. 15-6 Chronic progressive nephropathy, photomicrograph (100×). Note glomerulosclerosis and tubular distension with proteinaceous fluid.

Grade	Reference: Yu et al., 1982.	Grade	Reference: Coleman et al., 1977.	Grade	Reference: Rao et al., 1993.
1	Minimal severity -- primarily involves glomerular capillary basement membrane and mesangial matrix; an occasional hyaline cast.	1	Thickened glomerular capillary basement membranes and slight mesangial thickening in some glomeruli; a few cortical tubules shrunken, with thickened, wrinkled basement membranes, and lined by enlarged cells containing basophilic cytoplasm and large nuclei.	1	Minimal -- less than 20% of the cortex and outer medulla affected; a few small foci of tubular cell degeneration and regeneration in the cortex and outer medulla; occasional scattered tubules containing eosinophilic proteinaceous casts.
2	Mild severity – involves glomerular basement membrane and mesangial matrix; tubular proteinaceous casts invariably.	2	Grade 1 findings plus scattered dilated tubules lined by atrophic epithelium with occasional hyaline casts, particularly at the corticomedullary junction.	2	Mild -- approximately 20-50% of the cortex and outer medulla affected; focal areas of tubular cell regeneration and degeneration generally larger and more frequent than Grade 1; thickened peritubular basal lamina in some regenerative tubules; eosinophilic proteinaceous casts in cortical and outer medullary tubules, some tubular distention with casts.
3	Moderate severity -- Grade 2 findings but more extensive, plus thickening of Bowman's capsule, lymphocyte infiltration, mild interstitial fibrosis.	3	Grade 2 findings plus prominent protein casts within cortical, medullary, and papillary tubules; glomerular lesions more pronounced with atrophy of capillary tufts, sclerosis, thickening of Bowman's capsule, and tubular basement membranes.	3	Moderate -- approximately 50-75% of the cortex and outer medulla affected; tubular cell degeneration and regeneration, thickening of the basal lamina, and proteinaceous casts more prominent than in Grade 2; large number of tubulare lumen distended with eosinophilic protein casts; some glomeruli with mesangial proliferation and adhesions of the glomerular tuft to the wall.
4	Very severe -- Grade 3 findings but more marked, plus segmental or diffuse glomerular sclerosis; frequent adhesion of glomerular tuft to Bowman's capsule.	4	Grade 3 findings, but more pronounced; adhesions of glomerular tufts to Bowman's capsule; enlarged nuclei in the tunica media and proliferation of adventitial connective tissue in afferent arterioles of the severely affected glomeruli.	4	Marked -- more than 75% of the cortex and outer medulla affected; foci of tubular cell regeneration and degeneration more numerous and often merged; peritubular basal lamina thickened; frequent interstitial fibrosis and inflammation; many tubules distended with proteinaceous casts; glomerular lesions (adhesions, sclerosis and atrophy of glomerular tuft, etc.), more extensive than in Grade 3.
E	End-stage -- widespread glomerular sclerosis, obsolescence of glomeruli, diffuse interstitial fibrosis, frequent calcification, marked tubular dilation with numerous proteinaceous casts.	End-stage	No normal parenchyma remaining, with widespread glomerular sclerosis, marked tubular dilation, atrophy and hyaline cast formation, and interstitial fibrosis.		

Fig. 15-7   Three scales used to grade the histological appearance and severity of chronic progressive nephropathy in laboratory rats.

to several multifocal cross sections of basophilic tubules undergoing varying degrees of regeneration (Dixon et al., 1995). Glomerular hypertrophy and basement membrane thickening progresses until the basement membranes of the glomerulus, Bowman's capsule, and proximal tubule fracture and wrinkle; proximal tubules degenerate and collapse (Kohn and Barthold, 1984). Within affected glomerular tufts, there is mesangial proliferation, adhesion to Bowman's capsule, and segmental sclerosis (Percy and Barthold, 2001). Proximal tubules become dilated with copious eosinophilic casts, and eosinophilic, periodic acid-Schiff (PAS) positive droplets (lysosomes) are demonstrable in the tubular epithelial cells (Christensen and Madsen, 1978; Percy and Barthold, 2001). Tubules eventually atrophy, which may be accompanied by interstitial fibrosis and inflammatory infiltration (Kohn and Barthold, 1984).

Renal secondary hyperparathyroidism is frequently observed with chronic CPN. Itakura et al. (1977) described this sequela in 15.6% (24.4% males, 6.7% females) of aged Sprague-Dawley rats. All parathyroid glands in rats with end-stage CPN may be hyperplastic (Maeda et al., 1985), leading to osteodystrophy with fibroplasia, and osteoid formation of the femur, humerus, vertebrae, scapula, mandible, and parietal bone, and metastatic calcification in the kidney, gastrointestinal tract, lungs, and arterial media (Durand et al., 1964; Itakura et al., 1977; Owen and Heywood, 1986; Masoro et al., 1989; Percy and Barthold, 2001).

Treatment may be unrewarding and is palliative, such as hydration (Harkness and Wagner, 1995). A low-protein diet and anabolic steroid use have been advocated per other severe renal disorders (Donnelly, 1997). Prevention involves a careful balance of the nutritional requirements for sustaining health while adjusting for the deleterious effects of excess dietary calories and protein.

A high rate of PAN correlated with CPN in a mutant colony derived from Sprague-Dawley rats, termed *juxtaglomerular index* (JGI), that develop age-related hypertension, suggests a relationship between these age-related disorders (Rapp, 1973). Similarly, the Fawn-hooded rat, another mutant that develops hypertension at an early age, may develop CPN at 3 months (Alden and Frith, 1991), related most likely to the development of glomerular hypertension (Provoost, 1994). Systemic hypertension may not be necessary, however, as glomerular capillary pressure and its effects on the nephron are independent of systemic blood pressure (Anderson et al., 1986). As an additional correlation with a disorder affecting the cardiovascular system, the reduction of CPN following dietary restriction has also been shown to reduce cardiomyopathy (Keenan et al., 1995a). Interestingly, early lesions of CPN also resemble glomerular lesions associated with diabetic glomerulopathy (Velasquez et al., 1990). Thus, it is tempting to suggest an inter-relationship between two or more

of these processes in older rats. Alternatively, the rat kidney might simply be exquisitely sensitive to alterations in systemic and local physiology, with a variety of insults initiating a cascade of events resulting in lesions classically recognized as CPN.

B. Nephrocalcinosis

Nephrocalcinosis, or intranephronic calculosis, are terms used for renal mineralization, frequently seen in female rats (Nguyen and Woodard, 1980; Kohn and Barthold, 1984). Lesions occur in young rats as early as 3 to 5 weeks of age (Hall et al., 1992). The lesions' predilection for females suggests a hormonal influence. Ovariectomy reduces the development of mineralization while estrogen administration to either gender induces nephrocalcinosis (Cousins and Geary, 1966; Geary and Cousins, 1969). Some strain differences exist (Ritskes-Hoitinga et al., 1989; Du Bruyn, 1970), with one report indicating up to 77% of female Fischer-344 rats between 19 and 21 weeks of age affected (Dixon et al., 1995).

The appearance of nephrocalcinosis has historically been related to feeding semi-synthetic or purified diets, although rats fed standard laboratory feed also develop lesions (Hitchman et al., 1979; Nguyen and Woodard, 1980; Clapp et al., 1982; Ritskes-Hoitinga et al., 1989; Percy and Barthold, 2001). The severity of lesions is greater in casein-based semi-synthetic diets than in those using lactalbumin (Meyer et al., 1989) or soya protein (Anastasia et al., 1990). Diets with a low concentration of magnesium, high concentrations of either calcium or phosphorus, or a low calcium-to-phosphorus ratio may induce typical lesions (Ritskes-Hoitinga et al., 1989; Cockell et al., 2002). Feeding diets with high protein levels (Van Camp et al., 1990) or with calcium-to-phosphorus ratios approximating 1.2:1 (Clapp et al., 1982; Hoek et al., 1988; Phillips et al., 1986; Ritskes-Hoitinga et al., 1991) assist in preventing lesions. Mineral content and chemical formulations affecting nutritional availability, e.g., solubility and carrier binding, are also involved (Woodard, 1971; Clapp et al., 1982; Anastasia et al., 1990). Dietary chloride, water, lipid, and protein may also play a role (Nguyen and Woodard, 1980; Alden and Frith, 1991). There is some controversy on whether dietary changes can induce nephrocalcinosis regression, although reported inconsistencies might reflect analytical method differences and the approaches used to induce the disease (Grimm et al., 1991; Soeterboek et al., 1991; Beynen, 1992).

Microscopically, intratubular lithiasis appear as basophilic lamellar deposits of calcium phosphates, or hydroxyapatite (Nguyen and Woodard, 1980) typically occurring in proximal tubules at the corticomedullary junction (Clapp et al., 1982; Kohn and Barthold, 1984; Dixon et al.,

1995; Percy and Barthold, 2001). Since standard histological stains may not identify mineral deposition, specific stains such as Von Kossa's or Alizarin Red S are suggested (Alden and Frith, 1991). Tubular hyperplasia and fibrosis may also be seen (Ritskes-Hoitinga et al., 1989). Clinical evidence of renal insufficiency, such as albuminuria, may only be present in advanced cases of nephrocalcinosis (Ritskes-Hoitinga et al., 1989). Behavior and clinical index analysis in an experimental model suggests that the disorder is not overtly painful (Soeterboek et al., 1991).

C. Urolithiasis

The incidence of spontaneous calculi formation in the urinary tract of laboratory rats appears to be quite variable (Table 15-4). Urolithiasis may be more common in males because of the length and relative rigidity of the urethra (Harkness and Wagner, 1995).

Hematuria, hemorrhagic cystitis, and/or anuria may be associated with urinary stones (Sharp and LaRegina, 1998). Uroliths may be discovered in the urinary bladder, renal pelvis, ureter, and/or urethra, and can result in a thickened bladder wall, ureteral distention, and hydronephrosis (Kuhlmann and Longnecker, 1984; Paterson, 1979; Percy and Barthold, 2001; Bingel, 2003; Fig. 15-8). Ammonium magnesium phosphate (struvite) calculi ranged in size, from minute to over 5 g, in one report of Sprague-Dawley rats (Paterson, 1979). Surgical extraction of urinary bladder calculi and treatment for secondary cystitis may be curative in some cases (Harkness and Wagner, 1995).

Microscopic evidence of epithelial hyperplasia and inflammation may be seen in association with urolith

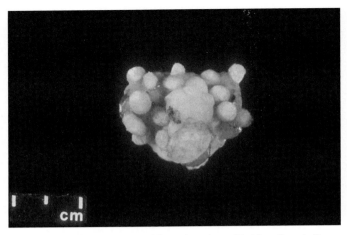

Fig. 15-8 Uroliths recovered from the urinary bladder of a rat. (*Image courtesy of Dr. S.W. Barthold*)

location (Kuhlmann and Longnecker, 1984). Urolithiasis and crystalluria may be involved in the pathogenesis of uroepithelial proliferative lesions (Boorman and Hollander, 1974; Alden and Frith, 1991). Conversely, epithelial hyperplasia was considered an initiating event in the struvite stone formation in a Sprague-Dawley rat colony, with degenerating epithelial cells providing a nidus for urolith formation (Magnusson and Ramsay, 1971).

Calculi of struvite, carbonate, phosphate, citrate, and oxalate, and mixtures of several of these minerals have been described. Struvite calculi in Lewis and Wistar rats followed dietary retinoid supplementation, which may have disrupted vitamin A metabolism (Kuhlmann and Longnecker, 1984). When combined with lactose as the primary carbohydrate source, vitamin A deficiency has been shown

TABLE 15-4

STOCK/STRAIN AND GENDER DISTRIBUTION OF UROLITHIASIS IN LABORATORY RATS

| Stock/Strain | Incidence (%) of Urolithiasis | | | Reference |
	Male	Female	Total (or Gender Not Specified)	
Fischer-344	0.2	0.2	—	Goodman et al., 1979
	0.3	0.3	—	Biology Databook Editorial Board, 1985
Osborne-Mendel	0.9	0.0	—	Biology Databook Editorial Board, 1985
Sprague-Dawley[1]	—	—	0.5	Paterson, 1979
Sprague-Dawley[2]	7.8	0.3		Biology Databook Editorial Board, 1985
Sprague-Dawley[3]	—	—	18.1	Anver et al., 1982
Sprague-Dawley[2]	31.8	18.2	25.0	Magnusson and Ramsay, 1971

[1]COBS.
[2]Designation not specified .
[3]Hap:(SD).

to predispose to weddellite (calcium oxalate dihydrate) or apatite (calcium phosphate) lithogenesis (Gershoff and McGandy, 1981). Other predisposing factors include genotype, metabolic or nutritional imbalances, and dehydration (Harkness and Wagner, 1995). Reportedly, germfree male rats are more susceptible to urolithiasis, with 50% of male and 2% of female germfree rats diagnosed with calcium citrate and calcium oxalate urinary bladder calculi; administering gut microflora eliminated the increased incidence of stone formation (Gustafsson and Norman, 1962). The authors attributed this predisposition to increased urinary calcium, citrate, and pH in germfree animals. Table 15-5 summarizes the composition and suspected predisposing factors involved in urolithiasis of laboratory rats.

It is important to distinguish between uroliths and copulatory plugs. Copulatory plugs, also known as urethral plugs, proteinaceous plugs, bladder plugs, mucoid calculi, or soft calculi, are frequently seen in the urethra or urinary bladder following retrograde ejaculation in male rats (Lee, 1986). These structures are generally regarded as an agonal change (Percy and Barthold, 2001); however, some regard them as normal structures in healthy rats and absent only in ill animals (Kunstýř et al., 1982). Proteinaceous casts were found within the urinary bladder in 1.8% of Sprague-Dawley males up to 12 months of age, with no mention of casts in Fischer-344 or Osborne-Mendel rats (Biology Databook Editorial Board, 1985). Although their relationship to urolith formation is unclear, calcium carbonate calculi may be found incorporated into copulatory plugs in the urinary bladder of male rats (Paterson, 1979).

D. Hydronephrosis

Renal pelvis dilation is not an uncommon necropsy finding in many laboratory rats and must be distinguished from polycystic kidneys, pyelonephritis, and renal papillary necrosis (Percy and Barthold, 2001). Hydronephrosis can be easily visualized radiographically via excretory urogram (Cohen et al., 1970; Lozzio et al., 1967). Analysis of breeding data and the great variation of incidence in different strains and stocks attest to its heritable nature (see Table 15-6). The condition is thought to be congenital and inherited as an autosomal polygenic trait with incomplete penetrance in Brown Norway (BN/Bi; Cohen et al., 1970), Sprague-Dawley (Van Winkle et al., 1988), and ACI (Cramer and Gill, 1975) rats, and as an autosomal dominant lethal gene when homozygous in Wistar-derived Gunn rats (Lozzio et al., 1967). A subline of Wistar rats was developed by Friedman and colleagues (1979) that developed hydronephrosis in 95% and 60% of males and females, respectively.

Hydronephrosis is typically considered incidental and producing little renal dysfunction (Cohen et al., 1970), although it can be lethal when bilateral (Percy and Barthold, 2001). Some reports indicate a predilection for the right kidney (Burton et al., 1979). One hypothesis suggested constriction of the right ureter by the right internal spermatic artery and vein, but surgical resection of these vessels fails to prevent occurrence (O'Donoghue and Wilson, 1977). Accompanying ureteral dilation is inconsistent (Fujikura, 1970; Burton et al., 1979). Hydronephrosis has been associated with pyelonephritis, renal papillary congestion and hemorrhage, uroepithelial

TABLE 15-5

SPONTANEOUS UROLITH COMPOSITION AND POSSIBLE PREDISPOSING FACTORS IN LABORATORY RATS*

Composition	Location (in Descending Order)	Suggested Predisposing Factors	References
Struvite (magnesium ammonium phosphate)	Urinary bladder, ureter, renal pelvis	Dietary retinoids, vitamin A disturbance	Kuhlmann and Longnecker, 1984
	Renal pelvis	Degenerating hyperplastic uroepithelium	Magnusson and Ramsay, 1971
	Urinary bladder, renal pelvis	None	Paterson, 1979
Calcium carbonate	Urinary bladder, renal pelvis	None	Paterson, 1979
Calcium phosphate	Renal pelvis	None	Paterson, 1979
Mixed: calcium carbonate and struvite	Urinary bladder, renal pelvis, ureter	None	Paterson, 1979
Mixed: calcium phosphate and struvite	Urinary bladder, renal pelvis	None	Bingel, 2003
Mixed: calcium carbonate, phosphate, and struvite	Urinary bladder	None	Paterson, 1979
Mixed: calcium citrate and calcium oxalate	Urinary bladder	Germfree status	Gustafsson and Norman, 1962
Mixed: calcium carbonate and calcium oxalate	Urinary bladder	None	Paterson, 1979

*Sprague-Dawley rats were described in each report except Kuhlmann and Longnecker (1984), which involved both Lewis and Wistar rats. Stock/strain not identified in Gustafsson and Norman (1962).

TABLE 15-6

STOCK/STRAIN AND GENDER DISTRIBUTION OF HYDRONEPHROSIS IN LABORATORY RATS

| Stock/Strain | Incidence (%) of Hydronephrosis | | | Reference |
	Male	Female	Total (or Gender Not Specified)	
Fischer-344	2.0	2.0	—	Solleveld and Boorman, 1986
	1.1	0.3	—	Biology Databook Editorial Board, 1985
Osborne-Mendel	22.0	10.0	—	Solleveld and Boorman, 1986
	0.0	4.5	—	Biology Databook Editorial Board, 1985
Sprague-Dawley[1]	—	—	3.9	Anver et al., 1982
Sprague-Dawley[2]	—	—	13.9	Anver et al., 1982
Sprague-Dawley[3]	—	—	2.0	Van Winkle et al., 1988
Sprague-Dawley[4]	4.8	4.4	—	Biology Databook Editorial Board, 1985
Sprague-Dawley[5]	1.0	0.4	—	Owen and Heywood, 1986
August Copenhagen Irish (ACI)	48	41	—	Solleveld and Boorman, 1986
	7.9	5.3	—	Cramer and Gill, 1975
August (A28807)	3.0	9.0	—	Solleveld and Boorman, 1986
Marshall (M520)	34.0	63.0	—	Solleveld and Boorman, 1986
Wistar [Hla:(WI)BR]	13.9	4.6	—	Burton et al., 1979
Brown Norway (BN/Bi)	36.8	21.8	—	Cohen et al., 1970
Lewis x Brown Norway (LBN)	14.8	4.7	—	Treloar and Armstrong, 1993

[1]Crl:COBS[R]CD[R] (SD).
[2]Hap:(SD).
[3]Crl:CD[R](SD) BR.
[4]designation not specified.
[5]CD.

proliferative lesions, and renal pelvic urolithiasis (Maronpot, 1986; Solleveld and Boorman, 1986), although urolithiasis does not appear to precede hydronephrosis (Van Winkle et al., 1988). There may be a relationship between hydronephrosis and benign intermittent hematuria as seen in Lewis and Brown Norway rat crosses subsequent to rupture of well-vascularized proliferative renal pelvis masses (Treloar and Armstrong, 1993).

Renal agenesis is associated with hydronephrosis in the ACI rat. Unilateral renal agenesis has been described in 12% to 21% of male and 14% to 16% of female ACI rats, with a predilection for right kidney and concurrent absence or malformation of the ipsilateral adrenal gland and genital organs (Cramer and Gill, 1975; Solleveld and Boorman, 1986). Interestingly, hydronephrosis occurs in this strain at similar rates, but involving the left kidney (Fujikura, 1970; Cramer and Gill, 1975).

E. Miscellaneous Conditions

Rats are particularly prone to proximal tubular and renal papillary necrosis following administration of NSAIDs (Alden and Frith, 1991), including salicylate (Kyle and Koscis, 1985) and acetaminophen (Beierschmitt et al., 1986). Congenital lesions described include persistent urachus in three female Wistar rats (Borrás, 1983)

and ectopic kidney in a female Sprague-Dawley rat (Gupta, 1975).

VII. ENDOCRINE SYSTEM

Several endocrine gland lesions have been described in rats, especially in older animals. Few, if any, have known clinical significance, but are described as an introduction to the spontaneous conditions of these organs.

Several lines of the laboratory rat, including the BB, Goto-Kakisaki (GK), Otsuka Long-Evans Tokushima Lean (OLETL), Cohen, obese/SHR, Wistar fatty, SHR/N-cp, BHE, and Zucker fatty rat, have been propagated to serve as models for human diabetes mellitus, (Velasquez et al., 1990; Ktorza et al., 1997; Cheta, 1998). Less is known about the occurrence of this disorder in standard stocks and strains. Beta-cell hyperplasia is frequently seen in rats even as young as 3 months of age; with aging, rats may develop islet fibrosis and hyperglycemia, especially in males (Dillberger, 1994). Glucose metabolism alterations can be identified in rats without islet disease, although dysfunction becomes more severe when morphological changes occur (Anver and Cohen, 1979). Islet changes have also been related to *ad libitum* feeding (Keenan et al., 1994; Keenan et al., 1995a), which also likely predisposes rats to the development of islet cell neoplasia (Dillberger, 1994).

Accessory adrenal cortical nodules are frequently seen in aged rats (Hamlin and Banas, 1990). Such nodules were found in 1.3% of female and 1.9% of male Fischer-344 rats aged 19 to 21 weeks (Dixon et al., 1995). Sinusoidal dilatation and thrombi occur frequently, with up to 32.4% of some Sprague-Dawley colonies (Crl:COBS[R]CD[R](SD)) affected (Anver et al., 1982). Other lesions include focal cortical degeneration and adenoma-like cells in the zona fasciculata (Anver and Cohen, 1979).

Ultimobranchial cysts are a congenital thyroid gland lesion thought to originate from the ultimobranchial body (Hardisty and Boorman, 1990). These are often filled with debris and develop in approximately 8.9% of male and 9.7% of female 19- to 21-week-old Fischer-344 rats (Dixon et al., 1995).

Neoplasia is the most significant abnormality involving the pituitary gland, which in some reports is cited as the most frequent cause of laboratory rat mortality (Ettlin et al., 1994; Keenan et al., 1995b). Colloid cysts are frequently seen in the pituitary gland of aging rats. The incidence of cysts varies with age, stock/strain, and specific colony, and occurs in approximately 2% of most strains, including Fischer-344 and Osborne-Mendel rats (Coleman et al., 1977; Goodman et al., 1979; Goodman et al., 1980; Biology Database Editorial Board, 1985). The reported incidence in Sprague-Dawley colonies ranges from 1.5% (Biology Database Editorial Board, 1985), to 12.5% to 23.7% (Anver et al., 1982; Cohen et al., 1978).

Parathyroid hyperplasia is frequently seen in older rats, primarily as a sequela to chronic progressive nephropathy (Anver et al., 1982). A brief discussion of this condition can be found in Chronic Progressive Nephropathy section of this chapter.

VIII. NEUROLOGICAL SYSTEM

A. Hydrocephalus

Internal hydrocephalus appears to be a relatively rare phenomenon in laboratory rats. Reported incidences have ranged from 0.4% in Fischer-344 (Biology Databook Editorial Board, 1985) to 1.4% in Sprague-Dawley rats (Anver et al., 1982). One subline of an unknown stock was reported to develop congenital hydrocephalus in 50% of the offspring secondary to poorly developed dural venous and pia-arachnoid tissues (Kohn et al., 1981).

B. Radiculoneuropathy

Radiculoneuropathy, also known as spinal nerve root degeneration or degenerative myelopathy, is a long-recognized degenerative disease of aging rats (Anver and Cohen, 1979). Lesions are not usually apparent until rats reach 18 to 20 months of age (Gilmore, 1972; Van der Kogel, 1977), but may affect 80% to 100% in rats surpassing 2 years of age (Berg et al., 1962; Anver et al., 1982). A gender predisposition is unclear (Berg et al., 1962; Burek et al., 1976; Gilmore, 1972). Examination of initial peripheral nerve degeneration suggests an earlier onset in males than females (Cotard-Bartley et al., 1981).

This condition has been reported in a number of stocks and strains, including Sprague-Dawley, Wistar, WAG/Rij, Brown Norway, and Fischer-344 colonies (Krinke et al., 1981; Biology Databook Editorial Board, 1985). Because of the number of clinical cases of posterior paresis and paralysis, Cohen and colleagues (1978) more carefully examined the nervous system of 30- to 39-month-old Sprague-Dawley [Crl:CD(SD)BR] rats, and discovered radiculoneuropathy in 12% and 95% of the cervical/thoracic and lumbar/sacral nerves, respectively. Although incidence rates are approximately 2% in inbred BN/Bi/Rij and WAG/Rij strains, the F1 hybrids of these strains develop lesions in 21% of the offspring (Burek et al., 1976).

The principal clinical signs of radiculoneuropathy are slowly advancing rear limb ataxia or paralysis, leading to gradual wasting of rear limb musculature, weight loss, and lethargy (Anver and Cohen, 1979; Witt and Johnson, 1990). Other clinical signs include urinary incontinence and constipation (Anver et al., 1982). Hind limb neurological assessment may reveal proprioceptive defects (Witt and Johnson, 1990).

Although lesions are generally difficult to appreciate grossly, the key histological feature is swelling of the myelin sheath and demyelination (Krinke et al., 1981). Early in the course of development, axonal atrophy is most commonly seen in distal nerve sections (Kazui and Fujisawa, 1988). Segmental demyelination and remyelination may be disseminated, occurring most regularly in the lumbar ventral spinal roots (Krinke et al., 1981; Anver et al., 1982; Kazui and Fujisawa, 1988). Lesions may be prominent in peripheral nerves such as the sciatic (Van Steenis and Kroes, 1971; Witt and Johnson, 1990) and tibial (Cotard-Bartley et al., 1981) nerves, and may also develop in the cauda equina, spinal column white matter, lower brainstem, and, rarely, the dorsal spinal nerve roots (Berg et al., 1962; Gilmore, 1972; Burek et al., 1976). Evidence of Wallerian-type degeneration and myelinated fiber loss is seen in the distal segments of affected nerves (Mitsumori et al., 1981; Mitsumori et al., 1986).

Theories regarding the pathogenesis of radiculoneuropathy include obesity-related pressure resulting in trauma, chronic progressive nephropathy, and hyperglycemia, although radiculoneuropathy can be found in rats not exhibiting any of these concurrent disorders (Krinke, 1983; Witt and Johnson, 1990). An association with vertebral disk degeneration, protrusion, and rupture, with subsequent spinal cord compression, was suggested by Burek et al. (1976), who also described vertebral bone aseptic necrosis in rats displaying posterior paralysis.

A case of intervertebral disk disease resulting in posterior paralysis was noted in a 12- to 18-month-old Fischer-344 rat (Coleman et al., 1977).

C. Traumatic and Toxic Diseases

A relatively high incidence (1.5%) of spontaneous clonic convulsions was described in a Wistar rat colony (Crl:(WI) BR) over 16 weeks of age (Nunn and Macpherson, 1995). Rats may present with neurological signs following trauma to the peripheral or central nervous system. Differential diagnoses include encephalitis, Rat Virus infection, otitis interna, arthritis, and pituitary adenoma or other lesions compressing nervous tissue (Harkness and Wagner, 1995).

Lethal organophosphate toxicity following accidental exposure to an agricultural insecticide was described in Long-Evans rats presenting with muscle fasciculations, depression, exophthalmos, and ptyalism (Gibson et al., 1987). Careful selection of bedding, with attention to vendor assurances with regard to contaminants such as organophosphates, chlorinated hydrocarbons, aflatoxins, and heavy metals is prudent (Kohn and Barthold, 1984).

IX. OCULAR SYSTEM

A. Glaucoma/Buphthalmos

Primary, or congenital, glaucoma is rare in rats (Goldblum and Mittag, 2002), but has been described in Wistar rats (Heywood, 1975). Secondary glaucoma occurs with regularity in association with hypermature cataract formation (Wegener et al., 2002). Persistent papillary membranes incited unilateral and bilateral buphthalmos in up to 15.3% of a WAG rat colony (Young et al., 1974). Glaucoma was a sequela to intraocular inflammatory disease in a report involving Sprague-Dawley rats (Heywood, 1975).

B. Conjunctival and Lacrimal Diseases

Chromodacryorrhea is an easily appreciated clinical sign in rats, and therefore frequently reported. It is characterized by reddish secretions or staining around the eyes and external nares (chromodacryorhinorrhea), may be incorrectly reported as a "bloody nose" or "bleeding around the eyes," and often accumulates on the forepaws during grooming. The pigments are porphyrins secreted from the harderian gland, and can be differentiated from blood by their orange-red fluorescence when viewed under ultraviolet light. Chromodacryorrhea may result from any condition producing distress in laboratory rats. Although not induced by epinephrine or corticosterone administration, stressors against which rats cannot cope, such as immobilization, overcrowding, and infection, can result in chromodacryorrhea within a few minutes (Harkness and Ridgway, 1980). Identification and treating the underlying cause is necessary for alleviation.

Harderian gland dacryoadenitis has also been described. In one study, marked glandular necrosis with cellular infiltrates and edema was noted after as little as 12 hours of exposure to high intensity (2500 lux) light, with the damage hypothesized to result from porphyrins' photodynamic properties (Kurisu et al., 1996). Earlier studies have also noted a causal link between uninterrupted light and harderian gland damage (Johnson et al., 1979; O'Steen et al., 1978).

Dacryoadenitis was described as a sequela to blood sample retrieval via orbital venous plexus puncture (McGee and Maronpot, 1979). Focal hemorrhages can be identified along the penetration path, including within the harderian gland, periocular musculature, and orbital periosteum; damage and subsequent contraction of ocular muscles can cause enophthalmos (Van Herck et al., 1992). Excessive blood loss via orbital plexus puncture can instigate retinal atrophy (Krinke et al., 1988).

Conjunctivitis in rats may result from infectious agents (e.g., Pasteurella pneumotropica and sialodacryoadenitis virus), or other noninfectious causes such as trauma, foreign bodies, and increased environmental ammonia (Harkness and Wagner, 1995). Epiphora may also be seen in these situations.

C. Corneal and Lens Lesions

Corneal dystrophy has been described in a variety of rat stocks and strains, including Sprague-Dawley, Wistar (Bellhorn et al., 1988), and Fischer-344 rats (Losco and Troup, 1988; Brunner et al., 1992; Whorton and Tilstead, 1984). Although it affects both sexes, Brunner et al. (1992) suggest1ed it may be more prevalent and severe in males. Wide incidence ranges (reports vary between 15% and 100%) might be due to the method of detection employed. The lesion typically appears as a superficial, punctate to linear opacity, located in the interpalpebral fissure. Microscopically there is focal to segmental thickening and disruption of the basement membrane with mineral deposition. The lesion may be seen in animals as young as 10 weeks of age but fails to progress or increase in severity with age. Because the etiology has not been established, corneal dystrophy is typically considered spontaneous. However, in one report, it was seen as a sequela to desiccation of the eye during a surgical procedure

(King et al., 1995). Corneal dystrophy in rats is very similar to calcific band keratopathy in humans and has been suggested as an animal model of this condition.

Cataracts are the most common lenticular disease reported in many different rat stocks and strains, including Sprague-Dawley (Durand et al., 2001), Wistar, SHR, Royal College of Surgeons (RCS) rats (Hess et al., 1985), and Brown Norway rats. Several rat strains, such as the Ihara cataract rat and the Shumiya cataract rat (SCR) serve as animal models of this disease (Ihara, 1983; Shumiya, 1995). Cataract formation in these lines ranges from 60% to 100% and shows no sex-based differences. Posterior subcapsular cataracts were described in 32% of Wistar rats over 2 years of age, and tended to occur more frequently in females (Wegener et al., 2002). In contrast, the maximum incidence in a Sprague-Dawley rat colony was 9.8% (Durand et al., 2001). In this study, the lenticular opacity was caused by focal epithelial proliferation, which failed to progress.

In addition to corneal changes, transient lens opacities, mydriasis, and proptosis can be induced by anesthesia, especially following xylazine administration (Kufoy et al., 1989; Calderone et al., 1986). Cataracts, as well as retinal degeneration, are a common sequela to diabetes mellitus, features that have been well-described in the spontaneously diabetic WBN/Kob strain (Miyamura and Amemiya, 1998). Similar to other degenerative disorders, calorie restriction delays the onset of cataract formation in Brown Norway rats (Wang and Sun, et al., 2004), although interestingly, not in albino Fischer-344 rats (Wolf et al., 2000).

D. Retinal Disorders

Phototoxic retinopathy is a retinal degeneration produced by light exposure first reported in albino rats in 1966 by Noell et al. It has been reported in multiple rat stocks and strains, affecting both sexes equally. Grossly, fundic exam reveals hyperreflectivity and narrowing of the retinal vessels (Everitt et al., 1987). Bilateral progressive loss of the outer retinal layers (photoreceptors and their nuclei) is seen in histological sections (Fig. 15-9). Factors affecting the development of phototoxic retinopathy include spectral wavelength of the light, photoperiod, illumination intensity, and body temperature (Lanum, 1978; Bellhorn, 1980). Most important are the synergetic effects of photoperiod and illumination intensity, with lower intensity light exposure for longer lengths of time producing equivalent phototoxicity to higher illumination levels for a shorter duration. Rats housed in areas of greater light intensity, such as in a rack's top row and outer columns, are therefore at greater risk (Rao, 1991). The ambient light intensity during rat development affects the damage-producing threshold later in life—an approximate 1.3 log unit increase is sufficient to induce phototoxicity (Semple-Rowland and Dawson, 1987). Methods to prevent phototoxic retinopathy include limiting light intensities to below 325 lux at 1 meter from the floor, rotating cage position on the rack, using opaque or translucent caging, and providing environmental enrichment devices or nesting material, allowing rats to self-regulate light level exposure.

Other than the RCS rat, which is a standard model of retinal dystrophy (Chader, 2002), spontaneous retinal degeneration not associated with light is relatively rare in rats. Lin and Essner (1987) describe lesions in rats housed under lighting conditions well below levels inducing phototoxicity. The lesions were similar to those produced by phototoxicity, but were found both unilaterally and bilaterally. Similarly, Lee and colleagues (1990) discovered unilateral lesions associated with optic nerve degeneration in approximately 16% of F-344 rats between 57 and 64 weeks of age; no association with diet, toxicity, or trauma was observed.

Other retinal conditions include proliferative retinopathy, which is characterized by the growth of new blood vessels on the retina or vitreous body similar to lesions seen in humans with diabetes mellitus (Matsuura et al., 1999). Preretinal arteriolar loops in which a branch of the central artery extends into the vitreous before descending back into the retina have also been described (Tanaka et al., 1994). Finally, in a large study of over 6,000 untreated Crl:CD(SD)BR rats, the two most common spontaneous retinal changes were linear focal retinopathy (3%) and coloboma (0.5%) (Hubert et al., 1994). Other changes such as retinal hemorrhage, retinal folds, optic disk aplasia, and saccular aneurysm of the retinal vessels were noted as rare occurrences.

X. MUSCULOSKELETAL SYSTEM

Many musculoskeletal problems seen in rats may be iatrogenic or traumatic in nature following improper handling, fighting, or procedures associated with experimental studies. For example, ketamine mixed with xylazine is a popular anesthetic cocktail used in rats. When given intramuscularly, this combination has been shown to produce edema, muscle necrosis, skin ulcerations, and lameness (Smiler et al., 1990).

Osteoarthritis affecting rats generally occurs in animals greater than one year of age and may be seen in the tibiotarsal joints, the medial femoral condyles, and the sternum (Berg, 1967; Sokoloff, 1967; Yamasaki and Inui, 1985). There is no apparent sex predilection, but when both Fischer-344 and Wistar rats were examined, the Fischer-344 rats were found to have a greater incidence and an

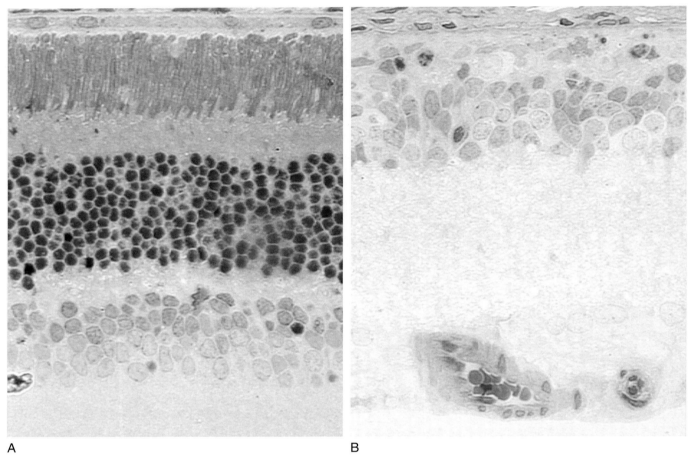

A   B

Fig. 15-9   Comparison of normal rat retina and phototoxic retinopathy, photomicrograph (400×). In the normal retina (**A**), note the distinct layers comprising (from top) the retinal pigmented epithelium, rod outer segments, rod inner segments, outer nuclear layer (dark-staining round nuclei), outer plexiform layer, inner nuclear layer, and inner plexiform layer. Structures not appearing in this image are the ganglion cell layer and inner limiting membrane. In the light-damaged retina (**B**), note the loss of essentially all layers caudal to the inner nuclear layer; in this image, the inner plexiform layer, ganglion cell layer (containing capillaries) and inner limiting membrane is visible. (*Images and description courtesy of Dr. D. Vaughan*)

increased severity of lesions, indicating strain differences (Smale et al., 1995). Gross lesions may not be apparent, but microscopically, osteoarthritis is characterized by erosion of articular cartilage and chondromucoid degeneration of the matrix. Cyst formation may also be observed.

Aseptic necrosis may be seen at the epiphyses of various bones, most commonly affecting the vertebrae, knee joint, and the sternum (Sokoloff, 1967; Burek et al., 1976; Jasty et al., 1986). Approximately 92% of Sprague-Dawley (Crl:CD(SD)BR) rats examined at 180 days of age contained aseptic necrosis of the intersternebral cartilage. No etiology was discovered (Jasty et al., 1986).

Osteochondrosis is a disease of young animals characterized by abnormal growth cartilage differentiation (Kato and Onodera, 1987). The incidence is higher in males than in females. Lesions typically begin to develop at around 6 weeks of age and may be found in areas such as the medial femoral condyle and the humeral head.

Microscopically, a thick deep zone with incomplete mineralization of the matrix and small cavity formation is seen. Viable chondrocytes are present around the edges of the cavities but there is a lack of blood vessel invasion from the subchondral bone (Kato and Onodera, 1987).

Proliferative lesions may be seen in histological sections as new bone formation both inside the marrow cavity and outside the cortex of the bones of the carpal, tarsal, and digital joints (Yamasaki and Houshuyama, 1994). Yamasaki and Anai (1989) described a small group of animals with a congenital bony lesion producing a kinked tail. In all animals affected by the trait, abnormal bony and cartilaginous tissue was found between the 18th and 19th vertebrae causing the tail to curve to the right. Hypertrophic osteopathy associated with congestive heart failure, multifocal granulomatous pneumonia, and both intra- and extra-thoracic neoplasia was described in a Wistar rat (WI/HicksCar) presenting with bilateral metatarsal swelling and hind limb paresis (Jackson et al., 1997).

Skeletal muscle degeneration is frequently observed in the hind limbs of older rats. Clinical signs generally occur in rats greater than 24 months of age and vary from posterior gait abnormalities to posterior paresis or paralysis. Urinary incontinence or a loss of posterior muscle mass with resulting change in conformation may also be seen. On gross exam, hind limb muscles appear brown, shrunken, and soft, especially in comparison to muscles from the anterior portion of the body. Histological appearance varies with disease chronicity. Early stages may be limited to a loss of muscle fiber diameter with or without loss of cross-striations (Berg, 1955b; Berg, 1967; Van Steenis and Kroes, 1971). Disease progression continues through fragmentation and disintegration of fibers until the terminal collapse of the sarcolemmal sheath. Inflammation is characteristically absent. The etiology of skeletal myodegeneration is unclear, but may be associated with neurogenic atrophy secondary to radiculoneuropathy (Van Steenis and Kroes, 1971; Burek et al., 1976). However, one study indicated the muscular atrophy, unlike radiculoneuropathy, could be delayed by calorie restriction (Berg, 1967).

XI. INTEGUMENTARY SYSTEM

A common change seen in the integumentary system of the rat is a roughened hair coat. While underlying causes such as stress, diarrhea, leaking water bottles, and parasitism should be ruled out, such a hair coat may be a product of aging and thus not indicative of disease *per se*. The development of pigmented scales along the dorsum, perineum, and tail is a common finding in older rats. While this change is dependent on age, strain, and gender, it does not appear to be related to any specific disease (Tayama and Shisa, 1994).

A. Traumatic and Husbandry-related Conditions

Various noninfectious causes of alopecia have been reported in the rat. While barbering is not as common in rats as it is in mice, it can occur (Bresnahan et al., 1983). Other causes of alopecia include excessive mutual grooming related to 20% dietary fat (Beare-Rogers and McGowan, 1973) and mechanical abrasion due to the use of feeding cups (Andrews, 1977).

A well-recognized cutaneous lesion in rats is the development of single or multiple annular constrictive rings on the tail, especially in suckling or pre-weanling animals. Edema is frequently seen distal to the constriction and may progress to a dry necrosis of the affected area (Fig. 15-10).

This syndrome, commonly known as "ringtail," is traditionally attributed to low environmental humidity (Njaa et al., 1957). Environmental temperature, dietary deficiencies, hydration status of the animal, and genetic susceptibility have also been suggested as predisposing factors (Percy and Barthold, 2001). Crippa and colleagues (2000) have detailed the histological findings associated with this condition.

The tail is frequently used during handling and restraint. Attempts to pick up large rats by the distal tail may lead to skin tears and a de-gloving injury (Kohn and Barthold, 1984).

While fighting may occur among some group-housed rats, traumatic cutaneous injuries are more likely self-induced. Self-trauma may play an important role in the pathogenesis of ulcerative dermatitis, which is also associated with *Staphylococcus aureus* infection (Fox et al., 1977); clipping the toenails or amputating the toes of the hind feet resulted in resolution of skin lesions (Wagner et al., 1977). Other causes of self-induced trauma may include ectoparasites or dermatophyte infection.

Nodular swellings, calluses, and decubital ulcers may also be seen on the plantar surface of the rat foot. These lesions are usually found near the carpus on animals housed in wire-bottomed cages, and can progress to granulomas, generalized edema, and regional lymphadenopathy of the affected limb (Honma and Kast, 1989). Foot lesion development is thought to be related to increased body weight, may show stock/strain predilection, and generally requires over one year to develop (Peace et al., 2001).

B. Auricular Chondritis

Characterized by a gross thickening of the pinnae, auricular chondritis (also called auricular chondropathy) is typically bilateral and ranges from nodular to diffuse.

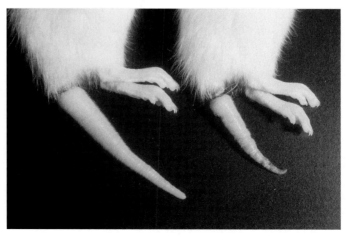

Fig. 15-10 Annular constrictions ("ringtail").

It has been described in Sprague Dawley (Chiu and Lee, 1984), Fawn-Hooded (Prieur et al., 1984), and Wistar rats (McEwen and Barsoum, 1990). Histological examination reveals a predominantly granulomatous infiltrate disrupting the normal cartilaginous structures and resulting in multifocal nodular proliferation of cartilaginous tissue (Linton et al., 1994). Differential diagnoses include fibrochondroma, chondroma, and chondrosarcoma (Chiu and Lee, 1984).

This condition has been proposed as a model for the suspected autoimmune-associated human disorder relapsing polychondritis. Similar lesions are also induced via immunization with Type II collagen, suggesting a similar pathogenesis (Cremer et al., 1981). There are differences, however. Rats immunized with Type II collagen develop arthritis (Cremer et al., 1981). A lack of responses to delayed hypersensitivity tests for Type II collagen in spontaneous auricular chondritis suggests that the autosensitization results from other pinnal cartilage proteins (Meingassner, 1991). Although lesions develop in rats without them, spontaneous auricular chondritis is often associated with metal ear tags. Unilateral lesions may occur within one week of ear tag placement and may be followed by lesions in the contralateral ear within four weeks (Kitagaki et al., 2002). Chronic irritation and inflammation caused by the tags may initiate the autoimmune response (Meingassner, 1991).

XII. REPRODUCTIVE SYSTEM

While not common, twinning has been noted in multiple strains of rat; both monoamniotic and conjoined twins have been described (Cockrell, 1972; Levinsky, 1973; Mutinelli et al., 1992; Kawaguchi et al., 1995). Other anomalies noted in the literature include supernumerary forelimb (Borrás, 1987) and placental fusion (Arora, 1977).

Various noninfectious causes of infertility have been described in the rat including high environmental temperatures (Pucak et al., 1977), vitamin E deficiency (Jager, 1972), protein deficiency (Brown et al., 1960), organophosphate poisoning (Timmons et al., 1975), prenatal stress (Herrenkohl, 1979), elevated humidity (Njaa et al., 1957), and constant light (Weihe, 1976).

Light cycles and total light exposure time can greatly effect the estrus cycle of rats. Constant environmental light may initiate polycystic ovaries, endometrial hypertrophy, hyperestrogenism, and persistent estrus (Percy and Barthold, 2001). Other spontaneous changes of the female reproductive tract include ovarian degeneration (Meites et al., 1978), cystic dilatation of the vaginal fornix (Yoshitomi, 1990), hydrometra and pyometra

(Franks, 1967), and granular cell aggregations (Markovits and Sahota, 2000). Both longitudinal and transverse vaginal septa preventing conception have also been described in the rat (Barbolt and Brown, 1989; De Schaepdrijver et al., 1995).

Many changes seen in the male reproductive system are associated with aging. These include seminiferous tubule degeneration, polyarteritis nodosa, interstitial cell hyperplasia, and testicular interstitial edema, as well as abscesses, hyperplasia, and concretion formation in the prostate. Of these, seminiferous tubule atrophy was found to be the most common testicular lesion in two-year-old Sprague-Dawley rats (James and Heywood, 1979). The age of onset of interstitial cell hyperplasia could be significantly delayed by food restriction. Preputial gland lesions are also common in the aging rat and include acinar cell atrophy, fibrosis, duct distension, and inflammation, affecting up to 68% of 112-week-old Fischer-344 rats (Reznik and Reznik-Schïller, 1980). In contrast to those conditions occurring primarily in older animals, epididymal granulomas may be found in 12-week-old Sprague-Dawley rats, possibly resulting in outflow obstruction of spermatozoa from the epididymis, producing distension, thinning, and eventually rupture of the ducts resulting in granuloma formation (Yamasaki, 1990).

ACKNOWLEDGMENTS

The authors would like to acknowledge the assistance of Sheila Carpenter, Melissa Carpenter, and Manon Gant, and are deeply indebted to Drs. James E. Artwohl, Stephen W. Barthold, Michael T. Fallon, J. Russell Lindsey, Daniel T. Organisciak, Mary L. Proctor, Trenton R. Schoeb, and Dana K. Vaughan for their assistance in obtaining images used in this manuscript.

REFERENCES

Abrass, C.K. (2000). The nature of chronic progressive nephropathy in aging rats. *Adv. Ren. Replace. Ther.* 7, 4–10.

Alden, C.L., and Frith, C.H. (1991). Urinary system. *In* "Handbook of Toxicologic Pathology" (W.M. Haschek and C.G. Rousseaux, eds.), pp. 315–387. Academic Press, San Diego.

Alison, R.H., Elwell, M.R., Jokinen, M.P., Dittrich, K.L., and Boorman, G.A. (1987). Morphology and classification of 96 primary cardiac neoplasms in Fischer 344 rats. *Vet. Pathol.* 24, 488–494.

Amouzadeh, H.R., Sangiah, S., Qualls, C.W., Cowell, R.L., and Mauromoustakos, A. (1991). Xylazine-induced pulmonary edema in rats. *Toxicol. Appl. Pharmacol.* 108, 417–427.

Amouzadeh, H.R., Qualls, C.W., Wyckoff, J.H., Dzata, G.K., Sangiah, S., Mauromoustakos, A., and Stein, L.E. (1993). Biochemical and morphological alterations in xylazine-induced pulmonary edema. *Toxicol. Pathol.* 21, 562–571.

Anastasia, J.V., Braun, B.L., and Smith, K.T. (1990). General and histopathological results of a two-year study of rats fed semi-purified diets containing casein and soya protein. *Food Chem. Toxicol.* **28**, 147–156.

Anderson, S., Rennke, H.G., and Brenner, B.M. (1986). Therapeutic advantage of converting enzyme inhibitors in arresting progressive renal disease associated with systemic hypertension in the rat. *J. Clin. Invest.* **77**, 1993–2000.

Andrews, E.J. (1977). Muzzle trauma in the rat associated with the use of feeding cups. *Lab. Anim. Sci.* **27**, 278.

Anver, M.R., and Cohen, B.J. (1979). Lesions associated with aging. *In* "The Laboratory Rat" (H.J. Baker, J.R. Lindsey, and S.H. Weisbroth, eds.), Vol. 1, pp. 377–399. Academic Press, New York.

Anver, M.R., Cohen, B.J., Lattuada, C.P., and Foster, S.J. (1982). Age-associated lesions in barrier-reared male Sprague-Dawley rats: a comparison between Hap:(SD) and Crl:COBS[R]CD[R](SC) stocks. *Exp. Aging Res.* **8**, 3–24.

Anversa, P., and Capasso, J.M. (1991). Cellular basis of aging in the mammalian heart. *Scanning Microsc.* **5**, 1065–1073.

Apostolou, A., Saidt, L., and Brown, W.R. (1976). Effect of overnight fasting of young rats on water consumption, body weight, blood sampling, and blood composition. *Lab. Anim. Sci.* **26**, 959–960.

Arora, K.L. (1977). A case of placental fusion in a Sprague Dawley rat. *Lab. Anim. Sci.* **27**, 377–379.

Aspinall, R., and Andrew, D. (2000). Immunosenescence: potential causes and strategies for reversal. *Biochem. Soc. Trans.* **28**, 250–254.

Ayers, K.M., and Jones, S.R. (1978). The cardiovascular system. *In* "Pathology of Laboratory Animals" (K. Benirschke, F.M. Garner, and T.C. Jones, eds.), Vol. 1, pp. 1–69. Springer-Verlag, New York.

Baczako, K., and Dolderer, M. (1997). Polyarteritis nodosa-like inflammatory vascular changes in the pancreas and mesentery of rats treated with streptozotocin and nicotinamide. *J. Comp. Pathol.* **116**, 171–180.

Baiocco, A., Boujon, C.E., Häfeli, W., Cherubini, Y.D., Bestetti, G.E., and Rossi, G.L. (1993). Megaoesophagus in rats: a clinical, pathological and morphometrical study. *J. Comp. Pathol.* **108**, 269–281.

Barbolt, T.A., and Brown, G.L. (1989). Vaginal septum in the rat. *Lab. Anim.* **18**, 47–48.

Barthold, S.W. (1979). Chronic progressive nephropathy in aging rats. *Toxicol. Pathol.* **7**, 1–6.

Beare-Rogers, J.L., and McGowan, J.E. (1973). Alopecia in rats housed in groups. *Lab. Anim.* **7**, 237–238.

Beaver, D.L., Ashburn, L.L., McDaniel, E.G., and Brown, N.D. (1963). Lipid deposits in the lungs of germ-free animals. *Arch. Pathol.* **76**, 565–570.

Beierschmitt, W.P., Keenan, K.P., and Weiner, M. (1986). Age-related susceptibility of male Fischer-344 rats to acetaminophen nephrotoxicity. *Life Sci.* **39**, 2335–2342.

Bellhorn, R.W. (1980). Lighting in the animal environment. *Lab. Anim. Sci.* **30**, 440–450.

Bellhorn, R.W., Korte, G.E., and Abrutyn, D. (1988). Spontaneous corneal degeneration in the rat. *Lab. Anim. Sci.* **38**, 46–50.

Bender, H.M. (1998). Pica behavior associated with buprenorphine administration in the rat (letter). *Lab. Anim. Sci.* **48**, 5.

Berg, B.N. (1955a). The electrocardiogram in aging rats. *J. Gerontol.* **10**, 420–423.

Berg, B.N. (1955b). Muscular dystrophy in aging rats. *J. Gerontol.* **11**, 134–139.

Berg, B.N. (1967). Longevity studies in rats. II. Pathology of ageing rats. *In* "Pathology of Laboratory Rats and Mice" (E. Cotchin and F.J.C. Roe, eds.), pp. 749–786, Blackwell Scientific Publications, Oxford and Edinburgh.

Berg, B.N., and Harmison, C.R. (1955). Blood pressure and heart size in aging rats. *J. Gerontol.* **10**, 416–419.

Berg, B.N., and Simms, H.S. (1960). Nutrition and longevity in the rat. II. Longevity and onset of disease with different levels of food intake. *J. Nutr.* **71**, 255–263.

Berg, B.N., Wolf, A., and Simms, H.S. (1962). Degenerative lesions of spinal roots and peripheral nerves in aging rats. *Gerontologia* **6**, 72–80.

Bertani, T., Zoja, C., Abbate, M., Rossini, M., and Remuzzi, G. (1989). Age-related nephropathy and proteinuria in rats with intact kidneys exposed to diets with different protein content. *Lab. Invest.* **60**, 196–204.

Beynen, A.C. (1992). Can pre-established nephrocalcinosis regress? (letter) *Lab. Anim.* **26**, 306–308.

Bilhun, C. (1997). Basic anatomy, physiology, husbandry, and clinical techniques: anatomic and physiologic features. *In* "Ferrets, Rabbits and Rodents: Clinical Medicine and Surgery" (E.V. Hillyer and K.E. Quesenberry, eds.), pp. 295–306, W.B. Saunders Company, Philadelphia.

Bingel, S.A. (2003). Spontaneous concretions in the urinary tract of rats. *Lab Anim. (NY).* **32**(8), 22–25.

Biology Databook Editorial Board. (1985). "Pathology of Laboratory Mice and Rats" (P.L. Altman, ed.), Pergamon Infoline, Inc., McLean.

Bishop, S.P. (1989). Animal models of vasculitis. *Toxicol. Pathol.* **17**, 109–117.

Blatherwick, N.R., and Medlar, E.M. (1937). Chronic nephritis in rats fed high protein diets. *Arch. Intern. Med.* **59**, 572–596.

Bolton, W.K., Benton, F.R., Maclay, J.G., and Sturgill, B.C. (1976). Spontaneous glomerular sclerosis in aging Sprague-Dawley rats. I. Lesions associated with mesangial IgM deposits. *Am. J. Pathol.* **85**, 277–302.

Boorman, G.A., and Hollander, C.F. (1974). High incidence of spontaneous urinary bladder and ureter tumors in the Brown Norway rat. *J. Natl. Cancer Inst.* **52**, 1005–1008.

Boorman, G.A., Zurcher, C., Hollander, C.F., and Feron, V.J. (1973). Naturally occurring endocardial disease in the rat. *Arch. Pathol.* **96**, 39–45.

Borrás, M. (1983). 3 cases of persistent urachus with umbilical abscess in Wistar rats. *Lab. Anim.* **17**, 55–58.

Borrás, M. (1987). A supernumerary forelimb in a Wistar rat. *Lab. Anim.* **21**, 223–225.

Bras, G. (1969). Age-associated kidney lesions in the rat. *J. Infect. Dis.* **120**, 131–135.

Bras, G., and Ross, M.H. (1964). Kidney disease and nutrition in the rat. *Toxicol. Appl. Pharmacol.* **6**, 247–262.

Brenner, B.M. (1985). Nephron adaptation to renal injury or ablation. *Am. J. Physiol.* **249**, F324–F337.

Bresnahan, J.F., Kitchell, B.B., and Wildman, M.F. (1983). Facial hair barbering in rats. *Lab. Anim. Sci.* **33**, 290–291.

Brockus, C.W., Rakich, P.M, King, C.S., and Broderson, J.R. (1999). Rhabdomyosarcoma associated with incisor malocclusion in a laboratory rat. *Contemp. Top. Lab. Anim. Sci.* **38**(6), 42–43.

Broderson, J.R., Lindsey, J.R., and Crawford, J.E. (1976). The role of environmental ammonia in respiratory mycoplasmosis of rats. *Am. J. Pathol.* **85**, 115–130.

Brors, O. (1987). Gastrointestinal mucosal lesions: a drug formulation problem. *Med. Toxicol.* **2**, 105–111.

Brown, A.M., Cook, M.J., Lane-Petter, W., Porter, G., and Tuffery, A.A. (1960). Influence of nutrition on reproduction in laboratory rodents. *Proc. Nutr. Soc.* **19**, 32–37.

Brown, A.P., Dinger, N., and Levine, B.S. (2000). Stress produced by gavage administration in the rat. *Contemp. Top. Lab. Anim. Sci.* **39** (1), 17–21.

Brunner, R.H., Keller, W.F., Stitzel, K.A., Sauers, L.J., Reer, P.J., Long, P.H., Bruce, R.D., and Alden, C.L. (1992). Spontaneous corneal dystrophy and generalized basement membrane changes in Fischer-344 rats. *Toxicol. Pathol.* **20**, 357–366.

Burek, J.D. (1978). "Pathology of Aging Rats." CRC Press, West Palm Beach.

Burek, J.D., Van der Kogel, A.J., and Hollander, C.F. (1976). Degenerative myelopathy in three strains of aging rats. *Vet. Pathol.* **13**, 321–331.

Burton, D.S., Maronpot, R.R., and Howard, F.L., III. (1979). Frequency of hydronephrosis in Wistar rats. *Lab. Anim. Sci.* **29**, 642–644.

Calderone, L., Grimes, P., and Shalev, M. (1986). Acute reversible cataract induced by xylazine and by ketamine-xylazine anesthesia in rats and mice. *Exp. Eye Res.* **42**, 331–337.

Chader, G.J. (2002). Animal models in research on retinal degenerations: past progress and future hopes. *Vision Res.* **42**, 393–399.

Cherry, C.P., Eisenstein, R., and Glïcksmann, A. (1967). Epithelial cords and tubules in the rat thymus: effects of age, sex, castration, of sex, thyroid and other hormones on their incidence and secretory activity. *Br. J. Exp. Pathol.* **48**, 90–106.

Chea, D. (1998). Animal models of type I (insulin-dependent) diabetes mellitus. *J. Pediatr. Endocrinol. Metab.* **11**, 11–19.

Chiu, T., and Lee, K.P. (1984). Auricular chondropathy in aging rats. *Vet. Pathol.* **21**, 500–504.

Christensen, E.I., and Madsen, K.M. (1978). Renal age changes: observations on the rat kidney cortex with special reference to structure and function of the lysosomal system in the proximal tubule. *Lab. Invest.* **39**, 289–297.

Clapp, M.J.L., Wade, J.D., and Samuels, D.M. (1982). Control of nephrocalcinosis by manipulating the calcium:phosphorus ratio in commercial rodent diets. *Lab. Anim.* **16**, 130–132.

Clark, J.A., Myers, P.H., Goelz, M.F., Thigpen, J.E., and Forsythe, D.B. (1995). Gastric distention associated with buprenorphine administration in the rat. *Contemp. Top Lab. Anim. Sci.* **34**(4), 61.

Clark, J.A., Jr., Myers, P.H., Goelz, M.F., Thigpen, J.E., and Forsythe, D.B. (1997). Pica behavior associated with buprenorphine administration in the rat. *Lab. Anim. Sci.* **47**, 300–303.

Cockell, K.A., L'Abbé, M.R., and Belonje, B. (2002). The concentrations and ratio of dietary calcium and phosphorus influence development of nephrocalcinosis in female rats. *J. Nutr.* **132**, 252–256.

Cockrell, B.Y. (1972). A case of monster twinning (thoracopagus) in the Sprague Dawley rat (Spartan strain). *Lab. Anim. Sci.* **22**, 102–103.

Cohen, B.J., DeBruin, R.W., and Kort, W.J. (1970). Heritable hydronephrosis in a mutant strain of Brown Norway rats. *Lab. Anim. Care* **20**, 489–493.

Cohen, B.J., Anver, M.R., Ringler, D.H., and Adelman, R.C. (1978). Age-associated pathological changes in male rats. *Fed. Proc.* **37**, 2848–2850.

Coleman, G.L., Barthold, S.W., Osbaldiston, G.W., Foster, S.J., and Jonas, A.M. (1977). Pathological changes during aging in barrier-reared Fischer 344 male rats. *J. Gerontol.* **32**, 258–278.

Cornwell, G.G., III, Thomas, B.P., and Snyder, D.L. (1991). Myocardial fibrosis in aging germ-free and conventional Lobund-Wistar rats: the protective effect of diet restriction. *J. Gerontol.* **46**, B167–B170.

Cotard-Bartley, M.P., Secchi, J., Glomot, R., and Cavanagh, J.B. (1981). Spontaneous degenerative lesions of peripheral nerves in aging rats. *Vet. Pathol.* **18**, 110–113.

Couser, W.G., and Stilmant, M.M. (1975). Mesangial lesions and focal glomerular sclerosis in the aging rat. *Lab. Invest.* **33**, 491–501.

Cousins, F.B., and Geary, C.P.M. (1966). A sex-determined renal calcification in rats. *Nature* **211**, 980–981.

Cramer, D.V., Gill, T.J., III. (1975). Genetics of urogenital abnormalities in ACI inbred rats. *Teratology* **12**, 27–32.

Cremer, M.A., Pitcock, J.A., Stuart, J.M., Kang, A.H., and Townes, A.S. (1981). Auricular chondritis in rats: an experimental model of relapsing polychondritis induced with Type II collagen. *J. Exp. Med.* **154**, 535–540.

Crippa, L., Gobbi, A., Ceruti, R.M., Clifford, C.B., Remuzzi, A., and Scanziani, E. (2000). Ringtail in suckling Munich Wistar Fromter rats: a histopathologic study. *Comp. Med.* **50**, 536–539.

Cutts, J.H. (1966). Vascular lesions resembling polyarteritis nodosa in rats undergoing prolonged stimulation with oestrogen. *Br. J. Exp. Pathol.* **47**, 401–404.

Danneman, P.J., Stein, S., and Walshaw, S.O. (1997). Humane and practical implications of using carbon dioxide mixed with oxygen for anesthesia or euthanasia of rats. *Lab. Anim. Sci.* **47**, 376–385.

Davis, H., Cox, N.R., and Lindsey, J.R. (1985). Diagnostic exercise: distended abdomens in rats. *Lab. Anim. Sci.* **35**, 392–394.

De Schaepdrijver, L.M., Fransen, J.L., Van der Eycken, E.S., and Coussement, W.C. (1995). Transverse vaginal septum in the specific-pathogen-free Wistar rat. *Lab. Anim. Sci.* **45**, 181–183.

Deerberg, F., and Pittermann, W. (1972). Megaösophagus als Todesursache bei BDE/HAN-Ratten. *Z. Versuchstierkd* **14**, 177–182.

Dillberger, J.E. (1994). Age-related pancreatic islet changes in Sprague-Dawley rats. *Toxicol. Pathol.* **22**, 48–55.

Dixon, D., and Jure, M.N. (1988). Diagnostic exercise: pneumonia in a rat. *Lab. Anim. Sci.* **38**, 727–728.

Dixon, D., Heider, K., and Elwell, M.R. (1995). Incidence of nonneoplastic lesions in historical control male and female Fischer-344 rats from 90-day toxicity studies. *Toxicol. Pathol.* **23**, 338–348.

Donnelly, T.M. (1997). Disease problems of small rodents. *In* "Ferrets, Rabbits and Rodents: Clinical Medicine and Surgery" (E.V. Hillyer and K.E. Quesenberry, eds.), pp. 307–327, W.B. Saunders Company, Philadelphia.

Du Bruyn, D.B. (1970). A comparison of certain rat strains with respect to susceptibility to nephrocalcinosis. *S. Afr. Med. J.* **44**, 1417–1418.

Durand, A.M.A., Fisher, M., and Adams, M. (1964). Histology in rats as influenced by age and diet. I. Renal and cardiovascular systems. *Arch. Pathol.* **77**, 268–277.

Durand, G., Hubert, M.-F., Kuno, H., Cook, W.O., Stabinski, L.G., Darbes, J., and Virat, M. (2001). Spontaneous polar anterior subcapsular lenticular opacity in Sprague-Dawley rats. *Comp. Med.* **51**, 176–179.

Elema, J.D., and Arends, A. (1975). Focal and segmental glomerular hyalinosis and sclerosis in the rat. *Lab. Invest.* **33**, 554–561.

el-Nassery, S.M., Rahmy, A.E., el Gebaly, W.M., and Sadaka, H.A. (1994). *Pneumocystis carinii*: recognition of the infection in albino rats using different strains. *J. Egypt. Soc. Parasitol.* **24**, 285–294.

Emily, P. (1991). Problems peculiar to continually erupting teeth. *J. Small Exotic Anim. Med.* **1**, 56–59.

Ettlin, R.A., Stirnimann, P., and Prentice, D.E. (1994). Causes of death in rodent toxicity and carcinogenicity studies. *Toxicol. Pathol.* **22**, 165–178.

Eustis, S.L., Boorman, G.A., Harada, T., and Popp, J.A. (1990). Liver. *In* "Pathology of the Fischer Rat" (G.A. Boorman, S.L. Eustis, M.R. Elwell, C.A. Montgomery, Jr., and W.F. MacKenzie, eds.), pp. 71–94, Academic Press, San Diego.

Everitt, A.V. (1958). The urinary excretion of protein, nonprotein nitrogen, creatinine and uric acid of ageing male rats. *Gerontologia* **2**, 33–46.

Everitt, J.I., McLaughlin, S.A., and Helper, L.C. (1987). Diagnostic exercise: eye lesions in rats. *Lab. Anim. Sci.* **37**, 203–204.

Ewaldsson, B., Fogelmark, B., Feinstein, R., Ewaldsson, L., and Rylander, R. (2002). Microbial cell wall product contamination of bedding may induce pulmonary inflammation in rats. *Lab. Anim.* **36**, 282–290.

Fairweather, F.A. (1967). Cardiovascular disease in rats. *In* "Pathology of Laboratory Rats and Mice." (E. Cotchin and F.J.C. Roe, eds.), pp. 213–228, Blackwell Scientific Publications, Oxford and Edinburgh.

Feldman, D.B., and Gupta, B.N. (1976). Histopathologic changes in laboratory animals resulting from various methods of euthanasia. *Lab. Anim. Sci.* **26**, 218–221.

Flecknell, P.A., Roughan, J.V., and Stewart, R. (1999). Use of oral buprenorphine ('buprenorphine jello') for postoperative analgesia in rats – a clinical trial. *Lab. Anim.* **33**, 169–174.

Fleischman, R.W., McCracken, D., and Forbes, W. (1977). Adynamic ileus in the rat induced by chloral hydrate. *Lab. Anim. Sci.* **27**, 238–243.

Fogo, A., and Ichikawa, I. (1989). Evidence for the central role of glomerular growth promoters in the development of sclerosis. *Semin. Nephrol.* **9**, 329–342.

Fox, J.G., Niemi, S.M., Murphy, J.C., and Quimby, F.W. (1977). Ulcerative dermatitis in the rat. *Lab. Anim. Sci.* **27**, 671–678.

Fox, M.H. (1969). Evidence suggesting a multigenic origin of membranous septal defects in rats. *Circ. Res.* **24**, 629–637.

Franks, L.M. (1967). Normal and pathological anatomy and histology of the genital tract of rats and mice. *In* "Pathology of Laboratory Rats and Mice." (E. Cotchin and F.J.C. Roe, eds.), pp. 469–499, Blackwell Scientific Publications, Oxford and Edinburgh.

Friedman, J., Hoyer, J.R., McCormick, B., and Lewy, J.E. (1979). Congenital unilateral hydronephrosis in the rat. *Kidney Int.* **15**, 567–571.

Fries, J.W.U., Sandstrom, D.J., Meyer, T.W., and Rennke, H.G. (1989). Glomerular hypertrophy and epithelial cell injury modulate progressive glomerulosclerosis in the rat. *Lab. Invest.* **60**, 205–218.

Frith, C.H., Farris, H.E., and Highman, B. (1977). Endocardial fibromatous proliferation in a rat. *Lab. Anim. Sci.* **27**, 114–117.

Fujikura, T. (1970). Kidney malformations in fetuses of AXC line 9935 rats. *Teratology* **3**, 245–249.

Gamble, M.R., and Clough, G. (1976). Ammonia build-up in animal boxes and its effect on rat tracheal epithelium. *Lab. Anim.* **10**, 93–104.

Geary, C.P., and Cousins, F.B. (1969). An oestrogen-linked nephrocalcinosis in rats. *Br. J. Exp. Pathol.* **50**, 507–515.

Germann, P-G., and Ockert, D. (1994). Granulomatous inflammation of the oropharyngeal cavity as a possible cause for unexpected high mortality in a Fischer 344 rat carcinogenicity study. *Lab. Anim. Sci.* **44**, 338–343.

Germann, P-G., Ockert, D., and Tuch, K. (1995). Oropharyngeal granulomas and tracheal cartilage degeneration in Fischer-344 rats. *Toxicol. Pathol.* **23**, 349–355.

Gershoff, S.N., and McGandy, R.B. (1981). The effects of vitamin A-deficient diets containing lactose in producing bladder calcula and tumors in rats. *Am. J. Clin. Nutr.* **34**, 483–489.

Ghione, S. (1996). Hypertension-associated hypalgesia. Evidence in experimental animals and humans, pathophysiological mechanisms, and potential clinical consequences. *Hypertension* **28**, 494–504.

Gibson, S.V., Besch-Williford, C., Raisbeck, M.F., Wagner, J.E., and McLaughlin, R.M. (1987). Organophosphate toxicity in rats associated with contaminated bedding. *Lab. Anim. Sci.* **37**, 789–791.

Gilmore, S.A. (1972). Spinal nerve root degeneration in aging laboratory rats: a light microscopic study. *Anat. Rec.* **174**, 251–258.

Goldblum, D., and Mittag, T. (2002). Prospects for relevant glaucoma models with retinal ganglion cell damage in the rodent eye. *Vision Res.* **42**, 471–478.

Goldman, D., Lee, S.C., Casadevall, A. (1994). Pathogenesis of pulmonary *Cryptococcus neoformans* infection in the rat. *Infect. Immun.* **62**, 4755–4761.

Goldstein, R.S., Tarloff, J.B., and Hook, J.B. (1988). Age-related nephropathy in laboratory rats. *FASEB J.* **2**, 2241–2251.

Goodman, D.G., Ward, J.M., Squire, R.A., Chu, K.C., and Linhart, M.S. (1979). Neoplastic and nonneoplastic lesions in aging F344 rats. *Toxicol. Appl. Pharmacol.* **48**, 237–248.

Goodman, D.G., Ward, J.M., Squire, R.A., Paxton, M.B. Reichardt, W.D., Chu, K.C., and Linhart, M.S. (1980). Neoplastic and nonneoplastic lesions in aging Osborne-Mendel rats. *Toxicol. Appl. Pharmacol.* **55**, 433–447.

Gray, J.E. (1986). Chronic progressive nephrosis, rat. *In* "Monographs on Pathology of Laboratory Animals: Urinary System." (T.C. Jones, U. Mohr, and R.D. Hunt, eds.), pp. 174–179, Springer-Verlag, New York.

Gray, J.E., Larsen, E.R., and Greenberg, H.S. (1982a). A breeding study to control chronic progressive nephrosis in rats. *Lab. Anim. Sci.* **32**, 609–612.

Gray, J.E., Van Zwieten, M.J., and Hollander, C.F. (1982b). Early light microscopic changes of chronic progressive nephrosis in several strains of aging laboratory rats. *J. Gerontol.* **37**, 142–150.

Grimm, P., Nowitzki-Grimm, S., Weidenmaier, W., Schumacher, K.A., and Classen, H.G. (1991). *In-vivo* monitoring of corticomedullary nephrocalcinosis in rats using computerized tomography. *Lab. Anim.* **25**, 354–359.

Gumprecht, L.A., Long, C.R., Soper, K.A., Smith, P.A., Hascheck-Hock, W.M., and Keenan, K.P. (1993). The early effects of dietary restriction on the pathogenesis of chronic renal disease in Sprague-Dawley rats at 12 months. *Toxicol. Pathol.* **21**, 528–537.

Gupta, B.N. (1975). Ectopic kidney in a rat. *Lab. Anim. Sci.* **25**, 361.

Gustafsson, B.E., and Norman, A. (1962). Urinary calculi in germfree rats. *J. Exp. Med.* **116**, 273–283.

Hall, W.C., Ganaway, J.R., Rao, G.N., Peters, R.L., Allen, A.M., Luczak, J.W., Sandberg, E.M., and Quigley, B.H. (1992). Histopathologic observations in weanling B6C3F1 mice and F344/N rats and their adult parental strains. *Toxicol. Pathol.* **20**, 146–154.

Hamlin, M.H., and Banas, D.A. (1990). Adrenal gland. *In* "Pathology of the Fischer Rat." (G.A. Boorman, S.L. Eustis, M.R. Elwell, C.A. Montgomery, Jr., and W.F. MacKenzie, eds.), pp. 501–518, Academic Press, Inc., San Diego.

Hardisty, J.F., and Boorman, G.A. (1990). Thyroid gland. *In* "Pathology of the Fischer Rat." (G.A. Boorman, S.L. Eustis, M.R. Elwell, C.A. Montgomery, Jr., and W.F. MacKenzie, eds.), pp. 519–536, Academic Press, Inc., San Diego.

Harkness, J.E., and Ferguson, F.G. (1979). Idiopathic megaesophagus in a rat (*Rattus norvegicus*). *Lab. Anim. Sci.* **29**, 495–498.

Harkness, J.E., and Ridgway, M.D. (1980). Chromodacryorrhea in laboratory rats (*Rattus norvegicus*): etiologic considerations. *Lab. Anim. Sci.* **30**, 841–844.

Harkness, J.E., and Wagner, J.E. (1995). "Biology and Medicine of Rabbits and Rodents." Williams and Wilkins, Baltimore, Maryland.

Hartmann, H.A., Miller, E.C., and Miller, J.A. (1959). Periarteritis in rats given single injection of 4'-fluoro-10–methyl-1,2–benzanthracene. *Proc. Soc. Exp. Biol. Med.* **101**, 626–629.

Herrenkohl, L.R. (1979). Prenatal stress reduces fertility and fecundity in female offspring. *Science* **206**, 1097–1099.

Hess, H.H., Knapka, J.J., Newsome, D.A., Westney, I.V., and Wartofsky, L. (1985). Dietary prevention of cataracts in the pink-eyed RCS rat. *Lab. Anim. Sci.* **35**, 47–53.

Heywood, R. (1975). Glaucoma in the rat. *Br. Vet. J.* **131**, 213–221.

Hitchman, A.J., Hasany, S.A., Hitchman, A., Harrison, J.E., and Tam, C. (1979). Phosphate-induced renal calcification in the rat. *Can. J. Physiol. Pharmacol.* **57**, 92–97.

Hoek, A.C., Lemmons, A.G., Mullink, J.W.M.A., and Beynen, A.C. (1988). Influence of dietary calcium: phosphorus ratio on mineral excretion and nephrocalcinosis in female rats. *J. Nutr.* **118**, 1210–1216.

Hollander, C.F. (1976). Current experience using the laboratory rat in aging studies. *Lab. Anim. Sci.* **26**, 320–328.

Honma, M., and Kast, A. (1989). Plantar decubitus ulcers in rats and rabbits. *Jikken Dobutsu* **38**, 253–258.

Hubert, M.F., Gillet, J.P., and Durand-Cavagna, G. (1994). Spontaneous retinal changes in Sprague Dawley rats. *Lab. Anim. Sci.* **44**, 561–567.

Ihara, N. (1983). A new strain of rat with an inherited cataract. *Experientia* **39**, 909–911.

Itakura, C., Iida, M., and Goto, M. (1977). Renal secondary hyperparathyroidism in aged Sprague-Dawley rats. *Vet. Pathol.* **14**, 463–469.

Iwasaki, K., Gleiser, C.A., Masoro, E.J., McMahan, C.A., Seo, E.-J., and Yu, B.P. (1988). The influence of dietary protein source on longevity and age-related disease processes in Fischer rats. *J. Gerontol.* **43**, B5–B12.

Jablonski, P., Howden, B.O., and Baxter, K. (2001). Influence of buprenorphine analgesia on postoperative recovery in two strains of rats. *Lab. Anim.* **35**, 213–222.

Jackson, T.A., Chrisp, C.E., Dysko, R.C., and Carlson, B.M. (1997). Spontaneous hypertrophic osteopathy in a Wistar rat. *Contemp. Top. Lab. Anim. Sci.* **36**(5), 68–70.

Jacobson, C. (2000). Adverse effects on growth rates in rats caused by buprenorphine administration. *Lab. Anim.* **34**, 202–206.

Jager, F.C. (1972). Long-term dose-response effects of vitamin E in rats. *Nutr. Metab.* **14**, 1–7

James, R.W., and Heywood, R. (1979). Age-related variations in the testes of Sprague-Dawley rats. *Toxicol. Lett.* **4**, 257–261.

Jasty, V., Bare, J.J., Jamison, J.R., Porter, M.C., Kowalski, R.L., Clemens, G.R., Jackson, G.E., Jr., and Hartnagel, R.E., Jr. (1986). Spontaneous lesions in the sternums of growing rats. *Lab. Anim. Sci.* **36**, 48–51.

Jimenez-Marin, D. (1971). Protein phenotypes of rat substrains showing a high incidence of heart defects. *Diss. Abstr. Int.* **32B**, 3117.

Johnson, D.D., Rudden, P.K., and O'Steen, W.K. (1979). Photically induced experimental exophthalmos: role of Harderian and pituitary glands. *Invest. Ophthalmol. Visual Sci.* **18**, 1280–1285.

Johnson, P.D., Dawson, B.V., and Goldberg, S.J. (1993). Spontaneous congenital heart malformations in Sprague Dawley rats. *Lab. Anim. Sci.* **43**, 183–188.

Kato, M., and Onodera, T. (1987). Early changes of osteochondrosis in medial femoral condyles from rats. *Vet. Pathol.* **24**, 80–86.

Kawaguchi, T., Matsubara, Y., Okada, F., Okuda, Y., Niwa, N., and Yamatsu, K. (1995). A case of spontaneous monoamniotic twins in a Sprague Dawley rat. *Contemp. Top. Lab. Anim. Sci.* **34**(5), 99–100.

Kazui, H., and Fujisawa, K. (1988). Radiculoneuropathy in aging rats: a quantitative study. *Neuropathol. Appl. Neurobiol.* **14**, 137–156.

Keenan, K.P., Smith, P.F., Hertzog, P., Soper, K., Ballam, G.C., and Clark, R.L. (1994). The effects of overfeeding and dietary restriction on Sprague-Dawley rat survival and early pathology biomarkers of aging. *Toxicol. Pathol.* **22**, 300–315.

Keenan, K.P., Soper, K.A., Hertzog, P.R., Gumprecht, L.A., Smith, P.F., Mattson, B.A., Ballam, G.C., and Clark, R.L. (1995a). Diet, overfeeding and moderate dietary restriction in control Sprague-Dawley rats: II. Effects on age-related proliferative and degenerative lesions. *Toxicol. Pathol.* **23**, 287–302.

Keenan, K.P., Soper, K.A., Smith, P.F., Ballam, G.C., and Clark, R.L. (1995b). Diet, overfeeding and moderate dietary restriction in control Sprague-Dawley rats: I. Effects on spontaneous neoplasms. *Toxicol. Pathol.* **23**, 269–286.

Keenan, K.P., Coleman, J.B., McCoy, C.L., Hoe, C.-M., Soper, K.A., and Laroque, P. (2000). Chronic nephropathy in *ad libitum* overfed Sprague-Dawley rats and its early attenuation by increasing degrees of dietary (caloric) restriction to control growth. *Toxicol. Pathol.* **28**, 788–798.

Kerns, W.D., Arena, E., Macia, R.A., Bugelski, P.J., Matthews, W.D., and Morgan, D.G. (1989). Pathogenesis of arterial lesions induced by dopaminergic compounds in the rat. *Toxicol. Pathol.* **17**, 203–213.

Kim, Ch., and Cervós-Navarro, J. (1991). Spontaneous saccular cerebral aneurysm in a rat. *Acta Neurochir. (Wien)* **109**, 63–65.

King, C.S., and Miller R.T. (1997). Fatal perforating intestinal ulceration attributable to flunixin meglumine overdose in rats. *Lab. Anim. Sci.* **47**, 205–208.

King, K.L., West, W.L., and Hunter, W.L. (1995). Diagnostic exercise: ocular lesions in hypophysectomized rats. *Contemp. Top. Lab. Anim. Sci.* **34**(3), 103–104.

Kitagaki, M., Suwa, T., Yanagi, M., and Shiratori, K. (2002). Auricular chondritis in young ear-tagged Crj:CD(SD)IGS rats *Lab. Anim.* **37**, 249–253.

Klahr, S., Buerkert, J., and Purkerson, M.L. (1983). Role of dietary factors in the progression of chronic renal disease. *Kidney Int.* **24**, 579–587.

Kohn, D.F., and Barthold, S.W. (1984). Biology and diseases of rats. *In* "Laboratory Animal Medicine." (J.G. Fox, B.J. Cohen, and F.M. Loew, eds.), pp. 91–122, Academic Press, Inc., San Diego.

Kohn, D.F., Chinookoswong, N., and Chou, S.M. (1981). A new model of congenital hydrocephalus in the rat. *Acta Neuropathol.* **54**, 211–218.

Koppel, J., Kuchár, S., Mozeš, Š., Petrušová, K., and Jasenovec, A. (1982). The changes in glycaemia after the administration of xylazine and adrenergic blockers in the rat. *Vet. Med. (Praha)* **27**, 113–118.

Krinke, G. (1983). Spinal radiculoneuropathy in aging rats: demyelination secondary to neuronal dwindling? *Acta Neuropathol.* **59**, 63–69.

Krinke, G., Suter, J., and Hess, R. (1981). Radicular myelinopathy in aging rats. *Vet. Pathol.* **18**, 335–341.

Krinke, A., Kobel, W., and Krinke, G. (1988). Does the repeated orbital sinus puncture alter the occurrence of changes with age in the retina, the lens, or the Harderian gland of laboratory rats? *Z. Versuchstierkd* **31**, 111–119.

Krugner-Higby L., Wolden-Hanson T., Gendron A., Atkinson R.L. (1996). High prevalence of gastric trichobezoars (hair balls) in Wistar-Kyoto rats fed a semi-purified diet. *Lab. Anim. Sci.* **46**, 635–639.

Ktorza, A., Bernard, C., Parent, V., Penicaud, L., Froguel, P., Lathrop, M., and Gauguier, D. (1997). Are animal models of diabetes relevant to the study of the genetics of non-insulin-dependent diabetes in humans? *Diabetes Metab.* **23**, 38–46.

Kufoy, E.A., Pakalnis, V.A., Parks, C.D., Wells, A., Yang, C.H., and Fox, A. (1989). Keratoconjunctivitis sicca with associated secondary uveitis elicited in rats after systemic xylazine/ketamine anesthesia. *Exp. Eye Res.* **49**, 861–871.

Kuhlmann, E.T., and Longnecker, D.S. (1984). Urinary calculi in Lewis and Wistar rats. *Lab. Anim. Sci.* **34**, 299–302.

Kunstýř, I., Küpper, W., Weisser, H., Naumann, S., and Messow, C. (1982). Urethral plug–a new secondary male sex characteristic in rat and other rodents. *Lab. Anim.* **16**, 151–155.

Kuper, C.F., Beems, R.B., and Hollanders, V.M.H. (1986). Spontaneous pathology of the thymus in aging Wistar (Cpb:WU) rats. *Vet. Pathol.* **23**, 270–277.

Kurisu, K., Sawamoto, O., Watanabe, H., and Ito, A. (1996). Sequential changes in the Harderian gland of rats exposed to high intensity light. *Lab. Anim. Sci.* **46**, 71–76.

Kyle, M.E., and Koscis, J.J. (1985). The effect of age on salicylate-induced nephrotoxicity in male rats. *Toxicol. Appl. Pharmacol.* **81**, 337–347.

Lalich, J.J., and J.R. Allen. (1971). Protein overload nephropathy in rats with unilateral nephrectomy. *Arch. Pathol.* **91**, 372–382.

Lanum, J. (1978). The damaging effects of light on the retina. Empirical findings, theoretical and practical implications. *Surv. Ophthalmol.* **22**, 221–249.

Lee, E.W., Render, J.A., Garner, C.D., Brady, A.N., and Li, L.C. (1990). Unilateral degeneration of retina and optic nerve in Fischer-344 rats. *Vet. Pathol.* **27**, 439–444.

Lee, F.T., Sproat, I.A., Hammond, T.G., and Naidu, S.G. (1994). Placement of intraperitoneal injections in rats. *Contemp. Top. Lab. Anim. Sci.* **33**(1), 64–65.

Lee, J.C., Karpeles, L.M., and Downing, S.E. (1972). Age-related changes of cardiac performance in male rats. *Am. J. Physiol.* **222**, 432–438.

Lee, K.P. (1986). Ultrastructure of proteinaceous bladder plugs in male rats. *Lab. Anim. Sci.* **36**, 671–677.

Levinsky, H.V. (1973). A case of conjoined twins in the Sprague-Dawley rat. *Lab. Anim. Sci.* **23**, 903–904.

Lewis, D.J. (1992). Nonneoplastic lesions in the cardiovascular system. *In* "Pathobiology of the Aging Rat, Vol. 1." (U. Mohr, D.L. Dungworth, and C.C. Capen, eds.), pp. 301–309, International Life Sciences Institute Press, Washington, D.C.

Liles, J.H., and Flecknell, P.A. (1992). The use of non-steroidal anti-inflammatory drugs for the relief of pain in laboratory rodents and rabbits. *Lab. Anim.* **26**, 241–255.

Lin, W., and Essner, E. (1987). An electron microscopic study of retinal degeneration in Sprague-Dawley rats. *Lab. Anim. Sci.* **37**, 180–186.

Linton, C.G., Gordon, B.E., and Richardson, J.A. (1994). Diagnostic exercise: thickened auricular pinnae in a Sprague Dawley rat. *Lab. Anim. Sci.* **44**, 69–70.

Lipman, N.S., Wardrip, C.L., Yuan, C.-S., Coventry, S., Bunte, R.M., and Li, X. (1998). Familial megacecum and colon in the rat: a new model of gastrointestinal neuromuscular dysfunction. *Lab. Anim. Sci.* **48**, 243–252.

Losco, P.E., and Troup, C.M. (1988). Corneal dystrophy in Fischer 344 rats. *Lab. Anim. Sci.* **38**, 702–710.

Lozzio, B.B., Chernoff, A.I., and Lozzio, C.B. (1967). Hereditary renal disease in a mutant strain of rats. *Science* **156**, 1742–1744.

Madsen, C. (1989). Squamous-cell carcinoma and oral, pharyngeal and nasal lesions caused by foreign bodies in feed. Cases from a long-term study in rats. *Lab. Anim.* **23**, 241–247.

Maeda, H., Gleiser, C.A., Masoro, E.J., Murata, I., McMahan, C.A., and Yu, B.P. (1985). Nutritional influences on aging of Fischer 344 rats: II. Pathology. *J. Gerontol.* **40**, 671–688.

Magnusson, G., and Ramsay, C.H. (1971). Urolithiasis in the rat. *Lab. Anim.* **5**, 153–162.

Magro, A.M., and Rudofsky, U.H. (1982). Plasma renin activity decrease precedes spontaneous focal glomerular sclerosis in aging rats. *Nephron.* **31**, 245–253.

Manninen, A., Anttila, S., Savolainen, H. (1988). Rat metabolic adaptation to ammonia inhalation. *Proc. Soc. Exp. Biol. Med.* **187**, 278–281.

Markovits, J.E., and Sahota, P.S. (2000). Granular cell lesions in the distal female reproductive tract of aged Sprague-Dawley rats. *Vet. Pathol.* **37**, 439–448.

Maronpot, R.R. (1986). Spontaneous hydronephrosis, rat. *In* "Monographs on Pathology of Laboratory Animals: Urinary System." (T.C. Jones, U. Mohr, and R.D. Hunt, eds.), pp. 268–271, Springer-Verlag, New York.

Masoro, E.J. (1991). Use of rodents as models for the study of "normal aging": conceptual and practical issues. *Neurobiol. Aging* **12**, 639–643.

Masoro, E.J., and Yu, B.P. (1989). Editorial: diet and nephropathy. *Lab. Invest.* **60**, 165–167.

Masoro, E.J., Iwasaki, K., Gleiser, C.A., McMahan, C.A., Seo, E-J., and Yu, B.P. (1989). Dietary modulation of the progression of

nephropathy in aging rats: An evaluation of the importance of protein. *Amer. J. Clin. Nutr.* **49**, 1217–1227.

Masoro, E.J., Shimokawa, I., and Yu, B.P. (1991). Retardation of the aging process in rats by food restriction. *Ann. N.Y. Acad. Sci.* **621**, 337–352.

Matsuura, T., Horikiri, K., Ozaki, K., and Narama, I. (1999). Proliferative retinal changes in diabetic rats (WBN/Kob). *Lab. Anim. Sci.* **49**, 565–569.

Mazzeo, R.S. (1994). The influence of exercise and aging on immune function. *Med. Sci. Sports Exerc.* **26**, 586–592.

McDonald, M.M., and Boorman, G.A. (1989). Pancreatic hepatocytes associated with chronic 2,6–dichloro-p-phenylenediamine administration in Fischer 344 rats. *Toxicol. Pathol.* **17**, 1–6.

McEwen, B.J., and Barsoum, N.J. (1990). Auricular chondritis in Wistar rats. *Lab. Anim.* **24**, 280–283.

McGee, M.A., and Maronpot, R.R. (1979). Harderian gland dacryoadenitis in rats resulting from orbital bleeding. *Lab. Anim. Sci.* **29**, 639–641.

Meihuizen, S.P., and Burek, J.D. (1978). The epithelial cell component of the thymuses of aged female BN/Bi rats: a light microscopic, electron microscopic, and autoradiographic study. *Lab. Invest.* **39**, 613–622.

Meingassner, J.G. (1991). Sympathetic auricular chondritis in rats: a model of autoimmune disease? *Lab. Anim.* **25**, 68–78.

Meites, J., Huang, H.H., and Simpkins, J.W. (1978). Recent studies on neuroendocrine control of reproductive senescence in rats. *In* "The Aging Reproductive System." (E.L. Schneider, ed.), pp. 213–235, Raven, New York.

Meyer, O.A., Kristiansen, E., and Wirtzen, G. (1989). Effects of dietary protein and butylated hydroxytoluene on the kidneys of rats. *Lab. Anim.* **23**, 175–179.

Mikhail, A.A., and Holland, H.C. (1966). A simplified method of inducing stomach ulcers. *J. Psychosom. Res.* **9**, 343–347.

Mitchell, D., Wells, C., Hoch, N., Lind, K., Woods, S.C., and Mitchell, L.K. (1976). Poison induced pica in rats. *Physiol. Behav.* **17**, 691–697.

Mitchell, D., Krusemark, M.L., and Hafner, E. (1977). Pica: a species relevant behavioral assay of motion sickness in the rat. *Physiol. Behav.* **18**, 125–130.

Mitsumori, K., Maita, K., and Shirasu, Y. (1981). An ultrastructural study of spinal nerve roots and dorsal root ganglia in aging rats with spontaneous radiculoneuropathy. *Vet. Pathol.* **18**, 714–726.

Mitsumori, K., Maita, K., Nakashima, N., and Shirasu, Y. (1986). Electronmicroscopy of sciatic nerves in aging rats with spontaneous radiculoneuropathy. *Jpn. J. Vet. Sci.* **48**, 219–226.

Miyamura, N., and Amemiya, T. (1998). Lens and retinal changes in the WBN/Kob rat (spontaneous diabetic strain). Electron-microscopic study. *Ophthalmic Res.* **30**, 221–232.

Mutinelli, F., Nani, S., and Zampiron, S. (1992). Conjoined twins (thoracopagus) in a Wistar rat (*Rattus norvegicus*). *Lab. Anim. Sci.* **42**, 612–613.

Neuhaus, O.W., and Flory, W. (1978). Age-dependent changes in the excretion of urinary proteins by the rat. *Nephron* **22**, 570–576.

Nguyen, H.T., and Woodard, J.C. (1980). Intranephric calculosis in rats: an ultrastructural study. *Am. J. Pathol.* **100**, 39–56.

Njaa, L.R., Utne, F., and Braekkan, O.R. (1957). Effect of relative humidity on rat breeding and ringtail. *Nature* **180**, 290–291.

Noell, W.K., Walker, V.S., Kang, B.S., and Berman, S. (1966). Retinal damage by light in rats. *Invest. Ophthalmol.* **5**, 450–473.

Novilla, M.N., Sandusky, G.E., Hoover, D.M., Ray, S.E., and Wightman, K.A. (1991). A retrospective survey of endocardial proliferative lesions in rats. *Vet. Pathol.* **28**, 156–165.

Nunn, G., and Macpherson, A. (1995). Spontaneous convulsions in Charles River Wistar rats. *Lab. Anim.* **29**, 50–53.

Nygård, G., Anthony, A., Piasecki, C., Trevethick, M.A., Hudson, M., Dhillon, A.P., Pounder, R.E., and Wakefield, A.J. (1994). Acute indomethacin-induced jejunal injury in the rat: early morphological and biochemical changes. *Gastroenterology* **106**, 567–575.

O'Donoghue, P.N., and Wilson, M.S. (1977). Hydronephrosis in male rats. *Lab. Anim.* **11**, 193–194.

Ogino, K., Hobara, T., Kobayashi, H., and Iwamato, S. (1990). Gastric mucosal injury induced by chloral hydrate. *Toxicol. Lett.* **52**, 129–133.

Opie, E.L., Lynch, C.J., and Tershakovec, M. (1970). Sclerosis of the mesenteric arteries of rats. Its relation to longevity and inheritance. *Arch. Pathol.* **89**, 306–313.

O'Steen, W.K., Kraeer, S.L., and Shear, C.R. (1978). Extraocular muscle and Harderian gland degeneration after exposure of rats to continuous fluorescent illumination. *Ophthalmol. Visual Sci.* **17**, 847–856.

Owen, R.A., and Heywood, R. (1986). Age-related variations in renal structure and function in Sprague-Dawley rats. *Toxicol. Pathol.* **14**, 158–167.

Palm, M. (1998). The incidence of chronic progressive nephrosis in young Sprague-Dawley rats from two different breeders. *Lab. Anim.* **32**, 477–482.

Paterson, M. (1979). Urolithiasis in the Sprague-Dawley rat. *Lab. Anim.* **13**, 17–20.

Peace, T.A., Singer, A.W., Niemuth, N.A., and Shaw, M.E. (2001). Effects of caging type and animal source on the development of foot lesions in Sprague Dawley rats (*Rattus norvegicus*) *Contemp. Top. Lab. Anim. Sci.* **40**(5), 17–21.

Percy, D.H., and Barthold, S.W. (2001). "Pathology of Laboratory Rodents and Rabbits, Second Edition." Iowa State University Press, Ames, IA.

Perry, S.W. (1965). Proteinuria in the Wistar rat. *J. Pathol. Bacteriol.* **89**, 729–733.

Peter, C.P., Burek, J.D., and van Zwieten, M.J. (1986). Spontaneous nephropathies in rats. *Toxicol. Pathol.* **14**, 91–100.

Petruska, J.M., Haushalter, T.M., Scott, A., and Davis, T.E. (2001). Diet restriction in rat toxicity studies: automated gravimetric dispensing equipment for allocating daily rations of powdered rodent diet into pouches and 7–day feeders. *Contemp. Top. Lab. Anim. Sci.* **40**(5), 37–43.

Phillips, J.C., Bex, C., Mendis, D., and Gangolli, S.D. (1986). Studies on the mechanism of diet-induced nephrocalcinosis: calcium and phosphorus metabolism in the female rat. *Food Chem. Toxicol.* **24**, 283–288.

Pollard, M., and Kajima, M. (1970). Lesions in aged germfree Wistar rats. *Am. J. Pathol.* **61**, 25–31.

Pollock, W.B., and Hagan, T.R. (1972). Two cases of torsion of the cecum and ileum in rats. *Lab. Anim. Sci.* **22**, 549–551.

Prieur, D.J., Young, D.M., and Counts, D.F. (1984). Auricular chondritis in fawn-hooded rats. A spontaneous disorder resembling that induced by immunization with type II collagen. *Am. J. Pathol.* **116**, 69–76.

Provoost, A.P. (1994). Spontaneous glomerulosclerosis: insights from the fawn-hooded rat. *Kidney Int.* **45**, S2–S5.

Pucak, G.J., Lee, C.S., and Zaino, A.S. (1977). Effects of prolonged high temperature on testicular development and fertility in the male rat. *Lab. Anim. Sci.* **27**, 76–77.

Rao, G. (1991). Light intensity-associated eye lesions of Fischer-344 rats on long-term studies. *Toxicol. Pathol.* **19**, 148–155.

Rao, G.N., Edmondson, J., and Elwell, M.R. (1993). Influence of dietary protein concentration on severity of nephropathy in Fischer-344 (F-344/N) rats. *Toxicol. Pathol.* **21**, 353–361.

Rao, M.S., Scarpelli, D.G., and Reddy, J.K. (1986). Transdifferentiated hepatocytes in rat pancreas. *Curr. Top. Dev. Biol.* **20**, 63–78.

Rapp, J.P. (1973). Age-related pathologic changes, hypertension, and 18–hydroxydeoxycorticosterone in rats selectively bred for high or low juxtaglomerular granularity. *Lab. Invest.* **28**, 343–351.

Reddy, J.K., Rao, M.S., Qureshi, S.A., Reddy, M.K., Scarpelli, D.G., and Lalwani, N.D. (1984). Induction and origin of hepatocytes in rat pancreas. *J. Cell Biol.* **98**, 2082–2090.

Reid, W.C., Carmichael, K.P., Srinivas, S., and Bryant, J.L. (1999). Pathologic changes associated with use of tribromoethanol (Avertin) in the Sprague Dawley rat. *Lab. Anim. Sci.* **49**, 665–667.

Reznik, G., and Reznik-Schïller, H. (1980). Pathology of the clitoral and prepucial glands in aging F344 rats. *Lab. Anim. Sci.* **30**, 845–850.

Richard, D., Bouley, G, and Boudène, C. (1978). Effects of ammonia gas continually inhaled by rats and mice. *Bull. Eur. Physiopathol. Respir.* **14**, 573–582.

Richardson, B., and H. Luginbïhl. (1976). The role of prolactin in the development of chronic progressive nephropathy in the rat. *Virchows Arch. A Pathol. Anat. Histol.* **370**, 13–19.

Ritskes-Hoitinga, J., and Beynen, A.C. (1988). Atherosclerosis in the rat. *Artery* **16**, 25–50.

Ritskes-Hoitinga, J., Lemmens, A.G., and Beynen, A.C. (1989). Nutrition and kidney calcification in rats. *Lab. Anim.* **23**, 313–318.

Ritskes-Hoitinga, J., Mathot, J.N.J.J., Danse, L.H.J.C., and Beynen, A.C. (1991). Commercial rodent diets and nephrocalcinosis in weanling female rats. *Lab. Anim.* **25**, 126–132.

Robinson, M., Hart, D., and Pigott, G.H. (1991). The effects of diet on the incidence of periodontitis in rats. *Lab. Anim.* **25**, 247–253.

Roe, F.J.C. (1994). Historical histopathological control data for laboratory rodents: valuable treasure or worthless trash? *Lab. Anim.* **28**, 148–154.

Roe, F.J.C., Lee, P.N., Conybeare, G., Tobin, G., Kelly, D., Prentice, D., and Matter, B. (1991). Risk of premature death and cancer predicted by body weight in early life. *Hum. Exp. Toxicol.* **10**, 285–288.

Roughan, J.V., and Flecknell, P.A. (2002). Buprenorphine: a reappraisal of its antinociceptive effects and therapeutic use in alleviating post-operative pain in animals. *Lab. Anim.* **36**, 322–343.

Ruben, Z., Rohrbacher, E., and Miller, J.E. (1983). Esophageal impaction in BHE rats. *Lab. Anim. Sci.* **33**, 63–65.

Rudofsky, U.H., and Magro, A.M. (1982). Spontaneous hypertension in fawn-hooded rats. *Lab. Anim. Sci.* **32**, 389–391.

Saito, M., Nakayama, K., Muto, T., and Nakagawa, M. (1982). Effects of gaseous ammonia on Mycoplasma pulmonis infection in mice and rats. *Exp. Anim.* **31**, 203–206.

Saito, N., and Kawamura, H. (1999). The incidence and development of periarteritis nodosa in testicular arterioles and mesenteric arteries of spontaneously hypertensive rats. *Hypertens. Res.* **22**, 105–112.

Saxton, J.A., Jr., and Kimball, G.C. (1941). Relation of nephrosis and other diseases of albino rats to age and to modifications of diet. *Arch. Pathol.* **32**, 951–965.

Schaerdel, A.D., White, W.J., Lang, C.M., Dvorchik, B.H., and Bohner, K. (1983). Localized and systemic effects of environmental ammonia in rats. *Lab. Anim. Sci.* **33**, 40–45.

Schoeb, T.R., Davidson, M.K., and Lindsey, J.R. (1982). Intracage ammonia promotes growth of Mycoplasma pulmonis in the respiratory tract of rats. *Infect. Immun.* **38**, 212–217.

Scipioni, R.L., Diters, R.W., Myers, W.R., and Hart, S.M. (1997). Clinical and clinicopathological assessment of serial phlebotomy in the Sprague Dawley rat. *Lab. Anim. Sci.* **47**, 293–299.

Semple-Rowland, S.L., and Dawson, W.W. (1987). Retinal cyclic light damage threshold for albino rats. *Lab. Anim. Sci.* **37**, 289–298.

Serrano, L.J. (1971). Carbon dioxide and ammonia in mouse cages: effects of cage covers, population, and activity. *Lab. Anim. Sci.* **21**, 75–85.

Shackelford, C.C., and Jokinen, M.P. (1991). Decreasing severity of chronic nephropathy in male F344/N rats gavaged with safflower oil. *Proc. Am. Coll. Vet. Pathologists (42nd Annual Meeting)*, Middleton, Wisconsin, USA. p. 48.

Sharp, P.E., and LaRegina, M.C. (1998). "The Laboratory Rat." CRC Press LLC, Boca Raton, FL.

Shibuya, K., Tajima, M., Yamate, J., Sutoh, M., and Kudow, S. (1986). Spontaneous occurrence of pulmonary foam cells in Fischer 344 rats. *Jpn. J. Vet. Sci.* **48**, 413–417.

Shibuya, K., Tajima, M., Yamate, J., Saitoh, T., and Sannai, S. (1991). Genesis of pulmonary foam cells in rats with diet-induced hyper beta-lipoproteinemia. *Int. J. Exp. Pathol.* **72**, 423–435.

Shibuya, K., Tajima, M., Saitoh, T., Yamate, J., and Nunoya, T. (1995). Suppressive effect of beta-carotene on the development of pulmonary foam cells in rats with hyper beta-lipoproteinemia. *Toxicol. Pathol.* **23**, 47–55.

Shibuya, K., Tajima, M., Yamate, J., Saitoh, T., and Nunoya, T. (1997). Carbon tetrachloride-induced hepatotoxicity enhances the development of pulmonary foam cells in rats fed a cholesterol-cholic acid diet. *Toxicol. Pathol.* **25**, 487–494.

Shimokawa, I., Higami, Y., Hubbard, G.B., McMahan, C.A., Masoro, E.J., and Yu, B.P. (1993). Diet and the suitability of the male Fischer 344 rat as a model for aging research. *J. Gerontol.* **48**, B27–B32.

Shinkai, S., Konishi, M., and Shephard, R.J. (1997). Aging, exercise, training, and the immune system. *Exerc. Immunol. Rev.* **3**, 68–95.

Short, B.G., and R.S. Goldstein. (1992). Nonneoplastic lesions in the kidney. *In* "Pathobiology of the Aging Rat, Vol. 1." (U. Mohr, D.L. Dungworth, and C.C. Capen, eds.), pp. 211–225. International Life Sciences Institute Press, Washington, D.C.

Shumiya, S. (1995). Establishment of the hereditary cataract rat strain (SCR) and genetic analysis. *Lab. Anim. Sci.* **45**, 671–673.

Silverman, J., and Muir, W.W. (1993). A review of laboratory animal anesthesia with chloral hydrate and chloralose. *Lab. Anim. Sci.* **43**, 210–216.

Skold, B.H. (1961). Chronic arteritis in the laboratory rat. *J. Am. Vet. Med. Assoc.* **138**, 204–207.

Smale, G., Bendele, A., and Horton, W.E., Jr. (1995). Comparison of age-associated degeneration of articular cartilage in Wistar and Fischer 344 rats. *Lab. Anim. Sci.* **45**, 191–194.

Smiler, K.L., Stein, S., Hrapkiewicz, K.L., and Hiben, J.R. (1990). Tissue response to intramuscular and intraperitoneal injections of ketamine and xylazine in rats. *Lab. Anim. Sci.* **40**, 60–64.

Smith, P.C., Stanton, J.S., Buchanan, R.D., and Tanticharoenyos, P. (1968). Intestinal obstruction and death in suckling rats due to sawdust bedding. *Lab. Anim. Care* **18**, 224–228.

Soeterboek, S.J.A.J., Ritskes-Hoitinga, J., Lemmens, A.G., and Beynen, A.C. (1991). Phosphorus-induced nephrocalcinosis in female rats: a study on regression and clinical abnormalities. *Lab. Anim.* **25**, 258–262.

Sokoloff, L. (1967). Articular and musculoskeletal lesions of rats and mice. *In* "Pathology of Laboratory Rats and Mice." (Cotchin, E., and Roe, F.J.C., eds.), pp. 373–390, Blackwell Scientific Publications, Oxford and Edinburgh.

Solleveld, H.A., and Boorman, G.A. (1986). Spontaneous renal lesions in five rat strains. *Toxicol. Pathol.* **14**, 168–174.

Spangler, E.L., and Ingram, D.K. (1996). Utilization of the rat as a model of mammalian aging: impact of pathology on behavior. *Gerontology* **42**, 301–311.

Spikes, S.E., Hoogstraten-Miller, S.L., and Miller, G.F. (1996). Comparison of five anesthetic agents administered intraperitoneally in the laboratory rat. *Contemp. Top. Lab. Anim. Sci.* **35**(2), 53–56.

Squire, R.A., and Levitt, M.H. (1975). Report of a workshop on classification of specific hepatocellular lesions in rats. *Cancer Res.* **35**, 3214–3215.

Stern, J.S., Gades, M.D., Wheeldon, C.M., and Borchers, A.T. (2001). Calorie restriction in obesity: prevention of kidney disease in rodents. *J. Nutr.* **131**, 913S–917S.

Stringer, S.K., and Seligmann, B.E. (1996). Effects of two injectable anesthetic agents on coagulation assays in the rat. *Lab. Anim. Sci.* **46**, 430–433.

Suzuki, T., Oboshi, S., and Sato, R. (1979). Periarteritis nodosa in spontaneously hypertensive rats–incidence and distribution. *Acta Pathol. Jpn.* **29**, 697–703.

Takeda, N., Hasegawa, S., Morita, M., and Matsunaga, T. (1993). Pica in rats is analogous to emesis: an animal model in emesis research. *Pharmacol. Biochem. Behav.* **45**, 817–821.

Tanaka, K., Inagaki, S., and Doi, K. (1994). Preretinal arteriolar loops in rats. *Lab. Anim. Sci.* **44**, 71–72.

Tanaka, M., Yanagi, M., Shirota, K., Une, Y., Nomura, Y., Masaoka, T., and Akahori, F. (1995). Eosinophil and foam cell accumulation in lungs of Sprague-Dawley rats fed purified, biotin-deficient diets. *Vet. Pathol.* **32**, 498–503.

Tapp, D.C., Wortham, W.G., Addison, J.F., Hammonds, D.N., Barnes, J.L., and Venkatachalam, M.A. (1989). Food restriction retards body growth and prevents end-stage renal pathology in remnant kidneys of rats regardless of protein intake. *Lab. Invest.* **60**, 184–195.

Tayama, K., and Shisa, H. (1994). Development of pigmented scales on rat skin: relation to age, sex, strain, and hormonal effect. *Lab. Anim. Sci.* **44**, 240–244.

Thomas, D.S., and Pass, D.A. (1997). Chronic ulcerative typhlocolitis in CBH-*rnu/rnu* (athymic nude) rats. *Lab. Anim. Sci.* **47**, 423–427.

Timmons, E.H., Chaklos, R.J., Bannister, T.M., and Kaplan, H.M. (1975). Dichlorvos effects on estrous cycle onset in the rat. *Lab. Anim. Sci.* **25**, 45–47.

Treloar, A.F., and Armstrong, A. (1993). Intermittent hematuria in a colony of Lewis x Brown Norway hybrid rats. *Lab. Anim. Sci.* **43**, 640–641.

Vachon, P., Faubert, S., Blais, D., Comtois, A., and Bienvenu, J.G. (2000). A pathophysiological study of abdominal organs following intraperitoneal injections of chloral hydrate in rats: comparison between two anaesthesia protocols. *Lab. Anim.* **34**, 84–90.

Van Camp, I., Ritskes-Hoitinga, J., Lemmens, A.G., and Beynen, A.C. (1990). Diet-induced nephrocalcinosis and urinary excretion of albumin in female rats. *Lab. Anim.* **24**, 137–141.

Van der Kogel, A.J. (1977). Radiation-induced nerve root degeneration and hypertrophic neuropathy in the lumbosacral spinal cord of rats: the relation with changes in aging rats. *Acta Neuropathol.* **39**, 139–145.

Van Herck, H., Baumans, V., Van der Craats, N.R., Hesp, A.P.M., Meijer, G.W., Van Tintelen, G., Walvoort, H.C., and Beynen, A.C. (1992). Histological changes in the orbital region of rats after orbital puncture. *Lab. Anim.* **26**, 53–58.

Van Steenis, G. and Kroes, R. (1971). Changes in the nervous system and musculature of old rats. *Vet. Pathol.* **8**, 320–332.

Van Winkle, T.J., Womack, J.E., Barbo, W.D., and Davis, T.W. (1988). Incidence of hydronephrosis among several production colonies of outbred Sprague-Dawley rats. *Lab. Anim. Sci.* **38**, 402–406.

Velasquez, M.T., Kimmel, P.L., and Michaelis, O.E., IV. (1990). Animal models of spontaneous diabetic kidney disease. *FASEB J.* **4**, 2850–2859.

Vessell, E.S. (1967). Induction of drug-metabolizing enzymes in liver microsomes of mice and rats by softwood bedding. *Science* **157**, 1057–1058.

Vesell, E.S., Lang, C.M., White, W.J., Passananti, G.T., and Tripp, S.L. (1973). Hepatic drug metabolism in rats: impairment in a dirty environment. *Science* **179**, 896–897.

Wagner, J.E., Owens, D.R., LaRegina, M.C., and Vogler, G.A. (1977). Self trauma and *Staphylococcus aureus* in ulcerative dermatitis of rats. *J. Am. Vet. Med. Assoc.* **171**, 839–841.

Wang, K., Li, D., and Sun, F. (2004). Dietary restriction may delay the development of cataract by attenuating the oxidative stress in the lenses of Brown Norway rats. *Exp. Eye Res.* **78**, 151–158.

Weber, K.T., Janicki, J.S., Pick, R., Capasso, J., and Anversa, P. (1990). Myocardial fibrosis and pathologic hypertrophy in the rat with renovascular hypertension. *Am. J. Cardiol.* **65**, 1G–7G.

Weber, K.T., Brilla, C.G., and Janicki, J.S. (1993). Myocardial fibrosis: functional significance and regulatory factors. *Cardiovasc. Res.* **27**, 341–348.

Wegener, A., Kaelger, M., and Stinn, W. (2002). Frequency and nature of spontaneous age-related eye lesions observed in a 2–year inhalation toxicity study in rats. *Ophthalmic Res.* **34**, 281–287.

Weihe, W.H. (1976). *In* "Control of the Animal House Environment." (T. Mcsheeby, ed.), Laboratory Animals Ltd., London.

Wexler, B.C. (1970). Co-existent arteriosclerosis, PAN, and premature aging. *J. Gerontol.* **25**, 373–380.

Wexler, B.C., McMurtry, J.P., and Iams, S.G. (1981). Histopathologic changes in aging male vs female spontaneously hypertensive rats. *J. Gerontol.* **36**, 514–519.

Whorton, J.A., and Tilstead, J. (1984). Cloudy corneas (what's your diagnosis?) *Lab Anim. (NY)* **13**(6), 21–23.

Wilens, S.L., and Sproul, E.E. (1938a). Spontaneous cardiovascular disease in the rat. I. Lesions of the heart. *Am. J. Pathol.* **14**, 177–200.

Wilens, S.L., and Sproul, E.E. (1938b). Spontaneous cardiovascular disease in the rat. II. Lesions of the vascular system. *Am. J. Pathol.* **14**, 201–221.

Wilhelmi, G. (1974). Species differences in susceptibility to the gastro-ulcerogenic action of anti-inflammatory agents. *Pharmacology* **11**, 220–230.

Will, L.A., Leininger, J.R., and Donham, K.J. (1979). Regurgitation and choke in rats. *Lab. Anim. Sci.* **29**, 360–363.

Witt, C.J., and Johnson, L.K. (1990). Diagnostic exercise: rear limb ataxia in a rat. *Lab. Anim. Sci.* **40**, 528–529.

Wolf, N.S., Li, Y., Pendergrass, W., Schmeider, C., and Turturro, A. (2000). Normal mouse and rat strains as models for age-related cataract and the effect of caloric restriction on its development. *Exp. Eye Res.* **70**, 683–692.

Wolman, M., Cervós-Navarro, J., Sampaolo, S., and Cardesa, A. (1993). Pathological changes in organs of rats chronically exposed to hypoxia. Development of pulmonary lipidosis. *Histol. Histopathol.* **8**, 247–255.

Woodard, J.C. (1971). A morphologic and biochemical study of nutritional nephrocalcinosis in female rats fed semipurified diets. *Am. J. Pathol.* **65**, 253–268.

Wright, J.R., Yates, A.J., Sharma, H.M., and Thibert, P. (1981). Spontaneous gastric erosions and ulcerations in BB Wistar rats. *Lab. Anim. Sci.* **31**, 63–66.

Yagil, Y., and Yagil, C. (1998). Genetic basis of salt-susceptibility in the Sabra rat model of hypertension. *Kidney Int.* **53**, 1493–1500.

Yamasaki, K. (1990). Spontaneous spermatic granuloma and related epididymal lesions in Sprague-Dawley rats. *Lab. Anim. Sci.* **40**, 533–534.

Yamasaki, K., and Anai, S. (1989). Kinked tail in Sprague-Dawley rats. *Lab. Anim. Sci.* **39**, 77–78.

Yamasaki, K., and Houshuyama, S. (1994). Proliferative bone lesions in Sprague Dawley rats. *Lab. Anim. Sci.* **44**, 177–179.

Yamasaki, K., and Inui, S. (1985). Lesions of articular, sternal and growth plate cartilage in rats. *Vet. Pathol.* **22**, 46–50.

Yamazaki, H., Ikeda, H., Ishizu, A., Nakamaru, Y., Sugaya, T., Kikuchi, K., Yamada, S., Wakisaka, A., Kasai, N., Koike, T., Hatanaka, M., and Yoshiki, T. (1997). A wide spectrum of collagen vascular and autoimmune diseases in transgenic rats carrying the env-pX gene of human T lymphocyte virus type I. *Int. Immunol.* **9**, 339–346.

Yang, Y.H. (1965). Polyarteritis nodosa in laboratory rats. *Lab. Invest.* **14**, 81–88.

Yang, Y.H., Yang, C.Y, and Grice, H.C. (1966). Multifocal histiocytosis in the lungs of rats. *J. Pathol. Bacteriol.* **92**, 559–561.

Yoshitomi, K. (1990). Cystic dilatation of the vaginal fornix in aged female Crj:F344/Du rats. *Vet. Pathol.* **27**, 282–284.

Young, C., Festing, M.F.W., and Barnett, K.C. (1974). Buphthalmos (congenital glaucoma) in the rat. *Lab. Anim.* **8**, 21–31.

Zaidi, I., Sullivan, D.J., and Seiden, D. (1982). Endocardial thickening in the Sprague-Dawley rat. *Toxicol. Pathol.* **10**, 27–32.

Chapter 16

Medical Management and Diagnostic Approaches

Glen Otto and Craig L. Franklin

THE LABORATORY RAT, 2ND EDITION

I. INTRODUCTION

This chapter reviews the basic principles of medical management of rat colonies and diagnostic approaches to detect infectious diseases of rats. As is the case with all other species, rats are susceptible to a variety of injuries and diseases that can cause distress, morbidity, or mortality. Any facility that houses rats must develop monitoring programs designed to rapidly identify health-related problems so they can be communicated to appropriate veterinary or animal care personnel to be resolved. These programs generally consist of multiple components, some of which are directed towards individual animals, and others that assess the health status of rat populations as a whole. Relevant aspects of medical management of rat colonies include individual animal monitoring and care; signs of illness and distress; colony health management; components of microbiological monitoring programs, including agents commonly targeted and sentinel programs; quarantine; biological material screening; diagnostic testing methodologies, including culture, serology, molecular diagnostic and histopathology; test profiles and interpretation; management of disease outbreaks; and treatment and prevention strategies for infectious agents.

II. INDIVIDUAL ANIMAL MONITORING AND CARE

A. Observation and Examination

Daily direct observation of rats for signs of abnormality is important, from a humane and ethical standpoint (Institute for Laboratory Animal Research-National Research Council, 1996a), and is vital in order to quickly identify problems caused by mechanical failure, trauma, pathogenic organisms, spontaneous disease, or research manipulations (Institute for Laboratory Animal Research-National Research Council, 1996b). The persons given the responsibility for this observation need to have had sufficient training or previous experience with rats in order to adequately detect abnormalities that might be present. Grossly visible trauma or lesions may not always be present in ill rats, but more subtle behavioral clues might be present. For this reason, it is important to be familiar with the normal range of behavioral patterns observed in healthy captive rats, which has been described (Saibaba et al., 1995).

Using appropriately sized equipment and methods similar to those used for larger species, it is possible to perform a relatively complete physical exam upon rats

when indicated (Sharp and LaRegina, 1998). The use of fabric or plastic restraint devices and/or protective gloves may be needed if the animal is fractious or the individual doing the examination is inexperienced, but most strains of rats are fairly amenable to gentle handling.

Careful observation and diagnostic evaluation is indicated for animals that show clinical signs when irradiated, exposed to corticosteroids or other immunosuppressive agents, or subjected to other types of significant stress, because latent infections might become symptomatic at these times (Small, 1984). Likewise, genetically immunodeficient animals (such as *rnu/rnu* rats) may manifest signs of disease from agents that are clinically silent in immunocompetent animals housed in the same area. Although the overall health status of most institutional rat colonies is monitored by routine screening of asymptomatic animals, it is important to realize that daily individual animal observation can sometimes identify an "index case" of a newly introduced disease that has not yet been revealed via routine scheduled testing.

B. Signs of Illness and Distress

Abnormal physical findings in rodents are not always useful in localizing an illness to a specific organ system. A very common constellation of findings indicative of pain or distress is piloerection, decreased activity, an ungroomed appearance, and often a hunched posture (Institute for Laboratory Animal Research-National Research Council, 1992). Weight loss is another nonspecific finding, but since weight determination is a simple, rapid, objective, and noninvasive technique, it is commonly used to assess the general health status of an animal placed under. It should be realized that stress is not always manifested as an absolute weight loss in a growing animal, so it may be necessary to take into account the normal weight gain of young rats in order to document a variation (Dymsza et al., 1963).

Table 16-1 describes signs of illness that can be seen in rats, along with possible diagnoses. This list is not meant to be an exhaustive summary, but it includes some of the more common clinical signs and suggests potential differential diagnoses.

C. Treatment of Disease

The majority of drugs administered to laboratory rats are provided prophylactically (for example, as part of perioperative care) or as a direct component of the research study. Because both the disease state and the use of xenobiotics (antibiotics) can affect the physiology of animals in a way that is difficult to control within the experimental design and could invalidate a study

TABLE 16-1

PHYSICAL FINDINGS

Abnormality	Potential diagnosis
Pale mucous membranes, extremities, or eyes	Anemia (if rat appears otherwise relatively normal)
	Excessive or too frequent blood collection
	Circulatory deficiency (if animal appears weak or depressed)
Alopecia with normal, intact skin	Physical abrasion from cage or feeder
	Self-grooming or barbering from cage mate
Alopecia with crusted, inflamed or ulcerated skin	Ectoparasites or dermatophytes
	Bacterial opportunists such as *Staphylococcus aureus*
	Pruritic syndromes
Dermal or subcutaneous masses	Tumors of skin or mammary origin
	Lymphadenopathy
	Abscess, granuloma, cyst
Nodular deformity of ears	Auricular chondritis
"Red" or "bloody" tears	Chromodacryorrhea caused by Harderian gland secretions (can also be seen on front paws and over the back from grooming)
	Frequent clinical sign of SDAV infection
Circumferential, annular constrictions on tail	Ringtail (generally seen in suckling animals under conditions of low humidity and a cool or poorly insulated environment)
Head tilt, circling or spinning when lifted by tail	Bacterial or mycoplasmal otitis interna/media
	Tumor or other space-occupying brain lesion
Hairless, swollen, or bleeding plantar lesions	"Sore hock" syndrome associated with large and/or aged rats kept on wire or mesh flooring
Salivation, weight loss, swollen oral tissues	Malocclusion
Fecal staining	Diarrhea
Dyspnea/rales/hyperventilation	*Mycoplasma pulmonis*, CAR bacillus, *C. kutscheri*, or *S. pneumoniae* infection
	Overheating
Facial swellings	Parotid and/or submandibular salivary gland swelling from coronavirus (RCV/SDAV) infection
	Abscess of lymph nodes (lymphadenitis)
	Zymbal gland tumor at the base of the ear
Abdominal distension (pot-bellied appearance)	Ascites
	Intestinal distension from toxicity (chloral hydrate)
	Enteritis (possibly megaloileitis associated with Tyzzer's disease)
	Obesity
	Abdominal mass (tumor, abscess)
	Pregnancy
Excessively wet hair coat and/or bedding	Diabetic polyuria
	Leaking bottle or automatic water system
	Behavioral water wastage from "playing"
	Overheating
Eye lesions	Blepharospasm, corneal opacities, keratitis due to coronaviral infection
	Cataracts (aging lesion)

RCV = rat corona virus; SDAV = Sialodacryoadenitis virus.

(Lipman and Perkins, 2002), ill rats are often euthanized rather than treated. However, the situation surrounding the incident should be carefully considered to determine whether it is prudent to gather appropriate antemortem diagnostic samples and to submit the carcass for necropsy evaluation even if the animal is terminated. Likewise, a process to monitor animal mortality records and to perform necropsy on animals whose death is suspicious is quite important, because in some cases such an evaluation can allow early detection of a problem that otherwise would reoccur and eventually affect a much larger group of animals.

In some situations, it certainly is useful to treat individual animals or larger groups if the animals are considered valuable to an ongoing study or are not being used to generate sensitive data. It is beyond the scope of this chapter to describe particular pharmaceutical dosages and treatment indications, but the reader can be directed elsewhere in this volume for disease-specific recommendations. Well-referenced and comprehensive formularies that include rat-specific drug dosages are also available, written for veterinarians in both the laboratory animal and "exotic" pet specialties (Carpenter et al., 1996; Hawk and Leary, 1999).

III. COLONY HEALTH MANAGEMENT

A. Need for Monitoring

Despite the fact that some infectious agents (for example, virulent strains of rat coronavirus) can induce readily identifiable abnormal physical signs, most infectious agents encountered in laboratory rat populations cause only subclinical disease, that is often signaled by interference with a research process. Such conditions can only be detected and identified via sensitive and comprehensive testing protocols (Institute for Laboratory Animal Research-National Research Council, 1991). Despite the lack of observable morbidity or mortality, these subclinical infections pose a significant risk to the research conducted with affected animals because they can alter the background physiology of experimental subjects or cause variation and alteration in specific experimental responses. These adverse effects have been summarized in a number of reviews (Institute for Laboratory Animal Research-National Research Council, 1991; Baker, 1998; Nicklas et al., 1999; Lipman and Perkins, 2002; Baker, 2003), and more recent studies continue to add to the list of potential adverse implications (Ball-Goodrich et al., 2002). Two of the physiologic processes that have specifically been shown to be altered by the presence of these infectious agents in rodents include immune function and neoplasia, which are quite relevant since immunogenetics, transplantation, and tumor biology are three of the disciplines that most heavily depend on the use of rats (Gill et al., 1989). As a result, one of the primary aims of a rodent colony health monitoring program is to document the presence or absence of particular infections agents irrespective of any observable disease state. A term which is generally synonymous with colony health monitoring and is often used to convey this emphasis is *microbiologic monitoring* (Fujiwara and Wagner, 1994; Waggie et al., 1994).

Although there has been great progress in defining and improving the overall health status of laboratory rodent colonies in recent decades, many of the agents of concern remain endemic in institutional colonies (Institute for Laboratory Animal Research-National Research Council, 1991; Jacoby and Lindsey, 1998; Livingston and Riley, 2003), or may be introduced via human contacts or feral rodent contamination. Significant risks of one or more of these agents being introduced still exist in contemporary colonies. Due to the fact that a comprehensive colony health monitoring program is so vital in protecting the validity and reproducibility of experimental research data, it must be devoted an appropriate priority in terms of budget, personnel, and other resources. The policies and practices should be defined in written plans, and agreement with the principles set forth should be secured by the scientific and administrative leadership of the institution, as well as by the veterinary and animal care group.

B. Health Status Terminology

The terms *axenic* and *gnotobiotic* refer to animals that harbor no cultivatable organisms or have a completely defined microbiological flora, respectively (see gnotobiology chapter); as a consequence, the health status of these animals regarding pathogenic or opportunistic agents is relatively easy to characterize. Terms that are less useful without detailed backup information include *specific pathogen-free* (SPF), and *conventional*. In general use, SPF refers to animals that are 1) considered to be free of major pathogens and some or all opportunists, 2) maintained under housing and use conditions designed to protect this high-quality status by excluding infectious agents, and 3) monitored closely to assure that there is no undetected introduction of excluded agents. Conventional animals are usually considered to be those that originate from uncontrolled colonies that are not subjected to routine health monitoring, or those in which some degree of monitoring occurs but there is no action taken if infectious agents are found. However, these terms are not really descriptive or representative enough to use when assigning risk to animals proposed for introduction to monitored, disease-free animal facilities. For example, animals from an institution that experienced an outbreak of rat coronavirus and decided not to undertake the steps needed to eliminate the agent from the facility can still be considered to be "SPF", since by definition this status is only defined by the particular list of agents of which the animals are specifically free. The term *conventional* also suffers from some ambiguity, since it can be used to refer to not only to animal health status but also to facility design. For example, a facility that allows direct staff entry into rat rooms without changing out of street clothes would be termed a conventional facility rather than a "barrier" in the purest sense, but if micro-isolators and high efficiency particulate air (HEPA)-filtered changing hoods have been used to successfully institute pathogen exclusion at the cage level, the animals themselves might possess a high-quality health status that is far from "conventional." In practice, it may be more effective to evaluate a specific panel of agents that animals have been tested for than to assign quality descriptions.

C. Colony Health Management Considerations

In contrast to a program designed to monitor individual animal health through the use of direct methods such as close observation and physical examination, a program created to monitor the overall health status of a colony population will often utilize more indirect methods. Routine testing of selected representative animals (even in the absence of any signs of illness or disease) can provide valuable information regarding the viral, parasitic, and bacterial agents that such animals are either currently harboring or have been exposed to in the past.

Risk analysis should be done by any institution planning on holding rodents, a process which should involve a discussion of the relative costs and benefits of the various options available for routine health monitoring as well as quarantine isolation and testing. Although the available expertise of trained veterinarians, colony managers, and other professionals must be utilized, the discussions should not completely exclude the primary research directors and institutional officials who are needed to support the program both financially and administratively. It is also important to establish good communications with those individuals utilizing the animals for research so that they can report abnormal physiological responses or other experimental variation. It is not uncommon for research personnel to identify a problem with rodent-derived experimental data that ultimately is found to be due to microbiological contamination (Small, 1986; McKisic et al., 1993).

Risk-based sampling strategies will take into account 1) the frequency of introduction of new animals, 2) the quality and reliability of the source of introduced animals, 3) the mode of transport, 4) the pattern of personnel traffic into and out of the room, 5) the frequency of animal transport into and out of the room as part of the research project, 6) the potential for cross-contamination from other rooms inherent in the facility design, 7) the housing system, 8) the facility configuration (barrier design), 9) the proportion of the animals that are irreplaceable, 10) the proportion that are immunocompromised due to genetic factors, chemotherapy, or experimental stress, 11) the prevalence of infectious pathogens within the animal facility among laboratory rats in general, and 12) the potential for introduction of pathogens through the use of biological materials. The development of genetically engineered rats will add additional factors to consider when assessing risks of infectious disease.

It is important to realize that the window of detection varies for different types of diagnostic tests, and this must be taken into account whenever vendor screening or quarantine test results are interpreted. For example, when an antibody detection method (serology) is performed upon arrival, it is often considered representative of the vendor's colony, while subsequent seroconversion evident in serum drawn 2 to 4 weeks after delivery may be indication of exposure during transport or shortly after arrival at the user facility. However, tests that directly identify components of the agent (such as an antigen detection assay or polymerase chain reaction, or PCR test) could theoretically be positive upon arrival due to either a pre-existing vendor problem or in-transit contamination.

D.   Specific Components of Microbiological Monitoring

The primary goal of health monitoring is to detect the presence of an organism in at least one animal in the sample population, provided the organism is present. Equally as important, such testing is the means by which any of a panel of agents may be confirmed as not present. The components of most colony health monitoring programs include 1) periodic routine assessment of resident animals via random screening or targeted testing of dedicated sentinel animals, 2) the assessment of incoming animals through the use of vendor screening and/or quarantine testing, and 3) the assessment of biological materials destined for use in rat experiments. However, as stated earlier, the program for individual animal monitoring should include an attempt to identify index cases of diseases in the early stages of outbreaks that have not yet been detected by overall colony monitoring.

There is great variety in rodent health surveillance programs, and no two designs are usually identical (Institute for Laboratory Animal Research-National Research Council, 1991). However, some authorities advocate a certain degree of standardization (Jacoby and Homberger, 1999), and there are regional organizations that provide detailed specific guidelines for institutions that wish to participate (Yamamoto et al., 2001; Nicklas et al., 2002). Ultimately, health monitoring program design should cater to the needs of the institution. Consideration should also be given to needs of other institutions that could receive rats from the home institution (for example, the sharing of genetically engineered rats).

1.   Random Testing of Resident Animals

Health monitoring is often performed on representative residents removed from the colony for specific testing. When selecting animals for screening, there are certain points to keep in mind. Animals to be sampled should be taken from rack and shelf locations spread throughout the room to maximize the possibility of detecting an isolated focus of contamination. If multiple stocks or strains are present, an attempt should be made to sample representative rats from each of these subcolonies. It is also desirable to test both young and old animals (avoiding geriatric animals) if they are available, since parasite burdens may be higher in the young (Institute for Laboratory Animal Research-National

Research Council, 1991), while the old would have had the best chance of seroconverting to agents that may have not yet affected younger animals. In a breeding colony, ideal choices might be retired breeders and surplus weanlings. Immunodeficient animals are a good choice for detecting parasites and bacterial contamination, since they may have a lowered resistance to such agents. However, it is important to remember that serology will be subject to false negative results if performed on an animal with a genetic or induced immunosuppression that may impair the antibody response. The purpose of health surveillance is generally not to accurately determine the specific *prevalence* of infection or disease in an area, but rather to accurately identify its *presence* by finding at least one positive animal in an endemic colony (Institute for Laboratory Animal Research-National Research Council, 1991). The minimum number of animals from a population that need to be tested in order to identify one positive animal can be viewed as a statistical exercise in random sampling. Probability theory can provide the equation necessary to determine the sampling size needed, based on the assumption that one is dealing with an ideal population (100 or more animals, where all animals have an equal opportunity for pathogen exposure) and calculated based on variables such as the prevalence of infection (often estimated at 30%) within the population and the degree of confidence required in the result (Clifford, 2001). These equations have been used to prepare charts that have been published to assist in the selection of sample size (Small, 1984; DiGiacomo and Koepsell, 1986; Institute for Laboratory Animal Research-National Research Council, 1991). For example, if an infectious agent affects 25% of the rats in a population, one would only need to test 15 randomly selected individuals in order to have a 99% probability of detecting the agent. These calculations have the most robust application when dealing with large populations of animals held under conditions that provide little or no barrier to cage-to-cage transmission (that is, an open shoe box or suspended caging with no filter tops) because, under those circumstances, agents are fairly uniformly distributed and would be expected to have a prevalence of 30% or higher. Under those circumstances, even if a room holds 1000 rats, it would still only be necessary to sample 8 of them to be 95% sure that an agent is not present. Many vendor quality assurance programs are based on this type of calculation.

2. Targeted Sentinel Programs

Alternatives to a random sampling approach are needed because a large percentage of rats in contemporary research colonies are housed under circumstances that do not result in a uniform distribution of transmissible agents, due to the popularity and utility of cubicles (segregating

fewer than 100 animals into functional groups) and/or systems that provide a barrier at the cage level, such as static microisolators or ventilated rack caging. These housing systems are beneficial in decreasing the likelihood of disease transmission, but they also make it harder to detect infectious agents based on random screening protocols, because the agent distribution is not uniform and the prevalence of infection may be far below 30%. Another problem with random screening techniques is the impracticality of selecting and testing animals from active research colonies without disrupting the ongoing research. For this reason, it is common to place *sentinel* animals into a colony for the sole purpose of health status testing. These animals are not assigned to any particular study, and under ideal circumstances they will be exposed to the same agents as the *principal* animals actually being used for research. Because they exist solely for the screening program, sentinels can be bled, sampled, or removed for nonsurvival testing at the discretion of the colony management, without interfering with ongoing experiments.

Sentinels should be immunocompetent young adult rats (6 to 8 weeks of age) (Koszdin and DiGiacomo, 2002). The use of aged rats should be avoided if possible, as these animals may be more prone to false positive seroreactivity (Wagner et al., 1991). Selection of a particular stock or strain of rat to be used as sentinels will vary, and there is no single correct choice. Using the same stock and source as the principals that are being monitored may be ideal because it eliminates the additional risk of contamination that would occur if animals were imported from another source specifically to be used as sentinels (Institute for Laboratory Animal Research-National Research Council, 1991). For closed breeding colonies, this can be done by setting aside some of the animals bred locally to be used as sentinels, and if animals are commercially obtained, extra animals can be ordered along with the principal shipment. However, this approach is not always practical in non-closed colonies, and it is common for facilities to specifically order sentinel animals from a reliable commercial source to be used as sentinels. In this case, a readily available outbred line is often chosen for sentinel use, since they are inexpensive and will generally mount a robust antibody response. Inbred lines can also be used, but it is important to consider any strain-specific limitations of disease susceptibility or immune responses, since these may affect their utility as sentinels. Occasionally, sentinels will be chosen so that they have a coat color that differs from the principal animals they are associated with to minimize the possibility that they will be mistaken for experimental animals.

Sentinels should be placed in physical proximity to the principal animals they are associated with to ensure that they are exposed to equivalent environmental contamination. It is desirable to place them in a consistent spot on each rack so that husbandry and research staff can anticipate their location. If a single cage is used, it is customary to place it on the bottom shelf, since it is assumed that the concentration of aerosolized agents and particulate fomites will be highest near the floor. There are no firm guidelines for the relative density of sentinels, but for logistical reasons at least one sentinel cage should be in place on each rack or in each cubicle. The placement of one sentinel cage on each standard 25–cage rack has historically worked well in most situations. Other approaches can be taken, such as allocating sentinel cages to each breeding or experimental subgroup, placing multiple cages on each rack to increase the theoretical sensitivity of the program, etc. Depending on the specific design of the program, each sentinel cage may contain either a single animal or a small group of rats. When multiple sentinels of the same age are kept in a cage, it is rarely useful to sample more than one at any time, since the microbiological status of cohabitating animals is generally uniform. The use of small groups offers an advantage: the other sentinels in the cohort can be used to confirm positive results found in the rat initially submitted for testing.

In housing situations where filter-topped cages are being monitored, it is a common practice to remove the lids from the cages used to hold sentinels, effectively keeping them in "open" cages. This is done to increase the exposure of the sentinels to environmental contamination that might be transmitted by either true aerosols or small particulate fomites that are generated and dispersed within the room as part of routine rodent care and use. However, it should be noted that, in this type of situation, the subpopulation of rats with the highest cumulative risk of becoming infected with an agent (for example, the sentinel cages receiving a constant flow of dirty bedding) are NOT being held with the same degree of cage-level containment as the principal animals, and if they do become infected, the amount of environmental contamination and subsequent cross-contamination to other cages in the home room or elsewhere may be increased. Since other open sentinel cages in the room would be at highest risk for secondary transmission, it also may become more difficult to determine the point of origin of an agent within a room if sentinels are becoming infected, not from their assigned principal cages, but from other sentinel cages, essentially giving a type of false positive result (Weisbroth et al., 1998). During the sentinel program planning process, the benefit of a potential increase in the sensitivity of open-caged sentinels to detect an agent needs to be balanced against these potential adverse effects.

The process of routinely transferring *soiled* bedding from principal rodent cages into sentinel cages will increase the sensitivity of a monitoring program and can decrease the duration of sentinel exposure needed to detect endemic agents (Thigpen et al., 1989). The specific procedures utilized for the collection of soiled bedding and the transfer to sentinel cages vary widely as a result of the different types of cage/rack/hood configurations that are used and because they must integrate with the specific procedural methods being used for overall cage changing. However, to ensure that the transfer of bedding is having a net positive impact on colony health (by aiding in the detection of excluded agents) rather than a net negative impact (by increasing the cage-to-cage transmission between principal cages) this practice should be standardized and incorporated into both written procedural descriptions and employee training programs. It should be realized that dirty bedding transfer may not reliably transmit all agents of concern in a rat colony (Dillehay et al., 1990; Artwohl et al., 1994; Cundiff et al., 1995).

The optimum time interval between the placement of sentinels and their screening is another factor that has not been definitively determined. The time it takes a sentinel to be exposed to endemic infection would be expected to vary depending on specifics such as 1) the relative density of sentinels, 2) the frequency of cage changing and soiled bedding transfer, 3) the percentage of principal cages that have bedding sampled at each change, 4) the caging system in place, 5) the prevalence and transmissibility of the infectious agent present, and 6) possibly the macro-environmental characteristics of the room, such as relative humidity and ventilation. Once a sentinel is exposed, there will be an additional delay until the development of an immune response ascends to levels that can be detected by serologic means. Experimentally, it has been shown that sensitive antibody determination tests can identify sero-conversion in a period as short as 1 week post-exposure for rats infected with agents such as the rat coronavirus/sialodacryoadenitis virus (RCVSDAV) (Smith, 1983) and the RV parvovirus (Ball-Goodrich et al., 2002). However, a more "average" timeframe is within the range of 2 to 3 weeks, and it is felt that the utility of testing results will be greatest if a period of 21 to 28 days is allowed for seroconversion. For this reason, sentinels should generally not be sampled before they have had at least 1 month of exposure. It cannot be assumed that an agent will make its way into a sentinel cage during the first week or two, so many programs allow for an exposure period longer than 1 month (for example, utilizing 2 to 3 months of exposure as part of a quarterly monitoring schedule).

3. Vendor Screening

Facilities wishing to verify the reports obtained from commercial colonies may establish formal vendor screening programs whereby a small group of rats are obtained specifically for diagnostic testing without the vendor's knowledge. Sampling of animals that are euthanized immediately upon arrival can provide confirmation of the health status of the animals as maintained by the vendor, although it should be recognized that serology would generally not be expected to consistently identify animals infected less than a week previously. If the intent is to fully evaluate the status of all animals delivered, this testing must be repeated for each breeding unit of animals accepted from the vendor, and it should also account for the fact that vendors may produce the same strain in multiple, physically distinct breeding or holding areas (Small, 1984). Such a program might be feasible for facilities with a very limited list of vendors and a small number of strains in use, but is often impractical for facilities that serve large, multidisciplinary institutions. In this situation, a more limited and targeted vendor surveillance program might be useful (for example, surveying animals when a new vendor is under consideration, or getting more information if there are specific concerns about the status of animals from a particular vendor for some reason). Occasionally, the status of the vendor's production colony is not in question, but possible contamination during transport and delivery is suspected. If that is the case, incoming vendor animals for testing should not be killed upon arrival, but should be placed in a quarantine facility that provides for not only containment but also exclusion of infectious agents (to eliminate confounding cross-contamination within the facility). They can then be given time to fully colonize with and/or seroconvert to agents they were exposed to in transport, and tested on a schedule similar to other animals subjected to quarantine.

4. Quarantine

In many cases, the relative risk to the existing colony from newly-acquired animals that are shipped from a high-quality vendor and arrive in intact, filtered shipping containers is small enough to allow direct introduction into the room (Small, 1986; Institute for Laboratory Animal Research-National Research Council, 1996b). The documented procedures for rodent receipt under these circumstances should include a careful inspection of the containers upon arrival, the rejection of those that are damaged, and careful handling and disinfection of the external surfaces to minimize the risks from superficial contamination of the crate.

In contrast, animals proposed for introduction from noncommercial sources are often bred, packed, and shipped under less stringent conditions, and the establishment of a quarantine program for this type of transfer is very important.

The type of health monitoring documentation available when animals are obtained from a university, research institute, or biotechnology/pharmaceutical company may

be quite variable, and should be carefully interpreted as plans are made to receive and quarantine rats. Terms such as *SPF* or *conventional* are useful in relaying the general status of a colony, or to contrast the differing characteristics of animals from different rooms/buildings/facilities (much the same as the terms *clean* and *dirty*) but much more specific information should be obtained from the sending institution. From a health monitoring perspective, the status of each cohort of imported animals must be defined individually, based on the recent and historical findings of specific health monitoring tests. When introducing animals into a disease-free facility and making decisions about the relative risk, all animals should be considered suspect until there is data to suggest otherwise.

It is vitally important to achieve functional segregation and isolation of animals during a quarantine period, not only to protect the health status of other rodents in the facility, but also to ensure the ability to accurately determine the actual source of any contamination identified during quarantine. Room-level isolation would be ideal, but often there are space constraints when dealing with small shipments of rodents, and the common procedure is to utilize flexible-film isolators, cubicles, or ventilated cabinets of some type to partition a quarantine room (Small, 1984). In contemporary colonies, the introduction of a novel, noncommercial rat strain is a much less frequent occurrence than the transfer of a mutant mouse line. However, if this activity increases in the future (as many in the field feel that it will) it may be necessary to consider programs similar to those described for mouse quarantine that group multiple shipments into a single cohort for batch testing (Rehg and Toth, 1998). The availability of microisolator-type caging, either as static units or within ventilated racks, has also allowed programs to be designed that are not all-in-all-out but still allow functional isolation and segregation of multiple shipments within the same room (Institute for Laboratory Animal Research-National Research Council, 1996b; Otto and Tolwani, 2002). Although this option will provide more flexibility and may reduce the space requirements for quarantine, proper operational procedures are extremely important, since the whole system is reliant upon proper technique.

5. Screening of Imported Biological Materials

All tissue cultures and tumors should be tested and approved as free of infective contaminants prior to use in rats (Sharp and LaRegina, 1998). Recent experiences have shown that even cell-free biologicals have the potential to introduce agents to rodent colonies when imported (Lipman et al., 2000). Similar to the procedures used for mouse tissues, a rat antibody production (RAP) bioassay can be performed, whereby naïve animals held in quarantine are inoculated with a representative aliquot of the suspect material and tested 4 to 6 weeks later for seroconversion to excluded agents (Small, 1984; Johnson, 1986). Alternatively, newer technology makes it possible to utilize various types of polymerase chain reaction (PCR) assays on the materials themselves to more directly assess them for the presence of infectious agents (Besselsen et al., 2002; Bootz and Sieber, 2002; Blank et al., 2004).

IV. DIAGNOSTIC TESTING

A. Agents to be Monitored

There should be a very specific justification for each agent tested, based on the potential for an adverse effect on animal health or research studies (Institute for Laboratory Animal Research-National Research Council, 1991). There is no firm agreement on exactly which agents should be eliminated from high-quality rat populations, but there is a general consensus on the organisms which have the most potential detrimental impact and thus are almost universally monitored for and excluded (Waggie et al., 1994; Institute for Laboratory Animal Research-National Research Council, 1996b; Nicklas et al., 2002). These agents are listed in Table 16-2.

There are a number of agents not on this list that also have the potential for significant impact and merit monitoring in colonies being carefully maintained. *Clostridium piliforme* (the agent of Tyzzer's disease) and the cilia-associated respiratory (CAR) bacillus have proven difficult to detect as part of routine screening, but innovative diagnostic tests are making this more easily done (Boivin et al., 1994; Franklin et al., 1999).

TABLE 16-2

Core Agents for Screening

Type of organism	Specific agent
Viruses	Sendai virus
	Rat corona virus (RCV or SDAV)
	Rat virus (RV or KRV)
	H-1 parvovirus
	Rat parvovirus (RPV)
	Pneumonia virus of mice (PVM)
	Hantaan virus
Bacteria	*Streptococcus pneumonia*
	Mycoplasma pulmonis
	Corynebacterium kutscheri
	Salmonella spp.
Parasites	*Syphacia muris*
	Radfordia ensifera

SDAV = Sialodacryoadenitis virus.

Members of the *Pasteurella pneumotropica* complex often colonize the murine respiratory tract and can cause opportunistic infections in rats as well as mice (Kohn and Clifford, 2002). Emerging pathogens such as newly-discovered parvoviruses and members of the bacterial genus *Helicobacter* have been studied and characterized primarily in mice, but since they can colonize rats, facilities should begin to develop plans for monitoring these agents and taking action should they be found (Riley et al., 1996; Ball-Goodrich et al., 1998; Haines et al., 1998; Goto et al., 2000; Wan et al., 2002). Although it would be rarely encountered in rat colonies maintained at a high health status, many facilities opt to monitor for the bacterial agent *Streptobacillus moniliformis*, because it can cause a zoonotic disease (rat bite fever) and also could be a marker for wild rat contamination. Other potential pathogens include the Theiler murine encephalomyelitis virus-like agent, reovirus-3, mouse adenovirus-1, and *Pneumocystis carinii*. The latter is likely carried by most rats, but its potential to cause clinical or histologic disease is primarily in immunodeficient rats.

Agent-specific frequencies of testing can be determined based on the perceived risks of infection, transmissibility of the agent, potential impact of the agent to the population and associated research, immunocompetence of the colony being screened, ubiquity of agent, and the requirements of the biomedical research community, but it must be recognized that economic considerations also will play a role (Institute for Laboratory Animal Research-National Research Council, 1991). Tests for agents that are considered to pose a similar risk can be grouped together. For example, commonly encountered agents such as coronavirus, parvovirus, *Mycoplasma pulmonis*, and pinworms might be tested for on a quarterly basis, while more infrequently-detected agents such as Hantaan and Sendai virus could be surveyed on an annual or semi-annual basis.

B. Tests Used in Health Monitoring

The monitoring of laboratory rats for infectious disease utilizes a variety of tests. These include those in the general categories of gross necropsy, examination of serum for antibodies to infectious agents (serology), culture of bacterial pathogens, molecular tools designed to amplify infectious agent genomes, microscopic examination for parasites, and histologic examination of tissues. For routine health monitoring, these tests are often packaged, depending on the institution's needs, into profiles that include one or more testing modalities. For additional reading, a number of excellent reviews are available (Weisbroth et al., 1998; Compton and Riley, 2001; Feldman, 2001; Livingston and Riley, 2003).

1. General Test Performance Guidelines

Determining which test or which battery of tests to use in detecting infectious disease requires some knowledge about test performance. An ideal test is one that in all cases clearly distinguishes between exposed and unaffected animals (Weisbroth et al., 1998). Diagnostic tests can be assessed via several parameters (Table 16-3; from Bellamy and Olexson, 2000). In general, diagnostic sensitivity and specificity are of greatest importance when designing a health monitoring program. Tests with high sensitivity will generate a very low percentage of false negative results, whereas tests with high specificity will generate a low percentage of false positive results. Tests with high (>90%) sensitivity and specificity (for example, serology) should be used when available. For those tests that lack sensitivity or specificity (for example, histology), results must be interpreted accordingly. Other parameters, such as positive and negative predictive values, may also be of value in interpreting results, however these parameters can be affected by agent prevalence. For example, when agent prevalence is low, the calculated positive predictive value may also be misleadingly low (Lipman and Homberger, 2003). While highly sensitive and specific tests are available, it should be realized that no test is 100% sensitive or 100% specific. To this end, all unexpected results should be confirmed either through the use of corroborative testing platforms, the testing of additional animals, or both. In no case, should a decision about colony status be made based on a single positive result.

As discussed earlier, daily observation is a critical component to any health monitoring program. Recognition of clinical signs is especially important in the early detection of outbreaks of disease and documenting emerging diseases. However, because most agents that infect rats cause subclinical disease, observation is a very insensitive means of screening for infectious disease. As a result, sentinel and colony monitoring programs have been developed. For health monitoring, animals may either be euthanized and a necropsy examination performed, or samples for testing may be collected from live animals such as blood for serology, feces for molecular diagnostics, or perianal tape test samples for pinworms.

2. Testing Methodologies

a. PRE-NECROPSY EXAMINATION. A pre-necropsy examination should include collection of important circumstantial and historical data that may be important to test interpretation. Historical data include, but are not limited to, housing and husbandry conditions, rat strains housed, the genetic and immune status of the rats to be screened, number of animals in the colony, and individual identification, including sex, approximate age, and pelage color (Weisbroth et al., 1998). Prior to necropsy, animals should

TABLE 16-3

PARAMETERS FOR ASSESSING DIAGNOSTIC TEST PERFORMANCE.

Test characteristics	Formula
Diagnostic Sensitivity—likelihood that an animal will be positive for a particular test given that the animal is truly infected with the agent	$TP/(TP+FN) \times 100\%$
Diagnostic Specificity—likelihood that an animal will be will be negative for a particular test given that the animal is truly free of that agent	$TN/(FP+TN) \times 100\%$
Positive Predictive Value—estimate of the likelihood that an animal with a positive test has an infection; provides an estimate of the percentage of animals that are likely to have an infection given that they are positive for a particular test	$TP/(TP+FP) \times 100\%$
Negative Predictive Value—estimate of the likelihood that an animal with a negative test is free of the infection; provides an estimate of the percentage of animals that are likely to be free of an infection given that they are negative for a particular test	$TN/(TN+FN) \times 100\%$
Diagnostic Accuracy—provides a measure of all results (positive and negative) that correctly classify infectious disease status	$(TP+TN)/(TP+FP+TN+FN) \times 100\%$
Prevalence—an estimate of the frequency of an infection in a population at a point in time	$(TP+FN)/(TP+FP+TN+FN) \times 100\%$

FN = false negative results; FP = false positive results; TN = total negative results; TP = total positive results.

be examined for normal activity, ambulation, posture, hair coat appearance, and the presence or absence of discharges (Weisbroth et al., 1998).

b. NECROPSY EXAMINATION. A gross necropsy examination of the rat is critical in situations where rats exhibit clinical signs or when increased mortality is noted (see Euthanasia and Necropsy chapter). For example, if multiple rats develop cervical swellings and/or excessive periocular and perinasal porphyrin accumulation, a gross necropsy may reveal swollen salivary glands that support a tentative diagnosis of SDAV coronavirus infection. As a result, colony management decisions can be implemented while confirmatory tests are being pursued. Gross necropsy examinations are also often a component of health monitoring screens. Such necropsy examinations usually include a thorough examination of major organ systems with sample collection dependent on gross findings. Gross necropsy may reveal a multitude of lesions including abscesses, pneumonia, developmental defects, urolithiasis, neoplasia, trauma, and malocclusion. Unfortunately, there are few gross lesions that are pathognomonic for specific infectious diseases. Moreover, in many infectious diseases, gross lesions are not evident and the gross necropsy serves to enable sample collection for other more sensitive assays.

c. SEROLOGY. Examination for antibodies produced during an infection is the most economical and efficient means of screening rats for infectious disease. Serology offers several advantages: 1) testing requires serum, which can be obtained from an either euthanized or anesthetized rat; 2) multiple tests can be performed on a single serum sample; 3) antibodies (IgM followed by IgG) are detectable 1 to 2 weeks following exposure to the infectious agent; 4) serum antibody is long lasting (months), so the organism does not need to persist in the host for the infection to be detected; and 5) the antigens used in serologic assays can be highly purified, rendering these tests very sensitive and specific (Livingston and Riley, 2003).

A variety of serologic assays have been developed; the enzyme-linked immunosorbent assay (ELISA) and the immunofluorescence assay (IFA) have emerged as the two most popular platforms. A new multiplex fluorescent immunoassay (MFI), that utilizes antigen coated beads, has recently been developed and may supplant ELISA in high-throughput diagnostic laboratories. Other methods such as hemagglutination inhibition, complement fixation, and serum neutralization are time consuming and not as sensitive as ELISA or IFA and are thus no longer routinely used in rat infectious disease diagnosis. Additional methods such as Western blot analysis are valuable adjunct tests for ELISA and IFA but are not commonly used as primary tests.

ELISAs are highly sensitive and can be highly specific depending on the choice of antigen. Because ELISA is adaptable to automation, a large number of samples can be rapidly screened, and this testing platform is relatively inexpensive. Indirect ELISA methodology utilizes antigen bound to a solid phase (96 well plates or beads; Kendall et al., 1999). Serum is added and if antibody to the antigen is present, it will bind in a specific manner. Antibodies not specific for the antigen are removed in subsequent washing steps. Following washing, enzyme-labeled antibodies that are specific for rat antibody (enzyme conjugated anti-immunoglobulin) are added. These bind to rat antibodies that were bound in the first step. The last step involves the addition of a substrate for the enzyme label. If specific antibody is present, the secondary enzyme labeled antibody will have bound and the enzyme will cleave the substrate resulting in color change; the latter can be measured spectrophotometrically to give a semiquantitative assessment of serum antibody to the specific antigen (usually measured in absorbance units).

Serum quality is critical in ELISA testing. Non-specific absorbance may also occur in serum from aged rodents (over 6 months old; Wagner et al., 1991), strains subject to autoimmunity, animals whose immune systems are non-specifically stimulated because of injury, neoplasm, other noninfectious disease processes, or other types of antigenic stimulation (Wagner et al., 1991; Weisbroth et al., 1998). Additionally, a variety of experimental manipulations of rodents may result in non-specific absorbance.

Antigens employed in ELISA testing vary in complexity from crude extracts containing multiple antigens and impurities to select recombinant proteins generated in viral vectors. The use of highly purified subunit antigens may increase specificity as cross-reactive impurities are not present. However, the use of these subunit antigens may negatively impact sensitivity (Compton and Riley, 2001). This happens because the host response is polyclonal, with many antibodies being produced to different epitopes on the infectious agent. Highly purified subunit antigens may lack the immunodominant epitopes to which antibodies have been produced and result in a test with decreased sensitivity when compared to one that utilizes crude protein preparations. Moreover, agents may express different epitopes during different stages of disease. Therefore, an ELISA that uses an antigen that is only expressed at certain stages may miss some infections. In practice, a balance is sought so that purified preparations of multiple antigens are used, resulting in very sensitive and specific assays.

Variations in ELISA methodology also exist. One such variation, often referred to as antigen capture ELISA, allows for the detection of antigen rather than antibody. These assays are particularly useful in detecting agents in feces or secretions and may ultimately serve as adjuncts or alternatives for more expensive molecular-based techniques.

IFA methodology is similar in principle to ELISA (Kendall et al., 1999). In contrast to ELISA, the antigen is supplied in the form of a virus or bacteria growing in culture. Virus-infected cells and uninfected cells (internal controls) are fixed to wells of a glass slide. Test serum is added to wells, and if antibody is present it binds to the antigen. The secondary antibody is labeled with a fluorescent molecule rather than an enzyme (fluorescent dye-conjugated anti-immunoglobulin). Sensitivity is similar to ELISA. Specificity is equal to or better than ELISA because patterns or location of fluorescence may provide additional information (granular or nuclear fluorescence may be consistent with certain viral infections as opposed to diffuse fluorescence, which may indicate a non-specific reaction). IFAs are relatively inexpensive, but more expensive than ELISA. The main disadvantages of IFA are that it is labor intensive, and interpretation is subjective and dependent on the expertise of the observer. An additional requirement is a specialized epifluorescence microscope. The choice between ELISA and IFA is based on personal preference of the laboratory. These tests are often used in combination, with the ELISA serving as the primary test and the IFA serving as a confirmatory test.

As described earlier, serology is ideal because detectable antibodies persist for months, allowing for a large window of opportunity to detect infections. This is very advantageous in health monitoring programs. Because of this large window of opportunity, serologic testing at a single time point cannot distinguish active from prior infections (for example, the infectious potential of a rat tested). Serology also allows for multiple tests to be run on a single serum sample. Although serology has few limitations, it is unreliable in the diagnosis of infections in immunodeficient rodents (Compton and Riley, 2001; Livingston and Riley, 2003) and, as mentioned earlier, cannot distinguish exposure from active infection. Both ELISA and IFA are subject to non-specific reactivity, which can lead to false positive results. This is especially true in bacterial ELISAs due to the complexity and abundance of potentially cross-reactive bacterial antigens. With improvements in antigen production and reagents, false positive results are uncommon. However, because of this possibility, a single positive should always be confirmed with additional testing.

d. CULTURE. Culture of bacteria, using a variety of media may be incorporated into health monitoring programs. Culture is especially useful when evidence of infection such as abscess formation or pneumonia is present. Culture is most effective during the height of infection, and prior to administration of antibiotics or the development of an immune response (Compton and Riley, 2001). Culture may also be used as a screening tool for pathogens or agents capable of causing opportunistic infections. In the latter scenario, mucosal sites of the intestinal tract (for example, the

cecum) and respiratory tract (for example, the nasopharynx) are cultured on broad spectrum or selective media. Bacterial speciation is based on colony morphology, Gram staining characteristics, organism morphology, biochemical tests, and growth on selective media or in selective conditions (Feldman, 2001; Livingston and Riley, 2003). Culture has the advantage of determining whether a live agent is present, as opposed to potentially nonviable DNA remnants or antibacterial antibodies from a past infection, in the animal. Culture and subsequent biochemical analyses are also very specific for most agents and can be supplemented with molecular techniques where speciation is desired. Culture does have some drawbacks. For example, agents colonizing the mucosal surface may be present in low numbers or sequestered in areas not accessed by routine procedures (for example, the deep recesses of the nasal turbinates). Moreover, fastidious organisms may not grow well unless conditions are optimized, or their growth may be hindered by the growth of more vigorous bacteria. Some agents may take several days to grow into identifiable colonies, while some agents such as CAR bacillus and *Clostridium piliforme* have yet to be cultivated on cell-free media. Collection of samples for submission to diagnostic laboratories may also be problematic in that some bacteria, notably the *Pasteurellaceae*, do not survive well in transfer media.

Culture of viruses is also possible using cell culture systems or embryonated eggs; however, procedures are time consuming, expensive, and require considerable expertise. Viral culture is important in the characterization of novel viral infections. However, because other means of detecting viral infection are readily available, viral culture is rarely used in rat health monitoring programs.

e. MOLECULAR DIAGNOSTICS. Molecular diagnostic techniques, primarily those based on the PCR technique are rapidly replacing traditional culture techniques (Compton and Riley, 2001). PCR utilizes specific oligonucleotide primers to exponentially amplify small amounts of target deoxyribonucleic acid (DNA) or ribonucleic acid (RNA) from a particular organism that is present in a clinical specimen. PCR offers exquisite sensitivity and specificity, detecting as few as 1 to 10 viral virions and 3 to 10 bacteria (Compton and Riley, 2001), and is readily adapted to the detection of bacterial, viral, parasite, and fungal agents.

Details of PCR technique can be found in a number of technique manuals. Briefly, PCR consists of repetitive cycles of a 3-step amplification procedure. Double-stranded sample DNA is denatured into two single strands. Oligonucleotide primers specific for the agent (complementary to the specific microorganisms genome and typically situated a few hundred base pairs apart) are added and allowed to anneal to target sequences in the sample DNA. A polymerase (for example, Taq polymerase), an enzyme that functions in DNA synthesis, is added along with nucleotide bases and new DNA strands of a specific size are created. After "n" cycles of this 3-step process of denaturation, annealing, and synthesis, the target sequence is amplified 2^{nth} times (30 cycles $= 2^{30} = 1,073,741,824$ copies of DNA). The product is then subjected to gel electrophoresis, and if a targeted sequence is amplified, it will migrate to a specific location based on its molecular weight. RNA (RNA viral genomes) may also be detected by PCR. In this case, reverse transcriptase PCR (RT-PCR) is utilized. With RT-PCR, RNA is converted to complementary DNA (cDNA) using the enzyme, reverse transcriptase. This cDNA becomes the template for PCR, as described earlier. Because RNA is very susceptible to degradation, additional protective steps in sample preparation and storage should be incorporated.

PCR offers superior sensitivity and specificity, and results can be obtained in a single working day. The main disadvantages of PCR are directly related to its advantages. The exquisite sensitivity renders contamination especially problematic, and false positives may occur if strict laboratory technique protocols are not in place or followed. The test is also relatively expensive due to the need for expensive equipment and its labor-intensiveness. This expense can be partially overcome by pooling of samples. In addition, costs will likely be lowered as technology allows for more automation of PCR. The expense of PCR also relates to the need for multiple samples from which multiple tests must be performed. This problem is being addressed by the development of high-throughput multiplexed assays where multiple agents can be tested in a single reaction. Lastly, PCR is often performed on biological samples, many of which contain inhibitors of components of the PCR reaction such as heme and plant products that contaminate feces (Panaccio and Lew, 1991; Al-Soud and Radstrom, 2001; Compton and Riley, 2001; Feldman, 2001). This possibility must be considered when testing these samples; however, the use of highly purified DNA can eliminate or sufficiently dilute inhibitors, so that accurate results are obtained.

Sampling for PCR requires knowledge of the pathogenesis of the agent, including tissue tropism and duration of infection (Compton and Riley, 2001). PCR is an ideal primary test for the detection of active or persistent infections (for example, infections by parvoviruses, LCMV, RCMV, *Mycoplasma pulmonis*, *Helicobacter* spp.) or those agents for which other diagnostic tests are of poor sensitivity (for example, the culture of *Helicobacter* spp.). In contrast, detection of infections where colonization is transient (many viral infections) is possible for only brief periods of time. In the latter case, PCR may serve as an adjunct test. In this scenario, infections may be detected by a primary test such as serology. To confirm infection, additional rats of appropriate target age (an age at which colonization or shedding is expected) are selected and target tissues are tested by PCR. This two-methodology approach provides very convincing evidence of infection. Moreover, although PCR cannot distinguish between live and dead organisms, results from PCR testing can provide valuable information about the current status (actively shedding, free of colonization) among individual animals or groups of animals.

Newer modifications of PCR, such as fluorogenic $5'$ nuclease PCR are also being developed (Feldman, 2001; Besselsen et al., 2002; Drazenovich et al., 2002; Besselsen et al., 2003; Uchiyama and Besselsen, 2003). These assays offer improved sensitivity and specificity in some cases, require no post-PCR processing, and can be used to quantify infectious agents. Moreover, other molecular techniques such as microarrays, which allow for the screening of thousands of agents simultaneously, will surely add to the arsenal of molecular techniques available to the diagnostician in the near future. Molecular techniques are also applicable in many other areas of rat medicine and biology, including the detection of contaminants in tissue culture material and monitoring of genetic purity of inbred strains or genetically engineered rats.

f. PARASITE SCREENING. Screening for parasites is usually accomplished by a subgross or microscopic examination of parasite niches. The three general classes of parasites that infect rats include ectoparasites (mites and lice), endoparasitic helminths (pinworms, other nematodes and cestodes), and endoparasitic protozoa. For ectoparasites, the pelage can be collected and examined for mite or louse infestation. Most protocols suggest allowing the sample to cool to encourage mites to venture to the tips of the hair shaft in search of a warmer host. Alternatively, scotch tape tests may be effective for detecting mites or mite eggs attached to hair shafts. The latter can also be utilized in the live animal.

Endoparasitic helminths may be detected by direct examination of the intestinal tract for adult worms. *Syphacia muris* pinworms usually inhabit the cecum and *Rodentolepis (Hymenolepis)* spp. tapeworms the small intestine. Detection of helminths in gross specimens may be enhanced by the use of a dissecting microscope. Incubation of a section of intestine in saline for a short period of time may also facilitate detection by allowing worms to migrate out of the dark fecal matter into the more transparent saline. Pinworms of the genera *S. muris* also deposit ova on the perineum and can thus be detected by perineal tape testing. For this test, a piece of clear cellophane tape is applied to the perineal skin, placed on a microscope slide, and examined for typical banana-shaped ova. This test offers the advantage of being usable in live animals. As an alternative, fecal flotation may be used to detect pinworm or cestode ova.

Endoparasitic protozoa are usually commensal organisms of questionable pathogenicity. These agents are generally detected by wet mount preparations of intestinal contents. Protozoa are readily identified based on motility, morphology, and intestinal locale. For example, *Spironucleus muris* is most often found in the small intestine and is characterized by its small tear drop shape with darting motility. *Giardia* sp. are also found in the small intestine but are larger, have a cup-shaped morphology with an "owl face" appearance, and a "falling leaf" motility. Other protozoa include trichomonads (lemon-shaped with undulating membrane and rolling motility), *Chilomastix* sp. (oval to bar-shaped with spiraling motility), and *Entamoeba* sp. (amoeboid shape with slow motility by pseudopod formation).

In general, patent infections by parasites are more readily detectable in young animals (Wagner et al., 1991; Weisbroth et al., 1998). Microscopic and gross examinations for parasites are advantageous in that they are relatively simple, straightforward techniques; some (tape tests, fecal flotation) can be performed on live animals; and they are relatively specific. The disadvantage of these tests is that they lack sensitivity, and ultimately the development of more sensitive techniques, such as PCRs, may be warranted.

g. SCREENING OF TISSUE BY HISTOLOGY FOR LESIONS OF INFECTIOUS DISEASE.   A variety of tissues may be screened for lesions indicative of infectious disease. While there are very few pathognomonic lesions of rodent infections, screening of tissues may provide presumptive diagnoses that can be confirmed by other means. The disadvantages of histology as a screening tool include the narrow window of opportunity to detect certain transient infections and the fact that many opportunistic pathogens do not cause histologic disease. Screening of tissues may be useful in several situations: 1) screening of target tissues for known lesions of infectious disease; 2) screening immunodeficient rats in which tests such as serology are not appropriate; 3) detecting disease early in its time course prior to the development of detectable antibody; 4) detecting bacterial agents that are difficult to cultivate (such as CAR bacillus and *Clostridium piliforme*); and 5) detecting emerging or previously unrecognized infectious diseases. The latter is exemplified by the recent discovery of rat respiratory virus. This agent was discovered after the recognition that chronic idiopathic interstitial pneumonia became prevalent in several colonies of rats (Simmons and Riley, 2002). In addition, it is only by histopathology that noninfectious degenerative conditions, such as renal and cardiac calcinosis, may be recognized.

The use of tissue screening for infectious disease relies upon the selection of certain target tissues. It is unrealistic to screen all tissues for signs of disease and many tissues are not common sites of infection. Most commonly, systems exposed to the external environment (for example, respiratory and enteric systems) are screened. Other tissues often screened are based on known disease pathogeneses. These include the Harderian and salivary glands, which are screened for lesions of rat coronavirus (RCV/SDAV), and the urinary bladder, which is screened for *Trichosomoides crassicauda* infections.

h. OTHER TESTING STRATEGIES.   Historically, other testing platforms were employed, including stress testing for *C. piliforme* (Fries and Ladefoged, 1979) or *P. carinii* (Armstrong et al., 1991). These tests may still be used as a diagnostic tool or for the characterization of a novel pathogen, but they are rarely if ever used in routine health monitoring. Moreover, certain strains of rats, such as gnotobiotic or axenic rats, may require additional tests, such as flora confirmation, that can be coupled with health monitoring.

Table 16-4, adapted from Livingston and Riley (Livingston and Riley, 2003), lists agents commonly tested for in rat health monitoring programs and methodologies used to test these agents.

C.   Testing Profiles

Tests used in the monitoring of rats for infectious disease are often packaged, depending on the institution's needs, into profiles that include one or more testing modalities. These profiles almost invariably include serologic examination for antibodies to viral and bacterial agents and may include gross necropsy examination, parasite examination, examination for enteric or respiratory pathogens using culture or molecular techniques, and histologic examination of target tissues. The design of these testing profiles requires consideration of several factors as outlined in earlier sections of this chapter. Tiered testing strategies are very economical and are becoming commonplace (Laber-Laird and Proctor, 1993). With these strategies, the most prevalent agents are tested for on a frequent basis via inexpensive high-throughput tests, while testing for agents of low prevalence or screening of animals for indication of emerging diseases occurs on a less frequent basis. For example, rats may be screened for pinworms, fur mites, *M. pulmonis*, RCV/SDAV, pneumonia virus of mice (PVM), and parvoviruses on a quarterly basis while less prevalent agents such as *Corynebacterium kutscheri, Streptococcus pneumoniae*, Sendai, REO3, LCM, MAD1, CAR bacillus and *Clostridium piliforme* are tested for once a year (Nicklas et al., 2002).

TABLE 16-4
COMMONLY USED TESTING METHODOLOGIES FOR RAT PATHOGENS

Agent (species)	Primary testing methodology (sample tested)	Confirmatory testing methodology
Viruses		
Hantaan (HTN)	Serology (serum)	PCR (kidney)
Lymphocytic choriomeningitis virus (LCMV)	Serology (serum)	PCR (kidney)
Mouse adenovirus 1 (MAD1)	Serology (serum)	PCR (lung)
Pneumonia virus of mice (PVM)	Serology (serum)	PCR (trachea, lung)
Rat coronavirus (RCV/SDAV)	Serology (serum)	PCR, Histology (salivary and Harderian glands)
Rat parvoviruses	Serology (serum)	PCR (mesenteric lymph node)
Reovirus type 3 (REO 3)	Serology (serum)	PCR (liver, lung feces)
Sendai virus (Sendai)	Serology (serum)	PCR (trachea, lung)
Theiler murine encephalomyelitis virus (TMEV)*	Serology (serum)	PCR (feces, intestine)
Rat respiratory virus	Histology (lung)	
Bacteria		
Cilia-associated respiratory (CAR) bacillus	Serology (serum)/PCR (trachea)	Histology (nasopharynx, trachea, lung)
Corynebacterium kutscheri	Culture (NP)	
Helicobacter spp.	PCR (feces)	Culture (feces)
Mycoplasma pulmonis	Serology (serum)/PCR (NP)	Culture (nasopharynx)
Pasteurella pneumotropica	Culture (NP)	PCR (nasopharynx)
Salmonella spp.	Culture (cecal contents, feces)	PCR (cecal contents, feces)
Streptococcus pneumoniae	Culture (NP)	PCR (nasopharynx)
Clostridium piliforme	ELISA (serum)	PCR, Histology (intestine, liver)
Parasites		
Radfordia ensifera	Direct exam (pelage)	
Rodentolepis (Hymenolepis) spp.	Direct exam (small intestine)	
Syphacia muris	Direct exam (cecal contents)	
	Direct exam (perianal tape test)	
Fungus		
Pneumocystis carinii **	PCR (lung)	Histology (lung)

ELISA = enzyme-linked immunosorbent assay; NP = nasopharynx; PCR = polymerase chain reaction.
*TMEV-like agent (see Virology chapter)
**Monitored only in immunodeficient rats.

D. Test Interpretation and Retesting

In many cases, interpretation of health monitoring results is straightforward. For example, when several rats with cervical swellings are found to be seropositive for RCV/SDAV, it is reasonable to determine that an outbreak of this infection is occurring. In other cases, the diagnosis is not so clear-cut, and test results require careful interpretation and follow-up testing. For example, in cases where a single animal is found to be seropositive for PVM, this may indicate either an early outbreak or a false positive result.

There are several approaches to test interpretation. First, results should be interpreted in the context of the entire colony and the health monitoring program. Decisions about rodent health should rarely if ever be made on a single positive result and the latter should be assumed to be a false positive until verified (Laber-Laird and Proctor, 1993; Compton and Riley, 2001; Livingston and Riley, 2003). Verification may include testing a sample (serum) using an alternative test platform, testing a second sample from the affected animals using an alternative test platform (for example, through PCR), or testing cohort animals (Weisbroth et al., 1998; Livingston and Riley, 2003). As discussed earlier, there are two primary serologic testing platforms: ELISA and IFA. These platforms can also be used as adjuncts for each other. In most diagnostic laboratories, the ELISA serves as the primary test, and borderline or solitary positive results are confirmed by IFA. Confirmatory testing may also involve the use of different testing platforms. For example, if a rat is found to be seropositive for CAR bacillus, the lungs may be examined with a silver stain to detect the presence of the organism. Often, only serum is collected for health monitoring, so samples for confirmatory testing by other platforms may not be possible. In these cases, testing of cohort animals may be warranted and a diagnostic plan to test with different testing platforms should be designed. For example, if a rat is found to be seropositive for a rat parvovirus, additional animals from that colony may be tested by serology and their mesenteric lymph nodes may concurrently be tested by PCR for rat parvovirus. Testing of cohort animals is also warranted in the case where very few animals are seropositive. This scenario may indicate an early outbreak or a false positive result. If an early outbreak is occurring, cohort animals will have additional time to seroconvert and the percentage of positives should increase.

V.   MANAGEMENT OF COLONY DISEASE OUTBREAKS

A.   Confirmation and Containment

As discussed earlier, when laboratory testing suggests a change in colony status for a particular agent, it is important to verify the information. Once confidence in the lab result is obtained, the positive sample should be tracked back to confirm its origin by comparing the date of testing, or cage or animal identification numbers, etc., to the monitoring schedule and sampling documentation. Based on this information, if an excluded agent appears to be present, the room should be quarantined to prevent further spread throughout the facility. Scheduled incoming shipments should be diverted to other areas, and transfers out of the room should be canceled. Changes to standard practices that have the potential to affect cross-contamination should be considered, such as the room entry order, the handling and transport of soiled cages, the protective clothing and disinfectants used in the room, and the amount of personnel traffic allowed. A follow-up plan should be implemented to establish whether the agent truly does exist within the room by testing remaining sentinels or principal animals. It may be useful to draft a generic initial response plan for suspected contamination events in advance, so that these initial steps can be instituted promptly and efficiently.

B.   Response Plan

When contamination has been confirmed, a plan of action should be developed by the veterinary and animal facility management groups in concert with others that are affected, such as the research groups holding animals in the area. It is also prudent to include key individuals from the administration or upper management (since there may be a significant fiscal impact) and the institutional animal care and use committee. In some cases there will be no question what the follow-up response to contamination will be (for example, eradication of the agent), but in others the potential costs and benefits of the available options may need to be considered. Regardless of the decisions made, the plan must be documented and distributed so that the goal is very clear to all involved and the sequence of events and projected timeline is evident.

C.   Eradication Options

There are a variety of methods that can be used to eliminate an infectious agent from an area, and careful professional judgment is needed to determine the most appropriate course of action. If the animals are replaceable and the primary consideration is to return the room to normal use, complete depopulation followed by environmental decontamination can be performed. Attempting partial depopulation by removing positive animals (via test and cull) is not a very productive approach for most rodent colony disease outbreaks due to the large numbers of animals often involved, the delay between exposure and seroconversion, and the possibility that the disease will be further spread during the handling and sampling procedures needed to test the entire population. As an alternative approach, if the agent does not establish persistent infection, it may be useful to test and remove the *negative* animals. Fully immune populations should pose little risk of shedding to naïve animals after infection by agents such as coronavirus or Sendai virus. By retaining only previously exposed seropositive animals to re-establish a breeding program, it is possible to produce seronegative offspring (Brammer et al., 1993). A related approach to break the chain of transmission and repopulate an area without determining the serologic status of each individual rat allowed to remain is the cessation of breeding method, also known as *burnout* or *stop-breeding* schemes. By eliminating the introduction of naïve animals from outside and eliminating all internal breeding for a period of time (6 to 8 weeks is recommended) coronavirus can be eliminated from a population (Bhatt and Jacoby, 1985; Jacoby and Gaertner, 1994). Reciprocal transfer of soiled bedding between all cages during the early weeks of a burnout period can help assure that all animals have had equivalent exposure to the agent. Of note, these techniques may not work in immunodeficient rats and should be used with caution in genetically engineered rats.

If rats having a valuable or irreplaceable genotype are involved in an outbreak, there are methods for rederiving the strain. Detailed description of techniques used for rat cesarean section rederivation with or without superovulation have been published (Rouleau et al., 1993; Sharp and LaRegina, 1998) and are described elsewhere in the Assisted Reproductive Technologies chapter of this text. Although the techniques may not be as well established or efficient in rats, superovulation and embryo transfer methods similar to those used in mice can be successful in rats (Robl and Heideman, 1994).

REFERENCES

Al-Soud, W.A., and Radstrom, P. (2001). Purification and characterization of PCR-inhibitory components in blood cells. *J. Clin. Microbiol.* **39**, 485–493.

Armstrong, M.Y., Smith, A.L., and Richards, F.F. (1991). Pneumocystis carinii pneumonia in the rat model. *J. Protozool.* **38**, 136S–138S.

Artwohl, J.E., Cera, L.M., Wright, M.F., Medina, L.V., and Kim, L.J. (1994). The efficacy of a dirty bedding sentinel system for detecting Sendai virus infection in mice: a comparison of clinical signs and seroconversion. *Lab. Anim. Sci.* **44**, 73–75.

Baker, D.G. (1998). Natural pathogens of laboratory mice, rats, and rabbits and their effects on research. *Clin. Microbiol. Rev.* **11**, 231–66.

Baker, D.G. (2003). "Natural Pathogens of Laboratory Animals Their Effects on Research." ASM Press, Washington.

Ball-Goodrich, L.J., Leland, S.E., Johnson, E.A., Paturzo, F.X., and Jacoby, R.O. (1998). Rat parvovirus type 1: the prototype for a new rodent parvovirus serogroup. *J. Virol.* **72**, 3289–3299.

Ball-Goodrich, L.J., Paturzo, F.X., Johnson, E.A., Steger, K., and Jacoby, R.O. (2002). Immune responses to the major capsid protein during parvovirus infection of rats. *J. Virol.* **76**, 10044–10049.

Bellamy, J.E.C., and Olexson, D.W. (2000). Evaluating Laboratory Procedures. In: "Quality Assurance Handbook for Veterinary Laboratories." pp. 61–77, Iowa State University Press, Ames, IA.

Besselsen, D.G., Wagner, A.M., and Loganbill, J.K. (2002). Detection of rodent coronaviruses by use of fluorogenic reverse transcriptase-polymerase chain reaction analysis. *Comp. Med.* **52**, 111–116.

Besselsen, D.G., Wagner, A.M., and Loganbill, J.K. (2003). Detection of lymphocytic choriomeningitis virus by use of fluorogenic nuclease reverse transcriptase-polymerase chain reaction analysis. *Comp. Med.* **53**, 65–69.

Bhatt, P.N., and Jacoby, R.O. (1985). Epizootiological observations of natural and experimental infection with sialodacryoadenitis virus in rats. *Lab. Anim. Sci.* **35**, 129–134.

Blank, W.A., Henderson, K.S., and White, L.A. (2004). Virus PCR assay panels: an alternative to the mouse antibody production test. *Lab. Anim.* **32**, 26–32.

Boivin, G.P., Hook, R.R., Jr., and Riley, L.K. (1994). Development of a monoclonal antibody-based competitive inhibition enzyme-linked immunosorbent assay for detection of Bacillus piliformis isolate-specific antibodies in laboratory animals. *Lab. Anim. Sci.* **44**, 153–158.

Bootz, F., and Sieber, I. (2002). Replacement of mouse and rat antibody production test; comparison of sensitivity between the in vitro and in vivo methods. [German] *ALTEX* **19 Suppl 1**, 76–86.

Brammer, D.W., Dysko, R.C., Spilman, S.C., and Oskar, P.A. (1993). Elimination of sialodacryoadenitis virus from a rat production colony by using seropositive breeding animals. *Lab. Anim. Sci.* **43**, 633–634.

Carpenter, J., Mashima, T., and Rupiper, D. (1996). "Exotic Animal Formulary." Greystone Publications, Manhattan, KS.

Clifford, C.B. (2001). Samples, sample selection, and statistics: living with uncertainty. *Lab. Anim.* **30**, 26–31.

Compton, S.R., and Riley, L.K. (2001). Detection of infectious agents in laboratory rodents: traditional and molecular techniques. *Comp. Med.* **51**, 113–119.

Cundiff, D.D., Riley, L.K., Franklin, C.L., Hook, R.R., Jr., and Besch-Williford, C. (1995). Failure of a soiled bedding sentinel system to detect cilia-associated respiratory bacillus infection in rats. *Lab. Anim. Sci.* **45**, 219–221.

DiGiacomo, R.F., and Koepsell, T.D. (1986). Sampling for detection of infection or disease in animal populations. *J. Am. Vet. Med. Assoc.* **189**, 22–23.

Dillehay, D.L., Lehner, N.D., and Huerkamp, M.J. (1990). The effectiveness of a microisolator cage system and sentinel mice for controlling and detecting MHV and Sendai virus infections. *Lab. Anim. Sci.* **40**, 367–370.

Drazenovich, N.L., Franklin, C.L., Livingston, R.S., and Besselsen, D.G. (2002). Detection of rodent Helicobacter spp. by use of fluorogenic nuclease polymerase chain reaction assays. *Comp. Med.* **52**, 347–353.

Dymsza, J., Miller, S., Maloney, J., and Foster, H. (1963). Equilibration of the laboratory rat following exposure to shipping stresses. *Lab. Anim. Care* **13**, 60–65.

Feldman, S.H. (2001). Diagnostic molecular microbiology in laboratory animal health monitoring and surveillance programs. *Lab. Anim.* **30**, 34–42.

Franklin, C.L., Pletz, J.D., Riley, L.K., Livingston, B.A., Hook, R.R., Jr., and Besch-Williford, C.L. (1999). Detection of cilia-associated respiratory (CAR) bacillus in nasal-swab specimens from infected rats by use of polymerase chain reaction. *Lab. Anim. Sci.* **49**, 114–117.

Fries, A.S., and Ladefoged, O. (1979). The influence of Bacillus piliformis (Tyzzer) infections on the reliability of pharmacokinetic experiments in mice. *Lab. Anim.* **13**, 257–261.

Fujiwara, K., and Wagner, J. (1994). Sendai virus. In "Manual of Microbiologic Monitoring of Laboratory Animals." (T. Nomura, ed.), pp. 107–109, NIH Publication No. 94–2498.

Gill, T.J., 3rd, Smith, G.J., Wissler, R.W., and Kunz, H.W. (1989). The rat as an experimental animal. *Science* **245**, 269–276.

Goto, K., Ohashi, H., Takakura, A., and Itoh, T. (2000). Current status of Helicobacter contamination of laboratory mice, rats, gerbils, and house musk shrews in Japan. *Curr. Microbiol.* **41**, 161–166.

Haines, D.C., Gorelick, P.L., Battles, J.K., Pike K.M., Anderson, R.J., Fox, J.G., Taylor, N.S., Shen, Z., Dewhirst, F.E., Anver, M.R., and Ward, J.M. (1998). Inflammatory large bowel disease in immunodeficient rats naturally and experimentally infected with Helicobacter bilis. *Vet. Pathol.* **35**, 202–208.

Hawk, C., and Leary, S. (1999). "Formulary for Laboratory Animals." Iowa State University Press, Ames, IA.

Institute for Laboratory Animal Research-National Research Council (1991). "Infectious Diseases of Mice and Rats." National Academy Press, Washington, DC.

Institute for Laboratory Animal Research-National Research Council (1992). "Recognition and Alleviation of Pain and Distress in Laboratory Animals." National Academy Press, Washington, DC.

Institute for Laboratory Animal Research-National Research Council (1996a). "Guide for the Care and Use of Laboratory Animals." National Academy Press, Washington, DC.

Institute for Laboratory Animal Research-National Research Council (1996b). "Laboratory Animal Management: Rodents." National Academy Press, Washington, DC.

Jacoby, R., and Gaertner, D. (1994). Rat coronavirus/sialodachryoadenitis virus. In "Manual of Microbiologic Monitoring of Laboratory Animals." (T. Nomura, ed.), pp. 87–91, NIH Publication No. 94–2498.

Jacoby, R.O., and Homberger, F.R. (1999). International standards for rodent quality. *Lab. Anim. Sci.* **49**, 230.

Jacoby, R.O., and Lindsey, J.R. (1998). Risks of infection among laboratory rats and mice at major biomedical research institutions. *ILAR J* **39**, 266–271.

Johnson, K. (1986). Hemorrhagic fever-Hantaan virus. *In* "Viral and Mycoplasmal Infections of Laboratory Rodents: Effects on Biomedical Research." (E. New, ed.), pp. 193–216, Academic Press, Orlando, FL.

Kendall, L.V., Steffen, E.K., and Riley, L.K. (1999). Indirect fluorescent antibody (IFA) assay. *Contemp. Top. Lab. Anim. Sci.* **38**, 68.

Kohn, D., and Clifford, C. (2002). Biology and diseases of rats. *In* "Laboratory Animal Medicine." (F. Quimby, ed.), pp. 121–165, Academic Press, San Diego.

Koszdin, K.L., and DiGiacomo, R.F. (2002). Outbreak: detection and investigation. *Contemp. Top. Lab. Anim. Sci.* **41**, 18–27.

Laber-Laird, K., and Proctor, M. (1993). An example of a rodent health monitoring program. *Lab. Anim.* **22**, 24–32.

Lipman, N.S., and Homberger, F.R. (2003). Rodent quality assurance testing: use of sentinel animal systems. *Lab. Anim.* **32**, 36–43.

Lipman, N.S., and Perkins, S. (2002). Factors that may influence animal research. *In* "Laboratory Animal Medicine." (F. Quimby, ed.), pp. 1143–1184, Academic Press, San Diego.

Lipman, N.S., Perkins, S., Nguyen, H., Pfeffer, M., and Meyer, H. (2000). Mousepox resulting from use of ectromelia virus-contaminated, imported mouse serum. *Comp. Med.* **50**, 426–435.

Livingston, R.S., and Riley, L.K. (2003). Diagnostic testing of mouse and rat colonies for infectious agents. *Lab. Anim.* **32**, 44–51.

McKisic, M.D., Lancki, D.W., Otto, G., Padrid, P., Snook, S., Cronin, D.C., 2nd, Lohmar, P.D., Wong, T., and Fitch, F.W. (1993). Identification and propagation of a putative immunosuppressive orphan parvovirus in cloned T cells. *J. Immunol.* **150**, 419–428.

Nicklas, W., Baneux, P., Boot, R., Decelle, T., Deeny, A.A., Fumanelli, M., and Illgen-Wilcke, B. (2002). Recommendations for the health monitoring of rodent and rabbit colonies in breeding and experimental units. *Lab. Anim.* **36**, 20–42.

Nicklas, W., Homberger, F.R., Illgen-Wilcke, B., Jacobi, K., Kraft, V., Kunstyr, I., Mahler, M., Meyer, H., and Pohlmeyer-Esch, G. (1999). Implications of infectious agents on results of animal experiments. Report of the Working Group on Hygiene of the Gesellschaft fur Versuchstierkunde—Society for Laboratory Animal Science (GV-SOLAS). *Lab. Anim.* **33 Suppl 1**, S39–87.

Otto, G., and Tolwani, R.J. (2002). Use of microisolator caging in a risk-based mouse import and quarantine program: a retrospective study. *Contemp. Top. Lab. Anim. Sci.* **41**, 20–27.

Panaccio, M., and Lew, A. (1991). PCR based diagnosis in the presence of 8% (v/v) blood. *Nucleic Acids Res.* **19**, 1151.

Rehg, J.E., and Toth, L.A. (1998). Rodent quarantine programs: purpose, principles, and practice. *Lab. Anim. Sci.* **48**, 438–447.

Riley, L.K., Franklin, C.L., Hook, R.R., Jr., and Besch-Williford, C. (1996). Identification of murine helicobacters by PCR and restriction enzyme analyses. *J. Clin. Microbiol.* **34**, 942–946.

Robl, J., and Heideman, J. (1994). Production of transgenic rats and rabbits. *In* "Transgenic Animal Technology." (C. Pinkert, ed.), pp. 265–277, Academic Press, San Diego.

Rouleau, A.M., Kovacs, P.R., Kunz, H.W., and Armstrong, D.T. (1993). Decontamination of rat embryos and transfer to specific pathogen-free recipients for the production of a breeding colony. *Lab. Anim. Sci.* **43**, 611–615.

Saibaba, P., Sales, G., Stodulski, G., and Hau, J. (1995). Behaviour of rats in their home cages: daytime variations and effects of routine husbandry procedures analyzed by time sampling techniques. *Lab. Anim.* **30**, 13–21.

Sharp, P., and LaRegina, M. (1998). "The Laboratory Rat." CRC Press, Boca Raton, FL.

Simmons, J.H., and Riley, L.K. (2002). Hantaviruses: an overview. *Comp. Med.* **52**, 97–110.

Small, J. (1986). Decision making, detection, prevention and control. *In* "Viral and Mycoplasmal Infections of Laboratory Rodents: Effects on Biomedical Research." (E. New, ed.), pp. 777–799, Academic Press, Orlando, FL.

Small, J.D. (1984). Rodent and lagomorph health surveillance: quality assurance. *In* "Laboratory Animal Medicine." (F. Loew, ed.), pp. 709–723, Academic Press, Orlando, FL.

Smith, A.L. (1983). An immunofluorescence test for detection of serum antibody to rodent coronaviruses. *Lab. Anim. Sci.* **33**, 157–160.

Thigpen, J.E., Lebetkin, E.H., Dawes, M.L., Amyx, H.L., Caviness, G.F., Sawyer, B.A., and Blackmore, D.E. (1989). The use of dirty bedding for detection of murine pathogens in sentinel mice. *Lab. Anim. Sci.* **39**, 324–327.

Uchiyama, A., and Besselsen, D.G. (2003). Detection of reovirus type 3 by use of fluorogenic nuclease reverse transcriptase polymerase chain reaction. *Lab. Anim.* **37**, 352–359.

Waggie, K., Kagiyama, N., Allen, A., and Nomura, T. (1994). "Manual of Microbiologic Monitoring of Laboratory Animals." NIH Publication No. 94–2498.

Wagner, J.E., Besch-Williford, C.L., and Steffen, E.K. (1991). Health surveillance of laboratory rodents. *Lab. Anim.* **20**, 40–45.

Wan, C.H., Soderlund-Venermo, M., Pintel, D.J., and Riley, L.K. (2002). Molecular characterization of three newly recognized rat parvoviruses. *J. Gen. Virol.* **83**, 2075–2083.

Weisbroth, S.H., Peters, R., Riley, L.K., and Shek, W. (1998). Microbiological assessment of laboratory rats and mice. *ILAR J* **39**, 272–290.

Yamamoto, H., Sato, H., Yagami, K., Arikawa, J., Furuya, M., Kurosawa, T., Mannen, K., Matsubayashi, K., Nishimune, Y., Shibahara, T., Ueda, T., and Itoh, T. (2001). Microbiological contamination in genetically modified animals and proposals for a microbiological test standard for national universities in Japan. *Exp. Anim.* **50**, 397–407.

Chapter 17

Occupational Health and Safety

Sanford H. Feldman and David N. Easton

THE LABORATORY RAT, 2ND EDITION

I. INTRODUCTION

Laboratory rats are one of the most commonly used species for biomedical research, second only to mice. Developing an adequate institutional animal care and use program requires that the institution's occupational health and safety program perform risk assessment at the interface between humans and rats. The human-rat interface must be characterized for the presence of risk agents that may adversely affect human health. Notable examples of risks that are intrinsic to the rat are its potential to inflict bite or scratch wounds, transmit zoonotic diseases, and shed allergens. Risk modifiers intrinsic to the human are determined by an individual's general health, immune competence, existing allergies, propensity to develop new allergies, and technical competence in handling laboratory rats. There are extrinsic factors introduced at the interface inherent to the experimental paradigm that may include administration of carcinogens, toxins, biological hazards, viral gene therapy vectors, radioisotopes and/or physical hazards (for example, electricity and laser light). There are ergonomic hazards for animal care personnel associated with the repetitive movements executed when performing animal husbandry leading to carpal tunnel syndrome and/or back strain. The design of buildings, animal facilities and animal caging can be engineered to mitigate the exposure of personnel to some risks inherent to working with laboratory rats.

Minimizing risks during human-rat interactions requires an overall assessment of risk agents and risk modifiers. Risk assessment requires determining the source and nature of the risk to humans, the frequency and duration of human exposure, and the intensity of the exposure. Risk assessment examines engineering controls and operational practices for their ability to minimize aerosol exposure and to limit the contact of humans to risk agents. Occupational risk of exposure is associated with provision of animal husbandry, cage washing activity, colony management, veterinary care, facility management and research activities. A comprehensive occupational health and safety program implements adequate controls at a variety of levels: administrative, physical plant engineering and maintenance, animal husbandry equipment, personal protective equipment and periodic assessment of employee health. Additionally, consideration must be given to second-hand exposure to non-research personnel to risk agents because of their job requirements (for example, facilities maintenance personnel, housekeeping personnel, administrative personnel) or by association with coworkers that passively transfer hazards on their soiled clothing, hands and equipment. The dynamics of how various administrative areas in an institution interact as part of a comprehensive rodent exposure program can be found elsewhere (National Research Council, 1997). The purpose of this chapter is to cover many of the risk agents to consider during development of an occupational health and safety program in a research facility that uses laboratory rats. It should be noted, however, that at this point in time specific pathogen-free laboratory rats are used in the vast majority of research in rats. Therefore, concerns regarding zoonotic disease transmission are almost exclusively associated with studies of feral rat populations or research purposefully inoculating rats with known biohazardous agents.

II. ALLERGY

Allergens associated with rat fur, urine and saliva are some of the most antigenic substances described in the literature, are responsible for the allergic reactions of many research and animal care personnel. Eighteen allergens in rat urine and saliva have been identified using immunoglobulin E from allergic humans (Gordon et al., 1993; Gordon et al., 1996). Three major allergenic proteins reacted with a high percentage of atopic human sera tested: a 44 kDa protein (100% sera), a 20.5 kDa protein (78% sera), and a 17 kDa protein (88% sera). This 17 kDa protein has been identified as alpha 2u-globulin (Rat nI), a protein determined to be a suitable environmental marker for airborne rat allergen (Gordon et al., 1993). Rat nI is found in two forms that together comprise the major protein fraction found in rat urine (Bayard et al., 1996). Rat nI is a lipocalin pheromone binding protein (Wood, 2001). Lipocalins are a family of proteins that transport small hydrophobic molecules such as steroids, bilins, retinoids, and lipids that share limited regions of sequence homology and a common tertiary structure of an eight stranded antiparallel beta-barrel enclosing in internal ligand binding. Many allergenic proteins found in the urine of various species of animals are lipocalins (Virtanen et al., 1999). Rat nI is the main allergenic constituent of the urine of adult male rats and is not excreted in significant amounts by female rats (Gordon, 2001). The secretion of rat allergenic proteins is similar in prepubescent male and female rats (Gordon et al., 1993). In adulthood, high molecular weight proteins secreted in rat urine increase dramatically in males and moderately in females; low molecular weight allergenic proteins increase moderately in both males and females (Gordon et al., 1993).

According to one study, animal technicians are exposed to approximately 32.4 $\mu g/m^3$ of rat aeroallergen when compared with office personnel in the same building whose exposure was 0.1 $\mu g/m^3$ (Nieuwenhuijsen, et al., 1994). The most intense exposure occurred to persons performing experiments on rats in laboratories (323 $\mu g/m^3$), postmortem examination of rats (199 $\mu g/m^3$), direct handling of rats (108 $\mu g/m^3$), and cage cleaning activity (74 $\mu g/m^3$) (Nieuwenhuijsen et al., 1994). In general, rat aeroallergens are associated with relatively large particles >5.8 μm in

diameter (Hollander et al., 1998) and perhaps >8 microns (Gordon et al., 1992). One study demonstrated that rat aeroallergen levels tend to be twice as high on Mondays due to the increased number of cage changes and cage washing activity associated with compensation for not having done this activity on weekends (Hollander et al., 1998).

There is a linear relationship between the number of rats in a volume of space and the level or aeroallergens present (Gordon et al., 1992; Hollander et al., 1998). Placing filter tops on static cages leads to a 6–17 fold reduction in the amount of allergen present in ambient air in one study (Hollander et al., 1998), a four-fold reduction in another (Taylor et al., 1994) and a 50-fold reduction in a third study (Platts-Mills et al., 2005). Animal holding racks that provided HEPA filtration of air exhausted from rodent cages decreased airborne allergen still further below levels associated with filter tops alone (Platts-Mills et al., 2005). Mean airborne allergen levels in rodent facilities where rodent cages were filter-topped was significantly lower than airborne cat or dog allergen levels in household living rooms (Platts-Mills et al., 2005). The type of rodent bedding dramatically affects aeroallergens in ambient air in a rat room: absorbent pads 2.47 $\mu g/m^3$, hardwood chip 6.16 $\mu g/m^3$ and sawdust 7.79 $\mu g/m^3$ (Gordon et al., 1992). The use of corncob bedding leads to a 57% reduction of aeroallergens elaborated by mice when compared with mice bedded on hardwood chip (Sakaguchi et al., 1990); no similar study has been done in rats.

In humans, clinical signs of allergy to rats manifest as contact dermatitis (angioedema) with or without urticaria, rhinitis and conjunctivitis, and/or asthma and respiratory compromise. Rat urine is a very potent sensitizer (Heederick et al., 1999). The incidence rate of allergy symptoms in laboratory animal research personnel range is 20% (Botham et al., 1987; Aoyama et al., 1992) ranging from 11% to 40% in the literature (Seward, 2001). There is a relationship between exposure to rodent allergens and the development of rodent allergy (Cullinan et al., 1994; Hollander et al., 1997; Heederick et al., 1999; Gordon, 2001). Personnel exposed to high levels of rat allergen from handling rats or the soiled bedding of rats, have twice the prevalence of positive skin testing as those with lower level exposures such as that resulting from infrequent work with rats or from work with rats housed in filter-topped caging (Hollander et al., 1997). A relationship between positive skin testing results to rat allergens and higher-level allergen exposure was found in personnel was associated with a prior history of atopy (Gautrin et al., 2001). Risk of allergy development increases rapidly at relatively low-level rat allergen exposure (Cullinan et al., 1994). Risk factors associated with the development of rat allergy are atopy, tobacco smoking, and other animal allergy (Krakowiak et al., 1996). Allergy testing can be performed by the skin prick test with Rat nI and by radioallergosorbent test (RAST) for circulating anti-rat IgE.

The Occupational Safety and Health Administration (OSHA) released a bulletin (National Institute for Occupational Health and Safety, 1998) pertaining to prudent practices for personnel exposed to animal allergens in the workplace. The bulletin recommends that animal handlers work within ventilated hoods or biological safety cabinets, wear dedicated work clothes and leave them in the workplace, keep animal facility areas clean, and use gloves and particulate respirators to reduce direct allergen contact. This OSHA bulletin recommends employers diminish aeroallergens using operational procedures based upon information obtained from research focusing on rodent allergens (Gorden et al., 1997, Harrison, 2001; Thulin et al., 2002), increase room ventilation rates and humidity, decrease animal density, use absorbent pads or corncob bedding, use less allergenic species, provide personal protective equipment, and provide training and health counseling for employees. The development of an adequate occupational health program involves the coordinated efforts of several specialists, including an occupational health physician, supervisory veterinary and animal husbandry staff, industrial hygienist, health physicists, and perhaps safety engineers. The occupational health physician can use a variety of tools as predictors of rat allergy. These include obtaining a comprehensive history to identify risk factors such as atopy, smoking, and preexisting animal allergies. Skin prick testing can be implemented to demonstrate the relative sensitivity of individuals to rat allergen, but cannot predict the type of clinical signs that will occur. RAST can be performed, however it has been shown to be positive in asymptomatic individuals and is lacking as a predictor of clinical symptom severity (Botham et al., 1987). A medical surveillance program for rat allergy should include annual evaluation by completion of a well-contrived questionnaire, general medical examination for rhinitis, conjunctivitis, urticaria, and asthma. If diminished pulmonary function is apparent in personnel after rat exposure, the annual exam may incorporate cross shift spirometry (before and after work) to determine if the diminished pulmonary function is due to workday exposure (Seward, 2001).

Minimizing allergy risk requires implementing preventive measures that fall into one of the following categories: substitution, engineering controls, administrative controls, and personal protective equipment. Substitution could involve using female instead of male rats, juvenile instead of adult rats, or a less allergenic species than rats. Engineering controls provide for high-volume directional ventilation, filter-topped caging with exhaust HEPA filtration, HEPA-filtered negatively pressurized bedding dump stations, and robotic cage dumping and washing systems. Administrative controls include adequate training of individuals regarding the hazards associated with occupational exposure to rat allergens. Animal care and research personnel training should include the importance of personal hygiene to remove allergens from superficial body surfaces and prevent the passive transfer of allergens outside the work environment. Personal protective equipment could include wearing respirators, eye protection, gloves, coveralls or lab coats, work-dedicated footwear, and other clothing to reduce respiratory, mucus membrane, and skin exposure. Respirators and other protective clothing can be difficult to wear throughout the entire workday making employee compliance problematic. If respirators are provided, they should be either the dust and mist type or HEPA cartridge, which are sufficient to remove large (5 to 8 microns) airborne particles. Only respirators approved by the United States National Institutes of Occupational Health and Safety (NIOSH) should be worn. A respiratory protection program includes pulmonary function testing and fit testing, and takes into account the employees work environment.

Safe levels for exposure to rat allergens have not been established, however 0.7 $\mu g/m^3$ has been suggested as a legally binding threshold limit in Germany (Bauer et al., 1998). There should be consideration for potential exposure to other allergens in the employees' environment, including other laboratory animals, latex rubber in gloves and medical products, fungal spores in feed and bedding, and chemicals used for disinfecting and cleaning. Irritating inhalation agents such as chlorine, acid or alkaline aerosols, peroxides, aldehydes, and alcohols may play a pivotal role in the induction of asthma associated with laboratory rats.

III. PHYSICAL AND CHEMICAL HAZARDS

A. Rat Bites

There are several retrospective studies of bite and scratch wounds inflicted by the Norway rat in urban dwellings, but comparable studies of the laboratory animal environment have not been done. One retrospective study of bite wounds over the 10-year period of 1974 to 1984 in Philadelphia identified several risk factors associated with feral rat bites (Hirschhorn and Hodge, 1999). A total of 622 rat bite cases were identified with the victims typically 5 years of age or younger. Most rat bites were inflicted on the face and hands and occurred between midnight and 8AM during summer months. The majority of rat bite victims in this study were black or Hispanic, 66% of the victims lived in single-family dwellings with poor sanitation and poor structural integrity, and 8% of the bites occurred in research laboratories. There are several reports of leptospirosis (*Leptospira icterohemorrhagica*) following the bite of feral rats (Luzzi et al., 1987; Cerny et al., 1992; Gollop et al., 1993), rat bite fever caused by *Streptobacillus moniliformis* (McHugh et al., 1985; Anglada et al., 1990; Rupp, 1992; Hagelskjaer et al., 1998; Schuurman et al., 1998) and a single case of sporotrichosis (*Sporothrix schenkii*) (Fishman et al., 1973). There is a single case report of near-fatal anaphylaxis due to rat bite in an allergic individual (Hesford et al., 1995). The most common bacterial isolates from rat bites are *Staphylococcus epidermidis*, *Bacillus subtilis*, diphtheroids and α-hemolytic streptococci (Hirschhorn and Hodge, 1999).

Rat bites, which commonly occur in laboratory animal research facilities, generally go unreported. Risk factors associated with rat bites include experience of the individual handling the rat, ambient noise levels during the handling procedure, prior experience of the animal in being handled during experiments, rat age, rat strain, rat gender, health and comfort level of the animal, and perhaps prior experimental manipulations. Rats assume easily recognizable postures prior to inflicting bites, typically raising the forelegs and turning the head with ears slightly turned backward and flattened. Training individuals in prudent rat handling practices can help prevent the occurrence of bite wounds. The training should include moving slowly and deliberately when grasping a rat, and stroking the animal while speaking in soft tones prior to actually grasping the animal. For particularly fractious animals, the rat can be grasped through a small towel or while wearing chain link gloves that prevent skin penetration by the rat's teeth.

B. Sharps

Hypodermic needles and scalpel blades are often used in research facilities using laboratory rats. Workers must be trained to handle and dispose of sharps in an appropriate fashion. Hypodermic needles should not be recapped, and once used should be disposed of in approved puncture resistant leak proof containers. State and local municipalities often regulate disposal of these sharps and their containers. The use of glass bottles is discouraged in most animal facilities. Metabolic rat cages are often purchased with glass bottles that are marked so that water consumption can be accurately recorded. Broken glass bottles should be disposed of properly according to acceptable institutional practices. Improper disposal of sharps and broken glass in regular waste containers can lead to injury of housekeeping staff with potential exposure to pathogenic organisms.

C. Pressurized Containers

Compressed gases are commonly used in research facilities. Oxygen used for inhalation anesthesia is provided at 2000 psi in green cylinders of various sizes. Carbon dioxide used for euthanasia and cell culture incubators is provided at 750 psi in a gas or liquid form in black cylinders or dewars. Nitrogen used for vibration-less tables, surgical air tools, and low oxygen tension experiments is provided at 2000 psi in a blue cylinder. Unsecured cylinders can become high-velocity projectiles if they fall and lose their valve stem. There are a variety of suitable securing mechanisms including stands, table edge clamps with straps, brackets with straps or chains for wall mounting, and cylinder carts.

Steam sterilizers are highly pressurized vessels that require personnel to routinely access the interior chamber. Steam sterilizers (autoclaves) are used to provide 100% kill of all microbial organisms associated with a material. Sterilizers are commonly used to sterilize all components of the caging, feed, and water in animal facilities that maintain rodents in filter topped caging. Sterilizers are also used to process surgical instruments and to decontaminate hazardous waste materials. Most sterilizers close with a radial arm clocking mechanism allowing the internal chamber to reach 121°C at 15 psi or 132°C at 30 psi (flash sterilization). The amount of time to sterilize a material depends on the type of cycle (prevacuum or gravity), the type of material being sterilized, the packing density of the chamber, and the steam permeability of wrapping used to enclose the material. The sterilizer door contains a gasket to allow the internal pressure to be maintained during the normal operating cycle. Hazards associated with the sterilizers include rapid release of superheated steam due to door or gasket failure, and the handling of very hot materials removed from the sterilizer. At the end of a sterilization cycle it is recommended that the door be opened a small crack for one minute to allow the escape of residual steam in a manner that directs it away from the operator. A plume of steam will escape from the chamber when opened at the end of a sterilization cycle if a dry phase is not used. Most large capacity sterilizers are installed with an overhead exhaust canopy to direct steam released when the sterilizer door is opened at the end of a cycle away from the operators face. The inadvertent exposure of the sterilizer operator to steam can occur if there is inadequate air velocity in the exhaust canopy, if canopy design is poor, and/or due to failure of the operator to crack the door for a short period of time before opening it fully.

D. Flammable Liquids

Substances that cause fires have been classified based upon their flammable and combustible characteristics into four categories by the National Fire Protection Association (National Fire Protection Association, 2003).

Class A materials that are found in animal facilities include bedding, paper gowns, plastic animal cages, paper towels, and paper bags. Class B materials are liquids (petroleum products and other organic compounds) categorized according to their flash points. Flammable liquids have flash points less than 100°F; combustible liquids have flash points greater than 100°F but less than 200°F and are more difficult to ignite at room temperatures. Class B liquids must be stored in solvent cabinets according to OSHA regulations (Occupational Safety and Health Administration, 1975). These cabinets are designed to contain and retard flame, and prevent permeation of solvent fumes into workspaces.

NFPA class C materials represent electrical fire hazard and include motors on ventilated cage racks, wet vacuums, cage washing equipment, lights, computers and printers, and commonly used electrical laboratory equipment. Ground fault interrupters (GFI) circuit breakers should be used in all areas where electrical circuits are likely to get wet. This would include areas near aquatic systems, cage washing areas, water mazes for behavioral testing, and procedure areas where electrical power strips are near sinks. All other electrical sockets within animal facilities should have waterproof covers.

NFPA class D substances are chemical compounds found in some laboratories, but not commonly found in animal facilities such as elemental sodium, magnesium and potassium. These class D compounds explode when combined with water.

E. Ionizing Radiation

The following radiation producing isotopes are those most commonly encountered in research using laboratory rats: ^{35}S, ^{32}P, ^{31}P, ^{125}I, ^{131}I, ^{57}Cr, ^{3}H, ^{99}Tc, ^{14}C, ^{18}F, ^{111}In, ^{85}Sr and ^{90}Yt. Radiation is classified as particulate or nonparticulate. Nonparticulate radiation includes x-rays and gamma rays. These are very short photons that are difficult to shield and are highly penetrating. To be classified as ionizing, the radiation must be able to displace electrons from atoms and create ions. Ionizing radiation is used for a variety of purposes in research involving laboratory rats. X-rays are the form of ionizing radiation produced by medical imaging equipment to visualize bones and soft tissues of laboratory rats. Gamma irradiators use a cesium or cobalt source that produce gamma ray photons during the process of nuclear decay. Irradiators allow one or more animals to be introduced into a small, shielded chamber. A timing device removes the shielding, exposing the animal to gamma rays. Gamma irradiation is used in laboratory rat facilities to treat solid tumors, as an immunosuppressive regimen prior to bone marrow transplantation, and for sterilization of materials that cannot be rendered sterile by steam or ethylene oxide. Radiology (x-ray) equipment and gamma irradiators are licensed and regulated by each state, and require periodic inspection and validation of radiation containment. The radiation source in gamma irradiators is licensed and regulated by the United States Nuclear Regulatory Commission. These are considered specialty equipment that require documented training to use them.

Particulate radiation can be either alpha particles, composed of two protons and two neutrons, or beta particles, which are electrons ejected at high energy. Alpha particles travel very short distances and cannot penetrate skin. Beta particles travel a distance dependent on their ejection energy, an inch for ^{3}H and many feet for ^{32}P. Beta particles penetrate skin and can cause serious skin and eye damage. Particulate radiation can be an internal hazard if radioisotopes are inhaled, ingested, or introduced into the skin contact or external hazard from close proximity to a high-energy radiation source. Some isotopes are concentrated in a particular organ, such as iodine (^{125}I, ^{131}I) in the thyroid gland. The United States Nuclear Regulatory Commission regulates use, handling, and disposal of radioactive material, including contaminated tissues and soiled bedding from laboratory rats used in radio-isotopic labeling experiments (Nuclear Regulatory Commission, 1978). Personnel using radioisotopes in research must receive specialized training in their parent institution; a radiation safety office must periodically inspect laboratories where radioisotopes are used. Personnel using radioisotopes that produce moderately to highly penetrating radiation must wear dosimetry badges, used to evaluate an individual's cumulative radiation exposure. Signage is required for identifying laboratories and animal facility spaces where radiation is used; the signage must be maintained until the area is validated free of radioactive contamination by testing environmental samples.

F. Ultraviolet Radiation

Ultraviolet radiation is used for a variety of purposes in facilities performing experimentation in laboratory rats. Ultraviolet light is classified as UV-A (320 to 400 nm), UV-B (280 to 320 nm) and UV-C (100 to 280 nm). UV-C is the most commonly used form of ultraviolet radiation due to its germicidal effects and for visualizing ethidium stained ribonucleic acids after electrophoresis in molecular biology laboratories. UV-C lamps can be found in biological safety cabinets, clean rooms, cell culture rooms, and in automatic watering systems used in animal facilities. UV-A in the form of a Wood's lamp is used to examine the fur of animals for apple-green fluorescence detected with in a high percentage of ringworm infections caused by *Microsporum canis*. Exposure of the skin or the eyes to

UV-B or UV-C can result in skin or corneal burns. Exposure of chlorinated hydrocarbons to UV-B or UV-C leads to production of phosgene gas, a potent respiratory irritant. Window glass is an effective shield to prevent exposure to UV-B and UV-C radiation. Only quartz glass will allow passage of these wavelengths.

G. Laser Radiation

Light amplification by stimulated emission of radiation is given the acronym *LASER*. The American National Standards Institute (ANSI) has categorized laser on the basis of power level and potential hazard (American National Standards Institute, 2000). Class I lasers do not emit a hazardous amount of radiation. Class II lasers are low power lasers that do not cause harm directly but will damage the retina if viewed for a prolonged period of time. Class III lasers are subdivided into IIIa, lasers that will cause harm if viewed through magnifying optics, and IIIb, lasers that can produce injury directly. Class IV lasers are similar to class III but additionally represent a fire hazard. Laboratory rats are used in the learning of surgical laser techniques in human medicine. Lasers can cause burns or lacerations and produce aerosols and fumes from tissue. The electrical power required to produce a useful surgical laser can represent a substantial electrical shock and fire hazard.

H. Chemical Hazards

Various chemical hazards are associated with the routine cleaning and sanitation within animal facilities, as well as with the experimentation performed on laboratory rats. Most chemical compounds are described in terms of their flammability, corrosiveness, toxicity, teratogenicity, and carcinogenicity. Compounds used in sanitation of animal facilities include detergents, quaternary ammonium compounds, phenolics, iodophors, chlorine dioxide generators, organic solvents, peroxides, alcohols, acids, and strong alkali solutions. Other typical sources of hazardous chemical exposure in animal facilities are pesticides, gas anesthetics, tissue fixatives, and carcinogens and toxins used in research experiments.

An excellent resource of hazardous chemical assessment can be found in the text *Prudent Practices in the Laboratory: Handling and Disposal of Chemicals* (National Research Council, 1995). An excellent resource for methods in decontamination of carcinogens used experimentally can be found in the text by Slein and Sansone (1980). Several occupational diseases, including cancer, induction of spontaneous abortion, and liver disease have been associated with exposure to waste anesthetic gases (Occupational Health and Safety Adminstration, 2005).

Periodic monitoring of waste anesthetic gases is a prudent portion of any occupational health and safety program involving research animals. Chemical burns or skin irritations are the most commonly reported chemical injuries in animal facilities. Appropriate precautions associated with any particular chemical can be found in its Material Safety Data Sheet (MSDS). Most animal facilities maintain a comprehensive book of all MSDS associated with chemical compounds used in the animal facility. The risk of illness or exposure to chemicals can be reduced with appropriate training for safe handling practices, chemical resistant gloves, respiratory protection, and use of goggles to prevent eye injury when warranted.

I. Ergonomic Hazards

The highly repetitive hand, wrist, and elbow movements associated with cage cleaning, scooping bedding, and moving animals can lead to cumulative stress injuries. These stress injuries include carpal tunnel syndrome, forearm tenosynovitis and tendonitis, and Osgood Slaughter disease. Activities required of personnel working in animal facilities include lifting 40 to 50lb bags of feed and bedding, lifting crates of animals, mopping and sweeping, pushing carts or flat beds of bedded cages, bending and reaching to service cages located on the top and bottom of racks, and pushing rack caging systems that can weigh several hundred pounds. To reduce injuries associated with repetitive motion, tasks should be varied to lessen the number of repetitions. Anyone lifting heavy loads should wear back support, avoid sudden movements, and use a two-handed lifting technique relying on the legs and not back muscles.

J. Noise and Machinery

Conveyor belts in tunnel washers, high-pressure sprayers in cage and rack washers, floor machines, and the handling of caging and equipment all add to the sound level of noise in animal facilities using laboratory rats. Loud noises are associated with cage cleaning and washing activity, and vocalization of large animals such as dogs and swine. Exposure to chronic high noise levels can lead to loss of hearing usually noted initially by a decline in sensitivity to frequencies above 2,000 Hz. OSHA limits the exposure of employees to noise 90 decibels or above over an eight hour time period. Employers that expose employees to 85 decibels or above over an 8-hour time period should provide a hearing conservation program. This program includes monitoring noise levels, audiometric testing of employees, provision of hearing protective equipment and record keeping (Occupational Safety and Health Administration, 1983). Chronic exposure of personnel to high levels of sound can lead to hearing loss, make speech difficult, cause loss of concentration and increase fatigue.

Conveyor belts, steam guns, floor machines, mechanics tools, sterilizers and cage and rack washers all have potential for causing injury. The most common types of injury are burns, crush injury, pinch points, and hearing impairment. Ramps for ingress and egress of heavy, wheeled racks into cage and rack washers are a frequent cause of crush injury. Each piece of equipment must be evaluated individually and standard procedures put into place that will prevent injury to the worker.

IV.   BACTERIAL DISEASES

The transmission of zoonotic diseases from laboratory rats to humans is uncommon. This is due to the concerted effort made by commercial and institutional producers of laboratory rats to continually verify that the rats are free of a detailed list of microbial organisms, including zoonotic disease agents (see chapter on rat colony health surveillance). Programs of veterinary care at research institutions continually survey for the causative agents of diseases in laboratory rats to validate the specific pathogen-free status of rats and identify new pathogens not previously been described. The following discussion is a comprehensive list of zoonotic diseases of rats. Emphasis is placed on those diseases that are more likely to occur. Other diseases are introduced to encompass research involving feral rats, including genera other than *Rattus*, which may harbor these uncommon agents.

A.   Rat Bite Fever

Streptobacillus moniliformis is the causative agent of rat bite fever (Andersen et al., 1983). This Gram-negative organism has been reported worldwide and can be present in the nasopharyngeal cavity of asymptomatic rats. This disease has largely been eradicated from laboratory animal colonies, however outbreaks in laboratory animals have been reported as recently as 1996 in an Australian colony of Swiss Webster mice (Glastonbury et al., 1996) and in a colony of rats in the Netherlands (Boot et al., 1996). *S. moniliformis* has been identified as a causative agent of otitis media in conventionally housed laboratory rats in a report from Germany (Wullenweber et al., 1992); reports of the organism being isolated in barrier-reared rat colonies still occur (Wullenweber, 1995). The most common mode of transmission of rat bite fever from rats to humans is a bite wound inoculating the organism. After a short incubation period of 2 to 10 days in the infected person, there is routine healing of the bite wound.

Clinical symptoms of the disease in humans manifest as fever, malaise, headache, sore throat, lymphadenitis, myalgia, and a petechial rash on the extremities. Septic arthritis occurs in 50% of cases usually affecting one or more large joints (Mandel, 1985). Complications of the disease include focal abscess, endocarditis, pneumonia, hepatitis, and pyelonephritis. Definitive diagnosis of Streptobacillosis requires isolation of this fastidious organism, whose growth characteristics require inoculation onto a specialized media. The most recent report of rat-bite fever in a laboratory research technician occurred in early 1980s (McHugh et al., 1985)

Sodoku is less common cause of rat bite fever, resulting from infection with the bacterium *Spirillum minus*. Sodoku is characterized by a longer incubation period (2 to 3 weeks) than Streptobacillosis. *S. minus* is a causative agent of conjunctivitis in rats but can be found in the blood and urine of asymptomatic rats. The organism is transmitted by bite wound inoculation. The clinical symptoms in humans that distinguish *S. minus* as the cause of rat bite fever, from *S. moniliformis*, is an indurated ulcer at the site of the original bite wound and the absence of development of septic arthritis. Other symptoms in humans infected with *S. minus* include fever, malaise, regional lymphadenitis, myalgia, and a macular rash. Reports of rat-bite fever caused by *S. minus* due to bites from feral rats have been reported as recently as 1992 in Kenya (Bhatt and Mirza, 1992) and Portugal (Hinrichsen et al., 1992).

B.   Salmonellosis

Organisms belonging to the genus *Salmonella* have a worldwide distribution among animals, but are rarely recovered from laboratory rats. It is possible that laboratory personnel infected with *Salmonella* or interacting with other species (nonhuman primate, cat, or dog) could inadvertently infect naïve rats. The most recent published investigation of an outbreak of salmonellosis in laboratory rats occurred in Amsterdam in 1987 (Seeps et al., 1987). *S. typhimurium* can be isolated from at low frequency from feral Norwegian rats (Seguin et al., 1986). Once infected, rats become chronic carriers episodically shedding the organisms in their feces (Naughton et al., 1996). The mode of transmission from rat to humans is fecal-oral due to poor hygiene practices of personnel working with infected laboratory rats. Clinical signs of salmonellosis in humans include fever, enterocolitis, diarrhea, septicemia, and focal infections of the liver that may lead to jaundice. The organism can be isolated on standard microbiological media. Rats naturally or experimentally infected with *Salmonella* spp. should be contained at Animal Biosafety Level 2 (Center for Disease Control-National Institutes of Health, 1999).

C. Leptospirosis

Leptospirosis (Weil's disease) in rats is most commonly caused by *Leptospira icterohemorrhagiae* and *L. ballum*. This infection is generally inapparent in rats. Endemic leptospirosis in laboratory animals was reported as recently as 1996 in a facility in Madras, India (Natrajaseeivasan and Ratnam, 1996) where two serovars were found, namely *L. javanica* in rats and *L. autumnalis* in mice and guinea pigs. Leptospires can be isolated from feral rats (Seguin et al., 1986).

After initial infection, rats undergo bacteremia with the organism settling in the kidneys and being shed continually in the urine. The disease spreads to humans by ingestion of infected urine, contamination of wounds with infectious urine, inhalation of infected aerosols, or less frequently by bite-wound. Clinical signs of leptospirosis in humans include headache, fever, myalgia, conjunctival suffusion, jaundice, hemolytic anemia, hepatorenal failure and mental confusion associated with encephalitis. Diagnosis of the disease is by serologic detection of antibodies against leptospires and by phase contrast microscopic examination of urine demonstrating the presence of leptospires. Containment of laboratory rats at animal biosafety level 2 is recommended for personnel working with naturally or experimentally infected animals (Centers for Disease Control-National Institutes of Health, 1999).

D. Plague

Yersinia pestis is the causative agent of bubonic and pneumonic plague (The Black Death). The disease is transmitted from rats to humans by rat fleas (*Xenopsylla cheopis* and *Nosopsyllus fasciatus*) leading to the bubonic form (bubos = abscessed regional lymph nodes). Focal outbreaks are recognized in feral rodent populations in the western third of the United States and elsewhere worldwide (Keeling and Gilligan, 2000). Pneumonic forms of the disease are caused by inhalation of aerosols of infectious material and perhaps ingestion.

Clinical signs in humans manifest as either a bubonic form, localized to the regional lymph node draining the region of the flea bite, or pneumonic form, which represents a far greater hazard to other humans due to the infective aerosols generated during coughing. The bubonic form can progress to septicemia, leading to the pneumonic form in some individuals. Diagnosis of *Yersinia pestis* infection is confirmed by bacterial isolation, direct fluorescent antibody testing of suspect tissues, or an antigen-capture ELISA. Prevention includes control of feral rodents in animal facilities coupled with prevention of ectoparasite infestation of laboratory rat colonies. Containment of laboratory rats at animal biosafety level 2 is recommended for personnel working with naturally or experimentally infected animals (Centers for Disease Control-National Institutes of Health, 1999).

E. Streptococcal Infection

Streptococcus pneumoniae (synonyms: Pneumococcus, Diplococcus) continues to be infrequently isolated as a commensal flora of commercially produced Fisher 344 rats in the United States (Fallon et al., 1988). The upper respiratory tract of older laboratory rats is most frequently colonized with serotype 35. There is no evidence of streptococcal transmission from rats to humans.

F. Staphylococcal Infection

Staphylococcal infections are the most common sequel to feral rat bites. The overall wound infection rate after rat bite is low, but when it occurs *Staphylococcus epidermidis* is isolated in 43% of the cases. The Centers for Disease Control estimates that bites from feral rats occur in major metropolitan areas at a rate of 10 per 100,000 people per year (Bjornson et al., 1968). In an epidemiologic survey of feral rats in Lyon, France, *S. aureus* was cultured from the oropharyngeal cavity of 53% of the rats sampled (Seguin et al., 1986). In one study the incidence of *S. aureus* in laboratory rats was found to be 100%, however the phage type varied, with type 35 being most common (Wullenweber et al., 1990). The clinical presentation of staphylococcal infection in humans is most often as a localized abscess. Diagnosis of staphylococcal infection is confirmed by bacterial isolation, and treatment by determination of antimicrobial administration after sensitivity testing of the isolate often coupled with establishing drainage of the infected wound.

G. Pseudotuberculosis

There are several reports demonstrating that feral rats act as a reservoir for *Yersinia pseudotuberculosis* outbreaks in humans (Borowski et al., 1970; Fukushmina et al., 1988; Ol'iakova and Antoniuk, 1989). *Y. pseudotuberculosis* is routinely isolated from feral rats trapped in abattoirs, barns, and zoos in Japan (Zen-Yoji et al., 1974; Kaneko et al., 1979; Kaneko and Hashimoto, 1983). *Y. pseudotuberculosis* was isolated from sewer and laboratory rats in Eastern Europe (Aldova et al., 1977). The organism is persistently excreted in the feces of infected rats. Humans infected with *Y. pseudotuberculosis* present with symptoms of gastroenteritis and mesenteric lymphadenitis (Press et al., 2001). The diagnosis of *Y. pseudotuberculosis* is confirmed by microbial isolation from stool or lymph nodes of infected people. *Y. pseudotuberculosis* is responsive to rational antimicrobial therapy.

H. Listeriosis

Listeria monocytogenes has been isolated from feral rats in the United Kingdom, Eastern Europe, Japan, Singapore, and the Sverdlosk region of the former Soviet Union (Zhukokva et al., 1966; Iida et al., 1991; Webster and MacDonald, 1995a; Webster, 1996; Chan et al., 2001). Feral rats from areas enzootically infected with this organism form an important reservoir of the organism and so represent a significant zoonotic hazard to humans living in these areas. Symptoms of listeriosis in humans include spontaneous abortion and gastroenteritis from contaminated food (Hof, 2001). Meningoencephalitis is a common clinical manifestation of listeriosis caused by subdural abscesses associated with the presence of this organisms, these abscesses leading to neurologic symptoms whose manifestation is dependent on the area of brain affected. Diagnosis of *L. monocytogenes* infection is confirmed by isolation of the organism from stool or cerebral spinal fluid using cold enrichment methods. Central nervous system involvement can lead to chronic seizures or fatality if not diagnosed and treated early with appropriate antimicrobial therapy (Chang et al., 2001).

I. Erysipelas

There is a single report of spontaneous fibrinopurulent polyarthritis in Sprague-Dawley rats due to *Erysipelothrix rhusiopathiae* (Feinstein and Eld, 1989). Additionally, this organism has been isolated from feral rats in Australia (Eamens et al., 1988), Siberia (Ol'iakova and Antoniuk, 1989) and Sverdlosk (Zhukova et al., 1966). Feral rats constitute a nidus of *E. rhusiopathiae* that potentially could transmit to humans. Infected humans often develop a localized cutaneous infection leading to full thickness skin necrosis of the infected area. The cutaneous infection can progress to bacteremia seeding most commonly the mitral valve of the heart causing valvular endocarditis (Hedirich et al., 2001). In humans, one uncommon sequel to cutaneous *Erysipelothrix rhusiopathiae* infection is the development of a localized acute osteomyelitis (Leveque et al., 2001).

J. Bordetellosis

The bacterium *Bordetella bronchiseptica* is infrequently isolated from laboratory rats in commercial production facilities (Boot et al., 1996); the incidence in Japanese laboratory rats by serologic monitoring was reported as high as 40% during the 1970s (Fujiwara et al., 1979). The bacterium colonizes the upper respiratory tract of laboratory rats (Brockmeier, 1999). *B. bronchiseptica* is an uncommon cause of pneumonia in children (Stefanelli et al., 1997), but not an uncommon cause of pneumonia in patients affected with human immunodeficiency virus (Dworkin et al., 1999). Zoonotic transmission from rabbits to an elderly woman has been reported (Gueirard et al., 1995), however no similar transmission from rats to humans has also been reported. *B. bronchiseptica* infection manifests as mild to moderate upper respiratory symptoms in immune competent humans to pneumonia in immunodeficient persons.

K. Bartonellosis

Forty six percent of sera collected from intravenous drug users in Central and East Harlem during 1997 and 1998 had antibodies to rodent-associated *Bartonella elizabethae*, commonly found in *Rattus norvegicus* (Comer et al., 2001). *Bartonella* infection of woodland rats is common in the United Kingdom (Birtles et al., 2001) and the United States (Ellis et al., 1999). In humans, bartonellosis is characterized by a localized caseous lymphadenitis developing near the site of a recent bite or scratch. This localized lymphadenitis can progress to parasitemia with endocarditis as a potential outcome. The *Bartonella* organisms are difficult to cultivate, therefore diagnosis is most often confirmed with serologic or molecular microbiologic tests.

V. RICKETTSIAL AND PIROPLASMOSIS DISEASES

A. Flea-Borne Typhus

This disease caused by *Rickettsia typhi* (*R. mooseri*) occurs worldwide, most commonly seen in the Americas. Feral rats (*R. norvegicus* and *R. rattus*) act as the reservoir for the disease in humans; infected rats do not exhibit symptoms of illness. The organism replicates in the Malpighian tubules (an excretory organ) of the flea, a process that damages the flea. The organism is shed in flea excrement. The rat flea (*Xenopsylla cheopis*) (Acha and Szyfres, 2003) transmits the disease from rat to humans by contamination of the flea bite wound with flea excrement. The flea *Nosopsyllus fasciatus* and possibly the louse *Polyplax spinulosa* can also act as vectors for transmission of *R. typhi* between rats. In areas of enzootic disease where *X. cheopis* is absent, opossums (*Didelphis virginiana*) act as the reservoir host of *R. typhi* and the cat flea

(*Ctenocephalides felis*) functions as the vector for the disease from opossums to humans. The incubation period in humans for flea-borne typhus is 1 to 2 weeks. Symptoms of flea-borne typhus in humans manifest as severe headache, myalgia, arthralgia, coughing, nausea, and vomiting, followed by macular eruption on the trunk and extremities but not hands or face.

B.   Scrub (Mite-borne) Typhus

This disease of the Pacific Islands, Southeast Asia, and Northern Australia is caused by *Rickettsia tsutsugamushi*. The reservoir is wild rodents, most commonly *Mus domesticus*, and the disease is most commonly transmitted to humans by the bite of *Trombicula akamushi* and *Leptotrombidium deliensis*. Less commonly, the disease is spread by the mites *L. pallidum* and *L. scutellare* (Acha and Szyfres, 2003). A black-scabbed ulcer is found at the site of the skin bite on Caucasians but rarely on persons of Asiatic origin. Initially, symptoms of scrub typhus in humans include congestion of the conjunctiva and generalized lymphadenopathy. These early symptoms are followed in 1 week by a generalized maculopapular eruption. Depending on the strain of rickettsia, the disease if untreated can progress to potentially fatal pulmonary, encephalitic, or cardiac forms.

C.   Rickettsialpox

Rickettsialpox is caused by *Rickettsia akari*. The condition was initially described in New York City in 1946, and later in Baltimore, Maryland (Comer et al., 1999; Comer et al., 2001), and in North Carolina (Krussell et al., 2002). Rickettsialpox has been reported in Korea, Africa, and the Ukraine region of Eastern Europe (Acha and Szyfres, 20031980). Feral rats (of the genus *Rattus*) and mice (of the genus *Mus*) act as reservoir hosts. The rodent mite *Liponyssoides sanguineus* transmits rickettsialpox between infected rodents and humans. Initially, the site of the infected mite bite develops a papular eruption, which is followed in 1 week by a fever. After 2 to 3 weeks, the initial lesion develops a central vesicle that is finally covered by a dark scab, leaving a scar. Symptoms of rickettsialpox in infected humans manifest as fever, sweating, headache, and myalgia. Between the first and fourth day of fever a maculopapular eruption occurs on many parts of the body except the palms of the hands and soles of the feet (Boyd, 1997). The eruption is nonpruritic and resolves without leaving scars.

D.   Babesiasis

Babesiasis occurs everywhere in the world where ticks are found. Rats harboring *Babesia microti* act as a reservoir for transmitting the disease to naïve feral rats and humans via the bite of the tick *Ixodes spinipalpis* (Burkot et al., 2001). Splenectomized humans can become symptomatic when infected. *B. microti* infection in humans manifests as fever, headache, myalgia, anemia, jaundice, and hemoglobinuria. There have been reports of disease in nonsplenectomized individuals on Nantucket Island (Acha and Szyfres, 2003).

VI.   VIRAL DISEASES

A.   Arenaviruses

1.   Lymphocytic Choriomeningitis Virus

The lymphocytic choriomeningitis virus (LCMV) can be found in the feral rat population worldwide, capable of acting as the source of human infections. A serologic study of people residing in the Baltimore, Maryland inner city between 1986 and 1988 reported that 4.7% of the adult human population possessed antibodies to LCMV demonstrating previous exposure to the virus (Childs et al., 1991). This viral infection is rare to nonexistent in laboratory rats; most outbreaks in humans are attributed to exposure to the urine of infected hamsters (Bowen et al., 1975) or mice (Sato and Miyata, 1986). LCMV disease in rats is characterized by retinitis (del Cerro et al., 1984), auditory and visual impairment (Campo et al., 1985), and cerebellar hypoplasia. LCMV inhibits the development of insulin-dependent diabetes mellitus in BB rats (Dryberg et al., 1988). Contacting tissues or urine from rodent infected can result in infection in humans. The disease in humans can be clinically inapparent, or, on rare occasion, manifest as a fatal meningoencephalitis. Most persons infected with LCMV experience symptoms similar to those caused by influenza virus. The incubation period for LCMV in humans is 1 to 2 weeks. Symptoms of LCMV infection in humans manifest as stiff neck, fever, headache, malaise, and myalgia. Samples of cerebrospinal fluid from LCMV-infected people contain 100 to 3000 lymphocytes per milliliter. Transplacental infection of rats leads to immune tolerance to the virus from birth; prenatal infection of rats with LCMV leads to lifelong infection with continual shedding of virus in urine.

2.   Lassa Fever Virus

Lassa fever virus is found in West and Central Africa, first isolated in 1969 in Lassa, Nigeria. The only rodent known to carry the virus is *Praomys (Masomys) natalensis* the multimammate rat (Green et al., 1978). Animals inoculated during the first 4 days of life are immune tolerant and develop persistent infection. Inoculation of late gestation *P. natalensis* females has no influence on litter size and the entire litter is persistently infected by 2 weeks of age. In Sierra Leone, the incidence of antibodies in humans to this disease is 11 percent. Humans that contact infected rodent urine or infected humans can contract the disease. The incubation period in humans is 1 to 3 weeks. Humans infected with Lassa fever virus manifest symptoms of fever, myalgia, headache, vomiting, diarrhea and abdominal pain, ulcerative pharyngitis, coughing, stertorous breath, and thoracic pain (Acha and Szyfres, 2003). Most Lassa fever virus-infected people are asymptomatic. If the disease persists, an infected person will eventually demonstrate apathy, vomiting and diarrhea, capillary hemorrhages, and a small percentage progress to fatal circulatory collapse. The disease can be transmitted between humans. The virus is isolated for the respiratory, gastrointestinal, and genitourinary tracts of infected individuals. Epidemics occur in hospitals from the rapid spread of Lassa fever from an initial case after hospitalization to other people before a definitive diagnosis of Lassa fever is made. Transmission is associated with infected blood, pharyngeal secretions, urine and sexual transmission. Case fatality rates for nocosomial hospital based infections are 10%.

3. Machupo Hemorrhagic Fever (Black Typhus)

Bolivian hemorrhagic fever (black typhus) caused by Machupo virus is a member of the family Arenaviridae known as the Tacaribe-LCMV complex that includes Lassa fever, Junin and lymphocytic choriomeningitis virus. Machupo virus is only found in Bolivia and was first described in 1959, although cases have been described in other neighboring Central American countries (Acha and Szyfres, 2003). Hospitalized cases lead to small nosocomial outbreaks similar to Lassa fever virus; horizontal transmission of black typhus between humans occurs. Experimental infection of neonatal Wistar rats and mice resulted in virus replication to high titer in the thymus, causing persistent infection without the development of neurologic symptoms (Calello et al., 1985). Rodents persistently infected with Machupo hemorrhagic fever virus manifest neurologic symptoms. In affected localities, the spiny rat (*Proechimys gyannensis*) was initially suspected as the reservoir of this disease, however the cricetid large vesper mouse (*Calomys callosus*) is now known to act as the reservoir of black typhus; Machupo virus can be isolated from 25% of *C. callosus* in enzootically infected areas. The virus spreads horizontally between rodents. Infection of *C. callosus* younger than

10 days of age leads to immune tolerance and persistent viremia. Humans contract the disease through inadvertent contact with the excreta of infected rodents or by consuming contaminated food and water supplies, or by contacting secretions and blood of infected symptomatic humans. The incubation period in humans is 2 weeks followed by the emergence of symptoms including fever, myalgia, headache, conjunctivitis, and petechial hemorrhage of the gums and gastrointestinal tract. As many as 50% of persons with Machupo virus infection exhibit neurologic symptoms, including tremor of the tongue and extremities, convulsion, and coma. Rarely, infected patients succumb to hypotensive shock. Leucopenia is a consistent clinical laboratory finding in infected humans.

B. Hepatitis E

Antibodies to Hepatitis E (an RNA-based alpha virus) are found in less than 1% of U.S. citizens. One study examining antibody titers to hepatitis E in the serum of feral rats trapped within the United States reported positive results in 77% of rats sampled in Maryland, 90% sampled in Hawaii, and 44% sampled in Louisiana (Kabrane-Lazizi et al., 1999). Domestic swine are also susceptible to hepatitis E infection, developing hepatic inflammation, shedding the virus in their stool (Halbur et al., 2001). Both swine and feral rats have been developed as animal models of hepatitis E (Purcell and Emerson, 2001). Clinical symptoms in humans infected with hepatitis E include gastrointestinal upset and jaundice. These symptoms resolve in most infected persons after several weeks; however, fatal disease occurs in 25% of pregnant women that contract infection with hepatitis E during the second or third trimester of pregnancy (Krawczynski et al., 2001).

C. Hemorrhagic Fever with Renal Syndrome

Hemorrhagic fever with renal syndrome (HFRS) is caused by members of the RNA virus family Bunyaviridae in the genus Hantavirus. These viruses include Seoul virus, Korean Hemorrhagic Fever, and Hantaan virus (LeDuc, 1989). By 1985, 126 human cases of laboratory-acquired HFRS were reported in Japan (Kawamata et al., 1987). Laboratory research staff was infected more frequently than animal caretakers, and one animal caretaker died of the disease. Inhalation of aerosolized virus and wound infection were the major modalities of transmission from laboratory rats to humans. The tropical rat mite *Ornithonyssus bacoti* can act as a vector for transmission of Hantaan and Seoul viruses between rats, as well as from rats to humans (Zhuge et al., 1998). Symptoms of hemorrhagic fever with renal syndrome in humans manifest initially as fever, bradycardia, and conjunctival congestion.

The disease progresses with the development of abdominal pain, vomiting, hemorrhages, and nephropathy. Symptoms of renal involvement are hematuria, oliguria, proteinuria, and, later, production of copious isosthenuric urine. Fatality can occur during the oliguric phase.

D. Hantavirus Pulmonary Syndrome

Sin Nombre virus, the prototypic member of this gorup in the genus Hantavirus, is the causative agent of hantavirus pulmonary syndrome (HPS) a disease with a 50% mortality rate. Humans infected with several other hantaviruses exhibit HPS. Deer mice (*Peromyscus maniculatus*) are the reservoir of Sin Nombre virus (Duchin et al., 1994). Other hantaviruses of the Americans causing HPS include Black Creek Canal virus, with cotton rats (*Sigmodon hispidus*) acting as the reservoir (Rakov et al., 1997); New York 1 virus, with the reservoir existing in white-footed mice (*P. leucopus*) (Gavrilovskaya et al., 1999); El Moro Canyon virus, with harvest mice (*Reithrodontomys megalotis*) acting as the reservoir (Mantooth et al., 2001); and Bayou virus, with rice rats (*Oryzomys palustris*) acting as the reservoir (Ksiazek et al., 1997; Torrez-Maartinez et al., 1998). Initial symptoms in humans of HPS manifest as a low-grade fever, headache, cough or dyspnea, nausea and vomiting, and myalgia. Clinical laboratory findings of thrombocytopenia and leucocytosis are found in all cases. Persons infected with hantavirus causing HPS exhibit poor pulmonary oxygen diffusion from severe pulmonary edema, often leading to a fatal cyanosis (Figueriredo et al., 2001).

E. Rabies (*Rhabdoviridae*)

Rodents do not act as a reservoir of rabies virus. Annually in the United States there are a small number of rodents examined by state departments of public health that demonstrate rabies virus in post-mortem examination of brain tissue by direct fluorescent antibody assay. There is a single report of a local outbreak of paralytic rabies in children in Surinam, attributed to the bites of feral rats (Verlinde et al., 1975).

F. Encephalomyocarditis Virus (*Picornaviridae*)

Early reports of this virus presented serologic evidence that enzootic infection was common in feral rats (*Rattus norvegicus* and *R. rattus*) in the United States and Canada (Acha and Szyfres, 2003). It has subsequently been shown that encephalomyocarditis virus (EMC) is more often associated with infected herds of swine that remain asymptomatic, or which have a herd history of sudden death and abortion (Kim et al., 1991). The reservoir of this virus is still unknown, and may involve wild rodents and free-ranging birds. One report of an epizootic of EMC in a baboon colony was attributed to infestation of the area with infected feral rodents (Hubbard et al., 1992). However, no rigorous testing of feral rodents occurred in that study. EMC is excreted in the urine and feces of infected animals. The D variant of this virus has been used to produce an experimental model of diabetes mellitus in rodents (Matsuzaki et al., 1989). Symptoms of EMC infection in humans are fever of a 2 to 3 week duration, headache, pharyngitis, stiff neck, and, infrequently, paralysis of the limbs.

VII. PROTOZOAL DISEASES

A. Amoebiasis

Historically, *Entamoeba histolytica* was reported as a common isolate from the gastrointestinal tract of feral rats (Fulton and Joiner, 1948; Neal, 1951); however, no recent reports in the literature support these earlier findings. Experimental oral inoculation of *E. histolytica* into the laboratory rat leads to replication of the organism in the gastrointestinal tract, but not in extra-intestinal sites (Tsutsumi, 1994). In humans, amoebiasis manifests as dysentery with severe gastrointestinal discomfort, fever and malaise, and mucus in the stool. Amoebic abscesses can occur in the liver or brain of infected humans with clinical signs associated with damage to these organs.

B. Encephalitozoonosis

Enzootic infection with *Encephalitozoon cuniculi* has been reported in laboratory rat colonies (Majeed and Zubaidy, 1982) and in their feral counterparts (Muller-Doblies et al., 2002). An isolate of *E. cuniculi* from feral rats was passaged in laboratory Brown Norway rats, and characteristically disseminated to liver and lung, inciting granulomatous lesions (Muller-Doblies et al., 2002). There are several reports of secondary infection with *E. cuniculi* in persons with acquired immunodeficiency syndrome (AIDS); however, the origin of these microsporidian infections is unclear. Humans infected with *E. cuniculi* demonstrate clinical symptoms associated with intestinal infection, as well as conjunctivitis, rhinitis, sinusitis, pneumonia, nephritis, hepatitis, and meningoencephalitis (Van Gool and Dankert, 1995; Fournier et al., 2000; Schotellius and da Costa, 2000; Del Aguila et al., 2001; Tosoni et al., 2002).

C. Cryptosporidiosis

Cryptosporidium parvum has been isolated from feral rats (and other wild rodents) in Spain and the United Kingdom (Quy et al., 1999). The prevalence of *C. parvum* infection in feral rats in the UK ranges from 24% to 63% (Webster and MacDonald, 1995b, Torres et al., 2000). Humans at risk for acquiring *C. parvum* infection are immunocompromised children and adults (especially those with AIDS), children in day care, cattle farmers, and travelers to areas of enzootic infection (Keusch et al., 1995; Bhattacharya et al., 1997). The primary target of *C. parvum* is the intestinal epithelium. Infection of humans results in clinical symptoms including watery diarrhea, although the gall bladder and lungs can be affected, with symptoms arising from dysfunction of these organs. Chronic diarrhea in up to 22% of AIDS patients has been attributed to intestinal infection with *C. parvum* (Beaugeri et al., 1998).

VIII.   HELMINTH INFESTATIONS

A. *Hymenolepis* Species

The rodent cestodes *Hymenolepis nana* and *H. diminuta* are capable of infecting humans and causing disease. These tapeworms are commonly found in feral rats. *H. nana* completes its life cycle within its definitive host, whereas the life cycle of *H. diminuta* requires an intermediate host, typically coprophagic arthropods. Humans acquire *H. nana* infection by ingesting infectious ova; infection with *H. diminuta* requires humans accidentally ingesting an arthropod intermediate host containing an *H. diminuta* cysticercoid. Human infection with *H. diminuta* occurs most often by ingestion of flour and cereal beetles in food contaminated with rodent feces (Acha and Szyfres, 2003). Humans can act as a reservoir of *H. nana* infection, since auto-reinfection occurs, and horizontal transmission between humans is via poor hygiene associated with the fecal-oral route (Buscher and Haley, 1972; Moon, 1976; Sahin, 1979). Symptoms of *Hymenolepis* infestation of humans include nausea, vomiting, abdominal pain, diarrhea, irritability, restless sleep, and allergic symptoms (anal and nasal itching). One-third of infected humans develop peripheral eosinophilia of greater than 5% on a white blood count differential as a consistent laboratory finding.

B. *Capillaria hepatica*

One survey of feral rats (*R. norvegicus* and *R. rattus*) in Marseille, France demonstrated a 44% incidence of *C. hepatica* (Davous et al., 1997). This nematode causes necropurulent hepatitis with fibrosis in infested rats. The eggs are passed in feces from infected animals and subsequently embryonate in the soil. Human infections with *C. hepatica* are rare since infection requires ingesting of soil contaminated with embryonated eggs. The disease in humans is similar to that seen in rodents. Hepatomegaly, high fever, nausea and vomiting, diarrhea, and peripheral blood eosinophilia are characteristic symptoms of *C. hepatica* infestation in humans (Bhattacharya et al., 1999). A total of 37 cases of *C. hepatica* in humans are reported in the literature (Juncker-Voss et al., 2000).

C. Pacific Basin Eosinophilic Meningitis

The principal etiologic agent of this disease is *Angiostrongylus cantonensis*, the rat lungworm. This disease, endemic to the islands of the Pacific, was initially described during World War II (Klicks and Palumbo, 1992). The life cycle of this nematode in rats begins as adult nematodes living in the airways of rats. Sexually mature *A. cantonensis* adults lay eggs in the airways that migrate up the trachea, are swallowed, and expelled to the environment in the stool of infested rats. Under favorable environmental conditions the eggs embryonate and are eaten by a mollusk or crustacean intermediate host. The rodent ingests the infested mollusk or crustacean that contains third stage larvae of *A. cantonensis*. Once ingested, the larvae enter the gut and penetrate into the circulatory system where they are carried to the right heart and eventually the left heart. The larvae are pumped to the brain where they accumulate and undergo two molts before leaving via the subarachnoid spaces, migrating back to the lungs. The presence of adult worms in the lungs of rats induces a modest pulmonary fibrosis. Humans that ingest infested mollusks or crustaceans undergo a similar release of larvae, which migrate to the brain but are incapable of completing their migration to the lung. The symptoms of Pacific basin eosinophilic meningitis in infested humans is characterized by clinical laboratory findings of a peripheral blood eosinophilia of greater than 10%, and symptoms of headache, neck stiffness, vomiting, and pain in the arms and shoulders.

IX.   ARTHROPOD INFESTATIONS

A. Fleas

In recent times, fleas have rarely been seen on laboratory rats, but are a common ectoparasite of feral rats. The oriental flea (*Xenopsylla cheopis*) is the vector for bubonic plague (*Yersinia pestis*) and flea-borne typhus

(*Rickettsia typhi*). The oriental flea, the mouse flea (*Leptosylla segnis*) and the northern rat flea (*Nosopsyllus fasciatus*) will bite humans. All three of these species of fleas can act as an intermediate host for *Hymenolepis nana* and *H. diminuta*. The northern rat flea is important in the transmission of *Trypanosoma lewisi*, which is pathogenic to animals but not man (Molyneux, 1969).

B. Mites

The tropical rat mite (*Ornithonyssus bacoti*), once a common ectoparasite of laboratory rats, is now rarely found in laboratory rats. This mite is a vector for transmission of hantavirus infections between rats and humans (Zhuge et al., 1998); Hantaan virus has been detected within infected *O. bacoti* when examined by *in situ* hybridization techniques (Wu et al., 1998). *O. bacoti* can support the survival of *Borrelia burgdorferi* (Lyme disease) however, it has not been shown to be an important vector for Lyme disease transmission (Lopatina et al., 1999). Tropical rat mite dermatitis of humans manifests as a transient pruritic disease that resolves 2 weeks after exposure (Engel et al., 1998).

Liponyssoides sanguineus is the vector for the transmission of rickettsialpox caused by *Rickettsia akari*. Persons contracting rickettsialpox manifest symptoms that include fever, headache, and a vesicular eruption over the trunk and extremities (Boyd, 1997). *L. sanguineus* can also be found on gerbils (Levine et al., 1984). *L. sanguineus* and *Ornithonyssus bacoti* are not resident on the host, but infest the rat only when feeding. *Trombicula akamushi*, *Leptotrombidium deliensis*, *L. pallidum* and *L. scutellare* are capable of transmitting scrub typhus to humans.

X. MYCOTIC INFECTIONS

A. Dermatophytosis

The dermatophyte most commonly infecting laboratory rats is *Trichophyton mentagrophytes* (Connole et al., 2000). This fungus is capable of causing ringworm lesions in humans. Infection of laboratory rats is rarely seen and there have been no studies over the last 40 years of the incidence of this organism in feral rat populations. Dermatophytosis is transmissible to humans from infected rats, and manifests as a localized ringworm lesions affecting areas of haired skin. Dermatophytes infect the epidermis and adnexal structures, including hair follicles and hair shafts, inducing pruritus, patchy alopecia, erythema, and crusting.

B. Pneumocystosis

Pneumocystis carinii is a fungus commonly found in the respiratory tract of rats. One can induce fulminate respiratory disease due to pneumocystosis in rats by administering immunosuppressive dosages of corticosteroids or cyclophosphamide (Weisbroth et al., 1999). Recent research has demonstrated that isolates of *P. carinii* are host specific even in immune deficient animals. Therefore, rat strains of *P. carinii* are unlikely to infect humans (Gigliotti et al., 1993; Durand-Joly et al., 2002), and conversely human strains of *P. carinii* are not likely to infect rats.

XI. EXPERIMENTS USING BIOHAZARDOUS AGENTS

A. Risk Assessment

Experimental procedures involving the deliberate administration of biohazardous agents to laboratory rats requires prior planning so that the appropriate containment practices and personal protective equipment are implemented for the handling of inoculated rats, procedures for processing soiled caging, and procedures for decontamination of rat rooms. The process of identifying components of hazardous activities and appropriate precautionary practices is referred to as *risk assessment*. Performing risk assessment is hardly a novel concept in the average individual's everyday life. Lifestyle decisions all have inherent risks or hazards, including such activities as alcohol or tobacco consumption, dietary indulgences, and driving an automobile. However, we are willing to accept risks associated with these activities and exercise caution to derive benefit from these activities. Although decision-making is more reliable when a quantitative means is used to determine the hazard, this is generally not possible in risk assessment and so decisions are made subjectively.

The lack of objective risk assessment is particularly apparent in research that requires the inoculation of biohazardous agents into laboratory animals. Unlike the quantitative assessment of the risk posed by a chemical exposure, at a known concentration for a specified time resulting in a calculated dose, the outcome of inoculation of a biohazardous agent into laboratory rats can is dependent upon a variety of qualitative factors. These qualitative factors are often called *risk factors,* and include the pathogenicity of the agent (it's virulence), the route(s) of transmission (airborne, parenteral, ingestion, or arthropod vector), the agent's concentration and stability in the environment, and the susceptibility of humans and the laboratory rat (Centers for Disease Control-National Institutes of Health, 1999).

B. Biosafety Principles and Practices

Several sources of information categorize risk factors associated with biohazardous agents that are helpful in determining relative risk. Appendix B of the *NIH Guidelines for Research Involving Recombinant DNA Molecules* (National Institutes of Health, 1994) contains the "Classification of Human Etiologic Agents on the Basis of Hazard." The agents listed in this appendix are assigned to "Risk Groupings" after consideration of qualitative risk factors associated with them. Biohazardous agents are classified into the following categories:

1. Risk Group 1 (RG1): Agents that are not associated with disease in healthy adult humans.
2. Risk Group 2 (RG2): Agents that are associated with human disease which is rarely serious and for which preventive or therapeutic interventions are often available.
3. Risk Group 3 (RG3): Agents that are associated with serious or lethal human disease for which preventive or therapeutic interventions *may* be available (high individual risk but low community risk).
4. Risk Group 4 (RG4): Agents that are likely to cause serious or lethal human disease for which preventive or therapeutic interventions are *not usually* available (high individual and high community risk).

After determining the risk group classification of an agent, one can consult the 4th Edition of *Biosafety in Microbiological and Biomedical Laboratories* (BMBL) (1999) to determine the appropriate containment level and associated signage, work practices, and facility engineering and design. For example, biosafety level 1 (BSL1) describes the standard and specialized practices, the safety equipment, and the specifications of the laboratory facility design deemed necessary to work safely with RG1 agents. As the risk grouping increase above RG1, the complexity of the appropriate practices and controls also increase.

Another excellent source of information regarding the safe handling and administration of biological agents may be accessed on the internet at http://www.hc-sc.gc.ca/ pphb-dgspsp/msds-ftss/index.html. This site maintained by Health Canada's Office of Laboratory Security contains material safety data sheets for biological agents. The individual summary sheets provide health hazard information, information on the biohazardous agent dissemination and viability, decontamination procedures, medical information, laboratory hazard data, recommended precautions, handling information, and spill response procedures.

C. Biosafety in Animal Care Facilities

While there are many comparable concerns and appropriate precautions that must be observed whether one is working in the laboratory (*in vitro*) or in a laboratory rat facility (*in vivo*), there are some additional hazards that are associated with the specific activities when working with laboratory rats. The normal activity of rats in cages with contact bedding generates aerosols (airborne exposure potential) that disseminate allergens and could transmit certain zoonotic diseases if they were enzootic in the rat colony. Rats can bite and scratch (parenteral exposure potential), another means whereby rats can transmit some of the zoonotic organisms described earlier in this chapter. Conducting a risk assessment for working safely with experimental agents in rats, therefore, must include these risk factors in addition to those listed previously.

After determining the risk grouping for a particular laboratory rat experiment, the selection of the appropriate practices, equipment, and facility design is very similar to requirements for laboratory operations. These biocontainment requirements are detailed in the section of the BMBL entitled, "Vertebrate Animal Biosafety Level (ABSL) Criteria," addressing physical separation and traffic flow from "clean" areas to "dirty" areas in order to avoid cross contamination. Accordingly, ABSL1 design specifications are appropriate for activities involving RG1, ABSL2 for RG2, and so on. In the BMBL, each succeeding animal biosafety level builds upon the standard practices, procedures, containment equipment, and facility requirements of the preceding section.

D. Viral and Non-viral Vector Gene Transfer

The *Guidelines for Experiments Involving Recombinant DNA Molecules* (National Institutes of Health, 1994) provide a risk assessment and biocontaminment framework for institutions that perform gene therapy experimentation in laboratory rats. *Gene therapy* is the method of experimentation for delivery of exogenous ribonucleic or deoxyribonucleic acid (RNA or DNA gene transfer), in a specific or nonspecific manner, into cells of rats altering cellular phenotype. The vehicle used for the delivery of exogenous genetic material is called the gene vector, implying unidirectional transfer of material. The gene vector can be viral or nonviral. Viral vectors can be replication-competent or replication-deficient (unable to replicate other than under special conditions). There can be amplification of replication-competent viral vectors increasing the vector concentration augmenting the potential for horizontal transmission, or passive shedding of the replication-deficient viral vectors. There has been no demonstration of horizontal transmission of nonviral gene vectors. When performing risk assessment of gene transfer experiments, consideration needs to be given to the vector, gene delivered, and method of delivery (Feldman, 2003).

The potential risk to laboratory personnel of gene therapy experiments is the unintended gene delivery in exposed persons and the effects this may cause.

Viral gene vectors demonstrate specific cell tropism. Expression of the transferred gene can be transient when the gene is transferred only to cells that are terminally differentiated or if the transferred gene is not stably integrated into a progenitor cell's genome. The viral vector titer (concentration) shed from the laboratory rat after vector delivery is a primary determinant of potential personnel exposure. Many viral vectors encode endogenous viral proteins, in addition to the gene to be transferred, which are concomitantly expressed as protein.

Specific DNA sequences upstream of the gene, and the therapeutic gene itself, comprise the gene therapy expression cassette. The therapeutic gene may encode a structural protein, an enzyme, a toxin, a cytokine, a hormone, etc. When assessing the relative risk associated with a therapeutic gene, the potential to produce adverse systemic effects is greatest for toxin genes, hormones, or cytokines. Far less potential for adverse systemic effects exist if the therapeutic gene encodes an enzyme, structural proteins, or reporter molecules. The promoter region of the expression cassette controls the temporal and spatial aspects of gene expression. The promoter determines whether gene expression is constitutive or inducible, whether gene expression is cell type specific or generalized. The level of gene expression can be above normal physiological limits, within normal levels of expression, or so low as to be inconsequential. The expression cassette can include additional DNA elements that promote the secretion of the gene product from the cell to the peripheral circulation systems, or traffic the expressed protein to a specific intracellular region.

The physical method of gene therapy delivery can be via hypodermic injection or pressurized through catheters, both methods increasing the potential for aerosol generation and airborne exposure of personnel to the gene vector. One ballistic, nonviral method of gene delivery utilizes high velocity gold or tungsten microprojectiles that are coated with the expression cassette (Johnston et al., 1988).

The most commonly used replication-deficient viral vectors are retroviruses and lentiviruses (Dull et al., 1998), human adenovirus types 2 and 5 (Graham and Prevec, 1995), adeno-associated virus (AAV) types 1 and 2 (Rolling and Samulski, 1995), sindbis virus (Johanning et al., 1995) and Simliki Forest virus (Olkkonen et al., 1994). Adenovirus, sindbis and Simliki Forest virus vector systems are RG 2 agents, whereas the replication deficient AAV and retroviral/lentiviral vectors have been relegated RG1 (Cornetta et al., 1991). Each research facility should have an Institutional Biosafety Committee that can change the risk group designation of a gene therapy study if warranted by the contents of the expression cassette.

Such a case can be made for increased biocontainment if the gene to be expressed is a potent toxin, or a protein with the potential to cause significant systemic effects. Horizontal transmission of replication-deficient viral vectors has not been well studied. However, prudence dictates containment of laboratory rats for a minimum of 72 hours at animal biosafety level 2 (ABSL2) after administration of viral vector due to documented low titer viral shedding in urine and feces during this post-inoculation period (Feldman and Hoyt, 1994; Feldman, 2003).

Common replication-competent viral gene vectors are poxviruses (vaccinia virus and canarypox) (Mackett and Smith, 1986), herpesvirus simplex type 1 (HSV-1), and pseudorabies virus. The poxviruses are most often used as vaccine vectors, whereas the herpesviruses are used for gene therapy targeting the central nervous system. There is documentation of shedding of vaccinia virus for 10 days or less after a scarification challenge, that lead to the development of anti-poxvirus antibodies in sentinel animals (Gaertner et al., 2003), however this was not a common occurence. In contrast, horizontal transmission to naïve cage mates did not occur after subcutaneous administration of vaccinia to guinea pigs (Holt et al., 2002). There was no evidence of horizontal transmission after oral vaccination of foxes and raccoons with a recombinant vaccinia rabies vaccine (Brochier et al., 1988; Brochier et al., 1989). No detectable shedding of Herpesvirus vectors has been reported after administration to laboratory rats. Contact with freshly dissected tissues from inoculated animals has the potential to lead to significant exposure of research personnel. The *Herpesvirus* vector pseudorabies virus is not a zoonotic agent, and most humans are latently infected with HSV-1 demonstrating prior immunity to HSV-1 based vectors. However, prudent practices dictates the use of ABSL-2 biocontainment for a minimum of ten days when providing husbandry for laboratory rats inoculated with vaccinia, and for 72 hours for HSV-1 vectors. If it is demonstrated that viral gene therapy vector shedding does not occur after inoculation, then biocontainment can be relaxed to ABSL-1 practices.

XII. CONCLUSIONS

This chapter covers occupational health risks to humans associated with research that use laboratory or feral rats. The risks associated with laboratory rats can be chemical, physical, biological, electrical, ergonomic, or a combination of these. Understanding the intrinsic and extrinsic risks associated with the species, experimental paradigm, and personnel handling rats, will assist the reader in determining the overall risk associated with the particular research activity. It is reasonable to assume that all activities in research have some risk inherent to them, and that the determination of overall risk is a subjective venture. Personnel examining these risks are scientists, research staff, animal care and use committee members, institutional biosafety committee members, radiation safety officials, chemical safety officials, veterinarians, animal care staff, environmental health and safety office, and the institutional occupational health physician. Interaction between these parties facilitates the development of prudent procedures and practices necessary to minimize experimental risk to research personnel using rats.

REFERENCES

Acha, P.N., and Szyfres, B. (2003). Volume II. Chlamydioses, Rickettsioses, and Viroses. In: "Zoonoses and Communicable Disease Common to Man and Animals." 3rd Edition, Scientific and Technical Publication Number 580, World Health Organization, Pan American Health Organization. Washington, D.C.

Aldova, E., Cerny, J., and Chmela, J. (1977). Findings of Yersinia in rats and sewer rats. Zentralbl. Bakteriol. H239, 208–212.

American National Standards Institute. (2000). "Safe Use of Lasers." ANSI Z136.1. Washington, D.C.

Andersen, L.C., Leary, S.L., and Manning, P.J. (1983). Rat-bite fever in animal research laboratory personnel. Lab. Anim. Sci. 33, 292–294.

Anglada, A., Comas, L., Euras, J.M., Sanmarti, R., Vilaro, J., and Brugues, J. (1990). Arthritis caused by Streptobacillus moniliformis: a case of fever induced by a rat bite. Med. Clin. (Barc). 94, 535–537.

Aoyama, K., Ueda, A., Manda, F., Matsushita, T., and Ueda, T. (1992). Allergy to laboratory animals: an epidemiologic study. Br. J. Ind. Med. 49, 41–47.

Bauer, X., Chen, Z., and Liebers, V. (1998). Exposure-response relationships of occupational inhalative allergens. Clin. Exp. Allergy 28, 537–544.

Bayard, C., Holmquist, L., and Vesterberg, O. (1996). Purification and identification of allergenic alpha (2u)-globulin species of rat urine. Biochim. Biophys. Acta. 1290, 129–134.

Beaugeri, L., Carbonnel, F., Carrat, F., Rached, A.A., Maslo, C., Gendre, J.P., Rozenbaum, W.W., and Cosnes, J. (1998). Factors of weight loss in patients with HIV and chronic diarrhea. J. Acquir. Immune. Defic. Syndr. Hum. Retrovirol. 19, 34–39.

Bhatt, K.M., and Mirza, N.B. (1992). Rat bite fever: a case report of a Kenyan. East Afr. Med. J. 69, 542–543.

Bhattacharya, D., Teka, T., Faruque, A.S., and Fuchs, G.J. (1997). Cryptosporidium infection in children in urban Bangladesh. J. Trop. Pediatr. 43, 282–286.

Bhattacharya, D., Patel, A.K., Das, S.C., and Sikdar, A. (1999). Capillaria hepatica, a parasite of zoonotic importance—a brief overview. J. Commun. Dis. 31, 267–269.

Birtles, R.J., Hazel, S. M., Bennett, M., Bown, K., Raoult, D., and Begon, M. (2001). Longitudinal monitoring of the dynamics of infections due to Bartonella species in UK woodland rodents. Epidemiol. Infect. 126, 323–329.

Bjornson, B.F., Pratt, H.D., and Littig, K.S. (1968). Control of domestic mice and rats. Atlanta, GA. U. S. Public Health Service Bulletin. Bureau of State Services. National Communicable Disease Center.

Boot, R., van Herck, H., and van der Logt, J. (1996). Mutual viral and bacterial infections after housing rats of various breeders within an experimental unit. Lab. Anim. 30, 42–45.

Borowski, J., Kupryanow-Wolfert, K., Kurasz, S., and Sokolewicz, E. (1970). "A minor epidemic" of Yersinia pseudotuberculosis infection. Pol. Med. Sci. Hist. Bull. 13, 165–168.

Botham, P.A., Davies, G.E., and Teasdale, E.L. (1987). Allergy to laboratory animals: A prospective study of its incidence and of the influence of atopy on its development. Br. J. Ind. Med. 44, 627–632.

Bowen, G.S., Calisher, C.H., Winkler, W.G., Kraus, A.L., Fowler, E.H., Garman, R.H., Fraser, D.W., and Hinman, A.R. (1975). Laboratory studies of a lymphocytic choriomeningitis virus outbreak in man and laboratory animals. Am. J. Epidemiol. 102, 233–240.

Boyd, A.S. (1997). Rickettsialpox. Dermatol. Clin. 15, 313–318.

Brochier, B.M., Blancou, J., Thomas, I., Languet, B., Artois, M., Kieny, M.P., Lecocq, J.P., Costy, F., Desmettre, P., and Cahppuis, G. (1989). Use of recombinant vaccinia-rabies glycoprotein virus for oral vaccination of wildlife against rabies: innocuity to several nontarget bait consuming species. J. Wildl. Dis. 25, 540–547.

Brochier, B.M., Languet, B., Blancou, J., Kieny, M.P., Lecocq, J.P., Costy, F., Desmettre, P., and Pastoret, P.P. (1988). Use of recombinant vaccinia-rabies virus for oral vaccination of fox cubs (Vulpes vulpes, L) against rabies. Vet. Microbiol. 18, 103–108.

Brockmeier, S.L. (1999). Early colonization of the rat upper respiratory tract by temperature modulated Bordetella bronchiseptica. FEMS Microbiol. Lett. 174, 225–229.

Burkot, T.R., Maupin, G.O., Schneider, B.S., Denatale, C., Happ, C.M., Rutherford, J.S., and Zeidner, N.S. (2001). Use of a sentinel host system to study the questing behaviors of Ixodes spinipalpis and its role in the transmission of Borrelia bissettis, human granulocytic ehrlichiosis, and Babesia microti. Am. J. Trop. Med. Hyg. 65, 293–299.

Buscher, H.N., and Haley, A.J. (1972). Epidemiology of Hymenolepis nana infections of Punjabi villagers in West Pakistan. Am. J. Trop. Med. Hyg. 21, 422–429.

Calello, M.A., Rabinovich, R.D., Boxaca, M.C., and Weissenbacher, M.C. (1985). Association of the infection of the thymus and bone marrow with the establishment of persistent infection with Junin virus in 2 genera of rodents. Rev. Argent. Microbiol. 17, 115–119.

Campo, A., del Cerro, M., Foss, J.A., Ison, J.R., Warren, J.L., and Monjan, A.A. (1985). Impairment in auditory and visual function follows perinatal viral infection in the rat. Int. J. Neurosci. 27, 85–90.

Centers for Disease Control-National Institutes of Health (1999). "Biosafety in Microbiological and Biomedical Laboratories," 4th Edition. HHS Publication 93–8395. U.S. Government Printing Office, Washington, D.C.

Cerny, A., Ettlin, D., Betschen, K., Berchtold, E., Hottinger, S., and Neftel, K.A. (1992). Weil's disease after a rat bite. Eur. J. Med. 1, 315–326.

Chan, Y.C., Ho, K.H., Tambyah, P.A., Lee, K.H., and Ong, B.K. (2001). Listeria meningoencephalitis: two cases and a review of the literature. Ann. Acad. Med. Singapore 30, 659–663.

Childs, J.E., Glass, G.E., Kasiazek, T.G., Rossi, C.A., Oro, J.G., and Leduc, J.W. (1991). Human-rodent contact and infection with lymphocytic choriomeningitis and Seoul viruses in an inner-city population. Am. J. Trop. Med. Hyg. 44, 117–121.

Comer, J.A., Biaz, T., Vlahov, D., Monterroso, E., and Childs, J.E. (2001). Evidence of rodent-associated Bartonella and Rickettsia infections among intravenous drug users from Central and East Harlem, New York City. Am. J. Trop. Med. Hyg. 65, 855–860.

Comer, J.A., Tzianabox, T., Flynn, C., Vlahov, D., and Childs, J.E. (1999). Serologic evidence of rickettsialpox (Rickettsia akari) infection among intravenous drug users in inner-city Baltimore, Maryland. Am. J. Trop. Med. Hyg. 60, 894–898.

Connole, M.D., Yamaguchi, H., Elas, D., Hasegawa, A., Segal, E., and Torres-Rodriguez, J.M. (2000). Natural pathogens of laboratory animals and their effects on research. Med. Mycol. 38 Suppl 1, 59–65.

Cornetta, K., Morgan, R.A., and Anderson, W.F. (1991). Safety issues related to retroviral-mediated gene transfer in humans. *Hum. Gene Ther.* **2,** 5–14.

Cullinan, P., Lowson, D., Nieuwenhuijsen, M.J., Gordon, S., Tee, R.D., Venables, K.M., McDonald, J.C., and Newman-Taylor, A.J. (1994). Work related symptoms, sensitization, and estimated exposure in workers not previously exposed to laboratory rats. *Eur. J. Resp. Med.* **51,** 589–592.

Davous, B., Boni, M., Branquet, D., ducos De Lahitte, J., and Martet, G. (1997). Research on three parasitic infestations in rats captured in Marseille: evaluation of the zoonotic risk. *Bull. Acad. Natl. Med.* **181,** 887–895.

Del Aguila, C., Moura, H., Fenoy, S., Navajas, R., Lopez-Velez, R., Ki, L., Xiao, L., Leitch, G.J., da Silva, A., Pienizek, N.J., Lal, A.A., and Visvesvara, G.S. (2001). In vitro culture, ultrastructure, antigenic, and molecular characterization of *Encephalitozoon cuniculi* isolated from urine and sputum samples from a Spanish patient with AIDS. *J. Clin. Microbiol.* **39,** 1105–1108.

del Cerro, M., Grover, D., Monjan, A.A., Pfau, C., and Dematte, J. (1984). Congenital retinitis in the rat following maternal exposure to lymphocytic choriomeningitis virus. *Exp. Eye Res.* **38,** 313–324.

Dryberg, T., Schwimmbeck, P.L., and Oldstone, M.B. (1988). Inhibition of diabetes in BB rats by virus infection. *J. Clin. Invest.* **81,** 928–931.

Duchin, J., Koster, F., Peters, C.J., Simpson, G.L., Temperst, B., Zaki, S.R., Ksaizek, T.G., Rollin, P.E., Nichol, S., and Umland, E.T., for The Hantavirus Study Group. (1994). Hantavirus pulmonary syndrome: a clinical description of 17 patients with a newly recognized disease. *N. Engl. J. Med.* **330,** 949–955.

Dull, T.R. Zufferey, M., Kelly, R.J., Mandel, M., Nguyen, D., and Naldini, L. (1998). A third-generation lentivirus vector with a conditional packaging system. *J. Virol.* **72,** 8463–8471.

Durand-Joly, I., Aliouat, el M., Recourt, C., Guyot, K., Francois, N., Wauquier, M., Camus, D., and Dei-Cas, E. (2002). *Pneumocystis carinii f. sp. hominis* is not infectious for SCID mice. *J. Clin. Microbiol.* **40,** 1862–1865.

Dworkin, M.S., Sullivan, P.S., Buskin, S.E., Harrington, R.D., Olliffe, J., MacArthur, R.D., and Lopez, C.E. (1999). *Bordetella bronchiseptica* infection in human immunodeficiency virus-infected patients. *Clin. Infect. Disl.* **28,** 1095–1099.

Eamens, G.J., Turner, M.J., and Catt, R.E. (1988). Serotypes of *Erysipelothrix rhusiopathiae* in Australian pigs, small ruminants, poultry, and captive wild birds and animals. *Aust. Vet. J.* **65,** 249–252.

Ellis, B.A., Regnery, R.L., Beati, L., Bacellar, F., Rood, M., Glass, G.G., Marston, E., Ksaizek, T.G., Jones, D., and Childs, J.E. (1999). Rats of the genus *Rattus* are reservoir hosts for pathogenic *Bartonella* species: an Old World origin for a New World disease? *J. Infect. Dis.* **180,** 220–224.

Engel, P.M., Welzel, J., Maass, M., Schramm, U., and Wolff, H.H. (1998). Tropical rat mite dermatitis: case report and review. *Clin. Infect. Disl.* **27,** 1465–1469.

Fallon, M.T., Reinhard, M.K., Gray, B.M., Davis, T.W., and Lindsey, J.R. (1988). Inapparent *Streptococcus pneumoniae* type 35 infections in commercial rats and mice. *Lab. Anim. Sci.* **38,** 129–132.

Feinstein, R.E., and Eld, K. (1989). Naturally occurring erysipelas in rats. *Lab. Anim.* **23,** 256–260.

Feldman, S.H., and Hoyt, R.F. (1994). Biosafety considerations for lab animal gene therapy studies. *Lab. Anim.* **23,** 28–30.

Feldman, S.H. (2003). Components of gene therapy experimentation that contribute to relative risk. *Comp. Med.* **53,** 147–158.

Figueiredo, L.T., Campos, G.M., and Rodridues, F.B. (2001). Hantavirus pulmonary and cardiovascular syndrome: epidemiology, clinical presentation, laboratory diagnosis and treatment. *Rev. Soc. Bras. Med. Trop.* **34,** 13–23.

Fishman, O., Alchorne, M.M., and Portugal, M.A. (1973). Human sporotrichosis following a rat bite. *Rev. Inst. Med. Trop. Sao Paulo.* **15,** 99–102.

Fournier, S., Liguory, O., Sarfati, C., David-Ouaknine, F., Derouin, F., Decazes, J.M., and Molina, J.M. (2000). Disseminated infection due to *Encephalitozoon cuniculi* in a patient with AIDS: case report and review. *HIV Med.* **1,** 155–161.

Fujiwara, K., Tanishima, Y., and Tanaka, M. (1979). Seromonitoring of laboratory mouse and rat colonies for common murine pathogens. *Jikken Dobutsu.* **28,** 297–306.

Fukushmina, H., Gomyoda, M., Shiozawa, K., Kaneko, S., and Tsubokura, M. (1988). *Yersinia pseudotuberculosis* infection contracted through water contaminated by a wild animal. *J. Clin. Microbiol.* **26,** 584–585.

Fulton, J.D., and Joiner, L.P. (1948). Natural amoebic infections of laboratory rodents. *Nature (London).* **161,** 66–68.

Gaertner, D.J., Batchelder, M., Herbst, L.H., and Kaufman, H.L. (2003). Administration of vaccinia virus to mice may cause contact or bedding sentinel mice to test positive for orthopoxvirus antibodies: case report and follow-up investigation. *Comp. Med.* **53,** 85–88.

Gautrin, D., Infante-Rivard, C., Ghezzo, H., and Malo, J.L. (2001). Incidence and host determinants of probable occupational asthma in apprentices exposed to laboratory animals. *Am. J. Respir. Crit. Care Med.* **163,** 899–904.

Gavrilovskaya, I., LaMonica, R., Fay, M.E., Hjelle, B., Schmaljohn, C., Shaw, R., and MacKow, E.R. (1999). New York 1 and Sin Nombre Viruses are serotypically distinct viruses associated with Hantavirus Pulmonary Syndrome. *J. Clin. Microbiol.* **37,** 122–126.

Gigliotti, F., Harmsen, A.G., Haidaris, C.G., and Haidaris, P.J. (1993). *Pneumocystis carinii* is not universally transmissible between mammalian species. *Infect. Immun.* **61,** 2886–2890.

Glastonbury, J.R., Morton, J.G., and Matthews, L.M. (1996). *Streptobacillus moniliformis* infection in Swiss white mice. *J. Vet. Diagn. Invest.* **8,** 202–209.

Gollop, J.H., Katz, A.R., Rudoy, R.C., and Saski, D.M. (1993). Rat-bite leptospirosis. *West. J. Med.* **159,** 76–77.

Gordon, S. (2001). Laboratory animal allergy: a British perspective on a global problem. *ILAR J.* **42,** 37–46.

Gordon, S., Tee, R.D., Lowson, D., Wallace, J., and Newman Taylor, A.J. (1992). Reduction of airborne allergenic urinary proteins from laboratory rats. *Br. J. Ind. Med.* **49,** 416–422.

Gordon, S., Tee, R.D., and Newman Taylor, A.J. (1996). Analysis of the allergenic composition of rat dust. *Clin. Exp. Allergy.* **26,** 533–541.

Gordon, S., Tee, R.D., Nieuwenhuijsen, M.J., Lowson, D., Harris, J., and Newman Taylor, A.J. (1994). Measurement of airborne rat urinary allergen in an epidemiological study. *Clin. Exp. Allergy* **24,** 1070–1077.

Gordon, S., Tee, R.D., Stuart, M.C., and Newman Taylor A.J. (2001). Analysis of allergens in rat fur and saliva. *Allergy* **56,** 563–567.

Gordon, S., Tee, R.D., and Taylor, A.J. (1993). Analysis of rat urine proteins and allergens by sodium dodecyl sulfate-polyacrylamide gel electrophoresis and immunoblotting. *J. Allergy Clin. Immunol.* **92,** 298–305.

Gordon, S., Wallace, J., Cook, A., Tee, R.D., and Newman Taylor, A.J. (1997). Reduction of exposure to laboratory animal allergens in the workplace. *Clin. Exp. Allergy.* **27,** 744–751.

Graham, F.L., and Prevec, L. (1995). Methods for construction of adenovirus vectors. *Mol. Biotechnol.* **3,** 207–220.

Green, C.A., Gordon, D.H., and Lyons, N.F. (1978). Biological species in *Praomys (Mastomys) natalensis* (Smith), a rodent carrier of Lassa virus and bubonic plague in Africa. *Am. J. Trop. Med. Hyg.* **27,** 627–629.

Gueirard, P., Weber, C., Le Coustumier, A., and Guiso, N. (1995). Human *Bordetella bronchiseptica* infection related to contact with infected animals: persistence of bacteria in host. *J. Clin. Microbiol.* **33,** 2002–2006.

Hagelskjaeer, L., Soresen, I., and Randers, E. (1998). *Streptobacillus moniliformis* infection: 2 cases and a literature review. *Scand. J. Infect. Dis.* **30,** 309–311.

Halbur, P. G., Kasorndorkbua, C., Gilbert, C., Guenette, D., Potters, M.B., Purcell, R.H., Emerson, S.U., Toth, T.E., and Meng, X.J. (2001). Comparative pathogenesis of infection of pigs with hepatitis E viruses recovered from a pig and a human. *J. Clin. Microbiol.* **39,** 918–923.

Harrison, D.J. (2001). Controlling exposure to laboratory animal allergens. *ILAR J.* **42,** 17–36.

Hedirich, J.P., Stahl, M., Dittmann, R., Maass, M., and Solbach, W. (2001) Mitral valve endocarditis caused by *Erysipelothrix rhusiopathiae*. *Dtsch. Med. Wochenschr.* **126,** 431–433.

Heederick, D., Venables, K.M., Malmberg, P., Hollander, A., Karlsson, A-S., Renström, S. Doekes, G., Nieuwenhuijsen, M.J., and Gordon, S. (1999). Exposure-response relationships for work-related sensitization in workers exposed to rat urinary allergens: results from a pooled study. *J. Allergy Clin. Immunol.* **103,** 142–149.

Hesford, J.D., Platts-Mills, T.A., and Edlich, R.F. (1995). Anaphylaxis after laboratory rat bite: an occupational hazard. *J. Emerg. Med.* **13,** 765–768.

Hinrichsen, S.L., Ferraz, S., Romeiro, M., Muniz Filho, M., Abath, A.H., Magalhaes, C., Damasceno, F., Araujo, C.M., Campos, C.M., and Lamprea, D.P. (1992). Sodoku–a case report. *Rev. Soc. Bras. Med. Trop.* **25,** 135–138.

Hirschhorn, R.B., and Hodge, R.R. (1999). Identification of risk factors in rat bite incidents involving humans. *Pediatrics* **104,** 1–6.

Hof, H. (2001). *Listeria monocytogenes*: a causative agent of gastroenteritis. *Eur. J. Clin. Microbiol. Infect. Dis.* **20,** 369–373.

Hollander, A., Heederik, D., and Doekes, G. (1997). Respiratory allergy to rats: exposure-response relationships in laboratory animal workers. *Am. J. Respir. Crit. Care Med.* **155,** 562–567.

Hollander, A., Heederik, D., Doekes, G., and Kronhout, H. (1998). Determinants of airborne rat and mouse urinary allergen exposure. *Scand. J. Work Environ. Health* **24,** 228–235.

Holt, R.K., Walker, B.K., and Ruff, A.J. (2002). Horizontal transmission of recombinant vaccinia virus in strain 13 guinea pigs. *Contemp. Top. Lab. Anim. Sci.* **41,** 57–60.

Hubbard, G.B., Soike, K.F., Butler, T.M., Carey, K.D., David, H., Butcher, W.I., and Gauntt, C.J. (1992). An encephalomyocarditis virus epizootic in a baboon colony. *Lab. Anim. Sci.* **42,** 233–239.

Iida, T., Kanzaki, M., Maruyama, T., Inoue, S., and Kaneuchi, C. (1991). Prevalence of *Listeria monocytogenes* in intestinal contents of healthy animals in Japan. *J. Vet. Med. Sci.* **53,** 873–875.

Johanning, F.W., Conry, R.M., LoBuglio, A.F., Sumerel, L.A., Pike, M.J., and Curiel, D.T. (1995). A sindbis virus mRNA polynucleotide vector achieves prolonged and high level heterologous gene expression in vivo. *Nucleic Acids Res.* **23,** 1495–1501.

Johnston, S.A., Anziano, P.Q., Shark, K., Sanford, J.C., and Butow, R.A. (1988). Mitochondrial transformation in yeast by bombardment with microprojectiles. *Science* **240,** 1538–1541.

Juncker-Voss, M., Prosl, H., Lussy, H., Enzenberg, U., Auer, H., and Nowotny, N. (2000). Serologic detection of *Capillaria hepatica* by indirect immunofluorescence assay. *J. Clin. Microbiol.* **38,** 431–433.

Kabrane-Lazizi, Y., Fine, J.B., Elm, J., Glass, G.E., Diwan, A., Gibbs, C.J., Jr., Meg, X.J., Emerson, S.U., and Purcell, R.H. (1999). Evidence for widespread infection of wild rats with hepatitis E virus in the United States. *Am. J. Trop. Med. Hyg.* **61,** 331–335.

Kaneko, K.I., Hamada, S., Kasai, Y., and Hashimoto, N. (1979). Smouldering epidemic of *Yersinia pseudotuberculosis* in barn rats. *Appl. Environ. Microbiol.* **37,** 1–3.

Kaneko, K., and Hashimoto, N. (1983). Acetone-treated agglutinin against *Yersinia pseudotuberculosis* in free-living rats. *Am. J. Vet. Res.* **44,** 511–515.

Kawamata, J., Yamanouchi, T., Dohmae, K., Miyamoto, H., Takahashki, M., Yamanishi, K., Kurata, T., and Lee, H.W. (1987). Control of laboratory acquired hemorrhagic fever with renal syndrome (HFRS) in Japan. *Lab. Anim. Sci.* **37,** 431–436.

Keeling, M.J., and Gilligan, C.A. (2000). Bubonic plague: a metapopulation model of a zoonosis. *Proc. R. Soc. Lond. B. Biol. Sci.* **267,** 2219–2230.

Keusch, G.T., Hamer, D., Joe, A., Kelley, M., Griffiths, J., and Ward, H. (1995). *Cryptosporidia*–who is at risk? *Schweiz. Med. Wochenschr.* **125,** 899–908.

Kim, H.S., Christianson, W.T., and Joo, H.S. (1991). Characterization of encephalomyocarditis virus isolated from aborted swine fetuses. *Am. J. Vet. Res.* **52,** 1649–1652.

Klicks, M.M., and Palumbo, N.E. (1992). Eosinophilic meningitis behind the pacific basin: the global dispersal of a peridomestic zoonosis caused by *Angiostrongylus cantonensis,* the nematode lungworm of rats. *Soc. Sci. Med.* **34,** 199–212.

Krakowiak, A., Szule, B., Palczynski, C., and Gorski, P. (1996). Laboratory animals as a cause of occupational allergy. *Med. Pr.* **47,** 523–531.

Krawczynski, K., Kamili, S., and Aggarwal, R. (2001). Global epidemiology and medical aspects of hepatitis E. *Forum* (Genova) **11,** 166–179.

Krussell, A., Comer, J. A., and Sexton, D. J. (2002). Rickettsialpox in North Carolina: a case report. *Emerg. Infect. Dis.* **8,** 727–728.

Ksiazek, T.G., Nichol, S.T., Mills, J.N., Groves, M.G., Wozniak, A., McAdams, S., Monroe, M.C., Johnson, A.M., Martin, M.L., Peters, C.J., and Rollin, P.E. (1997). Isolation, genetic diversity, and geographic distribution of Bayou virus (Bunyaviridae: hantavirus). *Am. J. Trop. Med. Hyg.* **57,** 445–458.

LeDuc, J.W. (1989). Epidemiology of hemorrhagic fever viruses. *Rev. Infect. Dis.* 11 **Suppl 4,** S730–S735.

Leveque, L. Piroth, L., Baulot, E., Dutronc, Y., Dalac, S., and Lambert, D. (2001). Acute osteomyelitis. A rare erysipelas differential diagnosis. *Ann. Dermatol. Venereol.* **128,** 1233–1236.

Levine, J.F., and Lage, A.L. (1984). House mouse mites infesting laboratory rodents. *Lab. Anim. Sci.* **34,** 393–394.

Lopatina, I.V., Vasil'eva, I.S., Gutova, V.P., Ershova, A.S., Burakova, O.V., Naumov, R.L., and Petrova, A.D. (1999). An experimental study of the capacity of the rat mite *Ornithonyssus bacoti* (Hirst, 1913) to ingest, maintain and transmit *Borrelia*. *Med. Parazitol.* **2,** 26–30.

Luzzi, G.A., Milne, L.M., and Waitkins, S.A. (1987). Rat-bite acquired leptospirosis. *J. Infect.* **15,** 57–60.

Mackett, M., and Smith, G.L. (1986). Vaccinia virus expression vectors. *J. Gen. Virol.* **67,** 2067–2082.

Majeed, S.K., and Zubaidy, A.J. (1982). Histopathological lesions associated with *Encephalitozoon cuniculi* (nosematosis) infection in a colony of Wistar rats. *Lab. Anim.* **16,** 244–247.

Mandel, D.R. (1985). Streptobacillary fever. An unusual cause of infectious arthritis. *Cleve. Clin. Q.* **52,** 203–205.

Mantooth, S.J., Milazzo, M.L., Bradley, R.D., Hice, C.K. Ceballos, G., Tesh, R.B., and Fulhorst, C.F. (2001). Geographical distribution of rodent-associated hantaviruses in Texas. *J. Vector Ecol.* **26,** 7–14.

Matsuzaki, H., Doi, K., Doi, C., Onodera, T., and Misuoka, T. (1989). Susceptibility of four species of small rodents to encephalomyocarditis (EMC) virus infection. *Jikken Dobutsu.* **38,** 357–361.

McHugh, T.P., Bartlett, R.L., and Raymond, J.I. (1985). Rat bite fever: a report of a fatal case. *Ann. Emerg. Med.* **14,** 1116–1118.

Molyneux, D.H. (1969). The attachment of *Trypanosoma lewisi* in the rectum of its vector flea *Nosopsyllus fasiatus*. *Trans. R. Soc. Trop. Med. Hyg.* **63,** 117.

Moon, J.R. (1976). Public health significance of zoonotic tapeworms in Korea. *Int. J. Zoonoses* **3**, 1–18.

Muller-Doblies, U., Herzog, K., Tanner, I., Mathis, A., and Deplazes, P. (2002). First isolation and characterization of *Encephalitozoon cuniculi* from a free-ranging rat (*Rattus norvegicus*). *Vet. Parasitol.* **107**, 279.

National Fire Protection Association. (2003). "NFPA1 Uniform Fire Code." National Fire Protection Association, Quincy, MA.

National Institute for Occupational Safety and Health. (1998). "Occupational Exposure to Animal Allergens." Bulletin 97–116. ONLINE. Available: http://www.cdc.gov/niosh/animalrt.html.

National Institutes of Health. (1994). "NIH Guidelines for Research Involving Recombinant DNA Molecules." 59 FR 34496 and amendments. ONLINE. Available: http://www4.od.nih.gov/oba/rac/guidelines/guidelines.html.

National Research Council. (1995). "Prudent Practices in the Laboratory: Handling and Disposal of Chemicals." National Academy Press. Washington, D.C.

National Research Council. (1997). "Occupational Health and Safety in the Care and Use of Research Animals." National Academic Press, Washington, D.C.

Natrajaseeivasan, K., and Ratnam, S. (1996). An investigation of leptospirosis in a laboratory animal house. *J. Commun. Dis.* **28**, 153–157.

Naughton, P. J., Gran, G., Spencer, R.J., Bardocz, S., and Pusztai, A. (1996). A rat model of infection by *Salmonella typhimurium* or *Salmonella enteritidis*. *J. Appl. Bacteriol.* **81**, 651–656.

Neal, R.A. (1951). The duration and epidemiological significance of *Entamoeba histolytica* infections in rats. *Trans. R. Soc. Trop. Med. Hyg.* **45**, 363–370.

Nieuwenhuijsen, M.J., Gordon, S., Tee, R.D., Venables, K.M., McDonald, J.C., and Newman Taylor, A.J. (1994). Exposure to dust and rat urinary aeroallergens in research establishments. *Occup. Environ. Med.* **51**, 593–596.

Nuclear Regulatory Commission. (1978). "Medical Uses of By-product Materials." 10 CFR 35. ONLINE. GPO Access. Available: http://ecfr.gpoaccess.gov.

Occupational Safety and Health Administration. (1975). "Occupational Safety and Health Standards." 29 CFR 1910.106 ONLINE. GPO Access. Available: http://ecfr.gpoaccess.gov.

Occupational Safety and Health Administration. (1983). "Guidelines for Noise Enforcement." 29 CFR 1910.95 (b) (1). ONLINE. GPO Access. Available: http://ecfr.gpoaccess.gov.

Occupational Safety and Health Adminstration (2005). Anesthetic Gases: Guidelines for Workplace Exposure. Available: http://www.osha.gov/dts/osta/anestheticgases/.

Ol'iakova, N.V., and Antoniuk, V.Ia. (1989). The gray rat (*Rattus norvegicus*) as a carrier of infectious causative agents in Siberia and the Far East. *Med. Parazitol.* **3**, 73–77.

Olkkonen, V.M., Dupree, P., Simons, K., Liljestrom, P., and Garoff, H. (1994). Expression of exogenous proteins in mammalian cells with Semliki Forest virus vector. *Methods Cell Biol.* **43**, 4353–4364.

Platts-Mills, J., Curtis, N., Kenney, A., Tsay, A., Chapman, M., Feldman, S., and Platts-Mills, T. (2005). The effects of cage design on airborne allergens and endotoxin in animal rooms: high-volume measurements with an ion-charging device. *Contemp. Top. Lab. An. Sci* **44**, 12–16.

Press, N., Fyfe, M., Bowie, W., and Kelly, M. (2001). Clinical and microbiological follow-up of an outbreak of *Yersinia pseudotuberculosis* serotype Ib. *Scand. J. Infect. Dis.* **33**, 523–526.

Purcell, R.H., and Emerson, S.U. (2001). Animal models of hepatitis A and E. *ILAR J.* **42**, 161–177.

Quy, R.J., Cowan, D.P., Haynes, P.J., Sturdee, A.P. Chalmers, R.M., Bodley-Tickell, A.T., and Bull, S.A. (1999). The Norway rat as a reservoir host of *Cryptosporidium parvum*. *J. Wildl. Dis.* **35**, 660–670.

Rakov, E.V., Nichol, S.T., and Compans, R.W. (1997). Polarized entry and release in epithelial cells of Black Creek Canal virus, a new world hantavirus. *J. Virol.* **71**, 1147–1154.

Rolling, F., and Samulski, R.J. (1995). AAV as a viral vector for human gene therapy. *Mol. Biotechnol.* **3**, 9–15.

Rupp, M.E. (1992). *Streptobacillus moniliformis* endocarditis: case report and review. *Clin. Infect. Disl.* **14**, 769–772.

Sahin, I. (1979). Parasitosis and zoonosis in mice and rats caught in and around Beytepe Village. *Mikrobioyol. Bull.* **13**, 283–290.

Sakaguchi, M., Inouye, S., Miyazawa, H., Kamimura, H., Kimura, M., and Yamazaki, S. (1990). Evaluation of counter-measures for reduction of mouse airborne allergens. *Lab. Anim. Sci.* **40**, 613–615.

Sato, H., and Miyata, H. (1986). Detection of lymphocytic choriomeningitis virus antibody in colonies of laboratory animals in Japan. *Jikken Dobutsu.* **35**, 189–192.

Schottelius, J., and da Costa, S.C. (2000). Microsporidia and acquired immunodeficiency syndrome. *Mem. Inst. Oswaldo Cruz.* **95 Suppl 1**, 133–139.

Schuurman, B., van Griethuysen, A.J., Marcelis, J.H., and Nijs, A.M. (1998). Rat bite fever after a bite from a tame rat. *Ned. Tijdschr. Geneeskd.* **142**, 2006–2009.

Seguin, B., Boucaud-maitre, Y., Quenin, P., and Lorgue, G. (1986). Epidemiologic evaluation of a sample of 91 rats (*Rattus norvegicus*) captured in the sewers of Lyon. *Zentralbl. Bakteriol. Microbiol. Hyg.* **261**, 539–546.

Seeps, S.L., Cera, L.M., Terese, S. C., Jr., and Vosicky, J. (1987). Investigations of the pathogenicity of *Salmonella enteritidis* serotype Amsterdam following a naturally occurring infection in rats. *Lab. Anim. Sci.* **37**, 326–330.

Seward, J.P. (2001). Medical surveillance of allergy in laboratory animal handlers. *ILAR J.* **42**, 47–54.

Slein, M.W., and Sansone, E.B. (1980). "Degradation of Chemical Carcinogens: An Annotated Bibliography." Van Nostrand, Reinhold, Co., New York.

Stefanelli, P., Matrantonio, P., Hausman, S.Z., Giuliano, M., and Burns, D.L. (1997). Molecular characterization of two *Bordetella bronchiseptica* strains isolated from children with coughs. *J. Clin. Microbiol.* **35**, 1550–1555.

Taylor, A.J., Gordon, S., and Tee, R.D. (1994). Influence of bedding, cage design, and stock density on rat urinary aeroallergen levels. *Am. J. Ind. Med.* **25**, 89.

Thulin, H., Björkdahl, M., Karlsson, A.S., and Rentröm, A. (2002). Reduction of exposure to laboratory animal allergens in a research laboratory. *Ann. Occup. Hyg.* **46**, 61–68.

Torres, J., Gracenea, M., Gomez, M.S., Arrizabalaga, A., and Gonzalez-Moreno, O. (2000). The occurrence of *Cryptosporidium parvum* and *C. muris* in wild rodents and insectivores in Spain. *Vet. Parasitol.* **92**, 253–260.

Torrez-Maartinez, N., Mausumi, B., Doade, D., Delurey, J., Moran, P., Hicks, B., Nix, B., Davis, J.L., and Hjelle, B. (1998). Bayou virus-associated Hantavirus Pulmonary Syndrome in eastern Texas: identification of the rice rat, *Oryzomys palustris*, as reservoir host. *Emerg. Infect. Dis.* **4**, 105–111.

Tosoni, A., Nebuloni, M., Ferri, A., Bonetto, S., Antinori, S., Scaglia, M., Xiao, L., Moura, H., Visvesvara, G.S., Vago, L., and Costanzi, G. (2002). Disseminated microsporidiosis caused by *Encephalitozoon cuniculi III* (dog type) in an Italian AIDS patient: a retrospective study. *Mod. Pathol.* **15**, 577–583.

Tsutsumi, V. (1994). In vivo experimental models of amebiasis. *Gac. Med. Mex.* **130**, 450–453.

Van Gool, T. and Dankert, J. (1995). Human microsporidiosis: clinical, diagnostic and therapeutic aspects of an increasing infection. *Clin. Microbiol. Infect.* **1**, 75–85.

Verlinde, J.D., Li-Fo-Sjoe, E., Versteeg, J., and Dekker, S.M. (1975). A local outbreak of paralytic rabies in Surinam children. *Trop. Geogr. Med.* **27**, 137–142.

Virtanen, T., Zeiler, T., and Mantyjarvi, R. (1999). Important animal allergens are lipocalin proteins: why are they allergenic? *Int. Arch. Allergy Immunol.* **120**, 247–258.

Webster, J.P. (1996). Wild brown rats (*Rattus norvegicus*) as a zoonotic risk on farms in England and Wales. *Commun. Dis. Rep. CDR Rev.* **6**, R46–R49.

Webster, J.P., and MacDonald, D.W. (1995a). Parasites of wild brown rats (*Rattus norvegicus*) on UK farms. *Parasitology* **111**, 247–255.

Webster, J.P., and MacDonald, D.W. (1995b). Cryptosporidiosis reservoir in wild brown rats (*Rattus norvegicus*) in the UK. *Epidemiol. Infect.* **115**, 207–209.

Weisbroth, S.H., Geistfeld, J., Weisbroth, S.P., Williams, B., Feldman, S.H., Linke, M.J., Orr, S., and Cushion, M.T. (1999). Latent *Pneumocystis carinii* infection in commercial rat colonies: comparison of inductive immunosuppressants plus histopathology, PCR, and serology as detection methods. *J. Clin. Microbiol.* **37**, 1441–1446.

Wood, R.A. (2001). Laboratory animal allergens. *ILAR J.* **42**, 12–16.

Wu, J., Meng, Y., Li, Y., Zhou, H., Zhuge, H., Lai, P., and Wang J. (1998). Detection of HFRSV in *Eulaelaps shanghaiensis* and *Ornithonyssus bacoti* by using in situ hybridization. *Zhongguo Ji Sheng Chong Bing Za Zhi.* **16**, 441–444.

Wullenweber, M. (1995). *Streptobacillus moniliformis*—a zoonotic pathogen. Taxonomic considerations, host species, diagnosis, therapy, geographical distribution. *Lab. Anim.* **29**, 1–15.

Wullenweber, M., Jonas, C., and Kunstyr, I. (1992). *Streptobacillus moniliformis* isolated from otitis media of conventionally kept laboratory rats. *J. Exp. Anim. Sci.* **35**, 49–57.

Wullenweber, M., Lenz, W., and Werhan, K. (1990). *Staphylococcus aureus* phage types in barrier-maintained colonies of SPF mice and rats. *Z. Versuchstierkd.* **33**, 57–61.

Zen-Yoji, H., Sakai, S., Maruyama, T., and Yanagawa, Y. (1974). Isolation of *Yersinia enterocolitica* and *Yersinia pseudotuberculosis* from swine, cattle and rats at an abattoir. *Jpn. J. Microbiol.* **18**, 103–105.

Zhuge, H., Meng, Y., Wu, J., Zhu, Z., Liang, W., and Yao, P. (1998). Studies on the experimental transmission of *Rattus*-borne Hantavirus by *Ornithonyssus bacoti*. *Zhongguo Ji Sheng Chong Bing Za Zhi.* **16**, 445–448.

Zhukova, L. N., Kon'shina, T.A., and Popugailo, V.M. (1966). Infection of rodents by the agents of listeriosis and erysipeloid in the Sverdlovsk region. *Zh. Microbiol. Epidemiol. Immunobiol.* **43**, 18–23.

Chapter 18

Experimental Modeling and Research Methodology

Michael A. Koch

THE LABORATORY RAT, 2ND EDITION

I. INTRODUCTION

The rat remains one of the most popular laboratory animals in contemporary biomedical research. Although the mouse is arguably the most widely used laboratory animal due to the explosion of transgenic, knock-out and other genetic research, the rat is still commonly used for areas as diverse as pharmaceutical discovery, toxicology, oncology, gerontology, neurophysiology, and cardiovascular physiology. A vast number of research techniques and methods have been developed and refined by scores of investigators from many disciplines over the past several decades. This chapter will describe some of the most common research techniques and methods developed by investigators to enhance the usefulness of the laboratory rat. Although the list of methods will be extensive, it is not within the scope of this chapter to describe all methods. The author wishes to acknowledge and thank Alan L. Kraus, the author of the corresponding chapter in the original edition of this text, whose previous work served as an excellent outline and reference for this edition.

II. HANDLING AND RESTRAINT

Investigators in all disciplines today recognize the impact of environmental, nutritional, health-related and other factors on biomedical research. Research animal facilities and programs are designed to help reduce variables. Animal caretakers and laboratory animal veterinarians have long recognized that proper handling and restraint of rats in the laboratory can help to reduce unwanted stress and variation. Rats become relatively more gentle and probably less prone to stress reactions when handled appropriately and frequently. Investigators should also learn to handle and restrain rats properly to reduce stress and experimental variation. The physiologic effects of restraint have been well characterized. Chronic restraint has been associated with development of gastric ulcers, cardiac necrosis, hypothermia and chromodacryorrhea, altered behavior and food intake, promotion of tumor growth, reduced reproductive performance, and impairment of memory in rats (Renaud, 1959; Hanson and Brodie, 1960; Harkness and Ridgway, 1980; Monteiro et al., 1989; Glockner and Karge, 1991; Schumacher et al., 1991; Tejwani et al., 1991; Marti et al., 1994; Ely et al., 1997; Sunanda et al., 2000). Current best practices involve minimizing restraint time and favor the use of tethered, telemetry or other systems when chronic restraint is needed (NRC Guide for the Care and Use of Laboratory Animals, 1996).

A. Manual Restraint

Most rats bred specifically for research are docile and can be handled easily. Some stocks or strains are relatively more docile than others. The Sprague-Dawley and Wistar stocks tend to be more docile while the Fischer 344 and Long-Evans and other hooded rats are more difficult to handle. Handling and restraint of wild rats is entirely different and will be discussed later in this section.

Rats should be handled with latex, nitrile or other similar gloves to minimize exposure to hazardous agents and urinary and other allergens (Occupational Health and Safety in the Care and Use of Research Animals, 1997). Rats that are not accustomed to being handled should be approached cautiously. The operator can initially grasp such a rat by the base of the tail and place it on an arm or elbow to allow the animal to be calmed. The animal may then be grasped confidently over the thorax with either the thumb and index finger or the index finger and second finger used to control the mandible to prevent biting

(Flecknell, 1991; Waynforth and Flecknell, 1992; Sharp and LaRegina, 1998; Fig. 18-1). The rat should be held firmly but not too tightly to prevent excessive struggling and respiratory compromise. Alternatively, for particular applications, the rat may be restrained by firmly grasping the loose skin behind the neck and thorax.

The use of heavy gloves or forceps to handle domestic rats is generally unnecessary and may cause the animals to be more fractious than they would otherwise be. Likewise, excessive handling by the tail may lead to animals that resist restraint. If it is necessary, the tail should be grasped near its base to avoid de-gloving injuries.

B. Restraint Apparatus

Many types of restraint devices have been developed. Laboratory animal science experts and research investigators alike recognize that prolonged restraint must be scientifically justified and approved by oversight groups

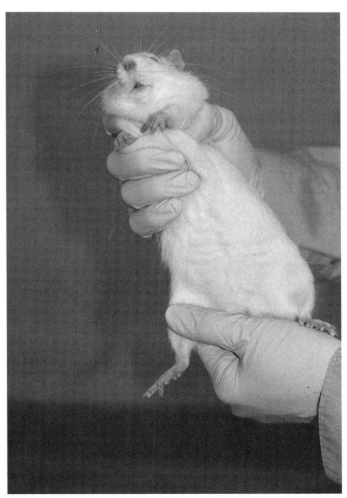

Fig. 18-1 A method of manual restraint of the rat.

such as institutional animal care and use committees
(NRC Guide for the Care and Use of Laboratory Animals,
1996). The earliest short-term restraint devices included
laboratory glassware and large syringes where rats could
be briefly immobilized for rectal and vaginal swabs, tail
vein blood collections and injections and other procedures
(Ganis, 1962). Other restraint devices have been fashioned
from readily available materials (Lawson et al., 1966;
Faulkner, 1975; Schumacher, 1983; Kirk, et al., 1997).
Commercially available plastic restraint devices have been
manufactured to facilitate withdrawal of samples and
injection of substances (Fig. 18-2).

Other methods for short-term restraint have been
developed more recently. Decapitation cones can be used
to immobilize rats for sample collections and injections.
A decapitation cone is a plastic sleeve that is tapered at
one end so that a rat can be inserted head first toward
the tapered end and restrained for various procedures such
as blood collection, injection of substances or decapitation (Reigle and Bukova, 1984; Sharp and LaRegina, 1998;
Fig. 18-3). Rats can also be wrapped in towels so that the
tail and caudal parts of the body are available for manipulation (Sharp and LaRegina, 1998). Devices for restraint
during electrocardiogram recording (Osborne, 1973;
Yagiela and Bilger, 1986), neurological recording (Prazma
and Kidwell et al.1980; Daunicht, 1985; Soltysik et al.,
1996), pharmacological sampling (Toon and Rowland,
1981; Lewis et al., 1989) and ophthalmic (Anderson et al.,
1991) and oral examinations (Johansen, 1952; Macedo-
Sobrinho et al., 1978) have been described. Most of these
devices are plastic restrainers that immobilize the animals
for procedures for short periods of time. Some are devices
that force the mouth open for examination of the oral
cavity and the Prazma device is a high platform that rats
will sit quietly on for neurological recording. Restraining
boards have also been developed for blood collection and
surgery in the rat (John, 1941).

As stated earlier, prolonged restraint is generally used
in contemporary biomedical research only if scientifically
justified and approved. Modern techniques for chronic
restraint generally allow rats to move freely within their
cage. Metabolism cages allow separation and collection of
urine and feces from conscious, freely moving rats. Tether
and jacket techniques allow chronic administration of
test materials or collection of bodily fluids or physiologic
measurements (Cocchetto and Bjornsson, 1983; Takeyama
et al., 1988; Parker and Martin, 1989; Mandavilli et al.,
1991; Jolly and Vezina, 1996; Van Wijk et al., 2000).
Rats are generally instrumented with in-dwelling catheters that are protected by jackets and tethers that exit
the cage through a swivel joint for connection to pumps
or collection bags and allow the animals full range of
motion (Fig. 18-4). Multi-channel, multipurpose sample
collection and drug delivery systems have recently been

Fig. 18-2   A commercially available plastic restraint device.

developed (Bonsall et al., 1998; Bohs et al., 2000).
Such systems completely eliminate the swivel and replaces
it with a powered turntable that allows the animals to move
freely while cannulated (Fig. 18-5). These techniques
obviously allow chronic collection of research samples or

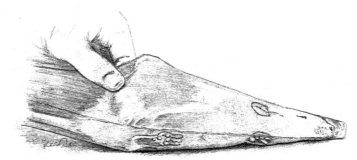

Fig. 18-3   Decapicone restraint of the rat.

bite while in these dark bags. Other devices designed to hold and manipulate wild rats include the Emlen sleeve and similar devices (Emlen, 1944; Carmichael et al., 1946; Evans et al., 1968). The Emlen device consists of a collapsible cone of wires connected to a short cloth sleeve. A wild rat is induced to enter the cloth sleeve and will

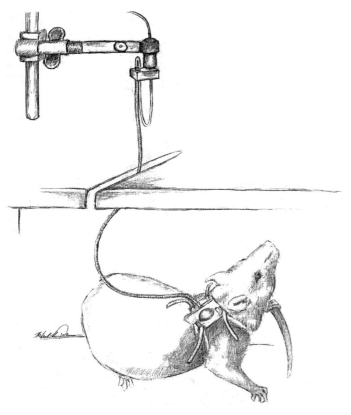

Fig. 18-4 Jacket and tether restraint of the rat.

administration of test substances while simultaneously addressing the welfare of the animals.

C. Handling and Restraint of Wild Rats

Wild rats are aggressive and much more difficult to handle than domesticated laboratory rats. These animals are used in contemporary research far less often than domestic animals due to zoonotic and colony health concerns. When wild rats are used, they must be handled and restrained in ways that are much different than those used with gentler, domesticated rats. Investigators have designed several devices to safely handle wild rats in the laboratory (Emlen, 1944; Carmichael et al., 1946; Evans et al., 1968; Andrews et al., 1971; Inglis and Hudson 1999). Dim red light can be used to calm wild rats and facilitate their handling and restraint in laboratory environments (Fall, 1974).

Wild rats can be safely removed from cages using heavy cloth bags that the animals readily enter. The open end of the bag is placed tightly over the cage door. The rat is induced to enter and the bag removed for manipulation of the animal (Andrews et al., 1971; Inglis and Hudson, 1999). Wild rats generally become quiet and tend not to

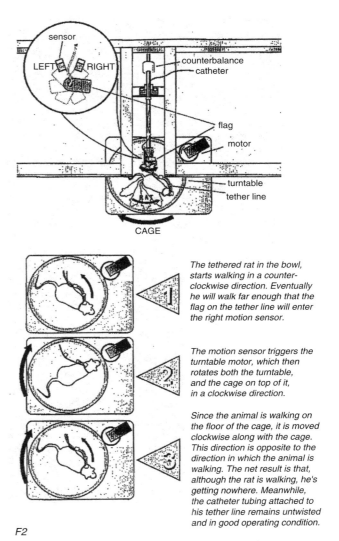

The tethered rat in the bowl, starts walking in a counter-clockwise direction. Eventually he will walk far enough that the flag on the tether line will enter the right motion sensor.

The motion sensor triggers the turntable motor, which then rotates both the turntable, and the cage on top of it, in a clockwise direction.

Since the animal is walking on the floor of the cage, it is moved clockwise along with the cage. This direction is opposite to the direction in which the animal is walking. The net result is that, although the rat is walking, he's getting nowhere. Meanwhile, the catheter tubing attached to his tether line remains untwisted and in good operating condition.

F2

An animal housed in either the Culex or Ratum system is free to move up and down, or left and right, at will. The catheter line(s) are attached to a tether wire which is in turn attached to a counterbalanced arm. The arm responds to the animal's vertical movements by pivoting up when the animal rears up, or down when he returns to a resting position. The arm also holds the sensor assembly which registers rotational behavior within the round cage. Clockwise, or counterclockwise rotations beyond 270° will trigger a sensor, which in turn starts a turntable under the cage rotating in an opposite direction. Because of these systems, the catheter tubing remains unstretched, untwisted, unchewed, and intact.

Fig. 18-5 Device for sampling blood from freely moving rats (reproduced from Bohs et al., 2000).

generally continue into the wire cone where it can be held and manipulated safely.

III. IDENTIFICATION METHODS

Identification of individual rats or groups of rats is often required in biomedical research studies. Cage cards may be all that is necessary for many studies but rigorous individual identification is often required for drug safety studies done according to Good Laboratory Practice (GLP) regulations and for other types of experiments. Individual identification may also help prevent problems associated with inadvertent switching of cage cards or accidents such as escape of similar colored animals. This section will briefly outline some of the methods available to identify individual rats.

A. Color Patterns

Color patterns of Long-Evans hooded rats can be used to identify the animals because individual patterns are unique (Roberts, 1961). The patterns can be sketched or photographed. Identification by color pattern is used infrequently with rats.

B. Ear Notching and Punching

Rats can be individually identified by marking, notching or punching the ear pinnae with commercially available ear punches (Koolhaas, 1999). Coding schemes have been described and can be customized by using location (left vs. right ear) and number of notches or punches to establish numbering systems (Fig. 18-6). This method of identification is inexpensive and easy to accomplish.

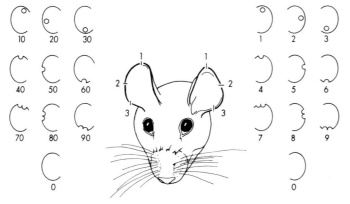

Fig. 18-6 A commonly used method of individual animal identification by using ear notching and punching. One ear is selected to represent tens, the other units.

C. Toe Clipping

Clipping of digits has been used to identify individual neonatal rats (Waynforth and Flecknell, 1992). This method may not be suitable for certain behavioral tests (Iwaki et al., 1989) and is no longer considered humane or acceptable in contemporary biomedical research (Sharp and LaRegina, 1998).

D. Ear Tags

Metal or plastic ear tags with unique identifying numbers are commercially available. Like ear notching, this technique is easy to perform and inexpensive but fighting or cannibalism could lead to loss of the tags. In addition, results from a carcinogenesis study suggested that inflammatory, proliferative and neoplastic lesions may be associated with long term exposure to metal tags (Waalkes et al., 1987).

E. Tattooing

Rats can also be identified by tattooing numbers onto the tail, ears, feet and other areas (Geller and Geller, 1966; Schoenborne et al., 1977; Iwaki et al., 1989; Koolhaas, 1999). Tattooing offers an effective and permanent means to identify individual adult and neonatal rats. The required equipment is inexpensive and personnel can be readily trained to effectively tattoo rats.

F. Dyes and Other Color Markings

Commercially available waterproof markers can be used to temporarily identify rats (Koolhaas, 1999). Codes for identification can be established using various markings or stripes. Alternatively, numbers or letters can be directly written on tails or fur. These markings will generally need to be renewed every 2 weeks due to skin and fur re-growth (Koolhaas, 1999).

G. Electronic Microchip Devices

Rats can be permanently identified with commercially available electronic microchips (Ball et al., 1991). Each chip is encoded with a unique number, implanted subcutaneously with a large bore needle and read by a special detector (Fig. 18-7). Implantation is simple and rarely requires chemical restraint of the animal. Some models can be programmed to read body temperature and capture other data in addition to the identification number. This technique is particularly useful for studies (e.g., drug

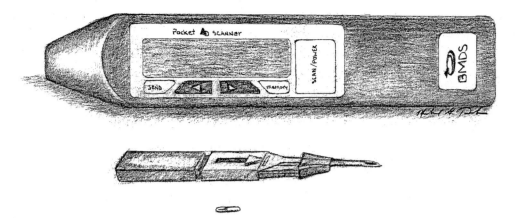

Fig. 18-7   An electronic microchip device used to identify rats.

safety studies done according to GLP regulations) where individual identification of animals is critical. However, animal identification microchip systems are expensive and the technology is imperfect (e.g., some chips may be difficult to read). In addition, foreign body reactions to the chips can occur and mesenchymal tumor formation has been associated with microchips (Elcock et al., 2000) during long-term studies.

IV.   SPECIMEN COLLECTION

Collection of bodily fluids and other specimens is an important part of experiments in many research disciplines. Toxicology, pharmacology, physiology, infectious disease and many other types of studies require collection of fluids for analysis. For example, blood collection is required in toxicology studies to determine test material exposure levels and to analyze clinical pathology for test material effects. Blood is also collected to determine the pharmacokinetics and distribution of new drug compounds during pre-clinical testing. Blood is a bodily fluid that travels through all tissues and, therefore, can serve as a window to the function of many organ systems. Bile, milk and urine can be collected to examine metabolism and excretion of test compounds. Other fluids or specimens, such as cerebrospinal fluid, bone marrow and saliva, are collected for specialized research projects and provide information about function of corresponding tissues. The following section will describe procedures for collection of these specimens from the laboratory rat.

A.   Blood

Since blood collection technique can impact physiologic parameters, research investigators should thoroughly

consider how the technique chosen might affect their experiments. A thorough review of the pertinent literature would obviously be helpful in considering the best technique for a particular study.

Many techniques are available for collection of blood from rats. The technique chosen depends on (1) the volume to be collected, (2) the frequency of collection, (3) whether anesthesia can be used, (4) the impact of the method on physiological measurements, (5) whether or not separate venous or arterial blood is required, and (6) whether or not aseptic technique is required. Techniques to be considered here include (1) cardiac puncture, (2) retroorbital plexus, (3) dorsal metatarsal vein, (4) saphenous vein, (5) tail vessels, (6) jugular vein, (7) carotid artery, (8) aorta and vena cava, (9) decapitation, and (10) other techniques.

The physiologic effects of blood collection techniques have been reviewed. Several systems can be affected by stress associated with collection technique. Hormones such as epinephrine, corticosterone, prolactin, growth hormone, insulin, and renin may be affected by the method of collection chosen (Brown and Martin, 1974; Oates and Stokes, 1974; Bellinger and Mendel, 1975). Sampling technique can impact serum enzyme levels (Friedel et al., 1974; Van Herck et al., 1998). Blood glucose levels reportedly can also be affected by the method of blood collection (Klinger et al., 1965; Beauzeville, 1968). Hemoglobin levels, hematocrit and erythrocyte and leukocyte counts may differ depending on the collection site chosen (Wright, 1970; Suber and Kodell, 1985; van Herck et al., 1998). Finally, repeated retroorbital collections can alter peripheral leukocyte counts (Golba et al., 1974) and are associated with higher creatine kinase and aspartate aminotransferase levels compared to collections from the sublingual vein (Mahl et al., 2000). These effects should be considered when planning experiments so that unwanted variables do not impact investigations.

The blood volume collected may impact investigations and should also be considered when designing experiments.

The volume collected at a single survival time point should not exceed 1.25 ml/100 g of body weight (Sharp and LaRegina, 1998) or roughly 10% to 15% of blood volume (Joint Working Group on Refinement, 1993; McGuill and Rowan, 1989). As much as 40% of blood volume may be collected without clinical signs of distress if multiple small samples are taken from rats over 24 hours (Scipioni et al., 1997). However, large volumes will likely elevate corticosterone levels (Waynforth and Flecknell, 1992), increase heart rate and create fluid shifts that could impact parameters being studied. Repeated withdrawal of large volumes over 24 hours leads to significant decreases in hematological parameters such as hemoglobin concentration and hematocrit (Nahas et al., 2000; Mahl et al., 2000). Recovery time for these hematological parameters is related to the magnitude of the withdrawal volume (Nahas et al., 2000). Excessive volume collection could eventually lead to hypovolemia, hypotension and clinical signs of shock. For repeated blood collections, a maximum of 1% of circulating blood volume can be safely removed every 24 hours (Joint Working Group on Refinement, 1993). Diehl et al. (2001) also listed recommendations for recovery periods for single and multiple blood sample studies. They recommended that animals be allowed to recover at least 1 week following removal of 7.5% of blood volume as part of either single sample (e.g., acute toxicity study) or multiple sample studies (e.g., toxicokinetic studies). They further recommended that 3 to 4 weeks recovery be allowed if larger volumes (e.g., 15% to 20%) are removed either as a single sample or as multiple samples. Hematocrit and serum proteins should be monitored if collections of larger volumes or frequent, excessive collections are planned.

1. Cardiac Puncture

Cardiac puncture is commonly used as a terminal procedure in contemporary biomedical research for collecting large volumes of blood from the anesthetized rat (Waynforth and Flecknell, 1992; Sharp and LaRegina, 1998). The adult rat is placed in dorsal recumbency and the apex beat palpated. A one inch 21 gauge needle attached to a syringe is introduced at a 30° angle under the skin just to the left of the xyphoid and advanced a short distance. The needle is advanced slowly and blood aspirated as the needle enters the heart. A momentary easing of the effort required to push through the heart wall will signal entrance into the heart. Several milliliters of blood can generally be obtained using this method in adult rats. Less experienced operators may need to use syringe vacuum and transmission of the apex beat through the needle to help determine the best approach to puncturing the left ventricle. Alternatively, cardiac puncture can be accomplished by penetration through the left side of the thorax (Fig. 18-8). The anesthetized rat is placed

in right lateral recumbency and the heartbeat is palpated. A needle is introduced perpendicular to the left chest wall at the point of greatest beat intensity and blood is aspirated. If little blood is obtained, the needle tip may be inappropriately positioned in the myocardium, atrium or adjacent to the heart. The needle may need to be gently re-positioned to ensure that it is in the left ventricle. Only 2 or 3 fresh attempts should be made since each attempt will likely cause bleeding in and around the heart. Cardiac tamponade can occur even with the best technique (Burhoe, 1940). Therefore, as stated earlier, this technique is generally used for terminal blood collection. If the rat is allowed to recover, the smallest gauge needle (e.g., 23 gauge) possible should be used and the volume minimized to help prevent complications (Waynforth and Flecknell, 1992).

Techniques are available for collection of blood by cardiac puncture in neonatal rats (Grazer, 1958; Gupta, 1973). Grazer's technique uses a 30-gauge needle and adapter attached to a length of polyethylene tubing and tuberculin syringe. The needle is introduced into the neonatal thorax at a 90° angle just to the left of the midline and just cranial to the costal margin. Gupta's technique uses a 21-gauge needle and Vacutainer system. The neonate is held with head down and the needle is introduced into the chest through the thoracic inlet. Small amounts of blood (<1 mL depending on the age of the neonate) can be slowly aspirated with these techniques.

2. Retroorbital Plexus

The retroorbital venous plexus can be used to collect small amounts of blood from rats. The technique was often used to collect blood serially and without anesthesia. However, corneal lesions, ocular discharge and other ocular problems can be caused by this technique, depending on the experience of the operator (van Herck et al., 1998). In addition, creatine kinase and aspartate aminotransferase activities are higher in retroorbital samples compared to samples collected from the sublingual vein,

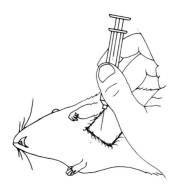

Fig. 18-8 Technique of cardiac puncture in the rat.

suggesting that tissue damage is greater when the retro-orbital plexus is sampled (Mahl et al., 2000). Thus, animal welfare concerns today dictate that rats are anesthetized for retroorbital sampling and the frequency of collections is limited (Waynforth and Flecknell, 1992). The technique has generally been replaced for welfare reasons by other methods such as metatarsal or jugular venipuncture.

When retroorbital bleeding is performed, customized and commercially available pipettes or capillary tubes are used. The technique is simple and operators can be trained quickly (Sharp and LaRegina, 1998). The rat is anesthetized and the non-dominant thumb and index finger used to pull the skin away from the orbit so that the eye protrudes slightly (Fig. 18-9). The thumb can be pressed behind the angle of the jaw to help impede venous return from the head so that blood will more freely flow from the plexus. A pipette is introduced into the medial canthus and the plexus is entered caudal and medial to the eye. Gentle rotation and retraction of the pipette tip will aid flow of blood from the plexus. The pipette is removed when sufficient volume is collected. Pressure on the re-positioned globe, through closed eyelids, must be maintained for a sufficient period of time to avoid post-orbital hematoma and exophthalmos following the procedure. The lateral canthus can also be used for this technique (Sorg and Bruckner, 1964). Timm (1989) suggested that the preferred site for retroorbital collection in the rat is through the conjunctiva dorsal to the eye.

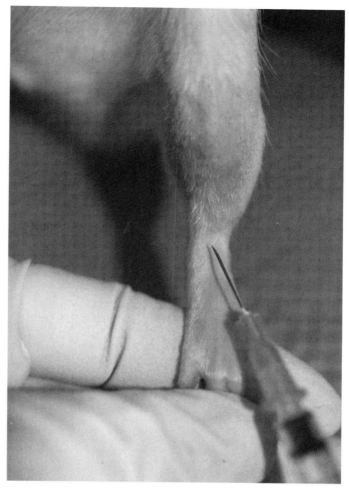

Fig. 18-10   Technique of bleeding using the dorsal metatarsal vein.

3. Dorsal Metatarsal Vein

The dorsal metatarsal vein can be used in the rat to collect small amounts of blood (Nobunaga et al., 1966). The vein may also be used to inject test materials. The technique is generally easier to perform if one person restrains the rat and another performs the venipuncture. The animal is restrained with the caudal portion of the body supported. The supporting thumb and index finger grasp and extend the leg so that the dorsal metatarsal vein

can be visualized. The dorsal surface of the leg should be shaved and swabbed with alcohol. The operator then grasps the foot, flexes the first phalangeal joint, slightly adducts the limb and inserts a small gauge needle (e.g., 23 gauge) into the vein (Fig. 18-10). 0.1 to 0.2 mL of blood can generally be collected with this technique.

4. Saphenous Vein

The medial (Rusher and Birch, 1975) or lateral (Hem et al., 1998) saphenous veins may be used to collect single or serial blood samples (Fig. 18-11). Each vein runs superficially over the medial and lateral aspects of the hind limb. Blood can be collected from either vein in conscious, manually restrained rats. The area to be punctured should be shaved and swabbed with alcohol to help visualize the vein. For the medial vein, the inguinal area should be compressed to engorge the vein and a 23-gauge needle used to puncture the vessel. On the opposite side, the lateral thigh area should be compressed and the vein

Fig. 18-9   Technique of bleeding from the retroorbital plexus.

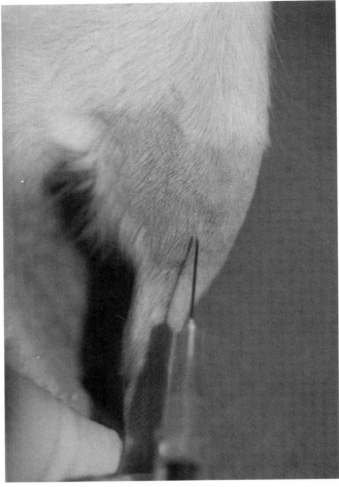

Fig. 18-11 Technique of bleeding using the saphenous vein.

punctured near the tarsal joint with an appropriately sized needle. Blood is then collected in a capillary tube as it forms a drop at the puncture site. The scab can easily be removed to allow serial sampling over 24 hours (Hem et al., 1998).

5. Tail

Many techniques for obtaining blood from the tail of rats have been described (Burhoe, 1940; Grice, 1964; Sandiford, 1965; Wright, 1970; Hurwitz, 1971; Agrelo and Miliozzi, 1974; Nerenberg and Zedler, 1975; Omaye et al., 1987; Conybeare et al., 1988; Sarlis, 1991; Waynforth and Flecknell, 1992; Sharp and LaRegina, 1998; Fluttert et al., 2000). The technique chosen will depend on the volume to be collected and the frequency of sampling.

a. VENIPUNCTURE. The tail has a ventral artery and dorsal and lateral veins that can be used for venipuncture. Collection of blood from the tail veins by needle and

syringe can be accomplished without anesthesia. Exposing the tail to warm water or a heat lamp or bulb (Grice, 1964) will help dilate the veins. Incubators warmed to 40° C can be used to house rats for several minutes to facilitate dilation of tail veins (Conybeare et al., 1988, Waynforth and Flecknell, 1992). The rat is then placed in a restraining device. The operator applies pressure to the base of the tail and punctures the lateral or dorsal vein with a 23-gauge or smaller needle (Sharp and LaRegina, 1998). If serial samples are required, the first sample should be taken as distal on the tail as possible. Subsequent punctures should occur progressively more proximal on the tail. Alternatively, a 21-gauge butterfly set has been used successfully for weekly sampling (Conybeare et al., 1988).

b. INCISION OF LATERAL TAIL VEIN. The tail veins of the rat can be incised to collect blood. In one report, the vein is incised longitudinally after the area is coated with petroleum jelly (Nerenberg and Zedler, 1975). Blood is collected after the tail is placed in a vacuum device fashioned from a test tube and other common laboratory materials. More recently, Fluttert et al. (2000) describes an alternative technique where rats are loosely restrained within a towel and a 2-mm incision is made in a tail vein. A drop of blood forms at the site and is collected in capillary tubes. The tail is gently stroked to facilitate blood flow. The authors consider the technique to be easy to learn and "animal friendly." However, serum enzymes such as creatine kinase were not evaluated to determine the extent of tissue trauma from this procedure. Either of these techniques can be used for serial collections since the scab that forms at the site can be easily removed.

c. PUNCTURE OF THE VENTRAL ARTERY OR VESSELS. Hurwitz (1971) originally described a technique that required anesthesia for puncture of the caudal artery. In this technique, a heparinized capillary tube is inserted into the hub of a 21-gauge needle. With the rat in dorsal recumbency, the needle is inserted into the artery as it courses midventrally in the tail. About 0.35 mL of blood can be collected in one tube. A modified version of this technique has been described (Omaye et al., 1987). The rat is restrained in a commercial restraint device without anesthesia. A special tourniquet is placed proximal to the puncture site and a 21-gauge needle used to puncture the midventral vein that is associated with the artery. Drops of blood are then collected in a vial. A particular advantage to this technique is that good blood flow is maintained without the need for suction or vacuum devices. In addition, the technique is safe, convenient and rapid.

d. AMPUTATION OR TAIL CLIPPING. Amputation of the tip of the tail has been used to collect small amounts of blood (Grice, 1964, Moreland, 1965). Less than 5 mm of the tail tip is transected. Tail warming, swabbing with xylol or gentle massage help to dilate vessels and facilitate blood flow. Generally, less than 0.5 mL of blood is collected by this method but some reports (Sarlis, 1991; Waynforth and Flecknell, 1992) indicate that a larger volume can be obtained. This technique is simple and rapid and can be used for repeated sampling (Liu et al., 1996) but has several disadvantages. Massaging the tail can cause a leukocytosis that artificially elevates white blood cell count in the sample (Waynforth and Flecknell, 1992). Additionally, the samples contain mixed venous, arterial and tissue fluids and the number of samples is limited since repeated cutting will result in trauma to cartilage and vertebrae, which is potentially painful and distressful to the rat (Waynforth and Flecknell, 1992; Weiss et al., 2000).

e. CANNULATION OF THE CAUDAL ARTERY OR LATERAL VEIN. The caudal artery or lateral tail vein can be cannulated for relatively short-term studies where test materials are infused for a short period or through a series of injections (Waynforth and Flecknell, 1992; Weiss et al., 2000). A 23- or 24-gauge over-the-needle catheter can be used (Waynforth and Flecknell, 1992). The tail vein is occluded proximally with digital pressure or a tourniquet and the vein punctured with the needle and catheter. The catheter is then advanced into the vein after verifying that it is in the vessel (the catheter should fill with blood). The catheter can be secured by taping or gluing (with skin glue) it to the tail. The cannulated vessels can be used to repeatedly collect blood samples and inject test substances. If large volumes of blood are required, cannulation allows blood to be transfused to offset the effects of the blood collections. Anesthesia may be required to place the cannula but the animal can be awakened and restrained for sampling. The technique is particularly useful for studies of 5 to 6 hours duration (Weiss et al., 2000).

6. Jugular Vein

The jugular vein has been used as a site for blood collection via in-dwelling catheters. However, the jugular vein can also be used to collect routine blood samples by venipuncture. Both techniques require considerable training, skill and experience to be used effectively.

Jugular venipuncture can be performed with or without anesthesia in rats (Waynforth and Flecknell, 1992; Sharp and LaRegina, 1998). The right jugular vein is generally preferred. The animal can be restrained by hand (Waynforth and Flecknell, 1992) or on a restraint board

(Phillips et al., 1973). The rat is held tightly by the loose skin behind the neck and restrained with the ventral side up. Alternatively, the rat is placed on a restraining board in dorsal recumbency and the legs are momentarily tied down. With the restraint board method, the head is flexed slightly to the left and the area bounded by the neck and shoulder is visualized. Shaving the area may enhance visualization of landmarks. A 21-gauge needle is used to penetrate the skin at the angle formed by the neck and point of the shoulder (Fig. 18-12). The needle is advanced until blood is aspirated and slowly withdrawn. The needle is removed and modest pressure is applied to the area momentarily to promote hemostasis. Bleeding or hematoma formation does not generally occur. Well-trained, experienced personnel can collect blood from about 60 rats in 1 hour using this technique.

Techniques for cannulation of the jugular vein for blood sampling are used for pharmacokinetic studies and experiments where anesthesia or restraint may influence the outcome. Cannulation techniques allow blood collection from conscious rats at several time points without confounding restraint and other variables. Surgical exposure of the jugular vein is required for cannulation. The vein is exposed in the ventral cervical area just lateral to the midline. Fat and other tissue is dissected from the vein and a forceps is passed under the vein. Suture is then passed and used to ligate the vessel cranially. The vein is entered with a small gauge catheter through a needle hole and the catheter is secured to muscle with suture. The end of the catheter is exteriorized between the scapulae or attached to a vascular access port. As with all chronic catheterization techniques, the catheter should be flushed routinely and filled with an anti-coagulant such as heparin-saline (20 units heparin per milliliter; Waynforth and Flecknell, 1992) to help maintain catheter patency.

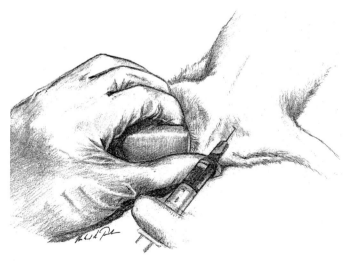

Fig. 18-12 Jugular venipuncture technique in the rat.

Other commonly used anti-coagulant solutions include heparinized glycerol, 50% dextrose and polyvinylpyrrolidone. Heparinized glycerol outperformed the other solutions in a recent study (Luo et al., 2000).

Improvements to these techniques have been described (Suzuki et al., 1997; Wiersma and Kastelijn, 1985). Wiersma and Kastelijn (1985) cannulated the jugular vein and exteriorized it through a steel tube attached to the skull. Blood samples were then collected through a 90-cm long cannula attached to the steel tube. The technique allowed relatively unrestricted movement and did not interfere with normal hormone secretion (when compared to decapitation). Suzuki et al., (1997) extended the connection tubing through a swivel mechanism that allowed the rats full range of motion while cannulated. Arginine vasopressin and corticosterone concentrations were lower when sampled with this technique compared to decapitation and restrained cannulation techniques. Thus, these improved techniques probably decrease stress associated with restraint and reduce variability.

The jugular cannulation technique is apparently less stressful than carotid or femoral artery techniques since post-surgical weight loss is minimal after jugular cannulation (Yoburn et al., 1984). However, the jugular cannula is also most likely to occlude after 14 days so the authors felt that femoral catheterization was preferable since patency can be maintained longer and weight loss was intermediate to the other techniques.

7. Carotid Artery

The carotid artery can be used for rapid blood collection due to high arterial blood pressure and flow. Access to the artery is generally accomplished via surgical placement of an in-dwelling catheter. Cannulation of the carotid artery was described by D'Amour and Blood (1954). Leal et al. (1988) described a microsampling technique where the carotid artery was catheterized and used for maternal blood collections. However, carotid artery catheterization causes greater post-surgical weight loss than femoral or jugular cannulation, suggesting that cannulation of the carotid artery may be more stressful than the other techniques (Yoburn et al., 1984). In addition, severe bleeding can occur if the catheter is dislodged, and thromboembolic events can occur and lead to interruption of blood flow to the brain.

8. Aorta and Vena Cava

Large volumes of blood can be easily obtained terminally by abdominal aortic (Lushbough and Moline et al., 1961) or caudal vena cava transection (Grice, 1964) or puncture (Waynforth and Flecknell, 1992). The anesthetized rat is laid in dorsal recumbency and the abdominal skin and muscle incised to expose the viscera. The intestinal loops are reflected and the vessels located. The aorta is entered just cranial to its bifurcation into the iliac arteries. The vena cava is entered at its widest point adjacent to the kidneys. Blood should be aspirated slowly from the vena cava to avoid collapsing the vessel. Winsett et al. (1985) described a technique for collection of serial samples from the caudal vena cava in conscious, restrained rats. A translumbar approach using a 25-gauge needle was used safely to collect multiple samples. The aorta or vena cava can also be chronically cannulated to collect blood from conscious animals. The aorta is generally accessed by placing a catheter surgically in peripheral arteries such as the carotid (Popovic and Popovic, 1960) or iliac (Carvalho et al., 1975). The vena cava can be cannulated directly by surgical exposure (Koeslag et al., 1984) or via the jugular or femoral vein (Moslen et al., 1988). The advantage to these techniques is that blood can be collected without the confounding effects of restraint and anesthesia. However, prolonged cannulation of the vena cava can lead to local endothelial changes and lung emboli (de Jong et al., 2001).

9. Decapitation

Decapitation can rapidly yield large volumes of mixed venous and arterial blood from anesthetized or unanesthetized rats when performed competently by well-trained personnel using appropriate equipment (Waynforth and Flecknell, 1992). Anesthesia is generally preferred for this technique if allowed by the experimental design. Some workers may find the technique to be aesthetically displeasing. The head is removed with a properly maintained, routinely sharpened guillotine and the neck placed quickly over a collecting vessel. Experienced personnel may restrain the rats by hand. Plastic cones are also available to facilitate handling during decapitation. These cones appear to reduce animal stress from handling, improve positioning of the animal in the guillotine and reduce the potential for operator injury (American Veterinary Medical Association Panel on Euthanasia, 2000).

10. Other Techniques

Several other techniques are available for collection or sampling of blood from rats. The axillary vessels of anesthetized rats can be severed and blood collected as it pools in a pocket formed from the incised skin (Waynforth and Flecknell, 1992). Serial blood samples (7 total samples within 24 hours) using a 23-gauge needle can be collected from the sublingual vein in isoflurane-anesthetized rats (Zeller et al., 1998). Horton et al. (1986) described a femoral venipuncture technique for use in conscious, restrained rats. The rat is restrained in dorsal recumbency and a 25-gauge needle used to enter the vein in the femoral triangle. Multiple samples can be collected in a few hours. Automated blood samplers are also available (Kissinger et al., 2003). Cannulated rats are housed in metabolic or other cages (depending on study type) and allowed to move freely while blood and other samples are collected for metabolism or other types of studies. Microdialysis systems for sampling blood parameters are also now available for use in rats (Jolly and Vezina, 1996). These systems involve cannulation of vessels, such as the jugular vein, with probes for sampling blood chemicals.

B. Urine

Urine is an important substance to collect for metabolism studies because many test compounds are excreted in urine. Urinalysis may also help determine if test compounds are nephrotoxic. Several techniques are available for collection of urine from rats (Fig. 18-13). Metabolism cages designed to separate feces and urine will be described in a later section. Techniques to be described here include (1) manual expression, (2) bladder centesis, (3) cystotomy, cannulation and urinary fistula, (4) free catch, (5) urethral catheterization, and (6) other methods.

1. Manual Expression

Urine can be collected by manually manipulating the urinary bladder in the caudal abdomen (Sharp and LaRegina, 1998). Gradually increasing, but not excessive, force should be applied to express urine (Fig. 18-13A). The urine is collected in capillary tubes (Hayashi and Sakaguchi, 1975) or other suitable collection devices. An average of 150 to 200 µl can be collected with this method (Weiss et al., 2000).

2. Bladder Centesis

Urine can easily be collected with needle and syringe at necropsy. The small urinary bladder of rats is difficult to locate percutaneously in conscious or anesthetized rats. Therefore, it is recommended that centesis be used only at necropsy in rats (Weiss et al., 2000).

3. Cystotomy and Cannulation

Urine can reportedly be collected by surgically opening the urinary bladder and placing a cannula within it (Fig. 18-13B). Hoy and Adolph (1956) first described a technique for creating a urinary bladder fistula in infant and adult rats. A short plastic tube was inserted in the bladder, exited through the body wall and attached to urine collection tubes. Samples were collected from

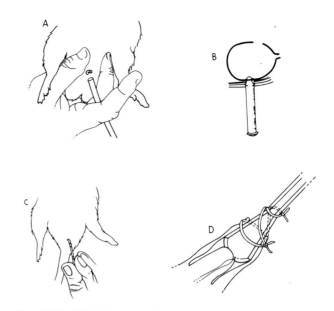

Fig. 18-13 Methods for obtaining urine specimens from rats. (A) Manual pressure and collection into capillary tube. (B) Urinary cystotomy technique. (C) Catheterization of female rats with a No. 4 coude catheter. Catheter itself is also depicted. (D) External drainage catheter used with male rats. [Redrawn from White, 1971.]

conscious rats using this technique. Serial urine samples can also be collected from anesthetized rats using a surgically implanted special silicon plastic catheter (Xu and Melethil, 1990). Urine flow can be maintained at 0.3 to 0.7 ml/hr using these techniques.

4. "Free-Catch"

Rats will often spontaneously urinate when handled, especially if they are not acclimated to being restrained. The handler simply positions the urethral opening near a collection device and "catches" small quantities of urine as it is voided.

5. Urethral Catheterization

Female rats may be catheterized with a 22-gauge catheter under general anesthesia, although it is essentially impossible to directly catheterize male rats due to curves in the male urethra (Sharp and LaRegina, 1998). Good aseptic technique should be followed to avoid contaminating the urinary bladder and increasing the risk of cystitis. A sterile catheter is used and the urethral opening disinfected prior to passing the catheter. Cohen and Oliver (1964) used a catheter with a curved tip to facilitate passing the catheter over the pubic symphysis (Fig. 18-13C). The anesthetized rat is placed in dorsal recumbency, the urethral meatus grasped with forceps and the catheter inserted and rotated, as needed. If performed with proper aseptic technique, the procedure can be repeated for serial sampling without infection or excessive trauma. Rosas-Arellano et al. (1988) described a surgical technique for cannulation of the female urethra. A catheter is secured within the urinary bladder and exteriorized through the urethral opening. The authors reported that urine could be collected intermittently for up to 6 weeks with this method.

Urine may be collected from anesthetized male rats by using a plastic catheter fitted over the tip of the penis (White, 1971). The catheter is fixed in place by suturing it to the prepuce (Fig. 18-13D). Urine flow of 6-10 ml/hr can be obtained from rats that are fluid "loaded" prior to collection.

6. Other Methods

Khosho et al. (1985) described a simple, reliable and efficient technique for urine collection in rats. A 5-mL polystyrene beaker is either attached to the perineal area with adhesive tape or held in place by hand. The back of the rat is stimulated cranial to the tail and urine collected in the beaker. Eighty percent of the animals on the study voided an average of 0.5 mL of urine.

Mandavilli et al. (1991) described a novel surgical catheterization technique for urine collection in conscious, unrestrained male rats. The urethra is ligated and the urinary bladder catheterized. The catheter is tunneled to the dorsal cervical area and exteriorized. A tether and swivel system allows the rats full range of motion while urine is collected in a specimen vessel. The technique was used to collect urine successfully from more than 100 rats. The technique is considered technically more demanding in females. The ureters can also be surgically cannulated for collection of urine in conscious, freely moving rats (Horst et al., 1988; Oz et al., 1989).

Finally, urine is probably most commonly collected from rats housed in metabolic cages. See the subsequent section on metabolic cages for further details.

C. Feces

Feces can be collected from rats on metabolism studies to determine if a test compound is excreted through the gastrointestinal tract. Fecal samples may be collected directly from cage pans under wire bottom cages or from bedding in plastic solid bottom cages. However, these samples will likely be contaminated with urine, dander, fur, bedding, drinking water and other substances.

Rats are coprophagic and will ingest feces directly from the anus whether housed in wire- or solid-bottom cages. Rats may consume as much as 50% to 65% of voided feces when housed in wire bottom cages (Barnes et al., 1957). Special procedures are required to prevent coprophagy and collect all voided feces.

Anal cups can be applied to the perineal area of rats to prevent coprophagy and collect uncontaminated samples (Ryer and Walker, 1971; Scheuermann and Lantzsch, et al., 1982; Waynforth and Flecknell, 1992). The cups are generally fashioned from small, plastic laboratory bottles. The upper portion of the bottle is cut to fit the perineal area of the rat and secured to the tail with adhesive tape. The cup is loosened slightly to allow collection of voided feces. The feces are best collected in the morning since defecation is heaviest during the dark cycle.

Armstrong and Softly (1966) described a leather jacket that, when applied to rats, could also prevent coprophagy. Jacketed rats were housed on wire floors so that all feces would drop through for collection. However, growth was significantly decreased in jacketed rats compared to rats allowed to consume their feces. It is known that coprophagy significantly impacts gut flora and nutrient utilization (Waynforth and Flecknell, 1992) and probably is required for normal growth and body maintenance in rats.

Small amounts of fecal material may be obtained by swabbing the rectum with a small cotton swab, plastic probe or other devices. This technique depends on feces being present in the rectum at the time of collection.

D. Metabolic Cages

Metabolic cages are primarily used to separate and collect urine and feces without contamination by fur, dander, feed, drinking water and other substances. A few commercially available designs also incorporate collection of inspired and expired gases but most simply separate urine and feces. Metabolic cages allow chronic collection of urine and feces while also allowing the rats some freedom of movement. Many different designs have been described (Weiss et al., 2000).

Two types of metabolic cages are generally available commercially. The plastic type is depicted in Fig. 18-14. The rat is housed in an upper enclosure on a suspended wire floor. The food hopper and water bottle are housed outside the enclosure and may be graduated for measurement purposes. The design helps to minimize contamination of urine and feces with food and water. Urine flows through the wire floor and drops onto the walls of the collecting chamber or the fecal deflector. Eventually, all urine flows through the collecting ring and into the graduated collection tube. Fecal pellets are deflected by an inverted cone and eventually drop vertically into the fecal collection vial. The stainless steel metabolic cage is depicted in Fig. 18-15. The apparatus used to separate urine and feces is similar but less sophisticated than the plastic cage.

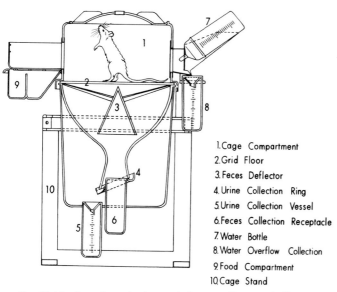

1. Cage Compartment
2. Grid Floor
3. Feces Deflector
4. Urine Collection Ring
5. Urine Collection Vessel
6. Feces Collection Receptacle
7. Water Bottle
8. Water Overflow Collection
9. Food Compartment
10. Cage Stand

Fig. 18-14 A modern plastic-metal, free-standing metabolism cage.

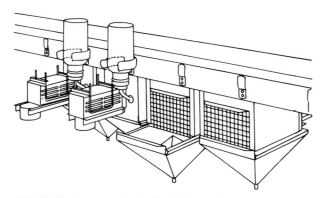

Fig. 18-15 A conventional all-stainless-steel rat metabolism cage.

E. Bone Marrow

Bone marrow can be collected to answer questions about regeneration of red cell mass or immune function. A number of techniques for collection of bone marrow from rats have been described. Specimens can be collected at necropsy or under anesthesia as a terminal procedure. The dissected femur is used to collect bone marrow by cracking it either longitudinally or transversely. The bone marrow specimen is collected with a fine brush and painted on to a microscope slide for examination or flushed from the bone for tissue culture (Dobson et al., 1999).

Bone marrow can be collected from anesthetized rats that are expected to recover. The site should be aseptically prepared. Biopsy needles (e.g., 18-gauge, 1 inch) can be used to obtain samples from the iliac crest. Alternatively, a low speed dental drill is used to create a hole in the tibia or femur. Marrow can then be aspirated with a needle

and syringe or collected with a brush, spatula tip or polyethylene tube (Sundberg and Hodgson, 1949; Burke et al., 1955; Sharp and LaRegina, 1998).

A technique for aspirating marrow from the sternebrae of mice can be adapted for use in rats. The animal is anesthetized and the area over the sternum prepared aseptically. The skin is then incised and the muscles overlying the sternum are bluntly dissected until the bone is visible. A hole in the bone is created with a dental drill and marrow is aspirated with a fine needle and syringe or pipette (Pilgrim, 1963).

F. Ocular Fluids

Collection of ocular fluids may be required for specialized studies such as ocular toxicity or metabolism studies. Aqueous humor may be collected easily from the anesthetized rat using a 26-gauge needle and tuberculin syringe. Vitreous humor may be collected from the dissected globe into a syringe after enucleation (Gershbein et al., 1975).

G. Lacrimal Secretion

Lacrimal fluid collection may be required for studies of lacrimal gland function. Lacrimal fluid can be collected from the medial canthus of the rat following treatment with parasympathomimetic drugs such as pilocarpine (Weiss et al., 2000). The fluid is generally collected in capillary tubes. Tear samples have also reportedly been collected from the lower lid margin using a folded 5-mm section of a Schirmer strip (Kapicioglu et al., 1998). Finally, the lacrimal gland excretory duct can be directly cannulated to collect lacrimal secretions from anesthetized rats (Thorig et al., 1984).

H. Peritoneal Cells or Ascitic Fluid

Peritoneal cells or ascitic fluid may be collected for immune function or antibody production protocols. The cells or fluid can be collected by (1) percutaneous aspiration using a needle and syringe in the conscious, anesthetized or euthanized rat, (2) abdominal incision in the anesthetized or euthanized rat, (3) using a peritoneal cell glass pipette or (4) surgical implantation of silicon tubes.

1. Percutanteous Collection

Ascitic fluid from hybridoma-inoculated rats can be collected percutaneously from anesthetized rats or at the time of necropsy with a 20-gauge needle and syringe

(Weiss et al., 2000). Gentle suction should be used to avoid occlusion of the needle with tissue.

2. Abdominal Incision

Ascitic fluid can also be collected from euthanized rats through a small abdominal incision using either a needle and syringe or a standard pipette attached to a bulb-type suction device. Alternatively, one can use a 12-mL plastic tube (Cooper and Stanworth, 1974). In the case of the latter, the tip of the tube is perforated several times with a hot needle. The peritoneal cavity is lavaged through the tube with 20 mL of sterile phosphate buffer as the rat is held vertically by the tail. The tube is then clamped, the rat laid horizontally and the abdomen massaged for approximately 90 seconds. Finally, the rat is held vertically, the tube unclamped and the fluid collected.

3. Peritoneal Cell Glass Pipette

Nashed (1975) described a technique to collect peritoneal cells from conscious rats. A glass pipette was designed and consisted of a mouth or syringe piece, egg-shaped main body and 15-gauge needle (Fig. 18-16). The needle is inserted in the lower right abdominal quadrant after the area is shaved and swabbed with alcohol. The abdomen is then lavaged with 30 to 35 mL of warmed Hank's solution. The solution is immediately aspirated to collect peritoneal cells. As many as 5 to 20 million cells can reportedly be collected using this method and multiple collections from a single rat are possible without apparent adverse clinical effects.

4. Surgical Implantation of Silicon Tubes

Nagel et al. (1989) described a technique for intraperitoneal injection of drugs and collection of ascitic fluid in conscious rats. Rats are anesthetized and two silicon tubes are surgically implanted in the abdomen. One tube is used for drug delivery while the other tube is used for ascitic fluid collection. Eighty percent of the implantations were successful and all rats tolerated the tubes for

24 hours. Approximately 9 to 14 mL of ascitic fluid could be recovered after instillation of 20 mL of saline.

I. Bile

Bile is an important substance to collect for pharmacokinetic and metabolism studies because many test compounds are transformed in the liver and excreted in bile. Bile can be collected easily from anesthetized rats for acute studies (Xu and Melethil, 1990). Numerous techniques for chronic bile duct cannulation have been described (Lambert, 1965; Cocchetto and Bjornsson, 1983; Weiss et al., 2000). A few general techniques will be outlined briefly in this section.

Anatomical features of the biliary and pancreatic systems should be considered when planning for chronic collection of bile. The rat, unlike many other species, lacks a gall bladder. The common bile duct carries both bile and pancreatic juices. Therefore, for collection of bile without contamination by pancreatic substances, the common bile duct must be catheterized near the liver and before the pancreatic ducts enter the common bile duct (Fig. 18-17).

Current techniques for chronic collection of bile samples generally involve either continuous bile collection and bile salt replacement or intermittent bile collection with re-circulation of bile when it is not being collected (Mulder et al., 1981; Cocchetto and Bjornsson; 1983). Balabaud et al. (1981) described a method to exteriorize the bile duct cannula from the subcutaneous area of the skull. A swivel and tether system is employed to allow the rats to move freely. Waynforth and Flecknell (1992) described a surgical technique where a needle is passed through the flank skin into the abdomen. The bile duct is catheterized and ligated near the hilum of the liver and the catheter is exteriorized through the needle. In addition, rats with bile duct cannulas are commercially available from rodent vendors.

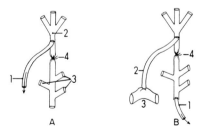

Fig. 18-17 Technique for collecting bile (A) and pancreatic juice (B). (A) Cannula (1) placed into common bile duct (2), proximal to entrance of pancreatic ducts (3), and ligature (4), distal to cannula and proximal to pancreatic ducts. (B) Cannula (1) into common duct distal to entrance of pancreatic ducts, second cannula (2) ("artificial bile duct") into proximal common duct and also into duodenum (3), ligature distal to "artificial bile duct" and proximal to pancreatic ducts (4).

Fig. 18-16   A specially constructed peritoneal cell glass pipette.

T-cannulae have been developed to allow undisturbed bile flow except when bile is being collected (Chipman and Cropper, 1977; Cocchetto and Bjornsson, 1983). A T-piece cannula is placed in the common bile duct with one arm directed toward the liver and the other toward the duodenum. Alternatively, a T-disc is used to receive cannulae that are implanted in the bile duct and directed toward the liver and the duodenum. Other current techniques involve the use of either external or internal collection reservoirs. For external reservoir procedures, one end of the cannula is implanted proximally toward the liver and the other end is placed distally toward the duodenum using a polyethylene loop (Rath and Hutchison, 1989). The loop is placed external to the abdomen and protected in a plastic housing. To continuously collect bile, the housing cap is removed, the loop disconnected and bile collected into an external reservoir from the proximal cannula while commercial bile salt solution is infused into the distal cannula. Bile can reportedly be collected at a rate of 1 to 3 ml/hr with this technique. Rolf et al. (1991) described a similar external loop procedure for measuring bile flow and biliary pressure. Their procedure could easily be adapted for continuous bile collection. Alternatively, internal reservoir procedures involve implantation of an externally accessible, continuous-loop cannula that is threaded through an internally implanted glass collection vial (Heitmeyer and Powers, 1992). After recovery from surgery, the loop is cut and bile diverted into the glass collection vial. The bile is then aspirated through a sampling port with flexible tubing attached to a syringe. Approximately 1 mL of bile per hour can be collected with this technique.

Jacket and tether systems are often utilized to allow conscious animals to move relatively freely about their cages. Catheterized rats can also be restrained in restraint cages and 10 to 15 mL of bile can be collected every 24 hours for several weeks.

J. Pancreatic Juice

Gastrointestinal physiology studies and studies of pancreatic enzyme function may require collection of pancreatic juice. As stated earlier, the common bile duct carries bile from the liver and pancreatic enzymes and other substances from the multiple, small pancreatic ducts (Fig. 18-17). Therefore, it is necessary to fully obstruct bile flow in order to collect uncontaminated pancreatic juice. For general reviews of the methodology associated with collection of pancreatic fluid, see Lambert (1965), Zabielski et al. (1997), and Weiss et al. (2000). Unfortunately, techniques for collection of pancreatic fluid have historically been plagued by clinical problems associated with obstructive jaundice because the bile duct has generally been ligated near

the liver. Although technique improvements have occurred over the years, there is still not an ideal model for chronic collection of pancreatic fluid (Zabielski et al., 1997).

Waynforth and Flecknell (1992) described a surgical technique for chronic collection of pancreatic fluid. The surgical approach is similar to the one for collection of bile. The common bile duct is ligated at its proximal and distal ends and a catheter is inserted between the ligatures. The catheter is exteriorized and the rat restrained in a cage or by a tether and jacket system. The flow rate of pancreatic secretion is variable but usually is about 0.5 to 1 ml/hr. However, rats survived this technique for only a few weeks due to obstructive jaundice.

Pancreatic secretions are, of course, critical for digestive processes and regulation of intestinal flora (Weiss et al., 2000). Therefore, recent improvements in collection methodology include efforts to maintain flow of pancreatic juice into the duodenum (Zabielski et al., 1997). Rats can be housed in metabolic cages and peristaltic pumps used to deliver pancreatic juices back to the animals at a slow rate (Onaga et al., 1993). The rate of re-introduction of juices should approximate the rate of endogenous secretions (Weiss et al., 2000). Bodziony et al. (1985) described a technique that diverted bile directly into the duodenum and allowed chronic collection and re-circulation of pancreatic juice in conscious rats. The proximal portion of the common bile duct is implanted micro-surgically into the duodenum. A continuous loop cannula is then placed in the distal common bile duct and looped into the duodenum. A subcutaneous stopper connects the loop and is punctured to collect pancreatic secretions. The authors concluded that their method allowed collection of pancreatic juice from conscious, unrestrained rats without bile obstruction and changes in the composition of the pancreatic secretion.

K. Female Reproductive Products

1. Ova and Embryos

Collection of ova or embryos may be required for fertility, embryology or other reproductive studies. Oocytes may be collected from normal cycling female rats or from rats treated with drugs to induce superovulation. Primordial oocytes may be collected from the developing fetus prior to sexual differentiation at 14 to 15 days of gestation.

Oocytes may be obtained by removing the ovaries from post-pubertal female rats. The ovaries can be collected from freshly euthanized animals or as part of a survival surgery. Waynforth and Flecknell (1992) described a dorsal midline skin incision technique for ovariectomy in the rat. Briefly, following appropriate aseptic technique, a dorsal midline skin incision is made, subcutaneous tissue is dissected and incisions in the muscle are made on each side of the midline. The ovaries are externalized and removed by incising associated fat and vessels. Ligation and hemostasis are rarely required and the incisions are closed routinely. The oocytes are then placed in warm saline and teased from the ovarian follicle with a needle or other sharp instrument. The oocytes are cleared from associated cumulus cells using trypsin solution (Engel and Kreutz, 1973), hyaluronidase (Akira et al., 1993) or by forcing the cell masses through a tapered pipette (Zeilmaker and Verhamme, 1974).

Investigators may also flush the oviducts of euthanized female rats on day 1 or 2 of the estrous cycle to obtain oocytes (Zeilmaker and Verhamme, 1974). Pre-pubertal rats may be used by superovulating them with 20 to 30 IU of pregnant mare serum gonadotropin (PMSG) followed 48 hours later by 20 to 30 IU of human chorionic gonadotropin (HCG). The rats are euthanized 15 to 18 hours later and the oviducts flushed to collect oocytes (Oikawa et al., 1975). Corbin and McCabe (2002) also demonstrated that the combination of PMSG (30 IU) and HCG (25 IU) effectively superovulated female rats and that outbred rats are generally more responsive than inbred or hybrid rats. Pre-implantation embryos can also be flushed with appropriate media from the oviducts of pregnant, euthanized rats (Ansai et al., 1994).

Early techniques for egg transfer in the rat may be consulted (Bennett and Vickery, 1972).

2. Vaginal Fluids and Cells

Exfoliative cytology techniques are well developed in modern human and veterinary medicine. These techniques have been used to study vaginal mucosal morphologic changes during the estrous cycle in the rat. Gill (1976) thoroughly described the principles and practices of cytopreparation. Long and Evans (1922) first correlated vaginal morphology with ovarian activity in the rat and proposed that the estrous cycle be divided into 5 stages. Early investigators such as Nicholas (1967) described methods of collecting vaginal fluids and recommended various staining techniques. Smooth, polished glass rods, moistened cotton tipped swabs or lavage can be used to collect vaginal cell specimens. Samples for cytology are spread on a glass microscope slide, allowed to dry, fixed with alcohol and stained with suitable stains.

Several related techniques for collection of vaginal fluid have been reported more recently. Vaginal fluid can be collected with blunt-ended pipettes (Flecknell, 1987) or calibrated eyedroppers (Atherton et al., 1986). Saline is introduced into the vagina with the pipette or dropper and the fluid is aspirated. A suspension of cellular debris is obtained and can be directly applied to a microscope

slide or centrifuged to form a cellular pellet. The cells are stained with methylene blue or some other stain to determine the stage of estrus. The supernatant can be frozen and used for analysis. Small, calibrated plastic loops can also be inserted into the vagina of the rat to collect fluid (Bernardis et al., 1994). The collected fluid can be divided and either applied to a slide for examination or is used to inoculate appropriate media for cell culture.

3. Milk

Milk is a possible route of excretion for many drug compounds. Thus, for certain new drug compounds, metabolism studies may include collection of milk for analysis. Several methods for collection of milk from lactating rats have been developed. Direct methods include manual ejection (stripping), cannulation of mammary tissue and machine milking (Brake, 1979; Skala et al., 1981; Rodgers, 1995; Weiss et al., 2000). Indirect methods include collecting milk from the stomachs of neonates (Skala et al., 1981).

Teats can be manually stripped to collect milk (Brake, 1979; Skala et al., 1981; Gouldsborough and Ashton, 1998). Repetitive stroking motions over the teat will stimulate milk release and milk is collected in suitable vessels. Milk yield is reportedly dependent on the time that the dam is separated from her pups and is enhanced by treatment with oxytocin (Brake, et al.1979). Electrolyte content of the milk can also be altered by the collection procedure. Prolonged separation from the pups or repeated milking increases milk yield but the samples may not be representative of milk delivered under normal nursing conditions (Gouldsborough and Ashton, 1998). These authors recommended that dams be milked once 2 hours after separation from the pups for best results. Up to 0.4 mL of milk can be collected from each teat in anesthetized rats.

Waynforth and Flecknell (1992) described a milking method using a homemade apparatus. Two teat cups are attached by way of plastic tubing to a capped collection beaker. The collection beaker is connected in series to a second beaker with a vacuum side port and vertical tube with stopcock. Vacuum is maintained at 25 to 28 cm of mercury and the stopcock rotated 25 times/min. Vacuum pulses are created that simulate the suckling action of rat pups. Treatment with 5 units of oxytocin stimulates milk ejection and removal of the dam's pups about 4 hours prior to collection enhances milk yield. The authors reported that 3 to 7 mL of milk can be obtained daily from dams that have been lactating for at least 7 days. Rodgers (1995) also described a machine for collection of milk from rats without anesthesia. The machine consists of a capped glass jar with a connected vacuum line and vacuum control hole. A silicone suction cup is attached to the teat and connected to flexible plastic tubing that enters the jar through the cap. The tubing extends to a tube where milk is collected. The vacuum is controlled at 200 mbar and pulsations are created by repeatedly moving a finger over the control hole. About 0.5 mL of milk can be collected per nipple and the dam can be immediately returned to her litter after milking. Reducing the litter size to 6 within 2 days of parturition can enhance milk yield. The litter should be removed about 5 minutes prior to collection to enhance milk yield. Longer periods of separation do not appreciably increase milk yield according to the author. Finally, milk production can be initially stimulated with 4 units of oxytocin per kg of body weight.

L. Male Reproductive Secretions

Male reproductive fluids may be used for studies of fertility, sperm morphology and sperm function and can be easily collected from epididymis, prostate and seminal vesicles during terminal surgery or at necropsy (Blandau and Jordan, 1941). Ryan et al. (1988) reported that the vas deferens of the rat can also be removed, cannulated and flushed to provide sperm samples to simply and accurately measure exocrine testicular function. Excision of the cauda epididymis and collection of sperm from the distal portion of this anatomic structure are also often used to analyze sperm motility and velocity (Weiss et al., 2000). Samples may be aspirated with a capillary tube from a small incision in the excised epididymis (Klinefelter et al., 1991). Alternatively, the epididymis can be placed in appropriate media, excised and sperm allowed to diffuse into the media (Klinefelter et al., 1991). See Klinefelter et al. (1991) for a more complete description of these epididymal procedures.

Sperm can be collected from various portions of the female reproductive tract following breeding (Carballada and Esponda, 1998). However, total numbers of sperm ejaculated cannot easily be quantified by this technique and samples cannot be used for artificial insemination because the ejaculate coagulates (Weiss et al., 2000). Rats will also reportedly ejaculate spontaneously and the specimen can be collected if the rats are prevented from orally grooming the genital area (Orbach, 1961). Finally, the ductus deferens can be surgically anastomosed to the urinary bladder for continuous collection of sperm from urine (Vreeburg et al., 1974).

Electroejaculation allows for repeated sampling of male reproductive secretions in the same rat over time. Semzcuk et al. (1977) reported an electroejaculation procedure that can be applied multiple times over several weeks to anesthetized rats without the need for removal of accessory glands. The stimulus energy is decreased relative to earlier reported techniques and 92.6% of rats ejaculated samples with mobile and lively spermatozoa. Finally, Hundal et al. (1978) claimed that stimulation at 618 cps and maximum voltage of 10 volts produced consistent samples every week for 4 weeks without the need for surgical removal of accessory sex glands.

Waynforth and Flecknell (1992) reported a technique for electroejaculation in the rat. They reported that removal of the coagulating glands is not necessary to prevent coagulation of the semen sample if conditions are carefully controlled. The conscious rat is restrained in sternal recumbency on a board and the hind legs raised. A bipolar electrode is inserted into the rectum and the following electrical sequence performed. An audio oscillator is set to 30 cycles per second and the voltage is increased to 2 volts over 1 to 2 seconds where it remains for about 5 seconds before being decreased to 0 volts over 1 to 2 seconds. Ten seconds of recovery are allowed before another cycle is repeated. This sequence is repeated until ejaculation occurs (usually within 5 to 6 repetitions). The specimen is collected in physiologic solution on a warmed glass plate or into a suitable container. Adherence to the described electrical sequence is critical to preventing coagulation of the sample. This method can be used safely many times in the same rat and about 60 million sperm may be collected.

M. Pulmonary Cells and Fluid

Cells and fluid collected from the respiratory tract can be used for diagnostic purposes or various laboratory procedures. Brain and Frank (1968) described a lavage technique for collection of pulmonary or alveolar macrophages from euthanized rats. The lungs are removed, rendered gas free by evacuating at a pressure of 20 mmHg for 10 minutes and the trachea cannulated with polyethylene tubing. After an hour, the lungs are flushed with saline 12 times for 3 minutes each time. Approximately 5 mL of saline per gram of lung tissue are used for each flush. Wash contents are collected after each flush. Kosugi et al. (1980) described another method for collection of tracheobronchial secretions from rats. Two polyethylene catheters are inserted into one of the bronchi via a tracheotomy in anesthetized rats. One is used to flush the lungs and the other to collect secretions. Modern techniques for collection of various types of pulmonary cells and fluids generally involve similar terminal procedures where rats are deeply anesthetized. The lungs are often perfused with balanced salt solution, and are cannulated and lavaged multiple times with phosphate-buffered saline or another salt solution. Finally, the lungs are minced to collect cells such as alveolar macrophages that are not readily available in lavage fluid. Pneumocytes (Lag et al., 1996), alveolar and interstitial macrophages (Zetterberg et al., 2000), lymphocytes (Struhar et al., 1989), neutrophils (Bassett et al., 2000), and lipids (Alam and Alam, 1984) can be collected using these techniques for various research disciplines.

Pulmonary cells can also be collected from tracheal washings in anesthetized rats as part of a terminal procedure (Schreiber and Nettesheim, 1972). The rat is anesthetized, placed on a slanted board with mouth fixed open and the trachea cannulated. A specially designed system delivers 2 mL of wash fluid for collection of cells. The system includes the tracheal cannula, an automatic injector, regulator, manometer and vacuum control. Varner et al. (1999) described serial segmental bronchoalveolar lavage in lightly anesthetized rats. A tracheal catheter is advanced to a wedge position through an endotracheal tube and 5 small (0.1 mL) volumes of buffer solution are instilled and withdrawn with gentle suction. The technique allowed airway samples to be collected longitudinally at 2 week intervals so that animals could remain alive and not be euthanized as part of the collection process.

N. Lymph

Lymph collection may be required for immunology, lipid physiology and other types of studies. Several techniques for collection of lymph have been described. Reinhardt (1945) first described thoracic duct cannulation for collection of lymph from anesthetized rats. Trypan blue is injected 30 minutes prior to cannulation to enhance visualization of the duct. Flow of lymph averaged 0.45 ml/hr following cannulation. Reich and Keller (1988) described several modifications to existing methods, including the use of heparinized saline rather than Evans blue to dilate the duct and a newly designed restrainer that allowed increased mobility. The authors reported lymph flow of 80 ml/day.

Waynforth and Flecknell (1992) provided a detailed description of thoracic and mesenteric duct cannulation techniques. Restraint cages can be used but these techniques can be enhanced to utilize jacket and tether methods for chronic collection of lymph from conscious, freely moving rats. The authors consider the thoracic duct procedure to be technically easier than the mesenteric duct method. Evans blue or trypan blue can be injected into the mesenteric lymph nodes to temporarily enhance visualization of the thoracic duct. Alternatively, olive oil or glycerol trioleate is given orally to cause white-colored lymph to flow through both the mesenteric and thoracic ducts. A dissecting microscope is also helpful during surgery because the vessels are delicate and transparent. Maintenance of good hydration is important for good

lymph flow, therefore, the authors recommend that intravenous fluids are administered and a sugar- and electrolyte-rich fluid is offered orally. The lymph flow rate is expected to be 0.13 to 0.7 ml/hr. Lymph clots can occur within the first few hours post-surgery and may inhibit lymph flow. This problem generally disappears after the first day. Frank et al. (1992) also described a technique for cannulation of the thoracic duct. The duct is approached through a left sub-costal abdominal incision and is cannulated with polyethylene tubing that is exteriorized through the left lateral body wall. The rat is restrained in a restraint cage while lymph is collected by gravity flow. The authors claimed that their surgical modifications minimized surgery time and maximized collection of lymph.

Finally, lymph can be collected terminally as part of acute experiments. Jakab et al. (1980) cannulated the cervical thoracic duct and reported higher lymph flow from this site than from the abdominal thoracic duct in anesthetized rats. Zawieja and Barber (1987) described procedures for micro-puncture of villus initial and mesenteric lymphatic vessels. Segments of intestine are isolated in anesthetized rats and glass micropipettes are used to puncture the mesenteric lymphatic and villus lymphatic vessels with the aid of a special micromanipulator. Fifteen to sixty seconds are required to collect these samples and collected volume varies from 2 to 500 nl.

O. Cerebrospinal Fluid

Collection of cerebrospinal fluid or injection of test materials into the cerebrospinal space may be required for neurochemistry, neurophysiology and neurotoxicity studies. This section will cover techniques to collect cerebrospinal fluid (CSF) or to inject substances into the cerebrospinal space. The techniques available include "free hand," stereotaxic and necropsy procedures. A review of the literature regarding these techniques is available (Weiss et al., 2000).

Necropsy techniques include a method to collect CSF directly from the third ventricle of the decapitated rat using a microcannula (Knigge and Joseph, 1974). Up to 0.5 µl of CSF can reportedly be collected using this technique. Perfusion techniques for collection of CSF from several sites have also been described (Dudzinski and Cutler et al., 1974; Franklin et al., 1975).

Several acute and chronic techniques for collection of CSF from a variety of sites have been developed. Cerebrospinal fluid can be continuously sampled with an in-dwelling cannula inserted in the third ventricle for acute pharmacokinetic studies (Meulemans et al., 1986). Waynforth and Flecknell (1992) provided a detailed description of a cisternal puncture technique for collection of CSF. The rat is anesthetized and placed on a special wooden stand with the head flexed. A 24-gauge needle attached to polyethylene tubing is used to puncture the atlanto-occipital membrane. As CSF begins to flow, the tubing should be held below the level of the head to promote flow. The procedure should be repeated no more than every 3 to 7 days to prevent contamination of samples with blood. The technique allows 0.1 to 0.15 mL to be collected in approximately 3 minutes. Other workers reported collecting 0.1 to 0.5 mL of CSF by cisternal puncture (Hudson et al., 1994; Weiss et al., 2000). Frankmann (1986) described a similar method for repeated sampling of CSF from anesthetized rats using stereotaxic equipment. Samples as large as 250 μl can be collected every 3 days for up to 2 weeks. Finally, the lateral ventricles or cisterna magna can be implanted with guide cannulae for chronic collection of CSF (Sanvitto et al., 1987; Westergren and Johansson, 1991; Consiglio and Lucion, 2000). The cannulae are implanted surgically and CSF is collected after recovery using sampling tubes or needles that are introduced into the guide cannulae. Investigators should be aware that numbers of cells and levels of albumin in CSF may be elevated when collected from guide cannulae compared to repeated sampling in the absence of permanent cannulae (Westergren and Johansson, 1991).

Microdialysis techniques for sampling artificial CSF for compounds such as caffeine and radiolabeled carbon dioxide have recently been developed (Nakazono et al., 1992; Zielke et al., 1998). Guide cannulae are generally placed surgically in cisterna magna, cerebral cortex, lateral ventricle or another structure of interest using stereotaxic equipment. A microdialysis probe inserted in the guide cannula then delivers artificial CSF and samples the fluid for the compound of interest.

Finally, injection of test substances can be accomplished with the Waynforth and Flecknell method for CSF collection. Techniques for injection of neurosphere cells, tracer agents or other substances into the fourth ventricle, cisterna magna or other areas of the brain using stereotaxic equipment have also been developed (Roeling et al., 1993; Stoodley et al., 1999; Wu et al., 2002).

P. Saliva

1. General

Collection of oral fluids from rats is important in research disciplines such as oral fluid immunology, circadian salivary flow physiology, oral electrolyte physiology and drug metabolism (Shannon and Suddick, 1975; Cocchetto and Bjornsson, 1983; Weiss et al., 2000). Researchers in these fields have described numerous techniques for collection of oral fluids from rats.

Research investigators have generally collected saliva from rats by one of the following four methods: (1) direct collection from the oral cavity, (2) cannulation of a salivary duct, (3) direct collection from the transected end of a surgically isolated salivary duct, and (4) removal of a salivary gland at necropsy. Consult Cheyne (1939) for a detailed description of the anatomy and extirpation of the salivary glands of the rat.

2. Stimulation of Salivation

Salivary flow can be stimulated chemically using sialogogues or by direct stimulation of secretory nerves supplying the salivary glands. The increased flow can facilitate collection of saliva. Pilocarpine is a parasympathomimetic drug that can be used to stimulate salivary flow (Cocchetto and Bjornsson, 1983). The drug also stimulates lacrimal, bronchial and nasal secretions that may accompany saliva in collected oral fluids. The submandibular salivary glands apparently secrete the majority of saliva stimulated by pilocarpine administration (Holloway and Williams, 1965). A dosage of 0.1 to 10 mg/kg promotes salivary flow in the rat (Holloway and Williams, 1965; Robinovitch and Sreebny, 1969; Cocchetto and Bjornsson, 1983; Weiss et al., 2000). Isopreterenol can also be used successfully at 10 to 250 mg/kg (Robinovitch and Sreebny, 1969; Menaker et al., 1974; Weiss et al., 2000). Salivary flow can also be enhanced by stimulating the secretory nerves supplying parotid and submandibular salivary glands (Hellekant and Hagstrom, 1974; Hellekant and Kasahara, 1973; Woodruff and Carpenter, 1973). Investigators should be aware that saliva collected after stimulation by pilocarpine may differ from saliva collected after secretory nerve stimulation (Schneyer and Hall, 1965). Amylase activity and total protein levels were shown to be higher in pilocarpine-stimulated saliva compared to nerve-stimulated saliva.

3. Methods

a. DIRECT COLLECTION FROM THE ORAL CAVITY. Saliva can be collected into a graduated tube as it drips directly from the mouth of the anesthetized rat (Holloway and Williams, 1965). Discs or strips of filter paper can also be placed directly in the oral cavity to collect saliva (Tatevossian and Wright, 1974; Damas, 1994; Guhad and Hau, 1996). Finally, to avoid contamination of saliva with nasal and lacrimal secretions in pilocarpine-stimulated rats, some investigators collect saliva directly from the orifice of a salivary duct using capillary tubes or special devices (Menaker et al., 1974; Wolf and Kakehashi, 1966; Woodruff and Carpenter, 1973).

b. CANNULATION TECHNIQUES. All three groups of salivary glands can be cannulated to collect saliva. A polyethylene tube or glass cannula is typically used to enter the orifice of the duct or the proximal end of a surgically isolated and transected salivary duct (Hellekant and Kasahara, 1973; Martinez et al., 1975; D'Amour et al., 1965; Stahlin et al., 1978; Pull and Reynolds, 1978; Takai et al., 1982; Martinez and Camden, 1983; Abe et al., 1987). The parotid salivary duct can be cannulated with a 30-gauge steel cannula attached to polyethylene tubing (Robinovitch and Sreebny, 1969). This duct can also be cannulated via its orifice using a stereomicroscope and a polyethylene catheter with inserted stylet (Ulmansky et al., 1971). The method is reportedly simple and not traumatic and can be repeated in the same animal since the duct is entered through its orifice (not through the transected duct). Each of the salivary gland ducts can be cannulated with blunted 25- to 26-gauge needles attached to polyethylene tubing (Drum, 1963). Volumes up to 0.8 mL of saliva can be collected from single submaxillary ducts and up to 0.2 mL from single parotid ducts using 200-g rats. Vissink et al. (1989) described collection of saliva from the parotid and submandibular salivary ducts using a miniaturized cup that is attached by silicon tubing to a vacuum pump. The cup is placed over the duct orifice in an anesthetized rat and saliva is collected in a plastic container. According to the authors, the method is non-invasive, causes no tissue damage and is appropriate for chronic studies of salivary composition and secretion in the rat. Finally, Scott and Berry (1989) also described intra-oral cannulation of the parotid duct using a blunted 25-gauge needle in anesthetized rats.

c. REMOVAL OF A SALIVARY GLAND AT NECROPSY. An entire salivary gland can be removed at necropsy and collected for weighing or secretory protein analysis (Sreebny and Johnson, 1969; Nagler et al., 1993).

V. SUBSTANCE ADMINISTRATION

A. General

The route of administration of drugs, tumor cells, carcinogens and other test materials obviously depends on the purpose of the administration. Consult Woodard (1965), Waynforth and Flecknell (1992), and Nebendahl (2000) for full discussions of the principles of drug or test substance administration.

B. Oral

This route is very commonly used for regulatory toxicity and safety testing of new drug compounds and for basic research studies in many disciplines. Investigators have developed numerous techniques for delivery of test materials through the oral cavity to the esophagus or stomach. The maximum volume to be administered depends on the type of solution and its properties but should be limited to 10 to 20 ml/kg (Wolfensohn and Lloyd, 1994; Hull, 1995; Sharp and LaRegina, 1998; Morton et al., 2001). Brown et al. (2000) recommended that the volume should not exceed 10 ml/kg due to potential aspiration, pulmonary injury or elicitation of a stress response (especially with oil-based vehicles such as corn oil).

The classic steel, ball-tipped 15- to 16-gauge hypodermic needle was developed for oral gavage of test materials (Ferrill, 1943). Only minor modifications of the original design have been reported over the years (Fig. 18-18). For example, Moreland (1965) recommended adding a 20 to 30° bend to the end of the needle to facilitate passage through the oral cavity and into the esophagus (Fig. 18-18C). The gavage needle can be readily passed in conscious rats using general handling and restraint techniques described earlier in this chapter. Nebendahl (2000) recommends that the gavage needle should be of suitable length (e.g., the needle should reach from the mouth to the end of the sternum) to help determine if the needle is properly placed. The rat is grasped around the thorax and the index finger and thumb are used to control the animal's mandible. Alternatively, the rat can be held firmly by the loose skin of the neck and back

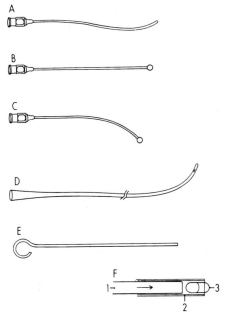

Fig. 18-18 Devices used for oral administration of materials. (A) Original flexible silk tube. (B) Straight ball-tipped needle. (C) Curved ball-tipped needle. (D) Flexible No. 8 French catheter. (E) and (F) "Balling gun" for gelatin capsules: (E) ram portion of device, (F) close up of ram (1) within outer sleeve (2) showing position of gelatin capsule (3).

(Waynforth and Flecknell, 1992). The gavage needle is passed through the space between the incisors and premolars, rotated slightly, moved over the tongue and into the esophagus (Fig. 18-19). The needle tip can usually be seen passing down the esophagus on the left side of the neck (Wolfensohn and Lloyd, 1994). If the tip is in the trachea, it may be possible to feel the tip rubbing against the cartilage rings (Baumans et al., 1993). However, accidental tracheal intubation is rare if the procedure is performed properly. The test material can be expelled into the esophagus or the needle tip can be further pushed into the stomach for direct gastric administration. The procedure is rapidly learned and mastered. Experienced technicians can reportedly gavage approximately 120 rats per hour (Ferrill, 1943; Moreland, 1965).

Plastic gavage needles are also available for delivery of fluids to the esophagus or stomach (Fig. 18-18A and D). These needles are more flexible and, thus, may cause less trauma to the esophagus than steel needles. However, rats

can accidentally bite and sever the needle during gavage and the needle can become a gastric foreign body.

Gelatin capsules can be administered to rats using a balling gun (Fig. 18-18E and F). The device was first developed for use in guinea pigs (Nelson and Hoar, 1969) but can be modified for use in rats. The capsule is deposited in the caudal oral cavity and the rat's mouth is held closed until the capsule is swallowed. Solid or liquid materials may be administered and the possibility of esophageal irritation is eliminated. Accidental tracheal instillation of test material and secondary aspiration pneumonia are unlikely to occur with this method and the possibility of contamination of the oral cavity with radioactive or toxic agents is minimized. Lax et al. (1983) described a modification using a stainless steel needle with a cup at the end for capsule dosing. The capsule fits snugly within the cup and the needle is introduced into the esophagus through the oral cavity. The capsule is ejected by air (Waynforth and Flecknell, 1992) or water from an attached syringe or by a steel rod. The rat's mouth may need to be held closed while the capsule is swallowed.

Simple mouth gags can be used to facilitate passage of stomach tubes in rats (Lehr, 1945; Kesel, 1964). Other devices were developed to facilitate examination of the oral cavity in rats and could be similarly useful (Johansen, 1952; Rizzo, 1959). Lehr's device is simply a hairpin-shaped steel wire that is held in a vise. The rat's incisors are hooked over the wire to hold the mouth open. More recently, Waynforth and Flecknell (1992) described an oval-shaped gag made of wood or plastic with central diameter of 2 cm and a central hole of 5-mm diameter. The upper and lower incisors are hooked over the gag to hold the mouth open while a flexible dosing catheter is passed.

Rats can also be administered fluids through a chronic gastric tube that bypasses the oral cavity (Epstein and Teitelbaum, 1962). The authors developed a swivel joint that allowed administration of materials into the stomach in moving animals. A chronic gastric fistula procedure described by Waynforth and Flecknell (1992) for collection of gastric secretions could probably also be used to chronically administer test materials.

Dietary administration of test materials can be used for toxicology and other types of research studies. There are several advantages to dietary administration of test materials. It is often simple to mix materials in food or water (although a commercial feed company or testing facility should be used for GLP-type studies). Rats generally require no acclimation to the procedures and the test material is provided relatively stress-free. However, this method may not be practical for every test material (since some may not mix well). See Nebendahl (2000) for a complete review of the principles associated with dietary administration of test materials.

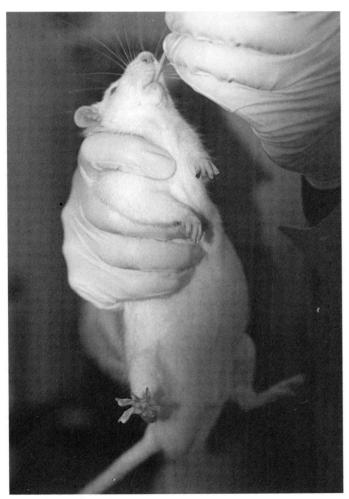

Fig. 18-19   Restraint and passage of a gavage needle.

Finally, neonatal rats weighing 5 to 6 grams can be gavaged orally with a blunt tip 30-gauge needle attached to polyethylene tubing (Gibson and Becker, 1967). Other techniques for delivery of materials to the stomach of neonatal rats have been described (Miller and Dymsza, 1973; Smith and Kelleher, 1973).

C. Intravenous

1. General

Intravenous injection of test materials is commonly used for toxicology and basic science studies and is usually performed in one of the superficial veins that are used to collect blood. Other peripheral veins are also available and will be described in this section. The choice of site for intravenous injection depends on several factors: (1) ease of venipuncture, (2) age, size and sex of the rat, (3) relative risk to the animal, (4) whether anesthesia is required, (5) whether surgical exposure is necessary, (6) whether chronic catheterization is required, and (7) whether the vein is part of the hepatic portal system.

Several standard techniques will help ensure that intravenous injection is successful. Clean technique using sterile syringe and needle is required to prevent blood-borne infection after percutaneous intravenous injection. Strict aseptic technique is required for chronic catheterization techniques or for procedures that involve surgically exposing the vein. Care should be taken to avoid introducing air when injecting test materials intravenously to prevent air embolism problems. Finally, if a peripheral vein will be injected multiple times, the first punctures should be as distal as possible so that plenty of normal vein is left for later punctures.

For drug metabolism studies, it is important to determine whether the vein to be injected is part of the hepatic portal system. Test materials administered orally or intraperitoneally are absorbed into mesenteric veins that form the hepatic portal vein. Test materials that are metabolized by the liver may undergo a marked "first pass" metabolism if given by the oral or intraperitoneal routes. Nightingale and Mouravieff (1973) showed, for example, that blood from the dorsal vein of the penis does not enter the portal blood flow and, thus, can be used as an injection site for certain types of studies.

The total volume of test material formulations injected should also be considered for humane and scientific reasons. Rapid injection of large volumes of material can possibly cause life-threatening complications such as pulmonary edema and can certainly impact the cardiovascular, pulmonary and renal systems in ways that may affect the research study. Sharp and LaRegina (1998) and Hull (1995) recommend that a single intravenous injection not exceed 10% of circulating blood volume.

Hull (1995) also recommends no more than 5 ml/kg be given intravenously by rapid bolus (if the fluid is not too viscous). Other authors recommend that rapid intravenous injections given to adult rats be limited to 1 to 5 mL (Wolfensohn and Lloyd, 1994; Morton et al., 2001). More volume may be given by slow injection or intravenous infusion (Hull, 1995). Up to 4 ml/kg/hr has been given to rats by continuous infusion for 28 days (Hull, 1995).

Injection of test materials may be made into the jugular, femoral, caudal, saphenous, lateral marginal, dorsal metatarsal, lingual and dorsal penile vein. Only the femoral, lingual and dorsal penile veins will be described in detail in this section since access to other veins was described in the section regarding blood collection. A few specialized techniques will also be described here.

2. Femoral Vein/Continuous Infusion

The femoral vein and others such as the jugular vein, vena cava and tail vein are most commonly used for catheterization and acute or chronic continuous infusion of test materials (Nebendahl, 2000; Nolan and Klein, 2002). Many techniques for periodic or continuous infusion have been described (Cocchetto and Bjornsson, 1983). Nishihira et al. (1994) described general improvements that allowed continuous infusion of parenteral nutritional solution simultaneously to the jugular and portal veins of rats. The authors used vascular clamps to control bleeding during catheter insertion and increased numbers of anchoring sutures to improve the technique. They were able to continuously infuse the jugular and portal vein of rats for 6 days. Commercially available femoral-cannulated rats are commonly used for modern toxicology and other studies. The surgically implanted catheter is generally exteriorized between the scapulae or attached to a vascular access port. The catheter can also be exteriorized in other ways such as through a plastic housing attached to the skull (Ohashi et al., 1985). Externalized catheters are disadvantageous because they disrupt skin integrity and leave the animal susceptible to local or systemic infections (Nolan and Klein, 2002). The externalized catheter is often protected by a jacket and tether system with a swivel joint that allows the rat to freely move while test material is continuously infused (Cocchetto and Bjornssson, 1983). Catheter tract infections can be reduced by using subcutaneous vascular access ports (Nolan and Klein, 2002). Access ports are placed between muscle fascia and skin and are composed of a chamber that communicates with the implanted catheter (Nolan and Klein, 2002). A hard silicone septum overlies the chamber and is penetrated with a special Huber needle. Careful aseptic technique should be used to minimize bacterial contamination of the port and catheter. The ports are useful for periodic injection or for continuous infusion over several hours using a jacket and tether system (Nolan and Klein, 2002). An infusion pump can be used to deliver exact quantities of test material. As a general rule, infusion of about 1% of the rat's blood volume per hour will not affect fluid physiology (Nebendahl, 2000).

3. Lingual Vein

The lingual veins are readily accessible and can be used for intravenous injections in anesthetized rats. The rat is placed in dorsal recumbency and the tongue is dried and pulled out to one side. A 25- to 27-gauge needle is used to penetrate the prominent sublingual veins that run from the tip to the root of the tongue (Fig. 18-20). A hematoma can develop easily so care should be taken to adequately control hemorrhage after withdrawing the needle (Anderson, 1963). This technique is easy in larger rats and may be preferable to caudal venipuncture since the skin of aged rats can be tough and difficult to penetrate. A suture placed through the tip of the tongue can be used to facilitate the procedure by pulling the tongue out and fully exposing the sublingual veins (Greene and Wade, 1967). Waynforth and Flecknell (1992) also described a technique for venipuncture of the lingual vein. The rat is anesthetized and placed in dorsal recumbency with the head toward the operator. The tongue is pulled out with forceps and held with thumb and forefinger. One of the two lingual veins is entered almost horizontal to its surface with a 25-gauge needle for weanling and older rats (30-gauge for smaller rats). Hemostasis is accomplished with the thumb initially and then with a small cotton-wool pledget that is pushed with the tongue back into the mouth. The pledget falls out of the mouth as the rat recovers from anesthesia. The authors claim that hematoma formation is rare and frequent injections can be performed if proper technique is used.

4. Dorsal Vein of the Penis

The dorsal vein of the penis can be used for intravenous injection in the rat, although authors differ regarding the usefulness of this procedure. For example, Waynforth and Flecknell (1992) argue that this route should be used only under special circumstances because the vein can be damaged. However, other authors indicated that the vein is readily visible and easily injected. The technique was originally described for mice but can be used in rats (Salem et al., 1963). The rat is restrained by an assistant who gently hyperextends the animal's vertebral column (Fig. 18-21). The tip of the penis is everted with thumb and index finger and the vein is penetrated with a 25- or 30-gauge needle (Waynforth and Flecknell, 1992). Aspiration of blood is apparently nearly impossible so slow injection to detect a subcutaneous bleb or injected material freely moving in the vein is required to confirm proper intravenous injection. Kononov et al. (1994) described a simple technique for continuous IV infusion

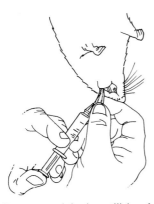

Fig. 18-20 Intravenous injection utilizing the lingual vein.

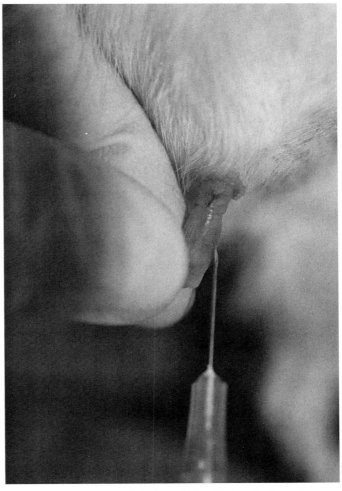

Fig. 18-21 Intravenous injection utilizing the dorsal vein of the penis.

of supportive fluids during small bowel transplantation surgery in rats. The penis is extruded with a hemostat and the vein is entered with a 24-gauge, 3/4 inch angiocatheter. The penis is re-positioned within the prepuce and the catheter left in place for the duration of the surgery. The authors were able to perform 74 successful surgeries using this technique for fluid support.

5. Other Intravenous Injection Techniques

Early workers described intracardiac injection (Postnikova, 1960) and injection of the superficial palpebral veins (Anderson et al., 1959) of newborn rats. The orbital plexus has also been used historically as an intravenous injection site (Pinkerton and Webber, 1964).

The lateral tail vein is approached for injection as it is for blood collection. A 23- to 25-gauge needle is directed at a very shallow angle to enter the vein. The tail should be bent down and entered at the point of the bend (Waynforth and Flecknell, 1992). A tourniquet may be useful to enhance visualization of the vein. The dorsal metatarsal vein is also approached like it is for blood collection. An assistant extends the leg and applies pressure to the vein proximally. The area over the vein is shaved and the operator holds the rat's foot. The vein is entered with a 25-gauge needle at a shallow angle. A thumb or finger can be used to control hemorrhage. The previous section on blood collection describes the approach to the saphenous vein.

D. Intraperitoneal

Intraperitoneal (IP) injection can be used for toxicology and other experiments where other routes are not considered suitable. Compounds absorbed IP will pass through hepatic circulation prior to distribution to other organs. Compounds delivered IP should be sterile and in a form that is readily absorbed to avoid chemical or septic peritonitis. The rat is restrained normally by an assistant to help control the rear legs. The abdomen is penetrated with a needle at a 20 to 45° angle just off the midline in the lower left or right quadrants (Waynforth and Flecknell, 1992; Sharp and LaRegina, 1998; Nebendahl, 2000). These quadrants are chosen because the intestinal mass, liver and spleen generally lie in other areas of the abdomen. Thus, inadvertent penetration of a hollow or solid organ is less likely to occur in these quadrants. Greener and Gillies (1983) and Murgas et al. (1988) described simple methods for continuous long-term intraperitoneal administration through surgically implanted polyethylene catheters. Greener and Gillies (1988) instilled materials 4 times daily for up to 94 days with 65% of the

catheters remaining patent during this time. The intraperitoneal route should be chosen carefully since many test material formulations can be irritating to the peritoneum. Sharp and LaRegina (1998), Hull (1995), and Morton et al. (2001) recommend that the volume administered be limited to 10 ml/kg while other authors recommend limiting the volume to 5 to 10 mL total (Wolfensohn and Lloyd, 1994; Nebendahl, 2000).

E. Subcutaneous

Subcutaneous injection is also often used for administration of test materials and is easily performed in the rat. The animal is restrained appropriately and a needle is used to penetrate the skin overlying the shoulders, flank or ventral body wall. The needle should enter at a 20 to 30° angle into a pocket formed beneath the tented skin. It may be helpful to pull the skin over the needle. Redirecting the needle may be necessary if large volumes are injected. The volume to be injected should be limited to 5 to 10 ml/kg or 5 to 10 mL total distributed over two to four sites (Wolfensohn and Lloyd, 1994; Hull, 1995; Sharp and LaRegina, 1998; Nebendahl, 2000; Morton et al., 2001).

F. Osmotic Pumps

Osmotic minipumps can be used for continuous long-term delivery of test substances by the subcutaneous or intraperitoneal routes (Morton et al., 2001). The pumps are surgically implanted in anesthetized rats using appropriate aseptic technique. The pumps are commercially available and designed to deliver various volumes of test material at different infusion rates. The surgical incision area should be shaved and scrubbed prior to surgery. Sterile instruments are used to incise skin and muscle. For subcutaneous placement, a 1-cm incision is made along the dorsal midline and blunt forceps are used to create a subcutaneus pocket for placement of the pump. The skin is closed with staples, sutures or skin glue. The pumps can also be implanted in the abdomen through a ventral midline incision. Appropriate post-surgical care should be provided. The minipumps allow continuous exposure to test compounds without restraint and repeated subcutaneous or intraperitoneal injections. However, there is likely some initial stress associated with surgical implantation. The minipumps should be implanted by experienced surgeons to help minimize surgical complications and associated stress (Morton et al., 2001). Only small volumes can be delivered by this technique so test material solutions may be very concentrated when administered.

G. Intramuscular

The intramuscular route is often used for administration of therapeutic agents and, therefore, may be used for toxicology or safety studies for new drug compounds. Small quantities of test material formulation can be administered in the muscles of rats. The quadriceps muscle group or the epaxial muscles are free of major nerves and blood vessels unlike the caudal thigh muscles. Thus, injections into these areas may not cause nerve or vessel damage and associated lameness or self-mutilation (Nebendahl, 2000). An assistant properly restrains the animal and the quadriceps muscle is immobilized with thumb and index finger. An appropriate sized needle is introduced, the syringe plunger pulled back to aspirate blood and the test material is injected. Several authors recommend 22- to 25-gauge needles for intramuscular injection. The volume injected should not exceed 0.05 ml/kg/site (Morton et al., 2001) or 0.1 to 0.5 mL per site with a maximum of two to four sites per animal (Wolfensohn and Lloyd, 1994; Hull, 1995; Sharp and LaRegina, 1998; Nebendahl, 2000).

VI. SPECIALIZED RESEARCH TECHNIQUES

A. Techniques in Cardiovascular Research

1. Blood Pressure Determination

Blood pressure determination can be an important component of pharmacology, toxicology, physiology and many other types of studies. Many techniques and modifications of techniques to determine arterial blood pressure in conscious and anesthetized rats have been published over the years. Reviews of the methods to determine blood pressure are available (Chodorowicz et al., 1973; Udenfriend et al., 1976; Sponer et al., 1988; Van Vliet et al., 2000).

a. DIRECT TECHNIQUES. Arterial blood pressure has been measured directly by placing a catheter in a major vessel such as the aorta, carotid artery or femoral artery. The exteriorized catheter is then connected to a pressure transducer for accurate measurement of blood pressure. This procedure can be performed easily in anesthetized rats for acute, terminal experiments but the physiologic effects of anesthesia must be accounted for. Thus, investigators have developed models for measurement of blood pressure in conscious, freely moving rats (Stanton, 1971; Waynforth and Flecknell, 1992; Gonzalez et al., 2000). Jacket and tether systems are available for measurement of blood pressure in conscious animals. Wireless telemetry systems are also now available (Schierok et al., 2000; Van Vliet et al., 2000) for direct measurement of arterial blood pressure in conscious, freely moving animals. A catheter is generally placed surgically in a major vessel such as the aorta and connected to an implanted transmitter. A remote receiver then records pressure data. Week to week variability of systolic, diastolic and mean arterial pressures was found to be minimal using telemetry and responses to a hypotensive agent using telemetry correlated well with a direct arterial catheterization system (Guiol et al., 1992). Thus, these systems offer chronic, high fidelity recording without the physiologic effects of restraint or anesthesia (Van Vliet et al., 2000).

b. INDIRECT TECHNIQUES. Indirect techniques are noninvasive but generally less accurate than direct methods because restraint and tail warming are generally required (Van Vliet et al., 2000). Restraint stress and warming can obviously impact arterial blood pressure and reduce the accuracy of data obtained by this method. These techniques are most advantageous where the investigator wishes to sample large numbers of rats without the expense and variables associated with surgery. O'Neill and Kaufman (1990) argued that repeated indirect measurements are preferable for studies that may be affected by alterations in nutrient intake and/or animal growth. An inflatable cuff and pulse detector are placed over a peripheral artery (most often the caudal tail artery) and connected to a commercially available blood pressure recorder (Waynforth and Flecknell, 1992; Van Vliet et al., 2000). Plethysmographic and piezoelectric crystal systems are also available (Widdop and Li, 1997). The cuff is inflated to a pressure above the systolic pressure of the rat and then is slowly deflated. The cuff pressure where the pulse can no longer be detected is considered comparable to the systolic pressure (Waynforth and Flecknell, 1992).

Investigators should carefully choose the instrument and calibration to be used and should consider the specific and controlled conditions under which blood pressure will be determined if meaningful and reproducible results are to be obtained.

2. Blood and Plasma Volume Determination

Accepted techniques for determination of blood volume (Bruckner-Kardoss and Wostmann, 1974) and plasma volume (Wunder, 1965) have been described. Regional blood volumes can now be quantified in rats by injecting contrast agent and mapping with magnetic resonance imaging (MRI; Schwarzbauer et al., 1993). Quantitative MRI maps of skeletal muscle, heart, liver and kidney were obtained in jugular cannulated rats 20 minutes after injection of contrast material. The intravascular and extravascular spaces in each organ are compared by complex MR formulae using intrinsic spin-lattice relaxation time. Blood volume values are derived from the formulae.

3. Electrocardiographic Techniques

ECG recordings can be valuable sources of data for certain types of studies such as toxicology experiments. Techniques for recording ECGs from anesthetized rats have been described (Sharp and LaRegina, 1998; Ueyama et al., 2000). The ECG can be recorded by standard limb electrode systems or by surgical implantation of electrodes for telemetry. Electrodes can be placed directly on the skin of the limbs but this method generally requires anesthesia or prolonged restraint with associated physiologic changes (e.g., increased heart rate and blood pressure). Investigators have generally used the bipolar limb lead system of Einthoven (leads I, II, III) with or without the augmented unipolar limb leads (aVR, aVL, aVF; Detweiler, 1981). Three-dimensional ECG has more recently been developed as a valuable research tool (Sharp and LaRegina, 1998). Telemetry techniques have also recently been developed for evaluation of ECGs in conscious, freely moving rats (Kuwahara et al., 1994; Baillard et al., 2000). Sgoifo et al. (1996) further showed that electrodes placed within the thorax for telemetry produced ECGs with less noise and more recognizable beats compared to electrodes placed subcutaneously.

B. Techniques in Neurophysiology

Several commonly used techniques are available for physiological evaluation of the nervous system. Consult Classen (2000) for a complete review of contemporary techniques in neurophysiology.

1. Electroencephalography

Several techniques for recording of EEGs in rats have been described for behavioral, epilepsy, learning, sleep, brain injury or brain electrophysiology studies (Sharp and LaRegina, 1998; Datta and Hobson, 2000; Kondo et al., 2001; Sawamura et al., 2001). The EEG generally measures integrated brain activity and is composed of electrical signals of different frequencies (Classen, 2000). Surface or epidural electrodes measure brain activity within a few millimeters of the electrodes. The EEG electrodes are placed surgically through the skull and anchored with dental acrylic at the coordinates of choice (Kaufmann et al., 1977; Troncoso et al., 1995; Datta and Hobson, 2000). Deeper structures can be evaluated by implanting electrodes near the structures of interest (Datta and Hobson, 2000; Kondo et al., 2001; Sawamura et al., 2001). The EEG can be monitored with tethered or telemetry systems so that rats can be freely moving or minimally restrained (Gohd et al., 1974; Troncoso et al., 1995; Sharp and LaRegina, 1998).

2. Peripheral Nerve Conduction Velocity

The speed of nerve signals along peripheral nerves can be measured in excised nerves, surgically exposed nerves, and in intact rats (Classen, 2000). This technique is particularly useful for studies of diabetic neuropathy (Stevens et al., 2000; Cameron et al., 2001). Stimulus electrodes are placed along a selected nerve segment of appropriate length and conduction velocity is measured. The temperature of the tissue or bath surrounding the nerve must be appropriately controlled since conduction velocity is temperature-dependent (Classen, 2000; Stevens et al., 2000; Cameron et al., 2001).

3. Evoked Potentials

Electrical potentials recorded from the nervous system after external stimulation are called evoked potentials (Classen, 2000). Investigators measure these potentials in studies of diabetic neuropathy, spinal cord injury, brain ischemia and neural circuit mapping. These potentials can be stimulated visually using light flashes or other visual stimuli or by implanted light emitting diodes (Szabo-Salfay et al., 2001). They are also generally elicited by auditory stimuli such as brief clicks or by electrical stimulation of peripheral nerves of the foot, tail or skin (Hayton et al., 1999; Winkler et al. 2000). A recording electrode is placed near the structure of interest (visual cortex, brain stem, cerebellum, spinal cord, etc.) in the anesthetized rat. Hayton et al. (1999) recommended fentanyl/fluanisone and midazolam as the combination anesthetic of choice in rats since the combination has fewer deleterious effects on somatosensory evoked potentials than other anesthetic regimens. A method for recording magnetic motor evoked potentials from the brain of conscious, restrained rats has also been described (Linden et al., 1999). Classen (2000) provides a more complete description of these techniques.

4. Brain Imaging

Modern imaging techniques are used not only for clinical diagnostic purposes but are being used increasingly for research in animals, including rats. Stroke and brain ischemia, brain injury, neurodegenerative disease, demyelination disease and basic brain physiology and metabolism are just a few of the research disciplines now utilizing these powerful tools (Kohno et al., 1995; Yamada et al., 1997; van Bruggen et al., 1998; Lester et al., 1999; Guzman et al., 2000; Kornblum et al., 2000; Rubens et al., 2001). Magnetic resonance imaging (MRI) and positron emission tomography (PET) are used to evaluate anatomical, physiological and chemical processes associated with certain disorders. With MRI, cross-sectional images are created using magnetic fields and radio-frequency signals (Paulus et al., 2001). The animal is placed inside a magnet scanner and exposed to brief radio-frequency pulses that disturb the alignment of hydrogen nuclei. The recovery of the nuclei, called spin-lattice relaxation, is used to map the animal's anatomy since tissues differ in their recovery rates. The resolution of small animal scanners is much greater than human clinical scanners. With PET, anatomy is imaged using radio-labeled compounds (Paulus et al., 2001). The animal is placed within a scanner that consists of a ring of detectors. The animal is exposed to a radio-labeled compound, such as carbon-11, that combines with electrons. Gamma rays are produced and measured by the scanner to construct an anatomical map. Small animal scanners have higher resolution than human clinical scanners. The use of these technologies is generally limited to large medical or academic centers where magnets and PET scanners are available.

C. Body Temperature Monitoring

Body temperature can be measured in rats, just as in other species, with ordinary mercury thermometers placed rectally. However, these thermometers require relatively lengthy periods of time to measure temperature and can be hazardous if accidentally broken. Modern electronic thermometers can be used conveniently (Kobayashi et al., 2000; Oprica et al., 2002) without the potential hazards associated with glass mercury thermometers. Larger probe sizes (2 to 4 mm in diameter) are available for large rats while smaller probes (1 to 2 mm diameter) can be used for small rats (Waynforth and Flecknell, 1992). Thermistor probes can also be connected to a chart recorder or computer for continuous real-time monitoring and data capture of body temperature (Halvorson et al., 1992). Finally, probes or sensors can now be implanted subcutaneously or intra-abdominally and connected by a tether system to a recorder or read by a telemetry unit (Waynforth and Flecknell, 1992; Malkinson and Pittman, 1997; Kamerman et al., 2002). Temperature measured by telemetry implants has been shown to be comparable to results obtained with rectal probes (Matthew, 1997). Thus, the telemetry method is particularly useful for measuring body temperature without the potential effects of restraint or anesthesia.

D. Antibody Production

Antibody production has become an important component of many modern research disciplines. Investigators must consider a complex array of issues when contemplating how to best generate antisera for a particular antigen. Issues to be considered in the immunization protocol should include the form of the antigen, quantity of antigen, route of injection, number and distribution of injections, particular adjuvant and its quantity and others (Hanly et al., 1995). For a complete review of the issues and procedures associated with antibody production, the reader is urged to consult Stills (1994) or volume 37, number 3 of the *Institute of Laboratory Animal Resources (ILAR) Journal* (1995).

The rabbit is the most used species for antibody production, with the rat being used less frequently because it does not offer significant advantages over the rabbit. However, the rat is particularly useful for producing IgE antibody (Hirano et al., 2001) and IgG antibody with restricted specificity to mouse proteins (Hanly et al., 1995). The general techniques for antibody production in rats do not differ from those used in other species such as the rabbit. Antigen is usually injected with an adjuvant and antiserum is subsequently collected. Typical injection routes include intravenous, intramuscular, subcutaneous, intraperitoneal and intradermal and the choice is dependent upon choice of adjuvant and character, quantity and volume of the antigen (Hanly et al., 1995). The volume of antigen-adjuvant mixture should be limited to 100 μl per injection site for the commonly used routes such as intradermal and subcutaneous. Immunization schedules involve priming and booster injections that are separated by varying numbers of weeks, depending on the type of antigen mixture injected. The choice of adjuvant is critical to the success of the immunization protocol and the welfare of the immunized animals. Most adjuvants commonly used for research act by creating an antigen depot effect and stimulating the immune system. Many also have surfactant properties that help immune cells and molecules to adhere and contact the antigen.

Freund's adjuvants have historically been the most widely used adjuvants in research immunization protocols. Freund's complete adjuvant (FCA) is a water-in-oil emulsion of mineral oil, a surfactant and heat killed tuberculous organisms or components of the organism (Jennings, 1995). FCA has been the adjuvant of choice in research immunization protocols because of its potent ability to stimulate both humoral and cell-mediated immunity. However, FCA is toxic because the mineral oil is not metabolizable and the tuberculous components can cause severe granulomatous reactions. FCA granulomas are known to be painful in human beings (Jennings, 1995). Freund's incomplete adjuvant (FIA) is less toxic because it lacks the tuberculous components. In general, due to animal welfare concerns, FCA should be used only for weak antigens and only for initial immunizations. FIA should be used for booster immunizations (Jennings, 1995). Alternatives to Freund's adjuvants should always be considered when preparing immunization protocols.

Many alternatives to Freund's adjuvants have been developed. The alternative adjuvants are generally less toxic than Freund's but vary with regard to their ability to enhance the immune response to antigens. The choice of adjuvant depends on the character of the antigen (size, net charge, hydrophilic or hydrophobic properties) and the species chosen (Hanly et al., 1995). The Ribi Adjuvant System® and Hunter's Titermax® are commonly used alternatives to Freund's adjuvants. The Ribi system is an oil-in-water adjuvant with refined mycobacterial components and metabolizable squalene oil. It is less toxic than FCA and is a satisfactory adjuvant for many purposes (Jennings, 1995). Titermax® is a water-in-oil emulsion with a co-polymer and squalene oil. A key feature of this adjuvant is that the co-polymer is stabilized by silica particles and, therefore, can contain a variety of antigens without large amounts of toxic emulsifying agents (Jennings, 1995). Titermax® generally stimulates antibody titers comparable to FCA with less toxicity (Bennett et al., 1992; Jennings, 1995).

Other less frequently used alternatives include Montanide ISA Adjuvants®, Syntex Adjuvant Formulation (SAF)®, GERBU™, aluminum salt adjuvants and liposome entrapped antigens. Montanide ISA Adjuvants® consist of different combinations of oil and surfactant that yield performance similar to FIA with less inflammation (Hanly et al., 1995). Leenaars et al. (1998) reported that Montanide ISA50 induced acceptable antibody titers with fewer pathological changes than Freund's adjuvant. SAF® is an oil-in-water emulsion stabilized by Tween 80 and a co-polymer. It contains squalene and a refined derivative of muramyl dipeptide that is potent with low toxicity. This adjuvant was developed as an alternative to Freund's adjuvants (Hanly et al., 1995). GERBU can induce polyclonal titers that are similar to Freund's adjuvant with less toxicity (Ferber et al., 1999). Other alternatives such as alum or liposome preparations are generally not as effective as other adjuvants and can be difficult to prepare (Hanly et al., 1995).

VII. SELECTION OF ANIMAL MODELS OF DISEASE

A model is "a small copy, an imitation or preliminary representation which serves as the plan from which a final larger object is constructed or formulated;" also, "a model is intended to be copied or followed" (Van Citters, 1973). Animals are used in contemporary biomedical research to model human biological and disease processes. Most animal models are only partially accurate in representing the human condition or disease process.

However, animal models have allowed investigators to learn much of what we know about the human and animal condition and will be required for the advancement of biomedical knowledge for the foreseeable future.

Leader and Padgett (1980) described the basic criteria for a good animal model system. The animal model system should accurately reproduce the disease being studied and should be available to multiple investigators. The model system should be exportable to other laboratories. The species used should be large enough for adequate sample collection and should be readily handled by investigators. The species should survive long enough to be useable and should "fit" into available laboratory animal facilities. A good animal model system should be available in multiple species. Finally, if the disease is genetic, the species chosen should have large litters. The laboratory rat meets these criteria for a vast number of disease model systems (some of which will be discussed in subsequent chapters in this text). Many model systems using rats accurately reproduce the disease under study. Many stocks and strains of laboratory rats are readily available from modern commercial breeders. Rats are large enough for collection of many biological samples and rats from most contemporary stocks and strains can be easily handled by investigators. Most modern animal facilities are well equipped to house large numbers of rats. Thus, the laboratory rat is obviously a good choice for an animal model system.

Animal models of disease can be classified two ways. A spontaneous disease of animals may be used to model a related human disease. Spontaneous models often resemble the counterpart human disease more closely than experimentally induced models. Animal models of disease can be experimentally induced. These models can be produced surgically, genetically, chemically or biologically. Chemical induction of diabetes mellitus and neoplasia has been classically used in the laboratory rat. The literature contains references too numerous to count regarding these sorts of models using the laboratory rat.

Perhaps the most important induced models of disease today involve transgenic and other genetic engineering applications. The mouse is the preferred species for genetic manipulations in contemporary biomedical research. The laboratory rat is currently rarely used for genetic manipulations. However, the rat has many desirable biological features (larger size, more human-like physiological responses to disease, etc.) that may warrant its use rather than a mouse model (Charreau et al., 1996; Bolon et al., 2000). Like mice, rats are also small enough and breed efficiently enough to be advantageous with regard to the cost and logistics of housing large numbers of genetically manipulated animals in modern animal facilities. Thus, the laboratory rat may be the next logical choice of species for genetic manipulation on a large scale.

VIII. STATISTICALLY APPROPRIATE ANIMAL NUMBERS

Federal regulations and guidelines require that appropriate numbers of animals be used in biomedical experiments. Where appropriate, the number of animals used should be determined using rigorous statistical procedures. Other factors such as ethical and economic considerations may also help determine the appropriate sample size. This section will briefly describe statistical and nonstatistical factors associated with determining appropriate animal sample sizes. It is not within the scope of this section to consider all statistical tests associated with calculating the appropriate numbers of animals to be used in a given experiment.

The number of animals to be used in a given experiment is often determined by statistical method. The appropriate sample size is dependent on the nature of the experiment and the test statistic chosen for data analysis. The statistical test of the data should ideally be chosen prior to conducting the work (Mann et al., 1991). Several parameters such as alpha error, beta error, effect size and population variability should be considered when determining appropriate sample size.

Alpha error is the error associated with concluding that a difference between experimental and control populations exists when there is no difference. It is the error made when rejecting the null hypothesis when it is, in fact, true (Jones, 2000; Norton and Strube, 2001). The alpha probability or p value is often set arbitrarily at 0.05 or 0.01 for biomedical experiments. In general, the smaller the alpha error, the larger the sample size must be to detect a difference between experimental and control populations (Mann et al., 1991). Beta error is the error associated with concluding that a difference between experimental and control populations does not exist when there is a difference. It is the error made when accepting the null hypothesis when it is, in fact, false (Jones, 2000; Norton and Strube, 2001). Beta error is related to the power of the study. Power = 1-β and is the probability that a difference that exists will be detected (Mann et al., 1991; Norton and Strube, 2001). In general, higher power studies require a larger sample size and this relationship is greater if the effect size is small. The effect size is the actual difference between experimental and control populations with regard to the parameters being measured (Mann et al., 1991). A larger effect size generally requires a smaller sample size and, conversely, a smaller effect size requires a larger sample size. Finally, some estimate of the variability in the populations (e.g., standard deviation) is required to determine the appropriate sample size. Increased sample size generally reduces standard deviation and, therefore, variability.

Minimum sample size can be estimated if the previously discussed parameters are known. As stated earlier, the alpha error is often arbitrarily set at 0.05 or less, since these values are generally accepted within the biomedical research community. Reasonable values for power are 0.7 to 0.9 (Mann et al., 1991). Effect size and variability can be estimated from previous, similar experiments or from pilot studies. Investigators should choose a clinically or biologically relevant effect size since not all statistically significant differences between experimental and control groups are meaningful. These parameters can then be used in published calculation formulas, sample size tables or nomograms, depending on the specific statistical test chosen.

Other factors may influence how the appropriate number of animals for a particular experiment is determined. Concerns for the welfare of the animals on study may dictate that the number of animals used is altered from the statistical estimate. For example, studies involving significant pain or distress may require reduction in sample size. However, to avoid wasting animals, investigators should ensure that the sample size is adequate to test the hypothesis. Economic issues may also influence the number of animals used. Rats are generally less expensive and easier to house than larger species and, therefore, may be used preferentially and in greater numbers than other larger animals. Finally, granting or regulatory agencies may rely on historical precedent to dictate the number of animals to be used.

Contemporary best practices in biomedical research include reducing the numbers of animals used to the minimum necessary. Investigators can reduce the sample size in a number of ways without reducing the power of studies (Mann et al., 1991). The effect size can be increased, if appropriate. Decreasing genetic and environmental variability within and between populations can also help to decrease sample size (Engeman and Shumake, 1993). The use of repeated sampling or sequential testing can also reduce sample size. Improvement of experimental design and focus and scientific collaboration among several investigators from different disciplines using the same set of animals can also help to reduce animal numbers (Engeman and Shumake, 1993). Finally, replacement of animals with *in vitro* or other methods would obviously reduce numbers of animals used.

ACKNOWLEDGMENTS

The author wishes to acknowledge and thank Mary Ballweg, Pat McDonough, and Sandra Thompson for their diligence and invaluable service with literature searches and collection of references for preparing this chapter. The author also wishes to recognize the illustrators from the original edition of this chapter and to thank Kristi Hall for photographic assistance and Michael Taschwer for preparation of new illustrations concerning Decapicone and tether restraint, electronic microchip identification and jugular venipuncture.

REFERENCES

Abe, K., Hidaka, S., Ishibashi, K., Yanabu, M., Kamogashira, K., Itoh, T., and Matsumoto, M. (1987). Developmental changes in the volumes, protein, and some electrolyte concentrations of male and female rat submandibular saliva secreted in response to methoxamine and pilocarpine. *J. Dent. Res.* **66,** 745–750.

Agrelo, C.E., and Miliozzi, J.O. (1974). A technique for repeated blood sampling with transfusion in the conscious rat. *J. Pharm. Pharmacol.* **26,** 207–208.

Akira, S., Sanbuissho, A., Lin, Y.C., and Araki, T. (1993). Acceleration of embryo transport in superovulated adult rats. *Life Sci.* **53,** 1243–1251.

Alam, S.Q., and Alam, B.S. (1984). Lung surfactant and fatty acid composition of lung tissue and lavage of rats fed diets containing different lipids. *Lipids* **19,** 38–43.

American Veterinary Medical Association Panel on Euthanasia. (2000). Report of the AVMA Panel on Euthanasia. *J. Am. Vet. Med. Assoc.* **218,** 669–696.

Anderson, G.W., Lawrence, W.B., Lee, J., and Young, M. (1991). A restraint for ophthalmic examination of unanesthetized rats. *Lab. Anim. Sci.* **41,** 288–289.

Anderson, J.M. (1963). Lingual vein injection in the rat. *Science* **140,** 195.

Anderson, N.F., Delorme, E.J., Woodruff, M.F.A., and Simpson, D.C. (1959). An improved technique for intravenous injection of new-born rats and mice. *Nature (London)* **184,** 1952–1953.

Andrews, R.V., Belknap, R.W., Southard, J., Hess, S., and Strohbehn, R. (1971). Capture and handling of wild Norway rats used for endocrine studies. *J. Mammal.* **52,** 820–823.

Ansai, T., Ikeda, A., Toyoda, Y., and Takahashi, M. (1994). The chronologically defined developmental process of rat preimplantation embryos. *J. Reprod. Dev.* **40,** 33–38.

Armstrong, B.K., and Softly, A. (1966). Prevention of coprophagy in the rat: a new method. *Br. J. Nutr.* **20,** 595–598.

Atherton, R.W., Culver, B., Seitz, J., Khatoon, S., and Gern, W. (1986). Vaginal fluid adenosine 3',5'-cyclic monophosphate (cAMP) in the rat: interaction with sperm cAMP-dependent protein kinase regulatory subunits. *Arch. Androl.* **16,** 215–226.

Baillard, C., Mansier, P., Ennezat, P., Mangin, L., Medigue, C., Swynghedauw, B., and Chevalier, B. (2000). Converting enzyme inhibition normalizes QT interval in spontaneously hypertensive rats. *Hypertension* **36,** 350–354.

Balabaud, C., Saric, J., Gonzalez, P., and Delphy, C. (1981). Bile collection in freely moving rats. *Lab. Anim. Sci.* **31,** 273–275.

Ball, D.J., Argentieri, G., Krause, R., Lipinski, M., Robison, R.L., Stoll, R.E., and Visscher, G.E. (1991). Evaluation of a microchip implant system used for animal identification in rats. *Lab. Anim. Sci.* **41,** 185–186.

Barnes, R.H., Fiala, G., McGehee, B., and Brown, A. (1957). Prevention of coprophagy in the rat. *J. Nutr.* **63,** 489–498.

Bassett, D.J.P., Elbon-Copp, C., Ishii, Y., Barraclough-Mitchell, H., and Yang, H. (2000). Lung tissue neutrophil content as a determinant of ozone-induced injury. *J. Toxicol. Environ. Health* **60,** 513–520.

Baumans, V., ten Berg, R., Bertens, A., Hackbarth, H., and Timmermann, A. (1993). Experimental procedures. *In* "Principles of Laboratory Animal Science." (L. Zutphen, V. Baumans, A. Beynen, eds.), pp. 299–318. Elsevier, Amsterdam.

Beauzeville, C. (1968). Catheterization of the renal artery of rats. *Proc. Soc. Exp. Biol. Med.* **129,** 932–936.

Bellinger, L.L., and Mendel, V.E. (1975). Hormone and glucose responses to serial cardiac puncture in rats. *Proc. Soc. Exp. Biol. Med.* **148,** 5–8.

Bennett, J.P., and Vickery, B.H. (1972). Rats and mice. *In* "Reproduction and Breeding Techniques for Laboratory Animals." (E.S.E. Hafez, ed.), pp. 299–315. Lea & Febiger, Philadelphia, Pennsylvania.

Bennett, B., Check, I.J., Olsen, M.R., and Hunter, R.L. (1992). A comparison of commercially available adjuvants for use in research. *J. Immunol. Methods* **153,** 31–40.

Bernardis, F.D., Molinari, A., Boccanera, M., Strinaro, A., Robert, R., Senet, J., Arancia, G., and Cassone, A. (1994). Modulation of cell surface-associated mannoprotein antigen expression in experimental candidal vaginitis. *Infect. Immun.* **62,** 509–519.

Blandau, R.J., and Jordan, E.S. (1941). A technique of artificial insemination in the white rat. *J. Lab. Clin. Med.* **26,** 1361–1362.

Bodziony, J., Scwille, P.O., and Szcurek, Z. (1985). A method for inducing exocrine atrophy and collecting juice from the nonatrophied pancreas in the rat. *Eur. Surg. Res.* **17,** 292–300.

Bohs, C., Cregor, M., Gunaratna, G., and Kissinger, C. (2000). Culex automated blood sampler part II: managing freely-moving animals and monitoring their activity. *Curr. Sep.* **18,** 147–151.

Bolon, B., Galbreath, E., Sargent, L., and Weiss, J. (2000). Genetic engineering and molecular technology. *In* "The Laboratory Rat." (G.J. Krinke, ed.), pp. 603–634. Academic Press, London.

Bonsall, R.W., Emery, M.S., and Weiss, J.M. (1998). Apparatus permitting tethered laboratory animals to move freely. U.S. Patent No. 5,832,878, Emory University, Atlanta, GA.

Brain, J.D., and Frank, N.R. (1968). Recovery of free cells from rat lungs by repeated washings. *J. Appl. Physiol.* **25,** 63–69.

Brake, S.C. (1979). Procedures for the collection of milk from the rat dam. *Physiol. Behav.* **22,** 795–797.

Brown, A.P., Dinger, N., and Levine, B.S. (2000). Stress produced by gavage administration in the rat. *Contemp. Top. Lab. Anim. Sci.* **39,** 17–21.

Brown, G.M., and Martin, J.B. (1974). Corticosterone, prolactin, and growth hormone responses to handling and new environment in the rat. *Psychosom. Med.* **36,** 241–247.

Bruckner-Kardoss, E., and Wostmann, B.S. (1974). Blood volume of adult germ-free and conventional rats. *Lab. Anim. Sci.* **24,** 633–635.

Burhoe, S.O. (1940). Methods of securing blood from rats. *J. Hered.* **31,** 445–448.

Burke, W.T., Brotherston, G., and Harris, C. (1955). An improved technique for obtaining bone marrow smears from the rat. *Am. J. Clin. Pathol.* **25,** 1226–1228.

Cameron, N.E., Tuck, Z., McCabe, L., and Cotter, M.A. (2001). Effects of the hydroxyl radical scavenger, dimethylthiourea, on peripheral nerve tissue perfusion, conduction velocity and nociception in experimental diabetes. *Diabetologia* **44,** 1161–1169.

Carballada, R., and Esponda, P. (1998). Binding of seminal vesicle proteins to the plasma membrane of rat spermatozoa in vivo and in vitro. *Int. J. Androl.* **21,** 19–28.

Carmichael, E.B., McBurney, R., and Cason, L.R. (1946). A trap with holder for handling vicious laboratory animals such as wild rats. *J. Lab. Clin. Med.* **31,** 365–368.

Carvalho, J.S., Shapiro, R., Hopper, P., and Page, L.B. (1975). Methods for serial study of renin-angiotensin system in the unanesthetized rat. *Am. J. Physiol.* **228,** 369–375.

Charreau, B., Tesson, L., Soulillou, J., Pourcel, C., and Anegon, I. (1996). Transgenesis in rats: technical aspects and models. *Transgenic Res.* **5,** 223–234.

Cheyne, V.D. (1939). A description of the salivary gland of the rat and a procedure for their extirpation. *J. Dent. Res.* **18,** 457–468.

Chipman, J.K., and Cropper, N.C. (1977). A technique for chronic intermittent bile collection from the rat. *Res. Vet. Sci.* **22,** 366–370.

Chodorowicz, M., Brzozowski, J., and Bojarski, R. (1973). Modifications of an apparatus for indirect measurement of blood pressure in the rat. *Acta Physiol. Pol.* **24,** 473–477.

Classen, W. (2000). Behaviour, neurology and electrophysiology. *In* "The Laboratory Rat." (G.J. Krinke, ed.), pp. 419–435. Academic Press, London.

Cocchetto, D.M., and Bjornsson, T.D. (1983). Methods for vascular access and collection of body fluids from the laboratory rat. *J. Pharm. Sci.* **72,** 465–492.

Cohen, A. E., and Oliver, H.M. (1964). Urethral catheterization of the rat. *Lab. Anim. Care* **14,** 471–473.

Consiglio, A.R., and Lucion, A.B. (2000). Technique for collecting cerebrospinal fluid in the cisterna magna of non-anesthetized rats. *Brain Res. Protoc.* **5,** 109–114.

Conybeare, G., Leslie, G.B., Angles, K., Barrett, R.J., Luke, J.S., and Gask, D.R. (1988). An improved simple technique for the collection of blood samples from rats and mice. *Lab. Anim.* **22,** 177–182.

Cooper, P.H., and Stanworth, D.R. (1974). A simple and reproducible method of isolating rat peritoneal mast cells in high yield and purity. *Prep. Biochem.* **4,** 105–114.

Corbin, T.J., and McCabe, J.G. (2002). Strain variation of immature female rats in response to various superovulatory hormone preparations and routes of administration. *Contemporary Topics* **41,** 18–23.

D'Amour, F.E., and Blood, F.R. (1954). "Manual for Laboratory Work in Mammalian Physiology." Univ. of Chicago Press, Chicago, IL.

D'Amour, F.E., Blood, F.R., and Belden, D.A. (1965). The secretion and digestive action of saliva. *In* "Manual for Laboratory Work in Mammalian Physiology." Univ. of Chicago Press, Chicago, IL.

Damas, J. (1994). Pilocarpine-induced salivary secretion, kinin system and nitric oxide in rats. *Arch. Int. Physiol. Biochim. Biophys.* **102,** 103–105.

Datta, S., and Hobson, J.A. (2000). The rat as an experimental model for sleep neurophysiology. *Behav. Neurosci.* **114,** 1239–1244.

Daunicht, W.J. (1985). Painless head holder for unitary recordings from alert rats. *J. Neurosci. Methods* **14,** 301–306.

de Jong, W.H., Timmerman, A., and van Raaij, T.M. (2001). Long-term cannulation of the vena cava of rats for blood sampling: local and systemic effects observed by histopathology after six weeks of cannulation. *Lab. Anim.* **35,** 243–248.

Detweiler, D.K. (1981). The use of electrocardiography in toxicological studies with rats. *In* "The Rat Electrocardiogram in Pharmacology and Toxicology." (R. Budden, D.K. Detweiler and G. Zbinden, eds.), pp. 83–115. Pergamon Press, New York, NY.

Diehl, K., Hull, R., Morton, D., Pfister, R., Rabemampianina, Y., Smith, D., Vidal, J., and van de Vorstenbosch, C. (2001). A good practice guide to the administration of substances and removal of blood, including routes and volumes. *J. Appl. Toxicol.* **21,** 15–23.

Dobson, K.R., Reading, L., Haberey, M., Marine, X., and Scutt, A. (1999). Centrifugal isolation of bone marrow from bone: an improved method for the recovery and quantitation of bone marrow osteoprogenitor cells from rat tibiae and femurae. *Calcif. Tissue Int.* **65,** 411–413.

Drum, D.E. (1963). Simple technique for direct cannulation of rat salivary ducts. *J. Dent. Res.* **42,** 892.

Dudzinski, D.S., and Cutler, R.W. (1974). Spinal subarachnoid perfusion in the rat: glycine transport from spinal fluid. *J. Neurochem.* **22,** 355–361.

Elcock, L.E., Stuart, B.P., Wahle, B.S., Hoss, H.E., Crabb, K., Millard, D.M., Mueller, R.E., Hastings, T.F., and Lake, S.G. (2000). Tumors in chronic rat studies associated with microchip animal identification devices. *Tox. Path.* **28,** 860.

Ely, D.R., Dapper, V., Marasca, J., Correa, J.B., Gamaro, G.D., Xavier, M.H., Michalowski, M.B., Catelli, D., Rosat, R., Ferreira, M.B., and Dalmaz, C. (1997). Effect of restraint stress on feeding behavior of rats. *Physiol. Behav.* **61,** 395–398.

Emlen, J.T. (1944). Device for holding live wild rats. *J. Wildl. Manage.* **8,** 264–265.

Engel, W., and Kreutz, R. (1973). Lactate dehydrogenase isoenzymes in the mammalian egg-investigations by micro disc electrophoresis in 15 species of the orders Rodentia, Lagomorpha, Carnivora, Artiodactyla and in man. *Humangenetik* **19,** 253–260.

Engeman, R.M., and Shumake, S.A. (1993). Animal welfare and the statistical consultant. *Am. Statistic.* **47,** 229–233.

Epstein, A.N., and Teitelbaum, P. (1962). A watertight swivel joint permitting chronic injection into moving animals. *J. Appl. Physiol.* **17,** 171–172.

Evans, C.S., Smart, J.L., and Stoddart, R.C. (1968). Handling methods for wild house mice and wild rats. *Lab. Anim.* **2,** 29–34.

Fall, M.W. (1974). The use of red light for handling wild rats. *Lab. Anim. Sci.* **24,** 686–687.

Faulkner, K.D. (1975). A restraining device for irradiating rats. *Aust. Dent. J.* **February,** 19–21.

Ferber, P.C., Ossent, P., Homberger, F.R., and Fischer, R.W. (1999). The generation of monoclonal antibodies in mice: influence of adjuvants on the immune response, fusion efficiency and distress. *Lab. Anim.* **33,** 334–350.

Ferrill, H.W. (1943). A simplified method for feeding rats. *J. Lab. Clin. Med.* **28,** 1624.

Flecknell, P.A. (1987). Non-surgical experimental procedures. *In* "Laboratory Animals: An Introduction for New Experimenters." (A.A. Tuffery, ed.), pp. 225–260. Wiley-Interscience, Chichester, West Sussex.

Flecknell, P.A. (1991). Small mammals. *In* "Practical Animal Handling." (R.S. Anderson and A.T. Edney, eds.), pp. 177–187. Pergamon Press, Oxford.

Fluttert, M., Dalm, S., and Oitzl, M.S. (2000). A refined method for sequential blood sampling by tail incision in rats. *Lab. Anim.* **34,** 372–378.

Frank, W.L., Stuhldreher, D., Muchnik, S., Ray, P., and Guinan, P. (1992). A new technique for the cannulation of the rat thoracic duct. *Lab. Anim. Sci.* **42,** 526–527.

Franklin, G.M., Dudzinski, D.S., and Cutler, R.W. (1975). Amino acid transport into the cerebrospinal fluid of the rat. *J. Neurochem.* **24,** 367–372.

Frankmann, S.P. (1986). A technique for repeated sampling of CSF from the anesthetized rat. *Physiol. Behav.* **37,** 489–493.

Friedel, R., Trautschold, T., Gartner, K., Helle-Feldman, M., and Gaudssuhm, D. (1974). Effects of blood sampling on enzyme activities in the serum of laboratory animals. *Z. Clin. Chem. Klin. Biochem.* **12,** 229.

Ganis, F.M. (1962). Convenient mouse and rat holder for use in various laboratory procedures. *J. Lab. Clin. Med.* **60,** 354–356.

Geller, L.M., and Geller, E.H. (1966). A simple technique for the permanent marking of newborn albino rats. *Psychol. Rep.* **18,** 221–222.

Gershbein, L.L., Dan, T.C., and Shurrager, P.S. (1975). Glycolytic enzyme levels of intraocular fluids and lens as compared to the respective sera of various animal species. *Enzyme* **20,** 165–177.

Gibson, J.E., and Becker, B.A. (1967). The administration of drugs to one day old animals. *Lab. Anim. Care* **17,** 524–527.

Gill, G.W. (1976). Principles and practice of cytopreparation. *In* "Handbook of Laboratory Animal Science." (E.C. Melby, and N. H. Altman, eds.), pp. 519–552. CRC Press, Cleveland, Ohio.

Glockner, R., and Karge, E. (1991). Influence of chronic stress before and/or during gestation on pregnancy outcome of young and old Uje:WIST rats. *J. Exp. Anim. Sci.* **34,** 93–98.

Gohd, R., Hubbard, J.E., and Pendleton, F.M. (1974). A simple prefabricated electrode connector unit for chronic rat EEG studies. *Physiol. Behav.* **12**, 1097–1099.

Golba, S., Golba, M., and Wilczok, T. (1974). The effect of trauma, in the form of intraperitoneal injections or puncture of the orbital venous plexus on peripheral white blood cell count in rats. *Acta Physiol. Pol.* **25**, 339–345.

Gonzalez, J.J., Cordero, J.J., Feria, M., and Pereda, E. (2000). Detection and sources of nonlinearity in the variability of cardiac R-R intervals and blood pressure in rats. *Am. J. Physiol.* **279**, H3040–H3046.

Gouldsborough, I., and Ashton, N. (1998). Milking procedure alters the electrolyte composition of spontaneously hypertensive and normotensive rat milk. *Physiol. Behav.* **63**, 883–887.

Grazer, F.M. (1958). Technic for intravascular injection and bleeding of newborn rats and mice. *Proc. Soc. Exp. Biol. Med.* **99**, 407–409.

Greene, F.E., and Wade, A.E. (1967). A technique to facilitate sublingual vein injection in the rat. *Lab. Anim. Care* **17**, 604–606.

Greener, Y., and Gillies, B.A. (1983). Chronic intraperitoneal administration of fluids in rats. *Lab. Anim. Sci.* **33**, 175–176.

Grice, H.C. (1964). Methods of obtaining blood and for intravenous injections in laboratory animals. *Lab. Anim. Care* **14**, 483–493.

Guhad, F.A., and Hau, J. (1996). Salivary IgA as a marker of social stress in rats. *Neurosci. Lett.* **216**, 137–140.

Gupta, B.N. (1973). Technic for collecting blood from neonatal rats. *Lab Anim. Sci.* **23**, 559–561.

Guiol, C., Ledoussal, C., and Surge, J-M. (1992). A radiotelemetry system for chronic measurement of blood pressure and heart rate in the unrestrained rat validation of the model. *J. Pharmacol. Toxicol. Meth.* **28**, 99–105.

Guzman, R., Lovblad, K.O., Meyer, M., Spenger, C., Schroth, G., and Widmer, H.R. (2000). Imaging rat brain on a 1.5 T clinical MR-scanner. *J. Neurosci. Methods.* **97**, 77–85.

Halvorson, I., Zimmerman, R., and Thornhill, J. (1992). Computer interface and software analysis for multi-channel, on-line temperature recording to any micro-computer. *Comput. Biol. Med.* **22**, 83–95.

Hanly, W.C., Artwohl, J.E., and Bennett, B.T. (1995). Review of polyclonal antibody production procedures in mammals and poultry. *ILAR J.* **37**, 93–118.

Hanson, H.M., and Brodie, D.A. (1960). Use of the restrained rat technique for study of the antiulcer effect of drugs. *J. Appl. Physiol.* **15**, 291–294.

Harkness, J.E., and Ridgway, M.D. (1980). Chromodacryorrhea in laboratory rats (Rattus norvegicus): etiologic considerations. *Lab. Anim. Sci.* **30**, 841–844.

Hayashi, S, and Sakaguchi, T. (1975). Capillary tube urinalysis for small animals. *Lab. Anim. Sci.* **25**, 781–782.

Hayton, S.M., Kriss, A., and Muller, D.P.R. (1999). Comparison of the effects of four anaesthetic agents on somatosensory evoked potentials in the rat. *Lab. Anim.* **33**, 243–251.

Heitmeyer, S.A., and Powers, J.F. (1992). Improved method for bile collection in unrestrained conscious rats. *Lab. Anim. Sci.* **42**, 312–315.

Hellekant, G., and Hagstrom, E.C. (1974). Efferent chorda tympani activity and salivary secretion in the rat. *Acta Physiol. Scand.* **90**, 533–543.

Hellekant, G., and Kasahara, Y. (1973). Secretory fibres in the trigeminal part of the lingual nerve to the mandibular salivary gland of the rat. *Acta Physiol. Scand.* **89**, 198–207.

Hem, A., Smith, A., and Solberg, P. (1998). Saphenous vein puncture for blood sampling of the mouse, rat, hamster, gerbil, guinea pig, ferret and mink. *Lab. Anim.* **32**, 364–368.

Hirano, T., Kawasaki, N., Miyataka, H., and Satoh, T. (2001). Wistar strain rats as the model for IgE antibody experiments. *Biol. Pharm. Bull.* **24**, 962–963.

Holloway, P.J., and Williams, R.A.D. (1965). A study of the oral secretions of rats stimulated by pilocarpine. *Arch. Oral Biol.* **10**, 237–244.

Horst, P., Bauer, M., Veelken, R., and Unger, T. (1988). A new method for collecting urine directly from the ureter in conscious unrestrained rats. *Renal Physiol. Biochem.* **11**, 325–331.

Horton, M.L., Olson, C.T., and Hobson, D.W. (1986). Femoral venipuncture for collection of multiple blood samples in the non-anesthetized rat. *Am. J. Vet. Res.* **47**, 1781–1782.

Hoy, P.A., and Adolph, E.F. (1956). Diuresis in response to hypoxia and epinephrine in infant rats. *Am. J. Physiol.* **187**, 32–40.

Hudson, L.C., Hughes, C.S., Bold-Fletcher, N.O., and Vaden, S.L. (1994). Cerebrospinal fluid collection in rats: modification of a previous technique. *Lab. Anim. Sci.* **44**, 358–361.

Hull, R.M. (1995). Guideline limit volumes for dosing animals in the preclinical stage of safety evaluation. *Hum. Exp. Toxicol.* **14**, 305–307.

Hundal, R.S., Kaur, C., and Mangat, H.K. (1978). Improved technique for electroejaculation in albino rats. *Indian J. Exp. Biol.* **16**, 1182–1184.

Hurwitz, A. (1971). A simple method for obtaining blood samples from rats. *J. Lab. Clin. Med.* **78**, 172–174.

Inglis, I.R., and Hudson, A. (1999). Wild rats and mice. *In* "The UFAW Handbook on the Care and Management of Laboratory Animals, Seventh Edition, Vol. I." (T. Poole, ed.), pp. 274–281. Blackwell Science Ltd, Oxford.

Iwaki, S., Matsuo, A., and Kast, A. (1989). Identification of newborn rats by tattooing. *Lab. Anim.* **23**, 361–364.

Jakab, F., Sugar, I., and Szabo, G. (1980). Cannulation of the cervical thoracic duct in the rat. *Lymph.* **13**, 184–185.

Jennings, V.M. (1995). Review of selected adjuvants used in antibody production. *ILAR J.* **37**, 119–125.

Johansen, E. (1952). A new technique for oral examination of rodents. *J. Dent. Res.* **31**, 361–365.

John, R. (1941). An improved electrically-heated operating board for small animals. *J. Physiol. (London)* **99**, 157–160.

Joint Working Group on Refinement. (1993). Removal of blood from laboratory mammals and birds. *Lab. Anim.* **27**, 1–22.

Jolly, D., and Vezina, P. (1996). In vivo microdialysis in the rat: low cost and low labor construction of a small diameter, removable, concentric-style microdialysis probe system. *J. Neurosci. Methods* **68**, 259–267.

Jones, J.B. (2000). Research fundamentals: statistical considerations in research design: a simple person's approach. *Acad. Emerg. Med.* **7**, 194–199.

Kamerman, P., Mitchell, D., and Laburn, H. (2002). Circadian variation in the effects of nitric oxide synthase inhibitors on body temperature, feeding and activity in rats. *Eur. J. Physiol.* **443**, 609–616.

Kapicioglu, Z., Kalyoncu, I.N., Deger, O., and Can, G. (1998). Effect of a somatostatin analogue (SMS 201–995) on tear secretion in rats. *Int. Ophthalmol.* **22**, 43–45.

Kaufmann, P.G., Bennett, P.B., and Farmer, J.C. (1977). Cerebellar and cerebral electroencephalogram during the high pressure nervous syndrome (HPNS) in rats. *Undersea Biomed. Res.* **4**, 391–402.

Kesel, H. (1964). A simple aid in the intubation of small animals. *Lab. Anim. Care* **14**, 499–500.

Khosho, F.K., Kaufmann, R.C., and Amankwah, K.S. (1985). A simple and efficient method for obtaining urine samples from rats. *Lab. Anim. Sci.* **35**, 513–514.

Kirk, K.W., Araujo, F.G., and Rich, S.T. (1997). Customized rat restraint. *Contemp. Topics* **36**, 75–76.

Kissinger, C., Hilt, R., Cadle, K., and Bohs, C. (2003). Animal housing options for the Culex® automated blood sampler. *Current Separations* **20**, 103–104.

Klinefelter, G.R., Gray, L.E., and Suarez, J.D. (1991). The method of sperm collection significantly influences sperm motion parameters following ethane dimethanesulfonate administration in the rat. *Reprod. Toxicol.* **5**, 39–44.

Klinger, W., Kersten, L., and Melhorn, G. (1965). The influence of blood taking technique on the high blood sugar level and the total liver content of glucose in Wistar rats and Agnes-Bluhm mice (Jena) fixation of normal values. *Z. Versuchstierkd.* **6**, 35–47.

Kobayashi, T., Tamura, M., Hayashi, M., Katsuura, Y., Tanabe, H., Ohta, T., and Komoriya, K. (2000). Elevation of tail skin temperature in ovariectomized rats in relation to menopausal hot flushes. *Am. J. Physiol.* **278**, R863–R869.

Kosugi, T., Morimitsu, T., Matsuo, O., and Mihara, H. (1980). Fibrinolytic enzyme in tracheobronchial secretion of rats: existence of plasminogen activator. *Laryngo.* **90**, 1045–1051.

Knigge, K.M., and Joseph, S.A. (1974). Thyrotrophin releasing factor (TRF) in cerebrospinal fluid of the third ventricle of rat. *Acta Endocrinol. Copenhagen* **76**, 209–213.

Koeslag, D., Humphreys, A.S., and Russell, J.C. (1984). A technique for long-term venous cannulation in rats. *J. Appl. Physiol.* **57**, 1594–1596.

Kohno, K., Back, T., Hoehn-Berlag, M., and Hossmann, K-A. (1995). A modified rat model of middle cerebral artery thread occlusion under electrophysiogical control for magnetic resonance investigations. *Magn. Reson. Imaging* **13**, 65–71.

Kondo, S., Najm, I., Kunieda, T., Perryman, S., Yacubova, K., and Luders, H.O. (2001). Electroencephalographic characterization of an adult rat model of radiation-induced cortical dysplasia. *Epilepsia* **10**, 1221–1227.

Kononov, A., Browne, E.Z., Alexander, F., and Porvasnik, S. (1994). Continuous rat intravenous infusion. *Microsurgery* **15**, 443–445.

Koolhaas, J.M. (1999). The laboratory rat. *In* "The UFAW Handbook on the Care and Management of Laboratory Animals, Seventh Edition, Vol. I." (T. Poole, ed.), pp. 313–330. Blackwell Science Ltd, Oxford.

Kornblum, H.I., Araujo, D.M., Annala, A.J., Tatsukawa, K.J., Phelps, M.E., and Cherry, S.R. (2000). In vivo imaging of neuronal activation and plasticity in the rat brain by high resolution positron emission tomography (microPET). *Nature Biotech.* **18**, 655–660.

Kuwahara, M., Yayou, K., Ishii, K., Hashimoto, S., Tsubone, H., and Sugano, S. (1994). Power spectral analysis of heart rate variability as a new method for assessing autonomic activity in the rat. *J. Electrocardiol.* **27**, 333–337.

Lag, M., Becher, R., Samuelsen, J.T., Wiger, R., Refsnes, M., Huitfeldt, H.S., and Schwarze, P.E. (1996). Expression of CYP2B1 in freshly isolated and proliferating cultures of epithelial rat lung cells. *Exp. Lung Res.* **22**, 627–649.

Lambert, R. (1965). Surgery of the bile and pancreatic ducts. In "Surgery of the Digestive System in the Rat." pp. 107–170. Charles C. Thomas, Springfield, Illinois.

Lawson, R.L., Barranco, S., and Sorensen, A.M. (1966). A device to restrain the mouse, rat, hamster, and chinchilla to facilitate semen collection and other reproductive studies. *Lab. Anim. Care* **16**, 72–79.

Lax, E.R., Militzer, K., and Trauschel, A. (1983). A simple method for oral administration of drugs in solid form to fully conscious rats. *Lab. Anim.* **17**, 50–54.

Leader, R.W., and Padgett, G.A. (1980). The genesis and validation of animal models. *Am. J. Pathol.* **101**, S11–S16.

Leal, M., Carson, J.H., Bidanset, J.H., Balkon, J., Barletta, M., and Hyland, M.D. (1988). A method to obtain maternal-fetal plasma samples using a microsampling technique in the rat: transplacental passage of cefoxitin. *Reprod. Toxicol.* **1**, 111–116.

Leenaars, P.P.A.M., Koedam, M.A., Wester, P.W., Baumans, V., Claassen, E., and Hendricksen, C.F.M. (1998). Assessment of side effects induced by injection of different adjuvant/antigen combinations in rabbits and mice. *Lab. Anim.* **32**, 387–406.

Lehr, D. (1945). Stomach tube feeding of small laboratory animals. *J. Lab. Clin. Med.* **30**, 977–980.

Lester, D.S., Lyon, R.C., McGregor, G.N., Engelhardt, R.T., Schmued, L.C., Johnson, G.A., and Johannessen, J.N. (1999). 3–dimensional visualization of lesions in rat brain using magnetic resonance imaging microscopy. *NeuroReport* **10**, 737–741.

Lewis, A., Long, A., and Griffiths, R. (1989). New versatile restraining device for use with rats in pharmacokinetic or pharmacological studies involving chronically implanted cannulae or electrodes. *J. Pharm. Meth.* **22**, 59–63.

Linden, R.D., Zhang, Y., Burke, D.A., Hunt, M.A., Harpring, J.E., and Shields, C.B. (1999). Magnetic motor evoked potential monitoring in the rat. *J. Neurosurg.* **91**, 205–210.

Liu, J.Y., Diaz, T.G., Vadgama, J.V., and Henry, J.P. (1996). Tail sectioning: a rapid and simple method for repeated blood sampling of the rat for corticosterone determination. *Lab. Anim. Sci.* **46**, 243–245.

Long, J.A., and Evans, H.M. (1922). The estrus in the rat and its associated phenomena. *Mem. Univ. Calif.* **6**, 1–148.

Luo, Y.S, Luo, Y.L., Ashford, E.B., Morin, R.R., White, W.J., and Fisher, T.F. (2000). Comparison of catheter lock solutions in rats. *Contemporary Topics* **39**, 83–84.

Lushbough, C.H., and Moline, S.W. (1961). Improved terminal bleeding method. *Proc. Anim. Care Panel* **11**, 305–308.

Macedo-Sobrinho, B., Roth, G., and Grellner, T. (1978). Tubular device for intraoral examination of rodents. *Lab. Anim.* **12**, 137–139.

Mahl, A., Heining, P., Ulrich, P., Jakubowski, J., Bobadilla, M., Zeller, W., Bergmann, R., Singer, T., and Meister, L. (2000). Comparison of clinical pathology parameters with two different blood sampling techniques in rats: retrobulbar plexus versus sublingual vein. *Lab. Anim.* **34**, 351–361.

Malkinson, T.J., and Pittman, Q.J. (1997). Temperature treck the next generation in data analysis. *Annals New York Acad. Sci.* **813**, 230–232.

Mandavilli, U., Schmidt, J., Rattner, D.W., Watson, W.T., and Warshaw, A.L. (1991). Continuous complete collection of uncontaminated urine in conscious rodents. *Lab. Anim. Sci.* **41**, 258–261.

Mann, M.D., Crouse, D.A., and Prentice, E.D. (1991). Appropriate animal numbers in biomedical research in light of animal welfare considerations. *Lab. Anim. Sci.* **41**, 6–14.

Marti, O., Marti, J., and Armario, A. (1994). Effects of chronic stress on food intake in rats: influence of stressor intensity and duration of daily exposure. *Phys. Behav.* **55**, 747–753.

Martinez, J.R., Adshead, P.C., Quissell, D.O., and Barbero, G.J. (1975). The chronically reserpinized rat as a possible model for cystic fibrosis. II. Comparison and cilioinhibitory effects of submaxillary saliva. *Pediatr. Res.* **9**, 470–475.

Martinez, J.R., and Camden, J. (1983). Volume and composition of pilocarpine- and isoproterenol-stimulated submandibular saliva of early postnatal rats. *J. Dent. Res.* **62**, 543–547.

Matthew, C.B. (1997). Telemetry augments the validity of the rat as a model for heat acclimation. *Ann. N. Y. Acad. Sci.* **813**, 233–238.

McGuill, M.W., and Rowan, A.N. (1989). Perspectives on animal use; biological effects of blood loss: implications for sampling volumes and techniques. *ILAR News* **31**, 5–18.

Menaker, L., Sheetz, J.H., Cobb, C.M., and Navia, J. (1974). Gel electrophoresis of whole saliva and associated with histologic changes in submandibular glands of isoproterenol-treated rats. *Lab. Invest.* **30**, 341–349.

Meulemans, A., Vicart, P., Mohler, J., Vulpillat, M., and Pocidalo, J.J. (1986). Continuous sampling for determination of pharmacokinetics in rat cerebrospinal fluid. *Antimicrob. Agents Chemo.* **30**, 888–891.

Miller, S.A., and Dymsza, H.A. (1973). Artificial feeding of neonatal rats. *Science* **141**, 517–518.

Monteiro, F., Abraham, M.E., Sahakari, S.D., and Mascarenhas, J.F. (1989). Effect of immobilization stress on food intake, body weight and weights of various organs in rat. *Ind. J. Physiol. Pharmac.* **33**, 186–190.

Moreland, A.F. (1965). Collection and withdrawal of body fluids and infusion techniques. *In* "Methods of Animal Experimentation" (W.I. Gay, ed.), Vol. I, pp. 1–42. Academic Press, New York.

Morton, D.B., Jennings, M, Buckwell, A., Ewbank, R., Godfrey, C., Holgate, B., Inglis, I., James, R., Page, C., Sharman, I., Verschoyle, R., Westfall, L., and Wilson, A.B. (2001). Refining procedures for the administration of substances. *Lab. Anim.* **35**, 1–41.

Moslen, M.T., Kanz, M.F., Bhatia, J., and Catarau, E.M. (1988). Two cannula method for parenteral infusion and serial blood sampling in the freely moving rat. *J. Parent. Ent. Nutr.* **12**, 633–638.

Mulder, G.J., Scholtens, E., and Meijer, D.K. (1981). Collection of metabolites in bile and urine from the rat. *In* "Methods in Enzymology: Detoxification and Drug Metabolism: Conjugation and Related Systems." (S.P. Colowick and N.O. Kaplan, eds.), Vol. 77, pp. 21–30. Academic Press, New York.

Murgas, K., Oprsalova, Z., and Dobrakovova, M. (1988). Stress free administration of drugs by intraperitoneal cannulation in small laboratory animals. *Endocrinologia Experimentalis* **22**, 281–284.

Nagel, J.D., Kort, W.J., Varossieau, F., and McVie, J.G. (1989). A new method of sampling ascitic fluid from rats. *Lab. Anim.* **23**, 197–199.

Nagler, R.M., Baum, B.J., and Fox, P.C. (1993). Effects of irradiation on the function of rat salivary glands at 3 and 40 days. *Rad. Res.* **136**, 392–396.

Nahas, K., Provost, J., Baneux, P., and Rabemampianina, Y. (2000). Effects of acute blood removal via the sublingual vein on haemotological and clinical parameters in Sprague-Dawley rats. *Lab. Anim.* **34**, 362–371.

Nakazono, T., Murakami, T., Sakai, S., Higashi, Y., and Yata, N. (1992). Application of microdialysis for study of caffeine distribution into brain and cerebrospinal fluid in rats. *Chem. Pharm. Bull.* **40**, 2510–2515.

Nashed, N. (1975). A technic for the collection of peritoneal cells from lab animals. *Lab. Anim. Sci.* **25**, 225–227.

National Research Council. (1996). "Guide for the Care and Use of Laboratory Animals." National Academy Press, Washington, D.C.

Nebendahl, K. (2000). Routes of administration. *In* "The Laboratory Rat." (G.J. Krinke, ed.), pp. 463–483. Academic Press, London.

Nelson, N.S., and Hoar, R.M. (1969). A small animal balling gun for oral administration of experimental compound. *Lab. Anim. Care* **19**, 871–872.

Nerenberg, S.T., and Zedler, P. (1975). Sequential blood samples from the tail vein of rats and mice obtained with modified Liebig condenser jackets and vacuum. *J. Lab. Clin. Med.* **85**, 523–526.

Nicholas, J.S. (1967). Experimental methods and rat embryos. *In* "The Rat in Laboratory Investigation" (J.Q. Griffith and E.J. Farris, eds.), pp. 51–67. Hafner, New York.

Nightingale, C.H., and Mouravieff, M. (1973). Reliable and simple method of intravenous injection into the laboratory rat. *J. Pharm. Sci.* **62**, 860–861.

Nishihira, T., Komatsu, H., Endo, Y., Shineha, R., Sagawa, J., Nakano, T., Hoshino, A., Yoshida, K., and Mori, S. (1994). New technology for continuous infusion via the central and portal veins in the rat. *Tohoku J. Exp. Med.* **173**, 275–282.

Nobunaga, T., Nabamura, K., and Imamichi, T. (1966). A method for intravenous injection and collection of blood from rats and mice without restraint and anesthesia. *Lab. Anim. Care* **16**, 40–49.

Nolan, T.E., and Klein, H.J. (2002). Methods in vascular infusion biotechnology in research with rodents. *ILAR J.* **43**, 175–182.

Norton, B.J., and Strube, M.J. (2001). Understanding statistical power. *J. Ortho. Sports Phys. Ther.* **31**, 307–315.

O'Neill, P.J., and Kaufman, L.N. (1990). Effects of indwelling arterial catheters or physical restraint on food consumption and growth patterns of rats: advantages of noninvasive blood pressure measurement techniques. *Lab. Anim. Sci.* **40**, 641–643.

Oates, H.F., and Stokes, G.F. (1974). Renin stimulation caused by blood collection techniques in the rat. *Clin. Exp. Pharmacol. Physiol.* **6**, 495–501.

Occupational Health and Safety in the Care and Use of Research Animals. (1997). National Research Council. National Academy Press, Washington, D.C.

Ohashi, H., Takami, T., and Yugari, Y. (1985). An improved procedure of chronic intravenous infusion in unrestrained rats. *Physiol. Behav.* **35**, 631–635.

Oikawa, T., Yanagimachi, R., and Nicolson, G.L. (1975). Species differences in the lactin binding sites on the zona pellucida of rodent eggs. *J. Reprod. Fertil.* **43**, 137–140.

Omaye, S.T., Skala, J.H., Gretz, M.D., Schaus, E.E., and Wade, C.E. (1987). Simple method for bleeding the unanaesthetized rat by tail venipuncture. *Lab. Anim.* **21**, 261–264.

Onaga, T., Zabielski, R., Mineo, H., and Kato, S. (1993). The temporal coordination of interdigestive pancreatic exocrine secretion and intestinal migrating myoelectric complex in rats. *XXXII Congress Int. Union Physiol. Sci.* **95** 1/P, Glasgow.

Oprica, N., Aronsson, A.F., Post, C., Eriksson, C., Ahlenius, S., Popescu, L.M., and Schultzberg, M. (2002). Effects of α-MSH on kainic acid induced changes in core temperature in rats. *Peptides* **23**, 143–149.

Orbach, J. (1961). Spontaneous ejaculation in rats. *Science* **134**, 1072–1073.

Osborne, B.E. (1973). A restraining device facilitating electrocardiogram recording in rats. *Lab. Anim.* **7**, 185–188.

Oz, M.C., Popilskis, S.J., Morales, A., and Nowygrod, R. (1989). Long term catheterization of the rat ureter. *Lab. Anim. Sci.* **39**, 349–350.

Parker, G.W., and Martin, D.G. (1989). Technique for cardiovascular monitoring in awake tethered rats. *Lab. Anim. Sci.* **39**, 463–467.

Paulus, M.J., Gleason, S.S., Easterly, M.E., and Foltz, C.J. (2001). A review of high-resolution x-ray computed tomography and other imaging modalities for small animal research. *Lab Anim.* **30**, 36–45.

Phillips, W.A., Stafford, W.W., and Stuut, J. (1973). Jugular vein technique for serial blood sampling and intravenous injection in the rat. *Proc. Soc. Exp. Biol. Med.* **143**, 733–735.

Pilgrim, H.I. (1963). Technic for sampling bone marrow from the living mouse. *Blood* **21**, 241–242.

Pinkerton, W., and Webber, M. (1964). A method of injecting small laboratory animals by the ophthalmic plexus route. *Proc. Soc. Exp. Biol. Med.* **116**, 959–961.

Popovic, V., and Popovic, P. (1960). Permanent cannulation of aorta and vena cava in rats and ground squirrels. *J. Appl. Physiol.* **15**, 727–728.

Postnikova, Z.A. (1960). Method of intracardiac injection of newborn mice and rats. *Vopr. Virusol.* **5**, 111–112.

Prazma, J., Orr, J.L., and Kidwell, S.A. (1980). Device for nonstressful restraining of rats and guinea pigs. *Phys. Behav.* **25**, 155–156.

Pull, S.L., and Reynolds, J.M. (1978). The collection of saliva in the new-born rabbit, guinea-pig and rat. *J. Physiol.* **289**, 1P–2P.

Rath, L., and Hutchison, M. (1989). A new method of bile duct cannulation allowing bile collection and re-infusion in the conscious rat. *Lab. Anim.* **23**, 163–168.

Reich, D.J., and Keller, S.E. (1988). An improvement on the Bollman method of restraining and collecting thoracic duct lymph from the rat. *J. Immun. Meth.* **110**, 179–181.

Reigle, R.D., and Bukva, N.F. (1984). A method of restraining rats for intravenous injection using a flexible plastic bag. *Lab. Anim. Sci.* **34**, 497–498.

Reinhardt, W.O. (1945). Rate of flow and cell count of rat thoracic duct lymph. *Proc. Soc. Exp. Biol. Med.* **58**, 123–124.

Renaud, S. (1959). Improved restraint-technique for producing stress and cardiac necrosis in rats. *J. Appl. Physiol.* **14**, 868–869.

Rizzo, A. (1959). A new mouth prop for oral examinations and operative procedures in rodents. *J. Dent. Res.* **38**, 830–832.

Roberts, W.H. (1961). Identification of multi-colored laboratory animals using electronic flash photography. *J. Anim. Tech. Assoc.* **12**, 11–14.

Robinovitch, M.R., and Sreebny, L.M. (1969). Separation and identification of some of the protein components of rat parotid saliva. *Arch. Oral Biol.* **14**, 935–949.

Rodgers, C.T. (1995). Practical aspects of milk collection in the rat. *Lab. Anim.* **29**, 450–455.

Roeling, T.A.P., Hekman, E., Helmer, J., and Veening, J.G. (1993). A new microcannula for injections in rat brains without disturbing social interactions. *Physiol. Behav.* **53**, 1007–1009.

Rolf, L.L., Bartels, K.E., Nelson, E.C., and Berlin, K.D. (1991). Chronic bile duct cannulation in laboratory rats. *Lab. Anim. Sci.* **41**, 486–492.

Rosas-Arellano, M.P., Vergara-Aragon, P., Cartas-Heredia, L., and Guevara-Rojas, A. (1988). Chronic bladder catheterization in the rat. *Physiol. Behav.* **43**, 127–128.

Rubens, D.J., Meadors, A.K., Yee, S., Melega, W.P., and Cherry, S.R. (2001). Evaluation of a stereotactic frame for repositioning of the rat brain in serial positron emission tomography imaging studies. *J. Neurosci. Methods.* **107**, 63–70.

Rusher, D.L., and Birch, R.W. (1975). A new method for rapid collection of blood from rats. *Physiol. Behav.* **14**, 377–378.

Ryan, P.C., Whelan, C.A., and Fitzpatrick, J.M. (1988). The vas deferens count: a new accurate method for experimental measurement of testicular exocrine function. *Eur. Urol.* **14**, 156–159.

Ryer, F.H., and Walker, D.W. (1971). An anal cup for rats in metabolic studies involving radioactive materials. *Lab. Anim. Sci.* **21**, 942–943.

Salem, H., Grossman, M.H., and Bilbey, D.J. (1963). Micro method for intravenous injection and blood sampling. *J. Pharm. Sci.* **52**, 794–795.

Sandiford, M. (1965). Some methods of collecting blood from small animals. *J. Anim. Tech. Assoc.* **16**, 9–14.

Sanvitto, G.L., Azambuja, N.A., and Marques, M. (1987). A technique for collecting cerebrospinal fluid using an intraventricular cannula in rats. *Physiol. Behav.* **41**, 523–524.

Sarlis, N.J. (1991). Chronic blood sampling techniques in stress experiments in the rat—a mini review. *Anim. Technol.* **42**, 51–59.

Sawamura, A., Hashizume, K., Yoshida, K., and Tanaka, T. (2001). Kainic acid-induced substantia nigra seizure in rats: behavior, EEG and metabolism. *Brain Res.* **911**, 89–95.

Scheuermann, S.E., and Lantzsch, H.J. (1982). An improved fecal trap to prevent coprophagy in rats. *Z. Versuchstierk.* **24**, 291–293.

Schierok, H., Markert, M., Pairet, M., and Guth, B. (2000). Continuous assessment of multiple vital physiological functions in conscious freely moving rats using telemetry and a plethysmography system. *J. Pharmacol. Toxicol. Meth.* **43**, 211–217.

Schneyer, C.A., and Hall, H.D. (1965). Comparison of rat saliva evoked by auriculotemporal and pilocarpine stimulation. *Am. J. Physiol.* **209**, 484–488.

Schoenborne, B.M., Schrader, R.E., and Canolty, N.L. (1977). Tattooing newborn mice and rats for identification. *Lab. Anim. Sci.* **27**, 110.

Schreiber, H., and Nettesheim, P. (1972). A new method for pulmonary cytology in rats and hamsters. *Cancer Res.* **32**, 737–745.

Schumacher, S.J. (1983). A device for restraining rats while administering intraperitoneal injections. *Lab. Anim.* **12**, 51.

Schumacher, S.J., Morris, M., and Riddick, E. (1991). Effects of restraint by tether jackets on behavior in spontaneously hypertensive rats. *Clin. Exper. Hyper.* **A13**, 875–884.

Schwarzbauer, C., Syha, J., and Haase, A. (1993). Quantification of regional blood volumes by rapid T1 Mapping. *Magn. Reson. Med.* **29**, 709–712.

Scipioni, R.L., Diters, R.W., Myers, W.R., and Hart, S.M. (1997). Clinical and clinicopathological assessment of serial phlebotomy in the Sprague-Dawley rat. *Lab. Anim. Sci.* **47**, 293–299.

Scott, J. and Berry, M.R. (1989). The effects of chronic ethanol administration on stimulated parotid secretion in the rat. *Alcohol and Alcoholism* **24**, 145–152.

Semczuk, M., Zrubek, H., and Glazowski, H.T. (1977). Some aspects of obtaining rat semen by electroejaculation method. *Acta Physiol. Pol.* **28**, 365–368.

Sgoifo, A., Stilli, D., Medici, D., Gallo, P., Aimi, B., and Musso, E. (1996). Electrode positioning for reliable ECG recordings during social stress in unrestrained rats. *Physiol. Behav.* **60**, 1397–1401.

Shannon, I.L., and Suddick, R.P. (1975). Current research on oral fluids. *J. Dent. Res.* **54**, B8–B18.

Sharp, P.E., and LaRegina, M.C. (1998). "The Laboratory Rat" (M.A. Suckow, ed.). CRC Press, Boca Raton, Florida.

Sholkoff, S.D., Glickman, M.G., and Powell, M.R. (1969). Restraint of small animals for radiopharmaceutical studies. *Lab. Anima. Care* **19**, 662–663.

Skala, J.P., Koldovsky, O., and Hahn, P. (1981). Cyclic nucleotides in breast milk. *Am. J. Clin. Nutr.* **34**, 343–350.

Smith, C.J., and Kelleher, P.C. (1973). A method for the intragastric feeding of neonatal rats. *Lab. Anim. Sci.* **23**, 682–684.

Soltysik, S., Jackson, R., and Jelen, P. (1996). Apparatus for studying behavior and learning in restrained rats. *Acta Neurobiol. Exp.* **56**, 697–701.

Sorg, D.A., and Bruckner, B.A. (1964). A simple method of obtaining venous blood from small laboratory animals. *Proc. Soc. Exp. Biol. Med.* **115**, 1131–1132.

Sponer, G., Schulz, L., and Bartsch, W. (1988). Methods for the measurement of blood pressure in conscious rats. *Contr. Nephrol.* **60**, 220–229.

Sreebny, L.B., and Johnson, D.A. (1969). Diurnal variation in secretory components of the rat parotid glands. *Arch. Oral Biol.* **14**, 397–405.

Stahlin, F.O., Schmid, G., Hempel, K., and Heidland, A. (1978). Technique of continuous collection of parotid saliva in the rat. *Res. Exp. Med.* **172**, 247–253.

Stanton, H.G. (1971). Experimental hypertension. *In* "Methods of Pharmacology." (A. Schwartz, ed.), pp. 125–150. Meredith Corp. New York.

Stevens, M.J., Obrosova, I., Cao, X., Van Huysen, C., and Greene, D.A. (2000). Effects of the DL-α-lipoic acid on peripheral nerve conduction, blood flow, energy metabolism, and oxidative stress in experimental diabetic neuropathy. *Diabetes* **49**, 1006–1015.

Stills, H.F. (1994). Polyclonal antibody production. *In* "The Biology of the Laboratory Rabbit." (P.J. Manning, D.H. Ringler and C.E. Newcomer, eds.), pp. 435–448. Academic Press, San Diego, CA.

Stoodley, M.A., Gutschmidt, B., and Jones, N.R. (1999). Cerebrospinal fluid flow in an animal model of noncommunicating syringomyelia. *Neurosurg.* **44**, 1065–1075.

Struhar, D., Harbeck, R.J., and Mason, R.J. (1989). Lymphocyte populations in lung tissue, broncheolar lavage fluid, and peripheral blood in rats at various times during the development of silicosis. *Am. Rev. Resp. Dis.* **139**, 28–32.

Suber, R.L., and Kodell, R.L. (1985). The effect of three phlebotomy techniques on hematological and clinical chemical evaluation in Sprague-Dawley rats. *Vet. Clin. Path.* **14**, 23–30.

Sunanda, B.S., Shankaranarayana, R., and Raju, T.R. (2000). Chronic restraint stress impairs acquisition and retention of spatial memory task in rats. *Curr. Sci.* **79**, 1581–1584.

Sundberg, R.D., and Hodgson, R.E. (1949). Aspiration of bone marrow in laboratory animals. *Blood* **4**, 557–561.

Suzuki, K., Koizumi, N., Hirose, H., Hokao, R., Takemura, N., and Motoyoshi, S. (1997). Blood sampling technique for measurement of plasma arginine vasopressin concentration in conscious and unrestrained rats. *Lab. Anim. Sci.* **47**, 190–193.

Szabo-Salfay, O., Palhalmi, J., Szatmari, E., Barabas, P., Szilagyi, N., and Juhasz, G. (2001). The electroretinogram and visual evoked potential of freely moving rats. *Brain Res. Bull.* **56**, 7–14.

Takai, N., Yoshido, Y., and Kakudo, Y. (1982). Technique for collection of saliva from the three major salivary glands of individual rats. *J. Osaka Dent. Univ.* **16**, 17–23.

Takeyama, H., Yura, J., Miyaike, H., Ishikawa, M., Mizuno, H., Taniguchi, M., Hanai, T., Mizuno, A., Shinagawa, N., and Kato, F. (1988). A new apparatus for chronic intravenous infusion in unrestrained rats. *J. Parent. Ent. Nutr.* **12**, 93–99.

Tatevossian, A., and Wright, W.G. (1974). The collection and analysis of resting rat saliva. *Arch. Oral Biol.* **19**, 825–827.

Tejwani, G.A., Gudehithlu, K.P., Hanissian, S.H., Gienapp, I.E., Whitacre, C.C., and Malarkey, W.B. (1991). Facilitation of dimethylbenz[*a*]anthracene-induced rat mammary tumorigenesis by restraint stress: role of β-endorphin, prolactin and naltrexone. *Carcinogenesis* **12**, 637–641.

Thompson, J.H. (1966). An adjustable restraining cage for rats. *Lab. Anim. Care* **16**, 520–522.

Thorig, L., van Haeringen, N.J., and Wijngaards, G. (1984). Comparison of enzymes of tears, lacrimal gland fluid and lacrimal gland tissue in the rat. *Exp. Eye Res.* **38**, 605–609.

Timm, K.I. (1989). Orbital venous anatomy of the Mongolian gerbil with comparison to the mouse, hamster and rat. *Lab. Anim. Sci.* **39**, 262–264.

Toon, S., and Rowland, M. (1981). A simple restraining device for chronic pharmacokinetic and metabolism studies in rats. *J. Pharm. Meth.* **5**, 321–323.

Troncoso, E., Rodriguez, M., and Feria, M. (1995). Light-induced arousal affects simultaneously EEG and heart rate variability in the rat. *Neurosci. Lett.* **188**, 167–170.

Udenfriend, S., Bumpus, F.M., Foster, H.L., Freis, E.D., Hansen, C.T., Lovenberg, W.M., and Yamori, Y. (1976). Spontaneously hypertensive rats: guidelines for breeding, care, and use. *ILAR News* **19**, G1–G20.

Ueyama, T., Yoshida, K., and Senba, E. (2000). Stress-induced elevation of the ST segment in the rat electrocardiogram is normalized by an adrenoceptor blocker. *Clin. Exp. Pharmacol. Physiol.* **27**, 384–386.

Ulmansky, M., Sela, J., Dishon, T., Rosenmann, E., and Boss, J.H. (1971). A technique for the intubation of the parotid duct in rats. *Arch. Oral Biol.* **17**, 609–612.

Van Bruggen, N., Busch, E., Palmer, J.T., Williams, S-P., and Crespigny, A.J. (1998). High-resolution functional magnetic resonance imaging of the rat brain: mapping changes in cerebral blood volume using iron oxide contrast media. *J. Cerebral Blood Flow Metab.* **18**, 1178–1183.

Van Citters, R. (1973). The role of animal research in clinical medicine. *In* "Research Animals in Medicine" (L.T. Harmison, ed.). *DHEW Publ. No.* (NIH) (*U.S.*) **72–333**, 3–8.

Van Herck, H., Baumans, V., Brandt, C., Boere, H., Hesp, A., van Lith, H., Schurink, M., and Beynen, A. (2001). Blood sampling from the retro-orbital plexus, the saphenous vein and the tail vein in rats: comparative effects on selected behavioural and blood variables. *Lab. Anim.* **35**, 131–139.

Van Herck, H., Baumans, V., Brandt, C., Hesp, A., Sturkenboom, J., van Lith, H., Tintelen, G., and Beynen, A. (1998). Orbital sinus blood sampling as performed by different animal technicians: the influence of technique and expertise. *Lab. Anim.* **32**, 377–386.

Van Vliet, B.N., Chafe, L.L., Antic, V., Schnyder-Candrian, S., and Montani, J-P. (2000). Direct and indirect methods used to study arterial blood pressure. *J. Pharmacol. Toxicol. Meth.* **44**, 361–373.

Van Wijk, H., Dick, A., Greenough, R.J., Oshodi, R.O., and Robb, D. (2000). Continuous intravenous infusion in athymic (nude) rats: an animal model for evaluating the efficacy of anti-cancer agents. *Lab. Anim.* **34**, 63–69.

Varner, A.E., Sorkness, R.L., Kumar, A., Kaplan, M.R., and Lemanske, R.F. (1999). Serial segmental bronchoalveolar lavage in individual rats. *J. Appl. Physiol.* **87**, 1230–1233.

Vissink, A., 's-Gravenmade, E.J., Konings, A.W.T., and Ligeon, E.E. (1989). An adaptation of the Lashley cup for use in rat saliva collection. *Arch. Oral Biol.* **34**, 577–578.

Vreeburg, J.T., Van Andel, M.V., Kort, W.J., and Westbroek, D.L. (1974). The effect of hemicastration on daily sperm output in the rat as measured by a new method. *J. Reprod. Fert.* **41**, 355–359.

Waalkes, M.P., Rehm, S., Kasprzak, K.S., and Issaq, H.J. (1987). Inflammatory, proliferative, and neoplastic lesions at the site of metallic identification ear tags in Wistar [Crl:(WI) BR] rats. *Cancer Res.* **47**, 2445–2450.

Warden, S.J., Bennell, K.L., McMeeken, J.M., and Wark, J.D. (2000). A technique for restraining rodents during hindlimb interventions. *Contemporary Topics* **39**, 24–27.

Waynforth, H.B., and Flecknell, P.A. (1992). "Experimental and Surgical Technique in the Rat." Academic Press, London.

Weiss, J., Taylor, G.R., Zimmermann, F., and Nebendahl, K. (2000). Collection of body fluids. *In* "The Laboratory Rat." (G.J. Krinke, ed.), pp. 485–510. Academic Press, London.

Westergren, I., and Johansson, B.B. (1991). Changes in physiological parameters of rat cerebrospinal fluid during chronic sampling: evaluation of two sampling methods. *Brain Res. Bull.* **27**, 283–286.

White, W.A. (1971). A technic for urine collection from anesthetized male rats. *Lab. Anim. Sci.* **21**, 401–402.

Widdop, R.E., and Li, X.C. (1997). A simple versatile method for measuring tail cuff systolic blood pressure in conscious rats. *Clin. Sci.* **93**, 191–194.

Wiersma, J., and Kastelijn, J. (1985). A chronic technique for high frequency blood sampling/transfusion in the freely behaving rat which does not affect prolactin and corticosterone secretion. *J. Endocrinol.* **107**, 285–292.

Winkler, T., Sharma, H.S., Stalberg, E., Badgaiyan, R.D., Westman, J., and Nyberg, F. (2000). Growth hormone attenuates alterations in spinal cord evoked potentials and cell injury following trauma to the rat spinal cord. *Amino Acids* **19**, 363–371.

Winsett, O.E., Townsend, C.M., and Thompson, J.C. (1985). Rapid and repeated blood sampling in the conscious rat: a new technique. *Am. J. Physiol.* **249**, G145–G146.

Wolf, R.O., and Kakehashi, S. (1966). Rat parotid saliva collection technique. *J. Dent. Res.* **45**, 979.

Wolfensohn, S., and Lloyd, M. (1994). Procedural data. *In* "Handbook of Laboratory Animal Management and Welfare," pp. 143–173. Oxford University Press, Oxford.

Woodard, G. (1965). Principles in drug administration. *In* "Methods of Animal Experimentation." (W.I. Gay, ed.), Vol. 1, pp. 343–359. Academic Press, New York.

Woodruff, C.R., and Carpenter, F.G. (1973). Actions of catecholamines and physostigmine on rat parotid salivary secretion. *Am. J. Physiol.* **225**, 1449–1453.

Wolf, R.O., and Kakehashi, S. (1966). Rat parotid saliva collection technique. *J. Dent. Res.* **45**, 979.

Wright, B.A. (1970). A new device for collecting blood from rats. *Lab. Anim. Care* **20**, 274.

Wu, S., Suzuki, Y., Kitada, M., Kataoka, K., Kitaura, M., Chou, H., Nishimura, Y., and Ide, C. (2002). New method for transplantation of neurosphere cells into injured spinal cord through cerebrospinal fluid in rat. *Neurosci. Letters* **318**, 81–84.

Wunder, C.C. (1965). Care and growth of animals during chronic centrifugation. *In* "Methods of Animal Experimentation" (W.I. Gay, ed.), Vol. 2. Academic Press, New York.

Xu, Z., and Melethil, S. (1990). Simultaneous sampling of blood, bile, and urine in rats for pharmacokinetic studies. *J. Pharmacol. Meth.* **24**, 203–208.

Yagiela, J.A., and Bilger, P.A. (1986). A custom restraining device for small animals. *Lab. Anim. Sci.* **36**, 303–305.

Yamada, K., Chen, C-J., Satoh, H., Hirota, T., Aoyagi, K., Enkawa, T., Ozaki, Y., Sekiguchi, F., and Furuhama, K. (1997). Magnetic resonance imaging of rat head with a high-strength (4.7 T) magnetic field. *J. Vet. Med. Sci.* **59**, 303–306.

Yoburn, B.C., Morales, R., and Inturrisi, C.E. (1984). Chronic vascular catheterization in the rat: comparison of three techniques. *Physiol. Behav.* **33**, 89–94.

Zabielski, R., Lesniewska, V., and Guilloteau, P. (1997). Collection of pancreatic juice in experimental animals: mini-review of materials and methods. *Reprod. Nutr. Dev.* **37**, 385–399.

Zawieja, D.C., and Barber, B.J. (1987). Lymph protein concentration in initial and collecting lymphatics of the rat. *Am. J. Physiol.* **252**, G602–G606.

Zeilmaker, G.H., and Verhamme, C.M. (1974). Observations on rat oocyte maturation in vitro: Morphology and energy requirements. *Biol. Reprod.* **11**, 145–152.

Zeller, W., Weber, H., Panoussis, B., Burge, T., and Bergmann, R. (1998). Refinement of blood sampling from the sublingual vein of rats. *Lab. Anim.* **32**, 369–376.

Zetterberg, G., Elmberger, G., Johansson, A., Lundahl, J., Lundborg, M., Skold, C.M., Tornling, G., Camner, P., and Eklund, A. (2000). Rat alveolar and interstitial macrophages in the fibrosing stage following quartz exposure. *Hum. Exp. Toxicol.* **19**, 402–411.

Zielke, H.R., Collins, R.M., Baab, P.J., Huang, Y., Zielke, C.L., and Tildon, J.T. (1998). Compartmentation of [C14] glutamate and [C14] glutamine oxidative metabolism in the rat hippocampus as determined by microdialysis. *J. Neurochem.* **71**, 1315–1320.

Chapter 19

Anesthesia and Analgesia

George A. Vogler

THE LABORATORY RAT, 2ND EDITION

I. INTRODUCTION

This chapter is a survey the most common anesthetic and analgesic drugs and techniques in the rat. It is necessarily limited in scope and depth, and the interested reader should consult an earlier book in this series, *Anesthesia and Analgesia in Laboratory Animals* (Kohn et al., 1997), as well as current anesthesia texts and published literature. While the methods and drugs used for anesthesia in the rat are similar to those used in human and veterinary clinical anesthesia, there are some important differences. Some venerable drugs, no longer used in clinical practice, persist in selected research disciplines because of their unique physiological properties. The need for efficient, brief anesthesia of large groups of animals is not unique to research, but the techniques used often are. For Scientific reasons, anesthesia and analgesia of research animals requires greater attention to short and long-term anatomical and physiological consequences than is frequently the case in clinical practice. Despite considerable progress, the problems of pain recognition, and of identifying drugs and methods of administration that are effective and feasible in a research setting, are formidable and persistent. The laboratory rat has been both the subject and beneficiary of more than a century of human and veterinary anesthetic research and practice.

II. PREOPERATIVE CONSIDERATIONS

A. Agent and Technique

There are no ideal anesthetic agents or perfect techniques. When choosing an anesthetic agent and method of administration, compromises are made to meet the needs of the animal and the research goals, and to accommodate available personnel, equipment and economic resources. Considerations include the anticipated duration of anesthesia, intra- and postoperative analgesic requirements, the training and technical expertise of the anesthesia provider, the health and age of the animal, and experimental constraints including the effect of anesthesia on the data.

1. Duration of Anesthesia

Short-acting or easily reversible anesthetics are used for minimally invasive brief procedures, lasting seconds to a few minutes. Inhalation methods are often the preferred choice because of their characteristic rapid onset and recovery from anesthesia, and relative simplicity if appropriate equipment is used. Alternatively, parenteral drugs with inherently short action or drug combinations for which reversal agents are available can also be used. For procedures of moderate duration inhalation anesthesia or any of several parenteral drugs or drug combinations may be satisfactory. Prolonged anesthesia might prompt consideration of inhalation anesthesia, or continuous infusion of parenteral agents, which offer the potential for improved stability of anesthesia, Either approach, or a combination of both, will tend to limit the excursions in drug levels and anesthetic depth that occur with intermittent bolus injections.

2. Procedure Invasiveness and Pain Potential

The invasiveness of the procedure and the expected severity and duration of resulting pain determine the need for additional analgesia before, during and following surgery. Many otherwise satisfactory anesthetics, including all current inhalation agents and many parenteral drugs, possess little or no inherent analgesia. Some analgesics, opioids prominent among them, will reduce the anesthetic dose needed to reach an adequate surgical plane. By reducing the dose of the primary anesthetic, analgesics can mitigate undesirable side effects, such as cardiopulmonary depression, that accompany the higher anesthetic doses required when no concurrent analgesic is used. Systemic analgesics and local anesthetics may also reduce analgesic requirements in the postoperative period. Because some analgesics have a delayed onset of action, early administration allows time for the drugs to act and to provide analgesia in the induction period, during surgery and into the critical recovery and the early postoperative period. The use of nonsteroidal analgesics (NSAIDs) and opioids during and after surgery is discussed below.

Drugs and procedures that would be unacceptable if the animal is to recover consciousness are sometimes suitable for acute, nonrecovery cases. Adequate anesthetic depth and stability, experimental necessity and, for toxic or carcinogenic drugs, personnel safety determine the choices.

3. Training and Expertise

Anesthetic and analgesic choices can be influenced by the expertise of the surgeon and the anesthetist. For the same procedure, the duration of anesthesia, the quality and speed of recovery, and the analgesic requirements can vary greatly among surgeons. For a very good surgeon, a relatively simple anesthetic and analgesic plan may suffice. A less adept surgeon will often require a more intensive anesthetic, analgesic and postoperative care regimen.

Similar factors are at work with respect to skill in anesthesia. An "optimum" anesthetic protocol can require knowledge, equipment and skills not readily available. The choice of anesthetics and methods of administration should fall within the abilities of the anesthetic provider. Older drugs and techniques in the hands of experienced personnel familiar with their effects will sometimes produce a better postoperative outcome than new methods using unfamiliar drugs, techniques and equipment.

In both instances, the importance of education and training is evident.

4. Age and Concurrent Disease

The physical condition and age of the patient also influence the choice of drugs and administration methods. By virtue of altered uptake, distribution, metabolism and elimination, anesthetics and doses that are well tolerated by healthy, naïve adults may be inadequate or lethal in the very young, the very old, and the debilitated patient. Safe anesthesia for ill patients requires knowledge of the physiological consequences of the disease, more intensive supportive care, and selection of agents and techniques that do not further exacerbate the condition. Preoperative steps include correction of fluid deficits and if time permits, providing high-density nutritional support before induction of anesthesia. To avoid undue delay during induction and surgery, the necessary preparations for induction, surgery, and monitoring and supportive measures should be completed before the patient is brought to the surgical area. Particular care should be taken to avoid hypothermia and stressful induction methods, and to provide analgesia, and these precautions should extend to the anesthetic recovery and postoperative periods. Techniques for neonatal rats are described in a following section.

5. Experimental Constraints

The influence of anesthetics on the proposed research must be considered because all anesthetics have effects that extend beyond the immediate goals of unconsciousness, muscle relaxation and analgesia. The problem is most evident when data is collected during the course of anesthesia, and effects on cardiopulmonary function, or other organ systems are immediately apparent. However, some anesthetics can produce persistent changes, including metabolic alterations, such as enzyme induction, or structural alterations in organ systems, such as the central nervous system, that can compromise or invalidate data collected subsequent to apparent recovery. Careful consultation with the researcher to gain a clear understanding of the research goals and methodology is essential to avoid, or to at least identify, such interactions. The point has been well made that a superficial consideration of a compound's effects, or reliance on published research methods, may not result in the optimum choice of anesthetic technique (Flecknell, 1996a). Consultation with colleagues who have experience managing similar studies, as well as with veterinary anesthesiologists experienced in biomedical research can be invaluable.

B. Preoperative Assessment and Preparation

1. Health Status

Spontaneous disease can alter the response to anesthesia and result in unexpected complications, even when a familiar, "safe" anesthetic protocol is used. In addition to confirming the identity and health history of the animals, a few moments spent assessing physical condition and behavior will often avert unpleasant surprises. In this context retrospective health reports are important, but are not a substitute for "real time" preanesthetic assessment. Unusual or unexpected findings should be clarified before the onset of anesthesia. Preanesthetic weight provides an objective guide to injectable drug dose and fluid supplementation, an additional indicator of health, and an essential baseline for subsequent evaluation and treatment.

Landi et al. 1992, have recommended providing a minimum of 3 days for acclimation of newly arrived animals. However, if the experimental protocol will involve radical departures from the usual husbandry routines, such as diet, caging, frequency or method of handling and restraint, acclimating the patient to these conditions before surgery is often beneficial and more time may be needed.

2. Fasting

Rats do not vomit, so preoperative fasting is generally unnecessary, and there is seldom any reason to restrict water. For gastrointestinal surgery, when fasting may be required, the duration should be limited to no longer than 12 hours and steps should also be taken to limit coprophagy (Wixson and Smiler, 1997; Flecknell, 1996b). Claassen, 1994 reviewed the varied and often substantial consequences of food deprivation, many of which affect response to anesthesia and surgery.

3. Preanesthetic Drugs

In larger species, analgesics and anxiolytics are routinely used to minimize the pain and distress associated with induction of anesthesia and to reduce doses of drugs needed for induction and maintenance. Anticholinergics are used to prevent bradycardia and reduce salivation and bronchial secretions caused by some anesthetic and analgesic drugs. In rats, these drugs are often given at the time of anesthetic induction to avoid the need for repeated handling and stress. However, many drugs, including analgesics, antibiotics, and sedative/tranquilizers do not have an immediate onset, and should be administered in advance if they are to be effective during induction and surgery.

For most surgical procedures in rats, prophylactic antibiotics are unnecessary if good surgical practices are followed. For those procedures where antibiotics are

medically indicated, the choice should be based on the efficacy of the drug against the bacteria most likely to cause complications.

4. Medical Records

Medical records are an essential component of adequate medical care. For many of the relatively simple procedures performed on rats, a group record is appropriate. Because procedures and anesthetic methods usually follow a standard protocol, the record need reflect only those factors which vary, such as animal identification, personnel involved, date, time, weight, drug doses administered, and adverse reactions or complications, if any. Depending on the number and intensity of required postprocedural observations, a single well-designed form can encompass all of these needs. More elaborate procedures, needing greater postoperative care, require appropriately more detail and room to record observations, treatments and progress. For common procedures, or those needing unusually intensive observations and specific care, it is sometimes useful to customize the record format, taking into account known complications, assessments and treatments.

In any case, a medical record should be clearly organized and provide sufficient information for appropriate therapeutic decisions. The responsibility for maintaining records will vary with the administrative structure of the organization, and is less important, so long as the records are clear, contain sufficient information to support medical care, and are quickly accessible. If institutional resources permit, computerized record systems may efficiently address all of these needs (American College of Laboratory Animal Medicine, 2004).

C. Intraoperative Monitoring and Support

1. Monitoring

Anesthetized patients should always be monitored. The responsibility for monitoring and the methods used are determined by the length, complexity and risks associated with the procedure. In many cases, the surgeon is responsible, but for complex and demanding procedures, it is often preferable for a well-trained assistant to assume the role of anesthetist.

a. OBSERVATIONAL METHODS. With experience and training, the senses are an important and reliable means of anesthetic monitoring, yielding information concerning respiration, cardiac function, temperature and anesthetic depth.

1. Respiration and Ventilation. An indication of respiratory function can be obtained by assessing respiratory rate, and pattern. These parameters are affected by the depth of anesthesia as well as by the specific anesthetic used, but with experience and careful

observation are useful indicators of function. It is more difficult to assess oxygenation by using the senses. Mucous membrane and extremity color are poor indicators of oxygenation, and can be rendered even less reliable by the tendency of α_2-adrenergics to produce peripheral vasoconstriction.

2. Cardiovascular Function. Heart rate can be assessed by palpation of the thoracic wall over the heart, however the high heart rate of rats usually defeats any attempt to count the pulse. With experience, however, it is possible to gain some indication of strength and relative rate.

3. Anesthetic Depth. Without elaborate equipment and training, anesthetic depth is best judged using physical means and observation. Steffey et al. (2003) found that pupil diameter correlated with anesthetic depth during inhalation anesthesia, but the method is probably not feasible for routine use. Loss of response to aversive stimulation, such as toe or tail-pinch, or a hemostat applied to the abdominal wall, is used as an indicator of adequate depth, as is loss of palpebral, corneal, pinnae and limb withdrawal reflexes. The most rigorous test of adequate anesthesia is absence of response to incision and surgical manipulation; with few exceptions, movement in response to these stimuli indicate inadequate anesthesia. All of these criteria of adequate anesthesia are defeated if paralytic drugs (muscle relaxants) are used.

b. INSTRUMENTAL METHODS. For more longer and more complex procedures, instrumental methods of monitoring offer continuous, specific information. Unlike busy operating room staff, instruments are not subject to distraction or fatigue. Most modern monitors provide data output for recording purposes, and subsequent or real-time analysis. Many, but not all, monitoring instruments made for human or veterinary medical use perform satisfactorily when used with rats. However, advertising claims and sales hyperbole are sometimes based more on extrapolation and anecdotal reports than on documented evidence of performance in rats. Vendors should be willing to provide specific references for users of their products. Ideally, a realistic on-site evaluation should be conducted before purchasing expensive equipment.

Instrumental monitoring is not a panacea. Simply connecting a monitor to the patient does not improve anesthetic care unless the data provided are reliable, correctly interpreted, and acted upon. Further, monitoring instruments are used to supplement, not replace, clinical observation and judgment (Gravenstein 1998). The value of monitoring equipment depends upon commitment, training and experience.

The instruments discussed below are grouped into categories corresponding to their principal applications, but the physiological parameters they measure are interrelated in complex ways, and are not independent, discrete functions. Impending heart failure, for example, is detected not only by ECG and blood pressure, but also by capnography and, finally, pulse oximetry. Numerous human and veterinary anesthesia texts describe the use of monitors in detail, and further information can be found in texts devoted to anesthesia equipment and monitoring (Dorsch and Dorsch, 1999).

i. Respiration and Ventilation. The methods usually used for monitoring in anesthesia are pulse oximetry, capnography, and blood gas determinations.

Pulse oximeters are used to monitor oxygen saturation and heart rate. The pulse oximeter probe contains two light emitting diodes and a photoreceptor to detect and display oxygen saturation as the percent ratio of oxygenated to deoxygenated hemoglobin. Other species of hemoglobin, such as carboxyhemoglobin and methemoglobin, are usually present only in very low levels and not directly measured, but significant elevations will cause inaccurate results. The instrument measures only the ratio of oxygenated to deoxygenated hemoglobin, not the absolute value, and is relatively insensitive to anemia. Thus, the highest possible value is 100%, and is not equivalent to arterial blood oxygen partial pressure (PaO_2). A value below 90% saturation is usually taken as a cause for concern because it correlates with a PaO_2 of approximately 60 mmHg, the point at which a further decrease in saturation can result in a large drop in the partial pressure in oxygen.

Most pulse oximeters have adjustable alarms to indicate acceptable upper and lower heart rates and oxygen saturation levels, as well as some indication of adequate signal quality. These alarms can be silenced for brief periods while the probe is connected and the patient positioned, but should not be permanently disabled if the monitor is to be useful. Pulse oximeters are sensitive to ambient light, dyes or pigmented skin, as well as patient motion and poor perfusion. The probe should be covered to prevent interference from room and surgical lights and, if possible, should be secured so that it is not dislodged during surgery. In some instances, motion sensitivity can be useful in a patient obscured by surgical drapes. If the probe is connected to the foot and the leg is not tightly secured, sudden disruption of the signal may indicate a return of response to noxious stimulation and an inadequate anesthetic depth. While occasionally useful, there are many other reasons for signal failure, and the instrument is not a dependable indicator of anesthetic depth. Drugs that cause vasoconstriction, such as α_2-adrenergics, can sometimes interfere with detection of pulsatile flow, as can restraints if applied too tightly.

The hind foot is usually a reliable site for probe application, although the front foot, tail and other sites are also useful depending upon the probe, and operator familiarity with the instrument. Probes are available in several styles, including reflectance probes, clip probes and wrap probes. Although clip probes are very popular because of their ease of use, they suffer from two disadvantages: they are easy to dislodge, and the spring pressure necessary to secure them will often slowly compress the capillary bed at the application site, resulting in loss of signal. Wrap probes, sometimes called pediatric or neonatal probes, are slightly more difficult to apply, but once they are properly secured generally perform well over time. Instruments designed for use in human patients are limited to an upper heart rate of 225 to 250 beats/min. Some veterinary instruments have upper limits of 450 beats/min or higher, and are better suited for rats. When used properly, pulse oximeters are simple, noninvasive monitors that warn of hypoxia during patient preparation, surgery and early recovery. Although pulse oximetry is often reserved for intraoperative use, unappreciated hypoxia occurs frequently during the induction and recovery periods, and pulse oximeters may be especially valuable at these times. The performance of a Nonin^{TM} 8600V pulse oximeter was recently evaluated in the rat (Bernard et al., 2004).

Capnography continuously measures inspired and expired carbon dioxide levels with each breath and displays a corresponding waveform. The concentrations are usually expressed as partial pressures (mmHg CO_2) in the United States, and as percent volume in Europe. Any condition that alters carbon dioxide production, transport and elimination, will affect the values and the waveform, including alterations in respiration, ventilation and cardiovascular function. Capnography requires endotracheal intubation in rats, limiting its use to those cases where the information is deemed critical. Two types of sampling systems are used, sidestream and mainstream. In sidestream analyzers, a continuous flow of gas is extracted via a small tube at the endotracheal tube connection and analyzed in the instrument. Some sidestream monitors also analyze the sample for inhaled and exhaled anesthetic agent concentration. Mainstream analyzers interpose the detector between the endotracheal tube and the breathing circuit connection, increasing the equipment dead space. At this time, equipment designed for human or small animal veterinary use is usually too large in terms of sample size, or equipment dead space for easy use in rats, although human clinical analyzers have been successfully used (Harper, 2000). A specialized instrument designed for rodent research purposes is available (Columbus Instruments, Columbus, Ohio).

Blood gas determinations are the gold standard for assessing ventilation and acid-base status. Arterial samples are preferred, necessitating an arterial catheter and, usually, a venous catheter for fluid or blood replacement. Compared with older models, modern blood gas analyzers have relatively small sample requirements making their routine use feasible in rats, once vascular access is established.

ii. Cardiovascular Function. Electrocardiography (ECG) and direct and indirect blood pressure are used to monitor cardiovascular function.

Electrocardiography is used to evaluate heart rate and rhythm. Instruments designed for experimental physiology are suitable for rats, as are some specialized small animal

veterinary monitors. Some instruments designed for clinical use in humans do not perform as well when used in rats. Needle electrodes or alligator clips are frequently used as electrodes, but unless they are well secured, needles are easily dislodged during surgery and both needles and alligator clips may result in unnecessary injury. For monitoring during anesthesia, the disposable adhesive human pediatric electrodes or small plate electrodes, both secured with tape, are often less traumatic and more secure.

Blood pressure can be monitored indirectly, using noninvasive methods, or directly, by means of an arterial catheter. Noninvasive methods usually rely on a pneumatic tail cuff with an oscillometric or Doppler flow transducer to detect the occlusion and resumption of pulsatile blood flow as the pressure in the cuff is periodically increased and released. In general, noninvasive monitoring is less accurate and reliable than invasive monitoring, but does not require additional surgery and can provide a reasonable determination of mean arterial pressure. For direct measurement, if suitable monitors and transducers are used, a complete waveform yielding diastolic, systolic and mean arterial pressure is obtained. Most invasive blood pressure monitors designed for veterinary or human patients will display a waveform, but the diastolic and systolic values may not be reliable.

iii. Temperature. Body temperature is best measured continuously using an electronic thermometer. Small probes are available for rectal use in the rat and for body surface temperature. Clinical thermometers designed for use in humans work equally well in the rat, and many have dual displays which can be used for a second animal, or for monitoring both core and surface temperature. Electronic thermometers not specifically designed for clinical monitoring are widely available, and may also perform well if care is taken to select the appropriate temperature range and probe.

2. Intraoperative Support

a. RESPIRATION AND VENTILATION. Ventilation is depressed by anesthesia. Anesthetized animals spontaneously breathing room air typically have elevated blood carbon dioxide levels (hypercapnia) and depressed blood oxygen levels (hypoxia). Hypoxia is the most immediately life-threatening of the two, and the easiest to address. When inhalation anesthesia is used, the carrier gas is usually oxygen, so hypoxia is less likely to be encountered. When injectable anesthetics are used, a mask or merely a plastic tube placed under the drape at the patient's head can be used to deliver a low flow of oxygen. Hypoxia can occur quickly in rats, and providing supplemental oxygen is an inexpensive safety measure, even for brief procedures.

When full respiratory support or control is needed, a ventilator is used. There are many types of ventilators and control options available, but a full discussion of this topic is beyond the scope of this chapter, and only the major features of the two most common designs are described.

Briefly, ventilators deliver a breath based on volume or pressure. The anesthetist controls the volume of gas for volume-limited ventilators, or the peak pressure for pressure-limited ventilators, and the respiration rate. Volume-limited ventilators deliver a fixed volume with each breath, regardless of airway pressure. Examples of this type of ventilator are the familiar mechanical piston designs long used in research. Pressure-limited ventilators deliver a breath until a set pressure limit is reached. Under these conditions, the actual tidal volume can vary with changes in lung compliance. For anesthesia use, both types of ventilators will work, but pressure-limited ventilators are somewhat more popular because they limit the risk of barotrauma, or lung damage due to excessive airway pressure. For safety, if a volume-limited ventilator is used, some means of monitoring airway pressure, such as a manometer or pressure transducer, should also be used. Endotracheal intubation is necessary for mechanical ventilation, and methods for oral endotracheal intubation are discussed later in this chapter. Connections between the endotracheal tube and the ventilator tubing must be kept as short as possible in order to minimize equipment dead space and limit rebreathing and the same caution applies if a gas analyzer is connected to the tube. When a ventilator is used with inhalation anesthetics, it is often possible to reduce the vaporizer settings somewhat compared to those used with spontaneous ventilation.

b. FLUID SUPPORT. Fluid administration is the primary means of supporting hydration and cardiovascular function during rodent surgery. For short procedures using anesthetics that result in rapid recovery, fluid support may not be necessary. When long surgeries, long anesthetic recovery periods, or blood loss are anticipated, however, fluid support is important. All fluids should be warmed before administration to avoid further depression of body temperature. For maintenance requirements during surgery, subcutaneous, intraperitoneal or intravenous routes of administration can be used. To replace blood loss, the intravenous route is preferred, because of the relatively slow absorption from subcutaneous and intraperitoneal spaces. Normal saline solution, lactated Ringer's solution, balanced electrolyte solutions (e.g., Normosol™) and other fluids are used, depending on the anticipated nature and severity of the fluid deficit and electrolyte disturbances. In the event of severe blood loss, transfusions are feasible. Under these conditions, there is rarely time for cross matching and, in most cases, a single transfusion of donor blood from an animal of the same strain is well tolerated (Wixson and Smiler, 1997).

c. TEMPERATURE SUPPORT. With the onset of anesthesia, normal thermoregulatory mechanisms are disrupted and heat moves from the core to the periphery of the body. The effect is accelerated with peripheral vasodilatation, as occurs with isoflurane, for example, and retarded to some extent by peripheral vasoconstriction, as seen with α_2-agonists such as xylazine or medetomidine. The small body mass and large relative surface area of the rat results in rapid cooling which is exacerbated by the necessary steps of shaving and surgical preparation. In the author's experience, virtually all rats are below normal body temperature by the time surgery begins.

Hypothermia is not benign. It delays recovery from anesthesia, increases the potency of many anesthetics, and places an added burden on the body to restore normal temperature during recovery (Grahn et al., 1996). Unless hypothermia is a deliberate goal, measures must be taken to reverse the process and support body temperature. Passive steps include providing additional insulation between the animal and a cold operating or preparation table. Synthetic fur, folded towels, and wrapping the animal in plastic bubble wrap are useful. These measures will slow cooling, but they generally do not reverse it in anesthetized patients. Instead, active warming is preferable. For this, heating pads, warming lights, and forced warm air are the usual methods. Plastic heating pouches containing saturated solutions of sodium acetate, or similar salts, release latent heat as the solute crystallizes and some brands can be re-used by immersing them in hot water to re-dissolve the solute. For occasional procedures of short to moderate duration, they can be satisfactory. Although household electric heating pads can be used, temperature control is uncertain, and hyperthermia and even burns are not unusual. Commercially produced rodent heating panels or operating tables are more expensive, but provide better control and safety. Some incorporate a patient temperature probe with control circuitry to regulate the heating pad and maintain body temperature within user-controlled limits. Circulating warm water pads are frequently chosen because they are safe, rarely causing hyperthermia or burns, and reasonably reliable. All heating pads are in contact with only a fraction of the body surface, and generally perform better if the patient is draped.

Forced-air warming air blankets have proven to be a very effective means of thermal support for human and veterinary patients, and are reported to be effective during rodent surgery and recovery as well (Rembert et al., 2004). The apparatus is somewhat bulky and adhesive drapes may be needed for optimum results. Heating lamps can also be useful, at the risk of tissue drying during surgery. For long procedures, when precise thermal control is desired, heating panels with integrated thermometers and control circuits may be the best, if most expensive, choice.

Continuous measurement of body temperature is not practical for many rodent surgical procedures, but should be used for long and very invasive surgeries. Further, steps intended to limit heat loss and maintain temperature should not be based on faith or authority, but should be assessed by directly measuring body temperature during patient preparation, surgery and recovery to insure the procedures are effective.

d. DRAPING AND POSITIONING.   Surgical drapes and sterile instrument fields are used to limit contamination of the surgical wound, the surgical instruments, and the surgeon's hands. To be effective, the drape should, at a minimum cover all, or nearly all, of the animal and enough of the surrounding area to avoid accidental contamination. If instrumental monitoring is not used, transparent plastic drapes, with or without adhesive, allow direct observation of the animal during surgery. Otherwise, both paper and cloth drapes are suitable. Cloth drapes are the easiest to position and use, but the initial purchase cost and continuing maintenance costs are high. Paper drapes are convenient, easy to customize and sterilize, and relatively inexpensive, but can be troublesome to keep in position during surgery. A simple means of securing paper drapes is to use three or four small pieces of poster putty to adhere the edges of the drape to the table.

For some surgeries, restraints are needed to hold the animal in position. Tape is probably the most flexible and least traumatic method, especially if the operating surface is smooth. During positioning, it is important to avoid unnecessary force and tension to prevent muscle or nerve injury, and restriction of circulation. Respiration should be checked following positioning to assure that it is not compromised.

e. MISCELLANEOUS MEASURES.   The eye of the rat frequently remains open or partially open during anesthesia. To prevent painful corneal drying and damage, bland ophthalmic ointment is applied during surgical preparation and re-applied as needed, and the eyes should be protected from direct contact with the operating table and drape.

D. Postoperative Care

1. Anesthesia Recovery

Following momentary or very brief anesthesia, especially with inhalant anesthetics, recovery is usually rapid and uncomplicated. The animal should be assessed to be sure that it is fully awake and that no complications have occurred before it is returned to its cage. Longer procedures generally entail longer recovery from anesthesia. Thus, many of the supportive measures used during anesthesia induction and surgery, including thermal support and oxygen enrichment, are maintained during recovery from anesthesia. The animal should be assessed to detect surgical complications, and the surgical site cleaned, if necessary. Reversal drugs are usually given at this time, as are fluids and analgesics if they have not been given in the preoperative or induction periods.

Thermal support should continue until the animal is fully conscious and active. Wixson and Smiler (1997) recommended an avian ICU cage as ideal for recovery of groups of rodents. These cages can be warmed, humidified, and connected to oxygen. Alternatively, a suitably large rodent housing cage can placed on a warm water circulating pad, and a plastic top fabricated to permit oxygen to be supplied. For individual patients, inhalation anesthesia induction chambers can similarly be used. Rembert et al. (2004) compared forced-air warmer blankets with warm water circulating pads and infrared heat emitters for surgical and postoperative warming and found the forced-air warming system to be superior for both purposes.

Unconscious or deeply sedated animals lack normal protective reflexes. It is safer to maintain them on a towel or similar soft surface in order to avoid airway compromise, corneal damage, and other mishaps attributable to cage bedding materials. If prolonged or intensive support is necessary, or in very busy surgical facilities, a separate rodent ICU area, equipped with appropriate caging, instruments and supplies may be worthwhile.

2. Surgical Recovery

Regular observation of surgical patients should continue until recovery is complete. This is generally characterized as return to preoperative weight, activity and behavior. In animals used to create surgical models of disease, the definition is obviously altered to take into account the intended consequences of the procedure. In addition to the physiological consequences of anesthesia and surgery, postoperative pain also has a major impact on the speed and quality of recovery. Recognition, assessment and treatment of pain are discussed elsewhere in this chapter.

The medical record for the animal or group of animals is used to record the weight, observations and treatments, forming an objective basis on which to assess recovery and provide further treatment, if indicated. The frequency of observation will vary with the severity of the procedure and local experience and professional judgment, but it is obvious that major surgical procedures and complications require more frequent observation and treatment than do minor ones. The historical documentation provided by medical records is invaluable in characterizing the expected postoperative course and the nature and incidence of complications. With this information, postoperative care protocols can be formulated on an objective basis and resources more efficiently allocated.

Body weight, fecal production, and appearance are used to gauge nutritional status. Many animals will return to eating rapidly following recovery from anesthesia. If this is not the case, nutritional support is needed. If the animal is unable to comfortably reach food or water, steps are taken to improve access by using bowls, crocks, jars or similar expedients. For access to water, extended sipper tubes are often helpful, as is transport, or shipping gel. A soft diet, such as moistened, ground chow, gelatin or agar can be used (Wixson and Smiler, 1997). Palatable nutritional supplements, such as Nutri-Cal® are also useful for short periods, until the animal is able to return to a standard diet. Hampshire et al. (2001) emphasized the importance of early postoperative fluid therapy, nutritional support and analgesia in improving postoperative recovery.

Skin turgor or elasticity is frequently suggested as a means of assessing hydration. While useful as a means of detecting severe dehydration, skin turgor is subjective, relatively insensitive, and difficult to assess in a struggling patient. In the postoperative rat, abrupt weight loss usually signals dehydration, invariably accompanied by inappetence. Substantial deficits are corrected using warmed subcutaneous or intraperitoneal fluids. With few exceptions, if there is doubt regarding hydration status, it is better to err on the side of fluid administration: subcutaneous or intraperitoneal fluids are unlikely to pose risks greater than those of progressive dehydration.

Implanted catheters or other exteriorized hardware are assessed, cleaned and flushed as necessary, and hair is clipped around the implant site to permit regular examination. Bandages, splints, jackets and Elizabethan collars are removed at frequent intervals to permit normal grooming and examination of the surgical site (Wixson and Smiler, 1997).

The operative site is cleaned, if required, and assessed to determine if healing is progressing as expected. With good surgical technique, wound infections are uncommon and wound healing uneventful. Delayed or absent healing attributable to poor wound closure technique, may respond to suture removal and conservative therapy, or require surgical debridement and wound revision. Skin sutures, Michelle clips, and staples are removed when healing is complete to prevent further irritation and subsequent complications.

If the surgery has resulted in long-term or permanent disability, additional care is continued as needed, often requiring changes in standard husbandry practices with respect to water, food, bedding, room temperature, etc., as well as a greater intensity of observation. Optimal care requires close cooperation among the investigator, the veterinarian, and the animal care staff. As the primary caregivers, dedicated animal care technicians are especially important, and their observations and suggestions often improve care and substantially contribute to success.

III. PAIN ASSESSMENT AND TREATMENT

The importance of this topic is apparent from the vast and rapidly growing literature, and the number of texts in both veterinary and human medicine devoted to the subject. Virtually every veterinary and laboratory animal meeting includes discussion of pain recognition and management in animals, and many professional organizations have issued position papers or standards. A thorough discussion of the pathophysiology of pain, analgesic pharmacology, and methods of clinical treatment is beyond the scope of this chapter, and the following section is intended only as brief survey of current approaches to the topic.

In laboratory animal medicine the difficulties in diagnosing painful or potentially painful injury or disease differ somewhat from those encountered in conventional clinical practice. For scientific reasons, considerable emphasis is placed on securing and maintaining healthy animals, or animals with well-defined genetically mediated disease. Surgical, pharmacological and genetic interventions are planned and documented, so the diagnostic challenges faced by the clinician are, in many cases, simplified by comparison to those faced in general clinical practice. Even so, identifying and assessing the intensity of pain in small mammals remains challenging. Many attempts have been made to identify specific signs of pain in rats and other animals (Soma, 1985; Morton and Griffiths, 1985; Flecknell, 1994; Stasiak et al., 2003; Gross et al., 2003; Roughan and Flecknell, 2003). However, clinical indices of physical condition, such as weight loss, unthriftiness, and reduced food and water intake, while useful for identifying the presence of abnormality and probable disease, are not specifically associated with pain. In some instances, such as abdominal surgery, more specific indices of pain have been identified, such as back arching, fall/stagger, writhe and poor gait described by Roughan and Flecknell (2003). These observations are further supported by their similarities to indices used in the classical writhing test and in other models of acute abdominal pain (Giamberdino et al., 2002). However, there is no comprehensive list of pain-specific behaviors or physical signs correlating with the many surgical procedures and disease-inducing manipulations encountered in laboratory animals. Anthropomorphic approaches are intuitively attractive and often may be useful, but they can also be misleading to the degree that they fail to take into account the anatomical, behavioral and physiological differences among species. Identification of pain depends upon physical examination, careful observation, a firm grasp of normal behavior in the animals being assessed, and knowledge of the history and nature of the surgical procedure or induced disease process.

Essential elements of a pain management scheme should include consultation with the investigator, who may be presumed to have some knowledge of the pathophysiologic consequences of the intervention; the animal caretaker and animal technicians, who have the most intimate knowledge of the normal and abnormal appearance and behavior of the animals; and the responsible clinician, who must assess and attempt to alleviate the pain. Using the information derived from each of these sources, it is possible to construct an assessment scheme, however imperfect, that will permit a more consistent and organized method of documenting the presence and severity of pain, and response to treatment. Once in use, assessment schemes are modified with experience and collaborative input to identify and focus more closely on those elements found to be specific and reliable indicators. To be successful, the personnel involved, especially caretakers, technicians and clinical veterinarians must have adequate administrative support in terms of training and time. Training is essential because of the often-subtle nature of the physical signs used. For the same reasons, consistency in care and excellent communication among those involved is also needed. Assessment schemes support rational clinical use of analgesics, identify animals in need of additional treatment and animals for which further treatment is not needed; and with refinement and training, need not be unduly onerous to use.

Once the presence and degree of pain are assessed, steps should be taken to alleviate it. While pharmacological methods are an important element of pain management, they should not be depended upon to obviate the consequences of poor surgical, technical, and nursing practices. Surgeons, technicians, and veterinarians require training and continuing education to assure a high level of proficiency, maintenance of skills, and exposure to improved methods and techniques. The effect of careful postoperative care in improving outcomes, regardless of analgesic treatments, cannot be overestimated. Early attention to a comfortable environment, easily accessible and palatable food, and adequate hydration will facilitate recovery and often reduce the need for postoperative analgesia. (Hampshire et al., 2001).

If possible, analgesics are used preemptively, in an attempt to prevent sensitization and amplification of postoperative pain. The primary analgesics used to treat of pain in rats are opioids and nonsteroidal antiinflammatory drugs (NSAIDs), and are discussed later in this chapter. These drugs differ in the means by which they contribute to pain relief. Because the nature and origin of postoperative pain varies over time, simultaneous use of different classes of analgesics may improve analgesia while lowering the needed dose of each component. Intraoperative use of local anesthetics can contribute to pain control in the immediate postoperative period. As with anesthesia, pain control measures are a part of project planning and, as with anesthesia, changes and adjustments may be needed as the project develops. Obviously, the intensity of observation and treatment will depend on the degree of trauma and anticipated pain. For many common procedures, only relatively brief analgesic treatment is indicated. In this context, a pain assessment scheme can support not only the need for continuing therapy, but also the decision to discontinue treatment. By contrast, in the event of severe, intractable pain leading to euthanasia, the record also documents regular assessment and professional care. Analgesics can reduce pain, rendering it more tolerable, but cannot realistically be expected to eliminate all of the physiological consequences of major surgery; recovery takes time, and the success of analgesic therapy cannot be judged solely on the basis of rapid return to normal function.

IV.   DRUG DOSES AND DOSE PREPARATION

A.   Dose Determination

Published dosages for anesthetics, anesthetic adjuvants, and analgesics are often derived from studies using a limited number of animals, of a specific age, sex, strain and health status. These studies are invaluable but they do not always predict the results seen when the setting, procedures and animals vary from the study conditions. In many cases, the definition of anesthesia, and the methods used to evaluate it, are not uniform, or not clearly described. In some reports, the variable assessed is not surgical anesthesia, but sleep time. Even when an attempt is made to duplicate published studies, the results often vary because of unconsidered or uncontrollable differences in procedures, environment, and the age, sex, health status and genetic background of the animal. Hence, published dose ranges tend to be wide, and adjustments are needed to achieve the best results.

Pilot studies are the best means to refine the anesthetic protocol when new methods of administration or unfamiliar drugs are used. A pilot study allows the adequacy and duration of anesthesia to be assessed, and technical and procedural problems to be identified and remedied. These advantages are not limited to new techniques and drugs, but apply equally to new surgical procedures and experimental conditions, where familiar anesthetic methods and doses may prove to be inadequate. A pilot study should mimic the experimental conditions as closely as possible and include the personnel who will be expected to use the anesthetic technique. Because surgeons and anesthetists improve with experience, and experimental procedures are modified over time, further adjustment of anesthetic and analgesic protocols can be anticipated until the procedures are optimized (see Table 19.1).

B. Dose Preparation

Parenteral anesthetics are frequently administered as a single dose, based on body weight. However, almost all anesthetic and analgesic drugs used in rats are, or once were, originally marketed for use in humans or for larger domestic animals. As a result, parenteral agents are often presented in concentrations best suited for much larger species, complicating accurate and convenient administration to small research animals. Small volume disposable syringes are helpful with precise measurement, but calculating and accurately drawing very small volumes of drugs is burdensome and prone to error. It is often possible to dilute a drug or a mixture of drugs so that the delivered volume is easily measured and conveniently calculated on the basis of weight. In rats, a typical dilution might yield a dose volume of 0.1 mL per 100 g of body weight. Assuming that preoperative weights are taken, dose calculation becomes a simple matter. Commonly used diluents include 0.9% saline, or water for injection, both of which are available as preservative-free preparations. Sterile, empty injection vials are also commercially available in a number of sizes, further simplifying the process. Drugs prepared in this manner eliminate the need for cumbersome calculations and, offer greater convenience and accuracy of dosing for small animals. Useful examples of such dilutions are have been published (Flecknell, 1996c).

However, while drug dilution is often useful, it is not always feasible. Due to solubility and stability issues, not all drugs can be diluted with saline or water. Safely compounding more complex solvent systems requires specialized knowledge and experience in parenteral formulation. The issue of expiration dates for such mixtures is also problematic, with little published guidance available. The accuracy of the dilution can be difficult to verify, although for some drugs the refraction of the mixture, determined with an ordinary clinical refractometer, can serve as a simple, though limited, quality control measure (Stabenow and Vogler, 2002).

V. ADMINISTRATION ROUTES AND EQUIPMENT

A. Oral Administration

Oral administration of anesthetics to rats is uncommon and offers little advantage over parenteral or inhalation

TABLE 19.1

SELECTED ANESTHETIC AND ANALGESIC DOSES

Drug	Dose	Route	Duration
TRANQUILIZERS			
Acepromazine	2.0–2.5 mg/kg	SC, IP	sedation
Droperidol	0.5–2.0 mg/kg	SC	sedation
Diazepam	2.5–15.0 mg/kg	SC	sedation
Midazolam	5.0 mg/kg	SC, IP	sedation
ANESTHETICS			
Barbiturates			
EMTU (Inactin)	80–100 mg/kg	IP	60–240 min
Methohexital	10–15 mg/kg	IV	5–10 min
Pentobarbital	30–60 mg/kg	IP	20–60 min
Thiopental	20–40 mg/kg	IV	5–10 min
Dissociative Agents & Combinations			
Ketamine + acepromazine	75 mg/kg 2.5 mg/kg	IP	20–30 min
Ketamine + medetomidine	60–75 mg/kg 0.25–0.50 mg/kg	IP	20–30 min
Ketamine + xylazine	40–75 mg/kg 5–10 mg/kg	IP	20–40 min
Neuroleptanesthetics			
Fentanyl/fluanisone (Hypnorm) + midazolam	2.7 ml/kg* 1.25 mg/kg	IP	20–40 min
Fentanyl + medetomidine	300 µg/kg 300 µg/kg	IP	60–70 min
Inhalant Agents			
CO_2/O_2	70%–80%/20%–30%	Inhalant	
Halothane	MAC ~1.0%, to effect	Inhalant	
Isoflurane	MAC ~1.4%, to effect	Inhalant	
Sevoflurane	MAC ~2.7%, to effect	Inhalant	
Other Agents			
Alpha-chloralose	55–65 mg/kg	IP	8–10 hr
Alphaloxone-alphadolone	10–15 mg/kg	IV	5–10 min
Chloral hydrate	300–450 mg/kg (4%–5%)	IP	60–120 min
Propofol	7.5–10 mg/kg	IV	5–10 min
Urethane	1000–1500 mg/kg	IP, SC	8–24 hr
ANALGESICS			
Buprenorphine	0.01–0.05 mg/kg	SC	8–12 hr
Butorphanol	2 mg/kg	SC	2–4 hr
Carprofen	5 mg/kg	SC	24? hr
Ketoprofen	5 mg/kg	SC	24? hr
Morphine	2.0–10.0 mg/kg	SC	2–4 hr
MISCELLANEOUS DRUGS			
Atropine	0.05 mg/kg	IP, SC	
Glycopyrrolate	0.50 mg/kg	IM	
Atipamezole	0.1–1.0 mg/kg, varies with α_2 dose	IP, SC	

EMTU = ethylmalonyl urea; IM = intramuscular; IP = intraperitoneal; IV = intravenous; MAC = minimum alveolar concentration; SC = subcutaneous.

*See Flecknell (1996a) for mixing instructions.

Dosages adapted with modifications from Wixson and Smiler (1997), and Flecknell (1996a).

methods. When the oral route is used, the most certain method is gavage, using a feeding needle or soft tube. In contrast to anesthetics, administration of analgesics in food, water or palatable treats is often advocated. While this route is convenient and avoids the need to handle and inject the animals, there are some disadvantages. For some analgesics, such as acetaminophen, the route appears to be ineffective (Wixson and Smiler, 1997; Cooper et al.,1995) while with others, e.g., buprenorphine, it is at least controversial (Martin et al., 2001; Roughan and Flecknell, 2002a; Thompson et al., 2004).

There are other problems with adding drugs to food or water. It can be difficult to attain uniform and stable drug dispersion, especially with suspensions of insoluble agents in water. Without extensive testing, chemical stability is uncertain. Adequate consumption depends on palatability and the condition of the animal, and may be decreased, increased or unchanged, but must be verified by some means if the goal is to administer a therapeutic dose. For analgesics known to be effective by the oral route, these issues may be addressed in the future by formulations in highly palatable, discrete dose forms. At this time, however, parenteral administration appears to be the most reliable means of alleviating moderate to severe acute pain.

B.   Parenteral Methods

For administration of anesthetic drugs, parenteral methods include intramuscular (IM), subcutaneous (SC) and intraperitoneal (IP), and intravenous (IV) injection. These methods have been described in an earlier chapter, and only intravenous access for anesthesia is described below.

While femoral or jugular vein catheters can be used for drug administration, both require surgical implantation. For the purposes of anesthesia, the tail of the rat presents three veins suitable for percutaneous puncture and is perhaps the most convenient site for use during anesthesia. The anatomy of the tail vasculature has been clearly illustrated and described by Stazyk et al. (2003). A needle and syringe or a butterfly set is adequate for brief, limited volume injections, but an indwelling catheter is more convenient and secure for intermittent or continuous infusion during anesthesia. A 24 gauge over the needle catheter is satisfactory for all but the smallest rats. Although the tail veins appear larger towards the base, the overlying skin is concomitantly tougher, and insertion is easier distally. Once the catheter is secured, an injection port and heparin lock will maintain a usable intravenous access site for the duration of anesthesia.

Intermittent injections are typically performed manually, but for continuous intravenous administration, a syringe pump is more reliable and consistent. Intravenous administration sets and drip controllers used for larger animals are poorly suited for the very low infusion rates and limited volumes used in rats. Syringe pumps provide accurate control of the infusion rate and those designed for anesthesia have some advantages over laboratory models. Anesthesia syringe pumps display the infusion rate, dose units and the total volume of drug infused and also permit the user to program bolus volume and rate. The latter feature is useful for quickly responding to inadequate anesthesia levels. Alarms warning of line occlusion and of an impending empty syringe further improve safety and reliability. For continuous intravenous anesthesia, these pumps are analogous to precision vaporizers in inhalation anesthesia.

C.   Inhalation Methods

For many years, methoxyflurane was the most common inhalation agent used in rodents, largely because the physical properties of this anesthetic made its use possible with a minimum of equipment. These same properties, however, were considered disadvantages in human and veterinary clinical practice, where newer and more inert agents displaced it. Manufacture ceased in the United States; at this time, the cost and inconvenience of importing methoxyflurane has led to sharply declining use. Newer inhalation anesthetics, such as halothane, isoflurane and sevoflurane, produce rapid induction and recovery and provide more responsive control of anesthetic depth. These agents are also characterized by relatively high vapor pressures and a narrow therapeutic index; safe use requires appropriate equipment to control inhaled concentrations, and to provide disposal of waste gas. While some aspects of anesthesia equipment specific to rodents are described below, a detailed discussion of the design, function and limitations of anesthesia equipment is beyond the scope of this chapter. The interested reader should consult relevant anesthesia texts (Dorsch and Dorsch 1999; Thurmon et al., 1996; Short, 1987).

1. Anesthesia Machines

The transition to modern inhalation agents has led to development of equipment specifically designed for rodent anesthesia. Precision calibrated vaporizers are used to control the concentration of anesthetic agent by setting a dial to the desired volume percent of inhaled agent. The vaporizers automatically compensate for changes in temperature, fresh gas flow rates, and back pressure. Modern vaporizers are agent specific, and should be used only for the agent for which they are designed. The maintenance requirements for anesthetic vaporizers vary with the make and model and range from annual service to "as required." Output can be verified using a specialized refractometer or agent analyzer.

Fresh gas flows are usually controlled by the operator by means of one or more flowmeters, calibrated in liters per minute. One machine design, however, uses flow restrictors to limit the flows to predetermined rates. The design trades flexibility for efficiency, safety and convenience during multiple simultaneous procedures (VetEquip Inc., Pleasanton, CA). The carrier gas is usually pure oxygen, but other gases, such as air or nitrous oxide, can also be used provided sufficient oxygen is present to avoid hypoxia. Most anesthesia machine manufacturers can add flowmeters for additional gases if desired. As with modern vaporizers, flowmeters are also agent, or gas, specific.

Several machines are available that split the vaporizer output into two or more circuits, for the efficient induction and maintenance of multiple animals for short procedures. With such designs each animal receives the same concentration of agent. This practice is efficient and usually safe for shorter procedures and is often the choice for these purposes in central facilities. When longer, more controllable anesthesia is needed for individual animals, as in an operating room or investigator laboratory, smaller machines are often more suitable. No single machine will serve all purposes equally well. When purchasing inhalation anesthesia equipment, the intended use and location of the equipment should be carefully defined. In the absence of central gas supplies, compressed gas tanks and regulators, and a means of safely securing them, are needed. Suitable arrangements for waste gas scavenging are also required, as well as additional equipment including induction chambers, delivery circuits, etc. These ancillary items should be evaluated to assure compatibility and ease of connection with the anesthetic machine.

2. Ancillary Equipment

a. INDUCTION CHAMBERS. With sufficient restraint, mask induction is feasible in the rat, but inhalation anesthesia is more often initiated using an induction chamber. Induction chambers should be made of clear plastic, so that the animal can be continuously observed during induction. The chamber should be large enough to comfortably hold the rat before and following loss of consciousness without compromising breathing, but an excessively large chamber will prolong induction and waste breathing gas and anesthetic agent. Induction can be hastened by increasing the fresh gas flow rate, the agent concentration, or both. An appropriate chamber size allows use of moderate flow rates and agent concentrations, lowering costs and improving safety. A chamber volume of 2 to 4 liters is usually adequate for rats. If the fresh gas flow in liters per minute equals the chamber volume in liters, the agent concentration in the chamber will closely approximate the vaporizer setting in about 3 minutes. For example, if a 3-liter chamber is used with the vaporizer set at 3% and the flowmeter set to deliver 3 L/m, the concentration of the anesthetic agent in the chamber will be nearly 3% at about 3 minutes. If the flow rate is doubled to 6 L/m, the time needed will be shortened to about 1.5 minutes.

The chamber lid should fit tightly enough to minimize leaks and to prevent the animal from dislodging it. Ideally, the chamber fresh gas inlet should fit the tubing from the anesthetic machine without further modification, and the outlet should be larger to avoid resistance to flow and leaks. Conventional induction chambers are major sources of waste gas pollution. Actively scavenged induction chambers are commercially available and are recommended to reduce waste gas exposure when the chamber is opened to remove the animal (VetEquip Inc., Pleasanton, CA; Harvard Apparatus, Holliston, MA).

For very brief anesthesia, if an anesthetic machine is not available, the volume of liquid anesthetic needed to produce a safe concentration in a chamber of known volume can be calculated and directly added to the chamber (Brunson, 1997). Volatile anesthetics are excellent organic solvents and will attack many of the plastics used for caging and induction chambers, so a secondary container resistant to the agent is needed to avoid damaging expensive equipment and to prevent contact between the liquid anesthetic and the animal. When the chamber is opened to remove the animal, the contents are mixed with ambient air and the remaining anesthetic concentration is no longer known, making repeated use less predictable. Alternatively, a method for diluting isoflurane in a nonvolatile solvent has been proposed as a means of reducing the delivered concentration of agent to safe levels (Refael et al., 2003). These techniques are useful on occasion, but for routine use, a small anesthetic machine is easier to use and more flexible.

b. BREATHING CIRCUITS. Anesthesia is most commonly maintained by mask, using simple, continuous flow, nonrecirculating circuits, without valves, carbon dioxide absorbers or reservoir bags. With these circuits, the fresh gas flow is used not only to supply oxygen and anesthetic, but also to remove expired gas. The most common examples of such circuits are Bain circuits, or other Mapleson D or E systems, used without a reservoir bag (Dorsch and Dorsch, 1999). The T piece designs used for stereotaxic stands operate similarly. These circuits are easy to configure and they impose very little resistance to breathing. However, if the fresh gas flow is not adequate, rebreathing will occur, with accumulation of carbon dioxide. The flow needed in any given instance depends upon a number of factors including the respiratory rate and depth, carbon dioxide production, and equipment and physiologic dead space. These variables are unpredictable, and precise adjustment of fresh gas flow requires measurement of expired or arterial carbon dioxide levels, so relatively high flow rates are usually recommended to reduce the risk of rebreathing. Typical recommendations for fresh gas flows are from 1.0 to 2.0 L/m, (Smith and Bolon, 2002), although with careful monitoring and attention to minimizing mask dead space, lower flows can successfully be used.

Other types of breathing circuits can be used. One equipment maker uses demand valves that open on inspiration and close on expiration, reducing anesthetic leakage (E-Z Anesthesia, Euthanex Corporation, Palmer, PA). Circle systems with carbon dioxide absorbers are feasible if the valves and circuits are designed to minimize breathing resistance and one such system is incorporated into an anesthesia ventilator (Model 2000 Small Animal Anesthesia Ventilator, Hallowell EMC, Pittsfield, MA).

Breathing gases are not recirculated and humidity is not retained with the circuits usually used for rodent anesthesia. When administered for prolonged periods, anesthetic gases can dry the airways, and impair epithelial ciliary function leading to lower airway obstruction. While this is not ordinarily a concern, humidification of breathing gas should be considered for prolonged anesthesia. A means of humidifying the gas in rodent breathing circuits was reported by Martenson, et al. (2005). Commercial devices intended to warm and humidify breathing gases for humans and larger animals should be used with caution in very small animals because of the danger of over-humidification and drowning.

c. MASKS. Conventional masks depend upon a tight seal between the patient and the circuit to prevent leaks. Perforated rubber septa or flared plastic tubing are used to achieve a tight fit. Defective seals or poorly fitting masks leak, resulting in unnecessary exposure of personnel to anesthetic gases. A sufficient leak may also dilute the anesthetic gas, requiring inappropriately high fresh gas flows or vaporizer settings and resulting in even greater pollution. Masks usually fit well when they are first applied, but tend to slip as the animal is moved during procedures, so frequent attention is needed to avoid leaks. Rubber seals deteriorate, and should be examined before use and replaced as needed. In addition to a tight seal, masks should also have a small internal volume and the connection to the breathing circuit should be as short as feasible to limit equipment dead space and reduce rebreathing. Masks designed for larger animals often have excessive internal volume and are poorly suited to rats. Specialized masks are available from stereotaxic equipment vendors, enabling the use of inhalation anesthesia for these procedures.

An alternative to the conventional masks described above is a coaxial mask that uses active scavenging and requires no septum (Glen et al., 1980; Levy et al., 1980; Hunter et al., 1984). In this design, fresh gas flows through an inner tube into which the rat's nose is placed. A larger coaxial tube surrounds the inner tube and vacuum, supplied by a central system or fan, is used to capture the expired gas and conduct it to a scavenging site. With proper construction and adequate scavenging flows, this arrangement is simple to use avoids many of the complications associated with conventional masks. Machines using these principles, with dedicated circuits and scavenging provisions, are commercially available from several sources.

3. Waste Gas Scavenging

Effective waste anesthetic gas scavenging is essential for both health and regulatory reasons. Multiple studies of the effects of trace exposure to halogenated anesthetics on human health have been carried out, in some cases with contradictory findings (Dorsch and Dorsch, 1999; Smith and Bolon, 2002). Controversy aside, there is general agreement that unnecessary exposure to trace anesthetic gases is undesirable. The National Institute for Occupational Safety and Health has issued recommendations that exposure to halogenated agents should not exceed 2 ppm, based on a time-weighted average (NIOSH, 1977).

Scavenging systems are classified as passive or active. Passive systems use fresh gas flow to push waste gas to the final scavenging site, or scavenging interface, while active systems use vacuum or fans to pull the waste gas to the disposal site. With either system, the ultimate fate of waste gas is either atmospheric dispersion or adsorption by an activated charcoal canister. Active scavenging systems are considered to be superior to passive systems (Gardner, 1989). The design of scavenging systems is described in more detail elsewhere (Dorsch and Dorsch, 1999b).

If there are no other suitable options, a charcoal canister can be used to remove the anesthetic agent from the waste gas stream. Most commercial brands of canisters claim to adsorb 50 g of halogenated anesthetic; they do not remove nitrous oxide. The canister is weighed and considered to be exhausted when it has gained 50 g. Considering the specific gravity of the agent, a canister will hold about 33 mL of isoflurane or sevoflurane, or somewhat less halothane. The time needed to reach saturation depends upon the concentration of anesthetic and fresh gas flow rates used.

Waste gas pollution and charcoal canister performance during rodent surgery has been studied (Smith and Bolon, 2002; Smith and Bolon, 2003). The performance of charcoal canisters was found to vary significantly among manufacturers. The authors also recommended active scavenging measures such as nonrecirculating hoods and actively scavenged masks, as well as cleaning chambers and circuits, and increasing room air turnover. Construction of a large capacity charcoal absorber system, and the results of testing were reported by Burkhart and Stobbe (1990). The inconvenience, expense and uncertainty of charcoal absorption as a means of waste gas scavenging are compelling arguments for installing active waste gas disposal systems during new construction or renovation.

Regardless of the scavenging system, poor work practices contribute to pollution. Avoiding excessive fresh gas flow rates, using appropriate chamber sizes, and assuring mask seal integrity will limit waste gas exposure. Other steps include turning off the fresh gas flow before opening the chamber and between patients, and filling vaporizers at the end of the work shift rather than at the beginning.

Downdraft or back draft tables, portable hoods and local directed suction devices are all useful to limit exposure, but in the author's opinion should be regarded as supplements, not replacements, for good work practices and the scavenging procedures described above. Work practices, environment, and equipment are uniquely combined in each setting: personnel should be periodically monitored in cooperation with the institutional safety officer to assess exposure and initiate corrective action as required.

4. Endotracheal Intubation

Endotracheal intubation is necessary in order to control ventilation. For nonsurvival procedures, a tracheotomy is suitable, and eliminates leaks around the tube. For survival procedures, oral intubation is a better approach. Many techniques have been published, including blind intubation, transillumination, direct laryngoscopy, fiber optic laryngoscopy, and the use of a stylet, a Seldinger guidewire, and various tables, stands, mouth gags, and other equipment. Of these various methods, blind intubation requires the least equipment but, possibly, the most skill (Stark et al., 1981). The methods of Yasaki and Dyck (1991) and of Cambron et al. (1995), using transillumination and a tilted board to hold the rat need little additional equipment, and offer improved vision and ergonomics. Tran and Lawson (1986) described a modified otoscope speculum to directly view the larynx. Their modification consists of a slit in the otoscope cone, creating a lighted laryngoscope with the added benefit of magnification, allowing the endotracheal tube to be introduced from the side of the mouth, as it is for larger species. A Seldinger guidewire can be used as an intubation guide (Proctor and Fernando, 1973; Weksler et al., 1994). More recently, a method using an intubation

wedge fabricated from a syringe barrel has been reported (Jou et al., 2000).

The author's preferred technique is a combination of the slanted table of Cambron, the modified otoscope of Tran and Lawson, and a short (15 cm) Seldinger guidewire. Briefly, the animal is anesthetized and placed supine on a slanted board. A cotton-tipped applicator used to withdraw the tongue while the modified otoscope cone is inserted and the cords are visualized. A drop of 2% lidocaine, applied with a curved ball-tipped feeding needle, is used to anesthetize the larynx and chords before intubation. The feeding needle is also useful to slightly depress the soft palate, to improve the view of the cords if needed. The guidewire is advanced into the trachea under direct vision, and the endotracheal tube is then advanced over the guidewire and into the trachea. The guidewire is removed, and the tube securely tied in place over the nose. With practice, the technique can be accomplished quickly, using inhalation anesthesia. Regardless of the anesthetic drug(s) used, however, a relatively deep plane of anesthesia is essential to reduce laryngeal trauma and to avoid fatal laryngospasm. Unless both the circuit and endotracheal tube are securely fixed in place during mechanical ventilation, the endotracheal tube will move with each breath, abrading the trachea.

Over-the-needle catheters are often used as endotracheal tubes. For rats of 200 g or greater, a 14-gauge, 1.5-inch catheter is adequate. Smaller animals may require a smaller catheter. DeLeonardis et al., (1995), recommended attaching a silicone tubing extension to the tip of the catheter in order to reduce tracheal trauma during anesthesia and ventilation, but the author has not found this to be necessary if the endotracheal tube is well secured. Connecting the Luer fitting of a catheter to a breathing circuit can be challenging. For ventilator connections, custom Y-connectors, often available from the ventilator manufacturer, can be used. Miniature plastic Y-connectors available from laboratory equipment suppliers can also be trimmed and smoothed to fit the Luer connection of the catheter. If a Mapleson-D or E circuit is used, endotracheal tube connectors that fit neonatal 2.0 or 2.5 mm endotracheal tubes will often work. If possible, low dead space connectors should be used (Vogler, 1997).

VI. INHALATION ANESTHETICS

Many older volatile anesthetics, such as chloroform, diethyl ether, and methoxyflurane now have limited use because they are toxic, difficult to administer safely, or suffer from limited availability. Readers interested in diethyl ether and methoxyflurane should consult *Anesthesia and Analgesia in Laboratory Animals* (Kohn et al., 1997) as well as veterinary and human anesthesia texts. Modern inhalation agents include halothane, isoflurane, sevoflurane, and desflurane. Of these, desflurane seems unlikely to become popular due to the expense of both the agent and the specialized heated vaporizer needed to use it. The use of the remaining agents, including carbon dioxide and nitrous oxide, is covered below. The pharmacology of these agents is discussed in greater detail by Brunson (1977), Steffey (1996), and in standard anesthesia and pharmacology texts.

The potency of inhaled agents is expressed as minimum alveolar concentration (MAC), the equivalent of the effective dose (ED_{50}) at which a single animal, or group of animals, responds or fails to respond to a standard noxious stimulus 50% of the time (Eger et al., 1965). When inhalant anesthetics are used alone, the concentration is increased to about 1.3 MAC to assure adequate anesthetic depth. MAC is affected by a number of factors, among them decreased body temperature, concurrent use of anesthetic drugs and adjuvants, circadian rhythms, age, and strain. In rats, the increase in anesthetic depth caused by hypothermia and the sparing effects of concurrently used anesthetics and analgesics are the most common concerns, but age and genetic background also affect the required dose.

Orliaguet et al. (2001) investigated the influence of age on MAC for halothane, isoflurane and sevoflurane. They determined MAC in 2-, 9-, and 30-day-old and 10 to 12-week-old (adult) Wistar rats, and found that for each of these agents, the MAC values increased significantly up to 9 days of age and then progressively declined to commonly reported values by 10 to 12 weeks of age, with differences exceeding 70% between the extremes for each agent.

Differences in MAC attributable to genetic background were confirmed by Gong et al. (1998) who determined the MAC of desflurane and nitrous oxide in Fischer, Brown Norway, Long-Evans, Sprague-Dawley, and Wistar rats. For desflurane differences in MAC varied from 8.3% for Brown Norway rats to 6.6% for Wistar rats. Using isoflurane, similar, though lower differences were reported between Sprague-Dawley, spontaneously hypertensive (SHR), and Wistar-Kyoto rats (Cole et al., 1990).

Differences in MAC and, hence, in vaporizer settings are easily obscured by variations in vaporizer accuracy, delivery systems, concurrent drug use, and intraoperative support measures. It is clear, however, that variations attributable to age and genetic background can be clinically relevant and should be taken into account.

A. Isoflurane

Isoflurane was introduced for widespread clinical use in humans in 1984 and was approved for veterinary use soon after. It has since become perhaps the most widely used inhalation agent for rats. Isoflurane is pungent, and was reported to be more aversive in rats during chamber inhalation than halothane, although markedly less so than carbon dioxide (Leach et al., 2002). About 0.2% of the inspired dose of isoflurane undergoes biotransformation, with only desflurane and nitrous oxide undergoing less degradation (Steffey, 1996). The MAC for isoflurane in the rat has been determined in numerous studies with slight differences in results, but an average value appears to be about 1.4% (Dardai and Heavner, 1987; Imai et al., 1999).

In common with other halogenated volatile anesthetics, isoflurane produces dose-dependent reductions in respiratory rate, mean arterial pressure and heart rate (Dardai and Heavner, 1987; Imai et al., 1999; Steffey et al., 2003). In these studies, respiration rates were similar for isoflurane and sevoflurane at 1.0 and 1.5 MAC, but lower than those for halothane at equivalent doses. Nevertheless, at equipotent doses of these agents, arterial PCO_2 and pH were comparable. Imai et al. (1999), using a Bain circuit to deliver isoflurane in oxygen, noted that the arterial PO_2, while adequate at levels of about 250 mmHg, was lower than expected, a finding attributed in part to inadequate attention to periodic lung inflation. However, these levels are probably representative of those seen in rats maintained on mask systems.

Isoflurane exerts a negative inotropic effect on the heart, although both isoflurane and sevoflurane generally support cardiac output at clinically used concentrations (Steffey, 1996). As with other inhalation anesthetics and opioids, isoflurane provides some protection against myocardial ischemia and reperfusion injury, possibly by reducing neutrophil adherence and increasing resistance to oxidant induced injury (Ludwig et al., 2003: Hu et al., 2004). Short-term protection against focal and forebrain ischemia also occurs with isoflurane, at least partly because of its ability to greatly reduce cerebral metabolic rate and, with hyperventilation, limit increases in intracranial pressure. However, in models of severe forebrain or focal cerebral ischemia, the protection does not persist after a 2- to 3-week recovery period (Elsersy et al., 2004; Kawaguchi et al., 2004). The success with which isoflurane can be used for central and peripheral nervous system electrophysiological studies is variable and appears to be specific to the experimental design. A careful review of the current literature is needed when selecting a suitable anesthetic for such studies. In general, isoflurane appears to depress function and signal acquisition to a greater degree than halothane or some injectable anesthetics (Hayton et al., 1999; Antunes et al., 2003; Villeneuve and Casanova, 2003).

The anesthetic dose of isoflurane is reduced by concurrent administration of opioids. Criado et al. (2000) examined the effects of intravenous buprenorphine and morphine on the MAC of isoflurane (MAC_{ISO}).

Intravenous doses of buprenorphine administered at 10, 30 and 100 μg/kg, resulted in approximately 15%, 30%, and 50% reductions in the required dose of isoflurane. Similar reductions were observed for IV morphine at doses of 1, 3, and 10 mg/kg. The duration of effect was greater for buprenorphine than for morphine, and both drugs markedly decreased blood pressure, heart rate and respiratory rates given by this route, with morphine causing up to 5 minutes of apnea in some animals at all doses. The authors suggested that SC or IP routes be used for clinical purposes. Fentanyl and remifentanil, administered by continuous IV infusion, reduced MAC_{ISO} by up to 60% at the highest doses used (Criado and Gomez de Segura, 2003). In this study, fentanyl resulted in apnea at each of three dose levels, and there was dose-related respiratory and cardiovascular depression. Tramadol, given IV at 10 mg/kg reduced MAC_{ISO} by approximately 11% (deWolff et al., 1999). In the same study, IV morphine at 1 mg/kg lowered MAC_{ISO} by approximately 15%, a value confirmed by Criado et al. (2000).

Nonsteroidal antiinflammatory drugs (NSAIDs) do not appear to reduce MAC_{ISO} when used alone, but Gomez de Segura et al. (1998) found that aspirin acted synergistically with morphine to further reduce MAC_{ISO} in rats. Under similar circumstances, however, meloxicam failed to produce additional MAC_{ISO} reduction, (Santos et al., 2004), suggesting that cyclooxygenase (COX) receptor specificity may play a role in the degree of MAC reduction caused by NSAIDs.

Concurrent use of other drugs, such as α_2-adrenergics and others, will also reduce the dose of isoflurane and other inhalants to variable, but often substantial, degrees.

B. Sevoflurane

Sevoflurane was first synthesized in 1968 but development was halted in the United States because of concern about toxicity. Following further extensive development and safety testing in Japan, sevoflurane was released for clinical use in that country in 1990 and in the United States in 1995, making it the newest of the inhaled agents to be marketed (Kronen, 2003). Sevoflurane is less pungent and, possibly, less aversive than isoflurane during mask or chamber induction, although neither agent causes coughing or breath-holding in rats. It is also less potent than isoflurane or halothane, and has a higher vapor pressure than either, but the precision vaporizers needed to use it are widely available and of conventional design. The low blood and tissue solubility of sevoflurane results in rapid induction and recovery, and more responsive control of anesthetic depth (Eger, 1994).

Sevoflurane undergoes 2% to 3% biotransformation in the liver, with formation of both hexaflouroisopropanol and inorganic fluoride (Eger, 1994; Karasch et al. 1995). Although this degree of biodegradation is greater than that of isoflurane, it is far less than that usually attributed to halothane, and the *in vivo* degradation products appear to have minimal renal or hepatic effects (Cook et al., 1975; Eger, 1994). Sevoflurane also undergoes degradation by common carbon dioxide absorbents, such as soda lime and others. Of particular importance is a degradation product called Compound A, which causes renal corticomedullary necrosis, proteinuria, and enzymuria in rats and can be lethal at sufficient concentrations (Stabernack et al., 2003). However, these authors found no histological evidence of renal toxicity in control rats receiving only sevoflurane in oxygen for four hours using a nonrecirculating delivery system. Circle systems and carbon dioxide absorbers are rarely used for inhalation anesthesia in the rat and their use is easily avoided, thus circumventing the problem. Even with traditional absorbent systems, sevoflurane has proven to be safe in clinical use in larger animals and in humans, and the development of nontraditional carbon dioxide absorbents may further alleviate concern.

Steffey et al. (2003) determined the MAC_{SEVO} in 16- to 23-week-old male Sprague-Dawley rats to be about 3%. Kashimoto et al. (1997) reported somewhat lower values of 2.7% for 9-week old rats and 2.3% for male Wistar rats older than 13 months. The differences in these values and others, e.g. Crawford et al. (1992), have been attributed to differences in animals and study design (Steffey et al., 2003).

In general, sevoflurane seems to share many of the characteristics of isoflurane although there are differences in both the mechanisms and magnitude of effects that might be of experimental importance. Sevoflurane administration causes dose-dependent respiratory and cardiovascular depression to about the same extent as isoflurane at 1 and 1.5 MAC (Imai et al., 1999; Steffey et al., 2003), but the mechanism of direct myocardial depression caused by sevoflurane differs somewhat from that caused by halothane and isoflurane (Davies et al., 2000). Crawford et al. (1992) concluded that sevoflurane produced minimal hemodynamic alterations when compared with halothane in spontaneously breathing rats. The authors reported that coronary artery resistance and blood flow were unchanged by sevoflurane and that cerebral blood flow and vascular resistance were markedly less affected when compared with halothane. Correa et al. (2001) examined the effects of sevoflurane on respiratory mechanics and lung histology in the rat. Compared with pentobarbital, sevoflurane caused increased stiffness and mechanical inhomogeneities of the lungs that could be partially, but not completely, mitigated by atropine.

Sevoflurane is slightly less stable than isoflurane *in vivo* and *in vitro* and remains considerably more expensive.

For most purposes, it appears to provide few compelling advantages over isoflurane in the rat. If speed of induction and recovery from anesthesia is paramount, however, sevoflurane may be preferred. The results of concurrent administration of analgesics with sevoflurane are very likely to parallel those seen with isoflurane.

C. Halothane

Halothane entered clinical use in 1956, the first in a series of halogenated compounds increasingly characterized by rapid uptake and elimination, minimal metabolic degradation, and favorable hemodynamics. With the exception of methoxyflurane, halothane is more potent, and somewhat more soluble in tissue and blood than newer agents. It is not a respiratory irritant and is usually well accepted during induction by mask or chamber. Halothane USP is a clear, colorless compound containing 0.01% thymol as preservative. The appearance of a yellowish color in the vaporizer sight glass indicates thymol accumulation, which can impair proper vaporizer function, and indicates the need for vaporizer service and cleaning. In their study of halothane and sevoflurane in the rat, Steffey et al. (2003) determined the MAC of halothane in the rat to be 1.0%, in good agreement with previous reports. In contrast to isoflurane and sevoflurane, about 20% to 25% of administered halothane undergoes biotransformation (Steffey, 1996). Under normal circumstances, the resulting metabolites are of little concern, but if P-450 enzyme activity has been induced by prior drug administration, hepatotoxicity can occur in rats (Kenna and Jones, 1995).

As with isoflurane and sevoflurane, halothane produces dose dependent depressions of mean arterial pressure and respiration. However, compared with isoflurane or sevoflurane at equivalent anesthetic doses, halothane causes greater direct myocardial depression and greater reduction of cardiac output. Thus, the drop in blood pressure associated with halothane is attributed more to its myocardial effects than to reduction in vascular resistance. Unlike other modern halogenated inhalation anesthetics, halothane is not an ether, and can sensitize the myocardium to epinephrine. Compared with sevoflurane at doses of 1 MAC, halothane produced greater reductions in coronary artery, renal, and total hepatic blood flow, and greater increases in cerebral blood flow (Crawford et al., 1992). Respiratory depth and frequency are reduced with increasing doses of halothane, as with isoflurane and sevoflurane, but at equivalent anesthetic doses, differences in arterial PCO_2 and pH are not pronounced. Antunes et al. (2003) noted that halothane causes less CNS depression than does isoflurane at equivalent doses and may be preferred if inhalation anesthesia is used for studies involving electroencephalography and auditory evoked potentials. Similar results were reported by Villeneuve and Casanova (2003) for single-cell recordings in the visual cortex of cats.

Morphine at dosages of 2, 4, and 8 mg/kg SC reduced the MAC_{HALO} by 22%, 33%, and 55%, respectively, with no adverse effects noted (Lake et al., 1985). In the same study, alfentanil given by IV infusion at rates up to 10 μg/min produced MAC reductions of about 27% without adverse effects, but higher doses produced an increasingly high frequency of chest wall and abdominal rigidity. Hecker et al. (1983), describe sufentanil infusions between 0.05 and 0.10 μg/kg/min as reducing the MAC_{HALO} from 24% to 86%, with respiratory depression.

Halothane is potent and safe when properly used and is the least expensive of modern inhalations anesthetics. After nearly 50 years of use, the body of experience and literature concerning it is vast. Even so, halothane has been displaced for most purposes by isoflurane and sevoflurane, both of which have better anesthetic characteristics, do not foul vaporizers, and are thought to present less risk to personnel. Manufacture of halothane has been discontinued in the United States, although it appears that production will continue elsewhere. It seems likely that the pattern of scarcity and increased cost already seen with methoxyflurane is about to be repeated.

D. Nitrous Oxide

Nitrous oxide is used as an adjuvant to reduce the dose of other inhaled or injectable anesthetics, but alone cannot produce anesthesia. While nitrous oxide does provide some analgesia, its potency in animals is only about of half that in humans. In the rat, the MAC for nitrous oxide is extrapolated to be between 136% and 235%, similar to most other mammals (Steffey, 1996). Thus, large doses of nitrous oxide, usually 70% to 75% of the breathing mixture, must be administered to provide a significant effect. Under these circumstances, particular care is necessary to ensure adequate oxygen is provided to avoid hypoxia. Nitrous oxide has a rapid onset and elimination with minimal metabolism, and relatively little effect on cardiovascular and respiratory parameters.

Nitrous oxide is often used in conjunction with muscle relaxants (paralytics) to provide "light anesthesia" following instrumentation, but the practice is questionable unless other agents are also used to assure unconsciousness and analgesia (Steffey and Eger II, 1985; Mahmoudi et al., 1989). When used in this way, pilot studies should be performed without paralytic drugs to assure adequate anesthesia is reliably attained.

E.   Carbon Dioxide

Carbon dioxide is used for euthanasia and for brief anesthesia of small laboratory animals. Anesthesia appears to be the result of acidosis, and presumably, subsequent electrolyte disturbances in the cerebrospinal fluid (Kohler et al., 1998). Using a precharged chamber, the onset of anesthesia following exposure to 70% to 80% carbon dioxide is rapid, with induction occurring in about 20 to 30 seconds, and anesthesia in about 1 to 3 minutes (Danneman et al., 1997a; Kohler et al., 1998). When used as an anesthetic, sufficient oxygen must be included to avoid fatal hypoxia. Although there are some discrepancies in recommended concentrations, an 80% CO_2/20% O_2 mixture is often used. Controversy exists regarding the best method for administration of CO_2/O_2. Danneman et al. (1997a), emphasize the aversive, and possibly painful, aspects of carbon dioxide and recommend using moderate flow rates into a nonprecharged chamber as the most humane technique. Kohler et al. (1998) urge use of high flow rates and a prefilled chamber in order to attain rapid induction and consistent anesthesia. Both authors, however, agree that recovery ensued within 1 minute following removal from the chamber.

Carbon dioxide/oxygen mixtures can be blended by the gas supplier for delivery in a compressed gas cylinder, or can be blended on site using a blending valve. The cost of premixed cylinders is high compared to the cost of cylinders of pure CO_2 and O_2, but adjustable gas mixing valves are also somewhat costly, so the choice of methods rests upon the anticipated rate of use. Users should remember that the mixture is used to produce unconsciousness and will work equally well for humans: it should be used in a well-ventilated area, or the effluent of the chamber should be scavenged.

Urbanski and Kelley (1991) reported minimal alteration of luteinizing hormone, follicle stimulating hormone, prolactin and corticosterone following carbon dioxide exposure. Similarly, Fowler et al. (1979) found no alterations in packed cell volume, total protein, alanine aminotransferase (ALT), and urea, but noted that glucose was elevated. Fomby et al. (2004) in a study using Fischer 344 rats advocate acclimation and administration of CO_2/O_2 in the home cages as a means of reducing baseline stress responses and obtaining reliable basal corticosterone concentrations. However, Nahas and Provost (2002) compared CO_2/O_2 and isoflurane to historical values for blood samples collected under ether anesthesia and noted that CO_2/O_2 resulted in elevated serum potassium levels, mild elevation of sodium and total protein and increases in several red cell parameters, including hemoglobin, red cell count, and packed cell volume (PCV). They further concluded that for the purposes of the clinical laboratory, a 3-minute exposure to isoflurane in oxygen was preferable to either ether or CO_2/O_2. Thus, carbon dioxide anesthesia does appear to alter at least some clinical chemistry and hematological values.

In their chapter on anesthesia and analgesia of rodents in *Anesthesia and Analgesia of Laboratory Animals* (1997), the authors conclude that there is "...no consensus on the usefulness and humane acceptability of carbon dioxide anesthesia" (Wixson and Smiler, 1997). Seven years following publication of that book, the status of carbon dioxide anesthesia remains controversial.

VII.   PARENTERAL AGENTS

A.   Preanesthetic Drugs and Anesthetic Adjuvants

1.   Anticholinergics

Atropine and glycopyrrolate are usually used in anesthesia to prevent or treat bradycardia, and to minimize salivation and respiratory secretions. Olsen et al. 1993 compared atropine at 0.05 mg/kg and glycopyrrolate at 0.5 mg/kg in the rat and found glycopyrrolate to be superior as a preanesthetic agent in terms of maintaining heart rate during ketamine/xylazine or ketamine/detomidine anesthesia. Anticholinergics are also used treat or prevent the oculocardiac and similar vagal reflexes in surgery of the head and neck (Hall et al., 2001a; Naff et al., 2004). Anticholinergic drugs are invaluable when indications for their use are present or reasonably predictable, but they are not wholly benign, and their indiscriminate use is not warranted. Unlike diethyl ether, modern inhalation agents are not respiratory irritants, so airway drying may be a greater concern than secretions. Further, the use of anticholinergics to prevent the bradycardia associated with α_2-adrenergic administration has also been questioned (Savola, 1989; Alibhai et al., 1996).

2.   Sedatives and Tranquilizers

These drugs are usually given during, rather than before anesthetic induction. In the rat, the distinctions between sedation, tranquilization and hypnosis, while real, are rarely made. If the goal is only to reduce anxiety without analgesia, benzodiazepines or phenothiazine tranquilizers may be appropriate. However, if some degree of analgesia is desirable, partial or mixed agonist opioids or α_2-agonists are often used alone or in combination with nonanalgesic drugs.

a.   PHENOTHIAZINE DERIVATIVES.   Acepromazine is the most commonly used representative of this large class of drugs, although chlorpromazine and promazine are occasionally encountered. In sufficient doses, these agents produce sedation, and some degree of antiemetic, antihistaminic, and antispasmodic activity but provide no significant analgesia. Despite many years of use, little specific information is available concerning the pharmacology of phenothiazine derivatives in rodents (Fish, 1997). Acepromazine is used in anesthetic cocktails, combined with ketamine, or with ketamine and xylazine, to prolong anesthesia and reduce the doses of ketamine required. Dose-dependent reduction of blood pressure results from vasodilatation, apparently mediated by blockade of peripheral α_1-adrenoreceptors (Hall et al., 2001a). Acepromazine alone is said to produce mild sedation in the rat at doses of 1.0 to 2.5 mg/kg, IM, IP, SC (Flecknell, P.A. 1996d).

b. BENZODIAZEPINE DERIVATIVES.   Diazepam, midazolam and zolazepam produce sedation, anxiolysis, and skeletal muscle relaxation, and potentiate the effects of anesthetics. Used alone, these agents usually produce only mild sedation, and provide no analgesia. Diazepam is poorly soluble in water, and is usually formulated in propylene glycol; when diluted in saline or water-soluble anesthetics it tends to precipitate. By contrast, midazolam is water soluble and much shorter acting than diazepam, properties that have advantages in anesthesia. For sedation in rats, doses of diazepam from 2.5 to 15.0 mg/kg IM, IP, SC, and midazolam at 5.0 mg/kg IP, have been suggested (Quinn et al., 1994; Flecknell, 1996d). Zolazepam is avaiable only in combination with tiletamine, marketed as Telazol®. Flumezenil, a specific benzodiazepine antagonist, is used as a reversal agent for benzodiazepines in the event of an overdose, or to hasten recovery.

c. BUTYROPHENONE DERIVATIVES.   For anesthesia purposes, droperidol is the single example of this class easily available in North America. In humans this class of drug is used as a major tranquilizer and antipsychotic, and sometimes as a preoperative drug to produce sedation with minimal cardiovascular and respiratory depression. Butyrophenone tranquilizers have been combined with opioids to produce neuroleptic anesthetics, including Innovar-Vet®, and Hypnorm®, discussed further below. The use of droperidol as a sedative in rats, given at 0.5 to 2.0 mg/kg SC, was described by Quinn et al. 1994.

d. α_2—AGONISTS.   Drugs in this group include detomidine, romifidine and several others, but the most frequently used in rats are xylazine and medetomidine. The pharmacology of these drugs has been reviewed in current anesthesia and pharmacology texts and, in the case of medetomidine, in a supplement to Acta Veterinaria Scandinavica, **85**, 1989, to which the interested reader is referred. Briefly, the ability of α_2-agonists to produce sedation and analgesia depends upon their affinity and specificity for receptors in the central nervous system. Because α_1-receptors mediate arousal, more selective α_2-agonists produce a greater anesthetic sparing effect (Vainio, 1997). However, adrenergic receptors are widely distributed in the body, and the cardiovascular and other effects of α_2-agonists are due to their activity in both central and peripheral sites. Xylazine was the first of these drugs widely used in veterinary medicine and, in combination with ketamine, remains one of the most commonly used drugs for anesthesia of the rat. Medetomidine is newer, more potent, and more selective for the α_2-adrenoreceptor than xylazine. Medetomidine is a racemic mixture of which only the dextrorotatory isomer, dexmedetomidine, is active. Purified dexmedetomidine is available and used in human anesthesia, but is considerably more expensive than medetomidine and there appears to be little advantage to its use in ordinary circumstances (Ramsay and Luterman, 2004).

Used alone, these drugs produce dose-dependent sedation and analgesia, but they are more often used in conjunction with other drugs to take advantage of their anesthetic sparing effects. Premedication with these drugs will significantly reduce the dose of concurrently administered inhalation and parenteral anesthetics. In the rat, sedative doses for xylazine range from 1 to 5 mg/kg and for medetomidine from 30 to 250 µg/kg, depending upon the degree of sedation required (Flecknell, 1996d; Flecknell, 1997: Hauptman et al., 2003; Macdonald et al., 1989). Both drugs cause dose-dependent hypothermia.

Cardiovascular effects include vasoconstriction with increased systemic vascular resistance, initial hypertension, and bradycardia. Following the initial hypertensive phase, mean arterial blood pressure falls to normal or below normal levels with reduced cardiac output. The magnitude of hypertension is affected by the route and speed of administration, rapid IV administration producing greatest rise in pressure. With medetomidine, premedication with atropine or glycopyrrolate delayed, but did not prevent bradycardia, and simultaneous administration produced persistent tachycardia and appeared to be related to unanticipated deaths (Savola, 1989; Vainio, 1989). Vasoconstriction and bradycardia can result in pale mucus membranes or cyanosis. It has been suggested that cyanosis does not represent a reduction in arterial oxygen content, but rather increased venous desaturation resulting from the slow movement of blood through the peripheral circulation (England and Clarke, 1989; Sinclair, 2003). In any event, vasoconstriction may be sufficient to interfere with pulse oximeters, depending upon the instrument, probe and monitoring site selected. Respiratory depression is usually moderate when α_2-adrenergic agents are used alone, but when they are combined with other drugs, such as potent opioids, it may become severe.

Alpha$_2$ agonists reduce both the central secretion of vasopressin and inhibit its renal effects, resulting in diuresis and sodium loss (Cabral et al., 1998). Hyperglycemia results from suppression of insulin release and stimulation of glucagon release and, if sufficiently great, may also contribute to diuresis. Thus, perioperative fluid support would seem particularly useful when these drugs are used. A unique side effect, the occurrence of acute, reversible cataracts, has been associated with xylazine use in the rat (Calderone et al., 1986). In the author's experience, the occurrence of cataracts is sporadic and transient, disappearing upon recovery from anesthesia. High doses of xylazine, in the range of 21 to 42 mg/kg have been reported to cause pulmonary edema in rats, attributed to increased permeability due to endothelial injury (Amouzadeh et al., 1991). Unexpected deaths and acute pulmonary edema were also reported by Roughan et al. (1999) in a study of medetomidine/ketamine anesthesia preceded by buprenorphine as a preanesthetic analgesic.

The ability to reverse the effects of α_2-adrenergic drugs provides the opportunity to shorten the anesthetic recovery period and to treat adverse effects. The antagonists most frequently used in rats are yohimbine and atipamezole. Yohimbine is a less specific antagonist, but has long been used to shorten recovery from xylazine/ketamine anesthesia, at doses of up to 1.0 to 2.1 mg/kg IP (Hsu et al., 1986; Komulainen & Olson, 1991). At these doses, recovery occurred in about 10 minutes with reversal of sedation, bradycardia, bradypnea and polyuria due to xylazine. Atipamezole is a much more selective α_2-antagonist, and doses from 0.1 to 1.0 mg/kg IM, IP, SC have been recommended, depending upon the dose of xylazine or medetomidine used (Flecknell, 1997).

Despite the cardiovascular and other effects described above, combinations of xylazine or medetomidine and ketamine are popular injectable anesthetics that have proven to be safe and reliable in healthy rats for surgical procedures of short to moderate duration. The concurrent use of α_2-adrenergic agonists with ketamine, fentanyl, sufentanil and other drugs are described in following sections.

3. Analgesics

Analgesics are used before, during and following surgery to reduce postoperative pain and to reduce the dose of anesthetic needed to provide adequate surgical anesthesia. While other drugs can provide analgesia and reduce anesthetic requirements, the opioids, nonsteroidal analgesics and local anesthetics are the drugs most commonly used to produce analgesia in the rat.

a. OPIOIDS.   Opioids act at μ, σ and κ receptors located in the central nervous system, and peripherally. They can be classified by their affinity for each receptor, or

by their activity as agonists, partial agonists, mixed agonist-antagonists, or antagonists at individual receptor types. Partial agonists exert an incomplete effect, regardless of dose, at the specified receptor. Some opioids, classified as mixed agonists-antagonists, can exert an agonist effect at one receptor and an antagonist effect at another (Nolan, 2000; Gutstein and Akil, 2001). The most effective analgesics appear to be full μ receptor agonists, although σ and κ receptors also mediate analgesia (Nolan, 2000; Gutstein and Akil, 2001). In humans, the μ receptor is associated with euphoria and the κ receptor with dysphoric effects. Peripheral receptors can interact with locally applied opioids, a fact sometimes used to advantage in larger species, but seldom if at all, in rats. However, systemic μ agonists are known to have antiinflammatory effects peripherally (Barber and Gottschlich, 1992).

Sex and strain related differences in response to opioids have been shown in rats using a thermal model for analgesiometry, with females generally requiring higher doses than males and Fischer 344 male rats requiring lower doses than Lewis males (Cook et al., 2000). Using a mechanical model for Fischer 344 rats, Barrett et al. (2002) reported that males had a higher threshold for nociception than females. By contrast, Kroin et al. (2003) found no differences between male and female Sprague-Dawley rats using an incisional pain model. Further indication of strain variability can be found in Mogil (1999). The differences in response seen using different noxious stimuli and means of assessment highlight the importance of clinical assessment as a basis for dose adjustment, rather than sole reliance on published dose ranges.

With few exceptions opioids are classified as narcotics with abuse potential and their acquisition and use is subject to legal controls in most jurisdictions.

i. Pure Agonists.   These drugs include a large number of morphine derivatives, often used for postoperative pain control, and other potent shorter-acting μ agonists more commonly used in conjunction with anesthesia. Thus, morphine and oxymorphone are more commonly used for perioperative analgesia, and fentanyl, sufentanil and other potent, shorter-acting drugs are more likely to be encountered as components of anesthesia.

Morphine is the archetypal full μ agonist, with much less affinity for σ and κ receptors, and is used more frequently as a perioperative analgesic than as an anesthetic component. In the rat, morphine is a potent analgesic, with some sedative effects, lasting approximately 2 to 4 hours at doses of 2 to 10 mg/kg (Wixson and Smiler, 1997; Liles et al., 1998; Gades et al., 2000). Wixson and Smiler (1997) suggest the higher dose of 10 mg/kg repeated every 2 to 4 hours for severe pain. However, in a study using midline ovariohysterectomy as a surgical pain model

Gonzalez et al. (2000), found that 3 mg/kg SC of morphine given 0.5 hours before surgery and 0.5 and 2 hours after surgery completely blocked abdominal signs of pain and delayed allodynia for up to 48 hours. They also noted that a single dose of morphine at 3.0 mg/kg was effective for only 1.5 hours.

Morphine, in common with other pure agonists, can result in bradycardia, respiratory depression, delayed gastrointestinal transit, urinary retention, and excitement, and some immunosuppression, among other unwanted effects, though at clinically used doses, these adverse effects are not ordinarily severe. If morphine is administered preoperatively, however, respiratory depression and cardiovascular effects should be taken into account when planning the anesthetic regimen. The use of morphine with isoflurane has been discussed earlier in this chapter. When used with injectable anesthetics, it may be preferable to administer morphine and similar drugs late in the procedure or during recovery to avoid prolonged recovery and respiratory depression (Dobromylskyj et al., 2000).

Oxymorphone is a morphine derivative that is more potent than morphine, but with similar effects. Gillingham et al. (2001) compared oxymorphone with buprenorphine for pain relief in rats following intestinal resection. Oxymorphone was given by intermittent injection at doses of 0.03 mg/kg hourly or 0.18 mg/kg every 6 hours, or by continuous intravenous infusion at 0.03 mg/kg/hr, or by intraperitoneal Alzet pump at the rate of 0.03 mg/kg; buprenorphine was administered at 0.5 mg/kg SC every 6 hours. The authors concluded that oxymorphone, given at 0.03 mg/kg hourly, was superior to buprenorphine in providing analgesia following intestinal resection in rats. An experimental liposome-encapsulated form of oxymorphone was reported by Krugner-Higby et al. (2003) to be effective for up to 24 hours following a single dose in rats undergoing intestinal resection. Smith et al. (2003) reported suppression of neuropathic pain in a sciatic nerve ligation model for up to 7 days following a single dose of the same liposome-encapsulated preparation. Unfortunately, while these results are exciting and hold promise for the future availability of effective long-acting analgesics, oxymorphone prepared in this way is not currently commercially available, and the method of preparation is not feasible in most facilities.

Fentanyl, sufentanil carfentanil, alfentanil and remifentanil are all short-acting potent full μ agonists, primarily used to provide analgesia during anesthesia as part of a balanced anesthesia technique. The dose-sparing effect of some of these agents on inhalation anesthetics is discussed earlier in this chapter. The potent analgesic effects of these opioids, and the ability to reverse their effects with weak or partial agonists or antagonists make them attractive as anesthetic components. Bradycardia caused by these drugs is greater than that seen with morphine and related drugs, but is responsive to anticholinergic drugs such as atropine and glycopyrrolate. As with other opioids, however, direct myocardial depression is minimal, and when used in conjunction with inhalation agents and some other anesthetics, they can provide excellent cardiovascular stability during anesthesia (Gutstein and Akil, 2001). Respiratory depression is also more prominent with these drugs, so that supplemental oxygen is always desirable and, at higher doses, mechanical ventilation is necessary. Of potential concern are reports of neurotoxicity in the rat related to fentanyl (Kofke et al., 1996a; Kofke et al., 1996b), alfentanil (Kofke et al., 1992), and remifentanil (Kofke et al., 2002), which may limit use of these agents in some neurological studies. Recent improvements in venous access, controlled intravenous infusion, endotracheal intubation and anesthetic monitoring in the rat combine to make the use of these drugs more attractive and feasible in selected cases.

Two fixed-dose combinations of fentanyl with butyrophenone tranquillizers have been marketed. Fentanyl was combined with droperidol as Innovar-Vet®, but is no longer manufactured, and the human version differs sufficiently in formulation that it is not a useful substitute in animals. Hypnorm™ is a combination of fluanisone (10 mg/mL) and fentanyl citrate (0.315 mg/mL) which, when administered IP with midazolam (1.25 mg/kg) or diazepam (2.5 mg/kg), becomes a very satisfactory anesthetic for rats, mice, rabbits and guinea pigs (Green, 1975). Fluanisone is available only as a component of this drug mixture. The combination provides good surgical anesthesia for 20 to 40 minutes in the rat, and anesthesia can be prolonged for up to 6 to 8 hours by intermittent IM injections (Flecknell and Mitchell, 1984). Respiratory depression occurs with this combination, and it has been recommended that supplemental oxygen should be provided (Whelan and Flecknell, 1994). Anesthesia recovery can be hastened by nalbuphine (1 mg/kg IP/SC) or butorphanol (2 mg/kg IP/SC), preserving some degree of analgesia. The use and preparation of Hypnorm™ and midazolam is discussed more fully by Wixson and Smiler (1997) and Flecknell (1996d). The original manufacturer discontinued production of Hypnorm™, but another company (Abbeyvet Ltd., Shelburn-in-Elmet, U.K.) acquired the license and it appears that the drug will remain available. Although widely used in Europe and elsewhere, Hypnorm™ remains unavailable in the United States without an FDA import license.

By using fentanyl with α_2-andrenergics, a completely reversible injectable anesthetic can be formulated. At least two such combinations of medetomidine with fentanyl and sufentanil have been evaluated in the rat. Hu et al. (1992) described the anesthetic combination of fentanyl and medetomidine and its reversal with atipamezole and nalbuphine in Wistar rats. Doses of 300 µg/kg of fentanyl and 200 to 300 µg/kg of medetomidine, given IP, were found to result in acceptable anesthesia. At the 300 µg/kg fentanyl + 300 µg/kg medetomidine dose level, the combination provided about 60 minutes of surgical anesthesia. Significant respiratory depression and prolonged recovery times were noted, but blood gases, blood pressure and heart rate were not reported. Providing supplemental oxygen during anesthesia seems prudent. Rapid recovery, with reversal of respiratory depression and sedation, followed administration of atipamezole (1 mg/kg SC/IP) and nalbuphine (1 to 2 mg/kg SC/IP). Fentanyl and medetomidine anesthesia has been characterized as safe and effective by Flecknell, (1996d), who also noted that allowing the animals to acclimate for 1 to 2 hours following movement to the procedure area greatly improves the quality of induction and recovery.

Sufentanil, a potent µ agonist, has also been used with medetomidine to produce surgical anesthesia in the rat (Hedenqvist et al., 2000). The increased potency of sufentanil compared with fentanyl allowed lower dose volumes to be used, both SC and IP. The authors found that SC administration was more reliable than IP, and suggested doses of 40 to 50 µg/kg of sufentanil and 150 µg/kg of medetomidine produced surgical anesthesia lasting from about 100 to about 120 minutes. However, they also noted profound respiratory depression, and blood gases confirmed severe hypoxemia and hypercapnia. It was suggested that the combination was suitable only for animals free of respiratory disease and that supplementary oxygen be used during anesthesia. The mixture can be reversed using butorphanol and atipamezole (0.2/0.5 mg/kg SC).

The effects of fentanyl/medetomidine or sufentanil medetomidine on heart rate and blood pressure were not reported in either of the two studies cited. However, the potential complications arising from simultaneous use of atropine or glycopyrrolate with medetomidine and other α_2-adrenergic drugs were discussed above, and their use with these combinations should be approached with circumspection.

Remifentanil is a novel ultra-short acting full µ agonist. The pharmacodynamics of remifentanil and its metabolite in the rat were reported by Cox et al. (1999). Criado and Gomez de Segura (2003) described its use with isoflurane in rats. In this study, remifentanil compared favorably with fentanyl in terms of respiratory depression and cardiovascular effects. The recovery characteristics of the combination were not studied, however. The very rapid onset and disappearance of remifentanil would seem to make it especially well suited for continuous infusion or for bolus administration to block brief, very painful procedures but no further reports of the clinical use of remifentanil in rats have appeared. Because of its rapid offset, persistent analgesia cannot be expected; postoperative pain must be addressed by other means.

Tramadol (Ultram^{TM}) is a synthetic codeine derivative and a weak µ receptor agonist. It is unusual in that a significant part of its analgesic effect is due to inhibition of norepinephrine and serotonin reuptake (Gutstein and Akil, 2001). Additionally, the principle µ opioid effects of tramadol appear to be due to one of its metabolites, M1, which has a 4- to 10-fold greater affinity for the µ receptor than the parent compound (de Wolff et al., 1999). Tramadol is available as 50 mg tablets. It is a racemic mixture in which the combined enantiomers are more effective that either alone (Gutstein and Akil, 2001). The MAC$_{ISO}$ sparing effects of 10 mg/kg of tramadol IV are moderate and approximately equal to that of 1 mg/kg of morphine IV, and the MAC sparing effects of both drugs are abolished by pretreatment with naloxone (de Wolff et al., 1999). The antiinflammatory effects of tramadol were examined by Bianchi et al. (1999) who reported significant reductions of inflammation and hyperalgesia in yeast-stimulated paw edema and subcutaneous carrageenan-induced inflammation. The antiinflammatory mechanism is not clearly established, but does not occur by direct action on cyclo-oxygenase (Biachi et al., 1999). Gaspani et al. (2002) reported that, in a natural killer cell sensitive tumor model in rats, tramadol blocked the enhanced metastasis associated with surgery and anesthesia and prevented suppression of natural killer cell activity, in contrast to morphine. Tramadol was also effective in suppressing hyperalgesia and producing analgesia in a sciatic nerve ligation model of neuropathic pain, although at the highest dose used, 10 mg/kg IP, responses approached baseline control values at 150 minutes (Apaydin et al., 2000). In a well-established model of colic pain produced by ureteral calculosis, tramadol injected IP twice daily over four days demonstrated dose-dependent reduction in the number and duration of ureteral crises (Affaitati et al., 2002). However, despite the encouraging and interesting results of these studies, there are no studies demonstrating the clinical utility or dosage of tramadol as an analgesic in laboratory rats.

ii. Partial Agonists.   The dominant drug in this group is buprenorphine, usually described as a partial µ agonist, with significant activity at the κ receptor. Following the publication by Flecknell (1984) suggesting buprenorphine as a useful, long acting analgesic in rats at doses of 0.1 to 0.5 mg/kg, it rapidly gained popularity and widespread use. Subsequent experience resulted in reduced dose recommendations to the level of 0.01 to 0.05 mg/kg SC currently suggested (Liles and Flecknell, 1992a). The same study documented reduced food intake in response to administration of buprenorphine, nalbuphine and butorphanol compared to unoperated rats, as well as increased locomotor activity attributable to buprenorphine. In a comparison of buprenorphine and bupivacaine, Liles and Flecknell (1993) again found that buprenorphine ameliorated decreases in weight loss, water consumption and food consumption in rats undergoing laparotomy with and without bile duct ligation. Hayes and Flecknell (1999) compared perioperative versus postoperative dosing of buprenorphine and found increased food intake with preoperative administration, but no significant differences in weight loss or water consumption between the groups over a 14-hour period. Jablonski et al. (2001), using a laparotomy model, assessed buprenorphine in two strains of rats, Sprague-Dawley and Dark Agouti. They reported that buprenorphine-treated groups lost less weight than untreated animals on the first postoperative day, but also found that weight losses in most animals receiving buprenorphine were greater than untreated animals over the following two days. They further noted apparent differences in dose-response between the strains, with Dark-Agouti requiring higher doses than Sprague-Dawley rats. Hedenqvist et al. (1999) and Roughan et al. (1999) reported unexpectedly high death rates following preanesthetic administration of buprenorphine with ketamine/medetomine anesthesia, and the authors recommended caution in the use of opioids with ketamine/medetomidine in rats. The duration of effect of buprenorphine is usually given as 6 to 12 hours (Flecknell, 1984; Gades et al., 2000). However, this can be expected to vary with the individual animal and the nature and degree of pain produced, and dosage intervals should be adjusted accordingly.

With increasing use, some undesirable effects of buprenorphine were reported. Clark et al. (1997) observed the occurrence of pica with bedding ingestion in rats undergoing laparotomy and partial hepatectomy, and treated with 0.3-mg/kg buprenorphine. In follow-up studies they demonstrated pica with dose dependent severity in all rats receiving buprenorphine at doses of 0.05 to 0.3 mg/kg. Jacobson (2000) also observed pica and occasional deaths in 2% to 3% of animals treated with 0.05 mg/kg buprenorphine following surgery, and recommended that animals receiving buprenorphine be kept on grid floors to avoid pica, and that the dose of buprenorphine be adjusted to the strain of animal and anticipated severity of pain. Both Jacobson (2000) who observed pica, and Jablonski et al. (2001), who did not, suggested that the gastrointestinal effects of buprenorphine contributed to adverse effects on postoperative weight gain. Thompson et al. (2004) also reported pica in Sprague-Dawley, but not Long-Evans rats. Overall, the frequency of pica is not well documented, but seems to be sporadic, and both strain and dose-related. The author has observed very low levels of pica in Sprague-Dawley male rats from one vendor and a virtually 100% incidence in Sprague-Dawley male rats from another under similar clinical conditions.

Peck et al. (2004) reported in a post-operative adhesion model, buprenorphine administered at 0.05 mg/kg SC during anesthesia induction and 0.3 mg/kg orally 6 hours later appeared to have adequate analgesia for 24 hours, but had a significantly increased incidence and severity of adhesions compared to saline-treated controls.

Pekow (1992) proposed an oral formulation of buprenorphine in gelatin as a suitable method of administration. Liles et al., 1998, reported that oral administration of buprenorphine in gelatin was effective in decreasing weight loss and maintaining water consumption in a rat laparotomy model, but also stressed the difficulty of behavioral assessments in evaluating the beneficial effects of buprenorphine. Martin et al. (2001) disputed the efficacy of oral buprenorphine. Using Long-Evans rats, they found that orally administered buprenorphine gelatin failed to produce measurable analgesia, when assessed by analgesiometry as opposed to 0.05 mg/kg buprenorphine injected SC., Roughan and Flecknell (2002a), in an extensive review of buprenorphine, questioned these conclusions, defending the efficacy of oral buprenorphine gelatin, based on previous studies and clinical experience, and suggested that failure to detect analgesia by Martin et al. (2001), might be due to technical errors in preparation of the buprenorphine gelatin. Thompson et al. (2004), repeated the findings of Martin et al. (2001). In that study, the authors also verified the availability of oral buprenorphine as prepared in the study by Martin et al. (2001), noted the appearance of strain-related pica, and suggested that the lowest effective parenteral dose of buprenorphine be used for postoperative analgesia. Thus, while analgesia following parenteral buprenorphine is not disputed, the efficacy of oral buprenorphine is contentious.

It seems likely at this time that there has been more experience and critical evaluation of buprenorphine as an analgesic in rats than for any other agent. This experience coincided with a period of increasing appreciation of the problem of pain in laboratory animals, and greater sophistication in clinical assessment and treatment, a situation in which buprenorphine, with its promise of effective durable analgesia, played a major role. Thus, it is hardly surprising that experience and critical appraisal of the drug has resulted in recognition of its disadvantages. While the role of buprenorphine will continue to change with the appearance of new analgesics and analgesic formulations, it remains a valuable drug in the rat when used with appropriate care and clinical judgment.

iii. Mixed Agonists. Butorphanol is a competitive μ antagonist and strong κ agonist. The analgesia produced by butorphanol stems from its activity at the κ receptor (Gutstein and Akil, 2001). Gades et al. (2000) compared morphine, butorphanol and buprenorphine in Sprague-Dawley rats using hot plate and tail-flick assays.

They found butorphanol to have the least duration, at 1-2 hours, and least analgesic effect, and recommended it be used for mild pain of short duration at a dose of 2 mg/kg every 1-2 hours. Similar recommendations are found in Dobromylskyj et al. (2000). Butorphanol can be used to reverse the effects of full μ agonists, such as fentanyl, while still providing some degree of analgesia (Hu et al., 1992; Flecknell, 1996d; Hedenqvist et al., 2000). In a study examining the antiinflammatory effects of opioids, Vachon and Moreau (2002) found that butorphanol reduced carrageenan-induced paw inflammation, rendering it unsuitable as a clinical analgesic in that experimental model.

Nalbuphine resembles butorphanol in its receptor activity. Analgesia is primarily mediated by the κ receptor, and is characterized by a ceiling effect such that increasing doses do not increase its analgesic effects. Like butorphanol, it is useful as a means of reversing μ-agonists. Flecknell and Liles (1991) found that 6 doses of nalbuphine at 1 mg/kg SC every 4 hours, but not 3 doses, ameliorated reductions in food and water consumption following nephrectomy in rats, when compared to saline-treated controls, but concluded that the beneficial effects were "...due to its stimulatory effect rather than a specific analgesic action." If used as an analgesic, recommended doses are 1 to 2 mg/kg every 4 hours (Dobromylskyj et al., 2000). DiFazio et al. (1981) describe the use of nalbuphine infusion as a component of balanced anesthesia, although more potent μ agonists such as fentanyl and sufentanil would be more likely choices at present.

iv. Antagonists. The use of butorphanol and nalbuphine to reverse μ agonist opioids is discussed above. Naloxone and naltrexone are both used to reverse or block the effects of opioids. Naltrexone is often used in drug treatment programs, and naloxone is generally used to treat opioid overdose. In rats, the dose for naloxone is 0.01 to 0.1 mg/kg, IM, SC, or IV (Dobromylskyj et al., 2000). Naloxone will reverse not only the respiratory depression and other unwanted effects of μ-agonists, but will also reverse analgesia. If circumstances permit, it is better to titrate an effective dose, rather than give a single bolus. Because naloxone is fairly short acting, the dose may need to be repeated if very high doses of μ agonists have been used, and for partial agonists, higher doses may be needed (Hall et al., 2001a).

b. NONSTEROIDAL ANTIINFLAMMATORY DRUGS. This large, chemically disparate, group of drugs includes aspirin, acetaminophen, ketoprofen, carprofen, flunixin, meloxicam, and many others. The pharmacology of NSAIDs is complex and the subject of much active research, and is reviewed in standard texts and current literature (Nolan, 2000; Roberts and Morrow, 2001; Warner and Mitchell, 2004). Briefly, these drugs act at to inhibit, with varying degrees of specificity, cyclooxygenase (COX) and subsequent production of prostaglandins and thromboxane. Two COX sites, referred to as COX-1 and COX-2, are the targets of the drugs, and both are involved in the production of prostaglandins and thromboxane (Warner and Mitchell, 2004). It is thought that the inhibition of COX-2 accounts for the antiinflammatory, antipyretic, and analgesics actions of NSAIDs, and inhibition of COX-1 results in undesirable side effects including gastric ulceration and renal damage (Roberts and Morrow, 2001). The distinction of COX-2 inhibition as good, and COX-1 inhibition as bad may be a useful generalization but is certainly an over simplification (Nolan, 2000). Of this very large group of drugs, only those NSAIDs commonly used in laboratory rats and, for which some data supporting their clinical use, are discussed below. In contrast to human and veterinary clinical medicine, controlled studies of the efficacy and safety of these agents for clinical use in rats and other small laboratory animals is sparse. The dosage recommendations are extrapolated, for the most part, from research, safety, and pharmacological studies that do not mimic clinical indications and use in the laboratory animal setting. Further, there are few published direct clinical comparisons of these agents in rats. The rapid proliferation of new NSAIDs coupled with aggressive commercial promotion and limited experience, combine to make rational selection of drugs and dosages difficult at best. While future research on the effectiveness and adverse reactions of these drugs in rats may clarify their role, at this point the choice and use of an appropriate NSAID must be based on limited data, experience, and uncertain clinical judgment.

By nature of their mechanisms of action, NSAIDs have the potential to interfere with a variety of research studies, and individual drugs vary in their specific effects. As a group, NSAIDs are prone to cause gastrointestinal irritation, nephrotoxicity, and in the case of acetaminophen, hepatotoxicity. Generally, gastrointestinal toxicity appears to be less frequently encountered with more COX-2 selective agents, although less selective agents have been safely used. Nephrotoxicity is associated with reduced renal blood flow, which can occur with anesthesia and dehydration (Nolan, 2000). Thus, it is important to maintain adequate renal perfusion if these drugs are used before or during anesthesia, and to maintain adequate hydration in any case. Hepatotoxicity caused by acetaminophen has been used a model for fulminate liver failure. However, large doses are used to produce this effect and, if overdosage is avoided, the drug is reasonably safe, if questionably effective. As a rule, NSAIDs should not be administered in conjunction with corticosteroids (Nolan, 2000).

i. Aspirin (acetylsalicylic acid). Aspirin is usually considered to be a minor analgesic, suitable for the relief of mild pain. Liles and Flecknell (1992b) suggested a dose extrapolated from algesiometric and toxicity data of 100 mg/kg PO for rats. In a study of Dark Agouti rats undergoing liver transplantation Jablonski and Bowen (2001) found a slight elevation in activity during the first dark cycle following surgery in rats receiving 100 mg/kg aspirin in gelatin compared to those receiving 0.5 mg/kg buprenorphine in gelatin, but on the following evening the results were reversed. They concluded that neither aspirin nor buprenorphine given for 48 hours had any adverse effect on liver transplantation, but that beneficial effects were difficult to demonstrate in this model.

ii. Acetaminophen. Acetaminophen is thought of as a mild analgesic and antipyretic suitable, at best, for mild to moderate pain. Its site of action has recently been identified as a COX-3 isoenzyme, a variant of the COX-1 enzyme (Chandrasekharan et al., 2002; Schwab et al., 2003; Ayoub et al., 2004). This discovery raises the possibility of developing more potent and selective drugs targeting the site. Acetaminophen has been advocated as an effective analgesic for rats, suitable for administration in the drinking water. However, Cooper et al. (1997) were unable to detect any analgesia following doses of 3.8 or 7.7 mg/mL of acetaminophen in drinking water. By contrast, acetaminophen administered at 200 mg/kg IP did produce analgesia. Speth et al. (2001) showed that both a suspension and an elixir of acetaminophen diluted to 6 mg/mL in drinking water reduced both food and water intake on the day of exposure and that water intake increased following return to normal drinking water; they attributed the results to neophobia. Bauer et al. (2003) found that continuous exposure of rats to a 6-mg/mL suspension of acetaminophen in drinking water for a week resulted in return to normal water intake after 24 hours, but that food consumption remained depressed throughout the exposure period. Neither of the latter two studies assessed the degree of analgesia produced, nor the blood levels of acetaminophen attained. Thus, regardless of the acceptability of acetaminophen in drinking water, it has not been shown be effective by that route. If acetaminophen is used as an analgesic in the rat, it appears that parenteral administration is preferable.

iii. Carprofen. Carprofen is a COX-2 selective propionic acid derivative, with relatively weak COX-1 activity. Its mechanism of action is not well understood, but carprofen has strong antiinflammatory and analgesic effects (Nolan, 2000). Flecknell et al. (1999) compared parenteral and oral carprofen and ketoprofen administered preoperatively to rats undergoing a 1 cm laparotomy for insertion of an osmotic pump. They found that both ketoprofen (5 mg/kg) and carprofen (5 mg/kg) given SC one hour before surgery prevented postoperative weight loss, maintained fluid intake, and minimized reductions in food intake when compared with oral administration and untreated controls. Oral administration was markedly less effective at the doses used. In a study designed to identify behavioral indicators of pain following laparotomy, Roughan and Flecknell (2000b) again used ketoprofen and carprofen as preoperative analgesics and found that at doses of 5 mg/kg SC 1 hour before surgery, the drugs provided 4 to 5 hours of analgesia assessed by behavioral methods. In this study, neither drug prevented weight loss over the 8-hour postsurgical assessment period, and both were equivalent in reducing behavioral indicators of pain. Roughan et al. (2004) also examined the effects of carprofen (5 mg/kg SC) and meloxicam (2 mg/kg SC) in ameliorating signs of chronic pain associated with growth of an experimental bladder tumor. In this study, animals were selected for treatment based on the appearance of clinical signs of chronic pain attributable to tumor growth. At the doses used, neither drug produced any improvement in clinical signs. The authors noted that the lack of activity in these rats before and following treatment rendered the behavioral means of analgesic assessment used inapplicable. They concluded that, although some benefit of treatment was probable, the extent could not be gauged. Carprofen is effective in improving signs of pain in dogs and cats following surgery, has a low incidence of gastric complications, and can be safely given preoperatively (Nolan, 2000). Doses of up to 5 mg/kg were administered by gavage to rats for 6 months without mortality or adverse effects (European Agency for the Evaluation of Medical Products, 1995). The recommended dose in the rat is 5 mg/kg SC once daily, although the studies cited above suggest that this may be inadequate for severe or enduring pain. Carprofen is available in tablet and injectable preparations.

Flunixin meglumine is a nonselective COX-1 and COX-2 inhibitor. Liles and Flecknell (1994) compared flunixin, buprenorphine and carprofen as postoperative analgesics in rats and found flunixin to be the least effective, a conclusion with which Wixson and Smiler et al. (1997) concurred. Doses of 2.5 mg/kg given once or twice daily have been recommended (Liles and Flecknell, 1992b). King and Miller (1997) published a report of fatal intestinal perforation in four rats attributed to an inadvertent overdose of 11 mg/kg every 12 hours. Nolan (2000) cautions that flunixin should be administered only postoperatively, to avoid renal toxicity. Flunixin is available in injectable and oral granular formulations.

iv. Ketoprofen. Ketoprofen, another propionic acid derivative, is a nonselective COX inhibitor, with good analgesic and antiinflammatory properties. Given preoperatively, 5 mg/kg ketoprofen SC, but not orally, was equivalent to 5 mg/kg carprofen in minimizing weight loss assessed 20 hours following surgery (Flecknell et al., 1999). In common with most NSAIDs, there is little published information available concerning the use of ketoprofen in rats.

v. Meloxicam.   Meloxicam, an enolic acid derivative, is COX-2 preferential, but does retain some COX-1 activity. Although approved for use in humans and animals in the United States and elsewhere, information concerning its clinical use in rats is sparse (Roughan et al., 2004). In a pharmacokinetic study in rats, Busch et al. (1998) found that meloxicam is well absorbed following oral dosing and that the plasma concentration-time profiles were comparable in rats, dogs and humans. They also noted that elimination of the drug was considerably slower in female than male rats and in albino versus hooded rats. The primary means of elimination was renal excretion. They also found that meloxicam appears in the milk and passes the placenta, with levels in the liver of day 18 fetuses being about one-third to one-fifth the levels in the maternal liver. Villegas et al. (2002) compared meloxicam with piroxicam at two dose levels and found equivalent levels of gastric ulceration. Meloxicam given at 7.5 mg/kg PO for 14 days, resulted in the death of 43% of the rats with weight loss and diarrhea as the primary clinical signs. Using a lower dose of 3.75 mg/kg PO for 28 days, no deaths were reported and hematological and liver enzymes, blood urea nitrogen (BUN) and creatinine were within normal limits with the exception of a mild elevation of aspartate aminotransferase (AST). The presence of gastric lesions within 9 hours following a single oral doses of meloxicam at 7.5 and 15 mg/kg was again confirmed by Villegas et al. (2004). It should be noted that the doses in these studies are far higher than those recommended for osteoarthritis in dogs, and at least 50% higher than the dose used by Roughan et al. (2004), thus their clinical relevance in terms of adverse effects in rats is questionable. Meloxicam is available in injectable and oral liquid formulations.

c. LOCAL ANESTHETICS.   Lidocaine is possibly the most common local anesthetic used in rats, typically as an anesthetic adjuvant and to prevent vasospasm during blood vessel cannulation. Lidocaine has a rapid onset and relatively short duration of effect in rats, certainly less than the 1 to 2 hours seen in larger species. It is most commonly given by multiple small injections along the line of incision, or by direct application to exposed tissues. A dose of 4 mg/kg (0.4 mL/kg of 1% lidocaine) is regarded as safe (Dobromylskyj et al., 2000).

Bupivacaine has a slower onset than lidocaine, but a longer duration of action. Again, the duration is likely to be considerably less than the 4 to 6 hours commonly cited for larger species. Bupivacaine is more potent and more toxic than lidocaine; 0.125% solutions are often used in the rat, and the total dose should be limited to 1 to 2 mg/kg (Dobromylskyj et al., 2000). Appropriate use of local anesthetics is particularly useful to limit the amount of general anesthetic needed and as a part of a comprehensive analgesic protocol.

Local anesthetics do not effectively penetrate intact skin. EMLA Cream^{TM}, a mixture of prilocaine and lidocaine, is designed to do so and is used in larger species and man to gain local anesthesia for venipuncture. However, it requires a 60-minute application time under an occlusive dressing in order to achieve local anesthesia. In rats, the thick skin of the tail defeats even the demanding measures needed to apply an occlusive dressing and maintain it for an hour; thus, EMLA Cream^{TM} does not work on rat tails (Flecknell et al., 1990). While other formulations of topical local anesthetics are available, the time required for effect and the need for an occlusive dressing in most cases, renders them cumbersome to use, and there are no reports of successful use in rats (Friedman et al., 2001: however, see Arevalo et al., 2004).

d. MISCELLANEOUS DRUGS.

i. Tricyclic antidepressants.   Tricyclic antidepressants, so-named because of their chemical structure, comprise a group of drugs which block neural uptake of norepinephrine or serotonin, or both, with varying degrees of selectivity (Baldessarini, 2001). These drugs are sometimes used in the treatment of chronic pain in humans. Wixson and Smiler et al. (1997) describe the successful use of amitriptyline in conjunction with methadone to prevent autotomy rats. In addition to its central effects, amitriptyline also has local anesthetic effects, as reported by Sudoh et al. (2003), who found that it produced a longer sciatic nerve blockade than bupivacaine.

ii. Gabapentin.   Gabapentin is an anticonvulsant which has recently been investigated as an antihyperalgesic agent. Using a plantar incision model in rats, Field et al. (1997) found that a single dose of gabapentin at 30 mg/kg given 1 hour before surgery in conjunction with 1 mg/kg morphine administered 30 minutes before surgery blocked the development of thermal hyperalgesia for 24 hours, and the development of tactile allodynia for up to 49 hours. Gilron et al. (2003), found that gabapentin partially restored the effects of morphine in a model of induced morphine tolerance. Administration of gabapentin in conjunction with low-dose morphine by means of a spinal catheter was effective in reducing behavioral indices of pain, as reported by Smiley et al. (2004). No reports of the clinical use of gabapentin in rats although the potential for reducing opioid dose and, possibly, reducing the development of tolerance to opioids is certainly of interest.

B. Barbiturates

The barbiturates used in rats are pentobarbital, thiopental, methohexital, and ethylmalonyl urea (EMTU; Inactin). All are sodium salts and, with the exception of pentobarbital, are presented as dry powders for dilution with water or saline. Solutions are alkaline and precipitate if mixed with acidic materials. All are controlled substances. While the pharmacokinetics and metabolism of these drugs differ, the quality of anesthesia produced at anesthetic doses is similar. Barbiturates are hypnotics, with no inherent analgesia, and produce respiratory and cardiovascular depression at anesthetic doses.

1. Pentobarbital

Pentobarbital was discovered in 1930 and, while no longer considered a synonym for "anesthetic" in rats, it remains in general use. Pentobarbital produces dose-related respiratory depression and cardiovascular depression. Wixson et al., (1987c) reported relatively stable heart rates and a 20% decline in mean arterial blood pressure from control levels over a 2-hour assessment period, following a 40 mg/kg dose of pentobarbital IP. Skolleborg et al., (1990) also found stable heart rates and declining mean arterial blood pressure over a 2-hour period following a 50 mg/kg dose. Both Wixson et al., (1987c) and Buelke-Sam et al., (1978) reported similar degrees of hypercapnia, mild acidosis and hypoxia in their studies. Skolleborg et al. (1990) compared midazolam/fentanyl/fluanisone to pentobarbital and found that, while heart rate and mean arterial pressures were higher with pentobarbital, cardiac output was considerably lower and muscle tissue perfusion was also decreased.

Using a toe-pinch withdrawal as a means of assessing anesthetic depth is problematic. Haberham et al. (1998) characterized toe-pinch withdrawal as being unreliable as the sole indicator of anesthetic depth with pentobarbital, inferring that it did not correlate well with level of consciousness. Field et al. (1993) suggested that attempting to abolish response to toe pinch in the presence of hypnotic agents can result in anesthetic overdose, and emphasized the use of multiple measures such as muscle relaxation, palpebral reflex and abdominal pinch, tail pinch and corneal reflex be used in that order as indicators of increasing anesthetic depth. These comments reflect the relatively narrow safety margin of hypnotic agents, and their lack of analgesia when used as monoanesthetics.

Pentobarbital is usually administered at doses of 30 to 60 mg/kg IP, (Buelke-Sam et al., 1978; Wixson et al., 1987a–d; Skollenborg et al., 1990; Haberham et al., 1998). Stock solutions of pentobarbital should be diluted to allow accurate dose measurement. Surgical anesthesia time is 15 to 60 minutes, with recovery taking from 2 to 4 hours, in most cases (Flecknell, 1996d). Repeated doses of pentobarbital can be given at rates of 20% to 25% of the initial dose to supplement or prolong anesthesia. However, intermittent bolus injections generally result in uneven levels of anesthesia and, in the author's experience, the mortality

rate correlates directly with the number of additional doses. Continuous intravenous infusion has been described (Seyde et al., 1985; Davis, 1992). An alternative method to provide prolong pentobarbital anesthesia is to administer an intragastric dose of pentobarbital following induction by IP injection (Zambricki and D'Alecy, 2004). With pentobarbital, general dose recommendations are merely guidelines and the doses must be optimized for age sex, strain and specific procedures. A further complication is that the formulation of pentobarbital may vary among manufacturers in composition and concentration. While reading the label can solve the issue of concentration differences, formulation differences are more subtle, and it is wise to evaluate an unfamiliar product before use. Roughan et al. (1999) reported that at equivalent doses of pentobarbital, female rats had greater respiratory depression than males. That observation was confirmed by Zambricki and D'Alecy (2004) who measured plasma pentobarbital levels in male and female Sprague-Dawley rats following equivalent doses and found higher levels and a slower rate of decline in females than in males. Wixson and Smiler (1997) note that rats that have eaten within 1 hour of injection have less satisfactory responses to pentobarbital in terms of induction time, depth of anesthesia and recovery time than do animals which are fasted.

Roughan et al. (1999) examined the effects of buprenorphine given at 0.05 mg/kg 1 hour before low (36 mg/kg), medium (48 mg/kg) and high (60 mg/kg) doses of pentobarbital, and found that buprenorphine increased surgical anesthesia and recovery time, and reduced respiratory rates in each case.

Pentobarbital is a useful drug, but has minimal analgesia at safe doses and a prolonged recovery period with impaired thermal regulation (Wixson et al., 1987c). Even animals that appear to have recovered well in a warmed cage may not be able to sustain body temperature when they are returned to their home cage. For that reason, animals recovering from pentobarbital should be reassessed within 1 or 2 hours following return to their home cages. For survival surgical procedures, many other anesthetics with better analgesic and recovery characteristics may be preferred.

2. Thiopental

Thiopental is commonly used for induction of anesthesia in larger animals and in man, usually given intravenously to allow for dose titration. Following a single dose, recovery in larger species is rapid. However, recovery results from redistribution of thiopental, not rapid metabolic elimination, thus multiple bolus doses result in prolonged recovery. In rats, doses of 30 to 100 mg/kg IP of 2.5% thiopental have been used, resulting in up to 4 hours of anesthesia with predictable dose-dependent respiratory depression, acidosis, hypercapnia, hypoxia and hypothermia (Kaczmarczyk and Reinhardt, 1975). However, given IV in rats at 30 mg/kg (1.25% solution), thiopental produces 5 minutes of surgical anesthesia with recovery in 15 minutes, according to Flecknell (1996d). There seems to be little advantage to using thiopental in preference to pentobarbital for long procedures.

3. Methohexital

Methohexital is very short-acting barbiturate with rapid clearance. In animals and humans, methohexital is more likely to cause muscle tremors and excitement during recovery. Wixson and Smiler et al. (1997) reported that 40 mg/kg IP produced 15 to 20 minutes of restraint, but insufficient analgesia for surgery. Given intravenously, a dose of 10 to 15 mg/kg will produce about 5 minutes of surgical anesthesia with rapid recovery (Flecknell, 1996d). With the advent of reversible injectable agents and inhalation equipment, there are fewer reasons to select either methohexital or thiopental for brief procedures.

4. Inactin (Ethylmalonyl Urea; EMTU)

Inactin is sometimes used to produce longer anesthesia than pentobarbital following a single dose. Buelke-Sam et al. (1978) reported stable anesthesia for 3 to 4 hours following a single IP dose of 80 to 100 mg/kg, and commented that the effects were similar to, but less variable, than pentobarbital. The author's experience is similar to that of Flecknell (1996d), in that EMTU was not reliable in producing longer anesthesia. Nevertheless, EMTU has a devoted following in selected research disciplines and appears to provide satisfactory anesthesia in experienced hands. It is supplied in vials as a dry powder under inert gas to prevent decomposition.

C. Dissociative Agents

Some compounds are referred to as dissociative agents because of the unique catoleptic state they produce in humans, characterized by analgesia, amnesia and altered consciousness (Korhs and Durlieux, 1998; Hall et al., 2001b; Evers and Crowder, 2002). Patients also display involuntary movement, rigidity and usually spontaneous respiration with intact airway reflexes. The parent drug for this group is phencyclidine, once used in human and veterinary medicine as Sernylan™, but ketamine and tiletamine are the only two currently available (legal) representatives. Ketamine, in various combinations, is by far the more commonly used of the two and, having displaced pentobarbital, is probably the most popular parenteral anesthetic for rats in the United States and other countries where Hypnorm™ is unavailable. In the United States, tiletamine is available only as Telazol™, a fixed-ratio combination of tiletamine and the benzodiazepine zolazepam.

1. Ketamine

Ketamine is a racemic mixture of two enantiomers, $S(+)$ and $R(-)$, of which $S(+)$ is more potent in terms of hypnotic effects and analgesia, but has fewer adverse effects (Kohrs and Durieux, 1998). The pharmacology of ketamine is discussed by Fish (1997), and in standard human and veterinary anesthesia and pharmacology texts. The mechanisms of action in the brain and spinal cord are not fully understood, but used alone ketamine produces strong analgesia with minimal respiratory depression, and increased blood pressure, heart rate and cardiac output (Fish, 1997; Evers and Crowder, 2001). Muscle relaxation does not occur and spontaneous involuntary movement is typical. Used alone, ketamine is not considered to provide sufficient analgesia or muscle relaxation for surgery in rats (Green et al., 1981; Wixson and Smiler, 1997). When ketamine is combined with other drugs for anesthetic use, the cardiovascular and respiratory support provided by ketamine may be compromised, depending upon the drugs and doses used, so that respiratory and cardiac depression may occur. Ketamine is well absorbed following IP and IM injection in rats, but IM injection is painful, irritating, and liable to result in tissue necrosis; thus, IP administration is preferred (Smiler et al., 1990; Sun et al., 2003). Unfortunately, there are no specific reversal agents for ketamine.

The subject of ketamine and ketamine cocktails is well reviewed by Wixson and Smiler (1997) in an earlier text in this series, *Anesthesia and Analgesia in Laboratory Animals* (Kohn et al., 1997). The following section is summarized from that review with the addition of the relatively little information added since that time.

Ketamine/xylazine combinations are widely used, with doses of ketamine ranging from 40 to 87 mg/kg and xylazine from 5 to 13 mg/kg. The mixture produces a rapid onset of anesthesia with generally good analgesia and muscle relaxation. Polyuria, hyperglycemia and bradycardia are dose-dependent effects attributable to the α_2-adrenergic component, xylazine. Under most circumstances doses of ketamine in the range of 40 to 60 mg/kg of ketamine and 5 to 10 mg/kg of xylazine are satisfactory for procedures of short to moderate duration (Smiler and Wixson, 1997a). In the author's opinion, if doses at the upper extremes for these drugs are found to be necessary, an alternative anesthetic should be considered to avoid prolonged recovery and cardiovascular and respiratory depression. Despite common practice, no single formulation of ketamine and xylazine is likely to be satisfactory for all situations, and doses and dose ratios should be adjusted to the potential of the procedure to cause pain and the duration of anesthesia required. The stability of mixtures of ketamine and xylazine, with or without further dilution, has not been documented although numerous anecdotal reports suggest that the mixture is stable for weeks to months.

Anesthesia can be prolonged by additional doses of ketamine at the rate of 1/4 to 1/3 of the original dose (Wixson and Smiler, 1997). Alternatively, Simpson (1997) described prolonged anesthesia for periods of up to 12 hours using ketamine/xylazine by continuous intravenous infusion. There are no reports of the use of preoperative buprenorphine with ketamine and xylazine mixtures. In light of the adverse effects noted by Roughan et al. (1999) and Hedenqvist et al. (1999) following preoperative administration of buprenorphine to rats anesthetized with ketamine/medetomidine (another α_2-agonist) such use should be approached with caution.

Recovery from ketamine and xylazine anesthesia can be hastened with yohimbine (1.0 to 2.1 mg/kg IP or SC) or atipamezole (0.1 to 1.0 mg/kg IP or SC). Thermal support is required during and after anesthesia with ketamine and xylazine, and warmed fluid supplementation is useful to compensate for the polyuria produced by this mixture. In the author's experience, every rat anesthetized with ketamine (50 to 55 mg/kg) and xylazine (7 mg/kg) is hypoxemic, or nearly so, when assessed by pulse oximetry, and supplemental oxygen is needed.

Ketamine and medetomidine are also used to produce anesthesia in rats similar in character, but of greater intensity and duration, than ketamine/xylazine. Doses of 60 to 75 mg/kg of ketamine and 0.25 to 0.5 mg/kg of medetomidine administered SC or IP are suggested (Nevalainen et al., 1989; Hedenqvist et al., 1999; Roughan et al., 1999). Repeated administration of ketamine/medetomidine with or without buprenorphine administered one hour before anesthetic induction was reported by Roughan et al. (1999) and Hedenqvist et al. (1999). While ketamine/medetomidine alone appeared to be safe when administered weekly for 6 weeks, the addition of buprenorphine resulted in unexpected deaths, and the authors concluded that opioids should not be used in conjunction with medetomidine/ketamine anesthesia in rats. Female rats are more sensitive to ketamine/medetomidine anesthesia than males (Nevalainen et al., 1989; Roughan et al., 1999).

Medetomidine in this combination is reversed using atipamezole, as previously described. The recommendations for intraoperative and postoperative care of rats receiving ketamine/medetomidine anesthesia are the same as for ketamine/xylazine although Hedenqvist et al. (1999) emphasized the value of oxygen supplementation with this mixture.

Ketamine is sometimes combined with acepromazine. While the mixture may provide muscle relaxation, it does not appear to enhance analgesia and should probably be limited to use as a restraint agent.

Diazepam and ketamine was assessed by Wixson et al. (1987a–d), at doses of 40 to 80 mg/kg ketamine and

5 to 10 mg/kg diazepam. The mixture resulted in poor to fair analgesia, with hyperacusia and minimal cardiovascular and respiratory depression. It appears that this mixture would be suitable for minor surgery.

2. Tiletamine/Zolazepam

Tiletamine/zolazepam (Telazol^{TM}) has been used as an anesthetic for rats. A mixture of tiletamine and zolazepam, it is presented as a powder for reconstitution. Used alone, at doses of 20 to 40 mg/kg, Telazol^{TM} produces 30 to 60 minutes of surgical anesthesia, but corneal, pedal and swallowing reflexes are retained, so the assessment of anesthetic depth is difficult. The addition of xylazine greatly improves the analgesia produced, but results in marked cardiovascular depression, and the addition of butorphanol also increases analgesia but at the cost of variable respiratory depression (Ward et al., 1974; Silverman et al., 1983; Wilson et al., 1993). Given the cost of Telazol^{TM}, its relatively short shelf life following reconstitution, and questionable analgesia there seems little reason to use it in preference to the ketamine mixtures described above.

D. Miscellaneous Agents

1. α-Chloralose

Alpha-chloralose alone or in combination with urethane is frequently used to achieve long-lasting, but light anesthesia with minimal depression of cardiovascular and respiratory depression and intact autonomic reflexes. Wixson and Smiler (1997) citing a study by Field (1988), report adverse effects including seizures, prolonged onset of anesthesia, hyperacusia, and viscous oral and nasal discharge, and further note that even at a dosage of 65 mg/kg, surgical anesthesia could not be obtained. The appearance of twitching, or myoclonic jerks, sometimes seen with α-chloralose has been attributed to contamination of the chemical with β-chloralose but, in an extensive review of the pharmacology of α-chloralose, Balis and Monroe (1964) regard that as unproven. Because α-chloralose has little, if any, analgesic properties, surgical procedures must be completed under conventional anesthesia. Even without surgical intervention, Flecknell (1996d) recommends induction of anesthesia with a short-acting anesthetic before administration of α-chloralose in order to minimize the complications associated with α-chloralose induction. If major surgery is performed, additional analgesia may be necessary to supplement the light anesthesia provided by α-chloralose.

2. Alphaxalone-alphadolone (Saffan®)

Alphaxalone-alphadolone is a mixture of two neurosteroids, alphaxalone and alphadolone solubilized in

Cremophor EL. Intravenous administration of alphaxalone-alphadolone to rats at 10 to 12 mg/kg produces brief anesthesia with recovery within about an hour, and administration of 25 to 30 mg/kg IP produces sedation and light anesthesia (Green et al., 1978). Because it has relatively little cumulative effect, alphaxalone-alphadolone is well suited to continuous IV infusion for prolonged anesthesia. Following an induction dose, alphaxalone-alphadolone is delivered continuously at rates of 0.25 to 0.45 mg/kg/min (Green et al., 1978), although Flecknell (1996d) suggests a somewhat broader range of 0.2 to 0.7 mg/kg/min. Both components of the mixture contribute to the sedation and analgesia when the drug is given intravenously. However, following IP administration in the rat, alphadolone failed to produce sedation, but did produce analgesia as assessed by noxious electrical current (Nadeson and Goodchild, 2001). The authors attributed the difference in the activity of alphadolone given IV versus IP to hepatic metabolism of alphadolone and conversion to a metabolite that is analgesic, but not anesthetic.

3. Chloral Hydrate

Chloral Hydrate is a sedative hypnotic that, in sufficient doses, results in general anesthesia. Following injection, chloral hydrate is quickly metabolized to form trichloroethanol, the active metabolite, which is subsequently eliminated by the kidney (Silverman and Muir, 1993). Typical doses for rats are 300 to 400 mg/kg IP, with additional doses of 40 mg/kg to maintain anesthesia (Silverman and Muir, 1993). Field et al. (1993) examined the effects chloral hydrate in male rats. Duration of anesthesia was dose-related with time from induction to recovery being 111 minutes at 300 mg/kg and 136 minutes at 400 mg/kg. Anesthesia was deeper and with greater suppression of reflexes and response to noxious stimulation at the 400 mg/kg dose. At that dose, chloral hydrate produced marked respiratory and cardiovascular depression with uncompensated acidosis. Recovery from chloral hydrate was noted to be rapid and abrupt. A major reported complication associated with chloral hydrate is the development of peritonitis and adynamic ileus (Fleischman et al., 1977; Davis et al., 1985). However, the complication appears to be related to the drug concentration used. Field et al. (1993) found no evidence of ileus 14 days after administration of 5% chloral hydrate, and Vachon et al. (2000) found minimal lesions at doses of 400 mg/kg using 4% concentrations.

Chloral hydrate is sometimes mixed with magnesium sulfate and pentobarbital to form an anesthetic called Equithesin, originally used in horses and other large animals, and currently still used in rats. It is no longer manufactured, but is reasonably simple to make from laboratory grade chemicals (n.b., in the United States both chloral hydrate and pentobarbital are controlled substances).

While there appear to be no studies regarding the use of this mixture in laboratory rats, its components, a mixture to two sedative hypnotics and a chemical that reduces muscle excitability, may serve as an indication of its effects.

4. Propofol

Propofol is a sedative hypnotic presented as an emulsion in a vehicle similar to Intralipid®. Propofol is effective only when delivered IV, and in general the cardiovascular and respiratory effects of propofol resemble those of barbiturates such as thiopental, but without the undesirable cumulative effects of those drugs. Original investigations by Glen (1980) and Glen and Hunter (1984) suggested useful dosages of 10 to 25 mg/kg IV in rats, or induction with 7.5 to 10 mg IV followed by continuous infusion of 44 to 55 mg/kg/hr, although Flecknell (1996d) recommends a slightly wider dose range of 30 to 60 mg/kg/hr. The influence of age and administration rate of propofol in male Sprague-Dawley rats was examined by Larson and Wahlström (1998) who determined that younger rats required a larger induction dose than older animals. Brammer et al. (1993) reported, in a nonsurvival study, that premedication with fentanyl-fluanisone (Hypnorm™) followed by a propofol infusion produced a smooth induction and stable hemodynamics for up to 3 hours.

5. Tribromoethanol

Tribromoethanol is another sedative-hypnotic, popular as an anesthetic in mice. Reid et al. (1999) administered tribromoethanol at three dosage levels to female Sprague-Dawley rats and reported deaths, dehydration, and the development of peritonitis and fibrous adhesions in all animals examined at 10 days. It appears that tribromoethanol is not a satisfactory anesthetic in rats intended to survive, and that there is little reason to use it in preference to other, safer drugs.

6. Urethane (Ethyl Carbamate)

Urethane (ethyl carbamate) is a mutagenic, carcinogenic and hepatotoxic anesthetic used to provide anesthesia adequate for surgical procedures lasting up to 8 or more hours (Maggi and Meli, 1986a–c; Field et al. 1993). Hara and Harris (2002) examined the effects of urethane on several neurotransmitter-gated ion channels, and concluded that urethane acts on both inhibitory and excitatory systems in a dose-dependent manner. Field et al. (1993) examined urethane at doses of 1200 to 1500 mg/kg IP in rats, and characterized the degree of analgesia produced as adequate for moderate to markedly painful procedures. However, they also found that, at doses of 1500 mg/kg, urethane produced a 25% mortality rate accompanied by severe metabolic acidosis. In an extensive review of the pharmacology of urethane, Maggi and Meli (1986a–c) note the toxic effects of IP urethane on the mesenteric vasculature, liver, spleen, and pancreas, as well as severe peritoneal effusion. Dose-dependent cardiovascular depression occurs as well, with significant depression of heart rate and blood pressure at doses of 1500 mg/kg IP. Administration of urethane by subcutaneous injection avoided or minimized most of these adverse effects, at the cost of less reliable anesthesia and slower onset times (Maggi and Meli, 1986 a–c). Both Field et al. (1993) and Maggi and Meli (1986 a–c), conclude that urethane is an effective anesthetic, but that the lowest doses that produce adequate anesthesia should be used to avoid or minimize physiological perturbations, a conclusion supported by Hara and Harris (2003). It seems unlikely that urethane differs from any other anesthetic in that the response can be expected to vary with the age, weight and strain of the animal, so a trial to determine the most effective dose in a given experimental setting is recommended.

Urethane is sometimes used in combination with α-chloralose. Hughes et al. (1982) described doses of 250 to 400 mg/kg urethane IP followed in 30 minutes by α-chloralose at 35 to 40 mg/kg IP, to produce stable anesthesia for up to 6 hours at the higher doses.

Urethane is considered suitable for use only in nonrecovery procedures because of its toxic and carcinogenic properties. The safety aspects of urethane in rats were reviewed by Field and Lang (1988). Because of the carcinogenic and toxic properties of urethane, solutions should be prepared under a chemical fume hood, and appropriate safety precautions used during administration to avoid skin or respiratory exposure.

VIII. SPECIAL ANESTHETIC CONSIDERATIONS

A. Anesthesia in Pregnancy

Pregnant rats can safely be anesthetized with a wide variety of anesthetics including ketamine/xylazine (Stickrod, 1979), and inhalation agents. In terms of maternal health, the anesthetic considerations already described apply. However, late term pregnant animals are more likely to experience respiratory embarrassment so careful attention to patient positioning and avoiding anesthetics prone to result in further respiratory depression is advisable. To avoid hypoglycemia, pregnant rats should not be fasted before anesthesia.

Adverse fetal effects have been shown, in varying degrees of subtlety and at varying time points of exposure, for almost every anesthetic investigated including ketamine, benzodiazepines, barbiturates, and inhalation agents (Wixson and Smiler, 1997l). Thus, it is impossible to offer general recommendations for anesthesia without specifying the criteria by which fetal outcomes will be assessed. When contemplating anesthesia of a pregnant rat in which subsequent fetal development is of experimental importance, close consultation with the investigator and a literature search directed to the specific research objectives will be needed to establish the optimum choice.

1. Anesthesia for Neonates

Hypothermia has been successfully used for anesthesia of early neonates (Phifer and Terry, 1986), and suggestions for refining the technique by enclosing the neonate in a latex glove finger were made by Danneman and Mandrell (1997). Using the glove finger refinement, Danneman and Mandrell (1997) reported loss of pedal reflexes in about 3 minutes following immersion in ice water, with return to righting at about 1 hour after 30 minutes of anesthesia. Wixson and Smiler (1997l) state that aggressive rewarming should be avoided following hypothermia, and recommend using an incubator at 33°C. Special adaptors with a well for alcohol/dry ice mixtures are available for maintaining hypothermia during stereotaxic surgery in neonates. Brunelli et al. (1994) note that distinctive ultrasonic vocalizations by rat pups recovering from hypothermic anesthesia served to direct search behavior by dams. Hypothermia is generally reserved for animals of less than 7 to 10 days of age and, despite controversy about the quality of anesthesia, currently remains an accepted method of anesthesia for neonatal rats.

Park et al. (1992) reported successful use of halothane by mask and fentanyl-droperidol (Innovar-Vet™) SC for eye surgery in 18- to 24-hour-old Long Evans neonates, with 55 of 57 rats surviving at 7 days following halothane anesthesia, and 16 of 16 rats surviving at 7 days following anesthesia with fentanyl-droperidol. Anesthesia was induced with 5% halothane and maintained at 2% at a fresh gas flow rate of 2 L/m for surgery lasting about 43 minutes. Danneman and Mandrell (1997) compared hypothermia, fentanyl-droperidol, ketamine, pentobarbital and methoxyflurane, and concluded that hypothermia and methoxyflurane were the most satisfactory. In contrast to Park et al. (1992), Danneman and Mandrell were unable to produce reliable anesthesia with fentanyl-droperidol IP in 1- to 3-day-old Sprague-Dawley neonates, with only 1 in 10 animals achieving surgical anesthesia. However, Clowry and Flecknell (1999) reported success using Hypnorm™, a mixture of fentanyl and fluanisone SC, in 7-day-old Wistar rat pups, although with a high level of spontaneous movement, and recommended SC administration of drug and reversal with nalbuphine. Differences in the dose of fentanyl, route of administration, and strain and age of the rat pups used in these studies make comparisons somewhat difficult. However, inhalation anesthesia was successful and safe in both instances where it was used. The author has successfully used isoflurane in oxygen for 24-hour-old Sprague-Dawley rat pups undergoing abdominal procedures lasting 15 to 20 minutes, with rapid induction and recovery and no mortalities attributable to anesthesia.

REFERENCES

Affaitati, G., Giamberardino, M.A., Lerza, R., Lapenna, D., De Laurentis, S., and Vecchiet, L. (2002). Effects of tramadol on behavioural indicators of colic pain in a rat model of ureteral calculosis. *Fundamental and Clinical Pharmacology*, p. 23–30.

Alibhai, H.I., Clark, K.W., Lee, Y.H., and Thompson, J. (1996). Cardiopulmonary effects of combinations of medetomidine hydrochloride and atropine sulphate in dogs. *Vet. Rec.* **138**, 11–13.

American College of Laboratory Animal Medicine (2004). Public statement: medical records for animals used in research, teaching, and testing. ONLINE: Available: http://www.aclam.org.

Amouzadeh, H.R., Sangiah, S., Qualls, Jr., C.W., Cowell, R.L., and Mauromoustakos, A. Xylaxine-induced pulmonary edema in rats. *Tox. Appl. Pharmacol.* **108**, 417–427.

Antunes, L.M., Golledge, H.D.R., Roughan, J.V., and Flecknell, P.A. (2003). Comparison of electroencephalogram activity and auditory evoked responses during isoflurane and halothane anaesthesia in the rat. *Vet. Anaesth. Analg.* **30**, 15–23.

Apaydin, S., Uyar, M., Karabay, N.U., Erhan, E., Yegul, I., Tuglular, I. (2000). The antinociceptive effect of tramadol on a model of neuropathic pain in rats. *Life Sciences*, p. 1627–1637.

Arevalo, M.I., Escribano. E., Calpena, A., Domenech, J., and Queralt, J. (2004). Rapid skin anesthesia using a new topical amethocaine formulation: a preclinical study. *Anesth. Analg.* **98**, 1407–1412.

Ayoub, S.S., Botting, R.M., Goorha, S., Colville-Nash, P.R., and Willoughby, D.A. (2004). Acetaminophen-induced hypothermia in mice is mediated by a prostaglandin endoperoxide synthase 1 gene-derived protein. *Proc. Natl. Acad. Sci. U.S.A.* **101**, 11165–11169.

Balis, G.U., and Monroe, R.L. (1964) The pharmacology of chloralose. *Psychopharmacologia* **6**, 1–30.

Barber, A., and Gottschlich, R. (1992). Opioid agonists and antagonists: an evaluation of their peripheral actions in inflammation. *Med. Res. Rev.* **12**, 525–562.

Baldessarini, R.J. (2001). Drugs and the treatment of psychiatric disorders. *In* "Goodman & Gilman's The Pharmacological Basis of Therapeutics." (J.G. Hardman, L.E. Limbird, and A.G. Goodman, eds.), pp. 451–470, McGraw-Hill, New York.

Barrett, A.C., Smith, E.S., and Picker, M.J. (2002). Sex-related differences in mechanical nociception and antinociception produced by μ- and κ-opioid receptor agonists in rats. *European Journal of Pharmacology*, p. 163–173.

Bauer, D.J., Christenson, T.J., Clark, K.R., Powell, S.K., and Swain, R.A. (2003). Acetaminophen as a postsurgical analgesic in rats: a practical solution to neophobia. *Contemp. Top. Lab. Anim. Sci.* 20–25.

Bernard, S.L., An, D., and Glenny, R.W. (2004). Validation of the Nonin 8600V pulse oximeter for heart rate and oxygen saturation measurements in the rat. *Contemp. Top. Lab. Anim. Sci.* **43**, 43–45.

Bianchi, M., Rossoni, G., Sacerdote, P., and Panerai, A.E. (1999). Effects of tramadol on experimental inflammation. *Undam. Clin. Pharmacol.* **13**, 220–225.

Brammer, A., West, C.D., and Allen, S.L. (1993). A comparison of propofol with other injectable anesthetics in a rat model for measuring cardiovascular parameters. *Lab. Anim.* **27**, 250–257.

Brunelli, S.A., Shair, H.N., and Hofer, M.A. (1994). Hypothermic vocalizations of rat pups (Rattus novegicus) elicit and direct maternal search behavior. *Journal of Comparative Psychology*, 298–303.

Brunson, D.B. (1997). Pharmacology of inhalation anesthetics. *In* "Anesthesia and Analgesia in Laboratory Animals." (D.H. Kohn, S.K. Wixson, W.J. White, and G.J. Benson, eds.), pp. 29–41, Academic Press, New York.

Buelke-Sam, J., Holson, J.F., Bazare, J.J., and Young, J. (1978). Comparative stability of physiological parameters during sustained anesthesia in rats. *Lab. Anim. Sci.* **28**, 157–162.

Burkhart, J.E., and Stobbe, T.J., (1990). Real-time measurement and control of waste anesthetic gases during veterinary surgeries. *Am. Ind. Hyg. Assoc. J.* **51**, 640–645.

Busch, U., Schmid, J., Heinzel, G., Schmaus, H., Baierl, J., Huber, C., and Roth, W. (1998). Pharmacokinetics of meloxicam in animals and the relevance to humans. *Drug Metab. and Disp.* **26**, 576–584.

Cabral A.M., Kapusta, D.R., Kenigs, V.A., and Varner, K. J. (1998). Central α_2-receptor mechanisms contribute to enhanced renal responses during ketamine-xylazine anaesthesia. *Am. J. Physiol.* **44**, R1867–R1874.

Calderone, L., Grimes, P., and Shalev, M. (1986). Acute reversible cataract induced by xylazine and by ketamine-xylazine anesthesia in rats and mice. *Exp. Eye Res.* **42**, 331–337.

Chandrasekharan, N.V., Dai, H., Roos, L.T., Evanson, N.K., Tomsik, J., Elton, T.S., and Simmons, D.L. (2002). COX-3, a cyclooxygenase-1 variant inhibited by acetaminophen and other analgesic/antipyretic drugs: cloning, structure, and expression. *Proc. Natl. Acad. Sci. U.S.A.* **99**, 13926–13931.

Cambron, H., Latulippe, J.F., Nguyen, T., and Cartier, R. (1995). Orotracheal intubation of rats by transillumination. *Lab. Anim. Sci.* **45**, 303–304.

Claassen V. (1994). Fasting. *In* "Neglected Factors in Pharmacology and Neuroscience Research." pp. 290–320, Elsevier, Amsterdam.

Clark, J.A., Myers, P.H., Goelz, M.F., Thigpen, J.E., and Forsythe, D.B. (1997). Pica behavior associated with buprenorphine administration in the rat. *Laboratory Animal Science*, p. 300–303.

Clowry, G.J., and Flecknell, P.A. (1999). The successful use of fentanyl/fluanisone ('Hypnorm') as an anaesthetic for intracranial surgery in neonatal rats. *Lab. Anim.* **34**, 260–264.

Cole, D.J., Marcantonio, S., and Drummond, J.C. (1990). Anesthetic requirement of isoflurane is reduced in spontaneously hypertensive and Wistar-Kyto rats. *Lab. Anim. Sci.* **40**, 506–509.

Cook, T.L., Beppu, W.J., Hitt, B.A., Kosek, J.C., and Mazze, R.I. (1975). A comparison of renal effects and metabolism of sevoflurane and methoxyflurane in enzyme-induced rats. *Anesth. Analg.* **54**, 929–935.

Cook, C.D., Barrett, A.C., Roach, E.L., Bowman, J.R., and Mitchell, J.P. (2000). Sex-related differences in the antinociceptive effects of opioids: importance of rat genotype, nociceptive stimulus intensity, and efficacy at the μ opioid receptor. *Psychopharmacology*, p. 430–442.

Cooper, D.M., De Long D., and Gillett C.S. (1995). Analgesic efficacy of acetaminophen elixir added to the drinking water of rats. *Contemp. Top. Lab. Anim. Sci.* **34**, 61.

Cooper, D.M., DeLong, D., and Gillett, C.S. (1997). Analgesic efficacy of acetaminophen and buprenorphine administered in the drinking water of rats. *Contemp. Top. Lab. Anim. Sci.* **36**, 58–62.

Correa, F.C.F., Ciminelii, P.B., Falcao, H., Alcantara, B.J.C., Contador, R.S., Medieros, A.S., Zin, W.A., and Rocco, P.R.M. (2001). Respiratory mechanics and lung histology in normal rats anesthetized with sevoflurane. *J. Appl. Physiol.* **91**, 803–810.

Cox, E.H., Langemeijer, M.W.E., Gubbens-Stibbe, J.M., Muir, K.T., and Danhof, M. (1999). The comparative pharmacodynamics or remifentanil and its metabolite, GR90291, in a rat electroencephalographic model. *Anesthesiology* **90**, 535–544.

Crawford, M.W., Lerman, J., Saldivia, V., and Carmichael, F.J. (1992). Hemodynamic and organ blood flow responses to halothane and sevoflurane anesthesia during spontaneous ventilation. *Aneasth. Analg.* **75**, 1000–1006.

Criado, A.B., Gomez de Segura, I.A., Tendillo, F.J., and Marsico, F. (2000). Reduction of isoflurane MAC with buprenorphine and morphine in rats. *Lab. Anim.* **34**, 252–259.

Criado, A.B., and Gomez de Segura, I.A. (2003). Reduction of isoflurane MAC by fentanyl or remifentanil in rats. *Vet. Anaesth. Analg.* **30**, 250–256.

Danneman, P.J., Stein, S., and Walshaw, S.O. (1997a). Humane and practical implications of using carbon dioxide mixed with oxygen for anesthesia or euthanasia of rats. *Lab. Anim. Sci.* **47**, 376–384.

Danneman, P.J., and Mandrell, T.D. (1997b). Evaluation of five agents/methods for anesthesia of neonatal rats. *Lab. Anim. Sci.* **47**, 836–895.

Dardai, E., and Heavner, J.E. (1987). Respiratory and cardiovascular effects of halothane, isoflurane and enflurane delivered via a Jackson-Reese breathing system in temperature controlled and temperature uncontrolled rats. *Meth. and Find. Explt. Clin. Pharmacol.* **9**, 717–720.

Davies, L.A., Gibson, C.N., Boyett, M.R., Hopkins, P.M., and Harris, S.M. (2000). Effects of isoflurane, sevoflurane, and halothane on myofilament Ca^{2+} sensitivity and sarcoplasmic reticulum Ca^{2+} release in rat ventricular myocytes. *Anesthesiology* **93**, 1034–1044.

Davis, H., Cox, N.R., and Lindsey, J.R. (1985). Diagnostic Exercise: Distended abdomens in rats. *Laboratory Animal Science*, p. 392–394.

Davis, S.C., (1992). A method of constant intravenous pentobarbital infusion in rats. *Cont. Top. Lab. Anim. Sci.* **31**, 50.

DeLeonardis, J.R. Clevenger, R. and Hoyt, Jr., R.F. (1995). Approaches in rodent intubation and endotracheal tube design. Poster abstract P27. *Contemp. Top.* **34**, 60.

deWolff, M.H., Leather, H.A., and Wouters, P.F. (1999). Effects of tramadol on minimum alveolar concentration (MAC) of isoflurane in rats. *Brit. J. Anaesth.* **83**, 780–783.

DiFazio, C.A., Moscicki, J.C., and Magruder, M.R. (1981). Anesthetic potency of nalbuphine and interaction with morphine in rats. *Anesth. Analg.* **60**, 629–633.

Dobromylskyj, P., Flecknell P.A., Lacelles, B.D., Pascoe, P.J., Taylor, P., and Waterman-Pearson, A. (2000). Management of postoperative and Other acute pain. *In* "Pain Management in Animals." (P. A. Flecknell, and A. Waterman-Pearson, eds.), pp. 81–145, W.B. Saunders, London.

Dorsch, J.A., and Dorsch, S.E. (1999). "Understanding Anesthesia Equipment, 4th Ed." Williams & Wilkins, Baltimore.

Dorsch, J.A., and Dorsch, S.E. (1999a). Mapleson breathing systems. *In* "Understanding Anesthesia Equipment, 4th ed.", pp. 207–208, Williams & Wilkins, Baltimore.

Dorsch J.A., and Dorsch S.E. (1999b). Controlling trace gas levels. *In* "Understanding Anesthesia Equipment, 4th ed.", pp. 363–373, Williams & Wilkins, Baltimore.

Eger II, E.I., Saidman L.J., and Brandstater, B. (1965). Minimal alveolar anesthetic concentration: a standard of anesthetic potency. *Anesthesiology* **26**, 756–763.

Eger, E.I. (1994). New inhaled anesthetics. *Anesthesiology* **80**, 906–922.

Elsersy, H., Sheng, H., Lynch, J.R., Moldovan, M., Pearlstein, R.D., and Warner, D.S. (2004). Effects of isoflurane versus fentanyl-nitrous oxide anesthesia on long-term outcome from severe forebrain ischemia in the rat. *Anesthesiology* **100**, 1160–1166.

European Agency for the Evaluation of Medicinal Products. (1999). Committee for Veterinary Medicinal Products Carprofen Summary Report (1). European Agency for the Evaluation of Medicinal Products.

England, G., and Clarke, K.W. (1989). The use of medetomidine/fentanyl combinations in dogs. *Acta. Vet. Scand.* **85**, 179–186.

Evers, A.S., and Crowder, C.M. (2002). General anesthetics. *In* "Goodman & Gilman's The Pharmacological Basis of Therapeutics." (J.G. Hardman, L.E. Limbird, and A.G. Goodman, eds.), p. 346, McGraw-Hill, New York.

Field, K.J., and Lang, C.M. (1988). Hazards of urethane (ethyl carbamate): a review of the literature. *Lab. Anim.* **22**, 255–262.

Field, K.J., White, W.J., and Lange, C.M. (1993). Anaesthetic effects of chloral hydrate, pentobarbitone and urethane in adult rats. *Lab. Anim.* **27**, 258–269.

Field, M.J., Holloman, E.F., McCleary, S., Hughes, J., and Singh, L. (1997). Evaluation of gabapentin and S-(+)-3-isobutylgaba in a rat model of postoperative pain. *J. Pharmacol. Exp. Ther.* **282**, 1442–1446.

Fish, R.E. (1997). Pharmacology of injectable anesthetics. *In* "Anesthesia and Analgesia in Laboratory Animals." (D.H. Kohn, S.K. Wixson, W.J. White, and G.J. Benson, eds.), pp. 1–28, Academic Press, New York.

Flecknell, P.A. (1984). The relief of pain in laboratory animals. *Lab. Anim.* **18**, 147–160.

Flecknell, P.A., and Mitchell, M. (1984). Midazolam and fentanyl-fluanisone: assessment of anesthetic effects in laboratory rodents and rabbits. *Lab. Anim.* **18**, 143–146.

Flecknell, P.A., Liles, J.H., and Williamson, H.A. (1990). The use of lignocaine-prilocaine local anesthetic cream for pain-free venipuncture in laboratory animals. *Lab. Anim.* **24**, 142–146.

Flecknell, P.A. and Liles, J.H. (1991). The effects of surgical procedures, halothane anaesthesia and nalbuphine on locomotor activity and food and water consumption in rats. *Lab. Anim.* **25**, 50–60.

Flecknell, P.A. (1994). Refinement of animal use–assessment and alleviation of pain and distress. *Lab. Anim.* **28**, 222–231.

Flecknell, P. A. (1996a). Anaesthesia. *In* "Laboratory Animal Anaesthesia, 2nd ed." pp. 15–73, Academic Press, San Diego.

Flecknell, P. A. (1996b). Pre-operative care. *In* "Laboratory Animal Anaesthesia, 2nd ed." pp. 1–5, Academic Press, San Diego.

Flecknell, P. A. (1996c). Appendix 3. *In* "Laboratory Animal Anaesthesia, 2nd ed." pp. 247–248, Academic Press, San Diego.

Flecknell, P. A. (1996d). Anaesthesia of common laboratory species. *In* "Laboratory Animal Anaesthesia, 2nd ed." pp. 160–223, Academic Press, San Diego.

Flecknell. P. (1997). Medetomidine and atipamezole: potential uses in laboratory animals. *Lab. Animal* **February,** 21–25.

Flecknell, P.A., Orr, H.E., Roughan, J.V., and Stewart, R. (1999). Comparison of the effects of oral or subcutaneous carprofen or ketoprofen in rats undergoing laparotomy. *Vet. Rec.* **144**, 65–67.

Fleischman, R.W., McCracken, D., and Forbes, W. (1977). Adynamic ileus in the rat induced by chloral hydrate. *Laboratory Animal Science*, p. 238–243.

Fomby, L.E., Wheat, T.M., Hartter, D.E., Tuttle, R.L., and Balck, C.A. (2004). Use of CO_2/O_2 anesthesia in the collection of samples for serum corticosterone analysis from Fischer 344 rats. *Contemp. Top. Lab. Anim. Sci.* **43**, 8–12.

Fowler, J.S.L., Brown, J.S., and Flower, E.W. (1979). Comparison between ether and carbon dioxide anaesthesia for removal of small blood samples from rats. *Lab. Anim.* **14**, 275–278.

Friedman, P.M., Mafong, E.A., Friedman, E.S. and Geronemus, R.G. (2001). Topical anesthetics update; EMLA and beyond. *Dermatol. Surg.* **27**, 1019–1026.

Gades, N.M., Danneman, P.J., Wixson, S.K., and Tolley, E.A. (2000). The magnitude and duration of the analgesic effect of morphine, butorphanol, and buprenorphine in rats and mice. *Cont. Top. Lab. Anim. Sci.* **39**, 8–13.

Gardner, R.J. (1989). Inhalation anaesthetics–exposure and control: a statistical comparison or personal exposures in operating theatres with and without scavenging systems. *Ann. Occup. Hyg.* **33**, 159–173.

Gaspani, L., Bianchi, M., Limiroli, E., Panerai, A.E., and Sacerdote, P. The analgesic drug tramadol prevents the effect of surgery on natural killer cell activity and metastatic colonization in rats. *J. Neuroimmunology* **129**, 18–24.

Giamberdino, M.A., Berkley, K.J., Affraitai, G., Lerza, R., Centurione, L., Lapenna, D., and Vecchiet, L. (2002). Influence of endometriosis on pain behaviors and muscle hyperalgesia induced by a ureteral calculosis in female rats. *Pain* **95**, 247–257.

Gillingham, M.B., Clark, M.D., Dahly, E.M., Krugner-Higby, L.A., and Ney, D.M. (2001). A comparison of two opioid analgesics for relief of visceral pain induced by intestinal resection in rats. *Contemp. Top.*, p. 21–26.

Gilron, I. Biederman, J., Jhamandas, K., and Hong, M. (2003). Gabapentin blocks and reverses antinociceptive morphine tolerance in the rat paw-pressure and tail-flick tests. *Anesthesiology* **98**, 1288–1292.

Glen, J.B. (1980). Animal studies of the anaesthetic activity of ICI 35868. *British Journal of Anaesthesia*, p. 731–741.

Glen, J.B., Cliff, G.S., and Jamieson, A. (1980). Evaluation of a scavenging system for use with inhalation anesthesia in rats. *Lab. Anim.* **14**, 207–211.

Glen, J.B., and Hunter, S.C. (1984). Pharmacology of an emulsion formulation of ICI 35868. *British Journal of Anaesthesia*, p. 617–625

Gomez de Segura, I.A., Criado, A.B., Santos, M., and Tendillo, F.J. (1998). Aspirin synergistically potentiates isoflurane minimum alveolar concentration reduction produced by morphine in the rat. *Anesthesiology* **89**, 1489–1494.

Gong, D., Fang, Z., Ionescu, P., Laster, M.J., Terrell, R.C., Ross, C., Eger, E.I. (1998). Rat strain minimally influences anesthetic and convulsant requirement of inhaled compounds in rats. *Anesth Analg*, p. 963–966.

Gonzalez, M.I., Field, M.J., Bramwell, S., McCleary, S. and Singh, L. (2000). Ovariohysterectomy in the rat: a model of surgical pain for evaluation of pre-emptive analgesia? *Pain* **88**, 79–88.

Grahn, D.A., Heller, M.C., Larkin, J.E., and Heller, H.C. (1996). Appropriate thermal manipulations eliminate tremors in rats recovering from halothane anesthesia. *J. Appl. Physiol.* **81**, 2547–2554.

Gravenstein, J.S. (1998). Monitoring with our good senses. *J. Clin. Monitor. Comput.* **14**, 451–453.

Green, C.J. (1975). Neuroleptanalgesics drug combinations in the anaesthetic management of small laboratory animals. *Lab. Anim.* **9**, 161–178.

Green, C.J., Halsey, M.J., and Precious, S. (1978). Alphaxalone-alphadolone anaesthesia in laboratory animals. *Lab. Anim.* **12**, 85–89.

Green, C.J., Knight, J., Precious, S., and Simpkin, S. (1981). Ketamine alone and combined with diazepam or xylazine in laboratory animals: a 10 year experience. *Lab. Anim.* **15**, 163–170.

Gross, D.R., Tranquilli, W.J., Greene, S.A., and Grimm, K.A. (2003). Critical anthropomorphic evaluation and treatment of postoperative pain in rats and mice. *J. Am. Vet. Med. Assoc.* **222**, 1505–1510.

Gutstein, H.B., and Akil, H., (2001). Opioid analgesics. *In* "Goodman & Gilman's The Pharmacological Basis of Therapeutics." (J. G. Hardman, L.E. Limbird, and A.G. Goodman, eds.), pp. 569–619, McGraw-Hill, New York.

Haberham, Z.L., van den Brom, W.E., Venker-van Haagen, A.J., Baumans, V., de Groot, H.N.M., and Hellebrakers. L.J. (1998). EEG evaluation of reflex testing as assessment of depth of pentobarbital anaesthesia in the rat. *Lab. Anim.* **33**, 47–57.

Hall, L.W., Clarke, K.W., and Trim, C.M. (2001a). Principles of sedation, analgesia and premedication. *In* "Veterinary Anaesthesia 10th ed.", pp. 75–112, W.B. Saunders, London.

Hall, L.W., Clarke, K.W., and Trim, C.M. (2001b). Injectable anesthetic agents. *In* "Veterinary Anaesthesia 10th ed.", pp. 101, W.B. Saunders, London.

Hampshire, V.A., Davis, J.A., McNickle, C.A., Williams, L., and Eskildson H. (2001). Retrospective comparison of rat recovery weights using inhalation and injectable anesthetics, nutritional and fluid supplementation for right unilateral neurosurgical lesioning. *Lab. Anim.* **35**, 223–229.

Hara, K., and Harris, R.A. (2002). The anesthetic mechanism of urethane: the effects on neurotransmitter-gated ion channels. *Anesth. Analg.* **94**, 313–318.

Harper, J.S. III., (2000). Personal communication.

Hauptman, K., Jekl, Jr., V., and Knotek, Z. (2003). Use of medetomidine for sedation in the laboratory rat (Rattus norvegicus). *Acta Vet. (Beogr)*, **72**, 583–591.

Hayes, J.H., and Flecknell, P.A. (1999). A comparison of pre- and post-surgical administration of bupivacaine or buprenorphine following laparotomy in the rat. *Lab. Anim.* **33**, 16–23.

Hayton, S.M., Kriss, A. and Muller, D.P.R. (1999). Comparison of the effects of four anaesthetic agents on somatosensory evoked potentials in the rat. *Lab. Anim.* **33**, 243–251.

Heavner, J.E. (1997). The pharmacology of analgesics. *In* "Anesthesia and Analgesia in Laboratory Animals." (D.H. Kohn, S.K. Wixson, W.J. White, and G.J. Benson, eds.), p. 43, Academic Press, New York.

Hecker, B.R., Lake, C.L., DiFazio, C.A., Moscicki, J.C., and Engle, J.S. (1983). The decrease of the minimum alveolar anesthetic concentration produced by sufentanil in rats. *Anesth. Analg.* **62**, 987–990.

Hedenqvist, P., Roughan, J.V., and Flecknell, P.A. (1999). Effects of repeated anaesthesia with ketamine/medetomidine and of pre-anaesthetic administration of buprenorphine in rats. *Lab. Anim.* **34**, 207–211.

Hedenqvist, P., Roughan, J.V., and Flecknell, P.A. (2000). Sufentanil and medetomidine anaesthesia in the rat and its reversal with atipamezole and butorphanol. *Lab. Anim.* **34**, 244–251.

Hsu, W.H., Bellin, S.I., Dellmann, H-D., and Hanson, C.E. (1989). Xylazine-ketamine-induced anesthesia in rats and its antagonism by yohimbine. *J. Am. Vet. Med. Assoc.* **189**, 1040–1043.

Hu, C., Flecknell, P.A., and Liles, J.H. (1992). Fentanyl and medetomidine anaesthesia in the rat and its reversal using atipamezole and either nalbuphine or butorphanol. *Lab. Anim.* **26**, 15–22.

Hu, G., Salem, M.R., and Crystal, G.J. (2004). Isoflurane and sevoflurane precondition against neutrophil-induced contractile dysfunction in isolated rat hearts. *Anesthesiology* **100**, 489–497.

Hughes, E.W., Martin-Body, R.L., Sarelius, I.H., and Sinclair, J.D. (1982). Effects of urethane-chloralose anaesthesia on respiration in the rat. *Clin. Exp. Pharmacol. Physiol.* **9**, 119–127.

Hunter, S.C., Glen, J.B., and Butcher, C.J. (1984). A modified anaesthetic vapour extraction system. *Lab. Anim.* **18**, 42–44.

Imai, A., Steffey, E.P., Farver, T.B., and Ilkiw, J.E. (1999). Assessment of isoflurane-induced anesthesia in ferrets and rats. *Am. J. Vet. Res.* **60**, 1577–1583.

Jablonski, P., and Bowen, B.O. (2001). Oral buprenorphine and aspirin analgesia in rats undergoing liver transplantation. *Lab. Anim.* **36**, 134–143.

Jablonski, P., Howden, B.O., and Baxter, K. (2001) Influence of buprenorphine analgesia on post-operative recovery in two strains of rats. *Lab. Anim.* **35**, 213–222.

Jacobson, C. (1999). Adverse effects on growth rates in rats caused by buprenorphine administration. *Laboratory Animals*, p. 202–206.

Jou, I-M., Tsai, C-L., Wu, M-H., Chan, H-Y., and Wang, N-S. (2002). Simplified rat intubation using a new oropharyngeal intubation wedge *J. App. Physiol.* **89**, 1766–1770.

Kaczmarczyk, G., and Reinhardt, H.W. (1975). Arterial blood gas tensions and acid-base status of Wistar rats during thiopental and halothane anesthesia. *Lab. Anim. Sci.* **25**, 184–189.

Karasch, E.D., Armstrong, A.S., Gunn, K. Artu, A., Cox, K., and Karol, M.D. (1995). Clinical sevoflurane metabolism and disposition. II. The role of cytochrome P450 2E1 in fluoride and hexaflourisopropanol formation. *Anesthesiology* **82**, 1379–1388.

Kashimoto, S., Furuya, A., Nonaka, A., Oguchi, T., Koshimizu, M., and Kumazawa, T. (1997). The minimum alveolar concentration of sevoflurane in rats. *European Journal of Anaesthesiology*, p. 359–361.

Kawaguchi, M., Drummond, J.C., Cole, D.J., Kelly, P.J., Spurlock, M.P., and Patel, P.M. (2004). Effect of isoflurane on neuronal apoptosis in rats subjected to focal cerebral ischemia. *Anesth. Analg.* **98**, 798–805.

Kenna, J.G., and Jones, R.M. (1995). The organ toxicity of inhaled anesthetics. *Anesth. Analg.* **81**, S51–S66.

King, C.S., and Miller, R.T. (1997). Fatal perforating intestinal ulceration attributable to flunixin meglumine overdose in rats. *Lab. Anim. Sci.* **47**, 265–208.

Kofke, W.W., Garman, R.H., Janosky, J., and Rose, M.E. (1996a). Opioid neurotoxicity; neuropathologic effects of different fentanyl congeners and effects of hexamethonioum-induced normotension. *Anesth. Analg.* **83**, 141–146.

Kofke, W.W., Garman, R.H., Stiller, R.L., Rose, M.E., and Garman, R. (1996b). Opioid neurotoxicity: fentanyl dose-response effects in rats. *Anesth. Analg.* **83**, 298–306.

Kofke, W.A., Garman, R.H., Tom, W.C., Rose, M.E., and Hawkins, R.A. (1992). Alfentanil-induced hypermetabolism, seizure, and histopathology in rat brain. *Anesth. Analg.* **75**, 953–964.

Kofke, W.A., Attaallah, A.F., Kuwabara, H., Garman, R.H., Sinz, E.H., Barbaccia, J., Gupta, N., and Hogg, J.P. (2002). The neuropathologic effects in rats and neurometabolic effects in humans of large-doe remifentanil. *Anesth. Analg.* **94**, 1229–1236.

Kohler, I., Meier, R., Busato, A., Neiger-Aeschbacher, and Schatzmann, U. (1998). Is carbon dioxide (CO_2) a useful short acting anaesthetic for small laboratory animals? *Lab. Anim.*, p. 155–161.

Kohn, D.S., Wixson, S.K., White, W.J., and Benson, G.J., eds. (1997). "Anesthesia and Analgesia in Laboratory Animals." Academic Press, New York.

Kohrs, R., and Durieux, M.E. (1998). Ketamine: teaching and old drug new tricks. *Anesth. Analg.* **87**, 1186–1193.

Komulainen, A., and Olson, M.E. (1991). Antagonism of ketamine-xylazine anesthesia in rats by administration of yohimbine, tolazoline, or 4–aminopyridine. *Am. J. Vet. Res.* **52**, 585–588.

Kroin, J.S., Buvanendran, A., Nagalla, S.K.S., and Tuamn, K.J. (2003). Postoperative pain and analgesic responses are similar in male and female Sprague-Dawley rats. *Can. J. Anesth.*, p. 904–908.

Kronen, P.W. (2003). Anesthetic management of the horse: inhalation anesthesia. *In* "Recent advances in anesthetic management of large domestic animals." (Steffey, E.P., ed.) International Veterinary Information Service. ONLINE. Available: *http://www. ivis.org*.

Krugner-Higby, L., Smith, L., Clark, M., Wendland, A., and Heath, T.D. (2003). A single dose of liposome-encapsulated oxymorphone or morphine provides long-term analgesia in an animal model of neuropathic pain. *Comp. Med.* **53**, 280–287.

Lake, C.L., DiFazia, C.A., Moscicki, J.C., and Engle, J.S. (1985). Reduction in halothane MAC: comparison of morphine and alfentanil. *Anesth. Analg.* **64**, 807–810.

Landi, M.S., Kreider, J.W., and Lang, C.M. (1982). Effects of shipping on the immune function of mice. *Am. J. Vet. Res.* **43**, 1654–1657.

Larsson, J.E., and Wahlström, G. (1998). The influence of age and administration rate on the brain sensitivity to propofol in rats. *Acta Anaesthesiologica Scandinavica*, p. 987–994.

Leach, M.C., Bowell, V.A., Allan, T.F., and Morton D.B. (2002). Degrees of aversion shown by rats and mice to different concentrations of inhalational anaesthetics. *Vet. Rec.* **150**, 180–189, 808–815.

Lee, F.T., Sproat, I.A., Hammond, T.G., and Naidu, S.G. (1994). Placement of intraperitoneal injections in rats. *Contemp. Top. Lab. Anim. Sci.* **33**, 64–65.

Levy, D.E., Zwies, A., and Duffy, T.E. (1980). A mask for delivery of gases to small laboratory animals. *Lab. Anim. Sci.* **30**, 868–870.

Liles, J.H., and Fecknell, P.A. (1992a). The effects of buprenorphine, nalbuphine and butorphanol alone or following halothane anaesthesia on food and water consumption and locomotor movement in rats. *Lab. Anim.* **26**, 180–189.

Liles, J.H., and Flecknell, P.A. (1992b). The use of non-steroidal anti-inflammatory drugs for the relief of pain in laboratory rodents and rabbits. *Lab. Anim.* **26**, 241–255.

Liles, J.H., and Flecknell, P.A. (1993). The influence of buprenorphine or bupivacaine on the post-operative effects of laparotomy and bile-duct ligation in rats. *Lab. Anim.* **27**, 374–380.

Liles, J.H., and Flecknell, P.A. (1994). A comparison of the effects of buprenorphine, carprofen and flunixin following laparotomy in rats. *J. Vet. Pharmacol.* **17**, 284–290.

Liles, J.H., Flecknell, P.A., Roughan, J., and Cruz-Madorran, I. (1998). Influence of oral buprenorphine, oral naltrexone or morphine on the effects of laparotomy in the rat. *Lab. Anim.* **32**, 149–161.

Ludwig, L.M., Patel, H.H., Gross, G.J., Kersten, J.R., Pagel, P.S., and Warltier, D.C. (2003). Morphine enhances pharmacological preconditioning by isoflurane. *Anesthesiology* **98**, 705–711.

Macdonald, E., Haapalinna, A., Virtanen, R., and Lammintausta, R. (1989). Effects of acute administration of medetomidine on the behaviour, temperature and turnover rates of brain biogenic amines in rodents and reversal of these effects by atipamezole. *Acta Vet. Scand.* **85**, 77–81.

Maggi, C.A., and Meli A. (1986a). Suitability of urethane anesthesia for physiopharmacological investigations in various systems. Part 1: general considerations. *Experientia* **42**, 109–210.

Maggi, C.A., and Meli A. (1986b). Suitability of urethane anesthesia for physiopharmacological investigations in various systems. Part 2: cardiovascular system. *Experientia* **42**, 292–297.

Maggi, C.A., and Meli A. (1986c). Suitability of urethane anesthesia for physiopharmacological investigations in various systems. Part 3; other systems and conclusions. *Experientia* **42**, 531–537.

Martin, L.B.E., Thompson, A.C., Martin, T., and Kristal. M.B. (2001). Analgesic efficacy of orally administered buprenorphine in rats. *Comp. Med.* **51**, 43–48.

Mahmoudi, N.W., Cole, D.J., and Shapiro, H.M. (1989). Insufficient anesthetic potency of nitrous oxide in the rat. *Anesthesiology* **70**, 345–349.

Mogil, J.S. (1999). The genetic mediation of individual differences in sensitivity to pain and its inhibition. *Proc. Nat. Acad. Sci. U.S.A.* **96**, 7744–7751.

Morton, D.B., and Griffiths, P.H.M. (1985). Guidelines on the recognition of pain, distress and discomfort in experimental animals and an hypothesis for assessment. *Vet. Rec.* **116**, 431–436.

Nadeson, R., and Goodchild, C.S. (2001). Antinociceptive properties of neurosteroids III: experiments with alphadolone given intravenously, intraperitoneally, and intragastrically. *Brit. J. Anesth.* **86**, 704–708.

Naff, K.A., Craig, S., and Gray, K. (2004). Intraoperative deaths in rats undergoing experimental ocular surgery. *Lab Animal* **33**, 22–25.

Nahas, K., and Provost, J-P. (2002). Blood sampling in the rat: current practices and limitations. *Comp. Clin. Path.* **11**, 14–37.

Nevalainen, T., Pyhälä, L., and Hanna-Maija, V. (1989). Evaluation of anesthetic potency of medetomidine-ketamine combination in rat, guinea pigs and rabbits. *Acta Vet. Scand.* **85**, 139–143.

National Institute for Occupational Safety and Health-U.S. Department of Health Education and Welfare. (1977). Criteria for a Recommended Standard: Occupational Exposure to Waste Anesthetic Gases and Vapors. Publication No. 77–140, U.S. Government Printing Office, Washington, D.C.

Nolan, A.M. (2000). Pharmacology of analgesic drugs. *In* "Pain Management in Animals." (P.A. Flecknell, and A. Watermen-Pearson, eds.), pp. 21–53, W.B. Saunders, London.

Olson, M.E., Vizzutti, D.W., and Cox, A.K. (1993). The parasympatholytic effects of atropine sulfate and glycopyrrolate in rats and rabbits. *Can. J. Vet. Res.* **57**, 254–258.

Orliaguet, G., Vivien, B., Langeron, O., Bouhemad, B., Coriat, P., and Riou, B. (2001). Minimum alveolar concentration of volatile anesthetics in rats during postnatal maturation. *Anesthesiology* **95**, 734–739.

Park, C.M., Clegg, K.E., Harvey-Clark, C.J., and Hollenberg, M.J. (1992). Improved techniques for successful neonatal rat surgery. *Lab. Anim. Sci.* **42**, 508–513.

Peck, L.S., Blendedn, S.E., Michelis, M., and Goldberg, E.P. (2004). An effective buprenorphine analgesic protocol increased post-operative adhesions in rats after laparotomy. Proceedings of the ACVA/IVAPM/AVAT Meeting, p. 64, Phoenix, AZ.

Pekow, C. (1992). Buprenorphine, Jell-O recipe for rodent analgesia. *Synapse* **25**, 35–36.

Phifer, C.B., and Terry, L.M. (1986). Use of hypothermia for general anesthesia in preweanling rodents. *Physiology & Behavior*, 887–890.

Proctor, E., and Fernando, A.R. (1973). Oro-tracheal intubation of the rat. *Br. J. Anaesth.* **45**, 139–142.

Quinn, R.H., Danneman, P.J., and Dysko, R.C. (1994). Sedative efficacy of droperidol and diazepam in the rat. *Lab. Anim. Sci.* **44**, 166–171.

Ramsay, M.A.E., and Lutermman, D.L. (2004). Dexmedetomidine as a total intravenous anesthetic agent. *Anesthesiology* **101**, 787–790.

Refael, I., Gitelman, I., and Davis, C. (2003). A replacement for methoxyflurane (Metofane) in open-circuit anaesthesia. *Lab. Anim.* **38**, 280–285.

Reid, W.C., Carmichael, K.P., Srinivas, S., and Bryant, J.L. (1999). Pathologic changes associated with use of tribromoethanol (Avertin) in the Sprague Dawley rat. *Laboratory Animal Science*, 665–667.

Rembert, M.S., Smith, J.A., and Hosgood, G. (2004). A comparison of a forced-air warming system to traditional thermal support for rodent microenvironments. *Lab. Anim.* **38**, 55–63.

Roberts II, L.J., and Morrow, J.D. (2001). Analgesic-antipyretic and antiinflammatory agents and drugs employed in the treatment of gout. *In* "Goodman & Gilman's The Pharmacological Basis of Therapeutics." (J. G. Hardman, L.E. Limbird, and A.G. Goodman, eds.), p. 687, McGraw-Hill, New York.

Roughan, J.V., Burzaco, Ojeda, O., and Flecknell, P.A. (1999). The influence of pre-anaesthetic administration of buprenorphine on the anesthetic effects of ketamine/medetomidine and pentobarbitone in rats and the consequences of repeated aneaesthesia. *Lab. Anim.* **33**, 234–232.

Roughan, J.V., and Flecknell, P.A. (2002a). Buprenorphine: a reappraisal of its antinociceptive effects and therapeutic use in alleviating post-operative pain in animals. *Lab. Anim.* **36**, 322–343.

Roughan, J.V., and Flecknell, P.A. (2002b). Behavioural effects of laparotomy and analgesic effects of ketoprofen and carprofen in rats. *Pain* **90**, 65–64.

Roughan, J.V., and Flecknell, P.A. (2003). Evaluation of a short duration behaviour-based post-operative pain scoring system in rats. *Europ. J. Pain* **7**, 397–406.

Roughan, J.V., Flecknell, P.A., and Davies, B.R. (2004). Behavioural assessment of the effects of tumor growth in rats and the influence of the analgesics carprofen and meloxicam. *Lab. Anim.* **38**, 286–296.

Santos, M., Kunkar V., Garcia-Iturralde, P., and Tendillo, F.J. (2004). Meloxicam, a specific COX-2 inhibitor, does not enhance the isoflurane minimum alveolar concentration reduction produced by morphine in the rat. *Anest. Analg.* **98**, 359–363.

Savola, J.-M., (1989). Cardiovascular actions of medetomidine and their reversal by atipamezole. *Acta. Vet. Scand.* **85**, 39–47.

Schwab, J.M., Schluesener, H.J., Meyermann, R., and Serhan, C.N. (2003). COX-3 the enzyme and the concept: steps towards highly specialized pathways and precision therapeutics. *Prostaglandins Leukotrienes and Essential Fatty Acids* **69**, 339–343.

Seyde, W.C., McGowan, L., and Land, N. (1985). Effects of anesthetics on regional hemo-dynamics in normovolemic and hemorrhaged rats. *Am. J. Physiol.* **249**(Pt. 2), H164–H173.

Short, C.E. (1987). "Principles and Practice of Veterinary Anesthesia." Williams & Wilkins, Baltimore.

Silverman, J., and Muir III, W.W. (1993). A review of laboratory animal anesthesia with chloral hydrate and chloralose. *Lab. Anim. Sci.* **43**, 210–216.

Silverman, J., Huhndorf. M., and Slater, G. (1983). Evaluation of a combination of tiletamine and zolazepam as an anesthetic for laboratory animals. *Lab. Anim. Sci.* **33**, 457–460.

Simpson, D.P. (1997). Prolonged (12 hours) intravenous anesthesia in the rat. *Lab. Anim. Sci.* **47**, 519–523.

Sinclair, M.D. (2003). A review of the physiological effects of α_2 agonists related to the clinical use of medetomidine in small animal practice. *Can. Vet. J.* **44**, 885–897.

Skolleborg, K.C., Grönbech, J.E., Grong, K., Åbyholm, F.E., and Lekven, J. (1990). Distribution of cardiac output during pentobarbital versus midazolam/fentanyl/fluanisone anaesthesia in the rat. *Lab. Anim.* **24**, 221–227.

Smiler, K.L., Stein, S., and Hrapkeiwicz, K.L. (1990). Tissue response to intramuscular and intraperitoneal injections of ketamine and xylazine in rats. *Lab. Anim. Sci.* **40**, 60–64.

Smiley, M.M., Lu, Y., Vera-Portocarrera, L.P., Zidan, A., and Westlund, K.N. (2004). Intrathecal gabapentin enhances the analgesic effects of subtherapeutic dose morphine in a rat experimental pancreatitis model. *Anesthesiology* **101**, 759–765.

Smith, L., Krugner-Higby, L., Clark, M., Heath, T.D., Dahly, E., Schiffman, B., Hubbard-VanStelle, S., Ney, D., and Wendland, A. (2003). Liposome-encapsulated oxymorphone hydrochloride provides prolonged relief of postsurgical visceral pain in rats. *Comp. Med.* **53**, 270–279.

Smith, J.C., and Bolon, B. (2002). Atmospheric waste isoflurane concentrations using conventional equipment and rat anesthesia protocols. *Contemp. Top. Lab. Anim. Sci.* **41**, 10–17.

Smith, J.C., and Bolon, B. (2003). Comparison of three commercially available activated charcoal canister for passive scavenging of waste isoflurane during conventional rodent anesthesia. *Contemp. Top. Lab. Anim. Sci.* **42**, 10–15.

Soma, L.R. (1985). Behavioral changes and the assessment of pain. *Proceedings of the 2nd International Congress of Veterinary Anaestheiology*, pp. 38–41.

Speth, R.C., Smith, M.S., and Brogan, R.S. (2001). Regarding the advisability of administering postoperative analgesics in the drinking water of rats *Contemp. Top. Lab. Anim. Sci.*, 15–17.

Stabenow, J.M., and Vogler, G.A. (2002). Refractometry for quality control of anesthetic drug mixtures. Poster abstract P40. *Contemp. Top. Lab. Anim. Sci.* **41**, 95.

Stabernack, C.R., Eger II, E.I., Warnken, U.H., Forster, H., Hanks, D.K., and Ferrell, L.D. (2003). Sevoflurane degradation by carbon dioxide absorbents may produce more than one nephrotoxic compound in rats. *Can. J. Anesth.* **50**, 249–252.

Stark, R.A., Nahrwold, M.L., and Cohen, P.J. (1981). Blind oral intubation of rats. *J. Appl. Physiol. Respirat. Environ. Exercise Physiol.* **51**, 1355–1356.

Stasiak, K.L., Maul, D., French, E., Hellyer, P.W., and Vandewoude, S. (2003). Species-specific assessment of pain in laboratory animals. *Contemp. Top. Lab. Anim. Sci.* **42**, 13–20.

Stazyk, C., Bohnet, W., Gasse, H., and Hackbarth, H. (2003). Blood vessels of the rat tail: a histological re-examination with respect to blood vessel puncture methods. *Lab. Anim.* **37**, 121–125.

Steffey, E.P. (1996). Inhalation anesthetics. *In* "Lumb & Jones' Veterinary Anesthesia, 3rd ed." (J.C. Thurmon, W. J. Tranquilli, and G.J. Benson, eds.), pp. 297–329, Williams & Wilkins, Baltimore.

Steffey, M.A., Brosnan, R.J., and Steffey, E.P. (2003). Assessment of halothane and sevoflurane anesthesia in spontaneously breathing rats. *Am. J. Vet. Res.* **64**(4), 470–474.

Steffey, E.P., and Eger II, E.I. (1985). Nitrous oxide in veterinary practice and animal research. *In* "Nitrous oxide/N₂O." (Eger II, E.I., ed.), p. 308, Elsevier, New York.

Stickrod, G. (1979). Ketamine/xylazine anesthesia in the pregnant rat. *J. Am. Vet. Med. Assoc.* **175**, 952–953.

Sudoh, Y., Cahoon, E.E., Gerner, P. and Wang, G.K. (2003). Tricyclic antidepressants as long-acting local anesthetics. *Pain* **103**, 49–55.

Sun, F.J., Wright, D.E., and Pinson, D.M. (2003). Comparison of ketamine versus combination of ketamine and medetomidine in injectable anesthetic protocols; chemical immobilization in macaques and tissue reaction in rats. *Contemp. Top. Lab. Anim. Sci.* **42**, 32–37.

Thompson, A.C., Kirstal, M.B., Sallaj, A., Acheson A., Martin, L.B.E., and Martin, T. (2004). Analgesic efficacy of orally administered buprenorphine: methodologic considerations. *Compar. Med.* **54**, 293–300.

Thurmon, J.C., Tranquilli, W.J., and Benson, G.J., eds. (1996). "Lumb and Jones' Veterinary Anesthesia, 3rd ed." Williams & Wilkins, Baltimore.

Tran, D.Q., and Lawson, D. (1986). Endotracheal ventilation and manual ventilation of the rat. *Lab. Anim. Sci.* **36**, 540–541.

Urbanski, H.F., and Kelley, S.T. (1991). Sedation by exposure to a gaseous carbon dioxide-oxygen mixture: application to studies involving small laboratory animal species. *Lab. Anim. Sci.* **41**, 80–81.

Vachon, P., Faubert, S., Blais, D., Comtois, A., and Bienvenu, J.G. (2000). A pathophysiological study of abdominal organs following intraperitoneal injections of chloral hydrate in rats: comparison between two anaesthesia protocols. *Lab. Anim.* **34**, 84–90.

Vachon, P., and Moreau, J.-P. (2002). Butorphanol decreases edema following carrageenan-induced paw inflammation in rats. *Cont. Top. Lab. Anim. Sci.* **41**, 15–17.

Vainio, O. (1989). Introduction to the clinical pharmacology of medetomidine. *Acta. Vet. Scand.* **85**, 85–88.

Vainio, O. (1997). α₂-Adrenergic agonists and antagonists. 6th Proceeding of the International Congress of Veterinary Anaesthesiology, Thessaloniki, Greece, pp. 75–77.

Villeneuve, M.Y., and Casanova, C. (2003). On the use of isoflurane in the study of visual response properties of single cells in the primary visual cortex. *J. Neurosci. Methods.* **129**, 19–31.

Villegas, I., de la Lastra, C.A., Matin, M.J., Motilva, V., and La Casa Garcia, C. (2002). Gastric damage induced by subchronic administration of preferential cyclooxygenase-1 and cyclooxygenase-2 inhibitors in rats. *Pharmacology* **66**, 68–75.

Villegas, I., LaCasa, C., de la Lastra, A., Motilva, V., Herrias, J.M., and Martin, M.J. (2004). Mucosal damage induced by preferential COX-1 and COX-2 inhibitors: role of prostaglandins ans inflammatory response. *Life Sci.* **74**, 873–884.

Vogler, G.A. (1997). Anesthesia equipment: types and uses. *In* "Anesthesia and Analgesia in Laboratory Animals." (D.H. Kohn, S.K. Wixson, W.J. White, and G.J. Benson, eds.), pp. 105–147, Academic Press, New York.

Ward, G.S., Johnson, D.O., and Roberts, C.R. (1974). The use of CI 744 as an anesthetic for laboratory rats. *Lab. Anim. Sci.* **24**, 737–742.

Warner, T.D., and Mitchell, J.A. (2004). Cyclooxygenases: new forms, new inhibitors, and lessons from the clinic. *FASEB J.* **18**, 790–804.

Weksler, B., Ng, B., Lenert, J., and Burt, M. (1994). A simplified method of endotracheal intubation in the rat. *J. Appl. Physiol.* **76**, 1823–1825.

Whelan, G., and Flecknell, P.A. (1994). The use of etorphine/methotrimeprazine and midazolam as an anesthetic technique in laboratory rats and mice. *Lab. Anim.* **28**, 70–77.

Wixson, S.K., White, W.J., Hughes Jr., H.C., Lang, C.M., and Marshall, W.K. (1987a). A comparison of pentobarbital, fentanyl-droperidol, and ketamine-xylazine and ketamine-diazepam anesthesia in adult male rats. *Lab. Anim. Sci.* **37**, 726–730.

Wixson, S.K., White, W.J., Hughes, Jr., H.C., Marshall, W.K., and Lang, C.M. (1987b). The effects of pentobarbital, fentanyl-droperidol, and ketamine-xylazine and ketamine-diazepam on noxious stimulus perception in adult male rats. *Lab. Anim. Sci.* **37**, 731–735.

Wixson, S.K., White, W.J., Hughes Jr., H.C., Lang, C.M., and Marshall, W.K. (1987c). The effects of pentobarbital, fentanyl-droperidol, and ketamine-xylazine and ketamine-diazepam on arterial blood pH, blood gases, mean arterial pressure and heart rate in adult rats. *Lab. Anim. Sci.* **37**, 736–749.

Wilson, R.P., Zagon, I.S., Larach, D.R., and Lang, C.M. (1993). Cardiovascular and respiratory effects of tiletamine-zolazepam. *Pharmacol. Biochem. Behaviour.* **44**, 1–8.

Wixson, S.K., and Smiler, K.L (1997). Anesthesia and Analgesia in Rodents. *In* "Anesthesia and Analgesia in Laboratory Animals" (D.H. Kohn, S.K. Wixson, W.J. White, and G.J. Benson, eds.), pp. 165–203, Academic Press, New York.

Wolff, M.H., Leather, H.A., and Wouters, P.F. (1999). Effects of tramadol on minimum alveolar concentration (MAC) of isoflurane in rats. *Brit. J. Anaesth.* **83**, 780–783.

Yasaki, S., and Dyck, P.J. (1991). A simple method for rat endotracheal intubation. *Lab. Anim. Sci.* **45**, 303–304.

Zambricki, E.A., and D'Alecy, L.G. (2004). Rat sex differences in anesthesia. *Comparative Medicine*, p. 49–53.

Chapter 20

Euthanasia and Necropsy

Jeffrey I. Everitt and Elizabeth A. Gross

I. INTRODUCTION

Recommendations for animal euthanasia are provided by the American Veterinary Medical Association (AVMA) Panel on Euthanasia (AVMA, 2001). Physical, inhalant, and pharmacologic methods of euthanasia all have their place in research studies that employ laboratory rats. Several factors are critically important to consider when choosing the method of euthanasia. The method must be humane and produce minimal discomfort, the time elapsed until unconsciousness and death should be as short

THE LABORATORY RAT, 2ND EDITION

as possible, the method must have a minimal impact on the experimental parameters being investigated, and the method should consider the safety factors and emotional effects on staff. Consideration of the act of euthanasia in conjunction with the experimental needs and objectives to be fulfilled by the necropsy procedure is a main emphasis of this chapter.

The necropsy examination, including tissue collection and preservation, represents one of the most critical phases for the evaluation of rats in experimental studies. Although there are many variations of necropsy technique, systematic examination to ensure that all lesions and specified tissues are collected and thoroughly examined macroscopically, as well as collected for subsequent experimental evaluation, is required regardless of specific design. The actual necropsy technique used depends on the objectives of the procedure. Most rat necropsies are conducted within the following categories: (1) diagnostic necropsy for determination of cause of clinical outcome, (2) rodent health surveillance, (3) complete necropsy for experimental purposes, and (4) target organ collection or evaluation. One example necropsy method is described in detail.

II. EUTHANASIA OF THE LABORATORY RAT

A. General Issues

Recommendations for animal euthanasia are provided by the AVMA Panel on Euthanasia (AVMA, 2001) and have been reviewed in a number of publications (Close et al., 1996a,b). Physical methods, including decapitation; inhalant methods such as the use of carbon dioxide (CO_2) asphyxiation and inhalant anesthetic overdose; and pharmacologic methods such as injectable anesthetic overdose all have their place in the research armamentarium for studies that employ laboratory rats. Several factors are critically important to consider when choosing the method of euthanasia. The method must be humane and produce minimal discomfort; the time elapsed until unconsciousness and death should be as short as possible, the method must have minimal impact on the experimental parameters being investigated, and the method should consider the emotional effects on personnel. There are other associated factors to consider such as minimizing exposure of laboratory personnel to physical and chemical hazards. In the United States, the euthanasia procedure requires review by the institutional animal care and use committee (IACUC) and involves careful examination of not only the method used but also the training and experience of personnel who conduct the procedure. In all cases, it should be recognized that for a rat to be humanely euthanized, the individual conducting the procedure must

feel comfortable handling this species. Gentle, compassionate animal handling is paramount for the minimization of distress during the procedure.

There are a variety of logistical issues that must be considered so that the act of euthanasia can be linked with the requirements for animal necropsy. For example, certain techniques that may be part of the rat necropsy, such as a whole-body or target organ vascular perfusion, may need to be treated as a terminal surgical procedure. There is an associated anesthesia knowledge requisite for the conduct of terminal surgical procedures, including knowledge of anesthetic monitoring. In certain instances, investigators desire to obtain blood and other clinical samples at the time of the terminal sacrifice, and this may necessitate certain methods of euthanasia so that obtained samples are compatible with those of other study time points.

B. Common Methods of Rat Euthanasia

The AVMA guidelines for euthanasia list a variety of techniques that are suitable for use in rats. One physical method used for euthanasia of laboratory rats is decapitation, usually using a rodent guillotine (Fig. 20-1). Rats weighing less than 200 g can be humanely killed by using cervical dislocation. The AVMA guidelines on euthanasia recommend sedation before decapitation and/or cervical dislocation. Many IACUCs make exceptions to this requirement if the investigator can demonstrate the possibility of deleterious effects of sedation on the experimental parameters being studied. Physical methods are most often utilized when there is a need to collect biological samples for biochemical and metabolism studies that would be compromised by exposure to xenobiotic agents. Although

Fig. 20-1 Guillotine used for rodent decapitation. This equipment should be kept very sharp and should be cleaned between animals.

Fig. 20-2 Plexiglas chamber for rodent euthanasia. Note that the chamber is clear to allow visualization of the animals and is easily disconnected from the gas line for cleaning.

physical methods of euthanasia are often effectively utilized for rats, there are a number of drawbacks. Decapitation of nonsedated rats entails danger to the operator of the rodent guillotine and can be difficult in fractious animals. Some individuals find that placement of animals in disposable plastic cones enhances the ability to carefully position the rat in the rodent guillotine and conduct the decapitation procedure. Both decapitation and cervical dislocation can be aesthetically displeasing and cause emotional impact to individuals conducting or viewing the procedure.

CO_2 or oxygen (O_2)/CO_2 inhalation is a commonly used method of rat euthanasia and is used in many types of studies (Danneman et al., 1997). Gas is administered via various forms of inhalation chambers hooked directly to gas sources (Fig. 20-2). Chambers used for rodent euthanasia should allow the animals to be seen during the process and should be easy to sanitize. In addition to CO_2 a variety of inhalant anesthetic gases have been used either alone or in combination with exsanguination or bilateral pneumothorax to accomplish euthanasia of rats (Blackshaw et al., 1988). Terminal surgical procedures in rats that are conducted before or as part of the necropsy process are often conducted by using various forms of inhalation anesthesia such as halogenated ethers (halothane, isoflurane, etc.) using various rodent anesthetic circuits. Care must be taken to scavenge waste anesthetic gases when these methods are used. Rats subjected to inhalant anesthesia in these terminal procedures are generally exsanguinated under deep general anesthesia or are killed via an anesthetic overdose.

Many laboratories that conduct pathology assessments of rats use parenteral anesthesia followed by exsanguination via cardiac puncture or via vena caval transection

through ventral midline celiotomy. Sodium pentobarbital is often the parenteral agent of choice owing to its effectiveness in inducing a reproducible deep surgical plane of general anesthesia and its relatively low cost. There are advantages to the investigator regarding tissue suitability for morphology and other types of studies when exsanguination is conducted as part of the necropsy procedure.

C. Controversies

Over the past several years, there have been a number of controversies concerning the euthanasia of rats. One such controversy concerns the humaneness of CO_2 asphyxia alone versus admixed O_2/CO_2 inhalation (Coenen et al., 1995; Hackbarth et al., 2000; Leach et al., 2002). This controversy stems in part from the finding that human volunteers exposed to 100% CO_2 judged the experience as "painful" (Danneman et al., 1997). There have been controversies concerning the optimal use of CO_2 inhalation chambers; these include flow rates of gas, precharging of chambers, loading of chambers with multiple animals, and schedules and methods for cleaning chambers between animals (Hewett et al., 1993; Smith and Harrap, 1997; Gos et al., 2002). Associated with the euthanasia issue is a general lack of scientific information concerning how rats in the necropsy room may perceive pheromone and other signals from rats that have previously undergone euthanasia and other procedures. Thus, how equipment is cleaned, maintained, and operated in the necropsy area is important for the conduct of the euthanasia procedure. One aspect of the necropsy room layout that may be important is the location of animals relative to other animals being handled, anesthetized, euthanized, and prosected (Fig. 20-3). Some studies suggest that this is not as important an issue as might be perceived (Sharp et al., 2002).

The use of short (30 seconds) high concentration (70%) CO_2 inhalation induces anesthesia and antinociception in the rat (Mischler et al., 1994). This method is commonly used for collection of blood samples before euthanasia by CO_2 asphyxiation. Although CO_2 is effective for euthanasia and for analgesia relative to minor procedures such as needle sticks, it should not be routinely used for use in more radical terminal surgical procedures such as ventral midline celiotomy, owing to how quickly the animals awaken when they are removed from the gas.

Controversy related to the humaneness of decapitation in rats has been recently addressed in the scientific literature, thereby ameliorating earlier concerns among IACUCs (Allred and Berntson, 1986; Vanderwolf et al., 1988; Derr, 1991). How decapitation is conducted in rats must take into account the experimental parameters being studied as well as the experience of the laboratory and associated staff. There are significant safety issues involved

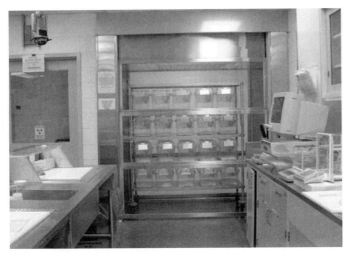

Fig. 20-3 Photograph of a rodent necropsy room equipped with an Illinois cubicle for rodent cage and rack isolation. A well-designed necropsy room such as the one shown here has well-lighted and ventilated areas with easily sanitized surfaces that allow for prosection (left) and areas for weighing tissues and animals (right).

in the use of rodent guillotines, as well as issues pertaining to physical restraint and associated stress.

The purpose of discussion in this chapter is not to review the methodology of rat euthanasia per se because this is an evolving process outlined in the AVMA euthanasia guidelines, which serve as the general reference on the topic. Instead, the focus of the present discussion is to emphasize the need for careful consideration of the act of euthanasia in conjunction with the experimental needs and objectives to be fulfilled by the necropsy procedure. In all instances, scientific personnel who conduct euthanasia should do so with a full understanding of how the method induces death and an understanding of any controversial issues that may pertain to the procedure.

D.   Influence of Euthanasia Method on Experimental Parameters

In general, there is a relative paucity of published data that discuss the influence of euthanasia methods on experimental parameters. Consideration of this topic is a critically important issue for every investigator. A number of recent publications have demonstrated varied results. Exposure to O_2/CO_2 as an inhalant anesthetic agent did not alter sperm motility (Slott et al., 1994), trophic hormone levels, or central nervous system acetyl cholinesterase and choline acetyl transferase (Berger-Sweeney et al., 1994) in the rat. CO_2 has been shown to affect the respiratory system (Fawell et al., 1972; Feldman and Gupta, 1976) and hepatic metabolism (Brooks et al., 1999) in rats. Experimental studies have demonstrated that euthanasia

produced by exposure to isoflurane for 3 minutes, as well as euthanasia by exposure to 2.5 minutes of CO_2 or O_2/CO_2, altered liver metabolism in rats in comparison to data obtained from rats killed by decapitation (Brooks et al., 1999). Studies have demonstrated that euthanasia technique affects vascular arachidonic acid metabolism in rats as well as vascular and intestinal smooth muscle contractility (Butler et al., 1990).

In addition to studies that have been conducted in rats, there are studies in mice that may be extrapolated to what might be expected in rats. Studies have demonstrated that euthanasia techniques, including those of CO_2 euthanasia, had effects on certain immunologic assays (Howard et al., 1990; Pecaut et al., 2000). The use of various inhalant and parenteral anesthetic techniques can dramatically affect endocrine parameters in both tissues and blood. In summary, in studies with biochemical, physiological, and functional endpoints, it is particularly important for investigators to carefully examine the potential effects of the euthanasia method on experimental parameters.

In addition to effects on experimental parameters, several of the most commonly used methods for euthanizing rats can result in significant alterations and artifacts that affect tissue morphology in pathology studies (Feldman and Gupta, 1976). Physical methods of euthanasia such as decapitation and cervical dislocation have profound effects on cervical structures and can affect the respiratory system through the aspiration of blood into the trachea and through hemorrhage into mediastinal regions. Alveolar hemorrhage and profound pulmonary congestion have been commonly noted with CO_2 asphyxiation (Fawell et al., 1972). Intraperitoneal and intrathoracic injection of euthanasia agents and parenteral anesthetics can produce artifacts on the serosal surfaces of visceral organs (Feldman and Seely, 1988).

Laboratories that conduct necropsy of rats and use organ weight data generally recommend exsanguination under anesthesia as part of the euthanasia procedure. Severance of the abdominal aorta caused rapid and relatively complete exsanguination in rats and results in reduction of the standard deviation of mean absolute and relative organ weights (Sullivan, 1985). Exsanguination of tissues also offers advantages for certain pathology procedures such as immunohistochemistry and usually is favored by pathologists conducting microscopic examination of tissues.

III.   NECROPSY OF THE RAT

A.   General Considerations

The term *necropsy* is derived from the Greek words *Nekros*, meaning corpse, and *Opsis*, meaning to view.

The necropsy examination, including tissue collection and preservation, represents one of the most important phases for the evaluation of rats on study. Proper conduct of the necropsy is critical because there is only a single opportunity to conduct the procedure. Although there are many variations of necropsy technique, one should use systematic examination to ensure that all lesions and specified target tissues for collection are thoroughly examined macroscopically, as well as collected for subsequent experimental evaluation such as histopathologic analysis. The actual necropsy technique used depends on the objectives of the procedure. Most rat necropsies are conducted within the following categories: (1) diagnostic necropsy for determination of cause of clinical outcome, (2) rodent health surveillance, (3) complete necropsy for experimental purposes, and (4) target organ collection or evaluation. The methods used for these differing purposes will vary quite substantially and have been reviewed (Feldman and Seely, 1988). In many studies the necropsy procedure will need to be customized to suit the experimental needs.

Generally, the function of the gross necropsy is to identify lesions present in the animal and to collect tissue in an orderly fashion for subsequent microscopic and other experimental evaluations. Only systematic necropsy technique can ensure that all lesions and specified tissues are examined macroscopically and collected for histologic and other analyses. Even the most careful microscopic preparation and examination cannot salvage tissues lost owing to autolysis, tissue mishandling, or the failure to recognize and select lesions. A tissue discarded at necropsy is lost forever and with it, potentially valuable information. Whenever possible, the rat necropsy should be guided by a standard operating procedure and conducted by trained prosectors. Although dissection of the rat and collection of tissues is often the goal, the quality of the necropsy depends on the entire procedure and highly depends on the degree of planning and preparation for the event (Table 20-1) (Black, 1986).

A properly conducted necropsy reduces the number of artifacts noted in tissues. Artifacts can be produced by a number of conditions, including excessive tissue manipulation, improper dissection technique, poor instrumentation, osmotically injurious moistening fluids, autolysis, and poor fixation. Excessive digital manipulation and improper use of instruments, such as the use of dull instruments, can easily lead to injury of fresh tissue with distortion of normal tissue architecture. This is particularly true with rats, in which small tissues are easily distorted by poor handling, such as crush injury resulting from improper scissor use. It is important to optimize dissection technique. For example, stretching the optic nerve during removal of the globe from the skull can lead to extensive artifact such as retinal detachment. Artifacts can result from the

TABLE 20-1

ISSUES TO CONSIDER IN LOGISTICAL PLANNING OF THE RAT NECROPSY

- What is the method of euthanasia, and is it compatible with experimental needs?
- What are expected "target organs," and are any special procedures warranted such as ancillary clinical pathology studies?
- What special prelabeling and identification requirements are needed?
- Do fixatives and/or clinical preservatives need to be prepared fresh?
- Are there randomization issues and/or special timing issues required between experimental groups?
- Are target organs to be weighed, and is exsanguination required?
- Do terminal clinical, clinical chemistry, urinalysis, or hematology specimens need to be collected and, if so, by what method? Is food restriction required before this collection?
- Are microbiology samples to be collected, and is aseptic technique required?
- Are there critical steps that require a "single individual" to collect target tissues for consistency purposes?
- Is a special tissue collection order required for target organ study owing to autolysis issues or biochemistry needs?
- Are special identification methods to be used for multiple or paired tissues?
- Are study protocols, standard operating procedures, and clinical records available to the prosectors?
- Are frozen specimens to be collected and, if so, by what method?
- Are morphometric techniques to be used, and does this require special tissue handling such as organ perfusion?
- Are there special postfixation needs for molecular, ultrastructural, histochemical, or immunohistochemical studies?
- What are the microbial, radiological, and chemical hazards posed by the necropsy procedures, and what precautions are to be used?

drying of tissues or the contact of tissues with fluids that are not osmotically compatible. These changes can be reduced by the use of physiological saline as a moistening fluid and the avoidance of tissue contact with water. Artifacts of autolysis can be minimized by the proper use of fixatives and by ensuring rapid mucosal contact with fixatives. The need for rapid fixation means that it is important to flush rat nasal passages with fixative and to infuse fixative intraluminally into bowel segments.

B.   Necropsy Preparation

The rat necropsy is a procedure that should be carefully planned to obtain the optimal results. During the logistical phase of the necropsy planning, research personnel should review critical questions that will impact the conduct of the procedure and make plans accordingly (Table 20-1). The purpose of the necropsy should be carefully reviewed and will dictate the subsequent course of action. One must carefully consider the ancillary tests that will be performed on samples taken at necropsy, and this may dictate the order of collection and the necropsy methods that are chosen. For example, animal health and sentinel rat necropsies used to determine the pathogen status of colonies might necessitate aseptic technique and methods to acquire proper tissue samples for microbiological analysis. Similarly, samples might need to be analyzed for the presence of macroscopically visible ecto- or endoparasites using special methods. There are many ancillary imaging and other specialized methods that might need to be used during a rat necropsy for experimental purposes, such as the use of radiographic methods for the detection of skeletal abnormalities.

If a large number of rats are to be subjected to complete necropsy using multiple prosectors and necropsy stations, one must determine whether single individuals should be responsible for certain technical aspects of the necropsy in order to maintain consistency of data. One such example would be tissue dissection for target organ weight determinations where trim methods might contribute to differences in outcome. There are situations in which individuals each having a particular necropsy task is preferable to the use of multiple stations in which each individual performs a complete necropsy. One must carefully determine the need for randomization of animals across study groups at the terminal sacrifice. Sometimes randomization is warranted; at other times it might be useful to first necropsy controls before experimental animals to allow easier determination of macroscopic observations. Whenever multiple prosectors are used, careful records need to be kept about who is responsible for individual prosections or individual organs. It is important to have individuals work across experimental groups to prevent operator-specific tissue handling artifacts from being confused with experimental outcomes. These types of decisions are best made during planning sessions before the actual conduct of the necropsy procedure.

Rats should be necropsied as soon as possible after death as postmortem changes begin to occur immediately. Microscopically, these autolytic changes closely mimic the histologic alterations that one might note in many types of acute experiments and have been well described (Seaman, 1987). If immediate necropsy is not possible, the rat should be refrigerated to slow autolysis. Carcasses should not be frozen if microscopic examination is to be performed because ice crystals cause disruption of cellular architecture.

C. Equipment

All equipment and supplies should be arranged before the beginning of the necropsy procedure. Personnel should wear personal protective equipment, including laboratory coats, protective gloves, and safety eyewear. In addition, individuals need to be protected from volatile anesthetics, fixatives, solvents, and the like; thus, chemical safety hoods, vented enclosures, down-draft necropsy tables, and scavenging devices are often present in rodent necropsy areas. Depending on the type of study, it might be necessary to have equipment such as biological safety cabinets and autoclaves present in the necropsy room to handle biological hazards. Modern rodent necropsy rooms are often equipped with computer stations for on-line collection of body/organ weight and necropsy data.

Prelabeling of equipment such as specimen containers, tissue cassettes and weigh boats is useful to save time and to prevent sample mix-ups. All specimen containers should be labeled with indelible inks on their sides (never on container tops); this will help to avoid specimen mix-up. A cutting surface is necessary to lay the rat upon, and various hard plastic or cork dissecting boards have been used. Some prosectors favor corkboards or disposable or Styrofoam boards because they allow the use of pushpins to fasten rodents to the surface. In general, the surface should be easily cleaned between animals. If cork is used it, must be covered with disposable paper, which is changed between animals, and the board must be thoroughly scrubbed at the end of the day. Lighting and magnification are important factors to consider during the design of the rat necropsy station as both tissues and lesions can be quite small.

Surgical instruments are satisfactory for the necropsy procedure, and instrumentation depends on personal preference but should include bone cutting forceps or rongeurs to penetrate the rat skull, as well as clean, sharp, well-maintained scissors, forceps, and scalpels. Poor choice of instruments or instruments that are dull and soiled will cause tissue artifacts despite otherwise good dissection technique. An example includes large scissors, 15 to 18 cm in length, with sharp/blunt points that are suitable for the initial cutting through the skin and opening up the body cavities. For dissection of internal organs, various smaller, 10 to 13-cm-long scissors, either straight or curved with blunt/sharp or sharp/sharp points, are adequate. Scalpels with no. 11 or no. 22 blades for trimming tissues are useful and may be used to prevent artifacts related to tissue compression from scissors. Various straight and curved forceps are used during the rodent necropsy, and "rat-toothed" forceps are helpful for gripping surrounding tissues, although great care must be used to prevent injury to target tissues. Surgical spatulas are helpful for lifting small organs such as the pituitary when forceps become impractical.

A variety of consumables and other supplies are needed for the rat necropsy procedure. Suture material or umbilical tape are used for ligation of tissues such as trachea/lungs and urinary bladder after infusion of fixative. Physiological saline is used to rinse tissues. Blunt-tip hypodermic needles/dosing needles and a 5 to 20-mL syringe are needed for fixative infusion. In addition, dyes for labeling lesions, gauze sponges, various bags, tubes, and containers are all necessary supplies to have readily available. Small tissues such as rat lymph nodes or adrenal glands should be placed into prelabeled tissue cassettes or embedding bags.

It is often important to weigh the carcass as well as target tissues and to have properly calibrated balances available for this purpose. In some experiments the brain is weighed to derive organ/brain weight ratios for assessment of parenchymal organ size. Body weight, although useful, can change with conditions such as obesity and make organ/body weight ratios misleading, although this parameter is commonly assessed in rodent studies. Rat organs and lesions are often measured, so calipers and other measuring devices such as metric rulers are useful to have available. A well-designed and located writing surface is useful to keep documents such as necropsy sheets away from the dissection work area.

D. Necropsy Procedure

There is no single best method for the conduct of the rat necropsy, and the method should be selected based on the objective of the experiment, use of the tissues to be sampled, and the experience of prosection and pathology personnel. The suggested background reading and references by the following should aid those that have an interest in the conduct of the rat necropsy: Greene (1963); Cook (1965); Olds and Olds (1979); Winegerd (1988); Popesko et al. (1990); Bono (1994); Feinstein (1994); Sharp and La Regina (1998); Bono et al. (2000); National Cancer Institute Veterinary Pathology (*http://home.neifcrf.gov/vetpath/necropsy.html*); The Registry Nomenclature Information System (*http://www.item.fraunhofer.de/reni/index.html*); and The Virtual Mouse Necropsy, (*http://www.geocities.com/virtualbiology/necropsy.html*). The general necropsy procedure described in this section has proved useful in our laboratory.

1. External Examination

The first step in the rodent necropsy is a thorough external examination. The animal is palpated for gross anatomical abnormalities or subcutaneous masses. The anus and urethral areas are examined for discharge, blood, or diarrhea. The oral cavity (Fig. 20-4a) is examined for malaligned or broken teeth, blood, or foreign object (food or bedding). The ears and nose are examined for discharge. If identification implants such as ear tags or microchip transponders are used, their associated tissues are examined for the presence of macroscopic lesions. Eyes are examined for discharge and corneal or lens opacity. The external genitalia are examined for developmental abnormalities. The color and texture of the fur is also examined for any abnormality. Previous clinical observations are

reviewed, and the animal is checked for any external finding to confirm observations, such as an injury. All external findings should be noted on a necropsy record form.

2. Exsanguination

Blood samples are often collected at the time of necropsy from the heart of the anesthetized rat via cardiac puncture, from the vena cava, or after decapitation from the body trunk. Even if blood samples are not required, it is desirable to completely exsanguinate the animal to obtain consistent organ weights, provide for better histopathology or immunohistochemistry, and prevent pooled blood in tissues from obscuring lesions during the prosection process. After collection of blood samples according to protocol, rapid exsanguination of rats can be accomplished by severing the abdominal aorta (Fig. 20-4b), vena cava, or femoral arteries or by decapitation.

3. Dissection

The order of collection for tissues during any necropsy depends on the purpose of the study, on endpoints to be studied, and on which target tissues are of most importance. For purposes of this chapter, a good general method for rat prosection with full screen tissue collection is presented.

Skin with mammary gland is collected first because it is easily forgotten when it is not a target tissue. If the protocol allows, the fur on the ventral surface of the rat can be moistened with saline or alcohol. This will prevent fur from getting into the other tissue samples. Two rows of teats can be located lateral to the midline from the cervical to the inguinal region (Fig. 20-4c). Grasping the fur, a section of skin that includes teat, skin, and mammary gland (Fig. 20-4d) is cut. This piece of tissue (fur side up) is then placed onto a piece of light cardboard or paper towel and into fixative.

If bone marrow is to be collected for impression smears, it is critically important to do so quickly because bone marrow cell morphology deteriorates rapidly postmortem.

The tissues from the head are dissected next because the brain autolyzes quickly after death. The eyes, with optic nerves attached and Harderian glands, are removed first by grasping the adnexa with forceps and cutting into the socket of the globe with a long narrow scalpel blade. (Fig. 20-4e). Once the globe has been removed, the remaining fur is trimmed from the head, taking care not to damage the external nasal tissue. The nasal bones and frontal and parietal bones of the calvarium (skull) are easily identified. The brain is exteriorized by first removing the calvarium without disturbing the bones that cover the nasal cavity and turbinates. Shallow cuts made with bone cutting forceps circumscribe the calvarium

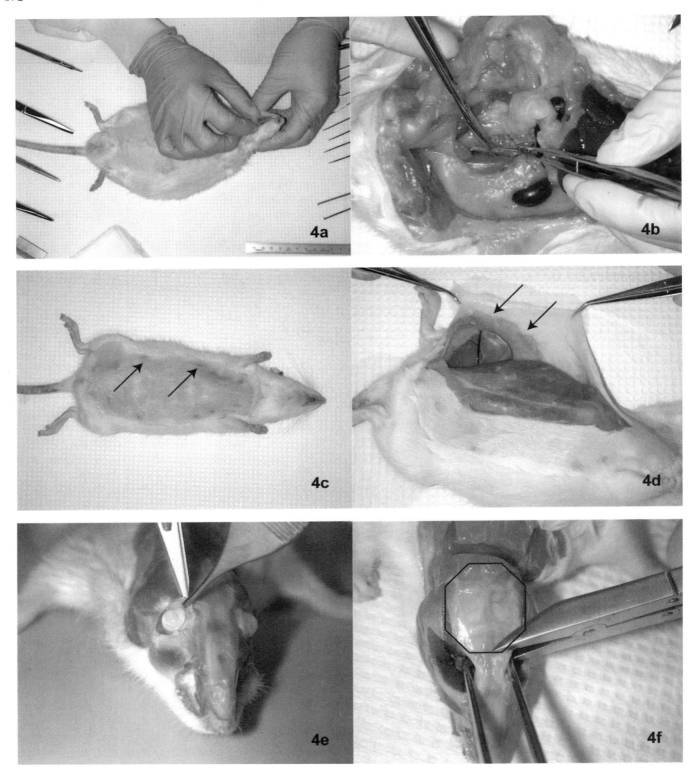

Fig. 20-4 (a) Photograph of rat necropsy showing the prosector examining the oral cavity before initiating dissection. (b) Rapid exsanguination can be accomplished by severing the abdominal aorta. Note the location deep in the abdominal cavity. Overlying fat has been pushed aside. (c) Photograph of the ventral surface of a rat that has had the fur moistened to facilitate observation and dissection. Arrows show the location of mammary gland teats. (d) The arrows in this figure show the location of the mammary glands on the underside of the teats seen in panel c. (e) Eyes are removed with the optic nerve and Harderian gland attached by grasping the adnexa with forceps and cutting into the ocular orbit with a long narrow scalpel blade. (f) The brain is exteriorized by removing the calvarium. Shallow cuts are made with bone cutting forceps along the lines indicated in this figure.

(Fig. 20-4f); it is then lifted off to expose the brain. Care is always taken to keep the cuts shallow enough to prevent damage to the brain tissue. The brain is removed with a small blunt spatula. The brainstem and optic nerves are cut so as not to stretch tissues, because stretching may cause artifact damage that will be noted by the pathologist during microscopic examination. Care is also taken to prevent damage to the pituitary gland that lies below the brain (Fig. 20-5a). This gland is generally removed after fixation.

The nasal cavity can be collected at this stage in the necropsy procedure or later with the remainder of the respiratory tract tissues. If it is collected at this stage, the lower jaw (with tongue) is separated from the head, and the head portion is cut from the carcass with scissors. The nasal cavity is flushed with fixative by inserting a blunt needle or cannula into the nasopharynx (Fig. 20-5b) and backflushing fixative slowly with a syringe until it exits the external nares.

At this stage in the necropsy, the abdominal and thoracic cavities (Fig. 20-5c,d) are opened with a ventral midline incision from the urethra to the jaw. The salivary glands and mandibular lymph nodes located in the cervical region are removed. The skin and musculature are reflected to expose the xyphoid cartilage. The thoracic cavity is opened by a cut through the diaphragm and ribs about 1 inch lateral to the midline. The lungs will collapse toward the spine and can be seen during the removal of the sternum, ribs, and muscle. Blunt scissors are suggested for this portion of the dissection.

The tissues of the abdominal and thoracic cavities and the ventral cervical region are observed for macroscopic lesions. It is important to observe each organ in relation to neighboring organs, especially if lesions involve more than one tissue or displace organs. Masses that were palpated during the in-life stage or at the start of the necropsy are located, observed, removed, measured, described, and recorded at this time.

Those organs of the abdominal cavity that are easily lost should be located and dissected first. The mesenteric lymph nodes (Fig. 20-5e) are removed before the remainder of the gastrointestinal (GI) tract. The liver is removed by cutting the diaphragm and using a piece of this attached tissue to manipulate the organ. The liver hilus (where the lobes come together) can also be grasped with forceps to manipulate the organ. The spleen is removed next. The pancreas may be left attached to the spleen, attached to the stomach, or removed separately. Because of high enzymatic activity, care should be taken to keep the pancreas from contacting other tissues, and instruments used to dissect the pancreas should be rinsed off before contacting other tissues.

The entire GI tract (stomach to rectum) may be removed as a single structure (Fig. 20-5f), filled with fixative and examined at a later time. Alternatively, each section of GI tract may be examined at necropsy. The stomach and cecum are opened along their greater curvatures to facilitate removal of ingesta and rinsing of the luminal surfaces with physiologic saline. The rat stomach has two regions: the nonglandular mucosa of the cardiac portion in the esophageal area, and the glandular epithelium of the pyloric part. The serosal and mucosal surfaces of the tubular GI organs are examined for lesions. These tissues may be placed on a card or paper to keep them flat during fixation. Adrenal glands are located cranial to the kidneys, embedded in perirenal fat (Fig. 20-6a). They are removed with some fat attached but later trimmed free after removal from the body. The kidneys with the ureters intact are examined for macroscopic lesions. If needed, kidneys can be individually identified as to left and right by a longitudinal nick in the greater curvature of the left kidney. The right kidney is cut in cross-section.

The thoracic viscera are removed in a single piece. This includes the tongue, larynx, trachea, thyroid gland, parathyroid glands, esophagus, aorta, heart, thymus, lungs, and lung-associated lymph nodes. Lungs are often infused with fixative by syringe (to a fixed volume) or by pressure infusion (to a set pressure, usually 30 cm) either in the thoracic cavity (Fig. 20-6b) or after removal. The photomicrographs of Fig. 20-7 show rat lung tissue in both the desirable inflated and undesirable collapsed state. After infusion with fixative, the trachea is ligated to keep the tissue inflated until it has fixed.

Once the other abdominal viscera have been removed or placed aside, the urogenital track can be easily observed and dissected. Figure 20-6c shows the male reproductive tract. To expose the testes in this manner, the epididymal fat is gently grasped so that the testes slide easily from the scrotal sac. The epididymis may be left attached to the testis or carefully dissected free. Testis tissue is under pressure and any nick in the wall will cause seminiferous tubules to extrude from the testis, changing the orientation of tubules in the sample used for histopathology. Bone cutting forceps or strong scissors are used to cut the pelvic girdle to remove the anus, penis, prostate, preputial glands, vas deferens, urinary bladder, and seminal vesicles in one piece. If the prostate is to be weighed, or certain lobes are to be weighed, it should be carefully trimmed and the lobes identified individually. Often the dorsal and lateral lobes are combined and weighed separately from the ventral lobes (Suwa et al., 2001). In standard studies these organs can be fixed as a unit. If it is not already filled with urine, the bladder is filled with formalin by injection.

Figure 20-6d shows the female urogenital system. Bone cutting forceps or strong scissors are used to cut the pelvic girdle to remove the anus, rectum, vagina, urinary bladder, uterine horns, cervix, body of the uterus, and both ovaries. As with the male, the bladder may be inflated with fixative if it is not full of urine. The ovaries are

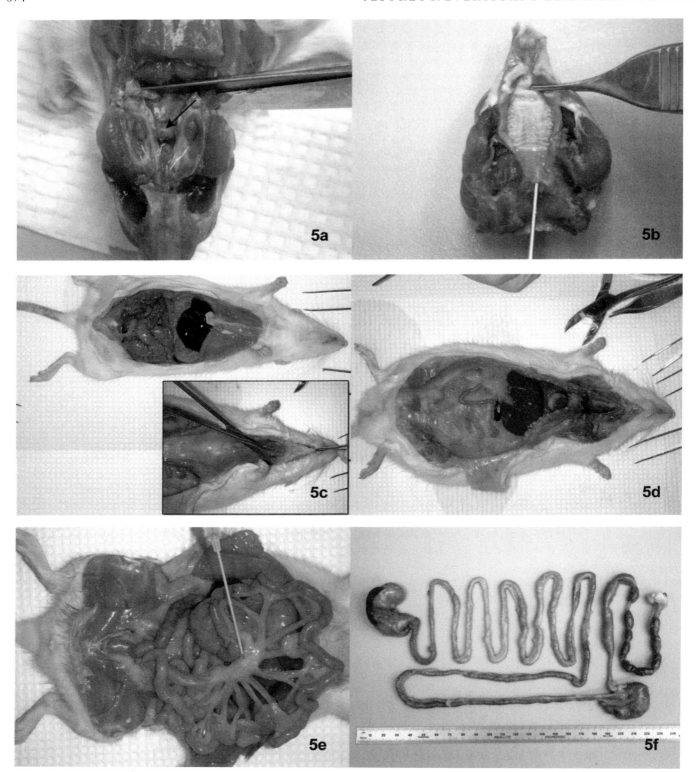

Fig. 20-5   (a) After the brain has been removed, the pituitary gland can be visualized (arrow). This gland is generally removed after fixation. (b) The nasal cavity can be flushed with fixative by inserting a blunt needle or cannula into the nasopharynx and gently instilling fixative until it exits the external nares. (c) Photograph of rat necropsy with abdominal cavity opened to view the organs *in situ*. The inset shows the removal of the salivary glands and mandibular lymph nodes. (d) Photograph of rat necropsy after removal of the ribs and sternum to show the organs of the thoracic cavity *in situ*. (e) The mesenteric lymph nodes (located at the tip of the pointer) can be easily lost during the dissection of the abdominal organs. They are best located and removed before the other abdominal organs. (f) The entire gastrointestinal tract (stomach to the rectum) can be removed as a single structure.

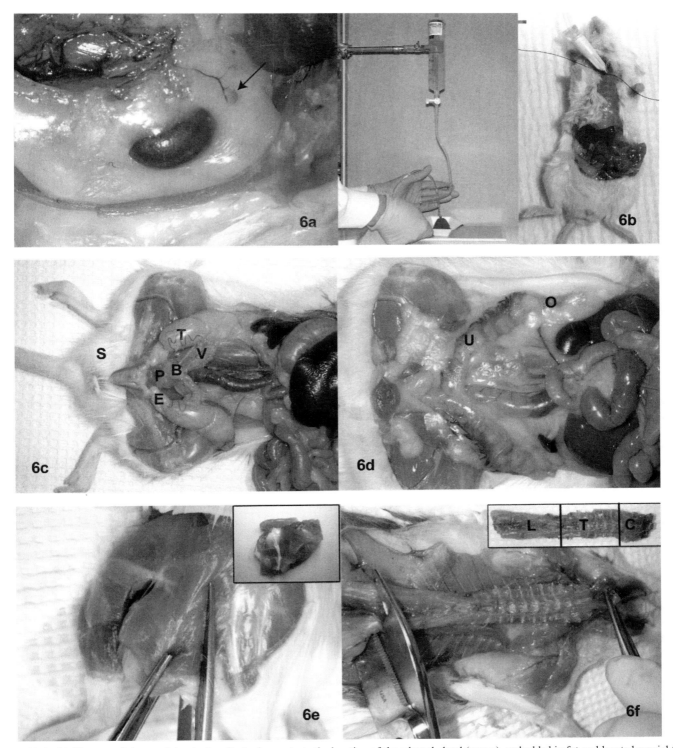

Fig. 20-6 (a) Close-up of the rat abdominal cavity to demonstrate the location of the adrenal gland (arrow), embedded in fat and located cranial to the kidney. (b) Lungs may be infused with fixative to a set pressure (usually 30 cm) with a simple pressure infusion apparatus made from a syringe barrel, tubing, stopcock, and pipette. This can be performed after removal of the lungs from the animal or as the lungs are in the thoracic cavity. (c) Organs of the male reproductive tract. The testes (T) have been pulled from the scrotal sac (S); epididymis (E), seminal vesicles (V), urinary bladder (B), and prostate (P) are also visible. (d) Organs of the female reproductive tract. The uterus (U) and ovary (O) are visible. (e) This photograph shows the location for cuts made in the biceps femoris muscle, perpendicular to the femur for the collection of a standard skeletal muscle sample. After the sample is excised and flipped over, the sciatic nerve (insert) is visible. (f) Spinal cord may be collected from the cervical (C), thoracic (T), or lumbar (L) regions. Strong scissors or bone cutting forceps are used to cut away the ribs and muscle.

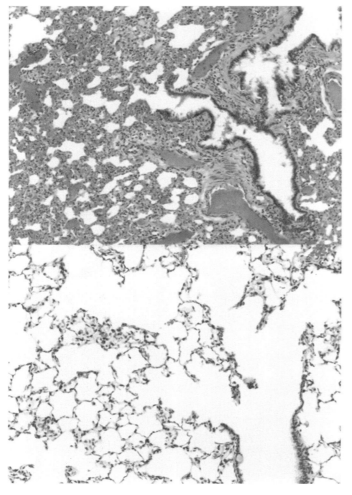

Fig. 20-7 These photomicrographs show lung tissue that has been properly pressure infused with fixative (bottom) versus the appearance of lung tissue that was immersion fixed in a collapsed state (top). Inflation is necessary for proper microscopic evaluation of alveolar septa.

dissected away from their surrounding fat and may be removed for weighing. As with the male, the reproductive viscera are generally fixed on paper or cardboard as a unit, but for studies with special emphasis on the female reproductive system, it may be preferable to dissect free some of these organs separately for weight determination and other endpoint analyses. If ovaries or other small tissues are dissected free, they may be placed in a tissue cassette to assure against loss.

Skeletal muscle is collected with sciatic nerve by skinning the dorsal surface of a hind limb. Two cuts are made into the muscle (biceps femoris) perpendicular to the femur about 1 inch apart, followed by one cut parallel to the femur, close to the bone (Fig. 20-6e). The piece of muscle is excised and flipped over to expose the sciatic nerve (Fig. 20-6e, insert). The remaining hind limb may be saved for bone marrow or knee joint collection.

Spinal cord for routine studies is collected from the thoracic region (Fig. 20-6f). Bone cutting forceps or strong scissors are used to cut away the ribs and muscle. At least two cuts are made into the vertebrae of each region to allow fixative to penetrate. If lumbar spinal cord is to be collected, prosectors should collect in the L1-L2 vertebral region to ensure a cord and not spinal nerve root sample.

IV. PATHOLOGY CONSIDERATIONS

A. Descriptive Pathology

It is critically important for the prosector to observe, describe, and record all macrocopic abnormalities that might influence the interpretation of subsequent studies such as histopathology (Ward and Reznik, 1983). To accomplish this task, it is important to understand pathology terminology and develop a standardized lexicon for use. Terms associated with medical and pathological diagnoses are complex and specialized. Terms associated with gross necropsy description are best kept simple and descriptive. There is no one standard set of terms or way to describe findings. Good working examples of systems have been reviewed (Feldman and Seely, 1988; Bono et al., 2000). Because a pathology diagnosis will be made from the collective microscopic and clinical data, a good description of the location, size, color, shape, texture, and severity of any abnormality seen is adequate to assure further examination. Some terms and examples are presented in Table 20-2.

TABLE 20-2

COMMON CRITERIA AND TERMS USED TO CHARACTERIZE NECROPSY FINDINGS

Morphology	Adhesion
	Reduction in size
	Discoloration
	Focus
	Mass
Location	The organ, limb, or cavity where the lesion is located, with directional or organ specific terms
Size	Use two or three dimensions (largest first)
Weight	In grams, for organs and lesions; record significant figures with consistency
Color	Use standard real colors (no "-ish") and patterns
Shape	Examples include nodular, spherical, ovoid, crateriform, polypoid
Texture (consistency)	Examples include soft, firm, hard, gritty
Severity	Not graded (0)
	Minimal (1)
	Mild (2)
	Moderate (3)
	Severe (4)

B. General Issues

1. Tissue Preparation

Fixation of tissues is one of the basic processing steps used in histology. Fixatives are chemicals or physical processes that are used to "fix" in time the relationships among cells, cellular components, and extracellular matrix (Jones, 2001). Unfortunately, each combination of fixation, processing and staining is a compromise as to the best representation of the living tissue. Thus, there is no ideal fixation protocol, and one must bear in mind the purpose and needs of the study and the endpoints being studied. Chemical fixation is a complex topic for which there are many excellent reviews (Eltoum et al., 2001; Jones, 2001). The advent of molecular pathology has increased the need for a prospective consideration of fixation protocols. One should always avoid "overfixation" in aldehyde-based fixatives such as the 10% neutral buffered formalin used as the most common fixative in the laboratory. Over-fixation can impair certain histological analyses such as immunohistochemistry. For routine histology, tissues specimens are fixed by immersion in 10% neutral buffered formalin solution for 24 to 48 hours. Many research necropsy procedures, such as those that involve nervous tissue examination or special target organ studies, will involve intravascular perfusions of fixative using flow and pressure control. The best way to avoid overfixation is to fix for a specified amount of time and then transfer tissues to 70% alcohol or buffer or to process quickly to paraffin.

For immersion fixation techniques, the volume of fixative should always exceed that of tissue specimens by at least 10-fold for aldehyde-based fixatives (20x for alcohol based fixatives). For adequate penetration of formalin fixative, tissues should not be more than 5 mm thick. Other commonly used fixatives for rats include Bouin's solution for reproductive and fetal tissues, Davidson's fixative for ocular and male reproductive tissues, and Zenker's solution for eyes. For ultrastructural studies using electron microscopy, glutaraldehyde and Karnovsky's fixative are commonly used on very small pieces of tissue. Many of these fixatives entail hazardous agent handling and special waste handling considerations. For example, formalin is a carcinogenic substance, Bouin's solution has picric acid that can be explosive under certain conditions, and Zenker's solution contains mercury, a toxic material that can be readily absorbed into the body.

Modern pathology combines traditional histology with molecular biology techniques and requires the scientist to give special consideration to the collection of tissues and specimens for techniques such as immunohistochemistry, *in situ* hybridization, and *in situ* polymerase chain reaction (Eltoum et al., 2001). This may necessitate specific tissue collection protocols that preserve biological targets in special ways. Rapid freezing of specimens with or without cryoprotectants and or special fixatives are some examples. RNA-based techniques may require special handling to prevent enzymatic destruction of targets. These include the use of RNAase-destroying solutions and the protection of all surfaces that contact tissues, such as the use of gloves and disposable cryostat blades. As the use of genetically modified rats grows, the need for collection of tissues for molecular analysis will grow.

2. Interfacing the Rat Necropsy with the Microscopic Evaluation of Tissues

After fixation, tissues need to be trimmed and then processed for embedment into paraffin or plastic media for subsequent microtomy and histologic preparation. The tissue trim procedure for rat tissues requires standardization so that the pathologist is presented with adequate specimens to evaluate. Good tissue trim involves uniform sample size, sampling from standard regions, maintaining natural borders, maintenance of orientation, and maintenance of tissue identification. Tissue trim methods for the rat have been described (Bahneman et al., 1995; Bono et al., 2000; *http://www.item.fraunhofer.de/reni/index.htm*). After tissue trim, rat tissues are generally multiple-embedded. A complete full screen tissue examination from a rat will generally involve approximately 60 tissues distributed on 18 to 20 hematoxylin/eosin-stained slides.

ACKNOWLEDGMENTS

We thank Sadie Leak for manuscript processing, formatting, and editing; David Weil and Delorise Williams for the prosection involved in staging the necropsy illustrations; and Donald Joyner and Otis Lyght for photography and photo-editing.

REFERENCES

Allred, J.B. and Berntson, G.G. (1986). Is euthanasia of rats by decapitation inhumane? *J. Nutr.* **116,** 1859–1861.

American Veterinary Medical Association [AVMA] (2001). 2000 report of the AVMA Panel on Euthanasia. *J. Am. Vet. Med. Assoc.* **218,** 669–696.

Bahnemann, R., Jacobs, M., Karbe, E., Kaufmann, W., Morawietz, G., Nolte, T., and Rittinghausen, S. (1995). RITA-registry of industrial toxicology animal data-guides for organ sampling and trimming procedures in rats. *Exp. Toxicol. Pathol.* **47,** 247–266.

Berger-Sweeney, J., Berger, U.V., Sharma, M., and Paul, C.A. (1994). Effects of carbon dioxide induced anesthesia on cholinergic parameters in rat brain. *Lab. Anim. Sci.* **44,** 369–371.

Black, H.E. (1986). A manager's view of the musts in a quality necropsy. *In* "Managing Conduct and Data Quality of Toxicology Studies" (B.K. Hoover, ed.), pp. 249–255. Princeton Scientific, Princeton, NJ.

Blackshaw, J.K., Fenwick, D.C., Beattie, A.W., and Allan, D.J. (1988). The behavior of chickens, mice and rats during euthanasia with chloroform, carbon dioxide and ether. *Lab. Anim.* **22,** 67–75.

Bono, C.D. (1994). "Necropsy of the Laboratory Rat." Bono & Associates, Vienna, VA.

Bono, C.D., Elwell, M.R., and Rogers, K. (2000). Necropsy techniques with standard collection and trimming of tissues. *In* "The Laboratory Rat" (G.J. Krinke, ed.), pp. 569–587. Academic Press, San Diego, CA.

Brooks, S.P., Lampi, B.J., and Bihun, C.G. (1999). The influence of euthanasia methods on rat liver metabolism. *Contemp. Top. Lab. Anim. Sci.* **38**, 19–24.

Butler, M.M., Griffey, S.M., Clubb Jr., F.J., Gerrity, L.W., and Campbell, W.B. (1990). The effect of euthanasia technique on vascular arachidonic acid metabolism and vascular and intestinal smooth muscle contractility. *Lab. Anim. Sci.* **40**, 277–283.

Close, B., Banister, K., Baumans, V., Bernoth, E.M., Bromage, N., Bunyan, J., Erhardt, W., Flecknell, P., Gregory, N., Hackbarth, H., Morton, D., and Warwick, C. (1996a). Recommendations for euthanasia of experimental animals: part 1. *DGXI of the European Commission Lab. Anim.* **30**, 293–316.

Close, B., Banister, K., Baumans, V., Bernoth, E.M., Bromage, N., Bunyan, J., Erhardt, W., Flecknell, P., Gregory, N., Hackbarth, H., Morton, D., and Warwick, C. (1996b). Recommendations for euthanasia of experimental animals: part 2. *Lab. Anim.* **31**, 1–32.

Coenen, A.M., Drinkenburg, W.H., Hoenderken, R., and van Luijtelaar, E.L. (1995). Carbon dioxide euthanasia in rats: oxygen supplementation minimizes signs of agitation and asphyxia. *Lab. Anim.* **29**, 262–268.

Cook, M. (1965). "The Anatomy of the Laboratory Mouse." Academic Press, San Diego, CA.

Danneman, P.J., Stein, S., and Walshaw, S.O. (1997). Humane and practical implications of using carbon dioxide mixed with oxygen for anesthesia or euthanasia of rats. *Lab. Anim. Sci.* **47**, 376–385.

Derr, R.F. (1991). Pain perception in decapitated rat brain. *Life Sci.* **49**, 1399–1402.

Eltoum, I., Fredenburgh, J., and Grizzle, W. E. (2001). Advanced concepts in fixation, 1: effects of fixation on immunohistochemistry, reversibility of fixation and recovery of proteins, nucleic acids, and other molecules from fixed and processed tissues. 2: developmental methods of fixation. *J. Histotech.* **24**, 201–210.

Fawell, J.K., Thomson, C., and Cooke, L. (1972). Respiratory artifact produced by carbon dioxide and pentobarbital sodium euthanasia in rats. *Lab. Anim.* **6**, 321–326.

Feinstein, R.E. (1994). Postmortem procedures. *In* "Handbook of Laboratory Animal Science, Volume 1: Selection and Handling of Animals in Biomedical Research" (P. Svendsen and J. Hau, eds.), pp. 373–396. CRC Press, Boca Raton, FL.

Feldman, D.B. and Gupta, B.N. (1976). Histopathologic changes in laboratory animals resulting from various methods of euthanasia. *Lab. Anim. Sci.* **26**, 218–221.

Feldman, D.B., and Seely, J.C. (1988). "Necropsy Guide: Rodents and the Rabbit." CRC Press, Inc., Boca Raton, FL.

Gos, T., Hauser, R., and Krzyzanowski, M. (2002). Regional distribution of glutamate in the central nervous system of rat terminated by carbon dioxide euthanasia. *Lab. Anim.* **36**, 127–133.

Greene, E.C. (1963). "Anatomy of the Rat." Hafner Publishing Company, New York, NY.

Hackbarth, H., Kuppers, N., and Bohnet, W. (2000). Euthanasia of rats with carbon dioxide-animal welfare aspects. *Lab. Anim.* **34**, 91–96.

Hewett, T.A., Kovacs, M.S., Artwohl, J.E., and Bennett, B.T. (1993). A comparison of euthanasia methods in rats, using carbon dioxide in prefilled and fixed flow rate filled chambers. *Lab. Anim. Sci.* **43**, 579–582.

Howard, H.L., McLaughlin-Taylor, E., and Hill, R.L. (1990). The effect of mouse euthanasia technique on subsequent lymphocyte proliferation and cell mediated lympholysis assays. *Lab. Anim. Sci.* **40**, 510–514.

Jones, M.L. (2001). To fix, to harden, to preserve-fixation: a brief history. *J. Histotech.* **24**, 155–162.

Leach, M.C., Bowell, V.A., Allan, T.F., and Morton, D.B. (2002). Aversion to gaseous euthanasia agents in rats and mice. *Comp. Med.* **52**, 249–257.

Maronpot, R. (1999). "Pathology of the Mouse." Cache River Press, St. Louis, MO.

Mischler, S.A., Alexander, M., Battles, A.H., Raucci, Jr., J.A., Nalwalk, J.W., and Hough, L.B. (1994). Prolonged antinociception following carbon dioxide anesthesia in the laboratory rat. *Brain Res.* **640**, 322–327.

National Cancer Institute Veterinary Pathology, *http://home.ncifcrf.gov/vetpath/necropsy.html.*

Olds, R.J. and Olds, J.R. (1979). "A Colour Atlas of the Rat: Dissection Guide." Wolfe Medical Publications, Ltd., London, UK.

Pecaut, M.J., Smith, A.L., Jones, T.A., and Gridley, D.S. (2000). Modification of immunologic and hematologic variables by method of CO_2 euthanasia. *Comp. Med.* **50**, 595–602.

Popesko, P., Rajtova, V., and Horak, J. (1990). "A Colour Atlas of Anatomy of Small Laboratory Animals." Wolfe Publishing Co., London, UK.

The Registry Nomenclature Information System, *http://www.item. fraunhofer.de/reni/index.htm.*

Seaman, W.J. (1987). "Postmortem Change in the Rat: A Histologic Characterization." Iowa State University Press, Ames, IA.

Sharp, J., Zammit, T., Azar, T., and Lawson, D. (2002). Does witnessing experimental procedures produce stress in male rats? *Contemp. Top. Lab. Anim. Sci.* **41**, 8–12.

Sharp, P.E. and La Regina, M.C. (1998). "The Laboratory Rat." CRC Press, New York.

Slott, V.L., Linder, R.E., and Dyer, C.J. (1994). Method of euthanasia does not affect sperm motility in the laboratory rat. *Reprod. Toxicol.* **8**, 371–374.

Smith, W. and Harrap, S.B. (1997). Behavioral and cardiovascular responses of rat to euthanasia using carbon dioxide gas. *Lab. Anim.* **31**, 337–346.

Sullivan, D.J. (1985). The effect of exsanguination on organ weight of rats. *Toxicol. Pathol.* **13**, 229–231.

Suwa, T., Nyska, A., Peckham, J.C., Hailey, J.R., Mahler, J.F., Haseman, J.K., and Maronpot, R.R. (2001). A retrospective analysis of background lesions and tissue accountability for male accessory sex organs in Fischer-344 rats. *Toxicol. Pathol.* **29**, 467–478.

Vanderwolf, C.H., Buzsaki, G., Cain, D.P., Cooley, R.K., and Robertson, B. (1988). Neocortical and hippocampal electrical activity following decapitation in the rat. *Brain Res.* **451**, 340–344.

The Virtual Mouse Necropsy, *http://www.geocities.com/virtualbiology/necropsy.html.*

Ward, J.M. and Reznik, G. (1983). Refinements of rodent pathology and the pathologist's contribution to evaluation of carcinogenesis bioassays. *Prog. Exp. Tumor. Res.* **26**, 266–291.

Winegerd, B. (1988). "Rat Dissection Manual." The Johns Hopkins University Press, Baltimore, MD.

Chapter 21

Integrating Biology with Rat Genomic Tools

Carol Moreno and Howard J. Jacob

I. BACKGROUND TO THE RAT GENOME AND TOOLS

For many decades, the rat has been the dominant model organism for physiological and pharmacological research.

In some cases, the rat is the unique model for certain human diseases (Gill et al., 1989) and is arguably the most significant model for human cardiovascular disease, as well as diabetes, arthritis, and many behavioral disorders. It is also extensively used in the study of the genetics of diseases, making it the ideal model for disease-based genetic research. Although the mouse has been the rodent model of choice for geneticists over the past century, the rat has progressively gained importance. This trend has increased with the recent release of the rat genome sequence (Gibbs et al., 2004).

There are many rat models that resemble human diseases, developed by phenotypic selection for certain traits, leading to creation of inbred strains (20 generations of brother-sister mating or 10 generations of backcrossing). These rat models were initially generated for physiological and pharmacological studies, and with the advent of the genomic era, they have become key models for gene discovery, proteomics, and, now, systems biology. Since the development of the first inbred rat strain by King in 1909, there have been 512 inbred rat strains developed for different phenotypes and diseases, ranging from biochemical to cardiovascular and diabetes. Selective breeding for isolation of a specific trait fixes the genes related to the trait in a particular rat strain, which allows further identification of those genes by strain comparison. Over the past decade, a number of valuable rat genetic resources have been developed. The National Institutes of Health (NIH) funded the Rat Genome Database (RGD) to "collect, consolidate, and integrate data generated from ongoing rat genetic and genomic research efforts and make the data widely available to the scientific community." Description of inbred rat strains, genetic maps and markers, and mapping tools can be found at RGD Web site (*http:// rgd.mcw.edu*). This site also contains curated information on rat quantitative trait loci (QTL) for different phenotypes, a genome browser, and a dynamic homology tool (VCMap) to compare rat genetic data to the homologous regions in mouse and human. These and other bioinformatics tools allow researchers to attach biological data to the rat genome. There are other sites available to the research community that provide additional information on rat genetics and genomics. Genome browsers, such as ENSEMBL and the ones from University of California Santa Cruz (UCSC) and the National Center for Biotechnology Information (NCBI), integrate genomic information from RGD with gene prediction tools and links to other databases and resources for gene expression and annotation, among others. Some other achievements in rat genomics in the past decade include the development of consomic rat strains, recombinant inbred (RI) strains, numerous congenics, and the assembly and publication of the rat genome sequence. With these and other genetic tools in hand, in the next years we expect to see a substantial increase in the identification of genes underlying complex rat phenotypes, and their link to human diseases.

There are several ways to link physiological data to the rat genome such as QTL mapping, congenic and consomic strains, genetic manipulation, and comparative mapping, to name a few. These tools not only facilitate gene identification and characterization but also contribute to the advance of systems biology at the whole organism level.

II. MAPPING TOOLS

Different approaches have been used in the search for the genes underlying simple and complex traits. Genetic linkage studies are used to locate the chromosomal position of genes, by the development of segregating crosses of rats. By this technique, hundreds of QTL for different traits have been mapped to the rat genome. A QTL is defined as a chromosomal region that segregates with a phenotype in a cross between rat strains at a defined statistical significance (Lander et al., 1987; Lander and Botstein, 1989). The first traits to be mapped were related to coat color (Bateson, 1903; MacCurdy et al., 1907). The segregating population is constructed by crossing two phenotypically and genetically different strains of rats, to produce a first filial (F1) generation. The best design is to use two inbred rat strains, so only two alleles are segregating, as an inbred strain is homozygous at all loci (genetic twins). The rats in the F1 population from two inbred strains are identical, as they have inherited one chromosome (50% of their genetic material) from each parental strain. The F1 population is then either intercrossed to generate a second filial (F2) generation or backcrossed to one of the parental strains (N2 generation). Each rat in the F2 generation is unique phenotypically and in their genotype, owing to recombination events during meiosis. Linkage analysis is then performed by computer software packages such as Mapmaker (Lander et al., 1987) or Map Manager (Manly et al., 2001), which first construct a genetic linkage map and then detect the loci related to the phenotype (for an example, see Fig. 21-1). Genome-wide scans are performed by genotyping genetic markers, evenly spaced along the chromosomes. All genetic markers are based on sequence variations. A microsatellite marker or simple sequence length polymorphism (SSLP) is a DNA sequence that contains repetitive sequence, generally of 2- to 4-bp repeats (e.g. CACACACACA), and that frequently differs in length between rat strains, thus generating a polymorphism. Currently, there are 10,035 rat SSLPs available to the scientific community (*http://rgd.mcw.edu*). To determine which genetic markers are likely to be polymorphic, 4200 of these SSLPs were screened in 48 different rat strains, yielding a large number of potential crosses that can be used for whole-genome scans. These data are publicly available at the RGD (*http://rgd.mcw.edu*), where information about all published rat QTLs can be found. The reason genetic mapping is so powerful is that it does not

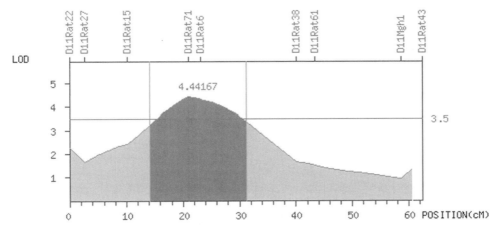

Fig. 21-1 MapMaker output analysis representing the log of the odds ratio (LOD) plot for susceptibility to glomerular injury in chromosome 11, in a female F2 cross between the hypertensive SS/JrHsdMcwi rat and the normotensive BN/SSNHsdMcwi rat. The 95% confidence interval is marked by the vertical bars and defines the quantitative trait loci. A LOD score above 3.2 is suggestive of linkage of the chromosomal region to the phenotype studied (Kruglyak and Lander, 1995). (From Moreno et al., 2003.)

require prior knowledge of causative genes, allowing for the discovery of novel genes.

Genetic linkage analysis studies require large populations and a very accurate phenotyping measurement in order to achieve enough statistical power for QTL detection. Interaction of the genetic background with the environment, gene-gene interaction, and incomplete penetrance (i.e., individuals who carry the gene but do not express the phenotype) can also have masking effects. Many of the biological traits mapped to the genome are quantitative (i.e., display a normal distribution)—they show a variation within individuals of a population—and have complex inheritance patterns, being controlled by several genes (polygenic). In addition, the chromosomal region that underlies a QTL generally includes hundred of genes, which makes gene discovery a very lengthy process (Fig. 21-1). QTLs span such large intervals because genetic mapping is not very precise, with the relatively small number of animals or genetic markers typically used for genome scans. Once a QTL has been identified by genetic linkage, the next step for gene discovery is the generation of congenic strains, the study of candidate genes (known as positional candidate genes) of that region, or a combination of both.

There are a few things to consider when performing a study using genetic linkage.

A. Selection of Rat Strains

When selecting the strains to perform genetic linkage for a quantitative trait, the difference between the control and the diseased strain for that phenotype should be greater than two standard deviations (otherwise the number of animals required is too large to be practical), and there needs to be evidence that the trait is genetic (Lander and Botstein, 1989). As genetic mapping can only detect differences, success is enriched by having the strains be as different as possible physiologically and genetically. The strains selected should be inbred and genetically divergent, having a high number of polymorphic genetic markers (with the onset of genomic sequencing, closely related strains can be used, but this is beyond the scope of this chapter).

B. Cross Design

Intercross studies (intercrosses of F1 population) should be chosen when the mode of inheritance is additive and there is a large number of independent traits being studied in parallel. An added advantage is that there are twice as many meioses that can be detected (recombinants in both F1 parents) than in a backcross; however, mapping accuracy is less because it can not be determined in which parent the recombination occurred. Backcross studies (i.e., F1 population is crossed with one of the parental strains) are chosen when the mode of inheritance is dominant or recessive. As the recombinations can only be detected from the F1, the mapping accuracy is higher but may require more animals. A second consideration is if a reciprocal cross design (two crosses, A × B and B × A, where the first strain designates the female) should be used. If experiments are performed by single-cross design, in which an affected male is typically crossed with an unaffected female, such as in hypertension studies as the Y-chromosome has been frequently implicated to carry hypertensive genes, they will fail to address the possibility that phenotypic differences between male and female animals are caused by sex-linked gene(s) or to a sex-specific effect. We recommend that a relatively large cohort of F1 animals, generated by reciprocal crosses, should also be studied to determine the mode of inheritance and, if there are any sex-specific phenotypic differences, to enable the investigator to estimate genetic contribution to each trait studied.

III. GENOMIC SEQUENCE AND COMPARATIVE MAPPING

The diploid chromosome number of the rat is 42 (20 autosomes and 2 sex chromosomes). The rat karyotype contains three large and three medium submetracentric chromosome pairs (chromosome 1–3 and 11–13), whereas seven autosomal pairs and the sex chromosome pair are large to medium telocentrics (chromosomes 4–10, X, Y), and seven pairs are medium to small metacentrics (chromosomes 14–20). Rat chromosomes are designated with the prefix RNO (for *Rattus novergicus*) followed by the chromosome number. Before the release of the first rat genetic linkage map, chromosome assignment for rat genes was performed cytogenetically (Szpirer et al., 1984, 1996; Levan et al., 1991; Yasue et al., 1991) or by using somatic cell hybrids. Maps evolved from being genetic based (Jacob et al., 1995; Pravenec et al., 1996) to being radiation hybrid based (Steen et al., 1999; Watanabe et al., 1999), to now being sequence based (Gibbs et al., 2004). With the genomic sequence of the rat now described, the reader may think that other techniques will no longer be required. However, the rat sequence is a draft sequence, meaning it contains errors. We anticipate that the genetic maps and the radiation maps, as well as newer maps based on single nucleotide polymorphisms (SNPs) and haplotypes will be standard for the next 5 to 10 years. A final methodology for analyzing chromosomes is *in situ* fluorescence hybridization (FISH), which uses fluorescent dyes to tag genes or other large segments of interest on a chromosome (Wiegant et al., 1991). This technique is particularly useful not only for gene mapping but also for detecting chromosomal patterns and abnormalities such as inversions, gene duplications, deficiencies, and translocations, many of which are very difficult to detect by sequencing. Most of these mapping tools have been used in the mouse and human and form the basis of comparative mapping (comparing genes, gene order, and other genomic features between species).

The availability of the rat genomic sequence along with the genomic sequence of many other mammals, other vertebrates, invertebrates, and diploids introduces the capability of comparative genome analysis at a nucleotide level rather than the ordering of large blocks of genes used previously. The utility of comparative analysis is based on the hypothesis that functionally important sequences will be conserved across species. Sequence comparison of rat, mouse, human, and partial data sets from additional mammals makes it possible to perform studies of mammalian evolution. It also allows drawing inferences about gene evolution. For example, it has been shown that human, mouse, and rat encode a similar number of genes, with a high degree of conservation not only of the coding regions but also of intronic structures (Cooper et al., 2004), suggesting these conserved regions are relevant to gene function.

The rat is the third mammalian genome to be sequenced, after human and mouse. The Rat Genome Sequencing Consortium (RGSC) was launched to generate a draft sequence of the rat genome, and the last assembly was released in June 2003 and published in 2004 (Gibbs et al., 2004). The sequencing strategy was a combination of whole shotgun sequencing and bacterial artificial chromosome hierarchical approach. The rat strain selected for sequencing was the BN/SsNHsd/Mcw (Brown Norway, BN) rat, and sequencing covered 90% of the rat genome. The rat genome is estimated to be 2.75 Gb long, distributed in 21 chromosomes (the Y chromosome has not been sequenced), and is predicted to encode for approximately 20,973 genes, with 28,516 transcripts and 205,623 exons (Gibbs et al., 2004). The exact number of genes and transcripts will take several more years to resolve, but the bulk of the data is available for investigators to use now.

The data from the rat genome sequence provide researchers with a precise knowledge of the rat gene content, essential for the advance of biomedical research. It also improves physical and genetic mapping, as chromosomal position no longer depends on recombination rates and statistical analysis. However, it must be noted that various lines of evidence will be required, including genetic linkage analysis or other forms of mapping, to ensure that the region of the genome under investigation has been constructed correctly.

As mentioned above, the next generation in genetic mapping will involve SNPs, which are single nucleotide sequence variants between strains. These SNPs will be used to construct haplotypes of each inbred strain. These haplotypes will then be used for genetic mapping and for helping to find sequence variants that are causally related to the phenotype of interest. The rat haplotype and SNP (pronounced *snip*) project will be analogous the human HapMap project (International HapMap Consortium, 2003). SNPs are the most common type of genetic variation between rat strains; it is estimated that on average there is 1 SNP every 1000 bp between any two strains. Obviously, some of these will be responsible for the phenotypic differences among strains. Identification of SNPs can be performed by applying different experimental techniques (Vignal et al., 2002) and by analyzing the wealth of sequence information already available in public databases. By this approach, Guryev et al. (2004) have identified more than 30,000 SNPs between rat species. The reader can anticipate that this form of mapping will be the dominant form of mapping over the next decade.

The information derived from various model organisms can be integrated into improving our understanding of disease in humans by the use of comparative genomics, facilitating the identification of genes and helping in the understanding of the biology of physiological and pathophysiological events. In this way, the information gained from one organism can be applied to others. One of the primary advantages of the rat is the amount of physiological and pharmacological information, and one of the key advantages of the mouse is the availability of gene knockout technology. Integration of functional data from the rat and gene knockout data from the mouse will translate to better understanding of gene biology and the pathogenesis of common, complex disease in humans. Comparative genomics will also facilitate the discovery of new genes or, at least, the function of new genes. Traditionally, it has been taught that sequences highly conserved between two distantly related mammals would correspond to the coding region of a gene, subjected to strong negative selection. However, several studies over the past 5 years have shown that there are large blocks of the genome that do not contain exons, but still have remarkable conservation (Bagheri-Fam et al., 2001; Nobrega et al., 2003), in some cases down to fugu (puffer fish) (Ahituv et al., 2004). Some of the elements in these regions appear to be important for gene regulation and tissue specificity. Identification of these regions by computational sequence analysis can lead to the identification of novel genes and regulatory regions in the genome.

IV. GENE EXPRESSION AND MICROARRAY

One of the most challenging tasks in genomics is the prediction of gene function and the study of the interactions between genes in the genome, what is called functional genomics. However, it should be noted that function has many different connotations, from regulation of gene expression to systems biology at the level of the whole organism. A common form of functional genomics is assessment of gene expression on a global scale by using a variety of techniques. They include serial analysis of gene expression (Velculescu et al., 1995), differential display (Liang and Pardee, 1992), oligonucleotide arrays (Lockhart et al., 1996), and microarrays (Scheena et al., 1995; Duggan et al., 1999). It appears that microarrays are the most widely used high-throughput gene expression technology. With the use of microarray techniques, one can assess the expression level of thousands of genes simultaneously, potentially the entire genome, allowing the examination of the dynamics of a whole biological system at one time. This capability has the potential to greatly enhance our

knowledge of the genes and pathways involved in the physiological responses to physiological stressors, drugs, and environmental stimuli and in pathogenesis of diseases. The use of high-throughput gene expression is referred to as "profiling" or "molecular finger printing," and it is gaining importance in the field of integrative physiology as a tool for understanding functional pathways and regulatory mechanism in the context of intact organisms.

Most microarrays consist of a solid support, typically a glass slide or nylon membrane, onto which DNA probes (synthetic oligonucleotides or cDNA fragments) are attached as thousands of microspots, using robotic arrayer pins (Shalon et al., 1996), an ink-jet printer (Cooley et al., 2001), or photolithography (Lipshutz et al., 1999). Then, target molecules derived from mRNA are labeled and hybridized to the probes and quantified for each microspot. Expression analysis using glass slide microarrays is typically done by the competitive hybridization of two targets. Fig. 21-2 illustrates the basic process of a typical cDNA-based microarray technique, utilized in our laboratories (Kita et al., 2002; Liang et al., 2003). For the cDNA probe arrays, the experiment requires a competition assay between two samples in order to make a comparison of the relative expression of the genes under different conditions; each sample RNA is reverse transcribed and the cDNA labeled with Cy3 and Cy5 fluorescent dyes, respectively. The labeled samples are pooled and hybridized to the microarray slide. Fluorescent intensities of Cy3 and Cy5 are quantified for each spot by using a laser scanner, and data are then analyzed to yield log-transformed ratios between Cy3 and Cy5 intensities in each spot, which reflect the relative mRNA expression levels of the corresponding genes for the compared samples. Oligonucleotide microarrays are usually made by in situ synthesis on glass, using a combination of photolithography and oligonucleotide chemistry (Pease et al., 1994), although some manufacturers make their arrays by spotting conventionally synthesized oligonucleotides. The use of light-directed synthesis has enabled the large-scale manufacture of arrays containing hundreds of thousands of oligonucleotide probe sequences on glass slides (McGall and Fidanza, 2001); this method is now used by Affymetrix (Santa Clara, CA) to produce the high-density GeneChip probe arrays (Lockhart et al., 1996). Oligonucleotide arrays allow the use reference samples that are not derived from mRNA, being instead a calibrated oligo reference containing sequences complementary to every spot on the array (Dudley et al., 2002); hence, RNA abundance is expressed as a ratio to the calibrated oligo reference, a sample that is easily reproduced and provides significant signal for every feature on the array. In addition, each labeled target is hybridized to a separate Affymetrix GeneChip array, in contrast to glass slide arrays.

The DNA microarray technology has introduced new opportunities in the field of integrative physiology, by

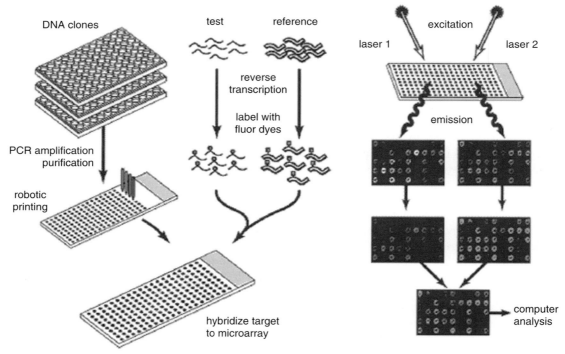

Fig. 21-2 Schema of cDNA microarray. Templates for genes of interest are printed on a slide. Total RNA from the test and reference samples is labeled fluorescently with cy3 and cy5, and hybridized to the clones on the array. Slides are scanned and images analyzed for calculation of Cy3/Cy5 ratio. (From Duggan et al., 1999.)

allowing the study of a global scale mRNA expression, which then can be integrated with the physiological data, protein expression, and regulatory data to compose a "biological atlas" (Vidal, 2001), in which the genome is linked to the transcriptome, proteome, and function. Microarray technology has been used in the rat in conjunction with other genetic strategies, such as QTL analysis, congenic mapping, or transgenic techniques, to accelerate the search for genes underlying various phenotypes (Aitman et al., 1999; Monti et al., 2001; Liang et al., 2003; Vitt et al., 2004).

Although the dominant application of microarrays is in measuring gene expression in different situations, other array applications include comparative genomic hybridization, chromatin immunoprecipitation, mutation and methylation detection, and genotyping.

V. GENETIC MANIPULATIONS

For more than a dozen years, genetics studies have followed two tracks: positional cloning from genetic mapping to gene identification, and transgenesis (random insertion of genes, knockouts, knockins, and conditional knockouts) to unravel gene function. Traditional rat transgenesis by pronuclear injection has been established

since 1990 (Mullins et al., 1990). The release of the rat genome sequence and the need to determine gene function in rats, as well as the need to validate genes cloned by position in rats by using a transgenic rescue (the phenotype is normalized via a transgene), has facilitated an explosion of these technologies in the rat. In mice, knockout of genes by homologous recombination in embryonic stem cells has become a standard technology that has driven functional genomics in mice. Unfortunately, at the writing of this chapter, embryonic stem cell manipulation is not technically possible in the rat. Transgenic rescue is being used when the trait shows a recessive mode of inheritance (Pravenec et al., 2001; Jacob and Kwitek, 2002). Recently, cloning of fertile adult rats has been achieved by nuclear transfer (Zhou et al., 2003), which opens the door for targeted gene manipulation techniques not possible until now for the rat, such as knockout technology. It will be some time before this can be done routinely as the efficiency is too low to be used as a general methodology. There are also several groups working to develop rat embryonic stem cells. In the interim there are other technologies.

A. Chemical Mutagenesis

Mutational analysis has become one of the most important tools in biological research, greatly contributing to the understanding of biological processes and gene function. Chemical mutagenesis has been widely used to induce novel phenotypes (Ohgaki et al., 1993) and to help identify gene function (Zan et al., 2003; Smits et al., 2004).

Random point mutations can be induced by chemical mutagens, such as ethyl methanesulfonate or N-ethyl-N-nitrosourea. Administration of a mutagen will induce genome-wide mutations randomly in the cells of the whole organism, including the gametes, which will carry the mutations to the next generation. Mutations are subsequently identified by screening of the genome by TILLING (targeting-induced local lesion in genomes) with cell digestion, denaturing high-performance liquid chromatography, yeast-based translational screening assays, or resequencing. The mutagenesis approach could be phenotype driven, in which animals are screened in the search of novel mutant phenotypes followed by the isolation of the underlying gene involved (Brown and Hardisty, 2003), or it could use a gene-driven approach, in which some genes are selected as candidates for a genetic disease and screened for mutations in the offspring, followed by breeding of the selected mutations to homozygosis and subsequent phenotyping of the mutated animals for phenotypic characterization.

B. Transgenic Rats

Transgenic rats have been very important for understanding gene function and for providing new animal models of human disease. A transgenic rat is one in which foreign DNA is incorporated into the genome, by pronuclear microinjection of a fertilized oocyte. To obtain transgenic rats, females are superovulated, mated, and sacrificed the next day for recovery of the oocytes. The transgene construct containing the gene, promoter, etc., of interest is then microinjected into the male pronucleus of the individual oocyte. The embryos are transplanted into a foster mother, and the pups are screened for the presence of the transgene. Those that carry the transgene are then bred to determine if the transgene is present in the germ-line. Once the transgene is germ-line, a strain can be produced that is homozygous for the transgene.

The microinjected DNA is randomly integrated into the chromosomal DNA, usually at a single site, but frequently the transgene goes in with multiple concatenated copies. Although transgenes often integrate into the genome without affecting the expression of endogenous genes, occasionally the integration alters expression of a gene (insertional mutation), producing a phenotype not related to the transgene expression itself. To control for effects of insertion, typically more than one strain of transgenic rat is made. This control also helps to determine if copy number is also relevant to the phenotype observed. If the transgene

contains a tissue-specific promoter coupled to the gene of interest, then it can be expressed in specific cells or tissue of the organism.

One of the great advantages of transgenic models is that a specific disease or phenotype model can be constructed by altering the expression of a chosen target gene. The result of a mutation is typically a gain of function or overexpression of the inserted gene. It must also be taken into account that the phenotype observed very often depends on the genetic background and the expression of modifier genes that interact with the transgene. Some transgenic rat strains are commercially available.

VI. ANIMAL MODEL SYSTEMS

Animal models provide an additional way to map phenotypes to the genome and facilitate the process of gene identification by narrowing the region where linkage to a phenotype resides and by fixing the effect of a gene in a homogeneous genetic background.

A. Congenic Strains

To validate the functional importance of a genomic region, initially identified by genetic linkage analysis, congenic techniques were originally developed to study the major histocompatibility complex in the mouse by Nobel prize winner Dr. Snell (1948) at the Jackson Laboratory (Bar Harbor, ME). This strategy remains a common way to study genes nearly 60 years later. By this approach, a chromosomal region containing the segregating phenotype from one strain is transferred to another strain. Confirmation of a QTL is achieved when substitution of the segregating region alters the phenotype of interest. Because congenic strains differ only in a short chromosomal segment from their background strain, it is possible to investigate the phenotypic effect of the locus isolated from other effects caused by other loci on the original genetic background. Generation of congenic strains is performed by backcrossing the control (donor) strain onto the susceptible (recipient) strain (or vice versa) for 10 generations, selecting for each round of backcrossing the animals that carry the target region from the donor by using genetic markers. Without recombination, each backcross reduces the donor (contaminant) genome and enriches the recipient genome by 50%, following an exponential decay model. Therefore, after 10 generations of backcrossing, the congenic is isogenic for the recipient genome background. The congenic region is maintained as a heterozygote throughout the process. After the 10 generations, the heterozygotes are intercrossed, yielding the homozygous strain. With recombination, on average there

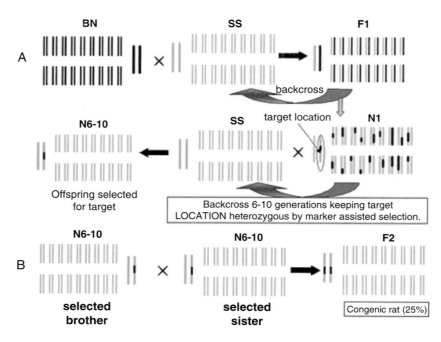

Fig. 21-3 Schematic representation of the generation of a congenic strain from two genetically different rat strains. (A) Parental strains Brown Norway (BN) and Dahl salt-sensitive (SS) are intercrossed for the generation of a heterozygous F1 population. The F1 is then crossed with the parental background of interest (in this example, the SS) to generate an N2 population. The N2 rats are then backcrossed 6 to 10 generations by using marker-assisted selection of the offspring, in order to substitute a selected genomic region from the BN rat. (B) A male and female rat, selected by genotyping for this specific target region containing the phenotype of interest, are then mated. Twenty-five percent of the offspring from this cross will be homozygous for this region. These rats are then inbred to produce a stable inbred congenic strain. (From Cowley et al., 2004.)

is still 50% fixation with each generation; however, with any average there are some animals above and below the mean (Fig. 21-3).

The development of congenic strains can be accelerated by genotyping the whole genome and selecting the breeders that, in addition to containing the target region of the donor strain, have a greater proportion of alleles from the recipient strain throughout the genome. This process of whole-genome marker-assisted selection, also called "speed congenics," can reduce the breeding time by half (Visscher, 1999). With this method, generation of a congenic strain is reduced from 4 to 5 years to 2 to 3 years. In complex diseases with multiple QTL determining a trait, generation of double or triple congenic strains is some times necessary to confirm a causative locus. These multiple congenics are typically constructed one at a time and then assembled onto multicongenics.

B. Consomic Strains

The process of generating congenic strains is very lengthy, because it takes many backcrosses to obtain an isogenic or homogeneous background. Consomic strains are rat strains in which a whole chromosome from one strain is transferred to the genomic background of another strain.

The methodology of creating consomic strains is the same as for creating congenic strains, with generations of backcrossing and marker-assisted selection, but in this case the full-length chromosome is transferred from one rat inbred strain into the isogenic background of another inbred strain. Generation of congenic rats from consomic rat strains will take, at most, three generations of breeding, following an intercross with the consomic and parental strain. Because all the alleles on the other chromosomes remain homozygous to the recipient strain, there is no need for backcrossing to "clean" the genome (Fig. 21-4). The Medical College of Wisconsin has assembled two complete panels of consomic strains, using the BN strain that was sequenced as the donor strain. In these consomic strains, a chromosome from the BN rat was substituted, one at a time, into the genetic background of the SS/JrHsdMcwi Dahl salt-sensitive (SS) and FHH/EurMcwi Fawn-hooded hypertensive (FHH) rats. The SS rat is a model for salt-sensitive hypertension (Rapp, 1982), insulin resistance (Kotchen et al., 1991), hyperlipidemia (Reaven et al., 1991), endothelial dysfunction (Luscher et al., 1987), cardiac hypertrophy (Ganguli et al., 1979), and glomerulosclerosis (Roman and Kaldunski, 1991). The FHH rat is a model for systolic hypertension, renal disease, pulmonary hypertension, bleeding disorder, alcoholism, and depression (Provoost, 1994). The two consomic panels developed

Rapid generation of congenics from consomics

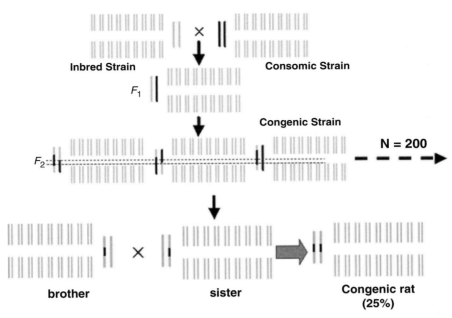

Fig. 21-4 Generation of congenic rats from consomic strains. The parental strain is crossed with the consomic strain to generate an F1 population with identical genetic background and a heterozygous target chromosome. These F1 rats are intercrossed to generate an F2 population of rats whose target chromosome will be congenic, owing to recombination events. Two similar F2 rats are selected (by genotype) and mated to fix the region of interest. (From Cowley et al., 2004.)

capture nearly 50% of the genetic variation in the rat, based on the Allele Characterization Project, which determined the average number of alleles for a given genetic marker is 6 when 48 rat strains are screened. These consomic panels provide a foundation of genetic resources for the study of disorders of heart, kidney, lung, and vasculature; given the 50% genetic variability, it is reasonable to assume that a similar level of biological variability can be expected.

There are additional applications for consomic rat strains, in addition to the rapid development of congenic strains for any region of the genome; for example, they can be used to assess the role of a genomic region in different backgrounds, assessing whether background effects modify gene function. Consomic rat strains can be also used to develop polygenic models to study gene-gene interactions. The use of a complete panel of consomic rats provides a much greater power for mapping traits in complex diseases. Gene-gene and gene-environment interactions reduce the power of identifying QTL with weak effects in segregating populations. Consomic strains have a fixed genetic background and, because they are inbred strains, provide the possibility of reproducing experiments in genetically identical animals, which allows identifying genes with small phenotypic effects. The Medical College of Wisconsin Programs for Genomic Applications (PhysGen), in an effort to dissect multigenic common diseases, has characterized each consomic strain for more than 300 heart, lung, and

blood phenotypes. The aim of the PhysGen project is to link biological functions of heart, lung, and blood systems to genomic data; develop a renewable national resource for investigators to study the impact of multiple disease genes on systems biology; and provide basic information that the investigators can use to understand the impact of genetic variations and their interactions with the environment on diseases that influence heart, lung, and blood systems, by mapping these traits to a particular chromosome. The data are publicly available at its Web site (*http://pga.mcw.edu*). Environmental stressors such as hypoxia, exercise, and high salt intake are being used to unmask deficiencies in normal homeostatic mechanisms and idiopathic mechanisms that contribute to disease. Toward this goal, the strains of consomic rats are being characterized by using 166 phenotypes specific to heart, lung, kidney, vasculature, and blood function in response to environmental stressors such as hypoxia, exercise, and salt intake. This information allows mapping of phenotypes to the chromosomes, as well as the likely determinants or intermediate components of diseases and of the responses to stress.

The contributions of genes on each chromosome to the observed traits can be assessed by phenotyping and by expression profiling. Comparisons between the consomic and parental strains provide valuable insights into the genomic pathways (clustered gene expression patterns) that differ between strains and how these differences might be connected to a particular pathogenic phenotype of the animal.

C. Recombinant Inbred Strains

RI strains provide an additional tool for mapping phenotypes to the genome. This strategy is based on the generation of different rat strains derived from an F2 population. After two inbred strains are crossed, the F1 hybrids are intercrossed to generate an F2 generation, in which each rat is genetically unique owing to the genetic recombination occurring during meiosis. Pairs of F2 animals are chosen randomly to serve as founders for new strains of rats, which are then brother-sister mated for 20 generations to become inbred, that is, homozygous for each genetic locus. This fixation process is random; therefore, each strain of RI will carry a unique pattern of chromosomal recombination. As each strain is unique, it takes a relatively large number of RI strains to provide genetic mapping power. The RI strains derived from the same progenitor strains constitute a RI strain panel. Genotyping of the RI strains will identify what genomic regions are inherited from each parental strain, the strain distribution pattern of genetic loci. Comparison of the strain distribution pattern of different RI strains allows genetic mapping and linkage analysis for identification of phenotype-related loci. The RI strategy allows development of lines that contain more than one QTL, thereby permitting the analysis of gene interaction and detection of weak loci. The RI strains are similar to the consomics in that they are isogenic and can be used to study many phenotypes by using the same strains. However, the presence of a unique genome background in each strain prevents their use in rapid generation of congenic animals. They also have the advantage and disadvantage of variable gene-gene interactions that are not currently known *a priori*.

One of the largest rodent RI panels, derived form the spontaneously hypertensive and the BN rat strains, is the reciprocal HXB/BXH RI strains (Pravenec et al., 1989; Kren et al., 1999). These strains are a great resource for genetic analysis of cardiovascular and metabolic phenotypes. It is noticeable that the BN rat is the strain used for the consomic and RI strain project and is also the rat strain used in the Rat Sequencing Project. However, the substrain used to make the RI strain is about 7% different from the sequenced BN (data not shown).

VII. BIOINFORMATIC TOOLS

In the past several years, there has been a rapid increase in rat genetic and genomic data. This explosion of information highlighted the need for a centralized database to efficiently and effectively collect, manage, and distribute a rat-centric view of these data to researchers around the world. The RGD (*http:rgd.mcw.edu*) is based at the Medical College of Wisconsin and works in collaboration with the Mouse Genome Database (Mouse Genome Informatics), NCBI, UCSC, European Bioinformatics Institute, RGSC, Baylor University, SWISS-PROT, Biomolecular Interaction Network Database, RatMap, and Program for Genomic Applications (de la Cruz et al., 2005). RGD coordinates all nomenclature for genes, strains, and QTLs with RatMap (*http://ratmap.gen.gu.se*), a European effort to manage rat related data, and the International Rat Genome and Nomenclature Committee (*http://rgnc.gen.gu.se*). RGD also curates and integrates rat genetic and genomic data and provides access to these data to support research using the rat as a genetic model for the study of human disease; it was created to serve as a repository of rat genetic and genomic data, as well as mapping, strain, and physiological information. It also facilitates investigators' research efforts by providing tools to search, mine, and analyze these data sets.

RGD includes curated data on rat genes, QTLs, microsatellite markers and expressed sequences tags, rat strains, genetic and radiation hybrid maps, sequence information, and ontologies. Ontologies provide standardized vocabularies for annotating molecular function, biological process, cellular component, phenotype, and disease associations. It also allows searching across genes, QTLs, and strains and provides a basis for cross-species comparisons. RGD also provides a variety of tools to search and analyze both the data within RGD and data generated from rat research in the laboratory. The list below identifies the major tools provided at RGD.

A. Virtual Comparative Map

Virtual Comparative Map (VCMap) is a dynamic sequence-based homology tool that allows researchers of rat, mouse, and human to view mapped genes and sequences and their locations in the other two organisms, an essential tool for comparative genomics. It is currently based on radiation hybrid maps but is being converted to sequence-based conservation between species that allows sequence-related information derived from one species to be mapped to the corresponding (syntenic) region in another. The VCMap tool (*http://rgd.mcw.edu/VCMAP/*) provides a visual representation of the chromosome of the rat and the syntenic regions from the other two species. The maps are clickable, allowing the view of information on markers, chromosomal regions, and mapped QTL.

B. Genome Scanner

Genome Scanner (*http://rgd.mcw.edu/GENOMESCAN-NER/*) combines data on allele size for more than 4000 genetic markers in 48 rat strains, with the position in genetic and radiation hybrid maps. By selecting the strain of interests, Genome Scanner will select polymorphic markers between the selected rat strains, which can be further used to perform genetic linkage studies or to increase mapping density in a specific chromosomal region.

C. ACP Haplotyper

ACP Haplotyper (*http://rgd.mcw.edu/ACPHAPLOTY-PER/*) allows visualization of chromosomal segments that are conserved between different inbred strains and provides a measurement of relatedness between these inbred strains. This tool creates a visual haplotype that can be used to identify conserved and nonconserved chromosomal regions between any of the 48 rat strains characterized as part of the Allele Characterization Project. For the selected chromosome and between the selected strains, the tool compares the allele size data for microsatellite markers on the selected genetic or RH map. Results are presented as PDF documents containing the interstrain identity matrices and as a visual haplotype showing conserved microsatellite marker allele sizes across the selected chromosome.

D. Gene Annotation Tool

The purpose of the Gene Annotation Tool (*http://rgd.mcw.edu/gatool/*) is to gather information about genes by parsing several databases available online—Entrez, SWISS-PROT, KEGG, and RGD—to provide the user with a comprehensive file of gene descriptions and annotations.

E. MetaGene

MetaGene (*http://rgd.mcw.edu/METAGENE/*) is a powerful gene prediction tool that analyses a submitted sequence to seven prediction engines simultaneously to obtain a comprehensive report on sequence features.

F. Genome Browser

The Genome Browser (*http://rgd.mcw.edu/sequence-resources/gbrowse.shtml*) allows rapid visualization of different types of information beneath genome coordinate positions. Information includes genes, genetic markers, expressed sequence tags, and mapped QTLs for different phenotypes. Other genome browsers, with numerous tracks and links, are listed in Table 21-1.

G. Genome Viewer Tool

Genome Viewer Tool (GViewer; *http://www.rgd.mcw.edu/gviewer/Gviewer.jsp*) provides users with a complete genome view of genes and QTLs annotated with function, biological process, cellular component, phenotype, disease,

TABLE 21-1

RAT GENOMIC RESOURCES

Rat Resources

- Genome Tool Suite
- The Programs for Genomic Applications (PGA)
- RatMap
- Baylor College of Medicine Rat Resources
- Rat Genome Sequencing Consortium (RGSC) chromosomal assemblies
- Rat EST Project at the University of Iowa
- National Center Biotechnology Information (NCBI) Rat Genome Resources
- National Institute of Arthritis and Musculoskeletal and Skin Diseases (NIAMS): Arthritis and Rheumatism Branch Rat Genetic Database
- Wellcome Trust Centre: yeast artificial chromosome sequence-tagged site-based (YAC/STS) framework for the rat genome
- Max Delbruck Center (MDC)
- Laboratory of Neuro Imaging (LONI) Rat Image Database
- University of New South Wales (UNSW) Embryology Rat Development
- Rat Resource and Research Center (RRRC)

Genome Resources

- Virtual Comparative Map (VCMap) Browser
- Rat Genome Database (RGD) Genome Browser
- University of California Santa Cruz (UCSC) Browser
- Emsembl Browser
- Ensmart: Data mining tool
- NCBI Browser
- Evolutionary Conserved Region (ECR) Comparative Genome Browser
- Regulatory VISTA (rVISTA): Regulatory Sequence Comparative Sequence Alignment Browser
- VISTA: Comparative Sequence Alignment Browser
- Softberry Genome Explorer
- PipMaker: Comparative "Percent Identity Plot" Tool

Bioinformatics Resources

- European Bioinformatics Institute (EBI)
- The Institute for Genomic Research (TIGR)
- International Mammalian Genome Society (IMGS)
- National Genetic Resources Program (NGRP)
- The Sanger Centre
- DNA Data Bank of Japan (DDBJ)
- National Heart, Lung, and Blood Institute (NHLBI): Programs for Genomic Applications
- The Genomics Website at Wiley

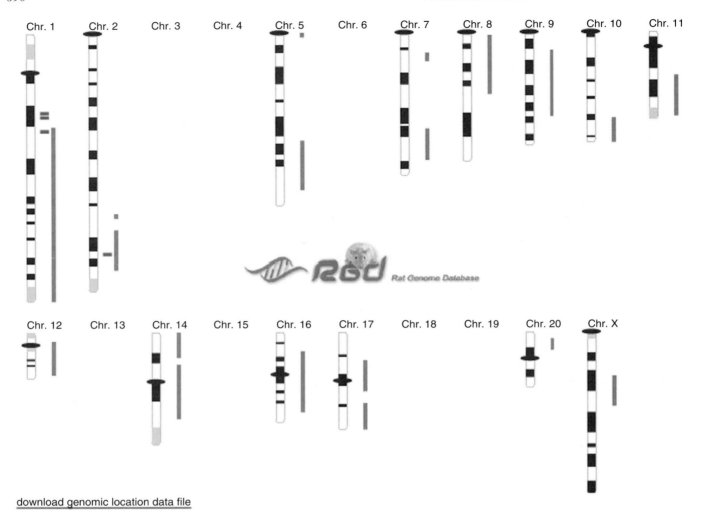

Fig. 21-5 Genome Viewer (Gviewer) Report for diabetes mellitus, non-insulin dependent. The Gviewer gives a genome view on an Rat Genome Database (RGD) ontology report page of records that match a query statement. The dynamic visual report allows the user to zoom in or out the image and view a single chromosome by clicking on one. Symbols are hyperlinked to RGD reports, and bars are hyperlinked to the region in the RGD genome browser. The example shows the different quantitative trait loci that have been mapped in the rat for non-insulin–dependent diabetes mellitus, according to the different curated publications.

or pathway information. Gviewer will search for terms from the Gene Ontology, Mammalian Phenotype Ontology, Disease Ontology, or Pathway Ontology (Fig. 21-5). In addition to the previous tools, there is a wealth of genomic information and bioinformatic tools for the rat, freely available on the Web. Table 21-1 lists some of the rat genomic sources.

become relatively standard in research using rats. We anticipate that these tools will help further refine our understanding of complex diseases and will be directly transferable to increase our knowledge of disease in humans via comparative genomics. Finally, rat models of the future will incorporate both physiological and genetic characterization.

VIII. SUMMARY

In 1989, there were seven linkage groups, cytogenics and a recent panel of somatic cell hybrids. Over the past 16 years there has been a phenomenal growth in the molecular genetic tool set for the rat. We expect this toolbox will

REFERENCES

Ahituv, N., Rubin, E.M., and Nobrega, M.A. (2004). Exploiting human-fish genome comparisons for deciphering gene regulation. *Hum. Mol. Genet.* **13(Spec No 2)**, R261–R266.

Aitman, T.J., Glazier, A.M., Wallace, C.A., Cooper, L.D., Norsworthy, P.J., Wahid, F.N., Al-Majali, K.M., Trembling, P.M., Mann, C.J., Shoulders, C.C., Graf, D., St Lezin, E., Kurtz, T.W., Kren, V.,

Pravenec, M., Ibrahimi, A., Abumrad, N.A., Stanton, L.W., and Scott, J. (1999). Identification of Cd36 (Fat) as an insulin-resistance gene causing defective fatty acid and glucose metabolism in hypertensive rats. *Nat. Genet.* **21,** 76–83.

Bagheri-Fam, S., Ferraz, C., Demaille, J., Scherer, G., and Pfeifer, D. (2001). Comparative genomics of the SOX9 region in human and *Fugu rubripes*: conservation of short regulatory sequence elements within large intergenic regions. *Genomics* **78,** 73–82.

Bateson, W. (1903). The present state of knowledge of colour-heredity in mice and rats. *Proc. Zool. Soc. (Lond)* **2,** 71–93.

Brown, S.D. and Hardisty, R.E. (2003). Mutagenesis strategies for identifying novel loci associated with disease phenotypes. *Semin. Cell. Dev. Biol.* **14,** 19–24.

Cooley, P., Hinson, D., Trost, H.J., Antohe, B., and Wallace, D. (2001). Ink-jet–deposited microspot arrays of DNA and other bioactive molecules. *Methods Mol. Biol.* **170,** 117–129.

Cooper, G.M., Brudno, M., Stone, E.A., Dubchak, I., Batzoglou, S., and Sidow, A. (2004). Characterization of evolutionary rates and constraints in three mammalian genomes. *Genome Res.* **14,** 539–548.

Cowley Jr., A.W., Roman, R.J., and Jacob, H.J. (2004). Application of chromosomal substitution techniques in gene-function discovery. *J. Physiol.* **554,** 46–55.

de la Cruz, N., Bromberg, S., Pasko, D., Shimoyama, M., Twigger, S., Chen, J., Chen, C.F., Fan, C., Foote, C., Gopinath, G.R., Harris, G., Hughes, A., Ji, Y., Jin, W., Li, D., Mathis, J., Nenasheva, N., Nie, J., Nigam, R., Petri, V., Reilly, D., Wang, W., Wu, W., Zuniga-Meyer, A., Zhao, L., Kwitek, A., Tonellato, P., and Jacob, H. (2005). The Rat Genome Database (RGD): developments towards a phenome database. *Nucleic Acids Res.* **33,** D485–D491.

Dudley, A.M., Aach, J., Steffen, M.A., and Church, G.M. (2002). Measuring absolute expression with microarrays with a calibrated reference sample and an extended signal intensity range. *Proc. Natl. Acad. Sci. USA* **99,** 7554–7559.

Duggan, D.J., Bittner, M., Chen, Y., Meltzer, P., and Trent, J.M. (1999). Expression profiling using cDNA microarrays. *Nat. Genet.* **21,** 10–14.

Ganguli, M., Tobian, L., and Iwai, J. (1979). Cardiac output and peripheral resistance in strains of rats sensitive and resistant to NaCl hypertension. *Hypertension* **1,** 3–7.

Gibbs, R.A., Weinstock, G.M., Metzker, M.L., et al. (2004). Genome sequence of the Brown Norway rat yields insights into mammalian evolution. *Nature* **428,** 493–521.

Gill 3rd, T.J., Smith, G.J., Wissler, R.W., and Kunz, H.W. (1989). The rat as an experimental animal. *Science* **245,** 269–276.

Guryev, V., Berezikov, E., Malik, R., Plasterk, R.H., and Cuppen, E. (2004). Single nucleotide polymorphisms associated with rat expressed sequences. *Genome Res.* **14,** 1438–1443.

International HapMap Consortium (2003). The International HapMap Project. *Nature* **426,** 789–796.

Jacob, H.J., Brown, D.M., Bunker, R.K., Daly, M.J., Dzau, V.J., Goodman, A., Koike, G., Kren, V., Kurtz, T., Lernmark, A. (1995). A genetic linkage map of the laboratory rat, *Rattus norvegicus. Nat. Genet.* **9,** 63–69.

Jacob, H.J., and Kwitek, A.E. (2002). Rat genetics: attaching physiology and pharmacology to the genome. *Nat. Rev. Genet.* **3,** 33–42.

Kita, Y., Shiozawa, M., Jin, W., Majewski, R.R., Besharse, J.C., Greene, A.S., and Jacob, H.J. (2002). Implications of circadian gene expression in kidney, liver and the effects of fasting on pharmacogenomic studies. *Pharmacogenetics* **12,** 55–65.

Kotchen, T.A., Zhang, H.Y., Covelli, M., and Blehschmidt, N. (1991). Insulin resistance and blood pressure in Dahl rats and in one-kidney, one-clip hypertensive rats. *Am. J. Physiol.* **261,** E692–E697.

Kren, V., Pravenec, M., Bila, V., Krenova, D., Zidek, V., Simakova, M., and Printz, M. (1999). Rat congenic and recombinant inbred strains: a genetic model for the study of quantitative trait loci. *Transplant Proc.* **31,** 1592–1593.

Kruglyak, L. and Lander, E.S. (1995). A nonparametric approach for mapping quantitative trait loci. *Genetics* **139,** 1421–1428.

Lander, E.S. and Botstein, D. (1989). Mapping mendelian factors underlying quantitative traits using RFLP linkage maps. *Genetics* **121,** 185–199.

Lander, E.S., Green, P., Abrahamson, J., Barlow, A., Daly, M.J., Lincoln, S.E., and Newburg, L. (1987). MAPMAKER: an interactive computer package for constructing primary genetic linkage maps of experimental and natural populations. *Genomics* **1,** 174–181.

Levan, G., Szpirer, J., Szpirer, C., Klinga, K., Hanson, C., and Islam, M.Q. (1991). The gene map of the Norway rat (*Rattus norvegicus*) and comparative mapping with mouse and man. *Genomics* **10,** 699–718.

Liang, M., Briggs, A.G., Rute, E., Greene, A.S., and Cowley Jr., A.W. (2003). Quantitative assessment of the importance of dye switching and biological replication in cDNA microarray studies. *Physiol. Genomics* **14,** 199–207.

Liang, M., Yuan, B., Rute, E., Greene, A.S., Olivier, M., and Cowley Jr., A.W. (2003). Insights into Dahl salt-sensitive hypertension revealed by temporal patterns of renal medullary gene expression. *Physiol. Genomics* **12,** 229–237.

Liang, P. and Pardee, A.B. (1992). Differential display of eukaryotic messenger RNA by means of the polymerase chain reaction. *Science* **257,** 967–971.

Lipshutz, R.J., Fodor, S.P., Gingeras, T.R., and Lockhart, D.J. (1999). High density synthetic oligonucleotide arrays. *Nat. Genet.* **21,** 20–24.

Lockhart, D.J., Dong, H., Byrne, M.C., Follettie, M.T., Gallo, M.V., Chee, M.S., Mittmann, M., Wang, C., Kobayashi, M., Horton, H., and Brown, E.L. (1996). Expression monitoring by hybridization to high-density oligonucleotide arrays. *Nat. Biotechnol.* **14,** 1675–1680.

Luscher, T.F., Raij, L., and Vanhoutte, P.M. (1987). Endothelium-dependent vascular responses in normotensive and hypertensive Dahl rats. *Hypertension* **9,** 157–163.

MacCurdy, H., Hansford, M., and Castle, W. (1907). Selection and cross-breeding in relation to the inheritance of coat-pigments and coat-patterns in rats and guinea-pigs. *Carnegie Inst. Wash. Publ.* **40.**

Manly, K.F., Cudmore Jr., R.H., and Meer, J.M. (2001). Map Manager QTX, cross-platform software for genetic mapping. *Mamm. Genome* **12,** 930–932.

McGall, G.H. and Fidanza, J.A. (2001). Photolithographic synthesis of high-density oligonucleotide arrays. *Methods Mol. Biol.* **170,** 71–101.

Monti, J., Gross, V., Luft, F.C., Franca Milia, A., Schulz, H., Dietz, R., Sharma, A.M., and Hubner, N. (2001). Expression analysis using oligonucleotide microarrays in mice lacking bradykinin type 2 receptors. *Hypertension* **38,** E1–E3.

Moreno, C., Dumas, P., Kaldunski, M.L., Tonellato, P.J., Greene, A.S., Roman, R.J., Cheng, Q., Wang, Z., Jacob, H.J., and Cowley Jr., A.W. (2003). Genomic map of cardiovascular phenotypes of hypertension in female Dahl S rats. *Physiol. Genomics* **15,** 243–257.

Mullins, J.J., Peters, J., and Ganten, D. (1990). Fulminant hypertension in transgenic rats harbouring the mouse Ren-2 gene. *Nature* **344,** 541–544.

Nobrega, M.A., Ovcharenko, I., Afzal, V., and Rubin, E.M. (2003). Scanning human gene deserts for long-range enhancers. *Science* **302,** 413.

Ohgaki, H., Vogeley, K.T., Kleihues, P., and Wechsler, W. (1993). neu mutations and loss of normal allele in schwannomas induced by N-ethyl-N-nitrosourea in rats. *Cancer Lett.* **70,** 45–50.

Pease, A.C., Solas, D., Sullivan, E.J., Cronin, M.T., Holmes, C.P., and Fodor, S.P. (1994). Light-generated oligonucleotide arrays for rapid DNA sequence analysis. *Proc. Natl. Acad. Sci. USA* **91,** 5022–5026.

Pravenec, M., Gauguier, D., Schott, J.J., Buard, J., Kren, V., Bila, V., Szpirer, C., Szpirer, J., Wang, J.M., Huang, H., St Lezin, E., Spence, M.A., Flodman, P., Printz, M., Lathrop, G.M., Vergnaud, G., and Kurtz, T.W. (1996). A genetic linkage map of the rat derived from recombinant inbred strains. *Mamm. Genome* **7**, 117–127.

Pravenec, M., Klir, P., Kren, V., Zicha, J., and Kunes, J. (1989). An analysis of spontaneous hypertension in spontaneously hypertensive rats by means of new recombinant inbred strains. *J. Hypertens.* **7**, 217–221.

Pravenec, M., Landa, V., Zidek, V., Musilova, A., Kren, V., Kazdova, L., Aitman, T.J., Glazier, A.M., Ibrahimi, A., Abumrad, N.A., Qi, N., Wang, J.M., St Lezin, E.M., and Kurtz, T.W. (2001). Transgenic rescue of defective Cd36 ameliorates insulin resistance in spontaneously hypertensive rats. *Nat. Genet.* **27**, 156–158.

Provoost, A.P. (1994). Spontaneous glomerulosclerosis: insights from the fawn-hooded rat. *Kidney Int. Suppl.* **45**, S2–S5.

Rapp, J.P. (1982). Dahl salt-susceptible and salt-resistant rats: a review. *Hypertension* **4**, 753–763.

Reaven, G.M., Twersky, J., and Chang, H. (1991). Abnormalities of carbohydrate and lipid metabolism in Dahl rats. *Hypertension* **18**, 630–635.

Roman, R.J. and Kaldunski, M. (1991). Pressure natriuresis and cortical and papillary blood flow in inbred Dahl rats. *Am. J. Physiol.* **261**, R595–R602.

Schena, M., Shalon, D., Davis, R.W., and Brown, P.O. (1995). Quantitative monitoring of gene expression patterns with a complementary DNA microarray. *Science* **270**, 467–470.

Shalon, D., Smith, S.J., and Brown, P.O. (1996). A DNA microarray system for analyzing complex DNA samples using two-color fluorescent probe hybridization. *Genome Res.* **6**, 639–645.

Smits, B.M., Mudde, J., Plasterk, R.H., and Cuppen, E. (2004). Target-selected mutagenesis of the rat. *Genomics* **83**, 332–334.

Snell, G. (1948). Methods for the study of histocompatibility genes. *J. Genet.* **49**, 87–108.

Steen, R.G., Kwitek-Black, A.E., Glenn, C., et al. (1999). A high-density integrated genetic linkage and radiation hybrid map of the laboratory rat. *Genome Res.* **9**, AP1-8, insert.

Szpirer, C., Szpirer, J., Klinga-Levan, K., Stahl, F., and Levan, G. (1996). The rat: an experimental animal in search of a genetic map. *Folia Biol. (Praha)* **42**, 175–226.

Szpirer, J., Levan, G., Thorn, M., and Szpirer, C. (1984). Gene mapping in the rat by mouse-rat somatic cell hybridization: synteny of the albumin and alpha-fetoprotein genes and assignment to chromosome 14. *Cytogenet. Cell. Genet.* **38**, 142–149.

Velculescu, V.E., Zhang, L., Vogelstein, B., and Kinzler, K.W. (1995). Serial analysis of gene expression. *Science* **270**, 484–487.

Vidal, M. (2001). A biological atlas of functional maps. *Cell* **104**, 333–339.

Vignal, A., Milan, D., SanCristobal, M., and Eggen, A. (2002). A review on SNP and other types of molecular markers and their use in animal genetics. *Genet. Sel. Evol.* **34**, 275–305.

Visscher, P.M. (1999). Speed congenics: accelerated genome recovery using genetic markers. *Genet. Res.* **74**, 81–85.

Vitt, U., Gietzen, D., Stevens, K., Wingrove, J., Becha, S., Bulloch, S., Burrill, J., Chawla, N., Chien, J., Crawford, M., Ison, C., Kearney, L., Kwong, M., Park, J., Policky, J., Weiler, M., White, R., Xu, Y., Daniels, S., Jacob, H., Jensen-Seaman, M.I., Lazar, J., Stuve, L., and Schmidt, J. (2004). Identification of candidate disease genes by EST alignments, synteny, and expression and verification of Ensembl genes on rat chromosome 1q43-54. *Genome Res.* **14**, 640–650.

Watanabe, T.K., Bihoreau, M.T., McCarthy, L.C., Kiguwa, S.L., Hishigaki, H., Tsuji, A., Browne, J., Yamasaki, Y., Mizoguchi-Miyakita, A., Oga, K., Ono, T., Okuno, S., Kanemoto, N., Takahashi, E., Tomita, K., Hayashi, H., Adachi, M., Webber, C., Davis, M., Kiel, S., Knights, C., Smith, A., Critcher, R., Miller, J., James, M.R., et al. (1999). A radiation hybrid map of the rat genome containing 5,255 markers. *Nat. Genet.* **22**, 27–36.

Wiegant, J., Ried, T., Nederlof, P.M., van der Ploeg, M., Tanke, H.J., and Raap, A.K. (1991). In situ hybridization with fluoresceinated DNA. *Nucleic Acids Res.* **19**, 3237–3241.

Yasue, M., Serikawa, T., and Yamada, J. (1991). Chromosomal assignments of 23 biochemical loci of the rat by using rat x mouse somatic cell hybrids. *Cytogenet. Cell. Genet.* **57**, 142–148.

Zan, Y., Haag, J.D., Chen, K.S., Shepel, L.A., Wigington, D., Wang, Y.R., Hu, R., Lopez-Guajardo, C.C., Brose, H.L., Porter, K.I., Leonard, R.A., Hitt, A.A., Schommer, S.L., Elegbede, A.F., and Gould, M.N. (2003). Production of knockout rats using ENU mutagenesis and a yeast-based screening assay. *Nat. Biotechnol.* **21**, 645–651.

Zhou, Q., Renard, J.P., Le Friec, G., Brochard, V., Beaujean, N., Cherifi, Y., Fraichard, A., and Cozzi, J. (2003). Generation of fertile cloned rats by regulating oocyte activation. *Science* **302**, 1179.

Chapter 22

Gnotobiotics

Philip B. Carter and Henry L. Foster

I. INTRODUCTION

A. History

The earliest description of research involving gnotobiotic or germfree (GF)[1] animals, *Thierisches Leben ohne Bakterien im Verdauungskanal* (Animal life without bacteria in the digestive tract), by George Nuttal and H. Thierfelder working in Berlin, dates back to the late 19th century. These animals used by Nuttal and Thierfelder, maintained in the most rudimentary devices (Fig. 22-1), were first guinea pigs (Nuttal and Thierfelder, 1895, 1896) and then chickens (Nuttal and Thierfelder, 1897); they had considered using chickens first but were concerned about reports of *in ovo* infections. Other mammals were used by later investigators. Significant advances in the production, use, and characterization of GF animals did not occur until the 1930s and was virtually simultaneous at the University of Notre Dame in Indiana by Art Reyniers and coworkers (Fig. 22-2) and at the University of Lund, Sweden (later moving to the Karolinska Institutet in Stockholm), by Bengt Gustafsson (Fig. 22-3); his professor, Güsta Glimstedt, and colleagues. These groups later reported the establishment of the first GF rat colonies (Gustaffson, 1948; Carter, 1971). Interest in gnotobiotic science and technology appeared later in Asia with the work of Masazumi Miyakawa (Fig. 22-4) and colleagues at Nagoya University, Japan.

Since that time, there have been major advances in methodology that have facilitated the breeding and utilization of gnotobiotic rats in a continually expanding spectrum of biomedical research. A literature survey for recent decades would yield thousands of references on GF rats and their gnotobiotic and disease-free descendants.

Fig. 22-2 Art Reyniers with colleagues Philip Trexler and Robert Ervin. (Courtesy University of Notre Dame.)

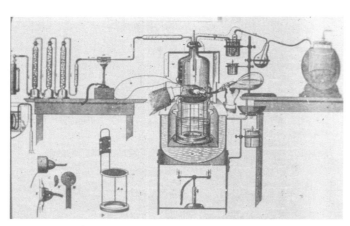

Fig. 22-1 Early isolator of the type used by Nuttal and Thierfelder. (Courtesy University of Notre Dame; attributed to Nuttal and Thierfelder, 1895.)

[1]Germfree (or germ-free; the latter is the more grammatically correct, but most authors use the former, some using both in different papers as can be observed in the References), gnotobiotic, and axenic will be discussed in the section"Terminology."

Fig. 22-3 Bengt Gustafsson. (Courtesy Gustafsson family.)

Fig. 22-4 Masazumi Miyakawa. (Courtesy B. Sakakibara and Nagoya University.)

Fig. 22-5 Stainless steel isolators. (Courtesy University of Notre Dame.)

The vast majority of these research reports are concerned with other experimental uses of these animals rather than their derivation, rearing, and establishment. The scientific literature contains reviews specific for the GF laboratory rat by Pollard (1971a), Maejima et al. (1974), and Miyakawa (1968), whereas the more current publication by Wostmann (1996) provides a general overview. The Web site of the Association for Gnotobiotics (2005) provides regularly updated information on methods, animal stocks, and other resources, as well as information on scientific and technical conferences. This useful site also provides links to the Web sites of allied organizations.

The quantum jump in recent decades in the use of these animals was facilitated in large part by the introduction of the Trexler flexible plastic film isolator system in 1957 (Trexler and Reynold, 1957). This innovation greatly reduced costs while simultaneously increasing design flexibility. Before the late 1950s, GF research was conducted in rigid isolators made of stainless steel or steel and glass (Fig. 22-5). The most commonly used type consisted of a stainless steel cylinder bolted together and gasketed at the joints. Many of these systems contained a steam autoclave as a pass-through lock. These isolator systems were heavy, cumbersome, and very costly and occupied a large amount of floor space, because their weight prohibited placing them in tiers. The advent of the plastic isolator not only reduced the cost to one-tenth to one-fifteenth of the cost of steel systems but allowed units to be stacked in tiers of two and three in a multitude of sizes and configurations to conserve floor space (Fig. 22-6). This marked the birth of a new era, permitting vast expansion in the production and availability of gnotobiotic animals. It also paved the way for the development of lightweight, disposable shipping units, which made the transport of gnotobiotic animals to distant locations a practical reality.

Fig. 22-6 Gnotobiotic facility utilizing flexible film isolators. (Courtesy Charles River Labs.)

B. Terminology

The word *gnotobiotic* is derived from the Greek words *gnotos* and *biota* meaning known flora or fauna. Therefore, when referring to gnotobiotes (GNs), one refers to an animal with a known flora or fauna. This term also is applicable when a microbial flora does not exist or is not detectable. In other words, gnotobiotic is the broad term encompassing axenic, GF, and defined flora/fauna-associated animals (Luckey, 1963).

The general review of gnotobiotics by Pleasants (1974) defines a gnotobiotic animal as follows:

> One of an animal stock or strain derived by aseptic caesarian section or sterile hatching of eggs that is reared and continuously maintained with GF technics under isolator conditions and in which the composition of an associated fauna and flora, if present, is fully defined by accepted and current methodology.

Axenic animals are GNs known to be free of all detectable microorganisms. This term is often used synonymously with GF, although the latter is more commonly used. The detection of leukemia virus particles in mice by Pollard (1966) raises the question of whether axenic animals exist at all, although these viruses have yet to be detected in cesarean-derived rats.

Germfree is the historical term that has been used over the longest period of time and is part of the colloquial scientific language, especially in North America. Its definition is the same as axenic, and although those working within the field prefer the term *axenic* as it is more accurate, *germfree* continues to remain the more popular term.

Specific pathogen-free (SPF) animals are those originally cesarean-derived within an axenic environment and subsequently associated with a definable microbial flora. These animals are then transferred from the rigidly controlled isolator system into a controlled larger physical space in which technicians wear protective clothing while working. It is designed to preclude the entrance of pathogenic organisms. This environment is most commonly termed a *barrier facility* (Foster, 1959a).

SPF is considered a controversial term in that there does not appear to be consensus as to its true definition. Some have argued that any animal free of one or more specific pathogens, for example, *Salmonella* or *Mycoplasma,* fits the definition. This controversy is confusing and is gradually being clarified by adherence to the original intent that cesarean derivation and confirmed axenic status is an essential precursor to the true SPF animal.

Pathogen-free is used loosely and interchangeably with SPF, because animals in both categories are implied to be free of pathogens. It is, however, theoretically possible to maintain animals free of pathogens through testing and eradication as well as through the use of broad-spectrum antibiotics (van der Waaij et al., 1971).

Cesarean-derived, cesarean-derived barrier-maintained, and *cesarean-originated barrier-maintained* or *sustained* are commonly used terms that imply an initial derivation of axenic animals and their subsequent association with a defined microflora (DF) followed by the continuing maintenance within a controlled barrier, where all materials entering the barrier are subjected to a procedure that removes or destroys pathogenic microorganisms.

Conventional (CV) animals are all other animals maintained under accepted husbandry practices but that do not fall within any of the previously described definitions (i.e., axenic, SPF, or pathogen-free). To some working in the field of gnotobiotics, animals are either GNs or CV animals. Today, the largest category of research animals falls in between gnotobiotic and noncontainment, conventionally raised animals in that commercial stocks are cesarean-derived, associated with a defined flora, and bred and maintained in barrier rooms to inhibit the introduction of additional microorganisms.

II. THE GERMFREE AND DEFINED FLORA LABORATORY RAT

One of the main advantages of using GF and DF laboratory rats in biomedical research is that the nutrition and physiology of many such colonies and strains have been well established. These rats have been used extensively, for example, in metabolic experiments. These animals are quite prolific in the isolator environment, notwithstanding the greatly enlarged cecum, which is thought to impair reproduction in GF guinea pigs.

There are many research areas in which the investigator, by using microbiologically sterile animals, can elicit information that cannot be obtained by using animals with normal flora. These research areas have included nutrition, immunology, infectious diseases, and dental caries studies. It is probably a lack of training and confidence in gnotobiotic technology on the part of investigators that limits more extensive use of GN animals. In reality and as a practical matter, the training of technicians in gnotobiology is a routine procedure and does not require special aptitude or formal education.

The other major uses and importance of GN rats are as nucleus seed stocks for the production of disease-free animals and as diagnostic tools for infectious disease studies, particularly in situations in which routinely used culture media are inadequate. In the mid-1950s, major laboratory animal breeders reported the use of GN rats as foster nursing stock for the rederivation of breeding colonies (Foster, 1959b). It became apparent to the laboratory animal breeding industry that testing, eradication, and selection techniques for the elimination of infectious diseases and parasites were often inconclusive, as well as tedious, and were not totally reliable. Therefore, as a natural evolution of the technology developed for the production of GN animals, the latter became the building blocks and nucleus stock for the production of microbially associated defined flora and disease-free animals. This was accomplished by introducing a known microflora to GF animals, placing them in a barrier, and maintaining them in an environment that precluded the entry of pathogenic organisms.

When a clinical syndrome or a set of pathological findings fails to elicit an etiologic agent by routine microbiologic techniques, the GF rat provides an excellent model for transmission and diagnostic studies. It provides the almost perfect model to establish Koch's postulates, because the use of this definable animal model frequently ensures valid results in the determination of specific etiologic agents, that is, the effect of a single organism can be evaluated, in the absence of other microbial forms.

A. Derivation Philosophy

The primary method of deriving GF rats is through surgical intervention of pregnancy at term. Another technique has been reported whereby the gut tract has been rendered sterile through the successive use of a variety of broad-spectrum antibiotics. Van der Waaij (1971) has reported that mice have been rendered GF within sterile isolator systems. The presumed advantages are the rapidity with which this regime can be accomplished as opposed to the traditional and proven method of cesarean derivation. For certain types of studies of short or medium duration, animals can be freed from microorganisms and maintained within isolator systems free of those organisms by using gnotobiotic techniques.

Cesarean delivery is the classic method of deriving axenic animals. Early workers delivered GF rats in stainless steel GF tanks in which visibility was possible only through small viewing ports. The surgical technician's movements were restricted by the rigid steel isolator, even though sufficient mobility remained to perform the cesarean section. Because the weakest component in any GF system is probably the rubber sleeves and gloves, in the 1950s it was established that flexible film polyethylene or polyvinyl chloride isolators afforded at least the same degree of microbiological security as the type of Neoprene sleeves and gloves (Trexler and Reynold, 1957; Trexler, 1959) used in the earlier rigid systems.

Therefore, since the late 1950s the most common procedure for cesarean delivery of GF rats is the use of a 1.5 × 0.6-m flexible film isolator fitted with at least two pairs of Neoprene (DuPont Dow Elastomers, Wilmington, DE) sleeves fitted to 22.9-cm glove ports attached to the isolator wall. In addition, standard surgical gloves are affixed to the wrists of the rubber sleeves to permit maximum tactile sensitivity. There is a 30.5- or 46-cm transfer port in the isolator for the introduction of supplies and instruments. An additional port is installed at one end, which is attached to a long, tapered sleeve or a rigid, clear plastic tube that is approximately 7.6 cm in diameter. This sleeve or tube terminates beneath the surface of a liquid germicide trap filled with a warm, 38°C, chlorine solution (a 0.525% solution of sodium hypochlorite, which can be prepared by a 1:9 dilution of classic bleach solution that most often comes as a 5.25% solution of sodium hypochlorite). A thermostatically controlled electric heating pad is placed between the exterior floor of the isolator and the rigid surface supporting the isolator. This provides warmth to the neonates after delivery. Within the sterile system, in addition to the required surgical instruments, sponges, and water, a plastic cage 35.5 cm long, 30.5 cm wide, and 14.5 cm deep is fitted with a taut Mylar membrane (a product of Du Pont, East Orange, NJ), which provides a work area for the surgeons and which can be replaced after each procedure with a new membrane. This arrangement permits the uterus and fetal membranes to drop to the floor of the plastic cage.

B. Cesarean Methods

Delivery of GF rats can be accomplished by a two-stage hysterectomy technique or by a single-stage hysterotomy procedure (Foster, 1959b; Wostmann, 1970; Pollard, 1971a). In the latter procedure, a plastic Mylar membrane in the floor of the isolator is sealed to the shaved and surgically prepared abdominal wall of the pregnant rat. The surgical technician performs the hysterotomy through the Mylar membrane window in the isolator floor. This method is more tedious than is the preferred and more rapidly performed hysterectomy. In addition to speed, the hysterectomy method permits an almost mass production routine, because the two stages can be performed simultaneously by separate surgical teams. One team is responsible for the extra-isolator phase, which consists of preparation, euthanasia, hysterectomy, and introduction of the uterus into the sterile surgical isolator. Another team of usually two technicians performs the actual cesarean, removing the fetuses from the uterus and its membranes. With the proper planning and coordination, the two surgical teams can perform 6 to 8 cesareans in 1 hour.

C. Derivation Procedure

A vital key to successful cesarean delivery of GF rats is the assurance that the pregnant rat has completed the normal gestation period of 20 to 21 days (Foster, 1959b). This is best accomplished by observed or timed matings that are confirmed by the presence of the spermatic plug in the vaginal opening. If the plug is not seen, confirmation can be made by vaginal smear for the presence of spermatozoa. Because a large percentage of natural births occur from 6 PM to 6 AM, normal parturition can be forestalled by the administration of progesterone or other drugs. This technique is used mainly for the convenience of the surgical team and, with a high degree of success, precludes premature natural parturition during the night and allows for better planning for cesareans in the morning.

The timed, gravid female is shaved along the ventral portion of the abdomen from the xiphoid cartilage to the genital opening. The use of a depilatory ensures complete removal of hair and a clean incision without the contamination of animal fur. The female is sacrificed usually by cervical fracture outside of, but close to, the surgical isolator. The surgical site is washed and disinfected. The abdomen is draped with a surgical shroud containing an elliptical opening through which a midline incision is made. Good surgical technique is required to prevent the accidental incision of the intestines with their abundance of microorganisms and parasites. The uterus, with the cervix clamped, is lifted from the abdomen onto sterile drapes. After severance from the maternal body, the uterus is lifted and removed to a 38°C primary germicide of 0.525% sodium hypochlorite solution, where it remains for 5 seconds. It is then placed into a perforated container, which is lowered via the sleeve or rigid tube by a nylon cord from inside the sterile isolator to beneath the surface of a 5% iodide germicide. After an additional 15 to 30 seconds, the container—still beneath the surface of the germicide—is guided into the mouth of a 10.4-cm-wide rigid clear plastic tube connected to the isolator wall. A surgical technician, with arms inside the isolator via the surgical sleeves and gloves, raises the uterus within the perforated container to the interior of the sterile isolator, allowing the germicide to drain down to the germicidal trap. Two technicians on opposite sides of the isolator rapidly remove the fetuses from the uterus and separate them from their fetal membranes by using a taut Mylar membrane secured to a cage top as a surgical table. The fetuses are quickly washed with surgical sponges and rendered clean from amniotic fluid and blood. This is essential because a foster mother might cannabalize them if body fluids and remnants of the fetal membranes remain. These procedures must be accomplished rapidly, because maternal support is lost on separation of the placenta from the uterus.

The neonates are dried and massaged to stimulate breathing, and the umbilical cord is separated by clamping and cutting or by electric cautery (Pollard, 1971a). If additional procedures are to be performed, the neonates are loosely wrapped in a small surgical towel and placed on the isolator floor above the warmth of the heating pad resting beneath the isolator floor. The Mylar membrane attached to the plastic cage is punctured, permitting the uterus and membranes to fall below to the floor of the cage. A new membrane is placed across the mouth of the cage and once again held in place by rubber bands. This procedure can now be repeated many times without breaching the integrity of the sterile system.

It is good practice to transfer the neonates to a rearing isolator, where they are foster-nursed on lactating GF mothers until of weaning age (Pollard, 1971a). This method has replaced the more cumbersome hand-rearing and feeding (Gustaffson, 1948; Griffiths and Barrow, 1972), principally because of the more readily available lactating GF mothers and the increased survival rate as compared with that of hand-feeding methods. One would only consider hand-rearing where minimal antigenic stimulation is

desirable, and in these instances, highly purified synthetic formulas are used. Pleasants (1974) reviews the surgical procedures, including equipment. Figures 22-7 through 22-10 show an adaptation of the procedure described above using tandem laminar flow hoods. Numerous references exist on the establishment and maintenance of breeding colonies of GF rats (Reyniers et al., 1946; Gustaffson, 1948; Foster and Pfau, 1963; Lev, 1963; Gordon et al., 1966; Reid and Gates, 1966; Miyakawa, 1968; Kappel, 1969; Kellog and Wostmann, 1969; Yale and Linsley, 1970; Coates, 1973; Maejima et al., 1974).

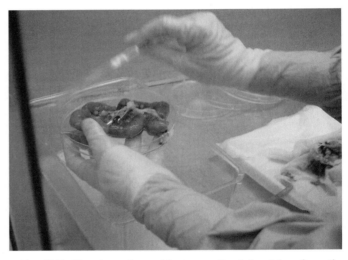

Fig. 22-7 The clamped, gravid uterus, after being taken from the donor dam following euthanasia, is placed into a plastic dish, immersed in disinfectant, and moved to a laminar flow hood. (Courtesy Charles River Labs.)

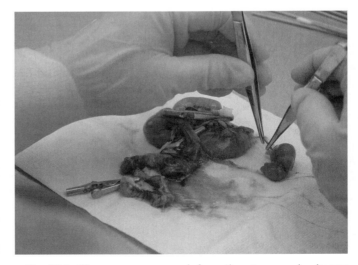

Fig. 22-8 The pups are removed from the uterus, each placenta detached, and the pups cleansed before being taken into an isolator. (Courtesy Charles River Labs.)

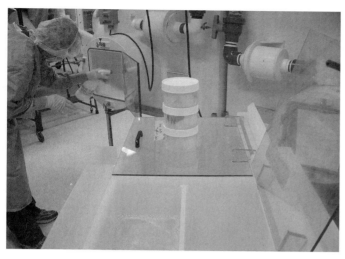

Fig. 22-9 The clean, active pups are placed into a sealed container that is sprayed with disinfectant and passed into an isolator. (Courtesy Charles River Labs.)

Fig. 22-10 The container of pups is introduced into the isolator and placed with a receptive foster mother. (Courtesy Charles River Labs.)

D. Production of Defined Microflora Rats

The elimination of the rat's normal gut microflora by cesarean derivation results in dramatic changes in the host's physiology, nutrition, tissue morphology, and defense against infectious agents. The most pronounced anomaly of the GF state in rats, as well as other species, is the enlargement of the cecum, which can lead to volvulus at the ileo-cecal-colonic junction and eventual death (Wostmann and Bruckner-Kardoss, 1959). The content of the cecum and intestines are fluid, and the animal is said to have a chronic mild diarrhea. In addition, aside from low levels of antigenic material in the feed and bedding, the immune system of the GF rat is unstimulated (Gordon and Pesti, 1971). The lamina propria is thin and almost devoid

of antibody-producing plasma cells, and lymph nodes are smaller (Gordon and Wostmann, 1960; Gordon et al., 1966). Also, because of the absence of its vitamin K-synthesizing gut flora, the GF rat must have this vitamin added to its food or it rapidly develops prolonged prothrombin times and hemorrhages (Gustaffson, 1959). GF rats are also much more susceptible to infections than are their CV counterparts, which is why they often die soon after being introduced into a CV colony (Luckey, 1963). This can be prevented if they are first colonized by at least several species of their normal gut microflora.

Gordon and Wostmann demonstrated that GF rats could be normalized by feeding them cecal contents of CV rats (Gordon and Wostmann, 1959). However, no attempt was made to determine which member(s) of the microflora was responsible for this phenomenon until Schaedler and Dubos' classic work describing the bacterial colonization of the gastrointestinal tract of mice, which subsequently became the cornerstone for much of the work that followed (Dubos et al., 1965; Schaedler et al., 1965a); they reported that soon after birth the entire gastrointestinal tract was populated by lactobacilli and a group N *Streptococcus*. During the second week of life, high concentrations of aerobic bacteria, such as enterococci and slow lactose-fermenting coliforms, were observed in the large intestine. Their numbers abruptly dropped during the third week of life when obligately anaerobic bacteria, such as *Bacteroides* species, colonized this organ. Throughout the adult life of mice, the obligately anaerobic bacteria remained at very high levels, and the aerobic component of the microflora remained suppressed at very low levels. The microflora of the rat has been found to closely resemble that of the mouse (Smith, 1965; Savage, 1969).

Schaedler et al. (1965b) then proceeded to colonize GF animals with a flora consisting of *Bacteroides*, lactobacilli, an anaerobic *Streptococcus*, and a slow lactose-fermenting coliform. This flora was able to drastically reduce the size of the cecum and, therefore, almost normalize the animals. Consequently, this flora, and variations of it, has been used extensively to colonize both GF mice and rats before their removal from an isolator into a new colony. A process for colonizing GF animals, commonly referred to as associating animals, merely consists of colonizing an initial isolator of GF rats with pure cultures of each of the individual members of the flora. Additional associations are then achieved by simply introducing an associated animal into a GF isolator and placing fecal pellets from the associated animal into the water bottles of the GF animals on two consecutive days. During the first day, the aerobic bacteria colonize the GF animals and lower the oxidation-reduction potential, so that on the second day the extremely oxygen-sensitive fusiform-shaped anaerobes are able to colonize the animals. It should be noted that these few bacteria represent a very small fraction of the gut microflora, and many additional members are necessary to normalize a GF animal completely (Syed et al., 1970).

E. Microbiological Testing

It is good practice to perform certain examinations on the euthanized dam. Historical data of the health status of the donor female provide excellent reference material should subsequent contaminations occur in the GF or SPF colony. Therefore, before losing the identification of a cesarean-delivered litter, examination of the dam for *Mycoplasma* and intestinal parasites and serological examination for murine viruses are recommended. Certainly, at the very least, careful culturing of the ovaries and uterus for *Mycoplasma* should be performed (Ganaway et al., 1973; Schultz et al., 1974), because there have been occasional reports of *Mycoplasma pulmonis* isolation from GF rats that may have resulted from *in utero* contamination of the dam (Kappel et al., 1969; Ganaway et al., 1973). Careful workers discard neonates as a precaution if the donor female exhibits positive *Mycoplasma* in the reproductive system.

Subclinical *Pasteurella pneumotropica* infections have been reported in GF rats (Ganaway et al., 1973; Pleasants, 1974), and these might be transmitted to GF progeny. Microbiological testing by fecal swabs of the neonates 24 to 48 hours after delivery ensures the asepsis of the surgical procedure as well as the sterility of the rearing isolator. Detailed methods are described elsewhere (Wostmann, 1970). The methods for gross observation and the detection of bacteria, fungi, and parasites are relatively simple and standardized. The methods for the detection of exposure to murine viruses are usually accomplished through serological tests for the specific antibodies such as the enzyme-linked immunosorbent assay (ELISA) or the indirect immunofluorescence assay. Wagner (1959) worked out detailed sterility testing procedures that are used as standard procedures in many laboratories. Usually when an isolator becomes contaminated with bacteria, the exhaust air loses its non-animal-like, almost sweet, odor to the more familiar odor of laboratory rats. In nearly all instances, contaminants can be observed in fecal wet mounts before routine culturing. Culturing 24 to 48 hours on appropriate media readily reveals typical contaminants at 37°C. Because 21°C, 37°C, and 55°C are standard incubation temperatures, molds and thermophilic organisms are also detected in the less common contaminations.

Intrauterine infection represents a potential hazard to the axenic integrity of GF rats (Asano, 1969; Altura et al., 1975), because vertical transplacental transmission of Kilham rat virus (rat parvovirus type 1) (Kajiwara et al., 1996; Jacoby et al., 2001), a leukemia virus (Pollard, 1966), and a still undefined virus from the submaxillary gland

(Ashe et al., 1965) have been noted, although the latter has not been reported in later literature. The confirmation of encephalitozoan in GF rabbits (Hunt et al., 1972) suggests that such vertical transmission is possible in rats and other species. With these limited reports as background, the examination of donor stock would be the only means currently available to help reduce the vertical transmission potential in newly derived colonies. Unfortunately, the job to search histologically for leukemia virus and the virus from the submaxillary gland is tedious, and negative results would not be conclusive. With regard to encephalitozoan, immunofluorescent and India dye tests are fairly straightforward procedures and provide a high degree of accuracy (Wosu et al., 1977a).

F. Anatomy, Reproduction, and Lifespan

Although strain, sex, age, and organs must be taken into account in making comparisons (Banasaz et al., 2000), the GF rat is an experimental animal that differs significantly from the CV rat in a number of characteristics. GF rats provide a uniform and relatively stable baseline of morphologic and physiologic activities, which in turn facilitate studies of superimposed changes. The earliest noted and most conspicuous effect of the GF status in rodents, including rats, is enlargement of the cecum. It becomes voluminous, usually five times larger than that of its CV counterpart on the same diet, and may approach 25% of the rat's body weight (Coates, 1973). This enlargement sometimes interferes with normal reproduction but can be significantly reduced with dietary manipulation of inorganic ions. The cecal wall is much thinner in GF rats, and the cecal contents more liquid than that in CV animals. This is owing to an excess of water and anionic-soluble mucins, the latter being degraded in the CV rat (Asano, 1967; Pleasants, 1974; Carlstedt-Duke et al., 1986). Cecectomy of GF rats restores most functional and metabolic parameters to within the CV range (Wostmann, 1975).

The enlarged cecum is associated with altered metabolic functions, particularly slower cholesterol and bile turnover (Einarsson et al., 1973), depressed reducing capacity of the cecal contents, and reduced cecal concentrations of chloride and carbonate ions (Thompson and Trexler, 1971). These animals also require less exogenous choline (Nagler et al., 1969), and there is a total absence of metabolism of flavonoid compounds in the gut (Griffiths and Barrow, 1972). If the GF rat's cecal contents are replaced with saline, then the water absorption capacity of the cecum becomes normal or greater than that in CV rats (Gordon, 1974). GF rats have been reported to accumulate a compound or compounds in their cecum that changes the tone and reactivity of mesenteric vascular smooth muscle to adrenaline (Baez and Gordon, 1971) and that is normally destroyed by the CV normal flora (Carlstedt-Duke et al., 1986). Current understanding suggests that alpha-pigment, prostaglandins, eicosanoids, fatty acids, and kallikrein all contribute to change in cecal muscle tone (Bruckner-Kardoss and Wostmann, 1974).

The weight and surface area of the small intestine and the associated lymphoid cells and tissues of GF rats are generally decreased (Gordon et al., 1966). Depending on the diet, the rate of peristalsis may be the same (Gustafsson and Norman, 1969) or slower (Sacquet et al., 1973) in GF than in CV rats. The rate of mucosal sloughing is generally half that of CV controls (Gordon et al., 1966; Reddy and Wostmann, 1966), and digestive enzymes, such as pro-teases, lipase, and amylase, persist longer and farther down the gut under GF conditions (Lepovsky et al., 1966; Reddy et al., 1969a; Norin et al., 1986). Urease appears to be absent under GF status (Delluva et al., 1968). The GF rat gut is more efficient in digestion and absorption, in part because the villi are longer and more even (Coates, 1973).

Nutrient requirements in the diet of GF rats are usu-ally higher than in CV rats but vary with experimental conditions (Pleasants, 1974). In general, there is a higher need for total food and water, for vitamin K and the B vitamins, and for choline to prevent liver cirrhosis. GF rats maintained on a diet without supplemental vitamin K rapidly develop a hemorrhagic condition, whereas CV rats on the same diet do not (Coates, 1973). Antagonism between vitamins A and K occurs only when vitamin A intake is 10 times above normal (Wostmann and Knight, 1965; Reddy and Wostmann, 1966). On the other hand, GF rat nutrient requirements are less than CV require-ments for vitamin A (Rogers et al., 1971; Coates, 1973), lysine, cysteine and vitamin E to prevent liver necrosis, protein (Pleasants, 1974), calcium, and magnesium and zinc (Smith et al., 1973). GF rats given vitamin A-deficient diets survived much longer than do CV rats (Coates, 1973). Assessment of the rat's nutritional requirements are often difficult because of coprophagy.

GF rat studies show unequivocally that the CV rat's microbial flora has a significant effect on the basal meta-bolism, on the response to adrenaline, cardiac output, and vascular distribution (Pleasants, 1974). The overall meta-bolic rate of GF rats has been reported to be one-fourth that of CV rats of the same strain (Wostmann et al., 1968). This undoubtedly results from reduced (one-third normal) cardiac output and oxygen consumption (Wostmann et al., 1968), reduced regional blood flow and distribution (Gordon et al., 1966), decreased heart weight (Gordon et al., 1966; Albrecht and Souhrada, 1971), decreased total blood volume (Bruckner, 1997), and decreased pulmonary partial pressure of carbon dioxide values (Schwartz, 1975). Aortas and portal veins of GF rats have an attenuated reactivity to angiotensin, vasopressin, and epinephrine (Altura et al., 1975). Production of short-chain fatty

acids are related to microbial activity (Høverstad and Midtvedt, 1986), and reduction of total body fats has been reported (Reina-Guerra et al., 1969). In GF rats the lymph nodes and Peyer's patches are small, lack germinal zones, and contain few, if any, plasma cells (Miyakawa et al., 1969; Carter, 1971; Balish et al., 1972). Serum globulin values are one-third those of CV rats, and GF rats have fewer total serum proteins (Balish et al., 1972). The decreased immunological stimulation of GF animals leads to very low titers of agglutinating antibodies for *Streptococcus fecalis, Proteus vulgaris, Pseudomonas aeru-ginosa, Lactobacillus acidiphilus,* and *Bacillus fragilis* (Balish et al., 1972). This is accompanied by decreased severity of conjunctival inflammation when rats are infected by bacteria (McMaster et al., 1967). In relation to the above, it has also been reported that the mucosa of the nasal cavity and middle ear have few lymphocytes and no inflammatory infiltrates (Giddens et al., 1971).

Tissue enzyme levels usually tend to be lower in GF rats. There is less mitochondrial succinate oxidase and glycerophosphate dehydrogenase activity in the liver (Sewell et al., 1975). Lower muramidase levels have also been reported (Ikari and Donaldson, 1970). However, some tissue enzyme activities are higher in GF rats, namely, peroxidase-mediated antibacterial activity of the salivary glands (Morioka et al., 1969), liver microsomal hydroxyla-tion of steroid hormones (Einarsson et al., 1974), and fatty acid synthetase and citrate lyase activity in the liver (Reddy et al., 1973; Wostmann, 1975).

In general, GF rats eat more, grow better, and are less subject to disease. They absorb saturated and unsaturated fats better, particularly palmitic and stearic acid (Demarne et al., 1973); have greater serum and liver cholesterol concentrations (Reina-Guerra et al., 1969); use more total fat in the diet (Nolen and Alexander, 1965); and have higher cholesterol turnover rates (Reina-Guerra et al., 1969). A report on experimental cholesterol synthesis in GF and CV rats (Ukai et al., 1976) indicated that there is an inverse proportionality between the log phase rate of hepatic cholesterol synthesis and liver cholesterol levels in GF rats. Therefore, the endogenous cholesterol synthesis in GF rats may not be responsible for the high cholesterol levels in plasma or in the liver. Liver cholesterol may play a major role in the regulation of hepatic cholesterogenesis in the GF rat by a mechanism similar to that in the CV rat.

There is total conjugation of bile acids in GF rats compared with almost total lack of conjugation in the cecum of CV rats (Madsen et al., 1976), which depends on clostridial species (Midtvedt and Gustafsson, 1981). The bile turnover rates are higher (Reina-Guerra et al., 1969) as is the pH of the cecal contents (Thompson and Trexler, 1971). In GF cecal contents, the colloid osmotic pressures are approximately 100 mm Hg. This results in a pressure gradient of 60 to 70 mm Hg between the gut lumen and the blood plasma, in contrast to a smaller gradient in CV rats (Gordon, 1974). GF rats have greater reabsorption of bile acids from the gastrointestinal tract and, therefore, have greater recirculation of bile (Einarsson et al., 1973). Use of the GF rat in studies of cholesterol metabolism are particularly concerned with the factors that influence the absorption of cholesterol from the gut and its elimination from the body as bile acids via the feces (Wostmann, 1973, 1975). The GF rat appears to be unable to decrease the reabsorption of bile acids in the lower gut, a function of normal microbial flora. Work indicates that differences in the histochemical nature of mucosaccharides depend on whether they are located in areas of normal bacterial flora in CV rats or in areas relatively free of intestinal flora (Yamada and Ukai, 1976).

Other metabolic parameters that tend to be higher or greater in GF animals compared with CV animals include pH of cecal contents (Thompson and Trexler, 1971), mean intracolonic oxygen pressure (Bornside et al., 1976), pulmonary arterio-venous oxygen values (Schwartz, 1975), plasma levels of some steroids (Einarsson et al., 1973), urinary citrate excretion (Gustafsson, 1948), and fasting blood glucose (Pleasants, 1974). In addition, aortas and portal veins have a higher total calcium content in GF animals (Altura et al., 1975), and the microvasculature is refractory to catecholamines (Gordon et al., 1966; Baez and Gordon, 1971).

The GF state has little influence on the functional respiration or oxidative phosphorylation of mitochondria isolated from the livers of adult rats (Sewell and Wostmann, 1975). Serum chemistry and hematological values are within the normal range except for the depressed leukocyte level (Burns et al., 1971). Minimal differences have also been reported in serum β-lysin (Ikari and Donaldson, 1970), fasting blood glucose and glucose tolerance (Sewell et al., 1976), metabolism of nicotinamide and nicotinic acid (Lee et al., 1972), carbon dioxide production in the gut (Rodkey et al., 1972), mean pulmonary arterial oxygen partical pressure (Schwartz, 1975), and activities of hepatic enzymes of urea synthesis (Norin et al., 1986). No differences were found in the histology of the eye of GF rats (McMaster et al., 1967).

It is evident that there are many basic physiological and morphological parameters of GF rats that have not yet been studied. Furthermore, one must keep in mind that often reports cannot be reliably compared because of variables of age, sex, and strain of rat as well as environmental conditions, diet, and unknown interactions among these factors. For example, it has been reported that differences in thyroid function and related hepatic enzymes tend to lessen with age of the animals (Sewell et al., 1975) and that the qualitative and quantitative composition of the bile acids varies considerably between male and female GF rats (Gustafsson et al., 1975).

G. Nutrition

It can be generally stated that the nutritional requirements of animals are inversely proportional to their biosynthetic capacity (Luckey, 1963). The need for special diets for GN animals has been reviewed by Wostmann (1975). Special diets are necessary principally because food sterilization methods usually require compensation for the loss of vitamins and the reduction of nutrient value of proteins resulting from heat sterilization (Weisburger et al., 1975). The dietary requirements for microbiologically synthesized vitamins are higher (Coates, 1973; Wostmann, 1975), because the lack of normal microbial flora affects the absorption, which is greatly enhanced in the GF rat and which leads to the formation of urinary calculi unless dietary levels of calcium are reduced (Gustafsson and Norman, 1962; Smith et al., 1973).

Diets tend to vary according to the specific GF research objectives (Luckey, 1963), that is, antigen-free diets for immune system studies or high-sugar diets in dental caries studies. Some diets have been found nutritionally adequate for short-term experiments even if autoclaved, as long as the diets are supplemented with filter-sterilized heat-labile vitamins (Oace, 1972). A canned, moist, presterilized (autoclaved) diet of known composition can be provided by spraying it into the isolator system (Foster and Pfau, 1963; Foster et al., 1967). Autoclavable diets are also available (Kellog and Wostmann, 1969; Oace, 1972; Pleasants, 1974), as are chemically and water-soluble ones. The latter can be filter-sterilized (Pleasants, 1974) and used as special purified diets for nutritional research (Wostmann and Kellog, 1967). Growth of GF rats on these diets is comparable to that of CV animals. Reddy et al. (1969b) grew GF rats from birth to maturity by using membrane-filtered, chemically defined, water-soluble diets based on amino acids and glucose. Diets sterilized by gamma irradiation have also been used in rearing GF rats (Paterson and Cook, 1971). Radiation sterilization using cobalt 60 irradiation is recommended for studies of cholesterol and bile acid metabolism in GF rats (Wostmann et al., 1975). Current diets suitable for studies in nutrition and metabolism of GF rats are listed in Wostmann's review (1975).

H. Strains and Stocks

1. Laboratories of Bacteriology of the University of Notre Dame

The Laboratories of Bacteriology of the University of Notre Dame (LOBUND) has maintained GF colonies of the Fischer strain and of the Wistar and Sprague Dawley rat stocks (Maejima et al., 1974), and Pleasants (1959) reported experiments with the Holtzmann and Lobund stocks of rats bred in closed colony. LOBUND has a distinguished history as a site for training, supplies, and animal stock, in addition to research, into the 1990s; this laboratory is now much reduced in size and scope and does not currently maintain GF animals.

2. University of Wisconsin

An internationally known laboratory was established in the 1960s by Balish and colleagues in Madison, Wisconsin, that created and maintained large colonies of GF rats, mice, and, uniquely, beagle dogs, for 3 decades. Unfortunately, similar to LOBUND, this resource is now much reduced in size and scope.

3. Karolinska Institutet

Gustafsson (1948) used the Long-Evans hooded stock, and other strains are currently used in Sweden.

4. Commercial Suppliers

Charles River Labs, Harlan-Sprague-Dawley, Taconic Farms, IFFA-Credo (Charles River, France) and a few others are commercial entities that maintain various gnotobiotic stocks and strains, generally not catalog items, as seed stocks for their commercially available pathogen-free animals.

5. Others

The Gifu hybrid has been produced in the GF state by Miyakawa (1968) in Japan, and Dajani et al. (1975) used the GF Agus strain.

Although there are many laboratories that actively use GN rats—such as those supported by the National Institutes of Health gastrointestinal disease center at North Carolina State University and the University of North Carolina at Chapel Hill or those at Tokai University in Japan and the University of Minas Gerais in Brazil, none is large enough to maintain stocks equivalent to those mentioned above that would provide more than a few breeding pairs of animals. There is a profound need for a central supplier, supported by a large, national research fund, to maintain a GF animal resource center, particularly for the major rat and mouse strains, for use by the world's research laboratories.

III. RESEARCH APPLICATIONS OF GNOTOBIOTIC RATS

A. Infectious Disease

The infectious and chronic diseases of CV rats are described by Tuffery and Innes (1963) and may serve as a basis for comparison of monoassociation and experimental infection studies. *M. pulmonis* is the primary pathogen in chronic respiratory disease (Sugiyama and Bruckner, 1975). Luckey (1969) has provided an extensive bibliography on the effects of bacterial species on the monoassociated rat. No differences were found in the susceptibility of GF rats to *Plasmodium berghei* primary infections via mosquito-borne sporozoites, nor were there any differences in the resulting pathology (Martin et al., 1966). Work in Balish's laboratory (Rogers and Balish, 1976) indicates that the GF rat can serve as an animal model of nephritis owing to *Candida* infections, because the yeasts multiply in the kidneys.

B. Cancer

Cancer development in GF rats can be related in part to the absence of microbial flora (Pollard and Teah, 1963; Pollard, 1972; Walburg, 1973; Pollard et al., 1985). Experimental cancer yields are lower in GF rats when the carcinogens tested are of the type necessitating enzymatic activation (Weisburger et al., 1975). In general, the oncogenic potential is the same as in CV rats, but tumor-related changes are more clearly defined in GF animals (Pollard et al., 1968). GF rats with either spontaneous or induced tumors have higher numbers of plasma cells but have no germinal zones in their lymph nodes (Pollard et al., 1968).

Gnotobiotic animals are particularly suitable for testing candidate viral carcinogens, because derivation by hysterectomy and gnotobiotic maintenance has been found to eliminate all known viruses from GF rats (Luckey, 1963; Pleasants, 1974). Nevertheless, GF rats have a very low rate of spontaneous neoplasm development compared with that of GF mice (Walburg, 1973). The most frequent spontaneous tumors in aged GF rats involve the mammary and pituitary glands (Pittermann and Deerberg, 1975).

1. Colon Cancer

Cycasin from cycad bean flour is carcinogenic for CV rats, whose microflora convert it to a carcinogen, whereas it does not induce tumors in GF rats (Laqueur et al., 1967; Luckey, 1969). If cycasin is first hydrolyzed to aglycone (methylazoxymethanol), it is then carcinogenic to GF rats (Laqueur et al., 1967). Spontaneous colon adenomas are twice as prevalent in GF rats (Weisburger et al., 1975). No difference in the incidence of adenocarcinoma has been reported after intracolonic exposure to nitrosoguanadine carcinogens (Weisburger et al., 1975), whereas others reported greater susceptibility of GF rats to these same direct-acting carcinogens (Shih et al., 1975). Results significantly depend on the route of administration, because oral administration of *N*-methyl-N-nitro-N-nitrosoguanadine produced very few adenocarcinomas in GF rats compared with CV animals (Miyakawa et al., 1975).

2. Breast and Prostate Cancers

GF rats are as susceptible to dimethylbenzanthracene (DMBA)-induced breast cancer as are CV rats (Pollard, 1966). Primarily through the efforts of Pollard (Pollard, 1973, 1977, 1992; Pollard and Luckert, 1987; Snyder et al., 1990; Pollard and Wolter, 2000), a spontaneous prostate tumor that appeared in GF LOBUND Wistar rats has been developed into an exquisite animal model for human prostate cancer. This has been further characterized and extended in conventionalized rats of the same line (Fig. 22-11). Although the Wistar rat is not inbred, the LOBUND line has been bred in a closed colony for so many generations that skin grafts are accepted among rats of that line.

3. Leukemia

A spontaneous, transplantable, lymphatic leukemia called Nova rat leukemia has been reported in aged GF Fischer rats (Sacksteder et al., 1973). Leukemia could not be induced by whole-body irradiation, but GF rats were found as susceptible to passage of gross A leukemia virus as were CV rats (Pollard, 1966).

4. Urethral Cancer

Urethral cancer is rare in CV rats but is relatively frequent in some older GF rat strains (Pollard, 1973).

5. Endocrine Cancer

Endocrine-related cancers of the nonleukemic type, such as thymomas and mammary neoplasms, occur in GF rats (Pollard, 1977).

C. Oral Pathology

Mechanisms of oral pathology have been clarified through GF rat studies. This is particularly true in caries research in which *Streptococcus* sp. in association with GN rats fed a carcinogenic diet produced carious lesions (Green et al., 1973), and in periodontal disease studies in which a number of streptococci, actinomycetes, and Gram-positive bacilli caused typical periodontal disease under conditions of monoassociation (Green et al., 1974).

Fig. 22-11 Spontaneous prostate carcinoma in an aged LOBUND (Laboratories of Bacteriology of the University of Notre Dame) Wistar rat. (Courtesy M. Pollard, University of Notre Dame)

D. Senescence and Wound Repair

In general, GF animals tend to live longer than do their CV counterparts. Premature death of GF animals may be caused by infection or by environmental factors. Delayed morbidity in 2- to 3-year-old GF rats is a common observation, which shows them to be virtually free of age-related kidney, heart, and lung changes (Pollard and Kajima, 1970; Pollard, 1971b). Postmortem differences include a minimum of odoriferous putrefactive changes and autolysis of the intestinal area by digestive enzymes; dead GF animals undergo drying and mummification if in a dry atmosphere (Luckey, 1963). GF rats are less sensitive by half to X irradiation as it affects the rate of wound closure (Donati et al., 1973).

E. Immunology

Immunological studies with GF or DF animals enable one to distinguish primary mediation lesions from those associated with microbial infections. From work on the biological effects of radiation, it has been determined that GF rats survive larger doses of total-body X irradiation for a longer time (Reyniers et al., 1956). In basic immunological studies, GF or DF rats provide information on the role of the microbial flora in stimulating humoral and cell-mediated immune responses. Immunity, as measured by opsonic activity, was depressed in GF rats infected with *P. aeruginosa* whereas the GF rat's responsiveness to H and O antigens of *Escherichia coli* and to sheep erythrocytes was increased (McClellan et al., 1974). Differences in phagocytosis depend on differences in opsonic activity rather than on functional differences at the cellular level. *In vitro* studies with ^{32}P-labeled *E. coli* opsonized with sera of GF rats indicated that cells from GF rats were slightly more active in ingesting capacity than were cells from CV rats. Thus, the opsonic activity of CV rat sera is higher than that of GF rat sera when tested *in vitro* (Trippestad and Midtvedt, 1971). GF rats have been reported to reject skin allografts more rapidly than do CV rats (Lev, 1963), whereas autografts of skin transplants on rat tails healed quickly (Ashman, 1975). The latter is postulated to be owing to genetic uniformity of histocompatibility factors in GF animals (Ashman, 1975). The GF allogenic radiation-induced chimera has a greatly reduced or absent T-cell response, whereas the B-cell response is almost normal (Bealmear et al., 1973).

F. Xenotransplantation

Xenotransplantation, focusing on the use of porcine organs in humans, is a topic of great current interest, but the first successful xenotransplant was actually performed by using bone marrow transplantation in GF rats and mice (Pollard et al., 1985; Wade et al., 1987). The immunological basis for the success of xenotransplantation in this system remains to be defined but gives hope that organ transplants between species may someday be successful.

G. Inflammatory Bowel Disease

Understanding the etiology and pathogenesis of human inflammatory bowel diseases, such as Crohn's disease and ulcerative colitis, has been thwarted by the lack of a suitable animal model. Toward the end of the 20th century, several potential mouse models were investigated (Elson et al., 1995), but none appeared to show the overall promise of a transgenic rat model (Sartor, 2000) that has allowed critical evaluation of the host response and the host's associated intestinal flora in the pathogenesis of chronic intestinal disease. The breakthrough was provided by the successful introduction of a human gene into Fischer rats, creating a transgenic rat line that reproduced and was cesarean-derived into the GF state by Balish (Taurog et al., 1994; Warner et al., 1996). Sartor and coworkers then used this transgenic line to show that spontaneous colitis did not develop in the GF host but did develop in both CV and monoassociated rats (Rath et al., 1996, 1999, 2001; Dieleman et al., 2003).

IV. CONCLUSIONS

At the New York Academy of Sciences Conference in 1959, "Germfree Vertebrates: Present Status," there were discussions on the state of the art, contemporary research

uses, and some practical applications of the GF animal in the development of disease-free breeding colonies. Much of these discussions have come to fruition through further study and expanded implementation of the technology.

Before the late 1950s equipment for GF animals was fabricated principally of stainless steel with various sizes and shapes of viewing areas. The early pioneers were convinced that the security of GF systems could be best achieved when rigid components were used that could withstand impact and be less vulnerable to breakage. Stainless steel cylinder-shaped GF isolators bolted and gasketed together were the most commonly used systems in the United States at the LOBUND Laboratory, the National Institutes of Health, and Walter Reed Army Institute for Research. In Sweden, at the Karolinska Institute, Gustafsson (1948) developed rectangular rigid steel isolators with glass tops maximizing the visibility inside. These units could be totally introduced into a steam autoclave for sterilization. In both the Reyniers cylindrical tank and the Gustafsson isolator, Neoprene < RM > sleeves and gloves were used for manipulation, constituting a potential weakness because they were subject to tear or puncture. Miyakawa (1968) in Japan operated remote mechanical arms and hands from outside a small sterile room similar to the equipment sometimes seen in radiation research laboratories. Even though there are very fine research reports from this period, the cost of equipment and its inefficient utilization of space limited the number of workers in the field. Also, the complexity of the fabrication and construction added further to discourage interested scientists. It was not until Trexler developed the low cost, lightweight flexible film isolator that GF research came within the budgetary and technical reach of the research community in general. In 1957, a standard flexible film $1.5 \times 0.6 \times 0.6$ m isolator complete with filtration, transfer port, exhaust trap, and flexible sleeves and gloves bore a price of $300 to $400 compared to $5000 for a typical stainless steel tank type isolator. Thus, technological advances produced a system that has been proved to be equally or more secure, light in weight to permit use in tiers, and made of clear plastic to allow complete visibility. An extension of the flexible film technology brought forth a lightweight, disposable flexible film shipping unit (Trexler and Reynold, 1957) weighting 5–6 kg compared to earlier units weighing 70–80 kg and which required battery-operated blower systems. Thus, both rearing and research units, plus shipping units, were readily obtainable to those interested in using and transporting GF and GN animals.

Diets which early workers developed were either mixed and formulated in their laboratories or prepared at significant cost by organizations that specialized in small batch mixing of complicated formulas. From the late 1950s onward, commercial feed manufacturers offered standard laboratory rodent diets pre-fortified with sufficient thermolabile nutrients to withstand sterilization and still support reproduction and growth. There have, from time to time, appeared in the marketplace canned sterile water and prepackaged sterilized bedding for those with limited sterilization capacity or capability.

In effect, it is entirely feasible to conduct a single GF research project without setting up a vast facility with expensive support, laboratories, and personnel. Also, when a single project is completed, the inflated flexible film isolator can be stored flattened in its deflated form for subsequent reuse. The initial set-up costs are nominal as are the operating costs.

Perhaps one of the single most important benefits of the GF and GN rats is the utilization of these animals as seed stock for new colonies. By deriving CV rats into the GF state, all microorganisms and parasites are eliminated except the few that are thought to be transplacental and thus vertically transmitted. Colonies plagued with external and intestinal parasites, chronic murine pneumonia, and one or more other bacterial or viral diseases can be rendered free of these infections by utilization of the gnotobiotic technology and deriving these animals into the GF state. It is then a routine procedure to associate these rats with a defined "bacterial cocktail" of gut flora before removal from isolator systems and introduction into some type of clean barrier facility. "Pathogen-free" is the commonly used terminology for rats derived in this manner and is the accepted practice by industry, government, and academia for providing healthy animals for research. Even though testing and eradication techniques can work, as does induction of a bacterial-free state through broad-spectrum antibiotics, the GN must be derived within a sterile isolator system.

Valuable genetic strains of rats are assured continuity by maintaining them GF, thereby greatly reducing and almost eliminating the possible loss through an epizootic. The National Cancer Institute for many years has maintained Rodent Genetic Centers under contract whereby valuable genetic strains and stocks have been maintained in the GN state. Through the technology herein described, genetically defined CV animals are cesarean-delivered into the axenic state then subsequently associated with a defined flora and maintained in isolators assuring microbial definition and uniformity. These rats can be sent with genetic and microbiological pedigree to laboratories for research utilization or breeding programs.

Like the pure or refined chemical reagents available to researchers, the laboratory rat can be obtained in the purest microbiological sense (axenic) or with an easily described and definable flora (DF) within the isolator. The barrier-reared animal which is an extension of the GN can be maintained in a controlled environment to preclude contamination by pathogens. Even though isolator systems break down on occasion, usually through human error,

their use is a giant step forward toward supplying defined rats as one of the basic tools of biomedical research. Compared to its CV counterpart, the SPF, cesarean-delivered, cesarean-originated, barrier-sustained (COBS), barrier-maintained rat has provided the research community with some point of reference in that at one time during the immediate past they were GNs. The CV animal usually is not reared in a barrier environment where all materials contacting the animals have undergone decontamination, pasteurization, or sterilization. The probability, therefore, is far greater that microbiological variability does occur in CV animals because of their undefined origin and their more loosely controlled environment.

The science of breeding and rearing GF, DF, and SPF rats is an attempt by professionals in laboratory animal science to keep pace with the rapidly evolving technology in the instrumentation field. What is the value of highly sophisticated instrumentation designed to make finite measurements of biological materials if the biological tool is uncontrolled and undefinable? The horizons of gnotobiotic research applications have expanded to include space travel and its potential and unpredictable effects on man and his biosphere. Other areas of research include oral pathology, cancer, wound repair, infectious diseases, and nutrition. The GF animal, therefore, offers a multitude of research opportunities.

Because of the technology developed through the years, the cost of such research using GF animals is within the scope of most budgets. However, there is still an inherent resistance to undertake research with GF rats principally as a carry over from other eras when the cost of this type of research was excessive and the technology was too highly specialized for the average laboratory setting. The contents of this chapter, in connection with the literature citations, should provide those desirous of conducting research on GN rats the necessary technical information with regard to methodology, characteristics, and utilization.

ACKNOWLEDGMENTS

We gratefully acknowledge the advice and critical review of the manuscript by Drs. Tore Midtvedt, Roger Orcutt, Morris Pollard, and Philip Trexler. The efforts of Dr. William White, Charles River Laboratories, in providing photographs of cesarean derivations are most appreciated.

REFERENCES

Albrecht, I. and Souhrada, J. (1971). Defining the laboratory rat for cardiovascular research. In: "Defining the Laboratory Animal" (H.A. Schneider, ed.), pp. 616–625. National Academy of Science, Washington, DC.

Altura, B.T., Altura, B.M., and Baez, S. (1975). Reactivity of aorta and portal vein in germfree rats. *Blood Vessels* **12**, 206–218.

Asano, T. (1967). Inorganic ions in cecal content of gnotobiotic rats. *Proc. Soc. Exp. Biol. Med.* **124**, 424–430.

Asano, T. (1969). Modification of cecal size in germfree rats by long-term feeding of anion exchange resin. *Am. J. Physiol.* **217**, 911–918.

Ashe, W.K., Scherp, H.W., and Fitzgerald, R.J. (1965). Previously unrecognized virus from submaxillary glands of gnotobiotic rats. *J. Bacteriol.* **90**, 1719–1729.

Ashman, R.B. (1975). A rapid method for skin grafting in germfree rats. *Lab. Anim. Sci.* **25**, 502–504.

Association for Gnotobiotics. (2004), *http://www.gnotobiotics.org.*

Baez, S. and Gordon, H.A. (1971). Tone and reactivity of vascular smooth muscle in germfree rat mesentery. *J. Exp. Med.* **134**, 846–856.

Balish, E., Yale, C.E., and Hong, R. (1972). Serum proteins of gnotobiotic rats. *Infect. Immun.* **6**, 112–118.

Banasaz, M., Alam, M., Norin, E., and Midtvedt, T. (2000). Gender, age and microbial status influence upon cell kinetics in a compartmentalised manner: an experimental study in germfree and conventional rats. *Microb. Ecol. Health Dis.* **12**, 208–218.

Bealmear, P.M., Loughman, B.E., Nordin, A.A., and Wilson, R. (1973). Evidence for graft vs. host reaction in the germfree allogeneic radiation chimera. *In* "Germfree Research" (J.B. Heneghan, ed.), pp. 471–475. Academic Press, New York.

Bornside, G.H., Donovan, W.E., and Myers, M.B. (1976). Intracolonic tensions of oxygen and carbon dioxide in germfree, conventional, and gnotobiotic rats. *Proc. Soc. Exp. Biol. Med.* **151**, 437–441.

Bruckner, G. (1997). How it started–and what is MAS? *In* "Gastrointestinal motility: 1995 Old Herborn University Seminar Monograph No. 9" (P.J. Heidt, V. Rusch, and D. van der Waijj, eds.), pp. 24–34. Herborn Litterae, Herborn-Dill, Germany.

Bruckner-Kardoss, E. and Wostmann, B.S. (1974). Blood volume of adult germfree and conventional rats. *Lab. Anim. Sci.* **24**, 633–635.

Burns, K.P., Timmons, E.H., and Poiley, S.M. (1971). Serum chemistry and hematological values for axenic (germfree) and environmentally associated inbred rats. *Lab. Anim. Sci.* **21**, 415–419.

Carlstedt-Duke, B., Midtvedt, T., Nord, C.E., and Gustafsson, B.E. (1986). Isolation and characterization of a mucin degrading strain of peptostreptococcus from rat intestinal tract. *Acta Pathol. Microbiol. Immunol. Scand. [B]* **94**, 292–300.

Carter, P.B. (1971). Host responses to normal intestinal microflora (Ph.D. thesis). University of Notre Dame, Notre Dame, IN.

Coates, M.E. (1973). Gnotobiotic animals in nutrition research. *Proc. Nutr. Soc.* **32**, 53–58.

Dajani, R.M., Gorrod, J.W., and Beckett, A.H. (1975). Reduction *in vivo* of (minus)-nicotine-l-N-oxide by germfree and conventional rats. *Biochem. Pharmacol.* **24**, 648–650.

Delluva, A.M., Markley, K., and Davies, R.E. (1968). The absence of gastric disease in germfree animals. *Biochim. Biophys. Acta* **151**, 646–650.

Demarne, Y., Flanzy, J., and Jacquet, E. (1973). The influence of gastrointestinal flora on digestive utilization of fatty acids in rats. *In* "Germfree Research" (J.B. Heneghan, ed.), pp. 553–560. Academic Press, New York.

Dieleman, L.A., Goerres, M.S., Arends, A., Sprengers, D., Torrice, C., Hoentjen, F., Grenther, W.B., and Sartor, R.B. (2003). Lactobacillus GG prevents recurrence of colitis in HLA-B27 transgenic rats after antibiotic treatment. *Gut* **52**, 370–376.

Donati, R.M., McLaughlin, M.M., and Stromberg, L.R. (1973). Combined surgical and radiation injury: the effect of the gnotobiotic state on wound closure. *Experientia* **29**, 1388–1390.

Dubos, R., Schaedler, R.W., Costello, R., and Hoet, P. (1965). Indigenous, normal, and autochthonous flora of the gastrointestinal tract. *J. Exp. Med.* **122**, 67–76.

Einarsson, K., Gustafsson, J.A., and Gustafsson, B.E. (1973). Differences between germfree and conventional rats in liver microsomal metabolism of steroids. *Biol. Chem.* **248**, 3623–3630.

Einarsson, K., Gustafsson, J.A., and Gustafsson, B.E. (1974). Liver microsomal hydroxylation of steroid hormones after establishing an indigenous microflora in germfree rats. *Proc. Soc. Exp. Biol. Med.* **145**, 48–52.

Elson, C.O., Sartor, R.B., Tennyson, G.S., and Riddell, R.H. (1995). Experimental models of inflammatory bowel disease. *Gastroenterology* **109**, 1344–67.

Foster, H.L. (1959a). Housing of disease-free vertebrates. *Ann. NY Acad. Sci.* **78**, 80–88.

Foster, H.L. (1959b). A procedure for obtaining nucleus stock for a pathogen-free animal colony. *Proc. Anim. Care* **9**, 135–142.

Foster, H.L. and Pfau, E.S. (1963). Gnotobiotic animal production at The Charles River Breeding Laboratories, Inc. *Lab. Anim. Care* **13**, 629–632.

Foster, H.L., Trexler, P.C., and Rumsey, G.L. (1967). A canned sterile shipping diet for small laboratory rodents. *Lab. Anim. Care* **17**, 400–405.

Ganaway, J.R., Allen, A.M., Moore, T.D., and Bohner, H.J. (1973). Natural infection of germfree rats with *Mycoplasma pulmonis. J. Infect. Dis.* **127**, 529–537.

Giddens Jr., W.E., Whitehair, C.K., and Carter, G.R. (1971). Morphological and microbiologic features of nasal cavity and middle ear in germfree, defined-flora, conventional, and chronic respiratory disease-affected rats. *Am. J. Vet. Res.* **32**, 99–114.

Gordon, H.A. (1974). Intestinal water absorption in young and old, germfree and conventional rats. *Experientia* **30**, 214–215.

Gordon, H.A., Bruckner-Kardoss, E., Staley, T.E., Wagner, M., and Wostmann, B.S. (1966). Characteristics of the germfree rat. *Acta Anat.* **64**, 367–389.

Gordon, H.A. and Pesti, L. (1971). The gnotobiotic animal as a tool in the study of host-microbial relationships. *Bacterial Rev.* **35**, 390–429.

Gordon, H.A. and Wostmann, B.S. (1959). Responses of the animal host to changes in the bacterial environment: Transition of the albino rat from germfree to the conventional state. *In* "Recent Progress in Microbiology" (G. Tunevau, ed.), pp. 336–339. Almqvist & Wiksell, Stockholm.

Gordon, H.A. and Wostmann, B.S. (1960). Morphological studies on the germfree albino rat. *Anat. Rec.* **137**, 65–70.

Green, R.M., Blackmore, D.K., and Drucker, D.B. (1973). The role of gnotobiotic animals in the study of dental caries. *Br. Dent. J.* **134**, 537–540.

Green, R.M., Drucker, D.B., and Blackmore, D.K. (1974). The reproducibility of experimental caries studies within and between two inbred strains of gnotobiotic rat. *Arch. Oral Biol.* **19**, 1049–1054.

Griffiths, L.A. and Barrow, A. (1972). Metabolism of flavonoid compounds in germfree rats. *Biochem. J.* **130**, 1161–1162.

Gustafsson, B. (1948). Germfree rearing of rats. *Acta Pathol. Microbiol. Scand.* **73(Suppl)**, 1–130.

Gustafsson, B.E. (1959). Vitamin K deficiency in germfree rats. *Ann. NY Acad. Sci.* **78**, 166–174.

Gustafsson, B.E., Cronholm, T., and Gustafsson, J.A. (1975). Relation of intestinal bile acids in rats to intestinal microflora and sex. *In* "Clinical and Experimental Gnotobiotics" (T.M. Fliedner, H. Heit, D. Neithammer, and H. Pflieger, eds.). Gustav Fischer, New York.

Gustafsson, B.E. and Norman, A. (1962). Urinary calculi in germfree rats. *J. Exp. Med.* **116**, 273–284.

Gustafsson, B.E. and Norman, A. (1969). Influence of the diet on the turnover of bile acids in germfree and conventional rats. *Br. J. Nutr.* **23**, 429–442.

Høverstad, T. and Midtvedt, T. (1986). Short-chain fatty acids in germfree mice and rats. *J. Nutr.* **116**, 1772–1776.

Hunt, R.D., King, N.W., and Foster, H.L. (1972). Encephalitozoonosis: evidence for vertical transmission. *J. Infect. Dis.* **126**, 212–224.

Ikari, N.S. and Donaldson, D.M. (1970). Serum beta-lysin and murami-dase levels in germfree and conventional rats. *Proc. Soc. Exp. Biol. Med.* **133,** 49–52.

Jacoby, R.O., Ball-Goodrich, L., Paturzo, F.X., and Johnson, E.A. (2001). Prevalence of rat virus infection in progeny of acutely or persistently infected pregnant rats. *Comp. Med.* **51,** 38–42.

Kajiwara, N., Ueno, Y., Takahashi, A., Sugiyama, F., Sugiyama, Y., and Yagami, K. (1996). Vertical transmission to embryo and fetus in maternal infection with rat virus. *Exp. Anim.* **45,** 239–244.

Kappel, H.K., Kappel, J.P., Weisbroth, S.H., and Kozma, C.K. (1969). Establishment of a hysterectomy-derived, pathogen-free breeding nucleus of Blu-(LE) rats. *Lab. Anim. Care* **19,** 738–741.

Kellogg, T.F. and Wostmann, B.S. (1969). Stock diet for colony production of germfree rats and mice. *Lab. Anim. Care* **19,** 812–814.

Laqueur, G.L., McDaniel, E.G., and Mateumoto, H. (1967). Tumor induction in germfree rats with methylazoxymethanol (MAM) and synthetic MAM acetate. *J. Natl. Cancer Inst.* **39,** 355–371.

Lee, Y.C., McKenzie, R.M., Gholson, R.K., and Raica, N. (1972). A comparative study of nicotinamide and nicotinic acid in normal and germfree rats. *Biochim. Biophys. Acta* **264,** 59–64.

Lepkovsky, S., Furuta, F., Ozone, K., and Koike, T. (1966). The proteases, amylase, and lipase of the pancreas and intestinal contents of germfree and conventional rats. *Br. J. Nutr.* **20,** 257–261.

Lev, M. (1963). Germfree animals. *In* "Animals for Research" (W.L. Lane-Petter, ed.), pp. 139–175. Academic Press, New York.

Luckey, T.D. (1963). "Germfree Life and Gnotobiology." Academic Press, New York.

Luckey, T.D. (1969). Gnotobiology and aerospace systems. *In* "Advances in Germfree Research and Gnotobiology" (M. Miyakawa and T.D. Luckey, eds.), pp. 317–353. Iliffe, London.

Madsen, D., Beaver, M., Chang, L., Bruckner-Kardoss, E., and Wostmann, B. (1976). Analysis of bile acids in conventional and germfree rats. *J. Lipid Res.* **17,** 107–111.

Martin, L.K., Einheber, A., Porro, R.F., Sadun, E.H., and Bauer, H. (1966). *Plasmodium berghei* infections in gnotobiotic mice and rats: parasitologic, immunologic, and histopathologic observations. *Mil. Med.* **131(Suppl),** 870–889.

McClellan, M.A., Hummel, R.P., and Alexander, J.W. (1974). Opsonic activity in germfree and monocontaminated rat sera. *Surg. Forum* **25,** 27–28.

McMaster, P.R., Aronson, S.B., and Bedford, M.J. (1967). Mechanisms of the host response in the eye, IV: the anterior eye in germfree animals. *Arch. Ophthalmol.* **77,** 392–399.

Maejima, K., Mitsuoka, T., Namioka, S., Nomura, T., Tajima, Y., and Yoshida, T. (1974). Bibliography of technology for germfree animal research. *Exp. Anim.* **24,** 229–253.

Midtvedt T. and Gustafsson, B.E. (1981). Microbial conversion of bilirubin to urobilins *in vitro* and *in vivo*. *Acta Pathol. Microbiol. Scand. [B]* **89,** 57–60.

Miyakawa, M. (1968). Studies of rearing germfree rats. In "Advances in Germfree Research and Gnotobiology" (M. Miyakawa and T.D. Luckey, eds.), pp. 48–62. Iliffe, London.

Miyakawa, M., Sumi, Y., Kanzaki, M., and Imaeda, F. (1975). Tumor induction by oral administration of A′-methyl-A′-nitro-/V-nitrosogua-nadine or aflatoxin Bl in germfree and conventional rats. In "Clinical and Experimental Gnotobiotics" (T.M. Fliedner, H. Heit, D. Neithammer, and H. Pflieger, eds.). Gustav Fischer, New York.

Miyakawa, M., Sumi, Y., Sakurai, K., Ukai, M., Hirabayashi, N., and Ito, G. (1969). Serum gammaglobulin and lymphoid tissue in the germfree rats. *Acta Haematologica Japonica* **32,** 501–518.

Morioka, T., Saji, S., Inoue, M., and Matsumura, T. (1969). Peroxidase-mediated antibacterial activity in the salivary gland of germfree and conventional rats. *Arch. Oral Biol.* **14,** 549–553.

Nagler, A.L., Seifter, E., Geever, E.F., Dettbarn, W.D., and Levenson, S.M. (1969). The nephropathy of acute choline deficiency in germfree conventional, and open animal room rats. *In* "Germfree Biology" (E.A. Mirand and N. Back, eds.), pp. 317–324. Plenum, New York.

Nolen, G.A. and Alexander, J.C. (1965). A comparison of the growth and fat utilization of caesarean-derived and conventional albino rats. *Lab. Anim. Care* **15,** 295–303.

Norin, K.E., Gustafsson, B.E., and Midtvedt, T. (1986). Strain differences in fecal tryptic activity of germfree and conventional rats. *Lab Anim.* **20,** 67–69.

Nuttal, G.H.F. and Thierfelder, H. (1895). Thierisches Leben ohne Bakterien im Verdauungskanal. *Hoppe-Seyler's Zeitschrift Physiologische Chemie* **21,** 109–121.

Nuttal, G.H.F. and Thierfelder, H. (1896). Thierisches Leben ohne Bakterien im Verdauungskanal, II: Mittheilung). *Hoppe-Seyler's Zeitschrift Physiologische Chemie* **22,** 62–73.

Nuttal, G.H.F. and Thierfelder, H. (1897). Thierisches Leben ohne Bakterien im Verdauungskanal, III: Mittheilung; Versuche an Hïhnern. *Hoppe-Seyler's Zeitschrift Physiologische Chemie* **23,** 231–235.

Oace, S.M. (1972). A purified soy protein diet for nutrition studies with germfree rats. *Lab. Anim. Sci.* **22,** 528–531.

Paterson, J.S. and Cook, R. (1971). Utilization of diets sterilized by gamma irradiation for germfree and specific-pathogen-free laboratory animals. *In* "Defining the Laboratory Animal" (H.A. Schneider, ed.), pp. 586–596. National Academy of Science, Washington, DC.

Pittermann, W. and Deerberg, F. (1975). Spontaneous tumors and lesions of the lung, kidney and gingiva in aged germfree rats. *In* "Clinical and Experimental Gnotobiotics" (T.M. Fliedner, H. Heit, D. Neithammer, and H. Pflieger, eds.). Gustav Fischer, New York.

Pleasants, J.R. (1959). Rearing germfree caesarian-born rats, mice, and rabbits through weaning. *Ann. NY Acad. Sci.* **78,** 116–126.

Pleasants, J.R. (1974). Gnotobiotics. *In* "Handbook of Laboratory Animal Science," Vol. I (E.C. Melby Jr. and N.H. Altman, eds.), pp. 119–174. CRC Press, Cleveland, Ohio.

Pollard, M. (1966). Leukemia in germfree rats. *Proc. Soc. Exp. Biol. Med.* **121,** 585–589.

Pollard, M. (1971a). The germfree rat. *Pathobiol. Annu.* **1,** 83–94.

Pollard, M. (1971b). Senescence in germfree rats. *Gerontologia* **17,** 333–338.

Pollard, M. (1972). Carcinogenesis in germfree animals. *Prog. Immunobiol. Stand.* **5,** 226–230.

Pollard, M. (1973). Spontaneous prostate adenocarcinomas in aged germfree Wistar rats. *J. Natl. Cancer Inst.* **51,** 1235–1241.

Pollard, M. (1977). Animal model of human disease: metastatic adenocarcinoma of the prostate. *Am. J. Pathol.* **86,** 277–80.

Pollard, M. (1992). The Lobund-Wistar rat model of prostate cancer. *J. Cell Biochem.* **16H,** 84–88.

Pollard, M., and Kajima, M. (1970). Lesions in aged germfree Wistar rats. *Am. J. Pathol.* **61,** 25–36.

Pollard, M., Kajima, J., and Lorans, G. (1968). Tissue changes in germfree rats with primary tumors. *Res. J. Reticuloendothel. Soc.* **5,** 147–160.

Pollard, M. and Luckert, P.H. (1987). Autochthonous prostate adeno-carcinomas in Lobund-Wistar rats: a model system. *Prostate* **11,** 219–227.

Pollard, M, Luckert, P.H., and Meshorer, A. (1985). Xenogeneic bone marrow chimerism in germfree rats. *Prog. Clin. Biol. Res.* **181,** 447–450.

Pollard, M. and Teah, B.A. (1963). Spontaneous tumors in germ-free rats. *J. Natl. Cancer Inst.* **31,** 457–465.

Pollard, M. and Wolter, W. (2000). Prevention of spontaneous prostate-related cancer in Lobund-Wistar rats by a soy protein isolate/isoflavone diet. *Prostate* **45,** 101–105.

Rath, H.C., Herfarth, H.H., Ikeda, J.S., Grenther, W.B., Hamm Jr., T.E., Balish, E., Taurog, J.D., Hammer, R.E., Wilson, K.H., and Sartor, R.B. (1996). Normal luminal bacteria, especially *Bacteroides* species, mediate chronic colitis, gastritis, and arthritis in HLA-B27/human β_2 microglobulin transgenic rats. *J. Clin. Invest.* **98**, 945–953.

Rath, H.C., Schultz, M., Freitag, R., Dieleman, L.A., Li, F., Linde, H.J., Scholmerich, J., and Sartor, R.B. (2001). Different subsets of enteric bacteria induce and perpetuate experimental colitis in rats and mice. *Infect. Immun.* **69**, 2277–2285.

Rath H.C., Wilson K.H., and Sartor R.B. (1999). Differential induction of colitis and gastritis in HLA-B27 transgenic rats selectively colonized with *Bacteroides vulgatus* or *Escherichia coli*. *Infect. Immun.* **67**, 2969–2974.

Reddy, B.S., Pleasants, J.R., and Wostmann, B.S. (1969a). Pancreatic enzymes in germfree and conventional rats fed chemically defined, water-soluble diet free from natural substrates. *J. Nutr.* **97**, 327–334.

Reddy, B.S., Pleasants, J.R., and Wostmann, B.S. (1973). Metabolic enzymes in liver and kidney of the germfree rat. *Biochim. Biophys. Acta* **320**, 1–8.

Reddy, B.S. and Wostmann, B.S. (1966). Intestinal disaccharidase activities in the growing germfree and conventional rat. *Arch. Biochem. Biophys* **113**, 609–616.

Reddy, B.S., Wostmann, B.S., and Pleasants, J.R. (1969b). Protein metabolism in germfree rats fed chemically defined, water-soluble diet and semi-synthetic diet. *In* "Germfree Biology" (E.A. Mirand and N. Back, eds.), pp. 301–305. Plenum, New York.

Reid Jr., L.C. and Gates Jr., A.S. (1966). A method of sterilizing supplies for germfree isolators in plastic bags. *Lab. Anim. Care* **16**, 246–254.

Reina-Guerra, M., Tennant, B., Harrold, D., and Goldman, M. (1969). The absorption of fat by germfree and conventional rats. *In* "Germfree Biology" (E.A. Mirand and N. Back, eds.), pp. 297–300. Plenum, New York.

Reyniers, J.A., Trexler, P.C., and Ervin, R.F. (1946). "Rearing Germfree Albino Rats: Lobund Report 1." University of Notre Dame Press, Notre Dame, IN.

Reyniers, J.A., Trexler, P.C., Scruggs, W., Wagner, M., and Gordon, H.A. (1956). Observations on germfree and conventional albino rats after total-body x-irradiation. *Radiat. Res.* **5**, 591.

Rodkey, F.L., Collison, H.A., and O'Neal, J.D. (1972). Carbon monoxide and methane production in rats, guinea pigs, and germfree rats. *J. Appl. Physiol.* **33**, 256–260.

Rogers, T. and Balish, E. (1976). Experimental *Candida* infection in conventional mice and germfree rats. *Infect. Immun.* **14**, 33–38.

Rogers Jr., W.E., Bieri, J.G., and McDaniel, E.G. (1971). Vitamin A deficiency in the germfree state. *Fed. Proc. Fed. Am. Soc. Exp. Biol.* **30**, 1773–1778.

Sacksteder, M., Kasza, L., Palmer, J., and Warren, J. (1973). Cell transformation in germfree Fischer rats. *In* "Germfree Research" (J.B. Heneghan, ed.), pp. 153–157. Academic Press, New York.

Sacquet, E., Lachkar, M., Mathis, C., and Raibaud, P. (1973). Cecal reduction in "gnotoxenic" rats. *In* "Germfree Research" (J.B. Heneghan, ed.), pp. 545–552. Academic Press, New York.

Sartor, R.B. (2000). Colitis in HLA-B27/β_2 microglobulin transgenic rats. *Int. Rev. Immunol.* **19**, 39–50.

Savage, D.C. (1969). Microbial interference between indigenous yeast and lactobacilli in the rodent stomach. *J. Bacteriol.* **98**, 1278.

Schaedler, R.W., Dubos, R., and Costello, R. (1965a). Association of germfree mice with bacteria isolated from normal mice. *J. Exp. Med.* **122**, 77–82.

Schaedler, R.W., Dubos, R., and Costello, R. (1965b). The development of the bacterial flora in the gastrointestinal tract of mice. *J. Exp. Med.* **122**, 59–66.

Schultz, K.D., Appel, K.R., Goeth, H., and Wilk, W. (1974). Experiments to establish a rat stock free of mycoplasma. *Z. Versuchstierk.* **16**, 105–112.

Schwartz, B.F. (1975). Pulmonary gas exchange in germfree and conventional rats. *In* "Clinical and Experimental Gnotobiotics"

(T.M. Fliedner, H. Heit, D. Neithammer, and H. Pflieger, eds.). Gustav Fischer, New York.

Sewell, D.L., Bruckner-Kardoss, E., Lorenz, L.M., and Wostmann, B.S. (1976). Glucose tolerance, insulin and catecholamine levels in germfree rats. *Proc. Soc. Exp. Biol. Med.* **152**, 16–19.

Sewell, D.L. and Wostmann, B.S. (1975). Thyroid function and related hepatic enzymes in the germfree rat. *Metab. Clin. Exp.* **24**, 695–701.

Sewell, D.L., Wostmann, B.S., Gairola, C., and Aleem, M.I.H. (1975). Oxidative energy metabolism in germfree and conventional rat liver mitochondria. *Am. J. Physiol.* **228**, 526–529.

Shih, C.N., Balish, E., Lower Jr., G.M., Yale, C.E., and Bryan, G.T. (1975). Induction of colonic tumors in germfree and conventional rats. *In* "Clinical and Experimental Gnotobiotics" (T.M. Fliedner, H. Heit, D. Neithammer, and H. Pflieger, eds.). Gustav Fischer, New York.

Smith, H.W. (1965). The development of the flora of the alimentary tract in young animals. *J. Pathol. Bacterial.* **90**, 495–513.

Smith Jr., J.C., McDaniel, E.G., and Doft, F.S. (1973). Urinary calculi in germfree rats alleviated by varying the dietary minerals. *In* "Germfree Research" (J.B. Heneghan, ed.), pp. 285–290. Academic Press, New York.

Snyder, D.L., Pollard, M., Wostmann, B.S., and Luckert, P. (1990). Life span, morphology, and pathology of diet-restricted germ-free and conventional Lobund-Wistar rats. *J. Gerontol.* **45**, B52–B58.

Sugiyama, T. and Bruckner, G.G. (1975). Mycoplasma-bacteria relationships in the pathogenesis of chronic respiratory disease in conventional rats. *In* "Clinical and Experimental Gnotobiotics" (T.M. Fliedner, H. Heit, D. Neithammer, and H. Pflieger, eds.). Gustav Fischer, New York.

Syed, S.A., Abrams, G.D., and Freter, R. (1970). Efficiency of various intestinal bacteria in assuming normal function of enteric flora after association with germfree mice. *Infect. Immun.* **2**, 376–386.

Taurog, J.D., Richardson, J.A., Croft, J.T., Simmons, W.A., Zhou, M., Fernandez-Sueiro, J.L., Balish, E., and Hammer, R.E. (1994). The germfree state prevents development of gut and joint inflammatory disease in HLA-B27 transgenic rats. *J. Exp. Med.* **180**, 2359–2364.

Thompson, G.R. and Trexler, P.C. (1971). Gastrointestinal structure and function in germfree or gnotobiotic animals. *Gut* **12**, 230–235.

Trexler, P.C. (1959). The use of plastic in the design of isolator systems. *Ann. NY Acad. Sci.* **78**, 29–36.

Trexler, P.C. and Reynold, L.I. (1957). Flexible film apparatus for the rearing and use of germfree animals. *Appl. Microbiol.* **5**, 6.

Trippestad, A. and Midvedt, T. (1971). The phagocytic activity of polymorphonuclear leucocytes from germfree and conventional rats. *Acta Pathol. Microbiol. Scand. [B]* **79**, 519–522.

Tuffery, A.A. and Innes, J.R.M. (1963). Diseases of laboratory mice and rats. *In* "Animals for Research" (W. Lane-Petter, ed.), pp. 48–109. Academic Press, New York.

Ukai, M., Tomura, A., and Ito, M. (1976). Cholesterol synthesis in germfree and conventional rats. *J. Nutr.* **106**, 1175–1183.

van der Waaij, D., Berghus-de Vries, J.M., and Lekkerkerk van der Wees, J.E.C. (1971). Colonization resistance of the digestive tract in conventional and antibiotic-treated mice. *J. Hyg.* **69**, 405–411.

Wade, A.C., Luckert, P.H., Tazume, S., Niedbalski, J.L., and Pollard, M. (1987). Characterization of xenogeneic mouse to rat bone marrow chimeras. *Transplantation* **44**, 88–92.

Wagner, M. (1959). Determination of germfree status. *Ann. NY Acad. Sci.* **78**, 89–102.

Walburg Jr., H.E. (1973). Carcinogenesis in gnotobiotic rodents. *In* "Germfree Research" (J.B. Heneghan, ed.), pp. 115–122. Academic Press, New York.

Warner, T.F., Madsen, J., Starling, J., Wagner, R.D., Taurog, J.D., and Balish, E. (1996). Human HLA-B27 gene enhances susceptibility of rats to oral infection by *Listeria monocytogenes*. *Am. J. Pathol.* **149**, 1737–1743.

Weisburger, J.H., Reddy, B.S. Narisawa, T., and Wynder, E. L. (1975). Germfree status and colon tumor induction by N-methyl-N-nitro-N-nitrosoguanidine. *Proc. Soc. Exp. Biol. Med.* **148,** 1119–1121.

Wostmann, B.S., ed. (1970). "Gnotobiotes: Standards and Guidelines for the Breeding, Care and Management of Laboratory Animals." National Research Council, National Academy of Science, Washington, DC.

Wostmann, B.S. (1973). Intestinal bile acids and cholesterol absorption in the germfree rat. *J. Nutr.* **103,** 982–990.

Wostmann, B.S. (1975). Nutrition and metabolisrn of the germfree mammal. *World Rev. Nutr. Diet* **22,** 40–92.

Wostmann, B.S. (1996). "Germfree and gnotobiotic animal models: Background and applications." CRC Press, Boca Raton, FL.

Wostmann, B.S., Beaver, M., and Modsen, D. (1975). Effect of diet sterilization on cholesterol and bile acid values of germfree rats. *In* "Clinical and Experimental Gnotobiotics" (T.M. Fliedner, H. Heit, D. Neithammer, and H. Pflieger, eds.). Gustav Fischer, New York.

Wostmann, B.S. and Bruckner-Kardoss, E. (1959). Development of cecal distention in germfree baby rats. *Am. J. Physiol.* **197,** 1345–1346.

Wostmann, B.S., Bruckner-Kardoss, E., and Knight Jr., P.L. (1968). Cecal enlargment, cardiac output, and O_2 consumption in germfree rats. *Proc. Soc. Exp. Biol. Med.* **128,** 137–1141.

Wostmann, B.S. and Kellogg, T.F. (1967). Purified starch-casein diet for nutritional research with germfree rats. *Lab. Anim. Care* **17,** 589–593.

Wostmann, B.S. and Knight, P.L. (1965). Antagonism between vitamins A and K in the germfree rat. *J. Nutr.* **87,** 155–160.

Wosu, N.J., Olsen, R., Shadduck, J.A., Koestner, A., and Pakes, S.P. (1977a). Diagnosis of experimental encephalitozoonosis in rabbits by complement fixation. *J. Infect. Dis.* **135,** 944–948.

Wosu, N.J., Shadduck, J.A., Pakes, S.P., Frenkel, J.K., Todd Jr., K.S., and Conroy, J.D. (1977b). Diagnosis of encephalitozoonosis in experimentally infected rabbits by intradermal and immunofluorescence test. *Lab. Anim. Sci.* **27,** 210–216.

Yale, C.E. and Linsley, J.G. (1970). A large efficient isolator holding germfree rats. *Lab. Anim. Care* **20,** 749–755.

Yamada, K. and Ukai, M. (1976). The histochemistry of mucosaccharides in some organs of germfree rats. *Histochemistry* **47,** 219–238.

Chapter 23

Spontaneous, Surgically and Chemically Induced Models of Disease

Dwight R. Owens

THE LABORATORY RAT, 2ND EDITION

I. INTRODUCTION

Historically, the rat has been a nuisance, curse, and bane to man. It has been known to be a carrier of pathogenic microorganisms and pests in the destruction of the world's food supply, as well as being involved in the spread of contagious diseases. However, the rat has become a useful tool in biomedical research, both for understanding disease processes and for developing and testing new drugs. The contributions of the rat to our basic knowledge and understanding of human health is substantial.

The use of the rat for the study of disease and for basic research dates back to the work done at the Wistar Institute in 1906 by Donaldson (Lindsay, 1979). Since then, rats have been used to study nutrition, physiology, and pathobiology of disease states, as well as the aging process. As a result, there is a wealth of knowledge available on the physiology of the rat (Jacob et al., 1995; James and Lindpaintner, 1997).

At present, more research is performed in the mouse than in the rat. This is due in part to the availability of reagents for the mouse and to its smaller size; as a result, the mouse requires less housing space and reduced amounts of study reagents the cost is lower. In addition, gene knockout lines for rats were not developed until Zan and coworkers (2003) reported the creation of two related gene knockout strains of rat for the breast tumor suppressor genes *Brca1* and *Brca2*. Zan et al. used the rat because tumors in the rat have a spectrum of hormonal responses similar to that of tumors in humans. Another deterrent to the use of the rat in research was the slow development of the rat gene map and the lack of polymorphic genetic markers. However, with the development of newer rat disease models and the emerging science of proteomics, the future for the rat as a model in biomedical research is promising. Furthermore, the recent sequencing of the Brown Norway genome will yield many insights into genes associated with disease, and almost all human genes associated with disease have orthologs in the rat (Rat Genome Sequencing Project Consortium, 2004).

Models of disease can be developed by (1) spontaneous heritable mutations that result in alterations of normal structures or functions; (2) alterations of normal functions by the administration of chemicals (pharmacologically induced model); (3) surgical manipulations in which functions or organs are altered; (4) genetic modification by molecular biological techniques (transgenic and knockouts) or by specific breeding methods to yield consomic (movement of one chromosome from one rat strain to another) or congenic (movement of one gene or region in the chromosome) animals; and (5) induced mutation, typically after administration of chemical mutagens.

This chapter is limited to models resulting from naturally occurring mutations, pharmacological induction, and surgical manipulation. Only examples of commonly used rat models will be discussed, as this chapter provides a general overview; a thorough review of all rat models is beyond the scope of a single chapter.

II. MODELS OF METABOLIC DISEASE

Recent work has shown the importance of the metabolic cluster of disease syndromes (insulin resistance, glucose intolerance, hyperinsulinemia, dyslipidemia, obesity and hypertension) as strong predictors of obesity-related complications, including type 2 diabetes, renal failure, and cardiovascular disease. It has been shown that resistance to insulin-stimulated glucose uptake is associated with a series of related metabolic variations termed metabolic "syndrome X," which cluster in the same individual animal and include glucose intolerance, disturbed plasma lipids, and high blood pressure (Reaven, 1988).

Obesity is a common disorder that is rapidly increasing worldwide. It is estimated that 60% of the population of the United States is overweight or obese, and the number of individuals with this disorder is increasing. Obesity appears to amplify the underlying problem of insulin resistance, leading this metabolic cluster of disorders into a clinical syndrome, which in turn can lead to type 2 diabetes, hypertension, arteriosclerosis, renal failure, strokes, and coronary heart disease. In this regard, the following models have been used for the study of the metabolic syndrome and obesity-related conditions.

A. Obesity

1. Genetic Obesity

ZUCKER OBESE (ZUCKER FATTY) RAT. The Zucker obese rat is outbred and multicolored, with four principal coat colors: (1) predominantly brown, (2) brown and white, (3) predominantly black, and (4) black and white. The first rat model of genetic obesity was described by Zucker and Zucker (1961, 1962, 1963; Zucker and Antoniades, 1972). The Zucker rat was developed from crosses between animals from the Sherman strain and the Merck stock 13 M strain (Kava et al., 1990). The leptin mutation (called "fa") that occurred in this strain of rat is a recessive trait and causes the rat's obesity. The mutation was determined to be a shortened leptin-receptor protein resulting from a single nucleotide substitution at position 880 of the leptin-receptor gene (Chua et al., 1996).

The outbred Zucker fatty rat (sometimes called the Zucker fa/fa) is probably the best known and most widely used model of the genetic obesity trait (Kava et al., 1990). This obese rat model is characterized by hyperlipidemia, hypercholesterolemia, and hyperinsulinemia and develops adipocyte hypertrophy and hyperplasia. It has been a valuable contributor to the study of early-onset hyperplastic-hypertrophic obesity.

Although the obese Zucker rat is hyperphagic compared with its lean littermates, the hyperphagia is not necessary for expression of the obesity syndrome. A study by Greenwood et al. (1982) showed that jejunoileal bypass surgery, a treatment that produces both decreased food intake and malabsorption of ingested nutrients, resulted in smaller, lighter Zucker fa/fa rats, although the obese body composition did not normalize. Some data suggest that the obese Zucker has a profoundly abnormal brain neuropeptide physiology and that this abnormality may show a fa gene-dose effect (Baskin et al., 1985). Insulin infusion into the third ventricle of the brain reduced food intake and body weight in the heterozygote +Zucker +/fa rat, but not in the homozygous obese Zucker fa/fa rat, suggesting that the genetically obese Zucker rats have reduced sensitivity to insulin in the central nervous system (Ikeda et al., 1986).

Although the outbred Zucker fa/fa model does not develop type 2 diabetes, it shows some variation in the occurrence and level of fasting hyperglycemia and glucose intolerance. Furthermore, this rat is hyperinsulinemic, hypertriglyceridemic, and non-hypertensive, making it useful for studies of the pre-diabetic state.

SPONTANEOUSLY HYPERTENSIVE OBESE RAT (KOLETSKY RAT). The spontaneously hypertensive obese (SHROB) rat is an inbred albino developed from a mating between a spontaneously hypertensive rat (SHR)/N female and a male Sprague-Dawley rat (Koletsky, 1973). After inbreeding several generations in the progeny, which developed hypertension, a new obese phenotype was noted. This new phenotype was unique because it not only expressed genetic obesity but also demonstrated hypertriglyceridemia, spontaneous hypertension, hyperinsulinemia, hyperlipidemia, proteinuria and renal disease (Koletsky, 1973; Koletsky et al., 2001).

The obese phenotype results from a single recessive trait, aTyr763 stop nonsense mutation, which leads to the absence of a leptin receptor (Takaya et al., 1996) and is designated fa^k, cp, or fa^{cp} (Yen et al., 1977). This mutation in the SHROB rat is located on the same allele as the fa in the outbred Zucker fa/fa rat. SHROB rats and their lean (Kol) littermates develop hypertension spontaneously, and the systolic blood pressure (SBP) reaches a hypertensive level > (more than 150 mm Hg) when the rat is about 3 months old, and the pressure increasing with age. Once hypertensive, the SHROB rat will maintain its hypertensive state until shortly before death (Koletsky et al., 2001). SHRKol rats have slightly lower blood pressures than do lean SHR littermates (Koletsky et al., 2001). Male and female rats have similar blood pressure levels.

The SHROB rat is useful for studies of metabolic disease because it has hyperinsulinemia and glucose intolerance, but with a fasting euglycemia. In addition, it is hypertensive and develops proteinuria at about 6 weeks of age, which is maintained as the animal develops renal disease. The renal disease in the SHROB rat is characterized by focal segmental glomerulosclerosis and nephrosclerosis, which resembles the disease found in the human diabetic (Koletsky et al., 2001). These characteristics make the SHROB rat a very useful model for the study of human metabolic "syndrome X". The absence of a fasting hyperglycemia would seem to suggest that additional factors are needed for the development of that characteristic as well as type 2 diabetes (Koletsky et al., 2001).

2. Nutritional Obesity

DIETARY-INDUCED OBESITY PRONE RAT. This dietary-induced obesity (DIO) prone rat model of obesity is a outbred, albino, and polygenic model that has been selectively bred to develop obesity associated with impaired glucose tolerance, dyslipidemia, and insulin resistance when fed a high-fat, high-sucrose, and high-caloric diet (Levin et al., 1997). The rat is a useful model for the study of obesity because the obesity is expressed only when the rats are fed a diet moderately high in energy and fat content, allowing control of the obesity rate and leptin production. Obesity in the DIO rat shares many characteristics of human obesity conditions, including polygenic inheritance, insulin resistance, reduced growth hormone secretion, and a propensity to oxidize carbohydrate preferentially over fat (Levin and Dunn-Meynell, 2000). Similar to many obese humans, DIO rats reduce their resting metabolic rate when calorically restricted and return to their previously high body weight when restriction is discontinued (Levin and Dunn-Meynell, 2000).

The DIO rat was developed by the selective breeding of a group of outbred CD (Sprague-Dawley rats). CD rats develop a bimodal obese and lean populations when fed a high energy diet. After feeding the rats a high-fat, high-energy diet (Research Diets D 122266B; Research Diets, New Brunswick, NJ) based on condensed milk, bimodal groups of animals were selected for further breeding to develop the DIO and diet-resistant lines. The DIO line was selected for high weight gain; the diet-resistant line, for low weight gain (Levin et al., 1997).

In weight-matched studies using regular laboratory diet (4% fat), DIO rats had 44% greater carcass fat than did diet-resistant rats having similar energy intake and feed efficiency. The basic insulin level of the DIO rats was 70% higher and blood glucose 14% greater than levels for the diet-resistant rats. It was further noted that DIO rats ate 25% more food compared with the intake of diet-resistant animals and gained 115% more body weight (Levin et al., 2000; Levin and Dunn-Meynell, 2002).

The DIO rat is a useful model for studies of non-leptin-deficient obesity and for investigations of glucose tolerance and insulin resistance, "metabolic syndrome X". The DIO rat develops significant obesity with normal levels of leptin, whereas most other models of obesity are leptin deficient. In addition, the DIO rat can be used to study the development of hypertension in the obese state; development of vascular and renal changes are common. In this regard, the DIO rat closely resembles obesity of humans (Dobrian et al., 2000).

OSBORNE-MENDEL RAT. The inbred, albino Osborne-Mendel (OM) rat is a model of diet-induced obesity, becoming obese on a high-fat (36% fat) diet (Schemmel et al., 1970). A high incidence of spontaneous tumors has been noted in the this rat, including thyroid carcinoma (33%), adrenal cortical tumors (94%), and pituitary tumors (8% to 18%) at 18 months of age (Lindsey et al., 1968; Hansen et al., 1973).

In addition to obesity, when fed a high-fat diet, the OM rat demonstrates several metabolic abnormalities, including insulin resistance and decreased expression of hypothalamic leptin receptor (Madiehe et al., 2000). This rat model is hyperinsulinemic but does not develop as a bimodal population distribution, as does the DIO rat. The elevated insulin levels observed in the obese rat are thought to be necessary to maintain normal glucose levels in the presence of impaired insulin action (Hall, 1994). Furthermore, the OM rat develops some of the cardiovascular and hormonal changes associated with human obesity, including cardiac and renal hypertrophy, a slight increase in mean arterial pressure, and hyperinsulinemia.

B. Diabetes

Diabetes is an important, worldwide disease with an increasing incidence. More than 17 million people in the United States currently have diabetes, and more than 200,000 will die each year due to the disease and its related complications. In adults, the incidence of type 2 diabetic disease has increased 49% since 1990 (Koplan, 2002).

There are two significant types of diabetic disease: type 1 and type 2. Individuals with type 1 disease have reduced or no insulin production by pancreatic beta cells. Type 1 diabetes accounts for 5% to 10% of all diabetic disease. Type 2 diabetes accounts for 90% to 95% of all diabetic disease and most often appears in individuals older than 40 years. The disease has occurred increasingly in younger individuals and is linked to diet, obesity, and physical inactivity. Type 2 diabetes is caused by decreased insulin production or by the inability of the body to utilize the insulin that is produced.

1. Spontaneous Models of Diabetes

a. MODEL OF TYPE 1 DIABETES MELLITUS. The inbred, albino Biomedical Research Models (BBDP) rat is a model of spontaneous autoimmune, insulin-dependent diabetes mellitus. A number of observations suggest that the condition in humans is the result of autoimmune disease; these include the presence of pancreatic insulitis, islet autoantibodies, reappearance of disease after syngeneic pancreatic allografts, induction of disease in bone marrow allograft recipients, successful immunoprophylaxis with cyclosporine, and association with the major histocompatibility complex. Analogous evidence suggests that the hyperglycemic syndrome found in the BBDP rat is a similar disorder (Mordes et al., 2001).

Inheritance of the disease in the BBDP rat is polygenic and depends on several loci. The model was first recognized in 1974 in a colony of outbred Wistar rats at the Bio-Breeding Laboratories in Ottawa, Ontario, Canada. In 1977 a program of inbreeding of the these rats was begun at the University of Massachusetts Medical Center at Worcester, Massachusetts, with 300 breeder rats purchased from the Bio-Breeding Laboratories (Nakhooda et al., 1977). The inbred line developed at the University of Massachusetts is the BBDP/Wor, and clinical, metabolic and pathological features of the model have been extensively characterized (Nakhooda et al., 1978; Guberski, 1993).

The BBDP/Wor rat is characterized by plasma glucose levels of 250 to 750 mg/dL and hypoinsulinemia, with insulin levels of less than 1 ng/mL (Nakhooda et al., 1978). Development of diabetes in the inbred line occurs between 60 and 120 days of age in both the male and female. Onset of the disease is characterized by weight loss, polyuria, polydipsia, glycosuria, hyperglycemia, hypoinsulinemia, and ketosis (Guberski, 1993). Animals that develop diabetes must be given exogenous insulin, as the animals will die within 4 to 7 days after onset of the disease without the administered insulin.

The BBDP/Wor rat is considered to be one of the best models of spontaneous autoimmune diabetes (MacMurray et al., 2002). Disease is characterized by neuropathy (Greene et al., 1984; Yagihashi and Sima 1985a,b, 1986), nephropathy (Chakrabarti et al., 1989), and retinal changes (Sima et al., 1985; Chakrabarti et al., 1991).

The inbred BBDP/Wor as well as its control, the BBDR/Wor as well as other inbred lines, are available from Biomedical Research Models, Inc., Worcester, MA. There are a number of lines and sublines available throughout the research community, but it should be noted that there are genotypic and phenotypic variation differences between the lines. When results from different studies are compared, it is important to identify the source of the BBPR rats used in the study.

b. MODELS OF TYPE 2 DIABETES MELLITUS.

i. Zucker Diabetic Fatty Rat. The inbred, hooded, black and white, Zucker diabetic fatty (ZDF) rat has become the primary rat model for the study of type 2 diabetes. In 1977, the first diabetic ZDF were observed in a colony of Zucker rats belonging to Dr. Walter Shaw of Eli Lilly and Company (Indianapolis, IN). The Shaw colony was subsequently transferred to Dr. Julia Clark's laboratory at Indiana University School of Medicine. In 1985, after several years of trying to perpetuate the disease, the colony was turned over to Dr. Richard Peterson, also of Indiana University School of Medicine (Peterson, 2001). Under his direction, animals were selected from the Clark colony, inbred, caesarean derived, and developed into a production colony in the late 1980s (Peterson, 1990).

The ZDF rat also carries the same recessive fa/fa mutation as does the outbred Zucker for the leptin membrane receptor, which was described by Zucker and Zucker in 1961. The result of the mutation is a shortened leptin-receptor protein that does not interact with leptin effectively (Chua et al., 1966; White et al., 1997).

The ZDF rat is characterized by development of diabetes between 7 and 12 weeks of age. As shown in Figure 23-1, non-fasting plasma glucose levels of 200 to 400 mg/dL are seen at 8 weeks, and levels of 400 to 600 mg/dL are seen at 12 weeks in the obese male rat fed a diet of Purina Rodent LabDiet 5008 (PMI Nutrition, Richmond, IN).

The female ZDF rat does not develop diabetes on the same diet (Purina Rodent LabDiet 5008) that induces diabetes in the male. Instead, a diet (Research Diets 13004; Research Diets, New Brunswick, NJ) that is higher in fat and lower in protein is necessary to induce the disease.

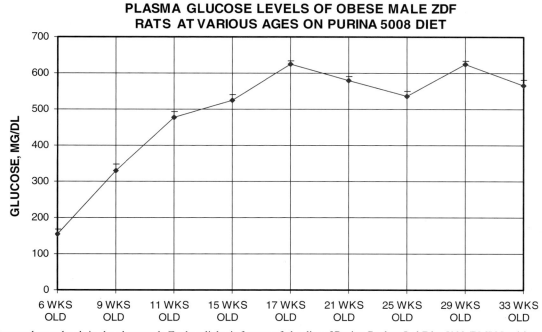

Fig. 23-1 Plasma glucose levels in the obese male Zucker diabetic fatty rat fed a diet of Purina Rodent LabDiet 5008 (PMI Nutrition, Richmond, IN).

The disease in the female becomes a good model for studying type 2 diabetes, because the diabetes induced by the high diet can be compared with controls of the same sex and strain maintained on the Purina 5008 diet. Approximately 4 weeks are required for the female obese ZDF to develop diabetes when placed on the high-fat diet, but once diabetes has developed and the obese female rat has been on the diet for 6 weeks, the diabetes is not reversible (Corsetti et al., 2000). The obese ZDF female is a useful model because of the prolonged period of time needed to develop diabetes; this translates into a longer period of insulin sensitivity, during which testing may be conducted. The female maintains a higher level of insulin (3 to 4 ng/mL) once the disease has developed than does the male (0.5 to 1 ng/mL).

The diabetic ZDF rat develops symptoms of type 2 diabetes that are characterized by (1) hyperglycemia that develops between 7 and 12 weeks of age, (2) early hyperinsulinenemia that begins to diminish as the beta-cells in the pancreas are destroyed, (3) fasting hyperglycemia, (4) abnormal glucose tolerance, (5) increased plasma triglycerides, and (6) a mild hypertension. The ZDF model has proven to be useful in studying type 2 diabetes as well as the influence of diabetes on wound healing, periodontal disease, and neuropathy (Peterson, 2001). This model is prone to hydronephrosis, which may interfere with renal studies.

ii. Cohen Diabetic Rat. The inbred, albino Cohen diabetic rat is a model of diet-induced type 2 diabetes that reproduces many of the features of the disease in humans (Cohen et al., 1993). This model has the following distinctive features: (1) expression of genetic susceptibility (either sensitive or resistant) to a carbohydrate-rich diet that is not present in other type 2 models of diabetes; (2) availability of two contrasting strains originating from the same parent strain; and (3) a nonobese model of diabetes that allows dissociation of the confounding obesity factor from other diabetogenic genes (Weksler-Zangen et al., 2001).

Although first described in 1972, the Cohen diabetic model has only been studied on a limited basis, primarily because it had never been systematically characterized in terms of phenotype or genotype (Welsler-Zangen et al., 2001). The Cohen model has a milder, more slowly developing hyperglycemia than is found in the male ZDF model. In contrast to the male ZDF rat fed a Purina 5008 diet, which has an average non-fasting glucose level of 450 mg/dL at 12 weeks of age, the Cohen model manifests glucose values of approximately 400 mg/dL at 6 months of age. The Cohen model has been shown to be hyperinsulinemic at a young age, and animals that were 6 months of age were unable to produce sufficient insulin to handle the glucose load (Weksler-Zangen et al., 2001).

The Cohen model is normotensive, but a hypertensive variant (Cohen-Rosenthal rat) has been developed by crossing the SHR rat to the Cohen diabetic rat (Cohen et al., 1993; Rosenthal et al., 1995).

iii. Goto-Kakizaki (GK) Rat. The inbred, albino Goto-Kakizaki (GK) rat was developed from an outbred Wistar background in 1973 (Goto et al., 1988a,b). It was developed by selecting animals from both sexes that demonstrated the highest blood glucose levels during a glucose tolerance test (Goto et al., 1976). The process for selecting the animals bred to develop the diabetic colony was repeated for five generations. Since the ninth generation, brother x sister matings have been performed (Suzuki et al., 1998). Unlike the ZDF, the GK rat is lean rather than obese. A neonatal deficiency in beta-cell number is the primary defect and leads to basal hyperglycemia, which is first detectable at 3 weeks of age (Tourrel et al., 2002).

The GK rat is characterized by a uniformly lean phenotype, development of diabetes in both sexes, mild insulin resistance, a non-fasting glucose level of 160 to 200 mg/dL by 12 weeks, a fasting hyperglycemia, and development of neuropathy (Goto et al., 1988a,b; Suzuki, 1998). There are also reports of retinopathy developing late in the life of the animal (Agardh et al., 1997, 1998). This model has been used to study diabetic type 2 end-stage renal disease (ESRD) (Vesely et al., 1999).

The GK rat demonstrates rather stable levels of glucose intolerance and an impaired glucose-induced insulin response, but there are several colonies around the world, and characteristics such as islet morphology, insulin content, and islet metabolism vary between substrain colonies (Ostenson, 2001).

iv. Otsuka Long-Evans Tokushima Fatty Rat. The inbred, black-hooded Otsuka Long-Evans Tokushima fatty (OLETF) rat was developed from a spontaneous diabetic rat that occurred in a colony of Long-Evans rats purchased from Charles River Canada by Tokushima Research Institute (Otsuka Pharmaceutical, Tokushima, Japan) in 1982 (Kawano et al., 2001).

OLETF rats are hyperphagic and develop mild obesity. There is a late onset of hyperglycemia, and high glucose levels do not develop until after 18 weeks of age. After 40 weeks of age, the pancreatic islets deteriorate to a point that a type 1 diabetic syndrome develops. The resulting disease is much like what is seen in the db/db mouse; this is in contrast to what is seen in ZDF rats or GK diabetic rats, which develop type 2 diabetes with accompanying polyphagic hyperinsulinemia and with no transformation into insulin-dependent diabetes mellitus (type 1 diabetes) (Kawano et al., 2001).

Characteristics of the OLETF rat include (1) late-onset hyperglycemia (after 18 weeks of age in the male,

70 weeks in the female), (2) maximum plasma glucose of 300 mg/dL in the oral glucose tolerance test at 30 weeks of age, (3) mild obesity, and (4) islet hyperplasia in the pancreas (Kawano et al., 2001). It is also reported that by 30 weeks of age the animals develop renal dysfunction and glomerular damage, simulating the end-stage kidney disease that is seen in advanced human nephropathy (Kawano et al., 2001).

v. Wistar Fatty Rat. The inbred, albino Wistar fatty rat was the first rat model of obese type 2 diabetes (Odaka et al., 2001). The model was established by crossing the obese outbred Zucker rat with the Wistar-Kyoto (WKY) rat (Kava et al., 1990). Only the male expresses type 2 diabetes, which is characterized by hyperglycemia, hyperlipidemia, glucose intolerance, hyperinsulinemia, and decreased whole-body insulin sensitivity. By 14 weeks of age, the obese male has non-fasting glucose levels of more than 400 mg/dL. The male obese Wistar fatty rat is infertile, and the obese animal is obtained by breeding lean heterozygotes. This rat is not currently available commercially and must be obtained from Takeda Chemical Industries (Osaka, Japan).

2. Chemically Induced Diabetes

a. ADULT STREPTOZOTOCIN-INDUCED MODEL OF DIABETES. Chemicals such as streptozotocin (STZ) have been used for more than 25 years to induce diabetes in rats and other animals for studies on the long-term effects of diabetes. STZ causes type 1 diabetes by inducing destruction of pancreatic beta-cells. Type 1 diabetes has been induced in many strains of rats by intravenous administration of 45 to 55 mg/kg of STZ (Farber et al., 1991; Kurthy et al., 2002) or by intraperitoneal administration of 100 mg/kg (Portha et al., 1974). This model of type 1 diabetes has been used to study a variety of sequelae to type 1 diabetes, including peripheral nerve degeneration (neuropathy) (Doss et al., 1997), alterations in erectile function (Rehman et al., 1997), and changes in contractile properties of the urinary bladder (Ozturk et al., 2002).

b. NEONATAL STREPTOZOTOCIN (n-STZ) MODEL. In contrast to the induction of type 1 diabetes in the adult rat, administration of STZ to neonatal animals (100 mg/kg via the saphenous vein) at the day of birth results in a disease that resembles type 2 diabetes (Portha et al., 1974). The specific time at which STZ is administered to the neonate is dictated by the target to be studied: (1) beta-cell lesions (day 0 n-STZ model), (2) insulin target cell lesions (day 2 and day 5 n-STZ models), (3) hepatic glucose output lesions (day 2 and day 5 n-STZ models), and (4) imbalance of the glucose fatty acid cycle (day 5 n-STZ model) (Portha et al., 2001).

This n-STZ model is useful for studying interactions between type 2 diabetes and other disease processes, such as hypertension. For example, an n-STZ SHR rat allows for the study of the interactions between hyperglycemia and hypertension, as well as related influences on the renal system (Iwase et al., 1987; Wakisaka et al., 1988).

3. Models of Retinopathy

Retinopathy is a common complication of diabetes. Rat models commonly used to study vascular abnormalities of the retina include the following strains: SHROB (Huang et al., 1995), SHR and WKY (More et al., 1986), ZDF fa/fa type 2 diabetic rat (Danis and Yang, 1993; Yang et al., 2000), GK type 2 diabetic rat (Agardh et al., 1997 et al., 1998), and the BB/Wor type 1 diabetic rat (Sima et al., 1985; Chakrabarti et al., 1991).

C. Hypertension

Hypertension has been classified as either primary (essential), when the cause of the disease is known, or as secondary, when the cause is not known. Hypertension is not a single disease but a syndrome with many causes (e.g., narrowing of the aorta, salt sensitivity, renal abnormalities, and diabetes). Several animal models are available for studies into the pathogenesis and treatment of hypertension.

1. Spontaneous Hypertension

a. SPONTANEOUSLY HYPERTENSIVE RAT. The inbred, albino SHR rat is probably the best known and most widely used of the rat hypertension models. This strain of rat was developed from an outbred colony of WKY rats by Okamoto and Aoki (1963). The line was established by breeding a male with mild hypertension to a female with marked hypertension (Okamoto and Aoki, 1963; Okamoto et al., 1974). They were selectively bred for high blood pressure without any provocative dietary or environmental stimuli (Okamoto and Aoki, 1963). A number of SHR sublines have been developed, some of which are stroke-prone (Okamoto et al., 1974). There is no evidence of sub-strain differentiation among the SHR stock from major commercial suppliers in the United States, with respect to both phenotype and DNA (Blizard et al., 1991). Inheritance of high blood pressure in the SHR rat is polygenic with at least three major genes involved (Tanase et al., 1970).

SHR rats develop SBP values of 150 mm Hg by 10 weeks of age. Because they express insulin resistance, SHR rats are suitable subjects for studies that need to model the insulin resistance and essential hypertension found in some non-obese humans (Swislocki and Tsuzuki, 1993). In general, the SHR rat is a good and suitable model for screening anti-hypertensive drugs for humans (Roba, 1976).

The use of the WKY rat as a normotensive control for the SHR rat has been questioned by many (Kurtz et al., 1989; Johnson et al., 1992). Although the WKY rat was developed from the same base population, it differs from the SHR rat in many genetic marker loci (Festing and Bender, 1984). The WKY rat was distributed to commercial suppliers in the United States before it was fully inbred, resulting in genetic differences between suppliers (Rapp, 2000). St. Lezin et al (1992) found that that only 50% of DNA fingerprint bands were shared between the WKY and the SHR rat. This led to the suggestion that other normotensive strains be used as controls for comparative studies.

b. GENETICALLY HYPERTENSIVE (GH) RAT. The inbred, albino, genetically hypertensive (GH) rat was developed at the University of Otago Medical School (Dunedin, New Zealand) from a colony of outbred Wistar rats that had been imported from England in 1930. The New Zealand strain of genetically hypertensive rats was the first experimental demonstration that it was possible to transmit hypertension from parent to offspring by a hereditary mechanism (Simpson et al., 1994). Smirk, in 1955, bred animals with marked hypertension to develop the inbred line (Heslop and Phelan, 1973).

The GH rat is characterized by hypertension, cardiac hypertrophy, and vascular disease (Phelan, 1968; Simpson et al., 1973). By 10 months of age, mesenteric arteritis is common (Gresson and Simpson, 1974). The heart rate is about 20% faster and heart weight about 50% greater than are values in normotensive strains. Abnormally high blood pressures are noted practically from birth and are associated with increased vascular resistance from at least 4 weeks of age. The GH rat is a useful model for study of the pathogenesis of spontaneously occurring hypertension, having analogies with human essential hypertension.

c. FAWN HOODED HYPERTENSIVE RAT. The inbred, Fawn hooded hypertensive (FHH) rat was derived from a colony of outbred Fawn hooded rats. The Fawn hooded rat originated at the University of Michigan, most likely as a spontaneous mutation in a strain of Long-Evans rats. Subsequently, Fawn hooded rats were brought to Europe in the 1980s. An inbreeding program was started by A.P. Provoost at Erasmus University Rotterdam (The Netherlands), to develop two strains that differed in the level of SBP (Kuijpers and de Jong, 1986). The tail cuff SBP of the FHH rat increases to about 150 to 160 mm Hg, whereas the SBP is 120 to 130 mm Hg in the normotensive Fawn hooded low blood pressure rat (Kuijpers et al., 1986; Simons et al., 1993a). The two strains differ not

only in the level of SBP but also in the development of renal damage. The FHH rat develops progressive proteinuria (mainly albuminuria) and focal segmental glomerulosclerosis at a relatively young age, whereas the normotensive model shows little renal damage (Kuijpers and Gruys, 1984; Kuijpers and de Jong, 1987). The FHH rat has elevated glomerular capillary pressure, which results from low afferent arteriolar resistance (Simons et al., 1993a), owing to an impaired myogenic responsiveness (Van Dokkum et al., 1999). In contrast, the SHR has an increased afferent arteriolar resistance, and major renal damage is absent (Dworkin and Feiner, 1986). Early treatment with an angiotensin-converting enzyme inhibitor prevents the development of hypertension and renal damage (Verseput et al., 1997). The severity of renal damage is influenced not only by the blood pressure level but also by unilateral nephrectomy (Simons et al., 1993b) and dietary protein intake (De Keijzer and Provoost, 1990). In FHH rats, blood pressure and susceptibility to renal damage are under separate genetic control (Brown et al., 1996). Five chromosomal loci influence the development of proteinuria, suggesting complex gene-gene interactions (Shiozawa et al., 2000). It should be noted that other Fawn hooded rat strains exist: the Tester Moriyama strain and others with names such as FH/Wjd, FH/IR, or FH/Har. All Fawn hooded strains have a hereditary bleeding disorder, owing to a platelet storage pool deficiency (Tschopp and Zucker, 1972). The bleeding disorder is pleiotropic with respect to coat color dilution. A gene relevant to the clotting disorder has recently been mapped in the FHH (Datta et al., 2003), and the Tester Moriyama rats (Oiso et al., 2004). FHH rats have been used to model pulmonary hypertension under mild hypoxic conditions (Sato et al., 1992; Le Cras et al., 2000), as well as alcoholism and depression (Rezvani et al., 2002).

d. LYON HYPERTENSIVE RAT. The inbred, albino Lyon hypertensive (LH) rat was developed from outbred Sprague-Dawley rats in 1969; this model of hypertension is characterized by increased body weight, hyperlipidemia, and salt sensitivity. The LH rat exhibits mild hypertension (165 mm Hg in males, and 140 mm Hg in females at 14 weeks of age). Hypertension without hyperinsulinemia is characteristic of the LH rat (Boulanger et al., 1997). Neonatal thymectomy prevents the development of hypertension (Bataillard et al., 1993), and the renin-angiotensin system, renal synthesis of prostaglandins, and secretion of vasopressin do not seem to be involved in the development of increased blood pressure (Vincent et al., 1984). Florin et al. (2001) demonstrated that LH rats are salt sensitive despite the lack of an active renin-angiotensin system. Dubay et al. (1993) showed a significant association between blood pressure and a microsatellite marker near the renin gene on chromosome 13, and between pulse pressure and the *carboxypeptidase-B* gene on chromosome 2.

e. SABRA HYPERTENSION-PRONE AND HYPERTENSION-RESISTANT RATS. The inbred Sabra hypertension-prone (SBH) rat is a hypertensive, salt sensitive and testosterone dependent model that was selectively bred on the basis of its blood pressure response to experimental manipulation with unilateral nephrectomy (Ben-Ishay et al., 1972). SBH rats have a greater average body weight, and their average blood pressure is about 20 mm Hg higher compared with that of the Sabra hypertension-resistant rat. Large differences in hypertension between the two strains are observed after treatment with hypertension-inducing agents such as excess dietary sodium chloride (Mekler et al., 1985). In addition, the SBH rat has a preference for drinking saline over water. Young SBH rats have enhanced norepinephrine turnover and an impaired baroreflex function, which precedes the development of hypertension (Ben-Ishay and Yagil, 1994).

f. MILAN HYPERTENSIVE STRAIN AND MILAN NORMOTENSIVE STRAIN RAT. The inbred, albino Milan hypertensive (MHS) strain was developed from a colony of outbred Wistar rats by mating brother to sister and then selecting animals for high SBP. As in the SHR rat, the MHS rat model was developed without special dietary or environmental challenges (Bianchi et al., 1984). At 3 to 4 weeks of age, the mean SBP of the MHS rat does not differ from the 120 mm Hg found in the Milan normotensive rat, but by 7 weeks, the SBP of the MHS rat increases to values of greater than 170 mm Hg. Hypertension can be "transplanted" from the MHS rat to the Milan normotensive rat by kidney transplantation (Barber et al., 1994). This model is also characterized by a genetic deficiency in calpastatin, the endogenous inhibitor of calpain (Muller et al., 1995). It has also been reported that the MHS rat has defects in the regulation of adenyl cyclase in vascular smooth muscle cell membranes (Clark et al., 1993).

The normotensive Milan strain spontaneously develops severe proteinuria, interstitial fibrosis, and focal glomerulosclerosis in the absence of hypertension by 10 months of age (Floege et al., 1997).

2. Salt-Sensitive Hypertension

The inbred, albino Dahl hypertensive-sensitive (SS)/Jr rat was developed by John Rapp to study the effects of an exogenous source of salt on hypertension (Rapp and Dene, 1985). The SS/Jr model was derived from a colony of salt-induced hypertensive, outbred Sprague-Dawley rats developed by L.K. Dahl at Brookhaven National Laboratories (Rapp, 1982). St. Lezin et al.

(1992) found that the SS/Jr and the control that is resistant to salt, the SR/Jr, have about 80% of DNA fingerprint bands in common. Another inbred strain from the same outbred source was developed (Iwai and Heine, 1986); this strain has been designated the Dahl-Iwai S for the salt-sensitive line, and Dahl-Iwai R for the salt-resistant line, but both are currently available only in Japan (Rapp, 2000).

Key features of the SS/Jr rat model are development of fulminating hypertension and marked vascular and renal lesions after being on a high salt (8% sodium chloride) diet for 3 to 4 weeks; typically, all SS/Jr individuals die within 8 weeks of being placed on the high-salt diet (Rapp and Dene, 1985). On a low-salt diet (4% sodium chloride), the animals survive and develop marked hypertension, but have a slower increase in blood pressure.

Young SS/Jr rats will uniformly manifest low-renin hypertension and develop hypertensive nephrosclerosis within weeks of being placed on the high-salt diet. The pathology is identical to human hypertensive renal disease and consists of arteriolosclerosis, glomerulosclerosis, and interstitial scarring with tubular cell dropout. Because of the reproducible in the development of disease, this is a very good model for studying low-renin hypertension.

3. Surgically-Induced Renovascular Hypertension

In 1934, Harry Goldblatt observed from results he obtained in an experiment he performed on dogs that he could induced high blood pressure by constricting either one or both main renal arteries with an adjustable clamp which reduced renal perfusion pressure and renal blood flow. This model led to the elucidation of the renin-angiotensin system and its relationship to many other regulatory mechanisms. In the dog, constriction of one main renal artery will produce elevated blood pressure, but this elevation is temporary and blood pressure returns to normal within a few weeks unless both of the renal arteries are constricted, a procedure referred to as the two-kidney, two-clamp model.

In the rat, constriction of only one main renal artery is sufficient to produce sustained high blood pressure and renovascular hypertension disease. This method is sometimes referred to as the two-kidney, one-clamp model or the "Goldblatt kidney," and is a high rennin model of hypertension. In contrast, the one-kidney, one-clip model, in which the unclamped kidney is extirpated, is associated with the inhibition of the renin-angiotensin system. No matter which system is used, the Goldblatt kidney is a non-strain-dependent model and can be performed without administration of a high-salt diet so that experimental studies on severe hypertrophy of small blood vessels can be performed (Sventek et al., 1996).

III. MODELS OF NEUROLOGICAL DISEASE

Animal models of neurological deficits are essential for research into the pathogenesis and treatment of such disease. It has been suggested that non-human primates rather than rats be used in neurological studies; however, a number of valuable rat models of neurological disease have been characterized in recent years. Rat models exhibiting akinesia, tremor, postural deficits, and dyskinesia—all of which are deficits relevant to Parkinson's disease—have been developed (Betarbet et al., 2002; Cenci et al., 2002). Studies conducted in the rat are an effective and inexpensive means to complement studies in non-human primates.

A. Alzheimer's Disease

Alzheimer's disease is a neurodegenerative disease characterized by progressive memory loss and dementia. It is characterized by deposition of β-amyloid peptide into senile plaques of the affected brain (Goodenough et al., 2003), as well as β-dystrophic neuritis, activated microglia, reactive astrocytes, and neuronal loss (Selkoe, 1991). The lack of an adequate animal model for Alzheimer's disease has been a significant barrier in the understanding of the disease and development of therapeutic drugs. To date, the development of transgenic rats for Alzheimer's disease has not been successful (Cole and Frautschy, 1997), and most studies inject β-amyloid into rat brain tissue to study the pathology, neurodegeneration, or effect of drugs (Weldon et al., 1998; Jin et al., 2004; Yagami et al., 2004).

B. Epilepsy

Epilepsy, a disease characterized by the tendency to experience recurrent seizures that originate in the brain, is a common neurological disorder that affects approximately 1% of the human population. Types of epilepsy include (1) general epilepsy and (2) absence epilepsy, in which there are brief behavioral absences with staring spells but no convulsive seizure. The occurrence of absence epilepsy is primarily in children (Sohal and Huguenard, 2001). Several commonly used rat models to study this disease are described here.

1. Genetically Epilepsy-Prone Rats

Both substrains of the inbred, albino genetically epilepsy-prone (GEPR/3 and GEPR/9) rats of this model were developed from outbred Sprague-Dawley rats by selecting for moderate susceptibility to audiogenic-induced seizures (Reigel et al., 1986). In response to a standard sound stimulus, the GEPR/3 rat demonstrates a single running phase that is followed by clonic convulsions and subsequent postictal rage, which includes vocalization, jumping, biting movements, and intermittent catatonia (Reigel et al., 1986). The GEPR/3 rat becomes susceptible to sound at approximately 15 day of age (Reigel et al., 1989). The GEPR/9 substrain is subject to seizures of greater severity and becomes sensitive to the sound stimulus at about 21 days of age. The GEPR/9 males show greater differences in expressivity and penetrance compared with that of GEPR-9 females; the sex-associated differences in expressivity and penetrance are smaller in GEPR/3 animals (Kurtz et al., 2001). Because seizure predisposition is determined by factors present at birth, the GEPR rat is a valuable model for the study of mechanisms influencing neonatal seizure susceptibility.

2. WAG/Rij Rat

The inbred, albino WAG/Rij rat was developed from Wistar stock by A.L. Bacharach at Glaxo in 1924 (Mouse Genome Database, 2004). There are several substrains available, and it is important that the specific substrain used be designated in reports. The WAG/Rij subtrain spontaneously develops a clinical phenomenon resembling human "petit mal" or absence epilepsy (van Luijtelaar and Coenen, 1988; Coenen et al., 1992). This model has been used for the evaluation of anticonvulsant drugs, and responses in the WAG/Rij rat generally correspond to those seen in human epilepsy patients (Peeters et al., 1988, 1994). The WAG/Rij rat is regarded as a well-validated animal model of childhood absence epilepsy (Schridde and van Luijtelaar, 2004).

3. Genetic Absence Epilepsy Rat from Strasbourg

The inbred, albino genetic absence epilepsy rat from Strasboug (GAERS) was originally derived from a colony of Wistar rats. The GAERS presents with generalized, non-convulsive epileptic seizures characterized by behavioral arrest and sometimes accompanied by rhythmic twitching of the vibrissae and facial muscles. During the epileptic episodes, the animals show spontaneous occurrence of spike and wave discharges (SWD) in the cortical electroencephalogram (Van Hese et al., 2003). In this regard, the GAERS is useful for the evaluation of anti-epileptic compounds, because the efficacy of a drug is directly related to an increase or decrease in the number of spike and wave discharges in the cortical electroencephalogram (Van Hese et al., 2003).

C. Multiple Sclerosis

The experimental autoimmune encephalomyelitis (EAE) rat is a useful model of multiple sclerosis, which is characterized by remittance and relapse of the disease and presence of demyelinating lesions in the central nervous system. EAE results from a Th1 inflammatory response directed at the central nervous system of susceptible laboratory animals (Swanborg, 1995). Two inbred strains of rats have been used extensively for the induction of EAE and other autoimmune diseases: the Lewis (LEW) rat and the Dark Agouti (DA) rat. The DA rat differs from the LEW rat in that it develops EAE when immunized with encephalitogenic myelin basic protein peptide (MBP63-81) in incomplete Freund's adjuvant. In contrast, the LEW rat will develop EAE when injected with myelin basic protein and complete Freund's adjuvant, but will develop a state of tolerance when immunized with the peptide (MBP 63-81) and incomplete Freund's adjuvant (Lenz et al., 1999).

D. Parkinson's Disease

A chemically-induced rat model has been used to study Parkinson's disease. The model is produced by dopamine depletion after the administration of 6-hydroxydopamine bilaterally into the *nucleus accumbens septi* or the caudate nucleus (Kelly et al., 1975).

In contrast, the Langston model, which is induced by the metabolites of adminstered 1-methyl-4-phenyl-1,2,3,6-tetrahydropyridine (MPTP) and is widely used in non-human primates, typically induces only minimal effects in the rat (Chiueh et al., 1984).

A recently described rat model of Parkinson's disease involves induction of the disease in rats by administration of the Japanese encephalitis virus (Ogata et al., 1997).

Recently, rotenone, a naturally occurring rotenoid pesticide derived from the roots of certain plants, has been found to reproduce in rats the substantia nigra nerve cell loss and the occurrence of Lewy bodies that are characteristic of Parkinson's disease (Betarbet et al., 2002). Rotenone appears to exert this effect by inhibition of complex I of the electron transfer chain, which catalyzes the transfer of electrons from nicotinamide adenine dinucleotide to ubiquinone and translocates protons from the mitochondrial matrix to the intermembrane space. In addition, this model of Parkinson's disease exhibits bradykinesia, postural instability, and an unsteady gait (Greenamyre et al., 2001).

E. Peripheral Neuropathy

Peripheral neuropathy is a generic term denoting functional disturbances and/or pathological changes in the peripheral nervous system. If the changes are restricted to a single nerve, the condition is commonly referred to as mononeuropathy; involvement of several nerves is referred to as mononeuritis multiplex; and diffuse, bilateral nerve involvement is referred to as polyneuropathy. Peripheral neuropathy is not a specific disease but rather a manifestation of many conditions that cause damage to the peripheral nerves.

Diabetes is one of the most common causes of neuropathy, and therefore, diabetic rat models have been used to study this condition. Peripheral neuropathy is a relatively common complication for diabetics, particularly in those who have had difficulty with controlling their blood glucose levels; in those with elevated levels of blood lipids, high blood pressure, and obesity; and in people over the age of 40. Peripheral neuropathy is characterized by numbness and sometimes pain and weakness in the extremities. Additional neuropathic complications may also occur in every organ system, including the digestive tract, heart, and sex organs.

Investigations into the pathogenesis of peripheral neuropathy have often used the STZ-induced type 1 diabetic rat model. Early work using this model has shown that nerve fiber size was diminished in the peroneal nerve in the diabetic rat (Jakobsen and Lundbaek, 1976). The rat STZ-induction model has also been used to examine the impairment of nitric oxide in neuropathy in the experimental type 1 diabetic rat model (Sakurai et al., 2002; Thomsen et al., 2002), as well as the interrelationships of metabolic, vascular, and nerve conduction in development of the neuropathy (Dall'Ago et al., 2002; Pop-Busui et al., 2002; Ulugo et al., 2002).

Studies of neuropathy in the ZDF type 2 diabetic rat model have been reported by Peterson et al. (1990), who demonstrated reduced nerve conduction velocities, nerve edema, and morphological abnormalities in the nerve fibers. Mathew et al. (1997) demonstrated that the metabolic abnormalities in the peripheral nerve fibers of the ZDF rat are consistent with other diabetic animal models. Other studies using the ZDF rat have investigated the effects of different potentially therapeutic compounds on peripheral neuropathy (Shibata et al., 2000; Shimoshige et al., 2000; Kurthy et al., 2002).

Peripheral neuropathy has also been demonstrated in the BB/Wor type 1 diabetic rat (Greene et al., 1984; Yagihashi and Sima, 1985a,b). Further limited studies have been conducted using the GK type 2 diabetes rat model (Wada et al., 1999) and the OLETF model of type 2 diabetes (Iwase et al., 2002).

IV. MODELS OF CANCER

Rats are susceptible to the development of spontaneous, virally induced and chemically induced tumors. There are several rat models that have been used to elucidate the pathogenesis and molecular mechanisms involved in cancer.

Information on models using the growth of xenografted tumor tissue in nude rats is included elsewhere in this book.

A. Prostate Cancer

1. Lobund-Wistar Rat

The albino Lobund-Wistar (LW) rat is used as a model of prostate cancer and was first described by Pollard (1973) in a colony of germ-free outbred Wistar rats. Although there are several transgenic mouse models of prostate cancer, this is the only known rat strain that is genetically predisposed to spontaneous metastasizing, sex hormone-influenced adenocarcinoma of the prostate (Pollard, 1998). Cancer of the LW rat model resembles the human disease in many respects, including inherited susceptibility, spontaneous development and progression to androgen independence, metastasis from the primary tumor, and resistance to a variety of therapeutic agents in the refractory state (Plata-Salaman et al., 1998).

Two models of autochthonous prostate adenocarcenoma have been described in the LW rat: (1) spontaneous prostate adenocarcinoma, which develops at a mean age of 26 months; and (2) chemically-induced prostate adenocarcinoma, which develops within a mean of 10.5 months after treatment (Pollard and Luckert, 1986, 1987). Pollard (1992) demonstrated that prostate adenocarcinoma is manifested spontaneously in 26% of aged LW rats and in 90% of younger animals that have been treated with a combination of N-methyl-N-nitrosourea/testosterone propionate. The "premaligant" stage of induced tumorgenesis is susceptible to intervention, but the later malignant stage resists therapeutic intervention.

2. Noble Rat

Another rat model which has been commonly used for the study of prostate tumor is the inbred, black-hooded Noble rat. In this strain, the tumor is characterized by hyperplasia and dysplasia of the dorsolateral prostate after implantation of capsules containing testosterone and 17-β-estradiol (Noble et al., 1975; Noble, 1977). On a 16-week treatment schedule of testosterone and 17-β-estradiol, intraductal dysplasia developed in the dorsolateral lobe of the prostate (Yu et al., 1993). More prolonged exposure to these agents results in prostatic carcinoma in up to 100% of treated individuals (Leav et al., 1988; Cavalieri, 2002).

B. Breast Cancer

Breast cancer is a major cause of cancer morbidity in women in the United States, and is expected to result in over 40,000 deaths in 2005. The rat is an extremely valuable model for studying susceptibility to breast cancer.

The characteristics of rat mammary cancer are remarkably similar to those of the human disease. For example, mammary cancer in the rat is responsive to hormone therapy (Gould, 1995).

Several rat models have been commonly used to study mammary carcinogenesis susceptibility and resistance. The Copenhagen (COP) strain carries a dominant-resistance gene; in contrast, the Wistar-Furth (WF) rat is susceptible to inducible carcinoma of the mammary gland (Shepel and Gould, 1995). By backcrossing F1 animals from a WF × COP mating to a WF rat, Hsu et al. (1994) were able to identify the first mammary carcinoma susceptibility locus (Mcs1) on rat chromosome 2. Gould et al. (1998) expanded that initial study and conducted a genome-wide screen of the original backcross as well as of the F2 intercross between the COP and the WF. This approach not only confirmed the Mcs 1 locus but also identified three additional loci that modify the susceptibility to 7,12-dimethylbenzanthracene-induced breast cancer (Martin and Weber, 2000).

V. MODELS OF RENAL DISEASE

A. End-Stage Renal Disease

End-stage renal disease (ESRD) owing to hypertension, diabetes and to polycystic kidney disease is of great clinical significance. Kidney disease resulting from diabetes and hypertension—two common causes of loss of renal function—is characterized by an increase in glomerular filtration rate, proteinuria, and glomerular lesions. Loss of renal function is a common complication of both type 1 and type 2 spontaneous diabetes mellitus in man.

As with humans, renal disease in rats tends to be characterized by hypoinsulinemia, hyperinsulinemia, and impaired glucose tolerance: the metabolic syndrome. To date, there is no rodent model that develops renal changes that are identical to those seen in human ESRD. For example, mesangial expansion, glomerular basement membrane thickening, and renal arteriolar hyalinosis are commonly found in the human disease, but these changes are not typically found in animal models, although such changes are found occasionally in the SHR/N-cp rat (Velasquez et al., 1990).

One rat model used for investigating the mechanism of diabetic nephropathy and for evaluating various drugs for treatment is the hybrid, black-hooded ZSF1 rat. This animal model was developed from a cross between a ZDF +/fa female rat and a spontaneously hypertensive heart failure (SHHR)/Mcc-fa^k male rat. A good model for studying kidney dysfunction, the obese male animal develops hypertension, type 2 diabetes (when fed Purina Lab Diet 5008), elevated blood apolipoprotein B levels, hyperlipidemia, hyperinsulinemia, and ESRD.

By 46 weeks of age, ZSF1 rats develop elevated blood levels of very low density lipoprotein, renal hypertrophy, and a three- to eight-fold decrease in creatinine clearance when compared with levels of SHHF, WKY, and SHR rats. There is a high degree of segmental and global glomerulosclerosis, as well as severe tubulointerstitial and vascular changes (Tofovic et al., 2000).

It has been suggested that lipotoxicity causes nephropathy in both the human and the rat. Hawes et al. (2001) found that in the ZSF1 rat lipotoxicity induced oxidant injury and altered the activity of hydroxyisobutyrate dehydrogenase, a mitochondrial enzyme; furthermore, they showed that nephropathy in type 2 diabetes is caused or enhanced by severely depressed levels of renal mitochondrial enzymes.

Additional rat models used to study the effect of hypertension and diabetes on kidney function and for the study of end-stage kidney disease include ZDF fa/fa, SHROB, SHR, stroke-prone spontaneously hypertensive (SHRSP), Fawn-hooded, GH, and Dahl salt-sensitive SS/Jr rats.

B. Polycystic Kidney Disease (Bright's Disease)

Polycystic kidney disease is the most common genetic disease of the kidney, with more than 600,000 people in the United States and 13 million worldwide being affected. The disease is characterized by multiple renal cysts that increase in size and number over time, resulting in increased kidney mass and eventual renal failure. There are two forms of the disease: autosomal dominant polycystic kidney disease (ADPKD) and autosomal recessive polycystic kidney disease, which is less common.

1. Han:Sprague-Dawley (Cy/+) Rat

A model of ADPKD, the outbred, albino Han:Sprague-Dawley (Han: SPRD) rat develops polycystic kidneys as well as hypertension and vascular disease (Wang et al., 1999). The homozygous Cy/Cy male rats of this strain develop a rapidly progressive polycystic kidney disease and typically die by 3 weeks of age. The heterozygous Cy/+ male rat shows signs of renal failure by 8 to 12 weeks of age and seldom live beyond one year. The female Cy/+ develops a slowly progressive form of the disease and will live into the second year of life (Cowley et al., 2001).

2. Polycystic Kidney Rat

A spontaneous mutation was observed in an inbred, albino rat that developed polycystic lesions in the liver as well as the kidneys and that has an autosomal recessive mode of inheritance (Lager et al., 2001). The mutation occurred in a colony of CRJ:CD (SD) BR rats in Japan,

and the rats develop congenital intrahepatic biliary dilatation associated with congenital hepatic fibrosis. The lesions found in this model are similar to those seen in humans with Caroli's disease (Sanzen et al., 2001). Genetic analysis has shown the polycystic kidney rat model to be orthologous to the autosomal recessive polycystic kidney disease that occurs in man (Harris, 2002).

C. Diabetes Insipidus

Diabetes insipidus is a disorder in which there is neurogenic deficiency of the antidiuretic hormone or nephrogenic insensitivity/resistance to the antidiuretic hormone. In both cases the kidney is unable to reabsorb the proper amount of water. Both types of the disease are characterized by the inability of the kidneys to respond to the antidiuretic hormone, arginine vasopressin. It is a rare disease and is often mistaken for diabetes mellitus because of the large amounts of urine that are produced.

1. Brattleboro Rat

The outbred, hooded Brattleboro rat is a model of hereditary hypothalamic diabetes insipidus, which demonstrates the characteristics of the neurogenic form of the disease (Sokol and Valtin, 1965; Valtin et al., 1965; Valtin, 1967; Saul et al., 1968). The principal characteristic of the disease in this model are high plasma rennin activity, high anglotensin II concentration, spontaneous hypertension and abnormally large volume of dilute urine, which is produced because of the lack of the hormone vasopressin.

2. M520/DI Rat

A inbred, albino, congenic strain lacking the ability to produce vasopressin owing to the presence of the "di" mutation from the Brattleboro rat, this model has also been used to study diabetes indisidus (Colombo et al., 1992).

VI. MODELS OF VASCULAR AND HEART DISEASE

A. Stroke-Prone Rat

This model of essential hypertension and cerebrovascular disease has been used to study factors that affect the outcome of cerebral ischemia and hemorrhage was developed from the A1 and A3 substrains of the SHR rat. They had been bred as parallel lines from generation F20 to F36. Animals from these lines were crossed, and further inbreeding was performed with the selection of offspring from parents that died of stroke (Okamoto et al., 1974; Yamori, 1984; Okamoto et al., 1986).

The A3 substrain was found to have a high incidence of cerebrovascular lesions and was designated by the National Institutes of Health in 1976 as the SHRSP/A3N line. A major characteristic of this model is high blood pressure, which increases to a level of greater than 180 mm Hg by 10 weeks of age; the blood pressure in this model is greater than that in the SHR rat of the same age. Transplant studies have shown that the kidneys of the SHRSP rat are intrinsically more susceptible than are those of the SHR rat to renal damage when exposed to similar blood pressures and metabolic environments (Churchill et al., 2002).

SHRSP rats possess a genetic predisposition to the development of cerebrovascular lesions. Stroke in the SHRSP rat has been reported to be similar to that seen in man (Yamori et al., 1976).

A higher incidence of stroke results in the SHRSP rat when the animals are fed a stroke-prone Japanese style diet (Yamori et al., 1984). Steir et al. (2002) found that placement of 7.5 weeks of age SHRSP rats on a 1% sodium chloride drinking solution in combination with feeding of a stroke-prone Japanese style diet (Zeigler Diet 52880000, Zeilger Brothers Inc., Gardners, PA) having reduced potassium and protein content will markedly accelerate the onset of microvascular damage. When this regimen is used, beginning at 53 days of age, SHRSP rats began to show increased protein in the urine by 15 weeks of age and to experience stroke at 110 days of age (Hernandez et al., 1994).

Cerebral hemorrhage or infarction are present in 82% of male SHRSP rats over 100 days of age and in 58% of females over 150 days of age. Signs of stroke include development of proteinuria, repetitive lifting and convulsion of the paws, paralysis of hind limbs, and hyperirritability.

B. Spontaneous Hypertensive Heart Failure Rat

The inbred, albino SHHF is a model of obesity, type 2 diabetes, and congestive heart failure (CHF). The stock for this colony was transferred to Dr. Sylvia McCune at the University of Chicago Medical School in 1983 from the laboratory of J.E. Miller at G.D. Searle and Company. Dr. McCune received the breeder stock from the seventh backcross of a Koletsky (SHROB) obese rat to a SHR/N rat. The animals from the backcross were inbred to fix the CHF trait; this trait was established beyond 21 generations in Dr. McCune's laboratory (McCune et al., 1990).

As a model of cardiomyopathy, the SHHF is characterized by development of hypertension and cardiac hypertrophy, which lead to CHF in all phenotypes, regardless of sex. The obese phenotypes of the rat are insulin resistant and hyperinsulinemic, and have elevated triglyceride levels and moderately elevated cholesterol levels. The female is insulin resistant; the obese male is hyperinsulinemic and insulin resistant until the development of type 2 diabetes at about 12 weeks of age. In this model, the influence of diabetes, hypertension, and obesity on the development of CHF can be evaluated in the male. The lean phenotype allows one to study the development of CHF without the influence of diabetes or obesity. CHF develops in the obese male SHHF at 10 to 14 months of age, in obese females and lean males at 16 to 18 months of age, and in the lean female at 20 months or older. The model has been used also to evaluate the role of tumor necrosis factor in heart failure (Bergman et al., 1999), pregnancy-associated hypertension (Sharkey et al., 1999), the effect of caffeine consumption on renal function (Tofovic and Jackson, 1999), A(1) receptor blockade (Jackson et al., 2001), the role of the renin-angiotension system in CHF (Radin et al., 2002; Tamura et al., 2002), and mechanical alternans and electromechanical restitution in the failing heart (Dumitrescu et al., 2002; Narayan et al., 1995).

C. JCR:LA-Corpulent Rat

After the identification of the *cp* gene mutation (Koletsky, 1973), rats having the mutation were initially bred into two standard strains: the LA/N and the SHR/N (Hansen, 1983). The progeny were backcrossed to the parental strains to produce two congenic strains: LA/N-cp and SHR/N-cp (Russell and Koeslag, 1990). The fully backcrossed, congenic strains of these rats have incorporated the corpulent gene and are defined, stable inbred lines. The partially backcrossed (five backcross) strain, JCR:LA-cp, which was sent to Dr. J.C. Russell at the University of Alberta (Edmonton, Alberta, Canada) in 1978 retains some genetic contributions from the original Koletsky colony, which are reflected in metabolism and disease expression (Russell and Koeslag, 1990).

Although the JCR:LA-cp strain is phenotypically indistinguishable from the congenic LA/N-cp, there are major metabolic differences. The JCR:LA-cp exhibits a unique pattern of atherosclerotic and myocardial lesions and is a close model of the human syndrome, which is characterized by obesity, hyperlipidemia, hyperinsulinemia, and insulin resistance and is at risk for cardiovascular disease (Russell and Koeslag, 1990; Russell and Graham, 2001).

The JCR:LA-cp rat is characterized by hyperphagia and high body weight, with males normally weighing 800 to 900 g by 9 months of age. In contrast, the lean male weighs 400 to 450 g at this age. The strain is insulin resistant, hyperlipidemic, and hypertriglyceridemic but is not hypertensive. The obese male exhibits spontaneous atherosclerosis and ischemic myocardial lesions developing from occlusive thrombi. Neither the obese nor the lean females develop such thrombi (Russell, 1993).

The atherosclerotic changes are evident by 12 weeks of age in the obese male and increase in severity with age. By 9 months of age, essentially all of the obese males have advanced intimal lesions of the aortic arch that are similar to those seen in the human and are characterized by accumulation of lipids, proteoglycan, collagen, macrophages, smooth muscle cells, and cellular debris in the intimal space (Russell, 1993; Russell and Graham, 2001).

D. Surgical Myocardial Infarction Model

Induction of myocardial infarction can be accomplished by permanently ligating the left anterior descending coronary artery. This procedure can be done on any rat strain. Many commercial rat vendors will perform this procedure for a fee.

E. Aortic Banding (Pressure Overload) Model of Heart Failure

A surgical model of heart failure, banding of the ascending and descending aorta in rats, produces a severe ventricular hypertrophy after about 12 weeks. The model shows a transition of left ventricular hypertrophy into cardiac failure (Weinberg et al., 1994). This model mimics heart failure in of patients with stenosis of the aortic valve (Muders and Elsner, 2000). Many commercial rat vendors will perform this procedure for a fee.

VII. MODELS OF MENOPAUSE

Menopause is the cessation of ovarian cyclicity resulting from the depletion of ovarian follicles by a natural process of attrition known as atresia. Follicular maturation in the ovary is a dynamic series of events in which primordial follicles provide a finite pool from which preovulatory follicles are selected for development and ovulation. The primordial follicle, the most immature stage of development, is formed in the ovary during fetal development. Because the oocyte is arrested in meiosis, this pool is nonregenerating after birth. In an on-going process, follicles continually progress from the primordial to ovulatory stages after puberty. However, the vast majority of the follicles do not develop to the ovulation stage but instead undergo cell death by atresia. As a result, the pool of primordial follicles gradually becomes depleted, and ultimately ovarian failure (menopause) occurs.

Until recently, the only method for inducing menopause has been surgical ovariectomy. However, this method alters ovarian steroid hormone production, a process that does not occur in natural menopause. In contrast, Smith et al. (1990) showed that intraperitoneal dosing with 4-vinycyclohexene for 30 days produced ovarian follicle loss in mice but not in rats. However, dosing with 4-vinyl-1-cyclohexene diepoxide (VCD) resulted in loss of ovarian follicles in both species (Smith et al., 1990). Flaws et al. (1994) found that VCD destroyed oocytes in small preantral (primordial and primary) follicles in the rat. Further, Hoyer et al. (2001) found that intraperitoneal dosing for 30 days with 80 mg/kg of VCD destroyed the majority of ovarian primordial follicles in rats.

Surgical ovariectomy mimics the loss of 17-β estradiol typical of menopause; however, it does not include the influence and the impact of secreted androgens that are present in the menopausal ovary. Although there is a decrease in circulating 17-β estradiol in menopause, there is a concomitant rise in the level of other gonadotropins, specifically follicle stimulating hormone and luteinizing hormone, owing to negative feedback from the ovary to the pituitary gland (Hoyer and Sipes, 1996).

VIII. MODELS OF RHEUMATOID ARTHRITIS

Rheumatoid arthritis is a chronic autoimmune disease that involves inflammation of the lining of the joints and/or other internal organs. Rheumatoid arthritis typically affects many different joints. The synovium can become inflamed, leading to damage of the bone and cartilage. The inflammatory cells cause thickening of the synovial lining, pannus formation, and release enzymes that erode bone and cartilage. The involved joint can lose its shape and alignment, resulting in pain and loss of movement.

There are two types of induced arthritis models typically used to study this disease: the collagen-induced arthritis model and the adjuvant induced arthritis model.

Autoimmune arthritis can be experimentally induced in rodents and non-human primates with type II collagen (Trentham et al., 1977), complete Freund's adjuvant (Audibert and Chedid, 1976), and incomplete Freund's adjuvant (Holmdahl et al., 1992), as well as with pristane (Vingsbo et al., 1996; Olofsson et al., 2002) and extracts of streptococcal cell walls (Wilder et al., 1982). After immunization with type II collagen, susceptible animals develop an autoimmune-mediated polyarthritis that shares many characteristics with the rheumatiod arthritis found in humans. The rat strains commonly used in experimentally-induced arthritis include the LEW rat, the DA rat, and the BB(DR)/Wor rat. The DA rat is highly susceptible to the development of experimentally induced arthritis with type II collagen, whereas the LEW rat is not; however, the LEW rat is susceptible to arthritis induced by streptococcal cell wall extracts. The DA rat was also found to be susceptible to arthritis induced by incomplete Freund's adjuvant alone. Susceptibility and severity of the disease varies among the inbred strains (Kawahito et al., 1998). The BB(DR)/Wor rat, which is resistant to the development of type 1 diabetes, is susceptible to collagen-induced arthritis (Watson, 1990). Experimental rheumatoid arthritis is influenced by genetic factors, especially those related to the major histocompatibility complex class II alleles; many of the loci are sex influenced (Furuya et al., 2000).

REFERENCES

Agardh, C.D., Ahardh, E., Hultberg, B., Qian, Y., and Ostenson, C.G. (1998). The glutathione levels are reduced in Goto-Kakizaki rat retina, but are not influenced by aminoguanidine treatment. *Curr. Eye Res.* **17**, 251–256.

Agardh, C.D., Agardh, E., Zhang, H., and Ostenson, C.G. (1997). Altered endothelial/pericyte ratio in Goto-Kakizaki rat retina. *J. Diabetes Complications* **11**, 158–162.

Audibert, F. and Chedid, L. (1976). Adjuvant disease induced by mycobacteria, determinants of arthritogenicity. *Agents Actions* **6**, 75–85.

Awai, J. and Heine, M. (1986). Dahl salt-sensitive and salt-resistant rats. *Hypertension* **9(Suppl I)**, I-18–I-20.

Barber, B.R., Ferrari, P., and Bianchi, G. (1994). The Milan hypertensive strain: a description of the model. *In* "Handbook of Hypertension, Vol 16: Experimental and Genetic Models of Hypertension." (D. Ganten and W. de Jong, eds.), pp 316–345, Elsevier, Amsterdam.

Baskin, D.G., Stein, J., Ikeda, H., Woods, S.C., Figlewicz, D.P., Porte, Jr., D., Greenwood, M.R.C., and Dorsa, D.M. (1985). Genetically obese Zucker rats have abnormally low brain insulin content. *Life Sci.* **36**, 627–633.

Bataillard, A., Sacquet, J., Julien, C., Vincent, M., Gomezsanchez, C., Touraine, J., and Sassard, J. (1993). Do thymic-neuroendocrine interactions play a role in the antihypertensive effect of neonatal thymectomy in Lyon hypertensive rats. *Am. J. Hypertens.* **6**, 407–412.

Ben-Ishay, D., Saiternik, D.R., and Welner, A. (1972) Separation of two strains of rats with inbred dissimilar sensitivity to DOCA-salt hypertension, *Experientia* **28**, 1321–1322.

Ben-Ishay, D. and Yagil, Y. (1994) The Sara hypertension-prone and -resistant strains. *In* "Handbook of Hypertension, Vol 16: Experimental and Genetic Models of Hypertension." (D. Ganten and W. de Jong, eds.), pp 272–299, Elsevier, Amsterdam.

Bergman, M., Kao, R., McCune, S., and Holycross, B. (1999). Myocardial tumor necrosis factor-alpha secretion in hypertensive and heart failure-prone rats. *Am J. Physiol.* **277(2 Pt 2)**, G543–G550.

Betarbet, R., Sherer, T., and Greenamyre, J. (2002). Animal models of Parkinson's disease. *Bioessays* **24**, 308–318.

Bianchi, G., Ferrari, P., and Barber, B.R. (1984) The Milan hypertensive strain. *In* "Handbook of Hypertension, Vol 4: Experimental and Genetic Models of Hypertension." (D. Ganten and W. de Jong, eds.), pp 328–349, Elsevier, Amsterdam.

Blizard, D.A., Peterson, W.N., and Adams, N. (1991). Dietary salt and accelerated hypertension: lack of sub-line differentiation in spontaneously hypertensive rat stocks from the United States. *J. Hypertens.* **9**, 1169–1175.

Boulanger, M., Duhault, J., Broux, O., Batalillard, A., and Sassard. K. (1997). Lack of insulin resistance in the Lyon hypertensive rat. *Fundam. Clin. Pharmacol.* **11**, 546–549.

Bray, G.A. and York, D.A. (1971). Genetically transmitted obesity in rodents. *Physiol. Rev.* **51**, 598–646.

Bray, G.A. and York, D.A. (1979). Hypothalamic and genetic obesity in experimental animals: an autonomic and endocrine hypothesis. *Physiol. Rev.* **59**, 719–809.

Brenner, B.M. (1993). Pathogenesis of glomerular injury in the fawn-hooded rat: early glomerular capillary hypertension predicts glomerular sclerosis. *J. Am. Soc. Nephrol.* **3**, 1775–1786.

Brown, D.M., Provoost, A.P., Daly, M.J., Lander, E.S., and Jacob, H.J. (1996). Renal disease susceptibility and hypertension are under independent genetic control in the fawn-hooded rat. *Nature Genet.* **12**, 44–51.

Butler, L., Guberski, D.L., and Like, A.A. (1983). Genetic analysis of the BB/W diabetic rat. *Canad. J. Genet. Cytol.* **25**, 7–15.

Butler, L., Guberski, D.L., and Like, A.A. (1988). Genetics of diabetes production in the Worcester colony of the BB rat. *In* "Frontiers in Diabetes Research: Lessons from Animal Diabetes II." (E. Shafrir and A.E. Renold, eds.), pp. 74–78. Smith Gordon & Company, London.

Cavalieri, E., Devanesan, P. Bosland, M., Badawi, A., and Rogan, E. (2002). Catechol estrogen metabolites and conjugates in different regions of the prostate of Noble rats treated with 4-hydroxyestradiol: implications for estrogen-induced initiation of prostate cancer. *Carcinogenesis* **23**, 29–33.

Cenci M.A., Whishaw I.Q., and Schallert T. (2002). Animal models of neurological deficits: how relevant is the rat? *Nat. Rev. Neurosci.* **3**, 574–579.

Chakrabarti, S., Ma, N., and Sima, A. (1989). Reduced number of anionic sites is associated with glomerular basement membrane thickening in the diabetic BB rat. *Diabetologia* **32**, 826–828.

Chakrabarti, S., Zhang, W., and Sima, A. (1991) Optic neuropathy in the diabetic BB rat. *Adv. Exp. Med. Biol.* **291**, 257–264.

Chiueh, C., Markey, S., Burns, R., Johannessen, J., Pert, A., and Kopin, I. (1984) Neurochemical and behavioral effects of systemic and intranigral administration of *N*-methyl-4-phenyl-1,2,3,6-tetrahydropyridine in the rat. *Eur. J. Pharmacol.* **100**, 189–194.

Chua, S., White, D., Wu-Peng, X., Liu, S., Okada, N., Kershaw, E., Chung, W., Power-Kehoe, L., Chua, M., Tartaglia, L., and Leibel, R. (1996). Phenotype of fatty due to Gln269Pro mutation in the leptin receptor (Lepr). *Diabetes* **45**, 1141–1143.

Churchill, P.C., Churchill, M.C., Griffin, K.A., Picken, M., Webb, R.C., Kurtz, T.W., and Bidani, A.K. (2002). Increased genetic susceptibility to renal damage in the stroke-prone spontaneously hypertensive rat. *Kidney Int.* **61**, 1794–1800.

Clark, C.J., Milligan, G., and Connell, J.M.C. (1993). Guanine-nucleotide regulatory protein alterations in the Milan hypertensive rat strain, *J. Hyperten.* **11**, 1161–1169.

Cohen, A.M., Rosenmann, E., and Rosenthal, T. (1993). The Cohen diabetic (non-insulin-dependent) hypertensive rat model: description of the model and pathologic findings. *Am. J. Hyperten.* **6**, 989–995.

Cole, G.M. and Frautschy, S.A. (1997). Animal models for Alzheimer's disease. *Alzheimer's Dis. Rev.* **2**, 2–10.

Coenen, A.M.L., Drinkenberg, W.H.I.M., Inoue, M., and Van Luijtelaar, E.L.J.M. (1992). Genetic models of absence epilepsy, with emphasis on the WAG/Rij strain of rats. *Epilepsy Res.* **12**, 75–86.

Columbo, G., Hansen, C., Hoffman, P., and Gran, K. (1992). Decreased performance in a delayed alternation task by rats genetically deficient in vasopressin. *Physiol. Behav.* **52**, 827–830.

Corsetti, J., Sparks, J., Peterson, R., Smith, R., and Sparks, C. (2000). Effect of dietary fat on the development of non-insulin dependent diabetes mellitus in obese Zucker diabetic fatty male and female rats. *Atherosclerosis* **148**, 231–241.

Cowley, Jr., B., Ricardo, S., Nagao, S., and Diamond, J. (2001). Increased renal expression of monocyte chemoattractant protein-1 and osteopontin in ADPKD in rats. *Kidney Int.* **60**, 2087–2096.

Dall'Ago, P., Silva, V., De Angelis, K., Irigoyen, M., Fazan, Jr., R., and Salgado, H. (2002). Reflex control of arterial pressure and heart rate in short-term streptozotocin diabetic rats. *Braz. J. Med. Biol. Res.* **35,** 843–849.

Danis, R. and Yang, Y. (1993). Microvascular retinopathy in the Zucker diabetic fatty rat. *Invest. Ophthal. Visual Sci.* **34,** 2367–2371.

Datta, Y.H., Wu, F.C., Dumas, P.C., Rangel-Filho, A., Datta, M.W., Ning, G., Cooley, B.C., Majewski, R.R., Provoost, A.P., and Jacob, H.J. (2003). Genetic mapping and characterization of the bleeding disorder in the fawn-hood hypertensive rat. *Thromb. Haemost.* **89,** 951–952.

Dobrian, A., Davies, M., Schriver, S., Lauterio, T., and Prewitt, R. (2001). Oxidative stress in a rat model of obesity-induced hypertension. *Hypertension* **37,** 554–560.

Doss, D.J., Kuruvilla, R., Bianchi, R., Peterson, R.G., and Eichberg, J. (1997). Effects of hypoxia and severity of diabetes on Na, K-ATPase activity and arachidonoyl-containing glycerophospholipid molecular species in nerve from streptozotocin diabetic rats. *J. Peripher. Nerv. Syst.* **2,** 155–163.

Dubay, C., Vincent, M., Samani, N.J., Hilbert, P., Kaiser, M.A., Beressi, J.P., Kotelevtsey, Y., Beckmann, J.S., Soubrier, F., Sassard, J., and Lathrop, G.M. (1993). Genetic-determinants of diastolic and pulse pressure map to different loci in Lyon Hypertensive Rats. *Nat. Genet.* **3,** 354–357.

Dumitrescu, C., Narayan, P., Efimov, I., Cheng, Y., Radin, M., McCune, S., and Altschuld, R. (2002). Mechanical alternans and restitution in failing SHHF rat left ventricles. *Am J. Physiol. Heart Circ. Physiol.* **282,** H1320–H1326.

Dworkin, L.D. and Feiner, H.D. (1986) Glomerular injury in uninephrectomized spontaneously hypertensive rats: a consequence of glomerular capillary hypertension. *J. Clin. Invest.* **77,** 797–809

Farber, S.D., Farber, M.O., Brewer, G., Magnes, C.J., and Peterson, R.G. (1991). Oxygen affinity of hemoglobin and peripheral nerve degeneration in experimental diabetes. *J. Neurol. Sci.* **101,** 204–207.

Flaws, J., Salyers, K., Sipes, I., and Hoyer, P. (1994). Reduced ability of rat preantral ovarian follicles to metabolize 4-Vinyl-1-cyclohexene diepoxide in vitro. *Toxicol. Appl. Phar.* **126,** 286–294.

Festing, M.F. and Bender, K. (1984). Genetic relationships between inbred strains of rats: an analysis based on genetic markers at 28 biochemical loci. *Genet. Res.* **44,** 271–281.

Floege, J., Hackmann, B., Kliem, V., Kriz, W., Alpers, C.E., Johnson, R.J., Kuhn, K.W., Koch, K.M., and Brunkhorst, R. (1997). Age-related glomerulosclerosis and interstitial fibrosis in Milan normotensive rats: a podocyte disease. *Kidney Int.* **51,** 230–243.

Florin, M., Lo, M., Liu, K., and Sassard, J. (2001). Salt sensitivity in genetically hypertensive rats of the Lyon strain. *Kidney Int.* **59,** 1865–1872.

Furuya, T., Salstrom, J., McCall-Vining, S., Cannon, G., Joe, B., Remmers, E., Griffiths, M., and Wilder, R. (2000). Genetic dissection of rat model for rheumatoid arthritis: significant gender influences on autosomal modifier loci. *Hum. Mol. Genet.* **9,** 2241–2250.

Goodenough, S., Schafer, M., and Behl, C. (2003) Estrogen-induced cell signaling in a cellular model of Alzheimer's disease. *J. Steroid Biochem. Mol. Biol.* **84,** 301–305.

Goto, Y., Kakizaki, M., and Masaki, N. (1976). Production of spontaneous diabetic rats by repetition of selective breeding. *Tokohu J. Exp. Med.* **119,** 85–90.

Goto, Y., Suzuki, K., Ono, T., Sasaki, M., and Toyota, T. (1988a). Development of diabetes in the non-obese NIDDM rat (GK rat). *Adv. Exp. Med. Biol.* **246,** 29–31.

Goto, Y., Suzukki, K., Sasaki, M., Ono, T., and Abe, S. (1988b). GK rat as a model of nonobese, nonisulin-dependent diabetes: selective breeding over 35 generations. *In* "Frontiers in Diabetes Research: Lessons from Animal Diabetes II." (E. Shafir and A.E. Renold, eds.), pp 301–303. Smith Gordon & Company, London.

Gould, M.N. (1995). Rodent models for the study of etiology, prevention and treatment of breast cancer. *Semin. Cancer Biol.* **6,** 147–152.

Greenamyre, J., Sherer, T., Betarbet, R., and Panov, A. (2001). Complex I and Parkinson's disease. *IUBMM Life* **52,** 135–141.

Greene, D., Yagihashi, S., Lattimer, S., and Sima, A. (1984). Nerve Na+-K+-ATPase, conduction and myo-inositol in the insulin-deficient BB rat. *Am. J. Physiol.* **247,** E534–E539.

Greenwood, M.R.C., Maggio, C.A., Koopmans, H.S., and Sclafani, A. (1982). Zucker fa/fa rats maintain their obese body composition ten months after jejunoileal bypass surgery. *Int. J. Obesity* **6,** 513–525.

Gresson, C.R. and Simpson, F.O. (1974) Increased plasma volume in hypertensive rats with polyarteritis. *Br. J. Exp. Pathol.* **55,** 519–523.

Guberski, D. (1993) Models of type 1 diabetes: part two. *ILAR News* **35,** 1–14.

Hall, J.E. (1994) Renal and cardiovascular mechanisms of hypertension in obesity. *Hypertension* **23,** 381–394.

Hansen, C. (1983). Two new congenic strains for nutrition and obesity research. *Fed. Proc.* **42,** 537.

Hansen, C.T., Judge, F.J., and Whitney, R.A. (1973). Catalogue of NIH Rodents. DREW Publ. No. 74-606. Department of Health, Education and Welfare, Washington, DC.

Harris, P. (2002). Molecular basis of polycystic kidney disease: PKD1 PKD2 and PKHD1. *Curr. Opin. Nephrol. Hypertens.* **11,** 309–314.

Hawes, J., Xu, W., Peterson, R., and Dominguez, J. (2001). Renal mitochondria injury in NIDDM-linked lipotoxic nephropathy. Abstract T1-0145. 2001 Annual ASN Meeting, San Francisco, CA.

Hernandez, N.E., MacDonall, J.S., Stier, Jr., C.T., Belmonte, A., Fernandez, R., and Karpiak, S.E. (1994). GM1 ganglioside treatment of spontaneously hypertensive stroke prone rats. *Exp. Neurol.* **126,** 95–100.

Heslop, B. and Phelan, E. (1973). The GH and AS hypertensive rat strains. *Lab. Anim.* **7,** 41–46.

Holmdahl, R., Goldschmidt, T., Kleinau, S., Kvick, C., and Jonsson, R. (1992). Arthritis induced in rats with adjuvant oil is a genetically restricted, alpha beta T-cell dependent autoimmune disease. *Immunology* **76,** 197–202.

Hoyer, P., Devine, P., Hu, X., Thompson, K., and Sipes, I. (2001). Ovarian toxicity of 4-vinylcyclohexene diepoxide: a mechanistic model. *Toxicol. Pathol.* **29,** 91–99.

Hoyer, P. and Sipes, I. (1996). Assessment of follicle destruction in chemical-induced ovarian toxicity. *Ann. Rev. Pharm. Toxicol.* **36,** 307–331.

Hsu, L.C., Kennan, W.S., Shepel, L.A., Jacob, H.J., Szpirer, C., Szpirer J., Landers, E.S., and Gould, M.N. (1994). Genetic identification of Mcs-1, a rat mammary carcinoma suppressor gene. *Cancer Res.* **54,** 2765–2770.

Huang, S., Khosrof, S., Koletsky, R., Benetz, B., and Ernsberger, P. (1995). Characterization of retinal vascular abnormalities in lean and obese spontaneously gypertensive rats. *Clin. Exp. Pharm. Physiol. Suppl.* **1,** S129–S131.

Ikeda, H., West, D.B., Pustek, J.J., Figlewicz, D.P., Greenwood, M.R.C., Porte, D., and Woods, S.C. (1986). Intraventricular insulin reduces food intake and body weight of lean but not obese Zucker rats. *Appetite* **7,** 381–386.

Iwai, J. and Heine, M. (1986). Dahl salt-sensitive rats and human essential hypertension, *J. Hyperten. Suppl.* **4,** S29–S31.

Iwase, M., Kikuchi, M., Nunoi, K., Wakisaka, M., Maki, Y., Sadoshima, S., and Fujishima M. (1987). Diabetes induced by neonatal streptozotocin treatment in spontaneously hypertensive and normotensive rats. *Metabolism* **36,** 654–657.

Iwase, M., Uchizono, Y., Tashiro, K.L., Goto, D., and Iida, M. (2002). Islet hyperperfusion during prediabetic phase in OLETF rats, a model of type 2 diabetes. *Diabetes* **51,** 2530–2535.

Jackson, E., Kost, Jr., C., Herzer, W., Smits, G., and Tofovic, S. (2001). A(1) receptor blockade induced natriuresis with a favorable renal hemodynamic profile in SHHF/Mcc-fa(cp) rats chronically treated with salt and furosemide. *J. Pharmacol. Exp. Ther.* **299**, 78–87.

Jacob, H., Brown, D., Bunker, R., Daly, M., Dzau, V., Goodman, A., Koike, G., Kren, V., Kurtz, T., Lernmark, A., Levan, G., Mao, Y., Pettersson, A., Pravenec, M., Simon, J., Szpirer, C., Szpirer, J., Trolliet, M., Winer, E., and Lander, E. (1995). A genetic linkage map of the laboratory rat, Rattus norvegicus. *Nat. Genet.* **9**, 63–69.

Jakobsen, J. and Lundbaek, K. (1976) Neuropathy in experimental diabetes: an animal model. *Br. Med. J.* **2**, 278–279.

James, M. and Lindpaintner, K. (1997). Why map the rat? *Trends Genet.* **13**, 171–173.

Jin, D.Q., Park, B.C., Lee, J.S., Choi, H.D., Lee, Y.S., Yang, J.H., and Kim, J.A. (2004). Mycelial extract of *Cordyceps ophioglossoides* prevents neuronal cell death and ameliorates beta-amyloid peptide-induced memory deficits in rats. *Biol. Pharmacol. Bull.* **27**, 1126–1129.

Johnson, M.L., Ely, D.L., and Turner, M.E. (1992). Genetic divergence between the Wistar-Kyoto rat and the spontaneously hypertensive rat. *Hypertension.* **19**, 425–427.

Kava, R., Peterson, R.G., West, D.B., and Greenwood, M.R.C. (1990). Wistar diabetic fatty rat, *ILAR News,* **32**, 9–13.

Kawahito, Y., Cannon, G., Gulko, P., Remmers, E., Longman, R., Reese, V., Wang, J., Griffiths, M., and Wilder, R. (1998). Localization of quantitative trait loci regulating adjuvant-induced arthritis in rats: evidence for genetic factors common to multiple autoimmune disease, *J. Immunol.* **161**, 4411–4419.

Kawano, K., Hirashima, T., Mori, S., Man, Z., and Natori, T. (2001). The OLETF rat. *In* "Animal Models of Diabetes: A Primer." (A.A.F. Sima and E. Shafrir, eds.), pp. 213–225. Harwood Academic Publishers, The Netherlands.

Kelly, P., Seviour, P., and Iversen, S. (1975). Amphetamine and apomorphine responses in the rat following 6-OHDA lesions of the nucleus accumbens septi and corpus striatum. *Brain Res.* **94**, 507–522.

Koletsky, S. (1973). Obese spontaneously hypertensive rats: a model for the study of atherosclerosis. *Exp. Mol. Pathol.* **19**, 53–60.

Koletsky, R.J., Friedman, J.E., and Ernsberger, P. (2001). The obese spontaneously hypertensive rat (SHROB, Koletsky rat): a model of metabolic syndrome X. *In* "Animal Models of Diabetes: A Primer." (A.A.F. Sima and E. Shafrir, eds.), pp. 143–158. Harwood Academic Publisher, The Netherlands.

Koplan, J.P. (2002). Diabetes Public Health Resource, CDC Diabetes Program Publication-Diabetes at-a-Glance, CDC, Atlanta, GA.

Kuijpers, M. and de Jong, W. (1986). Spontaneous hypertension in the fawn-hooded rat: a cardiovascular disease model. *J. Hypertens Suppl.* **4**, S41–S44.

Kuijpers, M. and de Jong, W. (1987). Relationship between blood pressure level, renal histopathological lesions and plasma rennin activity in fawn-hooded rats. *Br. J. Exp. Pathol.* **68**, 179–187.

Kuijpers, M. and Gruys, E. (1984). Spontaneous hypertension and hypertensive renal disease in the fawn-hooded rat. *Br. J. Exp. Pathol.* **65**, 181–190.

Kuijpers, M., Provoost, A., and de Jong, W. (1986). Development of hypertension and proteinuria with age in fawn-hooded rats. *Clin. Exp. Pharmacol. Physiol.* **13**, 201–209.

Kurthy, M., Mogyorosi, T., Nagy, K., Kukorelli, T., Jednakovits, A., Talosi, L., and Biro, M. (2002). Effect of BRX-220 against peripheral neuropathy and insulin resistance in diabetic rat models. *Ann. N.Y. Acad. Sci.* **967**, 482–489.

Kurtz, B.S., Lehman, J., Garlick, P., Amberg, J., Mishra, P.K., Dailey, J.W., Weber, R., and Jobe, P.C. (2001). Penetrance and expressivity of genes involved in the development of epilepsy in the genetically epilepsy-prone rat (GEPR). *J. Neurogenet.* **15**, 233–244.

Kurtz, T.W., Montano, M., Chan, L., and Kabra, P. (1989) Molecular evidence of genetic heterogeneity in Wistar-Kyoto rats: implications for research with the spontaneously hypertensive rat. *Hypertension* **13**, 188–192.

Lager, D., Qian, Q., Bengal, R., Ishibashi, M., and Torres, V. (2001). The pck rat: a new model that resembles human autosomal dominant polycystic kidney and liver disease. *Kidney Int.* **59**, 126–136.

Leav, I., Ho, S.M., Ofner, P., Merk, F., Kwan, P.L., and Damassa, D. (1988). Biochemical alterations in the sex hormone-induced hyperplasia and dysplasia of the dorsolateral prostates of Nobel rats. *J. Natl. Cancer Inst.* **80**, 1045–1053.

Le Cras, T.D., Kim, D.H., Markham, N.E., and Abman, S.H. (2000). Early abnormalities of pulmonary vascular development in the fawn-hooded rat raised at Denver's altitude. *Am. J. Physiol. Lung Cell Mol. Physiol.* **279**, L283–L291.

Lenz, D., Wolf, N., and Swanborg, R. (1999). Strain variation in autoimmunity: attempted tolerization of DA rats in the induction of experimental autoimmune encephalomyelitis. *J Immunol.* **163**, 1763–1768.

Levin, B. and Dunn-Meynell, A. (2002). Defense of body weight depends on dietary composition and palatability in rats with diet-induced obesity. *Am. J. Physiol. Reg. Integr. Comp Physiol.* **282**, R46–R54.

Levin, B., Dunn-Meynell, A., Balkan, B., and Keesey, R. (1997). Selective breeding for diet-induced obesity and resistance in Sprague-Dawley rats. *Am. J. Physiol.* **273**, R225–R230.

Levin, B., Michel, R., and Servatius, R. (2000). Differential stress responsivity in diet-induced obese and resistant rats. *Am. J. Physiol. Reg. Integr. Comp. Physiol.* **279**, R1357–R1364.

Lindsay, J.R. (1979). Historical foundations. *In* "The Laboratory Rat," Vol. 1. (H.J. Baker, J.R. Lindsay, and S.H. Weisbroth, eds.), pp 2–36, Academic Press, Orlando, FL.

Lindsey, S., Nichols, Jr., C.W., and Chaikoff, I.L. (1968). Naturally occurring thyroid carcinoma in the rat. Similarities to human medullary carcinoma. *Arch. Pathol.* **86**, 353–364.

MacMurray, A., Moralejo, D., Kwitek, A., Rutledge, E., Van Yserloo, B., Gohlke, P., Speros, S., Snyder, B., Schaefer, J., Bieg, S., Jiang, J., Ettinger, R., Fuller, J., Daniels, T., Pettersson, A., Orlebeke, K., Birren, B., Jacob, H., Lander, E., and Lernmark, A. (2002). Lymphopenia in the BB rat model of type 1 diabetes is due to a mutation in a novel immune-associated nucleotide (Ian)-related gene. *Genome Res.* **7**, 1029–1039.

Madiehe, A.M., Schaffhauser, A.O., Braymer, H.D., Bray, G.A., and York, D.A. (2000). Differential expression of leptin receptor in high and low-fat-fed Osborne-Mendel and S5B/PI rats. *Obes. Res.* **8**, 367–474,

Martin, A. and Weber, B. L. (2000). Genetic and hormonal risk factors in breast cancer. *J. Natl. Cancer Inst.* **92**, 1126–1135.

Mathew, J., Bianchi, R., McLean, W., Peterson, R., Roberts, R., and Savaresi, S. (1997). Phosphoinositide metabolism, Na, K-ATPase and protein kinase C are altered in peripheral nerve from Zucker fatty diabetic rats (ZDF/Gmi-fa). *Neurosci. Res. Comm.* **20**, 21–30.

McCune, S.A., Baker, P.B., and Stills, H.F. (1990). AHHF/Mcc-cp rat: model of obesity, non-insulin-dependent diabetes, and congestive heart failure. *ILAR News* **32**, 23–27.

Mekler, J., Yagil, Y., and Ben-Ishay, D. (1985) Renal response to acute saline loading in Sabra hypertension-prone and -resistant rats. *Experientia* **41**, 923–925.

Mouse Genome Database, Mouse Genome Informatics Web Site, The Jackson Laboratory, Bar Harbor, Maine. World Wide Web (*URL:http.www.informatics.jax.org*). June, 2004.

Mordes, J., Bortell, R., Groen, H., Guberski, D., Rossini, A., and Greiner, D. (2001). Autoimmune diabetes mellitus in the BB rat. *In* "Animal Models of Diabetes: A Primer." (A.A. Sima and E. Shafrir, eds.), pp. 1–41. Harwood Academic Publisher, Amsterdam.

More N.S., Rao N.A., and Preuss H.G. (1986). Early sucrose-induced retinal vascular lesions in spontaneously hypertensive rats (SHR) and Wistar Kyoto rats (WKY). *Ann. Clin. Lab. Sci.* **16**, 419–426.

Muders, F. and Elsner, D. (2000) Animal models of chronic heart failure. *Pharmacol. Res.* **41**, 605–612.

Muller, D., Molinari, I., Soldati, L., and Bianchi, G. (1995). A genetic deficiency in calpastatin and isovalerylcarnitine treatment is associated with enhanced hippocampal long-term potentiation. *Synapse* **19**, 37–45.

Nakhooda, A.F., Like, A.A., Chappel, C.L., Murray, F.T., and Marliss, E.B. (1977). The spontaneously diabetic Wistar rat metabolic and morphologic studies. *Diabetes* **26**, 100–112.

Nakhooda, A.F., Like, A.A., Chappel, C.I., Wei, C.N., and Marliss, E.B. (1978). The spontaneously diabetic Wistar rat (The "BB" Rat): studies prior to and during development of the overt syndrome. *Diabetologia* **14**, 199–207.

Narayan, P., McCune, S., Robitaille, P., Hohl, C., and Altschuld, R. (1995). Mechanical alternans and the force-frequency relationship in failing rat hearts. *J. Mol. Cell Cardiol.* **27**, 523–30.

Nobel, R. (1977). The development of prostatic adenocarcinoma in Nb rats following prolonged sex hormone administration. *Cancer Res.* **37**, 1929–1933.

Noble, R., Hochachka, B., and King, D. (1975). Spontaneous and estrogen-produced tumors in Nb rats and their behavior after transplantation. *Cancer Res.* **35**, 766–780.

Odaka, H., Sugiyama, Y., and Ikeda, H. (2001). Characteristics of the Wistar fatty rat. *In* "Animal Models of Diabetes: A Primer." (A.A.F. Sima and E. Shafrir, eds.), pp. 159–169. Harwood Academic Publishers, The Netherlands.

Ogata, A., Tashiro, K., Nukuzuma, S., Nagashima, K., and Hall, W. (1997). A rat model of Parkinson's disease induced by Japanese encephalitis virus, *J. Neurovirol.* **3**, 141–147.

Oiso, N., Riddle, S.R., Serikawa, T., Kuramoto, T., and Spritz, R.A. (2004). The rat Ruby <RM> locus is Rab38:identical mutations in Fawn-hooded and Tester-Moriyama rats derived from an ancestral Long Evans rat sub-strain. *Mamm. Genome* **15**, 307–314.

Okamoto, K. and Aoki, K. (1963) Development of a strain of spontaneously hypertensive rats. *Jpn. Circ. J.* **27**, 282–293.

Okamoto, K., Yamamoto, K., Morita, N., Ohta, Y., Chikugo, T., Higashizawa, T., and Suzuki, T. (1986). Establishment and use of the M strain of stroke-prone spontaneously hypertensive rat. *J. Hypertens. Suppl.* **4**, S21–S24.

Okamoto, K., Yamori, Y., and Nagaoka, A. (1974) Establishment of the stroke-prone spontaneously hypertensive rat (SHR). *Circ. Res.* **34(Suppl I)**, I-143–I-153

Olofsson, P., Nordquist, N., Vingsbo-Lundberg, C., Larsson, A., Falkenberg, C., Pettersson, U., Akerstrom, B., and Holmdahl, R. (2002). Genetic links between the acute-phase response and arthritis development in rats. *Arthritis Rheum.* **46**, 259–268.

Ostenson C.G., Sandberg-Nordqvist, A.C., Chen, J., Hallbrink, M., Rotin, D., Langel, U., and Efendic, S. (2002). Overexpression of protein-tyrosine phosphatase PTP sigma is linked to impaired glucose-induced insulin secretion in hereditary diabetic Goto-Kakizaki rats. *Biochem. Biophys. Res. Commun.* **291**, 945–950.

Ozturk, B., Cetinkaya, M., Gur, S., Uguriu, O., Oztekin, V., and Aki, T. (2002). The early effects of partial outflow obstruction on contractile properties of diabetic and non-diabetic rat bladder. *Urol. Res.* **30**, 178–184.

Peeters, B.W.M.M., Ramakers, G.M.J., Vossen, J.M.H., and Coenen, A.M.L. (1994). The WAG/Rij rat model for nonconvulsive absence

epilepsy-involvement of non-NMDA receptors. *Brain Res. Bull.* **33**, 709–713.

Peeters, B.W.M.M., Spooren, W.P.J.M., Van Luijtelaar, E.L.J.M., and Coenen, A.M.L. (1988). The WAG/Rij rat model for absence epilepsy: anticonvulsant drug evaluation. *Neurosci. Res. Commun.* **2**, 93–97.

Peterson, R. (2001). The Zucker diabetic fatty (ZDF) Rat. *In* "Animal Models of Diabetes: A Primer." (A.A.F. Sima and E. Shafrir, eds.), pp. 109–128. Harwood Academic Publishers, The Netherlands.

Peterson, R., Neel, M., Little, L., Kincaid, J., and Eicherg, J. (1990). Neuropathic complications in the Zucker diabetic fatty rat (ZDF/Drt-fa). *In* "Frontiers in Diabetes Research. Lessons from Animal Diabetes III." (E. Shafrir, ed.), pp. 456–458. Smith-Gordon, London.

Phelan, E. (1968) The New Zealand strain of rats with genetic hypertension. *New Zeal. Med. J.* **67**, 334–344.

Plata-Salaman, C.R., Ilyin, S.E., and Gayle, D. (1998). Brain cytokine mRNAs in anorectic rats bearing prostate adenocarcinoma tumor cells. *Am. J. Physiol.* **275**, R566–R573.

Pollard, M. (1973). Spontaneous prostate adenocarcinomas in aged germfree Wistar rats. *J. Natl. Cancer Inst.* **511**, 1235–1241.

Pollard, M. (1992). The Lobund-Wistar rat model of prostate cancer. *J. Cell Biochem. Suppl.* **16H**, 84–88.

Pollard, M. (1998). Lobund-Wistar rat model of prostate cancer in man. *Prostate* **37**, 1–4.

Pollard, M. and Luckert P.H. (1986). Production of autochthonous prostate cancer in Lobund-Wistar rats by treatments with *N*-nitroso-*N*-methylurea and testosterone. *J. Natl. Cancer Inst.* **77**, 583–587.

Pollard, M. and Luckert, P.H. (1987). Autochthonous prostate adenocarcinomas in Lobund-Wistar rats: a model system, *Prostate* **11**, 219–227.

Pop-Busui, R., Marinescu, V., Van Huysen, C., Li, F., Sullivan, K., Greene, D., Larkin, D., and Stevens, M. (2002). Dissection of metabolic, vascular, and nerve conduction interrelationships in experimental diabetic neuropathy by cyclooxygenase inhibition and acetyl-L-carnitine administration. *Diabetes* **51**, 2619–2628.

Portha, B., Giroix, M.H., Serradas, P., Movassat, J., Bailbe, D., and Kergoat, M. (2001). The neonatally streptozotocin induced (n-STZ) diabetic rats, a family of NIDDM models. *In* "Animal Models of Diabetes: A Primer." (A.A.F. Sima and E. Shafrir, eds.), pp. 247–281. Harwood Academic Publishers, The Netherlands.

Portha, B., Levacher, C., Picon, L., and Rosselin, G. (1974). Diabetogenic effects of streptozotocin during the perinatal period in the rat. *Diabetes* **23**, 889–895.

Radin, M., Holycross, B., Sharkey, L., Shiry, L., and McCune, S. (2002). Gender modulates activation of rennin-angiotensin and endothelin system in hypertension and heart failure. *J. Appl. Physiol.* **92**, 935–40.

Rapp J.P. (1982). Dahl salt-susceptible and salt-resistant rats: a review. *Hypertension* **4**, 753–763.

Rapp, J.P. (2000). Genetic analysis of inherited hypertension in the rat. *Physiol. Rev.* **80**, 135–172.

Rapp, J.P. and Dene, H. (1985). Development and characteristics of inbred strains of Dahl salt-sensitive and salt-resistant rats. *Hypertension* **7**, 340–349.

Rat Genome Sequencing Project Consortium (2004). Genome sequence of the brown Norway rat yields insight into mammalian evolution. *Nature* **493**, 493–521.

Reaven, G.E. (1988). Banting lecture 1988: role of insulin resistance in human disease. *Diabetes* **37**, 1595–1607.

Rehman, J., Chenven, E., Brink, P., Petereson, B., Walcott, B., Wen, Y.P., Melman, A., and Christ, G. (1997). Diminished neurogenic but not pharmacological erections in the 2-to 3-month experimentally diabetic F-344 rat. *Am. J. Physiol.* **272**, H1960–H1971.

Reigel, C.E., Dailey, J.W., and Jobe, P.C. (1986). The genetically epilepsy-prone rat: an overview of seizure-prone characteristics and responsiveness to anticonvulsant drugs. *Life Sci.* **39**, 763–774.

Reigel, C.E., Jobe, P.C., Dailey, J.W., and Savage, D.D. (1989). Ontogeny of sound-induced seizures in the genetically epilepsy-prone rat. *Epilepsy Res.* **4**, 63–71.

Rezvani, A.H., Parsian, A., and Overstreet, D.H. (2002). The Fawn-hooded (FH/Wjd) rat: a genetic animal model of comorbid depression and alcoholism. *Psychiatr. Genet.* **12**, 1–16.

Roba J.L. (1976). The use of spontaneously hypertensive rats for the study of anti-hypertensive agents. *Lab. Anim. Sci.* **26**, 305–319.

Rosenthal, T., Erlich, Y., Rosenmann, E., Grossman, E., and Cohen, A. (1995). Enalapril improves glucose tolerance in two rat models: a new hypertensive diabetic strain and a fructose-induced hyperinsulinaemic rat. *Clin. Exp. Pharmacol. Physiol. Suppl.* **22**, S353–S354.

Russell, J. (1993). Insulin resistance and cardiovascular disease: lessons from the *JCR:LA-cp and other strains of obese rat. In* "Lessons from Animal Diabetes IV." (E. Shafrir, ed.), pp. 137–144. Smith-Gordon, Great Britain.

Russell, J. and Graham, S. (2001). The JCR:LA-cp rat: an animal model of obesity and insulin resistance with spontaneous cardiovascular disease. *In* "Animal Models of Diabetes: A Primer." (A.A.F. Sima and E. Shafrir, eds.), pp. 227–245. Harwood Academic Publishers: The Netherlands.

Russell, J. and Koeslag, D. (1990). Jcr:LA-corpulent rat: a strain with spontaneous vascular and myocardial disease. *ILAR News* **32**, 27–32.

Sakurai, M., Higashide, T., Takeda, H., and Shirao, Y. (2002). Characterization and diabetes-induced impairment of nitric oxide synthase in rat choroids. *Curr. Eye Res.* **24**, 139–146.

Sanzen, T., Harada, K., Yasoshima, M., Kawamura, Y., Ishibashi, M., and Nakanuma, Y. (2001). Polycystic kidney rat is a novel animal model of Caroli's disease associated with congenital hepatic fibrosis. *Am J. Pathol.* **158**, 1605–1612.

Sato, K., Webb, S., Tucker, A., and Rabinovitch, M. (1992). Factors influencing the idiopathic development of pulmonary hypertension in the Fawn hooded rat. *Am. Rev. Respir. Dis.* **145**, 793–797.

Saul, G., Garrity, E., Benirschke, K., and Valtin, H. (1968). Inherited hypothalamic diabetes insipidus in the Brattleboro strain of rats. *J. Hered.* **59**, 113–117.

Schemmel, R., Mickelsen, O., and Gill, J.L. (1970). Dietary obesity in rats: Body weight and body fat accretion in seven strains of rats. *J. Nutr.* **100**, 1041–1048.

Schridde, U. and Van Luijtelaar, G. (2004). Corticosterone increases spike-wave discharges in a dose-and time-dependent manner in WAG/Rij rats. *Pharmacol. Biochem. Behav.* **78**, 369–375.

Selkoe, D.J. (1991). The molecular pathology of Alzheimer's disease. *Neuron* **6**, 487–498.

Sharkey, L., Holycross, G., Park, S., Shiry, L., Hoepf, T., McCune, S., and Radin, M. (1999). Effect of ovariectomy and estrogen replacement on cardiovascular disease in heart failure-prone SHHF/Mcc-fa cp rats. *J. Mol. Cell Cardiol.* **8**, 1527–1537.

Shepel, L.A. and Gould, M.N. (1999). The genetic components of susceptibility to breast cancer in the rat. *Prog. Exp. Tumor Res* **35**, 158–169.

Shibata, T., Takeuchi, S., Yokota, S., Kakimoto, K., Yonemori, F., and Wakitani, K. (2000). Effects of peroxisome proliferator-activated receptor-alpha and -gamma agonist, JTT-501, on diabetic complications in Zucker diabetic fatty rats. *Br. J. Pharmacol.* **130**, 495–504.

Shimoshige, Y., Ikuma, K., Yamamoto, T., Takakura, S., Kawamura, I., Seki, J.L., Mutoh, S., and Goto, T. (2000). The effects of zenarestat, an aldose reductase inhibitior, on peripheral neuropathy in Zucker diabetic fatty rats. *Metabolism* **49**, 1395–1399.

Shiozawa, M., Provoost, A.P., Van Dokkum, R.P.E., Majewski, R.R., and Jacob, H.J. (2000). Evidence of gene-gene interactions in the genetic susceptibility to renal impairment after unilateral nephrectomy. *J. Am. Soc. Nephrol.* **11**, 2068–2078.

Sima, A., Chakrabarti, S., Garcia-Salinas, R., and Basu, P. (1985). The BB rat: an authentic model of human diabetic retinopathy. *Curr. Eye Res.* **4**, 1087–1092.

Simons, J., Provoost, A., Anderson, S., Troy, J., Rennke, H., Sandstorm, D., and Brenner, B. (1993a). Pathogenesis of glomerular injury in the Fawn-hooded rat: early glomerular capillary hypertension predicts glomerular sclerosis. *J. Am. Soc. Nephrol.* **3**, 1775–1782.

Simons, J.L., Provoost, A.P., De Keijzer, M.H., Anderson, S., Rennke, H.G., and Brenner, B.M. (1993b). Pathogenesis of glomerular injury in the fawn-hooded rat: effect of unilateral nephrectomy. *J. Am. Soc. Nephrol.* **4**, 1362–1370.

Simpson, F. and Phelan, E. (1984). Hypertention in the genetically hypertensive strain. *In* "Handbook of Hypertension, Vol. 4: Experimental and Genetic Models of Hypertension." (E. de Jong, ed.), pp. 200–223. Elsevier, Amsterdam.

Simpson F.O., Phelan E.L., Clark D.W., Jones D.R., Gresson C.R., Lee D.R., and Bird D.L. (1973). Studies on the New Zealand strain of genetically hypertensive rats. *Clin. Sci. Mol. Med. Suppl.* **1**, 15s–21s.

Simpson, F.O., Phelan, E.L., Ledingham, J.M., and Millar, J.A. (1994). Hypertension in the genetically hypertensive (GH) strain. *In* "Handbook of Hypertension, Vol 16: Experimental and Genetic Models of Hypertension." (D. Ganten and W. de Jong, eds), pp. 228–271. Elsevier Science, Amsterdam.

Smith, B., Mattison, D., and Sipes, I. (1990). The role of epoxidation in 4-vinylcyclohexene-induced ovarian toxicity. *Toxicol. Appl. Pharmacol.* **105**, 372–381.

Sohal, V.S. and Huguenard, J.R. (2001). It takes T to tango. *Neuron* **31**, 3–7.

Sokol, H. and Valtin, H. (1965). Morphology of the neurosecretory system in rats homozygous and heterozygous for hypothalamic diabetes insipidus (Brattleboro strain). *Endocrinology* **77**, 692–700.

Steir, Jr., C.T., Chander, P.N., and Rocha, R. (2002). Aldosterone as a mediator in cardiovascular injury. *Cardiol. Rev.* **10**, 97–107.

St. Lezin, E., Simonet, L., Pravenec, M., and Kurtz, T.W. (1992). Hypertensive strains and normotensive 'control' strains. *Hypertension* **19**, 419–424.

Suzuki, S. and Toyota, T (1998). GK rat. *Nippon Rinsho* **56(Suppl. 3)**, 695–699.

Sventek, P., Turgeon, A., Garcia, R., and Schiffrin, E.L. (1996). Vascular and cardiac overexpression of endothelin-1 gene in one-kidney, one clip Goldblatt hypertensive rats but only in the late phase of two-kidney one clip Goldblatt hypertension. *J. Hypertens.* **14**, 57–64.

Swanborg, R. (1995). Experimental autoimmune encephalomyelitis in rodents as a model for human demyelinating disease. *Clin. Immunol. Immunmopathol.* **77**, 4–13.

Swislocki, A. and Tsuzuki, A. (1993). Insulin resistance and hypertension: glucose intolerance, hyperinsulinemia, and elevated free fatty acids in the lean spontaneously hypertensive rat. *Am. J. Med. Sci.* **306**, 282–286.

Takaya, K., Ogawa, Y., Hiraoka, J., Hosoda, K., Yamori, Y., Nakao, K., and Koletsky, R.J. (1996). Nonsense mutation of leptin receptor in the obese spontaneously hypertensive Koletsky rat. *Nat. Genet.* **14**, 130–131.

Tamura, T., Said, S., Anderson, S., McCune, S., Mochizuki, S., and Gerdes, A.M. (2002). Temporal regression of myocyte hypertrophy in hypertensive, heart failure-prone rats treated with an AT_1-receptor antagonist. *J. Cardiol. Fail.* **8**, 43–47.

Tanase, H., Suzuki, Y., Ooshima, A., Yamori, Y., and Okamoto, K. (1970). Genetic analysis of blood pressure in spontaneously hypertensive rats. *Jpn. Circ. J.* **34**, 1197–1212.

Thomsen, K., Rubin, I., and Lauritzen, M. (2002). NO- and non-NO-, non-prostanoid-dependent vasodilatation in rat sciatic nerve during maturation and developing experimental diabetic neuropathy. *J. Physiol.* **543**, 977–993.

Tofovic, S. and Jackson, E. (1999). Effects of long-term caffeine consumption on renal function in spontaneously hypertensive heart failure prone rats. *J. Cardiovasc. Pharmacol.* **33**, 360–366.

Tofovic, S., Kusaka, H., Kost, C., and Bastacky, S. (2000). Renal function and structure in diabetic, hypertensive, obese ZDF x SHHF hybrid rat. *Renal Failure* **22**, 387–406.

Tourrel, C., Bailbe, D., Lacorne, M., Meile, M.J., Kergoat, M., and Portha, B. (2002). Persistent improvement of type 2 diabetes in the Goto-Kakizaki rat model by expansion of the beta-cell mass during the prediabetic period with glucagons-like peptide-1 or exendin-4. *Diabetes* **51**, 1443–1452.

Trentham, D., Townes, A., and Kang, A. (1977). Autoimmunity to type II collagen: an experimental model of arthritis, *J. Exp. Med.* **146**, 857–868.

Tschopp, T.B. and Zucker, M.B. (1972). Heriditary defect in platelet function in rats. *Blood.* **40**, 217–226.

Ulugo, A., Karadag, H., Tamer, M., Firat, Z., Aslantas, A., and Dokmeci, I. (2002). Involvement of adenosine in the anti-allodynic effect of amitriptyline in streptozotocin-induced diabetic rats. *Neurosci. Lett.* **328**, 129–132.

Valtin, H. (1967). Hereditary hypothalamic diabetes insipidus in rats (Brattleboro strain): a useful experimental model. *Am. J. Med.* **42**, 814–827.

Valtin, H., Sawyer, W., and Sokol, H. (1965). Neurohypophysial principles in rats homozygous and heterozygous for hypothalamic diabetes insipidus (Brattleboro strain). *Endocrinology* **77**, 701–706.

Van Dokkum, R.P.E., Sun, C.W., Provoost, A.P., Jacob, J.J., and Roman, R.J. (1999). Altered renal hemodynamics and impaired myogenic responses in the fawn-hooded rat. *Am. J. Physiol.* **276**, R855–R863

Van Hese, P., Martens, J.P., Boon, P., Dedeurwaerdere, S., Lemahieu, I., and Van de Walle, R. (2003). Detection of spike and wave discharges in the cortical EEG of genetic absence epilepsy rats from Strasbourg. *Phys. Med. Biol.* **48**, 1685–1700.

Van Luijtelaar, E.L.J.M. and Coenen, A.M.L. (1988). Circadian rhythmicity in absence epilepsy in rats. *Epilepsy Res.* **2**, 331–336.

Velasquez, M., Kimmel, P., and Michaelis, O.I.V. (1990). Animal Models of Spontaneous Diabetic Kidney Disease. *FASEB J.* **4**, 2850–2859.

Verseput, G.H., Braam, B., Provoost, A.P., and Koomans, H.A. (1998). Tubuloglomerular feedback and prolonged ACE-inhibitor treatment in the hypertensive fawn-hooded rat. *Nephrol. Dial. Transplant.* **13**, 893–899.

Verseput, G.H., Provoost, A.P., Braam, B., Weening, J.J., and Kooamns, H.J. (1997). Angiotensin-converting enzyme inhibition in the prevention and treatment of chronic renal damage in the hypertensive fawn-hooded rat. *J. Am. Soc. Nephrol.* **8**, 249–259.

Vesely, D.L., Gower, Jr., W.R., Dietz, J.R., Overton, R.M., Clark, L.C., Antwi, E.K., and Farese, R.V. (1999). Elevated atrial natriuretic peptides and early renal failure in type 2 diabetic Goto-Kakizaki rats. *Metabolism* **48**, 771–778.

Vincent, M., Saquet, J., and Sassard, J. (1984). The Lyon strains of hypertensive normotensive and low-blood pressure rats. *In* "Handbook of Hypertension, Vol. 4: Experimental and Genetic Models of Hypertension." (W. de Jong, ed.), pp. 314–327. Elsevier, Amsterdam.

Vingsbo, C., Sahlstrand, P., Brun, J., Jonsson, R., Saxne, T., and Holmdahl, R. (1996). Pristane-induced arthritis in rats: a new model for rheumatoid arthritis with a chronic disease course influenced by both major histocompatibility complex and non-major histocompatibility complex genes. *Am. J. Pathol.* **149**, 1675–1683.

Wada, R., Koyama, M., Mizukami, H., Odaka, H., Ikeda, H., and Yagihashi, S. (1999). Effects of long-term treatment with alpha-glucosidase inhibitor on the peripheral nerve function and structure in Goto-Kakizaki rats: a genetic model for type 2 diabetes. *Diabetes Metab. Res. Rev.* **15**, 332–337.

Wakisaka, M., Nunoi, K., Iwase, M., Kikuchi, M., Maki, Y., Yamamoto, K., Sadoshima, S., and Fujishima, M. (1988). Early development of nephropathy in a new model of spontaneously hypertensive rat with non-insulin-dependent diabetes mellitus. *Diabetologia* **31**, 291–296.

Wang, D., Iversen, J., and Strandgaard, S. (1999). Contractility and endothelium-dependent relaxation of resistance vessels in polycystic kidney disease rats. *J. Vasc. Res.* **36**, 502–509.

Watson, W. (1990). Human HLA-DRb gene hypervariable region homology in the biobreeding BB rats: selection of the diabetic-resistant subline as a rheumatoid arthritis research tool to characterize the immunopathologic response to human type II collagen, *J. Exp. Med.* **155**, 1331–1339.

Weinberg, E.O., Schoen, F.J., George, D., Kagaya, Y., Douglas, P.S., Litwin, S.E., Schunkert, H., Benedict, C.R., and Lorell, B.H. (1994). Angiotensin-converting enzyme inhibition prolongs survival and modifies the transition to heart failure in rats with pressure overload hypertrophy due to ascending aortic stenosis. *Circulation* **90**, 1410–1422.

Weksler-Zangen, S., Yagil, C., Zangen, D.H., Ornoy, A., Jacob, H.J., and Yoram, Y. (2001). The newly inbred Cohen diabetic rat: a nonobese normolipidemic genetic model of diet-induced Type 2 diabetes expressing sex differences. *Diabetes* **50**, 2521–2529.

Weldon, D.T., Rogers, S.D., Ghilardi, J.R., Finke, M.P., Cleary, J.P., O'Hare, E., Esler, W.P., Maggio, J.E., and Mantyh, P.W. (1998). Fibrillar β-amyloid induces microglial phagocytosis, expression of inducible nitric oxide synthase, and loss of a select population of neurons in the rat CNS in vivo. *J. Neuroscience* **18**, 161–173.

White, D., Wang, D., Chua, S., Jr., Morgenstern, J., Leibel, R., Baumann, H., and Tartaglia, L. (1997). Constitutive and impaired signaling of leptin receptors containing the Gln→ Pro extracellular domain fatty mutation. *Proc. Natl. Acad. Sci. U.S.A.* **94**, 10657–10662.

Wilder, R., Calandra, G., Garvin, A., Wright, K., and Hansen, C. (1982). Strain and sex variation in the susceptibility to streptococcal cell wall-induced polyarthritis in the rat. *Arthritis Rheum.* **25**, 1064–1072.

Yagami, T., Takahara, Y., Ishibashi, C., Sakaguchi, G., Itoh, N., Ueda, K., Nakazato, H., Okamura, N., Hiramatsu, Y., Honma, T., Arimura, A., Sakaeda, T., and Katsuura, G. (2004). Amyloid beta protein impairs motor function via thromboxane (2) in the rat striatum. *Neurobiol. Dis.* **16**, 481–489.

Yagihashi, S. and Sima, A. (1985a). Diabetic autonomic neuropathy in the BB rat: ultrastructural and morphometric changes in sympathetic nerves. *Diabetes* **34**, 558–564.

Yagihashi, S. and Sima, A. (1985b). Diabetic autonomic neuropathy: the distribution of structural changes in sympathetic nerves of the BB rat. *Am. J. Pathol.* **121**, 138–147.

Yagihashi, S. and Sima, A. (1986). Neuroaxonal and dendritic dystrophy a diabetic autonomic neuropathy: classification and topographic distribution in the BB rat. *J. Neuropathol. Exp. Neurol.* **45**, 545–565.

Yamori, Y., Horie, R., Handa, H., Sato, M., and Fukase, M. (1976). Pathogenetic similarity of strokes in stroke-prone spontaneously hypertensive rats and humans. *Stroke* **7**, 46–53.

Yamori, Y., Horie, R., Tanase, H., Fujiwara, K., Nara, Y., and Lovenberg, W. (1984). Possible role of nutritional factors in the incidence of cerebral lesions in stroke-prone spontaneously hypertensive rats. *Hypertension* **6**, 49–53.

Yang, Y., Danis, R., Peterson, R., Dolan, P., and Wu, Y. (2000). Acarbose partially inhibits microvascular retinopathy in the Zucker diabetic fatty rat (ZDF/Gmi–fa), *J. Ocular Pharmacol.* **16**, 471–479.

Yen, T.T., Shaw, W.N., and Yu, P.L. (1977). Genetics of obesity of Zucker rats and Koletsky rats. *Heredity* **38,** 373–377.

Yu, M., Leav, B., Leav, I., Merk, F., Wolfe, H., and Ho, S. (1993). Early alterations in ras protooncogene mRNA expression in testosterone and estradiol-17β induced prostatic dysplasia of noble rats. *Lab Invest.* **68,** 33–44.

Zan, Y., Haag, J.D., Chen, K.S., Shepel, L.A., Wigington, D., Want, Y.R., Hu, R., Lopez-Guajardo, C.C., Brose, H.L., Porter, K.I., Leonard, R.A., Hitt, A.A., Schommer, S.L., Elegbede, A.F., and Gould M.N. (2003). Production of knockout rats using ENU mutagenesis and a yeast-based screening assay. *Nat. Biotechnol.* **21,** 645–651.

Zucker, L.M. and Antoniades, H.N. (1972). Insulin and obesity in the Zucker genetically obese rat "fatty." *Endocrinology* **90,** 1320–1330.

Zucker, L.M. and Zucker, T.F. (1961). Fatty: a new mutation in the rat. *J. Hered.* **52,** 275–278.

Zucker, T.F. and Zucker, L.M. (1962). Hereditary obesity in the rat associated with high serum fat and cholesterol. *Proc. Soc. Exp. Biol. Med.* **110,** 165–171.

Zucker, T.F. and Zucker, L.M. (1963). Fat accretion and growth in the rat. *J. Nutr.* **80,** 6–19.

Chapter 24

The Nude Rat

Martha A. Hanes

THE LABORATORY RAT, 2ND EDITION

I. INTRODUCTION

The congenitally athymic nude rat (rnu/rnu) is a genetic mutant found in *Rattus norvegicus* characterized by a severe alteration in the cell-mediated branch of immunity (CMI). Altered cellular immune responses are owing to a lack of appropriate development and education of functional T-cell (thymus-derived) lymphocytes. The size and robustness of the nude rat and the absence of major endocrinological abnormalities make it an appropriate experimental model for a variety of immunological, surgical, infectious, transplant-related, and oncological procedures.

II. HISTORY

The nude rat mutation was described in an outbred colony of hooded rats at the Rowett Research Institute in Aberdeen, Scotland, in 1953, and was lost shortly afterward. The phenotype reappeared more than 20 years later in two research facilities, once again in Aberdeen (1977) and separately in a random outbred albino rat in New Zealand (1979) (Festing et al., 1978, 1979; Schuurman et al., 1995).

III. GENETICS

The allele associated with the New Zealand mutation (rnu^{nz}, also known as nznu, rnu^{n} is recessive to the Rowett (rnu) (Marshall and Miller, 1981). The rnu alleles are located within chromosome 10 of the rat genome within the winged-helix domain (*whn*) (also known as forkhead domain) (Cash et al., 1993) family of transcription factors in the q24-q32 interval (Kuramoto et al., 1993; Hirasawa et al., 1998). The mutated transcripts translate into defective proteins with a deletion in the DNA-binding domain. This was the first described member of a class of genes where lack of DNA binding of the protein was the direct cause of the developmental defect in rats (Nehls et al., 1994; Segre et al., 1995; Schuddekopf et al., 1996). The rnu mutation involves a transposable element, a so-called jumping gene, as the result of the integration of an endogenous retrovirus sequence (Jones and Jesson, 1995). This is the first example of a transposon mutation in the rat (Huth et al., 1997). Linkage mapping for the rat homolog of mouse Mpo (myeloperoxidase) is assigned to chromosome 10 (Murakumo et al., 1995), as is the nitric oxide synthase gene (iNOS2) (Zha et al., 1995). The significance of this spatial relationship to chromosome 10 is unknown.

The autosomal recessive mutation has been backcrossed onto various commonly used background strains. A comparative evaluation of the immune status, focused on the T-cell system, was conducted by using rats from 11 institutions and sampling animals at ages 1.5 to 2 months and 6 months (Schuurman et al., 1992). Strains upon which the rnu gene have been backcrossed include the following: LEW/Ztm (Germany); LOU/Cnb (Belgium); LEW/Mol (Denmark); LEW/Iv (Germany); PVG/Bell (England); CBH/Arc (Australia); Orl:RNU (France); R/APfd (Belgium); WAG/RiijRiv (the Netherlands); F344/Wits (South Africa); and Han:RNU, Han:NZNU, Lew/Han, WAG/Han, and DA/Han (Germany). Congenic inbred homozygous female rats did not raise their offspring, whereas the Han: RNU rnu/rnu females were able to procreate successfully. The evaluated strains demonstrated similar histology of the basic architecture of the spleen, lymph nodes, and Peyer's patches. After ovalbumin immunization, none of the animals demonstrated anti-ovalbumin antibody in sera at day 10 or 24, except one, the WAG/rnu rat. Similarly, no significant delayed hypersensitivity response to ear pinnae challenge was noted. Body weight, spleen weight, spleen red pulp percentage, and spleen periarteriolar lymphocyte sheath percentage were compared between groups and with age, with significant differences for all variables tested. A cluster analysis placed the animals into four different clusters. The extent of $\alpha\beta$-TCR (T-cell receptor)-bearing cells in the splenic periarteriolar lymphocyte sheath differs between the various strains and increases with age.

IV. BASIC ASPECTS OF RNU/RNU BIOLOGY AND CARE

A. Physical Characteristics

The nude rat is characterized by lack of hair over most of the body. The albino New Zealand rnu^{nz} mutation reportedly has no hair, whereas the Rowett rnu/rnu hooded rats have cyclic hair growth. Curly vibrissae are present in both but are reduced in number. Short and thin hair may grow in patches around the face and feet or may cover the entire body. The Rowett rat has a hooded gene, so that the eyes, head, and part of the dorsal torso are pigmented (Fig. 24-1). The degree of hairlessness is not affected by background strain (Rolstad, 2001). Heterozygous (rnu/+) rats are fully haired and resemble their wild-type (+/+) siblings. Heterozygous rats exhibit a lower density of paracortical lymphocytes than does the wild type (Hougen, 1987). The rnu mutant has been used more extensively than has the rnunz; the majority of this chapter will refer to the rnu^{nz} unless otherwise noted.

Fig. 24-1 Young Rowett nude rat (rnu/rnu) demonstrates the Long Evans–associated hooded pigmentation and variable hair loss pattern with cyclical retention of vibrissae.

B. Histologic Features

The primary unique feature of the athymic nude rat is congenital hypoplasia, sometimes referred to as aplasia, of the thymus. At day 14 of fetal life, a thymic anlage is present in both euthymic and rnu rats. However, in the rnu mutant, the thymus fails to become appropriately populated with lymphocytes, macrophages, and dendritic cells (DCs). In an adult nude rat, the rudimentary or residual thymus may show variable amounts of cystic epithelial degeneration, some embryonic remnants, or secretory differentiation (Rolstad, 2001).

Histologically, the lack of secondary follicles in the spleen and lymph nodes results in a sparse lymphocyte population and in more prominent hemosiderin-bearing macrophages (Fig. 24-2) (Fossum, 1990). Lymph nodes may be ectatic and contain increased numbers of mast cells or more prominent high endothelial lined venules (Sainte-Marie and Peng, 1990; Sainte-Marie et al., 1991). The basic stromal structure of the T-cell-dependent areas of secondary lymphoid organs remains intact. The mucosal-associated lymphoid tissue, the interfollicular areas of Peyer's patches, cecal and colic mucosal/submucosal lymphoid follicles, and paracortical regions of the mesenteric lymph nodes demonstrate T-cell depletion. Interfollicular areas in bronchiolar-associated lymphoid tissue and nasal-associated tissues are lymphodepleted (Vos et al., 1980a,b). The formation of germinal centers for secondary follicles requires T-cell activity; thus secondary follicles are typically absent in rnu/rnu rats. The bone marrow and erythopoiesis appear unaffected. There is a histochemical heterogeneity of the dermal mast cells in euthymic verses rnu/rnu rats (Aldenborg and Enerback, 1988). The lack

of normal levels of functioning T cells in rnu/rnu rats provides a successful model system for studying immunologic function of T-cells, B-cells, natural killer (NK) cells, and Langhan's/DCs in a larger format than that of the laboratory mouse.

C. Clinical Pathology

The presence of the homozygous rnu gene demonstrates little effect on total leukocyte (granulocytes, monocytes/macrophages) counts. Although absolute counts of lymphocytes are reduced, B-cells and NK cells are usually within normal range. Rnu/rnu rats raised in isolators may yield absolute values of 2000 white blood cells/dL or less. The immunoglobulins (Igs) in rnu/rnu rats exhibit mild variation: IgA, IgG1, and IgG2 are elevated; IgG2a is lower; and IgM, IgD, and IgG2b values are similar to background strains. The thyroid function of athymic rats is intact (thyroid-stimulating hormone levels, 2.9 ± 0.6 ng/mL; T3, 2.6 ± 0.4 ng/mL; T4, 22.0 ± 5.6 ug/dL) (Hoang-Vu et al., 1992).

D. Growth and Lifespan

The size of adult nude rats is approximately 20% less than that of euthymic rats of congenic strains. Female rnu rats are smaller than are males. When raised in conventional conditions, the lifespan is severely reduced: rnu/rnu rats may live 9 months, and rnu^{nz}/rnu^{nz} may only live 4 months. When maintained under specific pathogen-free

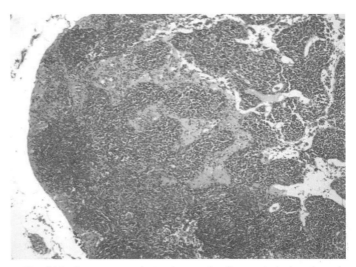

Fig. 24-2 Low power photomicrograph of mesenteric lymph node demonstrating lack of primary and secondary follicles associated with the rnu T-cell paucity. Intrahistiocytic hemosiderin, and erythrophagocytosis in lymph nodes are frequent findings in older nude rats (hematoxylin and eosin, 100×).

conditions or when raised and maintained in flexible film isolators or micro-isolators, athymic nude rats live longer than 2 years (Schuurman, 1995).

E. Reproduction

Female rnu/rnu are fertile, however the early perinatal loss is high, owing in part to poor development of the mammary glands related to the nude phenotype. Successful reproduction has been reported by using homozygous males and heterozygous females (Liang et al., 1997). This scheme is the most commonly used by breeders and may produce as low as 36% rnu/rnu, suggesting *in utero* losses but reportedly higher yields of rnu/rnu overall. Some production colonies, such as the one at Duke University Cancer Center Isolation Facility, use male and female heterozygotes (rnu/+) to produce mendelian ratios (3:1) of rnu/rnu animals. There is no difference in male and female homozygous survival. The production of rnu^{nz}/rnu^{nz} yields fewer pups weaned per litter.

F. Rearing Conditions

ABSL2 or ABSL3 rearing conditions are typically used for maintenance of athymic rats, with husbandry practices being similar to those used for immune-compromised mice. Procedures used include individually ventilated cages, filter hoods, laminar flow devices, flexible film or rigid isolators, micro-isolators, autoclaved or irradiated diets, and autoclaved water and bottles. Practices preventing contamination or elimination of contamination include caesarian rederivation, barrier maintenance, vermin control and escapee capture, limited personnel access, disinfection of handling forceps, sterile or freshly disinfectant-sprayed gloves and transfer forceps, protective gowns, masks, foot covers, eye protection, and training. Some colonies are maintained with rederived isolator-reared animals with defined flora, using a Schaedler's modified cocktail, with only non-urease fermenters, non-pathogenic enteric bacteria (*Lactobacillus* spp and others).

V. IMMUNOLOGY

A. Lymphocyte Markers

Peripheral lymphoid tissues are grossly deficient in mature and immature T-cells of nude rats. Lymphocyte recirculation appears to be mildly slowed. In studies with flow cytometry using cell suspensions from the thoracic duct, spleen, and lymph nodes, CD4 (T-lymphocyte helper-inducer and some macrophages) and CD8 (T-cell

suppressor-cytotoxic and some NKs) markers identified the presence of T-like cells. These T-like cells have an immunophenotype of immature cells. The expression of CD45 (common leukocyte marker, OX22) and RT6 antigens on CD4-positive or CD8-positive cells differs from that in euthymic rats, suggesting that the T-like cell carries different markers in rnu rats (Waite et al., 1996). After exposure to T-lymphocyte-specific antibodies for $\alpha\beta$-heterodimeric T-lymphocyte receptor (α/βTCR), cells with T-lymphocyte characteristics were identified. Current reasoning is that these T-like cells have not been properly educated in a functional thymic microenvironment, and they remain immature and poorly differentiated. These "uneducated" T-like cells are functionally defective owing in part to the defective nature of the receptor protein (Yang et al., 2000).

T-like cells can be stimulated by T cell mitogens (PHA and ConA) and by alloantigens to proliferate in mixed lymphocyte cultures *in vitro*. They can also develop into cytotoxic T lymphocytes that kill allogenic blast cells and produce interleukin-2 (Schuurman, 1995). In contrast, the T-like cells are unable to initiate skin allograft or kidney cell rejection without the addition of CD4+ cells from a euthymic donor. T-like cells in athymic animals rearrange their TCR genes, but there is no firm information about their TCR repertoire as compared with normal T lymphocytes. T-like cells may also function in non-antigen-specific reactivity, causing the emergence of germinal centers in secondary follicles that were not originally described (Schuurman, 1995).

Athymic nude rats accumulate CD3+ cells mostly expressing α/β-TCR, but also some γ/δ-TCR, in their peripheral lymphatic tissues with age. Several research groups have identified the phenotype (CD5+, CD2+, and CD43+) and CD4+ or CD8+ single positive cells in a ratio of 3:1. Rnu/rnu rats 8 to 12 months old express approximately 50% of the normal level (Bischof et al., 2000; Rolstad, 2001).

B. Age Differences in Immunity

The numbers of T-like cells increase with age in rnu/rnu rats. In one study, using the CD-5 (pan-T marker, antibody MRC monoclonal antibody OX19), less than 2% of the cells present in the spleen, lymph nodes, and peripheral blood were CD5 positive at 6 weeks, increasing to 19% at 17 months. This contrasted to euthymic rats with an age-independent value of 34%. T-like cells accumulate in the gut epithelium in rnu/rnu rats with age; absent before 2 months; these cells appear in variably increasing numbers after 6 months. There is a preponderance of γ/δ TCR cells, and a dominance of CD8+ cells in the intestine.

In addition to an increase of T-like cells during the aging process, functionality of these cells increases with age. With *in vivo* graft-versus-host (GVH) reactivity, alloreactivity in graft rejection and response to thymus-dependent antigens are reduced or absent in young rnu rats. Mixed lymphocyte culture alloreactivity is increased in adult animals. Aging rnu rats also gain the ability to reject skin allografts from MHC (RT1) incompatible donors. Lymphocytes from older rnu/rnu rats mount a proliferative response against different MHC haplotypes. However, these responses are reduced compared with that of euthymic controls. Aged rnu/rnu rats demonstrate a somewhat random allograft rejection *in vivo,* supporting a difference in individual clonal repertoire. Importantly, as an association between acceptance of xenografts and the age of the recipients has been demonstrated, rnu should be younger than 8 weeks when grafted with tumors or transplant tissues.

C. Natural Killer Cells

Increased non-thymus-dependent cytotoxicity, including markedly increased NK cell activity, is a proposed compensatory response to the T-cell defect. Studies in rnu rats have provided the foundation for exploration of these multifunctional cells. NK cells are CD3-lymphocytes that express a lectin-like receptor, NKR-P1. NKR-P1 is coded for by a genetic region termed the natural killer cell gene complex, located on chromosome 4. Rat NK cells are found mainly in the blood, spleen, lungs, and liver. A comparison of NK cells from euthymic rats and rnu rats has revealed identical marker expression. By using a fluorescent-activated cell sorted scan fluorocytometer, the cells positive for NK-RP1 are subtracted from all cells positive for CD8+ cells (NK and T cells) to obtain NK numbers. NK cells have large granules in the cytoplasm that contain perforins and granzymes that facilitate the killing of tumor or allogenic cells identified as foreign. NK cells also express ligands for inducing apoptosis, and kill cells expressing Fas through the FasL-dependent apoptotic pathway. NK cells produce interferon-γ, and tumor necrosis factor-α. NK cells respond to several well-known interleukins by activation, maturation, cytokine release, or migration. Rnu NK cells in the blood and spleen are more numerous and, paradoxically, kill tumor target cells better than euthymic cells. NK cells do not reach normal levels in the peripheral circulation until rats are 4 to 6 weeks of age, supporting the argument for increased resistance of xenograft success after this age (Cook et al., 1995). Several additional paradoxical findings of rnu NK cells have been encountered in GVH reactions after bone marrow transplantation, suggesting the existence of naturally occurring alloreactivity owing to receptors on the NK cells; NK receptors recognize the MHC-1 molecules (Rolstad, 2001).

D. Other Cell Types

B-cells appear relatively, but not absolutely, increased in numbers in lymphoid organs of rnu/rnu rats, and B-cell function is intact. Proliferation of eosinophils and intestinal mucosal mast cells, and cytokine production in response to parasites were within normal range in rnu/rnu rats (Koga et al., 1999).

DC migration has been studied extensively in the rat. Rnu DCs can be isolated from afferent lymph vessels and the spleen. An especially potent method to gain large numbers of relatively pure DC is to collect lymph from the thoracic duct of rnu rats that have had the mesenteric lymph nodes removed some weeks before duct canulation, and with use of Percoll gradient separation. DC are strongly OX-6 (MHC-Class II) and OX-62-positive. Some DCs have NK-like receptors (Rolstad, 2001).

VI. SPONTANEOUS DISEASE

A. Bacterial Disease

Athymic nude rats appear more resistant than nude mice to bacterial infection. Handling and treatment-related infections are reported less frequently and are generally less severe than those in nude mice.

In general, nude rats have impaired reactivity to infectious microorganisms that require functionally active thymus-dependent immunity. One study looked at the natural exposure of 15 strains of nude rats from 11 facilities and detected *Clostridium piliforme,* Kilham's rat virus (RV), and *Pasteurella pneumotropica* (Boot et al., 1996). Tyzzer's disease (*C. piliforme*) (Thuner et al., 1985) has been reported to cause widespread mortality within 3 weeks of exposure in young rnu/rnu rats. Natural infections of nude rats with *Mycoplasma pulmonis,* causing conjunctivitis and periorbital abscesses, have been reported.

Staphylococcus sp. *(S. xylosus, S. aureus, S. intermiditis)* have been isolated from surgical wounds, tumor implant sites, drug/device implant sites, and spontaneous skin lesions in nude rats (Fig. 24-3). Discrete multifocal to coalescing ulcerative dermatitis on the dorsum, perineal/ scrotum areas, appendages, or tail regions are typical. Microscopically, lesions are of variable duration, with chronic fibrosis, scar formation, and mild mixed inflammation in those of longer duration. Necroulcerative lesions, with abundant neutrophils admixed with coccobacillary bacteria in a serocellular crust, are seen in the more florid lesions. Such lesions are usually seen in animals on

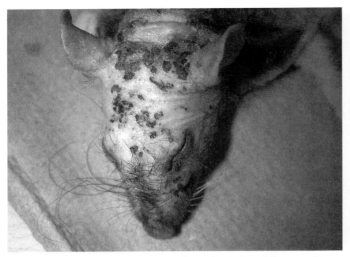

Fig. 24-3 Exudative dermatitis of the head found in a nude rat. Pruritic lesions can lead to an increase in the severity of the lesions owing to trauma. After they reached 150 g body weight, these (flexible-film gnotobiotic isolator-reared) animals were maintained under micro-isolator tops within conventional housing. A break in designated hygienic practices during cage change was suspected as the precipitating event of the causative *Staphylococcus* spp. infection.

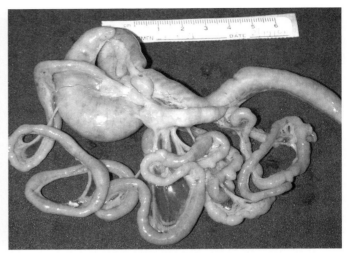

Fig. 24-4 Chronic inflammation of the small and large intestines is characterized by distended loops of bowel, thickened intestinal walls, and hyperplasia of the mesenteric lymph nodes. This lesion has been related to infection with *Helicobacter bilis*.

protocols involving tumor growth in rnu/rnu rats and are assumed to be associated with breaks in appropriate handling techniques. Attempts to isolate corynebacterial species have been unrewarding. Occasionally a concomitant *Candida* sp. infection is associated with such gross lesions. A study examining the effects of ampicillin, spiramycin, sulfadoxine/trimethoprim, tetracycline, lincomycin, and spectinomycin over a period of 30 weeks demonstrated the suppression of *P. pneumotropica,* the elimination of *S. aureus,* and appearance of resistant *Klebsiella pneumonia* and *Streptococcus* species (Hansen, 1995). Augmentin (amoxicillin and clavulinic acid) at a dosage of 0.35 mg/mL in drinking water has been used as a postsurgical treatment. A caveat to the use of prophylactic antibiotics in rnu/rnu rats is the probability that antibiotic-resistant bacteria will arise.

Chronic ulcerative typhlocolitis was diagnosed in a nude rat colony in Australia in 1994 (Thomas and Pass, 1997). At 8 to 23 weeks of age, clinical signs were noted, including pale-yellow-brown feces with mucus, intermittent diarrhea, and inflammation and ulceration of the perineum. Positive occult blood tests were a sensitive and specific clinical indicator. Gross necropsy lesions were confined to the cecum and colon, with thickening of the wall and dilation with gas (Fig. 24-4). Ileocecal lymph nodes were enlarged (Fig. 24-5). Microscopic lesions included ulceration and metaplasia of enterocytes. Subacutely, hyperplastic branched and dilated crypts containing debris, as well as extension of crypts into the submucosa, were seen (Fig. 24-6). The lamina propria was infiltrated with mixed inflammatory cells. Active

fibroplasia and giant cells were seen in chronic lesions. No lesions were seen in heterozygotes. In this epizootic, cultures and special stains failed to reveal bacteria. In another laboratory, large bowel disease was induced in Cr:NIH-rnu rats after detection of natural proliferative typhlitis in a retired breeder male rnu rat. In this case, Steiner's stain revealed spiral organisms within cecal crypts. Culture, polymerase chain reaction (PCR), and sequencing revealed *Helicobacter bilis* (Haines DC et al., 1998). Although not infectious, a wasting disease characterized by failure to normally gain weight followed by a rapid decline in body weight with low body fat has been associated with malocclusion (Fig. 24-7) and dental caries in nude rats.

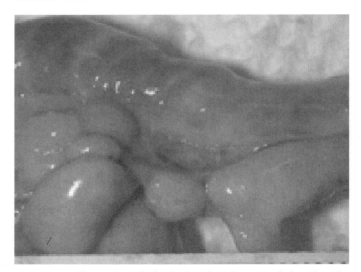

Fig. 24-5 Closer view of Fig. 24-4. Enlarged mesenteric lymph nodes owing to chronic inflammatory disease.

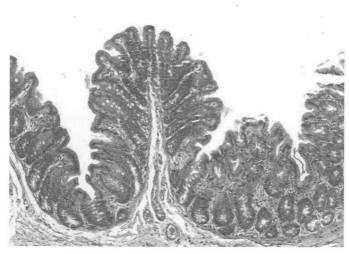

Fig. 24-6 Photomicrograph of cecal wall from rat in Figs. 24-4 and 24-5. Moderate hyperplasia of the cecal epithelium is present. *Helicobacter* sp. has been reported as a cause of this syndrome in nude rats.

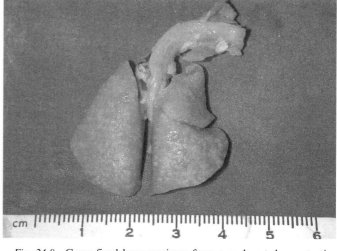

Fig. 24-8 Gross fixed lung specimen from a nude rat demonstrating pulmonary lesions suggestive of *Pneumocystis carnii*. Raised slightly pale lesions are noted diffusely scattered throughout the pulmonary lobes. In unfixed specimens, these foci may be associated with areas of hemorrhage, atelectasis, or emphysema.

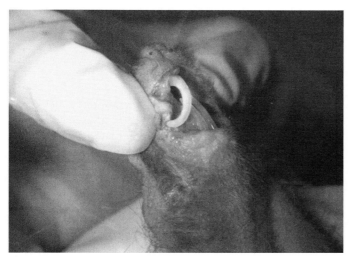

Fig. 24-7 Sporadically occurring lesions in closed colonies of nude rats may be seen. This malocclusion and tooth loss was not detected until weight loss was noted.

B. Fungal Disease

Epizootics owing to *Pneumocyctis carnii* in nude rat colonies are significant, and have been reported in several facilities (Zeifer et al., 1984; Furuta and Ueda, 1987; Deerberg et al., 1993: Furuta et al., 1993; Pohlmeyer and Deerberg 1993). Lungs that do not deflate and that have a red to white mottled pattern to the pulmonary parenchyma are characteristic gross lesions (Fig. 24-8). In contrast, no gross or histologic change is seen with PCR-positive rnu/+ heterozygous rats. Homozygous nude rats are very susceptible and spontaneously develop a severe pneumonia, with large numbers of intrahistiocytic and free fungal organisms in the alveolar airspace. Organisms are seen more easily with periodic acid-Schiff or Gomori's methenamine silver staining. The presence of alveolar foamy macrophages, pockets of neutrophils, and peri-bronchiolar lymphoid aggregates characterize chronic lesions (Fig. 24-9). Alveolar septal walls contain increased fibrous connective tissue. Areas of atelectasis and emphysema may appear, depending on the duration of the disease. Organisms have been determined to be specifically of rat, rather than of mouse or human origin, based on antisera generated against the 52-kDa antigens of *P. carinii* (Furuta, and Ueda, 1987). PCR diagnosis for *Pneumocystis* sp. using heterozygous littermates can be conducted by various commercial laboratories. Two new anti-pneumocystis drugs (azasordarins, GW471552 and GW471558) developed for humans were tested in nude rats and significantly reduced the number of *P. carinii* cysts per gram of lung homogenate with a therapeutic efficacy at 0.25 mg/kg twice a day for 10 days (Jimenez et al., 2002).

C. Viral Disease

Naturally occurring reports of viral disease include papovavirus, RV (parvovirus), and sialodacryoadenitis virus (SDAV). Wasting animals with suspected papovavirus infection developed pneumonia (Fig. 24-10) and

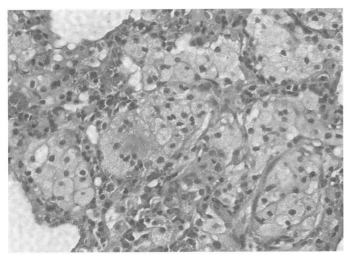

Fig. 24-9 Photomicrograph of lungs from a nude rat with *Pneumocystis carinii* infection. Microscopically, pulmonary alveoli are distended with abundant large foamy macrophages and a few lymphocytes and polymorphonuclear leukocytes. Depending on the time course, alveolar septa maybe altered by the presence of fibrous connective tissue, inflammatory cells, erythrocytes, and hypertrophic endothelial cells of pulmonary capillaries. Occasionally, type II pneumocyte hyperplasia and bronchial epithelial hyperplasia may be apparent (hematoxylin and eosin, 400×).

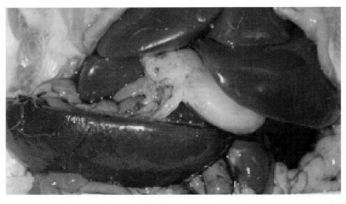

Fig. 24-10 Gross photograph of the opened abdomen at necropsy of a nude rat. Hepatomegaly and splenomegaly are characterized by increased tissue weight and volume, changes in organ hue with tense capsules, and rounded organ edges. Lymphocytic leukemia was diagnosed.

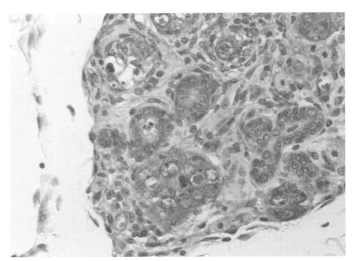

Fig. 24-11 Parotid salivary gland: Chronic lesions associated with papovavirus infection of the salivary glands are characterized by the fibrosis and atrophy of the normal glandular acini and by inflammation with macrophages and lesser numbers of lymphocytes. A cell bearing an intranuclear inclusion body is adjacent to several megakaryotic cells (hematoxylin and eosin, 400×).

seen. Virus shedding persisted over 6 months (Weir et al., 1990; Hajjar et al., 1991).

D. Neoplasia

In 2001, a previously unreported lymphocytic leukemia was noted in six BIG/NCI:rnu rats housed at Duke University Medical Center. The initial site of malignancy appeared to be the spleen (Figs. 24-12 and 24-13). The first clinical signs were related to submandibular swelling (Fig. 24-14). Hemograms obtained from one case revealed a leukogram of 498,800 with 100% lymphoid cells composed of lymphoblasts, with fewer numbers of small and reactive lymphocytes. A regenerative anemia with 2 nRBC/ 100 white blood cells, a hemoglobin value of 8.1 g/dL and a RBC count of 4.0 × 10⁶ cells was reported. Hepatic cytosolic enzymes were mildly increased, with no renal analyte alterations. Histopathologically, the spleen was severely affected by the loss of normal architecture and the presence of neoplastic lymphoblastic cells (Fig. 24-15). Likewise, the normal liver parenchyma was effaced by large numbers of neoplastic lymphocytes. Diagnosis was based on clinical signs of anemia, weight loss, and necropsy findings of splenomegally with histologic peripheral leukocytosis. Diffuse infiltration of other organs (mesenteric lymph nodes, Peyer's patches, and mandibular lymph nodes) was seen. Several studies have addressed the ability to transfer the large granulocytic leukemia commonly found in F344 rats to nude rats (Stromberg et al., 1992). Mammary fibroadenomas and hemangiosarcomas have been diagnosed in nude rats.

sialoadenitis. Intranuclear inclusions were seen in the parotid gland duct epithelium (Fig. 24-11). Papovavirus antigen found in kidney, laryngeal, and bronchial epithelium did not cross-react with K virus or polyoma virus (Ward et al., 1984). Naturally occurring and experimental RV causes persistent parvovirus infection in nude rats (Gaertner et al., 1989, 1991, 1995). SDAV was detected in a colony of nude rats with concomitant *P. pneumotropica, Pseudomonas aeruginosa,* and *Staphylococcus* sp. infections. Clinically, sneezing, loss of weight, and death were noted. At necropsy, rhinitis and bronchopneumonia were

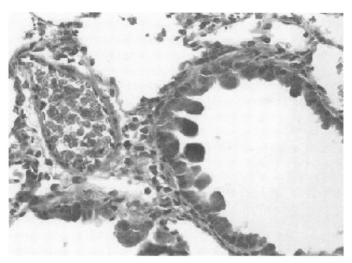

Fig. 24-12 Lung: Papavovirus infection of the bronchiolar epithelium in nude rats causes multinucleated giant cell bearing intranuclear inclusions (hematoxylin and eosin, 400×).

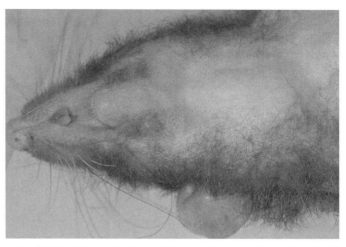

Fig. 24-13 Submandibular swelling seen at necropsy in a nude rat with leukemia. The cervical lymph nodes, spleen, liver, and mesenteric lymph nodes were involved. Common differential diagnoses might include salivary gland swelling owing to viral infections (e.g., sialodacryoadenitis virus, papovavirus), bacterial cellulitis and abscessation, or other tumors (mammary, hemangiosarcoma).

VII. EXPERIMENTAL DISEASE

Reduced and delayed responses in nude rats to tumors and pathogens indicate a primary role of T cells in many host defense mechanisms. Miscellaneous studies in which the nude rat has been used as a model include those studying the induction of autoimmune uveoretinitis (Caspi et al., 1982), function of T cells in atherosclerosis (Haraoka et al., 1997), vascular injury (Hansson et al., 1991), osteoporosis and bone formation (Kirkeby, 1991; Buchinsky et al., 1995), induction of autoimmune diabetes mellitus by adoptive transfer (Whalen et al., 1994), pyelonephritis (Miller, 1984), amyloidosis (Vargas and Stephens, 1983), and the role of metalloproteinases in erosive arthritis (Case et al., 1989).

A. Experimental Viral Disease

Nude rats have been infected with mouse hepatitis virus (JHM strain) as a model of demyelinating disease (Sorensen et al., 1987a,b). Interestingly, serial passage of tumors through rnu rats have been used to eliminate adventitious mouse viruses in tumor lines, especially mouse hepatitis virus (Rulicke et al., 1991). Experimental Sendai paramyxoviral infection caused interstitial pneumonia and

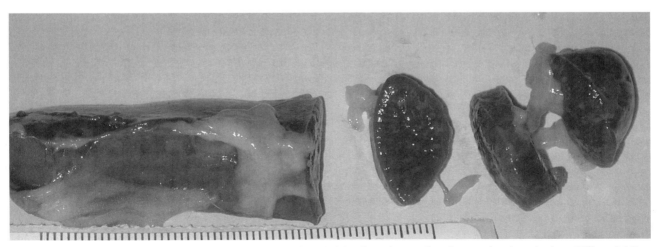

Fig. 24-14 Photograph of spleen taken at necropsy demonstrates tissue bulging from the cut edge of a severely enlarged spleen. Differential diagnoses would include leukemia/lymphoma and splenic extramedullary hematopoiesis (EMH).

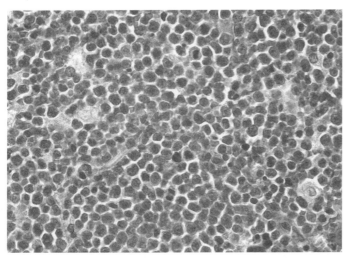

Fig. 24-15 High power photomicrograph of spleen from Fig. 24-14 demonstrating a monomorphic round cell neoplastic population characteristic of lymphocytic leukemia (hematoxylin and eosin, 400×).

epithelial sloughing. Antigen persisted for greater than 32 days, although no circulating antibody was detected. By use of rnu rats with human retinal transplants, infection with human cytomegalovirus was successful (Laycock et al., 1997). Genetically altered viruses (e.g., herpesvirus, G207) have been used successfully to introduce DNA into xenograft tissues and tumors (Carew et al., 1999; Kooby et al., 1999; McAuliffe et al., 2000).

B. Experimental Bacterial Disease

Studies with nude rats vaccinated against *Legionella pneumophila* or *Streptococcus mutans* demonstrated protection that was not seen in naïve rnu rats. The initial defense to *Listeria montocyogenes* that depends on nonspecific macrophage activity is increased in rnu rats, but the terminal elimination that requires intact CMI is impaired. Injection with bacille Calmette-Guerin and *Corynebacterium parvum* stimulate normal NK activity. *Salmonella typhimurium* infection has been developed for the study of passive immunization (Hougen et al., 1989, 1990).

A hypertensive strain of nude rat, SHR/NVrj-rnu, was developed that was more highly susceptible to *Mycobacterium leprae*. Massive nodular lesions developed at inoculation sites and un-inoculated forefeet and muco-cutaneous junctions. Macrophage proliferation was evident with mild lymphocyte proliferation. Congenic F344-rnu rats demonstrated milder nodules and greater neutrophil responses (Yogi et al., 1999).

In an effort to create a chronic pneumonia model of cystic fibrosis, *P. aeruginosa* was administered to nude rats; the rats mounted high IgA titers in response. When

rats were given ginseng extract at 25 mg/kg/day for 10 days, the bacterial load and the number of mast cells were reduced (Johansen et al., 1993; Song et al., 1997).

Mycoplasma arthritidis infection in nude rats resulted in progressive infection without development of protective antibodies (Binder et al., 1993).

C. Experimental Dental Disease

By use of adoptive transfer methods in studies with an oral pathogen, *Porphyromonas gingivalis,* it was determined that salivary IgA (Th2-cell dependent) and serum IgA (Th1-cell dependent) antibodies are generated when nude rats are given donor lymphocytes. Control rats had low IgM levels, and undetectable IgG and IgA (Katz and Michalek, 1998). Dental caries and higher bacterial counts resulted in nude rats given *Streptococcus sobrinus* strain 6715 compared with heterozygote rats (Stack et al., 1990). The role of T cells has been studied in the development of periapical lesions exposed to normal flora. Lesion evolution and severity were similar between rnu and conventional rats (Wallstrom et al., 1993). However, in another study, adoptive transfer of T-cells ameliorated periodontal lesions (Yoshie et al., 1985; Eastcott et al., 1994).

D. Experimental Fungal Disease

Studies with the skin fungus *Tricophyton metagrophytes* support the nude rat as a model for human dermatophytoses (Jones 1986; Green et al., 1987; Balish 1993). Fungal clearance took almost twice as long in nude rats as in euthymic rats (60 versus 32 days).

E. Experimental Protozoal Disease

Nude rats have been used in studies related to a variety of parasites. Infection of nude rats with, for example, *Entamoeba histolytica* resulted in shedding of organisms for more than 1 year (Owen 1987); cryptosporidial infection was established in nude rats, with prolonged shedding and defective immunity (Gardner et al., 1991). It was determined that there is no clearance of *Plasmodium bergei* and *Trypanosmoa cruzi* after experimental inoculation of nude rats; parasitic organisms were found in large numbers and death of the host animal occured within 4 weeks (Kamiyama et al., 1987). Infection with *Plasmodium chabaudi* resulted in persistent parasitemia of nude rats, which resolved with adoptive transfer, validating the role of T-cells in the immunity to malarial and other protozoal infections (Watier et al., 1992).

Nude rats can be infected with *Toxoplasma gondii* (including the RH tachyzoites and Prugniaud cysts) and therefore serve as a good animal model system for the study of the development of pulmonary, myocardial, and hepatic disease in an immunocompromised host (Foulet et al., 1994; Zenner et al., 1999). Nude rats have been used for the testing of anti-parasiticals such as epiroprim or antigenic peptides directed toward the *T. gondii* P30 antigen. Different subsets of T cells, macrophages, and NK cells have been determined to be involved in the intracerebral immune response to *T. gondii* in nude rats (Schluter et al., 1995).

F. Experimental Helminth Disease

Studies conducted in nude rats experimentally infected with *Nippostrongylus brasiliensis* have yielded information regarding IgE receptors, IgE occupancy, and secretory ability of mast cells (Smith et al., 1991; D'Inca et al., 1992; Chen and Enerback, 1994; Chen, 1995; Chen et al., 1995; McKay et al., 1995; Chen and Enerback, 1996; Holmes et al., 1996). Nude rats have also been used for study of a variety of other helminthes. For example, the survival of *Oncocera volvulus* and *O. lienalis* have been explored by using athymic rats (Rajan et al., 1992; Taylor et al., 1992); infection with *Taenia taeniaformis* caused gastric epithelial hyperplasia in both nude and euthymic rats, suggesting a T cell-independent mechanism in the development of this lesion (Abella et al., 1997); nude rats failed to develop significant resistance after vaccination with irradiated *Schistosoma mansoni* cercariae, although immunity could be induced with passive transfer from a rnu/+ donor (Ford et al., 1987); infection of nude rats with *Fasciola hepatica* generated a T cell-independent eosinophilia (Milbourne and Howell 1997; Topfer et al., 1995); and administration of *Trichenella spiralis* and *Brugia* sp. larvae to nude rats result in delayed or failed clearance and does not stimulate an antibody response (Vos et al., 1983; Lawrence and Denham, 1992). By use of nude rats administered eggs of *Rodentolepis (Hymenolepis) nana,* it was demonstrated that the rat host used thymus-independent, cortisone-sensitive resistance to an initial infection, which is the main component of the innate resistance and blocks the lumenal establishment of this parasite. The thymus-dependent resistance suppressed the established worms' fecundity and may be ascribed to acquired resistance to the ongoing infection (Ito and Kamiyama, 1987).

G. Oncology

Studies in nude rats have exploited impaired function of cell-mediated immunity, and used tumor and graft tolerance in many tissues. Xenograft, orthograft, and metastatic tumors have been studied with regards to tumor growth kinetics, chemotherapy, chemoprotection, and various imaging modalities in an attempt to discover new treatments and more accurate diagnostic predictors of outcomes (Table 24-1). Acceptance of xenotransplanted and allotransplanted tissues into athymic nude rats allows for continued elucidation of GVH and tolerance (Table 24-2).

The literature is replete with references attesting to the versatility of the nude rat model in xenograft transplantation of human tumors. Many cell lines have been tested in nude rats, capitalizing on the advantages that size and robustness of the nude rat provide. Suspensions of cells in the range of 10^5, or controlled dimensions of solid tumor blocks, may be surgically implanted in multiple subcutaneous sites. Orthotopic sites have been used to resemble the "normal" environment, (e.g., intrathecal, intra-bronchial, intra-esophageal, subcutaneous, subcapsular). Metastatic scenarios can be created with cell dispersal by way of intravenous, intra-arterial, intra-carotid, intrasplenic, intrahepatic, intracardiac, and intra-tibial injections of cells. Natural metastasis can be encouraged with surgical removal of the primary. Nude rats have been used as hosts for primary and metastatic malignant lymphoid diseases, sometimes incorporating adoptive transfer procedures. In-depth evaluation of these tumor models is beyond the scope of this chapter, however, several models that use rnu are discussed below.

Neoplasia of the central nervous system (CNS) presents multiple unique roadblocks to the effective control and elimination of primary and metastatic neoplastic cells. Glioma implants have been delivered stereotactically with co-ordinates based on small increments from the bregma (e.g., 1.4 mm anterior and 2 mm lateral to the intersection of the coronal and sagittal sutures). With techniques requiring intrathecal administration of antineoplastics, the nude rat is the model of choice for the testing of drugs that are normally ineffective owing to interference with the intact blood-brain barrier, or first-pass metabolism through the body. Treatment with an analog of cyclophosphamide placed intrathecally in nude rats with meningeal populations of TE-671 (human rhabdomyosarcoma cell line) or D-54 (human glioma cell line) resulted in an increase of median survival of 23%, demonstrating the efficacy and lack of toxicity at therapeutic level of this drug in an *in vivo* model (Friedman et al., 1994). Intrathecal therapy of microcrystalline temozolomide, an alkalating agent with proven efficacy against intraparenchymal malignant gliomas, was found to increase median survival time 132% over that of untreated nude rats (Heimberger et al., 2000). The use of angiogenesis inhibitor TNP470 against a neuroblastoma line (SH-SY5Y) suggested that by inhibiting angiogenesis, TNP470 induced metabolic stress; in turn, chromaffin differentiation and

TABLE 24-1

Tumor Cell Line Athymic Nude Rat Studies

Tumor line	Cell type/tumor	Origin	Route of administration	Treatment modality	Imaging modality	Reference
SK-Mel123	melanoma	human	SQ	^{76}Br-thiouracil	PET	Mars et al., 2000
U-251, SF-767, SF-129, GBM	glioblastoma	human	intracerebral	n/a	growth	Ozawa et al., 2002
A549	non-small cell	human	orthotopic lung	docetaxel	growth	Chan et al., 2002
U-87	glioblastoma	human	intracerebral	VEGF-antibody A4.6.1	MRI	Gossmann et al., 2002
U-87	glioblastoma	human	intracerebral	HPPH photodynamic	growth	Lobel et al., 2002
FPM1-V1AX	HTLV infected T-cell leukemia	rat	subcutaneous	vaccination, adaptive transfer	growth	Kannagi et al., 2001
OMC-1	fibrosarcoma with PTHrP production	human	subcutaneous	bisphosphonate, YM529	Ca^{2+} and vitamin D levels	Michigami et al., 2001
SKBR3-express HER2; MCF-7-not express HER2	mammary adenocarcinoma	human	lumbar placement meningeal metastasis model	antip185/HER2 monoclonal antibody, 4D5	growth, clinical signs	Bergman et al., 2001
PC-14	lung adenocarcinoma	human	subpleural or intrathoracic	anti-VEGF antibodies	formation of effusion and dissemination	Ohta et al., 2001
U-251MG, A1690	glioblastoma	human	intravascular	TALL-104 cytoxic T-cell line	fluorescence labeling	Geoerger et al., 2000
D341, D283, DAOY	medulloblastoma	human	intravascular	TALL-104 cytoxic T-cell line	fluorescence labeling	Geoerger et al., 2000
A431	epidermoid carcinoma	human	intravascular	TALL-104 cytoxic T-cell line	fluorescence labeling	Geoerger et al., 2000
PANC-1(CRL-1469), hs776t (HTB134), HTB147	pancreatic carcinoma	human	subcutaneous	replication competant multimutated herpesvirus G207	growth	McAuliffe et al., 2000
T.T, T.T-1	esophageal carcinoma	human	orthotopic esophagus	CD444 mediation	histology	Hori et al., 001
D54MG	glioma	human	intracerebral	temozolamide	survival	Heimberger et al., 2000
SKOV-3	ovarian	human	intraperitoneal	VEGF antibody	MRI	Gossmann et al., 2000
G55	glioblastoma	human	basal ganglia	anti-VEGF	histology	Rubenstein et al., 2000
MATLyLu	Dunning's prostate cell line	rat	subcutaneous	TGF-β	growth	Matthews et al., 2000
Unknown	osteosarcoma	rat	intratibial	growth hormone, somatostatin, carboplatin	growth	Conzemius et al., 2000
MT-1; MA-11	breast carcinoma	human	intracardiac, intratibial	Mab MOC31, BM7, 425.3-PE, MUC-1	growth	Engebraaten et al., 2002
NCI-H460	non-small cell lung carcinoma	human	orthotopic endobronchial	n/a	histology	Howard et al., 1999
D456MG	glioma	human	meningeal	busulfan	histology	Archer et al., 1999
L2987	adenocarcinoma lung	human	subcutaneous	BR96-doxorubicin	growth	Trail PA et al., 1999
RCA	colon carcinoma	human	subcutaneous	BR96-doxorubicin	growth	Trail et al., 1999
LS174T	colon carcinoma	human	subcutaneous	BR96-doxorubicin	growth	Trail et al., 1999
U87deltaEGFR	glioma	human	subcutaneous	MR-1 Pseudomonas exotoxin	survival	Archer et al., 1999
C18, C29, C85, C86, HCT8	colorectal carcinoma	human	subcutaneous	G207	histology	Kooby et al., 1999
Seven different lines	head and neck squamous cell carcinoma	human	subcutaneous	G207	growth	Carew et al., 1999
SKMEL-1	melanoma	human	meningeal	intraventricular Mab 3F8	histology	Bergman et al., 1999
NMB&	neuroblastoma	human	meningeal	intraventricular Mab 3F8	histology	Bergman et al., 1999
SH-SyY5Y	neuroblastoma	human	subcutaneous	TNP-470	differentiation	Wassberg et al., 1999

Cell line	Tumor type	Species	Site	Treatment	Measurement	Reference
HT-29, WiDr	colorectal carcinoma	human	liver metastasis	n/a	histology	Vogel et al., 998
A431	epidermoid carcinoma	human	meningeal	immunotoxin-1	survival	Bigner et al., 1995
H-81	stomach cancer	human	subcutaneous	S-1 (5FU)	growth	Fukushima et al., 1998
KM12C	colon cancer	human	subcutaneous	S-1 (5FU)	growth	Fukushima et al., 1998
H-31	breast cancer	human	subcutaneous	S-1 (5FU)	growth	et al., 1998
n/a	osteosarcoma	human	right leg	iron oxide	MRI, histology	Axmann et al., 1998
Benzanthrecene induced	malignant fibrous histiocytoma	rat	subcutaneous, orthotopic prostate	adriamycin	growth	Yakushiji et al., 1997
LNCaP	prostate	human		antisense nucelotide	PSA	Rubenstein et al., 1997
TE8, H69	squamous cell carcinoma	human	n/a	EFGR monoclonal and iron	MRI	Suwa et al., 1998
HeLa, GBM, mouse AM-12	cervical Ca; glioblastoma	human, mouse	subcutaneous	n/a	growth	Liang et al., 1998
GBM	glioblastoma	human	intracerebral	n/a	growth	Liang et al., 1998
CEM	T-cell lumphoma	human	meningeal	ricin immunotoxin	histology	Herrlinger et al., 1998
MHH-MED-1	medulloblastoma	human	meningeal	n/a	MRI, histology	Schabet et al., 1997
HT-29	colon carcinoma	human	liver metastasis	n/a	immuno-histochemistry	Ashraf et al., 1997
C6-2B	glioma	human	intracerebral	n/a	culture, histology	Tornatore et al., 1997
D456 MG	glioma	human	intracerebral	O6-benzylguanine	survival	Kurpad et al., 1997
U-87	glioma	human	intracerebral	photodynamic therapy	histology	Chopp et al., 1996
BHK-21	E1A oncogene sarcoma	hamster	newborn rats	n/a	cytolytic activity	Cook et al., 1995
n/a	gastrin receptor colon carcinoma	human	colon	gastrin receptor antagonist	metastasis	Chu et al., 1995
D-54MG, D-456MG	glioma	human	intracerebral	melphalan	survival	Kurpad et al., 1995
DAOY	medulloblastoma	human	meningeal	transferrin toxin conjugate	immunohistochemistry	Wen et al., 1995
U-87	glioma	human	intracerebral	Mab bFGF	angiogenesis	Stan et al., 1995
A431	epidermoid carcinoma	human	meningeal	LMB-7 *Pseudomonas* exotoxin	n/a	Pastan et al., 1995
RT4	transitional cell carcinoma	human	intrarenal, intravesicular	n/a	histology, growth	Oshinsky et al., 1995
NCI-H441	CEA expressing	human	subcutaneous	anti CEA	histology	Gridley et al., 1994
n/a	small cell lung carcinoma	human	intracerebral	Gd-DTPA	MRI	Roman-Goldstein et al., 1994
TE-671	rhabdo-myosarcoma	human	meningeal	melphalan/alpha immunoconjugate	survival	Friedman et al., 1994
LOX	melanoma	human	meningeal	Tfn-PE	histology	Hall et al., 1995
H33396	mammary adenocarcinoma	human	subcutaneous	BR96sFv-PE40	histology	Siegall et al., 1994
A172, A1207, A1235	glioma	human	intracerebral	a and b BGF	histology	Finkelstein et al., 1994
OHS, MHMX	osteosarcoma	human	intratibial	strontium89, ifosasfamide	99mTc labeling	Kjonniksen et al., 1994
FaDu	head and neck squamous cell carcinoma	human	vascular pedicle	cisplatin	histology	Moses et al., 1993
D-54MG, D-456MG	glioma	human	intracerebral	cyclophosphamide	survival	Schuster et al., 1993
SH-SY5Y	neuroblastoma	human	n/a	MIBG	histology	Nilsson et al., 1993
Bl6-F10	melanoma	mouse	meningeal	ACNU	growth, MRI	Schabet et al., 1992
LGL	large granular cell leukemia	rat	intraperitoneal	n/a	histology	Stromberg et al., 1992
AKG	mesothelioma	human	subcutaneous	cyclophosphamide, doxorubicin, etopside, vincristine, vindesine, cisplatin, carmustine, mitomycin c	growth	Linden 1990, 1991

(continued)

TABLE 24-1

Continued

Tumor line	Cell type/tumor	origin	Route of administration	Treatment modality	Imaging modality	Reference
S117	sarcoma	human	subcutaneous	pH, O₂, glucose, heat	pO₂	Roszinski et al., 1991
HSNLV	transplantable sarcoma	rat	n/a	n/a	IL-2 secretion	Russell et al., 1991
2PR-129, 102PR	prostate	rat	n/a	n/a	histology	Gershwin et al., 1982
GER	pancreas exocrine	human	n/a	n/a	histology	Davies et al., 1981
LOX, FEMX-1	melanoma	human	n/a	Fab fragments	Immuno-scintigraphy	Storeng et al., 1991
OHSX	osteosarcoma	human	n/a	Fab fragments	Immuno-scintigraphy	Storeng et al., 1991
HK-1, HK-2	large cell lung carcinoma	human	n/a	n/a	histology	Bonaldo et al., 1991

VEGF indicates vascular endothelial growth factor; BGF, basic growth factor; MRI, magnetic resonance imaging; HPPH, 2-[1-hexyloxyethyl]-2-devinyl pyropheophorbide-a; HTLV, human T-cell lymphotrophic virus; PTHrP, parathyroid hormone-related protein; TGF, transforming growth factor; PSA, periodic acid–Schiff; EFGR, epidermal growth factor receptor; CEA, carcinoembryonic antigen; DTPA, diethylenetriaminepenta acetic acid; MIBG, iobenguane sulfate ¹²³I; ACNU, nidran; IL, interleukin; n/a, not available.

TABLE 24-2

TRANSPLANTATION STUDIES USING THE ATHYMIC NUDE RAT

Organ	Form	Origin	Site	Result	Reference
Pancreas (islets)	fetal cells	human	kidney capsule	extendin 4 encouraged growth	Movassat et al., 2002
Eye (retina)	sheet	human	intraocular	good retention of architecture	Aramant and Seiler, 2002
Peripheral blood mononuclear cells and platelets	cells	human	ischemic limb	angiogenesis	Iba et al., 2002
Kidney	endothelial cells	human	ischemic kidney	increased function	Brodsky et al., 2002
Neurons/neuronal cells	striatal graft	human	striatum after damage	migration of neurons (6 months)	Hurelbrink et al., 2002
Pancreas (islets)	cells	pig	intraportal	nicotinamide reduces failure	Brandhorst D, Brandhorst H, Zwolinski A, Nahidi F, Bretzel RG. 2002
Vein	graft	rat	artery	used Sry gene to tag donor	Redwood and Tennant, 2001
Heart	organ	hamster	intra-abdominal	many studies, most recent used pretransplant blood infusion tolerance	Vriens et al., 2002
Osteoprogenitor cells	cloned bone marrow	mouse	lumbar spine	increased bone formation	Cui et al., 2002
Spinal cord	graft	human	spinal cavities	space occupying, compensatory function	Akesson E, Holmberg L, Jonhagen ME, Kjaeldgaard A, Falci S, Sundstrom E,Seiger A. 2002
Heart	angioblast cells	human	ischemic myocardium	good retention of architecture	Kocher et al., 2001
Ovary	graft	wombat	kidney capsule	follicle formation	Wolvekamp et al., 2001
Thymus	graft	hamster	kidney capsule	long-term tolerance	Xia et al., 2001
Aortic valve	graft	rat	infrarenal	T cell-mediated rejection	Legare et al., 2000
Pancreas (islets)	graft	canine	n/a	stimulation of nitric oxide production	Ketchum et al., 2000
Heart	organ	guinea pig	abdominal	delayed xenograph rejection	Ji et al., 2000
Nasal	graft	human	spinal cord, central nervous system	remyelination	Barnett et al., 2000
Kidney	fetal cells	human	kidney capsule	fetal tissue tolerance	Dekel et al., 2000
Ganglion, sympathetic T2	organ	human	medial forebrain bundle	temporary remission of Parkinson's symptoms	Liu et al., 1999
Skin	meshed	human	skin	growth factors stimulated repair	Smith et al., 2000
Skin	proliferative scar	human	skin	collagen synthesis	Wang et al., 1999
Parathyroid	organ	human	sternocleidomastoid muscle	successful	Walgenbach et al., 1999
Heart	organ	hamster	abdominal	rejection	Lin Y, Soares MP, Sato K, Takigami K, Csizmadia E, Anrather J, Bach FH. 1999
Bone matrix	demineralized	human	cranial	successful	Chesmel et al., 1998
Mesenchymal	stem cells	human	femur	bone regeneration	Bruder et al., 1998
Neurons/neuronal cells	cells	pig	intracerebral	successful	Jacoby et al., 1997
Schwann	cells	human	spinal cord, central nervous system	promote regeneration	Guest JD, Rao A, Olson L, Bunge MB, Bunge RP. 1997
Liver	organ allograph	rat (RNU)	Brown-Norway liver	10-day survival	Brami et al., 1997
Bone marrow	cells	rat	n/a	tolerance, chimerism	Cutler and Bell, 1996

(continued)

TABLE 24-2

CONTINUED

Organ	Form	Origin	Site	Result	Reference
Cornea	graft	guinea pig	eye	rejection	Larkin et al., 1995
Skin	graft	human	skin	tolerance	Harries et al., 1995
Eye (retina)	embryonic	human, rat	eye	synaptic junction	Aramant and Seiler, 1994
Hepatocyte	cell suspension	human	liver	organization	Fontaine et al., 1995
Thymus	cell subsets	rat	n/a	adaptive transfer	Yang and Bell, 1994
Pancreas (islets)	cell clusters	pig	capsule (kidney, liver, spleen)	only kidney capsule	Korsgren and Jansson, 1994
Spinal cord	fetal graft	human	eye	survival	Inoue et al., 1994
Brain	fetal graft	human	eye	HIV-1 and cytomegalovirus AD169 survival	Epstein et al., 1994
Thymus	cultured	rat	n/a	neuroendocrine studies	Martin-Fontecha et al., 1993
Bone	cortical graft	rat	n/a	healing	Kirkeby et al., 1992
Brain	fetal graft	rabbit	eye	comparison of strains	Hall et al., 1992
Hepatocyte	cell	human, dog, rat	n/a	athymic Gunn rat	Lahiri et al., 1991
Thymus	graft	rat	vascular anastomosis	successful	Kampinga et al., 1990
Skin	graft	human	skin	hair growth, cyclosporin A	Gilhar et al., 1990
Muscle	graft	human	n/a	n/a	Gulati et al., 1989
Thymus	graft	calf, rat	n/a	n/a	Hougen et al., 1987

n/a, not available.

apoptosis in the neuroblastoma occurred (Wassberg et al., 1999).

An orthotopic xenograft model using a transitional cell carcinoma cell line (RT4-a well differentiated papillary human bladder tumor cell line) was developed in the nude rat (Oshinsky et al., 1995). The upper urinary tract model of bladder tumor provided a highly reproducible rat model that easily allowed intravesicular (inside the bladder) treatment. Young rats (2 to 4 weeks of age) were surgically implanted with RT4 cells by open cystotomy; pre-irradiation, acid treatment, or cautery did not increase the implantation rate. A similar study used RT4 cells injected from bilateral flank incisions into the renal pelvis after temporarily clamping the ureter (Jarrett et al., 1995).

Orthotopic and metastatic cancers have been studied in nude rats with a variety of experimental tumors. Groups of male nude rats were given 500 rads of gamma radiation, and then 50 mg of NCI-H460 tumor fragments were endo-bronchially implanted in the right caudal lobe airway. High metastatic rates for mediastinal lymph nodes, kidney, bone, and brain were observed (Howard et al., 1999). By use of PC-14, lung adenocarcinoma cells were injected into the pleural space of nude rats, and lesions associated with disseminated cancer (pleural carcinomatosis) were generated (Ohta et al., 2001). The role of vascular endothelial growth factor was explored by using anti-vascular endothelial growth factor neutralizing antibody in hopes of resolving concerns associated with vascular permeability and the development of thoracic effusions with this model.

A human colon cancer model was developed by using HT-29 cells injected into the hepatic portal circulation of nude rats. After 6 to 12 weeks, tumor cells were detected in the liver, lung, and peritoneum. These tumors were detected by using PCR and immunohistochemistry methods for cytokeratin 20 (Vogel et al., 1998). In another study, PCR methodology was compared with histology as a diagnostic method by injecting nude rats with human colorectal cancer (LS174T) by various routes (subcutaneous, intrasplenic, intrahepatic, and intraportal). The PCR method was diagnostic at 10 cancer cells per milligram of liver, 100 to 1000-fold more sensitive than was histology (Stein and Berger, 1999).

Breast cancer studies with nude rats have exploited the generation of breast cancer cells with increased metastatic potential. MA-11 cells, an estrogen and progesterone receptor-negative human breast cancer cell line, progressed metastatically to the CNS when injected into the left chamber of the heart. MT-1 cells, with similar hormone traits, consistently demonstrated bone metastasis (Engebraaten et al., 1999). A novel DNA-binding protein of the helix-turn-helix type called candidate of metastasis-1, which is present in human breast tumors that typically metastasize to the CNS, has been studied in the nude rat xenograft model (Ree et al., 1999).

H. Organ Transplantation

Beyond the studies examining the roles of various compounds on T-cell-mediated rejection and NK cell stimulation with allografts, the nude rat has been used to study the ability of immunosuppressive compounds to increase tolerance of xenograph tissues. One popular model is the hamster-heart-to-nude-rat (Vriens et al., 2001). Miscellaneous studies have been conducted to explore the factors affecting tissue repair by using surgical wounds and xenografts in the nude rat. For example, keratinocyte growth factor has been evaluated for its role in wound healing by using the nude rat (Maggi et al., 1999; Soler et al., 2000). Nude rats have also been used to explore the response of transplanted human/rat skin flaps to percutaneous and transdermal pharmaceuticals and vasoactive drugs (Wojciechowski et al., 1987).

The nude rat has been used as a tool for many studies related to major organ transplantation. Studies conducted using cryopreserved human parathyroid grafted into microsurgically parathroidectomized 6-week-old nude rats showed that the tolerance of the nude rat for parathyroprival hypocalcemia increased the success of the study during the delay of human graft function (Walgenbach et al., 1999).

The feasibility of nerve transplantation and repair of demyelination has been explored by using the nude rat model. After creation of a focal area containing demyelinated axons in the dorsal funiculus of the spinal cord, irradiation with 40 Gy, and exposure to 0.1% ethidium bromide (XEB), it was demonstrated that the XEB model provided an area that was depleted of host derived glial cells and demyelinated axons that could be grafted with cultures derived from human olfactory glial cells (Barnett et al., 2000) or with human peripheral nerve Schwann cells derived from phrenic nerve or lumbosacral plexus that have been depleted of macrophages (Brierley et al., 2001).

Studies using whole-organ heart discordant xenografts of guinea pig (Ji et al., 2000) or hamster (Vriens et al., 2001) have been described by using the rnu rat as host. Hearts are grafted in an abdominal location, and the donors and recipients are treated with a variety of techniques intended to delay or suppress xenograft rejection. It has been determined that there are high titers of preexisting IgM xenoantibodies (xAbs) in the rnu rat that can be neutralized by the injection of a donor cell suspension, and reformation of the xAbs is suppressed by malononitriloamide (MNA 715) (Ji et al., 1999). Hyperacute rejection can be prevented by cobra venom factor, but delayed xenograft rejection still occurs. Rabbit antisialo-GM (100 μL) can be used to deplete NK cells (Ji et al., 2000).

Studies involving liver transplantation have been described in which polymer constructs for human hepatocytes injected intraperitoneally into nude rats (Moscioni et al., 1989; Fontaine et al., 1995). One study involved the use of a Gunn-athymic rat strain with hepatic xenographs from dog, human, and rat. Results of this investigation suggested that based on the ability of the transplanted liver to excrete bilirubin conjugates specific for the species, the transplanted cells retained the metabolic qualities of the donor (Lahiri et al., 1991). In studies investigating liver allograft tolerance, it was found that the peripheral blood and liver of nude rats contain 4 to 10 times fewer CD3+ cells than did wild-type rats. Further, the α/β TCR subset of CD3+ cells is not present in the transplanted liver, whereas the CD3+ γ/δ TCR levels are the same as in wild-type rats, suggesting that these T cells matured in extrathymic secondary lymphoid organs (Brami et al., 1997).

Transplantation of pancreatic islets is a possible treatment of insulin-dependent diabetes, and the nude rat has been used in research as a recipient for porcine, human fetal and adult, and genetically engineered islet cells. For this type of study, the nude rat is pretreated once with

85 mg/kg streptozotocin, the same dose used in euthymic models. Early failure of the intraportally delivered pancreatic islet grafts was accompanied by inflammation and the production of cytokines and nitric oxide (Brandhorst et al., 1998). Specific inducible nitric oxide synthase inhibitors (thiourea) in the transplanted nude rat have been used to explore the mode action of nitric oxide (Brandhorst et al., 2001).

I. Imaging Technologies

The size and predictability of the tumor models in nude rats make them superior subjects for the development of imaging technologies and methods (Table 24-3). Although gross necropsies and subsequent histopathologic examination remain the gold standard in terms of tumor evaluation, long-term evaluation is limited by euthanasia of the subject. The first use of repeat X-ray examinations for evaluation of the progress of tumors in visceral organs of

TABLE 24-3

IMAGING STUDIES: ATHYMIC NUDE RATS

Imaging modality	Contrast technique	References
MRI	Gd-DTPA	Roman-Goldstein et al., 1994
MRI	magnetite particles coated with Mab to EGFR	Suwa et al., 1998
MRI	iron oxide	Axmann et al., 1998
MRI	magnetic starch microspheres	Sundin et al., 1998
MRI	monocrystalline iron oxide nanoparticles	Remsen et al., 1996
MRI	contrast enhanced	Gossmann et al., 2002
MRI	Mab to VEGF loss of vascular permeability	Pham et al., 1998
MRI	diffusion weighted	Lang et al., 1998
MRI	Gd-DTPA	Schabet et al., 1997
MRI	dysprosium-enhanced	Wang et al., 1997
MRI	bone	Witzel et al., 1992
MRI	DMP disrupted BBB	Bulte et al., 1990
MRI	1H nuclear magnetic resonance	Bernsen et al., 1992
POSITRON EMISSION TOMOGRAPHY (PET)	FDG	Burt et al., 2001
PET	76Br 5-bromo-2-thiouracil	Mars et al., 2000
PET	herpes simplex thymidine kinase system	Hospers et al., 1998
PET	11C-colchicine for MDR	Levchenko et al., 2000
PET	Br76-Mab; FDG; 11C-methionine	Lovqvist et al., 1997a,b
PET	124I-Mab	Rubin et al., 1997
COMPUTED TOMOGRAPHY (CT)	iodinated lipid emulsion, iohexol	Bergman et al., 1997
CT	FP736-03, hepatocyte specific contrast	Bergman et al., 1997a,b,c
CT	iodonated hepatocyte emulsion	Sundin et al., 1994
CT	contrast enhanced	Sundin et al., 1992
RADIOGRAPHY	n/a	Le Pimpec-Barthes et al., 1999
RADIOGRAPHY	n/a	Zeligman et al., 1992
Bioluminencence	n/a	Mueller-Klieser et al., 1989
RAR	radio-labeled Mab	Ingvar et al., 1994
RAR	radiolabeled meta-iodobenzylguanidine	Nilsson et al., 1993
RAR	111In-labeled monoclonal antibody MOC-31	de Jonge et al., 1993

MRI indicates, magnetic resonance imaging; DTPA, diethylenetriaminepenta acetic acid; EFGR, epidermal growth factor receptor; VEGF, vascular endothelial growth factor; DMP, dextran-magnetite particles; BBB, blood-brain barrier; PET, positron emission tomography; FDG, 18Fl-labeled deoxyglucose; MDR, multidrug resistant; CT, computed tomography; RAR, autoradiography; n/a, not available.

a rodent host was performed using LOX human malignant melanoma cells in a nude rat model. Metastasis was noted in the lungs after intravenous injection of cells to 4-week-old rats (Kjonniksen et al., 1989).

Nude rats administered melanoma cells have been used in development of positron emission tomography (Mars et al., 2000). Those studies used [76]Br-labeled thiouracil and capitalized on the natural tendency of melanin to take up thiouracil. Studies designed to refine magnetic resonance imaging have used nude rats administered human ovarian cancer cells (Gossmann et al., 2000), glioma xenografts (Gossmann et al., 2002), and melanomameningeal melanomatosis (Martos et al., 1999). Magnetic resonance imaging research methods have used the absorption of iron particles or starch granules, normal mineral densities of bone, and contrast agents injected intravenously or intrathecally to provide suitable contrast for detection of metastasis. With the development of micromagnetic resonance imaging techniques, this method will continue to become more sensitive to microscopic metastases.

J. Cancer Pharmacology

Nude rats are uniquely capable of handling an increased tumor burden without overt distress, and the additional body size allows therapeutic intervention more easily than in a nude mouse. Therapeutic data for anti-neoplastic studies are frequently presented in terms of (1) increased lifespan reflected by the relative median survival time of treated verses control groups (%treated/control); (2) cures; or (3) relative median times (in days) for the treated and control groups to grow tumors of a predetermined target size. Also considered valuable in assessing effectiveness is tumor volume doubling time. The weight (in milligrams) of the tumor is generally estimated *in vivo* to be the measure of one diameter (width) times the square of the measure at 90° (length) (width x length^2).

A variety of classic antineoplastic agents have been tested in the nude rat—including thiouracil, docetaxel, carboplatin, cisplatin, mitomycin, and doxorubicin—as single agents and in variable combinations. Vitamin A analogs demonstrated significant apoptotic (programmed cell death) acceleration in nude rats (Ponthan et al., 2001).

Monoclonal antibodies, with and without attached killer components have been widely tested in cancer-bearing nude rats. Unfortunately, the specificity of antibodies is both the strength and weakness of this treatment modality, with the monoclonal antibodies being specific for the human cell line being studied but conflicting with the intact humoral branch of immunity in the nude rat host.

Some antimetastatic agents use the ability of matrix metalloproteinases inhibitors (e.g., batimastat) to complicate and slow the dissemination of metastatic tumor without toxicity. Other ancillary treatments have included bisphosphonates (Michigami et al., 2001). Nude rats have provided significant and reproducible contributions to our knowledge base for these treatment therapies.

K. Toxicology

Nude rats have been infrequently used in initial toxicologic investigations primarily owing to expense. Exceptions include studies concerned with oncology or autoimmunity, especially T-cell and NK cell activity. The assessment of the development of autoimmune inflammatory skin response to hexachlorobenzene in nude and euthymic rats was hampered by preexistent skin lesions in the nude rats (Michielsen et al., 1999). Hexachlorobenzene caused increased numbers of fibroblasts, histiocytes, and moderate extramedullary hematopoiesis in the red pulp of the spleen in nude and euthymic rats. In control and treated nude rats, brown scaly flakes appeared and disappeared over the dorsum. Histologically, these lesions demonstrated hyperplasia and loss of the epithelium, activated endothelial cells of the deeper dermal vessels, and mononuclear inflammation.

The function of T-cells in the generation of inflammation related to drug-induced neuron cell death was examined in nude rats exposed to guanethidine sulfate. The peripheral sympathetic neuron cell death was associated with the proliferation of T-cells in the euthymic rat, and non-functional CD5-positive T-cell-like cells in the nude rat (Juul et al., 1989; Thygesen et al., 1990). This model was also used to study the effects of cyclosporine, cyclophosphamide and methylpredisone on the mononuclear cell population (Hougen et al., 1992).

The function of T-cells in chronic pulmonary hypertension caused by monocrotaline (MCT) inhalation has been examined in nude and euthymic rats (Miyata et al., 2000). Histologically significant differences included increased interstitial edema in euthymic rats but not in nude rats exposed to MCT, as well as increased alveolar wall thickness in nude rats. The MCT-induced pulmonary hypertension was determined to be more severe in nude rats. An overall increase in the numbers of mast cells in MCT-exposed animals was a major contribution to the development and progression of the hypertensive lesions in nude rats.

VIII. SUMMARY

The nude rat model has demonstrated value for many areas of biomedical research. From basic cellular function of immune cells to intricate interactions of tumors, treatment, and imaging modalities, the nude rat has provided a robust model of suitable size and appropriate response. As the usefulness of this model will continue to increase, it is important that the researcher be familiar with the expected in order to analyze the unexpected.

REFERENCES

Abella, J.A., Oku, Y., Nonaka, N., Ito, M., and Kamiya, M. (1997). Role of host immune response in the occurrence of gastropathy in rats infected with larval *Taenia taeniaeformis*. *J. Vet. Med. Sci.* **59**, 1039–1043.

Akesson, E., Holmberg, L., Jonhagen, M.E., Kjaeldgaard, A., Falci, S., Sundstrom, E., and Seiger, A. (2001). Solid human embryonic spinal cord xenografts in acute and chronic spinal cord cavities: a morphological and functional study. *Exp. Neurol.* **170**, 305–316.

Aldenborg, F., and Enerback, L. (1988). Histochemical heterogeneity of dermal mast cells in athymic and normal rats. *Histochem. J.* **20**, 19–28.

Aramant, R.B. and Seiler, M.J. (1994). Human embryonic retinal cell transplants in athymic immunodeficient rat hosts. *Cell Transplant.* **3**, 461–474.

Aramant, R.B. and Seiler, M.J. (2002). Transplanted sheets of human retina and retinal pigment epithelium develop normally in nude rats. *Exp. Eye Res.* **75**, 115–125.

Archer, G.E., Sampson, J.H., Lorimer, I.A., McLendon, R.E., Kuan, C.T., Friedman, A.H, Friedman, H.S., Pastan, I.H., and Bigner, D.D. (1999). Regional treatment of epidermal growth factor receptor vIII-expressing neoplastic meningitis with a single-chain immunotoxin, MR-1. *Clin. Cancer Res.* **5**, 2646–2652.

Archer, G.E., Sampson, J.H., McLendon, R.E., Friedman, A.H., Colvin, O.M., Rose, M., Sands, H., McCullough, W., Fuchs, H.E., Bigner, D.D., and Friedman, H.S. (1999). Intrathecal busulfan treatment of human neoplastic meningitis in athymic nude rats. *J. Neurooncol.* **44**, 233–241.

Ashraf, S., Loizidou, M., Crowe, R., Turmaine, M., Taylor, I., and Burnstock, G. (1997). Blood vessels in liver metastases from both sarcoma and carcinoma lack perivascular innervation and smooth muscle cells. *Clin. Exp. Metastasis* **15**, 484–498.

Axmann, C., Bohndorf, K., Gellissen, J., Prescher, A., and Lodemann, K.P. (1998). Mechanisms of accumulation of small particles of iron oxide in experimentally induced osteosarcomas of rats: a correlation of magnetic resonance imaging and histology: preliminary results. *Invest. Radiol.* **33**, 236–245.

Balish, E. (1993). Experimental mycoses in congenitally immunodeficient rodents. *Arch. Med. Res.* **24**, 225–231.

Barnett, S.C., Alexander, C.L., Iwashita, Y., Gilson, J.M., Crowther, J., Clark, L., Dunn, L.T., Papanastassiou, V., Kennedy, P.G., and Franklin, R.J. (2000). Identification of a human olfactory ensheathing cell that can affect transplant-mediated remyelination of demyelinated CNS axons. *Brain* **123**, 1581–1588.

Bergman, I., Barmada, M.A., Griffin, J.A., and Slamon, D.J. (2001). Treatment of meningeal breast cancer xenografts in the rat using an anti-p185/HER2 antibody. *Clin. Cancer Res.* **7**, 2050–2056.

Bergman, I., Barmada, M.A., Heller, G., Griffin, J.A., and Cheung, N.K. (1999). Treatment of neoplastic meningeal xenografts by intraventricular administration of an antiganglioside monoclonal antibody, 3F8. *Int. J. Cancer* **82**, 538–548.

Bergman, A., Magnusson, A., and Sundin, A. (1997a). Relationship between contrast enhancement and diagnostic accuracy with the liver-specific CT contrast medium FP 736–04 in an experimental model of liver metastases. *Acad. Radiol.* **4**, 736–741.

Bergman, A., Sundin, A., and Magnusson, A. (1997b). An iodinated lipid emulsion for CT of the liver. Comparison with iohexol in the detection of experimental hepatic metastases. *Acta. Radiol.* **38**, 55–60.

Bergman, A., Sundin, A., and Magnusson, A. (1997c). CT with different doses of the hepatocyte-specific contrast medium FP 736–03: evaluation in a nude-rat model of experimental metastases. *Acta. Radiol.* **38**, 1003–1006.

Bernsen, H.J., Heerschap, A., van der Kogel, A.J., van Vaals, J.J., Prick, M.J., Poels, E.F., Meyer, J., and Grotenhuis, J.A. (1992). Image-guided 1H NMR spectroscopical and histological characterization of a human brain tumor model in the nude rat: a new approach to monitor changes in tumor metabolism. *J. Neurooncol.* **13**, 119–130.

Bigner, D.D., Archer, G.E., McLendon, R.E., Friedman, H.S., Fuchs, H.E., Pai, L.H., Herndon II, J.E., and Pastan, I.H. (1995). Efficacy of compartmental administration of immunotoxin LMB-1 (B3-LysPE38) in a rat model of carcinomatous meningitis. *Clin. Cancer Res.* **1**, 1545–1555.

Bonaldo, Mde. F., Pestano, C.B., Ribeiro, M.C., Machado-Santelli, G.M., Mori, L., and Oliveira, A.R. (1991). Comparative characterization of a human large cell lung carcinoma cell line and the xenograft derived cell line. *Cell Biol. Int. Rep.* **15**, 229–241.

Binder, A., Hedrich, H.J., Wonigeit, K., and Kirchhoff, H. (1993). The *Mycoplasma arthritidis* infection in congenitally athymic nude rats. *J. Exp. Anim. Sci.* **35**, 177–185.

Bischof, A., Park, J.H., and Hunig, T. (2000). Expression of T-cell receptor beta-chain mRNA and protein in γ/δ T-cells from euthymic and athymic rats: implications for T-cell lineage divergence. *Dev. Immunol.* **8**, 19–30.

Boot, R., van Herck, H., and van der Logt, J. (1996). Mutual viral and bacterial infections after housing rats of various breeders within an experimental unit. *Lab Anim.* **30**, 42–45.

Brami, F.C., Gambiez, L.P., Labalette, M., Sfeir, R., Dessaint, J.P., and Pruvot, F.R. (1997). Liver allograft tolerance: do donor thymus-independent T cells play a role? Preliminary results in a nude rat model. *Transplant Proc.* **29**, 2177–2178.

Brandhorst, D., Brandhorst, H., Zwolinski, A., Nahidi, F., and Bretzel, R.G. (2002). High-dosed nicotinamide decreases early graft failure after pig to nude rat intraportal islet transplantation. *Transplantation* **73**, 74–79.

Brierley, C.M., Crang, A.J., Iwashita, Y., Gilson, J.M., Scolding, N.J., Compston, D.A., and Blakemore, W.F. (2001). Remyelination of demyelinated CNS axons by transplanted human Schwann cells: the deleterious effect of contaminating fibroblasts. *Cell Transplant.* **10**, 305–315.

Brodsky, S.V., Yamamoto, T., Tada, T., Kim, B., Chen, J., Kajiya, F., and Goligorsky, M.S. (2002). Endothelial dysfunction in ischemic acute renal failure: rescue by transplanted endothelial cells. *Am. J. Physiol. Renal Physiol.* **282**, 1140–1149.

Bruder, S.P., Kurth, A.A., Shea, M., Hayes, W.C., Jaiswal, N., and Kadiyala, S. (1998). Bone regeneration by implantation of purified, culture-expanded human mesenchymal stem cells. *J. Orthop. Res.* **16**, 155–162.

Buchinsky, F.J., Ma, Y., Mann, G.N., Rucinski, B., Bryer, H.P., Paynton, B.V., Jee, W.S., Hendy, G.N., and Epstein, S. (1995). Bone mineral metabolism in T lymphocyte-deficient and replete strains of rat. *J. Bone Miner. Res.* **10**, 1556–1565.

Bulte, J.W., de Jonge, M.W., de Leij, L., The, T.H., Kamman, R.L., Blaauw, B., Zuiderveen, F., and Go, K.G. (1990). Passage of DMP across a disrupted BBB in the context of antibody-mediated MR imaging of brain metastases. *Acta. Neurochir. Suppl. (Wien)* **51**, 43–45.

Burt, B.M., Humm, J.L., Kooby, D.A., Squire, O.D., Mastorides, S., Larson, S.M., and Fong, Y. (2001). Using positron emission tomography with [(18)F]FDG to predict tumor behavior in experimental colorectal cancer. *Neoplasia.* **3**, 189–195.

Carew, J.F., Kooby, D.A., Halterman, M.W., Federoff, H.J., and Fong, Y. (1999). Selective infection and cytolysis of human head and neck squamous cell carcinoma with sparing of normal mucosa by a cytotoxic herpes simplex virus type 1 (G207). *Hum. Gene Ther.* **10**, 1599–1606.

Case, J.P., Sano, H., Lafyatis, R., Remmers, E.F., Kumkumian, G.K., and Wilder, R.L. (1989). Transin/stromelysin expression in the synovium of rats with experimental erosive arthritis: in situ localization and kinetics of expression of the transformation-associated metalloproteinase in euthymic and athymic Lewis rats. *J. Clin. Invest.* **84**, 1731–1740.

Cash, J.M., Remmers, E.F., Goldmuntz, E.A., Crofford, L.J., Zha, H., Hansen, C.T., and Wilder, R.L. (1993). Genetic mapping of the athymic nude (RNU) locus in the rat to a region on chromosome 10. *Mamm. Genome* **4**, 37–42.

Caspi, R.R., Chan, C.C., Fujino, Y., Najafian, F., Grover, S., Hansen, C.T., and Wilder, R.L. (1993). Recruitment of antigen-nonspecific cells plays a pivotal role in the pathogenesis of a T cell-mediated organ-specific autoimmune disease, experimental autoimmune uveoretinitis. *J. Neuroimmunol.* **47**, 177–188.

Chan, D.C., Earle, K.A., Zhao, T.L., Helfrich, B., Zeng, C., Baron, A., Whitehead, C.M., Piazza, G., Pamukcu, R., Thompson, W.J., Alila, H., Nelson, P., and Bunn, Jr., P.A. (2002). Exisulind in combination with docetaxel inhibits growth and metastasis of human lung cancer and prolongs survival in athymic nude rats with orthotopic lung tumors. *Clin. Cancer Res.* **8**, 904–912.

Chen, X.J. and Enerback, L. (1994). IgE receptors, IgE content and secretory response of mast cells in athymic rats. *Immunology* **83**, 595–600.

Chen, X.J. and Enerback, L. (1995). Surface expression of IgE receptors, IgE occupancy and histamine releasability of mast cells: influence of genetic and T cell factors. *Int. Arch. Allergy Immunol.* **107**, 132–134.

Chen, X.J. and Enerback, L. (1996). Immune responses to a *Nippostrongylus brasiliensis* infection in athymic and euthymic rats: surface expression of IgE receptors, IgE occupancy and secretory ability of mast cells. *Int. Arch. Allergy Immunol.* **109**, 250–257.

Chen, X.J., Wiedermann, U., Dahlgren, U., Hanson, L.A., and Enerback, L. (1995). T-cell–independent and T-cell–dependent IgE responses to the nematode *Nippostrongylus brasiliensis*: comparison of serum IgE and mast-cell-bound IgE. *Immunology* **86**, 351–355.

Chesmel, K.D., Branger, J., Wertheim, H., and Scarborough, N. (1998). Healing response to various forms of human demineralized bone matrix in athymic rat cranial defects. *J. Oral Maxillofac. Surg.* **56**, 857–863.

Chopp, M., Madigan, L., Dereski, M., Jiang, F., and Li, Y. (1996). Photodynamic therapy of human glioma (U87) in the nude rat. *Photochem. Photobiol.* **64**, 707–711.

Chu, M., Nielsen, F.C., Franzen, L., Rehfeld, J.F., Holst, J.J., and Borch, K. (1995). Effect of endogenous hypergastrinemia on gastrin receptor expressing human colon carcinoma transplanted to athymic rats. *Gastroenterology* **109**, 1415–1420.

Conzemius, M.G., Graham, J.C., Haynes, J.S., and Graham, C.A. (2000). Effects of treatment with growth hormone and somatostatin on efficacy of diammine [1,1-cyclobutane dicarboxylato (2-)-0,0′]-(SP-4-2) in athymic rats with osteosarcoma. *Am. J. Vet. Res.* **61**, 646–650.

Cook, J.L., Ikle, D.N., and Routes, B.A. (1995). Natural killer cell ontogeny in the athymic rat: Relationship between functional maturation and acquired resistance to E1A oncogene-expressing sarcoma cells. *J. Immunol.* **155**, 5512–5518.

Cui, Q., Ming Xiao, Z., Balian, G., and Wang, G.J. (2001). Comparison of lumbar spine fusion using mixed and cloned marrow cells. *Spine* **26**, 2305–2310.

Cutler, A.J. and Bell, E.B. (1996). Neonatally tolerant rats actively eliminate donor-specific lymphocytes despite persistent chimerism. *Eur. J. Immunol.* **26**, 320–328.

Davies, G., Duke, D., Grant, A.G., Kelly, S.A., and Hermon-Taylor, J. (1981). Growth of human digestive-tumour xenografts in athymic nude rats. *Br. J. Cancer* **43**, 53–58.

Dekel, B., Marcus, H., Herzel, B.H., Bocher, W.O., Passwell, J.H., and Reisner, Y. (2000). *In vivo* modulation of the allogeneic immune response by human fetal kidneys: the role of cytokines, chemokines, and cytolytic effector molecules. *Transplantation* **69**, 1470–1478.

Deerberg, F., Pohlmeyer, G., Wullenweber, M., and Hedrich, H.J. (1993). History and pathology of an enzootic *Pneumocystis carinii* pneumonia in athymic Han:RNU and Han:NZNU rats. *J. Exp. Anim. Sci.* **36**, 1–11.

de Jonge, M.W., Kosterink, J.G., Bin, Y.Y., Bulte, J.W., Kengen, R.A., Piers, D.A., The, T.H., and de Leij, L. (1993). Radio-immunodetection of human small cell lung carcinoma xenografts in the nude rat using ^{111}In-labelled monoclonal antibody MOC-31. *Eur. J. Cancer* **29**, 1885–1890.

D'Inca, R., Ernst, P., Hunt, R.H., and Perdue, M.H. (1992). Role of T lymphocytes in intestinal mucosal injury: inflammatory changes in athymic nude rats. *Dig. Dis. Sci.* **37**, 33–39.

Eastcott, J.W., Yamashita, K., Taubman, M.A., Harada, Y., and Smith, D.J. (1994). Adoptive transfer of cloned T helper cells ameliorates periodontal disease in nude rats. *Oral Microbiol. Immunol.* **9**, 284–289.

Engebraaten, O., Hjortland, G.O., Hirschberg, H., and Fodstad, O. (1999). Growth of precultured human glioma specimens in nude rat brain. *J. Neurosurg.* **90**, 125–132.

Engebraaten, O., Sivam, G., Juell, S., and Fodstad, O. (2000). Systemic immunotoxin treatment inhibits formation of human breast cancer metastasis and tumor growth in nude rats. *Int. J. Cancer* **88**, 970–976.

Epstein, L.G., Cvetkovich, T.A., Lazar, E.S., DiLoreto, D., Saito, Y., James, H., del Cerro, C., Kaneshima, H., McCune, J.M., and Britt, W.J. (1994). Human neural xenografts: progress in developing an in-vivo model to study human immunodeficiency virus (HIV) and human cytomegalovirus (HCMV) infection. *Adv. Neuroimmunol.* **4**, 257–260.

Festing, M.F., Lovell, D., and Sparrow, S. (1979). The athymic nude rat. *Folia. Biol. (Praha)* **24**, 402.

Festing, M.F., May, D., Conners, T.A., Lovell, D., and Sparrow, S. (1978). An athymic nude mutation in the rat. *Nature* **274**, 365–366.

Fontaine, M., Schloo, B., Jenkins, R., Uyama, S., Hansen, L., and Vacanti, J.P. (1995). Human hepatocyte isolation and transplantation into an athymic rat, using prevascularized cell polymer constructs. *J. Pediar. Surg.* **30**, 56–60.

Fossum, S. (1990). Differences between lymph node structure and function in normal and athymic rats. *Curr. Top. Pathol.* **84**, 65–83.

Ferns, G.A., Raines, E.W., Sprugel, K.H., Motani, A.S., Reidy, M.A., and Ross, R. (1991). Inhibition of neointimal smooth muscle accumulation after angioplasty by an antibody to PDGF. *Science* **253**, 1129–1132.

Ferns, G.A., Reidy, M.A., and Ross, R. (1991). Balloon catheter de-endothelialization of the nude rat carotid: response to injury in the absence of functional T lymphocytes. *Am. J. Pathol.* **138**, 1045–1057.

Finkelstein, S.D., Black, P., Nowak, T.P., Hand, C.M., Christensen, S., and Finch, P.W. (1994). Histological characteristics and expression of acidic and basic fibroblast growth factor genes in intra-cerebral xenogeneic transplants of human glioma cells. *Neurosurgery* **34**, 136–143.

Ford, M.J., Bickle, Q.D., and Taylor, M.G. (1987). Immunity to *Schistosoma mansoni* in congenitally athymic, irradiated and mast cell-depleted rats. *Parasitology* **94**, 313–326.

Foulet, A., Zenner, L., Darcy, F., Cesbron-Delauw, M.F., Capron, A., and Gosselin, B. (1994). Pathology of *Toxoplasma gondii* infection in the nude rat: an experimental model of toxoplasmosis in the immunocompromised host? *Pathol. Res. Pract.* **190**, 775–781.

Friedman, H.S., Archer, G.E., McLendon, R.E., Schuster, J.M., Colvin, O.M., Guaspari, A., Blum, R., Savina, P.A., Fuchs, H.E., and Bigner, D.D. (1994). Intrathecal melphalan therapy of human neoplastic meningitis in athymic nude rats. *Cancer Res.* **54,** 4710–4714.

Fukushima, M., Satake, H., Uchida, J., Shimamoto, Y., Kato, T., Takechi, T., Okabe, H., Fujioka, A., Nakano, K., Ohshimo, H., Takeda, S., and Shirasaka, T. (1998). Preclinical antitumor efficacy of S-1: a new oral formulation of 5-fluorouracil on human tumor xenografts. *Int. J. Oncol.* **13,** 693–698.

Furuta, T., Fujita, M., Machii, K., Kobayashi, K., Kojima, S., and Ueda, K. (1993). Fatal spontaneous pneumocystosis in nude rats. *Lab Anim. Sci.* **43,** 551–556.

Furuta, T. and Ueda, K. (1987). Intra- and inter-species transmission and antigenic difference of *Pneumocystis carinii* derived from rat and mouse. *Jpn. J. Exp. Med.* **57,** 11–17.

Gaertner, D.J., Jacoby, R.O., Johnson, E.A., Paturzo, F.X., and Smith, A.L. (1995). Persistent rat virus infection in juvenile athymic rats and its modulation by immune serum. *Lab Anim. Sci.* **45,** 249–253.

Gaertner, D.J., Jacoby, R.O., Paturzo, F.X., Johnson, E.A., Brandsma, J.L., and Smith, A.L. (1991). Modulation of lethal and persistent rat parvovirus infection by antibody. *Arch. Virol.* **118,** 1–9.

Gaertner, D.J., Jacoby, R.O., Smith, A.L., Ardito, R.B., and Paturzo, F.X. (1989). Persistence of rat parvovirus in athymic rats. *Arch. Virol.* **105,** 259–268.

Gardner, A.L., Roche, J.K., Weikel, C.S., and Guerrant, R.L. (1991). Intestinal cryptosporidiosis: pathophysiologic alterations and specific cellular and humoral immune responses in rnu/+ and rnu/rnu (athymic) rats. *Am. J. Trop. Med. Hyg.* **44,** 49–62.

Geoerger, B., Tang, C.B., Cesano, A., Visonneau, S., Marwaha, S., Judy, K.D., Sutton, L.N., Santoli, D., and Phillips, P.C. (2000). Antitumor activity of a human cytotoxic T-cell line (TALL-104) in brain tumor xenografts. *Neuro-oncology* **2,** 103–113.

Gershwin, M.E., Ruebner, B.H., and Ikeda, R.M. (1982). Transplantation and metastasis of NB rat prostate neoplasia in congenitally athymic (nude) mice, nude mice treated with antilymphocyte sera and congenitally athymic rats. *Exp. Cell Biol.* **50,** 145–154.

Gilhar, A., Etzioni, A., and Krueger, G.G. (1990). Hair growth in human split-thickness skin grafts transplanted onto nude rats: the role of cyclosporin. *Dermatologica* **181,** 117–121.

Gossmann, A., Helbich, T.H., Mesiano, S., Shames, D.M., Wendland, M.F., Roberts, T.P., Ferrara, N., Jaffe, R.B., and Brasch, R.C. (2000). Magnetic resonance imaging in an experimental model of human ovarian cancer demonstrating altered microvascular permeability after inhibition of vascular endothelial growth factor. *Am. J. Obstet. Gynecol.* **183,** 956–963.

Gossmann, A., Helbich, T.H., Kuriyama, N., Ostrowitzki, S., Roberts, T.P., Shames, D.M., van Bruggen, N., Wendland, M.F., Israel, M.A., and Brasch, R.C. (2002). Dynamic contrast-enhanced magnetic resonance imaging as a surrogate marker of tumor response to anti-angiogenic therapy in a xenograft model of glioblastoma multiforme. *J. Magn. Reson. Imaging* **15,** 233–240.

Green III, F., Weber, J.K., and Balish, E. (1987). Acquired immunity to *Trichophyton mentagrophytes* in thymus-grafted or peritoneal exudate cell-injected nude rats. *J. Invest. Dermatol.* **88,** 345–349.

Gridley, D.S., Smith, T.E., Liwnicz, B.H., and McMillan, P.J. (1994). Pilot study of monoclonal antibody localization in subcutaneous and intracranial lung tumor xenografts after proton irradiation. *Anticancer Res.* **14,** 2493–2500.

Guest, J.D., Rao, A., Olson, L., Bunge, M.B., and Bunge, R.P. (1997). The ability of human Schwann cell grafts to promote regeneration in the transected nude rat spinal cord. *Exp. Neurol.* **148,** 502–522.

Gulati, A.K., Rivner, M.H., Gulati, N.K., Swift, T.R., and Cole, G.P. (1989). Growth of human muscle in athymic rodents. *Muscle Nerve* **12,** 249–250.

Haines, D.C., Gorelick, P.L., Battles, J.K., Pike, K.M., Anderson, R.J., Fox, J.G., Taylor, N.S., Shen, Z., Dewhirst, F.E., Anver, M.R., and Ward, J.M. (1988). Inflammatory large bowel disease in immunodeficient rats naturally and experimentally infected with *Helicobacter bilis. Vet. Pathol.* **35,** 202–208.

Hajjar, A.M., DiGiacomo, R.F., Carpenter, J.K., Bingel, S.A., and Moazed, T.C. (1991). Chronic sialodacryoadenitis virus (SDAV) infection in athymic rats. *Lab Anim. Sci.* **41,** 22–25.

Hall, M., Wang, Y., Granholm, A.C., Stevens, J.O., Young, D., and Hoffer, B.J. (1992). Comparison of fetal rabbit brain xenografts to three different strains of athymic nude rats: electrophysiological and immunohistochemical studies of intraocular grafts. *Cell Transplant.* **1,** 71–82.

Hall, W.A., Myklebust, A., Godal, A., Nesland, J.M., and Fodstad, O. (1994). *In vivo* efficacy of intrathecal transferrin-*Pseudomonas* exotoxin An immunotoxin against LOX melanoma. *Neurosurgery* **34,** 649–655.

Hansen, A.K. (1995). Antibiotic treatment of nude rats and its impact on the aerobic bacterial flora. *Lab. Anim.* **29,** 37–44.

Hansson, G.K., Holm, J., Holm, S., Fotev, Z., Hedrich, H.J., and Fingerle, J. (1991). T lymphocytes inhibit the vascular response to injury. *Proc. Natl. Acad. Sci. USA* **88,** 10530–10534.

Haraoka, S., Shimokama, T., and Watanabe, T. (1997). Role of T lymphocytes in the pathogenesis of atherosclerosis: animal studies using athymic nude rats. *Ann. NY Acad. Sci.* **811,** 515–518.

Harries, R.H., Rogers, B.G., Leitch, I.O., and Robson, M.C. (1995). An *in vivo* model for epithelialization kinetics in human skin. *Aust. NZ J. Surg.* **65,** 600–603.

Heimberger, A.B, Archer, G.E., McLendon, R.E., Hulette, C., Friedman, A.H., Friedman, H.S., Bigner, D.D., and Sampson, J.H. (2000). Temozolomide delivered by intracerebral micro-infusion is safe and efficacious against malignant gliomas in rats. *Clin. Cancer Res.* **6,** 4148–4153.

Herrlinger, U., Schmidberger, H., Buchholz, R., Wehrmann, M., Vallera, D.A., and Schabet, M. (1998). Intrathecal therapy of leptomeningeal CEM T-cell lymphoma in nude rats with anti-CD7 ricin toxin A chain immunotoxin. *J. Neurooncol.* **40,** 1–9.

Hirasawa, T., Yamashita, H., and Makino, S. (1988). Genetic typing of the mouse and rat nude mutations by PCR and restriction enzyme analysis. *Exp. Anim.* **47,** 63–67.

Hoang-Vu, C., Dralle, H., Schroder, S., Oertel, M., Kohrle, J., Hesch, R.D., von zur Muhlen, A., and Brabant, G. (1992). The nude rat is established as a model for endocrine studies: regulation of thyroid function and xenotransplantation of a thyroid tumor cell line. *Lab Anim. Sci.* **42,** 164–167.

Holmes, B.J., Diaz-Sanchez, D., Lawrence, R.A., Bell, E.B., Maizels, R.M., and Kemeny, D.M. (1996). The contrasting effects of CD8+ T cells on primary, established and *Nippostrongylus brasiliensis*-induced IgE responses. *Immunology* **88,** 252–260.

Hori, T., Yamashita, Y., Ohira, M., Matsumura, Y., Muguruma, K., and Hirakawa, K. (2001). A novel orthotopic implantation model of human esophageal carcinoma in nude rats: CD44H mediates cancer cell invasion *in vitro* and *in vivo. Int. J. Cancer* **92,** 489–496.

Hospers, G.A., Calogero, A., van Waarde, A., Doze, P., Vaalburg, W., Mulder, N.H., and de Vries, E.F. (2000). Monitoring of herpes simplex virus thymidine kinase enzyme activity using positron emission tomography. *Cancer Res.* **60,** 1488–1491.

Hougen, H.P., Klausen, B., Stenvang, J.P., Kraemmer, J., Rygaard, J. (1987). Effects of xenogeneic, allogeneic and isogeneic thymus grafts on lymphocyte populations in peripheral lymphoid organs of the nude rat. *Lab. Anim.* **21,** 103–111.

Hougen, H.P., Jensen, E.T., and Klausen, B. (1989). Experimental *Salmonella typhimurium* infections in rats, I: influence of thymus on the course of infection. *APMIS* **97,** 825–832.

Hougen, H.P., Jensen, E.T., and Klausen, B. (1990). Experimental *Salmonella typhimurium* infections in rats, II: active and passive immunization as protection against a lethal bacterial dose. *APMIS* **98**, 30–36.

Hougen, H.P., Thygesen, P., Christensen, H.B., Rygaard, J., Svendsen, O., and Juul, P. (1992). Effect of immunosuppressive agents on the guanethidine-induced sympathectomy in athymic and euthymic rats. *Int. J. Immunopharmacol.* **14**, 1113–1123.

Howard, R.B., Mullen, J.B., Pagura, M.E., and Johnston, M.R. (1999). Characterization of a highly metastatic, orthotopic lung cancer model in the nude rat. *Clin. Exp. Metastasis* **17**, 157–162.

Hurelbrink, C.B., Armstrong, R.J., Dunnett, S.B., Rosser, A.E., and Barker, R.A. (2002). Neural cells from primary human striatal xenografts migrate extensively in the adult rat CNS. *Eur. J. Neurosci.* **15**, 1255–1266.

Huth, H., Schlake, T., and Boehm, T. (1997). Tranposon induced splicing defect in the nude rat gene. *Immunogenetics* **45**, 282–283.

Iba, O., Matsubara, H., Nozawa, Y., Fujiyama, S., Amano, K., Mori, Y., Kojima, H., and Iwasaka, T. (2002). Angiogenesis by implantation of peripheral blood mononuclear cells and platelets into ischemic limbs. *Circulation* **106**, 2019–2025.

Ingvar, C., Norrgren, K., Strand, S.E., and Jonsson, P.E. (1994). Heterogeneous distribution of radiolabelled monoclonal antibody: autoradiographic evaluation in the nude rat model. *Anticancer Res.* **14**, 141–146.

Inoue, H.K., Henschen, A., and Olson, L. (1994). Human fetal spinal cord xenografted to the eye of athymic nude rats: survival, ultrastructural differentiation, glial responses and vascular interactions. *J. Electron Microsc. (Tokyo)* **43**, 1–9.

Ito, A. and Kamiyama, T. (1987). Cortisone-sensitive, innate resistance to *Hymenolepis nana* infection in congenitally athymic nude rats. *J. Helminthol.* **61**, 124–128.

Jacoby, D.B., Lindberg, C., Ratliff, J., Wunderlich, M., Bousquet, J., Wetzel, K., Beaulieu, L., and Dinsmore, J. (1997). Fetal pig neural cells as a restorative therapy for neurodegenerative disease. *Artif. Organs* **21**, 1192–1198.

Jarrett, T.W., Chen, Y., Anderson, A.E., Oshinsky, G., Smith, A.D., and Weiss, G.H. (1995). Model of human transitional cell carcinoma: tumor xenografts in upper urinary tract of nude rat. *J. Endourol.* **9**, 1–7.

Ji, P., Xia, G.L., and Waer, M. (2000). Absence of cross-reactivity between xenoantibodies directed against concordant or discordant xenoantigens in rats. *Transplant Proc.* **32**, 861.

Jimenez, E., Martinez, A., Aliouat el, M., Caballero, J., Dei-Cas, E., and Gargallo-Viola, D. (2002). Therapeutic efficacies of GW471552 and GW471558, two new azasordarin derivatives, against pneumocystosis in two immunosuppressed-rat models. *Antimicrob. Agents Chemother.* **46**, 2648–2650.

Johansen, H.K., Espersen, F., Pedersen, S.S., Hougen, H.P., Rygaard, J., and Hoiby, N. (1993). Chronic *Pseudomonas aeruginosa* lung infection in normal and athymic rats. *APMIS* **101**, 207–225.

Jones, B. and Jesson, M.J. (1995). On the nature of the mutation in the nude rat. *Trends Genet.* **11**, 257–258.

Jones, H.E. (1986). Cell-mediated immunity in the immunopathogenesis of dermatophytosis. *Acta Derm. Venereol. Suppl.* **121**, 73–83.

Juul, A., Juul, P., and Christensen, H.B. (1989). Guanethidine-induced sympathectomy in the nude rat. *Pharmacol. Toxicol.* **64**, 20–22.

Kamiyama, T., Cortes, G.T., and Rubio, Z. (1987). The role of thymocytes and IgG antibody in protection against malaria in nude rats. *Zentralbl. Bakteriol. Mikrobiol. Hyg.* **264**, 496–501.

Kampinga, J., Schuurman, H.J., Pol, G.H., Bartels, H., Broekhuizen, R., Vaessen, L.M., Tielen, F.J., Rozing, J., Blaauw, E.H., and Roser, B. (1990). Vascular thymus transplantation in rats: technique, morphology, and function. *Transplantation* **50**, 669–678.

Kannagi, M., Ohashi, T., Hanabuchi, S., Kato, H., Koya, Y., Hasegawa, A., Masuda, T., and Yoshiki, T. (2001a). Immunological aspects of rat models of HTLV type 1-infected T lymphoproliferative disease. *AIDS Res. Hum. Retroviruses* **16**, 1737–1740.

Kannagi, M., Ohashi, T., Hanabuchi, S., Kato, H., Koya, Y., Hasegawa, A., Masuda, T., and Yoshiki, T. (2001b). Regression of human T-cell leukemia virus type I (HTLV-I)-associated lymphomas in a rat model: peptide-induced T-cell immunity. *J. Natl. Cancer Inst.* **93**, 1775–1783.

Katz, J., and Michalek, S.M. (1998). Effect of immune T cells derived from mucosal or systemic tissue on host responses to *Porphyromonas gingivalis*. *Oral Microbiol. Immunol.* **13**, 73–80.

Ketchum, R.J., Deng, S., Weber, M., Jahr, H., and Brayman, K.L. (2000). Reduced NO production improves early canine islet xenograft function: a role fornitric oxide in islet xenograft primary nonfunction. *Cell Transplant.* **9**, 453–462.

Kirkeby, O.J. (1991). Bone metabolism and repair are normal in athymic rats. *Acta. Orthop. Scand.* **62**, 253–256.

Kirkeby, O.J., Nordsletten, L., and Skjeldal, S. (1992). Healing of cortical bone grafts in athymic rats. *Acta Orthop. Scand.* **63**, 318–322.

Kjonniksen, I., Storeng, R., Pihl, A., McLemore, T.L., and Fodstad, O. (1989). A human tumor lung metastasis model in athymic nude rats. *Cancer Res.* **49**, 5148–5152.

Kjonniksen, I., Winderen, M., Bruland, O., and Fodstad, O. (1994). Validity and usefulness of human tumor models established by intratibial cell inoculation in nude rats. *Cancer Res.* **54**, 1715–1719.

Kocher, A.A., Schuster, M.D., Szabolcs, M.J., Takuma, S., Burkhoff, D., Wang, J., Homma, S., Edwards, N.M., and Itescu, S. (2001). Neovascularization of ischemic myocardium by human bone-marrow-derived angioblasts prevents cardiomyocyte apoptosis, reduces remodeling and improves cardiac function. *Nat. Med.* **7**, 430–436.

Koga, H., Aoyagi, K., Matsumoto, T., Iida, M., and Fujishima, M. (1999). Experimental enteropathy in athymic and euthymic rats: synergistic role of lipopolysaccharide and indomethacin. *Am. J. Physiol.* **276**, 576–582.

Kooby, D.A., Carew, J.F., Halterman, M.W., Mack, J.E., Bertino, J.R., Blumgart, L.H., Federoff, H.J., and Fong, Y. (1999). Oncolytic viral therapy for human colorectal cancer and liver metastases using a multi-mutated herpes simplex virus type-1 (G207). *FASEB J.* **13**, 1325–1334.

Korsgren, O., and Jansson, L. (1994). Discordant cellular xenografts revascularized in intermediate athymic hosts fail to induce a hyperacute rejection when transplanted to immuno-competent rats. *Transplantation* **57**, 1408–1411.

Kuramoto, T., Serikawa, T., Hayasaka, N., Mori, M., and Yamada, J. (1993). Regional mapping of the Rowett nude gene (RNU) to rat chromosome 10q24 → q32 by localizing linked SYB2 and GH loci. *Cytogenet. Cell Gen.t.* **63**, 107–110.

Kurpad, S.N., Dolan, M.E., McLendon, R.E., Archer, G.E., Moschel, R.C., Pegg, A.E., Bigner, D.D., and Friedman, H.S. (1997). Intraarterial O6-benzylguanine enables the specific therapy of nitrosourea-resistant intracranial human glioma xenografts in athymic rats with 1,3-bis(2-chloroethyl)-1-nitrosourea. *Cancer Chemother. Pharmacol.* **39**, 307–316.

Kurpad, S.N., Friedman, H.S., Archer, G.E., McLendon, R.E., Petros, W.M., Fuchs, H.E., Guaspari, A., and Bigner, D.D. (1995). Intraarterial administration of melphalan for treatment of intracranial human glioma xenografts in athymic rats. *Cancer Res.* **55**, 3803–3809.

Lahiri, P., Yerneni, P.R., Khan, Z., Demetriou, A.A., Moscioni, A.D., Shouval, D., Chowdhury, J.R., and Chowdhury, N.R. (1991). Bilirubin conjugates produced by human, dog, and rat hepatocytes transplanted into athymic Gunn rats. *J. Assoc. Acad. Minor Phys.* **2**, 58–63.

Lang, P., Wendland, M.F., Saeed, M., Gindele, A., Rosenau, W., Mathur, A., Gooding, C.A., and Genant, H.K. (1998). Osteogenic sarcoma: noninvasive *in vivo* assessment of tumor necrosis with diffusion-weighted MR imaging. *Radiology* **206**, 227–235.

Larkin, D.F., Takano, T., Standfield, S.D., and Williams, K.A. (1995). Experimental orthotopic corneal xenotransplantation in the rat: mechanisms of graft rejection. *Transplantation* **60**, 491–497.

Lawrence, R. and Denham, D.A. (1992). *Brugia pahangi* infections in immune-compromised rats demonstrate that separate mechanisms control adult worm and microfilarial numbers. *Parasite Immunol.* **14**, 371–384.

Laycock, K.A., Fenoglio, E.D., Hook, K.K., and Pepose, J.S. (1997). An *in vivo* model of human cytomegalovirus retinal infection. *Am. J. Ophthalmol.* **124**, 181–189.

Laycock, K.A., Kumano, Y., and Pepose, J.S. (1998). Reproduction of antiviral effect in an *in vivo* model of human cytomegalovirus retinal infection. *Graefes Arch. Clin. Exp. Ophthalmol.* **236**, 527–530.

Legare, J.F., Lee, T.D., Creaser, K., and Ross, D.B. (2000). T lymphocytes mediate leaflet destruction and allograft aortic valve failure in rats. *Ann. Thorac. Surg.* **70**, 1238–1245.

Le Pimpec-Barthes, F., Bernard, I., Abd Alsamad, I., Renier, A., Kheuang, L., Fleury-Feith, J., Devauchelle, P., Quintin Colonna, F., Riquet, M., and Jaurand, M.C. (1999). Pleuro-pulmonary tumours detected by clinical and chest X-ray analyses in rats transplanted with mesothelioma cells. *Br. J. Cancer* **81**, 1344–1350.

Levchenko, A., Mehta, B.M., Lee, J.B., Humm, J.L., Augensen, F., Squire, O., Kothari, P.J., Finn, R.D., Leonard, E.F., and Larson, S.M. (2000). Evaluation of 11C-colchicine for PET imaging of multiple drug resistance. *J. Nucl. Med.* **41**, 493–501.

Liang, S.C., Lin, S.Z., Yu, J.F., Wu, S.F., Wang, S.D., and Liu, J.C. (1997). F344-rnu/rnu athymic rats: breeding performance and acceptance of subcutaneous and intracranial xenografts at different ages. *Lab Anim. Sci.* **47**, 549–553.

Lin, Y., Soares, M.P., Sato, K., Takigami, K., Csizmadia, E., Anrather, J., and Bach, F.H. (1999). Rejection of hamster cardiac xenografts by rat CD4+ or CD8+ T cells. *Transplant Proc.* **31**, 959–960.

Linden, C.J. (1990). Response to doxorubicin and cyclophosphamide of a human pleural mesothelioma clinically and as a xenograft in nude rats. *In Vivo* **4**, 55–59.

Linden, C.J. (1991). Chemosensitivity spectrum of a human pleural mesothelioma xenograft grown in athymic rats. *In Vivo* **5**, 375–380.

Lovqvist, A., Sundin, A., Ahlstrom, H., Carlsson, J., and Lundqvist, H. (1997). Pharmacokinetics and experimental PET imaging of a bromine-76-labeled monoclonal anti-CEA antibody. *J. Nucl. Med.* **38**, 395–401.

Lovqvist, A., Sundin, A., Roberto, A., Ahlstrom, H., Carlsson, J., and Lundqvist, H. (1997). Comparative PET imaging of experimental tumors with bromine-76-labeled antibodies, fluorine-18-fluorodeoxyglucose and carbon-11-methionine. *J. Nucl. Med.* **38**, 1029–1035.

Liu, D.M., Lin, S.Z., Wang, S.D., Wu, M.Y., and Wang, Y. (1999). Xenografting human T2 sympathetic ganglion from hyperhidrotic patients providesshort-term restoration of catecholaminergic functions in hemiparkinsonian athymic rats. *Cell Transplant.* **8**, 583–591.

Lobel, J., MacDonald, I.J., Ciesielski, M.J., Barone, T., Potter, W.R., Pollina, J., Plunkett, R.J., Fenstermaker, R.A., and Dougherty, T.J. (2001). 2-[1-hexyloxyethyl]-2-devinyl pyropheophorbide-a (HPPH) in a nude rat glioma model: implications for photodynamic therapy. *Lasers Surg. Med.* **29**, 397–405.

Maggi, S.P., Soler, P.M., Smith, P.D., Hill, D.P., Ko, F., and Robson, M.C. (1999). The efficacy of 5% Sulfamylon solution for the treatment of contaminated explanted human meshed skin grafts. *Burns* **25**, 237–241.

Mars, U., Tolmachev, V., and Sundin, A. (2000). Positron emission tomography of experimental melanoma with [76Br]5-bromo-2-thiouracil. *Nucl. Med. Biol.* **27**, 845–849.

Marshall, E. and Miller, T. (1981). Characterization of the nude rat (rnunz): functional characteristics. *Aust. J. Exp. Biol. Med. Sci.* **59**, 287–296.

Martin-Fontecha, A., Broekhuizen, R., de Heer, C., Zapata, A., and Schuurman, H.J. (1993). The neuro-endocrine component of the rat thymus: studies on cultured thymic fragments before and after transplantation in congenitally athymic and euthymicrats. *Brain Behav. Immun.* **7**, 1–15.

Martos, J., Petersen, D., Klose, U., Requardt, H., Buchholz, R., Ohnseit, P., Schabet, M., and Voigt, K. (1992). MR imaging of experimental meningeal melanomatosis in nude rats. *J. Neurooncol.* **14**, 207–211.

Matthews, E., Yang, T., Janulis, L., Goodwin, S., Kundu, S.D., Karpus, W.J., and Lee, C. (2000). Down-regulation of TGF-beta1 production restores immunogenicity in prostate cancer cells. *Br. J. Cancer* **83**, 519–525.

McAuliffe, P.F., Jarnagin, W.R., Johnson, P., Delman, K.A., Federoff, H., and Fong, Y. (2000). Effective treatment of pancreatic tumors with two multi-mutated herpes simplex oncolytic viruses. *J. Gastrointest. Surg.* **4**, 580–588.

McKay, D.M., Benjamin, M., Baca-Estrada, M., D'Inca, R., Croitoru, K., and Perdue, M.H. (1995). Role of T lymphocytes in secretory response to an enteric nematode parasite: studies in athymic rats. *Dig. Dis. Sci.* **40**, 331–337.

Michielsen, C.C., Bloksma, N., Klatter, F.A., Rozing, J., Vos, J.G., and van Dijk, J.E. (1999). The role of thymus-dependent T cells in hexachlorobenzene-induced inflammatory skin and lung lesions. *Toxicol. Appl. Pharmacol.* **161**, 180–191.

Michigami, T., Yamato, H., Suzuki, H., Nagai-Itagaki, Y., Sato, K., and Ozono, K. (2001). Conflicting actions of parathyroid hormone-related protein and serum calcium as regulators of 25-hydroxyvitamin D(3)-1 α-hydroxylase expression in a nude rat model of humoral hypercalcemia of malignancy. *J. Endocrinol.* **171**, 249–257.

Milbourne, E.A. and Howell, M.J. (1997). Eosinophilia in nude rats and nude mice after infection with *Fasciola hepatica* or injection with its E/S antigens. *Int. J. Parasitol.* **27**, 1099–1105.

Miller, T. (1984). Pyelonephritis: the role of cell-mediated immunity defined in a congenitally athymic rat. *Kidney Int.* **26**, 816–822.

Miyata, M., Sakuma, F., Ito, M., Ohira, H., Sato, Y., and Kasukawa, R. (2000). Athymic nude rats develop severe pulmonary hypertension following monocrotaline administration. *Int. Arch. Allergy Immunol.* **121**, 246–252.

Moses, B.L., Chan, D.W., Hruban, R.H., Forastiere, A., and Richtsmeier, W.J. (1993). Comparison of intra-arterial and intravenous infusion of cisplatin for head and neck squamous cell carcinoma in a modified rat model. *Arch. Otolaryngol. Head Neck Surg.* **119**, 612–617.

Movassat, J., Beattie, G.M., Lopez, A.D., and Hayek, A. (2002). Exendin 4 up-regulates expression of PDX 1 and hastens differentiation and maturation of human fetal pancreatic cells. *J. Clin. Endocrinol. Metab.* **87**, 4775–4781.

Mueller-Klieser, W., Walenta, S., Kallinowski, F., Vaupel, P., and Fortmeyer, H.P. (1989). Metabolic imaging in human tumor xenografts in rnu/rnu-rats using bioluminescence. *Strahlenther Onkol.* **165**, 506–507.

Murakumo, Y., Takahashi, M., Hayashi, N., Taguchi, M., Arakawa, A., Sharma, N., Sakata, K., Saito, M., Amo, H., and Katoh, H. (1995). Linkage of the athymic nude locus with the myeloperoxidase locus in the rat. *Pathol. Int.* **45**, 261–265.

Nehls, M., Pheifer, D., Schorpp, M., Hedrich, H., and Boehm, T. (1994). New member of the winged-helix protein family disrupted in mouse and rat nude mutations. *Nature* **372**, 103–107.

Nilsson, S., Pahlman, S., Arnberg, H., Letocha, H., and Westlin, J.E. (1993). Characterization and uptake of radiolabelled meta-iodobenzylguanidine (MIBG) in a human neuroblastoma hetero-transplant model in athymic rats. *Acta Oncol.* **32,** 887–891.

Ohta, Y., Kimura, K., Tamura, M., Oda, M., Tanaka, M., Sasaki, T., and Watanabe, G. (2001). Biological characteristics of carcinomatosa pleuritis in orthotopic model systems using immune-deficient rats. *Int J. Oncol.* **18,** 499–505.

Oshinsky, G.S., Chen, Y., Jarrett, T., Anderson, A.E., and Weiss, G.H. (1995). A model of bladder tumor xenografts in the nude rat. *J. Urol.* **154,** 1925–1929.

Owen, D.G. (1987). The effect of orally administered *Entamoeba histolytica* on Wistar and athymic (rnu/ rnu) rats observed during a 12-month period. *Trans. R. Soc. Trop. Med. Hyg.* **81,** 621–623.

Ozawa, T., Wang, J., Hu, L.J., Bollen, A.W., Lamborn, K.R., and Deen, D.F. (2002). Growth of human glioblastomas as xenografts in the brains of athymic rats. *In Vivo* **16,** 55–60.

Pastan, I.H., Archer, G.E., McLendon, R.E., Friedman, H.S., Fuchs, H.E., Wang, Q.C., Pai, L.H., Herndon, J., and Bigner, D.D. (1995). Intrathecal administration of single-chain immunotoxin, LMB-7 [B3(Fv)-PE38], produces cures of carcinomatous meningitis in a rat model. *Proc. Natl. Acad. Sci. USA* **92,** 2765–2769.

Pham, C.D., Roberts, T.P., van Bruggen, N., Melnyk, O., Mann, J., Ferrara, N., Cohen, R.L., and Brasch, R.C. (1998). Magnetic resonance imaging detects suppression of tumor vascular permeability after administration of antibody to vascular endothelial growth factor. *Cancer Invest.* **16,** 225–230.

Pohlmeyer, G. and Deerberg, F. (1993). Nude rats as a model of natural *Pneumocystis carinii* pneumonia: sequential morphological study of lung lesions. *J. Comp. Pathol.* **109,** 217–230.

Ponthan, F., Borgstrom, P., Hassan, M., Wassberg, E., Redfern, C.P., and Kogner, P. (2001). The vitamin A analogues: 13–*cis* retinoic acid, 9-cis retinoic acid, and Ro 13–6307 inhibit neuroblastoma tumor growth *in vivo. Med. Pediatr. Oncol.* **36,** 127–131.

Rajan, T.V., Nelson, F.K., Cupp, E., Schultz, L.D., and Greiner, D. (1992). Survival of *Onchocerca volvulus* in nodules implanted in immunodeficient rodents. *J. Parasitol.* **78,** 160–163.

Redwood, A.J. and Tennant, M. (2001). Cellular survival in rat vein-to-artery grafts: extensive depletion of donor cells. *Cell Tissue Res.* **306,** 251–256.

Ree, A.H., Engebraaten, O., Hovig, E., and Fodstad, O. (2002). Differential display analysis of breast carcinoma cells enriched by immunomagnetic target cell selection: gene expression profiles in bone marrow target cells. *Int. J. Cancer* **97,** 28–33.

Ree, A.H., Tvermyr, M., Engebraaten, O., Rooman, M., Rosok, O., Hovig, E., Meza-Zepeda, L.A., Bruland, O.S., and Fodstad, O. (1999). Expression of a novel factor in human breast cancer cells with metastatic potential. *Cancer Res.* **59,** 4675–4680.

Remsen, L.G., McCormick, C.I., Roman-Goldstein, S., Nilaver, G., Weissleder, R., Bogdanov, A., Hellstrom, I., Kroll, R.A., and Neuwelt, E.A. (1996). MR of carcinoma-specific monoclonal antibody conjugated to monocrystalline ironoxide nanoparticles: the potential for noninvasive diagnosis. *AJNR Am. J. Neuroradiol.* **17,** 411–418.

Rolstad, B. (2001). The athymic nude rat: an animal experimental model to reveal novel aspects of innate immune responses? *Immunol. Rev.* **184,** 136–144.

Roman-Goldstein, S.M., Barnett, P.A., McCormick, C.I., Szumowski, J., Shannon, E.M., Ramsey, F.L., Mass, M., and Neuwelt, E.A. (1994). Effects of Gd-DTPA after osmotic BBB disruption in a rodent model: toxicity and MR findings. *J. Comput. Assist. Tomogr.* **18,** 731–736.

Roszinski, S., Wiedemann, G., Jiang, S.Z., Baretton, G., Wagner, T., and Weiss, C. (1991). Effects of hyperthermia and/or hyperglycemia on pH and pO$_2$ in well oxygenated xenotransplanted human sarcoma. *Int. J. Radiat. Oncol. Biol. Phys.* **20,** 1273–1280.

Rubenstein, J.L., Kim, J., Ozawa, T., Zhang, M., Westphal, M., Deen, D.F., and Shuman, M.A. (2000). Anti-VEGF antibody treatment of glioblastoma prolongs survival but results in increased vascular cooption. *Neoplasia* **2,** 306–314.

Rubenstein, M., Mirochnik, Y., Ray, V., and Guinan, P. (1997). Lack of toxicity associated with the systemic administration of antisense oligonucleotides for treatment of rats bearing LNCaP prostate tumors. *Med. Oncol.* **14,** 131–136.

Rubin, S.C., Kairemo, K.J., Brownell, A.L., Daghighian, F., Federici, M.G., Pentlow, K.S., Finn, R.D., Lambrecht, R.M., Hoskins, W.J., and Lewis, Jr., J.L. (1993). High-resolution positron emission tomography of human ovarian cancer in nude rats using ^{124}I-labeled monoclonal antibodies. *Gynecol. Oncol.* **48,** 61–67.

Rulicke, T., Hassam, S., Autenried, P., and Briner, J. (1991). The elimination of mouse hepatitis virus by temporary transplantation of human tumors from infected athymic nude mice into athymic nude rats (rnuN/rnuN). *J. Exp. Anim. Sci.* **34,** 127–131.

Russell, S.J., Eccles, S.A., Flemming, C.L., Johnson, C.A., and Collins, M.K. (1991). Decreased tumorigenicity of a transplantable rat sarcoma following transfer and expression of an IL-2 cDNA. *Int. J. Cancer* **47,** 244–251.

Sainte-Marie, G. and Peng, F.S. (1990). Mast cells and fibrosis in compartments of lymph nodes of normal, gnotobiotic, and athymic rats. *Cell Tissue Res.* **261,** 1–15.

Sainte-Marie, G. and Peng, F.S. (1991). Dilatation of high endothelial venules in compartments of rat lymph nodes with abundant cortical mast cells. *J. Anat.* **174,** 163–170.

Sainte-Marie, G., Peng, F.S., and Guay, G. (1997). Ectasias of the subcapsular sinus in lymph nodes of athymic and euthymic rats: a relation to immunodeficiency. *Histol. Histopathol.* **12,** 637–643.

Salinas-Carmona, M.C., Nussenblatt, R.B., and Gery, I. (1982). Experimental autoimmune uveitis in the athymic nude rat. *Eur. J. Immunol.* **12,** 480–484.

Schabet, M., Martos, J., Buchholz, R., and Pietsch, T. (1997). Animal model of human medulloblastoma: clinical, magnetic resonance imaging, and histopathological findings after intra-cisternal injection of MHH-MED-1 cells into nude rats. *Med. Pediatr. Oncol.* **29,** 92–97.

Schabet, M., Ohneseit, P., Buchholz, R., Santo-Holtje, L., and Schmidberger, H. (1992). Intrathecal ACNU treatment of B16 melanoma leptomeningeal metastasis in a new athymic rat model. *J. Neurooncol.* **14,** 169–175.

Schluter, D., Hein, A., Dorries, R., and Deckert-Schluter, M. (1995). Different subsets of T cells in conjunction with natural killer cells, macrophages, and activated microglia participate in the intracerebral immune response to *Toxoplasma gondii* in athymic nude and immunocompetent rats. *Am. J. Pathol.* **146,** 999–1007.

Schuddekopf, K., Schorpp, M., and Boehm, T. (1996). The whn transcription factor encoded by the nude locus contains an evolutiona-rily conserved and functionally indispensable activation domain. *Proc. Natl. Acad. Sci. USA* **93,** 9661–9664.

Schuster, J.M., Friedman, H.S., Archer, G.E., Fuchs, H.E., McLendon, R.E., Colvin, O.M., and Bigner, D.D. (1993). Intra-arterial therapy of human glioma xenografts in athymic rats using 4-hydroperoxycyclophosphamide. *Cancer Res.* **53,** 2338–2343.

Schuurman, H.J. (1995). The nude rat. *Hum. Exp. Toxicol.* **14,** 122–125.

Schuurman, H.J., Bell, E.B., Gartner, K., Hedrich, H.J., Hansen, A.K., Kruijt, B.C., de Very, P., Leyten, R., Maeder, S.J., Moutier, R., et al. (1992). Comparative evaluation of the immune status of congenitally athymic and euthymic rat strains bred and maintained at different institutes, 2: athymic rats. *J. Exp. Anim. Sci.* **35,** 33–48.

Schuurman, H.J., Hougen, H.P., and van Loveren, H. (1992). The rnu (Rowlett nude) and rnu N (nznu, New Zealand nude) rat: an update. *ILAR News* **34,** 3–12.

Segre, J.A., Nemhauser, J.L., Taylor, B.A., Nadeau, J.H., and Lander, E.S. (1995). Positional cloning of the nude locus: genetic, physical and transcriptional maps of the region and mutations in the mouse and rat. *Genomics* **28**, 549–559.

Siegall, C.B., Chace, D., Mixan, B., Garrigues, U., Wan, H., Paul, L., Wolff, E., Hellstrom, I., and Hellstrom, K.E. (1994). *In vitro* and *in vivo* characterization of BR96 sFv-PE40: a single-chain immunotoxin fusion protein that cures human breast carcinoma xenografts in athymic mice and rats. *J. Immunol.* **152**, 2377–2384.

Siegall, C.B., Liggitt, D., Chace, D., Tepper, M.A., and Fell, H.P. (1994). Prevention of immunotoxin-mediated vascular leak syndrome in rats with retention of antitumor activity. *Proc. Natl. Acad. Sci. USA* **91**, 9514–9518.

Smith, N.C., Ovington, K.S., and Bryant, C. (1991). Free radical generation and the course of primary infection with *Nippostrongylus brasiliensis* in congenitally athymic (nude) rats. *Parasite Immunol.* **13**, 571–581.

Smith, P.D., Polo, M., Soler, P.M., McClintock, J.S., Maggi, S.P., Kim, Y.J., Ko, F., and Robson, C.M. (2000). Efficacy of growth factors in the accelerated closure of interstices in explanted meshed human skin grafts. *J. Burn Care Rehabil.* **21**, 5–9.

Soler, P.M., Wright, T.E., Smith, P.D., Maggi, S.P., Hill, D.P., Ko, F., Jimenez, P.A., and Robson, M.C. (1999). *In vivo* characterization of keratinocyte growth factor-2 as a potential wound healing agent. *Wound Repair Regen.* **7**, 172–178.

Song, Z.J., Johansen, H.K., Faber, V., and Hoiby, N. (1997). Ginseng treatment enhances bacterial clearance and decreases lung pathology in athymic rats with chronic *P. aeruginosa* pneumonia. *APMIS* **105**, 438–444.

Sorensen, O., Saravani, A., and Dales, S. (1987a). *In vivo* and *in vitro* models of demyelinating disease, XVII: the infectious process in athymic rats inoculated with JHM virus. *Microb. Pathog.* **2**, 79–90.

Sorensen, O., Saravani, A., and Dales, S. (1987b). *In vivo* and *in vitro* models of demyelinating disease, XXIII: infection by JHM virus of athymic rats. *Adv. Exp. Med. Biol.* **218**, 383–390.

Stack, W.E., Taubman, M.A., Tsukuda, T., Smith, D.J., Ebersole, J.L., and Kent, R. (1990). Dental caries in congenitally athymic rats. *Oral Microbiol. Immunol.* **5**, 309–314.

Stan, A.C., Nemati, M.N., Pietsch, T., Walter, G.F., and Dietz, H. (1995). *In vivo* inhibition of angiogenesis and growth of the human U-87 malignant glial tumor by treatment with an antibody against basic fibroblast growth factor. *J. Neurosurg.* **82**, 1044–1052.

Stein, T.N. and Berger, M.R. (1999). Quantification of liver metastases from LS174T human colorectal cancer cells in nude rats by PCR. *Anticancer Res.* **19**, 3939–3945.

Stromberg, P.C., Rosol, T.J., Grants, I.S., and Mezza, L.E. (1992). Transplantation of large granular lymphocyte leukemia in congenitally athymic rats. *Vet. Pathol.* **29**, 216–222.

Sundin, A., Ahlstrom, H., Graf, W., and Magnusson, A. (1994). Computed tomography of experimental liver metastases using an iodinated hepatocyte-specific lipid emulsion: a correlative study in the nude rat. *Invest. Radiol.* **29**, 963–969.

Sundin, A., Graf, W., Glimelius, B., Ahlstrom, H., and Magnusson, A. (1992). Contrast-enhanced CT scanning *in vivo* for the quantification of hepatic metastases from a human colonic cancer in the nude rat. *Eur. J. Surg. Oncol.* **18**, 615–623.

Sundin, A., Wang, C., Ericsson, A., and Fahlvik, A.K. (1998). Magnetic starch microspheres in the MR imaging of hepatic metastases. A preclinical study in the nude rat. *Acta. Radiol.* **39**, 161–166.

Suwa, T., Ozawa, S., Ueda, M., Ando, N., and Kitajima, M. (1998). Magnetic resonance imaging of esophageal squamous cell carcinoma using magnetite particles coated with anti-epidermal growth factor receptor antibody. *Int. J. Cancer* **75**, 626–634.

Taylor, M.J., Van Es, R.P., Townson, S., and Bianco, A.E. (1992). Host strain, H-2 genotype and immunocompetence do not affect the survival or development of *Onchocerca lienalis* infective

larvae implanted within micropore chambers into mice or rats. *Parasitology* **105**, 445–451.

Thomas, D.S. and Pass, D.A. (1997). Chronic ulcerative typhlocolitis in CBH-rnu/rnu (athymic nude) rats. *Lab Anim. Sci.* **47**, 423–427.

Thuner, A.L., Jonas, L.C., Rahm, S., and Sickel, E.(1985). Transmission and course of Tyzzer's disease in euthymic and thymus aplastic nude HanRNU rats. *Z. Versuchstierk.* **27**, 241–248.

Thygesen, P., Hougen, H.P., Christensen, H.B., Rygaard, J., Svendsen, O., and Juul, P. (1990). Identification of the mononuclear cell infiltrate in the superior cervical ganglion of athymic nude and euthymic rats after guanethidine-induced sympathectomy. *Int. J. Immunopharmacol.* **12**, 327–330.

Tjuvajev, J.G., Avril, N., Oku, T., Sasajima, T., Miyagawa, T., Joshi, R., Safer, M., Beattie, B., DiResta, G., Daghighian, F., Augensen, F., Koutcher, J., Zweit, J., Humm, J., Larson, S.M., Finn, R., and Blasberg, R. (1998). Imaging herpes virus thymidine kinase gene transfer and expression by positron emission tomography. *Cancer Res.* **58**, 4333–4341.

Tjuvajev, J.G., Doubrovin, M., Akhurst, T., Cai, S., Balatoni, J., Alauddin, M.M., Finn, R., Bornmann, W., Thaler, H., Conti, P.S., and Blasberg, R.G. (2002). Comparison of radiolabeled nucleoside probes (FIAU, FHBG, and FHPG) for PET imaging of HSV1-tk gene expression. *J. Nucl. Med.* **43**, 1072–1083.

Tjuvajev, J.G., Finn, R., Watanabe, K., Joshi, R., Oku, T., Kennedy, J., Beattie, B., Koutcher, J., Larson, S., and Blasberg, R.G. (1996). Noninvasive imaging of herpes virus thymidine kinase gene transfer and expression: a potential method for monitoring clinical gene therapy. *Cancer Res.* **56**, 4087–4095.

Topfer, F., Lenton, L.M., Bygrave, F.L., and Behm, C.A. (1995). Importance of T-cell–dependent inflammatory reactions in the decline of microsomal cytochrome P450 concentration in the livers of rats infected with *Fasciola hepatica*. *Int. J. Parasitol.* **25**, 1259–1262.

Tornatore, C., Rabin, S., Baker-Cairns, B., Keir, S., and Mocchetti, I. (1997). Engraftment of C6–2B cells into the striatum of ACI nude rats as a tool for comparison of the *in vitro* and *in vivo* phenotype of a glioma cell line. *Cell Transplant.* **6**, 317–326.

Trail, P.A., Willner, D., Bianchi, A.B., Henderson, A.J., Trail-Smith, M.D., Girit, E., Lasch, S., Hellstrom, I., and Hellstrom, K.E. (1999). Enhanced antitumor activity of paclitaxel in combination with the anticarcinoma immunoconjugate BR96-doxorubicin. *Clin. Cancer Res.* **5**, 3632–3638.

Vargas, K.J. and Stephens, L.C. (1983). Resistance of Rowett athymic (nude) rats to casein-induced amyloidosis. *Am. J. Vet. Res.* **44**, 1597–1599.

Vogel, I., Shen, Y., Soeth, E., Juhl, H., Kremer, B., Kalthoff, H., and Henne-Bruns, D. (1998). A human carcinoma model in athymic rats creflecting solid and disseminated colorectal metastases. *Langenbecks Arch. Surg.* **383**, 466–473.

Vos, J.G., Berkvens, J.M., and Kruijt, B.C. (1980). The athymic nude rat, I: morphology of lymphoid and endocrine organs. *Clin. Immunol. Immunopathol.* **15**, 213–228.

Vos, J.G., Kreeftenberg, J.G., Kruijt, B.C., Kruizinga, W., and Steerenberg, P. (1980). The athymic nude rat, II: immunological characteristics. *Clin. Immunol. Immunopathol.* **15**, 229–237.

Vos, J.G., Ruitenberg, E.J., Van Basten, N., Buys, J., Elgersma, A., and Kruizinga, W. (1983). The athymic nude rat, IV: immunocytochemical study to detect T-cells, and immunological and histopathological reactions against *Trichinella spiralis*. *Parasite Immunol.* **5**, 195–215.

Vriens, P.W., Stoot, J.H., Hoyt, G., Scheringa, M., Bouwman, E., and Robbins, R.C. (2001). Pre-transplant blood transfusion induces tolerance to hamster cardiac xenografts in athymic nude rats. *Xenotransplantation* **8**, 247–257.

Waite, D.J., Appel, M.C., Handler, E.S., Mordes, J.P., Rossini, A.A., and Greiner, D.L. (1996). Ontogeny and immunohistochemical localization of thymus-dependent and thymus-independent RT6+ cells in the rat. *Am. J. Pathol.* **148**, 2043–2056.

Walgenbach, S., Rosniatowski, R., Bittinger, F., Schicketanz, K.H., Hafner, F., Hengstler, J., and Junginger, T. (1999). Modified cryopreservation and xenotransplantation of human parathyroid tissue. *Langenbecks Arch. Surg.* **384**, 277–283.

Wallstrom, J.B., Torabinejad, M., Kettering, J., and McMillan, P. (1993). Role of T cells in the pathogenesis of periapical lesions: a preliminary report. *Oral Surg. Oral Med. Oral Pathol.* **76**, 213–218.

Wang, X., Smith, P., Pu, L.L., Kim, Y.J., Ko, F., and Robson, M.C. (1999). Exogenous transforming growth factor β_2 modulates collagen I and collagen III synthesis in proliferative scar xenografts in nude rats. *J. Surg. Res.* **87**, 194–200.

Wang, C., Sundin, A., Ericsson, A., Bach-Gansmo, T., Hemmingsson, A., and Ahlstrom, H. (1997). Dysprosium-enhanced MR imaging for tumor tissue characterization: an experimental study in a human xenograft model. *Acta. Radiol.* **38**, 281–286.

Ward, J.M., Lock, A., Collins, Jr., M.J., Gonda, M.A., and Reynolds, C.W. (1984). Papovaviral sialoadenitis in athymic nude rats. *Lab. Anim.* **18**, 84–89.

Wassberg, E., Hedborg, F., Skoldenberg, E., Stridsberg, M., and Christofferson, R. (1999). Inhibition of angiogenesis induces chromaffin differentiation and apoptosis in neuroblastoma. *Am. J. Pathol.* **154**, 395–403.

Watier, H., Verwaerde, C., Landau, I., Werner, E., Fontaine, J., Capron, A., and Auriault, C. (1992). T-cell-dependent immunity and thrombocytopenia in rats infected with *Plasmodium chabaudi*. *Infect. Immun.* **60**, 136–142.

Weir, E.C., Jacoby, R.O., Paturzo, F.X., Johnson, E.A., and Ardito, R.B. (1990). Persistence of sialodacryoadenitis virus in athymic rats. *Lab Anim. Sci.* **40**, 138–143.

Wen, D.Y., Hall, W.A., Conrad, J., Godal, A., Florenes, V.A., and Fodstad, O. (1995). *In vitro* and *in vivo* variation in transferrin receptor expression on a human medulloblastoma cell line. *Neurosurgery* **36**, 1158–1163.

Whalen, B.J., Greiner, D.L., Mordes, J.P., and Rossini, A.A. (1994). Adoptive transfer of autoimmune diabetes mellitus to athymic rats: synergy of CD4+ and CD8+ T cells and prevention by RT6+ T cells. *J. Autoimmun.* **7**, 819–831.

Witzel, J.G., Bohndorf, K., Prescher, A., and Adam, G. (1992). Osteosarcoma of the nude rat: a model for experimental magne resonance imaging studies of bone tumors. *Invest. Radiol.* **27**, 205–210.

Wojciechowski, Z., Pershing, L.K., Huether, S., Leonard, L., Burton, S.A., Higuchi, W.I., and Krueger, G.G. (1987). An experimental skin sandwich flap on an independent vascular supply for the study of percutaneous absorption. *J. Invest. Dermatol.* **88**, 439–446.

Wolvekamp, M.C., Cleary, M.L., Cox, S.L., Shaw, J.M., Jenkin, G., and Trounson, A.O. (2001). Follicular development in cryopreserved common wombat ovarian tissue xenografted to nude rats. *Anim. Reprod. Sci.* **65**, 135–147.

Xia, G., Goebels, J., Rutgeerts, O., Vandeputte, M., and Waer, M. (2001). Transplantation tolerance and autoimmunity after xenogeneic thymus transplantation. *J. Immunol.* **166**, 1843–1854.

Yakushiji, T., Yonemura, K., Nishida, K., and Takagi, K. (1997). *In vivo* tumouricidal effect of T lymphocytes activated by liposomes containing adriamycin on chemically-induced rat malignant fibrous histiocytoma. *Anticancer Res.* **17**, 4347–4354.

Yang, C.P. and Bell, E.B. (1994). Thymic education curtailed: defective immune responses in nude rats reconstituted with immature thymocyte subsets. *Int. Immunol.* **6**, 569–577.

Yang, C.P., Shittu, E., and Bell, E.B. (2000). Specific B cell tolerance is induced by cyclosporin A plus donor-specific blood transfusion pretreatment: prolonged survival of MHC class I disparate cardiac allografts. *J. Immunol.* **164**, 2427–2432.

Yogi, Y., Banba, T., Kobayashi, M., Katoh, H., Jahan, N., Endoh, M., and Nomaguchi, H. (1999). Leprosy in hypertensive nude rats (SHR/NCrj-rnu). *Int. J. Lepr. Other Mycobact. Dis.* **67**, 435–445.

Yoshie, H., Taubman, M.A., Ebersole, J.L., Smith, D.J., and Olson, C.L. (1985). Periodontal bone loss and immune characteristics of congenitally athymic and thymus cell-reconstituted athymic rats. *Infect. Immun.* **50**, 403–408.

Zalutsky, M.R., McLendon, R.E., Garg, P.K., Archer, G.E., Schuster, J.M., and Bigner, D.D. (1994). Radioimmunotherapy of neoplastic meningitis in rats using an alpha-particle–emitting immuno-conjugate. *Cancer Res.* **54**, 4719–4725.

Zeifer, A., Jacobs, T., Hedrich, H.J., and Seitz, H.M. (1984) *Pneumocystis carnii* infections in immunosuppressed and thymus deficient rats. *Zbl. Bakt.* 258: 387 [abstract].

Zeligman, B.E., Howard, R.B., Marcell, T., Chu, H., Rossi, R.P., Mulvin, D., and Johnston, M.R. (1992). Chest roentgenographic techniques for demonstrating human lung tumour xenografts in nude rats. *Lab. Anim.* **26**, 100–106.

Zenner, L., Foulet, A., Caudrelier, Y., Darcy, F., Gosselin, B., Capron, A., and Cesbron-Delauw, M.F. (1999). Infection with *Toxoplasma gondii* RH and Prugniaud strains in mice, ratsand nude rats: kinetics of infection in blood and tissues to pathology in acute and chronic infection. *Pathol. Res. Pract.* **195**, 475–485.

Zha, H., Remmers, E.F., Du, Y., Cash, J.M., Goldmuntz, E.A., Crofford, L.J., and Wilder, R.L. (1995). The rat athymic nude (rnu) locus is closely linked to the inducible nitric oxide synthase gene (Nos2). *Mamm. Genome* **6**, 137–138.

Chapter 25

Gerontology and Age-Associated Lesions

Nancy L. Nadon

I. OVERVIEW

Gerontology research involves the study of normal aging processes and age-related diseases. Normal aging manifests itself as loss of cells and loss of function in many organ systems, including bone, muscle, and brain. It is likely a cumulative effect of changes in gene expression and environmental factors leading to cellular damage. Laboratory rodents provide important models for studies of the genetic and cellular basis of normal aging, mechanistic investigations of age-related diseases, and testing of therapeutic interventions for normal aging processes and age-related diseases.

The rat has long been an important rodent model for studies of aging and age-related disease. Its size, temperament, and the wealth of published physiological and biochemical data on young rats make it a good choice for studies of the changes that occur during normal aging. Recent advances in rat genomics, covered elsewhere in this book, contribute to the growing value of the rat in biogerontology. In addition, several rat models mimic human pathological conditions that have increased incidence with age. These include the spontaneously hypertensive rat (SHR), the spontaneously hypertensive heart failure (SHHF) rat, and Zucker obese rats (Anderson et al., 1999; Naruse et al., 2002). These disease-specific models are discussed in other chapters of this book.

THE LABORATORY RAT, 2ND EDITION

Although the incidence and types of lesions vary greatly from strain to strain, all strains of rats develop characteristic lesions with very old age that can confound interpretation of experimental data. A good rule to follow for gerontological research is to use animals at about their 50% survival point to avoid confounding lesions with incidences that increase with extreme old age. Even then, some strain-specific lesions, such as glomerulonephropathy in Fischer 344 (F344) rats, have very high incidence, making the choice of strain an important consideration in experimental design. Likewise, it is important to use young controls that have attained full adulthood, which is usually 4 to 6 months of age, although this can vary for different physiological systems. Issues of model choice specific to the field of biogerontology are presented in Miller and Nadon (2000).

A. Strains Used in Gerontology Research

The primary consideration for choice of a rat model for gerontology research is the goal of the study—studies aimed at delineating the progress of age-related diseases benefit from models that specifically recapitulate that disease with high incidence, whereas studies of normal aging are best served by models with long life-spans and few strain-specific lesions.

The genetic background (inbred, hybrid, or outbred) is another important consideration. Outbred rats are genetically diverse and more accurately reflect the genetics of the human population. However, there are several disadvantages to use of outbred rats (Lipman, 1997). Populations of outbred rats can vary from one another, particularly if derived from a small starting population. Even populations of outbred rats from different facilities of the same vendor can vary from one another, introducing a confounding factor into analysis and interpretation of experimental data. As a result, sample sizes must be large to accommodate the genetic diversity of the population. Outbred strains are also not suitable for genetic studies.

Inbred rats provide a more uniform genetic background, but they may be prone to strain-specific lesions. F1 hybrids are genetically identical to one another, facilitating comparison between individuals, yet their heterozygosity protects them from the high incidence of inbred strain-specific disorders. Hybrid rats also in general have an advantage over inbred rats in that they have longer life-spans and generally fewer background lesions than their parental strains.

The National Institute on Aging (NIA) has chosen the F344, Brown Norway (BN), and F344BN F1 hybrid (F344BN) strains to supply for biology of aging studies, as these three strains combine some genetic diversity with specific genetic and physiological advantages for aging

TABLE 25-1

MEDIAN LIFE EXPECTANCY AND BODY WEIGHTS OF RAT STRAINS

	Median survival (months)		Average peak body weight (g)	
Strain	Male	Female	Male	Female
Fischer 344 (F344)	24	26	480	320
Brown Norway (BN)	32	32	480	250
F344BN F1	34	29	630	380

(Data from Turturro et al., 1999.)

research. Lifespans (Table 25-1), food intake, and body weight over life were determined for these three genotypes as part of the NIA-supported Biomarkers of Aging program (Turturro et al., 1999). This chapter will discuss these three strains and the Sprague Dawley rat, an outbred stock used for some aging studies.

The most commonly used strain for aging research, the F344 rat, has advantages for biogerontology research, including a moderate body weight throughout life. Median survival of F344 rats is 24 months for males and 26 months for females. F344 are albino rats related to many of the other common laboratory strains as they have a Wistar background. The disadvantage of working with aged F344 rats is that they have a high incidence of certain diseases, including glomerulonephropathy and leukemia. Lipman et al. (1999) reported that by 24 months of age (50% survival), F344 had a greater than 50% incidence of kidney dysfunction and 25% incidence of large granular cell leukemia (Table 25-2).

BN rats have a long lifespan, and they have far fewer strain specific lesions with age. The nephropathy and leukemia so prevalent in F344 are rare in BN rats

TABLE 25-2

COMMON LESIONS IN AGED RATS

Lesion	Fischer 344[a]	Brown Norway[a]	F344BN F1[b]	Sprague Dawley[c]
Glomerulonephropathy (males)	56%	0	34%	na
Glomerulonephropathy (females)	22%	0	19%	na
Hydronephrosis (males)	0	62%	45%	na
Interstitial cell adenoma	48%	0	25%	0.82%
Testicular atrophy	3.25%	81%	18%	na
Pituitary adenoma (males)	12.9%	1.2%	19%	27%
Pituitary adenoma (females)	33.9%	16%	23%	48%
Leukemia (males)	32.3%	1.25%	na	na
Leukemia (females)	25.4%	0	na	na
Mammary gland tumors (females)	6.8%	2.3%	na	31%

na = data not available.
(Data from [a]Lipman et al., 1999; [b]Lipman et al., 1996; and [c]Chandra et al., 1992.)

(Table 25-2), making them suitable for many studies in which these diseases complicate interpretation of results. The mean lifespan for both male and female BN rats is 32 months, and lesions noted at necropsy cover a wide range, similar to the range seen in the human population. In addition, there are physiological similarities to humans that make BN rats useful models for human senescence, particularly in the male reproductive system. Interstitial pneumonia of undefined etiology has been observed in barrier-raised BN rats, including the NIA aging colonies, but it has a low incidence and is not age-related.

F344BN F1 hybrid rats have a mean lifespan of 34 months for males, slightly longer than the BN mean lifespan, and they develop few background lesions until extreme old age. Mean survival for females is 29 months; the basis for this gender effect is unknown.

Sprague Dawley are outbred rats that have been used in biogerontology research. These rats often become obese and suffer from high insulin resistance with age, which may complicate some experimental design and interpretation and be advantageous to others. Sprague Dawley rats have a median survival of about 24 months. Males commonly attain a body weight of more than 700 g and can reach weights of more than 1 kg (Levin and Dunn-Meynell, 2000; Skalicky et al., 2000; Narath et al., 2001). Weight gain in Sprague Dawley is influenced by many factors, including diet and housing conditions, and there is an extreme variation between individuals even under strictly controlled conditions (Skalicky et al., 2000; Levin and Dunn-Meynell, 2000). Exercise has also been shown to influence not only weight gain in Sprague Dawley rats but also longevity, with sedentary rats having a median lifespan several months shorter than that of exercising rats (Skalicky et al., 2000; Narath et al., 2001).

It is important to mention here reports that suggest that average body weights of both F344 and Sprague Dawley rats have increased over the past 2 to 3 decades. An extensive literature review by Keenan et al. (1998) indicated that Sprague Dawley rats have increased significantly in size since the 1970s. With an outbred stock, it is particularly difficult to make comparisons because colonies may vary greatly depending on the size of the original breeding colony, time since derivation of the breeding colony, and manner of breeding. But this is an essential consideration in aging research, because increased body weight appears to correlate with decreased lifespan and increased incidence of neoplasms and other diseases (Keenan et al., 1995a,b).

Rao et al. (1990) analyzed data from toxicological studies using almost 300 study groups of F344 rats. From 1971 through 1981, a trend of increasing peak body weight and decreasing mean survival was seen. A separate but overlapping analysis of 88 long-term toxicological studies showed a similar trend, with mean body weight increasing

17% from 1971 to 1983 and survival at 24 months of age decreasing from 85% to 61% over the same interval (Haseman and Rao, 1992). Decreased survival is likely tied to increased incidence of many diseases such as leukemia. The F344 rats used in these studies were from commercial vendors and derived from National Institutes of Health (NIH) stock.

The NIA Biomarkers program also produced weight curves for NIH-derived F344 rats in the mid-1990s, and the mean peak body weight for male F344 rats (480 g) was virtually the same as that reported for the early 1980s (477 g) (Rao et al., 1990; Turturro et al., 1999). There has been no increase in mean body weight documented in the NIA aging colonies, which are re-derived from NIH stock every 6-7 years to counter the effects of genetic drift. These reports illustrate the difficulties of comparing studies, particularly those performed in different laboratories at different times. It has been proposed that all long-term studies in rodents should be done under mild caloric restriction (CR) to prevent excessive weight gain (Allaben et al., 1996; Keenan et al., 1999). However, the expense associated with this protocol precludes it from general use by the majority of investigators. These concerns should be weighed when deciding on a model system to use for aging research.

Alliot et al. (2002) recently proposed the LOU/c/jall rat as a model for normal aging. Similar to the BN rat, the LOU/c/jall rat is characterized by a long lifespan and, similar to the F344, by lack of obesity. Derived from the Louvain/c strain, albino rats related to Wistar, LOU/c/jall are reported to be easy to handle. Males attain an average maximal body weight of slightly more than 300 g, whereas females reach an average weight of about 200 g. Females exhibit a very stable weight throughout life; males attain their maximal body weight by 10 months and remain relatively stable in weight until 2 years of age. Body weight and lean mass maintain a strong correlation throughout life. Blood glucose levels remain constant until about 26 months of age.

Mean life expectancy of LOU/c/jall is 28-30 months for males and 33-34 months for females. Maximal survival is 35 months for males and exceeds 37 months for females. Alliot et al. (2002) presented data on changes in muscle mass with age, showing atrophy in the second half of life, and on activity levels, showing decreasing motor activity with age. However, there were no data presented on lesion incidence at any age. The latter information will be essential in order to fully exploit the LOU/c/jall rat as a model for aging.

B. Weight and Body Composition Changes with Age

F344 and BN males reach an average maximal weight of just under 500 g at middle age, whereas the average for

F344BN F1 males is just over 600 g (Table 25-1) (Turturro et al., 1999). For all three strains, the maximal weight is attained shortly before the 50% survival point, with a significant decrease in weight during the last quadrant of life. Females of all strains are smaller and show a less pronounced weight gain through middle age. BN females in particular show a relatively stable weight throughout most of adult life, with an average maximal weight of about 250 g. F344 and F344BN F1 females show patterns of weight gain very similar to that of males, but maximal weight is approximately 350 g.

Changes in weight during aging are accompanied by changes in the lean mass to fat mass ratio. Matsumoto et al. (2000) reported a decrease in the lean mass to fat mass ratio in male BN rats as they aged. Total lean mass was the same at 15 and 29 months of age, whereas total fat mass increased. They suggested that weight gain is not directly linked to increased food intake as food intake progressively decreased with age. The NIA Biomarkers Program measured daily food intake in rats over their entire lifespan and found no decrease in food intake with age but rather a slight increase in food intake from middle age to old age in BN and F344BN F1 rats, as well as a relatively stable level of food intake during life in F344 rats (Turturro et al., 1999). Although food intake varied between the two studies, the Biomarkers study supports the suggestion that food intake and body weight are disassociated in middle and old age.

F344BN F1 rats also exhibit increased fat mass with age that does not correlate with food intake (Li et al., 1997; Scarpace and Tumer, 2001). Leptin, a peptide hormone that participates in the regulation of food intake and energy expenditure, was investigated as a potential mediator of the increased adiposity in old rats. Li et al. (1997) found that although leptin gene expression and serum leptin levels correlated well with adiposity, they did not correlate with food intake. The same correlation between serum leptin levels and adiposity is seen in F344 rats (Greenberg and Boozer, 1999; Mooradian and Chehade, 2000). This increase in both adiposity and serum leptin levels is also seen in humans, making the rat a good model for age-related obesity. One hypothesis put forth to explain the increase in fat mass in older humans and rats is that leptin resistance develops with age (Gabriely et al., 2002). To this end, when young rats were infused with leptin, food intake and visceral fat decreased and insulin response improved. In old rats, the same treatment had no effect (Ma et al., 2002).

II. AGE-ASSOCIATED LESIONS

A. Strain Differences in Lesion Incidence

There are many strain differences with respect to the incidence of common lesions found in aged rats. It must be

noted however that diet influences on tumorigenesis are also strong for some lesion types, making comparisons between studies difficult. Histological analysis of control rats in the NIA Biomarkers Program study underscored the differences between the three strains (Table 25-2 provides some examples) (Lipman et al., 1999). For example, the incidence of adrenal gland enlarged pale cell lesions was 72% in aged BN rats. Conversely, this lesion was absent in aged F344 rats and the incidence was greatly reduced in aged hybrid rats.

Twenty-two percent of F344 females and 56% of F344 males exhibited glomerulonephropathy, which was virtually absent in the other two strains (Lipman et al., 1999). However, 60-70% of BN rats exhibited hydronephrosis of the kidneys, a condition not seen at all in F344 rats. The same pattern was seen in an earlier study in which rats were necropsied after a battery of behavioral assessments (Spangler et al., 1994). Hydronephrosis with reduced urine flow, which in severe cases can cause total obstruction, is not age-related and is not progressive (Provoost et al., 1991). Hydronephrosis also exhibits a gender effect in hybrid rats, irrespective of the direction of the cross: this lesion was observed in about 43-45% of the kidneys from male hybrid rats but only 16% of the kidneys from female hybrid rats (Lipman et al., 1999). Although hydronephrosis in BN rats is not age-related, the high incidence could affect some studies and must be considered when choosing a strain.

Lipman et al. (1996) published a comprehensive histological study of BN, F344BN F1, and BNF344 F1 rats ranging in age from 3 months to 43 months. Most lesions occurred sporadically in few animals, but some lesions were very common. For example, chronic cardiomyopathy was observed in approximately 70% of old males and approximately 35% of old females, irrespective of strain. A gender effect was also observed for the average age of the afflicted rats—the average age was 27 months for females and 23 months for males, suggesting a younger age of onset in males. The basis for these gender effects is unknown.

Not surprisingly, a gender effect was seen for mammary gland hyperplasia—BNF344 F1 and F344BN F1 females exhibited 44% and 38% mammary gland hyperplasia, respectively, whereas males exhibited 16% and 13%. However, the average age of males showing mammary gland hyperplasia was 16 months, compared with 29 months for females, suggesting that the causes of this lesion may differ between sexes.

Lesions of the adrenal gland were very common in F344 rats and both F344BN F1 and BNF344 F1 hybrids, as was inflammation of the preputial gland (Lipman et al., 1996). Metaplasia of the Harderian gland was common in BN males (55%), less common in the BNF344 F1 and F344BN F1 males (20% and 14%, respectively), and even less prevalent in the hybrid females (9% and 11%, respectively).

Testicular atrophy is common in aged BN rats but rare in F344 rats. Lipman et al. (1996) reported a 56% incidence of testicular atrophy in aged BN rats and, in a second study in 1999, reported an 81% incidence in BN rats and a 3.2% incidence in aged F344 rats. Hybrid rats had an intermediate incidence of 8-16%, depending on the direction of the cross. Interstitial (Leydig) cell tumors show the opposite pattern, with a high incidence in F344 (48%) and a negligible incidence (2% or less) in BN rats (Lipman et al., 1996, 1999). Again, the hybrid rats showed an intermediate incidence of about 25%. It has been suggested that the testicular atrophy in BN rats may protect them from proliferative diseases, including interstitial cell adenomas. BN rats are used extensively in the study of age-related lesions of the prostate, as the aging process appears similar to that of humans (Zirkin and Chen, 2000). Prostate atrophy, inflammation, and concretions of the lumen are all prevalent in BN rats and their F1 hybrids.

In a study by Lipman et al. (1996), 19% of male BN rats developed late onset (average of 31 months) cataracts, providing a model for age-related cataracts in humans. F344 rats showed a similar pattern of age-dependent cataract development (Wolf et al., 2000). Cataracts were observed in far fewer hybrid male rats (5-7%) and even fewer female hybrid rats (1%), and the age of onset in the male hybrid rats (37-38 months) was significantly later than in BN rats (Lipman et al., 1996). Although the genetic diversity of the hybrid is clearly protective, these results also suggest a gender effect. The incidence of cataracts is particularly important to consider in behavioral studies that require a response to visual cues.

Sprague Dawley rats have a mean survival time of just less than 24 months, and aged Sprague Dawley rats frequently develop neoplasia. In a study documenting tumor lesions at death in 240 male and 240 female Sprague Dawley rats, greater than 70% of males and 80% of females exhibited at least one tumor at the time of death (Nakazawa et al., 2001). Similar to F344 and BN rats, lesions of the endocrine system were by far the most commonly observed. Pituitary adenomas were common in both genders, as were pheochromocytomas. Interstitial cell tumors were common in males, and mammary gland tumors were common in females. Although it was not quantitated in this study, the authors described chronic progressive nephropathy as a serious problem in Sprague Dawley rats, causing significant early death. Other reports indicate Sprague Dawley rats are extremely prone to chronic renal disease, which begins before they are a year old and can affect nearly every animal by 2 years of age (Bolton et al., 1976; Keenan et al., 2000).

Chandra et al. (1992) also documented neoplastic lesions in Sprague Dawley rats at time of death and found the same incidence of endocrine tumors in both genders. Pituitary adenomas were seen in 28% of males and 48% of females, a similar incidence to that seen in F344 rats. The other common lesion in females was mammary gland neoplasia, observed in 31% of aged Sprague Dawley females. In males, a wide variety of neoplastic lesions of low incidence were found, including pheochromocytomas (4%), keratoacanthomas (4%), pancreatic islet cell adenomas (3.7%) thyroid adenomas (3.6%), fibromas of the skin (2%), and adrenal adenomas (1.2%).

Stain/stock differences have also been noted in lung histology and function. Aged Sprague Dawley rats show fibrotic changes in the lung with age, resulting in changes in air exchange capacity, whereas aged F344 rats show no such changes (Kerr et al., 1990). Moreover, Stiles and Tyler (1988) reported that old Sprague Dawley rats responded to ozone exposure with a greater change in lung volume than did young Sprague Dawley rats.

B. Caloric Restriction

Caloric restriction (CR) is the only intervention proven to increase both lifespan and health span in rodents. Oxidative stress and the glucose-insulin system are both hypothesized to play roles in normal aging, and both are modulated by CR protocols (reviewed in Masoro, 2002). The rat was used for many of the early studies that demonstrated that tumorigenesis is lower in rats raised under life-long CR (Maeda et al., 1985; Thurman et al., 1994). CR is designed to maintain animals at a relatively constant adult weight (Turturro et al., 1999). Although most CR protocols are initiated early in adult life, a positive effect was seen in mice when CR was initiated in middle age (Weindruch and Walford, 1982). However, Lipman et al. (1998) examined F344BN F1 rats with CR initiated at either middle age (approximately 17 months) or old age (approximately 24 months) and found no significant differences in either longevity or tumor burden between CR and control rats. Conversely, there are other benefits of short-term CR feeding when it is initiated in old age. Dean and Cartee (1996) reported that even 20 days of CR feeding initiated at 24 months of age improved insulin responsiveness in skeletal muscle of F344 rats.

BN rats responded to a 40% CR protocol initiated at 4 months of age with a 25% increase in median lifespan for females and 21% increase for males (Turturro et al., 1999). F344 showed less of an increase (approximately 10%) under CR. F344BN F1 hybrid males on the CR protocol showed a 20% increase in median lifespan, whereas the median lifespan for hybrid females was increased almost 40% by the CR protocol.

The NIA Biomarkers Program analyzed age-related lesions in F344, BN, and F344BN F1 rats with CR initiated at 3 months of age (Lipman et al., 1999). Histology was performed on CR and *ad libitum* (AL) fed rats at 24 and 30 months of age. This study demonstrated a significant health benefit of the CR regimen that was evident by 12 months of age. By 30 months of age, the group fed AL exhibited twice the number of lesions per rat when compared to the CR diet group. Most lesion types were reduced in prevalence in the CR rats as compared to age-matched AL controls. For example, the nephropathy that is so prevalent in F344 rats was reduced when the rats were raised under a CR protocol (Masoro et al., 1989; Van Liew et al., 1992).

There are, however, strain differences in the effect of CR on lesion development (Table 25-3). One illustration of the differences between strains is the extent of atrophy of the pancreas. AL animals of all strains showed a significant level of atrophy by 30 months of age, ranging from 9% in F344 to 13% in BN. BN rats on the CR protocol showed an increase in pancreatic atrophy, whereas F344 and F344BN F1 rats on the CR diet showed a decrease. Spinal root degeneration was also commonly seen in all strains, and again, the prevalence increased in BN rats under the CR regimen whereas the prevalence decreased in F344BN F1 rats. Atrophy of the testes, a common lesion in old BN rats, increased in prevalence in the CR BN rats, but the prevalence of interstitial cell adenoma in F344 decreased under CR. Collectively, these findings indicate that some types of lesions show different strain-specific responses to CR protocols. These studies also highlighted that variation in the number of lesions per rat increased with age. The latter is an especially important consideration when studying a small group of very aged rats.

Other functions that decline with age can respond beneficially to CR with varying responses seen in different strains. For example, long-term potentiation, essential to memory development, declined with age, but in F344 rats the decline was abolished by CR (Eckles-Smith et al., 2000). In contrast, Markowska (1999a) reported that CR did not delay or reduce age-dependent loss of memory function in F344 rats. Sensorimotor function at 30 months of age was improved by CR in both BN and F344BN F1

rats, with the hybrid showing more significant improvement than did the BN rats (Markowska and Savonenko, 2002). The CR protocol also improved spatial learning in aged F344BN F1 rats but had no effect on spatial learning in BN rats.

The incidence of cataracts increases with age in most rodent strains, with strains differences evident in both onset and severity. In BN rats, CR delayed the onset of cataracts, but by 34 months of age there was no difference in incidence between CR and control rats (Wolf et al., 2000). F344 rats showed the opposite effect—the incidence was the same in CR and control rats through 25 months of age, after which CR rats had a reduced incidence of cataracts. Presbyacusis (age-related hearing loss) is another significant affliction of the elderly that can be studied in rats. In a rat model of presbyacusis, auditory brainstem response and loss of cochlear hair cells were both significantly reduced in F344 rats fed life-long a CR diet compared with age-matched AL controls (Seidman, 2000).

Sprague Dawley rats also respond to CR with increased lifespan and decreases in age-related degenerative disease and neoplasms (Keenan et al., 1995a,b; Laroque et al., 1997). Cardiomyopathy, a common cause of death in Sprague Dawley rats, was decreased in severity by a CR protocol that extended mean survival of Sprague Dawley rats by 20% (Kemi et al., 2000). Molon-Noblot et al. (2001) reported decreases in several age-related degenerative processes in Sprague Dawley rats, including degeneration of the endocrine pancreas and islet hyperplasia. The incidence of nephropathy was reduced in Sprague Dawley rats raised under CR protocols, with a clear correlation between the degree of CR and the reduction in nephropathy (Keenan et al., 2000). Moreover, evidence from the obese Zucker rat indicates that CR begun in middle age after the onset of nephropathy can attenuate the disease progress (Maddox et al., 2002).

CR is thus an important consideration in the design of many, if not all, biogerontology studies. Moreover, the study of CR rats can also model the beneficial effects of CR in humans, and investigations into the mechanisms of lifespan and health span extension by CR may define new targets for therapeutic interventions.

TABLE 25-3

Effects of Caloric Restriction on Prevalence of Selected Age-Related Lesions in Rat Strains

Lesion	Fischer 344	Brown Norway	BNF344 F1
Pancreatic exocrine atrophy	decreased	increased	no change
Interstitial cell adenoma	decreased	nd	nd
Testicular atrophy	increased	increased	nd

nd = lesion not detected or at very low prevalence.
(Data from Lipman *et al.*, 1999.)

III. HIGHLIGHTS OF THE RAT MODEL IN GERONTOLOGY RESEARCH

The rat has been used extensively for biochemical and physiological assays, providing a wealth of background data to use as the foundation for studies on the biology of aging. Rats are suitable for the modeling of several age-dependent declines in physiological parameters, including muscle strength, bone strength, memory, and motor function. One reason the rat has been so popular is that its size allows for far more manipulation that the mouse. Behavioral, biochemical, and electrophysiological data can all be collected on the same animals. Pellets implanted under the skin of rats to deliver hormones, drugs, and other compounds are tolerated well by rats, and stereotactic surgery allows for precise manipulations.

A. Neurological and Behavioral Research

Rats are easily trained to do basic tasks and respond to environmental stimuli, making them a suitable model for behavioral, cognitive, and neurological research. This has been exploited greatly by the aging research community, as cognitive decline and loss of memory are two especially devastating problems for the elderly. Many different experimental designs have been developed for measuring learning and memory in rats (Markowska and Brechler, 1999). The Morris water task requires a rat to use spatial cues to remember the location of a platform submerged in murky water (Morris, 1981). Land-based mazes can be used to accomplish the same goals, using food rewards as the motivator. The radial arm maze measures memory by providing visual cues on which to train the rats (Ward et al., 1999). A T-maze is a series of interlocking T's requiring a complex series of maneuvers to transit the entire maze, and has been used to measure more complex learning functions (Ingram et al., 1998b; Meyer et al., 1998; Umegaki et al., 2001). These assays are important tools for dissecting the changes in memory and learning ability that occur with age.

BN and Sprague Dawley rats are particularly docile and easy to handle. BN rats have been used for studies on age-related eye diseases, including cataracts and glaucoma, in part because of the ease of handling and performing eye exams without anesthesia (Wolf et al., 2000). BN rats are also very active, showing little aversion to entering new environments and open areas. By a large number of tests, F344 rats are more anxiety-prone, less exploratory, and less social than are BN rats (Rex et al., 1996, Ramos et al., 1997).

There are strain differences in behavior, learning, memory, and motor function. A comparison of F344, BN, and F344BN F1 hybrid rats showed age-related declines of several parameters in all strains, including motor function as assayed by rotarod performance, daily locomotor activity, and completion of a T-maze (Spangler et al., 1994). F344 rats fared the poorest on motor function and T-maze performance. Age-dependent decline in performance did not correlate with lesion ratings obtained from necropsy, suggesting that the poorer performance in aged rats was not due to underlying disease.

F344 showed a marked age-dependent deficit in spatial memory as assayed by the Morris water task (Markowska and Brechler, 1999). The decline began earlier in females (by 12 months) than males, although by 18 months of age the deficit was approximately equivalent in both sexes (Markowska, 1999b). The earlier decline in females may be related to changes in estrogen levels. By 12 months of age, only 30% of F344 females were cycling regularly, and by 18 months of age, 60% had irregular cycles and the other 40% were acyclic.

Motor function, another neurologic parameter that changes with age, is important in adaptation to the environment. Basic coordination can be assayed by measuring the length of time it takes a rat to turn around on an inclined screen or its ability to walk down a wide, flat bar without falling. Sensorimotor skills can be measured by mechanisms such as a rotating rod on which the rat must balance, a stationary round dowel on which the rat must balance and walk to a platform, or a runway made up of rods down which the rat must run quickly. F344 rats show impairment in all of these tests as they age (Frick et al., 1997; Markowska and Brechler, 1999). Bickford et al. (2000) illustrated the utility of these measurements in testing antioxidant-rich diets for their ability to improve motor learning in old rats. The runway they used could be configured with either regular or irregular spacing between the rods to test how fast the rats learned to navigate different courses. This study demonstrated the beneficial effects of antioxidants on motor learning in aged rats.

Neurobiology has been advanced greatly by experiments that measure activities of individual neurons in conscious rats. Moreover, stereotactic surgery to implant electrodes allows for electrophysiological measurements in ambulating rats. For example, Barnes et al. (2000) used a movable microdrive carrying both stimulating and recording electrodes to measure evoked responses in F344 rats performing the Morris water task. This laboratory also used a multiple electrode microdrive with 14 independently movable probes implanted over the hippocampus to record signals from multiple individual cells at the same time (Smith et al., 2000). They were able to record electrophysiological data while the rat rested, slept, and moved down a track toward a food reward. Such studies have helped elucidate the normal changes in neural cell activity and plasticity with age.

Electrophysiological measurements can also be performed on anesthetized rats, allowing testing of compounds directly administered to specific parts of the brain. Microdialysis probes can be implanted in specific regions of the brain to deliver controlled amounts of compounds, as demonstrated by Gerhardt and Maloney (1999) in their study of dopamine metabolites in young and old F344 rat brains. In another study, Hebert and Gerhardt (1999) investigated age-related changes in dopamine uptake in the brains of anesthetized rats, using a micropipette for delivery of precise volumes of drugs and recording electrodes to measure cell function.

The size of the rat also allows for stereotactic surgical manipulations to transplant cells or gene therapy vectors into very specific regions of the brain. For example, Hodges et al. (2000) described experiments that assessed whether stem cells can populate the aged brain and correct age-related deficits in learning and memory. Ingram et al. (1998a) tested an adenovirus gene therapy model to deliver dopamine D2 receptors to the brain, by injecting the vector into the rat brain striatum and looking for evidence of receptor expression. The rat has also been used as a surgical model for stroke that involves occlusion of cerebral arteries to cause ischemia. Dubal and Wise (2001) used such a model to demonstrate that estrogen protects old female rats from brain injury resulting from stroke.

The ability to combine behavioral and learning assays with biochemical assays in the same animal has also made the rat a good model for Alzheimer's disease (AD). For example, Bimonte et al. (2002) reported that young F344 rats showed a positive linear relationship between levels of soluble amyloid protein fragments (sAPP) and memory. Conversely, in old F344 rats a narrow range of sAPP levels correlated with normal memory, whereas both high and low levels of sAPP correlated with marked memory deficits. Such biochemical differences between young and old rats are important clues to understanding the devastating progression of AD.

B.   Insulin and Glucose Metabolism

Similar to humans, rodents exhibit increasing blood glucose and insulin levels with age (Barzilai and Rossetti, 1995). Insulin resistance is associated with Syndrome X, a constellation of conditions including obesity, hypertension, and type II diabetes mellitus that affects a significant proportion of the elderly population. The rat is an important model for investigating the etiology of Syndrome X and determining the respective roles of aging and lifestyle changes.

Most rat strains show an increase in body fat mass throughout adult life and a decrease at the very end of life, a pattern similar to that seen in the human population. Barzilai and Gupta (1999) hypothesized that changes in glucose metabolism are related to changes in body composition, specifically increasing obesity with age, and suggested changes in insulin responsiveness are related to changes in fat mass, not aging.

Larkin et al. (2001) used F344BN F1 rats to examine insulin responsiveness in aged muscles, choosing the hybrid rat because of the concerns that the nephropathy common

in F344 rats and the obesity common in Sprague Dawley rats might complicate interpretation of the results. They found that plasma insulin increased with age, insulin responsiveness decreased, and levels of muscle glut-4 glucose transporter decreased with age. These changes are similar to those seen in humans, indicating that the hybrid rat is a good model for age-related changes in glucose metabolism. Conversely, F344 rats show no changes in insulin responsiveness with age, a finding possibly related to their leaner body composition. This study by Larkin et al. (2001) also illustrates an advantage of the rat as an animal model. The carotid artery was cannulated to allow for collection of blood at regular intervals, and cannulae were implanted in the jugular vein to infuse insulin and glucose in precise amounts. Such an experiment could not have been done in a mouse.

C.   Musculoskeletal Biology

Rats have been used to model age-related loss of bone density, even though significant bone loss in rats is not accompanied by a propensity for fractures as it is in humans. To recapitulate the symptoms of post-menopausal bone loss in women, rats can be ovariectomized to produce estrogen-dependent decreases in bone mineral density. Several strains have been used in the ovariectomized model of post-menopausal bone loss, including F344, Sprague Dawley, and Wistar rats. This model works well for histological analysis, density measurements by peripheral quantitative computerized tomography, measurement of bone remodeling through tetracycline labeling, and mechanical strength testing of isolated bones (Mosekilde et al., 1998; Wang et al., 2001b,c). Bone strength is measured in vertebrae by compression analysis and in the femur by the bend-to-fracture stress analysis.

Age-related bone loss of males is harder to model in the rat. F344 male rats experience loss of cancellous bone mineral density at some sites, notably the tibia and femoral neck (Banu et al., 2002). However, cancellous bone loss in vertebrae occurs only late in life and there is no loss of cortical bone, a significantly different bone loss pattern than that observed in aged men. Therefore, the F344 rat does not appear to be a suitable model for age-related bone loss in males. Sprague Dawley rats, on the other hand, do appear to mirror the process in humans, losing cortical bone by 9 months of age (Wang et al., 2001a).

Sarcopenia is another common affliction of the elderly, and rats have been used to model the effects of exercise on this condition (Cartee, 1994). F344BN F1 rats in particular show age-related muscle loss similar to that seen in humans. Surgical ablation of the gastrocnemius muscle mimics resistance training by increasing resistance on the plantaris muscle (Blough and Linderman, 2000). Although such increased resistance resulted in increased plantaris muscle mass in young hybrid rats, there was no such response in old hybrid rats, suggesting changes in muscle plasticity with age. Mayhew et al. (1998) reported that in old muscle, there is an uncoupling of the DHPR and RyR1 receptors that are necessary for the excitation-contraction of skeletal muscle. CR reduces the uncoupling and preserves muscle function, suggesting a direct connection between the uncoupling of the receptors and the loss of muscle strength in aged rats. Studies such as these provide hope for therapeutic interventions for sarcopenia.

D. Cardiovascular Disease

The large size of the rat allows manipulations and measurements in aged animals that are impossible in smaller animals. Blood flow to tissues can be measured by implanting a catheter into the left ventricle of the heart via the carotid artery and infusing labeled microspheres. Catheters can be implanted in the caudal tail artery to measure blood pressure and withdraw blood samples. These procedures are done as survival surgeries so that measurements can be made on conscious rats (Delp et al., 1998; Salter et al., 1998). An in vivo model of coronary infarct involving surgical ligation of the coronary artery has also been used to study cardiovascular disease in aged rats (Raya et al., 1997).

The large size of rat organs also makes in vitro manipulations easier and more accurate. A model of myocardial infarction uses hearts dissected from the body and cannulated through the aorta for attachment to a perfusion apparatus (Gao et al., 2000). Lucas and Szweda (1999) used such a scheme to demonstrate age-dependent effects of cardiac reperfusion. Blood vessels were large enough for in vitro manipulations, allowing Muller-Delp et al. (2002) to measure vasodilatory responses in arterioles isolated from muscle. This study reported that blood vessels from slow-twitch and fast-twitch muscles of aged F344 rats showed differential dependence on endothelium-derived signals for the vasodilatory response.

Peripheral arterial occlusion disease is a common cardiovascular disease in humans that greatly increases risk of mortality from coronary heart disease. Physical activity has been shown to be therapeutic, but there is need for other interventions that non-ambulatory elderly can use. Rats can be trained to use a treadmill and are used for investigation of disease progression and intervention strategies. These investigations use femoral artery ligation to model arterial occlusion. Yang and Feng (2000) used such a model to deliver basic fibroblast growth factor via catheter. The carotid artery was catheterized to monitor blood pressure and heart rate and to infuse labeled microspheres to measure blood flow to the muscle. Measurements obtained while rats were running on the treadmill demonstrated improved blood flow in rats treated with basic fibroblast growth factor.

E. Reproductive Aging

The primary rat model for male reproductive aging is the BN rat. This strain shows declines in serum testosterone with age similar to those observed in human males. This decline in testosterone is believed to contribute to osteoporosis, sarcopenia, and decreased libido. Analyses in the BN rat demonstrated deficits in the testicular interstitial cells concomitant with decreased testosterone levels (Chen et al., 1994). Correlating increases in follicle stimulating hormone suggest the deficit is caused by changes in the testes, not the hypothalamic-pituitary axis. These findings again mirror changes seen in aging human males (Zirkin and Chen, 2000).

Another reason BN rats are used for male reproductive aging studies is that they rarely develop testicular tumors that would complicate analysis, whereas F344 rats exhibit a >50% incidence of interstitial cell adenomas by 24 months of age. The disadvantage of the BN rat model is that they develop testicular atrophy. Although other strains also develop this lesion, it is more common in BN rats than in F344 or BNF344 F1 rats. Moreover, although CR usually improves health span in rodents, it increases the incidence of testicular atrophy in BN and F344 rats by an unknown mechanism.

Reproductive aging in female rats is a function of changes in the hypothalamus and the ovary. The hypothalamus-pituitary-ovarian axis is a complex interaction of signals, and the rat model has contributed greatly to the understanding of this system. During aging, female rats transition from regular estrus cycles to a period of irregular cycling to cessation of cycling altogether. Even before loss of cycle regularity, there are changes in hormone levels, including the pre-ovulatory surge in luteinizing hormone (Rubin, 2000; Wise et al., 2002). Changes at the neuronal level, including the numbers and activity of specific neurons, are being determined through studies on the rat. Reproductive aging in female rats appears to result from both aging and estrogen exposure and may be rooted in the oxidative stress theory of aging (Hung et al., 2003).

IV. SUMMARY

The rat is a valuable research tool for investigating the biology of aging, but care must be exercised when choosing the strain. Inbred strains provide a uniform genetic background that facilitates many studies, but it must be understood that a finding in one strain will not necessarily hold true for other inbred strains. Hybrid rats in general have fewer strain-specific lesions than do inbred rats and provide a robust model for most studies. Outbred rats may more naturally reflect the varied genetics of the human population, but when small groups of outbred rats are analyzed, individual variation may be large. This makes comparisons between studies difficult. CR has been shown to reduce variability between groups of outbred rats, by reducing body weight and morbidity, and mild CR has been proposed as a basic component of experimental design in aging studies. All aspects of the rat—physiology and behavior, genetics, feeding protocols, and even housing—must be considered when developing the experimental paradigm.

ACKNOWLEDGMENT

The author is grateful to Dr. Huber Warner for critical review of the manuscript.

REFERENCES

Allaben, W.T., Turturro, A., Leakey, J.E.A., Seng, J.E., and Hart, R.W. (1996). FDA points-to-consider documents: the need for dietary control for the reduction of experimental variability within animal assays and the use of dietary restriction to achieve dietary control. *Toxicol. Pathol.* **24**, 776–781.

Alliot, J., Boghossian, S., Jourdan, D., Veyrat-Durebex, C., Pickering, G., Meydnial-Denis, D., and Guamet, N. (2002). The LOU/c/jall rats as an animal model of healthy aging? *J. Gerontol. A Biol. Sci. Med. Sci.* **57**, B312–B320.

Anderson, K.M., Eckhart, A.D., Willette, R.N., and Koch, W.J. (1999). The myocardial beta-adrenergic system in spontaneously hypertensive heart failure (SHHF) rats. *Hypertension* **33**, 402–407.

Banu, J., Wang, L., and Kalu, D.N. (2002). Age-related changes in bone mineral content and density in intact male F344 rats. *Bone* **30**, 125–130.

Barnes, C.A., Rao, G., and Orr, G. (2000). Age-related decrease in the schaffer collateral-evoked EPSP in awake, freely behaving rats. *Neural Plast.* **7**, 167–178.

Barzilai, N. and Gupta, G. (1999). Interaction between aging and syndrome X: new insights on the pathophysiology of fat distribution. *Ann. NY Acad. Sci.* **892**, 58–72.

Barzilai, N. and Rosetti, L. (1995). Relationship between changes in body composition and insulin responsiveness in models of the aging rat. *Am. J. Physiol.* **269**, E591–E597.

Bickford, P.C., Gould, T., Briederick, L., Chadman, K., Pollock, A., Young, D., Shukitt-Hale, B., and Joseph, J. (2000). Antioxidant-rich diets improve cerebellar physiology and motor learning in aged rats. *Brain Res.* **886**, 211–217.

Bimonte, H.A., Granholm, A.-C.E., Seo, H., and Isacson, O. (2002). Spatial memory testing decreases hippocampal amyloid precursor protein in young, but not aged, female rats. *Neurosci. Lett.* **298**, 50–54.

Blough, E.R. and Linderman, J.K. (2000). Lack of skeletal muscle hypertrophy in very aged male Fischer 344 × Brown Norway rats. *J. Appl. Physiol.* **88**, 1265–1270.

Bolton, W.K., Benton, F.R., Maclay, B.A., Sturgill, B.C. (1976). Spontaneous glomerular sclerosis in aging Sprague-Dawley rats. I. lesions associated with mesangial IgM deposits. *Am. J. Pathol.* **85**, 277–302.

Cartee, G.D. (1994). Influence of age on skeletal muscle glucose transport and glycogen metabolism. *Med. Sci. Sports Exerc.* **26**, 577–585.

Chandra, M., Riley, M.G.I., and Johnson, D. (1992). Spontaneous neoplasms in aged Sprague-Dawley rats. *Arch. Toxicol.* **66**, 496–502.

Chen, H., Hardy, M.P., Huhtaniemi, I., and Zirkin, B.R. (1994). Age-related decreased Leydig cell testosterone production in the Brown Norway rat. *J. Androl.* **15**, 551–557.

Dean, D.J. and Cartee, G.D. (1996). Brief dietary restriction increases skeletal muscle glucose transport in old Fisher 344 rats. *J. Gerontol. A Biol. Sci. Med. Sci.* **51**, B208–B213.

Delp, M.D., Evans, M.V., and Duan, C. (1998). Effects of aging on cardiac output, regional blood flow, and body composition in Fisher 344 rats. *J. Appl. Physiol.* **85**, 1813–1822.

Dubal, D.B. and Wise, P.M. (2001). Neoruoprotective effects of estradiol in middle-aged female rats. *Endocrinology* **142**, 43–48.

Eckles-Smith, K., Clayton, D., Bickford, P., and Browning, M.D. (2000). Caloric restriction prevents age-related deficits in LTP and in NMDA receptor expression. *Mol. Brain Res.* **78**, 154–162.

Frick, K.M., Price, D.L., Koliatsos, V.E., and Markowska, A.L. (1997). The effects of nerve growth factor on spatial recent memory in aged rats persists after discontinuation of treatment. *J. Neurosci.* **17**, 2543–2550.

Gabriely, I., Ma, X.H., Yang, X.M., Rossetti, L., and Barzilai, N. (2002). Leptin resistance during aging is independent of fat mass. *Diabetes* **51**, 1016–1021.

Gao, F., Christopher, T.A., Lopez, B.L., Friedman, E., Cai, G., and Ma, X.L. (2000). Mechanism of decreased adenosine protection in reperfusion injury of aging rats. *Am. J. Physiol.* **279**, H329–H338.

Gerhardt, G.A. and Maloney, Jr., R.E. (1999). Microdialysis studies of basal levels and stimulus-evoked overflow of dopamine and metabolites in the striatum of young and aged Fischer 344 rats. *Brain Res.* **816**, 68–77.

Greenberg, J.A. and Boozer, C.N. (1999). The leptin-fat ratio is constant, and leptin may be part of two feedback mechanisms for maintaining the body fat set point in non-obese male Fischer 344 rats. *Horm. Metab. Res.* **31**, 525–532.

Haseman, J.K. and Rao, G.N. (1992). Effects of corn oil, time-related changes, and inter-laboratory variability on tumor occurrence in control Fischer 344 (F344/N) rats. *Toxicol. Pathol.* **20**, 52–60.

Hebert, M.A. and Gerhardt, G.A. (1999). Age-related changes in the capacity, rate, and modulation of dopamine uptake within the striatum and nucleus accumbens of Fischer 344 rats: an in vivo electrochemical study. *J. Pharmacol. Exp. Ther.* **288**, 879–887.

Hodges, H., Veizovic, T., Bray, N., French, S.J., Rashid, T.P., Chadwick, A., Patel, S., and Gray, J.A. (2000). Conditionally immortal neuroepithelial stem cell grafts reverse age-associated memory impairments in rats. *Neuroscience* **101**, 945–955.

Hung, A.J., Stanbury, M.G., Shanabrough, M., Horvath, T.L., Garcia-Segura, L.M., and Naftolin, F. (2003). Estrogen, synaptic plasticity and hypothalamic reproductive aging. *Exp. Gerontol.* **38**, 53–59.

Ingram, D.K., Ikari, H., Umegaki, H., Chernak, J.M., and Roth, G.S. (1998a). Application of gene therapy to treat age-related loss of dopamine D2 receptor. *Exp. Gerontol.* **33**, 793–804.

Ingram, D.K., Spangler, E.L., Meyer, R.C., and London, E.D. (1998b). Learning in a 14-unit T-maze is impaired in rats following systemic treatment with Nomeganitro-L-argine. *Eur. J. Pharmacol.* **341**, 1–9.

Keenan, K.P., Ballam, G.C., Soper, K.A., Laroque, P., Coleman, J.B., and Dixit, R. (1999). Diet, caloric restriction, and the rodent bioassay. *Toxicol. Sci.* **52**, 24–34.

Keenan, K.P., Coleman, J.B., McCoy, C.L., Hoe, C.M., Soper, K.A., and Laroque, P. (2000). Chronic nephropathy in *ad libitum* overfed Sprague-Dawley rats and its early attenuation by increasing degrees of dietary (caloric) restriction to control growth. *Toxicol. Pathol.* **28**, 788–798.

Keenan, K.P., Laroque, P., and Dixit, R. (1998). Need for dietary control by caloric restriction in rodent toxicology and carcinogenicity studies. *J. Toxicol. Environ. Health B Crit. Rev.* **1**, 135–148.

Keenan, K.P., Soper, K.A., Hertzog, P.R., Gumprecht, L.A., Smith, P.F., Mattson, B.A., Ballam, G.C., and Clark, R.L. (1995a). Diet, over-feeding, and moderate dietary restriction in control Sprague-Dawley rats. II. effects on age-related proliferative and degenerative lesions. *Toxicol. Pathol.* **23**, 287–302.

Keenan, K.P., Soper, K.A., Smith, P.F., Ballam, G.C., and Clark, R.L. (1995b). Diet, overfeeding, and moderate dietary restriction in control Sprague-Dawley rats. I. effects on spontaneous neoplasms. *Toxicol. Pathol.* **23**, 269–286.

Kemi, M., Keenan, K.P., McCoy, C., Hoe, C.M., Soper, K.A., Ballam, G.C., and van Zwieten, J. (2000). The relative protective effects of moderate dietary restriction versus dietary modification on spontaneous cardiomyopathy in male Sprague-Dawley rats. *Toxicol. Pathol.* **28**, 285–296.

Kerr, J.S., Yu, S.Y., and Riley, D.J. (1990). Strain specific respiratory air space enlargement in aged rats. *Exp. Gerontol.* **25**, 563–574.

Larkin, L.M., Reynolds, T.H., Supiano, M.A., Kahn, B.B., and Halter, J.B. (2001). Effect of aging and obesity on insulin responsiveness and glut-4 glucose transporter content in skeletal muscle of Fischer 344 × Brown Norway rats. *J. Gerontol. A Biol. Sci. Med. Sci.* **56**, B486–B492.

Laroque, P., Kennan, K.P., Soper, K.A., Dorian, C., Gerin, G., Hoe, C.-M., and Duprat, P. (1997). Effect of early body weight and moderate dietary restriction on the survival of the sprague-dawley rat. *Exp. Toxic Pathol.* **49**, 459–465.

Levin, B.E. and Dunn-Meynell, A.A. (2000). Defense of body weight against chronic caloric restriction in obesity-prone and –resistant rats. *Am. J. Physiol.* **278**, R231–R237.

Li, H., Matheny, M., Nicolson, M., Tumer, N., and Scarpace, P.J. (1997). Leptin gene expression increases with age independent of increasing adiposity in rats. *Diabetes* **46**, 2035–2039.

Lipman, R. (1997). Pathobiology of aging rodents: inbred and hybrid models. *Exp. Gerontol.* **32**, 215–228.

Lipman, R.D., Chrisp, C.E., Hazzard, D.G., and Bronson, R.T. (1996). Pathologic characterization of Brown Norway, Brown Norway × Fischer 344, and Fischer 344 × Brown Norway rats with relation to age. *J. Gerontol. A Biol. Sci. Med. Sci.* **51**, B54–B59.

Lipman, R.D., Dallal, G.E., and Bronson, R.T. (1999). Effects of genotype and diet on age-related lesions in *ad libitum* fed and calorie-restricted F344, BN and BNF3F1 rats. *J. Gerontol. A Biol. Sci. Med. Sci.* **54**, B478–B491.

Lipman, R.D., Smith, D.E., Blumberg, J.B., and Bronson, R.T. (1998). Effects of caloric restriction or augmentation in adult rats: longevity and lesion biomarkers of aging. *Aging Clin. Exp. Res.* **10**, 463–470.

Lucas, D.T. and Szweda, L.I. (1999). Declines in mitochondrial respiration during cardiac reperfusion: age-dependent inactivation of alpha-ketoglutarate dehydrogenase. *Proc. Natl. Acad. Sci. USA* **96**, 6689–6693.

Ma, X.H., Muzumdar, R., Yang, X.M., Gabriely, I., Berger, R., and Barzilai, N. (2002). Aging is associated with resistance to effects of leptin on fat distribution and insulin action. *J. Gerontol. A Biol. Sci. Med. Sci.* **57**, B225–B231.

Maddox, D.A., Alavi, F.K., Santella, R.N., and Zawada, Jr., E.T. (2002). Prevention of obesity-linked renal disease: age-dependent effects of dietary food restriction. *Kidney Int.* **62**, 208–219.

Maeda, H., Gleiser, C.A., Masoro, E.J., Murata, I., McMahan, C.A., and Yu, B.P. (1985). Nutritional influences on aging in Fisher 344 rats: II. Pathology. *J. Gerontol.* **40**, 671–688.

Markowska, A.L. (1999a). Life-long diet restriction failed to retard cognitive aging in Fisher-344 rats. *Neurobiol. Aging* **20**, 177–189.

Markowska, A.L. (1999b). Sex dimorphisms in the rate of age-related decline in spatial memory: relevance to alterations in the estrous cycle. *J. Neurosci.* **19**, 8122–8133.

Markowska, A.L. and Breckler, S.J. (1999). Behavioral biomarkers of aging: illustration of a multivariate approach for detecting age-related behavioral changes. *J. Gerontol. A Biol. Sci. Med. Sci.* **54**, B549–B566.

Markowska, A.L. and Savonenko, A. (2002). Retardation of cognitive aging by life-long diet restriction: implications for genetic variance. *Neurobiol. Aging* **23**, 75–86.

Masoro, E.J. (2002) "Caloric Restriction: A Key to Understanding and Modulating Aging." Elsevier, New York.

Masoro, E.J., Iwasaki, K., Gleiser, C.A., McMahan, C.A., Seo, E.-J., and Yu, B.P. (1989). Dietary modulation of the progression of nephropathy in aging rats: an evaluation of the importance of protein. *Am. J. Clin. Nutr.* **49**, 1217–1227.

Matsumoto, A.M., Marck, B.T., Gruenewald, D.A., Wolden-Hanson, T., and Naai, M.A. (2000). Aging and the neuroendocrine regulation of reproduction and body weight. *Exp. Gerontol.* **35**, 1251–1265.

Mayhew, M., Renganathan, M., and Delbono. O. (1998). Effectiveness of caloric restriction in preventing age-related changes in rat skeletal muscle. *Biochem. Biophys. Res. Comm.* **251**, 95–99.

Meyer, R.C., Spangler, E.L., Patel, N., London, E.D., and Ingram, D.K. (1998). Impaired learning in rats in a 14-unit t-maze by 7-nitroindazole, a neuronal nitric oxide synthase inhibitor, is attenuated by the nitric oxide donor, molsidomine. *Eur. J. Pharmacol.* **341**, 17–22.

Miller, R.A. and Nadon, N.L. (2000). Principles of animal use for gerontological research. *J. Gerontol. A Biol. Sci. Med. Sci.* **55**, B117–B123.

Molon-Noblot, S., Keenan, K.P., Coleman, J.B., Hoe, C.-M., and Laroque, P. (2001). The effects of *ad libitum* overfeeding and moderate and marked dietary restriction on age-related spontaneous pancreatic islet pathology in Sprague-Dawley rats. *Toxicol. Pathol.* **29**, 353–362.

Mooradian, A.D. and Chehade, J.M. (2000). Serum leptin response to endogenous hyperinsulinemia in aging rats. *Mech. Aging Dev.* **115**, 101–106.

Morris, R.G.M. (1981). Spatial localization does not require the presence of local cues. *Learn. Motiv.* **12**, 239–260.

Mosekilde, L., Thomsen, J.S., Orhii, P.B., and Kalu, D.N. (1998). Growth hormone increases vertebral and femoral bone strength in osteopenic, ovariectomized, aged rats in a dose-dependent and site-specific manner. *Bone* **23**, 343–352.

Muller-Delp, J.M., Spier, S.A., Ramsey, M.W., and Delp, M.D. (2002). Aging impairs endothelium-dependent vasodilation in rat skeletal muscle arterioles. *Am. J. Physiol.* **283**, H1662–H1672.

Nakazawa, M., Tawaratani, T., Uchimoto, H., Kawaminami, A., Ueda, M., Ueda, A., Shinoda, Y., Iwakura, K., Kura, K., and Sumi, N. (2001). Spontaneous neoplastic lesions in aged Sprague Dawley rats. *Exp. Anim.* **50**, 99–103.

Narath, E., Skalicky, M., and Viidik, A. (2001). Voluntary and forced exercise influence the survival and body composition of ageing male rats differently. *Exp. Gerontol.* **36**, 1699–1711.

Naruse, M., Tanabe, A., Sato, A., Takagi, S., Tsuchiya, K., Imaki, T., and Takano, K. (2002). Aldosterone breakthrough during angiotensin II receptor antagonist therapy in stroke-prone spontaneously hypertensive rats. *Hypertension* **40**, 28–33.

Provoost, A.P., Van Aken, M., and Melenaar, J.C. (1991). Sequential renography and renal function in Brown Norway rats with congenital hydronephrosis. *J. Urol.* **146**, 588–591.

Ramos, A., Berton, O., Mormede, P., and Chaouloff, F. (1997). A multiple-test study of anxiety-related behaviours in six inbred rat strains. *Behav. Brain Res.* **85**, 57–69.

Rao, G.N., Haseman, J.K., Grumbein, S., Crawford, D.D., and Eustis, S.L. (1990). Growth, body weight, survival, and tumor trends in F344/N rats during an 11-year period. *Toxicol. Pathol.* **18**, 61–70.

Raya, T.E., Gaballa, M., Anderson, P., and Goldman, S. (1997). Left ventricular function and remodeling after myocardial infarction in aging rats. *Heart Circ. Physiol.* **42**, H2652–H2658.

Rex, A., Sondern, U., Voigt, J.P., Franck, S., and Fink, H. (1996). Strain differences in fear-motivated behavior of rats. *Pharmacol. Biochem. Behav.* **54**, 107–111.

Rubin, B.S. (2000). Hypothalamic alterations and reproductive aging in female rats: evidence of altered luteinizing hormone-releasing hormone neuronal function. *Biol. Reprod.* **63**, 968–976.

Salter, J.M., Cassone, V.M., Wilkerson, M.K., and Delp, M.D. (1998). Ocular and regional cerebral blood flow in aging Fischer-344 rats. *J. Appl. Physiol.* **85**, 1024–1029.

Scarpace, P.J. and Tumer, N. (2001). Peripheral and hypothalamic leptin resistance with age-related obesity. *Physiol. Behav.* **74**, 721–727.

Seidman, M.D. (2000). Effects of dietary restriction and antioxidants on presbyacusis. *Laryngoscope* **110**, 727–738.

Skalicky, M., Narath, E., and Viidik, A. (2000). *Ad libitum* fed rats can keep lean or grow fat when they age. *Exp. Gerontol.* **35**, 1093-1094.

Smith, A.C., Gerrard, J.L., Barnes, C.A., and McNaughton, B.L. (2000). Effect of age on burst firing characteristics of rat hippocampal pyramidal cells. *NeuroReport* **11**, 3865–3871.

Spangler, E.L., Waggie, K.S., Hengimihle, J., Roberts, D., Hess, B., and Ingram, D.K. (1994) Behavioral assessments of aging in male Fisher 344 and Brown Norway rat strains and their F1 hybrid. *Neurobiol. Aging* **15**, 319–328.

Stiles, J. and Tyler, W.S. (1988). Age-related morphometric differences in responses of rat lungs to ozone. *Toxicol. Appl. Pharmacol.* **92**, 274–285.

Thurman, J.D., Bucci, T.J., Hart, R.W., and Turturro, A. (1994). Survival, body weight, and spontaneous neoplasms in *ad libitum*-fed and food-restricted Fisher 344 rats. *Toxicol. Pathol.* **22**, 1–9.

Turturro, A., Witt, W.W., Lewis, S., Hass, B.S., Lipman, R.D., and Hart, R.W. (1999). Growth curves and survival characteristics of the animals used in the biomarkers of aging program. *J. Gerontol. A Biol. Sci. Med. Sci.* **54**, B492–B501.

Umegaki, H., Munoz, J., Meyer, R.C., Spangler, E.L., Yoshimura, J., Ikari, H., Iguchi, A., and Ingram, D.K. (2001). Involvement of dopamine D2 receptors in complex maze learning and acetylcholine release in ventral hippocampus of rats. *Neuroscience* **103**, 27–33.

Van Liew, J.B., Davis, F.B., Davis, P.J., Noble, B., and Bernardis, L.L. (1992). Calorie restriction decreases microalbuminuria associated with aging in barrier-raised Fischer 344 rats. *Am. J. Physiol.* **263**, F554–F561.

Wang, L., Banu, J., McMahan, C.A., and Kalu, D.N. (2001a). Male rodent model of age-related bone loss in men. *Bone* **29**, 141–148.

Wang, L., Orhii, P.B., Banu, J., and Kalu, D.N. (2001b). Bone anabolic effects of separate and combined therapy with growth hormone and parathyroid hormone on femoral neck in aged ovariectomized osteopenic rats. *Mech. Aging Dev.* **122**, 89–104.

Wang, L., Orhii, P.B., Banu, J., and Kalu, D.N. (2001c). Effects of separate and combined therapy with growth hormone and parathyroid hormone on lumbar vertebral bone in aged ovariectomized osteopenic rats. *Bone* **28**, 202–207.

Ward, M.T., Stoelzel, C.R., and Markus, E.J. (1999). Hippocampal dysfunction during aging II: deficits on the radial-arm maze. *Neurobiol. Aging* **20**, 373–380.

Weindruch, R. and Walford, R.L. (1982). Dietary restriction in mice beginning at 1 year of age: effect on life-span and spontaneous cancer incidence. *Science* **215**, 1415–1418.

Wise, P.M., Smith, M.J., Dubal, D.B., Wilson, M.E., Rau, S.W., Cashion, A.B., Bottner, M., and Rosewell, K.L. (2002). Neuroendocrine modulation and repercussions of female reproductive aging. *Recent Prog. Horm. Res.* **57**, 235–256.

Wolf, N.S., Li, Y., Pendergrass, W., Schmeider, C., and Turturro, A. (2000). Normal mouse and rat strains as models for age-related cataract and the effect of caloric restriction on its development. *Exp. Eye Res.* **70**, 683–692.

Yang, H.T. and Feng, Y. (2000). bFGF increases collateral blood flow in aged rats with femoral artery ligation. *Am. J. Physiol.* **278**, H85-H93.

Zirkin, B.R. and Chen, H. (2000). Regulation of leydig cell steroidegenic function during aging. *Biol. Reprod.* **63**, 977–981.

Chapter 26

Cardiovascular Research

Peter G. Anderson, Sanford P. Bishop and J. Thomas Peterson

THE LABORATORY RAT, 2ND EDITION

I. INTRODUCTION

The rat is widely used for many types of studies involving the cardiovascular system. Several inbred strains are readily available from commercial sources, providing homogeneous groups of animals and thus reducing the number required to obtain statistically significant results. Rats are easily handled and are of sufficient size to allow a variety of surgical procedures and functional studies of the cardiovascular system to be performed, yet they are considerably less expensive to maintain than are larger species commonly used such as the dog, pig, and sheep. The purpose of this chapter is to indicate those areas of cardiovascular research in which the rat has been particularly useful, to describe those features of the rat cardiovascular system that may influence experimental studies, to identify rat strains useful for cardiovascular research, and to discuss the advantages and disadvantages of the rat for specific research objectives.

II. MACROSCOPIC ANATOMY

A. Heart

The heart of the rat is a four-chambered organ, as in other mammals, without normal communications between the left and right chambers. The heart is enclosed within a thin, transparent pericardium that is attached to the major arteries and veins at the base of the heart. The topographic features of the heart are similar to those of other mammals, with a clearly defined atrio-ventricular (AV) groove separating the atria from the ventricles.

Anterior and posterior papillary muscles in the left ventricle—which have relatively long, slender, apical portions projecting into the ventricular lumen—anchor the chordae tendineae of the mitral valve. The anterior papillary muscle of the rat is located more laterally than that in the dog or man. Papillary muscles in the right ventricle are slender, elongated structures varying from two to five or more in number. The cardiac valves are similar to those in other species. The aortic and pulmonic valves have three leaflets; the mitral valve, an anterior and posterior leaflet; and the tricuspid, a posterior septal leaflet and two lateral leaflets on the free-wall portion of the right ventricular wall. There are multiple thin chordae tendineae connecting the mitral and tricuspid leaflets to the papillary muscles.

B. Heart Shape during Growth

Left ventricular morphology has been studied in detail by Grimm et al. (1973) in rats from 85 to 445 g (body weight) to evaluate the mechanical capabilities of the growing myocardium. During normal growth, the increasing radius of the left ventricular lumen was accompanied by a proportional increase in wall thickness and valve-to-apex measurement. Therefore, during growth the shape of the heart remained unchanged.

C. Heart Weight

The weight of the rat heart at various stages during growth is of interest because many studies are designed to alter normal heart weight (e.g., hypertrophy or atrophy models). Because the rat continues to gain weight through the first year of life, the heart also continually increases in weight. Therefore, studies designed to increase the heart weight must take into consideration not only the experimentally induced increase in heart weight but also the increase owing to normal growth. If the experimental design also alters normal body weight gain, this must be considered in evaluation of the results.

Unlike some larger species of mammals, in the rat heart weight does not keep pace with increasing body weight during normal growth. Rakusan et al. (1963) have studied heart weights in Sprague-Dawley rats from birth through old age, and Clubb et al. (1987) have studied neonatal Wistar-Kyoto (WKY) rats and spontaneously hypertensive rats (SHR) from birth through 28 days of age. Rakusan and Poupa (1963) found a total ventricular weight-to-body weight ratio of 5.20 in neonates, which decreased to 4.48 by 35 days of age (50 g body weight). Heart weight-to-body weight ratios continue to decrease for the growing rat throughout the first year of life. These investigators

and others (Grimm et al., 1963; Setnikar and Magistretti, 1965; Penny et al., 1974; Kawamura et al., 1976) report heart weight-to-body weight values for various size rats within the following ranges: 100 g body weight, 3.3 to 3.8 mg/g; 200 g, 2.9 to 3.5 mg/g; 300 g, 2.4 to 2.9 mg/g; 400 g, 2.4 to 2.8 mg/g; and 500 g, 2.25 to 2.6 mg/g. Regression equations calculated from data published by these several investigators give a Y intercept (heart weight in milligrams) of 165 to 235 and a slope of 1.9 to 2.6X (body weight in grams). Beznak (1955) reported a regression equation of $Y = 186.2 + 2.59X$.

Weights of the right and left ventricles and the combined atria have been determined for 289 male Sprague-Dawley rats weighing 23 to 600 g. These data with the computer-derived regression lines are shown in Fig. 26-1 A–D. These data illustrate the decreasing rate of growth relative to increasing body weight for the various cardiac chambers. Total heart weight-to-body weight data from some commonly used strains derived from our laboratories (S.P.B., P.G.A.) are shown in Table 26-1 and regional heart weight data from Sprague-Dawley rats in Table 26-2.

Differences in values reported by various investigators are owing to differing strains, environmental conditions during growth, and methods of dissection of the heart. Each investigator must establish a normal body weight and cardiac growth curve for each strain under specific environmental conditions.

D. Major Arteries and Veins Leaving and Entering the Heart

The anatomy of the rat circulatory system has been described by Greene and Halpern (Halpern, 1953, 1957; Greene, 1959). The aorta and pulmonary artery originate from the left and right ventricles at the base of the heart, respectively. The pulmonary artery divides into left and right branches in the usual manner. The left and right coronary artery ostia are located at the root of the ascending aorta. There are three major branches arising from the arch of the aorta in the rat: the innominate (brachiocephalic), the left common carotid, and the left subclavian arteries. The innominate artery gives rise to the right common carotid, right subclavian, and right internal mammary arteries. The left subclavian artery gives rise to the left internal mammary artery.

There are three major veins entering the right atrium of the rat heart (Halpern, 1953). The inferior vena cava enters the right atrium at the posterior right border, and the right superior vena cava enters cranially. In addition, the rat heart has a left superior vena cava that enters the coronary sinus region with the inferior vena cava after crossing the posterior surface of the heart from the left. The

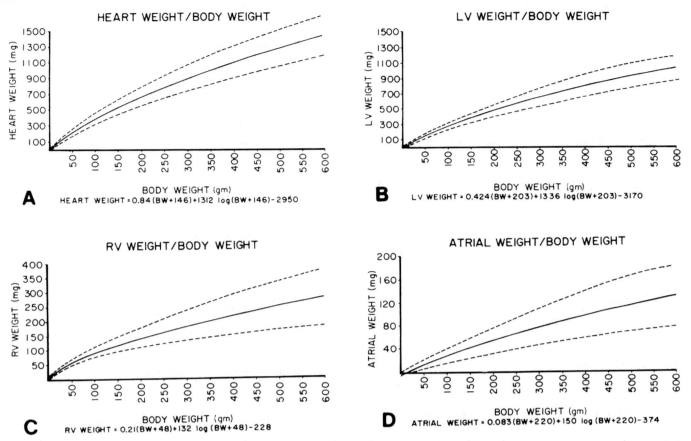

Fig. 26-1 Heart weight-to-body weight ratios determined on 289 male Sprague-Dawley rats. The solid line is the computer-derived mean for the group, with the regression equation given under each graph. The dashed lines were drawn to include 95% of all observations. LV, left ventricle; RV, right ventricle. (Data from David G. Penney, Department of Physiology, Wayne State University, Detroit, Michigan, and from a colony maintained at the University of Illinois, Chicago Circle.)

TABLE 26-1

HEART WEIGHT–TO–BODY WEIGHT RATIOS IN RATS DURING NORMAL GROWTH

Age	Sprague-Dawley		Fisher		WKY		SHR	
	BW (g)	HW/BW	BW (g)	HW/BW	BW (g)	HW/BW	BW (g)	HW/BW
Newborn			5 ± 0.3	7.01 ± 0.79	5 ± 0.6	7.83 ± 1.26	5.0 ± 0.5	9.52 ± 1.83
3 days			9 ± 1.1	7.75 ± 0.44	8 ± 0.4	7.70 ± 0.42	6.5 ± 1.4	7.77 ± 0.44
1 week			12 ± 1.9	8.13 ± 1.19	11.6 ± 2		8.9 ± 1.7	
2 weeks	25 ± 3.7	7.20 ± 0.74	16 ± 1.5	9.20 ± 1.32	21.8 ± 2		18.2 ± 3	
3 weeks	62 ± 4	4.99 ± 0.50	27 ± 0.6	9.87 ± 1.26	42.5 ± 4		39.2 ± 7	
4 weeks	107 ± 12	4.34 ± 0.50			60 ± 6	5.48 ± 0.87		
5 weeks	172 ± 14	3.87 ± 0.30			85 ± 10		71.6 ± 8	
6 weeks	204 ± 5	4.13 ± 0.12	135 ± 5	5.06 ± 0.54	117 ± 18	4.33 ± 0.28	100 ± 14	4.79 ± 0.25
8 weeks	296 ± 23	3.44 ± 0.22	177 ± 7	5.04 ± 0.49	167 ± 21	4.43 ± 0.36	172 ± 8	4.95 ± 0.76
10 weeks	337 ± 35	3.24 ± 0.12	252 ± 14	3.55 ± 0.28	190 ± 32		198 ± 15	
3 months	424 ± 42	3.10 ± 0.20	286 ± 10	3.78 ± 0.40	255 ± 35	3.51 ± 0.59	241 ± 15	4.51 ± 0.32
6 months	622 ± 69	2.62 ± 0.43			351 ± 4	2.99 ± 0.15	298 ± 5	4.25 ± 0.14
8 months	720 ± 93	2.29 ± 0.32	370 ± 47	3.20 ± 0.40	503 ± 38			
12 months			397 ± 5	2.94 ± 0.13	515 ± 40	2.78 ± 0.13	404 ± 15	4.35 ± 0.46
18 months					432 ± 45	2.71 ± 0.24	401 ± 30	5.14 ± 0.64
24 months					400 ± 54	2.63 ± 0.31	405 ± 38	6.21 ± 0.88

Values are mean ± SD; n = 6 to 40; BW indicates body weight (g); HW/BW, heart weight (mg)/body weight (g).

TABLE 26-2

REGIONAL HEART WEIGHTS IN SPRAGUE-DAWLEY RATS DURING
NORMAL GROWTH

Age	Body weight (g)	HW/BW	LV + S/BW	RV/BW
5 weeks	164 ± 30	4.25 ± 0.40	2.64 ± 0.23	0.89 ± 0.12
8 weeks	298 ± 23	3.44 ± 0.20	2.23 ± 0.20	0.78 ± 0.03
11 weeks	415 ± 58	3.60 ± 0.55	2.12 ± 0.13	0.62 ± 0.12
25 weeks	622 ± 89	2.62 ± 0.43	1.96 ± 0.21	0.57 ± 0.27

HW/BW indicates heart weight (mg)/body weight (g); LV + S/BW, left
ventricle plus septum (mg)/body weight (g); RV/BW, right ventricular free
wall weight (mg)/body weight (g).

Data from the laboratories of S.P.B. and P.G.A., published and
unpublished.

persistence of two superior venae cavae is normal in the
rat. The azygous vein is on the left and drains into the left
superior vena cava, quite the opposite to the situation in
higher mammals without a left superior vena cava. A
hemiazygous vein may be present, and it drains into the
azygous vein.

E. Blood Supply of the Heart

Both coronary and extracoronary sources supply blood
to the rat heart (Halpern, 1957). The major coronary arteries, in contrast to larger mammalian species, do not lie on
the surface of the myocardium but rather are covered by a
layer of myocardium throughout their course (Fig. 26-2.).

There may be two major arteries arising from one or
both aortic sinuses. The tendency for variant patterns is
more common in some strains, apparently as an inherited
trait (Bloor et al., 1967). The right coronary artery arises
from the aortic root and courses diagonally across the
lateral wall of the right ventricle. A large septal artery

commonly is given off near the origin of the right coronary
artery and is the major blood supply to the septum. Small
branches supply the right atrium adjacent to the AV sulcus.
The right coronary artery also sends a branch to the inter-
atrial septum and the region of the AV node.

The left coronary artery descends over the anterior
lateral wall of the left ventricle, parallel to the anterior longitudinal sulcus. It may give rise to a major septal artery near
its origin. Branches of the left coronary artery extend to the
pulmonary conus region and ramify laterally over the left
ventricular free wall, including the posterior wall of the left
ventricle. There is no clearly defined left circumflex coronary artery branch as in higher mammals, but a small
branch often courses through the left ventricle parallel to
the AV sulcus. Small branches reach the left atrium.

The left and right internal mammary arteries are the
origin of the left and right cardiacomediastinal arteries,
which are important extracardiac sources of blood to the
heart (Halpern, 1957). The right cardiacomediastinal artery
courses along the right superior vena cava to the right
atrium, where it supplies the sinus node, the bulk of the
right atrium, and portions of the left atrium. The left cardiacomediastinal artery follows the left superior vena cava
and supplies the remaining portion of the left atrium.

The atria, including the sinus node, receive their major
blood supply from a noncoronary artery source. This fact
must be recognized when the atria and ventricles of the
rat heart are compared, and may be useful for certain
types of vascular studies. The retention of a dual blood
supply in the rat heart is similar to the situation found
in lower animals, including reptiles and amphibians.
Although this dual blood supply is better developed in
the rat than in higher mammals, it is by no means
nonexistent in higher forms. Petelenz (1965) studied 100
human specimens and demonstrated branches of the

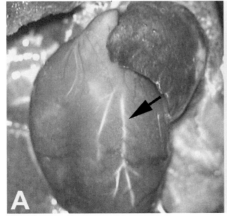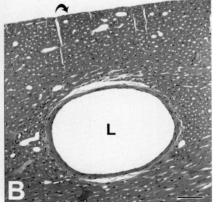

Fig. 26-2 (A) View of the anterior surface of the heart demonstrating the course of the left anterior descending artery (arrow). The coronary
arteries have been injected with white latex to illustrate the arterial bed. (B) Low-power histology section of the left ventricular anterior wall with the
left anterior descending coronary artery (L) of a perfusion-fixed rat (250 g body weight). A layer of myocardium covers the major coronary arteries
in the rat. Epicardium is at the top (curved arrow). Bar, 100 μm. Hematoxylin and eosin.

bronchial artery reaching the sinus node region in 44%, and in 26% the extent of vascularization was judged to be of functional significance.

III. MICROSCOPIC ANATOMY

A. Ventricular Myocardium

1. Myocyte Orientation

Cardiac muscle cells are arranged in parallel bundles joined to each other at the intercalated disc regions at the ends and along the sides of the cells. In general, orientation of muscle fibers within the left ventricular subendocardium is vertical, running from base to apex, with parallel fibers within the papillary muscles and trabecular muscles of both ventricles. Successively deeper layers in the left ventricle from the endocardium are oriented more horizontally, with fibers in the midwall region approximately 70° to 90° to those of the subendocardium. At the subepicardium, fibers have turned approximately 120° from those in the subendocardium. Thus, a full-wall thickness section cut horizontally through the midportion of the left ventricle will have fibers oriented in cross-section in the subendocardium and subepicardium, and in longitudinal section in the midwall region. The longitudinal orientation of subendocardial fibers in both ventricles generally follows the lines of blood flow.

2. Cellular Composition

The cell type, volume, and numerical composition of the myocardium are of importance in many studies utilizing heart tissue. Morphometric studies of adult rat myocardium have shown the volume composition to be 80% to 83% myocytes, with the remainder consisting of interstitial connective tissue, fibrocytes, capillary endothelial cells, and various other cell types (Anversa et al., 1976; Kawamura et al., 1976; Lund and Tomanke, 1978; Clubb et al., 1987; Engelmann et al., 1987), The capillary-to-myofiber ratio is approximately 1:1 in the mature rat (Rakusan and Poupa, 1963; Poole et al., 1992), as in other species (Wearn, 1928; Shipley et al., 1937). However, on a strictly numerical basis, heart muscle tissue in the 200-g rat consists of only 20% to 25% heart muscle cells, the bulk of cells consisting of the much smaller endothelial cells and other interstitial connective tissue cells (Morkin and Ashford, 1968; Grove et al., 1969). This fact is particularly important for studies utilizing cell number rather than cell volume, such as DNA synthesis or other nuclear-dependent methodologies.

3. Myocyte Dimensions

The cross-sectional diameter or area of heart muscle cells has been determined by several investigators. Some investigators have measured cell diameter by using longitudinally sectioned paraffin embedded tissue, but because of irregular cross-section, poorly defined cell borders, and shrinkage artifact, this method is unreliable. Measurement of cross-sectional area in frozen sections or methacrylate embedded tissues stained with silver, laminin, or other methods to clearly identify cell borders is preferable (Fig. 26-3). Several reports in which formalin-fixed, paraffin-embedded tissues were used indicated mean cross-sectional diameter of mature ventricular myocardial

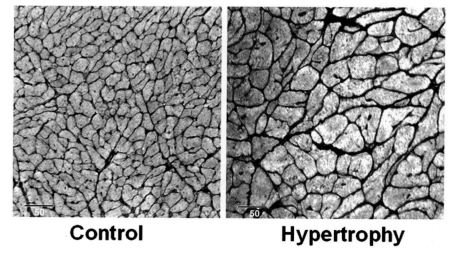

Control   **Hypertrophy**

Fig. 26-3 Cross-sectioned left ventricular myocardium from a control untreated 300-g rat and from a weight-matched rat with cardiac hypertrophy produced by banding of the ascending aorta at 4 weeks of age. Tissue was immersion fixed in formalin, embedded in glycol methacrylate, sectioned at 1-μm thickness, and stained with silver to outline cell borders. In addition to the increase in cross-sectional area of the hypertrophied myocytes, note the increased variability in cross-sectional size and shape. Bars, 50 μm.

cells to be 11 to 14 μm (Angelakos et al., 1964; Rakusan, 1971). However, other studies that used either fresh preparations of isolated myocytes, frozen sections, or methacrylate-embedded tissues (in which shrinkage artifact is essentially eliminated) report mean cross-sectional diameters of 17 to 20 μm (Anversa et al., 1976; Bishop et al., 1979). Cross-sectional area of myocytes in normal adult rats is in the range of 200 to 300 μm^2 (Gerdes et al., 1986; Smith et al., 1988; Campbell et al., 1993).

The length of heart muscle cells is difficult to obtain in histologic sections owing to orientation of the plane of section and cell irregularity. By use of isolated myocytes, cell length for adult rats has been found to be between 120 and 140 μm (Bishop et al., 1980; Gerdes et al., 1986, 1996; Smith et al., 1988; Clubb et al., 1989; Zimmer et al., 1990; Campbell et al., 1993).

Cell volume may be calculated from measurements of myocyte length and cross-sectional area, or by use of Coulter counter measurement of isolated myocytes. Because cell volume continues to increase with the increase in heart weight as the rats age, cell volume of adult rats from 250 to 500 g body weight of different strains ranges from 20,000 to 40,000 μm^3. Myocytes from the inner half of the left ventricle are slightly larger in both width and length than are those from the outer half of the left ventricular wall. In addition, right ventricular myocytes are significantly smaller than left ventricular myocytes (Gerdes et al., 1986; Smith et al., 1988; Clubb et al., 1989; Campbell et al., 1993).

The fractional organelle volume of ventricular myocytes determined by morphometric analysis of electron micrographs is 35% to 40% mitochondria, 45% to 56% myofibrils, 1.0% to 2.2% T tubular system, and 3.0% to 3.5% sarcoplasmic reticulum (Page and McCallister, 1973; Page et al., 1974; Anversa et al., 1976; Kawamura et al., 1976).

4. Myocyte Structural Changes during Growth

During the early neonatal period, cardiac myocytes change from an ovoid shape to a more elongated shape as the myocardium converts from the hyperplastic growth characteristic of the fetal and early neonatal period to growth by increase in cell size, or hypertrophy. Myocyte volume remains constant at approximately 1500 μm^3 during the first 4 to 6 days of postnatal life (hyperplastic growth), and myocytes are single-nucleated. By 6 days of age, conversion to hypertrophic cell growth has been started, and cells are becoming binucleated. By 12 days of age, 85% to 90% of myocytes are binucleated, as in adult rats, and all further growth is by cellular hypertrophy. The number of cardiac myocytes increases from approximately 10×10^6 at birth to the adult number of 20×10^6 to 30×10^6 (Clubb et al., 1987, 1989; Campbell et al., 1993; Li et al., 1996).

B. Atrial Myocardium

Myocytes in the atrium are smaller and more loosely arranged than are ventricular myocytes (Prakash, 1954; Bompiani et al., 1959; Jamieson and Palade, 1964). Although some reports indicate an absence of T tubules in atrial myocardium (Hibbs and Ferrans, 1969), studies with horseradish peroxidase tracer have identified two populations of atrial cells: one without T tubules and the other possessing T tubules. Atrial myocytes have been reported to have only a single nucleus with large mitochondrial collections at the nuclear poles (Jamieson and Palade, 1964). However, unpublished studies with isolated myocytes in one of our laboratories (S.P.B.) show that atrial myocytes have two nuclei, as in ventricular myocytes. Jamieson and Palade (1964) described the ultrastructure of the rat atrium and specifically drew attention to the electron-dense, membrane-bound, atrial-specific granules present not only in rat atrium but in all mammalian atrial myocytes. In the rat heart, atrial granules are larger and more numerous than those in larger mammals. These granules are now recognized to contain a specific natriuretic substance, atrial natriuretic factor, which plays a role in systemic fluid balance (DeBold, 1985; Brenner et al., 1990).

C. Conduction System

Descriptions of cardiac conduction tissue have been reviewed by a number of authors (Truex and Smythe, 1965; James and Sherf, 1971; James, 2001). The sinus node of the rat heart is described by Halpern (1955) and King (1954) as a relatively large horseshoe-shaped structure located at the junction of the right superior vena cava and the right atrium. A significant feature of the sinus node in the rat is the extracoronary origin of the sinus node artery described above (Halpern, 1957). Ultrastructurally, both sinus node and AV node cells are reported to always contain a single nucleus, are smaller in size, and have fewer and poorly organized contractile elements than do atrial or ventricular working myocytes (Melax and Leeso, 1970; James and Sherf, 1971). Sinus node cells contain large numbers of atrial-specific granules. The AV node cells also contain a single nucleus (Melax and Leeso, 1971) and have no distinctive features that are different from AV node cells of other species. T tubules are not present in sinus node, AV node, or Purkinje cells (Sommer and Johnson, 1968). AV node cells of the rat have abundant amounts of glycogen, as do the peripheral Purkinje cells. Ventricular Purkinje cells arborize over the left and right subendocardium with only minor penetration into the ventricular wall before connecting with working myocytes.

IV. METHODS FOR *IN VIVO* EVALUATION OF THE CARDIOVASCULAR SYSTEM

A. Electrocardiogram

Electrocardiograms can be obtained by using standard electrocardiographic equipment. If the animal is not anesthetized, a restraining device is required to hold the animal during the recording (Osborne, 1973). Electrocardiograms of isolated perfused hearts can also be accomplished with needle electrodes (Anderson et al., 1990b). Heart rate may be determined from the electrocardiogram.

B. Hemodynamic Measurements

Blood pressure can be obtained from the tail of conscious animals by using a specially designed cuff equipped with either an ultrasound microphone or a pressure transducer and an attached recording system (Williams et al., 1939; Fregly, 1963; Maistrello and Matscher, 1969; Pfeffer et al., 1971). Systems are also available for continuous recording of blood pressure in unrestrained animals by using chronically implanted catheters connected to recording devices by a swivel mechanism (Laffan et al., 1972; Averill et al., 1996; Peng et al., 1996), or by using radiotelemetry from implanted transducers (Anderson et al., 1999). In anesthetized animals, left and right ventricular and aortic blood pressure and rate of pressure change (dp/dt) may be recorded by cannulation of a peripheral artery, such as the femoral or carotid, with a small catheter tip transducer system (Zimmer et al., 1990). Hemodynamic data can also be obtained during echocardiographic (ECHO)/Doppler studies and nuclear magnetic imaging (see later in this chapter).

C. Large Vessel and Regional Blood Flow

Although the rat is relatively small compared with other animals commonly used for cardiovascular functional studies, small flow probes have been developed that may be placed on the aorta, renal artery, mesenteric artery, carotid artery, or others for measurement of total flow to body regions or organs (Van Orden et al., 1984; D'Aleida et al., 1996). Regional blood flow within an organ may be evaluated with either radioactive or colored microspheres (Nishiyama et al., 1978; Flaim et al., 1984; Hiller et al., 1996). Injection into the left ventricle has been demonstrated to provide adequate mixing of microspheres in the blood, eliminating the need for left atrial injection, as is commonly practiced in larger animals (Wicker and Tarzi, 1982).

D. Echocardiography

With the development of small ultrasonic probes, ECHO evaluation of the rat has become feasible and routine in many laboratories. Methods are available for recording ECHOs by use of M-mode, two-dimensional mode, and with Doppler mode for repeated imaging of cardiac and large-vessel morphology and function (Litwin et al., 1995; Wu et al., 1997; Douglas et al., 1998; Prahash et al., 2000; Prunier et al., 2002). An example of M-mode and two-dimensional imaging of a rat with pressure overload induced left ventricular enlargement is shown in Fig. 26-4.

E. Magnetic Resonance Imaging

Alterations in myocardial structure and function may be serially followed in the intact animal by the use of magnetic resonance imaging (Rudin et al., 1991; Crowley et al., 1997; Rehwald et al., 1997; Waller et al., 2001). With this non-invasive methodology, excellent images of the heart may be obtained at different phases of the cardiac cycle, allowing evaluation of serial changes in total ventricular mass and infarct size, changes in chamber size and shape, and alterations in function.

V. METHODS FOR *IN VITRO* STUDY OF MYOCARDIUM

A variety of *in vitro* techniques have been used to study the normal structure, function, and metabolism of rat myocardium. The rat has been the most widely used species for most of these *in vitro* techniques, because an adequate amount of tissue is available, function is easily maintained in isolated perfused hearts, and this well-characterized animal model is readily available and relatively inexpensive to procure and maintain.

A. Tissue Culture of Neonatal Myocytes

Neonatal myocytes in culture are commonly used for a wide variety of studies, using variations of trypsinization and mincing to prepare the cells. Neonatal cells have poorly developed control of membrane physiology and, owing to leakage of ions, continue to beat after isolation. In early studies by Harary and others (Harary and Farley, 1963; Desmond and Harary, 1972), heart muscle cells continued to beat for 9 weeks or more. During the fetal and early neonatal period, heart muscle cells from the rat are metabolically adapted to a highly anaerobic type of environment in contrast to the oxidative metabolism

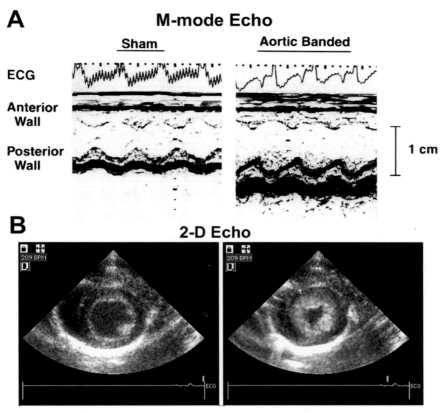

A   **M-mode Echo**

Sham   Aortic Banded

ECG

Anterior Wall

Posterior Wall

1 cm

B   **2-D Echo**

Fig. 26-4   (A) Tracings of M-mode echocardiograms obtained from a sham operated rat (left) and an aortic banded rat (right). Note the thickening of the ventricular wall in the aortic banded animals. (B) Two-dimensional echocardiograms showing a cross-sectional view at the mid portion of the ventricle of a sham-operated rat heart (left) and an aortic banded heart (right). The left ventricular chamber is significantly smaller, and the left ventricular walls are thickened in the hypertrophied heart. (Echocardiograms courtesy of Sheldon E. Litwin, M.D., Division of Cardiology, University of Utah School of Medicine.)

required by adult myocardial cells. In addition, fetal and neonatal cells retain the capability of growth by cellular division, or hyperplasia, which is lost a few days after birth (Clubb and Bishop, 1984; Li et al., 1996). Although these characteristics of the neonatal myocyte provide the possibility of maintaining beating heart muscle cells in culture, which are useful for a variety of studies, it must be kept in mind that such myocytes are structurally, functionally, and metabolically different from adult cardiac myocytes. Neonatal myocytes in tissue culture often form single layers of synchronously beating cells. Some cells will beat independently of others in the culture, whereas others will not beat unless in actual contact with beating cells (Harary and Farley, 1963).

Several investigators have also used non-myocyte cells derived from cardiac cell isolation techniques. These cells include cardiac fibroblasts and other cells thought to be smooth muscle cells, myofibroblasts, primitive mesenchymal cells, or stem cells. The role of cardiac fibroblasts in studies of neuohumeral influences on remodeling have been well characterized, and these cells have been utilized in studies to repair injured myocardium (Long and

Brown, 2002; Tian et al., 2003). And other cell types have been used to study structural proteins and other aspects of heart development and growth (Jones et al., 1979; de Clerck and Jones, 1980).

B. Isolated Myocytes from Adult Myocardium

Isolated adult cardiac myocytes have been useful both for studies of myocytes maintained in culture and for morphologic studies of myocytes during normal growth and with a variety of pathologic conditions. Cell isolation is performed by using aortic retrograde perfusion of an isolated heart with oxygenated calcium-free buffer mixtures containing collagenase and other optional enzymes (Berry et al., 1970; Powell, 1986). Cells isolated in calcium-free buffer and then gradually restored to buffer containing 1 to 2 mM calcium can be maintained in a quiescent state in culture for weeks to months, undergoing gradual dedifferentiation over time (Bugaisky, 1988; Decker et al., 1991). Because the majority of cells in the myocardium are non-myocytes, it is important to perform an enrichment step to separate the myocytes from the other cell types. This process usually entails plating cells on plastic culture dishes for 1 hour and then gently rinsing the dish (to dislodge the myocytes, which do not stick to the plastic) and decanting the myocyte enriched fraction. Other techniques for myocyte enrichment involve gentle centrifugation through a Ficoll gradient (Bugaisky and Zak, 1985; Bugaisky, 1988).

Morphologic evaluation of myocytes may be performed either on freshly prepared cells or after fixation of cells in suspension. Cell dimensions are determined by direct microscopic measurement, and cell volume is best evaluated with a Coulter counter (Bishop and Drummond, 1979; Smith et al., 1988; Gerdes et al., 1996). Freshly prepared isolated myocytes have been used to evaluate cellular function under *in vitro* conditions by using controlled stimulation of myocytes while recording contractile shortening (Frank et al., 1986; Ren et al., 2000) or measuring calcium function (Sumida et al., 1998; Balnave and Vaughan-Jones, 2000).

C. Organ Culture

Organ culture of fetal hearts was originally used with the mouse (Wildenthal, 1970) by suspending fetal mouse hearts on a mesh at the air-culture media interface; these hearts continue to beat for many days. Fetal rat hearts have been similarly cultured for use in studies of embryonic myocyte growth (Higo et al., 1987; Ban et al., 1996). A unique model for study of the developing embryonic heart was studied by Tucker and co-workers, using a 12-day embryonic rat heart implanted *in oculo* on the anterior surface of the iris (Tucker and Gist, 1986; Bishop et al., 1990; Tucker et al., 1992). With this *in oculo* model, the embryonic heart is perfused and innervated from ocular tissues, beats, and continues to grow over a several-week period.

D. Isolated Organelles

The rat myocardium has been used by a number of investigators for preparation of isolated cellular organelles, such as nuclei, mitochondria, ribosomes, or sarcoplasmic reticulum. One factor that must be considered when studying functions of the myocardium related to cellular organelles is that heart muscle is composed of several cell types. Although the bulk of myocardial mass is cardiac muscle tissue, on a numerical basis, myocytes are a definite minority. In the mature rat myocardium, only 20% to 30% of the total cell number is composed of cardiac myocytes (Morkin and Ashford, 1968; Grove et al., 1969), the largest number of cells being the much smaller endothelial cells and connective tissue cells such as fibrocytes. Therefore, studies measuring nuclear function, for example, would

principally detect endothelial cell or fibroblast nuclear function rather than myocyte nuclear function.

E. Isolated Heart Preparation

There are two basic procedures used today for isolated perfused heart preparations. The first is the aortic perfusion technique originally described by Professor Oscar Langendorff (1895) and used extensively by investigators since (Anderson et al., 1990b). With this system (the Langendorff preparation), the aorta is cannulated and retrograde flow closes the aortic valve, resulting in direct perfusion of the coronary arteries. Left ventricular functional assessments can be made with a left ventricular balloon. The balloon and catheter system usually consists of a latex balloon affixed to a double-bore catheter attached to a pressure transducer. Cardiac function is usually recorded as rate-pressure product (heart rate × peak left ventricular pressure). Because this is an isovolumic balloon system, it is important to keep the balloon volume to a minimum and keep left ventricular end diastolic pressure low. At higher end diastolic pressure, there is compression of the ventricular chamber with impediment of perfusion in the subendocardial region of the myocardium.

The second most common type of isolated perfused heart preparation (referred to as the working heart preparation) was described and characterized by Neely et al. (1967). This technique involves cannulation of both the aorta and the left atrium with perfusate flowing into the atrium at a constant pressure. This results in left ventricular filling and allows the normal systolic-diastolic cycling of cardiac contraction with ejection of fluid through the aortic valve against an afterload pressure. The coronary arteries are perfused normally by the ejection of perfusate out the aorta against this pressure head. Because perfusate is ejected, work is performed by the heart, thus the designation working heart preparation. Cardiac function can be quantitated by measuring the aortic output against a known pressure. The amount of work performed by this preparation can be changed by adjusting the preload, afterload, or both. Because coronary perfusion depends on left ventricular function, the preparation will convert to a Langendorff perfusion system if systolic left ventricular pressure falls below the aortic pressure.

The standard perfusate used in both types of isolated rat heart preparations is an asanguinous bicarbonate buffer, the most common of which is Krebs-Henseleit buffer (Anderson et al., 1989). Krebs buffer contains either glucose or dextrose as the sole metabolic substrate, with some investigators adding other constituents, for example, lactate, pyruvate, ketone bodies, and insulin (Morgan et al., 1984; Williamson and Kobayashi, 1984). Because these buffers do not contain plasma proteins, they have a low oncotic pressure; thus edema forms in the myocardial tissues soon after the onset of perfusion. The low oncotic pressure and low oxygen carrying capacity of these asanguinous buffers also produce coronary flow rates that are much higher than are normal physiologic levels. The normal *in vivo* coronary flow rate is approximately 1 mL/min/g wet heart weight. In hearts perfused with Krebs buffer at 100 mm Hg perfusion pressure, the coronary flow rate is 14 to 18 mL/min/g wet heart weight (Anderson et al., 1988, 1989). In an effort to more accurately duplicate the normal substrate of the *in vivo* heart, fatty acids and albumin can be added to Krebs buffer. The most commonly used substrate is palmitate (Lopaschuk et al., 1990). With this buffer, cardiac metabolism shifts to the preferred substrate of non-esterified fatty acids, and the respiratory quotient decreases, as would be expected. Another modification of these fatty acid containing buffers is the addition of washed red blood cells (Duvelleroy et al., 1976). In studies that used washed red blood cells from pigs or man at concentrations of 20% to 40%, coronary flow rates decreased to more physiologic levels.

Isolated heart preparations have an advantage over *in vivo* studies in that all neurogenic and hormonal influences are removed. Perfusion fluid volume and content are easily controlled in the isolated heart preparation, and metabolic substrates are conserved because only a small amount of tissue is being perfused compared with that of whole-animal preparations. Function of the heart is easily controlled and monitored. Light microscopic histology is unaltered except for mild interstitial edema after 1 to 3 hours at 37°C (Brown et al., 1968), and myocardial ultrastructure is maintained for at least 90 minutes at 32°C if glucose and insulin are present in the perfusion media (Weissler et al., 1968). Disadvantages of the isolated heart preparation are that hemodynamic and other functional conditions are altered from the *in vivo* situation, and after several hours, functional and structural alterations occur, leading to myocardial failure. Although function and structure appear stable for several hours, in fact, degenerative changes are probably occurring throughout the perfusion period.

VI. GENETIC MODELS OF HUMAN CARDIOVASCULAR DISEASE

The maintenance of numerous highly inbred strains of rats for many years has led to the development of several strains and substrains that have particular usefulness in the study of cardiovascular disease. The SHR is by far the most widely used, but several other strains useful for studies of hypertension and other aspects of cardiovascular

disease have been developed (Table 26.3). Most of these strains have been developed from baseline stocks of Wistar or Sprague-Dawley rats, both of the same origin, and thus have many genetic similarities. With genetic analysis, information is becoming available concerning specific genetic loci associated with various phenotypic traits such as hypertension (Rapp, 2000). Several recent reviews have described these various strains in detail and provide information on availability (Lovenberg, 1987; Ganten et al., 1994; Doggrell and Brown, 1998; Rapp, 2000).

A. Spontaneously Hypertensive Rats

1. Historical Development

A strain of rats with persistently elevated blood pressure was developed by selective brother–sister mating by Okamoto and Aoki (1963) and has been maintained by selective sibling matings. By 1969 these SHRs had reached the 21st generation, defining an inbred line. Unfortunately, the parent WKY rat, the line often used as a control, was distributed to commercial sources before sufficient inbreeding to develop a line. Consequently, there are various phenotypes relative to DNA fingerprint patterns, growth, blood pressure, and response to pharmacologic agents among different sources of Wistar rats (Kurtz and Morris, 1987; Kurtz et al., 1989a; St. Lezin et al., 1992), bringing to question the use of the WKY rat as a control for SHR. Further, WKY rats may share a number of genes responsible for hypertension with SHR (Louis and Howes, 1990).

2. Hypertension and Complicating Lesions

Hypertension is present in 100% of SHR. In many colonies, prehypertension (100 to 120 mm Hg) is present during the first month of life (Adams et al., 1989). Increased blood pressure (greater than 140 mm Hg) is present by 5 weeks of age in many animals and by 10 weeks of age is maintained at over 180 mm Hg, frequently exceeding 200 mm Hg. Minimal cardiac hypertrophy has been reported in neonatal SHR (Cutilletta et al., 1978), and approximately 30% left ventricular hypertrophy follows the development of hypertension. By 12 months of age, many animals develop focal areas of myocardial fibrosis, increasing with age, with congestive heart failure in some animals by 18 months of age (Pfeffer et al., 1979b; Bing et al., 1984; Engelmann et al., 1987; Boluyt et al., 1995). In a review of the model, Okamoto (1969) described complicating lesions associated with the hypertension in several organs in addition to the heart. Cerebral softening owing to infarction, subarachnoid hemorrhage, and microscopic cerebral hemorrhages occurred in 8%. Myocardial necrosis and fibrosis, mainly detected by microscopy, were

TABLE 26-3

GENETIC STRAINS OF RATS USEFUL IN CARDIOVASCULAR RESEARCH

Strain	Characteristics	Reference
New Zealand genetically hypertensive rats (GH)	Hypertension	Smirk and Hall, 1958
Dahl salt-sensitive (DS) and Dahl salt-resistant (DR) rats	Hypertension	Dahl et al., 1962b
Spontaneously hypertensive rats (SHR)	Hypertension	Okamoto and Aoki, 1963
Obese SHR (Koletsky)	Obese, hypertension, hypercholesterolemia, atherosclerosis	Koletsky, 1973
Spontaneously hypertensive rats-stroke prone (SPSHR)	Hypertension, cerebral hemorrhage	Okamoto et al., 1974
Israel DOCA salt-sensitive (SBH) and -resistant (SBN) rats	Hypertension	Ben-Ishay et al., 1972
Lyon hypertensive (LH), Lyon normotensive (LN), and Lyon low blood pressure (LL) rats	Hypertension, hypotension	Dupont et al., 1973
Milan hypertensive strain (MHS) and Milan normotensive (MNS) rats	Hypertension	Bianchi et al., 1974
Utrecht Fawn hooded hypertensive (FHH) and Fawn hooded low blood pressure (FHL) rats	Hypertension, hypotension	Kuijpers and Gruys, 1984
Prague hypertensive rat (PHR) and Prague normotensive rat (PNR)	Hypertension	Heller et al., 1993
Hypothalamic diabetes insipidus rats (Brattleboro strain)	Diabetes insipidus, hypertension	Hall et al., 1973
Cohen-Rosenthal diabetic hypertensive rats	Insulin-dependent diabetes, hypertension	Cohen et al., 1993
Obese Zucker rat	Non-insulin-dependent diabetes, hypertension, hypercholesterolemia	Zucker and Zucker, 1961
JCR:LA-cp rat	Obese, insulin resistant, hypertriglyceridemic	Russell et al., 1987
Otsuka Long-Evans Tokushima fatty rat (OLETF)	Non-insulin dependent diabetes, hypertension	Kawano et al., 1992
Hypertensive heart failure prone rats (SHHR/Mcc-cp)	Obese, hypertensive, diabetes, hypertrophy, heart failure	McCune et al., 1990a
Wistar fatty rat	Non-insulin-dependent diabetes, obese, hypertension	Ikeda et al., 1981

DOCA indicates deoxycorticosterone acetate.

found in 19%, nephrosclerosis in 17%, and vascular lesions (including fibrinoid necrosis and hyaline degeneration) in renal, pancreatic, hepatic and, less commonly, coronary arteries were present in 39%.

3. Mechanism of Hypertension in Spontaneously Hypertensive Rats

There is clearly a genetic basis for the hypertensive trait in SHR. Mating SHR with normotensive rats of the parent strain (WKY) produces offspring with moderately elevated blood pressure, suggesting a polygenic mode of inheritance for the hypertensive trait (Yamori et al., 1972). Strain-specific esterase isozymes of nonspecific α-naphthyl acetate esterase serve as a marker for the strain, are single gene-controlled, and are codominantly transmitted (Yamori and Okamoto, 1970). Although considerable advance has been made in genetic analysis in recent years and the rat renin gene has been localized to chromosome 13 (Pravenec et al., 1991), the specific genes responsible for increased blood pressure in SHR have not been identified (Rapp, 1991, 2000).

The renin angiotensin system clearly plays a role in development and maintenance of hypertension in SHR, although there have been conflicting reports on the level of plasma renin or tissue renin RNA. Plasma renin activity is reportedly increased in the young SHR, coinciding with increased systemic pressure, and then becomes suppressed once the hypertension has stabilized (Sen et al., 1972). Plasma angiotensin II levels were found to be not different in young SHR and WKY rats, but levels were lower in SHR over 28 weeks of age compared with levels in WKY. However, angiotensin II in left ventricular tissue was significantly increased in old SHR compared with old WKY rats (Dang et al., 1999).

4. Pathogenesis of Hypertrophy

The pathogenesis of the cardiac hypertrophy in SHR is not clear, but clearly is related to angiotensin II, because treatment with converting enzyme inhibitors prevents both the hypertension and the cardiac hypertrophy (Pfeffer et al., 1982; Limas et al., 1984). Treatment of SHR with nerve growth factor antiserum produced peripheral sympathectomy and prevented the development of hypertension, but not cardiac hypertrophy (Cutilletta et al., 1977). Hydralazine treatment, although normalizing blood pressure, results in increased plasma renin activity and persistence of cardiac hypertrophy (Sen et al., 1972; Limas et al., 1984). It has also been proposed that myocardial hypertrophy in SHR may develop as a result of a genetic cardiovascular abnormality that does not require increased systemic pressure for its expression (Cutilletta et al., 1977).

SHR have been widely used for studies of hypertension and are responsive to the same drugs that are useful in treating hypertension in humans. The model consistently results in both hypertension and cardiac hypertrophy, however, despite more than 30 years of study, the mechanism for hypertension remains unclear. The onset of both hypertension and cardiac hypertrophy early in life also differs from the situation in humans.

B. Substrains of Spontaneously Hypertensive Rats

1. Stroke-Prone Spontaneously Hypertensive Rats-

In a strain with severe hypertension derived from the parent SHR strain, the incidence of these complicating lesions was greatly increased, establishing a substrain of stroke-prone SHR (SPSHR) (Okamoto et al., 1972; Ooshima et al., 1972). In SPSHR, 80% of male rats over 100 days of age and 60% of females over 150 days of age develop cerebral hemorrhages. They develop severe hypertension of about 240 mm Hg. Signs of stroke include excitement, hyperirritability, paroxysm, coma, and death. Thromboses and cerebral hemorrhages are most common in the left occipital area of the cortex and subcortex. Lesions also occur in frontal and medial cortical areas and the basal ganglia. Death occurs during mid-adulthood, around 52 to 64 weeks of age (Nagura et al., 1995) and even earlier when the rats are salt loaded, in contrast to about 2 years in SHR or 3 years in WKY rats. Although this model is useful for studies of the cerebral lesions of stroke, the relatively young age of stroke development differs from the condition in humans.

2. Obese Spontaneously Hypertensive Rats (Koletsky)

The obese SHR substrain was developed from litters in which some animals were obese (Koletsky, 1973). By selective breeding of heterozygous parents, approximately 25% of the offspring were obese. The obese trait appears to be inherited as an autosomal recessive gene. Obese SHR have a marked increase in food consumption and a more rapid body weight gain than do nonobese SHR. Fat is deposited in subcutaneous tissues, retroperitoneally, and in the mesentery, although the face and paws are spared. The rats take on a rounded body shape and may weigh up to 1000 g. They have a high incidence of lipid-containing atherosclerotic plaques in the aorta, coronary arteries, and other vessels. There is marked hyperlipemia, detected as early as 6 weeks of age. There are marked increases in both serum cholesterol and, especially, triglycerides compared with levels in control nonobese SHR.

3. Hypertensive Heart Failure Prone Rats

A strain of heart failure-prone rats (SHHR) was developed by mating SHR with Koletsky obese rats heterogeneous for the corpulent gene (McCune et al., 1990a,b). Approximately 25% of rats are obese and manifest hyperinsulinemia and diabetes. The strain has been maintained for more than 27 generations and is hypertensive, and obese rats begin dying of congestive heart failure between 10 and 12 months of age for males and between 14 and 16 months of age for females; lean male SHHR develop hypertrophy by 5 months of age and die of congestive heart failure by 16 to 20 months of age, and lean females die between 22 to 26 months of age. The left ventricular chamber becomes dilated as demonstrated by ECHO (Haas et al., 1995), and the left ventricular myocytes are hypertrophied by increase in cell length, characteristic of dilated cardiomyopathy (Gerdes et al., 1996). Both SHR and SHHR exhibit cachexia; however, SHR typically exhibit thoracic effusion, ascites, and atrial thrombi much more frequently than do SHHR. As in humans with developing congestive heart failure, SHHR have increased plasma levels of plasma renin activity, atrial natriuretic peptide, and aldosterone with increasing age (Holycross et al., 1997). Thus, this strain with many similarities to congestive heart failure in humans has proven to be a useful model for studies of congestive heart failure owing to hypertension, hypertrophy, and diabetes.

C. Dahl/Rapp Salt-Sensitive and Salt-Resistant Rats

Dahl et al. (1962a,b) have developed two lines of rats by selective inbreeding that are either resistant (DR) or sensitive (DS) to the effects of a high (8%) salt diet in regard to development of hypertension. Neither of these lines has hypertension if salt is not included in the diet. The DS rats develop hypertension in excess of 180 mm Hg when 8% NaCl is added to the diet starting at 6 weeks of age, with left ventricular hypertrophy by 11 weeks and, finally, left ventricular dilation, cardiac failure, and death at 15 to 20 weeks of age (Inoko et al., 1994). Although these lines have been maintained since 1962, the lines do not breed true and must be maintained by continued selection for or against the development of high blood pressure when fed the high-salt diet. Inbred strains from Dahl's outbred lines have been developed and are designated SS/Jr and SR/Jr for the salt sensitive and salt resistant strains (Rapp, 1982; Rapp and Dene, 1985), respectively. Selective breeding studies have demonstrated a genetic basis for salt sensitivity in the Dahl rats (Knudsen et al., 1970), and the model is a useful one for studies involving both genetic and environmental factors. Different genetic loci apparently are involved in the Dahl rats than in the SHR of Okamoto.

The strain-specific aryl esterase isoenzymes found in the SHR (Yamori and Okamoto, 1970) were not present in either the DS or DR lines (Rapp and Dahl, 1972). In more recent studies, it has been suggested that modulation of nitric oxide production from L-arginine may be involved in the development of hypertension and cardiac hypertrophy in these salt-sensitive rats (He et al., 1997). Although such a high dietary content of salt is not common in humans, salt is implicated in the development of hypertension, and thus, this model is useful in studies of hypertension.

D.   Other Strains of Hypertensive Rats

Several additional strains of hypertensive rats have been developed by different groups of investigators around the world, but most are only available on a limited basis. The Sabra hypertension-prone (SBH) and Sabra hypertension-resistant (SBN) rats were developed in Israel by selecting animals with increased blood pressure response to unilateral nephrectomy, deoxycorticosterone acetate (DOCA) and 1% salt in drinking water (Ben-Ishay et al., 1972). The Lyon strains of high and low blood pressure were selected without special dietary or environmental challenges (Dupont et al., 1973). The Milan strains were also selected without special dietary or environmental challenges for high or low blood pressure (Bianchi et al., 1974), and are available in the United States and in Italy. Selective inbreeding of the Fawn-hooded rat produced a high blood pressure strain and a normotensive strain characterized by spontaneous hypertension and renal lesions with proteinuria (Kuijpers and Gruys, 1984). Another strain of high and normotensive rats has been developed in Prague (Dobesova et al., 1991; Heller et al., 1993).

E.   Obese Diabetic Rat Models

Adult onset type II non-insulin-dependent diabetes is characterized by obesity, hypertension, hypercholesterolemia, and cardiovascular lesions involving the heart and blood vessels. The Zucker diabetic obese rat model has hyperglycemia, hypercholesterolemia, progressive kidney damage, and mild hypertension compared with the lean Zucker rat, thus fulfilling several criteria for a model for diabetes (Zucker and Zucker, 1961; Kurtz et al., 1989b; Alonso-Galicia et al., 1996; Carlson et al., 2000). Other related models of obesity and non-insulin-dependent diabetes include the Otsuka Long-Evans Tokushima fatty rat with mild hypertension, renal lesions, and perivascular fibrosis (Yagi et al., 1997), the JCR:LA-cp rat (Russell and Amy, 1986), and the Wistar fatty rat, developed from the obese Zucker rat (Ikeda et al., 1981; Yamakawa et al., 1995).

F.   Hereditary Hypothalamic Diabetes Insipidus

The Brattleboro strain of rats with hereditary hypothalamic diabetes insipidus owing to a deficiency of arginine vasopressin has hypertension that responds to treatment with vasopressin (Hall et al., 1973; Valtin, 1982). This model has been widely used to study the effects of vasopressin on cardiac hypertrophy, hypertension, and renal function.

VII.   TRANSGENIC RAT MODELS OF HUMAN HEART DISEASE

The rat has had far fewer transgenic models produced than has the mouse, while it is likely that others will be introduced in the future. The rat appears to have several genes that play a role in the development of hypertension, although spontaneous hypertension is rare in the mouse, making the rat a suitable model for genetic studies of hypertension. The most widely used transgenic rat to date has been the TGR (mRen2)27 rat with overexpression of the mouse Ren2 gene (Mullins and Ganten, 1990; Lee et al., 1996). Hypertension is associated with high renin tissue activity, although plasma renin is low and, if not treated with angiotensin-converting enzyme inhibitors, is lethal by several months of age. Introduction of both renin and human angiotensinogen has resulted in severe hypertension and early mortality in rats (Mullins and Ganten, 1990; Bohlender et al., 1996, 1997, 2000; Lee et al., 1996). A number of other transgenic rat models have been developed for use in cardiovascular research, a few of which are listed in Table 26.4.

TABLE 26-4

Some Transgenic Rat Models for Cardiovascular Research

Model	Reference
TGR (mREN-2)27 mouse Ren2d	Mullins et al., 1990
TGR (hAOGEN)1623 human angiotensinogen	Ganten et al., 1992
Angiotensin II type 1 receptor	Hoffmann et al., 2001
Myocardial AT1 (alphaMHC-hAT1)594-17	Han et al., 2002
Human tissue kallikrein	Silva et al., 2000
Human α 2 Na, K-ATPase	Ruiz-Opazo et al., 1997
Cardiac troponin T deletion	Frey et al., 2000
Calmodulin-dependent calcium ATPase (PMCA)	Hammes et al., 1998
α1B-adrenergic receptors	Harrison et al., 1998
Human cholesteryl ester transfer protein	Herrera et al., 1999
Endothelin-B receptor-deficient	Matsumura et al., 2000
Neuropeptide Y	Michalkiewicz et al., 2001
Green fluorescent protein	Hakamata et al., 2001

VIII. EXPERIMENTALLY INDUCED MODELS FOR CARDIOVASCULAR RESEARCH

A. Systemic Hypertension

Systemic hypertension is produced in rats by a variety of techniques, the most common of which are related to stimulation of the renin-angiotensin system by manipulation of blood flow to the kidneys. The most commonly used systems use a modification of the classic model originally described by Goldblatt (1934). In the two-kidney Goldblatt model, partial occlusion of one renal artery is produced and the other kidney is left intact. In the one-kidney Goldblatt model, the renal artery to one kidney is stenosed and the second kidney is removed. Other variations of this model include removal of one kidney with cellophane wrapping of the other and complete occlusion of the aorta between the renal arteries, producing ischemic necrosis of one kidney and increased blood flow to the other (Latta et al., 1975). Acute hypertension has been produced by direct infusion of angiotensin (Wiener and Giacomelli, 1973). Systemic arterial pressure usually exceeds 150 mm Hg with these models and, in some, exceeds 200 mm Hg.

A frequently used model for induction of systemic hypertension is the uninephrectomized rat treated with DOCA and saline for drinking water. One kidney is removed at 4 to 6 weeks of age, and weekly subcutaneous injections of 10 mg DOCA are administered (Hall and Hall, 1965). Blood pressure gradually increases and by 3 weeks stabilizes at approximately 200 mm Hg. Heart weight-to-body weight ratios are increased compared with those of nontreated controls.

B. Pulmonary Hypertension

Pulmonary hypertension with resulting right ventricular hypertrophy has been produced in rats by administration of *Crotalaria spectabilis* seeds or extracted monocrotaline pyrrole and by chronic or intermittent exposure to hypobaric pressure.

1. Crotalaria

Severe and gradually increasing pulmonary hypertension is produced by intravenous or subcutaneous injection of the monocrotaline pyrrole from *C. spectabilis* seeds (Turner and Lalich, 1965; Chesney et al., 1974; Ishikawa et al., 1991). The pulmonary vessels become thickened by medial hyperplasia and intimal fibrosis, leading to increased pulmonary vascular resistance (Reindel et al., 1990; Wilson et al., 1992). Studies using monocrotaline treatment

have produced peak systolic right ventricular pressures of 91 mm Hg with no change in aortic pressure. These animals also had increased hemoglobin and hematocrit, as well as a two to threefold increase in right ventricular free-wall weight. Right ventricular weights increased by 40% at 2 weeks, 85% at 4 weeks, and up to two to threefold increase at later time points compared with weights in control rats (Turner and Lalich, 1965; Chesney et al., 1974; Ishikawa et al., 1991). These animals subsequently died with congestive heart failure.

2. Normobaric Hypoxia and High Altitude (hypobaric hypoxia)

Rats exposed to hypoxia (10% O_2) develop pulmonary hypertension, which resulted in pressure overload right ventricular hypertrophy (Chen et al., 1997; Tilton et al., 2000). Hypoxic exposure can be achieved by placing rats in a Plexiglass glove box that is gassed with N_2 to maintain an oxygen concentration of 10%. A carbon dioxide (CO_2) concentration of less than 0.2% can be maintained by using a baralym CO_2 scrubber (Chen et al., 1997). By use of this technique, Chen et al. (1997) reported that rats developed pulmonary pressures of 36 mm Hg after 2 weeks (compared with control values of 16 mm Hg) and 47 mm Hg by 6 weeks. This pressure change resulted in an 80% to 120% increase in right ventricular weight compared with that of controls, with no change in the weight of the left ventricle plus septum. There was a decrease in the rate of body weight gain in animals subjected to hypoxia.

Chronic exposure to high altitude (3000 to 4000 m or higher) or simulated high altitude in a hypobaric chamber results in pulmonary hypertension mediated by pulmonary vasoconstriction and the development of right ventricular hypertrophy. In one study, rats maintained for 5 weeks at a simulated altitude of 5500 m nearly doubled right ventricular weight from control weight of 216 ± 10 to 426 ± 22 mg (York et al., 1976). In many reported studies, however, severe weight loss during chronic exposure to hypobaric conditions has caused difficulty in interpretation of data. Despite this confounding variable, simulated high altitude has been used by many investigators to study various aspects of the cardiopulmonary system subjected to pulmonary hypertension (Mager et al., 1968; McGrath and Bullard, 1968; Bischoff et al., 1969).

C. Left Ventricular Hypertrophy and Heart Failure

A variety of stresses, generally resulting in an increased workload on the myocardium, have been used to produce an increase in mass of the left ventricle. The major methods used in the rat are summarized below.

1. Swimming

One of the most common methods of producing an increased work load on the heart is to subject the rats to swimming exercise several hours per day for a period of several weeks (Stevenson, 1966, 1967; Leon and Bloor, 1968). This swimming regimen produces a 10% to 20% increase in heart weight after several weeks of exercise. Rats have also been trained to exercise by running on a treadmill (Thomas and Miller, 1958; Van Liere et al., 1965). Both of these methods, however, require a great deal of time and produce only a mild degree of hypertrophy. In addition, stress factors associated with swimming or running may produce effects on the animal other than those due to exercise alone.

2. Banding of the Aorta

Pressure overload or increased afterload on the left ventricle has been produced by banding of the aorta. Beznak (1952; 1955) introduced this method by placing a suture on the aorta below the diaphragm. The degree of obstruction is controlled by placing a wire of known gauge against the aorta while making the tie. The wire is then removed, leaving a lumen the size of the wire. Although the method is easily performed without the necessity of thoracotomy, only mild cardiac hypertrophy is produced, often less than 10% increase in mass, and there is increased systemic pressure to the cranial portion of the animal. Modifications of this technique have been reported to produce up to 30% increase in cardiac mass (Schonekess et al., 1996).

Banding of the ascending aorta was developed by Nair et al. (1968) and has been modified and used by others (Bugaisky et al., 1983; Anderson et al., 1990a) to produce increased pressure load on the heart without increased pressure to the rest of the body. A metal clip is placed on the ascending aorta of young rats (approximately 50 g body weight) with specially designed forceps. This band is initially non-constricting, but as the animal grows, the aortic stenosis gradually increases. The procedure is rapidly completed through a small thoracotomy without the need for a respirator. There is a modest decrease in body weight gain for several days followed by normal weight gain. By 6 to 8 weeks after the band placement, there is a 60% to 80% increase in heart weight. This procedure can also be slightly modified if the band is placed on the aortic arch (Fig. 26-5). This procedure allows investigators to pass a catheter down the right carotid artery and into the left ventricle to measure left ventricular pressures.

3. Volume Overload

Volume overload of the heart occurs when the heart is required to pump a larger than normal amount of blood

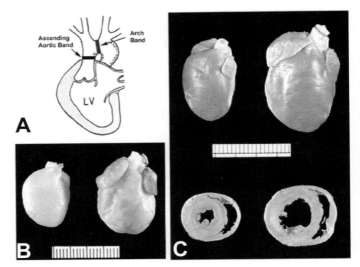

Fig. 26-5 Aortic banding procedure for producing left ventricular pressure overload hypertrophy. (A) Diagram of a rat heart demonstrating placement of aortic coarctation on the ascending aorta or on the aortic arch. Placement of the band on the aortic arch allows for cannulation of the left ventricle via the right carotid artery. (B) Hearts from age-matched rats sacrificed 6 weeks after placement of the aortic arch band (right) versus a heart from a sham-operated control (left). (C) Hearts (top) from age-matched rats 8 weeks after placement of an ascending aortic band. The banded heart, right, is significantly larger than that of the sham-operated control (left). Cross-sections (bottom) of hearts demonstrate the left ventricular wall thickening and the slight dilation of the left ventricular chamber in the aortic banded heart (right).

in order to meet the requirements of the body. Surgical preparations creating AV shunts, although technically difficult, have been used in the rat with considerable success (Su et al., 1999; Wei et al., 2003).

One commonly used method involves creation of an infrarenal abdominal aorta-to-vena cava fistula. For this procedure a ventral abdominal laparotomy is performed to expose the aorta and caudal vena cava approximately 1.5 cm below the renal arteries. Via blunt dissection, the overlying adventitia is removed and the vessels are exposed, with care taken not to disrupt the tissue that connects the vessels. Both vessels are then occluded proximal and distal to the intended puncture site, and an 18-gauge needle is inserted into the exposed abdominal aorta and advanced through the medial wall into the vena cava to create the fistula. The needle is withdrawn, and the ventral aortic puncture is sealed with cyanoacrylate. Creation of the aorta-to-vena cava fistula can be visualized by the pulsatile flow of oxygenated blood into the vena cava.

In studies by Wei et al. (2003), there was a 45% increase in total heart weight-to-body weight ratio compared to sham at 5 days after creation of the aorta-to-vena cava fistula. The left ventricular weight-to-body weight ratio was increased 30% compared with sham animals ($P < 0.05$), and right ventricular weight-to-body weight ratio was also increased. As expected, there was a significant decrease in mean arterial pressure owing to the aorta-to-vena cava fistula but left ventricular end-diastolic pressure and peak positive left ventricular dp/dt was not different from that of sham animals.

Iron deficiency anemia in weaning rats has also been used as a model of volume overload cardiac hypertrophy (Korecky et al., 1964; Korecky and French, 1967). After the weaning, rats are fed a milk and sugar diet, which is deficient in iron and copper, Hemoglobin concentration drops to 3.5 to 5.0 g/dL blood, and control rats with iron and copper added to the milk and sugar diet maintain hemoglobin of 13 to 16 g/dL blood. Heart weights are increased by more than 100% by 11 to 19 weeks; both left and right ventricles are hypertrophied.

4. Cardiac Hypertrophy Owing to Carbon Monoxide Exposure

Exposure of neonatal rats to 500 ppm carbon monoxide has been shown to result in approximately a 70% increase in heart weight after 35 days of exposure (Penney et al., 1974a). In this model approximately 40% of the hemoglobin is in the form of carboxy-hemoglobin, and hemoglobin values are increased to 17 g/dL compared with 13 g/dL in nonexposed controls. Although the mechanism for production of cardiac hypertrophy in carbon monoxide toxicity is not clear, it is probably related to increased blood viscosity with secondarily increased vascular resistance and to the increased volume of blood, which must be pumped by the heart owing to its decreased oxygen carrying capacity. This appears to be an excellent model for induction of cardiac hypertrophy in young animals.

5. Thyroxine

Cardiac hypertrophy of approximately 38% increase in total heart weight may be produced by daily administration of thyroxine (T_4) subcutaneously over a 10-week period (Sandler and Wilson, 1959; Beznak, 1962; Beznack et al., 1969; Bedotto et al., 1989). There is complete regression of the thyroxine-induced cardiac hypertrophy by 2 weeks after cessation of thyroxine administration (Beznack et al., 1969; Van Liere and Sizemore, 1971).

6. Isoproterenol

Cardiac hypertrophy has also been produced in the rat by administration of low, subnecrotizing doses of isoproterenol to the rat (Rona et al., 1959; Stanton et al., 1969; Wood et al., 1971). Administration of large doses of isoproterenol, 80 to 100 mg/kg, results in myocardial necrosis, but repeated administration of 5 mg/kg body weight results in stimulation of RNA synthesis in the myocardium and development of cardiac hypertrophy (Gordon et al., 1969). Administration of 175 mg/kg of isoproterenol on two

successive days has been reported to induce rapid heart failure (Teerlink et al., 1994). However, in our hands (S.P.B.) this technique produces an initial mortality of 64% and substantial left ventricular dilation but relatively little change in left ventricular dp/dt or peak developed pressure, and extremely variable pattern of subendocardial necrosis with half the rats showing a transmural apical scar. The high initial mortality, variability in necrosis, and small reductions in left ventricular function do not make this an attractive model of heart failure.

7. Nitric Oxide Synthase Inhibition

Nitric oxide is an important regulator of vascular tone and myocardial contractile function (Dai et al., 2001). Nitric oxide produced by the constitutive endothelial form of NO-synthase (eNOS or NOS3) can modulate cardiomyocyte energetics, contractility, blood flow, and platelet aggregation (Loke et al., 1999; Brookes et al., 2001). Inducible nitric oxide synthase (NOS2) cosegregates with the increased blood pressure in the DS rat, but endothelial cell nitric oxide synthase (NOS3) seems uninvolved in human essential hypertension (Bonnardeaux et al., 1995; Deng and Rapp, 1995). Nitric oxide synthesis can be blocked by inhibitors such as L-NAME (N'-nitro-L-arginine methyl ester) and nitro-L-arginine. Chronic administration of L-NAME increased systolic blood pressure and heart weight (Bernatova et al., 2000), whereas other investigators report hypertension with no cardiac hypertrophy (Bartunek et al., 2000). Although the role of decreased nitric oxide production in human hypertension is unclear, this model of experimental nitric oxide inhibition is technically easy and physiologically relevant.

D. Atherosclerosis

The normotensive rat is very resistant to development of severe atheromatous lesions when placed on high lipid-containing diets. Arterial lesions containing extensive lipid deposits are difficult to produce, and complicated lesions with hemorrhage, thrombosis, and resulting myocardial infarction are very rare. Therefore, the rat is seldom used for studies of atherosclerosis.

A notable exception to the resistance of the rat to development of atherosclerosis is found in the SHR. Okamoto and co-workers (Okamoto and Aoki, 1963; Okamoto et al., 1972) have demonstrated that certain substrains of the SHR are particularly susceptible to development of lipid-containing lesions in the arteries and show a pronounced susceptibility to cholesterol-containing diets. These investigators have concluded that certain substrains have genetic differences in addition to the hypertension that account for the susceptibility to development of atheromatous lesions. In general, the rat is an atherosclerosis-resistant species that, unlike humans, does not have plasma cholesteryl ester transfer protein (CETP). In addition, rats use high-density lipoprotein as the major carrier of plasma cholesterol. Hyperlipidemia in itself is insufficient to produce atherogenesis in rats. Atherogenesis can be induced in rats by adding cholic acid and thiouracil to a high-cholesterol/high-fat diet (Joris et al., 1983). Cholic acid increases cholesterol absorption and produces a concomitant suppression of cholesterol 7-α-hydroxlyase activity that results in decreased cholesterol excretion. Thiouracil induces hypothyroidism, which results in a decrease in low density lipoprotein receptor activity and hypercholesterolemia. Several strains of rats have been identified with heritable hyperlipidemia. Some of these are associated with a mild to moderate atherogenesis (St. John and Bell, 1991; Russell et al., 1993; Chinellato et al., 1994).

Transgenic overexpression of human CETP (hCETP) in DS hypertensive rats (Tg[hCETP]DS) has been shown to produce a more aggressive atherosclerosis than transgenic CETP mice exhibit (Herrera et al., 1999). Tg[hCETP]DS rats have hypertriglyceridemia, hypercholesterolemia, and decreased high-density lipoprotein levels when fed regular rat chow, and this lipid profile worsens with age. Rats in the two "high expresser" lines had coronary artery disease that was proportional to lipoprotein abnormalities. The transgenic rat line expressing the highest level of hCETP, Tg53, had coronary artery disease and myocardial infarction at 6 months and a significantly higher rate of mortality compared with that of age-matched nontransgenic control DS rats.

Exposure to allylamine (3-aminopropene), a specific cardiovascular toxin, induces myocardial necrosis and atherosclerotic-like vascular lesions characterized by vascular smooth muscle hyperplasia and necrotizing vasculitis in rats (Guzman et al., 1961; Will et al., 1971; Lalich et al., 1972; Boor et al., 1979; Boor and Hysmith, 1987). In this model, myocardial necrosis occurs within 24 hours after allylamine treatment, whereas vascular lesions requires weeks to months of treatment depending on the route of administration and dose (Boor et al., 1979; Boor and Hysmith, 1987). A combination of coronary artery vasospasm (Lalich et al., 1972; Boor et al., 1979; Kato et al., 1995), increased oxidant stress and lipid peroxidation (Awasthi and Boor, 1993; Misra et al., 1995), and possibly calcium overload mediate cardiovascular injury in this model. Isolated thoracic aortic rings from allylamine-treated rats exhibit substantial phasic tension oscillations, supporting the concept that allylamine-induced myocardial necrosis is triggered by prolonged coronary artery contraction (Conklin and Boor, 1998). Several studies indicate that the vascular enzyme semicarbazide-sensitive amine oxidase metabolizes allylamine to acrolein, hydrogen peroxide, and ammonia (Ramos et al., 1988; Awasthi and Boor, 1993),

and increase basal tension as well as generate agonist-induced hypercontractility (Conklin et al., 2001). These reactive intermediates alter vascular contractility, decrease responsivity to catecholaminergic stimulation, and ultimately result in myocardial and vascular smooth muscle toxicity. The vasculitis induced by allylamine may provide a useful alternative atherosclerosis model in rats.

E. Ischemic Heart Disease

The rat has been used extensively to study ischemic heart disease by ligation of a major coronary artery (Bryant et al., 1958; Selye et al., 1960; MacLean et al., 1975; Litwin et al., 1991) and was instrumental in identifying the therapeutic potential of the angiotensin-converting enzyme inhibitors in heart failure (Pfeffer et al., 1985, 1987). The left coronary artery is accessible through a left thoracotomy for ligation (Selye et al., 1960). As discussed above, the left coronary artery does not usually have a large circumflex branch. Ligation of the proximal left coronary will result in 20% to 60% necrosis of the left ventricle (Fig. 26-6). The procedure is usually reported to have a 40% or greater mortality, although Bajusz and Jasmin (1964) reported no mortality after occlusion of the left coronary artery 1 mm below its origin. This model produces compensatory hypertrophy of the surviving myocardium and eventually leads to congestive heart failure (Litwin et al., 1991, 1994). Liu et al. (1997) reported that they are able to produce more consistent large myocardial infarcts with lower mortality by use of Lewis rather than Sprague-Dawley rats. This result is owing to greater uniformity in left anterior descending artery branching patterns in Lewis versus Sprague-Dawley rats. Progressive deterioriation of left ventricular function becomes evident between 8 to 26 weeks after infarct. A difficulty in using this model is that the onset of congestive heart failure is difficult to predict and occurs over a 3 to 4 month timeframe, making assessment of drug efficacy problematic. We have observed that the time between clinical manifestations of congestive heart failure (i.e., edema and/or tachypnea) and death is usually 1 to 2 weeks in rodents (data not shown). Although myocardial infarct size has been reported to be important in determining drug effect (Goldman and Raya, 1995), myocardial infarct size in our hands does not strongly correlate with progressive left ventricular dysfunction as measured by dp/dt ($r < -0.33$) or peak developed pressure ($r < -0.32$) at either 8 or 16 weeks after-infarct in contrast to strong relationship reported by Pfeffer (1979a, 1984, 1991). If the relationship between ischemic injury and the rate of left ventricular remodeling and dysfunction is variable, then understanding the factor(s) underlying this variability may provide novel

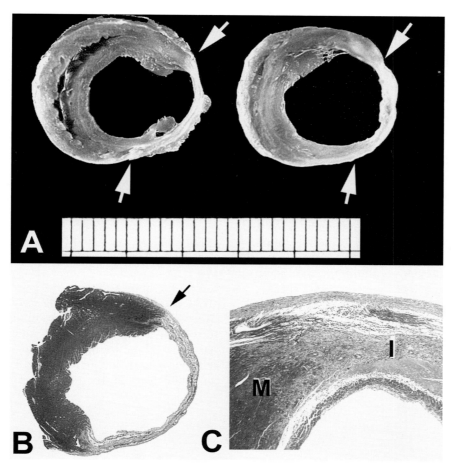

Fig. 26-6 (A) Gross photograph of cross-section slices of left ventricle from a rat that has undergone coronary artery ligation to produce a large transmural myocardial infarction. The borders of the infarction are highlighted with arrows. (B) Low-power photomicrograph of hematoxylin and eosin–stained cross-section of the heart in panel A. Note the sharp line of demarcation (arrow) between the surviving myocardium and the scar that has formed 4 weeks after the coronary ligation. (C) Higher power of the region below the arrow in panel B. There is compensatory hypertrophy in the surviving myocardial tissue (M) adjacent to the area of infarction (I).

pharmacological targets for treating progressive ischemic heart disease.

Ischemic myocardial injury has been extensively studied by using the isolated heart preparation described in "Isolated Heart Preparation." By use of this isolated heart technique, global ischemia or hypoxia can be initiated for specific time periods with exquisite control of temperature, perfusate composition, metabolism, pharmacologic intervention and, importantly, reperfusion condition (Apstein et al., 1978; Anderson et al., 1987, 1990b; Allard et al., 1994).

F. Adriamycin (Doxorubicin)

The common cancer chemotherapeutic agent adriamycin is associated with cardiotoxicity that can lead to congestive heart failure and death. Rats administered adriamycin develop cardiotoxicity and heart failure (Doggrell and Brown, 1998). Treatment of rats with adriamycin, 2 mg/kg/week, for 12 weeks results in a decreased cardiac output, decreased blood pressure, pleural effusions, ascites, and hepatic congestion. There is no change in heart weight, but there are pathological changes in the myocytes such as cytoplasmic vacuolation, disorganisation of myofibrils, and some necrosis (Ambler et al., 1993).

G. Alcoholic Heart Disease

Chronic ingestion of ethanol by humans is often associated with alcoholic cardiomyopathy. Male rats given oral ethanol (30% in their drinking water for 8 months) developed myocardial damage, especially multiple areas of myocyte loss and replacement fibrosis, and ventricular wall remodelling resulting in ventricular dysfunction (Segel et al., 1979; Kino et al., 1981; Capasso et al., 1991).

Chronic ethanol treatment in rats produced moderate hypertrophy with significantly decreased responses to phenylephrine, glucagon, and dobutamine (Segel et al., 1979). Although some investigators show little impact of ethanol on cardiac function or structure (Hepp et al., 1984), differences in dose and duration are likely the cause for this discrepancy.

H.   Endocarditis

Bacterial endocarditis may be produced in the rat experimentally by introduction of bacteria into the animal either as the sole agent or in combination with some other stress, such as cold exposure or tumbling. Under these conditions, bacteria will localize on the mitral and aortic valves, producing a fibrinothrombotic lesion that resembles spontaneous bacterial endocarditis in humans (Clawson, 1950). Nonbacterial thrombotic endocarditis has been extensively studied in the rat by Angrist and co-workers (Angrist et al., 1960; Oka et al., 1966). The condition is induced by stress situations, such as cold, trauma, or corticosteroid administration, without concomitant introduction of pathogenic bacteria.

VIII.   USE OF THE RAT FOR MISCELLANEOUS STUDIES

A.   Carotid Balloon Injury Model

The rat carotid balloon injury model has been used extensively to further our understanding of the vascular response to injury (Anderson et al., 2001). In this model a deflated Fogarty balloon is inserted into the carotid artery, and the balloon is inflated and gently removed from the artery. This process results in dilation of the vessel wall and mechanical denudation of endothelial cells as the balloon is removed. The sequence of events after balloon injury, as initially described by Fishman (1975) and expounded on by others (Fingerle et al., 1990; Majesky et al., 1992), includes platelet adhesion and smooth muscle cell proliferation to form a neointima composed of almost 100% smooth muscle cells (Fig. 26-7). The initial sequences of platelet adhesion and degranulation result in thrombus formation and the release of numerous growth factors. At 24 to 72 hours after balloon injury, smooth muscle cells begin to divide and migrate from the media into the intima (Clowes et al., 1983). Approximately 50% of the

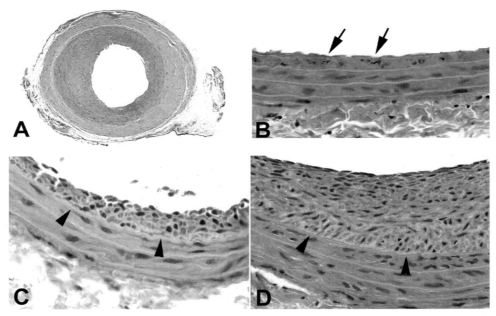

Fig. 26-7 Morphologic changes associated with the balloon injury model in the rat carotid artery. (A) Low-power photomicrograph demonstrating the concentric neointimal response with preservation of the media in this rat carotid artery 3 weeks after balloon injury. (B) High-power view of carotid artery wall 24 hours after balloon injury. Note the absence of endothelial cells and the multiple breaks in the internal elastic membrane. There is also nuclear fragmentation in smooth muscle cells along the outermost layer of the vessel. (C) High-power view of carotid artery wall 4 days after balloon injury. Note the early formation of a neointima on the luminal side of the internal elastic membrane (arrowheads). (D) At 3 weeks after balloon injury, the neointima is well formed and has increased in thickness. Some of the smooth muscle cells have begun to develop a fusiform shape and line up circumferentially perpendicular to the direction of blood flow. Although difficult to appreciate at this magnification, the luminal surface of the neointima is covered with endothelial cells. The arrowheads outline the internal elastic membrane. (Reprinted with permission from Anderson et al., 2001.)

cells that migrate into the intima do not divide; rather they change to the secretory phenotype (loss of myofilaments and increase of rough endoplasmic reticulum) and begin to produce extracellular matrix material. The cell division in the neointima continued for up to 8 weeks after injury in areas that were re-endothelialized, and cell division in the neointima continues up to 12 weeks in areas where endothelial cells did not re-grow.

It is evident that endothelial cell denudation plays a critical role in restenosis even without concurrent injury to the media. If a small region of the artery is denuded with a fine nylon thread that does not injure the underlying media, endothelial cells quickly grow over the denuded region, and the proliferative process is greatly diminished (Fingerle et al., 1990).

With the rat carotid artery balloon injury model, the neointimal proliferation is reproducible, and accurate quantitation can be used to characterize the restenosis phenomenon. There are no breaks in the internal elastic membrane, thus the neointima is easily measured by morphometric techniques. This well-characterized model system can be used to study specific pathogenic mechanisms in the restenosis process. This model system does have several caveats, however. First, the arteries are normal before the experimental manipulation. This is very different than what is seen in atherosclerotic human coronary arteries that undergo angioplasty or other interventions. A more disturbing finding is that many of the treatments that have been successful in preventing restenosis in the rat model have not been successful in large animal species and in humans. Thus, there appear to be differences between rats and larger species that make it difficult to predict from rat studies the outcome of pharmacologic interventions on restenosis in man.

B. Thrombosis Models for Anticoagulation Studies

A wide variety of rat models has been developed to assess *in vivo* efficacy and safety of antithrombotic agents. The rat has been used as a primary screen because of its small size, which reduces drug requirements, animal costs, and time in surgical preparation. These advantages are counterbalanced by the amount of blood that can be withdrawn from rodents, which can limit the study of biomarkers and pharmacokinetics, and by cross-species differences in inhibitor potency between rodent and human coagulation enzymes (e.g., factor Xa). The therapeutic window in these models is assessed as the ratio of dose or plasma concentration required for clot weight reduction versus that inducing a substantial increase in bleeding (template, tail transection, and/or hydrogen chloride-induced gastric hemorrhage). Rat thrombosis models can generally be categorized by the type of clot: platelet dependent (arterial clot); platelet and fibrin dependent, also called a "mixed" (venous clot); and disseminated intravascular coagulation or fibrin dependent.

Rat arterial thrombosis models include carotid artery ferrous or ferric chloride injury (Kurz et al., 1990; Lockyer and Kambayashi, 1999; Heran et al., 2000), electrolytic injury of the carotid artery with and without stasis (Philip et al., 1978; Massad et al., 1987; Guarini, 1996), photochemical-induced injury of the vascular endothelium by using rose bengal (Matsuno et al., 1991; Hirata et al., 1995; Ikeda and Umemura, 2001), mechanical crushing (Butler et al., 1992; Chen et al., 1996; Hupkens and Cooley, 1996), heating (Lenfors et al., 1993), carotid de-endothelialization (Cooley et al., 1992), and aortic de-endothelialization and partial constriction (Hladovec, 1986; Sato et al., 1998; Henke et al., 2000). The usual endpoints in arterial injury studies are time to occlusion, quality of perfusion, and clot weight.

Antithrombotic efficacy against venous clots has been assessed by several techniques: isolating a section of the inferior vena cava to produce a zone of stasis, which results in a thrombosis (Sato et al., 1998; Henke et al., 2000); vena caval de-endothelialization (Wong et al., 1996); venous grafting (Hirigoyen et al., 1995; Hupkens and Cooley, 1996); and laser-induced vascular lesions (Giedrojc et al., 1999). Venous thrombus formation can be accelerated by administering a thrombogenic substance such as factor Xa or tissue factor directly into the stasis region. Another commonly used rodent thrombosis model that produces a "mixed" thrombus (platelet and fibrin-dependent) is the AV shunt model in which a tube containing a cotton or silk thread (Peters et al., 1991; Wong et al., 1996), copper wire (Morishima et al., 1997), or other thrombogenic surface is placed in-line between the carotid artery and the jugular vein (Freund et al., 1993; Herbert et al., 1996; Wong et al., 1996; Sato et al., 1998). After a specified time, the thrombus is removed, and the clot size is determined by weight or by a protein assay.

Another widely used coagulation model is based on disseminated intravascular coagulation. Systemic thrombosis is induced by tissue factor, endotoxin (lipopolysaccharide), or factor Xa (Yamazaki et al., 1994; Herbert et al., 1996; Sato et al., 1998). After systemic administration of a thrombogenic stimulus, fibrin deposition occurs. Some studies determine antithrombotic efficacy by measuring changes in survival (Herbert et al., 1996). If stasis is used in addition to a thrombogenic stimulus, then the endpoint is typically thrombus mass. When stasis is not used, endpoints are usually biochemical (e.g., fibrin degradation products, fibrinogen, platelet count, prothrombin time, and activated partial thromboplastin time).

C. Other Studies

Embryologic and fetal development of the vascular system has been extensively studied in the rat, including a variety of morphologic, biochemical, and functional studies. The fetal rat is an ideal species for studies of teratogenic effects of drugs.

A major application of the rat has been its use in determining cardiotoxicity of a variety of toxic agents. Notable examples include such agents as rapeseed oil and other brominated oils (Munro et al., 1972) and cobalt, the toxic agent in beer drinker cardiomyopathy (Grice et al., 1969).

Our understanding of aortic aneurysm formation and the role of collagen cross-linking has been significantly enhanced by studies of lathyrism (angiolathyrism) in the rat. The role of collagen and elastin in the development of blood vessels, damage to vessels, and the healing process have used rat models extensively (Hosoda and Iri, 1966; Barrow et al., 1974; Siegel, 1979; Langford et al., 1999, 2002).

IX. SPECIAL CONSIDERATIONS FOR USE OF THE RAT IN CARDIOVASCULAR RESEARCH

A. Anatomic Differences

There are several anatomic features of the rat heart that differ from the human heart or larger species of animals commonly used for cardiovascular research. The rat normally has a left superior vena cava in addition to the usual right superior vena cava found in most larger mammals (the rabbit also normally has both left and right superior venae cavae). Although because of its small size, the rat is seldom used for hemodynamic studies or measurements of blood flow within the heart, it should be kept in mind that coronary sinus blood is composed of blood draining from the left superior vena cava as well as from the coronary vein.

The rat heart has a dual blood supply, the atria, including the sinus node region, receive blood supply from the cardiacomediastinal arteries, which arise from the internal mammary arteries. This fact would have special significance in certain types of preparations; for example, the isolated heart perfused retrograde through the aorta presumably would not have the sinus node region directly perfused.

Coronary artery distribution is different in the rat compared with higher mammals. The main branches of the right and left coronary arteries descend over the free-wall portions of the respective ventricles rather than follow the AV grooves or the anterior interventricular sulcus as in higher mammals.

The male rat continues to increase in body weight throughout most of its normal life span, and the heart weight increases at a somewhat lesser rate than does body weight, resulting in a constantly decreasing heart weight-to-body weight ratio. This is particularly important when conducting studies related to changes in heart weight, because heart weight-to-body weight ratios from animals of different body weight cannot be compared without reference to a set of normal values for animals of various weights and from the same colony.

B. Electrocardiography and Electrophysiology

Evaluation of the electrocardiogram of the rat is difficult owing to the very rapid heart rate (350 to 450 beats/min), and poorly defined wave form of the body surface electrocardiogram. The S-T segment is markedly slurred in the rat electrocardiogram, with poorly defined S and T waves, making interpretation of changes in these waves difficult. The action potential recorded from rat myocardial cells is known to have different characteristics from that of other mammals. The membrane action potential has essentially no plateau as in other mammals. Further, it has been demonstrated that stretch-induced excitation is different from that of other mammals, behaving in a graded and asynchronous manner (Spear and Moore, 1972). Stretch appears to induce a partial conduction block in isolated rat papillary muscle, producing delayed activation, decremental conduction, subthreshold electrotonic potentials, and other electrical and mechanical abnormalities not found in papillary muscle preparations from other species.

C. Pharmacological Differences

It is well known that the rat is very resistant to digitalis glycosides, being many times more resistant to its effects than is the dog or man. Detweiler (1967) has reviewed the comparative pharmacology of the cardiac glycosides and notes a variety of differences between the rat and other species in the handling of digitalis glycosides. The rat, for example, has a very high level of $Na^+ K^+$-ATPase compared with other species, which has been postulated as a major reason for the species difference in glycoside action (Repke and Portius, 1963).

X. CONCLUSIONS

The rat has been extensively used for a large variety of cardiovascular studies despite a number of peculiarities of this species compared with other mammals. In selecting an animal species for use in a particular research problem, the investigator must carefully weigh the advantages and disadvantages of the species for the particular type of research. The economic advantages or availability of background data for the species must be weighed against the possible disadvantages of differences in anatomy, function, or metabolism compared with other species.

The rat has been particularly useful for studies of myocardial structure and metabolism. The response of the myocardial cell to a variety of physiologic and pathologic stimuli has been studied in both *in vivo* and *in vitro* situations and thus provided basic information on the response of the myocardial cell to such stimuli. Another particularly promising area of research using the rat is the study of hypertension, an area that has been greatly facilitated by the development of the SHR and its various substrains. Major areas in which the rat has been less useful include electrophysiology of the myocardium and atherosclerosis. Selection of the rat as a species for cardiovascular research, therefore, must be based on consideration of the peculiarities of this species.

REFERENCES

Adams, M.A., Bobik, A., and Korner, P.I. (1989). Differential development of vascular and cardiac hypertrophy in genetic hypertension: relation to sympathetic function. *Hypertension* **14**, 191–202.

Allard, M.F., Emanuel, P.G., Russell, J.A., Bishop, S.P., Digerness, S.B., and Anderson, P.G. (1994). Preischemic glycogen reduction or glycolytic inhibition improves postischemic recovery of hypertrophied rat hearts. *Am. J. Physiol.* **267**, H66–H74.

Alonso-Galicia, M., Brands, M.W., Zappe, D.H., and Hall, J.E. (1996). Hypertension in obese Zucker rats: role of angiotensin II and adrenergic activity. *Hypertension* **28**, 1047–1054.

Ambler, G.R., Johnston, B.M., Maxwell, L., Gavin, J.B., and Gluckman, P.D. (1993). Improvement of doxorubicin induced cardiomyopathy in rats treated with insulin-like growth factor I. *Cardiovasc Res.* **27**, 1368–1373.

Anderson, N.H., Devlin, A.M., Graham, D., Morton, J.J., Hamilton, C.A., Reid, J.L., Schork, N.J., and Dominiczak, A.F. (1999). Telemetry for cardiovascular monitoring in a pharmacological study: new approaches to data analysis. *Hypertension* **33**, 248–255.

Anderson, P.G., Allard, M.F., Thomas, G.D., Bishop, S.P., and Digerness, S.B. (1990a). Increased ischemic injury but decreased hypoxic injury in hypertrophied rat hearts. *Circ. Res.* **67**, 948–959.

Anderson, P.G., Bishop, S.P., and Digerness, S.B. (1987). Transmural progression of morphologic changes during ischemic contracture and reperfusion in the normal and hypertrophied rat heart. *Am. J. Pathol.* **129**, 152–167.

Anderson, P.G., Bishop, S.P., and Digerness, S.B. (1988). Coronary vascular function and morphology in hydralazine treated DOCA salt rats. *J. Mol. Cell. Cardiol.* **20**, 955–967.

Anderson, P.G., Bishop, S.P., and Digerness, S.B. (1989). Vascular remodeling and improvement of coronary reserve after hydralazine treatment in spontaneously hypertensive rats. *Circ. Res.* **64**, 1127–1136.

Anderson, P.G., Digerness, S.B., Sklar, J.L., and Boor, P.J. (1990b). Use of the isolated perfused heart for evaluation of cardiac toxicity. *Toxicol. Pathol.* **18**, 497–510.

Anderson, P.G., Ruben, Z., and Wagner, B.M. (2001). Pathobiology of the vascular response to injury. *In* "Cardiovasuclar Toxicology," 3rd Ed. (D. Acosta, Jr., ed.), pp 525–5556, Taylor and Francis, New York.

Angelakos, E.T., Bernardini, P., and Barrett, Jr., W.C. (1964). Myocardial fiber size and capillary fiber ratio in the right and left ventricles of the rat. *Anat. Rec.* **149**, 671–676.

Angrist, A.A., Oka, M., Nakao, K., and Marquiss, J. (1960). Studies in experimental endocarditis, I: production of valvular lesions by mechanisms not involving infection or sensitivity factors. *Am. J. Pathol.* **36**, 181–200.

Anversa, P., Loud, A.V., and Vitali-Mazza, L. (1976). Morphometry and autoradiography of early hypertrophic changes in the ventricular myocardium of adult rat: an electron microscopic study. *Lab Invest* **35**, 475–548.

Apstein, C.S., Deckelbaum, L., Hagopian, L., and Hood, Jr., W.B. (1978). Acute cardiac ischemia and reperfusion: contractility, relaxation, and glycolysis. *Am. J. Physiol.* **235**, H637–H648.

Averill, D.B., Matsumura, K., Ganten, D., and Ferrario, C.M. (1996). Role of area postrema in transgene hypertension. *Hypertension* **27**, 591–597.

Awasthi, S. and Boor, P.J. (1993). Semicarbazide protection from *in vivo* oxidant injury of vascular tissue by allylamine. *Toxicol. Lett.* **66**, 157–163.

Bajusz, E. and Jasmin, G. (1964). Histochemical studies on the myocardium following experimental interference with coronary circulation in the rat, I: occlusion of coronary artery. *Acta Histochem.* **18**, 222–237.

Balnave, C.D. and Vaughan-Jones, R.D. (2000). Effect of intracellular pH on spontaneous Ca^{2+} sparks in rat ventricular myocytes. *J. Physiol.* **528**, 25–37.

Ban, Y., Nakatsuka, T., and Matsumoto, H. (1996). Effects of calcium channel blockers on cultured rat embryos. *J. Appl. Toxicol.* **16**, 147–151.

Barrow, M.V., Simpson, C.F., and Miller, E.J. (1974). Lathyrism: a review. *Q. Rev. Biol.* **49**, 101–128.

Bartunek, J., Weinberg, E.O., Tajima, M., Rohrbach, S., Katz, S.E., Douglas, P.S., and Lorell, B.H. (2000). Chronic N^G-nitro-L-arginine methyl ester-induced hypertension: novel molecular adaptation to systolic load in absence of hypertrophy. *Circulation* **101**, 423–429.

Bedotto, J.B., Gay, R.G., Graham, S.D., Morkin, E., and Goldman, S. (1989). Cardiac hypertrophy induced by thyroid hormone is independent of loading conditions and beta adrenoceptor blockade. *J. Pharmacol. Exp. Ther.* **248**, 632–636.

Ben-Ishay, D., Saliternik, R., and Welner, A. (1972). Separation of two strains of rats with inbred dissimilar sensitivity to DOCA-salt hypertension. *Experientia* **28**, 1321–1322.

Bernatova, I., Pechanova, O., Pelouch, V., and Simko, F. (2000). Regression of chronic L-NAME-treatment–induced left ventricular hypertrophy: effect of captopril. *J. Mol. Cell. Cardiol.* **32**, 177–185.

Berry, M.N., Friend, D.S., and Scheuer, J. (1970). Morphology and metabolism of intact muscle cells isolated from adult rat hearts. *Circ. Res.* **26**, 679–687.

Beznak, M. (1955). The effect of different degrees of subdiaphragmatic aortic constriction on heart weight and blood pressure of normal and hypophysectomized rats. *Can. J. Biochem. Physiol.* **33**, 985–994.

Beznak, M. (1962). Cardiovascular effects of thyroxine treatment in normalrats. *Can. J. Biochem.* **40**, 1647–1654.

Beznak, M. (1952). The effect of the adrenals and the pituitary on blood pressure and cardiac hypertrophy of rats. *J. Physiol. (London)* **116**, 219–227.

Beznak, M., Korecky, B., and Thomas, G. (1969). Regression of cardiac hypertrophies of various origin. *Can. J. Physiol. Pharmacol.* **47**, 579–586.

Bianchi, C., Fox, U., and Impasciati, E. (1974). The development of a new strain of spontaneously hypertensive rats. *Life Sci.* **14**, 339–347.

Bing. O.H.L., Sen, S., Conrad, C.H., and Brooks, W.W. (1984). Myocardial function, structure and collagen in the spontaneously hypertensive rat: progression from compensated hypertrophy to haemodynamic impairment. *Eur. Heart. J.* **5**, 43–52.

Bischoff, M.B., Dean, W.D., Bucci, T.J., and Fries, L.A. (1969). Ultrastructural changes in myocardium of animals after five months residence at 14,100 feet. *Fed. Proc. Fed. Am. Soc. Exp. Biol.* **28**, 1268–1673.

Bishop, S.P., Anderson, P.G., and Tucker, D.C. (1990). Morphological development of the rat heart growing in oculo in the absence of hemodynamic work load. *Circ. Res.* **66**, 84–102.

Bishop, S.P., Dillon, D., Naftilan, J., and Reynolds, R. (1980). Surface morphology of isolated cardiac myocytes from hypertrophied hearts of aging SHR. *Scan. Electron. Microsc.* **2**, 193–199.

Bishop, S.P. and Drummond, J.L. (1979). Surface morphology and cell size measurement of isolated rat cardiac myocytes. *J. Mol. Cell. Cardiol.* **11**, 423–433.

Bishop, S.P., Oparil, S., Reynolds, R.H., and Drummond, J.L. (1979). Regional myocyte size in normotensive and spontaneously hypertensive rats. *Hypertension* **1**, 378–383.

Bloor, C., Leon, A.S., and Pitt, B. (1967). The inheritance of coronary artery anatomic patterns in rats. *Circulation* **36**, 771–776.

Bohlender, J., Fukamizu, A., Lippoldt, A., Nomura, T., Dietz, R., Menard, J., Murakami, K., Luft, F.C., and Ganten, D. (1997). High human renin hypertension in transgenic rats. *Hypertension* **29**, 428–434.

Bohlender, J., Ganten, D., and Luft, F.C. (2000). Rats transgenic for human renin and human angiotensinogen as a model for gestational hypertension. *J. Am. Soc. Nephrol.* **11**, 2056–2061.

Bohlender, J., Menard, J., Wagner, J., Luft, F.C., and Ganten, D. (1996). Human renin-dependent hypertension in rats transgenic for human angiotensinogen. *Hypertension* **27**, 535–540.

Boluyt, M.O., Bing, O.H., and Lakatta, E.G. (1995). The ageing spontaneously hypertensive rat as a model of the transition from stable compensated hypertrophy to heart failure. *Eur. Heart J.* **16(Suppl N)**, 19–30.

Bompiani, G.D., Rouiller, C., and Hatt, P.Y. (1959). Le tissu de conduction de coeur chez le rat. Etude au microscope electronique, I: le tranc commun du faisceau de His et les cellules claires de l'oreillette droite. *Arch. Ma lCoeur Vaiss.* **52**, 1257–1274.

Bonnardeaux, A., Nadaud, S., Charru, A., Jeunemaitre, X., Corvol, P., and Soubrier, F. (1995). Lack of evidence for linkage of the endothelial cell nitric oxide synthase gene to essential hypertension. *Circulation* **91**, 96–102.

Boor, P.J. and Hysmith, R.M. (1987). Allylamine cardiovascular toxicity. *Toxicology* **44**, 129–145.

Boor, P.J., Moslen, M.T., and Reynolds, E.S. (1979). Allylamine cardiotoxicity, I: sequence of pathologic events. *Toxicol. App. Pharmacol.* **50**, 581–592.

Brenner, B.M., Ballermann, B.J., Gunning, M.E., and Zeidel, M.L. (1990). Diverse biological actions of atrial natriuretic peptide. *Physiol. Rev.* **70**, 665–699.

Brookes, P.S., Zhang, J., Dai, L., Zhou, F., Parks, D.A., Darley-Usmar, V.M., and Anderson, P.G. (2001). Increased sensitivity of mitochondrial respiration to inhibition by nitric oxide in cardiac hypertrophy. *J. Mol. Cell. Cardiol.* **33**, 69–82.

Brown, J.W., Cristian, D., and Paradise, R.R. (1968). Histological effects of procedural and environmental factors on isolated rat heart preparations. *Proc. Soc. Exp. Biol. Med.* **129**, 455–462.

Bryant, R.E., Thomas, W.A., and O'Neal, R.M. (1958). An electron micrscopic study of myocardial ischema in the rat. *Circ. Res.* **6**, 699–709.

Bugaisky, L.B. (1998). Differentiation in culture of adult and neonatal cardiac myocytes. *Circulation* **78**, 179.

Bugaisky, L.B., Siegel, E., and Whalen, R.G. (1983). Myosin isozyme changes in the heart following constriction of the ascending aorta of a 25-day old rat. *FEBS Lett.* **161**, 230–234.

Bugaisky, L.B. and Zak, R. (1985). Adult rat cardiac myocytes in culture. *J. Cell. Biol.* **101**, 39a.

Butler, K.D., Ambler, J., Dolan, S., Giddings, J., Talbot, M.D., and Wallis, R.B. (1992). A non-occlusive model of arterial thrombus formation in the rat and its modification by inhibitors of platelet function, or thrombin activity. *Blood. Coag. Fibrinol.* **3**, 155–165.

Campbell, S.E., Turek, Z., Rakusan, K., and Kazda, S. (1993). Cardiac structural remodelling after treatment of spontaneously hypertensive rats with nifedipine or nisoldipine. *Cardiovasc. Res.* **27**, 1350–1358.

Capasso, J.M., Li, P., Guideri, G., and Anversa, P. (1991). Left ventricular dysfunction induced by chronic alcohol ingestion in rats. *Am. J. Physiol.* **261**, H212–H219.

Carlson, S.H., Shelton, J., White, C.R., and Wyss, J.M. (2000). Elevated sympathetic activity contributes to hypertension and salt sensitivity in diabetic obese Zucker rats. *Hypertension* **35**, 403–408.

Chen, L.E., Seaber, A.V., and Urbaniak, J.R. (1996). Thrombosis and thrombolysis in crushed arteries with or without anastomosis: a new microvascular thrombosis model. *J. Reconst. Microsurg.* **12**, 31–38.

Chen, S.J., Chen, Y.F., Opgenorth, T.J., Wessale, J.L., Meng, Q.C., Durand, J., DiCarlo, V.S., and Oparil, S. (1997). The orally active nonpeptide endothelin A–receptor antagonist A-127722 prevents and reverses hypoxia-induced pulmonary hypertension and pulmonary vascular remodeling in Sprague-Dawley rats. *J. Cardiovasc. Pharmacol.* **29**, 713–725.

Chesney, C.F., Allen, J.R., and Hsu, I.C. (1974). Right ventricular hypertrophy in monocrotaline pyrrole treated rats. *Exp. Mol. Pathol.* **20**, 257–268.

Chinellato, A., Ragazzi, E., Petrelli, L., Paro, M., Mironov, A., and Aliev, G. (1994). Effect of cholesterol-supplemented diet in heritable hyperlipidemic Yoshida rats: functional and morphological characterization of thoracic aorta. *Atherosclerosis* **106**, 51–63.

Clawson, B.J. (1950). Experimental endocarditis with fibrinoid degeneration in the heart valves of rats. *Arch. Pathol.* **50**, 68–74.

Clowes, A.W., Reidy, M.A., and Clowes, M.M. (1983). Mechanisms of stenosis after arterial injury. *Lab. Invest.* **49**, 208–215.

Clubb, Jr., F.J., Bell, P.D., Kriseman, J.D., and Bishop, S.P. (1987). Myocardial cell growth and blood pressure development in neonatal spontaneously hypertensive rats. *Lab. Invest.* **56**, 189–197.

Clubb, Jr., F.J. and Bishop, S.P. (1984). Formation of binucleated myocardial cells in the neonatal rat: an index for growth hypertrophy. *Lab. Invest.* **50**, 571–577.

Clubb, Jr., F.J., Penney, D.G., and Bishop, S.P. (1989). Cardiomegaly in neonatal rats exposed to 500 ppm carbon monoxide. *J. Mol. Cell. Cardiol.* **21**, 945–956.

Cohen, A.M., Rosenmann, E., and Rosenthal, T. (1993). The Cohen diabetic (non-insulin-dependent) hypertensive rat model: description of the model and pathologic findings. *Am. J. Hyperten.* **6**, 989–995.

Conklin, D.J. and Boor, P.J. (1998). Allylamine cardiovascular toxicity: evidence for aberrant vasoreactivity in rats. *Toxicol. Appl. Pharmacol.* **148**, 245–251.

Conklin, D.J., Boyce, C.L., Trent, M.B., and Boor, P.J. (2001). Amine metabolism: a novel path to coronary artery vasospasm. *Toxicol. Appl. Pharmacol.* **175**, 149–159.

Cooley, B.C., Li, X., Dzwierzynski, W., Gruel, S.M., Hall, R.L., Wright, R.R., O'Brien, E.M., Fagan, D., Hanel, D.P., and Gould, J.S. (1992). The de-endothelialized rat carotid arterial graft: a versatile experimental model for the investigation of arterial thrombosis. *Thromb. Res.* **67**, 1–14.

Crowley, J.J., Huang, C.L., Gates, A.R., Basu, A., Shapiro, L.M., Carpenter, T.A., and Hall, L.D. (1997). A quantitative description of dynamic left ventricular geometry in anaesthetized rats using magnetic resonance imaging. *Exp. Physiol.* **82**, 887–904.

Cutilletta, A.F., Benjamin, M., Culpepper, W.S., and Oparil, S. (1978). Myocardial hypertrophy and left ventricular performance in the absence of hypertension in spontaneously hypertensive rats. *J. Mol. Cell. Cardiol.* **10,** 689–703.

Cutilletta, A.F., Erinoff, L., Heller, A., Low, J., and Oparil, S. (1977). Development of left ventricular hypertrophy in young spontaneously hypertensive rats after peripheral sympathectomy. *Circ. Res.* **40,** 428–434.

Dahl, L.K., Heine, M., and Tassinari, L. (1962a). Effects of chronic excess salt ingestion: evidence that genetic factors play an important role in susceptibility to experimental hypertension. *J. Exp. Med.* **115,** 1173–1190.

Dahl, L.K., Heine, M., and Tassinari, L. (1962b). Role of genetic factors in susceptibility to experimental hypertension due to chronic excess salt ingestion. *Nature (London)* **194,** 480–482.

Dai, L., Brookes, P.S., Darley-Usmar, V.M., and Anderson, P.G. (2001). Bioenergetics in cardiac hypertrophy: mitochondrial respiration as a pathological target of NO*. *Am. J. Physiol. Heart Circ. Physiol.* **281,** H2261–H2269.

D'Almeida, M.S., Cailmail, S., and Lebrec, D. (1996). Validation of transit-time ultrasound flow probes to directly measure portal blood flow in conscious rats. *Am. J. Physiol.* **271,** H2701–H2709.

Dang, A., Zheng, D., Wang, B., Zhang, Y., Zhang, P., Xu, M., Liu, G., and Liu, L. (1999). The role of the renin-angiotensin and cardiac sympathetic nervous systems in the development of hypertension and left ventricular hypertrophy in spontaneously hypertensive rats. *Hypertens. Res. Clin. Exp.* **22,** 217–221.

DeBold, A.J. (1985). Atrial natriuretic factor: a hormone produced by the heart. *Science* **230,** 767–770.

Decker, M.L., Behnke-Barclay, M., Cook, M.G., LaPres, J.J., Clark, W.A., and Decker, R.S. (1991). Cell shape and organization of the contractile apparatus in cultured adult cardiac myocytes. *J. Mol. Cell. Cardiol.* **23,** 817–832.

de Clerck, Y.A. and Jones, P.A. (1980). The effect of ascorbic acid on the nature and production of collagen and elastin by rat smooth-muscle cells. *Biochem. J.* **186,** 217–225.

Deng, A.Y. and Rapp, J.P. (1995). Locus for the inducible, but not a constitutive, nitric oxide synthase cosegregates with blood pressure in the Dahl salt-sensitive rat. *J. Clin. Invest.* **95,** 2170–2177.

Desmond, W.J. and Harary, I. (1972). *In vitro* studies of beating heart cells in culture. *Arch. Biochem. Biophys.* **151,** 285–294.

Detweiler, D.K. (1967). Comparative pharmacology of cardiac glycosides. *Fed. Proc. Fed. Am. Soc. Exp. Biol.* **26,** 1119–1124.

Dobesova, Z., Zicha, J., Heller, J., and Kunes, J. (1991). The lack of cardiac hypertrophy in newborn Prague hypertensive rats. *Physiol. Res.* **40,** 373–376.

Doggrell, S.A. and Brown, L. (1998). Rat models of hypertension, cardiac hypertrophy and failure. *Cardiovasc. Res.* **39,** 89–105.

Douglas, P.S., Katz, S.E., Weinberg, E.O., Chen, M.H., Bishop, S.P., and Lorell, B.H. (1998). Hypertrophic remodeling: gender differences in the early response to left ventricular pressure overload. *J. Am. Coll. Cardiol.* **32,** 1118–1125.

Dupont, J., Dupont, J.C., Froment, A., Milon, H., and Vincent, M. (1973). Selection of three strains with spontaneously different levels of blood pressure. *Biomedicine* **19,** 36–41.

Duvelleroy, M.A., Duruble, M., Martin, J.L., Teisseire, B., Droulez, J., and Cain, M. (1976). Blood-perfused working isolated rat heart. *J. Appl. Physiol.* 41, 603–607.

Engelmann, G.L., Vitullo, J.C., and Gerrity, R.G. (1987). Morphometric analysis of cardiac hypertrophy during development, maturation, and senescence in spontaneously hypertensive rats. *Circ. Res.* **60,** 487–494.

Fingerle, J., Au, Y.P., Clowes, A.W., and Reidy, M.A. (1990). Intimal lesion formation in rat carotid arteries after endothelial denudation in absence of medial injury. *Arteriosclerosis* **10,** 1082–1087.

Fishman, J.A., Ryan, G.B., and Karnovsky, M.J. (1975). Endothelial regeneration in the rat carotid artery and the significance of endothelial denudation in the pathogenesis of myointimal thickening. *Lab. Invest.* **32,** 339–351.

Flaim, S.F., Nellis, S.H., Toggart, E.J., Drexler, H., Kanda, K., and Newman, E.D. (1984). Multiple simultaneous determinations of hemodynamics and flow distribution in conscious rat. *J. Pharmacol. Methods* **11,** 1–39.

Frank, J.S., Brady, A.J., Farnsworth, S., and Mottino, G. (1986). Ultrastructure and function of isolated myocytes after calcium depletion and repletion. *Am. J. Physiol.* **250,** H265–H275.

Fregly, M.J. (1963). Factors affecting indirect determination of systolic blood pressure of rats. *J. Lab. Clin. Med.* **62,** 223–230.

Freund, M., Mantz, F., Nicolini, P., Gachet, C., Mulvihill, J., Meyer, L., Beretz, A., and Cazenave, J.P. (1993). Experimental thrombosis on a collagen coated arterioarterial shunt in rats: a pharmacological model to study antithrombotic agents inhibiting thrombin formation and platelet deposition. *Thromb. Haemost.* **69,** 515–521.

Frey, N., Franz, W.M., Gloeckner, K., Degenhardt, M., Muller, M., Muller, O., Merz, H., and Katus, H.A. (2000). Transgenic rat hearts expressing a human cardiac troponin T deletion reveal diastolic dysfunction and ventricular arrhythmias. *Cardiovasc. Res.* **47,** 254–264.

Ganten, D., Schmidt, S., and Paul. M. (1994). Genetics of primary hypertension. *J. Cardiovasc. Pharmacol.* **24,** S45–S50.

Ganten, D., Wagner, J., Zeh, K., Bader, M., Michel, J.B., Paul, M., Zimmermann, F., Ruf, P., Hilgenfeldt, U., and Ganten, U. (1992). Species specificity of renin kinetics in transgenic rats harboring the human renin and angiotensinogen genes. *Proc. Natl. Acad. Sci. USA* **89,** 7806–7810.

Gerdes, A.M., Moore, J.A., Hines, J.M., Kirkland, P.A., and Bishop, S.P. (1986). Regional differences in myocyte size in normal rat heart. *Anat. Rec.* **215,** 420–426.

Gerdes, A.M., Onodera, T., Wang, X., and McCune, S.A. (1996). Myocyte remodeling during the progression to failure in rats with hypertension. *Hypertension* **28,** 609–614.

Giedrojc, J., Klimiuk, M., Radziwon, P., Kloczko, J., Bielawiec, M., and Breddin, H.K. (1999). Comparative study on the *in vitro* and *in vivo* activities of heparinoids derivative investigated on the animal model. *J. Cardiovasc. Pharmacol.* **34,** 340–345.

Goldblatt, H., Lynch, J., Hanzal, R.F., and Summerville, W.W. (1934). Studies on experimental hypertension, I: the production of persistant elevation of systolic blood pressure by means of renal ischemia. *J. Exp. Med.* **59,** 347–381.

Goldman, S. and Raya, T.E. (1995). Rat infarct model of myocardial infarction and heart failure. *J. Cardiol. Fail.* **1,** 169–177.

Gordon, A.L., Inchiosa, Jr., M.A., and Lehr, D. (1969). Myocardial actomyosin and total protein in isoproterenol-induced hypertrophy. *Bull. NY Acad. Med.* 2: 98.

Greene, E.C. (1959). "Anatomy of the Rat." Hafner, New York.

Grice, H.C., Goodman, T., Munro, I.C., Wiberg, G.S., and Morrison, A.B. (1969). Myocardial toxicity of cobalt in the rat. *Ann. NY Acad. Sci.* **156,** 189–194.

Grimm, A.F., Katele, K.V., Klein, S.A., and Lin, H.L. (1973). Growth of the rat heart. Left ventricular morphology and sarcomere lengths. *Growth* **37,** 189–206.

Grimm, A.F., Kubota, R., and Whitehorn, W.V. (1963). Properties of myocardium in cardiomegaly. *Circ. Res.* **12,** 118–124.

Grove, D., Nair, K.G., and Zak, R. (1969). Biochemical correlates of cardiac hypertrophy, III: changes in DNA content; the relative contributions of polyploidy and mitotic activity. *Circ. Res.* **25,** 463–471.

Guarini, S. (1996). A highly reproducible model of arterial thrombosis in rats. *J. Pharmacol. Toxicol. Methods* **35,** 101–105.

Guzman, R.J., Loquvam, G.S., Kodama, J.K., and Hine, C.H. (1961). Myocarditis produced by allylamines. *Arch. Environ. Health* **2,** 62–73.

Haas, G.J., McCune, S.A., Brown, D.M., and Cody, R.J. (1995). Echocardiographic characterization of left ventricular adaptation in a genetically determined heart failure rat model. *Am. Heart J.* **130**, 806–811.

Hakamata, Y., Tahara, K., Uchida, H., Sakuma, Y., Nakamura, M., Kume, A., Murakami, T., Takahashi, M., Takahashi, R., Hirabayashi, M., Ueda, M., Miyoshi, I., Kasai, N., and Kobayashi, E. (2001). Green fluorescent protein-transgenic rat: a tool for organ transplantation research. *Biochem. Biophys. Res. Comm.* **286**, 779–785.

Hall, C.E., Ayachi, S., and Hall, O. (1973). Spontaneous hypertension in rats with hereditary hypothalmic diabetes insipidus (Brattleboro strain). *Tex. Rep. Biol. Med.* **31**, 471–487.

Hall, C.E. and Hall, O. (1965). Hypertension and hypersalimentation, II: deoxycorticosterone hypertension. *Lab. Invest.* **14**, 1727–1735.

Halpern, M.H. (1953). Extracoronary cardiac veins in the rat. *Am. J. Anat.* **92**, 307–328.

Halpern, M.H. (1955). The sino-atrial node of the rat heart. *Anat. Rec.* **123**, 425–436.

Halpern, M.H. (1957). The dual blood supply of the rat heart. *Am. J. Anat.* **101**, 1–16.

Hammes, A., Oberdorf-Maass, S., Rother, T., Nething, K., Gollnick, F., Linz, K.W., Meyer, R., Hu, K., Han, H., Gaudron, P., Ertl, G., Hoffmann, S., Ganten, U., Vetter, R., Schuh, K., Benkwitz, C., Zimmer, H.G., and Neyses, L. (1998). Overexpression of the sarcolemmal calcium pump in the myocardium of transgenic rats. *Circ. Res.* **83**, 877–888.

Han, H., Hoffmann, S., Hu, K., and Ertl, G. (2002). Angiotensin II subtype 1 (AT₁) receptors contribute to ischemic contracture and regulate chemomechanical energy transduction in isolated transgenic rat (αMHC-hAT1)594-17 hearts. *Eur. J. Heart Fail.* **4**, 131–137.

Harary, I. and Farley, B. (1963). *In vitro* studies on single beating rat heart cells, I: growth and organization. *Exp. Cell. Res.* **29**, 451–465.

Harrison, S.N., Autelitano, D.J., Wang, B.H., Milano, C., Du, X.J., and Woodcock, E.A. (1998). Reduced reperfusion-induced Ins(1,4,5)P3 generation and arrhythmias in hearts expressing constitutively active α₁B-adrenergic receptors. *Circ. Res.* **83**, 1232–1240.

He, H., Kimura, S., Fujisawa, Y., Tomohiro, A., Kiyomoto, K., Aki, Y., and Abe, Y. (1997). Dietary L-arginine supplementation normalizes regional blood flow in Dahl-Iwai salt-sensitive rats. *Am. J. Hyperten.* **10**, 89S–93S.

Heller, J., Hellerova, S., Dobesova, Z., Kunes, J., and Zicha, J. (1993). The Prague hypertensive rat: a new model of genetic hypertension. *Clin. Exp. Hypertens.* **15**, 807–818.

Henke, P.K., DeBrunye, L.A., Strieter, R.M., Bromberg, J.S., Prince, M., Kadell, A.M., Sarkar, M., Londy, F., and Wakefield, T.W. (2000). Viral IL-10 gene transfer decreases inflammation and cell adhesion molecule expression in a rat model of venous thrombosis. *J. Immunol.* **164**, 2131–2141.

Hepp, A., Rudolph, T., and Kochsiek, K. (1984). Is the rat a suitable model for studying alcoholic cardiomyopathy? Hemodynamic studies at various stages of chronic alcohol ingestion. *Basic Res. Cardiol.* **79**, 230–237.

Heran, C., Morgan, S., Kasiewski, C., Bostwick, J., Bentley, R., Klein, S., Chu, V., Brown, K., Colussi, D., Czekaj, M., Perrone, M., and Leadley, Jr., R. (2000). Antithrombotic efficacy of RPR208566, a novel factor Xa inhibitor, in a rat model of carotid artery thrombosis. *Eur. J. Pharmacol.* **389**, 201–207.

Herbert, J.M., Bernat, A., Dol, F., Herault, J.P., Crepon, B., and Lormeau, J.C. (1996). DX 9065A a novel, synthetic, selective and orally active inhibitor of factor Xa: *in vitro* and *in vivo* studies. *J. Pharmacol. Exp. Ther.* **276**, 1030–1038.

Herrera, V.L., Makrides, S.C., Xie, H.X., Adari, H., Krauss, R.M., Ryan, U.S., and Ruiz-Opazo, N. (1999). Spontaneous combined hyperlipidemia, coronary heart disease and decreased survival in Dahl salt-sensitive hypertensive rats transgenic for human cholesteryl ester transfer protein. *Nat. Med.* **5**, 1383–1389.

Hibbs, R.G. and Ferrans, V.J. (1969). An ultrastructural and histochemical study of rat myocardium. *Am. J. Anat.* **124**, 251–280.

Higo, H., Higo, K., Satow, Y., and Lee, J.Y. (1987). Induction of an HSP70-like protein during organ culture of rat embryonic heart. *Biochem. Int.* **15**, 727–733.

Hiller, K.H., Adami, P., Voll, S., Roder, F., Kowallik, P., Bauer, W.R., Haase, A., and Ertl, G. (1996). *In vivo* colored microspheres in the isolated rat heart for use in NMR. *J. Mol. Cell. Cardiol.* **28**, 571–577.

Hirata, Y., Umemura, K., Uematsu, T., and Nakashima, M. (1995). An experimental myocardial infarction model in the rat and its properties. *Jpn. J. Pharmacol.* **67**, 51–57.

Hirigoyen, M.B., Zhang, W.X., and Weinberg H. (1995). A standardized model of microvenous thrombosis. *J. Reconstruc. Microsurg.* **11**, 455–459.

Hladovec, J. (1986). A new model of arterial thrombosis. *Thromb. Res.* **41**, 659–664.

Hoffmann, S., Krause, T., van Geel, P.P., Willenbrock, R., Pagel, I., Pinto, Y.M., Buikema, H., van Gilst, W.H., Lindschau, C., Paul, M., Inagami, T., Ganten, D., and Urata, H. (2001). Overexpression of the human angiotensin II type 1 receptor in the rat heart augments load induced cardiac hypertrophy. *J. Mol. Med.* **79**, 601–608.

Holycross, B.J., Summers, B.M., Dunn, R.B., and McCune, S.A. (1997). Plasma renin activity in heart failure-prone SHHF/Mcc-facp rats. *Am. J. Physiol.* **273**, H228–H233.

Hosoda, Y. and Iri, H. (1966). Angiolathyrism, I; a histological and histochemical study on successive changes of the lathyritic rat aorta. *Acta Pathol. Jpn.* **16**, 239–252.

Hupkens, P. and Cooley, B.C. (1996). Comparison of arterial and venous patency in a rat model of subendothelium-stimulated thrombosis. *Microsurgery* **17**, 226–229.

Ikeda, H., Shino, A., Matsuo, T., Iwatsuka, H., and Suzuoki, Z. (1981). A new genetically obese-hyperglycemic rat (Wistar fatty). *Diabetes* **30**, 1045–1050.

Ikeda, Y. and Umemura, K. (2001). A characterization of myocardial infarction induced by thrombotic occlusion and a comparison with mechanical ligation of the rat coronary artery. *Methods Find. Exp. Clin. Pharmacol.* **23**, 23–28.

Inoko, M., Kihara, Y., Morii, I., Fujiwara, H., and Sasayama, S. (1994). Transition from compensatory hypertrophy to dilated, failing left ventricles in Dahl salt-sensitive rats. *Am. J. Physiol.* **267**, H2471–H2482.

Ishikawa, S., Honda, M., Yamada, S., Goto, Y., Kuzuo, H., Morioka, S., and Moriyama, K. (1991). Changes in contractile and noncontractile protein metabolisms in both ventricles in monocrotaline-treated rats. *J. Cardiovasc. Pharmacol.* **17(Supp 2)**, S119–S121.

James, T.N. (2001). The internodal pathways of the human heart. *Prog. Cardiovasc. Dis.* **43**, 495–535.

James, T.N. and Sherf, L. (1971). Specialized tissues and preferential conduction in the atria of the heart. *Am. J. Cardiol.* **28**, 414–427.

Jamieson, J.D. and Palade, G.E. (1964). Specific granules in atrial muscle cells. *J. Cell. Biol.* **23**, 151–172

Jones, P.A., Scott-Burden, T., and Gevers, W. (1979). Glycoprotein, elastin, and collagen secretion by rat smooth muscle cells. *Proc. Natl. Acad. Sci. USA* **76**, 353–357.

Joris, I., Zand, T., Nunnari, J.J., Krolikowski, F.J., and Majno, G. (1983). Studies on the pathogenesis of atherosclerosis, I; adhesion and emigration of mononuclear cells in the aorta of hypercholesterolemic rats. *Am. J. Pathol.* **113**, 341–358.

Kato, K., Nakazawa, M., Masani, F., Izumi, T., Shibata, A., and Imai, S. (1995). Ethanol ingestion on allylamine-induced experimental subendocardial fibrosis. *Alcohol* **12**, 233–239.

Kawamura, K., Kashii, C., and Imamura, K. (1976). Ultrastructural changes in hypertrophied myocardium of spontaneously hypertensive rats. *Jpn. Circ. J.* **40**, 1119–1145.

Kawano, K., Hirashima, T., Mori, S., Saitoh, Y., Kurosumi, M., and Natori, T. (1992). Spontaneous long-term hyperglycemic rat with diabetic complications Otsuka Long-Evans Tokushima fatty (OLETF) strain. *Diabetes* **41**, 1422–1428.

King, T.S. (1954). Sinuatrial node in the rat. *Am. Heart J.* **48**, 785–786.

Kino, M., Thorp, K.A., Bing, O.H., and Abelmann, W.H. (1981). Impaired myocardial performance and response to calcium in experimental alcoholic cardiomyopathy. *J. Mol. Cell. Cardiol.* **13**, 981–989.

Knudsen, K., Dahl, L.K., Thompson, K., Iwai, J., Heine, M., and Leitl, G. (1970). Effects of chronic salt ingestion: inheritance of hypertension in the rat. *J. Exp. Med.* **132**, 967–1000.

Koletsky, S. (1973). Obese spontaneously hypertensive rats: a model for study of aterosclerosis. *Exp. Mol. Pathol.* **19**, 53–60.

Korecky, B. and French, I.W. (1967). Nucleic acid synthesis in enlarged hearts of rats with nutritional anemia. *Circ. Res.* **21**, 635.

Korecky, B., Rakusan, K., and Poupa, O. (1964). The effect of anemia due to iron deficiency during early postnatal development of the rat on growth and body composition later in life. *Physiol. Bohemoslov.* **13**, 72–77

Kuijpers, M.H. and Gruys, E. (1984). Spontaneous hypertension and hypertensive renal disease in the fawn-hooded rat. *Br. J. Exp. Pathol.* **65**, 181–190.

Kurtz, T.W., Montano, M., Chan, L., and Kabra, P. (1989a). Molecular evidence of genetic heterogeneity in Wistar-Kyoto rats: implications for research with the spontaneously hypertensive rat. *Hypertension* **13**, 188–192.

Kurtz, T.W. and Morris, Jr., R.C. (1987). Biological variability in Wistar-Kyoto rats. Implications for research with the spontaneously hypertensive rat. *Hypertension* **10**, 127–131.

Kurtz, T.W., Morris, R.C., and Pershadsingh, H.A. (1989b). The Zucker fatty rat as a genetic model of obesity and hypertension. *Hypertension* **13**, 896–901.

Kurz, K.D., Main, B.W., and Sandusky, G.E. (1990). Rat model of arterial thrombosis induced by ferric chloride. *Thromb. Res.* **60**, 269–280.

Laffan, R.J., Peterson, A., Hitch, S.W., and Jeunelot, C. (1972). A technique for prolonged, continuous recording of blood pressure of unrestrained rats. *Cardiovasc. Res.* **6**, 319–324.

Lalich, J.J., Allen, J.R., and Paik, W.C. (1972). Myocardial fibrosis and smooth muscle cell hyperplasia in coronary arteries of allylamine-fed rats. *Am. J. Pathol.* **66**, 225–240.

Langendorff, O. (1895). Untersuchungen am uberlebender Saugetierherzen. *Pfluegers Arch Gesamte. Physiol. Menschen Tiere* **61**, 291–332.

Langford, S.D., Trent, M.B., Balakumaran, A., and Boor, P.J. (1999). Developmental vasculotoxicity associated with inhibition of semicarbazide-sensitive amine oxidase. *Toxicol. Appl. Pharmacol.* **155**, 237–244.

Langford, S.D., Trent, M.B., and Boor, P.J. (2002). Semicarbazide-sensitive amine oxidase and extracellular matrix deposition by smooth-muscle cells. *Cardiovasc. Toxicol.* **2**, 141–150.

Latta, H., White, F.N., Osvaldo, L., and Johnston, W.H. (1975). Unilateral renovascular hypertension in rats: Measurements of medullary granules, juxtaglomerular and cellularity, and areas of adrenal zones. *Lab. Invest.* **33**, 379–390.

Lee, M.A., Bohm, M., Paul, M., Bader, M., Ganten, U., and Ganten, D. (1996). Physiological characterization of the hypertensive transgenic rat TGR(mREN2)27. *Am. J. Physiol.* **270**, E919–E929.

Lenfors, S., Marberg, L., Wikstrom, S., Jonsson, U., Eriksson, A.W., and Gustafsson, D. (1993). A new rat model of arterial thrombosis with a platelet-rich head and an erythrocyte-rich tail: thrombolysis experiments with specific thrombin inhibition. *Blood Coag. Fibrinol.* **4**, 263–271.

Leon, A.S. and Bloor, C.M. (1968). Effects of exercise and its cessation on the heart and its blood supply. *J. Appl. Physiol.* **24**, 485–490.

Li, F., Wang, X., Capasso, J.M., and Gerdes, A.M. (1996). Rapid transition of cardiac myocytes from hyperplasia to hypertrophy during postnatal development. *J. Mol. Cell. Cardiol.* **28**, 1737–1746.

Limas, C., Westrum, B., and Limas, C.J. (1984). Comparative effects of hydralazine and captopril on the cardiovascular changes in spontaneously hypertensive rats. *Am. J. Pathol.* **117**, 360–371.

Litwin, S.E., Katz, S.E., Morgan, J.P., and Douglas, P.S. (1994). Serial echocardiographic assessment of left ventricular geometry and function after large myocardial infarction in the rat. *Circulation* **89**, 345–354.

Litwin, S.E., Katz, S.E., Weinberg, E.O., Lorell, B.H., Aurigemma, G.P., and Douglas, P.S. (1995). Serial echocardiographic-Doppler assessment of left ventricular geometry and function in rats with pressure-overload hypertrophy: chronic angiotensin-converting enzyme inhibition attenuates the transition to heart failure. *Circulation* **91**, 2642–2654.

Litwin, S.E., Raya, T.E., Anderson, P.G., Litwin, C.M., Bressler, R., and Goldman, S. (1991). Induction of myocardial hypertrophy after coronary ligation in rats decreases ventricular dilatation and improves systolic function. *Circulation* **84**, 1819–1827.

Liu, Y.H., Yang, X.P., Nass, O., Sabbah, H.N., Peterson, E., and Carretero, O.A. (1997). Chronic heart failure induced by coronary artery ligation in Lewis inbred rats. *Am. J. Physiol.* **272**, H722–H727.

Lockyer, S. and Kambayashi, J. (1999). Demonstration of flow and platelet dependency in a ferric chloride-induced model of thrombosis. *J. Cardiovasc. Pharmacol.* **33**, 718–725.

Loke, K.E., McConnell, P.I., Tuzman, J.M., Shesely, E.G., Smith, C.J., Stackpole, C.J., Thompson, C.I., Kaley, G., Wolin, M.S., and Hintze, T.H. (1999). Endogenous endothelial nitric oxide synthase-derived nitric oxide is a physiological regulator of myocardial oxygen consumption. *Circ. Res.* **84**, 840–845.

Long, C.S. and Brown, R.D. (2002). The cardiac fibroblast, another therapeutic target for mending the broken heart? *J. Mol. Cell. Cardiol.* **34**, 1273–1278.

Lopaschuk, G.D., Spafford, M.A., Davies, N.J., and Wall, S.R. (1990). Glucose and palmitate oxidation in isolated working rat hearts reperfused after a period of transient global ischemia. *Circ. Res.* **66**, 546–553.

Louis, W.J. and Howes, L.G. (1990). Genealogy of the spontaneously hypertensive rat and Wistar-Kyoto rat strains: implications for studies of inherited hypertension. *J. Cardiovasc. Pharmacol.* **16(Suppl 7)**, S1–S5.

Lovenberg, W. (1987). Animal models for hypertension research. *Prog. Clin. Biol. Res.* **229**, 225–240.

Lund, D.D. and Tomanek, R.J. (1978). Myocardial morphology in spontaneously hypertensive and aortic-constricted rats. *Am. J. Anat.* **152**, 141–152.

MacLean, D., Fishbein, M.C., Maroko, P.R., and Braunwald, E. (1975). Hyaluronidase-induced reductions in myocardial infarct size. *Science* **194**, 199–200.

Mager, M., Blatt, W., Natale, P.J., and Blatteis, C.M. (1968). Effect of high altitude on lactic dehydrogenase isozymes of neonatal and adult rats. *Am. J. Physiol.* **215**, 8–13.

Maistrello, I. and Matscher, R. (1969). Measurement of systolic blood pressures of rats: comparison of intraarterial and cuff values. *J. Appl. Physiol.* **26**, 188–193.

Majesky, M.W., Giachelli, C.M., Reidy, M.A., and Schwartz, S.M. (1992). Rat carotid neointimal smooth muscle cells reexpress a developmentally regulated mRNA phenotype during repair of arterial injury. *Circ. Res.* **71**, 759–768.

Massad, L., Plotkine, M., Capdeville, C., and Boulu, R.G. (1987). Electrically induced arterial thrombosis model in the conscious rat. *Thromb. Res.* **48**, 1–10.

Matsumura, Y., Kuro, T., Kobayashi, Y., Konishi, F., Takaoka, M., Wessale, J.L., Opgenorth, T.J., Gariepy, C.E., and Yanagisawa, M. (2000). Exaggerated vascular and renal pathology in endothelin-B receptor-deficient rats with deoxycorticosterone acetate-salt hypertension. *Circulation* **102**, 2765–2773.

Matsuno, H., Uematsu, T., Nagashima, S., and Nakashima, M. (1991). Photochemically induced thrombosis model in rat femoral artery and evaluation of effects of heparin and tissue-type plasminogen activator with use of this model. *J. Pharmacol. Methods* **25**, 303–317.

McCune, S.A., Baker, P.B., and Stills, Jr., H.F. (1990a). SHHF/Mcc-cp rat: model of obesity, non-insulin dependent diabetes and congestive heart failure. *ILAR News* **32**, 23–27.

McCune, S.A., Jenkins, J.E., Stills, H.F., Park, S., Radin, M.J., Jurin, R.R., and Hamlin, R.E. (1990b). "Renal and heart function in the SHHF/Mcc-fa cp rat." Smith-Gordon, London.

McGrath, J.J. and Bullard, R.W. (1968). Altered myocardial performance in response to anoxia after high altitude exposure. *J. Appl. Physiol.* **25**, 761–764.

Melax, H. and Leeso, T.S. (1970). Fine structure of the impulse conducting system in rat heart. *Can. J. Zool.* **48**, 837–839.

Michalkiewicz, M., Michalkiewicz, T., Kreulen, D.L., and McDougall, S.J. (2001). Increased blood pressure responses in neuropeptide Y transgenic rats. *Am. J. Physiol.* **281**, R417–R426.

Misra, P., Srivastava, S.K., Singhal, S.S., Awasthi, S., Awasthi, Y.C., and Boor, P.J. (1995). Glutathione S-transferase 8-8 is localized in smooth muscle cells of rat aorta and is induced in an experimental model of atherosclerosis. *Toxicol. Appl. Pharmacol.* **133**, 27–33.

Morgan, H.E., Neely, J.R., and Kira, Y. (1984). Factors determining the utilization of glucose in isolated rat hearts. *Basic Res. Cardiol.* **79**, 292–299.

Morishima, Y., Tanabe, K., Terada, Y., Hara, T., and Kunitada, S. (1997). Antithrombotic and hemorrhagic effects of DX-9065a, a direct and selective factor Xa inhibitor: comparison with a direct thrombin inhibitor and antithrombin III–dependent anticoagulants. *Thromb. Haemost.* **78**, 1366–1371.

Morkin, E. and Ashford, T.P. (1968). Myocardial DNA synthesis in experimental cardiac hypertrophy. *Am. J. Physiol.* **215**, 1409–1413.

Mullins, J.J. and Ganten, D. (1990). Transgenic animals: new approaches to hypertension research. *J. Hyperten. Suppl.* **8**, S35–S37.

Mullins, J.J., Peters, J., and Ganten, D. (1990). Fulminant hypertension in transgenic rats harbouring the mouse Ren-2 gene. *Nature* **344**, 541–544.

Munro, I.C., Hasnain, S., Salem, F.A., Goodman, T., Grice, H.C., and Heggtveit, H.A. (1972). Carditoxicity of brominated vegetable oils. *Recent Adv. Stud. Cardiac Struct. Metab.* **1**, 588–595.

Nagura, J., Hui, C., Yamamoto, M., Yasuda, S., Abe, M., Hachisu, M., and Konno, F. (1995). Effect of chronic treatment with ME3221 on blood pressure and mortality in aged stroke-prone spontaneously hypertensive rats. *Clin. Exp. Pharmacol. Physiol. Suppl.* **22**, S363–S365.

Nair, K.G., Cutilletta, A.F., Zak, R., Koide, T., and Rabinowitz, M. (1968). Biochemical correlates of cardiac hypertrophy, I: experimental model; changes in heart weight, RNA content, and nuclear RNA polymerase activity. *Circ. Res.* **23**, 451–462.

Neely, J.R., Liebermeister, H., Battersby, E.J., and Morgan, H.E. (1967). Effect of pressure development on oxygen consumption by isolated rat heart. *Am. J. Physiol.* **212**, 804–814.

Nishiyama, K., Nishiyama, A., Pfeffer, M.A., and Frohlich, E.D. (1978). Systemic and regional flow distribution in normotensive and spontaneously hypertensive young rats subjected to lifetime β-adrenergic receptor blockade. *Blood Vessels* **15**, 333–347.

Oka, M., Girerd, R.J., Brodie, S.S., and Angrist, A.A. (1966). Cardio valve and aortic lesions in beta-aminopropionitrile fed rats with and without high salt. *Am. J. Pathol.* **48**, 45–64.

Okamoto, K. (1969). Spontaneous hypertension in rats. *Int. Rev. Exp. Pathol.* **7**, 227–270.

Okamoto, K. and Aoki, K. (1963). Development of a strain of spontaneously hypertensive rats. *Jpn. Circ. J.* **27**, 282.

Okamoto, K., Yamori, Y., and Nagaoka, A. (1974). Establishment of the stroke-prone spontaneously hypertensive rat. *Circ. Res.* **34(suppl)**, I-143–I-153.

Okamoto, K., Yamori, Y., Ooshima, A., and Tanaka, T. (1972). Development of substrains in spontaneously hypertensive rats: geneoloiosozymes and effect of hypercholesterolemic diet. *Jpn. Circ. J.* **36**, 461–470.

Ooshima, A., Yamori, Y., and Okamoto, K. (1972). Cardiovascular lesions in the selectivel-bred group of spontaneously hypertensive rats with severe hypertension. *Jpn. Circ. J.* **36**, 797–812.

Osborne, B.E. (1973). A restraining device facilitating electrocardiogram recording in rats. *Lab. Anim.* **7**, 185–188.

Page, E., Early, J., and Power, B. (1974). Normal growth of ultrastructures in rat left ventricular myocardial cells. *Circ. Res.* **34–35**, 12–16.

Page, E. and McCallister, L.P. (1973). Quantitative electron microscopic description of heart muscle cells: application to normal, hypertrophied and thyroxin-stimulated hearts. *Am. J. Cardiol.* **31**, 172–181.

Peng, N., Meng, Q.C., Oparil, S., and Wyss, J.M. (1996). Acute saline infusion decreases norepinephrine release in the anterior hypothalamic area. *Hypertension* **27**, 578–583.

Penney, D., Dunham, E., and Benjamin, M. (1974a). Chronic carbon monoxide exposure: time course of hemoglobin, heart weight and lactate dehydrogenase isozyme changes. *Toxicol. Appl. Pharmacol.* **28**, 493–497.

Penney, D.G., Sakai, J., and Cook, K. (1974b). Heart growth: Interacting effects of carbon monoxide and age. *Growth* **38**, 321–328.

Petelenz, T. (1965). Extracoronary blood supply of the sinu-atrial (Keith-Flack's) node. *Cardiology* **47**, 57–67.

Peters, R.F., Lees, C.M., Mitchell, K.A., Tweed, M.F., Talbot, M.D., and Wallis, R.B. (1991). The characterisation of thrombus development in an improved model of arterio-venous shunt thrombosis in the rat and the effects of recombinant desulphatohirudin (CGP 39393), heparin, and iloprost. *Thromb. Haemost.* **65**, 268–274.

Pfeffer, J.M., Pfeffer, M., and Frohlich, E.D. (1971). Validity of an indirect tail-cuff method for determining systolic arterial pressure in unanesthitized normotensive and spontaneously hypertensive rats. *J. Lab. Clin. Med.* **78**, 957–962.

Pfeffer, J.M., Pfeffer, M.A., and Braunwald, E. (1985). Influence of chronic captopril therapy on the infarcted left ventricle of the rat. *Circ. Res.* **57**, 84–95.

Pfeffer, J.M., Pfeffer, M.A., and Braunwald, E. (1987). Hemodynamic benefits and prolonged survival with long-term captopril therapy in rats with myocardial infarction and heart failure. *Circulation* **75**, I-149–I-155.

Pfeffer, J.M., Pfeffer, M.A., Fishbein, M.C., and Frohlich, E.D. (1979b). Cardiac function and morphology with aging in the spontaneously hypertensive rat. *Am. J. Physiol.* **6**, H461–H468.

Pfeffer, J.M., Pfeffer, M.A., Fletcher, P.J., and Braunwald, E. (1991). Progressive ventricular remodeling in rat with myocardial infarction. *Am. J. Physiol.* **260**, H1406–1414.

Pfeffer, J.M., Pfeffer, M.A., Fletcher, P.J., and Braunwald, E. (1982). Regression of LVH and prevention of LV dysfunction by captopril in SHR. *Proc. Natl. Acad. Sci. USA* **79**, 3310–3314.

Pfeffer, J.M., Pfeffer, M.A., Fletcher, P.J., and Braunwald, E. (1984). Ventricular performance in rats with myocardial infarction and failure. *Am. J. Med.* **76**, 99–103.

Pfeffer, M.A., Pfeffer, J.M., Fishbein, M.C., Fletcher, P.J., Spadaro, J., Kloner, R.A., and Braunwald, E. (1979a). Myocardial infarct size and ventricular function in rats. *Circ. Res.* **44**, 503–512.

Philp, R.B., Francey, I., and Warren, B.A. (1978). Comparison of antithrombotic activity of heparin, ASA, sulfinpyrazone and VK 744 in a rat model of arterial thrombosis. *Haemostasis* **7**, 282–293.

Poole, D.C., Batra, S., Mathieu-Costello, O., and Rakusan, K. (1992). Capillary geometrical changes with fiber shortening in rat myocardium. *Circ. Res.* **70**, 697–706.

Powell, T. (1986). Methods of isolation of cardiac myocytes. *J. Mol. Cell. Cardiol.* **18**, 30–30.

Prahash, A.J., Gupta, S., and Anand, I.S. (2000). Myocyte response to β-adrenergic stimulation is preserved in the noninfarcted myocardium of globally dysfunctional rat hearts after myocardial infarction. *Circulation* **102**, 1840–1846.

Prakash, R. (1954). The heart of the rat with special reference to the conducting system. *Am. Heart J.* **47**, 241–251

Pravenec, M., Simonet, L., Kren, V., Kunes, J., Levan, G., Szpirer, J., Szpirer, C., and Kurtz, T. (1991). The rat renin gene: assignment to chromosome 13 and linkage to the regulation of blood pressure. *Genomics* **9**, 466–472.

Prunier, F., Gaertner, R., Louedec, L., Michel, J.B., Mercadier, J.J., and Escoubet, B. (2002). Doppler echocardiographic estimation of left ventricular end-diastolic pressure after MI in rats. *Am. J. Physiol. Heart Circ. Physiol.* **283**, H346–H352.

Rakusan, K. (1971). Quantitative morphology of capillaries of the heart: number of capillaries in animal and human hearts under normal and pathological conditions. *Methods Achiev. Exp. Pathol.* **5**, 272–286.

Rakusan, K., Korecky, B., Roth, Z., and Poupa, O. (1963). Development of the ventricular weight of the rat heart with special reference to the early phases of postnatal ontogenesis. *Physiol. Bohemoslov.* **12**, 518–525.

Rakusan, K. and Poupa, O. (1963). Changes in the diffusion in the rat heart muscle during development. *Physiol. Bohemoslov.* 220–227.

Ramos, K., Grossman, S.L., and Cox, L.R. (1998). Allylamine-induced vascular toxicity *in vitro*: prevention by semicarbazide-sensitive amine oxidase inhibitors. *Toxicol. Appl. Pharmacol.* **95**, 61–71.

Rapp, J.P. (1982). Dahl salt-susceptible and salt-resistant rats: a review. *Hypertension* **4**, 753–763.

Rapp, J.P. (1991). Dissecting the primary causes of genetic hypertension in rats. *Hypertension* **18(Suppl. I)**, I-18–I-28.

Rapp, J.P. (2000). Genetic analysis of inherited hypertension in the rat. *Physiol. Rev.* **80**, 135–172.

Rapp, J.P. and Dahl, L.K. (1972). Arylesterase isoenzymes in rats bred for susceptibility and resistance to the hypertensive effect of salt. *Proc. Soc. Exp. Biol. Med.* **139**, 349–352.

Rapp, J.P. and Dene, H. (1985). Development and characteristics of inbred strains on Dahl salt-sensitive and salt-resistant rats. *Hypertension* **7**, 340–349.

Rehwald, W.G., Reeder, S.B., McVeigh, E.R., and Judd, R.M. (1997). Techniques for high-speed cardiac magnetic resonance imaging in rats and rabbits. *Magn. Reson. Med.* **37**, 124–130.

Reindel, J.F., Ganey, P.E., Wagner, J.G., Slocombe, R.F., and Roth, R.A. (1990). Development of morphologic, hemodynamic, and biochemical changes in lungs of rats given monocrotaline pyrrole. *Toxicol. Appl. Pharmacol.* **106**, 179–200.

Ren, J., Walsh, M.F., Jefferson, L., Natavio, M., Ilg, K.J., Sowers, J.R., and Brown, R.A. (2000). Basal and ethanol-induced cardiac contractile response in lean and obese Zucker rat hearts. *J. Biomed. Sci.* **7**, 390–400.

Repke, K. and Portius, H.J. (1963). Uber die Identitat der Ionenpumpen-ATPase in der Zellmembran des Herzmuskels mit einen Digitalis-Receptorenzym. *Experientia* **19**, 452–458.

Rona, G., Chappel, C.I., Balazs, T., and Gaudry, R. (1959). An infarct-like myocardial lesion and other toxic manifestations produced by isoproternol in the rat. *Arch. Pathol.* **67**, 443–455.

Rudin, M., Pedersen, B., Umemura, K., and Zierhut, W. (1991). Determination of rat heart morphology and function *in vivo* in two models of cardiac hypertrophy by means of magnetic resonance imaging. *Basic Res. Cardiol.* **86**, 165–174.

Ruiz-Opazo, N., Xiang, X.H., and Herrera, V.L. (1997). Pressure-overload deinduction of human α 2 Na,K-ATPase gene expression in transgenic rats. *Hypertension* **29**, 606–612.

Russell, J.C., Ahuja, S.K., Manickavel, V., Rajotte, R.V., and Amy, R.M. (1987). Insulin resistance and impaired glucose tolerance in the atherosclerosis-prone LA/N corpulent rat. *Arteriosclerosis* **7**, 620–626.

Russell, J.C. and Amy, R.M. (1986). Myocardial and vascular lesions in the LA/N-corpulent rat. *Can. J. Physiol. Pharmacol.* **64**, 1272–1280.

Russell, J.C., Koeslag, D.G., Dolphin, P.J., and Amy, R.M. (1993). Beneficial effects of acarbose in the atherosclerosis-prone JCR:LA-corpulent rat. *Metabolism* **42**, 218–223.

Sandler, C. and Wilson, G.M. (1959). The production of cardiac hypertrophy with thyroxine in the rat. *Q. J. Exp. Physiol. Cogn. Med. Sci.* **44**, 282–289.

Sato, K., Kawasaki, T., Hisamichi, N., Taniuchi, Y., Hirayama, F., Koshio, H., and Matsumoto, Y. (1998). Antithrombotic effects of YM-60828, a newly synthesized factor Xa inhibitor, in rat thrombosis models and its effects on bleeding time. *Br. J. Pharmacol.* **123**, 92–96.

Schonekess, B.O., Allard, M.F., and Lopaschuk, G.D. (1996). Recovery of glycolysis and oxidative metabolism during postischemic reperfusion of hypertrophied rat hearts. *Am. J. Physiol.* **271**, H798–H805.

Segel, L.D., Rendig, S.V., and Mason, D.T. (1979). Left ventricular dysfunction of isolated working rat hearts after chronic alcohol consumption. *Cardiovasc. Res.* **13**, 136–146.

Selye, H., Bajusz, E., Grasso, S., and Mendell, P (1960). Simple techniques for the surgical occlusion of coronary vessels in the rat. *Angiology* **11**, 398–407.

Sen, S., Smeby, R.R., and Bumpus, F.M. (1972). Renin in rats with spontaneous hypertension. *Circ. Res.* **31**, 876–880.

Setnikar, I. and Magistretti, M.J. (1965). Relationships between organ and body weight in the male rat. *Arzneimittelforschung* **15**, 1042–1048.

Shipley, R.A., Shipley, L.J., and Wearn, J.T. (1937). The capillary supply in normal and hypertrophied hearts of rabbits. *J. Exp. Med.* **65**, 29–44.

Siegel, R.C. (1979) Lysyl oxidase. *Int. Rev. Connect. Tissue. Res.* **8**, 73–118.

Silva, Jr., J.A., Araujo, R.C., Baltatu, O., Oliveira, S.M., Tschope, C., Fink, E., Hoffmann, S., Plehm, R., Chai, K.X., Chao, L., Chao, J., Ganten, D., Pesquero, J.B., and Bader, M. (2000). Reduced cardiac hypertrophy and altered blood pressure control in transgenic rats with the human tissue kallikrein gene. *FASEB J.* **14**, 1858–1860.

Smirk, F.H., and Hall, W.H. (1958). Inherited hypertension in rats. *Nature* **182**, 727–728.

Smith, S.H., McCaslin, M.D., Sreenan, C., and Bishop, S.P. (1998). Regional myocyte size in two-kidney, one clip renal hypertension. *J. Mol. Cell. Cardiol.* **20**, 1035–1042.

Sommer, J.R. and Johnson, E.A. (1968). Cardiac muscle: a comparative study of Purkinje fibers and ventricular fibers. *J. Cell. Biol.* **36**, 497–526.

Spear, J.F. and Moore, E.N. (1972). Stretch-induced excitation and conduction disturbances in the isolated rat myocardium. *J. Electrocardiol.* **5**, 15–24.

St. John, L.C. and Bell, F.P. (1991). Arterial lipid biochemistry in the spontaneously hyperlipidemic Zucker rat and its similarity to early atherogenesis. *Atherosclerosis* **86**, 139–144.

St. Lezin, E., Simonet, L., Pravenec, M., and Kurtz, T.W. (1992). Hypertensive strains and normotensive control strains. How closely are they related? *Hypertension* **19**, 419–424.

Stanton, H.C., Brenner, G., and Mayfield, Jr., E.D. (1969). Studies on isoproterenol-induced cardiomegaly in rats. *Am. Heart J.* **77**, 72–80.

Stevenson, J.A.F. (1966). Bouts of exercise and food intake in the rat. *J. Appl. Physiol.* **21**, 118–122.

Stevenson, J.A.F. (1967). Exercise, food intake and health in experimental animals. *Can. Med. Assoc. J.* **96**, 862–867.

Su, X., Brower, G., Janicki, J.S., Chen, Y.F., Oparil, S., and Dell'Italia, L.J. (1999). Differential expression of natriuretic peptides and their receptors in volume overload cardiac hypertrophy in the rat. *J. Mol. Cell. Cardiol.* **31**, 1927–1936.

Sumida, E., Nohara, M., Muro, A., Sumida, E., Kaku, H., Koga, Y., Toshima, H., and Imaizumi, T. (1998). Altered calcium handling in compensated hypertrophied rat cardiomyocytes induced by pressure overload. *Jap. Circ. J.* **62**, 36–46.

Teerlink, J.R., Pfeffer, J.M., and Pfeffer, M.A. (1994). Progressive ventricular remodeling in response to diffuse isoproterenol-induced myocardial necrosis in rats. *Circ. Res.* **75**, 105–113.

Thomas, B.M. and Miller, Jr., A.T. (1958). Adaptation to forced exercise in the rat. *Am. J. Physiol.* **193**, 350–354.

Tian, B., Liu, J., Bitterman, P., and Bache, R.J. (2003). Angiotensin II modulates nitric oxide-induced cardiac fibroblast apoptosis by activation of AKT/PKB. *Am. J. Physiol. Heart Circ. Physiol.* **285**, H1105–H1112.

Tilton, R.G., Munsch, C.L., Sherwood, S.J., Chen, S.J., Chen, Y.F., Wu, C., Block, N., Dixon, R.A., and Brock, T.A. (2000). Attenuation of pulmonary vascular hypertension and cardiac hypertrophy with sitaxsentan sodium, an orally active ET(A) receptor antagonist. *Pulm. Pharmacol. Ther.* **13**, 87–97.

Truex, R.C. and Smythe, M.Q. (1965) Comparative morphology of the cardiac conduction tissue in animals. *Ann. NY Acad. Sci.* **127**, 19–33.

Tucker, D.C., Askari, M., and Bishop, S.P. (1992). Effects of sympathetic innervation on size of myocytes in embryonic rat heart cultured in oculo. *J. Mol. Cell. Cardiol.* **24**, 925–935.

Tucker, D.C. and Gist, R. (1986). Sympathetic innervation alters growth and intrinsic heart rate of fetal rat atria maturing in oculo. *Circ. Res.* **59**, 534–544.

Turner, J.H. and Lalich, J.J. (1965). Experimental cor pulmonale in the rat. *Arch. Pathol.* **79**, 409–418.

Valtin, H. (1982). The discovery of the Brattleboro rat, recommended nomenclature, and the question of proper controls. *Ann. NY Acad. Sci.* **394**, 1–9.

Van Liere, E.J., Krames, B.B., and Northup, D.W. (1965). Differences in cardiac hypertrophy in exercise and hypoxia. *Circ. Res.* **16**, 244–248.

Van Liere, E.J. and Sizemore, D.A. (1971). Regression of cardiac hypertrophy following experimental hyperthyroidism in rats. *Proc. Soc. Exp. Biol. Med.* **136**, 645–648.

Van Orden, D.E., Farley, D.B., Fastenow, C., and Brody, M.J. (1984). A technique for monitoring blood flow changes with miniaturized Doppler flow probes. *Am. J. Physiol.* **247**, H1005–H1009.

Waller, C., Hiller, K.H., Kahler, E., Hu, K., Nahrendorf, M., Voll, S., Haase, A., Ertl, G., and Bauer, W.R. (2001). Serial magnetic resonance imaging of microvascular remodeling in the infarcted rat heart. *Circulation* **103**, 1564–1569.

Wearn, J.T. (1928). The extent of the cappillary bed of the heart. *J. Exp. Med.* **47**, 273–292.

Wei, C.C., Lucchesi, P.A., Tallaj, J., Bradley, W.E., Powell, P.C., and Dell'Italia, L.J. (2003). Cardiac interstitial bradykinin and mast cells modulate pattern of LV remodeling in volume overload in rats. *Am. J. Physiol. Heart Circ. Physiol.* **285**, H784–H792.

Weissler, A.M., Kruger, F.A., Baba, N., Scarpelli, D.G., Leighton, R.F., and Gallimore, J.K. (1968). Role of anaerobic metabolism in th preservation of functional capacity and structure of anoxic myocardium. *J. Clin. Invest.* **47**, 403–416.

Wicker, P. and Tarazi, R.C. (1982). Importance of injection site for coronary blood flow determinations by microspheres in rats. *Am. J. Physiol.* **242**, H94–H97.

Wiener, J. and Giacomelli, F. (1973). The cellular pathology of experimental hypertension, VII: structure and permeability of the mesenteric vasculature in angiotensin-induced hypertension. *Am. J. Pathol.* **72**, 221–231.

Wildenthal, K. (1970). Factors promoting the survival and beating of intact foetal mouse hearts in organ culture. *J. Mol. Cell. Cardiol.* **1**, 101–104.

Will, J.A., Rowe, G.G., Olson, C., and Crumpton, C.W. (1971). A chemically induced acute model of myocardial damage in intact calves. *Res. Comm. Chem. Pathol. Pharmacol.* **2**, 61–66.

Williams, J.R., Harrison, T.R., and Grollman, A. (1939). A simple method for determining the systolic blood pressure of the unanesthetized rat. *J. Clin. Invest.* **18**, 373–376.

Williamson, J.R. and Kobayashi, K. (1984). Use of the perfused rat heart to study cardiac metabolism: retrospective and prospective views. *Basic Res. Cardiol.* **79**, 283–291.

Wilson, D.W., Segall, H.J., Pan, L.C., Lame, M.W., Estep, J.E., and Morin, D. (1992). Mechanisms and pathology of monocrotaline pulmonary toxicity. *Crit. Rev. Toxicol.* **22**, 307–325.

Wong, P.C., Crain, Jr., E.J., Nguan, O., Watson, C.A., and Racanelli, A. (1996). Antithrombotic actions of selective inhibitors of blood coagulation factor Xa in rat models of thrombosis. *Thromb. Res.* **83**, 117–126.

Wood, W.G., Lindenmayer, G.E., and Schwartz, A. (1971). Myocardial synthesis of ribonucleic acid, I: stimulation by isoproterenol. *J. Mol. Cell. Cardiol.* **3**, 127–138.

Wu, C.C., Feldman, M.D., Mills, J.D., Manaugh, C.A., Fischer, D., Jafar, M.Z., and Villanueva, F.S. (1997). Myocardial contrast echocardiography can be used to quantify intramyocardial blood volume: new insights into structural mechanisms of coronary autoregulation. *Circulation* **96**, 1004–1011.

Yagi, K., Kim, S., Wanibuchi, H., Yamashita, T., Yamamura, Y., and Iwao, H. (1997). Characteristics of diabetes, blood pressure, and cardiac and renal complications in Otsuka Long-Evans Tokushima fatty rats. *Hypertension* **29**, 728–735.

Yamakawa, T., Tanaka, S., Tamura, K., Isoda, F., Ukawa, K., Yamakura, Y., Takanashi, Y., Kiuchi, Y., Umemura, S., and Ishiiu, M. (1995). Wistar fatty rat is obese and spontaneously hypertensive. *Hypertension* **25**, 146–150.

Yamazaki, M., Asakura, H., Aoshima, K., Saito, M., Jokaji, H., Uotani, C., Kumabashiri, I., Morishita, E., Ikeda, T., and Matsuda, T. (1994). Effects of DX-9065a, an orally active, newly synthesized and specific inhibitor of factor Xa, against experimental disseminated intravascular coagulation in rats. *Thromb. Haemost.* **72**, 392–396.

Yamori, Y. and Okamoto, K. (1970). Zymogram analyses of various organs from spontaneously hypertensive rats: a genetico-biochemical study. *Lab. Invest.* **22**, 206–211.

Yamori, Y., Ooshima, A., and Okamoto, K. (1972). Genetic factors involved in spontaneous hypertension in rats: an analysis of F-2 segregate generation. *Jpn. Circ. J.* **36**, 561–568.

York, J.W., Penney, D.G., Weeks, T.A., and Stagno, P.A. (1976). Lactate dehydrogenase changes following several cardiac hypertrophic stresses. *J. Appl. Physiol.* **40**, 923–926.

Zimmer, H.G., Gerdes, A.M., Lortet, S., and Mall, G. (1990). Changes in heart function and cardiac cell size in rats with chronic myocardial infarction. *J. Mol. Cell. Cardiol.* **22**, 1231–1243.

Zucker, L.M. and Zucker, T.F. (1961). Fatty, a new mutation in the rat. *J. Heredity* **52**, 275–278.

Chapter 27

Toxicology

Thomas E. Hamm, Jr., Angela King-Herbert, and Mary Ann Vasbinder

I. INTRODUCTION

It is impossible in a single chapter to cover the enormous body of information that has resulted from the use of the rat in toxicology. Rather, this chapter will focus on major points concerning the use of the rat in toxicology and provide key references that will allow the reader to delve further into the subject. In any discussion of toxicology, it should be emphasized that the toxicity of any compound can be easily influenced by the design and conduct of the experiment. For example, results of an experiment using one strain or stock of rats, in which experimental variables (diet, water quality, intestinal flora, caging, bedding, etc.) are not defined or controlled cannot be extrapolated with absolute confidence to other species, including humans. Moreover, putative toxicities to humans are most validly defined when a compound causes similar findings in multiple experiments.

II. HISTORY

Toxicology as an organized discipline is relatively new. The first journal expressly dedicated to experimental toxicology, *Archiv fur Toxikologie,* started publication in Europe in 1930 (Amdur et al., 1991), and the preeminent toxicology organization, the Society of Toxicology, was not founded until 1961 (Hays, 1986). The reader is referred to excellent reviews of the history of toxicology (Gallo and Doull, 1991) and regulatory toxicology (Gad, 2001).

Animals have been used in experiments and testing for hundreds of years based on the assumption that they react similarly to experimental conditions as do humans. The use of animals in toxicology was formalized in the 1930s with the passage of the Food Drug and Cosmetic Act in 1938. Basis for this act included a 1933 report describing a severe allergic reaction to Lash-Lure, a synthetic aniline dye for eyebrows and eyelashes (Parascandola, 1991), and a rash of deaths associated with administration of the new wonder drug sulfanilamide that were later attributed to the vehicle of the drug, ethylene glycol. The Food, Drug, and Cosmetic Act gave the U.S. Food and Drug Administration (FDA) the authority to require toxicology tests necessary to assess the safety of food, drugs, medical devices, and cosmetics.

In 1970, the U.S. Environmental Protection Agency (EPA) was created. It was given a wide variety of regulatory duties: (1) pesticides are regulated under authority of the Federal Insecticide, Fungicide, and Rodenticide Act; (2) chemicals in commercial production are regulated under the Toxic Substances Control Act; (3) hazardous wastes are regulated under the Resource Conservation and Recovery Act; (4) water pollution is regulated under the Federal Water Pollution Control Act; (5) drinking water is regulated under the Safe Drinking Water Act; and (6) air pollution is regulated under the Clean Air Act (CAA).

Two other major governmental agencies were also charged with consumer safety in the United States. The Occupational Safety and Health Administration (OSHA) including the Consumer Products Safety Commission, relied mainly on already established toxicology data; and The National Institutes of Occupational Safety and Health (NIOSH) collected and analyzed data for OSHA. The FDA, EPA and the Organization for Economic Cooperation and Development (OECD) have worked together to define standard toxicology tests. These standards are constantly evolving, and the latest standards, which can be found by contacting respective organizations or visiting their Web sites, must be considered when designing a test that might be used in a regulatory approval process.

III. HUMANE USE OF ANIMALS IN TOXICOLOGY

Many events in the 19th (French, 1975) and 20th centuries lead to current concepts in animal use in toxicology, including the concepts of humane endpoints and the use of alternatives. For example, any experiment, including those in toxicology, should be preceded by a review of the three R's—reduction, refinement and replacement—that were first outlined in *The Principles of Humane Experimental Technique* (Russell and Burch, 1959). The atrocities of World War II lead to the Nuremberg Code of Ethics of 1947. The Nuremberg Code included the following provision: "The experiment should be so designed and based on the results of animal experimentation and a knowledge of the natural history of the disease or other problem under study, that the anticipated results will justify the performance of the experiment" (Silverman, 1985). In 1964, the World Medical Association adopted the Declaration of Helsinki, a formal code of ethics for doctors in clinical research. This code was extended in 1973 by the 29th World Medical Assembly. This new code, Helsinki II, included the following provision: "Biomedical Research involving human subjects must conform to generally accepted scientific principles and should be based on adequately performed laboratory and animal experimentation and on a thorough knowledge of the scientific literature" (Silverman, 1985). These ethical standards established that animal experimentation must be conducted before human experimentation when possible.

IV. THE USE OF THE RAT IN PHARMACOLOGIC TESTING: AN OVERVIEW

Drug discovery is a complex process that can take more than 10 years and require hundreds of millions of dollars for the development of a single product. All of the testing before clinical trials requires both *in vivo* and *in vitro* testing. The rat is the major rodent species used for *in vivo* testing during drug development. Drug discovery is rigorous and carefully staged, beginning with the selection of a disease on which to focus. This decision is based on needs in the patient population, expertise and technology within the company, and potential competition and commercial considerations. Once marketing and commercial studies are completed, a commitment can be made to search for a particular "product type." The desired characteristics of that product are outlined, including efficacy, safety, route of administration, frequency of dosing, pricing, and needs of the patient populations. Scientists are then asked to consider the types of biological "targets," which are molecules that may be addressed by drugs to produce a therapeutic effect. A decision is then made to select a particular target class or family of molecules that will be investigated for drug development. A hypothesis is formed, and targets are selected with the hope that a new drug molecule with the proper characteristics can be created.

The route of drug administration must be determined in the preclinical phase. Oral formulations include capsules, tablets, or oral gavage. Solutions may be administered intravenously, intraperitoneally, or intramuscularly, but this requires a more limited application to the general human population. Inhalational administration can be accomplished by nebulization or by powder-form or metered-dose inhalers. Creams, ointments, lotions, and sprays are the options for topical application. Lastly, preparations for nasal application can be made as solutions or suspensions. Once the route of administration is determined, the vehicle or carrier is decided. Important considerations for selection of vehicles include pH, solubility, viscosity, osmolality, sterility, and biocompatibility with the compound. The volume administered is also critical, and several authors have reviewed these issues (Diel, 2000; Morton, 2000).

The next phases require setting up screening systems that are robust and conducive to high-volume, automated tests. The new compounds are analyzed by medicinal chemists to identify molecules that may be useful as a candidate compound. Selected compounds are then expanded into a series of compounds that contain variations of the starting molecule. Eventually, a compound is selected based on *in vitro* potency, selectivity, and the drug-like properties of the compound.

Synthesis and testing of molecules to create an optimal balance of potency, selectivity, and safety are part of drug optimization. Simultaneously, pharmacological screening for efficacy and potential toxicity are carried out by using *in vitro* testing. Once chemicals are determined to have characteristics that make them interesting as potential drugs, they are called drug candidates and are slated for *in vivo* testing.

Drug metabolism and pharmacokinetic studies are aimed at understanding the pharmacokinetic and pharmacodynamic nature of compounds. The initial testing typically consists of a single oral dose, administered to rats, with evaluation over a short period of time (less than 72 hours). Samples such as urine, feces, and blood are collected temporally and analyzed for drug or drug metabolites. After the studies are accomplished in rodents, the study is repeated in a second species such as the dog. These studies are the first effort at characterizing the nature of the drug in animals.

Target validation studies investigate efficacy and physiologic response of the drug molecule on the target. This work may also require a basic understanding of the mechanism of action of the target. Studies that provide supportive data for the pharmacologic usefulness of a molecule are often called "proof of concept" studies.

Safety parameters must be well defined for each chemical entity. Dose-range-finding studies, typically in the rodent and dog, provide a range of safe and efficacious dosages for compounds. Studies describing toxic doses and description of the toxicity are required for drug approval.

All of the data collected are then presented to an internal review process for approval, and a request is made for review by an external ethical review committee. These approvals are necessary to gain permission to initiate clinical evaluation in humans. Next the data is submitted to the FDA who must approve the Investigational New Drug (IND), and the plan to initiate clinical studies. Safety studies using animals continue during phase I clinical trials, including chronic toxicity tests, reproductive toxicity, oncogenicity studies, and supplementary pharmacology work. This is a dynamic process and occurs as data are collected and compared across species and the need for information becomes apparent (Marzo, 1997; Roberts, 2001). After successful phase III trials, the new drug application (NDA) is submitted to the FDA with the request for approval to market the new drug.

V. TOXICOLOGICAL STUDIES

The goal of toxicology is to minimize the adverse effects of compounds in the human population and the environment. Assessment of chemicals, pharmaceuticals, devices, etc., in animals is the best measure of interactions between complex organ system functions and the test article. Evaluation of compounds in an animal, at doses that approach a maximally tolerated dose, provides an understanding of the consequences of physiologic, pharmacological, and toxicological actions.

The FDA has written guidelines for toxicity testing (FDA, 1993b) and international harmonization of tests has been described by the OECD (OECD, 1993), but these guidelines have not yet been adopted. Regulations for both Europe and Japan have also been published (EEC, 1992; PAFSC, 1997). Toxicology studies are conducted as described by *Good Laboratory Practice* in the United States (FDA, 1978) or by similar standards in other countries.

A. Acute Toxicity Testing

In acute toxicity testing, a single dose of a test compound is given once to ascertain the adverse effects on the animal. The compound may be administered orally, parenterally, and dermally or by inhalation (Echobichon, 1992). The route of exposure in the test animals usually depends on the usual route of exposure in humans. Typically, rats and mice are the species of choice for acute toxicity testing, with test protocols requiring the use of both sexes. The OECD has established guidelines that recommend that in acute oral toxicity testing, only female rats should be used when there is no indication that males are more sensitive to the compound being tested (OECD, 1998). Range-finding tests with a small number of animals are usually used to determine the relative lethality of a compound. Then larger numbers of animals are used to determine relative toxicity. LD_{50} is a dose that is lethal to half of the animals on study. The ED_{50} is a measure of the dose for which half of the animals exhibit an effect and half of the animals exhibit no effect. In LD_{50} testing, several dose levels of the test compound are used, with one dose per group of animals, and the testing is performed in replicates (Ballantyne, 1999). Typical LD_{50} tests require 100 animals or more. In addition to providing the median lethal concentration of a test compound, LD_{50} determinations can provide information in terms of the dose-response relationship, clinical signs of toxicity, and gross and histopathologic findings (Hayes, 2001).

LD_{50} testing for acute lethality is no longer required by the FDA in an effort to reduce the number of animals necessary for study and lessen the number of animals that experience pain and distress. Alternatives to the LD_{50} have been described (Sass, 2000; Botham, 2002). The FDA specifically refers to the limit test as an alternative to LD_{50} testing. The limit test is used to determine if the toxicity of a compound is above or below a specified dose (FDA, 1993b). Five to 10 animals of each sex are exposed to the test compound, and toxic responses are recorded in a given time. Further testing is pursued based on the animal response.

Other alternatives to the LD_{50} test include the up and down procedure (UPD), fixed dose method (FDM), and the acute toxic class method (ATC). These alternative tests provide information about the toxicity of the compound while minimizing the number of animals on study and number of animal deaths. The UPD test gives a point estimation of the LD_{50} using six or seven animals. This test starts with a fixed dose and adjusts the subsequent testing based on the toxic response of the initial test. The UPD is composed of three tests: the primary test, the limit test, and a final test for estimation of the slope and confidence interval for the dose response curve. The FDM avoids using death of animals as an endpoint and relies on observed signs of toxicity at one of a series of fixed dose levels (Botham, 2004). The ATC uses fixed dose levels, but animal mortality serves as an endpoint. Details of these testing strategies can be found on the OECD Web site (*http://oecd.org/ehs/test/testlist.htm*).

B. Chronic Studies

Various longer term tests may be conducted. Carcinogenicity, mutagenicity, reproductive toxicology, developmental toxicology, and neurotoxicology will be discussed in the following sections.

1. Carcinogenicity Testing

Carcinogenicity studies are performed to assess the potential of a compound to cause cancer. The primary species for this testing is the rat; the mouse is used secondarily. The experimental approach to demonstration of carcinogenic potential of a drug must be flexible, but a scheme usually includes one long-term rodent study and either another rodent study in an alternate species or shorter *in vivo* studies. These shorter investigational studies are performed to provide information about carcinogenesis, including mechanism or insight into induction and promotion of the cancer. Dose dependency and the relationship to carcinogenicity should be evaluated in these studies. This may be accomplished in a number ways. For example, tissues can be collected and evaluated for morphologic and histochemical changes. Biochemical analysis of systemic or tissue hormone levels, enzymes, and growth factors may also be important information in these studies (ICH, 1997).

Animal strain, food type, and housing must be taken into consideration when setting up long-term rodent studies. The species must have longevity that meets the criterion for the duration of the study. The species must not have background disease or cancers that confound the study. The diet can significantly impact the manifestation of cancer and the longevity of the animals.

The National Cancer Institute has recommended guidelines for long-term carcinogenicity studies (NCI, 1976), and a review of the carcinogenesis bioassay test has been published (Hamm, 1985). Carcinogenicity studies in rodents are 2-year studies incorporating exposure to the compound for a significant portion of the lifespan of the animal. Usually, both sexes are used with the route of exposure the same as would be found for human exposure. Groups of approximately 50 randomly assigned animals per exposure concentration (high, mid, and low dosage) are used, along with untreated controls. A complete necropsy and subsequent microscopic examinations are performed on all exposed and control animals on the study, including those that are euthanized for humane reasons before the end of the study.

Evaluation of the carcinogenic potential of a drug is multifold. The decision should be based on tumor incidence and latency in the long-term studies, the pharmacokinetics of the compound in the animal compared with humans, and information gained from the ancillary data that may shed light on human disease. This information together forms a "weight of evidence" approach to the scientific interpretation of the data (ICH, 1997).

2. Mutagenicity Studies

These studies are designed to determine if compounds induce genetic damage. Both *in vitro* and *in vivo* tests are used to assess genetic damage directly or indirectly, by various mechanisms. *In vivo* tests for genotoxicity are important test batteries because they allow examination of the whole animal, including tissue distribution, absorption, metabolism, and excretion of chemicals and their metabolites (Dean et al., 1999). The use of animal models allows investigation into the mechanisms of DNA damage, the relationship between gene structure and cell and tissue function, and the relationship between sequence alteration and disease.

Damage to DNA can include frameshifts, insertions, and deletions of DNA. Recombinatorial and numerical chromosome changes have the potential to induce mutagenesis. These changes can be manifested by inheritable disease or cancer (Derelanko and Hollinger, 2002). The International Conference on Harmonization of Technical Requirements for Registration of Pharmaceuticals for Human Use drafted two key guidance documents for determination of genotoxicity in pharmaceutical testing. The first recommendation, *Guidance of Specific Aspects of Regulatory Genotoxicity Tests for Pharmaceuticals* (ICH, 1995) provides recommendations for *in vitro* and *in vivo*

testing and makes recommendations for interpretation of test results. The second document, *Genotoxicity: A Standard Battery for Genotoxicity Testing of Pharmaceuticals* (ICH, 1997), establishes a standard set of tests for potential genotoxicity and provides guidance on the test limitations. The two primary *in vivo* tests recommended are the micronucleus and metaphase analysis assays. The micronucleus test screens for chemicals that cause spindle formation and micronuclei (FDA, 2000), which are clumps of cytoplasmic chromatin that are formed when chromosomes are fragmented or fail to be incorporated into the cell nuclei during anaphase of cell division. This leads to abnormal chromosomes, spindle abnormalities, and micronuclear fragmentation. Clastogens are agents that cause double-strand breaks in the chromosomes, which result in the formation of micronuclei. This assay is typically performed in healthy, young adult rats or mice. The micronucleus test is performed by exposing the animal to a chemical by a predetermined route, typically orally or intraperitoneally, over a predetermined time course. At necropsy, bone marrow or peripheral blood is collected and stained. Positive results indicate that a substance is the cause of the formation of micronuclei (Ashby, 1995; FDA, 2000). Another test less frequently used for *in vivo* testing of compounds is the metaphase analysis assay. Mice or rats are administered a dose of test compound. At predetermined time intervals, colchicine is given to arrest bone marrow cell division in the metaphase stage. A necropsy is performed, and bone marrow cells are evaluated microscopically for chromosomal damage (FDA, 2000).

Some of the most promising newer test systems are the transgenic rodent mutation models. Early transgenic animal models were created in the early 1990s, by inserting bacteriophage lambda shuttle vector systems of either Lac I or Lac Z, for creation of the Muta Mouse and the Big Blue assay systems, respectively (Kohler et al., 1991; Provost et al., 1993). The Big Blue assay system has been developed for both rats and mice. This recoverable reporter gene allows the detection of *in vivo* mutations in various target tissue locations (Turner et al., 2001; Yu et al., 2002). Animal models have also been developed that provide the ability to monitor, recover, and sequence mutations that arise in all mammalian tissues (MacGregor et al., 1998).

3. Reproductive Studies

The purpose of reproductive toxicology studies is to discover any possible substances that impact reproduction (OECD, 1996). Reproductive toxicology studies are used to identify chemicals or physical agents that impact reproduction for example by interfering with fertility in both sexes (Neubert, 2002). The basic unit of study is the mother, father, and the offspring (Hoyer, 2001). Toxicants may affect reproduction as a primary target, causing damage to the reproductive organs, germ cells, or the developing embryo/fetus. Alternatively, indirect toxicity may impact reproduction by mimicking or eliminating sex hormones or affecting hormone receptors. Rats are the preferred rodents used for studying reproductive toxicology. They are suited for this purpose for several reasons. Rats are small and highly fertile and have a short gestation period (21 days) and estrus cycle (4 to 5 days). They reach puberty at 37 to 67 days and are spontaneous ovulators that can reproduce all year. This comparatively short reproductive cycle is desirable for multigenerational studies. The average litter size is six to nine pups, and several litters may be born to one dam in a year. The large litter size allows for intra- and inter-litter comparisons that are more statistically powerful. It is also helpful that there exists a large database of knowledge on rat toxicology (Ecobichon, 2002). Guidelines for reproductive toxicity studies have been published by the FDA, (FDA, 1993a, 2000) and by other regulatory bodies (EPA, 1996; OECD, 1996). These flexible guidelines provide a framework for initial testing. Rats of age 5 to 9 weeks are acclimated and randomized by body weight. The goal is to obtain 20 males and 20 pregnant females per group, with high-dose, intermediate-dose, low-dose, and control groups. Males for the trial should be dosed at least 10 weeks before mating to enable examination of the effect of the compound on a full cycle of spermatogenesis. Dosing of females begins at the same time and continues through parturition and weaning of the litter. Offspring should be dosed throughout the postnatal life.

Successful mating is determined by daily examination for a copulatory plug and/or vaginal smears. After mating, the male and female pairs are separated. Studies typically continue for two generations, so at least one male and one female from each litter should be randomly selected to produce the next generation (FDA, 2000). Clinical observations are made twice daily, and abnormalities in condition or behavior are recorded. Body weights are recorded weekly, and vaginal smears can provide information regarding estrus cycle length and regularity. Litters are examined, and the number of live, dead, or abnormal pups are noted. Sex and weight of the pups is recorded (OECD, 1996). Necropsies are conducted on any dead or anomalous pups that are found during the study. Pups that are not to be used for breeding are euthanized, and a necropsy examination is performed. Reproductive organs and other tissue are collected for gross and histologic evaluation. Sperm motility, morphology, and number are assessed at necropsy. Review of female reproductive performance is made by a series of indexes. The female fertility, gestation, live-born, weaning, and sex ratio indexes provide population-based data for reproductive performance (FDA, 2000).

4. Developmental Studies

The purpose of developmental toxicological studies (teratology) is to evaluate compounds for their effect on the developing fetus. The rat and rabbit are the preferred species for these studies. These tests may be integrated into the multigenerational tests during evaluation of reproductive toxicity. The test compound is administered to pregnant animals from implantation (day 6 for rats) to the day before parturition (day 20 or 21). Manifestations of toxicity to the fetus are categorized as death, structural anomaly, altered or retarded growth, or functional deficiencies. Studies are monitored in a manner similar to that of the reproductive studies, with clinical observations of the dam conducted twice daily. At necropsy, reproductive tracts and fetuses are collected for counts of live/dead fetus and for gross and histologic examination of the fetus and reproductive organs. Special stains may be used to facilitate understanding of the mechanism of toxicity (FDA, 2000).

5. Neurotoxicology Studies

Neurotoxicity has been defined as an adverse change in the structure or function of the nervous system that results from exposure to a chemical, biological, or physical agent. Damage to the nervous system can occur by many different mechanisms. The chemical may affect the nervous system as a primary target or may affect the nervous system indirectly by toxicity to other target organs. Disease of the nervous system may be categorized as neuronopathies, axonopathies, myelinopathies, or neurotransmitter effects (Mandella, 2002). The goals of neurologic testing should include detection of neurotoxic potential and definition of dose range and time-course, as well as give a profile for effect to focus future testing (Moser, 1991). Assessment of neurotoxicity must include the severity, type, and reversibility of the effects. Various governmental bodies have described neurotoxic testing guidelines. The FDA has not identified any formal protocol requirement for systematic neurotoxicity testing for pharmaceutics; requirements are made on a case-by-case basis (Mandella, 2002). The FDA has taken this approach to accommodate the broad spectrum of behavioral, structural, and biochemical changes and to lend flexibility to the testing strategy (Ross, 2002). The guidelines require that neurobehavioral and neuropathologic testing be assessed to describe toxicity and risk.

The FDA has published draft guidelines for food additives in the *Redbook* (FDA, 2000). An ad hoc expert panel described a two-tiered method of testing for assessment of neurotoxic potential of pharmaceuticals. Tier-one testing requires assessment of neurobehavioral changes through clinical observations and testing of motor activity (Irwin, 1968). The second tier of testing is based on findings from the primary screenings and is to be tailored to evaluate specific toxicities. The OECD and EPA have published other guidelines (EPA, 1998; Haggarty, 1989; OECD, 2000b). The first-tier studies typically follow EPA study guidelines by using 20 (10 male, 10 female), young-adult rats (Mandella, 2002; Ross, 2002). Half of these animals (10) are used for neuropathology. The standard toxicity studies use at least three dose groups and a control. The highest dose is usually the maximum tolerated dose; mid- and low-dose groups are calculated from this dose. Route of exposure may depend on a number of factors, including the human indication for dosing or optimal bioavailability. Acute, single-dose studies are followed by 2- to 4-week, 90-day, and 12-month chronic studies. (EPA, 1998; FDA, 2000).

The functional observational battery (FOB) is a testing strategy that was created by the EPA to make quantitative and qualitative measurements of clinical changes. These observations are made in the home-cage, by open-field activity monitoring and by interactive assessments of the rat behavior and neurological function (EPA, 1998). The observers must be well trained to critically evaluate the animals in a consistent manner. It is desired that the same observer be used for evaluation across the course of a study and that observation should be randomized. Variables such as sound, lighting, humidity, temperature, odor, and time of day should be minimized during observations. Defined scales for measuring observation response are to be used. The goal is to describe the test animal's appearance, behavior, and functional integrity. Tests proceed from the least to most interactive tests. Most of these tests do not rely on learned responses or require training or conditioning of the animal. Measurements that may be taken for the FOB are as follows.

Autonomic functions can be assessed: ranking of lacrimation and salivation, presence or absence of piloerection and exophthalmos, count of frequency of urination and defecation, pupillary size and responsiveness, degree of palpebral closure, presence, intensity and incidence of seizures, ranking of response to general stimuli, ranking of activity level, assessment of posture and gait, fore-and hind-limb strength, and quantitative measure of foot splay. Sensorimotor responses to stimuli can be assessed by gross deficits. Tail pinch, tail-flick, or hot-plate procedures may test deficits or exaggeration in pain perception. Gross assessments such as body weight, body temperature, vocalization, quality of respiration, and righting ability may also be documented (EPA, 1998).

In the evaluation of some compounds, enhanced clinical observations or detailed clinical observations are necessary. This would include all of the observational testing of the FOB without the manipulative aspects (i.e., grip strength, landing foot splay, and motor activity). Motor activity can be monitored by automated sampling devices. This is typically achieved with video cameras or activity measurement by photodetector beams. The device must be capable of detecting increases and decreases in activity, and each animal must be tested individually.

Second-tier testing is generally determined after the initial screening, to characterize the nature of a chemical's toxicity. The tests performed depend on the questions that are generated by the first-tier testing. Typically second-tier tests further study sensorimotor deficits, evaluate cognitive behaviors related to learning and memory, and assess performance of complex tasks. Third-tier tests are usually mechanistic studies that attempt to describe the toxicity on the cellular or molecular level. These studies provide information to understand the basis of the neurotoxicity (Harry, 1995).

Routine neuropathologic testing for standard toxicology studies, in which abnormalities in the FOB have not been discovered, includes fixation of tissues with buffered formalin, staining with hematoxylin and eosin stain and examination with light microscopy. Tissue samples taken for evaluation should represent all major regions of the nervous system. When there is reason to believe that neurological damage has occurred, systemic perfusion with fixatives is recommended for tissue preparation. Paraffin and/or plastic embedment is recommended to better preserve tissue for areas of suspected local damage and to improve the resolution of the images. Special staining methods, such as Bodian's or Bielchowsky's silver methods, or immunohistochemistry may be used to better characterize disease processes (O'Donoghue, 1989, EPA, 1998).

C. Inhalation Toxicology

Inhalation exposure allows the controlled delivery of an airborne compound to the respiratory tract. Inhalation exposures can be acute, in which the duration of exposure is very short; subchronic, a varying interval usually less than 90 days; or chronic, exposure to the airborne compound several hours a day for 1 to 2 years. There are several kinds of exposure systems that facilitate the delivery of the airborne compound. These include the following: whole-body exposure, head-only exposure, nose-only exposure, lung-only exposure (tracheostomy, airway catheters, or intubation), and partial-lung exposure. Each type of exposure system has advantages and disadvantages. Whole-body exposure requires the use of inhalation chambers, which can be quite expensive. Continuous airflow is maintained in the chamber during exposure. Rats are usually housed in cages within the chambers. Although whole-body exposure affords minimal restraint of the animal, inhalation of the aerosolized compound may be decreased owing to animal avoidance (i.e., rats burying their nose in their fur, thereby decreasing the amount of compound being inhaled). Whole-body exposure also allows exposure to the aerosolized compounds by other methods, such as orally and topically, therefore adding other variables to research results. Head-only exposure provides a more efficient delivery of the aerosolized compound because the animal cannot avoid inhalation other than limiting the number or depth of respirations. The head of the rat is firmly restrained, which may be quite stressful. With nose-only exposure, rats are restrained in tubes that are open at one end to the aerosolized compound. As in the head-only exposure system, the delivery of the aerosolized compound is very direct. Restraint within these tubes is stressful. Rats subjected to even mild restraint have been reported to have increased baseline serum corticosterone concentrations (Pitman, 1988). Rats immobilized in nose-only tubes have been reported to gain significantly less weight than did unrestrained controls (McConnell, 1994). Large control groups and historical data from animals in the restraint device are needed to assist in the interpretation of this data. Lung-only and partial-lung-only exposure bypasses the upper respiratory system and directly instills the aerosolized compound. These types of exposure systems require the rats to be anesthetized during exposure, which may affect the animal's breathing pattern. The anesthetic may also have some interactions with the test compound. Intratracheal instillation has become sufficiently widely used that the Inhalation Specialty Section of the Society of Toxicology has recommendations and guidelines when using this exposure method (Driscoll et al., 2000).

D. Pharmacokinetics, Pharmacodynamics, and Biometabolism Studies

Pharmacokinetics is the study of the absorption, distribution, biotransformation, and excretion of compounds (Ross and Gillman, 1990). Pharmacokinetic parameters such as peak plasma concentrations (Cmax), the time of peak absorption (Tmax), the rate of clearance (CL), biological half-life (T1/2), and the extent of distribution in the body (V) are determined from these studies. Area under the time-concentration curve (AUC) is calculated by plotting the plasma concentration against time and measuring the plotted area. These measurements are used to establish absorption and exposure in circulation. Linearity of exposure, relative to dose, is a key parameter used to define appropriate dose ranges for a drug.

Pharmacodynamics is the study of the mechanisms of action and the relationship between concentration and effect (Ross and Gillman, 1990). Maximum serum concentrations and duration in circulation are used as an estimate of drug dosages for humans. Characteristics, including oral bioavailability, and plasma elimination half-life are also assessed. Single-dose and repeated-dose studies are examined to predict steady-state plasma concentrations (Marzo, 1997; White, 2000).

Rats with jugular catheters or with jugular and femoral catheters are a primary model for pharmacokinetic and pharmacodynamic studies. Compounds are administered by mouth or through a venous catheter, and blood is collected over time. If rats are housed in metabolic cages, compound excretion may also be characterized by collection of feces and/or urine. Provided that the half-life and clearance rates are acceptable, the animals may be used repeatedly to reduce the number of animals used for studies.

Automated blood sampling machines have been created to reduce the manpower required for blood collection and decrease the stress on the animals used in these studies. Computer-driven systems withdraw the blood from a catheter, replace the blood volume with saline or another appropriate solution, and then refrigerate the stored samples for future analysis. The frequency and timing of samples can be programmed for optimal time intervals. These systems incorporate elaborate mechanisms to prevent unwanted twisting or crimping of the catheter, increasing the comfort of the animal while increasing the time of patency of the line (Roberts, 2001).

Simultaneous dosing of multiple compounds to a single animal is a technique called N-in-one dosing or cassette dosing. N represents some number of drugs, typically between 1 and 10 that are dosed to a single animal. This method requires fewer animals and results in fewer samples for analysis. This technique, although highly desirable, has inherent limitations. It can be difficult to dose a meaningful quantity of several compounds in one oral administration, and it requires that the chemistry of the compounds be compatible and assumes there are not interactions in the co-administered compounds. The initial assumption is the dose will not be toxic or overwhelm the metabolic capacity of the animal (White, 2000). Positive results must be confirmed in traditional tests.

Compounds are evaluated for dosage range, regimen, and metabolic profiles. A study series known as ADME (absorption, distribution, metabolism, and excretion) is used to answer questions about excretion route, plasma and tissue concentrations, extent of metabolism, and administered/excreted balance. Excretion-balance studies using radioactive isotopes allow characterization of compound distribution and excretion over time. After administration of a dose of a radio-labeled compound, urine, feces, and respired air (if it is a suspected route of elimination) are collected. In mass balance studies, the first goal is to determine where the entire compound goes relative to the dose administered. Blood samples are taken periodically during the study, and the exposure to the compound (radioactivity) is determined. Radioactivity in the carcass can be quantified to determine if compound resides in the tissues during or at the end of the study.

Autoradiography is another method used to study whole-body distribution and organ concentration of compounds and their metabolites. Rats are administered a radioactive dose and sacrificed over a pre-selected time course. The sections taken over the time course demonstrate uptake and subsequent decrease over time as the compound is cleared. Rats are used as the species of choice for this work primarily because of their size and the ease of obtaining samples (Marzo, 1997).

Bile duct-cannulated rats are frequently used to study biliary excretion and enterohepatic recirculation of compounds and metabolites. Bile can be collected at predetermined time intervals from a biliary cannula. Urine, blood, and feces are collected at the same time interval, allowing for examination of excretion balances and flow. Mass balance of drug can be assessed by adding blood, bile, urine, feces, and final carcass content for total radioactivity (White, 2000).

VI. CONSIDERATION OF VARIABLES IN TOXICOLOGICAL TESTING

Well-designed toxicology experiments always include appropriate control animals. In addition to the main independent variable of the project (compound), many other variables should be considered in experimental design. This is especially relevant to toxicology experiments as interactions between test variables and environmental or animal variables may be present. For example, a specific environmental variable may interact with the test article but have no effect in control animals. Thereby, a variable can increase or decrease toxicity and may cause the approval or disapproval of a compound based on erroneous experimental data. Thus, attempts to control every possible variable are critical to toxicology experiments. Variables to consider include strain or stock of rat, water, diet, intestinal flora, caging and bedding, and disease status.

A. Rat Strain or Stock

The rat has been and is a major animal in toxicology testing. Historically, specific strains and stocks of rats have been used for various tests. Because interpretation of any toxicological test often requires historical control information for comparison, introduction of new strains or stocks should be carefully considered. Moreover, when

working on a specific test, the literature on that test should be reviewed to determine the strain or stock that has been used in the past. To ensure genetic stability, strains or stocks of rats used in toxicology testing should be genetically monitored and bred in a system that minimizes genetic drift over time. The breeding colony should be maintained in a maximum barrier under rigidly controlled conditions of husbandry (see below). Efforts are ongoing to create international standards for the production and monitoring of rats for use in toxicology and pharmacology to minimize variability in test animals. These include those of the Global Alliance for Laboratory Animal Standardization (GALAS, 1993), Charles Rivers Laboratories International Genetic Standard (Charles River, 1999) and the Rat Resource and Research Center (RRRC, 2003).

B. Water

Water contains numerous and constantly changing impurities and can be an important source of variability in experiments. Monitoring the quality is difficult because each analysis can only account for a small number of possible contaminants and only represents the water at the site and on the day the sample was collected. Water can be purified using reverse osmosis systems (Raynor et al., 1984). The reverse osmosis water can then be monitored periodically to ensure the system is operating properly. Purified water can be corrosive so consideration should also be given to water lines, and equipment and systems should be flushed regularly. Systems, which automatically flush lines, are available and recommended.

C. Diet

Diet can be an extremely important source of variability in the results, and standardization of diets has been discussed for many years (Newberne et al., 1978; Coates, 1982). Toxicology experiments have usually been conducted by using diets made from natural ingredients that are either closed formula or open formula. A closed-formula diet can contain one of several possible sources of ingredients. Ingredients in these diets are usually those that cost the least or are more readily available at the time of preparation. An open-formula diet is prepared following an established and invariable list of ingredients. Although there is more variability in the ingredients in a closed-formula diet than in an open-formula diet, an open-formula diet still has variability because the ingredients are natural products that may vary from batch to batch. It is generally recommended that rats in toxicology experiments be fed an open-formula diet to minimize dietary variability in experiments (NCI, 1976; AIN, 1977; Newberne et al., 1978; Feron et al., 1980; Hamm, 1980;

Coates, 1982). The National Institutes of Health has used open-formula diets since 1972; the National Cancer Institute/National Toxicology Program, for all bioassays since 1976; and the Chemical Industry Institute of Toxicology Centers for Health Research, since 1981. The American Institute of Nutrition has recommended that the NIH-07 open-formula diet be used for rat studies (AIN, 1977). NIH-31 was developed as an autoclavable alternative to NIH-07; however, it should be noted that NIH-31 is not an autoclavable NIH-07, it is a unique diet.

Experiments that require that a particular component of the diet be controlled (e.g., a vitamin deficiency) can be conducted by using purified diets. These diets are composed of refined proteins, carbohydrates, and fats with added vitamin and mineral mixtures. For experiments that require a purified diet, the AIN-93 purified diet (AIN, 1993; Reeves et al., 1993) was developed. The use of purified diets should be carefully pursued as there are reports of complications associated with their use (Bieri, 1980), and these complications can increase with the length of the experiment (Nguyen and Woodard 1980; Hamm et al., 1982; Medinsky et al., 1982). Moreover, the use of different purified diets may affect the toxicity of the compound being studied. As an example, Goldsworthy and colleagues (1986) showed that feeding NIH-07 diet containing 2,6-dinitrotoluene resulted in liver tumors in rats, whereas the same chemical fed in the AIN-76a diet was not carcinogenic. The only known variable in these experiments was the diet. One mechanism for this phenomenon was the effect of diet on intestinal flora (see next section). Collectively, these studies highlight the principle that dietary changes should be made with care. The use of dietary restriction, pair feeding, and other feeding methods also require careful experimental design. The reader is referred to an excellent review of the implications of dietary restriction for the design and interpretation of toxicity and carcinogenicity studies (Hart, 1995).

D. Intestinal Flora

The influence of intestinal flora on toxicology experiments is one variable that is easily overlooked. An excellent review of this topic is available (Rowland, 1988). The intestinal flora may metabolize xenobiotics more than the host, and different intestinal floras can result in different toxicological endpoints in experiments. The diet may also affect the intestinal flora and subsequently affect compound toxicity. For example, deBethizy et al. (1983) showed that Fischer-344 rats fed Purina 5002 or NIH-07 diets had 135% and 150% higher hepatic covalent binding of 2,6-dinitrotoluene, respectively, compared with that of rats fed a AIN-76A purified diet or AIN-76A diet with added pectin. Changes in betaglucuronidase and

nitroreductase activities, enzymes implicated in the activation of 2,6-dinitrotoluene to a toxicant, correlated with the total number of cecal microflora that were associated with feeding of each diet. Goldstein et al. (1984) studied the influence of dietary pectin on intestinal microfloral metabolism and toxicity of nitrobenzene. They found that nitrobenzene induced methemoglobinemia in male Fischer-344 rats fed NIH-07 and was not detectable in rats fed AIN-76a when both groups were exposed to 200 mg/kg of nitrobenzene. Inclusion of 5% pectin in the AIN-76a diet resulted in methemoglobin concentrations comparable to those found in rats fed NIH-07.

To control and study the importance of the microflora on a toxic endpoint, one may consider the use of axenic or gnotobiotic rats. Gnotobiotic technology has been used in experimentation for many years (Trexler et al., 1999). More information can be obtained in the gnotobiology chapter of this textbook and from the Association for Gnotobiotics (*http://www.gnotobiotics.org/*). As an example of the use of gnotobiotic rats in toxicology experiments, Mirsalis and colleagues (1982) used axenic Fischer 344 rats maintained in sterile flexible isolators to show the role of the intestinal flora in the genotoxicity of dinitrotoluene. They showed that there was extensive dinitrotoluene-induced DNA repair in rats with an intestinal flora but not in rats that had no intestinal flora, indicating that metabolism by the intestinal flora is a necessary step in the genotoxicity of this compound.

E. Caging and Bedding

Historically, rats were housed in either wire-bottom or solid-bottom caging. Solid-bottom cages with filter caps changed in hoods and individually ventilated filter-top cages are now commonplace. Filter-top cages exclude infectious organisms, allowing experiments to be completed without this confounding factor. The cages also contain the test chemical and the animal byproducts, resulting in greatly increased worker safety. Use of these cages also allows the use of bedding to absorb the urine, reducing the exposure of the animals to ammonia and other odors. Rats often consume and possibly inhale some of their bedding. The type of bedding that produces the least variability is heat-treated hardwood chips. The hardwood should be manufactured exclusively for this use and should not be a treated byproduct of some other product. Dust and contamination should be minimized. The toxicologist may also wish to analyze for specific contaminants if the test compound is likely to be affected by a specific contaminant in the bedding. Wire floor inserts for solid-bottom cages minimize the access of the animals to the bedding below the insert and have been used in some experiments to minimize ingestion of bedding

(Raynor et al., 1983). The *Guide for the Care and Use of Laboratory Animals* (NRC, 1996) recommends solid-bottom caging with bedding for rodents. Therefore, the use of wire-bottom cages or the use of wire inserts in solid-bottom caging should be carefully justified and approved by the institutional animal care and use committee.

F. Disease Status

Because it is usually impossible to determine the effect of any disease on the outcome of a particular toxicology study, it is imperative that disease-free animals be used and maintained as disease free for the entire study. A plan to accomplish this in a toxicology laboratory has been published (Hamm et al., 1986).

G. Environmental Enrichment Considerations

The recent International Symposium on Regulatory Testing and Animal Welfare (Richmond et al., 2002a) recommended that rodents be housed in compatible social groups and not housed singly. They also recommended that rodents be provided with environmental enrichment. Whenever an enrichment device for a rat in a toxicology experiment is considered, the possible effect of the device on the results must be taken into account. Anything the rat ingests can affect metabolism; therefore, only devices that cannot be ingested or inhaled should be considered. Careful use of controls and historical data should be incorporated whenever a new variable such as an enrichment device is added to an experiment.

VII. HUMANE ENDPOINTS AND ROLE OF ALTERNATIVES IN TOXICOLOGY TESTING

Historically, the endpoint of many toxicologic tests was death. Over the course of the past 20 years, an abundance of effort has been placed on the development of humane endpoints and alternatives to death as an endpoint. A thorough review of endpoints in toxicologic testing is beyond the scope of this chapter, but the reader is referred to a number of excellent reviews and position papers including the following: the Society of Toxicology's position paper on the LD_{50} and acute eye and skin irritation tests (SOT 1989); the American College of Toxicology's policy statement on the care of animals used in toxicology (ACT, 1992); a review of death as an endpoint (Hamm 1995); the OECD's guidelines on the recognition and assessment of pain and distress and informed decision making for euthanasia (OECD, 2000a); and the Institute for Laboratory Animal Research's journal issues devoted

to humane endpoints in biomedical research and testing (ILAR, 2000) and to regulatory testing and animal welfare (ILAR, 2002). Moreover, the Interagency Coordinating Committee on the Validation of Alternative Methods and its supporting center, the National Toxicology Program Interagency Center for the Evaluation of Alternative Toxicological Methods (*http://iccvam.niehs.nih.gov/home.htm*), are coordinating the development, validation, acceptance, and harmonization of alternative toxicological test methods.

VIII. SAFETY IN TOXICOLOGY EXPERIMENTS

The reader is referred to several recent texts and manuscripts regarding safety in toxicology experiments: The text *Occupational Health and Safety in the Care and Use of Research Animals* (National Research Council, 1997) is an excellent source of information that can be used to develop a safety program. Reviews of chemical hazard control are also provided in *Prudent Practices in the Laboratory: Handling and Disposal of Chemicals* (NRC, 1995), and chapter 24 of the *Laboratory Animal Medicine* textbook (Hamm, 2002) is devoted to worker safety in the animal facility. The Institute for Laboratory Animal Research has devoted a recent issue of its journal to laboratory safety (ILAR, 2003).

All projects should minimize worker exposure and emphasize the use of personnel protective equipment when applicable. Four activities involving animals in toxicology testing are potentially hazardous and, as a result, should have procedures developed to minimize risks. These areas include the following: manipulating the animal's enclosure, mixing of test chemicals into feed, filling food containers with dosed feed, and washing the cage. A discussion of animal enclosure is presented in a previous section. Precautions to consider when mixing chemicals into feed include placement of equipment such as balances and feed blenders into enclosures, such as chemical fume hoods or negative pressure enclosures, that prevent worker exposure. Food containers can be filled in hoods designed to minimize exposure to the personnel, and cage waste can be removed before washing in "dump stations" that connected to exhaust systems.

IX. CONCLUSION

The rat has been one of the most important animal models in toxicology. This chapter overviews the types of toxicological studies and provides information about variables that must be considered in designing toxicology experiments such as rat strain, diet, water quality, intestinal flora, and environmental enrichment.

REFERENCES

Amdur, M.O., Doull, J., and Klaassen, C.D. (1991). "Casarett and Doull's Toxicology the Basic Science of Poisons," 4th Ed. Pergamon Press, New York.

American College of Toxicology [ACT] (1992). Policy statement: care and use of animals in toxicology. *J. Am. Coll. Toxicol.* **11**, 1.

American Institute of Nutrition [AIN] (1977). Report of the American Institute of Nutrition ad hoc committee on standards for nutritional studies. *J. Nutr.* **107**, 1340–1348.

American Institute of Nutrition [AIN] (1993). Final report of the AIN ad hoc writing committee on the reformulation of the AIN-76A rodent diet. *J. Nutr.* **123**, 1939–1951.

Ashby, J. (1995). Rodent germ cell mutagens and their activity in the rodent bone marrow micronucleus assay. *Mutat. Res.* **328**, 239–241.

Ballantyne, B., Marrs, T., and Syversen, T. (1999). "General and Applied Toxicology," 2nd Ed. Grove's Dictionaries, Inc., New York.

Bieri, J.G. (1980). Second report of the ad hoc committee on standards for nutritional studies. *J. Nutr.* **110**, 1726.

Botham, P.A. (2002). Acute systemic toxicity. *ILAR J.* **43**, S27–S30.

Botham, PA. (2004). Acute systemic toxicity-prospects for tiered testing strategies. *Toxicol. In Vitro* **18**, 227–230.

Charles River Laboratories' International Genetic Standard (1999). Charles River Reference Paper, Volume 11, #1.

Coates, M. (1982). Workshop on laboratory animal nutrition. *ICLAS Bull.* **50**, 4–11.

Dean, S.W., Brooks, T.M., Burlinson, B., Mirsalis, J., Myhr, B., Recio, L., and Thybaud, V., (1999). Transgenic mouse mutation assay systems can play an important role in regulatory mutagenicity testing *in vivo* for the detection of site-of-contact mutagens. *Mutagenesis* **14**, 141–151.

deBethizy, J.D., Sherrill, M., Rickert, D.E., and Hamm, Jr., T.E. (1983). Effects of pectin-containing diets on hepatic macromolecular covalent binding of 2,6-Dinitro[3H]toluene in Fischer-344 rats. *Toxicol. Appl. Pharmicol.* **69**, 369–376.

Derelanko, M.J. and Hollinger, M.A. (2002). "Handbook of Toxicology." CRC Press, New York.

Diehl, K.H., Hull, R., Morton, D., Pfister, R., Rabemampianina, Y., Smith, D., Vidal, J.M., and Vorstenbosch, C. (2001). A practice guide for the administration of substances and removal of blood, including routes and volumes. *J. Appl. Toxicol.* **21**, 15–23.

Driscoll, K.E., Costa, D.L., Hatch, G., Henderson, R., Oberdorster, G., Salem, H., and Schlesinger, R.B. (2000). Intratracheal instillation as an exposure technique for the evaluation of respiratory tract toxicity: uses and limitations. *Toxicol. Sci.* **55**, 24–35.

Ecobichon, D.J. (1992). "The Basis of Toxicity Testing." CRC Press, Boca Raton, FL.

Ecobichon, D.J. (2002). Reproductive toxicology. *In* "Handbook of Toxicology," Vol. 1 (M. Derelanko and M. Hollinger, eds.), pp. 371–388. CRC Press, Boca Raton, FL.

European Economic Community [EEC] (1992). Council Directive 92/32/EEC, April 30, 1993, amending for the seventh time Directive 67/548/EEC on the approximation of the laws, regulations and administrative provisions relating to the classification, packaging and labeling of dangerous substances. *Official Journal of the European Communities* **35(no. L-154)**, pp. 1–29.

Feron, V.J., Grice, H.C., Griesemer, R., Peto, R., Agthe, C., Althoff, J., Arnold, D.L., Blumenthal, H., Cabral, J.R., Della Porta, G., Ito, N.,

Kimmerle, G., Kroes, R., Mohr, U., Napalkov, N.P., Odashima, S., Page, N.P., Schramm, T., Steinhoff, D., Sugar, J., Tomatis, L., Uehleke, H., and Vouk, V. (1980). Basic requirements for long-term assays for carcinogenicity. *IARC Monogr. Eval. Carcinog. Risk Chem. Hum. Suppl.* **2 Suppl**, 21–83.

French, R.D. (1975). "Antivivisection and Medical Science in Victorian Society." Princeton University Press, Princeton, NJ.

Gad, S.C (2001). "Regulatory Toxicology," Taylor and Francis, New York.

Gallo, M.A., and Doull, J. (1991). History and scope of toxicology. *In* "Casarett and Doull's Toxicology," 4th Ed. (M.A. Amdur, J. Doull, and C.D. Klaassen, eds.), chapter 1. Pergamon Press, NY.

Global Alliance for Laboratory Animal Standardization [GALAS] (1993), *http://www.galas.org.*

Goldstein, R.S., Chism, J.P., Sherrill, J.M., and Hamm, Jr., T.E. (1984). Influence of dietary pectin on intestinal microfloral metabolism and toxicity of nitrobenzene. *Toxicol. Appl. Pharmicol.* **75**, 547–553.

Goldworthy, T.L., Hamm, Jr., T.E., Rickert, D.E., and Popp, J.A. (1986). The effect of diet on 2,6-dinitrotoluene hepatocarcinogenesis. *Carcinogenesis* **7**, 1909–1915.

Haggarty, G.C. (1989). Development of tier 1 neurobehavioral testing capabilities for incorporation into pivotal rodent safety assessment studies. *J. Am. Coll. Toxicol.* **8**, 53.

Hamm, Jr., T.E. (1980). *Proceedings of a Workshop on the Optimal use of Facilities for Carcinogenicity/Toxicity Testing.* NTP/NCI Carcinogenesis Testing Program, Boiling Springs, PA.

Hamm, Jr., T.E. (1985). Design of the carcinogenesis bioassay. *In* "Handbook of Carcinogen Testing" (H.A. Milman and E.K. Weisburger, eds.), pp. 252–267, Noyes Publications, Park Ridge, NJ.

Hamm, Jr., T.E. (1995). Proposed IACUC guidelines for death as an endpoint in rodent studies. *Contempt. Top. Lab. Anim. Med.* **34**, 69–71.

Hamm, Jr., T.E. (2002). Control of biohazards associated with the use of experimental animals. *In* "Laboratory Animal Medicine," 2nd Ed. (J.G. Fox, L.C. Anderson, F.M. Loew and F.W. Quimby, eds.), pp 1047–1056, Academic Press, New York.

Hamm, Jr., T.E., Raynor, T., and Caviston, T. (1982). Unsuitability of the AIN-76A diet for Male F-344 and CD rats and improvement by substituting starch for sucrose. *Lab Anim. Sci.* **32**, 414–415.

Hamm, Jr., T.E., Raynor, T.H., and Sherrill, M. (1986). Animal Management Procedures to Avoid or Minimize Disease Problems. *In* "Complications of Viral and Mycoplasma Infections in Rodents to Toxicology Research and Testing" (T.E. Hamm, Jr., ed.), pp. 175–184, Hemisphere Publishing Corporation, Washington, DC.

Harry, J., O'Donoghue, J.L., and Goldberg, A.M. (1995). Neurotoxicity. *In* "Screening and Testing Chemicals in Commerce: OTA-BP-ENV-166 Background Paper for U.S. Congress." Office of Technology Assessment, Washington, DC.

Hart, R.W., Neumann, D.A., and Robertson, R.T. (1995). "Dietary Restriction Implications for the Design and Interpretation of Toxicity and Carcinogenicity Studies," ILSI Press, Washington, DC.

Hayes, A.W. (2001). "Principles and Methods of Toxicology," 4th Ed. Taylor & Francis, Philadelphia.

Hays, H.W. (1986). "Society of Toxicology History." Society of Toxicology, Washington, DC.

Hoyer, P.B. (2001). Reproductive toxicology: current and future directions. *Biochem. Pharmicol.* **62**, 1557–1564.

Institute of Laboratory Animal Research [ILAR] (2000). Humane endpoints for animals used in biomedical research and testing. *ILAR J.* **41(2)**.

Institute of Laboratory Animal Research [ILAR] (2002). Regulatory testing and animal welfare. *ILAR J.* **43(Suppl)**.

Institute of Laboratory Animal Research [ILAR] (2003). Occupational safety and health in biomedical research in the laboratory. *ILAR J.* **44 (1)**.

International Conference on Harmonization [ICH] (1995). "Guidance on Specific Aspects of Regulatory Genotoxicity Tests for Pharmaceuticals." International Conference on Harmonization of Technical Requirements for Registration of Pharmaceuticals for Human Use, Geneva, Switzerland.

International Conference on Harmonization [ICH[(1997). "Genotoxicity: A Standard Battery for Genotoxicity Testing for Pharmaceuticals." International Conference on Harmonization of Technical Requirements for Registration of Pharmaceuticals for Human Use, Geneva, Switzerland.

Irwin, S. (1968). Comprehensive observational assessment: a systematic, quantitative procedure for assessing the behavioral and physiologic state of the mouse. *Psychopharmacologia* **13**, 222–257.

Kohler, S.W., Provost, G.S., Fieck, A., Kretz, P.L., Bullock, W.O., Putman, D.L., Sorge, J.A., and Short, J.M. (1991). Spectra of spontaneous and induced mutations using a lambda ZAP/*lacI* shuttle vector. *Environ. Mol. Mutagen.* **18**, 316–321.

MacGregor, J.T. (1998). Transgenic animal models for mutagenesis studies: role in mutagenesis research and regulatory testing. *Environ. Mol. Mutagen.* **32**, 106–109.

Mandella, R.C. (2002). Applied neurotoxicology. *In* "Handbook of Toxicology," Vol. 1 (M. Derelanko and M. Hollinger, eds.), pp. 371–388. CRC Press, Bocca Raton, FL.

Marzo, A. (1997). Clinical pharmacokinetic registration file for NDA and ANDA procedures. *Pharmacol. Res.* **36**, 425–50.

McConnell, E.E., Kamstrup, O., Mussellman, R., Hesterberg, T.W., Chevalier, J., Miiller, W.C., and Thevenaz, P. (1994). Chronic inhalation study of size-separated rock and slag wool insulation fibers in Fischer 344/N rats. *Inhal. Toxicol.* **6**, 571–614.

Medinsky, M.A., Popp, J.A., Hamm, Jr., T.E., and Dent, J.G. (1982). Development of hepatic lesions in male Fisher-344 rats fed AIN-76A purified diet. *Toxicol. Appl. Pharmacol.* **62**, 111–120

Milman, H.A. and Weisburger, E.K. (1985). "Handbook of Carcinogen Testing." Noyes Publications, Park Ridge, NJ.

Mirsalis, J.C., Hamm, Jr., T.E., Sherrill, J.M., and Butterworth, B.E. (1982). The role of intestinal flora in the genotoxicity of dinitrotoluene in rats. *Nature* **295**, 322–323.

Morton, D.B., Jennings, M., Buckwell, A., Ewbank, R., Godfrey, C., Holgate, B., Inglis, I., James, R., Page, C., Sarman, I.L., Verschoyle, R., Westall, L., and Wilson, A.B. (2001). Refining procedures for the administration of substances. *Lab. Anim.* **31**, 1–41.

Moser, V.C. (1991). Applications of a neurobehavioral screening battery. *J. Am. Coll. Toxicol.* **10**, 661–669.

National Cancer Institute [NCI] (1976). "Guidelines for Carcinogen Bioassay in Small Rodents, National Cancer Institute Technical Report Series, No. 1." National Academy Press, Bethesda, MD.

National Research Council [NRC] (1995). "Prudent Practices in the Laboratory Handling and Disposal of Chemicals." National Academy Press, Washington, DC.

National Research Council [NRC] (1996): "Guide for the Care and Use of Laboratory Animals." National Academy Press, Washington, DC.

National Research Council [NRC] (1997): "Occupational Health and Safety in the Care and Use of Research Animals." National Academy Press, Washington, DC.

Neubert, D. (2002). Reproductive toxicology: the science today. *Teratog. Carcinog. Mutagen.* **22**, 159–174.

Newberne, P.M., Dieri, J.G., Briggs, G.M., and Nesheim, M.C. (1978). "Control of Diets in Laboratory Animal Experimentation." Institute of Laboratory Animal Resources, Assembly of Life Sciences, National Research Council, National Academy of Sciences, Washington, DC.

Nguyen, H.T. and Woodard, T.C. (1980). Intranephronic calcinosis in rats. *Am. J. Pathol.* **100**, 39–55.

O'Donoghue, J.L. (1989). Screening for neurotoxicity using a neurologically based examination and neuropathology. *J. Am. Coll. Toxicol.* **8**, 97–115.

Organization for Economic Co-operation and Development [OECD] (1993). "Guidelines for the Testing of Chemicals." OECD, Paris.

Organization for Economic Co-operation and Development [OECD] (1996). "Guidelines for Testing of Chemicals: Proposal for Updating Guideline 414: Prenatal Developmental Toxicity Study. Draft Document." OECD, Paris.

Organization for Economic Co-operation and Development [OECD] (1998). "Guideline 425: Acute Oral Toxicity-Modified Up and Down Procedure." OECD, Paris.

Organization for Economic Co-operation and Development [OECD] (2000a). "Guideline 19: Guidance Document on the Recognition, Assessment, and Use of Clinical Signs as Humane Endpoints for Experimental Animals Used in Safety Evaluation," OECD, Paris.

Organization for Economic Co-operation and Development [OECD] (2000b). "Guideline 20: Regulatory Guidelines for Neurotoxicity Testing, Series on Testing and Assessment." OECD, Paris.

Parascandola, J. (1991). The development of the Draize test for eye toxicity. *Pharm. Hist.* **33**, 111–117.

Pitman, D.L., Ottenweller, J.E., and Natelson, B.H. (1988). Plasma corticosterone concentrations during repeated presentation of two intensities of restraint stress: chronic stress and habituation. *Physiol. Behav.* **4**, 47–55.

Pharmaceutical Affairs and Food Sanitation Council [PAFSC] (1997). Articles 12, 2214, 19–2, 5 and 24, Ministry of Health, Labour and Welfare, Tokyo, Japan. *http://www.mhlw.go.jp/english/index*.

Provost, G.S., Kretz, P.L., Hamner, R.T., Matthews, C.D., Rogers, B.J., Lundberg, K.S., Dycaico, M.M., and Short, J.M. (1993). Transgenic systems for *in vivo* mutation analysis. *Mutat. Res.* **288**, 133–149.

Rat Resource and Research Center [RRRC] (2003). *www.nrrc.missouri.edu.*

Raynor, T.H., Steinhagen ,W.H., and Hamm, Jr., T.E. (1983). Differences in the microenvironment of a polycarbonate caging system: bedding vs. raised wire floors. *Lab. Anim.* **17**, 85–89.

Raynor, T.H., White, E.L., Cheplen, J.M., Sherrill J.M., and Hamm, Jr., T.E. (1984). An evaluation of a water purification system for use in animal facilities. *Lab. Anim.* **18**, 45–51.

Reeves, P.G., Rossow, K.L., and Lindlauf J. (1993). Development and testing of the AIN-93 purified diets for rodents: results on growth, kidney calcification and bone mineralization in rats and mice. *J. Nutr.* **123**, 1923–1931.

Richmond, J., Fletch, A., and Van Tongerloo, R. (2002a): The International Symposium on Regulatory Testing and Animal Welfare: recommendations on best scientific practices for animal care in regulatory toxicology. *ILAR J.* **43 Suppl**, S129–S132.

Roberts, S.A.(2001). High-throughput screening approaches for investigating drug metabolism and pharmacokinetics. *Xenobiotica* **31**, 557–589.

Ross, E.M. and Gilman, A.G. (1990). Pharmacodynamics: Mechanism of drug action and the relationship between drug concentration and effect. *In* "The Pharmacological Basis of Therapeutics," 8th Ed. (A.G. Gilman, L.S. Goodman, T.W. Rall, and F. Murad, eds.), chapter 2. Macmillan Publishing Company, New York.

Ross, J.F. (2002). Tier 1 neurological assessment in regulated animal safety studies. *In* "Handbook of Neurology," Vol. 2 (J. Massaro, ed.), pp. 461–505. Humana Press, Totowa, NJ.

Rowland, I.R. (1988). "Role of the Intestinal Flora in Toxicity and Cancer." Academic Press, San Diego, CA.

Russell, W.M.S. and Burch, R.L. (1959). "The Principles of Humane Experimental Technique". Charles C. Thomas, Springfield IL.

Sass, N. (2000). Humane endpoints and acute toxicity testing. *ILAR J.* **41**, 114–123.

Silverman, W.A. (1985). "Human Experimentation: A Guided Step into the Unknown." Oxford University Press, New York.

Society of Toxicology [SOT] (1989). SOT position paper: comments on the LD_{50} and acute eye and skin irritation tests. *Fundam. Appl. Toxicol.* **134**, 621–623.

Technical Committee on Alternatives to Animal Testing [TCAAT] (1996). Replacing the Draize eye irritation test: scientific background and research needs. *J. Toxicol. Cutaneous Ocul. Toxicol.* **15**, 211–234.

Trexler, P.C. and Orcutt, R.P. (1999). Development of gnotobiotics and contamination control in laboratory animal science. *In* "Fifty Years of Laboratory Animal Science," (C.W. McPherson and S.F. Mattingly, eds.), chapter 16. Sheridan Books, Chelsea, MI.

Turner, S.D., Tinwell, H., Piegorsch, W., Schmezer, P., and Ashby, J. (2001). The male rat carcinogens limonene and sodium saccharin are not mutagenic to male Big Blue rats. *Mutagenesis* **16**, 329–332.

U.S. Environmental Protection Agency [EPA] (1996). "Health Effects Test Guidelines: Prenatal Developmental Toxicity study, Document OPPTS 870.3700. U.S. Government Printing Office, Washington, DC.

U.S. Environmental Protection Agency [EPA] (1998). "Health Effects Test Guidelines: Neurotoxicity Screening Battery," Document OPPTS 870.6200. U.S. Government Printing Office, Washington, DC.

U.S. Food and Drug Administration [FDA] (1978). "Good Laboratory Practice Regulations for Nonclinical Laboratory Studies." U.S. Code of Federal Regulations, Title 21, Part 8.

U.S. Food and Drug Administration [FDA] (1993a). International Conference on Harmonization; guideline on detection of toxicity to reproduction for medicinal products. *Fed. Reg.* **59(183)**, 48746–48752.

U.S. Food and Drug Administration [FDA] (1993b). "New drug evaluation guidance document, refusal to file, July 12, 1993." FDA, Center for Drug Evaluation and Research, Rockville, MD.

U.S. Food and Drug Administration [FDA] (2000). "Toxicological Principles for the Safety of Food and Ingredients, Redbook 2000." FDA, Center for Food Safety and Applied Nutrition, Rockville, MD.

White, R.E. (2000). High-throughput screening in drug metabolism and pharmacokinetic support of drug discovery. *Annu. Rev. Pharmacol. Toxicol.* **40**, 133–157.

Yu, M., Jones, M.L., Gong, M., Sinha, R., Schut, H., and Snyderwine, E.G. (2002). Mutagenicity of 2-amino-1-methyl-6-phenylimidazo[4,5]pyridine in the mammary gland of Big Blue rats on high-and-low-fat diets. *Carcinogenesis* **23**, 877–884.

Chapter 28

Embryology and Teratology

Carol Erb

I. INTRODUCTION

Rodents, specifically the rat, have been used extensively to study the development of the mammalian embryo, and the rat is one of the species of choice for the regulatory assessment of developmental toxicology. The purposes of this chapter are to present the normal embryology of the rat and the methods used in experimental teratology.

II. EMBRYOLOGY

A. Preimplantation Embryo

1. Fertilization

Fertilization is a process by which two gametes fuse together to create a new individual. Fertilization functions to allow genes to be transferred from parent to offspring and the process of development to begin. The rat is an excellent model for the study of fertilization because the timing of events has been well established. In the female rat, the onset of behavioral estrus, or heat, is a suitable reference point to begin the timing of the estrous cycle. Females respond to manual manipulation of the pudendal region with a typical lordosis, which is accompanied by a definite ear quiver at the onset of heat (Blandau et al., 1941; Odor and Blandau, 1951), and it has been established that ovulation usually occurs 11 to 13 hours after the onset of heat (Odor and Blandau, 1951) but may occur as early as 6.5 hours (Boling et al., 1941).

In the female, the ovum forms the first polar body in the ovarian follicle, although the polar body is lost soon after its formation. The second maturation division begins immediately after completion of the first division. The appearance of the second maturation spindle in the monaster stage of the second maturation division marks the end of the ovarian ova maturation. After ovulation, the second maturation division of the ova does not progress beyond the metaphase stage unless sperm penetration occurs (Odor and Blandau, 1951). The zona pellucida is laid down during the follicular development of the egg and appears to be a secretory product of the granulosa cells. The granulosa cells immediately around the egg are shed with the egg at ovulation and are carried down the fallopian tube into the ampullar portion of the oviduct. They form a loose cluster of cells around the egg (corona radiata) that must be penetrated by the sperm before it can come into contact with the plasma membrane of the egg (Davies and Hesseldahy, 1971). The corona radiata are shed shortly after fertilization. In the ampullar portion of the oviduct, the recently discharged ova are clumped together (Huber, 1915b).

In the male, spermatozoa from the testes are stored in the epididymis, where they mature and acquire the ability to move. At the time of ejaculation, sperm still lack the capacity to bind to and fertilize an egg. This final stage of sperm maturation (capacitation) occurs after the sperm are inside the female reproductive tract. Spermatozoa have been found disseminated throughout the uterine cornua and already migrating into the uterine segments of the oviducts within 15 minutes after ejaculation. At 30 minutes after ejaculation, spermatozoa were found in 88% of the oviducts examined, and 100% of the examined oviducts had spermatozoa present 1 hour after ejaculation. Blandau and Money (1944) and Huber (1915b) found that the number of spermatozoa entering the oviduct is exceedingly small compared with the number present in the uterine cornua. It is estimated that the lifespan of spermatozoa in the genital tract is only about 10 hours in the albino rat (Huber, 1915b).

Fertilization takes place in the ampullae of the oviducts (Huber, 1915b). When the fertilizing spermatozoon comes into contact with the vitelline membrane, the cell membranes of both the spermatozoon and the egg rupture in the area of contact and fuse with each other as the spermatozoon passes into the cytoplasm of the egg (Austin, 1961). The zona pellucida undergoes a change (zona reaction) after the entry of the first spermatozoon that tends to exclude other spermatozoa. The zona reaction is thus a mechanism that helps to prevent the occurrence of polyspermic fertilization. In the rat, the mean time to complete the zona reaction has been estimated to be not less than 10 minutes or more than 1.5 to 2 hours (Austin, 1961). Odor and Blandau (1951) found that 47% of ova observed 11 hours after the onset of heat contained fertilizing sperm, whereas 90% of ova examined after 13 hours contained sperm. After sperm penetration occurs, the second maturation division of the ova continues in the oviduct. The sequence of events that lead to the formation of the two-cell embryo are discussed in detail by Odor and Blandau (1951). The second polar body is formed and extruded by the ova, and the sperm head undergoes enlargement during the first 4 hours after sperm penetration in the oviduct. From 5 to 8 hours after sperm penetration, the female pronucleus forms but remains close to the extruded second polar body. The typical hook-shaped sperm head is transformed into an elongated slipper-shape pronucleus that remains in the vicinity of the proximal end of the middle-piece of the sperm flagellum. There is a rapid increase in the size of the pronuclei and in the number of their nucleoli and a gradual migration of the pronuclei toward the center of the ovum during the interval of 9 through 19 hours after sperm entry (Fig. 28-1). The first segmentation spindle is commonly found 21 to 23 hours after sperm penetration. Fertilization may be considered complete with the condensation of the chromosomes in

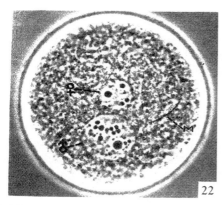

Fig. *28-1* Living ovum in which the pronuclei are of maximum size. Note the presence of the flagellum. ×645. (Data from Odor and Blandau, 1951.)

the male and female pronuclei and the coming together of the two groups of chromosomes to form a single chromosome group. These events constitute the prophase of the first cleavage mitosis (Austin, 1961). By the end of the first day, the fertilized ova have traveled about one-fourth the length of the oviduct and are found lying free it its lumen (Huber, 1915a,b).

2. Cleavage/Morula Formation

The chromosomes proceed immediately thereafter to become arranged at the metaphase plate of the first cleavage spindle (Austin, 1961). The first segmentation occurs during the early part of the second day after insemination, and the resulting two-cell stage lasts for a period of about 24 hours. By the end of the second day after insemination, the two-cell embryo has traversed a little over one-half the length of the oviduct. One of the cells usually divides before the other, resulting in a three-cell stage. By the end of the third day, only four-cell embryos are found in the oviduct. The first three segmentations are spaced at intervals of about 18 to 24 hours, and by the end of the fourth day, the 12- to 16-cell embryos pass from the oviducts to the uterine horns. In the albino rat, during the first 4 days of the development, there is only a very slight increase on the size of the cell mass compared with the unsegmented ovum with two pronuclei (Huber, 1915a,b).

During the initial cleavage stages, at least in the mouse for which the most data are available, there are relatively minor changes in the wet weight, dry weight, volume, or protein content of the embryo (Wales, 1975). Cleavage is the initial phase of cell division after fertilization and is similar to mitotic division except that the individual blastomeres do not undergo a period of growth between successive cleavages, as with other cells. As a result, the blastomeres become progressively smaller as cleavage continues (Gulyas, 1975). Cell lineage studies in early mammalian embryos have been confined largely to the mouse.

During the early cleavage stages up to the eight-cell stage, all blastomeres appear morphologically identical, and each is competent to contribute to descendants of both trophectoderm and inner cell mass cells (Fleming, 1987). The mouse inner cell mass is established by cells that are allocated to internal positions after the eight-cell stage (Pedersen et al., 1986). Inner cell mass pluripotency is preserved until at least the late 32- or early 64-cell stage (Dyce et al., 1987). Initially, both turnover of energy substrates and synthetic activity are low, but they increase significantly over the first 3 or 4 days when glycolysis and tricarboxylic acid cycle activity increase significantly. Increasing macromolecular synthesis leads to the appearance of new RNA, DNA, protein, and polysaccharides (Wales, 1975). In the mouse and rabbit, RNA synthesis is low during the early cleavage stages and then increases rapidly at the early morula stage. The rate of protein synthesis is low during preimplantation development in the mouse and rabbit, increasing rapidly in the eight-cell stage in the mouse; however, the rate does not increase until the blastocyst stage in the rabbit (Biggers and Borland, 1976). During the preimplantation period, development follows a program determined partly by the transcription of the maternal genome during oogenesis and partly by reading of the new genome after fertilization. Biggers and Borland (1976) concluded that the mammalian embryo is not completely self-contained but undergoes important exchanges with the dam. These exchanges are of three types: exchanges concerned with homeostatic functions that preserve the internal environment of the embryo (e.g., pH and osmotic pressure), exchanges involving the uptake of substances that can be metabolized, (e.g., glucose and amino acids), and specific signals that pass between dam and embryo.

Early on the fifth day, the 24- to 32-cell morulae are found lying free in the lumen of the uterus, spaced in a manner similar to how they will implant. Krehbiel and Plagge (1962) concluded that the random scattering of rat preimplantation embryos at the time of implantation determines the spacing of the implanted embryos. No predetermined sites for blastocyst attachment have been found (Abrahamsohn and Zorn, 1993).

3. Blastocyst

Formation of the blastocoele by cavitation of morula masses composed of 30 to 32 cells begins mid day 5 (Huber, 1915a). The blastocyst is a roughly spherical epithelial structure that ranges in size from 60 to 85 μm and is surrounded by the 2.5 to 3-μm-thick zona pellucida.

It consists of a hollow sphere of cuboidal trophoblast cells with their apical surface facing on the zona pellucida (Schlafke and Enders, 1968), with an inner eccentrically placed group of large (7–12 μm), roughly spherical, and loosely associated inner cell mass cells, which occupies about one-quarter of the volume of the sphere. The trophectoderm cells have descendants only in placental structures (chorion, ectoplacental cone, and trophoblast giant cells). The trophoblast is continuous, with a massing of epithelial cells at one pole of the blastocyst, the inner cell mass, from which the embryo and its immediate membranes are developed (amnion, yolk sac, allantois, and mesodermal components of the placenta). The blastocysts shed their zonae during the afternoon of day 5 in the uterus before implantation (Dickmann and Noyes, 1961; Enders, 1971; Lundkvist and Nilsson, 1984) when the pregnant rat is under the influence of both progesterone and estrogen (Fig. 28-2) (Dickmann, 1969). The shedding of the zona pellucida depends on the stage of development of the ovum (Dickmann and Noyes, 1961), and this is important

because it marks the beginning of the first direct contact between the embryonic and maternal tissues (Davies and Hesseldahy, 1971). In the rat, the loss of the zona pellucida is probably the last event that precedes the beginning of implantation. During the sixth day, the size of the blastodermic vesicles increase, partly as a result of flattening of the roof cells, and partly as a result of re-arranging and flattening of the cells constituting the floor of the vesicle. By the end of day 6, the blastodermic vesicle consists of a discoidal area, the germ disc, and the remainder of the vesicle wall, a single layer of very flattened cells (Huber, 1915a,b). During the early hours of day 7, cell proliferation, rearrangement, and enlargement of cells takes place in the region of the germinal disc, initiating the phenomena known as inversion of the germ layers or entypy of the germ layers. The anlage of the ectoplacental cone, or *träger*, can be recognized (Huber, 1915a,b).

The preimplantation period is usually regarded as a period during which toxic insult generally results in embryonic death or absence of an effect because of the

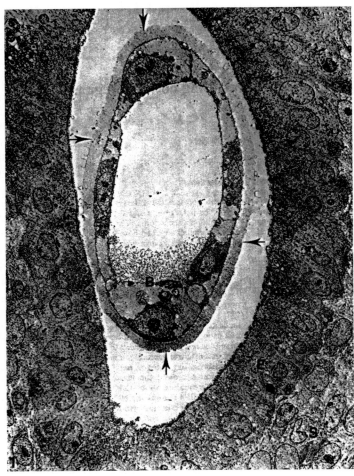

Fig. 28-2 Blastocyst-containing site at 06.00 h on Day 5 of pregnancy. The blastocyst (*B*) is encased in an intact zona pellucida (*arrows*). *E*, uterine luminal epithelium; *S*, uterine stroma. ×3,900. (Data from Lundkvist and Nilsson, 1984.)

regenerative powers of the pleuripotent cells of the embryo at this stage. However, preimplantion exposure to actinomycin D and methotrexate has been shown to affect growth and development of the embryo (Christian, 2001), whereas dichloro-diphenyl-trichloroethane, nicotine, or methylmethane sulfonate results in body and/or brain weight deficits or death (Fabro, 1973; Fabro et al., 1974). Malformations have been caused by administration of methylnitrosourea, cyproterone acetate, and medroyprogesterone acetate during the preimplantation period (Eibs et al., 1982; Takeuchi, 1984).

4. Yolk Sac Placenta Development

Brunschwig (1927) and Everett (1933) suggested that the yolk sac epithelium of the rat is physiologically a placenta that functions as an organ for maternal and embryo exchange. Over much of its surface, Reichert's membrane is directly bathed by circulating maternal blood that allows for diffusion of materials into the yolk sac cavity, from which they are absorbed into the embryo (Everett, 1935). Therefore, in the rat there are two placentas that serve as organs for maternal embryo exchange. The first to develop is a villiary highly vascularized yolk sac placenta, which will be followed on day 11.5 to 12.5 by the chorioallantoic placenta (Beck et al., 1967b; Jollie, 1990; Abdulrazzaq et al., 2001). The two placentas are present together throughout most of gestation.

The formation of the yolk sac placenta begins on day 7 with the proliferation at the periphery of the inner cell mass of the hypoblast, which will line the blastocoel, converting this chamber into a primitive yolk sac cavity lined with a bilaminar structure of trophoblast and a single layer of hypoblast (endodermal cells); (Fig. 28-3) (Jollie, 1990). Subsequent development consists of two separate processes: an inversion of the disc, and an inversion of the yolk sac that lies peripheral to it. The inversion of the yolk sac or entypy is a fundamental phenomenon observed in rodents; it is associated with a precocious development of the inner cell mass that proliferates rapidly and becomes invaginated into the yolk sac cavity. The inversion of the germ layers involves a process that temporarily brings embryonic hypoblast (endoderm) external to the embryonic epiblast (ectoderm) as a result of the conversion of the inner cell mass to a bilaminar embryonic disc by the cavitation of the inner cell mass to form an amniotic cavity. By day 8, the amniotic cavity has separated the ectoplacental cone from the inner cell mass. Three cavities eventually form within the ectoderm of the egg cylinder: the amniotic cavity (most ventral cavity), the extraembryonic cavity (middle), and the ectoplacental cavity that is just under the ectoplacental cone and is transitory. The membrane separating the extraembryonic cavity from the ectoplacental cavity is the chorion, and the membrane between the extraembryonic coelom and the amniotic cavity is the amnion (Beaudoin, 1980). The visceral (adembryonal) portion of the yolk sac becomes trilaminar by an interposition of extraembryonic mesoderm between epiblast and hypoblast and will soon undergo angiogenesis. The vascular adembryonal portion or visceral wall of the yolk sac membrane, with its endodermal epithelium directed outward, completely surrounds the embryo external to the amnion. The more peripheral portion or the parietal wall of the yolk sac membrane remains bilaminar, but the ectoderm (trophoblast) is separated from the yolk sac endodermal epithelium by a thick basement membrane, Reichert's membrane. The space between the two walls of the yolk sac membrane constitutes the yolk sac cavity. By day 10, the visceral yolk sac mesoderm has split to form an extraembryonic coelom, undergoes angiogenesis, and becomes vascularized by the peripheral vitelline circulation. The yolk sac is thus the first hematopoietic organ. Division of the yolk sac into visceral and parietal walls is complete in the rat by gestation day 11 to 12. Meanwhile, the visceral wall of the yolk sac cavity also differentiates into the villous portion, a more highly vascular and more absorptive area, and a smooth portion that is less vascular and presumably a less absorptive area. In rodents, the visceral yolk sac villi are generally found close to the fetal surface of the chorioallantoic placenta. At this stage of development, the yolk sac is the only placental exchange site (Jollie, 1990). Inversion occurs as a result of the apposition of the inner (visceral) wall of the yolk sac, which is vascular, to the parietal wall of the yolk sac. The parietal wall of the yolk sac and with its associated trophoblast then disappears about two-thirds of the way through gestation in rodents. This brings the absorptive cells of the visceral yolk sac into contact with the uterine epithelium and its secretion.

B. Implantation and Chorioallantoic Placenta Development

Implantation is a complex sequence of events that requires proper synchrony between embryonic and endometrial development. It has been shown that implantation requires strictly timed hormonal conditioning, consisting of continuous preparation of the endometrium by progesterone for at least 48 hours, and then a brief intervention of minute amounts of estrogen on the fourth day. A delay in the presence of estrogen results in the postponement of implantation (Psychoyos, 1966). Between day 5 and 7 of gestation, the implantation chamber develops from a flattened shallow crypt to an elongated tube (Schlafke et al., 1985). On day 5, the blastocyst is still free in the lumen but is pressed between the enlarged cells of the superficial epithelium of the antimesometrial

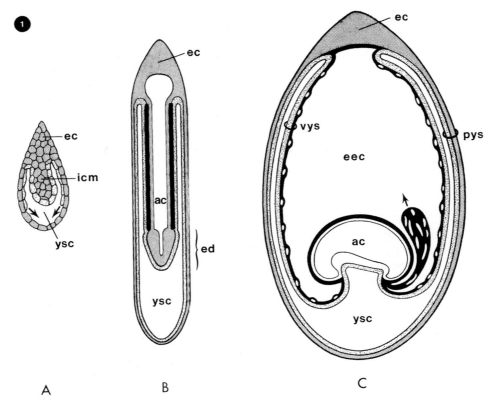

Fig. 28-3 **1.** Diagrams of an implanting rat conceptus that illustrate the mode of formation of the yolk sac in myomorphic laboratory rodents. *A:* At 7 days postcoiturn hypoblast has delaminated from inner cell mass. At the periphery of the inner cell mass, hypoblast proliferates to line the blastocoel, converting this chamber into a yolk-sac cavity. *B:* At 8 days an amnionic cavity has separated the ectoplacental cone from the inner cell mass: and the cells of the latter are arranged to form a bilaminar embryonic disc. The adembryonal (i.e., visceral) portion of the yolk sac has become trilaminar by an interposition of extraembryonic mesoderm between epiblast (ectoderm) and hypoblast (endoderm); the parietal yolk-sac endodermimural trophoblast complex, however, remains two-layered, viz., the "bilaminar omphalopleure." *C:* By 10 days visceral yolk-sac mesoderm has split to form an extraembryonic coelom and undergoes angiogenesis to vascularize this portion of the yolk sac. The visceral yolk sac almost encloses the conceptus/amnionic cavity complex; and an allantoic diverticulum, also angiogenic, has differentiated. **2.** By 11½ days, anatomic relationships of rat fetal membranes essentially are definitive. The distal portion of the allantoic diverticulum is invading the ectoplacental cone (which, at this stage, already contains maternal blood sinuses) to form with it a definitive chorioallantoic placenta. Note that the allantoic stalk (or umbilical cord) attaches to this placenta at a point around which visceral and parietal walls of the yolk sac join and that, as a consequence, at this time the conceptus and its amnionic cavity are enclosed completely by both yolk-sac membranes, except at the root of the umbilical cord and at the yolk stalk. (Tone Code: cross-hatching, epiblast (or ectoderm); stippling, hypoblast (or endoderm); solid black, mesoderm) (Data from Jollie, 1990.).

wall (Psychoyos, 1966). This first association between the blastocyst and the uterus establishes a definitive position for the blastocyst in relation to the uterus and is frequently referred to as attachment (Schlafke and Enders, 1975). This attachment stage can be further divided into an appositional stage and a subsequent adhesion stage. It is during the apposition stage of implantation that the first structural interaction occurs between embryonic and maternal cells. Apposition is characterized by increasing contact between the microvilli on the surface of the trophoblastic and uterine epithelial cells, leading to their interdigitation (Abrahamsohn and Zorn, 1993). Contact between the embryo and the surface of the uterine epithelium increases gradually at the beginning of implantation and is probably caused by a combination of blastocyst swelling and closure of the uterine lumen (Abrahamsohn and Zorn, 1993).

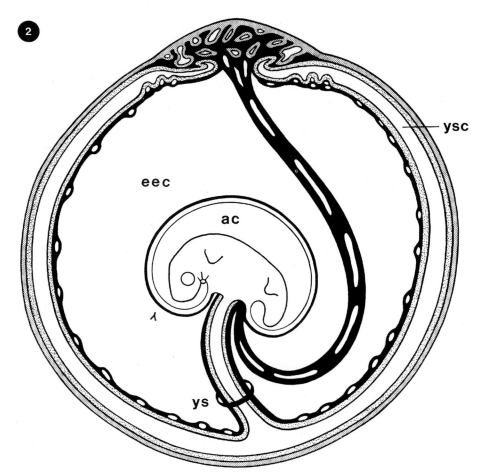

Figure 28.3 Continued.

Abbreviations
ac   amnionic cavity
ad   allantoic diverticulum
bl   basal lamina(e)
cp   chorioallantoic placenta
ec   ectoplacental cone
ed   embryonic disk
eec   extraembryonic coelom
icm   inner cell mass
mt   mural trophoblast
pee   parietal endodermal epithelium
pys   parietal wall of the yolk sac
Rm   Reichert's membrane
uc   umbilical cord
vcl   vitelline capillary lumen
vee   visceral endodermal epithelium
vys   visceral wall of the yolk sac
ys   yolk stalk
ysc   yolk-sac cavity

Fig. 28-3 Continued

The attachment reaction that obliterates the uterine lumen occurs when the uterine surfaces approximate each other (Schlafke and Enders, 1975). By the seventh day, the decidual growth that has occluded the uterus has isolated each segment of the uterus that contains an embryo (Bridgman, 1948). This encapsulating uterine tissue underlying the ectoplacental cone becomes the maternal part of the definitive placenta, the decidual basalis. Implantation proceeds into the adhesion stage when the surfaces of trophoblastic and luminal epithelial cells have lost their microvilli and run parallel to one another. A lack of estrogen which occurs during lactation or after ovariectomy will induce a state of delayed implantation, and development of the embryo will stop at the stage of apposition (Psychoyos, 1966; Abrahamsohn and Zorn, 1993). The rat blastocyst is in contact with the uterine surface for nearly a day between initial uterine vascular response and the first invasion of the luminal epithelium of the uterus by the trophoblast (Enders et al., 1980). According to Schlafke et al. (1985), the decidual cells are principally responsible for the formation of the implantation chamber and may assist in the expansion of the blastocyst by initiating disruption of the residual uterine luminal epithelial basal lamina. During the seventh day, the blastodermic vesicles become definitely oriented in the decidual crypts, which are lined by slightly flatted cubic epithelium and directed toward the antimesometrial border (Huber, 1915b; Abrahamsohn and Zorn, 1993; De Rijk et al., 2002). In the rat, embryo implantation normally occurs at the anti-mesometrial side of the uterus. The inner cell mass always faces toward the mesometrial border, and the final placenta is formed at the mesometrial side (Fig. 28-4) (Abrahamsohn and Zorn, 1993; De Rijk, et al., 2002). Uterine luminal epithelium is present both mesometrial and antimesometrial to the blastocyst, but it is discontinuous in a central band on the walls of the implantation chamber, indicating that these are the initial areas of cellular interaction between embryonic and maternal tissues. In this area, proximal mural trophoblasts are seen adjacent to the decidualized stromal cells of the primary decidual zone (Schlafke et al., 1985). In the rat, the interaction of trophoblast with the uterine epithelium is a displacement penetration in which the uterine luminal epithelium is readily dislodged from the basal lamina and the trophoblast comes to lie on the basal lamina (Abrahamsohn and Zorn, 1993). The trophoblast, which remains cellular, not only phagocytizes the sloughed cells but also sends projections extending mesometrially between the basal lamina and the overlying uterine cells at the margin of the advancing embryonic pole of the blastocyst. Invasion of the endometrial connective tissue, the subepithelial region of which has now been transformed into decidual tissue by a population of trophoblast giant cells, occurs next in the rat.

The outermost giant cells penetrate between the decidual cells, which form a narrow avascular layer, the primary decidual, around the embryo. The giant cells also extent to blood vessels outside the primary deciduas and penetrate between endothelial cells. Maternal blood now begins to circulate through spaces in the network of giant cells. All of these events happen in the antimesometrial endometrium at the site of initial embryonic attachment (Schlafke and Enders, 1975; Abrahamsohn and Zorn, 1993). Implantation in the rat is termed interstitial because the blastocyst sinks beneath the level of the endometrium and becomes enclosed in a recess of the uterine cavity. There is evidence for extensive endocytic activity by the blastocyst before and during implantation. Lysosomal enzymes are present and are involved in autolysis and phagolysis in trophoblast, and in addition, proteolytic activity has been found in the uterine lumen (Schlafke and Enders, 1975).

In the rat, the chorioallantoic placental development is a strictly timed and synchronized process defined by the appearance of five different layers. The rat placenta is classified as chorioallantoic, based on the origin of the placental membranes. The rat placenta has been further defined as hemotrichorial because it has three layers of trophoblasts separating maternal from fetal endothelium (De Rijk et al., 2002). The chorioallantoic placenta results from the growth of the ectoplacental cone, differentiation of its cells, and the formation of an association with cells of allantoic origin. The placenta will form on the mesometrial side of the uterus.

On day 8, the ectoplacental cone, or ä träger, is at least one-fourth of the length of the egg cylinder (Huber, 1915a), and it is penetrated by maternal vascular channels that are without endothelium but limited peripherally by a layer of chorionic giant cells that extend around the entire blastocyst. A transitory ectoplacental cavity exits within the ectoplacental cone (Davies and Glasser, 1968). By day 9.5 the rat allantois first appears as a small cluster of cells attached to the caudal region of the embryo in the angle between the amnion and the yolk sac (Ellington, 1985).

An allantoic diverticulum that is angiogenic differentiates and invaginates into the extraembryonic coelom during day 9 through 10. Beginning about day 10, the allantois will come into contact and fuse with the chorionic mesoderm (Ellington, 1985). The allantois differentiates into fetal mesenchyme and vascular channels lined by endothelium, which will become confluent at its proximal end with the visceral yolk sac vessels. The vasculature of the allantois extends deeply into the ectoplacental trophoblast between the pre-existing maternal blood channels. The allantois and chorionic mesoderm, together with the ectoplacental cone, will develop to form the fetal component of the chorioallantoic placenta beginning at day 11.5 to 12.5 (Beck et al., 1967a; Jollie 1990; Abdulrazzaq et al., 2001).

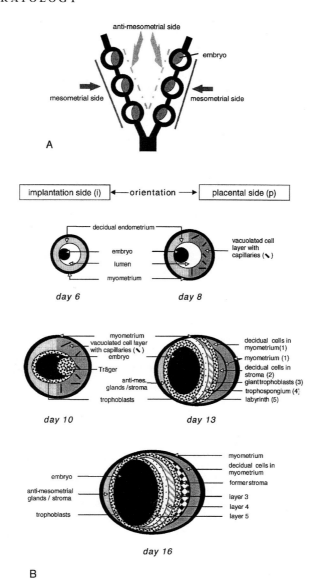

Fig. 28-4 Schematic drawing of implantation and placental development. **A:** Orientation of the implantation of rat fetus. **B:** The development of the different placental layers and structures on day 6, 8, 10, 13, and 16 of rat pregnancy. (Data from De Rijk et al., 2002.)

This is the labyrinth or zone of combined fetal and maternal vascularization (Davies and Glasser, 1968). The allantoic stalk gives rise to the umbilical cord with its root centered on the fetal surface of the chorioallantoic placenta. In the rat, a portion of the avascular parietal wall of the yolk sac membrane covers this organ.

As a result in the rat, all embryonic tissues (intraembryonic and extraembryonic) except the chorioallantoic placenta, are enclosed within the bilaminar yolk sac, where the chorionic ectoderm (trophoblast) is separated from yolk sac endodermal epithelium by a singular thick basement membrane, called Reichert's membrane, that has

been shown to be the basal lamina of parietal yolk sac endodermal epithelium. By day 12, the trophoblastic ectoplacenta on the mesometrial side is vascularized by allantoic capillaries to form a definitive chorioallantoic placenta, with maternal blood sinuses near its fetal surface that lie just external to the placental portion of the parietal yolk sac.

Implantation in rodents is accompanied by marked changes in the stromal cells of the endometrium connective tissue that lead to the formation of the deciduas. Decidualization begins in the stromal region immediately surrounding the mucosal crypt where the embryo is implanting. Decidualization then spreads antimesometrially toward the myometrium, and a few days later, it can also be observed in the mesometrial endometrium in the narrow layer immediately surrounding the embryo. This is the primary decidual zone, and it appears to be avascular. A region of fully transformed decidual cells surrounds this primary decidual zone. In the rat decidua, populations of granulated cells called granulated metrial gland cells increase around day 8 or 9 and migrate toward the mesometrium, where they will form the metrial gland.

A number of giant trophoblast with phagocytotic features invaded the stromal tissue from the ä *träger* to the lateral and antimesometrial direction between days 8 and 10 (De Rijk et al., 2002). Beginning about day 10, actively angioblastic chorioallantoic mesoderm appears in relation to the superficial part of the ectoplacental cone, which differentiates into fetal mesenchyme and vascular channels lined by endothelium. It extends into the ectoplacental trophoblast between the pre-existing maternal blood channels to establish the labyrinth (Davies and Glasser, 1968). By day 11, a clear single layer of giant trophoblasts covers the embryonic outer membrane (De Rijk et al., 2002). By day 9, maternal blood lacunae appear between the cells of the ectoplacental cone (Bridgman, 1948). From day 12 and throughout pregnancy, five distinct morphologic layers are present in the placenta. Starting from the myometrium, there is a strongly decidualized myometrium with many large metrial cells. A stromal layer consisting of moderately large decidual cells is followed by a layer of giant trophoblasts that separates the tissue of maternal origin from that of embryonic origin. A narrow layer follows, which is called the trophospongium (also called the basal zone, reticular zone, or junctional zone), that consists of highly proliferative cells with rather large syncytiotrophoblasts and lacunae with mature blood cells. Last is a layer called labyrinth, which contains many lacunae; some filled with maternal blood and some with embryonic blood. The last two layers comprise the fetal placenta and occupy about one-third of the thickness of the uterine mesometrial wall (Davies and Glasser, 1968). The labyrinthine trophoblast of the rat is trilaminar; the layer in direct contact with the maternal blood is cellular and fenestrated; and the intermediate and basal layers are syncytial, with the third layer in immediate contact with the trophoblastic basement membrane and fetal mesenchyme (Davies and Glasser, 1968). These lacunae will develop later into capillaries with separate vascular channels for maternal and fetal blood. Fetal red blood cells are 100% nucleated on days 12 through 15 of pregnancy, but the number of nucleated red blood cells decreases to 50% by day 16 and 5% to 10% as the end of pregnancy nears (De Rijk et al., 2002).

On about day 14, the uterine cavity is re-established ventral (antimesometrial) to the embryo, creating a capsule of decidua over the implantation site, the decidua capsularis. The remaining uterine lining is the decidua parietalis (Beaudoin, 1980).

C. Gastrulation

The period of gastrulation (days 8.5 to 9.5 in the rat) covers the period of development of the conceptus from a bilaminar germ disc of epiblast and hypoblast that arises from the inner cell mass through the formation of the primitive streak to the trilaminar ectoderm/mesoderm/endoderm of the embryo. Only in two areas is the original bilaminar condition retained: the prochordal plate, or buccopharyngeal membrane, and the cloacal membrane (Van Mierop, 1979). Structures formed during gastrulation include the notochord and neural plate (gestation day 9.5). The notochord appears as a cranial outgrowth of Hensen's node, situated in an axial position between the ectoderm and the endoderm (Gajović et al., 1989). The first segregation of cells within the inner cell mass involves the formation of the hypoblast (primitive endoderm) layer at the blastocoel side as discussed previously. These cells will also line the blastocoel cavity, where they will form the yolk sac endoderm. No derivatives of these cells form the embryo. The remaining inner cell mass cells are referred to as the epiblast. Derivatives of these cells will form all embryonic tissues and the amniotic ectoderm. During the middle and latter part of the ninth day, active cell proliferation in the embryonic ectoderm in the region of the future caudal end of the embryo leads to a distinct thickening of the embryonic ectoderm of this region. This thickening constitutes the primitive streak region and defines unequivocally the anteroposterior axis of the embryo. Within the primitive streak, a short axial groove called the primitive groove develops. During gastrulation, first the definitive endodermal cells, which will replace the primitive endodermal or hypoblast cells, and then the mesodermal cells leave the primitive ectoderm and become placed in their positions in the definitive germ layers (Švajger et al., 1986). Cells remaining in the topographic position of the primitive ectoderm become the definitive ectoderm and will, at the head-fold stage, undergo neurulation. Mac Auley et al. (1993) showed that during gastrulation, most cells of the ectoderm and mesoderm had a cell cycle time of 7 to 7.5 hours, but the cells of the primitive streak divided every 3 to 3.5 hours. By the end of day 9 or at the beginning of day 10, the mesoderm extends so as to form a distinct layer situated between the two primary germ layers (Huber, 1915a).

With gastrulation and the displacement of cell groups, new spatial relationships, interactive events, and cell commitments occur that lead to regional restrictions of developmental potencies.

Grafting of isolated rat embryo germ layers under the kidney capsule have shown that the primitive embryonic ectoderm of the pre-primitive streak stage gives rise to tissue derivatives of all three definitive germ layers. At the head-fold stage, embryonic ectoderm differentiates only into ectodermal and mesodermal tissues, whereas the definitive endoderm, as an isograft, shows no capacity for differentiation. Definitive endoderm, together with mesoderm, differentiates into a variety of derivatives of the primitive gut. Isolated mesoderm at the head-fold stage differentiates only into brown adipose tissue. However, in combination with the definitive endoderm, it develops into adipose tissue, cartilage, bone, and muscle (Levak-Švajger and Švajger, 1974; Švajger et al., 1986).

Smith (1985) noted in the mouse that the anterior-posterior axis of the blastocyst was determined before streak formation and before implantation. In mice the hypoblast forms on the side of the inner cell mass that is exposed to the blastocyst fluid. As development proceeds, the notochord maintains dorsal-ventral polarity by inducing specific dorsal-ventral patterns of gene expression in the neural tube (Goulding et al., 1993). Sidedness of asymmetric body structures is determined during the early neural plate stage, which is before the 5- to 8-somite stage when the bulboventricular loop is formed from a single heart tube and when morphological signs of body asymmetry first appear. The embryonic axis starts to rotate from a concave to a convex dorsal curvature, where the tail takes up a position on the right side of the anterior-posterior body axis at the 9- to 10-somite stage. This rotation is completed at the 17- to 18-somite stage (Fujinaga and Baden, 1993).

Most defects of gastrulation are probably incompatible with survival.

It is important to stage embryos as there is considerable variation in embryonic development by gestation day 9, ranging from embryos at early primitive streak stage to late neural plate stage with a foregut pocket (Fujinaga and Baden, 1992), and by day 11, embryos may be up to 12 hours apart in development (Fujinaga and Baden, 1991). Causes for this difference in development include length of mating period, time required for sperm to reach the ova, actual time of ovulation and implantation, and nutritional access of implants owing to differences in uterine site (Holson, et al., 1976; Fujinaga and Baden, 1992). Two staging systems are in common use: Theiler's system (Theiler, 1972) for mice and Witschi's system (Witschi, 1962) for rats. Both systems use consecutive numbered stages and cover all of the gestation period; however, neither system precisely defines the stages during the early postimplantation period. Fujinaga et al. (1992) proposed modifying Theiler's system by making subdivisions of some of the stages to permit a more precise staging of early postimplantation embryos. For the somite period, embryo staging is usually based on the number of somite pairs, but this can be unsatisfactory because the number of somites varies by as much as six pairs at any one stage of development and because the total number cannot be counted throughout development as the somites gradually become obliterated. Edwards (1968) gives a table that correlates gestational age in days to the development of external features in the rat.

D. Organogenesis

1. Neurulation and Early Central Nervous System Development

Neurulation is the process of neural tube formation that forms the future brain and much of the length of the spinal cord. The caudal portion of the spinal cord forms during secondary neurulation. The formation of the nervous system depends on cell movements and cell-cell interactions that begin during gastrulation. It has also been shown that many of the *Pax* genes are expressed during early neurogenesis, where they exhibit distinct regional patterns of transcriptional activity (Goulding et al., 1993). The induction of the neural plate has been studied principally in amphibian embryos, and little is known about neural induction in mammals. Based on presumed parallels with amphibian embryos, it is believed that this migration of cells brings the mesodermal cell population into a close relationship with the overlying ectoderm and sets up the initial embryonic inductive interaction between the medial portion of the mesoderm and the overlying prospective neural plate. Numerous microsurgical and transfilter experiments have shown that the subjacent notochord is a requisite for neural induction to occur. The development of the notochord in the rat takes place between day 8.5 and 9 (Florez-Cossio, 1975). Induction is followed by elevation of the lateral margins of the neural plate to form neural folds. On elevation, the neural folds first become bi-convex, then beome concave, and finally meet and fuse in the midline. During this time, neuroepithelial cells change in shape from low columnar to high columnar and then to wedge shaped.

The initial neural tube fusion occurs in the cervical region of the second and third somites at the seven-somite stage, extending upward into the hindbrain. At the early 10-somite stage, a separate region of fusion begins at the midbrain/forebrain junction immediately above the developing cranial flexure. By the 14-somite stage, final closure of the spindle-shaped opening between these two areas of fusion is complete (Morriss-Kay, 1981). The last portion of the cephalic neural fold to close, the anterior neuropore, is located in the future forebrain and closes by the 16-somite stage. This is followed by progressive fusion from the cervical region caudad (Morriss and New, 1979; Campbell et al., 1986; Morriss-Kay and Tucket, 1991). The posterior neuropore is found at the caudal end of the embryo in the future lumbosacral region and remains open until about the 21-somite stage. In the rat, the neuropores close between days 10.5 and 11, with the anterior neuropore closing first (Long and Burlingame, 1938; Hoar and Monie, 1981). The final processes of neural tube closure are migration of the neural crest cells and individual fusion of the surface ectoderm and neural ectoderm (Campbell et al., 1986).

In the mouse and other tailed mammals, including the rat, the secondary neural tube forms from a process that starts within the tail bud just after the posterior neuropore closure has been completed. The tail bud is a mass of undifferentiated mesenchyme, which consist of cells derived from the remnants of the primitive streak and Hensen's node, the posterior end of the neural groove and the posterior coelomic mesodermal epithelium. During the elongation of the tail, new cells are added by proliferation of those at the proximal end of the tail bud. The neural tube within the tail originates by a process of secondary neurulation by simultaneous processes of extension of the central canal of the primary neural tube and recruitment or mitosis of cells that become radially arranged around the extending lumen. This lumen formed by cavitation is always directly continuous cranially with the lumen of the fully formed neural tube (Hughes and Freeman, 1974; Schoenwolf, 1984; Campbell et al., 1986; Gajović et al., 1989). The notochord in the tail of the rat develops on day 12 and 13 from a common mass of condensed mesenchymal cells located ventrally to the secondary neural tube, which subsequently splits to form a thin cord (which becomes the notochord) and a thick portion (which gives rise to the tail gut) (Gajović et al., 1989).

Neural crest cells differentiate at the lateral margins of the neural plate and are derived solely from the neural epithelium from the six-somite stage onward (Tan and Morriss-Kay, 1985). They will migrate away from the neural fold or roof of the incipient neural tube and are induced by structures with which they come in contact to form many cell types, including neurons and glial cells of the sensory, sympathetic, and parasympathetic nervous systems; the medulla cells of the adrenal gland; adrenergic paraganglia; calcitonin-producing cells; melanocytes of the epidermis; endothelium of the aortic arch arteries; septum between the aorta and pulmonary artery; and skeletal and connective tissues of the face and anterior neck region (Adelman, 1925; Le Douarin and Ziller, 1993).

Neural crest migration occurs in the following sequence: midbrain, rostral hindbrain, caudal otic hindbrain, postotic, and finally rostral otic hindbrain, with a general shift in migratory intensity from rostral to caudal regions (Tan and Morriss-Kay, 1985). Fukiishi and Morriss-Kay (1992) showed that in the rat, neural crest cells from the midbrain levels migrate to the maxillary process and the mandibular arch and contribute to the development of the face, and neural crest cells from the rostral hindbrain migrate to the first and second pharyngeal arches. Occipital neural crest cells migrate to the third, fourth, and sixth pharyngeal arches and are essential for the division of the outflow tract of the heart into aorta and pulmonary trunk. Clonal analysis of the developmental potentialities of individual neural crest cells has shown most of the migrating cells exhibit various levels of pluripotency, but some are fully committed at the time of migration (Le Douarin and Zeller, 1993). Inhibition or errors in neural crest migration produces anomalies of facial development, defects of the aorticopulmonary septum in the heart, small or missing dorsal root ganglia, and errors in pigmentation.

In the anterior region, the neural tube balloons into three primary vesicles as the anterior neuropore closes at gestation day 10.5: forebrain (prosencephalon), midbrain (mesencephalon), and hindbrain (rhombencephalon). The optic vesicles extend laterally from each side of the developing forebrain. On gestation day 12, the forebrain becomes subdivided into the anterior telencephalon (cerebral hemispheres) and the more caudal diencephalons (thalami, posterior lobe of the pituitary gland, pineal body and optic stalk, and optic cup, which later generate the optic nerve and retina). The rhombencephalon becomes subdivided into a posterior myelencephalon (medulla oblongata) and a more anterior metencephalon (pons and cerebellum). The cerebellum is formed in the rat on day 14. The remainder of the neural tube forms the spinal cord. Expression profiles of *Wnt* genes in the mouse suggest that *Wnt* signaling plays a major role in early central nervous system development in vertebrates (Parr et al., 1993). Neurogenesis in the rat spinal cord occurs in a temporal gradient, with neurons in the ventral spinal cord becoming postmitotic before those in the dorsal cord. Most motor neurons are generated between days 11 and 13 (Goulding et al., 1993). Neurons form in the fetal rat brain principally between days 13 and 20—except in the cerebellum, where they do not appear until about day 19—and continue to form during the early postnatal weeks (Hoar and Monie, 1981; Gilbert, 1997, 2000).

Naturally occurring neural tube defects have been extensively studied in several strains of mice (curly tail, splotch mutant, loop tail, and those trisomy for chromosomes 12 and 14). Campbell et al. (1986) lists teratogens that have been administered to rats to interrupt the process of neurulation. Defective closure of the neural folds results in the following malformations: anencephaly, exencephaly, rachischisis, encephalocele, or spina bifida.

2. Somite Formation

The mesoderm is found on day 9 extending on each side of the notochordal plate as two lateral wings, and it

TABLE 28-1

NUMBER OF SOMITES AND CORRESPONDING GESTATIONAL AGE

Number of Somites	Age (Days from Mating)
1	9
5	9.5
10	9–10
20	10–11
25	11–12
30	11–11.5
35	12
40	12–12.5
45	13.5
50	13
60	14.75

divides peripherally into the somatic and splanchnic layers. Butcher (1929) describes somite formation in the rat. The somite divides into three parts. The cells of the most medial part, the sclerotome, will migrate around the notochord and neural tube and will form connective tissue, cartilage, and bone (vertebrae and intervertebral discs of the vertebral column). Immediately lateral to the sclerotome are the cells that form the myotome, which will later become skeletal muscle. The most lateral part of the somite forms the dermatome that will form the dermal connective tissue. The first somite will later involute, and other somites develop in a head-to-tail sequence until there are 65 complete somites and a mesodermal remnant. The number of somites present (Table 28-1) has been used to classify the developmental stage of rat embryos (Schneider and Norton, 1979). Each succeedingly appearing somite is more advanced structurally than the one just anterior to it. The size of the successive somites increases from the 1st to the 31st, and then the somites are successively smaller.

3. Cardiovascular System

The heart is the first functional embryonic organ and is required by the embryo for efficient growth (Christoffels et al., 2000; Linask, 2003). The cardiogenic mesoderm forms by anterior-lateral migration from the primitive streak and constitutes a single primordium, which will organize into an epithelia. The heart forms in the anterior margins of the embryonic disc in front of the neural plate. The pericardial cavity also originates in the cardiogenic mesoderm, first appearing as small spaces that coalesce to form larger single closed right and left cavities, which in turn unite into a single horseshoe-shaped cavity. Because of the rapid cephalad growth of the forebrain and cephalocaudal folding of the embryo, the prochordal plate

(buccopharyngeal membrane) and pericardial cavity are carried further under the rising head of the embryo. Cells that form the heart initially have a bilateral distribution, but they gradually become localized in the cardiac crescent lying in front of the prechordal plate and stretching along the future neural plate at the level of the first two paired somitomeres. Subsequently, after the lateral body folding, the two sides of the crescent meet in the midline to form the single heart tube (Baldwin et al., 1991; Suzuki et al., 1995). An epithelial myocardium is established and forms a tubular heart in the rat on day 9 to 9.5, and contractions can be observed on day 9.5 to 10 (Goss, 1938; Hebel and Stromberg, 1986; Clark, 1989; Baldwin et al., 1991). In a complex morphogenetic progression, the linear tube is transformed into a synchronously contracting four-chambered heart (Christoffels et al., 2000). Both myosin and actin filaments become identifiable for the first time in the day-10 myocardium of the C-shaped tubular heart (Chacko, 1976). The splanchnopleuric mesoderm that surrounds the endothelial heart tube differentiates into three layers: an inner layer of cardiac jelly surrounding the endothelium, an early myocardium that acquires contractile elements, and an outer layer (the epicardium). The last two layers are referred to as the myoepicardial mantle.

Cranially at the arterial pole of the bulboventricular tube, the first pair of aortic arches arises and is continuous with the dorsal aortae. Caudally, the bulboventricular tube receives blood from the yolk sac by means of the paired sinus venosus and atrium. From this, the bulboventricular tube will develop into the embryonic ventricle and bulbus cordis, and later into the ventricle and their outflow tracts. As it grows, the bulboventricular tube forms a loop bending right in the shape of a *C*. Looping results in the correct spatial orientation for subsequent modeling of the four chambers of the heart (Linask, 2003). It has been shown by Christoffels et al. (2000) that the onset of the transcriptional program for cardiac genes specifically associated with the formation of the ventricular and atrial chamber myocardium is restricted to the ventral side of the linear heart and the outer curvature of the looped heart. The atrial portion of the heart dilates and fuses to form a large common atrium, which, as a result of growth, moves craniad from its original caudal position to lie behind the bulboventricular loop. The paired sinus venosus also partly fuse to form the right and left sinus horns that receive the vitelline, allantoic, and cardinal veins. Dilatations develop in the bulboventricular loop, whereas the atrioventricular junction and the junction of the ventricle and proximal bulbus cordis remain relatively narrow.

The embryonic ventricle becomes the primitive left ventricle, and the proximal one-third of the bulbus cordis becomes the primitive right ventricle. Alternatively, it has been proposed that the right and left ventricular chambers "balloon out" from the outer curvature of the heart tube (Christoffels et al., 2000). The conus cordis or outflow portions of both ventricles will be formed from the middle one-third of the bulbus. The remaining one-third of the bulbus after partitioning develops into the truncus arteriosus. The cardiac jelly in the atrioventricular canal area and in the conus cordis and truncus arteriosus becomes invaded by cells derived from the endothelium and is referred to as endocardial cushion tissue. The atrioventricular and outflow tract cushions function as primitive valves in the early heart tube, ensuring directional blood flow (Barnett and Desgrosllier, 2003). Growth of the right and left ventricles results in the formation of the muscular ventricular septum that begins on day 11.5 to 12 in the rat (Hoar and Monie, 1981; Hebel and Stromberg, 1986). The atrioventricular canal shifts position to the right, allowing blood to enter both ventricles.

In the rat, the division of the atrium is accomplished by the formation of the septum primum, which contains the foramen ovale, and begins on day 11.5 to 12 (Hoar and Monie, 1981; Hebel and Stromberg, 1986). Opposing pairs of endocardial cushions appear in the atrioventricular canal, the conus cordis, and truncus arteriosus and then meet and fuse. This divides the atrioventricular canal into right and left ostia and divides the right anterior half of the conus cordis into the right ventricular outflow tract, and the posteromedial half becomes absorbed into the primitive left ventricle to become its outflow portion. The truncus arteriosus is divided into pulmonary and aortic channels. Conotruncal septation involves morphogenic movement of mesenchymal tissue from the branchial arches, neural crest cells, and endothelium and also involves the conotruncal cushions. As a result of shifts in the forth and sixth aortic arches, the sixth arches become aligned with the pulmonary channel; the fourth arches, with the aortic channel. Endocardial cushion tissue derived from the atrioventricular canal and conal septa complete the ventricular septum.

Additional pairs of opposing masses of endocardial cushion tissue in the atrioventricular canal area along with the atrioventricular canal septum act as provisional valves. Most of the atrioventricular valves (tricuspid and mitral valves) form by the increased diverticularization and undermining of the interior of the ventricular walls that later become fibrous and form the valve cusps and their chordae tendineae. Additional pairs of endocardial cushion tissue in the truncus arteriosus eventually develop into the anterior cusp of the pulmonary valve and the posterior cusp of the aortic valve. The remaining cusps of the pulmonary and aortic valve develop from the endocardial cushion tissue of the truncus septum (Burlingame and Long, 1939; Moffat, 1959; Van Mierop, 1979; deVries, 1981; Clark, 1989; Barnett and Desgrosellier,

2003). After septation, these swellings thin and become "excavated" on the distal side to become leaflets (Rothenburg et al., 2003).

The heartbeat is initiated and coordinated by a heterogeneous set of tissues that are collectively referred to as the pacemaking and conduction system (Gourdie et al., 2003). The sinus node, which determines the intrinsic rhythm of the heart, develops at the junction of the right common cardinal vein and the right sinus horn and can be observed on day 12 in the rat and is completely developed by day 15 (Domenech-Mateu and Orts-Llorca, 1976; Hebel and Stromberg, 1986; Gourdie et al., 2003). After initiation of a cardiac action potential within the node, activation is propagated through the muscular tissue of the atria, eventually focusing into the atrioventricular node. The atrioventricular node, located at the junction of the atria and ventricles, functions as part of a mechanism for generating a momentary delay in the propagation of the cardiac action potential. This separates the activation of the atrial chambers from that of the ventricles (Gourdie et al., 2003).

After exiting from the atrioventricular node, the action potential rapidly propagates along the His bundle and its distal branches, finally activating the ventricular chambers via the Purkinje fibers. The His-Purkinje system is a network of fast conduction tissues that are responsible for coordinating ventricular contraction. The fast conduction system of the ventricles is the last element of the pacemaking and conduction system to develop (Gourdie et al., 2003). The atrioventricular node and bundle of His develops from a specialized reticular segment of the myocardium that initially surrounds the atrioventricular canal (Van Mierop, 1979).

The pulmonary veins grow as a bud from the lung buds and connect with the left atrium. Of the six pairs of aortic arches, the first and second arches, which appear by the 12-somite stage, regress, and the fifth arch disappears. In embryos with crown-rump length of between 3 and 4 mm, the first aortic arch has broken down, but the second and third arches are present and join the aortic sac to the dorsal aorta. In the rat, there is a temporary communication that forms between the fourth and sixth arches, at sometime after the sixth arches are fully formed, as a result of the marked caudally directed convexity of the fourth arch. The third arches, which appear between the 15- to 18-somite stages form the proximal parts of the internal carotids, and the sixth arches form the pulmonary arteries and, on the left, the ductus arteriosus. The left fourth arch forms part of the aortic arch, which lies proximal to the origin of the left common carotid artery, and the right fourth arch forms part of the innominate artery (Moffat, 1959). The common carotid artery is formed as a result of the elongation of the cranial end of the aortic sac. Coronary arteries that arise as buds from the ascending aorta grow down across the heart surface and into the myocardium by day 15.5 in the rat (Hoar and Monie, 1981). The coronary veins form about the same time and connect with the left sinus venosus later to form the coronary sinus (Clark, 1989). Development of both the coronary arteries and capillaries continues after birth (Hew and Keller, 2003).

Structures peculiar to the fetal circulation include the foramen ovale, ductus arteriosus, umbilical vessels, and the ductus venosus. The foramen ovale permits blood to flow directly from the right atrium to the left atrium, bypassing the lungs; the ductus arteriosus allows blood to flow from the pulmonary arteries to aorta, again bypassing the lung. The ductus venosus allows blood in the umbilical vein to bypass the liver. The umbilical vein carries oxygen-rich blood from the placenta to the fetus, while oxygen-poor blood is returned to the placenta through the umbilical arteries. Two umbilical veins form in young rat embryos, but the right umbilical vein disappears on day 13. In the rat the umbilical arteries arise from the internal iliac arteries, but there is only a single umbilical artery within the cord. In the rat embryo, there is a communication between the vitelline and umbilical arteries until day 11 (Hoar and Monie, 1981). The transition from prenatal to postnatal circulation involves three primary steps: (1) cessation of unbilical circulation, (2) the transfer of gas exchange function from the placenta to the lungs, and (3) closure of the prenatal shunts. In addition, the heart function changes from acting as two parallel pumps to acting as two pumps performing in series. The foramen ovale is diminished in the first two days after birth and is completely closed by day 3. Functional closure of the ductus arteriosus is complete 2 to 5 hours after birth in the rat. The ductus venosus narrows rapidly after birth and closes completely in 2 days (Hew and Keller, 2003). Hew and Keller (2003) report heart rate and blood pressure values for the early postnatal period in the rat.

Heart malformations attributed to aberrant neural crest cell migration include aorticopulmonary septal defects (persistent truncus arteriosus, transposition of the great vessels, truncus arteriosus communis). Malformations attributed to extracellular matrix abnormalities include persistent atrioventricular canal and tricuspid atresia. Malformations related to abnormal blood flow include pulmonary valvular atresia, coarctation of the aorta, and aortic valvular stenosis or atresia. Aberrant cell death can cause muscular ventricular septal defects. Left-sided looping of the developing heart will result in dextrocardia (Kirby et al., 1983; DeSesso, 1993). Overriding aorta is a malalignment defect in which wedging of the aorta between the mitral valve and tricuspid atrioventricular valves fails to occur because the outflow track does not lengthen normally during looping. This leads to a double-outlet right ventricle and tetralogy of Fallot (Hutson and Kirby, 2003).

4. Urogenital System

The urogenital system is actually two separate systems united through inductive interactions during development. The urinary system develops within the intermediate mesodermal ridge, which is along the peritoneal aspect of the dorsal body wall. The rat, similar to other mammals, develops three different but overlapping kidneys systems, which appear during days 10 to 12 of gestation: pronephros, mesonephros, and metanephros. The first two will regress in utero in the rat. The excretory system of the rat begins as a nephrogenic ridge that runs from the 9th to the 12th somites.

The pronephros develops at the cervical somite levels in the intermediate mesoderm and migrates caudally. The anterior region of the pronephric duct induces the adjacent mesenchyme to form the pronephric kidney tubules. This is a non-functional kidney. The pronephric tubules and the anterior portion of the pronephric duct degenerate, but the more caudal portion of the pronephric duct persists and is referred to as the Wolffian duct. In the rat, the Wolffian duct appears at the end of day 10 (Hebel and Stromberg, 1986).

The mesonephric duct (formerly the caudal portion of the pronephric duct) initiates over time a new set of vesicles, beginning at the level of somite 12. These vesicles develop in a cranial to caudal pattern and will become the second kidney or mesonephros. One end of each elongating vesicle connects to the mesonephric duct, and the other end is invaginated by capillaries from the dorsal aorta to form the functional mesonephric glomerulus. The tubular components of the mesonephric nephron are found between the glomerulus and the mesonephric duct and consist of a secretory segment and a collecting segment. 24- to 30-somite rat embryos have from 2 to 13 mesonephric tubules in various stages of differentiation, with the principle of antero-posterior development roughly applying, and a pair of Wolffian ducts with varying degrees of canalization, which run toward the cloaca but terminate at the lateral wall and do not empty into the cloacal cavity. On day 12, the mesonephros is at the height of its development, with 15 to 18 tubules between the levels of somites 12 and 18. The distal ends of the mature tubules are dilated to form a Bowman's capsule. Some of the more anterior tubules share a common basal junction instead of opening directly into the Wolffian duct. On day 12, the mesonephric tubules and the Wolffian duct develop a lumen that opens into the cloaca. The Wolffian duct bends rather sharply at the level of the 26th somite before it opens into the cloaca. It is at this bend that the ureteric bud forms (Hebel and Stromberg, 1986). The mesonephric nephrons regress with only the three most anterior tubules being retained. These three comprise the future epigonadal system (Torrey, 1943; Arey, 1974; Gilbert, 1997).

The permanent kidney or metanephros forms from an evagination (the ureteric or metanephric bud) that appears just proximal to where the mesonephric duct enters the cloaca on day 13 (Hebel and Stromberg, 1986). These buds eventually separate from the nephric duct to become the ureters that will take the urine from the kidney to the bladder. The ureteric bud elongates and migrates cranially and projects into the mesenchyme, which then is induced to condense around the buds and differentiate into the secretory part of the nephrons of the metanephros. This nephron-forming tissue then induces further branching of the ureteric bud to form the major and minor calyces of the metanephros. Outgrowths from the minor calyces develop and project into the condensed mesenchyme or metanephric blastema to form the collecting duct portion of the nephron. The metanephros consists of three cell lineages: the epithelial cells of the Wolffian-duct-derived-ureter bud, the mesenchymal cells of the nephric blastema, and the endothelial cells of the capillaries.

Kidney development requires a set of inductive interactions between the metanephric mesenchyme and the ureteric bud, where first the metanephric mesenchyme induces a bud to form on the nephric duct. This is followed by two reciprocal inductive interactions, in which the mesenchyme causes the ureteric bud to grow and bifurcate, forming the collecting duct system, whereas the bud induces the mesenchyme to differentiate into nephrons by first causing the mesenchyme to differentiate into stem cells located at the periphery of the growing kidney. Groups of these stem cells then aggregate, epithelialize, and undergo morphogenetic events to produce comma-then S-shaped bodies, which later unwind into elongated tubular nephrons that interact with capillaries at their proximal ends to form glomeruli, convolute and differentiate to become the selective reabsorption apparatus, and finally fuse to the collecting ducts at the distal end. The connecting tubules of the juxtamedullary and midcortical nephrons form arcades; however the subcapsular nephrons directly join the collecting duct. Nephrons continue to be initiated at the periphery of the kidney over the fetal period so that the less mature nephrons are located in the kidney periphery. The first primitive loop of Henle appears at day 17 in the rat, and the last loops mature by approximately postnatal day 15. The thin ascending limb of Henle begins to differentiate, beginning in the tip of the papilla, and progresses outward, with all the loops differentiated at postnatal day 8 (Arey, 1974; Kazimierczak, 1980; Hoar and Monie, 1981; Neiss, 1982a and 1982b; Saxén and Sariola, 1987; Evan et al., 1994; Bard et al., 1996; Gilbert, 1997).

The primordial germ cells are identifiable in the yolk sac epithelium on day 9 to 10 in the rat and begin to migrate on day 10 or 11 (Beaudoin, 1980). The gonadal ridge containing primordial germ cells has developed by day 13 (Maeda et al., 2000). The mammalian gonad first develops through an indifferent or bisexual state, with the default sex being the female. During the indifferent gonadal stage, both the Wolffian and Müllerian ducts are present. The Wolffian (mesonephric) duct is in close association with the gonad and empties into the urogenital sinus. The Müllerian (paramesonephric) ducts are also in close association to the developing gonad being induced within the intermediate mesodermal ridges by the mesonephric ducts, beginning on day 14 (Hebel and Stromberg, 1986). Depending on the sex of the embryo, one of these two duct systems completes its development and the other disappears almost completely.

The gonadal primordium or genital ridge can be observed in the rat at day 11 at the level of somite 16. In day 13 embryos, the gonadal blastema blends with the mesonephric stroma and makes contact with the first three mesonephric tubules. It is in this area that the rete forms. The junction of the rete cords with the mesonephric tubules takes place in both male and female rats on day 17. The epithelium of the genital ridge proliferates into the mesenchymal tissue above it to form the sex cords that will surround the germ cells that migrate from the yolk sac to form the gonad. Two pairs of genital ducts are established and present at the indifferent-gonad stage. The Müllerian ducts begin to degenerate in the male, and Wolffian ducts degenerate in the female on day 18 in the rat. The Müllerian (paramesonephric) ducts, which can form the oviducts and uterus, are also in close association to the developing gonad being induced within the intermediate mesodermal ridges by the mesonephric ducts beginning on day 14.

In the male, the genital tract develops primarily from two embryonic anlagen: the Wolffian ducts and the urogenital sinus. The formation of the male phenotype involves the secretion of the hormones testosterone, secreted from the fetal testicular Leydig cells that stimulate the mesonephric (Wolffian) duct and urogenital sinus to differentiate, and anti-Müllerian duct hormone (Müllerian-inhibiting substance) that destroys the Müllerian duct over days 18 to 22 in the rat. The Wolffian duct, whose epithelium is mesodermal in origin, gives rise to the epididymis, ductus deferens, seminal vesicle, and ejaculatory duct. The urogenital sinus, whose epithelium is derived from endoderm, gives rise to the prostate, bulbourethral glands, urethra, and periurethral glands. The mesonephric tubules give rise to the efferent ducts. The testes can be observed toward the end of day 13 or beginning of day 14, when primary sex cords are seen at right angles to the surface of the testis. An area of unorganized blastema found between the peripheral ends of the primary cords and the celomic epithelium is the tunica albuginea. The mesenchyme-like cells found in the narrow angles between some of the cords are the progenitors of the interstitial tissues of the testis. Testicular secretion of testosterone begins on day 15 and is a requirement for differentiation of the reproductive organs that develop from the urogenital sinus, Wolffian ducts, and perineal tissue (Timms et al., 2002). By day 16, the testis cords have become elongated and tortuous, and Sertoli cells are first observed. Leydig cells are found among the stromal cells around day 17. At the anterior end of the testis, the cords are beginning to radiate toward the mesonephros and abut on the developing rete cords. After forming a junction with the first three mesonephric tubules on day 17, the rete cords in the male begin to cavitate late on day 18. By day 20, most of the volume of the cord is occupied by germ cells. By postnatal day 1, the rete tubules are broad, branching, and open into the mesonephric tubules that comprise the efferent ducts that connect to the Wolffian duct or vas deferens. On about postnatal day 15, the cavitation of the sex cords begins to form the seminiferous tubules. The seminal vesicles appear as diverticula just proximal to the ejaculatory duct, and the prostate gland develops from the pelvic portion of the urogenital sinus on day 19 (Torrey, 1945, 1947; Merchant-Larios, 1976; Hebel and Stromberg, 1986; Cunha et al., 1992; Gilbert, 1997).

If the Y chromosome is absent, the gonadal primordial develops into ovaries that secrete estrogenic hormones and enable the development of the Müllerian duct into the uterus, oviduct, and upper end of the vagina. The mesonephric (Wolffian) duct disappears almost completely. Ovarian differentiation is first observed in the rat at day 16 as a pronounced clumping of cells and the beginning of cord formation within the blastema. The rete cells extend through the mesovarium into the stroma of the mesonephros. They are continuous with the primary sex cords and are in contact with the distal ends of the mesonephric tubules that will become the tubules of the epoöphoron. The cords become tightly packed in the interior of the ovary, and no tunica albuginea separates the cords from the celomic epithelium. Secondary cords that originate in the epithelium form by day 17. The Wolffian duct disappears, and the rete break up into sharply delineated cords. Cavitation of the rete cords usually begins during postnatal day 3. At parturition, the central two-thirds of the ovary is divided into rectangular columns of sex cords by connective tissue septa. Degeneration of the medullary cords and formation of the primary follicles occurs during postnatal development (Torrey, 1945, 1947; Hoar and Monie, 1981; Gilbert, 1997).

5. Craniofacial Development

Proper temporal and spatial patterns of growth are critical for normal development of the craniofacial region, which includes externally the face and head (excluding the caudal pharyngeal arches) and internally the brain; skull, jaws, and facial skeleton; teeth; and associated soft tissues. The face develops from the growth and fusion of five facial prominences that surround the stomodeum: the frontonasal prominence and the paired maxillary and mandibular prominences (Sulik and Schoenwolf, 1985). The cells that form the skeletal and connective tissues of the face, with the exception of tooth enamel, are derived from cranial neural crest cells that migrate into the first and second pharyngeal arches (Johnson and Sulik, 1980). The frontonasal prominence is composed of the tissue that surrounds the forebrain. The olfactory placodes located on the lateral aspects of the frontonasal prominence will invaginate to form the widely separated nasal pits. Cells from the nasal pits will line the nasal cavities and later form the sensory epithelium for olfaction. Midbrain crest cells migrate into the frontonasal region and then form the lateral nasal prominence on the rim of the nasal placodes. The median portion of the face develops from the frontonasal processes that separate the nasal pits. Forebrain crest cells also migrate into the frontonasal region and then contribute to the formation of the medial nasal prominence that surrounds the nasal pit closer to the midline (Osumi-Yamashita et al., 1997; Hall and Miyake, 2000).

The first and second pharyngeal arches are evident on days 11 to 12, the time the anterior neuropore closes (Hebel and Stromberg, 1986; Ross and Persaud, 1990). The first pharyngeal arch has both the maxillary and mandibular prominence that contributes to the upper and lower jaw. The maxillary process is first distinguishable at day 12 (Smith and Monie, 1969). At day 11, the mandibular arches meet in the midline but do not commence fusion (Symons and Moxham, 2002). The maxillary processes emerge laterally from the proximal portion of the mandibular processes at the lateral edges of the stomodeum and extend upward toward the nasal region and reach the lateral nasal process by day 11.5 to 12. The ends of the lateral and medial nasal processes come into contact below the nasal pit and begin to fuse with the result that the nasal pit becomes deeper and narrower. Around the 31-somite stage, the medial nasal process and the maxillary process come into contact, and the groove between the maxillary process and the lateral nasal process becomes the naso-lachrymal groove. The fusion of the mandibular arches begins at their caudal margin and extends cranially and is complete around day 13 to form the floor of the mouth (Christie, 1964; Smith and Monie, 1969). By day 13, the upper jaw of the rat is forming. The maxillary processes extend around the developing mouth, and the lateral and medial nasal processes surround the developing nasal pits. These processes fuse by the merging of the mesenchyme to form the intermaxillary segment (Lejour, 1970; Symons and Moxham, 2002).

The medial nasal prominences merge in the midline and extend caudally to the oral aperture forming the columella. Lateral extensions of the tips of each nasal medial process extend cranially between the nasolateral and maxillary processes (Smith and Monie, 1969). By day 13.5 a swelling of the medial borders of the maxillary processes begins to form the definitive or primary palatal processes. By day 14, the lower jaw extends about halfway along the maxillary process (Smith and Monie, 1969). On day 14, fusion occurs between the ventral end of the palatal proliferation of the maxillary process and that of the frontonasal process to form the roof of the anterior portion of the primitive oral cavity, as well as the initial separation between the oral and nasal cavities.

Later, derivatives of the primary palate form portions of the upper lip, anterior maxilla, and teeth (Johnson and Sulik, 1980). In the caudal portion of the oral cavity on day 14, new outgrowths from the medial edges of the maxillary processes form the shelves of the secondary palate. The anterior four-fifths of each palatine process grow downward on either side of the tongue, and on day 15 the tip of the tongue extends beneath the posterior portion of the primary palate. The posterior one-fifth of each palatal shelf (future soft palate) grows horizontally above the dorsum of the tongue from the beginning and never elevates (Ferguson, 1977, 1978). The vertical palatal shelves elevate to a horizontal position between the tongue and the nasal septum by day 16, with the shelves taking an active role in their elevation (Srivastava and Rao, 1979). After the shelves elevate, the shape of the tongue changes from a highly arched structure that fills the oronasal cavity to a broad flat structure. On day 16, the tip of the tongue protrudes from the mouth, but by day 17, the mandibular growth rate has exceeded that of the tongue, and the tongue returns to the oral cavity (Wragg et al., 1970). Within 2 hours of elevation, the anterior shelves begin to fuse, leaving a Y-shaped gap in the palate in front of the region of contact and a long straight gap behind it. The anterior gap is closed later by a combination of backward growth of the nasal septum and primary palate and forward growth of the palate shelves. The more posterior shelves grow rapidly toward each other, and epithelial fusion proceeds, with the major part of the future hard palate being fused within 5 hours (Ferguson, 1978). The future soft palate is still widely separated and does not fuse until day 17.5 (Ferguson, 1977). Fusion of the palate is complete by day 18, and the characteristic pattern of rugae is present.

Cleft palate can be induced at several developmental stages (days 9, 11, or 15) in the rat by various agents that include high doses of vitamin A, glucocorticoids, retinoids, and β-aminoproprionitrile (Kochhar and Johnson, 1965; Vig et al., 1984; Granström et al., 1992). Cleft palate may result from disturbances at any stage of palate development by defective palatal shelf growth, delayed or failed shelf elevation, defective fusion, or post-fusion rupture (Kerrigan et al., 2000). The Small eye rat (rSey) lacks eyes and nose owing to an impairment in which the frontonasal ectoderm fails to induce lens and olfactory placodes. At day 13, the rSey/rSey homozygous embryo lacks eye and nasal pits. In addition, the nasal prominence appears to be missing and the frontonasal prominence protrudes, making an appearance at the medial nasal prominence (Fujiwara et al., 1994).

6. Limb Development

Limb development begins when mesenchymal cells proliferate from the somatic layer of the lateral plate mesoderm (skeletal precursors) and the somites (Gilbert, 1997). Progenitor cells of skeletal muscle migrate from the lateral edge of limb level somites into the limb bud of the rat on day 14 and colonize in the dorsoproximal region of the limb bud. After the establishment of myogenic masses, these cells continue migrating within the limb (Hayashi and Ozawa, 1995; Rahman et al., 1997). Limb development involves a three dimensional pattern of development: proximal-distal (shoulder-finger or hip to toe), anterior-posterior (digit 1 to digit 5), and dorsal-ventral (knuckle-palm). In the rat, the anterior limb bud is apparent on day 11 as a lateral swelling of the lateral plate mesoderm between about the 6th and 10th somite (Christie, 1964; Hoar and Monie, 1981; Hebel and Stromberg, 1986).

The middle of the forelimb buds in mice is found at the most anterior expressing region of Hoxc-6 gene, the position of the first thoracic vertebra (Burke et al., 1995; Gilbert, 1997). Retinoic acid appears to be essential for the initiation of limb bud outgrowth because blocking its synthesis inhibits limb bud formation (Stratford et al., 1996). The hindlimb bud is first observed at day 12 (34-somite stage) opposite somites 23 to 28 (Long and Burlingame, 1938; Christie, 1964; Hoar and Monie, 1981; Hebel and Stromberg, 1986). There is a dynamic expression pattern of Hoxa and Hoxd genes during limb development. The initial phase of Hoxd gene expression begins at the time of the initial outgrowth of the limb (Nelson et al., 1996). In both the chick and mouse limbs, expression domains along both proximal-distal and anterior-posterior axes give the mesenchyme cells unique positional addresses in terms of their patterns of Hox gene expression. Experimental evidence suggests that the particular combinations of expression of homeobox genes may relate to the type of digit or other structure that forms, determining whether the structure is posterior or anterior, and proximal or distal (Hinchliffe, 1994).

The proximal-distal growth and differentiation of the limb bud require interactions between the limb bud mesenchyme and the ectoderm. The mesodermal cells induce the overlying ectoderm to form a structure called the apical ectodermal ridge (AER) that will be responsible for the sustained outgrowth and development of the limb. The AER runs along the distal margin of the limb bud (Paulsen, 1994; Gilbert, 1997). The formation of the AER may involve an interaction between the secretions of fibroblast growth factor 8 or other fibroblast growth factors from the mesoderm and the boundary of expression of the secreted protein called radical fringe along the dorsal-ventral ectoderm border (Gilbert, 1997). Two delineated areas of cell death appear in the early AER in the postaxial and preaxial areas. Later just before the outgrowth of digital buds transforms the circular contour of the marginal border into a polygonal one, a second wave of ectodermal cell death starts in the postaxial portion of the AER facing the rudiment of digit V (Milaire and Rooze, 1983). The limb bud elongates by the proliferation of the mesenchyme cells in the progress zone that extends for about 200 μm under the AER.

It is hypothesized that the AER promotes growth by secreting fibroblast growth factors (Gilbert, 1997). When the mesenchyme cells leave the progress zone, they will differentiate such that the first cells to leave will form more proximal structures (humerus). Phase-specific *Hox* gene expression corresponds to each of the three proximodistal segments of the limb (Nelson et al., 1996). There are three sites of cell death that successively affect the mesoderm of the limb bud: (1) the deep preskeletal mesoderm located opposite the preaxial extremity of the AER, (2) an area located preaxially from digit I to III (forelimb) or II (hindlimb), and (3) an area postaxially from digit V to the postaxial border of digit IV. The last areas of physiological cell death involved in limb morphogenesis take place in the interdigital mesoderm that separates the proximal portions of the digits (Milaire and Rooze, 1983). During day 16, the forelimb digits fully separate (Christie, 1964).

The anterior-posterior axis is specified by a small block of mesodermal tissue near the posterior junction of the limb bud and the body wall. This region of mesoderm has been called the zone of polarizing activity (ZPA) and provides positional information for the specification of the digits (Tickle et al., 1975; Fallon and Crosby, 1977; Summerbell, 1979). The expression of the gene *Sonic Hedgehog* is localized exactly in the region of the limb bud shown to contain the highest ZPA activity (Riddle et al., 1993). The ZPA has been proposed to be the source of a morphogen, the concentration of which sets the identity

of digits; tissues that sense a high level of the ZPA morphogen develop into digits of posterior character and tissues exposed to lower levels develop into more anterior digits (Johnson et al., 1994). Although *Sonic Hedgehog* appears to define the ZPA, it is not the soluble morphogen responsible for specifying digits but may specify the pattern by acting in conjunction with signals from the AER (López-Martínez et al., 1995). Retinoic acid, which can convert anterior limb bud tissue into tissue with polarizing activity, also induces *Sonic Hedgehog* expression in the anterior limb bud. ZPA tissue, sonic hedgehog protein, and retinoic acid all have been shown to produce mirror-image duplications of digits if implanted in an anterior portion of the limb bud, and both *Sonic Hedgehog* and retinoic acid will regulate the expression of pattern-related *Hox* genes (Riddle et al., 1993; Paulson, 1994; Nelson et al., 1996; Gilbert, 1997). By day 14 in the rat, the forelimb has formed digital rays (Hoar and Monie, 1981; Hebel and Stromberg, 1986).

MacCabe et al. (1974) demonstrated that the dorsal-ventral axis is determined by the ectoderm encasing it. The *Wnt*-7a gene is expressed selectively in the dorsal but not ventral ectoderm and is required for the establishment of a normal dorsal pattern in the limb mesoderm of mice, whereas *Wnt*-5a is expressed in the ventral ectoderm of limb buds (Parr et al., 1993). *Wnt*-7a gene induces the *Lmx*1 gene that is expressed selectively by the dorsal mesenchyme that encodes a transcription factor that appears to be essential for specifying the dorsal cell fates in the limb (Riddle et al., 1995).

The forelimb and hindlimb follow the same rules of pattern formation, yet their patterns differ. Common processes between forelimbs and hindlimbs include the local accumulation of fibronectin that promotes precartilage mesenchymal condensation, and distinguishing processes include effects of growth and differentiation factors such as retinoids and their receptors and *Hox* gene products, which vary in concentration across limb buds and between forelimbs and hindlimbs (Downie and Newman, 1994). Transcription factors have also been shown to be differentially expressed in forelimbs and hindlimb buds (Gibson-Brown et al., 1996; Nelson, et al., 1996).

III. EXPERIMENTAL TERATOLOGY

A. Historical Overview

Experimental teratology in the modern sense began in the 1940s when Warkany and his associates first called attention to the fact that maternal dietary deficiencies and X-rays could adversely affect the *in utero* development of mammals (Wilson, 1977). Three human tragedies resulting from *in utero* exposure to drugs or chemicals led to the development and revision of testing guidelines. The thalidomide tragedy of 1961 resulted in the birth of infants with rare limb malformations, amelia (absence of the limbs), or various degrees of phocomelia (absence of the proximal portion of a limb) to mothers who took this sedative. Additional malformations seen with thalidomide treatment included defects of the external ears, facial hemangioma, atresia of the esophagus or duodenum, tetralogy of Fallot, and renal agenesis (Shepard, 1995b). This completely changed the perception regarding the vulnerability of the intrauterine embryo to outside influences.

In the early 1970s, adenocarcinoma of the vagina began occurring in young women whose mothers had been treated with diethylstilbestrol during the first trimester of pregnancy. This raised concerns regarding adverse effects that were not evident until after puberty. The third event was the epidemic of organomercurial poisoning that resulted from the dumping of metallic mercury into Minimata Bay, Japan, and its conversion by aquatic plant life into methylmercury. The concern for the potential to affect postnatal development resulted in additional testing for postnatal behavioral changes (Christian, 2001).

Wilson (1977) formulated six principles of teratology and they remain valid today:

1. Susceptibility to teratogenesis depends on the genotype of the conceptus and the manner in which this interacts with environmental factors.
2. Susceptibility to teratogenic agents varies with the developmental stage at the time of exposure. Wilson (1965a) graphically illustrated this concept, showing periods at which specific organs are more susceptible to teratogenic insult (Fig. 28-5).
3. Teratogenic agents act in specific ways (mechanisms) on developing cells and tissues to initiate abnormal embryogenesis (pathogenesis).
4. The final manifestations of abnormal development are death, malformation, growth retardation, and functional disorder.
5. The access of adverse environmental influences to developing tissues depends on the nature of the influences (agent).
6. Manifestations of deviant development increase in degree as dosage increases, from the no-effect to the totally lethal level.

In addition, Wilson (1977) listed eight mechanisms of teratogenesis: mutation, chromosomal nondisjunction and breaks, mitotic interference, altered nucleic acid integrity or function, lack of precursors and substrates needed for biosynthesis, altered energy sources, enzyme inhibitions, osmolar imbalance, and altered membrane characteristics. Schardein (2000) and Shepard (1995a) regularly update compilations of animal teratogens that provide overviews of the effects of test agents. After the thalidomide tragedy, the U.S. Food and Drug Administration (FDA) issued more extensive testing protocols (segments I, II, and III) (FDA, 1966). In 1994 as a result of the International Conference of Harmonisation of Technical Requirements for Registration of Pharmaceuticals for Human Use (ICH), a new internationally accepted set

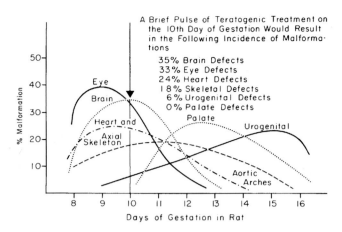

Fig. 28-5 Hypothetical representation of how the syndrome of malformations produced by a given agent might be expected to change when treatment is given at different times. The percentage of animals affected as well as the incidence rank of the various types of malformations would be somewhat different from that shown for the tenth day if treatment were given instead on the twelfth or the fourteenth day, for example. (Data from Beaudoin, 1980.)

of guidelines was issued that the FDA, the European Union, and Japan accept (ICH, 1994). These guidelines cite recommendations for flexible study designs, kinetics, requirements for mechanistic studies, and essentially equal emphasis on all endpoints (death, malformation, and growth). Additional guidelines for evaluating chemicals for teratogenic potential include those by the U.S. Environmental Protection Agency (EPA, 1998) and the Organization for Economic and Co-operation and Development (OECD): Guidelines for Testing of Chemicals (section 4, nos. 414, Teratogencity, and 422, Combined Repeated Dose Toxicity Study with the Reproduction/Developmental Toxicity Screening Test) (OECD, 1981, 1996). Harmonization efforts are ongoing between the EPA and OECD (Christian, 2001). Christian (2001) gives a listing of the majority of the regulatory guidelines currently in use.

The most common treatment period to evaluate teratogenicity requires exposure from implantation to closure of the hard palate or from gestation day 6 through 17 in the rat. Studies to evaluate effects on embryo-fetal development are required in two species: one rodent and one non-rodent, usually the rat and the rabbit. The ICH guidelines state "reasons for using rats as the predominant rodent species are practicality, comparability with other results obtained in this species, and the large amount of background knowledge accumulated."

The rat has the following characteristics that make it a suitable species for evaluating effects on embryo-fetal development: it is relatively disease resistant, withstands operative procedures well, and has a short reproductive cycle (estrus stage can be determined) and a high breeding rate with a relatively short gestation period and good litter size. The fetuses have a low spontaneous malformation rate and are of sufficient size (Beaudoin, 1980). Historical data are available in the scientific literature and in a compilation of historical control data from 1992–1994 (MARTA/MTA, 1995), but the most relevant data are control data generated in the specific laboratory using the same animal strains, scientific nomenclature, and technical conditions. An internationally developed glossary of terms for structural developmental abnormalities, which gives developmental findings, synonyms, and definitions, has also been published (Wise, et al., 1997).

B. Breeding

Pregnant females used for embryo-fetal development studies may be obtained in one of three ways: (1) by purchasing timed pregnant females from a commercial supplier, (2) by purchasing male and females from a commercial supplier and breeding them in house, and (3) maintaining a breeding colony. The ICH guidelines (1994) state "that within and between studies, animals should be of comparable age, weight, and parity at the start: the easiest way to fulfill these criteria is to use animals that are young, mature adults at the time of mating with the females being virgin." Strains with low fecundity should not be used. The amount of background data available and the experience of the laboratory with specific rat strains will determine the final choice; however, the ICH guidelines (1994) states that "it is generally desirable to use the same species and strain as in other toxicological studies."

If mating is done in house, the stage of the estrus cycle (proestrus, estrus, metestrus, or diestrus) can be determined by evaluating vaginal smears of cells from the vaginal epithelium (Beaudoin, 1980; Taylor, 1986; Christian, 2001). Rats in the proestrus stage are placed for overnight breeding. Rats in estrus are paired with a male if bred for short periods (2 to 4 hours) because ovulation in the rat occurs spontaneously near the end of the estrous period, the period when the female is receptive to the male. The shorter mating period is used to reduce interlitter variability. Males and females are usually cohabited in a 1:1 ratio, but males may be cohabitated with up to five females. Insemination is confirmed by the presence of spermatozoa in a vaginal smear or the presences of a vaginal plug. The vaginal plug is a coagulated mass of semen, which may be found lying on the bottom of the cage (Beaudoin, 1980; Taylor, 1986). The day of insemination is usually considered gestation day 0, but day 1 is also used. The gestation period in the rat is 22 days (Hoar and Monie, 1981; Barrow, 2000). Care should be taken to avoid excessive stimulation of the vagina when evaluating the female for the estrous cycle stage or for the presence of sperm in the vagina, because this can upset the estrous cycle and produce a pseudopregnancy with a duration of 12 to 16 days before the resumption of the normal cycle (Beaudoin, 1980).

C. Embryo-Fetal Development Study Design

The ICH guidelines (1994) for the testing of drugs require treatment from implantation (day 6) of the embryo until closure of the hard palate (day 17), and group sizes that will yield between 16 to 20 litters to evaluate. The latest EPA (1998) guidelines and OECD drafts for chemicals modify the traditional study design and require treatment to include the fetal period (gestation days 17 to 20). The extended duration of treatment is intended to improve the sensitivity of the test to detect toxic influences on fetal development (Barrow, 2000). Satellite animals that are not included in the fetal evaluations of the experiment are generally needed for toxicokinetic evaluations in the rat because of the volume of blood that is collected and the possible stress induced in the animal has the potential to affect pregnancy outcomes.

1. Dose Selection

Selection of the correct doses is a critical issue. Usually there are at least three test agent treatment groups and appropriate control groups. Vehicle controls are normally included; however, in some cases an untreated or sham-treated control group and a positive control group may be included. When different treatment volumes are used for the different treatment groups, the control group is normally given the volume administered to the highest treated group. The selection of the high dose should be based on data from all available studies (pharmacology, acute and repeat dose toxicity, and kinetic studies), and doses should be used that are similar to those used in general toxicity studies, thus allowing for comparison. Dose selection is usually based on results of a dose range-finding study in pregnant animals with about five or six animals per group. The same in-life and caesarean examinations as for the embryo-fetal development study are used except that fetuses are only evaluated externally. According to the ICH guidelines, the high dose is expected to induce minimal maternal toxicity, as indicated by reductions in body weight gain, specific target organ toxicity, changes in hematology or clinical chemistry parameters, and/or

TABLE 28-2

RECOMMENDED DOSING VOLUMES

Gavage	Intravenous (Bolus)[a]	Intraperitoneal	Subcutaneous	Intramuscular (mL/injection site)	Intranasal (mL)	Intradermal (mL per site)
10–30	5–10	5–10	5–10	0.1–0.3	0.1	0.05–0.1

[a]Bolus injections are generally given over approximately 1 minute. The maximum volume for slow intravenous infusions lasting up to 20 minutes is 4 mL/kg/min. Maximum volumes for continuous intravenous infusion are 4 mL/kg/hr in rats.

an exaggerated pharmacological response. However, maternal effects should not be overly severe so as to limit the number of fetuses to be evaluated.

The kinetics of the drug, both systemic exposures (area under the concentration-time curve [AUC]) and peak plasma levels (C_{max}) also should be considered as both can influence developmental effects. Ideally, the doses selected should provide systemic exposures with an adequate margin above the estimated human therapeutic exposure. It should be noted that for some compounds, particularly cytotoxic oncolytics, a "safety" margin may not be achieved. The physico-chemical properties of the test substance may impose practical limitations on the amount of drug that can be administered, and under most circumstances, a dose of 1 g/kg/day should be an adequate limit dose. The ICH guidelines (1994) state "lower dosages should be selected in a descending sequence, the intervals depending on kinetic and other toxicity factors. While it is desirable to be able to determine a no observed adverse effect level, priority should be given to setting dosage intervals close enough to reveal any dosage-related trends that may be present." The ICH further cautions that any dose response may be steep.

2. Routes of Administration, Frequency of Dosing, and Dose Volume

For drugs, the proposed human route of administration is preferred for embryo-fetal development studies, provided this route provides an adequate exposure compared with the estimated human therapeutic exposure. For chemicals, the principal route of human exposure should be used (Barrow, 2000). Gavage administration is the preferred method of oral treatment for embryo-fetal development studies and has the advantage of allowing the exact predefined dose to be administered to each animal. The daily dose should be adjusted if possible according to the most recent body weight. Administration by admixture in the diet or by dissolution in the drinking water is the preferred method of oral administration for multigenerational studies with chemicals. Maternal food consumption and body weight increase during gestation; therefore, it is necessary to vary the concentration in the diet to maintain a constant dose level relative to body weight. Pregnant rats have particular nutritional

requirements that can have a significant influence on pregnancy outcomes (Wilson, 1977); therefore, the percentage of drug in the diet must be considered or vitamin deficiencies may result. When the route of exposure is by inhalation, particularly with pregnant rats, whole-body exposure is preferred because of the stress associated with the restraint necessary in nose only exposure (Astroff, et al., 2000). Other acceptable methods of drug administration for the rat include intraperitoneal, intramuscular, subcutaneous, intravenous, intra-arterial, or intradermal injections. Acceptable volumes for the rat are given in Table 28-2. The first value in the range is considered to be an optimal dose volume for most toxicology protocols, taking into account both animal tolerance and the characteristics of the dosing preparation. All volumes are expressed in milliliters per kilogram unless otherwise noted. Special consideration should also be given to pH, irritation potential, osmolality, and pharmacologic and toxicologic effects of the drugs to be administered.

3. *In Vivo* Observations

The dams are monitored daily for clinical signs and mortality. Clinical observations may be increased to include a predose and postdose time point during the treatment period. The time of expected maximum plasma concentration (t_{max}) is often used as the postdose time point.

Body weights are measured at least twice weekly, and body weight change is determined. The ICH guidelines state "daily weighing of pregnant females during treatment can provide useful information." Food consumption is usually measured concomitant with body weight; however, the ICH guidelines (1994) only require weekly measurements. Water consumption can be measured if there are scientific reasons to do so, but it is not normally done.

4. Toxicokinetic Parameters

Developmental effects depend on the gestational stage at the time of the exposure and the exposure level. Some effects seem to relate to the peak plasma levels achieved (C_{max}), whereas other effects occur at lower levels related to AUC and sustained exposures. Nau (1985) showed in rodents (mice) that with valproic acid, a high maternal plasma concentration at specific stages of development resulted in exencephaly in the fetuses, whereas sustained exposure to lower levels resulted in embryonic resorptions. Therefore, it is important to consider both the expected C_{max} and AUC when selecting doses for an embryo-fetal development study.

Many of the physiological changes that occur during pregnancy may potentially influence the toxicokinetic and pharmacodynamic properties of the test substance and its metabolites. These changes include reduced gastric secretion, increased transit time, increased plasma and extracellular fluid volume, increased body fat, decreased xenobiotic hepatic transformation, increased kidney function, and differences in protein binding (Clarke, 1993). In rats, many blood parameters change considerably during pregnancy, including drastic decreases in red blood cell counts, hemoglobin concentration, and hematocrit, that strongly suggest that hemodilution occurs in rat pregnancy. Total protein, albumin, glucose, cholesterol, high-density lipoprotein cholesterol, triglycerides, and phospholipids change significantly during pregnancy. Electrolyte values change during pregnancy in the rat as shown by significant declines in sodium and chloride levels (De Rijk et al., 2002).

5. Dam Necropsy and Collection of Fetal and Placental Data

In embryo-fetal development studies, dams are typically euthanized on gestation day 20 or 21 by carbon dioxide inhalation and are given a detailed macroscopic necropsy to detect lesions in the maternal organs. The uterus and ovaries are excised (Taylor, 1986; Barrow, 2000; Christian, 2001). The number of corpora lutea in each ovary is counted and recorded. The uterus of female rodents that do not appear to be pregnant can be examined either by pressing the uterus between glass plates and examining it for the presence of implantation sites or by staining it with ammonium sulfide (Kopf et al., 1964; Salewski, 1964). The weight of the gravid uterus may be recorded (required for chemical testing), and this value can be used to calculate the corrected body weight (terminal body weight minus the gravid uterine weight), which can be useful for evaluation of maternal body weight independent of uterine effects. The uterine horns are cut open, and the chorionic and amniotic sacs are opened to expose the fetuses. The number and location of each implantation site are recorded. The status of each implant site (live and dead fetus, early or late resorption) is recorded. The number of implantations is compared to the number of corpora lutea to determine preimplantation loss. An early resorption is a conceptus in which only placental remnants (metrial gland) are visible, and a late

resorption has both placental and fetal remnants visible. A live fetus is pink in color and will respond to stimulation. A dead fetus is often pale to tan in color and is not responsive to stimulation but does not demonstrate marked autolysis. Fetuses with marked autolysis are considered to be late resorptions (Taylor, 1986; Barrow, 2000; Christian, 2001). Individually identified fetuses are separated from the placenta by cutting the umbilical cord, and any membranes are removed. The fetuses are then blotted dry and individually weighed. The individually identified placentas from live fetuses are then removed from the uterus, and after any remaining membranes are removed, they are individually examined for abnormalities in appearance and weighed. Placental weights normally correlate with fetal weights. Abnormal placentas can be evaluated microscopically. According to the ICH guidelines (1994), it must be possible to relate all findings in a single fetus; therefore, accurate fetal identification is very important.

6. Fetal Evaluations

A. EXTERNAL EVALUATIONS. Each fetus is carefully examined for any external abnormalities, including palatine abnormalities, and the sex is determined by inspection of the anal-genital distance. Taylor (1986) gives a detailed method for conducting an external evaluation. Fetal examinations are singled out in the EPA guidelines as requiring blinded examination in which the technician is unaware of the identity of the fetus with regard to treatment group or control at the time of the examination (Barrow, 2000).

B. VISCERAL EVALUATIONS. The ICH guidelines (1994) state that a minimum of 50% of fetuses (every other fetus in the uterus) from each litter should be examined for visceral or soft-tissue alterations. Examinations can be performed by serial sectioning (Wilson, 1965b) or by a combination of serial sectioning of the head and fresh gross microdissection of the thorax and abdomen (Barrow and Taylor, 1967; Staples, 1974; Stuckhardt and Poppe, 1984; Barrow, 1990; Christian, 2001). In the microdissection technique, the thoracic and abdominal cavities are carefully opened under a microscope, and the organs and major vessels of the heart are examined *in situ*. The heart is cut by using scissors, with the first cut beginning to the right of the ventral midline surface at the apex and extending anteriorly and ventrally into the pulmonary artery. The incision is opened, and the papillary muscles, tricuspid valve, and the three cusps of semilunar valve of the pulmonary artery are evaluated. The interventricular septum is examined for any defects. The second cut is made starting to the left of the ventral midline surface at the apex and extending through the left ventricle into the ascending aorta. The mitral valve (bicuspid valve) and the three cusps of the semilunar valve of the aorta and papillary muscles are observed in addition to the interventricular septum. Igarashi (1993) published a gelatin-embedding-slice method to evaluate cardiac malformations. The kidneys are sectioned to allow for an examination of the renal papillae. A frozen-sectioning method for the evaluation of the head permits the complete visceral examination to be completed at the time of cesarean section (Astroff et al., 2002).

The fetuses are eviscerated to aid in clearing and staining of the skeleton if the skeleton is to be evaluated on the same fetus. The fetuses must be skinned if double staining of the skeleton for bone and cartilage is to be done (Young et al., 2000). Fetuses need not be skinned if only ossification sites are stained using alizarin red S (Kawamura et al., 1990).

C. SKELETAL EVALUATIONS. Before skeletal evaluation can begin, the fetal carcasses that were fixed in 95% ethanol must be dehydrated, macerated, stained with alizarin red S, and cleared. The stained specimens can be stored indefinitely in glycerol. This results in specimens in which all ossified parts of the skeleton are stained red and are strikingly visible through the other tissues that have become completely transparent. Alternatively, the cartilage can also be stained with alcian blue (Inouye, 1976; Kimmel and Trammell, 1981; Whitaker and Dix, 1979; Young et al., 2000). Fetal rat specimens, previously stained only for bone, can be overstained for cartilage according to the methods of Yamada (1991). Automatic processors are available (Rousseaux, 1985 and Conway and Erb, 1995).

The EPA guidelines for chemicals suggest an examination of the fetal cartilage in addition to the ossified bones (Barrow, 2000). After staining, the ossified skeleton can be examined and checked for the presence, size, and shape of every bone (Taylor, 1986; Barrow, 1990; Christian, 2001). Also, if the maceration step with potassium hydroxide is carefully controlled and kept short, the cartilaginous areas of the skeleton remain partially visible as light shadowy regions (Barrow, 2000; Christian, 2001); this allows for an accurate distinction to be made between agenesis of a bone, a malformation, and a developmental delay of ossification, a variation.

An atlas of the fetal skeleton on gestation day 20 for the Crl:CD rat is available (Menegola et al., 2001), and Yasuda and Yuki (1996) published a comparable atlas of the skeleton of the gestation day 21 Wistar rat. Data on the sequence of fetal ossification in the Long-Evans strain from gestation day 14 to 21 are described by Wright et al. (1958). X-ray imaging has also been used for skeletal examination of fetuses. Counting and recording numbers of selected ossification sites in each skeleton provides a method to identify delays in ossification and analyze them statistically for significance.

7. Evaluation Criteria

Developmental toxicity is manifested by death, structural abnormalities, and developmental delay. Malformations are structural defects that are incompatible with life or of major physiological consequence, represent gross structural changes, and are rare in occurrence. Variations are structural alterations with no significant biological effect. Ossification retardation that represents a delay in the normal amount of ossification, or decreases in ossification sites are classified as variations. Teratogenicity is characterized by a significant increase in: (1) the percentage of malformed fetuses per litter, (2) number and percentage of litters with malformed fetuses, and (3) number of fetuses or litters with a particular malformation that appears to increase with dose. However, the level of concern regarding these findings is increased in the absence of causal maternal toxicity. Significant dose-related increases in the incidence of variations or reduction in fetal weights are considered an indication of developmental toxicity. A significant increase in the number of dead or resorbed embryos/fetuses is indicative of embryo/fetal lethality.

Maternal toxicity is assessed by the evaluation of morbidity, body weight gain, food and water consumption, and pathologic lesions in the dam. Material toxicity may directly or indirectly influence the development of the embryos and should be considered when evaluating developmental toxicity. Maternal body weight gain during the second half of gestation in the rat is influenced by the size and rate of growth of the litter. Litter influences on maternal weight can be assessed by subtracting the weight of the gravid uterus (determined at necropsy) from the terminal body weight of the dam. Treatment-related effects on gestation length have an influence on the relative state of development of the fetus at the time of examination or pups at birth and need to be considered in the evaluation of fetus or pup weight and fetal or postnatal development.

In the presentation of data in report form, the entire spectrum of observed effects needs to be presented. These findings must be analyzed in relation to the concurrent controls in addition to historical controls (from the specific laboratory if possible). The conclusion should concisely state the effect of the compound on the fetuses and should include any maternal toxicity at each of the specific dose levels.

Risk assessment is defined as consisting of hazard identification, dose-response assessment, exposure assessment, and risk characterization (Christian, 2001). Guidelines for developmental toxicity risk assessment were issued by the EPA in 1996 and provide guidance for interpreting, analyzing, and using the data from developmental toxicity studies. The FDA considers risk on a risk/benefit basis, and the conclusions regarding this risk/benefit consideration are published on the label (package insert) of the drug or biologic, which given a pregnancy label category of A, B, C, D, or X (Christian, 2001). At present, the FDA is reviewing this procedure and considering a change from categories to text that would include each of the seven potentially affected reproductive and developmental endpoints. Conclusions from animal data are to be based on all pharmacology, toxicology, and pharmacokinetic data. It is to be performed using "the wedge," a tool that is designed to assess each endpoint and the strength of its effect, which ultimately results in a quantification of the risk for human use of the agent (Christian, 2001).

8. Statistical Analysis

When using statistical procedures, the litter should be the experimental unit of comparison (Jensh et al., 1970; Haseman and Hogan, 1975; ICH, 1994). According to the ICH guidelines (1994), the use of statistics should support the interpretation of the results based on biological significance. Statistical analysis can be conducted on maternal body weight, maternal body weight change, and food and water consumption over periods of interest; that is, before dosing, during the dosing period, and after the dosing period. Fetal body weights by sex, sex ratio, and placental weights may be analyzed by using appropriate statistical tests. Statistical analysis can be conducted on the data collected at necropsy, including number of corpora lutea, implant sites, live and dead fetuses, and resorptions (early and late may be individually analyzed). The calculated values of the percentage of preimplantation loss ([number of corpora lutea minus number of implant sites divided by number of corpora lutea] × 100) and postimplantation loss ([number of implant sites minus number of viable fetuses divided by number of implant sites] × 100) can also be analyzed. Numbers of specific sites of ossification can also be analyzed statistically as a measure of developmental delay. The percent fetal incidence and litter incidence of external, visceral or skeletal malformations and variations can also be evaluated.

D. *In Vitro* Methods

Many different culture systems—including cell culture, tissue culture, organ culture, and whole-embryo culture—have been proposed as *in vitro* alternatives to existing *in vivo* assays for developmental toxicity. Most have been developed and tested in only a single laboratory or lack large-scale validations (Walmod et al., 2002). Teratogenesis screening systems can provide a rapid, cost-effective method of assessing potential teratogens that requires only small amounts of test material. However, *in vitro* assays for developmental toxicity are considered a useful method to screen structurally related compounds for teratogenic potential when at least one compound from the series has been tested *in vivo*.

New et al. (1971, 1976) described the technique for the culture of postimplantation rodent embryos from the early head-fold stage (9.5 days of gestation) for 48 hours (at the approximately 25-somite stage). They found that the rates of growth and differentiation of all the embryos grown for up to 48 hours in culture under optimum culture conditions were similar to those of littermates *in vivo* (New et al., 1976). Brown and Fabro (1981) developed and Van Maele-Fabry et al. (1990) modified an objective scoring system that provides a precise measure of morphological development, aides in the detection of retardation or dysmorphogenesis, and permits a quantitative comparison of growth and development. Studies have shown that whole-embryo cultures have a high predictive capacity for teratogens (New, 1990), and an interlaboratory evaluation of embryo toxicity that uses whole-embryo culture has been published (Piersma et al., 1995).

Assay systems using artificial hydra "embryos" have been used to evaluate developmental toxicity. In this system, a differential assessment of toxicity of a compound to the adult and to regenerating artificial hydra embryos permits the determination of a developmental hazard index determined by calculating the minimal affective concentrations in the adult hydra and developing artificial hydra embryo (A/D ratio) (Johnson et al., 1982, 1988; Mayura et al., 1991).

The frog embryo teratogenesis assay *Xenopus* (FETAX) has been used to identify the potential developmental toxicity of compounds. FETAX is a 96-hour whole-embryo test that uses the embryos of the frog *Xenopus laevis*. The embryos are cultured with and without an exogenous metabolic activation system. The endpoints measured include mortality, malformation, and growth. The 96-hour median lethal (LC_{50}) and teratogenic (EC_{50}) concentrations are determined. From these values, a teratogenic index (IT) is determined (TI = 96-hour LC_{50}/96-hour EC_{50}). The teratogenic index provides a measurement of the separation between mortality and malformation concentration-response curves. The separation between the two curves is used to indicate significant teratogenic hazard. The length data are used to calculate a minimum concentration to inhibit growth (Fort et al., 1992, 1996; Dresser et al., 2000).

Embryonic development is highly sensitive to changes in cell death, growth, migration, and differentiation, and by using comparable endpoints with cell culture systems, it may be possible to predict developmental toxicity. Cell culture techniques have been developed to evaluate teratogenic potential based on the inhibition of differentiation by using limb buds, measuring chondrogenesis (Kistler, 1987; Tsuchiya et al., 1987), and midbrain cells, determining the number of neuron foci (Flint and Orton, 1984; Tsuchiya et al., 1991).

Cultures of pluripotent embryonal stem cells derived from blastocysts that are able to differentiate into a variety of embryonal tissues, retain the euploid chromosome constitution, and proliferate rapidly have been developed and used to evaluate cytotoxicity; such systems appear to hold promise as an *in vitro* teratogenesis screen if differentiation of these pluripotent cells can be quantified (Laschinski et al., 1991). Organ cultures of explanted palates have been used to study normal palate formation and the effect of compounds/drugs on palatogenesis *in vitro* (Shiota et al., 1990; Mino et al., 1994).

REFERENCES

Abdulrazzaq, Y.M., Padmanabhan, R., Bastaki, S.M.A., Ibrahim, A., and Bener, A. (2001). Placental transfer of vigabatrin (γ-vinyl GABA) and its effect on concentration of amino acids in the embryo of TO mice. *Teratology* **63**, 127–133.

Abrahamsohn, P.A. and Zorn, T.M.T. (1993). Implantation and decidualization in rodents. *J Exp. Zool.* **266**, 603–628.

Adelman, H.B. (1925). The development of the neural folds and cranial ganglia of the rat. *J. Comp. Neurol* **39**, 19–171.

Arey, L.B. (1974). The urinary system. *In* "Developmental Anatomy." pp 295–314, W.B. Saunders., Philadelphia.

Astroff, A.B., Ray, S.E., Rowe, L.M., Hilbish, K.G., Linvelle, A.L., Stutz, J.P., and Breslin, W.J. (2002). Frozen-sectioning yields similar results as traditional methods for fetal cephalic examination in the rat. *Teratology* **66**, 77–84.

Astroff, A.B., Sturdivant, D.W., Lake, S.G., Shiotsuka, R.N., Simon, G.S., and Andrews, L.S. (2000). Developmental toxicity of 1,6-hexamethylene diisocyanate (HDI) in the Sprague-Dawley rat. *Teratology* **62**, 205–213.

Austin, C.R. (1961). "The Mammalian Egg." Blackwell Scientific Publication, Oxford.

Baldwin, H.S., Jensen, K.L., and Solursh, M. (1991). Myogenic cytodifferentiation of the precardiac mesoderm in the rat. *Differentiation* **47**, 163–172.

Bard, J.B.L., Davies, J.A., Karavanova, I., Lehtonen, E., Sariola, H., and Vainio, S. (1996). Kidney development: the inductive interactions. *Semin. Cell. Dev. Biol.* **7**, 195–202.

Barnett, J.V. and Desgrosellier, J.S. (2003). Early events in valvulogenesis: a signaling perspective. *Birth Defects Research Part C* **69**, 58–72.

Barrow, M.V. and Taylor, W.J. (1967) A rapid method for detecting malformations of rat foetuses. *J. Morphol.* **127**, 291–306.

Barrow, P. (1990). Technical procedures in reproduction toxicology: fetal pathology investigations. *In* "Laboratory Animal Handbooks." pp 31–40, Royal Society of Medicine, London.

Barrow, P. (2000). Reproductive and developmental toxicology safety studies. *In* "The Laboratory Rat" (G. J. Krinke, Ed.), pp 199–225, Academic Press, New York.

Beaudoin, A.R. (1980). Embryology and teratology. *In* "The Laboratory Rat," Vol. 2. (H.J. Baker, J.R. Lindsey, and S.H. Weisbroth, eds.), pp 75–101, Academic Press, New York.

Beck, F., Lloyd, J.B., and Griffiths, A. (1967a). A histochemical and biochemical study of some aspects of placental function in the rat using maternal injection of horseradish peroxidase. *J. Anat.* **101**, 461–478.

Beck, F., Lloyd, J.B., and Griffiths, A. (1967b). Lysosomal enzyme inhibition by trypan blue: a theory of teratogenesis. *Science* **157**, 1180–1182.

Biggers, J.D. and Borland, R.M. (1976). Physiological aspects of growth and development of the preimplantation mammalian embryo. *Annu. Rev. Physiol.* **38**, 95–119.

Blandau, R.J., Boling, J.L., and Young, W.C. (1941). The length of heat in the albino rat as determined by the copulatory response. *Anat. Rec.* **79**, 453–463.

Blandau, R.J. and Money, W.L. (1944). Observations on the rate of transport of spermatozoa in the female genital tract of the rat. *Anat. Rec.* **90**, 255–260.

Boling, J.L., Blandau, R.J., Soderwall, A.L., and Young, W.C. (1941). Growth of the graafian follicle and the time of ovulation in the albino rat. *Anat. Rec.* **79**, 313–331.

Bridgman, J. (1948). A morphological study of the development of the placenta of the rat, I: an outline of the development of the placenta of the white rat. *J. Morphol.* **83**, 61–85.

Brown, N.A. and Fabro, S. (1981). Quantitation of rat embryonic development in vitro: a morphological scoring system. *Teratology* **24**, 65–78.

Brunschwig, A.E. (1927). Notes on experiments in placental permeability. *Anat. Rec.* **34**, 237–244.

Burke, A.C., Nelson, C.E., Morgan, B.A., and Tabin, C. (1995). Hox genes and the evolution of vertebrate axial morphology. *Development* **121**, 333–346.

Burlingame, P.L. and Long, J.A. (1939). The development of the heart in the rat. *Univ. Calif. Publ. Zool.* **43**, 249–319.

Butcher, E.O. (1929). The development of the somites in the white rat (*Mus norvegicus albinus*) and the fate of the myotomes, neural tube, and gut in the tail. *Am. J. Anat.* **44**, 381–439.

Campbell, L.R., Dayton, D.H., and Sohal, G.S. (1986). Neural tube defects: a review of human and animal studies on the etiology of neural tube defects. *Teratology* **34**, 171–187.

Chacko, K.J. (1976). Observations on the ultrastructure of developing myocardium of rat embryos. *J. Morphol.* **150**, 681–709.

Christian, M.S. (2001). Test methods for assessing female reproductive and developmental toxicology. *In* "Principles and Methods of Toxicology," 4th Ed. (A.W. Hayes, ed.), pp 1301–1381, Taylor & Francis, Philadelphia.

Christie, G.A. (1964). Developmental stages in somite and post-somite rat embryos, based on external appearance, and including some features of the macroscopic development of the oral cavity. *J. Morph.* **114**, 263–286.

Christoffels, V.M., Habets, P.E.M.H., Franco, D., Campione, M., de Jong, F., Lamers, W.H., Bao, Z., Palmer, S., Biben, C., Harvey, R.P., and Moorman, A.F.M. (2000). Chamber formation andmorphogenesis in the developing mammalian heart. *Dev. Biol.* **223**, 266–278.

Clark, E.B. (1989). Growth, morphogenesis and function: the dynamics of heart development. *In* "Fetal, Neonatal and Infant Heart Disease." (J.H. Moller, W.A. Neal, and J.E. Lock, eds.), pp 1–17, Appleton-Centurey-Crofts, New York.

Clarke, D.O. (1993). Pharmacokinetic studies in developmental toxicology: practical considerations and approaches. *Toxicol. Meth.* **3**, 223–251.

Conway, M.J. and Erb, C.A. (1995). Double staining fetal rat and rabbit skeletons in an automated processor. *Teratology* **51**, 182.

Cunha, G.R., Alarid, E.T., Turner, T., Donjacour, A.A., Boutin, E.L., and Foster, B.A. (1992). Normal and abnormal development of the male urogenital tract: role of androgens, mesenchymal-epithelial interactions, and growth factors. *J. Androl.* **13**, 465–475.

Davies, J. and Glasser, S.R. (1968). Histological and fine structural observations on the placenta of the rat. *Acta. Anat.* **69**, 542–608.

Davies, J. and Hesseldahy, H. (1971). Comparative embryology of mannalian blastocysts. *In* "The Biology of the Blastocyst." (R.J. Blandau, ed.), pp 27–48, University of Chicago Press, Chicago.

De Rijk, E.P.C.T., Van Esch, E., and Flik, G. 2002. Pregnancy dating in the rat: placental morphology and maternal blood parameters. *Toxicol. Pathol.* **30**, 271–282.

deVries, P.A. (1981). Evolution of precardiac and splanchnic mesoderm in relationship to the infundibulum and truncus. *In* "Perspectives in Cardiovascular Research," Vol. 5. (T. Pexieder, ed.), pp 31–48, Raven Press, New York.

DeSesso, J. (1993). Cardiovascular development. *In* "Embryology for Teratologist: A Refresher Couse Conducted by the Education Committee of the Teratology Society." pp C1–C45.

Dickmann, Z. (1969). Shedding of the zona pellucida. *Adv. Reprod. Physiol.* **4**, 187–206.

Dickmann, Z. and Noyes, R.W. (1961). The zona pellucida at the time of implantation. *Fertil. Steril.* **12**, 310–318.

Domenech-Mateu, J.M. and Orts-Llorca, F. (1976). Arterial vascularization of the sinuatrial node in the embryonic rat heart. *Acta. Anat.* **94**, 343–355.

Downie, S.A. and Newman, S.A. (1994). Morphogenetic differences between fore and hind limb precartilage mesenchyme: relation to mechanisms of skeletal pattern formation. *Dev. Biol.* **162**, 195–208.

Dresser, T.H., Rivera, E.R., Hoffmann, F.J., and Finch, R.A. (1992). Teratogenic assessment of four solvents using the frog embryo teratogenesis assay *Xenopus* (FETAX). *J. Appl. Toxicol.* **12**, 49–56.

Dyce, J., George, M., Goodall, H., and Fleming, T.P. (1987). Do trophectoderm and inner cell mass cells in the mouse blastocyst maintain discrete lineages? *Development* **100**, 685–698.

Edwards, J.A. (1968). The external development of the rabbit and rat embryo. *Adv. Teratol.* **3**, 239–263.

Eibs, H.G., Spielman, H., and Hägele M. (1982). Teratogenic effects of cyproterone acetate and medroyprogesterone treatment during the pre- and postimplantation period of mouse embryos. *Teratology* **25**, 27–36.

Ellington, S.K.L. (1985). A morphological study of the development of the allantois of rat embryos *in vivo*. *J. Anat.* **142**, 1–11.

Enders, A.C. (1971). The fine structure of the blastocyst. *In* "The Biology of the Blastocyst." (R.J. Blandau, ed.), pp 72–88, University of Chicago Press, Chicago.

Enders, A.C. (1975). The implantation chamber, blastocyst and blastocyst imprint of the rat: a scanning electron microscope study. *Anat. Rec.* **82**, 137–149.

Enders, A.C., Schlafke, S., and Welsh, A.O. (1980). Trophoblastic and uterine luminal epithelial surfaces at the time of blastocyst adhesion in the rat. *Am. J. Anat.* **159**, 59–72.

Evan, A.P., Gattone II, V.H., and Blomgren, P.M. (1984). Application of scanning electron microscopy to kidney development and nephron maturation. *Scanning Electron Microscopy* 455–473.

Everett, J.W. (1933). Structure and function of the yolk-sac placenta in *Mus norvegicus albinus*. *Proc. Soc. Exp. Biol. Med.* **31**, 77–79.

Everett, J.W. (1935). Morphological and physiological studies of the placenta in the albino rat. *J. Exp. Zool.* **70**, 243–285.

Fabro, S. (1973). Passage of drugs and other chemicals into the uterine fluids and preimplantation blastocyst. *In* "Fetal Pharmacology." (L. Boreus, ed.), pp 443–461, Raven Press, New York.

Fabro, S., McLachlan, J.A., and Dames, N.M. (1974). Chemical exposure of embryos during the preimplantation stages of pregnancy: mortality rate and intrauterine development. *Am. J. Obstet. Gynecol.* **148**, 929–938.

Fallon, J.F. and Crosby, G.M. (1977). Polarising zone activity in limb buds of amniotes. *In* "Vertebrate Limb and Somite Morphogenesis." (D.A. Ede, J.R. Hinchliffe, and M. Balls, eds.) pp 55–69, Cambridge University Press, New York.

Ferguson, M.W.J. (1977). The mechanism of palatal shelf elevation and the pathogenesis of cleft palate. *Virchows Arch. A Pathol. Anat. Histol.* **375**, 97–113.

Ferguson, M.W.J. (1978). Palatal shelf elevation in the Wistar rat fetus. *J. Anat.* **125**, 555–577.

Fleming, T.P. (1987). A quantitative analysis of cell allocation to trophectoderm and inner cell mass in the mouse blastocyst. *Exp. Biol.* **119**, 520–531.

Flint, O.P. and Orton, T.C. (1984). An in vitro assay for teratogens with cultures of rat embryo midbrain and limb bud cells. *Toxicol. Appl. Pharmacol.* **76**, 383–395.

Florez-Cossio, T.J. (1975). Studies on the development of the albino rat notochord. *Anatomischer Anzeiger* **137**, 35–55.

Fort, D.J., Stover, E.L., Farmer, D.R., and Lemen, J.K. (2000). Assessing the predictive validity of frog embryo teratogenesis assay: *Xenopus* (FETAX). *Teratogen. Carcinog. Mutagen.* **20**, 87–98.

Fort, D.J., Stover, E.L., Propst, T., Hull, M.A., and Bantle, J.A. (1996). Evaluation of the developmental toxicity of theophylline, dimethyuric acid, and methylxanthine metabolites using *Xenopus Drug Chem. Toxicol.* **19**, 267–278.

Fujinaga, M. and Baden, J.M. (1991). Critical period of rat development when sidedness of asymmetric body structures is determined. *Teratology* **44**, 453–462.

Fujinaga, M. and Baden, J.M. (1992). Variation in development of rat embryos at the presomite period. *Teratology* **45**, 661–670.

Fujinaga, M. and Baden, J.M. (1993). Microsurgical study on the mechanisms determining sidedness of axial rotation in rat embryos. *Teratology* **47**, 585–593.

Fujinaga, M., Brown, N.A., and Baden, J.M. (1992). Comparison of staging systems for the gastrulation and early neurulation period in rodents: a proposed new system. *Teratology* **46**, 183–190.

Fujiwara, M., Uchida, T., Osumi-Yamashita, N., and Eto, K. (1994). Uchida rat (rSey): a new mutant rat with craniofacial abnormalities resembling those of the mouse Sey mutant. *Differentiation* **57**, 31–38.

Fukiishi, Y. and Morriss-Kay, G.M. (1992). Migration of cranial neural crest cells to the pharyngeal arches and heart in rat embryos. *Cell Tissue Res.* **268**, 1–8.

Gajović, S., Kostović-Knežević, L., and Svajger, A. (1989). Origin of the notochord in the rat embryo tail. *Anat. Embryol.* **179**, 305–310.

Gibson-Brown, J.J., Agulnik, S.I., Chapman, D.L., Alexiou, M., Garvey, N., Silver, L.M., and Papaioannou, V.E. (1996). Evidence of a role for T-box genes in the evolution of limb morphogenesis and the specification of forelimb/hindlimb identity. *Mech. Dev.* **56**, 93–101.

Gilbert, S.F. (1997). "Developmental Biology," 5th Ed. Sinauer Associates, Sunderland.

Gilbert, S.F. (2000). "Developmental Biology," 6th Ed. Sinauer Associates, Sunderland.

Goulding, M.D., Lumsden, A., and Gruss, P. (1993). Signals from the notochord and floor plate regulate the region-specific expression of two *Pax* genes in the developing spinal cord. *Development* **117**, 1001–1016.

Gourdie, R.G., Harris, B.S., Bond, J., Justus, C., Hewett, K.W., O'Brien, T.X., Thompson, R.P., and Sedmera, D. (2003). Development of the cardiac pacemaking and conduction system. *Birth Defects Res. C* **69**, 46–57.

Granström, G., Jacobsson, C., and Kirkeby, S. (1992). Diagnosis of induced cleft palate in rats from defective patterns of differentiation of isoenzymes. *Scand. J. Plast. Reconstr. Hand Surg.* **26**, 13–20.

Gross, C.M. (1938). The first contractions of the heart in rat embryos. *Anat. Rec.* **70**, 505–524.

Gulyas, B.J. (1975). A reexamination of cleavage patterns in eutherian mammalian eggs: rotation of blastomere pairs during second cleavage in the rabbit. *J. Exp. Zool.* **193**, 235–248.

Hall, B.K. and Miyake, T. (2000). Craniofacial development of avian and rodent embryos. *Methods Mol. Biol.* **135**, 127–137.

Haseman, J.K. and Hogan, M.D. (1975). Selection of the experimental unit in teratology studies. *Teratology* **12**, 165–172.

Hayashi, K. and Ozawa, E. (1995). Myogenic cell migration from somites is induced by tissue contact with medial region of the presumptive limb mesoderm in chick embryos. *Development* **121**, 661–669.

Hebel, R. and Stromberg, M.W. (1986). Embryology. *In* "Anatomy and Embryology of the Laboratory Rat." pp 231–257, BioMed Verlag Wörthsee, Federal Republic of Germany.

Hew, K.W. and Keller, K.A. (2003). Postnatal anatomical and functional development of heart: a species comparison. *Birth Defects Res. B* **68**, 309–320.

Hinchliffe, J.R. (1994). Evolutionary developmental biology of the tetrapod limb. *Development* **1994(Suppl.)**, 163–168.

Hoar, R.M. and Monie, I.W. (1981). Comparative development of specific organ systems. *In* "Developmental Toxicology." (C.A. Kimmel and J. Buelke-Sam, eds.), pp 13–33, Ravin Press, New York.

Holson, J.F., Scott, W.J., Gaylor, D.W., and Wilson, J.G. (1976). Reduced interlitter variability in rats resulting from a restricted mating period, and reassessment of the "litter effect." *Teratology* **14**, 135–141.

Huber, G.C. (1915a). The development of the albino rat: from the end of the first to the tenth day after insemination. *Anat. Rec.* **9**, 84–88.

Huber, G.C. (1915n). The development of the albino rat, *Mus Norvegicus Albinus*, I: from the pronuclear stage to the stage of mesoderm anlage: end of the first to the end of the ninth day. *J. Morphol.* **26**, 247–357.

Hughes, A.F. and Freeman, R.B. (1974). Comparative remarks on the development of the tail cord among higher vertebrates *J. Embryol. Exp. Morph.* **32**, 355–363.

Hutson, M.R. and Kirby, M.L. (2003). Neural crest and cardiovascular development: a 20-year perspective. *Birth Defects Res. C* **69**, 2–13.

Igarashi, E. (1993). New method for the detection of cardiovascular malformations in rat fetuses: gelatin-embedding-slice method. *Teratology* **48**, 329–333.

Inouye, M. (1976). Differential staining of cartilage and bone in fetal mouse sleleton by alcian blue and alizarin red S. *Congenit. Anom. (Kyoto)* **16**, 171–173.

International Conference of Harmonisation [ICH]. (1994). Guideline on Detection of Toxicity to Reproduction for Medicinal Products. *Fed. Reg.* **59**(183), 48746–48752.

Jensh, R.P., Brent, R.L., and Barr, Jr., M. (1970). The litter effect as a variable in teratologic studies of the albino rat. *Am. J. Anat.* **128**, 185–192.

Johnson, E.M., Gorman, R.M., Gabel, B.E.G., and George, M.E. (1982). The *Hydra attenuate* system for detection of teratogenic hazards. *Teratogen. Carcinog. Mutagen.* **2**, 263–276.

Johnson, E.M., Newman, L.M., Gabel, B.E.G., Boerner, T.F., and Dansky, L.A. (1988). An analysis of the hydra assay's applicability and reliability as a developmental toxicity prescreen. *J. Am. Toxicol.* **7**, 111–126.

Johnson, M.C. and Sulik, K.K. (1980). Development of face and oral cavity. *In* "Orban's Oral Histology and Embryology." (S.N. Ghaskar, ed.), pp 1–23. Mosby, St. Louis.

Johnson, R.L., Riddle, R.D., Laufer, E., and Tabin, C. (1994). Sonic hedgehog: a key mediator of anterior-posterior patterning of the limb and dorso-ventral patterning of axial embryonic structures. *Biochem. Soc. Trans.* **22**, 569–574.

Jollie, W. (1964). Fine structural changes in placental labyrinth of the rat with increasing gestational age. *J. Ultrastruct. Res.* **10**, 27–47.

Jollie, W.P. (1990). Development, morphology, and function of the yolk-sac placenta of laboratory rodents. *Teratology* **41**, 361–381.

Kawamura, S., Hirohashi, A., Kato, T., and Yasuda, M. (1990). Bone-staining technique for fetal rat specimens without skinning and removing adipose tissue. *Congenit. Anom. (Kyoto)* **30**, 93–95.

Kazimierczak, J. (1980). A study by scanning (SEM) and transmission (TEM) electron microscopy of the glomerular capillaries in developing rat kidney. *Cell Tissue Res.* **212**, 241–255.

Kerrigan, J.J., Mansell, J.P., Sengupta, A., Brown, N., and Sandy, J.R. (2000). Palatogenesis and potential mechanisms for clefting. *J. R. Coll. Surg.* **45**, 351–358.

Kimmel, C.A. and Trammell, C. (1981). A rapid procedure for routine double staining of cartilage and bone in fetal and adult animals. *Stain Technol.* **56**, 271–273.

Kirby, M.L., Bale, T.F., and Stewart, D.E. (1983). Neural crest cells contribute to normal aorticopulmonary septation. *Science* **220**, 1059–1061.

Kistler, A. (1987). Limb bud cell cultures for estimating the teratogenic potential of compounds. Validation of the test system with retinoids. *Arch. Toxicol.* **60**, 403–414.

Kochhar, D.M. and Johnson, E.M. (1965). Morphological and auto-radiographic studies of cleft palate induced in rat embryos by maternal hypervitaminosis A. *J. Embryol. Exp. Morph.* **14**, 223–238.

Kopf, R., Lorenz, D., and Salewski, E. (1964). Influence of thalidomide on the fertility of rats in a generation study over two generations. *Naunyn-Schmiedebergs Arch. Exp. Pharmacol.* **247**, 121–135.

Krehbiel, R.H. and Plagge, J.C. (1962). Distribution of ova in the rat uterus. *Anat. Rec.* **143**, 239–241.

Laschinski, G., Vogel, R., and Spielmann, H. (1991). Cytotoxicity test using blastocyst-derived euploid embryonal stem cells: a new approach to in vitro teratogenesis screening. *Reprod. Toxicol.* **5**, 57–64.

Le Douarin, N.M. and Ziller, C. (1993). Plasticity in neural crest cell differentiation. *Curr. Opin. in Cell Biol.* **5**, 1036–1043.

Lejour, M. (1970). Cleft lip induced in the rat. *Cleft Palate J.* **7**, 169–186.

Levak-Švajger, B. and Švajger, A. (1974). Investigation on the origin of the definitive endoderm in the rat embryo. *J. Embryol. Exp. Morph.* **32**, 445–459.

Linask, K.K. (2003). Regulation of heart morphology: current molecular and cellular perspectives on the coordinated emergence of cardiac form and function. *Birth Defects Res. C* **69**, 14–24.

Long, J.A. and Burlingame, P.L. (1938). The development of the external form of the rat, with some observations on the origin of the extraembryonic coelom and foetal membranes. *Univ. Calif. Publ. Zool.* **43**, 143–183.

López-Martínez, A., Chang, D.T., Chiang, C., Porter, J.A., Ros, M.A., Simandl, B.K., Beachy, P.A., and Fallon, J.F. (1995). Limb-patterning activity and restricted posterior localization of the animo-terminal product of sonic hedgehog cleavage. *Curr. Biol.* **5**, 791–796.

Lundkvist, Ö. and Nilsson, B.O. (1984). Ultrastructural studies of the temporal relationship between loss of zona pellucida and appearance of blastocyst-induced stroma changes during normal pregnancy in rats. *Anat. Embryol.* **170**, 45–49.

Mac Auley, A., Werb, Z., and Mirkes, P.E. (1993). Characterization of the unusually rapid cell cycles during rat gastrulation. *Development* **117**, 873–883.

MacCabe, J.A., Errick, K., and Saunders, Jr., J.W. (1974). Ectodermal control of dorsovental axis in leg bud of chick embryo. *Dev. Biol.* **39**, 69–82.

Maeda, K., Ohkura, S., and Tsukamura, H. (2000). Physiology of reproduction. *In* "The Laboratory Rat." (G.J. Krinke, ed.), pp 145–176, Academic Press, New York.

Mayura, K., Smith, E.E., Clement, B.A., and Phillips, T.D. (1991). Evaluation of the developmental toxicity of chlorinated phenols utilizing Hydra attenuata and postimplantation rat embryos in culture. *Toxicol. Appl. Pharmacol.* **108**, 253–266.

Menegola, E., Broccia, M.L., and Giavini, E. (2001). Atlas of rat fetal skeleton double stained for bone and cartilage. *Teratology* **64**, 125–133.

Merchant-Larios, H. (1976). The onset of testicular differentiation in the rat: an ultrastructural study. *Am. J. Anat.* **145**, 319–330.

Middle Atlantic Reproduction and Teratology Association Midwest Teratology Association [MARTA/MTA]. (1995). "Historical control data (1992–1994) for developmental and reproductive studies using the CRL:CD BR rat." Charles River Laboratories.

Milaire, J. and Rooze, M. (1983). Hereditary and induced modifications of the normal necrotic patterns in the developing limb buds of the rat and mouse: facts and hypotheses. *Arch. Biol.* **94**, 459–490.

Mino, Y., Mizusawa, H., and Shiota, K. (1994). Effects of anticonvulsant drugs on fetal mouse palates cultured in vitro. *Reprod. Toxicol.* **8**, 225–230.

Moffat, D.B. (1959). Developmental changes in the aortic arch system of the rat. *Am. J. Anat.* **105**, 1–35.

Morriss, G.M. and New, D.A.T. (1979). Effect of oxygen concentration on morphogenesis of cranial neural folds and neural crest in cultured rat embryos. *J. Embryol. Exp. Morphol.* **54**, 17–35.

Morriss-Kay, G. (1981). Growth and development of pattern in the cranial neural epithelium of rat embryos during neurulation. *J. Embryol. Exp. Morph.* **65(Suppl.)**, 225–241.

Morriss-Kay, G. and Tucket, F. (1991). Early events in mammalian craniofacial morphogenesis. *J. Craniofac. Genet. Dev. Biol.* **11**, 181–191.

Nau, H. (1985). Teratogenic valproic acid concentrations: infusion by implanted minipumps vs conventional injection regimen in the mouse. *Toxicol. Appl. Pharmacol.* **80**, 243–250.

Neiss, W.F. (1982a). Histogenesis of the loop of Henle in the rat kidney. *Anat. Embryol.* **164**, 315–330.

Neiss, W.F. (1982b). Morphogenesis and histogenesis of the connecting tubule in the rat kidney. *Anat. Embryol.* **165**, 81–95.

Nelson, C.E., Morgan, B.A., Burke, A.C., Laufer, E., DiMambro, E., Murtaugh, L.C., Gonzales, E., Tessarollo, L., Parada, L.F., and Tabin, C. (1996). Analysis of *Hox* gene expression in the chick limb bud. *Development* **122**, 1449–1466.

New, D.A.T. (1971). Methods for the culture of postimplantation embryos of rodents. *In* "Methods in Mammalian Embryology." (J.C. Daniel, Jr., ed.), pp 305–319, W.H. Freeman, San Francisco.

New, D.A.T. (1990). Whole embryo culture, teratogenesis, and the estimation of teratogenic risk. *Teratology* **42**, 635–642.

New, D.A.T., Coppola, P.T., and Cockroft, D.L. (1976). Comparison of growth in vitro and in vivo of post-implantation rat embryos. *J. Embryol. Exp. Morph.* **36**, 133–144.

Odor, D.L. and Blandau, R.J. (1951). Observations on fertilization and the first segmental division in rat ova. *Am. J. Anat.* **89**, 29–62.

Organization for Economic and Cooperation and Development [OECD]. (1981). Guidelines for testing of chemicals, section 4, no. 414: teratogenicity, Adopted 12 May 1981.

Organization for Economic and Cooperation and Development [OECD]. (1996). OECD guidelines for testing of chemicals, section 4, no. 422: combined repeated dose toxicity study with the reproduction/developmental toxicity screening test. Adopted 22 March 1996.

Osumi-Yamashita, N., Ninomiya, Y., and Eto, K. (1997). Mammalian craniofacial embryology in vitro. *Int. J. Dev. Biol.* **41**, 187–194.

Parr, B.A., Shea, M.J., Vassileva, G., and McMahon, A.P. (1993). Mouse *Wnt* genes exhibit discrete domains of expression in the early embryonic CNS and limb buds. *Development* **119**, 247–261.

Paulsen D.F. (1994). Retinoic acid in limb bud outgrowth: review and hypothesis. *Anat. Embryol.* **190**, 399–415.

Pedersen, R.A., Wu, K., and Balakier, H. (1986). Origin of the inner cell mass in mouse embryos: cell lineage analysis by microinjection. *Dev. Biol.* **117**, 581–595.

Piersma, A.H., Attenon, P., Bechter, R., Govers, M.J.A.P., Krafft, N., Schmid, B.P., Stadler, J., Verhoef, A., and Verseil, C. (1995). Interlaboratory evaluation of embryotoxicity in the postimplantation rat embryo culture. *Reprod. Toxicol.* **9**, 275–280.

Psychoyos, A. (1966). Recent researches on egg implantation. *In* "Egg Implantation." (G.E.W. Wolstenholme, and M.L. Connor, eds.), pp 4–28, Little Brown, Boston.

Rahman, M.M., Iida, H., and Shibata, Y. (1997). Expression and localization of annexin V and annexin VI durig limb bud formation in the rat fetus. *Anat. Embryol.* **195**, 31–39.

Riddle, R.D., Ensini, M., Nelson, C., Tsuchida, T., Jessell, T.M., and Tabin, C. (1995). Induction of the LIM homeobox gene *Lmx1* by WNT7a establishes dorsoventral pattern in the vertebrate limb. *Cell* **83**, 631–640.

Riddle, R.D., Johnson, R.L., Laufer, E., and Tabin, C. (1993). Sonic hedgehog mediates the polarizing activity of the ZPA. *Cell* **75**, 1401–1416.

Ross, C.P. and Persaud, T.V.N. (1990). Craniofacial and limb development in early rat embryos following in utero exposure to ethanol and caffeine. *Anat. Anz. Jena.* **170**, 9–14.

Rothenberg, F., Fisher, S.A., and Watanabe, M. (2003). Sculpting the cardiac outflow tract. *Birth Defects Res. C* **69**, 38–45.

Rousseaux, C.G. (1985). Automated differential staining for cartilage and bone in whole mount preparations of vertebrates. *Stain Technol.* **60**, 295–297.

Saxén, L. and Sariola, H. (1987). Early organogenesis of the kidney. *Pediat. Nephrol.* **1**, 385–392.

Salewski, E. (1964). Färbemethode zum makroskopischen nachweis von implantationsstellen am uterus der ratte. *Arch. Pathol. Exp. Pharmakol.* **247**, 367.

Schardein, J.L. (2000). "Chemically Induced Birth Defects." 3rd Ed. Marcel Dekker, New York.

Schlafke, S. and Enders, A.C. (1968). Observations on the fine structure of the rat blastocyst. *J. Anat.* **97**, 353–360.

Schlafke, S. and Enders, A.C. (1975). Cellular basis of interaction between trophoblast and uterus at implantation. *Biol. Reprod.* **12**, 41–65.

Schlafke, S., Welsh, A.O., and Enders, A.C. (1985). Penetration of the basal lamina of the uterine luminal epithelium during implantation in the rat. *Anat. Rec.* **212**, 47–56.

Schneider, B.F. and Norton, S. (1979). Equivalent ages in rat, mouse and chick embryos. *Teratology* **19**, 273–278.

Schoenwolf, G.C. (1984). Histological and ultrastructural studies of secondary neurulation in mouse embryos. *Am. J. Anat.* **169**, 361–376.

Shepard, T.H. (1995a). "Catalog of Teratogenic Agents." 8th Ed. The Johns Hopkins University Press, Baltimore.

Shepard, T.H. (1995b). Thalidomide. *In* "Catalog of Teratogenic Agents," 8th Ed. (T.H. Shepard, ed.), pp 407–409, Johns Hopkins University Press, Baltimore.

Shiota, K., Kosazuma, T., Klug, S., and Neubert, D. (1990). Development of the fetal mouse palate in suspension organ culture. *Acta Anat.* **137**, 59–64.

Smith, L.J. (1985). Embryonic axis orientation in the mouse and its correlation with blastocyst relationships to the uterus, II: relationships from 41/4 to 91/2 days. *J. Embryol. Exp. Morph.* **89**, 15–35.

Smith, S.C. and Monie, I.W. (1969). Normal and abnormal nasolabial morphogenesis in the rat. *Teratology* **2**, 1–12.

Srivastava, H.C. and Rao, P.P. (1979). Movement of palatal shelves during secondary palate closure in rat. *Teratology* **19**, 87–104.

Staples, R.E. (1974). Detection of visceral alterations in mammalian fetuses. *Teratology* **9**, A37–A38.

Stratford, T., Horton, C., and Maden, M. (1996). Retinoic acid is required for the initiation of outgrowth in the chick limb bud. *Curr. Biol.* **6**, 1124–1133.

Stuckhardt, J.L. and Poppe, S.M. (1984). Fresh visceral examination of rat and rabbit fetuses used in teratogenicity testing. *Teratogen. Carcinog. Mutagen.* **4**, 181–188.

Sulik, K.K. and Schoenwolf, G.C. (1985). Highlights of craniofacial morphogenesis in mammalian embryos, as revealed by scanning electron microscopy. *Scanning Electron Microscopy* **4**, 1735–1752.

Summerbell, D. (1979). The zone of polarizing activity: evidence for a role in abnormal chick limb morphogenesis. *J. Embryol. Exp. Morphol.* **50**, 217–233.

Suzuki, H.R., Solursh, M., and Baldwin, H.S. (1995). Relationship between fibronectin expression during gastrulation and heart formation in the rat embryo. *Dev. Dyn.* **204,** 259–277.

Švajger, A., Levak-Švajger, B., and Škrub, N. (1986). Rat embryonic ectoderm as renal isograft. *J. Embryol. Exp. Morph.* **94,** 1–27.

Symons, D. and Moxham, B.J. (2002). Ultrastructural characterization of the mesenchyme of the facial processes during development of the rat intermaxillary segment. *Connective Tissue Res.* **43,** 238–244.

Takeuchi, I.K. (1984). Teratogenic effects of methylnitrosourea on pregnant mice before implantation. *Experientia* **40,** 879–881.

Tan, S.S. and Morriss-Kay, G. (1985). The development and distribution of the cranial neural crest in the rat embryo. *Cell Tissue Res.* **240,** 403–416.

Taylor, P. (1986). "Practical Teratology." Academic Press, New York.

Theiler, K. (1972). "The House Mouse: Development and Normal Stages from Fertilization to 4 Weeks of Age." Springer-Verlag, New York.

Tickle, C., Summerbell, D., and Wolpert, L. (1975). Positional signalling and specification of digits in chick limb morphogenesis. *Nature* **254,** 199–202.

Timms, B.G., Peterson, R.E., and vom Saal, F.S. (2002). 2,3,7,8-Tetrachlorodibenzo-p-dioxin interacts with endogenous estradiol to disrupt prostate gland morphogenesis in male rat fetuses. *Toxicol. Sci.* **67,** 264–274.

Torrey, T.W. (1943). The development of the urinogenital system of the albino rat, I: the kidney and its ducts. *Am. J. Anat.* **72,** 113–147.

Torrey, T.W. (1945). The development of the urinogenital system of the albino rat, II: the gonads. *Am. J. Anat.* **76,** 375–397.

Torrey, T.W. (1947). The development of the urinogenital system of the albino rat, III: the urogenital union. *Am. J. Anat.* **81,** 139–158.

Tsuchiya, T., Bïrgin, H., Tsuchiya, M., Winternitz, P., and Kistler, A. (1991). Embryolethality of new herbicides is not detected by the micromass teratogen tests. *Arch. Toxicol.* **65,** 145–149.

Tsuchiya, T., Hisano, T., Tanaka, A., and Takahashi, A. (1987). Effects of benzimidazoles on mouse and rat limb bud cells in culture. *Toxicol. Lett.* **38,** 97–102.

U.S. Environmental Protection Agency [EPA]. (1996). *Fed. Reg.* **61,** no. 212.

U.S. Environmental Protection Agency [EPA]. (1997). *Fed. Reg.* **40,** 799, 9370.

U.S. Environmental Protection Agency [EPA]. (1998). "Health effects test guidelines: prenatal developmental toxicity study. Office of Prevention. Pesticides and Toxic Substances" (OPPTS), 870.3700.

U.S. Food and Drug Administration [FDA]. (1966). "Guidelines for Reproduction Studies for Safety Evaluation of Drugs for Human Use." FDA, Rockville, MD.

Van Maele-Fabry, G., Delhaise, F., and Picarrd, J.J. (1990). Morphogenesis and quantification of the development of postimplantation mouse embryos. *Toxicol. In Vitro* **4,** 149–156.

Van Mierop, L.H.S. (1979). Morphological development of the heart. *In* "Handbook of Physiology Section 2," Vol. 1. (R.M. Berne, ed.), pp 1–28, American Physiological Society, Bethesda.

Vig, K.W.L., Millicovsky, G., and Johnson, M.C. (1984). Craniofacial development: the possible mechanisms for some malformations. *Brit. J. Orthodont.* **11,** 114–118.

Wales, R.G. (1975). Maturation of the mammalian embryo: biochemical aspects. *Biol. Reprod.* **12,** 66–81.

Walmod, P.S., Berezin, A., Gallagher, H.C., Gravemann, U., Lepekhin, E.A., Belman, V., Bacon, C.L., Nau, H., Regan, C.M., Berezin, V., and Bock, E. (2002). Automated *in vitro* screening of teretogens. *Toxicol. Appl. Pharmacol.* **181,** 1–15.

Whitaker, J. and Dix, K.M. (1979). Double staining technique for rat foetus skeletons in teratological studies. *Lab. Anim.* **13,** 309–310.

Wilson, J.G. (1965a). Embryological considerations in teratology. *In* "Teratology Principles and Techniques." (J.G. Wilson and J. Warkany, eds.), pp 251–261, University of Chicago Press, Chicago.

Wilson, J.G. (1965b). Methods for administering agents and detecting malformations in experimental animals. *In* "Teratology Principles and Techniques" (J.G. Wilson and J. Warkany, eds.), pp 262–277, University of Chicago Press, Chicago.

Wilson, J.G. (1977). Current status of teratology: general principles and mechanisms derived from animal studies. *In* "Handbook of Teratology," Vol. 1. (J.G. Wilson and F.C. Fraser, eds.), pp 47–74, Plenum Press, New York.

Wise, L.D., Beck, S.L., Beltrame, D., Beyer, B.K., Chahoud, I., Clark, R.L., Clark, R., Druga, A.M., Feuston, M.H., Guittin, P., Henwood, S.M., Kimmel, C.A., Lindstrom P., Palmer A.K., Petrere, J.A., Solomon, H.M., Yasuda, M., and York, R.G. (1997). Terminology of developmental abnormalities in common laboratory mammals (Version 1). *Teratology* **55,** 249–292.

Witschi, E. (1962). Development: Rat. *In* "Growth Including Reproduction and Morphological Development." (P.L. Altman and D.S. Dittmer, eds.), pp 304–314, Federal American Society of Experimental Biology, Washington, DC.

Wragg, L.E., Klein, M., Steinvorth, G., and Warpeha, R. (1970). Facial growth accommodating secondary palate closure in rat and man. *Archs. Oral Biol.* **15,** 705–719.

Wright, H.V., Asling, C.W., Dougherty, H.L., Nelson, M.M., and Evans H.M. (1958). Prenatal development of the skeleton in Long-Evans rats. *Anat. Rec.* **130,** 659–672.

Yamada T. (1991). Selective staining methods for cartilage of rat fetal specimens previously treated with alizarin red S. *Teratology* **43,** 615–619.

Yasuda, M. and Yuki, T. (1996). "Color Atlas of Fetal Skeleton of the Mouse, Rat, and Rabbit." Ace Art, Osaka.

Young, A.D., Phipps, D.E., and Astroff, A.B. (2000). Large-scale double-staining of rat fetal skeletons using Alizarin red S and alcian blue. *Teratology* **61,** 273–276.

Chapter 29

Immunology

Veronica M. Jennings and Dirck L. Dillehay

I. INTRODUCTION

Rats have been the backbone of many types of research, including immunology immunogenetics, and transplantation. For example, Brown Norway (BN) rats are used as models of respiratory track allergy, and BN and Lewis (LEW) rats are used as models of several autoimmune diseases (Pat Happ et al., 1988; Caspi et al., 1992; Beijleveld et al., 1996; White et al., 2000; Warbrick et al., 2002). The rat is second to the mouse as an animal model for immunological research, and the numbers of available monoclonal antibodies to rat cell surface antigens and soluble factors have rapidly grown over the past 20 years. Many homologs of human cluster of differentiation (CD) antigens have been cloned in the rat, further enhancing the importance of this species as an immunology research tool. Also, because rats are larger than mice, they lend themselves to a variety of well-established and novel microsurgical techniques that have proven to be valuable in addressing certain fundamental immunological questions. Moreover, the size of rats makes them the premier rodent choice for transplantation studies.

A wide range of genetically well-defined inbred laboratory rat strains are available, and a variety of mutant, congenic, and recombinant inbred rat strains of immunological interest have been developed. Rats are also frequently used for hybridoma and monoclonal antibody production (Bazin 1990). Indeed, immunological experimentation in rats has been critical in biomedical research and will continue to be for many years to come. A thorough review of immunology is beyond the scope of a single chapter, and we refer the reader to a number of excellent immunology textbooks that are frequently updated, including, but not limited to, Paul's *Fundamental Immunology,* Janeway's *Immunobiology,* and Abbas and Lichtman's *Cellular and Molecular Immunology.* This chapter will review highlights of specific aspects of the immune system and transplant immunology of the laboratory rat, *Rattus norvegicus.*

II. CELLULAR COMPONENTS OF THE IMMUNE SYSTEM

The immune response can be divided into early non-specific innate immunity and later highly antigen-specific adaptive immunity. Adaptive immunity can be further divided into cell-mediated and humoral components. Innate immunity is mediated by epithelial barriers; immune cells, including neutrophils, macrophages, and NK cells; circulating proteins such as those of the complement system; and cytokines. Cell-mediated immunity is mediated by T lymphocytes and their products, cytokines, and is important in defense against intracellular pathogens. Humoral immunity is mediated by B lymphocytes and their products, antibodies, and functions primarily in defense against extracellular pathogens. Immune responses often, if not always, involve participation by and cooperation of innate and adaptive immunity and have both cell-mediated and humoral components.

Both B and T lymphocytes arise from a common precursor in the bone marrow. B cells develop in the bone marrow, whereas T-cell precursors migrate to and mature in the thymus. After maturation, B and T lymphocytes migrate to and populate peripheral lymphoid organs (lymph nodes, spleen, and mucosal-associated lymphoid tissues). Naïve B and T-cells are mature lymphocytes that have not been stimulated by antigen. When they encounter an antigen, they differentiate into effector lymphocytes or memory lymphocytes. The latter are responsible for rapid and enhanced response to subsequent antigen exposures. Effector B lymphocytes are antibody-secreting plasma cells, whereas effector T lymphocytes include cytokine-secreting $CD4^+$ cells and $CD8^+$ cytotoxic lymphocytes.

A. B Lymphocytes

Rat bone marrow has a high content of lymphoid cells, especially B-cell progenitors in various stages of differentiation. For example, 25% of nucleated cells isolated from one femur belonged to the B-cell lineage, as determined by reactivity with monoclonal antibody HIS24 (Deenen et al., 1987). Bazin et al. (1985) demonstrated that the entire peripheral B-cell pool can be reconstituted within 8 days of B-cell depletion. B cells mature in the bone marrow and are defined by the expression of B-cell receptor complex (BCR). This complex is formed by membrane-bound antibody molecules (immunoglobulin [Ig] M and IgD) complexed with the Igα (CD79a) and Igβ (CD79b) molecules. Although membrane-bound antibody binds antigen, CD79a and CD79b participate in BCR signal transduction. On activation, B cells may ultimately differentiate into antibody-producing plasma cells, the effector cells of the humoral immune response.

B cells also play a vital role in cell-mediated immunity responses because they function as antigen-presenting cells for T-cell activation. Mature B cells in rats lack Thy-1/CD90 (Soares et al., 1996) and can be distinguished from other leukocytes either by their expression of surface Ig or CD45 isoforms unique to B cells, which include CD45R and CD45RA. Once fully matured, B cells have a lifespan of months with a turnover rate of 1% to 2% per day (Deenen and Kroese, 1993). B cells also express Fcγ receptors (FcγRs). The major FcγRs in rats, humans, and mice include FcγRI (CD64), FcγII (CD32), and FcγIII

(CD16). All of these receptors recognize and bind the Fc portion of IgG and function primarily to help clear immune complexes. CD64 binds monomeric IgG, whereas CD32 and CD16 bind aggregated IgG.

1. B-1 Cells

One major difference between rat B-cells and those in the mouse is that in the rat, a B-1 (also knows as Ly-1 B-cells or $CD5^+$ B-cells) subset of B-cells has not yet been detected (De Boer, et al., 1994; Vermeer, et al., 1994). In other species, $CD5^+$ B-cells are involved in autoreactive antibody production (Kipps, 1989). Although B-1 cells are absent in adult rats, B cells in the neonatal spleen express low levels of CD5.

B. T Lymphocytes

T cells are critical for immune-mediated cytotoxicity, generation of the cell-mediated arm of an immune responses, and coordination of humoral immune responses. These cells mature in the thymus and can be differentiated from other leukocytes based on their function and expression of the T-cell receptor (TCR) complex, which consists of an antigen-binding dimeric hypervariable TCR, and an associated invariant CD3 complex involved in TCR signal transduction (see below). Mature T-cells can be divided into $CD4^+$ (T helper or TH cells) and $CD8^+$ (T cytotoxic cells). TH cells can be further subdivided into T_H1 or T_H2 cells, depending on their pattern of cytokine secretion (see below). Another T-cell classification is based on the specific chains that make up the TCR (α/β versus γ/δ). T-cell rate of formation is constant, and the mobilized pool is proportional to body weight and calculated to be 7.8×10^6 lymphocytes per gram of body weight.

Intra-thymic T-cell development in the rat is similar to that in the mouse in terms of TCR, CD2, CD4, and CD8 expression (Aspinall et al., 1991). However, there is discrepancy with mouse data as to whether or not interleukin (IL)-2 is an essential growth factor for T-cell development in the $CD4^-8^-$ double-negative preselectional stage of differentiation (Aspinall et al., 1991; Kroemer and Martinez, 1991). This is based on the fact that $TCR^-/CD4^-8^-$ thymocytes do not express IL-2Rα (CD25) or other IL-2 binding proteins (Paterson and Williams 1987; Takacs et al., 1988). Also, unlike mice and humans, most immature rat $TCR^{low}/CD4^+8^+$ thymocytes express little or no CD28, and $TCR^{intermediate}$ and TCR^{high} thymocytes express high levels of CD28 (Tacke et al., 1995). Unlike mice, the majority of rat thymocytes express CD157, with spontaneously hypertensive rats (SHR) and Wistar-Kyoto (WKY) rats being the exception, having

unusually low levels (similar to those of mice) of $CD157^+$ cells (Milicevic et al., 2001).

With post-thymic T-cell development, most rat peripheral T-cells demonstrate mutually exclusive expression of CD4 and CD8 (Brideau et al., 1980). These subsets can be further subdivided according to the types of cytokines the T-cells secrete (see below). Furthermore, Thy-1, RT-6, and CD45RC have been used substantially in rats to study phenotypic peripheral T-cell development. Expression of these markers represents not only different stages of development but also different functional properties. Thy-1 in adult rats is expressed in a subset of bone marrow cells, all thymocytes, and a small subset of T-lymphocytes (Crawford and Goldschneider, 1978; Ritter et al., 1978). $Thy-1^+$ T-cells function in delayed allograft rejection (Yang and Bell, 1992).

Depending on the age and strain of the rat, RT6 is expressed in 50% to 85% of peripheral $CD4^+$ and $CD8^+$ T-cells (Mojcik et al., 1988, 1991), a subset of NK cells (Wonigeit et al., 1996), and a subset of intraepithelial lymphocytes (Fangmann et al., 1990). $RT6^+$ T-cells play an important regulatory role in the prevention of autoimmunity. CD45RC in the rat is expressed by all B-cells, approximately 70% of $CD4^+$ T cells, and 90% of $CD8^+$ T-cells (Spickett et al., 1983). $CD4^+/CD45RC^+$ T-cells provide primary B-cell help *in vivo,* produce high amounts of IL-2, and interferon (IFN)-γ and are associated with cellular autoimmunity (Fowell et al., 1991) and prevention of humoral autoimmunity.

1. TH1/TH2 Cells

T_H1 and T_H2 cell differential cytokine production is known to play a crucial role in immune response regulation. In general, T_H1 T-cells produce IFN-γ, whereas T_H2 cells produce IL-4, IL-5, IL-10, and/or IL-13 (Paul, 1993). Rat strains may differ in their capacity to produce T_H1 or T_H2 responses. For example, LEW and BN strains tend to mount different cytokine profiles that dictate their susceptibility to either T_H1 or T_H2-mediated autoimmune disease, respectively. T_H1-responder LEW rats are susceptible to experimental autoimmune encephalitis induction primarily through the actions of IFN-γ and tumor necrosis factor (TNF)-α, whereas T_H2-responder BN rats are highly susceptible to immune complex glomerulonephritis, primarily through the action of IL-2 and IL-4 (Westermann et al., 1989; and Groen et al., 1993). Recent advances have determined that the $CD8^+$ T-cell compartment plays a dominant role in the inability of BN rats to mount a proper T_H1 immune response (Cautain et al., 2002). However, not all factors that make certain rat strains susceptible or resistant to autoimmune diseases have been determined.

2. T-Cell Receptor

As with other genes, rat TCR genes demonstrate many similarities to those of mice. The TCR complex consists of an antigen binding dimeric hypervariable TCR and an invariant CD3 complex that is involved in TCR signal transduction. αβ T-cells comprise the majority of T-cells in the adult rat and are found in peripheral lymphoid organs and blood. Frequencies of this T-cell population and CD4/CD8 ratios are similar to those of mice and humans (Hunig et al., 1989). γδT-cells are found in low numbers in the thymus, peripheral lymphoid organs, and blood. Intestinal intraepithelial lymphocytes may also include γδ T-cells, but their frequency is lower that that of the mouse (Lawetzky et al., 1990; Kuhnlein et al., 1994, 1995). Similar to mice and humans, the components of the TCR signal transduction complex—mainly CD2, CD3, CD4, CD8, CD5, p56lck, and p59fyn–are found in rats (Tanaka et al., 1989; Beyers et al., 1992; Kuhnlein et al., 1994).

C. Natural Killer Cells

Natural killer (NK) cells are instrumental in innate immune responses against certain classes of viruses, bacteria, and parasites and against tumor growth and metastasis (Kim et al., 2000). NK cells are large, granular lymphocytes that do not express receptors and function via non-major histocompatibility complex (MHC)-restricted cell mediated cytotocicity. In the rat, these cells can be distinguished by either high levels of NKR-P1A expression or by the fact that most NK cells express CD8a, but not CD3 (Lanier et al., 2000). NK cells do not require prior sensitization or coating of target cell with antibody. However, NK cells can act as effector cells for antibody-dependent cell-mediated cytotoxicity. This involves NK cells binding to antibody-coated target cells, degranulating, and releasing perforin and granzymes, which kill target cells. These functions place NK cells at a borderline between innate and adaptive immune responses (Long et al., 1999).

Virally induced αIFN-α and IFN-β, and to a certain extent IL-12 (Biron et al., 1999) and IL-18 (Liu et al., 2000), lead to activation of NK cells, which in turn induce NK cell production of IFN-γ and other cytokines that orchestrate adaptive immune responses toward invading pathogens. NK cells rely on several non-rearranging receptors for their cytolytic activity and cytokine production. Some of the receptors implicated in rat NK cell activation include CD2, CD16, CD28, Ly-6, CD69, NKR-P1, CD244, NKG2D, and CD94/NKG2A,C,E (Lanier et al., 2000). Also, a synergistic type of interaction between multiple activating receptors is required for successful initiation of effector function.

D. Monocytes/Macrophages

Monocytes and macrophages compose another class of leukocytes that plays a major role in the immune response. Monocytes, monocyte precursors, and macrophages are present in low numbers in healthy rat bone marrow. The latter two await homing signals to be recruited to become various types of tissue-associated macrophages. Within tissues, macrophages have several important roles, including phagocytosis, production and secretion of cytokines and chemokines, and antigen presentation. In delayed-type hypersensitivity reactions, macrophages are attracted and activated by cytokines to participate in this lymphocyte-mediated disease. Rat monocytes in particular can be identified by expression of CD4 without co-expression of CD3. They comprise about 10% of circulating white blood cells (Valli et al., 1983). In contrast to mice, except under pathological conditions, no cells from the myeloid series (such as polymorphonuclear cells) are formed in the rat spleen, and stem cell content in the spleen is negligible.

III. HUMORAL COMPONENTS OF THE IMMUNE SYSTEM

A. Immunoglobulin Classes and Nomenclature

Table 29-1 gives the different nomenclatures employed for rat Ig (sub)isotypes (subclasses). According to the World Health Organization (WHO) recommendations (WHO, 1973), the "Ig" sign should be adopted and all the letters should be written on the line, that is IgG1 and not IgG_1. The division of the IgG classes is based on electrophoretic mobility and is maintained for historical reasons.

Most of the biological and physicochemical properties of the five Ig isotypes (IgM, IgD, IgG, IgA, and IgE) are common to all mammalian species in which they have been described (Bazin et al., 1990). However, even today, it continues to be difficult to precisely correlate functions of rat IgG subclasses with those of human or mouse subclasses. Rat light chains, as in other species, are designated κ and λ (Hood et al., 1967). These were first characterized by Querinjean et al., (1972, 1975) using the LOU/Ws1 rat strain. As do other laboratory rodents, rats produce both IgM and IgG (sub)classes of antibodies. Depending on the antigen and the physiological status of the rat, the dose of antigen required to obtain precipitating antibodies varies, but doses between 100 μg and 1 mg with complete Freund's adjuvant are generally adequate. In all cases, the immunological response seems to be under the influence of genetic factors (Gill and Kunz et al., 1970;

TABLE 29-1

HISTORICAL NOMENCLATURE OF RAT IMMUNOGLOBULIN (SUB)CLASSES

IgM	IgA	IgD	IgE	IgG1	IgG2a	IgG2b	IgG2c	References
(γM, γ1M, 2M)	(β2A, γ1A, γA)	(γD)	(γE)		(7Sγm γss ir γG)			
IgM				IgX	IgG			Arnason et al., 1963
IgM				IgA	IgG			Arnason et al., 1964
IgM				IgA	IgGa	IgGb		Austin et al., 1968
	IgA							Nash et al., 1969
γM	γA			7Sγ1	7Sγ2			Jones et al., 1969
			IgE					Stechschulte et al., 1970
IgM	IgA		IgE	IgG1	IgG2a	IgG2b	IgG2c	Bazin et al., 1973
		IgD						Ruddick et al., 1977

Ladoulis et al., 1973; Armerding et al., 1974; Tada et al., 1974; Kunz et al., 1975; Ruscetti, et al., 1976).

B. History of Rat Immunoglobulin Nomenclature

Grabar and Courcon (1958) and Escribano and Grabar (1962) published the first data on rat serum proteins, obtained by immunoelectrophoretic analysis. These investigators distinguished two precipitation lines in the γ-globulin area. Some years later, Arnason and colleagues (1963, 1964) identified three proteins with antibody activities in rat serum. The authors referred to them as IgM, IgX, and IgG.

1. *IgM*. This class of antibody functions primarily as a naïve B-cell antigen receptor and in complement fixation. Arnason et al. (1963, 1964) characterized this class by the cysteine degradation technique and later by its molecular weight.
2. *IgG*. This class of antibody has multiple functions, including opsonization, complement fixation, and antibody-dependent cell-mediated cytotoxicity. As in the mouse, the nomenclature for rat IgG antibodies has changed over its history. For example, the present IgG1 isotype was originally misnamed IgA. At present, it is known that IgA (or IgX) and IgG classes distinguished by Arnason et al. (1963, 1964) are both IgG (sub)classes. The various IgG subisotypes were correctly identified as IgG1 by Jones (1969) and as IgG2a, IgG2b, and IgG2c by Bazin (1974a). Functionally, rat IgG1 and IgG2a antibodies bind complement. This is in contrast to the mouse, where only the IgG2a subtype binds complement (Ovary et al., 1966; Medgyesi et al., 1974).
3. *IgA*. This Ig class functions primarily in mucosal immunity. It was identified by Arnason et al., (1963, 1964) and named IgX and later IgA because, similar to human IgA, it was found in saliva, and a serum homolog was identified. Concurrently, Nash et al. (1969) identified a putative rat IgA in the milk, using rabbit anti-mouse IgA serum. Antiserum raised against this rat Ig, allowed for the demonstration of a large number of IgA containing cells in rat lamina propria. Nash et al. called this Ig class IgA and demonstrated that its anodal mobility was much faster than the previously termed "IgA," which was, in fact, the IgG1 class. Orlans and Feinstein (1971) later identified cross-reactivity between human and rat IgA.
4. *IgE*. Reaginic (mediate immediate hypersensitivity) antibodies have been shown to exist in the rat (Mota et al., 1963; Binaghi et al., 1964) and to have properties similar to those of man (Jones et al., 1967; Bloch et al., 1968). Stechschulte et al. (1970) and Jones and Edwards (1971) published initial reports of a rat Ig equivalent to human IgE. The existence of a rat IgE was confirmed by Bazin et al. (1974a,c) through the use of an IgE myeloma protein from the LOU rat strain.
5. *IgD*. This antibody class serves primarily as a naïve B-cell antigen receptor. By using chicken antiserum to human IgD, Ruddick and Leslie (1977) demonstrated an antigenically cross-reactive homolog to human IgD on rat lymphocyte membranes. Moreover, Bazin et al. (1978) fully characterized rat IgD with LOU rat IgD monoclonal Igs.

In rats, the discovery of LOU ileocecal malignant immunocytomas contributed greatly to the development of knowledge on Igs of this species (Bazin et al., 1972, 1973). Between 80% and 90% of these tumors secreted monoclonal Igs with or without Bence-Jones proteins, and most were transplantable in histocompatible rats. Study of this rat also lead to the discovery of the first monoclonal IgE (Bazin et al., 1974a,c) and IgD (Bazin et al., 1978) isotypes in a species other than humans. Furthermore, it facilitated development of the current nomenclature for rat Igs and the discovery of the IgG2c subisotype in rats (Bazin et al., 1974a,c).

C. Allotypes of Rat Immunoglobulins

Molecules of an Ig isotype (or subisotype) do not always bear identical antigenic determinants in all individuals from the same species because they may have different amino acid sequences. They can present two or more antigenic forms, each of them known as allotypes according to the Oudin terminology (1956), and four haplotypes have been indentified in rats. In this case, subjects from a unique species may be divided into as many allotypic groups as there are antigenic forms. Each individual expresses one allotype because each Ig-producing cell expresses only one of the paired allelic genes. Allotypic specificities are supported by structural differences determined by genes, the inheritance of which is mendelian. Ig genes are codominant; that is, both alleles are phenotypically expressed equally in heterozygotes.

A common nomenclature was adopted in 1983 for rat Ig allotypes (Gutman et al., 1983), and it defines an allotype called IgK-1 and two alleles, IgK-1a (reference strain, LOU/C) and IgK-1b (reference strain, DA). Gutman et al. (1975) described the differences between the two alleles as one sequence gap and many amino acid substitutions between the κ chains of the LEW and Dark Agouti (DA) rat strains. Another type located on the rat α heavy-chain of Igs was described by Bazin et al., (1974b). It is now called IgH-1 with two alleles: IgH-1a (reference strain, LOU/C) and IgH-non1a (reference strain, OKA). The rat γ2b heavy-chain (IgH-2) with two alleles IgH-2a (reference strain, LOU/C) and IgH-2b (reference strain, OKA) was described by Beckers and Bazin (1975). IgH-3 is also on the rat Ig heavy-chain and has been associated with IgG2c subisotype (Leslie et al., 1984). It has two alleles: IgH-3a in COP, Fischer 344 (F344), BN, and DA rat strains and IgH-3b in Wistar-Furth (WF), LOU, and SHR rat strains.

IV. THE MAJOR HISTOCOMPATIBILITY COMPLEX

In 1960, two independent groups of investigators published reports on the production of MHC-specific, hemagglutinating antibodies by inbred rats after tissue transplantation. The first report by Stark and colleagues referred to this rat MHC as RtH-1 (Frenzl et al., 1960), whereas the second report used the term Ag-B (Bogden and Aptekman, 1960). In 1979, the recommendation was made at the Second International Workshop of Alloantigenic Systems in the Rat that the rat MHC be designated as RT1. There are multiple genes involved in controlling histocompatibility differences between allogeneic individuals. These genes are usually divided into major and minor histocompatibility loci, the latter being responsible for weak transplantation reactions. These histocompatibility loci encode for cell surface antigens, which influence the success or failure of transplanted tissue. In most species, the MHC is composed of multiple, closely linked genetic loci, controlling not only tissue rejection but a number of other important immunological functions. For example, one of the most important characteristics of the MHC complex is MHC-restriction for self/nonself discrimination of the immune system (Zinkernagel et al., 1997). Moreover, several rat autoimmune models demonstrate an almost entire association with certain RT1 haplotypes.

A. The RT1 Complex

The rat MHC (RT1) is similar to that of mice (Gill et al., 1995) and resides on chromosome 20 (Locker et al., 1990).

Class I antigens are found on all nucleated cells and red blood cells, are encoded for by the RT1.A loci, and have a molecular weight (45,000 Da) similar to that of other species (Blankenhorn et al., 1978). Class II antigens are found on B-lymphocytes, macrophages, antigen-presenting cells, and dendritic cells. Their molecular weight ranges from 57,000 to 63,000 Da (Fukumoto et al., 1982), and they are encoded for by the RT1.B, RT1.D, and RT1.H loci. RT1.B and RT1.D loci also encode for I-A and I-E-like (based on mouse MHC nomenclature) class II molecules (Blankenhorn et al., 1983), and RT2 and RT3 encode for blood groups (Pinto et al., 1983).

V. RAT CLUSTER OF DIFFERENTIATION ANTIGENS

An overview of this field can never be complete because data are constantly being gathered and updated for these antigens. Some rat antigens differ from those in humans in terms of cDNA sequence, protein homology, antigen distribution, and function, whereas others have homologs in humans and other species. The reader is referred to the chart on rat CD antigens published by Lai et al. (1997) and the Rat Genome Database at *http://ratmap.gen.gu.se* for further information.

VI. CYTOKINES AND THEIR RECEPTORS

Cytokines are pleiotropic molecules that mediate numerous biological effects through the transmission of stimulatory and inhibitory signals between various types of cells. Differential cytokine production plays a major role in orchestrating which type of immune response (i.e., T_H1 versus T_H2) is induced. In general a T_H1 response is associated with production of IL-12 and IFN-γ. In contrast, a T_H2 response is associated with production of IL-4, IL-10, and IL-13 (Paul, 1993). The development and function of proper immune responses also depend on a well-orchestrated production of and response to cytokines. Tables 29-2 and 29-3 show a concise source of information on rat cytokines, specific cell-surface receptors, and references where more detailed information can be obtained.

Despite a growing interest in rat cytokines, literature on rat cytokines is incomplete. For some *in vivo* studies, better-characterized mouse cytokines may be used in rat systems; however, when homologous (rat) cytokines are available, these are preferable as they avoid possible antigenic stimulation induced by heterologous cytokines. Human cytokines show little biological activity in animal systems and should therefore be avoided in rat studies.

TABLE 29-2

RECOMBINANT RAT CYTOKINES

TABLE 29-2

RECOMBINANT RAT CYTOKINES

Cytokine	Reference(s)
IFN-α	Dijkema et al., 1984; van der Meide et al., 1986a
IFN-β	Ruuls et al., 1996
IFN-γ	Dijkema et al., 1985; van der Meide et al., 1986b
IL-1-α	Nishida et al., 1989
IL-2	McKnight et al., 1989; McKnight and Classon, 1992
IL-3	Cohen et al., 1986; Kamegai et al., 1990; Gebicke-Haerter et al., 1994
IL-4	Leitenberg and Feldbush, 1988; Richter et al., 1990; McKnight et al., 1991; McKnight and Classon, 1992
IL-5	Uberla et al., 1991
IL-6	Fey et al., 1989; Northemann et al., 1989; Frorath et al., 1992
IL-10	Goodman et al., 1992; Feng et al., 1993
IL-13	Lakkis and Cruet, 1993
TNF-α	Chung and Benveniste, 1990; Rothe et al., 1992; Appel et al., 1995a
SCF	Martin et al., 1990; Zsebo et al., 1990
TGF-β3	Wang et al., 1995
TGF-α	Blasband et al., 1990
EGF	Saggi et al., 1992; Abraham et al., 1993
IGF-II	Frunzio et al., 1986; Soares et al., 1986; Chiariotti et al., 1988

TABLE 29-3

RECOMBINANT RAT CYTOKINE RECEPTORS

Cytokine receptor	Reference(s)
TGF-βIIR	Tsuchida et al., 1993
IL-4R	Richter et al., 1995
IL-2R-α	Page and Dallman, 1991
IL-2R-β	Page and Dallman, 1991; Park et al., 1996
IL-3R-β	Appel, 1995b
GPR14	Marchese et al., 1995
IL-1R	Hart et al., 1993; Sutherland et al., 1994
IL-1RII	Bristulf et al., 1994

VII. RATS IN IMMUNOLOGY RESEARCH

Rats have been instrumental in immunology research for decades. For example, rats were critical to our understanding of lymphocyte recirculation and homing. In 1959, Gowans et al. demonstrated that small lymphocytes in the rat recirculate from blood to lymph, and this was later confirmed by Shorter and Bollman (1960) and Caffrey et al., (1962). By 1964, Gowans and Knight published a benchmark paper in which they observed that transfused small lymphocytes homed rapidly to the lymph nodes, splenic white pulp, and Peyer's patches. Radiolabeled lymphocytes were observed to transverse postcapillary venules located in the paracortex of lymph nodes. Because the cells forming the endothelium in these venules were exceptionally tall, they were referred to as "high" endothelium. By the use of electron microscopy, Marchesi and

Gowans (1964) were able to observe lymphocytes actually traversing the endothelium of postcapillary venules. Schoefl (1972) later presented evidence to indicate that lymphocytes traverse the endothelium by squeezing between the endothelial cells rather than passing through them. A high incidence of lymphocyte clusters suggested that particularly weak spots in the cohesion of the endothelial cell sheet existed, which favored the passage of lymphocytes. It was then postulated that the high endothelium was an adaptation in lymphoid tissues, excluding the thymus, which allowed sustained cell traffic without excessive fluid loss. These studies highlight the historical importance of rats in basic immunology research. The remaining sections of this chapter will overview the current use of rats in two important areas of immunology: immunodeficiency and transplantation immunology.

A. Rat Models of Immunodeficiency

This section reviews major rat models of natural immunodeficiency, which can range from partial to complete absence of certain immune functions. For a discussion of immunodeficient nude rats, refer to the chapter on spontaneous and induced models.

1. Diabetes-Prone, Bio-Breeding Rats

In 1974, the spontaneously bio-breeding diabetes-prone (BBDP) rat was discovered in a colony of outbred Wistar-derived rats at the Bio-Breeding Laboratories, Ottawa, Canada (Nakhooda et al., 1977). A diabetes-resistant (DR) line was derived from BBDP forbearers in the fifth generation of inbreeding (Butler et al., 1991). BBDP rats develop spontaneous diabetes and thyroiditis at high frequency between 60 and 120 days old. Although all BBDP rats are descendants of the original Ottawa litters, differences in incidence and severity of diabetes are noted not only among colonies but also within sublines maintained within the same colony (Butler et al., 1991). These rats are severely lymphopenic and immunocompromised; therefore, it is recommended that they be housed under virus-antibody–free or specific-pathogen–free conditions. The clinical manifestation of diabetes is acute and characterized by hyperglycemia, weight loss, hypoinsulinemia, and ketonuria. Exogenous insulin therapy (PZI U-40, 1.0 U/100g body weight/day) is necessary to prevent death. Like et al. (1986) also developed a line of BB rats, the nonlymphopenic diabetic Wor-BB rat line, that become diabetic but are nonlymphopenic with normal numbers of T-cells.

The mechanism of diabetes in BBDP rats involves, at least in part, their deficiency in RT6[+] T-cells

(Crisa et al., 1992). RT6.1 is an alloantigen expressed on 60% to 80% of peripheral CD4$^+$ and CD8$^+$ Tcells (Greiner et al., 1987). RT6$^+$ cells were found to have immuno-regulatory properties (Greiner et al., 1987) in depletion studies that used rat monoclonal antibodies directed against RT6.1, beginning at 30 days of age. Moreover, depletion of this cell population in young nonlymphopenic nondiabetic conventionally housed BBDR/Wor rats induced diabetes in more than 60% of rats (Thomas et al., 1991). Collectively, these results suggest that the lack of this immunoregulatory population is critical to the pathogenesis of diabetes.

2. Long-Evans Cinnamon Rats

Long-Evans Cinnamon (LEC) mutant rats have a maturational arrest in the development of CD4$^+$ T-cells, but not CD8$^+$ T-cells, from CD4$^+$CD8$^+$ T-cells (Agui et al., 1990, 1991). This results in a significant decrease in the number of peripheral CD4$^+$ T-cells; however, these cells are present in peripheral lymphoid organs and are most likely generated by an alternative pathway. The deficiency of CD4$^+$ T-cells is owing a single recessive gene system, *thid* (T-helper immunodeficiency) (Yamada et al., 1991). This mutation is found in bone marrow-derived cells but not in thymic stromal cells (Agui et al., 1991). CD4$^+$ T-cells that are present show functional abnormalities (Sakai et al., 1995). For example, there is no IL-2 production after Con A stimulation. Interestingly, the presence of the allele *thid* does not prevent the development of CD4$^+$ intra-epithelial lymphocytes since LEC rats possess normal numbers of both CD4$^+$CD8$^+$ and CD4$^+$8$^-$ intra-epithelial lymphocytes (Sakai et al., 1994).

3. Brown Norway Rats

BN rats are highly susceptible to T_H2-mediated auto-immune disease such as mercury (HgCl$_2$)-induced auto-immune glomerulonephritis (Aten et al., 1988). Rats of this inbred strain are generally considered to be poor immunological responders in various *in vivo* and *in vitro* assays (Rozing et al., 1979). CD4$^+$ T-cells in BN rats are polarized toward a T_H2 type of immune response, as they are prone to produce cytokines IL-4, IL-6, and IL-10 and antibody isotypes IgG1 and IgE upon antigenic stimula-tion (Mathieson et al., 1992). Their vigorous T_H2 immune response may explain their low response in IL-2-dependent assays and their high sensitivity to antibody-mediated forms of autoimmunity (Rozing et al., 1979; Aten et al., 1988). For example, BN rats are high IgE responders and are used in models of respiratory tract allergy (Haczku et al., 1995). Although both CD4$^+$ and CD8$^+$ T-cells are involved in the BN rats inability to develop a robust T_H1 immune response, CD8$^+$ T-cells appear to play the major role. This defect in T_H1/T_H2 immunity is not driven by MHC haplotype or high IL-4 expression (Cautain et al., 2002).

4. Lewis Rats

LEW rats are susceptible to T_H1-mediated autoimmune disease such as experimental autoimmune encephalomye-litis, experimental autoimmune uveoretinitis, and cyclo-sporine A-induced autoimmunity (Pat Happ et al., 1988; Caspi et al., 1992 Beijleveld et al., 1996). These suscept-ibilities are likely owing, at least in part, to the LEW rats' strong cellular immune responses and associated polarized production of T_H1 cytokines, including IL-2 and IFN-γ (Groen et al., 1993; Beijleveld et al., 1996).

B. Rat Models in Transplantation Immunology

The rat has played an integral role in understanding the immunologic process of transplant rejection, and stud-ies using rats have led to important breakthroughs in the-rapy for transplant rejection. The following paragraphs highlight transplantation research of three organs, the heart, kidney, and liver, in which the rat has served as a critical animal model.

1. Cardiac Transplantation

A reproducible model of cardiac transplant rejection has been developed in rats and is widely used to mimic human disease. This model uses an abdominal heart trans-plant, made by anastomosis between the aorta of the donor heart and the abdominal aorta of the recipient rat and between the pulmonary artery of the donor heart and the inferior vena cava of the recipient rat (Alexis et al., 2003). This allograft model, coupled with use of minor and major histocompatibility-mismatched rat strains, has been used to study the pathogenesis of and therapy for accele-rated graft arteriosclerosis (AGA) (Ono and Lindsay, 1969; Bujan et al., 2001; Suzuki et al., 2002). AGA is a clinically significant lesion that may develop within months after human cardiac transplantation (Qian et al., 2001). Lesions are characterized by smooth muscle cell proliferation that is preceded by inflammation, suggesting a critical inflam-matory component to disease pathogenesis (Adams et al., 1993). Models of AGA mimic the intimal thickening that occurs in arteries and arterioles of chronically rejected transplanted human hearts (Kouwenhoven et al., 2000; Bujan et al., 2001).

Several different strains of rats have been used in cardiac allograft studies. Donor strains have included WF, BN, LEW, and PVG, whereas recipient strains have included ACI, LEW, and F344 rats. When crossing MHC class I and II barriers, immunosuppression with cyclosporine is necessary to prevent acute graft rejection (Bujan et al., 2001). To study chronic rejection without the use of immunononsuppression, models have been developed that cross only minor histocompatability barriers (Lacha et al., 1995). One of the most common cardiac transplant models avoids immunosuppression by using LEW and F344 rat strains, which differ only in their non-MHC histocompatibility antigens (Gunther and Walter, 2001). This model involves transplanting LEW heart grafts into F344 recipients (Adams et al., 1993). These allografts can survive up to 120 days without immunosuppression.

The cellular inflammatory component of the intimal lesion consists of macrophages (Yang et al., 1998; Kanno et al., 2001; Miller et al., 2001), helper CD4$^+$ (Yamamoto et al., 2000), and cytotoxic CD8$^+$ lymphocytes (Yamamoto et al., 2000; Bujan et al., 2001) and neutrophils (Suzuki et al., 2002). Cytokines (Adams et al., 1993; Coito et al., 1995; Russell et al., 1995; Yang et al., 1998), macrophage-produced nitric oxide (Miller et al., 2001), chemokines (Rollins 1991; Paul et al., 1992; Russell et al., 1995; Horiguchi et al., 2002) and platelet-derived growth factor (Adams et al., 1993) also contribute to lesion development, likely through the promotion of smooth muscle hyperplasia within the intima resulting in arteriosclerosis (Adams et al., 1993). The precise pathogenic mechanism of smooth muscle cell proliferation is, however, unclear.

Another important immunologic component in the pathogenesis of AGA is the complement system (Qian et al., 1999, 2001). The complement system consists of a number of serum and cell surface proteins that interact with one another in a regulated cascade. Antibody-endothelium or antibody-antigen complexes deposited in vessels initiate this cascade, and ultimately membrane attack complexes (MACs) consisting of components C5b to C9 are formed. These MACs are inserted into endothelial cells, resulting in damage that may contribute to progression of AGA. The importance of the MAC was demonstrated in cardiac allograft studies that utilized complement component 6 (C6)-deficient and C6-sufficient PVG congenic rat strains. Allografts in C6-deficient recipients developed AGA in a lower percentage of arteries than did allografts in C6-sufficient recipients (Qian et al., 2001). These findings suggest that MACs are an important component of the pathogenesis of AGA in long-term cardiac allografts.

Rats have also been important in the study of the relative importance of different inflammatory cells in AGA pathogenesis. In the LEW-to-F344 allograft model, the early-stage (less than 30 days) has a predominance of macrophages in the neointima with few T lymphocytes (CD4$^+$ and CD8$^+$); from 30 to 90 days, smooth muscle cells arise in the neointima, and after 90 days the latter cells are the predominant cell type in this vessel wall (Bujan et al., 2001). A recent study found that incidence of AGA

with participation of ED1$^+$ macrophages, CD8$^+$ lymphocytes, and α-actin$^+$ cells (smooth muscle cells) was decreased in transplant recipients treated with cyclosporine, an immunosuppressive drug that inhibits the earliest phase of T-cell activation (Bujan et al., 2001). In this study, it appeared that CD8$^+$ lymphocytes were the predominant cell in affected vessel walls, suggesting that these cells may be primary contributors to disease progression. However, in another study, anti-CD4 antibody therapy induced tolerance in a WF-to-LEW rat cardiac transplant model (Yamamoto et al., 2000), suggesting that CD4$^+$ T-cells are critical to disease pathogenesis. Furthermore, depletion of recipient CD4$^+$ but not CD8$^+$ T lymphocytes prevented the development of cardiac allograft vasculopathy (Szeto et al., 2002). Thus, it is likely that both populations contribute to pathogenesis of arteriosclerosis, but their relative importance remains unclear.

Rat models have been critical to the development of therapies for AGA. Therapies studied in rats include treatment with donor liver lymphocytes (Sfeir et al., 2000), donor immature myeloid dendritic cells (Buonocore et al., 2003; DePaz et al., 2003), and anti-lymphocyte antibodies (Yamamoto et al., 2000; Dong et al., 2002; Yuan et al., 2002). Drugs that have been used to inhibit the inflammatory response associated with the graft rejection process include tacrolimus and malononitrilamides (Ebbs et al., 2002; Bilolo et al., 2003); FTY720 (Koshiba et al., 2003); LF15-0195, a deoxyspergualine that decreases IL-2 and IFN-γ in the heart graft (Chiffoleaui et al., 2002); CNI 1493 (Yang et al., 1998); cyclosporine (Bujan et al., 2001); sulfasalazine (Feeley et al., 1999); and methotrexate (Sakuma et al., 2002). Recently, the vascular growth factor, angiopoietin-1, has been shown in a rat cardiac allograft model to inhibit the production of arteriosclerosis through its anti-inflammatory properties (Nyakanen et al., 2003). Similar anti-inflammatory effects have been observed with granulocyte colony-stimulating factor and tacrolimus injections, which suppressed acute cardiac graft rejection in DA-to-LEW transplant rats by downregulating IL-2 and IL-12 production (Egi et al., 2002). Furthermore, dietary studies have shown that a high isoflavone diet and intravenous genistein delay rejection of cardiac allografts. This effect is enhanced with additional treatment with cyclosporine (O'Connor et al., 2002). Rats have also been important in the assessment of potential side effects associated with drug-induced immunosuppression including hypercholesterolemia, diabetes, hypertension, infection, and kidney disease, which could lead to accelerated arteriosclerosis (Bujan et al., 2001). Many studies have shown that cyclosporine plays an important role and is necessary in the prevention of acute cardiac graft rejection; however, renal toxicity can occur (Adams et al., 1993).

These examples highlight the important role that rat models of cardiac transplantation have played in study of transplant rejection. Future investigations using rat allograft models for chronic cardiac rejection will continue to be critical in expanding our knowledge of the immunopathogenesis of AGA and aiding in the development of novel therapies to prevent cardiac transplant rejection.

2. Kidney Transplantation

Genetically inbred rodents have been essential to studies of both immunological and non-immunological mechanisms associated with progressive renal allograft rejection (Paul, 1993). Transplantation of kidneys from F344 to LEW rats has been used to investigate the chronic renal rejection process. These strains are MHC compatible, and thus, this model requires no immunosuppression (Gill et al., 1990; Azuma et al., 1994). By 6 to 8 months, rats develop glomerular basement membrane thickening, glomerulosclerosis, arterial intimal thickening, interstitial nephritis, and fibrosis and proteinuria. This disease eventually progresses to chronic renal failure (Paul, 1995). Another rat kidney transplant model uses DA donors and WF recipients. These strains are strongly histoincompatible, and thus, acute rejection occurs in a few weeks. Immunosuppression with cyclosporine for 2 weeks can be used with this model to study chronic rejection lesions (Hayry et al., 1993).

The rejection process in rats mimics that observed in human kidney transplant recipients. This process involves three immunological components: humoral immunity, cellular immunity, and cytokine production (Jindal and Hariharan, 1999; Grau et al., 2000). Humoral components include antigen-antibody complex formation in transplanted tissue, anti-endothelial basement membrane antibodies (IgG and IgM), and fixation of complement (Paul, 1995; Pratt et al., 2000). Cellular components critical to the rejection process include lymphocytes and macrophages and their products (cytokines) (Azuma et al., 1994; Le Meur et al., 2002). Acute rejection involves expression of cytokines from $CD8^+$ lymphocytes, whereas immunological injury in chronic stages of rejection is associated with $CD4^+$ cells and macrophages (Paul, 1993, 1995). Cytokines associated with a T_H1 response, including IL-1, TNF-α, and IFN-γ are produced in acute rejection, whereas the chronic rejection process resembles a T_H2 response with production of cytokines that promote B lymphocyte/plasma cell production of antibody (Paul, 1995). These findings suggest that T_H1 or T_H2 responses are important in the process of graft rejection, and that the pathogenesis of rejection varies depending on the stage of disease (Paul, 1993; Engstrand et al., 1999). Interestingly, it has recently been shown that IL-17 expression in a renal graft from BN to LEW rats was a possible early predictor for subclinical renal allograft

rejection (Loong et al., 2000; Hsieh et al., 2001; Loong et al., 2002).

Non-cellular factors are also involved in the graft rejection process. For example, intravascular leukocytes in renal grafts are predominantly monocytes and many of them display an activated phenotype and produce TNF-α, nitric oxide synthase, and tissue factor (Grau et al., 2001). Other non-cellular factors include platelet-derived growth factor, monocyte-colony stimulating factor, IL-1, and IL-6, factors that are produced by many different cell types, especially macrophages (Kouwenhoven et al., 2000). Furthermore, it has been shown that the expression of CC chemokines, MCP-1, MIP-1α, and RANTES are up-regulated during the kidney rejection process (Grau et al., 2001). Lastly, the complement component, C3, is produced by renal tubular epithelial cells during acute renal rejection along with the upregulation of TNF-α and IL-1 within the graft (Pratt et al., 2000).

Rat models have been key tools in development of treatments for kidney transplant rejection. Therapy for renal rejection has relied on immunosuppressive drugs such as cyclosporine, tacrolimus, and daclizumab (Cianco et al., 2002). Recently, monoclonal antibody interference with the T-cell costimulatory molecule, CD28, significantly prolonged and normalized renal function in renal allografts in a F344-to-LEW graft model (Laskowski et al., 2002). Similarly, antibody-based inhibition of inducible costimulator (ICOS), a molecule with homology to the T-cell costimulatory molecules, CD28 and CTLA4, prolongs kidney allograft survival (Guo et al., 2002). Collectively, these results highlight the importance of T-cell activation in graft rejection and that modulating activation via interference with co-stimulatory molecules may have therapeutic benefit.

The transfusion of different types of lymphocytes into the graft kidney or graft recipient is a new type of experimental therapy. Transfusion of donor rat B lymphocytes at the time of kidney transplantation into a fully MHC mismatched rat induced allograft acceptance long term. This appeared to be owing to the lack of induction of a Th2 cytokine immune response (Yan et al., 2001). Another novel therapy investigated in a rat model involves the use of a promising *ex vivo* genetic manipulation technique. In this study, transduction of the donor kidney with the *CTLA4Ig* gene via an adenovirus vector blocked T-cell activation and induced donor-specific renal allograft tolerance (Tomasoni et al., 2000).

Other new immunosuppressive drug therapies that inhibit chronic renal transplant rejection include tacrolimus and malononitrilamides (Bilolo et al., 2003; Vu et al., 2003), mycophenolate mofetil and rapamycin (Jolicoeur et al., 2003), sirolimus, leflunomide, and dietary supplementation with omega-3 fatty acids and thromboxane synthase inhibitors (Jindal and Hariharan, 1999).

Moreover, a combination treatment regimen of tacrolimus, mycophenolate mofetil, and steroids resulted in significantly lower incidence in acute rejection at 6 months in human patients in comparison to the incidence in single drug treatments (Ciancio et al., 2002). However, complete inhibition of renal rejection over the life of the transplant has not been achieved (Jindal and Hariharan, 1999).

These examples highlight how rat models have been and will continue to be critical tools in the investigation of the pathogenesis of renal graft rejection and in the development of novel therapies.

3. Liver Transplantation

The liver is a unique immunologically privileged organ that often survives transplantation without immunosuppression (Kamada, 1992). Many different rat strains have been use for orthotopic liver transplantation, including those that cross both class I and class II MHC barriers. Commonly, portal veins are anastomosed, and the bile duct is surgically connected to the duodenum. These orthotopic liver allografts may survive indefinitely and, interestingly, can also protect subsequent allografts of the same strain from rejection (Kamada, 1992). Orthotopic transplantation techniques have resulted in a 96% to 98% successful liver graft tolerance in some crosses of rats (Kamada, 1992). Because the liver is uniquely immunologically privileged, studies involving this organ allow investigations that are not possible with other organ systems. For example, researchers may study the weak rejection process that occurs in comparison to other organs; this includes studies of graft survival across MHC barriers. Moreover, the unique phenomenon, whereby liver transplantation results in graft tolerance of other systems, is an area of great interest.

Many different strains of rat have been used for liver transplants, in which rejection can take place anywhere from 11 to 50 days or grafts can survive indefinitely through the induction of tolerance. One of the most common rat liver allograft models involves orthotopic transplantation of a DA liver into a LEW rat (Kamada, 1992; Savier et al., 1994). Orthotopic liver grafts of allogeneic nonrejection combinations also include DA into PVG, and BN into LEW. Allogeneic rejector combinations include PVG into LEW, DA into WAG, and LEW into BN. DA rats accept livers from a wider range of allogeneic donors than does any other strain (Kamada, 1992).

The cellular basis for liver-induced transplantation tolerance likely involves several different immunologic cell types (Savier et al., 1994; Shintaku et al., 2000; Okabe et al., 2001). It has been suggested that with liver transplantation, a host microchimerism develops in which passenger leukocytes, such as T-lymphocytes, induce tolerance (Adachi et al., 2000). Others have shown that CD4$^+$

TH cells regulate alloantigen recognition and graft rejection and that tolerance occurs when CD4$^+$ T-cells produce Th2 cytokines (Gassel et al., 2002). However, the precise cellular mechanisms for liver transplantation tolerance remain unknown.

Donor leucocytes transferred with the liver appear to be responsible for both liver acceptance and the abortive activation of the recipient T-cells. Although T-cells in the recipient lymphoid tissues are activated to express IL-2 and IFN-γ mRNA, cells within the graft are not adequately activated. Subsequently the activated T-cells die, leading to specific clonal deletion of liver-donor-reactive T-cells. Promotion of this "death by neglect phenomenon" warrants further study as it may have therapeutic potential (Bishop et al., 2002).

Other cell types also likely contribute to liver transplant tolerance. For example, NKT-cells are a category of T-cells that express both NK and TCRs and are found primarily in bone marrow and liver (Ohkawa et al., 1999). A study of LEW rats transplanted with DA livers suggested that NKT-cells within the allograft produce IL-4, and the associated T$_H$2 cell response results in allograft tolerance (Ohkawa et al., 1999). Another immunologic cell type with a role in liver allograft tolerance is the CD45RC$^+$ $\gamma\delta$ T-cell. CD45RC cell surface glycoprotein, which is expressed on many leukocytes in different isoforms, has been proposed to be a marker for memory T-cells. With transfusion of donor-specific blood before or just after liver allografting, CD45RC$^+$ $\gamma\delta$ T-cells infiltrate into rat liver allografts and induce tolerance. These lymphocytes produce T$_H$2 type cytokines, which are associated with immunologic unresponsiveness (Okabe et al., 2001).

Cytokines are integral to immunological responses in transplantation. In liver grafts, it has been shown that reduction of the T$_H$1 cytokines, IL-12, IL-18, and IFN-γ, resulted in prolonged liver graft survival after donor-specific transfusion (Yamaguchi et al., 2000). Another cytokine involved in inducing graft tolerance is the T$_H$2 cytokine, IL-10, which suppresses the alloimmune response to allograft transplantation. A rat orthotopic liver transplantation model treated with an adenovirus vector expressing human IL-10 demonstrated local and effective immunosuppression (Shinozaki et al., 1999). This IL-10, and the associated T$_H$2 response within the graft, induced graft survival for more than 100 days compared with 11 days for controls. Furthermore, rat liver allografts, transduced to produce IL-10, develop an anti-inflammatory monocytosis that inhibits donor antigen-stimulated secretion of IFN-γ and IL-2 and results in improved graft survival (Fan et al., 2001; Hayamizu et al., 200). Similarly, the T$_H$2 cytokine, IL-13, has been found to be expressed in CD4$^+$ cells in transplanted liver that are not rejected (Dresske et al., 2002). Collectively, the previous paragraphs highlight the importance of a T_H2 immune response in liver graft tolerance.

A monoclonal anti-CD4 antibody was shown to induce donor-specific tolerance in a high-responder liver transplant in LEW rats receiving livers from DA rats (Kohlhaw et al., 2001), suggesting that CD4$^+$ T-cells may also contribute to rejection. This observation was only seen when the monoclonal antibody therapy was started before the recipient contacted the donor alloantigens. This suggests that the anti-CD4 antibody is more immunomodulative than immunosuppressive (Kohlhaw, et al., 2001). This phenomenon has also been seen in rat modles of renal (Lacha et al., 1995) and cardiac (Arima et al., 1997) transplantation.

Immunosuppressive treatment is not often necessary for survival of liver allografts; however, cyclosporine and other immunosuppressive agents have been used. FTY720 (FTY) is a new immunosuppressive agent that elicits its effects by reducing the peripheral blood lymphocytes. It has produced allograft survival in animal transplantation models of the heart, liver, and kidney (Suzuki, 1999). In a DA-to-LEW transplantation model, treatment with low dosages of FTY and tacrolimus resulted in a prolonging liver allograft survival associated with FTY-induced apoptosis of the liver-infiltrating lymphocytes (Tamura et al., 1999). The synergistic effect of this combination therapy at very low doses is a very promising result for the field of clinical liver transplantation.

In a study that used a DA-to-LEW transplantation model, short-term selective immunosuppression combining monoclonal antibodies directed against CD25 and CD54, and low doses of cyclosporine prevented both acute and chronic graft rejection and induced donor-specific tolerance (Gassel et al., 2000). A short course of cyclosporine has also been shown to inhibit rejection but not tolerance in the rat liver transplant rat models of PVG to DA (tolerates grafts) and PVG to LEW (rejects grafts) (Huang et al., 2003).

Similarly, ICOS, a T-cell costimulatory molecule that has a homology with CD28 and CTLA4 may be an important therapeutic target. When anti-rat ICOS antibody was given intravenously to LEW recipients of DA liver transplants, T-cell activation was inhibited, resulting in significant prolongation of graft survival. The graft-infiltrating cells, both CD4$^+$ and CD8$^+$ T-cells, were not completely depleted; however, the expression of ICOS was significantly reduced (Guo et al., 2002). Thus, modulation of these lymphocyte molecules with monoclonal antibodies is an effective method for modulating transplantation immunity (Hayamizu et al., 2001).

Studies using rat transplantation models of heart, kidney, and liver have provided significant breakthroughs in our knowledge about transplantation immunology. This better understanding has lead to the development of more effective therapy for graft rejection. Future investigations using rat models of transplantation will continue to be critical in expanding our knowledge of the immunopathogenesis of grant rejection and grant tolerance and in aiding in the development of novel therapies to prevent transplant rejection.

REFERENCES

Abraham, J.A., Damm, D., Bajardi, A., Miller, J., Klagsbrun, M., and Ezekowitz, R.A.B. (1993). Heparin-binding EGF-like growth factor: characterization of rat and mouse cDNA clones, protein domain conservation across species, and transcript expression in tissues. *Biochem. Biophys. Res. Commun.* **190**, 125–133.

Adachi, K., Tamura, A., Sugioka, A., Morita, M., Yan, H., Li, X-K., Kitazawa, Y., Ammiya, H., Suzuki, S., Miyata, M., and Kimura, H. (2000). Evidence of regulatory T lymphocytes constitutes peripheral blood microchimerism following rat liver transplantation. *Transplant. Proc.* **32**, 2297–2299.

Adams, D.H., Russell, M.E., Hancock, W.W., Sayegh, M.H., Wyner, L.R., and Karnmovsky, M.J. (1993). Chronic rejection in experimental cardiac transplantation: Studies in the Lewis-F344 model. *Immunol. Rev.* **134**, 5–19.

Agui, T., Oka, M., Yamada, T., Sakai, T., Izumi, K., Ishida, Y., Himeno, K., and Matsumoto, K. (1990). Maturational arrest from CD4$^+$8$^-$ to CD4$^+$8$^-$ thymocytes in a mutant strain (LEC) of rats. *J. Exp. Med.* **172**, 1615–1624.

Agui, T., Sakai, T., Himeno, K., and Matsumoto, K. (1991). Bone marrow-derived progenitor T-cells convey the origin of maturational arrest from CD4$^+$CD8$^+$ to CD4$^+$CD8$^-$ thymocytes in LEC mutant rats. *Eur. J. Immunol.* **21**, 2277–2280.

Alexis, J.D., Pyo, R.T., and Chereshnev, I. (2003). Immunologic factors in transplant arteriopathy insight from animal models. *Mt. Sinai J. Med.* **70**, 191–195.

Appel, K., Buttini, M., Sauter, A., and Gebicke-Haerter, P.J. (1995a). Cloning of rat interleukin-3 receptor β-subunit from cultured microglia and its mRNA expression *in vivo*. *J. Neurosci.* **15**, 5800–5809.

Appel, K., Honegger, P., Gebicke-Haerter, P.J. (1995b). Expression of interleukin-3 and tumor necrosis factor-β mRNAs in cultured microglia. *J. Neuroimmunol.* **60**, 83–91.

Arima, T., Lehmann, M., and Flye, W.M. (1997). Induction of donor specific transplantation tolerance to cardiac allografts following treatment with nondepleting (RIB 5/2) or depleting (OX-38) anti-CD4 mAb plus intrathymic or intravenous donor alloantigen. *Transplantation* **63**, 284–292.

Armerding, D., Katz, D.H., and Benacerraf, B. (1974). Immune response genes in inbred rats, I: analysis of responder status to synthetic polypeptides and low doses of bovine serum albumin. *Immunogenetics* **1**, 329–339.

Arnason, B.G., St. deVaux, C.C., and Grabar, P. (1963). Immunoglobulin abnormalities of the thymectomized rats. *Nature (London)* **199**, 1199–1200.

Arnason, B.G., St. deVaux, C.C., and Relvveld, E.H. (1964). Role of the thymus in immune reactions in rats. IV. Immunoglobulins and antibody formation. *Int. Arch. Allergy Appl. Immunol.* **25**, 206–224.

Aspinall, R., Kampinga, J., and van Den Bogaerde, J. (1991). T-cell development in the fetus and the invariant series hypothesis. *Immunol. Today* **12**, 7–10.

Aten, J., Bosman, C.B., Rozing, J., Stijnen, T., Hoedemaeker, P.J., and Weening, J.J. (1988). Mercuric chloride-induced autoimmunity in the Brown Norway rat: cellular kinetics and major histocompatibility complex antigen expression. *Am. J. Pathol.* **133**, 127–138.

Austin, C.M. (1968). Patterns of migration of lymphoid cells. *Aust. J. Exp. Biol. Med. Sci.* **46**, 581–593.

Azuma, H. and Tilney, N.L. (1994). Chronic graft rejection. *Curr. Opin. Immunology* **6**, 770–776.

Bazin, H., Beckers, A., Deckers, C., and Moriame, M. (1973). Transplantable immunoglobulin-secreting tumors in rats, V. Monoclonal immunoglobulins secreted by 250 ileocecal immunocytomas of the LOU/Wsl rats. *J. Natl. Cancer Inst.* **51**, 1359–1361.

Bazin, H., Beckers, A., and Querinjean, P. (1974a). Three classes and four (sub)classes of rat immunoglobulins: IgM, IgA, IgE, and IgG1, IgG2a, IgG2b, IgG2c. *Eur. J. Immunol.* **4**, 44–48.

Bazin, H., Beckers, A., Urbain-Vansanten, G., Pauwels, R., Bruyns, C., Tilkin, A.F., Platteau, B., and Urbain, J. (1978). Transplantable IgD immunoglobulin-secreting tumours in rats. *J. Immunol.* **121**, 2077–2082.

Bazin, H., Beckers, A., Vaerman, J.P., and Heremans, J.F. (1974b). Allotypes of rat immunoglobulins, I: an allotype at the α-chain locus. *J. Immunol.* **112**, 1035–1041.

Bazin, H., Deckers, C., Beckers, A., and Heremans, J.F. (1972). Transplantable immunoglobulin-secreting tumors in rats, I: general features of LOU/Wsl strain rat immunocytomas and their monoclonal proteins. *Int. J. Cancer* **10**, 568–580.

Bazin, H., Platteau B., MacLennan, I.C.M., and Johnson, G.D. (1985). B-cell production and differentiation in adult rats. *Immunology* **54**, 79–88.

Bazin, H., Querinjean, P., Beckers, A., Heremans, J.F., and Dessy, F. (1974c). Transplantable immunoglobulin-secreting tumors in rats, IV: sixty-three IgE-secreting immunocytoma tumors. *Immunology* **26**, 713–723.

Bazin, H., Rousseaux, J., Rousseaux-Prevost, R., Platteau, B., Querinjean, P., Malache, J.M., and Delaunay, T. (1990). Rat immunoglobulins. *In* "Rat Hybridomas and Rat Monoclonal Antibodies." (H. Bazin, ed.), pp 5–42, CRC Press, Boca Raton.

Beckers, A. and Bazin, H. (1975). Allotypes of rat immunoglobulins, III: an allotype of the γ2b chain locus. *Immunochemistry* **12**, 671–675.

Beijleveld, L.J.J., Groen, H., Broeren, C.M., Klatter, F.A., Kampinga, J., Damoiseaux, J.G.M.C., and van Brenda Vriesman, P.J.C. (1996). Susceptibility to clinically manifest cyclosporin A (CsA)-induced autoimmune disease is associated with interferon-gamma producing CD45RC+RT6−T helper cells. *Clin. Exp. Immunol.* **105**, 486–496.

Beyers, A.D., Spruyt, L.L., and Williams, A.F. (1992). Molecular associations between the T-lymphocyte antigen receptor complex and the surface antigens CD2, CD4, or CD8 and CD5. *Proc. Natl. Acad. Sci. USA* **89**, 2945–2949.

Bilolo, K.K., Ouyang, J., Wang, X., Zhu, S., Jiang, W., Qi, S., Xu, D., Herbert, M.J., Bekersky, I., Fitzimmons, W.E., and Chen, H. (2003). Synergistic effects of malononitrilamides (FK778, FK779) with tacrolimus (FK506) in prevention of acute heart and kidney allograft rejection and reversal of ongoing heart allografts rejection in the rat. *Transplantation* **75**, 1881–1887.

Binaghi, R.A. and Benacerraf, B. (1964). Production of anaphylactic antibody in the rat. *J. Immunol.* **92**, 920–926.

Biron, C.A., Nguyen, K.B., Pien, G.C., Cousens, L.P., and Salazar-Mather, T.P. (1999). Natural killer cells in antiviral defense: function and regulation by innate cytokines. *Annu. Rev. Immunol.* **17**, 189–220.

Bishop, G.A., Wang, C., Sharland, A.F., and McCaughan, G. (2002). Spontaneous acceptance of liver transplants in rodents: evidence that liver leucocytes induce recipient T-cell death by neglect. *Immunol. Cell Biol.* **80**, 93–100.

Blankenhorn, E.P., Cecka, J.M., Gotze, D., and Hood, L. (1978). Partial N-terminal amino acid sequences of rat transplantation antigens. *Nature* **274**, 90–92.

Blankenhorn, E.P., Symington, F.W., and Cramer, D.V. (1983). Biochemical characterization of Ia antigens encoded by the RT1.B and RT1.D loci in the rat MHC. *Immunogenetics* **17**, 475–484.

Blasband, A.J., Rogers, K.T., Chen, X., Azizkhan, J.C., and Lee, D.C. (1990). Characterization of the rat transforming growth factor α gene and identification of promoter sequences. *Mol. Cell. Biol.* **10**, 2111–2121.

Bloch, K.J. and Wilson, R.J.M. (1968). Homocytotropic antibody response in the rat infected with the nematode *Nippostrongylus brasiliensis*, III: characteristics of the antibody. *J. Immunol.* **100**, 629–636.

Bogden, A.E. and Aptekman, P.M. (1960). The R-1 factor: a histocompatibility antigen in rat. *Cancer Res.* **20**, 1372.

Brideau, R.J., Carter, P.B., McMaster, W.R., Masson, D.W., and Williams, A.F. (1980). Two subsets of rat T lymphocytes defined with monoclonal antibodies. *Eur. J. Immunol.* **10**, 609–615.

Bristulf, J., Gatti, S., Malinowsky, D., Bjork, L., Sundgren, A.K., and Bartfai, T. (1994). Interleukin-1 stimulates the expression of type I and type II interleukin-1 receptors in the rat insulinoma cell line Rinm5F: sequencing a rat type II interleukin-1 receptor cDNA. *Eur. Cytokine Netw.* **5**, 319–330.

Bujan, J., Jurado, F., Gimeno, M.J., Rodriguez, M., and Bellon, J.M. (2001). Function of inflammatory cells and neoral cyclosporin-A in heart transplant-associated coronary vasculopathy. *Histol. Histopathol.* **16**, 297–203.

Buonocore, S., Flamand, V., Goldman, M., and Braun, M.Y. (2003). Bone marrow–derived immature dendritic cells prime *in vivo* alloreactive T-cells for interleukin-4–dependent rejection of major histocompatibility complex class II antigen-disparate cardiac allograft. *Transplantation* **75**, 407–413.

Butler, L., Guberski, D.L., and Like, A.A. (1991). *In* "Frontiers in Diabetes Research: Lessons from Animal Diabetes III." (E. Shafrir, ed.), pp 50–53, Smith-Gordon, London.

Caffrey, R.W., Rieke, W.O., and Everett, N.B. (1962). Radioautographic studies of small lymphocytes in the thoracic duct of the rat. *Acta Haematol.* **28**, 145–154.

Caspi, R.R., Chan, C.-C., Fujino, Y., Oddo, S., Najafian, F., Bahmanyar, S., Heremans, H., Wilder, R.L., and Wiggert, B. (1992). Genetic factors in susceptibility and resistance to experimental autoimmunity uveoretinitis. *Curr. Eye Res.* **11(Suppl.)**, 81–86.

Cautain, B., Damoiseaux, J., Bernard, I., Xystrakis, E., Fournie, E., van Breda Vriesman, P., Druet, P., and Saoudi, A. (2002). The CD8 T-cell compartment plays a dominant role in the deficiency of Brown-Norway rats to mount a proper type 1 immune response. *J. Immunol.* **168**, 162–170.

Chiariotti, L., Brown, A.L., Frunzio, R., Clemmons, D.R., Rechler, M.M., and Bruni, C.B. (1988). Structure of the rat insulin-like growth factor II transcriptional unit: heterogeneous transcripts are generated from two promoters by use of multiple polyadenylation sites and differential ribonucleic acid splicing. *Mol. Endocrinol.* **2**, 1115–1126.

Chiffoleau, E., Beriou, G., Dutartre, P., Usal, C., Soulillou, J.P., and Cuturi, M.C. (2002). Induction of donor-specific allograft tolerance by short-term treatment with LF15-0195 after transplantation: evidence for a direct effect on T-cell differentiation. *Am. J. Transplant.* **2**, 745–754.

Chung, I.Y. and Benveniste, E.N. (1990). Tumor necrosis factor-α production by rat astrocytes: induction by lipopolysaccharide, IFN-γ and IL-1β. *J. Immunol.* **144**, 2999–3007.

Ciancio, G., Burke, G.W., Suzart, K., Roth, D., Kupin, W., Rosen, A., Olson, L., Esquenazi, V., and Miller, J. (2002). Daclizumab induction, tacrolimus, mycophenolate mofetil and steroids as an immunosuppression regimen for primary kidney transplant recipients. *Transplantation* **73**, 1100–1106.

Cohen, D.R., Hapel, A.J., and Young, I.G. (1986). Cloning and expression of the rat interleukin-3 gene. *Nucleic Acids Res.* **14**, 3641–3658.

Coito, A.J., Binder, J., Brown, L.F., de Sousa, M., Van De Water, L., and Kupiec-Weglinski, J.W. (1995). Anti-TNF α treatment down-regulates the expression of fibronectin and decreases cellular infiltration of cardiac allografts in rats. *J. Immunol.* **154**, 2949–2958.

Crawford, J.M. and Goldschneider, I. (1980). Thy-1 antigen, and B-lymphocyte differentiation in the rat. *J. Immunol.* **124**, 969–976.

Crisa, L., Mordes, J.P., and Rossini, A.A. (1992). Autoimmune diabetes mellitus in the BB rat. *Diab. Metab. Rev.* **8**, 9–37.

De Boer, N.K., Meedendorp, B., Ammerlaan, W.A.M., Nieuwenhuis, P., and Kroese, F.G.M. (1994). B cells specific for bromelain treated erythrocytes are not derived from adult rat bone marrow. *Immunobiology* **190**, 105–115.

Deenen, G.J., Hunt, S.V., and Opstelten, D. (1987). A stathmokinetic study of B lymphocytopoiesis in rat bone marrow: proliferation of cells containing cytoplasmic μ-chains, terminal deoxynucleotidyl transferase and carrying HIS24 antigen. *J. Immunol.* **139**, 702–710.

Deenen, G.J. and Kroese, F.G. (1993). Kinetics of B cell subpopulations in peripheral lymphoid tissues: evidence for the presence of phenotypically distinct short-lived and long-lived B cell subsets. *Int. Immunol.* **5**, 735–741.

DePaz, H.A., Oluwole O.O., Adeyeri, A.O., Witkowski, P., Jin M.X., Hardy, M.A., and Oluwole, S.F. (2003). Immature rat myeloid dendritic cells generated in low-dose granulocyte macrophage-colony stimulating factor prolong donor-specific rat cardiac allograft survival. *Transplantation* **75**, 521–528.

Dijkema, R., Pouwels, P., de Reus, A., and Schellekens, H. (1984). Structure and expression in *Escherichia coli* of a cloned rat interferon-α gene. *Nucleic Acids Res.* **12**, 1227–1242.

Dijkema, R., van der Meide, P.H., Pouwels, P.H., Caspers, M., Dubbeld, M., and Schellekens, H. (1985). Cloning and expression of the chromosomal immune interferon gene of the rat. *EMBO J.* **4**, 761–767.

Dong, V.M., Yuan, X., Coito, A.J., Waaga, A.M., Sayegh, M.H., and Chandraker, A. (2002). Mechanisms of targeting CD-28 by signaling, monoclonal antibody in acute and chronic allograft rejection. *Transplantation* **73**, 1310–1317.

Dresske, B., Lin, X., Huang, D.S., Zhou, X., and Fandrich, F. (2002). Spontaneous tolerance: experience with the rat liver transplant model. *Hum. Immunol.* **63**, 853–886.

Ebbs, A., Pan, F., Wynn, C., Erickson, L., Kobayashi, M., and Jiang, H. (2002). Tacrolimus treats ongoing allograft rejection by inhibiting interleukin-10 mediated functional cytotoxic cell infiltration. *Transplant. Proc.* **34**, 1378–1381.

Egi, H., Hayamizu, K., Kitayama, T., Ohmori, I., Okajima, M., and Ashara, T. (2002). Downregulation of both interleukin-12 and interleukin-2 in heart allografts by pretransplant host treatment with granulocyte colony-stimulating factor and tacrolimus. *Cytokine* **18**, 164–167.

Engstrand, M., Johnsson, C., Korsgren, O., and Tufveson, G. (1999). *Ex vivo* propagation and characterization of lymphocytes from rejecting rat-kidney allograft. *Transpl. Immunol.* **7**, 189–196.

Escribano, M.J. and Grabar, P. (1962). L'analyse immunoelectrophoretique du serum de rat normal. *C. R. Hebd. Seances Acad. Sci.* **255**, 206–208.

Fan, X.H., Hayamizu, K., Yahata, H., Shinozaki, K., Tashiro, H., Sakaguchi, T., Ito, H., and Asahara, T. (2001). Emergence of anti-inflammatory monocytes in long-term surviving hosts of IL-10 transduced liver allografts. *Cytokine* **13**, 183–187.

Fangmann, J., Schwinzer, R., Winkler, M., and Wonigeit, K. (1990). Expression of RT6 alloantigens and the T-cell receptor on intestinal intraepithelial lymphocytes of the rat. *Transplant. Proc.* **22**, 2543–2544.

Feeley, B.T., Park, A.K., Hoyt, E.G., and Robbins, R.C. (1999). Sulfasalazine inhibits reperfusion injury and prolongs allograft survival in rat cardiac transplants. *J. Heart Lung Transplant.* **18**, 1088–1095.

Feng, L., Tang., W.L., Chang, J.C., and Wilson, C.B. (1993). Molecular cloning of rat cytokine synthesis inhibitory factor (IL-10) DNA and expression in spleen and macrophages. *Biochem. Biophys. Res. Commun.* **192**, 452–458.

Fey, G.H., Hattori, M., Northemann, W., Abraham, L.J., Baumann, M., Braciak, T.A., Fletcher, R.C., Gauldie, J., Lee, F., Reymond, M.F. (1989). Regulation of rat liver acute phase gene by interleukin-6 and production of hepatocyte stimulating factors by rat hepatome cells. *Ann. NY Acad. Sci.* **557**, 317–331.

Fowell, D., McKnight, A.J., Powrie, F., Dyke, R., and Mason, D. (1991). Subsets of CD4+ T-cells and their roles in the induction and prevention of autoimmunity. *Immunol. Rev.* **123**, 37–64.

Frenzl, B., Kren, V., and Stark, O. (1960). Attempt to determine blood groups in rats. *Folia Biologica Prag.* **6**, 121–126.

Frorath, B., Abney, C.C., Berthold, H., Scanarini, M., and Northemann, W. (1992). Production of recombinant rat interleukin-6 in *Escherichia coli* using a novel highly efficient expression vector pGEX-3T. *BioTechniques* **12**, 558–563.

Frunzio, R., Chiariotti, L., Brown, A.L., Graham, D.E., Rechler, M.M., and Bruni, C.B. (1986). Structure and expression of the rat insulin-like growth factor II (rIGF-II) gene. rIGF-II RNAs are transcribed from two promoters. *J. Biol. Chem.* **261**, 17138–17149.

Fukumoto, T., McMaster, W.R., and Williams, A.F. (1982). Mouse monoclonal against rat major histocompatibility antigens: two Ia antigens and expression of Ia and class I antigens in rat thymus. *Eur. J. Immunol.* **12**, 237–243.

Gassel, H.J, Kauczock, J., Martens, N., Steger, U., Timmermann, W., Ulrichs, K., and Otto, C. (2002). Tolerance induction following orthotopic rat liver transplantation: cytokine production by CD4+ T-cells determine the immunological response. *Transplant. Proc.* **34**, 1429–1430.

Gassel, H.J., Otto, C., Gassel, A.M., Meyer, D., Steger, U., Timmermann, W., Ulrichs, K., and Thiede, A. (2000). Tolerance of rat liver allografts induced by short-term selective immunosuppression combining monoclonal antibodies directed against CD25 and CD54 with subtherapeutic cyclosporine. *Transplantation* **69**, 1058–1067.

Gebicke-Haerter, P.J., Appel, K., Taylor, G.D., Schobert, A., Rich, I.N., Northoff, H., and Berger, M. (1994). Rat microglial interleukin-3. *J. Neuroimmunol.* **50**, 203–214.

Gill III, T.J. and Kunz, H.W. (1970). Genetic and cellular factors in the immune response. II. Evidence for the polygenic control of the antibody response from further breeding studies and from pedigree analysis. *J. Immunol.* **106**, 980–992.

Gill III, T.J., Misra, D.N., Vardimon, D., Kunz, H.W., Rushton, J., Kiristis, M.J., Locker, J., and Cortese Hassett, A.L (1990). Structure of the major histocompatibility complex in the rat. *Transplant. Proc.* **22**, 2508.

Gill III, T.J., Natori, T., Salgar, S.K., and Kunz, H.W. (1995). Current status of the major histocompatibility complex in the rat. *Transplant. Proc.* **27**, 1495–1500.

Goodman, R.E., Oblak, J., and Bell, R.G. (1992). Synthesis and characterization of rat interleukin-10 (IL-10) cDNA clones from the RNA of cultured OX8⁻OX22⁻thoracic duct T-cell. *Biochem. Biophys. Res. Commun.* **189**, 1–7.

Gowans, J.L. (1959). The recirculation of lymphocytes from blood to lymph in the rat. *J. Physiol. (London)* **146**, 54–69.

Gowans, J.L. and Knight, E.J. (1964). The route of recirculation of lymphocytes in the rat. *Proc. R. Soc. Lond. B* **159**, 257–282.

Grabar, P. and Courcon, J. (1958). Etude des serum de cheval, lapin, rat er souris par l'analyse immunoelectrophoretique. *Bull. Soc. Chim. Biol.* **40**, 1993–2003.

Grau, V., Gemsa, D., Steinger, B., and Garn, H. (2000). Chemokine expression during acute rejection of rat kidneys. *Scand. J. Immunol.* **51**, 435–440.

Grau, V., Stehling, O., Garn, H., and Steiniger, B. (2001). Accumulating monocytes in the vasculature of rat renal allografts: phenotype, cytokine, inducible no synthase, and tissue factor mRNA expression. *Transplantation* **71**, 37–46.

Greiner, D.L., Mordes, J.P., Handler, E.S., Angelillo, M., Nakamura, N., and Rossini, A.A. (1987). Depletion of RT6.1$^+$ T lymphocytes induces diabetes in resistant BioBreeding/Worcester (BB/W) rats. *J. Exp. Med.* **166**, 461–475.

Groen, H., Klatter, F.A., van Petersen, A.S., Pater, J.M., Nieuwenhuis, P., and Kampinga, J. (1993). Composition of rat CD4$^+$ resting memory T-cell pool is influenced by major histocompatibility complex. *Transplant. Proc.* **25**, 2782–2783.

Gunther, E. and Walter, L. (2001). The major histocompatibility complex of the rat *(Rattus norvegicus)*. *Immunogenetics* **53**, 520–542.

Guo, L., Li, X.K., Funeshima, N., Fujino, M., Nagata, Y., Kimura, H., Ameniya, H., Enosawa, S., Tsuji, T., Harihara, Y., Mukuuchi, M., and Suzuki, S. (2002). Prolonged survival in rat liver transplantation with mouse-monoclonal antibody against an inducible costimulator (ICOS). *Transplantation* **73**, 1027–1032.

Gutman, G.A., Bazin, H., Rocklin, C.V., and Nezlin, R.S. (1983). A standard nomenclature for rat immunoglobulin allotypes. *Transplant. Proc.* **15**, 1685–1686.

Gutman, G.A., Loh, E., and Hood, L. (1975). Structure and regulation of immunoglobulins: kappa allotypes in the rat have multiple amino acid differences in the constant region. *Proc. Natl. Acad. Sci.* **72**, 5046–5050.

Haczku, A., Chung, K.F., Sun, J., Bames, P.J., Kay, A.B., and Moqbel, R. (1995). Airway hyperresponsiveness, elevation of serum-specific IgE and activation of T-cells following allergen exposure in sensitized Brown-Norway rats. *Immunology* **85**, 598–603.

Hart, R.P., Liu, C., Shadiack, A.M., McCormack, R.J., and Jonakait, G.M. (1993). An mRNA homologous to interleukin-1 receptor type I is expressed in cultured rat sympathetic ganglia. *J. Neuroimmunol.* **44**, 49–56.

Hayamizu, K., Shinozaki, K., Fan, X., Yahata, H., Tashiro, H., and Asahara, T. (2001). IL-10 transduction of liver allografts induces antiinflammatory monocytes in long-term-surviving hosts. *Transplant. Proc.* **33**, 355.

Hayry, P., Isoniemi, H., Yilmaz, S., Mennander, A., Lemstrom, K., Raisanen-Sokolowski, A., Koskinen, P., Ustinov, J., Lautenschlager, I., and Taskinen, E. et al. (1993). Chronic allograft rejection. *Immunol. Rev.* **134**, 33–81.

Hood, L., Gray, W.R., Sanders, B.G., and Dreyer, W.J. (1967). Light chain evolution. *Cold Spring Harbor Symp. Quant. Biol.* **22**, 133.

Horiguchi, K., Kitagawa-Sakakida, S., Sawa, Y., Li, Z.Z., Fukushima, N., Shirakura, R., and Matsuda, H. (2002). Selective chemokine and receptor gene expression in allografts that develop transplant vasculopathy. *J. Heart Lung Transplant.* **21**, 1090–1100.

Hsieh, H.G., Loong, C.C., Lui, W.Y., Chen, A.L., and Lin, C.Y. (2001). IL-17 expression as a possible predictive parameter for subclinical renal allograft rejection. *Transpl. Int.* **14**, 287–298.

Huang, W.H., Yan, Y., De Boer, B., Bishop, G.A., and House, A.K. (2003). A short course of cyclosporine immunosuppression inhibits rejections but not tolerance of rat liver allografts. *Transplantation* **75**, 368–374.

Hunig, T., Wallny, H.-J., Hartley, J.K., Lawetzky, A., and Tiefenthaler, G. (1989). A monoclonal antibody to a constant determinant of the rat T-cell antigen receptor that induces T-cell activation. *J. Exp. Med.* **169**, 73–86.

Jindal, R.M. and Hariharan, S.J. (1999). Chronic rejection in kidney transplants. *Nephron* **83**, 13–24.

Jolicoeur, E.M., Qi, S., Xu, D., Dumont, L., Daloze, P., and Chen, H. (2003). Combination therapy of mycophenolate mofetil and rapamycin in prevention of chronic renal allograft rejection in the rat. *Transplantation* **75**, 54–59.

Jones, V.E. (1969). Rat 7S immunoglobulins: characterization of γ2- and γ1-anti-hapten antibodies. *Immunology* **16**, 589–599.

Jones, V.E. and Edwards, A.J. (1971). Preparation of an antiserum specific for rat reagin (rat gamma E?). *Immunology* **21**, 383–385.

Jones, V.E. and Ogilvie, B.M. (1967). Reaginic antibodies and immunity to *Nippostrongylus brasiliensis* in the rat, II: some properties of the antibodies and antigens. *Immunology* **12**, 583–597.

Kamada, N. (1992). Animal models of hepatic allograft rejection. *Semin. Liver Dis.* **12**, 1–15.

Kamegai, M., Niijima, K., Kunishita, T., Nishizawa, M., Ogawa, M., Araki, M., Ueki, A., Konishi, Y., and Tabira, T. (1990). Interleukin-3 as a trophic factor for central cholinergic neurons *in vitro* and *in vivo*. *Neuron* **2**, 429–436.

Kanno, S., Wu, Y.J., Lee, P.C., Dodd, S.J., Williams, M., Griffith, B.P., and Ho, C. (2001). Macrophage accumulation associated with rat cardiac allograft rejection detected by magnetic resonance imaging with ultrasmall superparamagnetic iron oxide particles. *Circulation* **104**, 934–938.

Kim, S., Iizuka, K., Aguila, H.L., Weissman, I.L., and Yokoyama, W.M. (2000). *In vivo* natural killer cell activities revealed by natural killer cell–deficient mice. *Proc. Natl. Sci. Acad. USA* **97**, 2731–2736.

Kipps, T.J. (1989). The CD5 B cell. *Adv. Immunol.* **47**, 117–185.

Kohlhaw, K., Sack, U., Lehmann, I., Drews, G., Schwarz, R., Hartwig, T., Tannapfel, A., Berr, F., Oertel, M., Marx, U., Lehmann, M., Witterkind, C., Hauss, J., and Emmrich, F. (2001). Anti-CD4–induced long-term acceptance of rat liver allografts in a high-responder model is not based on pure immunosuppression. *Transplant. Proc.* **33**, 2173–2175.

Koshiba, T., Van Damme, B., Rutgeerts, O., Waer, M., and Pirenne, J. (2003). FTY720, an immunosuppressant that alters lymphocyte trafficking, abrogates chronic rejection in combination with cyclosporine A. *Transplantation* **75**, 945–952.

Kouwenhoven, E.A., IJzermans, J.N., and de Bruin, R.W. (2000). Etiology and pathophysiology of chronic transplant dysfunction. *Transpl. Int.* **13**, 385–401.

Kroemer, G. and Martinez, A. C. (1991). Invariant involvement of IL-2 in thymocyte differentiation. *Immunol. Today* **12**, 246.

Kuhnlein, P., Park, H.J., Hermann, T., Elbe, A., and Hunig, T. (1994). Identification and characterization of rat Δ/γ T-lymphocytes in peripheral lymphoid organs, small intestine and skin with a monoclonal antibody to a constant determinant of the Δ/γ T-cell receptor. *J. Immunol.* **153**, 979.

Kuhnlein, P., Vicente, A., Varas, A., Huing, T., and Zapata, A. (1995). γ/Δ T-cells in fetal, neonatal and adult rat lymphoid organs. *Dev. Immunol.* **4**, 181–188.

Kunz, H.W., Gill III, T.J., Hansen, C.T., and Poloskey, P.E. (1975). Genetic studies in inbred rats, III: histocompatibility type and immune response of some mutant stocks. *J. Immunogenet.* **2**, 51–54.

Lacha, J., Chadimoval, M., Havlickova, J., Brok, J., Matl, I., and Volk, H.D. (1995). A short course of cyclosporine A combined with anti-CD4 and/or anti-TCR MAb treatment induces long term acceptance of kidney allografts in the rat. *Transplant. Proc.* **27**, 125–126.

Ladoulis, C.T., Shonnard, J.W., Kunz, H.W., and Gill III, T.J. (1973). Genetic control of the induction of the antibody response. *Protides Biol. Fluids. Proc. Colloq.* **21**, 283–289.

Lai, L., Alaverdi, N., Chen, Z., Kroese, F.G.M., Bos, N.A., and Huang, E.C.M. (1997). Monoclonal antibodies to human, mouse, and rat cluster of differentiation antigens. *In* "The Handbook of Experimental Immunology," 5th Ed. (L.A.H.D. Weir and L.A. Herzenberg, eds.), pp 61.1–61.37, Blackwell Science Inc., Cambridge.

Lakkis, F.G. and Cruet, E.N. (1993). Cloning of rat interleukin-13 (IL-13) cDNA and analysis of IL-13 gene expression in experimental glomerulonephritis. *Biochem. Biophys. Res. Commun.* **197**, 612–618.

Lanier, L.L. (2000). Turning on natural killer cells. *J. Exp Med* **191**, 1259–1262.

Laskowski, I.A., Pratschke, J., Wilhelm, M.J., Dong, V.M., Beato, F., Taal, M., Gasser, M., Hancock, W.W., Sayegh, M.H., and Tilney, N.L. (2002). Anti-CD28 monoclonal antibody therapy prevents chronic rejection of renal allografts in rats. *J. Am. Soc. Nephrol.* **13**, 519–527.

Lawetzky, A., Tiefenthaler, G., Kubo, R., and Huing, T. (1990). Identification and characterization of rat T-cell subpopulations expressing α/β and Δ/γ T-cell receptors. *Eur. J. Immunol.* **20**, 343–349.

Leitenberg, D. and Feldbush, T.L (1988). Lymphokine regulation of surface Ia expression on rat B cells. *Cell. Immunol.* **111**, 451–460

Le Meur, Y., Jose, M.D., Mu, W., Atkins, R.C., and Chadban, S.J. (2002). Macrophage colony-stimulating factor expression and macrophage accumulation in renal allograft rejection. *Transplantation* **73**, 1318–1324.

Leslie, G.A. (1984). Allotypic determinants (IgH-3) associated with the IgG2c subclass of rat immunoglobulins. *Mol. Immunol.* **21**, 577–580.

Like, A.A., Guberski, D.L., and Butler, L. (1986). Diabetic BB/Wor rats need not be lymphopenic. *J. Immunol.* **136**, 3254–3258.

Liu, C.C., Perussia, B., and Young, J.D. (2000). The emerging role of IL-15 in NK cell development. *Immunol. Today* **21**, 113–116.

Locker, J., Gill III, T.J., Kraus, J.P., Ohura, T., Swarop, M., Riviere, M., Islam, M.Q., Levan, G., and Szpirer, C. (1990). The rat MHC and cystathione beta-synthase gene are syntenic on chromosome 20. *Immunogenetics* **31**, 271–274.

Long, E.O. (1999). Regulation of immune responses through inhibitory receptors. *Annu. Rev. Immunol.* **17**, 875–904.

Loong, C.C., Hsieh, H.G., Lui, W.Y., Chen, A., and Lin, C.Y. (2002). Evidence for the early involvement of interleukin 17 in human and experimental renal allograft rejection. *J. Pathol.* **197**, 322–332.

Loong, C.C., Lin, C.Y., and Lui, W.Y. (2000). Expression of interleukin-17 as a predictive parameter in acute renal allograft rejection. *Transplant. Proc.* **32**, 1773.

Marchese, A., Heiber, M., Nguyen, T., Heng, H.H.Q., Saldivia, V.R., Cheng, R., Murphy, P.M., Tsui, L.-C., Shi, X., Gregor, P., George, S.R., O'Dowd, B.F., and Docherty, J.M. (1995). Cloning and chromosomal mapping of three novel genes, GPR9, GPR10 and GPR14, encoding receptors related to interleukin-8, neuropeptide Y, and Somatostatin receptors. *Genomics* **29**, 335–344.

Marchesi, V.T. and Gowan, J.L. (1964). The migration of lymphocytes through the endothelium of venules in lymph nodes: an electron microscopic study. *Proc. R. Soc. Lond. B* **159**, 283–290.

Martin, F.H., Suggs, S.V., Langley, K.E., Lu, H.S., Ting, J., Okino, K.H., Morris, C.F., McNiece, I.K., Jacobson, F.W., Mendiaz, E.A., and Birkett, N.C. et al. (1990). Primary structure and functional expression of rat and human stem cell factor DNAs. *Cell* **63**, 203–211.

Mathieson, P.W. (1992). Mercuric chloride induced autoimmunity. *Autoimmunity* **13**, 243–247.

McKnight, A.J., Barclay, A.N., and Mason, D.W. (1991). Molecular cloning of rat interleukin-4 cDNA and analysis of the cytokine repertoire of subsets of CD4⁺ T-cells. *Eur. J. Immunol.* **21**, 1187–1194.

McKnight, A.J. and Classon, B.J. (1992). Biochemical and immunological properties of rat recombinant interleukin-2 and interleukin-4. *Immunology* **75**, 286–292.

McKnight, A.J., Mason, D.W., and Barclay, A.N. (1989). Sequence of a rat MHC class II–associated invariant chain cDNA clone containing a 64 amino acid thyroglobulin-like domain. *Nucleic Acids Res.* **17**, 3983–3984.

Medgyesi, G.A., Fust, G., Bazin, H., Ujhelyi, E., and Gergely, J. (1974). Interactions of rat immunoglobulins with complement. *FEBS Meet.* **86**, 123–128.

Miller, L.W., Granville, D.J., Narula, J., and McManus, B.M. (2001). Apoptosis in cardiac transplant rejection. *Cardiol. Clin.* **19**, 141–154.

Mojcik, C.F., Greiner, D.L., Goldschneider, I. (1991). Characterization of RT6-bearing rat lymphocytes, II: developmental relationships of RT6 and TR6⁺ T-cells. *Dev. Immunol.* **1**, 191–201.

Mojcik, C.F., Greiner, D.L., Medlock, E.S., Komschlies, K.L., and Goldschneider, I. (1988). Characterization of RT6 bearing rat lymphocytes, I: ontogeny of the RT6⁺ subset. *Cell. Immunol.* **114**, 336–346.

Mota, I. (1963). Mast cells and anaphylaxis. *Ann. NY Acad. Sci.* **103**, 264–277.

Nakhooda, A.F., Like, A.A., Chappel, C.I., Murray, F.T., and Marliss, E.B. (1977). The spontaneously diabetic Wistar rat: metabolic and morphologic studies. *Diabetes* **26**, 100–112.

Nash, D.R., Vaeman, J.P., Bazin, H., and Heremans, J.F. (1969). Identification of IgA in rat serum secretions. *J. Immunol.* **103**, 145–148.

Nishida, T., Nishino, N., Takano, M., Sekiguchi, Y., Kawai, K., Mizumo, K., Nakai, S., Masai, Y., and Hirai, Y. (1989). Molecular cloning and expression of rat interleukin-1 α cDNA. *J. Biochem.* **105**, 351–357.

Northemann, W., Braciak, T.A., Hattori, M., Lee, F., and Fey, G.H. (1989). Structure of the rat interleukin-6 gene and its expression in macrophage-derived cells. *J. Biol. Chem.* **264**, 16072–16082.

O'Connor, T.P., Liesen, D.A., Mann, P.C., Rolando, L., and Banz, W.J. (2002). A high isovlavone soy protein diet and intravenous genistein delay rejection of rat cardiac allografts. *J. Nutr.* **132**, 2283–2287.

Ohkawa, A., Ito, T., Yumiba, T., Maeda, A., Tori, M., Sawai, T., Kiyomoto, T., Akamaru, Y., and Matsuda, H. (1999). Immunological characteristics of intragraft NKR-P1⁺ TCR α β⁺ T (NKT) cells in rat hepatic allografts. *Transplant. Proc.* **31**, 2699–2700.

Okabe, K., Yamaguchi, Y., Takai, E., Ohshiro, H., Zhang, J.L., Hidaka, H., Ishihara, K., Uchino, S., Furuhashi, T., Yamada, S., Mori, K., and Ogawa, M. (2001). CD45RC γΔ T-cell infiltration is associated with immunologic unresponsiveness induced by prior donor-specific blood transfusion in rat hepatic allografts. *Hepatology* **33**, 877–886.

Ono, K. and Lindsay, E.S. (1969). Improved technique of heart transplantation in rats. *J. Thorac. Cardiovasc. Surg.* **57**, 225–229.

Orlans, E. and Feinstein, A. (1971). Detection of α, κ and λ chains in mammalian immunoglobulins using fowl antisera to human IgA. *Nature (London)* **233**, 45–47.

Oudin, J. (1956). "L'allotypie" de certains antigenes proteidiques du serum. *C. R. Hebd. Seances Acad. Sci.* **242**, 2606–2608.

Ovary, Z. (1966). The structure of various immunoglobulins and their biologic activities. *Ann. NY Acad. Sci.* **129**, 776–786.

Page, T.H. and Dallman, M.J. (1991). Molecular cloning of cDNAs for the rat interleukin 2 receptor α and β chain genes: differentially regulated gene activity in response to mitogenic stimulation. *Eur. J. Immunol.* **21**, 2133–2138.

Park, J.H., Hanke, T., and Hunig, T. (1996). Identification and cellular distribution of the rat interleukin-2 receptor β chain: induction of the IL-2Rα⁻β⁺ phenotype by major histocompatibility complex class I recognition during T-cell development *in vivo* and by T-cell receptor stimulation of CD4⁺8⁺ immature thymocytes *in vitro*. *Eur. J. Immunol.* **26**, 2371–2375.

Pat Happ, M., Wettstein, P., Dietzschold, B., and Herber-Katz, E. (1988). Genetic control of the development of experimental allergic encephalomyelitis in rats. *J. Immunol.* **141**, 1489–1494.

Paterson, D.J. and Williams, A.F. (1987). An intermediate cell in thymocyte differentiation that expresses CD8 but not CD4 antigen. *J. Exp. Med.* **166**, 1603–1608.

Paul, L.C. (1993). Animal models of chronic heart and kidney allograft rejection. *Transplant. Proc.* **25**, 2080–2081.

Paul, L.C. (1995). Experimental models of chronic renal allograft rejection. *Transplant. Proc.* **27**, 2126–2128.

Paul, L.C., Grothman, G.T., Benediktsson, H., Davidoff, A., and Rozing, J. (1992). Macrophage subpopulations in normal and transplanted heart and kidney tissues in the rat. *Transplantation* **53**, 157–162.

Pinto, M., Gill III, T.J., Kunz, H.W., and Dixon-McCarthy, B.D. (1983). The relative roles of MHC and non-MHC antigens in bone marrow transplantation in rats: graft acceptance and antigenic expression on donor red blood cells. *Transplantation* **35**, 607–611.

Pratt, J.R., Abe, K., Miyazaki, M., Zhou, W., and Sachs, S.H. (2000). *In situ* localization of C3 synthesis in experimental acute renal allograft rejection. *Am. J. Pathol.* **157**, 825–831.

Qian, Z., Hu, W., Liu, J., Sanfilippo, F., Hruban, R.H., and Baldwin III, W.M. (2001). Accelerated graft arteriosclerosis in cardiac transplants: complement activation promotes progression of lesions from medium to large arteries. *Transplantation* **72**, 900–906.

Qian, Z., Wasowska, B.A., Behrens, E., Cangello, D.L., Brody, J.R., Kadkol, S.S., Horwitz, L., Liu, J., Lowenstein, C., Hess, A.D., Sanfilippo, F., and Baldwin III, W.M. (1999). C6 produced by macrophages contributes to cardiac allograft rejection. *Am. J. Pathol.* **155**, 1293–1302.

Querinjean, P., Bazin, H., Beckers, A., Deckers, C., Heremans, J.F., and Milstein, C. (1972). Transplantable immunoglobulin-secreting tumors in rats. Purification of chemical characterization of four κ chains from LOU/Wsl rats. *Eur. J. Biochem.* **31**, 354–359.

Querinjean, P., Bazin, H., Kehoe, J.M., Capra, J.D., (1975). Transplantable immunoglobulin-secreting tumors in rats, VI: N-terminal sequence variability in LOU/Wsl rat monoclonal heavy chains. *J. Immunol.* **114**, 1375–1378.

Richter, G., Blankenstein, T., and Diamantstein, T. (1990). Evolutionary aspects, structure and expression of the rat interleukin 4 gene. *Cytokine* **2**, 221–228.

Richter, G., Hein, G., Blankenstein, T., and Diamantstein, T. (1995). The rat interleukin 4 receptor: co-evolution of ligand and receptor. *Cytokine* **7**, 237–241.

Ritter, M.A., Gordon, L.K., and Goldschneider, I. (1978). Distribution and identity of Thy-1 bearing cells during ontogeny in rat hemopoietic and lymphoid tissues. *J. Immunol.* **121**, 2463–2471.

Rollins, B.J. (1991). JE/MCP-1: An early-response gene encodes a monocyte-specific cytokine. *Cancer Cells* **3**, 517–524.

Rothe, H., Schuller, I., Richter, G., Jongeneel, C.V., Kiesel, U., Diamantstein, T., Blankenstein, T., and Kolb, H. (1992). Abnormal TNF production in prediabetic BB rats is linked to defective CD45 expression. *Immunology* **77**, 1–6.

Rozing, J. and Vaessen, L.M.B. (1979). Mitogen responsiveness in the rat. *Transpl. Proc.* **11**, 1657–1659.

Ruddick, J.H. and Leslie, G.A. (1977). Structure and biological functions of human IgD, XI: identification and ontogeny of a rat lymphocyte immunoglobulin having antigenic cross-reactivity with human IgD. *J. Immunol.* **118**, 1025–1031.

Ruscetti, S.K, Gill III, T.J., and Kunz, H.W. (1976). The genetic control of the antibody response in inbred rats. *Mol. Cell. Biochem.* **7**, 145–156.

Russell, M.E., Wallace, A.F., Hancock, W.W., Sayegh, M.H., Adams, D.H., Sibinga, N.E.S., Wyner, L.R., and Karnovsky, M.J. (1995). Upregulation of cytokines associated with macrophage activation in the Lew-to-F344 rat transplantation model of chronic cardiac rejection. *Transplantation* **59**, 572–578.

Ruuls, S.R., de Labie, M.C.D., Weber, K.S., Botman, C.A.D., Groenestein, R.J., Dijkstra, C.D., Olsson, T., and van der Meide, P.H. (1996). The length of treatment determines whether IFN-β prevents or aggravates experiment autoimmune encephalomyelitis in Lewis rats. *J. Immunol.* **157**, 5721–5731.

Saggi, S.J., Safirstein, R., and Price, P.M. (1992). Cloning and sequencing of the rat preproepidermal growth factor cDNA: comparison with mouse and human sequence. *DNA Cell Biol.* **11**, 481–487.

Sakai, T., Agui, T., and Matsumoto, K. (1994). Intestinal intraepithelial lymphocytes in LEC mutant rats. *Immunol. Lett.* **41**, 185–189.

Sakai, T., Agui, T., and Matsumoto, K. (1995). Abnormal CD45RC expression and elevated CD45 protein tyrosine phosphatase activity in LEC rat peripheral CD4$^+$ T-cells. *Eur. J. Immunol.* **25**, 1399–1404.

Sakuma, Y., Xiu D., Uchida, H., Hakamata, Y., Takahashi, M., Murakami, T., Nagai, H., and Kobayashi, E. (2002). Short-course methotrexate and long-term acceptance of fully allogeneic rat cardiac grafts: a possible mechanism of tolerance. *Transpl. Immunol.* **10**, 49–54.

Savier, E., Lemasters, J.J., and Thurman, R. (1994). Kupffer cells participate in rejection following liver transplantation in the rat. *Transpl. Int.* **7**, 183–186.

Schoefl, G.I. (1972). The migration of lymphocytes across the vascular endothelium in lymphoid tissue: a reexamination: with an appendix by R.E. Miles. *J. Exp. Med.* **136**, 568–588.

Seki, M., Fairchild, S., Rosenwasser, O.A., Tada, N., and Tomonari, K. (2001). An immature rat lymphocyte marker CD157: striking differences in the expression between mice and rats. *Immunobiology* **203**, 725–742.

Sfeir, R., Gambiez, L., Labalette, M., Brami, F., Lecomte, M., Dessaint, J.P., and Pruvot, F.R. (2000). Prolongation of cardiac allograft survival by selective injection of donor liver leukocytes in non-immunosuppressed rats. *Eur. Surg. Res.* **32**, 274–278.

Shinozaki, K., Yahata, H., Tanji, H., Sakaguchi, T., Ito, H., and Dohi, K. (1999). Allograft transduction of IL-10 prolongs survival following orthotopic liver transplantation. *Gene Ther.* **6**, 816–822.

Shintaku, S., Tashiro, H., Yamamoto, H., Fudaba, Y., Shibata, S., Okimasa, S., Mizunuma, K., Noriyuki, T., Fukuda, Y., Asahara, T., and Dohi, K. (2000). Predictive value of CD45RC mRNA for acute rejection in hepatic allografts in rats. *Transplant. Proc.* **32**, 2360–2362.

Shorter, R.G. and Bollman, J.L. (1960). Experimental transfusion of lymphocytes. *Am. J. Physiol.* **198**, 1014–1018.

Soares, M., Havaux, X., Nisol, F., Bazin, H., and Latinne, D. (1996). Modulation of rat B-cell differentiation *in-vivo* by the administration of an anti-mu monoclonal antibody. *J. Immunol.* **156**, 108–118.

Soares, M.B., Turken, A., Ishii, D., Mills, L., Episkopou, V., Cotter, S., Zeitlin, S., and Efstratiadis, A. (1986). Rat insulin-like growth factor II gene: a single gene with two promoters expressing a multitranscript family. *J. Mol. Biol.* **192**, 737–752.

Spickett, G.P., Brandon, M.R., Mason, D.W., Williams, A.F., Woollett, G.R. (1983). MRC OX-22, a monoclonal antibody that labels a new subset of T lymphocytes and reacts with the high molecular weight form of the leukocyte-common antigen. *J. Exp. Med.* **158**, 795–810.

Stechschulte, D.J., Orange, R.P., and Austen, K.F. (1970). Immuno-chemical and biologic properties of rat IgE, I: immunochemical identification of rat IgE. *J. Immunol.* **104**, 1082–1086.

Sutherland, D.B., Varilek, G.W., and Neil, G.A. (1994). Identification and characterization of the rat intestinal epithelial cell (IEC-18) interleukin-1. *Am. J. Physiol.* **266**, C1198–C1203.

Suzuki, K., Murtuza, B., Smolenski, R.T., Suzuki, N., and Yacoub, M.H. (2002). Development of an *in vivo* ischemia-reperfusion model in heterotopically transplanted rat hearts. *Transplantation* **73**, 1398–1402.

Suzuki, S. (1999). FTY720: mechanisms of action and its effect on organ transplantation (review). *Transplant. Proc.* **31**, 2779–2782.

Szeto, W.Y., Krasinskas, A.M., Kreisel, D., Krupnick, A.S., Popma, S.H., and Rosengard, B.R. (2002). Depletion of recipient CD4$^+$ but not CD8$^+$ T lymphocytes prevents the development of cardiac allograft vasculopathy. *Transplantation* **73**, 1116–1122.

Tacke, M., Clark, G.J., Dallman, M.J., Hunig, T. (1995). Cellular distribution and costimulatory function of rat CD28. Regulated expression during thymocyte maturation and induction of cyclosporin A sensitivity of costimulated T-cell responses by phorbol ester. *J. Immunol.* **154**, 5121–5127.

Tada, N., Itakura, K., and Aizawa, M. (1974). Genetic control of the antibody response in inbred rats. *J. Immunogenet.* **1**, 265–275.

Takacs, L., Ruscetti, F.W., Kovacs, E.J., Rocha, B., Brocke, S., Diamantstein, T., and Mathieson, B.J. (1988). Immature, double negative (CD4-, CD8-) rat thymocytes do not express IL-2 receptors. *J. Immunol.* **141**, 3810–3818.

Tamura, A., Li, X.K., Funeshima, N., Enosawa, S., Amemiya, H., and Suzuki, S. (1999). Combination effect of tacrolimus and FTY720 in liver transplantation in rats. *Transplant. Proc.* **31**, 2785–2786.

Tanaka, T., Masuko, T., Yagita, H., Tamura, T., and Hashimoro, Y. (1989). Characterization of a CD3-like rat T-cell surface antigen recognized by a monoclonal antibody. *J. Immunol.* **142**, 2791–2795.

Thomas, V.A., Woda, B.A., Handler, E.S., Greiner, D.L., Mordes, J.P., and Rossini, A.A. (1991). Altered expression of diabetes in BB/Wor rats by exposure to viral pathogens. *Diabetes* **40**, 255–258.

Tomasoni, S., Azzollini, N., Casiraghi, F., Capogrossi, M.C., Remuzzi, G., and Benigni, A. (2000). CTLA4Ig gene transfer prolongs survival and induces donor-specific tolerance in a rat renal allograft. *J. Am. Soc. Nephrol.* **11**, 747–752.

Tsuchida, K., Lewis, K.A., Mathews, L.S., and Vale, W.W. (1993). Molecular characterization of rat transforming growth factor-β type II receptor. *Biochem. Biophys. Res. Commun.* **191**, 790–795.

Uberla, K., Li, W., Qin, Z., Richter, G., Raabe, T., Diamantstein, T., and Blankenstein, T. (1991). The rat interleukin-5 gene: characterization and expression by retroviral gene transfer and polymerase chain reaction. *Cytokine* **3**, 72–81.

Valli, V.E., Villeneuve, D., Reed, B., Barsoum, N., and Smith, G. (1983). Evaluation of blood and bone marrow, rat. *In* "Hemopoietic Systems." (T.C. Jones, J.M. Ward, and R.D. Hunt, eds.), pp 9–26, Springer Verlag, Berlin.

van der Meide, P.H., Dijkema, R., Caspers, M., Vijyerberg, K., and Schellekens, H. (1986a). Clonning, expression and purification of rat IFN-α1. *Meth. Enzymol.* **119**, 441–453.

van der Meide, P.H., Dubbeld, M., Vijverberg, K., Kos, T., and Schellekens, H. (1986b). The purification and characterization of rat gamma interferon by use of two monoclonal antibodies. *J. Gen. Virol.* **67**, 1059–1071.

Vermeer, L.A., De Boer, N.K., Bucci, C., Bos, N.A., Kroese, F.G.M., and Alberti, S. (1994). MRC OX 19 recognizes the rat CD5 surface glycoprotein, but does not provide evidence for a population of CD5bright B cells. *Eur. J. Immunol.* **24**, 585–592.

Vu, M.D., Qi, S., Wang, X., Jiang, W., Ma, A., Xu, D., Bekersky, I., Fitzsimmons, W.E., Wu, J., and Chen, H. (2003). Combination therapy of malononitrilamide FK778 with tacrolimus on cell proliferation assays and in rats receiving renal allografts. *Transplantation* **75**, 1455–1459.

Wang, J., Kuliszewski, M., Yee, W., Sedlackova, L., Xu, J., Tseu, I., and Post, M. (1995). Cloning and expression of glucocorticoid-induced genes in fetal rat lung fibroblasts: transforming growth factor-beta 3. *J. Biol. Chem.* **270**, 2722–2728.

Warbrick, E.V., Dearman, R.J., and Kimber, I. (2002). Induced changes in total serum IgE concentration in the Brown Norway rat: potential for identification of chemical respiratory allergens. *J. Appl. Toxicol.* **22**, 1–11.

Westermann, J., Ronneberg, S., Fritz, F.J., and Pabst, R. (1989). Proliferation of lymphocyte subsets in the adult rat: a comparison of different lymphoid organs. *Eur. J. Immunol.* **19**, 1087–1093.

White, K.L., David, D.W., Butterworth, L.F., and Klykken, P.C. (2000). Assessment of autoimmunity-inducing potential using the Brown Norway rat challenge model. *Toxicol. Lett.* **112–113**, 443–451.

Wonigeit, K., Dinkel, A., Fangmann, J., and Thude, H. (1996). Expression of the ectoenzyme RT6 is not restricted to resting peripheral T-cells and is differently regulated in normal peripheral T-cells, intestinal IEL, and NK cells. *Adv. Exp. Med. Biol.* **419**, 257–264.

World Health Organization [WHO]. (1973). Nomenclature of human immunoglobulins. *Bull. WHO* **48**, 373.

Yamada, T., Natori, T., Izumi, K., Sakai, T., Agui, T., and Matsumoto, K. (1991). Inheritance of T helper immunodeficiency (thid) in LEC mutant rats. *Immunogenetics* **33**, 216–219.

Yamaguchi, Y., Matsumura, F., Liang, J., Akizuki, E., Matsuda, T., Okabe, K., Ohshiro, H., Ishihara, K., Yamada, S., Mori, K., and Ogawa, M. (2000). Reduced interleukin-12, interleukin-18, and interferon-gamma production with prolonged rat hepatic allograft survival after donor-specific blood transfusion. *Dig. Dis. Sci.* **45**, 2429–2435.

Yamamoto, T., Yamaguchi, J., Nakayama, E., and Kanematsu, T. (2000). Anti-CD4 induced rat heart tolerance: no presence of primed T-cells and regulatory mechanisms for cytotoxic T-cells. *Transpl. Immunol.* **8**, 101–107.

Yan, Y., Shastry, S., Richards, C., Wang, C., Bowen, D.G., Sharland, A.F., Painter, D.M., McCaughan, G.W., and Bishop, G.A. (2001). Posttransplant administration of donor leukocytes induces long-term acceptance of kidney or liver transplants by an activation-associated immune mechanism. *J. Immunol.* **166**, 5258–5264.

Yang, C.P. and Bell, E.B. (1992). Functional maturation of recent thymic emigrants in the periphery; development of alloreactivity correlates with the cyclic expression of CD45RC isoforms. *Eur. J. Immunol.* **22**, 2261–2269.

Yang, X., Szabolcs, M., Minanov, O., Ma, N., Sciacca, R.R., Bianchi, M., Tracey, K.J., Michler, R.E., and Cannon, P.J. (1998). CNI-1493 prolongs survival and reduces myocyte loss, apoptosis, and inflammation during rat cardiac allograft rejection. *J. Cardiovasc. Pharmacol.* **32**, 146–155.

Yuan, X., Dong, V.M., Coito, A.J., Waaga, A., Salama, A.D., Benjamin, C.D., Sayegh, M., and Chandraker, A. (2002). A novel CD154 monoclonal antibody in acute and chronic rat vascularized cardiac allograft rejection. *Transplantation* **73**, 1736–1742.

Zinkernagel, R.M. (1997). The discovery of MHC restriction. *Immunol. Today* **18**, 14–17.

Zsebo, K.M., Wypych, J., McNiece, I.K., Lu, H.S., Smith, K.A., Karkare, S.B., Sachdev, R.K., Yuschenkoff, V.N., Birkett, N.C., Williams, L.R., Satyagal, V.N., Tung, W., Bosselman, R.A., Mendiaz, E.A., and Langley, K.E. (1990). Identification, purification and biological characterization of hematopoietic stem cell factor from Buffalo rat liver conditioned medium. *Cell* **63**, 195–201.

Chapter 30

Wild and Black Rats

Marc S. Hulin and Robert Quinn

I. INTRODUCTION

Wild and black rats maintained for biomedical research purposes have had an extremely diverse and interesting background over the past 60 years. Taxonomic classification of these rodents has been among the more interesting aspects. All of the animals described in this chapter clearly belong in the order Rodentia, but the "finer" classification for some of these animals becomes a challenge as to whether they should be designated as rats, mice, gerbils, hamsters, or even "relatives" of guinea pigs (hystricomorphs). Determining the real distinction between mice and rats is particularly challenging. For example, *Zygodontomys*; in different literature references has been referred to as both the "cane rat" and the "cane mouse." Williams (1980) makes anecdotal reference to size as being a potential distinguishing factor in the scientific distinction between mice and rats. However, literature references have described wild rats down to the size of 25 g

(*Oryzomys* spp.) and up to the size of 500 g (*Rattus* spp.). Taxonomic reclassification of some species also impedes a clear understanding of a true designation. For example, the rice rat was originally classified in the Cricetidae family (Walker, 1975), but recent references (Levin et al., 1995; Moscarella and Aguilera, 1999) indicate the rice rat is in the Sigmodontidae family. In addition, the multimammate rat has had two genus names (*Mastomys* and *Praomys*) for more than a decade.

The uses of wild rats in biomedical research range the spectrum of human disease syndromes, including oncology, virology, parasitology, and metabolic diseases (obesity and diabetes). There are also several specific animal models that have been described in the literature as extremely important species in our understanding of the basic biology and mechanisms of disease. For example, the sand rat has been used extensively to study non-insulin-dependent diabetes mellitus and obesity. There is a plethora of literature on the usage of *Psammomys obesus*; in diabetes, covering the gamut of leptin receptor effects, exercise, beta cells, and diabetic complications (vascular, renal, and ocular) to name a few. Another important animal model has been the cotton rat, which has become one of the most commonly used species to study the pathogenesis of human respiratory viruses.

There is also a major focus in wild and black rats to study how their close proximity to the daily lives of mankind impacts the transmission of zoonotic diseases such as hantavirus, viral hemorrhagic fevers, and bacterial infections. Wild rats are the reservoirs for several deadly diseases that affect human health. Therefore, pest control and environmental contamination are significant portions of many literature reviews on wild rats. The future use of wild rats in biomedical research is likely to increase in the next decade owing to the increased awareness and potential threat of agroterrorism and bioterrorism.

II. LABORATORY MANAGEMENT OF CAPTIVE WILD RATS

Before the advances made in husbandry systems over the past 20 to 50 years, investigators using wild-caught rats often had to devise their own housing system to meet both the needs of the experiment and the need to keep the animals contained. Wild-caught rats tend to be more aggressive than their purpose-bred counterparts, and this often correlates with an increased tenacity for finding means of escape. The previously used, common wood and wire containment systems often would not suffice to hold these animals, and the investigators would have to devise more creative ways of securing them.

Current technology using polycarbonate or polysulfone cages with interlocking wire lids appears to be adequate for housing all of the wild rodents discussed in this chapter. The direct bedding used in these systems allows somewhat for the natural burrowing behavior, which is characteristic for most of these species. If nesting material (facial tissue or Nestlets®) is added to the cage, the animals will often use it to enhance their ability to conceal themselves, which appears to provide security and comfort.

Although current recommendations (National Research Council, 1996) call for the group housing of social species, caution should be used when attempting to group house wild-caught rats. Because, in most cases, these animals will already be adults and unfamiliar with each other, there is a very high likelihood that they will display aggression toward each other if pairing or group housing is attempted. Individual housing should be considered initially with gradual, well-monitored attempts at socialization made at a later time.

Handling these animals in a manner safe for both the handler and the animal is critical to working with captive wild rodents. These are animals that have not been selectively bred to be docile and often present a unique challenge when it comes time for experimental manipulation or even simple cage changes. Some of these animals will attack aggressively when approached by human hands. Fortunately, many types of safety gloves exist that are made of materials (e.g., Kevlar®) that will resist rodent bites while still allowing adequate flexibility and tactile sensation. If animals are too active or anxious for routine handling with gloves, other means of transfer, such as gently "dumping" the animal into a secondary container, may be required. Jumping animals constitute a major escape hazard during cage access, and it may be necessary to use a secondary containment system (such as a deep sink, large garbage can, or escape-proof room) whenever the cage is opened.

Improvements in anesthesia systems for rodents have also simplified the experimental manipulation of wild animals. Inhalant systems utilizing volatile agents (such as isoflurane) allow for rapid, safe anesthetic induction with minimal animal handling. Whole-body induction chambers require only minimal manipulation to transfer the animal to the chamber, and systems can even be devised to completely eliminate the need to handle the animal. In addition, the plastic top of a shoebox cage can be modified to make the home cage an anesthetic induction chamber for particularly anxious species.

Very little information is available concerning the specific nutrient requirements of most of these species. Although it would appear that most can adequately maintain, grow, and even reproduce on standard rodent chow, it is unknown if these diets are truly the most nutritionally complete for each species. More studies detailing the specific requirements of each species are needed.

In general, these species adapt well to the use of water bottles. Although there are limited references to the use of automatic watering systems with wild rodent species, it is assumed that they would also be adequate. With either watering system, close monitoring for signs of clinical dehydration and/or loss of body weight must be performed initially to ensure that new animals are able to obtain adequate amounts of water.

One very important aspect to the husbandry of wild-caught rodents is the consideration of the number and variety of human pathogens for which they serve as reservoirs. This will become obvious in reading through the details of each species' experimental use, and serves to emphasize the need to treat all wild-caught rodents as potential zoonotic hazards. Standard practices in working with these animals should include gloves, dedicated clothing, and other personal protective devices as necessary for the species and type of research being conducted.

Although, the transmission of zoonotic disease is rare in the animal research setting, laboratory animal personnel must be always cognizant for the potential severe hazards that can develop from working with wild-caught rodents. In particular, institutions should thoroughly review proposed activities with wild-caught rodents. Species, genus, and location from which a wild-caught rodents are trapped should drive the decision-making process for the institution. Depending on the expertise of the investigator or institution, knowledge and experience from specialists or consultants may be needed to ensure that practices and procedures are conducted in a safe manner.

III.   EXAMPLES OF SPECIES USED IN RESEARCH

A.   Norway Rat (*Rattus norvegicus*)

Brown Norway rats are classified in the family Muridae. They exist wild throughout the world, and the vast majority of laboratory rat strains have been derived from mutations or selective breeding of these animals. The wild-type Brown Norway rat has been inbred and is available commercially from several sources. The inbred strains are more commonly used for biomedical research than are wild-caught rats. The body weight range for wild-caught animals has been reported as 200 to 500 g for adults (Williams, 1980).

The most common uses of wild-caught Norway rats are either for the monitoring of environmental contamination (Ceruti et al., 2002) or for assessment of zoonotic transmission (Sire et al., 2001). The largest use of wild-caught rats is in testing rodenticides and other pest control mechanisms (Jakel et al., 1996). Rats often

cohabitate with and subsist off the byproducts of human production, and therefore, they serve as an excellent monitor of our environment as well as a likely reservoir for disease.

Wild rats have been used as environmental monitors for lead contamination (Mouw et al., 1975; Ceruti et al., 2002), general environmental contamination (Kapu et al., 1991; Fouchecourt and Riviere, 1996; Eckl and Riegler, 1997), and even nuclear waste (Decat and Leonard, 1985). They often consume the same foods that we produce for ourselves, thereby potentially concentrating any contaminants to a degree reflective of human levels.

To an even greater extent, wild rats are studied as potential reservoirs of human disease (Webster et al., 1995; Webster, 1996). This is particularly true in the field of parasitology, in which they have been used to study infection with *Leishmania* (Mukhtar et al., 2000), *Schistosoma* (Imbert-Establet, 1982; Alarcon de Noya et al., 1997), *Toxoplasma* (Fan et al., 1998; Webster et al., 1994), *Capillaria* (Brown et al., 1975), *Paragonimus*, (Cabrera, 1977), *Strongyloides* (Fisher and Viney, 1998), *Angiostrongylus* (Uchikawa et al., 1984), and other parasites (Abd el-Wahed et al., 1999).

Wild rats are an important reservoir for human pathogens such as hepatitis virus. Studies have implicated rats in the transmission and/or monitored for infection levels of hepatitis (Kabrane-Lazizi et al., 1999), hantavirus (Lee et al., 1982; Glass et al., 1988; Ibrahim et al., 1996), rickettsia (Ibrahim et al., 1999), *Bartonella* (Heller et al., 1998), *Coxiella* (Webster et al., 1995), *Listeria* (Inoue et al., 1992), *Yersinia* (Suzuki and Hotta, 1979), and *Pneumocystis* (Palmer et al., 2000).

Another active area of research involving wild rats is comparative genetics and biology. Because the majority of inbred strains have been derived from wild *Rattus norvegicus* it is logical that there are numerous studies comparing these inbred strains to their wild-type "ancestors" (Gregson et al., 1979; Clark and Price, 1981; Cramer et al., 1988; Nagabuchi et al., 1993; Yamada et al., 1993; Kloting et al., 1997, 2001; Voigt et al., 1997, 2000; van Den Brandt et al., 1999; Heeley et al., 1998; Kloting and Kovacs, 1998; Yan et al., 1999; van Den Brandt et al., 2000).

Behavior has also been an area of active study using wild rats. This research has focused on determining central control over defense mechanisms (Mitchell, 1976; Hewitt et al., 1981; Blanchard et al., 1986; Popova et al., 1998) or on understanding and/or controlling aggressive behavior (Boice, 1981; Nikulina et al., 1986; Blanchard et al., 1989; Nikulina, 1991; Hammer et al., 1992; Lucion et al., 1994, 1996; Plyusnina and Oskina, 1997). Much of this behavioral research has been in the context of understanding the normal behavior of wild rats in order to devise effective mechanisms of population control.

Very little work has been conducted on the subject of normative biology or natural disease of wild rats. It is presumed that much of the data derived from laboratory strains can be extrapolated to wild rats. Cramer et al. (1978) compared the different phenotypes of the major histocompatibility complex in wild rats from different regions, and there have been two reports describing filamentous respiratory bacteria in wild rats (MacKenzie et al., 1981; Brogden et al., 1993).

B. Black Rat (*Rattus rattus*)

The black rat (also known as the roof rat) is usually found in tropical to subtropical regions along waterways. In the United States, it is primarily found in the southern and Pacific coastal regions. The reported adult body weight has a wide range of variation (115 to 350 g) (Williams, 1980). It is not used in biomedical research to a significant extent in the United States, but appears to be used more in Europe, India, and Japan. This assessment is skewed by that fact that there also appears to be some confusion of nomenclature within the literature. There are a significant number of publications that, although the investigators stated they were using *Rattus rattus*, they were probably using *R. norvegicus*.

The vast majority of black rat use is in the study of pest control. These animals commonly infest shipyards and can cause significant economic loss owing to damage of product shipments during storage and/or shipping. This work has centered on the mechanism of warfarin resistance (Sugano et al., 2001) and the use of predator odors as repellants (Burwash et al., 1998; Bramley and Waas, 2001).

As an offshoot of pest control studies, a significant body of literature exists concerning natural behavior (Bhardwaj and Prakash, 1981, 1982; Eilam and Golani, 1988) and mechanisms of behavioral control (Barnett and Sanford, 1982; Blanchard et al., 1988, 1989; Kemble et al., 1990) in this species. One study evaluated the effect of the cage environment on stereotypic behavior in captive *R. rattus*; (Callard et al., 2000), and Barnett et al. (1982) conducted a comparative analysis of social postures between several species of *Rattus,* including *R. rattus*.

Other areas of research in *R. rattus,* as in *R. norvegicus*, are the monitoring of toxic environmental contaminants (Pahwa and Chatterjee, 1988; Eckl and Riegler, 1997) and as reservoirs of human disease (Torres et al., 2000; Wang et al., 2000). This species has not been used in these areas nearly as extensively as has *R. norvegicus*, probably because the latter has a much wider geographical distribution.

A small body of literature exists concerning normal anatomy and biology of this species. Emura et al. (1999) conducted an electron microscopic study of the parathyroid glands; Thiele et al. (1997) examined the expression profiles of lymphocyte surface markers; Kaltwasser (1990) studied acoustic signaling; Cavagna et al. (2002) characterized genomic variability using chromosome painting; and Singh and Sabnis (1980) measured hematological changes during gestation.

C. Polynesian Rat (*Rattus exulans*)

Also known as the Burmese house rat, this species is primarily found in the equatorial Pacific islands. Adult body weights range from 70 to 150 g (Williams, 1980).

Polynesian rats are rarely used in biomedical research, but they are captured primarily for studies on pest control (Bramley and Waas, 2001) and for studies on the role of this species as a reservoir for human pathogens such as hantaviruses (Nitatpattana et al., 2000). There are also a pair of studies looking at the behavior of the Polynesian rat (Brooks and Htun, 1980; Gander and Lewis, 1983), and one novel use of DNA analysis from animals captured in different regions is to assess prehistoric human migration in Polynesia (Matisoo-Smith et al., 1998).

D. Tree Rat (*Thamnomys surdaster*)

The tree rat is a very small rat (55 to 65 g) that is also classified in the family Muridae (Williams, 1980). Although traditionally used as a propagation host for *Plasmodium* spp., the causative agent of malaria (Cambie et al., 1990), this use has diminished over the past 30 years owing to the development of alternative species (mice) as hosts. Other species of *Thamnomys* have been used in studies of chromosomal variability (Viegas-Pequignot et al., 1983; Civitelli et al., 1989).

E. Multimammate Rat (*Mastomys [Praomys] coucha*)

This African rodent is also in the family Muridae and used to be included in the genus *Rattus* (Williams, 1980). Its small size (40 to 80 g) has caused it to also be known as the multimammate mouse. Several inbred, specific pathogen-free strains currently exist in various laboratories throughout the world (Yamamoto et al., 1999). Although *Mastomys* is currently considered the most appropriate genus designation, these animals are also referred to by the genus name *Praomys* in some literature (Yamamoto et al., 1999). There are seven different species, with *M. coucha* and *M. natalensis* being used the most frequently in research. The taxonomy is gradually being refined based on genetic analysis (Hisatomi et al., 1994; Fieldhouse et al., 1997; Volobouev et al., 2001).

Mastomys has been used most extensively as a model of gastric neoplasia (Nilsson et al., 1992; Gilligan et al., 1995). One study reported that this species had an abnormally high occurrence of gastric adenocarcinoma (Oettle, 1957). Further studies have determined these masses to be gastric carcinoids (Snell and Stewart, 1969). This spontaneous carcinoid formation can be greatly accelerated (from 2 years down to 4 months) by H2 blockade or administration of proton pump inhibitors, either of which result in an acid-inhibition-induced hypergastrinemia (Tang and Modlin, 1996). As a result, this species is actively used in the study of mechanisms of gastric hyperplasia and carcinoid formation (Kidd et al., 2000).

Mastomys has been promoted as a model of other types of neoplasia as well. Madarame et al. (1995) reported on spontaneous rhabdomyosarcomas, and Pruthi et al. (1983) described intracutaneous cornifying epitheliomas. Papillomavirus has been used to experimentally induce keratoacanthomas and squamous cell carcinomas, although the classification of these lesions has been questioned (Rudolph and Busse, 1981; Rudolph et al., 1981). A histopathological survey of aged *Mastomys* displayed a wide range of spontaneous neoplastic and preneoplastic lesions (Solleveld et al., 1982).

Recently, this species was investigated as a possible model for testing parvoviral-mediated antitumor therapies, but the species was found to be more acutely susceptible to parvoviral disease than were standard laboratory rats and mice (Haag et al., 2000). It has also been touted as a model of autoimmune thyroiditis (Solleveld et al., 1985) and as a potentially useful model for reproductive biology with the successful development of an *in vitro* fertilization procedure (Nohara et al., 1998).

Studies on normative biology of *Mastomys* are rather limited, yet still under active investigation. Yamamoto et al. (1999) documented normal serum biochemistry values for two inbred strains, and Schares and Zahner (1994) characterized the immunoglobulin (Ig) IgG system. This species (along with other desert rodents) has been used to investigate mechanisms of body temperature regulation (Shukla et al., 1997). Mammary gland growth and hormonal response have been compared between *Mastomys* and C3H/He mice (Nagasawa et al., 1989). Relative to standard laboratory rats, *Mastomys* appear to be more resistant to the hepatotoxic affects of aflatoxins (Kumagai et al., 1998) and the nephrotoxic effects of mercury II chloride (Holmes et al., 1996). One study reported on stereotypic behavior in laboratory-housed animals (Gulatti et al., 1988).

Several studies have described anatomical peculiarities within this species. Both males and females possess a well-developed prostate gland, which has previously been exploited to develop an experimental model of infectious prostatitis (Jantos et al., 1990). The submaxillary salivary glands are considered one of the richest natural sources of nerve growth factor (Aloe et al., 1981; Burcham et al., 1991). Females of the wild-colored inbred strain, MWC, have a unique adrenal border zone between the zona fasciculata and the zona reticularis (Tanaka et al., 1996). The MCC strain was observed to have lysosomal glycolipid storage within the renal proximal tubular epithelium (Fujimura et al., 1996). Another active area of research utilizing *Mastomys* is in the study of microfilarial infections and development of antifilarial agents (Tripathi et al., 2000).

Previously, these animals were used extensively in pesticide research (Gill and Redfern, 1979) because they constitute a significant agricultural pest in Africa. This use appears to have diminished in recent years.

F. Cotton Rat (*Sigmodon hispidus*)

The cotton rat is a New World rodent that is distributed from the southern United States to the northern regions of South America. Seven species of *Sigmodon*; have been identified, but *Sigmodon hispidus* (Fig. 30-1) has the widest geographical distribution. Cotton rats are classified in the family Cricetidae, and although there are only minor phenotypic differences between species, there are significant karyotypic differences (ranging from 22 to 52 chromosomes) (Prince, 1994). Taxonomy continues to be an active area of research (Pfau et al., 1999). Body weight of adults has been reported from 90 to 200 g (Williams, 1980).

The handling of cotton rats can be a challenge because they are quick, can jump quite high, and often attempt to

Fig. 30-1 Adult female cotton rat (*Sigmodon hispidus*) with two pups. (Photograph courtesy of P.R. Wyde, Baylor College of Medicine, Houston, Texas.)

bite when handled (Ward, 2001). Implementation of the suggested handling techniques listed in the general husbandry section for this chapter should alleviate some of the difficulties. In addition, a unique device for the handling of adult animals has been described (Niewiesk et al., 1997).

Several historical uses led to the development of cotton rats as a laboratory species. From the 1940s through the 1970s, cotton rats were used extensively in polio research (Prince, 1994). Before the development of susceptible inbred mouse strains, the cotton rat was the only rodent identified that developed paralytic disease after experimental infection. A similar historical use was in vaccine development for scrub typhus by the British during World War II (Worth and Rickard, 1951). In addition, cotton rats have been used extensively in the study of filarial infections (Pringle and King, 1968). *Litomosoides carinii* is an endogenous pathogen of this species and has been used to investigate both filarial-host interactions and potential therapeutic modalities, although this use appears to be waning.

Currently, the most common biomedical uses of cotton rats are in the study of respiratory viral pathogenesis and the use of viral agents as vectors for gene therapy. These animals are susceptible to respiratory infections with respiratory syncytial virus (Piazza et al., 1993; Curtis et al., 2002), measles virus (Niewiesk, 1999), influenza virus (Langley et al., 1998), parainfluenza virus (Langley et al., 1998), vaccinia virus (Weidinger et al., 2001), and human adenovirus. Infection with adenovirus has been particularly exploited as a means of developing potential gene therapies or recombinant vaccines (Breker-Klassen et al., 1995; Papp et al., 1997).

A limiting factor preventing the more widespread use of these animals in respiratory viral pathogenesis studies is a lack of commercially-available immunological reagents, as are currently available for both laboratory mice and rats. This deficiency is being addressed by a variety of laboratories (Houard et al., 2000). There is also only limited information concerning normal hematological values (Katahira and Ohwada, 1993; Robel et al., 1996).

In addition to the respiratory viruses listed above, cotton rats have also recently been promoted as a superior model (over mice) of herpes simplex virus type 1 infection and latency (Lewandowski et al., 2002).

Other uses of *Sigmodon* include their potential as a reservoir for the Lyme disease agent *Borrelia burgdorferi* (Oliver et al., 1995; Clark et al., 2001); their role in the pathogenesis and perpetuation of hantavirus infections (Hutchinson et al., 1998); their use as a model of genetic diversity in different populations (Pfau et al., 2001), an indicator system for toxicity associated with environmental contamination (McMurry et al., 1995; Rafferty et al., 2000; Kim et al., 2001), or a rodent model of

Bartonella spp. bacteremia (Kosoy et al., 2000); their use in the study of dietary affects on immune function (Lochmiller et al., 1998) or as a potential model for gastric adenocarcinoma (Kawase and Ishikura, 1995); and their use in rodenticide trials (Gill and Redfern, 1979).

Several sporadic reports of normal anatomy and biology exist, including a detailed study of the pineal gland (Sakai et al., 1996), a calculation of body surface area (Ohwada and Katahira, 1993), and an assessment of the gastrointestinal flora (Itoh et al., 1989).

An extensive review of cotton rats and their use in biomedical research was compiled by Faith et al. (1997).

G. Wood Rat (*Neotoma* spp.)

Wood rats are also known as pack rats or trade rats. Similar to cotton rats, they also belong to the family Cricetidae, yet they are significantly larger (adult body weight 200 to 430 g) (Williams, 1980). They are native to North America.

Neotoma have very limited use in biomedical research. They are more commonly studied from an ecological (Reichman, 1988; Sakai and Noon, 1997; Matocq, 2002) or taxonomic (Castleberry et al., 2000) perspective. Current biomedical use focuses on their being a reservoir for arenavirus infection (Kosoy et al., 1996; Fulhorst et al., 2002) and Lyme disease (Lane and Brown, 1991; Burkot et al., 1999; DeNatale et al., 2002).

Other research uses of *Neotoma* have included mechanisms of biotransformation of ingested plant compounds (Lamb et al., 2001), hearing (Heffner and Heffner, 1985), hormonal and sensory control of scent marking (Clarke, 1975; Fleming and Tambosso, 1980), and thermoregulation (Schwartz and Bleich, 1976), as well as studies relating to pest control, such as taste preference (Harriman, 1978).

Several studies have centered on characterization of the anatomy and biology of this species, including the ocular lens (Ramos and Smith, 1975), thyroid gland (Capeheart and Burns, 1976), and serum proteins (Zimmerman and Nejtek, 1975). Detailed anatomy has been described previously (Howell, 1926). One noteworthy anatomical characteristic is the tendency for the incisors of nursing pups to splay outward for the first 2 weeks of life (Hamilton, 1953).

H. Cane Rat (*Zygodontomys brevicauda*)

The genus *Zygodontomys* has been referred to as cane rats or cane mice. Further increasing the confusion on the designation of this animal, some references have referred to the cane rat as the "grasscutter," with the scientific nomenclature of *Thryonomys swinderianus*

(Baptist and Mensah, 1986; McCoy et al., 1997). The cane rat is classified as a New World rat belonging to the Cricetidae family (Williams, 1980), thus the name of cane mouse or grasscutter is inappropriate. The cane rat can be found in South and Central America and appears to have a combination of phenotypic characteristics found in mice, rats, and hamsters. The average body weight is around 100 g or less; the tail is about three-quarters the length of the body; and the eyes, ears, and face are almost identical to several species in the *Mus* genus.

The most commonly used species of cane rat described in biomedical research is *Zygodontomys brevicauda*. The husbandry for this species varies depending on the sources referenced, but the standards for most rodents, as based on the *Guide for the Care and Use of Laboratory Animals* (National Research Council, 1996), seem to maintain the species. The urinary and fecal output of cane rats appears to be similar to that of gerbils, and as a result, the cages can be changed and cleaned once every 7 to 14 days (Voss et al., 1992). A variety of diets have been used, including a commercial fox diet, laboratory rat chow, and commercial mouse chow supplemented with dry cat food and green vegetables (Worth, 1967; Williams, 1980; Voss et al., 1992). There are some conflicting reports on whether these animals are aggressive in captivity; however, one consistent finding in the literature is that cane rats are very agile and "jumpy" and will try to escape from their cages with minimal stimulation.

The reproductive biology of the cane rat has been nicely reviewed and summarized by Donnelly and Quimby (2002). In short, the gestation period (25 to 28 days), weaning age (16 days of age), and litter size are similar, with slight variations, to those of many other rodents.

Cane rats have been used primarily in biomedical research to determine if they are a source of viruses of the Arenaviridae family known to produce viral hemorrhagic fevers in man. In particular, a recent report by Fulhorst et al. (1999) indicated that cane rats are the natural reservoir host of Guanarito virus, which is known to produce Venezuelan hemorrhagic fever in humans. Guanarito virus has been shown to be non-pathogenic in cane rats, and studies indicated persistent viremia and shedding of the virus from saliva and urine without the development of morbidity or mortality (Fulhorst et al., 1999).

In addition, older references describe the use of cane rats to study various viral infections such as Cocal (Jonkers et al., 1964), Nariva (Beare, 1975), Venezuelan equine encephalomyelitis (Downs et al., 1962), and yellow fever (Bates and Weir, 1944).

Voss et al. (1992) maintained a colony of cane rats to study basic genetics. Analysis and comparison of evolutionary genetic divergence is easier in *Z. brevicauda* because of the short gestation, diploid chromosome number, and hybrid crosses available compared with other species.

Wild-caught cane rats are particularly sensitive to stress-induced self-mutilation. McCoy et al. (1997) determined that the long-acting neuroleptic tranquilizer pipothiazine palmitate (25 mg/kg) provided a significant calming effect of over 39 days on male and female cane rats after a single administration.

I. Rice Rat (*Oryzomys* spp.)

There are numerous species in the genus *Oryzomys*, which are commonly known as rice rats. The taxonomy of the rice rat is confusing, with several different references placing this rodent to be in either the Cricetidae or Sigmodontidae family (Williams, 1980; Levin et al., 1995; Moscarella and Aguilera, 1999). Rice rats are native to several countries in South America. Williams (1980) reports the weight range to be 25 to 150 g, whereas Wolfe (1982) indicates the body weight is around 40 to 80 g.

The most commonly used species in biomedical research is *Oryzomys palustris,* sometimes known as the marsh rat (Edmonds and Stetson, 1995; Edmonds et al., 2003). The rice rat may be a reservoir host for Lyme disease (Levin et al., 1995). *O. palustris* is easily infected with *B. burgdorferi* via infected tick vectors (*Ixodes scapularis*), and maintains infectivity of the spirochaete for greater than 2 months.

Marsh/rice rats have been used as an animal model to study peridontal disease (Gotcher and Jee, 1981; Shklair and Ralls, 1988; Cohen and Meyer, 1993). Rice rats spontaneously develop peridontal disease, and feeding of a high-sucrose diet accelerates the disease progression. Various investigators have tried to determine the etiology for periodontitis and bone loss, but they have had little success in identifying a primary etiologic agent.

Edmonds and Stetson (1993, 1994, 1995) studied the effects of the pineal gland on reproductive physiology and development in male rice rats. Their work suggested a clear connection between melatonin release and pineal gland function. In addition, length of photoperiod and availability of food had multiple effects on the reproductive development of juvenile and adult marsh rice rats (Edmonds et al., 2003).

O. albigularis (Moscarella and Aguilera, 1999); *Oligoryzomys flavescens,* the yellow pigmy rice rat (Delfraro et al., 2003); and *Oligoryzomys longicaudatus,* the long-tail rice rat (Murua, 2003), have been used in various ways as described in the literature. *O. albiguris* have been monitored under laboratory conditions to compare their growth and reproductive development to that of other sigmodontine rodents. The yellow pigmy and long-tail rice rats have been investigated as reservoir hosts for hantavirus in South America (Milazzo et al., 2002; Delfraro et al., 2003; Murua et al., 2003). Hantavirus pulmonary syndrome is a potentially fatal human disease that develops after zoonotic transmission from infected rodents. Rice rats appear to be the main reservoir host in several South American countries, and a newly discovered hantavirus (Maporal virus) has been recently isolated from arboreal rice rats (Murua et al., 2003).

J. Climbing Rat (*Tylomys nudicaudus*)

Tylomys nudicaudus have been infrequently used in biomedical research, beginning in the early to mid-1970s. The neotropical climbing rat belongs to the family Cricetidae and is native to Central America. Williams (1980) indicates that these rats easily adapt to the research laboratory setting, are relatively tame, and can be handled by the tail. Restraining these animals by holding around the thorax should be avoided because they have a tendency to bite. Climbing rats can be housed in rat shoebox cages on contact bedding and fed a standard laboratory rodent diet supplemented with vegetables and fruits. Helm and Dalby (1975) reported that *Tylomys* females reach sexual maturity around 3 months of age and that the average gestation length is 40 days. The typical litter size ranges from 1 to 4 pups, with an average of 2.3 offspring per female. Young climbing rats stay closely attached to their mothers for the first 3 weeks of life, and the weaning age is around 3 to 4 weeks of age.

K. African White-Tailed Rat (*Mystromys albicaudatus*)

The white-tailed rat represents the only species in the genus *Mystromys* that belongs in the family Cricetidae. This animal is indigenous to dry sandy regions of South Africa and has a similar body size and shape to that of *Mesocricetus auratus* (Syrian hamster). The long white tail of *M. albicaudatus* is the source of its common name and helps to distinguish this animal from the hamster. One unique reproductive biologic characteristic is that female white-tailed rats have a rudimentary prostate gland (Hall et al., 1967).

The husbandry of these rats is very similar to that of other rodents. The animals appear to adapt to the laboratory setting and should be housed in the standard size plastic rodent shoebox cage with contact bedding (Hall et al., 1967; Williams, 1980). Feeding these animals a commercially available standard rodent diet *ad libitum* is sufficient to sustain normal growth and maintenance. Their reproductive physiology appears to be similar to

that of climbing rats, with an average litter size of three, a 38-day gestation period, and a 3-week weaning age (Hall et al., 1967; Hallett and Meester, 1971; Williams, 1980; Donnelly and Quimby, 2002).

Mystromys albicaudatus have been used in biomedical research to study the development of spontaneous diabetes mellitus. Diabetic white-tailed rats develop the classic characteristics of the disease (hyperglycemia, glycosuria, ketonuria, and polyuria) but apparently do not develop obesity (Little et al., 1982; Clark, 1984). These animals have also been used to study a variety of other human diseases and pathophysiological processes such as dental caries (Larson and Fitzgerald, 1968), hemorrhagic fevers (Shepard et al., 1989), thermoregulation (Downs and Perrin, 1995), carcinogenicity (Roebuck and Longnecker, 1979), and gastrointestinal disorders (Mahida and Perrin, 1994). In addition, *M. albicaudatus* has been used as an experimental host for a variety of protozoal species that cause leishmaniasis, such as *Leishmania braziliensis, L. donovani,* and *L. mexicani* (McKinney and Hendricks, 1980; Sayles et al., 1981; Franke et al., 1985).

L. Sand rat (*Psammomys obesus*)

The sand rat has also been referred to as "fat" sand rat and the "Israeli" sand rat (Collier et al., 1997; Lerbowitz et al., 2001). *P. obesus* is classified in the family Cricetidae and is in the same subfamily as the gerbil (Gerbillinae). These rats are natural residents of the of the desert habitats of the Middle East, which accounts for their dietary needs and nutritional physiology. Laboratory colonies of sand rats have been well established, with successful breeding programs (Frenkel et al., 1972; Gruber, et al., 2002). *P. obesus* adapt well to the laboratory environment and may be housed in standard shoebox cages with contact bedding. Williams (1980) reports that sand rats are solitary in nature, and fighting may develop if breeding pairs are housed together when the female is not in estrus. Female sand rats have a similar gestation period (24 days) and average litter size (3.6) to that of the gerbil, *Meriones unguiculatus.*

Dietary factors are critical because of the potential development of spontaneous diabetes mellitus in this species (Marquie et al., 1984). Sand rats should be maintained on a low-energy diet supplemented with desert salt bush (Moskowitz et al., 1990). Kalman et al. (2001) reported that in their natural environment, *P. obesus* has a metabolism geared toward low energy, which helps accumulate fat stores for maintenance and breeding during periods of nutritional drought. However, this low-energy metabolism switches off with consumption of an increased nutrient diet, and the animal's metabolism is shifted to a diabetic-prone state.

P. obesus seems to be the perfect animal model to study non-insulin-dependent diabetes mellitus, the so-called adult-onset type II diabetes mellitus (Leibowitz et al., 2001). Adult sand rats spontaneously develop diabetes with moderate obesity when fed a standard laboratory rodent or high-energy diet (Pertusa et al., 2002; Zoltowska et al., 2003). One colony of sand rats became obese and developed hyperglycemia when fed a commercial diet supplemented with natural low-calorie plants containing a high salt concentration (Marquie, 1984). The diabetic syndrome in sand rats is characterized by hyperinsulinemia with rebound hypoinsulinemia, which eventually leads to insulin resistance and obesity (Cerasi et al., 1997). Researchers have investigated the role of hyperglycemia in vascular contractility (Zoltowska et al., 2003), diabetic nephropathy secondary to renal hypertrophy (Raz et al., 2003), collagen synthesis (Bouguerra et al., 2001), cataract development (Borenshtein et al., 2001), obesity phenotypes (Walder et al., 2000), and genetic and dietary factors (Nesher et al., 1999). Several investigations have focused on the role of leptin, dyslipidemia, and hypothalamic gene expression in obesity and its close association with the diabetic metabolic syndrome (Collier et al., 1997; Walder et al., 2002).

Moskowitz et al. (1990) began using aged sand rats to study the effects of aging on vertebral disc degeneration. Initial work with these animals revealed that as early as 3 months of age, sand rats have signs of disc space narrowing and sclerosis of end plates. Following these initial studies, *P. obesus* has become a well-characterized and reliable model to study degenerative disc disease. Gruber et al. (2002, 2003) has worked extensively on the characterization of sand rats as animal models by establishing radiologic progression of the disease. Bone mineral density increases in the end plates of lumbar vertebrae of aging males, and radiographic changes are evident as early as 2 month of age. Wilson et al. (2003) have established a computer-aided analytical system to digitize radiographic images of disc degeneration in sand rats. In addition, fat sand rats may be a valuable model for autologous disc cell implantation (Gruber et al., 2002).

P. obesus has also been used to study otic cholesteatoma (Feinmesser et al., 1988), dental abnormalities (Ulmansky et al., 1984), and surgical anatomy of the ear (Sichel et al., 1999). Sand rats have been proposed as an animal model for Müllerian mixed tumors since Czernobilsky et al. (1982) reported a high incidence of spontaneous uterine neoplasms in a colony of female sand rats. Fifty-three percent of the spontaneous tumors were adenofibromas, which led investigators to propose sand rats might be a useful animal model for Müllerian mixed tumors.

Gruber et al. (2001) found tapeworm infestation (*Rodentolepis nana*) in a research-breeding colony; the clinical diagnosis was unexpected in this colony because of the large size of the tapeworms.

M. Kangaroo Rat (*Dipodomys* spp.)

There are two species of kangaroo rats that have been studied in biomedical research: *Dipodomys merriami* (Merriam's kangaroo rat) and *D. spectabilis* (banner-tailed kangaroo rat). However, other species of kangaroo rats, *D. ordii* (Ord's) and *D. deserti* (Desert), have been used to study reproductive physiology (Jollie, 1956) and breeding dynamics (Butterworth, 1961). Kangaroo rats are taxonomically classified in the family Heteromyidae and live almost exclusively in the desert areas of the southwestern United States and Mexico (Hall, 1981; Servin et al., 2003). Nowak (1999) has extensively described the kangaroo rat and, in particular, notes the strength and conformation of the hindlegs, which lend to its characteristic common name. *Dipodomys* are true desert animals and have an incredible water conservation system. This system allows the rats to gain most of their water supply from their diet. Williams (1980) states that kangaroo rats rarely drink water if freely accessible, but Donnelly and Quimby (2002) contend that these animals will drink water readily if offered in captivity. Although their diet in the wild appears to be grains, in the laboratory they should be fed seeds and grains, supplemented with lettuce (Williams, 1980).

A single report of a gastric trichobezoar in a banner-tailed kangaroo rat suggested that consumption of fur during grooming may result in anorexia and wasting in this species (Suckow et al., 1996).

Kangaroo rats have external cheek pouches to store food and long tails prone to degloving injury if restrained, and they must have access to dust baths to maintain a healthy fur coat (Nowak, 1999). Because these animals cannot be handled by the tail, the best restraint technique is to scruff the animal by the nape of the neck (Fine et al., 1986). With persistence and patience, kangaroo rats can be successfully bred and raised in the laboratory setting (Eisenberg and Isaac, 1963; Daly et al., 1984; Donnelly and Quimby, 2002). *Dipodomys* have an average gestation length of 31 days, and the average litter size is two.

Several investigators have used kangaroo rats to study the mechanisms of water conservation and physiology of the kidney (Schmidt-Nielsen, 1964; Stallone and Braun, 1988). Kangaroo rats have also been used to study human decompression sickness (Hills and Butler, 1978), thyroxine-induced basal metabolism (Banta and Holcombe, 2002), osteoporosis (Muths and Reichman, 1996), and neuroanatomy (Jacobs and Spencer, 1994).

Winters and Waser (2003) studied the inbreeding tendencies of banner-tailed kangaroo rats. The philopatric nature of this animal led to the belief that they were closely inbred. However, detailed genetic analysis revealed genetic variability. Merriam's and Ord's kangaroo rats have been studied in the field to better characterize their dietary habits (Sipos et al., 2002) and burrowing thermal tolerance to classic desert conditions (Tracy and Walsberg, 2002).

N. Degu (*Octodon degus*)

This hystricomorph desert rodent is commonly referred to as the "degu," but it has also been called the "trumpet-tailed rat" because of the prominent black hair at the tip of its tail. *Octodon degus* belongs to the family Octodontidae and naturally resides in the Andes Mountains of Chile (Woods and Boraker, 1975). The genus name for this species was derived from the unusual shape of the molar teeth. Degus have a strong social structure and live in small colonies, in which females often raise their offspring together in a common burrow (White et al., 1982). These animals readily adapt to the research laboratory and exhibit good social acceptance of strange males and females into the social organization (Altmann et al., 1994; Donnelly and Quimby, 2002). An adult female degu is shown in Fig. 30-2, and a weanling degu is shown in Fig. 30-3.

Female degus have reproductive traits similar to guinea pigs, such as a vaginal closure membrane (Williams, 1980), long gestation period (90 days), and birth of precocious young (Weir, 1970). A unique trait of this hystricomorph rodent is that the females are induced ovulators (Weir, 1970). The average litter size is 5 to 6 pups and weaning age averages 3 to 4 weeks. The 3-month gestation period, 1-month weaning age, and induced ovulation result in the female having a maximum of two litters per year.

Fig. 30-2   Adult female degu (*Octodon degus*). (Photograph courtesy of Erik Stortz, Charles River Laboratories, Wilmington, MA.)

Fig. 30-3   Weanling male degu (*Octodon degus*). (Photograph courtesy of Erik Stortz, Charles River Laboratories, Wilmington, MA.).

Several investigators have reported that degus are gentle, social rodents that respond positively to gentle handling, but they may bite handlers who are unfamiliar with the species (Williams, 1980). Restraining these animals by the tail must be done cautiously because the skin may deglove if the animals fight the restraint technique (Woods and Boraker, 1975; Williams, 1980). If degu pups are gently handled from birth, they become familiar with their handlers and can be efficiently restrained by the tail or scooped up in two hands (Altmann et al., 1994).

O. degus can be group housed in large opaque shoebox cages with contact bedding. These animals should be fed a commercial rodent diet with vegetable supplements (e.g., potatoes and carrots). However, caretakers should use caution in the amount of supplemental food provided, because degus will tend to exclusively hoard vegetables. Obesity is likely to occur if degus are fed a commercial rodent diet *ad libitum*.

Degus have some unique characteristics that have made it a very popular animal model in the past 10 to 15 years. *O. degus* are naturally diurnal and have a similar circadian rhythm to humans. This feature makes this caviomorph rodent an excellent animal model to study human circadian rhythm patterns and sleep behavior (Goel and Lee, 1995, 1996, 1997; Kas and Edgar, 1999; Jechura et al., 2003; Lee, 2004). Social organization of the degu colony has led to several investigations into the development of neurologic behavior of offspring in response to maternal vocalizations (Poeggel and Braun, 1996; Braun et al., 2003), visual adaptation (Jacobs et al., 2003), and neonatal separation and isolation (Poeggel et al., 2000; Ziabreva et al., 2003). In addition, the female degu has a labyrinthine hemomonochorial placenta, which has been examined as an potential model for the human placenta (Bosco, 1997; Kertschanska et al., 1997).

Investigators have also used *O. degus* to understand hepatic drug metabolism and resistance. Degus demonstrate natural resistance to morphine and greater tolerance to pentobarbital compared with that of other rodents (Letelier et al., 1984, 1985; Pelisser et al., 1989; Gaule et al., 1990). Letelier et al. (1984, 1985) determined that the degus' tolerances to morphine and pentobarbital were related to higher levels of mixed function oxidases and cytochrome P-450 in hepatic microsomes.

This desert rodent has also been used in investigations in renal pathology and toxicology (Bosco et al., 1997; Cadillac et al., 2003), colonic water absorption (Gallardo et al., 2002), age-related degenerative neuropathology (Ostroff et al., 2002), and developmental physiology and behavior (Reynolds and Wright, 1979; White et al., 1982; Wilson, 1982).

REFERENCES

Abd el-Wahed, M.M., Salem, G.H., and el-Assaly, T.M. (1999). The role of wild rats as a reservoir of some internal parasites in *Qalyobia governorate. J. Egypt Soc. Parasitol.* **29**, 495–503.

Alarcon de Noya, B., Pointier, J.P., Colmenares, C., Theron, A., Balzan, C., Cesari, I.M., Gonzalez, S., and Noya, O. (1997). Natural *Schistosoma mansoni* infection in wild rats from Guadeloupe: parasitological and immunological aspects. *Acta Tropica* **68**, 11–21.

Aloe, L., Cozzari, C., and Levi-Montalcini, R. (1981). The submaxillary salivary glands of the African rodent *Praomys (Mastomys) natalensis* as the richest available source of the nerve growth factor. *Exp. Cell Res.* **133**, 475–480.

Altmann, D., Schwendenwein, I., and Wagner, K. (1994). Zu Biologie, Haltung, ernahrung und Erkrankungen des Degus (*Octodon degus*). *Erkrank. Zoot.* **36**, 277–292.

Banta, M.R. and Holcombe, D.W. (2002). The effects of thyroxine on metabolism and water balance in a desert-dwelling rodent, Merriam's kangaroo rat (*Dipodomys merriami*). *J. Comp. Physiol.* **172**, 17–25.

Baptist, R. and Mensah, G.A. (1986). The cane rat: farm animal of the future. *World Anim. Rev.* **60**, 2–6.

Barnett, S.A., Fox, I.A., and Hocking, W.E. (1982). Some social postures of five species of *Rattus. Austral. J. Zool.* **30**, 581–601.

Barnett, S.A. and Sandford, M.H. (1982). Decrement in "social stress" among wild *Rattus rattus* treated with antibiotic. *Physiol. Behav.* **28**, 483–487.

Bates, M. and Weir, J.M. (1944). The adaptation of a cane rat (*Zygodontomys*) to the laboratory and its susceptibility to the virus of yellow fever. *Am. J. Trop. Med. Hyg.* **24**, 35–37.

Beare, A.S. (1975). Myxoviruses. *Dev. Biol. Stand.* **28**, 3–17.

Bhardwaj, D. and Prakash, I. (1981). Movements of *Rattus rattus* in an artificial environment. *Indian J. Exp. Biol.* **19**, 794–796.

Bhardwaj, D. and Prakash, I. (1982). Discrimination between harmless and harmful stimuli by the house rat *Rattus rattus rufescens* (Gray). *Indian J. Exp. Biol.* **20**, 302–304.

Blanchard, D.C., Hori, K., Rodgers, R.J., Hendrie, C.A., and Blanchard, R.J. (1989). Attenuation of defensive threat and attack in wild rats (*Rattus rattus*) by benzodiazepines. *Psychopharmacology* **97**, 392–401.

Blanchard, D.C., Rodgers, R.J., Hendrie, C.A., and Hori, K. (1988). Taming of wild rats (*Rattus rattus*) by 5HT1A agonists buspirone and gepirone. *Pharmacol. Biochem. Behav.* **31**, 269–278.

Blanchard, R.J., Flannelly, K.J., and Blanchard, D.C. (1986). Defensive behavior of laboratory and wild *Rattus norvegicus. J. Comp. Psychol.* **100**, 101–107.

Boice, R. (1981). Behavioral comparability of wild and domesticated rats. *Behav. Genetics* **11**, 545–553.

Borenshtein, D., Ofri, R., Werman, M., Stark, A., Tritschler, H.J., Moeller, W., and Madar, Z. (2001). Cataract development in diabetic sand rats treated with alpha-lipoic acid and its gamma-linolenic acid conjugate. *Diabetes Metab. Res. Rev.* **17**, 44–50.

Bosco, C. (1997). Ultrastructure of the degu term placental barrier (*Octodon degus*): a labyrinthine hemomonochorial placental model. *Med. Sci. Res.* **25**, 15–18.

Bosco, C., Rodrig, R., Diaz, S., and Borax, J. (1997). Renal effects of chronic exposure to malathion in *Octodon degus. Comp. Biochem. Physiol.* **118**, 247–253.

Bouguerra, S.A., Bourdillon, M.C., Dahmani, Y., and Beckkhoucha, E. (2001). Non-insulin dependent diabetes in sand rat (*Psammomys obesus*) and production of collagen in cultured aortic smooth muscle cells influence of insulin. *Int. J. Exp. Diabetes Res.* **2**, 37–46.

Bramley, G.N. and Waas, J.R. (2001). Laboratory and field evaluation of predator odors as repellents for kiore (*Rattus exulans*) and ship rats (*R. rattus*). *J. Chem. Ecol.* **27**, 1029–1047.

Braun, K., Kremz, P., Wetzel, W., Wagner, T., and Poeggel, G. (2003). Influence of parental deprivation on the behavioral development in *Octodon degus*: modulation by maternal vocalizations. *Dev. Psychobiol.* **42**, 237–245.

Breker-Klassen, M.M., Yoo, D., Mittal, S.K., Sorden, S.D., Haines, D.M., and Babiuk, L.A. (1995). Recombinant type 5 adenoviruses expressing bovine parainfluenza virus type 3 glycoproteins protect *Sigmodon hispidus* cotton rats from bovine parainfluenza virus type 3 infection. *J. Virol.* **69**, 4308–4315.

Brogden, K.A., Cutlip, R.C., and Lehmkuhl, H.D. (1993). Cilia-associated respiratory bacillus in wild rats in central Iowa. *J. Wildl. Dis.* **29**, 123–126.

Brooks, J.E. and Htun, P.T. (1980). Early post-natal growth and behavioral development in the Burmese house rat, *Rattus exulans. J. Zool.* **190**, 125–136.

Brown, R.J., Carney, W.P., Van Peenen, P.F., Cross, J.H., and Saroso, J.S. (1975). Capillariasis in wild rats of Indonesia. *Southeast Asian J. Trop. Med. Public Health* **6**, 219–222.

Burcham, T.S., Sim, I., Bolin, L.M., and Shooter, E.M. (1991). The NGF complex from the African rat *Mastomys natalensis. Neurochem. Res.* **16**, 603–612.

Burkot, T.R., Clover, J.R., Happ, C.M., DeBess, E., and Maupin, G.O. (1999). Isolation of *Borrelia burgdorferi* from *Neotoma fuscipes, Peromyscus maniculatus, Peromyscus boylii*, and *Ixodes pacificus* in Oregon. *Am. J. Trop. Med. Hyg.* **60**, 453–457.

Burwash, M.D., Tobin, M.E., Woolhouse, A.D., and Sullivan, T.P. (1998). Laboratory evaluation of predator odors for eliciting an avoidance response in roof rats (*Rattus rattus*). *J. Chem. Ecol.* **24**, 49–66.

Butterworth, B.B. (1961). The breeding of *Dipodomys deserti* in the laboratory. *J. Mamm.* **42**, 413–414.

Cabrera, B.D. (1977). Studies on Paragonimus and paragonimiasis in the Philippines, II: prevalence of pulmonary paragonimiasis in wild rats, *Rattus norvegicus* in Jaro, Leyte, Philippines. *Int. J. Zoonoses* **4**, 49–55.

Cadillac, J.M., Rush, H.G., and Sigler, R.E. (2003). Polycystic and chronic kidney disease in a young degu (*Octodon degus*). *Contemp. Topics Lab. Anim. Sci.* **42**, 43–45.

Callard, M.D., Bursten, S.N., and Price, E.O. (2000). Repetitive backflipping behaviour in captive roof rats (*Rattus rattus*) and the effects of cage environment. *Anim. Welfare* **9**, 139–152.

Cambie, G., Landau, I., and Chabaud, A.G. (1990). Timing niches of three species of *Plasmodium* coexisting in a rodent in Central Africa. *Comp. Rendus l'Academie des Sci. Series III, Sciences de la Vie* **310**, 183–188.

Capeheart, J.R. and Burns, J.M. (1976). Effects of temperature, photoperiod, and thiourea on the thyroid of the woodrat (*Neotoma micropus*). *J. Mamm.* **57**, 567–569.

Castleberry, S.B., King, T.L., Wood, P.B., and Ford, W.M. (2000). Microsatellite DNA markers for the study of Allegheny woodrat (*Neotoma magister*) populations and cross-species amplification in the genus *Neotoma*. *Mol. Ecol* **9**, 824–826.

Cavagna, P., Stone, G., and Stanyon, R. (2002). Black rat (*Rattus rattus*) genomic variability characterized by chromosome painting. *Mamm. Genome* **13**, 157–163.

Cerasi, E., Kaiser, N., and Gross, D.J. (1997). From sand rats to diabetic patients: is non-insulin diabetes mellitus a disease of the beta cell? *Diabetes Metab.* **23**, 47–51.

Ceruti, R., Ghisleni, G., Ferretti, E., Cammarata, S., Sonzogni, O., and Scanziani, E. (2002). Wild rats as monitors of environmental lead contamination in the urban area of Milan, Italy. *Environ. Pollution* **117**, 255–259.

Civitelli, M.V., Consentino, P., and Capanna, E. (1989). Inter- and intra-individual chromosome variability in *Thamnomys (Grammomys) gazellae* (Rodentia, Muridae) B-chromosomes and structural heteromorphisms. *Genetica* **79**, 93–105.

Clark, B.R. and Price, E.O. (1981). Sexual maturation and fecundity of wild and domestic Norway rats (*Rattus norvegicus*). *J. Reprod. Fertil.* **63**, 215–220.

Clark, J.D. (1984). Biology and diseases of other rodents. *In* "Laboratory Animal Medicine," 1st Ed. (J.G. Fox, B.J. Cohen, and F.M. Loew, eds.), pp 183–206, Academic Press, San Diego.

Clark, J.W. (1975). Androgen control of the ventral scent gland in *Neotoma floridana*. *J. Endocrinol.* **64**, 393–394.

Clark, K.L., Oliver, Jr., J.H., Grego, J.M., James, A.M., Durden, L.A., and Banks, C.W. (2001). Host associations of ticks parasitizing rodents at *Borrelia bergdorferi* enzootic sites in South Carolina. *J. Parasitol.* **87**, 1379–1386.

Cohen, M.E. and Meyer, D.M. (1993). Effect of dietary vitamin E supplementation and rotational stress on alveolar bone loss in rice rats. *Arch. Oral Biol.* **38**, 601–606.

Collier, G.R., Collier, F.M., Sanigorski, A., Walder, K., Cameron-Smith, D., and Sinclair, A.J. (1997). Non-insulin dependent diabetes mellitus in *Psammomys obesus* is independent of changes in tissue fatty acid composition. *Lipids* **32**, 317–322.

Collier, G.R., De Silva, A., Sanigorski, A., and Zimmet, P. (1997). Development of obesity and insulin resistance in the Israeli sand rat (*Psammomys obesus*): does leptin play a role? *Ann. NY Acad. Sci.* **827**, 50–63.

Collier, G.R., Walder, K., Lewandowski, P., Sanigorski, A., and Zimmet, P. (1997). Leptin and the development of obesity and diabetes in *Psammomys obesus*. *Obesity Res.* **5**, 455–458.

Cramer, D.V., Chakravarti, A., Arenas, O., Humprieres, J., and Mowery, P.A. (1988). Genetic diversity within and between natural populations of *Rattus norvegicus*. *J. Hered.* **79**, 319–324.

Cramer, D.V., Davis, B.K., Shonnard, J.W., Stark, O., and Gill III, T.J. (1978). Phenotypes of the major histocompatibility complex in wild rats of different geographic origins. *J. Immunol.* **120**, 179–187.

Curtis, S.J., Ottolini, M.G., Porter, D.D., and Prince, G.A. (2002). Age-dependent replication of respiratory syncytial virus in the cotton rat. *Exp. Biol. Med.* **227**, 799–802.

Czernobilsky, B., Ungar, H., and Adler, J.H. (1982). Spontaneous uterine neoplasms in the fat sand rat (*Psammomys obesus*). *Lab. Anim.* **16**, 285–289.

Daly, M.M., Wilson, I., and Behrends, P. (1984). Breeding of captive kangaroo rats, *Dipodomys merriami* and *D. microps*. *J. Mamm.* **65**, 338–341.

Decat, G. and Leonard, A. (1985). Chromosome analysis of the lymphocytes of wild rats living close to waterways contaminated with effluents from nuclear plants. *Comp. Rendus Sci. Societe de Biologie Ses Filiales* **179**, 480–486.

Delfrano, A., Clara, M., Tome, L., Achaval, F., Levis, S., Calderon, G., Enria, D., Lozano, M., Russi, J., and Arbiza, J. (2003). Yellow pigmy rice rat (*Oligoryzomys flavescens*)and hantavirus pulmonary syndrome in Uruguay. *Emerg Infect Dis.* **9**, 846–852.

DeNatale, C.E., Burkot, T.R., Schneider, B.S., and Zeidner, N.S. (2002). Novel potential reservoirs for *Borrelia sp.* and the agent of human granulocytic ehrlichiosis in Colorado. *J. Wildl. Dis.* **38**, 478–482.

Donnelly T.M. and Quimby, F.W. (2002). *In* "Laboratory Animal Medicine," 2nd Ed. (J.G. Fox, L.C. Anderson, F.M. Loew, and F.W. Quimby, eds.), pp 247–307, Academic Press, San Diego.

Downs, C.T. and Perrin, M.R. (1995). The thermal biology of the white-tailed rat *Mystromys albicaudatus,* a cricetine relic in southern African grassland. *Comp. Biochem. Physiol. A Comp. Physiol. A* **110**, 65–69.

Downs, W.G., Spence, L., and Aitken, T.H (1962). Studies on the virus of Venezuelan equine encephalomyelitis in Trinidad, W.I. III: reisolation of virus. *Am. J. Trop. Med. Hyg.* **11**, 841–843.

Eckl, P.M. and Riegler, D. (1997). Levels of chromosomal damage in hepatocytes of wild rats living within the area of a waste disposal plant. *Sci. Total Environ.* **196**, 141–149.

Edmonds, K.E., Riggs L., and Stetson, M.H. (2003). Food availability and photoperiod affect reproductive development and maintenance in the marsh rice rat (*Oryzomys palustris*). *Physiol. Behav.* **78**, 41–49.

Edmonds, K.E. and Stetson, M.H. (1993). Effect of photoperiod on gonadal maintenance and development in the marsh rice rat (*Oryzomys palustris*). *Gen. Comp. Endocrinol.* **92**, 281–291.

Edmonds, K.E. and Stetson, M.H. (1994). Photoperiod and melatonin affect testicular growth in the marsh rice rat (*Oryzomys palustris*). *J. Pineal Res.* **17**, 86–93.

Edmonds, K.E. and Stetson, M.H. (1995) Effects of photoperiod on pineal melatonin in the marsh rice rat (*Oryzomys palustris*). *J. Pineal Res.* **18**, 148–153.

Eilam, D. and Golani, I. (1988). The ontogeny of exploratory behavior in the house rat (*Rattus rattus*): the mobility gradient. *Dev. Psychobiol.* **21**, 679–710.

Eisenberg, J.F. and Isaac, D.E. (1963). The reproduction of heteromyid rodents in captivity. *J. Mamm.* **44**, 61–67.

Emura, S., Tamada, A., Hayakawa, D., Chen, H., and Shoumura, S. (1999). Electron microscopic study of the parathyroid gland of *Rattus rattus*. *Okajimas Folia Anat. Jpn.* **76**, 71–80.

Faith, R.E., Montgomery, C.A., Durfee, W.J., Aguilar-Cordova, E., and Wyde, P.R. (1997). The cotton rat in biomedical research. *Lab. Anim. Sci.* 47, 337–345.

Fan, C.K., Su, K.E., Chung, W.C., Tsai, Y.J., Chiol, H.Y., Lin, C.F., Su, C.T., and Chao, P.H. (1998). Seroepidemiology of *Toxoplasma gondii* infection among Atayal aborigines and local animals in Nan-ao district, Ilan county and Jen-ai district, Nan-tou county, Taiwan. *Kaohsiung J. Med. Sci.* **14**, 762–769.

Feinmesser, R., Ungar, H., and Adler, J. (1988). Otic cholesteatoma in the sand rat (*Psammomys obesus*). *Am. J. Otol.* **9**, 409–411.

Fieldhouse, D., Yazdani, F., and Golding, G.B. (1997). Substitution rate variation in closely related rodent species. *Heredity* **79**, 21–31.

Fine, J., Quimby, F.W., and Greenhouse, D.D. (1986). Annotated bibliography on uncommonly used laboratory animals: mammals. *ILAR News* **24**, 3–38.

Fisher, M.C. and Viney, M.E. (1998). The population genetic structure of the facultatively sexual parasitic nematode *Stongyloides ratti* in wild rats. *Proc R. Soc Lond. B Biol. Sci.* **265**, 703–709.

Fleming, A.S. and Tambosso, L. (1980). Hormonal and sensory control of scent-marking in the desert woodrat (*Neotoma lepida lepida*). *J Comp. Physiol. Psychol.* **94**, 564–578.

Fouchecourt, M.O. and Riviere, J.L. (1996). Activities of liver and lung cytochrome P450-dependent monooxygenases and antioxidant enzymes in laboratory and wild Norway rats exposed to reference and contaminated soils. *Arch. Environ. Contam. Toxicol.* **30**, 513–522.

Franke, E.D., McGreevy, P.B., Katz, S.P., and Sacks, D.L. (1985). Growth cycle-dependent generation of complement-resistant *Leishmania promastigotes*. *J. Immunol.* **134**, 2713–2718.

Frenkel, G., Shaham, Y., and Kraicer, P.F. (1972). Establishment of conditions for colony-breeding of the sand rat *Psammomys obesus*. *Lab. Anim. Sci.* **22**, 40–47.

Fujimura, H., Ogura, A., Asano, T., Noguchi, Y., Mochida, K., and Takimoto, K. (1996). Lysosomal glycolipid storage in the renal tubular epithelium in mastomys (*Praomys coucha*). *Histol. Histopathol.* **11**, 171–174.

Fulhorst, C.F., Ksiazek, T.G., Peters, C.J., and Tesh, R.B. (1999). Experimental infection of the can mouse *Zygodontomys brevicauda* (family Muridae) with Guanarito virus (Arenaviridae), the etiologic agent of Venezuelan hemorrhagic fever. *J. Infect. Dis.* **180**, 966–969.

Fulhorst, C.F., Milazzo, M.L., Carroll, D.S., Charrel, R.N., and Bradley, R.D. (2002). Natural host relationships and genetic diversity of Whitewater Arroyo virus in southern Texas. *Am. J. Trop. Med. Hyg.* **67**, 114–118.

Gallardo, P., Olea, N., and Sepulveda, F.V. (2002). Distribution of aquaporins in the colon of *Octodon degus*, a South American desert rodent. *Am. J. Physiol.* **283**, 779–788.

Gander, P.H. and Lewis, R.D. (1983). Phase-resetting action of light on the circadian activity rhythm of *Rattus exulans*. *Am. J. Physiol.* **245**, R10–R17.

Gaule, C., Vega, P., Sanchez, E., and Del Villar, E. (1990). Drug metabolism in *Octodon degus*: low inductive effect of phenobarbital. *Comp. Biochem. Physiol.* **96**, 217–222.

Gill, J.E., and Redfern, R. (1979). Laboratory test of seven rodenticides for the control of *Mastomys natalensis*. *J. Hyg.* **83**, 345–352.

Gilligan, C.J., Lawton, G.P., Tang, L.H., West, A.B., and Modlin, I.M. (1995). Gastric carcinoid tumors: the biology and therapy of an enigmatic and controversial lesion. *Am. J. Gastroenterol.* **90**, 338–352.

Glass, G.E., Childs, J.E., Korch, G.W., and LeDuc, J.W. (1988). Association of intraspecific wounding with hantaviral infection in wild rats (*Rattus norvegicus*). *Epidemiol. Infect.* **101**, 459–472.

Goel, N. and Lee, T.M. (1995). Sex differences and effects of social cues on daily rhythms following phase advances in *Octodon degus*. *Physiol. Behav.* **58**, 205–213.

Goel, N. and Lee, T.M. (1995). Social cues accelerate reentrainment of circadian rhythms in diurnal female *Octodon degus* (Rodentia-Octodontidae) *Chronobiol. Int.* **12**, 311–323.

Goel, N. and Lee, T.M. (1996). Relationship of circadian activity and social behaviors to reentrainment rates in diurnal *Octodon degus* (Rodentia). *Physiol. Behav.* **59**, 817–826.

Gotcher, J.E. and Jee, W.S. (1981). The progression of the peridontal syndrome in the rice rat, I: morphometric and autoradiographic studies. *J. Periodont. Res.* **16**, 275–291.

Gregson, R.L., Davey, M.J., and Prentice, D.E. (1979). Bronchus-associated lymphoid tissue (BALT) in the laboratory-bred and wild rat, *Rattus norvegicus*. *Lab. Anim.* **13**, 239–243.

Gruber, H.E., Gordon, B.E., Williams, C., James, N.H., and Hanley, E.N. (2003). Bone mineral density of lumbar vertebral end plates in the aging male sand rat spine. *Spine* **16**, 1766–1772.

Gruber, H.E., Johnson, T.L., Kinsella, J.M., Greiner, E.C., and Gordon, B.E. (2001). Tapeworm identification in the fat sand rat (*Psammomys obesus obesus*). *Contemp. Topics Lab. Anim. Sci.* **40**, 22–24.

Gruber, H.E., Johnson, T.L., Leslie, K., Ingran J.A., Martin, D., Hoelscher, G., Banks, D., Phieffer, L., Coldham, G., and Hanley, E.N. (2002). Autologous intervertebral disc cell implantation: a model using *Psammomys obesus*, the sand rat. *Spine* **27**, 1626–1633.

Gulati, A., Srimal, R.C., and Dhawan, B.N. (1988). An analysis of stereotyped behaviour in *Mastomys natalensis*. *Naunyn-Schmiedebergs Arch. Pharmacol.* **337**, 572–575.

Haag, A., Wayss, K., Rommelaere, J., and Cornelis J.J. (2000). Experimentally induced infection with autonomous parvoviruses, minute virus of mice and H-1, in the African multimammate mouse (*Mastomys coucha*). *Comp. Med.* **50**, 613–621.

Hall, A.D., Persing, R.L., White, D.C., and Rickets, R.T. Jr. (1967). *Mystromys albicaudatus* (the African white tailed rat) as a laboratory species. *Lab. Anim. Care* **17**, 180–188.

Hall, E.R. (1981). "The Mammals of North America," 2nd Ed. Wiley, New York.

Hallett, A.F. and Meester, J. (1971). Early postnatal development of the South African hamster. *Zool. Afr.* **6**, 221–228.

Hamilton, Jr., W.J. (1953). Reproduction and young of the Florida wood rat *Neotoma f. floridana* (Ord). *J. Mammal.* **34**, 180–189.

Hammer, Jr., R.P., Hori, K.M., Blanchard, R.J., and Blanchard, D.C. (1992). Domestication alters 5-HT1A receptor binding in rat brain. *Pharmacol. Biochem. Behav.* **42**, 25–28.

Harriman, A.E. (1978). Concordance of taste preferences by *Neotoma micropus* with those of other rodents in a grassland community. *Percept. Motor Skills* **46**, 703–708.

Heeley, R.P., Gill, E., van Zutphen, B., Kenyon, C.J., and Sutcliffe, R.G. (1998). Polymorphisms of the glucocorticoid receptor gene in laboratory and wild rats: steroid binding properties of trinucleotide CAG repeat length variants. *Mamm. Genome* **9**, 198–203.

Heffner, H.E. and Heffner, R.S. (1985). Hearing in two cricetid rodents: wood rat (*Neotoma floridana*) and grasshopper mouse (*Onychomys leucogaster*). *J. Comp. Psychol.* **99**, 275–288.

Heller, R., Riegal, P., Hansmann, Y., Delacour, G., Bermond, D., Dehio, C., Lamarque, F., Monteil, H., Chomel, B., and Piemont, Y. (1998). *Bartonella tribocorum* sp. nov., a new Bartonella species isolated from the blood of wild rats. *Int. J. Syst. Bacteriol.* **48**, 1339–1339.

Helm, J.D. and Dalby, P.L. (1975). Reproductive biology and postnatal development of the neotropical climbing rat, *Tylomys*. *Lab. Anim. Sci.* **25**, 741–747.

Hewitt, J.K., Fulker, D.W., and Broadhurst, P.L. (1981). Genetics of escape-avoidance conditioning in laboratory and wild populations of rats: a biometrical approach. *Behav. Genet.* **11**, 533–544.

Hills, B.A. and Butler, B.D. (1978). The kangaroo rat as a model for type 1 decompression sickness. *Undersea Biomed. Res.* **5**, 309–321.

Hisatomi, H., Miura, R., Shiota, K., Ogawa, T., Sayama, K., Tanaka, S., Matsuzawa, A., and Kanoh, Y. (1994). Phylogenetic relationships of mastomys to mouse and rat deduced from satellite DNA sequences. *Jikken Dobutsu Exp. Anim.* **43**, 403–408.

Holmes, E., Bonner, F.W., and Nicholson, J.K. (1996). Comparative biochemical effects of low doses of mercury II chloride in the F344 rat and the multimammate mouse (*Mastomys natalensis*). *Comp. Biochem. Physiol. C* **114**, 7–15.

Houard, S., Jacquet, A., Haumont, M., Daminet, V., Milican, F., Glineur, F., and Bollen, A. (2000). Cloning expression and purification of recombinant cotton rat interleukin-5. *Gene* **257**, 149–155.

Howell, A.B. (1926). "Anatomy of the Wood Rat." Williams & Wilkins, Baltimore.

Hutchinson, K.L., Rollin, P.E., and Peters, C.J. (1998). Pathogenesis of a North American hantavirus, Black Creek Canal virus, in experimentally infected *Sigmodon hispidus*. *Am. J. Trop. Med. Hyg.* **59,** 58–65.

Ibrahim, I.N., Okabayashi, T., Ristiyanto, Lestari, E.W., Yanase, T., Muramatsu, Y., Ueno, H., and Morita, C. (1999). Serosurvey of wild rodents for Rickettsioses (spotted fever, murine typhus, and Q fever) in Java Island, Indonesia. *Eur. J. Epidemiol.* **15,** 89–93.

Ibrahim, I.N., Sudomo, M., Morita, C., Uemura, S., Muramatsu, Y, Ueno, H., and Kitamura, T. (1996). Seroepidemiological survey of wild rats for Seoul virus in Indonesia. *Jpn. J. Med. Sci. Biol.* **49,** 69–74.

Imbert-Establet, D. (1982). Natural infestation of wild rats by *Schistosoma mansoni* in Guadeloupe: quantitative data on the development and fertility of the parasite. *Ann. Parasitol. Hum. Comp.* **57,** 573–585.

Inoue, S., Tanikawa, T., Kawaguchi, J., Iida, T., and Morita, C. (1992). Prevalence of Listeria (spp.) in wild rats captured in the Kanto area of Japan. *J. Vet. Med. Sci.* **54,** 461–463.

Itoh, K., Tamura, H., and Mitsuoka, T. (1989). Gastrointestinal flora of cotton rats. *Lab. Anim.* **23,** 62–65.

Jacobs, G.H., Calderone, J.B., Fenwick, J.A., Krogh, K., and Williams, G.A. (2003). Visual adaptations in a diurnal rodent, *Octodon degus. J. Comp. Physiol.* **189,** 347–361.

Jacobs, L.F. and Spencer, W.D. (1994). Natural space use patterns and hippocampal size in kangaroo rats. *Brain Behav. Evol.* **44,** 125–132.

Jantos, C., Altmannsberger, M., Weidner, W., and Schiefer, H.G. (1990). Acute and chronic bacterial prostatitis due to *E. coli*: description of an animal model. *Urol. Res.* **18,** 207–211.

Jechura, T.J., Walsh, J.M., and Lee, T.M. (2003). Testosterone suppress circadian responsiveness to social cues in the diurnal rodent *Octodon degus. J. Biol. Rhythms* **18,** 43–50.

Jollie, W.P. (1956). Rearing the pallid Ord kangaroo rat in the laboratory. In "Symposium on Ecology of Disease Transmission in Native Animals." pp 54–56, Army Chemical Corps, Dugway, UT.

Jonkers, A.H., Spence, L., Downs, W.G., and Worth, C.B. (1964). Laboratory studies with wild rodents and viruses native to Trinidad, II: studies with the Trindad Caraparu-like agent TRVL 34053-1. *Am. J. Trop. Med. Hyg.* **13,** 728–733.

Kabrane-Lazizi, Y., Fine, J.B., Elm, J., Glass, G.E., Higa, H., Diwan, A., Gibbs, Jr., C.J., Meng, X.J., Emerson, S.U., and Purcell, R.H. (1999). Evidence for widespread infection of wild rats with hepatitis E virus in the United States. *Am. J. Trop. Med. Hyg.* **61,** 331–335.

Kalman, R., Ziv, E., Shafrir, E., Bar-On, H., and Perez, R. (2001). *Psammomys obesus* and the albino rat: two different models of nutritional insulin resistance, representing two different types of human populations. *Lab. Anim.* **35,** 346–352.

Kaltwasser, M.T. (1990). Acoustic signaling in the black rat (*Rattus rattus*). *J. Comp. Psychol.* **104,** 227–232.

Kapu, M.M., Akanya, H.O., Ega, R.A., Olofu, E.O., Balarabe, M.L., Chafe, U.M., and Schaeffer, D.J. (1991). Concentrations of trace and other elements in the organs of wild rats and birds from the northern Guinea savanna of Nigeria. *Bull. Environ. Contam. Toxicol.* **46,** 79–83.

Kas, M.J. and Edgar, D.M. (1999). Circadian timed wakefullness at dawn opposes compensatory sleep responses after sleep deprivation in *Octodon degus. Sleep* **22,** 1045–1053.

Katahira, K. and Ohwada, K. (1993). Hematological standard values in the cotton rat (*Sigmodon hispidus*). *Jikken Dobutsu Exp. Anim.* **42,** 653–656.

Kawase, S. and Ishikura, H. (1995). Female-predominant occurrence of spontaneous gastric adenocarcinoma in cotton rats. *Lab. Anim. Sci.* **45,** 244–248.

Kemble, E.D., Blanchard, D.C., and Blanchard, R.J. (1990). Effects of regional amygdaloid lesions on flight and defensive behaviors of wild black rats (*Rattus rattus*). *Physiol. Behav.* **48,** 1–5.

Kertschanska, S., Schroder, H., and Kaufmann, P. (1997). The ultrastructure of the trophoblstic layer of the degu (*Octodon degus*) placenta: a re-evaluation of the channel problem. *Placenta* **18,** 219–225.

Kidd, M., Tang, L.H., Modlin, I.M., Zhang, T., Chin, K., Holt, P.R., and Moss, S.F. (2000). Gastrin-mediated alterations in gastric epithelial apoptosis and proliferation in a mastomys rodent model of gastric neoplasia. *Digestion* **62,** 143–151.

Kim, S., Stair, E.L., Lochmiller, R.L., Lish, J.W., and Qualls, Jr., C.W. (2001). Evaluation of myelotoxicity in cotton rats (*Sigmodon hispidus*) exposed to environmental contaminants, I: *in vitro* bone-marrow progenitor culture. *J. Toxicol. Environ. Health A* **62,** 83–96.

Kloting, I. and Kovacs, P. (1998). Crosses between diabetic BB/OK and wild rats confirm that a third gene is essential for diabetes development. *Acta Diabet.* **35,** 109–111.

Kloting, I., Kovacs, P., and van den Brandt, J. (2001). Quantitative trait loci for body weight, blood pressure, blood glucose, and serum lipids: linkage analysis with wild rats (*Rattus norvegicus*). *Biochem. Biophys. Res. Comm.* **284,** 1126–1133.

Kloting, I., Voigt, B., and Kovacs, P. (1997). Comparison of genetic variability at microsatellite loci in wild rats and inbred strains (*Rattus norvegicus*). *Mamm. Genome* **8,** 589–591.

Kosoy, M.Y., Elliott, L.H., Ksiazek, T.G., Fulhorst, C.F., Rollin, P.E., Childs, J.E., Mills, J.N., Maupin, G.O., and Peters, C.J. (1996). Prevalence of antibodies to arenaviruses in rodents from the southern and western United States: evidence for an arenavirus associated with the genus Neotoma. *Am. J. Trop. Med. Hyg.* **54,** 570–576.

Kosoy, M.Y., Saito, E.K., Green, D., Marston, E.L., Jones, D.C., and Childs, J.E. (2000). Experimental evidence of host specificity of Bartonella infection in rodents. *Comp. Immunol. Microbiol. Infect. Dis.* **23,** 221–238.

Kumagai, S., Sugita-Konishi, Y., Hara-Kudo, Y., Ito, Y., Noguchi, Y., Yamamoto, Y., and Ogura, A. (1998). The fate and acute toxicity of aflatoxin B1 in the *Mastomys* and rat. *Toxiconcology* **36,** 179–188.

Lamb, J.G., Sorensen, J.S., and Dearing, M.D. (2001). Comparison of detoxification enzyme mRNAs in woodrats (*Neotoma lepida*) and laboratory rats. *J. Chem. Ecol.* **27,** 845–857.

Lane, R.S. and Brown, R.N. (1991). Wood rats and kangaroo rats: potential reservoirs of the Lyme disease spirochete in California. *Jf. Med. Entomol.* **28,** 299–302.

Langley, R.J., Prince, G.A., and Ginsberg, H.S. (1998). HIV type-1 infection of the cotton rat (*Sigmodon fulviventer* and *S. hispidus*). *Proc. Natl. Acad. Sci. USA* **95,** 14355–14360.

Larson, R.H. and Fitzgerald, R.J. (1968). Caries development in the African white-tailed rat (*Mystromys albicaudatus*) infected with streptococcus of human origin. *J. Dent. Res.* **47,** 746–749.

Lee, H.W., Baek, L.J., and Johnson, K.M. (1982). Isolation of Hantaan virus: the etiologic agent of Korean hemorrhagic fever, from wild urban rats. *J. Infect. Dis.* **146,** 638–644.

Lee, T.M. (2004). *Octodon degus*: A diurnal, social, and long-lived rodent. *ILAR J.* **45,** 14–24.

Leibowitz, G., Yuli, M., Donath, M.Y., Nesher, R., Melloul, D., Cerasi, E., Gross, D.J., and Kaiser, N. (2001). Beta-cell glucitoxicity in the *Psammomys obesus* model of type 2 diabetes. *Diabetes* **50,** 113–117.

Letelier, M.E., Del Villar, E., and Sanchez, E. (1984). Enhanced metabolism of morphine in *Octodon degus* compared to Wistar rats. *Gen. Pharmacol.* **15,** 403–406.

Letelier, M.E., Del Villar, E., and Sanchez, E. (1985). Drug tolerance and detoxicating enzymes in *Octodon degus* and Wistar rats: a comparative study. *Comp. Biochem. Physiol.* **80,** 195–198.

Levin, M., Levine, J.F., Apperson, C.S., Norris, D.E., and Howard, P.B. (1995). Reservoir competence of the rice rat (Rodentia: Cricetidae) for *Borrelia burgdoferi. J. Med. Entomol.* **32,** 138–142.

Lewandowski, G., Zimmerman, M.N., Denk, L.L., Porter, D.D., and Prince, G.A. (2002). Herpes simplex type 1 infects and establishes latency in the brain and trigeminal ganglia during primary infection of the lip in cotton rats and mice. *Arch. Virol.* **147**, 167–179.

Little, R.R., Parker, K.M., England, J.D., and Goldstein, D.E. (1982). Glycosylated hemoglobin in *Mystromys albicaudatus*: A diabetic animal model. *Lab. Anim. Sci.* **32**, 44–47.

Lochmiller, R.L., Sinclair, J.A., and Rafferty, D.P. (1998). Tumoricidal activity of lymphokine-activated killer cells during acute protein restriction in the cotton rat (*Sigmodon hispidus*). *Comp. Biochem. Physiol. C* **119**, 149–155.

Lucion, A.B., De-Almeida, R.M., and Da-Silva, R.S. (1996). Territorial aggression, body weight, carbohydrate metabolism and testosterone levels of wild rats maintained in laboratory colonies. *Brazil J. Med. Biol. Res.* **29**, 1657–1662.

Lucion, A.B., De-Almeida, R.M., and De-Marques, A.A. (1994). Influence of the mother on development of aggressive behavior in male rats. *Physiol. Behav.* **55**, 685–689.

MacKenzie, W.F., Magill, L.S., and Hulse, M. (1981). A filamentous bacterium associated with respiratory disease in wild rats. *Vet. Pathol.* **18**, 836–839.

Madarame, H., Kashimoto, Y., Kawamoto, T., Toyonaga, S., and Hasegawa, Y. (1995). Spontaneous rhabdomyosarcomas in aged Mastomys (*Praomys coucha*). *Lab. Anim.* **29**, 464–469.

Mahida, H. and Perrin, M.R. (1994). The effect of different diets on the amount of organic acid produced in the digestive tract of *Mystromys albicaudatus*. *Acta Theriol.* **39**, 21–27.

Marquie, G., Duhault, J., and Jacotot, B. (1984). Diabetes mellitus in sand rats (*Psammomys obesus*): metabolic pattern during development of the diabetic syndrome. *Diabetes* **33**, 438–443.

Matisoo-Smith, E., Roberts, R.M., Irwin, G.J., Allen, J.S., Penny, D., and Lambert, D.M. (1998). Patterns of prehistoric human mobility in polynesia indicated by mtDNA from the Pacific rat. *Proc. Natl. Acad. Sci. USA* **95**, 15145–15150.

Matocq, M.D. (2002). Phylogeographical structure and regional history of the dusky-footed woodrat, *Neotoma fuscipes*. *Mol. Ecol.* **11**, 229–242.

McCoy, J., Jori, F., and Sterm, C. (1997). Tranquilization of cane rats (*Thryonomys swinderianus*) with depot neuroleptic (pipothiazine palmitate). *J. Vet. Pharmacol. Therap.* **20**, 233–239.

McKinney, L.A. and Hendricks, L.D. (1980). Experimental infection of *Mystromys albicaudatus* with Leishmania braziliensis: pathology. *Am. J. Trop. Med. Hyg.* **29**, 753–760.

McMurry, S.T., Lochmiller, R.L., Chandra, S.A.M., and Qualls, Jr., C.W. (1995). Sensitivity of selected immunological, hematological, and reproductive parameters in the cotton rat (*Sigmodon hispidus*) to subchronic lead exposure. *J. Wildl. Dis.* **31**, 193–204.

Milazzo, M.L., Eyzaguirre, E.J., Molina, C.P., and Fulhorst, C.F. (2002). Maporal viral infection in the Syrian golden hamster: a model of hantavirus pulmonary syndrome. *J. Infect. Dis.* **186**, 1390–1395.

Mitchell, D. (1976). Experiments on neophobia in wild and laboratory rats: a reevaluation. *J. Comp. Physiol. Psychol.* **90**, 190–197.

Moscarella, R.A. and Aguliera, M. (1999). Growth and reproduction of *Oryzomys albigularis* (Rodentia: Sigmondontinae) under laboratory conditions. *Mammalia* **63**, 349–362.

Moskowitz, R.W., Ziv, I., Denko, C.W., Boja, B., Jones, P.K., and Adler, J.H. (1990). Spondylosis in sand rats: a model of intervertebral disc degeneration and hyperostosis. *J. Orthop. Res.* **8**, 401–411.

Mouw, D., Kalitis, K., Anver, M., Schwartz, J., Constan, A., Hartung, R., Cohen, B., and Ringler, D. (1975). Lead: possible toxicity in urban vs. rural rats. *Arch. Environ. Health* **30**, 276–280.

Mukhtar, M.M., Sharief, A.H., el Saffi, S.H., Harith, A.E., Higazzi, T.B., Adam, A.M., and Abdalla, H.S. (2000). Detection of antibodies to *Leishmania donovani* in animals in a kala-azar endemic region in eastern Sudan: a preliminary report. *Trans. R. Soc. Trop. Med. Hyg.* **94**, 33–36.

Murua, R., Gonzalez, L.A., and Lima, M. (2003). Population dynamics of rice rats (a hantavirus reservoir) in southern Chile: feedback structure and non-linear effects of climatic oscillations. *OIKOS* **102**, 137–145.

Muths, E. and Reichman, O.J. (1996). Kangaroo rat bone compared to white rat bone after short-term disuse and exercise. *Comp. Biochem. Physiol. A. Physiol.* **114**, 355–361.

Nagabuchi, M., Kawamoto, Y., Nishikawa, T., and Nishimura, M. (1993). Polymorphism of transferring found in the laboratory rat and wild rats in Japan. *Biochem. Genet.* **31**, 147–154.

Nagasawa, H., Koshimizu, U., Watanabe, M., Tokuda, K., Sumita, H., and Kano, Y. (1989). Mammary gland growth and response to hormones in *Mastomys* compared with mice. *Lab. Anim. Sci.* **39**, 313–317.

National Research Council. (1996). "Guide for the Care and Use of Laboratory Animals." National Academy Press, Washington, DC.

Nesher, R., Gross, D.J., Donath, M.Y., Cesari, E., and Kaiser, N. (1999). Interaction between genetic and dietary factors determines beta-cell function in *Psammomys obesus*: an animal model of type 2 diabetes. *Diabetes* **48**, 731–737.

Niewiesk, S. (1999). Cotton rats (*Sigmodon hispidus*): an animal model to study the pathogenesis of measles virus infection. *Immunol. Lett.* **65**, 47–50.

Niewiesk, S., Voelp, F., and Meulen, V. (1997). A maintenance and handling device for cotton rats (*Sigmodon hispidus*). *Lab. Anim.* **26**, 32–33.

Nikulina, E.M. (1991). The effect of the S1A-receptor agonist ipsapirone on behavior types in wild and domesticated rats. *Zh. Vysshei Nerv. Deiatelnosti Imeni. I.P. Pavlova* **41**, 1149–1153.

Nikulina, E.M., Borodin, P.M., and Popova, N.K. (1986). Change in certain forms of aggressive behavior and monamine content in the brain during selection of wild rats for taming. *Neurosci. Behav. Physiol.* **16**, 466–471.

Nilsson, O., Wangberg, B., Johansson, L., Modlin, I.M., and Ahlman, H. (1992). *Praomys (Mastomys) natalensis*: a model for gastric carcinoid formation. *Yale J. Biol. Med.* **65**, 741–751.

Nitatpattana, N., Chauvancy, G., Dardaine, J., Poblap, T., Jumronsawat, K., Tangkanakul, W., Poonsuksombat, D., Yoksan, S., and Gonzalez, J.P. (2000). Serological study of hantavirus in the rodent population of Nakhon Pathom and Nakhon Ratchasima Provinces Thailand. *Southeast Asian J. Trop. Med. Pub. Health* **31**, 277–282.

Nohara, M., Hirayama, T., Ogura, A., Hiroi, M., and Araki, Y. (1998). Partial characterization of the gametes and development of a successful in vitro fertilization procedure in the mastomys (*Praomys coucha*): a new species for reproductive biology research. *Biol. Reprod.* **58**, 226–233.

Nowak, R.M. (1999). "Walker's Mammals of the World," 6th Ed. Johns Hopkins University Press, Baltimore.

Oettle, A.G. (1957). Spontaneous carcinoma of the glandular stomach in *Rattus (Mastromys) natalensis*, an African rodent. *Br. J. Cancer* **11**, 415–433.

Ohwada, K. and Katahira, K. (1993). Indirect measurement for body surface area of cotton rats. *Jikken Dobutsu. Exp. Anim.* **42**, 635–637.

Oliver, Jr., J.H., Chandler, Jr., F.W., James, A.M., Sanders, Jr., F.H., Hutcheson, H.J., Huey, L.O., McGuire, B.S., and Lane, R.S. (1995). Natural occurrence and characterization of the Lyme disease spirochete, *Borrelia burgdorferi*, in cotton rats (*Sigmodon hispidus*) from Georgia and Florida. *J. Parasitol.* **81**, 30–36.

Ostroff, R., Trimbur, M., Aldinger, K., Sadasivan, S., Kormhauser, J., and Tate, B. (2002). The degu (*Octodon degus*) models aspects of human age-related neuropathology. *Neurobiol. Aging* **23**, 412.

Pahwa, R. and Chatterjee, V.C. (1988). The toxicity of Indian *Calotropis procera* RBr latex in the black rat, *Rattus rattus Linn*. *Vet. Hum. Toxicol.* **30**, 305–308.

Palmer, R.J., Settnes, O.P., Lodal, J., and Wakefield, A.E. (2000). Population structure of rat-derived *Pneumocystis carinii* in Danish wild rats. *App. Environ. Microbiol.* **66**, 4954–4961.

Papp, Z., Middleton, D.M., Mittal, S.K., Babiuk, L.A., and Baca-Estrada, M.E. (1997). Mucosal immunization with recombinant adenoviruses: induction of immunity and protection of cotton rats against respiratory bovine herpesvirus type 1 infection. *J. Gen. Virol.* **78**, 2933–2943.

Pelissier, T., Saavedra, H., Bustamante, D., and Paeile, C. (1989). Further studies on the understanding of *Octodon degus* natural resistance to morphine: a comparative study with the Wistar rat. *Comp. Biochem. Physiol.* **92**, 319–322.

Pertusa J.G., Nesher, R., Kaiser, N., Cerasi, E., Henquin, J.C., Jonas, J.C. (2002). Increased glucose sensitivity of stimulus-secretion coupling in islets from *Psammomys obesus* after diet induction of diabetes. *Diabetes* **51**, 2552–2560.

Pfau, R.S., Van Den Bussche, R.A., and McBee, K. (2001). Population genetics of the hispid cotton rat (*Sigmodon hispidus*): patterns of genetic diversity at the major histocompatibility complex. *Mol. Ecol.* **10**, 1939–1945.

Pfau, R.S., Van Den Bussche, R.A., McBee, K., and Lochmiller, R.L. (1999). Allelic diversity at the Mhc-DQA locus in cotton rats (*Sigmodon hispidus*) and a comparison of DQA sequences within the family muridae (Mammalia: Rodentia). *Immunogen* **49**, 886–893.

Piazza, F.M., Johnson, S.A., Darnell, M.E.R., Porter, D.D., Hemming, V.G., and Prince, G.A. (1993). Bovine respiratory syncytial virus protects cotton rats against human respiratory syncytial virus infections. *J. Virol.* **67**, 1503–1510.

Plyusnina, I. and Oskina, I. (1997). Behavioral and adrenocortical responses to open-field test in rats selected for reduced aggressiveness towards humans. *Physiol. Behav.* **61**, 381–385.

Poeggel, G. and Braun, K. (1996). Early auditory filial learning in degus (*Octodon degus*): behavioral and autoradiographic studies. *Brain Res.* **743**, 162–170.

Poeggel, G., Haase, C., Gulyaeva, N., and Braun, K. (2000). Quantitative changes in reduced nicotinamide adenine dinucleotide phosphate-diaphorase-reactive neurons in the brain of *Octodon degus* after periodic maternal separation and early social isolation. *Neuroscience* **99**, 381–387.

Popova, N.K., Avgustinovich, D.F., Kolpakov, V.G., and Plyusnina, I.Z. (1998). Specific [3H]8-OH-DPAT binding in brain regions of rats genetically predisposed to various defense behavior strategies. *Pharmacol. Biochem. Behav.* **59**, 793–797.

Prince, G.A. (1994). The cotton rat in biomedical research. *AWIC Newslett.* **5**, 1–6.

Pringle, G., and King, D.F. (1968). Some developments in techniques for the study of the rodent filarial parasite, I: a preliminary comparison of the host efficiency of the multimammate rat, *Praomys (Mastromy) natalensis*, with that of the cotton rat, *Sigmodon hispidus*. *Ann. Trop. Med. Parasitol.* **62**, 462–468.

Pruthi, A.K., Kharole, M.U., Gupta, R.K.P., and Kumar, B.B. (1983). Occurrence and pathology of intracutaneous cornifying epitheliomas in *Mastomys natalensis*, the multimammate mouse. *Res. Vet. Sci.* **35**, 127–129.

Rafferty, D.P., Lochmiller, R.L., Kim, S., Qualls, C.W., Schroder, J., Basta, N., and McBee, K. (2000). Fluorosis risks to resident hispid cotton rats on land-treatment facilities for petrochemical wastes. *J. Wildl. Dis.* **36**, 636–645.

Ramos, F. and Smith, A.C. (1975). Protein differences in the lens nucleus from the desert woodrat (*Neotoma lepida*). *Psychol. Rep.* **37**, 219–222.

Raz, I., Wexler, I., Weiss, O., Flyvbjerg, A., Segev, Y., Rauchwerger, A., Raz, G., and Khamaisi, M. (2003). Role of insulin and the IGF system in renal hypertrophy in diabetic *Psammomys obesus* (sand rat). *Nephrol. Dial. Transplant.* **18**, 1293–1298.

Reichman, O.J. (1988). Caching behaviour by eastern woodrats, *Neotoma floridana*, in relation to food perishability. *Anim. Behav.* **36**, 1525–1532.

Reynolds, T.J. and Wright, J.W. (1979). Early postnatal physical and behavioural development of degus (*Octodon degus*). *Lab. Anim.* **13**, 93–99.

Robel, G.L., Lochmiller, R.L., McMurry, S.T., and Qualls, Jr., C.W. (1996). Environmental, age, and sex effects on cotton rat (*Sigmodon hispidus*) hematology. *J. Wildl. Dis.* **32**, 390–394.

Roebuck, B.D. and Longnecker, D.S. (1979). Response to two rodents, *Mastomys natalensis* and *Mystromys albicaudatus*, to the pancreatic carcinogen azaserine. *J. Natl. Cancer Inst.* **62**, 1269–1271.

Rudolph, R. and Busse, C. (1981). Electron microscopical studies on spontaneous and experimentally induced neoplasms and on normal skin of *Mastomys natalensis* (strains GRA Giessen and WSA Giessen). *Zentralbl. Veterinar. Reihe. B* **24**, 312–326.

Rudolph, R.L., Muller, H., Reinacher, M., and Thiel, W. (1981). Morphology of experimentally induced so-called keratoacanthomas and squamous cell carcinomas in two inbred lines of *Mastomys natalensis*. *J. Comp. Pathol.* **91**, 123–124.

Sakai, H.F. and Noon, B.R. (1997). Between-habitat movement of dusky-footed woodrats and vulnerability to predation. *J. Wildl. Manag.* **61**, 343–350.

Sakai, Y., Hira, Y., and Matsushima, S. (1996). Regional differences in the pineal gland of the cotton rat, *Sigmodon hispidus*: light microscopic, electron microscopic, and immunohistochemical observations. *J. Pineal Res.* **20**, 125–137.

Sayles, P.C., Hunter, K.W., Stafford, E.E., and Hendricks, L.D. (1981). Antibody response to Leishmania mexicana in African white-tailed rats (*Mystromys albicaudatus*). *J. Parasitol.* **67**, 585–586.

Schares, G. and Zahner, H. (1994). IgG subclasses of the multimammate rat, *Mastomys coucha*: isolation and characterization of IgG1 and IgG2. *J. Ex.p Anim. Sci.* **36**, 55–69.

Schmidt-Nielsen, K. (1964). "Desert Animals." Oxford University Press, London.

Schwartz, O.A. and Bleich, V.C. (1976). The development of thermo-regulation in two species of woodrats, *Neotoma lepida* and *Neotoma albigula*. *Comp. Biochem. Physiol. A Comp. Physiol.* **54**, 211–213.

Servin, J., Chacon, E., Alonso-Perez, N., and Huxley, C. (2003). New records of mammals from Durango, Mexico. *Southwest Nat.* **48**, 136–138.

Shepard, A.J., Leman, P.A., and Swanepoel, R. (1989). Viremia and antibody response of small african and laboratory animals to Crimean-Congo hemorrhagic fever virus infection. *Am. J. Trop. Med. Hyg.* **40**, 541–547.

Shklair, I.L. and Ralls, S.A. (1988). Periodontopathic micro-organisms in the rice rat (Oryzomys palustris). *Microbios* **55**, 25–31.

Shukla, R., Srimal, R.C., and Prasad, C. (1997). Cyclo (His-Pro) modulation of body temperature at hot ambient temperature in the desert rat (*Mastomys natalensis*). *Peptides* **18**, 689–693.

Sichel, J.Y., Plotnik, M., Cherny, L. Sohmer, H., and Elidan, J. (1999). Surgical anatomy of the ear of the fat sand rat. *J. Otolaryngol.* **28**, 217–222.

Singh, V.H. and Sabnis, J.H. (1980). Haematological changes during gestation in the rat, *Rattus rattus*. *Comp. Physiol. Ecol.* **5**, 211–214.

Sipos, M.P., Andersen, M.C., Whitford, W.G., and Gould, W.R. (2002). Graminivory by *Dipodomys ordii* and *Dipodomys merriami* on four species of perennial grasses. *Southwest Nat.* **47**, 276–281.

Sire, C., Durand, P., Pointier, J.P., and Theron, A. (2001). Genetic diversity of *Schistosoma mansoni* within and among individual hosts (*Rattus rattus*): infrapopulation differentiation at microspatial scale. *Int. J. Parasitol.* **31**, 1609–1616.

Snell, K.C. and Stewart, H.L. (1969). Malignant argyrophilic gastric carcinoids of *Praomys (Mastomys) natalensis*. *Science* **163**, 470.

Solleveld, H.A., Coolen, J., Haaijman, J.J., Hollander, C.F., and Zurcher, C. (1985). Autoimmune thyroiditis: spontaneous autoimmune thyroiditis in *Praomys (Mastomys) coucha*. *Am. J. Pathol.* **119**, 345–349.

Solleveld, H.A., van Zwieten, M.J., Zurcher, C., and Hollander, C.F. (1982). A histopathological survey of aged *Praomys (Mastomys) natalensis*. *J. Gerontol.* **37**, 656–665.

Stallone, J.N. and Braun, E.J. (1988). Regulation of plasma antidiuretic hormone in the dehydrated kangaroo rat (*Dipodomys spectabilis* M.). *Gen. Comp. Endocrinol.* **69**, 119–127.

Suckow, M.A., Terril-Robb, L.A., and Grigdesby, C.F. (1996). Gastric trichobezoar in a banner-tailed kangaroo rat (*Dipodomys spectabilis*). *Lab. Anim.* **30**, 383–385.

Sugano, S., Kobayashi, T., Tanikawa, T., Kawakami, Y, Kojima, H., Nakamura, K., Uchida, A., Morishima, N., and Tamai, Y. (2001). Suppression of CYP3A2 mRNA expression in the warfarin-resistant roof rat, *Rattus rattus*: possible involvement of cytochrome P450 in the warfarin resistance mechanism. *Xenobiotica* **31**, 399–407.

Suzuki, S. and Hotta, S. (1979). Antiplague antibodies against *Yersinia pestis* fraction-I antigen in serum from rodents either experimentally infected or captured in harbor areas of Japan. *Microbiol. Immunol.* **23**, 1157–1168.

Tanaka, S., Nozaki-Ukai, M., Kitoh, J., and Matsuzawa, A. (1996). Genetic regulation of border zone formation in female mastomys (*Praomys coucha*) adrenal cortex. *J. Hered.* **87**, 70–74.

Tang, L.H. and Modlin, I.M. (1996). Somatostatin receptor regulation of gastric carcinoid tumours. *Digestion* **57**, 11–14.

Thiele, H.G., Haag, F., Nolte, F., Lischke, C., and Bauschus, S. (1997). Expression profiles of RT6 and other T lymphocyte surface markers in the black rat (*Rattus rattus*). *Transplant Proc.* **29**, 1697–1698.

Torres, J., Gracenea, M., Gomez, M.S., Arrizabalaga, A., and Gonzalez-Moreno, O. (2000). The occurrence of *Cryptosporidium parvum* and *C. muris* in wild rodents and insectivores in Spain. *Vet. Parasitol.* **92**, 253–60.

Tracy, R. and Walsberg, G.E. (2002). Kangaroo rats revisited: re-evaluating a classic case of desert survival. *Oecologia* **133**, 449–457.

Tripathi, R.P., Tripathi, R., Bhaduri, A.P., Singh, S.N., Chatterjee, R.K., and Murthy, P.K. (2000). Antifilarial activity of some 2H-1-benzopyran-2-ones (coumarins). *Acta Tropica.* **76**, 101–106.

Uchikawa, R., Takagi, M., Matayoshi, S., and Sato, A. (1984). The presence of *Angiostrongylus cantonensis* in Viti Levu, Fiji. *J. Helminthol.* **58**, 231–234.

Ulmansky, M., Ungar, H., and Adler, J.H. (1984). Dental abnormalities in the aging sand rats (*Psammomys obesus*). *J. Oral. Pathol.* **13**, 366–372.

van den Brandt, J., Kovacs, P., and Kloting, I. (1999). Blood pressure, heart rate and motor activity in six inbred rat strains and wild rats (*Rattus norvegicus*): a comparative study. *Exp. Anim.* **48**, 235–240.

van den Brandt, J., Kovacs, P., and Kloting, I. (2000). Metabolic variability among disease-resistant inbred rat strains and in comparison with wild rats (*Rattus norvegicus*). *Pharmacol. Physiol.* **27**, 793–795.

Viegas-Pequignot, E., Dutrillaux, B., Prod'Homme, M., and Petter, F. (1983). Chromosomal phylogeny of Muridae: a study of 10 genera. *Cytogen. Cell Genet.* **35**, 269–278.

Voigt, B., Kitada, K., Kloting, I., and Serikawa, T. (2000). Genetic comparison between laboratory rats and Japanese and German wild rats. *Mamm. Genome* **11**, 789–790.

Voigt, B., Kovacs, P., Vogt, L., and Kloting, I. (1997). How heterozygous are wild rats (*Rattus norvegicus*)? *Transplant Proc.* **29**, 1772–1773.

Volobouev, V.T., Hoffmann, A., Sicard, B., and Granjon, L. (2001). Polymorphism and polytypy for pericentric inversions in 38-chromosome *Mastomys* (Rodentia, Murinae) and possible taxonomic implications. *Cytogenet. Cell Genet.* **92**, 237–242.

Voss, R.S., Heideman, P.D., Mayer, V.L., and donnelly, T.M. (1992) Husbandry, reproduction, and postnatal development of the neotropical muroid rodent *Zygodontomys brevicauda*. *Lab Anim.* **26**, 38–46.

Walder, K., Fahey, R.P., Morton, G.J., Zimmet, P.Z., and Collier, G.R. (2000). Characterization of obesity phenotypes in *Psammomys obesus* (Israeli sand rats). *Int. J. Exp. Diabetes Res.* **1**, 177–184.

Walder, K., Oakes, N., Fahey, R.P., Cooney, G., Zimmet, P.Z., and Collier, G.R. (2002). Profile of dyslipidemia in *Psammomys obesus*, an animal model of the metabolic syndrome. *Endocr. Regul.* **36**, 1–8.

Walder, K., Ziv, E., Kalman, R., Whitecross, K., Shafrir, E., Zimmet, P.Z., and Collier, G.R. (2002). Elevated hypothalamic beacon gene expression in *Psammomys obesus* prone to develop obesity and type 2 diabetes. *Int. J. Obesity Rel. Metab. Dis.* **26**, 605–609.

Walker, E.P. (1975). "Mammals of the World," 3rd Ed., Vol. 2. Johns Hopkins, Baltimore.

Wang, H., Yoshimatsu, K., Ebihara, H., Ogino, M., Araki, K., Kariwa, H., Wang, Z., Luo, Z., Li, D., Hang, C., and Arikawa, J. (2000). Genetic diversity of hantaviruses isolated in China and characterization of novel hantaviruses isolated from *Niviventer confucianus* and *Rattus rattus*. *Virology* **278**, 332–345.

Ward, L.E. (2001). Handling the cotton rat for research. *Lab. Anim.* **30**, 45–50.

Webster, J.P. (1996). Wild brown rats (*Rattus norvegicus*) as a zoonotic risk on farms in England and Wales. *Comm. Dis. Rep. CDR Rev.* **6**, R46–R49.

Webster, J.P., Brunton, C.F., and MacDonald, D.W. (1994). Effect of *Toxoplasma gondii* upon neophobic behaviour in wild brown rats, *Rattus norvegicus*. *Parasitology* **109**, 37–43.

Webster, J.P., Lloyd, G., and MacDonald, D.W. (1995). Q fever (*Coxiella burnetii*) reservoir in wild brown rat (*Rattus norvegicus*) populations in the UK. *Parasitology* **110**, 31–35.

Weidinger, G., Ohlmann, M., Schlereth, B., Sutter, G., and Niesiesk, S. (2001). Vaccination with recombinant modified vaccinia virus Ankara protects against measles virus infection in the mouse and cotton rat model. *Vaccine* **19**, 2764–2768.

Weir, B.J. (1970). The management and breeding of some more hystricomorph rodents. *Lab. Anim.* **4**, 83–97.

White, P., Fischer, R., and Meunier, G. (1982). The lack of recognition of lactating females by infant *Octodon degus*. *Physiol. Behav.* **28**, 623–625.

Williams, C.S.F. (1980). Wild rats in research. *In* "The Laboratory Rat, Vol 2: Research Applications" (H.J. Baker, J.R. Lindsey, and S.H. Weisbroth, eds.), pp 245–256, Academic Press, San Diego.

Wilson, C., Brown, D., Najarian, K., Hanley, E.N., Gruber, H.E. (2003). Computer aided vertebral visualization and analysis: a methodology using the sand rat, a small animal model of disc degeneration. *BMC Musculoskel. Dis. Electronic Resource* **4**, 4.

Wilson, S.C. (1982). Contact-promoting behavior, social development, and relationship with parents in sibling juvenile degus (*Octodon degus*). *Dev. Psychobiol.* **15**, 257–268.

Winters, J.B. and Waser, P.M. (2003). Gene dispersal and outbreeding in a philopatric mammal. *Mol. Ecol.* **12**, 2251–2259.

Wolfe, J.L. (1982). *Oryzomys palustris*. *Mamm. Spec.* **176**, 1–5.

Woods, C.A. and Boraker, D. (1975). *Octodon degus*. *Mamm. Spec.* **67**, 1–5.

Worth, C.B. (1967). Reproduction, development, and behavior of captive *Oryzomys laticeps* and *Zygodontomys brevicauda* in Trinidad. *Lab. Anim. Care* **17**, 355–361.

Worth, C.B. and Rickard, E.R. (1951). Evaluation of the efficiency of the common cotton rat ectoparasites in the transmission of murine typhus. *Am. J. Trop. Med.* **31**, 295–298.

Yamada, T., Moralejo, D., Tsuchiya, K., Agui, T., and Matsumoto, K. (1993). Biochemical polymorphisms in wild rats (*Rattus norvegicus*) captured in Oita city, Japan. *J. Vet. Med. Sci.* **55**, 673–675.

Yamamoto, Y., Noguchi, Y., Noguchi, A., Nakayama, K., Mochida, K., Takano, K., Koura, M., and Ogura, A. (1999). Serum biochemical values in two inbred strains of Mastomys (*Praomys coucha*). *Exp. Anim.* **48**, 293–295.

Yan, Y., Todaka, N., Yamamura, K., Hirano, H., Gotoh, S., Katoh, T., Higashi, K., Arai, S., Murata, Y., Higashi, T., and Jirtle, R.L. (1999). The occurrence of polymorphism of mannose 6-phosphate/insulin-like growth factor 2 receptor gene in laboratory and wild rats. *Sangyo Ika Daigaku Zasshi* **21**, 199–208.

Ziabreva, I., Poeggel, G., Schnabel, R., and Braun, K. (2003). Separation-induced receptor changes in the hippocampus and amygdala of *Octodon degus*: influence of maternal vocalizations. *J. Neurosci.* **23**, 5329–5336.

Ziabreva, I., Schnabel, R., Poeggel, G., and Braun, K. (2003). Mother's voice "buffers" separation-induced receptor changes in the prefrontal cortex of *Octodon degus*. *Neuroscience* **119**, 433–441.

Zimmerman, E.G. and Nejtek, M.E. (1975). The hemoglobins and serum albumins of three species of wood rats (Neotoma Say and Ord). *Comp. Biochem. Physiol. B Comp. Biochem.* **50**, 275–278.

Zoltowska, M., St.-Louis, J., Ziv, E., Sicotte, B., Delvin, E.E., and Levy, E. (2003). Vascular responses to alpha-adrenergic stimulation and depolarization are enhanced in insulin-resistant and diabetic *Psammomys obesus*. *Can. J. Physiol. Pharmacol.* **81**, 704–710.

Appendix 1

Selected Normative Data

Henry J. Baker, J. Russell Lindsey, and Steven H. Weisbroth

Adult weight	
Male	300–400 gm
Female	250–300 gm
Life Span	
Usual	2.5–3 years
Maximum reported	4 yrs. 8 mo.
Surface area	0.03–0.06 cm^2
Chromosome number (diploid)	42
Water consumption	80–110 ml/kg/day
Food consumption	100 gm/kg/day
Body temperature	99.5°F, 37.5°C
Puberty	
Male	50 ± 10 days
Female	50 ± 10 days
Breeding season	None
Gestation	21–23 days
Litter size	8–14 pups
Birth weight	5–6 gm
Eyes open	10–12 days
Weaning	21 days
Heart rate	330–480 beats/min
Blood pressure	
Systolic	88–184 mm Hg
Diastolic	58–145 mm Hg
Cardiac output	50 (10–80) ml/min
Blood volume	
Plasma	40.4 (36.3–45.3) ml/kg
Whole blood	64.1 (57.5–69.9) ml/kg
Respiration frequency	85.5 (66–114)/min
Tidal volume	0.86 (0.60–1.25) ml

(*continued*)

(Continued)

Minute volume	0.073 (0.05–0.101) ml
Stroke volume	1.3–2.0 ml/beat
Plasma	
pH	7.4 ± 0.06
CO_2	22.5 ± 4.5 mM/liter
CO_2 pressure	40 ± 5.4 mm Hg
Leukocyte counts	
Total	$14\ (5.0–25.0) \times 10^3/\mu$l
Neutrophils	22 (9–34)%
Lymphocytes	73 (65–84)%
Monocytes	2.3 (0–5)%
Eosinophils	2.2 (0–6)%
Basophils	0.5 (0.15)%
Platelets	$1240\ (1100–1380) \times 10^3/\mu$l
Packed cell volume	46%
Red blood cells	$7.2–9.6 \times 10^6/\text{mm}^3$
Hemoglobin	15.6 gm/dl
Maximum volume of single bleeding	5 ml/kg
Urine	
pH	7.3–8.5
Specific gravity	1.04–1.07

Subject Index

Page numbers followed by *f* indicate figures; page numbers followed by *t* indicate tables.

A

Abdominal cavity, dissection at necropsy, 673, 674*f*
Abducens nerve, 119*t*
Acanthocytes, 136
Accelerated graft arteriosclerosis (AGA)
 complement system in, 855
 inflammatory response in, 854, 855
 therapies for, 855
Accessory organs, 105
Accessory sex glands, tumors of, 503–504
Acepromazine, 646
 dose determination, 638*t*
Acetaminophen, 652
Acid-base abnormalities, 140
Acid-base balance
 phosphate in, 233
Acid-base status
 intraoperative, blood gas determinations of, 632
Acid secretion, stomach, 102
Acinar cell tumor, pancreatic, 493
ACI strain, genealogy of, 34
ACP haplotype, 689
ACP strain genealogy, 34
Actinobacillus
 coinfection with *P. pneumotropica*, 378
 phylogenetic relationship with
  P. pneumotropica, 378
Activated partial thromboplastin time
  (APTT), 131
Acute-phase responses, 142, 142*t*
Adenocarcinoma
 of kidney, 498
 of mammary glands, 501–502, 502*f*
 ovarian, 500
 of pancreas, 139
 of preputial gland, 504
 of salivary glands, 490
Adenofibroma, 502
Adenoma
 adrenal cortical, 495–496, 496*f*

bronchiolar/alveolar, 487, 487*f*
 of colon, 491, 491*f*
 endometrial, 500, 501
 of glandular stomach, 490
 hepatocellular, 491
  vs. focal hyperplasia, 491–492, 492*f*
 of kidney, 498
 mammary gland, 501, 502
 mixed acinar-islet-cell, 497, 497*f*
 ovarian, 500
 pancreatic, 493
 pancreatic islet cell, 497
 parathyroid, 495, 495*f*
 pituitary, 494
  prolactin-secreting, 494
 of preputial gland, 504
 of prostate, 504
 of salivary glands, 490
 sebaceous gland, 484
 of thyroid
  C-cell, 494–495, 495*f*
  follicular cell, 494, 495*f*
Adenovirus, 424*t*–425*t*, 435
 papovavirus, 424*t*–425*t*, 435
Adipose, 98
Adjuvant-induced arthritis, 725–726
Adjuvants, in antibody production, 615–616
Ad libitum feeding, cancer and, 481
Administration route(s)
 intramuscular, 613
 intraperitoneal, 612
 intravenous, 610–612
 oral, 608–610
 osmotic pumps, 612
 subcutaneous, 612
Adrenal cortex
 accessory nodules of, 533
 in Donyru rats, 496*t*
 tumors of, 495–496
 in Wistar rats, 483*t*
Adrenalectomy, in albino rats, 2, 20
Adrenal gland, 110

age and lesions of, 764
 dissection at necropsy, 673, 675*f*
 morphophysiology of, 116, 116*f*
 tumors of, 495–496
Adrenal medulla, tumors of, 496
Aeroallergen exposure, 310
 caging systems for, 310
Aeroallergens, 567–568
Aerobic exercise capacity
 genetic component to, 108
Aflatoxin, 266, 267
 ionizing radiation detoxification of, 272
African white-tailed rat
  (*Mystromus albicaudatus*), 872
Agar diets, 274
Age
 and age-related disease
  rat models of, 761–762
 changes in body weight and composition
  and, 763–764
 immunity and, 736–737
 incidence of lesions and
  age and strain differences in, 764–765
 skeletal muscle function decline in,
  95–96
Aggression
 patterns of, 209
 purpose of, 209
Alanine aminotransferase (ALT)
 in hepatocellular disease, 136
Albany strain, 32
Albendazole, 454
Albino rats
 Hatai and, 3, 5
 King Albino, 6–7
 origin of, Geneva, 3
 at Wistar Institute, 5, 75
ALB strain, genealogy of, 34
Alcoholic heart disease, 791–792
Alimentary system
 parasitic infections of, 464–470
 schematic of, 104*f*